18

He 2	
helium	
4.00	
$1s^2$	

	Metal
	Metalloid
	Nonmetal

13	**14**	**15**	**16**	**17**	
B 5	**C** 6	**N** 7	**O** 8	**F** 9	**Ne** 10
boron	carbon	nitrogen	oxygen	fluorine	neon
10.81	**12.01**	**14.01**	**16.00**	**19.00**	**20.18**
$2s^22p^1$	$2s^22p^2$	$2s^22p^3$	$2s^22p^4$	$2s^22p^5$	$2s^22p^6$
Al 13	**Si** 14	**P** 15	**S** 16	**Cl** 17	**Ar** 18
aluminum	silicon	phosphorus	sulfur	chlorine	argon
26.98	**28.09**	**30.97**	**32.06**	**35.45**	**39.95**
$3s^23p^1$	$3s^23p^2$	$3s^23p^3$	$3s^23p^4$	$3s^23p^5$	$3s^23p^6$

10	**11**	**12**					
Ni 28	**Cu** 29	**Zn** 30	**Ga** 31	**Ge** 32	**As** 33	**Se** 34	**Br** 35
nickel	copper	zinc	gallium	germanium	arsenic	selenium	bromine
58.69	**63.55**	**65.41**	**69.72**	**72.64**	**74.92**	**78.96**	**79.90**
$3d^84s^2$	$3d^{10}4s^1$	$3d^{10}4s^2$	$4s^24p^1$	$4s^24p^2$	$4s^24p^3$	$4s^24p^4$	$4s^24p^5$

Kr 36 / krypton / **83.80** / $4s^24p^6$

Pd 46	**Ag** 47	**Cd** 48	**In** 49	**Sn** 50	**Sb** 51	**Te** 52	**I** 53	**Xe** 54
palladium	silver	cadmium	indium	tin	antimony	tellurium	iodine	xenon
106.42	**107.87**	**112.41**	**114.82**	**118.71**	**121.76**	**127.60**	**126.90**	**131.29**
$4d^{10}$	$4d^{10}5s^1$	$4d^{10}5s^2$	$5s^25p^1$	$5s^25p^2$	$5s^25p^3$	$5s^25p^4$	$5s^25p^5$	$5s^25p^6$

Pt 78	**Au** 79	**Hg** 80	**Tl** 81	**Pb** 82	**Bi** 83	**Po** 84	**At** 85	**Rn** 86
platinum	gold	mercury	thallium	lead	bismuth	polonium	astatine	radon
195.08	**196.97**	**200.59**	**204.38**	**207.2**	**208.98**	**(209)**	**(210)**	**(222)**
$5d^96s^1$	$5d^{10}6s^1$	$5d^{10}6s^2$	$6s^26p^1$	$6s^26p^2$	$6s^26p^3$	$6s^26p^4$	$6s^26p^5$	$6s^26p^6$

Ds 110	**Rg** 111	**Cn** 112	113	**Fl** 114	115	**Lv** 116	117	118
darmstadtium	roentgenium	copernicium		flerovium		livermorium		
(281)	**(280)**	**(285)**		**(289)**		**(293)**		
$6d^87s^2$	$6d^{10}7s^1$	$6d^{10}7s^2$		$7s^27p^2$		$7s^27p^4$		

Eu 63	**Gd** 64	**Tb** 65	**Dy** 66	**Ho** 67	**Er** 68	**Tm** 69	**Yb** 70	**Lu** 71
europium	gadolinium	terbium	dysprosium	holmium	erbium	thulium	ytterbium	lutetium
151.96	**157.25**	**158.93**	**162.50**	**164.93**	**167.26**	**168.93**	**173.04**	**174.97**
$4f^76s^2$	$4f^75d^16s^2$	$4f^96s^2$	$4f^{10}6s^2$	$4f^{11}6s^2$	$4f^{12}6s^2$	$4f^{13}6s^2$	$4f^{14}6s^2$	$5d^16s^2$

Am 95	**Cm** 96	**Bk** 97	**Cf** 98	**Es** 99	**Fm** 100	**Md** 101	**No** 102	**Lr** 103
americium	curium	berkelium	californium	einsteinium	fermium	mendelevium	nobelium	lawrencium
(243)	**(247)**	**(247)**	**(251)**	**(252)**	**(257)**	**(258)**	**(259)**	**(262)**
$5f^77s^2$	$5f^76d^17s^2$	$5f^97s^2$	$5f^{10}7s^2$	$5f^{11}7s^2$	$5f^{12}7s^2$	$5f^{13}7s^2$	$5f^{14}7s^2$	$6d^17s^2$

FREQUENTLY USED TABLES AND FIGURES

		Page

CHEMICAL PRINCIPLES

THE QUEST FOR INSIGHT

SEVENTH EDITION

PETER ATKINS

Oxford University

LORETTA JONES

University of Northern Colorado

LEROY LAVERMAN

University of California, Santa Barbara

w.h.
freeman

Macmillan Learning

New York

Publisher: Kate Ahr Parker

Acquisitions Editor: Alicia Brady

Developmental Editor: Heidi Bamatter

Marketing Manager: Maureen Rachford

Marketing Assistant: Cate McCaffery

Media Editor: Amy Thorne

Media Producer: Jenny Chiu

Photo Editor: Robin Fadool

Photo Licensing Editor: Richard Fox

Senior Project Editor: Elizabeth Geller

Cover Designer: Blake Logan

International Edition Cover Design: Dirk Kaufman

Text Designer: Marsha Cohen

Art Manager: Matthew McAdams

Illustrations: Peter Atkins and Leroy Laverman

Production Manager: Susan Wein

Composition: Aptara

Printing and Binding: RR Donnelley

Cover Image: © Ted Kinsman/Alamy

Library of Congress Control Number: 2015951706

ISBN-13: 978-1-4641-8395-9

ISBN-10: 1-4641-8395-3

Printed in the United States of America

First printing

1007740928

W. H. Freeman and Company
One New York Plaza
Suite 4500
New York, NY 10004-1562
www.whfreeman.com

MAJOR TECHNIQUES (Online Only)
http://macmillanhighered.com/chemicalprinciples7e

CONTENTS IN BRIEF

CONTENTS

v

Chemical Principles

The central theme of this text is to challenge students to think and question, while providing a sound foundation in the principles of chemistry. Students of all levels also benefit from assistance in learning how to think, pose questions, and approach problems. We show students how to build models, refine them systematically in the light of experimental input, and express them quantitatively. To that end, *Chemical Principles: The Quest for Insight*, Seventh Edition, aims to build understanding and offer students a wide array of pedagogical support.

New Overall Organization

In this seventh edition, we have implemented a new organization. The content is presented as a series of 85 short *Topics* arranged into 11 thematic groups called *Focuses*. Our aim is twofold: to present reader and instructor with maximum flexibility and digestibility. We had a particular structure in mind when writing this edition, but instructors might have different ideas. Although the content is arranged along the lines of an atoms first approach, the division of Topics allows the instructor not only to tailor the text within the time constraints of the course, as it will be much easier to omit selected Topics, but also to take a path through the text that matches individual teaching and learning objectives. We have carefully avoided language that suggests the Topics should be read in the order they appear in the book. The student should also find the Topics easy to absorb and review, as each Topic is organized into smaller, more manageable sections. As such, since the Focuses are of very different lengths, instructors should target Topics, and not necessarily entire Focuses, when assigning content in their syllabi.

Each Focus begins with a brief discussion of how its Topics share a theme and how that theme links to others in the book. This contextual relationship is also captured visually by the "Road Map" that prefaces each Focus. We wanted to convey the intellectual structure of the subject, while leaving open the order of presentation.

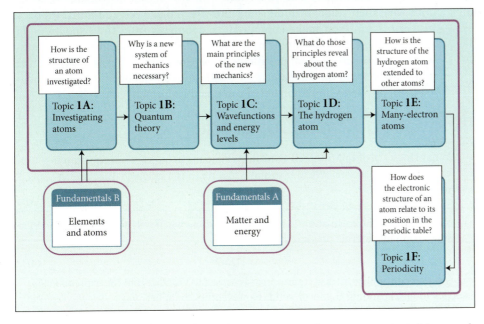

Our core motivation is to help students to master the course content. Thus, each Topic opens with two questions a student typically faces: "Why do you need to know this material?", and "What do you need to know already?" The answers to the second question point to other Topics that we consider appropriate to have studied in advance of the Topic at hand. We listened to the thoughtful advice of our reviewers and have

Why Do You Need to Know This Material? Ionic bonding is one of the principal forms of bonding between atoms. Understanding how bonds form between ions allows you to predict the formulas of ionic compounds and to estimate how strongly the ions are held together.

What Do You Need to Know Already? You need to know about electron configurations of many-electron atoms (Topic 1E), the concept of potential energy, and the nature of the Coulomb interaction between charges (*Fundamentals* A). You need to be familiar with ionic radii and the ionization energy and electron affinity of elements (Topic 1F).

ensured that this new organization guides and supports instructors and students through the individual paths they choose, to provide an improved classroom experience. Even the Road Map is designed to be an encouragement to learn, because we show how each Topic is inspired by a conceptual question.

How Is That Explained...

...using kinetics?

The *kinetic interpretation* of equilibrium is based on a comparison of competing rates, in this instance, the rates of evaporation and condensation. Vapor forms as molecules leave the surface of the liquid through evaporation. However, as the number of molecules in the vapor increases, more of them are available to condense, that is, to strike the surface of the liquid, stick to it, and become part of the liquid again. Eventually, the rate of molecules returning to the liquid matches the rate escaping (**FIG. 5A.2**). The vapor is now condensing as fast as the liquid is vaporizing, and so the equilibrium is *dynamic* in the sense that both the forward and reverse processes are still occurring but now their rates are equal. The **dynamic equilibrium** between liquid water and its vapor is denoted

$$H_2O(l) \rightleftharpoons H_2O(g)$$

Wherever the symbol $\rightleftharpoons$ appears, it means that the species on both sides of it are in dynamic equilibrium with each other. With this picture in mind, the **vapor pressure** of a liquid (or a solid) can be defined as the pressure exerted by its vapor when the vapor and the liquid (or the solid) are in dynamic equilibrium with each other.

...using thermodynamics?

In the *thermodynamic interpretation* of equilibrium, the condensed and vapor phases of a substance are in equilibrium, denoted

$$H_2O(l) \rightleftharpoons H_2O(g)$$

when there is no change in Gibbs free energy, $\Delta G = 0$ for the phase change process. In short, neither the forward nor the reverse process is spontaneous at equilibrium. The **vapor pressure** of a liquid (or a solid) is the pressure exerted by its vapor when the vapor and the liquid (or the solid) are in equilibrium with each other.

New to this edition, and specifically to Focus 5, is a new two-column approach for presenting derivations from both a kinetic and a thermodynamic viewpoint. This innovation aims to accommodate instructors who approach equilibrium from differing viewpoints and allows the instructors to take either path or to include both perspectives in their instruction.

Finally, we have collected all the Major Techniques in one group. These technique sections have been placed online for convenient access from laboratories or classroom, on our textbook catalog page: http://macmillanhighered.com/chemicalprinciples7e.

Reviewing the Basics

The *Fundamentals* sections are identified by green-edged pages. These sections provide a streamlined overview of the basics of chemistry. This material can be used either to provide a useful, succinct review of elementary material to which students can refer for extra help as they progress through the course, or as a concise survey of material before starting on the main text.

To support the *Fundamentals* sections pedagogically, we continue to provide the *Fundamentals Diagnostic Test*. This test allows instructors to determine what their students understand and where they need additional support. Instructors can then make appropriate assignments from the *Fundamentals*. The test includes 5 to 10 problems for each *Fundamentals* section. The diagnostic test was created by Cynthia LaBrake at the University of Texas, Austin. More information about the Fundamentals Diagnostic test can be found on our catalog page: http://macmillanhighered.com/chemicalprinciples7e.

Innovative Math Coverage

- **What Does This Equation Tell You?** helps students to interpret an equation in physical and chemical terms. We aim to show that math is a language that reveals aspects of reality.

The result of the calculation is that the work done when a system expands by ΔV against a constant external pressure P_{ex} is

$$w = -P_{ex}\Delta V \tag{3}$$

This expression applies to all systems. A gas is easiest to visualize, but the expression also applies to an expanding liquid or solid. However, Eq. 3 applies *only when the external pressure is constant* during the expansion.

What Does This Equation Tell You? When the system expands, ΔV is positive. Therefore the minus sign in Eq. 3 tells you that the internal energy of the system decreases when the system expands. The factor P_{ex} tells you that more work is done for a given change in volume when the external pressure is high. The factor ΔV tells you that, for a given external pressure, more work is done the greater the change in volume.

- **How Is That Done?** The text is designed so that mathematical derivations are set apart from the body of the text, making it easy for instructors to avoid or assign this material. This feature, which is structured in a way that encourages students to appreciate the power of math (by showing that vital progress depends on it), sets off derivations of key equations from the rest of the text. Virtually all the calculus in the text is confined to this feature, so it can be avoided if appropriate. For instructors who judge that their students can cope with this material and who want their students to realize the power that math puts into their hands, these derivations provide that encouragement. A selection of end-of-Focus exercises that make use of calculus is provided and marked with an icon: $\boxed{\int_{dx}^{C}}$. Some derivations that we consider to be beyond this level but are useful as a resource, are located on the website.

How Is That Done?

To calculate the fraction of occupied space in a close-packed structure, consider a ccp structure. First, look at how the cube is built from the spheres representing the atoms. **FIGURE 3H.18** shows that eight spheres lie at the corners of the cubes. Only $\frac{1}{8}$ of each of these spheres projects into the cube, so the eight corner spheres collectively contribute $8 \times \frac{1}{8} = 1$ sphere to the cube. Half a sphere on each of the six faces projects into the cube, so the spheres on each face contribute $6 \times \frac{1}{2} = 3$ spheres, giving four spheres in all within the cube. The length of the diagonal of the face of the cube shown in Fig. 3H.18 is $4r$, where r is the radius of the sphere. Each of the two corner spheres contributes r and the sphere at the center of the face contributes $2r$. According to the Pythagorean theorem, the length of the side of the face, a, is related to the diagonal by $a^2 + a^2 = (4r)^2$, or $2a^2 = 16r^2$, and so $a = 8^{1/2}r$. The volume of the cube is therefore $a^3 = 8^{3/2}r^3$. The volume of each sphere is $\frac{4}{3}\pi r^3$, so the total volume of the spheres inside the cube is $4 \times \frac{4}{3}\pi r^3 = \frac{16}{3}\pi r^3$. The ratio of this occupied volume to the total volume of the cube is therefore

$$\frac{\text{Total volume of spheres}}{\text{Total volume of cube}} = \frac{(16/3)\pi r^3}{8^{3/2}r^3} = \frac{16\pi}{3 \times 8^{3/2}} = 0.74\ldots$$

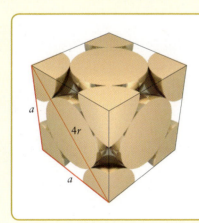

a

$4r$

a

FIGURE 3H.18 The relation of the dimensions of a face-centered cubic unit cell to the radius, r, of the spheres. The spheres are in contact along the face diagonals.

- **Annotated equations** help students interpret an equation and see the connection between symbols and numerical values. We consider the correct use of units an important part of a student's vocabulary, not only because it is a part of the international language of chemistry but also because it encourages a systematic approach to calculations; in more complicated or unfamiliar contexts, we also use annotations to explain the manipulation of units.

$$w = -\underbrace{(0.100\ \text{mol})}_{n} \times \underbrace{(8.3145\ \text{J·K}^{-1}\text{·mol}^{-1})}_{R} \times \underbrace{(298\ \text{K})}_{T} \times \ln\underbrace{\frac{2.00\ \text{L}}{1.00\ \text{L}}}_{V_2/V_1} = -172\ \text{J}$$

Emphasis on Problem Solving

- **Notes on Good Practice** encourage conformity to the language of science by setting out the language and procedures adopted by the International Union of Pure and Applied Chemistry (IUPAC). In many cases, they identify common mistakes and explain how to avoid them.

> **A Note on Good Practice:** A property y is said to "vary linearly with x" if the relation between y and x can be written $y = b + mx$, where b and m are constants. A property y is said to be "proportional to x" if $y = mx$ (that is, $b = 0$).

- **Anticipate/Plan/Solve/Evaluate Strategy.** This problem-solving approach encourages students to anticipate or predict what a problem's answer should be qualitatively and to map out the solution before trying to solve the problem quantitatively. Following the solution, the original anticipation is evaluated. Students are often

---

EXAMPLE 6B.1   Calculating a pH from a concentration

You are working in a medical laboratory monitoring the recovery of patients in intensive care. The pH of their blood must be carefully monitored and controlled because even small deviations from normal levels can be fatal. What is the pH of (a) human blood, in which the concentration of H_3O^+ ions is 4.0×10^{-8} mol·L^{-1}; (b) 0.020 M HCl(aq); (c) 0.040 M KOH(aq)?

ANTICIPATE The concentration of H_3O^+ ions in blood is lower than in pure water, so you should expect pH > 7; in HCl(aq), an acid, you should expect pH < 7, and in KOH(aq), a base, pH > 7.

PLAN The pH is calculated from Eq. 1b. For strong acids, the molar concentration of H_3O^+ is equal to the molar concentration of the acid. For strong bases, first find the concentration of OH^-, then convert that concentration into $[H_3O^+]$ by using $[H_3O^+][OH^-] = K_w$ in the form $[H_3O^+] = K_w/[OH^-]$.

What should you assume? Assume that any strong acid (HCl here) is fully deprotonated in solution and any ionic compound (KOH here) is fully dissociated in solution.

SOLVE

(a) From pH $= -\log[H_3O^+]$,

$$pH = -\log(4.0 \times 10^{-8}) = 7.40$$

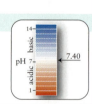

(b) HCl is a strong acid, so it is treated as completely deprotonated in water.

$$[H_3O^+] = [HCl] = 0.020 \text{ mol·L}^{-1}$$

From pH $= -\log[H_3O^+]$,

$$pH = -\log 0.020 = 1.70$$

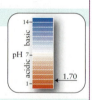

(c) Because KOH is assumed to dissociate completely in solution each formula unit provides one OH^- ion,

$$[OH^-] = [KOH] = 0.040 \text{ mol·L}^{-1}$$

Find $[H_3O^+]$ from $[H_3O^+][OH^-] = K_w$ in the form $[H_3O^+] = K_w/[OH^-]$.

$$[H_3O^+] = \frac{K_w}{[OH^-]} = \frac{1.0 \times 10^{-14}}{0.040} = 2.5\ldots \times 10^{-13}$$

From pH $= -\log[H_3O^+]$,

$$pH = -\log(2.5\ldots \times 10^{-13}) = 12.60$$

EVALUATE The calculated pH values are in line with what was anticipated.

Self-test 6B.1A Calculate the pH of (a) household ammonia, in which the OH^- concentration is about 3×10^{-3} M·L^{-1}; (b) 6.0×10^{-5} M HClO$_4$(aq).

[***Answer:*** (a) 11.5; (b) 4.22]

Self-test 6B.1B Calculate the pH of 0.077 M NaOH(aq).

Related Exercises: 6B.3, 6B.4

puzzled about what they should assume in a calculation; many worked examples now include an explicit statement about what should be assumed. Because students process information in different ways, many steps in the worked examples are broken down into three components: a *qualitative* statement about what is being done, a *quantitative* explanation with the mathematics worked out, and a *visual* representation to aid with interpreting each step.

- **Real-world contexts for Worked Examples.** We want to motivate students and encourage them to see that the calculations are relevant to all kinds of careers and applications. With that aim in mind, we pose the problem in a context in which such calculations might occur.

- **Self-Tests** are provided as pairs throughout the book. They enable students to test their understanding of the material covered in the preceding section or worked example. The answer to the first self-test is provided immediately, and the answer to the second can be found at the back of the book.

- **Thinking Points** encourage students to speculate about the implications of what they are learning and to transfer their knowledge to new situations. This edition now provides instructors with suggested answers to the Thinking Points online on the textbook's catalog page: http://macmillanhighered.com/chemicalprinciples7e.

THINKING POINT

By what factor does the unique average reaction rate change if the coefficients in a chemical equation are doubled?

- **Toolboxes** show students how to tackle major types of calculations and demonstrate how to connect concepts to problem solving. The Toolboxes are designed as learning aids and handy summaries of key material. Each summarizes the conceptual basis of the following steps, because we are concerned that students understand what they are doing as well as be able to do it. Each Toolbox is followed immediately by one or more related Examples; these Examples apply the problem-solving strategy outlined in the Toolbox and illustrate each step of the procedure explicitly.

Toolbox 6H.2 HOW TO CALCULATE THE pH DURING A TITRATION OF A WEAK ACID OR A WEAK BASE

CONCEPTUAL BASIS

The pH is governed by the major solute species present in solution. As strong base is added to a solution of a weak acid, a salt of the conjugate base of the weak acid is formed. This salt affects the pH and needs to be taken into account. **TABLE 6H.1** outlines the regions encountered during a titration and the primary equilibrium to consider in each region.

PROCEDURE

The procedure is like that in Toolbox 6H.1, except that an additional step is required to calculate the pH from the proton transfer equilibrium. First use reaction stoichiometry to find the amount of excess acid or base. Begin by writing the chemical equation for the reaction, then:

Step 1 Calculate the amount of weak acid or base in the original analyte solution. Use $n_J = V_{analyte}[J]$.

Step 2 Calculate the amount of OH^- ions (or H_3O^+ ions if the titrant is an acid) in the volume of titrant added. Use $n_J = V_{titrant}[J]$.

Step 3 Use reaction stoichiometry to calculate the following amounts:

- Weak acid–strong base titration: the amount of conjugate base formed in the neutralization reaction, and the amount of weak acid remaining.

- Weak base–strong acid titration: the amount of conjugate acid formed in the neutralization reaction, and the amount of weak base remaining.

Calculate the concentrations.

Step 4 Find the "initial" molar concentrations of the conjugate acid and base in solution after neutralization, but before any proton transfer equilibrium with water is taken into account. Use $[J] = n_J/V$, where V is the total volume of the solution, $V = V_{analyte} + V_{titrant}$.

Calculate the pH.

Step 5 Use the expression for K_a or K_b to find the H_3O^+ concentration in a weak acid or the OH^- concentration in a weak base. Alternatively, if the concentrations of conjugate acid and base calculated in step 4 are both large relative to the concentration of hydronium ions, use them in the Henderson–Hasselbalch equation, Eq. 2 of Topic 6G, pH $\approx$ pK_a + log([base]$_{initial}$/[acid]$_{initial}$), to determine the pH. In each case, if the pH is less than 6 or greater than 8, assume that the autoprotolysis of water does not significantly affect the pH. If necessary, convert between K_a and K_b by using $K_a \times K_b = K_w$.

This procedure is illustrated in Example 6H.3.

- **"The skills you have mastered are the ability to:"** are checklists of key concepts provided at the end of each Topic. These checklists not only are a reminder of the subjects with which students should feel comfortable by the end of the topic but also offer a satisfying opportunity to check off the items that they consider they have grasped.

The skills you have mastered are the ability to:

☐ **1.** Determine the activation energy from the experimental temperature dependence of reaction rate constants (Example 7D.1).

☐ **2.** Predict the rate constant for a reaction at a new temperature if the activation energy and rate constant at one temperature are known (Example 7D.2).

☐ **3.** Discuss the Arrhenius parameters, A and E_a, in terms of models of reactions (Sections 7D.2 and 7D.3).

- **Margin Notes** are brief asides, placed in the margin right next to the relevant text, that provide an extra note of help to clarify concepts or usage or to make a historical point.

The VSEPR model was first proposed by the British chemists Nevil Sidgwick and Herbert Powell and has been developed by the Canadian chemist Ronald Gillespie.

Lewis structures (Topics 2B and 2C) show only how the atoms are connected and how the electrons are arranged around them. The **valence-shell electron-pair repulsion model** (VSEPR model) extends Lewis's theory of bonding by adding rules that account for bond angles and molecular shapes:

- **NEW! Interludes** describe a number of contemporary applications of chemistry by showing how chemistry is being used in a variety of modern contexts. New for this edition, there are five interludes, placed between various Focuses.
- **NEW! Topic- and Focus-Specific Exercises** give students the opportunity to practice solving problems that draw upon one Topic (these appear at the end of every Topic) and exercises that include and combine concepts from the entire Focus (these appear at the end of each Focus).

Topic 3I Exercises

3I.1 Estimate the relative density (compared to pure aluminum) of magnalium, a magnesium–aluminum alloy in which 30.0% of the aluminum atoms have been replaced by magnesium atoms without distortion of the crystal structure.

3I.2 Estimate the relative density (compared to pure copper) of aluminium bronze, an alloy that is 8.0% by mass aluminium. Assume no distortion of the crystal structure.

3I.9 A unit cell for the calcite structure can be found at http://webmineral.com. From this structure, identify (a) the crystal system and (b) the number of formula units present in the unit cell.

3I.10 Consult http://webmineral.com and examine the unit cells of calcite and dolomite. (a) In what respects are these two structures the same? (b) In what respect are they different? (c) Where are the magnesium and calcium ions located in dolomite?

The following Example and Exercises draw on material from throughout Focus 3.

FOCUS 3 Online Cumulative Example

Some of the earliest mortars were *nonhydraulic cements*, which harden by reaction with CO_2 rather than with water. These cements are prepared by heating calcite, $CaCO_3(s)$, strongly to drive off CO_2 gas and form quicklime, $CaO(s)$. The resulting solid is mixed with water to give a paste of slaked lime, $Ca(OH)_2$, to which sand or volcanic ash is added to form lime mortar. The Roman Colosseum and Pantheon were constructed with this type of mortar and have endured the ages. You are investigating ancient building methods and want to understand the chemistry of these materials.

(a) Write the balanced chemical equations for (i) the conversion of calcite to quicklime, (ii) the reaction of quicklime with water to form slaked lime, and (iii) the reaction of slaked lime with CO_2 to form calcium carbonate.

(b) Preparing quicklime releases the greenhouse gas carbon dioxide. If 1.000 t (1 t $= 10^3$ kg) of $CaCO_3$ is placed in a kiln and heated to 850 °C, what volume of $CO_2(g)$ is formed at 850 °C and 1 atm?

(c) If the $CO_2(g)$ from part (b) is cooled to room temperature of 22 °C what volume would it occupy?

(d) Calcium oxide has the cubic structure shown in (**1**). The length of each edge is 481.1 pm. All the atoms are on an edge, face, or corner of the cube with one O atom in the center of the cube. Use this information and the density of $CaCO_3(s)$, 2.711 g·cm^{-3}, to calculate the change in volume of the solid as CO_2 is driven off from 1.0 t of $CaCO_3$.

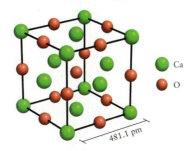

○ Ca
○ O

481.1 pm

1   Calcium oxide, CaO

(e) From the results in part (d), suggest a reason why buildings constructed of bricks held together with lime mortar might collapse during a fire.

 The online Cumulative Example solution can be found at http://macmillanhighered.com/chemicalprinciples7e

FOCUS 3 Exercises

3.1 The drawing below shows a tiny section of a flask containing two gases. The orange spheres represent neon atoms and the blue spheres represent argon atoms. (a) If the partial pressure of neon in this mixture is 420. Torr, what is (a) the partial pressure of argon; (b) the total pressure?

3.2 The four flasks below were prepared with the same volume

Given that the partial pressure of carbon ⊘ sphere is 0.26 Torr and that the temperatu the volume of air at 1.0 atm needed to prod

3.4 Roommates fill ten balloons for a part and five with helium. After the party the hy lost one-fifth of their hydrogen due to effus of the balloons. What fraction of helium w have lost at that same time?

3.5 Suppose that 200. mL of hydrogen chl and 20. °C is dissolved in 100. mL of water. T

- **NEW! Online worked examples**. Each Focus ends with a Cumulative Example that challenges students to combine their understanding of concepts from several parts of the Focus. Full solutions presented in the same format as the worked examples in the text are available to students on the book's online catalog page: http://macmillanhighered.com/chemicalprinciples7e.

Improved Illustration Program

- **NEW!** All the line art has been redrawn or refreshed for this edition using a new and more vibrant color palette.

- We have replaced many of the photographs with more revealing and often more relevant images.

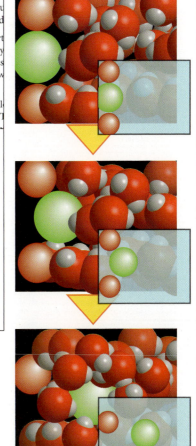

FIGURE 5D.1 The events that take place at the interface of a solid ionic solute and a solvent (water). Only the surface layer of ions is shown. When the ions at the surface of the solid become hydrated, they move off into the solution. The insets at the right show the ions alone.

Contemporary Chemistry for All Students

Chemistry has an extraordinary range of applications, and we have sought to be inclusive and extensive in our discussion and use of examples. The brief contextual remarks in the worked examples help to illustrate this range. So too do some of the end-of-Focus exercises and the boxes that illustrate modern applications that occur throughout the text. We have kept in mind that engineers need a knowledge of chemistry, that biologists need a knowledge of chemistry, and that anyone anticipating a career in which materials are involved needs chemistry. Specific points relevant to the study of green chemistry are noted with an icon: . An important aspect of chemistry is that it provides transferable skills that can be deployed in a wide variety of careers; we have kept that in mind throughout, by showing readers how to think systematically, to build models based on observation, to be aware of magnitudes, to express qualitative ideas, concepts, and models quantitatively, and to interpret mathematical expressions physically.

Media and Supplements

For Students

We believe a student needs to interact with a concept several times in a variety of scenarios in order to obtain a thorough understanding. With that in mind, Macmillan Learning has developed a comprehensive package of student learning resources.

Printed Resources

Student Study Guide, by John Krenos, Rutgers University
ISBN: 1-319-01755-X

The Student Study Guide helps students to improve their problem-solving skills, avoid common mistakes, and understand key concepts. After a brief review of each Topic's critical ideas, students are taken through worked-out examples, try-it-yourself examples, and quizzes, all structured to reinforce the text's objectives and build problem-solving techniques.

Student Solutions Manual, by Laurence Lavelle, University of California, Los Angeles; Yinfa Ma, Missouri University of Science and Technology; and Christina Johnson, University of California, San Diego
ISBN: 1-319-01756-8

The *Student Solutions Manual* follows the problem-solving structure set out in the main text and includes detailed solutions to all odd-numbered exercises in the text.

Media Resources

The *Chemical Principles* student resources at http://macmillanhighered.com/chemical-principles7e provide a range of tools for problem solving and chemical explorations.

- **Solutions to Cumulative Examples.** Each Focus ends with a Cumulative Example that combines concepts from several parts of the Focus. Full solutions, presented in the same format as the worked examples in the text, are available to students on the catalog page.

- **Major Techniques** have been placed online for convenient access.

- **Media Tools:**
  - **Living Graphs** allow the user to control the parameters.
  - **Animations** from the Vischem group are once again available to students and instructors.
  - **Lab Videos** are connected to figures in the text and demonstrate a laboratory experiment.

- **Molecule Database** links to *ChemSpider*, a free database of chemical structures, providing students access to information on over 35 million structures from hundreds of data sources. ChemSpider ID numbers have been provided in selected exercises to help students find the correct structures.

- **ChemCasts** replicate the face-to-face experience of watching an instructor work a problem. Using a virtual whiteboard, these video tutors show students the steps involved in solving key worked examples, while explaining the concepts along the way. They are easy to view on a computer screen or to download to a tablet or other media player.

- **Key Equations**, a compilation of key equations from the text

- **Interactive Periodic Table of Elements** links to www.Ptable.com, a dynamic periodic table with extensive information about each of the elements.

For Instructors

Whether you are teaching the course for the first time or the hundredth time, the Instructor Resources to accompany *Chemical Principles* provide the resources you need to make teaching preparation efficient.

Media Resources

Instructors can access valuable teaching tools through the *Chemical Principles* catalog page, http://macmillanhighered.com/chemicalprinciples7e. These resources are designed to aid the instructor throughout the teaching experience. They include:

- **Instructor's Solutions Manual,** by Laurence Lavelle, University of California, Los Angeles; Yinfa Ma, Missouri University of Science and Technology; and Christina Johnson, University of California, San Diego, which contains full, worked-out solutions to all even-numbered exercises in the text.

- **Updated Illustrations from the textbook** are offered as high-resolution .jpeg files and in PowerPoint format.

- **Newly Updated Lecture PowerPoints with Integrated Clicker Questions** have been developed to minimize preparation time for new users of the book. These files offer suggested lectures, including key illustrations, summaries, and clicker questions that instructors can adapt to their teaching styles.

- **Test Bank,** by Robert Balahura, University of Guelph, and Mark Benvenuto, University of Detroit, Mercy, which offers over 1400 multiple-choice, fill-in-the-blank, and essay questions and is available exclusively on the book's catalog page.

Online Learning Environment

Sapling Learning
www.saplinglearning.com

Developed by educators with both online expertise and extensive classroom experience, Sapling Learning provides highly effective interactive homework and instruction that improve student learning outcomes for the problem-solving disciplines. Sapling Learning offers an enjoyable teaching and effective learning experience that is distinctive in three important ways:

- **Ease of Use:** Sapling Learning's easy-to-use interface keeps students engaged in problem solving, not struggling with software.

- **Targeted Instructional Content:** Sapling Learning increases student engagement and comprehension by delivering immediate feedback and targeted instructional content.

- **Unsurpassed Service and Support:** Sapling Learning makes teaching more enjoyable by providing a dedicated Masters- or Ph.D.-level colleague to serve instructors' unique needs throughout the course, including help with content customization.

We offer bundled packages that include Sapling Learning Online Homework with all versions of our text.

Lab Resources

Available stand-alone or bundled with the text for a nominal charge.
ACS Molecular Structure Model Set, by Maruzen Company, Ltd.
ISBN: 0-7167-4822-3

Molecular modeling helps students understand physical and chemical properties by providing a way to visualize the three-dimensional arrangement of atoms. This model set uses polyhedra to represent atoms and plastic connectors to represent bonds (scaled to correct bond length). Plastic plates representing orbital lobes are included for indicating lone pairs of electrons, radicals, and multiple bonds—a feature unique to this set.

Acknowledgments

We are grateful to the many instructors, colleagues, and students who have contributed their expertise to this edition and to all the preceding editions. We would like above all to thank those who evaluated the seventh edition so carefully and helped us to develop the new organization:

Natalya Bassina, *Boston University*
Charles Carraher, *Florida Atlantic University*
Patricia Christie, *Massachusetts Institute of Technology*
Gregory M. Ferrence, *Illinois State University*
David Finneran, *Miami Dade College*
James Fisher, *Imperial Valley College*
Teresa Garrett, *Vassar College*
Dawit Gizachew, *Purdue University Calumet*
Susan Green, *Macalester College*
P. Shiv Halasyamani, *University of Houston*
Vlad M. Iluc, *University of Notre Dame*
Elon Ison, *North Carolina State University*
Adam Johnson, *Harvey Mudd College*
Humayun Kabir, *Oglethorpe University*
James I. Lankford, *St. Andrews University*
Susan Maleckar, *University of Pittsburgh*
Lynn Mandeltort, *Auburn University*
David W. Millican, *Guilford College*
Apryl Nenortas, *Clovis Community College*

Brian Northrop, *Wesleyan University*
John W. Overcash, *University of Illinois*
Pat Owens, *Winthrop University*
Rene Rodriguez, *Idaho State University*
Michael P. Rosynek, *Texas A&M University*
Suzanne Saum, *Washington University*
Carlos Simmerling, *Stony Brook University*
Thomas Speltz, *DePaul University*
Melissa Strait, *Alma College*
John Straub, *Boston University*
Hal Van Ryswyk, *Harvey Mudd College*
Kirk Voska, *Rogers State University*
Dunwei Wang, *Boston College*
Kim Weaver, *Southern Utah University*
Scott Weinert, *Oklahoma State University*
Carl T. Whalen, *Central New Mexico Community College*
Kenton H. Whitmire, *Rice University*
Burke Scott Williams, *Claremont McKenna*

The contributions of the reviewers of the first, second, third, fourth, fifth, and sixth editions remain embedded in the text, so we also wish to renew our thanks to:

Rebecca Barlag, *Ohio University*
Thomas Berke, *Brookdale Community College*
Amy Bethune, *Albion College*
Lee Don Bienski, *Blinn Community College*
Simon Bott, *University of Houston*
Luke Burke, *Rutgers University—Camden*
Rebecca W. Corbin, *Ashland University*
Charles T. Cox, Jr., *Stanford University*
Irving Epstein, *Brandeis University*
David Esjornson, *Southwest Oklahoma State University*

Theodore Fickel, *Los Angeles Valley College*
David K. Geiger, *State University of New York—Geneseo*
John Gorden, *Auburn University*
Amy C. Gottfried, *University of Michigan*
Myung Woo Han, *Columbus State Community College*
James F. Harrison, *Michigan State University*
Michael D. Heagy, *New Mexico Tech*
Michael Hempstead, *York University*
Byron Howell, *Tyler Junior College*
Gregory Jursich, *University of Illinois at Chicago*

Jeffrey Kovac, *University of Tennessee*

Evguenii Kozliak, *University of North Dakota Main Campus*

Richard Lavallee, *Santa Monica College*

Laurence Lavelle, *University of California, Los Angeles*

Hans-Peter Loock, *Queens University*

Yinfa Ma, *Missouri University of Science and Technology*

Marcin Majda, *University of California, Berkeley*

Diana Mason, *University of North Texas*

Thomas McGrath, *Baylor University*

Shelly Minteer, *University of Utah*

Nixon Mwebi, *Jacksonville State University*

Maria Pacheco, *Buffalo State College*

Hansa Pandya, *Richland College*

Gregory Peters, *Wilkes University*

Britt Price, *Grand Rapids Community College*

Robert Quant, *Illinois State University*

Christian R. Ray, *University of Illinois at Urbana-Champaign*

William Reinhardt, *University of Washington*

Michael P. Rosynek, *Texas A&M*

George Schatz, *Northwestern University*

David Shaw, *Madison Area Technical College*

Conrad Shiba, *Centre College*

Lothar Stahl, *University of North Dakota*

John B. Vincent, *University of Alabama*

Kirk W. Voska, *Rogers State University*

Joshua Wallach, *Old Dominion University*

Meishan Zhao, *University of Chicago*

Thomas Albrecht-Schmidt, *Auburn University*

Matthew Asplund, *Brigham Young University*

Matthew P. Augustine, *University of California, Davis*

Yiyan Bai, *Houston Community College System Central Campus*

David Baker, *Delta College*

Alan L. Balch, *University of California, Davis*

Maria Ballester, *Nova Southeastern University*

Mario Baur, *University of California, Los Angeles*

Robert K. Bohn, *University of Connecticut*

Paul Braterman, *University of North Texas*

William R. Brennan, *University of Pennsylvania*

Ken Brooks, *New Mexico State University*

Julia R. Burdge, *University of Akron*

Paul Charlesworth, *Michigan Technological University*

Patricia D. Christie, *Massachusetts Institute of Technology*

William Cleaver, *University of Vermont*

Henderson J. Cleaves, II, *University of California, San Diego*

David Dalton, *Temple University*

J. M. D'Auria, *Simon Fraser University*

James E. Davis, *Harvard University*

Walter K. Dean, *Lawrence Technological University*

Ivan J. Dmochowski, *University of Pennsylvania*

Jimmie Doll, *Brown University*

Ronald Drucker, *City College of San Francisco*

Jetty Duffy-Matzner, *State University of New York, Cortland*

Christian Ekberg, *Chalmers University of Technology, Sweden*

Robert Eierman, *University of Wisconsin*

Bryan Enderle, *University of California, Davis*

David Erwin, *Rose-Hulman Institute of Technology*

Kevin L. Evans, *Glenville State College*

Justin Fermann, *University of Massachusetts*

Donald D. Fitts, *University of Pennsylvania*

Lawrence Fong, *City College of San Francisco*

Regina F. Frey, *Washington University*

Dennis Gallo, *Augustana College*

P. Shiv Halasyamani, *University of Houston*

David Harris, *University of California, Santa Barbara*

Sheryl Hemkin, *Kenyon College*

Michael Henchman, *Brandeis University*

Geoffrey Herring, *University of British Columbia*

Jameica Hill, *Wofford College*

Timothy Hughbanks, *Texas A&M University*

Paul Hunter, *Michigan State University*

Keiko Jacobsen, *Tulane University*

Alan Jircitano, *Penn State, Erie*

Robert C. Kerber, *State University of New York, Stony Brook*

Robert Kolodny, *Armstrong Atlantic State University*

Lynn Vogel Koplitz, *Loyola University*

Petra van Koppen, *University of California, Santa Barbara*

Mariusz Kozik, *Canisius College*

Julie Ellefson Kuehn, *William Rainey Harper College*

Cynthia LaBrake, *University of Texas, Austin*

Brian B. Laird, *University of Kansas*

Gert Latzel, *Riemerling, Germany*

Nancy E. Lowmaster, *Allegheny College*

Yinfa Ma, *Missouri University of Science and Technology*

Paul McCord, *University of Texas, Austin*

Alison McCurdy, *Harvey Mudd College*

Charles W. McLaughlin, *University of Nebraska*

Matthew L. Miller, *South Dakota State University*

Clifford B. Murphy, *Boston University*

Maureen Murphy, *Huntingdon College*

Patricia O'Hara, *Amherst College*

Noel Owen, *Brigham Young University*

Donald Parkhurst, *The Walker School*

Enrique Peacock-Lopez, *Williams College*

LeRoy Peterson, Jr., *Francis Marion University*

Montgomery Pettitt, *University of Houston*

Joseph Potenza, *Rutgers University*

Wallace Pringle, *Wesleyan University*

Philip J. Reid, *University of Washington*

Tyler Rencher, *Brigham Young University*

Michael Samide, *Butler University*

Gordy Savela, *Itasca Community College*

Barbara Sawrey, *University of California, San Diego*

George Schatz, *Northwestern University*

Paula Jean Schlax, *Bates College*

Carl Seliskar, *University of Cincinnati*

Robert Sharp, *University of Michigan, Ann Arbor*

Peter Sheridan, *Colgate University*

Jay Shore, *South Dakota State University*

Herb Silber, *San Jose State University*

Lori Slavin, *College of Saint Catherine*

Lee G. Sobotka, *Washington University*

Mike Solow, *City College of San Francisco*

Michael Sommer, *Harvard University*

Nanette A. Stevens, *Wake Forest University*

John E. Straub, *Boston University*

Laura Stultz, *Birmingham-Southern College*

Tim Su, *City College of San Francisco*

Peter Summer, *Lake Sumter Community College*

Sara Sutcliffe, *University of Texas, Austin*

Larry Thompson, *University of Minnesota, Duluth*

Dino Tinti, *University of California, Davis*

Sidney Toby, *Rutgers University*

David Vandenbout, *University of Texas, Austin*

Deborah Walker, *University of Texas, Austin*

Lindell Ward, *Franklin College*

Thomas R. Webb, *Auburn University*

Peter M. Weber, *Brown University*

David D. Weis, *Skidmore College*

Ken Whitmire, *Rice University*

James Whitten, *University of Massachusetts*

Lowell David W. Wright, *Vanderbilt University*

Gang Wu, *Queen's University*

Mamudu Yakubu, *Elizabeth City State University*

Meishan Zhao, *University of Chicago*

Zhiping Zheng, *University of Arizona*

Marc Zimmer, *Connecticut College*

Martin Zysmilich, *Massachusetts Institute of Technology*

Some contributed in substantial ways. Roy Tasker, Purdue University, contributed to the website for this book and designed related animations. Kent Gardner (Thundercloud Consulting) redesigned the living graphs on the website for this book. Michael Cann, University of Scranton, opened our eyes to the world of green chemistry in a way that has greatly enriched this book. We would also like to thank Nathan Barrows, Grand Valley State University, for contributing to the Self-Test answers and for generating the ChemCast problem-solving videos. The supplements authors, especially John Krenos, Laurence Lavelle, Yinfa Ma, and Christina Johnson have offered us a great deal of useful advice. Valerie Keller, University of Chicago, provided careful checking of all the solutions. Many others wrote to us with advice, and reviewers were particularly helpful and influential. We are grateful to them all.

We are also grateful to the staff at W. H. Freeman and Company, who understood our vision and helped to bring it to fruition. Among so many we could mention, our special thanks go to Alicia Brady, chemistry editor, who offered guidance and support; Heidi Bamatter, our development editor, who brought keen insight and conscientious oversight to many aspects of this edition; Liz Geller, senior project editor, who guided the complex process through production; Marjorie Anderson, our copyeditor, who polished our text; Robin Fadool and Richard Fox, our photo and licensing editors; Marsha Cohen and Blake Logan, who provided sparkling designs; Susan Wein, who supervised composition and printing; and Amy Thorne, who directed the development and production of the media supplements. We also thank the Aptara staff for turning our manuscript into a finished product. The authors could not have wished for a better or more committed team.

Welcome to chemistry! You are about to embark on a remarkable journey that will take you to the center of science. Looking in one direction, toward physics, you will see how the principles of chemistry are based on the behavior of atoms and molecules. Looking in another direction, toward biology, you will see how chemists contribute to an understanding of that most awesome property of matter, life. Eventually, you will be able to look at an everyday object, see in your mind's eye its composition in terms of atoms, and understand how that composition determines its properties.

Introduction and Orientation

Chemistry is the science of matter and the changes it can undergo. The world of chemistry therefore embraces everything material around us—the stones you stand on, the food you eat, the flesh you are made of, and the silicon in your computers. There is nothing material beyond the reach of chemistry, be it living or dead, vegetable or mineral, on Earth or in a distant star.

Chemistry and Society

In the earliest days of civilization, when the Stone Age gave way to the Bronze Age and then to the Iron Age, people did not realize that they were doing chemistry when they changed the material they found as stones—they would now be called *minerals*—into metals (**FIG. 1**). The possession of metals gave them a new power over their environment, and treacherous nature became less brutal. Civilization emerged as skills in transforming materials grew: glass, jewels, coins, ceramics, and, inevitably, weapons became more varied and effective. Art, agriculture, and warfare became more sophisticated. None of this would have happened without chemistry.

The development of steel accelerated the profound impact of chemistry on society. Better steel led to the Industrial Revolution, when muscles gave way to steam and giant enterprises could be contemplated. With improved transport and greater output from

FIGURE 1 Copper is easily extracted from its ores and was one of the first metals worked. The Bronze Age followed the discovery that adding some tin to copper made the metal harder and stronger. These four bronze swords date from 1250 to 850 BCE, the Late Bronze Age, and are from a collection in the Naturhistorisches Museum, Vienna, Austria. From bottom to top, they are a short sword, an antenna-type sword, a tongue-shaped sword, and a Liptau-type sword. (*Erich Lessing/Art Resource, NY.*)

FIGURE 2 Cold weather triggers chemical processes that reduce the amount of the green chlorophyll in leaves, allowing the colors of various other pigments to show. *(David Q. Cavagnaro/Photolibrary/Getty Images.)*

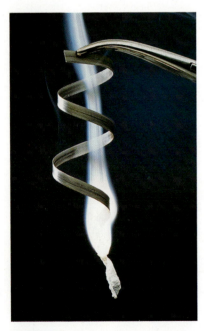

FIGURE 3 When magnesium burns in air, it gives off a lot of heat and light. The gray-white powdery product looks like smoke. *(©1991 Richard Megna–Fundamental Photographs.)*

LAB VIDEO FIGURE 3

factories came more extensive trade, and the world became simultaneously a smaller but busier place. None of this would have happened without chemistry.

With the twentieth century, and now the twenty-first, came enormous progress in the development of the chemical industry. Chemistry transformed agriculture. Synthetic fertilizers provided the means of feeding the enormous, growing population of the world. Chemistry transformed communication and transportation. Today chemistry provides advanced materials, such as polymers for fabrics, ultrapure silicon for computers, and glass for optical fibers. It is producing more efficient renewable fuels and the tough, light alloys that are needed for modern aircraft and space travel. Chemistry has transformed medicine, substantially extended life expectancy, and provided the foundations of genetic engineering. The deep understanding of life that we are developing through molecular biology is currently one of the most vibrant areas of science. None of this progress would have been achieved without chemistry.

However, the price of all these benefits has been high. The rapid growth of industry and agriculture, for instance, has stressed the Earth and damaged our inheritance. There is now widespread concern about the preservation of our extraordinary planet. It will be up to you and your contemporaries to draw on chemistry—in whatever career you choose—to build on what has already been achieved. Perhaps you will help to start a new phase of civilization based on new materials, just as semiconductors transformed society in the twentieth century. Perhaps you will help to reduce the harshness of the impact of progress on our environment. To do that, you will need chemistry.

Chemistry: A Science at Three Levels

Chemistry can be understood at three levels. At one level, chemistry is about matter and its transformations. This is the level at which you can see the changes, as when a leaf changes color in the fall (**FIG. 2**) or magnesium burns brightly in air (**FIG. 3**). This level is the **macroscopic level,** the level dealing with the properties of large, visible objects. However, there is an underworld of change, a world that you cannot see directly. At this deeper, **microscopic level,** chemistry interprets these phenomena in terms of the rearrangements of atoms (**FIG. 4**). The third level is the **symbolic level,** the expression of chemical phenomena in terms of chemical symbols and mathematical equations. A chemist thinks at the microscopic level, conducts experiments at the macroscopic level, and represents both symbolically. These three aspects of chemistry can be mapped as a triangle (**FIG. 5**). As you read further in this text, you will find that sometimes the topics and explanations are close to one vertex of the triangle, sometimes to another. Because it is helpful in understanding chemistry to make connections among these levels, in the worked examples in this book you will find drawings of the molecular level as well as graphical interpretations of equations. As your understanding of chemistry grows, so will your ability to travel easily within the triangle as you connect, for example, a laboratory observation to the symbols on a page and to mental images of atoms and molecules.

How Science Is Done

Scientists pursue ideas in an ill-defined but effective way called the **scientific method.** There is no strict rule of procedure that will lead you from a good idea to a Nobel Prize or even to a publishable discovery. Some scientists are meticulously careful; others are highly creative. The best scientists are probably both careful and creative. Although there are various scientific methods in use, a typical approach consists of a series of steps (**FIG. 6**). The first step is often to collect **data,** the record of observations and measurements. These measurements are usually made on small **samples** of matter, representative pieces of the material being studied.

Scientists are always on the lookout for patterns. When a pattern is observed in the data, it can be stated as a scientific **law,** a succinct summary of a wide range of observations. For example, water was found to have eight times the mass of oxygen as it has of hydrogen, regardless of the source of the water or the size of the sample. One of the earliest laws of chemistry summarized those types of observations as the **law of constant composition,** which states that a compound has the same composition regardless of the source of the sample.

Formulating a law is just one way, not the only way, of summarizing data. There are many properties of matter (such as superconductivity, the ability of a few cold solids to conduct electricity without any resistance) that are currently at the forefront of research but are not described by grand "laws" that embrace hundreds of different compounds. A major current puzzle, which might be resolved in the future either by finding the appropriate law or by detailed individual computation, is what determines the shapes of protein molecules such as those that govern almost every aspect of life, including serious diseases such as Alzheimer's, Parkinson's, and cancer.

Once they have detected patterns, scientists may develop **hypotheses,** possible explanations of the laws—or the observations—in terms of more fundamental concepts. Observation requires careful attention to detail, but the development of a hypothesis requires insight, imagination, and creativity. In 1807, John Dalton interpreted experimental results to propose his **atomic hypothesis,** that matter consists of atoms. Although Dalton could not see individual atoms, he was able to imagine them and formulate his hypothesis. Dalton's hypothesis was a monumental insight that helped others to understand the world in a new way. The process of scientific discovery never stops. With luck and application, you may acquire that kind of insight as you read through this text, and one day you may make your own extraordinary and significant hypotheses.

After formulating a hypothesis, scientists design further **experiments**—carefully controlled tests—to verify it. Designing and conducting good experiments often requires ingenuity and sometimes good luck. If the results of repeated experiments—often in other laboratories and sometimes by skeptical coworkers—support the hypothesis, scientists may go on to formulate a **theory,** a formal explanation of a law. Quite often the theory is expressed mathematically. A theory originally envisioned as a **qualitative** concept—a concept expressed in words or pictures—is converted into a

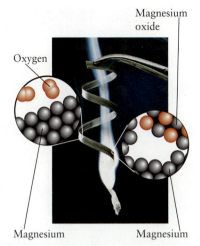

FIGURE 4 When a chemical reaction takes place, atoms exchange partners, as in Fig. 3, where magnesium and oxygen atoms form magnesium oxide. As a result, two forms of matter (left inset) are changed into another form of matter (right inset). Atoms are neither created nor destroyed in chemical reactions. *(Photo: ©1991 Richard Megna–Fundamental Photographs.)*

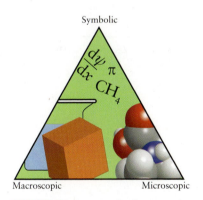

FIGURE 5 This triangle illustrates the three modes of scientific inquiry used in chemistry: macroscopic, microscopic, and symbolic. Sometimes chemists work more at one corner than at the others, but it is important to be able to move from one approach to another inside the triangle.

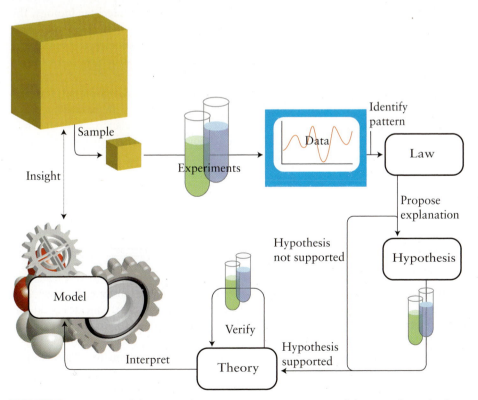

FIGURE 6 A summary of the principal activities in a common version of the scientific method. The ideas proposed must be tested and possibly revised at each stage.

FIGURE 7 Scientific research today often requires sophisticated equipment and computers. These scientists are using a using a portable gamma spectrometer to measure gamma radiation levels near Quezon City in the Philippines. *(Bullit Marquez/AP Photo.)*

quantitative form—the same concept expressed in terms of mathematics. After a concept has been expressed quantitatively, it can be used to make numerical predictions and is subjected to rigorous experimental confirmation. You will have plenty of practice with the quantitative aspects of chemistry while working through this text.

Scientists commonly interpret a theory in terms of a **model,** a simplified version of the object of study that they can use to make predictions. Like hypotheses, theories and models must be subjected to experiment and revised if experimental results do not support them. For example, the current model of the atom has gone through many formulations and progressive revisions, starting from Dalton's vision of an atom as an uncuttable solid sphere to the current, much more detailed model, which is described in Focus 1. One of the goals of this text is to show you how chemists build models, turn them into a testable form, and then refine them in the light of additional evidence.

The Branches of Chemistry

Chemistry is more than test tubes and beakers. New technologies have transformed chemistry dramatically in the past 50 years, and new areas of research have emerged (**FIG. 7**). Traditionally, the field of chemistry has been organized into three main branches: **organic chemistry,** the study of compounds of carbon; **inorganic chemistry,** the study of all the other elements and their compounds; and **physical chemistry,** the study of the principles of chemistry.

New areas of study have developed as information has been acquired in specialized areas or as a result of the use of particular techniques. They include biochemistry, analytical chemistry, theoretical chemistry, computational chemistry, chemical engineering, medicinal chemistry, and biological chemistry. Various interdisciplinary branches of knowledge with roots in chemistry have also arisen, including **molecular biology,** the study of the chemical and physical basis of biological function and diversity; **materials science,** the study of the chemical structure and composition of materials; and **nanotechnology,** the study of matter on the scale of nanometers, at which structures consisting of a small number of atoms can be manipulated.

 A newly emerging concern of chemistry is **sustainable development,** the economical utilization and renewal of resources coupled with hazardous waste reduction and concern for the environment. This sensitive approach to the environment and our planetary inheritance is known colloquially as **green chemistry.** When it is appropriate to draw your attention to this important development, we display the small icon shown here.

All sciences, medicine, and many fields of commercial activity draw on chemistry. You can be confident that whatever career you choose in a scientific or technical field, it will make use of the concepts discussed in this text. Chemistry is truly central to science.

Mastering Chemistry

You might already have a strong background in chemistry. These introductory pages with colored edges will provide you with a summary of a number of basic concepts and techniques. Your instructor will advise you how to use these sections to prepare yourself for the Topics in the text itself.

If you have little experience of chemistry, these pages are for you, too. They contain a brief but systematic summary of the basic concepts and calculations of chemistry that you should know before studying the Topics in the text. You can return to them as needed. If you need to review the mathematics required for chemistry, especially algebra and logarithms, Appendix 1 has a brief review of the important procedures.

A Matter and Energy

Whenever you touch, pour, or weigh something, you are working with matter. Chemistry is concerned with the properties of matter and particularly the conversion of one form of matter into another kind. But what is matter? Matter is in fact difficult to define precisely without drawing on advanced ideas from elementary particle physics, but a straightforward working definition is that **matter** is anything that has mass and takes up space. Thus, gold, water, and flesh are forms of matter; electromagnetic radiation (which includes light) and justice are not.

One characteristic of science is that it uses common words from everyday language but gives them a precise meaning. In everyday language, a "substance" is just another name for matter. However, in chemistry, a **substance** is a *single, pure form of matter*. Thus, gold and water are distinct substances. Flesh is a mixture of many different substances, and, in the technical sense used in chemistry, it is not a "substance." Air is matter, but, because it is a mixture of several gases, it is not a substance in the technical sense.

Substances, and matter in general, can take different forms, called **states of matter.** The three most common states of matter are solid, liquid, and gas.

A **solid** is a form of matter that retains its shape and does not flow.

A **liquid** is a fluid form of matter that has a well-defined surface; it takes the shape of the part of the container it occupies.

A **gas** is a fluid form of matter that fills any vessel containing it.

The term **vapor** denotes the gaseous form of a substance that is normally a solid or liquid. For example, water exists as solid (ice), liquid, and vapor (steam).

FIGURE A.1 shows the different arrangements and mobilities of atoms and molecules in these three states of matter. In a solid, such as copper metal, the atoms are packed together closely; the solid is rigid because the atoms cannot move past one another. However, the atoms in a solid are not motionless: they oscillate around their average locations, and the oscillation becomes more vigorous as the temperature is raised. The atoms (and molecules) of a liquid are packed together about as closely as they are in a solid, but they have enough energy to move past one another readily. As a result, a liquid, such as water or molten copper, flows in response to a force, such as gravity. In a gas, such as air (which is mostly nitrogen and oxygen) and water vapor, the molecules have achieved almost complete freedom from one another: they fly through empty space at close to the speed of sound, colliding when they meet and immediately flying off in another direction.

A.1 Symbols and Units

Chemistry is concerned with the **properties** of matter, its distinguishing characteristics. A **physical property** of a substance is a characteristic that can be observed or measured without changing the identity of the substance. For example, two physical properties of a sample of water are its mass and its temperature. Physical properties include characteristics such as melting point (the temperature at which a solid turns into a liquid), hardness, color, state of matter (solid, liquid, or gas), and density. When a substance undergoes a **physical change,** the identity of the substance does not change; only its physical properties are different. For example, when water freezes, the solid ice is still water. A **chemical property** refers to the ability of a substance to be changed into another substance. For example, a chemical property of the gas hydrogen is that it reacts with (burns in) oxygen to produce water; a chemical property of the metal zinc is that it reacts with acids to produce hydrogen gas. When a substance undergoes a **chemical change,** it is transformed into a different substance, such as hydrogen changing to water.

A measurable physical property is represented by an italic or sloping letter (thus, *m* for mass, not m). The result of the measurement, the "value" of a physical property, is reported as a multiple of a **unit,** such as reporting a mass as 15 kilograms, which is understood to be 15 times the unit "1 kilogram." Scientists have reached international agreement on the units to use when reporting measurements, so their results can be used with

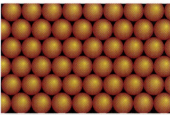

(a)

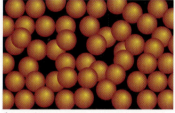

(b)

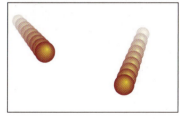

(c)

FIGURE A.1 Molecular representations of the three states of matter. In each case, the spheres represent particles that may be atoms, molecules, or ions. (a) In a solid, the particles are packed tightly together and held in place, but they continue to oscillate. (b) In a liquid, the particles are in contact, but they have enough energy to move past one another. (c) In a gas, the particles are far apart, move almost completely freely, and are in ceaseless random motion.

confidence and checked by people anywhere in the world. You will find most of the symbols used in this textbook together with their units in Appendix 1.

A Note on Good Practice: All units are denoted by Roman letters, such as m for meter and s for second, which distinguishes them from the physical quantity to which they refer (such as *l* for length and *t* for time).

The **Système International** (SI) is the internationally accepted form and elaboration of the metric system. It defines seven **base units** in terms of which all measureable physical properties can be expressed. At this stage all you need are

1 meter, 1 m	1 **meter,** the unit of length
1 kilogram, 1 kg	1 **kilogram,** the unit of mass
1 second, 1 s	1 **second,** the unit of time

All the units are defined in Appendix 1B. Each unit may be modified by a prefix that represents a multiple of 10 (and typically 10^3 or $1/10^3$). The full set is given in Appendix 1B; some common examples are

Prefix	Symbol	Factor	Example
kilo-	k	10^3 (1000)	$1\ km = 10^3\ m$ (1 kilometer)
centi-	c	10^{-2} (1/100, 0.01)	$1\ cm = 10^{-2}\ m$ (1 centimeter)
milli-	m	10^{-3} (1/1000, 0.001)	$1\ ms = 10^{-3}\ s$ (1 millisecond)
micro-	μ	10^{-6} (1/1 000 000, 0.000 001)	$1\ \mu g = 10^{-6}\ g$ (1 microgram)
nano-	n	10^{-9} (1/1 000 000 000, 0.000 000 001)	$1\ nm = 10^{-9}\ m$ (1 nanometer)

Units may be combined into **derived units** to express a property that is more complicated than mass, length, or time. For example, **volume,** *V*, the amount of space occupied by a substance, is the product of three lengths; therefore, the derived unit of volume is (meter)^3, denoted m^3. Similarly, **density,** the mass of a sample divided by its volume, is expressed in terms of the base unit for mass divided by the derived unit for volume—namely, kilogram/(meter)^3, denoted kg/m^3 or, equivalently, kg·m^{-3}.

A Note on Good Practice: The SI convention is that a power, such as the 3 in cm^3, refers to the base unit and its prefix. That is, cm^3 should be interpreted as (cm)^3 or 10^{-6} m^3, *not* as c(m^3) or 10^{-2} m^3.

It is often necessary to convert measurements from another set of units into SI units. For example, when converting a length measured in inches (in.) into centimeters (cm), it is necessary to use the relation 1 in. = 2.54 cm. Relations between common units can be found in Table 5 of Appendix 1B. They are used to construct a **conversion factor** of the form

$$\text{Conversion factor} = \frac{\text{units required}}{\text{units given}}$$

which is then used as follows:

$$\text{Information required} = \text{information given} \times \text{conversion factor}$$

When using a conversion factor, treat the units just like algebraic quantities: they can be multiplied or canceled in the normal way.

EXAMPLE A.1   Converting units

Suppose you are in a store—perhaps in Canada or Europe—where paint is sold in liters. You know you need 1.7 qt of a particular paint. What is that volume in liters?

ANTICIPATE A glance at Table 5 in Appendix 1B shows that 1 L is slightly more than 1 qt, so you should expect a volume of slightly less than 1.7 L.

PLAN Identify the relation between the two units from Table 5 of Appendix 1B:

$$1 \text{ qt} = 0.946\,3525 \text{ L}$$

Then set up the conversion factor from the units given (qt) to the units required (L).

SOLVE

Form the conversion factor as (units required)/(units given).

$$\text{Conversion factor} = \frac{0.946\,3525 \text{ L}}{1 \text{ qt}}$$

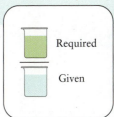

Convert the measurement into the required units.

$$\text{Volume (L)} = (1.7 \text{ qt}) \times \frac{0.946\,3525 \text{ L}}{1 \text{ qt}} = 1.6 \text{ L}$$

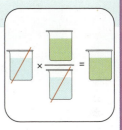

EVALUATE As expected, you need slightly less than 1.7 L. The answer has been rounded to two digits, as explained in Appendix 1.

Self-test A.1A Express the height of a person 6.00 ft tall in centimeters.

[**Answer:** 183 cm]

Self-test A.1B Express the mass in ounces of a 250.-g package of breakfast cereal.

Related Exercises A.13, A.14, A.31, A.32

It is often necessary to convert a unit that has been raised to a power (including negative powers). In such cases, the conversion factor is raised to the same power. For example, to convert a density, d, of $11\,700 \text{ kg·m}^{-3}$ into grams per centimeter cubed (g·cm^{-3}), use the two relations

$$1 \text{ kg} = 10^3 \text{ g and } 1 \text{ cm} = 10^{-2} \text{ m}$$

as follows:

$$d = (11\,700 \text{ kg·m}^{-3}) \times \frac{10^3 \text{ g}}{1 \text{ kg}} \times \left(\frac{1 \text{ cm}}{10^{-2} \text{ m}}\right)^{-3}$$

$$= (11\,700 \text{ kg·m}^{-3}) \times \frac{10^3 \text{ g}}{1 \text{ kg}} \times \frac{10^{-6} \text{ m}^3}{1 \text{ cm}^3}$$

$$= 11.7 \frac{\text{g}}{\text{cm}^3} = 11.7 \text{ g·cm}^{-3}$$

Self-test A.2A Express a density of 6.5 g·mm^{-3} in micrograms per nanometer cubed ($\mu\text{g·nm}^{-3}$).

[**Answer:** $6.5 \times 10^{-12} \text{ } \mu\text{g·nm}^{-3}$]

Self-test A.2B Express an acceleration of 9.81 m·s^{-2} in kilometers per hour squared.

As remarked above, units are treated like algebraic quantities and are multiplied and canceled just like numbers. One consequence is that a quantity like $m = 5 \text{ kg}$ could also

Answers to all B self-tests are in the back of this book.

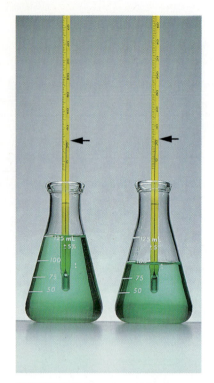

FIGURE A.2 Mass is an extensive property, but temperature is intensive. These two samples of iron(II) sulfate solution were taken from the same well-mixed supply; they have different masses but the same temperature. *(W.H. Freeman photo by Ken Karp.)*

Units for physical properties and temperature scales are discussed in Appendix 1B.

be reported as $m/kg = 5$ by dividing both sides by kg. Likewise, the answer in the density conversion could have been reported as $d/(g \cdot cm^{-3}) = 11.7$.

Properties can be classified according to their dependence on the size of a sample:

An **extensive property** is a property that depends on the size ("extent") of the sample.

An **intensive property** is independent of the size of the sample.

More precisely, if a system is divided into parts and it is found that the property of the complete system has a value that is the sum of the values of the property of all the parts, then that property is extensive. If that is not the case, then the property is intensive. Volume is an extensive property: 2 kg of water occupies twice the volume of 1 kg of water. Temperature is an intensive property, because whatever the size of the sample taken from a uniform bath of water, it has the same temperature (**FIG. A.2**). The importance of the distinction is that different substances can be identified by their intensive properties. Thus, a sample can be recognized as water by noting its color, density ($1.00 \text{ g} \cdot cm^{-3}$), melting point (0 °C), boiling point (100 °C), and the fact that it is a liquid.

Some intensive properties are ratios of two extensive properties. For example, density is a ratio of the mass, m, of a sample divided by its volume, V:

$$\text{Density} = \frac{\text{mass}}{\text{volume}} \quad \text{or} \quad d = \frac{m}{V} \tag{1}$$

The density of a substance is independent of the size of the sample because doubling the volume also doubles the mass, so the ratio of mass to volume remains the same. Density is therefore an intensive property and can be used to identify a substance. Most properties of a substance depend on its state of matter and conditions, such as the temperature and pressure. For example, the density of water at 0 °C is $1.000 \text{ g} \cdot cm^{-3}$, but at 100 °C it is $0.958 \text{ g} \cdot cm^{-3}$. The density of ice at 0 °C is $0.917 \text{ g} \cdot cm^{-3}$, but the density of water vapor at 100 °C and atmospheric pressure is nearly 2000 times less, at $0.597 \text{ g} \cdot L^{-1}$.

THINKING POINT

When you heat a gas at constant pressure, it expands. Does the density of a gas increase, decrease, or stay the same as it expands?

Self-test A.3A The density of selenium is $4.79 \text{ g} \cdot cm^{-3}$. What is the mass of 6.5 cm^3 of selenium?

[*Answer:* 31 g]

Self-test A.3B The density of helium gas at 0 °C and 1.00 atm is $0.176 \, 85 \text{ g} \cdot L^{-1}$. What is the volume of a balloon containing 10.0 g of helium under the same conditions?

Chemical properties involve changing the identity of a substance; physical properties do not. Extensive properties depend on the size of the sample; intensive properties do not.

A.2 Accuracy and Precision

All measured quantities have some uncertainty associated with them; in science it is important to convey the degree to which you are confident about not only the values you report but also the results of calculations using those values. Notice that in Example A.1 the result of multiplying 1.7 by 0.946 3525 is written as 1.6, not 1.608 799 25. The number of digits reported in the result of a calculation must reflect the number of digits known from the data, not the entire set of digits the calculator might provide.

The number of **significant figures** in a numerical value is the number of digits that can be justified by the data:

When reporting the results of multiplication and division, identify the number of digits in the least precise value and retain that number of digits in the answer.

Thus, the measurement 1.7 qt has two significant figures (2 sf) and 0.946 3525 has seven (7 sf), so in Example A.1 the result is limited to 2 sf.

When reporting the results of addition or subtraction, identify the quantity with the least number of digits following the decimal point and retain that number of digits in the answer.

For instance, two very precise measurements of length might give 55.845 mm and 15.99 mm, and the total length would be reported as

$$55.845 \text{ mm} + 15.99 \text{ mm} = 71.83 \text{ mm}$$

with the precision of the answer governed by the number of digits in the data (shown here in red). The full set of rules for counting the number of significant figures and determining the number of significant figures in the result of a calculation is given in Appendix 1C, together with the rules for rounding numerical values.

An ambiguity may arise when dealing with a whole number ending in a zero, because the number of significant figures in the number may be less than the number of digits. For example, does 400 mean 4×10^2 (1 sf), 4.0×10^2 (2 sf), or 4.00×10^2 (3 sf)? To avoid ambiguity, in this book, when all the digits in a number ending in zero are significant, the number is followed by a decimal point. Thus, the number 400. has 3 sf. In the "real world," this helpful convention only rarely is adopted.

To make sure of their data, scientists usually repeat their measurements several times, report the average value, and assess the precision and accuracy of their measurements:

The **precision** of a measurement is an indication of how close repeated measurements are to one another.

The **accuracy** of a series of measurements is the closeness of their average value to the true value.

The illustration in **FIG. A.3** distinguishes precision from accuracy. As the illustration suggests, even precise measurements can give inaccurate values.

More often than not, measurements are accompanied by two kinds of error. A **systematic error** is an error that is present in every one of a series of repeated measurements. Systematic errors in a series of measurements always have the same sign and magnitude. For instance, a laboratory balance might not be calibrated correctly and all recorded masses will be reported as either too high or too low. If you are using that balance to measure the mass of a sample of silver, then even though you might be justified in reporting your measurements to a precision of five significant figures (such as 5.0450 g), the reported mass of the sample will be inaccurate. In principle, systematic errors can be discovered and corrected, but they often go unnoticed and in practice may be hard to identify. A **random error** is an error that varies in both sign and magnitude and can average to zero over a series of observations. An example is the effect of drafts of air from an open window moving a balance pan either up or down a little, decreasing or increasing the mass measurements randomly. Scientists attempt to minimize random error by making many observations and taking the average of the results.

THINKING POINT

What are some means that scientists can use to identify and eliminate systematic errors?

The precision of a measurement is an indication of how close together repeated measurements are; the accuracy of a measurement is its closeness to the true value.

A.3 Force

Speed, v, is the rate of change of a body's position and is reported (in SI units) in meters per second ($\text{m} \cdot \text{s}^{-1}$). **Velocity** is closely related to speed but takes into account the direction of motion as well as its rate. Thus, a particle moving in a circle at a constant speed has a constantly changing velocity. **Acceleration,** a, is the rate of change of velocity: a particle moving in a straight line at a constant speed is not accelerating (its speed and direction of travel is unchanging), but a particle moving at a constant speed in a curved path accelerates because although its speed is constant its velocity is changing (**FIG. A.4**). In SI units, acceleration is reported in meters per second squared ($\text{m} \cdot \text{s}^{-2}$).

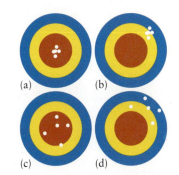

FIGURE A.3 The holes in these targets represent measurements that are (a) precise and accurate, (b) precise but inaccurate, (c) imprecise but accurate on average, and (d) both imprecise and inaccurate.

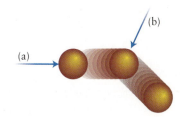

FIGURE A.4 (a) When a force acts along the direction of travel, the speed (the magnitude of the velocity) changes, but the direction of motion does not. (b) The direction of travel can be changed without affecting the speed if the force is applied in an appropriate direction. Both changes in velocity correspond to acceleration.

A force, F, is an influence that changes the state of motion of an object. For instance, you exert a force to open a door—to start the door swinging open—and you exert a force on a ball when you hit it with a bat. According to Newton's second law of motion, when an object experiences a force, it is accelerated in proportion to the force that it experiences:

$$\text{Acceleration} \propto \text{force} \quad \text{or} \quad a \propto F$$

and specifically

$$a = \frac{F}{m}$$

where m is the mass of the body. Thus, for a given force, a heavy body undergoes less acceleration than a light body. This relation, which is called **Newton's second law of motion,** is normally written

$$\text{Force} = \text{mass} \times \text{acceleration} \quad \text{or} \quad F = ma \qquad (2)$$

It follows from this relation that the derived SI unit of force, with mass in kilograms and acceleration in meters per second squared, is 1 kg·m·s^{-2}. This derived unit occurs so often that it is given a special name, 1 newton, and symbol, 1 N. That is, $1 \text{ N} = 1 \text{ kg·m·s}^{-2}$. A force of 1 N is approximately the same as the gravitational pull on a small apple (100 g) hanging from a tree.

Acceleration, the rate of change of velocity, is proportional to applied force.

A.4 Energy

Work is the process of moving an object against an opposing force, and its magnitude is the product of the strength of the opposing force and the distance through which the object is moved:

$$\text{Work done} = \text{force} \times \text{distance}$$

With force in newtons and distance in meters, the SI unit of work is 1 N·m, or $1 \text{ kg·m}^2\text{·s}^{-2}$. This combination of base units is even more common in chemistry than the newton itself and is called 1 joule, with the symbol 1 J. That is, $1 \text{ J} = 1 \text{ kg·m}^2\text{·s}^{-2}$. Each beat of the human heart does about 1 J of work.

Energy is the capacity to do work. Thus, energy is needed to do the work of raising a weight a given height or the work of forcing an electric current through a circuit. The greater the energy of an object, the greater is its capacity to do work. To raise this book (of mass close to 2.0 kg) from the floor to a tabletop about 0.97 m above the floor requires about 19 J (**FIG. A.5**). Because energy changes in chemical reactions tend to be of the order of thousands of joules for the amounts usually studied, it is more common in chemistry to use the kilojoule (kJ, where $1 \text{ kJ} = 10^3 \text{ J}$).

A Note on Good Practice: Names of units derived from the names of people are always lowercase (as for joule), but their abbreviations are always uppercase (as in J for joule).

Three contributions to energy are important in chemistry: kinetic energy, potential energy, and electromagnetic energy. **Kinetic energy,** E_k, is the energy that a body possesses due to its motion. For a body of mass m traveling at a speed v, the kinetic energy is

$$E_k = \tfrac{1}{2}mv^2 \qquad (3)$$

A heavy body traveling rapidly has a high kinetic energy. A body at rest (stationary, $v = 0$) has zero kinetic energy.

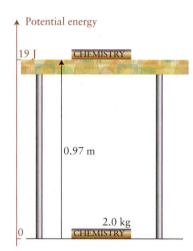

FIGURE A.5 The energy required to raise the book that you are now reading from the floor to the tabletop is approximately 19 J. The same energy would be released if the book fell from the tabletop to the floor.

Potential energy

19 J   CHEMISTRY

0.97 m

2.0 kg

0   CHEMISTRY

EXAMPLE A.2   Calculating kinetic energy

Athletes can expend a lot of energy in a race, not only while moving but also in the process of starting to move. Suppose you are working as a sports physiologist and are studying the energetics of cycling. You would need to know the energy involved in each phase of a race. How much energy does it take to accelerate a person and a bicycle of total mass 75 kg to 20. mph (8.9 m·s^{-1}), starting from rest and ignoring friction and wind resistance?

PLAN A stationary cyclist has zero kinetic energy. You need to decide how much energy must be supplied to reach the kinetic energy of the cyclist traveling at the final speed.

SOLVE

From $E_k = \frac{1}{2}mv^2$,

$$E_k = \frac{1}{2} \times (75 \text{ kg}) \times (8.9 \text{ m·s}^{-1})^2$$
$$= 3.0 \times 10^3 \underbrace{\text{kg·m}^2\text{·s}^{-2}}_{\text{J}} = 3.0 \text{ kJ}$$

EVALUATE A minimum of 3.0 kJ is needed. More energy is needed to achieve that speed when friction and wind resistance are taken into account.

Self-test A.4A Calculate the kinetic energy of a ball of mass 0.050 kg traveling at 25 m·s^{-1}.

[**Answer:** 16 J]

Self-test A.4B Calculate the kinetic energy of a 1.5-kg book just before it lands on your foot after falling off a table, when it is traveling at 3.0 m·s^{-1}.

Related Exercises A.35, A.36

The **potential energy,** E_p, of an object is the energy that it possesses on account of its position in a field of force. There is no single formula for the potential energy of an object, because the potential energy depends on the nature of the force that it experiences. However, two simple cases are important in chemistry: gravitational potential energy (for a particle in a gravitational field) and Coulomb potential energy (for a charged particle in an electrostatic field).

A body of mass m at a height h above the surface of the Earth has a gravitational potential energy

$$E_p = mgh \tag{4}$$

relative to its potential energy on the surface itself (**FIG. A.6**), where g is the **acceleration of free fall** (and, commonly, the "acceleration of gravity"). The value of g depends on location, but in most inhabited locations on the surface of the Earth g has close to its "standard value" of 9.81 m·s^{-2}, which is used in all calculations in this book. Equation 4 shows that the greater the height of an object above the surface of the Earth, the greater is its gravitational potential energy. For instance, a book on a table has a greater potential energy on the table than on the floor.

A Note on Good Practice: You will sometimes see kinetic energy denoted KE and potential energy denoted PE. Modern practice is to denote all physical quantities by a single (italic) letter (accompanied, if necessary, by subscripts).

Potential energy is also commonly denoted V. A *field* is a region where a force acts.

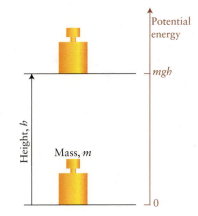

FIGURE A.6 The potential energy of a mass m in a gravitational field is proportional to its height h above a point (here, the surface of the Earth), which is taken to correspond to zero potential energy.

EXAMPLE A.3   Calculating the gravitational potential energy

A skier of mass 65 kg boards a ski lift at a resort in eastern British Columbia and is lifted 1164 m above the starting point. What is the change in potential energy of the skier?

ANTICIPATE When a mass of 1 kg is raised by 1 m on the surface of the Earth, it gains nearly 10 J of potential energy. In this example, 65 kg is raised over 1000 m, so you should expect the gain in potential energy to be greater than 650 kJ.

PLAN To calculate the change, suppose that the potential energy of the skier at the bottom of the lift is zero, then calculate the potential energy at the height of the top of the lift.

SOLVE The potential energy of a skier at the top of the lift relative to the bottom of the lift is

From $E_p = mgh$,

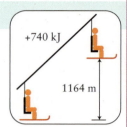

$$E_p = (65 \text{ kg}) \times (9.81 \text{ m·s}^{-2}) \times (1164 \text{ m})$$
$$= 7.4 \times 10^5 \text{ kg·m}^2\text{·s}^{-2} = +740 \text{ kJ}$$

EVALUATE As expected, the potential energy difference is greater than 650 kJ.

Self-test A.5A What is the gravitational potential energy of this book (mass 2.0 kg) when it is on a table of height 0.82 m, relative to its potential energy when it is on the floor?

[*Answer:* 16 J]

Self-test A.5B How much energy has to be expended to raise a can of soda (mass 0.350 kg) to the top of the Willis Tower in Chicago (height 443 m)?

Related Exercises A.37–A.39

The energy due to attractions and repulsions between electric charges is of great importance in chemistry, which deals with electrons, atomic nuclei, and ions, all of which are charged. The **Coulomb potential energy** of a particle of charge Q_1 at a distance r from another particle of charge Q_2 is proportional to the two charges and inversely proportional to the distance between them:

$$E_p = \frac{Q_1 Q_2}{4\pi\varepsilon_0 r} \tag{5}$$

In this expression, which applies when the two charges are separated by a vacuum, ε_0 (epsilon zero) is a fundamental constant called the *vacuum permittivity;* its value is $8.854 \times 10^{-12} \text{ J}^{-1}\text{·C}^2\text{·m}^{-1}$. The Coulomb potential energy is obtained in joules when the charges are in coulombs (C, the SI unit of charge) and their separation is in meters (m). The charge on an electron is $-e$, with $e = 1.602 \times 10^{-19}$ C, the "fundamental charge."

What Does This Equation Tell You? The Coulomb potential energy approaches zero as the distance between two particles approaches infinity. If the particles have the same charge—if they are two electrons, for instance—then the numerator, $Q_1 Q_2$, and therefore E_p itself, is positive, and the potential energy *rises* (becomes more strongly positive) as the particles approach each other (r decreases). If the particles have opposite charges—an electron (which has a negative charge) and an atomic nucleus (which has a positive charge), for instance—then the numerator, and therefore E_p, is negative and the potential energy *decreases* (in this case, becomes more negative) as the particles approach each other (**FIG. A.7**).

The "electromagnetic energy" mentioned at the beginning of this section is the energy of the **electromagnetic field,** such as the energy carried through space by radio waves, light waves, and x-rays. An electromagnetic field is generated by the acceleration of charged particles and consists of an oscillating **electric field** and an oscillating **magnetic field** (**FIG. A.8**). The crucial distinction is that an electric field affects charged particles whether they are stationary or moving, whereas a magnetic field affects only moving charged particles. The electromagnetic field is discussed in more detail in Topic 1A.

The **total energy,** E, of a particle is the sum of its kinetic and potential energies:

$$\text{Total energy} = \text{kinetic energy} + \text{potential energy} \quad \text{or} \quad E = E_k + E_p \tag{6}$$

A very important feature of the total energy of an object is that, provided there are no outside influences, it is constant. This observation is summarized by saying that *energy is conserved.* Kinetic energy and potential energy can change into each other, but their

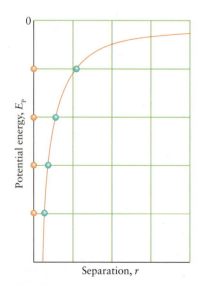

FIGURE A.7 The variation of the Coulomb potential energy of two opposite charges (one represented by the red sphere, the other by the green sphere) with their separation. Notice that the potential energy decreases as the charges approach each other.

FIGURE A.8 An electromagnetic field oscillates in time and space. The magnetic field (shown in blue) is perpendicular to the electric field (shown in red). The length of an arrow at any point represents the strength of the field at that point, and its orientation denotes its direction. Both fields are perpendicular to the direction of travel of the radiation.

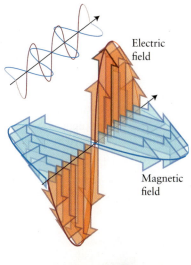

Electric field

Magnetic field

sum for a given object, whether as large as a planet or as tiny as an atom, is constant. For instance, a ball thrown up into the air initially has high kinetic energy and zero potential energy. At the top of its flight, it has zero kinetic energy and high potential energy. However, as it returns to Earth, its kinetic energy rises and its potential energy approaches zero again. At each stage, its *total* energy is the same as it was when it was launched (**FIG. A.9**). When it strikes the Earth, the ball is no longer isolated, and its energy is dissipated as **thermal motion,** the chaotic, random motion of atoms and molecules. If all the kinetic and potential energies of these atoms and molecules were added together, you would find that the total energy of the Earth had increased by exactly the same amount as that lost by the ball. No one has ever observed any exception to the **law of conservation of energy,** the observation that energy can be neither created nor destroyed. One region of the universe—an individual atom, for instance—can lose energy, but another region must gain that energy.

Chemists often refer to two other kinds of energy. The term **chemical energy** is used to refer to the change in energy when a chemical reaction takes place, as in the combustion of a fuel. "Chemical energy" is not a special form of energy: it is simply a shorthand name for the sum of the potential and kinetic energies of the substances participating in the reaction, including the potential and kinetic energies of their electrons. The term **thermal energy** is another shorthand term. In this case it is shorthand for the sum of the potential and kinetic energies arising from the thermal motion of atoms, ions, and molecules.

> *Work is motion against an opposing force. Energy is the capacity to do work. Kinetic energy results from motion, potential energy from position. An electromagnetic field carries energy through space.*

What have you learned in Fundamentals A?

You have learned how to use and report measurements, how to manipulate units, and how to report results of calculations with the correct number of significant figures. You have also been introduced to force and energy and have learned how to distinguish kinetic and potential energy.

The skills you have mastered are the ability to:

- ☐ **1.** Identify properties as physical or chemical and extensive or intensive.
- ☐ **2.** Convert between units (Example A.1).
- ☐ **3.** Calculate the kinetic energy of an object (Example A.2).
- ☐ **4.** Calculate the gravitational potential energy of an object (Example A.3).
- ☐ **5.** Write the expression for the Coulomb potential energy between electric charges.

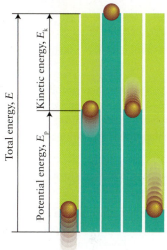

FIGURE A.9 Kinetic energy (represented by the height of the light green bar) and potential energy (the dark green bar) are interconvertible, but their sum (the total height of the bar) is a constant in the absence of external influences, such as air resistance. A ball thrown up from the ground loses kinetic energy as it slows, but it gains potential energy. The reverse happens as it falls back to Earth.

Fundamentals A Exercises

A.1 Laws and hypotheses are distinguished in the *Introduction and Orientation*. Which of the following statements are best described as laws? (a) The volume of a gas increases as it is heated at constant pressure. (b) Sodium reacts with chlorine to produce sodium chloride. (c) The universe is infinite in extent. (d) All animal life on Earth requires water to survive. (e) The burning of coal is a cause of global warming.

A.2 Which of the statements in Exercise A.1 are best described as hypotheses?

A.3 Classify the following properties as chemical or physical: (a) objects made of silver become tarnished; (b) the red color of rubies is due to the presence of chromium ions; (c) the boiling point of ethanol is 78 °C.

A.4 A chemist investigates the transparency, boiling point, and flammability of hexane, a component of mineral spirits. Which of these properties are physical properties and which are chemical properties?

A.5 Identify all the physical properties and changes in the following statement: "The camp nurse measured the temperature of the injured camper and ignited a propane burner; when the water began to boil, some of the water vapor condensed on the cold window."

A.6 Identify all the chemical properties and changes in the following statement: "Copper is a red-brown element obtained from copper sulfide ores by heating them in air, which forms copper oxide. Heating the copper oxide with carbon produces impure copper, which is purified by electrolysis."

A.7 In the containers below, the green spheres represent atoms of one element, the red spheres the atoms of a second element. In each case, the pictures show either a physical or chemical change; identify the type of change.

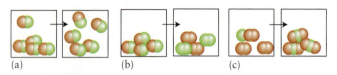
(a)   (b)   (c)

A.8 Which of the containers in Exercise A.7 shows a substance that could be a gas?

A.9 State whether the following properties are extensive or intensive: (a) the temperature at which water boils; (b) the color of copper; (c) the humidity of the atmosphere; (d) the intensity of light emitted by glowing phosphorus.

A.10 State whether the following properties are extensive or intensive: (a) the heat produced when gasoline burns; (b) the volume of gasoline; (c) the cost of gasoline; (d) the air pressure in a tire.

A.11 The following units may sound strange, but they were actually used in ancient times. Suppose that they have been adopted into the SI system. Rewrite each value with the appropriate SI prefix. (a) 1000 grain; (b) 0.01 batman; (c) 1×10^6 mutchkin.

A.12 The following units may sound strange, but they were actually used by the Romans. Suppose that they have been adopted into the SI system. Rewrite each value with the appropriate SI prefix. (a) 1×10^{-12} stadium; (b) 1×10^2 pertica (perch); (c) 1×10^{-1} gradus (step).

A.13 Express the volume in milliliters of a 1.00 cup sample of milk, given that 2 cups = 1 pint, 2 pints = 1 quart.

A.14 The ångström unit (1 Å = 10^{-10} m) is still widely used to report measurements of the dimensions of atoms and molecules. Express the following data in ångströms: (a) the radius of a sodium atom, 180 pm (2 sf); (b) the wavelength of yellow light, 550 nm (2 sf). (c) Write a (single) conversion factor between ångströms and nanometers.

A.15 Two wavelengths (the distance between neighboring peaks in a wave) were measured in a laboratory. Without using a calculator, decide which wavelength is longer: (a) 5.4×10^2 μm; (b) 1.3×10^9 pm.

A.16 The densities of two substances were measured in a laboratory. Without using a calculator, decide which substance has the greater density: (a) 2.4 g·cm^{-3}; (b) 2.4 kg·m^{-3}.

A.17 When a piece of unreactive metal of mass 11.23 g is dropped into a graduated cylinder containing 2.34 mL of water, the water level rises to 2.93 mL. What is the density of the metal (in grams per cubic centimeter)?

A.18 When a piece of unreactive metal of mass 17.32 g is dropped into a graduated cylinder containing 3.24 mL of water, the water level rises to 4.98 mL. What is the density of the metal (in grams per cubic centimeter)?

A.19 The density of diamond is 3.51 g·cm^{-3}. The international (but non-SI) unit for weighing diamonds is the carat (1 carat = 200 mg exactly). What is the volume of a 0.750-carat diamond?

A.20 Use data from Appendix D to calculate the volume of 1.00 ounce of gold.

A.21 A flask weighs 43.50 g when it is empty and 105.50 g when it is filled with water. When the same flask is filled with another liquid, the mass is 96.75 g. What is the density of the second liquid?

A.22 What volume (in cubic centimeters) of lead (of density 11.3 g·cm^{-3}) has the same mass as 215 cm^3 of a piece of redwood (of density 0.38 g·cm^{-3})?

A.23 Spacecraft are commonly clad with aluminum to provide shielding from radiation. Adequate shielding requires that the cladding provide 20. g of aluminum per square centimeter. Use data from Appendix D to calculate how thick the aluminum cladding must be to provide adequate shielding.

A.24 Assume that the entire mass of an atom is concentrated in its nucleus, a sphere of radius 1.5×10^{-5} pm. (a) If the mass of a carbon atom is 2.0×10^{-23} g, what is the density of a carbon nucleus? The volume of a sphere of radius r is $^4/_3\pi r^3$. (b) What would the radius of the Earth be if its matter were compressed to the same density as that of a carbon nucleus? (The Earth's average radius is 6.4×10^3 km, and its average density is 5.5 g·cm^{-3}.)

A.25 Express the answer to the following calculation to the correct number of significant figures:

$$\frac{51.875 \times 1.700}{50.4 + 207.2}$$

A.26 Express the answer to the following calculation to the correct number of significant figures:

$$\frac{64\,500 \times 0.001\,962}{3.02 - 1.007}$$

A.27 Express the answer to the following calculation to the correct number of significant figures:

$$\frac{0.082\,06 \times (27.015 + 1.2)}{3.25 \times 7.006}$$

A.28 Express the answer to the following calculation to the correct number of significant figures:

$$\frac{(604.01 + 0.53) \times 321.81 \times 0.001\,80}{3.530 \times 10^{-3}}$$

A.29 Use the conversion factors in Appendix 1B and inside the back cover to express the following measurements in the designated units: (a) 4.82 nm to pm; (b) 1.83 mL·min^{-1} to mm^3·s^{-1}; (c) 1.88 ng to kg; (d) 2.66 g·cm^{-3} to kg·m^{-3}; (e) 0.044 g·L^{-1} to mg·cm^{-3}.

A.30 Use the conversion factors in Appendix 1B and inside the back cover to express the following measurements in the designated units: (a) 36 L to m^3; (b) 45 $g \cdot L^{-1}$ to $mg \cdot mL^{-1}$; (c) 1.54 $mm \cdot s^{-1}$ to $nm \cdot \mu s^{-1}$; (d) 7.01 $cm \cdot s^{-1}$ to $km \cdot h^{-1}$; (e) \$3.50/gallon to peso/liter (use 1 dollar $\approx$ 13.1 peso).

A.31 The density of a metal was measured by two different methods. In each case, calculate the density. Indicate which measurement is more precise. (a) The dimensions of a rectangular block of the metal were measured as 1.10 cm $\times$ 0.531 cm $\times$ 0.212 cm. Its mass was found to be 0.213 g. (b) The mass of a cylinder of water filled to the 19.65-mL mark was found to be 39.753 g. When a piece of the metal was immersed in the water, the level of the water rose to 20.37 mL and the mass of the cylinder with the metal was found to be 41.003 g.

A.32 A chemist at Trustworthy Labs determined in a set of four experiments that the density of magnesium metal was 1.68 $g \cdot cm^{-3}$, 1.67 $g \cdot cm^{-3}$, 1.69 $g \cdot cm^{-3}$, 1.69 $g \cdot cm^{-3}$. A chemist at Righton Labs repeated the measurements but found the following values: 1.72 $g \cdot cm^{-3}$, 1.63 $g \cdot cm^{-3}$, 1.74 $g \cdot cm^{-3}$, 1.86 $g \cdot cm^{-3}$. The accepted value for the density of magnesium metal is 1.74 $g \cdot cm^{-3}$. Compare the precision and accuracy of the two chemists' data.

A.33 The reference points on a newly proposed temperature scale expressed in °X are the freezing and boiling points of water, set equal to 50. °X and 250. °X, respectively. (a) Derive a formula for converting temperatures on the Celsius scale to the new scale. (b) Comfortable room temperature is 22 °C. What is that temperature in °X?

A.34 When Anders Celsius first proposed his scale, he took 100 as the freezing point of water and 0 as the boiling point. (a) To what temperature would 25 °C correspond on that proposed scale? (b) To what temperature would 98.6 °F correspond on the proposed scale?

A.35 The maximum ground speed of a chicken is 14 $km \cdot h^{-1}$. Calculate the kinetic energy of a chicken of mass 4.2 kg crossing a road at its maximum speed.

A.36 Mars orbits the Sun at 25 $km \cdot s^{-1}$. A spaceship attempting to land on Mars must match its orbital speed. If the mass of the spaceship is 3.6×10^5 kg, what is its kinetic energy when its speed has matched that of Mars?

A.37 A vehicle of mass 2.8 t slows from 100. $km \cdot h^{-1}$ to 50. $km \cdot h^{-1}$ as it enters a city. How much energy could have been recovered instead of being dissipated as heat? To what height, neglecting friction and other losses, could that energy have been used to drive the vehicle up a hill?

A.38 What is the minimum energy that a football player must expend to kick a football of mass 0.51 kg over a goalpost of height 3.0 m?

A.39 It has been said in jest that the only exercise some people get is raising a fork to their lips. How much energy is expended to raise a loaded fork of mass 40.0 g a height of 0.50 m, 30 times in the course of a meal?

A.40 Calculate the energy released when an electron is brought from infinity to a distance of 53 pm from a proton. (That distance is the most probable distance of an electron from the nucleus in a hydrogen atom.) The actual energy released when an electron and a proton form a hydrogen atom is 13.6 electronvolts (eV; 1 eV = 1.602×10^{-19} J). Account for the difference.

A.41 The expression $E_p = mgh$ applies only close to the surface of the Earth. The general expression for the potential energy of a mass m at a distance R from the center of the Earth (of mass m_E) is $E_p = -Gm_Em/R$. Write $R = R_E + h$, where R_E is the radius of the Earth; show that, when $h \ll R_E$, this general expression reduces to the special case, and find an expression for g. You will need the expansion $(1 + x)^{-1} = 1 - x + ...$.

A.42 The expression for the Coulomb potential energy is very similar to the expression for the general gravitational potential energy given in Exercise A.41. Is there an expression resembling $E_p = mgh$ for the change in potential energy when an electron a long way from a proton is moved through a small distance h? Find the expression of the form $E_p = egh$, with an appropriate expression for g, using the same procedure as in Exercise A.41.

B Elements and Atoms

Science is a quest for simplicity. Although the complexity of the world appears boundless, this complexity springs from an underlying simplicity that science attempts to discover. Chemistry shows how everything around us—mountains, trees, people, computers, brains, concrete, oceans—is in fact made up of a handful of simple entities. The ancient Greeks had much the same idea. They supposed that there were four elements—earth, air, fire, and water—that could produce all other substances when combined in the right proportions. Their concept of an element is similar to the modern view, but, on the basis of experiments it is now known that there are actually more than 100 elements, which—in various combinations—make up all the matter on Earth (**FIG. B.1**).

B.1 *Atoms*

B.2 *The Nuclear Model*

B.3 *Isotopes*

B.4 *The Organization of the Elements*

B.1 *Atoms*

The Greeks wondered what would happen if they continued to cut matter into ever smaller pieces. Is there a point at which they would have to stop because the pieces no longer had the same properties as the whole, or could they go on cutting forever? It is now known that there is a point at which the cutting has to stop. That is, matter consists of

FIGURE B.1 Samples of common elements. Clockwise from the red-brown liquid bromine are the silvery liquid mercury and the solids iodine, cadmium, red phosphorus, and copper. *(W.H. Freeman photo by Ken Karp.)*

FIGURE B.2 John Dalton (1766–1844), the English schoolteacher who used experimental measurements to argue that matter consists of atoms. *(Photos. com/Getty Images.)*

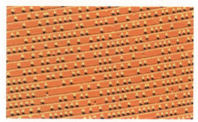

FIGURE B.3 Individual atoms can be seen as bumps on the surface of a solid by the technique called scanning tunneling microscopy (STM). This image shows how information can be stored in individual atoms. The yellow spheres are silicon atoms arranged on a surface of gold and silicon atoms in an arrangement that can be read by an STM microscope. These tiny arrays can lead to very-high-density storage media. *(Franz Himpsel/ University of Wisconsin/Science Source.)*

almost unimaginably tiny particles that cannot be further divided simply by cutting. The smallest particle of an element that can exist is called an **atom.** The story of the development of the modern model of the atom is an excellent illustration of how scientific models are developed and is continued in Focus 1.

The first convincing argument for atoms, one based on experiment rather than speculation, was made in 1807 by the English schoolteacher and chemist John Dalton (**FIG. B.2**). He made many measurements of the ratios of the masses of elements that combine together to form the substances now called "compounds" and found that the ratios formed patterns. For example, he found that, in every sample of water, there is 8 g of oxygen for every 1 g of hydrogen and that, in another compound of the two elements (hydrogen peroxide), there is 16 g of oxygen for every 1 g of hydrogen. These and related data led Dalton to develop his **atomic hypothesis:**

> The name *atom* comes from the Greek word for "not cuttable."

1. All the atoms of a given element are identical.
2. The atoms of different elements have different masses.
3. A compound is a specific combination of atoms of more than one element.
4. In a chemical reaction, atoms are neither created nor destroyed; they exchange partners to produce new substances.

> Dalton's first assumption later had to be modified, because it is now known that the atoms of an element are not all exactly the same, for they can differ slightly in mass (Section B.3).

Modern instrumentation provides much more direct evidence of atoms than was available to Dalton (**FIG. B.3**). There is no longer any doubt that atoms exist and that they are the units that make up the elements. In fact, chemists now use the existence of atoms as the definition of an element: an **element** is a substance composed of only one kind of atom. By 2015, 114 elements had been discovered or created, in some cases in only very small amounts. For instance, when element 110 was made, only two atoms were produced and they lasted less than a millisecond before disintegrating. Claims that several other new elements have been made are currently being assessed.

> Appendix 2D lists the names and chemical symbols of the elements and gives the origins of their names.

All matter is made up of various combinations of the simple forms of matter called the chemical elements. An element is a substance that consists of only one kind of atom.

B.2 The Nuclear Model

According to the current **nuclear model** of the atom, an atom consists of a small positively charged **nucleus,** which is responsible for almost all its mass, surrounded by negatively charged **electrons** (denoted e^-). Compared with the size of the nucleus (about 10^{-14} m in diameter), the space occupied by the electrons is enormous (about 10^{-9} m in diameter, or a hundred thousand times as great). If the nucleus of an atom were the size of a fly at the

TABLE B.1   The Properties of Subatomic Particles Relevant to Chemistry

Particle	Symbol	Charge/e*	Mass/kg
electron	e^-	-1	9.109×10^{-31}
proton	p	$+1$	1.673×10^{-27}
neutron	n	0	1.675×10^{-27}

*Charges are given as multiples of the fundamental charge, which in SI units is $e = 1.602 \times 10^{-19}$ C (see Appendix 1B).

center of a baseball field, then the space occupied by the surrounding electrons would be about the size of the entire stadium (**FIG. B.4**).

The negative charge of the electrons exactly cancels the positive charge of the central nucleus. As a result, an atom is electrically neutral (uncharged). Because each electron has a single negative charge, a nucleus can be regarded as containing one positively charged particle for each surrounding electron (which experiments confirm). These positively charged particles are called **protons** (denoted p); their properties are given in **TABLE B.1**. Note that a proton is nearly 2000 times as heavy as an electron.

The number of protons in an element's atomic nucleus is called the **atomic number**, Z, of that element. The nucleus of a hydrogen atom has one proton, so its atomic number is $Z = 1$; the nucleus of a helium atom has two protons, so its atomic number is $Z = 2$. Henry Moseley, a young British scientist, was the first to determine atomic numbers unambiguously, shortly before he was killed in action in World War I. Moseley knew that when elements are bombarded with rapidly moving electrons, they emit x-rays. He found that the properties of the x-rays emitted by an element depend on its atomic number; by studying the x-rays of many elements, he was able to determine the values of Z for them. Scientists have since determined the atomic numbers of all the known elements (see the list of elements inside the back cover).

Technological advances in electronics early in the twentieth century led to the invention of the **mass spectrometer**, a device for determining the mass of an atom (**FIG. B.5**). Mass spectrometry has been used to determine the masses of the atoms of all the elements. It is now known, for example, that the mass of a hydrogen atom is 1.67×10^{-27} kg and that of a carbon atom is 1.99×10^{-26} kg. Even the heaviest atoms have masses of only about 5×10^{-25} kg. Once the mass of an individual atom is known, the number of those atoms in a given mass of element can be determined simply by dividing the mass of the sample by the mass of one atom.

FIGURE B.4 Think of a fly at the center of this stadium: that is the relative size of the nucleus of an atom if the atom were magnified to the size of the stadium. (*Walter Schmid/The Image Bank/Getty Images.*)

EXAMPLE B.1   Calculating the number of atoms in a sample

Suppose you are preparing a sample of carbon for use as a substrate in a study of molecular electronics, in which organic molecules serve as components in an electronic circuit. You might need to know how many atoms are in your sample. How many atoms are there in a lump of carbon of mass 10.0 g?

ANTICIPATE Because atoms are very tiny, you should expect a very large number.

PLAN Divide the mass of the sample by the mass of one atom.

SOLVE The mass of one carbon atom is 1.99×10^{-26} kg (given in the text).

From $N = $ (mass of sample)/(mass of one atom),

$$N = \frac{\overset{\text{10.0 g}}{\overbrace{1.00 \times 10^{-2} \text{ kg}}}}{1.99 \times 10^{-26} \text{ kg}} = 5.03 \times 10^{23}$$

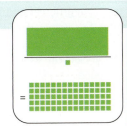

EVALUATE As anticipated, the number of atoms, 5.03×10^{23}, is very large.

A Note on Good Practice: Notice how the given units (grams, for the mass of this sample) have been converted into units that cancel (here, kilograms). It is often prudent to convert all units to SI base units.

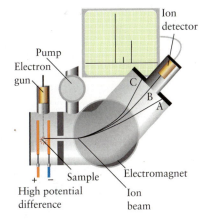

FIGURE B.5 A mass spectrometer is used to measure the masses of atoms. Electrons emerge from the electron gun, are accelerated through a potential difference, and pass through a magnetic field. A pump is used to remove air. As the strength of the magnetic field is changed, the path of the accelerated ions moves from A to C. When the path is at B, the ion detector sends a signal to the recorder. The mass of the ion is proportional to the strength of the magnetic field needed to move the beam into position to strike the detector.

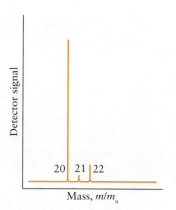

FIGURE B.6 The mass spectrum of neon. The locations of the peaks on the x-axis tell us the masses of the atoms as multiples of the atomic mass constant, m_u, and the intensities tell us the relative numbers of atoms that have each mass.

In the nuclear model of the atom, the positive charge and almost all of the mass is concentrated in the tiny nucleus, and the negatively charged electrons take up most of the surrounding space. The atomic number is the number of protons in the nucleus.

B.3 Isotopes

As happens so often in science, a new and more precise technique of measurement led to a major discovery. When scientists first used mass spectrometers, they found—much to their surprise—that not all the atoms of a single element have the same mass, so Dalton's model is not quite right. In a sample of perfectly pure neon, for example, most of the atoms have mass 3.32×10^{-26} kg, which is about 20 times as great as the mass of a hydrogen atom. Some neon atoms, however, are found to be about 22 times as heavy as hydrogen. Others are about 21 times as heavy (**FIG. B.6**). All three types of atoms have the same atomic number, so they are definitely atoms of neon, but contrary to Dalton's view, they are not identical.

The observation that atoms of a single element can have different masses helped scientists refine the nuclear model still further. They realized that an atomic nucleus must contain subatomic particles other than protons and proposed that it also contains electrically neutral particles called **neutrons** (denoted n). Because neutrons have no electric charge, their presence does not affect the nuclear charge or the number of electrons in the atom. However, they have approximately the same mass as protons, and so they add substantially to the mass of the atom. Therefore, different numbers of neutrons in a nucleus give rise to atoms of different masses, even though the atoms are of the same element. Apart from their charge, neutrons and protons are very similar (Table B.1); they are jointly known as **nucleons.**

Another, better, name for the mass number is *nucleon number.*

When the term "nuclide" was first introduced it referred to the bare nucleus; in its modern usage it refers to the entire atom.

The name *isotope* comes from the Greek words for "same place."

Neon-20   Neon-21   Neon-22
($^{20}_{10}$Ne)   ($^{21}_{10}$Ne)   ($^{22}_{10}$Ne)

FIGURE B.7 The nuclei of different isotopes of the same element have the same number of protons but different numbers of neutrons. These three diagrams show the composition of the nuclei of the three isotopes of neon. On this scale, the atom itself would be about 1 km in diameter. These diagrams make no attempt to show how the protons and neutrons are arranged inside the nucleus.

The total number of protons and neutrons in a nucleus is called the **mass number,** A, of the atom. A nucleus of mass number A is about A times as heavy as a hydrogen atom, which has a nucleus that consists of a single proton. Therefore, if you know that an atom is a certain number of times as heavy as a hydrogen atom, you can infer the mass number of the atom. For example, because mass spectrometry shows that the three varieties of neon atoms are 20, 21, and 22 times as heavy as a hydrogen atom, you know that the mass numbers of the three types of neon atoms are 20, 21, and 22. Because for each of them $Z = 10$, these neon atoms must contain 10, 11, and 12 neutrons, respectively (**FIG. B.7**).

An atom with a specified atomic number and mass number is called a **nuclide.** Thus oxygen-16 ($Z = 8$, $A = 16$) is one nuclide and neon-20 ($Z = 10$, $A = 20$) is another. Atoms with the same atomic number (that is, atoms of the same element) but with different mass numbers are called **isotopes** of the element. All isotopes of an element have exactly the same atomic number, hence they have the same number of protons and electrons but different numbers of neutrons. An isotope is named by writing its mass number after the name of the element, as in neon-20, neon-21, and neon-22. Its symbol is obtained by writing the mass number as a superscript to the left of the chemical symbol of the element, as in ^{20}Ne, ^{21}Ne, and ^{22}Ne. You will occasionally see the atomic number included as a subscript on the lower left, as in the symbol $^{22}_{10}$Ne used in Fig. B.7.

Because isotopes of the same element have the same number of protons and therefore the same number of electrons, they have essentially the same chemical and physical properties. However, the mass differences between isotopes of hydrogen are comparable to the masses of the atoms themselves, leading to noticeable differences in some physical properties and slight variations in some of their chemical properties. Hydrogen has three isotopes (**TABLE B.2**). The most common (^{1}H) has no neutrons; its nucleus is a lone proton. The other two isotopes are less common but nevertheless are so important in

TABLE B.2   Some Isotopes of Common Elements

Element	Symbol	Atomic number, Z	Mass number, A	Abundance/%
hydrogen	^{1}H	1	1	99.985
deuterium	^{2}H or D	1	2	0.015
tritium	^{3}H or T	1	3	—*
carbon-12	^{12}C	6	12	98.90
carbon-13	^{13}C	6	13	1.10
oxygen-16	^{16}O	8	16	99.76

*Radioactive, short-lived.

chemistry and nuclear physics that they are given special names and symbols. The isotope with one neutron (^{2}H) is called *deuterium* (D), and the one with two neutrons (^{3}H) is called *tritium* (T).

Self-test B.2A How many protons, neutrons, and electrons are present in (a) an atom of nitrogen-15; (b) an atom of iron-56?

[*Answer:* (a) 7, 8, 7; (b) 26, 30, 26]

Self-test B.2B How many protons, neutrons, and electrons are present in (a) an atom of oxygen-16; (b) an atom of uranium-236?

Isotopes of an element have the same atomic number but different mass numbers. Their nuclei have the same number of protons but different numbers of neutrons.

B.4 The Organization of the Elements

Each element has a name and a unique chemical symbol made up of one or two letters. Many of the symbols are the first one or two letters of the element's name; some have symbols formed from the first letter of the name and a later letter:

carbon C	nitrogen N	aluminum Al	nickel Ni
magnesium Mg	chlorine Cl	zinc Zn	plutonium Pu

Notice that the first letter of a symbol is always uppercase and the second letter is always lowercase (for example, He, not HE). Other symbols are taken from the element's name in Latin, German, or Greek. For example, the symbol for iron is Fe, from its Latin name *ferrum*. Appendix 2D lists the names and chemical symbols of all the elements and gives the origins of their names.

Self-test B.3A Give the symbols of (a) rhenium and (b) boron. Name (c) Hg and (d) Zr.

[*Answer:* (a) Re; (b) B; (c) mercury; (d) zirconium]

Self-test B.3B Give the symbols of (a) tin and (b) sodium. Name (c) I and (d) Y.

There are currently (in 2015) 114 known and confirmed elements, of which only 88 occur in significant quantities on Earth and are considered to be naturally occurring. At first sight, the prospect of learning their properties might seem overwhelming. The task is made much easier—and much more interesting—by one of the most important discoveries in the history of chemistry. As is explained in more detail in Topic 1F, it has been found that when the elements are listed in order of their atomic number and arranged in rows of certain lengths, they form families that show regular trends in properties. The arrangement of elements that shows their family relationships is called the **periodic table** (it is printed inside the front cover of this book and repeated schematically in **FIG. B.8**).

The vertical columns of the periodic table are called **groups.** These groups identify the principal families of elements. The taller columns (Groups 1, 2, and 13 through 18) are called the **main groups** of the table. The horizontal rows are called **periods** and are

The story of the discovery of periodic relationships by Dmitri Mendeleev can be found in Topic 1F.

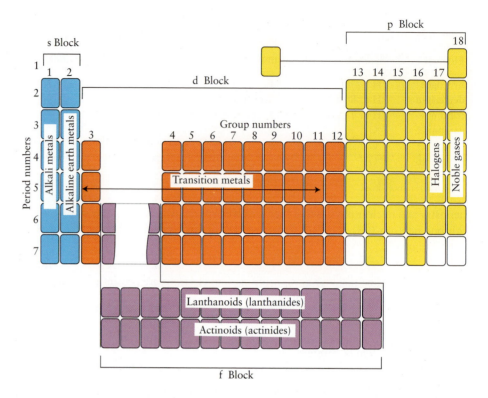

In some versions of the periodic table, you will see a different notation for groups, with the noble gases belonging to Group VIII or VIIIA. These alternatives are given in the table printed inside the front cover.

In some versions of the periodic table, the lanthanoids begin with lanthanum (element 57) and the actinoids begin with actinium (element 89).

FIGURE B.9 All metals can be deformed by hammering. Gold can be hammered into a sheet so thin that light can pass through it. Here, it is possible to see the light of a flame through the sheet of gold. (*Chip Clark/Fundamental Photographs, NYC.*)

numbered from the top down. The four rectangular regions of the periodic table are called **blocks** and, for reasons related to atomic structure (Topic 1D), are labeled s, p, d, and f. The members of the d block, with the exception of the elements in Group 12 (the zinc group), are called **transition metals.** The name indicates that they are "transitional" in character between the vigorously reactive metals in the s block and the less reactive metals of the p block. The members of the f block, which is shown below the main table (to save space), are the **inner transition metals.** The upper row of this block, following lanthanum (element 57) in Period 6, consists of the **lanthanoids** (more commonly and traditionally, the "lanthanides"), and the lower row, following actinium (element 89) in Period 7, consists of the **actinoids** (more commonly, the "actinides").

Some of the main groups have special names:

Group 1: the **alkali metals**

Group 2: (more precisely, calcium, strontium, and barium): the **alkaline earth metals**

Group 17: the **halogens**

Group 18: the **noble gases**

At the head of the periodic table, standing alone, is hydrogen. Some tables place hydrogen in Group 1, others place it in Group 17, and yet others place it in both groups. In this text it is treated as a very special element and is placed in none of the groups.

Most of the elements are solid metals. Only two elements (mercury and bromine) are liquids at ordinary temperatures, and only 11 are gases. The elements are classified as metals, nonmetals, and metalloids:

A **metal** conducts electricity, has a luster, and is malleable and ductile.

A **nonmetal** does not conduct electricity and is neither malleable nor ductile.

A **metalloid** has the appearance of a metal but can behave chemically like a metal or a nonmetal, depending on conditions.

A *malleable* substance (from the Latin word *malleus*, meaning "hammer") is one that can be hammered into thin sheets, such as copper (**FIG. B.9**). A *ductile* substance (from the Latin word *ductire*, meaning "to draw out") is one that can be drawn out into wires. Many nonmetals are brittle and break when hammered. The distinctions between metals and metalloids and between metalloids and nonmetals are not very precise (and not always made),

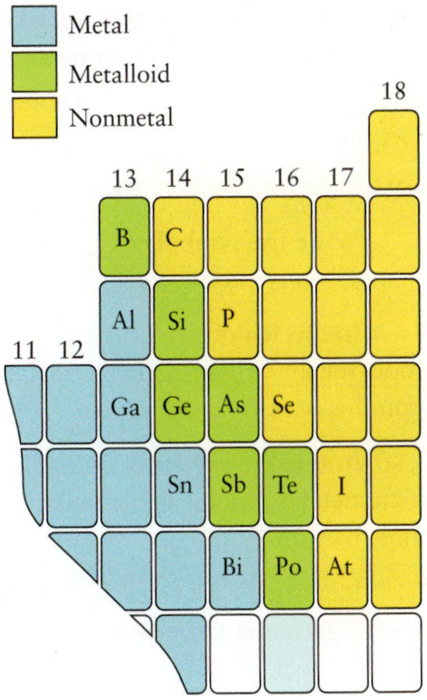

FIGURE B.10 The location of the seven elements commonly regarded as metalloids; these elements have characteristics of both metals and nonmetals. Other elements, notably beryllium and bismuth, are sometimes included in the classification. Boron (B), although not resembling a metal in appearance, is included because it resembles silicon (Si) chemically.

but the metalloids are often taken to be the seven elements shown in **FIG. B.10** on a diagonal band between the metals on the left and the nonmetals on the right.

The periodic table is an arrangement of the elements by atomic number that reflects their family relationships; members of the same group typically show a smooth trend in properties.

What have you learned in Fundamentals B?

You have learned about the structure of the nuclear atom and how it was discovered. You have also learned how to convert from the mass of a sample to the number of atoms it contains, and you have seen that there is a pattern of repeating properties when the elements are organized into the periodic table.

The skills you have mastered are the ability to:

☐ **1.** Describe the structure of an atom.

☐ **2.** Find the number of atoms in a sample of an element of given mass (Example B.1).

☐ **3.** State the numbers of neutrons, protons, and electrons in a nuclide (Self-Test B.2).

☐ **4.** Write the symbols of elements (Self-Test B.3).

☐ **5.** Describe the organization of the periodic table and the characteristics of elements in different regions of the table.

Fundamentals B Exercises

B.1 The mass of an atom of beryllium is 1.50×10^{-26} kg. How many beryllium atoms are present in a beryllium film of mass 0.210 g used as a window on an x-ray tube?

B.2 The mass of an atom of fluorine is 3.16×10^{-26} kg. How many fluorine atoms are present in a tank of fluorine gas of volume 0.970 L? The density of the fluorine gas in the tank is 0.777 g·L^{-1}.

B.3 Give the number of protons, neutrons, and electrons in an atom of (a) boron-11; (b) ^{10}B; (c) phosphorus-31; (d) ^{238}U.

B.4 Give the number of protons, neutrons, and electrons in an atom of (a) ^{40}K; (b) ^{58}Co; (c) tantalum-180; (d) ^{210}At.

B.5 Identify the nuclide that has atoms with (a) 117 neutrons, 77 protons, and 77 electrons; (b) 12 neutrons, 10 protons, and 10 electrons; (c) 28 neutrons, 23 protons, and 23 electrons.

B.6 Identify the nuclide that has atoms with (a) 44 neutrons, 42 protons, and 42 electrons; (b) 40 neutrons, 32 protons, and 32 electrons; (c) 101 neutrons, 70 protons, and 70 electrons.

B.7 Complete the following table:

Element	Symbol	Protons	Neutrons	Electrons	Mass number
	^{36}Cl				
		30			65
			20	20	
lanthanum			80		

B.8 Complete the following table:

Element	Symbol	Protons	Neutrons	Electrons	Mass number
osmium					190
	^{120}Sn				
		74			184
			30	25	

B.9 (a) What characteristics do atoms of argon-40, potassium-40, and calcium-40 have in common? (b) In what ways are they different? (Consider the numbers and types of subatomic particles.)

B.10 (a) What characteristics do atoms of manganese-55, iron-56, and nickel-58 have in common? (b) In what ways are they different? (Consider the numbers and types of subatomic particles.)

B.11 Determine the fraction of the total mass of a ^{56}Fe atom that is due to (a) neutrons; (b) protons; (c) electrons. (d) What is the mass of neutrons in an automobile of mass 1.000 t? Assume that the total mass of the vehicle is due to ^{56}Fe.

B.12 (a) Determine the total number of protons, neutrons, and electrons in one phosphorus pentachloride molecule, PCl_5, assuming that all atoms are the most common isotopes of that element. (b) What is the total mass of protons, of neutrons, and of electrons in one PCl_5 molecule? (Calculate three masses.)

B.13 Name each of the following elements: (a) Sc; (b) Sr; (c) S; (d) Sb. List their group numbers in the periodic table. Identify each as a metal, a nonmetal, or a metalloid.

B.14 Name each of the following elements: (a) Tc; (b) Te; (c) Ti; (d) Tm. List their group numbers in the periodic table, and identify each as a metal, a nonmetal, or a metalloid.

B.15 Write the symbol of (a) strontium; (b) xenon; (c) silicon. Classify each as a metal, a nonmetal, or a metalloid.

B.16 Write the symbol of (a) ytterbium; (b) manganese; (c) selenium. Classify each as a metal, a nonmetal, or a metalloid.

B.17 In this list of elements identify the (a) alkali metal; (b) transition metal; (c) lanthanoid: cerium, cadmium, radium, radon, bromine, barium.

B.18 In this list of elements identify the (a) halogen; (b) alkaline earth metal; (c) noble gas: cerium, cadmium, radium, radon, bromine, barium.

B.19 Identify the block of the periodic table to which each of the following elements belongs: (a) zirconium; (b) As; (c) Ta; (d) barium; (e) Si; (f) cobalt.

B.20 Identify the block of the periodic table to which each of the following elements belongs: (a) phosphorus; (b) No; (c) Po; (d) Mo; (e) osmium; (f) krypton.

B.21 Write the symbol for each of the following elements, state its group and period in the periodic table, and indicate whether the element is a metal, a nonmetal, or a metalloid: (a) an element with 118 neutrons and mass number 200; (b) an element with 78 neutrons and mass number 133.

B.22 Write the symbol for each of the following elements, state its group and period in the periodic table, and indicate whether the element is a metal, a nonmetal, or a metalloid: (a) an element with 67 neutrons and mass number 116; (b) an element with 22 neutrons and mass number 40.

C Compounds

The small number of elements that make up our world combine to produce matter in a seemingly limitless variety of forms. You have only to look at the vegetation, flesh, landscapes, fabrics, building materials, and other things around you to appreciate the wonderful variety of the material world. A part of chemistry is **analysis**: the discovery of which elements have combined together to form a substance. Another aspect of chemistry is **synthesis**: the process of combining elements to produce compounds or converting one compound into another. If the elements are the alphabet of chemistry, then the compounds are its plays, its poems, and its novels.

C.1 What Are Compounds?

A **compound** is an electrically neutral substance that consists of two or more different elements with their atoms present in a definite ratio. A **binary compound** consists of only two elements. For example, water is a binary compound of hydrogen and oxygen, with two hydrogen atoms for each oxygen atom. Whatever the source of the water, it has exactly the same composition; indeed, a substance with a different ratio of atoms would not be water. Hydrogen peroxide (H_2O_2), for instance, has one hydrogen atom for each oxygen atom.

Compounds are classified as either organic or inorganic. **Organic compounds** contain the element carbon and usually hydrogen, too. There are millions of organic compounds, including fuels such as methane and propane, sugars such as glucose and sucrose, and most medicines. These compounds are called *organic* because it was once believed, incorrectly, that they could be formed only by living organisms. **Inorganic compounds** are all the other compounds; they include water, calcium sulfate, ammonia, silica, hydrochloric acid, and many, many more. In addition, some very simple carbon compounds, particularly carbon dioxide and the carbonates, which include chalk (calcium carbonate), are treated as inorganic compounds.

The elements in a compound are not just mixed together. Their atoms are actually joined, or *bonded,* to one another in a specific way due to a chemical change. The result is a substance with chemical and physical properties different from those of the elements that form it. For example, when sulfur is ignited in air, it combines with oxygen from the air to form the compound sulfur dioxide. Solid yellow sulfur and odorless oxygen gas produce a colorless, pungent, and poisonous gas (**FIG. C.1**).

Chemists have found that atoms can bond together to form molecules or can be present in compounds as ions.

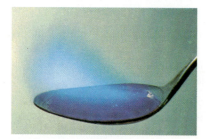

FIGURE C.1 Elemental sulfur burns in air with a blue flame and produces the dense gas sulfur dioxide, a compound of sulfur and oxygen. *(©1983 Chip Clark–Fundamental Photographs.)*

A **molecule** is a discrete group of atoms bonded together in a specific arrangement.

An **ion** is a positively or negatively charged atom or molecule.

The prefixes *cat-* and *an-* come from the Greek words for "down" and "up." Oppositely charged ions travel in opposite directions in an electric field.

A positively charged ion is called a **cation,** and a negatively charged ion is called an **anion.** For instance, a positively charged sodium atom is a cation and is denoted Na^+; a negatively charged chlorine atom is an anion and is denoted Cl^-. An example of a *polyatomic* (many-atom) cation is the ammonium ion, NH_4^+, and an example of a polyatomic anion is the carbonate ion, CO_3^{2-}; note that the latter has two negative charges. An **ionic compound** consists of ions in a ratio that results in overall electrical neutrality; a **molecular compound** consists of electrically neutral molecules.

Binary compounds of two nonmetals are typically molecular, whereas binary compounds formed by a metal and a nonmetal are typically ionic. Water (H_2O) is an example of a binary molecular compound; sodium chloride (NaCl) is an example of a binary ionic compound. These two types of compounds have characteristic properties, and knowing the type of compound that you are studying can be a source of insight into its properties.

Compounds are combinations of elements in which the atoms of the different elements are present in a characteristic, constant ratio. A compound is classified as molecular if it consists of molecules and as ionic if it consists of ions.

C.2 Molecules and Molecular Compounds

The **chemical formula** of a compound represents its composition in terms of chemical symbols. Subscripts show the numbers of atoms of each element present in the smallest unit that is representative of the compound. For molecular compounds, it is common to give the **molecular formula,** a chemical formula that shows how many atoms of each type of element are present in a single molecule of the compound. For instance, the molecular formula for water is H_2O: each molecule contains one O atom and two H atoms. The molecular formula for estrone, a female sex hormone, is $C_{18}H_{22}O_2$, showing that a single molecule of estrone consists of 18 C atoms, 22 H atoms, and 2 O atoms. A molecule of a male sex hormone, testosterone, differs by only a few atoms: its molecular formula is $C_{19}H_{28}O_2$. Think of the consequences of that tiny difference!

Some elements also exist in molecular form. Except for the noble gases, all the elements that are gases at ordinary temperatures are found as diatomic (two-atom) molecules and, in a small number of cases, triatomic (three-atom) molecules. For example, molecules of hydrogen gas contain two hydrogen atoms bonded together and are represented as H_2. The most common form of oxygen consists of diatomic molecules and is known formally as dioxygen, O_2; a less common form is ozone, O_3. Solid sulfur exists as S_8 molecules and phosphorus as P_4 molecules. Nitrogen and all the halogens exist as diatomic molecules: N_2, F_2, Cl_2, Br_2, and I_2.

1   Methanol, CH_3OH

A **structural formula** indicates how the atoms are linked together, but not their actual three-dimensional arrangement in space. For instance, the molecular formula of methanol (wood spirit or wood alcohol) is CH_4O, and its structural formula is shown in (**1**): each line represents a chemical bond (the link between two atoms) and each symbol an atom. Structural formulas contain more information than the chemical formula, but they are cumbersome. So chemists condense them and write, for instance, CH_3OH to represent the structure of methanol. This "condensed" structural formula indicates the groupings of the atoms and summarizes the full structural formula. In most cases, symbols with subscripts represent atoms connected to the preceding element in the formula. A group of atoms attached to another atom in the molecule is set off with parentheses. For example, methylpropane (**2**) has a *methyl* group ($-CH_3$) attached to the central atom in a chain of three carbon atoms, and the condensed version of its structural formula is written $CH_3CH(CH_3)CH_3$ or $HC(CH_3)_3$.

2   Methylpropane, $CH_3CH(CH_3)CH_3$

The richness of organic chemistry arises in part from the fact that, although carbon atoms nearly always form four bonds, they can form chains and rings of almost limitless variety. Another source of richness is that the carbon atoms can be linked together by single bonds, indicated by a single line (C—C); double bonds, represented by a double line (C=C); and triple bonds, represented by a triple line (C≡C). A carbon atom can form four single bonds, two double bonds, or any combination that results in four bonds, such as one single and one triple bond.

(a)

(b)

3 2-Chlorobutane, CH₃CHClCH₂CH₃

OH

O

4 Testosterone, C₁₉H₂₈O₂

Organic chemists have found a way to draw complex molecular structures in a simplified way, by not showing the C and H atoms explicitly. A **line structure** represents a chain of carbon atoms by a zigzag line, where each short line indicates a bond and the end of each line represents a carbon atom. Atoms other than C and H are shown by their symbols. Because carbon almost always forms four bonds in organic compounds, there is no need to show the C—H bonds explicitly; just fill in the correct number of hydrogen atoms mentally: compare the line structure of 2-chlorobutane, $CH_3CHClCH_2CH_3$ (**3a**), with its structural form (**3b**). Line structures are particularly useful for complex molecules, such as testosterone (**4**).

Another important aspect of a molecular compound is its shape. The pictorial representations of molecules that most accurately show their shapes are images based on computation or software that represents atoms by spheres of various sizes. An example is the **space-filling model** of an ethanol molecule shown in **FIG. C.2a**. The atoms are represented by colored spheres (they are not the actual colors of the atoms) that fit into one another. Another representation of the same molecule, called a **ball-and-stick model,** is shown in Fig. C.2b. Each ball represents the location of an atom, and the sticks represent the bonds. Although this kind of model does not represent the actual molecular shape as well as a space-filling model does, it shows bond lengths and angles more clearly. It is also easier to draw and interpret.

THINKING POINT

Which type of molecular representation would be more useful for studying (a) bond lengths, (b) molecular volume?

A molecular formula shows the composition of a molecule in terms of the numbers of atoms of each element present. Different styles of molecular models are used to emphasize different molecular characteristics.

C.3 Ions and Ionic Compounds

To imagine what ionic compounds look like, you need to visualize a huge number of cations and anions stacked together in a regular three-dimensional formation and held in place by the attraction between their opposite charges. Each crystal of sodium chloride, for example, consists of an orderly array of a vast number of alternating Na^+ and Cl^- ions (**FIG. C.3**). When you take a pinch of salt, each crystal that you are picking up consists of more ions than there are stars in the visible universe.

The nuclear model of the atom readily explains the existence of **monatomic ions** (single-atom ions). When an electron is removed from a neutral atom, the charge of the remaining electrons no longer fully cancels the positive charge of the nucleus (**FIG. C.4**). Because an electron has one unit of negative charge, each electron removed from a neutral atom leaves behind a cation with one additional unit of positive charge. For example, a

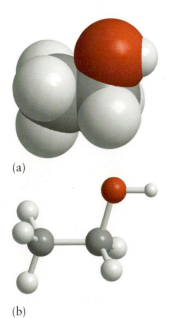

(a)

(b)

FIGURE C.2 Two representations of an ethanol molecule: (a) space-filling, (b) ball-and-stick.

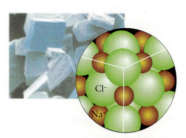

FIGURE C.3 An ionic solid consists of an array of cations and anions stacked together. This illustration shows the arrangement of sodium cations (Na^+) and chlorine anions (chloride ions, Cl^-) in a crystal of sodium chloride (common table salt). The faces of the crystal are where the stacks of ions come to an end. *(Photo: Andrew Syred/Science Source.)*

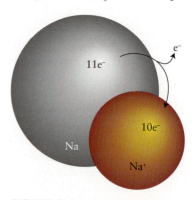

FIGURE C.4 A neutral sodium atom (left) consists of a nucleus that contains 11 protons and is surrounded by 11 electrons. When 1 electron is lost, the remaining 10 electrons cancel only 10 of the proton charges, and the resulting ion (right) has one overall positive charge.

FIGURE C.5 A neutral fluorine atom (left) consists of a nucleus that contains 9 protons and is surrounded by 9 electrons. When the atom gains 1 electron, the 9 proton charges cancel all but 1 of the 10 electron charges, and the resulting ion (right) has one overall negative charge.

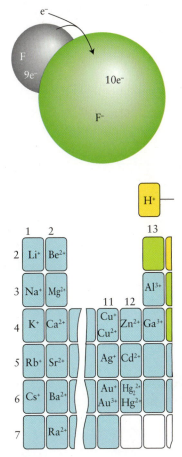

sodium cation, Na^+, is a sodium atom that has lost one electron. When a calcium atom loses two electrons, it becomes the doubly positively charged calcium ion, Ca^{2+}.

Each electron gained by an atom increases the negative charge by one unit (**FIG. C.5**). So, when a fluorine atom gains an electron, it becomes the singly negatively charged fluoride ion, F^-. When an oxygen atom gains two electrons, it becomes the doubly charged oxide ion, O^{2-}. When a nitrogen atom gains three electrons, it becomes the triply charged nitride ion, N^{3-}.

The periodic table can help you decide what type of ion an element forms and what charge to expect the ion to have. One important pattern is that metallic elements—those toward the left of the periodic table—typically form cations by electron loss. Nonmetallic elements—those toward the right of the table—typically form anions by gaining electrons. Thus, the alkali metals form cations, and the halogens form anions.

FIGURE C.6 shows another pattern in the charges of monatomic cations. For elements in Groups 1 and 2, for instance, the charge of the ion is equal to the group number. Thus, cesium in Group 1 forms Cs^+ ions; barium in Group 2 forms Ba^{2+} ions. Figure C.6 also shows that atoms of the d-block elements can form cations with different charges. An iron atom, for instance, can lose two electrons to become Fe^{2+} or three electrons to become Fe^{3+}. Copper can lose either one electron to form Cu^+ or two electrons to become Cu^{2+}. Some of the heavier metals of Groups 13 and 14 also form cations with different charges.

Some of the anions present in compounds are listed in **FIG. C.7**. You should be able to identify another important pattern: a main-group element on the right side of the table forms an anion with a negative charge equal to its separation from the noble-gas element to its right. Oxygen is two groups away from the noble gases and forms the oxide ion, O^{2-}; phosphorus, which is three groups away, forms the phosphide ion, P^{3-}. In general, if the group number of a main-group element is N (in the 1-to-18 system), the charge of the anions that it forms is $N - 18$.

The pattern of ion formation by main-group elements can be summarized by a single rule: for atoms toward the left or right of the periodic table, atoms lose or gain electrons until they have the same number of electrons as the nearest noble-gas atom. That is:

- Elements in Groups 1, 2, and 3 lose electrons until they have the same number of electrons as the noble-gas atom at the end of the previous period.
- Elements in Groups 14–17 gain electrons until they have the same number of electrons as the noble-gas atom at the end of their period.

Thus, a magnesium atom can lose two electrons and become Mg^{2+}, which has the same number of electrons as an atom of neon. A selenium atom can gain two electrons and become Se^{2-}, which has the same number of electrons as a krypton atom.

Metallic elements typically form cations, and nonmetallic elements typically form anions; the charge of a monatomic ion of a main-group element is related to its group in the periodic table.

FIGURE C.6 The typical cations formed by a selection of elements in the periodic table. The transition metals (Groups 3–11) form a wide variety of cations; only a few are shown here.

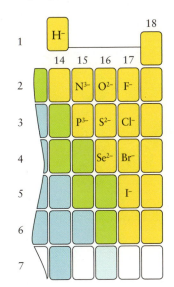

FIGURE C.7 The typical monatomic anions formed by a selection of elements in the periodic table. Notice how the charge on each ion depends on its group number. Only the nonmetals form monatomic anions under common conditions.

EXAMPLE C.1   Identifying the likely charge of a monatomic ion

If you work in materials science, you might study elements that can be used to prepare new ionic ceramic compounds; in that case, you would need to keep in mind the types of ions that these elements form. What ions are (a) nitrogen and (b) calcium likely to form?

ANTICIPATE Nitrogen is a nonmetal, so you should expect it to form an anion; calcium is a metal, so you should expect it to form a cation.

PLAN Identify the group to which each element belongs and decide whether it is likely to lose or gain electrons in order to reach a nearby noble-gas electron count. Typically, a nonmetal forms an anion of charge $N - 18$ and a metal in Group 1 or 2 forms a cation of charge N, where N is the group number.

SOLVE

(a) Nitrogen (N) is in Group 15 and a nonmetal.

$N = 15$; so $N - 18 = 15 - 18 = -3$; nitrogen forms N^{3-}

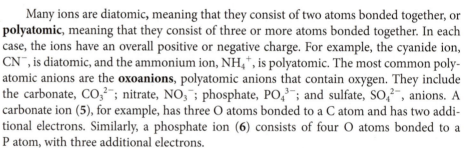

(b) Calcium (Ca) is in Group 2 and a metal.

$N = 2$; so calcium forms Ca^{2+}

Self-test C.1A What ions are (a) iodine and (b) aluminum likely to form?

[*Answer:* (a) I^-; (b) Al^{3+}]

Self-test C.1B What ions are (a) potassium and (b) sulfur likely to form?

Related Exercises C.7, C.8

Many ions are diatomic, meaning that they consist of two atoms bonded together, or **polyatomic**, meaning that they consist of three or more atoms bonded together. In each case, the ions have an overall positive or negative charge. For example, the cyanide ion, CN^-, is diatomic, and the ammonium ion, NH_4^+, is polyatomic. The most common polyatomic anions are the **oxoanions**, polyatomic anions that contain oxygen. They include the carbonate, CO_3^{2-}; nitrate, NO_3^-; phosphate, PO_4^{3-}; and sulfate, SO_4^{2-}, anions. A carbonate ion (**5**), for example, has three O atoms bonded to a C atom and has two additional electrons. Similarly, a phosphate ion (**6**) consists of four O atoms bonded to a P atom, with three additional electrons.

The formulas of ionic compounds have a different meaning from those of molecular compounds. Each crystal of sodium chloride may have a different total number of cations and anions from any other crystal. The numbers of ions present as the formula of this ionic compound cannot be specified, because each crystal would have a different formula and the subscripts would be enormous numbers. However, the *ratio* of the number of cations to the number of anions is the same in all the crystals, and the chemical formula shows this ratio. In sodium chloride, there is one Na^+ ion for each Cl^- ion, so its formula is NaCl. Sodium chloride is an example of a **binary ionic compound,** a compound formed from the ions of two elements. Another binary ionic compound, $CaCl_2$, is formed from Ca^{2+} and Cl^- ions in the ratio 1:2, which is required for electrical neutrality.

The formulas of compounds containing polyatomic ions follow similar rules. In sodium carbonate, there are two Na^+ (sodium) ions per CO_3^{2-} (carbonate) ion, so its formula is Na_2CO_3. When a subscript has to be added to a polyatomic ion, the ion is written within parentheses, as in $(NH_4)_2SO_4$, where $(NH_4)_2$ means that there are two NH_4^+ (ammonium) ions for each SO_4^{2-} (sulfate) ion in ammonium sulfate. In each case, the ions combine in such a way that the positive and negative charges cancel: *all compounds are electrically neutral overall.*

A group of ions having the number of atoms given by the formula is called a **formula unit.** For example, the formula unit of sodium chloride, NaCl, consists of one Na^+ ion and one Cl^- ion; and the formula unit of ammonium sulfate, $(NH_4)_2SO_4$, consists of two NH_4^+ ions and one SO_4^{2-} ion.

You can often decide whether a substance is an ionic compound or a molecular compound by examining its formula. Binary molecular compounds are typically formed

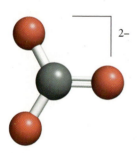

5 Carbonate ion, CO_3^{2-}

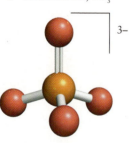

6 Phosphate ion, PO_4^{3-}

Clusters of ions, such as Na^+Cl^-, do exist in the gas phase and in concentrated aqueous solutions, in which case they are commonly termed *ion pairs* rather than molecules.

from two nonmetals (such as hydrogen and oxygen, the elements in water). Ionic compounds are typically formed from the combination of a metallic element with nonmetallic elements (such as the combination of potassium with sulfur and oxygen to form potassium sulfate, K_2SO_4). Ionic compounds typically contain one metallic element; the principal exceptions are compounds containing polyatomic cations such as the ammonium ion (for example, ammonium nitrate), which are ionic even though all the elements present are nonmetallic.

EXAMPLE C.2   Writing the likely formula of a binary ionic compound

The binary ionic compound formed by magnesium and phosphorus is flammable in air, and the fire produced is difficult to extinguish. If you are preparing training materials for firefighters, you might need to list the formulas of flammable compounds. Write the formula of this compound.

ANTICIPATE Since magnesium is a metal and appears on the left side of the periodic table, you should anticipate that Mg can easily lose electrons and form cations. Because a compound must be electrically neutral, you should expect phosphorus to form anions.

PLAN Identify the charges of the cation and anion, and then choose numbers of each so that the total charge is zero.

SOLVE

Find the charge on the cation from the periodic table.

Mg is in Group 2; so it forms ions with charge +2.

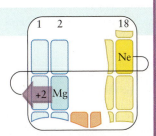

Find the charge on the anion from the periodic table.

P is in Group 15 and forms ions with charge $15 - 18 = -3$.

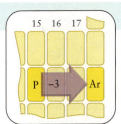

Combine the ions so that their charges cancel.

3 Mg^{2+} ions have a charge of +6.
2 P^{3-} ions have a charge of −6.
The formula is Mg_3P_2.

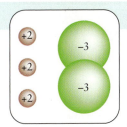

EVALUATE Magnesium forms Mg^{2+} cations as anticipated, whereas phosphorus forms the expected anion, P^{3-}.

Self-test C.2A Write the formula of the binary ionic compound formed by (a) barium and bromine; (b) aluminum and oxygen.

[*Answer:* (a) $BaBr_2$; (b) Al_2O_3]

Self-test C.2B Write the formula of the binary ionic compound formed by (a) lithium and nitrogen; (b) strontium and bromine.

Related Exercises C.13, C.14

The chemical formula of an ionic compound shows the ratio of the numbers of atoms of each element present in one formula unit. A formula unit of an ionic compound is a group of ions with the same number of atoms of each element as appears in its formula.

What have you learned in Fundamentals C?

You have learned that atoms of different elements can combine to form molecular or ionic compounds. You have also seen how to predict the charge on the monatomic ion that an atom is likely to form, how to interpret chemical formulas, and how to predict the formulas of ionic compounds.

The skills you have mastered are the ability to:

- ☐ **1.** Distinguish between atoms, molecules, and ions.
- ☐ **2.** Identify compounds as organic or inorganic and as molecular or ionic.
- ☐ **3.** Use various means of representing molecules.
- ☐ **4.** Predict the anion or cation that a main-group element is likely to form (Example C.1).
- ☐ **5.** Interpret chemical formulas in terms of the number of each type of atom present.
- ☐ **6.** Predict the formulas of binary ionic compounds (Example C.2).

Fundamentals C Exercises

C.1 Each of the containers pictured below holds a mixture, a single compound, or a single element. The blue spheres represent atoms of one element, the green spheres the atoms of a second element. In each case, identify the type of contents.

(a)   (b)

C.2 Each of the containers pictured below holds a mixture, a single compound, or a single element. The blue spheres represent atoms of one element, the green spheres the atoms of a second element. In each case, identify the type of contents.

(a)   (b)

C.3 The compound xanthophyll, also known as lutein, is a yellow compound found in egg yolks, green leaves, bird feathers, and flowers. It is thought to be important for maintaining good eyesight. Xanthophyll contains atoms of carbon, hydrogen, and oxygen in the ratio 20:28:1. Its molecules each have two oxygen atoms. Write the molecular formula of xanthophyll.

C.4 Casimiroedine is an alkaloid extracted from the seeds of the zapote blanco plant, which is native to Mexico. Because it has been found to lower blood pressure in guinea pigs, medical researchers are studying it. Casimiroedine contains atoms of carbon, hydrogen, nitrogen, and oxygen in the ratio 7:9:1:2. Its molecules each have 6 oxygen atoms. Write the molecular formula of casimiroedine.

C.5 In the following ball-and-stick molecular structures, dark gray indicates carbon; red, oxygen; white, hydrogen; blue, nitrogen; and green, chlorine. Write the chemical formula of each structure.

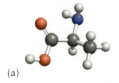 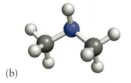

(a)   (b)

C.6 Write the chemical formula of each ball-and-stick structure. See Exercise C.5 for the color code.

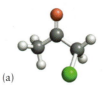 

(a)   (b)

C.7 State whether each of the following elements is more likely to form a cation or an anion, and write the formula for the ion most likely to be formed: (a) cesium; (b) iodine; (c) selenium; (d) calcium.

C.8 State whether each of the following elements is more likely to form a cation or an anion, and write the formula for the ion most likely to be formed: (a) nitrogen; (b) bismuth; (c) sulfur; (d) magnesium.

C.9 How many protons, neutrons, and electrons are present in (a) $^{10}Be^{2+}$; (b) $^{17}O^{2-}$; (c) $^{80}Br^{-}$; (d) $^{75}As^{3-}$?

C.10 How many protons, neutrons, and electrons are present in (a) $^{66}Zn^{2+}$; (b) $^{150}Sm^{3+}$; (c) $^{133}Cs^{+}$; (d) $^{127}I^{-}$?

C.11 Write the symbol of the ion that has (a) 9 protons, 10 neutrons, and 10 electrons; (b) 12 protons, 12 neutrons, and 10 electrons; (c) 52 protons, 76 neutrons, and 54 electrons; (d) 37 protons, 49 neutrons, and 36 electrons.

C.12 Write the symbol of the ion that has (a) 11 protons, 13 neutrons, and 10 electrons; (b) 13 protons, 14 neutrons, and 10 electrons; (c) 34 protons, 45 neutrons, and 36 electrons; (d) 24 protons, 28 neutrons, and 22 electrons.

C.13 Write the formula of a compound formed by combining (a) Al and Te; (b) Mg and O; (c) Na and S; (d) Rb and I.

C.14 Write the formula of a compound formed by combining (a) Ca and C; (b) Al and O; (c) Li and N; (d) Sr and S.

C.15 A main-group element in Period 3 forms the following ionic compounds: EBr_3 and E_2O_3. (a) To which group does the element E belong? (b) Write the name and symbol of element E.

C.16 A main-group element in Period 4 forms the molecular compound H_2E and the ionic compound Na_2E. (a) To which group does the element E belong? (b) Write the name and symbol of element E.

C.17 Determine the fraction of the total mass of a ^{48}Ti atom that is due to (a) neutrons; (b) protons; (c) electrons. (d) What is the mass of protons in a titanium bicycle frame of mass 30. kg? Assume the total mass of the bicycle frame is due to ^{48}Ti. *Note:* The masses of free protons and neutrons are slightly higher than the masses of these particles in atoms, so the answer is only an approximation.

C.18 (a) Determine the total number of protons, neutrons, and electrons in one water molecule, H_2O, assuming that only the most common isotopes, ^{1}H and ^{16}O, are present. (b) What are the total masses of protons, neutrons, and electrons in this water molecule? (c) What fraction of your own mass is due solely to the electrons in your body, assuming that you consist primarily of water made from this type of molecule? *Note:* The masses of free protons and neutrons are slightly higher than the masses of these particles in atoms, so the answer is only an approximation.

C.19 The following compounds contain polyatomic ions. (a) Write the formula for the compound formed from sodium ions and hydrogen phosphite ions (HPO_3^{2-}). (b) Write the formula for ammonium carbonate, which is formed from ammonium ions (NH_4^+) and carbonate ions (CO_3^{2-}). (c) The formula for magnesium sulfate is $MgSO_4$; what is the charge of the cation in $CuSO_4$? (d) The formula for potassium phosphate is K_3PO_4; what is the charge of the cation in $Sn_3(PO_4)_2$?

C.20 The following compounds contain polyatomic ions. (a) Write the formula for the compound formed from sodium ions and chromate ions (CrO_4^{2-}). (b) Write the formula for ammonium sulfite, which is formed from ammonium ions (NH_4^+) and sulfite ions (SO_3^{2-}). (c) The formula for manganese(II) sulfate is $MnSO_4$; what is the charge of the cation in $Co_2(SO_4)_3$? (d) The formula for potassium hydrogen phosphate is K_2HPO_4; what is the charge of the cation in $Pb_3(PO_4)_4$?

C.21 Identify the following substances as elements, molecular compounds, or ionic compounds: (a) HCl; (b) S_8; (c) CoS; (d) Ar; (e) CS_2; (f) $SrBr_2$.

C.22 Identify the following substances as elements, molecular compounds, or ionic compounds: (a) F_2; (b) BF_3; (c) BaF_2; (d) NO; (e) Ca; (f) CaO.

D   Nomenclature

D.1	**Names of Cations**
D.2	**Names of Anions**
D.3	**Names of Ionic Compounds**
D.4	**Names of Inorganic Molecular Compounds**
D.5	**The Nomenclature of Some Common Organic Compounds**

Many compounds were given informal, **common names** before their compositions were known. Common names include water, salt, sugar, ammonia, and quartz. A **systematic name,** on the other hand, reveals which elements are present and, in some cases, the arrangement of atoms. The systemic naming of compounds, which is called **chemical nomenclature,** follows the simple rules described here.

D.1 Names of Cations

The name of a monatomic cation is the same as the name of the element forming it, with the addition of the word "ion," as in "sodium ion" for Na^+. When an element can form more than one kind of cation, such as Cu^+ and Cu^{2+} from copper, use the **oxidation number,** in this context the charge of the cation, written as a Roman numeral in parentheses immediately following the name of the element. Thus, Cu^+ is a copper(I) ion and Cu^{2+} is a copper(II) ion. Similarly, Fe^{2+} is an iron(II) ion and Fe^{3+} is an iron(III) ion. As shown in Fig. C.6, most transition metals form more than one kind of ion; so unless you are given other information, it is important to include the oxidation number in the names of their compounds.

Some older systems of nomenclature are still in use. For example, some cations were once denoted by the endings *-ous* and *-ic* for the ions with lower and higher charges, respectively. To make matters worse, these endings were in some cases added to the Latin form of the element's name. Thus, iron(II) ions were called ferrous ions and iron(III) ions were called ferric ions (see Appendix 3C). Although that system is not used in this text, you will sometimes come across it and should be aware of it.

> *The name of a monatomic cation is the name of the element plus the word "ion"; for elements that can form more than one type of cation, the oxidation number, a Roman numeral indicating the charge, is included.*

D.2 Names of Anions

Monatomic anions, such as the Cl^- ions in sodium chloride and the O^{2-} ions in *quicklime* (calcium oxide, CaO), are named by adding the suffix *-ide* and the word "ion" to the first part of the name of the element (the *stem* of its name), as shown in **TABLE D.1**; thus, S^{2-} is a sulfide ion and O^{2-} is an oxide ion. There is usually no need to specify the charge, because most elements that form monatomic anions form only one kind of ion. The ions formed by the halogens are collectively called *halide* ions and include fluoride (F^-), chloride (Cl^-), bromide (Br^-), and iodide (I^-) ions.

TABLE D.1   Common Anions and Their Parent Acids

Anion	Parent acid	Anion	Parent acid
fluoride ion, F^-	hydrofluoric acid,* HF (hydrogen fluoride)	nitrite ion, NO_2^-	nitrous acid, HNO_2
		nitrate ion, NO_3^-	nitric acid, HNO_3
chloride ion, Cl^-	hydrochloric acid,* HCl (hydrogen chloride)	phosphate ion, PO_4^{3-}	phosphoric acid, H_3PO_4
bromide ion, Br^-	hydrobromic acid,* HBr (hydrogen bromide)	hydrogen phosphate ion, HPO_4^{2-} dihydrogen phosphate ion, $H_2PO_4^-$	
iodide ion, I^-	hydroiodic acid,* HI (hydrogen iodide)	sulfite ion, SO_3^{2-} hydrogen sulfite ion, HSO_3^-	sulfurous acid, H_2SO_3
oxide ion, O^{2-} hydroxide ion, OH^-	water, H_2O	sulfate ion, SO_4^{2-} hydrogen sulfate ion, HSO_4^-	sulfuric acid, H_2SO_4
sulfide ion, S^{2-} hydrogen sulfide ion, HS^-	hydrosulfuric acid,* H_2S (hydrogen sulfide)	hypochlorite ion, ClO^-	hypochlorous acid, HClO
cyanide ion, CN^-	hydrocyanic acid,* HCN (hydrogen cyanide)	chlorite ion, ClO_2^-	chlorous acid, $HClO_2$
		chlorate ion, ClO_3^-	chloric acid, $HClO_3$
acetate ion, $CH_3CO_2^-$	acetic acid, CH_3COOH	perchlorate ion, ClO_4^-	perchloric acid, $HClO_4$
carbonate ion, CO_3^{2-} hydrogen carbonate (bicarbonate) ion, HCO_3^-	carbonic acid, H_2CO_3		

*The name of the aqueous solution of the compound. The name of the compound itself is in parentheses

Polyatomic ions (see *Fundamentals* C) include the *oxoanions*, which are ions that contain oxygen (see Table D.1). If only one oxoanion of an element exists, its name is formed by adding the suffix *-ate* to the stem of the name of the element, as in the carbonate ion, CO_3^{2-}. Some elements can form two types of oxoanions, with different numbers of oxygen atoms, so names are needed that distinguish them. Nitrogen, for example, forms both NO_2^- and NO_3^-. In such cases:

- The ion with the larger number of oxygen atoms is given the suffix *-ate*.
- The ion with the smaller number of oxygen atoms is given the suffix *-ite*.

Thus, NO_3^- is nitrate, and NO_2^- is nitrite. The nitrate ion is an important source of nitrogen for plants and is included in some fertilizers (such as ammonium nitrate, NH_4NO_3). Some elements—particularly the halogens—form more than two kinds of oxoanion:

Hypo comes from the Greek word for "under."

- The name of the oxoanion with the smallest number of oxygen atoms is formed by adding the prefix *hypo-* to the *-ite* form of the name, as in the *hypo*chlor*ite* ion, ClO^-.
- The oxoanion with the most oxygen atoms is named with the prefix *per-* added to the *-ate* form of the name. An example is the *per*chlor*ate* ion, ClO_4^-.

Per is the Latin word for "all over," suggesting that the element's ability to combine with oxygen is finally satisfied.

The rules for naming polyatomic ions are summarized in Appendix 3A, and common examples are listed in Table D.1.

Hydrogen is present in some anions. Two examples are HS^- and HCO_3^-. The names of these anions begin with "hydrogen." Thus, HS^- is the hydrogen sulfide ion, and HCO_3^- is the hydrogen carbonate ion. You might also see the name of the ion written as a single word, as in hydrogencarbonate ion, which is the more modern style. In an older system of nomenclature, which is still quite widely used, an anion containing hydrogen is named with the prefix *bi-*, as in bicarbonate ion for HCO_3^-. If two hydrogen atoms are present in an anion, as in $H_2PO_4^-$, then the ion is named as a dihydrogen anion, in this case as dihydrogen phosphate.

Self-test D.1A   Write (a) the name of the IO^- ion and (b) the formula of the hydrogen sulfite ion.

[*Answer:* (a) Hypoiodite ion; (b) HSO_3^-]

Self-test D.1B   Write (a) the name of the $H_2AsO_4^-$ ion and (b) the formula of the chlorate ion.

Names of monatomic anions end in -ide. Oxoanions are anions that contain oxygen. The suffix -ate indicates a greater number of oxygen atoms than the suffix -ite within the same series of oxoanions. In series of three or more related oxoanions, the prefix per- indicates the maximum number of oxygen atoms; the prefix hypo- indicates the least number of oxygen atoms.

D.3 Names of Ionic Compounds

An ionic compound is named with the cation name first, followed by the name of the anion; the word *ion* is omitted in each case. The oxidation number of the cation is given if more than one charge is possible. However, if the cation comes from an element that exists in only one charge state (as listed in Fig. C.6), then the oxidation number is omitted. Typical names include potassium chloride (KCl), a compound containing K^+ and Cl^- ions; and ammonium nitrate (NH_4NO_3), which contains NH_4^+ and NO_3^- ions. The cobalt chloride that contains Co^{2+} ions ($CoCl_2$) is called cobalt(II) chloride; $CoCl_3$ contains Co^{3+} ions and is called cobalt(III) chloride. Notice that the number of chloride ions is determined by the need for charge balance.

Some ionic compounds form crystals that incorporate a definite proportion of molecules of water as well as the ions of the compound itself. These compounds are called **hydrates.** For example, copper(II) sulfate normally exists as blue crystals of composition $CuSO_4 \cdot 5H_2O$ (**FIG. D.1**). The raised dot in this formula separates the "waters of hydration" from the rest of the formula, and the number before the H_2O indicates how many water molecules are present in each formula unit. Hydrates are named by first giving the name of the compound, then adding *-hydrate* with a Greek prefix indicating how many molecules of water are found in each formula unit (**TABLE D.2**). For example, the name of $CuSO_4 \cdot 5H_2O$, the common blue form of this compound, is copper(II) sulfate pentahydrate.

TABLE D.2 Prefixes Used for Naming Compounds

Prefix	Meaning	Prefix	Meaning
mono-	1	hepta-	7
di-	2	octa-	8
tri-	3	nona-	9
tetra-	4	deca-	10
penta-	5	undeca-	11
hexa-	6	dodeca-	12

FIGURE D.1 Blue crystals of copper(II) sulfate pentahydrate ($CuSO_4 \cdot 5H_2O$) lose water above 150 °C and form the white anhydrous powder ($CuSO_4$) seen in this petri dish. The color is restored when water is added; in fact, anhydrous copper(II) sulfate has such a strong attraction for water that it is usually colored a very pale blue as a result of its reaction with the water in air. *(W.H. Freeman photo by Ken Karp.)*

 LAB VIDEO FIGURE D.1

Toolbox D.1 HOW TO NAME IONIC COMPOUNDS

CONCEPTUAL BASIS

The aim of chemical nomenclature is to be simple but unambiguous. Ionic and molecular compounds use different procedures, so it is important first to identify the type of compound. This Toolbox describes the procedure for naming ionic compounds. The nomenclature of molecular compounds is introduced in Toolbox D.2.

PROCEDURE

To name an ionic compound, name the ions present and then combine the names of the ions.

Step 1 Identify the cation and the anion (see Table D.1 or Appendix 3A, if necessary). To determine the oxidation number of the cation, decide what cation charge is required to cancel the total negative charge of the anions.

Step 2 Name the cation. If the metal can have more than one oxidation number (most transition metals and some metals in Groups 12 through 15), give its charge as a Roman numeral in parentheses.

Step 3 Name the anion. If the anion is monatomic, change the ending of the element's name to *-ide*. Combine the names of the ions with the cation written first.

For an oxoanion:

(a) For elements that form two oxoanions, give the ion with the larger number of oxygen atoms the suffix *-ate* and that with the smaller number of oxygen atoms the suffix *-ite*.

(b) For elements that form a series of four oxoanions, add the prefix *hypo-* to the name of the oxoanion with the smallest number of oxygen atoms. Add the prefix *per-* to the oxoanion with the highest number of oxygen atoms.

For other polyatomic anions, find the name of the ion in Table D.1 or Appendix 3A. If hydrogen is present, add *hydrogen* to the name of the anion. If two hydrogen atoms are present, add *dihydrogen* to the name of the anion.

Step 4 If water molecules appear in the formula, the compound is a hydrate. Add *-hydrate* with a Greek prefix corresponding to the number in front of H_2O.

Example D.1 shows how these rules are applied.

EXAMPLE D.1   Naming ionic inorganic compounds

When working in science, you will need to know both the names and formulas of compounds because sometimes only one is on the label of a container. Name (a) the green solid $CrCl_3 \cdot 6H_2O$, which is used to synthesize some organic compounds, and (b) $Ba(ClO_4)_2$, which gives a green color to fireworks.

PLAN Use the rules in Toolbox D.1.

SOLVE

Step 1 Identify the cation and anion.

   (a) $CrCl_3 \cdot 6H_2O$: Cr^{3+}, Cl^-   **(b) $Ba(ClO_4)_2$: Ba^{2+}, ClO_4^-**

Step 2 Name the cation, giving the charge of the transition-metal cation as a Roman numeral in parentheses. Note that in (a) there are three Cl^- ions, so the charge on Cr must be $+3$.

   chromium(III)   barium

Step 3 Name the anion and combine the names of the ions with the cation first. Because chlorine forms a series of four oxoanions, add the prefix *per-* to the oxoanion in (b).

   chromium(III) chloride   barium perchlorate

Step 4 If H_2O is present, add the word *hydrate* with a Greek prefix corresponding to the number in front of H_2O.

   chromium(III) chloride hexahydrate

Self-test D.2A Name the compounds (a) $NiCl_2 \cdot 2H_2O$; (b) AlF_3; (c) $Mn(IO_2)_2$.

[***Answer:*** (a) Nickel(II) chloride dihydrate; (b) aluminum fluoride; (c) manganese(II) iodite]

Self-test D.2B Name the compounds (a) $AuCl_3$; (b) CaS; (c) Mn_2O_3.

Related Exercises D.7, D.8

Ionic compounds are named by starting with the name of the cation (with its oxidation number if more than one charge is possible), followed by the name of the anion; hydrates are named by adding the suffix -hydrate, preceded by a Greek prefix indicating the number of water molecules in the formula unit.

TABLE D.3	Common Names for Some Simple Molecular Compounds
Formula*	**Common name**
NH_3	ammonia
N_2H_4	hydrazine
NH_2OH	hydroxylamine
PH_3	phosphine
NO	nitric oxide
N_2O	nitrous oxide
C_2H_4	ethylene
C_2H_2	acetylene

*For historical reasons, the molecular formulas of binary hydrogen compounds of Group 15 elements are written with the Group 15 element first.

D.4 Names of Inorganic Molecular Compounds

Many simple inorganic molecular compounds are named by using the Greek prefixes in Table D.2 to indicate the number of each type of atom present. A prefix is not normally used if only one atom of an element is present; for example, NO_2 is nitrogen dioxide. An important exception to this rule is carbon monoxide, CO. When naming common binary molecular compounds—molecular compounds built from two elements—name the element that occurs farther to the right in the periodic table second, with its ending changed to *-ide*:

   phosphorus trichloride, PCl_3 dinitrogen oxide, N_2O
   sulfur hexafluoride, SF_6 dinitrogen pentoxide, N_2O_5

Some exceptions to these rules are the phosphorus oxides and compounds that are generally known by their common names. The phosphorus oxides are distinguished by the oxidation number of phosphorus, which is written as though phosphorus were a metal and the oxygen present as O^{2-}. Thus, P_4O_6 is named phosphorus(III) oxide as though it were $(P^{3+})_4(O^{2-})_6$, and P_4O_{10} is named phosphorus(V) oxide as though it were $(P^{5+})_4(O^{2-})_{10}$. These compounds, though, are molecular. Some binary molecular compounds, such as NH_3 and H_2O, have widely used common names (**TABLE D.3**).

In both the names and the molecular formulas of compounds formed between hydrogen and nonmetals in Group 16 or 17, hydrogen is written first; for example, the formula for hydrogen chloride is HCl and that for hydrogen sulfide is H_2S. Note that, when these compounds are dissolved in water, most act as acids and are named as acids. Binary acids are named by adding the prefix *hydro-* and changing the ending of the name of the second element to *-ic acid*, as in hydrochloric acid for HCl in water and hydrosulfuric acid for H_2S in water. An **aqueous solution,** a solution in water, is indicated by (aq) after the formula. Thus, HCl, the compound itself, is hydrogen chloride, and HCl(aq), its aqueous solution, is hydrochloric acid.

An **oxoacid** is an acidic molecular compound that contains oxygen. Oxoacids are the parents of the oxoanions in the sense that an oxoanion is formed by removing one or more hydrogen ions from an oxoacid molecule (see Table D.1). In general, *-ic* oxoacids are the parents of *-ate* oxoanions and *-ous* oxoacids are the parents of *-ite* oxoanions. For example, H_2SO_4, sulfuric acid, is the parent acid of the sulfate ion, SO_4^{2-}. Similarly, the parent acid of the sulfite ion, SO_3^{2-}, is the molecular compound H_2SO_3, sulfurous acid.

Toolbox D.2 HOW TO NAME SIMPLE INORGANIC MOLECULAR COMPOUNDS

CONCEPTUAL BASIS

The name of a compound should be unambiguous and as simple as possible. The systematic name of a molecular compound typically specifies the elements present and the numbers of atoms of each element in the molecule.

PROCEDURE

Determine the type of compound and then apply the corresponding rules.

Binary molecular compounds other than acids

The compound is generally not an acid if its formula does not begin with H.

Step 1 Write the name of the first element, followed by the name of the second, with its ending changed to *-ide*.

Step 2 Add Greek prefixes to indicate the number of atoms of each element; *mono-* is usually omitted.
Exceptions are the phosphorus oxides, which are named as if they were ionic compounds.

See Example D.2(a) and (b).

Acids

An inorganic acid has a formula that typically begins with H; oxoacids have formulas that begin with H and end in O. Distinguish between binary hydrides, such as HX, which are not named as acids, and their aqueous solutions, HX(aq), which are.

If the compound is a binary acid in solution, add *hydro . . . ic acid* to the root of the element's name.

See Example D.2(c).

If the compound is an oxoacid, derive the name of the acid from the name of the polyatomic ion that it produces, as in Toolbox D.1. In general,

-ate ions come from *-ic acids*
-ite ions come from *-ous acids*

Retain any prefix, such as *hypo-* or *per-*. See Appendix 3A for a systematic method for naming oxoacids.

See Example D.2(d).

EXAMPLE D.2   Naming inorganic molecular compounds

Give the systematic names of (a) N_2O_4, once studied as a rocket fuel; (b) ClO_2, which is used to disinfect water; (c) HI(aq), which is used in the synthesis of methamphetamine; and (d) HNO_2, which is formed in the atmosphere during lightning strikes.

PLAN Proceed as in Toolbox D.2.

SOLVE

Compounds (a) and (b) are not acids.

Step 1 Write the name of the first element, followed by the name of the second, with its ending changed to *-ide*.

(a) __Nitrogen __oxide; (b) __Chlorine __oxide

Step 2 Add Greek prefixes to indicate the number of atoms of each element. "Mono-" is usually omitted.

(a) The molecule N_2O_4 has two nitrogen atoms and four oxygen atoms.

Dinitrogen tetroxide

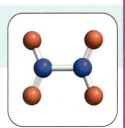

(b) A ClO_2 molecule has one chlorine atom and two oxygen atoms.

Chlorine dioxide

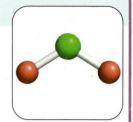

Compounds (c) and (d) are acids.

(c) Because the compound is a binary acid in solution, add "*hydro . . . ic acid*" to the root of the element's name. HI(aq) is a binary acid formed when hydrogen iodide dissolves in water.

Hydroiodic acid (The molecular compound HI is hydrogen iodide.)

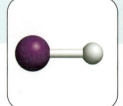

(d) Because the compound is an oxoacid, derive the name of the acid from the name of the polyatomic ion that it produces, as in Toolbox D.1.

HNO_2 is an oxoacid, the parent acid of the nitrite ion, NO_2^-.

Nitrous acid

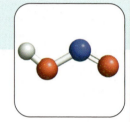

Self-test D.3A Name (a) HCN(aq); (b) BCl_3; (c) IF_5.

[**Answer:** (a) Hydrocyanic acid; (b) boron trichloride; (c) iodine pentafluoride]

Self-test D.3B Name (a) PCl_3; (b) SO_3; (c) HBr(aq).

Related Exercises D.9–D.12

EXAMPLE D.3   Writing the formula of a binary compound from its name

Write the formulas of (a) cobalt(II) chloride hexahydrate, a pink compound used to indicate humidity; and (b) diboron trisulfide, which is used to make some lithium-ion batteries.

PLAN First check to see whether the compounds are ionic or molecular. Many compounds that contain a metal are ionic. Write the symbol of the metal first, followed by the symbol of the nonmetal. The charges on the ions are determined as shown in Examples C.1 and C.2. Subscripts are chosen to balance charges. Compounds of two nonmetals are normally molecular. Write their formulas by listing the symbols of the elements in the same order as in the name, with subscripts corresponding to the Greek prefixes used.

SOLVE

(a) Determine if the compound is ionic or molecular.

Metal and nonmetal, so ionic.

Determine the charge on the cation.

The II in cobalt(II) indicates a +2 charge, so Co^{2+}.

Determine the charge on the anion.

Cl is in Group 17, so has the charge $17 - 18 = -1$, so Cl^-.

Balance charges.

2 Cl^- ions are required for each Co^{2+} ion, so $CoCl_2$.

Add waters of hydration.

Hexahydrate, so 6 water molecules are attached: $CoCl_2 \cdot 6H_2O$.

(b) Determine if the compound is ionic or molecular.

Two nonmetals, so molecular.

Convert Greek prefixes into subscripts.

di = 2; tri = 3, so B_2S_3.

Self-test D.4A Write the formulas for (a) vanadium(V) oxide; (b) magnesium carbide; (c) germanium tetrafluoride; (d) dinitrogen trioxide.

[*Answer:* (a) V_2O_5; (b) Mg_2C; (c) GeF_4; (d) N_2O_3]

Self-test D.4B Write the formulas for (a) cesium sulfide tetrahydrate; (b) manganese(VII) oxide; (c) hydrogen cyanide (a poisonous gas); (d) disulfur dichloride.

Related Exercises D.3, D.4, D.15, D.16

Binary molecular compounds are named by using Greek prefixes to indicate the number of atoms of each element present; the element named second has its ending changed to -ide.

D.5 The Nomenclature of Some Common Organic Compounds

There are millions of organic compounds; many consist of highly intricate molecules, so their names can be very complicated. However, for most of this text, you will need to know only a few simple organic compounds, and this section will introduce some of them.

Compounds of hydrogen and carbon are called **hydrocarbons.** They include methane, CH_4 (**1**); ethane, C_2H_6 (**2**); and benzene, C_6H_6 (**3**). Hydrocarbons that have no carbon–carbon multiple bonds are called **alkanes.** Thus, methane and ethane are both alkanes. The unbranched alkanes with up to 12 carbon atoms are listed in **TABLE D.4**. Notice that Greek prefixes are used to name all the alkanes with five or more carbon atoms. Hydrocarbons with double bonds are called **alkenes.** Ethene, $CH_2{=}CH_2$, is the simplest example of an alkene. It used to be (and still widely is) called ethylene. Benzene is a hydrocarbon with double bonds, but it has such distinctive properties that it is regarded as the parent hydrocarbon of a separate class of compounds called **aromatic**

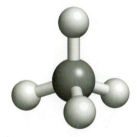

1 Methane, CH_4

In industrial chemistry alkenes are commonly called *olefins* from the French word *oléfiant*, meaning "oil-forming."

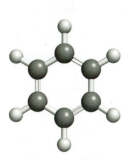

2 Ethane, C_2H_6

3 Benzene, C_6H_6

TABLE D.4 Alkane Nomenclature

Number of carbon atoms	Formula	Name of alkane	Name of alkyl group
1	CH_4	methane	methyl
2	CH_3CH_3	ethane	ethyl
3	$CH_3CH_2CH_3$	propane	propyl
4	$CH_3(CH_2)_2CH_3$	butane	butyl
5	$CH_3(CH_2)_3CH_3$	pentane	pentyl
6	$CH_3(CH_2)_4CH_3$	hexane	hexyl
7	$CH_3(CH_2)_5CH_3$	heptane	heptyl
8	$CH_3(CH_2)_6CH_3$	octane	octyl
9	$CH_3(CH_2)_7CH_3$	nonane	nonyl
10	$CH_3(CH_2)_8CH_3$	decane	decyl
11	$CH_3(CH_2)_9CH_3$	undecane	undecyl
12	$CH_3(CH_2)_{10}CH_3$	dodecane	dodecyl

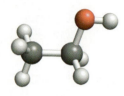

4 Ethanol, CH_3CH_2OH

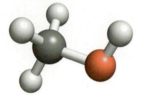

5 Methanol, CH_3OH

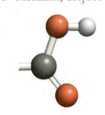

6 Carboxyl group, —COOH

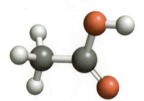

7 Acetic acid, CH_3COOH

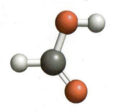

8 Formic acid, HCOOH

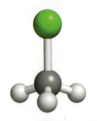

9 Chloromethane, CH_3Cl

10 Trichloromethane, $CHCl_3$

compounds. The hexagonal benzene ring is exceptionally stable and can be found in many important compounds. Specific groups of atoms derived from hydrocarbons, such as —CH_3, methyl, and —CH_2CH_3, ethyl, are named by replacing the ending of the parent hydrocarbon's name by -yl.

The hydrocarbons are the basic framework for all organic compounds. Different classes of organic compounds have one or more of the hydrogen atoms replaced by other atoms or groups of atoms. All you need to be aware of at this stage are the three classes of compounds known as alcohols, carboxylic acids, and haloalkanes:

An **alcohol** is a type of organic compound that contains an —OH group.

Ethanol, CH_3CH_2OH (**4**), the "alcohol" of beer and wine, is an ethane molecule in which one H atom has been replaced by an —OH group; and CH_3OH (**5**) is the toxic alcohol called methanol, or wood alcohol.

A **carboxylic acid** is a compound that contains the carboxyl group, —COOH (**6**).

The most common example is acetic acid, CH_3COOH (**7**), the acid that gives vinegar its sharp taste. Formic acid, HCOOH (**8**), is the acid of ant venom.

A **haloalkane** is an alkane in which one or more H atoms have been replaced by halogen atoms.

Haloalkanes include chloromethane, CH_3Cl (**9**), and trichloromethane, $CHCl_3$ (**10**). The latter compound has the common name chloroform and was an early anesthetic. Notice that the names are formed by shortening the name of the halogen to fluoro-, chloro-, bromo-, or iodo- and adding a Greek prefix indicating the number of halogen atoms.

Self-test D.5A (a) Name CH_2BrCl. (b) What type of compound is $CH_3CH(OH)CH_3$?

[*Answer:* (a) Bromochloromethane; (b) alcohol]

Self-test D.5B (a) Name $CH_3CH_2CH_2CH_2CH_3$. (b) What type of compound is CH_3CH_2COOH?

The names of organic compounds are based on the names of the parent hydrocarbons; alcohols contain —OH groups, carboxylic acids contain —COOH groups, and haloalkanes contain halogen atoms.

What have you learned in Fundamentals D?

You have learned to write the names of chemical compounds from their formulas and to write the formulas of compounds from their names. You have also learned to identify some simple organic compounds.

The skills you have mastered are the ability to:

☐ **1.** Name ions and ionic compounds with common polyatomic ions, and hydrates (Self-Test D.1, Toolbox D.1, and Example D.1).

☐ **2.** Name binary inorganic molecular compounds and oxoacids (Toolbox D.2 and Example D.2).

☐ **3.** Write the formulas of binary inorganic compounds from their names (Example D.3).

☐ **4.** Name simple hydrocarbons and substituted methane (Section D.5).

☐ **5.** Identify alcohols, carboxylic acids, and haloalkanes from their formulas (Self-Test D.5).

Fundamentals D Exercises

D.1 Write (a) the name of the BrO_2^- ion and (b) the formula of the hydrogen sulfite ion.

D.2 Write (a) the name of the CN^- ion and (b) the formula of the periodate ion.

D.3 Write the formula of (a) manganese(II) chloride; (b) calcium phosphate; (c) aluminum sulfite; (d) magnesium nitride.

D.4 Write the formula of (a) barium hydroxide; (b) cobalt(II) phosphite; (c) manganese(II) bromide; (d) chromium(III) sulfide.

D.5 Name each of the following compounds: (a) PF_5; (b) IF_3; (c) OF_2; (d) B_2Cl_4; (e) $CoSO_4 \cdot 7H_2O$; (f) $HgBr_2$; (g) $Fe_2(HPO_4)_3$; (h) W_2O_5; (i) $OsBr_3$.

D.6 Name each of the following compounds: (a) P_2S_5; (b) SO_3; (c) SiO_2; (d) S_2F_4; (e) $Co_3(PO_4)_2 \cdot 8H_2O$; (f) Cr_2O_3; (g) $Ir(HCO_3)_3$; (h) $Sr(BrO)_2$; (i) $MoBr_4$.

D.7 Name the following ionic compounds. Write both the old and the modern names wherever appropriate. (a) $Ca_3(PO_4)_2$, the major inorganic component of bones; (b) SnS_2; (c) V_2O_5; (d) Cu_2O.

D.8 The following ionic compounds are commonly found in laboratories. Write the modern name of (a) $NaHCO_3$ (baking soda); (b) Hg_2Cl_2 (calomel); (c) $NaOH$ (lye); (d) ZnO (calamine).

D.9 Name each of the following binary molecular compounds: (a) SF_6; (b) N_2O_5; (c) NI_3; (d) XeF_4; (e) $AsBr_3$; (f) ClO_2.

D.10 The following compounds are often found in chemical laboratories. Name each compound: (a) SiO_2 (silica); (b) SiC (carborundum); (c) N_2O (a general anesthetic); (d) P_4O_{10} (a drying agent for organic solvents); (e) CS_2 (a solvent); (f) SO_2 (a bleaching agent); (g) NH_3 (a common reagent).

D.11 The following aqueous solutions are common laboratory acids. What are their names? (a) $HCl(aq)$; (b) $H_2SO_4(aq)$; (c) $HNO_3(aq)$; (d) $CH_3COOH(aq)$; (e) $H_2SO_3(aq)$; (f) $H_3PO_4(aq)$.

D.12 Name each of the following acids: (a) $H_2SeO_4(aq)$; (b) $HClO_2(aq)$; (c) $HIO_4(aq)$; (d) $H_3PO_3(aq)$; (e) $HBrO(aq)$; (f) $H_2S(aq)$.

D.13 Write the formula of (a) perchloric acid; (b) hypochlorous acid; (c) hypoiodous acid; (d) hydrofluoric acid; (e) phosphorous acid; (f) periodic acid.

D.14 Write the formula of (a) nitrous acid; (b) carbonic acid; (c) hydroselenic acid; (d) bromous acid; (e) iodic acid; (f) telluric acid.

D.15 Write the formula of (a) titanium dioxide; (b) silicon tetrachloride; (c) carbon disulfide; (d) sulfur tetrafluoride; (e) lithium sulfide; (f) antimony pentafluoride; (g) dinitrogen pentoxide; (h) iodine heptafluoride.

D.16 Write the formula of (a) dinitrogen pentoxide; (b) hydrogen iodide; (c) oxygen difluoride; (d) phosphorus trichloride; (e) sulfur trioxide; (f) carbon tetrabromide; (g) bromine trifluoride.

D.17 Write the formula for the ionic compound formed from (a) zinc and fluoride ions; (b) barium and nitrate ions; (c) silver and iodide ions; (d) lithium and nitride ions; (e) chromium(III) and sulfide ions.

D.18 Write the formula for the ionic compound formed from (a) calcium and bromide ions; (b) ammonium and phosphite ions; (c) cesium and oxide ions; (d) gallium and sulfide ions; (e) lithium and nitride ions.

D.19 (a) Write the formula for and name the compound formed when the element in Group 2, Period 6, of the periodic table combines with the element in Group 17, Period 3. (b) Is the compound molecular or ionic?

D.20 (a) Write the formula for and name the compound formed when the element in Group 13, Period 3, of the periodic table combines with the element in Group 16, Period 2. (b) Is the compound molecular or ionic?

D.21 Name each of the following compounds: (a) Na_2SO_3; (b) Fe_2O_3; (c) FeO; (d) $Mg(OH)_2$; (e) $NiSO_4 \cdot 6H_2O$; (f) PCl_5; (g) $Cr(H_2PO_4)_3$; (h) As_2O_3; (i) $RuCl_2$.

D.22 Name each of the following compounds: (a) $CrBr_2 \cdot 6H_2O$; (b) $Co(NO_3)_3 \cdot 6H_2O$; (c) $InCl_3$; (d) BrF; (e) CrO_3; (f) $Cd(NO_2)_2$; (g) $Ca(ClO_3)_2$; (h) $Ni(ClO_2)_2$; (i) V_2O_5.

D.23 Each of the following compounds is named incorrectly. Correct each of the mistakes. (a) $CuCO_3$, copper(I) carbonate; (b) K_2SO_3, potassium sulfate; (c) $LiCl$, lithium chlorine.

D.24 Each of the following compounds has an incorrect formula. Correct each of the mistakes. (a) sodium sulfate, $NaSO_4$; (b) magnesium chlorite, $Mg(ClO)_2$; (c) phosphorus(V) oxide, P_2O_{10}.

D.25 Name each of the following organic compounds: (a) $CH_3CH_2CH_2CH_2CH_2CH_2CH_3$; (b) $CH_3CH_2CH_3$; (c) $CH_3CH_2CH_2CH_2CH_3$; (d) $CH_3CH_2CH_2CH_3$.

D.26 Name each of the following organic compounds: (a) CH_4; (b) CH_3F; (c) CH_3Br; (d) CH_3I.

D.27 You come across some old bottles in a storeroom that are labeled (a) cobaltic oxide monohydrate, (b) cobaltous hydroxide. Using Appendix 3C as a guide, write their modern names and chemical formulas.

D.28 The formal rules of chemical nomenclature result in a certain compound used for electronic components being called barium titanate(IV), in which the oxidation number of titanium is $+4$. See if you can work out its likely chemical formula. When you have identified the rules for naming oxoanions, suggest a formal name for H_2SO_4.

D.29 A main-group element E in Period 3 forms the molecular compound EH_4 and the ionic compound Na_4E. Identify element E and write the names of the compounds described.

D.30 A main-group element E in Period 5 forms the ionic compounds EBr_2 and EO. Identify element E and write the names of the compounds described.

D.31 The names of some compounds of hydrogen are exceptions to the usual rules of nomenclature. Look up the following compounds, write their names, and identify them as ionic or molecular: (a) $LiAlH_4$; (b) NaH.

D.32 The names of some compounds of oxygen are exceptions to the usual rules of nomenclature. Look up the following compounds, write their names, and identify them as ionic or molecular: (a) KO_2; (b) Na_2O_2; (c) CsO_3.

D.33 Name each of the following compounds, using analogous compounds with phosphorus and sulfur as a guide: (a) H_2SeO_4; (b) Na_3AsO_4; (c) $CaTeO_3$; (d) $Ba_3(AsO_4)_2$; (e) H_3SbO_4; (f) $Ni_2(SeO_4)_3$.

D.34 Name each of the following compounds, using analogous compounds with phosphorus and sulfur as a guide: (a) AsH_3; (b) H_2Se; (c) Cu_2TeO_4; (d) $Ca_3(AsO_3)_2$; (e) NaH_2SbO_4; (f) $BaSeO_3$.

D.35 What type of organic compound is (a) $CH_3CH_2CH_2OH$; (b) $CH_3CH_2CH_2CH_2COOH$; (c) CH_3F?

D.36 What type of organic compound is (a) $CH_3CH_2CH_3$; (b) CH_3CH_2Br; (c) CH_3CH_2COOH?

E.1   *The Mole*

E.2   *Molar Mass*

E Moles and Molar Masses

Astronomically large numbers of molecules are present even in a tiny sample: even 1 mL of water contains 3×10^{22} H_2O molecules, more molecules than there are stars in the visible universe. How can you find out the numbers that are present, and how do you report them in a simple, clear way? It is difficult to imagine and inconvenient to refer to large numbers such as 3×10^{22} molecules, just as wholesalers find it inconvenient to report individual items instead of dozens (12) or gross (144). To keep track of the enormous numbers of atoms, ions, or molecules in a sample, you need an efficient way of reporting them.

E.1 *The Mole*

Chemists report numbers of atoms, ions, and molecules in terms of a unit called a *mole*. A mole is the analog of the wholesaler's dozen, which is defined as 12 objects. In the case of a mole,

> 1 mole of objects contains the same number of objects as there are atoms in exactly 12 g of carbon-12 (**FIG. E.1**).

FIGURE E.1 The definition of the mole: exactly 12 g of carbon-12 corresponds to exactly 1 mol of carbon-12 atoms. There is exactly an Avogadro's number of atoms in the pile.

How is that number determined? First, you should note that a "dozen" could be defined in a roundabout way as the number of soda cans in a "twelve pack" carton. Even if you could not open the carton to count the number of cans inside, you could find out how many cans are inside it by weighing the carton and dividing its mass by the mass of one can. Similarly, to "count" the atoms in 12 g of carbon-12, you could divide the mass of 12 g of carbon-12 by the mass of one atom. The mass of a carbon-12 atom has been found by mass spectrometry to be $1.992\ 65 \times 10^{-23}$ g. It follows that the number of atoms in exactly 12 g of carbon-12 is

$$\text{Number of C atoms} = \frac{\overbrace{12\ \text{g [exactly]}}^{\text{mass of sample}}}{1.992\ 65 \times 10^{-23}\ \text{g}} = 6.0221 \times 10^{23}$$

Because the mole is equal to this number, the definition can be applied to *any* object, not just carbon atoms (**FIG. E.2**):

> **1 mole** of objects means 6.0221×10^{23} of those objects.

Therefore, 1 mole of atoms of any element, 1 mole of ions, and 1 mole of molecules each contain 6.0221×10^{23} atoms, ions, and molecules, respectively. The symbol for the unit mole is mol; that is, mol is to mole as g is to gram.

Just as 1 g and 1 m are units for physical properties, so too is the unit 1 mol. The mole is the unit for the physical property formally called the **amount of substance,** n. However, that name has found little favor among chemists, who colloquially refer to n as "the number of moles." A compromise, which is gaining acceptance, is to refer to n as the "chemical amount" or simply "amount" of entities present in a sample. Thus, 1.0000 mol of hydrogen atoms (6.0221×10^{23} hydrogen atoms), which is written 1.0000 mol H, is the chemical amount of hydrogen atoms in the sample. Take the advice of your instructor on whether to use the formal term. Like any SI unit, the mole can be used with prefixes. For example, 1 mmol = 10^{-3} mol and 1 μmol = 10^{-6} mol. Chemists encounter such small quantities when dealing with rare or expensive natural products and pharmaceuticals.

The number of objects per mole, 6.0221×10^{23} mol^{-1}, is called **Avogadro's constant,** N_A, in honor of the nineteenth-century Italian scientist Amedeo Avogadro (**FIG. E.3**),

The name *mole* comes from the Latin word for "massive heap." Coincidentally, the animal of the same name makes massive heaps of soil on lawns.

FIGURE E.2 Each sample consists of 1 mol of atoms of the element. Clockwise from the upper right are 32 g of sulfur, 201 g of mercury, 207 g of lead, 64 g of copper, and 12 g of carbon. (© 1991 Chip Clark–Fundamental Photographs.)

who helped to establish the existence of atoms. Avogadro's constant is used to convert between the chemical amount (the number of moles) and the number of atoms, ions, or molecules in that amount:

Number of objects = amount in moles × number of objects per mole
= amount in moles × Avogadro's constant

When the number of objects is denoted N and the amount of substance (in moles) is denoted n, this relation becomes

$$N = nN_A \qquad \qquad (1)$$

A Note on Good Practice: Avogadro's constant is a constant with units, not a pure number. You will often hear people referring to *Avogadro's number*: this is the pure number 6.0221×10^{23}.

To be unambiguous when using moles, you should be specific about which entities (that is, which atoms, molecules, formula units, or ions) you mean. For example, hydrogen occurs naturally as a gas, with each molecule built from two atoms, which is why it is denoted H_2. Write 1 mol H if you mean hydrogen atoms or 1 mol H_2 if you mean hydrogen molecules; do not write "1 mol hydrogen," because that is ambiguous. Note that 1 mol H_2 corresponds to 2 mol H.

FIGURE E.3 Lorenzo Romano Amedeo Carlo Avogadro, Count of Quaregna and Cerreto (1776–1856). (SPL/Science Source.)

EXAMPLE E.1   Converting number of atoms into an amount in moles

 Nanotechnology researchers have developed a hydrogen storage device capable of storing large quantities of hydrogen. This kind of research is vitally important for finding ways to transport hydrogen safely and economically in "green" vehicles. Suppose you have developed a hydrogen storage device that can store 1.29×10^{24} hydrogen atoms. What is the chemical amount (in moles) of hydrogen atoms that can be stored?

ANTICIPATE Because the number of atoms in the sample is greater than 6×10^{23}, you should anticipate that more than 1 mol of hydrogen atoms is present.

PLAN Rearrange Eq. 1 into $n = N/N_A$ and then substitute the data.

SOLVE

From $n = N/N_A$,

$$n = \frac{1.29 \times 10^{24} \text{ H}}{6.0221 \times 10^{23} \text{ mol}^{-1}} = 2.14 \text{ mol H}$$

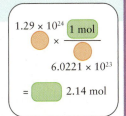

EVALUATE There is indeed more than 1 mol of hydrogen atoms present in the device, as anticipated.

Self-test E.1A A sample of a drug extracted from a fruit used by the Achuar Jivaro people of Peru to treat fungal infections is found to contain 2.58×10^{24} oxygen atoms. What is the amount (in moles) of O atoms in the sample?

[**Answer:** 4.28 mol O]

Self-test E.1B A double espresso contains 3.14 mol H_2O. What is the number of H atoms present in the espresso?

Related Exercises E.7, E.8, E.18

The amounts of atoms, ions, or molecules in a sample are expressed in moles, and Avogadro's constant, N_A, is used to convert between numbers of these particles and amount in moles.

E.2 Molar Mass

How can you determine the amount of atoms present if you can't count the atoms directly? The amount of substance can be found if you know the mass of the sample and the **molar mass**, M, the mass per mole of particles:

The molar mass of an *element* is the mass per mole of its *atoms*.

The molar mass of a *molecular* compound is the mass per mole of its *molecules*.

The molar mass of an *ionic compound* is the mass per mole of its *formula units*.

The units of molar mass in each case are grams per mole ($g \cdot mol^{-1}$). The mass of a sample is the amount (in moles) multiplied by the mass per mole (the molar mass),

$$\text{Mass of sample} = \text{amount in moles} \times \text{mass per mole}$$

Therefore, if the mass of the sample is denoted by m, then

$$m = nM \tag{2}$$

It follows that $n = m/M$. That is, to find the amount in moles, n, divide the mass, m, of the sample by the molar mass of the species present.

EXAMPLE E.2   Calculating the amount of atoms in a sample

Fluorine gas is so reactive that it reacts explosively with almost every other element. If you are working with fluorine and making quantitative observations, it is very important to know how much material you have. Calculate (a) the amount of F_2 and (b) the number of F atoms in 22.5 g of F_2. The molar mass of fluorine molecules is 38.00 $g \cdot mol^{-1}$, or, more specifically, 38.00 $g \cdot (mol\ F_2)^{-1}$.

ANTICIPATE Because the mass of fluorine present in the sample is less than the mass of 1 mol F_2, you should expect to find that less than 1 mol F_2 is present.

PLAN To find the amount in moles, divide the mass of the sample by the molar mass.

SOLVE

(a) From $n = m/M$,

$$n(F) = \frac{22.5\ g}{38.00\ g \cdot (mol\ F_2)^{-1}} = \frac{22.5}{38.00}\ mol\ F_2 = 0.592\ mol\ F_2$$

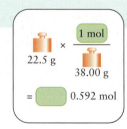

EVALUATE As expected, less than 1 mol F_2 is present.

A Note on Good Practice: To avoid ambiguity, specify the entities (in this case, fluorine molecules, F_2, and not fluorine atoms, F) in the units of the calculation.

(b) To calculate the actual number, N, of atoms in the sample, multiply the amount (in moles) by Avogadro's constant:

From $N = nN_A$,

$$N = (0.592 \text{ mol } F_2) \times (6.0221 \times 10^{23} \text{ mol}^{-1})$$
$$= 3.57 \times 10^{23} \text{ } F_2$$

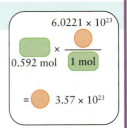

Because each F_2 molecule contains 2 F atoms,

$$N = 3.57 \times 10^{23} \text{ } F_2 \times \frac{2 \text{ F}}{1 \text{ } F_2} = 7.13 \times 10^{23} \text{ F}$$

That is, the sample contains 7.13×10^{23} fluorine atoms.

Self-test E.2A The mass of a copper coin is 3.20 g. Suppose it were pure copper. (a) How many moles of Cu atoms would the coin contain, given a molar mass of Cu of 63.55 g·mol^{-1}? (b) How many Cu atoms are present?

[***Answer:*** 0.0504 mol Cu; 3.03×10^{22} Cu atoms]

Self-test E.2B In one day, 5.4 kg of aluminum was collected from a recycling bin. (a) How many moles of Al atoms did the bin contain, given that the molar mass of Al is 26.98 g·mol^{-1}? (b) How many Al atoms were collected?

Related Exercises E.9, E.10, E.17

The molar masses of elements are determined by using mass spectrometry to measure the masses of the individual isotopes and their abundances. The mass per mole of atoms is the mass of an individual atom multiplied by Avogadro's constant (the number of atoms per mole):

$$M = m_{atom}N_A \tag{3a}$$

The greater the mass of an individual atom, the greater is the molar mass of the substance. However, most elements exist in nature as a mixture of isotopes. As explained in *Fundamentals B*, for instance, neon exists as three isotopes, each with a different mass. In chemistry, you almost always deal with natural samples of elements, which have the natural abundance of isotopes. The average atom mass is determined by calculating the weighted average, the sum of the product of the mass of each isotope, $m_{isotope}$, multiplied by its fractional abundance in a natural sample, $f_{isotope}$.

Σ means "take the sum of the following"

$$m_{atom, average} = \sum_{isotopes} f_{isotope}m_{isotope} \tag{3b}$$

The corresponding average molar mass is

$$M = m_{atom, average}N_A \tag{3c}$$

All molar masses quoted in this text refer to these average values. Their values are given in Appendix 2D. They are also included in the periodic table inside the front cover and in the alphabetical list of elements inside the back cover.

EXAMPLE E.3   Evaluating an average molar mass

Chlorine, which is used to sanitize water supplies and swimming pools, has two naturally occurring isotopes: chlorine-35 and chlorine-37. The mass of an atom of chlorine-35 is 5.807×10^{-23} g, and that of an atom of chlorine-37 is 6.139×10^{-23} g. In a typical natural sample of chlorine, 75.77% of the sample is chlorine-35 and 24.23% is chlorine-37. What is the molar mass of a typical sample of chlorine?

ANTICIPATE Because the more abundant isotope is chlorine-35, you should expect to find that the molar mass of a typical sample will be only slightly greater than 35 g·mol^{-1}.

PLAN First calculate the average atomic mass of the isotopes by adding together the individual masses, each multiplied by the fraction that represents its abundance. Then obtain the molar mass, the mass per mole of atoms, by multiplying the average atomic mass by Avogadro's constant.

SOLVE

From Eq. 3b, $m_{atom, average} = f_{chlorine-35}m_{chlorine-35} + f_{chlorine-37}m_{chlorine-37}$,

$$m_{atom, average} = 0.7577 \times (5.807 \times 10^{-23}\ g) + 0.2423 \times (6.139 \times 10^{-23}\ g)$$

$$= 5.887 \times 10^{-23}\ g$$

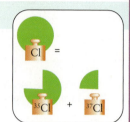

It follows from Eq. 3c that the molar mass of a typical sample of chlorine atoms is

From $M = m_{atom, average}N_A$,

$$M = (5.887 \times 10^{-23}\ g) \times (6.0221 \times 10^{23}\ mol^{-1})$$

$$= 35.45\ g \cdot mol^{-1}$$

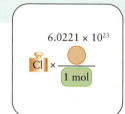

EVALUATE As expected, the molar mass is just over 35 g·mol^{-1}.

Self-test E.3A In a typical sample of magnesium, 78.99% is magnesium-24 (with atomic mass 3.983×10^{-23} g), 10.00% is magnesium-25 (4.149×10^{-23} g), and 11.01% is magnesium-26 (4.315×10^{-23} g). Calculate the molar mass of a typical sample of magnesium, given the atomic masses (in parentheses).

[**Answer:** 24.31 g·mol^{-1}]

Self-test E.3B Calculate the molar mass of copper, given that a natural sample typically consists of 69.17% copper-63, which has a molar mass of 62.94 g·mol^{-1}, and 30.83% copper-65, which has a molar mass of 64.93 g·mol^{-1}. (Note that the molar mass of an isotope is proportional to its atomic mass, so the procedure in Example E.3 can be followed, but there is no need to convert from atomic mass to molar mass at the end.)

Related Exercises E.11, E.12, E.14

To calculate the molar masses of compounds, use the molar masses of the elements present: *the molar mass of a compound is the weighted sum of the molar masses of the elements that make up the molecule or the formula unit*. It is necessary to note how many atoms or ions are present in the molecular formula or the formula unit of the ionic compound. For example, 1 mol of the ionic compound $Al_2(SO_4)_3$ contains 2 mol Al, 3 mol S, and 12 mol O. Therefore, the molar mass of $Al_2(SO_4)_3$ is

$$M(Al_2(SO_4)_3) = 2M(Al) + 3M(S) + 12M(O)$$

$$= 2(26.98\ g \cdot mol^{-1}) + 3(32.06\ g \cdot mol^{-1}) + 12(16.00\ g \cdot mol^{-1})$$

$$= 342.14\ g \cdot mol^{-1}$$

Self-test E.4A Calculate the molar mass of (a) ethanol, C_2H_5OH; (b) copper(II) sulfate pentahydrate.

[**Answer:** (a) 46.07 g·mol^{-1}; (b) 249.69 g·mol^{-1}]

Self-test E.4B Calculate the molar mass of (a) phenol, C_6H_5OH; (b) sodium carbonate decahydrate.

Two terms used throughout the chemical literature are *atomic weight* and *molecular weight*:

For the most authoritative and up-to-date listing of atomic weights, see the IUPAC periodic table at iupac.org/reports/periodic_table/.

The **atomic weight** of an element is the numerical value of its molar mass.

The **molecular weight** of a molecular compound or the **formula weight** of an ionic compound is the numerical value of its molar mass.

For instance, the atomic weight of hydrogen (molar mass 1.0079 $g \cdot mol^{-1}$) is 1.0079, the molecular weight of water (molar mass 18.02 $g \cdot mol^{-1}$) is 18.02, and the formula weight of sodium chloride (molar mass 58.44 $g \cdot mol^{-1}$) is 58.44. These three terms are traditional and deeply ingrained in chemical conversations, even though the numbers are not "weights." The mass of an object is a measure of the *quantity of matter* that it contains, whereas the weight of an object is a measure of the *gravitational pull* that it experiences. Mass and weight are proportional to each other, but they are not identical. An astronaut would have the same mass (contain the same quantity of matter) but different weights on Earth and on Mars.

Once you know the molar mass of a compound, you can use the same technique that was used for elements to determine how many moles of molecules or formula units are in a sample of a given mass.

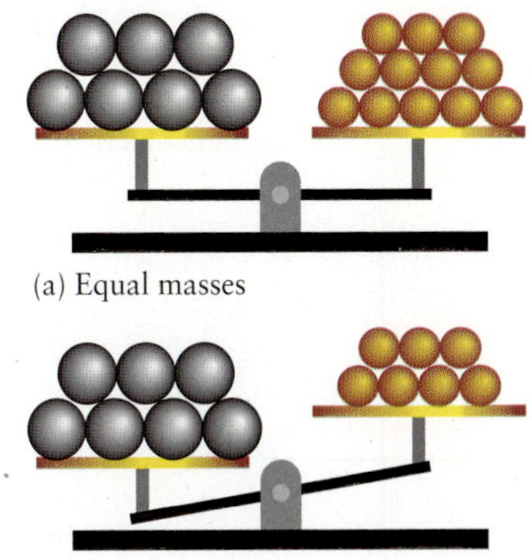

> **Self-test E.5A** Calculate the amount of urea molecules, $OC(NH_2)_2$, in 2.3×10^5 g of urea, which is used in facial creams and, on a bigger scale, as an agricultural fertilizer.
>
> [***Answer:*** 3.8×10^3 mol]
>
> **Self-test E.5B** Calculate the amount of $Ca(OH)_2$ formula units in 1.00 kg of slaked lime (calcium hydroxide), which is used to adjust the acidity of soils.

Molar mass is important when you want to know the number of atoms in a sample. It would be impossible to count out 6×10^{23} atoms of an element, but it is easy to measure out a mass equal to the molar mass of the element in grams. Each of the samples shown in Fig. E.2 was obtained in this way: each sample contains the same number of atoms of the element (6.022×10^{23}), but the masses vary because the masses of the atoms are different (**FIG. E.4**). The same rule applies to compounds. Because the molar mass of sodium chloride is 58.44 $g \cdot mol^{-1}$, if you measure out 58.44 g of sodium chloride, you obtain a sample that contains 1.000 mol NaCl formula units (**FIG. E.5**).

In practice, chemists rarely try to measure out a specific mass. Instead, they estimate the mass required and spoon out that mass approximately. Then they measure the mass of the sample precisely and convert it into an amount in moles (by using Eq. 2, $n = m/M$) to find the precise amount that they have obtained.

THINKING POINT

Why do chemists typically not measure out very precise, predetermined masses when making up solutions?

FIGURE E.4 (a) These two samples have the same mass but, because the atoms on the right are lighter than those on the left, more of them are required to balance the sample on the left. (b) The two samples contain the same number of atoms but, because the atoms on the right are lighter than those on the left, the mass of the sample on the right is the smaller of the two. Equal amounts (equal numbers of moles) of atoms do not necessarily have the same mass.

(a) Equal masses

(b) Equal amounts

FIGURE E.5 Each sample contains 1 mol of formula units of an ionic compound. From left to right are 58 g of sodium chloride (NaCl), 100 g of calcium carbonate ($CaCO_3$), 278 g of iron(II) sulfate heptahydrate ($FeSO_4 \cdot 7H_2O$), and 78 g of sodium peroxide (Na_2O_2). *(Chip Clark/ Fundamental Photographs, NYC.)*

EXAMPLE E.4   Calculating the mass of a sample corresponding to a given amount in moles

Potassium permanganate, $KMnO_4$, is very reactive and is used by film studios to stain new materials to make them look old. Suppose you are a film studio technician and are examining how different concentrations of the compound have different effects. You need about 0.10 mol $KMnO_4$ to prepare a solution. What mass (in grams) of the compound do you need?

ANTICIPATE You should expect the mass of $KMnO_4$ to be just a few grams, as only a tenth of a mole is required.

PLAN To find the mass of a stated amount of compound, multiply the amount in moles by the molar mass of the compound.

SOLVE To find the mass of $KMnO_4$ that corresponds to 0.10 mol $KMnO_4$, note that, because the molar mass of the compound is 158.04 g·mol⁻¹,

From $m = nM$,

$$m = (0.10 \text{ mol}) \times (158.04 \text{ g·mol}^{-1}) = 16 \text{ g}$$

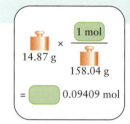

Therefore, you need to measure out about 16 g of $KMnO_4$.

EVALUATE As expected, the mass is small. If, when you weigh the sample, you find that its actual mass is 14.87 g, the amount of $KMnO_4$ actually weighed out is

From $n = m/M$,

$$n(KMnO_4) = \frac{14.87 \text{ g}}{158.04 \text{ g·}(\text{mol } KMnO_4)^{-1}}$$
$$= 0.094\,09 \text{ mol } KMnO_4$$

Self-test E.6A What mass of anhydrous sodium hydrogen sulfate should you weigh out to obtain about 0.20 mol $NaHSO_4$?

[**Answer:** About 24 g]

Self-test E.6B What mass of acetic acid should you weigh out to obtain 1.5 mol CH_3COOH?

Related Exercises E.25, E.26

The molar mass of a compound, the mass per mole of its molecules or formula units, is used to convert between the mass of a sample and the amount of molecules or formula units that it contains.

What have you learned in Fundamentals E?

You have learned how to use the concept of chemical amount, the meaning of the unit 1 mole (1 mol) and Avogadro's constant, and how to convert between mass in grams and amount in moles by using the molar mass. You know that an "atomic weight" (and similarly "molecular weight" and "formula weight") is a unitless average of atomic masses for the natural isotopic composition of a sample.

The skills you have mastered are the ability to:

☐ **1.** Use Avogadro's constant to convert between amount in moles and the number of atoms, molecules, or ions in a sample (Examples E.1 and E.2).

☐ **2.** Calculate the molar mass of an element, given its isotopic composition (Example E.3).

☐ **3.** Calculate the molar mass of a compound, given its chemical formula.

☐ **4.** Convert between mass and amount in moles by using the molar mass (Example E.4).

Fundamentals E Exercises

E.1 The field of nanotechnology offers some intriguing possibilities, such as the creation of fibers one atom wide. Suppose you were able to string together 1.00 mol Ag atoms, each of radius 144 pm, into one of these fibers by encapsulating them in carbon nanotubes. How long would the fiber extend?

E.2 If you won 1 mol of dollars in a lottery the day you were born, and spent 1 billion dollars a second for the rest of your life, what percentage of the prize money would remain, if any, when you decide to retire from spending, exhausted, at 90 years of age?

E.3 In a nanotechnology lab you might have the capability to manipulate individual atoms. The atoms on the left are gallium atoms (molar mass 70 g·mol^{-1}), and those on the right are atoms of astatine (molar mass 210 g·mol^{-1}). How many astatine atoms would the pan on the right have to contain for the masses on the two pans to be equal?

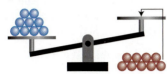

E.4 In a nanotechnology lab you might have the capability to manipulate individual atoms. The atoms on the left are silicon atoms (molar mass 28 g·mol^{-1}), and those on the right are atoms of lithium (molar mass 7 g·mol^{-1}). How many lithium atoms would the pan on the right have to contain for the masses on the two pans to be equal?

E.5 (a) The approximate population of Earth is 7.0 billion people. How many moles of people inhabit Earth? (b) If all people were pea pickers, how long would it take for the entire population of Earth to pick 1 mol of peas at the rate of one pea per second, working 24 hours per day, 365 days per year?

E.6 (a) One thousand tonnes (1000 t, 1 t = 10^3 kg) of sand contains about a trillion (10^{12}) grains of sand. How many tonnes of sand are needed to provide 1 mol of grains of sand? (b) Assuming the volume of a grain of sand is 1.0 mm^3 and the land area of the continental United States is 3.6 × 10^6 square miles, how deep would the sand pile over the United States be if this area were evenly covered with 1.0 mol of grains of sand?

E.7 A molecule of human DNA was found to contain 2.1 × 10^9 atoms of carbon. Calculate the chemical amount (in moles) of carbon atoms in the molecule of DNA.

E.8 "Artificial fossils" consisting of thin sheets of TiO_2 are created by depositing a film of TiO_2 on filter paper, which is later removed. The TiO_2 film retains the imprint of the texture of the filter paper surface. In one preparation, 2.4 × 10^{21} TiO_2 formula units were deposited. Calculate the chemical amount (in moles) of TiO_2 in the film produced.

E.9 Epsom salts consist of magnesium sulfate heptahydrate. Write its formula. (a) How many atoms of oxygen are in 5.15 g of Epsom salts? (b) How many formula units of the compound are present in 5.15 g? (c) How many moles of water molecules are in 5.15 g of Epsom salts?

E.10 Copper metal can be extracted from a copper(II) sulfate solution by electrolysis. If 45.20 g of copper(II) sulfate pentahydrate, $CuSO_4·5H_2O$, is dissolved in 100 mL of water and all the copper is electroplated out, what mass of copper will be obtained?

E.11 The nuclear power industry extracts ^{6}Li but not ^{7}Li from natural samples of lithium. As a result, the molar mass of commercial samples of lithium is increasing. The current abundances of the two isotopes are 7.42% and 92.58%, respectively, and the

masses of their atoms are 9.988 × 10^{-24} g and 1.165 × 10^{-23} g, respectively. (a) What is the current molar mass of a natural sample of lithium? (b) What will the molar mass be when the abundance of ^{6}Li is reduced to 5.67%?

E.12 Calculate the molar mass of sulfur, S, in a sample that consists of 93.0% ^{32}S (molar mass 31.97 g·mol^{-1}), 1.2% ^{33}S (molar mass 32.97 g·mol^{-1}), and 5.8% ^{34}S (molar mass 33.97 g·mol^{-1}).

E.13 The molar mass of boron atoms in a natural sample is 10.81 g·mol^{-1}. The sample is known to consist of ^{10}B (molar mass 10.013 g·mol^{-1}) and ^{11}B (molar mass 11.093 g·mol^{-1}). What are the percentage abundances of the two isotopes?

E.14 Calculate the molar mass of the noble gas krypton, Kr, in a natural sample, which is 0.3% ^{78}Kr (molar mass 77.92 g·mol^{-1}), 2.3% ^{80}Kr (molar mass 79.91 g·mol^{-1}), 11.6% ^{82}Kr (molar mass 81.91 g·mol^{-1}), 11.5% ^{83}Kr (molar mass 82.92 g·mol^{-1}), 56.9% ^{84}Kr (molar mass 83.91 g·mol^{-1}), and 17.4% ^{86}Kr (molar mass 85.91 g·mol^{-1}).

E.15 The molar mass of the metal hydroxide $M(OH)_2$ is 74.10 g·mol^{-1}. What is the molar mass of the sulfide of this metal?

E.16 The molar mass of the metal oxide M_2O is 231.74 g·mol^{-1}. What is the molar mass of the chloride of this metal?

E.17 Which sample in each of the following pairs contains the greater number of moles of atoms? (a) 75 g of indium or 80 g of tellurium; (b) 15.0 g of P or 15.0 g of S; (c) 7.36 × 10^{27} atoms of Ru or 7.36 × 10^{27} atoms of Fe.

E.18 Calculate the mass, in micrograms, of (a) 3.27 × 10^{16} Hg atoms; (b) 963 nmol Hf atoms; (c) 5.50 μmol Gd atoms; (d) 6.02 × 10^{25} Sb atoms.

E.19 A report stated that the Sudbury Neutrino Observatory in Sudbury, Canada, uses 1.00 × 10^3 tonne (1 t = 10^3 kg) of heavy water, D_2O, in a spherical tank of diameter 12 m to detect neutrinos. The density of normal water (H_2O) at the temperature of the tank is 1.00 g·cm^{-3}. (a) Using a molar mass for deuterium of 2.014 g·mol^{-1}, calculate the molar mass of D_2O. (b) Assuming that the volume occupied by a D_2O molecule is the same as that occupied by an H_2O molecule, calculate the density of heavy water. (c) Calculate the volume of the tank in cubic meters from your density data and the given mass of the heavy water, then compare this volume with the volume of the tank calculated from its reported diameter. (d) Is the reported mass of heavy water accurate? (e) Is the assumption you made in part (b) reasonable? Explain your reasoning. (The volume of a sphere of radius r is $V = \frac{4}{3}\pi r^3$.)

E.20 A chemist orders 1.000 kg of D_2O from an isotope supplier, who claims that the compound is 98% pure (that is, it contains no more than 2% H_2O by mass). To check this claim, the chemist measures the density of the D_2O and finds it to be 1.10 g·cm^{-3}. The density of normal water (H_2O) at the temperature of the test is 1.00 g·cm^{-3}. Is the D_2O as pure as advertised? (Molar mass of deuterium is 2.014 g·mol^{-1}.) Make the same assumption as in Exercise E.19(b).

E.21 Calculate the amount (in moles) and the number of molecules and formula units (or atoms, if indicated) in (a) 10.0 g of alumina, Al_2O_3; (b) 25.92 mg of hydrogen fluoride, HF; (c) 1.55 mg of hydrogen peroxide, H_2O_2; (d) 1.25 kg of glucose, $C_6H_{12}O_6$; (e) 4.37 g of nitrogen as N atoms and as N_2 molecules.

E.22 Convert each of the following masses into amount (in moles) and into number of molecules (and atoms, if indicated): (a) 3.60 kg of H_2O; (b) 91 kg of benzene (C_6H_6); (c) 350.0 g of phosphorus, as P atoms and as P_4 molecules; (d) 1.2 g of CO_2; (e) 0.37 g of NO_2.

E.23 Calculate the amount (in moles) of (a) Cu^{2+} ions in 3.00 g of $CuBr_2$; (b) SO_3 molecules in 7.00×10^2 mg of SO_3; (c) F^- ions in 25.2 kg of UF_6; (d) H_2O in 2.00 g of $Na_2CO_3 \cdot 10H_2O$.

E.24 Calculate the amount (in moles) of (a) CN^- in 4.00 g of NaCN; (b) O atoms in 4.00×10^2 ng of H_2O; (c) $CaSO_4$ in 4.00 kg of $CaSO_4$; (d) H_2O in 4.00 mg of $Al_2(SO_4)_3 \cdot 8H_2O$.

E.25 (a) Determine the number of KNO_3 formula units in 0.750 mol KNO_3. (b) What is the mass (in milligrams) of 2.39×10^{20} formula units of Ag_2SO_4? (c) Estimate the number of $NaHCO_2$ formula units in 3.429 g of $NaHCO_2$, sodium formate, which is used in dyeing and printing fabrics.

E.26 (a) How many CaH_2 formula units are present in 6.177 g of CaH_2? (b) Determine the mass of 8.75×10^{21} formula units of $NaBF_4$, sodium tetrafluoroborate. (c) Calculate the amount (in moles) of 8.15×10^{20} formula units of CeI_3, cerium(III) iodide, a bright yellow, water-soluble solid.

E.27 (a) Calculate the mass, in grams, of one water molecule. (b) Determine the number of H_2O molecules in 1.00 kg of water.

E.28 Octane, C_8H_{18}, is typical of the molecules found in gasoline. (a) Calculate the mass of one octane molecule. (b) Determine the number of C_8H_{18} molecules in 1.00 mL of octane, the mass of which is 0.82 g.

E.29 A chemist measured out 8.61 g of copper(II) chloride tetrahydrate, $CuCl_2 \cdot 4H_2O$. (a) How many moles of $CuCl_2 \cdot 4H_2O$ were measured out? (b) How many moles of Cl^- ions are present in the sample? (c) How many H_2O molecules are present in the sample? (d) What fraction of the total mass of the sample is due to oxygen?

E.30 Anhydrous copper(II) sulfate is difficult to dry completely. What mass of copper(II) sulfate would remain after removing 90% of the water from 360. g of $CuSO_4 \cdot 5H_2O$?

E.31 Suppose you had bought 2.5 kg of $Na_2CO_3 \cdot 10H_2O$ for $175, instead of 2.5 kg of Na_2CO_3 for $195. (a) How much water did you buy, and how much did you pay per liter of water? (The mass of 1 L of water is 1 kg.) (b) What would have been a fair price for the hydrated compound, valuing water at zero cost?

E.32 A chemist wants to extract the gold from 13.62 g of gold(III) chloride dihydrate, $AuCl_3 \cdot 2H_2O$, from an aqueous solution. What mass of gold can be obtained from the sample?

E.33 Fluoridated toothpaste can slow the decay of teeth. The fluoride ions convert the hydroxyapatite, $Ca_5(PO_4)_3OH$, in tooth enamel into fluorapatite, $Ca_5(PO_4)_3F$. If all the hydroxyapatite is converted into fluorapatite, by what percentage does this conversion increase the mass of the enamel?

E.34 The antibiotic tetracycline has the formula $C_{22}H_{24}N_2O_8$. Safe dosage of the drug is 0.24 μmol per kilogram (of body mass) per day. If tetracycline is administered in four equally divided doses per day to a child weighing 45 lb, what mass of tetracycline should be present in each dose?

F The Determination of Composition

One way that scientists discover new drugs is by extracting a biologically active compound from a natural source (**FIG. F.1**). They then try to identify its molecular structure so that it can be improved or manufactured in greater quantities. The first steps in identifying the molecular structure are the determination of the "empirical formula" and the "molecular formula" of the compound.

The **empirical formula** of a compound expresses the *relative* numbers of atoms of each element present in it. For example, the empirical formula of glucose, which is CH_2O, shows that carbon, hydrogen, and oxygen atoms are present in the ratio 1:2:1. The elements are present in these proportions regardless of the size of the sample. The **molecular formula** shows the actual numbers of atoms of each element in a molecule. The molecular formula for glucose, which is $C_6H_{12}O_6$, shows that each glucose molecule consists of 6 carbon atoms, 12 hydrogen atoms, and 6 oxygen atoms (**1**). Because the empirical formula gives only the *ratios* of the numbers of atoms of each element, different compounds with different molecular formulas can have the same empirical formula. For example, formaldehyde, CH_2O (**2**), the preservative in formalin solution), acetic acid, $C_2H_4O_2$ (the acid in vinegar), and lactic acid, $C_3H_6O_3$ (the acid in sour milk), all have the same empirical formula (CH_2O) as that of glucose but are different compounds with different properties.

F.1 *Mass Percentage Composition*

To determine the empirical formula of a compound, the mass of each element present in a sample is measured. The result is usually reported as the **mass percentage composition**—that is, the mass of each element expressed as a percentage of the total mass:

$$\text{Mass percentage of element} = \frac{\text{mass of element in sample}}{\text{total mass of sample}} \times 100\% \qquad (1)$$

1 α-D-Glucose, $C_6H_{12}O_6$

2 Formaldehyde, CH_2O

FIGURE F.1 A marine zoologist surveys sea sponges that may contain compounds of medicinal value. Compounds found to have antifungal or antiviral properties are subjected to the kinds of analyses described in this section. (© Images & Stories/Alamy.)

Because the mass percentage composition is independent of the size of the sample—in the language of *Fundamentals* A, it is an intensive property—every sample of the substance has that same composition. A principal technique for determining the mass percentage composition of an unknown organic compound is *combustion analysis,* which is described in *Fundamentals* M.

Self-test F.1A Australian Aborigines used the leaves of the eucalyptus tree to alleviate sore throats and other pains. The primary active ingredient has been identified and named eucalyptol. An analysis of a sample of eucalyptol of total mass 3.16 g gave its composition as 2.46 g of carbon, 0.373 g of hydrogen, and 0.329 g of oxygen. Determine the mass percentages of carbon, hydrogen, and oxygen in eucalyptol.

[***Answer:*** 77.8% C, 11.8% H, 10.4% O]

Self-test F.1B The compound α-pinene, a natural antiseptic found in the resin of the piñon tree, has been used since ancient times by Zuni healers. A 7.50-g sample of α-pinene contains 6.61 g of carbon and 0.89 g of hydrogen. What are the mass percentages of carbon and hydrogen in α-pinene?

If the chemical formula of a compound is already known, its mass percentage composition can be obtained from that formula.

EXAMPLE F.1   Calculating the mass percentage of an element in a compound

Suppose you are working on the design of a photovoltaic cell that you hope will be able to use sunlight to split water into its elements in an economical manner. As a preliminary step, you need to know what mass of hydrogen can be produced from a given mass of water. What is the mass percentage of hydrogen in water?

ANTICIPATE Even though there are two hydrogen atoms in each H_2O molecule, because hydrogen atoms are much lighter than oxygen atoms they contribute only a small mass to each molecule and you should therefore expect only a small mass percentage contribution.

PLAN To calculate the mass percentage of hydrogen in water, find the mass of H atoms in 1 mol H_2O by noting that there are 2 mol H in 1 mol H_2O, dividing that mass by the mass of 1 mol H_2O, and multiplying by 100%.

SOLVE

From

$$\text{Mass percentage of H} = \frac{\text{mass of H atoms}}{\text{mass of } H_2O \text{ molecules}} \times 100\%$$

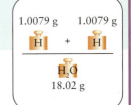

$$\text{Mass percentage of H} = \frac{(2 \text{ mol}) \times (1.0079 \text{ g·mol}^{-1})}{(1 \text{ mol}) \times (18.02 \text{ g·mol}^{-1})} \times 100\%$$

$$= 11.19\%$$

EVALUATE As expected, the mass percentage of hydrogen in water, 11.19%, is quite small.

Self-test F.2A Calculate the mass percentage of Cl in NaCl.

[***Answer:*** 60.66%]

Self-test F.2B Calculate the mass percentage of Ag in $AgNO_3$.

Related Exercises F.1, F.2, F.5, F.6, F.23

Mass percentage composition is found by calculating the fraction of the total mass contributed by each element present in a compound and expressing the fraction as a percentage.

F.2 Determining Empirical Formulas

To convert the mass percentages into an empirical formula, convert the mass percentage of each type of atom into the relative number of atoms of that element. The simplest procedure is to imagine that the mass of the sample is exactly 100 g, for then the mass percentage composition gives the mass in grams of each element. Then the molar mass of each element is used to convert these masses into amounts in moles and finally to the relative numbers of moles of each type of atom.

EXAMPLE F.2   Determining the empirical formula from mass percentage composition

In some cases, you might want to test a compound to see if it is the expected product. A first step is to determine its empirical formula. Suppose you sent a sample of an unknown compound, which you suspect to be vitamin C, to a laboratory for combustion analysis. The laboratory reported a composition of 40.9% carbon, 4.58% hydrogen, and 54.5% oxygen. Might it be vitamin C?

ANTICIPATE Because the molar mass of carbon is slightly less than that of oxygen but 12 times that of hydrogen, you should expect to find about the same number of atoms of each element in the compound.

PLAN Divide each mass percentage by the molar mass of the corresponding element to obtain the number of moles of that element found in exactly 100 g of the compound. Divide the number of moles of each element by the smallest number of moles. If fractional numbers result, then multiply by the factor that gives the smallest whole numbers of moles. Denote the amount of a substance J by $n(J)$.

SOLVE

The mass of each element X, $m(X)$, in exactly 100 g of the compound is equal to its mass percentage in grams.

$$m(C) = 40.9 \text{ g}$$
$$m(H) = 4.58 \text{ g}$$
$$m(O) = 54.5 \text{ g}$$

Convert each mass to an amount of atoms, $n(J)$, in moles by using the molar mass, $M(J)$, of the element, $n(J) = m(J)/M(J)$.

$$n(C) = \frac{40.9 \text{ g}}{12.01 \text{ g} \cdot (\text{mol C})^{-1}} = 3.41 \text{ mol C}$$

$$n(H) = \frac{4.58 \text{ g}}{1.0079 \text{ g} \cdot (\text{mol H})^{-1}} = 4.54 \text{ mol H}$$

$$n(O) = \frac{54.5 \text{ g}}{16.00 \text{ g} \cdot (\text{mol O})^{-1}} = 3.41 \text{ mol O}$$

Divide each amount by the smallest amount (3.41 mol).

$$\text{Carbon: } \frac{3.41 \text{ mol}}{3.41 \text{ mol}} = 1.00$$

$$\text{Hydrogen: } \frac{4.54 \text{ mol}}{3.41 \text{ mol}} = 1.33$$

$$\text{Oxygen: } \frac{3.41 \text{ mol}}{3.41 \text{ mol}} = 1.00$$

Because a compound has only whole numbers of atoms, multiply by the smallest factor that gives a whole number for each element.

Because 1.33 is $\frac{4}{3}$, multiply all the numbers by 3 to obtain the ratios 3.00:3.99:3.00, or approximately 3:4:3. The empirical formula is $C_3H_4O_3$.

EVALUATE As expected, there are similar numbers of atoms of each element in the formula.

Self-test F.3A Use the mass percentage composition of eucalyptol calculated in Self-Test F.1A to determine its empirical formula.

[**Answer:** $C_{10}H_{18}O$]

Self-test F.3B The mass percentage composition of the compound thionyl difluoride is 18.59% O, 37.25% S, and 44.16% F. Calculate its empirical formula.

Related Exercises F.9–F.12, F.15, F.16

The empirical formula of a compound is determined from the mass percentage composition and the molar masses of the elements present.

F.3 Determining Molecular Formulas

One more piece of information, the molar mass, is needed if you want to find the molecular formula of a molecular compound, because to find the molecular formula you need to decide how many empirical formula units are needed to account for the observed molar mass.

EXAMPLE F.3   Determining the molecular formula from the empirical formula

You are continuing your investigation of the compound from Example F.2 that you suspect is vitamin C and await results from a mass spectrometry laboratory. Mass spectrometry shows that the molar mass of the unknown sample is 176.12 $g \cdot mol^{-1}$. Given its empirical formula of $C_3H_4O_3$, what is the molecular formula of the compound? Can you be confident that it is indeed vitamin C?

ANTICIPATE You can estimate the molar mass of the empirical formula unit by noting that it contains three atoms of carbon, four of hydrogen, and three of oxygen, to give $(3 \times 12) + (4 \times 1) + (3 \times 16)$ $g \cdot mol^{-1} = 88$ $g \cdot mol^{-1}$, which is half the reported molar mass of the compound. Therefore, the molecular formula is expected to have twice as many atoms of each element.

PLAN To find the number of empirical formula units needed to account for the observed molar mass of a compound, divide the molar mass of the compound by the molar mass of the empirical formula unit.

SOLVE

The molar mass of a $C_3H_4O_3$ formula unit is:

$$\begin{aligned}
\text{Molar mass of } C_3H_4O_3 = &\; 3 \times (12.01 \text{ g} \cdot mol^{-1}) \\
&+ 4 \times (1.008 \text{ g} \cdot mol^{-1}) \\
&+ 3 \times (16.00 \text{ g} \cdot mol^{-1}) \\
\hline
= &\; 88.06 \text{ g} \cdot mol^{-1}
\end{aligned}$$

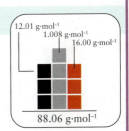

Divide the molar mass of the compound by the molar mass of the empirical formula unit:

$$\frac{\text{Molar mass of compound}}{\text{Molar mass of empirical formula unit}} = \frac{176.14 \text{ g} \cdot mol^{-1}}{88.06 \text{ g} \cdot mol^{-1}} = 2.000$$

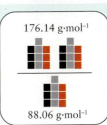

Multiply the coefficients in the empirical formula by the factor 2 to obtain the molecular formula.

$$2 \times (C_3H_4O_3), \text{ or } C_6H_8O_6$$

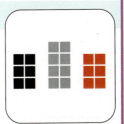

EVALUATE As expected, there are twice as many atoms of each element in the formula. The molecular formula is the same as that of vitamin C, so you can be reasonably confident about your identification but not certain because several different compounds might have the same molecular formula.

Self-test F.4A The molar mass of styrene, which is used in the manufacture of the plastic polystyrene, is 104 $g \cdot mol^{-1}$, and its empirical formula is CH. Deduce its molecular formula.

[*Answer:* C_8H_8]

Self-test F.4B The molar mass of oxalic acid, an acid present in rhubarb, is 90.0 $g \cdot mol^{-1}$, and its empirical formula is CHO_2. What is its molecular formula?

Related Exercises F.17–F.22

The molecular formula of a compound is found by determining how many empirical formula units are needed to account for the measured molar mass of the compound.

What have you learned in Fundamentals F?

You have learned how to determine the empirical and molecular formulas of a compound from its mass percentage composition.

The skills you have mastered are the ability to:

☐ **1.** Calculate the mass percentage of an element in a compound from a formula (Example F.1).

☐ **2.** Calculate the empirical formula of a compound from its mass percentage composition (Example F.2).

☐ **3.** Determine the molecular formula of a compound from its empirical formula and its molar mass (Example F.3).

Fundamentals F Exercises

F.1 Citral is a fragrant component of lemon oil that is used in colognes. It has the molecular structure shown. (a) Write the molecular formula of citral. (b) Calculate its mass percentage composition (dark gray = C, white = H, red = O).

F.2 The compound mainly responsible for the odor of musk produced by musk deer is muscone, which has the molecular structure shown. (a) Write the molecular formula of muscone. (b) Calculate its mass percentage composition (dark gray = C, white = H, red = O).

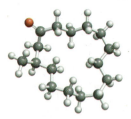

F.3 (a) Write the formula for nitric acid. (b) Without doing a calculation, estimate which element in nitric acid occurs with the greatest mass percentage.

F.4 (a) Write the formula for lithium sulfide. (b) Without doing a calculation, estimate which element in lithium sulfide occurs with the greater mass percentage.

F.5 What is the mass percentage composition of L-carnitine, $C_7H_{15}NO_3$, a compound that is taken as a dietary supplement to reduce muscle fatigue?

F.6 What is the mass percentage composition of the hormone testosterone, $C_{19}H_{28}O_2$?

F.7 A metal M forms an oxide with the formula M_2O, for which the mass percentage of the metal is 88.8%. (a) What is the molar mass of the metal? (b) Write the name of the compound.

F.8 A metal M forms an oxide with the formula M_2O_3, for which the mass percentage of the metal is 69.9%. (a) What is the identity of the metal? (b) Write the name of the compound.

F.9 Vanillin is found in vanilla extracted from Mexican orchids. A report on the analysis of vanillin listed the mass percentage composition as 63.15% C, 5.30% H, and 31.55% O. In what atom ratios are the atoms present in vanillin?

F.10 Cadaverine is formed in decomposing flesh. A report on the analysis of cadaverine listed the mass percentage composition as 58.77% C, 13.81% H, and 27.42% N. In what atom ratios are the atoms present in cadaverine?

F.11 Determine the empirical formulas from the following analyses. (a) The mass percentage composition of cryolite, a compound used in the production of aluminum, is 32.79% Na, 13.02% Al, and 54.19% F. (b) A compound used to generate O_2 gas in the laboratory has mass percentage composition 31.91% K and 28.93% Cl, the remainder being oxygen. (c) A fertilizer is found to have the following mass percentage composition: 12.2% N, 5.26% H, 26.9% P, and 55.6% O.

F.12 Determine the empirical formula of each of the following compounds from the data given. (a) Talc (used in talcum powder) has mass percentage composition 19.2% Mg, 29.6% Si, 42.2% O, and 9.0% H. (b) Saccharin, a sweetening agent, has mass percentage composition 45.89% C, 2.75% H, 7.65% N, 26.20% O, and 17.50% S. (c) Salicylic acid, used in the synthesis of aspirin, has mass percentage composition 60.87% C, 4.38% H, and 34.75% O.

F.13 In an experiment, 4.14 g of phosphorus combined with chlorine to produce 27.8 g of a white solid compound. (a) What is the empirical formula of the compound? (b) Assuming that the empirical and molecular formulas of the compound are the same, what is its name?

F.14 A chemist found that 4.69 g of sulfur combined with fluorine to produce 15.81 g of a gas. (a) What is the empirical formula of the gas? (b) Assuming that the empirical and molecular formulas of the compound are the same, what is its name?

F.15 Diazepam, a drug used to treat anxiety, has the mass percentage composition 67.49% C, 4.60% H, 12.45% Cl, 9.84% N, and 5.62% O. What is the empirical formula of the compound?

F.16 The compound fluoxetine is sold as the antidepressant Prozac when combined with HCl. Fluoxetine has the mass percentage composition 66.01% C, 5.87% H, 18.43% F, 4.53% N, and 5.17% O. What is the empirical formula of fluoxetine?

F.17 Osmium forms a molecular compound with mass percentage composition 15.89% C, 21.18% O, and 62.93% Os. (a) What is the empirical formula of this compound? (b) From the mass spectrum of the compound, the molecule was determined to have a molar mass of 907 g·mol^{-1}. What is its molecular formula?

F.18 Paclitaxel, which is extracted from the Pacific yew tree *Taxus brevifolia*, has antitumor activity for ovarian and breast cancer. It is sold under the trade name Taxol. Upon analysis, its mass percentage composition is 66.11% C, 6.02% H, and 1.64% N, with the balance being oxygen. (a) What is the empirical formula of paclitaxel? (b) The molar mass of paclitaxel is 853.91 g·mol^{-1}. What is its molecular formula?

F.19 Caffeine, a stimulant in coffee and tea, has a molar mass of 194.19 g·mol^{-1} and a mass percentage composition of 49.48% C, 5.19% H, 28.85% N, and 16.48% O. What is the molecular formula of caffeine?

F.20 A sample of mass 2.00 mg from a pungent compound was extracted from skunk spray. Upon analysis, the sample was found to have the following composition: 1.09 mg C, 0.183 mg H, and 0.727 mg S. The molar mass of the compound was found to be 88.17 g·mol^{-1}. What is the molecular formula of this compound?

F.21 In 1978, scientists extracted a compound with antitumor and antiviral properties from marine animals in the Caribbean Sea. A sample of the compound didemnin-A of mass 1.78 mg was analyzed and found to have the following composition: 1.11 mg C, 0.148 mg H, 0.159 mg N, and 0.363 mg O. The molar mass of didemnin-A was found to be 942 g·mol^{-1}. What is the molecular formula of didemnin-A?

F.22 Etomidate is an anesthetic used for outpatient procedures. A sample of etomidate of mass 5.80 mg was found to have the following composition: 3.99 mg C, 0.383 mg H, 0.665 mg N, and 0.760 mg O.

The molar mass of etomidate is 244.29 g·mol^{-1}. What is the molecular formula of etomidate?

F.23 The CO_2 produced by the combustion of hydrocarbons contributes to climate change. Rank the following fuels according to increasing mass percentage of carbon: (a) ethene, C_2H_4; (b) propanol, C_3H_7OH; (c) heptane, C_7H_{16}.

F.24 Dolomite is a mixed carbonate of calcium and magnesium. Calcium and magnesium carbonates both decompose upon heating to produce the metal oxides (MgO and CaO) and carbon dioxide (CO_2). If 5.12 g of residue consisting of MgO and CaO remains when 10.04 g of dolomite is heated until decomposition is complete, what percentage by mass of the original sample was $MgCO_3$?

F.25 In the following ball-and-stick molecular structures, dark gray indicates carbon; white, hydrogen; blue, nitrogen; and green, chlorine. Write the empirical and molecular formulas of each structure. *Hint:* It may be easier to write the molecular formula first.

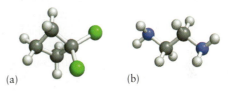

(a)   (b)

F.26 Write the empirical and molecular formulas of each ball-and-stick structure. Dark gray indicates carbon; white, hydrogen; red, oxygen; and green, chlorine. *Hint:* It may be easier to write the molecular formula first.

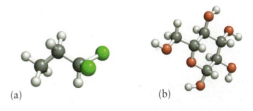

(a)   (b)

F.27 A mixture of $NaNO_3$ and Na_2SO_4 of mass 5.37 g contains 1.61 g of sodium. What is the percentage by mass of $NaNO_3$ in the mixture?

F.28 A mixture of KBr and K_2S of mass 6.14 g contains 2.50 g of potassium. What is the percentage by mass of KBr in the mixture?

G Mixtures and Solutions

Most materials are neither pure elements nor pure compounds, and so they are not "substances" in the technical sense of the term (*Fundamentals* A): they are **mixtures** with one substance mingled with another. For example, air, blood, and seawater are mixtures. A medicine, such as a cough syrup, is often a mixture of various ingredients that has been formulated to achieve an overall biological effect. Much the same can be said of a perfume.

G.1 *Classifying Mixtures*

A compound has a fixed composition but the composition of a mixture may vary. There are always two H atoms for each O atom in a sample of the compound water, but sugar and sand, for instance, can be mixed in any proportions. Because the components of a mixture are merely mingled with one another, they retain their own chemical properties in the mixture. In contrast, a compound has chemical properties that differ from those of its component elements. The formation of a mixture is a *physical* change, whereas the

FIGURE G.1 This piece of granite is a heterogeneous mixture of several substances. *(De Agostini/A. Rizzi/Getty Images.)*

TABLE G.1   Differences Between Mixtures and Compounds

Mixture	Compound
components can be separated with physical techniques.	components cannot be separated with physical techniques.
composition is variable.	composition is fixed.
properties of its components are retained.	properties of its components are are lost.

formation of a compound requires a *chemical* change. The differences between mixtures and compounds are summarized in **TABLE G.1**.

Some mixtures have component particles that are so large that they can be seen with an optical microscope or even the unaided eye (**FIG. G.1**). Such a patchwork of different substances is called a **heterogeneous mixture**. Many of the rocks that form the landscape are heterogeneous mixtures of crystals of different minerals.

In some mixtures, the molecules or ions of the components are so well mixed that the composition is the same throughout, no matter how small the sample. Such a mixture is called a **homogeneous mixture** (**FIG. G.2**) or **solution**. In a solution, there is typically one dominant substance, the **solvent;** the other substance or substances present are called the **solutes.** Filtered seawater is a solution of common table salt (sodium chloride) and many other substances in water. There are also **solid solutions,** in which the solvent is a solid. An example is a common form of brass that can be regarded as a solution of zinc in copper. Although a solution may look uniform in composition, the components that form it retain their identities. Forming a solution is a physical change, not a chemical change. In practice, gaseous mixtures are not thought of as solutions, even though one gas might be dominant (such as nitrogen in the atmosphere).

Crystallization is the process in which a solute slowly comes out of solution as crystals, perhaps as the solvent evaporates. For example, salt crystals that form when water evaporates line the shores of the Great Salt Lake in Utah. In **precipitation,** a solute comes out of solution so rapidly that a single crystal does not have time to form. Instead, the solute forms a finely divided powder (a collection of very tiny crystals) called a precipitate. Precipitation is often almost instantaneous, occurring as soon as two solutions are mixed (**FIG. G.3**).

Beverages and seawater are examples of **aqueous solutions,** solutions in which the solvent is water. Aqueous solutions are very common in everyday life and in chemical laboratories; for that reason, most of the solutions mentioned in this text are aqueous. **Nonaqueous solutions** are solutions in which the solvent is not water. Although they are less common than aqueous solutions, nonaqueous solutions have important uses. In "dry cleaning," the grease and dirt on fabrics are dissolved in a nonaqueous liquid solvent such as tetrachloroethene, C_2Cl_4.

> *In mixtures, the properties of the constituents are retained; they differ from compounds, as summarized in Table G.1. Mixtures are classified as homogeneous or heterogeneous; solutions are homogeneous mixtures of two or more substances.*

FIGURE G.3 Precipitation occurs when an insoluble substance is formed. Here lead(II) iodide, PbI_2, which is an insoluble yellow solid, precipitates when we mix solutions of lead(II) nitrate, $Pb(NO_3)_2$, and potassium iodide, KI. *(© 1995 Chip Clark–Fundamental Photographs.)*

 **LAB VIDEO FIGURE G.3**

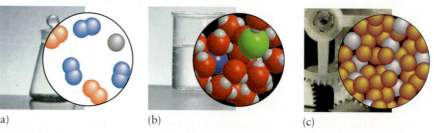

(a)   (b)   (c)

FIGURE G.2 Three examples of homogeneous mixtures. (a) Air is a homogeneous mixture of many gases, including the nitrogen, oxygen, and argon depicted here. (b) Table salt dissolved in water consists of sodium ions and chloride ions among water molecules. (c) Many alloys are solid homogeneous mixtures of two or more metals. The insets show that in a solution mixing is uniform at the molecular level. *(Photos: W. H. Freeman photos by Ken Karp.)*

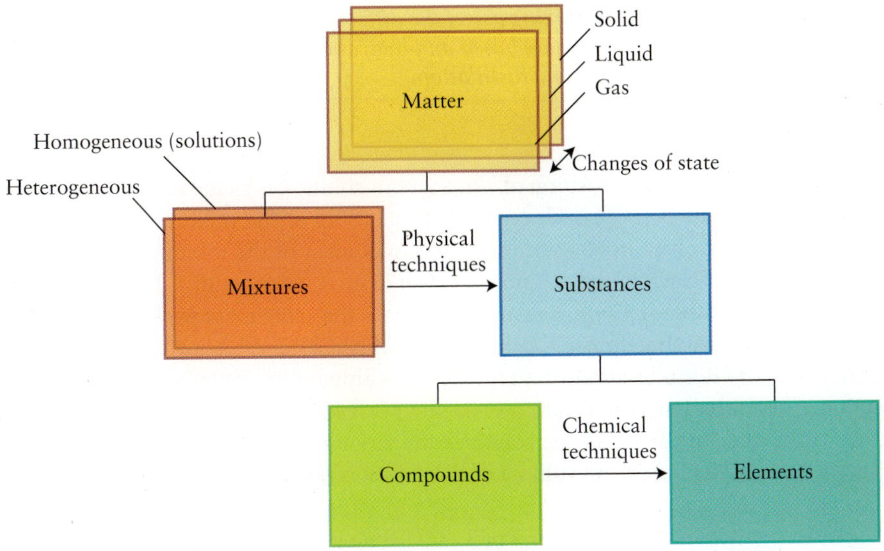

FIGURE G.4 The hierarchy of materials: matter, whether solid, liquid, or gas, consists of either mixtures or substances; substances consist of either compounds or elements. Physical techniques are used to separate mixtures into pure substances. Chemical techniques are used to separate compounds into elements.

G.2 Separation Techniques

To analyze the composition of any sample that you suspect is a mixture, it is first separated into its components by physical means and then each individual substance present is identified (**FIG. G.4**). Common physical separation techniques include decanting, filtration, chromatography, and distillation.

Decanting makes use of differences in density. One liquid floats on another liquid or lies above a solid and is poured off. **Filtration** is used to separate substances when there is a difference in the ability to dissolve in a solvent. The sample is stirred with a liquid and then poured through a fine mesh filter. Components of the mixture that have dissolved in the liquid pass through the filter, whereas the filter captures solid components that do not dissolve. The technique can be used to separate sugar from sand because sugar is soluble in water but sand is not. A related technique, one of the most sensitive techniques available for separating a mixture, is **chromatography,** which relies on the different abilities of substances to **adsorb,** or stick, to surfaces (**FIG. G.5**). The dried support showing the separated components of a mixture is called a **chromatogram.** Chromatography is discussed in detail in *Major Technique* 4.

Distillation makes use of differences in boiling points to separate mixtures. When a solution is distilled, the components of the mixture boil away at different temperatures and condense in a cooled tube called a *condenser* (**FIG. G.6**). Distillation can be used to remove water from table salt (sodium chloride); the solid salt is left behind when the water evaporates.

The first known distillation apparatus, assembled in the first century CE, is attributed to Mary the Jewess, an alchemist who lived in the Middle East.

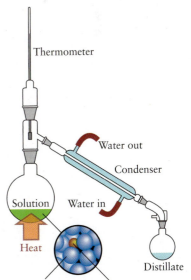

FIGURE G.6 The technique of distillation, which is used to separate a low-boiling liquid from a dissolved solid or a liquid with a much higher boiling point. When the solution is heated, the low-boiling liquid boils off, condenses in the water-jacketed tube (the condenser), and is collected as the distillate.

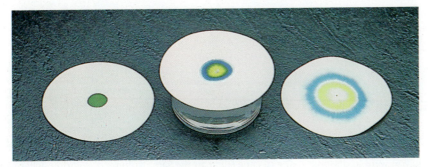

FIGURE G.5 In paper chromatography, the components of a mixture are separated by washing them along a paper—the support—with a solvent. A primitive form of the technique is shown here. On the left is a dry filter paper to which a drop of green food coloring was applied. Solvent was then poured onto the center of the filter paper. The blue and yellow dyes that were combined to make the green color begin to separate. The filter paper on the right was allowed to dry after the solvent had spread out to the edges of the paper, carrying the two dyes to different distances as it spread. (*W. H. Freeman photo by Ken Karp.*)

Mixtures are separated by making use of the differences in physical properties of the components; common techniques based on physical differences include decanting, filtration, chromatography, and distillation.

G.3 Concentration

One way to express the composition of a mixture or solution is as the **mass percentage** of each component, the mass of each component in 100 g of the mixture. For example, if 15 g of NaCl is dissolved in 60. g of water, the total mass of the mixture is 75 g and the mass percentage of NaCl in the solution is (15 g/75 g) × 100% = 20.% NaCl. If a sample containing 30. g of that solution is taken, it will have the same mass percentage composition, 20.% NaCl, and therefore will contain 6.0 g of NaCl.

In chemistry it is often important to know the amount of solute in a given volume of solution. The **molar concentration,** *c*, of a solute in a solution, which is widely but unofficially called the "molarity" of the solute, is the amount of solute molecules or formula units (in moles) divided by the volume of the solution (in liters):

The formal name for molarity is the "amount of substance concentration."

$$\text{Molarity} = \frac{\text{amount of solute (in moles)}}{\text{volume of solutions (in liters)}}, \text{ or } c = \frac{\overbrace{n}^{\text{mol}}}{\underbrace{V}_{\text{L}}} \qquad (1)$$

The units of molarity are moles per liter (mol·L^{-1}), often denoted M:

$$1\,\text{M} = 1\,\text{mol·L}^{-1}$$

The symbol M is often read "molar"; it is not an SI unit. Chemists working with very low concentrations of solutes also report molar concentrations as millimoles per liter (mmol·L^{-1}) and micromoles per liter (μmol·L^{-1}).

EXAMPLE G.1   Calculating the molar concentration of a solute

Suppose you dissolved 10.0 g of cane sugar in enough water to make 200. mL of solution, which you might do (with less precision) if you were making a glass of lemonade, and want to report its concentration. Cane sugar is sucrose ($C_{12}H_{22}O_{11}$, molar mass 342 g·mol^{-1}). What is the molar concentration of the sucrose in the resulting solution?

ANTICIPATE The mass of sugar is only about 3% of the mass of 1 mol of sugar molecules; therefore, although only 0.200 L is being prepared, you should expect a molar concentration of less than 1 mol·L^{-1}.

PLAN The definition of molar concentration (molarity) is $c = n/V$; you first need to convert the mass of solute to an amount in moles (by using $n = m/M$) and then substitute that amount into this expression for c.

SOLVE

From $c = n/V$ and $n = m/M$,

$$c = \frac{\overbrace{(10.0\,\text{g})/(342\,\text{g·mol}^{-1})}^{n=m/M}}{\underbrace{0.200\,\text{L}}_{V}}$$

$$= \frac{10.0\,\text{g}}{(342\,\text{g·mol}^{-1}) \times (0.200\,\text{L})} = 0.146\,\text{mol·L}^{-1}$$

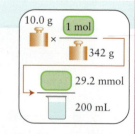

EVALUATE The molar concentration, which is reported as 0.146 M $C_{12}H_{22}O_{11}$(aq), is less than 1 mol·L^{-1}, as expected.

Self-test G.1A If, instead of 10.0 g, you were to dissolve 20.0 g of cane sugar in the same volume of solution, what would be the molarity of the sugar in the solution?

[***Answer:*** 0.292 M $C_{12}H_{22}O_{11}$(aq)]

Self-test G.1B What is the molarity of sodium sulfate in a solution prepared by dissolving 15.5 g of sodium sulfate in enough water to make 350. mL of solution?

Related Exercises G.5, G.6

Because molarity is defined in terms of the *volume of the solution,* not the volume of solvent used to prepare the solution, the volume must be measured after the solutes have been added. The usual way to prepare an aqueous solution of a solid substance of given molarity is to transfer a known mass of the solid into a **volumetric flask,** a flask calibrated to contain a specified volume, dissolve the solute in a little water, fill the flask up to the mark with water, and then mix the solution thoroughly by tipping the flask end over end (**FIG. G.7**).

The molar concentration of a solute is used to calculate the amount of solute in a given volume of solution:

$$\overset{\text{mol}}{\widehat{n}} = \overset{\text{mol·L}^{-1}}{\widehat{c}} \times \overset{\text{L}}{\widehat{V}} \qquad \text{(2a)}$$

where c is the molarity, V is the volume, and n is the amount. This formula is also used to estimate the mass of solute needed to make up a given volume of solution of known concentration. In that calculation, the molar mass of the solute is used to convert the amount into mass. First, write $n = m/M$, then Eq. 2a becomes $m/M = cV$, and therefore, after multiplying both sides by M,

$$m = cMV \qquad \text{(2b)}$$

EXAMPLE G.2   Determining the mass of solute required for a given concentration

Very dilute solutions of $CuSO_4$ are used to control algal growth in fish tanks. Suppose you are investigating the optimum concentration that will control the algae but not harm the fish. You are asked to prepare 250. mL of a solution that is approximately 0.0380 M $CuSO_4(aq)$ from solid copper(II) sulfate pentahydrate, $CuSO_4·5H_2O$. What mass of the solid do you need?

ANTICIPATE   Because the solution is dilute, even though a relatively large volume is needed, you should expect the required mass to be only a few grams.

PLAN   Use Eq. 2b to find the mass corresponding to the specified volume and molar concentration.

SOLVE   You need to know the amount of $CuSO_4$ in 250. mL (0.250 L) of solution. Because 1 mol $CuSO_4·5H_2O$ contains 1 mol $CuSO_4$, the amount of $CuSO_4·5H_2O$ you need to supply is the same as the amount of $CuSO_4$ needed to prepare the solution. That is,

$$n(CuSO_4·5H_2O) = n(CuSO_4)$$

Then, because the molar mass of copper(II) sulfate pentahydrate is 249.6 g·mol^{-1}, calculate the mass of the pentahydrate required as follows.

From $m = cMV$,

$$m(CuSO_4·5H_2O) = (0.0380 \text{ mol·L}^{-1}) \times (249.6 \text{ g·mol}^{-1})$$
$$\times (0.250 \text{ L}) = 2.37 \text{ g}$$

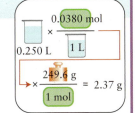

You will need about 2.37 g of copper(II) sulfate pentahydrate.

EVALUATE   As expected, the mass required is small.

A Note on Good Practice: The laboratory procedure is to take the approximate amount of solute needed, weigh it accurately, then make up the solution, calculating the actual concentration of solute from the mass that was used and the final volume of the solution. For example, you might find that you had measured out 2.403 g, in which case the molar concentration would be 0.0385 M $CuSO_4(aq)$.

Self-test G.2A   Calculate the mass of glucose needed to prepare 150. mL of 0.442 M $C_6H_{12}O_6(aq)$.

[***Answer:*** 11.9 g]

Self-test G.2B   Calculate the mass of oxalic acid needed to prepare 50.00 mL of 0.125 M $C_2H_2O_4(aq)$.

Related Exercises   G.7–G.10

FIGURE G.7   The steps in making up a solution of known molarity of solute. A known mass of the solute is dispensed into a volumetric flask (top). Some water is added to dissolve it (center). Finally, water is added up to the mark on the stem of the flask (bottom). The bottom of the solution's meniscus (the curved top surface of the liquid) should be level with the mark. (©1992 Richard Megna–Fundamental Photographs.)

The molarity is also used to calculate the volume of solution, V, that contains a given amount of solute, by rearranging Eq. 2a:

$$\overset{\text{L}}{V} = \frac{\overset{\text{mol}}{n}}{\underset{\text{mol}\cdot\text{L}^{-1}}{c}} \tag{3}$$

and then substituting the data.

EXAMPLE G.3   Calculating the volume of solution that contains a given amount of solute

Many reagents in chemistry stockrooms are prepared as aqueous solutions. Suppose you want to measure out 0.760 mmol CH_3COOH, acetic acid, an acid found in vinegar and often used in the laboratory, and you have available 0.0560 M CH_3COOH(aq). What volume of solution should you use?

ANTICIPATE   Because the amount of acetic acid required is very small, even though the solution is dilute, you should anticipate that only a few milliliters of the solution will be required.

PLAN   Rearrange $c = n/V$ into $V = n/c$. To keep the units straight, it is best to convert the amount required from millimoles (mmol) to moles (mol).

SOLVE

From $V = n/c$,

$$V = \frac{0.760 \times 10^{-3}\ \text{mol}}{0.0560\ \text{mol}\cdot\text{L}^{-1}} = 0.0136\ \text{L}$$

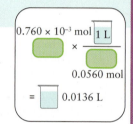

You should therefore transfer 13.6 mL of the acetic acid solution into a flask by using a buret or a pipet (**FIG. G.8**). The flask will then contain 0.760 mmol CH_3COOH.

EVALUATE   As expected, the volume of solution required is small.

Self-test G.3A   What volume of 1.25×10^{-3} M $C_6H_{12}O_6$(aq) contains 1.44 μmol of glucose molecules?

[**Answer:** 1.15 mL]

Self-test G.3B   What volume of 0.358 M HCl(aq) contains 2.55 mmol HCl?

Related Exercises   G.11, G.12

The molarity (molar concentration) of a solute in a solution is the amount of solute in moles divided by the volume of the solution in liters.

G.4 Dilution

A common space-saving practice in chemistry is to store a solution in a concentrated form called a **stock solution** and then to add solvent to **dilute** it, or reduce the concentration of the solute, to whatever value is needed. Chemists use techniques such as dilution when they need very precise control over the amounts of the substances that they are handling, especially when those amounts are very small. For example, pipetting 25.0 mL of 1.50×10^{-3} M NaOH(aq) corresponds to transferring only 37.5 μmol NaOH or 1.50 mg of the compound. A mass this small is difficult to weigh out accurately but the volume is easy to dispense accurately.

To dilute a stock solution to a desired concentration, a pipet is used to transfer the appropriate volume of stock solution to a volumetric flask. Then enough solvent is added to increase the volume of the solution to its final value. **Toolbox G.1** shows how to calculate the initial volume of stock solution required for a given volume of diluted solution.

FIGURE G.8   A buret is calibrated so that the volume of liquid delivered can be measured. *(Martyn F. Chillmaid/Science Source.)*

Toolbox G.1 HOW TO CALCULATE THE VOLUME OF STOCK SOLUTION REQUIRED FOR A GIVEN DILUTION

CONCEPTUAL BASIS

This procedure is based on a simple idea: although you might add more solvent to a given volume of solution, you change only the concentration, not the amount of solute (**FIG. G.9**). After dilution, the same amount of solute simply occupies a larger volume of solution.

PROCEDURE

The procedure has two steps:

Step 1 Calculate the amount of solute, n, in the final, dilute solution of volume, V_{final}. (This is the amount of solute to be transferred into the volumetric flask.)

$$n = c_{final}V_{final}$$

Step 2 Calculate the volume, $V_{initial}$, of the initial stock solution, of molarity $c_{initial}$, that contains this amount of solute. (This is the volume of stock solution that contains the amount of solute calculated in step 1.)

$$V_{initial} = \frac{n}{c_{initial}}$$

This procedure is illustrated in Example G.4.

Because the amount of solute, n, is the same in these two expressions, they can be combined into

$$V_{initial} = \frac{c_{final}V_{final}}{c_{initial}}$$

and rearranged into an easily remembered form:

$$c_{initial}V_{initial} = c_{final}V_{final} \tag{4}$$

In Eq. 4, the amount of solute in the final solution (the product on the right) is the same as the amount of solute in the initial volume of solution (the product on the left), $n_{final} = n_{initial}$.

EXAMPLE G.4   Calculating the volume of stock solution to dilute

Sodium hydroxide solution is used in the recycling of newspaper; it causes the paper fibers to swell, allowing the ink to be removed. Suppose you are working in the laboratory of a newsprint company and are investigating how cellulose fibers are affected by sodium hydroxide solutions with different concentrations. You need to prepare 250. mL of 1.25×10^{-3} M NaOH(aq) and will use a 0.0270 M NaOH(aq) stock solution. How much stock solution do you need?

ANTICIPATE Because the stock solution is about 22 times more concentrated than the diluted solution, you should expect to use about 1/22 of 250. mL, or just over 12 mL.

PLAN Proceed as in Toolbox G.1.

SOLVE

Step 1 Calculate the amount of solute, n, in the final, dilute solution of volume V_{final}.

From $n = c_{final}V_{final}$,

$$n = (1.25 \times 10^{-3}\ mol \cdot L^{-1}) \times (0.250\ L)$$
$$= (1.25 \times 10^{-3} \times 0.250)\ mol$$

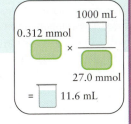

Step 2 Calculate the volume, $V_{initial}$, of the initial stock solution of molarity $c_{initial}$ that contains this amount of solute.

From $V_{initial} = n/c_{initial}$,

$$V_{initial} = \frac{(1.25 \times 10^{-3} \times 0.250)\,mol}{0.0270\ mol \cdot L^{-1}} = 1.16 \times 10^{-2}\ L$$

or 11.6 mL.

EVALUATE The volume of stock solution required is 11.6 mL, close to the expected value. This volume should be measured into a 250.-mL volumetric flask (by using a buret) and water should then be added up to the mark (**FIG. G.10**).

FIGURE G.9 When a solution is diluted, the same number of solute molecules occupy a larger volume. Therefore, the same volume (as depicted by the cube) will contain fewer molecules than in the concentrated solution.

A Note on Good Practice: Note that, to minimize rounding errors, carry out the calculation in a single step. However, to help guide you through the calculation, the examples will often show intermediate numerical results with unrounded values left as *n.nnn...* .

Self-test G.4A Calculate the volume of 0.0155 M HCl(aq) you should use to prepare 100. mL of 5.23×10^{-4} M HCl(aq).

[*Answer:* 3.37 mL]

Self-test G.4B Calculate the volume of 0.152 M $C_6H_{12}O_6$(aq) required to prepare 25.00 mL of 1.59×10^{-5} M $C_6H_{12}O_6$(aq).

Related Exercises G.15, G.16, G.18, G.27

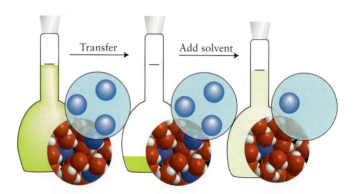

Transfer   Add solvent

FIGURE G.10 The steps in dilution. A small sample of the original solution is transferred to a volumetric flask, and then solvent is added up to the mark.

When a solution is diluted to a larger volume, the total amount of solute in the solution does not change, but the concentration of solute is reduced.

What have you learned in Fundamentals G?

You have learned how to classify mixtures and to prepare, dilute, and use a solution of given concentration.

The skills you have mastered are the ability to:

☐ **1.** Distinguish between heterogeneous and homogeneous mixtures and describe methods of separation (Section G.1 and G.2).

☐ **2.** Calculate the molar concentration (molarity) of a solute in a solution, volume of solution, and mass of solute, given the other two quantities (Examples G.1–G.3).

☐ **3.** Determine the volume of stock solution needed to prepare a dilute solution of a given molar concentration of solute (Toolbox G.1 and Example G.4).

Fundamentals G Exercises

G.1 Indicate whether each statement is true or false. If it is false, explain what is wrong with the statement. (a) The components of a compound can be separated from each other by physical means. (b) The composition of a solution can be varied. (c) The properties of a compound are the same as those of the elements that compose it.

G.2 Indicate whether each statement is true or false. If it is false, explain what is wrong with the statement. (a) A nonaqueous solution is one in which the solvent is water. (b) Decanting makes use of differences in boiling points to separate the components of a mixture. (c) In chromatography, the components of a mixture are separated according to their ability to be adsorbed on a surface.

G.3 Identify the following mixtures as homogeneous or heterogeneous, and suggest a technique for separating their components: (a) oil and vinegar; (b) chalk and table salt; (c) salt water.

G.4 Identify the following mixtures as homogeneous or heterogeneous, and suggest a technique for separating their components: (a) lemonade; (b) salad oil and vinegar; (c) salt and ground pepper.

G.5 A student prepared a solution of sodium carbonate by adding 2.111 g of the solid to a 250.0-mL volumetric flask and adding water to the mark. Some of this solution was transferred to a buret. What volume of solution should the student transfer into a flask to obtain (a) 2.15 mmol Na^+; (b) 4.98 mmol CO_3^{2-}; (c) 50.0 mg Na_2CO_3?

G.6 (a) A chemist prepares a solution by dissolving 1.734 g of $NaNO_3$ in enough water to make 250.0 mL of solution. What molar concentration of sodium nitrate should appear on the label? (b) If the chemist mistakenly uses a 500.0-mL volumetric flask instead of the 250.0-mL flask in part (a), what molar concentration of sodium nitrate will the chemist actually prepare?

G.7 You need to prepare 510. g of an aqueous solution containing 5.45% KNO_3 by mass. Describe how you would prepare the solution and what mass of each component you would use.

G.8 You need to prepare a sample containing 0.453 g of $CuSO_4$ from a solution that is 5.16% $CuSO_4$ by mass. What mass of solution do you need?

G.9 A chemist studying the properties of photographic emulsions needed to prepare 500.0 mL of 0.179 M $AgNO_3(aq)$. What mass of silver nitrate must be placed into a 500.0-mL volumetric flask, dissolved, and diluted to the mark with water?

G.10 What mass (in grams) of anhydrous solute is needed to prepare each of the following solutions? (a) 1.00 L of 0.125 M $K_2SO_4(aq)$; (b) 375 mL of 0.015 M $NaF(aq)$; (c) 500. mL of 0.35 M $C_{12}H_{22}O_{11}(aq)$.

G.11 A medical researcher investigating the properties of intravenous solutions prepared a solution containing 0.278 M $C_6H_{12}O_6$ (glucose). What volume of solution should the researcher use to provide 4.50 mmol $C_6H_{12}O_6$?

G.12 A student investigating the properties of solutions containing carbonate ions prepared a solution containing 8.124 g of Na_2CO_3 in a flask of volume 250.0 mL. Some of the solution was transferred to a buret. What volume of solution should be dispensed from the buret to provide (a) 5.124 mmol Na_2CO_3; (b) 8.726 mmol Na^+?

G.13 To prepare a fertilizer solution, a florist dilutes 1.0 L of 0.20 M $NH_4NO_3(aq)$ by adding 3.0 L of water. The florist then adds 100. mL of the diluted solution to each plant. How many moles of nitrogen atoms will each plant receive? Solve this exercise without using a calculator.

G.14 To prepare a nutrient solution, a nurse dilutes 1.0 L of 0.30 M $C_6H_{12}O_6(aq)$ by adding 4.0 L of water. The nurse then adds 100. mL of the diluted solution to an intravenous (IV) bag. How many moles of carbon atoms will the IV bag contain? Solve this exercise without using a calculator.

G.15 (a) What volume of 0.778 M $Na_2CO_3(aq)$ should be diluted to 150.0 mL with water to reduce its concentration to 0.0234 M $Na_2CO_3(aq)$? (b) An experiment requires the use of 60.0 mL of 0.50 M $NaOH(aq)$. The stockroom assistant can only find a reagent bottle of 2.5 M $NaOH(aq)$. How can the 0.50 M $NaOH(aq)$ be prepared?

G.16 A chemist dissolves 0.033 g of $CuSO_4 \cdot 5H_2O$ in water and dilutes the solution to the mark in a 250.0-mL volumetric flask. A 2.00-mL sample of this solution is then transferred to a second 250.0-mL volumetric flask and diluted. (a) What is the molarity of $CuSO_4$ in the final solution? (b) To prepare the final 250.0-mL solution directly, what mass of $CuSO_4 \cdot 5H_2O$ would need to be weighed out?

G.17 (a) Determine the mass of anhydrous copper(II) sulfate that must be used to prepare 250 mL of 0.20 M $CuSO_4(aq)$.

(b) Determine the mass of $CuSO_4 \cdot 5H_2O$ that must be used to prepare 250 mL of 0.20 M $CuSO_4(aq)$.

G.18 The ammonia solution that is purchased for a stockroom has a molarity of 15.0 mol·L^{-1}. (a) Determine the volume of 15.0 M $NH_3(aq)$ that must be diluted to 250. mL to prepare 0.720 M $NH_3(aq)$. (b) An experiment requires 0.050 M $NH_3(aq)$. The stockroom manager estimates that 8.10 L of the base is needed. What volume of 15.0 M $NH_3(aq)$ will be required for the preparation?

G.19 (a) A sample of 1.345 M $K_2SO_4(aq)$ of volume 12.56 mL is diluted to 250.0 mL. What is the molar concentration of K_2SO_4 in the diluted solution? (b) A sample of 0.366 M $HCl(aq)$ of volume 25.00 mL is drawn from a reagent bottle with a pipet. The sample is transferred to a flask of volume 125.00 mL and diluted to the mark with water. What is the molar concentration of the dilute hydrochloric acid solution?

G.20 To prepare a very dilute solution, it is advisable to perform successive dilutions of a single prepared reagent solution, rather than to weigh out a very small mass or to measure a very small volume of stock chemical. A solution was prepared by transferring 0.661 g of $K_2Cr_2O_7$ to a 250.0-mL volumetric flask and adding water to the mark. A sample of this solution of volume 1.000 mL was transferred to a 500.0-mL volumetric flask and diluted to the mark with water. Then 10.0 mL of the diluted solution was transferred to a 250.0-mL flask and diluted to the mark with water. (a) What is the final concentration of $K_2Cr_2O_7$ in solution? (b) What mass of $K_2Cr_2O_7$ is in this final solution? (The answer to the last question gives the amount that would have had to have been weighed out if the solution had been prepared directly.)

G.21 A solution is prepared by dissolving 0.500 g of KCl, 0.500 g of K_2S, and 0.500 g of K_3PO_4 in 500. mL of water. What is the concentration in the final solution of (a) potassium ions; (b) sulfide ions?

G.22 Describe the preparation of each of the following solutions, starting with the anhydrous solute and water and using the indicated size of volumetric flask: (a) 75.0 mL of 5.0 M $NaOH(aq)$; (b) 5.0 L of 0.21 M $BaCl_2(aq)$; (c) 300. mL of 0.0340 M $AgNO_3(aq)$.

G.23 In medicine it is sometimes necessary to prepare solutions with a specific concentration of a given ion. A lab technician has made up 100.0 mL of a solution containing 0.50 g of NaCl and 0.30 g of KCl, as well as glucose and other sugars. What is the concentration of chloride ions in the solution?

G.24 When a sample of iron ore of mass 2.016 g is treated with 50.0 mL of hydrochloric acid, the iron dissolves in the acid to form a solution of $FeCl_3$. The $FeCl_3$ solution is diluted to 100.0 mL and the concentration of Fe^{3+} ions is determined by spectrometry to be 0.145 mol·L^{-1}. What is the mass percentage of iron in the ore?

G.25 Practitioners of the branch of alternative medicine known as homeopathy claim that very dilute solutions of substances can have an effect. Is the claim plausible? To explore this question, suppose that you prepare a solution of a supposedly active substance, X, with a molar concentration of 0.10 mol·L^{-1}. Then you dilute 10. mL of that solution by doubling the volume, doubling it again, and so on, for 90 doublings in all. How many molecules of X will be present in 10. mL of the final solution? Comment on the possible health benefits of the solution.

G.26 Refer to Exercise G.25. How many successive tenfold dilutions of the original solution will result in one molecule of X being left in 10 mL of solution?

G.27 Concentrated hydrochloric acid is 37.50% HCl by mass and has a density of 1.205 g·cm^{-3}. What volume (in milliliters) of concentrated hydrochloric acid must be used to prepare 10.0 L of 0.7436 M HCl(aq)?

G.28 You need 500. mL of 0.10 M AgNO$_3$(aq). You have on hand 100. mL of 0.30 M AgNO$_3$(aq), 1.00 L of 0.050 M AgNO$_3$(aq), and lots of distilled water. Describe how you would prepare the desired solution and what volume of each solution you would use.

G.29 The concentration of toxic chemicals in the environment is often measured in parts per million (ppm) or even parts per billion (ppb). A solution in which the concentration of the solute is 3 ppb by mass has 3 g of the solute for every billion grams (1000 t) of the solution. The World Health Organization has set the acceptable standard for lead in drinking water at 10 ppb. You need to analyze some tap water for lead concentration, but your equipment is able to detect lead in concentrations only as low as 1×10^{-8} mol·L^{-1}. Is your equipment satisfactory? You can assume that in such dilute solutions the density of the solution is 1.00 g·cm^{-3}. Explain your reasoning.

G.30 Refer to Exercise G.29. In 1992 the tap water in one-third of the homes in Chicago was found to have a concentration of about 10 ppb lead by mass. If you lived in one of these homes and drank 2 L of tap water at home each day, what is the total mass of lead you would have ingested per year?

H Chemical Equations

The growth of a child, the production of polymers from petroleum, and the digestion of food are all the outcome of **chemical reactions,** processes by which one or more substances are converted into other substances. This type of process is a chemical change (*Fundamentals* A). The starting materials are called the **reactants** and the substances formed are called the **products.** The chemicals typically available in a laboratory are called **reagents.** Only when a reagent is being used in a particular reaction is it called a reactant. In *Fundamentals* H you will see how to express the outcome of a chemical reaction in terms of symbols, a fundamental part of the language of chemistry.

H.1 Representing Chemical Reactions

A chemical reaction is symbolized by an arrow, as in:

$$\text{Reactants} \longrightarrow \text{products}$$

For example, sodium is a soft, shiny metal that reacts vigorously with water. If a small lump of sodium metal is dropped into a container of water, hydrogen gas forms rapidly and sodium hydroxide is produced in solution (**FIG. H.1**). This reaction can be described in words:

$$\text{Sodium} + \text{water} \longrightarrow \text{sodium hydroxide} + \text{hydrogen}$$

and by using chemical formulas:

$$\text{Na} + \text{H}_2\text{O} \longrightarrow \text{NaOH} + \text{H}_2$$

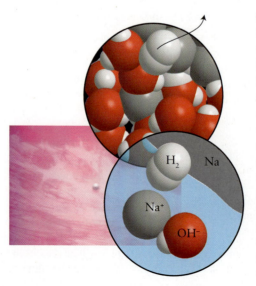

FIGURE H.1 When a small piece of sodium is dropped into water, a vigorous reaction takes place. Hydrogen gas and sodium hydroxide are formed, and the heat released is enough to melt the sodium, which forms a sphere. The pink color is due to the presence of a dye that changes color in the presence of sodium hydroxide. The balanced chemical equation shows that when two sodium atoms give rise to two sodium ions, two water molecules give rise to one hydrogen molecule (which escapes as a gas) and two hydroxide ions. There is a rearrangement of partners, not a creation or annihilation of atoms. The unreacted water molecules are not shown in the lower inset. (*Photo: ©2012 Chip Clark–Fundamental Photographs.*)

This expression is called a **skeletal equation** because it shows the "bare bones" of the reaction (the identities of the reactants and products) in terms of chemical formulas. A skeletal equation is a *qualitative* summary of a chemical reaction.

To summarize reactions *quantitatively*, note that atoms are neither created nor destroyed in a chemical reaction: they simply change their partners. The principal evidence for this conclusion is that there is no overall change in mass when a reaction takes place in a sealed container. The observation that the total mass is constant during a chemical reaction is called the **law of conservation of mass.** Because atoms are neither created nor destroyed, the chemical formulas in a skeletal equation must be multiplied by factors that result in the same numbers of atoms of each element on both sides of the arrow. The resulting expression is said to be **balanced** and is called a **chemical equation.** For example, there are two H atoms on the left of the preceding skeletal equation but three H atoms on the right. So, the balanced equation is

$$2\,Na + 2\,H_2O \longrightarrow 2\,NaOH + H_2$$

Now there are four H atoms, two Na atoms, and two O atoms on each side, and the equation conforms to the law of conservation of mass. The number multiplying an *entire* chemical formula in a chemical equation (for example, the 2 multiplying H_2O) is called the **stoichiometric coefficient** of the substance. A coefficient of 1 (as for H_2) is not written explicitly.

The somewhat awkward word stoichiometric *is derived from the Greek words for "element" and "measure."*

A Note on Good Practice: Be careful to distinguish coefficients from subscripts. Subscripts in a formula show how many atoms of that element are present in one molecule. Coefficients show how many formula units or molecules are present.

A chemical equation typically also shows the physical state of each reactant and product by using a **state symbol:**

(s): solid   (l): liquid   (g): gas   (aq): aqueous solution

For the reaction between solid sodium and liquid water, the complete, balanced chemical equation is therefore

$$2\,Na(s) + 2\,H_2O(l) \longrightarrow 2\,NaOH(aq) + H_2(g)$$

When it is important to emphasize that a reaction requires a high temperature, the Greek letter Δ (delta) is written over the arrow. For example, the conversion of limestone into quicklime takes place at about 800 °C:

$$CaCO_3(s) \xrightarrow{\Delta} CaO(s) + CO_2(g)$$

Sometimes a **catalyst,** a substance that increases the rate of a reaction but is not itself consumed in the reaction (Topic 7E), is needed. For example, vanadium pentoxide, V_2O_5, is a catalyst in one step of the industrial process for the production of sulfuric acid. The presence of a catalyst is indicated by writing the formula of the catalyst above the reaction arrow:

$$2\,SO_2(g) + O_2(g) \xrightarrow{V_2O_5} 2\,SO_3(g)$$

An important interpretation of a chemical equation is as follows. First, note that the equation for the reaction of sodium with water ($2\,Na + 2\,H_2O \rightarrow 2\,NaOH + H_2$) shows that

- when any 2 *atoms* of sodium react with 2 *molecules* of water, they produce 2 *formula units* of NaOH and 1 *molecule* of hydrogen.

You can multiply through by the number of entities in a mole (6.0221×10^{23}, *Fundmentals* E), and conclude that

- when 2 *moles* of Na atoms react with 2 *moles* of H_2O molecules, they produce 2 *moles* of NaOH formula units and 1 *mole* of H_2 molecules.

In other words, the stoichiometric coefficients multiplying the chemical formulas in any balanced chemical equation show the relative number of moles of each substance that reacts or is produced in the reaction.

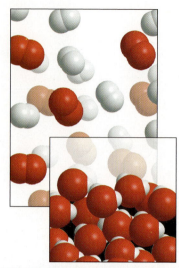

FIGURE H.2 A representation of the reaction between hydrogen and oxygen, with the production of water. No atoms are created or destroyed; they simply change partners. For every two hydrogen molecules that react (background), one oxygen molecule (shown in red) is consumed and two water molecules are formed (foreground).

A balanced chemical equation symbolizes both the qualitative and the quantitative changes that take place in a chemical reaction. The stoichiometric coefficients summarize the relative amounts (numbers of moles) of reactants and products taking part in the reaction.

H.2 Balanced Chemical Equations

In some cases, the stoichiometric coefficients needed to balance an equation are easy to determine. For example, consider the reaction in which hydrogen and oxygen gases combine to form water. This qualitative information is summarized as a skeletal equation:

$$H_2 + O_2 \longrightarrow H_2O$$ ⚠

The international *Hazard!* road sign, ⚠, is used (in this text) to warn you that a skeletal equation is not balanced. Then the hydrogen and oxygen atoms are balanced:

$$2 H_2 + O_2 \longrightarrow 2 H_2O$$

There are four H atoms and two O atoms on each side of the arrow. Finally, the state symbols are attached to the formulas:

$$2 H_2(g) + O_2(g) \longrightarrow 2 H_2O(l)$$ **(A)**

FIGURE H.2 is a molecular-level representation of this reaction.

An equation must never be balanced by changing the subscripts in the chemical formulas. That change would imply that different substances were taking part in the reaction. For example, changing H_2O to H_2O_2 in the skeletal equation and writing

$$H_2 + O_2 \longrightarrow H_2O_2$$

certainly results in a balanced equation. However, it is a summary of a different reaction—the formation of hydrogen peroxide (H_2O_2) from its elements. Nor should you write

$$H_2 + O \longrightarrow H_2O$$

Although this equation is balanced, it summarizes the reaction between hydrogen molecules and oxygen atoms, not the oxygen molecules that are the actual starting materials. For the same reason, adding unattached atoms to balance an equation is wrong. Writing

$$H_2 + O_2 \longrightarrow H_2O + O$$

would indicate that free oxygen atoms are being formed along with the water, but this does not happen.

Normally the coefficients in a balanced chemical equation are the smallest possible whole numbers, as in the equation describing the reaction of hydrogen and oxygen (reaction **A**). However, a chemical equation can be multiplied through by a factor and still be a valid equation. At times it is convenient to use fractional coefficients; for example, reaction **A** could be multiplied by ½ to give

$$H_2(g) + ½ O_2(g) \longrightarrow H_2O(l)$$

if you wanted the equation to correspond to the consumption of 1 mol H_2.

FIGURE H.3 When methane burns, it forms carbon dioxide and water. The blue color is due to the presence of C_2 molecules in the flame. If the oxygen supply is inadequate, these carbon molecules can stick together and form soot, thereby producing a smoky flame. Note that one carbon dioxide molecule and two water molecules are produced for each methane molecule that is consumed. The two hydrogen atoms in each water molecule do not necessarily come from the same methane molecule: the illustration depicts the overall outcome, not the specific outcome of the reaction of one molecule. The excess oxygen remains unreacted. *(Photo: SPL/Science Source.)*

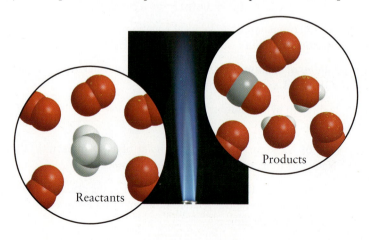

Reactants   Products

A good strategy for more complicated equations is to balance one element at a time, starting with one that appears in the fewest formulas and balancing uncombined elements last. For example, suppose you need to balance the equation for the combustion of methane. **Combustion** refers to burning in air, specifically to reaction with molecular oxygen. The products in this case are carbon dioxide and water (**FIG. H.3**). First, write the skeletal equation:

$$CH_4 + O_2 \longrightarrow CO_2 + H_2O$$

It is easier to balance carbon and hydrogen first and oxygen last. Then, once the equation is balanced, specify the states. If, under the conditions of the experiment, water is produced as a vapor, the equation is written:

$$CH_4(g) + 2\,O_2(g) \longrightarrow CO_2(g) + 2\,H_2O(g)$$

EXAMPLE H.1 Writing and balancing a chemical equation

Chemists are always on the lookout for new and more effective fuels, especially as concerns increase about the availability of fossil fuels. If you become involved in this search, you will need to study combustion reactions. Write and balance the chemical equation for the combustion of liquid hexane, C_6H_{14}, to gaseous carbon dioxide gas and gaseous water.

ANTICIPATE Because hexane contains six C atoms and fourteen H atoms, you can expect each molecule to give rise to six CO_2 molecules and seven H_2O molecules, so the balanced equation will be of the form $C_6H_{14} + ?\,O_2 \rightarrow 6\,CO_2 + 7\,H_2O$ or a multiple of it.

PLAN First write the skeletal equation, using the rules for writing formulas in *Fundamentals* D if necessary. Balance the element that occurs in the fewest formulas, then balance the remaining elements. If a stoichiometric coefficient turns out to be fractional, it is common practice to multiply through by a factor to produce whole-number coefficients. Finally, specify the physical state of each reactant and product.

SOLVE

Write the skeletal equation.

$$C_6H_{14} + O_2 \longrightarrow CO_2 + H_2O$$

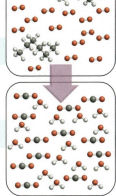

Balance carbon and hydrogen.

$$C_6H_{14} + O_2 \longrightarrow 6\,CO_2 + 7\,H_2O$$

Next balance oxygen. In this case, a fractional stoichiometric coefficient is needed.

$$C_6H_{14} + {}^{19\!/2}\,O_2 \longrightarrow 6\,O_2 + 7\,H_2O$$

The equation is balanced at this point. However, multiply by 2 to clear the fraction and to obtain the smallest whole number coefficients.

$$2\,C_6H_{14} + 19\,O_2 \longrightarrow 12\,CO_2 + 14\,H_2O$$

Finally, add the physical states.

$$2\,C_6H_{14}(g) + 19\,O_2(g) \longrightarrow 12\,CO_2(g) + 14\,H_2O(g)$$

EVALUATE As anticipated, the balanced equation, $2\,C_6H_{14}(g) + 19\,O_2(g) \rightarrow 12\,CO_2(g) + 14\,H_2O(g)$, is a multiple (by a factor of 2) of an equation of the form $C_6H_{14} + ?\,O_2 \rightarrow 6\,CO_2 + 7\,H_2O$.

Self-test H.1A When aluminum is melted and heated with solid barium oxide, a vigorous reaction takes place, and elemental molten barium and solid aluminum oxide are formed. Write the balanced chemical equation for the reaction.

[***Answer:*** $2\,Al(l) + 3\,BaO(s) \xrightarrow{\Delta} Al_2O_3(s) + 3\,Ba(l)$]

Self-test H.1B Write the balanced chemical equation for the reaction of solid magnesium nitride with aqueous sulfuric acid to form aqueous magnesium sulfate and aqueous ammonium sulfate.

Related Exercises H.7, H.8, H.13–H.21

A chemical equation expresses a chemical reaction in terms of chemical formulas; the stoichiometric coefficients are chosen to show that atoms are neither created nor destroyed in the reaction.

What have you learned in Fundamentals H?

You have learned how to express a chemical reaction symbolically and how to ensure that it is balanced. You also know how to interpret the stoichiometric coefficients in a balanced equation.

The skills you have mastered are the ability to:

☐ **1.** Explain the role of stoichiometric coefficients (Section H.1).

☐ **2.** Write, balance, and label a chemical equation on the basis of verbal information (Example H.1).

Fundamentals H Exercises

H.1 It appears that balancing the chemical equation $Cu + SO_2 \rightarrow CuO + S$ would be simple if we could just add another O atom to the product side: $Cu + SO_2 \rightarrow CuO + S + O$. (a) Why is that balancing procedure not allowed? (b) Balance the equation correctly.

H.2 Indicate which of the following are conserved in a chemical reaction: (a) mass; (b) number of atoms; (c) number of molecules; (d) number of electrons.

H.3 The first box below represents the reactants for a chemical reaction and the second box the products that form if all the reactant molecules shown react. Use the following key to write a balanced equation for the reaction. Assume that if two atoms are touching, they are bonded together. Key: ● oxygen; ○ hydrogen; ◆ silicon.

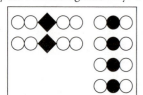

H.4 The first box below represents the reactants for a chemical reaction and the second box the products that form if all the reactant molecules shown react. Use the following key to write a balanced equation for the reaction using the smallest whole-number coefficients. Assume that if two atoms are touching, they are bonded together. Key: ● oxygen; ○ hydrogen; ☐ nitrogen.

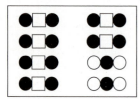

H.5 Balance the following skeletal chemical equations:

(a) $NaBH_4(s) + H_2O(l) \rightarrow NaBO_2(aq) + H_2(g)$
(b) $Mg(N_3)_2(s) + H_2O(l) \rightarrow Mg(OH)_2(aq) + HN_3(aq)$
(c) $NaCl(aq) + SO_3(g) + H_2O(l) \rightarrow Na_2SO_4(aq) + HCl(aq)$
(d) $Fe_2P(s) + S(s) \rightarrow P_4S_{10}(s) + FeS(s)$

H.6 Balance the following skeletal chemical equations:

(a) $KClO_3(s) \xrightarrow{\Delta} KCl(s) + O_2(g)$
(b) $KClO_3(l) \xrightarrow{\Delta} KCl(s) + KClO_4(g)$
(c) $N_2H_4(aq) + I_2(aq) \rightarrow HI(aq) + N_2(g)$
(d) $P_4O_{10}(s) + H_2O(l) \rightarrow H_3PO_4(l)$

H.7 Write a balanced chemical equation for each of the following reactions. (a) Calcium metal reacts with water to produce hydrogen gas and aqueous calcium hydroxide. (b) The reaction of solid sodium oxide, Na_2O, and water produces aqueous sodium hydroxide. (c) Hot solid magnesium metal reacts in a nitrogen atmosphere to produce solid magnesium nitride, Mg_3N_2. (d) The reaction of ammonia gas with oxygen gas at high temperatures in the presence of a copper metal catalyst produces the gases water and nitrogen dioxide.

H.8 Write a balanced chemical equation for each of the following reactions. (a) In the first step of recovering copper metal from ores containing $CuFeS_2$, the ore is heated in air. During this "roasting" process, molecular oxygen reacts with the $CuFeS_2$ to produce solid copper(II) sulfide, iron(II) oxide, and sulfur dioxide gas. (b) The diamondlike abrasive silicon carbide, SiC, is made by reacting solid silicon dioxide with elemental carbon at 2000 °C to produce solid silicon carbide and carbon monoxide gas. (c) The reaction of elemental hydrogen and nitrogen gases is used for the commercial production of gaseous ammonia in the Haber process. (d) Under acidic conditions, oxygen gas can react with aqueous hydrobromic acid to form liquid water and liquid bromine.

H.9 A shortcut you can use to balance reactions in which polyatomic ions remain intact is to treat the ions as if they were elements. Use that shortcut to balance the following reactions:

(a) $Pb(NO_3)_2(aq) + Na_3PO_4(aq) \rightarrow$
$$Pb_3(PO_4)_2(s) + NaNO_3(aq)$$
(b) $Ag_2CO_3(s) + NaBr(aq) \rightarrow AgBr(s) + Na_2CO_3(aq)$

H.10 A shortcut you can use to balance reactions in which polyatomic ions remain intact is to treat the ions as if they were elements. Use that shortcut to balance the following reactions:

(a) $H_3PO_4(aq) + Ca(OH)_2(aq) \rightarrow Ca_3(PO_4)_2(s) + H_2O(l)$
(b) $Cr_2(SO_4)_3(aq) + HClO_2(aq) \rightarrow$
$$Cr(ClO_2)_3(aq) + H_2SO_4(aq)$$

H.11 In one stage in the commercial production of iron metal in a blast furnace, the iron(III) oxide, Fe_2O_3, reacts with carbon monoxide to form solid Fe_3O_4 and carbon dioxide gas. In a second stage, the Fe_3O_4 reacts further with carbon monoxide to produce solid elemental iron and carbon dioxide. Write the balanced equation for each stage in the process.

H.12 The bacteria-catalyzed oxidation of ammonia in waste water takes place in two steps. In the first step, aqueous ammonia reacts with oxygen gas to form aqueous nitrous acid and water. In the second step, the nitrous acid reacts with additional oxygen to form aqueous nitric acid. Write the balanced equation for each stage in the process.

H.13 When nitrogen and oxygen gases react in the hot exhaust environment of an automobile engine, nitric oxide gas, NO, is formed. After it escapes into the atmosphere with the other exhaust gases, the nitric oxide reacts with oxygen to produce nitrogen dioxide gas, one of the precursors of acid rain. Write the two balanced equations for the reactions leading to the formation of nitrogen dioxide.

H.14 The reaction of boron trifluoride, $BF_3(g)$, with sodium borohydride, $NaBH_4(s)$, leads to the formation of sodium tetrafluoroborate, $NaBF_4(s)$, and diborane gas, $B_2H_6(g)$. The diborane reacts with the oxygen in air, forming boron oxide, $B_2O_3(s)$, and water. Write the two balanced equations leading to the formation of boron oxide.

H.15 Hydrofluoric acid is used to etch glass because it reacts with the silica, $SiO_2(s)$, in glass. The products of the reaction are aqueous silicon tetrafluoride and water. Write a balanced equation for the reaction.

H.16 The volatile liquid pentaborane, B_5H_9, once studied as a rocket fuel, is known as the "green dragon" because it burns with a hot, bright, green flame. (a) In jet engines it reacts with oxygen gas to produce $B_2O_3(s)$ and liquid water. Write a balanced equation for the reaction. (b) Because it is highly toxic and unstable, pentaborane is no longer used. Pentaborane that had been stored at the Redstone Arsenal in Alabama was safely decomposed in the "dragonslayer" process, in which it reacts with liquid water to produce hydrogen gas and an aqueous solution of boric acid, H_3BO_3. Write a balanced equation for the dragonslayer reaction.

H.17 Write a balanced equation for the complete combustion (reaction with oxygen gas) of liquid heptane, C_7H_{16}, a component typical of the hydrocarbons in gasoline, to carbon dioxide gas and water vapor.

H.18 Write a balanced equation for the incomplete combustion (reaction with oxygen gas) of liquid heptane, C_7H_{16}, to carbon monoxide gas and water vapor.

H.19 Aspartame, $C_{14}H_{18}N_2O_5$, is a solid used as an artificial sweetener. Write the balanced equation for its combustion to carbon dioxide gas, liquid water, and nitrogen gas.

H.20 Dimethazan, $C_{11}H_{17}N_5O_2$, is a solid antidepressant drug. Write the balanced equation for its combustion to carbon dioxide gas, liquid water, and nitrogen gas.

H.21 The psychoactive drug methamphetamine ("speed"), which is sold as the prescription medication Desoxyn, $C_{10}H_{15}N$, undergoes a series of reactions in the body; the net result of these reactions is the oxidation of solid methamphetamine by oxygen gas to produce carbon dioxide gas, liquid water, and an aqueous solution of urea, CH_4N_2O. Write the balanced equation for this net reaction.

H.22 The psychoactive street drug sold as MDMA ("ecstasy"), $C_{11}H_{15}NO_2$, undergoes a series of reactions in the body; the net result of these reactions is the oxidation of aqueous MDMA by oxygen gas to produce carbon dioxide gas, liquid water, and an aqueous solution of urea, CH_4N_2O. Write the balanced equation for this net reaction.

H.23 Sodium thiosulfate, which as the pentahydrate $Na_2S_2O_3 \cdot 5H_2O$ forms the large white crystals used as "photographer's hypo," can be prepared by bubbling oxygen through a solution of sodium polysulfide, Na_2S_5, in alcohol and adding water. Sulfur dioxide gas is formed as a byproduct. Sodium polysulfide is made by the action of hydrogen sulfide gas on a solution of sodium sulfide, Na_2S, in alcohol, which, in turn, is made by the reaction of hydrogen sulfide gas, H_2S, with solid sodium hydroxide. Write the three chemical equations that show how hypo is prepared from hydrogen sulfide and sodium hydroxide. Use (alc) to indicate the state of species dissolved in alcohol.

H.24 The first stage in the production of nitric acid by the Ostwald process is the reaction of ammonia gas with oxygen gas, producing nitric oxide gas, NO, and liquid water. The nitric oxide reacts with oxygen to produce nitrogen dioxide gas, which, when dissolved in water, produces nitric acid and nitric oxide. Write the three balanced equations that lead to the production of nitric acid.

H.25 Phosphorus and oxygen react to form two different phosphorus oxides. The mass percentage of phosphorus in one of these oxides is 43.64%; in the other, it is 56.34%. (a) Write the empirical formula of each phosphorus oxide. (b) The molar mass of the former oxide is 283.33 $g \cdot mol^{-1}$ and that of the latter is 219.88 $g \cdot mol^{-1}$. Determine the molecular formula and name of each oxide. (c) Write a balanced chemical equation for the formation of each oxide.

H.26 One step in the refining of titanium metal is the reaction of $FeTiO_3$ with chlorine gas and carbon. Balance the equation for the reaction: $FeTiO_3(s) + Cl_2(g) + C(s) \rightarrow TiCl_4(l) + FeCl_3(s) + CO(g)$.

FIGURE I.1 When a solution of yellow K_2CrO_4 is added to a colorless solution of $AgNO_3$, a red precipitate of silver chromate, Ag_2CrO_4, forms. (*©1998 Richard Megna–Fundamental Photographs.*)

In aqueous solution hydrogen ions exist as H_3O^+ ions, as described in *Fundamentals* J.

I Precipitation Reactions

When two solutions are mixed the result may be simply a new solution that contains both solutes. In some cases, however, the solutes can react with each other. For instance, when a colorless aqueous solution of silver nitrate is mixed with a clear yellow aqueous solution of potassium chromate, a powdery red solid forms, indicating that a chemical reaction has occurred (**FIG. I.1**).

I.1 *Electrolytes*

A **soluble substance** is one that dissolves to a significant extent in a specified solvent. When solubility is mentioned without indicating a solvent, it normally means "soluble in water." An **insoluble substance** is one that does not dissolve significantly in a specified solvent; substances are often regarded as "insoluble" if they do not dissolve to more than about 0.1 mol·L^{-1}. Unless otherwise specified, the term *insoluble* in this text normally means "insoluble in water." For instance, calcium carbonate, $CaCO_3$, which makes up limestone and chalk, dissolves to form a solution that contains only 0.01 g·L^{-1} (corresponding in this case to $1 \times 10^{-4} \text{ mol·L}^{-1}$), and $CaCO_3$ is regarded as insoluble. This insolubility is important for landscapes: chalk hills and limestone buildings do not wash away in natural rainwater.

A solute may be present as ions or as molecules. You can find out if a solute is present as ions by noting whether the solution conducts an electric current. Because a current is a flow of electric charge, only solutions that contain ions conduct electricity. There is such a tiny concentration of ions in pure water (about $10^{-7} \text{ mol·L}^{-1}$) that pure water itself does not conduct electricity significantly.

An **electrolyte** is a substance that conducts electricity by the migration of ions. Solutions of ionic solids are electrolytes because the ions become free to move when they dissolve (**FIG. I.2**). The term **electrolyte solution** is commonly used to emphasize that the electrolyte medium is in fact a solution. Some compounds, such as acids, form ions when they dissolve and hence produce an electrolyte solution even though no ions are present before they dissolve. For example, hydrogen chloride is a gas of HCl molecules, but when it dissolves in water it reacts with the water to form hydrochloric acid, and the solution consists of hydrogen ions, H^+, and chloride ions, Cl^-.

A **nonelectrolyte** is a substance that does not conduct electricity, even in solution. A **nonelectrolyte solution** is a solution that, because no ions are present, does not conduct

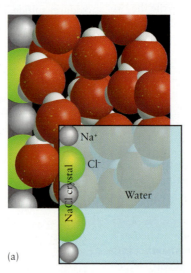
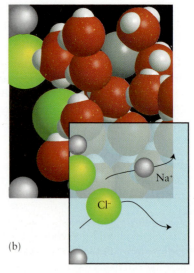

FIGURE I.2 Sodium chloride consists of sodium ions and chloride ions. When sodium chloride comes in contact with water (left), the ions are separated by the water molecules, and they spread throughout the solvent (right). The solution consists of water, sodium ions, and chloride ions. There are no NaCl molecules present. In the insets, water is represented by the blue background.

ANIMATION FIGURE I.2

electricity. Aqueous solutions of acetone (**1**) and the sugar ribose (**2**) are nonelectrolyte solutions. Except for acids and bases, most organic compounds that dissolve in water form nonelectrolyte solutions. If you could see the individual molecules in a nonelectrolyte solution, you would see the intact solute molecules dispersed among the solvent molecules (**FIG. I.3**).

A **strong electrolyte** is a substance that is present almost entirely as ions in solution. Three types of solutes are strong electrolytes: strong acids and strong bases, which are described in more detail in *Fundamentals* J, and soluble ionic compounds. Hydrochloric acid is a strong electrolyte; so are sodium hydroxide and the salt sodium chloride. A **weak electrolyte** is a substance that is incompletely ionized in solution; in other words, most of the molecules remain intact. Acetic acid is a weak electrolyte: in aqueous solution at normal concentrations, only a small fraction of CH_3COOH molecules separate into hydrogen ions, H^+, and acetate ions, $CH_3CO_2^-$. One way to distinguish strong and weak electrolytes is to measure the abilities of their solutions to conduct electricity: for the same molar concentration of solute, a solution of strong electrolyte is a better conductor than a solution of a weak electrolyte (**FIG. I.4**).

1 Acetone, C_3H_6O

2 D-Ribose

(a)   (b)   (c)

FIGURE I.4 (a) Pure water is a poor conductor of electricity, as shown by the unlit bulb in the circuit on the left. (b) When ions are present, as in this weak electrolyte solution, the solution has a low ability to conduct electricity and a dim glow is seen. That ability is significant when the solute is a strong electrolyte (c), even when the solute concentration is the same in each instance. (©1970 *George Resch–Fundamental Photographs.*)

 LAB VIDEO FIGURE I.4

Self-test I.1A Identify each of the following substances as an electrolyte or a nonelectrolyte and predict which will conduct electricity when dissolved in water: (a) NaOH; (b) Br_2.

[*Answer:* (a) ionic compound, so a strong electrolyte, conducts electricity; (b) molecular compound and not an acid, so a nonelectrolyte, does not conduct electricity]

Self-test I.1B Identify each of the following substances as an electrolyte or a nonelectrolyte and predict which will conduct electricity when dissolved in water: (a) ethanol, $CH_3CH_2OH(aq)$; (b) $Pb(NO_3)_2(aq)$.

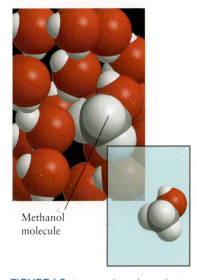

FIGURE I.3 In a nonelectrolyte solution, the solute remains as intact molecules and does not break up into ions. Methanol, CH_3OH, is a nonelectrolyte and is present as intact molecules when it is dissolved in water. The inset shows the methanol molecule alone.

The solute in an aqueous strong electrolyte solution is present as ions that can conduct electricity through the solution. The solutes in nonelectrolyte solutions are present as molecules. Only a small fraction of the solute molecules in weak electrolyte solutions are present as ions.

I.2 Precipitates

Consider what happens when a solution of sodium chloride (a strong electrolyte) is poured into a solution of silver nitrate (another strong electrolyte). A solution of sodium chloride contains Na^+ cations and Cl^- anions. Similarly, a solution of silver nitrate,

$AgNO_3$, contains Ag^+ cations and NO_3^- anions. When these two aqueous solutions are mixed, a white **precipitate,** a cloudy, finely divided solid deposit, forms immediately. Analysis shows that the precipitate is silver chloride, AgCl, an insoluble white solid. The colorless solution remaining above the precipitate contains dissolved Na^+ cations and NO_3^- anions. These ions remain in solution because sodium nitrate, $NaNO_3$, is soluble in water.

In a **precipitation reaction,** an insoluble solid product forms when two electrolyte solutions are mixed. When an insoluble substance is formed in water, it immediately precipitates. In the chemical equation for a precipitation reaction, (aq) is used to indicate substances that are dissolved in water and (s) to indicate the solid that has precipitated:

$$AgNO_3(aq) + NaCl(aq) \longrightarrow AgCl(s) + NaNO_3(aq)$$

A precipitation reaction takes place when solutions of two electrolytes are mixed and react to form an insoluble solid.

I.3 Ionic and Net Ionic Equations

The **complete ionic equation** for a precipitation reaction shows all the species as they actually exist in solution; because dissolved ionic compounds exist as separate aqueous ions, the ions are shown separately. For example, the complete ionic equation for the silver chloride precipitation reaction shown in **FIG. I.5** is

$$Ag^+(aq) + NO_3^-(aq) + Na^+(aq) + Cl^-(aq) \longrightarrow AgCl(s) + Na^+(aq) + NO_3^-(aq)$$

Because the Na^+ and NO_3^- ions appear as both reactants and products, they play no direct role in the reaction. They are **spectator ions,** ions that are present while the reaction takes place but do not participate in it, like spectators at a sports event. Because spectator ions remain unchanged, they can be canceled on each side of the arrow to simplify the ionic equation:

$$Ag^+(aq) + \cancel{NO_3^-(aq)} + \cancel{Na^+(aq)} + Cl^-(aq) \longrightarrow AgCl(s) + \cancel{Na^+(aq)} + \cancel{NO_3^-(aq)}$$

Canceling the spectator ions leaves the **net ionic equation** for the reaction, the chemical equation that displays only the net change taking place in the reaction:

$$Ag^+(aq) + Cl^-(aq) \longrightarrow AgCl(s)$$

The net ionic equation shows that Ag^+ ions combine with Cl^- ions to precipitate as solid silver chloride, AgCl (see Fig. I.5).

FIGURE I.5 (a) Silver chloride precipitates immediately when sodium chloride solution is added to a solution of silver nitrate. (b) If we imagine the removal of the spectator ions from the complete ionic reaction (top), we can focus on the essential process, the net ionic reaction (bottom). *(Part a: ©1995 Richard Megna–Fundamental Photographs.)*

 **ANIMATION FIGURE I.5**

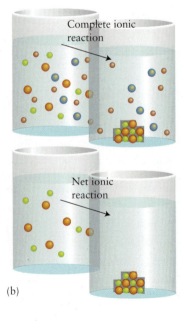

(a)   (b)

EXAMPLE I.1   Writing a net ionic equation

Suppose you are working on water purification and need to know how much barium is present in a water sample. You can isolate barium ions (Ba^{2+}) by reacting them with another substance to form a precipitate. When concentrated ammonium iodate solution, $NH_4IO_3(aq)$, is added to an aqueous solution of barium nitrate, $Ba(NO_3)_2(aq)$, insoluble barium iodate, $Ba(IO_3)_2(s)$, forms. The chemical equation for the precipitation reaction is

$$Ba(NO_3)_2(aq) + 2\,NH_4IO_3(aq) \longrightarrow Ba(IO_3)_2(s) + 2\,NH_4NO_3(aq)$$

Write the net ionic equation for the reaction.

PLAN First, write and balance the complete ionic equation, showing all the dissolved ions separately. Insoluble solids are shown as complete compounds. Next, cancel the spectator ions, the ions that appear on both sides of the arrow.

SOLVE

The complete ionic equation, which shows all the dissolved ions, is

$$Ba^{2+}(aq) + 2\,NO_3^-(aq) + 2\,NH_4^+(aq) + 2\,IO_3^-(aq) \longrightarrow Ba(IO_3)_2(s) + 2\,NH_4^+(aq) + 2\,NO_3^-(aq)$$

Now cancel the spectator ions NH_4^+ and NO_3^-:

$$Ba^{2+}(aq) + 2\,\cancel{NO_3^-(aq)} + 2\,\cancel{NH_4^+(aq)} + 2\,IO_3^-(aq) \longrightarrow Ba(IO_3)_2(s) + 2\,\cancel{NH_4^+(aq)} + 2\,\cancel{NO_3^-(aq)}$$

and obtain the net ionic equation:

$$Ba^{2+}(aq) + 2\,IO_3^-(aq) \longrightarrow Ba(IO_3)_2(s)$$

Self-test I.2A Write the net ionic equation for the reaction in Fig. I.1, in which aqueous solutions of colorless silver nitrate and yellow potassium chromate react to give a precipitate of red silver chromate.

[**Answer:** $2\,Ag^+(aq) + CrO_4^{2-}(aq) \rightarrow Ag_2CrO_4(s)$]

Self-test I.2B The mercury(I) ion, Hg_2^{2+}, consists of two Hg^+ ions joined together. Write the net ionic equation for the reaction in which colorless aqueous solutions of mercury(I) nitrate, $Hg_2(NO_3)_2$, and potassium phosphate, K_3PO_4, react to give a white precipitate of mercury(I) phosphate.

Related Exercises I.5, I.6, I.15, I.16

A complete ionic equation expresses a reaction in terms of the ions that are present in solution; a net ionic equation is the chemical equation that remains after the cancellation of the spectator ions.

I.4 Putting Precipitation to Work

One of the many uses for precipitation reactions is to make a compound by preparing two solutions that, when mixed, give a precipitate of the desired insoluble compound. Then the insoluble compound can be separated from the reaction mixture by filtration. Precipitation reactions are also used in chemical analysis. In **qualitative analysis**—the identification of the substances present in a sample—the formation of a precipitate is used to confirm the identity of certain ions. In **quantitative analysis**, the aim is to determine the amount of each substance or element present. In particular, in **gravimetric analysis**, which is used in environmental monitoring, the amount of substance present is determined by measurements of mass. In this application, an insoluble compound is precipitated, the precipitate is filtered off and weighed, and from its mass the amount of a substance in one of the original solutions is calculated (**FIG. I.6**).

Topic 6J describes the use of precipitation in qualitative analysis in more detail.

You will find an example of how to use this technique in Fundamentals L.

TABLE I.1 summarizes the solubility patterns of common ionic compounds in water. Notice that all nitrates and all common compounds of the Group 1 metals are soluble, so they make useful starting solutions for precipitation reactions. Any spectator ions can be used, provided that they remain in solution and do not otherwise react. For example, Table I.1 shows that mercury(I) iodide, Hg_2I_2, is insoluble. It is formed as a precipitate when solutions containing Hg_2^{2+} ions and I^- ions are mixed:

$$Hg_2^{2+}(aq) + 2\,I^-(aq) \longrightarrow Hg_2I_2(s)$$

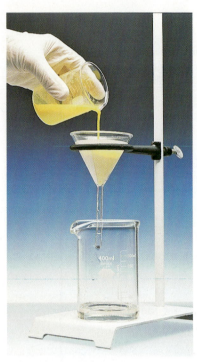

FIGURE I.6 A step in gravimetric analysis. To determine how much of a particular type of ion was present in a solution, the ion was precipitated as a salt and the precipitate is being filtered. The filter paper, which has a known mass, will be dried and weighed, thereby allowing the mass of the precipitate to be determined. *(Richard Megna–Fundamental Photographs.)*

TABLE I.1   Solubility Rules for Inorganic Compounds

Soluble compounds	Insoluble compounds
compounds of Group 1 elements	carbonates (CO_3^{2-}), chromates (CrO_4^{2-}), oxalates ($C_2O_4^{2-}$), and phosphates (PO_4^{3-}), **except** those of the Group 1 elements and NH_4^+
ammonium (NH_4^+) compounds	
chlorides (Cl^-), bromides (Br^-), and iodides (I^-), **except** those of Ag^{++}, Hg_2^{2+}, and Pb^{2+}*	sulfides (S^{2-}), **except** those of the Group 1 and 2 elements and $NH_4^{+‡}$
nitrates (NO_3^-), acetates ($CH_3CO_2^-$), chlorates (ClO_3^-), and perchlorates (ClO_4^-)	hydroxides (OH^-) and oxides (O^{2-}), **except** those of the Group 1 elements and Group 2 elements later than Period 2§
sulfates (SO_4^{2-}), **except** those of Ca^{2+}, Sr^{2+}, Ba^{2+}, Pb^{2+}, Hg_2^{2+}, and $Ag^{+†}$	

*$PbCl_2$ is slightly soluble.
† Ag_2SO_4 is slightly soluble.
‡Group 2 sulfides react with water to form the hydroxide and H_2S.
§$Ca(OH)_2$ and $Sr(OH)_2$ are sparingly (slightly) soluble; $Mg(OH)_2$ is only very slightly soluble.

Because the spectator ions are not shown, the net ionic equation will be the same when any soluble mercury(I) compound is mixed with any soluble iodide.

EXAMPLE I.2   Predicting the outcome of a precipitation reaction

If you intend to use a precipitation reaction to produce a new compound, you need to be able to predict the products of the reaction and whether any of them will precipitate as insoluble solids. Predict the precipitate likely to be formed, if any, when aqueous solutions of sodium phosphate and lead(II) nitrate are mixed. Write the net ionic equation for the reaction.

PLAN Decide which ions are present in the mixed solutions and consider all possible combinations. Use the solubility rules in Table I.1 to decide which combination corresponds to an insoluble compound and write the net ionic equation to match.

SOLVE

The mixed solution will contain Na^+, PO_4^{3-}, Pb^{2+}, and NO_3^- ions. All nitrates and all compounds of Group 1 metals are soluble, but phosphates of other elements are generally insoluble.

Hence, Pb^{2+} and PO_4^{3-} ions will form an insoluble compound and lead(II) phosphate, $Pb_3(PO_4)_2$, will precipitate.

Now write the net ionic equation. The Na^+ and NO_3^- ions are spectators, and so they are omitted.

$$3\,Pb^{2+}(aq) + 2\,PO_4^{3-}(aq) \longrightarrow Pb_3(PO_4)_2(s)$$

Self-test I.3A Predict the identity of the precipitate that forms, if any, when aqueous solutions of ammonium sulfide and copper(II) sulfate are mixed, and write the net ionic equation for the reaction.

[**Answer:** Copper(II) sulfide; $Cu^{2+}(aq) + S^{2-}(aq) \rightarrow CuS(s)$]

Self-test I.3B Suggest two solutions that can be mixed to prepare strontium sulfate, and write the net ionic equation for the reaction.

Related Exercises I.11–I.14

The solubility rules in Table I.1 can be used to predict the outcomes of precipitation reactions.

What have you learned in Fundamentals I?

You have learned that, in a precipitation reaction, two solutions are mixed and form a precipitate of an insoluble solid. You have also learned the difference between electrolytes and nonelectrolytes and have seen how to write a net ionic equation for a reaction.

The skills you have mastered are the ability to:

☐ **1.** Identify electrolytes or nonelectrolytes on the basis of the formulas of the solutes (Self-Test I.1).

☐ **2.** Construct balanced complete ionic and net ionic equations for reactions involving ions (Example I.1).

☐ **3.** Use the solubility rules to select appropriate solutions that, when mixed, produce a desired precipitate (Section I.4).

☐ **4.** Identify any precipitate that may form upon mixing two given solutions (Example I.2).

Fundamentals I Exercises

I.1 The solution on the left pictured below contains 0.50 M $CaCl_2$(aq), and the solution on the right contains 0.50 M Na_2SO_4(aq). Suppose the contents of the two solutions are mixed. Draw a picture of the resulting products.

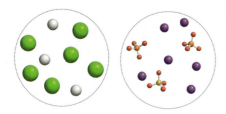

I.2 The solution on the left pictured below contains 0.50 M $Hg_2(NO_3)_2$(aq), and the solution on the right contains 0.50 M K_3PO_4(aq). Suppose the contents of the two solutions are mixed. Draw a picture of the resulting products.

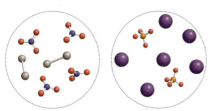

I.3 Classify each of the following substances as a strong electrolyte or a nonelectrolyte: (a) CH_3OH; (b) $BaCl_2$; (c) KF.

I.4 Classify each of the following substances as a strong electrolyte, a weak electrolyte, or a nonelectrolyte: (a) H_2SO_4; (b) KOH; (c) $CH_3CH_2CH_2COOH$.

I.5 Write the balanced overall, complete ionic, and net ionic equations corresponding to each of the following reactions:
(a) $BaBr_2(aq) + Li_3PO_4(aq) \rightarrow Ba_3(PO_4)_2(s) + LiBr(aq)$
(b) $NH_4Cl(aq) + Hg_2(NO_3)_2(aq) \rightarrow NH_4NO_3(aq) + Hg_2Cl_2(s)$
(c) $Co(NO_3)_3(aq) + Ca(OH)_2(aq) \rightarrow$
$$Co(OH)_3(s) + Ca(NO_3)_2(aq)$$

I.6 Write the balanced overall, complete ionic, and net ionic equations corresponding to each of the following reactions:
(a) $MgBr_2(aq) + Na_3PO_4(aq) \rightarrow Mg_3(PO_4)_2(s) + NaBr(aq)$
(b) $CsI(aq) + Hg_2(HSO_3)_2(aq) \rightarrow Hg_2I_2(s) + CsHSO_3(aq)$
(c) $K_2C_2O_4(aq) + Co(NO_3)_3(aq) \rightarrow$
$$Co_2(C_2O_4)_3(s) + KNO_3(aq)$$

I.7 Use the information in Table I.1 to classify each of the following ionic compounds as soluble or insoluble in water:
(a) potassium phosphate, K_3PO_4; (b) lead(II) chloride, $PbCl_2$;
(c) cadmium sulfide, CdS; (d) barium sulfate, $BaSO_4$.

I.8 Use the information in Table I.1 to classify each of the following ionic compounds as soluble or insoluble in water:(a) potassium phosphate, K_3PO_4; (b) lead(II) chloride, $PbCl_2$; (c) cadmium sulfide, CdS; (d) barium sulfate, $BaSO_4$.

I.9 What are the principal solute species present in an aqueous solution of (a) NaI; (b) Ag_2CO_3; (c) $(NH_4)_3PO_4$; (d) $FeSO_4$?

I.10 What are the principal solute species present in an aqueous solution of (a) $CoCO_3$; (b) $LiNO_3$; (c) K_2CrO_4; (d) Hg_2Cl_2?

I.11 (a) When aqueous solutions of iron(III) sulfate and sodium hydroxide were mixed, a precipitate formed. Write the formula of the precipitate. (b) Does a precipitate form when aqueous solutions of silver nitrate, $AgNO_3$, and potassium carbonate are mixed? If so, write the formula of the precipitate. (c) If aqueous solutions of lead(II) nitrate and sodium acetate are mixed, does a precipitate form? If so, write the formula of the precipitate.

I.12 (a) Solid calcium nitrate and solid sodium carbonate were mixed, placed in water, and stirred. What is observed? If a precipitate is present, write its formula. (b) Solid nickel(II) sulfate and solid copper(II) chloride were placed in water, and the mixture was stirred. Is the formation of a precipitate expected? If so, write the formula of the precipitate. (c) Aqueous solutions of sodium phosphate and barium chloride are mixed. What is observed? If a precipitate is present, write its formula.

I.13 When the solution in Beaker 1 is mixed with the solution in Beaker 2, a precipitate forms. Using the following table, write the net ionic equation describing the formation of the precipitate, and then identify the spectator ions.

Beaker 1	Beaker 2
(a) $FeCl_2$(aq)	Na_2S(aq)
(b) $Pb(NO_3)_2$(aq)	KI(aq)
(c) $Ca(NO_3)_2$(aq)	K_2SO_4(aq)
(d) Na_2CrO_4(aq)	$Pb(NO_3)_2$(aq)
(e) $Hg_2(NO_3)_2$(aq)	K_2SO_4(aq)

I.14 The contents of Beaker 1 are mixed with those of Beaker 2. If a reaction takes place, write the net ionic equation and indicate the spectator ions.

Beaker 1	Beaker 2
(a) K_2SO_4(aq)	$AgNO_3$(aq)
(b) H_3PO_4(aq)	$SrBr_2$(aq)
(c) Na_2S(aq)	NH_4NO_3(aq)
(d) $CdSO_4$(aq)	$(NH_4)_2CO_3$(aq)
(e) H_2SO_4(aq)	Hg_2Cl_2(aq)

I.15 Each of the following procedures results in the formation of a precipitate. For each reaction, write the chemical equations describing the formation of the precipitate: the overall equation, the complete ionic equation, and the net ionic equation. Identify the spectator ions.
(a) $(NH_4)_2CrO_4(aq)$ is mixed with $BaCl_2(aq)$.
(b) $CuSO_4(aq)$ is mixed with $Na_2S(aq)$.
(c) $FeCl_2(aq)$ is mixed with $(NH_4)_3PO_4(aq)$.
(d) Potassium oxalate, $K_2C_2O_4(aq)$, is mixed with $Ca(NO_3)_2(aq)$.
(e) $NiSO_4(aq)$ is mixed with $Ba(NO_3)_2(aq)$.

I.16 Each of the following procedures results in the formation of a precipitate. For each reaction, write the chemical equations describing the formation of the precipitate: the overall equation, the complete ionic equation, and the net ionic equation. Identify the spectator ions.
(a) $AgNO_3(aq)$ is mixed with $Na_3PO_4(aq)$.
(b) $Hg_2(NO_3)_2(aq)$ is mixed with $NH_4I(aq)$.
(c) $BaCl_2(aq)$ is mixed with $Na_2SO_4(aq)$.
(d) $K_2S(aq)$ is mixed with $Bi(NO_3)_3(aq)$.
(e) Barium acetate, $Ba(CH_3CO_2)_2(aq)$, is mixed with $Li_2CO_3(aq)$.

I.17 For each of the following reactions, suggest two soluble ionic compounds that, when mixed together in water, result in the net ionic equation given.
(a) $2\,Ag^+(aq) + CrO_4^{2-}(aq) \rightarrow Ag_2CrO_4(s)$
(b) $Ca^{2+}(aq) + CO_3^{2-}(aq) \rightarrow CaCO_3(s)$ the reaction responsible for the deposition of chalk hills and sea urchin spines
(c) $Cd^{2+}(aq) + S^{2-}(aq) \rightarrow CdS(s)$, a yellow substance used to color glass

I.18 For each of the following reactions, suggest two soluble ionic compounds that, when mixed together in water, result in the net ionic equation given:
(a) $2\,Ag^+(aq) + CO_3^{2-}(aq) \rightarrow Ag_2CO_3(s)$
(b) $Mg^{2+}(aq) + 2\,OH^-(aq) \rightarrow Mg(OH)_2(s)$, the suspension present in milk of magnesia
(c) $3\,Ca^{3+}(aq) + 2\,PO_4^{3-}(aq) \rightarrow Ca_3(PO_4)_2(s)$, gypsum, a component of concrete

I.19 How would you use the solubility rules in Table I.1 to separate the following pairs of ions? In each case, indicate what reagent you would add and write the net ionic equation for the precipitation reaction: (a) lead(II) and copper(II) ions; (b) ammonium and magnesium ions.

I.20 How would you use the solubility rules in Table I.1 to separate the following pairs of ions? In each case, indicate what reagent you would add and write the net ionic equation for the precipitation reaction: (a) cesium and zinc ions; (b) nickel(II) and barium ions.

I.21 Write the net ionic equation for the formation of each of the following insoluble compounds in aqueous solution: (a) silver sulfate, Ag_2SO_4; (b) mercury(II) sulfide, HgS, used as an electrolyte in some primary batteries; (c) calcium phosphate, $Ca_3(PO_4)_2$, a component of bones and teeth. (d) Select two soluble ionic compounds that, when mixed in solution, form each of the insoluble compounds in parts (a), (b), and (c). Identify the spectator ions.

I.22 Write the net ionic equation for the formation of each of the following insoluble compounds in aqueous solution: (a) lead(II) chromate, $PbCrO_4$, the yellow pigment that has been used for centuries in oil paints; (b) aluminum phosphate, $AlPO_4$, used in cements and as an antacid; (c) iron(II) hydroxide, $Fe(OH)_2$. (d) Select two soluble ionic compounds that, when mixed in solution, form each of the insoluble compounds in parts (a), (b), and (c). Identify the spectator ions.

I.23 You are given a solution and asked to analyze it for the cations Ag^+, Ca^{2+}, and Zn^{2+}. You add hydrochloric acid, and a white precipitate forms. You filter out the solid and add sulfuric acid to the solution. Nothing appears to happen. Then you add hydrogen sulfide. A black precipitate forms. Which ions should you report as being present in your solution?

I.24 You are given a solution and asked to analyze it for the cations Ag^+, Ca^{2+}, and Hg^{2+}. You add hydrochloric acid. Nothing appears to happen. You then add dilute sulfuric acid, and a white precipitate forms. You filter out the solid and add hydrogen sulfide to the solution that remains. A black precipitate forms. Which ions should you report as being present in your solution?

I.25 Suppose that 40.0 mL of 0.100 M $NaOH(aq)$ is added to 10.0 mL of 0.200 M $Cu(NO_3)_2(aq)$. (a) Write the chemical equation for the precipitation reaction, the complete ionic equation, and the net ionic equation. (b) What is the concentration of Na^+ ions (in moles per liter) in the final solution?

I.26 Suppose that 2.50 g of solid $(NH_4)_3PO_4$ is added to 50.0 mL of 0.125 M $CaCl_2(aq)$. (a) Write the chemical equation for the precipitation reaction and the net ionic equation. (b) What is the concentration of each spectator ion (in moles per liter) after the reaction is complete? Assume a final volume of 70.0 mL.

The pH scale is discussed in detail in Topic 6B.

J Acids and Bases

Early chemists applied the term *acid* to substances that had a sharp or sour taste. Vinegar, for instance, contains acetic acid, CH_3COOH. Aqueous solutions of substances that they called *bases* or **alkalis** were recognized by their soapy feel. Fortunately, there are less hazardous ways of recognizing acids and bases. For instance, acids and bases change the color of certain dyes known as **indicators** (**FIG. J.1**). One of the best-known indicators is litmus, a vegetable dye obtained from a lichen. Aqueous solutions of acids turn litmus red; aqueous solutions of bases turn it blue. The electronic instrument known as a pH meter provides a rapid way of identifying a solution as acidic or basic:

A pH *lower* than 7 (pH < 7) is characteristic of an **acidic solution.**

A pH *higher* than 7 (pH > 7) is characteristic of a **basic solution.**

FIGURE J.1 The acidities of various household products can be demonstrated by adding an indicator (an extract of red cabbage, in this case) and noting the resulting color. Red indicates an acidic solution; blue, basic. From left to right, the household products are stomach acid, lemon-lime soda, tap water, detergent, and a solution of lye. The yellow seen in the lye solution is a sign that the lye is so basic that it has destroyed some of the dye. *(Andrew Lambert Photography/Science Source.)*

J.1 Acids and Bases in Aqueous Solution

Chemists debated the concepts of acid and base for many years before precise definitions emerged. Among the first useful definitions was the one proposed by the Swedish chemist Svante Arrhenius in about 1884. He defined an "acid" as a compound that contains hydrogen and reacts with water to form hydrogen ions and a "base" as a compound that produces hydroxide ions in water. Compounds that conform to these definitions are called **Arrhenius acids and bases.** For instance, HCl is an Arrhenius acid, because it releases a hydrogen ion, H^+ (a proton), when it dissolves in water; CH_4 is not an Arrhenius acid, because it does not release hydrogen ions in water. Sodium hydroxide is an Arrhenius base, because OH^- ions go into solution when it dissolves; ammonia is an Arrhenius base, because it produces OH^- ions by reacting with water:

$$NH_3(aq) + H_2O(l) \longrightarrow NH_4^+(aq) + OH^-(aq) \qquad \textbf{(A)}$$

Sodium metal, although it produces OH^- ions when it reacts with water, is not an Arrhenius base, because it is an element, not a compound, as the definition requires.

The problem with the Arrhenius definitions is that they are specific to one particular solvent, water. When chemists studied nonaqueous solvents, such as liquid ammonia, they found that a number of substances showed the same pattern of acid–base behavior. A major advance in understanding what it means to be an acid or a base came in 1923, when two chemists working independently, Johannes Brønsted in Denmark and Thomas Lowry in England, came up with the same idea. Their insight was to realize that the key process responsible for the properties of acids and bases was the transfer of a proton (a hydrogen ion) from one substance to another. The **Brønsted–Lowry definition** of acids and bases is as follows:

- An **acid** is a proton donor.
- A **base** is a proton acceptor.

Such substances are called "Brønsted acids and bases" or just plain "acids and bases" because the Brønsted–Lowry definition is the one commonly accepted today and the one used throughout this text.

When a molecule of an acid dissolves in water, it donates a hydrogen ion, H^+, to one of the water molecules and forms a **hydronium ion,** H_3O^+ (**1**). For example, when hydrogen chloride, HCl, dissolves in water, it releases a hydrogen ion to water, and the resulting solution consists of hydronium ions and chloride ions:

$$HCl(aq) + H_2O(l) \longrightarrow H_3O^+(aq) + Cl^-(aq)$$

Notice that because H_2O accepts the hydrogen ion to form H_3O^+, water is acting as a Brønsted base.

How can acids be identified from their formulas? A Brønsted acid contains an **acidic hydrogen atom,** a hydrogen atom that can be released as a proton. An acidic hydrogen atom is often written as the first element in the molecular formula of inorganic acids. For

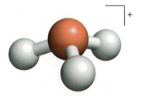

1 Hydronium ion, H_3O^+

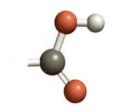

2 Carboxyl group, —COOH

example, hydrogen chloride, HCl, and nitric acid, HNO_3, are Brønsted acids. The molecules of both compounds contain hydrogen atoms that they can donate as protons to other substances. The formula of an organic acid is different in that the acidic hydrogen atom is given at the end, as part of the carboxyl group, —COOH (**2**). The carboxyl group, —COOH, is written explicitly, which makes it easier to remember that the H atom in this group of atoms is the acidic one. Acetic acid, CH_3COOH, releases *one* hydrogen ion (from the hydrogen atom of the carboxyl group) to water and any other Brønsted bases present in the solution. When a carboxyl group loses a proton, it becomes a **carboxylate anion.** In the case of acetic acid, the anion formed is the acetate ion, $CH_3CO_2^-$. Methane, CH_4, and ammonia, NH_3, are not Brønsted acids because, although they contain hydrogen, they do not normally donate protons to other substances.

Like HCl and HNO_3, acetic acid is a **monoprotic acid,** an acid that can donate only one proton from each molecule. Sulfuric acid, H_2SO_4, can release both its hydrogen atoms as ions—the first one much more readily than the second—and so it is an example of a **polyprotic acid,** an acid that can donate more than one proton from each molecule.

You can use these guidelines to recognize from their formulas that HCl, H_2CO_3 (carbonic acid), H_2SO_4 (sulfuric acid), and HSO_4^- (hydrogen sulfate) are acids in water but that CH_4, NH_3 (ammonia), and $CH_3CO_2^-$ (the acetate ion) are not. The common oxoacids, acids containing oxygen, are introduced in *Fundamentals* D and are listed in Table D.1.

Hydroxide ions are bases, because they accept protons from acids to form molecules of water:

$$OH^-(aq) + CH_3COOH(aq) \longrightarrow H_2O(l) + CH_3CO_2^-(aq)$$

Ammonia is a base because, as you can see from reaction **A,** it accepts protons from water and forms NH_4^+ ions. Notice that because water donates a hydrogen ion, it is acting as a Brønsted acid in that reaction.

A Note on Good Practice: In the Arrhenius system, sodium hydroxide is a base; however, from the Brønsted point of view it simply *provides* the base OH^-. Chemists commonly slip into using the Arrhenius definition.

Self-test J.1A Which of the following compounds are Brønsted acids or bases in water? (a) HNO_3; (b) C_6H_6; (c) KOH; (d) C_3H_5COOH.

[***Answer:*** (a) and (d) are acids; (b) is neither an acid nor a base; (c) *supplies* the base OH^-]

Self-test J.1B Which of the following compounds are Brønsted acids or bases in water? (a) KCl; (b) HClO; (c) HF; (d) $Ca(OH)_2$.

Acids are molecules or ions that are proton donors. Bases are molecules or ions that are proton acceptors.

J.2 Strong and Weak Acids and Bases

Electrolytes are classified as strong or weak according to the extent to which they are present as ions in solution (*Fundamentals* I). Acids and bases are classified in a similar fashion according to their extent of **deprotonation,** the loss of a proton (for acids), or **protonation,** the gain of a proton (for bases):

The terms *ionized* and *dissociated* are commonly used instead of "deprotonated."

A **strong acid** is completely deprotonated in solution.

A **weak acid** is incompletely deprotonated in solution.

A **strong base** is completely protonated in solution.

A **weak base** is incompletely protonated in solution.

"Completely deprotonated" means that almost *every* acid molecule or ion has lost its acidic hydrogen atom by transferring it as a hydrogen ion (a proton) to a solvent molecule. Similarly, "completely protonated" means that almost *every* base molecule or ion has acquired a proton. "Incompletely deprotonated" means that only a fraction (usually a very tiny fraction) of the

TABLE J.1   Common Aqueous Strong Acids and Bases

Strong acids	Strong bases
hydrobromic acid, HBr(aq)	Group 1 hydroxides
hydrochloric acid, HCl(aq)	alkaline earth metal hydroxides*
hydroiodic acid, HI(aq)	Group 1 and Group 2 oxides
nitric acid, HNO_3	
perchloric acid, $HClO_4$	
chloric acid, $HClO_3$	
sulfuric acid, H_2SO_4 (to form HSO_4^-)	

*$Ca(OH)_2$, $Sr(OH)_2$, $Ba(OH)_2$.

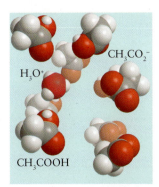

FIGURE J.2 Acetic acid, like all carboxylic acids, is a weak acid in water because when it dissolves most of it remains as acetic acid molecules, CH_3COOH; however, a small proportion of these molecules donate a hydrogen ion to a water molecule to form hydronium ions, H_3O^+, and acetate ions, $CH_3CO_2^-$.

acid molecules or ions have lost acidic hydrogen atoms as protons. "Incompletely protonated" means that only a tiny fraction of the base molecules or ions have acquired protons.

To understand the distinction between strong and weak acids, compare hydrochloric acid and acetic acid. Hydrogen chloride dissolves in water to form hydrochloric acid, which is a strong acid. This solution contains hydronium ions, chloride ions, and virtually no HCl molecules. The proton transfer reaction goes to completion:

$$HCl(g) + H_2O(l) \longrightarrow H_3O^+(aq) + Cl^-(aq)$$

Acetic acid, on the other hand, is a weak acid in water. Only a small fraction of its molecules undergo deprotonation in this reaction:

$$CH_3COOH(aq) + H_2O(l) \longrightarrow H_3O^+(aq) + CH_3CO_2^-(aq)$$

Therefore, the solute consists principally of CH_3COOH molecules (**FIG. J.2**). In fact, 0.1 M CH_3COOH(aq) contains only about one $CH_3CO_2^-$ ion for every hundred acetic acid molecules used to make a solution.

TABLE J.1 lists all the common substances that act as strong acids in water. They include three acids that are often found as reagents in laboratories—hydrochloric acid, nitric acid, and sulfuric acid (with respect to the loss of one proton from each H_2SO_4 molecule). Most other acids are weak in water. All carboxylic acids are weak in water.

The common strong bases are oxide ions and hydroxide ions, which are provided by the alkali metal and alkaline earth metal oxides and hydroxides, such as calcium oxide (see Table J.1). When an oxide dissolves in water, the oxide ions, O^{2-}, accept protons to form hydroxide ions:

$$O^{2-}(aq) + H_2O(l) \longrightarrow 2\,OH^-(aq)$$

Hydroxide ions, such as those provided by sodium hydroxide and calcium hydroxide, are also strong bases in water:

$$H_2O(l) + OH^-(aq) \longrightarrow OH^-(aq) + H_2O(l)$$

Even though a hydroxide ion is a strong base and is protonated in water, it effectively survives, because the H_2O molecule that donates a proton to OH^- becomes a hydroxide ion itself and takes its place!

All other common bases are weak in water. For example, ammonia (**3**) is a weak base in water, and the reaction

$$NH_3(aq) + H_2O(l) \longrightarrow NH_4^+(aq) + OH^-(aq)$$

produces only a small amount of OH^- ions. That is, in its aqueous solutions, ammonia exists almost entirely as NH_3 molecules, with just a small proportion—usually fewer than one in a hundred molecules at normal concentrations—of NH_4^+ cations and OH^- anions. Other common weak bases are the amines, the pungent compounds that are derived from ammonia by replacement of one or more of its hydrogen atoms by an organic group. For example, the replacement of one hydrogen atom in NH_3 by a methyl group, $-CH_3$ (**4**), results in methylamine, CH_3NH_2 (**5**). The replacement of all three hydrogen atoms in NH_3 by methyl groups results in trimethylamine, $(CH_3)_3N$ (**6**), a substance found in decomposing fish and on unwashed dogs. When amines become protonated, they form

3   Ammonia, NH_3

4   Methyl group, $-CH_3$

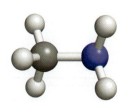

5   Methylamine, CH_3NH_2

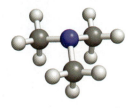

6   Trimethylamine, $(CH_3)_3N$

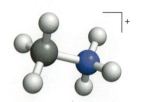

7   Methylammonium ion, $CH_3NH_3^+$

substituted ammonium cations. In the case of methylamine, the cation formed is the methyl ammonium ion, $CH_3NH_3^+$ (7).

Strong acids (the acids listed in Table J.1) are completely deprotonated in solution; weak acids (most other acids) are not. Strong bases (the metal oxides and hydroxides listed in Table J.1) are completely protonated in solution. Weak bases (ammonia and its organic derivatives, the amines) are only partially protonated in solution.

J.3 Neutralization

The reaction between an acid and a base is called a **neutralization reaction,** and the ionic compound produced in the reaction is called a **salt.** The general form of a neutralization reaction of a strong acid and a metal hydroxide that provides the hydroxide ion, a strong base, in water is

$$\text{Acid + metal hydroxide} \longrightarrow \text{salt + water}$$

The name *salt* is taken from ordinary table salt, sodium chloride, the ionic product of the reaction between hydrochloric acid and sodium hydroxide:

$$HCl(aq) + NaOH(aq) \longrightarrow NaCl(aq) + H_2O(l)$$

In the neutralization reaction between an acid and a metal hydroxide, the cation of the salt is provided by the metal hydroxide, such as Na^+ from NaOH, and the anion is provided by the acid, such as Cl^- from HCl. Another example is the reaction between nitric acid and barium hydroxide:

$$2\,HNO_3(aq) + Ba(OH)_2(aq) \longrightarrow Ba(NO_3)_2(aq) + 2\,H_2O(l)$$

The barium nitrate remains in solution as Ba^{2+} and NO_3^- ions.

The net chemical change of a neutralization reaction is clarified by writing its net ionic equation (*Fundamentals* I). For example, the complete ionic equation for the neutralization reaction between nitric acid and barium hydroxide in water is

$$2\,H^+(aq) + 2\,NO_3^-(aq) + Ba^{2+}(aq) + 2\,OH^-(aq) \longrightarrow Ba^{2+}(aq) + 2\,NO_3^-(aq) + 2\,H_2O(l)$$

The ions common to both sides now cancel,

$$2\,H^+(aq) + 2\,\cancel{NO_3^-(aq)} + \cancel{Ba^{2+}(aq)} + 2\,OH^-(aq) \longrightarrow \cancel{Ba^{2+}(aq)} + 2\,\cancel{NO_3^-(aq)} + 2\,H_2O(l)$$

and the net ionic equation of this reaction is therefore

$$2\,H^+(aq) + 2\,OH^-(aq) \longrightarrow 2\,H_2O(l)$$

which simplifies to

$$H^+(aq) + OH^-(aq) \longrightarrow H_2O(l)$$

The same net outcome is obtained for any neutralization reaction between a strong acid and a strong base in water: water is formed from hydrogen ions and hydroxide ions.

When the net ionic equation is written for the neutralization of a weak acid or a weak base, the formula for the molecular form of the weak acid or base is used, because intact acid molecules are the dominant species in solution. For example, the net ionic equation for the reaction of the weak acid HCN with the strong base NaOH in water (**FIG. J.3**) is written as

$$HCN(aq) + OH^-(aq) \longrightarrow H_2O(l) + CN^-(aq)$$

Similarly, the net ionic equation for the reaction of the weak base ammonia with the strong acid HCl in water is

$$NH_3(aq) + H^+(aq) \longrightarrow NH_4^+(aq)$$

Although hydrogen ions in aqueous solution are always attached to water molecules as hydronium ions, H_3O^+, to simplify the appearance of equations it is sometimes convenient to write them more simply as $H^+(aq)$.

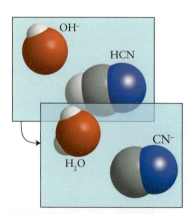

FIGURE J.3 The net ionic equation for the neutralization of HCN, a weak acid, by the strong base, NaOH, tells us that the hydroxide ion extracts a hydrogen ion from an acid molecule.

Self-test J.2A What acid and base solutions could you use to prepare rubidium nitrate? Write the chemical equation for the neutralization.

[**Answer:** $HNO_3(aq) + RbOH(aq) \rightarrow RbNO_3(aq) + H_2O(l)$]

Self-test J.2B Write the chemical equation for a neutralization reaction in which calcium phosphate is produced.

In a neutralization reaction in water, an acid reacts with a base to produce a salt (and water if the base is strong); the net outcome of the reaction between solutions of a strong acid and a metal hydroxide is the formation of water from hydrogen ions and hydroxide ions.

What have you learned in Fundamentals J?

You have learned that, according to the Brønsted definition, acids and bases are defined by the ability of a substance to donate or accept a proton. You are also now aware of the distinction between strong and weak acids and bases.

The skills you have mastered are the ability to:

☐ **1.** Describe the chemical properties of acids and bases (Section J.1).

☐ **2.** Classify substances as acids or bases (Self-Test J.1).

☐ **3.** Identify common strong acids and bases (Table J.1).

☐ **4.** Distinguish between strong and weak acids, and between strong and weak bases (Sections J.2 and J.3).

☐ **5.** Predict the outcome of neutralization reactions and write their chemical equations (Self-Test J.2).

Fundamentals J Exercises

J.1 Identify each compound as either a Brønsted acid or a Brønsted base: (a) NH_3; (b) HBr; (c) KOH; (d) H_2SO_3; (e) $Ca(OH)_2$.

J.2 Classify each compound as either a Brønsted acid or a Brønsted base: (a) H_2SeO_4; (b) $CH_3CH_2NH_2$, a derivative of ammonia; (c) $HCOOH$; (d) $CsOH$; (e) HIO_4.

J.3 Five compounds are being studied in a research lab: HCl, KOH, glucose ($C_6H_{12}O_6$, a sugar), CH_3COOH, and NH_3. A technician has prepared an aqueous solution of one of the compounds but forgot to label it and has gone home. You need to identify the solution, so you test it with litmus paper and with a conductivity meter. The solution turns the litmus paper pink, and it is found to be only weakly conducting compared to a standard NaCl solution. Which of the compounds is in the solution?

J.4 Five compounds are being studied in a research lab: HNO_3, NaOH, methanol (CH_3OH, an alcohol), $HCOOH$, and CH_3NH_3. A technician has prepared an aqueous solution of one of the compounds but forgot to label it and has gone home. You need to identify the solution, so you test it with litmus paper and with a conductivity meter. The solution turns the litmus paper blue, and it is found to conduct electricity as well as a standard NaCl solution. Which of the compounds is in the solution?

J.5 Complete the overall equation, and write the complete ionic equation and the net ionic equation for each of the following neutralization reactions. If the substance is a weak acid or base, leave it in its molecular form in the equation.
(a) $HF(aq) + NaOH(aq) \rightarrow$
(b) $(CH_3)_3N(aq) + HNO_3(aq) \rightarrow$
(c) $LiOH(aq) + HI(aq) \rightarrow$

J.6 Complete the overall equation, and write the complete ionic equation and the net ionic equation for each of the following neutralization reactions. If the substance is a weak acid or base, leave it in its molecular form in the equation.
(a) $H_3AsO_4(aq) + NaOH(aq) \rightarrow$

(Arsenic acid, H_3AsO_4, is a triprotic acid. Write the equation for complete reaction with NaOH.)
(b) $Sr(OH)_2(aq) + HClO_4(aq) \rightarrow$
(c) $Ca(OH)_2(s) + HBrO(aq) \rightarrow$

J.7 Select an acid and a base for a neutralization reaction that results in the formation of (a) potassium bromide; (b) zinc nitrite; (c) calcium cyanide, $Ca(CN)_2$; (d) potassium phosphate. Write the balanced equation for each reaction.

J.8 Select an acid and a base for a neutralization reaction that results in the formation of (a) lead(II) iodide; (b) cadmium sulfide; (c) magnesium sulfite; (d) iron(II) hypochlorite. Write the balanced equation for each reaction.

J.9 Identify the salt that is produced from the acid–base neutralization reaction between (a) potassium hydroxide and acetic acid, CH_3COOH; (b) ammonia and phosphoric acid; (c) calcium hydroxide and bromous acid; (d) sodium hydroxide and hydrosulfuric acid, H_2S (both H atoms react). Write the complete ionic equation for each reaction.

J.10 Identify the salt that is produced from the neutralization reaction between (a) sodium hydroxide and propanoic acid, CH_3CH_2COOH; (b) ammonia and nitrous acid; (c) cobalt(III) hydroxide and hydrosulfuric acid, H_2S (both H atoms react); (d) barium hydroxide and chloric acid. Write the complete ionic equation for each reaction.

J.11 Hydrochloric acid is a strong acid. Which of the following images best represents a solution of hydrochloric acid? The green spheres represent chlorine and the white spheres, hydrogen.

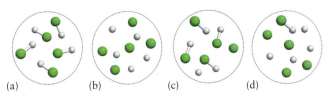

(a)   (b)   (c)   (d)

J.12 Hydrofluoric acid is a weak acid. Which of the following images best represents a solution of hydrofluoric acid? The blue spheres represent fluorine and the white spheres, hydrogen.

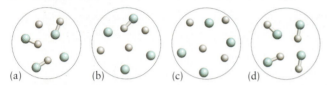

(a)   (b)   (c)   (d)

J.13 Identify the acid and the base in each of the following reactions.
(a) $CH_3NH_2(aq) + H_3O^+(aq) \rightarrow CH_3NH_3^+(aq) + H_2O(l)$
(b) $CH_3NH_2(aq) + CH_3COOH(aq) \rightarrow$
$$CH_3NH_3^+(aq) + CH_3CO_2^-(aq)$$
(c) $2\,HI(aq) + CaO(s) \rightarrow CaI_2(aq) + H_2O(l)$

J.14 Identify the acid and the base in each of the following reactions.
(a) $HBrO_3(aq) + NaHCO_3(aq) \rightarrow H_2CO_3(aq) + NaBrO_3(aq)$
(b) $(CH_3)_3N(aq) + HCl(aq) \rightarrow (CH_3)_3NH^+(aq) + Cl^-(aq)$
(c) $O^{2-}(aq) + H_2O(l) \rightarrow 2\,OH^-(aq)$

J.15 You are asked to identify compound X, which was extracted from a plant seized by customs inspectors. You run a number of tests and collect the following data. Compound X is a white, crystalline solid. An aqueous solution of X turns litmus red and conducts electricity poorly, even when X is present at appreciable concentrations. When you add sodium hydroxide to the solution, a reaction takes place. A solution of the products of the reaction conducts electricity well. An elemental analysis of X shows that the mass percentage composition of the compound is 26.68% C and 2.239% H, with the remainder being oxygen. A mass spectrum of X yields a molar mass of 90.0 g·mol^{-1}. (a) Write the empirical formula of X. (b) Write the molecular formula of X. (c) Write the balanced chemical equation and the net ionic equation for the reaction of X with sodium hydroxide. (Assume that X has two acidic hydrogen atoms.)

J.16 (a) White phosphorus, which has the formula P_4, burns in air to form compound A, in which the mass percentage of phosphorus is 43.64%, with the remainder oxygen. The mass spectrum of A yields a molar mass of 283.9 g·mol^{-1}. Write the molecular formula of compound A. (b) Compound A reacts with water to form compound B, which turns litmus red and has a mass percentage composition of 3.087% H and 31.60% P, with the remainder oxygen. The mass spectrum of compound B yields a molar mass of 97.99 g·mol^{-1}. Write the molecular formula of compound B. (c) Compound B reacts with an aqueous solution of calcium hydroxide to form compound C, a white precipitate. Write balanced chemical equations for the reactions in parts (a), (b), and (c).

J.17 In each of the following salts, either the cation or the anion is a weak acid or a weak base in water. Write the chemical equation for the proton transfer reaction of this cation or anion with water: (a) NaC_6H_5O; (b) $KClO$; (c) C_5H_5NHCl; (d) NH_4Br.

J.18 In each of the following salts, either the cation or the anion is a weak acid or a weak base in water. Write the chemical equation for the proton transfer reaction of this cation or anion with water: (a) $NaCH_2ClCO_2$; (b) C_2H_5NHBr; (c) $KBrO_2$; (d) $(CH_3)_2NH_2Cl$.

J.19 $C_6H_5NH_3Cl$ is a chloride salt with an acidic cation. (a) If 40.0 g of $C_6H_5NH_3Cl$ is dissolved in water to make 210.0 mL of solution, what is the initial concentration (in moles per liter) of the cation? (b) Write the chemical equation for the proton transfer reaction of the cation with water. Identify the acid and the base in this reaction.

J.20 The food preservative sodium benzoate, $NaC_6H_5CO_2$, is a salt with a basic anion. (a) If 25.0 g of sodium benzoate is dissolved in water to make 150.0 mL of solution, what is the initial molar concentration (in moles per liter) of the anion? (b) Write the chemical equation for the proton transfer reaction of the anion with water. Identify the acid and the base in this reaction.

J.21 The anion in Na_3AsO_4 is a weak base that can accept more than one proton. (a) Write the chemical equations for the sequential proton transfer reactions of the anion with water. Identify the acid and the base in each reaction. (b) If 35.0 g of Na_3AsO_4 is dissolved in water to make 250.0 mL of solution, how many moles of sodium cations are in the solution?

J.22 The anion in potassium sulfite, K_2SO_3, is a weak base that can accept more than one proton. (a) Write the chemical equations for the sequential proton transfer reactions of the anion with water. Identify the acid and the base in each reaction. (b) If 0.054 g of K_2SO_3 is dissolved in water to make 200.0 mL of solution, what amount (in moles) of potassium cations is in the solution?

J.23 The oxides of nonmetallic elements are called acidic oxides because they form acidic solutions in water. Write the balanced chemical equations for the reaction of 1 mol of formula units of each acidic oxide with 1 mol of water molecules to form an oxoacid and name the acid formed: (a) CO_2; (b) SO_3.

J.24 The oxides of metallic elements are called basic oxides because they form basic solutions in water. Write the balanced chemical equations for the reaction of 1 mol of each basic oxide with 1 mol of water molecules to form a metal hydroxide: (a) BaO; (b) Li_2O.

K Redox Reactions

Many common reactions, such as combustion, corrosion, photosynthesis, the metabolism of food, and the extraction of metals from their ores, appear to be completely different. However, when these changes are considered at the molecular level with a chemist's eye, they are found to be examples of a single type of process.

K.1 Oxidation and Reduction

Consider the reaction between magnesium and oxygen, which produces magnesium oxide (**FIG. K.1**); this reaction is used in fireworks to produce incandescent white sparks. It is also used, less agreeably, in tracer ammunition and incendiary devices. The reaction is a classic example of an *oxidation reaction*, in the original sense of the term: "reaction with oxygen." In the course of the reaction, Mg atoms in the solid magnesium lose electrons to form Mg^{2+} ions, and O atoms in the molecular oxygen gain electrons to form O^{2-} ions:

$$2\,Mg(s) + O_2(g) \longrightarrow 2\,Mg^{2+}(s) + 2\,O^{2-}(s), \text{ as } 2\,MgO(s)$$

A similar reaction takes place when magnesium reacts with chlorine to produce magnesium chloride:

$$Mg(s) + Cl_2(g) \longrightarrow 2\,Mg^{2+}(s) + 2\,Cl^{-}(s), \text{ as } MgCl_2(s)$$

Because the *pattern* of reaction is the same, the second reaction is also regarded as an "oxidation" of magnesium, even though no oxygen takes part. In each case, the common feature is the loss of electrons from magnesium and their transfer to another reactant. The loss of electrons from one species to another is now recognized as the essential event in an oxidation, and chemists now define **oxidation** as the loss of electrons, regardless of the species to which the electrons migrate.

You can often recognize loss of electrons by noting the increase in the charge of a species. This rule also applies to anions, as in the oxidation of bromide ions (charge -1) to molecular bromine (charge 0) in a reaction such as the one used commercially to make bromine (**FIG. K.2**):

$$2\,NaBr(s) + Cl_2(g) \longrightarrow 2\,NaCl(s) + Br_2(l)$$

Here, the bromide ion (as sodium bromide) is oxidized to bromine by the chlorine gas.

The name *reduction* originally referred to the extraction of a metal from its oxide, often by reaction with hydrogen, carbon, or carbon monoxide. One example is the reduction of iron(III) oxide by carbon monoxide in the manufacture of steel:

$$Fe_2O_3(s) + 3\,CO(g) \longrightarrow 2\,Fe(l) + 3\,CO_2(g)$$

In the reduction of iron(III) oxide, Fe^{3+} ions present in Fe_2O_3 are converted into uncharged Fe atoms when they gain electrons to neutralize their positive charges. This pattern is common to all reductions: in a **reduction**, an atom *gains* electrons from another reactant. Whenever the charge on a species is decreased (as from Fe^{3+} to Fe), reduction has taken place. The same rule applies if the charge is negative. For example, when chlorine is converted into chloride ions in the reaction

$$2\,NaBr(s) + Cl_2(g) \longrightarrow 2\,NaCl(s) + Br_2(l)$$

the charge decreases from 0 (in Cl_2) to -1 (in Cl^-), and we conclude that in this reaction chlorine has been reduced.

Self-test K.1A Identify the species that have been oxidized or reduced in the reaction $3\,Ag^+(aq) + Al(s) \rightarrow 3\,Ag(s) + Al^{3+}(aq)$.

[*Answer:* Al(s) is oxidized, $Ag^+(aq)$ is reduced]

Self-test K.1B Identify the species that have been oxidized or reduced in the reaction $2\,Cu^+(aq) + I_2(s) \rightarrow 2\,Cu^{2+}(aq) + 2\,I^-(aq)$.

Electrons are real particles and they cannot just be "lost." Therefore, *whenever a species is oxidized, another species must be reduced.* Oxidation or reduction taken separately is like one hand clapping: for reaction to take place, one process must occur in conjunction with the other. For instance, in the reaction between chlorine and sodium bromide, the bromide ions are oxidized and the chlorine molecules are reduced. Because oxidation is always accompanied by reduction, chemists speak of **redox reactions,** *redu*ction–*ox*idation reactions, rather than simply reduction reactions or oxidation reactions.

FIGURE K.1 An example of an oxidation reaction: magnesium burning brightly in air. Magnesium is so easily oxidized that it also burns brightly in water and carbon dioxide; consequently, magnesium fires are very difficult to extinguish. (*W.H. Freeman photo by Ken Karp.*)

LAB VIDEO FIGURE K.1

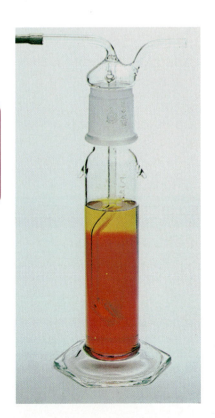

FIGURE K.2 When chlorine is bubbled through a solution of bromide ions, it oxidizes the ions to bromine, which colors the solution reddish-brown. (*©2012 Chip Clark–Fundamental Photographs.*)

Oxidation is electron loss; reduction is electron gain. Reduction and oxidation occur together in redox reactions.

K.2 Oxidation Numbers

The loss or gain of electrons is easy to identify for monatomic ions, because the charges of the species are easy to monitor. Thus, when Br^- ions are converted into bromine atoms (which go on to form Br_2 molecules), each Br^- ion must have lost an electron and therefore has been oxidized. When O_2 forms oxide ions, O^{2-}, each oxygen atom must have gained two electrons and therefore it has been reduced. Because electrons are so intimately involved in bond formation, their transfer often results in atoms being dragged from one reactant molecule to another, often making identification of redox reactions quite difficult. For example, is chlorine gas, Cl_2, oxidized or reduced when it is converted into hypochlorite ions, ClO^-? Oxygen has been added, suggesting oxidation, but the appearance of the negative sign arising from an additional electron suggests reduction.

Chemists have found a way to keep track of electrons by assigning an "oxidation number," N_{ox}, to each element (*Fundamentals* D).

- Oxidation corresponds to an *increase* in oxidation number.
- Reduction corresponds to a *decrease* in oxidation number.

A redox reaction, therefore, is any reaction in which there is a change in the oxidation numbers of one or more elements.

The oxidation number of the element as a monatomic ion is the same as its charge. For example, the oxidation number of magnesium is $+2$ when it is present as Mg^{2+} ions, and the oxidation number of chlorine is -1 when it is present as Cl^- ions. The oxidation number of the elemental form of an element is 0; so magnesium metal has oxidation number 0 and chlorine in the form of Cl_2 molecules also has oxidation number 0. When magnesium combines with chlorine, the oxidation numbers change as follows:

$$\overset{0}{Mg}(s) + \overset{2(0)}{Cl_2}(g) \longrightarrow \overset{+2\ 2(-1)}{MgCl_2}(s)$$

Thus, magnesium has been oxidized and chlorine has been reduced. Similarly, in the reaction between sodium bromide and chlorine,

$$\overset{2(+1-1)}{2\ NaBr}(s) + \overset{2(0)}{Cl_2}(g) \longrightarrow \overset{2(+1-1)}{2\ NaCl}(s) + \overset{2(0)}{Br_2}(l)$$

In this reaction, bromine has been oxidized and chlorine has been reduced, but the sodium ions remain unchanged as Na^+.

You will hear chemists speaking of both "oxidation number" and "oxidation state." An **oxidation number** is the number assigned according to the rules in **Toolbox K.1**. An **oxidation state** is the actual condition of a species with a specified oxidation number. Thus, an element *has* a certain oxidation number and *is in* the corresponding oxidation state. For example, Mg^{2+} is the $+2$ oxidation state of magnesium; and, in that state, magnesium has oxidation number $+2$. In practice, though, chemists often use the terms interchangeably.

Toolbox K.1 HOW TO ASSIGN OXIDATION NUMBERS

CONCEPTUAL BASIS

To assign an oxidation number, imagine that each atom in a molecule, formula unit, or polyatomic ion is present in ionic form (which it might not be). The oxidation number is then taken to be the charge on each "ion." The "anion" is usually oxygen as O^{2-} or the element farthest to the right in the periodic table. Then assign to the other atoms charges that balance the charge on the "anions." The method based on this approach is set out in Procedure 1. If you are familiar with the concept of electronegativity (it is introduced in Topic 2D), then you might find Procedure 2 more appropriate.

PROCEDURE 1

Rules for assigning an oxidation number $N_{ox}(E)$ to an element E:

1. The oxidation number of an element uncombined with other elements is 0.
2. The sum of the oxidation numbers of all the atoms in a species (an ion or molecule) is equal to its total charge.

The oxidation numbers of elements in most of the compounds in this text are assigned by using these two rules along with the following specific values:

- The oxidation number of hydrogen is $+1$ in combination with nonmetals and -1 in combination with metals.
- The oxidation number of elements in Groups 1 and 2 is equal to their group number.
- The oxidation number of all the halogens is -1 unless the halogen is in combination with oxygen or another halogen higher in the group. The oxidation number of fluorine is -1 in all its compounds.
- The oxidation number of oxygen is -2 in most of its compounds. Exceptions are its compounds with fluorine (in which case, the previous statement takes precedence) and its occurrence as peroxides (O_2^{2-}), superoxides (O_2^-), and ozonides (O_3^-).

This procedure is illustrated in Example K.1.

PROCEDURE 2

If you are already familiar with electronegativity (see Fig. 2D.2), then you can follow the first two rules with this more general one:

3. The oxidation number of each element is the charge when the more electronegative atom is imagined to be present as an ion typical of the element (such as O^{2-} for O).

This procedure is illustrated in Example K.2.

EXAMPLE K.1   Assigning oxidation numbers I

One of the most important reactions in industry is the conversion of sulfur dioxide, SO_2, into sulfate ion, SO_4^{2-}. Suppose you are working on reactions like this; you would have to know whether to oxidize or reduce the starting compound. Is the conversion of SO_2 to SO_4^{2-} oxidation or reduction?

ANTICIPATE   Even though the product has acquired a negative charge, suggesting reduction, it has also acquired two more O atoms, so you should anticipate that overall the conversion is an oxidation.

PLAN   Assign the oxidation numbers of sulfur in SO_2 and SO_4^{2-}, and then compare them. The process is oxidation if the oxidation number of sulfur increases, reduction if it decreases. In each case, represent the oxidation number of sulfur by $N_{ox}(S)$ and solve for $N_{ox}(S)$ after using the rules in Toolbox K.1. The oxidation number of oxygen is -2 in both compounds.

SOLVE

SO_2: By rule 2, the sum of the oxidation numbers of all the atoms in the molecule is equal to its total charge, 0: $N_{ox}(S) + 2N_{ox}(O) = 0$.

$$N_{ox}(S) + [2(-2)] = 0$$

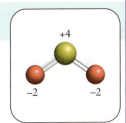

Therefore, the oxidation number of sulfur in SO_2 is $+4$.

SO_4^{2-}: By rule 2, the sum of oxidation numbers of the atoms in the ion is -2; so $N_{ox}(S) + 4N_{ox}(O) = -2$.

$$N_{ox}(S) + [4(-2)] = -2$$

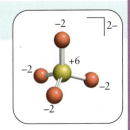

Therefore, the oxidation number of sulfur in SO_4^{2-} is $+6$.

EVALUATE   Sulfur is more highly oxidized in the sulfate ion than in sulfur dioxide. Therefore, as suspected, the conversion of SO_2 to SO_4^{2-} is an oxidation.

Self-test K.2A   Find the oxidation numbers of sulfur and phosphorus in (a) H_2S; (b) P_4O_6.

[**Answer:** (a) -2; (b) $+3$]

Self-test K.2B   Find the oxidation numbers of sulfur, nitrogen, and chlorine in (a) SO_3^{2-}; (b) NO_2^-; (c) $HClO_3$.

Related Exercises   K.1, K.2

EXAMPLE K.2 Assigning oxidation numbers II

The use of differences in electronegativity to assign oxidation number provides a very quick method in certain cases. What is the oxidation number of (a) sulfur in SF_6 and (b) nitrogen in N_2O_4?

ANTICIPATE The elements to the right in the periodic table (F and O) are likely to have negative oxidation numbers, and so S and N should have positive oxidation numbers.

PLAN Use Procedure 2 in Toolbox K.1. Refer to Fig. 2D.2 for electronegativities.

SOLVE

(a) SF_6. By rule 2, the sum of the oxidation numbers of all the atoms in the neutral molecule is

$$N_{ox}(S) + 6N_{ox}(F) = 0$$

To apply rule 3, note that the electronegativities of S and F are 2.6 and 4.0, respectively.

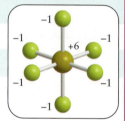

Because F is more electronegative than S, F is assigned an oxidation number of -1 (the typical charge for a fluoride ion).

As there are six fluorine atoms, each with oxidation number –1, the oxidation number of the single S atom is:

$$N_{ox}(S) = -6N_{ox}(F) = -6(-1) = +6$$

(b) N_2O_4. By rule 2, the sum of the oxidation numbers of all the atoms in the ion is:

$$2N_{ox}(N) + 4N_{ox}(O) = 0$$

To apply rule 3, note that the electronegativities of N and O are 3.0 and 3.4, respectively; so oxygen is assumed to be present as an O^{2-} ion. Therefore, the oxidation number of oxygen is -2. Because $2N_{ox}(N) + 4N_{ox}(O) = 0$, and $N_{ox}(O) = -2$, you can conclude that

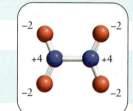

$$N_{ox}(N) = \frac{-4N_{ox}(O)}{2} = \frac{-4(-2)}{2} = +4$$

EVALUATE As expected, F and O have negative oxidation numbers.

Self-test K.3A Find the oxidation numbers of (a) nitrogen in N_2O_4 and (b) chlorine in ClO^-.

[*Answer:* (a) $+4$, (b) $+1$]

Self-test K.3B Find the oxidation numbers of (a) nitrogen in N_2S_4 and (b) bromine in BrO_3^-.

Related Exercises K.3, K.4

Oxidation increases the oxidation number of an element; reduction decreases the oxidation number. Oxidation numbers are assigned by using the rules in Toolbox K.1.

K.3 Oxidizing and Reducing Agents

The species that *causes* oxidation is called the *oxidizing agent* (or *oxidant*). When an oxidizing agent acts, it accepts the electrons released by the species being oxidized. In other words, the oxidizing agent contains an element that undergoes a *decrease* in oxidation number (**FIG. K.3**). That is,

* The **oxidizing agent** (or **oxidant**) in a redox reaction is the species that causes oxidation and, in the process, gets reduced.

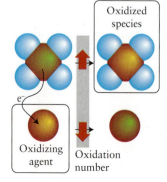

FIGURE K.3 The oxidizing agent (bottom left) is the species containing the element that undergoes a decrease in oxidation number. The color change shows how the oxidation number of the species on the top is driven upward as the oxidizing agent gains electrons. The more green the color the greater the number of electrons present.

For instance, oxygen removes electrons from magnesium. Because oxygen accepts those electrons, its oxidation number decreases from 0 to -2 (a reduction). Oxygen is therefore the oxidizing agent in this reaction. Oxidizing agents can be elements, ions, or compounds.

The species that brings about reduction is called the *reducing agent* (or *reductant*). Because the reducing agent supplies electrons to the species being reduced, the reducing agent loses electrons. That is, the reducing agent contains an element that undergoes an increase in oxidation number (**FIG. K.4**). In other words,

- The **reducing agent** (or **reductant**) in a redox reaction is the species that causes reduction and, in the process, gets oxidized.

For example, when magnesium metal supplies electrons to oxygen (reducing the oxygen atoms), the magnesium atoms lose electrons and the oxidation number of magnesium increases from 0 to $+2$ (an oxidation). It is the reducing agent in the reaction of magnesium and oxygen.

To identify the oxidizing and reducing agents in a redox reaction, you need to compare the oxidation numbers of the elements before and after the reaction to see which have changed. The reactant that contains an element that is reduced in the reaction is the oxidizing agent, and the reactant that contains an element that is oxidized is the reducing agent. For example, when a piece of zinc metal is placed in a copper(II) solution (**FIG. K.5**), the reaction is

$$\overset{0}{Zn}(s) + \overset{+2}{Cu^2}(aq) \longrightarrow \overset{+2}{Zn^{2+}}(aq) + \overset{0}{Cu}(s)$$

The oxidation number of zinc changes from 0 to $+2$ (an oxidation), whereas that of copper decreases from $+2$ to 0 (a reduction). Therefore, because zinc is oxidized, zinc metal is the reducing agent in this reaction. Conversely, because copper is reduced, copper(II) ions are the oxidizing agent.

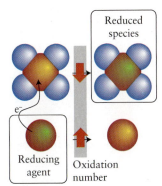

FIGURE K.4 The reducing agent (bottom left) is the species containing the element that undergoes an increase in oxidation number. The color change shows how the oxidation number of the species on the top is driven downward as it gains the electrons lost by the reducing agent. The more green the color the greater the number of electrons present.

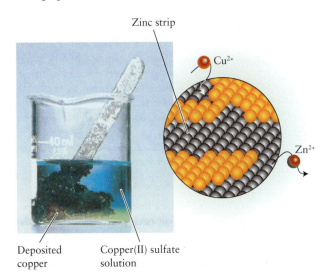

Zinc strip

Cu²⁺

Zn²⁺

Deposited copper

Copper(II) sulfate solution

FIGURE K.5 When a strip of zinc is placed in a solution that contains Cu^{2+} ions, the blue solution slowly becomes colorless and copper metal is deposited on the zinc. The inset shows that, in this redox reaction, the zinc metal is reducing the Cu^{2+} ions to copper metal and the Cu^{2+} ions are oxidizing the zinc metal to Zn^{2+} ions. *(W.H. Freeman photo by Ken Karp.)*

EXAMPLE K.3   Identifying oxidizing agents and reducing agents

An environmental chemist needs to determine the amount of Fe^{2+} ions in a sample of polluted water from a rusting industrial plant. One procedure is to analyze the sample by using a solution of sodium dichromate, $Na_2Cr_2O_7$. Identify the oxidizing agent and the reducing agent in the following reaction:

$$Cr_2O_7^{2-}(aq) + 6\,Fe^{2+}(aq) + 14\,H^+(aq) \longrightarrow 6\,Fe^{3+}(aq) + 2\,Cr^{3+}(aq) + 7\,H_2O(l)$$

ANTICIPATE   A species with many O atoms can be suspected to act as an oxidizing agent, so you should expect the dichromate ion ($Cr_2O_7^{2-}$) to be the oxidizing agent.

PLAN   The oxidation numbers of H and O have not changed, and so we concentrate on Cr and Fe.

SOLVE

Determine the oxidation numbers of chromium.

As a reactant (in $Cr_2O_7{}^{2-}$): let the oxidation number of Cr be $N_{ox}(Cr)$, then

$$\underbrace{2N_{ox}(Cr)}_{Cr_2O_7{}^{2-}} + \underbrace{[7 \times (-2)]}_{Cr_2O_7{}^{2-}} = \underbrace{-2,}_{Cr_2O_7{}^{2-}} \text{ or } 2N_{ox}(Cr) - 14 = -2$$

The oxidation number of Cr in $Cr_2O_7{}^{2-}$ is $+6$.

As a product (as Cr^{3+}): the oxidation number is $+3$.

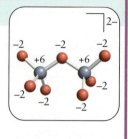

Decide whether Cr is oxidized or reduced.

$$\overset{+6}{Cr_2O_7^{2-}} \longrightarrow 2 \overset{+3}{Cr^{3+}}$$

The oxidation number of Cr decreases from $+6$ to $+3$, so Cr is reduced and the dichromate ion is the oxidizing agent.

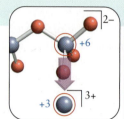

Determine the oxidation numbers of iron.

In Fe^{2+} (reactant): the oxidation number is $+2$.
In Fe^{3+} (product): the oxidation number is $+3$.

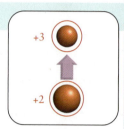

Decide whether Fe is oxidized or reduced.

The oxidation number of Fe increases from $+2$ to $+3$, so Fe is oxidized. Therefore, the Fe^{2+} ion is the reducing agent.

EVALUATE As anticipated, the dichromate ion is the oxidizing agent. The dichromate ion in acidic solution is a common laboratory oxidizing agent.

Self-test K.4A In the Claus process for the recovery of sulfur from natural gas and petroleum, hydrogen sulfide reacts with sulfur dioxide to form elemental sulfur and water: $2\,H_2S(g) + SO_2 \rightarrow 3\,S(s) + 2\,H_2O(l)$. Identify the oxidizing agent and the reducing agent.

[*Answer:* SO_2 is the oxidizing agent; H_2S is the reducing agent]

Self-test K.4B When sulfuric acid reacts with sodium iodide, sodium iodate and sulfur dioxide are produced. Identify the oxidizing and reducing agents in this reaction.

Related Exercises K.7–K.10, K.15

Oxidation is brought about by an oxidizing agent, a species that contains an element that undergoes reduction. Reduction is brought about by a reducing agent, a species that contains an element that undergoes oxidation.

K.4 Balancing Simple Redox Equations

Because electrons can be neither lost nor created in a chemical reaction, all the electrons lost by the species being oxidized must be transferred to the species being reduced. Because electrons are charged, the total charge of the reactants must be the same as the total charge of the products. Therefore, when balancing the chemical equation for a redox reaction, the charges as well as the atoms must balance.

For example, consider the net ionic equation for the oxidation of copper metal to copper(II) ions by silver ions (**FIG. K.6**):

$$Cu(s) + Ag^+(aq) \longrightarrow Cu^{2+}(aq) + Ag(s)$$

FIGURE K.6 (a) A silver nitrate solution is colorless. (b) Some time after the insertion of a copper wire, the solution shows the blue color of the copper(II) ion and the formation of crystals of silver metal on the surface of the wire. (©1986 Peticolas/Megna–Fundamental Photographs.)

(a)

(b)

 ANIMATION FIGURE K.6

At first glance, the equation appears to be balanced, because it has the same number of each kind of atom on each side. However, the total charge of the products is different from that of the reactants. Each copper atom has lost two electrons, whereas each silver atom has gained only one. To balance the electrons, the charge must be balanced too. Therefore, you need to write

$$Cu(s) + 2\,Ag^+(aq) \longrightarrow Cu^{2+}(aq) + 2\,Ag(s)$$

Self-test K.5A When tin metal is placed in contact with a solution of Fe^{3+} ions, it reduces the iron to iron(II) and is itself oxidized to tin(II) ions. Write the net ionic equation for the reaction.

[**Answer:** $Sn(s) + 2\,Fe^{3+}(aq) \rightarrow Sn^{2+}(aq) + 2\,Fe^{2+}(aq)$]

Self-test K.5B In aqueous solution, cerium(IV) ions oxidize iodide ions to solid diatomic iodine and are themselves reduced to cerium(III) ions. Write the net ionic equation for the reaction.

Some redox reactions, particularly those involving oxoanions, have complex chemical equations that require special balancing procedures. Examples and procedures are given in Topic 6K.

When balancing the chemical equation for a redox reaction involving ions, the total charge on each side must be balanced.

What have you learned in Fundamentals K?

You have learned that in a redox reaction, electrons (in some cases accompanied by atoms) are transferred from one species to another. You now know how to identify oxidizing and reducing agents by monitoring the oxidation numbers of elements.

The skills you have mastered are the ability to:

☐ **1.** Determine the oxidation number of an element (Toolbox K.1 and Examples K.1 and K.2).

☐ **2.** Identify the oxidizing and reducing agents in a reaction (Example K.3).

☐ **3.** Write and balance chemical equations for simple redox reactions (Self-Test K.5).

Fundamentals K Exercises

K.1 Determine the oxidation number of the italicized element in each of the following ions: (a) $Zn(OH)_4^{2-}$; (b) $PdCl_4^{2-}$; (c) UO_2^{2+}; (d) SiF_6^{2-}; (e) IO^-.

K.2 Determine the oxidation number of the italicized element in each of the following compounds: (a) $ClFO_3$; (b) SOF_2; (c) N_2O_5; (d) P_4S_3; (e) $HAsO_3$; (f) XeF_4.

K.3 Determine the oxidation number of the italicized element in each of the following compounds: (a) H_4SiO_4; (b) SnO_2; (c) N_2H_4; (d) P_4O_{10}; (e) S_2Cl_2; (f) P_4.

K.4 Determine the oxidation number of the italicized element in each of the following ions: (a) BF_4^-; (b) NO_2^-; (c) $S_2O_8^{2-}$; (d) VO^{2+}; (e) BrF_2^+.

K.5 When nickel is added to $CuCl_2(aq)$, Ni^{2+} ions and copper form. When iron is added to $NiCl_2(aq)$, Fe^{2+} ions and nickel form. What will happen if you add iron to $CuCl_2(aq)$? Explain your answer.

K.6 When lead is added to $AgCl(aq)$, Pb^{2+} ions and silver form. When zinc is added to $PbCl_2(aq)$, Zn^{2+} ions and lead form. What will happen if you add silver to $ZnCl_2(aq)$? Explain your answer.

K.7 For each of the following redox reactions, identify the substance oxidized and the substance reduced by the change in oxidation numbers.

(a) $CH_3OH(aq) + O_2(g) \rightarrow HCOOH(aq) + H_2O(l)$

(b) $2\,MoCl_5(s) + 5\,Na_2S(s) \rightarrow$
$$2\,MoS_2(s) + 10\,NaCl(s) + S(s)$$

(c) $3\,Tl^+(aq) \rightarrow 2\,Tl(s) + Tl^{3+}(aq)$

K.8 In each of the following reactions, use oxidation numbers to identify the substance oxidized and the substance reduced.
(a) The production of iodine from seawater:
$Cl_2(g) + 2 I^-(aq) \rightarrow I_2(aq) + 2 Cl^-(aq)$
(b) A reaction to prepare bleach:
$Cl_2(g) + 2 NaOH(aq) \rightarrow NaCl(aq) + NaOCl(aq) + H_2O(l)$
(c) A reaction that destroys ozone in the stratosphere:
$NO(g) + O_3(g) \rightarrow NO_2(g) + O_2(g)$

K.9 Identify the oxidizing agent and the reducing agent in each of the following reactions:
(a) $Zn(s) + 2 HCl(aq) \rightarrow ZnCl_2(aq) + H_2(g)$, a simple means of preparing H_2 gas in the laboratory
(b) $2 H_2S(g) + SO_2(g) \rightarrow 3 S(s) + 2 H_2O(l)$, a reaction used to produce sulfur from hydrogen sulfide, the "sour gas" in natural gas
(c) $B_2O_3(s) + 3 Mg(s) \rightarrow 2 B(s) + 3 MgO(s)$, a preparation of elemental boron

K.10 Identify the oxidizing agent and the reducing agent in each of the following reactions:
(a) $2 Al(l) + Cr_2O_3(s) \xrightarrow{\Delta} Al_2O_3(s) + 2 Cr(l)$, an example of a thermite reaction used to obtain some metals from their ores
(b) $6 Li(s) + N_2(g) \rightarrow 2 Li_3N(s)$, a reaction that shows the similarity of lithium and magnesium
(c) $2 Ca_3(PO_4)_2(s) + 6 SiO_2(s) + 10 C(s) \rightarrow P_4(g) + 6 CaSiO_3(s) + 10 CO(g)$, a reaction for the preparation of elemental phosphorus

K.11 The *Sabatier process* has been used to remove CO_2 from artificial atmospheres, such as those in submarines and spacecraft. An advantage is that it produces methane, CH_4, which can be burned as a fuel, and water, which can be reused. Balance the equation for the process and identify the type of reaction: $CO_2(g) + H_2(g) \rightarrow CH_4(g) + H_2O(l)$.

K.12 The industrial production of sodium metal and chlorine gas makes use of the Downs process, in which molten sodium chloride is electrolyzed. Write a balanced equation for the production of the two elements from molten sodium chloride. Which element is produced by oxidation and which by reduction?

K.13 Write a balanced equation for each of the following skeletal redox reactions:
(a) $NO_2(g) + O_3(g) \rightarrow N_2O_5(g) + O_2(g)$
(b) $S_8(s) + Na(s) \rightarrow Na_2S(s)$
(c) $Cr^{2+}(aq) + Sn^{4+}(aq) \rightarrow Cr^{3+}(aq) + Sn^{2+}(aq)$
(d) $As(s) + Cl_2(g) \rightarrow AsCl_3(l)$

K.14 Write a balanced equation for each of the following skeletal redox equations:
(a) $Sb_2S_3(s) + Fe(s) \rightarrow Sb(s) + FeS(s)$
(b) $BrO^-(aq) \rightarrow BrO_3^-(aq) + Br^-(aq)$
(c) $Cr_2O_3(s) + C(s) \rightarrow Cr_3C_2(s) + CO(g)$
(d) $PbS(s) + O_2(g) \rightarrow PbO(s) + SO_2(g)$

K.15 Identify the oxidizing and reducing agent in each of the following reactions.
(a) The production of tungsten metal from its oxide,
$WO_3(s) + 3 H_2(g) \rightarrow W(s) + 3 H_2O(l)$
(b) The generation of hydrogen gas in the laboratory,
$Mg(s) + 2 HCl(aq) \rightarrow H_2(g) + MgCl_2(aq)$
(c) The production of metallic tin from tin(IV) oxide,
$SnO_2(s) + 2 C(s) \xrightarrow{\Delta} Sn(l) + 2 CO(g)$

(d) A reaction used to propel rockets,
$2 N_2H_4(g) + N_2O_4(g) \rightarrow 3 N_2(g) + 4 H_2O(g)$

K.16 Identify the oxidizing and reducing agent in each of the following reactions.
(a) A reaction used to remove nitrate ions from wastewater:
$5 CH_3OH(aq) + 6 NO_3^-(aq) + 6 H^+(aq) \rightarrow$
$$5 CO_2(g) + 3 N_2(g) + 13 H_2O(l).$$
(b) A step in the production of an alternative fuel from coal:
$CH_4(g) + H_2O(g) \rightarrow CO(g) + 3 H_2(g).$
(c) A reaction used to set off fireworks: $4 KNO_3(s) \rightarrow$
$2 K_2O(s) + 2 N_2(g) + 5 O_2(g).$

K.17 Balance the following equations and identify the oxidizing and reducing agents in each.
(a) $Cl_2(g) + H_2O(l) \rightarrow HClO(aq) + HCl(aq)$
(b) $NaClO_3(aq) + SO_2(g) + H_2SO_4(aq, dilute) \rightarrow$
$$NaHSO_4(aq) + ClO_2(g)$$
(c) $CuI(aq) \rightarrow Cu(s) + I_2(s)$

K.18 Balance the following equations and identify the oxidizing and reducing agents in each.
(a) $CO(g) + H_2O(g) \rightarrow CO_2(g) + H_2(g)$
(b) $ClO_2(g) + O_3(g) \rightarrow Cl_2O_6(l) + O_2(g)$
(c) $Cl_2(g) + F_2(g) \rightarrow ClF_3(g)$

K.19 Write a balanced equation for each of the following redox reactions.
(a) The displacement of copper(II) ion from solution by magnesium metal:
$Mg(s) + Cu^{2+}(aq) \rightarrow Mg^{2+}(aq) + Cu(s)$
(b) The formation of iron(III) ion in the following reaction:
$Fe^{2+}(aq) + Ce^{4+}(aq) \rightarrow Fe^{3+}(aq) + Ce^{3+}(aq)$
(c) The synthesis of hydrogen chloride from its elements:
$H_2(g) + Cl_2(g) \rightarrow HCl(g)$
(d) The formation of rust (a simplified equation):
$Fe(s) + O_2(g) \rightarrow Fe_2O_3(s)$

K.20 The following redox reactions are important in the refining of certain elements. Balance the equations and, in each case, write the name of the source compound of the element (in bold type) and the oxidation state in that compound of the element that is being extracted (in red):
(a) $\mathbf{SiCl_4}(l) + H_2(g) \rightarrow Si(s) + HCl(g)$
(b) $\mathbf{SnO_2}(s) + C(s) \xrightarrow{1200\,°C} Sn(l) + CO_2$
(c) $\mathbf{V_2O_5}(s) + Ca(l) \xrightarrow{\Delta} V(s) + CaO(s)$
(d) $\mathbf{B_2O_3}(s) + Mg(s) \rightarrow B(s) + MgO(s)$

K.21 Some compounds of hydrogen and oxygen are exceptions to the common observation that H has oxidation number $+1$ and O has oxidation number -2. Assuming that each metal has the oxidation number of its most common ion, find the oxidation numbers of H and O in each of the following compounds: (a) KO_2; (b) $LiAlH_4$; (c) Na_2O_2; (d) NaH; (e) KO_3.

K.22 Nitrogen in the air of spaceships is likely to be gradually lost through leakage and must be replaced. One method is to store the nitrogen in the form of hydrazine, $N_2H_4(l)$, from which nitrogen is readily obtained by heating. The ammonia also produced can be processed further to obtain even more nitrogen: $N_2H_4(l) \rightarrow NH_3(g) + N_2(g)$. (a) Balance the equation. (b) Give the oxidation number of nitrogen in each compound. (c) Identify the oxidizing and reducing agent. (d) Given that 28 g of nitrogen gas occupies 24 L at room temperature and pressure, what volume of nitrogen gas can be obtained from 1.0 L of hydrazine? (The density of hydrazine is 1.004 g·cm^{-3} at room temperature.)

K.23 For each of the following incomplete reactions, would you choose an oxidizing agent or a reducing agent to make the conversion?
(a) $ClO_4^-(aq) \rightarrow ClO_2(g)$
(b) $SO_4^{2-}(aq) \rightarrow SO_2(g)$

K.24 For each of the following incomplete reactions, would you choose an oxidizing agent or a reducing agent to make the conversion?
(a) $H_3PO_3(aq) \rightarrow P_4O_{10}(g)$
(b) $CH_3CH_2OH(ethanol) \rightarrow CH_3CH_2COOH(acetic\ acid)$

K.25 Classify each of the following reactions as precipitation, acid–base neutralization, or redox. If it is a precipitation reaction, write a net ionic equation; if it is a neutralization reaction, identify the acid and the base; if it is a redox reaction, identify the oxidizing agent and the reducing agent.
(a) Reaction used to measure the concentration of carbon monoxide in a gas stream:
$5\ CO(g) + I_2O_5(s) \rightarrow I_2(s) + 5\ CO_2(g)$

(b) Test for the amount of iodine in a sample:
$I_2(aq) + 2\ S_2O_3^{2-}(aq) \rightarrow 2\ I^-(aq) + S_4O_6^{2-}(aq)$
(c) Test for bromide ions in solution:
$AgNO_3(aq) + Br^-(aq) \rightarrow AgBr(s) + NO_3^-(aq)$
(d) Heating of uranium tetrafluoride with magnesium, one stage in the purification of uranium metal:
$UF_4(g) + 2\ Mg(s) \rightarrow U(s) + 2\ MgF_2(s)$

K.26 Classify each of the following reactions as precipitation, acid–base neutralization, or redox. If it is a precipitation reaction, write a net ionic equation; if it is a neutralization reaction, identify the acid and the base; if it is a redox reaction, identify the oxidizing agent and the reducing agent.
(a) Burning of ammonia in air:
$4\ NH_3(g) + 3\ O_2(g) \rightarrow 2\ N_2(g) + 6\ H_2O(g)$
(b) Formation of silver oxide in solution:
$2\ AgNO_3(aq) + 2\ NaOH(aq) \rightarrow$
$$Ag_2O(s) + 2\ NaNO_3(aq) + H_2O(l)$$
(c) A reaction used to generate hydrogen in the laboratory:
$Mg(s) + 2\ HCl(aq) \rightarrow H_2(g) + MgCl_2(aq)$
(d) The reaction of sulfur trioxide with water vapor:
$SO_3(g) + 2\ H_2O(g) \rightarrow H_2SO_4(aq)$

L Reaction Stoichiometry

Sometimes you need to know how much product to expect from a reaction, or how much reactant you need to make a desired amount of product. To perform this kind of calculation, you use the quantitative aspect of chemical reactions called **reaction stoichiometry** in which the stoichiometric coefficients in a balanced chemical equation are interpreted in terms of the relative amounts that react or are produced. Thus, the stoichiometric coefficients in

$$N_2(g) + 3\ H_2(g) \longrightarrow 2\ NH_3(g)$$

indicate that if 1 mol N_2 reacts, then 3 mol H_2 will be consumed and 2 mol NH_3 will be produced. The relative amounts of reactants and products involved in a chemical reaction are summarized as **stoichiometric relations**

$$1\ mol\ N_2 \mathbin{\widehat{=}} 3\ mol\ H_2 \qquad 1\ mol\ N_2 \mathbin{\widehat{=}} 2\ mol\ NH_3$$

The sign $\widehat{=}$ is read "is chemically equivalent to." In general, different reactions give rise to different stoichiometric relations.

L.1 *Mole-to-Mole Predictions*

Stoichiometry has important practical applications, such as predicting how much product can be formed in a reaction. For example, in some fuel cells used to generate electricity, oxygen reacts with hydrogen to produce water. On the space shuttle the water generated was used for life support (**FIG. L.1**). Let's look at the calculation space shuttle engineers would have had to do to find out how much water is formed when 0.25 mol O_2 reacts with hydrogen.

The chemical equation for the reaction is

$$2\ H_2(g) + O_2(g) \longrightarrow 2\ H_2O(l)$$

The information that 1 mol O_2 reacts to form 2 mol H_2O is summarized by writing the stoichiometric relation between oxygen (the given substance) and water (the required substance):

$$1\ mol\ O_2 \mathbin{\widehat{=}} 2\ mol\ H_2O$$

Next, this stoichiometric relation is used to set up a conversion factor relating the given substance to the required substance:

$$\frac{Substance\ required}{Substance\ given} = \frac{2\ mol\ H_2O}{1\ mol\ O_2}$$

FIGURE L.1 A technician studies a lightweight, efficient hydrogen–oxygen fuel cell of the type used in the space shuttle. The three fuel cells on the space shuttle provided life-support electricity and drinking water. Because they had no moving parts, they had a long working life. (*Pasquale Sorrentino/Science Source.*)

Stoichiometric coefficients are exact numbers; therefore, they do not limit the significant figures of stoichiometric calculations (see Appendix 1C).

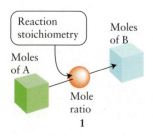

1

This ratio, which is commonly called the **mole ratio** for the reaction, allows the amount of O_2 molecules consumed to be related to the amount of H_2O molecules produced. It is used in the same way that a conversion factor is used to convert units (*Fundamentals* A):

$$\text{Amount of H}_2\text{O produced (mol)} = (0.25 \text{ mol O}_2) \times \frac{2 \text{ mol H}_2\text{O}}{1 \text{ mol O}_2}$$

$$= 0.50 \text{ mol H}_2\text{O}$$

Note that the unit mol and the species (in this case, O_2 molecules) cancel. The general strategy for this type of calculation is summarized in diagram (**1**).

Self-test L.1A What amount (in moles) of NH_3 can be produced from 2.0 mol H_2 in the reaction $N_2(g) + 3 H_2(g) \rightarrow 2 NH_3(g)$?

[***Answer:*** 1.3 mol NH_3]

Self-test L.1B What is the maximum amount of Fe atoms (in moles) that can be extracted from 25 mol Fe_2O_3?

The balanced chemical equation for a reaction is used to set up the mole ratio, a factor that is used to convert the amount of one substance into the amount of another.

L.2 Mass-to-Mass Predictions

To find the mass of a product that can be formed from a known mass of a reactant, the mass of reactant (in grams) is converted into an amount (in moles) by using its molar mass. The mole ratio from the balanced equation is then used to predict the amount of product (in moles), and finally that amount of product is converted to a mass (in grams) by using its molar mass, as described in **Toolbox L.1**:

$$\text{Mass of reactant} \xrightarrow{\text{Molar mass}} \text{amount of reactant} \xrightarrow{\text{Mole ratio}} \text{amount of product} \xrightarrow{\text{Molar mass}} \text{mass of product}$$

Toolbox L.1   HOW TO CARRY OUT MASS-TO-MASS CALCULATIONS

CONCEPTUAL BASIS

A chemical equation expresses the relations between the amounts (in moles) of each reactant and product. By using the molar masses as conversion factors, you can express these relations in terms of masses.

PROCEDURE

The general procedure for mass-to-mass calculations, summarized in diagram (**2**), requires that first you write the balanced chemical equation for the reaction. Then perform the following calculations:

Step 1 Convert the given mass of one substance (A) in grams into amount in moles by using its molar mass.

$$n_A = \frac{m_A}{M_A}$$

If necessary, first convert the units of mass to grams.

Step 2 Use the mole ratio derived from the stoichiometric coefficients in the balanced chemical equation to convert from the amount of one substance (A) into the amount in moles of the other substance (B).

For $a\,A \rightarrow b\,B$ or $a\,A + b\,B \rightarrow c\,C$, use

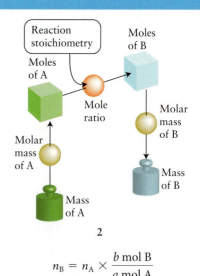

2

$$n_B = n_A \times \frac{b \text{ mol B}}{a \text{ mol A}}$$

Step 3 Convert from amount in moles of the second substance to mass (in grams) by using its molar mass.

$$m_B = n_B M_B$$

This procedure is illustrated in Example L.1.

EXAMPLE L.1   Calculating the mass of product that can be obtained from a given mass of reactant

Suppose you are involved in the design of a steel plant and are concerned about its environmental impact. You need to know not only the amount of iron you can extract from the ore supplied to the plant, but also the amount of carbon dioxide that will be produced in that extraction. You decide to conduct laboratory experiments to investigate this process. (a) What mass of iron(III) oxide, Fe_2O_3, present in iron ore is needed to produce 10.0 g of iron when it is reduced by carbon monoxide gas to metallic iron and carbon dioxide gas? (b) What mass of carbon dioxide is released when 10.0 g of iron is produced?

ANTICIPATE   Because not all the matter in Fe_2O_3 is iron, expect to use more than 10.0 g of the ore to obtain 10.0 g of iron. To answer part (b) there is no shortcut way to estimate the amount; you have to do the calculation.

PLAN   Follow the steps in Toolbox L.1.

SOLVE   The balanced chemical equation is

$$Fe_2O_3(s) + 3\,CO(g) \longrightarrow 2\,Fe(s) + 3\,CO_2(g)$$

which implies that

$$2\ mol\ Fe \mathrel{\widehat{=}} 1\ mol\ Fe_2O_3 \text{ and } 2\ mol\ Fe \mathrel{\widehat{=}} 3\ mol\ CO_2$$

(a) The molar mass of iron is 55.85 g·mol^{-1} and that of iron(III) oxide is 159.69 g·mol^{-1}.

Step 1 Convert the given mass of iron into amount of Fe atoms by using its molar mass ($n = m/M$).

$$\text{Amount of Fe (mol)} = \frac{10.0\ g}{55.85\ g\cdot(mol\ Fe)^{-1}}$$
$$= \frac{10}{55.85}\ mol\ Fe = 0.179\ldots\ mol\ Fe$$

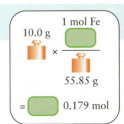

Step 2 Use the mole ratio to convert from the amount of Fe atoms to the amount of iron(III) oxide formula units.

$$\text{Amount of Fe}_2O_3\ (mol) = \frac{10}{55.85}\ mol\ Fe \times \frac{1\ mol\ Fe_2O_3}{2\ mol\ Fe}$$
$$= \frac{10}{55.85 \times 2}\ mol\ Fe_2O_3 = 0.0895\ldots\ mol\ Fe_2O_3$$

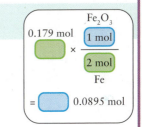

Step 3 Convert from amount of Fe_2O_3 into mass of iron(III) oxide by using its molar mass ($m = nM$).

$$\text{Mass of Fe}_2O_3\ (g) = \frac{10}{55.85 \times 2}\ mol\ Fe_2O_3 \times 159.69\ g\cdot(mol\ Fe_2O_3)^{-1}$$
$$= 14.3\ g$$

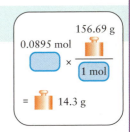

(b) The molar mass of carbon dioxide is 44.01 g·mol^{-1}.

Step 1 As above, convert the given mass of iron into amount of Fe atoms by using its molar mass ($n = m/M$).

$$\text{Amount of Fe (mol)} = \frac{10.0\ g}{55.85\ g\cdot(mol\ Fe)^{-1}}$$
$$= \frac{10}{55.85}\ mol\ Fe = 0.179\ldots\ mol\ Fe$$

Step 2 Use the mole ratio to convert from the amount of Fe atoms to the amount of CO_2 molecules.

$$\text{Amount of CO}_2\ (mol) = 0.179\ldots\ mol\ Fe \times \frac{3\ mol\ CO_2}{2\ mol\ Fe}$$
$$= 0.179\ldots \times \tfrac{3}{2}\ mol\ CO_2$$

Step 3 Use the molar mass to convert from amount of CO_2 molecules into mass of carbon dioxide ($m = nM$).

$$\text{Mass of } CO_2(g) = 0.179\ldots \times \tfrac{3}{2} \text{ mol } CO_2 \times 44.01 \text{g} \cdot (\text{mol } CO_2)^{-1}$$
$$= 11.8 \text{ g}$$

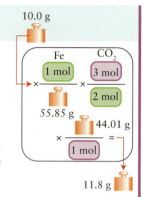

EVALUATE As anticipated, more than 10 g of the ore is needed to obtain 10 g of iron.

A Note on Good Practice: Although, as usual, rounding the numerical value has been left to a single, final step, the same is not true for the units: canceling units does not introduce rounding errors and clarifies each step.

Self-test L.2A Calculate the mass of potassium metal needed to react with 0.450 g of hydrogen gas to produce solid potassium hydride, KH.

[*Answer:* 17.5 g]

Self-test L.2B Carbon dioxide can be removed from power plant exhaust gases by combining it with an aqueous slurry of calcium silicate: $2\,CO_2(g) + H_2O(l) + CaSiO_3(s) \rightarrow SiO_2(s) + Ca(HCO_3)_2(aq)$. What mass of $CaSiO_3$ (of molar mass $116.17 \text{ g} \cdot \text{mol}^{-1}$) is needed to react completely with 0.300 kg of carbon dioxide?

Related Exercises L.3–L.8, L.11, L.12

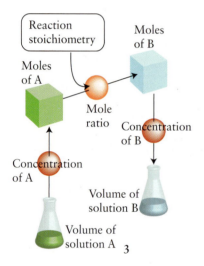

The stoichiometric point is also called the *equivalence point*.

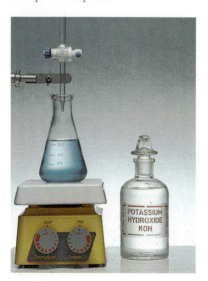

In a mass-to-mass calculation, convert the given mass into amount in moles, apply the mole-to-mole conversion factor to obtain the amount required, and finally convert amount in moles into mass of the unknown substance.

L.3 Volumetric Analysis

A common laboratory technique for determining the concentration of a solute is **titration** (**FIG. L.2**). Titrations are usually either **acid–base titrations**, in which an acid reacts with a base, or **redox titrations**, in which the reaction is between a reducing agent and an oxidizing agent. Titrations are widely used to monitor water purity and blood composition and for quality control in the food industry.

In a titration, one solution is gradually added to another until reaction is complete. A known volume of the solution being analyzed, which is called the **analyte**, is transferred into a flask. Then a solution containing a known concentration of the other reactant is measured into the flask from a buret until all the analyte has reacted. The solution in the buret is called the **titrant**, and the difference between the initial and final volume readings of the buret is the volume of titrant that has been added to the flask. The determination of concentration or amount by measuring volume is called **volumetric analysis**.

In an acid–base titration, the analyte is a solution of a base and the titrant is a solution of an acid or vice versa. An indicator, a water-soluble dye (*Fundamentals J*), is used to detect the **stoichiometric point**, the stage at which the volume of titrant added is exactly that required by the stoichiometric relation between titrant and analyte. For example, if hydrochloric acid containing a few drops of the indicator phenolphthalein is being titrated, the solution is initially colorless. After the stoichiometric point, when excess base is present, the solution in the flask is basic and the indicator is pink. Because the indicator color change is sudden, it is easy to detect the stoichiometric point (**FIG. L.3**). **Toolbox L.2** shows how to interpret a titration; the procedure is summarized in diagram (**3**), where A is the solute in the titrant and B is the solute in the analyte.

FIGURE L.2 The apparatus typically used for a titration: magnetic stirrer; flask containing the analyte; clamp; buret containing the titrant—in this case, potassium hydroxide. (*W.H. Freeman photo by Ken Karp.*)

Toolbox L.2 HOW TO INTERPRET A TITRATION

CONCEPTUAL BASIS

In a titration, one reactant (the titrant) is added gradually in solution to another (the analyte) until reaction is complete. The goal is to determine the concentration of the reactant in the analyte or its mass.

PROCEDURE

Step 1 Calculate the amount ($n_{titrant}$, in moles) of titrant species added from the volume of titrant ($V_{titrant}$) and its molarity ($c_{titrant}$).

$$\underbrace{n_{titrant}}_{\text{mol titrant}} = \underbrace{c_{titrant}}_{\text{mol titrant}\cdot L^{-1}} \times \underbrace{V_{titrant}}_{L}$$

Step 2 Calculate the amount of analyte: (a) Write the chemical equation for the reaction, (b) identify the mole ratio between the titrant species and the analyte species, and (c) use it to convert the amount of titrant into amount of analyte ($n_{analyte}$).

$$n_{analyte} = n_{titrant} \times \text{mole ratio}$$

Step 3 Calculate the initial molarity of the analyte ($c_{analyte}$) by dividing the amount of analyte species by the initial volume, $V_{analyte}$, of the analyte.

$$\underbrace{c_{analyte}}_{(\text{mol analyte})\cdot L^{-1}} = \dfrac{\overbrace{n_{analyte}}^{\text{mol analyte}}}{\underbrace{V_{analyte}}_{L}}$$

This procedure is illustrated in Example L.2.

If the mass of the analyte is required, instead of step 3, use the molar mass of the analyte to convert moles into grams.

This procedure is demonstrated in Example L.3.

A Note on Good Practice: Because it is important to keep track of the substances, be sure to indicate the exact species and concentration units, by writing, for example, 1.0 (mol HCl)·L^{-1} or 1.0 M HCl(aq).

EXAMPLE L.2   Determining the molarity of an acid by titration

A common laboratory procedure is the determination of the concentration of an acid solution by titration. Suppose that 25.00 mL of a solution of oxalic acid, $H_2C_2O_4$ (**4**), which has two acidic protons, is titrated with 0.100 M NaOH(aq) and that the stoichiometric point for reaction with both protons is reached when 38.0 mL of the solution of base is added. Find the molarity of the oxalic acid solution.

ANTICIPATE If the acid were monoprotic, then because more than 25 mL of the alkali is needed to neutralize it (about 1.5 times the volume of the acid), the molarity of the acid must be greater than the molarity of the base. However, the acid is in fact diprotic, so because each molecule of acid provides two acidic protons, its molarity will be half that of a monoprotic acid, so you should expect a molarity close to $\frac{1}{2} \times 1.5 \times 0.1$ M ≈ 0.08 M.

PLAN Proceed as in Toolbox L.2.

SOLVE

Step 1 Calculate the amount ($n_{titrant}$, in moles) of titrant species (NaOH) added from the volume of titrant ($V_{titrant}$) and its molarity ($c_{titrant}$).

$$n_{NaOH} = (38.0 \times 10^{-3}\,L) \times 0.100\,(\text{mol NaOH})\,L^{-1}$$
$$= 0.0038\ldots \text{ mol NaOH}$$

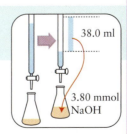

Step 2 (a) Write the chemical equation. (b) Identify the mole ratio. (c) Calculate the amount of acid present.

(a) $H_2C_2O_4(aq) + 2\,NaOH(aq) \rightarrow Na_2C_2O_4(aq) + 2\,H_2O(l)$

(b) 2 mol NaOH $\stackrel{\frown}{=}$ 1 mol $H_2C_2O_4$

(c) $n(H_2C_2O_4) = (0.0038\ldots \text{ mol NaOH}) \times \dfrac{1 \text{ mol } H_2C_2O_4}{2 \text{ mol NaOH}}$

$$= 0.0019\ldots \text{ mol } H_2C_2O_4$$

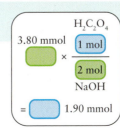

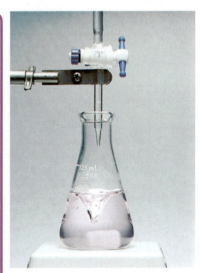

FIGURE L.3 An acid–base titration at the stoichiometric point. The indicator is phenolphthalein. *(W.H. Freeman photo by Ken Karp.)*

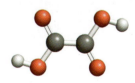

4 Oxalic acid, $(COOH)_2$

Step 3 Calculate the initial molarity of the analyte, $c(H_2C_2O_4)$, by dividing the amount of analyte species by the initial volume of the analyte.

$$c(H_2C_2O_4) = \frac{0.0019\ldots \text{ mol } H_2C_2O_4}{25.00 \times 10^{-3}\,L}$$
$$= 0.0760 \text{ (mol } H_2C_2O_4)\cdot L^{-1}$$

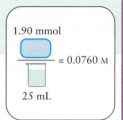

1.90 mmol

$\dfrac{}{25\text{ mL}} = 0.0760\text{ M}$

EVALUATE The solution is 0.0760 M $H_2C_2O_4$(aq). As expected, the acid is less concentrated than the base and has a molarity close to 0.08 M.

Self-test L.3A A student used a sample of hydrochloric acid that was known to contain 0.020 mol HCl in 500.0 mL of solution to titrate 25.0 mL of a solution of calcium hydroxide. The stoichiometric point was reached when 15.1 mL of acid had been added. What was the molarity of the calcium hydroxide solution?

[*Answer:* 0.012 M Ca(OH)$_2$(aq)]

Self-test L.3B Many abandoned mines have exposed nearby communities to the problem of acid mine drainage. Certain minerals, such as pyrite (FeS_2), decompose when exposed to air, forming solutions of sulfuric acid. The acidic mine water then drains into lakes and creeks, killing fish and other animals. At a mine in Colorado, a 16.45-mL sample of mine water was completely neutralized with 25.00 mL of 0.255 M KOH(aq). What is the molar concentration of H_2SO_4 in the water?

Related Exercises L.13–L.16

EXAMPLE L.3   Determining the purity of a sample by means of a redox titration

An experimental method to determine the purity of iron ore is to titrate a sample with a solution of purple potassium permanganate, $KMnO_4$. The ore is first dissolved in hydrochloric acid; the iron is then chemically reduced to iron(II) ions, which react with MnO_4^- ions:

$$5\,Fe^{2+}(aq) + MnO_4^-(aq) + 8\,H^+(aq) \longrightarrow 5\,Fe^{3+}(aq) + Mn^{2+}(aq) + 4\,H_2O(l)$$

The stoichiometric point is reached when all the Fe^{2+} has reacted and is detected when the purple color of the permanganate ion persists. This method assumes that the ore contains no impurities that might also react with permanganate ion. A sample of ore of mass 0.202 g was dissolved in hydrochloric acid, and the resulting solution required 16.7 mL of 0.0108 M $KMnO_4$(aq) to reach the stoichiometric point. (a) What mass of iron(II) ions is present? (b) What is the mass percentage of iron in the ore sample?

ANTICIPATE The mass of iron present must be less than the mass of ore used in the titration. The mass percentage of iron in the ore must also be no greater than that in pure Fe_2O_3, which is 70%.

PLAN (a) To obtain the amount of iron(II) in the analyte, follow the first two steps of the procedure in Toolbox L.2. Then convert the amount of Fe^{2+} ions into mass by using the molar mass of Fe^{2+}; because the mass of electrons is so small, use the molar mass of elemental iron for the molar mass of iron(II) ions. (b) To find the mass percentage, divide the mass of iron by the mass of the ore sample and multiply by 100%.

SOLVE

(a) Find the mass of iron present in the sample. In step 1, calculate the amount of titrant species (MnO_4^-) added from the volume of titrant and its concentration by using $n = cV$:

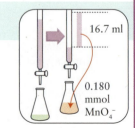

16.7 ml

0.180 mmol MnO_4^-

$$n(MnO_4^-) = \frac{\overbrace{0.0108 \text{ mol } MnO_4^-}^{\text{concentration}}}{1\,L} \times \overbrace{0.0167\,L}^{\text{volume}}$$
$$= 0.180\ldots \text{ mmol } MnO_4^-$$

In step 2, identify the mole ratio between the titrant species and the analyte species from the chemical equation (given above):

$$5 \text{ mol } Fe^{2+} \simeq 1 \text{ mol } MnO_4^-$$

Use it to convert the amount of titrant into amount of analyte:

$$n(\text{Fe}^{2+}) = \overset{\text{amount MnO}_4^-}{\overbrace{0.180\ldots \text{ mmol MnO}_4^-}} \times \overset{\text{mole ratio}}{\overbrace{\frac{5 \text{ mmol Fe}^{2+}}{1 \text{ mmol MnO}_4^-}}}$$

$$= 0.902\ldots \text{ mmol Fe}^{2+}$$

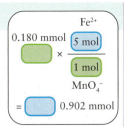

Find the mass of iron from $m = nM$:

$$m(\text{Fe}^{2+}) = \overset{\text{amount Fe}^{2+}}{\overbrace{0.902\ldots \text{ mmol Fe}^{2+}}} \times \overset{\text{molar mass}}{\overbrace{\frac{55.85 \text{ mg}}{\text{mmol Fe}^{2+}}}} = 50.4 \text{ mg}$$

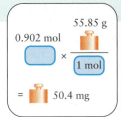

(b) Calculate the mass percentage of iron in the ore from mass percentage = {(mass of iron)/(mass of sample)} × 100%:

$$\text{Mass percentage of iron} = \frac{0.0504 \text{ g}}{0.202 \text{ g}} \times 100\% = 25.0\%$$

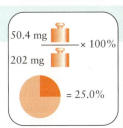

EVALUATE The sample contains 0.0504 g of iron (as anticipated, that is necessarily smaller than the mass of ore) and is 25.0% iron by mass, which is less than in pure Fe_2O_3 (70.0%).

Self-test L.4A A sample of clay of mass 20.750 g for use in making ceramics was analyzed to determine its iron content. The clay was washed with hydrochloric acid, and the iron converted into iron(II) ions. The resulting solution was titrated with cerium(IV) sulfate solution:

$$\text{Fe}^{2+}(\text{aq}) + \text{Ce}^{4+}(\text{aq}) \longrightarrow \text{Fe}^{3+}(\text{aq}) + \text{Ce}^{3+}(\text{aq})$$

In the titration, 13.45 mL of 1.340 M $Ce(SO_4)_2$(aq) was needed to reach the stoichiometric point. What is the mass percentage of iron in the clay?

[*Answer: 4.85%*]

Self-test L.4B The amount of arsenic(III) oxide in a mineral can be determined by dissolving the mineral in acid and titrating the solution with potassium permanganate:

$$24 \text{ H}^+(\text{aq}) + 5 \text{ As}_4\text{O}_6(\text{s}) + 8 \text{ MnO}_4^-(\text{aq}) + 18 \text{ H}_2\text{O}(\text{l}) \rightarrow 8 \text{ Mn}^{2+}(\text{aq}) + 20 \text{ H}_3\text{AsO}_4(\text{aq})$$

A sample of industrial waste was analyzed for the presence of arsenic(III) oxide by titration with 0.0100 M $KMnO_4$. It took 28.15 mL of the titrant to reach the stoichiometric point. What mass of arsenic(III) oxide did the sample contain?

Related Exercises L.25, L.26

The stoichiometric relation between analyte and titrant species, together with the molarity of the titrant, is used in titrations to determine the molarity of the analyte.

When an annotation above an expression refers to an entire fraction the term is printed in blue.

What have you learned in Fundamentals L?

You have learned how to use the stoichiometric coefficients in a balanced chemical equation to predict the quantitative outcome of a reaction in terms of the amounts or masses of the reactants and products.

The skills you have mastered are the ability to:

☐ **1.** Carry out stoichiometric calculations for any two species taking part in a chemical reaction (Toolbox L.1 and Example L.1).

☐ **2.** Calculate the molar concentration (molarity) of a solute from titration data (Toolbox L.2 and Example L.2).

☐ **3.** Calculate the mass of a solute from titration data (Toolbox L.2 and Example L.3).

Fundamentals L Exercises

L.1 Without using a calculator, estimate what amount of Br_2 (in moles) can be obtained from 0.30 mol ClO_2 in the reaction $6 \, ClO_2(g) + 2 \, BrF_3(l) \rightarrow 6 \, ClO_2F(s) + Br_2(l)$.

L.2 Without using a calculator, estimate what amount of Cl_2 (in moles) can be obtained from 0.75 mol N_2H_4 in the reaction $3 \, N_2H_4(g) + 4 \, ClF_3(g) \rightarrow 12 \, HF(g) + 3 \, N_2(g) + 2 \, Cl_2(g)$.

L.3 Compounds that can be used to store hydrogen in vehicles are being actively sought. One reaction being studied for hydrogen storage is $Li_3N(s) + 2 \, H_2(g) \rightarrow LiNH_2(s) + 2 \, LiH(s)$. (a) What amount (in moles) of H_2 is needed to react with 1.5 mg of Li_3N? (b) Calculate the mass of Li_3N that will produce 0.650 mol LiH.

L.4 In research on the synthesis of superconductors, the reaction $Tl_2O_3(l) + 2 \, BaO(s) + 3 \, CaO(s) + 4 \, CuO(s) \rightarrow Tl_2Ba_2Ca_3Cu_4O_{12}(s)$ is being studied. (a) What amount (in moles) of BaO is needed to react with 5.0 g of CaO? (b) Calculate the mass of CuO required to produce 0.24 mol of product.

L.5 The solid fuel in the booster stage of the space shuttle is a mixture of ammonium perchlorate and aluminum powder. Upon ignition, the reaction that takes place is $6 \, NH_4ClO_4(s) + 10 \, Al(s) \rightarrow 5 \, Al_2O_3(s) + 3 \, N_2(g) + 6 \, HCl(g) + 9 \, H_2O(g)$. (a) What mass of aluminum should be mixed with 1.325 kg of NH_4ClO_4 for this reaction? (b) Determine the mass of Al_2O_3 (alumina, a finely divided white powder that is produced as billows of white smoke) formed in the reaction of 3.500×10^3 kg of aluminum.

L.6 The compound diborane, B_2H_6, was at one time considered for use as a rocket fuel. Its combustion reaction is

$$B_2H_6(g) + 3 \, O_2(l) \longrightarrow 2 \, HBO_2(g) + 2 \, H_2O(l)$$

The fact that HBO_2, a reactive compound, was produced rather than the relatively inert B_2O_3 was a factor in the discontinuation of the investigation of diborane as a fuel. (a) What mass of liquid oxygen (LOX) would be needed to burn 257 g of B_2H_6? (b) Determine the mass of HBO_2 produced from the combustion of 106 g of B_2H_6.

L.7 The camel stores the fat tristearin, $C_{57}H_{110}O_6$, in its hump. As well as being a source of energy, the fat is also a source of water because, when it is used, the reaction $2 \, C_{57}H_{110}O_6(s) + 163 \, O_2(g) \rightarrow 114 \, CO_2(g) + 110 \, H_2O(l)$ takes place. (a) What mass of water is available from 1.00 pound (454 g) of this fat? (b) What mass of oxygen is needed to oxidize this amount of tristearin?

L.8 Potassium superoxide, KO_2, is used in a closed-system breathing apparatus to remove carbon dioxide and water from exhaled air. The removal of water generates oxygen for breathing by the reaction $4 \, KO_2(s) + 2 \, H_2O(l) \rightarrow 3 \, O_2(g) + 4 \, KOH(s)$. The potassium hydroxide removes carbon dioxide from the apparatus by the reaction $KOH(s) + CO_2(g) \rightarrow KHCO_3(s)$. (a) What mass of potassium superoxide generates 63.0 g of O_2? (b) What mass of CO_2 can be removed from the apparatus by 75.0 g of KO_2?

L.9 When a hydrocarbon burns, water is produced as well as carbon dioxide. The density of gasoline is $0.79 \, g \cdot mL^{-1}$. Assume gasoline to be represented by octane, C_8H_{18}, for which the combustion reaction is $2 \, C_8H_{18}(l) + 25 \, O_2(g) \rightarrow 16 \, CO_2(g) + 18 \, H_2O(l)$.

Calculate the mass of water produced from the combustion of 1.0 gallon (3.8 L) of gasoline.

L.10 The Sabatier reaction uses hydrogen to generate water from waste carbon dioxide on the space station: $CO_2(g) + 4 \, H_2(g) \rightarrow CH_4(g) + 2 \, H_2O(l)$. The hydrogen is stored as a liquid with a density of $0.070 \, g \cdot cm^3$. What mass of water can be produced from the reaction of 2.0 L $H_2(l)$?

L.11 The stomach uses HCl to digest food. However, excess stomach acid can cause problems and must sometimes be neutralized by chewing antacid tablets containing a base such as $Mg(OH)_2$. Carbonates such as $CaCO_3$ are also used, because they can react as bases: $CaCO_3(s) + 2 \, HCl(aq) \rightarrow CaCl_2(aq) + CO_2(g) + H_2O(l)$. A popular antacid tablet contains 400. mg of $CaCO_3$ and 150. mg of $Mg(OH)_2$. What mass of HCl can it neutralize?

L.12 You have used up all the antacid tablets, so you decide to use sodium bicarbonate as an antacid, because it can act as a base: $NaHCO_3(s) + HCl(aq) \rightarrow NaCl(aq) + H_2O(l) + CO_2(g)$. What mass of $NaHCO_3$ will you need to neutralize the same mass of acid neutralized by the antacid tablet in Exercise L.11?

L.13 A solution of $Ca(OH)_2$ of volume 25.00 mL was titrated to the stoichiometric point with 12.15 mL of 0.144 M $HNO_3(aq)$. What was the initial concentration of $Ca(OH)_2$ in the solution?

L.14 A solution of HCl of volume 10.00 mL was titrated to the stoichiometric point with 13.35 mL of 0.0152 M KOH(aq). What was the initial concentration of HCl in the solution?

L.15 A solution of sodium hydroxide of volume 15.00 mL was titrated to the stoichiometric point with 17.40 mL of 0.234 M HCl(aq). (a) What was the initial concentration (in moles per liter) of NaOH in the solution? (b) Calculate the mass of NaOH in the solution.

L.16 A solution of oxalic acid, $H_2C_2O_4$ (with two acidic protons), of volume 25.17 mL was titrated to the stoichiometric point with 26.72 mL of 0.327 M NaOH(aq). (a) What was the molarity of the initial oxalic acid solution? (b) Determine the mass of oxalic acid in the solution.

L.17 A sample of barium hydroxide of mass 9.670 g was dissolved and diluted to the mark in a 250.0-mL volumetric flask. It was found that 11.56 mL of this solution was needed to reach the stoichiometric point in a titration of 25.0 mL of a nitric acid solution. (a) Calculate the molarity of the HNO_3 solution. (b) What mass of HNO_3 was in the initial sample?

L.18 Suppose that 10.0 mL of 3.0 M KOH(aq) is transferred to a 250.0-mL volumetric flask and diluted to the mark. It is found that 38.5 mL of this diluted solution is needed to reach the stoichiometric point in a titration of 10.0 mL of a phosphoric acid solution according to the reaction $3 \, KOH(aq) + H_3PO_4(aq) \rightarrow K_3PO_4(aq) + 3 \, H_2O(l)$. (a) Calculate the concentration (in moles per liter) of H_3PO_4 in the solution. (b) What mass of H_3PO_4 was in the initial sample?

L.19 In a titration, 3.25 g of an acid, HX, requires 68.8 mL of 0.750 M NaOH(aq) for complete reaction. What is the molar mass of the acid?

L.20 Suppose that 14.56 mL of 0.115 M NaOH(aq) is required to titrate 0.2037 g of an unknown acid, HX. What is the molar mass of the acid?

L.21 Excess NaI was added to 50.0 mL of aqueous $AgNO_3$ solution, and 1.76 g of AgI precipitate was formed. What was the molar concentration of $AgNO_3$ in the original solution?

L.22 An excess of $AgNO_3$ reacts with 30.1 mL of 4.2 M K_2CrO_4(aq) to form a precipitate. What is the precipitate, and what mass of precipitate is formed?

L.23 A solution of hydrochloric acid was prepared by measuring 10.00 mL of the concentrated acid into a 1.000-L volumetric flask and adding water up to the mark. Another solution was prepared by adding 0.832 g of anhydrous sodium carbonate to a 100.0-mL volumetric flask and adding water up to the mark. Then, 25.00 mL of the carbonate solution was pipetted into a flask and titrated with the diluted acid. The stoichiometric point was reached after 31.25 mL of acid had been added. (a) Write a balanced equation for the reaction of HCl(aq) with Na_2CO_3(aq). (b) What was the molarity of the original concentrated hydrochloric acid?

L.24 A tablet of vitamin C was analyzed to determine whether it did in fact contain, as the manufacturer claimed, 1.0 g of the vitamin. One tablet was dissolved in water to form 100.00 mL of solution, and 10.0 mL of that solution was titrated with iodine (as potassium triiodide). It required 10.1 mL of 0.0521 M I_3^-(aq) to reach the stoichiometric point in the titration. Given that 1 mol I_3^- reacts with 1 mol vitamin C in the reaction, is the manufacturer's claim correct? The molar mass of vitamin C is 176 g·mol^{-1}.

L.25 The copper content of certain high-temperature superconductors can be determined by dissolving a sample of the superconductor in dilute acid and titrating a sample with iodide ion, which is oxidized to the triiodide ion by the copper:

$$6 Cu^{2+}(aq) + 15 I^-(aq) \longrightarrow 6 CuI(aq) + 3 I_3^-(aq)$$

The amount of triiodide reacted is then determined by further reaction with thiosulfate ion. In one experiment, 1.10 g of a superconductor required 24.4 mL of 0.0010 M I_3^-(aq) to reach the stoichiometric point. What was the mass percentage of copper in the superconductor?

L.26 The uranium content of ores can be determined by dissolving the ore in acid and converting all the uranium to the U^{4+} ion, which is then titrated with potassium permanganate in acidic solution until all the U^{4+} has reacted, at which point the purple color of the permanganate ion persists:

$$5 U^{4+}(aq) + 2 MnO_4^-(aq) + 2 H_2O(aq) \longrightarrow$$
$$5 UO_2^{2+}(aq) + 2 Mn^{2+}(aq) + 4 H^+(aq)$$

A sample of uranium ore of mass 11.020 g was dissolved in acid and the uranium converted to U^{4+}. The resulting solution required 25.8 mL of 0.538 M $KMnO_4$(aq) to reach the stoichiometric point. What was the mass percentage of uranium in the ore?

L.27 Iodine is a common oxidizing agent, often used as the triiodide ion, I_3^-. Suppose that, in the presence of HCl(aq), 25.00 mL of 0.120 M aqueous triiodide solution reacts completely with 30.00 mL of a solution containing 19.0 g·L^{-1} of an ionic compound containing tin and chlorine. The products are iodide ions and another compound of tin and chlorine. The reactant compound is 62.6% tin by mass. Write a balanced equation for the reaction.

L.28 A forensic laboratory is analyzing a mixture of two solids, calcium chloride dihydrate, $CaCl_2 \cdot 2H_2O$, and potassium chloride, KCl. The mixture is heated to drive off the water of hydration: $CaCl_2 \cdot 2H_2O(s) \xrightarrow{\Delta} CaCl_2(s) + 2 H_2O(g)$. A sample of the mixture weighed 2.543 g before heating. After heating, the resulting mixture of anhydrous $CaCl_2$ and KCl weighed 2.312 g. Calculate the mass percentage of each compound in the original sample.

L.29 Thiosulfate ions ($S_2O_3^{2-}$) "disproportionate" in acidic solution to give solid sulfur (S) and hydrogen sulfite ion (HSO_3^-):

$$2 S_2O_3^{2-}(aq) + 2 H_3O^+(aq) \longrightarrow$$
$$2 HSO_3^-(aq) + 2 H_2O(l) + 2 S(s)$$

A disproportionation reaction is a type of oxidation–reduction reaction. Which species is oxidized and which is reduced? (b) If 10.1 mL of 55.0% HSO_3^- by mass is obtained in the reaction, what mass of $S_2O_3^{2-}$ was present initially, assuming the reaction went to completion? The density of the HSO_3^- solution is 1.45 g·cm^{-3}.

L.30 Suppose that 25.0 mL of 0.50 M K_2CrO_4(aq) reacts with 15.0 mL of $AgNO_3$(aq) completely. What mass of NaCl is needed to react completely with 45.0 mL of the same $AgNO_3$ solution?

L.31 The compound $XCl_2(NH_3)_2$ can be formed by reacting XCl_4 with NH_3. Suppose that 3.571 g of XCl_4 reacts with excess NH_3 to give Cl_2 and 3.180 g of $XCl_2(NH_3)_2$. What is the element X?

L.32 The reduction of iron(III) oxide to iron metal in a blast furnace is another source of atmospheric carbon dioxide. The reduction takes place in these two steps:

$$2 C(s) + O_2(g) \longrightarrow 2 CO(g)$$
$$Fe_2O_3(s) + 3 CO(g) \longrightarrow 2 Fe(l) + 3 CO_2(g)$$

Assume that all the CO generated in the first step reacts in the second. (a) How many C atoms are needed to react with 600 Fe_2O_3 formula units? (b) What is the maximum volume of carbon dioxide (taken to have a density of 1.25 g·L^{-1}) that can be generated in the production of 1.0 t of iron (1 t = 10^3 kg)? (c) Assuming a 67.9% yield of iron, what volume of carbon dioxide is released to the atmosphere in the production of 1.0 t of iron? (d) How many kilograms of O_2 are required for the production of 5.00 kg of Fe?

L.33 Barium bromide, $BaBr_x$, can be converted into $BaCl_2$ by treatment with chlorine. It is found that 3.25 g of $BaBr_x$ reacts completely with an excess of chlorine to yield 2.27 g of $BaCl_2$. Determine the value of x and write the balanced chemical equation for the production of $BaCl_2$ from $BaBr_x$.

L.34 Sulfur is an undesirable impurity in coal and petroleum fuels. The mass percentage of sulfur in a fuel can be determined by burning the fuel in oxygen and dissolving the SO_3 produced in water to form aqueous sulfuric acid. In one experiment, 8.54 g of a fuel was burned, and the resulting sulfuric acid was titrated with 17.54 mL of 0.100 M NaOH(aq). (a) Determine the amount (in moles) of H_2SO_4 that was produced. (b) What is the mass percentage of sulfur in the fuel?

L.35 Sodium bromide, NaBr, which is used to produce AgBr for use in photographic film, can itself be prepared as follows:

$$Fe + Br_2 \longrightarrow FeBr_2$$

$$FeBr_2 + Br_2 \longrightarrow Fe_3Br_8$$

$$FeBr_2 + Na_2CO_3 \longrightarrow NaBr + CO_2 + Fe_3O_4$$

What mass of iron, in kilograms, is needed to produce 2.50 t of NaBr? Note that these equations must first be balanced!

L.36 Silver nitrate is an expensive laboratory reagent that is often used for quantitative analysis of chloride ion. A student preparing to conduct a particular analysis needs 100.0 mL of 0.0750 M $AgNO_3(aq)$, but finds only about 60 mL of 0.0500 M $AgNO_3(aq)$. Instead of making up a fresh solution of the exact concentration desired (0.075 M $AgNO_3(aq)$), the student decides to pipet 50.0 mL of the existing solution into a 100.0-mL flask, then add enough pure solid $AgNO_3$ to make up the difference and enough water to bring the volume of the resulting solution to exactly 100.0 mL. What mass of solid $AgNO_3$ must be added in the second step?

L.37 (a) How would you prepare 1.00 L of 0.50 M $HNO_3(aq)$ from "concentrated" (16 M) $HNO_3(aq)$? (b) How many milliliters of 0.20 M NaOH(aq) could be neutralized by 100. mL of the diluted solution?

L.38 You have been given a sample of an unknown diprotic acid. (a) Analysis of the acid shows that a 10.0-g sample contains 0.224 g of hydrogen, 2.67 g of carbon, and the rest oxygen. Determine the empirical formula of the acid. (b) A 0.0900-g sample of your unknown acid is dissolved in 30.0 mL of water and titrated to the endpoint with 50.0 mL of 0.040 M NaOH(aq). Determine the molecular formula of the acid. (c) Write a balanced chemical equation for the neutralization of the unknown acid with NaOH(aq).

L.39 A 1.50-g sample of metallic tin was placed in a 26.45-g crucible and heated until all the tin had reacted with the oxygen in air to form an oxide. The crucible and product together were found to weigh 28.35 g. (a) What is the empirical formula of the oxide? (b) Write the name of the oxide.

L.40 A 1.27-g sample of metallic copper was placed in a 26.32-g crucible and heated until all the copper had reacted with the oxygen in air to form an oxide. The crucible and product together were found to weigh 27.75 g. (a) What is the empirical formula of the oxide? (b) Write the name of the oxide.

L.41 A chemist is titrating a solution of KOH of unknown concentration with 0.0101 M HCl(aq); a number of events occur during the titration. Which of the following events will affect the reported concentration? If there is an effect, indicate whether the reported KOH concentration will be too high or too low.
(a) The flask used for the KOH solution was not dried and so contains a small amount of distilled water.
(b) The buret used for the HCl solution was not dried and so contains a small amount of distilled water.
(c) The walls of the buret were slightly oily and some water droplets stick to the sides as the level falls.
(d) The chemist misread the concentration of HCl as 0.0110 M HCl(aq).

L.42 A chemist is determining the mass percentage of iron in a sample of ore, using the method in Example L.3. After dissolving the ore in hydrochloric acid, it is discovered that some of the solid does not dissolve. Which of the following additional pieces of information will the chemist need to determine the mass percentage of iron in the ore?
(a) The mass of the insoluble solid.
(b) The color of the insoluble solid.
(c) Knowledge that the insoluble solid does not contain any iron.
(d) None of these pieces of information is necessary.

M Limiting Reactants

Stoichiometric calculations of the amount of product formed in a reaction are based on an ideal view of the world. They suppose, for instance, that all the reactants react exactly as described in the chemical equation. In practice, that might not be so. Some of the starting materials may be consumed in a **competing reaction,** a reaction taking place at the same time as the one in which you are interested and using some of the same reactants. Another possibility is that the reaction might not be complete at the time when the measurements were made. A third possibility is that many reactions do not go to completion. They appear to stop once a certain proportion of the reactants has been consumed. Therefore, the actual amount of product may be less than that calculated from the reaction stoichiometry.

M.1 Reaction Yield

The **theoretical yield** of a reaction is the *maximum* quantity (amount, mass, or volume) of product that can be obtained from a given quantity of reactant. The **percentage yield** is the fraction of the theoretical yield actually produced, expressed as a percentage:

$$\text{Percentage yield} = \frac{\text{actual yield}}{\text{theoretical yield}} \times 100\% \qquad (1)$$

EXAMPLE M.1   Calculating the percentage yield of a product

Incomplete combustion of the fuel in a poorly tuned engine can produce toxic carbon monoxide along with the usual carbon dioxide and water. You might be working for an engine manufacturer and asked to assess the efficiency of an off-road motorcycle engine by determining the percentage yield of carbon dioxide. In one test the engine burns 1.00 L of octane (of mass 702 g) and produces 1.84 kg of carbon dioxide. What is the percentage yield of carbon dioxide? Octane is C_8H_{18}.

ANTICIPATE   Apart from knowing that the percentage yield cannot exceed 100%, it is not possible to anticipate the actual value in this case.

PLAN   Begin by writing the chemical equation for the desired reaction. Then calculate the theoretical yield (in grams) of product by using the procedure in Toolbox L.1. To avoid rounding errors, do all the numerical work at the end of the calculation (in the sense of carrying unrounded intermediate values forward). To obtain the percentage yield, divide the actual mass produced by the theoretical mass of product and multiply by 100%.

SOLVE

The chemical equation is

$$2\,C_8H_{18}(l) + 25\,O_2(g) \longrightarrow 16\,CO_2(g) + 18\,H_2O(l)$$

Step 1   Convert the given mass of octane in grams into amount in moles by using its molar mass, $n = m/M$. The molar mass of C_8H_{18} is 114.2 g·mol^{-1}.

$$n(C_8H_{18}) = \frac{702\,\text{g}}{114.2\,\text{g}\cdot(\text{mol }C_8H_{18})^{-1}} = \frac{702}{114.2}\,\text{mol }C_8H_{18}$$

$$= 6.14...\,\text{mol }C_8H_{18}$$

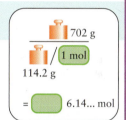

Step 2   Use the mole ratio derived from the stoichiometric coefficients in the balanced chemical equation, 2 mol $C_8H_{18} \triangleq$ 16 mol CO_2, to convert from the amount of C_8H_{18} molecules into the amount in moles of CO_2 molecules.

$$n(CO_2) = (6.14...\,\text{mol }C_8H_{18}) \times \frac{16\,\text{mol }CO_2}{2\,\text{mol }C_8H_{18}} = 49.1...\,\text{mol }CO_2$$

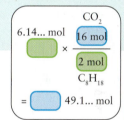

Step 3   Convert from amount in moles of CO_2 to mass (in grams and then into kilograms) by using its molar mass, which is 44.01 g·mol^{-1}, and $m = nM$:

$$m(CO_2) = (49.1...\,\text{mol }CO_2) \times 44.01\,\text{g}\cdot(\text{mol }CO_2)^{-1}$$

$$= 49.1... \times 44.01\,\text{g} = \underline{2.16 \times 10^3\,\text{g}}$$

2.16 kg

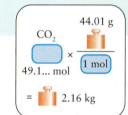

Finally, calculate the percentage yield of carbon dioxide from the fact that only 1.84 kg rather than 2.16 kg was produced.

$$\text{Percentage yield of }CO_2 = \frac{1.84\,\text{kg}}{2.16\,\text{kg}} \times 100\% = 85.2\%$$

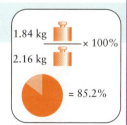

EVALUATE   The theoretical yield of 2.16 kg corresponds to about 8 kg of CO_2 per gallon of fuel. Although there are concerns about rising levels of CO_2, in this case the 85.2% yield of CO_2 is not a step in the right direction! It means that the fuel is being used inefficiently and that a lot of undesirable CO is being produced.

Self-test M.1A   When 24.0 g of potassium nitrate was heated with lead, 13.8 g of potassium nitrite was formed in the reaction $Pb(s) + KNO_3(s) \rightarrow PbO(s) + KNO_2(s)$. Calculate the percentage yield of potassium nitrite.

[*Answer:* 68.3%]

Self-test M.1B   Reduction of 15 kg of iron(III) oxide in a blast furnace produced 8.8 kg of iron in the reaction $Fe_2O_3(s) + 3\,CO(g) \rightarrow 2\,Fe(s) + 3\,CO_2(g)$. What is the percentage yield of iron?

Related Exercises M.1–M.4

The theoretical yield of a product is the maximum quantity that can be expected on the basis of the stoichiometry of a chemical equation. The percentage yield is the percentage of the theoretical yield that is actually achieved.

M.2 The Limits of Reaction

The **limiting reactant** in a reaction is the reactant that governs the maximum yield of product. A limiting reactant is like a part in short supply in a motorcycle factory. Suppose there are eight wheels and seven motorcycle frames. Because each frame requires two wheels, there are enough wheels for only four motorcycles, so the wheels play the role of the limiting reactant. When all the wheels have been used, three frames remain unused, because they were present in excess.

In some cases, the limiting reactant is not obvious and has to be decided by calculation. For example, to decide which reactant is the limiting one in the reaction

$$N_2(g) + 3 H_2(g) \longrightarrow 2 NH_3(g)$$

for which 1 mol $N_2 \mathrel{\widehat{=}} 3$ mol H_2, it is necessary to compare the number of moles of each reactant supplied with the stoichiometric coefficients. Thus, suppose you had available 1 mol N_2 but only 2 mol H_2. Because this amount of hydrogen is less than is required by the stoichiometric relation, hydrogen is the limiting reactant, even though it is present in the larger amount. Once you have identified the limiting reactant, you can calculate the amount of product that can be formed. You can also calculate the amount of excess reactant that remains at the end of the reaction.

Toolbox M.1 HOW TO IDENTIFY THE LIMITING REACTANT

CONCEPTUAL BASIS

The limiting reactant is the reactant that will be completely used up if the reaction goes to completion. All other reactants are in excess. Because the limiting reactant is the one that limits the amounts of products that can be formed, the theoretical yield is calculated from the amount of the limiting reactant.

PROCEDURE

There are two ways of determining which reactant is the limiting one.

Method 1

In this approach, use the mole ratio from the chemical equation to determine whether there is enough of one reactant to react with another.

Step 1 Convert the mass of each reactant into an amount in moles, if necessary, by using the molar masses of the substances.

Step 2 Choose one of the reactants and use the stoichiometric relation to calculate the theoretical amount of the second reactant needed for complete reaction with the first.

Step 3 If the actual amount of the second reactant is greater than the amount needed (the value calculated in step 2), then

the second reactant is present in excess; in this case, the first reactant is the limiting reactant. If the actual amount of the second reactant is less than that calculated, then all of it will react; so it is the limiting reactant and the first reactant is in excess.

This method is used in Example M.2.

Method 2

Calculate the theoretical yield of one of the products for each reactant separately, by using the procedure in Toolbox L.1. The reactant that would produce the smallest amount of product is the limiting reactant. This method is a good one to use when there are more than two reactants.

Step 1 Convert the mass of each reactant into an amount in moles, if necessary, by using the molar masses of the substances.

Step 2 For each reactant, calculate how many moles of the product it can form.

Step 3 The reactant that can produce the least product is the limiting reactant.

This method is used in Example M.3.

EXAMPLE M.2   Identifying the limiting reactant

Calcium carbide, CaC_2, reacts with water to form calcium hydroxide and the flammable gas ethyne (acetylene) in the reaction $CaC_2(s) + 2 H_2O(l) \rightarrow Ca(OH)_2(aq) + C_2H_2(g)$. This reaction was once used for fuel in lamps on bicycles and miners' helmets, because the reactants are easily transported and acetylene burns with a bright flame (due to the incandescent carbon particles that form). (a) Which is the limiting reactant when 100. g of water reacts with 100. g of calcium carbide? (b) What mass of ethyne can be produced? (c) What mass of excess reactant remains after reaction is complete? Assume that the reactants are pure and that the reaction goes to completion.

ANTICIPATE Because the molar mass of CaC_2 is so much greater than that of H_2O, you might suspect that only a small amount of it is present and therefore that it will be the limiting reactant. On the other hand, the mole ratio for the reaction indicates that the amount of H_2O required is twice that of the CaC_2, so you have to be cautious about that prediction.

PLAN Follow the procedure in Method 1 of Toolbox M.1. Molar masses are calculated by using the information in the periodic table inside the front cover or the alphabetical list of elements inside the back cover.

SOLVE

(a) *Step 1* Convert the mass of each reactant into an amount in moles by using molar masses: for CaC_2, 64.10 g·mol^{-1}; for H_2O, 18.02 g·mol^{-1}.

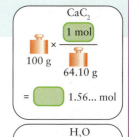

$$n(CaC_2) = \frac{100.\ g}{64.10\ g\cdot(mol\ CaC_2)^{-1}}$$
$$= \frac{100.}{64.10}\ mol\ CaC_2 = 1.56...\ mol\ CaC_2$$

$$n(H_2O) = \frac{100.\ g}{18.02\ g\cdot(mol\ H_2O)^{-1}}$$
$$= \frac{100.}{18.02}\ mol\ H_2O = 5.55...\ mol\ H_2O$$

Step 2 Select CaC_2 and use the stoichiometric relation 1 mol $CaC_2 \stackrel{\frown}{=}$ 2 mol H_2O to calculate the theoretical amount of H_2O needed for complete reaction with the given amount of CaC_2.

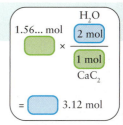

$$n(H_2O) = (1.56...\ mol\ CaC_2) \times \frac{2\ mol\ H_2O}{1\ mol\ CaC_2}$$
$$= 1.56... \times 2\ mol\ H_2O = 3.12\ mol\ H_2O$$

Step 3 Determine which reactant is the limiting reactant.

Because 3.12 mol H_2O is required and 5.55 mol H_2O is supplied, all the calcium carbide can react; so the calcium carbide is the limiting reactant and water is present in excess.

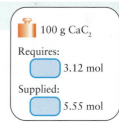

(b) Because C_2H_2 is the limiting reactant and 1 mol $CaC_2 \stackrel{\frown}{=}$ 1 mol C_2H_2, the mass of ethyne (of molar mass 26.04 g·mol^{-1}) that can be produced is

$$m(CaC_2) = (1.56...\ mol\ CaC_2) \times \frac{1\ mol\ C_2H_4}{1\ mol\ CaC_2} \times 26.04\ g\cdot(mol\ C_2H_4)^{-1}$$
$$= 40.6\ g$$

(c) The reactant in excess is water. Because 5.55 mol H_2O was supplied and 3.12 mol H_2O was consumed, the amount of H_2O remaining is 5.55 − 3.12 mol = 2.43 mol. Therefore, the mass of the excess reactant remaining at the end of the reaction is

$$Mass\ of\ H_2O\ remaining = (2.43\ mol) \times (18.02\ g\cdot mol^{-1}) = 43.8\ g$$

EVALUATE The CaC_2 is the limiting reactant (as anticipated).

Self-test M.2A (a) Identify the limiting reactant in the reaction $6\ Na(l) + Al_2O_3(s) \rightarrow 2\ Al(l) + 3\ Na_2O(s)$ when 5.52 g of sodium is heated with 5.10 g of Al_2O_3. (b) What mass of aluminum can be produced? (c) What mass of excess reactant remains at the end of the reaction?

[*Answer:* (a) Sodium; (b) 2.16 g Al; (c) 1.02 g Al_2O_3]

Self-test M.2B (a) What is the limiting reactant for the preparation of urea from ammonia in the reaction $2\ NH_3(g) + CO_2(g) \rightarrow OC(NH_2)_2(s) + H_2O(l)$ when 14.5 kg of ammonia is available to react with 22.1 kg of carbon dioxide? (b) What mass of urea can be produced? (c) What mass of excess reactant remains at the end of the reaction?

Related Exercises M.7, M.8, M.11, M.12

EXAMPLE M.3   Calculating the percentage yield from a limiting reactant

An important step in the refining of aluminum metal is the manufacture of cryolite, Na_3AlF_6, from ammonium fluoride, sodium aluminate, and sodium hydroxide in aqueous solution:

$$6\,NH_4F(aq) + NaAl(OH)_4(aq) + 2\,NaOH(aq) \longrightarrow Na_3AlF_6(s) + 6\,NH_3(aq) + 6\,H_2O(l)$$

Unfortunately, by-products can form, reducing the yield. You might be asked to investigate the efficiency of the process. You mix 100.0 g of NH_4F with 82.6 g of $NaAl(OH)_4$ and 80.0 g of NaOH and collect 75.0 g of Na_3AlF_6. What was the percentage yield of the reaction?

ANTICIPATE Apart from knowing that the percentage yield cannot exceed 100%, it is not possible to anticipate the actual value in this case.

PLAN First, the limiting reactant must be identified (Toolbox M.1). This limiting reactant determines the theoretical yield of the reaction, and so you need to use it to calculate the theoretical amount of product by Method 2 in Toolbox M.1. The percentage yield is the ratio of the mass produced to the theoretical mass times 100%.

SOLVE

Step 1 Convert the mass of each reactant into an amount in moles by using the molar masses of the substances, which are NH_4F: 37.04 g·mol^{-1}, NaOH: 40.00 g·mol^{-1}, $NaAl(OH)_4$: 118.00 g·mol^{-1}, and Na_3AlF_6: 209.95 g·mol^{-1}, and by using $n = m/M$.

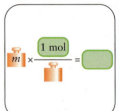

$$n(NH_4F) = \frac{100.0\ \text{g}}{37.04\ \text{g}\cdot(\text{mol}\,NH_4F)^{-1}} = 2.700\ \text{mol}\ NH_4F$$

$$n(NaAl(OH)_4) = \frac{82.6\ \text{g}}{118.00\ \text{g}\cdot(\text{mol}\,NaAl(OH)_4)^{-1}}$$
$$= 0.700\ \text{mol}\ NaAl(OH)_4$$

$$n(NaOH) = \frac{80.0\ \text{g}}{40.00\ \text{g}\cdot(\text{mol}\,NaOH)^{-1}} = 2.00\ \text{mol}\ NaOH$$

Step 2 For each reactant, calculate how many moles of product (Na_3AlF_6) it can form. Use the following mole ratios for the reaction:

$$6\ \text{mol}\ NH_4F \simeq 1\ \text{mol}\ Na_3AlF_6$$
$$1\ \text{mol}\ NaAl(OH)_4 \simeq 1\ \text{mol}\ Na_3AlF_6$$
$$2\ \text{mol}\ NaOH \simeq 1\ \text{mol}\ Na_3AlF_6$$

From NH_4F:

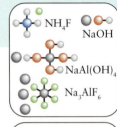

$$n(Na_3AlF_6) = 2.700\ \text{mol}\ NH_4F \times \frac{1\ \text{mol}\ Na_3AlF_6}{6\ \text{mol}\ NH_4F}$$
$$= 0.450\ \text{mol}\ Na_3AlF_6$$

From $NaAl(OH)_4$:

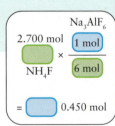

$$n(Na_3AlF_6) = 0.700\ \text{mol}\ NaAl(OH)_4 \times \frac{1\ \text{mol}\ Na_3AlF_6}{1\ \text{mol}\ NaAl(OH)_4}$$
$$= 0.700\ \text{mol}\ Na_3AlF_6$$

From NaOH:

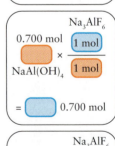

$$n(Na_3AlF_6) = 2.00\ \text{mol}\ NaOH \times \frac{1\ \text{mol}\ Na_3AlF_6}{2\ \text{mol}\ NaOH}$$
$$= 1.00\ \text{mol}\ Na_3AlF_6$$

NH_4F can produce only 0.450 mol Na_3AlF_6. Therefore, it is the limiting reactant.

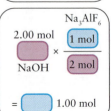

Calculate the theoretical yield as mass of Na_3AlF_6 from $m = nM$.

$$m(Na_3AlF_6) = 0.450 \text{ mol} \times 209.95 \text{ g·mol}^{-1} = 94.5 \text{ g}$$

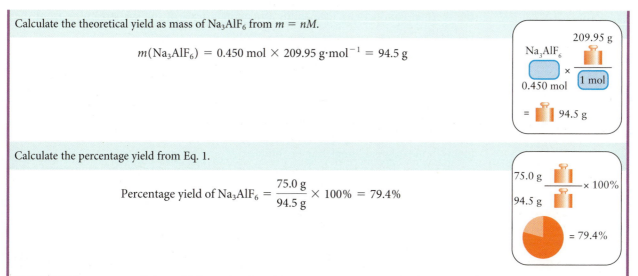

Calculate the percentage yield from Eq. 1.

$$\text{Percentage yield of } Na_3AlF_6 = \frac{75.0 \text{ g}}{94.5 \text{ g}} \times 100\% = 79.4\%$$

EVALUATE As expected, the yield (79.4%) is less than 100% on account of the formation of unwanted by-products.

Self-test M.3A In the synthesis of ammonia, what is the percentage yield of ammonia when 100. kg of hydrogen reacts with 800. kg of nitrogen to produce 400. kg of ammonia in the reaction $N_2(g) + 3 H_2(g) \rightarrow 2 NH_3(g)$?

[*Answer:* 71.0%]

Self-test M.3B Suppose that 28 g of NO_2 and 18 g of water are allowed to react to produce nitric acid and nitrogen monoxide in the reaction $3 NO_2(g) + H_2O(l) \rightarrow 2 HNO_3(aq) + NO(g)$. If 22 g of nitric acid is produced in the reaction, what is the percentage yield?

Related Exercises M.15, M.17, M.18, M.24

The limiting reactant in a reaction is the reactant supplied in an amount smaller than that required by the stoichiometric relation between the reactants.

M.3 Combustion Analysis

One technique used in modern chemical laboratories to determine the empirical formulas of organic compounds is combustion analysis (*Fundamentals* F). A sample is ignited and allowed to burn in a tube while a plentiful supply of oxygen passes through (**FIG. M.1**). The use of excess oxygen ensures that the sample is the limiting reactant. All the hydrogen in the compound is converted into water and all the carbon is converted into carbon dioxide. In the modern version of the technique, the product gases are separated chromatographically and their relative amounts determined by measuring

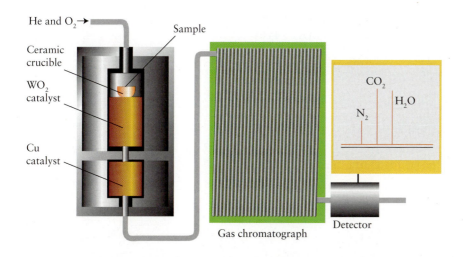

FIGURE M.1 The apparatus used for a combustion analysis. An oxygen–helium mixture is passed over a ceramic crucible containing the sample, which is oxidized. Product gases are passed through two filters. The WO_3 catalyst ensures that any CO produced is oxidized to CO_2; the Cu removes the excess oxygen. The masses of nitrogen, carbon dioxide, and water produced are obtained by separating the product gases and measuring their thermal conductivities.

the thermal conductivity (the ability to conduct heat) of the gas streaming from the apparatus.

In the presence of excess oxygen, each carbon atom in the compound ends up in one molecule of carbon dioxide. Therefore,

$$\text{1 mol C in sample} \doteq \text{1 mol } CO_2 \text{ as product}$$

or, more briefly, 1 mol C $\doteq$ 1 mol CO_2. Hence, by measuring the mass of carbon dioxide produced and converting that mass into amount of C atoms, the number of moles of C atoms in the original sample can be found.

Similarly, in the presence of excess oxygen, each hydrogen atom in a compound contributes to a water molecule when the compound burns:

$$\text{2 mol H in sample} \doteq \text{1 mol } H_2O \text{ as product}$$

or, more briefly, 2 mol H $\doteq$ 1 mol H_2O. Therefore, from the mass of water produced when the compound burns in plenty of oxygen, the amount of H atoms (in moles) in the sample can be determined.

Many organic compounds also contain oxygen. Provided the compound contains only carbon, hydrogen, and oxygen, the mass of oxygen originally present can be calculated by subtracting the masses of carbon and hydrogen in the sample from the original mass of the sample. That mass of oxygen can be converted into amount of O atoms (in moles) by using the molar mass of oxygen atoms (16.00 g·mol^{-1}).

EXAMPLE M.4   Determining an empirical formula by combustion analysis

If you become an organic chemist, you will almost certainly use combustion analysis at one stage of your work. Suppose you carry out a combustion analysis on 1.621 g of a newly synthesized compound, which is known to contain only C, H, and O. The masses of water and carbon dioxide produced are 1.902 g and 3.095 g, respectively. What is the empirical formula of the compound?

ANTICIPATE Only if you had carried out the synthesis with a specific product in mind could you anticipate the correct empirical formula.

PLAN Use the stoichiometric relations given earlier to find the amounts of carbon and hydrogen atoms in the sample and then convert those moles into masses. The mass of oxygen in the sample is obtained by subtracting the total mass of carbon and hydrogen from the mass of the original sample. That mass difference is due to oxygen and so needs to be converted into the number of moles of O atoms. Finally, the relative numbers of atoms are expressed as an empirical formula.

SOLVE The molar masses you need are

$$\text{C: } 12.01 \text{ g·mol}^{-1} \quad CO_2\text{: } 44.01 \text{ g·mol}^{-1} \quad \text{H: } 1.008 \text{ g·mol}^{-1} \quad H_2O\text{: } 18.02 \text{ g·mol}^{-1}$$

The mole ratio for the production of CO_2 is 1 mol C $\doteq$ 1 mol CO_2, and for the production of H_2O it is 1 mol $H_2O \doteq$ 2 mol H.

Convert mass of CO_2 produced into amount of C in the sample.

$$n(C) = \frac{3.095 \text{ g}}{44.01 \text{ g·(mol } CO_2)^{-1}} \times \frac{1 \text{ mol C}}{1 \text{ mol } CO_2} = \frac{3.095}{44.01} \text{ mol C}$$

$$= 0.070\ldots \text{ mol C}$$

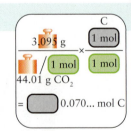

Calculate the mass of carbon in the sample from $m = nM$.

$$m(C) = \left(\frac{3.095}{44.01} \text{ mol C} \right) \times 12.01 \text{ g·(mol C)}^{-1} = 0.8446\ldots \text{ g}$$

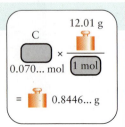

Convert mass of H_2O produced into amount of H in the sample.

$$m(H) = \frac{1.902 \text{ g}}{18.02 \text{ g} \cdot (\text{mol } H_2O)^{-1}} \times \frac{2 \text{ mol H}}{1 \text{ mol } H_2O}$$

$$= \frac{1.902 \times 2}{18.02} \text{ mol H} = 0.2111\ldots \text{ mol H}$$

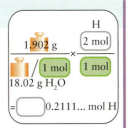

Calculate the mass of H in the sample.

$$m(H) = (0.2111\ldots \text{ mol H}) \times 1.008 \text{ g} \cdot (\text{mol H})^{-1} = 0.2128 \text{ g}$$

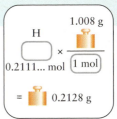

Find the total mass of C and H.

$$0.8446 \text{ g} + 0.2128 \text{ g} = 1.0574 \text{ g}$$

Calculate the mass of oxygen in the sample from the difference between sample mass and the masses of C and H.

$$m(O) = 1.621 \text{ g} - 1.0574 \text{ g} = 0.564 \text{ g}$$

Convert mass of oxygen into amount of O atoms.

$$n(O) = \frac{0.564 \text{ g}}{16.00 \text{ g} \cdot (\text{mol O})^{-1}} = 0.0352 \text{ mol O}$$

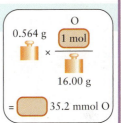

Write the ratios of the amount of each element in the sample. This ratio is the same as the relative numbers of atoms.

$$C:H:O = 0.070\,32:0.2111:0.0352$$

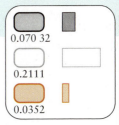

Divide by the smallest number (0.0352).

$$C:H:O = \frac{0.070\,32}{0.0352} : \frac{0.2111}{0.0352} : \frac{0.0352}{0.0352}$$

$$= 2.00:6.00:1.00$$

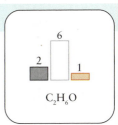

EVALUATE You can conclude that the empirical formula of the new compound is C_2H_6O.

Self-test M.4A When 0.528 g of sucrose (a compound of carbon, hydrogen, and oxygen) is burned, 0.306 g of water and 0.815 g of carbon dioxide are formed. Deduce the empirical formula of sucrose.

[**Answer:** $C_{12}H_{22}O_{11}$]

Self-test M.4B When 0.236 g of aspirin (a compound of carbon, hydrogen, and oxygen) is burned in oxygen, 0.519 g of carbon dioxide and 0.0945 g of water are formed. Deduce the empirical formula of aspirin.

Related Exercises M.19–M.22

In a combustion analysis, the amounts of C, H, and O atoms in a sample of a compound, and thus its empirical formula, are determined from the masses of carbon dioxide and water produced when the compound burns in excess oxygen.

What have you learned in Fundamentals M?

You have learned that a stoichiometric relation can be used to predict the theoretical yield of a reaction, but that not all yields reach that theoretical maximum; in some cases, the amount of one reactant limits the amount of product. You have also seen how to use the concept of theoretical yield to determine the empirical formula of a compound by combustion analysis.

The skills you have mastered are the ability to:

☐ **1.** Calculate the theoretical and percentage yields of the products of a reaction (Example M.1).

☐ **2.** Identify the limiting reactant of a reaction.

☐ **3.** Use the limiting reactant to calculate the yield of a product and the amount of excess reactant remaining after reaction is complete (Toolbox M.1 and Examples M.2 and M.3).

☐ **4.** Determine the empirical formula of an organic compound containing carbon, hydrogen, and oxygen by combustion analysis (Example M.4).

Fundamentals M Exercises

M.1 Hydrazine, N_2H_4, is an oily liquid used as a rocket fuel. It can be prepared in water by oxidizing ammonia with hypochlorite ions: $2\,NH_3(g) + ClO^-(aq) \rightarrow N_2H_4(aq) + Cl^-(aq) + H_2O(l)$. When 35.0 g of ammonia reacted with an excess of hypochlorite ion, 25.2 g of hydrazine was produced. What is the percentage yield of hydrazine?

M.2 The metal vanadium can be extracted from its oxide by heating it in the presence of calcium: $V_2O_5(s) + 5\,Ca(l) \xrightarrow{\Delta} 2\,V(s) + 5\,CaO(s)$. When 150.0 kg of V_2O_5 was heated with calcium, 36.7 kg of vanadium was produced. What is the percentage yield of vanadium?

M.3 When limestone, which is principally $CaCO_3$, is heated, carbon dioxide and quicklime, CaO, are produced by the reaction $CaCO_3(s) \xrightarrow{\Delta} CaO(s) + CO_2(g)$. If 17.5 g of CO_2 is produced from the thermal decomposition of 42.73 g of $CaCO_3$, what is the percentage yield of the reaction?

M.4 Phosphorus trichloride, PCl_3, is produced from the reaction of white phosphorus, P_4, and chlorine: $P_4(s) + 6\,Cl_2(g) \rightarrow 4\,PCl_3(g)$. A sample of PCl_3 of mass 180.5 g was collected from the reaction of 49.91 g of P_4 with excess chlorine. What is the percentage yield of the reaction?

M.5 Solve this exercise without using a calculator. The reaction $6\,ClO_2(g) + 2\,BrF_3(l) \rightarrow 6\,ClO_2F(s) + Br_2(l)$ is carried out with 12 mol ClO_2 and 5 mol BrF_3. (a) Identify the excess reactant. (b) Estimate how many moles of each product will be produced and how many moles of the excess reactant will remain.

M.6 Solve this exercise without using a calculator. The reaction $3\,N_2H_4(g) + 4\,ClF_3(g) \rightarrow 12\,HF(g) + 3\,N_2(g) + 2\,Cl_2(g)$ is carried out with 12 mol N_2H_4 and 12 mol ClF_3. (a) Identify the excess

reactant. (b) Estimate how many moles of each product will be produced and how many moles of the excess reactant will remain.

M.7 Solid boron can be extracted from solid boron oxide by reaction with magnesium metal at a high temperature. A second product is solid magnesium oxide. (a) Write a balanced equation for the reaction. (b) What mass of boron can be produced when 125 kg of boron oxide is heated with 125 kg of magnesium?

M.8 The antiperspirant aluminum chloride is made by reacting solid aluminum oxide, solid carbon, and chlorine gas. Carbon monoxide gas is also produced in the reaction. (a) Write a balanced equation for the reaction. (b) What mass of aluminum chloride can be produced when 185 kg of aluminum oxide is heated with 25 kg of carbon and 100. kg of chlorine?

M.9 Copper(II) nitrate reacts with sodium hydroxide to produce a precipitate of light blue copper(II) hydroxide. (a) Write the net ionic equation for the reaction. (b) Calculate the maximum mass of copper(II) hydroxide that can be formed when 2.00 g of sodium hydroxide is added to 80.0 mL of 0.500 M $Cu(NO_3)_2(aq)$.

M.10 Cobalt(III) nitrate reacts with sodium sulfide to form solid cobalt(III) sulfide. (a) Write the net ionic equation for the reaction. (b) Calculate the maximum mass of cobalt(III) sulfide that can be formed when 3.00 g of sodium sulfide is added to 65.0 mL of 0.620 M $Co(NO_3)_3(aq)$.

M.11 A reaction vessel contains 5.77 g of white phosphorus and 5.77 g of oxygen. The first reaction to take place is the formation of phosphorus(III) oxide, P_4O_6: $P_4(s) + 3\,O_2(g) \rightarrow P_4O_6(s)$. If enough oxygen is present, the oxygen can react further with this oxide to produce phosphorus(V) oxide, P_4O_{10}: $P_4O_6(s) + 2\,O_2(g) \rightarrow P_4O_{10}(s)$. (a) What is the limiting reactant for the formation of

P_4O_{10}? (b) What mass of P_4O_{10} is produced? (c) How many grams of the excess reactant remain in the reaction vessel?

M.12 A mixture of 12.375 g of iron(II) oxide and 6.144 g of aluminum metal is placed in a crucible and heated in a high-temperature oven, where a reduction of the oxide takes place: $3\,FeO(s) + 2\,Al(l) \rightarrow 3\,Fe(l) + Al_2O_3(s)$. (a) What is the limiting reactant? (b) Determine the maximum amount of iron (in moles of Fe) that can be produced. (c) Calculate the mass of excess reactant remaining in the crucible.

M.13 Polychlorinated biphenyls (PCBs) were once widely used industrial chemicals but were found to pose a risk to health and the environment. PCBs contain only carbon, hydrogen, and chlorine. Aroclor 1254 is the trade name for a PCB with molar mass $360.88\ g \cdot mol^{-1}$. Combustion of 1.52 g of Aroclor 1254 produced 2.224 g of CO_2, and combustion of 2.53 g produced 0.2530 g of H_2O. How many chlorine atoms does an Aroclor 1254 molecule contain?

M.14 In the reaction of hydrogen gas (H_2) and oxygen gas (O_2) to form water vapor, which is the limiting reactant in each situation? What is the maximum quantity of water vapor that can be produced in each case? Report your answer using the units in parentheses. (a) 1.0 g of hydrogen gas and 1.0 mol $O_2(g)$ (in moles of H_2O); (b) 100 H_2 molecules and 30 O_2 molecules (in numbers of H_2O molecules).

M.15 Aluminum metal reacts with chlorine gas to produce aluminum chloride. In one preparation, 255 g of aluminum is placed in a container holding 535 g of chlorine gas. After reaction ceases, it is found that 300. g of aluminum chloride has been produced. (a) Write the balanced equation for the reaction. (b) What mass of aluminum chloride can be produced by these reactants? (c) What is the percentage yield of aluminum chloride?

M.16 A mixture consisting of 4.94 g of 85.0% phosphine, PH_3, and 0.110 kg of $CuSO_4 \cdot 5H_2O$ (of molar mass $249.68\ g \cdot mol^{-1}$) is placed in a reaction vessel. (a) Balance the chemical equation for the reaction that takes place, given the skeletal form. $CuSO_4 \cdot 5H_2O(s) + PH_3(g) \rightarrow Cu_3P_2(s) + H_2SO_4(aq) + H_2O(l)$. (b) Name each reactant and product. (c) Determine the limiting reactant. (d) Calculate the mass (in grams) of Cu_3P_2 (of molar mass $252.56\ g \cdot mol^{-1}$) produced, given that the percentage yield of the reaction is 6.31%.

M.17 The acid HA (where A stands for an unknown group of atoms) has molar mass $231\ g \cdot mol^{-1}$. HA reacts with the base XOH (molar mass $125\ g \cdot mol^{-1}$) to produce H_2O and the salt XA. In one experiment, 2.45 g of HA reacts with 1.50 g of XOH to form 2.91 g of XA. What is the percentage yield of the reaction?

M.18 The acid H_2A (where A stands for an unknown group of atoms) has molar mass $168\ g \cdot mol^{-1}$. H_2A reacts with the base XOH (molar mass $125\ g \cdot mol^{-1}$) to produce H_2O and the salt X_2A. In one experiment, 1.20 g of H_2A reacts with 1.00 g of XOH to form 0.985 g of X_2A. What is the percentage yield of the reaction?

M.19 A stimulant in coffee and tea is caffeine, a substance of molar mass $194\ g \cdot mol^{-1}$. When 0.376 g of caffeine was burned,

0.682 g of carbon dioxide, 0.174 g of water, and 0.110 g of nitrogen were formed. Determine the empirical and molecular formulas of caffeine, and write the equation for its combustion.

M.20 Nicotine, a stimulant in tobacco, causes a very complex set of physiological effects in the body. It is known to have a molar mass of $162\ g \cdot mol^{-1}$. When a sample of mass 0.385 g was burned, 1.072 g of carbon dioxide, 0.307 g of water, and 0.068 g of nitrogen were produced. What are the empirical and molecular formulas of nicotine? Write the equation for its combustion.

M.21 A compound found in the nucleus of a human cell was found to be composed of carbon, hydrogen, oxygen, and nitrogen. A combustion analysis of 1.35 g of the compound produced 2.20 g of CO_2 and 0.901 g of H_2O. When a separate sample of mass 0.500 g of the compound was analyzed for nitrogen, 0.130 g of N_2 was produced. What is the empirical formula of the compound?

M.22 A compound produced as a by-product in an industrial synthesis of polymers was found to contain carbon, hydrogen, and iodine. A combustion analysis of 1.70 g of the compound produced 1.32 g of CO_2 and 0.631 g of H_2O. The mass percentage of iodine in the compound was determined by converting the iodine in a sample of mass 0.850 g of the compound into 1.15 g of lead(II) iodide. What is the empirical formula of the compound? Could the compound also contain oxygen? Explain your answer.

M.23 When aqueous solutions of calcium nitrate and phosphoric acid are mixed, a white solid precipitates. (a) What is the formula of the solid? (b) How many grams of the solid can be formed from 206 g of calcium nitrate and 150. g of phosphoric acid?

M.24 Small amounts of chlorine gas can be generated in the laboratory from the reaction of manganese(IV) oxide with hydrochloric acid: $4\,HCl(aq) + MnO_2(s) \rightarrow 2\,H_2O(l) + MnCl(s) + Cl_2(g)$. (a) What mass of Cl_2 can be produced from 42.7 g of MnO_2 with an excess of $HCl(aq)$? (b) What volume of chlorine gas (of density $3.17\ g \cdot L^{-1}$) will be produced from the reaction of 300. mL of $0.100\ M$ $HCl(aq)$ with an excess of MnO_2? (c) Suppose that only 150. mL of chlorine was produced in the reaction in part (b). What is the percentage yield of the reaction?

M.25 In addition to determining the elemental composition of pure unknown compounds, combustion analysis can be used to determine the purity of known compounds. A sample of 2-naphthol, $C_{10}H_7OH$, which is used to prepare antioxidants to incorporate into synthetic rubber, was found to be contaminated with a small amount of LiBr. The combustion analysis of this sample gave the following results: 77.48% C and 5.20% H. Assuming that the only species present are 2-naphthol and LiBr, calculate the percentage purity by mass of the sample.

M.26 An organic compound with the formula $C_{14}H_{20}O_2N$ was recrystallized from 1,1,2,2-tetrachloroethane, $C_2H_2Cl_4$. A combustion analysis of the compound gave the following data: 68.50% C, 8.18% H by mass. Because the data were considerably different from that expected for pure $C_{14}H_{20}O_2N$, the sample was examined and found to contain a significant amount of 1,1,2,2-tetrachloroethane. Assuming that only these two compounds are present, what is the percentage purity by mass of the $C_{14}H_{20}O_2N$?

M.27 Tu-jin-pi is a root bark used in traditional Chinese medicines for the treatment of "athlete's foot." One of the active ingredients in tu-jin-pi is pseudolaric acid A, which is known to contain carbon, hydrogen, and oxygen. A chemist wanting to determine the molecular formula of pseudolaric acid A burned 1.000 g of the compound in an elemental analyzer. The products of the combustion were 2.492 g of CO_2 and 0.6495 g of H_2O. (a) Determine the empirical formula of the compound. (b) The molar mass was found to be 388.46 $g \cdot mol^{-1}$. What is the molecular formula of pseudolaric acid A?

M.28 An industrial by-product consists of C, H, O, and Cl. When 0.100 g of the compound was analyzed by combustion analysis, 0.0682 g of CO_2 and 0.0140 g of H_2O were produced. The mass percentage of Cl in the compound was found to be 55.0%. What are the empirical and molecular formulas of the compound?

ATOMS

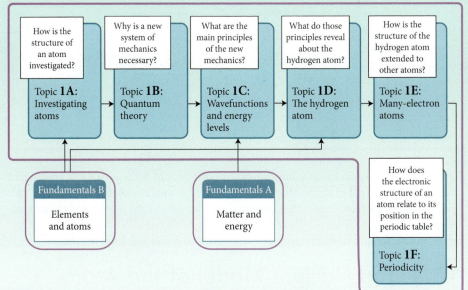

Atoms are the currency of chemistry. Almost all explanations in chemistry refer to them, and little of chemistry can be understood except in their terms. It is therefore essential to understand their characteristics.

This group of Topics provides the foundations for understanding atomic structure. First, in **TOPIC 1A** the experiments that led to the current nuclear model of the atom are reviewed and you will see how spectroscopy reveals information about the energies of the electrons around the nucleus of an atom. Then, in **TOPIC 1B**, the experiments that led to the replacement of classical mechanics by quantum mechanics are described. With the basic ideas established, in **TOPIC 1C**, the important concept of a wavefunction is introduced and some of its central features are illustrated by considering a very simple but instructive system.

TOPIC 1D turns to atoms and begins with the simplest atom of all, the hydrogen atom. To construct an acceptable model of the hydrogen atom, it is necessary to treat the electron as a wave and to consider the implications of the Schrödinger equation. The discussion of the hydrogen atom provides a foundation for the discussion of atoms with more than one electron in **TOPIC 1E**. That discussion in turn leads to understanding of the structure of the periodic table of the elements in **TOPIC 1F**, which you will come to see is central to learning chemistry.

Topic 1A Investigating Atoms

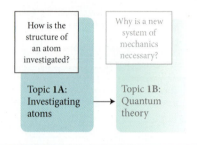

How is the structure of an atom investigated?

Why is a new system of mechanics necessary?

Topic **1A**: Investigating atoms → Topic 1B: Quantum theory

Why Do You Need to Know This Material? Atoms are central to chemical explanations, and it is therefore important to know how information about their structure is obtained and what it reveals.

What Do You Need to Know Already? You need to be familiar with the introductory discussion of the nuclear model of the atom (*Fundamentals* B).

John Dalton pictured atoms as featureless spheres, like billiard balls. Today, it is known that atoms have an internal structure: they are built from even smaller **subatomic particles.** This book deals with the three major subatomic particles: the electron, the proton, and the neutron. By understanding the internal structure of atoms, you will come to see how one element differs from another and how their properties are related to the structures of their atoms.

1A.1 The Nuclear Model of the Atom

The first experimental evidence for the internal structure of atoms was obtained in 1897. The British physicist J. J. Thomson (**FIG. 1A.1**) was investigating "cathode rays," the rays that are emitted when a high potential difference (a high voltage) is applied between two metal electrodes in an evacuated glass tube (**FIG. 1A.2**). By noting the direction of deflection of the beam caused by an applied electric field, Thomson showed that cathode rays are streams of negatively charged particles coming from inside the atoms that made up the negatively charged electrode, the *cathode*. He found that the charged particles, which came to be called **electrons,** were the same regardless of the metal used for the cathode and concluded that they are part of the makeup of all atoms.

Thomson was able to measure the value of e/m_e, the ratio of the magnitude of the electron's charge e to its mass m_e. However, the values of e and m_e themselves were not known until later workers, most notably the American physicist Robert Millikan, carried out experiments that enabled them to determine the value of e alone. Millikan designed an ingenious apparatus in which he could observe tiny electrically charged oil droplets (**FIG. 1A.3**). From the strength of the electric field required to overcome the pull of gravity on the droplets, he determined the values of the charges on the particles. Because each oil droplet contained more than one additional electron, he took the smallest increment of charge

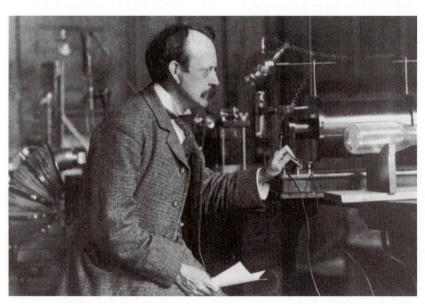

FIGURE 1A.1 Joseph John Thomson (1856–1940), with the apparatus that he used to discover the electron. (*© Pictorial Press Ltd/Alamy.*)

2

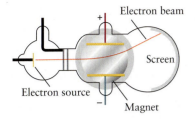

FIGURE 1A.2 The apparatus used by Thomson to investigate the properties of electrons. An electric field is set up between the two yellow plates and a magnetic field is applied perpendicular to the electric field.

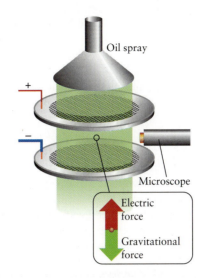

FIGURE 1A.3 A schematic diagram of Millikan's oil-drop experiment. Oil is sprayed as a fine mist into a chamber containing a charged gas, and the location of an oil droplet is monitored by using a microscope. Charged particles (ions) are generated in the gas by exposing it to x-rays. The fall of the charged droplet is balanced by the electric field.

between droplets to be the charge on one electron. The modern value, which is obtained in a much more sophisticated way, is $e = 1.602 \times 10^{-19}$ C. A charge of $-e$ is considered to be "one unit" of negative charge, and e itself, which is called the **fundamental charge,** is taken to be "one unit" of positive charge. The mass of the electron was calculated by combining this charge with the value of e/m_e measured by Thomson; its modern value is 9.109×10^{-31} kg.

Although electrons have a negative charge, an atom has zero charge: it is electrically "neutral." Therefore, an atom must contain enough positive charge to cancel the negative charge. But where is the positive charge? Thomson suggested a "plum pudding" model of an atom as a blob of a positively charged, jellylike material, with the electrons suspended in it like raisins in pudding. However, this model was overthrown in 1908 by another experimental observation. Ernest Rutherford (**FIG. 1A.4**) knew that some elements, including radon, emit streams of positively charged particles, which he called **α particles** (alpha particles). He asked two of his students, Hans Geiger and Ernest Marsden, to shoot α particles toward a piece of platinum foil only a few atoms thick (**FIG. 1A.5**). If atoms were indeed like blobs of positively charged jelly, then all the α particles would pass through the diffuse positive charge of the foil, with only occasional slight deflections in their paths.

Geiger and Marsden's observations astonished everyone. Although almost all the α particles did pass through and were deflected only very slightly, about 1 in 20 000 was deflected through more than 90°, and a few α particles bounced straight back in the direction from which they had come. "It was almost as incredible," said Rutherford, "as if you had fired a 15-inch shell at a piece of tissue paper and it had come back and hit you."

As explained in Fundamentals A, C stands for coulomb, the SI unit of electric charge.

FIGURE 1A.4 Ernest Rutherford (1871–1937), who was responsible for many discoveries about the structure of the atom and its nucleus. *(Prof. Peter Fowler/Science Source.)*

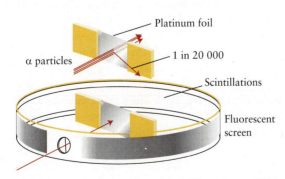

FIGURE 1A.5 Part of the experimental arrangement used by Geiger and Marsden. The α particles came from a sample of the radioactive gas radon. They were directed through a hole into a cylindrical chamber with a zinc sulfide coating on the inside. The α particles struck the platinum foil mounted inside the cylinder, and their deflections were measured by observing flashes of light (scintillations) where they struck the screen. About 1 in 20 000 α particles was deflected through very large angles; most went through the thin foil with almost no deflection.

They used platinum foil in their initial experiments; later they changed the foil to gold.

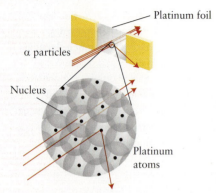

FIGURE 1A.6 Rutherford's model of the atom explains why most α particles pass almost straight through the platinum foil, whereas a very few—those scoring a direct hit on the nucleus—undergo very large deflections. Most of the atom is nearly empty space thinly populated by the atom's electrons. The nuclei are much smaller relative to their atoms than shown here.

The results of the Geiger–Marsden experiment suggested the **nuclear model** of the atom, in which there is a tiny, dot-like, heavy center of positive charge, the **nucleus,** surrounded by a large volume of mostly empty space in which the electrons are located. Rutherford reasoned that when a positively charged α particle scored a direct hit on one of the minute but heavy platinum nuclei, the α particle was strongly repelled by the positive charge of the nucleus and deflected through a large angle, like a tennis ball bouncing off a stationary cannonball (**FIG. 1A.6**). Later work by nuclear physicists showed that the nucleus of an atom contains particles called **protons,** each of which has a charge of $+e$, that are responsible for the positive charge, and **neutrons,** which are uncharged particles with almost the same mass as protons. Because protons and neutrons are of similar mass and electrons have a much smaller mass (*Fundamentals* B), the mass of an atom is almost entirely due to its nucleus. The number of protons in the nucleus is different for each element and is called the **atomic number,** Z, of the element. The total charge on an atomic nucleus of atomic number Z is $+Ze$ and, for the atoms to be electrically neutral, there must be Z electrons around it to provide a total negative charge of $-Ze$.

In the nuclear model of the atom, all the positive charge and almost all the mass is concentrated in the tiny nucleus, and the negatively charged electrons surround the nucleus. The atomic number is the number of protons in the nucleus.

1A.2 Electromagnetic Radiation

The question that scientists struggled with for years is how those Z electrons are arranged around the nucleus. To investigate the internal structures of objects as small as atoms, scientists observe them indirectly through the properties of the light the atoms emit when stimulated by heat or an electric discharge. The analysis of the light emitted or absorbed by substances is called **spectroscopy.**

Light is a form of **electromagnetic radiation,** which consists of oscillating (time-varying) electric and magnetic fields that travel through empty space at about 3×10^8 m·s^{-1}, or at just over 670 million miles per hour. This speed is denoted c and called the "speed of light." Visible light, radio waves, microwaves, and x-rays are all types of electromagnetic radiation. All these forms of radiation transfer energy from one region of space to another. The warmth you feel from the Sun is carried to you through space as electromagnetic radiation.

As a light ray passes an electron, its electric field pushes the electron first in one direction and then the opposite direction, over and over again. That is, the field oscillates in both direction and strength (**FIG. 1A.7**). The number of cycles (complete reversals of direction away from and back to the initial strength and direction) per second is called the **frequency,** ν (the Greek letter nu), of the radiation. The unit of frequency, 1 hertz (1 Hz), is defined as 1 cycle per second: $1\ \text{Hz} = 1\ \text{s}^{-1}$. Electromagnetic radiation of frequency 1 Hz pushes a charge in one direction, then the opposite direction, and returns to the original direction once per second. The frequency of the electromagnetic radiation that you see as visible light is close to 10^{15} Hz, and so its electric field returns to its initial direction about a thousand trillion (10^{15}) times a second as it travels past a given point.

An instantaneous snapshot of a wave of electromagnetic radiation spread through space would look like the one shown in Fig. 1A.7. The wave is characterized by its amplitude and wavelength. The **amplitude** is the height of the wave above the centerline. The square of the amplitude determines the **intensity,** or brightness, of the radiation. The **wavelength,** λ (the Greek letter lambda), is the peak-to-peak distance. Now imagine the wave in Fig. 1A.7 zooming along at its actual speed, the speed of light, c. If the wavelength of the light is very short, very many complete oscillations pass a given point in a second (**FIG. 1A.8a**). If the wavelength is long, fewer complete oscillations pass the point in a second (**FIG. 1A.8b**). A short wavelength therefore corresponds to high-frequency radiation and a long wavelength corresponds to low-frequency radiation. The precise relation is

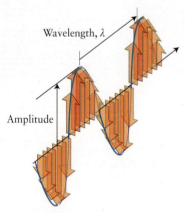

FIGURE 1A.7 The electric field of electromagnetic radiation oscillates in space and time. This diagram represents a "snapshot" of an electromagnetic wave at a given instant. The length of an arrow at any point represents the strength of the force that the field exerts on a charged particle at that point. The distance between the peaks is the wavelength of the radiation, and the height of the wave above the center line is the amplitude.

$$\underset{\lambda}{\underbrace{\text{wavelength}}} \times \underset{\nu}{\underbrace{\text{frequency}}} = \underset{c}{\underbrace{\text{speed of light}}}$$

(1)

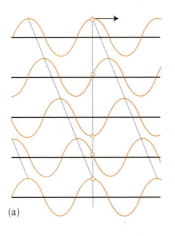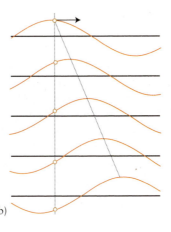

(a)   (b)

FIGURE 1A.8 (a) Short-wavelength radiation: the vertical line shows how the electric field changes markedly at five successive instants, from top to bottom. (b) For the same five instants, the electric field of the long-wavelength radiation changes much less. The horizontal arrows in the uppermost images show that in each case the wave has traveled the same distance. Short-wavelength radiation has a high frequency, whereas long-wavelength radiation has a low frequency.

EXAMPLE 1A.1 Calculating the wavelength of light of a known frequency

Rainbows form when the wavelengths of sunlight are refracted (bent) through different angles. When sunlight passes through droplets of water, the shorter the wavelength of the light, the greater the angle through which its path is bent. Which color of light has the shorter wavelength, red light of frequency 4.3×10^{14} Hz or blue light of frequency 6.4×10^{14} Hz?

ANTICIPATE Because long-wavelength waves result in fewer oscillations as they pass a given point, long wavelength is associated with low frequency. Therefore, because red light has a lower frequency than blue light, you should expect it to have the longer wavelength.

PLAN Use Eq. 1 to convert from frequency to wavelength.

SOLVE

For red light: from $\lambda \nu = c$ written as $\lambda = c/\nu$,

$$\lambda = \frac{\overset{c}{\overbrace{2.998 \times 10^8 \text{ m·s}^{-1}}}}{\underset{\text{Hz}}{\underbrace{4.3 \times 10^{14} \text{ s}^{-1}}}} = \frac{2.998 \times 10^8}{4.3 \times 10^{14}} \text{ m} = 7.0 \times 10^{-7} \text{ m}$$

(or 700 nm, to 2 sf)

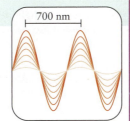
700 nm

For blue light: from $\lambda \nu = c$ written as $\lambda = c/\nu$,

$$\lambda = \frac{\overset{c}{\overbrace{2.998 \times 10^8 \text{ m·s}^{-1}}}}{\underset{\text{Hz}}{\underbrace{6.4 \times 10^{14} \text{ s}^{-1}}}} = \frac{2.998 \times 10^8}{6.4 \times 10^{14}} \text{ m} = 4.7 \times 10^{-7} \text{ m}$$

(or 470 nm, to 2 sf)

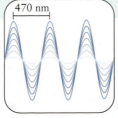
470 nm

EVALUATE As predicted, red light has a longer wavelength (700 nm) than blue light (470 nm).

Self-test 1A.1A Calculate the wavelengths of the light from traffic signals as they change. Assume that the lights emit the following frequencies: green, 5.75×10^{14} Hz; yellow, 5.15×10^{14} Hz; red, 4.27×10^{14} Hz.

[*Answer:* Green, 521 nm; yellow, 582 nm; red, 702 nm]

Self-test 1A.1B What is the wavelength of the signal from a radio station transmitting at 98.4 MHz?

Related Exercises: 1A.7–1A.10

Answers to all B self-tests are in the back of this book.

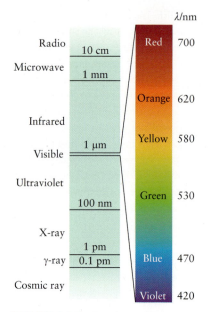

λ/nm

FIGURE 1A.9 The electromagnetic spectrum and the names of its regions. The region called "visible light" occupies a very narrow range of wavelengths. The regions are not drawn to scale.

Although 500 nm is only half of one-thousandth of a millimeter (so you might *just* be able to imagine it), it is much longer than the diameters of atoms, which are typically close to 0.2 nm.

Modern theories suggest that our concept of space breaks down on a scale of 10^{-34} m, so that may constitute a lower bound to the wavelength of electromagnetic radiation.

The components of different frequency or wavelength are called *lines* because, in the early spectroscopic experiments, the radiation from the sample was passed through a slit and then through a prism; the image of the slit was then focused on a photographic plate, where it appeared as a line.

TABLE 1A.1   Color, Frequency, and Wavelength of Electromagnetic Radiation*

Radiation type	Frequency/ $(10^{14}$ Hz)	Wavelength[†]/nm	Energy of photon/$(10^{-19}$ J)
x-rays and γ-rays	$\geq 10^3$	≤ 3	$\geq 10^3$
ultraviolet	8.6	350	5.7
visible light			
violet	7.1	420	4.7
blue	6.4	470	4.2
green	5.7	530	3.8
yellow	5.2	580	3.4
orange	4.8	620	3.2
red	4.3	700	2.8
infrared	3.0	1000	2.0
microwaves and radio waves	$\leq 10^{-3}$	$\geq 3 \times 10^6$	$\leq 10^{-3}$

* The values listed here are representative values within the range corresponding to each region.
† 2 sf

Different wavelengths of electromagnetic radiation correspond to different regions of the spectrum (**TABLE 1A.1**). The wavelengths of visible light are close to 500 nm. Human eyes detect electromagnetic radiation with wavelengths in the range from 700 nm (red light) to 400 nm (violet light). Radiation in this range is called **visible light,** and the frequency of visible light determines its color. White light is a mixture of all wavelengths of visible light. It is not known if there is a lower limit to the wavelength of electromagnetic radiation (**FIG. 1A.9**). **Ultraviolet radiation** has a higher frequency than violet light; its wavelength is less than about 400 nm. This component of solar radiation is responsible for sunburn and tanning, but it would destroy all living things on Earth if it were not largely prevented from reaching the surface of the Earth by the ozone layer. **Infrared radiation,** the radiation you experience as heat, has a lower frequency and longer wavelength than red light; its wavelength is greater than about 800 nm. Microwaves, which are used in radar and microwave ovens, have even longer wavelengths, in the millimeter-to-centimeter range.

The color of light depends on its frequency and wavelength; long-wavelength radiation has a lower frequency than short-wavelength radiation.

1A.3 Atomic Spectra

When an electric current is passed through a low-pressure sample of hydrogen gas, the gas emits light. Hydrogen gas itself does not conduct electricity, but a strong electric field strips off electrons from the H_2 molecules. As a result, they fall apart to form a "plasma" of H^+ ions and electrons, which conduct the current. The electrons almost immediately reattach to the H^+ ions to form energetically excited hydrogen atoms. These atoms quickly discard their excess energy by giving off electromagnetic radiation; then they recombine to form H_2 molecules again.

When white light is passed through a prism, a continuous spectrum of light results because white light consists of all wavelengths of visible radiation (**FIG. 1A.10a**). However, when the light emitted by excited hydrogen atoms is passed through a prism, the radiation is found to consist of a number of distinct components, or **spectral lines** (**FIG. 1A.10b**). The brightest line (at 656 nm) is red, and the excited atoms in the gas glow with this red light. Excited hydrogen atoms also emit ultraviolet and infrared radiation, which are invisible to the eye but can be detected electronically and photographically.

The first person to identify a pattern in the lines of the visible region of the hydrogen spectrum was Johann Balmer, a Swiss schoolteacher. In 1885, he noticed that the wavelengths of all the lines then known could be generated by the expression

$$\lambda \propto \frac{n^2}{n^2 - 4} \qquad n = 3, 4, \ldots$$

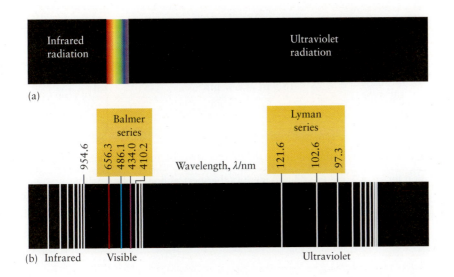

(a)

Balmer series

Lyman series

954.6 | 656.3 | 486.1 | 434.0 | 410.2 | Wavelength, λ/nm | 121.6 | 102.6 | 97.3

(b) Infrared   Visible   Ultraviolet

FIGURE 1A.10 (a) The infrared, visible, and ultraviolet spectrum. (b) The complete spectrum of atomic hydrogen. The spectral lines have been assigned to various groups called series, two of which are shown with their names.

Shortly after, the Swedish spectroscopist Johannes Rydberg suggested a form of this expression that was to prove much more revealing:

$$\frac{1}{\lambda} \propto \frac{1}{2^2} - \frac{1}{n^2} \qquad n = 3, 4, \ldots$$

This expression was easily extended to account for other series of lines that were subsequently discovered, simply by replacing 2^2 by 3^2, 4^2, and so on. The modern form of the more general expression is often written in terms of the frequency $\nu = c/\lambda$ as

$$\nu = \mathcal{R}\left\{\frac{1}{n_1^2} - \frac{1}{n_2^2}\right\} \qquad n_1 = 1, 2, \ldots, \qquad n_2 = n_1 + 1, \quad n_1 + 2, \ldots \qquad (2)$$

Here $\mathcal{R}$ is an empirical (experimentally determined) constant now known as the **Rydberg constant**; its value is 3.29×10^{15} Hz. The **Balmer series** consists of the lines with $n_1 = 2$ (and $n_2 = 3, 4, \ldots$). The **Lyman series** is a set of lines in the ultraviolet region of the spectrum with $n_1 = 1$ (and $n_2 = 2, 3, \ldots$).

EXAMPLE 1A.2   Identifying a line in the hydrogen spectrum

You can imagine Rydberg's excitement, just after he had identified his formula and found that it worked for all the known lines in the spectrum of atomic hydrogen. Calculate the wavelength of the radiation emitted by a hydrogen atom for $n_1 = 2$ and $n_2 = 3$. Identify the spectral line in Fig. 1A.10b.

ANTICIPATE Because $n_1 = 2$ the wavelength should match one of the lines in the Balmer series.

PLAN The frequency is given by Eq. 2. Convert frequency into wavelength by using Eq. 1.

SOLVE

From Eq. 2 with $n_1 = 2$ and $n_2 = 3$,

$$\nu = \mathcal{R}\left\{\frac{1}{2^2} - \frac{1}{3^2}\right\} = \frac{5}{36}\mathcal{R}$$

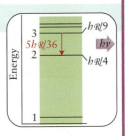

From $\lambda\nu = c$,

$$\lambda = \frac{c}{\nu} = \frac{c}{5\mathcal{R}/36} = \frac{36c}{5\mathcal{R}}$$

Now substitute the data:

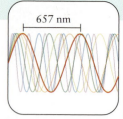

$$\lambda = \frac{36 \times \overbrace{(2.998 \times 10^8 \text{ m·s}^{-1})}^{c}}{5 \times \underbrace{(3.29 \times 10^{15} \text{ s}^{-1})}_{\mathcal{R}}} = 6.57 \times 10^{-7} \text{ m} = 657 \text{ nm}$$

EVALUATE This wavelength, 657 nm, corresponds to the red line in the Balmer series of lines in the spectrum, as expected.

Self-test 1A.2A Repeat the calculation for $n_1 = 2$ and $n_2 = 4$ and identify the spectral line in Fig. 1A.10b.

[**Answer:** 486 nm; blue line]

Self-test 1A.2B Repeat the calculation for $n_1 = 2$ and $n_2 = 5$ and identify the spectral line in Fig. 1A.10b.

Related Exercises 1A.13–1A.16

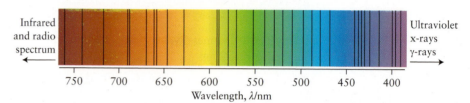

FIGURE 1A.11 When white light shines through a vapor, radiation is absorbed by the atoms at frequencies that correspond to their excitation energies. In this small section of the spectrum of the Sun, it is possible to identify which atoms in its outer layers are absorbing the radiation from the incandescence below. Many of the lines have been ascribed to hydrogen, showing that hydrogen is present in the cooler outer layers of the Sun.

If you passed white light through a gas composed of hydrogen atoms and analyzed the light you would see its **absorption spectrum,** a series of dark lines on an otherwise continuous spectrum (**FIG. 1A.11**). The absorption lines have the same frequencies as the lines in the emission spectrum and suggest that an atom can *absorb* radiation only of those same frequencies. Astronomers are able to use absorption spectra to identify elements in the outer layers of stars, because each element has a characteristic absorption spectrum.

The presence of spectral lines in an emission spectrum is explained if it is supposed that when an electron is part of an atom it can exist only with energies of discrete values, called **energy levels,** and that a line arises from a **transition,** a change of state, between two of the allowed energy levels. The difference in the energies of the two levels is carried away by the electromagnetic radiation emitted by the atom. If that is the case, then Rydberg's formula suggests that the permitted energies are proportional to $\mathcal{R}/n^2$, for then the difference in energy between the two states involved in the transition would be given by an expression that resembles the right-hand side of Rydberg's formula (Eq. 2). But why should the frequency of the emitted radiation be proportional to that energy difference? Furthermore, why does the Rydberg constant have the value that is observed? These important questions are addressed in Topic 1B.

> *The observation of discrete spectral lines suggests that an electron in an atom can have only certain energies.*

What have you learned in this Topic?

You have seen that scattering experiments show that an atom consists of a tiny dotlike central, massive nucleus surrounded by Z electrons, where Z is the atomic number of the element. You have also seen that electromagnetic radiation is a wave with characteristic frequency and wavelength that travels through space at the speed c. Atomic spectroscopy is the analysis of the frequencies of the light emitted or absorbed by atoms, and the observation of spectral lines strongly suggests that atoms can exist with only certain energies.

The skills you have mastered are the ability to:

☐ **1.** Describe the experiments that led to the formulation of the nuclear model of the atom (Section 1A.1).

☐ **2.** Calculate the wavelength or frequency of light from the relation $\lambda\nu = c$ (Example 1A.1).

☐ **3.** Calculate the wavelength of a transition in a hydrogen atom from the Rydberg formula, Example 1A.2.

Topic 1A Exercises

1A.1 At the time that J. J. Thomson conducted his experiments on cathode rays, the nature of the electron was in doubt. Some considered it to be a form of radiation, like light; others believed the electron to be a particle. Some of the observations made on cathode rays were used to advance one view or the other. Explain how each of the following properties of cathode rays supports either the wave or the particle model of the electron. (a) They pass through metal foils. (b) They travel at speeds slower than that of light. (c) If an object is placed in their path, they cast a shadow. (d) Their path is deflected when they are passed between electrically charged plates.

1A.2 J. J. Thomson originally referred to the rays produced in his apparatus (Fig. 1A.2) as "canal rays." The canal ray is deflected within the region between the poles of a magnet and strikes the phosphor screen. The ratio Q/m (where Q is the charge and m the mass) of the particles making up the canal rays is found to be 2.410×10^7 C·kg^{-1}. The cathode and anode of the apparatus are made of lithium, and the tube contains helium. Use the information inside the back cover to identify the particles (and their charges) that make up the canal rays. Explain your reasoning.

1A.3 Which of the following happens when the frequency of electromagnetic radiation decreases? Explain your reasoning.
(a) The speed of the radiation decreases.
(b) The wavelength of the radiation decreases.
(c) The extent of the change in the electrical field at a given point decreases.
(d) The energy of the radiation increases.

1A.4 Which of the following statements about the electromagnetic spectrum is true? Explain your reasoning.
(a) X-rays travel faster than infrared radiation because they have higher energy.
(b) The wavelength of visible radiation decreases as its color changes from blue to green.
(c) The frequency of infrared radiation, which has a wavelength of 1.0×10^3 nm, is half that of radio waves, which have a wavelength of 1.0×10^6 nm.
(d) The frequency of infrared radiation, which has a wavelength of 1.0×10^3 nm, is twice that of radio waves, which have a wavelength of 1.0×10^6 nm.

1A.5 Arrange the following types of photons of electromagnetic radiation in order of increasing energy: γ-rays, visible light, ultraviolet radiation, microwaves, x-rays.

1A.6 Arrange the following types of photons of electromagnetic radiation in order of increasing frequency: visible light, radio waves, ultraviolet radiation, infrared radiation.

1A.7 (a) The frequency of violet light is 7.1×10^{14} Hz. What is the wavelength (in nanometers) of violet light? (b) When an electron beam strikes a block of copper, x-rays with a frequency of 2.0×10^{18} Hz are emitted. What is the wavelength (in picometers) of these x-rays?

1A.8 (a) Radio waves for the FM station "Rock 99 at 99.3 on the FM dial" are generated at 99.3 MHz. What is the wavelength of this station? (b) Radioastronomers use 1420.-MHz waves to look at interstellar clouds of hydrogen atoms. What is the wavelength of this radiation?

1A.9 A college student recently had a busy day. Each of the student's activities on that day (reading, getting a dental x-ray, making popcorn in a microwave oven, and acquiring a suntan) involved radiation from a different part of the electromagnetic spectrum. Complete the following table and match each type of radiation to the appropriate event:

Frequency	Wavelength	Energy of photon	Event
8.7×10^{14} Hz			
		3.3×10^{-19} J	
300 MHz			
	2.5 nm		

1A.10 A college student encountered a variety of types of electromagnetic radiation when going to a restaurant for lunch (watching a red traffic light change, listening to the car radio, being struck by a stray γ-ray from outer space while entering the restaurant, and taking food from a serving table heated with an infrared lamp). Complete the following table and match each type of radiation to the appropriate event:

Frequency	Wavelength	Energy of photon	Event
		2.7×10^{-19} J	
	999 nm		
5×10^{19} Hz			
	155 cm		

1A.11 In the spectrum of atomic hydrogen, several lines are generally classified together as belonging to a series (for example, Balmer series or Lyman series, as shown in Fig. 1A.10). What is common to the lines within a series that makes grouping them together logical?

1A.12 A photon generated as a result of which of the following transitions in the hydrogen atom will have the greatest energy? Explain your answer. (a) From $n = 6$ to $n = 5$; (b) from $n = 4$ to $n = 3$; (c) from $n = 2$ to $n = 1$.

1A.13 Use the Rydberg formula for atomic hydrogen to calculate the wavelength of radiation generated by the transition from $n = 2$ to $n = 1$. (b) What is the name given to the spectroscopic series to which this transition belongs? (c) Use Table 1A.1 to determine the region of the spectrum in which the transition takes place.

1A.14 (a) Use the Rydberg formula for atomic hydrogen to calculate the wavelength for the transition from $n = 3$ to $n = 1$. (b) What is the name given to the spectroscopic series to which this transition belongs? (c) Use Table 1A.1 to determine the region of the spectrum in which the transition takes place.

1A.15 In the ultraviolet spectrum of atomic hydrogen, a line is observed at 102.6 nm. Determine the values of n for the initial and final energy levels of the electron during the emission of energy that leads to this spectral line.

1A.16 A violet line is observed at 434 nm in the spectrum of atomic hydrogen. Determine the values of n for the beginning and ending energy levels of the electron during the emission of energy that leads to this spectral line.

1A.17 The energy levels of hydrogenlike one-electron ions of atomic number Z differ from those of hydrogen by a factor of Z^2. Predict the wavelength of the transition from $n = 2$ to $n = 1$ in He^+.

1A.18 Some lasers work by exciting atoms of one element and letting these excited atoms collide with atoms of another element and transfer their excitation energy to those atoms. The transfer is most efficient when the separation of energy levels matches in the two species. Given the information in Exercise 1A.17, are there any transitions of He^+ (including transitions from its excited states) that could be excited by collision with an excited hydrogen atom with an electron in the $n = 2$ level?

Topic 1B Quantum Theory

1B.1 Radiation, Quanta, and Photons
1B.2 The Wave–Particle Duality of Matter
1B.3 The Uncertainty Principle

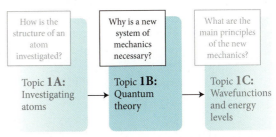

Toward the end of the nineteenth century, scientists became increasingly perplexed as they gathered more information about electromagnetic radiation that could not be explained by classical mechanics, and the lines in the spectrum of hydrogen discussed in Topic 1A remained deeply puzzling. Then, from 1900 on, a series of imaginative suggestions were made. By 1927 the puzzles had been resolved, only to be replaced by new and even more intriguing puzzles.

1B.1 Radiation, Quanta, and Photons

Important clues to the nature of electromagnetic radiation came from observations of objects as they are heated. At high temperatures an object begins to glow, the phenomenon called **incandescence.** As the object is heated to higher temperatures, it glows more brightly and the color of light it gives off changes from red through orange and yellow toward white. Those are *qualitative* observations. To understand what the color changes mean, scientists had to study the effect *quantitatively.* They measured the intensity of radiation at each wavelength and repeated the measurements at a variety of different temperatures. These experiments led to one of the greatest revolutions that has ever occurred in science.

FIGURE 1B.1 shows some of the experimental results. The "hot object" is known as a **black body** (even though it might be glowing white hot!). The name signifies that the object does not favor one wavelength over another in the sense of absorbing a particular wavelength preferentially or emitting one preferentially. The curves in Fig. 1B.1 show the intensity of **black-body radiation,** the radiation emitted at different wavelengths by a heated black body, for a series of temperatures. Notice that as the temperature rises, the maximum intensity of the radiation emitted occurs at shorter and shorter wavelengths.

A black body emits and absorbs a broad range of wavelengths because the atoms and their electrons behave collectively and because numerous transitions overlap in energy, rather than exhibit the discrete transitions of individual atoms.

THINKING POINT

Why does a heated metal object first glow red hot and then white hot?

Two crucial pieces of experimental information about black-body radiation had been discovered in the late nineteenth century. In 1879, Josef Stefan investigated the increasing brightness of a black body as it is heated and discovered that the total intensity of radiation emitted over all wavelengths increases as the fourth power of the absolute temperature (**FIG. 1B.2**). This quantitative relation is now called the **Stefan–Boltzmann law** and is usually written

$$\text{Total intensity} = \text{constant} \times T^4 \tag{1a}$$

The experimental value of the constant is 5.67×10^{-8} W·m^{-2}·K^{-4}, where W denotes watts (1 W = 1 J·s^{-1}). A few years later, in 1893, Wilhelm Wien examined the shift in color of black-body radiation as the temperature increases and discovered that the wavelength corresponding to the maximum in the intensity, $\lambda_{\max}$, is inversely proportional to the absolute temperature, $\lambda_{\max} \propto 1/T$ (that is, as T increases the wavelength of maximum

The name of the Stefan–Boltzmann law recognizes Ludwig Boltzmann's theoretical contribution.

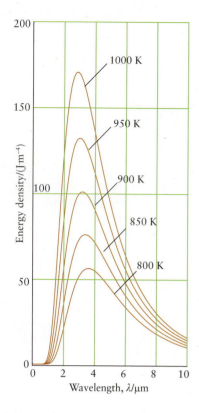

FIGURE 1B.1 The intensity of radiation emitted by a heated black body as a function of wavelength. As the temperature increases, the total energy emitted (the area under the curve) increases sharply, and the maximum intensity of emission moves to shorter wavelengths. To obtain the energy in a volume V and at wavelengths λ and $\lambda + \Delta\lambda$, multiply the energy density by V and $\Delta\lambda$.

intensity decreases), and therefore that $\lambda_{max} \times T$ is a constant (**FIG. 1B.3**). This quantitative result is now called **Wien's law** and is normally written

$$T\lambda_{max} = \text{constant} \qquad \text{(1b)}$$

The empirical value of the constant in this expression is 2.9 mm·K.

EXAMPLE 1B.1   Determining temperature from black-body radiation

Astronomers are often very interested in the temperatures of stars (including the Sun) because that gives a clue to the star's size, composition, and age. The maximum intensity of solar radiation occurs at 490. nm. What is the temperature of the surface of the Sun?

ANTICIPATE You should be aware that objects glowing white hot are at temperatures of several thousand degrees.

PLAN Use Wien's law in the form $T = \text{constant}/\lambda_{max}$.

SOLVE

From $T = \text{constant}/\lambda_{max}$,

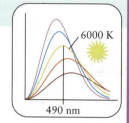

$$T = \overbrace{\frac{2.9 \times 10^{-3}\,\text{m·K}}{\underbrace{4.90 \times 10^{-7}\,\text{m}}_{490\,\text{mm}}}}^{2.9\,\text{mm·K}} = \frac{2.9 \times 10^{-3}}{4.90 \times 10^{-7}}\,\text{K} = 5.9 \times 10^{3}\,\text{K}$$

EVALUATE The surface temperature of the Sun is about 5900 K, in line with the value expected.

Self-test 1B.1A In 1965, electromagnetic radiation with a maximum at 1.05 mm (in the microwave region) was discovered to pervade the universe. What is the temperature of "empty" space?

[**Answer:** 2.76 K]

Self-test 1B.1B A red giant is a late stage in the evolution of a star. The average wavelength maximum at 700. nm shows that a red giant cools as it dies. What is the surface temperature of a red giant?

Related Exercises 1B.11–1B.14

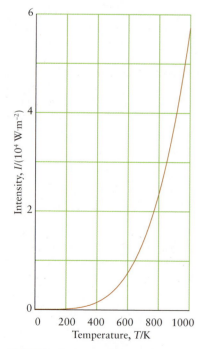

FIGURE 1B.2 The total intensity of radiation emitted by a heated black body increases as the fourth power of the temperature, so a body at 1000 K emits more than 120 times as much energy as is emitted by the same body at 300 K.

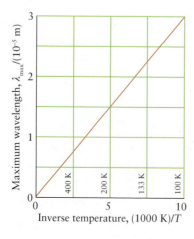

FIGURE 1B.3 As the temperature is raised ($1/T$ decreases), the wavelength of maximum emission shifts to smaller values.

For nineteenth-century scientists, the only way to account for the laws of black-body radiation was to use classical physics, the highly successful theory of motion devised by Newton two centuries previously, to derive its characteristics. However, much to their dismay, they found that the characteristics they deduced did not match their observations. Worst of all was the **ultraviolet catastrophe**: classical physics predicted that any hot body should emit intense ultraviolet radiation and even x-rays and γ-rays! According to classical physics, a hot object would devastate the countryside with high-frequency radiation. Even a human body at 37 °C would glow in the dark. There would, in fact, be no darkness.

The suggestion that resolved the problem came in 1900 from the German physicist Max Planck. He proposed that the exchange of energy between matter and radiation occurs in **quanta**, or packets, of energy. Planck focused his attention on the hot, rapidly oscillating electrons and atoms of the black body. His central idea was that a charged particle oscillating at a frequency ν (nu) can exchange energy with its surroundings by generating or absorbing electromagnetic radiation only in discrete packets of energy of magnitude

The word quantum *comes from the Latin for amount—literally, "How much?"*

$$E = h\nu \qquad (2)$$

The constant h, now called **Planck's constant**, has the value 6.626×10^{-34} J·s. If the oscillating atom releases a packet of energy of magnitude E into the surroundings, then radiation of frequency $\nu = E/h$ will be detected.

THINKING POINT

Why is ultraviolet radiation much more harmful to living tissues than infrared radiation?

Planck's hypothesis implies that radiation of frequency ν is generated only when an oscillator of that frequency has acquired the minimum energy required to start oscillating and then ejects it as a packet of electromagnetic radiation of energy $h\nu$. At low temperatures, there is not enough energy available to stimulate oscillations at very high frequencies, and so the object cannot generate high-frequency, ultraviolet radiation. As a result, the intensity curves in Fig. 1B.1 die away at high frequencies (short wavelengths) and the ultraviolet catastrophe is avoided. In contrast, in classical physics it was assumed that an oscillator could oscillate with any energy and therefore, even at low temperatures high-frequency oscillators could contribute to the emitted radiation. Planck's hypothesis is *quantitatively* successful, too, because not only was he able to use his proposal to derive the Stefan–Boltzmann and Wien laws, but he was also able to calculate the variation of intensity with wavelength and to obtain a curve that matched the experimental data almost exactly.

In fact, Planck struggled for years to keep classical mechanics alive, believing that his introduction of the quantum was just a mathematical trick. Einstein is often credited with establishing the physical reality of quantization but remained highly suspicious of the theory.

To achieve this successful theory, Planck had discarded classical physics, which puts no restriction on how small an amount of energy may be transferred from one object to another. He had proposed instead that energy is transferred in discrete packets. To justify such a dramatic revolution, more evidence was needed. That evidence came from the **photoelectric effect**, the ejection of electrons from a metal when its surface is exposed to ultraviolet radiation (**FIG. 1B.4**). The experimental observations were as follows:

1. No electrons are ejected unless the radiation has a frequency above a certain threshold value that is characteristic of the metal.
2. Electrons are ejected immediately, however low the intensity of the radiation.
3. The kinetic energy of the ejected electrons increases linearly with the frequency of the incident radiation.

A Note on Good Practice: A property y is said to "vary linearly with x" if the relation between y and x can be written $y = b + mx$, where b and m are constants. A property y is said to be "proportional to x" if $y = mx$ (that is, $b = 0$).

Albert Einstein found an explanation of these observations and, in the process, profoundly changed how scientists thought about the electromagnetic field. He proposed that electromagnetic radiation consists of particles, which were later called **photons.** Each photon can be regarded as a packet of energy, and the energy of a single photon is related

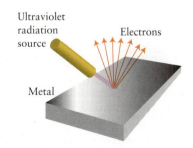

FIGURE 1B.4 When a metal is illuminated with ultraviolet radiation, electrons are ejected, provided the frequency is above a threshold frequency that is characteristic of the metal.

to the frequency of the radiation by Eq. 2 ($E = h\nu$). For example, ultraviolet photons are more energetic than photons of visible light, which has lower frequencies. According to this photon model of electromagnetic radiation, a beam of red light can be visualized as a stream of photons, each having the same energy, yellow light as a stream of photons, each of higher energy, and green light as a stream of photons, each of higher energy still. It is important to note that the *intensity* of radiation is an indication of the *number* of photons present, whereas $E = h\nu$ is a measure of the *energy* of each individual photon.

EXAMPLE 1B.2   Calculating the energy of a photon

Many chemical reactions are brought about by light: think of photosynthesis for the creation of carbohydrates, tanning caused by the ultraviolet component of solar radiation, and the dramatic molecular events that take place in the upper atmosphere. To understand these processes, chemists need to know how much energy is transferred to a molecule when a photon collides with it. What is (a) the energy of a single photon of blue light of frequency 6.4×10^{14} Hz; (b) the energy per mole of photons of the same frequency?

ANTICIPATE From Table 1A.1 you can see that the energy of a photon of blue light should be about 4×10^{-19} J.

PLAN (a) Use Eq. 2 to find the energy of light of a given frequency. (b) Multiply the energy of one photon by the number of photons per mole, which is Avogadro's constant (*Fundamentals* E).

SOLVE

(a) From $E(1\text{ photon}) = h\nu$,

$$E(1\text{ photon}) = (6.626 \times 10^{-34}\text{ J·s}) \times (6.4 \times 10^{14}\text{ Hz}) = 4.2 \times 10^{-19}\text{ J}$$

(b) From $E(\text{per mole of photons}) = N_A E(1\text{ photon})$,

$$E(\text{per mole of photons}) = (6.022 \times 10^{23}\text{ mol}^{-1}) \times (4.2 \times 10^{-19}\text{ J})$$
$$= 2.5 \times 10^5\text{ J·mol}^{-1}\text{, or } 250\text{ kJ·mol}^{-1}$$

To derive the energy in part (a), we have used 1 Hz = 1 s^{-1}, so 1 J·s $\times$ 1 Hz = 1 J·s $\times$ s^{-1} = 1 J.

EVALUATE The energies are in line with the values expected from Table 1A.1.

Self-test 1B.2A What is the energy of a photon of yellow light of frequency 5.2×10^{14} Hz?

[***Answer:*** 3.4×10^{-19} J]

Self-test 1B.2B What is the energy of a photon of orange light of frequency 4.8×10^{14} Hz?

Related Exercises 1B.5–1B.8

Be careful to distinguish the symbol for speed, v (from velocity), and the symbol for frequency, ν (the Greek letter nu).

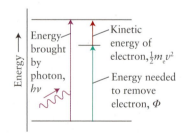

FIGURE 1B.5 In the photoelectric effect, a photon with energy $h\nu$ strikes the surface of a metal and its energy is absorbed by an electron. If the energy of the photon is greater than the work function, Φ, of the metal, the electron absorbs enough energy to break away from the metal. The kinetic energy of the ejected electron is the difference between the energy of the photon and the work function: $\frac{1}{2}m_e v^2 = h\nu - \Phi$.

The characteristics of the photoelectric effect are easy to explain if electromagnetic radiation is regarded as consisting of a stream of photons. If the incident radiation has frequency ν, it consists of photons of energy $h\nu$. When these photons collide with the electrons in the metal, the electrons absorb some of their energy. The energy required to remove an electron from a metal is called the **work function** of the metal and denoted Φ (uppercase phi). If the energy of a photon is less than the energy required to remove an electron from the metal, then an electron will not be ejected, regardless of the intensity of the radiation (which affects the rate at which the photons arrive). However, if the energy of the photon, $h\nu$, is greater than Φ, then an electron is ejected with a kinetic energy, $E_k = \frac{1}{2}m_e v^2$, equal to the difference between the energy of the incoming photon and the work function: $E_k = h\nu - \Phi$ (**FIG. 1B.5**). It follows that

$$\underbrace{\frac{1}{2}m_e v^2}_{\substack{\text{kinetic energy}\\\text{of an ejected}\\\text{electron}}} = \underbrace{h\nu}_{\substack{\text{energy}\\\text{supplied}\\\text{by a photon}}} - \underbrace{\Phi}_{\substack{\text{energy required}\\\text{to eject an electron,}\\\text{(the work function)}}} \tag{3}$$

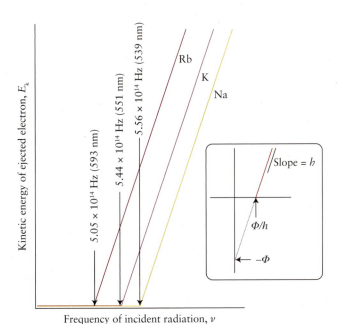

Frequency of incident radiation, v

What Does This Equation Tell You? Because the kinetic energy of the ejected electrons varies linearly with frequency, a plot of the kinetic energy against the frequency of the supplied radiation should look like the graph in **FIG. 1B.6**, a straight line of slope h, the same for all metals, and have an extrapolated intercept with the vertical axis at $-\Phi$, different for each metal. The intercept with the horizontal axis (corresponding to zero kinetic energy of the ejected electron) is at Φ/h in each case.

Einstein's theory provides the following interpretation of the photoelectric effect:

1. An electron can be driven out of the metal only if it receives from the photon during the collision at least a certain minimum energy equal to the work function, Φ. Therefore, the frequency of the radiation must have a certain minimum value if electrons are to be ejected. This minimum frequency depends on the work function and hence on the identity of the metal (as shown in Fig. 1B.6).

2. Provided a photon has enough energy, a collision results in the immediate ejection of an electron.

3. The kinetic energy of the electron ejected from the metal increases linearly with the frequency of the incident radiation according to Eq. 3.

EXAMPLE 1B.3   Analyzing the photoelectric effect

Suppose you are developing a radiation detector to be used on a spacecraft and decide to use a thin layer of metallic potassium to detect certain ranges of electromagnetic radiation. You need to make some estimates of the physical properties involved. The speed of an electron emitted from the surface of a sample of potassium by a photon is 668 km·s^{-1}. (a) What is the kinetic energy of the ejected electron? (b) The work function of potassium is 2.29 eV. What is the wavelength of the radiation that caused photoejection of the electron? (c) What is the longest wavelength of electromagnetic radiation that could eject electrons from potassium?

ANTICIPATE You should anticipate only that the wavelength of the radiation used (part b) must be less than or equal to the longest wavelength of radiation that can eject electrons from potassium (part c), as the longest wavelength corresponds to photons with the minimum energy required for the ejection.

PLAN (a) Find the kinetic energy of the ejected electron from $E_k = \frac{1}{2}m_e v^2$. To use SI base units (which is often a sensible strategy in calculations), first convert the speed to meters per second. (b) The energy of the ejected electron is equal to the difference in energy of the incident radiation and the work function (Eq. 3). The photon needs to provide enough energy to eject the electron from the metal (the work function) at a speed of 668 km·s^{-1}. Convert the value of the work function into joules and use Eq. 2 to determine the value of $h\nu$ for the photon. Then use $\lambda\nu = c$ to convert that energy to wavelength. Conversion factors and fundamental constants can be found inside the back cover of the book. (c) The longest wavelength of radiation that can eject electrons from a substance is the wavelength that results in the ejected electron having zero kinetic energy.

SOLVE

(a) From $E_k = \frac{1}{2} m_e v^2$,

$$E_k = \frac{1}{2} \times \overbrace{(9.109 \times 10^{-31} \text{ kg})}^{m_e} \times \left(\overbrace{6.68 \times 10^5 \text{ m·s}^{-1}}^{v} \right)^2$$

$$= 2.03\ldots \times 10^{-19} \underbrace{\text{kg·m}^2 \text{·s}^{-2}}_{J}$$

(b) Convert the work function from electronvolts to joules.

$$2.29 \text{ eV} \times \frac{1.602 \times 10^{-19} \text{ J}}{1 \text{ eV}} = 3.67\ldots \times 10^{-19} \text{ J}$$

From $\frac{1}{2} m_e v^2 = h\nu - \Phi$, $h\nu = \Phi + \frac{1}{2} m_e v^2 = \Phi + E_k$

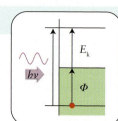

$$h\nu = \overbrace{3.67\ldots \times 10^{-19} \text{ J}}^{\Phi} + \overbrace{2.03\ldots \times 10^{-19} \text{ J}}^{E_k}$$

$$= 5.70\ldots \times 10^{-19} \text{ J}$$

so

$$\nu = \frac{5.70\ldots \times 10^{-19} \text{ J}}{h}$$

$$= \frac{5.70\ldots \times 10^{-19} \text{ J}}{6.626 \times 10^{-34} \text{ J·s}} = \frac{5.70\ldots \times 10^{-19}}{6.626 \times 10^{-34}} \text{ s}^{-1}$$

$$= 8.60\ldots \times 10^{14} \text{ s}^{-1}$$

Now use $\lambda = c/\nu$:

$$\lambda = \frac{2.998 \times 10^8 \text{ m·s}^{-1}}{\underbrace{8.60\ldots \times 10^{14} \text{ s}^{-1}}_{\nu}}$$

$$= 3.48 \times 10^{-7} \text{ m (or 348 nm)}$$

(c) To find the longest wavelength of radiation able to eject an electron, set $E_k = 0$ in Eq. 3, so $h\nu = \Phi$, and therefore $\lambda = ch/\Phi$.

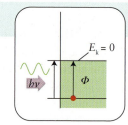

$$\lambda = \frac{(2.998 \times 10^8 \text{ m·s}^{-1}) \times (6.626 \times 10^{-34} \text{ J·s})}{3.67 \times 10^{-19} \text{ J}}$$

$$= \frac{(2.998 \times 10^8) \times (6.626 \times 10^{-34})}{3.67 \times 10^{-19}} \text{ m}$$

$$= 5.41 \times 10^{-7} \text{ m (or 541 nm)}$$

EVALUATE As expected, the wavelength of the radiation used (part b) was less than the longest wavelength of radiation that can eject electrons from potassium (part c).

Self-test 1B.3A The work function of zinc is 3.63 eV. What is the longest wavelength of electromagnetic radiation that could eject electrons from zinc?

[*Answer:* 342 nm]

Self-test 1B.3B The speed of an electron that is emitted from the surface of a sample of zinc by a photon is 785 km·s^{-1}. (a) What is the kinetic energy of the ejected electron? (b) The work function of zinc is 3.63 eV. What is the wavelength of the radiation that caused photoejection of the electron?

Related Exercises 1B.15, 1B.16

The existence of photons and the relation between the energy and the frequency of a photon help to answer one of the questions posed by the spectrum of atomic hydrogen. At the end of Topic 1A, the view that a spectral line arises from a transition between two energy levels started to form. Now you can see that if the energy difference is carried away as a photon, then the frequency of an individual line in a spectrum is related to the energy difference between two energy levels involved in the transition (**FIG. 1B.7**):

$$h\nu = E_{upper} - E_{lower} \qquad (4)$$

This relation is called the **Bohr frequency condition**. If the energies on the right of this expression are each proportional to $h\mathcal{R}/n^2$, then this relation has accounted for Rydberg's formula. Why the energies have this form still needs to be explained, but some progress has been made.

> *Studies of black-body radiation led to Planck's hypothesis of the quantization of electromagnetic radiation. The photoelectric effect provides evidence of the particulate nature of electromagnetic radiation.*

1B.2 The Wave–Particle Duality of Matter

The observation and interpretation of the photoelectric effect strongly supports the view that electromagnetic radiation consists of photons that behave like particles. However, there is plenty of evidence to show that electromagnetic radiation behaves like waves! The most compelling evidence is the observation of **diffraction**, the pattern of high and low intensities generated by an object in the path of a ray of light (**FIG. 1B.8**). A **diffraction pattern** results when the peaks and troughs of waves traveling along one path interfere with the peaks and troughs of waves traveling along another path. If the peaks coincide, the amplitude of the wave (its height) is enhanced; this enhancement is called **constructive interference** (**FIG. 1B.9a**). If the peaks of one wave coincide with the troughs of another wave, the amplitude of the wave is diminished by **destructive interference** (**FIG. 1B.9b**). This effect is the basis of a number of useful techniques for studying matter. For example, x-ray diffraction is one of the most important tools for studying the structures of molecules (*Major Technique* 3 on the website for this book).

You can appreciate why scientists were puzzled! The results of some experiments (the photoelectric effect) compelled them to the view that electromagnetic radiation is particlelike. The results of other experiments (diffraction) compelled them equally firmly to the view that electromagnetic radiation is wavelike! Thus you are brought to the heart of modern physics. Experiments oblige us to accept the **wave–particle duality** of electromagnetic radiation, in which the concepts of waves and particles blend together.

- In the wave model, the intensity of the radiation is proportional to the square of the amplitude of the wave.
- In the particle model, intensity is proportional to the number of photons present at each instant.

Now here is an interesting thought. If electromagnetic radiation, which for a long time had been regarded as wavelike, has a dual character, could it be that matter, which since Dalton's day had been regarded as consisting of particles, also has wavelike properties? In 1924, the French scientist Louis de Broglie proposed that *all* particles should be regarded as

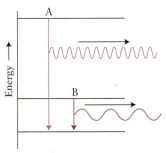

FIGURE 1B.7 When an atom undergoes a transition from a state of higher energy to one of lower energy, it loses energy that is carried away as a photon. The greater the energy difference, the higher the frequency (and the shorter the wavelength) of the radiation emitted. Compare the high frequency of the emission during a transition from a high-energy state (A) to the ground state (the lowest state) to that from a low-energy state (B) to the ground state.

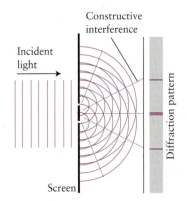

FIGURE 1B.8 In this illustration, colored lines represent the peaks of the waves of electromagnetic radiation. When radiation coming from the left (the vertical lines) passes through a pair of closely spaced slits, circular waves are generated at each slit. These waves interfere with each other. Where they interfere constructively (as indicated by the positions of the dotted lines), a bright line is seen on the screen behind the slits; where the interference is destructive, the screen is dark.

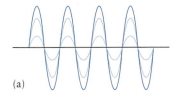 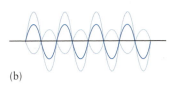

FIGURE 1B.9 (a) Constructive interference. The two component waves (pale blue) are "in phase" in the sense that their peaks and troughs coincide. The resultant (dark blue) has an amplitude that is the sum of the amplitudes of the components. The wavelength of the radiation is not changed by interference, only the amplitude is changed. (b) Destructive interference. The two component waves are "out of phase" in the sense that the troughs of one coincide with the peaks of the other. The amplitude of the resultant is much lower than in the case of constructive interference of either component.

having wavelike properties. He went on to suggest that the wavelength associated with the "matter wave" is inversely proportional to the particle's mass, m, and speed, v, and that

$$\lambda = \frac{h}{mv} \tag{5a}$$

The product of mass and speed is called the **linear momentum,** p, of a particle, and so this expression is more simply written as the **de Broglie relation:**

$$\lambda = \frac{h}{p} \tag{5b}$$

EXAMPLE 1B.4   Calculating the wavelength of a particle

Now suppose you were de Broglie and you had just devised your formula. A friend points out that the world obviously isn't wavelike. Maybe you should check whether your formula has worrying consequences for everyday objects. Calculate the wavelength of a particle of mass 1 g traveling at $1\ \text{m·s}^{-1}$.

ANTICIPATE Because the particle is much heavier than any subatomic particle, you should expect a very short wavelength.

PLAN Use Eq. 5a to find the wavelength of a particle of known mass.

SOLVE

From $\lambda = h/mv$,

$$\lambda = \frac{\overset{\displaystyle h}{\overbrace{6.626 \times 10^{-34}\ \overset{\text{kg·m}^2\text{·s}^{-2}}{\overbrace{\text{J}}}\text{·s}}}}{\underbrace{\left(1 \times 10^{-3}\ \text{kg}\right)}_{m} \times \underbrace{\left(1\ \text{m·s}^{-1}\right)}_{v}} = \frac{6.626 \times 10^{-34}}{1 \times 10^{-3}}\ \frac{\text{kg·m}^2\text{·s}^{-2}\text{·s}}{\text{kg·m·s}^{-1}}$$

$$= 7 \times 10^{-31}\ \text{m}$$

EVALUATE As expected, this wavelength is very—in fact, undetectably—small; the same is true for any macroscopic (visible) object traveling at normal speeds.

> **A Note on Good Practice:** Notice how, in the calculations, all the units are kept, writing them separately, and then canceling and multiplying them like ordinary numbers. We did not simply "guess" that the wavelength would turn out in meters. This procedure helps you to detect errors and ensures that your answer has the correct units.

Self-test 1B.4A Calculate the wavelength of an electron traveling at 1/1000 the speed of light (see the inside back cover of this book for the mass of an electron).

[*Answer:* 2.43 nm]

Self-test 1B.4B Calculate the wavelength of a rifle bullet of mass 5.0 g traveling at twice the speed of sound (the speed of sound is $331\ \text{m·s}^{-1}$).

Related Exercises 1B.21–1B.24

THINKING POINT

What is your wavelength when you are standing perfectly still? Should you be puzzled?

The wavelike character of electrons was confirmed by showing that they could be diffracted. The experiment was first performed in 1925 by two American scientists, Clinton Davisson and Lester Germer, who directed a beam of fast electrons at a single crystal of nickel. The regular array of atoms in the crystal, with centers separated by 250 pm, acts as a grid that diffracts waves; and a diffraction pattern was observed (**FIG. 1B.10**). Since then, heavier particles, such as molecules, have also been shown to undergo diffraction, and there is no doubt that particles have a wavelike character. Indeed, electron diffraction is now an important technique for determining the structures of molecules and exploring the structures of solid surfaces.

Electrons (and matter in general) have both wavelike and particlelike properties.

1B.3 The Uncertainty Principle

The discovery of wave–particle duality not only changed scientists' understanding of electromagnetic radiation and matter, it also swept away the foundations of classical physics. In classical mechanics, a particle has a definite **trajectory,** or path, on which location and linear momentum are specified at each instant. Think of the trajectory of a ball: in principle you could state its location and momentum at every moment of its flight. However, you cannot specify the precise location of a particle if it behaves like a wave: think of a wave in a guitar string, which is spread out all along the string, not localized at a precise point. A particle with a precise linear momentum has a precise wavelength; but, because it is meaningless to speak of the location of a wave, it follows that you cannot specify the location of a particle that has a precise linear momentum. The wave–particle duality of matter means that the electron in a hydrogen atom cannot be described as orbiting the nucleus with a definite trajectory. The popular picture of an electron in orbit around the nucleus is just plain wrong.

The difficulty will not go away. Wave–particle duality denies the possibility of specifying the location if the linear momentum is known, and so you cannot specify the trajectory of any particle exactly. The uncertainty is negligible for heavy particles, but for subatomic particles it can be huge. Thus, if you know that a subatomic particle is *here* at one instant, you can say nothing about where it will be an instant later! The impossibility of knowing the precise position if the linear momentum is known precisely is an aspect of the **complementarity** of location and momentum—if one property is known the other cannot be known simultaneously. The **Heisenberg uncertainty principle,** which was formulated by the German scientist Werner Heisenberg in 1927, expresses this complementarity quantitatively. It states that, if the location of a particle is known to within an uncertainty Δx, then the linear momentum, p, parallel to the x-axis can be known simultaneously only to within an uncertainty Δp, where

$$\underbrace{\Delta p}_{\substack{\text{uncertainty in} \\ \text{momentum}}} \times \underbrace{\Delta x}_{\substack{\text{uncertainty in} \\ \text{position}}} \geq \tfrac{1}{2}\hbar \tag{6}$$

The symbol $\hbar$, which is read "h bar," means $h/2\pi$, a useful combination that occurs widely in quantum mechanics.

What Does This Equation Tell You? The product of the uncertainties in two simultaneous measurements cannot be less than a certain constant value. Therefore, if the uncertainty in position is very small (Δx is very small), then the uncertainty in linear momentum must be large, and vice versa (**FIG. 1B.11**).

The uncertainty principle has negligible practical consequences for macroscopic objects, but it is of profound importance for very precise measurements dealing with subatomic particles, such as the locations and momenta of electrons in atoms, and the interpretation of their properties.

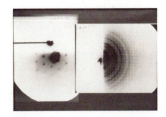

FIGURE 1B.10 Clinton Davisson and Lester Germer showed that electrons produce a diffraction pattern when reflected from a crystal. G. P. Thomson, working in Aberdeen, Scotland, showed that they also produce a diffraction pattern when they pass through a very thin gold foil. The latter is shown here. G. P. Thomson was the son of J. J. Thomson, who identified the electron (Section 1B.1). Both received Nobel prizes: J. J. for showing that the electron is a particle and G. P. for showing that it is a wave. (*Science & Society Picture Library/Getty Images.*)

Wide range of locations

(a)   Narrow range of momenta

Narrow range of locations

(b)   Wide range of momenta

FIGURE 1B.11 A representation of the uncertainty principle. (a) The location of the particle is ill defined, and so the momentum of the particle (represented by the arrow) can be specified reasonably precisely. (b) The location of the particle is well defined, and so the momentum cannot be specified very precisely.

EXAMPLE 1B.5   Using the uncertainty principle

To what extent does the Heisenberg uncertainty principle affect your ability to specify the properties of objects you can see? Can you be confident about their location? Estimate the minimum uncertainty in (a) the position of a marble of mass 1.0 g given that its speed is known to within ± 1.0 mm·s^{-1} and (b) the speed of an electron confined to an atom with the diameter 200. pm.

ANTICIPATE You should expect the uncertainty in the position of an object as heavy as a marble to be very small but the uncertainty in the speed of an electron, which has a very small mass and is confined to a small region, to be very large.

PLAN (a) The uncertainty Δp is equal to $m\Delta v$, where Δv is the uncertainty in the speed; use Eq. 6 to estimate the minimum uncertainty in position, Δx, along the direction of the travel of the marble from $\Delta p \Delta x = \tfrac{1}{2}\hbar$ (the minimum value of the product of uncertainties). (b) Assume Δx to be the diameter of the atom and use Eq. 6 to estimate Δp; use the mass of the electron inside the back cover and find Δv from $\Delta p = m\Delta v$.

SOLVE (a) First convert mass and speed into SI base units. The mass, m, is 1.0×10^{-3} kg, and the uncertainty in the speed, Δv, is $2 \times (1.0 \times 10^{-3} \text{ m·s}^{-1})$. The minimum uncertainty in position, Δx, is then:

From $\Delta p \Delta x = \frac{1}{2}\hbar$ and $\Delta p = m\Delta v$,

$$m\Delta v \Delta x = \frac{1}{2}\hbar \quad \text{and therefore} \quad \Delta x = \frac{\hbar}{2m\Delta v}$$

From $\Delta x = \hbar/2m\Delta v$,

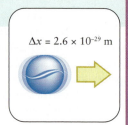

$\Delta x = 2.6 \times 10^{-29}$ m

$$\Delta x = \frac{\overbrace{1.054\,57 \times 10^{-34} \text{ J·s}}^{\hbar}}{2 \times \underbrace{(1.0 \times 10^{-3} \text{ kg})}_{1.0\text{ g}} \times \underbrace{(2.0 \times 10^{-3} \text{ m·s}^{-1})}_{2.0\text{ mm·s}^{-1}}}$$

$$= \frac{1.054\,57 \times 10^{-34}}{2 \times 1.0 \times 10^{-3} \times 2.0 \times 10^{-3}} \frac{\overbrace{\text{J}}^{\text{kg·m}^2\text{·s}^{-2}} \cdot \text{s}}{\text{kg·m·s}^{-1}}$$

$$= 2.6 \times 10^{-29} \frac{\text{kg·m}^2\text{·s}^{-2}\text{·s}}{\text{kg·m·s}^{-1}} = 2.6 \times 10^{-29} \text{ m}$$

EVALUATE As expected, this uncertainty is very small. Measurements of the location of a moving marble can be made with confidence.

A Note on Good Practice: Notice that to manipulate the units we have expressed derived units (J in this case) in terms of base units. Notice too that we are using the more precise values of the fundamental constants given inside the back cover (rather than the less precise values quoted in the text) to ensure reliable results.

(b) The mass of the electron is given inside the back cover; the diameter of the atom is $200. \times 10^{-12}$ m, or 2.00×10^{-10} m. The uncertainty in the speed, Δv, is equal to $\Delta p/m$:

From $\Delta p \Delta x = \frac{1}{2}\hbar$ and $\Delta p = m\Delta v$,

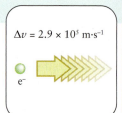

$\Delta v = 2.9 \times 10^{5}$ m·s^{-1}

$$\Delta v = \frac{\Delta p}{m} = \frac{\overbrace{\hbar}^{\hbar/2\Delta x}}{2m\Delta x}$$

$$\Delta v = \frac{\overbrace{1.054\,57 \times 10^{-34} \text{ J·s}}^{\hbar}}{2 \times \underbrace{(9.109\,39 \times 10^{-31} \text{ kg})}_{m_e} \times \underbrace{(2.00 \times 10^{-10} \text{ m})}_{\Delta x}}$$

$$= \frac{1.054\,57 \times 10^{-34}}{2 \times 9.109\,39 \times 10^{-31} \times 2.00 \times 10^{-10}} \frac{\overbrace{\text{J}}^{\text{kg·m}^2\text{·s}^{-2}} \cdot \text{s}}{\text{kg·m}}$$

$$= 2.89 \times 10^{5} \frac{\text{kg·m}^2\text{·s}^{-2}\text{·s}}{\text{kg·m}} = 2.89 \times 10^{5} \text{ m·s}^{-1}$$

EVALUATE As predicted, the uncertainty in the speed of the electron is very large, nearly ± 150 km·s^{-1}.

Self-test 1B.5A A proton is accelerated in a cyclotron to a very high speed that is known to within 3.0×10^2 km·s^{-1}. What is the minimum uncertainty in its position?

[*Answer:* 0.11 pm]

Self-test 1B.5B The police are monitoring an automobile of mass 2.0 t (1 t = 10^3 kg) speeding along a highway. They are certain of the location of the vehicle only to within 1 m. What is the minimum uncertainty in the speed of the vehicle? Can you use as your defense the argument that the uncertainly principle prevents the police from being able to measure your speed accurately?

Related Exercises 1B.25–1B.28

The location and momentum of a particle are complementary; that is, the location and the momentum cannot both be known simultaneously with arbitrary precision. The quantitative relation between the precision of each measurement is described by the Heisenberg uncertainty principle.

What have you learned in this Topic?

You have seen that not all classical concepts are applicable to subatomic particles and you now know that the concepts of waves and particles blend together. You have learned that one consequence of this blending is that it is impossible to specify the trajectory of a particle with arbitrary precision.

The skills you have mastered are the ability to:

☐ **1.** Use Wien's law to estimate a temperature (Example 1B.1).

☐ **2.** Use the relation $E = h\nu$ to calculate the energy, frequency, or number of photons emitted from a light source (Example 1B.2).

☐ **3.** Use the photoelectric effect to calculate the work function of a metal (Example 1B.3).

☐ **4.** Estimate the wavelength of a particle (Example 1B.4).

☐ **5.** Estimate the uncertainty in the location or speed of a particle (Example 1B.5).

Topic 1B Exercises

1B.1 Consider the following statements about electromagnetic radiation and decide whether they are true or false. If they are false, correct them. (a) The total intensity of radiation emitted from a black body at absolute temperature T is directly proportional to the temperature. (b) As the temperature of a black body increases, the wavelength at which the maximum intensity is found decreases. (c) Photons of radiofrequency radiation are higher in energy than photons of ultraviolet radiation.

1B.2 Consider the following statements about electromagnetic radiation and decide whether they are true or false. If they are false, correct them. (a) Photons of ultraviolet radiation have less energy than photons of infrared radiation. (b) The kinetic energy of an electron ejected from a metal surface when the metal is irradiated with ultraviolet radiation is independent of the frequency of the radiation. (c) The energy of a photon is inversely proportional to the wavelength of the radiation.

1B.3 From the following list of observations, select the one that best supports the idea that electromagnetic radiation has the properties of particles. Explain your reasoning.
(a) Black-body radiation.
(b) Electron diffraction.
(c) Atomic spectra.
(d) The photoelectric effect.

1B.4 From the following list of observations, select the one that best supports the idea that particles have wave properties. Explain your reasoning.
(a) Scattering of α particles by gold foil.
(b) Electron diffraction.
(c) Cathode rays.
(d) The photoelectric effect.

1B.5 The γ-ray photons emitted by the nuclear decay of a technetium-99 atom used in radiopharmaceuticals have an energy of 140.511 keV. Calculate the wavelength of these γ-rays.

1B.6 Neon lights glow with orange light and they also emit radiation of wavelength 865 nm. Calculate the energy change resulting from the emission of 1.00 mol of photons at this wavelength.

1B.7 Sodium vapor lamps, used for public lighting, emit yellow light of wavelength 589 nm. How much energy is emitted by (a) an excited sodium atom when it generates a photon; (b) 5.00 mg of sodium atoms emitting light at this wavelength; (c) 1.00 mol of sodium atoms emitting light at this wavelength?

1B.8 When an electron beam strikes a block of copper, x-rays with a frequency of 1.2×10^{17} Hz are emitted. How much energy is emitted at this wavelength by (a) an excited copper atom when it generates an x-ray photon; (b) 2.00 mol of excited copper atoms; (c) 2.00 g of copper atoms?

1B.9 A lamp rated at 32 W (1 W = 1 J·s^{-1}) emits violet light of wavelength 420 nm. How many photons of violet light can the lamp generate in 2.0 s? How many moles of photons are emitted in that time interval?

1B.10 A lamp rated at 40. W (1 W = 1 J·s^{-1}) emits blue light of wavelength 470 nm. How many photons of blue light can the lamp generate in 2.0 s? How many moles of photons are emitted in that time interval?

1B.11 The star Antares emits light with maximum intensity at 850 nm. What is the temperature at the surface of Antares?

1B.12 The very hot star Spica has a surface temperature of 23 kK (2.3×10^4 K). At what wavelength does Spica emit the maximum intensity of light?

1B.13 The temperature of molten iron can be estimated by using Wien's law. If the melting point of iron is 1540 °C, what will be the wavelength (in nanometers) corresponding to maximum intensity when a piece of iron melts? In what region of the electromagnetic spectrum is this light?

1B.14 An astronomer discovers a new red star that emits light with maximum intensity at 632 nm. What is the temperature at the surface of the star?

1B.15 The velocity of an electron that is emitted from a metallic surface by a photon is 3.6×10^3 km·s^{-1}. (a) What is the wavelength of the ejected electron? (b) No electrons are emitted from the surface of the metal until the frequency of the radiation reaches 2.50×10^{16} Hz. How much energy is required to remove the electron from the metal surface? (c) What is the wavelength of the radiation that caused photoejection of the electron? (d) What kind of electromagnetic radiation was used?

1B.16 The work function for chromium metal is 4.37 eV. What wavelength of radiation must be used to eject electrons with a velocity of 1.5×10^3 km·s^{-1}?

1B.17 Who has the shorter wavelength when running at the same speed: a person weighing 60 kg or a person weighing 80 kg? Explain your reasoning.

1B.18 (a) Calculate the wavelength of a hydrogen atom traveling at 10. m·s^{-1}. (b) What would cause the wavelength of the atom to decrease, accelerating the atom to higher speeds or slowing it down? Explain your reasoning.

1B.19 Protons and neutrons have nearly the same mass. How different are their wavelengths? Calculate the wavelength of each particle when traveling at 2.75×10^5 m·s^{-1} in a particle accelerator and report the difference as a percentage of the wavelength of the neutron.

1B.20 What is the wavelength of an electron when the distance it travels in 1 s is equal to its wavelength?

1B.21 A baseball must weigh between 5.00 and 5.25 ounces (1 ounce = 28.3 g). What is the wavelength of a 5.15-ounce baseball thrown at 92 mph?

1B.22 A certain automobile of mass 1531 kg travels on a German autobahn at 175 km·h^{-1}. What is the wavelength of the automobile?

1B.23 What is the velocity of a neutron of wavelength 100. pm?

1B.24 The average speed of a helium atom at 25 °C is 1.23×10^3 m·s^{-1}. What is the average wavelength of a helium atom at this temperature?

1B.25 What is the minimum uncertainty in the speed of an electron confined within a lead atom of diameter 350. pm? Model the atom as a one-dimensional box with a length equal to the diameter of the actual atom.

1B.26 What is the minimum uncertainty in the position of a hydrogen atom in a particle accelerator given that its speed is known to within ±5.0 m·s^{-1}?

1B.27 A bowling ball of mass 8.00 kg is rolled down a bowling alley lane at 5.00 ± 5.0 m·s^{-1}. What is the minimum uncertainty in its position?

1B.28 The uncertainty principle has negligible consequences for macroscopic objects. However, the properties of nanoparticles, which have dimensions ranging from a few to several hundred nanometers, may be different from those of larger particles. (a) Calculate the minimum uncertainty in the speed of an electron confined in a nanoparticle with a diameter of 2.00×10^2 nm. (b) Calculate the minimum uncertainty in the speed of a mobile Li$^+$ ion confined to a nanoparticle of the same size. (c) Which could be specified more accurately in a nanoparticle, the speed of an electron or the speed of a Li$^+$ ion?

Topic 1C Wavefunctions and Energy Levels

1C.1 The Wavefunction and Its Interpretation
1C.2 The Quantization of Energy

Why is a new system of mechanics necessary?	What are the main principles of the new mechanics?	What do those principles reveal about the hydrogen atom?
Topic 1B: Quantum theory	Topic 1C: Wavefunctions and energy levels	Topic 1D: The hydrogen atom

The observation of the spectra of atomic hydrogen was an important landmark in science. But in order to account for it, scientists of the early twentieth century had to revise the nineteenth-century description of matter to take into account wave–particle duality. This change transformed the modern description and understanding of chemistry, and its consequences will be found throughout this book.

1C.1 The Wavefunction and Its Interpretation

Because particles have wavelike properties, they cannot be expected to behave like minute "point-like" objects moving along precise trajectories. The Austrian scientist Erwin Schrödinger (**FIG. 1C.1**) devised a new approach in 1927. He replaced the precise trajectory of a particle by a **wavefunction,** ψ (the Greek letter psi), a mathematical function with values that vary with position. Don't be put off by the thought that wavefunctions are mysteriously complicated mathematical expressions: some wavefunctions are very simple; shortly you will meet one that is simply $\sin x$.

The German physicist Max Born suggested how the wavefunction should be interpreted physically. The **Born interpretation** of the wavefunction is that *the probability of finding the particle in a region is proportional to the value of ψ^2 in that region* (**FIG. 1C.2**). To be precise, ψ^2 is a **probability density,** the probability that the particle will be found in a small region divided by the volume of the region. "Probability density" is the analog of the more familiar "mass density," the mass of a region divided by the volume of the region. To calculate the mass of a region, its mass density is multiplied by the volume of the region. Likewise, to calculate the probability that a particle is in a region, the probability density is multiplied by the volume of the region. For instance, if $\psi^2 = 0.1 \text{ pm}^{-3}$ at a point, then the probability of finding the particle in a region of volume 2 pm^3 located at that point would be $(0.1 \text{ pm}^{-3}) \times (2 \text{ pm}^3) = 0.2$, or 1 chance in 5. Wherever ψ^2 is large, the particle has a high probability density; wherever ψ^2 is small, the particle has only a low probability density.

A Note on Good Practice: Distinguish between *probability* and *probability density*: the former is unitless and lies between 0 (certainly not there) and 1 (certainly there), but the latter has the dimensions of 1/volume. To go from probability density to probability, multiply it by the volume of the region of interest.

Because the square of any real number is positive, you don't have to worry about ψ having a negative sign in some regions of space (as a function such as $\sin x$ has): probability density is never negative. Wherever ψ, and hence ψ^2, is zero, the particle has zero probability density. A location where ψ passes *through* zero (not just reaching zero) is called a **node** of the wavefunction; a particle has zero probability density wherever the wavefunction has nodes.

Schrödinger's great contribution, the **Schrödinger equation,** is used to calculate the wavefunction for any particle. Although the equation is not used directly in this text (you

Why Do You Need to Know This Material? Whenever you are dealing with quantum mechanics you have to consider the properties of wavefunctions and the information they contain.

What Do You Need to Know Already? This Topic makes use of the properties of sine functions ($\sin x$). It assumes that you are familiar with the concept of duality and the de Broglie relation between momentum and wavelength (Topic 1B).

If you go on in quantum mechanics, you will see that wavefunctions may be "complex" in the technical sense of involving $i = \sqrt{-1}$. We ignore that possibility here.

FIGURE 1C.1 Erwin Schrödinger (1887–1961). (© Bettmann/Corbis.)

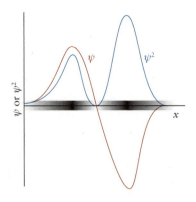

The Schrödinger equation is a "differential equation," an equation that relates the "derivatives" of a function (in this case, a second derivative of ψ, $d^2\psi/dx^2$) to the value of the function at each point. Derivatives are reviewed in Appendix 1F.

FIGURE 1C.2 The Born interpretation of the wavefunction. The probability density (the blue line) is given by the square of the wavefunction and depicted by the density of shading in the band beneath. Note that the probability density is zero at a node. A node is a point where the wavefunction (the orange line) passes through zero, not merely approaches zero.

will need to know only the form of some of its solutions, not how those solutions are found), it is of such central importance that it is appropriate at least to see what it looks like. For a particle of mass m moving in one dimension in a region where the potential energy is $V(x)$, the equation is

$$-\underbrace{\frac{\hbar^2}{2m}\frac{d^2\psi}{dx^2}}_{\text{kinetic energy}} + \underbrace{V(x)\psi}_{\substack{\text{potential} \\ \text{energy}}} = \underbrace{E\psi}_{\substack{\text{total} \\ \text{energy}}} \tag{1a}$$

The term $d^2\psi/dx^2$ can be thought of as a measure of how sharply the wavefunction is curved. The left-hand side of the Schrödinger equation is commonly written $H\psi$, where H is called the **hamiltonian** for the system:

$$\underbrace{-\frac{\hbar^2}{2m}\frac{d^2\psi}{dx^2} + V(x)\psi}_{H\psi} = E\psi$$

then the equation takes the deceptively simple form

$$H\psi = E\psi \tag{1b}$$

The Schrödinger equation is used to calculate both the wavefunction ψ and the corresponding energy E. To understand what is involved, consider one of the simplest systems, a single particle of mass m confined in a one-dimensional box between two rigid walls a distance L apart, a so-called **particle in a box** (**FIG. 1C.3**). The equation can be solved quite easily for this system and introduces a number of important concepts that recur throughout science. The idea to keep in mind is that a particle acts like a wave and only certain wavelengths can exist in the box, just as a stretched string can support only certain wavelengths. Think of a guitar string: because it is tied down at each end, it can support only shapes like the ones shown in Fig. 1C.3, which have zero displacement at each end. The shapes of the wavefunctions for the particle in the one-dimensional box are the same as the displacements of a vibrating string. Their mathematical form is

$$\psi_n(x) = \left(\frac{2}{L}\right)^{1/2}\sin\left(\frac{n\pi x}{L}\right) \quad n = 1, 2, \ldots \tag{2}$$

The integer n labels the wavefunctions and is called a "quantum number." In general, a **quantum number** is an integer (or, in some cases, as explained in Topic 1D, a half-integer such as ½) that labels a wavefunction, specifies a state, and can be used to calculate the value of a property of the system.

The probability density for a particle at a location is proportional to the square of the wavefunction at that point; the wavefunction is found by solving the Schrödinger equation for the particle.

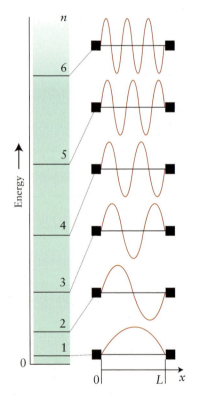

FIGURE 1C.3 The arrangement known as "a particle in a box," in which a particle of mass m is confined in one dimension between two impenetrable walls a distance L apart. The first six wavefunctions and their energies are shown. The numbers on the left are values of the quantum number n.

LIVING GRAPH FIGURE 1C.3

1C.2 The Quantization of Energy

An important feature of the particle in a box is that the particle has zero potential energy inside the box and infinite potential energy outside. The walls trap the particle inside the box. These constraints impose **boundary conditions** on the wavefunctions, values that it must have at certain locations. The boundary conditions for the particle in a box are that the wavefunctions must be zero at the walls and outside the box, and must be taken into account.

--- **How Is That Done?** ---

The kinetic energy of a particle of mass m is related to its speed, v, by $E_k = \frac{1}{2}mv^2$. This energy is related to the wavelength of the particle by noting that the linear momentum is $p = mv$ and then using the de Broglie relation (Eq. 5b of Topic 1B, $p = h/\lambda$):

$$E_k = \frac{1}{2}mv^2 = \frac{\overbrace{(mv)}^{p}{}^2}{2m} = \frac{\overbrace{(p)}^{h/\lambda}{}^2}{2m} = \frac{(h/\lambda)^2}{2m} = \frac{h^2}{2m\lambda^2}$$

The potential energy of the particle is taken to be zero everywhere inside the box, and so the total energy, E, is given by the expression for E_k alone.

At this point, you need to recognize that (like a guitar string) only whole-number multiples of half-wavelengths can fit into the box (see Fig. 1C.3; the waves have one bulge, two bulges, three bulges, and so forth, with each "bulge" a half-wavelength). That is, the wavelengths possible for a particle in a box of length L must meet the condition that:

$$L = \frac{1}{2}\lambda, \frac{2}{2}\lambda, \frac{3}{2}\lambda, \ldots = n \times \frac{1}{2}\lambda, \text{ with } n = 1, 2, \ldots$$

Therefore, the allowed wavelengths are

$$\lambda = \frac{2L}{n} \text{ with } n = 1, 2, \ldots$$

When this expression for λ is inserted into the expression for the energy, it becomes

$$E_n = \frac{h^2}{2m\lambda^2} = \frac{h^2}{2m(2L/n)^2} = \frac{n^2h^2}{8mL^2}$$

A subscript n has been attached to E as a reminder that the energy depends on the value of n.

> A more general method of finding the energy levels of a particle in a box is to use calculus to solve the Schrödinger equation: see the website for this book.

The calculation shows that the allowed energies of a particle of mass m in a one-dimensional box of length L are

$$E_n = \frac{n^2h^2}{8mL^2} \quad n = 1, 2, \ldots \tag{3}$$

What Does This Equation Tell You? Because the mass, m, of the particle appears in the denominator, for a given length of box, the energy levels lie at lower values for heavy particles than for light particles. Because the length of the box appears in the denominator, as the walls become more confining (L smaller), the energy levels are effectively squeezed upward.

A striking implication of Eq. 3 is that because n can take only integer values, the energy of the particle is **quantized,** or restricted to a series of discrete values, called **energy levels.** According to classical mechanics, an object can have any total energy—high, low, and anything in between. For example, a particle in a box could bounce from wall to wall with any speed and hence any kinetic energy. According to quantum mechanics, however, because only certain wavelengths fit into the box and each wavelength corresponds to a different energy, *energy is quantized.* The difference between the classical and quantum descriptions of energy is like the difference between the bulk and molecular descriptions of water: when water is poured from a bucket, the water seems to be a continuous fluid that can be transferred in any amount, no matter how small; however, the smallest amount of water that can be transferred is one H_2O molecule, one "quantum" of water.

As you saw in the derivation of Eq. 3, energy quantization stems from the boundary conditions on the wavefunction, the constraints that the wavefunction must satisfy at different points of space (such as fitting into a container correctly). You can now begin to see the origin of the quantized energy levels of an atom: *because an electron in an atom has a wavefunction that must satisfy certain constraints in three dimensions, only some solutions of the Schrödinger equation and their corresponding energies are acceptable.*

Equation 3 can be used to calculate the energy separation between two neighboring levels with quantum numbers n and $n + 1$:

$$E_{n+1} - E_n = \overbrace{\frac{(n+1)^2 h^2}{8mL^2}}^{\text{energy of level } n+1} - \overbrace{\frac{n^2 h^2}{8mL^2}}^{\text{energy of level } n}$$

$$= \{(n+1)^2 - n^2\}\frac{h^2}{8mL^2} = (2n+1)\frac{h^2}{8mL^2} \tag{4}$$

You can see that, as L (the length of the box) or m (the mass of the particle) increases, the separation between neighboring energy levels decreases (**FIG. 1C.4**). That is why no one noticed that energy is quantized until they investigated very small systems such as an electron in a hydrogen atom: the separation between levels is so small for ordinary particles in ordinary-sized vessels that it is completely undetectable. You can, in fact, ignore the quantization of the motion of the atoms of a gas in a typical flask. However, in very small systems, such as atoms or even nanoparticles, quantization is important (**BOX 1C.1**).

Box 1C.1 NANOCRYSTALS

Nanoscale particles of some semiconductor materials ranging from 1 to 100 nm in diameter are said to be *quantum confined*; that is, excited electrons in these materials behave like a particle in a box. Electrons that are excited by absorbing visible radiation are trapped inside the particle. When an electron returns to a lower energy state, a photon is emitted. The energy of the emitted photon depends on the size of the particle. The energy of a particle trapped in a one-dimensional box decreases as the size of the box increases. Extending this model to three dimensions gives analogous results: larger particles emit lower-energy photons.

Semiconductor materials that exhibit quantum confinement are called *quantum dots* or *nanocrystals* and are the foundation of the current interest in *nanotechnology*. Nanocrystals of CdSe that have been coated with a shell of ZnS are layered with a polymer and can be covalently linked to a wide variety of antibodies, forming a quantum dot–antibody conjugate (first illustration). When these conjugate materials are incubated with cells, the antibody binds tightly to very specific antigens located on proteins or other macromolecules of interest in the cell. The localization of these conjugates to specific regions allows scientists to know exactly where the antigens are located within the cell itself. Because the color of light emitted depends on the size of the quantum dot, it is possible to develop multicolor micrographs that highlight a variety of cell components by attaching different-size quantum dots to various antibodies. The second illustration shows a fluorescence micrograph of HeLa cells, a line of immortal human cervical cancer cells used in research that were cultured from Henrietta Lacks in 1951. These cells continue to be used in medical research today and were instrumental in the development of the polio vaccine.

Further Reading
- For a historical perspective of HeLa cells and their impact on medicine, see "*Henrietta Lacks' 'Immortal' Cells,*" Smithsonian.com, January 22, 2010, http://www.smithsonianmag.com/science-nature/Henrietta-Lacks-Immortal-Cells.html.

CdSe quantum dots with a shell of ZnS and polymer are linked to an antibody. The antibody binds to specific antigens within the cell.

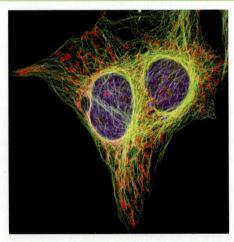

Multicolor fluorescence micrograph of HeLa cells showing the location of the cytoskeleton (yellow and green) and the cell nucleus (purple). (*Dr. Gopal Murti/Science Source.*)

EXAMPLE 1C.1 Calculating the energies of a particle in a box

Scientists often find it helpful to use very simple expressions to estimate the order of magnitude of a property without doing a detailed calculation. Treat a hydrogen atom as a one-dimensional box of length 150. pm (the approximate diameter of the atom) containing an electron and predict the wavelength of the radiation emitted when the electron falls to the lowest energy level from the next higher energy level.

ANTICIPATE According to Topic 1A, transitions in the hydrogen atom have wavelengths of the order of 100 nm, so expect a similar value.

PLAN The lowest energy level has $n = 1$, and so you can use Eq. 4 with $n = 1$ and $m = m_e$, the mass of the electron. The energy difference is carried away as a photon of radiation; so set the energy difference equal to $h\nu$ and express ν in terms of the corresponding wavelength by using Eq. 1 of Topic 1A ($\lambda = c/\nu$). The mass of the electron is given inside the back cover.

SOLVE

From Eq. 4 with $n = 1$, $2n + 1 = 3$,

$$E_2 - E_1 = \frac{3h^2}{8m_e L^2}$$

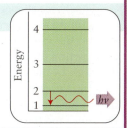

From $E_2 - E_1 = h\nu$,

$$h\nu = \frac{3h^2}{8m_e L^2} \quad \text{so} \quad \nu = \frac{3h}{8m_e L^2}$$

From $\lambda = c/\nu$,

$$\lambda = \frac{c}{3h/8m_e L^2} = \frac{8m_e c L^2}{3h}$$

24.7 nm

Now substitute the data:

$$\lambda = \frac{8 \times (9.109\,39 \times 10^{-31}\,\text{kg}) \times (2.998 \times 10^8\,\text{m·s}^{-1}) \times (1.50 \times 10^{-10}\,\text{m})^2}{3 \times (6.626 \times 10^{-34})\,\underbrace{\text{J·s}}_{\text{kg·m}^2\cdot\text{s}^{-2}}}$$

$$= 2.47 \times 10^{-8}\,\text{m (or 24.7 nm)}$$

A Note on Good Practice: Note how the complicated collection of units is treated: arriving at the correct units for the answer is a sign that you have set up the equation correctly. It is good practice to go as far as possible symbolically and then to insert numerical values at the last stage.

EVALUATE This result, 24.7 nm, is much shorter than the observed value, 122 nm, for this transition, but not completely out of line. The discrepancy is due to treating the atom in such a simple manner. An atom does not have the hard boundaries that confine a particle in a box, and is three-dimensional. The fact that the predicted wavelength has nearly the same order of magnitude as the actual value suggests that a quantum theory of the atom, based on a more realistic three-dimensional model, should give good agreement.

Self-test 1C.1A Use the same model for helium but suppose that the box is 100. pm long, because the atom is smaller. Estimate the wavelength of the same transition.

[*Answer:* 11.0 nm]

Self-test 1C.1B Use the same model for hydrogen and estimate the wavelength for the transition from the $n = 3$ energy level to the $n = 2$ level.

Related Exercises 1C.1, 1C.2

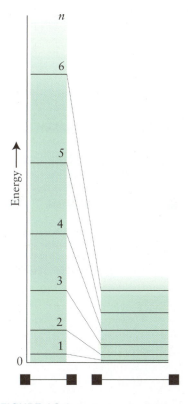

FIGURE 1C.4 As the length of the box increases (compare boxes on left and right), the energy levels fall and move closer together.

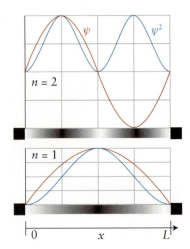

FIGURE 1C.5 The two lowest energy wavefunctions (ψ, orange) for a particle in a box and the corresponding probability densities (ψ^2, blue). The probability densities are also shown by the density of shading of the bands below each wavefunction.

Another surprising implication of Eq. 3 is that *a particle in a container cannot have zero energy*. Because the lowest allowed value of n is 1 (corresponding to a wave of one-half wavelength fitting into the box), the lowest energy is $E_1 = h^2/8mL^2$. This lowest possible energy is called the **zero-point energy.** The existence of a zero-point energy means that, according to quantum mechanics, a particle can never be perfectly still when it is confined between two walls: it must always possess an energy—in this case, a kinetic energy of at least $h^2/8mL^2$. This result is consistent with the uncertainty principle. When a particle is confined between two walls, the uncertainty in its position cannot be larger than the distance between the two walls. Because the position is not completely uncertain, the linear momentum must be uncertain, too, and so the particle cannot be said to be completely still and must therefore have some kinetic energy. The zero-point energy is a purely quantum mechanical phenomenon and is very small for macroscopic systems. For example, a billiard ball on a pool table has a completely negligible zero-point energy of about 10^{-67} J.

Finally, the shapes of the wavefunctions of a particle in a box also reveal some interesting information. Let's look at the two lowest energy wavefunctions, corresponding to $n = 1$ and $n = 2$. **FIGURE 1C.5** shows, by the density of shading, the likelihood of finding a particle (more formally, its probability density): you can see that when a particle is described by the wavefunction ψ_1 (and has energy $h^2/8mL^2$), then it is most likely to be found in the center of the box. Conversely, if the particle is described by the wavefunction ψ_2 (and has energy $h^2/2mL^2$), then it is most likely to be found in regions between the center and the walls and is unlikely to be found in the middle of the box. Remember that the wavefunction itself does not have any direct physical significance: you have to take the square of ψ before you can interpret it in terms of the probability of finding a particle somewhere.

> *When the Schrödinger equation is solved subject to the appropriate boundary conditions, it is found that the particle can possess only certain discrete energies.*

What have you learned in this Topic?

You now know that the location and properties of a particle are expressed by a wavefunction, the square of which expresses the probability (as a probability density) that the particle will be found in each region of space. You also know that a wavefunction is found by solving the Schrödinger equation and that one consequence of the wavefunction having to fit into a region of space is that a particle confined to a region can have only certain discrete energies.

The skills you have mastered are the ability to:

☐ **1.** Describe the origin and shapes of the wavefunctions of a particle in a box.

☐ **2.** Calculate the energies of a particle in a box (Example 1C.1) and explain how they depend on the length of the box and the mass of the particle.

☐ **3.** Explain what is meant by zero-point energy and account for its origin.

Topic 1C Exercises

Exercises labeled $\int_{dx}^{C}$ require calculus.

1C.1 (a) Using the particle-in-the-box model for the hydrogen atom and treating the atom as an electron in a one-dimensional box of length 150. pm, predict the wavelength of radiation emitted when the electron falls from the level with $n = 5$ to that with $n = 4$. (b) Repeat the calculation for the transition from $n = 4$ to $n = 3$.

1C.2 (a) Use the particle-in-the-box model for the helium atom and treat the electron as a particle in a one-dimensional box of length 100. pm to predict the wavelength of radiation emitted when the electron falls from the level with $n = 4$ to that with $n = 1$. (b) Repeat the calculation for the transition from $n = 4$ to $n = 2$.

1C.3 The energy levels of a particle of mass m in a two-dimensional square box of side L are given by $(n_1^2 + n_2^2)h^2/8mL^2$. Do any of these levels have the same energy? If so, find the values of the quantum numbers n_1 and n_2 for the first three cases.

1C.4 Refer to Exercise 1C.3. If one side of the box is twice that of the other, the energy levels are given by $(n_1^2/L_1^2 + n_2^2/L_2^2) \times h^2/8m$.

Can any of these levels have identical energies? If so, find the values of the quantum numbers n_1 and n_2 for the lowest levels having identical energies.

1C.5 (a) Use the Living Graphs on the Web site for this book to plot the particle-in-a-box wavefunction for $n = 2$ and $L = 1$ m. (b) How many nodes does the wavefunction have? Where do these nodes occur? (c) Repeat parts (a) and (b) for $n = 3$. (d) What general conclusion can you draw about the relation between n and the number of nodes present in a wavefunction? (e) Convert the $n = 2$ plot to a probability density distribution: at what values of x is it most likely to find the particle? (f) Repeat part (e) for $n = 3$.

1C.6 Verify the conclusion in part (d) of Exercise 1C.5 by plotting the wavefunction for $n = 4$ and determining the number of nodes.

1C.7 The wavefunction for a particle in a one-dimensional box is given in Eq. 2. Confirm that the probability of finding the particle in the left half of the box is $\frac{1}{2}$ regardless of the value of n.

1C.8 The wavefunction for a particle in a one-dimensional box is given in Eq. 2. Does the probability of finding the particle in the left-hand one-third of the box depend on n? If so, find the probability. *Hint:* The indefinite integral of $\sin^2 ax$ is $\frac{1}{2}x - (1/4a) \sin(2ax) + $ constant.

Topic 1D The Hydrogen Atom

What are the main principles of the new mechanics?

What do those principles reveal about the hydrogen atom?

How is the structure of the hydrogen atom extended to other atoms?

Topic **1C**: Wavefunctions and energy levels → Topic **1D**: The hydrogen atom → Topic **1E**: Many-electron atoms

Why Do You Need to Know This Material? The hydrogen atom is the simplest atom of all and is used to discuss the structures of all atoms. It is therefore central to many explanations in chemistry.

What Do You Need to Know Already? You need to be familiar with the nuclear model of the atom and the general layout of the periodic table (*Fundamentals* B). You also need to be familiar with the concepts of wavefunctions and energy levels (Topic 1C) that are implied by quantum mechanics.

The task of this Topic is to use the fact that an electron has wavelike properties and is described by a wavefunction to construct a quantum mechanical model of the atom consistent with experimental observations.

1D.1 Energy Levels

An electron in an atom is like a particle in a box (Topic 1C) in the sense that it is confined within the atom, not by walls but by the pull of the nucleus. The electron's wavefunctions can therefore be expected to obey certain boundary conditions, like the constraints encountered when fitting a wave between the walls of a one-dimensional box. As for a particle in a box, these constraints result in the quantization of energy and the existence of discrete energy levels. Even at this early stage, you can expect the electron to be confined to certain energies, just as the spectroscopic observations summarized in Topic 1A require.

> **THINKING POINT**
>
> The state of a particle in a box is defined by one quantum number. How many quantum numbers do you think will be needed to specify the wavefunction of an electron in a hydrogen atom?

To find the wavefunctions and energy levels of an electron in a hydrogen atom, it is necessary to solve the appropriate Schrödinger equation. To set up this equation, which allows for motion in three dimensions, the expression for the potential energy of an electron of charge $-e$ at a distance r from a nucleus of charge $+e$ is used. As explained in *Fundamentals* A, this "Coulomb" potential energy is

The *vacuum permittivity*, ε_0, is a fundamental constant. Its value can be found inside the back cover.

$$V(r) = \frac{\overbrace{(-e)}^{\text{charge of electron}} \times \overbrace{(+e)}^{\text{charge of proton}}}{4\pi\varepsilon_0 \underbrace{r}_{\substack{\text{distance of electron} \\ \text{from the proton}}}} = -\frac{e^2}{4\pi\varepsilon_0 r} \tag{1}$$

Solving the Schrödinger equation for a particle with this potential energy is difficult, but Schrödinger himself achieved it in 1927. He found that the allowed energy levels for an electron in a hydrogen atom are

$$E_n = -\frac{h\mathcal{R}}{n^2} \quad \text{with} \quad \mathcal{R} = \frac{m_e e^4}{8h^3\varepsilon_0^2} \quad n = 1, 2, \ldots \tag{2a}$$

These energy levels have exactly the form suggested spectroscopically (Topic 1A), but now the Rydberg constant, $\mathcal{R}$, is related to a number of fundamental constants. In science, it is always satisfying to realize that an experimental quantity is related to a combination of fundamental constants. When the fundamental constants are inserted, the value obtained is $\mathcal{R} = 3.29 \times 10^{15}$ Hz, the same as the experimental value. This agreement is a triumph for Schrödinger's theory and for quantum mechanics; it is easy to understand the thrill that Schrödinger must have felt when he arrived at this result. A very similar expression applies to other one-electron ions, such as He^+ and even C^{5+}, with atomic number Z:

$$E_n = -\frac{Z^2 h\mathcal{R}}{n^2} \quad n = 1, 2, \ldots \tag{2b}$$

What Does This Equation Tell You? First, note that all the energies are negative, meaning that the electron has a lower energy in the atom than when it is far from the nucleus. Next, note that the n is a quantum number, like the quantum number that occurs for a particle in a box, and the fact that it takes only integer values means that the energies have only discrete values. Now note that because n appears in the denominator, as n increases, the energies of successive levels increase (become less negative) and approach zero, at which point the electron is on the point of escaping from the atom. Finally, note that because Z appears in the numerator of Eq. 2b, the greater the value of the nuclear charge, the more tightly the electron is bound to a nucleus.

A skill to develop is to think about why a property depends on various parameters in a particular way. In this case you might wonder why the energy depends on Z^2 rather than on Z itself. The reason can be traced to two cooperating factors. First, a nucleus of atomic number Z and charge Ze gives rise to a field that is Z times stronger than that of a single proton. Second, the electron is drawn in by the higher charge and is Z times closer to the nucleus than it is in hydrogen. The two factors work together to give an overall energy lowering that is proportional to Z^2.

FIGURE 1D.1 shows the energy levels calculated from Eq. 2a. All the energies are negative, because they are measured relative to the energy of the free electron. Each level is labeled by the integer n, which is called the **principal quantum number,** from $n = 1$ for the first (lowest, most negative) level, $n = 2$ for the next higher energy level, continuing to infinity, for the highest energy level of zero, when the electron is on the point of escaping from the nucleus.

The lowest, most negative, energy possible for an electron in a hydrogen atom is obtained when $n = 1$ and is $-h\mathcal{R}$. This state of lowest energy is called the **ground state** of the atom. A hydrogen atom is normally found in its ground state, with its electron in the level with $n = 1$. When the bound electron is excited by absorbing a photon or by being bombarded by other particles, its energy is increased to a level with a higher value of n. It reaches $E = 0$, when n reaches infinity. At that point, the electron has effectively left the atom in the process called **ionization.** The **ionization energy** itself, which is discussed in more detail in Topic 1F, is the minimum energy needed to achieve ionization starting from the ground state. Any further supply of energy beyond the ionization energy simply adds to the kinetic energy of the liberated electron.

The energy levels of a hydrogen atom are defined by the principal quantum number, $n = 1, 2, \ldots$, and form a converging ladder, as shown in Fig. 1D.1.

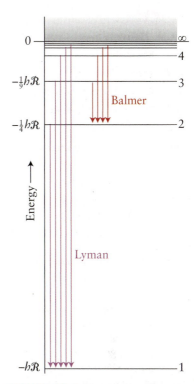

FIGURE 1D.1 The permitted energy levels of a hydrogen atom as calculated from Eq. 2a. The levels are labeled with the quantum number n, which ranges from 1 (for the lowest state) to infinity (for the separated, stationary proton and electron). Notice how the energy levels become closer together as n increases.

1D.2 Atomic Orbitals

The wavefunctions of electrons in atoms are called **atomic orbitals.** The name was chosen to suggest something less definite than an "orbit" of an electron around a nucleus and to take into account the wave nature of the electron. The mathematical expressions for atomic orbitals—which are obtained as solutions of the Schrödinger equation—are more complicated than the sine functions for the particle in a box described in Topic 1C, but their essential features are straightforward. Moreover, you should never lose sight of their interpretation, that the *square* of a wavefunction tells you the probability density of an electron at each point. To visualize this probability density, imagine a cloud centered on the nucleus. The density of the cloud at each point represents the probability of finding an electron there. The densest regions of the cloud represent locations where the electron is most likely to be found.

To write down the form of an atomic orbital, it is necessary to have a way to specify the location of each point around a nucleus, and then to give the value of the wavefunction at that point. Because an atom is like a little sphere, it is most convenient to describe

Geographical latitudes are measured from the equator, not the north pole. The technical name for θ is the colatitude.

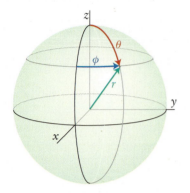

FIGURE 1D.2 The spherical polar coordinates: r is the radius, which gives the distance from the center, θ is the colatitude, which gives the angle from the z-axis, and ϕ, the "longitude," is the azimuth, which gives the angle from the x-axis.

these locations in terms of **spherical polar coordinates,** in which each point is labeled with three coordinates r, θ, and ϕ, where

- r is the distance from the nucleus;
- θ (theta) is the angle from the positive z-axis (the "north pole"), which can be thought of as playing the role of the geographical "latitude";
- ϕ (phi) is the angle about the z-axis, the geographical "longitude."

These coordinates are shown in **FIG. 1D.2**.

Each wavefunction, which in general varies with position, depends on the values of the three coordinates and is therefore denoted $\psi(r,\theta,\phi)$. It turns out, however, that all the wavefunctions can be written as the product of a function that depends only on r and another function that depends only on the angles θ and ϕ. That is,

$$\psi(r,\theta,\phi) = \underbrace{R(r)}_{\substack{\text{radial} \\ \text{wavefunction}}} \times \underbrace{Y(\theta,\phi)}_{\substack{\text{angular wave-} \\ \text{function}}} \tag{3}$$

The function $R(r)$ is called the **radial wavefunction;** it expresses how the wavefunction varies on moving away from the nucleus. The function $Y(\theta,\phi)$ is called the **angular wavefunction;** it expresses how the wavefunction varies as the angles θ and ϕ change.

The expressions for a number of atomic orbitals are shown in **TABLE 1D.1a** (for R) and Table 1D.1b (for Y). The full form of the wavefunctions can look quite fearsome at

TABLE 1D.1   Hydrogenlike Wavefunctions* (Atomic Orbitals), $\psi = RY$

(a) Radial wavefunctions			(b) Angular wavefunctions		
n	l	$R_{nl}(r)^{\dagger}$	l	"m_l"‡	$Y_{lm_l}(\theta,\phi)$
1	0	$2\left(\dfrac{Z}{a_0}\right)^{3/2} e^{-Zr/a_0}$	0	0	$\left(\dfrac{1}{4\pi}\right)^{1/2}$
2	0	$\dfrac{1}{2\sqrt{2}}\left(\dfrac{Z}{a_0}\right)^{3/2}\left(2-\dfrac{Zr}{a_0}\right)e^{-Zr/2a_0}$	1	x	$\left(\dfrac{3}{4\pi}\right)^{1/2}\sin\theta\cos\phi$
2	1	$\dfrac{1}{2\sqrt{6}}\left(\dfrac{Z}{a_0}\right)^{3/2}\left(\dfrac{Zr}{a_0}\right)e^{-Zr/2a_0}$	1	y	$\left(\dfrac{3}{4\pi}\right)^{1/2}\sin\theta\sin\phi$
3	0	$\dfrac{2}{9\sqrt{3}}\left(\dfrac{Z}{a_0}\right)^{3/2}\left(3-\dfrac{2Zr}{a_0}+\dfrac{2Z^2r^2}{9a_0^2}\right)e^{-Zr/3a_0}$	1	z	$\left(\dfrac{3}{4\pi}\right)^{1/2}\cos\theta$
3	1	$\dfrac{2}{9\sqrt{6}}\left(\dfrac{Z}{a_0}\right)^{3/2}\left(\dfrac{Zr}{a_0}\right)\left(2-\dfrac{Zr}{3a_0}\right)e^{-Zr/3a_0}$	2	xy	$\left(\dfrac{15}{16\pi}\right)^{1/2}\sin^2\theta\sin 2\phi$
3	2	$\dfrac{4}{81\sqrt{30}}\left(\dfrac{Z}{a_0}\right)^{3/2}\left(\dfrac{Zr}{a_0}\right)^2 e^{-Zr/3a_0}$	2	yz	$\left(\dfrac{15}{4\pi}\right)^{1/2}\cos\theta\sin\theta\sin\phi$
			2	zx	$\left(\dfrac{15}{4\pi}\right)^{1/2}\cos\theta\sin\theta\cos\phi$
			2	x^2-y^2	$\left(\dfrac{15}{16\pi}\right)^{1/2}\sin^2\theta\cos 2\phi$
			2	z^2	$\left(\dfrac{5}{16\pi}\right)^{1/2}(3\cos^2\theta-1)$

*For example, a $2p_x$ orbital ($n = 2$, $l = 1$, "m_l" = x, of hydrogen ($Z = 1$) is

$$\psi(r,\theta,\phi) = R_{2,1}(r)\,Y_{1,x}(\theta,\phi) = \frac{1}{2\sqrt{6}}\left(\frac{1}{a_0}\right)^{3/2}\frac{r}{a_0}e^{-r/2a_0} \times \left(\frac{3}{4\pi}\right)^{1/2}\sin\theta\cos\phi$$

$$= \frac{1}{(32\pi a_0^5)^{1/2}}re^{-r/2a_0}\sin\theta\cos\phi$$

and $1/(32\pi a_0^5)^{1/2} = 4.9 \times 10^{-6}\ \text{pm}^{-5/2}$.

† In each case, $a_0 = 4\pi\varepsilon_0\hbar^2/m_e e^2$, which has a value close to 52.9 pm; for hydrogen itself, $Z = 1$.
‡ In all cases except $m_l = 0$, the orbitals designated as x, y, etc. are sums and differences (linear combinations) of orbitals with equal but opposite values of m_l (such as +1 and −1).

first glance, but in fact their form is reasonably simple (and certainly need not be remembered in detail). For instance, the wavefunction corresponding to the ground state of the hydrogen atom ($n = 1$) is

$$\psi(r,\theta,\phi) = \left(\frac{1}{\pi a_0^3}\right)^{1/2} e^{-r/a_0}$$

The quantity a_0 is called the **Bohr radius** and has the value 52.9 pm. In this case, the wavefunction is **spherically symmetric,** which means it is independent of the angles θ and ϕ and that for a given radius, its value is the same in all directions. The wavefunction decays exponentially toward zero as r increases, which means that the probability density is highest close to the nucleus (at $r = 0$, using $e^0 = 1$). The key point to remember is that the ground-state wavefunction falls off exponentially with distance from the nucleus.

In Bohr's model of the lowest energy state of the hydrogen atom, the electron was viewed as traveling in a circular orbit of radius a_0 around the nucleus.

> *The distribution of an electron in an atom is described by a wavefunction known as an atomic orbital.*

1D.3 Quantum Numbers, Shells, and Subshells

When the Schrödinger equation is solved for a three-dimensional atom, it turns out that *three* quantum numbers are needed to label each wavefunction. The three quantum numbers are designated n, l, and m_l:

- n is related to the *size* and *energy* of the orbital.
- l is related to its *shape.*
- m_l is related to its *orientation* in space.

You have already encountered n, the principal quantum number, which specifies the energy of the orbital in a one-electron atom (see Eq. 2). In a one-electron atom, all atomic orbitals with the same value of the principal quantum number n have the same energy and are said to belong to the same **shell** of the atom. This name reflects the fact that as n increases, the region of greatest probability density is like a nearly hollow shell of increasing radius. The mean distance of an electron from the nucleus increases with the value of n.

The second quantum number needed to specify an orbital is l, the **orbital angular momentum quantum number.** This quantum number can take the values

$$l = 0, 1, 2, \ldots, n - 1$$

There are n different values of l for a given value of n. For instance, when $n = 3$, l can have any of the three values 0, 1, and 2. The orbitals of a shell with principal quantum number n therefore fall into n **subshells,** groups of orbitals that have the same value of l. There is only one subshell in the $n = 1$ level ($l = 0$), two in the $n = 2$ level ($l = 0$ and 1), three in the $n = 3$ level ($l = 0, 1,$ and 2), and so on. All orbitals with $l = 0$ are called **s-orbitals,** those with $l = 1$ are called **p-orbitals,** those with $l = 2$ are called **d-orbitals,** and those with $l = 3$ are called **f-orbitals:**

The labels come from the fact that spectroscopic lines were once classified as sharp, principal, diffuse, and fundamental.

Value of l	0	1	2	3
Orbital type	s	p	d	f

Although higher values of l (corresponding to g-, h-, . . . orbitals) are possible, the lower values of l (0, 1, 2, and 3) are the only ones that chemists need in practice.

Just as the value of n can be used to calculate the energy of an electron, the value of l can be used to calculate another physical property. As its name suggests, l tells us the **orbital angular momentum** of the electron, a measure of the rate at which (in classical terms) the electron circulates round the nucleus:

$$\text{Orbital angular momentum} = \{l(l + 1)\}^{1/2}\, \underset{h/2\pi}{\hbar} \tag{4}$$

TABLE 1D.2   Quantum Numbers for Electrons in Atoms

Name	Symbol	Values	Specifies	Indicates
principal	n	$1, 2, \ldots$	shell	size
orbital angular momentum*	l	$0, 1, \ldots, n-1$	subshell: $l = 0, 1, 2, 3, 4, \ldots$ s, p, d, f, g, $\ldots$	shape
magnetic	m_l	$l, l-1, \ldots, -l$	orbitals of subshell	orientation
spin magnetic	m_s	$+\frac{1}{2}, -\frac{1}{2}$	spin state	spin direction

*Also called the azimuthal quantum number.

An electron in an s-orbital (an "s-electron"), for which $l = 0$, has zero orbital angular momentum. That means that you should imagine it not as circulating around the nucleus but simply as distributed spherically around it. An electron in a p-orbital ($l = 1$) has nonzero orbital angular momentum (of magnitude $2^{1/2}\hbar$), so unlike an s-electron it can be thought of as circulating around the nucleus. An electron in a d-orbital ($l = 2$) has a higher orbital angular momentum ($6^{1/2}\hbar$), one in an f-orbital ($l = 3$) has an even higher angular momentum ($12^{1/2}\hbar$), and so on.

An important feature of the hydrogen atom (but not for atoms with more than one electron) is that all the orbitals of a given shell have the same energy, regardless of the value of their orbital angular momentum (you can see from Eq. 2 that l does not appear in the expression for the energy). The orbitals of a shell in a hydrogen atom are said to be **degenerate,** which means that they all have the same energy. This degeneracy of orbitals with the same value of n but different values of l is true only of the hydrogen atom and one-electron ions (such as He^+ and C^{5+}).

The third quantum number required to specify an orbital is m_l, the **magnetic quantum number,** which distinguishes the individual orbitals within a subshell. This quantum number can take the values

$$m_l = l, l - 1, \ldots, -l$$

There are $2l + 1$ different values of m_l for a given value of l and therefore $2l + 1$ orbitals in a subshell of quantum number l. For example, for a p-orbital, $l = 1$ and $m_l = +1, 0, -1$; so there are three p-orbitals in a given shell. Alternatively, a subshell with $l = 1$ can be said to consist of three orbitals. Orbitals with the same value of l but different values of m_l are degenerate even in many-electron atoms. Thus, the three p-orbitals of a shell are degenerate both in hydrogen and in many-electron atoms.

A Note on Good Practice: Because the magnetic quantum number can have a positive or negative value, always write the $+$ sign explicitly for positive values of m_l; for instance, write $m_l = +1$, not $m_l = 1$.

The magnetic quantum number specifies the orientation of the orbital motion of the electron. Specifically, it implies that the orbital angular momentum around an arbitrary axis is equal to $m_l\hbar$, the rest of the orbital motion (to make up the full amount of $\{l(l+1)\}^{1/2}\hbar$) being around other axes. For instance, if $m_l = +1$, then the orbital angular momentum of the electron around the arbitrary axis is $+\hbar$, whereas, if $m_l = -1$, then the orbital angular momentum of the electron around the same arbitrary axis is $-\hbar$. The difference in sign simply means that the direction of motion is opposite, the electron in one state circulating clockwise around the chosen axis and an electron in the other state circulating counterclockwise. If $m_l = 0$, then the electron is not circulating around the selected axis but, at a given radius, is evenly distributed around it. The shells and subshells are organized as shown in **FIG. 1D.3** and **TABLE 1D.2.**

Atomic orbitals are designated by the quantum numbers n, l, and m_l and fall into shells and subshells.

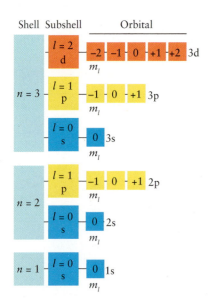

FIGURE 1D.3 A summary of the arrangement of shells, subshells, and orbitals in an atom and the corresponding quantum numbers. Note that the quantum number m_l is an alternative label for the individual orbitals: in chemistry, it is more common to use x, y, and z instead, as shown in Figs. 1D.9 through 1D.11.

1D.4 The Shapes of Orbitals

Each possible combination of the three quantum numbers specifies an individual orbital and acts as the "address" of the electron that "occupies" it, that is, has a distribution given by that wavefunction. For example, an electron in the ground state of a hydrogen atom has $n = 1$, $l = 0$, $m_l = 0$. Because $l = 0$, the ground state wavefunction is an example of an s-orbital and is denoted 1s. Each shell has one s-orbital, and the s-orbital in the shell with quantum number n is called an **ns-orbital.**

All s-orbitals are independent of the angles θ and ϕ, so are said to be spherically symmetrical (**FIG. 1D.4**). The probability density of an electron at the point (r, θ, ϕ) when it is in a 1s-orbital is given by the square of the corresponding wavefunction (which was given earlier):

$$\psi^2(r,\theta,\phi) = \frac{1}{\pi a_0^3}e^{-2r/a_0} \tag{5}$$

In this case, the probability density is independent of angle and, for simplicity, is commonly written simply $\psi^2(r)$. In principle, the cloud representing the probability density never thins to exactly zero, no matter how large the value of r. So you could think of an atom as being bigger than the Earth! However, there is virtually no chance of finding an electron farther from the nucleus than about 250 pm, and so atoms are for all practical purposes very small. As you can see from the high density of the cloud at the nucleus in Fig. 1D.4, an electron in an s-orbital has a nonzero probability of being found right at the nucleus: because $l = 0$, there is no orbital angular momentum to fling the electron away.

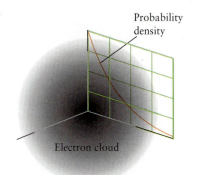

Probability density

Electron cloud

FIGURE 1D.4 The three-dimensional electron cloud corresponding to an electron in a 1s-orbital of hydrogen. The density of shading represents the probability of finding the electron at any point. The superimposed graph shows how the probability density varies with the distance of the point from the nucleus along any radius.

THINKING POINT

How would you expect a 1s-orbital in He$^+$ to differ from the same orbital in H?

EXAMPLE 1D.1   Calculating the probability of finding an electron at a certain location

Knowledge of the probable locations of electrons in atoms (and in molecules) is fundamental to understanding their properties, and it is important to be able to interpret their wavefunctions. Suppose the electron is in a 1s-orbital of a hydrogen atom. What is the probability of finding the electron in a small region a distance a_0 from the nucleus relative to the probability of finding it in the same small region located right at the nucleus?

ANTICIPATE You should expect a lower probability because the wavefunction decays exponentially with distance from the nucleus.

PLAN Compare the probability densities at the two locations. To do that, take the ratio of the squares of the wavefunction at the two locations. Because the probability density is independent of angle when $l = 0$, as remarked in the text, write the wavefunction simply as $\psi(r)$ rather than as the more complete expression $\psi(r,\theta,\phi)$.

SOLVE

The ratio of the probability that the electron is found at the nucleus or at $r = a_0$ is:

$$\frac{\text{Probability density at } r = a_0}{\text{Probability density at } r = 0} = \frac{\psi^2(a_0)}{\psi^2(0)}$$

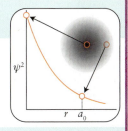

From $\psi^2(r) = (1/\pi a_0^3)e^{-2r/a_0}$, noting that the $(1/\pi a_0^3)$ cancel,

$$\frac{\psi^2(a_0)}{\psi^2(0)} = \frac{\overbrace{e^{-2a_0/a_0}}^{e^{-2}}}{\underbrace{e^0}_{1}} = e^{-2} = 0.14$$

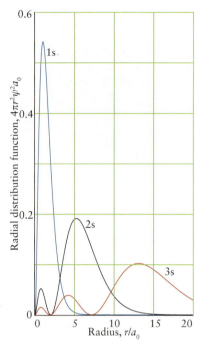

FIGURE 1D.5 The radial distribution function tells us the probability density for finding an electron at a given radius summed over all directions. The graph shows the radial distribution function for the 1s-, 2s-, and 3s-orbitals in hydrogen. Note how the most probable radius (corresponding to the greatest maximum) increases as n increases.

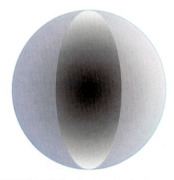

FIGURE 1D.6 The 1s-orbital. The simplest way of drawing an atomic orbital is as a boundary surface, a surface within which there is a high probability (typically 90%) of finding the electron. We shall use blue to denote s-orbitals, but that color is only an aid to their identification. The shading within the boundary surface is an approximate indication of the electron density at each point. The darker the shading, the higher the probability of finding the electron at that distance from the nucleus.

ANIMATION FIGURE 1D.6

EVALUATE As expected, the probability of finding the electron in a small region at a distance a_0 from the nucleus is lower than at the nucleus itself: the probability is only 14% of that of finding the electron in a region of the same volume located at the nucleus.

Self-test 1D.1A Calculate the same ratio but for the more distant point at $r = 2a_0$, twice as far from the nucleus.

[**Answer:** 0.018]

Self-test 1D.1B Calculate the same ratio but for a point at $3a_0$ from the nucleus.

Related Exercises 1D.3, 1D.4

The value of ψ^2 lets you predict the probability of finding the electron in a given region at a distance r from the nucleus; but suppose you want to know the total probability of finding the electron at a distance r at all possible orientations. To calculate this probability you need to use the **radial distribution function,** P. Specifically, the probability that the electron will be found anywhere in a thin shell around the nucleus, with radius r and thickness δr is given by $P(r)\delta r$ (**FIG. 1D.5**), with

$$P(r) = r^2 R^2(r) \qquad \text{(6a)}$$

For s-orbitals, $\psi = RY = R/2\pi^{1/2}$, so $R^2 = 4\pi\psi^2$, and this expression is then the same as

$$P(r) = 4\pi r^2 \psi^2(r) \qquad \text{(6b)}$$

This is the form that you will normally see; however, it applies only to s-orbitals, whereas Eq. 6a applies to any kind of orbital.

It is important to distinguish the radial distribution function from the wavefunction and its square, the probability density:

- The wavefunction itself tells us, through $\psi^2(r,\theta,\phi)\delta V$, the probability of finding the electron in the small volume δV at a particular location specified by r, θ, and ϕ.
- The radial distribution function tells us, through $P(r)\delta r$, the probability of finding the electron between r and $r + \delta r$ summed over all values of θ and ϕ.

The radial distribution function for the population of the Earth, for instance, is zero up to about 6400 km from the center of the Earth, rises sharply, and then falls back to almost zero (the "almost" takes into account the small number of people who are on mountains or flying in airplanes).

Note that for *all* orbitals, not just s-orbitals, P is zero at the nucleus simply because the region in which the electron is being sought has shrunk to zero size. (The probability density for an s-orbital is nonzero at the nucleus, but here it is being multiplied by a volume, $4\pi r^2 \delta r$, which becomes zero at the nucleus, at $r = 0$.) As r increases, the value of $4\pi r^2$ increases (the shell is getting bigger), but, for a 1s-orbital, the square of the wavefunction, ψ^2, falls toward zero as r increases; as a result, the product of $4\pi r^2$ and ψ^2 starts off at zero, goes through a maximum, and then declines to zero. The value of P is a maximum at a_0, the Bohr radius. Therefore, the Bohr radius corresponds to the radius at which an electron in a 1s-orbital in a hydrogen atom is most likely to be found.

THINKING POINT

How will the radial distribution function for an electron in He$^+$ differ from that in H itself?

Instead of drawing the s-orbital as a cloud, chemists usually draw a **boundary surface,** a smooth surface that encloses most of the cloud. However, although the boundary surface is easier to draw, it does not give the best picture of an atom; an atom has indefinite, or "fuzzy" edges and is not as smooth as the boundary surface might suggest. Despite this limitation, the boundary surface is useful because an electron is likely to be found only inside it. Keep in mind that the probability density inside the boundary surface is not uniform. An s-orbital has a spherical boundary surface (**FIG. 1D.6**) because the electron cloud is spherical. s-Orbitals with higher energies have spherical boundary surfaces of greater diameter. They also have a more complicated radial variation, with radial nodes at locations that can be found by examining the wavefunctions (**FIG. 1D.7**).

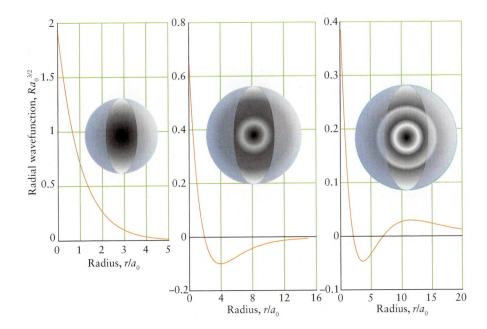

FIGURE 1D.7 The radial wavefunctions of the first three s-orbitals of a hydrogen atom. Note that the number of radial nodes increases (as $n - 1$), as does the average distance of the electron from the nucleus (compare with Fig. 1D.5). Because the probability density is given by ψ^2, all s-orbitals have a nonzero probability density at the nucleus.

LIVING GRAPH FIGURE 1D.7

The boundary surface of a p-orbital has two lobes (**FIG. 1D.8**). These lobes are marked + and − to signify that the wavefunction has different signs in the two regions. For instance, the $2p_z$-orbital is proportional to $\cos\theta$, and as θ increases from 0 to π on traveling around the atom from "pole" to "pole," $\cos\theta$ changes from +1 (at $\theta = 0$, the "north pole") through 0 (at the equator, when $\theta = \pi/2$) to −1 (at the "south pole", when $\theta = \pi$). Thus, in one hemisphere the wavefunction is positive and in the other it is negative, as depicted on the boundary surface in Fig. 1D.8. The two lobes of a p-orbital are separated by a **nodal plane,** which cuts through the nucleus and on which $\psi = 0$: the wavefunction changes sign on passing through this nodal plane. A **p-electron,** an electron in a p-orbital, will never be found at the nucleus because the wavefunction is zero there. This difference from s-orbitals occurs because an electron in a p-orbital has non-zero orbital angular momentum, which flings it away from the nucleus.

THINKING POINT

Does a 2p-orbital have a radial node? *Hint*: Think carefully about the definition of node in Topic 1C.

There are three p-orbitals in each subshell, corresponding to the quantum numbers $m_l = +1, 0, -1$. However, chemists commonly refer to the orbitals according to the axes along which the lobes lie; hence, they refer to p_x-, p_y- and p_z-orbitals (**FIG. 1D.9**).

A subshell with $l = 2$ consists of five d-orbitals. Each d-orbital has four lobes, except for the orbital designated d_{z^2}, which has a more complicated shape (**FIG. 1D.10**). A subshell with $l = 3$ consists of seven f-orbitals with even more complicated shapes (**FIG. 1D.11**).

As described in Topic 1C, as the value of n for a particle in a box is increased, the number of nodes in the wavefunction also increases: each wavefunction has $n - 1$ nodes. The same is also true for atomic orbitals, which also have $n - 1$ nodes: an orbital with quantum numbers n and l has l angular nodes and $n - l - 1$ radial nodes. Additional nodes appear as the principal quantum number for a given value of l is increased.

The total number of orbitals in a shell with principal quantum number n is n^2. To confirm this rule, recall that l has n integer values from 0 to $n - 1$ and that the number of orbitals in a subshell for a given value of l is $2l + 1$. For instance, for $n = 4$, there are

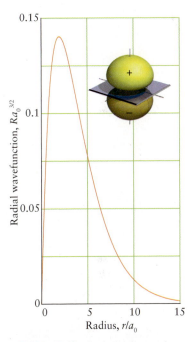

FIGURE 1D.8 The boundary surface and the radial variation of a 2p-orbital along the (vertical) z-axis. All p-orbitals have boundary surfaces with similar shapes, including one nodal plane. Note that the wavefunction has opposite signs (as depicted by the intensity of the color) on each side of the nodal plane.

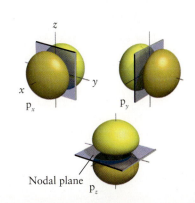

FIGURE 1D.9 There are three p-orbitals of a given energy, and they lie along three perpendicular axes. We shall use yellow to indicate p-orbitals: dark and light shades denote different signs of the wavefunction.

ANIMATION FIGURE 1D.9

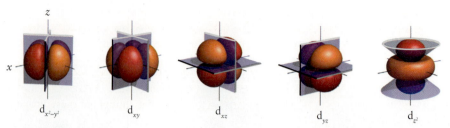

FIGURE 1D.10 The boundary surface of a d-orbital is more complicated than that of an s- or a p-orbital. There are, in fact, five d-orbitals of a given energy; four of them have four lobes, one is slightly different. In each case, an electron that occupies a d-orbital will not be found at the nucleus. We shall use orange to indicate d-orbitals: dark and light shades denote different signs of the wavefunction. The blue surfaces represent nodal planes.

ANIMATION FIGURE 1D.10

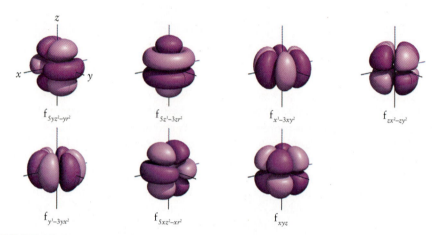

FIGURE 1D.11 The seven f-orbitals ($l = 3$) of a shell have a very complex appearance. Their detailed form will not be used again in this text. However, their existence is important for understanding the periodic table, the presence of the lanthanoids and actinoids (the "lanthanides" and "actinides"), and the properties of the later d-block elements. Dark and light shades denote different signs of the wavefunction.

four subshells with $l = 0, 1, 2,$ and 3, consisting of one s-orbital, three p-orbitals, five d-orbitals, and seven f-orbitals, respectively. There are therefore $1 + 3 + 5 + 7 = 16$, or 4^2, orbitals in the shell with $n = 4$ (**FIG. 1D.12**).

The shape of an atomic orbital is depicted by its boundary surface, and the average distance of an electron from the nucleus is calculated from the radial distribution function.

1D.5 Electron Spin

Schrödinger's calculation of the energies of the hydrogen orbitals was a milestone in the development of modern atomic theory. Yet the observed spectral lines did not have exactly the frequencies he predicted. In 1925 (before Schrödinger's work but after Bohr had developed an earlier model of the atom), two Dutch–American physicists, Samuel Goudsmit and George Uhlenbeck, proposed an explanation for the tiny deviations that had been observed. They suggested that an electron behaves in some respects like a spinning sphere, something like a planet rotating on its axis. This property is called **spin.** Schrödinger's theory did not account for spin, but it emerged naturally when the British physicist Paul Dirac found a way (in 1928) to combine Einstein's theory of relativity with Schrödinger's approach.

According to quantum mechanics, an electron has two spin states, represented by the arrows ↑ (up) and ↓ (down) or the Greek letters α (alpha) and β (beta). You can think of an electron as being able to spin counterclockwise at a certain rate (the ↑ state) or clockwise at exactly the same rate (the ↓ state). These two spin states are distinguished

4f	−3	−2	−1	0	+1	+2	+3
4d		−2	−1	0	+1	+2	
4p			−1	0	+1		
4s				0			16

FIGURE 1D.12 There are 16 orbitals in the shell with $n = 4$, each of which can hold two electrons (see Section 1D.6), for a total of 32 electrons.

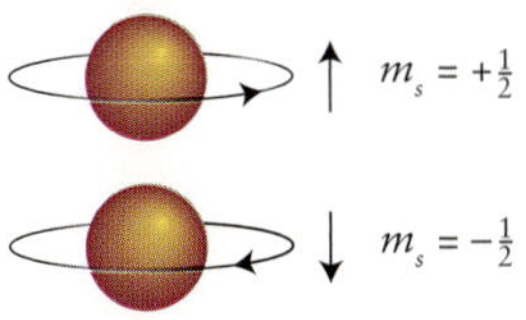

FIGURE 1D.13 The two spin states of an electron can be represented as clockwise or counterclockwise rotation around an axis passing through the electron. The two states are identified by the quantum number m_s and depicted by the arrows shown on the right.

$\uparrow \quad m_s = +\frac{1}{2}$

$\downarrow \quad m_s = -\frac{1}{2}$

by a fourth quantum number, the **spin magnetic quantum number,** m_s. This quantum number can have only two values: $+\frac{1}{2}$ indicates an $\uparrow$ electron and $-\frac{1}{2}$ indicates a $\downarrow$ electron (**FIG. 1D.13**). **BOX 1D.1** describes an experiment that confirmed these properties of electron spin.

> *An electron has the property of spin; the spin is described by the quantum number $m_s = \pm\frac{1}{2}$.*

Box 1D.1 HOW DO WE KNOW . . . THAT AN ELECTRON HAS SPIN?

Electron spin was first detected experimentally by two German scientists, Otto Stern and Walter Gerlach, in 1922. Because a moving electric charge generates a magnetic field, they predicted that a spinning electron should behave like a tiny bar magnet.

In their experiment (see the illustration), Stern and Gerlach removed all the air from a container and set up a highly nonuniform magnetic field across it. They then sent a narrow stream of silver atoms through the container toward a detector. A silver atom has one unpaired electron, with its remaining 46 electrons

Collection plate

Magnet

Atom beam

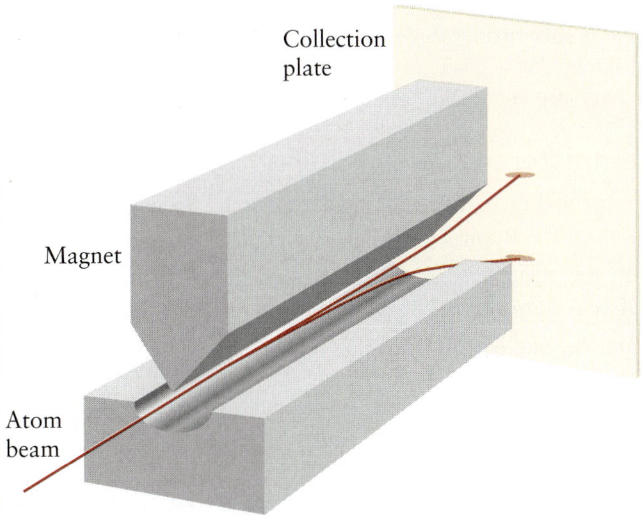

A schematic representation of the apparatus used by Stern and Gerlach. In the experiment, a stream of atoms splits into two as it passes between the poles of a magnet. The atoms in one stream have an odd $\uparrow$ electron, and those in the other an odd $\downarrow$ electron.

paired. The atom therefore behaves like a single unpaired electron riding on a heavy platform, the rest of the atom (Topic 1E).

If a spinning electron behaved like a spinning ball, the axis of spin could point in any direction. The electron would behave like a bar magnet that could have any orientation relative to the applied magnetic field. In that case, a broad band of silver atoms should appear at the detector, because the field would push the silver atoms by different amounts according to the orientation of the spin. Indeed, that is exactly what Stern and Gerlach observed when they first carried out the experiment.

However, the first results were misleading. The experiment is difficult because the atoms collide with one another in the beam. An atom moving in one direction might easily be knocked by its neighbors into a different direction. When Stern and Gerlach repeated their experiment, they used a much less dense beam of atoms, thereby reducing the number of collisions between the atoms. They now saw only two narrow bands. One band consisted of atoms flying through the magnetic field with one orientation of their spin; the other band consisted of the atoms with opposite spin. The two narrow bands confirmed not only that an electron has spin but also that it can have only two orientations.

Electron spin is the basis of the experimental technique called *electron paramagnetic resonance* (EPR), which is used to study the structures and motions of molecules and ions that have unpaired electrons. This technique is based on detecting the energy needed to flip an electron between its two spin orientations. Like Stern and Gerlach's experiment, it works only with ions or molecules that have an unpaired electron.

Further Reading
• "Stern and Gerlach: How a Bad Cigar Helped Reorient Atomic Physics," *Physics Today*, December 2003, p. 53. Online at: http://ptonline.aip.org/journals/doc/PHTOAD-ft/vol_56/iss_12/53_1.shtml.

1D.6 The Electronic Structure of Hydrogen

In the ground state, the electron is in the lowest energy level, the state with $n = 1$. The only orbital with $n = 1$ is the 1s-orbital; the electron is said to **occupy** a 1s-orbital or to be a "1s-electron." The electron in the ground state of a hydrogen atom is described by the following values of the four quantum numbers:

$$n = 1 \qquad l = 0 \qquad m_l = 0 \qquad m_s = +\tfrac{1}{2} \text{ or } -\tfrac{1}{2}$$

The electron can have either spin state.

When the atom acquires enough energy (by absorbing a photon of radiation, for instance) for its electron to reach the shell with $n = 2$, it can occupy any of the four orbitals in that shell. There are one 2s- and three 2p-orbitals in this shell; in hydrogen, they all have the same energy. When an electron is described by one of these wavefunctions, it is said to occupy a 2s-orbital or one of the 2p-orbitals or that it is a "2s-electron" or a "2p-electron," respectively. The average distance of an electron from the nucleus when it occupies any of the orbitals in the shell with $n = 2$ is greater than when $n = 1$, and so you should think of the atom as swelling up as it is excited energetically.

When the atom acquires even more energy, the electron moves into the shell with $n = 3$; the atom is now even larger. In this shell, the electron can occupy any of nine orbitals (one 3s-, three 3p-, and five 3d-orbitals). More energy moves the electron still farther from the nucleus to the $n = 4$ shell, where 16 orbitals are available (one 4s-, three 4p-, five 4d-, and seven 4f-orbitals). Eventually, enough energy is absorbed so that the electron can escape the pull of the nucleus and leave the atom.

> **Self-test 1D.2A** The three quantum numbers for an electron in a hydrogen atom in a certain state are $n = 4$, $l = 2$, and $m_l = -1$. In what type of orbital is the electron located?
>
> [*Answer:* 4d]
>
> **Self-test 1D.2B** The three quantum numbers for an electron in a hydrogen atom in a certain state are $n = 3$, $l = 1$, and $m_l = -1$. In what type of orbital is the electron located?

The state of an electron in a hydrogen atom is defined by the four quantum numbers n, l, m_l, and m_s; as the value of n increases, the size of the atom increases.

What have you learned in this Topic?

You have learned that electrons in atoms are described by wavefunctions that are called atomic orbitals and that each orbital is specified by three quantum numbers: n, l, and m_l. You now know that the shape and energy of a given orbital is found by solving the appropriate Schrödinger equation. You have also encountered the property of "electron spin."

The skills you have mastered are the ability to:

☐ **1.** Assess the relative probability of finding an electron at a given distance from the nucleus of an atom (Example 1D.1).

☐ **2.** Name and explain the relation of each of the four quantum numbers to the properties and relative energies of atomic orbitals (Sections 1D.1–1D.4).

☐ **3.** Describe the properties of electron spin (Section 1D.5).

☐ **4.** Describe the structure of a hydrogen atom in its ground state and its excited states (Section 1D.6).

Topic 1D Exercises

Exercises labeled $\int_{dx}^{C}$ **require calculus.**

1D.1 Which of the following increase when the electron in a hydrogen atom undergoes a transition from the 1s-orbital to a 2p-orbital? (a) Energy of the electron. (b) Value of n. (c) Value of l. (d) Radius of the atom.

1D.2 Which of the following increase when the electron in a hydrogen atom undergoes a transition from the 2s-orbital to a 2p-orbital? (a) Energy of the electron. (b) Value of n. (c) Value of l. (d) Radius of the atom.

1D.3 Evaluate the probability of finding an electron in a small region of a hydrogen 1s-orbital at a distance $0.55a_0$ from the nucleus relative to finding it in the same small region located at the nucleus.

1D.4 Evaluate the probability of finding an electron in a small region of a hydrogen 1s-orbital at a distance $0.83a_0$ from the nucleus relative to finding it in the same small region located at the nucleus.

1D.5 Show that the electron distribution is spherically symetrical for an atom in which an electron occupies each of the three p-orbitals of a given shell.

1D.6 Show that, if the radial distribution function is defined as $P = r^2R^2$, then the expression for P for an s-orbital is $P = 4\pi^2r^2\psi^2$.

1D.7 What is the probability of finding an electron anywhere inside a sphere of radius (a) a_0 or (b) $2a_0$ in the ground state of a hydrogen atom?

1D.8 At what distance from the nucleus is the electron most likely to be found if it occupies (a) a 3d-orbital or (b) a 3s-orbital in a hydrogen atom?

1D.9 (a) Sketch the shape of the boundary surfaces corresponding to 1s-, 2p-, and 3d-orbitals. (b) What is meant by a node? (c) How many radial nodes and angular *nodal surfaces* are there in each orbital? (d) Predict the number of nodal planes expected for a 4f-orbital.

1D.10 Describe the difference in orientation of the d_{xy}- and the $d_{x^2-y^2}$-orbitals with respect to the reference Cartesian axes. You might wish to refer to the interactive atomic orbitals found on the Web site for this text.

1D.11 How many orbitals are in subshells with l equal to (a) 0; (b) 2; (c) 1; (d) 3?

1D.12 (a) How many *subshells* are there with principal quantum number $n = 6$? (b) Identify the subshells in the form 6s, etc. (c) How many orbitals are there in the shell with $n = 6$?

1D.13 (a) How many values of the quantum number l are possible when $n = 7$? (b) How many values of m_l are allowed for an electron in a 6d-subshell? (c) How many values of m_l are allowed for an electron in a 3p-subshell? (d) How many subshells are there in the shell with $n = 4$?

1D.14 (a) How many values of the quantum number l are possible when $n = 6$? (b) How many values of m_l are allowed for an electron in a 5f-subshell? (c) How many values of m_l are allowed for an electron in a 2s-subshell? (d) How many subshells are there in the shell with $n = 3$?

1D.15 What are the principal and orbital angular momentum quantum numbers for each of the following orbitals: (a) 6p; (b) 3d; (c) 2p; (d) 5f?

1D.16 What are the principal and orbital angular momentum quantum numbers for each of the following orbitals: (a) 3s; (b) 4p; (c) 5d; (d) 6f?

1D.17 For each orbital listed in Exercise 1D.15, give the possible values of the magnetic quantum number.

1D.18 For each orbital listed in Exercise 1D.16, give the possible values of the magnetic quantum number.

1D.19 How many orbitals are present in the (a) 4p-subshell; (b) 3d-subshell; (c) 1s-subshell; (d) 4f-subshell of an atom?

1D.20 How many orbitals are present in a subshell with $l =$ (a) 0; (b) 1; (c) 2; (d) 3?

1D.21 Write the subshell notation (3d, for instance) and the number of orbitals having the following quantum numbers: (a) $n = 5$, $l = 2$; (b) $n = 1$, $l = 0$; (c) $n = 6$, $l = 3$; (d) $n = 2$, $l = 1$.

1D.22 Assume that each orbital can be occupied by a maximum of two electrons. Write the subshell notation (3d, for instance) and the number of orbitals that can have the following quantum numbers: (a) $n = 4$, $l = 1$; (b) $n = 5$, $l = 0$; (c) $n = 6$, $l = 2$; (d) $n = 7$, $l = 3$.

1D.23 How many orbitals can have the following quantum numbers in an atom: (a) $n = 2$, $l = 1$; (b) $n = 4$, $l = 2$, $m_l = -2$; (c) $n = 2$; (d) $n = 3$, $l = 2$, $m_l = +1$?

1D.24 How many orbitals can have the following quantum numbers in an atom: (a) $n = 3$, $l = 1$; (b) $n = 5$, $l = 3$, $m_l = -1$; (c) $n = 2$, $l = 1$, $m_l = 0$; (d) $n = 7$?

1D.25 Which of the following subshells cannot exist in an atom: (a) 2d; (b) 4d; (c) 4g; (d) 6f?

1D.26 Which of the following subshells cannot exist in an atom: (a) 1p; (b) 5f; (c) 5g; (d) 6g?

Topic 1E Many-Electron Atoms

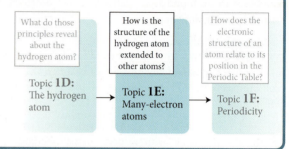

A neutral atom other than a hydrogen atom has more than one electron and is known as a **many-electron atom.** This Topic builds on the description of the hydrogen atom in Topic 1D to describe how the presence of more than one electron affects the energies of atomic orbitals and how they are occupied. The resulting electronic structures are the key to the periodic properties of the elements and the abilities of atoms to form chemical bonds. This material therefore underlies almost every aspect of chemistry.

1E.1 Orbital Energies

The electrons in a many-electron atom occupy orbitals like those of hydrogen. However, the energies of these orbitals are not the same as those for a hydrogen atom. The nucleus of a many-electron atom is more highly charged than the hydrogen nucleus, and the greater charge attracts electrons more strongly and hence lowers their energy. However, the electrons also repel one another; this repulsion opposes the nuclear attraction and raises the energies of the orbitals. In a helium atom, for instance, with two electrons, the charge of the nucleus is $+2e$ and the total potential energy is given by three terms:

$$V \propto \overbrace{-\frac{2e^2}{r_1}}^{\substack{\text{attraction of} \\ \text{electron 1 to} \\ \text{the nucleus}}} \quad \overbrace{-\frac{2e^2}{r_2}}^{\substack{\text{attraction of} \\ \text{electron 2 to} \\ \text{the nucleus}}} \quad \overbrace{+\frac{e^2}{r_{12}}}^{\substack{\text{repulsion} \\ \text{between the} \\ \text{two electrons}}} \tag{1}$$

where r_1 is the distance of electron 1 from the nucleus, r_2 is the distance of electron 2 from the nucleus, and r_{12} is the distance between the two electrons. The two terms with negative signs (indicating that the potential energy *decreases* as r_1 or r_2 decreases when the electron approaches the nucleus) represent the attractions between the nucleus and the two electrons. The term with a positive sign (indicating an *increase* in potential energy as r_{12} decreases when the two electrons approach each other) represents the repulsion between the two electrons. The Schrödinger equation based on this potential energy is impossibly difficult to solve exactly, but highly accurate numerical solutions can be obtained by using computers.

The hydrogen atom, with one electron, has no electron–electron repulsions and all the orbitals of a given shell are degenerate (have the same energy). For instance, as remarked in Topic 1D, the 2s-orbital and all three 2p-orbitals have the same energy. In many-electron atoms, however, the results of spectroscopic experiments and calculations show that electron–electron repulsions cause the energy of a 2p-orbital to be higher than that of a 2s-orbital. Similarly, in the $n = 3$ shell, the three 3p-orbitals lie higher than the 3s-orbital, and the five 3d-orbitals lie higher still (**FIG. 1E.1**). How can these energy differences be explained?

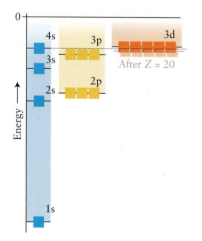

FIGURE 1E.1 The relative energies of the shells, subshells, and orbitals in a many-electron atom. Each of the boxes can hold at most two electrons. Note the change in the order of energies of the 3d- and 4s-orbitals after $Z = 20$.

As well as being attracted to the nucleus, each electron in a many-electron atom is repelled by the other electrons present. As a result, it is less tightly bound to the nucleus than it would be if those other electrons were absent. Each electron is said to be **shielded** from the full attraction of the nucleus by the other electrons in the atom. The shielding effectively reduces the pull of the nucleus on an electron. The **effective nuclear charge,** $Z_{eff}e$, experienced by the electron is always less than the actual nuclear charge, Ze, because the electron–electron repulsions work against the pull of the nucleus. A *very* approximate form of the energy of an electron in a many-electron atom is a version of Eq. 2b of Topic 1D ($E_n = -Z^2 h \mathcal{R}/n^2$) in which the true atomic number is replaced by the effective atomic number:

$$E_n = -\frac{Z_{eff}^2 h \mathcal{R}}{n^2} \tag{2}$$

Note that the other electrons do not "block" the influence of the nucleus; they simply provide additional repulsive coulombic interactions that partly counteract the pull of the nucleus. For example, the pull of the nucleus on an electron in the helium atom is less than its charge of $+2e$ would exert but greater than the net charge of $+e$ that we would expect if each electron balanced one positive charge exactly.

An s-electron of any shell can be found very close to the nucleus (remember that ψ^2 for an s-orbital is nonzero at the nucleus) and is said to **penetrate** through the inner shells. A p-electron penetrates much less because its orbital angular momentum prevents it from approaching close to the nucleus (**FIG. 1E.2**). In Topic 1D, it is pointed out that its wavefunction vanishes at the nucleus, and so there is zero probability density for finding a p-electron there. Because a p-electron penetrates less than an s-electron through the inner shells of the atom, it is more effectively shielded from the nucleus and hence experiences a smaller effective nuclear charge than an s-electron does. In other words, an s-electron is bound more tightly than a p-electron and has a slightly lower (more negative) energy. A d-electron is bound less tightly than a p-electron of the same shell because its orbital angular momentum is higher and it is therefore even less able to approach the nucleus closely. That is, d-electrons are higher in energy than p-electrons of the same shell, which are in turn higher in energy than s-electrons of that shell.

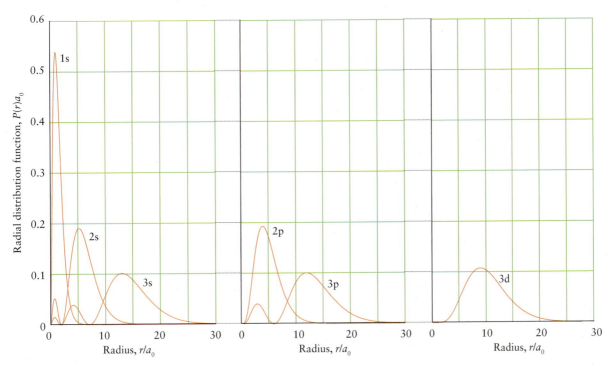

FIGURE 1E.2 The radial distribution functions for s-, p-, and d-orbitals in the first three shells of a hydrogen atom. Note that the probability maxima for orbitals of the same shell are close to each other; however, note that an electron in an ns-orbital has a higher probability of being found close to the nucleus than does an electron in an np-orbital or an nd-orbital.

The effects of penetration and shielding can be large. A 4s-electron generally has a much lower energy than that of a 4p- or 4d-electron; it may even have lower energy than that of a 3d-electron of the same atom (see Fig. 1E.1). The precise ordering of orbitals depends on the number of electrons in the atom, as explained in the next section.

In a many-electron atom, because of the effects of penetration and shielding, the order of orbital energies in a given shell is s < p < d < f.

1E.2 The Building-Up Principle

The electronic structure of an atom determines its chemical properties, and so you need to be able to describe that structure. To do so, you write the **electron configuration** of the atom—a list of all its occupied orbitals, with the numbers of electrons that occupy each one. In the ground state of a many-electron atom, the electrons occupy atomic orbitals in such a way that the total energy of the atom is a minimum. At first sight, it might seem that an atom should have its lowest energy when all its electrons are in the lowest energy orbital (the 1s-orbital). However, except for hydrogen and helium, which have no more than two electrons, that can never happen. In 1925, the Austrian scientist Wolfgang Pauli discovered a general and very fundamental rule about electrons and orbitals that is now known as the **Pauli exclusion principle:**

No more than two electrons may occupy any given orbital. When two electrons do occupy one orbital, their spins must be paired.

The spins of two electrons are said to be **paired** if one is ↑ and the other ↓ (**FIG. 1E.3**). Paired spins are denoted ↑↓, and electrons with paired spins have spin magnetic quantum numbers of opposite sign. Because an atomic orbital is designated by three quantum numbers (n, l, and m_l) and the two spin states are specified by a fourth quantum number, m_s, another way of expressing the Pauli exclusion principle for atoms is:

No two electrons in an atom can have the same set of four quantum numbers.

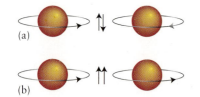

(a)

(b)

FIGURE 1E.3 (a) Two electrons are said to be paired if they have opposite spins (one clockwise, the other counterclockwise). (b) Two electrons are classified as having parallel spins if their spins are in the same direction.

The hydrogen atom in its ground state has one electron in the 1s-orbital. To show this structure, a single arrow is drawn in the 1s-orbital in a "box diagram," which shows each orbital as a box that can be occupied by no more than two electrons (see **1**, which is a fragment of Fig. 1E.1). Its electron configuration, the statement of the orbitals that are occupied, is then reported as 1s¹ ("one s one"). In the ground state of a helium (He) atom ($Z = 2$), both electrons are in a 1s-orbital, which is reported as 1s² ("one s two"). As you can see in (**2**), the two electrons are paired. At this point, the 1s-orbital and the shell with $n = 1$ are fully occupied. The helium atom in its ground state has a **closed shell,** a shell containing the maximum number of electrons allowed by the exclusion principle.

A Note on Good Practice: When a single electron occupies an orbital, write 1s¹, for instance, not simply 1s.

Lithium ($Z = 3$) has three electrons. Two electrons occupy the 1s-orbital and complete the $n = 1$ shell. The third electron must occupy the next available orbital up the ladder of energy levels, the 2s-orbital (see Fig. 1E.1). The ground state of a lithium (Li) atom is therefore 1s²2s¹ (**3**). You should think of the electronic structure of an atom as having an inner **core** consisting of the electrons in filled orbitals, surrounded by the **valence electrons,** the electrons in the outermost shell. In the case of lithium the core is made up of the inner heliumlike closed shell, the 1s² core, which is denoted [He]. The core is surrounded by an outer shell containing a higher-energy 2s-electron. Therefore, the electron configuration of lithium is [He]2s¹. In general, only valence electrons can be lost in chemical reactions because core electrons are in lower-energy orbitals and so are too tightly bound. Thus, lithium loses only one electron when it forms compounds; it forms Li⁺ ions, rather than Li²⁺ or Li³⁺ ions.

The element with $Z = 4$ is beryllium (Be), with four electrons. The first three electrons form the configuration 1s²2s¹, like lithium. The fourth electron pairs with the

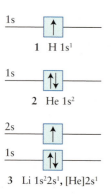

3 Li 1s²2s¹, [He]2s¹

The outermost electrons are used in the formation of chemical bonds (Topic 2A), and the theory of bond formation is called valence theory; hence the name of these electrons.

2s-electron, giving the configuration $1s^2 2s^2$, or more simply [He]$2s^2$ (**4**). A beryllium atom therefore has a heliumlike core surrounded by a valence shell containing two paired electrons. Like lithium—and for the same reason—a Be atom can lose only its valence electrons in chemical reactions. Thus, it loses both 2s-electrons to form a Be^{2+} ion.

Boron ($Z = 5$) has five electrons. Two enter the 1s-orbital and complete the $n = 1$ shell. Two enter the 2s-orbital. The fifth electron occupies an orbital of the next available subshell, which Fig. 1E.1 shows is a 2p-orbital. This arrangement of electrons is reported as the configuration $1s^2 2s^2 2p^1$ or [He]$2s^2 2p^1$ (**5**), showing that boron has three valence electrons around a heliumlike core.

Another decision has to be made at carbon ($Z = 6$): does the sixth electron join the one already in the 2p-orbital or does it enter a different 2p-orbital? (Remember, there are three p-orbitals in the subshell, all of the same energy.) To answer this question, you need to note that electrons are farther from each other and repel each other less when they occupy different p-orbitals than when they occupy the same orbital. So the sixth electron goes into an empty 2p-orbital, and the ground state of carbon is $1s^2 2s^2 2p_x^1 2p_y^1$ (**6**). The individual orbitals are written out like this only when it is necessary to emphasize that electrons occupy different orbitals within a subshell. In most cases, it is sufficient to write the shorter form, [He]$2s^2 2p^2$. Note that in the orbital diagram the two 2p-electrons have been drawn with **parallel spins** ($\uparrow\uparrow$), indicating that they have the same spin magnetic quantum numbers. For reasons based in quantum mechanics, electrons with parallel spins tend to avoid each other. Therefore, this arrangement has slightly lower energy than that of a paired arrangement. However, it is allowed only when the electrons occupy different orbitals.

The procedure is called the **building-up principle.** It can be summarized by two rules. To predict the ground-state configuration of a neutral atom of an element with atomic number Z with its Z electrons:

1. Add Z electrons, one after the other, to the orbitals in the order shown in **FIG. 1E.4** but with no more than two electrons in any one orbital.

2. If more than one orbital in a subshell is available, add electrons with parallel spins to different orbitals of that subshell rather than pairing two electrons in one of the orbitals.

The first rule takes into account the Pauli exclusion principle. The second rule is called **Hund's rule,** for the German spectroscopist Friedrich Hund, who first proposed it. This procedure gives the configuration of the atom that corresponds to the lowest total energy, which maximizes the attraction of the electrons to the nucleus and minimizes their repulsion by one another. An atom with electrons in energy states higher than predicted by the building-up principle is said to be in an **excited state.** For example, the electron configuration [He]$2s^1 2p^3$ represents an excited state of a carbon atom. An excited state is

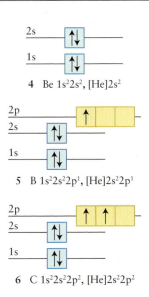

4   Be $1s^2 2s^2$, [He]$2s^2$

5   B $1s^2 2s^2 2p^1$, [He]$2s^2 2p^1$

6   C $1s^2 2s^2 2p^2$, [He]$2s^2 2p^2$

The building-up principle is also commonly called the *Aufbau principle,* from the German word for "building up."

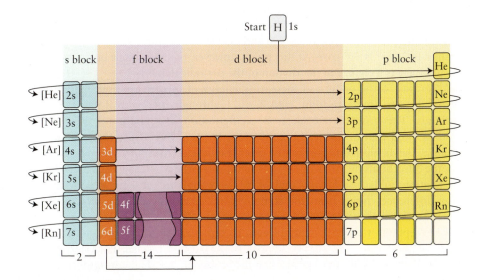

FIGURE 1E.4 The names of the blocks of the periodic table indicate the last subshell being occupied according to the building-up principle. The numbers of electrons that each type of orbital can accommodate are shown by the numbers across the bottom of the table. The colors of the blocks match the colors used for the corresponding orbitals.

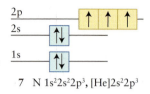

7 N $1s^2 2s^2 2p^3$, $[He]2s^2 2p^3$

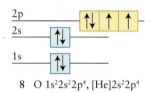

8 O $1s^2 2s^2 2p^4$, $[He]2s^2 2p^4$

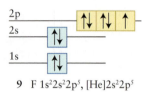

9 F $1s^2 2s^2 2p^5$, $[He]2s^2 2p^5$

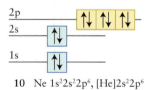

10 Ne $1s^2 2s^2 2p^6$, $[He]2s^2 2p^6$

unstable and emits a photon as the electron returns to an orbital that restores the atom to a lower energy.

In general, you should think of an atom of any element as having a noble-gas core surrounded by a number of electrons in the valence shell, the outermost occupied shell. The **valence shell** is the occupied shell with the largest value of n.

The underlying organization of the periodic table described in *Fundamentals* B now begins to unfold. All the atoms of the main-group elements in a given period have a valence shell with the same principal quantum number, which is equal to the period number. For example, the valence shell of elements in Period 2 (from lithium to neon) is the shell with $n = 2$. Thus all the atoms in a given *period* have the same type of core and different numbers of valence electrons with the same principal quantum number. Thus, the atoms of Period 2 elements all have a heliumlike $1s^2$ core, denoted [He], and those of Period 3 elements have a neonlike $1s^2 2s^2 2p^6$ core, denoted [Ne]. All the atoms of a given *group* (in the main groups, particularly) have analogous valence electron configurations that differ only in the value of n. For instance, all the members of Group 1 have the valence configuration ns^1; and all the members of Group 14 have the valence configuration $ns^2 np^2$. These similar electron configurations give the elements in a group similar chemical properties, as illustrated in *Fundamentals* B.

With these points in mind, let's continue building up the electron configurations across Period 2. Nitrogen has $Z = 7$ and one more electron than carbon, giving $[He]2s^2 2p^3$. Each p-electron occupies a different orbital, and the three have parallel spins (**7**). Oxygen has $Z = 8$ and one more electron than nitrogen; therefore, its configuration is $[He]2s^2 2p^4$ (**8**) and two of its 2p-electrons are paired. Similarly, fluorine, with $Z = 9$ and one more electron than oxygen, has the configuration $[He]2s^2 2p^5$ (**9**), with only one unpaired electron. Neon, with $Z = 10$, has one more electron than fluorine. This electron completes the 2p-subshell, giving $[He]2s^2 2p^6$ (**10**). According to Figs. 1E.1 and 1E.4, the next electron enters the 3s-orbital, the lowest-energy orbital of the next shell. The configuration of sodium is therefore $[He]2s^2 2p^6 3s^1$, or more briefly, $[Ne]3s^1$, where [Ne] denotes the neonlike core.

Self-test 1E.1A Predict the ground-state configuration of a magnesium atom.

[*Answer:* $1s^2 2s^2 2p^6 3s^2$, or $[Ne]3s^2$]

Self-test 1E.1B Predict the ground-state configuration of an aluminum atom.

The s- and p-orbitals of the shell with $n = 3$ are full at argon, $[Ne]3s^2 3p^6$, which is a colorless, odorless, unreactive gas resembling neon. Argon completes the third period. From Fig. 1E.1, you can see that the energy of the 4s-orbital is slightly lower than that of the 3d-orbitals. As a result, instead of electrons entering the 3d-orbitals, the fourth period now begins by filling the 4s-orbitals (see Fig. 1E.1 again). Hence, the next two electron configurations are $[Ar]4s^1$ for potassium and $[Ar]4s^2$ for calcium, where [Ar] denotes the argonlike $1s^2 2s^2 2p^6 3s^2 3p^6$ core. At this point, however, the 3d-orbitals begin to be occupied, and there is a change in the rhythm of the periodic table.

According to the pattern of increasing energy of the orbitals (see Fig. 1E.1), the next 10 electrons (for scandium, with $Z = 21$, through zinc, with $Z = 30$) enter the 3d-orbitals. The ground-state electron configuration of scandium, for example, is $[Ar]3d^1 4s^2$, and that of its neighbor titanium is $[Ar]3d^2 4s^2$. Note that, beginning at scandium, the 4s-electrons are written after the 3d-electrons: once they contain electrons, the 3d-orbitals lie lower in energy than the 4s-orbital. (Recall Fig. 1E.1, where the same relation holds true for nd- and $(n + 1)$s-orbitals in subsequent periods. These energy differences result from the complications caused by electron–electron repulsions). Successive electrons are added to the d-orbitals as Z increases. However, there are two exceptions: the experimental electron configuration of chromium is $[Ar]3d^5 4s^1$ instead of $[Ar]3d^4 4s^2$, and that of copper is $[Ar]3d^{10} 4s^1$ instead of $[Ar]3d^9 4s^2$. This apparent discrepancy occurs because the half-complete subshell configuration d^5 and the complete subshell configuration d^{10} turn out, for quantum mechanical reasons, to have a lower energy than simple theory suggests. As a result, a lower total energy may be achieved if an electron enters a 3d-orbital instead of the expected 4s-orbital, if that arrangement completes a half-subshell or a full subshell. Other exceptions to the building-up principle can be found in the complete listing of electron configurations in Appendix 2C and in the periodic table inside the front cover.

As you should be able to anticipate from the structure of the periodic table (see Fig. 1E.4), electrons occupy 4p-orbitals once the 3d-orbitals are full. The configuration of germanium, $[Ar]3d^{10}4s^24p^2$, for example, is obtained by adding two electrons to the 4p-orbitals outside the completed 3d-subshell. Arsenic has one more electron and the configuration $[Ar]3d^{10}4s^24p^3$. The fourth period of the table contains 18 elements, because the 4s- and 4p-orbitals together can accommodate a total of 8 electrons and the 3d-orbitals can accommodate 10. Period 4 is the first **long period** of the periodic table.

THINKING POINT

At approximately what value of atomic number might an "extra-long period," corresponding to the filling of g-orbitals appear, and how long would it be?

Next in line for occupation at the beginning of Period 5 is the 5s-orbital, followed by the 4d-orbitals. As in Period 4, the energies of the 4d-orbitals fall below that of the 5s-orbital after two electrons have been accommodated in the 5s-orbital. A similar effect is seen in Period 6, but now another set of inner orbitals, the 4f-orbitals, begins to be occupied. Cerium, for example, has the configuration $[Xe]4f^15d^16s^2$. Electrons then continue to occupy the seven 4f-orbitals, which are complete after 14 electrons have been added, at ytterbium, $[Xe]4f^{14}6s^2$. Next, the 5d-orbitals are occupied. The 6p-orbitals are occupied only after the 6s-, 4f-, and 5d-orbitals are filled at mercury; thallium, for example, has the configuration $[Xe]4f^{14}5d^{10}6s^26p^1$. You may notice in Appendix 2C several apparent disruptions in the order in which the 4f orbitals are filled. The apparent exceptions result because the 4f- and 5d-orbitals are very close in energy. In fact, closely spaced energy levels account for the fact that about 25% of all elements have electron configurations that deviate in some way from these rules. However, for the majority of elements the rules are useful guidelines and a good starting point for them all; **Toolbox 1E.1** outlines a procedure for writing the electron configuration of a heavy element.

Toolbox 1E.1 HOW TO PREDICT THE GROUND-STATE ELECTRON CONFIGURATION OF AN ATOM

CONCEPTUAL BASIS

Electrons occupy orbitals in such a way as to minimize the total energy of an atom by maximizing attractions and minimizing repulsions in accord with the Pauli exclusion principle and Hund's rule.

PROCEDURE

Use the following rules of the building-up principle to assign a ground-state configuration to a neutral atom of an element with atomic number Z:

1 Note from the periodic table in which period and group the element is found. Its core configuration will be that of the preceding noble-gas configuration together with any completed d- and f-subshells. The period number gives the value of the principal quantum number of the valence shell and the group number is used to find the number of valence electrons.

2 Add Z electrons, one after the other, to the orbitals in the order shown in Figs. 1E.1 and 1E.4 but with no more than two electrons in any one orbital (the Pauli exclusion principle).

3 If more than one orbital in a subshell is available, add electrons to different orbitals of the subshell before doubly occupying any of them (Hund's rule).

4 Write the labels of the orbitals in order of increasing energy, with a superscript that gives the number of electrons in that orbital. The configuration of a filled shell is represented by the symbol of the noble gas having that configuration, as in [He] for $1s^2$.

In most cases this procedure gives the ground-state electron configuration of an atom. Any arrangement other than the ground state corresponds to an excited state of the atom. Note that the structure of the periodic table can be used to predict the electron configurations of most elements by noting which orbitals are being filled in each block of the periodic table (see Fig. 1E.4).

Example 1E.1 shows how these rules are applied.

EXAMPLE 1E.1   Predicting the ground-state electron configuration of a heavy atom

In order to develop interesting new compounds inorganic chemists often start by noting the location of an element in the periodic table and its valence electron configuration. Suppose you are working in a laboratory developing catalysts for an industrial process. Predict the ground-state electron configuration of (a) a vanadium atom and (b) a lead atom.

ANTICIPATE Because vanadium is a member of the d-block, you should expect its atoms to have a partially filled set of d-orbitals. Because lead is in the same group as carbon, you should expect the configuration of its valence electrons to be similar to that of carbon (ns^2np^2).

PLAN Follow the procedure in Toolbox 1E.1.

SOLVE

(a) *Step 1* Find the period number, which gives the value of the principal quantum number of the valence shell, and the group number, which gives the number of valence electrons.

Vanadium is in Period 4 and Group 5, and so it has an argon core with five valence electrons.

Step 2 Add Z electrons, one after the other, to the orbitals in the order shown in Figs. 1E.1 and 1E.4 but with no more than two electrons in any one orbital.
Step 3 Add two electrons to the 4s-orbital, and the last three electrons to three separate 3d-orbitals.
Step 4 Write the labels of the orbitals in order of increasing energy.

$$[Ar]3d^34s^2$$

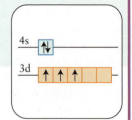

(b) *Step 1* As before, find the period and group of the element in the periodic table.

Lead belongs to Period 6 and Group 14. It has a xenon core with complete 5d- and 4f-subshells and four additional valence electrons.

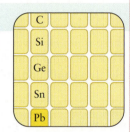

Step 2 Add Z electrons, one after the other, to the orbitals in the order shown in Figs. 1E.1 and 1E.4 but with no more than two electrons in any one orbital.
Step 3 Lead has two valence electrons in a 6s-orbital and two in different 6p-orbitals.
Step 4 Write the labels of the orbitals in order of increasing energy.

$$[Xe]4f^{14}5d^{10}6s^26p^2$$

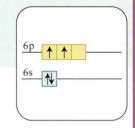

EVALUATE As expected, vanadium has an incomplete set of d-electrons and the valence-shell configuration of lead is analogous to that of carbon.

Self-test 1E.2A Write the ground-state configuration of a bismuth atom.

[*Answer:* $[Xe]4f^{14}5d^{10}6s^26p^3$]

Self-test 1E.2B Write the ground-state configuration of an arsenic atom.

Related Exercises 1E.11–1E.14

The ground-state electron configuration of an atom is predicted by using the building-up principle in conjunction with Fig. 1E.1, the Pauli exclusion principle, and Hund's rule.

What have you learned in this Topic?

You have learned that the structures of many-electron atoms are explained by the systematic occupation of orbitals by electrons, with the order determined by the effects of penetration and shielding in conjunction with the Pauli exclusion principle. The building-up principle is reflected in, and in a sense accounts for, the general structure of the periodic table.

The skills you have mastered are the ability to:

☐ **1.** Describe the factors affecting the energy of an electron in a many-electron atom (Section 1E.1).

☐ **2.** Write the ground-state electron configuration for an element (Toolbox 1E.1 and Example 1E.1).

Topic 1E Exercises

1E.1 Which of the following increase when an electron in a lithium atom undergoes a transition from the 1s-orbital to a 2p-orbital? (a) Energy of the electron. (b) Value of n. (c) Value of l. (d) Radius of the atom. Which answers would be different for a hydrogen atom and in what way would they be different?

1E.2 Which of the following increase when an electron in a lithium atom undergoes a transition from the 2s-orbital to a 2p-orbital? (a) Energy of the electron. (b) Value of n. (c) Value of l. (d) Radius of the atom. Which answers would be different for a hydrogen atom and in what way would they be different?

1E.3 (a) Write an expression for the total coulombic potential energy for a lithium atom. (b) What does each individual term represent?

1E.4 (a) Write an expression for the total coulombic potential energy for a beryllium atom. (b) If Z denotes the number of electrons present in an atom, write a general expression to represent the total number of terms that will be present in the total coulombic potential energy expression.

1E.5 Which of the following statements are true for many-electron atoms? If false, explain why. (a) The effective nuclear charge $Z_{eff}e$ is independent of the number of electrons present in an atom. (b) Electrons in an s-orbital are more effective than those in other orbitals at shielding other electrons from the nuclear charge because an electron in an s-orbital can penetrate to the nucleus of the atom. (c) Electrons having $l = 2$ are better at shielding than electrons having $l = 1$. (d) $Z_{eff}e$ for an electron in a p-orbital is lower than for an electron in an s-orbital in the same shell.

1E.6 For the electrons on a carbon atom in the ground state, decide which of the following statements are true. If false, explain why. (a) $Z_{eff}e$ for an electron in a 1s-orbital is the same as $Z_{eff}e$ for an electron in a 2s-orbital. (b) $Z_{eff}e$ for an electron in a 2s-orbital is the same as $Z_{eff}e$ for an electron in a 2p-orbital. (c) An electron in the 2s-orbital has the same energy as an electron in the 2p-orbital. (d) The electrons in the 2p-orbitals have spin quantum numbers m_s of opposite sign. (e) The electrons in the 2s-orbital have the same value of the quantum number m_s.

1E.7 Determine whether each of the following electron configurations represents the ground state or an excited state of the atom given.

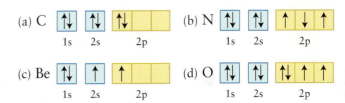

(a) C   1s   2s   2p   (b) N   1s   2s   2p
(c) Be   1s   2s   2p   (d) O   1s   2s   2p

1E.8 Each of the following *valence-shell configurations* is possible for a neutral atom of a certain element. What is the element and which configuration represents the ground state?

(a)   4s   4p   (b)   4s   4p
(c)   4s   4p   (d)   4s   4p

1E.9 Of the following sets of four quantum numbers $\{n, l, m_l, m_s\}$, identify the ones that are forbidden for an electron in an atom and explain why they are invalid:

(a) $\{4, 2, -1, +\frac{1}{2}\}$; (b) $\{5, 0, -1, +\frac{1}{2}\}$; (c) $\{4, 4, -1, +\frac{1}{2}\}$.

1E.10 Of the following sets of four quantum numbers $\{n, l, m_l, m_s\}$, identify the ones that are forbidden for an electron in an atom and explain why they are invalid:

(a) $\{2, 2, -1, +\frac{1}{2}\}$; (b) $\{6, 6, 0, +\frac{1}{2}\}$; (c) $\{5, 4, +5, +\frac{1}{2}\}$.

1E.11 Write the ground-state electron configuration for each of the following atoms: (a) sodium; (b) silicon; (c) chlorine; (d) rubidium.

1E.12 Write the ground-state electron configuration for each of the following atoms: (a) titanium; (b) chromium; (c) europium; (d) krypton.

1E.13 Write the ground-state electron configuration for each of the following atoms: (a) silver; (b) beryllium; (c) antimony; (d) gallium; (e) tungsten; (f) iodine.

1E.14 Write the ground-state electron configuration for each of the following atoms: (a) germanium; (b) cesium; (c) iridium; (d) tellurium; (e) thallium; (f) plutonium.

1E.15 Which elements are predicted to have the following ground-state electron configurations of their atoms: (a) $[Kr]4d^{10}5s^25p^4$; (b) $[Ar]3d^34s^2$; (c) $[He]2s^22p^2$; (d) $[Rn]7s^26d^2$?

1E.16 Which elements are predicted to have the following ground-state electron configurations of their atoms: (a) $[Ar]3d^{10}4s^24p^1$; (b) $[Ne]3s^1$; (c) $[Kr]5s^2$; (d) $[Xe]4f^76s^2$?

1E.17 For each of the following ground-state atoms, predict the type of orbital (1s, 2p, 3d, 4f, etc.) from which an electron will be removed to form the +1 ion: (a) Ge; (b) Mn; (c) Ba; (d) Au.

1E.18 For each of the following ground-state atoms, predict the type of orbital (1s, 2p, 3d, 4f, etc.) from which an electron will be removed to form the +1 ion: (a) Zn; (b) Cl; (c) Al; (d) Cu.

1E.19 Predict the number of valence electrons present in each of the following atoms (include the outermost d-electrons): (a) N; (b) Ag; (c) Nb; (d) W.

1E.20 Predict the number of valence electrons present in each of the following atoms (include the outermost d-electrons): (a) Ta; (b) Tc; (c) Te; (d) Tl.

1E.21 How many *unpaired* electrons are predicted for the ground-state configuration of each of the following atoms: (a) Bi; (b) Si; (c) Ta; (d) Ni?

1E.22 How many *unpaired* electrons are predicted for the ground-state configuration of each of the following atoms: (a) Pb; (b) Ir; (c) Y; (d) Cd?

1E.23 The elements Ga, Ge, As, Se, and Br lie in the same period in the periodic table. Write the electron configuration expected for the ground-state atoms of these elements and predict how many unpaired electrons, if any, each atom has.

1E.24 The elements N, P, As, Sb, and Bi belong to the same group in the periodic table. Write the electron configuration expected for the ground-state atoms of these elements and predict how many unpaired electrons, if any, each atom has.

1E.25 Give the notation for the valence-shell configuration (including the outermost d-electrons) of (a) the alkali metals; (b) Group 15 elements; (c) Group 5 transition metals; (d) the "coinage" metals (Cu, Ag, Au).

1E.26 Give the notation for the valence-shell configuration (including the outermost d-electrons) of (a) the halogens; (b) the chalcogens (the Group 16 elements); (c) the transition metals in Group 5; (d) the Group 14 elements.

Topic 1F Periodicity

How is the structure of the hydrogen atom extended to other atoms? Topic **1E:** Many-electron atoms → How does the electronic structure of an atom relate to its position in the Periodic Table? Topic **1F:** Periodicity

Why Do You Need to Know This Material? The periodic table summarizes trends in the properties of the elements, and the ability to predict the properties of an element from its location in the periodic table is a central skill of a chemist.

What Do You Need to Know Already? You need to be familiar with the structure of the periodic table and its relation to the structures of many-electron atoms, as implied by the building-up principle (Topic 1E). You need to be aware of the definition of ionization energy (Topic 1D).

The formulation of the periodic table is one of the most notable and useful achievements in chemistry because it helps to organize what would otherwise be a bewildering array of properties of the elements. However, the fact that its structure corresponds to the electronic structure of atoms was unknown to its discoverers. The periodic table was developed solely from a consideration of physical and chemical properties of the elements.

In 1869 two scientists, Lothar Meyer, a German, and Dmitri Mendeleev, a Russian (**FIG. 1F.1**), discovered independently that the elements fell into families with similar properties when they were arranged in order of increasing atomic mass. Mendeleev called this observation the **periodic law.** One problem with Mendeleev's table, however, was that some elements seemed to be out of place. For example, when argon was isolated, it did not seem to have the correct mass for its location. Its atomic weight of 40 (that is, its molar mass of 40 g·mol^{-1}) is almost the same as that of calcium, but argon is an inert gas and calcium is a reactive metal. Such anomalies led scientists to question the use of atomic weight as the basis for organizing the elements. In the early twentieth century, Henry Moseley examined x-ray spectra of the elements produced by bombarding a sample with a beam of electrons. He realized that he could infer the atomic number itself by noting how the frequencies of the x-rays depended on the nuclear charge, and therefore on Z. It was soon discovered that elements fall into the uniformly repeating pattern of the periodic table if they are organized according to atomic number rather than atomic weight.

1F.1 The General Structure of the Periodic Table

At the time the periodic table was formulated, the reason for the periodicity of the elements was a mystery. Now, however, its organization is understood in terms of the electron configurations of the elements. The table is divided into **blocks,** named for the last subshell that is occupied according to the building-up principle (s-, p-, d-, and f-blocks), as shown in Fig. 1E.4, which is repeated here as **FIG. 1F.2**. Two elements are exceptions. Because it has two 1s-electrons, helium ought to lie in the s-block, but it is shown in the p-block because of its properties: it is a gas with properties closely matching those of the noble gases in Group 18, rather than the reactive metals in Group 2. Its place in Group 18 is justified because it has a filled valence shell, like all the other Group 18 elements. Hydrogen occupies a unique position in the periodic table. It has one s-electron, and so it belongs in Group 1; but it is also one electron short of a noble-gas configuration, and so it can act like a member of Group 17. Because hydrogen has such a unique character, in this text it is not ascribed to any group; however, you will often see it placed in Group 1 or Group 17, and sometimes in both.

THINKING POINT

What are the arguments for and against including He in Group 2, above beryllium?

FIGURE 1F.1 (a) Dmitri Ivanovitch Mendeleev (1834–1907) and (b) Lothar Meyer (1830–1895). ((a) RIA Novosti/ Science Source. (b) Science Source.)

FIGURE 1F.2 The names of the blocks of the periodic table indicate the last subshell being occupied according to the building-up principle. The colors of the blocks match the colors used for the corresponding orbitals.

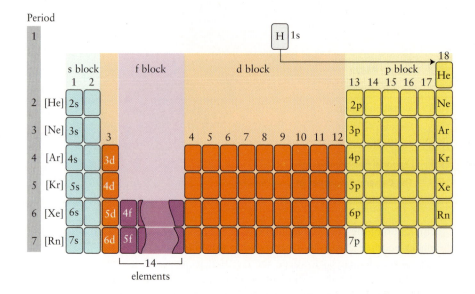

The s- and p-blocks form the **main groups** of the periodic table. The similar valence-shell electron configurations for the elements in the same main group are the reason for the similar properties of these elements. The group number indicates how many valence-shell electrons are present. In the s-block, the group number (1 or 2) is the same as the number of valence electrons. This relation is also true for all main groups when the older practice of using Roman numerals (I–VIII) to label the groups is used. However, when the modern group labels 1–18 are used, in the p-block subtract 10 from the group number to find the number of valence electrons. For example, fluorine in Group 17 (old notation: Group VII) has seven valence electrons.

Each new period corresponds to the occupation of a shell with a higher principal quantum number. This correspondence explains the different lengths of the periods:

- Period 1 consists of only two elements, H and He, in which the single 1s-orbital of the $n = 1$ shell is being filled with its two electrons.
- Period 2 consists of the eight elements Li through Ne, in which the one 2s- and three 2p-orbitals are being filled with eight more electrons.
- In Period 3 (Na through Ar), the 3s- and 3p-orbitals are being occupied by eight additional electrons.
- In Period 4, not only are the eight electrons of the 4s- and 4p-orbitals being added, so are the ten electrons of the 3d-orbitals. Hence there are 18 elements in Period 4.
- Period 5 elements add another 18 electrons as the 5s-, 4d-, and 5p-orbitals are filled.
- In Period 6, a total of 32 electrons are added, because 14 electrons are also being added to the seven 4f-orbitals.

The f-block elements have very similar chemical properties, because their electron configurations differ only in the population of inner f-orbitals, and electrons in these orbitals do not participate much in bond formation.

The periodic table can be used to predict a wide range of properties, many of which are crucial for understanding chemistry. The variation of effective nuclear charge, $Z_{eff}e$, through the periodic table plays an important role in the explanation of periodic trends because it influences the energies and locations of the electrons in the valence shells of the atoms. **FIGURE 1F.3** shows the variation for the first three periods. The effective charge increases from left to right across a period and falls back sharply upon going to the next period.

THINKING POINT

Before reading further, predict how the effective nuclear charge might affect some atomic properties, such as the size of an atom or the ease with which an outer electron can be removed.

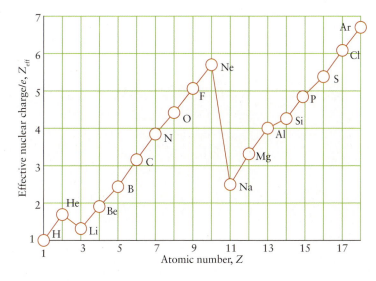

FIGURE 1F.3 The variation of the effective nuclear charge for the outermost valence electron with atomic number. Notice that the effective nuclear charge increases from left to right across a period but drops when the outer electrons occupy a new shell. (The effective nuclear charge is actually $Z_{eff}e$, but Z_{eff} itself is commonly referred to as the charge.)

The blocks of the periodic table are named for the last orbital to be occupied according to the building-up principle. The periods are numbered according to the principal quantum number of the valence shell.

1F.2 Atomic Radius

Electron clouds do not have sharp boundaries and so the exact radius of an atom cannot be measured. However, the electron density of many-electron atoms falls off very sharply at the "edge" of the atom, and when atoms pack together in solids and bond together to form molecules, their centers are found at definite distances from one another. The **atomic radius** of an element is defined as half the distance between the centers of neighboring atoms (**1**). Then:

- If the element is a metal, its atomic radius is taken to be half the distance between the centers of neighboring atoms in a solid sample.

For instance, because the distance between neighboring nuclei in solid copper is 256 pm, the atomic radius of copper is 128 pm.

- If the element is a nonmetal or a metalloid, half the distance between the nuclei of atoms joined by a chemical bond is used; this radius is also called the **covalent radius** of the element, for reasons explained in Topic 2D.

For instance, the distance between the nuclei in a Cl_2 molecule is 198 pm, and so the covalent radius of chlorine is 99 pm.

- If the element is a noble gas, the **van der Waals radius** is used, which is half the distance between the centers of neighboring atoms in a sample of the solidified gas.

The atomic radii of the noble gases listed in Appendix 2D are all van der Waals radii. Because the atoms in a sample of a noble gas are not chemically bonded together, van der Waals radii are generally much larger than covalent radii and are best not included in the discussion of trends.

FIGURE 1F.4 shows the atomic radii of some main-group elements and FIG. 1F.5 shows the variation in atomic radius with atomic number. Note the periodic, sawtooth pattern. A feature to note is that:

- Atomic radius generally decreases from left to right across a period and increases down a group.

The increase in atomic radius down a group, such as that from Li to Cs, makes sense: with each new period, the outermost electrons occupy shells with increasing principal quantum number and therefore lie farther from the nucleus. The decrease across a period, such as that from Li to Ne, might be surprising at first sight because the number of

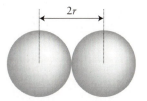

1 Atomic radius

FIGURE 1F.4 The atomic radii (in picometers) of the main-group elements. The radii decrease from left to right in a period and increase down a group. Atomic radii, including those of the d-block elements, are listed in Appendix 2D.

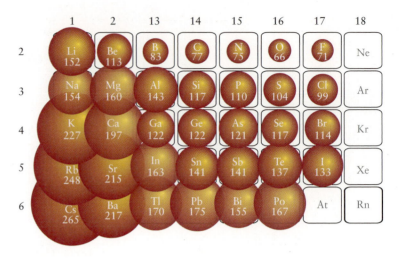

FIGURE 1F.5 The periodic variation in the atomic radii of the elements. The decrease across a period can be explained in terms of the effect of increasing effective nuclear charge and the increase down a group can be explained by the occupation of shells with increasing principal quantum number.

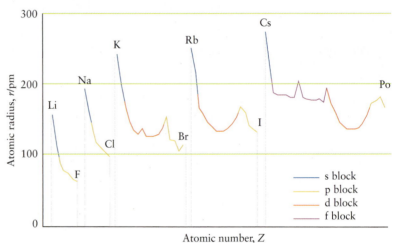

electrons is increasing along with the number of protons. The explanation is that the new electrons are in the same shell of the atom and about as close to the nucleus as other electrons in the same shell. However, because they are spread out in the shell, the electrons do not shield one another well from the nuclear charge; so the effective nuclear charge increases across the period. The increasing effective nuclear charge draws the electrons in. As a result, the atom is more compact and there is a trend for atomic radii to increase diagonally from the upper right of the periodic table to the lower left.

THINKING POINT

Which currently known element has the biggest atoms?

Atomic radii generally decrease from left to right across a period as the effective atomic number increases, and they increase down a group as successive shells are occupied.

1F.3 Ionic Radius

The radii of ions differ markedly from the radii of their parent atoms. As described in *Fundamentals* C, each ion in an ionic solid is surrounded by ions with the opposite charge. The **ionic radius** of an element is its share of the distance between neighboring ions in an ionic solid (**2**). The distance between the centers of a neighboring cation and anion is the sum of the two ionic radii. In practice, the radius of the oxide ion is taken to be 140. pm and the radii of other ions are calculated on the basis of that value. For example, because the distance between the centers of neighboring Mg^{2+} and O^{2-} ions in magnesium oxide is 212 pm, the radius of the Mg^{2+} ion is reported as 212 pm − 140. pm = 72 pm.

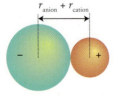

2 Ionic radius

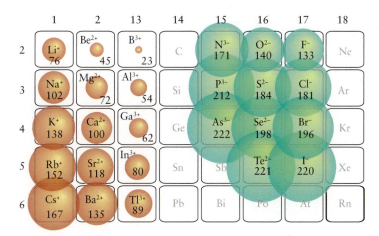

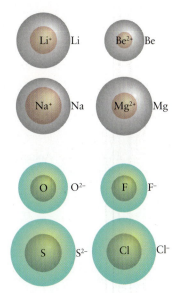

FIGURE 1F.6 The ionic radii (in picometers) of the ions of the main-group elements. Note that cations are typically smaller than their parent atoms, whereas anions are larger—in some cases, very much larger.

FIGURE 1F.7 The relative sizes of some cations and anions compared with their parent atoms. Note that cations (pink) are smaller than their parent atoms (gray), whereas anions (green) are larger.

FIGURE 1F.6 illustrates the trends in ionic radii, and **FIG. 1F.7** shows the relative sizes of some ions and their parent atoms. All cations are smaller than their parent atoms, because the atom loses its valence electrons to form the cation and exposes its core, which is generally much smaller than the parent atom. For example, the atomic radius of Li, with the configuration $1s^2 2s^1$, is 152 pm, but the ionic radius of Li^+, the bare heliumlike $1s^2$ core of the parent atom, is only 76 pm. This size difference is comparable to that between a cherry and its pit. Atoms in the same main group tend to form ions with the same charge. Like atomic radii, the radii of these ions increase down each group because the core electrons occupy shells with higher principal quantum numbers.

Figure 1F.7 shows that anions are larger than their parent atoms. The reason can be traced to the increased number of electrons in the valence shell of the anion and the repulsive effects exerted by electrons on one another. The variation in radii of anions shows the same diagonal trend as that for atoms and cations, with the smallest at the upper right of the periodic table, close to fluorine:

- Cations are smaller than their parent atoms, whereas anions are larger.

Atoms and ions with the same number of electrons are called **isoelectronic.** For example, Na^+, F^-, and Mg^{2+} are isoelectronic. All three ions have the same electron configuration, $[He]2s^2 2p^6$, but their radii differ because they have different nuclear charges (see Fig. 1F.3). The Mg^{2+} ion has the largest nuclear charge; so it has the strongest attraction for the electrons and therefore the smallest radius. The F^- ion has the lowest nuclear charge of the three isoelectronic ions and, as a result, it has the largest radius.

EXAMPLE 1F.1 Deciding the relative sizes of ions

Mineralogists and geologists often need to identify the relative sizes of atoms to judge whether one mineral might be modified by the inclusion of "alien" ions. For example, the different colors of some gemstones result from this type of insertion. Arrange each of the following pairs of ions in order of increasing ionic radius: (a) Mg^{2+} and Ca^{2+}; (b) O^{2-} and F^-.

PLAN The smaller member of a pair of isoelectronic ions in the same period will be an ion of an element that lies farther to the right in a period, because that ion has the greater effective nuclear charge. If the two ions are in the same group, the smaller ion will be the one that lies higher in the group, because its outermost electrons are closer to the nucleus.

SOLVE

(a) Mg lies above Ca in Group 2.

Mg^{2+} has the smaller ionic radius.

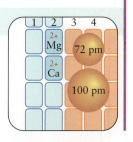

(b) F lies to the right of O in Period 2.

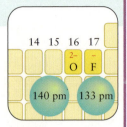

F^- has the smaller ionic radius.

EVALUATE Appendix 2C shows that the actual values are (a) 72 pm for Mg^{2+} and 100 pm for Ca^{2+}; (b) 133 pm for F^- and 140 pm for O^{2-}.

Self-test 1F.1A Arrange each of the following pairs of ions in order of increasing ionic radius: (a) Mg^{2+} and Al^{3+}; (b) O^{2-} and S^{2-}.

[***Answer:*** (a) $r(Al^{3+}) < r(Mg^{2+})$; (b) $r(O^{2-}) < r(S^{2-})$]

Self-test 1F.1B Arrange each of the following pairs of ions in order of increasing ionic radius: (a) Ca^{2+} and K^+; (b) S^{2-} and Cl^-.

Related Exercises 1F.3, 1F.4

Ionic radii generally increase down a group and decrease from left to right across a period. Cations are smaller than their parent atoms and anions are larger.

1F.4 Ionization Energy

As explained in Topic 2A, the formation of a bond in an ionic compound depends on the removal of one or more electrons from one atom and the transfer of those electrons to another atom. The energy needed to remove electrons from atoms is therefore of central importance for understanding their chemical properties. As remarked in Topic 1D, the ionization energy, I, is the minimum energy needed to remove an electron from an atom in the gas phase. Specifically:

> By referring to the minimum energy you don't have to worry about the kinetic energy of the electron: it is assumed to be stationary. Ionization can be achieved by using a higher energy, but then the electron would carry away the excess energy as kinetic energy.

$$J(g) \longrightarrow J^+(g) + e^-(g) \quad I = E(J^+) - E(J) \tag{1}$$

where $E(J)$ is the energy of species J. Ionization energies are reported either as molar quantities in kilojoules per mole ($kJ \cdot mol^{-1}$) or in **electronvolts** (eV), the change in energy of an electron when it moves through a potential difference of 1 volt ($1\ eV = 1.602 \times 10^{-19}$ J). The **first ionization energy,** I_1, is the minimum energy needed to remove an electron from a neutral atom in the gas phase. For example, for copper,

$$Cu(g) \longrightarrow Cu^+(g) + e^-(g) \quad \text{energy required} = I_1\ (7.73\ eV, 746\ kJ \cdot mol^{-1})$$

The **second ionization energy,** I_2, of an element is the minimum energy needed to remove an electron from a singly charged gas-phase cation. For copper,

$$Cu^+(g) \longrightarrow Cu^{2+}(g) + e^-(g) \quad \text{energy required} = I_2\ (20.29\ eV, 1958\ kJ \cdot mol^{-1})$$

Because ionization energy is a measure of how difficult it is to remove an electron, elements with low ionization energies can be expected to form cations readily and to conduct electricity (which requires that some electrons be free to move) in their solid (and liquid) forms. Elements with high ionization energies are unlikely to form cations and are unlikely to conduct electricity.

THINKING POINT

Why is the second ionization energy of an atom always higher than its first ionization energy?

As you can see from **FIG. 1F.8**:

- First ionization energies typically decrease down a group.
- First ionization energies generally increase across a period.

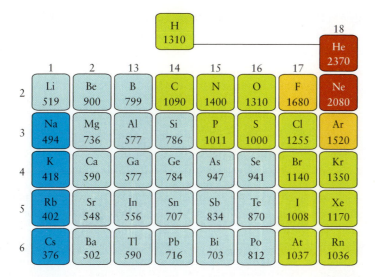

The decrease down a group can be explained by the finding that, in successive periods, the outermost electron occupies a shell that is farther from the nucleus and is therefore less tightly bound. Therefore, it takes less energy to remove an electron from a cesium atom, for instance, than from a sodium atom.

With few exceptions, the first ionization energy rises from left to right across a period (FIG. 1F.9). This trend can be traced to the increase in effective nuclear charge across a period. The small departures from this trend arise from repulsions between electrons, particularly electrons occupying the same orbital. For example, the ionization energy of oxygen is slightly lower than that of nitrogen because in a nitrogen atom each p-orbital has one electron, but in oxygen the eighth electron is paired with an electron already occupying an orbital. The repulsion between the two electrons in the same orbital raises their energy and makes one of them easier to remove from the atom than if the two electrons had been in different orbitals.

FIGURE 1F.10 shows that the second ionization energy of an element is always higher than its first ionization energy. It takes more energy to remove an electron from a positively charged ion than from a neutral atom. For the Group 1 elements, the second ionization energy is considerably larger than the first; in Group 2, however, the two ionization energies have similar values. This difference makes sense, because the Group 1 elements have an ns^1 valence-shell electron configuration. Although the removal of the first electron

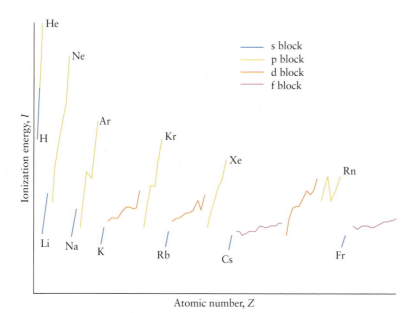

FIGURE 1F.9 The periodic variation of the first ionization energies of the elements.

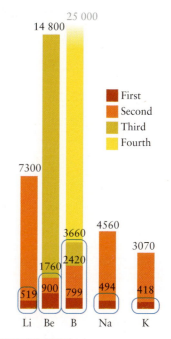

FIGURE 1F.10 The successive ionization energies of a selection of main-group elements. Note the great increase in energy required to remove an electron from an inner shell. In each case, the blue outline denotes ionization from the valence shell.

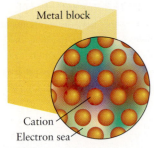

FIGURE 1F.11 A block of metal consists of an array of cations (the spheres) surrounded by a sea of electrons. The charge of the electron sea cancels the charges of the cations. The electrons of the sea are mobile and can move past the cations quite easily and hence conduct an electric current.

 ANIMATION FIGURE 1F.11

requires only a small input of energy, the second electron must come from the noble-gas core. The core electrons have lower principal quantum numbers and are much closer to the nucleus. They are strongly attracted to it and a lot of energy is needed to remove them.

Self-test 1F.2A Account for the slight decrease in first ionization energy between beryllium and boron.

[*Answer:* Boron loses an electron more easily from a higher-energy subshell than beryllium does.]

Self-test 1F.2B Account for the large decrease in third ionization energy between beryllium and boron.

The low ionization energies of elements at the lower left of the periodic table account for their metallic character. A block of metal consists of a collection of cations of the element surrounded by a sea of valence electrons that the atoms have lost (**FIG. 1F.11**). Only elements with low ionization energies—the members of the s-block, the d-block, the f-block, and the lower left of the p-block—can form metallic solids, because only they can lose electrons easily.

The elements at the upper right of the periodic table have high ionization energies, so they do not readily lose electrons and are therefore not metals. Note that your knowledge of electronic structure has helped you to understand a major feature of the periodic table—in this case, why the metals are found toward the lower left and the nonmetals are found toward the upper right.

The first ionization energy is highest for elements close to helium and is lowest for elements close to cesium. Second ionization energies are higher than first ionization energies (of the same element) and very much higher if the electron is to be removed from a closed shell. Metals are found toward the lower left of the periodic table because these elements have low ionization energies and can readily lose their electrons.

1F.5 Electron Affinity

To predict some chemical properties, you need to know how the energy changes when an electron attaches to an atom. The **electron affinity,** E_{ea}, of an element is the energy released when an electron is added to a gas-phase atom. A positive electron affinity means that energy is released when an electron attaches to an atom. A negative electron affinity means that energy must be *supplied* to push an electron onto an atom. This convention matches the everyday meaning of the term "affinity." More formally, the electron affinity of an element X is defined as

$$X(g) + e^-(g) \longrightarrow X^-(g) \qquad E_{ea}(X) = E(X) - E(X^-) \qquad (2)$$

where $E(X)$ is the energy of a gas-phase X atom and $E(X^-)$ is the energy of the gas-phase anion. For instance, the electron affinity of chlorine is the energy released in the process

$$Cl(g) + e^-(g) \longrightarrow Cl^-(g) \qquad \text{energy released} = E_{ea} \text{ (3.62 eV, 349 kJ·mol}^{-1})$$

Because the electron has a lower energy when it occupies one of the atom's orbitals, the difference $E(Cl) - E(Cl^-)$ is positive and the electron affinity of chlorine is positive. Like ionization energies, electron affinities are reported either in electronvolts for a single atom or in joules per mole of atoms.

In some books, you will see electron affinity defined with an opposite-sign convention. Those values are actually the electron-gain enthalpies (Topic 4C).

FIGURE 1F.12 shows the variation in electron affinity in the main groups of the periodic table. It is much less periodic than variations in radius and ionization energy. However, one broad trend is clearly visible. With the exception of the noble gases:

• Electron affinities are highest toward the right of the periodic table.

This trend is particularly true in the upper right, close to oxygen, sulfur, and the halogens. In these atoms, the incoming electron occupies a p-orbital close to a nucleus with a high effective charge and can experience its attraction quite strongly. The noble gases have negative electron affinities because any electron added to them must occupy an orbital outside a closed shell and far from the nucleus: this process requires energy, and so the electron affinity is negative.

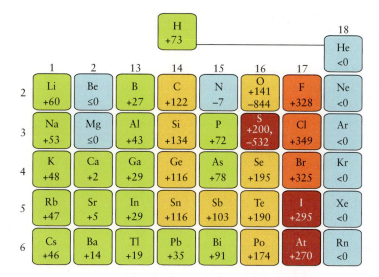

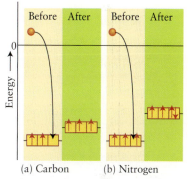

FIGURE 1F.12 The variation in electron affinity in kilojoules per mole of the main-group elements. Where two values are given, the first refers to the formation of a singly charged anion and the second is the additional energy needed to produce a doubly charged anion. The negative signs of the second values indicate that energy is required to add an electron to a singly charged anion. The variation is less systematic than that for ionization energy, but high values tend to be found close to fluorine (except for the noble gases).

Once an electron has entered the single vacancy in the valence shell of a Group 17 atom, the shell is complete and any additional electron would have to begin a new shell. In that shell, it not only would be farther from the nucleus but would also feel the repulsion of the negative charge already present. As a result, the second electron affinity of fluorine is strongly negative, meaning that a lot of energy has to be expended to form F^{2-} from F^-. Ionic compounds of the halogens are therefore built from singly charged ions such as F^- and never from doubly charged ions such as F^{2-}.

A Group 16 atom, such as O or S, has two vacancies in its valence-shell p-orbitals and can accommodate two additional electrons. The first electron affinity is positive because energy is released when an electron attaches to O or S. However, attachment of the second electron *requires* energy because of the repulsion by the negative charge already present in O^- or S^-. Unlike that of a halide ion, however, the valence shell of the O^- anion has only seven electrons and thus can accommodate an additional electron. Therefore, less energy will be needed to make O^{2-} from O^- than to make F^{2-} from F^-, where no such vacancy exists. In fact, 141 $kJ \cdot mol^{-1}$ is released when the first electron adds to the neutral atom to form O^-, but 844 $kJ \cdot mol^{-1}$ must be supplied to add a second electron to form O^{2-}; so the total energy required to make O^{2-} from O is +703 $kJ \cdot mol^{-1}$. As explained in Topic 2A, this energy can be achieved in chemical reactions, and O^{2-} ions are common in metal oxides.

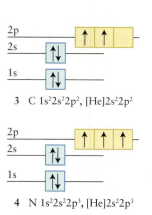

FIGURE 1F.13 The energy changes taking place when an electron is added to a carbon atom and a nitrogen atom. (a) A carbon atom can accommodate an additional electron in an empty p-orbital. (b) When an electron is added to a nitrogen atom it must pair with an electron in a p-orbital. The incoming electron experiences so much repulsion from those already present in the nitrogen atom that the electron affinity of nitrogen is less than that of carbon and is in fact negative.

EXAMPLE 1F.2   Predicting trends in electron affinity

Organic chemists need to think about the distribution of electrons in molecules, because electron-rich regions might prove to be centers of attack for reagents. One guide to where electrons are likely to accumulate is the electron affinity of the element. The electron affinity of carbon is greater than that of nitrogen; indeed, the latter is negative. Suggest a reason for this observation.

PLAN When a periodic trend is different from what is expected, you should examine the electron configurations of all the species to look for clues to the observed behavior.

SOLVE More energy is expected to be released when an electron enters the N atom, because an N atom is smaller than a C atom and its nucleus is more highly charged: the effective nuclear charges for the outermost electrons *of the neutral atoms* are 3.8 for N and 3.1 for C. However, the opposite is observed, and so the effective nuclear charges experienced by the valence electrons in the anions must also be considered (**FIG. 1F.13**). When C^- forms from C, the additional electron occupies an empty 2p-orbital (**3**). The incoming electron is well separated from the other p-electrons, and so it experiences an effective nuclear charge close to 3.1. When N^- forms from N, the additional electron must occupy a 2p-orbital that is already half full (**4**). The effective nuclear charge experienced by this electron is therefore much less than 3.8; so energy is required to form N^-, and the electron affinity of nitrogen is lower than that of carbon.

3   C $1s^2 2s^2 2p^2$, $[He]2s^2 2p^2$

4   N $1s^2 2s^2 2p^3$, $[He]2s^2 2p^3$

FIGURE 1F.14 When tin(II) oxide is heated in air, it becomes incandescent as it reacts to form tin(IV) oxide. Even without being heated, it smolders and can ignite. *(W. H. Freeman photo by Ken Karp.)*

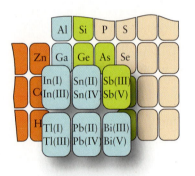

FIGURE 1F.15 The typical ions (more precisely, the oxidation states) formed by the heavy elements in Groups 13 through 15 show the influence of the inert pair—the tendency to form compounds in which the oxidation numbers differ by 2.

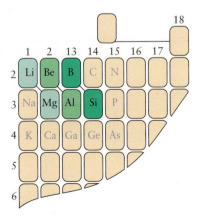

FIGURE 1F.16 The pairs of elements represented by similarly colored boxes show a strong diagonal relationship to each other.

Self-test 1F.3A Account for the large decrease in electron affinity between lithium and beryllium.

[**Answer:** The additional electron enters a 2s-orbital in Li but a 2p-orbital in Be, and a 2s-electron is more tightly bound than a 2p-electron.]

Self-test 1F.3B Account for the large decrease in electron affinity between fluorine and neon.

Related Exercises: 1F.11, 1F.12

Elements with the highest electron affinities are those in Groups 16 and 17.

1F.6 The Inert-Pair Effect

Although both aluminum and indium are in Group 13, aluminum forms Al^{3+} ions, whereas indium forms both In^{3+} and In^+ ions. The tendency to form ions two units lower in charge than expected from the group number is called the **inert-pair effect.** Another example of the inert-pair effect is found in Group 14: tin forms tin(IV) oxide when heated in air, but the heavier lead atom loses only its two p-electrons and forms lead(II) oxide. Tin(II) oxide can be prepared, but it is readily oxidized to tin(IV) oxide (**FIG. 1F.14**). Lead exhibits the inert-pair effect more strongly than tin.

The inert-pair effect is due in part to the relative energies of the valence p- and s-electrons. In the later periods of the periodic table, valence s-electrons are very low in energy because of their good penetration and the low shielding ability of the d-electrons. The valence s-electrons may therefore remain attached to the atom during ion formation. The inert-pair effect is most pronounced among the heaviest members of a group, where the difference in energy between s- and p-electrons is greatest (**FIG. 1F.15**). Even so, the pair of s-electrons can be removed from the atom under sufficiently vigorous conditions. An inert pair would be better called a "lazy pair" of electrons.

The inert-pair effect is the tendency to form ions two units lower in charge than expected from the group number; it is most pronounced for heavy elements in the p-block.

1F.7 Diagonal Relationships

A **diagonal relationship** is a similarity in properties among diagonal neighbors in the main groups of the periodic table (**FIG. 1F.16**). A part of the reason for this similarity can be seen in Fig. 1F.8 by concentrating on the colors that show the general trends in ionization energy. The colored bands of similar values lie in diagonal stripes across the table. Because these characteristics affect the chemical properties of an element, it is not surprising to find that the elements within a diagonal band show similar chemical properties. Diagonal relationships are helpful for making predictions about the properties of elements and their compounds.

The diagonal band of metalloids dividing the metals from the nonmetals (*Fundamentals* B) is one example of a diagonal relationship. So is the chemical similarity of lithium and magnesium and of beryllium and aluminum. For example, both lithium and magnesium react directly with nitrogen to form nitrides. Like aluminum, beryllium reacts with both acids and bases. Many examples of this diagonal similarity are encountered in Focus 8, in which the properties of the main-group elements are discussed.

Diagonally related pairs of elements often show similar chemical properties.

1F.8 The General Properties of the Elements

You are now at the point where you can begin to predict, in at least a general way, the properties of elements. For example, an s-block element has a low ionization energy, which means that its outermost electrons can easily be lost. An s-block element is therefore likely to be a reactive

TABLE 1F.1   Characteristics of Metallic and Nonmetallic Elements

Metallic	Nonmetallic
Physical properties	
good conductors of electricity	poor conductors of electricity
malleable	not malleable
ductile	not ductile
lustrous	not lustrous
typically: solid; high melting point; good conductors of heat	typically: solid, liquid, or gas; low melting point; poor conductors of heat
Chemical properties	
react with acids	do not react with acids
form basic oxides (which react with acids)	form acidic oxides (which react with bases)
form cations	form anions
form ionic halides	form covalent halides

metal with all the characteristics that the name "metal" implies (**TABLE 1F.1**, **FIG. 1F.17**). They are all soft, lustrous metals that melt at low temperatures and produce hydrogen when they come in contact with water. Because ionization energies are lowest at the bottom of each group and the elements there lose their valence electrons most easily, cesium and barium react most vigorously of all s-block elements and have to be stored out of contact with air and water.

Elements on the left of the p-block, especially the heavier elements, have ionization energies that are low enough for these elements to have some of the metallic properties of the members of the s-block. However, the ionization energies of the p-block metals are quite high, and they are less reactive than those in the s-block (**FIG. 1F.18**).

Elements at the right of the p-block (with the exception of the noble gases) have characteristically high electron affinities: they tend to gain electrons to complete closed shells. The elements in Group 18, the noble gases, have completed shells and are so unreactive that at one time they were called the "inert gases." Except for the metalloids tellurium and polonium, the members of Groups 16 and 17 are nonmetals (**FIG. 1F.19**). They typically form molecular compounds with one another.

All d-block elements are metals and are often referred to as the "d-metals" (**FIG. 1F.20**). Their properties are transitional between the s- and the p-block elements, which (with the exception of the members of Group 12) accounts for their alternative name, the "transition metals."

When a d-element atom loses electrons to form a cation, it first loses its outer s-electrons. However, most d-block elements form ions with different oxidation states,

FIGURE 1F.17 All the alkali metals are soft, reactive, silvery metals. Sodium is kept under mineral oil to protect it from air, and a freshly cut surface soon becomes covered with the oxide. (©1983 Chip Clark–Fundamental Photographs.)

FIGURE 1F.18 The Group 14 elements. From left to right: carbon (as graphite), silicon, germanium, tin, and lead. (©1984 Chip Clark–Fundamental Photographs.)

LAB VIDEO FIGURE 1F.17

FIGURE 1F.19 The Group 16 elements. From left to right: oxygen, sulfur, selenium, and tellurium. Note the trend from nonmetal to metalloid. *(©1989 Chip Clark–Fundamental Photographs.)*

FIGURE 1F.20 The elements in the first row of the d-block. Top row (left to right): scandium, titanium, vanadium, chromium, and manganese. Bottom row: iron, cobalt, nickel, copper, and zinc. *(©1989 Chip Clark–Fundamental Photographs.)*

because the d-electrons have similar energies and a variable number can be lost when they form compounds. Iron, for instance, forms Fe^{2+} and Fe^{3+}; copper forms Cu^+ and Cu^{2+}. Although copper is like potassium in having a single outermost s-electron, potassium forms only K^+. The reason for this difference can be understood by comparing their second ionization energies, which are $1958 \text{ kJ} \cdot \text{mol}^{-1}$ for copper and $3051 \text{ kJ} \cdot \text{mol}^{-1}$ for potassium. To form Cu^{2+}, an electron is removed from the d-subshell of $[Ar]3d^{10}$; but, to form K^{2+}, the electron would have to be removed from potassium's argonlike core.

The availability of d-orbitals and the similarity of the atomic radii of the d-block elements have a significant impact on many areas of our lives. The availability of d-orbitals is in large measure responsible for the action of d-block elements and their compounds as catalysts (substances that accelerate reactions but are not themselves consumed) throughout the chemical industry. Thus, iron is used in the manufacture of ammonia, nickel in the conversion of vegetable oils into shortening, platinum in the manufacture of nitric acid, vanadium(V) oxide in the manufacture of sulfuric acid, and titanium compounds in the manufacture of polyethylene. The ability to form ions with different charges is important for facilitating the very subtle changes that take place in organisms. For instance, iron is present as iron(II) in hemoglobin, the oxygen-transport protein in mammalian blood; copper is present in the proteins responsible for electron transport; and manganese is present in the proteins responsible for photosynthesis. The similarity of their atomic radii is largely responsible for the ability of transition metals to form the mixtures known as *alloys*, especially the wide variety of steels that make modern construction and engineering possible.

Difficulties in separating and isolating the lanthanoids (colloquially, the "lanthanides") delayed their widespread use in technology. However, today they are studied intensely, because superconducting materials often contain lanthanoids. All the actinoids (likewise, the "actinides") are radioactive. None of the elements following plutonium occurs naturally on Earth in any significant amount. Because they can be made only in nuclear reactors or particle accelerators, they are available only in small quantities.

All elements in the s-block are reactive metals that form basic oxides. The p-block elements tend to gain electrons to complete closed shells; they range from metals through metalloids to nonmetals. All d-block elements are metals with properties between those of s-block and p-block metals. Many d-block elements form cations in more than one oxidation state.

What have you learned in this Topic?

Many properties of the elements, especially their periodic variation, can be predicted through inspection of the periodic table and by considering the concept of effective nuclear charge.

The skills you have mastered are the ability to:

☐ **1.** Account for periodic trends in atomic radii, ionization energies, and electron affinities (Examples 1F.1 and 1F.2).

☐ **2.** Describe the inert-pair effect and its origin (Section 1F.6).

☐ **3.** Describe diagonal relationships and their origin (Section 1F.7).

☐ **4.** Summarize in a general way the properties of the elements in relation to their location in the periodic table (Section 1F.8).

Topic 1F Exercises

1F.1 Arrange the elements in each of the following sets in order of *decreasing* atomic radius: (a) sulfur, chlorine, silicon; (b) cobalt, titanium, chromium; (c) zinc, mercury, cadmium; (d) antimony, bismuth, phosphorus.

1F.2 Arrange the elements in each of the following sets in order of *decreasing* atomic radius: (a) bromine, chlorine, iodine; (b) gallium, selenium, arsenic; (c) calcium, potassium, zinc; (d) barium, calcium, strontium.

1F.3 Place the following ions in order of *increasing* ionic radius: S^{2-}, Cl^-, P^{3-}.

1F.4 Which ion of each of the following pairs has the *larger* radius: (a) Ga^{3+}, In^{3+}; (b) P^{3-}, S^{2-}; (c) Pb^{2+}, Pb^{4+}?

1F.5 Which member of each pair has the *smaller* first ionization energy: (a) Ca or Mg; (b) Mg or Na; (c) Al or Na?

1F.6 Which member of each pair is likely to have the *smaller second* ionization energy: (a) Ca or Mg; (b) Mg or Na; (c) Al or Na?

1F.7 Place each of the following sets of elements in order of *decreasing* ionization energy. Explain your choices. (a) Selenium, oxygen, tellurium; (b) gold, tantalum, osmium; (c) lead, barium, cesium.

1F.8 (a) Generally, the first ionization energies of elements in the same *period* increase upon going to higher atomic number. Why? (b) Examine the data for the p-block elements given in Fig. 1F.9. Note any exceptions to the rule given in part (a). How are these exceptions explained?

1F.9 The first and second ionization energies of phosphorus, sulfur, and chlorine atoms are listed in the following table. Explain why the first ionization energies of phosphorus and sulfur are nearly the same, whereas the second ionization energy of sulfur is much greater than that of phosphorus.

	$I_1/(kJ \cdot mol^{-1})$	$I_2/(kJ \cdot mol^{-1})$
P	1011	1903
S	1000	2251
Cl	1255	2296

1F.10 The first and second ionization energies of phosphorus, sulfur, and chlorine atoms are listed in the table in Exercise 1F.9. Explain why the first ionization energy of chlorine is much greater than that of sulfur, whereas their second ionization energies are nearly the same.

1F.11 Which element of each of the following pairs has the *higher* electron affinity: (a) tellurium or iodine; (b) beryllium or magnesium; (c) oxygen or sulfur; (d) gallium or indium?

1F.12 Which element of each of the following pairs has the *higher* electron affinity: (a) germanium or selenium; (b) boron or carbon; (c) phosphorus or arsenic?

1F.13 (a) What is the inert-pair effect? (b) Why is the inert-pair effect observed only for heavy elements?

1F.14 Identify which of the following elements experience the inert-pair effect and write the formulas for the ions that they form: (a) Sb; (b) As; (c) Tl; (d) Ba.

1F.15 (a) What is a diagonal relationship? (b) How does it arise? (c) Give two examples to illustrate the concept.

1F.16 Use Appendix 2D to find the values for the atomic radii of germanium and antimony, as well as the ionic radii for Ge^{2+} and Sb^{3+}. What do these values suggest about the chemical properties of these two ions?

1F.17 Which of the following pairs of elements exhibit a diagonal relationship: (a) Li and Mg; (b) Ca and Al; (c) F and S?

1F.18 Which of the following pairs of elements do not exhibit a diagonal relationship: (a) Be and Al; (b) As and Sn; (c) Ga and Sn?

1F.19 Why are s-block metals typically more reactive than p-block metals?

1F.20 Which of the following elements are transition metals: (a) radium; (b) radon; (c) hafnium; (d) niobium; (e) cadmium?

1F.21 Identify the following elements as metals, nonmetals, or metalloids: (a) lead; (b) sulfur; (c) zinc; (d) silicon; (e) antimony; (f) cadmium.

1F.22 Identify the following elements as metals, nonmetals, or metalloids: (a) aluminum; (b) carbon; (c) germanium; (d) arsenic; (e) selenium; (f) tellurium.

The following Example and Exercises draw on material from throughout Focus 1.

FOCUS 1 Online Cumulative Example

You are working in a laboratory investigating the properties of semiconductor nanomaterials. In your research you must synthesize CdSe nanocrystals by the reaction of CdO with Se in high-temperature solutions. A solution of Se is prepared by dissolving 152.6 mg of selenium metal in 25.0 mL of 1-octadecene solvent. In a separate flask, 64.2 mg of CdO is dissolved in 3.00 mL of oleic acid and 50.0 mL of 1-octadecene at 225 °C.

(a) Write the electron configurations of Cd and Se.

(b) Based on the data in Appendix 2D, which element has (i) the higher ionization energy (I_1); (ii) the greater electron affinity (E_{ea}); (iii) the larger atomic radius (r)?

(c) CdSe can be considered a binary ionic compound. Based on the electron affinities you found in (b), which element is more likely to form an anion in this compound? Predict the charge on the ions each element is likely to form.

(d) The final product requires a 1:1 mole ratio of Cd:Se. What volume of the selenium solution must be added to the CdO solution?

(e) A sample of the material you prepared was found to emit light at 546 nm when it is excited with UV radiation. An electron in a nanocrystal can be treated as an electron trapped in a one-dimensional box and the emitted light is due to an electronic transition from the $n = 2$ to the $n = 1$ energy level. What is the diameter of the nanocrystals in your sample? Note that the effective mass (m_e^*) of an electron in CdSe (the mass to use in the energy expression) is $m_e^* = 0.090m_e$. The mass of the electron is given in the table inside the back cover of this book.

(f) What is the wavelength of radiation required to excite an electron in the nanocrystals from $n = 1$ to $n = 3$?

 The Online Cumulative Example solution can be found at http://macmillanhighered.com/chemicalprinciples7e

FOCUS 1 Exercises

1.1 Lines in the Balmer series of the hydrogen spectrum are observed at 656.3, 486.1, 434.0, and 410.2 nm. What is the wavelength of the next line in the series?

1.2 Suppose that in a certain experiment the electron in a hydrogen atom could be excited no higher than the shell with $n = 5$. (a) How many different lines could appear in the spectrum as the excited atom falls back to lower states? (b) What would be the range of wavelengths emitted? (*Hint:* To answer this question, find the wavelength of the highest-energy transition and that of the lowest-energy transition.)

1.3 In each second, a certain lamp produces 2.4×10^{21} photons with a wavelength of 633 nm. How much power (in watts) is produced as radiation at this wavelength ($1\,W = 1\,J \cdot s^{-1}$)?

1.4 In one trial on Millikan's oil drop experiment, each droplet observed by technicians contained an even number of electrons. If the technicians were unaware of this limitation, how would it affect their report of the charge of an electron?

1.5 Wavefunctions corresponding to states of different energy of a particle in a box are mutually "orthogonal" in the sense that, if the two wavefunctions are multiplied together and then integrated over the length of the box, the outcome is zero. (a) Confirm that the wavefunctions for $n = 1$ and $n = 2$ are orthogonal. (b) Demonstrate, without doing a calculation, that all wavefunctions with even n are orthogonal to all wavefunctions with odd n. (*Hint:* Think about the area under the product of any two such functions.)

1.6 Wavefunctions are "normalized" to 1. This term means that the total probability of finding an electron in the system is 1. Verify this statement for a particle-in-the-box wavefunction (Eq. 2 in Topic 1C).

1.7 The intensity of a transition between the states n and n' of a particle in a box is proportional to the square of the integral $\mu_{nn'}$, where

$$\mu_{nn'} = -e\int_0^L \psi_n x \psi_{n'} \, dx$$

(a) Can there be a transition between states with quantum numbers 3 and 1? (b) Consider the transition between states with quantum numbers 2 and 1. Does the intensity decrease or increase as the box lengthens?

1.8 Millikan measured the charge of the electron in *electrostatic units,* esu. The data that he collected included the following series of charges found on oil drops: 9.60×10^{-10} esu, 1.92×10^{-9} esu, 2.40×10^{-9} esu, 2.88×10^{-9} esu, and 4.80×10^{-9} esu. (a) From this series, find the likely charge on the electron in electrostatic units. (b) Predict the number of electrons on an oil drop with the charge 6.72×10^{-9} esu. (c) The actual charge (in coulombs) of an electron is 1.602×10^{-19} C. What is the relation between esu and coulombs?

1.9 An electron in a hydrogen atom is excited to a $2p_x$-orbital. What is the probability that the electron will be found in the region of space for which the wavefunction has a positive sign?

1.10 In the f-block there are numerous exceptions to the regular order of orbital occupation predicted by the building-up principle. Suggest why so many exceptions are noted for these elements.

1.11 Photoelectron spectroscopy (PES, see Topic 1B) can be used to determine the energies of atomic orbitals by measuring the energies required to remove electrons from them. The following peaks were observed in the photoelectron spectra for two elements.

Identify the elements, write their electron configurations, and explain your reasoning:
(a) 75.7 eV (7.30 MJ·mol^{-1}) and 5.38 eV (0.519 MJ·mol^{-1})
(b) 153 eV (14.8 MJ·mol^{-1}) and 9.33 eV (0.90 MJ·mol^{-1})

1.12 The following peaks were observed in the photoelectron spectra for two elements (see Exercise 1.11). Identify the elements, write their electron configurations, and explain your reasoning:
(a) 257 eV (24.8 MJ·mol^{-1}), 25.2 eV (2.43 MJ·mol^{-1}), and 8.29 eV (0.800 MJ·mol^{-1})
(b) 301 eV (29.0 MJ·mol^{-1}), 47.8 eV (4.61 MJ·mol^{-1}), and 11.4 eV (1.10 MJ·mol^{-1})

1.13 Ionization energies usually increase on going from left to right across the periodic table. The ionization energy for oxygen, however, is lower than that of either nitrogen or fluorine. Explain this anomaly.

1.14 Thallium is the heaviest stable member of Group 13. Aluminum is also a member of that group, and its chemical properties are dominated by the +3 oxidation state. Thallium, however, is found most usually in the +1 oxidation state. Examine this difference by plotting the first, second, and third ionization energies for the Group 13 elements against atomic number (see Appendix 2D or the periodic-table data found on the Web site for this text). Explain the trends you observe.

1.15 The German physicist Lothar Meyer observed a periodicity in the physical properties of the elements at about the same time that Mendeleev was working on their chemical properties. Some of Meyer's observations can be reproduced by examining the molar volume for the solid element as a function of atomic number. Calculate the molar volumes for the elements in Periods 2 and 3 from the densities of the elements found in Appendix 2D and the following solid densities (in g·cm^{-3}): nitrogen, 0.88; fluorine, 1.11; neon, 1.21. Plot your results against atomic number and explain any variations that you observe.

1.16 In the spectroscopic technique known as photoelectron spectroscopy (PES), ultraviolet radiation is directed at an atom or a molecule. Electrons are ejected from the valence shell, and their kinetic energies are measured. Because the energy of the incoming ultraviolet photon is known and the kinetic energy of the outgoing electron is measured, the ionization energy, I, can be deduced from the fact that the total energy is conserved. (a) Show that the speed v of the ejected electron and the frequency n of the incoming radiation are related by $hv = I + \frac{1}{2}m_ev^2$. (b) Use this relation to calculate the ionization energy of a rubidium atom, given that radiation of wavelength 58.4 nm produces electrons with a speed of 2450 km·s^{-1}; note that 1 J = 1 kg·m^2·s^{-2}.

1.17 Apparent anomalies in the filling of electron orbitals in atoms occur in chromium and copper. In these elements an electron expected to occupy an s-orbital occupies a d-orbital instead. (a) Explain why these anomalies occur. (b) Similar anomalies are known to occur in seven other elements of the d-block. Using Appendix 2C, identify those elements and indicate for which ones the explanation used to rationalize the chromium and copper electron configurations is valid. (c) Explain why there are no elements in which electrons fill $(n + 1)$s-orbitals instead of np-orbitals.

1.18 The electron in a hydrogen atom is excited to a 4d-orbital. Calculate the energy of the photon released if the electron were then to move to each of the following orbitals: (a) 1s; (b) 2p; (c) 2s; (d) 4s. (e) If the outermost electron in a potassium atom were excited to a 4d-orbital and then fell to the same orbitals, describe qualitatively how the emission spectrum would differ from that of hydrogen (do not do any calculations). Explain your answer.

1.19 The following properties are observed for an unknown element. Identify the element from its properties. (a) The neutral atom has two unpaired electrons with $l = 2$. (b) The previous noble gas in the periodic table is krypton.

1.20 The electron affinity of thulium has been measured by a technique called *laser photodetachment electron spectroscopy*. In this technique a gaseous beam of the anions of an element is bombarded with photons from a laser. The photons eject electrons from some of the anions and the energies of the ejected electrons are detected. The incident radiation had a wavelength of 1064 nm and the ejected electrons had an energy of 0.137 eV. Although the analysis is somewhat more complicated, an estimate of the electron affinity can be obtained as the difference in energy between the photons and the energy of the ejected electrons. What is the electron affinity of thulium in electronvolts and in kilojoules per mole?

1.21 Francium is thought to be the most reactive of the alkali metals. Because it is radioactive and available in only very small amounts, it is difficult to study. However, its properties can be predicted based on its location in Group 1 of the periodic table. Estimate the following properties of francium: (a) atomic radius; (b) ionic radius of the +1 cation; (c) ionization energy.

1.22 Below is pictured the reaction between an atom of magnesium and an atom of oxygen. Identify each element and the ions formed and explain your reasoning.

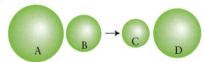

1.23 Below is pictured the reaction between an atom of sodium and an atom of chlorine. Identify each element and the ions formed and explain your reasoning.

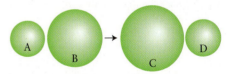

1.24 This plot shows the radial distribution function of the 3s- and 3p-orbitals of a hydrogen atom. Identify each curve and explain how you made your decision.

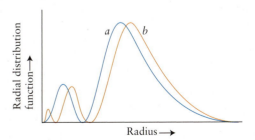

1.25 Suppose that in some other universe a rule corresponding to the Pauli exclusion principle reads "as many as two electrons in the same atom may have the same set of four quantum numbers." Suppose further that all other factors affecting electron configurations are unchanged. (a) Give the electron configuration of the element in the other universe that has five protons. (b) What is the most likely charge on the ion of this element? (c) Give the value of Z for the second inert gas in the other universe. Explain your reasoning.

1.26 Imagine a four-dimensional world. In that world, atoms would have one s-orbital and four p-orbitals in a given shell. (a) Describe the form of the periodic table for the first 24 elements. (b) Which elements would be the first two noble gases (use the names from our world that correspond to the atomic numbers).

1.27 A team of scientists recently developed a series of nanowires (very tiny wires) that can act as miniature lasers. (a) One nanowire was found to emit light of frequency 6.27×10^{14} Hz. What was the wavelength of the light emitted? (b) A second nanowire was found to emit light with a wavelength of 421 nm. What was the frequency of the light from the second wire?

1.28 A team of scientists recently developed a series of nanowires (very tiny wires) that can act as miniature lasers. (a) One nanowire was found to emit light of frequency 7.83×10^{14} Hz. What was the wavelength of the light radiated? (b) A second nanowire was found to emit light with a wavelength of 452 nm. What was the frequency of the light from the second wire?

1.29 Geiger counters can detect radioactivity because nuclear radiation consists of particles or radiation of sufficiently high energy to eject electrons from atoms. Consequently, this type of radiation is called "ionizing radiation." What is the longest wavelength of radiation that can be detected by a Geiger counter using neon gas as the ionizing medium?

1.30 Many fireworks mixtures depend on the highly energetic combustion of magnesium, in which the heat causes the oxide to become incandescent and to give out a bright white light. The color of the light can be changed by including nitrates and chlorides of elements that have emissions in the visible region of their spectra. Barium nitrate is often added to produce a yellow-green color. The excited barium ions generate light of wavelength 487 nm, 524 nm, 543 nm, 553 nm, and 578 nm. For each case, calculate (a) the change in energy (in electronvolts) of a barium atom and (b) the molar change in energy (in kilojoules per mole).

1.31 In a recent suspense film, two secret agents must penetrate a criminal's stronghold monitored by a lithium photomultiplier cell that is continually bathed in light from a laser. If the beam of light is broken, an alarm sounds. The agents want to use a hand-held laser to illuminate the cell while they pass in front of it. They have two lasers, a high-intensity red ruby laser (694 nm) and a low-intensity violet GaN laser (405 nm), but they disagree on which one would be better. Determine (a) which laser they should use and (b) the kinetic energy of the electrons emitted. The work function of lithium is 2.93 eV.

1.32 Clouds of hot, glowing interstellar hydrogen gas can be found in some parts of our galaxy. In some of the hydrogen atoms, electrons may be excited to quantum levels with $n = 100$ or more. (a) Calculate the wavelength that would be observed on Earth if electrons fell from the level with $n = 100$ to that with $n = 2$. (b) In what series would this transition be found? (c) Some of these high-energy electrons fall to intermediate states, such as $n = 90$. Would the wavelengths of a transition from a state with $n = 100$ to one with $n = 90$ be longer or shorter than those in the Balmer series? Explain your reasoning.

FOCUS 1 Cumulative Exercises

1.33 Electrons in molecules are described by wavefunctions that extend over more than one atom. Consider an electron that is described by a wavefunction that extends over two adjacent carbon atoms. The electron can move freely between the two atoms. The internuclear C—C distance is 139 pm.

(a) Using the one-dimensional particle-in-the-box model, calculate the energy required to promote an electron from the $n = 1$ to the $n = 2$ level, assuming that the length of the box is the same as the distance between the two carbon atoms.

(b) To what wavelength of radiation does this correspond?

(c) If each atom in a linear chain of 10 carbon atoms contributes one electron, what is the minimum number of wavefunctions required to account for all the electrons?

(d) Repeat the calculation in (a) for a linear chain of 10 carbon atoms with the same internuclear C—C distance (139 pm) but in which the transition takes place from the uppermost filled level to the one above.

(e) To what wavelength of radiation does the transition in part (d) correspond?

(f) A certain compound with a long chain of carbon atoms is found to require light of 696 nm to promote an electron from the $n = 6$ to the $n = 7$ level. How long is the chain of carbon atoms in this molecule?

1.34 "Green" chemistry methods, which use nontoxic chemicals, are replacing elemental chlorine for the bleaching of paper pulp. Chlorine causes problems because it is a strong oxidizing agent that reacts with organic compounds to form toxic byproducts such as furan and dioxins.

(a) Write the electron configuration of a chlorine atom in its ground state. How many unpaired electrons are present in the atom? Write the electron configuration you expect a chloride ion to have. The electron configuration of the chloride ion is identical to that of a neutral atom of what other element?

(b) When a chlorine atom is excited by heat or light, one of its valence electrons may be promoted to a higher energy level. Predict the most likely electron configuration for the lowest possible excited state for an excited chlorine atom.

(c) Estimate the wavelength (in nm) of the energy that needs to be absorbed for the electron to reach the excited state in part (b). To make this estimate, use Eq. 2 from Topic 1E and take the effective nuclear charge from Fig. 1F.3.

(d) What is the value of the energy required in part (c) in kilojoules per mole and electronvolts?

(e) The proportion of ^{37}Cl in a typical sample is 75.77%, with the remainder being ^{35}Cl. What would the molar mass of a sample of chlorine atoms be if the proportion of ^{37}Cl were reduced to half its current value? The mass of an atom of ^{35}Cl is 5.807×10^{-23} g and that of an atom of ^{37}Cl is 6.139×10^{-23} g.

(f) What are the oxidation numbers of chlorine in the bleaching agents ClO_2 and $NaClO$?

(g) What are the oxidation numbers of chlorine in the oxidizing agents $KClO_3$ and $NaClO_4$?

(h) Write the names of the compounds in parts (f) and (g).

MOLECULES

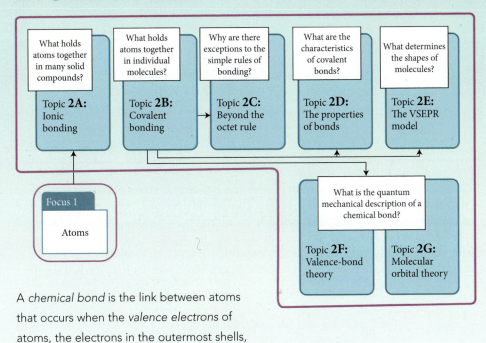

A *chemical bond* is the link between atoms that occurs when the *valence electrons* of atoms, the electrons in the outermost shells, move to new locations and settle into lower energy configurations. If the lowest energy can be achieved by the *complete transfer* of one or more electrons from each atom of one element to those of another, then ions form and the compound is held together by the electrostatic attraction between them. This attraction, which is described in **TOPIC 2A**, is called an *ionic bond*. We explain the ions that elements form by using a simple rule based on *Lewis symbols*. If the lowest energy can be achieved by *sharing* electrons, then the atoms link through a *covalent bond* and discrete molecules are formed, as described in **TOPIC 2B**. The pattern of bonds in molecules is commonly expressed by using some simple rules to draw a *Lewis structure*.

All descriptions of bonding are models. **TOPIC 2C** shows how the descriptions of both basic types of bond are improved and exceptions to the simple rules are encountered. **TOPIC 2D** shows how the properties of bonds, such as their strengths and lengths, can be explained and values transferred between molecules, and **TOPIC 2E** shows how the three-dimensional shape of a molecule can be predicted on the basis of a simple model based on the electrostatic (coulombic) interaction between pairs of electrons.

None of those descriptions draw on quantum theory directly. Modern theories of molecular structure are based on the wave nature of electrons and **TOPICS 2F** and **2G** introduce two rival theories that describe the distribution of electrons in terms of the occupation of orbitals. These models introduce a language that is used throughout chemistry and they help to explain the simpler models.

Topic 2A Ionic Bonding

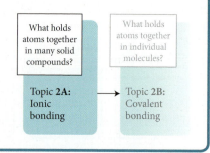

What holds atoms together in many solid compounds?	What holds atoms together in individual molecules?
Topic 2A: Ionic bonding	→ Topic 2B: Covalent bonding

Why Do You Need to Know This Material? Ionic bonding is one of the principal forms of bonding between atoms. Understanding how bonds form between ions allows you to predict the formulas of ionic compounds and to estimate how strongly the ions are held together.

What Do You Need to Know Already? You need to know about electron configurations of many-electron atoms (Topic 1E), the concept of potential energy, and the nature of the Coulomb interaction between charges (*Fundamentals* A). You need to be familiar with ionic radii and the ionization energy and electron affinity of elements (Topic 1F).

The **ionic model** is the description of chemical bonding in terms of ions. It is particularly appropriate for describing binary compounds formed between metallic and nonmetallic elements. An **ionic solid** is an assembly of cations and anions stacked together in a regular array. In sodium chloride, for instance, sodium ions alternate with chloride ions in all three dimensions (**FIG. 2A.1**). Ionic solids are examples of **crystalline solids,** or solids that consist of atoms, molecules, or ions stacked together in a regular pattern.

2A.1 The Ions That Elements Form

When an atom of a metallic element in the s-block forms a cation, it loses electrons down to its noble-gas core (**FIG. 2A.2**). That core typically has an eight-electron ns^2np^6 outer electron configuration, which is called an **octet** of electrons. For example, a sodium atom ($[Ne]3s^1$) loses its 3s-electron to form Na^+, which has the same electron configuration as a neon atom, $[Ne]$ or $1s^12s^22p^6$. The Na^+ ions cannot lose more electrons in a chemical reaction because the ionization energies of core electrons are so high. There are three exceptions to octet formation early in the periodic table. Hydrogen loses its only electron to form a bare proton. Lithium ($[He]2s^1$) and beryllium ($[He]2s^2$) atoms lose their 2s-electrons, leaving a heliumlike **duplet,** a pair of electrons with the heliumlike configuration $1s^2$, when they become Li^+ and Be^{2+} ions. Some typical electron configurations of atoms and the ions they form are shown in **TABLE 2A.1**.

When the atoms of metallic elements on the left of the p-block in Periods 2 and 3 lose their valence electrons, they form ions with the electron configuration of the preceding noble gas. Aluminum, $[Ne]3s^23p^1$, for instance, forms Al^{3+} with the same configuration as neon. When the metallic p-block elements in Period 4 and later periods lose their s- and p-electrons, they leave a noble-gas core surrounded by an additional, complete

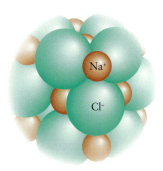

FIGURE 2A.1 This tiny fragment of sodium chloride is an example of an ionic solid. The sodium ions are represented by red spheres and the chloride ions by green spheres. An ionic solid consists of an array of enormous numbers of cations and anions stacked together to give the lowest-energy arrangement. The pattern shown here is repeated throughout the crystal.

ANIMATION FIGURE 2A.1

TABLE 2A.1	Electron Configurations of Some Atoms and the Ions They Form		
Atom	**Configuration**	**Ion**	**Configuration**
Li	$[He]2s^1$	Li^+	$[He]$ ($= 1s^2$)
Be	$[He]2s^2$	Be^{2+}	$[He]$
Na	$[Ne]3s^1$	Na^+	$[Ne]$ ($= [He]2s^22p^6$)
Mg	$[Ne]3s^2$	Mg^{2+}	$[Ne]$
Al	$[Ne]3s^23p^1$	Al^{3+}	$[Ne]$
N	$[He]2s^22p^3$	N^{3-}	$[Ne]$
O	$[He]2s^22p^4$	O^{2-}	$[Ne]$
F	$[He]2s^22p^5$	F^-	$[Ne]$
S	$[Ne]3s^23p^4$	S^{2-}	$[Ar]$ ($= [Ne]3s^23p^6$)
Cl	$[Ne]3s^23p^5$	Cl^-	$[Ar]$

subshell of d-electrons. For instance, gallium forms the ion Ga^{3+} with the configuration $[Ar]3d^{10}$. The d-electrons of the p-block atoms are gripped tightly by the nucleus and, in most cases, are not lost.

Many metallic elements, such as those in the p- and d-blocks, have atoms that can lose a variable number of electrons and so exhibit **variable valence.** As described in Topic 1F, the inert-pair effect implies that the elements listed in Fig. 1F.15, which is repeated here as **FIG. 2A.3**, can lose either their valence p-electrons alone or all their valence p- and s-electrons. These elements and the d-block metals can form compounds with different oxidation numbers, such as tin(II) oxide, SnO, and tin(IV) oxide, SnO_2, for tin. Many elements of the d-block also display variable valence by losing different numbers of d-electrons after their valence s-electrons have been removed. In the d-block, the ns-electrons are lost first, followed by a variable number of $(n-1)$d-electrons. For example, to obtain the Fe^{2+} ion, two 4s-electrons are removed from the Fe atom, which is $[Ar]3d^64s^2$, to give the configuration $[Ar]3d^6$, and then a third electron is removed from the 3d-subshell, giving Fe^{3+} with the configuration $[Ar]3d^5$.

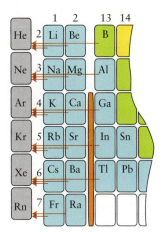

FIGURE 2A.2 When a main-group metal atom forms a cation, it loses its valence s- and p-electrons and acquires the electron configuration of the preceding noble-gas atom. The heavier atoms in Groups 13 and 14 retain their complete subshells of d-electrons.

The formulas of some common cations are shown in *Fundamentals* Fig. C.6.

EXAMPLE 2A.1   Writing the electron configurations of cations

Salts of indium(III) are used in some nutritional supplements, despite known side effects, because it is claimed that they can enhance the speed of memory, balance hormones, and reduce the need for sleep. However, indium(I) salts are unstable in water and so cannot be used in the diet. If you were working in a pharmaceutical laboratory, you would need to be able to distinguish the properties of these two ions. Write the electron configurations of (a) In^+ and (b) In^{3+}.

ANTICIPATE Because indium, In, is in Group 13, with the "generic" electron configuration $[core]s^2p^1$, you should expect successive loss of the p-electron and then the two s-electrons to give the configurations $[core]s^2$ and $[core]$, respectively.

PLAN Identify the configuration of the neutral atom from its position in the periodic table. Remove electrons from the valence-shell p-orbitals first, then from the s-orbitals, and finally, if necessary, from the d-orbitals in the next-lower shell, until the number of electrons removed equals the charge on the ion.

SOLVE

Identify the configuration of the neutral atom.

Indium is in Group 13, Period 5. Its ground-state configuration is therefore $[Kr]4d^{10}5s^25p^1$.

(a) Remove the outermost electron (from the 5p orbital).

In^+ is $[Kr]4d^{10}5s^2$

(b) Remove the next two electrons (from the 5s orbital).

In^{3+} is $[Kr]4d^{10}$

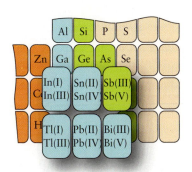

FIGURE 2A.3 The typical ions formed by the heavy elements in Groups 13 through 15 show the influence of the inert pair—the tendency to form compounds in which the oxidation numbers differ by 2.

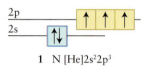

1 N [He]2s²2p³

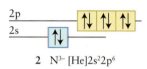

2 N³⁻ [He]2s²2p⁶

The formulas of some common anions are shown in *Fundamentals* Fig. C.7.

Nonmetals rarely lose electrons in chemical reactions because their ionization energies are too high. However, an atom of a nonmetallic element can acquire enough electrons to complete its valence shell and form an anion with an octet corresponding to the configuration of the next noble gas (1s² in the case of the hydride ion, H⁻), **FIG. 2A.4**. When the electron affinity of the atom is positive, energy is released in this step, but in some cases the electron affinity is negative and this process requires energy (as in the formation of O²⁻ from O). An O atom does not gain more electrons than that, because any additional electrons would have to be accommodated in a higher-energy shell and the energy demands would be too great. It follows that to write the formula for a monatomic anion, you need to add enough electrons to complete the valence shell. For example, nitrogen has five valence electrons (**1**); so adding three more electrons results in a noble-gas configuration, that of neon. Therefore, the nitride ion will be N³⁻ (**2**), which has the electron configuration of neon, the next noble gas.

To predict the electron configuration of a monatomic cation, remove the outermost electrons in the order np, ns, and (n − 1)d; for a monatomic anion, add electrons until the next noble-gas configuration has been reached. The transfer of electrons results in the formation of an octet (or duplet) of electrons in the valence shell on each of the atoms: metal atoms achieve an octet (or duplet) by electron loss and nonmetal atoms achieve it by electron gain.

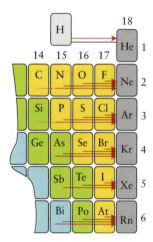

FIGURE 2A.4 When nonmetal atoms acquire electrons and form anions, they do so until they have reached the electron configuration of the next noble gas.

2A.2 Lewis Symbols

Many of the basic ideas about the chemical bond were proposed by G. N. Lewis in the early years of the twentieth century. Lewis devised a simple way to keep track of valence electrons when atoms form ionic bonds. He represented each valence electron as a dot and arranged the dots around the symbol of the element. A single dot represents an electron alone in an orbital; a pair of dots represents two paired electrons sharing an orbital. Examples of the **Lewis symbols** of atoms are

$$H\cdot \quad He: \quad :\overset{\cdot}{N}\cdot \quad \cdot\overset{\cdot\cdot}{O}: \quad :\overset{\cdot\cdot}{C}l\cdot \quad K\cdot \quad Mg:$$

The Lewis symbol for nitrogen, for example, represents the valence electron configuration 2s²2pₓ¹2pᵧ¹2p_z¹ (see structure **1**), with two electrons paired in a 2s-orbital and three unpaired electrons in different 2p-orbitals. The Lewis symbol is a visual summary of the valence-shell electron configuration of an atom or ion.

To work out the formula of an ionic compound by using Lewis symbols:

- Represent the cation by removing the appropriate number of dots from the symbol for the atom of the metallic element.
- Represent the anion by transferring those dots to the Lewis symbol for the atom of the nonmetallic elements to complete its valence shell.

- If necessary, adjust the numbers of atoms of each kind so that all the dots removed from the atom of the metallic element are accommodated by the atom of the non-metallic element.
- Write the charge of each ion as a superscript in the normal way.

A simple example is the formula of calcium chloride. The calcium atom loses its two valence electrons when it forms the Ca^{2+} ion. Because each chlorine atom has one vacancy, two are required to accept the two electrons from the calcium atom:

$$:\ddot{C}l\cdot \; + \; Ca: \; + \; :\ddot{C}l\cdot \longrightarrow \; :\ddot{C}l:^- \;\; Ca^{2+} \;\; :\ddot{C}l:^-$$

The ratio of two chloride ions for each calcium ion results in the formula $CaCl_2$. However, note that this is just a formula unit (*Fundamentals* E). There are no individual $CaCl_2$ molecules: a crystal of $CaCl_2$ consists of huge numbers of these ions in three-dimensional arrays.

Self-test 2A.3A   Draw the formula unit of lithium nitride using Lewis symbols.

[*Answer:* $Li^+ \; Li^+ \; :\ddot{N}:^{3-} \; Li^+$]

Self-test 2A.3B   Draw the formula unit of magnesium bromide using Lewis symbols.

Formulas of compounds consisting of the monatomic ions of main-group elements can be predicted by assuming that cations have lost all their valence electrons and anions have gained electrons in their valence shells until each ion has an octet of electrons, or a duplet in the case of H, Li, and Be.

2A.3 The Energetics of Ionic Bond Formation

To understand why a crystal of sodium chloride, an ionic compound, has a lower energy than widely separated sodium and chlorine atoms, you can use a strategy chemists often find useful: they analyze a complex process by breaking it down into simpler, often hypothetical steps. In this case, the formation of the solid is imagined as taking place in three hypothetical steps:

1.   Gaseous sodium atoms release electrons.
2.   These electrons attach to gaseous chlorine atoms.
3.   The resulting gaseous cations and anions clump together as a solid crystal.

Sodium is in Group 1 of the periodic table and can be expected to form a +1 ion. However, the valence electron is tightly held by the effective nuclear charge—it does not just fall off. In fact, the ionization energy of sodium is 494 kJ·mol^{-1} (see Fig. 1F.8), and so that much energy must be supplied to form the cations:

$$Na(g) \longrightarrow Na^+(g) + e^-(g) \qquad \text{energy required} = 494 \text{ kJ·mol}^{-1}$$

The electron affinity of chlorine atoms is $+349 \text{ kJ·mol}^{-1}$ (see Fig. 1F.12), and so 349 kJ·mol^{-1} of energy is *released* when electrons attach to chlorine atoms to form anions:

$$Cl(g) + e^-(g) \longrightarrow Cl^-(g) \qquad \text{energy released} = 349 \text{ kJ·mol}^{-1}$$

At this stage, the net change in energy (energy required − energy released) is $494 - 349 \text{ kJ·mol}^{-1} = +145 \text{ kJ·mol}^{-1}$, an *increase* in energy. A gas of widely separated Na^+ and Cl^- ions has a higher energy than a gas of neutral Na and Cl atoms.

Now consider what happens when gaseous Na^+ and Cl^- ions come together to form a crystalline solid. The difference in energy between the ions of a compound widely separated as a gas and packed together in a solid is the **lattice energy,** which is usually very large. This energy is released when the solid forms:

$$Na^+(g) + Cl^-(g) \longrightarrow NaCl(s) \qquad \text{energy released} = 787 \text{ kJ·mol}^{-1}$$

A positive electron affinity signifies a release of energy when an electron attaches to a gas-phase atom or ion (Topic 1F).

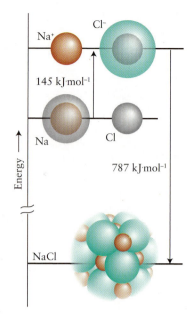

FIGURE 2A.5 Considerable energy is needed to produce cations and anions from neutral atoms; the ionization energy of the metal atoms is only partly recovered from the electron affinity of the nonmetal atoms. The overall lowering of energy that drags the ionic solid into existence arises from the strong attraction between cations and anions that occurs in the solid. This diagram does not show how the actual chemical reaction between Na(s) and $Cl_2(g)$ occurs; it simply illustrates the energies of Na(g) and Cl(g) atoms relative to the energy of NaCl(s).

FIGURE 2A.6 This sequence of images illustrates why ionic solids are brittle. (a) The original solid consists of an orderly array of cations and anions. (b) A hammer blow can push ions with like charges into adjacent positions; this proximity of like charges raises strong repulsive forces (as depicted by the double-headed arrows). (c) As the result of these repulsive forces, the solid breaks apart into fragments. (d) The smooth faces of this calcite crystal result from the regular arrangement of the calcium and carbonate ions. (e) The blow of a hammer has shattered the crystal, leaving flat, regular surfaces consisting of planes of ions. *(Parts (d) and (e) © 2009 Paul Silverman–Fundamental Photographs.)*

Therefore, the net change in energy for the overall process Na(g) + Cl(g) → NaCl(s) is $145 - 787$ kJ·mol^{-1} = -642 kJ·mol^{-1} (**FIG. 2A.5**), a huge *decrease* in energy. That is, a solid composed of Na$^+$ and Cl$^-$ ions has a lower energy than does a collection of widely separated Na and Cl atoms.

In summary, there is a net lowering of energy below that of the individual atoms, provided the net attraction between ions is greater than the energy needed to make them. The major contribution to the energy input is normally the ionization energy of the element that forms the cation. Although some of this energy may be recovered from the electron affinity of the nonmetal when the anion is formed, in some cases energy is also needed to make the anion. That energy must also be recovered from the interactions between ions. Typically, *only metallic elements have ionization energies that are low enough for the formation of ionic bonds to be energetically feasible.*

The energy lowering accompanying the formation of ionic bonds is due largely to the attraction between oppositely charged ions.

2A.4 Interactions Between Ions

The preceding discussion has shown that a key contribution to the formation of ionic bonds is the strength of the interaction between ions in a solid: it must be strong enough to overcome the energy investment needed to make the ions. However, a very important point is that an ionic solid is not held together by bonds between specific pairs of ions: *all* the cations interact to a greater or lesser extent with *all* the anions, *all* the cations repel one another, and *all* the anions repel one another. An ionic bond is a "global" interaction characteristic of the entire crystal, a net lowering of energy of the entire crystal relative to widely separated neutral atoms.

The strong electrostatic interactions between ions account for the typical properties of ionic solids, such as their high melting points and brittleness. A high temperature is required before the ions are able to move past one another and the solid melts to form a liquid. When an ionic solid is struck, ions with like charges come into contact, repel one another, and the solid shatters into fragments (**FIG. 2A.6**).

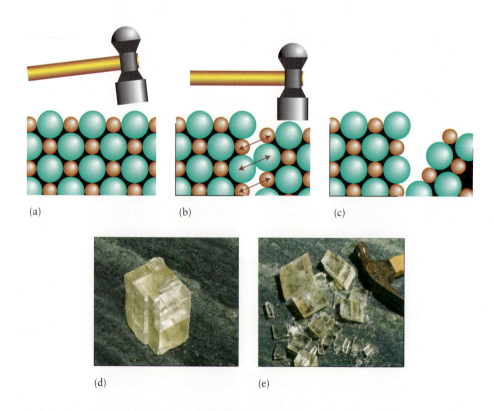

(a)   (b)   (c)

(d)   (e)

The starting point for understanding the interaction between ions in a solid is the expression for the Coulomb potential energy of the interaction of two individual ions (*Fundamentals* A):

$$E_{p,12} = \frac{\overbrace{(z_1 e)}^{\substack{\text{charge of}\\\text{ion 1}}} \times \overbrace{(z_2 e)}^{\substack{\text{charge of}\\\text{ion 2}}}}{4\pi\varepsilon_0 \underbrace{r_{12}}_{\text{separation}}} = \frac{z_1 z_2 e^2}{4\pi\varepsilon_0 r_{12}} \tag{1}$$

In this expression, e is the fundamental charge (the absolute value of the charge of an electron), z_1 and z_2 are the charge numbers of the two ions, r_{12} is the distance between the centers of the ions, and ε_0 ("epsilon zero") is the vacuum permittivity (see inside the back cover for the value of this fundamental constant).

A Note on Good Practice: The charge number, z, is positive for cations and negative for anions, and the charge of an ion is ze. However, chemists almost always refer to z itself as the charge and speak of a charge of $+1$, -1, and so on.

Each ion in a solid experiences attractions from all the other oppositely charged ions and repulsions from all the other ions with like charges. The total potential energy is the sum of all these contributions. Each cation is surrounded by anions, and there is a large negative (energy-lowering) contribution from the attraction of the opposite charges. Beyond those nearest neighbors, there are cations that contribute a positive (repulsive, energy-raising) term to the total potential energy of the central cation. There is also a negative contribution from the anions beyond those cations, a positive contribution from the cations beyond them, and so on, to the edge of the solid. These repulsions and attractions become progressively weaker as the distance from the central ion increases, but because the nearest neighbors of an ion give rise to a strong attraction, the net outcome of all these contributions is a lowering of energy. How far the energy is lowered can be assessed by using Eq. 1.

How Is That Done?

Consider a simple model consisting of a single line of uniformly spaced alternating cations and anions, with d the distance between their centers, the sum of the ionic radii (**FIG. 2A.7**). If the charge numbers of the ions have the same magnitude ($+1$ and -1, or $+2$ and -2, for instance), then $z_1 = +z$, $z_2 = -z$, and $z_1 z_2 = -z^2$. The potential energy of the central ion is calculated by summing all the Coulomb potential energy terms, with negative terms representing attractions to oppositely charged ions and positive terms representing repulsions from like-charged ions. For the interaction arising from ions extending in a line to the right of the central ion, the total potential energy of the central ion is

$$E_p = \frac{e^2}{4\pi\varepsilon_0} \times \left(\overbrace{-\frac{z^2}{d}}^{\text{attraction}} \overbrace{+\frac{z^2}{2d}}^{\text{repulsion}} \overbrace{-\frac{z^2}{3d}}^{\text{attraction}} \overbrace{+\frac{z^2}{4d}}^{\text{repulsion}} - \cdots \right)$$

$$= -\frac{z^2 e^2}{4\pi\varepsilon_0 d}\left(1 - \frac{1}{2} + \frac{1}{3} - \frac{1}{4} + \cdots \right) = -\frac{z^2 e^2}{4\pi\varepsilon_0 d} \times \ln 2$$

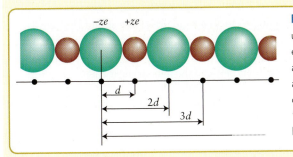

−ze +ze

d

$2d$

$3d$

FIGURE 2A.7 The arrangement used to calculate the potential energy of an ion in a line of alternating cations (red spheres) and anions (green spheres). We concentrate on one ion, the "central" ion denoted by the longest vertical line.

The last step used the relation $1 - \frac{1}{2} + \frac{1}{3} - \frac{1}{4} + \ldots = \ln 2$. Next, multiply E_p by 2 to obtain the total energy arising from interactions with ions on both sides of the central ion. Then multiply by Avogadro's constant, N_A, to obtain the potential energy per mole of ions. At this point, you know the total potential energy per mole of ions of one kind (cations, for instance):

$$E_p(\text{cations}) = -2 \ln 2 \times \frac{N_A z^2 e^2}{4\pi\varepsilon_0 d}$$

The same expression applies to the energy per mole of the anions present:

$$E_p(\text{anions}) = -2 \ln 2 \times \frac{N_A z^2 e^2}{4\pi\varepsilon_0 d}$$

However, the two expressions cannot simply be added together to get the total potential energy, for that would count each interaction twice: first ion with second, then second with first again. Therefore, the total energy per mole of pairs of ions is half this sum, which is

$$E_p = \tfrac{1}{2}\{E_p(\text{cations}) + E_p(\text{anions})\} = -2 \ln 2 \times \frac{N_A z^2 e^2}{4\pi\varepsilon_0 d}$$

with $d = r_{\text{cation}} + r_{\text{anion}}$ the distance between the centers of neighboring ions.

The calculation shows that the molar potential energy of a one-dimensional crystal in which cations and anions of equal but opposite charge alternate along a line has the form

$$E_p = -A \times \frac{N_A z^2 e^2}{4\pi\varepsilon_0 d} \tag{2}$$

with $A = 2 \ln 2$ (or 1.386) for this model system.

What Does This Equation Tell You? Because the potential energy is negative, there is a net lowering of energy, which means that the attraction between opposite charges overcomes the repulsion between like charges. The potential energy is strongly negative when the ions are highly charged (large values of z) and the separation between them is small (small values of d), which is the case when the ions themselves are small.

The calculation that led to Eq. 2 can be extended to more realistic three-dimensional arrays of ions with different charges. The result has the same form but with different values of A and $|z_1 z_2|$ (that is, the absolute value of $z_1 z_2$, its value without the negative sign) in place of z^2. The factor A is a numerical coefficient called the **Madelung constant**; its value depends on how the ions are arranged about one another. In all cases, the energy lowering that occurs when an ionic solid forms is greatest for small, highly charged ions. For example, there is a strong interaction between the Mg^{2+} and the O^{2-} ions in magnesium oxide, MgO, because the ions have high charges and small radii (so their centers are close together). This strong interaction is one reason why magnesium oxide survives at such high temperatures that it can be used for furnace linings. It is an example of a "refractory" material, a substance that can withstand high temperatures. You can also now see why Nature has adopted an ionic solid, calcium phosphate, for our skeletons: the doubly charged small Ca^{2+} ions and the triply charged PO_4^{3-} ions attract one another very strongly and clump together tightly to form a rigid, insoluble solid (**FIG. 2A.8**).

Self-test 2A.4A The ionic solids CaO and KCl crystallize to form structures of the same type. In which compound are the interactions between the ions stronger and what factors lead to that difference?

[*Answer:* CaO, higher charges and smaller radii]

Self-test 2A.4B The ionic solids KBr and KCl crystallize to form structures of the same type. In which compound are the interactions between ions stronger?

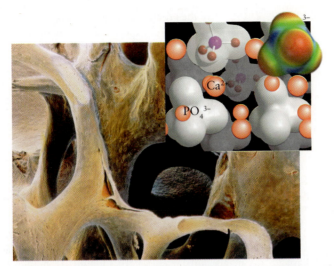

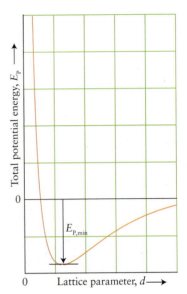

FIGURE 2A.8 A micrograph of bone, which owes its rigidity to calcium phosphate. The overlay shows part of the crystal structure of calcium phosphate. Phosphate ions are polyatomic ions; however, as shown in the inset, they are nearly spherical and fit into crystal structures in much the same way as monatomic ions of charge −3. (Prof. P. Motta/Science Source.)

FIGURE 2A.9 The potential energy of interaction between two close neighboring ions in an ionic solid, taking into account the Coulomb interaction of the ions and the increase in their repulsion when they are in contact.

The potential energy in Eq. 2 becomes more negative (that is, the favorable interaction between ions strengthens) as the separation d decreases. However, a collection of ions does not collapse to a point because repulsive effects between neighbors become important as soon as they come into contact, and the energy quickly rises again. To take the repulsion effects between close neighbors in an ionic solid into account, it is commonly supposed that the repulsive contribution to the potential energy rises exponentially with decreasing separation and therefore has the form $E_p{}^* \propto e^{-d/d^*}$ with d^* a constant (it is commonly taken to be 34.5 pm). The total potential energy is the sum of E_p and $E_p{}^*$ and passes through a minimum as the ions approach each other but then rises sharply again when they are very close (FIG. 2A.9).

Ionic solids typically have high melting points and are brittle. The lattice energy of an ionic solid is large when the ions are small and highly charged.

What have you learned in this Topic?

You have learned that in ionic bonding, electrons are transferred from one atom to another and that the patterns of ionic bond formation are represented by formula units based on Lewis symbols. You have seen that the greater the charge and the smaller the ion, the stronger is the energy of the ionic lattice.

The skills you have mastered are the ability to:

☐ **1.** Write the electron configuration for an ion (Example 2A.1 and Self-Test 2A.2).

☐ **2.** Account for the formation of ions in terms of ionization energy, electron affinity, and lattice energy (Section 2A.3).

☐ **3.** Predict the chemical formula of an ionic compound and draw its formula unit by using Lewis symbols (Self-Test 2A.3).

☐ **4.** Account for the origin and magnitude of the lattice energy (Section 2A.4).

Topic 2A Exercises

2A.1 Give the number of valence electrons (including d electrons) for each of the following elements: (a) Sb; (b) Si; (c) Mn; (d) B.

2A.2 Give the number of valence electrons (including d electrons) for each of the following elements: (a) V; (b) Fe; (c) Cd; (d) I.

2A.3 Give the ground-state electron configuration expected for each of the following ions: (a) S^{2-}; (b) As^{3+}; (c) Ru^{3+}; (d) Ge^{2+}.

2A.4 Give the ground-state electron configuration expected for each of the following ions: (a) V^{4+}; (b) Fe^{3+}; (c) Cd^{2+}; (d) I^-.

2A.5 Give the ground-state electron configuration expected for each of the following ions: (a) Cu^+; (b) Bi^{3+}; (c) Ga^{3+}; (d) Tl^{3+}.

2A.6 Give the ground-state electron configuration expected for each of the following ions: (a) Zr^{4+}; (b) Os^{3+}; (c) Cs^+; (d) P^{3-}.

2A.7 The following species have the same number of electrons: Cd, In^+, and Sn^{2+}. (a) Write the electron configurations for each species. Explain any differences. (b) How many unpaired electrons, if any, are present in each species? (c) What neutral atom, if any, has the same electron configuration as that of In^{3+}?

2A.8 The following species have the same number of electrons: Ca, Ti^{2+}, and V^{3+}. (a) Write the electron configurations for each ion. Explain any differences. (b) How many unpaired electrons, if any, are present in each species? (c) What neutral atom, if any, has the same electron configuration as that of Ti^{3+}?

2A.9 Which M^{2+} ions (where M is a metal) are predicted to have the following ground-state electron configurations: (a) $[Ar]3d^7$; (b) $[Ar]3d^6$; (c) $[Kr]4d^4$; (d) $[Kr]4d^3$?

2A.10 Which E^{3+} ions (where E is an element) are predicted to have the following ground-state electron configurations: (a) $[Xe]4f^{14}5d^8$; (b) $[Xe]4f^{14}5d^5$; (c) $[Kr]4d^{10}5s^25p^2$; (d) $[Ar]3d^{10}4s^2$?

2A.11 Which M^{3+} ions (where M is a metal) are predicted to have the following ground-state electron configurations: (a) $[Ar]3d^6$; (b) $[Ar]3d^5$; (c) $[Kr]4d^5$; (d) $[Kr]4d^3$?

2A.12 Which M^{2+} ions (where M is a metal) are predicted to have the following ground-state electron configurations: (a) $[Ar]3d^4$; (b) $[Kr]4d^9$; (c) $[Ar]3d^{10}$; (d) $[Xe]4f^{14}5d^{10}5s^2$?

2A.13 For each of the following ground-state atoms, predict the type of orbital (1s, 2p, 3d, 4f, etc.) from which an electron will need to be removed to form the +1 ions: (a) Zn; (b) Cl; (c) Al; (d) Cu.

2A.14 For each of the following ground-state ions, predict the type of orbital (1s, 2p, 3d, 4f, etc.) from which an electron will need to be removed to form the ions of one greater positive charge: (a) Mo^{3+}; (b) P^{3-}; (c) Bi^{2+}; (d) Mn^+.

2A.15 Write the most likely charge for the ions formed by each of the following elements: (a) S; (b) Te; (c) Rb; (d) Ga; (e) Cd.

2A.16 Write the most likely charge for the ions formed by each of the following elements: (a) Cs; (b) O; (c) Ca; (d) N; (e) I.

2A.17 Predict the number of valence electrons present for each of the following ions: (a) Mn^{4+}; (b) Rh^{3+}; (c) Co^{3+}; (d) P^{3+}.

2A.18 Predict the number of valence electrons present for each of the following ions: (a) In^+; (b) Tc^{2+}; (c) Ta^{2+}; (d) Re^+.

2A.19 Give the ground-state electron configuration and number of unpaired electrons expected for each of the following ions: (a) Sb^{3+}; (b) Sn^{4+}; (c) W^{2+}; (d) Br^-; (e) Ni^{2+}.

2A.20 Give the ground-state electron configuration and number of unpaired electrons expected for each of the following ions: (a) Sc^{3+}; (b) Co^{2+}; (c) Sr^{2+}; (d) Se^{2-}.

2A.21 Give the ground-state electron configuration and number of unpaired electrons expected for each of the following ions: (a) Ca^{2+}; (b) In^+; (c) Te^{2-}; (d) Ag^+.

2A.22 Give the ground-state electron configuration and number of unpaired electrons expected for each of the following ions: (a) Fe^{3+}; (b) Bi^{3+}; (c) Si^{4+}; (d) I^-.

2A.23 On the basis of the expected charges on the monatomic ions, give the chemical formula of each of the following compounds: (a) magnesium arsenide; (b) indium(III) sulfide; (c) aluminum hydride; (d) hydrogen telluride; (e) bismuth(III) fluoride.

2A.24 On the basis of the expected charges on the monatomic ions, give the chemical formula of each of the following compounds: (a) manganese(II) telluride; (b) barium arsenide; (c) silicon nitride; (d) lithium bismuthide; (e) zirconium(IV) chloride.

2A.25 On the basis of the expected charges of the monatomic ions, draw the formula unit of each of the following compounds using Lewis symbols: (a) thallium(III) chloride; (b) aluminum sulfide; (c) barium oxide.

2A.26 On the basis of the expected charges of the monatomic ions, draw the formula unit of each of the following compounds using Lewis symbols: (a) strontium iodide; (b) potassium phosphide; (c) magnesium nitride.

2A.27 Use data in Appendix 2D to predict which of the following pairs of ions would have the greatest coulombic attraction in a solid compound: (a) K^+, O^{2-}; (b) Ga^{3+}, O^{2-}; (c) Ca^{2+}, O^{2-}.

2A.28 Use data from Appendix 2D to predict which of the following pairs of ions would have the greatest coulombic attraction in a solid compound: (a) Mg^{2+}, S^{2-}; (b) Mg^{2+}, Se^{2-}; (c) Mg^{2+}, O^{2-}.

2A.29 Explain why the lattice energy of lithium chloride (861 $kJ \cdot mol^{-1}$) is greater than that of rubidium chloride (695 $kJ \cdot mol^{-1}$), given that they have similar arrangements of ions in the crystal lattice. See Appendix 2D.

2A.30 Explain why the lattice energy of silver bromide (903 $kJ \cdot mol^{-1}$) is greater than that of silver iodide (887 $kJ \cdot mol^{-1}$), given that they have similar arrangements of ions in the crystal lattice. See Appendix 2D.

Topic 2B Covalent Bonding

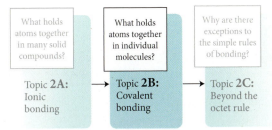

What holds atoms together in many solid compounds?

What holds atoms together in individual molecules?

Why are there exceptions to the simple rules of bonding?

Topic 2A: Ionic bonding → Topic 2B: Covalent bonding → Topic 2C: Beyond the octet rule

The nature of bonds between atoms of nonmetals, which have ionization energies that are too high for ionic bonding to be feasible (Topic 2A), puzzled scientists until 1916 when G. N. Lewis published his explanation. With brilliant insight, and before anyone knew about quantum mechanics or orbitals, Lewis proposed that a **covalent bond** is a pair of electrons shared between two atoms (**1**). A shared electron pair is denoted by a line (—). For example, the hydrogen molecule formed when two H· atoms share an electron pair (H : H) is represented by the symbol H—H. The **valence** of an element is the number of bonds that its atoms can form by sharing electron pairs, so hydrogen typically has a valence of 1.

2B.1 Lewis Structures

Lewis viewed a covalent bond as resulting from the sharing of pairs of electrons. He noted that atoms share electrons until they reach a noble-gas configuration, a principle that he called the **octet rule**:

> In covalent bond formation, atoms go as far as possible toward completing their octets by sharing electron pairs.

For instance, a fluorine atom has seven valence electrons and can achieve an octet by sharing an electron supplied by another atom, such as another fluorine atom:

$$:\ddot{F}\cdot \;+\; \cdot\ddot{F}: \;\longrightarrow\; :\ddot{F}\!\!:\!\!\ddot{F}: \quad \text{or} \quad :\ddot{F}\!-\!\ddot{F}:$$

The circles drawn around each F atom show how each one gets an octet by sharing one electron pair. The valence of fluorine is therefore **1**, the same as that of hydrogen. A hydrogen atom, as always, is anomalous: it tends to complete its duplet rather than octet.

As well as a bonding pair of electrons, a fluorine molecule also possesses three "lone pairs" of electrons on each atom: a **lone pair** is a pair of valence electrons that does not participate in bonding. The lone pairs on one F atom repel the lone pairs on the other F atom, and this repulsion is almost enough to overcome the favorable attractions of the bonding pair that holds the atoms together. This repulsion is one of the reasons why fluorine gas is so reactive: the atoms are bound together as F_2 molecules only very weakly. Among the common diatomic molecules, only H_2 has no lone pairs.

Just as Lewis devised a way to show the valence electron configurations of atoms (Topic 2A), he also devised a way to depict the arrangement of shared and lone pairs of electrons in molecules. The **Lewis structure** of a molecule shows atoms by their chemical symbols, covalent bonds by lines, and lone pairs by pairs of dots. For example, the Lewis structure of HF is H—$\ddot{F}$:. A Lewis structure does not in general portray the three-dimensional shape of a molecule; it simply shows how the atoms are bonded together and whether atoms have lone pairs. However, Lewis structures do help to explain the properties of molecules, including their shapes and their reactions.

Why Do You Need to Know This Material? A major type of bonding is covalent bonding, a concept that occurs throughout chemistry and which it is essential to master if you are to understand the properties and reactions of matter.

What Do You Need to Know Already? You need to be aware of the electron configurations of many-electron atoms (Topic 1F). You will find it helpful to be familiar with the various types of compounds and their nomenclature (*Fundamentals* C and D) and oxidation numbers (*Fundamentals* K).

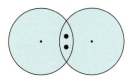

1 Shared electron pair

It is sometimes necessary to write a Lewis structure at the end of a sentence or clause: be careful to distinguish electron dots from periods and colons!

As an illustration of how to write the Lewis structure of a polyatomic molecule, consider methane, CH_4.

- Count the valence electrons available from all the atoms in the molecule. For methane, the Lewis symbols of the atoms are

$$:\!\overset{.}{C} \quad H\!\cdot \quad H\!\cdot \quad H\!\cdot \quad H\!\cdot$$

and so there are eight valence electrons.
- Arrange the dots representing the electrons so that the C atom has an octet, each H atom has a duplet, and the atoms share electron pairs.
- Draw the arrangement shown on the left in (**2**); the Lewis structure of methane is then drawn as shown on the right.

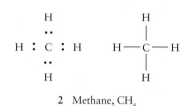

2 Methane, CH_4

Because the carbon atom is linked by four bonds to other atoms, carbon is said to be *tetravalent*: it has a valence of 4.

In many cases, neighboring atoms achieve their octets by sharing more than one pair of electrons. A single shared pair of electrons is called a **single bond.** Two electron pairs shared between two atoms constitute a **double bond,** and three shared electron pairs constitute a **triple bond.** A double bond, such as C::O, is written C=O in a Lewis structure. Similarly, a triple bond, such as C:::C, is written C≡C. Double and triple bonds are collectively called **multiple bonds.** The **bond order** is the number of bonds that link a specific pair of atoms. The bond order in H_2 is 1; in the group C=O it is 2; and for C≡C in a molecule such as ethyne, C_2H_2, the bond order is 3.

The general procedure for constructing the Lewis structure of any molecule or ion is set out in **Toolbox 2B.1.** In each case, you need to know which atoms are linked together in a molecule. A "terminal" atom is bonded to only one other atom; an H atom in methane is an example. Except in the unusual compounds called boranes (Topic 8E), an H atom is always a terminal atom. A "central" atom is bonded to at least two other atoms. Two examples are the O atom in a water molecule, HOH, and the C atom in methane. The arrangement of atoms in a molecule and the identity of the central atom are often known in advance (for instance, it is easy to remember the arrangements of atoms in CH_4, NH_3, and H_2O). If there is doubt, a good rule of thumb for molecules other than compounds of hydrogen is to *choose as the central atom the element with the lowest ionization energy.* This arrangement often results in the lowest energy because an atom in the central position shares more of its electrons than does a terminal atom. Atoms with higher ionization energies are more reluctant to share and are more likely to hold on to their electrons as lone pairs.

Another rule of thumb for predicting the structure of a molecule is to *arrange the atoms symmetrically around the central atom.* For instance, SO_2 is OSO, not SOO. One common exception to this rule is dinitrogen monoxide, N_2O (nitrous oxide), which has atoms in the asymmetrical arrangement NNO. Another clue for writing the correct arrangement of atoms is that, in simple chemical formulas, the central atom is often written first, followed by the atoms attached to it. For example, in the compound with the chemical formula OF_2, the arrangement of the atoms is actually FOF, not OFF; and, in SF_6, the S atom is surrounded by six F atoms. Acids are exceptions to this rule, because the H atoms are written first, as in H_2S, which has the arrangement HSH. If the compound is an oxoacid, then the acidic hydrogen atoms are attached to oxygen atoms, which in turn are

Another and more common way to express this rule is to say that the central atom is usually the element with lowest electronegativity (Topic 2D).

The formal charges of the atoms in a molecule can help to decide between alternative structures (Section 2B.3).

attached to the central atom. For example, the arrangement of atoms in sulfuric acid, H_2SO_4, is $(HO)_2SO_2$ (**3**), and in hypochlorous acid, with the formula HClO, the atoms are actually connected as HOCl.

The same general procedure is used to determine the Lewis structures of polyatomic ions, except that electrons are added or subtracted to account for the charge on the ion, as outlined in Toolbox 2B.1. As for neutral molecules, it is essential to know the general arrangement of atoms in the ion. For oxoanions, it is usually the case that (except for H) the atom written first in the chemical formula is the central atom. In CO_3^{2-}, for instance, the C atom is surrounded by three O atoms. Each atom provides the number of dots (electrons) equal to the number of electrons in its valence shell, but the total number of dots must be adjusted to represent the overall charge. For a cation, subtract one dot for each positive charge. For an anion, add one dot for each negative charge. The cation and the anion must be treated separately: they are individual ions and are not linked by shared pairs. The Lewis structure of ammonium carbonate, $(NH_4)_2CO_3$, for instance, is written as three bracketed ions (**4**).

Note that the entire Lewis structure is not always shown. At times it is convenient to draw a structure with only the electrons of interest displayed to emphasize a point related to a structure or a reaction.

4

Toolbox 2B.1 HOW TO WRITE THE LEWIS STRUCTURE OF A POLYATOMIC SPECIES

CONCEPTUAL BASIS

A valid Lewis structure accommodates all the valence electrons to give each atom, if possible, an octet (or duplet).

PROCEDURE

Step 1 Count the number of valence electrons on each atom; for ions, adjust the number of electrons to account for the charge. Divide the total number of valence electrons in the molecule by 2 to obtain the number of electron pairs.

Step 2 Write down the most likely arrangements of atoms by using common patterns and the clues indicated in the text.

Step 3 Place one electron pair between each pair of bonded atoms.

Step 4 Complete the octet (or duplet, in the case of H) of each atom by placing any remaining electron pairs around the atoms. If there are not enough electron pairs, form multiple bonds in place of one or more single bonds.

Step 5 Represent each bonding electron pair by a line.

To check on the validity of a Lewis structure, verify that each atom has an octet or (for hydrogen) a duplet. As explained in Topic 2C, a common exception to this rule arises when the central atom is an atom of an element in Period 3 or later. Such an atom can accommodate more than eight electrons in its valence shell. Consequently, the lowest energy Lewis structure may be one in which the central atom has more than eight electrons.

This procedure is illustrated in Examples 2B.1 and 2B.2.

EXAMPLE 2B.1   Writing the Lewis structure of a molecule or an ion

When you are assessing the properties of a compound, such as its ability to take part in a reaction, you need to know whether it is likely to have single or multiple bonds. Write the Lewis structures of (a) water, H_2O; (b) methanal, H_2CO; and (c) the chlorite ion, ClO_2^-. Use the rules in Toolbox 2B.1; note that you must add one electron for the negative charge of ClO_2^-.

ANTICIPATE It is hard to anticipate Lewis structures when you are starting out in your studies, but as you gain experience, you will be able to write them down without going through the systematic procedure used here.

PLAN Go through the steps set out in Toolbox 2B.1.

SOLVE

	(a) H₂O	(b) H₂CO	(c) ClO₂⁻

Step 1 Count the valence electrons and adjust the number for charges on ions.

$1 + 1 + 6 = 8$ $1 + 1 + 4 + 6 = 12$ $7 + 6 + 6 + 1 = 20$

Count the electron pairs.

4   6   10

Step 2 Arrange the atoms.

H   O   H

C   O
(with H above and H below C)

O   Cl   O

Step 3 Place one electron pair between each pair of bonded atoms.

H : O : H

C : O (with H above and H below)

O : Cl : O

Step 4 Count electron pairs not yet located.

: : (2)   : : : (3)   : : : : : : : : (8)

Complete the octets with lone pairs. If there are not enough electrons to give each atom an octet or duplet with single bonds, use multiple bonds.

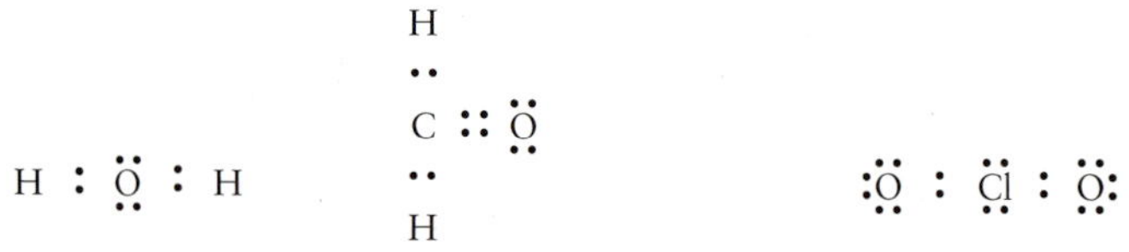

H : Ö : H

C :: Ö (with H atoms)

:Ö : Cl : Ö:

Step 5 Represent the bonds by lines and indicate charges.

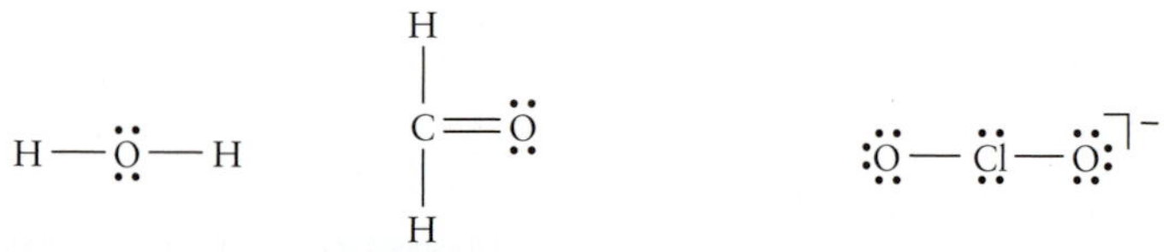

$$H—\overset{..}{\underset{..}{O}}—H$$

$$\overset{\displaystyle H}{\underset{\displaystyle H}{|}}C{=}\overset{..}{\underset{..}{O}}$$

$$\left[:\overset{..}{\underset{..}{O}}—\overset{..}{\underset{..}{Cl}}—\overset{..}{\underset{..}{O}}:\right]^{-}$$

EVALUATE Note that H₂CO did not have enough electrons to achieve complete octets without including a double bond.

Self-test 2B.2A Write a Lewis structure for the cyanate ion, CNO⁻ (the C atom is in the center).

[***Answer:*** $\left[\overset{..}{\underset{..}{N}}{=}C{=}\overset{..}{\underset{..}{O}}\right]^{-}$]

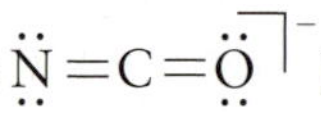

Self-test 2B.2B Write a Lewis structure for NH₃.

Related Exercises 2B.1–2B.6, B.9, B.10

EXAMPLE 2B.2   Writing Lewis structures for molecules with more than one "central" atom

When organic chemists think about how molecules undergo change during a reaction, they commonly find it helpful to write down a scheme that includes Lewis structures. Write the Lewis structure for acetic acid, CH₃COOH, which is found in vinegar. The formula of acetic acid suggests that the molecule consists of a CH₃— group and a —COOH group. In the —COOH group, both O atoms are attached to the same C atom, and one of them is bonded to the final H atom. The two C atoms are bonded to each other.

ANTICIPATE You should anticipate that the CH₃— group, by analogy with methane, will consist of a C atom joined to three H atoms by single bonds.

PLAN Apply the procedure in Toolbox 2B.1.

SOLVE

CH₃COOH

Step 1 Count the valence electrons and hence obtain the number of electron pairs:

$$4 + (3 \times 1) + 4 + 6 + 6 + 1 = 24, \; 12 \text{ pairs}$$

: : : : : :
: : : : : :

12 pairs

Step 2 Arrange the atoms (linked atoms are indicated by the blue rectangles).

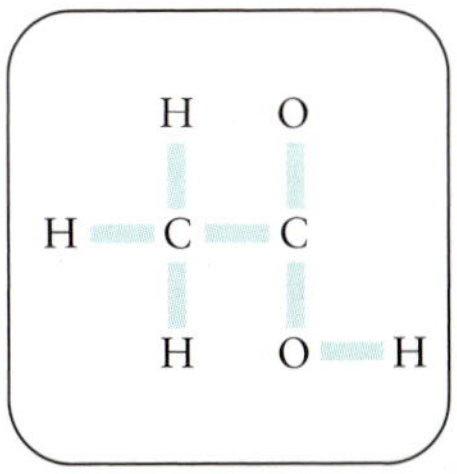

Step 3 Connect the atoms with bonding electron pairs.

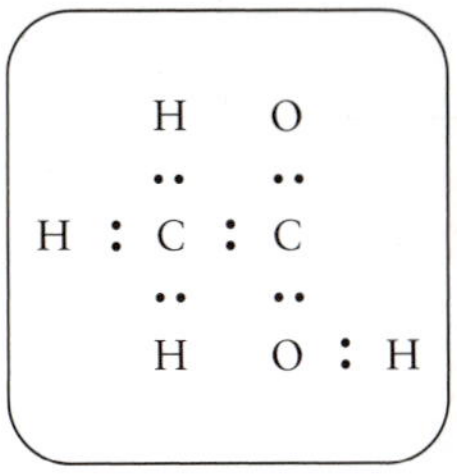

Step 4 Count electron pairs not yet located.

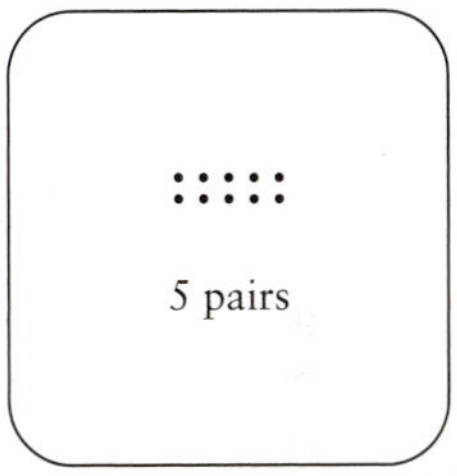

Complete the octets.

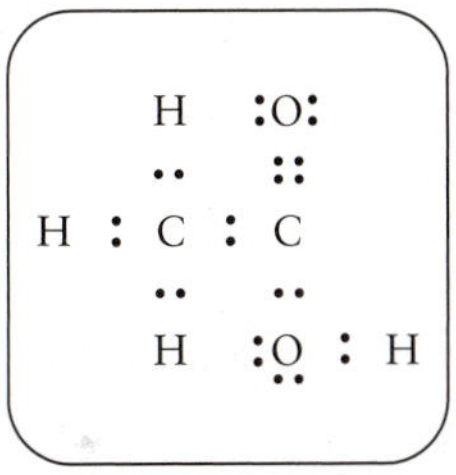

Step 5 Represent the bonds.

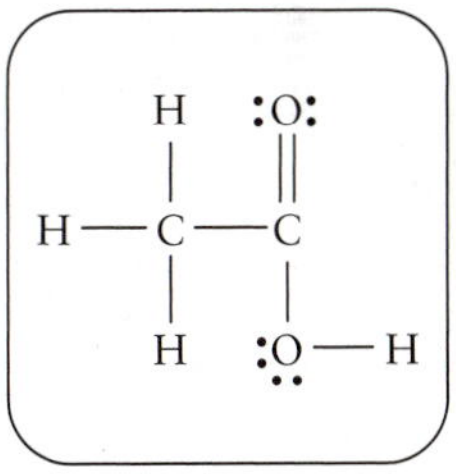

EVALUATE As expected, the methyl group has an arrangement similar to that of methane. With practice, you will begin to recognize common arrangements of atoms in many organic molecules.

Self-test 2B.3A Write a Lewis structure for the urea molecule, $(NH_2)_2CO$.

[*Answer:* See (**5**).]

Self-test 2B.3B Write a Lewis structure for hydrazine, H_2NNH_2.

Related Exercises: 2B.11, 2B.12, 2.59, 2.60.

> *In the Lewis structure of a polyatomic species all the valence electrons are used to complete the octets (or duplets) of the atoms present by forming single or multiple bonds; some electrons may be left as lone pairs.*

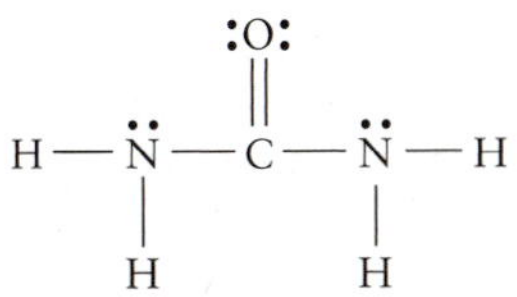

2B.2 Resonance

Some molecules are not represented adequately by a single Lewis structure. For example, consider the nitrate ion, NO_3^-, which, as potassium nitrate, is used in fireworks (to provide oxygen) and fertilizers (to provide nitrogen). The three Lewis structures shown in (**6**) differ only in the position of the double bond. All are valid structures and have exactly the same energy. If one of the pictured structures were correct, the nitrate ion would be

5 Urea, $(NH_2)_2CO$

6

7   Nitrate ion, NO_3^-

Bond length, the distance between the centers of bonded atoms, is discussed in more detail in Topic 2D.

expected to have two long single bonds and one short double bond, because a double bond between two atoms is shorter than a single bond between the same types of atoms. However, detailed measurements have shown that the bond lengths in a nitrate ion are all the same. At 124 pm, they are longer than a typical N=O double bond (120 pm) but shorter than a typical N—O single bond (140 pm). The bond order in the nitrate ion lies between 1 (a single bond) and 2 (a double bond).

Because all three bonds are identical, a better model of the nitrate ion is a *blend* of all three Lewis structures, with each bond intermediate in properties between a single and a double bond. This blending of structures, which is called **resonance,** is depicted in (7) by double-headed arrows. The actual structure is a resonance hybrid of the contributing Lewis structures. A molecule does not flicker between different structures: a **resonance hybrid** is a merging of structures, just as a mule is a hybrid of a horse and a donkey, not a creature that flickers between the two.

Electrons that are shown in different positions in a set of resonance structures are said to be **delocalized.** Delocalization means that a shared electron pair is distributed over several pairs of atoms and cannot be identified with just one pair of atoms. The three resonance structures in (7) do not exist as actual molecules; they are simply a way of showing that the electrons are spread across the molecule. As well as delocalizing electrons over the atoms, resonance also lowers the energy below that of any single contributing structure and helps to stabilize the molecule. This lowering of energy occurs for quantum mechanical reasons. Broadly speaking, the wavefunction that describes the resonance structure is a more accurate description of the electronic structure of the molecule than the wavefunction for any single structure alone, and the more accurate the wavefunction, the lower the corresponding energy.

The following points will help you to write appropriate resonance structures and to predict which contribute the most to the actual structure:

- In each contributing structure, the nuclei are in the same positions; only the locations of lone pairs and bonding pairs are changed.
- Structures with the same energy (so called "equivalent structures") contribute equally to the resonance.
- Low-energy structures contribute more to the resonance mixture than high-energy structures.

For example, although you could write the two hypothetical structures NNO and NON for the dinitrogen oxide (nitrous oxide) molecule, there is no resonance between them because the atoms lie in different locations.

EXAMPLE 2B.3   Writing a resonance structure

Stratospheric ozone, O_3, protects life on Earth from harmful ultraviolet radiation from the Sun. Suppose you are an atmospheric chemist; to understand the spectroscopic and structural properties of ozone, you would need to know how its electrons are arranged. Suggest two Lewis structures that contribute equally to the resonance structure for the O_3 molecule. Experimental data show that the two bond lengths are the same.

ANTICIPATE   You should expect to be able to write structures that differ only in the position of a multiple bond.

PLAN   Write a Lewis structure for the molecule by using the method outlined in Toolbox 2B.1. Decide whether there is another equivalent structure that results from the interchange of a single bond and a double or triple bond. Write the actual structure as a resonance hybrid of these Lewis structures.

SOLVE

Count the valence electrons.

Oxygen is a member of Group 16; so each atom has six valence electrons: 6 + 6 + 6 = 18 electrons.

Draw a Lewis structure for the molecule.

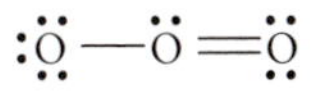

Draw a second Lewis structure by exchanging the bonds.

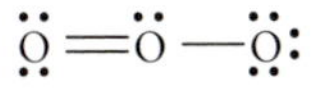

Draw the resonance hybrid as the two resonance structures connected with a double-headed arrow.

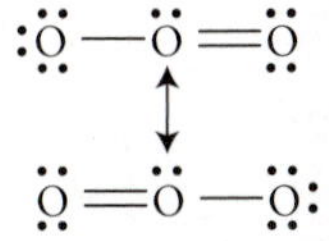

$:\ddot{O}—\ddot{O}=\ddot{O}:$

$\ddot{O}=\ddot{O}—\ddot{O}:$

$:\ddot{O}—\ddot{O}=\ddot{O}$
⇕
$\ddot{O}=\ddot{O}—\ddot{O}:$

EVALUATE As expected, the resonance in ozone can be described by writing two structures that differ only in the position of the double bond (of course, the lone pairs of electrons are also distributed differently).

Self-test 2B.4A Write Lewis structures contributing to the resonance hybrid for the acetate ion, $CH_3CO_2^-$. The structure of CH_3COOH is described in Example 2B.2; the acetate ion has a similar structure, except that it has lost the final H atom while keeping both electrons from the OH bond.

[*Answer:* See (8).]

Self-test 2B.4B Write Lewis structures contributing to the resonance hybrid for the nitrite ion, NO_2^-.

Related Exercises 2B.13–2B.18

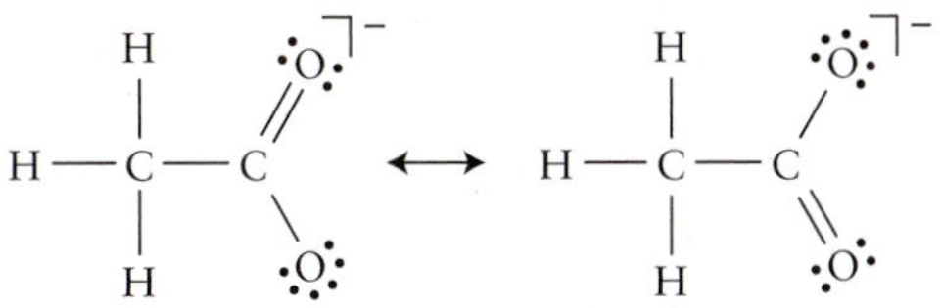

8 Acetate ion, $CH_3CO_2^-$

Benzene, C_6H_6, is another molecule best described as a resonance hybrid. It consists of a planar hexagonal ring of six carbon atoms, each having a hydrogen atom attached to it. One Lewis structure that contributes to the resonance hybrid is shown in (9); it is called a **Kekulé structure.** The structure is normally written as a line structure (see *Fundamentals* C), a simple hexagon with alternating single and double lines (10).

The difficulty with a single Kekulé structure is that it does not fit all the experimental evidence:

- *Reactivity:* Benzene does not undergo reactions typical of compounds with double bonds.

For example, when a solution of red–brown bromine is mixed with an alkene such as 1-hexene, $CH_2=CHCH_2CH_2CH_2CH_3$, the color due to bromine is lost as the Br_2 molecule reacts with the molecule at the double bond to produce $CH_2Br—CHBrCH_2CH_2CH_2CH_3$ (**FIG. 2B.1**). However, benzene does not react with liquid bromine.

- *Bond lengths:* All the carbon–carbon bonds in benzene are the same length.

A Kekulé structure suggests that benzene should have two different bond lengths: three longer single bonds (154 pm) and three shorter double bonds (134 pm). Instead, all the bonds are found experimentally to have the same intermediate length (139 pm).

- *Structural evidence:* Only one 1,2-dichlorobenzene (in which the chlorine atoms are attached to two adjacent carbon atoms) exists.

If the Kekulé structure were correct, there would be two distinct 1,2-dichlorobenzenes (11), one in which the carbon atoms are joined by a single bond and one with a double bond. In fact, only one 1,2-dichlorobenzene is known.

The concept of resonance explains these characteristics of the benzene molecule. There are two Kekulé structures with exactly the same energy: they differ only in the positions of the double bonds. As a result of resonance between these two equivalent structures

The German chemist Friedrich Kekulé first proposed (in 1865) that benzene has a cyclic structure with alternating single and double bonds.

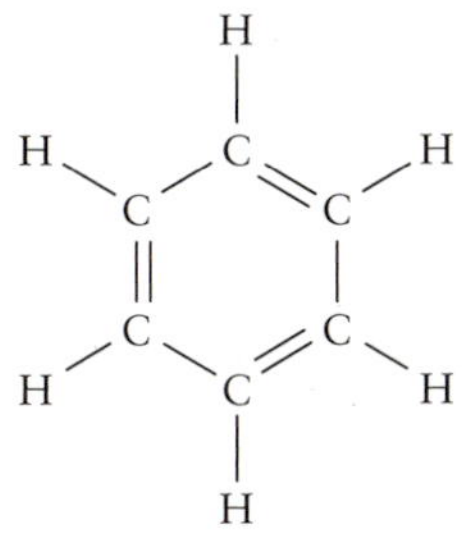

9 Kekulé structure

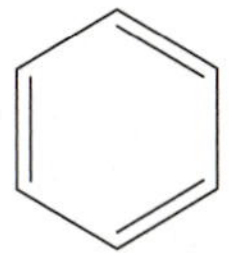

10 Kekulé structure, line structure

Other arrangements may be drawn, but they differ from the one shown only by a rotation of the molecule.

FIGURE 2B.1 When bromine dissolved in a solvent (the brown liquid) is mixed with an alkene (the colorless liquid), the bromine atoms add to the alkene molecule at the double bond, resulting in colorless products. (W. H. Freeman photo by Ken Karp.)

13   Benzene, C_6H_6

14   Dichlorobenzene, $C_6H_4Cl_2$

15

16

Some compounds, such as CO, are exceptions to the formal charge rules, but the rules are reliable in most common compounds.

11   Dichlorobenzene, $C_6H_4Cl_2$ **12**   Benzene resonance

(**12**), the electrons shared in the C=C double bonds are delocalized over the whole molecule, thereby giving each bond a length intermediate between that of a single and that of a double bond. Resonance makes all six C—C bonds identical; this equivalence is implied by representing the double bonds in the resonance hybrid with a circle (**13**). It can be seen from (**14**) why there can be only one 1,2-dichlorobenzene. Finally, as we have remarked, an important consequence of resonance is that it stabilizes a molecule by lowering its total energy. This stabilization makes benzene less reactive than expected for a molecule with three carbon–carbon double bonds.

Resonance is a blending of structures with the same arrangement of atoms but different arrangements of electrons. It spreads multiple-bond character over a molecule and results in a lower energy.

2B.3 Formal Charge

Nonequivalent Lewis structures—Lewis structures that do not correspond to the same energy—do not in general make the same contribution to a resonance structure. One way to decide which structures are likely to make the major contribution is to compare the number of valence electrons distributed around each atom in a structure with the number of valence electrons on the free atoms. The smaller these differences are for a structure, the lower is its energy and the greater is its contribution to a resonance hybrid.

A measure of the redistribution of electrons is the **formal charge** on an atom in a given Lewis structure—the charge it would have if the bonding were perfectly covalent in the sense that the atom had exactly a half-share in each pair of bonding electrons. That is, the formal charge takes into account the number of electrons that an atom can be regarded as "owning" in a molecule. It "owns" all its lone pairs and half of each shared bonding pair. The difference between this number and the number of valence electrons in the free atom is the formal charge:

$$\text{Formal charge} = V - (L + \tfrac{1}{2}B) \tag{1}$$

where V is the number of valence electrons in the free atom, L is the number of electrons present on the bonded atom as lone pairs, and B is the number of bonding electrons on the atom. If the atom has more electrons in the molecule than when it is a free, neutral atom, then the atom has a negative formal charge, like a monatomic anion. If the assignment of electrons leaves the atom with fewer electrons than when it is free, then the atom has a positive formal charge, as if it were a monatomic cation.

Formal charge can be used to predict the most favorable arrangement of atoms in a molecule and the most likely Lewis structure for that arrangement:

- A Lewis structure in which the formal charges of the individual atoms are closest to zero typically represents the lowest energy arrangement of the atoms and electrons.

A low formal charge indicates that an atom has undergone only a small redistribution of electrons relative to the free atom. The structure with formal charges closest to zero typically has the lowest energy of all possible structures. For example, the formal charge rule suggests that the structure OCO is more likely for carbon dioxide than COO, as shown in (**15**). Similarly, it also suggests that the structure NNO is more likely than NON for dinitrogen monoxide, as shown in (**16**).

Toolbox 2B.2 HOW TO USE FORMAL CHARGE TO IDENTIFY THE MOST LIKELY LEWIS STRUCTURE

CONCEPTUAL BASIS

Formal charge is assigned by establishing the "ownership" of the valence electrons of each atom in a molecule and comparing that ownership with the number of valence electrons on the free atom. An atom owns one electron of each bonding pair attached to it and owns its lone pairs completely. The most plausible Lewis structure will be the one in which the formal charges of the atoms are lowest.

PROCEDURE

Step 1 Count the number of valence electrons (V) possessed by each free atom by noting the number of its group in the periodic table. For ions, adjust the number of electrons according to its charge.

Step 2 Draw the Lewis structures.

Step 3 For each bonded atom, count both electrons in its lone pairs (L), plus one electron from each of its bonding pairs ($\frac{1}{2}B$, where B is the number of bonding electrons).

Step 4 For each bonded atom, subtract the total number of electrons it "owns" from V, as in Eq. 1.

Each equivalent atom (the same element, the same number of bonds and lone pairs) has the same formal charge. A check on the calculated formal charges is that their sum is equal to the overall charge of the molecule or ion. For an electrically neutral molecule, the sum of the formal charges is zero. Compare the formal charges of each possible structure. The structure with the lowest formal charges represents the least disturbance of the electronic structures of the atoms and is the most likely (lowest energy) structure.

This procedure is illustrated in Example 2B.4.

EXAMPLE 2B.4　Selecting the most likely atom arrangement

If you were an analytical chemist, one test you might use for the presence of iron(III) ions is to add a solution of potassium thiocyanate, KSCN; a blood-red color indicates that a compound of iron and the thiocyanate ion has formed. Write three Lewis structures with different atomic arrangements for the thiocyanate ion and select the most likely structure by identifying the structure with formal charges closest to zero. For simplicity, consider only structures with double bonds between neighboring atoms.

ANTICIPATE The element with the lowest ionization energy (after you have read Topic 2D, you will find it is more appropriate to think "lowest electronegativity") of the three is carbon, so you should expect it to be the central atom and for the structure to be NCS$^-$.

PLAN Proceed as set out in **Toolbox 2B.2**.

SOLVE

	NCS$^-$	CNS$^-$	CSN$^-$
Step 1 Count the valence electrons V and, for ions, adjust for the charge.	C: 4, N: 5, S: 6 Charge: −1 (16 electrons)	C: 4, N: 5, S: 6 Charge: −1 (16 electrons)	C: 4, N: 5, S: 6 Charge: −1 (16 electrons)
Step 2 Draw Lewis structures.	N̈=C=S̈	C̈=N=S̈	C̈=S=N̈
Step 3 For each bonded atom, count each electron of its lone pairs (L), plus one electron from each of its bonding pairs ($\frac{1}{2}B$).	6　4　6 N̈=C=S̈	6　4　6 C̈=N=S̈	6　4　6 C̈=S=N̈
Step 4 For each bonded atom, subtract the total number of electrons it "owns" from V, as in Eq. 1.	−1　0　0 N̈=C=S̈	−2　+1　0 C̈=N=S̈	−2　+2　−1 C̈=S=N̈

EVALUATE The individual formal charges are closest to zero in the first column; the atom arrangement NCS$^-$ is therefore the most likely one, as anticipated.

Self-test 2B.5A Suggest a likely structure for the poisonous gas phosgene, $COCl_2$. Write its Lewis structure and formal charges. C is the central atom.

[**Answer:** See (**17**).]

Self-test 2B.5B Suggest a likely structure for the oxygen difluoride molecule. Write its Lewis structure and formal charges.

Related Exercises 2B.21, 2B.22

0
:O:
‖
0 :C̈l —— C —— C̈l: 0
0

17 Phosgene, $COCl_2$

Although formal charge and oxidation number (*Fundamentals* K) both provide information about the number of electrons around an atom in a compound, they are distinct in origin and significance, are determined by different methods, and often have different values:

- Formal charge exaggerates the covalent character of bonds by assuming that the electrons are shared equally.
- Oxidation number exaggerates the ionic character of bonds. It represents the atoms as ions, and *all* the electrons in a bond are assigned to the more electronegative atom (the atom with the greater attraction for electrons, Topic 2D).

Thus, although the formal charge of C in structure **15** for CO_2 is 0, its oxidation number is +4 because all the bonding electrons are assigned to the oxygen atoms to give a structure that could be represented as $O^{2-}C^{4+}O^{2-}$. Formal charges depend on the particular Lewis structure you write; oxidation numbers do not.

The formal charge gives an indication of the extent to which atoms have gained or lost electrons in the process of covalent bond formation; atom arrangements and Lewis structures with the lowest formal charges are likely to have the lowest energy.

What have you learned in this Topic?

You have learned that a covalent bond consists of a shared electron pair and that atoms tend to acquire a complete octet (or duplet). The patterns of electron-pair sharing in covalent compounds are represented by drawing Lewis structures. You have seen that in some cases it is necessary to represent a molecule as a resonance hybrid that spreads multiple-bond character over the entire molecule. You have seen that the lowest energy Lewis structure can often be identified by calculating the formal charges on the atoms.

The skills you have mastered are the ability to:

☐ **1.** Draw the Lewis structures of molecules (Toolbox 2B.1 and Examples 2B.1 and 2B.2).

☐ **2.** Write the resonance structures for a molecule (Example 2B.3).

☐ **3.** Use formal charge calculations to select the most likely atom arrangement (Toolbox 2B.2 and Example 2B.4).

Topic 2B Exercises

2B.1 Draw the Lewis structure of (a) CCl_4; (b) $COCl_2$; (c) ONF; (d) NF_3.

2B.2 Draw the Lewis structure of (a) SCl_2; (b) $AsFr_3$; (c) SiH_4; (d) $InCl_3$.

2B.3 Draw the Lewis structure of (a) OF_2; (b) NHF_2; (c) SiO_2; (d) BrF_3.

2B.4 Draw the Lewis structure of (a) Cl_2O; (b) N_2F_2; (c) SO_3; (d) BrF_4^-.

2B.5 Draw the Lewis structure of (a) tetrahydridoborate ion, BH_4^-; (b) hypobromite ion, BrO^-; (c) amide ion, NH_2^-.

2B.6 Draw the Lewis structure of (a) nitronium ion, ONO^+; (b) chlorite ion, ClO_2^-; (c) peroxide ion, O_2^{2-}; (d) formate ion, HCO_2^-.

2B.7 The following Lewis structure was drawn for a Period 3 element. Identify the element.

:O:
‖
:C̈l —— E —— C̈l:
|
:C̈l:

2B.8 The following Lewis structure was drawn for a Period 4 element. Identify the element.

:O:
‖
:C̈l —— E ═ Ö
|
:C̈l:

2B.9 Draw the complete Lewis structure for each of the following compounds: (a) ammonium chloride; (b) potassium phosphide; (c) sodium hypochlorite.

2B.10 Draw the complete Lewis structure for a formula unit of each of the following compounds: (a) barium hydroxide; (b) cesium nitrite; (c) ammonium sulfide.

2B.11 Draw the complete Lewis structure for each of the following compounds: (a) formaldehyde, HCHO, which as its aqueous solution, "formalin," is used to preserve biological specimens; (b) methanol, CH_3OH, the toxic compound also called wood alcohol; (c) glycine, $H_2C(NH_2)COOH$, the simplest of the amino acids, the building blocks of proteins.

2B.12 Draw the Lewis structure of each of the following organic compounds: (a) urea, $H_2N-CO-NH_2$, a fertilizer that is also found in hand lotions (the O is a terminal atom); (b) CH_3SH, methanethiol, a compound found in breath odor; (c) $HOOC-COOH$, oxalic acid, a toxic compound found in rhubarb leaves.

2B.13 Anthracene has the formula $C_{14}H_{10}$. It is similar to benzene but has three six-membered rings that share common C—C bonds, as shown below. Complete the structure by drawing in multiple bonds to satisfy the octet rule at each carbon atom. Resonance structures are possible. Draw as many as you can find.

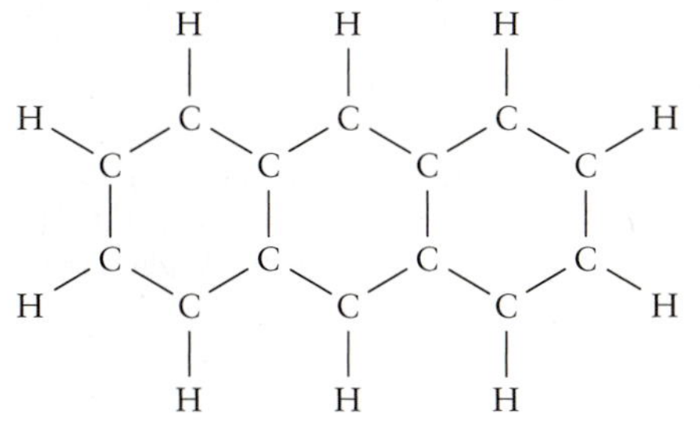

2B.14 Draw the Lewis structures that contribute to the resonance hybrid of the guanadinium ion, $C(NH_2)_3{}^+$.

2B.15 Draw the Lewis structures that contribute to the resonance hybrid of nitryl chloride, $ClNO_2$ (N is the central atom).

2B.16 Draw the Lewis structures that contribute to the resonance hybrid of chloric acid, $HClO_2$ (Cl is the central atom and the H is attached to one of the O atoms).

2B.17 In the compound cyclobutadiene, C_4H_4, the carbon atoms form a square, with a hydrogen atom attached to each one. Draw two resonance structures for C_4H_4.

2B.18 In the ion $Se_4{}^{2+}$, the selenium atoms form a square. Draw two resonance structures for $Se_4{}^{2+}$.

2B.19 Draw the Lewis structure and determine the formal charge on each atom in (a) NO^+; (b) N_2; (c) CO; (d) $C_2{}^{2-}$; (e) CN^-.

2B.20 Using only Lewis structures that obey the octet rule, draw the Lewis structures and determine the formal charge on each atom in (a) SO_2; (b) SO_3; (c) $SO_3{}^{2-}$.

2B.21 Hypochlorous acid, HClO, is found in white blood cells, where it helps to destroy bacteria. Draw two Lewis structures with different atom arrangements for HClO and select the most likely structure by identifying the structure with formal charges closest to zero. Consider only structures with single bonds.

2B.22 Nitrosyl fluoride, NOF, is an oxidizing agent used as a rocket fuel. Draw three Lewis structures with different atom arrangements for NOF and select the most likely structure by identifying the structure with formal charges closest to zero. Consider only structures with one single bond and one double bond.

2B.23 Determine the formal charge on each atom in the following molecules. Identify the structure of lower energy in each pair.

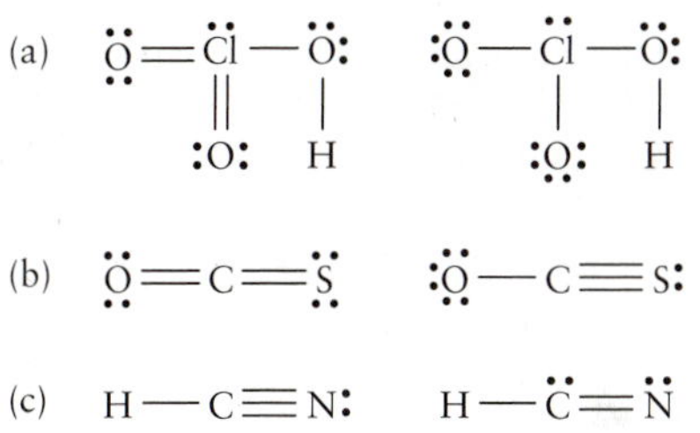

(a)

(b)

(c)

2B.24 Determine the formal charge on each atom in the following ions. Identify the structure of lowest energy in each case.

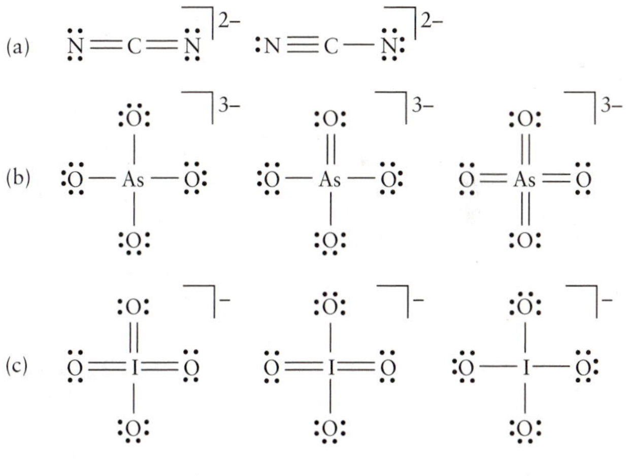

(a)

(b)

(c)

Topic 2C Beyond the Octet Rule

2C.1 Radicals and Biradicals
2C.2 Expanded Valence Shells
2C.3 Incomplete Octets

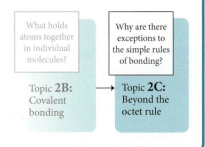

What holds atoms together in individual molecules?

Topic **2B:** Covalent bonding

→ Topic **2C:** Beyond the octet rule

Why are there exceptions to the simple rules of bonding?

Why Do You Need to Know This Material? Although the octet rule is a good starting point for the discussion of covalent bonding, there are many exceptions and subtle ways in which it is deployed. You need to be aware of these refinements if you are to understand the structures of all types of molecules.

What Do You Need to Know Already? You need to know how to draw and interpret Lewis structures (Topic 2B), be familiar with the concept of resonance, and be able to assign formal charges to atoms in a Lewis structure (Topic 2B).

The older, still widely used term for radicals is *free radicals*.

·CH₃
1

$\cdot \ddot{N} = \ddot{O}$
2

H   H   H   H   H
| | | | |
·C — C — C — C — C·
| | | | |
H   H   H   H   H

3 A biradical

The role of radicals in the depletion of stratospheric ozone is described in Box 7D.1.

H
|
:Ö — Ö·

4 Hydrogenperoxyl, HO_2·

The octet rule (Topic 2B) accounts for the valences of many of the elements and the structures of many compounds, especially those of the elements in Period 2 (specifically carbon, nitrogen, oxygen, and fluorine). However, there are several types of exception:

- A molecule might have an odd number of electrons, so octet formation is numerically impossible.
- Atoms of certain elements might be able to accommodate more than eight electrons in their valence shells.
- An atom might form compounds with incomplete octets.

2C.1 Radicals and Biradicals

Some species have an odd number of valence electrons, and so at least one of their atoms cannot have an octet. Species having electrons with unpaired spins are called **radicals.** Two examples, with their Lewis structures, are the methyl radical (**1**) and nitric oxide, NO (**2**).

Radicals are generally highly reactive and except in special cases cannot be stored: most have only a fleeting existence. The methyl radical, ·CH₃, occurs in the flames of burning hydrocarbon fuels. The single unpaired electron is indicated by the dot on the C atom in ·CH₃. Radicals are of crucial importance for the chemical reactions that take place in the upper atmosphere, where they contribute to the formation and decomposition of ozone. They also play a role in our daily lives, sometimes a destructive one. They are responsible for the rancidity of foods and the degradation of plastics in sunlight. Damage from radicals can be delayed by an additive called an **antioxidant,** which reacts rapidly with radicals before the radicals have a chance to do their damage. Human aging is believed to be partly due to the action of radicals, and antioxidants such as vitamins C and E may delay the process (see **BOX 2C.1**). Nitric oxide plays an important role in the body as a neurotransmitter and vasodilator. Because it is a radical, NO is very reactive and can be eliminated within a few seconds. Because it is small, the NO molecule can move easily throughout the body. These properties allow NO to play several roles, including blood pressure regulation and helping to fight infection during immune response.

A **biradical** is a molecule with two unpaired electrons. The unpaired electrons are usually on different atoms, as depicted in (**3**). In that biradical, one unpaired electron is on one carbon atom of the chain and the second is on another carbon atom several bonds away. In some cases, though, both electrons are on the same atom. One of the most important examples is the oxygen atom itself. Its electron configuration is $[He]2s^2 2p_x^2 2p_y^1 2p_z^1$ and its Lewis symbol is ·Ö·. The O atom has two unpaired electrons, and so it can be regarded as a special type of biradical.

Self-test 2C.1A Write a Lewis structure for the hydrogenperoxyl radical, HOO·, which is active in atmospheric chemistry and which, in the body, has been implicated in the degeneration of neurons.

[*Answer:* See (**4**).]

Self-test 2C.1B Write a Lewis structure for nitrogen dioxide, NO_2.

Box 2C.1 WHAT HAS THIS TO DO WITH . . . STAYING ALIVE?

CHEMICAL SELF-PRESERVATION

Nearly every pharmacy, supermarket, and health food store carries antioxidants and antioxidant-rich natural products, such as fish oils, *Gingko biloba* leaves, and wheat grass. These dietary supplements are intended to help the body control its population of radicals and, as a result, slow aging and degenerative diseases, such as heart failure and cancer.

Radicals occur naturally in the body, partly as a byproduct of metabolism. They serve many useful functions but can cause trouble if they are not eliminated when they are no longer needed. They often contain oxygen atoms and oxidize the *lipid* (fat) molecules that make up cell membranes and other vital tissues. This oxidation changes the structures of the lipid molecules and hence affects the function of the membranes. Cell membranes damaged in this way may not be able to protect cells against disease, and heart and nerve cells may lose their function. Some research suggests that radical damage to living cells may be an important factor in the aging process and the spread of diseases like cancer.

Antioxidants are molecules that react readily with radicals before the radicals can cause damage in the body. Many common foods, such as green leafy vegetables, orange juice, and dark chocolate, contain antioxidants, as do coffee and tea. The body maintains an antioxidant network consisting of vitamins A, C, and E, antioxidant enzymes, and a group of related compounds called coenzyme Q, for which the general formula is shown below. The n represents the number of times that a particular group is repeated; it can be 6, 8, or 10.

Molecular structure of coenzyme Q, an antioxidant used by the body to control the level of radicals.

Harmful environmental conditions, such as ultraviolet radiation, ozone in the air we breathe, poor nutrition, and tobacco smoke can cause *oxidative stress*, a condition in which the concentration of radicals has become so high that the body's natural antioxidant network cannot cope. Premature aging of skin that has been overexposed to sunlight and lung cancer in smokers are two possible results. Herbal medicines containing certain *phytochemicals*, chemicals derived from plant sources and fish oils, are being investigated as antioxidants that can supplement the diet to protect against damage from free radicals. These same chemicals are also being studied for their ability to prolong life beyond the currently expected lifespan.

Related Exercises 2C.5, 2C.6, 2.17

Further Reading D. Harman, "The aging process," *Proceedings of the National Academy of Sciences*, **78** (2004), 7124. M. W. Moyer, "The myth of antioxidants," *Scientific American*, vol. 308 (Feb., 2013), pp. 62–67. F. L. Muller, M. S. Lustgarten, Y. Jang, A. Richardson, and H. Van Remmen, "Trends in oxidative aging theories," *Free Radical Biology & Medicine*, 43 (2007), 477–503. G. Critser, *Eternity Soup: Inside the Quest to End Aging*, New York: Harmony Books (2010).

Leaves of the *Gingko biloba* tree, which originated in China. Extracts of these leaves are thought to have antioxidant properties. Some also believe that the extracts improve thinking ability by increasing oxygen flow to the brain. (© Rob Walls/Alamy.)

A radical is a species with an unpaired electron; a biradical has two unpaired electrons on either the same or different atoms.

2C.2 Expanded Valence Shells

The octet rule states that electron sharing continues until eight electrons fill the outer shell of an atom to give a noble-gas ns^2np^6 valence-shell configuration. However, when the central atom in a molecule has empty d-orbitals close in energy to the valence orbitals, it may be able to accommodate 10, 12, or even more electrons and possess an **expanded valence shell.**

This expansion can arise in either of two ways (and sometimes both):

- More atoms might attach to a central atom than is allowed by the octet rule.
- The number of atoms is the same as that allowed by the octet rule, but some single bonds are replaced by double bonds.

5   Phosphorus pentachloride, PCl_5

6a   6b

7   Phosphorus trichloride, PCl_3

A compound that contains an atom with more atoms attached to it than is permitted by the octet rule (the first of these possibilities) is called a **hypervalent compound,** as in the formation of PCl_5 (**5**). Hypervalence is often associated with **variable covalence,** the formation of compounds with different numbers of attached atoms, as in PCl_3 and PCl_5. The second possibility is commonly associated with the ability to write different Lewis structures for a molecule, with different arrangements of electron pairs, as in structures (**6a**) and (**6b**) for the chlorite ion, ClO_2^-.

A Note on Good Practice: Although "expanded valence shell" is the more logical term, most chemists still use the term *expanded octet.*

Only p-block atoms in Period 3 or later periods can expand their valence shells. Atoms of these elements have empty d-orbitals in the valence shell. Another factor—possibly the main factor—in determining whether more atoms than allowed by the octet rule can bond to a central atom is the size of that atom. A P atom is big enough for as many as six Cl atoms to fit comfortably around it, and PCl_5 is a common laboratory chemical. An N atom, though, is too small, and NCl_5 is unknown.

The variable valence of phosphorus shows itself in an interesting way. It reacts directly with a limited supply of chlorine to form the toxic, colorless liquid phosphorus trichloride: $P_4(s) + 6\,Cl_2(g) \rightarrow 4\,PCl_3(l)$, and the product, PCl_3 obeys the octet rule (**7**). However, when phosphorus trichloride reacts with more chlorine (**FIG. 2C.1**), phosphorus pentachloride, a pale yellow crystalline solid, is produced in the reaction $PCl_3(l) + Cl_2(g) \rightarrow PCl_5(s)$. This compound is an ionic solid consisting of PCl_4^+ cations and PCl_6^- anions; but at 160 °C, it vaporizes to a gas of PCl_5 molecules. The Lewis structures of the polyatomic ions are shown in (**8**) and the molecule in (**5**). In the PCl_6^- anion, the P atom has expanded its valence shell to 12 electrons, by making use of two of its 3d-orbitals. In PCl_5, the P atom has expanded its valence shell to 10 electrons by using one 3d-orbital.

8   Phosphorus pentachloride, $PCl_5(s)$

EXAMPLE 2C.1   Writing a Lewis structure with an expanded valence shell

The fluoride SF_4 is used in the pharmaceutical industry to synthesize fluorocarbons, some of which are used as anesthetics. Write the Lewis structure of sulfur tetrafluoride and give the number of electrons in the expanded valence shell.

ANTICIPATE Sulfur, in Group 16, has six valence electrons; if each fluorine atom provides one electron for the bond it forms, you should expect there to be $4 + 6 = 10$ electrons around the S atom.

PLAN Elements in Period 3 and later periods can expand their valence shells to accept additional electrons. After assigning all the valence electrons to bonds and lone pairs to give each atom an octet, assign any remaining electrons to the central atom as lone pairs.

SOLVE

Count the number of valence electrons.	
6 from sulfur ($\cdot \overset{\cdot\cdot}{S} \cdot$)	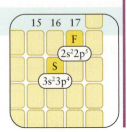
7 from each fluorine atom ($:\overset{\cdot\cdot}{F}\cdot$)	

FIGURE 2C.1 Phosphorus trichloride is a colorless liquid. When it reacts with chlorine (the pale yellow-green gas in the flask), it forms the very pale yellow solid phosphorus pentachloride (at the bottom of the flask). (*W. H. Freeman photo by Ken Karp.*)

Cl_2

PCl_3

PCl_5

Find the number of electron pairs.

There are $6 + (4 \times 7) = 34$ electrons, or 17 electron pairs.

Construct the Lewis structure.

Give each F atom 3 lone pairs and 1 bonding pair shared with the central S atom; place the 2 extra electrons on the S atom.

EVALUATE As expected, in this structure sulfur has 10 electrons in its expanded valence shell.

Self-test 2C.2A Write the Lewis structure for xenon tetrafluoride, XeF_4, and give the number of electrons in the expanded valence shell.

[***Answer:*** See (**9**); 12 electrons.]

Self-test 2C.2B Write the Lewis structure for the I_3^- ion and give the number of electrons in the expanded valence shell.

Related Exercises 2C.3, 2C.4, 2C.9–2C.12

9 Xenon tetrafluoride, XeF_4

When different resonance structures are possible, some assigning the central atom in a compound an octet and some an expanded valence shell (as for the chlorite ion in **6**), the dominant resonance structure is identified by assessing the formal charges on the atoms (Topic 2B). The dominant, most probable, structure is likely to be the one with the lowest formal charges. However, there are many exceptions, and the selection of the best structure often depends on a careful analysis of experimental data.

EXAMPLE 2C.2   Selecting the dominant resonance structure for a molecule

The sulfate ion, SO_4^{2-}, is present in a number of important minerals, including gypsum ($CaSO_4 \cdot 2H_2O$), which is used in cement, and Epsom salts ($MgSO_4 \cdot 7H_2O$), which is used as a purgative. Determine the dominant resonance structure of a sulfate ion from the three shown in (**10a–10c**) by calculating the formal charges on the atoms in each structure.

ANTICIPATE As you gain experience, you will be able to recognize the most favorable structure from the patterns of bonds and lone pairs, but until then the only way forward is to calculate the formal charges.

PLAN Follow the procedure in Toolbox 2B.2. You need do only one calculation for equivalent atoms, such as the oxygen atoms in the first diagram, because they all have the same arrangement of electrons and hence the same formal charge.

SOLVE Draw up the following table, in accord with Toolbox 2B.2.

	10a	10b	10c
Step 1 Count the valence electrons (V).		O: 6 S: 6 Total: 30 electrons, which provide 15 pairs of electrons, plus two electrons from the charge of -2.	
Step 2 Draw the Lewis structure.			

10a   10b   10c

Step 3 Assign electron ownership, $(L + \frac{1}{2}B)$.

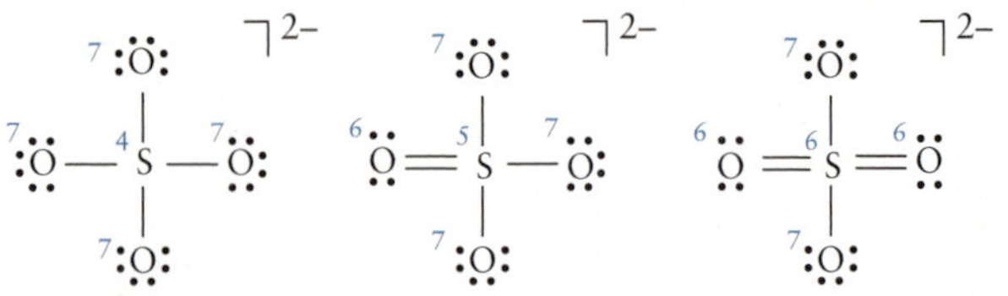

Step 4 Evaluate the formal charge, $V - (L + \frac{1}{2}B)$.

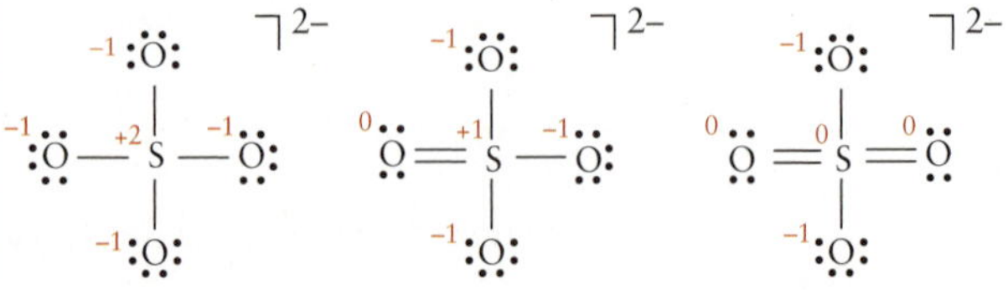

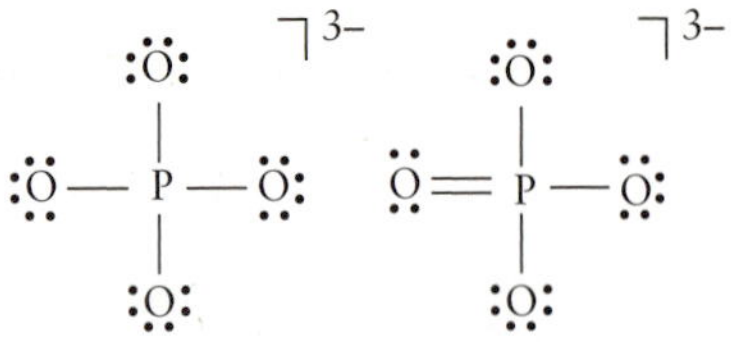

11

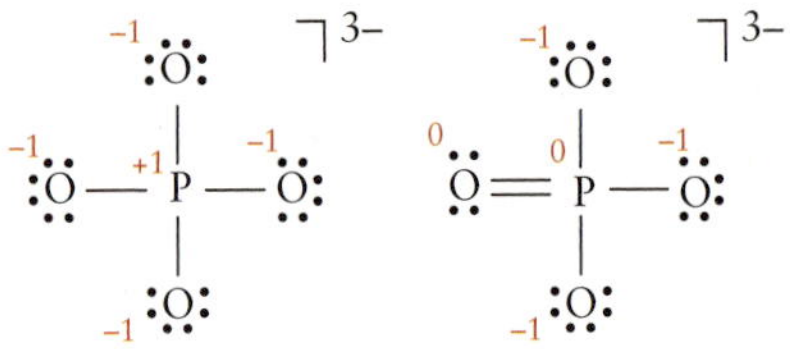

12

EVALUATE The individual formal charges are closest to zero in structure (**10c**), so the structure with two double bonds is likely to make the biggest contribution to the resonance hybrid even though the valence shell on the S atom has expanded to hold 12 electrons. This conclusion is consistent with experimental measurements of the S—O bond length in the sulfate ion, which is 149 pm, whereas the length of the single S—OH bond in sulfuric acid is 157 pm.

Self-test 2C.3A Calculate the formal charges for the two Lewis structures of the phosphate ion shown in (**11**).

[***Answer:*** See (**12**).]

Self-test 2C.3B Calculate the formal charges for the three oxygen atoms in one of the Lewis structures of the ozone resonance structure (see Example 2B.3).

Related Exercises 2C.15–2C.18

Expansion of the valence shell to more than eight electrons occurs in elements of Period 3 and later periods. These elements can exhibit variable covalence and be hypervalent. Formal charge helps to identify the dominant resonance structure.

2C.3 Incomplete Octets

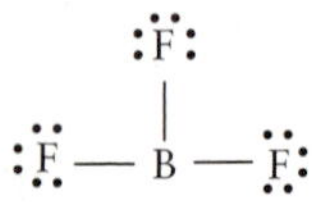

Some atoms exist in compounds with an **incomplete octet**. Boron is the prime example. One Lewis structure of the colorless gas boron trifluoride, BF_3 (**13**) leaves it with a valence shell of only six electrons. The boron atom might be expected to complete its octet by sharing more electrons with fluorine, as depicted in (**14**), but fluorine has such a high ionization energy that it is unlikely to exist with a positive formal charge. Experimental evidence, such as the short B—F bond lengths, suggests that the true structure of BF_3 is a resonance hybrid of both types of Lewis structures, with the singly bonded structure making the major contribution.

There is a special way for boron and similar atoms to complete their octets: an additional atom or ion with a lone pair of electrons might form a bond by providing *both* electrons. A bond in which both electrons come from one of the atoms is called a **coordinate covalent bond**. For example, the tetrafluoroborate anion, BF_4^- (**15**), forms when boron trifluoride is passed over a metal fluoride. Notice that both bonding electrons are provided by the fluoride ion. Another example of a coordinate covalent bond is that formed when boron trifluoride reacts with ammonia:

$$BF_3(g) + NH_3(g) \longrightarrow NH_3BF_3(s)$$

The Lewis structure of the product, a white molecular solid, is shown in (**16**). In this reaction, the lone pair on the nitrogen atom of ammonia, $:NH_3$, completes boron's octet in BF_3 by forming a coordinate covalent bond.

Another way in which coordinate covalent bonds form to complete an otherwise incomplete octet is by the formation of **dimers** (linked pairs of molecules). Aluminum

13 Boron trifluoride, BF_3

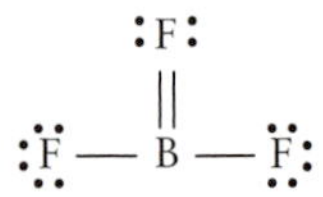

14 Boron trifluoride, BF_3

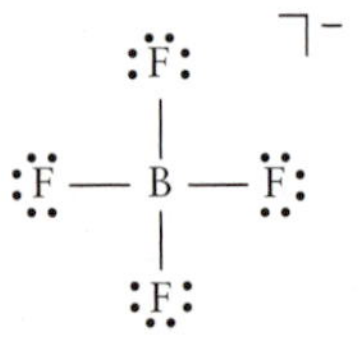

15 Tetrafluoroborate, BF_4^-

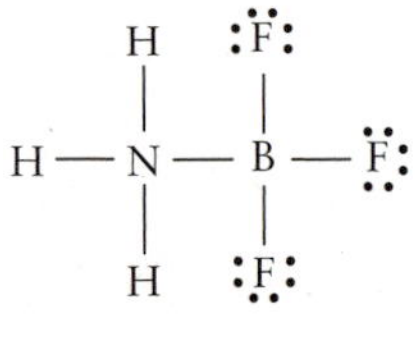

16 NH_3BF_3

chloride, for instance, is a volatile white solid that vaporizes at 180 °C to a gas of Al_2Cl_6 molecules. These molecules survive in the gas up to about 200 °C and only then fall apart into $AlCl_3$ molecules. The Al_2Cl_6 molecule exists because a Cl atom in one $AlCl_3$ molecule uses one of its lone pairs to form a coordinate covalent bond to the Al atom in a neighboring $AlCl_3$ molecule (**17**). This arrangement can occur in aluminum chloride but not boron trichloride because the atomic radius of Al is bigger than that of B and the Cl atom can get close enough to the Al atom to form a bridging bond.

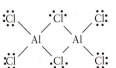

17 Aluminum chloride, Al_2Cl_6

Compounds of boron and aluminum may have unusual Lewis structures in which boron and aluminum have incomplete octets or halogen atoms that act as bridges.

What have you learned in this Topic?

You have learned that molecules with an unpaired electron are called radicals. These compounds are normally very reactive. When drawing Lewis structures for molecules that contain Period 3 and later elements, you have seen that it may be necessary to allow for the expansion of the valence shell.

The skills you have mastered are the ability to:

☐ **1.** Draw the Lewis structures of molecules and ions with expanded and incomplete valence shells (Example 2C.1).

☐ **2.** Use formal charge calculations to evaluate alternative Lewis structures (Example 2C.2).

Topic 2C Exercises

2C.1 Which of the following species are radicals? (a) NO_2^-; (b) CH_3; (c) OH; (d) CH_2O.

2C.2 Which of the following species are radicals? (a) NO_3; (b) ICl_2^+; (c) CH_2O; (d) HOCO.

2C.3 Draw the Lewis structure, including typical contributions to the resonance structure (where appropriate, allow for the possibility of octet expansion, including double bonds in different positions), for (a) periodate ion; (b) hydrogen phosphate ion; (c) chloric acid; (d) arsenate ion.

2C.4 Draw the Lewis structure, including typical contributions to the resonance structure (where appropriate, allow for the possibility of octet expansion), for (a) formate ion (HCO_2^-); (b) hydrogen phosphite ion; (c) bromate ion; (d) selenate ion.

2C.5 Draw the Lewis structure of each of the following reactive species, all of which are found to contribute to the destruction of the ozone layer, and indicate which are radicals: (a) chlorine monoxide, ClO; (b) dichloroperoxide, Cl—O—O—Cl; (c) chlorine nitrate, $ClONO_2$ (the central O atom is attached to the Cl atom and to the N atom of the NO_2 group).

2C.6 Draw the Lewis structure of each of the following species and indicate which are radicals: (a) the triiodide ion, I_3^-; (b) the methyl cation, CH_3^+; (c) the methyl anion, CH_3^-; (d) chlorine dioxide, ClO_2.

2C.7 Determine the numbers of electron pairs (both bonding and lone pairs) on the iodine atom in (a) ICl_2^+; (b) ICl_4^-; (c) ICl_3; (d) ICl_5.

2C.8 Determine the numbers of electron pairs (both bonding and lone pairs) on the phosphorus atom in (a) PCl_3; (b) PCl_5; (c) PCl_4^+; (d) PCl_6^-.

2C.9 Draw the Lewis structure for each of the following molecules or ions and give the number of electrons about the central atom: (a) SF_6; (b) XeF_2; (c) AsF_6^-; (d) $TeCl_4$.

2C.10 Draw the Lewis structure for each of the following molecules or ions and give the number of electrons about the central atom: (a) SiO_4^{4-}; (b) SCl_2; (c) BrF_5; (d) ICl_2^-.

2C.11 Draw the Lewis structure and state the number of lone pairs on xenon, the central atom of each of the following molecules: (a) $XeOF_2$; (b) XeF_4; (c) $XeOF_4$.

2C.12 Draw the Lewis structure and state the number of lone pairs on the central atom of each of the following molecules: (a) IF_5; (b) AsF_5; (c) H_2SO_3 (S is the central atom and each H atom is bonded to an O atom).

2C.13 In each of these compounds, an atom violates the octet rule. Identify the atom and explain the deviation from the octet rule: (a) $BeCl_2$; (b) ClO_2.

2C.14 In each of these compounds, an atom violates the octet rule. Identify the atom and explain the deviation from the octet rule: (a) SF_6; (b) BH_3.

2C.15 Two contributions to the resonance structure are shown below for each species. Determine the formal charge on each atom and then, if possible, identify the Lewis structure of lower energy for each species.

(a)

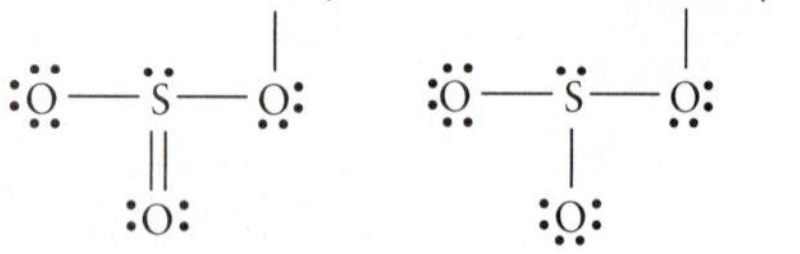

(b)

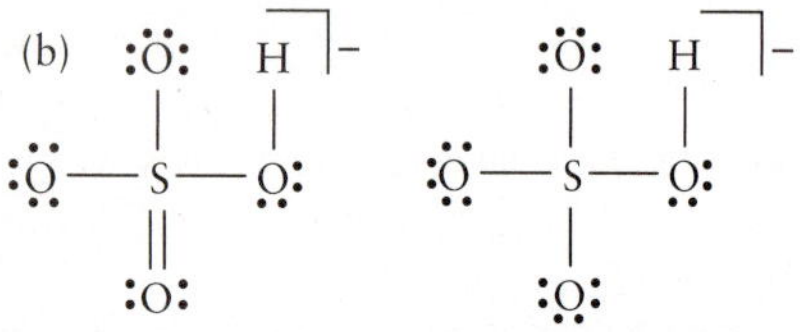

2C.16 Two Lewis structures are shown below for each species. Determine the formal charge on each atom and then, if appropriate, identify the Lewis structure of lower energy for each species.

(a)

(b)

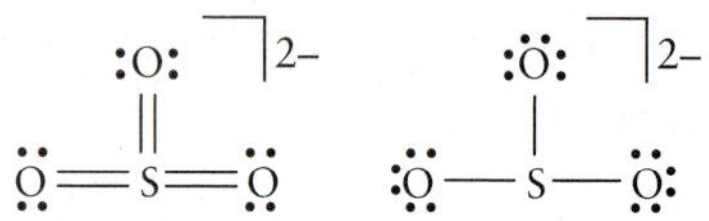

2C.17 Select from each of the following pairs of Lewis structures the one that is likely to make the dominant contribution to a resonance hybrid. Explain your selection.

(a)

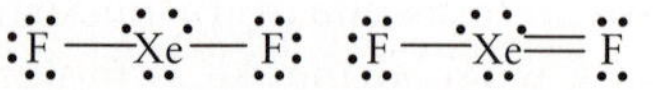

(b)

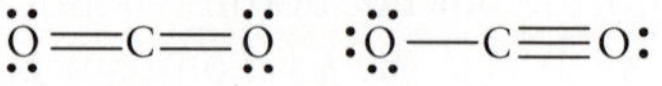

2C.18 Select from each of the following pairs of Lewis structures the one that is likely to make the dominant contribution to a resonance hybrid. Explain your selection.

(a)

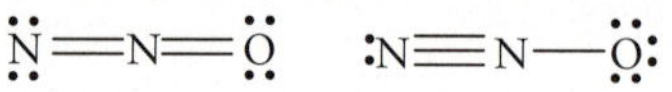

(b)

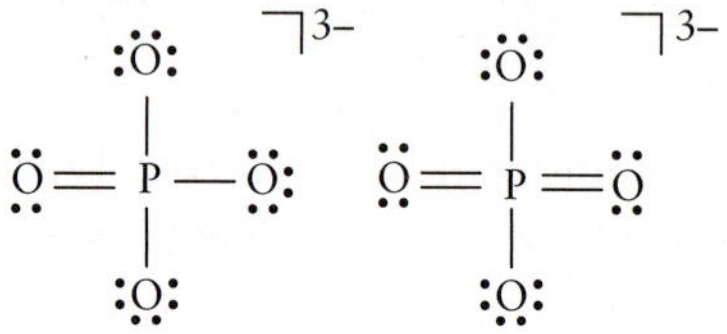

Topic 2D The Properties of Bonds

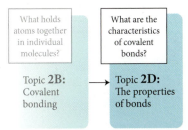

Why Do You Need to Know This Material? The properties of covalent compounds can vary widely, partly due to the way electrons are shared in covalent bonds. These variations, especially in bond strength, are used to explain the physical and chemical properties of molecules.

What Do You Need to Know Already? You need to know about trends in periodic properties (Topic 1F) as well as the role of electron-pair sharing in covalent bonding (Topic 2B).

The characteristics of a covalent bond between two atoms are due mainly to the properties of the atoms themselves and vary only a little with the identities of the other atoms present in a molecule. Consequently, some characteristics of a bond can be predicted with reasonable certainty once the identities of the two bonded atoms are known, regardless of the other atoms in the molecule. For instance, provided the bond order is the same, the length and strength of an A—B bond are approximately the same regardless of the molecule in which it is found. Thus, to understand the properties of a large molecule, such as how DNA replicates in our cells and transmits genetic information, the character of $C{=}O$ and N—H bonds can be studied in much simpler compounds, such as formaldehyde, $H_2C{=}O$, and ammonia, NH_3.

A further issue is that ionic and covalent bonding (Topics 2A and 2B) are two extreme models of the chemical bond. Most actual bonds lie somewhere between purely ionic and purely covalent. To describe bonds between nonmetals, covalent bonding is a good model. When a metal and nonmetal are present in a simple compound, ionic bonding is a good model. However, the bonds in many compounds seem to have properties between the two extreme models of bonding. Can these bonds be described more accurately by improving the two basic models?

2D.1 Correcting the Covalent Model: Electronegativity

A single Lewis structure is only the initial step in formulating a description of a covalently bonded molecule. That description can be improved by using resonance to blend together alternative contributions. All molecules can be viewed as resonance hybrids of purely covalent and purely ionic structures, even if only to a small extent. For example, the structure of a Cl_2 molecule can be described as

$$:\!\ddot{\underset{..}{Cl}}\!:^- \ \ddot{\underset{..}{Cl}}\!:^+ \longleftrightarrow :\!\ddot{\underset{..}{Cl}}\!-\!\ddot{\underset{..}{Cl}}\!: \longleftrightarrow Cl^+ :\!\ddot{\underset{..}{Cl}}\!:^-$$

In this case, the ionic structures (here, only a local pair of ions, not a conglomerate of many) make only a small contribution to the resonance hybrid; the bond is almost purely covalent. Moreover, the two ionic structures have the same energy and make equal contributions to the hybrid; so the average charge on each atom is zero. However, in a molecule composed of different elements, such as HCl, the resonance

$$H\!:^- \ \ddot{\underset{..}{Cl}}\!:^+ \longleftrightarrow H\!-\!\ddot{\underset{..}{Cl}}\!: \longleftrightarrow H^+ :\!\ddot{\underset{..}{Cl}}\!:^-$$

has unequal contributions from the two ionic structures. Because the chlorine atom has a greater attraction for electrons than does the hydrogen atom, the structure with the negative charge on the Cl atom, $H^+ :\!\ddot{\underset{..}{Cl}}\!:^-$, makes a bigger contribution than $H\!:^- \ \ddot{\underset{..}{Cl}}\!:^+$. As a result, there is a small net negative charge on the Cl atom and a small net positive charge on the H atom.

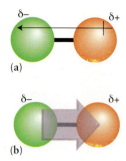

(a)

(b)

1

The debye is named for the Dutch chemist Peter Debye, who carried out important studies of dipole moments.

The SI unit of dipole moment is 1 C·m (1 coulomb meter). It is the dipole moment of a charge of 1 C separated from a charge of −1 C by a distance of 1 m; 1 D = 3.336×10^{-30} C·m.

Dissociation energies, a measure of the strengths of bonds, are discussed in Section 2D.3.

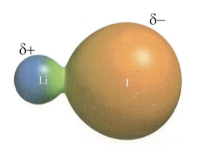

FIGURE 2D.1 The electronegativity of an element is its electron-pulling power when it is part of a compound. The bond in lithium iodide is approximately 50% covalent and 50% ionic in character. The electron density of LiI is shown here, with red denoting electron-rich regions and blue denoting electron-poor regions. The outcome of the tug-of-war is that the more electronegative atom (I) has a greater share in the electron pair of the covalent bond.

The net charges, the average outcome of the resonance, on the atoms in HCl are called **partial charges** and are shown by writing $^{\delta+}$H—Cl$^{\delta-}$. A bond in which there are nonzero partial charges on the atoms is called a **polar covalent bond.** All bonds between atoms of different elements are polar to some extent. The bonds in homonuclear (same-element) diatomic molecules and ions are nonpolar because, although ionic structures contribute to various extents, the net charges on the atoms are zero.

The partial charges on the two atoms in a polar covalent bond form an **electric dipole,** a partial positive charge next to an equal but opposite partial negative charge. In the original convention, a dipole was represented by an arrow that points toward the negative partial charge (**1a**); in the modern convention, the arrow points toward the positive partial charge (**1b**). Clearly, it is important to know which convention is being used! This text uses the modern convention. The size of an electric dipole—which is a measure of both the magnitude of the partial charges and their separation—is reported as the **electric dipole moment,** μ (the Greek letter mu), in non-SI units called **debye** (D). The debye is defined so that a single negative charge (an electron) separated by 100. pm from a single positive charge (a proton) has a dipole moment of 4.80 D. The dipole moment associated with a Cl—H bond is about 1.1 D. This dipole moment can be thought of as arising from a partial charge of about 23% of an electron's charge on the Cl atom and an equivalent positive charge on the H atom.

A covalent bond is polar if one atom has a greater attraction for electrons than the other atom, because then the electron pair is more likely to be drawn closer to the former atom. In 1932, the American chemist Linus Pauling proposed a quantitative measure of this attracting ability. The electron-pulling power of an atom when it is part of a molecule is called its **electronegativity.** Electronegativities are denoted χ (the Greek letter chi, pronounced "kye"). An atom of the element with the higher electronegativity has a stronger pulling power on electrons and tends to pull them away from the atom of the element with lower electronegativity (**FIG. 2D.1**). Pauling based his scale on the dissociation energies, D, of the A—A, B—B, and A—B bonds, measured in electronvolts. He defined the difference in electronegativity of the two elements A and B as

$$|\chi_A - \chi_B| = \{D(A-B) - \tfrac{1}{2}[D(A-A) + D(B-B)]\}^{1/2} \qquad (1)$$

He then assigned individual values of electronegativities that were consistent with this expression.

A different way of setting up a scale of electronegativities was devised by another American chemist, Robert Mulliken, in 1934. In his approach, the electronegativity is the average of the ionization energy (I) and electron affinity (E_a) of the element (both expressed as the numerical value in electronvolts):

$$\chi = \tfrac{1}{2}(I + E_a) \qquad (2)$$

The Mulliken definition makes sense because an atom gives up an electron reluctantly if the ionization energy is high. In addition, if the electron affinity of the atom is high, then attaching an electron to it is energetically favorable. Elements with high values of both these properties are reluctant to lose their electrons (in the sense that electron loss is energy intensive) and tend to gain them (in the sense that the energy is lowered if they succeed); hence, they are classified as highly electronegative. Conversely, if the ionization energy and the electron affinity are both low, then it takes very little energy for the element to give up its electrons and it has little tendency to gain more; hence, the electronegativity is low. The numerical values obtained in this way are broadly in step with the Pauling values, which are used in this text.

FIGURE 2D.2 shows the variation in electronegativity for the main-group elements of the periodic table. Because ionization energies and electron affinities are highest at the top right of the periodic table (close to fluorine, with the exception of the noble gases), it is not surprising to find that nitrogen, oxygen, bromine, chlorine, and fluorine are the elements with the highest electronegativities.

When the electronegativity difference between the two atoms in a bond is small, the partial charges are also small. As the difference in electronegativities increases, so do the partial charges. If the difference in electronegativities is large, then one atom can acquire

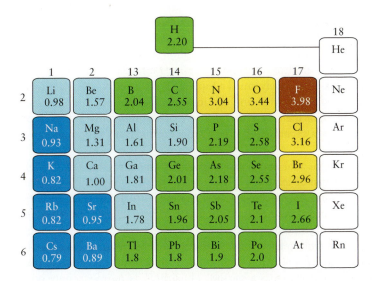

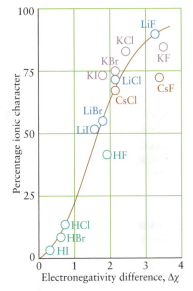

the lion's share of the electron pair, and the corresponding ionic structure makes a large contribution to the resonance. Because it has largely robbed the other atom of its share of the electrons, the highly electronegative element resembles an anion and the other atom resembles a cation. Such a bond is said to have considerable **ionic character.**

There is no sharp dividing line between ionic and covalent bonding. However, a good rule of thumb is that an electronegativity difference of about 2 means that the bond has so much ionic character that it is best regarded as ionic (**FIG. 2D.3**). For electronegativity differences smaller than about 1.5, a covalent description of the bond is reasonable. For example, the electronegativities of carbon and oxygen are 2.55 and 3.44, an electronegativity difference of 0.89, and C—O bonds are best regarded as polar covalent. However, there are exceptions to these guidelines. For instance, the electronegativity of magnesium is 1.31, and Mg—Cl bonds, with an electronegativity difference of 1.85, are considered ionic.

FIGURE 2D.3 The dependence of the percentage ionic character of the bond on the difference in electronegativity, $\Delta \chi$, between two bonded atoms for a number of halides.

Self-test 2D.1A In which of the following compounds do the bonds have greater ionic character: (a) P_4O_{10} or (b) PCl_3?

[*Answer:* (a)]

Self-test 2D.1B In which of the following compounds do the bonds have greater ionic character: (a) CO_2 or (b) NO_2?

Electronegativity is a measure of the pulling power of an atom on the electrons in a bond. A polar covalent bond is a bond between two atoms with partial electric charges arising from their difference in electronegativity. The presence of partial charges gives rise to an electric dipole moment.

2D.2 Correcting the Ionic Model: Polarizability

Now consider the alternative starting point of ionic bonding and how that description can be improved. All ionic bonds have some covalent character. To see how covalent character can arise, consider a monatomic anion (such as Cl^-) next to a cation (such as Na^+). As the cation's positive charge pulls on the anion's electrons, the spherical electron cloud of the anion becomes distorted in the direction of the cation. You can think of this distortion as the tendency of the electron density of the bond to occupy preferentially the region between the two nuclei, resulting in a bond with covalent character (**FIG. 2D.4**). The greater the distortion of the electron cloud on the anion, the greater is the covalent character of the bond.

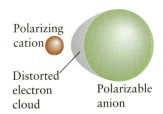

Polarizing cation

Distorted electron cloud

Polarizable anion

FIGURE 2D.4 When a small, highly charged cation is close to a large anion, the electron cloud of the anion is distorted in the process we call polarization. The green sphere represents the shape of an anion in the absence of a cation. The gray shadow shows how the shape of the sphere is distorted by the positive charge of the cation.

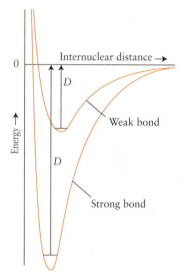

FIGURE 2D.5 The variation of the energy of a diatomic molecule with internuclear separation for weak and strong bonds. The dissociation energy is a measure of the depth of the well. (In practice, we have to take into account the small zero-point energy of the vibrating molecule, and so the dissociation energy is slightly less than the depth of the well.)

Atoms and ions with electron clouds that readily undergo a large distortion are said to be highly **polarizable.** An anion can be expected to be highly polarizable if it is large, such as an iodide ion, I^-. In such a large anion, the nucleus exerts only weak control over its outermost electrons because the effective nuclear charge experienced by the valence electrons is relatively small (Topic 1E). As a result, the electron cloud of the large anion is easily distorted and the ion is highly polarizable. Cations, which have fewer electrons than their parent atoms, are not significantly polarizable because the remaining electrons experience a greater effective nuclear charge and as a consequence are held tightly.

Atoms and ions that can cause large distortions are said to have high **polarizing power.** A cation can be expected to have high polarizing power if it is small and highly charged, such as an Al^{3+} cation. A small radius means that the center of charge of a highly charged cation can get very close to the anion, where it can exert a strong pull on the anion's electrons. Compounds composed of a small, highly charged cation and a large, polarizable anion tend to have bonds with considerable covalent character.

Cations become smaller, more highly charged, and hence more strongly polarizing, from left to right across a period. Thus, Be^{2+} is more strongly polarizing than Li^+, and Mg^{2+} is more strongly polarizing than Na^+. On the other hand, cations become larger and hence less strongly polarizing down a group. Thus, Na^+ is less strongly polarizing than Li^+, and Mg^{2+} is less strongly polarizing than Be^{2+}. Because polarizing power increases from Li^+ to Be^{2+} but decreases from Be^{2+} to Mg^{2+}, it follows that the polarizing power of the diagonal neighbors Li^+ and Mg^{2+} should be similar. Such similarities in the properties of other diagonally related neighbors can be expected and underlie the diagonal relationships in the periodic table introduced in Topic 1F.

Self-test 2D.2A In which of the compounds NaBr and $MgBr_2$ do the bonds have greater covalent character?

[*Answer:* $MgBr_2$]

Self-test 2D.2B In which of the compounds CaS and CaO do the bonds have greater covalent character?

Compounds composed of highly polarizing cations and highly polarizable anions have significant covalent character in their bonding.

2D.3 Bond Strengths

The strength of a chemical bond is measured by its **dissociation energy,** D, the energy required to separate the bonded atoms completely. On a plot of the potential energy of a diatomic molecule against the internuclear distance, the dissociation energy is the depth of the well below the energy of the separated atoms (**FIG. 2D.5**). In this type of bond breaking each atom retains one of the electrons from the bond. An example is $H—Cl(g) \longrightarrow \cdot H(g) + \cdot Cl(g)$. A high dissociation energy indicates a deep potential energy well and therefore a strong bond that requires a lot of energy to break. The strongest known bond between two nonmetal atoms is the triple bond in carbon monoxide, for which the dissociation energy is $1062 \ kJ\cdot mol^{-1}$. One of the weakest bonds is that between the iodine atoms in molecular iodine, for which the dissociation energy is only $139 \ kJ\cdot mol^{-1}$.

TABLES 2D.1 and **2D.2** list a selection of typical dissociation energies. The values given in Table 2D.1 are the actual dissociation energies of a variety of bonds in diatomic molecules. The values in Table 2D.2 are *average* dissociation energies for a number of different polyatomic molecules in which those bonds appear. The actual strength of a bond depends on the other atoms present in a polyatomic molecule, but the dependence is typically not too great and it can be useful for general discussions to have average values

TABLE 2D.1   Bond Dissociation Energies of Diatomic Molecules

Molecule	Bond dissociation energy/$(kJ\cdot mol^{-1})$
H_2	424
N_2	932
O_2	484
CO	1062
F_2	146
Cl_2	230
Br_2	181
I_2	139
HF	543
HCl	419
HBr	354
HI	287

TABLE 2D.2 Average Bond Dissociation Energies

Bond	Average bond dissociation energy/(kJ·mol⁻¹)	Bond	Average bond dissociation energy/(kJ·mol⁻¹)
C—H	412	C—I	238
C—C	348	N—H	388
C=C	612	N—N	163
C∴C*	518	N=N	409
C≡C	837	N—O	210
C—O	360	N=O	630
C=O	743	N—F	270
C—N	305	N—Cl	200
C—F	484	O—H	463
C—Cl	338	O—O	157
C—Br	276		

*In benzene.

that are typical of the bond in whatever molecule it occurs. For instance, the strength quoted for a C—O single bond is the average strength of such bonds in a selection of organic molecules, such as methanol (CH_3—OH), ethanol (CH_3CH_2—OH), and dimethyl ether (CH_3—O—CH_3). The values should therefore be regarded as typical rather than as accurate values for a particular molecule.

The trends in bond strengths shown in the tables are explained in part by the Lewis structures of the molecules. Consider, for example, the diatomic molecules of nitrogen, oxygen, and fluorine (**FIG. 2D.6**). Note the decline in bond strength as the bond order decreases, from 3 in N_2 to 1 in F_2. The triple bond in nitrogen is the origin of its chemical inertness. A multiple bond is almost always stronger than a single bond because more electrons bind the multiply bonded atoms. A triple bond between two atoms is always stronger than a double bond between the same two atoms, and a double bond is always stronger than a single bond between the same two atoms. However, a double bond between two carbon atoms is not twice as strong as a single bond, and a triple bond is a lot less than three times as strong. For example, the average dissociation energy of a C=C double bond is 612 kJ·mol⁻¹, whereas it takes 696 kJ·mol⁻¹ to break two C—C single bonds; similarly, the average dissociation energy of a C≡C triple bond is 837 kJ·mol⁻¹, but it takes 1044 kJ·mol⁻¹ to break three C—C single bonds (**FIG. 2D.7**). The origin of these differences is in part the repulsions between the electron pairs in a multiple bond, so each pair is not quite as effective at bonding as a pair of electrons in a single bond.

The values in Table 2D.2 show how resonance affects the strengths of bonds. For example, the strength of a carbon–carbon bond in benzene is intermediate between that of a single and that of a double bond. Resonance spreads multiple-bond character over the bonds between atoms; as a result, what were single bonds are strengthened and what were double bonds are weakened. The net effect overall is a stabilization of the molecule.

The presence of lone pairs may influence the strengths of bonds. Lone pairs repel each other; and, if they are on neighboring atoms, that repulsion can weaken the bond. This repulsion between lone pairs helps to explain why the bond in F_2 is weaker than the bond in H_2: the latter molecule has no lone pairs.

Trends in bond strengths correlate with trends in atomic radii. If the nuclei of the bonded atoms cannot get very close to the electron pair lying between them, the two atoms will be only weakly bonded together. For example, the bond strengths of the hydrogen halides decrease from HF to HI, as shown in **FIG. 2D.8**. The strength of the bond between hydrogen and a Group 14 element also decreases down the group (**FIG. 2D.9**). This weakening of the bond correlates with a decrease in the stability of the hydrides down the group. Methane, CH_4, can be kept indefinitely in air at room

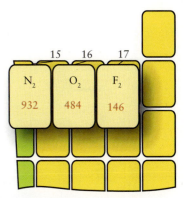

FIGURE 2D.6 The bond dissociation energies, in kilojoules per mole of N_2, O_2, and F_2 molecules. Note how the bonds weaken in the change from a triple bond in N_2 to a single bond in F_2.

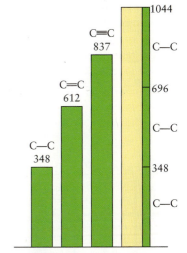

FIGURE 2D.7 The strengths (in kilojoules per mole) of single and multiple bonds between two carbon atoms. Note that, for bonds between carbon atoms, a double bond is less than twice as strong as a single bond and a triple bond is less than three times as strong as a single bond, as shown by the fourth column.

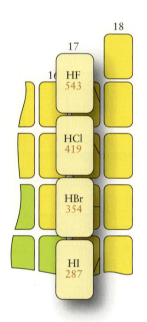

2 ATP

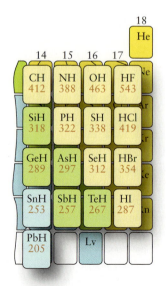

FIGURE 2D.8 The bond dissociation energies of the hydrogen halide molecules in kilojoules per mole of molecules. Note how the bond weakens as the halogen atom becomes larger.

FIGURE 2D.9 The dissociation energies for bonds between hydrogen and the p-block elements. The bond strength decreases down each group as the atoms increase in size.

temperature. Silane, SiH_4, bursts into flame on contact with air. Stannane, SnH_4, decomposes into tin and hydrogen. Plumbane, PbH_4, has never been prepared, except perhaps in trace amounts.

The relative strengths of bonds are important for understanding the way that energy is used in bodies to power our brains and muscles. For instance, adenosine triphosphate, ATP (**2**), is found in every living cell. The triphosphate part of this molecule is a chain of three phosphate groups. One of the phosphate groups is removed in a reaction with water. The P—O bond in ATP requires only 276 kJ·mol^{-1} to break, whereas the new P—O bond formed in $H_2PO_4^-$ releases 350 kJ·mol^{-1} when it forms. As a result, the conversion of ATP to adenosine diphosphate, ADP, in the reaction

$$\text{wwwO}-\overset{\overset{O}{\|}}{\underset{\underset{O^-}{|}}{P}}-O-\overset{\overset{O}{\|}}{\underset{\underset{O^-}{|}}{P}}-O-\overset{\overset{O}{\|}}{\underset{\underset{O^-}{|}}{P}}-O^- + H_2O \longrightarrow \text{wwwO}-\overset{\overset{O}{\|}}{\underset{\underset{O^-}{|}}{P}}-O-\overset{\overset{O}{\|}}{\underset{\underset{O^-}{|}}{P}}-O^- + H_2PO_4^-\text{(aq)}$$

(where the wiggly line indicates the rest of the molecule) can release energy that is used to power energy-demanding processes in cells.

Closely related to the strength of a bond is its stiffness (its resistance to stretching and compressing), with strong bonds typically being stiffer than weak bonds. The stiffness of bonds is measured by infrared (IR) spectroscopy and is used to identify compounds, as described in *Major Technique* 1 on the website for this book.

The strength of a bond between two atoms is measured by its dissociation energy: the greater the dissociation energy, the stronger the bond. Bond strength increases as the multiplicity of a bond increases, decreases as the number of lone pairs on neighboring atoms increases, and decreases as the atomic radii increase.

2D.4 Bond Lengths

A **bond length** is the distance between the centers of two atoms joined by a covalent bond. It corresponds to the internuclear distance at the potential-energy minimum for the two atoms (see Fig. 2D.5). Bond lengths affect the overall size and shape of a molecule. The transmission of hereditary information in DNA, for instance, depends on bond lengths because the two strands of the double helix must fit together like pieces of a jigsaw puzzle (Topic 11E). Bond lengths are also crucial to the action of enzymes because only a molecule of the right size and shape will fit into the active site of the enzyme molecule (Topic 7E). Bond lengths are determined experimentally by using either spectroscopy (*Major Technique* 1 on the website for this book) or x-ray diffraction (*Major Technique* 3 on the website for this book).

As **TABLE 2D.3** shows, the lengths of bonds between Period 2 elements typically lie in the range from 100 pm to 150 pm. Bonds between heavy atoms tend to be longer than those between light atoms because heavier atoms have larger radii than lighter ones (**FIG. 2D.10**). *Multiple bonds are shorter than single bonds between the same two elements,* because the additional bonding electrons attract the nuclei more strongly and pull the atoms closer together; compare the lengths of the various carbon–carbon bonds in Table 2D.3. The averaging effect of resonance can also be seen: the length of the carbon–carbon bond in benzene is intermediate between the lengths of the single and double bonds of a Kekulé structure (but closer to that of a double bond).

Some useful correlations can be drawn from these data. For example, for bonds between atoms of the same two elements, *the stronger the bond, the shorter it is.* Thus, a C≡C triple bond is both stronger and shorter than a C=C double bond. Similarly, a C=O double bond is both stronger and shorter than a C—O single bond.

TABLE 2D.3 Average and Actual Bond Lengths

Bond	Average bond length/pm		Molecule	Actual bond length/pm
C—H	109		H_2	74
C—C	154		N_2	110
C=C	134		O_2	121
C∷C*	139		F_2	142
C≡C	120		Cl_2	199
C—O	143		Br_2	228
C=O	112		I_2	268
O—H	96			
N—H	101			
N—O	140			
N=O	120			

*In benzene.

Just as bond strengths are approximately transferrable between molecules, so atomic radii are similar in whatever molecule they form. Thus, each atom makes a characteristic contribution, called its **covalent radius**, to the length of a bond (**FIG. 2D.11**). A bond length is approximately the sum of the covalent radii of the two atoms (**3**). The O—H bond length in ethanol, for example, is the sum of the covalent radii of H and O, 37 + 66 pm = 103 pm. The precise value to use depends, however, on the bond order: Fig. 2D.11 shows that the covalent radius of an atom taking part in a multiple bond is smaller than that for a single bond of the same atom.

Covalent radii typically decrease from left to right across a period. The reason is the same as for atomic radii (Topic 1F): the increasing effective nuclear charge draws in the electrons and makes the atom more compact. Like atomic radii, covalent radii increase down a group because, in successive periods, the valence electrons occupy shells that are more distant from the nucleus and are better shielded by the inner core of electrons.

> *The covalent radius of an atom is the contribution it makes to the length of a covalent bond; covalent radii are added together to estimate the lengths of bonds in molecules.*

What have you learned in this Topic?

You have learned that an important property of an element is its electronegativity, which enables you to assess which atom in a bond has the greater share of the electron pair. Chemical bonds show a range of character, from completely covalent to completely ionic. You have seen that bond strengths are approximately transferrable between molecules and that atoms make characteristic contributions to the lengths of bonds.

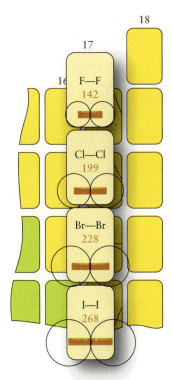

FIGURE 2D.10 Bond lengths (in picometers) of the diatomic halogen molecules. Notice how the bond length increases down the group as the atomic radii become larger.

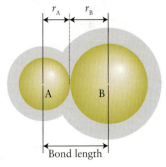

3 Covalent radii

FIGURE 2D.11 Covalent radii of hydrogen and the p-block elements (in picometers). Where more than one value is given, the values refer to single, double, and triple bonds. Covalent radii tend to become smaller toward fluorine. A bond length is approximately the sum of the covalent radii of the two participating atoms.

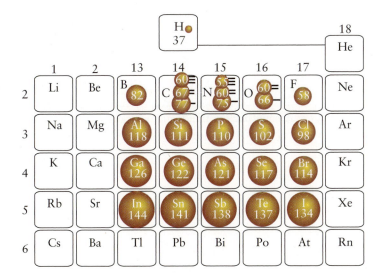

The skills you have mastered are the ability to:

☐ **1.** Explain the concept of electronegativity and use it to assess the polarity of a bond (Section 2D.1).

☐ **2.** Explain how resonance is used to improve the description of a covalent bond by introducing ionic character into it (Section 2D.1).

☐ **3.** Estimate relative ionic or covalent character (Self-Tests 2D.1 and 2D.2).

☐ **4.** Explain how the concept of polarizability is used to improve the description of an ionic bond (Section 2D.2).

☐ **5.** Predict and explain periodic trends in the polarizability of anions and the polarizing power of cations (Section 2D.2).

☐ **6.** Predict and explain relative bond strengths and bond lengths (Sections 2D.3−2D.4).

Topic 2D Exercises

2D.1 Place the following elements in order of increasing electronegativity: antimony, tin, selenium, and indium.

2D.2 Place the following elements in order of increasing electronegativity: iodine, sulfur, bromine, oxygen, and arsenic.

2D.3 Which of these compounds has bonds that are primarily ionic? (a) BBr_3; (b) $BaBr_2$; (c) $BeBr_2$.

2D.4 Which of these compounds has bonds that are primarily covalent? (a) $CaCl_2$; (b) $CsCl$; (c) CCl_4.

2D.5 For each pair, determine which compound has bonds with greater ionic character: (a) HCl or HI; (b) CH_4 or CF_4; (c) CO_2 or CS_2.

2D.6 For each pair, determine which compound has bonds with greater ionic character: (a) PH_3 or NH_3; (b) SO_2 or NO_2; (c) SF_6 or IF_5.

2D.7 Compounds having bonds with a high covalent character tend to be less soluble in water than similar compounds that have low covalent character. Use electronegativities to predict which of the following compounds is the more soluble in water: (a) $AlCl_3$ or KCl; (b) MgO or BaO.

2D.8 Use electronegativities to predict which of the following compounds is the more soluble in water (see Exercise 2D.7): (a) LiI or MgI_2; (b) CaS or CaO.

2D.9 Arrange the cations Rb^+, Be^{2+}, and Sr^{2+} in order of increasing polarizing power. Explain your reasoning.

2D.10 Arrange the cations K^+, Mg^{2+}, Al^{3+}, and Cs^+ in order of increasing polarizing power. Explain your reasoning.

2D.11 Arrange the anions Cl^-, Br^-, N^{3-}, and O^{2-} in order of increasing polarizability and give reasons for your decisions.

2D.12 Arrange the anions N^{3-}, O^{2-}, Cl^-, and Br^- in order of increasing polarizability and give reasons for your decisions.

2D.13 Place the following molecules or ions in order of *decreasing* bond length: (a) the CO bond in CO, CO_2, CO_3^{2-}; (b) the SO bond in SO_2, SO_3, SO_3^{2-}; (c) the CN bond in HCN, CH_2NH, CH_3NH_2. Explain your reasoning.

2D.14 Place the following molecules or ions in order of *decreasing* bond order: (a) NO bond in NO, NO_2, NO_3^-; (b) CC bond in C_2H_2, C_2H_4, C_2H_6; (c) CO bond in CH_3OH, CH_2O, CH_3OCH_3. Explain your reasoning.

2D.15 Which do you predict to have the strongest CX bond, where X is a halogen: (a) CF_4, (b) CCl_4, or (c) CBr_4? Explain.

2D.16 Which do you predict to have the strongest CN bond: (a) $NHCH_2$, (b) NH_2CH_3, or (c) HCN? Explain.

2D.17 Use the information in Fig. 2D.11 to estimate the length of (a) the CO bond in CO_2; (b) the CO and CN bonds in urea, $OC(NH_2)_2$; (c) the OCl bond in HClO; (d) the NCl bond in NOCl.

2D.18 Use the information in Fig. 2D.11 to estimate the length of (a) the CO bond in formaldehyde, H_2CO; (b) the CO bond in dimethyl ether, CH_3OCH_3; (c) the CO bond in methanol, CH_3OH; (d) the CS bond in methanethiol, CH_3SH.

2D.19 Use the covalent radii in Fig. 2D.11 to calculate the bond lengths in the following molecules. Account for the trends in your calculated values: (a) CF_4; (b) SiF_4; (c) SnF_4.

2D.20 Use the covalent radii in Fig. 2D.11 to calculate the lengths of the bonds between nitrogen atoms in the following molecules. Account for the trends in your calculated values: (a) hydrazine, H_2NNH_2; (b) nitrogen hydride, HNNH; (c) N_3^-.

Topic 2E The VSEPR Model

2E.1 The Basic VSEPR Model
2E.2 Molecules with Lone Pairs on the Central Atom
2E.3 Polar Molecules

What holds atoms together in individual molecules?	What determines the shapes of molecules?
Topic **2B**: Covalent bonding	→ Topic **2E**: The VSEPR model

Why Do You Need to Know This Material? The shapes of molecules determine their odors, their tastes, and their actions as drugs, and play an essential role in the reactions that are necessary for life. They also affect the properties of materials, including their physical states and their solubilities.

What Do You Need to Know Already? The VSEPR model extends the concept of Lewis structures (Topic 2B) and the discussion of polar molecules develops the material on polar bonds described in Topic 2D.

Lewis structures show the linkages between atoms and the presence of lone pairs, but except in very simple cases do not show how the atoms are arranged in space. In this Topic, Lewis's ideas are developed in order to predict the shapes of simple molecules.

2E.1 The Basic VSEPR Model

Consider a molecule that consists of one central atom to which all the other atoms are attached. Many molecules of this kind have the shapes of the geometrical figures shown in **FIG. 2E.1**; thus, CH_4 (**1**) is tetrahedral, SF_6 (**2**) is octahedral, and PCl_5 (**3**) is trigonal

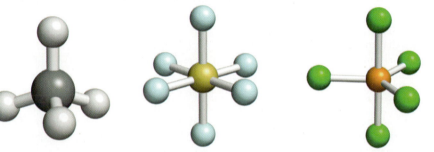

1 Methane, CH_4 2 Sulfur hexafluoride, SF_6 3 Phosphorus pentachloride, PCl_5

bipyramidal. In a number of these cases the **bond angles,** the angles between neighboring bonds (regarded as straight lines that join the atom centers), are fixed by the symmetry of the molecule; these bond angles are indicated in Fig. 2E.1. Thus, the HCH angle in CH_4 is 109.5° (the "tetrahedral angle"), the FSF angles in SF_6 are 90° and 180°, and the ClPCl

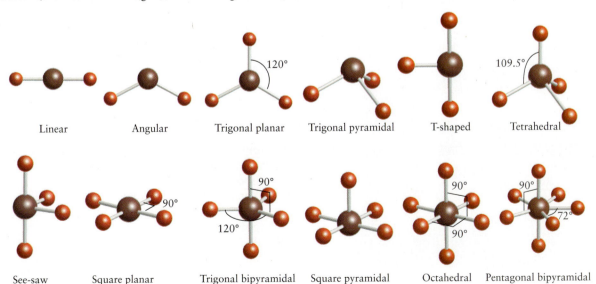

Linear   Angular   Trigonal planar   Trigonal pyramidal   T-shaped   Tetrahedral

See-saw   Square planar   Trigonal bipyramidal   Square pyramidal   Octahedral   Pentagonal bipyramidal

FIGURE 2E.1 The names of the shapes of simple molecules and their bond angles. Lone pairs of electrons are not shown because they are not included when identifying molecular shapes.

angles in PCl_5 are 90°, 120°, and 180°. The bond angles of molecules that are not fixed by symmetry must be determined experimentally. The HOH bond angle in the angular H_2O molecule, for instance, has been found to be 104.5°, and the HNH angle in the trigonal pyramidal NH_3 molecule is measured as 107°. The principal technique for determining bond angles in small molecules is spectroscopy, especially rotational and vibrational spectroscopy; x-ray diffraction is used for larger molecules. Knowledge of the shapes of molecules and their bond angles can be critical in computer-aided drug design (BOX 2E.1).

Box 2E.1 FRONTIERS OF CHEMISTRY: DRUGS BY DESIGN AND DISCOVERY

The search for new drugs relies not only on the skills of chemists but also on biologists, ethnobotanists, and medical researchers. Because there are so many millions of compounds, it would take too long to start with the elements, combine them in different ways, and then test their effects. Instead, chemists usually start either by *drug discovery,* the identification and possibly the modification of promising medicines that already exist, or by *rational drug design,* the identification of characteristics of a target enzyme, virus, bacterium, or parasite and the design of new compounds to react with it.

In drug discovery, a chemist usually begins by investigating compounds that have already shown medicinal value. A fruitful path is to find a *natural product,* an organic compound found in nature, that has been shown to have healing characteristics. Nature is the best of all synthetic chemists, with billions of chemicals that fulfill as many different needs. The challenge is to find compounds that have curative powers. These substances are found in different ways: random or "blind" collection of samples that are then tested, or collection of specific samples identified by native healers as being medically effective.

Observation of the properties of plants and animals can help to guide a random search. For example, if certain types of fruits remain fresh whereas others rot, we might expect the former to contain antifungal agents. An example of this type of collection is the gathering of tunicates ("sea squirts") and sponges in the Caribbean. The chemists harvest the samples by diving from research vessels. The samples are tested for antiviral and antitumor activity in a chemical laboratory on the ship. The antiviral drug didemnin-C and the anticancer drug bryostatin 1 were discovered in marine organisms.

The guided route requires fewer samples for testing because the chemist works with a native healer, the ancient lore guiding

the modern chemistry. Often an ethnobotanist, a specialist in plants used for native healing, joins the team. This approach saves time for the scientists and can provide an economic benefit for the healers and their nations as well. Drugs that have been discovered in this way include a variety of anticancer and antimalarial drugs, blood-clotting agents, antibiotics, and medicines for the heart and digestive system.

Once the empirical and molecular formulas of the active compounds are determined, then their structural formulas are sought. At that point, synthetic work can begin. The chemist can identify compounds in the material that have medicinal value and find a way to *synthesize* them, or prepare them in the laboratory, so that they can be made available in large quantities.

In rational drug design, the chemist begins with the tumor or organism that the drug is intended to eradicate. Virtually all processes in living cells depend on specific *enzymes,* types of proteins with very large molecules that have specific shapes. Usually there is an *active site* on the enzyme into which only specific molecules can fit and react. If the enzymes that control the growth of parasites or bacteria can be identified and their shapes known, compounds that fit into the active sites and block the reactions can be designed. The chemist taking this route begins by identifying key enzymes in the bacterium or parasite. Then the molecular structure of the enzyme is determined. A computer program is used to design molecules with structures that fit into the active site. Structure–activity relationships (SAR) are investigated to find out which aspects of the structure are important to the activity of the drug. The new compounds are synthesized, and their effects and side effects are tested.

HOW MIGHT YOU CONTRIBUTE?

Despite all the medicines in a modern pharmacy, there is still a great need for specific chemotherapeutic agents with few side effects. In addition, emerging drug-resistant strains of bacteria may require the development of new methods of drug discovery.

Related Exercises 2.29, 2.44

Further Reading

I. Chopra, "The 2012 Garrod Lecture: Discovery of antibacterial drugs in the 21st century," *Journal of Antimicrobial Chemotherapy,* vol. 68, 2013, pp. 496–505. W. H. Gerwick and B. S. Moore, "Lessons from the past and charting the future of marine natural products drug discovery and chemical biology," *Chemistry & Biology,* vol. 19, 2012, pp. 85–98. J. W.-H. Li and J. C. Vederas, "Drug discovery and natural products: End of an era or an endless frontier?" *Science,* vol. 325, pp. 161–165 [July 10, 2009]. D. Camp, "Discovery and development of natural compounds into medicinal products," *Drugs of the future,* vol. 38, 2013, pp. 245–256.

A field biologist examines a plant in a South American rainforest. The plant contains compounds that will be investigated for their medicinal value. (*Gary Retherford/Science Source.*)

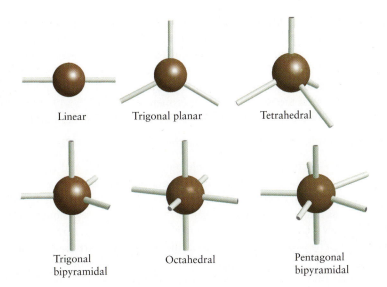

Linear   Trigonal planar   Tetrahedral

Trigonal bipyramidal   Octahedral   Pentagonal bipyramidal

FIGURE 2E.2 The positions that two to seven regions of high electron concentration (atoms and lone pairs) take around a central atom. The regions are denoted by the straight lines sticking out of the central atom. Use this diagram to identify the electron arrangement of a molecule, and then use Fig. 2E.1 to identify the shape of the molecule from the locations of its atoms. The seven-region pentagonal bipyramidal arrangement is not unique: several other arrangements have about the same energy.

Lewis structures (Topics 2B and 2C) show only how the atoms are connected and how the electrons are arranged around them. The **valence-shell electron-pair repulsion model** (VSEPR model) extends Lewis's theory of bonding by adding rules that account for bond angles and molecular shapes:

Rule 1 Regions of high electron concentration (bonds and lone pairs on the central atom) repel one another and, to minimize their repulsions, these regions move as far apart as possible while maintaining the same distance from the central atom (**FIG. 2E.2**).

These "most distant" locations give the **electron arrangement** of the molecule. Once this arrangement has been identified, the locations of the atoms are noted and the shape of the molecule is identified by referring to Fig. 2E.1. Note that, when naming the molecular shape, only the positions of atoms are considered, not any lone pairs that may be present on the central atom, even though they affect the shape.

A molecule with only two atoms attached to the central atom is $BeCl_2$. The Lewis structure is $:\!\ddot{C}l\!-\!Be\!-\!\ddot{C}l\!:$, and there are no lone pairs on the central atom. To be as far apart as possible, the two bonding pairs lie on opposite sides of the Be atom, and so the electron arrangement is linear. The Cl atoms lie on opposite sides of the Be atom and so the VSEPR model predicts a linear shape for the $BeCl_2$ molecule, with a bond angle of 180° (**4**). That shape is confirmed by experiment.

A boron trifluoride molecule, BF_3, has the Lewis structure shown in (**5**). There are three bonding pairs attached to the central atom and no lone pairs. To be as far apart as possible, the three bonding pairs must lie at the corners of an equilateral triangle. The electron arrangement is trigonal planar. Because an F atom is attached to each bonding pair, the BF_3 molecule is also trigonal planar (**6**), and all three FBF angles are 120°, as confirmed experimentally.

Methane, CH_4, has four bonding pairs on the central atom. To be as far apart as possible, the four pairs must take up a tetrahedral arrangement around the C atom. Because the electron arrangement is tetrahedral and an H atom is attached to each bonding pair, the VSEPR model predicts the molecule to be tetrahedral (as in **1**), with bond angles of 109.5°. That is the shape found experimentally.

In a phosphorus pentachloride molecule, PCl_5 (**7**), there are five bonding pairs and no lone pairs on the central atom. According to the VSEPR model, the five pairs and the atoms that they carry are farthest apart in a trigonal bipyramidal arrangement. In this arrangement, three atoms (the "equatorial" atoms) lie at the corners of an equilateral triangle with bond angles of 120° and the other two atoms (the "axial" atoms) lie above and below the plane of the triangle, each at a 90° angle from the equatorial atoms (as in **3**). The molecular shape is therefore predicted to be trigonal bipyramidal, and that is confirmed experimentally.

The VSEPR model was first proposed by the British chemists Nevil Sidgwick and Herbert Powell and has been developed by the Canadian chemist Ronald Gillespie.

4 Beryllium chloride, $BeCl_2$

$$:\!\ddot{F}\!: \\ | \\ :\!\ddot{F}\!-\!B\!-\!\ddot{F}\!:$$

5 Boron trifluoride, BF_3

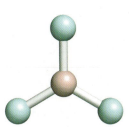

6 Boron trifluoride, BF_3

$$:\!\ddot{C}l\!: \\ | \\ :\!\ddot{C}l\!-\!P\!-\!\ddot{C}l\!: \\ :\!\ddot{C}l\!: \quad :\!\ddot{C}l\!:$$

7 Phosphorus pentachloride, PCl_5

8 Sulfur hexafluoride, SF_6

All six terminal atoms are equivalent in a regular octahedral molecule.

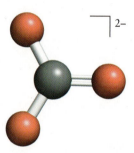

9 Carbon dioxide, CO_2

10 Carbonate ion, CO_3^{2-}

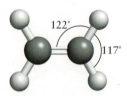

11 Carbonate ion, CO_3^{2-}

12 Ethene, C_2H_4

13 Ethene, C_2H_4

A sulfur hexafluoride molecule, SF_6, has six atoms attached to the central S atom and no lone pairs on that atom (**8**). According to the VSEPR model, the electron arrangement is octahedral, with four pairs at the corners of a square on the equator and the remaining two pairs above and below the plane of the square (see Fig. 2E.2). An F atom is attached to each electron pair, and so the molecule is predicted to be octahedral, with bond angles of 90° and 180°. That too is confirmed experimentally.

The second rule of the VSEPR model concerns the treatment of multiple bonds:

Rule 2 A multiple bond is treated as a single region of high electron concentration.

That is, the two electron pairs in a double bond stay together and repel other bonds or lone pairs as a unit. The three electron pairs in a triple bond also stay together and act like a single region of high electron concentration. For instance, a carbon dioxide molecule, $O=C=O$, has a linear structure similar to that of $BeCl_2$, even though both bonds are double bonds (**9**). One of the Lewis resonance structures of a carbonate ion, CO_3^{2-}, is shown in (**10**). The two pairs of electrons in the double bond are treated as a unit, and the resulting shape (**11**) is trigonal planar. Because each bond, whether single or multiple, acts as a single unit, to count the number of regions of high electron concentration and determine the electron arrangement of an ion or molecule simply count the number of atoms attached to the central atom and add the number of lone pairs.

When there is more than one "central" atom, the bonding about each atom is treated independently. For example, to predict the shape of an ethene (ethylene) molecule, $CH_2=CH_2$, each carbon atom is considered separately. From the Lewis structure (**12**) note that each carbon atom has three regions of high electron concentration (two single bonds and one double bond). The electron arrangement around each carbon atom is therefore trigonal planar. There are two H atoms and one C atom attached to each "central" C atom. You can therefore predict that the HCH and HCC angles will both be close to 120° (**13**); this prediction is confirmed experimentally (the actual bond angles being 117° and 122°, respectively). Experimental observations also show that all six atoms lie in a plane.

EXAMPLE 2E.1   Predicting the shape of a simple molecule

Knowledge of the shapes of molecules is particularly important in molecular biology, where the shape of a molecule can determine its function. The shapes of large biomolecules can often be predicted by first studying patterns in simple molecules, such as methanal (formaldehyde, $H_2C=O$), which was once used in preservatives for biological samples. Predict the shape of a methanal molecule.

ANTICIPATE Carbon atoms typically form four bonds. The central C atom in methanal has four bonds, so you should not expect there to be any lone pairs of electrons on the central atom.

PLAN Write the Lewis structure and identify the electron arrangement around the central atom (the C atom, in this case). Treat each multiple bond as a single unit. Then identify the overall shape of the molecule (refer to Fig. 2E.2 if necessary).

SOLVE

Write the Lewis structure of the molecule.

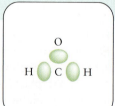

Identify the electron arrangement around the central atom.

Trigonal planar: The electron arrangement consists of three regions of high electron concentration and, to minimize repulsions, adopts a trigonal planar configuration.

Identify the arrangement of atoms around the C atom.

Trigonal planar: Since there are no lone pairs of electrons, the overall shape is described as trigonal planar.

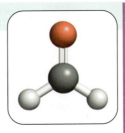

EVALUATE As expected, this molecule has no lone pairs of electrons on the central atom. From now on, you can predict

$$\overset{Z}{\underset{X\diagdown\underset{\displaystyle C}{}\diagup Y}{\parallel}}$$

that wherever you come across a X   Y group in a molecule, however complicated, you can anticipate that the structure about the C atom is trigonal planar.

Self-test 2E.1A Predict the shape of an arsenic pentafluoride molecule, AsF_5.

[**Answer:** Trigonal bipyramidal]

Self-test 2E.1B Predict the shape of an ethyne (acetylene) molecule, $HC{\equiv}CH$.

Related Exercises 2E.3, 2E.4

Because single bonds and multiple bonds are treated as equivalent in the VSEPR model, it does not matter which of the Lewis structures contributing to a resonance structure you consider. For example, because the nitrate ion is described by resonance involving three bonds about the central atom (Topic 2B), you can expect a trigonal planar structure (**14**).

Modern computational techniques provide additional detail about the distribution of electrons in molecules. A special kind of molecular representation shows the shape of the calculated electron distribution and is called a **density isosurface**. A combination of the density isosurface with electrostatic potential calculations results in an **electrostatic potential surface** (an "elpot" surface), in which the net electric potential is calculated at each point on the surface and is depicted by different colors (**15**): a blue tint at a point indicates a relative positive potential due to a positively charged nucleus; a red tint indicates the opposite. As you can see, although the electron distribution in NO_3^- is quite complicated, all three bonding regions are the same.

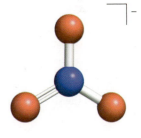

14 Nitrate ion, NO_3^-

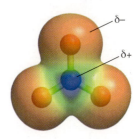

15 Nitrate ion, NO_3^-

THINKING POINT

Can you think of a way to present VSEPR theory without having to deal with Lewis structures?

According to the VSEPR model, regions of high electron concentration take up positions that maximize their separations; electron pairs in a multiple bond are treated as a single unit. The shape of the molecule is then identified from the relative locations of its atoms.

2E.2 Molecules with Lone Pairs on the Central Atom

It is sometimes helpful to use the generic "VSEPR formula" AX_nE_m to identify the different combinations of atoms and lone pairs attached to the central atom: A represents a central atom, X an attached atom, and E a lone pair. Molecules with the same VSEPR formula have essentially the same electron arrangement and the same shape; so, by recognizing the formula, you can immediately predict the shape (but not necessarily the precise numerical values of any bond angles that are not governed by symmetry). For example, the BF_3 molecule, with three attached fluorine atoms and no lone pairs on B, is an example of an AX_3 species, as is the NO_3^- ion (**14**). The sulfite ion, SO_3^{2-} (**16**), which has one lone pair on the S atom, is an example of an AX_3E species, as is NH_3.

$$\begin{array}{c} {:}\ddot{O}{:} \\ | \\ {:}\ddot{O}-\underset{\cdot\cdot}{S}-\ddot{O}{:} \end{array}\Bigg]^{2-}$$

16 Sulfite ion, SO_3^{2-}

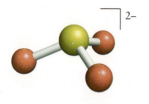

17 Sulfite ion, SO_3^{2-}

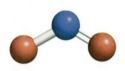

18 Nitrogen dioxide, NO_2

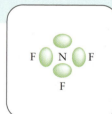

19 Nitrogen dioxide, NO_2

If there are no lone pairs on the central atom (an AX_n molecule), each region of high electron concentration corresponds to an atom, and so the molecular shape is the same as the electron arrangement. That was the case in the examples in Section 2E.1 ($BeCl_2$, BF_3, CH_4, PCl_5, and SF_6). However, if lone pairs are present, the molecular shape differs from the electron arrangement, because only the positions of the atoms are considered when naming the shape. For example, the four regions of high electron concentration in SO_3^{2-} are farthest apart if they adopt a tetrahedral arrangement (Fig. 2E.2). However, the *shape* of the ion is described by the locations of the atoms, not the lone pair. Because only three of the tetrahedral locations are occupied by atoms, the shape of an SO_3^{2-} ion is trigonal pyramidal (**17**). The rule to remember is

> **Rule 3** All regions of high electron concentration, lone pairs and bonds, are included in a description of the electronic arrangement, but only the positions of atoms are considered when identifying the shape of a molecule.

A single unpaired electron on the central atom is treated as a region of high electron concentration and acts like a lone pair when determining molecular shape. For example, radicals such as NO_2 have a single nonbonding electron, a "lone half-pair." Thus, NO_2 (**18**) has a trigonal planar electron arrangement (including the unpaired electron on N), but its shape, the arrangement of its atoms, is angular (**19**).

EXAMPLE 2E.2   Predicting the shape of a molecule with lone pairs on the central atom

The shapes of simple molecules are often relevant to their physical properties, and as you progress through your study of chemistry you will see how important it is to know whether a molecule is planar or not. Predict the electron arrangement and the shape of a nitrogen trifluoride molecule, NF_3.

ANTICIPATE The formula NF_3 resembles that of ammonia, NH_3, which is trigonal pyramidal; so you should suspect that NF_3 is also trigonal pyramidal.

PLAN For the electron arrangement, draw the Lewis structure and then use the VSEPR model to decide how the bonding pairs and lone pairs are arranged around the central atom (consult Fig. 2E.2 if necessary). Identify the molecular shape from the layout of atoms, as in Fig. 2E.1.

SOLVE

Draw the Lewis structure.

$$:\ddot{F} - \ddot{N} - \ddot{F}:$$
$$|$$
$$:\ddot{F}:$$

Count the bonds and lone pairs on the central atom.

The central N atom has one electron pair and three bonds, corresponding to four regions of high electron density.

Assign the electron arrangement.

Tetrahedral

Identify the shape considering only atoms.

The three atoms bonded to N form a trigonal pyramid.

Note that lone pairs are not normally shown explicitly when depicting molecular shape.

EVALUATE Spectroscopic measurements confirm the prediction of a trigonal pyramidal shape for NF_3. The value of the bond angle found experimentally is 102°.

Self-test 2E.2A Predict (a) the electron arrangement and (b) the shape of an IF_5 molecule.

[**Answer:** (a) Octahedral; (b) square pyramidal]

Self-test 2E.2B Predict (a) the electron arrangement and (b) the shape of an SO_2 molecule.

Related Exercises 2E.5–2E.16

So far lone pairs have been treated as equivalent to bonds in their effect on shape, but is that really the case? The electron arrangement of an SO_3^{2-} ion is tetrahedral, and so OSO angles of 109.5° might be expected. However, experimental observations have shown that, although the sulfite ion is indeed trigonal pyramidal, its bond angle is only 106° (**20**). Such experimental evidence implies that the VSEPR model as described so far is incomplete and needs to be refined.

To account for the smaller than expected bond angles in molecules with lone pairs, the VSEPR model treats lone pairs as having a more strongly repelling effect than do electrons in bonds. That is, the lone pairs push the atoms bonded to the central atom closer together. One possible rationalization for this effect is that the electron cloud of a lone pair can spread over a larger volume than a bonding pair can, because a bonding pair (or several bonding pairs in a multiple bond) is held in place by two atoms, not one (**FIG. 2E.3**). In summary, the following rule leads to reasonably reliable predictions for the VSEPR model:

> **Rule 4** The strengths of repulsions are in the order lone pair–lone pair > lone pair–atom > atom–atom.

Because of their strong mutual repulsion, the lowest energy is achieved when lone pairs are as far from each other as possible. The energy is also lowest if the atoms bonded to the central atom are far from lone pairs, even though that might bring the atoms closer together.

The improved model helps to account for the bond angle of the AX_3E sulfite ion. The atoms and the lone pair adopt a tetrahedral arrangement around the S atom. However, the lone pair exerts a strong repulsion on the O atoms, forcing them to move together slightly, reducing the OSO angle from the 109.5° of a regular tetrahedron to 106°. Note that, although the VSEPR model can predict the direction of the distortion, it cannot predict its extent. You can expect that, in any AX_3E species, the XAX angle will be less than 109.5°, but its actual value cannot be predicted: it must be measured experimentally or calculated by solving the Schrödinger equation numerically.

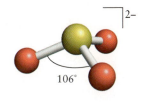

20 Sulfite ion, SO_3^{2-}

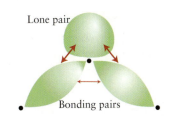

FIGURE 2E.3 A possible explanation of why lone pairs have a greater repelling effect than that of bonding electrons. A lone pair is less restrained than bonding pairs and so takes up more space; the bonding pairs (with their atoms) move away from the lone pair in an attempt to lower the repulsion that they experience, thus compressing the bond angle slightly.

Self-test 2E.3A (a) Give the VSEPR formula of an NH_3 molecule. Predict (b) its electron arrangement and (c) its shape.

[**Answer:** (a) AX_3E; (b) tetrahedral; (c) trigonal pyramidal (**21**, LP is lone pair), HNH angle less than 109.5°]

Self-test 2E.3B (a) Give the VSEPR formula of a ClO_2^- ion. Predict (b) its electron arrangement and (c) its shape.

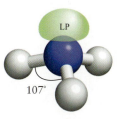

21 Ammonia, NH_3

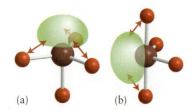

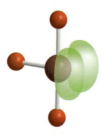

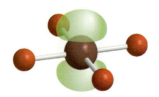

FIGURE 2E.4 (a) A lone pair in an axial position is close to three equatorial atoms. (b) In an equatorial position, a lone pair is close to only two atoms, a more favorable arrangement.

FIGURE 2E.5 Two lone pairs in an AX$_3$E$_2$ molecule adopt equatorial positions and move away from each other slightly. As a result, the molecule is approximately T-shaped.

FIGURE 2E.6 The square planar arrangement of atoms taken up in AX$_4$E$_2$ molecules: the two lone pairs are farthest apart when they are on opposite sides of the central atom.

Rule 4 allows you to predict the position in which a lone pair will be found. For example, the electron arrangement in an AX$_4$E molecule or ion, such as IF$_4^+$, is trigonal bipyramidal, but there are two different possible locations for the lone pair:

- An **axial lone pair** lies on the axis of the molecule, where it strongly repels the electron pairs in the three equatorial bonds, which lie at 90° angles to the axial position.
- An **equatorial lone pair** lies on the molecule's equator, on the plane perpendicular to the molecular axis, where it strongly repels only the electron pairs in the two axial bonds (**FIG. 2E.4**).

Therefore, the lowest energy is achieved when a lone pair is equatorial, producing a see-saw-shaped molecule. An AX$_3$E$_2$ molecule, such as ClF$_3$, also has a trigonal bipyramidal arrangement of electron pairs, but two of the pairs are lone pairs. These two pairs are farthest apart if they occupy two of the three equatorial positions, which are 120° from each other, but move away from each other slightly. The result is a T-shaped molecule (**FIG. 2E.5**). If the lone pairs had taken the axial positions, they would have been at 90° angles from the equatorial positions, resulting in greater repulsion. Now consider an AX$_4$E$_2$ molecule, which has an octahedral arrangement of electron pairs, two of which are lone pairs. The two lone pairs are farthest apart when they lie opposite each other, and so the molecule is square planar (**FIG. 2E.6**).

All molecules with the same VSEPR formula have the same general shape, although their bond angles generally differ slightly. For example, O$_3$ is an AX$_2$E species; it has a trigonal planar electron arrangement and an angular molecular shape (**22**). The bond angle in O$_3$ is 116.8°, which is slightly less than the predicted 120° for a trigonal planar molecule. The nitrite ion, NO$_2^-$, has the same general VSEPR formula (with a bond angle of 116°) and the same shape (**23**); so too does sulfur dioxide, SO$_2$ (**24**), with a bond angle of 119.5°. Exceptions sometimes arise when the energy difference between two possible structures is small and the central atom is so large that its lone pairs have little effect on the shape of the molecule. For example, the SeCl$_6^{2-}$ ion is octahedral, even though the Se atom contains a lone pair as well as bonds to six atoms.

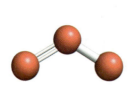

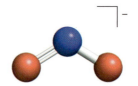

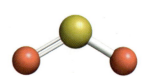

22 Ozone, O$_3$ **23** Nitrite ion, NO$_2^-$ **24** Sulfur dioxide, SO$_2$

Toolbox 2E.1   HOW TO USE THE VSEPR MODEL

CONCEPTUAL BASIS

Regions of high electron concentration—bonds and lone pairs attached to a central atom in a molecule—arrange themselves in such a way as to minimize mutual repulsions.

PROCEDURE

The general procedure for predicting the shape of a molecule follows the four VSEPR rules for predicting the shape of a molecule:

Step 1 Decide how many atoms and lone pairs are present on the central atom by writing a Lewis structure for the molecule.

Step 2 Identify the electron arrangement, including lone pairs and atoms, and treating a multiple bond as equivalent to a single bond (see Fig. 2E.2).

Step 3 Locate the atoms and identify the molecular shape (according to Fig. 2E.1). The molecular shape describes only the positions of the atoms, not the lone pairs.

Step 4 Allow the molecule to distort so that lone pairs are as far from one another and from bonding pairs as possible. The repulsions are in the order

Lone pair–lone pair > lone pair–atom > atom–atom

Example 2E.3 shows how this procedure is used.

EXAMPLE 2E.3 Predicting a molecular shape

You can't always guess a molecular shape from its chemical formula; even quite simple molecules can have surprising shapes. Predict the shape of a sulfur tetrafluoride molecule, SF_4.

ANTICIPATE You might guess that SF_4 has a tetrahedral geometry. However, you need to be aware that sulfur is capable of having an expanded valence shell (Topic 2C). You need to follow the steps in Toolbox 2E.1 to predict the shape of SF_4.

PLAN Use the procedure in Toolbox 2E.1.

SOLVE

Step 1 Decide how many atoms and lone pairs are present on the central atom by writing a Lewis structure for the molecule.

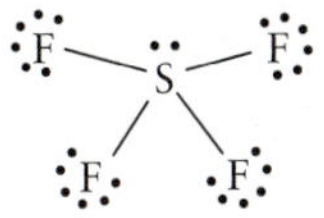

Each F atom should have three lone pairs of electrons, leaving four atoms and one lone pair on the central S atom.

Step 2 Identify the electron arrangement around the central atom, including lone pairs and atoms.

There are five regions of high electron density (four bonds and one lone pair) around the S atom, so the arrangement is trigonal bipyramidal.

Step 3 Locate the atoms and identify the molecular shape: AX_4E.

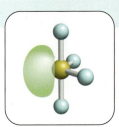

To minimize electron pair repulsions, the lone pair occupies an equatorial location. At this stage, SF_4 appears to have a seesaw shape.

Step 4 Allow the molecule to distort so that lone pairs are as far from one another and from bonding pairs as possible.

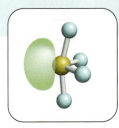

The atoms move slightly away from the lone pair.

A Note on Good Practice: All the lone pairs of electrons are shown here, but when predicting the shape of a molecule, it is necessary to draw only those on the central atom.

EVALUATE The slightly bent seesaw shape shown is the one found experimentally.

Self-test 2E.4A Predict the shape of an I_3^- ion.

[**Answer:** Linear]

Self-test 2E.4B Predict the shape of a xenon tetrafluoride molecule, XeF_4.

Related Exercises 2E.17–2E.24

In a molecule that has lone pairs or a single nonbonding electron on the central atom, the valence electrons contribute to the electron arrangement about the central atom, but only bonded atoms are considered in the identification of the shape. Lone pairs distort the shape of a molecule so as to reduce lone pair–bonding pair repulsions.

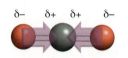

25   Carbon dioxide, CO_2

26   Carbon dioxide, CO_2

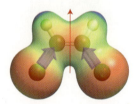

27   Water, H_2O

28   *cis*-Dichloroethene, $C_2H_2Cl_2$

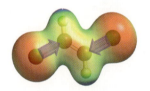

29   *trans*-Dichloroethene, $C_2H_2Cl_2$

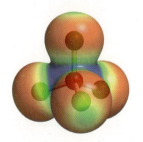

30   Tetrachloromethane, CCl_4

2E.3 Polar Molecules

A polar covalent *bond* in which electrons are not equally shared by the bonded atoms has a nonzero dipole moment (Topic 2B). A **polar molecule** is a *molecule* with a nonzero dipole moment. A diatomic molecule is polar if its bond is polar. An HCl molecule, with its polar covalent bond ($^{\delta+}H—Cl^{\delta-}$), is a polar molecule. Its dipole moment of 1.1 D is typical of polar diatomic molecules (**TABLE 2E.1**). All diatomic molecules composed of atoms of two different elements are at least slightly polar. A **nonpolar molecule** is a molecule that has no electric dipole moment. A **homonuclear diatomic molecule,** a diatomic molecule built from two atoms of the same element, such as O_2, N_2, and Cl_2, is nonpolar, because its only bond is nonpolar.

A polyatomic molecule may be nonpolar even if its bonds are polar. For example, the two $^{\delta+}C—O^{\delta-}$ dipole moments in carbon dioxide, a linear molecule, point in opposite directions, and so they cancel each other (**25**). Consequently, CO_2 is a nonpolar molecule. The electrostatic potential diagram (**26**) illustrates this conclusion. In contrast, the two $^{\delta+}H—O^{\delta-}$ dipole moments in H_2O lie at 104.5° to each other and do not cancel, and so H_2O is a polar molecule (**27**). This polarity is part of the reason why water is such a good solvent for ionic compounds.

THINKING POINT

Can you justify that last remark?

As you have seen by comparing CO_2 and H_2O, the shape of a polyatomic molecule affects whether or not it is polar. The same is true of more complicated molecules. For instance, the atoms and bonds are the same in *cis*-dichloroethene (**28**) and *trans*-dichloroethene (**29**); but in the latter, the C—Cl bonds point in opposite directions and the bond dipoles (which point along the C—Cl bonds) cancel. Thus, whereas *cis*-dichloroethene is polar, *trans*-dichloroethene is nonpolar.

If the four atoms attached to the central atom in a tetrahedral molecule are the same, as in tetrachloromethane (carbon tetrachloride), CCl_4 (**30**), the dipole moments cancel and the molecule is nonpolar. However, if one or more of the atoms are replaced by

TABLE 2E.1	Dipole Moments of Selected Molecules		
Molecule	Dipole moment/D	Molecule	Dipole moment/D
HF	1.91	PH_3	0.58
HCl	1.08	AsH_3	0.20
HBr	0.80	SbH_3	0.12
HI	0.42	O_3	0.53
CO	0.12	CO_2	0
ClF	0.88	BF_3	0
NaCl*	9.00	CH_4	0
CsCl*	10.42	*cis*-CHCl=CHCl	1.90
H_2O	1.85	*trans*-CHCl=CHCl	0
NH_3	1.47		

*For pairs of ions in the gas phase, not the bulk ionic solid.

different atoms, as in trichloromethane (chloroform), $CHCl_3$, or by lone pairs, as in NH_3, then the dipole moments associated with the bonds do not cancel. Thus, the $CHCl_3$ molecule is polar (**31**).

FIGURE 2E.7 summarizes the shapes of simple molecules that result in their being polar or nonpolar.

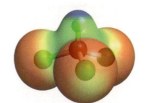

31 Trichloromethane, $CHCl_3$

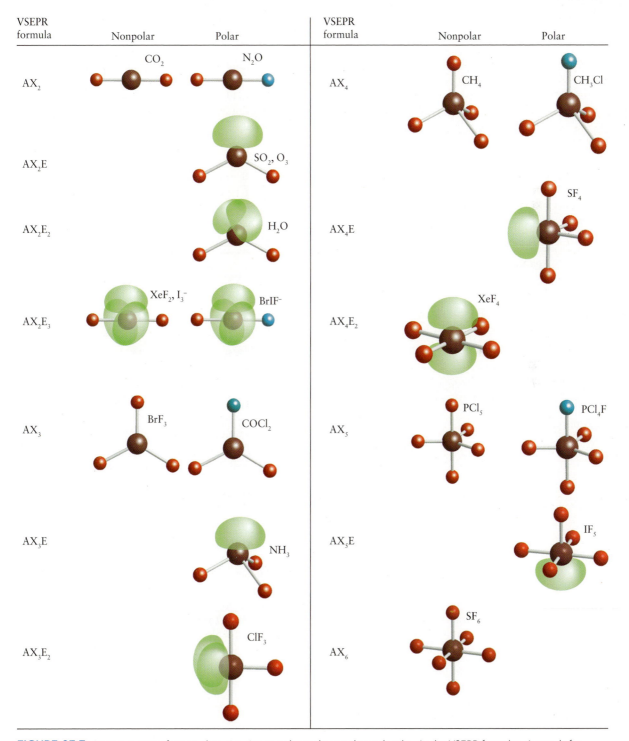

FIGURE 2E.7 Arrangements of atoms that give rise to polar and nonpolar molecules. In the VSEPR formulas, A stands for a central atom, X for an attached atom, and E for a lone pair. Identical atoms are the same color; attached atoms colored differently belong to different elements. The green lobes represent lone pairs of electrons.

EXAMPLE 2E.4 Predicting the polar character of a molecule

Because the polarity of a molecule affects its physical properties, you need to know whether a molecule is polar or nonpolar when predicting how it might interact with other molecules. Predict whether (a) a boron trifluoride molecule, BF_3, and (b) an ozone molecule, O_3, are polar.

ANTICIPATE The bonds in BF_3 are very polar, but the bonds in O_3 are not, so you might expect BF_3 to be polar and O_3 to be nonpolar. However, the shape of a molecule determines the polarity of the molecule, so the shapes must be identified before you can make a prediction.

PLAN In each case, predict the shape of the molecule by using the VSEPR model and then decide whether the dipole moments associated with the bonds cancel as a result of the symmetry of the molecule. If necessary, refer to Fig. 2E.7.

SOLVE

Draw the Lewis structure.

Assign the electron arrangement.

Identify the VSEPR formula.
Name the molecular shape.

Identify the polarity.

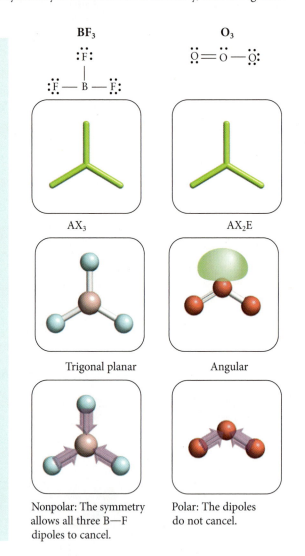

BF₃

O₃

AX_3

AX_2E

Trigonal planar

Angular

Nonpolar: The symmetry allows all three B—F dipoles to cancel.

Polar: The dipoles do not cancel.

EVALUATE This example shows that a homonuclear polyatomic molecule (O_3) can be polar. In this case, the electron density associated with the central O atom is different from that on the outer two O atoms: the central atom is bonded to two O atoms, whereas the outer atoms are bonded to only one O atom. This example also shows that, despite having very polar B—F bonds, BF_3 is nonpolar, because the individual dipoles cancel.

Self-test 2E.5A Predict whether (a) SF_4, (b) SF_6 is polar or nonpolar.

[*Answer:* (a) Polar; (b) nonpolar]

Self-test 2E.5B Predict whether (a) PCl_5, (b) IF_5 is polar or nonpolar.

Related Exercises 2E.25–2E.30.

A diatomic molecule is polar if its bond is polar. A polyatomic molecule is polar if it has polar bonds arranged in space in such a way that the dipole moments associated with the bonds do not cancel.

What have you learned in this Topic?

The shapes of simple molecules (and the shapes in localized regions of more complex ones) can be predicted by assessing the repulsions between regions of high electron density. You have seen that lone pairs of electrons are more strongly repelling than the electrons in bonds. You have learned that multiple bonds are treated like single bonds when predicting molecular shape and that the shape of a molecule is identified from the locations of its atoms (not its lone pairs). You have seen that lone pairs often distort the shape of a molecule in a predictable way. You have also seen how shape plays an important role in determining whether a polyatomic molecule is polar.

The skills you have mastered are the ability to:

☐ **1.** Explain the basis of the VSEPR model of bonding in terms of repulsions between electron pairs (Section 2E.1).

☐ **2.** Use the VSEPR model to predict the electron arrangement and shape of a molecule or polyatomic ion from its formula (Toolbox 2E.1 and Examples 2E.1, 2E.2, and 2E.3).

☐ **3.** Predict the polar character of a molecule (Example 2E.4).

Topic 2E Exercises

2E.1 Below are ball-and-stick models of two molecules. In each case, indicate whether there must be, may be, or cannot be one or more lone pairs of electrons on the central atom.

(a)   (b)

2E.2 Below are ball-and-stick models of two molecules. In each case, indicate whether there must be, may be, or cannot be one or more lone pairs of electrons on the central atom.

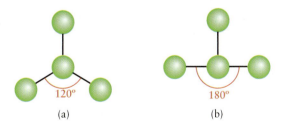

(a)   (b)

2E.3 Draw the structures and name the shapes of the following molecules: (a) HCN; (b) CH_2F_2.

2E.4 Draw the structures and name the shapes of the following molecules: (a) SeF_6; (b) IF_7.

2E.5 (a) What is the shape of a ClO_2^+ ion? (b) What is the expected OClO bond angle?

2E.6 (a) What is the shape of a ClO_2^- ion? (b) What value is expected for the OClO bond angle?

2E.7 (a) What is the shape of a thionyl chloride molecule, $SOCl_2$? Sulfur is the central atom. (b) How many different OSCl bond

angles are there in this molecule? (c) What values are expected for the OSCl and ClSCl bond angles?

2E.8 (a) What is the shape of a XeF_4 molecule? (b) What values are expected for the FXeF bond angles?

2E.9 (a) What is the shape of an ICl_3 molecule (iodine is the central atom)? (b) What value is expected for the ClICl bond angle?

2E.10 (a) What is the shape of a GeF_4 molecule? (b) What value is expected for the FGeF bond angle?

2E.11 Use Lewis structures and the VSEPR model to give the VSEPR formula for each of the following species and predict its shape: (a) sulfur tetrachloride; (b) iodine trichloride; (c) IF_4^-; (d) xenon trioxide.

2E.12 Use Lewis structures and the VSEPR model to give the VSEPR formula for each of the following species and predict its shape: (a) PF_4^-; (b) ICl_4^+; (c) phosphorus pentafluoride; (d) xenon tetrafluoride.

2E.13 Draw the Lewis structure, VSEPR formula, molecular shape, and bond angles for each of the following species: (a) I_3^-; (b) $POCl_3$; (c) IO_3^-; (d) N_2O.

2E.14 Draw the Lewis structure, VSEPR formula, molecular shape, and bond angles for each of the following species: (a) $SiCl_4$; (b) PF_5; (c) SBr_2; (d) ICl_2^+.

2E.15 Draw the Lewis structure and the VSEPR formula, list the shape, and predict the approximate bond angles of (a) CF_3Cl; (b) $TeCl_4$; (c) COF_2 (C is the central atom); (d) CH_3^-.

2E.16 Draw the Lewis structure and the VSEPR formula, list the shape, and predict the approximate bond angles of (a) PCl_3F_2; (b) SnF_4; (c) SnF_6^{2-}; (d) IF_5; (e) XeO_4.

2E.17 Predict the bond angles at the central atom of the following molecules and ions: (a) ozone, O_3; (b) azide ion, N_3^-; (c) cyanate ion, CNO^-; (d) hydronium ion, H_3O^+.

2E.18 Predict the bond angles at the central atom of the following molecules and ions: (a) OF_2; (b) ClO_2^-; (c) NO_2^-; (d) $SeCl_2$.

2E.19 Predict the shapes and estimate the bond angles of (a) the thiosulfate ion, $S_2O_3^{2-}$; (b) $(CH_3)_2Be$; (c) BH_2^-; (d) $SnCl_2$.

2E.20 For each of the following molecules or ions, draw the Lewis structure, list the number of lone pairs on the central atom, identify the shape, and estimate the bond angles: (a) PBr_5; (b) $XeOF_2$; (c) SF_5^+; (d) IF_3; (e) BrO_3^-.

2E.21 Draw the Lewis structure and give the approximate bond angles of (a) C_2H_4; (b) $ClCN$; (c) $OPCl_3$; (d) N_2H_4.

2E.22 Draw the Lewis structure and predict the shape of (a) ClF_5; (b) SbF_5; (c) IO_5^{3-}; (d) IO_6^{5-}.

2E.23 Draw the Lewis structure and predict the shape of (a) $OSbCl_3$; (b) SO_2Cl_2; (c) $IO_2F_2^-$. The atom in boldface red type is the central atom.

2E.24 Draw the Lewis structure and predict the shape of (a) AsO_4^{3-}; (b) OSF_4; (c) F_3IO_2. The atom in boldface red type is the central atom.

2E.25 Draw the Lewis structure and predict whether each of the following molecules is polar or nonpolar: (a) CH_2Cl_2; (b) CCl_4; (c) CS_2; (d) SF_4.

2E.26 Draw the Lewis structure and predict whether each of the following molecules is polar or nonpolar: (a) H_2Se; (b) AsF_5; (c) SiO_2; (d) NF_3.

2E.27 Predict whether each of the following molecules is likely to be polar or nonpolar: (a) C_5H_5N (pyridine, a molecule like benzene except that one —CH— group is replaced by a nitrogen atom); (b) C_2H_6 (ethane); (c) $CHCl_3$ (trichloromethane, also known as chloroform, a common organic solvent and once used as an anesthetic).

2E.28 Predict whether each of the following molecules is likely to be polar or nonpolar: (a) CH_3SH (methanethiol, found in breath odor and skunks); (b) CH_3NH_2 (methylamine, a drug precursor); (c) CH_3OCH_3 (dimethyl ether, used as an aerosol propellant).

2E.29 There are three isomers of dichlorobenzene, $C_6H_4Cl_2$, which differ in the relative positions of the chlorine atoms on the benzene ring. (a) Which of the three forms are polar? (b) Which has the largest dipole moment?

2E.30 There are three isomers of difluoroethene, $C_2H_2F_2$, which differ in the locations of the fluorine atoms. (a) Which of the forms are polar? (b) Which has the largest dipole moment?

Topic 2F Valence-Bond Theory

2F.1 Sigma and Pi Bonds

2F.2 Electron Promotion and the Hybridization of Orbitals

2F.3 Other Common Types of Hybridization

2F.4 Characteristics of Multiple Bonds

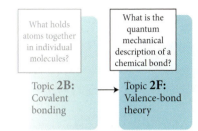

What holds atoms together in individual molecules?

What is the quantum mechanical description of a chemical bond?

Topic **2B:** Covalent bonding

Topic **2F:** Valence-bond theory

Valence-bond theory (VB theory) is the description of covalent bonding in terms of atomic orbitals. It was devised by Walter Heitler, Fritz London, John Slater, and Linus Pauling in the late 1920s. This theory is a quantum mechanical description of the distribution of electrons in bonds that goes beyond Lewis's theory and the VSEPR model by providing a way of calculating the numerical values of bond angles and bond lengths. The concepts and the language it introduces are used throughout chemistry.

The Lewis model of the covalent bond (Topic 2B) assumes that each bonding electron pair is located between the two bonded atoms—it is a *localized electron model*. However, the location of an electron in an atom cannot be described in terms of a precise position, but only in terms of the *probability* of finding it somewhere in a region of space defined by its orbital (Topic 1C). Valence-bond theory takes this wave nature of electrons into account.

2F.1 Sigma and Pi Bonds

First consider the formation of H_2, the simplest molecule of all. Each H atom in its ground state has one electron in a 1s-orbital (Topic 1D). Valence-bond theory adopts the view that, as two H atoms come together, their 1s-electrons pair (denoted ↑↓) and the two atomic orbitals merge together (**FIG. 2F.1**). The resulting sausage-shaped distribution of electrons, with an accumulation of electron density between the nuclei, is called a "σ-bond" (a sigma bond). More formally:

- A **σ-bond** is cylindrically symmetrical (the same in all directions around the long axis of the bond), with no nodal planes containing the internuclear axis.

The merging of the two atomic orbitals is called the **overlap** of orbitals. A general point to keep in mind throughout this Topic is that, the greater the extent of orbital overlap, the stronger is the bond.

Much the same kind of σ-bond formation ("σ-bonding") occurs in the hydrogen halides. For example, before an H and F atom combine to form hydrogen fluoride, the unpaired electron on the fluorine atom occupies a $2p_z$-orbital, and the unpaired electron on the hydrogen atom occupies a 1s-orbital (**1**). These two electrons pair to form a bond when the orbitals that they occupy overlap and merge into a cloud that spreads over both atoms (**FIG. 2F.2**). When viewed from the side, the resulting bond has a more complicated shape than that of the σ-bond in H_2. However, the σ-bond in HF, which shares many of the characteristics of the σ-bond in H_2, has cylindrical symmetry when viewed along the internuclear (z) axis and no nodal planes containing the internuclear axis; hence it too is a σ-bond. All *single* covalent bonds are σ-bonds.

A nitrogen molecule, N_2, has a different type of bond. There is a single electron in each of the three 2p-orbitals on each atom (**2**). However, only one of the three orbitals on

Why Do You Need to Know This Material? The valence-bond theory of the electronic structures of molecules gives insight into the quantum mechanical nature of the covalent bond, multiple bonds, and the shapes of molecules. It introduces language that is used throughout chemistry.

What Do You Need to Know Already? You need to be aware of the description of atomic structure in terms of the occupation of orbitals (Topics 1D and 1E), the notion of a wavefunction (Topic 1C), and the concept of electron spin (Topic 1D).

The Greek letter sigma, σ, is the equivalent of the letter s. It is a reminder that, looking along the internuclear axis, the electron distribution resembles that of an s-orbital.

By convention, the bond direction in a linear molecule defines the *z*-axis.

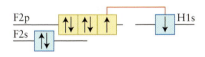

F2p   H1s

F2s

1 Hydrogen fluoride, HF

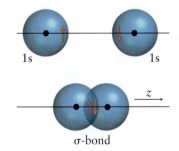

FIGURE 2F.1 When electrons with opposite spins (depicted as ↑ and ↓) in two hydrogen 1s-orbitals pair and the s-orbitals overlap, they form a σ-bond. The cloud has cylindrical symmetry around the internuclear axis and spreads over both nuclei. In the illustrations in this book, σ-bonds are usually colored blue.

ANIMATION FIGURE 2F.1

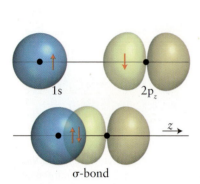

FIGURE 2F.2 A σ-bond can also be formed when electrons in 1s- and 2p$_z$-orbitals pair (where z is the direction along the internuclear axis). The two electrons in the bond are generally found spread over the entire region of space enclosed by the boundary surface.

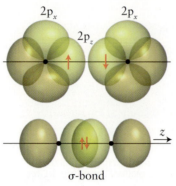

FIGURE 2F.3 A σ-bond is formed by the pairing of electron spins in two 2p$_z$-orbitals on neighboring atoms. At this stage, ignore the interactions of any 2p$_x$- (and 2p$_y$-) orbitals that also contain unpaired electrons, because they cannot form σ-bonds. Notice that the nodal plane of each p$_z$-orbital survives in the σ-bond.

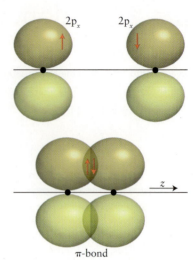

π-bond

FIGURE 2F.4 A π-bond is formed when electrons in two 2p-orbitals pair and overlap side by side. The bottom diagram shows the corresponding boundary surface. Even though the bond has a complicated shape, with two lobes, it is occupied by one pair of electrons and counts as one bond. In this text, π-bonds are usually colored a shade of yellow.

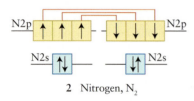

2   Nitrogen, N$_2$

The Greek letter pi, π, is the equivalent of the letter p. When you imagine looking along the internuclear axis, a π-bond resembles a pair of electrons in a p-orbital.

There are a few exceptions to the rule below about double bonds: in a very few instances, both bonds of a double bond are π-bonds.

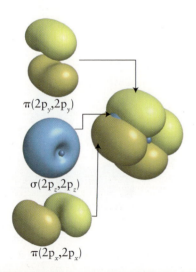

π(2p$_y$,2p$_y$)

σ(2p$_z$,2p$_z$)

π(2p$_x$,2p$_x$)

each atom, which is designated the 2p$_z$-orbital, can overlap end-to-end to form a σ-bond (**FIG. 2F.3**). Two of the 2p-orbitals on each atom (2p$_x$ and 2p$_y$) are perpendicular to the internuclear axis, and each one contains an unpaired electron (**FIG. 2F.4**, top). When the electrons in the p-orbitals on one N atom pair with the p-electrons on the other N atom, the orbitals can overlap only in a side-by-side arrangement. This overlap results in a "π-bond," a bond in which the two electrons lie in two lobes, one on each side of the internuclear axis (Fig. 2F.4, bottom). More formally:

- A **π-bond** has a single nodal plane containing the internuclear axis.

Although a π-bond has electron density on each side of the internuclear axis, it is only one bond, with the electron cloud in the form of two lobes, just as a p-orbital is one orbital with two lobes. In a molecule with two π-bonds, such as N$_2$, the electron densities in the two perpendicular π-bonds merge, and the two atoms appear to be surrounded by a cylinder of electron density (**FIG. 2F.5**).

THINKING POINT

Can you justify that last remark? How do you think a δ-bond (a delta bond) might be formed?

In general, as in these examples, valence-bond theory can be used to describe covalent bonds as follows:

A **single bond** is a σ-bond.
A **double bond** is a σ-bond plus one π-bond.
A **triple bond** is a σ-bond plus two π-bonds.

Self-test 2F.1A How many σ-bonds and how many π-bonds are there in (a) CO$_2$ and (b) CO?
[*Answer:* (a) Two σ, two π; (b) one σ, two π]

Self-test 2F.1B How many σ-bonds and how many π-bonds are there in (a) NH$_3$ and (b) HCN?

FIGURE 2F.5 The bonding pattern in a nitrogen molecule, N$_2$. The two atoms are held together by one σ-bond (blue) and two perpendicular π-bonds (yellow). Although they are shown separately here, in fact the two π-bonds merge to form a long doughnut-shaped cloud surrounding the σ-bond cloud; so the overall structure resembles a cylindrical hot dog.

According to valence-bond theory, bonds form when electrons in valence-shell atomic orbitals pair; the atomic orbitals overlap end to end to form σ-bonds or side by side to form π-bonds.

2F.2 Electron Promotion and the Hybridization of Orbitals

Difficulties are encountered when VB theory is applied to methane. A carbon atom has the configuration $[He]2s^22p_x^12p_y^1$ with four valence electrons (**3**). However, two valence electrons are already paired, and only the two half-filled 2p-orbitals appear to be available for bonding. It looks as though a carbon atom should have a valence of 2 and form two perpendicular bonds, but in fact it almost always has a valence of 4 (it is commonly "tetravalent") and in CH_4 has a tetrahedral arrangement of bonds.

A carbon atom has four unpaired electrons available for bonding when an electron is **promoted**—that is, relocated to a higher-energy orbital. When a 2s-electron is promoted into an empty 2p-orbital, the carbon atom has the configuration $[He]2s^12p_x^12p_y^12p_z^1$ (**4**) and can form four bonds.

The characteristic tetravalence of carbon is due to the small promotion energy of a carbon atom. The promotion energy is small because a 2s-electron is transferred from an orbital that it shares with another electron to an empty 2p-orbital. Although the promoted electron enters an orbital of higher energy, it experiences less repulsion from other electrons than before it was promoted. As a result, only a little energy is needed to promote the electron. Nitrogen, carbon's neighbor in the periodic table, cannot use promotion to increase the number of bonds it can form, because it has no empty p-orbitals (**5**). The same is true of oxygen and fluorine. Promotion of an electron is possible if the overall change, taking account of all contributions to the energy and especially the greater number of bonds that can thereby be formed, is toward lower energy. Boron, $[He]2s^22p^1$, like carbon, is an element in which promotion can lead to the formation of more bonds (three in boron's case), and boron does typically form three bonds.

At this stage, it looks as though electron promotion should result in two different types of bonds in methane, one bond from the overlap of a hydrogen 1s-orbital and a carbon 2s-orbital, and three more bonds from the overlap of hydrogen 1s-orbitals with each of the three carbon 2p-orbitals. The overlap with the 2p-orbitals should result in three σ-bonds at 90° to one another. However, this arrangement is inconsistent with the known tetrahedral structure of methane and its four equivalent bonds.

The model is improved by noting that s- and p-orbitals can be thought of as waves of electron density centered on the nucleus of an atom. Imagine that the four orbitals interfere with one another and produce new patterns where they intersect, like waves in water. Where the wavefunctions are all positive or all negative, the amplitudes are increased by this interference; where the wavefunctions have opposite signs, the overall amplitude is reduced and might even be canceled completely. As a result, the interference between the atomic orbitals results in new patterns called **hybrid orbitals.** Each of the four hybrid orbitals, designated h_i, is formed from a linear combination of the four atomic orbitals:

$$h_1 = s + p_x + p_y + p_z \qquad h_2 = s - p_x - p_y + p_z$$
$$h_3 = s - p_x + p_y - p_z \qquad h_4 = s + p_x - p_y - p_z$$

For instance, in h_1 the s- and p-orbitals all have their usual signs, and their amplitudes add together where they are all positive. In h_2, however, the signs of p_x and p_y are reversed, and so the resulting interference pattern is different. These four hybrid orbitals are called **sp^3 hybrids** because they are formed from one s-orbital and three p-orbitals. They differ only in their orientation, with one pointing toward each corner of a tetrahedron (**FIG. 2F.6**); in all other respects, they are identical.

In an orbital-energy diagram the hybridization is represented as the formation of four orbitals of equal energy. This energy lies between the energies of the s- and p-orbitals

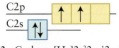

3 Carbon, $[He]2s^22p_x^12p_y^1$

Carbon monoxide, CO, is the only common exception to the tetravalence of carbon.

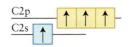

4 Carbon, $[He]2s^12p_x^12p_y^12p_z^1$

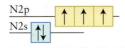

5 Nitrogen, $[He]2s^22p_x^12p_y^12p_z^1$

FIGURE 2F.6 This shading indicates the amplitude of the sp^3 hybrid orbital wavefunction in a plane that bisects it and passes through the nucleus. Each sp^3 hybrid orbital points toward the corner of a tetrahedron.

 **ANIMATION FIGURE 2F.6**

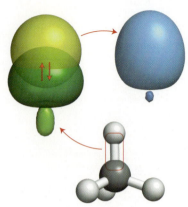

FIGURE 2F.7 Each C—H bond in methane is formed by the pairing of an electron in a hydrogen 1s-orbital and an electron in one of the four sp³ hybrid orbitals of carbon. Thus, valence-bond theory predicts four equivalent σ-bonds in a tetrahedral arrangement, which is consistent with experimental results.

 **ANIMATION FIGURE 2F.7**

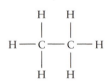

6 sp³ hybridized carbon

7 Ethane, CH₃CH₃

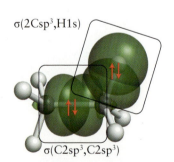

σ(2Csp³,H1s)

σ(C2sp³,C2sp³)

FIGURE 2F.8 The valence-bond description of bonding in an ethane molecule, C_2H_6. The boundary surfaces of only two of the bonds are shown. Each pair of neighboring atoms is linked by a σ-bond formed by the pairing of electrons in either H1s-orbitals or Csp³ hybrid orbitals. All the bond angles are close to 109.5° (the tetrahedral angle).

 **ANIMATION FIGURE 2F.8**

from which the hybrid orbitals are constructed (**6**). The hybrids are colored green as a reminder that they are a blend of (blue) s-orbitals and (yellow) p-orbitals. An sp³ hybrid orbital has two lobes, but one lobe extends farther than those of the contributing p-orbitals and the other lobe is shortened. The fact that hybrid orbitals have their amplitudes concentrated on one side of the nucleus allows them to overlap more effectively with other orbitals, and as a result the bonds that they form are stronger than in the absence of hybridization.

The bonding in methane can now be explained. In the promoted, hybridized atom, each of the electrons in the four sp³ hybrid orbitals can pair with an electron in a hydrogen 1s-orbital. Their overlapping orbitals form four σ-bonds that point toward the corners of a tetrahedron (**FIG. 2F.7**). The valence-bond description is now consistent with the molecule's known shape.

When there is more than one "central" atom in a molecule, the hybridization of each central atom is matched to the shape at that atom predicted by VSEPR. For example, in ethane, C_2H_6 (**7**), the two carbon atoms are both "central" atoms. According to the VSEPR model, the four electron pairs around each carbon atom take up a tetrahedral arrangement. This arrangement suggests the same sp³ hybridization of the carbon atoms as in methane (see Fig. 2F.7). Each C atom has one unpaired electron in each of its four sp³ hybrid orbitals and can therefore form four σ-bonds that point toward the corners of a regular tetrahedron. The C—C bond is formed by spin-pairing of the electrons in one sp³ hybrid orbital of each C atom. This bond is labeled σ(C2sp³,C2sp³) to show its composition: C2sp³ denotes an sp³ hybrid orbital composed of 2s- and 2p-orbitals on a carbon atom, and the parentheses show which orbitals on each atom overlap (**FIG. 2F.8**). Each C—H bond forms when an electron in one of the remaining sp³ hybrid orbitals pairs with an electron in a 1s-orbital of an H atom (denoted H1s). These bonds are denoted σ(C2sp³,H1s).

These ideas can be extended to molecules, such as ammonia (NH_3), that have a lone pair of electrons on the central atom. According to the VSEPR model, the four electron pairs in NH_3 take up a tetrahedral electron arrangement, so the nitrogen atom is described as forming four sp³ hybrid orbitals. Because nitrogen has five valence electrons, one of these hybrid orbitals is already doubly occupied (**8**). The three unpaired electrons in the remaining sp³ hybrid orbitals pair with the 1s-electrons of the three hydrogen atoms to form three N—H σ-bonds. *Whenever an atom of a nonmetallic element in a molecule has a tetrahedral electron arrangement, it is considered to be sp³ hybridized.*

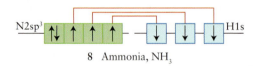

8 Ammonia, NH_3

The promotion of electrons will occur if, overall, it leads to a lowering of energy by permitting the formation of more bonds. Hybrid orbitals are constructed on an atom to reproduce the electron arrangement that is characteristic of the experimentally determined shape of a molecule.

2F.3 Other Common Types of Hybridization

Different hybridization schemes are used to describe other arrangements of electron pairs (**FIG. 2F.9**). For example, to explain a trigonal planar electron arrangement, like that in BF_3 and each carbon atom in ethene, one s-orbital and two p-orbitals are said to mix and so produce three sp² hybrid orbitals:

$$h_1 = s + 2^{1/2}p_y$$
$$h_2 = s + \left(\tfrac{3}{2}\right)^{1/2}p_x - \left(\tfrac{1}{2}\right)^{1/2}p_y$$
$$h_3 = s - \left(\tfrac{3}{2}\right)^{1/2}p_x - \left(\tfrac{1}{2}\right)^{1/2}p_y$$

FIGURE 2F.9 Three common hybridization schemes shown as boundary surfaces of the wavefunction and in terms of the orientations of the hybrid orbitals. (a) An s-orbital and a p-orbital hybridize into two sp hybrid orbitals that point in opposite directions, forming a linear molecular shape. (b) An s-orbital and two p-orbitals can blend together to give three sp² hybrid orbitals that point to the corners of an equilateral triangle. (c) An s-orbital and three p-orbitals can blend together to give four sp³ hybrid orbitals that point to the corners of a tetrahedron.

 ANIMATION FIGURE 2F.9

These three orbitals, which are identical apart from their orientation in space (and are not normalized) all lie in the same plane and point toward the corners of an equilateral triangle.

A linear arrangement of electron pairs requires two hybrid orbitals, and so an s-orbital is described as mixing with a p-orbital to obtain two sp hybrid orbitals:

$$h_1 = s + p \quad h_2 = s - p$$

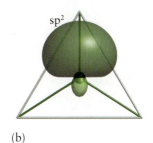

(b)

These two (unnormalized) sp hybrid orbitals point away from each other at $180°$ and result in a linear molecule. This is the arrangement in CO_2.

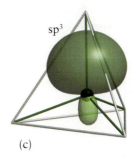

Self-test 2F.2A Suggest a structure in terms of hybrid orbitals for BF_3.

[**Answer:** Three σ-bonds formed from $F2p_z$-orbitals and $B2sp^2$ hybrids in a trigonal planar arrangement]

Self-test 2F.2B Suggest a structure in terms of hybrid orbitals for each carbon atom in ethyne, C_2H_2.

Some of the elements in Period 3 and later periods can accommodate five or more electron pairs, as in PCl_5. These types of bonds are described by a hybridization scheme that uses the d-orbitals of the central atom. To account for a trigonal bipyramidal arrangement of five electron pairs, one d-orbital is said to mix with all the valence s- and p-orbitals of the atom. The resulting five orbitals are called **sp³d hybrid orbitals** (**FIG. 2F.10**).

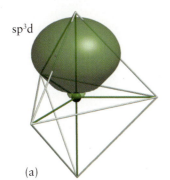

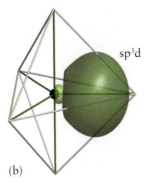

FIGURE 2F.10 Two of the five sp³d hybrid orbitals: (a) shows an axial orbital and (b) an equatorial orbital. The five sp³d orbitals account for the trigonal bipyramidal arrangement of electron pairs. The sp³d hybridization scheme can be used only when d-orbitals are available on the central atom.

 ANIMATION FIGURE 2F.10

Six orbitals are required to accommodate six electron pairs around an atom in an octahedral arrangement, as in SF_6 and XeF_4, and so two d-orbitals are said to mix with the valence s- and p-orbitals to form six **sp³d² hybrid orbitals** (**FIG. 2F.11**). These identical orbitals point toward the six corners of a regular octahedron.

TABLE 2F.1 summarizes the relation between electron arrangement and hybridization type. Note that the number of hybrid orbitals is always the same as the number of atomic orbitals used in their construction:

N atomic orbitals always produce *N* hybrid orbitals.

FIGURE 2F.11 One of the six sp³d² hybrid orbitals that may be formed when d-orbitals are available and it is necessary to account for an octahedral arrangement of electron pairs.

 ANIMATION FIGURE 2F.11

TABLE 2F.1 Hybridization and Molecular Shape*

Electron arrangement	Number of atomic orbitals	Hybridization of the central atom	Number of hybrid orbitals
linear	2	sp	2
trigonal planar	3	sp^2	3
tetrahedral	4	sp^3	4
trigonal bipyramidal	5	sp^3d	5
octahedral	6	sp^3d^2	6

*Other combinations of s-, p-, and d-orbitals can give rise to the same or different shapes, but the combinations in the table are the most common.

So far, terminal atoms, such as the Cl atoms in PCl_5, have not been regarded as hybridized. Spectroscopic data and calculations suggest that both s- and p-orbitals of terminal atoms take part in bond formation, and so it is reasonable to suppose that their orbitals are hybridized. The simplest model is to suppose that the three lone pairs and the bonding pair are arranged tetrahedrally and therefore that the chlorine atoms bond to the phosphorus atom by using sp^3 hybrid orbitals.

EXAMPLE 2F.1   Assigning a hybridization scheme

Identification of the hybridization of an atom in a molecule can often act as a good guide to the kind of chemical reactions it can undergo and help in the interpretation of its spectroscopic observations. What is the hybridization of phosphorus in PF_5?

ANTICIPATE To form five bonds, five hybrid orbitals are needed, which suggests that a d-orbital should be involved as well as the four s- and p-orbitals of a valence shell.

PLAN Choose the hybridization scheme to match the electron arrangement predicted by the VSEPR model, using N atomic orbitals to form N hybrid orbitals.

SOLVE

Draw the Lewis structure.

Determine the electron arrangement about the central atom.

Five bonds and no lone pairs; thus, trigonal bipyramidal

Identify the molecular shape.

Trigonal bipyramidal

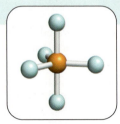

Select the same number of atomic orbitals as there are hybrid orbitals.

Construct the hybrid orbitals, starting with the s-orbital, and proceeding to the p- and d-orbitals.

$$sp^3d$$

EVALUATE As anticipated, a d-orbital is required to accommodate all the valence electrons.

Self-test 2F.3A Describe (a) the electron arrangement, (b) the molecular shape, and (c) the hybridization of the central chlorine atom in chlorine trifluoride.

[*Answer:* (a) Trigonal bipyramidal; (b) T-shaped; (c) sp^3d]

Self-test 2F.3B Describe (a) the electron arrangement, (b) the molecular shape, and (c) the hybridization of the central atom in BrF_4^-.

Related Exercises 2F.5–2F.10.

A hybridization scheme is adopted to match the electron arrangement of the molecule. Valence-shell expansion requires the use of d-orbitals.

2F.4 Characteristics of Multiple Bonds

Atoms of the Period 2 elements C, N, and O readily form double bonds with one another, with themselves, and (especially for oxygen) with atoms of elements in later periods. However, double bonds are rarely found between atoms of elements in Period 3 and later periods, because the atoms are so large and bond lengths consequently so great that it is difficult for their p-orbitals to take part in effective side-by-side overlap.

The pattern seen in ethene, $CH_2\!=\!CH_2$, is used to describe other carbon–carbon double bonds. It is known from experimental data that all six atoms in ethene lie in the same plane, with bond angles very close to 120°. Bond angles of 120° suggest a trigonal planar electron arrangement and sp^2 hybridization for each C atom (**9**). Each of the three hybrid orbitals on the C atom has one electron available for bonding; the fourth valence electron of each C atom occupies the unhybridized 2p-orbital, which is perpendicular to the plane formed by the hybrids. The two carbon atoms form a σ-bond by overlap of an sp^2 hybrid orbital on each atom. The H atoms form σ-bonds with the remaining lobes of the sp^2 hybrids. The electrons in the two unhybridized 2p-orbitals form a π-bond through side-by-side overlap. **FIGURE 2F.12** shows how the electron density in the π-bond lies above and below the axis of the C—C σ-bond.

In benzene, the C atoms and their attached H atoms all lie in the same plane, with the C atoms forming a hexagonal ring. To describe the bonding in the Kekulé structures of benzene (Topic 2B) in terms of VB theory, the hybrid orbitals on the C atoms must match the 120° bond angles of the hexagonal ring. Therefore, each carbon atom is taken to be sp^2 hybridized, as in ethene (**FIG. 2F.13**). The three hybrid orbitals each have one electron, with the fourth valence electron in an unhybridized 2p-orbital perpendicular to the plane of the hybrids. Two sp^2 hybrid orbitals on each carbon atom overlap with those of their neighbors, which results in six σ-bonds between them. The remaining sp^2 hybrid orbital on each carbon atom overlaps with a hydrogen 1s-orbital, resulting in six carbon–hydrogen bonds. Finally, the side-by-side overlap of the 2p-orbitals on each C atom results in a π-bond between each carbon atom and

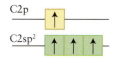

9 sp^2 hybridized carbon

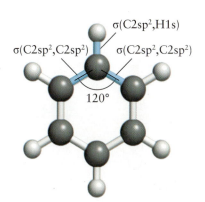

FIGURE 2F.13 The framework of σ-bonds in benzene. Each carbon atom is sp^2 hybridized, and the array of hybrid orbitals matches the bond angles (of 120°) in the hexagonal molecule. The bonds around only one carbon atom are labeled; all the others are the same.

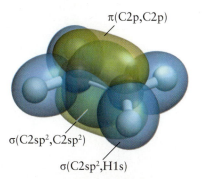

FIGURE 2F.12 A view of the bonding pattern in ethene (ethylene), showing the framework of σ-bonds and the single π-bond formed by side-to-side overlap of unhybridized C2p-orbitals. The double bond is torsionally rigid because twisting would reduce the overlap between the two C2p-orbitals and weaken the π-bond. Here the bonding structure is superimposed over a ball-and-stick model.

 **ANIMATION FIGURE 2F.12**

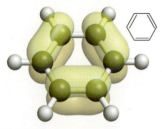

FIGURE 2F.14 Unhybridized carbon 2p-orbitals can form a π-bond with either of their immediate neighbors. Two arrangements are possible, each one corresponding to a different Kekulé structure. One Kekulé structure and the corresponding π-bonds are shown here.

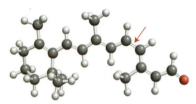

10   *cis*-Retinal

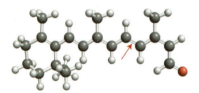

11   *trans*-Retinal

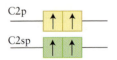

12   sp hybridized carbon

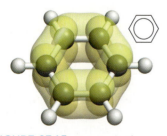

FIGURE 2F.15 As a result of resonance between two structures like the one shown in Fig. 2F.14 (corresponding to resonance of the two Kekulé structures), the π-electrons form a double doughnut-shaped cloud above and below the plane of the ring.

ANIMATION FIGURE 2F.15

one of its neighbors (**FIG. 2F.14**). The resulting pattern of π-bonds matches either of the two Kekulé structures, and the overall structure is a resonance hybrid of the two. This resonance ensures that the electrons in the π-bonds are spread around the entire ring (**FIG. 2F.15**).

The presence of a carbon–carbon double bond strongly influences the shape of a molecule because it prevents one part of a molecule from rotating relative to another part. The double bond of ethene, for example, holds the entire molecule flat. Figure 2F.12 shows that the two 2p-orbitals overlap best if all six atoms of the two CH_2 groups lie in the same plane. In order for the molecule to rotate about the double bond, the π-bond would need to break and re-form.

Double bonds and their influence on molecular shape are vitally important for living organisms. Vision depends on the shape of the molecule retinal in the retina of the eye. *cis*-Retinal is held rigid by its double bonds (**10**). When light enters the eye, it excites an electron out of the π-bond marked by the arrow. The double bond is now weaker, and the molecule is free to rotate about the remaining σ-bond. When the excited electron falls back, the molecule has rotated about the double bond and is now trapped in its trans shape (**11**). This change in shape triggers a signal along the optic nerve and is interpreted by the brain as the sensation of vision.

Now consider the alkynes, hydrocarbons with carbon–carbon triple bonds. The Lewis structure of the linear molecule ethyne (acetylene) is H—C≡C—H. A linear molecule has two equivalent orbitals at 180° from each other: this is sp hybridization. Each C atom has one electron in each of its two sp hybrid orbitals and one electron in each of its two perpendicular unhybridized 2p-orbitals (**12**). The electrons in one of the sp hybrid orbitals on each carbon atom pair and form a carbon–carbon σ-bond. The electrons in the two remaining sp hybrid orbitals pair with hydrogen 1s-electrons to form two carbon–hydrogen σ-bonds. The electrons in the two perpendicular sets of 2p-orbitals pair with a side-by-side overlap, forming two π-bonds at 90° to each other. As in the N_2 molecule, the electron density in the π-bonds forms a cylinder about the C—C bond axis. The resulting bonding pattern is shown in **FIG. 2F.16**.

Valence bond theory explains why a carbon–carbon double bond is stronger than one carbon–carbon single bond but weaker than the sum of two single bonds (Topic 2D) and why a carbon–carbon triple bond is weaker than the sum of three carbon–carbon single bonds. A single C—C bond is a σ-bond, but the additional bonds in a multiple bond are π-bonds. One reason for the difference in strength is that, because the p-orbitals overlap in a side-by-side fashion, the overlap in a π-bond is smaller and weaker than the end-to-end overlap that results in a σ-bond. The side-by-side overlap also explains why double bonds are rarely formed between elements in periods later than Period 3. The atoms are too large for the overlap to be strong enough to form a bond.

EXAMPLE 2F.2   Accounting for the structure of a molecule with multiple bonds

Formic acid is an important starting material for some polymers. Its structure makes it useful for the synthesis of strong polymer chains. Account for the structure of a formic acid molecule, HCOOH, in terms of hybrid orbitals, bond angles, and σ- and π-bonds. The C atom is attached to an H atom, a terminal O atom, and an —OH group.

ANTICIPATE Because the C atom is attached to three atoms, you should anticipate that its hybridization scheme is sp^2 and that one unhybridized p-orbital remains.

PLAN Use the VSEPR model to identify the shape of the molecule and then assign the hybridization consistent with that shape. All single bonds are σ-bonds, and multiple bonds are composed of a σ-bond and one or more π-bonds. Finally, form σ- and π-bonds by allowing the orbitals to overlap.

SOLVE

Draw the Lewis structure.

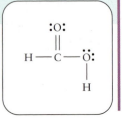

Use the VSEPR model to identify the electron arrangements around the central C and O atoms.

The C atom is bonded to three atoms and has no lone pairs; therefore, it has a trigonal planar arrangement. The O atom in the —OH group has two single bonds and two lone pairs; thus, it has a tetrahedral electron arrangement.

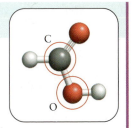

Identify the hybridization and bond angles.

C atom: trigonal planar, so 120° bond angle; sp^2 hybridized.
O atom of the —OH group: tetrahedral, so bond angles close to 109.5°; sp^3 hybridized.

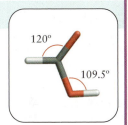

Form the bonds.

A π-bond forms by the overlap of the p-orbital on the C atom with the p-orbital on the terminal O atom.

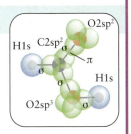

EVALUATE As anticipated, the hybridization scheme of the C atom is sp^2, leaving one unhybridized p-orbital.

Self-test 2F.4A Describe the structure of the carbon suboxide molecule, C_3O_2, in terms of hybrid orbitals, bond angles, and σ- and π-bonds. The atoms lie in the order OCCCO.

[**Answer:** Linear; bond angles all 180°; each C atom is sp hybridized and forms one σ-bond and one π-bond to each adjacent C or O atom.]

Self-test 2F.4B Describe the structure of the propene molecule, CH_3—CH=CH_2, in terms of hybrid orbitals, bond angles, and σ- and π-bonds.

Related Exercises 2F.17, 2F.18.

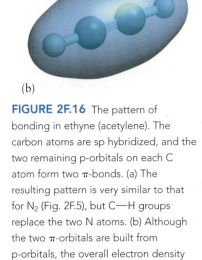

FIGURE 2F.16 The pattern of bonding in ethyne (acetylene). The carbon atoms are sp hybridized, and the two remaining p-orbitals on each C atom form two π-bonds. (a) The resulting pattern is very similar to that for N_2 (Fig. 2F.5), but C—H groups replace the two N atoms. (b) Although the two π-orbitals are built from p-orbitals, the overall electron density has cylindrical symmetry.

🔲 **ANIMATION FIGURE 2F.16**

Multiple bonds are formed when an atom forms a σ-bond by using an sp or sp^2 hybrid orbital and one or more π-bonds by using unhybridized p-orbitals. The side-by-side overlap that forms a π-bond makes a molecule resistant to twisting, results in bonds that are weaker than σ-bonds, and prevents atoms with large radii from forming multiple bonds.

What have you learned in this Topic?

You have learned that according to valence-bond theory a covalent bond forms when electrons in atomic orbitals pair their spins and the orbitals overlap. You have also learned that there are two types of bonds (σ and π) and that electron promotion occurs if the increased number of bonds that can form repays the energy investment. Finally, you have encountered the concept of hybridization, which enables the description of bond formation to be matched to the observed molecular shape.

The skills you have mastered are the ability to:

- ☐ **1.** Describe the difference between σ- and π-bonds and identify the types of bonds that make up double and triple bonds (Section 2F.1).
- ☐ **2.** Account for the occurrence of promotion.
- ☐ **3.** Describe the formation of hybrid orbitals from the mixing of atomic orbitals.
- ☐ **4.** Account for the structure of a molecule in terms of hybrid orbitals and σ- and π-bonds (Examples 2F.1 and 2F.2).

Topic 2F Exercises

Exercises labeled $\int_{dx}^{C}$ require calculus.

2F.1 State the relative orientations of each of the following hybrid orbitals: (a) sp^3; (b) sp; (c) sp^3d^2; (d) sp^2.

2F.2 The orientation of bonds on a central atom of a molecule that has no lone pairs can be one of the following. What is the hybridization of the orbitals used by each central atom for its bonding pairs: (a) tetrahedral; (b) trigonal bipyramidal; (c) octahedral; (d) linear?

2F.3 How many σ-bonds and how many π-bonds are there in (a) H_2S and (b) SO_2?

2F.4 How many σ-bonds and how many π-bonds are there in (a) NO and (b) N_2O?

2F.5 State the hybridization of the atom in boldface red type in each of the following molecules and ions: (a) **Be**Cl_2; (b) **B**H_3; (c) **B**H_4^-; (d) **Si**F_4.

2F.6 State the hybridization of the atom in boldface red type in each of the following molecules and ions: (a) **S**F_6; (b) **Cl**O_3^-; (c) **N**O_3^-; (d) O**C**Cl_2.

2F.7 Identify the hybrid orbitals used by the atom in boldface red type in each of the following species: (a) **B**F_3; (b) **As**F_3; (c) **Br**F_3; (d) **Se**F_3^+.

2F.8 Identify the hybrid orbitals used by the atom in boldface red type in each of the following molecules: (a) CH_3**C**CCH_3; (b) CH_3**N**NCH_3; (c) $(CH_3)_2$**C**$C(CH_3)_2$; (d) $(CH_3)_2$**N**$N(CH_3)_2$.

2F.9 Identify the hybrid orbitals used by the phosphorus atom in each of the following species: (a) PCl_4^+; (b) PCl_6^-; (c) PCl_5; (d) PCl_3.

2F.10 Identify the hybrid orbitals used by the atom in boldface red type in each of the following molecules: (a) H_2**C**CCH_2; (b) H_3**C**CH_3; (c) CH_3**N**NN; (d) CH_3**C**OOH.

2F.11 White phosphorus, P_4, is so reactive that it bursts into flame in air. The four atoms in P_4 form a tetrahedron in which each P atom is connected to three other P atoms. (a) Assign a hybridization scheme to the P_4 molecule. (b) Is the P_4 molecule polar or nonpolar?

2F.12 In the vapor phase, phosphorus can exist as P_2 molecules, which are highly reactive, whereas N_2 is relatively inert. Use valence-bond theory to explain this difference.

2F.13 Acrylonitrile, CH_2CHCN, is used in the synthesis of acrylic fibers (polyacrylonitriles), such as Orlon. Draw the Lewis structure of acrylonitrile and describe the hybrid orbitals on each carbon atom. What are the approximate values of the bond angles?

2F.14 Xenon forms XeO_3, XeO_4, and XeO_6^{4-}, all of which are powerful oxidizing agents. Draw their Lewis structures and state their bond angles and the hybridization of the xenon atom. Which would be expected to have the longest Xe—O distances? Explain your answer.

2F.15 Noting that the bond angle of an sp^3 hybridized atom is 109.5° and that of an sp^2 hybridized atom is 120°, do you expect the bond angle between two hybrid orbitals to increase or decrease as the s-character of the hybrids is increased?

2F.16 Both NH_2^- and NH_2^+ are angular species, but the bond angle in NH_2^- is less than that in NH_2^+. (a) What is the reason for this difference in bond angles? (b) Take the x-axis as lying perpendicular to the plane of the molecule. Does the $N2p_x$ orbital participate in the hybridization for either species? Briefly explain your answer.

2F.17 Describe the structure of the formaldehyde molecule, CH_2O, in terms of hybrid orbitals, bond angles, and σ- and π-bonds. The C atom is the central atom to which the other three atoms are attached.

2F.18 Describe the structure of the formamide molecule, $HCONH_2$, in terms of hybrid orbitals, bond angles, and σ- and π-bonds. The C atom is bonded to one H atom, a terminal O atom, and the N atom. The N atom is also bonded to two H atoms.

$\int_{dx}^{C}$ **2F.19** Given that the atomic orbitals used to form hybrids are normalized to 1 and mutually orthogonal, (a) show that the two tetrahedral hybrids $h_1 = s + p_x + p_y + p_z$ and $h_3 = s - p_x + p_y - p_z$ are orthogonal. (b) Construct the remaining two tetrahedral hybrids that are orthogonal to these two hybrids. *Hint:* Two wavefunctions are orthogonal if $\int \psi_1 \psi_2 d\tau = 0$, where $\int \ldots d\tau$ means "integrate over all space."

$\int_{dx}^{C}$ **2F.20** The hybrid orbital $h_1 = s + p_x + p_y + p_z$ referred to in Exercise 2F.19 is not normalized. Find the normalization factor N, given that all the atomic orbitals are normalized to 1. A wavefunction ψ can be normalized by writing it as $N\psi$ and finding the factor N which ensures that the integral over all space of $(N\psi)^2$ is equal to 1.

2F.21 The composition of hybrids can be discussed quantitatively. The outcome is that, if two equivalent hybrids composed of an s-orbital and two p-orbitals make an angle θ to each other, then the hybrids can be regarded as sp^λ, with $\lambda = -\cos\theta/\cos^2(\frac{1}{2}\theta)$. What is the hybridization of the O orbitals that form the two O—H bonds in H_2O?

2F.22 Given the information in Exercise 2F.21, plot a graph showing how the hybridization depends on the angle between two hybrids formed from an s-orbital and two p-orbitals, and confirm that it ranges from 90° when no s-orbital is included in the mixture to 120° when the hybridization is sp^2.

Topic 2G Molecular Orbital Theory

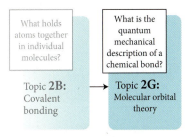

Molecular orbital theory, which was introduced in the late 1920s by Robert Mulliken, Friedrich Hund, John Slater, and John Lennard-Jones, has proved to be the most successful theory of the chemical bond: it overcomes all the deficiencies of Lewis's theory (Topic 2B) and is easier to use in computer calculations than valence-bond theory (Topic 2F).

2G.1 Molecular Orbitals

In **molecular orbital theory** (MO theory), electrons are described by wavefunctions called **molecular orbitals** that spread throughout the entire molecule. In other words, whereas in the Lewis and valence-bond models of molecular structure the electrons are localized on atoms or between pairs of atoms, in molecular orbital theory all valence electrons are delocalized over the whole molecule, not confined to individual bonds or atoms.

Molecular orbitals are built from atomic orbitals belonging to the valence shells of the atoms in the molecule. For example, a molecular orbital for H_2 is

$$\psi = \psi_{A1s} + \psi_{B1s} \tag{1}$$

where ψ_{A1s} is a 1s-orbital centered on one atom (A) and ψ_{B1s} is a 1s-orbital centered on the other atom (B). The technical terms for adding together wavefunctions (sometimes with different weighting coefficients) is called "forming a linear combination," and the molecular orbital in Eq. 1 is called a **linear combination of atomic orbitals** (LCAO). A molecular orbital formed from a linear combination of atomic orbitals on different atoms is called an **LCAO-MO.** Note that at this stage there are no electrons in the molecular orbital: a molecular orbital is just a combination—in this case, a simple sum—of wavefunctions. Like atomic orbitals, the molecular orbital in Eq. 1 is a well-defined mathematical function that can be evaluated at each point in space and pictured in three dimensions.

The LCAO-MO in Eq. 1 has a lower energy than either of the atomic orbitals used in its construction. Before the molecular orbital is formed, the two atomic orbitals are like waves centered on different nuclei. When the molecular orbital forms, the waves interfere constructively with each other between the nuclei in the sense that the total amplitude of the wavefunction is increased where the atomic orbitals overlap (**FIG. 2G.1**). The increased amplitude in the internuclear region means that there is an enhanced probability density there. Any electron that occupies that molecular orbital, therefore, is attracted to both nuclei and so has a lower energy than when it is confined to an atomic orbital on one atom. Moreover, because the electron now occupies a greater volume than when it is confined to a single atom, it also has a lower kinetic energy, just like a particle confined to a bigger box (Topic 1C). A combination of atomic orbitals that results in an overall lowering of energy, like that in Eq. 1, is called a **bonding orbital.**

FIGURE 2G.1 When two 1s-orbitals overlap in the same region of space in such a way that their wavefunctions have the same signs in that region, their wavefunctions (red lines) interfere constructively and give rise to a region of enhanced amplitude between the two nuclei (blue line).

Why Do You Need to Know This Material? The most common quantum mechanical approach to the description of the electronic structure of molecules is molecular orbital theory. It is used throughout chemistry and is essential for understanding the properties of modern materials.

What Do You Need to Know Already? This Topic makes use of atomic orbitals (Topic 1D). You also need to understand the Born interpretation of wavefunctions (Topic 1C) and how the building-up principle is used to describe the electronic structures of many-electron atoms (Topic 1E).

Whereas hybrid orbitals are linear combinations of atomic orbitals on the same atom, molecular orbitals are formed from linear combinations of atomic orbitals that may be on different atoms.

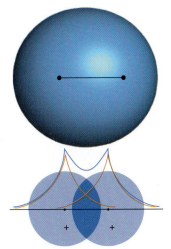

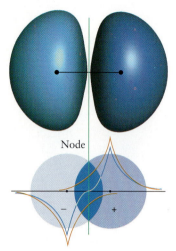

FIGURE 2G.2 When two 1s-orbitals overlap in the same region of space in such a way that their wavefunctions have opposite signs, the wavefunctions interfere destructively and give rise to a region of diminished amplitude and a node between the two nuclei (vertical blue line).

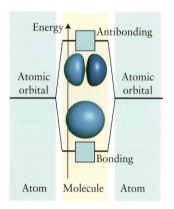

FIGURE 2G.3 A molecular orbital energy-level diagram for the bonding and antibonding molecular orbitals that can be built from two s-orbitals. Different signs of the s-orbitals (reflecting how they are combined to form the molecular orbital) are depicted by the different shades of blue.

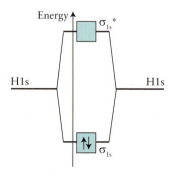

FIGURE 2G.4 The two electrons in an H_2 molecule occupy the lower-energy (bonding) molecular orbital and result in a stable molecule.

An important feature of MO theory is that

N molecular orbitals can be constructed from N atomic orbitals.

In molecular hydrogen, where LCAO-MOs are built from two atomic orbitals, *two molecular orbitals are formed.* The second molecular orbital has the form

$$\psi = \psi_{A1s} - \psi_{B1s} \tag{2}$$

The negative sign indicates that the amplitude of ψ_{B1s} *subtracts* from the amplitude of ψ_{A1s} where they overlap (**FIG. 2G.2**), and there is a nodal surface where the atomic orbitals cancel completely. In a hydrogen molecule, this nodal surface is a plane that lies halfway between the two nuclei. If an electron occupies this orbital, it is largely excluded from the internuclear region and consequently has a higher energy than when it occupies one of the atomic orbitals alone. A combination of atomic orbitals that results in a higher energy than that of the original atomic orbitals, like that in Eq. 2, is called an **antibonding orbital.**

A Note on Good Practice: What the negative sign in Eq. 2 really represents is that the sign of ψ_{B1s} is reversed everywhere (so a peak becomes a trough and vice versa), then the resulting wavefunction is added to (superimposed on) ψ_{A1s}.

The relative energies of the original atomic orbitals and the bonding and antibonding molecular orbitals are shown in a **molecular orbital energy-level diagram** like that in **FIG. 2G.3**. The increased energy of an antibonding orbital is about equal to or a little greater than the lowering of the energy of the corresponding bonding orbital.

Molecular orbitals are built from linear combinations of atomic orbitals: when atomic orbitals interfere constructively, they give rise to bonding orbitals; when they interfere destructively, they give rise to antibonding orbitals. N atomic orbitals combine to give N molecular orbitals.

2G.2 Electron Configurations of Diatomic Molecules

In the molecular orbital description of homonuclear diatomic molecules, all possible molecular orbitals are built from the available valence-shell atomic orbitals. Then the valence electrons are accommodated in these molecular orbitals by using the same procedure used in the building-up principle for atoms (Topic 1E). That is,

1. Electrons are accommodated in the lowest-energy molecular orbital, then in orbitals of increasingly higher energy.
2. According to the Pauli exclusion principle, each molecular orbital can accommodate up to two electrons. If two electrons are present in one orbital, they must be paired ($\uparrow\downarrow$).
3. If more than one molecular orbital of the same energy is available, the electrons enter them singly and adopt parallel spins (Hund's rule).

In H_2, two 1s-orbitals (one on each atom) merge to form two molecular orbitals. The bonding orbital is denoted σ_{1s} and the antibonding orbital σ_{1s}^*. The 1s in the notation shows the atomic orbitals from which the molecular orbitals are formed. The σ indicates that a "σ-orbital," a sausage-shaped orbital, has been built. More formally:

- A **σ-orbital** is a molecular orbital that has cylindrical symmetry and no nodal plane that contains the internuclear axis.

Two electrons, one from each H atom, are available. Both occupy the bonding orbital (the lower-energy orbital) and result in the configuration σ_{1s}^2 (**FIG. 2G.4**). Because only the bonding orbital is occupied, the energy of the molecule is lower than that of the separate atoms, and hydrogen exists as H_2 molecules. Two electrons in a σ-orbital form a σ-bond, like the σ-bond in VB theory. However, even a single electron may be able to hold two atoms together with about half the strength of an electron pair, and so—in contrast to Lewis's theory and VB theory—an electron pair is not required for a bond. A pair is simply the maximum number of electrons allowed by the Pauli exclusion principle to occupy any one molecular orbital.

Now these ideas are extended to other homonuclear diatomic molecules of Period 2 elements. The first step is to build up the molecular orbital energy-level diagram from the valence-shell atomic orbitals provided by the atoms. Because Period 2 atoms have 2s- and 2p-orbitals in their valence shells, molecular orbitals are formed from the overlap of these atomic orbitals. There are a total of eight atomic orbitals (one 2s- and three 2p-orbitals on each atom), so eight molecular orbitals can be formed. The two 2s-orbitals overlap to form two s-orbitals, one bonding (the σ_{2s}-orbital) and the other antibonding (the σ_{2s}^{*}-orbital); these orbitals resemble the σ_{1s}- and σ_{1s}^{*}-orbitals in H_2. The six 2p-orbitals (three on each neighboring atom) form the remaining six molecular orbitals. They can overlap in two distinct ways. The two 2p-orbitals that are directed toward each other along the internuclear axis form a bonding σ-orbital (σ_{2p}) and an antibonding σ*-orbital (σ_{2p}^{*}) where they overlap (**FIG. 2G.5**). The two 2p-orbitals on each atom that are perpendicular to the internuclear axis overlap side by side to form bonding and antibonding "π-orbitals" (**FIG. 2G.6**). More formally:

- A **π-orbital** is a molecular orbital with one nodal plane that contains the internuclear axis.

There are two 2p-orbitals on each atom perpendicular to the internuclear axis, and so four molecular orbitals—two bonding π_{2p}-orbitals and two antibonding π_{2p}^{*}-orbitals—are formed by their overlap.

Detailed calculation shows that there are some small differences in the order of energy levels from molecule to molecule (**BOX 2G.1**). **FIGURE 2G.7** shows the order

In precise computational work, molecular orbitals are built from core orbitals as well as valence orbitals.

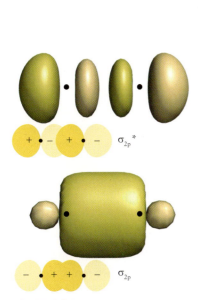

FIGURE 2G.5 Two p-orbitals can overlap to give bonding (lower) and antibonding (upper) σ-orbitals. Note that the antibonding combination has a node between the two nuclei. Both σ-orbitals have nodes passing through the nuclei, but no nodes along the axis of the bond.

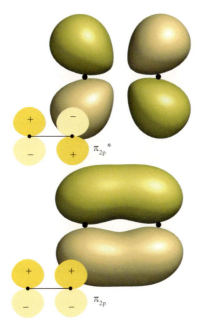

FIGURE 2G.6 Two p-orbitals can overlap side by side to give bonding (below) and antibonding (above) π-orbitals. Note that the latter has a nodal plane between the two nuclei. Both orbitals have a nodal plane through the two nuclei and look like p-orbitals when viewed along the internuclear axis.

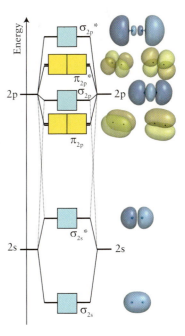

FIGURE 2G.7 A typical molecular orbital energy-level diagram for the homonuclear diatomic molecules Li_2 through N_2. Each box represents one molecular orbital and can accommodate up to two electrons.

Box 2G.1   HOW DO WE KNOW . . . THE ENERGIES OF MOLECULAR ORBITALS?

The energies of orbitals are calculated by finding solutions to the Schrödinger equation with computer software. The commercial software available is now so sophisticated that this approach can be as easy as typing in the name of the molecule or drawing it on a screen. However, these values are theoretical. How are the orbital energies determined experimentally?

One of the most direct methods is photoelectron spectroscopy (PES), an adaptation of the photoelectric effect (Topic 1B). A photoelectron spectrometer contains a source of high-frequency radiation (see the first illustration). Ultraviolet radiation is used most often for molecules, but x-rays are used to explore orbitals buried deeply inside solids. Photons in both frequency ranges have so much energy that they can eject electrons from the molecular orbitals they occupy.

Suppose that the frequency of the radiation is ν (nu), so each photon has an energy $h\nu$. An electron occupying a molecular orbital lies at an energy, $E_{orbital}$, below the zero of energy (corresponding to an electron far removed from the molecule). A photon that collides with the electron can eject it from the molecule if the photon can supply at least that much energy. The remaining energy of the photon, $h\nu - E_{orbital}$, then appears as the kinetic energy, E_k, of the ejected electron:

$$h\nu - E_{orbital} = E_k$$

The frequency, ν, of the radiation being used to bombard the molecules is known; so, if the kinetic energy of the ejected electron, E_k, is measured, this expression can be used to find the orbital energy, $E_{orbital}$.

The kinetic energy of an ejected electron depends on its speed, v, because $E_k = \frac{1}{2}m_e v^2$ (*Fundamentals* A). A photoelectron spectrometer acts like a mass spectrometer in that it measures v for electrons just as a mass spectrometer can measure v for ions (*Fundamentals* B). In this method, the electrons pass through a region of electric or magnetic field, which deflects their path. As the strength of the field is changed, the paths of the electrons change, too, until they fall on a detector and generate a signal. From the strength of the field required to obtain a signal, the speed of electrons ejected from a given orbital can be determined. From the speed, the kinetic energy of the electrons can be calculated, allowing the energy of the orbital from which they came to be obtained.

The photoelectron spectrum in the second illustration is that of nitrogen. Each peak corresponds to electrons being ejected from an orbital of different energy. A detailed analysis shows that the spectrum is a good portrayal of the qualitative orbital structure of nitrogen (as depicted in **1**).

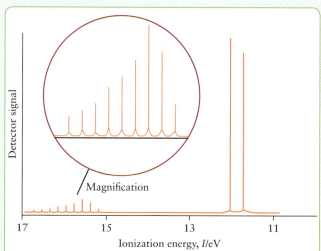

The photoelectron spectrum of nitrogen (N_2) has several peaks, a pattern indicating that electrons can be found in several energy levels in the molecule. Each main group of lines corresponds to the energy of a molecular orbital. The additional "fine structure" on some of the groups of lines is due to the excitation of molecular vibration when an electron is expelled.

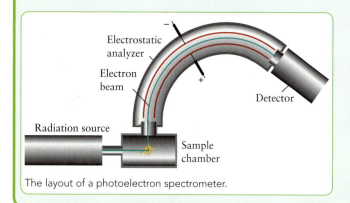

The layout of a photoelectron spectrometer.

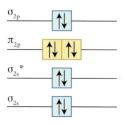

1 Nitrogen, N_2

for the Period 2 elements, with the exception of O_2 and F_2, which lie in the order shown in **FIG. 2G.8**. The order of energy levels is easy to explain for these two molecules. First, because each O and F atom has many electrons that contribute to shielding, the 2s-orbitals lie well below the 2p-orbitals and σ-orbitals can be built from the two sets of atomic orbitals separately. However, because the atoms of elements earlier in the period have fewer electrons, their 2s- and 2p-orbitals have more similar energies than in O and F. As a result, it is no longer possible to think of a σ-orbital as being formed from either the 2s-orbitals or the $2p_z$-orbitals separately, and all four of these orbitals must be used to build the four σ-orbitals. It is then hard to predict without detailed calculation where these four orbitals will lie, and it turns out that they in fact lie where they are shown in Fig. 2G.7.

Once it is known what molecular orbitals are available, the ground-state electron configurations of the molecules can be deduced by using the building-up principle. For example, consider N_2. Because nitrogen belongs to Group 15, each atom supplies five valence electrons. A total of ten electrons must therefore be assigned to the eight molecular orbitals shown in Fig. 2G.7. Two fill the σ_{2s}-orbital. The next two fill the σ_{2s}^*-orbital. Next in line for occupation are the two π_{2p}-orbitals, which can hold a total of four electrons. The last two electrons then enter the σ_{2p}-orbital. The ground configuration is therefore

$$N_2: \sigma_{2s}^2 \sigma_{2s}^{*2} \pi_{2p}^4 \sigma_{2p}^2$$

This configuration is shown as (**1**), where the boxes represent molecular orbitals.

At first sight, the molecular orbital description of N_2 looks quite different from the Lewis description ($:N\equiv N:$). However, it is, in fact, very closely related, as becomes clear when *bond order* is considered. In molecular orbital theory, **bond order,** b, is defined as the net number of bonds, allowing for the cancellation of bonds by antibonds:

$$\text{Bond order} = \tfrac{1}{2} \times (\text{number of electrons in bonding orbitals} -$$
$$\text{number of electrons in antibonding orbitals})$$

$$b = \tfrac{1}{2} \times (N_e - N_e^*) \qquad (3)$$

Here N_e is the number of electrons in bonding molecular orbitals and N_e^* is the number of electrons in antibonding molecular orbitals. In N_2, there are eight electrons in bonding orbitals and two in antibonding orbitals, and so the bond order is $\tfrac{1}{2}(8-2) = 3$. Because its bond order is 3, N_2 effectively has three bonds between the N atoms, just as the Lewis structure suggests.

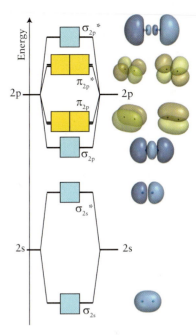

FIGURE 2G.8 The molecular orbital energy-level diagram for the homonuclear diatomic molecules on the right-hand side of Period 2, specifically O_2 and F_2.

Toolbox 2G.1 HOW TO DETERMINE THE ELECTRON CONFIGURATION AND BOND ORDER OF A HOMONUCLEAR DIATOMIC SPECIES

CONCEPTUAL BASIS

When N valence atomic orbitals overlap, they form N molecular orbitals. The ground-state electron configuration of a molecule is deduced by using the building-up principle to accommodate all the valence electrons in the available molecular orbitals. The bond order is the net number of bonds in the molecule.

PROCEDURE

Step 1 Identify *all* the atomic orbitals in the valence shells, ignoring how many electrons they contain.

Step 2 Use matching valence-shell atomic orbitals to build bonding and antibonding molecular orbitals and draw the

resulting molecular orbital energy-level diagram (see Figs. 2G.7 and 2G.8).

Step 3 Note the total number of electrons present in the valence shells of the two atoms. If the species is an ion, adjust the number of electrons to account for the charge.

Step 4 Accommodate the electrons in the molecular orbitals according to the building-up principle.

Step 5 To determine the bond order, subtract the number of electrons in antibonding orbitals from the number in bonding orbitals and divide the result by 2 (Eq. 3).

This procedure is illustrated in Example 2G.1.

EXAMPLE 2G.1 Deducing the ground-state electron configuration and bond order of a homonuclear diatomic molecule or ion

One way to understand the reactivities of compounds is to identify the strengths of their bonds, because weak bonds can often be broken easily. A guide to bond strength is bond order. Deduce the ground-state electron configuration of the fluorine molecule and calculate its bond order.

ANTICIPATE Since the Lewis structure of F_2 is $:\ddot{F}-\ddot{F}:$, you should anticipate that the bond order is 1.

PLAN Set up the molecular orbital energy-level diagram and use the building-up principle to accommodate the valence electrons as described in Toolbox 2G.1. Then calculate the bond order from the resulting configuration.

SOLVE

Step 1 Identify *all* the atomic orbitals in the valence shells, ignoring how many electrons they contain.

Each atom contributes a 2s-orbital and three 2p-orbitals, for a total of eight orbitals.

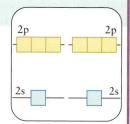

Step 2 Use matching valence-shell atomic orbitals to build bonding and antibonding molecular orbitals and draw the resulting molecular orbital energy-level diagram.

See Fig. 2G.8.

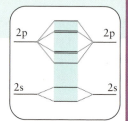

Step 3 Note the total number of electrons present in the valence shells of the two atoms.

$$2 \times 7 = 14$$

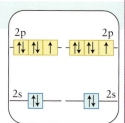

Step 4 Accommodate the electrons in the molecular orbitals according to the building-up principle:

$$\sigma_{2s}^{\,2} \sigma_{2s}^{*2} \sigma_{2p}^{\,2} \pi_{2p}^{\,4} \pi_{2p}^{*4}$$

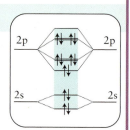

Step 5 To determine the bond order, subtract the number of electrons in antibonding orbitals from the number in bonding orbitals and divide the result by 2 (Eq. 3):

$$b = \tfrac{1}{2} \times [(2 + 2 + 4) - (2 + 4)] = 1$$

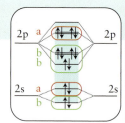

EVALUATE As expected, F_2 is a singly bonded molecule, in agreement with the Lewis structure. Notice that the first ten electrons repeat the N_2 configuration (apart from the change in order of the σ_{2p}- and π_{2p}-orbitals).

Self-test 2G.1A Deduce the electron configuration and bond order of the ion C_2^{2-}.

[***Answer:*** $\sigma_{2s}^{\,2} \sigma_{2s}^{*2} \pi_{2p}^{\,4} \sigma_{2p}^{\,2}$, $b = 3$]

Self-test 2G.1B Suggest a configuration for the O_2^{+} ion and state its bond order.

Related Exercises 2G.1–2G.4, 2G.11, 2G.12.

To find the ground-state electron configuration of O_2 add its 12 valence electrons (six from each atom) to the molecular orbitals shown in Fig. 2G.8. The first 10 electrons repeat the F_2 configuration, as in Example 2G.1. According to the building-up principle, the last two electrons occupy the two separate π_{2p}^*-orbitals and do so with parallel spins. The configuration is therefore

$$O_2:\ \sigma_{2s}^{\,2} \sigma_{2s}^{*2} \sigma_{2p}^{\,2} \pi_{2p}^{\,4} \pi_{2p}^{*1} \pi_{2p}^{*1}$$

as shown in (**2**). This conclusion is a minor triumph for molecular orbital theory, for there is an important experimental test of this configuration. You need to know that substances can be classified according to their behavior in a magnetic field: a **diamagnetic substance** is one that tends to move out of a magnetic field, and a **paramagnetic substance** is one that tends to move into a magnetic field. *Diamagnetism* indicates that all the electrons in a molecule are paired; *paramagnetism* indicates that a molecule possesses unpaired electrons (**BOX 2G.2**). According to both Lewis's approach and valence-bond theory, O_2 should be diamagnetic. In fact, it is found to be paramagnetic (**FIG. 2G.9**), just as MO theory implies, because the two π^* electrons are not paired. The bond order of O_2 is

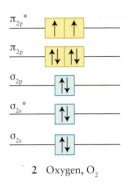

2 Oxygen, O_2

$$b = \tfrac{1}{2}\{\overbrace{(2 + 2 + 4)}^{N_e}\overbrace{(2 + 1 + 1)}^{N_e{}^*}\} = 2$$

The double bond in this molecule is actually a σ-bond plus two "*half* π-bonds," each "half bond" consisting of a pair of bonding electrons and one antibonding electron.

Box 2G.2 HOW DO WE KNOW. . . THAT ELECTRONS ARE NOT PAIRED?

Most common materials are diamagnetic. The tendency of a diamagnetic substance to move out of a magnetic field can be detected by hanging a long, thin sample from the arm of a balance and letting it lie between the poles of an electromagnet. This arrangement, which was once the primary technique used to measure the magnetic properties of samples, is called a *Gouy balance*. When the electromagnet is turned on, a diamagnetic sample tends to move upward, out of the field, so it appears to weigh less than in the absence of the field. The diamagnetism arises from the effect of the magnetic field on the electrons in the molecule: the field forces the electrons to circulate through the nuclear framework. Because electrons are charged particles, this circulation corresponds to an electric current circulating within the molecule. That current gives rise to its own magnetic field, which opposes the applied field. The sample tends to move out of the field so as to minimize this opposing field.

Compounds with unpaired electrons are *paramagnetic*. They tend to move into a magnetic field and can be identified because they seem to weigh more in a Gouy balance when a magnetic field is applied than when it is absent. Paramagnetism arises from the electron spins, which behave like tiny bar magnets that tend to line up with the applied field. The more that can line up in this way, the greater is the lowering of energy and the more the sample is drawn into the magnetic field, making it appear heavier. Oxygen is a paramagnetic substance because it has two unpaired electrons: this property is used to detect the concentration of

oxygen in incubators. All radicals are paramagnetic. Many compounds of the d-block elements are paramagnetic because they have various numbers of unpaired d-electrons.

The modern approach to measuring magnetic properties is to use a *superconducting quantum interference device* (a SQUID), which is highly sensitive to small magnetic fields and can be used to make very precise measurements on small samples.

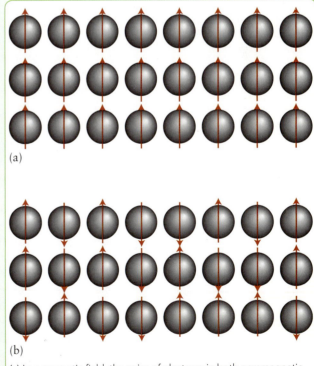

(a)

(b)

(a) In a magnetic field, the spins of electrons in both paramagnetic and ferromagnetic (Topic 3J) substances are aligned with the field to a large extent. (b) The spins of electrons in a paramagnetic substance return to a random orientation after an applied magnetic field is removed. However, the spins of electrons in a ferromagnetic substance remain aligned after a magnetic field is removed.

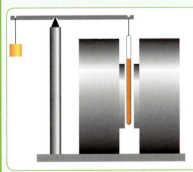

A Gouy balance is used to observe the magnetic character of a sample by detecting the extent to which it is driven out of (diamagnetic substances) or drawn into (paramagnetic substances) a magnetic field.

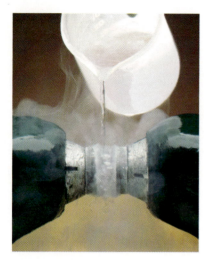

FIGURE 2G.9 The paramagnetic properties of oxygen are evident when liquid oxygen is poured between the poles of a magnet. The liquid sticks to the magnet instead of flowing past it. (© 1992 Richard Megna–Fundamental Photographs.)

THINKING POINT

Can you use molecular orbital theory to explain why fluorine is such a highly reactive gas?

The ground-state electron configurations of diatomic molecules are deduced by forming molecular orbitals from all the valence-shell atomic orbitals of the two atoms and adding the valence electrons to the molecular orbitals in order of increasing energy, in accord with the building-up principle.

2G.3 Bonding in Heteronuclear Diatomic Molecules

The bond in a **heteronuclear diatomic molecule,** a diatomic molecule built from atoms of two different elements, is polar, with the electrons shared unequally by the two atoms. Equation 1 then becomes the linear combination

$$\psi = c_A\psi_A + c_B\psi_B \tag{4}$$

where the coefficients c_A and c_B are not equal. Because the squares of wavefunctions are used to interpret them in terms of probabilities, if $c_A{}^2$ is large, then the molecular orbital looks more like the atomic orbital of A and the electron density is greater near A; if $c_B{}^2$ is large, the molecular orbital looks more like the atomic orbital of B and the electron density is greater near B. In general, the atom with the atomic orbitals of lower energy dominates the molecular orbitals and the electron density of a bonding orbital is greater on that atom. The relative values of $c_A{}^2$ and $c_B{}^2$ determine the type of bond:

- In a *nonpolar covalent bond,* $c_A{}^2 = c_B{}^2$ and the electron pair is shared equally between the two atoms.
- In an *ionic bond,* the coefficient belonging to one ion, the cation, is nearly zero because the other ion, the anion, captures almost all the electron density.
- In a *polar covalent bond,* the atomic orbital belonging to the more electronegative atom has the lower energy, and so it makes the larger contribution to the lowest-energy molecular orbital.

The last point refers to electrons in bonding orbitals. For electrons in antibonding orbitals, the opposite is true: the greater contribution comes from the *less* electronegative atom. This distinction is illustrated in **FIG. 2G.10**.

To find the ground-state electron configurations of heteronuclear diatomic molecules, use the same approach as for homonuclear diatomic molecules, but first modify the energy-level diagrams. For example, consider the HF molecule. The σ-bond in this molecule consists of an electron pair in a σ-orbital built from all the atomic orbitals with the appropriate symmetry: these are the $F2p_z$- and $F2s$-orbitals and the $H1s$-orbital. Because the electronegativity of fluorine is 3.98 and that of hydrogen is 2.20, you can expect the bonding σ-orbital to be mainly composed of fluorine orbitals and the antibonding σ^*-orbital to be mainly hydrogen in character. These expectations are confirmed by calculation. Because the two electrons in the bonding orbital are more likely to be found in the fluorine orbitals than in the $H1s$-orbital, there is a partial negative charge on the F atom and a partial positive charge on the H atom.

The molecular orbital energy-level diagrams of heteronuclear diatomic molecules are much harder to predict qualitatively (but are easily calculated using widely available software). Each energy level must be calculated explicitly because the atomic orbitals contribute differently to each one. Moreover, the diagrams have to show how all atomic orbitals of the appropriate symmetry contribute to a given molecular orbital. This requirement means that, as for HF, σ-orbitals are formed by combining both s- and p_z-orbitals on the atoms. When both atoms are in Period 2, each contributes an s- and a p_z-orbital, for four atomic orbitals in total, and four σ-orbitals can be formed. The same is true of π-orbitals: there are two sets of suitable atomic orbitals on each atom (the p_x- and p_y-orbitals); thus,

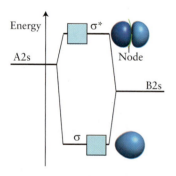

FIGURE 2G.10 A typical σ molecular orbital energy-level diagram for a heteronuclear diatomic molecule AB. The relative contributions of the atomic orbitals to the molecular orbitals are represented by the relative sizes of the contributing atomic orbitals and the horizontal position of the boxes. In this case, A is the more electronegative of the two elements.

four π-orbitals can be formed. **FIGURE 2G.11** shows the calculated scheme typically found for CO and NO. This diagram can be used to state the electron configuration by using the same procedure as for homonuclear diatomic molecules.

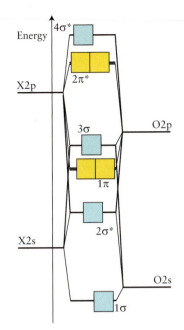

FIGURE 2G.11 The molecular orbital schemes typical of those calculated for a diatomic oxide molecule, XO (where X = C for CO and X = N for NO). Note that the σ-orbitals are formed from mixtures of s- and p_z-orbitals on both atoms; they are labeled simply 1σ, 2σ, etc., in order of increasing energy.

EXAMPLE 2G.2   Writing the configuration of a heteronuclear diatomic molecule or ion

Molecules take part in reactions and form compounds in ways that depend on the arrangement of electrons around their atoms. For instance, it is possible in this way to understand why carbon monoxide causes suffocation by attaching strongly to the iron atom of hemoglobin, thus preventing the attachment of oxygen molecules. Write the electron configuration of the ground state of the carbon monoxide molecule.

ANTICIPATE Carbon monoxide is known to have a triple bond, so you should expect to write a configuration with a bond order of three.

PLAN Assign the electrons to the orbitals shown in Fig. 2G.11, in accordance with the building-up principle.

SOLVE There are $4 + 6 = 10$ valence electrons to accommodate (4 from C, 6 from O). The resulting configuration is shown in Fig. 2G.11, and is

$$CO: 1\sigma^2 2\sigma^{*2} 1\pi^4 3\sigma^2$$

EVALUATE The bond order is $b = \frac{1}{2} \times \{(2 + 4 + 2) - 2\} = 3$, as anticipated.

Self-test 2G.2A Write the configuration of the ground state of the nitric oxide (nitrogen monoxide) molecule.

[**Answer:** $1\sigma^2 2\sigma^{*2} 1\pi^4 3\sigma^2 2\pi^{*1}$]

Self-test 2G.2B Write the configuration of the ground state of the cyanide ion, CN⁻, assuming that its molecular orbital energy-level diagram is the same as that for CO.

Related Exercises 2G.13, 2G.14.

Bonding in heteronuclear diatomic molecules involves an unequal sharing of the bonding electrons. The more electronegative element contributes more strongly to the bonding orbitals, whereas the less electronegative element contributes more strongly to the antibonding orbitals.

2G.4 Orbitals in Polyatomic Molecules

The molecular orbital theory of polyatomic molecules follows the same principles as those outlined for diatomic molecules, but the molecular orbitals spread over *all* the atoms in the molecule. An electron pair in a bonding orbital helps to bind together the *whole* molecule, not just an individual pair of atoms. The energies of molecular orbitals in polyatomic molecules can be studied experimentally by using ultraviolet and visible spectroscopy (see *Major Technique* 2 on the website for this book).

An important polyatomic molecule is benzene, C_6H_6, the parent of the aromatic compounds. In the molecular orbital description of benzene, all 30 C2s-, C2p-, and H1s-orbitals contribute to molecular orbitals spreading over all 12 atoms (six C plus six H). The orbitals in the plane of the ring (the C2s-, $C2p_x$-, and $C2p_y$-orbitals on each carbon atom and all six H1s-orbitals) form delocalized σ-orbitals that bind the C atoms together and link the H atoms to the C atoms. The six $C2p_z$-orbitals, which are perpendicular to the ring, contribute to six delocalized π-orbitals that spread all the way around the ring.

Chemists (other than those carrying out detailed calculations, who use a pure MO scheme) commonly mix MO and VB descriptions when discussing organic molecules. Because the language of hybridization and orbital overlap is well suited to the description of σ-bonds, chemists typically express the σ-framework of molecules in VB terms (Topic 2F). Thus, they think of the σ-framework of benzene as formed by the overlap of sp² hybridized

The z-axis is commonly chosen to lie along the dominant rotational axis of a molecule.

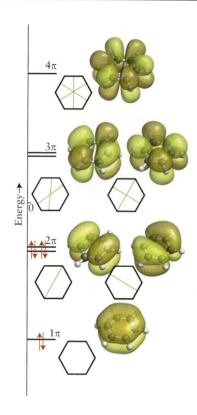

The six π-orbitals of benzene; locations of nodes are shown with line drawings. Note that the orbitals range from fully bonding (no internuclear nodes) to fully antibonding (six internuclear nodes). The zero of energy corresponds to the total energy of the separated atoms. The three orbitals with negative energies have net bonding character.

orbitals on neighboring atoms. Then, because delocalization is such an important feature of the π-bonding component of **conjugated double bonds,** in which single and double bonds alternate along a chain or around a ring as in —C=C—C=C—C=C—, they treat such π-bonds in terms of MO theory.

In the VB description of benzene each C atom is sp^2 hybridized, with one electron in each hybrid orbital. Each C atom has a p_z-orbital perpendicular to the plane defined by the hybrid orbitals, and it contains one electron. Two sp^2 hybrid orbitals on each C atom overlap and form σ-bonds with similar orbitals on the two neighboring C atoms, forming the 120° internal angle of the benzene hexagon. The third, outward-pointing sp^2 hybrid orbital on each C atom forms a σ-bond with a hydrogen atom. The resulting σ-framework is the same as that illustrated in Fig. 2F.13.

> **THINKING POINT**
>
> Suppose you wanted a fully MO description of the benzene molecule. How many σ-orbitals would you need to construct? How many would be occupied?

In the MO part of the description the six $C2p_z$-orbitals form six delocalized π-orbitals; their shapes are shown in **FIG. 2G.12** and their energies are shown in **FIG. 2G.13**. The character of the orbitals changes from net bonding to net antibonding as the number of internuclear nodes increases from none (fully bonding) to six (fully antibonding).

Each carbon atom provides one electron for the π-orbitals. Two electrons occupy the lowest-energy, most bonding orbital, and the remaining four electrons occupy the next higher-energy orbitals (two orbitals of the same energy). Figure 2G.13 shows one of the reasons for benzene's great stability: the π-electrons occupy only orbitals with a net bonding effect; none of the destabilizing antibonding orbitals is occupied.

The delocalization of electrons provides another test of MO theory. Lewis's theory fails to account for the structure of the molecules of diborane, B_2H_6, a colorless gas that bursts into flame on contact with air. The problem is that diborane has only 12 valence electrons (three from each B atom, one from each H atom); but, for a Lewis structure, it needs at least seven bonds, and therefore 14 electrons, to bind the eight atoms together! Diborane is an example of an **electron-deficient compound,** a compound with too few valence electrons to be assigned a valid Lewis structure. Valence-bond theory can account for the structures of electron-deficient compounds in terms of resonance, but the explanation is not straightforward. Molecular orbital theory accounts for the existence of electron-deficient molecules in a very straightforward way. Because the bonding influence of an electron pair is spread over all the atoms in the molecule, there is no need to provide one pair of electrons for each pair of atoms. A smaller number of pairs of electrons spread throughout the molecule may be able to bind all the atoms together, particularly if the nuclei are not highly charged and so do not repel one another strongly.

Another mystery solved by molecular orbital theory is the existence of hypervalent compounds, compounds in which a central atom forms more bonds than allowed by the octet rule (Topic 2C). In valence-bond theory, hybridization schemes are required to make sense of these compounds. For example, the expanded valence shell of Period 3 elements in compounds such as SF_6 is explained as sp^3d^2 hybridization. However, the d-orbitals of sulfur lie at relatively high energies and might not be accessible for bonding. In molecular orbital theory a bonding scheme can be devised for SF_6 that does not involve d-orbitals. The four valence orbitals provided by the sulfur atom and the six orbitals of the fluorine atoms that point toward the sulfur atom, a total of 10 atomic orbitals, result in 10 molecular orbitals with the energies shown in **FIG. 2G.14**. The 12 electrons occupy

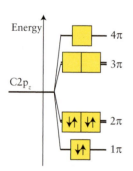

FIGURE 2G.13 The molecular orbital energy-level diagram for the π-orbitals of benzene. In the ground state of the molecule, only the net bonding orbitals are occupied.

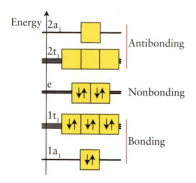

FIGURE 2G.14 The molecular orbital energy-level diagram for SF_6 and the occupation of the orbitals by the 12 valence electrons of the atoms. Note that no antibonding orbitals are occupied and that there is a net bonding interaction even though no d-orbitals are involved.

the lowest six orbitals, which are either bonding or nonbonding, and so bind all the atoms together without needing to use d-orbitals. Because four bonding orbitals are occupied, the average bond order of each of the six S—F links is $\frac{2}{3}$.

> *According to molecular orbital theory, the delocalization of electrons in a polyatomic molecule spreads the bonding effects of electrons over the entire molecule.*

What have you learned in this Topic?

You have learned that according to molecular orbital theory, electrons are described by wavefunctions that spread over all the atoms in a molecule and each of which can be occupied by up to two electrons. You now know about the existence of σ- and π-orbitals and bonding and antibonding orbitals. Their systematic occupation according to the building-up principle is used to predict the ground-state electron configuration of a molecule. You have seen how MO theory accounts for the paramagnetism of molecules and the existence of electron-deficient and hypervalent molecules.

The skills you have mastered are the ability to:

☐ **1.** Construct and interpret a molecular orbital energy-level diagram for homonuclear diatomic molecules (Section 2G.2).

☐ **2.** Deduce the ground-state electron configurations of Period 2 diatomic molecules (Toolbox 2G.1 and Examples 2G.1 and 2G.2).

☐ **3.** Define and use bond order as an assessment of the number of bonds between pairs of atoms (Example 2G.1).

☐ **4.** Account for the existence of hypervalent and electron-deficient compounds in terms of delocalized orbitals (Section 2G.2).

☐ **5.** Use molecular orbital theory to describe bonding in polyatomic molecules such as the benzene molecule (Section 2G.4).

Topic 2G Exercises

Exercises labeled $\int_{dx}^{C}$ require calculus.

2G.1 Draw a molecular orbital energy-level diagram and determine the bond order expected for each of the following diatomic species: (a) Li_2; (b) Li_2^+; (c) Li_2^-. State whether each molecule or ion will be paramagnetic or diamagnetic. If it is paramagnetic, give the number of unpaired electrons.

2G.2 Draw a molecular orbital energy-level diagram and determine the bond order expected for the following diatomic species: (a) B_2; (b) B_2^-; (c) B_2^+. State whether each molecule or ion will be paramagnetic or diamagnetic. If it is paramagnetic, give the number of unpaired electrons.

2G.3 (a) On the basis of the configuration of the neutral molecule F_2, write the molecular orbital configuration of the valence molecular orbitals for (i) F_2^-; (ii) F_2^+; (iii) F_2^{2-}. (b) For each species, give the expected bond order. (c) Which are paramagnetic, if any? (d) Is the highest-energy orbital that contains an electron σ or π in character?

2G.4 (a) On the basis of the configuration for the neutral molecule N_2, write the molecular orbital configuration of the valence molecular orbitals for (i) N_2^+; (ii) N_2^{2+}; (iii) N_2^{2-}. (b) For each species, give the expected bond order. (c) Which are paramagnetic, if any? (d) Is the highest-energy orbital that contains an electron σ or π in character?

2G.5 The ground-state electron configuration of the ion C_2^{n-} is $\sigma_{2s}^2\sigma_{2s}^{*2}\pi_{2p}^4\sigma_{2p}^2$. What is the charge on the ion, and what is its bond order?

2G.6 The ground-state electron configuration of the ion B_2^{n-} is $\sigma_{2s}^2\sigma_{2s}^{*2}\pi_{2p}^3$. What is the charge on the ion, and what is its bond order?

2G.7 Would you expect the HeH^- ion to exist? What would be its bond order? Which has the lower energy, HeH^- or HeH^+? Explain.

2G.8 Molecular orbital theory predicts that O_2 is paramagnetic. What other Period 2 homonuclear diatomic molecules are also expected to be paramagnetic?

2G.9 (a) Draw the molecular orbital energy-level diagram for N_2 and label the energy levels according to the type of orbitals from which they are made, whether they are σ- or π-orbitals, and whether they are bonding or antibonding. (b) The orbital structure of the heteronuclear diatomic ion NO^+ is similar to that of N_2. How will the fact that the electronegativity of N differs from that of O affect the molecular orbital energy-level diagram of NO^+ compared with that of N_2? Use this information to draw the energy-level diagram for NO^+. (c) In the molecular orbitals, will the electrons have a higher probability of being at N or at O? Why?

2G.10 (a) How does molecular orbital theory make it possible to explain both ionic *and* covalent bonding? (b) The degree of ionic character in bonding is related to electronegativity in Topic 2D. How does electronegativity affect the molecular orbital diagram so that bonds become ionic?

2G.11 Write the valence-shell electron configurations and bond orders of (a) B_2; (b) Be_2; (c) F_2.

2G.12 Write the valence-shell electron configurations and bond orders of (a) O_2^{2-}; (b) N_2^-; (c) C_2^-.

2G.13 Give the valence-shell electron configurations and bond orders for CO and CO^+. Use that information to predict which species has stronger bonds.

2G.14 Give the valence-shell electron configurations and bond orders for NO and NO^+. Use that information to predict which species has stronger bonds.

2G.15 Which of the following species are paramagnetic: (a) B_2; (b) B_2^-; (c) B_2^+? If the species is paramagnetic, how many unpaired electrons does it possess?

2G.16 Which of the following species are paramagnetic: (a) N_2^-; (b) F_2^+; (c) O_2^{2+}? If the species is paramagnetic, how many unpaired electrons does it possess?

2G.17 Determine the bond orders and use them to predict which species of each of the following pairs has the stronger bond: (a) F_2 or F_2^-; (b) B_2 or B_2^+.

2G.18 Determine the bond orders and use them to predict which species of each of the following pairs has the stronger bond: (a) C_2 or C_2^-; (b) N_2 or N_2^-.

2G.19 Based on their valence-shell electron configurations, which of the following species would you expect to have the lowest ionization energy: (a) C_2^+; (b) C_2; (c) C_2^-?

2G.20 Based on their valence-shell electron configurations, which of the following species would you expect to have the greatest electron affinity: (a) Be_2; (b) F_2; (c) B_2^+, (d) C_2^+?

2G.21 It is usually convenient to deal with wavefunctions that are "normalized," which means that the integral $\int \psi^2 \, d\tau = 1$. The bonding orbital in Eq. 1 is not normalized. A wavefunction ψ can be normalized by writing it as $N\psi$ and finding the factor N which ensures that the integral over $(N\psi)^2$ is equal to 1. Find the factor N that normalizes the bonding orbital in Eq. 1, given that the individual atomic orbitals are each normalized. Express your answer in terms of the "overlap integral" $S = \int \psi_{A1s} \psi_{B1s} \, d\tau$.

2G.22 The antibonding orbital in Eq. 2 is not normalized (see Exercise 2G.21). Find the factor that normalizes it to 1, given that the individual atomic orbitals are each normalized. Express your answer in terms of the overlap integral $S = \int \psi_{A1s} \psi_{B1s} \, d\tau$. Confirm that the bonding and antibonding orbitals are mutually orthogonal—that is, that the integral over the product of the two wavefunctions is zero.

The following Example and Exercises draw on material from throughout Focus 2.

FOCUS 2 Online Cumulative Example

You are working in a materials chemistry laboratory and are investigating hybrid organic–inorganic materials for use in optical memory devices. One potentially useful compound is 2,4,6-trimethylpyridinium dihydrogen phosphate. You want to know more about the structure of this material.

(a) Draw the Lewis structure for the dihydrogen phosphate anion, $H_2PO_4^-$ (P is the central atom and the H atoms are attached to two of the O atoms).

(b) What is the shape of the anion (disregarding the H atoms)? Estimate the O—P—O and HO—P—OH bond angles.

(c) The cation (**1**) has a structure analogous to benzene (Topic 2F), with nitrogen replacing one of the carbon atoms and three methyl (—CH_3) groups replacing hydrogen atoms. Draw its Lewis structure and indicate the hybridization of the carbon and nitrogen atoms.

1 2,4,6-trimethylpyridinium ion, $C_5H_{12}N$

(d) Use the valence bond model to sketch the π-bonds in the 2,4,6-trimethylpyridinium ion. Indicate which atomic orbitals are used (take the z-axis as perpendicular to the plane of the molecule).

(e) Refer to the π-molecular orbitals of benzene in Fig. 2G.12. Sketch the lowest energy π-orbital of the 2,4,6-trimethylpyridinium ion. How is it different from the corresponding orbital of benzene? *Hint:* Think about the relative electronegativities of carbon and nitrogen.

(f) Is the 2,4,6-trimethylpyridinium ion polar? If it is polar, draw the direction of the dipole moment.

The Online Cumulative Example solution can be found at http://macmillanhighered.com/chemicalprinciples7e

FOCUS 2 Exercises

2.1 Draw the Lewis structure, including resonance structures where appropriate, for (a) the oxalate ion, $C_2O_4^{2-}$ (there is a C—C bond with two oxygen atoms attached to each carbon atom); (b) BrO^+; (c) the acetylide ion, C_2^{2-}. Assign formal charges to each atom.

2.2 Draw three Lewis structures that follow the octet rule (including the most important structure) for the isocyanate ion, CNO^-. State which of the three Lewis structures is the most important and explain why.

2.3 Which do you expect to have the greater lattice energy, iron(III) chloride or iron(II) chloride? Explain your answer.

2.4 Explain the trend of decreasing lattice energies of the chlorides of the Group 2 metals down the group.

2.5 Draw resonance structures for the trimethylenemethane anion $C(CH_2)_3^{2-}$ in which a central carbon atom is attached to three CH_2 groups (CH_2 groups are referred to as methylene).

2.6 Show how resonance can occur in the following organic ions: (a) acetate ion, $CH_3CO_2^-$; (b) enolate ion, $CH_2COCH_3^-$, which has one resonance structure with a C=C double bond and an —O^- group on the central carbon atom; (c) allyl cation, $CH_2CHCH_2^+$; (d) amidate ion, CH_3CONH^- (the O and the N atoms are both bonded to the second C atom).

2.7 In 1999, Karl Christe synthesized and characterized a salt that contained the N_5^+ cation, in which the five N atoms are connected

in a chain. This cation is the first new all-nitrogen species to be isolated in more than 100 years. Draw the most important Lewis structure for this ion, including all equivalent resonance structures. Calculate the formal charges on all atoms.

2.8 Draw the most important Lewis structure for each of the following ring molecules (which have been drawn without showing the locations of the double bonds). Show all lone pairs and nonzero formal charges. If there are equivalent resonance forms, draw them.

(a) (b) (c) (d)

2.9 An important principle in chemistry is the *isolobal analogy*, which states that chemical fragments with similar valence-orbital structures can replace one another in molecules. For example, $\cdot\overset{\cdot}{C}$—H and $\cdot\overset{\cdot}{Si}$—H are isolobal fragments, each having three electrons with which to form bonds in addition to the bond to H. An isolobal series of molecules would be HCCH, HCSiH, HSiSiH. Similarly, a lone pair of electrons can be used to replace a bond so that $\cdot\overset{\cdot\cdot}{N}$: is isolobal with $\cdot\overset{\cdot}{C}$—H, with the lone pair taking the place of the $\overset{\cdot}{C}$—H bond. The isolobal set here is HCCH, HCN, NN. (a) Draw the Lewis structures for the molecules HCCH, HCSiH,

139

HSiSiH, HCN, and NN. (b) Use the isolobal principle to draw Lewis structures for molecules based on the structure of benzene, C_6H_6, in which one or more CH groups are replaced with N atoms.

2.10 The framework for the tropylium cation, $C_7H_7^+$, is a seven-membered ring of carbon atoms with a hydrogen atom attached to each carbon atom. Complete the structural drawing by adding the multiple bonds as appropriate. Resonance structures are possible. Draw as many as you can find. Determine the C—C bond order.

2.11 Quinone, $C_6H_4O_2$, is an organic molecule with the structure shown below; it can be reduced to the anion $C_6H_4O_2^{2-}$. (a) Draw the Lewis structure of the reduced product. (b) On the basis of formal charges derived from the Lewis structure, predict which atoms in the molecule are most negatively charged. (c) If two protons are added to the reduced product, where are they most likely to bond?

2.12 One of the following compounds does not exist. Use Lewis structures to identify that compound. (a) C_2H_2; (b) C_2H_4; (c) C_2H_6; (d) C_2H_8.

2.13 Draw the most important Lewis structure for each of the following molecules. Show all lone pairs and formal charges. Draw all equivalent resonance forms: (a) HONCO; (b) H_2CSO; (c) H_2CNN; (d) ONCN. Atoms are linked in the order shown.

2.14 Determine the formal charges of the atoms in (a) $BeCl_2$; (b) H_2NOH; (c) NO_2^-.

2.15 (a) Confirm that lattice energies are inversely proportional to the distance between the centers of the ions in MX (M = alkali metal, X = halide ion) by plotting the lattice energies of KF, KCl, and KI against the internuclear distances d_{M-X}. The lattice energies of KF, KCl, and KI are 826, 717, and 645 kJ·mol^{-1}, respectively. Use the ionic radii in Appendix 2D to calculate d_{M-X}. How good is the correlation? You should use a standard graphing program (for instance, with a spreadsheet program) to make the plot that will generate an equation for the line and calculate a correlation coefficient for the fit. (b) Estimate the lattice energy of KBr from your graph. (c) Find an experimental value for the lattice energy of KBr in the chemical literature and compare that value with the value that you calculated in part (b). How well do they agree?

2.16 (a) Explore whether the lattice energies of the alkali metal iodides are inversely proportional to the distances between the centers of the ions in MI (M = alkali metal) by plotting the lattice energies given below against the internuclear distances d_{M-I}.

	Lattice energy, $E/(\text{kJ·mol}^{-1})$
LiI	759
NaI	700
KI	645
RbI	632
CsI	601

Use the ionic radii in Appendix 2D to calculate d_{M-I}. How good is the correlation? Is a better fit obtained by plotting the lattice energies against $(1 - d^*/d)/d$, as suggested theoretically with $d^* = 34.5$ pm? You should use a standard graphing program to make the plot that will generate an equation for the line and calculate a correlation coefficient for the fit. (b) From the ionic radii given in Appendix 2D and the plot given in part (a), estimate the lattice energy of silver iodide. (c) Compare your results from part (b) with the experimentally determined value of 886 kJ·mol^{-1}. If they do not agree, provide an explanation for the deviation.

2.17 A common biologically active radical is the pentadienyl radical, RCHCHCHCHCHR′, where the carbons form a chain, with R and R′, which can be a number of different organic groups, at each end. Draw three resonance structures for this compound that maintain carbon's valence of four.

2.18 Predict the qualitative molecular potential energy curves for the NN bond by making sketches on one graph for N_2H_4, N_2, and N_3^-. Explain why the energy of the lowest point on each curve is not the same.

2.19 Thallium and oxygen form two compounds with the following characteristics:

	Compound I	Compound II
Mass percentage Tl	89.49%	96.23%
Melting point	717 °C	300 °C

Determine the chemical formulas of the two compounds. (b) Determine the oxidation number of thallium in each compound. (c) Assume that the compounds are ionic and write the electron configuration for each thallium ion. (d) Use the melting points to decide which compound has more covalent character in its bonds. Is your finding consistent with what you would predict from the polarizing abilities of the two cations?

2.20 How close are the Mulliken and Pauling electronegativity scales? (a) Use Eq. 2 in Topic 2D to calculate the Mulliken electronegativities of C, N, O, and F. Use the values in kilojoules per mole from Figs. 1F.8 and 1F.12 and divide each value by 230 kJ·mol^{-1} for this comparison. The Pauling values are those in Fig. 2D.2. (b) Plot both sets of electronegativities against atomic number on the same graph. (c) Which scale depends more consistently on position in the periodic table?

2.21 The perchlorate ion, ClO_4^-, is described by resonance structures. (a) Draw the Lewis structures that contribute to the resonance hybrid and identify the most plausible Lewis structures by using formal charge arguments. (b) The average length of a single Cl—O bond is 172 pm and that of a double Cl=O bond can be estimated at 140 pm. The Cl—O bond length in the perchlorate ion is found experimentally to be 144 pm for all four bonds. Identify the most plausible Lewis structures of the perchlorate ion from these experimental data. (c) What is the oxidation number of chlorine in the perchlorate ion? Identify the most plausible Lewis structure by using the oxidation number, assuming that lone pairs belong to the atom to which they are attached but that all electrons shared in a bond belong to the atom of the more negative element. (d) Are these three approaches consistent? Explain why or why not.

2.22 In the solid state, sulfur is commonly found in rings of eight atoms, but rings consisting of six atoms of sulfur were identified in 1958. (a) Draw a valid Lewis structure for S_6. (b) Is resonance possible in S_6? If so, draw one of the resonance structures.

2.23 *Structural isomers* are molecules that have the same composition but a different pattern of connectivity. Two isomers of disulfur difluoride, S_2F_2, are known. In each, the two S atoms are bonded to each other. In one isomer, each of the S atoms is bonded to an F atom. In the other isomer, both F atoms are attached to one of the S atoms. (a) In each isomer, the S—S bond length is approximately 190 pm. Are the S—S bonds in these isomers single bonds, or do they have some double bond character? (b) Draw two resonance structures for each isomer. (c) Determine for each isomer which structure is favored by formal-charge considerations. Are your conclusions consistent with the S—S bond lengths in the compounds?

2.24 Ionic compounds typically have higher boiling points and lower vapor pressures than covalent compounds. Predict which compound in the following pairs has the lower vapor pressure at room temperature: (a) Cl_2O or Na_2O; (b) $InCl_3$ or $SbCl_3$; (c) LiH or HCl; (d) $MgCl_2$ or PCl_3.

2.25 Which bond is longer: (a) the CN bond in HCN or in H_3CNH_2? (b) The NF bond in NF_3 or the PF bond in PF_3?

2.26 Which of the following bonds do you expect to be longer: (a) the BrO bond in BrO^- or in BrO_2^-; (b) the CH bond in CH_4 or the SiH bond in SiH_4?

2.27 (a) Draw a Lewis structure for each of the following species: CH_3^+; CH_4; CH_3^-; CH_2; CH_2^{2+}; CH_2^{2-}. (b) Identify each as a radical or not. (c) Rank them in order of increasing HCH bond angles. Explain your choices.

2.28 The halogens form compounds among themselves. These compounds, called the *interhalogens*, have the formulas X′X, X′X_3, and X′X_5, where X′ is the heavier halogen atom. (a) Predict their structures and bond angles. (b) Which of them are polar? (c) Why is the lighter halogen atom not the central atom of such molecules?

2.29 An organic compound distilled from wood was found to have a molar mass of 32.04 g·mol^{-1} and the following composition by mass: 37.5% C, 12.6% H, and 49.9% O. (a) Write the Lewis structure of the compound and determine the bond angles about the carbon and oxygen atoms. (b) Give the hybridization of the carbon and oxygen atoms. (c) Predict whether the molecule is polar or not.

2.30 Draw Lewis structures for each of the following species and predict the hybridization at each carbon atom: (a) H_2CCH^+; (b) $H_2CCH_3^+$; (c) $H_3CCH_2^-$.

2.31 (a) Draw the bonding and antibonding orbitals that correspond to the σ-bond in H_2. (b) Repeat this procedure for HF. (c) How do these orbitals differ?

2.32 (a) Consider the hypothetical species HeH. What charge (magnitude and sign), if any, should be present on this combination of atoms to produce the most stable molecule or ion possible? (b) What is the maximum bond order that such a molecule or ion could have? (c) If the charge on this species were increased or decreased by 1, what would be the effect on the bonding in the molecule?

2.33 (a) Order the following molecules according to increasing C—F bond length: CF^+, CF, CF^-. (b) Which of these species is diamagnetic, if any? Explain your reasoning.

2.34 Describe as completely as you can the structure and bonding in a carbamate ion, $H_2NCO_2^-$. The C—O bond lengths are both 128 pm, and the C—N bond length is 136 pm.

2.35 Borazine, $B_3N_3H_6$, a compound that has been called "inorganic benzene" because of its similar hexagonal structure (but with alternating B and N atoms in place of C atoms), is the basis of a large class of boron–nitrogen compounds. Write its Lewis structure and predict the composition of the hybrid orbitals used by each B and N atom.

2.36 Given that carbon has a valence of 4 in nearly all its compounds and can form chains and rings of C atoms, (a) draw any two of the three possible structures for C_3H_4; (b) determine all bond angles in each structure; (c) determine the hybridization of each carbon atom in the two structures; (d) ascertain whether the two structures are resonance structures and explain your reasoning.

2.37 The two atomic orbitals that contribute to the antibonding orbital in Eq. 2 in Topic 2G are each proportional to e^{-r/a_0}, where r is the distance of the point from its parent nucleus. Confirm that there is a nodal plane lying halfway between the two nuclei.

2.38 Show that a molecule with configuration π^4 has a cylindrically symmetrical electron distribution. *Hint:* Take the π-orbitals to be equal to xf and yf, where f is a function that depends only on distance from the internuclear axis.

2.39 In addition to forming σ- and π-types of bonds similar to the way that neighboring p-orbitals overlap, neighboring d-orbitals may overlap to form δ-bonds. (a) Draw overlap diagrams showing three different ways in which d-orbitals can combine to form bonds. (b) Place the three types of d—d bonds (σ, π, and δ) in the likely order of strongest to weakest.

2.40 An s-orbital and a p-orbital on neighboring atoms are being considered as contributing to bond formation in a molecule. Draw diagrams to show how one combination can contribute to σ-bonding but another combination cannot.

2.41 (a) Describe the changes in bonding that would occur in benzene if two electrons were removed from the HOMO (highest occupied molecular orbital). This removal would correspond to an oxidation of benzene to $C_6H_6^{2+}$. (b) Describe the changes in bonding that would occur if two electrons were added to the LUMO (lowest unoccupied molecular orbital). This addition would correspond to a reduction of benzene by adding two electrons to give $C_6H_6^{2-}$. Do you expect these ions to be diamagnetic or paramagnetic?

2.42 (a) Use trigonometry to confirm that the dipoles of the three bonds in a trigonal pyrimidal AB_3 molecule do not cancel, resulting in a polar molecule. (b) Show that the dipoles of the four bonds in a tetrahedral AB_4 molecule cancel and the molecule is nonpolar.

2.43 Benzyne, C_6H_4, is a highly reactive molecule that survives only at low temperatures. It is related to benzene in that it has a six-membered ring of carbon atoms; but, instead of three double bonds, the structure is normally drawn with two double bonds and a triple bond. (a) Draw a Lewis structure of the benzyne molecule. Indicate on the structure the hybridization at each carbon atom. (b) On the basis of your understanding of bonding, explain why this molecule might be highly reactive.

2.44 The Lewis structure of caffeine, $C_8H_{10}N_4O_2$, a common stimulant, is given below. (a) Give the hybridization of each atom other than hydrogen. (b) On the basis of your answers in part (a), estimate the bond angles around each carbon and nitrogen atom. (c) Search the chemical literature for the structure of caffeine, and compare the observed structural parameters with your predictions.

2.45 Consider the bonding in H_2C=$CHCHO$. (a) Draw the most important Lewis structure. Include all nonzero formal charges. (b) Identify the composition of the bonds and the hybridization of each lone pair—for example, by writing $\sigma(H1s,C2sp^2)$.

2.46 Consider the molecules H_2CCH_2, H_2CCCH_2, and H_2CCCCH_2. (a) Draw Lewis structures for these molecules. (b) What is the hybridization at each C atom? (c) What type of bond connects the carbon atoms (single, double, etc.)? (d) What are the HCH, CCH, and CCC angles in these molecules? (e) Do all the hydrogen atoms lie in the same plane? (f) A generalized formula for molecules of this type is $H_2C(C)_xCH_2$, where x is 0, 1, 2, etc. What can be said, if anything, about the relative orientation of the H atoms at the ends of the chain as a function of x?

2.47 The reaction between SbF_3 and CsF produces, among other products, the anion $Sb_2F_7^-$. This anion has no F—F bonds and no Sb—Sb bonds. (a) Propose a Lewis structure for the ion. (b) Assign a hybridization scheme to the Sb atoms.

2.48 The following molecules are bases that are part of the nucleic acids involved in the genetic code. Identify (a) the hybridization of each C and N atom, (b) the number of σ- and π-bonds, and (c) the number of lone pairs of electrons in the molecule.

2.49 Just as $AlCl_3$ forms dimers (Topic 2C), in the $Bi_2Cl_4^{2-}$ ion two of the Cl atoms form "bridges" between the two Bi atoms. Propose a Lewis structure for the $Bi_2Cl_4^{2-}$ ion.

2.50 Germanium forms a series of anions called "germides." In the germide ion, Ge_4^{n-}, the four Ge atoms form a tetrahedron in which each atom is bonded to the other three and each atom has a lone pair of electrons. What is the value of n, the charge on this anion? Explain your reasoning.

2.51 One form of the polyatomic ion I_5^- has an unusual V-shaped structure: one I atom lies at the point of the V, with a linear chain of two I atoms extending on each side. The bond angles are 88° at the central atom and 180° at the two atoms in the side chains. Draw a Lewis structure for I_5^- that explains its shape and indicate the hybridization you would assign to each nonterminal atom.

2.52 Molecules and ions, like atoms, can be isoelectronic. That is, they can have the same number of electrons. For example, CH_4 and NH_4^+ are isoelectronic. Therefore, they have the same molecular shape. Identify a molecule or ion that is isoelectronic with each of the following species and verify that the members of each pair have the same shape: (a) CO_3^{2-}; (b) O_3; (c) OH^-.

2.53 What changes in bond order, bond distance, and magnetic properties are expected to occur in the following ionization processes? (a) $C_2 \rightarrow C_2^+ + e^-$; (b) $N_2 \rightarrow N_2^+ + e^-$; (c) $O_2 \rightarrow O_2^+ + e^-$.

2.54 (a) Draw the Lewis structure of N_2O_3. (b) What is the N—N bond order? (c) Explain why N_2O_3 adopts this structure, whereas the analogous compound for phosphorus is P_4O_6.

2.55 Complexes of d- and f-block metals can be described in terms of hybridization schemes, each associated with a particular shape. Bearing in mind that the number of atomic orbitals hybridized must be the same as the number of hybrid orbitals produced, match the hybrid orbitals sp^3d, sp^3d^3, and sp^3d^3f to the following shapes: (a) pentagonal bipyramidal; (b) cubic; (c) square planar.

2.56 (a) The formula for bromate ion (BrO_3^-) is similar to that of chlorate ion (ClO_3^-); however, the formulas of the nitrate and phosphate ions have different numbers of oxygen atoms, as do the formulas of the nitrite and phosphite ions. Draw the Lewis structures of all four ions and explain why the formulas differ. (b) Energy is supplied to muscles in the body through changes in ATP, a molecule that contains chains of phosphate ions ($-PO_2-O-PO_2-O-PO_2-O-$). Explain how phosphate ions can connect to one another in this way, but nitrate ions cannot.

2.57 Acetonitrile, CH_3CN, is used as a solvent in the pharmaceutical industry. Describe the structure of the CH_3CN molecule in terms of hybrid orbitals, bond angles, and σ- and π-bonds. The N atom is a terminal atom.

2.58 Nitrosamines are potent carcinogens that are formed in tobacco smoke and in meat that is cooked at a high temperature. The basic formula for a nitrosamine is R_2NNO, in which the R represents an organic group. The atoms are connected as indicated in the formula with the two organic groups on the first nitrogen atom. In the simplest nitrosamine the R groups are methyl groups, $—CH_3$. Draw the Lewis structure of the simplest nitrosamine.

2.59 Draw the Lewis structure for each of the following compounds: (a) methanethiol, CH_3SH, one of the compounds found in bad breath and some cheeses; (b) carbon disulfide, CS_2, which is used in the manufacture of rayon; (c) dichloromethane, CH_2Cl_2, a common solvent.

2.60 Draw the Lewis structure for each of the following compounds: (a) urea, $OC(NH_2)_2$, a compound formed in the body when proteins are metabolized; (b) phosgene, Cl_2CO, a deadly gas once used in warfare; (c) trinitramide, $N(NO_3)_3$, an oxidizing agent used in rocket fuel.

2.61 Methane, the most abundant hydrocarbon in the atmosphere and a potent greenhouse gas, is slowly oxidized in the air to carbon dioxide. An intermediate in the oxidation of methane to carbon dioxide is HOCO. (a) Draw the Lewis structure for this compound. (b) Decide whether the compound is a radical.

2.62 The compound 2,4-pentanedione (also known as acetylacetone and abbreviated to acac) is acidic and can be deprotonated. The anion forms complexes with metals that are used in gasoline additives, lubricants, insecticides, and fungicides. (a) Estimate the bond angles marked with arcs and lowercase letters in 2,4-pentanedione and in the acac ion. (b) What are the differences, if any?

Acetylacetone   Acetylacetonate ion

2.63 Estimate the bond angles marked with arcs and lowercase letters in peroxyacetylnitrate, an eye irritant in smog:

Peroxyacetylnitrate

2.64 Hydroxylamine, $HONH_2$, is used to remove hair from animal hides and as a photoresist stripper in the electronics industry. Describe the structure of the hydroxylamine molecule in terms of hybrid orbitals, bond angles, and σ- and π-bonds. The O atom is bonded to the N atom and an H atom. The N atom is also bonded to two H atoms.

FOCUS 2 Cumulative Exercises

2.65 The nitrogen oxides are common pollutants generated by internal combustion engines and power plants. They not only contribute to the respiratory distress caused by smog but, if they reach the stratosphere, they also threaten the ozone layer that protects Earth from harmful radiation.

(a) The bond energy in NO is 632 kJ·mol^{-1} and that of each N—O bond in NO_2 is 469 kJ·mol^{-1}. Use Lewis structures and the average bond energies in Table 2D.2 to explain the difference in bond energies between the two molecules and the fact that the bond energies of the two bonds in NO_2 are the same.

(b) The bond length in NO is 115 pm. Use Fig. 2D.11 to predict the length of a single bond and a double bond between nitrogen and oxygen. Use Table 2D.3 to estimate the length of a triple bond between nitrogen and oxygen. Predict the bond order in NO from its bond length and explain any difference from the calculated values.

(c) When the NO in smog reacts with NO_2, a bond forms between the two N atoms. Draw the Lewis structure of each reactant and the product and indicate the formal charge on each atom.

(d) The NO_2 in smog also reacts with NO_3 to form a product with an O atom between the two N atoms. Draw the

Lewis structure of the most likely product and indicate the formal charge on each atom.

(e) Write the balanced chemical equation for the reaction of the product from part (d) with water to produce an acid. The acid produced acts as a secondary pollutant in the environment. Name the acid.

(f) If 4.05 g of the product from part (d) reacts with water as in part (e) to produce 1.00 L of acidic solution, what will be the molar concentration of the acid?

(g) Determine the oxidation number of nitrogen in NO, NO_2, and the products in parts (c) and (d). Which of these compounds would you expect to be the most potent oxidizing agent?

2.66 Hydrogen peroxide, H_2O_2, is a nontoxic bleaching agent being used as a replacement for chlorine in industry and home laundries. The bleaching process is an oxidation, and when hydrogen peroxide acts as a bleaching agent, the only waste it generates is H_2O.

(a) Draw the Lewis structure of hydrogen peroxide and determine the formal charge on each atom. What is the oxidation number of oxygen in hydrogen peroxide? Which is more useful in predicting the ability of H_2O_2 to act as an oxidizing agent, formal charge or oxidation number? Explain your reasoning.

(b) Predict the bond angles at each O atom in H_2O_2. Are all the atoms in the same plane? Is the molecule polar or nonpolar? Explain your reasoning.

(c) Write the valence electron configuration of (i) O_2; (ii) O_2^-; (iii) O_2^+; (iv) O_2^{2-}. For each species, give the expected bond order and indicate which, if any, are paramagnetic.

(d) The following bond lengths have been reported: (i) O_2, 121 pm; (ii) O_2^-, 134 pm; (iii) O_2^+, 112 pm; (iv) O_2^{2-}, 149 pm. Suggest a reason for the differences based on the configurations in part (c).

(e) One reaction in which H_2O_2 acts as an oxidizing agent is $Fe^{2+}(aq) + H_2O_2(aq) + H^+(aq) \rightarrow Fe^{3+}(aq) + H_2O(l)$. Balance the equation and determine what mass of iron(II) can be oxidized to iron(III) by 43.2 mL of 0.200 M $H_2O_2(aq)$. *Hint:* Make sure you balance charge as well as the elements.

(f) Hydrogen peroxide can also act as a reducing agent, as in $Fe^{3+}(aq) + H_2O_2(aq) + OH^-(aq) \rightarrow Fe^{2+}(aq) + H_2O(l) + O_2(g)$. Balance the equation and determine what mass of iron(III) can be reduced to iron(II) by 41.8 mL of 0.200 M $H_2O_2(aq)$. *Hint:* Make sure you balance charge as well as the elements.

(g) Hydrogen peroxide must be kept in brown bottles because, in the presence of light, it can *disproportionate,* which means that it oxidizes and reduces itself in the reaction $2 H_2O_2(aq) \rightarrow 2 H_2O(l) + O_2(g)$. How many electrons are transferred in the reaction represented by this equation?

(h) Although peroxides do not generate hazardous waste, they can cause problems in the atmosphere. For example, if a hydrogen peroxide molecule makes its way to the stratosphere, it can break into two ·OH radicals, which threaten the ozone layer that protects Earth from harmful radiation. Use data in Table 2D.2 to calculate the minimum frequency and corresponding wavelength of light needed to break the HO—OH bond.

STATES OF MATTER

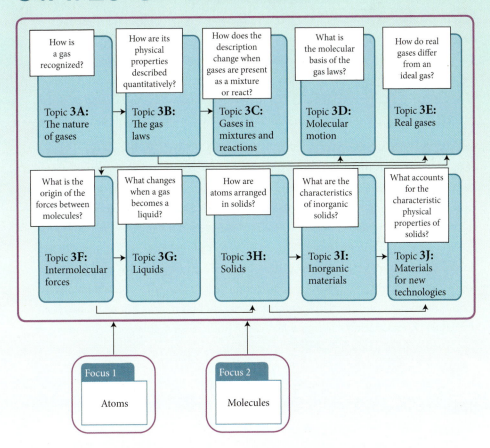

How is a gas recognized?

Topic **3A**: The nature of gases

How are its physical properties described quantitatively?

Topic **3B**: The gas laws

How does the description change when gases are present as a mixture or react?

Topic **3C**: Gases in mixtures and reactions

What is the molecular basis of the gas laws?

Topic **3D**: Molecular motion

How do real gases differ from an ideal gas?

Topic **3E**: Real gases

What is the origin of the forces between molecules?

Topic **3F**: Intermolecular forces

What changes when a gas becomes a liquid?

Topic **3G**: Liquids

How are atoms arranged in solids?

Topic **3H**: Solids

What are the characteristics of inorganic solids?

Topic **3I**: Inorganic materials

What accounts for the characteristic physical properties of solids?

Topic **3J**: Materials for new technologies

Focus 1 — Atoms

Focus 2 — Molecules

The materials of the twenty-first century are changing how we live. Whether designing a building, crafting artificial limbs, or developing new means of communication, engineers, medical researchers, architects, and scientists must all understand the chemical basis of materials. Materials not yet imagined will be developed—perhaps by you—as our ability to fabricate new forms of matter increases and thereby transform society.

This Focus deals with the properties of "bulk matter," matter that consists of huge numbers of particles. The simplest kind of bulk matter is a gas. Some of the earliest quantitative experiments in chemistry were made on gases and **TOPIC 3A** introduces a characteristic property of a gas: the pressure it exerts. **TOPIC 3B** shows how experiments on the dependence of pressure on volume and temperature led to an equation of state and the concept of an ideal gas. In **TOPIC 3C,** the relation of the volume of an ideal gas to the amount of gas present is then used to extend the stoichiometric relations for predicting the amounts of reactants consumed or products formed in chemical reactions to include the volume of any gas involved. **TOPIC 3D** then shows how the properties of an ideal gas can be understood in terms of a simple model in which widely separated molecules are in ceaseless chaotic motion. One outcome of this model is a quantitative understanding of the range of molecular speeds in a gas.

An ideal gas is a very important abstraction and is used throughout chemistry to formulate expressions for the properties of bulk matter, but it is an idealization, and real gases show deviations from ideal gas behavior. **TOPIC 3E** takes up the challenge of improving the model to take into account the interactions of molecules with one another. How they interact is discussed in **TOPIC 3F**, where the origin of the forces between molecules is traced to their structures. Intermolecular forces are weak in gases because the molecules spend so little time very close together, but they are of major importance for the existence and properties of liquids and solids. **TOPIC 3G** shows how liquids arise from the cohesion of molecules and **TOPIC 3H** shows how solids form when almost all molecular motion is lost and the molecules lie together in characteristic arrays.

Many of the past and current advances in technology have come from an understanding of the properties of solids. **TOPIC 3I** summarizes the characteristic properties of a wide range of inorganic solids and shows how those characteristics can be related to the location of the elements in the periodic table and the types of interactions between the constituent atoms and ions. A major aspect of some solid materials is their ability to conduct an electric current, and **TOPIC 3J** introduces the concepts that underlie the distinction between conductors, insulators, and—of huge importance to today's technologies—semiconductors and superconductors. It also shows how electrons contribute to the optical and magnetic properties of bulk matter. It then rounds out this discussion by introducing nanomaterials, which hold considerable promise and are at the center of much current research.

Topic 3A The Nature of Gases

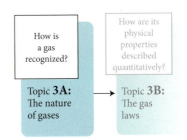

How is a gas recognized?

How are its physical properties described quantitatively?

Topic **3A**: The nature of gases → Topic **3B**: The gas laws

Why Do You Need to Know This Material? An important aspect of chemistry is its ability to explain the properties of bulk matter in terms of the behavior of single molecules. A gas is the simplest state of matter, and so the connections between the properties of individual molecules and those of bulk matter are relatively easy to identify.

What Do You Need to Know Already? You need to be familiar with the concept of force and the procedure for converting between units (*Fundamentals* A).

The most important gas on the planet is the atmosphere, a thin layer of gas held by gravity to the surface of the Earth. If you were to look from a point where the Earth appears to be the size of a basketball, the atmosphere would appear to be only 1 mm thick (**FIG. 3A.1**). Yet this delicate layer is vital to life: it shields us from harmful radiation and supplies substances needed for life, such as oxygen, nitrogen, carbon dioxide, and water.

Eleven elements are gases at room temperature and pressure (**FIG. 3A.2**). So are many compounds with low molar masses, such as carbon dioxide, hydrogen chloride, and organic compounds such as methane, CH_4. All substances that are gases at ordinary temperatures are molecular, except the six noble gases, which are monatomic (consist of single atoms).

3A.1 Observing Gases

Samples of gases large enough to study are examples of **bulk matter,** forms of matter that consist of large numbers of molecules. The properties of bulk matter emerge from the collective behavior of vast numbers of its individual particles. In the case of a gas, for example, when you push on a bicycle pump, you can see that air is **compressible**—that is, it can be confined into a smaller volume. The act of reducing the volume of a sample of gas is called **compression.** The observation that gases are more compressible than solids and liquids suggests that there is a lot of space between the molecules of gases.

You also know from everyday experience—by releasing air from an inflated balloon, for instance—that a gas expands rapidly to fill the space available to it. This observation suggests that the molecules are always moving rapidly and hence can respond quickly to changes in the space available to them. Because the pressure in a balloon is the same in all directions, it can also be inferred that the motion of the molecules is chaotic, not favoring any single direction. A primitive picture of a gas could therefore be as a collection of widely spaced molecules hurtling past each other in ceaseless rapid chaotic motion and changing speed and direction only when they happen to collide.

The fact that gases are readily compressible and immediately fill the space available to them suggests that molecules of gases are widely separated and in ceaseless chaotic motion.

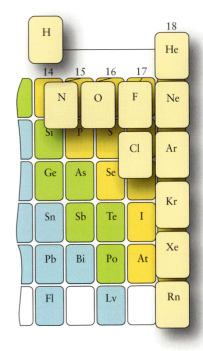

FIGURE 3A.2 The 11 elements that are gases under normal conditions. Note how they lie toward the upper right of the periodic table.

FIGURE 3A.1 The delicate film of the Earth's atmosphere as seen from space. (*Pete Turner/Iconica/Getty Images.*)

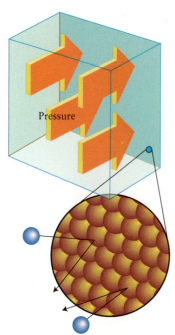

FIGURE 3A.3 The pressure of a gas arises from the collisions that its molecules make with the walls of the container. The storm of collisions, shown in the inset, exerts an almost steady force on the walls.

ANIMATION FIGURE 3A.3

The word barometer comes from the Greek word for weight, referring to the "weight" of the atmosphere.

3A.2 Pressure

If you have ever pumped up a bicycle tire or squeezed an inflated balloon, you have experienced an opposing force arising from the confined air. The **pressure,** P, of a gas is the force, F, exerted by the gas, divided by the area, A, on which the force is exerted:

$$P = \frac{F}{A} \tag{1}$$

The pressure that a gas exerts on the walls of its container results from the collisions of its molecules with the container (**FIG. 3A.3**). The more vigorous the storm of molecules colliding with a surface, the stronger the force and hence the higher the pressure.

The SI unit of pressure is the **pascal** (Pa):

$$1 \text{ Pa} = 1 \text{ kg·m}^{-1}\text{·s}^{-2}$$

A pascal is a rather small unit of pressure: the atmosphere exerts about 100 000 Pa (100 kPa) at sea level; 1 Pa is the pressure that would be exerted by a layer of water only 0.1 mm thick; the pressure of the atmosphere at sea level is equivalent to a layer of water about 10 m deep. Any object on the surface of the Earth stands in an invisible storm of molecules that beat on it incessantly and exert a force all over its surface. Even on an apparently calm day, you are in the midst of a molecular storm.

The pressure exerted by the atmosphere can be measured by various means. When a pressure gauge is used to measure the pressure of air in a tire, the "gauge pressure" is actually the difference between the pressure inside the tire and the atmospheric pressure. A flat tire registers a gauge pressure of zero, because the pressure inside the tire is the same as the pressure of the atmosphere. A laboratory pressure gauge attached to a scientific apparatus, however, measures the *actual* pressure in the apparatus.

The pressure of the atmosphere is measured with a **barometer.** The earliest form of a barometer was invented in the seventeenth century by the Italian scientist Evangelista Torricelli, a student of Galileo. Torricelli (whose name coincidentally means "little tower" in Italian) formed a little tower of liquid mercury. He sealed a long glass tube at one end, filled it with mercury, and inverted it into an open container partially filled with mercury (**FIG. 3A.4**). The column of mercury fell until the pressure that it exerted at its base matched the pressure exerted by the atmosphere. To interpret measurements with a barometer, it is necessary to know how the height of the column of mercury depends on the atmospheric pressure.

How Is That Done?

To find the relation between the height, h, of the column of mercury in a barometer and the atmospheric pressure, P, suppose that the cross-sectional area of the cylindrical column is A. The volume of mercury in the column is the height of the cylinder times this area, $V = hA$. The mass, m, of this volume of mercury is the product of mercury's density, d, and the volume; so $m = dV = dhA$. The mercury is pulled down by the force of gravity, and the total force that its mass exerts at its base is the product of the mass and the acceleration of free fall (the acceleration due to gravity), g: $F = mg$. The pressure at the base of the column is the force divided by the area.

From $P = F/A$, $F = mg$, and $m = dhA$:

$$P = \frac{F}{A} = \frac{\overset{F}{\overbrace{mg}}}{A} = \frac{\overset{m}{\overbrace{dhA}}g}{A} = dhg$$

Area, A
Mass, $m = dhA$
Force, $F = mg$
h

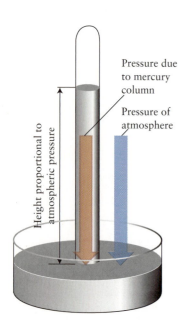

Height proportional to atmospheric pressure

Pressure due to mercury column

Pressure of atmosphere

FIGURE 3A.4 A mercury barometer has traditionally been used to measure the pressure of the atmosphere. The pressure of the atmosphere is balanced by the pressure exerted by the column of mercury, which falls to the appropriate height, leaving a near vacuum above it. The height of the column is proportional to the atmospheric pressure.

The calculation shows that the pressure of air that supports a column of liquid of height h and density d is

$$P = dhg \qquad (2)$$

where g is the acceleration of free fall. The pressure is obtained in pascals when the density, height, and value of g are expressed in SI units. This equation shows that the pressure, P, exerted by a column of mercury is proportional to the height of the column. Therefore, the height of the column can be used as a measure of atmospheric pressure.

THINKING POINT

Could you use a mercury barometer to measure pressure on the International Space Station?

EXAMPLE 3A.1 Calculating atmospheric pressure from the height of a column of mercury

Meteorologists monitor changes in atmospheric pressure to help predict future weather patterns. A drop in pressure is often a sign of an approaching storm system. Suppose the height of the column of mercury in a barometer is 760. mm at 15 °C. What is the atmospheric pressure in pascals? At 15 °C, the density of mercury is 13.595 g·cm^{-3} (corresponding to 13 595 kg·m^{-3}) and the standard acceleration of free fall at the surface of the Earth is 9.806 65 m·s^{-2}.

ANTICIPATE As noted earlier, atmospheric pressure is close to 100 kPa, so you should expect a value close to that figure.

PLAN Substitute the data into Eq. 2 and recognize that $1 \text{ kg·m}^{-1}\text{·s}^{-2} = 1$ Pa.

SOLVE

From $P = dhg$,

$$P = \overbrace{13\,595 \text{ kg·m}^{-3}}^{d} \times \overbrace{0.760 \text{ m}}^{h} \times \overbrace{9.806\,65 \text{ m·s}^{-2}}^{g}$$

$$= 1.01 \times 10^5 \underbrace{\text{kg·m}^{-1}\text{·s}^{-2}}_{\text{Pa}} = 1.01 \times 10^5 \text{ Pa}$$

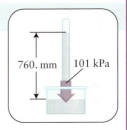

760. mm   101 kPa

EVALUATE This pressure can be reported as 101 kPa. This value is close to the anticipated value. The acceleration of free fall varies over the surface of the Earth and depends on the altitude; however, in all calculations in this text, g is assumed to have the "standard" value used here.

Self-test 3A.1A What is the atmospheric pressure in kilopascals when the height of the mercury column in a barometer at 15 °C is 756 mm?

[*Answer:* 100. kPa]

Self-test 3A.1B The density of water at 20 °C is 0.998 g·cm^{-3}. What height would the column of liquid in a water barometer at 20 °C reach when the atmospheric pressure corresponds to 760. mm of mercury?

Related Exercises 3A.7, 3A.8

Although an electronic pressure gauge is more commonly used to measure the pressure inside a laboratory vessel, a **manometer** is sometimes used (**FIG. 3A.5**). It consists of a U-shaped tube connected to the experimental system. The other end of the tube may be either open to the atmosphere or sealed. For an open-tube manometer (like that shown in Fig. 3A.5a), the pressure in the system is equal to that of the atmosphere when the levels of the liquid in each arm of the U-tube are the same. If the level of mercury on the system side of an open manometer is above that of the atmosphere side, the pressure in the system is lower than the atmospheric pressure. In a closed-tube manometer (like that shown in Fig. 3A.5b), one side is connected to a closed flask (the system) and the other side is vacuum. The difference in heights of the two mercury columns is proportional to the pressure in the system.

The word manometer comes from the Greek word for "thin," referring to a "thin" (low-pressure) atmosphere.

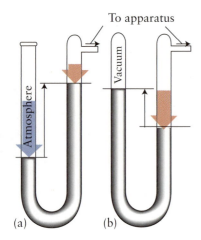

FIGURE 3A.5 (a) An open-tube manometer. The pressure inside the apparatus to which the narrow horizontal tube is connected pushes against the external pressure. In this instance, the pressure inside the system is lower than the atmospheric pressure to an extent proportional to the difference in heights of the liquid in the two arms. (b) A closed-tube manometer. The pressure in the adjoining apparatus is proportional to the difference in heights of the liquid in the two arms. The space inside the closed end is a vacuum.

Self-test 3A.2A What is the pressure in kilopascals in a system when the mercury level in the system-side column in an open-tube mercury manometer is 25 mm lower than the mercury level in the atmosphere-side column and the atmospheric pressure corresponds to 760. mm of mercury at 15 °C?

[*Answer:* 105 kPa]

Self-test 3A.2B What is the pressure in pascals inside a system when a *closed* mercury manometer shows a height difference of 10. cm at 15 °C?

The pressure of a gas, the force that it exerts divided by the area subjected to the force, arises from the impacts of its molecules.

3A.3 Alternative Units of Pressure

Although the SI unit of pressure is the pascal (Pa), several other units are commonly used. Normal atmospheric pressure is close to 100 kPa, and it is useful to have a unit, the **bar,** to denote exactly 100 kPa:

$$1 \text{ bar} = 10^5 \text{ Pa}$$

Weather charts commonly report pressures in millibars (1 mbar = 10^{-3} bar = 10^2 Pa), with typical atmospheric pressures close to 1000 mbar. The **standard pressure** for reporting data is now defined as 1 bar exactly and denoted P°.

A pressure of 1 bar is close to a traditional and still widely used unit based on the pressure typically exerted by the atmosphere at sea level and known as the **atmosphere** (atm). This unit is now defined by the exact relation

$$1 \text{ atm} = 1.013\ 25 \times 10^5 \text{ Pa}$$

Note that 1 atm = 1.013 25 bar, and so 1 atm is slightly greater than 1 bar. The use of mercury barometers in the past led to the use of the height of a mercury column (in millimeters) being used to report pressure, and expressed in **millimeters of mercury** (mmHg). This unit is defined in terms of the pressure exerted by a column of mercury exactly 1 mm high under certain conditions (15 °C and in a standard gravitational field), but it has now been largely displaced by a unit of very similar magnitude, the **torr** (Torr). This unit is defined by the exact relation

$$1 \text{ Torr} = \frac{1}{760} \text{ atm, so 1 atm} = 760 \text{ Torr}$$

It is normally safe to use mmHg and Torr interchangeably (so you will often see 1 atm reported as 760 mmHg), but in very precise work you should be aware that they are not exactly the same (they differ by less than 1 part in a million).

A Note on Good Practice: The name of the unit torr, like all names derived from the names of people, has a lowercase initial letter, as in the unit pascal. However, the symbol for the unit, like all symbols derived from proper names, has an uppercase initial letter (Torr, Pa).

Pressure units are summarized in **TABLE 3A.1**. It is important to be familiar with them and to be able to make conversions between them. For example, a pressure of 746 Torr is converted into a pressure in pascals by using a conversion factor derived from Table 3A.1:

$$P = 746 \text{ Torr} \times \frac{133.322 \text{ Pa}}{1 \text{ Torr}} = 9.95 \times 10^4 \text{ Pa or } 99.5 \text{ kPa}$$

The actual pressure exerted by the atmosphere varies with altitude and weather. The pressure of the atmosphere at the cruising height of a commercial jetliner (10 km) is only about 200 Torr (about 0.3 atm), and so airplane cabins must be pressurized. A region of very low atmospheric pressure, such as an area of low pressure on the weather chart in **FIG. 3A.6**, typically has a pressure of about 0.98 atm at sea level. A typical region of high pressure is about 1.03 atm.

TABLE 3A.1 Pressure Units*

SI unit: pascal (Pa)
1 Pa = 1 kg·m^{-1}·s^{-2} = 1 N·m^{-2}

Conventional units
1 bar = 10^5 Pa = **100** kPa
1 atm = **1.013 25 × 10^5** Pa
  = **101.325** kPa

1 Torr = **1/760** atm = 133.322... Pa[†]
1 mmHg = 133.322... Pa ≈ 1 Torr
1 atm = 14.7 lb·in.$^{-2}$ (psi)

*Figures in **bold** are exact. See inside back cover for more relations. N denotes newton (1 N = 1 kg·m·s^{-2}).
[†]1 mmHg is the same as 1 Torr to within 1 part in 10^7.

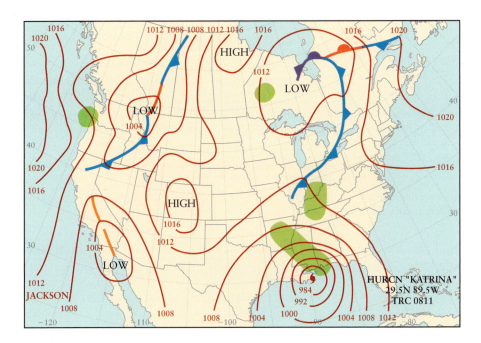

FIGURE 3A.6 A weather map of North America during Hurricane Katrina in 2005. The curves are contours of constant atmospheric pressure called *isobars*. Regions of low pressure (seen over southern Louisiana) are called *cyclones*, and regions of high pressure (seen over Colorado) are called *anticyclones*. All pressures are in millibars. The lowest pressure on the map is 984 mbar, at the center of the hurricane. Before making landfall, the pressure at the center of the hurricane had fallen to as low as 902 mbar. (*NOAA*).

Self-test 3A.3A The U.S. National Hurricane Center reported that the pressure in the eye of Hurricane Katrina (2005) fell to as low as 902 mbar. What is that pressure in atmospheres?
[***Answer:*** 0.890 atm]

Self-test 3A.3B The atmospheric pressure in Denver, Colorado, on a certain day was 630. Torr. Express this pressure in pascals.

The principal units for reporting pressure are torr, atmosphere, bar, and (the SI unit) pascal. Pressure units are interconverted by using the information in Table 3A.1.

What have you learned in this Topic?

You have learned that the properties of gases suggest that their molecules are in constant random motion and that the pressure exerted by gases arises from the impacts of the molecules with the walls of their containers.

The skills you have mastered are the ability to:

☐ **1.** Calculate the pressure at the base of a vertical column of liquid (Example 3A.1).

☐ **2.** Interpret a manometer reading (Self-Test 3A.2).

☐ **3.** Convert between the various units of pressure (Self-Test 3A.3)

Topic 3A Exercises

3A.1 The pressure needed to make synthetic diamonds from graphite is 8×10^4 atm. Express this pressure in (a) Pa; (b) kbar; (c) Torr; (d) lb·in.$^{-2}$.

3A.2 The pressure recorded for an argon gas cylinder is 35.0 lb·in.$^{-2}$. Convert this pressure into (a) kPa; (b) Torr; (c) bar; (d) atm.

3A.3 People sometimes experience altitude sickness when they arrive at mountain ski resorts at high altitudes as a result of the lower pressure of the air. Why is the pressure of the atmosphere lower at higher altitudes?

3A.4 How would the height of a mercury barometer be different on the planet Mars? Explain your reasoning.

3A.5 A student attaches a glass bulb containing neon gas to an open-tube manometer (like the one in Fig. 3A.5) and calculates the pressure of the gas to be 0.890 atm. (a) If the atmospheric pressure is 762 Torr, what height difference between the two sides of the mercury in the manometer did the student find? (b) Which side is higher, the side of the manometer attached to the bulb or the side open to the atmosphere? (c) If the student mistakenly switches the numbers for the sides of the manometer when recording the data in the laboratory notebook, what would be the reported pressure in the gas bulb?

3A.6 A reaction is performed in a vessel attached to a closed-tube manometer. Before the reaction, the levels of mercury in the two sides of the manometer were at the same height. As the reaction proceeds, a gas is produced. At the end of the reaction, the height of the mercury column on the vacuum side of the manometer has risen by 24.32 cm and the height on the side of the manometer connected to the flask has fallen by the same amount. What is the pressure in the apparatus at the end of the reaction expressed in (a) Torr; (b) atm; (c) Pa; (d) bar?

3A.7 Suppose you were marooned on a tropical island and had to use seawater (density 1.10 g·cm^{-3}) to make a primitive barometer. What height would the water reach in your barometer when a mercury barometer would reach 74.7 cm? The density of mercury is 13.6 g·cm^{-3}.

3A.8 An unknown liquid is used to fill a closed-tube manometer. The atmosphere is found to produce a height difference of 6.14 m in this manometer at the same time that a mercury manometer gives a displacement of 758.7 mm. What is the density of the unknown liquid?

3A.9 Assume that the width of your body (across your shoulders) is 20. in. and the depth of your body (chest to back) is 10. in. If atmospheric pressure is 14.7 lb·in.$^{-2}$, what mass of air does your body support when you are in an upright position?

3A.10 Low-pressure gauges in research laboratories are occasionally calibrated in inches of water (inH$_2$O). Given that the density of mercury at 15 °C is 13.6 g·cm^{-3} and the density of water at that temperature is 1.0 g·cm^{-3}, what is the pressure (in Torr) inside a gas cylinder that reads 8.9 inH$_2$O at 15 °C?

Topic 3B The Gas Laws

3B.1 The Experimental Observations
3B.2 Applications of the Ideal Gas Law
3B.3 Molar Volume and Gas Density

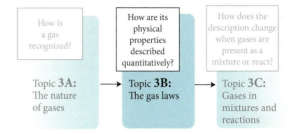

How is a gas recognized?

How are its physical properties described quantitatively?

How does the description change when gases are present as a mixture or react?

Topic **3A:** The nature of gases

Topic **3B:** The gas laws

Topic **3C:** Gases in mixtures and reactions

The Anglo-Irish scientist Robert Boyle made the first reliable measurements of the properties of gases in 1662 when he examined the effect of pressure on volume. A century and a half later, a new pastime, hot-air ballooning, motivated two French scientists, Jacques Charles and Joseph-Louis Gay-Lussac, to formulate additional gas laws. They measured how the temperature of a gas affects its pressure, volume, and density. Later in the nineteenth century, the Italian scientist Amedeo Avogadro made a further contribution that established the relation between the volume and the number of molecules in the sample and thereby helped to establish belief in the reality of atoms. Their findings have been combined into one simple, yet very useful, equation.

Why Do You Need to Know This Material? The equations in this Topic are used in many areas of chemistry, both for practical calculations and in the development of thermodynamics and its application to chemical equilibrium.

What Do You Need to Know Already? You need to be familiar with the concepts of pressure (Topic 3A) and the amount of substance (*Fundamentals* E).

3B.1 The Experimental Observations

Boyle took a long tube of glass curved into a J-shape, with the short end sealed (**FIG. 3B.1**). He then poured mercury into the tube, trapping air in the short end of the J. The more mercury he added, the more the air was compressed. He concluded that at constant temperature the volume of a fixed amount of gas (the air in this case) decreases as the pressure on it increases. **FIGURE 3B.2** shows a graph of the dependence. The curve shown there is called an **isotherm,** which is a general term for a constant-temperature plot. An **isothermal change** is one that takes place at constant temperature. Scientists often look for ways of plotting experimental data in a manner that gives straight lines, because such graphs are easier to identify, analyze, and interpret. Boyle's data give a straight line when the pressure is plotted against 1/volume (**FIG. 3B.3**). This result implies

> **Boyle's law:** For a fixed amount of gas at constant temperature, volume is inversely proportional to pressure.

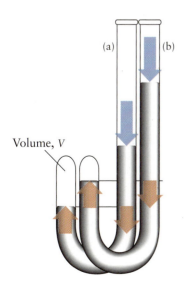

FIGURE 3B.1 (a) In Boyle's experiment, a gas was trapped by mercury inside the closed end of a J-shaped tube. (b) The volume of the trapped gas decreased as the pressure on it was increased by adding more mercury to the open end of the tube.

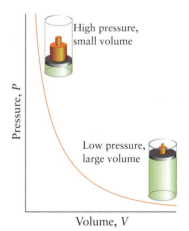

FIGURE 3B.2 Boyle's law summarizes the effect of pressure on the volume of a fixed amount of gas at constant temperature. As the pressure of a gas sample is increased (as depicted by the increasing size of the weight pressing on the piston), the volume of the gas decreases.

LIVING GRAPH FIGURE 3B.2

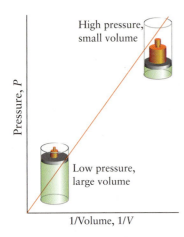

High pressure, small volume

Low pressure, large volume

Pressure, P

1/Volume, 1/V

FIGURE 3B.3 When the pressure is plotted against 1/volume, a straight line is obtained. Boyle's law breaks down at high pressures, and a straight line is not obtained in these regions (not shown).

Boyle's law is written

$$\text{Volume} \propto \frac{1}{\text{pressure}} \quad \text{or} \quad V = \frac{\text{constant}}{P} \quad \text{(at constant } n \text{ and } T)$$

Equivalently, by multiplying both sides by P,

$$PV = \text{constant (at constant } n \text{ and } T) \tag{1a}$$

Suppose the pressure and volume of a fixed amount of gas at the start of an experiment are P_1 and V_1, then at the start $P_1V_1 = $ constant. At the end of the experiment the pressure and volume have become P_2 and V_2, but P_2V_2 is equal to the same constant (provided the temperature has not changed). It follows that yet another form of Boyle's law is

$$P_2V_2 = P_1V_1 \text{ (at constant } n \text{ and } T) \tag{1b}$$

Self-test 3B.1A A sample of neon of volume 10.0 L at 300. Torr is allowed to expand isothermally into an evacuated tube with a volume of 20.0 L. What is the final pressure of the neon in the tube?

[***Answer:*** 150. Torr]

Self-test 3B.1B In a petroleum refinery, a container of volume 750. L containing ethylene gas at 1.00 bar was compressed isothermally until the pressure rose to 5.00 bar. What was the final volume of the container?

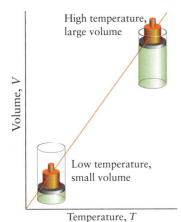

High temperature, large volume

Low temperature, small volume

Volume, V

Temperature, T

FIGURE 3B.4 When the temperature of a gas is increased and its volume is free to change at constant pressure (as depicted by the constant weight acting on the piston), the volume increases. A plot of volume against temperature is a straight line.

LIVING GRAPH FIGURE 3B.4

The Kelvin scale of temperature is described in Appendix 1B.

Charles and Gay-Lussac carried out a number of experiments with the hope of improving the performance of their balloons. They found that, provided the pressure is kept constant, for a fixed amount of gas the volume of a gas increases as its temperature is raised. In this case, a straight-line graph is obtained when the volume is plotted against the temperature (**FIG. 3B.4**). This result implies

Charles's law: For a fixed amount of gas under constant pressure, the volume varies linearly with the temperature.

Gay-Lussac's name is sometimes associated with the law, but "Charles's law" is now more common.

Charles's law has a very important implication. When the straight lines obtained by plotting volume against temperature for different gases and at various pressures are extrapolated (that is, extended beyond the range of the data), it is found that each line reaches zero volume at the same temperature, specifically −273.15 °C (**FIG. 3B.5**). Extrapolation is essential here because this point cannot be reached in practice: no real gas has zero volume and all real gases condense to a liquid before such low temperatures are reached. Because a volume cannot be negative, −273.15 °C must be the lowest possible temperature. It is the value corresponding to zero on the **Kelvin scale.** The SI unit of temperature is the **kelvin** (K, not °K). The sizes of the gradations on the Kelvin and Celsius scales are the same, so a change of 1 K is equal to a change of 1 °C. To convert from the Celsius scale to the Kelvin scale, add 273.15 to the temperature on the Celsius scale. It follows that, when the absolute temperature, T, is used, Charles's law becomes

$$\text{Volume} \propto \text{absolute temperature} \quad \text{or} \quad V = \text{constant} \times T \text{ (at constant } n \text{ and } P) \tag{2a}$$

Wherever the temperature is denoted T in this book, it denotes the absolute temperature, with the lowest attainable value at $T = 0$. A similar expression summarizes the linear

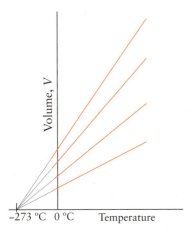

Volume, V

−273 °C   0 °C   Temperature

FIGURE 3B.5 The extrapolation of data like those in Fig. 3B.4 for a number of gases suggests that the volume of each gas should become 0 at $T = 0$ (−273 °C). Extrapolated data are shown as dotted lines. In practice, all gases condense to liquids before that temperature is reached.

variation in pressure of a sample of gas when it is heated in a container of fixed volume. The pressure is found experimentally to extrapolate to zero pressure at −273.15 °C (**FIG. 3B.6**). Therefore, as Gay-Lussac was able to confirm experimentally,

Pressure ∝ absolute temperature   or $P = \text{constant} \times T$ (at constant n and V)   **(2b)**

It follows that doubling the absolute temperature doubles the pressure of a gas, provided the amount and volume are constant.

> **A Note on Good Practice:** Note that for a fixed amount of gas the volume (or the pressure) doubles when the temperature is doubled on the absolute (Kelvin) scale, not when it is doubled on the Celsius scale. An increase from 20 °C to 40 °C corresponds to an increase from 293 K to 313 K, an increase of only 7%.

> **Self-test 3B.2A** A rigid oxygen tank stored outside a building has a pressure of 20.00 atm at 6:00 AM, when the temperature is 10 °C. What will be the pressure in the tank at 6:00 PM, when the temperature is 30. °C?
>
> [***Answer:*** 21.4 atm]
>
> **Self-test 3B.2B** A sample of hydrogen gas at 760. Torr and 20. °C is heated to 300. °C in a container of constant volume. What is the final pressure of the sample?

A further contribution was made by the Italian scientist Amedeo Avogadro:

Avogadro's principle: Under the same conditions of temperature and pressure, a given number of gas molecules occupy the same volume regardless of their chemical identity.

Avogadro's principle is commonly expressed in the terms of the **molar volume,** V_m, the volume occupied per mole of molecules:

$$\text{Molar volume} = \frac{\text{volume}}{\text{amount}} \quad \text{or} \quad V_m = \frac{V}{n} \tag{3a}$$

which can be expressed as:

$$V = nV_m \tag{3b}$$

The molar volumes of gases are all close to 22 L·mol⁻¹ at 0 °C and 1 atm (**FIG. 3B.7**).

> **Self-test 3B.3A** A helium weather balloon was filled at −20. °C and a certain pressure to a volume of 2.5×10^4 L with 1.2×10^3 mol He. What is the molar volume of helium under those conditions?
>
> [***Answer:*** 21 L·mol⁻¹]
>
> **Self-test 3B.3B** A large natural-gas storage tank contains 200. mol $CH_4(g)$ at 1.20 atm. An additional 100. mol $CH_4(g)$ is pumped into the tank at constant temperature. What is the final pressure in the tank?

All the properties that have been introduced so far are consistent with the model of a gas as a collection of widely spaced molecules in ceaseless motion. Boyle's law is consistent with this model because compression increases the number of molecules in a given region of the sample and hence increases the number of collisions that the molecules make with the walls. As a result, the pressure the molecules exert increases (**FIG. 3B.8**). The effect of temperature on the pressure of a gas in a constant-volume container suggests a new feature of a gas: *as the temperature of a gas is raised, the average speed of the molecules increases.* As a result of this increase in average speed, each molecule strikes the walls more often and with greater force. Therefore, the gas exerts a greater pressure as the temperature is increased at constant volume. The same model of a gas can be used to explain the effect of temperature on the volume when the pressure is held constant. To counteract the increase in pressure as the temperature is raised and molecular speed

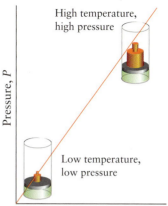

FIGURE 3B.6 The pressure of a fixed amount of gas in a vessel of constant volume is proportional to the absolute temperature. Notice the increase in the pressure of the gas in the cylinder (represented by the increasing mass of the weight on the piston) as the temperature increases. The pressure extrapolates to 0 at $T = 0$.

 LIVING GRAPH FIGURE 3B.6

Avogadro's principle is not a law because it is based not on observation alone but also on a model of matter—namely, that matter consists of molecules. Even though there is no longer any doubt that matter consists of atoms and molecules, it remains a principle rather than a law.

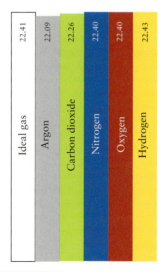

FIGURE 3B.7 The molar volumes (in liters per mole) of various gases at 0 °C and 1 atm. The values are all very similar and close to the molar volume of an ideal gas under these conditions, 22.41 L·mol⁻¹.

increases, the space available to the gas must increase so that fewer molecules are available to strike the walls in a given time interval. Finally, Avogadro's principle is consistent with the model: as more molecules are added to a container, in order for the pressure to remain constant the size of the container must increase.

The three properties of a gas expressed by Eqs. 1, 2, and 3 can be combined into a single expression relating pressure (P), volume (V), temperature (T), and amount (n) of a gas:

$$PV = \text{constant} \times nT$$

Thus, if the temperature and amount are constant, PV is a constant (Boyle's law); if the amount and pressure are constant, V is proportional to T (Charles's law); and if the pressure and temperature are constant, the volume is proportional to n (Avogadro's principle). When the constant of proportionality is written as R, this expression becomes the **ideal gas law:**

$$PV = nRT \tag{4}$$

The constant R is called the **gas constant** and is "universal" in the sense that it has the same value for all gases. Its value can be found by measuring P, V, n, and T and substituting their values into $R = PV/nT$. When SI units (pressure in pascals, volume in meters cubed, temperature in kelvins, and amount in moles) are used, R is obtained in joules per kelvin per mole: $R = 8.3145\ \text{J·K}^{-1}\text{·mol}^{-1}$. **TABLE 3B.1** lists the values of R when the volume and pressure are reported in other units.

The ideal gas law is an example of an **equation of state,** an expression showing how the pressure of a substance—in this case, a gas—is related to its temperature, volume, and amount of substance. A hypothetical gas that obeys the ideal gas law under all conditions is called an **ideal gas.** All real gases are found to obey Eq. 4, with increasing accuracy as the pressure is reduced toward zero (written $P \to 0$). Therefore, the ideal gas law is an example of a **limiting law,** a law that is strictly valid only in some limit—in this case, as $P \to 0$. Although the ideal gas law is a limiting law, it is in fact reasonably reliable at atmospheric pressures, and so it can be used to describe the behavior of most gases under normal laboratory conditions.

The ideal gas law, PV = nRT, summarizes the relation between the pressure, volume, temperature, and amount of molecules of an ideal gas and is used to assess the effect of changes in any of these properties; it is an example of a limiting law.

3B.2 Applications of the Ideal Gas Law

The individual gas laws can be used to make predictions when only one variable is changed, such as heating a fixed amount of gas at constant volume. The ideal gas law enables predictions to be made when two or more variables are changed.

To predict the result of changing more than one variable, note that, if the initial conditions of a gas are n_1, P_1, V_1, and T_1, then from Eq. 4, $P_1V_1 = n_1RT_1$ or $P_1V_1/n_1T_1 = R$. After the change, the conditions become n_2, P_2, V_2, and T_2; because the ideal gas law still applies, $P_2V_2/n_2T_2 = R$. Because R is a constant, P_1V_1/n_1T_1 and P_2V_2/n_2T_2 can be equated, to give

$$\underbrace{\frac{P_1V_1}{n_1T_1}}_{\text{initial conditions}} = \underbrace{\frac{P_2V_2}{n_2T_2}}_{\text{final conditions}} \tag{5}$$

This expression is called the **combined gas law.** However, it is a direct consequence of the ideal gas law and is not a new law.

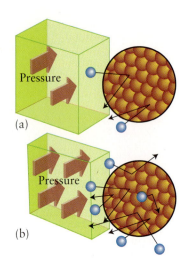

FIGURE 3B.8 (a) The pressure of a gas arises from the impact of its molecules on the walls of the container. (b) When the volume of the sample is decreased, there are more molecules in a given volume and so there are more collisions with the same area of the wall in a given time interval. Because the impact on the walls is now greater, so is the pressure.

 **ANIMATION FIGURE 3B.8**

TABLE 3B.1 The Gas Constant, R*
$8.205\ 74 \times 10^{-2}\ \text{L·atm·K}^{-1}\text{·mol}^{-1}$
$8.314\ 46 \times 10^{-2}\ \text{L·bar·K}^{-1}\text{·mol}^{-1}$
$8.314\ 46\ \text{L·kPa·K}^{-1}\text{·mol}^{-1}$
$8.314\ 46\ \text{J·K}^{-1}\text{·mol}^{-1}$
$62.364\ \text{L·Torr·K}^{-1}\text{·mol}^{-1}$

* The gas constant is related to Boltzmann's constant, k, by $R = N_A k$, where N_A is Avogadro's constant.

EXAMPLE 3B.1   Calculating the pressure of a given sample

In a plasma display panel (PDP), an electrical discharge causes the ionization of gases, producing UV radiation that strikes a phosphor, which generates red, green, or blue light depending on which phosphor coats the cell. Millions of such cells are combined to create an image. Estimate the pressure (in atmospheres) inside a single cell of a PDP, given that the volume of the cell is $0.030\ \text{mm}^3$, its temperature is $34\ ^\circ\text{C}$, and it contains $9.6\ \text{ng}$ of neon gas.

PLAN Use the ideal gas law, Eq. 4, to solve for pressure.

SOLVE

Step 1 Express the temperatures in kelvins, the amount in moles, and the volume in liters. Convert mass to amount using $n = m/M$ and temperature from degrees Celsius to kelvins by adding 273.15.

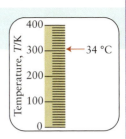

$$n = \frac{\overset{9.6 \text{ ng}}{\overbrace{9.6 \times 10^{-9} \text{ g}}}}{20.18 \text{ g·mol}^{-1}} = 4.7\ldots \times 10^{-10} \text{ mol}$$

$$T = (34 + 273.15) \text{ K} = 307 \text{ K}$$

$$V = 0.030 \text{ mm}^3 \times \frac{1 \text{ L}}{10^6 \text{ mm}^3} = 3.0 \times 10^{-8} \text{ L}$$

Step 2 Rearrange the equation $PV = nRT$ to give the desired quantity on the left and all other quantities on the right,

$$P = \frac{nRT}{V}$$

Step 3 Substitute the data, using R in $L \cdot atm \cdot K^{-1} \cdot mol^{-1}$.

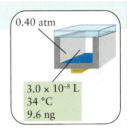

$$P = \frac{\overset{n}{\overbrace{4.7\ldots \times 10^{-10} \text{ mol}}} \times \overset{R}{\overbrace{8.205\,74 \times 10^{-2} \text{ L·atm·K}^{-1}\text{·mol}^{-1}}} \times \overset{T}{\overbrace{307 \text{ K}}}}{\underset{V}{\underbrace{3.0 \times 10^{-8} \text{ L}}}}$$

$$= 0.40 \text{ atm}$$

Self-test 3B.4A Calculate the pressure (in kilopascals) exerted by 1.0 g of carbon dioxide in a flask of volume 1.0 L at 300 °C.
[**Answer:** 1.1×10^2 kPa]

Self-test 3B.4B An idling, badly tuned automobile engine can release as much as 1.00 mol CO per minute into the atmosphere. At 27 °C, what volume of CO, adjusted to 1.00 atm, is emitted per minute?

Related Exercises 3B.5, 3B.6, 3B.25–3B.28

EXAMPLE 3B.2   Using the combined gas law when one variable is changed

In an automobile engine, the piston compresses a mixture of gasoline vapor and air before ignition occurs. The combined gas law can be used to calculate the new pressure. Assume that, when the piston is pushed in, the volume inside the chamber is decreased from 100. cm³ to 20. cm³ before ignition. Suppose that the compression is isothermal; estimate the final pressure of the compressed gas mixture, given an initial pressure of 1.00 atm.

ANTICIPATE The volume is reduced by a factor of 5 at constant temperature, so you should expect a fivefold increase in pressure.

PLAN Use the combined gas law, Eq. 5. Only the pressure and volume change, so all other variables cancel, resulting in Boyle's law.

SOLVE

Step 1 Rearrange $P_1V_1/n_1T_1 = P_2V_2/n_2T_2$ to find P_2 by multiplying both sides by n_2T_2/V_2. Set $n_2 = n_1$ (no change in the amount) and $T_2 = T_1$ (no change in temperature) and cancel these quantities.

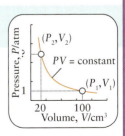

$$P_2 = \frac{P_1V_1}{n_1T_1} \times \frac{\overset{n_2 = n_1}{\overset{T_2 = T_1}{n_2T_2}}}{V_2} \xrightarrow{\hspace{1cm}} P_2 = \frac{P_1V_1}{V_2}$$

Step 2 Substitute the data:

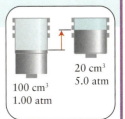

$$P_2 = \frac{(1.00 \text{ atm}) \times (100 \text{ cm}^3)}{20 \text{ cm}^3} = 5.0 \text{ atm}$$

EVALUATE The final pressure is higher by a factor of 5 (more precisely, 5.0), as expected.

Self-test 3B.5A A sample of argon gas of volume 10.0 mL at 200. Torr is allowed to expand isothermally into an evacuated tube with a volume of 0.200 L. What is the final pressure of the argon in the tube?

[*Answer:* 10.0 Torr]

Self-test 3B.5B A sample of dry air in the cylinder of a test engine at 80. cm^3 and 1.00 atm is compressed isothermally until the pressure has risen to 3.20 atm by pushing a piston into the cylinder. What is the final volume of the sample?

Related Exercises 3B.9–3B.14, 3B.23, 3B.24

EXAMPLE 3B.3   Using the combined gas law when two variables are changed

If you were designing an air-conditioning system in which a cooling effect is achieved by the expansion of a gas, you might need to assess the pressure changes of the gas as it underwent changes in both temperature and volume. A sample of gas of volume 500. mL at 28.0 °C was found to exert a pressure of 92.0 kPa. What pressure will the sample exert when it is compressed to 300. mL and cooled to −5.0 °C?

ANTICIPATE Your problem is to judge whether the compression, which will increase the pressure, dominates the cooling, which will decrease the pressure. On the Kelvin scale, the change in temperature is quite small, so you should expect compression to dominate.

PLAN Use the combined gas law, Eq. 5, noting that only the amount is the same in the two states.

SOLVE

Step 1 Rearrange $P_1V_1/n_1T_1 = P_2V_2/n_2T_2$ to find P_2, as in Example 3B.2, but only set $n_1 = n_2$ and cancel. Express the temperatures in kelvins.

$$P_2 = \frac{P_1V_1}{n_1T_1} \times \frac{n_2T_2}{V_2} \xrightarrow{n_1 = n_2} P_2 = \frac{P_1V_1T_2}{T_1V_2}$$

$$T_1 = (273.15 + 28.0) \text{ K} = 301.2 \text{ K}$$

$$T_2 = (273.15 - 5.0) \text{ K} = 268.2 \text{ K}$$

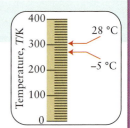

Step 2 Substitute the data:

$$P_2 = \frac{(92.0 \text{ kPa}) \times (500. \text{ mL}) \times (268.2 \text{ K})}{(301.2 \text{ K}) \times (300. \text{ mL})} = 137 \text{ kPa}$$

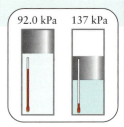

EVALUATE The net outcome is an increase in pressure, so, in this instance, compression has a greater effect than cooling, as suspected.

Self-test 3B.6A A parcel of air (the technical term in meteorology for a small region of the atmosphere) of volume 1.00 × 10^3 L at 20. °C and 1.00 atm rises up the side of a mountain range. At the summit, where the pressure is 0.750 atm, the parcel of air has cooled to −10. °C. What is the volume of the parcel at that point?

[*Answer:* 1.20 × 10^3 L]

Self-test 3B.6B A weather balloon is filled with helium gas at 20. °C and 1.00 atm. The volume of the balloon is 250. L. When the balloon rises to a layer of air where the temperature is −30. °C, it has expanded to 800. L. What is the pressure of the atmosphere at that point?

Related Exercises 3B.17, 3B.18, 3B.29, 3B.30

The combined gas law describes how a gas responds to changes in conditions.

3B.3 Molar Volume and Gas Density

The ideal gas law can be used to predict the molar volume of an ideal gas under any conditions of temperature and pressure. To do so, Eq. 3a ($V_m = V/n$) and Eq. 4 (as $V = nRT/P$) are combined by writing

$$V_m = \frac{V}{n} = \frac{nRT/P}{n} = \frac{RT}{P} \tag{6}$$

At **standard ambient temperature and pressure** (SATP), which means exactly 25 °C (298.15 K) and exactly 1 bar, the conditions commonly used to report data in chemistry, the molar volume of an ideal gas is 24.79 L·mol^{-1}, which is about the volume of a cube 1 ft on a side (**FIG. 3B.9**). The common expression **standard temperature and pressure** (STP) means 0 °C and 1 atm (both exactly), the conditions formerly used to report data and still widely used in some calculations. At STP, the molar volume of an ideal gas is 22.41 L·mol^{-1}. Note the slightly smaller value: the temperature is lower and the pressure is slightly higher, and so the same amount of gas molecules occupy a smaller volume than at SATP.

   TABLE 3B.2 gives values of the molar volume of an ideal gas under a variety of common conditions. To obtain the volume of a known amount of gas at a specified temperature and pressure, simply multiply the molar volume *at that temperature and pressure* by the amount in moles ($V = nV_m$, with $n = m/M$).

FIGURE 3B.9 The transparent cube is the volume (25 L) occupied by 1 mol of ideal gas molecules at 25 °C and 1 bar. (© 2001 Richard Megna–Fundamental Photographs.)

Self-test 3B.7A Calculate the volume occupied by 1.0 kg of hydrogen at 25 °C and 1.0 atm.
[***Answer:*** 1.2 × 10⁴ L]

Self-test 3B.7B Calculate the volume occupied by 2.0 g of helium at 25 °C and 1.0 atm.

TABLE 3B.2 Molar Volume of an Ideal Gas, V_m/(L·mol^{-1})

Temperature	Pressure	
	1 atm	1 bar
0 °C	22.4140	22.7110
25 °C	24.4654	24.7896

At $T = 0$, $V_m = 0$.

Because at low pressures any gas tends to follow the ideal gas law, that law can be used to calculate the density of a gas or to determine its molar mass from a known density. As described in *Fundamentals* G, the molar concentration of a substance is the amount of molecules (*n*, in moles) divided by the volume that they occupy (*V*). It follows from the ideal gas law that, for a gas behaving ideally (so that $n = PV/RT$ is valid),

$$\text{Molar concentration} = \frac{\text{amount}}{\text{volume}} = \frac{n}{V} = \frac{PV/RT}{V} = \frac{P}{RT} \overset{\text{Eq.6}}{=} \frac{1}{V_m} \tag{7}$$

The mass density, *d*, of a gas, commonly simply "density," like that of any substance, is the mass of the sample divided by its volume, $d = m/V$, and (for gases) is typically reported in grams per liter. The density of air, for instance, is about 1.6 g·L^{-1} at SATP. The density is inversely proportional to the molar volume and, at a given temperature, is proportional to the pressure:

$$\text{Density} = \frac{\text{mass}}{\text{volume}} = \frac{m}{V} = \frac{nM}{nV_m} = \frac{M}{V_m} = \frac{MP}{RT} \tag{8}$$

Equation 8 shows that:

- For a given pressure and temperature, the greater the molar mass of a gas, the greater is its density.
- At constant temperature, the density of a gas increases with pressure (the pressure is increased either by adding more of the gas or by reducing the volume).
- Raising the temperature of a gas that is free to expand at constant pressure increases the volume occupied by the gas and therefore reduces its density.

Equation 8 is the basis for using density measurements to determine the molar mass of a gas or vapor.

THINKING POINT

Why do hot-air balloons float in air?

EXAMPLE 3B.4   Calculating the molar mass of a gas from its density

Many compounds used in the fragrance industry are derived from plant extracts. One step in identifying a desired compound is the determination of its molar mass. The volatile organic compound geraniol is a component of oil of roses. The density of the vapor at 260. °C and 103 Torr is 0.480 g·L^{-1}. What is the molar mass of geraniol?

ANTICIPATE Because the compound is volatile, you should anticipate that it will have a moderately low molar mass.

PLAN List the information given and convert the temperature into an absolute value in kelvins. Then rearrange Eq. 8 into an expression for M, select a value of R with appropriate units, and substitute the data.

SOLVE

Assemble the data:

$$d = 0.480 \text{ g·L}^{-1}$$
$$P = 103 \text{ Torr}$$
$$T = (273.15 + 260.) \text{ K} = 533 \text{ K}$$

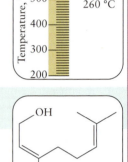

Rearrange $d = MP/RT$ into $M = dRT/P$ and substitute the data, selecting a value of R expressed in torr and liters:

$$M = \frac{\overbrace{0.480 \text{ g·L}^{-1}}^{d} \times \overbrace{62.36 \text{ L·Torr·K}^{-1}\text{·mol}^{-1}}^{R} \times \overbrace{533 \text{ K}}^{T}}{\underbrace{103 \text{ Torr}}_{P}} = 155 \text{ g·mol}^{-1}$$

154.25 g·mol^{-1}

EVALUATE This value, 155 g·mol^{-1}, is close to the value that could be calculated from the molecular formula (154.25 g·mol^{-1}, see the illustration).

Self-test 3B.8A The pharmaceutical chrysarobin was isolated from plants used by Zuni healers to treat skin diseases. At 213 °C and 64.5 Torr, a sample of chrysarobin vapor had a density of 0.511 g·L^{-1}. Calculate the molar mass of chrysarobin.
[*Answer:* 240. g·mol^{-1}]

Self-test 3B.8B The *Codex Ebers*, an Egyptian medical papyrus, describes the use of garlic as an antiseptic. Chemists have now verified that the oxide of diallyl disulfide (the volatile compound responsible for garlic odor) is a powerful antibacterial agent. At 177 °C and 200. Torr, a sample of diallyl disulfide vapor has a density of 1.04 g·L^{-1}. What is the molar mass of diallyl disulfide?

Related Exercises 3B.38, 3B.39, 3B.41

Standard ambient temperature and pressure (SATP) is 25 °C (298.15 K) and 1 bar; standard temperature and pressure (STP) is 0 °C (273.15 K) and 1 atm. The molar concentrations and densities of gases increase as they are compressed but decrease as they are heated. The density of a gas depends on its molar mass.

What have you learned in this Topic?

You have learned that the pressure of a gas is directly proportional to the temperature and amount but inversely proportional to its volume. You have started to see that the behavior of gases at low pressures obeys the ideal gas law, $PV = nRT$. You have seen how to calculate the molar volume and density of an ideal gas by using the ideal gas law and have learned that the density of a gas at a given temperature and pressure is directly proportional to its molar mass.

The skills you have mastered are the ability to:

☐ **1.** Use the gas laws to calculate P, V, T, or n for given conditions (Example 3B.1).

☐ **2.** Use the gas laws to calculate P, V, T, or n after a change in conditions (Examples 3B.2 and 3B.3).

☐ **3.** Use the ideal gas law to calculate the molar volume and density of a gas under given conditions (Section 3B.3).

Topic 3B Exercises

3B.1 Robert Boyle measured pressure in inches of mercury (inHg). On a day when the atmospheric pressure was 29.85 inHg, he trapped some air in the tip of a J-tube (**1**) and measured the difference in height of the mercury in the two arms of the tube (*h*). When *h* = 12.0 in., the height of the gas in the tip of the tube was 32.0 in.. Boyle then added additional mercury and the level rose in both arms of the tube so that *h* = 30.0 in. (**2**). (a) What was the height of the air space (in inches) in the tip of the tube in (**2**)? (b) What was the pressure of the gas in the tube in (**1**) and in (**2**) in inches of mercury?

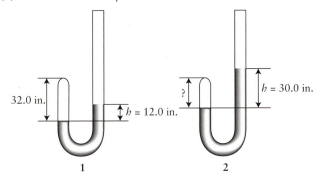

3B.2 Boyle continued to add mercury to the apparatus in Exercise 3B.1 until the height of the trapped air had been reduced to 6.85 in.. Assuming that atmospheric pressure had not changed, what was the pressure of the trapped air at that point (in inHg)?

3B.3 The following data were collected for a gas sample consisting of 1.00 mol of molecules in a rigid container. (a) Determine the volume of the sample. (b) Plot the data either by hand or using a spreadsheet. Now suppose you add an additional 1.00 mol of gas molecules into the same volume. Plot the corresponding line for the second sample. (c) At what temperature (in kelvins) do the two lines intersect?

Temperature/°C	0	20	40	60	80
Pressure, *P*/atm	1.12	1.20	1.28	1.36	1.45

3B.4 The following data were collected for a gas sample consisting of 5.00 mol of molecules in a rigid container. (a) Determine the volume of the sample. (b) Plot the data either by hand or using a spreadsheet. Now suppose you add another 5.00 mol of gas molecules into the same volume. Plot the corresponding line for the second sample. (c) At what temperature (in kelvins) do the two lines intersect?

Temperature/°C	0	20	40	60	80
Pressure, *P*/atm	4.48	4.80	5.14	5.46	5.80

3B.5 (a) A flask of volume 350. mL contains 0.1500 mol Ar at 24 °C. What is the pressure of the gas in kilopascals? (b) You are told that 23.9 mg of bromine trifluoride exerts a pressure of 10.0 Torr at 100 °C. What is the volume of the container in milliliters? (c) A flask of volume 100.0 L contains sulfur dioxide at 0.77 atm and 30 °C. What mass of gas is present? (d) A storage tank of volume 6.00×10^3 m^3 contains methane at 129 kPa and 14 °C. How many moles of CH_4 are present?

3B.6 (a) A flask of volume 125 mL contains argon at 1.85 atm and 86 °C. What amount of Ar is present (in moles)? (b) A flask of volume 250.0 mL contains 6.5 µg of O_2 at 17 °C. What is the pressure (in Torr)? (c) A flask of volume 50.0 L at 203 K and 20. Torr

contains nitrogen. What mass of nitrogen is present (in grams)? (d) A 0.0387-g sample of krypton exerts a pressure of 2.00×10^2 mTorr at 65 °C. What is the volume of the container (in liters)?

3B.7 (a) Use the Living Graphs on the Web site for this book to prepare a plot of the pressure against temperature for 1.00 mol of gas molecules, showing the curves at volumes ranging from 0.01 L to 0.05 L in increments of 0.01 L for *T* = 0 to 400 K. (b) What is the expression for the slope of each of these lines? (c) What is each intercept? Give your answers to two decimal places.

3B.8 (a) Use the Living Graphs on the Web site for this book to prepare a plot of the volume against temperature graph for 1.00 mol of gas molecules, showing the curves at pressures ranging from 11 000 atm to 15 000 atm in increments of 1000 atm for *T* = 0 to 400 K. (b) What is the expression for the slope of each of these lines? (c) What is each intercept? Give your answers to two decimal places.

3B.9 Determine the final pressure when (a) 7.50 mL of krypton gas at 2.00×10^5 kPa is transferred to a vessel of volume 1.0 L; (b) 54.2 cm^3 of oxygen gas at 643 Torr is compressed to 7.8 cm^3. Assume constant temperature.

3B.10 (a) Suppose that 5.00 L of methane at 620. Torr is transferred to a vessel of volume 2.50 L at the same temperature. What is the final pressure of methane? (b) A fluorinated organic gas in a cylinder is compressed from an initial volume of 654 mL at 152 Pa to 218 mL at the same temperature. What is the final pressure?

3B.11 A helium balloon has a volume of 12.4 L when the pressure is 0.885 atm and the temperature is 22 °C. The balloon is cooled at a constant pressure until the temperature is −18 °C. What is the volume of the balloon at this stage?

3B.12 A chemist prepares a sample of hydrogen bromide and finds that it occupies 500. mL at 45 °C and 120. Torr. What volume would it occupy at 0 °C at the same pressure?

3B.13 A chemist prepares 0.100 mol Ne(g) at a certain pressure and temperature in an expandable container. Another 0.010 mol Ne(g) is then added to the same container. How must the volume be changed to keep the pressure and temperature the same?

3B.14 A chemist prepares a sample of 0.0120 mol He(g) at a certain pressure, temperature, and volume and then adds another 0.0240 mol He(g). How must the temperature be changed to keep the pressure and volume the same?

3B.15 A sample of methane gas, CH_4, was slowly heated at a constant pressure of 0.90 bar. The volume of the gas was measured at a series of different temperatures and a plot of volume against temperature was constructed. The slope of the line was 2.88×10^{-4} L·K^{-1}. What was the mass of the sample of methane?

3B.16 A sample of butane gas, C_4H_{10}, was slowly heated at a constant pressure of 0.80 bar. The volume of the gas was measured at a series of different temperatures and a plot of volume against temperature was constructed. The slope of the line was 0.0208 L·K^{-1}. What was the mass of the sample of butane?

3B.17 A sample of xenon of volume 35.5 mL exerts a pressure of 0.255 atm at −45 °C. (a) What volume does the sample occupy at 1.00 atm and 298 K? (b) What pressure would it exert if it were transferred to a flask of volume 12.0 mL at 20 °C? (c) Calculate the temperature needed for the xenon to exert a pressure of 5.00×10^2 Torr in the flask.

3B.18 A lungful of air (332 cm³) is exhaled into a machine that measures lung capacity. If the air is exhaled from the lungs at a pressure of 1.08 atm at 37 °C but the machine is at ambient conditions of 0.964 atm and 25 °C, what is the volume of air measured by the machine?

3B.19 What is the molar volume of an ideal gas at 1.00 atm and (a) at 500. °C; (b) at the normal boiling point of liquid nitrogen (−196 °C)?

3B.20 What is the molar volume of an ideal gas at 1.00 atm and (a) at 212 °F; (b) at the normal sublimation point of dry ice (−78.5 °C)?

3B.21 Originally at −20. °C and 759 Torr, a sample of air of volume 1.00 L is heated to 235 °C. Next, the pressure is increased to 765 Torr. Then it is heated to 1250. °C, and finally the pressure is decreased to 252 Torr. What is the final volume of the air?

3B.22 A domestic water-carbonating kit uses steel cylinders of carbon dioxide having a volume of 250. mL. Each cylinder weighs 1.04 kg when full and 0.74 kg when empty. (a) What is the pressure of gas (in bar) in a full cylinder at 20. °C? (b) What is the pressure of the gas when the cylinder weighs 0.87 kg?

3B.23 The effect of high pressure on organisms, including humans, is studied to gain information about deep-sea diving and anesthesia. A sample of air occupied 1.00 L at 25 °C and 1.00 atm. What pressure (in atm) is needed to compress it to 239 cm³ at this temperature?

3B.24 To what temperature must a sample of helium gas be cooled from 115.0 °C to reduce its volume from 7.20 L to 0.325 L at constant pressure?

3B.25 The "air" in the space suit of astronauts is actually pure oxygen supplied at a pressure of 0.30 bar. Each of the two tanks on a space suit has a volume of 3980. cm³ and an initial pressure of 5860. kPa. Assuming a tank temperature of 16 °C, what mass of oxygen is contained in the two tanks?

3B.26 A balloon vendor has a helium tank of volume 18.0 L at 170 atm and 25 °C. How many balloons of volume 2.40 L can be filled at 1.0 atm and 25 °C by the helium in this tank?

3B.27 Nitrogen monoxide, NO(g), has been found to act as a neurotransmitter. To prepare to study its effect, a sample was collected in a container of volume 250.0 mL. At 19.5 °C, its pressure in this container is found to be 24.5 kPa. What amount (in moles) of NO has been collected?

3B.28 Volcanic eruptions can be a significant source of air pollution. The Kilauea volcano in Hawaii typically emits 200–300 t of sulfur dioxide gas (SO_2) each day (1 t = 10^3 kg). On one day, the volcano emitted 248 t of SO_2. If the gas was emitted at 800. °C and 1.00 atm, what volume of SO_2 was emitted that day?

3B.29 At sea level, where the pressure was 104 kPa and the temperature 21.1 °C, a certain mass of air occupied 2.0 m³. To what volume will the air mass expand when it has risen to an altitude

where the pressure and temperature are (a) 52 kPa, −5.0 °C; (b) 880. Pa, −52.0 °C?

3B.30 A meteorological balloon had a radius of 1.0 m when released at sea level at 20 °C. It expanded to a radius of 3.0 m when it had risen to its maximum altitude, where the temperature was −20 °C. What was the pressure inside the balloon at that altitude?

3B.31 Without doing a calculation, order the following gases according to increasing mass density: N_2H_4; N_2; NH_3. The temperature and pressure are the same for all three samples.

3B.32 Without doing a calculation, order the following gases according to increasing mass density: NO, NO_2, N_2O. The temperature and pressure are the same for all three samples.

3B.33 A sample of argon of mass 2.00 mg is confined to a vial of volume 0.0500 L at 20 °C; 2.00 mg of krypton is confined to a different vial of the same volume. What must the temperature of the krypton be if it is to have the same pressure as the argon?

3B.34 What mass of ammonia will exert the same pressure as 12 mg of hydrogen sulfide, H_2S, in the same container under the same conditions?

3B.35 What is the density (in g·L⁻¹) of chloroform, $CHCl_3$, vapor at (a) 2.00×10^2 Torr and 298 K; (b) 100. °C and 1.00 atm?

3B.36 What is the density (in g·L⁻¹) of ammonia at (a) 1.00 atm and 298 K; (b) 32.0 °C and 0.865 atm?

3B.37 A gaseous fluorinated methane compound has a density of 8.0 g·L⁻¹ at 2.81 atm and 300. K. (a) What is the molar mass of the compound? (b) What is the formula of the compound if it is composed solely of C, H, and F? (c) What is the density of the gas at 1.00 atm and 298 K?

3B.38 The density of a gaseous compound of phosphorus is 0.943 g·L⁻¹ at 420. K when its pressure is 727 Torr. (a) What is the molar mass of the compound? (b) If the compound remains gaseous, what would be its density at 1.00 atm and 298 K?

3B.39 A compound used in the manufacture of Saran wrap is 24.7% C, 2.1% H, and 73.2% Cl by mass. The storage of 3.557 g of the gaseous compound in a vessel of volume 755 mL at 0 °C results in a pressure of 1.10 atm. What is the molecular formula of the compound?

3B.40 The analysis of a hydrocarbon revealed that it was 85.7% C and 14.3% H by mass. When 1.77 g of the gas was stored in a flask of volume 1.500 L at 17 °C, it exerted a pressure of 508 Torr. What is the molecular formula of the hydrocarbon?

3B.41 The density of a gaseous compound was found to be 0.943 g·L⁻¹ at 298 K and 53.1 kPa. What is the molar mass of the compound?

3B.42 A sample of eugenol (the compound responsible for the odor of cloves) of mass 115 mg was placed in an evacuated flask of volume 500.0 mL at 280.0 °C. The pressure that eugenol exerted in the flask under those conditions was found to be 48.3 Torr. In a combustion experiment, 18.8 mg of eugenol burned to give 50.0 mg of carbon dioxide and 12.4 mg of water. What is the molecular formula of eugenol?

Topic 3C Gases in Mixtures and Reactions

3C.1   Mixtures of Gases

3C.2   The Stoichiometry of Reacting Gases

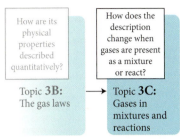

Topic **3B**:
The gas laws → Topic **3C**:
Gases in mixtures and reactions

In many chemical reactions a gas is a reactant or product. Knowledge of the ideal gas law allows the amount of gas produced or consumed in a reaction to be followed by monitoring its temperature, pressure, and volume. The calculations can be used regardless of whether the gas is the sole gas filling a container or is present as a component of a mixture.

3C.1 Mixtures of Gases

Many of the gases encountered in everyday life—and in the chemical laboratory—are mixtures. The atmosphere, for instance, is a mixture of nitrogen, oxygen, argon, carbon dioxide, and many other gases (TABLE 3C.1). Many gaseous anesthetics are carefully controlled mixtures. The description of an ideal gas needs to be extended to include mixtures of gases.

At low pressures all gases respond in the same way to changes in pressure, volume, and temperature. Therefore, for typical calculations on the physical properties of gases, it does not matter whether all the molecules in a sample are the same. *A mixture of gases that do not react with one another behaves like a single pure gas.*

John Dalton was the first to show how to calculate the pressure of a mixture of gases. To understand his reasoning, imagine that a certain amount of oxygen is introduced into a container so that the pressure is 0.60 atm. Then the container is evacuated to empty it of all gas. Now enough nitrogen gas is introduced into the container to give a pressure of 0.40 atm at the same temperature. Dalton wondered what the total pressure would be if these same amounts of the two gases were present in the container simultaneously. From some fairly crude measurements, he concluded that the total pressure resulting from the pressure of both gases in the same container would be 1.00 atm, the sum of the individual pressures.

Dalton reported his observations in terms of the **partial pressure** of each gas, the pressure that the gas would exert if it occupied the container alone. In the present example, the partial pressures of oxygen and nitrogen in the mixture are 0.60 atm and 0.40 atm, respectively, because those are the pressures that the gases exert when each one is in the container alone. Dalton summarized his observations in his **law of partial pressures**:

The total pressure of a mixture of gases is the sum of the partial pressures of its components.

Why Do You Need to Know This Material? Chemists are often concerned with mixtures of gases and with reactions that consume or produce gases. The ideal gas law provides a way to treat the production and consumption of gases quantitatively.

What Do You Need to Know Already? You need to be familiar with the ideal gas law (Topic 3B). This Topic extends the techniques of reaction stoichiometry (*Fundamentals* L and M) to gases.

This is the same Dalton whose contribution to atomic theory is introduced in *Fundamentals* B.

TABLE 3C.1 The Typical Composition of Dry Air at Sea Level		
Constituent	Molar mass,* $M/(\text{g·mol}^{-1})$	Mass percentage composition
N_2	28.02	75.52
O_2	32.00	23.14
Ar	39.95	1.29
CO_2	44.01	0.05

*The average molar mass of molecules in dry air is 28.97 g·mol^{-1}. The percentage of water vapor in air varies with the humidity.

163

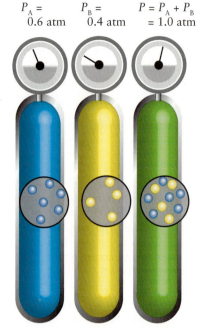

$$P_A = \qquad P_B = \qquad P = P_A + P_B$$
0.6 atm   0.4 atm   = 1.0 atm

FIGURE 3C.1 A representation of the experiment that Dalton performed on a gas mixture. According to Dalton's law, the total pressure, P, of a mixture of ideal gases is the sum of the partial pressures P_A and P_B of gases A and B. Each partial pressure is the pressure that one of the gases would exert if it were the sole gas in the container (at the same temperature).

If the partial pressures of the gases A, B, . . . are denoted P_A, P_B, . . . and the total pressure of the mixture is denoted P, Dalton's law can be written as

$$P = P_A + P_B + \cdots \tag{1}$$

The law is illustrated in **FIG. 3C.1**. It is exactly true only for gases that behave ideally, but it is a good approximation for nearly all gases under normal conditions (room temperature and pressure).

Dalton's law is consistent with the model of a gas outlined in Topic 3A and adds a little more information. The total pressure of a gas arises from the battering of the molecules on the walls of the container (Topic 3A). That battering is due to all the molecules in a mixture. The molecules of gas A strike the walls, as do the molecules of gas B, and if each impact is independent of the other, the resulting total pressure is the sum of the individual pressures, as Dalton's law asserts.

Partial pressures are used to describe the composition of a humid gas. For example, the total pressure of the damp air in your lungs is

$$P = P_{\text{dry air}} + P_{\text{water vapor}}$$

In a closed container, to which a lung is a good first approximation, water vaporizes until its partial pressure has reached a certain value, called its "vapor pressure." The vapor pressure of water at normal body temperature is 47 Torr. The partial pressure of the air itself in your lungs is therefore

$$P_{\text{dry air}} = P - P_{\text{water vapor}} = P - 47 \text{ Torr}$$

On a typical day, the total pressure at sea level is 760. Torr; so at sea level the pressure in your lungs due to all the gas except the water vapor is 760. − 47 Torr = 713 Torr.

THINKING POINT

Is damp air more dense or less dense than dry air under the same conditions?

Self-test 3C.1A As a sample of oxygen was collected over water at 24 °C and 745 Torr, it became saturated with water vapor. At this temperature, the vapor pressure of water is 24.38 Torr. What is the partial pressure of the oxygen?

[*Answer:* 721 Torr]

Self-test 3C.1B Students collecting hydrogen and oxygen gases by electrolysis of water failed to separate the two gases. If the total pressure of the dry mixture is 720. Torr, what is the partial pressure of each gas? *Hint:* Think about the relative amounts of each gas produced.

A useful way to express the relation between the total pressure of a mixture and the partial pressures of its components is to use the **mole fraction,** x, of each component A, B, ... , the fraction it contributes to the total amount of molecules in the mixture. If the total amount of gas molecules present is n and the amount of molecules of each gas A, B, etc., present are n_A, n_B, and so forth, the mole fraction of A is

$$x_A = \frac{n_A}{n} = \frac{n_A}{n_A + n_B + \cdots} \tag{2}$$

and likewise for the mole fractions of the other components. In a binary (two-component) mixture of gases A and B,

$$x_A + x_B = \frac{n_A}{n_A + n_B} + \frac{n_B}{n_A + n_B} = \frac{n_A + n_B}{n_A + n_B} = 1 \tag{3}$$

When $x_A = 1$, the mixture is pure A; when $x_B = 1$, the mixture is pure B. When $x_A = x_B = 0.50$, half the molecules are A and half are B (**FIG. 3C.2**). These definitions and the ideal gas law can be used to express the partial pressure of a gas in terms of its mole fraction in a mixture.

FIGURE 3C.2 A mole fraction, x, is the fraction of molecules of a particular kind in a mixture of two or more kinds of molecules. Here, the mole fraction of molecules shown in red is given below each mixture. The mixture can be solid or liquid as well as gaseous.

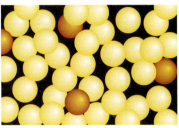

$x_{RED} = 0.1$

$x_{RED} = 0.5$

$x_{RED} = 0.9$

How Is That Done?

To express the partial pressure of a gas A in a mixture in terms of its mole fraction, first use the ideal gas law to express the partial pressure, P_A, of the gas in terms of the amount of A molecules present, n_A, the volume, V, occupied by the mixture, and the temperature, T:

$$P_A = \frac{n_A RT}{V}$$

Because $n_A = nx_A$ (where n is the total amount of all the molecules) and $P = nRT/V$,

$$P_A = \frac{nx_A RT}{V} = x_A \frac{nRT}{V} = x_A P$$

The result of the calculation is

$$P_A = x_A P \tag{4}$$

where P is the total pressure and x_A is the mole fraction of A in the mixture.

An important but subtle point is that, whereas Dalton defined the partial pressure as the pressure that gas would occupy if alone in the container, the modern approach is to use Eq. 4 as the definition of partial pressure for ideal *and* real gases. Equation 1 is then true for all gases, both ideal and real. For instance, for a binary mixture of any kinds of gases,

$$P_A + P_B = x_A P + x_B P = (x_A + x_B)P = P$$

However, according to this modern view, the partial pressures calculated from Eq. 4 can be interpreted as the pressures that each gas exerts when it is alone only if the gases are ideal.

EXAMPLE 3C.1 Calculating partial pressures

Air is a source of reactants for many chemical processes, such as the synthesis of ammonia. To determine how much air is needed for these reactions, it is useful to know the partial pressures of the components. A certain sample of dry air of total mass 1.00 g consists almost entirely of 0.76 g of nitrogen and 0.24 g of oxygen. Calculate the partial pressures of these gases when the total pressure is 0.87 atm.

ANTICIPATE The molar masses of N_2 and O_2 are very similar, so very approximately you should expect the amounts of N_2 and O_2 to be in the ratio of the masses present, which is 0.76:0.24. The partial pressures should be in the same ratio, or about 3:1.

PLAN To use Eq. 4, the total pressure (given) and the mole fraction of each component are needed. The first step is to calculate the amount (in moles) of molecules of each gas present and the total amount (in moles). Then calculate the mole fractions from Eq. 2. To obtain the partial pressures of the gases, multiply the total pressure by the mole fractions of the gases in the mixture (Eq. 4)

What should you assume? There is no need to assume that the gases are ideal because Eqs. 4 and 1 are valid for any kind of gas.

SOLVE

Use the molar masses of N_2 and O_2 to find the amounts (in moles) of each type of gas molecules.

$$n_{N_2} = \frac{0.76 \text{ g}}{28.02 \text{ g·mol}^{-1}} = \frac{0.76}{28.02} \text{ mol} = 0.027... \text{ mol}$$

$$n_{O_2} = \frac{0.24 \text{ g}}{32.00 \text{ g·mol}^{-1}} = \frac{0.24}{32.00} \text{ mol} = 0.0075... \text{ mol}$$

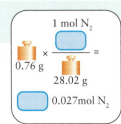

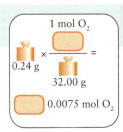

Find the total amount of gas molecules from $n_{total} = n_A + n_B$:

$$n_{N_2} + n_{O_2} = 0.027... + 0.0075... \text{ mol} = 0.035... \text{ mol}$$

Calculate the mole fractions from $x_A = n_A/n_{total}$:

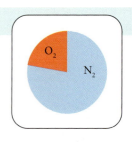

$$x_{N_2} = \frac{\overbrace{0.027...}^{n_{N_2}}}{\underbrace{0.035...}_{n_{N_2} + n_{O_2}}} = 0.78...$$

$$x_{O_2} = \frac{0.0075...}{\underbrace{0.035...}_{n_{N_2} + n_{O_2}}} = 0.22...$$

Multiply each mole fraction by the total pressure, $P_A = x_A P$

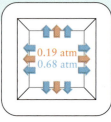

$$P_{N_2} = 0.78... \times (0.87\ atm) = 0.68\ atm$$
$$P_{O_2} = 0.22... \times (0.87\ atm) = 0.19\ atm$$

EVALUATE As expected, P_{N_2} is about three times P_{O_2}. A check of the calculation against given information shows that $0.68 + 0.19$ atm $= 0.87$ atm, as in the data.

Self-test 3C.2A A baby with a severe bronchial infection is in respiratory distress. The anesthetist administers heliox, a mixture of helium and oxygen with 92.3% by mass O_2. What is the partial pressure of oxygen being administered to the baby if the atmospheric pressure is 730 Torr?

[***Answer:*** 4.4×10^2 Torr]

Self-test 3C.2B Divers exploring a shipwreck and wishing to avoid the narcosis associated with breathing nitrogen under high pressure switch to a neon–oxygen gas mixture containing 141.2 g of oxygen and 335.0 g of neon. The pressure in the gas tanks is 50.0 atm. What is the partial pressure of oxygen in the tanks?

Related Exercises 3C.2–3C.4, 3C.7, 3C.8

The partial pressure of an ideal gas is the pressure that it would exert if alone in the container; the total pressure of a mixture of gases is the sum of the partial pressures of the components; the partial pressure of a gas is related to the total pressure by the mole fraction: $P_A = x_A P$.

3C.2 The Stoichiometry of Reacting Gases

Suppose you need to know the volume of carbon dioxide produced when a fuel burns or the volume of oxygen needed to react with a given mass of hemoglobin in red blood cells. To answer this kind of question, you can combine the mole-to-mole calculations of the type described in *Fundamentals* L and M with the conversion of moles of gas molecules into the volume that they occupy. The diagram in (**1**) extends stoichiometric strategies (*Fundamentals* L) to include the volume of a gas.

1

EXAMPLE 3C.2   Calculating the mass of reagent needed to react with a specified volume of gas

The carbon dioxide generated by the personnel in submarines and spacecraft must be removed from the air and the oxygen recovered. Submarine design teams have investigated the use of potassium superoxide, KO_2, as an air purifier because it reacts with carbon dioxide and releases oxygen (**FIG. 3C.3**):

$$4\ KO_2(s) + 2\ CO_2(g) \longrightarrow 2\ K_2CO_3(s) + 3\ O_2(g)$$

Calculate the mass of KO_2 needed to react with 50. L of carbon dioxide at 25 °C and 1.0 atm.

ANTICIPATE A volume of 50 L under normal conditions corresponds to about 2 mol CO_2, and the stoichiometry of the equation indicates that about 4 mol KO_2 would be needed. Because the molar mass of KO_2 is about 70 g·mol⁻¹, you should suspect that the answer will be close to 280 g.

PLAN Convert from the given volume of gas into amount of CO_2 molecules (by using the molar volume), then into the amount of KO_2 formula units (by using a mole ratio), and then into the mass of KO_2 (by using its molar mass). If the molar volume at the stated conditions is not available, then calculate the amount of gas molecules from the ideal gas law: $n = PV/RT$.

SOLVE

Find the molar volume under the stated conditions from a table or calculation.

Table 3B.2 gives $V_m = 24.47$ L·mol^{-1} (to two decimal places) under these conditions.

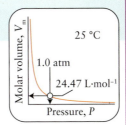

Find the stoichiometric relation between CO_2 and KO_2 from the chemical equation.

$$2 \text{ mol } CO_2 \; \hat{=} \; 4 \text{ mol } KO_2 \text{ or } 1 \text{ mol } CO_2 \; \hat{=} \; 2 \text{ mol } KO_2$$

Find the molar mass of KO_2 (*Fundamentals* E).

$$39.10 + 2(16.00) \text{ g·mol}^{-1} = 71.10 \text{ g·mol}^{-1}$$

Convert from volume of CO_2 to mass of KO_2.

$$\text{Mass of } KO_2 = 50. \text{ L} \times \frac{1 \text{ mol } CO_2}{24.47 \text{ L}} \times \frac{2 \text{ mol } KO_2}{1 \text{ mol } CO_2} \times \frac{71.10 \text{ g}}{1 \text{ mol } KO_2}$$

$$= 2.9 \times 10^2 \text{ g}$$

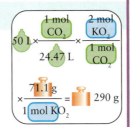

EVALUATE The answer, 290 g, is close to the 280 g anticipated.

Self-test 3C.3A Calculate the volume of carbon dioxide, adjusted to 25 °C and 1.0 atm, that plants need to make 1.00 g of glucose, $C_6H_{12}O_6$, by photosynthesis in the reaction 6 CO_2(g) + 6 H_2O(l) → $C_6H_{12}O_6$(s) + 6 O_2(g).

[*Answer:* 0.81 L]

Self-test 3C.3B The reaction of H_2 and O_2 gases to produce liquid H_2O was used in fuel cells on the space shuttles to provide electricity. What mass of water can be produced in the reaction of 100.0 L of oxygen stored at 25 °C and 1.00 atm?

Related Exercises 3C.9–3C.14

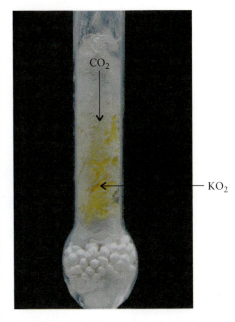

FIGURE 3C.3 When carbon dioxide is passed over potassium superoxide (the yellow solid), it reacts to form colorless potassium carbonate (the white solid coating the walls of the tube) and oxygen gas. The reaction is used to remove carbon dioxide from the air in a closed-system breathing environment. (*W. H. Freeman photo by Ken Karp.*)

There may be a considerable increase in volume when liquids or solids react to form a gas. Molar volumes of gases are close to 25 L·mol^{-1} under normal conditions, whereas liquids and solids occupy only about a few tens of milliliters per mole. The molar volume of liquid water, for instance, is only 18 mL·mol^{-1} at 25 °C. In other words, 1 mol of gas molecules at 25 °C and 1 atm occupies as much as 1000 times the volume of 1 mol of molecules in a typical liquid or solid.

The increase in volume as gaseous products are formed in a chemical reaction is even greater if several gas molecules are produced from each reactant molecule, such as the formation of CO and CO_2 from a solid fuel (**FIG. 3C.4**). Lead(II) azide, $Pb(N_3)_2$, which is used as a detonator for explosives, rapidly releases a large volume of nitrogen gas when it is struck and undergoes the reaction

$$Pb(N_3)_2(s) \longrightarrow Pb(s) + 3 N_2(g)$$

A reaction of the same kind but using sodium azide, NaN_3, is used in air bags in automobiles (**FIG. 3C.5**). The explosive release of nitrogen is detonated electrically when the vehicle decelerates abruptly in a collision.

FIGURE 3C.4 An explosion caused by the ignition of coal dust. A shock wave is created by the tremendous expansion of volume as large numbers of gas molecules form. (© *RIA Novosti/The Image Works*.)

FIGURE 3C.5 The rapid decomposition of sodium azide, NaN_3, results in the formation of a large volume of nitrogen gas. The reaction is triggered electrically in this air bag. (*Benelux Press BV/Science Source*.)

The molar volume (at the specified temperature and pressure) is used to convert the amount of a reactant or product in a chemical reaction into a volume of gas.

What have you learned in this Topic?

You have learned that each ideal gas in a mixture has a partial pressure equal to the pressure it would exert if it were the only gas in the container. You have also seen how to use the molar volume of a gas as a measure of its amount in a stoichiometric calculation and to predict the volume of a gas consumed or produced.

The skills you have mastered are the ability to:

☐ **1.** Express the composition of a mixture in terms of the mole fractions of the components (Section 3C.1).

☐ **2.** Calculate the partial pressures of gases in a mixture and the total pressure of the mixture (Example 3C.1).

☐ **3.** Calculate the mass or volume of reactant needed to react with a specified volume of gas (Example 3C.2).

Topic 3C Exercises

3C.1 A sample of hydrogen chloride gas, HCl, is being collected by bubbling it through liquid benzene. Assume that the molecules pictured as spheres show a representative sample of the mixture of HCl and benzene vapor (● represents an HCl molecule and ○ a benzene molecule). (a) Use the figure to determine the mole fractions of HCl and benzene vapor in the gas inside the container. (b) What are the partial pressures of HCl and benzene in the container when the total pressure inside the container is 0.80 atm?

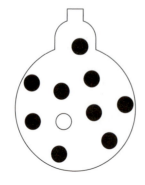

3C.2 A piece of sodium metal was added to a flask of water on a day when the atmospheric pressure was 757.5 Torr. The sodium reacted completely with the water to produce 137.0 mL of a

mixture of hydrogen gas and water vapor at 24 °C, at which temperature the vapor pressure of water is 22.38 Torr. (a) What is the partial pressure (in Torr) of hydrogen in the collection flask? (b) Write a balanced equation for the reaction of sodium with water. (c) What mass of sodium metal reacted?

3C.3 A vessel of volume 22.4 L contains 2.0 mol $H_2(g)$ and 1.0 mol $N_2(g)$ at 273.15 K. Calculate (a) their partial pressures and (b) the total pressure.

3C.4 A gas mixture being used to simulate the atmosphere of another planet consists of 376 mg of methane, 154 mg of argon, and 252 mg of nitrogen. The partial pressure of nitrogen at 300. K is 21.3 kPa. Calculate (a) the total pressure of the mixture and (b) the volume of the sample.

3C.5 A flask of volume 1.00 L contains nitrogen gas at a temperature of 15 °C and a pressure of 0.50 bar. 0.10 mol $O_2(g)$ is added to the flask and allowed to mix. Then a stopcock is opened to allow 0.020 mol of molecules to escape. What is the partial pressure of oxygen in the final mixture?

3C.6 An apparatus consists of a flask of volume 4.0 L containing nitrogen gas at 25 °C and 803 kPa, joined by a valve to a flask of volume 10.0 L containing argon gas at 25 °C and 47.2 kPa. The valve is opened and the gases mix. (a) What is the partial pressure of each gas after mixing? (b) What is the total pressure of the gas mixture?

3C.7 In the course of the electrolysis of water, hydrogen gas was collected at one electrode over water at 20. °C when the external pressure was 756.7 Torr. The vapor pressure of water at 20. °C is 17.54 Torr. The volume of the gas was measured to be 0.220 L. (a) What is the partial pressure of hydrogen? (b) The other product of the electrolysis of water is oxygen gas. Write a balanced equation for the electrolysis of water into H_2 and O_2. (c) What mass of oxygen was also produced in the reaction?

3C.8 Dinitrogen monoxide, N_2O, gas was generated from the thermal decomposition of ammonium nitrate and collected over water. The wet gas occupied 126 mL at 21 °C when the atmospheric pressure was 755 Torr. What volume would the same amount of *dry* dinitrogen monoxide have occupied if collected at 755 Torr and 21 °C? The vapor pressure of water is 18.65 Torr at 21 °C.

3C.9 The Haber process for the synthesis of ammonia is one of the most significant industrial processes for the well-being of humanity. It is used extensively in the production of fertilizers as well as polymers and other products. (a) What volume of hydrogen at 15.00 atm and 350. °C must be supplied to produce 1.0 t (1 t = 10^3 kg) of NH_3? (b) What volume of hydrogen is needed in part (a) if it is supplied at 376 atm and 250. °C?

3C.10 Nitroglycerin is a shock-sensitive liquid that detonates by the reaction

$$4\,C_3H_5(NO_3)_3(l) \longrightarrow$$
$$6\,N_2(g) + 10\,H_2O(g) + 12\,CO_2(g) + O_2(g)$$

Calculate the total volume of product gases at 88.5 kPa and 175 °C from the detonation of 1.00 lb (454 g) of nitroglycerin.

3C.11 Which starting condition would produce the larger volume of carbon dioxide by combustion of $CH_4(g)$ with an excess of oxygen gas to produce carbon dioxide and water: (a) 2.00 L of $CH_4(g)$; (b) 2.00 g of $CH_4(g)$? Justify your answer. The system is maintained at a temperature of 75 °C and 1.00 atm.

3C.12 Which starting condition would produce the larger volume of carbon dioxide by combustion of $C_2H_4(g)$ with an excess of oxygen gas to produce carbon dioxide and water: (a) 1.00 L of $C_2H_4(g)$; (b) 1.20 g of $C_2H_4(g)$? Justify your answer. The system is maintained at a temperature of 45 °C and 2.00 atm.

3C.13 The interhalogen compounds can be prepared by direct reaction of the elements. The following syntheses were carried out at 298 K and 1.00 atm. In each case, give the volume of product at the same temperature and pressure that can be produced from 2.00 mol F_2 and an excess of Cl_2: (a) $Cl_2(g) + F_2(g) \rightarrow 2\,ClF(g)$; (b) $Cl_2(g) + 3\,F_2(g) \rightarrow 2\,ClF_3(g)$; (c) $Cl_2(g) + 5\,F_2(g) \rightarrow 2\,ClF_5(g)$.

3C.14 The naturally occurring compound urea, $CO(NH_2)_2$, was first synthesized by Friedrich Wöhler in Germany in 1828 by heating ammonium cyanate. This synthesis was a significant event because it was the first time that an organic compound had been produced from an inorganic substance. Urea may also be made by the reaction of carbon dioxide and ammonia:

$$CO_2(g) + 2\,NH_3(g) \longrightarrow CO(NH_2)_2(s) + H_2O(g)$$

What volumes of CO_2 and NH_3 at 160. atm and 440. °C are needed to produce 1.50 kg of urea, assuming the reaction goes to completion?

3C.15 A sample of ammonia gas of volume 15.0 mL at 100. Torr and 30. °C is mixed with 25.0 mL of hydrogen chloride gas at 150. Torr and 25 °C, and the following reaction takes place:

$$NH_3(g) + HCl(g) \longrightarrow NH_4Cl(s)$$

(a) Calculate the mass of NH_4Cl that forms. (b) Identify the gas in excess and determine the pressure of the excess gas at 27 °C after the reaction is complete (in the combined volume of the original two flasks).

3C.16 A sample of ethene gas, C_2H_4, of volume 1.00 L at 1.00 atm and 298 K is burned in 4.00 L of oxygen gas to form carbon dioxide gas and liquid water at the same pressure and temperature. Ignore the volume of water, and determine the final volume of the reaction mixture (including any excess reactant) at 1.00 atm and 298 K if the reaction goes to completion.

Topic 3D Molecular Motion

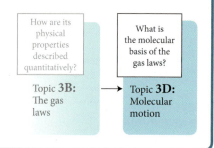

How are its physical properties described quantitatively?

What is the molecular basis of the gas laws?

Topic **3B:** The gas laws → Topic **3D:** Molecular motion

Why Do You Need to Know This Material? Knowledge of how the properties of gases arise from the behavior of the gas molecules builds an understanding that can be used to make predictions about gases. This Topic illustrates an important aspect of science: how qualitative models are used to derive quantitative, testable expressions.

What Do You Need to Know Already? You need to be familiar with the ideal gas law (Topic 3B) and the concepts of force and kinetic energy (*Fundamentals* A).

The empirical results summarized by the gas laws suggest a model of an ideal gas in which widely spaced (most of the time), noninteracting (except during collisions) molecules undergo ceaseless random motion, with average speeds that increase with temperature (Topics 3A and 3B). The model is refined in two steps in this Topic. First, experimental measurements of the rate at which gases spread from one region to another are used to gather information about the *average* speeds of molecules. Then this information is used to set up a quantitative version of the model.

3D.1 Diffusion and Effusion

Observations of two processes, diffusion and effusion, show how the average speeds of gas molecules are related to molar mass and temperature. **Diffusion** is the gradual dispersal of one substance through another substance, such as krypton dispersing through a neon atmosphere (**FIG. 3D.1**). Diffusion explains the spread of perfumes and pheromones (chemical signals exchanged by animals) through air. It also helps to keep the composition of the atmosphere approximately uniform. **Effusion** is the escape of a gas through a small hole into a vacuum or region of lower pressure (**FIG. 3D.2**). Effusion occurs whenever a gas is separated from a vacuum by a porous barrier—a barrier that contains microscopic holes—or even a single pinhole. A gas escapes through a pinhole because there are more "collisions" with the hole on the high-pressure side than on the low-pressure side, and so more molecules pass from the high-pressure region into the low-pressure region than pass in the opposite direction. Effusion is examined here; but similar remarks apply to diffusion too.

The nineteenth-century Scottish chemist Thomas Graham carried out a series of experiments on the rates of effusion of gases. He found that:

> At constant temperature, the rate of effusion of a gas is inversely proportional to the square root of its molar mass:

$$\text{Rate of effusion} \propto \frac{1}{\sqrt{\text{molar mass}}} \quad \text{or} \quad \text{Rate of effusion} \propto \frac{1}{\sqrt{M}} \quad \text{(1a)}$$

This observation is now known as **Graham's law of effusion.** The rate of effusion (in terms of the numbers or amounts of molecules) is proportional to the average speed of the molecules in the gas because that speed determines the rate at which molecules approach the hole. Hence, it can also be concluded that

$$\text{Average speed} \propto \frac{1}{\sqrt{M}} \quad \text{(1b)}$$

If Graham's law is written for two gases A and B with molar masses M_A and M_B, and then one equation is divided by the other, the constants of proportionality cancel and the result is

$$\frac{\text{Rate of effusion of A molecules}}{\text{Rate of effusion of B molecules}} = \frac{1/\sqrt{M_A}}{1/\sqrt{M_B}} = \sqrt{\frac{M_B}{M_A}} \quad \text{(2a)}$$

Because the times required for the same amount (in numbers or moles of molecules) of two substances to effuse through a small hole are inversely proportional to the rates at which they effuse, the *time* it takes for a given amount of substance to effuse through the

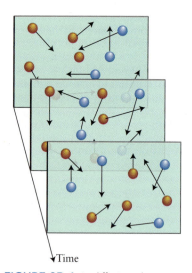

↓Time

FIGURE 3D.1 In diffusion, the molecules of one substance spread into the region occupied by molecules of another substance in a series of random steps, undergoing collisions as they move.

hole is *directly* proportional to the square root of its molar mass. Therefore, an equivalent statement to Eq. 2a is

$$\frac{\text{Time for A to effuse}}{\text{Time for B to effuse}} = \sqrt{\frac{M_A}{M_B}} \qquad \text{(2b)}$$

This relation can be used to estimate the molar mass of a substance by comparing the time required for a given amount of the unknown substance to effuse with that required for the same amount of a substance with a known molar mass.

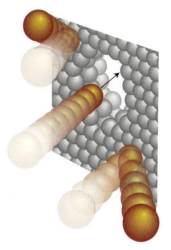

THINKING POINT

Why do heavier molecules diffuse more slowly than lighter molecules at the same temperature?

Self-Test 3D.1A It takes 40. s for 30. mL of argon to effuse through a porous barrier. The same volume of vapor of a volatile compound extracted from Caribbean sponges takes 120. s to effuse through the same barrier under the same conditions. What is the molar mass of this compound?

[*Answer:* $3.6 \times 10^2 \text{ g·mol}^{-1}$]

Self-Test 3D.1B It takes 10. s for a certain amount of helium atoms to effuse through a porous barrier. How long does it take the same amount of methane, CH_4, molecules to effuse through the same barrier under the same conditions?

FIGURE 3D.2 In effusion, the molecules of one substance escape through a small hole in a barrier into a vacuum or a region of low pressure.

Equation 1b shows that the average speed of molecules in a gas is inversely proportional to the square root of their molar mass. In effusion experiments at different temperatures, the rate of effusion increases as the square root of the temperature,

$$\frac{\text{Rate of effusion at } T_2}{\text{Rate of effusion at } T_1} = \sqrt{\frac{T_2}{T_1}} \qquad \text{(3a)}$$

Because the rate of effusion is directly proportional to the average speed of the molecules, it can be inferred that:

The average speed of molecules in a gas is directly proportional to the square root of the temperature,

$$\text{Average speed} \propto \sqrt{T} \qquad \text{(3b)}$$

This very important relation begins to reveal the significance of one of the most elusive concepts in science: the nature of temperature. When referring to a gas, the temperature is an indication of the average speed of the molecules: the higher the temperature, the higher is their average speed.

Equations 1 and 3 can be combined. Because the average speed of molecules in a gas is proportional to the square root of the temperature (Eq. 3b) and inversely proportional to the square root of the molar mass (Eq. 1b), it follows that

$$\text{Average speed} \propto \sqrt{\frac{T}{M}} \qquad \text{(4)}$$

That is, the higher the temperature and the lower the molar mass, the higher is the average speed of molecules in a gas.

The average speed of molecules in a gas is directly proportional to the square root of the temperature and inversely proportional to the square root of the molar mass.

3D.2 The Kinetic Model of Gases

The **kinetic model** of a gas, which is also known as the "kinetic molecular theory" (KMT), is a model of an ideal gas that accounts for the gas laws and effusion behavior and can be used to make numerical predictions. It is based on five assumptions (**FIG. 3D.3**):

1. A gas consists of a collection of molecules in ceaseless random motion.
2. Gas molecules are infinitesimally small points.

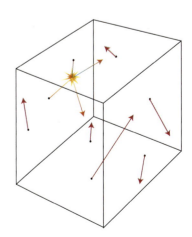

FIGURE 3D.3 In the kinetic model of gases, the molecules are regarded as infinitesimal points that travel in straight lines until they undergo instantaneous collisions.

3.   The molecules move in straight lines until they collide.

4.   The molecules do not influence one another except during collisions.

5.   The collisions are elastic.

In this context, "molecules" includes all types of particles, whether atoms, ions, or molecules.

The fourth assumption means that this model requires that there are no attractive or repulsive forces between ideal gas molecules except during the instantaneous collisions. A collision is "elastic" if the total kinetic energy of the colliding molecules is the same before and after the collision.

In the kinetic model of gases, molecules are pictured as widely separated for most of the time and in ceaseless random motion. They zoom from place to place, always in straight lines, changing direction only when they collide with a wall of the container or another molecule. The collisions change the speed and direction of the molecules, just like balls in a three-dimensional molecular game of pool. The kinetic model of a gas allows the quantitative relation between pressure and the speeds of the molecules to be derived.

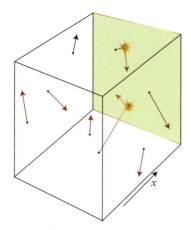

FIGURE 3D.4 In the kinetic model of gases, the pressure arises from the force exerted on the walls of the container when the impacting molecules are deflected.

How Is That Done?

The following calculation of the pressure of a gas based on the kinetic model may seem complicated at first, but it breaks down into a series of small steps, each of which can be interpreted physically.

Consider the image in **FIG. 3D.4** and the force that a single molecule exerts when it strikes the wall. Because Newton's second law of motion states that force is equal to the rate of change of momentum (*Fundamentals* A), the first task is to calculate the change in momentum. Momentum is the product of mass and velocity; so, if a molecule of mass m is traveling with a component of velocity v_x parallel to x, then its linear momentum before it strikes the wall on the right is mv_x. Immediately after the collision, when the direction of motion has reversed but the speed is the same (remember, the collision is elastic) the momentum of the molecule is $-mv_x$. That is:

The change in momentum of one molecule due to a collision with the wall is

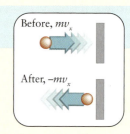

Before, mv_x

After, $-mv_x$

$$\Delta mv_x = \overbrace{mv_x}^{\text{initial}} - \overbrace{(-mv_x)}^{\text{final}} = 2mv_x$$

The total force exerted by all the molecules in the container is the total rate of change of momentum. Consider an interval Δt and the number of molecules that can strike the wall in that interval. Because a molecule that is traveling at v_x parallel to x travels a distance $v_x\Delta t$ in that interval:

Any molecule within a distance $v_x\Delta t$ of the wall and traveling toward it will strike the wall during the interval Δt.

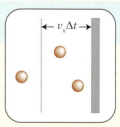

$\leftarrow v_x\Delta t \rightarrow$

If the wall has area A, all the particles in a volume $Av_x\Delta t$ will reach the wall if they are traveling toward it.

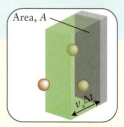

Area, A

$v_x\Delta t$

How many molecules are in that volume? If the total number of particles in the container is N and the volume of the container is V, the number of molecules in the volume $Av_x\Delta t$ is obtained as follows:

The number of molecules in the volume $Av_x\Delta t$ is that fraction of the total volume V, multiplied by the total number of molecules:

$$\text{Number of molecules} = \frac{\overbrace{Av_x\Delta t}^{\substack{V\\(\text{shown in green})}}}{\underbrace{V}_{\text{total volume}}} \times N = \frac{NAv_x\Delta t}{V}$$

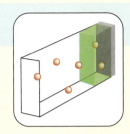

Half the molecules in the green box are moving toward the wall on the right and half are moving away from that wall. Therefore:

The average number of collisions with the wall during the interval Δt is half the number in the volume $Av_x\Delta t$:

$$\text{Number of collisions} = \frac{NAv_x\Delta t}{2V}$$

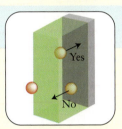

At this stage the change in momentum of one molecule in a single collision and the number of collisions made by all the molecules during the interval Δt have been calculated. Now the two parts of the calculation are brought together. The total momentum change of the molecules in that interval is the change $2mv_x$ that an individual molecule undergoes multiplied by the total number of collisions:

$$\text{Total momentum change} = \frac{NAv_x\Delta t}{2V} \times 2mv_x = \frac{NmAv_x^2\Delta t}{V}$$

Now recall that the force is given by the rate of change of momentum. The rate of change is the total momentum change in the interval Δt divided by the length of the time interval:

From rate of change of momentum = (total momentum change)/Δt,

$$\text{Rate of change of momentum} = \frac{NmAv_x^2\Delta t}{V\Delta t} = \frac{NmAv_x^2}{V}$$

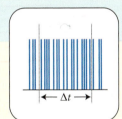

From Newton's second law,

$$\text{Force} = \text{rate of change of momentum} = \frac{NmAv_x^2}{V}$$

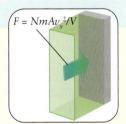

From pressure = force/area,

$$\text{Pressure} = \frac{NmAv_x^2}{VA} = \frac{Nmv_x^2}{V}$$

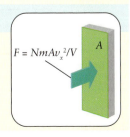

Not all the molecules in the sample are traveling at the same speed. To obtain the observed pressure, P, it is necessary to use the *average* value of v_x^2 in place of v_x^2 for an individual molecule. Averages are commonly denoted by angular brackets, so

$$P = \frac{Nm\langle v_x^2 \rangle}{V}$$

where $\langle v_x^2 \rangle$ is the average value of v_x^2 for all the molecules in the sample.

The calculation is nearly complete. To complete it, note that $\langle v_x^2 \rangle$ can be related to a quantity called the **root mean square speed**, $v_{rms} = \sqrt{\langle v_x^2 \rangle}$, which is the square root of the average of the squares of the molecular speeds (this quantity is explained in more detail in the text following this derivation). First, note that the speed, v, of a single molecule is related to the components of velocity parallel to the x, y, and z directions:

From the Pythagorean theorem,

$$v^2 = v_x^2 + v_y^2 + v_z^2$$

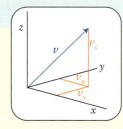

Therefore, the mean square speed is given by

$$v_{rms}^2 = \langle v^2 \rangle = \langle v_x^2 \rangle + \langle v_y^2 \rangle + \langle v_z^2 \rangle$$

However, because the particles are moving randomly in all directions, the average of v_x^2 is the same as the average of v_y^2 and the average of $\langle v_z^2 \rangle$, the analogous quantities in the y and z directions. Because $\langle v_x^2 \rangle = \langle v_y^2 \rangle = \langle v_z^2 \rangle$, then $\langle v^2 \rangle = 3\langle v_x^2 \rangle$; therefore, $\langle v_x^2 \rangle = \frac{1}{3}v_{rms}^2$. It follows that

$$P = \frac{Nmv_{rms}^2}{3V}$$

The final result can be expressed in terms of the amount (in moles) of molecules rather than the actual number. The total number of molecules, N, is the product of the amount, n, and Avogadro's constant, N_A ($N = nN_A$) so

$$P = \frac{\overbrace{nN_A}^{N}mv_{rms}^2}{3V} = \frac{n\overbrace{M}^{mN_A}v_{rms}^2}{3V}$$

where m is the mass of one molecule and $M = mN_A$ is the molar mass of the molecules.

The calculation has shown that from the reasonable assumptions of the kinetic molecular theory, the pressure of a gas and its volume are related by

$$PV = \tfrac{1}{3}nMv_{rms}^2 \tag{5}$$

where n is the amount (in moles) of gas molecules, M is their molar mass, and v_{rms} is the root mean square speed of the molecules (the square root of the average value of the squares of the molecular speeds). If there are N molecules in the sample and the speeds of these molecules at some instant are $v_1, v_2, ..., v_N$, then the root mean square speed is

The importance (and physical significance) of the root mean square speed stems from the fact that v_{rms}^2 is proportional to the average kinetic energy of the molecules, $\langle E_k \rangle = \frac{1}{2}m\langle v^2 \rangle = \frac{1}{2}mv_{rms}^2$.

$$v_{rms} = \sqrt{\frac{v_1^2 + v_2^2 + \cdots v_N^2}{N}} \tag{6}$$

THINKING POINT

What is the average *velocity* (not speed) of the molecules?

The ideal gas law can now be used to calculate the root mean square speed of the molecules of a gas. Because $PV = nRT$ for an ideal gas, the right-hand side of Eq. 5 can be set equal to nRT. The resulting expression, $\frac{1}{3}nMv_{rms}^2 = nRT$, can then be rearranged after canceling the n into

$$v_{rms} = \sqrt{\frac{3RT}{M}} \qquad \text{(7a)}$$

This important equation is used to find the root mean square speed of the molecules of any gas at any temperature (**FIG. 3D.5**). It can also be rewritten to emphasize that, for a gas, the temperature is a measure of mean molecular speed. From $v_{rms}^2 = 3RT/M$, it follows that

$$T = \frac{Mv_{rms}^2}{3R} \qquad \text{(7b)}$$

That is,

The temperature of a gas is proportional to the mean square speed of its molecules.

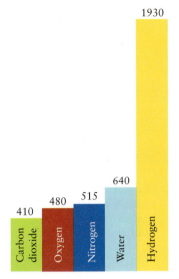

FIGURE 3D.5 The root mean square speeds of five gases at 25 °C in meters per second. Most of the gases are components of air; hydrogen is included to show that the root mean square speed of light molecules is much greater than that of heavier molecules.

EXAMPLE 3D.1 Calculating the root mean square speed of gas molecules

Air molecules are constantly striking your face. How fast are they moving when they collide with you? Calculate the root mean square speed of nitrogen molecules at 20. °C.

ANTICIPATE Because a sound wave is propagated by the movement of molecules, you should expect that typical speeds are about the same as the speed of sound in air. That is about 300 m·s^{-1}.

PLAN For the actual calculation, substitute the data into Eq. 7a. In calculations of basic physical properties such as speed, use R in SI units and molar masses in SI base units, namely, kilograms per mole.

SOLVE The temperature is 293 K and the molar mass of N_2 is 28.02 g·mol^{-1} (corresponding to 2.802×10^{-2} kg·mol^{-1}).

From $v_{rms} = (3RT/M)^{1/2}$,

$$v_{rms} = \left(\frac{3 \times \overbrace{8.3145\ \widehat{J}\cdot K^{-1}\cdot mol^{-1}}^{\substack{R \\ kg\cdot m^2\cdot s^{-2}}} \times \overbrace{293\ K}^{T}}{\underbrace{2.802 \times 10^{-2}\ kg\cdot mol^{-1}}_{M\ =\ 28.02\ g\cdot mol^{-1}}} \right)^{1/2} = 511\ m\cdot s^{-1}$$

The cancellation of units made use of the relation 1 J = 1 kg·m^2·s^{-2}.

EVALUATE This result, $v_{rms} = 511$ m·s^{-1}, is close to the speed of sound in air, as anticipated. It means that nitrogen molecules are colliding with your head at just over 1140 miles per hour.

Self-test 3D.2A Estimate the root mean square speed of water molecules in the vapor above boiling water at 100. °C.

[**Answer:** 719 m·s^{-1}]

Self-test 3D.2B Estimate the root mean square speed of CH_4 molecules at 25 °C.

Related Exercises 3D.9–3D.12

The kinetic model of gases is consistent with the ideal gas law and provides an expression for the root mean square speed of the molecules. Root mean square speeds of gas molecules are proportional to the square root of the temperature.

3D.3 The Maxwell Distribution of Speeds

Useful as it is, Eq. 7a gives only the root mean square speed of gas molecules. Like cars in traffic, individual molecules have speeds that vary over a wide range. Moreover, like a car in a head-on collision, a molecule might be brought almost to a standstill when it collides with another. In the next instant (but now unlike a colliding car), it might be struck by another molecule and move off at the speed of sound. An individual molecule might undergo several billion changes of speed and direction each second.

The formula for calculating the fraction of gas molecules having a given speed, v, at any instant was first derived from the kinetic model by the Scottish scientist James Clerk Maxwell. He derived the expression

$$\Delta N = Nf(v)\,\Delta v \qquad \text{with } f(v) = 4\pi\left(\frac{M}{2\pi RT}\right)^{3/2} v^2 e^{-Mv^2/2RT} \qquad (8)$$

where ΔN is the number of molecules with speeds in the narrow range between v and $v + \Delta v$, N is the total number of molecules in the sample, M is the molar mass, and R is the gas constant. This expression for $f(v)$ is the **Maxwell distribution of speeds** (BOX 3D.1).

The Maxwell distribution is sometimes called the Maxwell–Boltzmann distribution, to recognize Ludwig Boltzmann's theoretical contribution to its formulation.

What Does This Equation Tell You?   The exponential factor (which falls rapidly toward zero as v increases) means that very few molecules have very high speeds. The factor v^2 that multiplies the exponential factor goes to zero as v goes to zero, so it means that very few molecules have very low speeds. The factor $4\pi(M/2\pi RT)^{3/2}$ simply ensures that the total probability of a molecule having a speed between zero and infinity is 1.

Box 3D.1 HOW DO WE KNOW . . . THE DISTRIBUTION OF MOLECULAR SPEEDS?

The distribution of molecular speeds in a gas can be determined experimentally in a molecular beam apparatus. In this technique, the gas is heated to the required temperature in an oven. The molecules then stream out of the oven through a small hole into an evacuated region. To ensure that the molecules form a narrow beam, they also pass through a series of slits.

The molecular beam passes through a series of rotating disks (see the first illustration). Each disk contains a slit that is offset by a certain angle from its neighbor. A molecule that passes through the first slit will pass through the slit in the next disk only if the time that it takes to pass between the disks is the same as the time required for the slit in the second disk to move into the orientation originally occupied by the first disk. The two times must continue to match if the same molecule is to pass through subsequent disks. Therefore, a particular rate of rotation of the disks allows the passage of only those molecules having the speed that carries them through the slits. To determine the

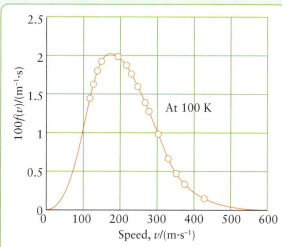

The points represent a typical result of a molecular speed distribution measurement. They are superimposed on the theoretical curve (Eq. 8). To obtain the fraction of molecules with speeds in the a narrow range from v to $v + \Delta v$, multiply $f(v)$ by Δv.

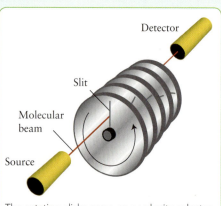

The rotating disks serve as a velocity selector in a molecular beam apparatus. Only molecules that are traveling with a specific speed can pass through the succession of slits.

distribution of molecular speeds, the intensity of the beam of molecules arriving at the detector for different rotational rates of the disks is measured. A typical result is shown in the graph in the second illustration, and shows a good match to the theoretical Maxwell expression.

There is a subtle point. The Maxwell distribution is for a sample of gaseous molecules free to move in three dimensions, not just the apparent one-dimensional arrangement characteristic of a beam. However, collisions occur within the beam, and on a molecular scale the beam is three-dimensional. As a result, the velocity along the direction of the beam has the characteristic distribution of a three-dimensional sample.

FIGURE 3D.6 The range of molecular speeds for three gases, as given by the Maxwell distribution. All curves correspond to the same temperature (300 K). The greater the molar mass, the lower is the average speed and the narrower the spread of speeds. To obtain the fraction of molecules with speeds in the range from v to $v + \Delta v$, multiply $f(v)$ by Δv.

LIVING GRAPH FIGURE 3D.6

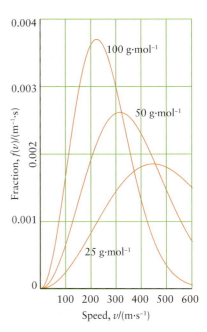

FIGURE 3D.6 shows a plot of the Maxwell distribution against speed for several different gases. You can see that heavy molecules (of molar mass, 100 g·mol^{-1}, for instance) travel with speeds close to their average values. Light molecules (of 20 g·mol^{-1}, for instance) not only have a higher average speed but also a wider range of speeds. Some molecules of gases with low molar masses have such high speeds that they can escape from the gravitational pull of small planets and go off into space. As a consequence, hydrogen molecules and helium atoms, which are both very light, are rare in the Earth's atmosphere, although they are abundant on massive planets such as Jupiter.

A plot of the Maxwell distribution for the same gas at several different temperatures shows that the average speed increases as the temperature is raised (**FIG 3D.7**), as expected from observations on diffusion and effusion; the curves also show that the spread of speeds widens as the temperature increases. At low temperatures, most molecules of a gas have speeds close to the average speed. At high temperatures, a high proportion of molecules have speeds that are very different from their average speed. Because the kinetic energy of a molecule in a gas is proportional to the square of its speed, the distribution of molecular kinetic energies follows the same trends.

The molecules of all gases have a wide range of speeds. As the temperature increases, the root mean square speed and the range of speeds both increase. The range of speeds is described by the Maxwell distribution, Eq. 8.

What have you learned in this Topic?

You have seen that effusion data show how the average speeds of molecules depend on the molar mass of the molecules and the temperature. You then saw how to use this information together with a model in which the pressure exerted by gases arises from the impact of the molecules on the walls of their container, to arrive at an equation that resembles the ideal gas law and that gives an expression for the root mean square speeds of molecules in a gas. You have also learned about the characteristic spread of speeds as expressed by the Maxwell distribution, which is broad at high temperatures and for gases with small molar masses.

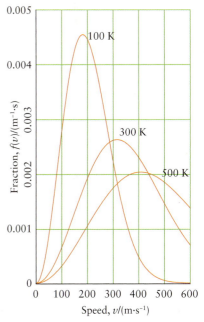

FIGURE 3D.7 The Maxwell distribution of a single substance (of molar mass 50 g·mol^{-1}) at different temperatures. The higher the temperature, the higher is the average speed and the broader the spread of speeds.

LIVING GRAPH FIGURE 3D.7

The skills you have mastered are the ability to:

☐ **1.** Use Graham's law to account for relative rates of effusion (Self-Test 3D.1).

☐ **2.** Describe the effect of temperature on average speed (Section 3D.1).

☐ **3.** List and explain the assumptions of the kinetic model of gases (Section 3D.2).

☐ **4.** Calculate the root mean square speed of molecules in a sample of gas (Example 3D.1).

☐ **5.** Describe the effect of molar mass and temperature on the Maxwell distribution of molecular speeds (Section 3D.3).

Topic 3D Exercises

Exercises labeled  require calculus.

3D.1 Do all the molecules of a gas strike the walls of their container with the same force? Justify your answer on the basis of the kinetic model of gases.

3D.2 How does the frequency of collisions of the molecules of a gas with the walls of the container change as the volume of the gas is decreased at constant temperature? Justify your answer on the basis of the kinetic model of gases.

3D.3 What is the molecular formula of a compound of empirical formula CH that diffuses 1.24 times more slowly than krypton at the same temperature and pressure?

3D.4 What is the molar mass of a compound that takes 2.7 times as long to effuse through a porous plug as it did for the same amount of XeF_2 at the same temperature and pressure?

3D.5 A sample of argon gas effuses through a porous plug in 147 s. Calculate the time required for the same amount of (a) CO_2, (b) C_2H_4, (c) H_2, and (d) SO_2 to effuse under the same conditions of pressure and temperature.

3D.6 A sample of neon gas effuses through a porous plug in 62.1 s. Calculate the time required for the same amount of (a) CO, (b) NH_3, (c) Kr, and (d) NO to effuse under the same conditions of pressure and temperature.

3D.7 A hydrocarbon of empirical formula C_2H_3 takes 349 s to effuse through a porous plug; under the same conditions of temperature and pressure, it took 210. s for the same number of molecules of argon to effuse. What is the molar mass and molecular formula of the hydrocarbon?

3D.8 The composition of a compound used to make polyvinyl chloride (PVC) is 38.4% C, 4.82% H, and 56.8% Cl by mass. It took 7.73 min for a given volume of the compound to effuse through a porous plug, but it took only 6.18 min for the same amount of Ar to diffuse at the same temperature and pressure. What is the molecular formula of the compound?

3D.9 Calculate the root mean square speeds of (a) methane, (b) ethane, and (c) propane molecules, all at −20. °C.

3D.10 Calculate the root mean square speed of molecules of (a) fluorine, (b) chlorine, and (c) bromine, all at 220. °C.

3D.11 The root mean square speed of methane molecules, CH_4, at a certain temperature was found to be 550. $m \cdot s^{-1}$. What is the root mean square speed of krypton atoms at the same temperature?

3D.12 In an experiment on gases, you are studying a sample of hydrogen gas of volume 1.00 L at 25 °C and 0.150 atm. You heat the gas until the root mean square speed of the molecules of the sample has increased by a factor of three. What will be the final pressure of the gas?

3D.13 An atmospheric probe reported that on a certain day the root mean square speed of ozone molecules, O_3, in the stratosphere over Fiji was 375 $m \cdot s^{-1}$. What was the temperature of that region of the stratosphere?

3D.14 In a study of turbulence caused by jet engines, the root mean square speed of nitrogen molecules in the air in a wind chamber was 495 $m \cdot s^{-1}$. What was the temperature in the wind chamber?

3D.15 A bottle contains 1.0 mol He(g) and a second bottle contains 1.0 mol Ar(g) at the same temperature. At that temperature, the root mean square speed of He is 1477 $m \cdot s^{-1}$ and that of Ar is 467 $m \cdot s^{-1}$. What is the ratio of the number of He atoms in the first bottle to the number of Ar atoms in the second bottle having these speeds? Assume that both gases behave ideally.

3D.16 A scientist is investigating a sample of 1.00 mol He(g) in a large cylinder at 200. K. Calculate by what factor each of the following variables would change the pressure of the gas and its root mean square speed: (a) The gas is compressed into a third of its original volume. (b) The temperature is reduced to 100. K. (c) The helium is replaced by the same amount of xenon atoms.

3D.17 (a) From the graph of Maxwell distribution of speeds in Fig. 3D.6, find the location that represents the most probable speed of the molecules at each temperature. (b) What happens to the fraction of molecules having a speed in the narrow range Δv centered on the most probable speed, v_{mp}, as the temperature increases (Eq. 8)?

3D.18 (a) From the graph of Maxwell distribution of speeds in Fig. 3D.6, estimate the most probable speed of the molecules for each molar mass. (b) What happens to the fraction of molecules having a speed in the narrow range Δv centered on the most probable speed, v_{mp}, as the molar mass of the gas increases (Eq. 8)?

3D.19 The number of molecules in a gas sample that have a speed in the narrow range Δv centered on the most probable speed (v_{mp}) at a temperature T is one-fourth the number of the same type of molecules that have a speed in the same range centered on the most probable speed at 200. K. What is the temperature?

3D.20 The number of molecules in a gas sample that have a speed in the narrow range Δv centered on the most probable speed (v_{mp}) at a temperature T is one-half the number of the same type of molecules that have a speed in the same range centered on the most probable speed at 300. K. What is the temperature?

Topic 3E Real Gases

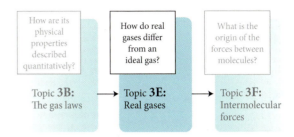

Topic 3B:
The gas laws

→ Topic 3E:
Real gases

→ Topic 3F:
Intermolecular
forces

Why Do You Need to Know This Material? Equations and models developed for limiting conditions, such as the ideal gas law, might not work well for real systems, and it is important to be able to account for departures from ideal behavior.

What Do You Need to Know Already? You need to be familiar with the nature of gases (Topic 3A), the ideal gas law (Topic 3B), and the kinetic model of gases (Topic 3D).

Under some conditions the ideal gas law fails. In industry and in many research laboratories, gases must be used under high pressures, but the ideal gas law is a *limiting law*, valid only as $P \rightarrow 0$. All actual gases, which are called **real gases,** have properties that differ from those predicted by the ideal gas law. These differences can be used to suggest modifications of the description of an ideal gas so that it applies at higher pressures.

3E.1 Deviations from Ideality

The effect of intermolecular forces can be assessed quantitatively by comparing the behavior of real gases with that expected of an ideal gas. One of the best ways of exhibiting these deviations is to measure the **compression factor,** Z, the ratio of the actual molar volume of the gas to the molar volume of an ideal gas under the same conditions:

$$Z = \frac{V_m}{V_m^{\text{ideal}}} \tag{1}$$

The compression factor of an ideal gas is 1, and so deviations from $Z = 1$ are a sign of nonideality. **FIGURE 3E.1** shows the experimental variation of Z for a number of gases. You can see that all gases deviate from $Z = 1$ as the pressure is raised. A refined model of gases must account for this behavior.

Deviations from ideal behavior can be explained by the presence of **intermolecular forces,** the attractions and repulsions between molecules. Topic 3F describes the origin of intermolecular forces. Here, all you need to know is that all molecules attract one another when they are a few molecular diameters apart but (provided they do not react) repel one another when their electron clouds come into contact. **FIGURE 3E.2** shows how the potential energy of a molecule varies with its distance from a second molecule. At moderate separations, its potential energy is lower than when it is infinitely far away: attractions always lower the potential energy of an object. As the molecules come into contact and repel each other, the potential energy starts to rise, because repulsions always increase the potential energy of an object.

The presence of attractive intermolecular forces accounts for the observation that, at low enough temperatures, gases condense to liquids when they are compressed. Compression brings the molecules closer together, and neighboring molecules can be captured by one another's attraction, provided that they are traveling slowly enough (that is, the sample is cool enough). The low compressibility of liquids and solids is consistent with the presence of strong repulsive forces when molecules are in contact. Another way

FIGURE 3E.1 A plot of the compression factor, Z, against pressure for a variety of gases. An ideal gas has $Z = 1$ for all pressures. For most gases, at low pressures the attractive forces are dominant and $Z < 1$ (see inset). At high pressures, repulsive forces become dominant and $Z > 1$ for all gases. For a few real gases with very weak intermolecular attractions, such as H_2, Z is always greater than 1.

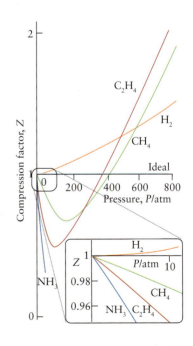

🠶 **LIVING GRAPH FIGURE 3E.1**

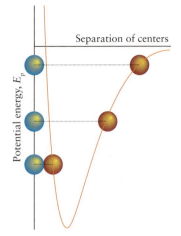

FIGURE 3E.2 The variation in potential energy of a molecule as it approaches another molecule. The potential energy rises sharply once the two molecules come into direct contact.

LIVING GRAPH FIGURE 3E.2

The name "virial" comes from the Latin word meaning "force."

to describe the repulsive forces between molecules is to say that molecules have definite volumes. When you touch a solid object, you feel its size and shape because your fingers cannot penetrate into it. The resistance to compression offered by the solid is due to the repulsive forces exerted by its atoms on the atoms in your fingers.

The presence of intermolecular forces also accounts for the variation in the compression factor. Thus, for gases under conditions of pressure and temperature such that $Z > 1$, the repulsions are more important than the attractions. Their molar volumes are greater than expected for an ideal gas because repulsions tend to drive the molecules apart. For example, a hydrogen molecule has so few electrons that its molecules are only very weakly attracted to one another. For gases under conditions of pressure and temperature such that $Z < 1$, attractions are more important than repulsions and the molar volume is smaller than for an ideal gas because attractions tend to draw molecules together. To improve the kinetic model of a gas, the effects of the attractive and repulsive forces that the molecules of a real gas exert on one another need to be included.

Real gases consist of atoms or molecules with intermolecular attractions and repulsions. Attractions have a longer range than repulsions. The compression factor is a measure of the strength and type of intermolecular forces. When $Z > 1$, intermolecular repulsions are dominant; when $Z < 1$, attractions are dominant.

3E.2 Equations of State of Real Gases

The behavior of real gases that do not obey the ideal gas law needs to be described mathematically so that more accurate predictions can be made. A common procedure in chemistry is to suppose that the term on one side of an equation (such as nRT in $PV = nRT$ of the ideal gas law) is just the leading (and predominant) term of a more complicated expression. Thus, to extend the ideal gas law to real gases, one procedure is to write

$$PV = nRT\left(1 + \frac{B}{V_m} + \frac{C}{V_m^2} + \cdots\right) \tag{2}$$

This expression is called the **virial equation.** The coefficients $B, C, \ldots$ are called the **second virial coefficient, third virial coefficient,** and so on. The virial coefficients, which depend on the temperature, are found by fitting experimental data to this equation.

Although the virial equation can be used to make accurate predictions about the properties of a real gas, provided the virial coefficients for the temperature of interest are known, it is not a source of much insight without a lot of advanced analysis. An equation that is less accurate but easier to interpret was proposed by the Dutch scientist Johannes van der Waals. The **van der Waals equation** is

$$\left(P + a\frac{n^2}{V^2}\right)(V - nb) = nRT \tag{3}$$

The temperature-independent **van der Waals parameters** a and b are unique for each gas and are determined experimentally (**TABLE 3E.1**). Parameter a represents the role of *attractions,* so it is relatively large for molecules that attract each other strongly and for large molecules with many electrons. Notice that the values of a in Table 3E.1 for gases with polar molecules, such as water and ammonia, are greater than those for nonpolar molecules and atoms with similar or greater molar masses, such as neon and oxygen. Parameter b represents the role of *repulsions;* it can be thought of as representing the volume of an individual molecule (more precisely, the volume per mole of molecules), because it is the repulsive forces between molecules that prevent one molecule from occupying the space already occupied by another molecule. You can see in Table 3E.1 that the values of b for the halogens increase as the size of the halogen atom increases.

One way to demonstrate the different roles of the two parameters is to express the compression factor in terms of them.

TABLE 3E.1 Van der Waals Parameters*

Gas	a/(bar·L²·mol⁻²)	b/(L·mol⁻¹)	Gas	a/(bar·L²·mol⁻²)	b/(L·mol⁻¹)
Noble gases			**Polar inorganic gases and vapors**		
helium	0.0346	2.38×10^{-2}	ammonia	4.225	3.71×10^{-2}
neon	0.208	1.67×10^{-2}	water	5.537	3.05×10^{-2}
argon	1.355	3.20×10^{-2}	carbon monoxide	1.472	3.95×10^{-2}
krypton	5.193	1.06×10^{-2}	hydrogen sulfide	4.544	4.34×10^{-2}
xenon	4.192	5.16×10^{-2}	**Nonpolar organic gases and vapors**		
Halogens			methane	2.303	4.31×10^{-2}
fluorine	1.171	2.90×10^{-2}	ethane	5.507	6.51×10^{-2}
chlorine	6.343	5.42×10^{-2}	propane	9.39	9.05×10^{-2}
bromine	9.75	5.91×10^{-2}	benzene	18.57	11.93×10^{-2}
Nonpolar inorganic gases					
hydrogen	0.2452	2.65×10^{-2}			
oxygen	1.382	3.19×10^{-2}			
carbon dioxide	3.658	4.29×10^{-2}			

*Within each category substances are arranged by increasing molar mass.

How Is That Done?

First, express the compression factor in terms of the pressure by writing $V_m = V/n$ and $V_m^{ideal} = RT/P$, so that

$$Z = \frac{V/n}{RT/P} = \frac{PV}{nRT}$$

Next, rearrange the van der Waals expression into an expression for P by dividing each side by $V - nb$ and then subtracting an^2/V^2 from both sides:

$$P = \frac{nRT}{V - nb} - a\frac{n^2}{V^2} \tag{4}$$

Now substitute this expression for P into the preceding expression for Z:

$$Z = \frac{V}{nRT} \times \left(\frac{nRT}{V - nb} - a\frac{n^2}{V^2} \right) = \frac{V}{V - nb} - \frac{an}{RTV}$$

Finally, divide the numerator and denominator of the term in blue by V. The result is given below as Eq. 5.

The expression for Z in terms of the parameters a and b that has been derived is

$$Z = \frac{1}{1 - nb/V} - \frac{an}{RTV} \tag{5}$$

What Does This Equation Tell You? For an ideal gas, a and b are both zero, and $Z = 1$. When the attractive contribution (a) is small, the second term on the right can be neglected. When the repulsive contribution (b) is appreciable, the denominator in the first term is less than 1 and the first term itself is greater than 1. As a result, when repulsions are dominant, $Z > 1$. Conversely, when the repulsive contribution is weak (b is small) and the attractive interaction is strong (a is large), the first term on the right is close to 1 and the second term on the right subtracts significantly from it. As a result, $Z < 1$ when attractions are dominant.

The values of a and b for a gas are found experimentally by fitting this expression for Z to curves like those in Fig. 3E.1. Once the parameters have been determined, they can be used in the van der Waals equation to predict the pressure of the gas under the conditions of interest.

EXAMPLE 3E.1 Estimating the pressure of a real gas

Investigators are studying the physical properties of a gas to be used as a refrigerant in an air-conditioning unit. Because refrigerants are used at low temperature and high pressure, they cannot be treated as ideal gases. A table of van der Waals parameters shows that for a certain refrigerant $a = 16.4$ bar·L^2·mol^{-2} and $b = 8.4 \times 10^{-2}$ L·mol^{-1}. Estimate the pressure of the gas when 1.50 mol occupies 5.00 L at 0 °C.

ANTICIPATE Fig. 3E.1 suggests that in most gases under normal conditions, the attractive interaction dominates the repulsive. So, you should suspect that the calculated pressure will be less than that based on an ideal gas, but the difference will be small, because real gases typically show only small deviations from ideal behavior.

PLAN Substitute the data into Eq. 4 after converting the temperature into the Kelvin scale. Use R in units that match those given.

SOLVE

From Eq. 4, $P = nRT/(V - nb) - an^2/V^2$,

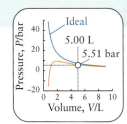

$$P = \frac{(1.50 \text{ mol}) \times (8.3145 \times 10^{-2} \text{ L·bar·K}^{-1}\text{·mol}^{-1}) \times (273 \text{ K})}{5.00 \text{ L} - (1.50 \text{ mol}) \times (8.4 \times 10^{-2} \text{ L·mol}^{-1})}$$

$$- (16.4 \text{ bar·L}^2\text{·mol}^{-2}) \times \frac{(1.50 \text{ mol})^2}{(5.00 \text{ L})^2}$$

$$= \frac{1.50 \times (8.3145 \times 10^{-2} \text{ bar}) \times 273}{5.00 - 1.50 \times 8.4 \times 10^{-2}} - (16.4 \text{ bar}) \times \frac{(1.50)^2}{(5.00)^2}$$

$$= 5.51 \text{ bar}$$

EVALUATE An ideal gas under the same circumstances exerts a pressure of 6.81 bar, so the "real" pressure (at least, as calculated from the van der Waals equation) of 5.51 atm is smaller, as anticipated.

Self-test 3E.1A A tank of volume 10.0 L containing 25 mol O_2 is stored in a diving supply shop at 25 °C. Use the data in Table 3E.1 and the van der Waals equation to estimate the pressure in the tank.

[*Answer:* 58.7 bar]

Self-test 3E.1B The properties of carbon dioxide gas are well known in the bottled beverage industry. In an industrial process, a tank of volume 100. L at 20. °C contains 20. mol CO_2. Use the data in Table 3E.1 and the van der Waals equation to estimate the pressure in the tank.

Related Exercises 3E.5, 3E.6, 3E.9, 3E.10

The virial equation is a general equation for describing real gases. The van der Waals equation is an approximate equation of state for a real gas; the parameter a represents the role of attractive forces and the parameter b represents the role of repulsive forces.

3E.3 The Liquefaction of Gases

At low temperatures, molecules of a gas move so slowly that, if it is a real gas, intermolecular attractions may result in one molecule being captured by others and sticking to them instead of moving freely. When the temperature falls below the boiling point of the substance, the gas condenses to a liquid (**FIG. 3E.3**).

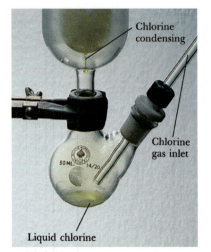

FIGURE 3E.3 Chlorine can be condensed to a liquid at atmospheric pressure by cooling it to −35 °C or lower. Here chlorine gas is introduced to the lower flask. The upper flask contains a "cold finger," a smaller tube filled with dry ice in acetone at −78 °C, on which the chlorine condenses. (*W. H. Freeman photo by Ken Karp.*)

FIGURE 3E.4 Cooling by the Joule–Thomson effect can be visualized as a slowing of the molecules (here shown blue) as they climb away from each other (in this case, from the one shown red) against the force of attraction between them.

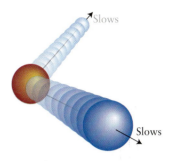

Gases can also be liquefied by making use of the relation between temperature and molecular speed. Because lower average speed corresponds to lower temperature, slowing the molecules is equivalent to cooling the gas. The molecules of a real gas can be slowed by making use of the attractions between them and allowing the gas to expand: the molecules have to climb away from one another's attractive forces when a gas expands, rather like a ball rising up above the surface of the Earth against the pull of gravity (**FIG. 3E.4**). Therefore, allowing the gas to occupy a larger volume, and hence increasing the average separation of the molecules, results in the molecules having a lower average speed. In other words, provided attractive effects are dominant, a real gas cools as it expands. This observation is called the **Joule–Thomson effect,** in honor of the scientists who first studied it, James Joule and William Thomson (later to become Lord Kelvin), the inventor of the absolute temperature scale.

Exceptions are helium and hydrogen, which have very weak attractive interactions and relatively strong repulsive interactions; at room temperature, they get warmer as they expand.

The Joule–Thomson effect is used in some commercial refrigerators to liquefy gases. The gas to be liquefied is compressed and then allowed to expand through a small hole, called the throttle. The gas cools as it expands, and the cooled gas circulates past the incoming high-pressure gas (**FIG. 3E.5**). This contact cools the incoming gas before it expands and cools still further. As the gas is continually recompressed and recirculated, its temperature progressively falls until finally it condenses to a liquid. If the gas is a mixture, such as air, then the liquid that it forms can later be distilled to separate its components. This technique is used for harvesting nitrogen, oxygen, neon, argon, krypton, and xenon from the atmosphere.

Many gases can be liquefied by making use of the Joule–Thomson effect, cooling brought about by expansion.

What have you learned in this Topic?

You have seen that deviations from ideal behavior are due to the attractions between molecules and their nonzero size. You now know that the virial equation adjusts the ideal gas law to provide a better description of the behavior of real gases and the van der Waals equation is an equation of state for a model real gas that uses two parameters to take into account the effect of molecular attractions and repulsions. You have also seen how intermolecular forces are used to liquefy a gas by the Joule–Thomson effect.

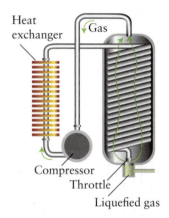

FIGURE 3E.5 A Linde refrigerator for liquefying gases. The compressed gas gives up heat to the surroundings in the heat exchanger (left) and passes through the coil (right). The gas is cooled still further by the Joule–Thomson effect as it emerges through the throttle. This gas cools the incoming gas and is recirculated through the system. Eventually, the temperature of the incoming gas is so low that it condenses to a liquid.

The skills you have mastered are the ability to:

☐ **1.** Explain how real gases differ from ideal gases (Section 3E.1).

☐ **2.** Use the van der Waals equation to estimate the pressure of a gas (Example 3E.1).

☐ **3.** Describe how the Joule–Thomson effect is used to cool gases (Section 3E.2).

Topic 3E Exercises

3E.1 The pressure of a sample of hydrogen fluoride is lower than expected and, as the temperature is increased, rises more quickly than the ideal gas law predicts. Suggest an explanation.

3E.2 Under what conditions of temperature and pressures would you expect a real gas to have (a) $Z < 1$; (b) $Z > 1$?

3E.3 Two identical flasks are each filled with a gas at 20 °C and 100 atm. One flask contains NH_3 and the other an equal amount of H_2. (a) In which flask do the molecules have the greater root mean square speed? (b) In which flask are the deviations from ideality stronger? Explain your reasoning.

3E.4 Two identical flasks are each filled with a gas at 20 °C and 100 atm. One flask contains NH_3 and the other an equal amount of H_2. (a) In which flask is the collision rate with the walls of the container greater? (b) For each flask, predict whether the pressure will be higher or lower than predicted by the ideal gas law. Explain your reasoning.

3E.5 Use the ideal gas equation to calculate the pressure at 298 K exerted by 1.00 mol $CO_2(g)$ when confined in a volume of (a) 15.0 L; (b) 0.500 L; (c) 50.0 mL. Repeat these calculations using the van der Waals equation. What do these calculations indicate about how the reliability of the ideal gas law depends on the pressure?

3E.6 Use the ideal gas equation to calculate the pressure at 298 K exerted by 1.00 mol $H_2(g)$ when confined in a volume of

(a) 30.0 L; (b) 1.00 L; (c) 50.0 mL. Repeat these calculations using the van der Waals equation. What do these calculations indicate about how the reliability of the ideal gas law depends on the pressure?

3E.7 The following table lists the van der Waals a parameters for HCl, CH_3CN, Ne, and CH_4. Without referring to Table 3E.1, use your knowledge of the factors governing the magnitudes of a to assign each of these four gases to its a value. Explain your reasoning.

$a/(\text{bar}\cdot\text{L}^2\cdot\text{mol}^{-2})$	17.813	3.700	2.303	0.208
Substance	?	?	?	?

3E.8 The following table lists the van der Waals b parameters for Br_2, N_2, He, and SF_6. Without referring to Table 3E.1, use your knowledge of the factors governing the magnitudes of b to assign each of these four gases to its b value. Explain your reasoning.

$100b/(\text{L}\cdot\text{mol}^{-1})$	8.79	5.91	3.87	2.38
Substance	?	?	?	?

3E.9 Calculate the pressure exerted by 1.00 mol H_2S, behaving as (a) an ideal gas; (b) a van der Waals gas when it is confined under the following conditions: (i) at 273.15 K in 22.414 L; (ii) at 800. K in 60.0 L.

3E.10 Calculate the pressure exerted by 1.00 mol C_2H_6(ethane), behaving as (a) an ideal gas; (b) a van der Waals gas when it is confined under the following conditions: (i) at 273.15 K in 1.121 L; (ii) at 1.00×10^3 K in 40.0 L.

3E.11 Plot pressure against volume for 1 mol of (a) ideal gas molecules, (b) ammonia molecules, and (c) oxygen molecules, for the range $V = 0.05$ L to 1.0 L at 298 K. Use the van der Waals equation to calculate the pressures of the two real gases. Use a spreadsheet or the graphical plotter on the Web site for this book. Compare the plots and explain the origin of any differences you find.

3E.12 Plot pressure against volume for 1 mol of (a) ideal gas molecules, (b) carbon dioxide molecules, (c) ammonia molecules, and (d) benzene molecules for the range $V = 0.1$ L to 1.0 L at 375 K. Use the van der Waals equation to determine the pressures of the real gases. Use a spreadsheet or the graphical plotter on the Web site for this book. Compare the shapes of the curves and explain the differences.

3E.13 (a) The van der Waals parameters for helium are $a = 3.46 \times 10^{-2}$ bar·L^2·mol^{-2} and $b = 2.38 \times 10^{-2}$ L·mol^{-1}. Calculate the apparent volume (in pm^3) and radius (in pm) of a helium atom as determined from the van der Waals parameters. (b) Estimate the volume of a helium atom on the basis of its atomic radius (Appendix 2D). (c) How do these quantities compare? Should they be the same? Discuss.

3E.14 Show that the van der Waals parameter b is related to the molecular volume V_{mol}, the volume occupied by one molecule, by $b = 4N_A V_{mol}$. Treat the molecules as spheres of radius r, so that $V_{mol} = \frac{4}{3}\pi r^3$. The closest that the centers of two molecules can approach is $2r$.

Topic 3F Intermolecular Forces

Molecules attract one another. Important consequences spring from that simple fact. Without forces between molecules, your flesh would drip off your bones and the oceans would be gas. Less dramatically, the forces between molecules govern the physical properties of bulk matter. They explain why CO_2 is a gas whereas SiO_2 is a solid, and why ice floats on water.

Molecules also repel one another. When pressed together, molecules resist further compression. As a result, you do not sink into the floor, and solid objects have a specific size and shape. Even in a gas, repulsions are important because they mean that molecules do not simply pass through one another but undergo actual collisions.

In gases, intermolecular forces lead to small deviations from ideal behavior (Topic 3E). In liquids and solids, on the other hand, the forces that hold molecules together are of great importance and control their physical properties. Individual water molecules, for instance, neither freeze nor boil, but bulk water does, because molecules stick together in the process of freezing and form a rigid array and in boiling they separate from one another and form a gas.

3F.1 The Origin of Intermolecular Forces

Intermolecular forces are responsible for the existence of several different "phases" of matter. A **phase** is a form of matter that is uniform throughout in both chemical composition and physical state. The phases of matter include the three common physical states, solid, liquid, and gas (or vapor), introduced in *Fundamentals* A. Many substances have more than one solid phase, each phase having different arrangements of its atoms or molecules. For instance, carbon has several solid phases; one is the hard, transparent diamond used in jewelry and cutting tools and another is the soft, black graphite used in common pencil "lead" and as a lubricant. Most solids melt to a liquid when heated. Both solids and liquids are called **condensed phases** because when a gas is cooled it condenses (becomes more compact) to a liquid or a solid. The temperature at which it condenses depends on the strength of the attractive forces between its molecules.

All interionic and almost all intermolecular interactions can be traced to the Coulomb interaction between charges (Topic 2A). The following discussion of intermolecular interactions is based on Eq. 5 of *Fundamentals* A, the expression for the potential energy E_p of two charges Q_1 and Q_2 separated by a distance r:

$$E_p = \frac{Q_1 Q_2}{4\pi\varepsilon_0 r} \tag{1}$$

This expression applies directly to ions, but it underlies all intermolecular interactions, even those between neutral molecules.

FIGURE 3F.1 shows a **molecular potential energy curve,** a graph that shows how the potential energy of a pair of molecules changes with the distance between their centers. In one case the molecules react and a chemical bond forms; in the other case a bond does not form. In each case, their energy falls initially as the widely separated molecules come

Why Do You Need to Know This Material? The formation and properties of liquids and solids depends on the presence of attractive and repulsive forces between molecules, and it is essential to understand their origin.

What Do You Need to Know Already? This Topic uses the concepts of potential energy (*Fundamentals* A), Coulomb interactions (Topic 2A), and polarizability (Topic 2D). You need to be aware of the existence of partial charges and dipole moments of polar molecules (Topic 2E), and intermolecular forces in gases (Topic 3E).

The vacuum permittivity, ε_0 (epsilon zero) is a fundamental constant with a value of 8.854×10^{-12} $J^{-1} \cdot C^2 \cdot m^{-1}$.

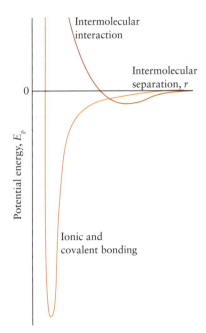

FIGURE 3F.1 The distance dependence of the potential energy of the interaction between atoms that form a bond (orange, lower line) and those that do not form bonds (red, upper line). Notice the deep energy well that is a sign of bond formation. The dip in the upper line is an indication that even when atoms or molecules do not form a bond, attractive forces lower the energies of the particles.

TABLE 3F.1   The Interactions Between Ions and Molecules*

Type of interaction	E_p dependence	Typical energy $E_p/(\text{kJ}\cdot\text{mol}^{-1})$	Interacting species
ion–ion	$-\lvert z\rvert^2/r$	250	ions
ion–dipole	$-\lvert z\rvert\mu/r^2$	15	ions and polar molecules
dipole–dipole	$-\mu_1\mu_2/r^3$	2	stationary polar molecules
	$-\mu_1\mu_2/r^6$	0.3	rotating polar molecules
dipole–induced-dipole	$-\mu_1^2\alpha_2/r^6$	2	molecules, at least one must be polar
London (dispersion)[†]	$-\alpha_1\alpha_2/r^6$	2	all types of molecules and ions
hydrogen bonding[‡]		20	molecules containing an N—H, O—H, or F—H bond

*The total interaction experienced by a species is the sum of all the interactions in which it can participate. In the expressions, r is the distance between the centers of the interacting particles, z is the charge number of an ion, μ is the electric dipole moment, and α is the polarizability of a molecule.

[†]Also known as the induced-dipole–induced-dipole interaction. Interactions that are proportional to $1/r^6$ are commonly regarded as van der Waals interactions.

[‡]The hydrogen bond is best regarded as a contact interaction.

closer together, indicating that attractive forces dominate. However, as their separation becomes very small, repulsions dominate and the potential energy climbs sharply to high values. The shapes of the two curves are similar, but the minimum is much shallower and at a greater distance for nonbonding interactions than for bond formation.

There are several types of interaction between ions, between ions and neutral molecules, and between neutral moleces. They and their typical strengths are summarized in **TABLE 3F.1** and discussed in more detail in the following sections. Notice that the interactions between ions are much stronger than those between molecules or dipoles. Covalent bonds are also much stronger than intermolecular interactions (compare Table 3F.1 with Table 2D.2).

Condensed phases form when attractive forces between molecules pull them together; repulsions dominate at short separations.

3F.2 Ion–Dipole Forces

When an ionic solid is added to water, a number of water molecules surround each ion on the surface of the solid, separate it from the other ions, and progressively dissolve it. The partial charges of the surrounding water molecules replace the charges of the ionic neighbors in the solid, so the ions slip into solution with little change of energy.

The attachment of water molecules to solute particles, particularly but not only ions, is called **hydration.** Hydration of ions is due to the polar character of the H_2O molecule (**1**). The partial negative charge on the O atom is attracted to cations, and the partial positive charges of the H atoms are repelled by them. Therefore, water molecules can be expected to cluster around a cation with the O atom pointing inward and the H atoms pointing outward (**FIG. 3F.2**, left). The reverse arrangement is found around an anion: the H atoms have partial positive charges, and so they are attracted to the anion's negative charge (Fig. 3F.2, right). Because hydration arises from the interaction between an ion and the partial charges on polar water molecules, it is an example of an **ion–dipole interaction.**

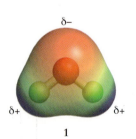

1

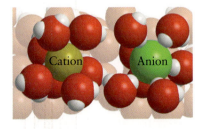

FIGURE 3F.2 In water, ions are hydrated. On the left, a cation is surrounded by water molecules, oriented with their partially negatively charged oxygen atoms facing the ion. On the right, an anion is surrounded by water molecules that direct their partially positively charged hydrogen atoms toward the ion.

The potential energy of interaction between the full charge of an ion and the two partial charges of a polar molecule is proportional to $-|z|\mu/r^2$ (Table 3F.1), where z is the charge number of the ion and μ is the dipole moment of the polar molecule. The negative sign means that the potential energy of the ion and its surrounding solvent molecules is *lowered* by their interaction, which signifies attraction.

THINKING POINT

What other solvents might be able to form strong ion–dipole interactions analogous to those formed by water?

The ion–dipole interaction is significant only when polar molecules are very close to an ion. Even then the interaction is still weaker than the attraction between two ions, because the dipole moment of a polar molecule arises only from partial charges. In addition, an ion attracted to the partial charge at one end of a polar molecule is repelled by the opposite charge at the other end and the two effects partly cancel. At large distances, the two partial charges are at almost the same distance from the ion, and the cancellation is nearly complete: that is why the potential energy of interaction between a point charge and a dipole falls off more quickly with distance (as $1/r^2$) than the interaction of two point charges does (as $1/r$) (**FIG. 3F.3**).

Both the size of the ion and its charge control the extent of hydration. The strength of the ion–dipole interaction is greater the closer the dipole can come to the ion. Consequently, small cations are hydrated more extensively than large cations. In fact, lithium and sodium commonly form hydrated salts, whereas potassium, rubidium, and cesium, with their larger cations, do not. The effect of charge on the extent of hydration can be seen by comparing barium and potassium cations, which have similar radii (135 pm for Ba^{2+} and 138 pm for K^+). In the solid state potassium salts are not hydrated to any appreciable extent, but barium salts are often hydrated. The difference can be traced to the barium ion's higher charge.

Ion–dipole interactions are strong for small, highly charged ions; one consequence is that small, highly charged cations are often hydrated in compounds.

3F.3 Dipole–Dipole Forces

Now consider the interactions between polar molecules. An example of a polar molecule is chloromethane, CH_3Cl, with a partial negative charge on the Cl atom and a partial positive charge spread over the H atoms (**2**). In solid chloromethane, a low energy is achieved when the Cl atom on one CH_3Cl molecule is next to the CH_3 end of the next molecule (**FIG. 3F.4**). The interaction between dipoles, and specifically between their partial charges, is called the **dipole–dipole interaction,** and the resulting potential energy for stationary polar molecules in a solid is proportional to $-\mu^2/r^3$ if the molecules are identical and to $-\mu_1\mu_2/r^3$ if they are different (Table 3F.1). Note that the greater the polarity of the molecules, the stronger is the interaction between them. The strength of the interactions depends more strongly on distance (as $1/r^3$) than the ion–dipole interaction, and doubling the distance between the molecules reduces the strength of their interaction by a factor of $2^3 = 8$. The reason for the very rapid falling off of the dipole–dipole interaction is that as the distance between molecules increases, the opposite partial charges on *each* molecule appear to merge and cancel, whereas in the interaction between a point charge and a dipole, only the partial charges on the dipole appear to merge.

Now imagine the same polar molecules in a gas in which all the molecules are rotating rapidly (**FIG. 3F.5**). For perfectly free rotation, the attractions between opposite partial charges and the repulsions between like partial charges cancel and there is no net interaction between neighboring molecules. However, in reality the rotating neighbors linger slightly in energetically favorable orientations (their oppositely charged ends adjacent), and so the attractive interactions between opposite partial charges slightly outweigh the repulsive interactions between like partial charges. As a result, there is a weak net attraction between rotating neighboring polar molecules in the gas phase. It turns out

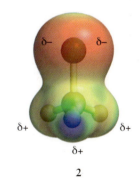

FIGURE 3F.3 The distance dependence of the potential energy of the interaction between ions (blue), ions and dipoles (green), stationary dipoles (orange), and rotating dipoles (red).

LIVING GRAPH FIGURE 3F.3

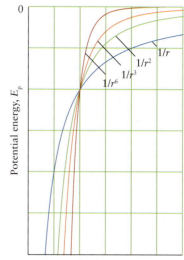

2

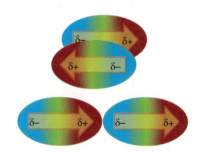

FIGURE 3F.4 Polar molecules attract each other by the interaction between the partial charges of their electric dipoles (represented by the arrows). Both orientations shown (end to end or side by side) result in a lower energy than a random distribution of orientations.

It takes about 1 ps for a small molecule to make a complete revolution in the gas phase.

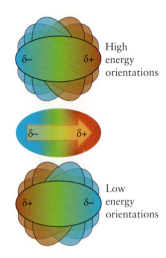

High energy orientations

Low energy orientations

FIGURE 3F.5 A polar molecule rotating near another polar molecule spends more time in the low-energy orientations (bottom) that maximize attractions, and so the net interaction is attractive but not as strong as it would be if the molecules were not rotating.

that the potential energy is proportional to $1/r^6$ (Table 3F.1). Doubling the separation of polar molecules reduces the strength of the interaction by a factor of $2^6 = 64$, and so dipole–dipole interactions between rotating molecules have a significant effect only when the molecules are very close.

You can now start to understand why the kinetic model accounts for the properties of gases so well (Topic 3D): gas molecules rotate almost freely and are far apart for most of the time, so any interactions between their molecules are very weak. Molecules also rotate in liquids (they do so in jerky steps rather than almost freely) but they are much closer together than in the gas phase and the dipole–dipole interactions are correspondingly stronger. Because considerable energy is required to separate molecules that are strongly attracted to one another, substances with strong intermolecular interactions have high boiling points.

3 *p*-Dichlorobenzene

4 *o*-Dichlorobenzene

5 *cis*-Dichloroethene

6 *trans*-Dichloroethene

7 **1,1**-Dichloroethene

EXAMPLE 3F.1   Predicting relative boiling points on the basis of dipole–dipole interactions

You are working on an organic synthesis and need to change a solvent to one that has a higher boiling point. Which would you expect to have the higher boiling point, *p*-dichlorobenzene (**3**) or *o*-dichlorobenzene (**4**)?

PLAN When two compounds have different dipole moments but are otherwise very similar, the molecules with the larger dipole moment interact more strongly. Therefore, assign the higher boiling point to the more strongly polar compound. To decide whether a molecule is polar, determine from the molecular structures whether the dipole moments of the bonds cancel each other, as explained in Topic 2E.

What should you assume? Assume that the polarity of a C—H group is much less than that of the a C—Cl group and therefore has little effect on the molecular polarity and can be ignored.

SOLVE

Because the two C—Cl bonds in *p*-dichlorobenzene lie directly across the ring, their dipole moments cancel and the molecule as a whole is nonpolar.

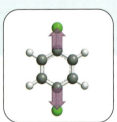

An *o*-dichlorobenzene molecule is polar because the two C—Cl bond dipoles do not cancel. Consequently, *o*-dichlorobenzene is predicted to have a higher boiling point than the nonpolar *p*-dichlorobenzene.

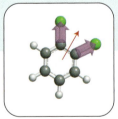

EVALUATE The experimental values are found to be 180 °C for *o*-dichlorobenzene and 174 °C for *p*-dichlorobenzene.

Self-test 3F.1A Which will have the higher boiling point, *cis*-dichloroethene (**5**) or *trans*-dichloroethene (**6**)?

[***Answer:*** *cis*-Dichloroethene]

Self-test 3F.1B Which will have the higher boiling point, 1,1-dichloroethene (**7**) or *trans*-dichloroethene?

Related Exercises 3F.5–3F.8

Polar molecules take part in dipole–dipole interactions. Dipole–dipole interactions are weaker than forces between ions and fall off rapidly with distance, especially in the liquid and gas phases, where the molecules rotate.

3F.4 London Forces

Attractive interactions are found even between nonpolar molecules. The evidence for these interactions includes the fact that the noble gases—which, because they are monatomic, are necessarily nonpolar—can be liquefied, and many nonpolar compounds, such as the hydrocarbons that make up gasoline, are liquids.

At first sight, there seems to be no means by which two nonpolar molecules could attract each other. The explanation lies in how electrons are distributed in a molecule. First, realize that representations of electron distributions and charge distributions (for example, structures **1** and **2**) show *average* values. In a nonpolar molecule or a single atom, the electrons appear to be symmetrically distributed. In fact, at any instant, the electron clouds of atoms and molecules are not uniform (**FIG. 3F.6**). If you could take a snapshot of a molecule at a given instant, the electron distribution would look like a swirling fog. At a given instant, electrons may accumulate in one part of a molecule, leaving a nucleus elsewhere partly exposed and in the next instant the accumulation is elsewhere. As a result, one region of the molecule will have a fleeting partial negative charge and another region will have a fleeting partial positive charge; in the next moment—about 10^{-16} s later—the charges may be reversed or in other locations. Even a nonpolar molecule may have an **instantaneous dipole moment,** a momentary dipolar separation of charge (**FIG. 3F.7**).

An instantaneous dipole moment on one molecule distorts the electron cloud on a neighboring molecule and induces a temporary dipole moment on that molecule; the two instantaneous dipoles attract each other. An instant later, the swirling electron cloud of the first molecule will give rise to a dipole moment in a different direction, but that new dipole moment induces a new dipole moment in the second molecule, and still the two molecules attract each other. That is, although there is a ceaseless flickering of the dipole moment of one molecule from one orientation to another, the induced dipole moment of the second molecule follows it faithfully, and there is a net attraction between the two molecules. This attractive interaction is called the **London dispersion interaction** or, more simply, the **London interaction.** It acts between *all* molecules and atoms and is the only interaction between nonpolar molecules and in monatomic gases.

The strength of the London interaction depends on the **polarizability,** α (alpha), of the molecules, the ease with which their electron clouds can be distorted. As described in Topic 2D, highly polarizable molecules are those in which the nuclear charges have little control over the surrounding electrons, perhaps because the atoms are large and the distance between electrons and the nucleus is great or because the valence electrons are well shielded by the inner electrons. There can then be considerable fluctuations in electron density, and hence highly polarizable molecules can have large instantaneous dipole moments and strong London interactions.

Detailed calculations show that the potential energy of the London interaction varies as $-\alpha^2/r^6$ when the molecules are identical and as $-\alpha_1\alpha_2/r^6$ when they are different (Table 3F.1). Like the potential energy of dipole–dipole interactions between rotating polar molecules, the potential energy of the London interaction decreases very rapidly with distance. The strength of the interaction increases with the polarizability of the interacting molecules. Because a large molecule with many electrons is typically more polarizable than a small molecule with only a few electrons, a large molecule is likely to have stronger London interactions than a smaller one (**FIG. 3F.8**).

FIGURE 3F.8 These hydrocarbons show how London forces increase in strength with molar mass. Pentane is a mobile liquid (left); pentadecane, $C_{15}H_{32}$, a viscous liquid (middle); and octadecane, $C_{18}H_{38}$, a waxy solid (right). To some extent, the effect of increasing intermolecular forces is enhanced by the ability of long-chain molecules to tangle with one another. (*W. H. Freeman photo by Ken Karp.*)

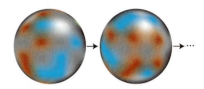

FIGURE 3F.6 The electron cloud around an atom resembles a constantly moving fog with instantaneous regions of enhanced or diminished electron density.

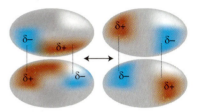

FIGURE 3F.7 Rapid fluctuations in the electron distribution in two neighboring molecules result in two instantaneous electric dipole moments that attract each other. The fluctuations flicker into different positions, but each new arrangement in one molecule induces an arrangement in the other that results in mutual attraction.

The London dispersion interaction is named after the German-American physicist Fritz London, who established the existence of dispersion forces between atoms of noble gases.

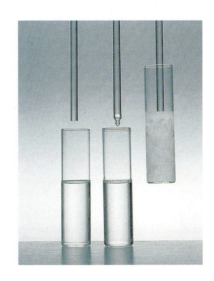

TABLE 3F.2 Melting and Boiling Points of Substances*

Substance	Melting point/°C	Boiling point/°C	Substance	Melting point/°C	Boiling point/°C
Noble gases			**Inorganic substances with small molecules**		
He	−270 (3.5 K)[†]	−269 (4.2 K)	H_2	−259	−253
Ne	−249	−246	N_2	−210	−196
Ar	−189	−186	O_2	−218	−183
Kr	−157	−153	H_2O	0	100
Xe	−112	−108	H_2S	−86	−60
Halogens			NH_3	−78	−33
F_2	−220	−188	CO_2	—	−78s
Cl_2	−101	−34	SO_2	−76	−10
Br_2	−7	59	**Organic compounds**		
I_2	114	184	CH_4	−182	−162
Hydrogen halides			CF_4	−150	−129
HF	−93	20	CCl_4	−23	77
HCl	−114	−85	C_6H_6	6	80
HBr	−89	−67	CH_3OH	−94	65
HI	−51	−35	glucose	142	d
			sucrose	184d	—

*Abbreviations: s, solid sublimes; d, solid decomposes.
[†] Under pressure.

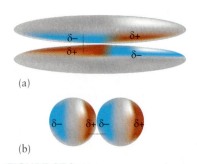

8 Pentane, C_5H_{12}

9 2,2-Dimethylpropane, $C(CH_3)_4$

(a)

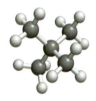

(b)

FIGURE 3F.9 (a) Instantaneous dipole moments in two neighboring rod-shaped molecules tend to be close together and interact strongly over a relatively broad region of the molecule, as indicated by the vertical lines. (b) Those on neighboring spherical molecules tend to be farther apart and interact weakly over only a small region of the molecule.

You can now see why the halogens range from gases (F_2 and Cl_2), through a liquid (Br_2), to a solid (I_2) at room temperature: the molecules have an increasing number of electrons, so their polarizabilities and hence their London interactions increase on going down the group. The increase in strength of London forces is striking when heavier atoms are substituted for hydrogen atoms. Methane boils at −16 °C, but tetrachloromethane (carbon tetrachloride, CCl_4) has many more electrons and is a liquid that boils at 77 °C (**TABLE 3F.2**). Tetrabromomethane, CBr_4, has even more electrons: it is a solid at room temperature which melts at 94 °C and boils at 190 °C.

The effectiveness of London forces also depends on the shapes of molecules. Both pentane (**8**) and 2,2-dimethylpropane (**9**), for instance, have the molecular formula C_5H_{12}, and so they each have the same number of electrons. However, they have different boiling points (36 °C and 10 °C, respectively). Molecules of pentane are relatively long and rod shaped. The instantaneous partial charges on adjacent rod-shaped molecules can be in contact at several points, leading to strong interactions. In contrast, the instantaneous partial charges on more spherical molecules such as 2,2-dimethylpropane cannot get so close to one another, because only a small region of each molecule can be in contact (**FIG. 3F.9**). As a result of the strong dependence on distance, the London interactions between rod-shaped molecules are more effective than those between spherical molecules with the same number of electrons.

Closely related to the London interaction is the **dipole–induced-dipole interaction,** in which a polar molecule interacts with a nonpolar molecule (for example, when oxygen dissolves in water). Like the London interaction, the dipole–induced-dipole interaction arises from the ability of one molecule to induce a dipole moment in the other. However, in this case, the molecule that induces the dipole moment has a permanent dipole moment. The potential energy of the interaction is once again inversely proportional to the sixth power of the separation.

As summarized in Table 3F.1, the potential energies of the dipole–dipole interaction of rotating polar molecules in the gas phase, the London interaction, and the dipole–induced-dipole interaction all depend on the inverse sixth power of the separation. They are known collectively as **van der Waals interactions,** after Johannes van der Waals, the Dutch scientist who studied them extensively.

EXAMPLE 3F.2   Accounting for a trend in boiling points

Chemists often gain insight into physical properties through examination of trends in closely related molecules. Explain the trend in the boiling points of the hydrogen halides: HCl, $-85\,°C$; HBr, $-67\,°C$; HI, $-5\,°C$.

PLAN Stronger intermolecular forces result in higher boiling points. The dipole moments, and therefore the strength of the dipole–dipole interactions, increase with the polarity of the H—X bond and therefore with the differences in electronegativity between hydrogen and the halogens. The strength of London forces increases with the number of electrons. Use the data to identify the dominant effect.

SOLVE The data in Fig. 2D.2 show that electronegativity differences decrease from HCl to HI, and so the dipole moments decrease as well. Therefore, dipole–dipole forces decrease too, a trend suggesting that the boiling points should decrease from HCl to HI. This prediction conflicts with the data, so the London forces must be considered. The number of electrons in a molecule increases from HCl to HI, and so the strength of the London interaction increases, too. Therefore, the boiling points should increase from HCl to HI, in accord with the data. This analysis suggests that London forces dominate dipole–dipole interactions for these molecules.

Self-test 3F.2A Account for the trend in boiling points of the noble gases, which increase from helium to xenon.

[**Answer:** The strength of the London interaction increases as the number of electrons increases.]

Self-test 3F.2B Suggest a reason why trifluoromethane, CHF_3, has a higher boiling point than tetrafluoromethane, CF_4.

Related Exercises 3F.15, 3F.16, 3F.19

The London interaction arises from the attraction between instantaneous dipoles on neighboring molecules and acts between all types of molecules; its strength increases with the number of electrons and occurs in addition to any dipole–dipole interactions. Polar molecules also attract nonpolar molecules by weak dipole-induced-dipole interactions.

3F.5 Hydrogen Bonding

The London interaction is "universal" in the sense that it applies to all molecules regardless of their chemical identity. Similarly, the dipole–dipole interaction depends only on the polarity of the molecule regardless of its chemical identity. However, there is another very strong interaction between molecules that is specific to molecules containing hydrogen atoms bonded to certain elements.

A plot of the boiling points of the common binary hydrogen compounds of the elements in Groups 14 to 17 suggests the presence of a special type of interaction (**FIG. 3F.10**). The trend in Group 14 is what you should expect for similar compounds that differ in their number of electrons—the boiling points increase down the group because the strength of the London interaction increases. However, ammonia, water, and hydrogen fluoride all show anomalous behavior. Their exceptionally high boiling points suggest that there are unusually strong attractive forces between their molecules.

The strong interaction responsible for the high boiling points of these substances and certain others is the **hydrogen bond,** an intermolecular interaction in which a hydrogen

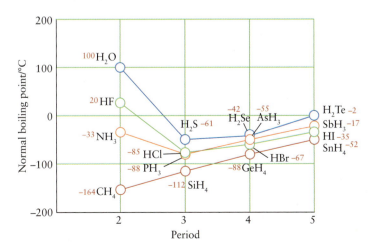

FIGURE 3F.10 The boiling points of most of the molecular hydrides of the p-block elements show a smooth increase with molar mass in each group. However, the boiling points of three compounds— ammonia, water, and hydrogen fluoride— are strikingly out of line.

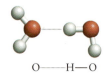

10　Hydrogen bond (in water)

11　Hydrogen fluoride (HF)$_n$

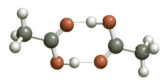

12　Acetic acid dimer

atom bonded to a small, strongly electronegative atom, specifically N, O, or F, is attracted to a lone pair of electrons on another N, O, or F atom (**10**).

To understand how it forms, imagine what happens when one water molecule comes close to another. Each O—H bond is polar. The electronegative O atom exerts a strong pull on the electrons in the bond, and the proton of the H atom is almost completely unshielded. Because it is so small, the hydrogen atom, with its partial positive charge, can get very close to one of the lone pairs of electrons on the O atom of another water molecule. The lone pair and the partial positive charge attract each other strongly and form a hydrogen bond. Hydrogen bonding is strongest when the hydrogen atom is on a straight line between the two oxygen atoms. A hydrogen bond is denoted by a dotted line, so the hydrogen bond between two O atoms is denoted O—H···O. The O—H bond length is 101 pm and the H···O distance is somewhat longer; in ice it is 175 pm. Hydrogen bonding is possible for HF and for any molecule that contains an N—H or O—H bond.

> **Self-test 3F.3A** Which of the following intermolecular links can be made by hydrogen bonds: (a) CH_3NH_2 to CH_3NH_2; (b) CH_3OCH_3 to CH_3OCH_3; (c) HBr to HBr?
> [**Answer:** Only in (a) is H bonded directly to N, O, or F.]
>
> **Self-test 3F.3B** Which of the following molecules can take part in hydrogen bonding with other molecules of the same compound: (a) CH_3OH; (b) PH_3; (c) HClO (which has the structure Cl—O—H)?

When it can form, a hydrogen bond is so strong—about 10% of typical covalent bond strengths—that it dominates all the other types of intermolecular interactions. Hydrogen bonding is strong enough to survive even in the vapor of some substances. Liquid hydrogen fluoride, for instance, contains zigzag chains of HF molecules (**11**), and the vapor contains short fragments of the chains and (HF)$_6$ rings. The vapor of acetic acid, CH_3COOH, contains **dimers,** pairs of identical molecules; here they are linked by two hydrogen bonds (**12**).

Hydrogen bonds play a vital role in maintaining the shapes of biological molecules. The overall shape of a protein molecule is governed largely by hydrogen bonds; once the bonds are broken, the delicately organized protein molecule loses its function. When you cook an egg, the clear albumen becomes milky white, because as the temperature rises the molecules move more rapidly and the more vigorous motion breaks the hydrogen bonds in the protein molecules, which then collapse into a random jumble. Trees are held upright by hydrogen bonds (**FIG. 3F.11**). Cellulose molecules (which have many —OH groups) can form many hydrogen bonds with one another, and the strength of wood is due in large part to the strength of the hydrogen bonds between neighboring ribbonlike cellulose molecules. Hydrogen bonding also links the two chains of a DNA molecule together, and so hydrogen bonding is a key to understanding reproduction (see Topic 11E). The hydrogen bonds are strong enough to keep the two chains of a DNA molecule together but so much weaker than typical covalent bonds that they readily give way in cell division without affecting the covalent bonds within the strands of the DNA.

> *Hydrogen bonding, which occurs when hydrogen atoms are bonded to oxygen, nitrogen, or fluorine atoms, is the strongest type of intermolecular interaction.*

3F.6 Repulsions

When molecules (or atoms that do not form bonds) are very close together, they repel one another. This repulsion, which is responsible for the steep rise in potential energy on the left of Fig. 3F.1, can be traced to the Pauli exclusion principle (Topic 3E). To see why this is so, think about two He atoms approaching each other. At short separations, their 1s atomic orbitals overlap and form a bonding and an antibonding molecular orbital. Two of the four electrons supplied by the two atoms occupy the lower, bonding orbital; the Pauli exclusion principle then requires the remaining two electrons to occupy the antibonding orbital. An antibonding orbital increases the energy slightly more than a bonding orbital lowers the energy (Topic 2G), so the net effect is an increase in energy as the two

FIGURE 3F.11 Vegetation, even huge trees such as these, is held upright by the powerful intermolecular hydrogen bonds that exist between the ribbonlike cellulose molecules that form much of its bulk. Without hydrogen bonding, these trees would collapse. (*J.A. Kraulis/Masterfile.*)

atoms merge into each other. The effect increases sharply as the distance decreases because the overlap between the atomic orbitals increases very rapidly as they approach. The same repulsion occurs for all molecules in which the atoms have filled shells, even though the details of the bonding and antibonding orbitals they form may be much more complicated.

The electron density in all atomic orbitals and the molecular orbitals they form decreases exponentially toward zero at large distances from the nucleus, so the overlap between orbitals on neighboring molecules, and therefore the repulsions, can be expected to depend exponentially on their separation. As a result, the repulsions are effective only when the two molecules are very close together, but once they are close, they increase sharply. This sharp dependence on separation is the underlying reason why solid objects have definite, well-defined shapes.

> *Intermolecular repulsions arise from the overlap of orbitals on neighboring molecules and the requirements of the Pauli exclusion principle.*

What have you learned in this Topic?

You have learned that the attractive forces between the permanent or transient partial charges of molecules pull them together, resulting in condensed phases. You also encountered the hydrogen bond, which is the strongest of the nonbonding interactions. You have seen that when molecules or atoms are very close together, the potential energy increases sharply, resulting in repulsive forces.

The skills you have mastered are the ability to:

☐ **1.** Predict the relative strengths of ion–dipole and dipole–dipole interactions (Sections 3F.2 and 3F.3).

☐ **2.** Explain how London forces arise and how they vary with the polarizability of an atom and the size and shape of a molecule (Section 3F.4).

☐ **3.** Predict the relative order of the boiling points of two substances from the strengths of their intermolecular forces (Examples 3F.1 and 3F.2).

☐ **4.** Identify molecules that can take part in hydrogen bonding (Self-Test 3F.3).

☐ **5.** Explain why solid objects have a definite size and shape (Section 3F.6).

Topic 3F Exercises

Exercises labeled $\int_{dx}^{C}$ require calculus.

3F.1 Identify the types of attractive intermolecular interactions that might arise between molecules of each of the following substances: (a) NH_2OH; (b) CBr_4; (c) H_2SeO_4; (d) SO_2.

3F.2 Identify the types of attractive intermolecular interactions that might arise between molecules of each of the following substances: (a) H_2S; (b) SiH_4; (c) N_2H_4; (d) CHF_3.

3F.3 For which of the following molecules will dipole–dipole interactions be important: (a) CH_4; (b) CH_3Cl; (c) CH_2Cl_2; (d) $CHCl_3$; (e) CCl_4?

3F.4 For which of the following molecules will dipole–dipole interactions be important: (a) O_2; (b) O_3; (c) CO_2; (d) SO_2?

3F.5 Suggest, giving reasons, which substance in each of the following pairs is likely to have the higher normal melting point (Lewis structures may help your arguments): (a) HCl or NaCl; (b) $C_2H_5OC_2H_5$ (diethyl ether) or C_4H_9OH (butanol); (c) CHI_3 or CHF_3; (d) C_2H_4 or CH_3OH.

3F.6 Suggest, giving reasons, which substance in each of the following pairs is likely to have the higher normal boiling point:

(a) H_2S or H_2Se; (b) NaCl or CH_3Cl; (b) NH_3 or PH_3; (d) SiH_4 or SiF_4.

3F.7 Use the VSEPR model (Topic 2E) to predict the shapes of each of the following molecules and identify the member of each pair that forms a liquid with the higher boiling point: (a) PBr_3 or PF_3; (b) SO_2 or O_3; (c) BF_3 or BCl_3.

3F.8 Use the VSEPR (Topic 2E) model to predict the shapes of each of the following molecules and identify the member of each pair that forms a liquid with the higher boiling point: (a) BF_3 or ClF_3; (b) SF_4 or CF_4; (c) *cis*-CHCl=CHCl or *trans*-CHCl=CHCl. (See structures **5** and **6**.)

3F.9 Place the following types of molecular and ion interactions in order of increasing strength: (a) ion–dipole; (b) induced-dipole–induced-dipole; (c) dipole–dipole in the gas phase; (d) ion–ion; (e) dipole–dipole in the solid phase.

3F.10 In each pair, indicate which substance has the stronger intermolecular forces and explain your reasoning: (a) CO, CO_2; (b) SiF_4, Si_2F_2; (c) O_2, O_3; (d) CH_3SH, CH_3OH.

3F.11 Which of the following molecules are likely to form hydrogen bonds: (a) PH_3; (b) HBr; (c) C_2H_4; (d) HNO_2?

3F.12 Which of the following molecules are likely to form hydrogen bonds: (a) CH_3OCH_3; (b) CH_3COOH; (c) CH_3CH_2OH; (d) CH_3CHO?

3F.13 Identify the arrangement (I, II, or III; all molecules are CH_2Cl_2) that should possess the strongest intermolecular attractions, and justify your selection.

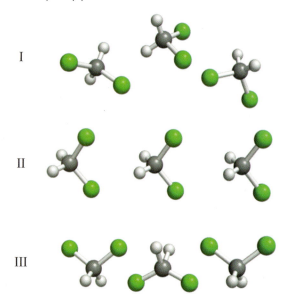

I

II

III

3F.14 Identify the arrangement (I, II, or III; all molecules are NH_3) that should possess the strongest intermolecular attractions, and justify your selection.

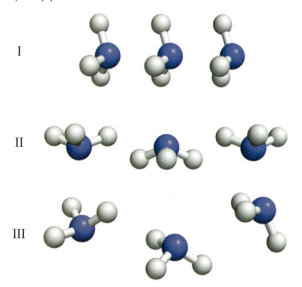

I

II

III

3F.15 Explain the difference in the boiling points of AsF_3 (63 °C) and AsF_5 (−53 °C).

3F.16 Explain the difference in the boiling points of NO_2 (21 °C) and N_2O (−88 °C).

3F.17 Calculate the ratio of the potential energies for the interaction of a water molecule with an Al^{3+} ion and with a Be^{2+} ion. Take the center of the dipole to be located at $r_{ion} + 100.$ pm. Which ion will attract a water molecule more strongly?

3F.18 Calculate the ratio of the potential energies for the interaction of a water molecule with a Li^+ ion and with a K^+ ion. Take the center of the dipole to be located at $r_{ion} + 100.$ pm. Which ion will attract a water molecule more strongly?

3F.19 Account for the following observations in terms of the type and strength of intermolecular forces. (a) The melting point of solid xenon is −112 °C and that of solid argon is −189 °C. (b) The vapor pressure of diethyl ether ($C_2H_5OC_2H_5$) is greater than that of water. (c) The boiling point of pentane, $CH_3(CH_2)_3CH_3$, is 36.1 °C, whereas that of 2,2-dimethylpropane (also known as neopentane), $C(CH_3)_4$, is 9.5 °C.

3F.20 Two students are assigned two pure compounds to investigate. They observe that compound A boils at 37 °C and compound B at 126 °C. However, they run out of time to take any additional measurements, so they make the following guesses about the two compounds. In each case, indicate if the guess can be justified by the data, if it is in error, or if it could be either true or false. Justify your answers. (a) Compound B has the higher molar mass. (b) Compound A is the more viscous. (c) Compound B has stronger intermolecular forces. (d) Compound B has the higher surface tension.

3F.21 The terms "intermolecular interaction" and "intermolecular force" are sometimes used interchangeably. However, it is important to distinguish the force from the potential energy of interaction. In classical mechanics, the strength of the force, F, is related to the distance dependence of the potential energy, E_p, by $F = -dE_p/dr$. How does the intermolecular force depend on separation for a typical intermolecular interaction that varies as $1/r^6$?

3F.22 Would you expect the energy of interaction of two rotating polar molecules to depend on the temperature? If so, should the strength of the interaction increase or decrease as the temperature is raised?

Topic 3G Liquids

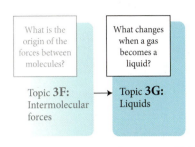

Why Do You Need to Know This Material? Liquids are all around us and many chemical reactions take place in liquid solutions, so it is essential to understand how their properties arise from their molecules.

What Do You Need to Know Already? This Topic draws on the concepts of intermolecular forces (Topic 3F).

The molecules in a liquid are held in close contact with their neighbors by attractive intermolecular forces, but they have enough energy to move past and tumble over one another (**FIG. 3G.1**). When you imagine a liquid, think of a jostling crowd of molecules, each one constantly changing places with its neighbors. A liquid at rest is like a crowd of people milling about in a stadium; a flowing liquid is like a crowd of people leaving a stadium.

3G.1 Order in Liquids

A crystalline solid has **long-range order.** In other words, its atoms or molecules lie in an orderly arrangement that is repeated over long distances. When the crystal melts, that long-range order is lost. In the liquid, the kinetic energy of the molecules can partly overcome the intermolecular forces, and so the molecules are able to move past one another. However, the molecules still experience strong attractions to one another. In water, for instance, only about 10% of the hydrogen bonds are broken upon melting. The rest are continually broken and re-formed with different water molecules. At any given moment, the immediate vicinity of a molecule in the liquid is like that in the solid, but the ordering does not extend very far past the nearest neighbors. This local ordering is called **short-range order.** In water, each molecule is surrounded by an approximately tetrahedral arrangement of other molecules as a result of hydrogen bonds. These interactions are ceaselessly changing from one moment to the next, but they are strong enough to affect the local structure of molecules up to the boiling point. Once the liquid has vaporized, the molecules move with almost complete freedom because the forces between them are almost negligible.

Molecules in a liquid have short-range but not long-range order.

3G.2 Viscosity and Surface Tension

The **viscosity** of a liquid is its resistance to flow: the higher the viscosity of the liquid, the more sluggish is the flow. A liquid with a high viscosity, such as molasses at room temperature or molten glass, is said to be "viscous." The viscosity of a liquid is an indication of the strength of the forces between its molecules: strong intermolecular forces hold molecules together and do not let them move past one another easily (**FIG. 3G.2**). However, the prediction of viscosity is very difficult because it depends not only on the strength with which molecules interact with one another but also on the ease with which

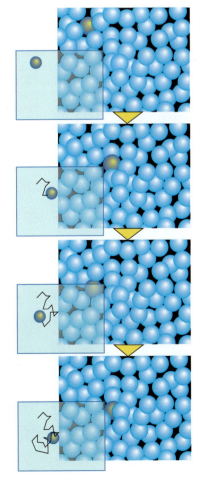

FIGURE 3G.1 The structure of a liquid. Although the molecules (represented by the spheres in this series of diagrams) remain in contact with their neighbors, they can move away from one another and have enough energy to push through to a new neighborhood. Consequently, the entire substance is fluid. One sphere is slightly darker so that you can follow its motion. Its path is shown in the insets.

ANIMATION FIGURE 3G.1

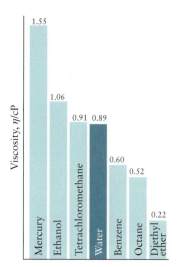

FIGURE 3G.2 The viscosities of several liquids. Liquids composed of molecules that cannot form hydrogen bonds are generally less viscous than those that can form hydrogen bonds. Mercury is an exception: its atoms stick together by a kind of metallic bonding with pronounced covalent character, and its viscosity is very high.

TABLE 3G.1	Surface Tensions of Liquids at 25 °C
Liquid	**Surface tension, $\gamma/(mN \cdot m^{-1})$**
benzene	28.88
carbon tetrachloride	27.0
ethanol	22.8
hexane	18.4
mercury	472
methanol	22.6
water	72.75
	58.0 at 100 °C

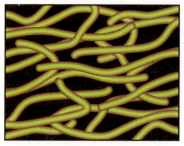

FIGURE 3G.3 The long threadlike molecules of heavy hydrocarbon oils tend to get tangled together like a plate of cooked spaghetti. As a result, the molecules do not move past one another very readily and the liquid is very viscous.

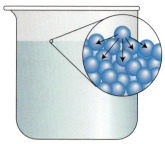

FIGURE 3G.4 Surface tension arises from the attractive forces acting on the molecules at the surface. The inset shows that a molecule within the liquid experiences attractive forces from all directions, but a molecule at the surface experiences a net inward force.

they can take up new positions as the fluid flows. The strong hydrogen bonds in water give it a viscosity greater than that of benzene: benzene molecules can easily slip past one another in the liquid but, for water molecules to move, hydrogen bonds have to be wrenched apart. Nevertheless, the viscosity of water is not very high because an H_2O molecule can quickly readjust the pattern of hydrogen bonds that it forms with its neighbors. Another factor is the entanglement of long hydrocarbon chains. The molecules of hydrocarbon oils and greases are nonpolar and so participate only in London forces. However, they have long chains that tangle together like cooked spaghetti (**FIG. 3G.3**), and the molecules move past one another only with difficulty.

The viscosity of liquids decreases as the temperature rises. The higher kinetic energy of molecules at high temperatures enables them to overcome the attractive intermolecular forces and wriggle past their neighbors more readily. The viscosity of water at 100 °C, for instance, is only one-sixth of its value at 0 °C, and so the same amount of water flows through a tube six times as fast at the higher temperature.

The surface of a liquid is smooth because intermolecular forces tend to pull the molecules together and inward (**FIG. 3G.4**); the **surface tension** is the net inward pull. Once again, liquids composed of molecules with strong intermolecular forces can be expected to have high surface tensions because the inward pull at the surface will be strong. Indeed, the surface tension of water is about three times as high as that of most other common liquids, because of its strong hydrogen bonds (**TABLE 3G.1**). The surface tension of mercury is higher still—more than six times that of water. Its high surface tension suggests that the bonds between mercury atoms in the liquid are very strong: in fact, they are partially covalent in character.

Surface tension accounts for a number of everyday phenomena. For example, a droplet of liquid suspended in air or on a waxy surface is spherical because the surface tension pulls the molecules into the most compact shape, a sphere (**FIG. 3G.5**). The attractive forces between water molecules are greater than those between water and wax, which is a mixture of nonpolar hydrocarbons. Surface tension decreases as the temperature rises and the interactions between molecules are overcome by the increased molecular motion.

Water has strong interactions with paper, wood, and cloth because it can form hydrogen bonds with the molecules in their surfaces. As a result, water maximizes its contact with these materials by spreading over them; in other words, water *wets* them by forming hydrogen bonds to them.

THINKING POINT

Why does adding soap or detergent to water decrease the surface tension? The structure of a typical soap is shown in (**1**)

FIGURE 3G.5 The nearly spherical shapes of these beads of water on the waxy surface of a leaf arise from the effect of surface tension. (*Nigel Cattlin/Science Source.*)

Capillary action, the rise of liquids up narrow tubes, occurs when there are favorable attractions between the molecules of the liquid and the tube's inner surface. These attractions are forces of **adhesion,** forces that bind a substance to a surface, as distinct from the forces of **cohesion,** the forces that bind the molecules of a substance together to form a bulk material. An indication of the relative strengths of adhesion and cohesion is the formation of a **meniscus,** the curved surface of a liquid, in a narrow tube (**FIG. 3G.6**). The meniscus of water in a glass capillary is curved upward at the edges (forming a concave shape) because the adhesive forces between water molecules and the oxygen atoms and —OH groups that are present on a typical glass surface are comparable in strength to the cohesive forces between water molecules. The water therefore tends to spread over the greatest possible area of glass. In contrast, the meniscus of mercury curves downward in glass (forming a convex shape). This shape is a sign that the cohesive forces between mercury atoms are stronger than the forces between mercury atoms and the glass, and so the liquid tends to reduce its contact with the glass.

The greater the viscosity of a liquid, the more slowly it flows. The viscosity of liquids decreases with increasing temperature. Surface tension arises from the net inward pull of intermolecular forces at the surface of a liquid. Capillary action arises from the imbalance of cohesive forces within a liquid and adhesive forces between the liquid and its container.

3G.3 Liquid Crystals

One type of material that has transformed electronic displays is neither a solid nor a liquid, but something intermediate between the two. **Liquid crystals** are substances that flow like viscous liquids but their molecules lie in a moderately orderly array, like those in a crystal. They are examples of a **mesophase,** a state of matter in which the molecules have an intermediate degree of order between a completely ordered crystal and a disordered liquid. Liquid crystalline materials have found many applications in the electronics industry because they are responsive to changes in temperature and electric fields.

A typical liquid-crystal molecule, such as *p*-azoxyanisole, is long and rodlike (**2**). Their rodlike shape enables the molecules to stack together like dry, uncooked spaghetti: they lie parallel to one another but are free to slide past one another along their long axes. Liquid crystals are anisotropic because of this ordering. **Anisotropic** materials have properties that depend on the direction of measurement. The viscosity of liquid crystals is least in the direction parallel to the long axis of the molecules: it is easier for the long rod-shaped molecules to slip past one another along their axes than to move sideways. **Isotropic** materials have properties that do not depend on the direction of measurement. Ordinary liquids, like water, are isotropic: their viscosities, for instance, are the same in every direction.

There are three classes of liquid crystals, which differ in the arrangement of their molecules. In the **nematic phase,** the molecules lie together, all in the same direction but staggered, like cars on a busy multilane highway (**FIG. 3G.7**). In the **smectic phase,** the molecules line up like soldiers on parade and form layers (**FIG. 3G.8**). Cell membranes are composed mainly of smectic liquid crystals. In the **cholesteric phase,** the molecules form ordered layers, but neighboring layers have molecules at different angles and so the liquid crystal has a helical arrangement of molecules (**FIG. 3G.9**).

Liquid crystals are also classified by their manner of preparation. **Thermotropic liquid crystals** are made by melting the solid phase. The highly viscous liquid-crystal phase exists over a short temperature range between the solid and liquid states. Thermotropic liquid crystals become isotropic liquids when they are heated above a characteristic temperature, because then the molecules have enough energy to overcome the attractions that restrict their movement. *p*-Azoxyanisole can form a thermotropic liquid crystal between 118 °C and 137 °C. Thermotropic liquid crystals are used in applications such as digital watches, computer screens, and thermometers. **Lyotropic liquid crystals** are layered structures that result from the action of a solvent on a solid or liquid. Examples are cell membranes and aqueous solutions of detergents and lipids (fats). These molecules, like the detergent sodium lauryl sulfate, have long, nonpolar hydrocarbon chains attached to polar heads (**3**). When the lipids that form cell membranes are mixed

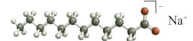

1   Sodium laurate (a soap)

FIGURE 3G.6 When the adhesive forces between a liquid and glass are stronger than the cohesive forces within the liquid, the liquid curves up to maximize contact with the glass, forming the meniscus shown here for water in glass (left). When the cohesive forces are stronger than the adhesive forces (as they are for mercury in glass), the edges of the surface curve downward to minimize contact with the glass (right). (©1990 Chip Clark–Fundamental Photographs.)

Nematic comes from the Greek word for "thread"; smectic comes from the Greek word for "soapy"; cholesteric is related to the word cholesterol, which comes from the Greek words for "bile solid."

2   *p*-Azoxyanisole

3   Sodium lauryl sulfate

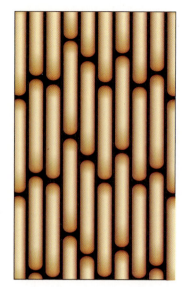

FIGURE 3G.7 A representation of the nematic phase of a liquid crystal. The long molecules lie parallel to one another but are staggered along their long axes.

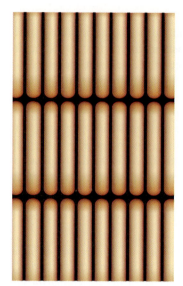

FIGURE 3G.8 The smectic phase of a liquid crystal. Not only do the molecules lie parallel to one another, they also line up next to one another to form sheets.

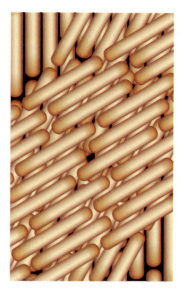

FIGURE 3G.9 The cholesteric phase of a liquid crystal. In this phase, sheets of parallel molecules are rotated relative to their neighbors and form a helical structure.

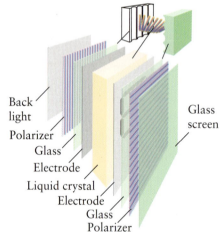

FIGURE 3G.10 The structure of a backlit LCD screen. The inset shows how the liquid-crystal molecules twist from one orientation to another as determined by the grooves in the transparent plates. When a potential difference is applied between the electrodes, this twist is lost.

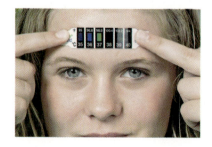

with water, they spontaneously form sheets in which the molecules are aligned in rows, forming a double layer, with their polar heads facing outward on each side of the sheet. These sheets form the protective membranes of the cells that make up living tissues.

Electronic displays make use of the anisotropy of the nematic phase and the changes in orientation of liquid-crystal molecules caused by an electric field. In an LCD (liquid-crystal display) television or computer monitor, layers of a liquid crystal in a nematic phase lie between the surfaces of two glass or plastic plates. The long axis of the molecules twists from the orientation governed by grooves in one plate to the perpendicular orientation governed by the grooves in the second plate (**FIG. 3G.10**). The light from the lamp is polarized, and as it passes through the twisted liquid crystal its plane of polarization changes too, and it is able to pass through the second polarizer. However, where a potential difference is applied between the electrodes (the second being in the shape of the characters it is intended to display, such as a simple numerical display), the twist is lost, and so too is the change of polarization of the light. As a result, a dark spot forms on the screen. In a "super-twist" LCD screen, the twist of the resting liquid crystal has more than one turn.

Cholesteric liquid crystals are also of interest because the helical structure unwinds slightly as the temperature is changed. Because the twist of the helical structure affects the optical properties of the liquid crystal, such as its color, these properties change with temperature. The effect is used in liquid-crystal thermometers (**FIG. 3G.11**).

Liquid crystals have a degree of order characteristic of solid crystals, but they can flow like viscous liquids. They are mesophases, intermediate between solids and liquids; their properties can be modified by electric fields and changes in temperature.

3G.4 Ionic Liquids

Liquid solvents are heavily used in industry to extract substances from natural products and to promote the synthesis of desired compounds. Because many of these solvents have high vapor pressures and so give off hazardous fumes, liquids have been sought that have low vapor pressures but dissolve organic compounds.

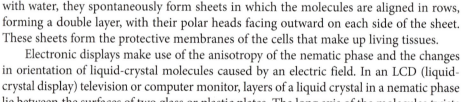

FIGURE 3G.11 The green rectangle on this liquid-crystal thermometer indicates a normal body temperature of 37 °C. (*Martyn F. Chillmaid/Science Source.*)

A new class of solvents called **ionic liquids** has been developed to meet this need. A typical ionic liquid has a relatively small anion, such as BF_4^-, and a relatively large, organic cation, such as 1-butyl-3-methylimidazolium (**4**). Because the cation has a large nonpolar region and is often asymmetrical, the compound does not crystallize easily and so is liquid at room temperature. Ionic liquids such as these can dissolve nonpolar organic compounds. However, the attractions between the ions reduce the vapor pressure to about the same as that of an ionic solid, thereby reducing air pollution. Because different cations and anions can be used, solvents can be designed for specific applications. For example, one formulation can dissolve the rubber in old tires so that it can be reused. Other ionic liquid solvents can be used to extract radioactive waste from groundwater.

4 1-Butyl-3-methylimidazolium ion

Ionic liquids are compounds in which one of the ions is a large, organic ion that prevents the liquid from crystallizing at ordinary temperatures. The low vapor pressures of ionic liquids make them desirable solvents that reduce pollution.

What have you learned in this Topic?

You have learned that the attractive forces between the molecules of a liquid result in its viscosity and surface tension and that strong adhesive forces between a liquid and its container can result in capillary action. The stronger the cohesive forces in a liquid the greater is the surface tension and, in many cases, the viscosity too. You have also learned that liquid crystals have properties intermediate between liquids and solids and that ionic substances in which one of the ions is organic can be liquid at room temperature.

The skills you have mastered are the ability to:

- ☐ **1.** Describe the structure of a liquid (Section 3G.1).
- ☐ **2.** Distinguish adhesive and cohesive forces (Section 3G.2).
- ☐ **3.** Explain how viscosity and surface tension vary with temperature and the strength of intermolecular forces (Section 3G.2).
- ☐ **4.** Distinguish the different types of liquid crystals (Section 3G.3).
- ☐ **5.** Explain why ionic liquids have low vapor pressures (Section 3G.4).

Topic 3G Exercises

3G.1 Predict how each of the following properties of a liquid varies as the strength of intermolecular forces increases and explain your reasoning: (a) boiling point; (b) viscosity; (c) surface tension.

3G.2 Predict how each of the following properties of a liquid changes as the temperature increases and explain your reasoning: (a) boiling point; (b) viscosity; (c) surface tension.

3G.3 Predict which liquid in each of the following pairs is likely to have the greater surface tension: (a) *cis*-dichloroethene or *trans*-dichloroethene (see structures **5** and **6** in Topic 3F); (b) benzene at 20 °C or benzene at 60 °C.

3G.4 Predict which substance in each of the following pairs is likely to have the greater viscosity in its liquid form at 20 °C: (a) methanol, CH_3OH, or ethanol, CH_3CH_2OH; (b) hexane, $CH_3CH_2CH_2CH_2CH_2CH_3$, or 1-pentanol, $CH_3CH_2CH_2CH_2CH_2OH$.

3G.5 Rank the following molecules in order of increasing viscosity at 50 °C: C_6H_5SH, C_6H_5OH, C_6H_6.

3G.6 Rank the following liquids in order of increasing viscosity at 25 °C: C_6H_6, CH_3CH_2OH, $CH_2OHCHOHCH_2OH$, CH_2OHCH_2OH, and H_2O. Explain your ordering.

3G.7 The following boiling points correspond to the substances listed. Match the boiling points to the substances by considering the relative strengths of their intermolecular forces. b.p./°C: −162, −88.5, 28, 36, 64.5, 78.3, 82.5, 140, 205, 290; substance: CH_4, $CH_3CHOHCH_3$, $C_6H_5CH_2OH$ (has a benzene ring), CH_3CH_3, C_5H_9OH (cyclic), $(CH_3)_2CHCH_2CH_3$, CH_3OH, $HOCH_2CHOHCH_2OH$, $CH_3(CH_2)_3CH_3$, CH_3CH_2OH. *Hint*: The boiling point of $(CH_3)_2CHCH_2CH_3$ is 28 °C and that of CH_3OH is 64.5 °C.

3G.8 The following surface tensions (in millinewtons per meter, $mN \cdot m^{-1}$, at 20 °C) correspond to the liquids listed. Match the surface tension to the substance. Surface tension: 18.43, 22.75, 27.80, 28.85, 72.75; compound: H_2O, $CH_3(CH_2)_4CH_3$, C_6H_6, CH_3CH_2OH, CH_3COOH.

3G.9 Explain why water forms a concave meniscus in a narrow glass tube but a convex meniscus in a narrow plastic tube.

3G.10 In an aqueous solution, solute molecules or ions require a certain amount of time to migrate through the solution. The rate of this migration sets an upper limit on how fast reactions can take place, because no reaction can take place faster than the ions can be supplied. This limit is known as the *diffusion-controlled rate*. It has been found that in aqueous solution the diffusion-controlled rate for hydronium ions (H_3O^+, *Fundamentals* J) is about three times as fast as that for other ions. Explain why this is so.

3G.11 The height, h, of a column of liquid in a capillary tube can be estimated by using $h = 2\gamma/gdr$, where γ is the surface tension, d is the density of the liquid, g is the acceleration of free fall ($g = 9.806$ m·s^{-2}), and r is the radius of the tube. Which will rise higher in a tube that is 0.15 mm in diameter at 25 °C, water or ethanol? The density of water is 0.997 g·cm^{-3} and that of ethanol is 0.79 g·cm^{-3}. See Table 3G.1.

3G.12 The expression for the capillary rise in Exercise 3G.11 assumes that the tube is vertical. How will the expression be modified when the tube is held at an angle θ (theta) to the vertical?

3G.13 Why do long hydrocarbon molecules that do not have multiple bonds, such as decane, $CH_3(CH_2)_8CH_3$, not form liquid crystals?

3G.14 *p*-Azoxyanisole (**2**) has a liquid crystalline range of approximately 117 °C to 137 °C. How might this molecule be modified to lower the melting point of the solid and make it more suitable for lower-temperature applications (near room temperature, for example)?

3G.15 Two solutes were used to study diffusion in liquids, methylbenzene, which is a small molecule that can be approximated as a sphere, and a liquid crystal that is long and rodlike. The two solutes were found to move and rotate in all directions to the same extent in benzene. In a liquid-crystal solvent the methylbenzene again moved and rotated to the same extent in all directions, but the liquid-crystal solute moved much more rapidly along the long axis of the molecule than it did in a "sideways" mode, perpendicular to the long axis. It also was found to rotate more rapidly around the long axis than perpendicular to it. Explain this behavior.

3G.16 The molecules of many common liquid crystals are long and rodlike. In addition, they contain polar groups. Explain how both characteristics of liquid crystals contribute to their anisotropic bulk properties.

3G.17 Which substance would be the better choice as an ionic liquid solvent, (a) $C_5H_6N^+Cl^-$; (b) $CH_3NH_3^+Cl^-$? Explain your selection.

3G.18 Which substance would be the better choice as an ionic liquid solvent (a) $C_{11}H_{23}Br$; (b) $C_{11}H_{14}N^+Br^-$? Explain your selection.

Topic 3H Solids

How are atoms arranged in solids?

What are the characteristics of inorganic solids?

Topic **3H**: Solids → Topic **3I**: Inorganic materials

When the temperature is so low that the molecules of a substance do not have enough energy to escape from their neighbors, the substance solidifies. The nature of the solid depends on the types of forces that hold the atoms, ions, or molecules together. An understanding of solids in terms of the properties of their atoms will help you to understand why, for instance, metals can be pounded into different shapes but salt crystals shatter, and why diamonds are so hard.

Why Do You Need to Know This Material? In order to understand the properties of solids you need to know how individual particles contribute to the structures and properties of bulk materials.

What Do You Need to Know Already? This Topic uses the concepts of ionic and covalent bonding (Topics 2A and 2B) and specifically ionic radius (Topic 1F).

3H.1 Classification of Solids

In a **crystalline solid** the atoms, ions, or molecules lie in an orderly array (**FIG. 3H.1**). A crystalline solid has long-range order. An **amorphous solid** is one in which the atoms, ions, or molecules lie in a random jumble, as in butter, rubber, and glass (**FIG. 3H.2**). An amorphous solid has a structure like that of a frozen instant in the life of a liquid, with only short-range order. Crystalline solids typically have flat, well-defined planar surfaces called **crystal faces,** which lie at definite angles to one another. These faces are formed by orderly layers of atoms (**BOX 3H.1**). Amorphous solids do not have well-defined faces unless they have been molded or cut. The arrangement of atoms, ions, and molecules within a crystal is determined by x-ray diffraction (*Major Technique* 3 on the website of this book).

FIGURE 3H.1 Crystalline solids have well-defined faces and an orderly internal structure. Each face of the crystal is formed by the top plane of an orderly stack of atoms, molecules, or ions. (*Chip Clark/Fundamental Photographs, NYC.*)

FIGURE 3H.2 (Left) Quartz is a crystalline form of silica, SiO_2, with the atoms in an orderly network, represented here in two dimensions. (Right) When molten silica solidifies in an amorphous arrangement, it becomes glass with the atoms in a disorderly network. (*Photos: Steven Smale (left); W. H. Freeman photo by Ken Karp (right).*)

Box 3H.1 HOW DO WE KNOW … WHAT A SURFACE LOOKS LIKE?

Even the most powerful optical microscopes cannot reveal the individual atoms of a solid surface. However, the new field of nanotechnology, the development and study of devices only nanometers in size (Box 1C.1), requires the ability to resolve surfaces on an atomic scale. A technique that allows individual atoms to be visualized, *scanning tunneling microscopy* (STM), is an important tool for nanotechnology.* The technique produces images like those on this page. The first illustration shows a tiny sodium iodide crystal lying on a copper surface.

An experimenter attempted to create a two-dimensional sodium iodide crystal on a copper surface, but the ions spontaneously rearranged themselves into a tiny three-dimensional crystal. (*Hopkinson, Lutz & Eigler/IBM.*)

The key idea behind STM is that, because electrons have wavelike properties (Topic 1B), they can penetrate into and through regions where classical mechanics would forbid them to be. This penetration is called *tunneling*. The effect is used in STM (hence the "tunneling" in the name) by bringing a fine tip up to a surface and monitoring the current that flows through the gap between the tip and the surface. The magnitude of the current, like the tunneling itself, is very sensitive to the separation of the tip and the surface, and even atomic-scale variations affect it.

To obtain an STM image, a very fine tip is moved back and forth across the surface in a series of closely spaced lines (hence the "scanning" in the name). The tip ends in a single atom (see the second figure). As the tip moves across the surface at a constant height, the tunneling ebbs and flows, and the current varies correspondingly through the circuit. The image is a portrayal of the current measured on each scan.

In a modification, the current is maintained at a constant level by varying and monitoring the height of the tip above the surface. That height is controlled by using a *piezoelectric* substance to support the tip, a substance that changes dimensions according to the electrical potential difference applied to it. By measuring the potential difference that must be applied to the

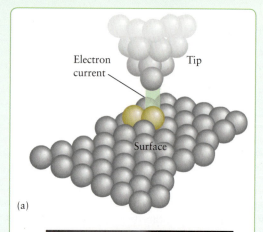

(a)

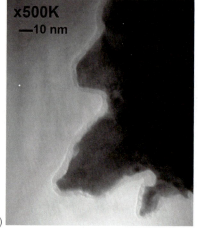

(b)

(a) The tip of a scanning tunneling microscope above a surface. Because the tip is too close to the surface for other molecules to interfere, STM devices can be used in a gaseous atmosphere or even in liquids. (b) This transmission electron microscopy (TEM) image of an STM tip prepared from a Pt/Ir alloy shows that the tip diameter is less than 20 nm. (*Part (b) Reifenberger Nanophysics Laboratory, Dept. of Physics, Purdue University.*)

piezoelectric support to maintain a constant current through the tip, the height of the tip can be inferred and plotted.

Yet another modification is *atomic force microscopy* (AFM), in which a fine tip attached to a cantilever (a tiny flexible beam), is scanned across the surface. The atom at the end of the tip experiences a force that pulls it toward or pushes it away from the atoms on the surface. The type of force between the tip and sample varies depending on the sample type and any coatings applied to the tip. For example, in a variant called *magnetic force microscopy* (MFM) the tip may be coated with a

---

*The 1986 Nobel Prize for Physics was awarded to Ernst Ruska and Gerd Binnig of Germany and Heinrich Rohrer of Switzerland for their invention of the scanning tunneling microscope.

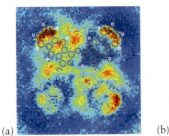 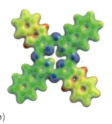

(a)   (b)

(a) Local contact potential difference (LCPD) image of naphthalocyanine. The circles represent the calculated locations of atoms. (b) Electrostatic potential isodensity surface showing the relative electrostatic potential in naphthalocyanine. Note that in these images positive potential is shown as red and less positive is shown as blue. (*Part (a) Fabian Mohn, Leo Gross, Nikolaj Moll & Gerhard Meyer, Imaging the charge distribution within a single molecule, Nature Nanotechnology (Feb. 2012), Vol. 7, pp. 227–231, courtesy of IBM Research–Zurich.*)

magnetic material to investigate the magnetic properties of a sample, such as the surface of a disk in a computer hard-drive. The deflection of the cantilever, which shows the shape of the surface, is monitored by using light from a laser. An advantage of AFM is that it can image biological surfaces that cannot conduct an electrical current. For example, AFM can be used to study the shapes of human chromosomes and to understand how carcinogenic (cancer-causing) substances promote the development of tumors by interfering with the reproduction of the DNA molecule. Recently, a new technique has been developed in which an AFM tip is modified with a single molecule of CO, allowing the electric field created by the charge distribution within a molecule to be observed. The third image shows the charge distribution for a single molecule of naphthalocyanine compared with the calculated electrostatic potential (elpot) isodensity surface.

The images of surfaces in this box, which have been obtained by increasingly sensitive techniques, have opened our eyes to the appearance of surfaces in the most extraordinary ways.

Crystalline solids are classified according to the types of bonds that hold their atoms, ions, or molecules in place:

Molecular solids are assemblies of discrete molecules held in place by intermolecular forces.
Network solids consist of atoms covalently bonded to their neighbors throughout the extent of the solid.
Metallic solids, also simply called *metals*, consist of cations held together by a sea of electrons.
Ionic solids consist of ions held in place by their mutual attraction.

TABLE 3H.1 lists examples of each type of solid and their typical physical characteristics. Solids are dense forms of matter because their atoms, ions, and molecules are packed closely together. Network solids (such as diamond) have very high melting points because their covalent bonds are so strong. Metals also have high melting points and many are very dense because their atoms are packed closely together. The metallic bond is relatively strong. As a result, most metals have high melting points and serve as tough, strong materials for construction. Ionic solids typically have higher melting points than molecular solids because interionic forces are much stronger than intermolecular forces.

Crystalline solids have a regular internal arrangement of atoms or ions; amorphous solids do not. Solids are classified as molecular, network, metallic, or ionic.

TABLE 3H.1 Typical Characteristics of Solids

Class	Examples	Physical characteristics
metallic	s- and d-block elements	malleable, ductile, lustrous, electrically and thermally conducting
ionic	$NaCl$, KNO_3, $CuSO_4 \cdot 5H_2O$	hard, rigid, brittle; high melting and boiling points; those soluble in water give conducting solutions
network	B, C, black P, BN, SiO_2	hard, rigid, brittle; very high melting points; insoluble in water
molecular	$BeCl_2$, S_8, P_4, I_2, ice, glucose, naphthalene	relatively low melting and boiling points; brittle if pure

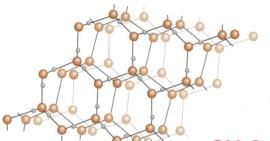

FIGURE 3H.3 Ice is made up of water molecules that are held together by hydrogen bonds in an open structure. Each O atom is surrounded tetrahedrally by four hydrogen atoms, two of which are σ-bonded to it and two of which are hydrogen-bonded to it. To show the structure more clearly, only the hydrogen atoms in the front layer are shown.

🔷 ANIMATION FIGURE 3H.3

3H.2 Molecular Solids

Molecular solids consist of molecules held together by intermolecular forces (Topic 3F), and their physical properties depend on the strengths of those forces. Amorphous molecular solids may be soft, like paraffin wax, which is a mixture of long-chain hydrocarbons. These molecules lie together in a disorderly way, and the forces between them are so weak that they can be pushed past one another very easily. Many other molecular solids have crystalline structures and strong intermolecular forces that make them brittle and hard. For example, sucrose molecules, $C_{12}H_{22}O_{11}$ (**1**), are held together by hydrogen bonds between their numerous —OH groups. Hydrogen bonding between sucrose molecules is so strong that, by the time the melting point has been reached (at 184 °C), the molecules themselves have started to decompose. The mixture of partly decomposed products, called caramel, is used to add flavor and color to food. Some molecular solids are very tough. For example, "ultrahigh-density polyethylene" consists of long hydrocarbon chains that lie densely together like close-packed cylinders; the resulting material is smooth yet so tough that it is used to make bullet-proof vests and joint replacements for the human body.

Because molecules have such diverse shapes, they stack together in a wide variety of different ways. In ice, for example, each O atom is surrounded by four H atoms in a tetrahedral arrangement. Two of these H atoms are linked covalently to the O atom through σ-bonds. The other two belong to neighboring H_2O molecules and are linked to the O atom by hydrogen bonds. As a result, the structure of ice is an open network of H_2O molecules held in place by hydrogen bonds (**FIG. 3H.3**). Some of the hydrogen bonds break when ice melts; as the orderly arrangement collapses, the molecules pack less uniformly but more densely (**FIG. 3H.4**). The openness of the network in ice compared with the closer arrangement of molecules in the liquid explains why it has a lower density than liquid water (0.92 g·cm⁻³ and 1.00 g·cm⁻³, respectively, at 0 °C). Solid benzene and solid tetrachloromethane, in contrast, have higher densities than their liquids (**FIG. 3H.5**). Their molecules are held in place by London forces, which are much less directional than hydrogen bonds, and so they can pack together more closely in the solid than in the liquid.

Molecular solids are typically soft and commonly melt at low temperatures.

3H.3 Network Solids

Whereas molecular solids consist of molecules held together by relatively weak intermolecular forces, the atoms in network solids are joined to their neighbors by strong covalent bonds that form a framework extending throughout the crystal. In order to break apart a crystal of a network solid, covalent bonds must be broken. Therefore, network solids are very hard, rigid materials with high melting and boiling points.

Diamond and graphite are examples of elemental network solids. These two forms of carbon are **allotropes,** meaning forms of an element that differ in the way in which

1 Sucrose, $C_{12}H_{22}O_{11}$

FIGURE 3H.4 As a result of its open structure, ice is less dense than liquid water and floats in it (left). Solid benzene is denser than liquid benzene and "benzenebergs" sink in liquid benzene (right). (©1988 Chip Clark–Fundamental Photographs.)

FIGURE 3H.5 Variation in the densities of water and tetrachloromethane with temperature. Note that ice is less dense than liquid water at its freezing point and that water has its maximum density at 4 °C.

🔷 ANIMATION FIGURE 3H.5

the atoms are linked. Each C atom in diamond is covalently bonded to four neighbors through sp^3-hybrid σ-bonds (**FIG. 3H.6**). The tetrahedral framework extends throughout the solid like the steel framework of a large building. This structure accounts for the great hardness of the solid. Diamond is a rigid, transparent solid. It is the hardest substance known and the best conductor of heat, being about five times better than copper. Diamond's high heat conductivity makes thin films of diamond ideal for use as a base for some integrated circuits and coatings on cutting tools so that they do not overheat.

In nature, diamond is found embedded in a soft rock called *kimberlite*. This rock rises in columns from deep in the Earth, where the diamonds are formed under intense pressure. One method used to make synthetic industrial diamonds re-creates the geological conditions that produce natural diamonds by compressing graphite at pressures greater than 80 kbar and temperatures above 1500 °C (**FIG. 3H.7**). Small amounts of metals such as chromium and iron are added to the graphite. The molten metals are thought to dissolve the graphite and then, as they cool, to deposit crystals of diamond, which are less soluble than graphite in the molten metal. Another, more common, method for producing synthetic diamonds is by thermal decomposition of methane. In this technique, the carbon atoms settle on a cool surface as both graphite and diamond. However, because hydrogen atoms produced in the decomposition react more quickly with graphite, forming volatile hydrocarbons, more diamond than graphite survives.

Graphite, the most important component of pencil "lead," is a black, lustrous, electrically conducting solid that vaporizes at 3700 °C. It consists of flat sheets of sp^2-hybridized carbon atoms bonded covalently into hexagons like chicken wire (**FIG. 3H.8**). There are also weak bonds between the sheets. Electrons spread through the delocalized π-network that extends across the plane. This delocalization accounts for graphite being a black, lustrous, electrically conducting solid; indeed, graphite is used as an electrical conductor in industry and as electrodes in electrochemical cells and batteries. Electrons can move within the sheets of graphite but much less readily from one sheet to another. Hence, graphite conducts electricity better parallel to the sheets than perpendicular to them.

Graphite is actually a moderately hard material, but in its commercially available forms many impurities, such as nitrogen and oxygen from the air, are trapped between the sheets; these impurities weaken the already weak intersheet bonds and allow the sheets of atoms to slide over one another easily. Thus, impure graphite is slippery and is used as a dry lubricant. In pencil "lead," the graphite is mixed with clay. The mark left on the paper by a pencil consists of rubbed-off layers of graphite.

THINKING POINT

Why is graphite not a suitable lubricant for use in outer space?

Each chickenwire-like sheet of carbon atoms in graphite is called a *graphene* sheet. Graphene itself, a single sheet of graphite, is a new material with exceptional promise in the electronics industry. Sheets of graphene can be prepared in a very pure state, and then stacked together with water molecules that act as a kind of glue between them. The result is a very strong, flexible, but vanishingly thin and sometimes nearly transparent paperlike material that conducts electricity but is tougher than diamond. Absorption of gas molecules changes its electrical properties, making graphene a good sensor for gases.

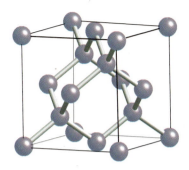

FIGURE 3H.6 The structure of diamond. Each sphere represents the location of the center of a carbon atom. Each atom is at the center of a tetrahedron formed by the sp^3 hybrid covalent bonds to each of its four neighbors.

ANIMATION FIGURE 3H.6

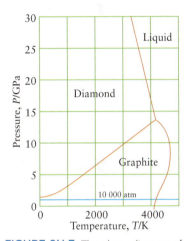

FIGURE 3H.7 The phase diagram of carbon, showing the regions of phase stability.

The 2010 Nobel Prize for Physics was awarded to Andre Geim and Konstantin Novoselov for their work on graphene.

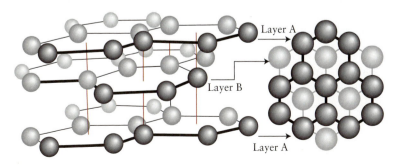

FIGURE 3H.8 Graphite consists of layers of hexagonal rings of sp^2 hybridized carbon atoms. The slipperiness of graphite results from the ease with which the layers can slide over one another when there are impurity atoms lying between the planes.

ANIMATION FIGURE 3H.8

Ceramic materials are typically noncrystalline inorganic oxides with a network structure that is produced by heat treatment of a powder. They include many silicate minerals, such as quartz (silicon dioxide, which has the empirical formula SiO_2), and high-temperature superconductors (see the Interlude following Focus 3). Ceramic materials have great strength and stability, because their ionic bonds have a great deal of covalent character and must be broken to cause any deformation in the crystal. As a result, ceramic materials under physical stress tend to shatter rather than bend.

Network solids are typically hard and rigid due to the covalent bonds holding them together; they have high melting and boiling points.

3H.4 Metallic Solids

A metallic crystal consists of an array of cations that are bound together by their interaction with the sea formed by their mobile electrons (recall Fig. 1F.11). For instance, metallic silver consists of Ag^+ ions held together by electrons that spread throughout the solid, with one electron for each cation. The characteristic luster of metals is due to the mobility of the electrons that form the sea. When an incident light wave strikes the surface of a metal, the electric field of the radiation pushes the mobile electrons backward and forward. These oscillating electrons radiate light, and it is seen as a luster—essentially a re-emission of the incident light (**FIG. 3H.9**). The electrons oscillate in step with the incident light, and so they generate light of the same frequency. In other words, red light reflected from a metallic surface is red and blue light is reflected as blue. That is why an image in a mirror—a thin metallic coating on glass—is a faithful portrayal of the reflected object.

The mobility of its electrons also explains a metal's **malleability,** its ability to be hammered into shape, and its **ductility,** its ability to be drawn into wires. Because the cations are surrounded by an electron sea, metallic bonding has very little directional character.

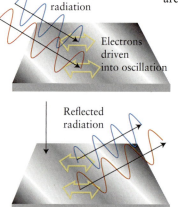

FIGURE 3H.9 (a) When light of a particular color shines on the surface of a metal, the electrons at the surface oscillate in step. This oscillating motion gives rise to an electromagnetic wave perceived as the reflection of the source. (b) Each of these solar mirrors at Sandia National Laboratories in California is positioned at the best angle to reflect sunlight into a collector that uses the incident energy to generate electricity. (*Part (b) Sandia National Laboratories/NREL (National Renewable Energy Laboratory).*)

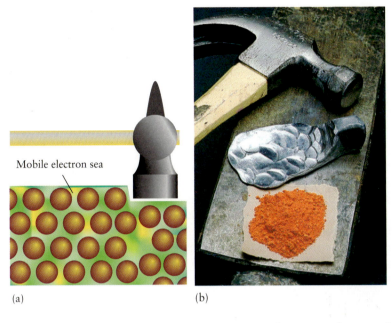

Mobile electron sea

(a)   (b)

FIGURE 3H.10 (a) When a blow from a hammer displaces a metal's cations, the mobile electrons can immediately respond and follow the cations to their new positions, and consequently the metal is malleable. (b) This piece of lead has been flattened by a hammer, whereas the orange crystals of lead(II) oxide, an ionic solid, have shattered. (*Part (b) ©1985 Chip Clark–Fundamental Photographs.*)

 ANIMATION FIGURE 3H.10

As a result, a cation can be pushed past its neighbors in any direction without much effort. A blow from a hammer can drive large numbers of cations past their neighbors. The electron sea immediately adjusts, so the atoms move relatively easily into their new positions (**FIG. 3H.10**).

Because the interaction between the ions and the electrons is the same in all directions, the arrangement of the cations in a metal can be modeled as hard spheres stacked together without favoring a particular direction. A bonding model that explains the structures and properties of many metals is a **close-packed structure,** in which spheres representing the cations stack together with the least waste of space, like oranges in a display (**FIG. 3H.11**).

FIGURE 3H.12 shows how identical spheres stack together to give a close-packed structure. In the first layer (A) each sphere lies at the center of a hexagon of other spheres. The spheres of the second (upper) layer (B) lie in the dips of the first layer (**FIG. 3H.13**). The third layer of spheres will lie in the dips of the second layer, with the pattern repeating over and over again.

The third layer of spheres may be added in either of two ways. There are two types of dips between the green spheres of the second layer in Fig. 3H.13: one type lies over the spheres of the first layer and the other type lies over the gaps in the first layer. If the spheres in the third layer lie in the dips that are directly above the spheres of the first layer (**FIG. 3H.14**), the third layer duplicates layer A, the next layer then duplicates B, and so on. This procedure results in an ABABAB . . . pattern of layers called a **hexagonal close-packed structure** (hcp). The hexagonal pattern in the arrangement of atoms can be seen in **FIG. 3H.15**. Notice that each sphere has three nearest neighbors in the plane below, six in its own plane, and three in the plane above, giving 12 in all. This arrangement

FIGURE 3H.11 The close-packed stacks of apples, oranges, and other produce in a grocery display illustrate how atoms stack together in metals to form crystals with flat faces. (© *Craig Lovell/Eagle Visions Photography/Alamy.*)

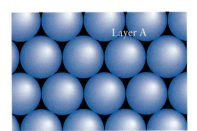

FIGURE 3H.12 A close-packed structure can be built up in stages. The first layer (A) is laid down with minimum waste of space.

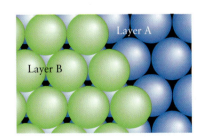

FIGURE 3H.13 The second layer (B) lies in the dips—the depressions—between the spheres of the first layer. Each sphere is touching six other spheres in its layer, as well as three in the layer below and three in the layer above.

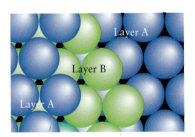

FIGURE 3H.14 When the spheres in the third layer lie directly above the spheres of the first layer, an ABABAB . . . structure is formed.

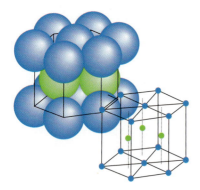

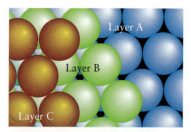

FIGURE 3H.16 In an alternative to the scheme shown in Fig. 3H.14, the spheres of the third layer can lie in the dips in the second layer that are directly above the dips in the first layer, to give an ABCABC . . . arrangement of layers.

The Pythagorean theorem states that the square of the hypotenuse of a right-angled triangle is equal to the sums of the squares of the other two sides. That is, if the hypotenuse is c and the other two sides are a and b, then $a^2 + b^2 = c^2$. In this calculation, $a = b$.

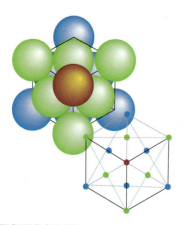

FIGURE 3H.17 A fragment of the structure constructed as described in Fig. 3H.16. This fragment shows the origin of the names "cubic close-packed" or "face-centered cubic" for this arrangement. The layers A, B, and C can be seen along the diagonals of the faces of the cube and are indicated by the different colors of the atoms.

FIGURE 3H.15 A fragment of the structure formed as described in Fig. 3H.14 shows the hexagonal symmetry of the arrangement—and the origin of its name, "hexagonal close-packed."

is reported by saying that the **coordination number,** the number of nearest neighbors of each atom, in the solid is 12. It is impossible to pack identical spheres together with a coordination number greater than 12. Magnesium and zinc are examples of metals that crystallize in hcp arrays.

In the second arrangement, the spheres of the third layer lie in the dips of the second layer that lie over the gaps of the first layer (**FIG. 3H.16**). If the third layer is labeled C, the resulting structure has an ABCABC . . . pattern of layers to give a **cubic close-packed structure** (ccp, **FIG. 3H.17**), because the atoms form cubes when viewed at an angle to the layers. A ccp structure can be thought of as many of these tiny cubes repeating over and over again in all directions. The coordination number is also 12: each sphere has three nearest neighbors in the layer below, six in its own layer, and three in the layer above. Aluminum, copper, silver, and gold are examples of metals that crystallize in ccp arrays.

Even in a close-packed structure, hard spheres do not fill all the space in a crystal. The gaps—the interstices—between the atoms are called "holes." To determine just how much space is occupied, you need to calculate the fraction of the total volume of a crystal that is occupied by the spheres.

How Is That Done?

To calculate the fraction of occupied space in a close-packed structure, consider a ccp structure. First, look at how the cube is built from the spheres representing the atoms. **FIGURE 3H.18** shows that eight spheres lie at the corners of the cubes. Only $\frac{1}{8}$ of each of these spheres projects into the cube, so the eight corner spheres collectively contribute $8 \times \frac{1}{8} = 1$ sphere to the cube. Half a sphere on each of the six faces projects into the cube, so the spheres on each face contribute $6 \times \frac{1}{2} = 3$ spheres, giving four spheres in all within the cube. The length of the diagonal of the face of the cube shown in Fig. 3H.18 is $4r$, where r is the radius of the sphere. Each of the two corner spheres contributes r and the sphere at the center of the face contributes $2r$. According to the Pythagorean theorem, the length of the side of the face, a, is related to the diagonal by $a^2 + a^2 = (4r)^2$, or $2a^2 = 16r^2$, and so $a = 8^{1/2}r$. The volume of the cube is therefore $a^3 = 8^{3/2}r^3$. The volume of each sphere is $\frac{4}{3}\pi r^3$, so the total volume of the spheres inside the cube is $4 \times \frac{4}{3}\pi r^3 = \frac{16}{3}\pi r^3$. The ratio of this occupied volume to the total volume of the cube is therefore

$$\frac{\text{Total volume of spheres}}{\text{Total volume of cube}} = \frac{(16/3)\pi r^3}{8^{3/2}r^3} = \frac{16\pi}{3 \times 8^{3/2}} = 0.74\ldots$$

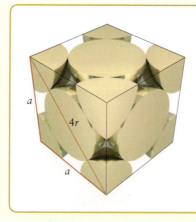

FIGURE 3H.18 The relation of the dimensions of a face-centered cubic unit cell to the radius, r, of the spheres. The spheres are in contact along the face diagonals.

The calculation shows that 74% of the crystal's space is occupied by atoms and 26% is considered to be empty. The hcp structure has the same coordination number of 12 and thus is packed as densely, with the same fraction of occupied space.

If a dip between three atoms is directly covered by another atom, the result is a **tetrahedral hole,** because it is formed by four atoms at the corners of a regular tetrahedron

FIGURE 3H.19 The locations of octahedral and tetrahedral holes. There are twice as many tetrahedral holes as octahedral holes. Note that both types of holes are defined by two neighboring close-packed layers, so they are present with equal abundance in both hcp and ccp structures.

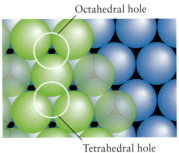

Octahedral hole

Tetrahedral hole

(**FIG. 3H.19**). There are two tetrahedral holes per atom in a close-packed lattice. When a dip in a layer coincides with a dip in the next layer, the result is an **octahedral hole,** because it is formed by six atoms at the corners of a regular octahedron, as shown in Fig. 3H.19. There is one octahedral hole for each atom in the lattice. Note that, because holes are formed by two adjacent layers and because neighboring close-packed layers have identical arrangements in hcp and ccp structures, the numbers of holes are the same for both close-packed structures. The holes in the close-packed structure of a metal can be filled with smaller atoms to form alloys (Topic 3I).

Which close-packed structure—if either—a metal adopts depends on which structure has the lower energy, and that in turn depends on details of its electronic structure. In fact, in the next section, you will see that some elements achieve a lower energy by adopting a different arrangement altogether.

Many metals have close-packed structures, with the atoms stacked in either a hexagonal or a cubic arrangement; close-packed atoms have a coordination number of 12. Close-packed structures have one octahedral and two tetrahedral holes per atom.

3H.5 Unit Cells

A lattice structure can be represented by a small repeating region of the crystal. The small region illustrated in Fig. 3H.17 is an example of a **unit cell,** the smallest unit that, when stacked together repeatedly without any gaps and without rotations, can reproduce the entire crystal (**FIG. 3H.20**).

A cubic close-packed unit cell like that in Fig. 3H.20 has an atom at each corner and one at the center of each face of the unit cell; for this reason, it is also called a **face-centered cubic structure** (fcc). In a **body-centered cubic structure** (bcc), a single atom lies at the center of a cube formed by eight other atoms (**FIG. 3H.21**). This structure is not close packed, and metals that have a body-centered cubic structure can often be forced under pressure into a close-packed form. Iron, sodium, and potassium are examples of metals that crystallize with bcc structures. A **primitive cubic structure** has an atom at each corner of a cube. The spheres representing the atoms are in contact along the edges (**FIG. 3H.22**). This structure is known for only one element, polonium: covalent forces, which favor certain directions, are so strong in this metalloid that they overcome the tendency toward close packing characteristic of metallic bonding. Unit cells are drawn by representing each atom by a dot that marks the location of the atom's center (**FIG. 3H.23**).

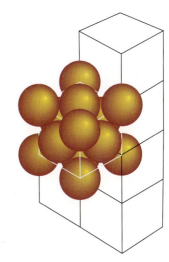

FIGURE 3H.20 The entire crystal structure is constructed from a single type of unit cell by stacking the cells together without any gaps.

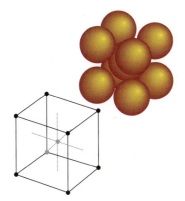

FIGURE 3H.21 The body-centered cubic (bcc) structure. This structure is not packed as closely as the face-centered cubic and hexagonal close-packed structures. It is less common among metals than close-packed structures. Some ionic structures are based on this model.

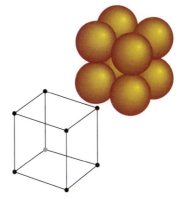

FIGURE 3H.22 A primitive cubic unit cell has an atom at each corner. It is rarely found in metals.

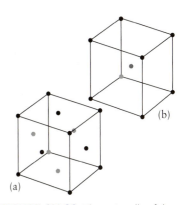

FIGURE 3H.23 The unit cells of the (a) ccp (fcc) and (b) bcc structures, in which the locations of the centers of the spheres are marked by dots.

FIGURE 3H.24 The 14 Bravais lattices. P denotes primitive; I, body-centered; F, face-centered; C, with a lattice point on two opposite faces; and R, rhombohedral (a rhomb is an oblique equilateral parallelogram).

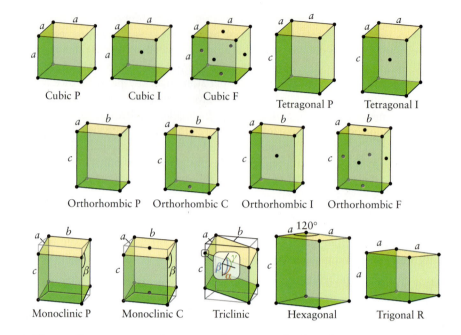

Cubic P   Cubic I   Cubic F   Tetragonal P   Tetragonal I

Orthorhombic P   Orthorhombic C   Orthorhombic I   Orthorhombic F

Monoclinic P   Monoclinic C   Triclinic   Hexagonal   Trigonal R

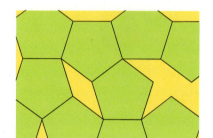

FIGURE 3H.25 A plane surface cannot be covered by regular pentagons without leaving gaps. The same is true of regular heptagons.

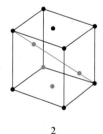

2

All crystal structures can be expressed in terms of only 14 basic patterns of unit cells called **Bravais lattices** (**FIG. 3H.24**). Regular pentagonal shapes are absent from the Bravais lattices: regular pentagons (pentagons with sides of the same length and all angles the same) cannot cover space without gaps (**FIG. 3H.25**). Similarly, regular heptagonal (seven-sided) and higher regular polygonal shapes cannot be stacked together to cover all space; hence they too do not occur among the Bravais lattices.

The number of atoms in a unit cell is counted by noting how they are shared between neighboring cells:

- An atom at the center of a cell belongs entirely to that cell, and counts as one atom.
- An atom on a face is shared between two cells and counts as one-half an atom.
- An atom at a corner is shared among eight cells and counts as one-eighth an atom.

As noted earlier, for an fcc structure, the eight corner atoms contribute $8 \times \frac{1}{8} = 1$ atom to the cell. The six atoms at the centers of faces contribute $6 \times \frac{1}{2} = 3$ atoms (**FIG. 3H.26**). The total number of atoms in an fcc unit cell is therefore $1 + 3 = 4$, and the mass of the unit cell is four times the mass of one atom. For a bcc unit cell (like that in Fig. 3H.23b), the center atom is entirely within the cell. Therefore, the atom at the center counts 1 and each of the eight corner atoms counts $\frac{1}{8}$, giving $1 + \left(8 \times \frac{1}{8}\right) = 2$ overall.

Self-test 3H.2A How many atoms are there in a primitive cubic cell (see Fig. 3H.22)?

[*Answer:* 1]

Self-test 3H.2B How many atoms are there in the structure made up of unit cells like the one shown in (2), which has an atom at each corner, two on opposite faces, and two inside the cell on a diagonal?

The best way to determine the type of unit cell adopted by a metal is x-ray diffraction, which gives a characteristic diffraction pattern for each type of unit cell (*Major Technique 3* on the website of this book). However, a simpler procedure that can be used to distinguish

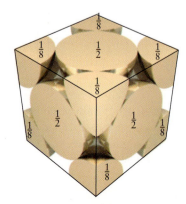

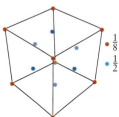

FIGURE 3H.26 The calculation of the net number of atoms in a face-centered cubic cell.

between close-packed and other structures is to measure the density of the metal, then calculate the densities of the candidate unit cells and decide which structure better accounts for the observed density. Density is an intensive property, which means that it does not depend on the size of the sample (*Fundamentals* A). Therefore, it is the same for a unit cell and a bulk sample. Hexagonal and cubic close packing cannot be distinguished in this way, because they have the same coordination numbers and therefore the same densities (for a given element).

EXAMPLE 3H.1 Deducing the structure of a metal from its density

One reason that copper is so valuable is its high malleability, which is related to its structure. Suppose you have been asked to examine how different types of heat treatment affect the crystal structure of copper. One way to detect a change in structure is from the density. After one heat treatment the density of copper is 8.93 g·cm^{-3}. Is the metal (a) body-centered cubic or (b) close-packed? The atomic radius of copper is 128 pm.

PLAN Calculate the density of the metal by assuming first that its structure is bcc and then that it is ccp (fcc). The structure with the density closer to the experimental value is more likely to be the actual structure. The mass of a unit cell is the sum of the masses of the atoms that it contains. The mass of each atom is equal to the molar mass of the element divided by Avogadro's constant. The volume of a cubic unit cell is the cube of the length of one of its sides. That length is obtained from the radius of the metal atom, the Pythagorean theorem, and the geometry of the cell.

SOLVE (a) To calculate the density of a bcc unit cell, first find the length of the side of the cube, a. Let the length of the diagonal of a face of the cell be f and the length of the diagonal through the body of the cell be b. Then, from **FIG. 3H.27b** and the Pythagorean theorem, $a^2 + f^2 = b^2 = (4r)^2$. The Pythagorean theorem also implies that $f^2 = 2a^2$, and so

$$a^2 + f^2 = a^2 + 2a^2 = 3a^2$$

It follows that $3a^2 = (4r)^2$ and therefore that $a = 4r/3^{1/2}$. Each unit cell contains one sphere at each of the eight corners and one sphere at the center for a total of $8 \times \frac{1}{8} + 1 = 2$ spheres; so the total mass of a body-centered cubic unit cell is $2M/N_A$. Therefore,

From $d = m/a^3$, $m = 2M/N_A$, and $a = 4r/3^{1/2}$,

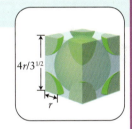

$$d = \underbrace{\frac{2M/N_A}{\underbrace{(4r/3^{1/2})^3}_{\text{volume}}}}_{\text{}} = \frac{3^{3/2} \times 2M}{N_A(4r)^3} = \frac{3^{3/2}M}{32N_Ar^3}$$

A radius of 128 pm corresponds to 1.28×10^{-8} cm, and the molar mass of copper (from the periodic table on the inside front cover) is 63.55 g·mol^{-1}. The predicted density is therefore

From $d = 3^{3/2}M/(32N_Ar^3)$,

$$d = \frac{3^{3/2} \times (63.55 \text{ g·mol}^{-1})}{32 \times (6.022 \times 10^{23} \text{ mol}^{-1}) \times (1.28 \times 10^{-8} \text{ cm})^3} = 8.17 \text{ g·cm}^{-3}$$

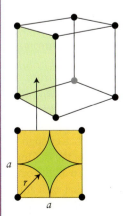

(a) Primitive cubic

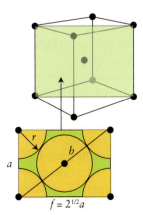

(b) Body-centered cubic

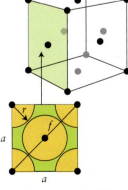

(c) Face-centered cubic

FIGURE 3H.27 The geometries of three cubic unit cells, showing the relation of the dimensions of each cell to the radius, r, of a sphere representing an atom or ion. The side of a cell is a, the diagonal of the body of a cell b, and the diagonal of a face f.

(b) The length, a, of the side of an fcc unit cell composed of spheres of radius r is $a = 8^{1/2}r$ (Section 3H.4). The volume of the unit cell is a^3 (Fig. 3H.27c). Because there are four atoms in the cell, the mass, m, of one unit cell is four times the mass of one atom (M/N_A). The density, d, is therefore

From $d = m/a^3$, $m = 4M/N_A$, and $a = 8^{1/2}r$,

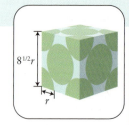

$$d = \underbrace{\frac{4M/N_A}{(8^{1/2}r)^3}}_{\text{volume}}^{\overbrace{}^{\text{mass}}} = \frac{4M}{8^{3/2}N_A r^3}$$

From $d = 4M/(8^{3/2}N_A r^3)$,

$$d = \frac{4 \times (63.55\ \text{g·mol}^{-1})}{8^{3/2} \times (6.022 \times 10^{23}\ \text{mol}^{-1}) \times (1.28 \times 10^{-8}\ \text{cm})^3} = 8.90\ \text{g·cm}^{-3}$$

EVALUATE The value for the body-centered structure, 8.17 g·cm^{-3}, is further from 8.93 g·cm^{-3}, the experimental value, than that for a close-packed structure, 8.90 g·cm^{-3}. This difference suggests that copper has a close-packed structure. In fact, x-ray diffraction shows that copper is cubic close-packed under normal conditions.

Self-test 3H.3A The atomic radius of silver is 144 pm and its density is 10.5 g·cm^{-3}. Is the structure close-packed or body-centered cubic?

[*Answer:* Close-packed]

Self-test 3H.3B The atomic radius of iron is 124 pm and its density is 7.87 g·cm^{-3}. Is this density consistent with a close-packed or a body-centered cubic structure?

Related Exercises: 3H.13, 3H.14

All crystal structures are derived from the 14 Bravais lattices. The atoms in a unit cell are counted by determining what fraction of each atom resides within the cell. The type of unit cell adopted by a metal can be identified by measuring its density.

3H.6 Ionic Solids

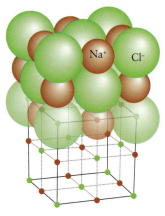

FIGURE 3H.28 The arrangement of ions in the rock-salt structure. Above is the unit cell, showing the packing of the individual ions, and below a representation of the same structure by spheres that identify the centers of the ions.

🅰 **ANIMATION FIGURE 3H.28**

The structures of ionic solids are based on the same kinds of arrays of spheres as elemental metals but are complicated by the need to take into account the presence of ions of opposite charges and different sizes. Sodium chloride, for instance, is modeled by stacking together positively charged spheres of radius 102 pm, representing the Na$^+$ ions, and negatively charged spheres of radius 181 pm, representing the Cl$^-$ ions, in such a way as to achieve the lowest possible energy. Because the crystal is electrically neutral overall, each unit cell must reflect the stoichiometry of the compound and itself be electrically neutral.

A helpful starting point is one of the close-packed structures. Because anions are usually larger than their accompanying cations, the anions can be visualized as forming a slightly expanded version of a close-packed structure with the smaller cations occupying some of the enlarged holes in the expanded lattice. A slightly enlarged tetrahedral hole is still relatively small and can accommodate only small cations. Octahedral holes are larger and can accommodate somewhat bigger cations.

The **rock-salt structure** is a common ionic structure that takes its name from the mineral form of sodium chloride. In it, the Cl$^-$ ions lie at the corners and in the centers of the faces of a cube, forming a face-centered cube (**FIG. 3H.28**). This arrangement is like an expanded ccp arrangement: the expansion keeps the anions out of contact with one another, thereby reducing their repulsion, and opens up holes that are big enough to accommodate the Na$^+$ ions. These ions fit into the octahedral holes between the Cl$^-$ ions. There is one octahedral hole for each anion in the close-packed array, and so all the octahedral holes are occupied. If you look carefully at the structure, you can see that each anion is surrounded by six cations and each cation is surrounded by six anions. The pattern

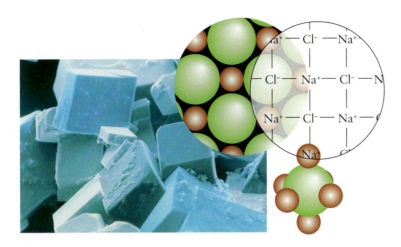

FIGURE 3H.29 Billions of unit cells stack together to recreate the smooth faces of the crystal of sodium chloride seen in this micrograph. The first inset shows some of the stacked unit cells. The second inset identifies the individual ions. The third inset (lower right) illustrates the coordination of an anion to its six cation neighbors. (*Photo: Andrew Syred/Science Source.*)

repeats over and over, with each ion surrounded by six other ions of the opposite charge (FIG. 3H.29). A crystal of sodium chloride is a three-dimensional array of a vast number of these little cubes.

In an ionic solid, the "coordination number" means the number of ions of *opposite* charge immediately surrounding a specific ion. In the rock-salt structure, the coordination numbers of the cations and the anions are both 6, and the structure overall is described as having *(6,6)-coordination*. In this notation, the first number is the cation coordination number and the second is that of the anion. The rock-salt structure is found for a number of other minerals having ions of the same charge number, including KBr, RbI, CaO, and AgCl. It is common whenever the cations and anions have very different radii, in which case the smaller cations can fit into the octahedral holes in a face-centered cubic array of anions. The **radius ratio**, ρ (rho), which is defined as

$$\text{Radius ratio} = \frac{\text{radius of smaller ion}}{\text{radius of larger ion}} \quad \text{or} \quad \rho = \frac{r_{\text{smaller}}}{r_{\text{larger}}} \tag{1}$$

is a guide to the type of structure to expect. Although there are many exceptions, a rock-salt structure can be expected when the radius ratio is in the range from 0.4 to 0.7. For example, the radius of the Mg^{2+} ion is 72 pm and that of the O^{2-} ion is 140. pm. Therefore, for MgO,

$$\rho = \frac{\overbrace{72 \text{ pm}}^{\text{radius of } Mg^{2+}}}{\underbrace{140. \text{ pm}}_{\text{radius of } O^{2-}}} = 0.51$$

This ratio is consistent with the rock-salt structure, which is indeed the one observed in MgO crystals.

When the radii of the cations and anions are similar and $\rho > 0.7$, more anions can fit around each cation. Now the ions may adopt the **cesium-chloride structure** typified by cesium chloride itself, CsCl (FIG. 3H.30). The radius of a Cs^+ ion is 167 pm and that of a Cl^- ion is 181 pm, giving a radius ratio of 0.923; so the two ions are almost the same size. In this structure, the anions form an expanded primitive cubic array, with a Cl^- ion at each of the eight corners of each cubic unit cell. There is a large "cubic" hole at the center of the cell, and the Cs^+ ion fits into it. Equivalently, each Cl^- ion can be regarded as being at the center of a cubic unit cell with eight Cs^+ ions at its corners (FIG. 3H.31). The coordination number of each type of ion is 8, and overall the structure has *(8,8)-coordination*. The cesium-chloride structure is much less common than the rock-salt structure, but it is also found for CsBr, CsI, TlCl, and TlBr.

When the radius ratio of an ionic compound is less than about 0.4, corresponding to cations that are significantly smaller than the anion, the small tetrahedral holes may be occupied. An example is the **zinc-blende structure** (which is also called the *sphalerite structure*), named after a form of the mineral ZnS (FIG. 3H.32). This structure is based on an expanded cubic close-packed lattice of big S^{2-} anions, with small Zn^{2+} cations occupying

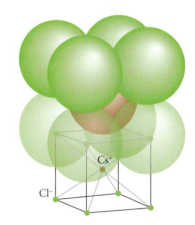

FIGURE 3H.30 The cesium chloride structure: above is the unit cell and below a second unit cell showing the location of the centers of the ions.

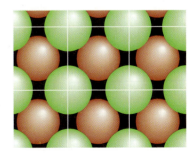

FIGURE 3H.31 The repetition of the cesium-chloride unit cell creates the entire crystal. This view is from one side of the crystal and shows several unit cells stacked together.

ANIMATION FIGURE 3H.31

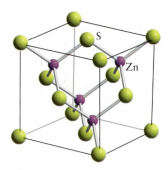

S

Zn

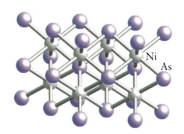

Ni
As

FIGURE 3H.33 The structure of nickel arsenide, NiAs. Atypical structures such as this one are often found when the covalent character of the bonding is important and the ions have to take up specific positions relative to one another to maximize their bonding.

FIGURE 3H.32 The zinc-blende (sphalerite) structure. The four zinc ions (purple) form a tetrahedron within a face-centered cubic unit cell composed of sulfide ions (yellow). The zinc ions occupy half the tetrahedral holes between the sulfide ions. Each zinc ion is surrounded by four sulfide ions; each sulfide ion is similarly surrounded by four zinc ions.

half the tetrahedral holes. Each Zn^{2+} ion is surrounded by four S^{2-} ions, and each S^{2-} ion is surrounded by four Zn^{2+} ions; so the zinc-blende structure has *(4,4)-coordination*.

> **Self-test 3H.4A** Predict (a) the likely structure and (b) coordination type of solid ammonium chloride. Assume that the ammonium ion can be approximated as a sphere with a radius of 151 pm.
>
> [*Answer:* (a) Cesium-chloride structure; (b) (8,8)-coordination]
>
> **Self-test 3H.4B** Predict (a) the likely structure and (b) coordination type of solid calcium sulfide.

The basic model of an ionic solid, as a collection of hard spheres of the appropriate radii stacked together in the arrangement that has the lowest total energy, may break down if the bonding is not purely ionic. In such a case the model must be adapted to include other arrangements. When bonding has a significant degree of covalent character certain orientations are favored over others and the ions will lie in specific positions around one another. An example is nickel arsenide, NiAs. In this solid, the small Ni^{3+} cations polarize the big As^{3-} anions (as described in Topic 2D), and the bonds have some covalent character. The ions pack together in an arrangement that is quite different from that of a purely ionic, sphere-packing model (**FIG. 3H.33**). Once the structure of an ionic compound is known, its density can be estimated by using an approach similar to that used for metals.

EXAMPLE 3H.2   Estimating the density of an ionic solid

Cesium chloride is being studied as an alternative treatment for cancer. Its medicinal effect is thought to be related to the large size of its cation. Estimate the density of cesium chloride from its crystal structure.

ANTICIPATE The densities of ionic compounds are typically a few grams per cubic centimeter, so expect a density of a similar magnitude.

PLAN The density of the bulk solid is the same as that of the unit cell. The volume of the cesium-chloride unit cell is the cube of the length of one of its sides, and that length can be worked out from the Pythagorean theorem. The mass of the unit cell is obtained by counting the net number of ions of each kind and using the mass of each ion (its molar mass divided by Avogadro's constant; the molar mass of an ion is negligibly different from that of its parent element). The density is obtained by dividing the mass by the volume. Assume that cations and anions touch either along diagonals (as in cesium chloride) or along edges, as in rock salt.

SOLVE

The radius of the Cs^+ ion is 167 pm and that of Cl^- is 181 pm. Therefore, the length of the diagonal of the unit cell is $b = r(Cl^-) + 2r(Cs^+) + r(Cl^-)$, so

$$b = \overbrace{181}^{r(Cl^-)} + \overbrace{2(167)}^{r(Cs^+)} + \overbrace{181}^{r(Cl^-)} \text{ pm} = 696 \text{ pm}$$
$$\text{or } 6.96 \times 10^{-8} \text{ cm (Fig. 3H.27b)}$$

The length of the side, a, is related to b by $a = b/3^{1/2}$ (recall Example 3H.1).

The volume of the unit cell is therefore $a^3 = (b/3^{1/2})^3$.

Each body-centered unit cell contains one Cs^+ ion (of molar mass 132.91 g·mol^{-1}) and one Cl^- ion (of molar mass 35.45 g·mol^{-1}). The total mass of the unit cell is the sum of these two masses divided by Avogadro's constant, N_A.

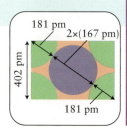

402 pm
181 pm
2×(167 pm)
181 pm

From $d = M/N_A(b/3^{1/2})^3$,

$$d = \frac{(132.91 + 35.45) \text{ g·mol}^{-1}}{(6.022 \times 10^{23} \text{ mol}^{-1}) \times (6.96 \times 10^{-8} \text{ cm/}3^{1/2})^3} = 4.31 \text{ g·cm}^{-3}$$

EVALUATE The calculated value is in line with typical values of ionic compounds, but the experimental value is 3.99 g·cm^{-3}. The discrepancy is probably due to the assumption that the ions behave like hard spheres.

Self-test 3H.5A Estimate the density of sodium chloride from its crystal structure.

[**Answer:** 2.14 g·cm^{-3}; experimental: 2.17 g·cm^{-3}]

Self-test 3H.5B Estimate the density of cesium iodide from its crystal structure.

Related Exercises 3H.31, 3H.32

Ions stack together in the regular crystalline structure corresponding to lowest energy. The structure adopted depends on the radius ratio of the cation and anion. Covalent character in an ionic bond imposes a directional character on the bonding.

What have you learned in this Topic?

You have seen that solids are classified as molecular, network, metallic, or ionic and that crystalline structures can be described in terms of their unit cells. You have also learned that an important way to model crystal structures is as collections of hard spheres packed together in characteristic ways that depend on their relative sizes and charges.

The skills you have mastered are the ability to:

☐ **1.** Summarize the properties and structures of molecular, network, metallic, and ionic solids (Sections 3H.1–3H.4).

☐ **2.** Calculate the fraction of occupied space in a given crystal lattice (Section 3H.4).

☐ **3.** State the coordination number of an atom or ion in a given crystal lattice (Section 3H.4).

☐ **4.** Count the number of atoms or ions in a given unit cell (Self-Test 3H.2).

☐ **5.** Infer the crystal structure of a metal from its density (Example 3H.1).

☐ **6.** Describe the structure and estimate the density of an ionic solid (Example 3H.2).

☐ **7.** Predict the structure of an ionic solid from the relative radii of the ions (Section 3H.6).

Topic 3H Exercises

3H.1 Glucose ($C_6H_{12}O_6$) and benzophenone ($C_6H_5COC_6H_5$) are examples of compounds that form molecular solids. The structures of glucose and benzophenone are given here. (a) What types of forces hold these molecules in a molecular solid? (b) Which of the two solids has the higher melting point?

Glucose   Benzophenone

3H.2 Chloromethane (CH_3Cl) and acetic acid (CH_3COOH) form molecular solids. (a) What types of forces hold these molecules in a molecular solid? (b) Which of the two liquids has the higher freezing point?

3H.3 Classify each of the following solids as ionic, network, metallic, or molecular: (a) quartz, SiO_2; (b) limestone, $CaCO_3$; (c) dry ice, CO_2; (d) sucrose, $C_{12}H_{22}O_{11}$; (e) polyethylene, a polymer with molecules consisting of chains of thousands of repeating $-CH_2CH_2-$ units.

3H.4 Classify each of the following solids as ionic, network, metallic, or molecular: (a) iron pyrite (fool's gold), FeS_2; (b) octane (a component of gasoline), C_8H_{18}; (c) cubic boron nitride (a compound with a structure similar to that of diamond, but with alternating boron and nitrogen atoms), BN; (d) calcium sulfate (gypsum), $CaSO_4$; (e) the chromium plating on a motorcycle.

3H.5 Three unknown substances were tested in order to classify them. The following table shows the results of the tests. Use Table 3H.1 to classify substances A, B, and C as metallic, ionic, network, or molecular solids.

Substance	Appearance	Melting point/°C	Electrical conductivity	Solubility in water
A	hard, white	800	only when dissolved in water	soluble
B	lustrous, malleable	1500	high	insoluble
C	soft, yellow	113	none	insoluble

3H.6 Three unknown substances were tested in order to classify them. The following table shows the results of the tests. Use Table 3H.1 to classify substances X, Y, and Z as metallic, ionic, network, or molecular solids.

Substance	Appearance	Melting point/°C	Electrical conductivity	Solubility in water
X	brittle, white	146	none	soluble
Y	very hard, colorless	1600	none	insoluble
Z	hard, orange	398	only when dissolved in water	soluble

3H.7 Iron crystallizes in a bcc structure. The atomic radius of iron is 124 pm. Determine (a) the number of atoms per unit cell; (b) the coordination number of the lattice; (c) the length of the side of the unit cell.

3H.8 The metalloid polonium crystallizes in a primitive cubic structure, with an atom at each corner of a cubic unit cell. The atomic radius of polonium is 167 pm. Sketch the unit cell and determine (a) the number of atoms per unit cell; (b) the coordination number of an atom of polonium; (c) the length of the side of the unit cell.

3H.9 Calculate the density of each of the following metals from the data given: (a) aluminum, fcc structure, atomic radius 143 pm; (b) potassium, bcc structure, atomic radius 227 pm.

3H.10 Calculate the density of each of the following metals from the data given: (a) nickel, fcc structure, atomic radius 125 pm; (b) rubidium, bcc structure, atomic radius 248 pm.

3H.11 Calculate the atomic radius of each of the following elements from the data given: (a) platinum, fcc structure, density 21.45 g·cm^{-3}; (b) tantalum, bcc structure, density 16.65 g·cm^{-3}.

3H.12 Calculate the atomic radius of each of the following elements from the data given: (a) silver, fcc structure, density 10.50 g·cm^{-3}; (b) chromium, bcc structure, density 7.19 g·cm^{-3}.

3H.13 The density of rhodium is 12.42 g·cm^{-3} and its atomic radius is 134 pm. Is the metal close-packed or body-centered cubic?

3H.14 The density of molybdenum is 10.22 g·cm^{-3} and its atomic radius is 136 pm. Is the metal close-packed or body-centered cubic?

3H.15 One form of silicon has density of 2.33 g·cm^{-3} and crystallizes in a cubic lattice with a unit cell edge of 543 pm. (a) What is the mass of each unit cell? (b) How many silicon atoms does one unit cell contain?

3H.16 Krypton crystallizes with a face-centered cubic unit cell of edge 559 pm. (a) What is the density of solid krypton? (b) What is the atomic radius of krypton? (c) What is the volume of one krypton atom? (d) What percentage of the unit cell is empty space if each atom is treated as a hard sphere?

3H.17 What percentage of space is occupied by close-packed cylinders of length *l* and radius *r*?

3H.18 Calculate the radius of the cavity formed by three circular disks of radius *r* that lie in a close-packed planar arrangement. Check your answer experimentally by using three compact disks and measuring the radius of the cavity that they form.

3H.19 Compare the hybridization and structure of carbon in diamond and graphite. How do these features explain the physical properties of the two allotropes?

3H.20 Ordinary "hexagonal" graphite has a structure that repeats the ABAB. . . alternation of layers; "rhombohedral graphite" has the repetition ABCABC. . . , with the C layer displaced from the other two. Sketch the structure of rhombohedral graphite.

3H.21 Sheets of graphene one atom thick were first prepared in the laboratory of Andre Geim and Kostya Novoselov at the University of Manchester. The scientists picked up flecks of graphite with adhesive tape and then pulled the layers apart with another piece of tape until one layer remained. Suppose you repeated this process with a piece of tape 2.0 cm wide until one layer remained that completely covered 1.0 cm of the tape. Estimate (a) the number of carbon atoms that remain on the tape and (b) the amount in moles.

3H.22 A layer of graphene 12 atoms thick was deposited on a circular silicon wafer 1.2 cm in diameter. Estimate (a) the number of carbon atoms deposited and (b) the amount in moles.

3H.23 Indium arsenide crystallizes in the zinc-blende (sphalerite) structure (Fig. 3H.32). (a) What are the coordination numbers of the indium and arsenide ions? (b) What is the formula of indium arsenide?

3H.24 Depending on the temperature, rubidium chloride can exist in either the rock-salt (Fig. 3H.25) or cesium-chloride structure (Fig. 3H.30). (a) What are the coordination numbers of the rubidium and chloride ions in each structure? (b) In which of these structures does the rubidium ion occupy the larger "hole"?

3H.25 Calculate the number of cations, anions, and formula units in a unit cell of each of the following solids: (a) the cesium-chloride unit cell shown in Fig. 3H.30; (b) the rutile (TiO$_2$) unit cell shown here. (c) What are the coordination numbers of the ions in rutile?

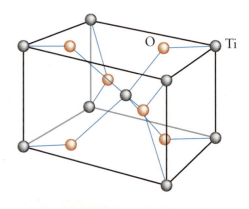

3H.26 Calculate the number of cations, anions, and formula units per unit cell in each of the following solids: (a) the rock-salt unit cell shown in Fig. 3H.28; (b) the fluorite (CaF_2) unit cell shown here. (c) What are the coordination numbers of the ions in fluorite?

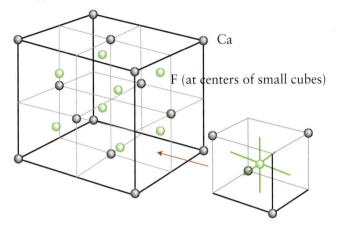

Ca

F (at centers of small cubes)

3H.27 An oxide of rhenium crystallizes with a cubic unit cell that has a rhenium cation at each corner and an oxide ion at the center of each edge of the crystal. (a) Determine the coordination numbers of the two ions. (b) Write the formula of the oxide.

3H.28 When an oxide of uranium crystallizes, the uranium cations form an expanded cubic close-packed array with an oxide ion in each tetrahedral hole. (a) Determine the coordination numbers of the two ions. (b) What is the formula of the oxide?

3H.29 Use radius ratios to predict the coordination number of the cation in (a) RbF; (b) MgO; (c) NaBr. (See Fig. 1F.6 for radius data.)

3H.30 Use radius ratios to predict the coordination number of the cation in (a) CsF; (b) NaI; (c) CaS. (See Fig. 1F.6 for radius data.)

3H.31 Estimate the density of each of the following solids from the ionic radii given in Fig. 1F.6: (a) calcium oxide (rock-salt structure, Fig. 3H.28); (b) cesium bromide (cesium-chloride structure, Fig. 3H.30).

3H.32 Calculate the density of each of the following solids: (a) magnesium oxide (rock-salt structure, Fig. 3H.28), distance between centers of Mg^{2+} and O^{2-} ions is 212 pm; (b) calcium sulfide (cesium-chloride structure, Fig. 3H.30), the distance between the centers of the Ca^{2+} and S^{2-} ions is 284 pm.

3H.33 Graphite forms extended two-dimensional layers (see Fig. 3H.8). (a) Draw the smallest possible rectangular unit cell for a layer of graphite. (b) How many carbon atoms are in that unit cell? (c) What is the coordination number of carbon in a single layer of graphite?

3H.34 The ammonium ion can be represented by a sphere of radius 151 pm. Use radius ratios to predict the type of lattice structure of (a) NH_4F; (b) NH_4I.

3H.35 If the edge length of an fcc unit cell of RbI is 732.6 pm, how long would the edge of a cubic single crystal of RbI be that contains 1.00 mol RbI formula units?

3H.36 The edge length of the fcc unit cell of NaCl is 562.8 pm. (a) How many unit cells are present in a single crystal of NaCl (table salt) that is a cube with edges of length 1.00 mm? (b) What amount (in moles) of NaCl formula units is present in this crystal?

Topic 3I Inorganic Materials

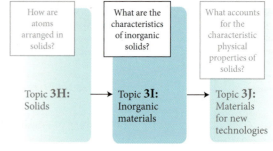

Why Do You Need to Know This Material? To select the best material for a purpose or to design a new one, you need to know what properties to expect from various aggregates of atoms or molecules.

What Do You Need to Know Already? This Topic assumes a familiarity with types of chemical bonding (Topics 2A and 2B) and the description of the structures of solids (Topic 3H).

Materials used in technology, medicine, and construction are classified somewhat loosely as either "hard" or "soft." **Hard matter** can withstand strong forces without deforming; **soft matter** responds more readily to an applied force. In general, hard matter is typically inorganic in composition, whereas soft matter is typically organic.

3I.1 Alloys

Alloys are created by mixing two or more molten metals and allowing them to cool. Some common alloys are listed in **TABLE 3I.1**. Their properties are affected by their composition, their crystalline structure, and the size and texture of the individual grains. In a **homogeneous alloy,** atoms of the different elements are distributed uniformly, like true compounds. Examples are some varieties of brass and bronze and the alloys used for coinage. A **heterogeneous alloy** consists of mixtures of crystalline phases with different compositions (**FIG. 3I.1**). Examples are tin–lead solder and the mercury–silver amalgams once used to fill teeth. Unlike pure metals, which have a distinct melting point, alloys typically melt and solidify over a range of temperatures.

An alloy in which atoms of one metal are substituted for atoms of another metal is called a **substitutional alloy** (**FIG. 3I.2**). Elements that can form substitutional alloys have atoms with atomic radii that differ by no more than about 15% (**FIG. 3I.3**). An example is the copper–zinc alloy used for some "copper" coins. Because zinc atoms are nearly the same size as copper atoms (their radii are 133 pm and 128 pm, respectively), zinc atoms can take the place of some of the copper atoms in the crystal. Because there are slight differences in size and electronic structure, the less abundant atoms in a substitutional alloy distort the shape of the lattice of the more abundant atoms of the host metal and hinder the flow of electrons and the spread of thermal motion. A substitutional alloy therefore typically has lower electrical and thermal conductivity than the pure element. Because the lattice is distorted, it is more difficult for one plane

FIGURE 3I.1 This micrograph shows some of the separate phases in a heterogeneous alloy consisting of 80% bismuth and 20% tin. The light regions have a high concentration of bismuth and the dark regions a high concentration of tin. However, each region is homogeneous. (*M. Charles and A. Cockburn/University of Cambridge.*)

TABLE 3I.1	Compositions of Typical Alloys
Alloy	**Mass percentage composition**
brass	up to 40% zinc in copper
bronze	a metal other than zinc or nickel in copper (casting bronze: 10% Sn and 5% Pb)
cupronickel	nickel in copper (coinage cupronickel: 25% Ni)
pewter	6% antimony and 1.5% copper in tin
solder	tin and lead
stainless steel*	more than 12% chromium in iron

*For more detailed information on steels, see Table 9B.2.

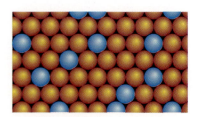

FIGURE 3I.2 In a substitutional alloy, the positions of some of the atoms of one metal are taken by atoms of another metal. The two elements must have similar atomic radii.

FIGURE 31.3 The relative metallic radii of the first row d-block metals (in picometers).

of atoms to slip past another. Consequently, a substitutional alloy is harder and stronger than the pure metal.

Self-test 31.1A Estimate the relative density of a brass (compared to pure copper) in which 20.00% of copper atoms have been replaced by zinc atoms without distortion of the crystal structure.

[*Answer:* 1.006]

Self-test 31.1B Estimate the relative density of a brass (compared to pure copper) in which 50.00% of copper atoms have been replaced by zinc atoms without distortion of the crystal structure.

Steel (Topic 9B) is a homogeneous alloy based on iron and containing up to 2% by mass of carbon. Carbon atoms (77 pm) are much smaller than iron atoms (124 pm), and do not substitute for iron in the crystal lattice. They are so small that they can fit into the **interstices** (the holes) in the iron lattice. The resulting material is called an **interstitial alloy** (**FIG. 31.4**). For two elements to form an interstitial alloy, the atomic radius of the solute element must be less than about 60% of the atomic radius of the host metal. The interstitial atoms interfere with electrical conductivity and with the movement of the atoms forming the lattice. This restricted motion makes the alloy harder and stronger than the pure host metal.

One of the oldest alloys is bronze. Pure copper melts at a 1083 °C and so is difficult to work with in a hot charcoal fire. Tin melts at 232 °C, and so cannot be used to make cooking pots. Both metals are rather soft and so do not make good tools. However, bronze melts at temperatures between the melting points of copper and tin, and the higher the proportion of copper, the higher is the melting point. In addition, bronze is much harder than either copper or tin and is more resistant to corrosion than either pure metal.

Brass is a substitutional homogeneous alloy of copper and zinc that has good corrosion resistance. The percentage of zinc in brass varies, but it is generally around 30%. Tin, arsenic, and antimony may be added to brass to improve its corrosion resistance, and iron may be added to increase its hardness.

Bismuth and cadmium form a heterogeneous alloy. When a molten mixture of bismuth and cadmium solidifies, the solid is a mixture of small crystals of pure bismuth and pure cadmium. As a bismuth-rich melt is cooled, bismuth is deposited and the composition of the remaining liquid changes, becoming richer in cadmium. As a cadmium-rich melt is cooled, solid cadmium is deposited and the composition of the remaining liquid changes, becoming richer in bismuth. Because of the changing compositions, the alloys freeze and melt over a temperature range. However, there is one composition at which the entire sample freezes and melts at a single fixed temperature that is lower than the melting point of either pure metal (**FIG. 31.5**). A mixture that behaves in this way is called a **eutectic** (from the Greek words for "easily melted").

Self-test 31.2A Suppose that the illustration in Fig. 31.5 corresponds to a mixture of lead (metal A) and tin (metal B). Estimate the mass percentage composition of the eutectic mixture.

[*Answer:* 51% Pb, 0.49% Sn]

Self-test 31.2B Suppose instead that the illustration in Fig. 31.5 corresponds to a mixture of silver (metal A) and nickel (metal B). Estimate the mass percentage composition of the eutectic mixture.

Alloys of metals tend to be stronger and have lower electrical conductivity than pure metals. In substitutional alloys, atoms of the solute metal take the place of some atoms of a metal of similar atomic radius. In interstitial alloys, atoms of the solute element fit into the interstices in a lattice formed by atoms of a metal with a larger atomic radius.

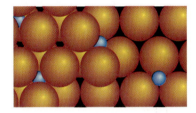

FIGURE 31.4 In an interstitial alloy, the atoms of one metal lie in the gaps between the atoms of another metal. The two elements need to have markedly different atomic radii.

In steel, carbon is so important and widely used that it is regarded as an "honorary metal" even though it is actually a nonmetal.

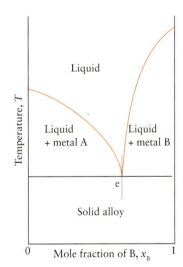

FIGURE 31.5 This temperature–composition plot shows how the melting point of an alloy changes with temperature. The vertical line through e is at the eutectic composition of the system. Note that it melts at the lowest temperature possible for the alloy.

(a)

(b)

(c)

FIGURE 31.6 Three common forms of silica (SiO_2): (a) quartz; (b) quartzite; and (c) cristobalite. The black parts of the sample of cristobalite are obsidian, a volcanic rock that contains silica. Sand consists primarily of small pieces of impure quartz. (*Field Museum of Natural History, Chicago/Getty Images.*)

31.2 Silicates

Silica, SiO_2, is a hard, rigid network solid (Topic 3H). It is insoluble in water and occurs naturally as *quartz* and sand, which consists of small fragments of quartz, usually colored golden brown by iron oxide impurities (**FIG. 31.6**). Minerals based on silica and silicates, such as sandstone and granite, are used when a strong, durable, corrosion-resistant construction material is required.

Silica gets its strength from its covalently bonded network structure. In silica itself, each Si atom is at the center of a tetrahedron of O atoms, and each corner O atom is shared by two Si atoms (**1**). Hence, each tetrahedron contributes one Si atom and $4 \times \frac{1}{2} = 2$ O atoms to the solid, which has the empirical formula SiO_2. Quartz has a complicated structure; it is built from helical chains of SiO_4 units wound around one another, giving a net composition of SiO_2 when the sharing of O atoms between units is taken into account. When it is heated to about 1500 °C, it changes into another arrangement, that of the mineral *cristobalite* (**FIG. 31.7**). This structure is easier to describe: its Si atoms are arranged like the C atoms in diamond, but, in cristobalite, an O atom lies between each pair of neighboring Si atoms.

1 An SiO_4 unit

THINKING POINT

Why do CO_2 and SiO_2 have such different properties despite being neighbors in the same group of the periodic table?

There are many different silicates, which have various arrangements of tetrahedral oxoanions of silicon. The Si—O bond has considerable covalent character. The differences in properties between the various silicates are related to the number of negative charges on each tetrahedron, the number of corner O atoms shared with other tetrahedra, and the manner in which chains and sheets of the linked tetrahedra lie together. Most glasses are primarily mixtures of silicates and ionic compounds (see the Interlude after this focus). Quartz itself is useful in spectroscopy because it is transparent to both ultraviolet and visible radiation.

The simplest silicates, the *orthosilicates*, are built from SiO_4^{4-} ions. They are not very common but include the mineral *zircon*, $ZrSiO_4$, which is used as a substitute for diamond in costume jewelry. The *pyroxenes* consist of chains of SiO_4 units in which two corner O atoms are shared by neighboring units (**FIG. 31.8**); the repeating unit is the metasilicate ion, SiO_3^{2-}. Electrical neutrality is provided by cations regularly spaced along the chain. The pyroxenes include the gemstone *jade*, $NaAl(SiO_3)_2$.

Chains of silicate units can link together to form ladderlike structures that include *tremolite*, $Ca_2Mg_5(Si_4O_{11})_2(OH)_2$. Tremolite is one of the fibrous minerals called *asbestos*, which can withstand high temperatures (**FIG. 31.9**). Their fibrous quality is due to the way in which the ladders of SiO_4 units lie together but can easily be torn apart. Because of their resistance to fire, asbestos fibers were once widely used for heat insulation in buildings. However, these fibers can lodge in lung tissue, where fibrous scar tissue forms around them, giving rise to asbestosis and a susceptibility to lung cancer. In some minerals, the SiO_4 tetrahedra link together to form sheets. An example is *talc*, a hydrated magnesium silicate, $Mg_3(Si_2O_5)_2(OH)_2$. Talc is soft and slippery because the silicate sheets can slide past one another.

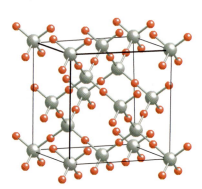

FIGURE 31.7 The structure of cristobalite is like that of diamond except that an O atom (red) lies between each pair of Si atoms (gray). The arrangement about each Si atom is shown in structure **1**.

More complex (and more common) structures result when some of the silicon(IV) in silicates is replaced by aluminum(III) to form the *aluminosilicates*. The missing positive charge is made up by the presence of extra cations. These cations account for the difference in properties between the silicate talc and the aluminosilicate *mica*. One form of mica is the mineral *phlogopite*, $KMg_3(Si_3AlO_{10})(OH)_2$. In this mineral, the sheets of tetrahedra are held together by extra K^+ ions. Although it cleaves neatly into transparent layers when the sheets are torn apart, mica is not slippery like talc (**FIG. 31.10**). Because they are more heat resistant than glass, sheets of mica are used for windows in furnaces.

The *feldspars* are aluminosilicates in which as much as half the silicon(IV) has been replaced by aluminum(III). A typical feldspar has the formula $KAlSi_3O_8$. Feldspars are the most abundant silicate materials on Earth and are a major component of *granite*, a compressed mixture of mica, quartz, and feldspar that serves as one of the most valued and attractive building materials (**FIG. 31.11**).

Silicate structures are based on SiO_4 tetrahedral units with different negative charges and different numbers of shared O atoms.

31.3 Calcium Carbonate

Ionic compounds of calcium are often used as structural materials in organisms, buildings, and civil engineering, on account of the rigidity of their structures. This rigidity stems from the strength with which the small, highly charged Ca^{2+} cation interacts with its neighbors. Because the carbonate ion is also doubly charged, calcium carbonate, $CaCO_3$, has a relatively high lattice energy (Topic 2A). Consequently, one of the earliest building materials was limestone, an impure form of calcium carbonate, which takes its honey color from impurities such as Fe^{2+} ions. In its compressed and harder form, it is *marble*; in its less compacted form it is *chalk*.

The two most common forms of crystalline calcium carbonate are *calcite* and *aragonite*. Although aragonite is harder and denser than calcite, it is less abundant and less stable. In addition, it converts to calcite at high temperatures.

Nature has made extensive use of the ability of calcium to form rigid structures. It is found as the calcium carbonate of the shells of shellfish and the calcium phosphate of bone. In fact, large deposits of limestone have resulted from seashells and microorganisms that accumulated on ocean floors millions of years ago. Both seashells and bones are much stronger than the pure calcium salts because the carbonate or phosphate is imbedded in a tough matrix.

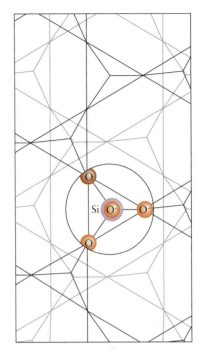

FIGURE 31.8 The basic structural unit of the minerals called pyroxenes. Each tetrahedron is an SiO_4 unit (like that in structure **1**), and a shared corner represents a shared O atom (red in the inset, where one O atom lies directly over an Si atom). The two unshared O atoms each carry a negative charge.

FIGURE 31.9 The minerals commonly called asbestos (from the Greek words meaning "not burning") are fibrous because they consist of long chains based on SiO_4 tetrahedra linked through shared oxygen atoms. (©1984 Chip Clark–Fundamental Photographs.)

FIGURE 31.10 The aluminosilicate mica cleaves into thin transparent sheets with high melting points. These properties allow it to be used for windows in furnaces. (©1989 Chip Clark–Fundamental Photographs.)

FIGURE 31.11 The mineral granite is a compressed mixture of mica, quartz, and feldspar. (*Phillip Hayson/Science Source.*)

Portland cement was named by Joseph Aspdin in 1824 in recognition of its similarity to a type of limestone mined on the Isle of Portland off the coast of England.

FIGURE 31.12 An electron micrograph of the surface of mortar, showing the growth of tiny interlocking crystals as carbon dioxide reacts with calcium oxide and silica. (*National Institute of Standards and Technology.*)

Self-test 31.3A What is the change in the length of a side of a 1.0 cm^3 cube of aragonite (density 2.83 g·cm^{-3}) when it converts to calcite (density 2.71 g·cm^{-3})?

[*Answer:* +0.14 mm]

Self-test 31.3B What is the change in the length of a side of a 1.0 cm^3 cube of diamond (density 3.51 g·cm^{-3}) when it converts to graphite (density 2.27 g·cm^{-3})?

Calcium compounds are common in structural materials because the small, highly charged Ca^{2+} ion results in rigid structures.

31.4 Cement and Concrete

The building blocks used in construction are commonly held together by binders called *mortars*. *Portland cement mortar* consists of about one part cement and three parts sand (largely silica, SiO$_2$). Lime mortar is a fast-setting mortar that contains Portland cement, lime (calcium hydroxide), and sand. It sets to a hard mass as the lime reacts with the carbon dioxide of the air to form the carbonate (**FIG. 31.12**). *Concrete* is also widely used in construction for the floors and walls of buildings. Concrete consists of a binder and a filler, usually gravel, and the binder is cement, usually Portland cement.

Portland cement is made by heating a mixture of crushed limestone, clay or shale, sand, and oxides such as iron ore in a kiln. As shown in **TABLE 31.2**, different types of Portland cement have been developed to meet different requirements. Clays are primarily aluminosilicates (Section 31.2), which are composed of layers of ions separated by water molecules. When they are heated with calcium carbonate, the water molecules are driven off. The hard pellets that result, called "clinkers," are a mixture of mainly calcium oxide, calcium silicates, and calcium aluminum silicates. These pellets are ground together with *gypsum*, CaSO$_4$·2H$_2$O, into a powder that sets into a hard mass when mixed with water. The water reacts with the mixture to produce hydrates and hydroxides. The reactions are complex, but a representative equation is

$$(Al_2O_3)\cdot(CaO)_3(s) + 3\ CaSO_4\cdot2H_2O(s) + 26\ H_2O(l) \longrightarrow$$
$$(Al_2O_3)\cdot(CaO)_3\cdot(CaSO_4)_3\cdot32H_2O(s)$$

Because the particles in Portland cement are finely ground and mixed together, the hydrates bind the salts together into an intricate three-dimensional network, forming a hard, strong material as the cement dries and other reactions take place, such as:

$$6\ (SiO_2)\cdot(CaO)_3(s) + 18\ H_2O(l) \longrightarrow (SiO_2)_6\cdot(CaO)_5\cdot5H_2O(s) + 13\ Ca(OH)_2(s)$$

TABLE 31.2   Compositions of Common Portland Cement Clinkers

Type of cement	Typical composition*
1. General use	50–70% (SiO$_2$)·(CaO)$_3$, 15–30% (SiO$_2$)·(CaO)$_2$, 5–10% (Al$_2$O$_3$)·(CaO)$_3$, 5–15% (4 CaO)·Al$_n$Fe$_{2-n}$O$_3$.
2. Moderate sulfate resistance	Resists attack by sulfates. Generates heat slowly and so sets slowly. Low (Al$_2$O$_3$)·(CaO)$_3$ content.
3. High early strength	Has more (SiO$_2$)·(CaO)$_3$ than Type 1 and has been ground finer to set faster.
4. Low heat	Used when hydration heat must be minimized. Has about half the percentage of (SiO$_2$)·(CaO)$_3$ and (Al$_2$O$_3$)·(CaO)$_3$ and double the abundance of (SiO$_2$)·(CaO)$_2$ of Type 1.
5. High sulfate resistance	Highly resistant to reaction with sulfates. Very low percentage of (Al$_2$O$_3$)·(CaO)$_3$ and high percentage of (SiO$_2$)·(CaO)$_2$.

*Small, variable amounts of other oxides are present in all Portland cements.

Self-test 3I.4A What mass of water is needed to react completely with 1.0 kg of $(Al_2O_3) \cdot (CaO)_3(s)$?

[*Answer:* 1.7 kg]

Self-test 3I.4B What mass of water is needed to react completely with 1.0 kg of $(SiO_2) \cdot (CaO)_3(s)$?

Portland cement forms when a mixture of limestone, clay, and other substances is heated to a high temperature. It sets when water is added, forming a network of hydrates.

What have you learned in this Topic?

You have learned how chemical principles are applied to a wide variety of problems relating to the development of inorganic materials, including alloys and building materials such as silicates, carbonates, and cements.

The skills you have mastered are the ability to:

☐ **1.** Distinguish the main types of alloys and explain how their properties differ from those of pure metals (Section 3I.1).

☐ **2.** Distinguish the principal silicate structures and describe their properties (Section 3I.2).

☐ **3.** Explain why calcium carbonate is the basis of so many structural materials (Section 3I.3).

Topic 3I Exercises

3I.1 Estimate the relative density (compared to pure aluminum) of magnalium, a magnesium–aluminum alloy in which 30.0% of the aluminum atoms have been replaced by magnesium atoms without distortion of the crystal structure.

3I.2 Estimate the relative density (compared to pure copper) of aluminium bronze, an alloy that is 8.0% by mass aluminium. Assume no distortion of the crystal structure.

3I.3 How do the physical properties of alloys differ from the pure metals from which they are made?

3I.4 What is the difference between homogeneous and heterogeneous alloys? Give examples of each type.

3I.5 When iron surfaces are exposed to ammonia at high temperatures, "nitriding"—the incorporation of nitrogen into the iron lattice—occurs. The atomic radius of iron is 124 pm. (a) Is the alloy interstitial or substitutional? Justify your answer. (b) How do you expect nitriding to change the properties of iron?

3I.6 Silicon can be doped with small amounts of phosphorus to create a semiconductor used in transistors. (a) Is the alloy interstitial or substitutional? Justify your answer. (b) How do you expect the properties of the doped material to differ from those of pure silicon?

3I.7 Calculate the relative number of atoms of each element contained in each of the following alloys: (a) coinage cupronickel, which is 25% Ni by mass in copper; (b) a type of pewter that is about 7% antimony and 3% copper by mass in tin.

3I.8 Calculate the relative number of atoms of each element contained in each of the following alloys: (a) Rose metal, which is a low-melting-point alloy used to solder decorative ironworks and is often 28% lead and 22% tin by mass in bismuth; (b) duralumin AA2024, an alloy that hardens aluminum for use in aircraft and is 4.4% copper, 1.5% magnesium, and 0.6% manganese in aluminum.

3I.9 A unit cell for the calcite structure can be found at http://webmineral.com. From this structure, identify (a) the crystal system and (b) the number of formula units present in the unit cell.

3I.10 Consult http://webmineral.com and examine the unit cells of calcite and dolomite. (a) In what respects are these two structures the same? (b) In what respect are they different? (c) Where are the magnesium and calcium ions located in dolomite?

3I.11 Iron pyrite (FeS_2) is known as fool's gold because of its resemblance to gold metal. However, it can easily be distinguished from gold by the difference in their densities. The density of gold is 19.28 $g \cdot cm^{-3}$ and that of fool's gold is 5.01 $g \cdot cm^{-3}$. What volume of fool's gold would have the same mass as a piece of gold of volume 4.0 cm^3?

3I.12 Mica, with a density of 1.5 $g \cdot cm^{-3}$, can be expanded into vermiculite, which is used as a low-density soil amendment. The vermiculite used in soils has an average density of 0.10 $g \cdot cm^{-3}$. Estimate the volume of the vermiculite obtained by expanding 52.0 cm^3 of mica.

3I.13 Draw a Lewis structure for the orthosilicate anion, SiO_4^{4-}, and deduce the formal charges and oxidation numbers of the atoms. Use the VSEPR model to predict the shape of the ion.

3I.14 Use the VSEPR model to estimate the Si—O—Si bond angle in silica.

3I.15 Describe the structures of a silicate in which the silicate tetrahedra share (a) one O atom; (b) two O atoms.

3I.16 What is the empirical formula of a potassium silicate in which the silicate tetrahedra share (a) two O atoms and form a chain or (b) three O atoms and form a sheet? In each case, there are single negative charges on the unshared O atoms.

Topic 3J Materials for New Technologies

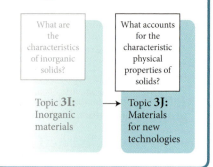

Why Do You Need to Know This Material? Modern technology relies heavily on the electrical, optical, and magnetic properties of matter, and knowledge of the origin of these properties is essential for the development of new materials.

What Do You Need to Know Already? This Topic draws on the properties of a particle in a box (Topic 1C) and extends the molecular orbital theory of bonding (Topic 2G) to solids.

Chemistry provides the material infrastructure of the modern world. It provides the substances that enable electronic material to be miniaturized so that a single computer chip can contain billions of components. It provides material that generates light and is used to display information of all kinds in compact, power-efficient devices. And, it provides magnetic materials that are used to store information in a compact, readily accessible form. On the grander scale, it also provides materials that can conduct electricity without resistance, which hold the promise—not yet realized—of enormous savings in the cost of transporting electricity and in the environmental impact of its generation.

3J.1 Electrical Conduction in Solids

An **electric current** is the flow of electric charge. In **electronic conduction,** the charge is carried by electrons. Electronic conduction is the mechanism of conduction in metals and graphite. In **ionic conduction,** the charge is carried by ions. A **solid electrolyte** is an ionic conductor. Ionic conduction is the mechanism of electrical conduction in a molten salt or an electrolyte solution. Because ions are too bulky to travel easily through most solids, the flow of charge through solids is in most cases a result of electronic conduction. However, solid electrolytes that allow ions to move through their lattices do exist and are important components of rechargeable batteries. An **insulator** is a substance that has such a high resistance that it does not conduct electricity.

Solids are classified according to their electrical resistance and how their electrical resistance varies with temperature (**FIG. 3J.1**):

A **metallic conductor** is an electronic conductor with a resistance that *increases* as the temperature is raised.

A **semiconductor** is an electronic conductor with a resistance that *decreases* as the temperature is raised.

A **superconductor** is an electronic conductor that conducts electricity with zero resistance, usually at very low temperatures.

In most cases, a metallic conductor has a much lower resistance than a semiconductor, but it is the temperature dependence of the resistance that distinguishes the two types of conductors, not the magnitude of the conductivity.

Molecular orbital theory explains the electrical properties of electronic conductors, semiconductors, and insulators by treating them as one huge molecule and supposing that their valence electrons occupy delocalized orbitals that spread throughout the solid. When N atomic orbitals merge together in a molecule, they form N molecular orbitals (Topic 2G). The same is true of a solid; but in that case N is enormous (about 10^{23} for 10 g of copper, for example). Instead of the few molecular orbitals with widely spaced energies typical of small molecules, the huge number of molecular orbitals in a solid are so close together in energy that they form a nearly continuous band (**FIG. 3J.2**). As a guide to the

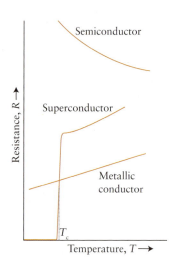

FIGURE 3J.1 The resistance of a metallic conductor increases with temperature. That of a semiconductor decreases with increasing temperature. A superconductor is a substance that has zero resistance below a certain temperature (T_c). An insulator behaves like a semiconductor with a very high resistance.

type of energy separation to expect, the separation of neighboring energy levels of a particle of mass m in a one-dimensional box of length L (Topic 1C) is

$$\overbrace{E_{n+1}}^{(n+1)^2 h^2/8mL^2} - \overbrace{E_n}^{n^2 h^2/8mL^2} = (2n+1)\frac{h^2}{8mL^2}$$

For an electron ($m = m_e$) in a box of length 1.0 cm the value of $h^2/8mL^2$ is 7.0×10^{-42} J, so even when n is enormous, the separation is still very tiny. Note that if two electrons occupy each level, then the quantum number of the uppermost filled level is $n = \frac{1}{2}N$.

> **Self-test 3J.1A** Estimate the value of n for the uppermost filled level in a one-dimensional line of sodium atoms of length 1.0 cm. *Hint:* Remember the Pauli exclusion principle, and take the radius of an Na atom as 154 pm.
>
> [**Answer:** $n \approx 1.6 \times 10^7$]
>
> **Self-test 3J.1B** Estimate the value of n for the uppermost filled level in a one-dimensional line of calcium atoms of length 1.0 cm. *Hint:* Take the radius of a Ca atom as 197 pm.

Now consider a molecular orbital treatment. For instance, in sodium each atom contributes one valence orbital (the 3s-orbital in this case) and one valence electron. If there are N atoms in the sample, then the N 3s-orbitals merge to form a band of N molecular orbitals, of which half are net bonding and half are net antibonding. The orbitals have *net* bonding or antibonding character because, in general and as for benzene (Topic 2G), a molecular orbital is bonding between some neighbors and antibonding between others, depending on where its internuclear nodes lie; only the molecular orbital of lowest energy, with no internuclear nodes, is bonding between all neighboring atoms. The N electrons contributed by the N atoms occupy the orbitals according to the building-up principle. Because two electrons can occupy each orbital, the N electrons occupy the lowest $\frac{1}{2}N$ orbitals.

An empty or incompletely filled band of molecular orbitals is called a **conduction band.** Because neighboring orbitals in a solid lie so close together in energy, it takes very little additional energy to excite the electrons from the topmost filled molecular orbitals of the band to its empty orbitals. As a result, electrons can move freely through the solid and form an electric current. The resistance of the metal increases with temperature because, when the metal is heated, its atoms vibrate more vigorously and impede the migration of the electrons.

THINKING POINT

Would you expect a single line of calcium atoms that use only their 4s-orbitals to form bands to be a metallic conductor?

In an insulator, the valence electrons fill all the available molecular orbitals to give a full band called a **valence band.** There is a substantial **band gap,** a range of energies for which there are no orbitals, before the next band, the conduction band composed of the empty orbitals, begins (**FIG. 3J.3**). The electrons in the valence band can be excited into the conduction band only by a very large injection of energy. Because the valence band is full and because the conduction band is separated from it by a large energy gap, the electrons are not mobile and the substance is an insulator.

> *Bonding in solids can be described in terms of bands of molecular orbitals. In metals, the conduction bands are incompletely filled orbitals that allow electrons to flow. In insulators, the valence bands are full and the large band gap prevents the promotion of electrons to empty orbitals.*

3J.2 Semiconductors

Semiconductors have revolutionized the electronics industry because very tiny semiconducting devices can be used to control the flow of an electric current. In an **intrinsic semiconductor,** an empty conduction band lies close in energy to a full valence

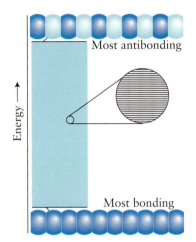

FIGURE 3J.2 A line of atoms gives rise to an almost continuous band of molecular orbital energies. At the lower edge of the band, the molecular orbitals are fully bonding; at the upper edge, the molecular orbitals are fully antibonding. The enlargement shows that, although the band of allowed energies appears to be continuous, it is in fact composed of discrete, closely spaced levels.

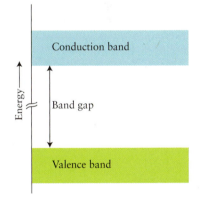

FIGURE 3J.3 In a typical insulating solid, a full valence band is separated by a substantial energy gap from the empty conduction band. Note the break in the vertical scale.

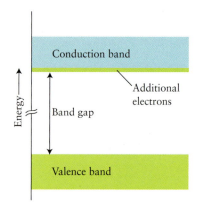

FIGURE 3J.4 In an n-type semiconductor, the additional electrons supplied by the electron-rich dopant atoms enter the conduction band (forming the green band at the bottom of the conduction band), where they can act as carriers for the current.

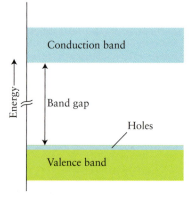

FIGURE 3J.5 In a p-type semiconductor, the electron-poor dopant atoms effectively remove electrons from the valence band, and the "holes" that result (the blue band at the top of the valence band) enable the remaining electrons to become mobile and conduct electricity through the valence band.

band. At ordinary temperatures, although most electrons are in the valence band, some occupy the nearby conduction band and are mobile. As the solid is warmed, more electrons are excited from the valence band into the conduction band. Hence, the resistance of a semiconductor decreases as its temperature is raised. In a formal sense, all solid insulators are in fact intrinsic semiconductors, but the band gap in an insulator is so large that its electrical conductivity remains negligible at all normal temperatures.

The ability of a semiconductor to carry an electric current can also be enhanced by adding electrons to the conduction band or by removing some from the valence band. These changes are brought about chemically by **doping** the solid, or spreading small amounts of impurities throughout it, and results in an **extrinsic semiconductor.** In one example, a minute amount of a Group 15 element such as arsenic is added to very pure silicon. The arsenic increases the number of electrons in the solid: each Si atom (Group 14) has four valence electrons, whereas each As atom (Group 15) has five. The additional electrons enter the upper, normally largely empty conduction band of silicon and allow the solid to conduct (**FIG. 3J.4**). This type of material is called an **n-type semiconductor** because it contains excess *n*egatively charged electrons. When silicon (Group 14) is doped with indium (Group 13) instead of arsenic, the solid has fewer valence electrons than does pure silicon; so the valence band is no longer completely full (**FIG. 3J.5**). The valence band is now said to contain "holes." Because the valence band is no longer full, it has been turned into a conduction band, and an electric current can flow. This type of semiconductor is called a **p-type semiconductor,** because the absence of negatively charged electrons is equivalent to the presence of *p*ositively charged holes.

Solid-state electronic devices such as diodes, transistors, and integrated circuits contain **p–n junctions** in which a p-type semiconductor is in contact with an n-type semiconductor (**FIG. 3J.6**). The structure of a p–n junction allows an electric current to flow in only one direction. When the electrode attached to the p-type semiconductor has a negative charge, the holes in the p-type semiconductor are attracted to it, the electrons in the n-type semiconductor are attracted to the other (positive) electrode, and current does not flow. When the polarity is reversed, with the negative electrode attached to the n-type semiconductor, electrons flow from the n-type semiconductor through the p-type semiconductor toward the positive electrode.

Self-test 3J.2A Which type of semiconductor is germanium doped with arsenic?

[*Answer:* n-type]

Self-test 3J.2B Which type of semiconductor is antimony doped with tin?

In semiconductors, empty levels are close in energy to filled levels.

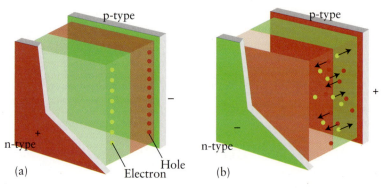

FIGURE 3J.6 The structure of a p–n junction allows an electric current to flow in only one direction. (a) Reverse bias: the negative electrode is attached to the p-type semiconductor and current does not flow. (b) Forward bias: the electrodes are reversed to allow charge carriers to be regenerated.

FIGURE 3J.7 Superconductors have the ability to levitate vehicles with embedded magnets. This picture shows an experimental zero-friction train in Japan, built to use helium-cooled metal superconductors. (*Andy Crump/Science Source.*)

3J.3 Superconductors

Superconductivity may occur when a substance is cooled below a characteristic transition temperature (T_c). Superconductors have enormous technological potential, in principle at least, as they would greatly reduce losses in power transmission. Moreover, because they trap magnetic fields and can levitate objects, energy-efficient transportation is also a possibility (**FIG. 3J.7**).

Conventional superconductivity rises from the ability of electrons to form a pair by making use of lattice vibrations. The electron pairs, which are called **Cooper pairs** after the scientist who first proposed the mechanism, can travel almost freely through the lattice, rather like the way oxen yoked together are less deflected by obstacles than individual oxen. A Cooper pair forms when the presence of one electron distorts the cations of the lattice in its vicinity, and that region of distorted lattice attracts a second electron into its vicinity (**FIG. 3J.8**). The two electrons are attracted together weakly in this way; the pair—and the resulting superconductivity—survives only if the temperature is so low that the pair is not shaken apart by lattice vibrations. Superconductivity was first observed in 1911 in mercury, for which $T_c = 4$ K. Over the years, many other metallic superconductors were identified, some having transition temperatures as high as 23 K. However, low-temperature superconductors need to be cooled with liquid helium, which is very expensive.

In 1986, a record-high transition temperature of 35 K was observed, not for a metal but for a ceramic material, a lanthanum–copper oxide doped with barium and which is a metallic conductor at room temperature. Then, early in 1987, a new record T_c of 93 K was set with a yttrium–barium–copper oxide and a series of related compounds. In 1988, two more oxide series of bismuth–strontium–calcium–copper and thallium–barium–calcium–copper exhibited transition temperatures of 110 K and 125 K, respectively. By 2015, the highest T_c attained at 1 atm was 138 K. These temperatures can be reached by cooling the materials with liquid nitrogen, which is much cheaper than liquid helium. Almost all these **high-temperature superconductors** (HTSCs) are hard, brittle ceramic oxides that have sheets of copper and oxygen atoms sandwiched between layers of either cations or a combination of cations and oxide ions, and all are derived from their respective parent insulators by doping (**FIG. 3J.9**). Because of their layered structure, their electrical and magnetic properties are strongly anisotropic (see Topic 3G). Because of their layered structure, the electric current flows easily along the planes of the copper–oxygen sheets, but only weakly perpendicular to them.

A major challenge to the use of HTSCs for transmitting electrical power is the difficulty of making electrical wires from a brittle ceramic material. One solution has been to deposit the superconducting material on the surface of wire or tape made of a metal such as silver. The metal is usually prepared with a textured surface that helps to align the crystal grains in the desired direction (**FIG. 3J.10**).

Superconductors conduct electricity with no resistance at low temperatures. Some metals and ceramic materials behave as superconductors.

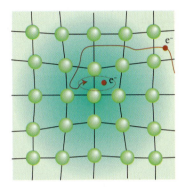

FIGURE 3J.8 The formation of a Cooper pair. As one electron distorts the crystal lattice, the energy of a second electron is lowered as it enters the same region. These electron–lattice interactions bind the two electrons into a pair.

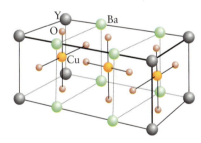

FIGURE 3J.9 The structure of a "123 superconductor," a ceramic that has the variable formula $YBa_2Cu_3O_{6.5-7.0}$. The numbers 1, 2, and 3 refer to the implied or specified subscripts on the first three elements in the formula.

FIGURE 3J.10 Twenty-five kilograms of the experimental superconducting wire on the right can carry as much current as 1800 kg of the bulky copper cable on the left. (*Courtesy American Superconductor (AMSC).*)

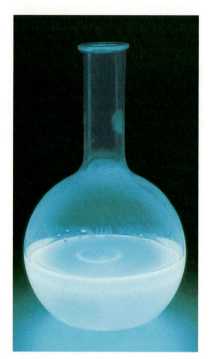

FIGURE 3J.11 Chemiluminescence, the emission of light as the result of a chemical reaction, occurs when hydrogen peroxide is added to a solution of the organic compound perylene. Although hydrogen peroxide itself can fluoresce, in this case the light is emitted by the perylene. (*W. H. Freeman photo by Ken Karp.*)

 **LAB VIDEO FIGURE 3J.11**

The 2014 Nobel Prize in Physics was awarded to Isamu Akasaki, Hiroshi Amano, and Shuji Nakamura for their development of the blue LED.

3J.4 Luminescent Materials

Incandescence is light emitted by a hot body, such as the filament in a lamp or the particles of hot soot in a candle flame. **Luminescence** is the emission of light by a process other than incandescence. For example, when hydrogen peroxide reacts with chlorine, the O_2 formed by the oxidation of H_2O_2 is produced in an energetically excited state and emits light as it returns to the ground state. This process is an example of **chemiluminescence**, the emission of light from products formed in energetically excited states in a chemical reaction (**FIG. 3J.11**). The light sticks used for emergency lighting glow with light from a chemiluminescent process. *Bioluminescence* is a form of chemiluminescence generated by a living organism. For example, the enzyme luciferase catalyzes the oxidation of luciferin in fireflies and some bacteria, producing oxyluciferin in an excited state.

Fluorescence and phosphorescence are the emission of light from molecules excited by radiation of higher frequency, such as visible light being emitted when a substance is illuminated with ultraviolet radiation. In **fluorescence,** emission of light typically persists for only a few nanoseconds after the illumination ceases. In **phosphorescence,** the emission persists, sometimes for seconds or much longer, as in the element phosphorus, from which the phenomenon gets its name (Topic 8G). The crucial difference in mechanism is that fluorescence involves the retention of the relative spin orientation of the excited electron, whereas in phosphorescence the electron becomes unpaired and it takes time for its spin to reverse again.

Fluorescent materials are very important in the electronics industry, as narrow fluorescent light tubes about the size of a pencil are used to provide the backlight for LCD screens in laptop computers and flat-screen televisions. Ultraviolet radiation generated in the tube excites a fluorescent material coating the inside surface of the tube, which illuminates the screen. Because fluorescent materials can be activated by radioactivity, they are also used in scintillation counters to measure radiation (see Box 10B.2).

The cathode-ray tubes that were once widely used for television and computer displays, and the plasma screens that (together with liquid-crystal displays) have replaced them, all make use of **phosphors,** phosphorescent materials that glow when they are excited by the impact of electrons or ultraviolet radiation.

In a **light-emitting diode,** LED, a luminescent material generates light when an electric current is applied to a p–n junction (**FIG. 3J.12**). The circuit attached to an LED is arranged so that electrons from the power source flow into the conduction band of the n-type side and are pushed to the conduction band of the p-type side. Once the electrons are in the higher-energy band of the p-type side, they fall into the lower-energy band and release the difference in energy as light. Compounds used to make different colors of light vary, but commonly aluminum gallium arsenide is used to make red LEDs, indium gallium nitride green LEDs, and zinc selenide blue LEDs. When tiny LEDs of these three colors are clustered tightly together on a display screen, any color can be generated depending on the relative brightness of the three colors. White LEDs are commonly formed from yellow and blue LEDs mixed in various proportions.

Organic light-emitting diodes (OLEDs) use a film consisting of an organic polymer that conducts electricity to generate light of different colors. Although LCD screens, which use liquid crystals (see Topic 3G), are more common, LED displays do not require

FIGURE 3J.12 A p-n junction (a) without bias, (b) with bias, resulting in emission of light as the incoming electrons migrate into the conduction band of the p-type semiconductor and then fall into the vacancies in its valence band.

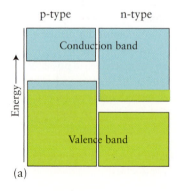
(a)

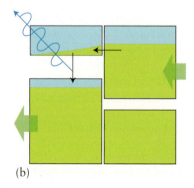
(b)

backlighting and so can be made much thinner. LED lights are now being introduced for many applications, as they use much less power than incandescent lights and last many times longer.

Self-test 3J.3A Explain how fluorescent materials can be used to detect radioactivity.
[*Answer:* The fluorescent materials absorb energy from the radiation and release it as light.]

Self-test 3J.3B Explain the difference between chemiluminescence and phosphorescence.

Luminescent materials release energy as light as they return from excited states to lower energy states.

3J.5 Magnetic Materials

Paramagnetism is the tendency of a substance to move into a magnetic field (Box 2G.2). It arises when an atom or molecule has at least one unpaired electron, which aligns with the applied field. However, because the spins on neighboring atoms or molecules are aligned almost randomly, paramagnetism is very weak and the alignment of electron spins is lost when the magnetic field is removed. In some d-metals, however, when the unpaired electrons of many neighboring atoms align with one another in an applied magnetic field, the much stronger effect of **ferromagnetism** arises. The regions of aligned spins, which are called **domains** (**FIG. 3J.13**), survive even after the applied field is removed.

Ferromagnetism is much stronger than paramagnetism, and ferromagnetic materials are used to make permanent magnets. They are also used in the coatings of computer hard drives. The electromagnetic recording heads align large numbers of spins as the platter rotates underneath, and the spin alignment in the domains remains for years. In an **antiferromagnetic material,** neighboring spins are locked into an *antiparallel* arrangement; so the magnetic moments cancel. Manganese is antiferromagnetic. In a **ferrimagnetic material,** the spins on neighboring atoms are different, and although they are locked together in an antiparallel arrangement, the two magnetic moments do not completely cancel. Ferromagnetism also occurs in alloys such as alnico and some compounds of the d-metals, such as the oxides of iron and chromium.

Ferrofluids are ferromagnetic liquids that are suspensions of finely powdered magnetite, Fe_3O_4, in a viscous, oily liquid (such as mineral oil) that contains a soap or detergent. The iron oxide particles do not settle out because they are attracted to the polar ends of the detergent molecules, which form compact clusters (a type of *micelle*, Topic 5D) around the particles. The nonpolar ends of the detergent molecules point outward, allowing the micelles to form a colloidal suspension in the oil. When a magnet is brought near a ferrofluid, the particles in the liquid tend to align with the magnetic field, but they are kept in place by the oil (**FIG. 3J.14**). As a result, it is possible to control the flow and position of the ferrofluid by means of an applied magnetic field. One application of ferrofluids is in the braking systems of exercise machines. The stronger the applied magnetic field, the greater is the resistance to motion.

Magnetic materials can be paramagnetic, ferrimagnetic, ferromagnetic, or antiferromagnetic. In ferromagnetic materials, large domains of electron spins are locked together in the same orientation.

3J.6 Nanomaterials

A new area of research with the potential, among other things, to revolutionize medical diagnosis and treatment is **nanoscience,** the study of the properties of nanomaterials, and its application, **nanotechnology,** the procedures used to manipulate matter on this scale. **Nanomaterials** are materials with particle sizes in the range 1 to 100 nm; they are larger

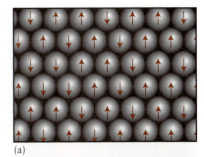
(a)

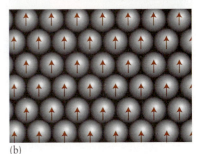
(b)

FIGURE 3J.13 Ferromagnetic materials include iron, cobalt, and the iron oxide mineral magnetite. They consist of crystals in which the electrons of many atoms spin in the same direction and give rise to a strong magnetic field. (a) Before magnetization, when the spins are almost randomly aligned; (b) after magnetization. The arrows represent the electron spins.

FIGURE 3J.14 When a magnet is pulled up from this viscous ferrofluid, the particles of Fe_3O_4 align themselves with the magnetic field. Because strong attractions exist between the particles and the detergent molecules in the oil, the liquid is pulled into the field along with the particles. (S. Odenbach, ZARM, University of Bremen, Germany.)

Nano is derived from the Greek word for dwarf.

FIGURE 3J.15 The different colors of these fluorescent CdSe quantum dot suspensions indicate the different sizes of the quantum dots that they contain. The greater the radius of the quantum dot, the longer is the wavelength of the color emitted. (*SPL/Science Source.*)

than individual molecules but too small to exhibit most bulk properties. Nanotechnology has the promise to provide new materials such as biosensors that monitor and even repair bodily processes, microscopic computers, artificial bone, and lightweight, remarkably strong materials.

In a block of metal the separation of neighboring energy levels is infinitesimal. However, the separation is significant in a cluster of atoms with dimensions of up to 100 nm or so. A cluster of atoms can act as a trap for electrons, and to a good approximation the cluster can be treated as a potential well. The properties of electrons trapped in potential wells can be used to estimate the allowed energy levels. The difference between the one-dimensional well treated in Topic 1C (with the energy levels $E_n = n^2h^2/8m_eL^2$) and the present case is that the well is three-dimensional and typically regarded as spherical. A spherical well is more difficult to treat mathematically than a rectangular well, but the results have some similarities. For instance, the spherically symmetrical energy levels ($l = 0$, corresponding to the s-orbitals of a hydrogen atom) of an electron in a spherical cavity of radius r are given by

$$E_n = \frac{n^2h^2}{8m_e r^2} \tag{1}$$

As usual, n is the quantum number, h is Planck's constant, and m_e is the mass of an electron. There are also solutions similar to p-, d-, etc., orbitals (corresponding to $l = 1, 2, \ldots$) that lie in a complicated pattern, the first four levels being

$E/(h^2/8m_e r^2)$	1	2.046	3.366	4
l	0	1	2	0

What Does This Equation Tell You? Equation 1 shows that as the radius of the well decreases, the separation of energy levels increases (as $1/r^2$), and so the wavelength of light (from $\Delta E = h\nu = hc/\lambda$) that can cause excitation decreases. That is, as the size of the cavity varies, so does the color of the material.

Three-dimensional crystals of semiconducting materials, such as cadmium selenide (CdSe), containing 10^2 to 10^5 atoms are called **quantum dots.** They can be made in solution or by depositing atoms on a surface. The variation of color with the radius of the quantum dot is readily observed in suspensions of CdSe quantum dots of different sizes (**FIG. 3J.15**).

Some quantum dots emit light when an excited electron falls back to a lower level in the dot. One application of this phenomenon is to monitor processes going on in biological cells. For example, a CdSe quantum dot can be attached to the surface of a cell, perhaps to a protein or another component, through an organic linking molecule and studied with a microscope.

THINKING POINT

The physicist Richard Feynman is credited with foreseeing the nanotechnology revolution in a talk called "There Is Plenty of Room at the Bottom." What did he mean?

In nanomaterials, such as quantum dots, differences between energy levels may give rise to transitions in the visible region of the spectrum.

3J.7 Nanotubes

In 1991, scientists identified a previously unknown form of carbon when they sealed two graphite rods inside a container of helium gas and passed an electric discharge between them. Much of one rod evaporated, but some surprising structures resulted

(**FIG. 3J.16**). A **nanotube** can be thought of as a graphene sheet of millions of carbon atoms rolled into a cylinder only 1–3 nm in diameter.

Carbon nanotubes conduct electricity through the extended network of delocalized π-bonds that runs from one end of the tube to the other. The conductivity of nanotubes depends on the way the tubes are rolled (**FIG. 3J.17**). If the points are aligned perpendicular to the long axis (the "armchair" configuration), the conductivity of the nanotubes is as high as in a metal. If the points of the hexagons are aligned along the long axis of the tube (the "zigzag" and "chiral" orientations), the nanotubes may behave as either metallic conductors or semiconductors, depending on the angle of wrap and the diameter of the tube. Like diamonds, carbon nanotubes are good conductors of heat. Their interesting electrical conductivity and high thermal conductivity make carbon nanotubes good candidates for the development of miniature integrated circuits.

The tensile strength of a carbon nanotube parallel to its axis is the greatest of any material that has been measured. Because they have a very low density, their strength-to-mass ratio is 40 times that of steel. The rigidity of nanotubes allows them to be used as minute molds or templates for nanostructures of other elements. For example, they can be filled with molten lead to create lead nanowires one atom in diameter. They can also serve as tiny "test tubes" that hold individual molecules in place and can be used as sensors. The very large surface area of nanotubes means that atoms of gases are readily adsorbed on the inner surfaces of the tubes. Nanotubes carrying hydrogen molecules could therefore become a storage medium for hydrogen-powered vehicles.

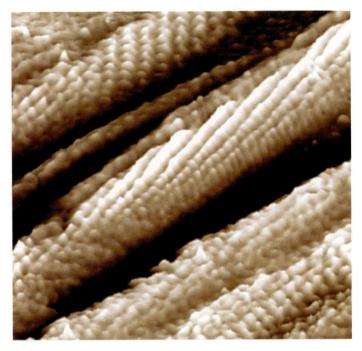

FIGURE 3J.16 A cluster of carbon nanotubes that has formed a ropelike structure. (*Eye of Science/Science Source.*)

Nanotubes are tiny cylinders only a few nanometers in diameter that are constructed of rolled sheets of a graphene-like array of atoms.

What have you learned in this Topic?

You have encountered the classification of materials according to the temperature dependence of their electrical resistance and the explanation of conductivity in terms of the occupation of bands of molecular orbitals spreading through the solid. You have seen that there are several different ways in which materials can emit light and that magnetism arises from the presence of unpaired electrons. You have also seen how quantum phenomena give rise to the special optical properties of nanomaterials.

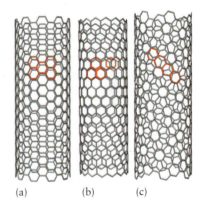

FIGURE 3J.17 Three versions of single-walled carbon nanotubes: (a) "armchair," (b) zigzag, and (c) chiral. Most tubes have ends capped by hemispheres of atoms. The electrical conducting properties of the nanotubes depend on their structures.

The skills you have mastered are the ability to:

☐ **1.** Use molecular orbital theory to account for the differences between metals, insulators, semiconductors, and superconductors (Sections 3J.1–3J.3).

☐ **2.** Explain the properties of luminescent materials (Section 3J.4).

☐ **3.** Distinguish the principal types of magnetism (Section 3J.5).

☐ **4.** Describe the nature of nanomaterials and how they differ from other materials (Sections 3J.6 and 3J.7).

Topic 3J Exercises

3J.1 How does the change in electrical resistance of a semiconductor differ from that of a metallic conductor as temperature is increased?

3J.2 Normally, in electrically conducting materials, the current is treated as being carried by electrons as they move through a solid.

In semiconductors, it is also common to treat the current as being carried by the "holes" in the valence band. (a) Explain how holes move through a solid material. (b) If, in a p-type semiconductor device, electric current is moving from left to right, in which direction will the holes be moving?

3J.3 Estimate the value of the quantum number n for the uppermost filled level in a one-dimensional line of silver atoms of length 2.4 mm. *Hint:* Take the radius of an Ag atom as 144 pm.

3J.4 Estimate the value of the quantum number n for the uppermost filled level in a one-dimensional line of copper atoms of length 1.0 mm. *Hint:* Take the radius of a Cu atom as 128 pm.

3J.5 If small amounts of the elements In, P, Sb, and Ga are present as impurities in germanium, which of them will make germanium into (a) a p-type semiconductor; (b) an n-type semiconductor?

3J.6 If you want to modify gallium arsenide by replacing a small amount of the arsenic with an element to produce an n-type semiconductor, which element would you choose: Se, P, or Si? Why?

3J.7 Distinguish between fluorescence and phosphorescence (a) observationally, (b) mechanistically.

3J.8 In fluorescence, how does the frequency of the emitted radiation compare with the frequency of the exciting radiation?

3J.9 Use Internet resources to identify the magnetic character at room temperature (diamagnetic, paramagnetic, ferromagnetic, antiferromagnetic) of as many d-block elements as you can.

3J.10 Consider the elements identified in Exercise 3J.9. What is the common feature of the electron configurations of paramagnetic elements?

3J.11 The Curie temperature, T_C, is the temperature at which a substance capable of being ferromagnetic actually becomes ferromagnetic. Would you expect ferromagnetism to occur above or below the Curie temperature?

3J.12 Describe the observational differences between diamagnetism, paramagnetism, ferromagnetism, and antiferromagnetism and the origin of these properties.

3J.13 A sample of CaS quantum dots is prepared by mixing 4.0 mL of 0.0015 M $CaCl_2$(aq) with 4.0 mL of 0.0015 M $(NH_4)_2S$(aq). Each quantum dot contains an average of 168 formula units of CaS. What is the molar concentration of the quantum dots in the final suspension?

3J.14 A sample of PbS quantum dots is prepared by mixing 7.0 mL of 0.0013 M $Pb(NO_3)_2$(aq) with 7.0 mL of 0.0013 M $(NH_4)_2S$(aq). Each quantum dot contains an average of 145 formula units of PbS. What is the molar concentration of the quantum dots in the final suspension?

3J.15 An electron trapped within a nanoparticle can be approximated as a particle of mass m_e confined to a cubic box of side L. The energy levels of the electron are

$$E = \frac{h^2}{8m_eL^2}(n_x^2 + n_y^2 + n_z^2)$$

Write expressions for the energies of the three lowest levels. Which of these levels are degenerate? For those levels that are degenerate, give the quantum numbers corresponding to each degenerate level.

3J.16 Buckminsterfullerene is a form of carbon having nearly spherical molecules that are composed of 60 carbon atoms (Topic 8F). The interior of a C_{60} molecule is about 0.7 nm in diameter and is being considered as a container for various atoms and molecules. Suppose that buckminsterfullerene were used to transport molecular hydrogen. The cavity in a single molecule can be approximated as an empty cube with a side of 0.7 nm. If a hydrogen molecule is represented by a point mass, how much energy does a hydrogen molecule inside a C_{60} molecule need to be excited from the lowest energy level to (a) the second energy level; (b) the third energy level? (See Exercise 3J.15.)

The following Example and Exercises draw on material from throughout Focus 3.

FOCUS 3 Online Cumulative Example

Some of the earliest mortars were *nonhydraulic cements,* which harden by reaction with CO_2 rather than with water. These cements are prepared by heating calcite, $CaCO_3(s)$, strongly to drive off CO_2 gas and form quicklime, $CaO(s)$. The resulting solid is mixed with water to give a paste of slaked lime, $Ca(OH)_2$, to which sand or volcanic ash is added to form lime mortar. The Roman Colosseum and Pantheon were constructed with this type of mortar and have endured the ages. You are investigating ancient building methods and want to understand the chemistry of these materials.

(a) Write the balanced chemical equations for (i) the conversion of calcite to quicklime, (ii) the reaction of quicklime with water to form slaked lime, and (iii) the reaction of slaked lime with CO_2 to form calcium carbonate.

(b) Preparing quicklime releases the greenhouse gas carbon dioxide. If 1.000 t (1 t = 10^3 kg) of $CaCO_3$ is placed in a kiln and heated to 850 °C, what volume of $CO_2(g)$ is formed at 850 °C and 1 atm?

(c) If the $CO_2(g)$ from part (b) is cooled to room temperature of 22 °C what volume would it occupy?

(d) Calcium oxide has the cubic structure shown in (**1**). The length of each edge is 481.1 pm. All the atoms are on an edge, face, or corner of the cube with one O atom in the center of the cube. Use this information and the density of $CaCO_3(s)$, 2.711 g·cm^{-3}, to calculate the change in volume of the solid as CO_2 is driven off from 1.0 t of $CaCO_3$.

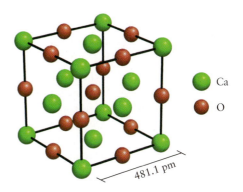

1 Calcium oxide, CaO

(e) From the results in part (d), suggest a reason why buildings constructed of bricks held together with lime mortar might collapse during a fire.

The online Cumulative Example solution can be found at http://macmillanhighered.com/chemicalprinciples7e

FOCUS 3 Exercises

3.1 The drawing below shows a tiny section of a flask containing two gases. The orange spheres represent neon atoms and the blue spheres represent argon atoms. (a) If the partial pressure of neon in this mixture is 420. Torr, what is (a) the partial pressure of argon; (b) the total pressure?

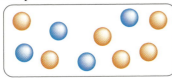

3.2 The four flasks below were prepared with the same volume and temperature. Flask I contains He atoms, Flask II contains Cl_2 molecules, Flask III contains Ar atoms, and Flask IV contains NH_3 molecules. Which flask has (a) the largest number of atoms; (b) the highest pressure; (c) the greatest (mass) density; (d) molecules or atoms with the highest root mean square speed; (e) molecules or atoms with the highest molar kinetic energy?

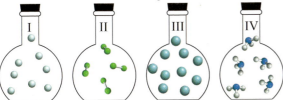

3.3 Through a series of photochemical and enzymatic steps, carbon dioxide and water participate in photosynthesis to produce glucose and oxygen according to the equation

$$6\,CO_2(g) + 6\,H_2O(l) \longrightarrow C_6H_{12}O_6(s) + 6\,O_2(g)$$

Given that the partial pressure of carbon dioxide in the troposphere is 0.26 Torr and that the temperature is 25 °C, calculate the volume of air at 1.0 atm needed to produce 10.0 g of glucose.

3.4 Roommates fill ten balloons for a party, five with hydrogen and five with helium. After the party the hydrogen balloons have lost one-fifth of their hydrogen due to effusion through the walls of the balloons. What fraction of helium will the other balloons have lost at that same time?

3.5 Suppose that 200. mL of hydrogen chloride gas at 690. Torr and 20. °C is dissolved in 100. mL of water. The solution is titrated to the stoichiometric point with 15.7 mL of a sodium hydroxide solution. What is the molar concentration of NaOH in the solution used for the titration?

3.6 When people dive into deep water, the pressure exerted by the water above can become great enough to collapse their lungs. Hence, scuba gear must sense the external pressure and increase the pressure of the breathing mixture provided to the diver until it equals the external pressure. A diver exploring a sunken ship uses an oxygen-enriched breathing mixture of 36% oxygen and 64% nitrogen by mass. At a depth of 33 ft, the pressure on the diver is 2.0 bar. What is the partial pressure of oxygen in the breathing mixture at that depth?

3.7 A flask of volume 5.00 L is evacuated and 43.78 g of solid dinitrogen tetroxide, N_2O_4, is introduced at −196 °C. The sample is then warmed to 25 °C, during which time the N_2O_4 vaporizes and some of it dissociates to form brown NO_2 gas. The pressure slowly increases until it stabilizes at 2.96 atm. (a) Write a balanced equation for the reaction. (b) If the gas in the flask at 25 °C were all N_2O_4, what would the pressure be? (c) If all the gas in the flask converted into NO_2, what would the pressure be? (d) What are the mole fractions of N_2O_4 and NO_2 once the pressure stabilizes at 2.96 atm?

3.8 When 0.40 g of impure zinc reacted with an excess of hydrochloric acid, 127 mL of hydrogen was collected over water at 10. °C. The external pressure was 737.7 Torr. (a) What volume would the dry hydrogen occupy at 1.00 atm and 298 K? (b) What amount (in moles) of H_2 was collected? (c) What is the percentage purity of the zinc, assuming that all the zinc present reacted completely with HCl and that the impurities did not react with HCl to produce hydrogen? The vapor pressure of water at 10. °C is 9.21 Torr.

3.9 Suppose that 0.473 g of an unknown gas that occupies 200. mL at 1.81 atm and 25 °C was analyzed and found to contain 0.414 g of nitrogen and 0.0591 g of hydrogen. (a) What is the molecular formula of the compound? (b) Draw the Lewis structure of the molecule. (c) If 0.35 mmol NH_3 effuses through a small opening in a glass apparatus in 15.0 min at 200. °C, what amount of the gas being studied will effuse through the same opening in 25.0 min at 200. °C?

3.10 When 2.36 g of phosphorus was burned in chlorine, the product was 10.5 g of a phosphorus chloride. Its vapor took 1.77 times as long to effuse as the same amount of CO_2 under the same conditions of temperature and pressure. What is the molar mass and molecular formula of the phosphorus chloride?

3.11 Determine the ratio of the number of molecules in a gas having a speed ten times as great as the root mean square speed to the number having a speed equal to the root mean square speed. Is this ratio independent of temperature? Explain your reasoning.

3.12 You are told that 2.55 g of a gaseous hydrocarbon occupies a vessel of volume 3.00 L at 0.950 atm and 82.0 °C. Draw the Lewis structure of this hydrocarbon.

3.13 A finely powdered solid sample of an osmium oxide (which melts at 40. °C and boils at 130. °C) with a mass of 1.509 g is placed into a cylinder with a movable piston that can expand against the atmospheric pressure of 745 Torr. Assume that the amount of residual air initially present in the cylinder is negligible. When the sample is heated to 200. °C, it is completely vaporized and the volume of the cylinder expands by 235 mL. What is the molar mass of the oxide? Assuming that the oxide is OsO_x, find the value of x.

3.14 How does the root mean square speed of gas molecules vary with temperature? Illustrate this relationship by plotting the root mean square speed of N_2 as a function of temperature from $T = 100$ K to $T = 300$ K.

3.15 The following plot shows the distribution of speeds for N_2 at 300, 500, and 1000 K. (a) Identify the temperature of the gas for each curve. (b) What is the root mean square speed of N_2 molecules at 227 °C?

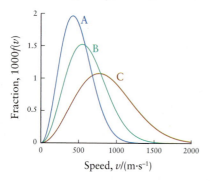

3.16 An expression for the root mean square speed, v_{rms}, of gas molecules is derived in Topic 3D. Using the Maxwell distribution of speeds, the mean speed and the most probable (mp) speed of a collection of molecules can be calculated and are $v_{mean} = (8RT/\pi M)^{1/2}$ and $v_{mp} = (2RT/M)^{1/2}$. These values have a fixed relation to each other. (a) Place these three quantities in order of increasing magnitude. (b) Show that the relative magnitudes are independent of the molar mass of the gas. (c) Using the smallest speed as the reference point, rank them by magnitude and determine the ratios of the larger values to the smallest.

3.17 A sample of chlorine gas of volume 1.00 L at 1.00 atm and 298 K reacts completely with 1.00 L of nitrogen gas and 2.00 L of oxygen gas at the same temperature and pressure. There is a single gaseous product, which fills a 2.00-L flask at 1.00 atm and 298 K. Use this information to determine the following characteristics of the product: (a) its empirical formula; (b) its molecular formula; (c) the most favorable Lewis structure based on formal charge arguments (the central atom is an N atom); (d) the molecular shape.

3.18 A sample of gaseous arsine, AsH_3, in a 500.0-mL flask at 280. Torr and 223 K is heated to 473 K, at which temperature arsine decomposes to solid arsenic and hydrogen gas. The flask is then cooled to 273 K, at which temperature the pressure in the flask is 485 Torr. Calculate the percentage of arsine molecules that have decomposed.

3.19 The van der Waals equation can be arranged into the following cubic equation:

$$V^3 + n\left(\frac{RT + bP}{P}\right)V^2 + \left(\frac{n^2a}{P}\right)V - \frac{n^3ab}{P} = 0$$

Using this equation, calculate the volume occupied by 0.505 mol $NH_3(g)$ at 25 °C and 95.0 atm. The van der Waals parameters for NH_3 are $a = 4.225$ bar·L^2·mol^{-2} and $b = 3.71$ L·mol^{-1}. (b) Do the attractive or the repulsive forces dominate at this temperature and pressure?

3.20 Photochemical smog is formed, in part, by the action of light on nitrogen dioxide. The wavelength of the radiation absorbed by NO_2 in this reaction is 197 nm.

$$NO_2 \xrightarrow{h\nu} NO + O$$

Draw the Lewis structure for NO_2 and sketch its π molecular orbitals. (b) When 1.56 mJ of energy is absorbed by 3.0 L of air at 20. °C and 0.91 atm, all the NO_2 molecules in this sample are dissociated by the reaction just shown. Assume that each photon absorbed results in the dissociation (into NO and O) of one NO_2 molecule. What is the proportion, in parts per million, of NO_2 molecules in this sample? Assume that the sample is behaving ideally.

3.21 The reaction of solid dimethylhydrazine, $(CH_3)_2N_2H_2$, and liquefied dinitrogen tetroxide, N_2O_4, has been investigated as a rocket fuel; the reaction produces gaseous carbon dioxide (CO_2), nitrogen (N_2), and water vapor (H_2O), which are ejected as exhaust gases. In a controlled experiment, solid dimethylhydrazine reacted with excess nitrogen tetroxide and the gaseous products were collected in a closed vessel until a pressure of 2.50 atm and a temperature of 400.0 K were reached. (a) What are the partial pressures of CO_2, N_2, and H_2O? (b) When the CO_2 is removed by chemical reaction, what are the partial pressures of the remaining gases?

3.22 In the pastime of hot-air ballooning, the air inside a balloon—called the envelope—is heated until its density is lower than that of the surrounding atmosphere, allowing the envelope and an attached basket to be lifted into the air. One team is preparing its balloon for a flight on a dry day at 16 °C. The envelope has a volume of 3125 m³ when inflated and, excluding the mass of the air, the total mass of the envelope and the basket loaded with fuel, heater, passengers, and sandwiches, is 586 kg. (a) What would be the mass of air in the envelope if it were at 16 °C? (b) To what temperature must the air in the envelope be heated to lift it and the loaded basket? (c) What is the mass of the air inside the envelope at that temperature? Assume that the volume of the loaded basket is negligible and that the molar mass of the air is 28.97 g·mol⁻¹.

3.23 Air bags in automobiles contain sodium azide, NaN₃, which decomposes rapidly during a collision to give nitrogen gas and sodium metal. The nitrogen gas liberated by this process instantly inflates the air bag. Assume that the nitrogen gas liberated behaves as an ideal gas and that any solid produced has a negligible volume (which may be ignored). (a) Calculate the mass (in grams) of sodium azide required to generate enough nitrogen gas to fill a 57.0-L air bag at 1.37 atm and 25.0 °C. (b) What is the root mean square speed of the N₂ gas molecules that are generated?

3.24 When long surfactant molecules having a polar headgroup and a nonpolar "tail" are added to water, *micelles* are formed in which the nonpolar tails aggregate, with the polar headgroups pointing out toward the solvent. *Inverse micelles* are similar but have the nonpolar regions pointing outward. How can inverse micelles be produced?

3.25 Draw the Lewis structure of (a) NI₃ and (b) BI₃, identify the shape of each molecule, and indicate whether it can participate in dipole–dipole interactions.

3.26 Draw the Lewis structure of (a) PF₃, (b) PF₅, identify the shape of the molecule, and indicate whether it can participate in dipole–dipole interactions.

3.27 (a) Calculate the surface areas of the isomers 2,2-dimethylpropane and pentane. Assume that 2,2-dimethylpropane is spherical with a radius of 254 pm and that pentane can be approximated by a rectangular prism with dimensions 295 pm × 692 pm × 766 pm. (b) Which has the larger surface area? (c) Which do you expect to have the higher boiling point?

3.28 (a) Calculate the ratio of potential energies for an ion–ion interaction of Li⁺ and of K⁺ with the same anion. (b) Repeat the ratio calculation, but for an ion–dipole interaction between Li⁺ and K⁺ and a water molecule. (c) What do these numbers indicate about the relative importance of the hydration of salts of lithium compared with those of potassium?

3.29 All noble gases except helium crystallize with ccp structures at very low temperatures. Find an equation relating the atomic radius to the density of a ccp solid of given molar mass and apply it to deduce the atomic radius of each of the following noble gases, given the density of each (in g·cm⁻³): Ne, 1.20; Ar, 1.40; Kr, 2.16; Xe, 2.83; Rn, 4.4 (estimated).

3.30 All the alkali metals crystallize with bcc structures. (a) Find a general equation relating the metallic radius to the density of a bcc solid of an element in terms of its molar mass and use it to deduce the atomic radius of each of the following elements, given the density of each (in g·cm⁻³): Li, 0.53; Na, 0.97; K, 0.86; Rb, 1.53; Cs, 1.87. (b) Find a factor for converting the density of a bcc element into the density that it would have if it crystallized in a ccp structure. (c) Calculate what the densities of the alkali metals

would be if they were ccp. (d) Which, if any, would float on water while reacting with it?

3.31 Metals with bcc structures, such as tungsten, are not close packed. Therefore, their densities would be greater if they were to change to a ccp structure (under pressure, for instance). What would the density of tungsten be if its structure were ccp rather than bcc? Its actual density is 19.3 g·cm⁻³.

3.32 An oxide of niobium has a cubic unit cell in which there are oxide ions at the middle of each edge and niobium ions at the center of each face. What is the empirical formula of this oxide?

3.33 A number of metal oxides are known to form "nonstoichiometric" compounds, in which the ratios of atoms that make up the compound cannot be expressed in small whole numbers. In the crystal structure of a nonstoichiometric compound, some of the lattice points where one would have expected to find atoms are vacant. Transition metals most easily form nonstoichiometric compounds because of the number of oxidation states that they can have. For example, a titanium oxide with formula TiO₁.₁₈ is known. (a) Calculate the average oxidation number of titanium in this compound. (b) If the compound has both Ti²⁺ and Ti³⁺ ions present, what fraction of the titanium ions will be in each oxidation state?

3.34 Uranium dioxide, UO₂, can be further oxidized to give a nonstoichiometric compound UO₂₊ₓ, where 0 < x < 0.25. (See Exercise 3.33 for a definition of nonstoichiometric compounds.) (a) What is the average oxidation number of uranium in a compound with composition UO₂.₁₇? (b) If you assume that the uranium exists in either the +4 or the +5 oxidation state, what is the fraction of uranium ions in each?

3.35 Are the following statements true or false? (a) If there is an atom present at the corner of a unit cell, there must be the same type of atom at all the corners of the unit cell. (b) A unit cell must be defined so that there are atoms at the corners. (c) If one face of a unit cell has an atom in its center, then the face opposite that face must also have an atom at its center. (d) If one face of a unit cell has an atom in its center, all the faces of the unit cell must also have atoms at their centers.

3.36 Are the following statements true or false? (a) Because cesium chloride has chloride ions at the corners of the unit cell and a cesium ion at the center of the unit cell, it is classified as having a body-centered unit cell. (b) The density of the unit cell must be the same as the density of the bulk material. (c) When x-rays are passed through a single crystal of a compound, the x-ray beam will be diffracted because it interacts with the electrons in the atoms of the crystal. (d) All the angles of a unit cell must be equal to 90°.

3.37 For each of the two-dimensional arrays shown here, draw a unit cell that, when repeated, generates the entire two-dimensional lattice.

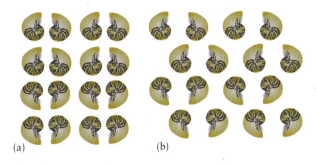

(a)   (b)

3.38 For each of the two-dimensional arrays shown here, draw a unit cell that, when repeated, generates the entire two-dimensional lattice.

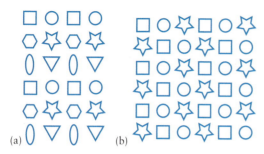

(a) (b)

3.39 Buckminsterfullerene is an allotrope of carbon in which the carbon atoms form nearly spherical molecules of 60 atoms each (Topic 8F). In the pure compound, the spheres pack in a cubic close-packed array. (a) The length of a side of the face-centered cubic cell formed by buckminsterfullerene is 142 pm. Use this information to calculate the radius of the buckminsterfullerene molecule treated as a hard sphere. (b) The compound K_3C_{60} is a superconductor at low temperatures. In this compound the K^+ ions lie in holes in the C_{60}^{3-} face-centered cubic lattice. Considering the radius of the K^+ ion and assuming that the radius of C_{60}^{3-} is the same as for the C_{60} molecule, predict in what type of holes the K^+ ions lie (tetrahedral, octahedral, or both) and indicate what percentage of those holes are filled.

3.40 Real gases and vapors deviate from ideal behavior on account of intermolecular interactions. One equation of state for a real gas is the van der Waals equation, which is expressed in terms of two parameters, a and b. (a) For each of the following pairs of gases, decide which substance has the larger van der Waals a parameter: (i) He and Ne; (ii) Ne and O_2; (iii) CO_2 and H_2; (iv) H_2CO and CH_4; (v) C_6H_6 and $CH_3(CH_2)_{10}CH_3$. (b) For each of the following pairs of gases, decide which substance has the larger van der Waals b parameter: (i) F_2 and Br_2; (ii) Ne and F_2; (iii) CH_4 and $CH_3CH_2CH_3$; (iv) N_2 and Kr; (v) CO_2 and SO_2.

3.41 A commonly occurring mineral has a cubic unit cell in which the metal cations, M, occupy the corners and face centers. Inside the unit cell, anions, A, occupy all the tetrahedral holes created by the cations. What is the empirical formula of the M_mA_a compound?

3.42 Tetrahedral and octahedral holes are formed by the vacancies left when anions pack in a ccp array. (a) Which hole can accommodate the larger ions? (b) What is the radius ratio of the largest metal cation that can occupy an octahedral hole to the largest that can occupy a tetrahedral hole while maintaining the close-packed nature of the anion lattice? (c) If half the tetrahedral holes are occupied, what is the empirical formula of the compound M_xA_y, where M represents the cations and A the anions?

3.43 The molar heat capacity of a substance is the heat required to raise the temperature by 1 °C per mole of molecules or formula units of the substance. For a given substance, the more varieties of movement—such as vibrations—that the atoms in a lattice undergo, the greater the heat capacity. In 1819 the chemists Pierre Dulong and Thérèse Petit claimed (in modern language) that the molar heat capacity of a crystalline solid per atom is the same for all crystals. (a) Verify this claim for metals by making a table of the molar heat capacities of Cu, Fe, Pb, and Zn (see Appendix 2A). (b) Verify this claim for ionic crystals by making a table of the heat

capacity per mole of atoms in the compound for CuO, FeS, $PbBr_2$, and ZnO. (c) Calculate the heat capacity per mole of atoms for $CuSO_4$ and $PbSO_4$. Do the atoms in a polyatomic ion act independently, or does the entire ion behave as one particle in the lattice? Use your results to support your conclusion.

3.44 Does the law of Dulong and Petit (Exercise 3.43) apply to network or molecular solids? (a) To check network solids, compare the molar heat capacities of graphite, diamond, and SiO_2 (see Appendix 2A). Use your findings to determine if the atoms in a network solid behave as independently as atoms in an ionic crystal. Justify your conclusion. (b) To check molecular solids, compare the molar heat capacities of benzoic acid, urea, and glycine (see Appendix 2A). Use your findings to determine whether the atoms in molecules vibrate completely independently or whether they move to some extent as an entire molecule. Use the data to justify your conclusion.

3.45 (a) If a pure element crystallizes with a primitive cubic lattice, what percentage of the unit cell is empty space? (b) How does this percentage compare with that of empty space in an fcc unit cell? Treat the atoms as hard spheres.

3.46 Many ionic compounds are considered to pack in such as way that the anions form a close-packed lattice in which the metal cations fill holes or interstitial sites left between the anions. These lattices, however, may not necessarily be as tightly packed as the label "close-packed" implies. The radius of an F^- ion is approximately 133 pm. The edge distances of the cubic unit cells of LiF, NaF, KF, RbF, and CsF, all of which pack in the rock-salt structure, are 568 pm, 652 pm, 754 pm, 796 pm, and 850 pm, respectively. Which of these lattices, if any, can be thought to be based on close-packed arrays of F^- ions treated as hard spheres? Justify your conclusions.

3.47 As a result of being able to form strong hydrogen bonds, in the vapor hydrogen fluoride is found as short chains and rings. Draw the Lewis structure of an $(HF)_3$ chain and indicate the approximate bond angles.

3.48 Calculate the Coulomb potential energy of an electron at a point a distance r from a dipole composed of charges Q and $-Q$ separated by a distance l in the arrangement shown in the diagram. Use the fact that $l \ll r$ to expand the expression $1/(1 + x)$ into $1 - x + \dots$ and identify the magnitude of the dipole moment as $\mu = Ql$. Show that the potential energy is proportional to $1/r^2$, as for an ion-dipole interaction (see Eq. 1 in Topic 3F and Table 3F.1).

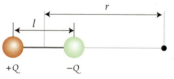

3.49 One of the most destructive natural processes in civilization is the corrosion of iron in the presence of oxygen to form rust, which for simplicity can be taken to be iron(III) oxide. If a cubic block of iron of side 1.5 cm reacts with 15.5 L of oxygen at 1.00 atm and 25 °C, what is the maximum mass of iron(III) oxide that can be produced? Iron metal has a bcc structure, and the atomic radius of iron is 124 pm. The reaction takes place at 298 K and 1.00 atm.

3.50 What mass of carbon is required for complete conversion of 14.0 kg of SiO_2 to SiC?

3.51 Some d-block metals, such as titanium, vanadium, zirconium, niobium, hafnium, and tantalum, react with nitrogen at high temperatures to form nitrides with the formula MN (in which M is the metal). The compounds are very hard, have high melting points, and have high electrical and thermal conductivities. The atoms are arranged in the same way as in NaCl (Topic 3H). (a) Would you classify these compounds as alloys, salts, molecular solids, or network solids? Justify your answer. (b) What is the evidence for covalent character in the bonding of these compounds? (c) Use electron configurations to explain why metals in Groups 11 and 12 do not form these types of nitrides.

3.52 The electrical and thermal conductivities of the Group 11 and Group 12 elements, along with manganese, were measured (see table). Use the data to answer the following: (a) The electrical conductivity of a metal is related to the ability of electrons to move freely through the lattice. Can the same be said of the thermal conductivity? Justify your answer by plotting the thermal conductivity against electrical conductivity. (b) Explain why the electrical and thermal conductivities of Group 11 metals are high while those of Group 12 metals are low. (c) Explain why the electrical and thermal conductivities of manganese are so low.

Element	Electrical conductivity/$(MS \cdot m^{-1})$, at 273 K	Thermal conductivity/ $(10^{-8} \, W \cdot K^{-1} \cdot m^{-1})$, at 300 K
Mn	0.694	7.82
Cu	60.7	401
Ag	63.0	429
Au	45.2	317
Zn	0.169	116
Cd	1.47	96.8
Hg	1.06	8.34

3.53 The unit cell of a high-temperature superconductor is shown here. What is its formula?

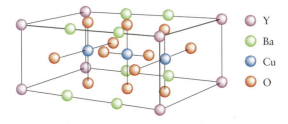

- Y
- Ba
- Cu
- O

3.54 The unit cell of the mineral perovskite, which has a structure similar to that of some high-temperature superconductors, is shown here. What is its formula?

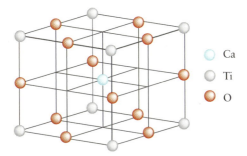

- Ca
- Ti
- O

3.55 Solid silicon carbide reacts with molten sodium hydroxide and oxygen gas at high temperatures to form solid Na_2SiO_3, gaseous water, and carbon dioxide. Write a balanced equation for the reaction.

3.56 Aluminum carbide is considered a covalent carbide. However, it reacts with water as ionic carbides do to produce solid aluminum hydroxide and methane gas, CH_4. Write a balanced equation for the reaction.

3.57 Below are two absorption plots. One was obtained from a solution of an organic dye and the other from a quantum dot suspension. Which plot was obtained from which solution? Explain your reasoning.

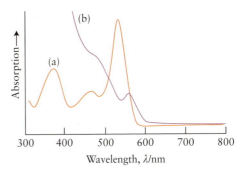

3.58 What is the total mass of (a) the electrons; (b) the (nuclear) protons in a block of the ceramic $BaTiO_3$ of mass 4.72 kg?

3.59 Zinc oxide is a semiconductor. Its conductivity increases when it is heated in a vacuum but decreases when it is heated in oxygen. Account for these observations.

The following nine exercises are based on the information in the Interlude following Focus 3.

3.60 Draw a simple chemical picture to show how the removal of water helps to make aluminosilicates into rigid ceramics.

3.61 What is the oxidation number of (a) phosphorus in $Li_7P_3S_{11}$, which forms in some ceramic electrolytes; (b) titanium in $BaTiO_3$?

3.62 Assuming that the oxidation number of aluminum is +3 in both compounds, what is the oxidation number of silicon in (a) clay, $Al_2Si_2O_5(OH)_4$; (b) mica, $KMg_3(Si_3AlO_{10})(OH)_2$?

3.63 Select the physical properties you would expect to differ between the glassy and crystalline phases of a substance and explain your choices: (a) ability to cleave along a plane; (b) rigidity; (c) sharp melting point; (d) transparency.

3.64 Select the type of substance (metal or ceramic) you would expect to have the greater value of each property and explain your choices: (a) ductility; (b) brittleness; (c) resistance to oxidation; (d) coefficient of thermal expansion (degree of expansion when heated).

3.65 What is the oxidation number of silicon in the anion formed when HF is used to etch glass?

3.66 In a typical procedure for etching glass, the glass surface is covered by a mask (a protective coating). The mask is then removed from the areas that are to be etched and a paste made of fluorite and sulfuric acid is placed on the surface. From standard reference sources, determine the chemical formula of fluorite and describe the chemical reactions that take place to achieve the etching.

3.67 Why is it unwise to store metal fluorides in glass bottles?

3.68 Solutions of strong bases stored in glass bottles react slowly with their containers. Write the balanced chemical equations for four possible reactions between OH^- and SiO_2.

FOCUS 3 Cumulative Exercises

3.69 Iron pyrite, FeS_2, is the form in which much of the sulfur exists in coal. In the combustion of coal, oxygen reacts with iron pyrite to produce iron(III) oxide and sulfur dioxide, which is a major source of air pollution and a substantial contributor to acid rain.

(a) Write a balanced equation for the burning of FeS_2 in air to produce iron(III) oxide and sulfur dioxide.

(b) Calculate the mass of Fe_2O_3 that is produced from the reaction of 75.0 L of oxygen at 2.33 atm and 150. °C with an excess of iron pyrite.

(c) If the sulfur dioxide that is generated in part (b) is dissolved to form 5.00 L of aqueous solution, what is the molar concentration of H_2SO_3 in the resulting solution?

(d) What mass of SO_2 is produced in the burning of 1.00 t ($1\ t = 10^3\ kg$) of high-sulfur coal, if the coal is 5% pyrite by mass?

(e) What is the volume of the SO_2 gas generated in part (d) at 1.00 atm and 25 °C?

(f) One way to remove SO_2 from exhaust gases is to remove it in the reaction $CaO(s) + SO_2(g) \rightarrow CaSO_3(s)$. In a test of this reaction, a mixture of sulfur dioxide and nitrogen gases is prepared at 25 °C in a vessel of volume 500. mL at 1.09 atm. The mixture is passed over warm calcium oxide powder, which removes the sulfur dioxide, and is then transferred to a vessel of volume 150. mL, where the pressure is 1.09 atm at 50. °C. (i) What was the partial pressure of the SO_2 in the initial mixture? (ii) What mass of SO_2 was present in the initial mixture?

(g) The van der Waals parameters for SO_2 are $a = 6.865\ L^2 \cdot atm \cdot mol^{-1}$ and $b = 0.0568\ L \cdot mol^{-1}$. For $SO_2(g)$ confined in a 1.00-L vessel at 27 °C, calculate the pressure of the gas by using the ideal gas law and the van der Waals equation for 0.100 mol to 0.500 mol SO_2 at 0.100-mol increments.

(h) Calculate the percentage deviation of the ideal value from the value calculated from the van der Waals equation at each point in part (g).

(i) Under the conditions in part (g), which term has the larger effect on the pressure of SO_2, the intermolecular attractions or the repulsions?

(j) If those gases are considered to be ideal for which the observed pressure differs by no more than 5% from the ideal value, at what pressure does SO_2 become a "real" gas?

3.70 Ethylammonium nitrate, $CH_3CH_2NH_3NO_3$, was the first ionic liquid to be discovered. Its melting point of 12 °C was reported in 1914, and it has since been used as a nonpolluting solvent for organic reactions and for facilitating the folding of proteins.

(a) Draw the Lewis structure of each ion in ethylammonium nitrate and indicate the formal charge on each atom (in the cation, the carbon atoms are attached to the N atom in a chain: C—C—N).

(b) Assign a hybridization scheme to each C and N atom.

(c) Ethylammonium nitrate cannot be used as a solvent for some reactions because it can oxidize some compounds. Which ion is more likely to be the actual oxidizing agent, the cation or the anion? Explain your answer.

(d) Ethylammonium nitrate can be prepared by the reaction of gaseous ethylamine, $CH_3CH_2NH_2$, and aqueous nitric acid. Write the chemical equation for the reaction. What type of reaction is this?

(e) In an experiment, 2.00 L of ethylamine at 0.960 atm and 23.2 °C was bubbled into 250.0 mL of 0.240 M $HNO_3(aq)$, and 4.10 g of ethylammonium nitrate was produced. What were the theoretical and percentage yields of the salt?

(f) Suggest ways in which the forces that hold ethylammonium nitrate ions together in the solid state differ from those that hold together salts such as sodium chloride or sodium bromide.

(g) Low-melting salts in which the cation is inorganic and the anion organic have been prepared. Explain the trend in melting point seen in the following series: sodium acetate ($NaCH_3CO_2$), 324 °C; sodium propanoate ($NaCH_3CH_2CO_2$), 285 °C; sodium butanoate ($NaCH_3CH_2CH_2CO_2$), 76 °C; and sodium pentanoate ($NaCH_3CH_2CH_2CH_2CO_2$), 64 °C.

INTERLUDE Ceramics and Glasses

Many of the materials used in advanced technologies are based on materials characteristic of the oldest technologies: sand and clay. *China clay* contains primarily *kaolinite*, an aluminosilicate of composition $Al_2Si_2O_5(OH)_4$, that can be obtained reasonably free of the iron impurities that make many clays appear reddish brown, and so it is white. However, other clays contain the iron oxides that cause the characteristic orange color of terra cotta tiles and flowerpots.

The appearance of a flake of clay reflects its internal structure, which is something like an untidy stack of papers (**FIG. 1**). Sheets of tetrahedral silicate units (Topic 3I) or octahedral units of aluminum or magnesium oxides are separated by layers of water molecules that serve to bind the layers of the flake together. Each flake of clay is surrounded by a double layer of ions that separates the flakes by repelling the like charges on the other flakes. This repulsion allows the flakes to slide past one another and gives the clay the ability to flow in response to stress. As a result, clays can be easily molded.

FIG. 1 The layers of clay particles can be seen in this micrograph. Because the surfaces of these layers have like charges, they repel one another and easily slide past one another, making clay soft and malleable. (© *The Natural History Museum/Alamy.*)

When clay is baked at high temperatures in a kiln, it forms the hard, tough *ceramic* materials used in firebricks, tiles, and pots. As the water is driven out during the firing process, strong chemical bonds form between the flakes. Large amounts of china clay, which is used to make ceramics such as porcelain and china, are applied in the coating of paper (such as this page) to give a smooth, nonabsorbent surface. Clay was the first substance to be made into a **ceramic,** an inorganic material that has been hardened by heating to a high temperature. Today a wide variety of compounds, often oxides, are used to create ceramics with specific properties. For example, bone china, which is strong enough to be used for thin, lightweight dishes, is made from china clay with bone ash as a binder to strengthen the ceramic.

A ceramic material is typically very hard, insoluble in water, and stable to corrosion and high temperatures. These characteristics are the key to their technological value. Although many ceramics are brittle, they can be used at high temperatures without failing, and they resist deformation. Ceramics are often the oxides of elements that lie on the border between metals and nonmetals, but d-metal oxides and some compounds of boron and silicon with carbon and nitrogen are also ceramic materials. Most ceramics are electrical insulators; however, some d-metal oxides, such as barium titanate, $BaTiO_3$, and zinc oxide, ZnO, are semiconductors. In addition, some of the most promising high-temperature superconductors are ceramics.

Aluminum oxide, Al_2O_3, is a ceramic oxide found in different solid forms. As α-alumina, it is the very hard, stable, crystalline substance *corundum*; impure microcrystalline corundum is the purple-black abrasive known as *emery*. Corundum makes up about 80% of the advanced ceramics used in high-technology applications. Its hardness, rigidity, thermal conductivity, stability at high temperatures, electrical insulating ability, and low cost make it suitable for a wide range of applications, including acting as the substrate for microchips. Corundum is prepared from powdered Al_2O_3 dispersed in liquid. Granules that form in the dispersion are compressed in a mold and sintered. Single-crystal forms of aluminum oxide are known, and large single crystals of aluminum oxide called sapphires, which derive their color from iron and titanium impurities, are grown for specialized applications such as acoustic microscopes and as heat-resistant windows on heat-seeking missiles.

The challenge presented by corundum and other ceramics is to find a way to overcome their brittleness. One route used with some silicon dioxide ceramics is the *sol–gel process*. In this process, an organic silicon compound is dissolved in a solvent such as an alcohol; water is then added to create hydrate cross-links as the compound is polymerized into a network structure. These cross-links form a rigid, strong matrix that has few of the tiny cracks that can initiate breakage in brittle ceramics. If the solvent is removed at high temperatures and low pressures, an *aerogel*, a solid foam that has a density about the same as that of air but is a good insulator, is formed. Aerogels are sometimes called "frozen smoke" because of their low density and translucency. They are used to insulate skylights and were used to insulate the Mars Rover. Despite their ephemeral appearance, aerogels are very strong and can support heavy weights (**FIG. 2**).

The stability of ceramic materials at high temperatures has made them useful as furnace liners and has led to interest in ceramic automobile engines, which could survive overheating. Currently, a typical automobile contains about 35 kg of ceramic

FIG. 2 An aerogel is a ceramic foam. Its low density and low thermal conductivity, combined with great strength, make it an ideal insulating material. Here a nearly transparent piece supports a large brick. (*NASA/JPL.*)

materials such as spark plugs, pressure and vibration sensors, brake linings, catalytic converters, and thermal and electrical insulation.

A **glass** is an ionic solid with an amorphous structure resembling that of a liquid, generally created by freezing a melt quickly so that crystals cannot form. Glass has a network structure based on a nonmetal oxide, usually silica, SiO_2, that has been melted together with metal oxides acting as "network modifiers," which alter the arrangement of bonds in the solid. Glass was first used in ancient times and has played an important role in the development of modern architecture. Glasses commonly shatter into fragments when struck with sufficient force because their atoms form an intricate network and bonds do not lie in orderly arrays. However, some glasses have been engineered with extraordinary toughness, flexibility, and scratch resistance for use in touchscreen devices such as cell phones and tablet computers (**FIG. 3**); a variant of this glass has been developed with antimicrobial properties. The commercial product known as "Gorilla Glass" is a glass that has been strengthened by immersion in a molten alkali salt. In this process, sodium ions are replaced by potassium ions. When the glass cools, the potassium ions are compressed, hardening and strengthening the material.

To make glass, silica in the form of sand is heated to about 1600 °C. Metal oxides, such as MO (where M is a metal cation, such as Na^+, Pb^{2+}, or Zn^{2+}), are added to the silica. As the mixture melts, many of the Si—O bonds break and the orderly structure of the individual crystals is lost. When the melt cools, the Si—O bonds re-form but are prevented from forming a crystalline lattice because some of the silicon atoms bond with the O^{2-} ions of the metal oxide to give —Si—O—M^{n+} groups in place of some of the —Si—O—Si— links present in pure silica. Because of the strong covalent character of the Si—O bonds, a short-range order is preserved in the glass; however, long-range order is lost. Silicate glasses are generally transparent and durable and can be formed into flat sheets, blown into bottles, or molded into desired shapes.

Almost 90% of all manufactured glass combines sodium and calcium oxides with silica to form *soda-lime glass*. This glass, which is used for windows and bottles, is about 12% by mass Na_2O prepared by the action of heat on sodium carbonate (the soda) and 12% CaO (the lime). When the proportions of soda and lime are reduced and 16% B_2O_3 is added, a *borosilicate glass*, such as Pyrex, is produced. Because borosilicate glasses do not expand much when heated, they survive rapid heating and cooling and are used for ovenware and laboratory beakers.

THINKING POINT

Why don't borosilicate glasses expand much when heated?

Because crystalline regions are not present in glass, light is not scattered by the tiny crystallites that make some minerals opaque, but instead can pass easily through the glass, just as it does through water. In general, silicate-based glasses are brittle, hard, and optically transparent. These properties make them desirable for use in optical fibers. Optical fibers are made by drawing a thin fiber from an optically pure glass rod heated until it softens. The fiber is then coated with plastic (**FIG. 4**). Optical fibers allow the transmission of information much faster than metal wires and are an important component of wideband telecommunication networks.

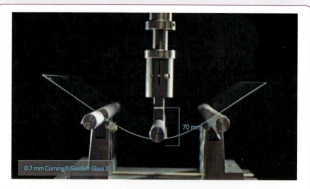

FIG. 3 A sheet of glass 0.7 mm thick used for touchscreen devices is tested for flexibility. (*Courtesy Corning Incorporated.*)

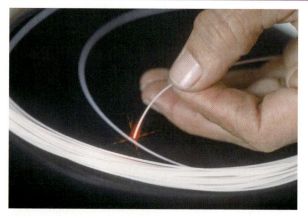

FIG. 4 Glass fibers, such as this one, about the diameter of a human hair, are used in communication networks when large amounts of information must be transmitted. (*David R. Frazier/Science Source.*)

THERMODYNAMICS

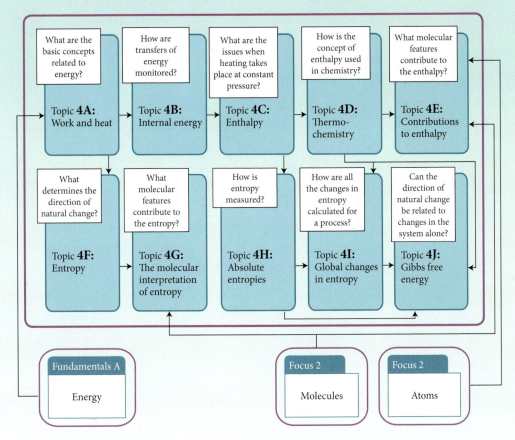

Energy lies at the heart of chemistry. No reaction takes place without energy being involved in one form or another, and almost all the physical properties of matter can be traced to the energy of interaction between and within atoms and molecules. Everyday phenomena, such as heat and work, are manifestations of energy. The study of the transformations of energy and specifically their role in chemistry is the field of **thermodynamics,** which is explored in this Focus.

What is energy? **TOPIC 4A** addresses this question by examining the roles of work and heat in changing the quantity of energy in a region and establishing the **first law of thermodynamics.** Thermodynamics distinguishes the system, essentially the region of interest, from the surroundings, the rest of the world. It then uses observations in the surroundings to monitor changes in the precisely defined "internal energy" of the system. Internal energy is explored in more detail in **TOPIC 4B**, where it is shown to behave like the reserves of a bank, with transactions taking place as either work or heat. An important point is that although thermodynamics is independent of any models about the structure of matter (whether it is atomic, and so on), this Topic shows how insight into it is immeasurably enriched by looking for explanations in terms of the behavior of atoms.

Thermodynamics, being a precise science, takes into account all the changes of energy that accompany a process. One that might be missed is the work done when a reaction generates or consumes a gas and has to drive back the atmosphere. This almost invisible contribution to the change in energy of a system is taken into account automatically in a property called "enthalpy," which is introduced in **TOPIC 4C**. So important is this property that it occurs throughout chemistry, especially, as shown in **TOPIC 4D**, in **thermochemistry,** the study of the heat transactions accompanying chemical reactions. In **TOPIC 4E**, insight into the factors contributing to the enthalpy of a system is presented by examining processes taking place in atoms and molecules.

Some things happen naturally; some don't. Decay is natural; construction requires work. Water flows downhill naturally; it has to be pumped uphill. Anyone who thinks about the world may wonder what determines the *natural* direction of change. What drives events forward? **TOPIC 4F** shows that a single quantity, the "entropy," in conjunction with the **second law of thermodynamics,** provides an elegant and quantitative answer to all these questions. Although entropy is a thermodynamic concept in the sense that it is a property of bulk samples of matter, **TOPIC 4G** shows that it has a very straightforward molecular interpretation and **TOPIC 4H** explains that a substance can be assigned an absolute entropy by appealing to the **third law of thermodynamics.**

Entropy changes in the surroundings of a system need to be taken into account as well, as shown in **TOPIC 4I**. Chemists have devised a clever way of doing that, as explained in **TOPIC 4J**, which introduces the very important Gibbs free energy.

Topic 4A Work and Heat

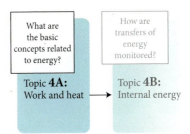

Thermodynamics is the study of the transformations of energy. Two of its fundamental concepts are *heat* and *work*. People once thought that heat was a material fluid called *caloric*, which flowed from a hot substance to a cooler one. The French engineer Sadi Carnot (**FIG. 4A.1**), who helped to lay the foundations of thermodynamics, believed that work resulted from the flow of caloric, just as the flow of water does work by turning a water wheel. Some of Carnot's conclusions survive, but it is now known that there is no such substance as caloric. About 25 years after Carnot proposed his ideas in the early nineteenth century, the English physicist James Joule (**FIG. 4A.2**) showed that heat and work are simply two ways in which energy can be transferred.

4A.1 Systems and Surroundings

To keep track of energy changes in thermodynamics, the world is divided into two parts. The region of interest, such as a flask of gas, a reaction mixture, or a muscle fiber, is called a **system** (**FIG. 4A.3**). Everything else, such as the water bath in which a reaction mixture may be immersed, is called the **surroundings.** The surroundings are where observations are made of the energy transferred into or out of the system. The system and the surroundings jointly make up the **universe.** However, often the only part of the universe significantly affected by a process consists of the sample, a flask, and a water bath; in such cases, which include most of the processes in this book, only changes in the immediate surroundings of the sample need be monitored.

There are three kinds of system (**FIG. 4A.4**):

- An **open system** can exchange both matter and energy with the surroundings.
- A **closed system** has a fixed amount of matter, but it can exchange energy with the surroundings.
- An **isolated system** can exchange neither matter nor energy with the surroundings.

Why Do You Need to Know This Material? Thermodynamics provides explanations for much of chemistry. Its central concept is energy, and to understand how the energy of a system can be changed you need to understand the concepts of work and heat.

What Do You Need to Know Already? This Topic assumes that you are familiar with the concepts of force and work (*Fundamentals* A), stoichiometry (*Fundamentals* L), and the ideal gas law (Topic 3B).

FIGURE 4A.1 Nicolas Leonard Sadi Carnot (1796–1832). (*Boyer/Roger Viollet/Getty Images.*)

FIGURE 4A.2 James Prescott Joule (1818–1889). (© *Corbis.*)

FIGURE 4A.3 The system is the sample or reaction mixture of interest. Outside the system are the surroundings. The system plus its surroundings is sometimes called the universe.

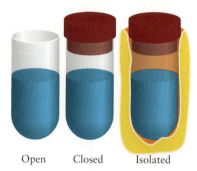

Open   Closed   Isolated

FIGURE 4A.4 A system is classified according to its interactions with its surroundings. An open system can exchange matter and energy with its surroundings. A closed system can exchange energy but not matter. An isolated system can exchange neither matter nor energy.

Examples of open systems are automobile engines and the human body. An example of a closed system is a cold pack used for treating athletic injuries. An isolated system can be thought of as having sealed, rigid, thermally insulating walls. A good approximation to an isolated system is a hot liquid inside a sealed thermos bottle.

In thermodynamics, the universe consists of a system and its surroundings. An open system can exchange both matter and energy with the surroundings; a closed system can exchange only energy; an isolated system can exchange nothing.

4A.2 Work

The most fundamental property in thermodynamics is **work**, the process of achieving motion against an opposing force (*Fundamentals* A). All forms of work can be thought of as equivalent to the work of raising a weight against the pull of gravity. The chemical reaction in a battery does work when it pushes an electric current through a circuit: the current can be used to drive an electric motor that raises a weight. Gas in a cylinder—such as the hot gas mixture in the cylinder of an automobile engine—does work when it expands and pushes back a piston: the piston may be connected to a pulley that raises a weight.

The work required to move an object a certain distance against an opposing force is calculated from

$$\text{Work} = \text{opposing force} \times \text{distance moved} \qquad (1)$$

As pointed out in *Fundamentals* A, the unit in which work is reported is the *joule,* J, with $1\ \text{J} = 1\ \text{kg·m}^2\text{·s}^{-2}$. Equation 1 is consistent with this definition, because force is measured in newtons ($1\ \text{N} = 1\ \text{kg·m·s}^{-2}$), so the units of force × distance are $\text{kg·m·s}^{-2} \times \text{m} = \text{kg·m}^2\text{·s}^{-2}$, or J.

Energy is the capacity of a system to do work (and, ultimately, to lift a weight). If a system can do a lot of work, it is said to possess a lot of energy. A hot compressed gas can do more work than the same gas after it has expanded and cooled, so it possesses more energy initially. A wound spring can do more work than an unwound spring, so the wound spring possesses more energy. When a system does work on the surroundings, its capacity to do work is reduced, so its energy has decreased. If work is done *on* a system, such as by winding a spring, then its capacity to do work is increased and therefore its energy has increased.

The total store of energy in a system is called its **internal energy,** U. The absolute value of the internal energy of a system cannot be measured, because that value includes the energies of all the atoms, their electrons, and the components of their nuclei. The best that can be done is to measure *changes* in internal energy. For instance, if a system does 15 J of work (and no other changes take place), then it has used up some of its store of energy, and its internal energy is said to have fallen by 15 J and we write $\Delta U = -15$ J. Throughout thermodynamics, the symbol ΔX means a difference in a property X:

$$\Delta X = X_{\text{final}} - X_{\text{initial}} \qquad (2)$$

A *negative* value of ΔX, as in $\Delta U = -15$ J, signifies that the value of X has decreased.

A Note on Good Practice: Except in the special cases we shall specify, always show the sign of ΔU (and of other ΔX) explicitly, even if it is positive. Thus, if the internal energy increases by 15 J during a change, write $\Delta U = +15$ J, not simply $\Delta U = 15$ J.

The symbol w is used to denote the energy transferred to a system by doing work and, *provided that no other type of transfer of energy is taking place,* $\Delta U = w$. When energy is transferred *to* a system by doing work, the internal energy of the system increases and w is positive. When energy *leaves* the system as it does work, the internal energy of the system decreases and w is negative. For example, when a system does 40 J of work, $w = -40$ J and $\Delta U = -40$ J.

TABLE 4A.1   Varieties of Work

Type of work	w	Comment	Units*
expansion	$-P_{ex}\Delta V$	P_{ex} is the external pressure	Pa
		ΔV is the change in volume	m^3
extension	$f\Delta l$	f is the tension	N
		Δl is the change in length	m
raising a weight	$mg\Delta h$	m is the mass	kg
		g is the acceleration of free fall	m·s^{-2}
		Δh is the change in height	m
electrical	$\mathcal{V}\Delta Q$	$\mathcal{V}$ is the electrical potential	V
		ΔQ is the change in charge	C
	$Q\Delta\mathcal{V}$	Q is the charge	C
		$\Delta\mathcal{V}$ is the potential difference	V
surface expansion	$\gamma\Delta A$	γ is the surface tension	N·m^{-1}
		ΔA is the change in area	m^2

*For work in joules (J). Note that 1 N·m = 1 J and 1 V·C = 1 J.

Work is the transfer of energy to a system by a process that is equivalent to raising or lowering a weight. For work done on a system, w is positive; for work done by a system, w is negative. The internal energy of a system may be changed by doing work: in the absence of other changes, $\Delta U = w$.

4A.3 Expansion Work

A system can do two kinds of work. **Expansion work** is the work arising from a change in the volume of a system. **Nonexpansion work** is work that does not involve a change in volume of the system. A chemical reaction in a battery can do nonexpansion work by causing an electrical current to flow, and your body does nonexpansion work when you move about. **TABLE 4A.1** lists some of the kinds of work that a system can do.

The simplest example of expansion work is the work done by a gas in a cylinder fitted with a piston. The external pressure acting on the piston provides the force opposing expansion. First, assume that the external pressure is constant, as when the piston is pressed on by the atmosphere (**FIG. 4A.5**). The challenge is to find an expression for the work done when the volume of the system changes by ΔV and the external pressure is P_{ex}.

─────────── **How Is That Done?** ───────────

To derive an expression for the work, you need to know the force opposing the expansion. Because pressure is force divided by the area to which the force is applied, $P = F/A$ (Topic 3A), which rearranges to $F = PA$. The force opposing expansion is the product of the pressure acting on the outside of the piston, P_{ex}, and the area of the piston, A: $F = P_{ex}A$. The work needed to drive the piston out through a distance d is therefore

From work = force × distance,

$$\text{Work} = P_{ex}A \times d$$

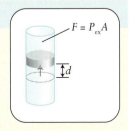

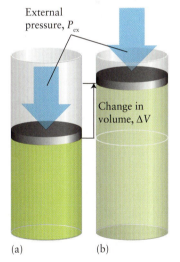

FIGURE 4A.5 A system does work when it expands against an external pressure. (a) A gas in a cylinder with a piston held in place. (b) The piston is released and (provided the pressure of the gas is greater than the external pressure, P_{ex}) the gas expands and pushes out against P_{ex}. The work done is proportional to P_{ex} and the change in volume, ΔV, that the system undergoes.

However, the product of area and distance moved is equal to the change in volume (ΔV) of the sample.

From volume = area × height,

$$A \times d = \Delta V$$

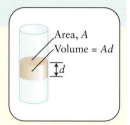

Therefore, the work done by the expanding gas is $P_{ex}\Delta V$. At this point, you need to adopt the sign convention. When a system expands, it loses energy as work; so, when ΔV is positive (an expansion), w is negative. Therefore,

From the sign convention,

$$w = -P_{ex}\Delta V$$

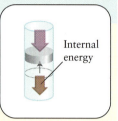

The result of the calculation is that the work done when a system expands by ΔV against a constant external pressure P_{ex} is

$$w = -P_{ex}\Delta V \tag{3}$$

This expression applies to all systems. A gas is easiest to visualize, but the expression also applies to an expanding liquid or solid. However, Eq. 3 applies *only when the external pressure is constant* during the expansion.

What Does This Equation Tell You? When the system expands, ΔV is positive. Therefore the minus sign in Eq. 3 tells you that the internal energy of the system decreases when the system expands. The factor P_{ex} tells you that more work is done for a given change in volume when the external pressure is high. The factor ΔV tells you that, for a given external pressure, more work is done the greater the change in volume.

In SI units, the external pressure is expressed in pascals (1 Pa = 1 kg·m^{-1}·s^{-2}, Topic 3A) and the change in volume is expressed in cubic meters (m^3). The SI units of work can therefore be found from the product of 1 Pa and 1 m^3:

$$1 \text{ Pa·m}^3 = \overbrace{1 \text{ kg·m}^{-1}\text{·s}^{-2}}^{\text{Pa}} \times \text{m}^3 = \overbrace{1 \text{ kg·m}^2\text{·s}^{-2}}^{\text{J}} = 1 \text{ J}$$

Therefore, if you perform calculations in pascals and cubic meters, the work is obtained in joules, as you should expect for a change in energy of the system. However, it is often more convenient to express pressure in atmospheres and the volume in liters. In this case, you might need to convert the answer (in liter-atmospheres) into joules. The conversion factor is obtained by noting that 1 L = 10^{-3} m^3 and 1 atm = 101 325 Pa exactly; therefore,

$$1 \text{ L·atm} = \overbrace{10^{-3}\text{m}^3}^{1 \text{ L}} \times \overbrace{101\,325 \text{ Pa}}^{1 \text{ atm}} = \overbrace{101.325 \text{ Pa·m}^3}^{\text{J}} = 101.325 \text{ J (exactly)}$$

There is one final point. If the external pressure is zero ($P_{ex} = 0$, a vacuum), it follows from Eq. 3 that $w = 0$; that is, *a system does no expansion work when it expands into a vacuum*, because there is no opposing force. No work is done by pushing if there is nothing to push against. Expansion against zero pressure is called **free expansion**.

EXAMPLE 4A.1   Calculating the work done when a gas expands

In the internal-combustion engine in an automobile, hot compressed gas expands against a piston, turning the driveshaft that moves the vehicle along the road. If you are investigating the performance of such an engine, you might need to assess the work that each stroke of the piston can achieve. Suppose a gas expands by 500. mL (0.500 L) against the opposing force due to the transmission, which is modeled as a pressure of 1.20 atm and no heat is exchanged with the surroundings during the expansion. (a) How much work is done in the expansion? (b) What is the change in internal energy of the system?

ANTICIPATE Because work is done by the system, expect w and therefore ΔU to be negative, signifying that energy has been lost from the system.

PLAN Use Eq. 3 to calculate the work, and then convert liter-atmospheres into joules.

What should you assume? Assume that the only energy exchanged with the surroundings is the work of expansion.

SOLVE

(a) From $w = -P_{ex}\Delta V$,

$$w = -(1.20 \text{ atm}) \times (0.500 \text{ L}) = -0.600 \text{ L·atm}$$

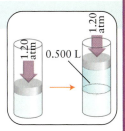

Convert into joules by using 1 L·atm = 101.325 J.

$$w = -(0.600 \text{ L·atm}) \times \frac{101.325 \text{ J}}{1 \text{ L·atm}} = -60.8 \text{ J}$$

(b) Because no energy is transferred as heat, $w = \Delta U$.

$$\Delta U = -60.8 \text{ J}$$

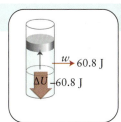

EVALUATE The negative sign in $w = -60.8$ J (which was anticipated) means that the internal energy of the gas decreases by 60.8 J when it expands and does 60.8 kJ of work on its surroundings (and there are no other changes).

Self-test 4A.1A Water expands when it freezes. How much work does 100. g of water do when it freezes at 0 °C and pushes back the metal wall of a pipe that exerts an opposing pressure of 1070 atm? The densities of water and ice at 0 °C are 1.00 g·cm^{-3} and 0.92 g·cm^{-3}, respectively.

[*Answer:* $w = -0.86$ kJ]

Self-test 4A.1B The gases in the four cylinders of an automobile engine expand from a total volume of 0.22 L to 2.2 L during one ignition cycle. Assuming a steady pressure of 9.60 atm on the gases, how much work can the engine do in one cycle?

Related Exercises 4A.3, 4A.4

To calculate the work done by a gas that expands against a *changing* external pressure, you need to know how that pressure changes in the course of the expansion.

First, you need to understand the term "reversible." In everyday language, a reversible process is one that can take place in either direction. This common usage is refined in science: in thermodynamics, a **reversible process** is one that can be reversed by an *infinitely small* change in a variable (an "infinitesimal" change). For example, if the external pressure exactly matches the pressure of the gas in the system, then the piston moves in neither direction. If the external pressure is increased infinitesimally, then the piston moves in. If, instead, the external pressure is reduced infinitesimally, the piston moves out. Expansion against an external pressure that differs by a measurable amount from the pressure of the system is an irreversible process in the sense that an infinitesimal change

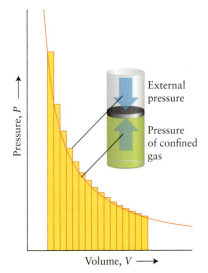

FIGURE 4A.6 When a gas expands reversibly, the external pressure is matched to the pressure of the gas at every stage of the expansion. This arrangement (when the steps corresponding to the increase in volume are infinitesimal) achieves maximum work.

in the external pressure does not reverse the direction of travel of the piston. For instance, if the pressure of the system is 2.0 atm at some stage of the expansion and the external pressure is 1.0 atm, then a tiny increase in the latter does not convert expansion into compression. Reversible processes are of considerable importance in thermodynamics because, as you will see, the work a system can do is greatest in a reversible process.

THINKING POINT

How could you ensure that an electric battery produced an electric current reversibly?

The simplest kind of reversible change to consider is the reversible, isothermal (constant-temperature) expansion of an ideal gas. The constant temperature of the gas is maintained by ensuring that the system is in thermal contact with a constant-temperature water bath at all stages of the expansion. In an isothermal expansion, the pressure of the gas falls as it expands (by Boyle's law, Topic 3B); therefore, to achieve reversible expansion, the external pressure must be reduced in step with the change in volume so that at every stage the external pressure is the same as the pressure of the gas (**FIG. 4A.6**). To calculate the work, the gradual reduction in external pressure, and therefore the changing matching opposing force, must be taken into account.

How Is That Done?

To calculate the work of reversible, isothermal expansion of a gas, think of the expansion as taking place in a series of infinitesimal steps, each one at a slightly lower pressure from the one before. The starting point is therefore Eq. 3 written for an infinitesimal change in volume, dV:

$$dw = -P_{ex}dV$$

The external pressure is matched to the pressure, P, of the gas at each stage of a reversible expansion, so $P_{ex} = P$, and

$$dw = -PdV$$

At each stage of the expansion, the pressure of the gas is related to its volume by the ideal gas law, $PV = nRT$, so P can be replaced by nRT/V,

$$dw = -\overbrace{\frac{nRT}{V}}^{P}dV$$

The total work done is the sum (integral) of these infinitesimal contributions as the volume changes from its initial value V_1 to its final value V_2. That is, the work is obtained by integrating dw from the initial to the final volume with nRT a constant (because the change is isothermal and there is a fixed amount of gas).

From $w = \int dw$, with nRT a constant,

$$w = -nRT\int_{V_1}^{V_2}\frac{dV}{V} = -nRT\ln\frac{V_2}{V_1}$$

The final step used the standard integral

$$\int\frac{dx}{x} = \ln x + \text{constant}$$

and then $\ln x - \ln y = \ln(x/y)$.

The result of the calculation is that the work of reversible, isothermal expansion of an ideal gas from V_1 to V_2 is

$$w = -nRT\ln\frac{V_2}{V_1} \tag{4}$$

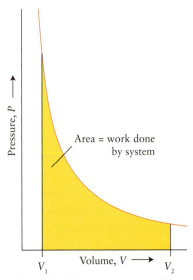

FIGURE 4A.7 The magnitude of the work done by an ideal gas is equal to the area below the curve of the graph of pressure plotted against volume (in this case, for isothermal expansion).

where n is the amount of gas molecules (in moles) in the system and T is the (absolute) temperature. An important result from calculus is that a definite integral of a function between two limits is equal to the area under the graph of the function and lying between the two limits. In this case, the magnitude of the work done is equal to the area under the plot of nRT/V against V and lying between the initial and final volumes (**FIG. 4A.7**).

What Does This Equation Tell You? For a given initial and final volume, more expansion work is done when the temperature is high than when it is low. For a given volume and amount of gas-phase molecules, a higher temperature corresponds to a higher gas pressure, and so the expansion takes place against a stronger opposing force and therefore must do more work. More expansion work is also done if the final volume is very much greater than the initial volume.

If the external pressure were to be increased even infinitesimally at any stage of the expansion, the piston would move in instead of out. Therefore, the work done during a reversible expansion of a gas is the maximum expansion work possible. This is a very important general point:

The greatest work is done for a process that takes place reversibly.

EXAMPLE 4A.2   Calculating the work of isothermal expansion

An engineer studying piston engines wishes to know the difference between the maximum work a system can achieve and the work obtained when expanding against a constant opposing pressure. A piston confines 0.100 mol Ar(g) in 1.00 L at 25 °C. Two experiments are performed. (a) The gas is allowed to expand through an additional 1.00 L against a constant pressure of 1.00 atm. (b) The gas is allowed to expand reversibly and isothermally to the same final volume. Which process does more work?

ANTICIPATE Provided the initial and final states are the same, a change carried out reversibly always does more work than a change carried out irreversibly, so you should expect the second path to produce more work and therefore correspond to a more negative value of w (because more energy is lost from the system).

PLAN For expansion against constant external pressure use Eq. 3 and for reversible, isothermal expansion use Eq. 4.

What should you assume? Assume that the gas is ideal and that it is immersed in a water bath to maintain the constant temperature.

SOLVE

(a) Irreversible path: From Eq. 3, $w = -P_{ex}\Delta V$, and then convert to joules,

$$w = -(1.00\ \text{atm}) \times (1.00\ \text{L}) = -1.00 \times 1.00\ \text{L·atm} \times \frac{101.325\ \text{J}}{1\ \text{L·atm}} = -101\ \text{J}$$

(b) Reversible isothermal path: From Eq. 4, $w = -nRT \ln(V_2/V_1)$,

$$w = -\underbrace{(0.100\ \text{mol})}_{n} \times \underbrace{(8.3145\ \text{J·K}^{-1}\text{·mol}^{-1})}_{R} \times \underbrace{(298\ \text{K})}_{T} \times \ln\underbrace{\frac{2.00\ \text{L}}{1.00\ \text{L}}}_{V_2/V_1} = -172\ \text{J}$$

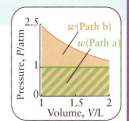

EVALUATE The gas does more work in the reversible process, as anticipated.

Self-test 4A.2A A cylinder of volume 2.00 L contains 0.100 mol He(g) immersed in a constant-temperature water bath held at 30 °C. In which process does the system do more work, expanding the gas isothermally to 2.40 L with a constant external pressure of 1.00 atm, or expanding it reversibly and isothermally to the same final volume?

                                                                         [Answer: Reversible expansion]

Self-test 4A.2B A cylinder of volume 2.00 L contains 1.00 mol He(g) immersed in a constant-temperature water bath held at 30 °C. Which process does more work on the surroundings, allowing the gas to expand isothermally to 4.00 L against a constant external pressure of 1.00 atm or allowing it to expand reversibly and isothermally to the same final volume?

Related Exercises 4A.5, 4A.6

A reversible process is a process that can be reversed by an infinitesimal change in a variable. The greatest work is done for a process that takes place reversibly.

FIGURE 4A.8 The thermite reaction is so exothermic that it melts the metal that it produces and has been used to weld railroad tracks together. Here, aluminum metal is reacting with iron(III) oxide, Fe_2O_3, causing a shower of molten iron sparks. (*W. H. Freeman photo by Ken Karp.*)

LAB VIDEO FIGURE 4A.8

FIGURE 4A.9 The endothermic reaction between ammonium thiocyanate, NH_4SCN, and barium hydroxide octahydrate, $Ba(OH)_2 \cdot 8H_2O$, absorbs a lot of heat and can cause water vapor in the air to freeze on the outside of the beaker. (*W. H. Freeman photo by Ken Karp.*)

LAB VIDEO FIGURE 4A.9

"Adiabatic" comes from the Greek words for "not passing through."

4A.4 Heat

The internal energy of a system, its capacity to do work, can also be changed by transferring energy to or from the surroundings as heat. For instance, a high-temperature gas might cool and, as a result, be able to do less work. "Heat" is a familiar term, but in thermodynamics it has a specific meaning. In thermodynamics, **heat** is the energy transferred as a result of a temperature difference. Energy flows as heat from a high-temperature region to a low-temperature region. Therefore, provided that the walls of the system are not thermally insulating, if the system is cooler than its surroundings, energy flows into the system from the surroundings.

The energy transferred to a system as heat is denoted q. When the internal energy of the system is changed by transferring energy as heat (and no other processes, including expansion or contraction, take place), $\Delta U = q$. If energy enters a system as heat, the internal energy of the system increases and q is positive; if energy leaves the system as heat, the internal energy of the system decreases and q is negative. Thus, if 10 J enters the system as heat, write $q = +10$ J and (provided no other processes take place) $\Delta U = +10$ J. Likewise, if 10 J leaves the system as heat, write $q = -10$ J and (if no other processes take place) $\Delta U = -10$ J.

As these examples show, energy transferred as heat (like any energy) is measured in joules, J. However, a unit still widely used in biochemistry and related fields is the *calorie*, cal. Originally, 1 cal was defined as the energy needed to raise the temperature of 1 g of water by 1 °C. The modern definition is

$$1 \text{ cal} = 4.184 \text{ J (exactly)}$$

This exact relation defines the calorie in terms of the joule; the joule is the fundamental unit. The **nutritional calorie,** Cal, is actually 1 kilocalorie (1 kcal), and so it is important to note which unit is being used when assessing the energy content of food.

THINKING POINT

How does heat transfer occur between hot and cold objects on the atomic level?

A process that releases heat into the surroundings is called an **exothermic process**. Most common chemical reactions—and all combustions, such as those that power transport and heating—are exothermic (**FIG. 4A.8**). Less familiar are chemical reactions that absorb heat from the surroundings. A process that absorbs heat is called an **endothermic process** (**FIG. 4A.9**). A number of common physical processes are endothermic. For instance, vaporization is endothermic because heat must be supplied to drive molecules of a liquid apart from one another. The dissolution of ammonium nitrate in water is endothermic; in fact, this process is used in instant cold packs for treating sports injuries.

Heat is the transfer of energy as a result of a temperature difference. When energy is transferred as heat and no other processes occur, $\Delta U = q$. When energy enters a system as heat, q is positive; when energy leaves a system as heat, q is negative.

4A.5 The Measurement of Heat

There are two types of boundaries between a system and its surroundings:

- An **adiabatic** wall does not permit the transfer of energy as heat even though there might be a temperature difference between the system and the surroundings.
- A **diathermic** wall allows the transfer of energy as heat between the system and surroundings.

Adiabatic walls are thermally insulating. The walls of a vacuum flask are a good approximation because the vacuum between them does not allow energy to be conducted from one wall to another by molecules and their silvered surfaces cut down the transfer of energy by radiation. A system with adiabatic walls is not necessarily isolated: the walls may be flexible, and energy may be transferred to or from the system as expansion work. In the case of a diathermic container, provided the system does not lose energy as work, an influx of energy through its walls typically raises the temperature of the system. Therefore, monitoring the temperature change is a way to monitor the heat transferred and therefore to infer the change in internal energy.

To convert a temperature change into an amount of energy transferred as heat, you need to know the **heat capacity,** C, the ratio of the heat supplied to the rise in temperature produced:

$$\text{Heat capacity} = \frac{\text{heat supplied}}{\text{temperature rise produced}}, \quad \text{that is,} \quad C = \frac{q}{\Delta T} \tag{5a}$$

Note the "typically." There is no change in temperature when the transfer takes place at the boiling point or freezing point of a liquid.

A high heat capacity means that a given supply of heat produces only a small rise in temperature (from $\Delta T = q/C$). A low heat capacity means that even a small transfer of energy as heat will produce a large rise in temperature. If the heat capacity of a system is known, the observed change in temperature, ΔT, of the system can be used to calculate how much heat has been supplied by using Eq. 5a in the form

$$q = C\Delta T \tag{5b}$$

Notice that the energy that has been supplied as heat is directly proportional to the temperature change, with C as the proportionality constant.

The heat capacity of most substances varies with temperature. However, when the change in temperature is small, its variation can be ignored.

The transfer of energy as heat occurs reversibly (in the thermodynamic sense) if the temperatures of the system and its surroundings are the same. Increasing the temperature of the surroundings infinitesimally would lead to energy flowing into the system; lowering their temperature infinitesimally would result in energy flowing out of the system into the surroundings. When the temperatures are the same energy flows as heat in both directions at the same rate. The system and its surroundings are then said to be in **thermal equilibrium.** A very important point is that equilibrium does not mean "dead." The equilibrium is *dynamic* in the sense that the molecular processes continue, but at equal rates. The great importance of an equilibrium being dynamic is that it is responsive to changes in the surroundings. All the equilibria encountered in chemistry are dynamic in this sense: they are living, responsive equilibria (Focus 5).

THINKING POINT

What is the heat capacity of water at its boiling point?

Heat capacity is an extensive property: the bigger the sample, the more heat is required to raise its temperature by a given amount (**FIG. 4A.10**). It is therefore common to report

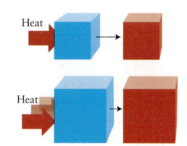

- The **specific heat capacity** (often called just "specific heat"), C_s, which is the heat capacity divided by the mass of the sample ($C_s = C/m$).
- The **molar heat capacity,** C_m, the heat capacity divided by the amount (in moles) of the sample ($C_m = C/n$).

FIGURE 4A.10 The heat capacity of an object determines the change in its temperature brought about by the quantity of energy transferred as heat. Heat capacity is an extensive property, so a large object (bottom) has a larger heat capacity than a small object (top) made of the same material.

For example, the specific heat capacity of liquid water at room temperature is $4.18 \, \text{J} \cdot (°\text{C})^{-1} \cdot \text{g}^{-1}$, or $4.18 \, \text{J} \cdot \text{K}^{-1} \cdot \text{g}^{-1}$, and its molar heat capacity is $75 \, \text{J} \cdot \text{K}^{-1} \cdot \text{mol}^{-1}$. **TABLE 4A.2** lists the specific and molar heat capacities of some common substances.

The heat capacity of a sample of a substance is calculated from its mass and its specific heat capacity by using $C = m \times C_s$. If you know the mass of a substance, its specific heat capacity, and the temperature rise it undergoes in a process, then the energy supplied as heat to the substance is

$$q = C\Delta T = mC_s\Delta T \tag{6a}$$

A similar expression, $C = n \times C_m$, is used if you know the molar heat capacity of a substance. Then you write

$$q = nC_m\Delta T \tag{6b}$$

These expressions may be rearranged to calculate the specific or molar heat capacity from the measured temperature rise caused by a known quantity of heat. One point to note is

TABLE 4A.2   Specific and Molar Heat Capacities of Common Materials*

Material	Specific heat capacity $C_s/[\text{J}\cdot(°\text{C})^{-1}\cdot\text{g}^{-1}]$	Molar heat capacity $C_m/(\text{J}\cdot\text{K}^{-1}\cdot\text{mol}^{-1})$
air	1.01	—
benzene	1.05	136
brass	0.37	—
copper	0.38	33
ethanol	2.42	111
glass (Pyrex)	0.78	—
granite	0.80	—
marble	0.84	—
polyethylene	2.3	—
stainless steel	0.51	—
water: solid	2.03	37
liquid	4.184	75
vapor	2.01	34

*More values are available in Appendices 2A and 2D; values assume constant pressure. Specific heat capacities commonly use Celsius degrees in their units, whereas molar heat capacities commonly use kelvins. All values except that for ice are for 25 °C.

that the specific heat capacity of a dilute solution is normally taken to be the same as that of the pure solvent (which is commonly water).

THINKING POINT

The molar heat capacity of lead is much greater than that of diamond at room temperature. Can you think of a reason for the difference?

EXAMPLE 4A.3   Calculating the heat required to bring about an increase in temperature

Dehydrated meals carried on camping trips are reconstituted with hot water. The energy used to heat the water is typically supplied by burning fuel in a camp stove. How much energy is required to heat the water? Calculate the heat necessary to increase the temperature of (a) 100. g of water, (b) 2.00 mol $H_2O(l)$ from 20. °C to 100. °C.

ANTICIPATE Because the molar mass of water is about 18 g·mol^{-1}, 100 g of water is greater than 2.00 mol H_2O, so you should expect more heat to be required for (a) than for (b).

PLAN The heat required is given by Eq. 6. For (a), use the specific heat capacity and for (b), use the molar heat capacity. In each case, $\Delta T = +80.$ K. Note that a heat capacity expressed in $\text{J}\cdot(°\text{C})^{-1}\cdot\text{g}^{-1}$ is numerically the same when expressed in $\text{J}\cdot\text{K}^{-1}\cdot\text{g}^{-1}$.

What should you assume? Assume that no energy is lost to the container or the surroundings during the heating and that the water is thoroughly mixed, so that it is at the same temperature throughout.

SOLVE

(a) From $q = mC_s\Delta T$,

$$q = (100.\,\text{g}) \times (4.18\,\text{J}\cdot\text{K}^{-1}\cdot\text{g}^{-1}) \times (80.\,\text{K}) = +33\,\text{kJ}$$

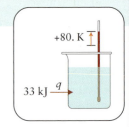

(b) From $q = nC_m\Delta T$,

$$q = (2.00\,\text{mol}) \times (75\,\text{J}\cdot\text{K}^{-1}\cdot\text{mol}^{-1}) \times (80.\,\text{K}) = +12\,\text{kJ}$$

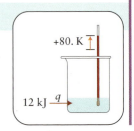

EVALUATE As anticipated, more heat is needed for (a) than for (b).

Self-test 4A.3A Potassium perchlorate, $KClO_4$, is used as an oxidizer in fireworks. Calculate the heat required to raise the temperature of 10.0 g of $KClO_4$ from 25 °C to an ignition temperature of 900. °C. The specific heat capacity of $KClO_4$ is 0.8111 $J·K^{-1}·g^{-1}$.

[**Answer:** 7.10 kJ]

Self-test 4A.3B Calculate the heat necessary to increase the temperature of 3.00 mol $CH_3CH_2OH(l)$, ethanol, by 15.0 °C from room temperature (see Table 4A.2).

Related Exercises 4A.7, 4A.8

Transfers of energy as heat are often measured with a **calorimeter,** a device in which energy transferred as heat is monitored by recording the change in temperature produced by a process taking place within it. A calorimeter can be as simple as a thermally insulated open container equipped with a thermometer (**FIG. 4A.11**). A more sophisticated version is a *bomb calorimeter* (**FIG. 4A.12**). The reaction takes place inside a sturdy sealed metal vessel of constant volume (the "bomb"), which is immersed in water, and the rise in temperature of the entire assembly is monitored. Its heat capacity is first measured by supplying a known quantity of heat and noting the resulting rise in temperature. This process is called "calibrating" the calorimeter.

It is important to remember that the heat lost by a reaction is gained by the calorimeter; that is, $-q = q_{cal}$ (so if $q = -15$ kJ, $q_{cal} = +15$ kJ). The heat gained by the calorimeter is found using $q_{cal} = C_{cal}\Delta T$, where C_{cal} is the heat capacity of the calorimeter (which is sometimes called the "calorimeter constant"). By combining these two results the heat lost (or gained) by a reaction can be related to the change in temperature of the calorimeter:

$$q = -C_{cal}\Delta T$$

Notice if the ΔT is positive, indicating that the temperature of the calorimeter has increased, then q is negative, indicating that energy has been released as heat by the reaction.

> "Calibrate" has an interesting etymology: the verb derives from the noun "caliber," the size of a bullet, and ultimately from the Arabic *qalib*, a mold for making bullets.

EXAMPLE 4A.4   Determining the internal energy change accompanying a reaction

Neutralization reactions occurring when acids and bases are mixed can be very exothermic. Suppose you are investigating how the heat released in various neutralization reactions is related to the structures of certain acids. A constant-volume calorimeter was calibrated by carrying out a reaction known to release 1.78 kJ of heat in 0.100 L of solution in the calorimeter ($q = -1.78$ kJ), resulting in a temperature rise of 3.65 °C. In a subsequent experiment, 50.0 mL of 0.200 M HCl(aq) and 50.0 mL of 0.200 M NaOH(aq) were mixed in the same calorimeter and the temperature rose by 1.26 °C (that is, $\Delta T = +1.26$ K). What is the change in the internal energy of the neutralization reaction?

ANTICIPATE The rise in temperature in the actual experiment is about one-third the rise in the calibration, so you can estimate that the heat released by the reaction is about one-third of 1.78 kJ, or about 0.6 kJ.

PLAN The calculation has two steps. First, calibrate the calorimeter by calculating its heat capacity with data from the first reaction, $C_{cal} = q_{cal}/\Delta T$ with $q_{cal} = -q$. Second, use that value of C_{cal} to find the energy change of the neutralization reaction. For the second step, use $q = -C_{cal}\Delta T$ but with ΔT now the change in temperature observed during the reaction. Note that the calorimeter contains the same volume of liquid in both cases, so that the temperature change is due only to a difference in the reaction that takes place. Finally, note that $\Delta U = q$.

What should you assume? Because dilute aqueous solutions have approximately the same heat capacities as pure water, assume that the heat capacity for the calorimeter during the reaction is the same as in the calibration. Also, assume that there is no change in volume on mixing, so the final volume is 0.100 L.

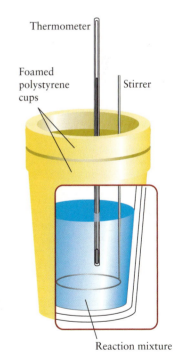

FIGURE 4A.11 The energy released or absorbed as heat by a reaction at constant pressure can be measured by noting the temperature change of this simple calorimeter. The outer polystyrene cup acts as an extra layer of insulation to ensure that no heat enters or leaves the inner cup.

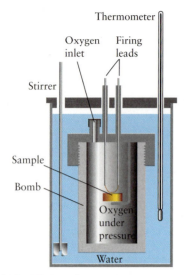

FIGURE 4A.12 A bomb calorimeter is used to measure heat transfers at constant volume. The sample in the central rigid container, called the bomb, is ignited electrically with a fuse wire. Once combustion has begun, energy released as heat spreads through the walls of the bomb into the water. The heat released is proportional to the temperature change of the entire assembly.

SOLVE

Calibration: From $-q = q_{cal}$ and $C_{cal} = q_{cal}/\Delta T$,

$$q_{cal} = -q = -(-1.78\text{ kJ}) = +1.78\text{ kJ}$$

$$C_{cal} = \frac{1.78\text{ kJ}}{3.65\text{ °C}} = \frac{1.78}{3.65}\text{ kJ·(°C)}^{-1} = 0.487...\text{ kJ·K}^{-1}$$

This step has used the fact that the degree Celsius is the same size as the kelvin, so "per °C" is the same as "per K."

Application: From $q = -C_{cal}\Delta T$ and $\Delta U = q$,

$$q = -(0.487...\text{ kJ·K}^{-1}) \times (1.26\text{ K}) = -0.614\text{ kJ}$$
$$\Delta U = -0.614\text{ kJ}$$

EVALUATE Because the temperature rises, the process is exothermic, energy leaves the system as heat, and $\Delta U = -0.614$ kJ, in agreement with expectation.

Self-test 4A.4A A small piece of calcium carbonate was placed in the same calorimeter as described in the preceding worked example, and 0.100 L of dilute hydrochloric acid was poured over it. The temperature of the calorimeter rose by 3.57 °C. What is the value of ΔU for the reaction of these quantities of hydrochloric acid and calcium carbonate?

[**Answer:** −1.74 kJ]

Self-test 4A.4B A calorimeter was calibrated by mixing two aqueous solutions together, each of volume 0.100 L. The heat output of the reaction that took place was known to be 4.16 kJ, and the temperature of the calorimeter rose by 3.24 °C. Calculate the heat capacity of this calorimeter when it contains 0.200 L of water.

Related Exercises 4A.13, 4A.14

THINKING POINT

Why does a hot pan inside a heated oven scald our hands quickly when we touch it, but the air inside does not?

The heat capacity of an object is the ratio of the heat supplied to the temperature rise produced. Heat transfers are measured by using a calibrated calorimeter.

What have you learned in this Topic?

You have seen that systems may be of three kinds (open, closed, and isolated) and that walls may be adiabatic or diathermic. You have learned that energy can be transferred to or from a system as either work or heat and have learned the sign convention for w and q. You also know how to relate the work done when a system expands (or contracts) to the change in volume and how maximum work is done if the change is reversible. You also know how to measure the transfer of energy as heat by monitoring the change in temperature and using the heat capacity of a substance.

The skills you have mastered are the ability to:

☐ **1.** Calculate the work done by a gas due to expansion against constant pressure (Example 4A.1).

☐ **2.** Calculate the work done by an ideal gas expanding reversibly and isothermally (Example 4A.2).

3. Use the heat capacity of a substance to calculate the heat required to raise its temperature by a given amount (Example 4A.3).

4. Determine the change in internal energy of a system that accompanies a chemical reaction (Example 4A.4).

5. Explain the terms "reversible" and "dynamic equilibrium" as used in thermodynamics (Sections 4A.3 and 4A.5).

Topic 4A Exercises

4A.1 Identify the following systems as open, closed, or isolated: (a) coffee in a very-high-quality thermos bottle; (b) coolant in a refrigerator coil; (c) a bomb calorimeter in which benzene is burned; (d) gasoline burning in an automobile engine; (e) mercury in a thermometer; (f) a living plant.

4A.2 (a) Describe three ways in which you could increase the internal energy of an open system. (b) Which of these methods could you use to increase the internal energy of a closed system? (c) Which, if any, of these methods could you use to increase the internal energy of an isolated system?

4A.3 Air in a bicycle pump is compressed by pushing in the handle. The inner diameter of the pump is 3.0 cm and the pump is depressed 20. cm with a pressure of 2.00 atm. (a) How much work is done in the compression? (b) Is the work positive or negative with respect to the air in the pump? (c) What is the change in internal energy of the system?

4A.4 Each of the four cylinders of a new type of combustion engine has a displacement of 3.60 L. (The volume of the cylinder expands by 3.60 L each time the fuel is ignited and a piston moves out.) (a) If each piston in the four cylinders is displaced against a pressure of 1.80 kbar and each cylinder is ignited once per second, how much work can the engine do in 1.00 min? (b) Is the work positive or negative with respect to the engine and its contents? (c) What is the change in internal energy of the system?

4A.5 A piston confines 0.200 mol Ne(g) in 1.20 L at 25 °C. Two experiments are performed. (a) The gas is allowed to expand through an additional 1.20 L against a constant pressure of 1.00 atm. (b) The gas is allowed to expand reversibly and isothermally to the same final volume. Which process does more work?

4A.6 A piston confines 0.250 mol He(g) in 1.50 L at 25 °C. Two experiments are performed. (a) The gas is allowed to expand through an additional 1.00 L against a constant pressure of 2.00 atm. (b) The gas is allowed to expand reversibly and isothermally to the same final volume. Which process does more work?

4A.7 (a) Calculate the heat that must be supplied to a copper kettle of mass 400.0 g containing 300.0 g of water to raise its temperature from 20.0 °C to the boiling point of water, 100.0 °C. (b) What percentage of the heat is used to raise the temperature of the water? (See Table 4A.2.)

4A.8 (a) Calculate the heat that must be supplied to a stainless steel vessel of mass 400.0 g containing 300.0 g of water to raise its temperature from 20.0 °C to the boiling point of water, 100.0 °C. (b) What percentage of the heat is used to raise the temperature of the water? (c) Compare these answers with those of Exercise 4A.7. (See Table 4A.2.)

4A.9 A piece of copper of mass 20.0 g at 100.0 °C is placed in a vessel of negligible heat capacity but containing 50.7 g of water at 22.0 °C. Calculate the final temperature of the water. Assume that no energy is lost to the surroundings.

4A.10 A piece of metal of mass 18.0 g at 100.0 °C is placed in a calorimeter containing 50.2 g of water at 22.0 °C. The final temperature of the mixture is 24.8 °C. What is the specific heat capacity of the metal? Assume that no energy is lost to the surroundings.

4A.11 A calorimeter was calibrated with an electric heater, which supplied 22.5 kJ of energy as heat to the calorimeter and increased the temperature of the calorimeter and its water bath from 22.45 °C to 23.97 °C. What is the heat capacity of the calorimeter?

4A.12 The heat released in the combustion of benzoic acid, C_6H_5COOH, which is often used to calibrate calorimeters, is -3228 kJ·mol^{-1}. When 1.685 g of benzoic acid was burned in a calorimeter, the temperature increased by 2.821 °C. What is the heat capacity of the calorimeter?

4A.13 A constant-volume calorimeter was calibrated by carrying out a reaction known to release 3.50 kJ of heat in 0.200 L of solution in the calorimeter ($q = -3.50$ kJ), resulting in a temperature rise of 7.32 °C. In a subsequent experiment, 100.0 mL of 0.200 M HBr(aq) and 100.0 mL of 0.200 M KOH(aq) were mixed in the same calorimeter and the temperature rose by 2.49 °C. What is the change in the internal energy of the reaction mixture as a result of the neutralization reaction?

4A.14 A constant-volume calorimeter was calibrated by carrying out a reaction known to release 0.90 kJ of heat in 0.60 L of solution in the calorimeter ($q = -0.90$ kJ), resulting in a temperature rise of 2.85 °C. In a subsequent experiment, 30.0 mL of 0.20 M $HClO_2$(aq) and 30.0 mL of 0.20 M NaOH(aq) were mixed in the same calorimeter and the temperature rose by 1.31 °C. What is the change in the internal energy of the reaction mixture as a result of the neutralization reaction?

Topic 4B Internal Energy

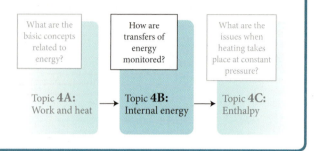

Topic **4A:**
Work and heat

Topic **4B:**
Internal energy

Topic **4C:**
Enthalpy

Why Do You Need to Know This Material? The internal energy underlies the first law of thermodynamics and much of the application of thermodynamics to chemistry.

What Do You Need to Know Already? This Topic assumes that you are familiar with the discussion of work and heat in Topic 4A and how they contribute to changes in the internal energy of a system. You also need to know how to express the kinetic energy of a moving body (*Fundamentals* A).

In Topic 4A, the transfers of energy as work or heat are considered separately, each process being a way to change the internal energy of a system. However, in many processes, the internal energy changes as a result of both work and heat. For example, when a spark ignites the mixture of gasoline vapor and air in the engine of an automobile, the vapor burns and the gaseous product mixture expands, transferring energy to its surroundings as both work and heat.

4B.1 The First Law

In Topic 4A, it is established that when a system does only work, w, the change in internal energy (the total energy of the system) is $\Delta U = w$ and that when a system exchanges energy only as heat, q, then $\Delta U = q$. In general, the change in the internal energy of a closed system is the net result of both kinds of transfers, and

$$\Delta U = q + w \tag{1}$$

This expression summarizes the experimental fact that both heat and work are means of transferring energy and thereby changing the internal energy of a system. A central point is that *work and heat are equivalent* in the sense that they are both modes of energy transfer; provided q and w have the same numerical value (+15 kJ, for instance), there is no distinction in the change in internal energy they bring about. A system is like an energy bank, with its reserves measured as internal energy and deposits and withdrawals made in currency of either work or heat.

> **Self-test 4B.1A**   When a certain automobile engine does 520. kJ of work it also loses 220. kJ of energy as heat. What is the change in the internal energy of the engine treated as the system?
> [***Answer:*** −740. kJ]
>
> **Self-test 4B.1B**   A system was heated by using 300. J of heat, yet it was found that its internal energy decreased by 150. J (so $\Delta U = -150$. J). Calculate w. Was work done on the system, or did the system do work?

It is an experimental fact—a fact supported by thousands of experiments—that a system cannot do work, be left isolated for a while, and then found to have its internal energy restored to its original value and ready to provide the same amount of work again. Despite the great amount of effort that has been spent trying to build a "perpetual motion machine," a device that would be an exception to this rule by producing work without using fuel, no one has ever succeeded in building one. In other words, Eq. 1 is a *complete* statement of how changes in internal energy may be achieved in a closed system of constant composition: the only way to change the internal energy of such a system is to transfer energy into it as heat or as work. If the system is isolated, then even that ability is eliminated, and the internal energy cannot change at all. This conclusion is known as the **first law of thermodynamics,** which states:

The internal energy of an isolated system is constant.

The first law is closely related to the conservation of energy (*Fundamentals* A) but goes beyond it: the concept of heat does not apply to the single particles treated in classical mechanics.

Work and heat are equivalent ways of changing the internal energy of a system; the first law of thermodynamics states that the internal energy of an isolated system is constant.

4B.2 State Functions

According to the first law, if an isolated system has a certain internal energy at one instant and then is inspected again later, it will be found to have exactly the same internal energy. Even if the system goes through a series of changes, as long as it is restored to its original state it will be found to have the same internal energy as it had originally. These statements are summarized by saying that the internal energy is a **state function,** a property that depends only on the current state of the system and is independent of how that state was prepared. The pressure, volume, temperature, and density of a system are also state functions.

The importance of state functions in thermodynamics is that, *because a state function depends only on the current state of the system, any change in its value is independent of how the change in state was brought about.* A state function is like altitude on a mountain (**FIG. 4B.1**). Any number of different paths might lie between two huts on the mountain, but the change in altitude between the two huts is the same regardless of the path. Similarly, if you took 100 g of water at 25 °C and raised its temperature to 60 °C, its internal energy would change by a certain amount. However, if you took the same mass of water at 25 °C and heated it to boiling, vaporized it, condensed the vapor, and then allowed the water to cool to 60 °C, the net change in the internal energy of the water would be exactly the same as when you had heated it to 60 °C in one step.

The work done by a system is *not* a state function: it depends on how a change is being brought about. For example, you could let a gas at 25 °C expand at constant temperature (by keeping it in contact with a water bath) through 100 cm³ by two different paths. In the first experiment, the gas might push on a piston and do a certain amount of work against an external force. In the second experiment, the gas might push a piston into a vacuum; it then does no work, because there is no external opposing force (**FIG. 4B.2**). The change in the state of the gas is the same in each case, but the work done by the system is different: in the first case, w is nonzero; in the second case, $w = 0$. Indeed, even everyday language suggests that work is not a state function, because we never speak of a system as possessing a certain amount of "work."

Similarly, heat is not a state function. The energy transferred as heat during a change in the state of a system depends on how the change is brought about. For example, suppose you want to raise the temperature of 100 g of water from 25 °C to 30 °C. One way to raise the temperature would be to supply energy as heat by using an electric heater. The heat required can be calculated from the specific heat capacity of water: $q = \{4.18\ \mathrm{J\cdot(°C)^{-1}\cdot g^{-1}}\} \times (100\ \mathrm{g}) \times (5\ °C) = +2\ \mathrm{kJ}$. Another way to raise the temperature would be to stir the water vigorously with paddles until 2 kJ of work has been done. In the

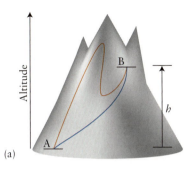

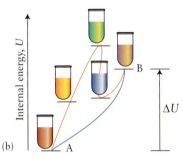

FIGURE 4B.1 (a) The altitude of a location on a mountain is like a thermodynamic state function: no matter what route is taken between points A and B, the net change in altitude is the same. (b) Internal energy is a state function: if a system changes from state A to state B (as depicted diagrammatically here), the net change in internal energy is the same whatever the route—the sequence of chemical or physical changes—between the two states.

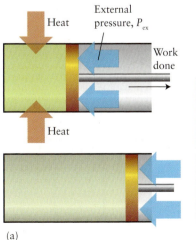

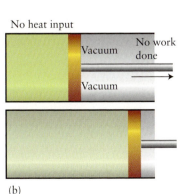

FIGURE 4B.2 Two different paths between the same initial and final states of an ideal gas. (a) The gas does work as it expands isothermally against an applied, matching pressure. Because energy flows in as heat to restore the energy lost as work, the temperature remains constant. (b) The gas does no work as it expands isothermally into a vacuum. Because internal energy is a state function, the change in internal energy is the same for both processes: $\Delta U = 0$ for the isothermal expansion of an ideal gas on any path. However, the exchange of heat and work is different in each case.

latter case, all the energy required is transferred as work; none is supplied as heat. So, in the first case, $q = +2$ kJ; and, in the second case, $q = 0$. However, the final state of the system is the same in each case. Because heat is not a state function, we should not speak of a system as possessing a certain amount of "heat" (but in everyday language we often do: science often refines language).

Because internal energy is a state function, *any convenient path* can be chosen between the given initial and final states of a system and ΔU can then be calculated for that path. The result will have the same value of ΔU as the actual path between the same two states, even if the actual path is so complicated that ΔU cannot be calculated directly. In some cases you can use insight about the behavior of molecules to identify the change in internal energy for a process without doing a calculation. For instance, when an ideal gas expands isothermally, its molecules continue to move at the same average speed, so their total kinetic energy remains the same. Because there are no forces between the molecules, their total potential energy also remains the same even though their average separation has increased. Because neither the total kinetic energy nor the total potential energy changes, the internal energy of the gas is unchanged too. That is,

$\Delta U = 0$ for the isothermal expansion (or compression) of an ideal gas.

It follows that when the volume of a sample of an ideal gas changes by *any* path between two states, then, provided the initial and final states have the same temperature, you know at once that $\Delta U = 0$.

EXAMPLE 4B.1   Calculating the work, heat, and change in internal energy accompanying the expansion of an ideal gas

Engineers designing new piston engines and turbines need to understand how work and heat are involved in various compression and expansion cycles. Suppose that 1.00 mol of ideal gas molecules at an initial pressure of 3.00 atm and 292 K expands against a constant external pressure of 0.20 atm from 8.00 L to 20.00 L by two different paths. (a) Path A is an isothermal, reversible expansion. (b) Path B, a hypothetical alternative to path A, has two steps. In step 1, the gas is cooled at constant volume until its pressure has fallen to 1.20 atm. In step 2, it is heated and allowed to expand against a constant pressure of 1.20 atm until its volume is 20.00 L and $T = 292$ K. Determine for each path the work done (w), the heat transferred (q), and the change in internal energy (ΔU).

ANTICIPATE You should expect the value of w to be less negative (less energy lost as work) in the irreversible path and therefore q to be less positive as less energy transferred as heat is needed to maintain the temperature.

PLAN It is a good idea to begin by sketching a plot of each process (**FIG. 4B.3**). (a) For an isothermal, reversible expansion, use Eq. 4 of Topic 4A ($w = -nRT \ln(V_2/V_1)$) to calculate w. (b) In step 1, the volume does not change, and so no work is done ($w = 0$). Step 2 is a constant-pressure process, so use Eq. 3 of Topic 4A ($w = -P_{ex}\Delta V$) to calculate w. Because internal energy is a state function and because the initial and final states are the same in both paths, ΔU for path B is the same as for path A. Because $\Delta U = 0$ for an isothermal expansion of an ideal gas, in each case, find q for the overall path from $\Delta U = q + w$ with $\Delta U = 0$. Use 1 L·atm = 101.325 J (Topic 4A) to convert liter-atmospheres into joules.

SOLVE

(a) From $w = -nRT \ln(V_2/V_1)$,

$$w = -(1.00 \text{ mol}) \times (8.3145 \text{ J·K}^{-1}\text{·mol}^{-1}) \times (292 \text{ K}) \times \ln \frac{20.00 \text{ L}}{8.0 \text{ L}}$$

$$= -2.22 \times 10^3 \text{ J} = -2.22 \text{ kJ}$$

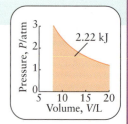

From $\Delta U = q + w = 0$,

$$q = -w = +2.22 \text{ kJ}$$

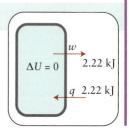

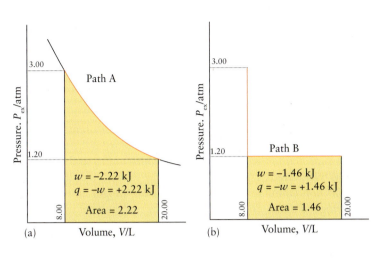

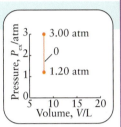

FIGURE 4B.3 (a) On the reversible path, the work done on the gas in Example 4B.1 is −2.22 kJ (negative, because the gas is doing the work); because the change in internal energy is zero, heat flows in to maintain constant temperature and constant internal energy. Therefore, $q = +2.22$ kJ. (b) On the irreversible path, the work done is also equal to the negative of the area beneath the curve and is less for this path ($w = −1.46$ kJ). The net heat input, allowing for outflow of heat in the cooling step and influx of heat in the expansion, is $q = +1.46$ kJ.

(b) **Step 1** Cool at constant volume, $\Delta V = 0$, so no work is done.

$$w = 0$$

Step 2 Heat and allow to expand. From $w = -P_{ex}\Delta V$,

$$w = -(1.20 \text{ atm}) \times (20.00 - 8.00)\text{L} = -14.4 \text{ L·atm}$$

Convert liter-atmospheres to joules.

$$w = -(14.4 \text{ L·atm}) \times \left(\frac{101.325 \text{ J}}{1 \text{ L·atm}}\right) = -1.46 \times 10^3 \text{ J} = -1.46 \text{ kJ}$$

Calculate total work for path B.

$$w = 0 + (-1.46 \text{ kJ}) = -1.46 \text{ kJ}$$

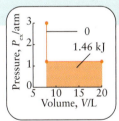

From $\Delta U = q + w = 0$,

$$q = -w = +1.46 \text{ kJ}$$

In summary,

	q	w	ΔU
For the reversible path:	+2.22 kJ	−2.22 kJ	0
For the irreversible path:	+1.46 kJ	−1.46 kJ	0

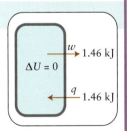

EVALUATE As expected, less work is done in the irreversible path and less energy has to enter the system as heat to maintain its temperature.

Self-test 4B.2A Suppose that 2.00 mol CO_2 at 2.00 atm and 300. K is compressed isothermally and reversibly to half its original volume before being used to produce soda water. Calculate w, q, and ΔU by treating the CO_2 as an ideal gas.
[**Answer:** $w = +3.46$ kJ, $q = −3.46$ kJ, $\Delta U = 0$]

Self-test 4B.2B Suppose that 1.00 kJ of energy is transferred as heat to oxygen in a cylinder fitted with a piston; the external pressure is 2.00 atm. The oxygen expands from 1.00 L to 3.00 L against this constant pressure. Calculate w and ΔU for the entire process by treating the O_2 as an ideal gas.

Related Exercises 4B.13, 4B.14

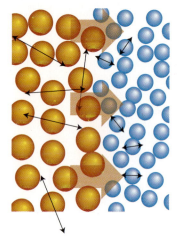

FIGURE 4B.4 On an atomic scale, the transfer of energy as heat can be pictured as a process in which the vigorous thermal motion of atoms in the system jostle the less vigorously moving atoms of the surroundings and transfer some of their energy to them. The double-headed arrows represent the motion of the atoms; the large pink arrows represent the direction of heat transfer.

A state function is a property that depends only on the current state of a system. The change in a state function between two states is independent of the path between them. Internal energy is a state function, but work and heat are not.

4B.3 A Molecular Interlude

Although heat and work are equivalent in the sense that the change in internal energy of a system is blind to the mode employed, there is a difference between them at a molecular level. That difference is related to the orderliness with which atoms in the surroundings move. When energy is transferred as work, the system moves molecules in the surroundings in a definite direction (think of the atoms of a rising weight all moving upward simultaneously). During the transfer of energy as heat, the molecules of the surroundings are moved in random directions (think of the atoms of a hot object jostling the atoms of the surroundings into more vigorous random motion, **FIG. 4B.4**). That is:

- Work makes use of orderly motion of atoms in the surroundings.
- Heat makes use of disorderly motion of atoms in the surroundings.

A Note on Good Practice: Be careful not to confuse the terms "heat" and "thermal energy." Heat is energy in transit as a result of a temperature difference. Thermal energy—or, better, the energy of thermal motion—is the energy associated with the chaotic motion of molecules at temperatures above absolute zero.

Internal energy is energy stored in a system as kinetic energy and potential energy. It includes all the energies of interaction of the fundamental particles that make up atoms and the energy stored as motion. Molecules in a gas can move in a variety of different ways, and each mode of motion can act as a store of energy (**FIG. 4B.5**):

- **Translational energy** is the energy of an atom or molecule due to its motion through space (this energy is kinetic energy).
- **Rotational energy** is the energy due to the rotational motion of a molecule (this energy is also kinetic; individual atoms do not rotate).
- **Vibrational energy** is the energy stored by a molecule as the oscillation of its atoms relative to one another; this contribution is the sum of kinetic and potential contributions.

Most molecules are not vibrationally excited at room temperature, and so that mode can be ignored for now. Kinetic energy is energy due to motion (*Fundamentals* A) and the faster a molecule travels or rotates, the greater is its kinetic energy. When the temperature of a gas is raised, either by doing work on a gas in an adiabatic container or by heating a gas, the average speed of translation and rotation of the molecules increases. That increase corresponds to an increase in the total kinetic energy of the molecules and therefore to an increase in the internal energy of the gas. It is always the case that, provided no chemical reaction occurs, *a system at high temperature has a greater internal energy than the same system at a lower temperature.* These remarks can be expressed quantitatively by using the **equipartition theorem:**

> The average value of each quadratic contribution to the energy of a molecule in a sample at a temperature T is equal to $\frac{1}{2}kT$.

A "quadratic contribution" to the energy is one that depends on the square of a velocity or a displacement, as in $\frac{1}{2}mv^2$ for translational kinetic energy. In this expression, k is Boltzmann's constant, a fundamental constant with the value 1.381×10^{-23} J·K^{-1}. Boltzmann's constant is related to the gas constant by $R = N_A k$, where N_A is Avogadro's constant. The name "equipartition" simply means that the available energy is shared (partitioned) equally over all the available modes. For example, the energy for translational kinetic energy is along a single direction. The total translational kinetic energy is the sum

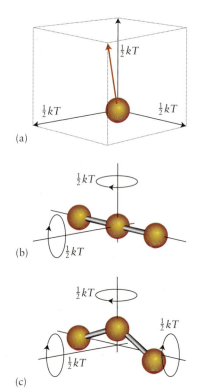

FIGURE 4B.5 The translational and rotational modes of atoms and molecules and the corresponding average energies of each mode at a temperature T. (a) An atom or molecule can undergo translational motion in three dimensions. (b) A linear molecule can also rotate about two axes perpendicular to the line of atoms. (c) A nonlinear molecule can rotate about three perpendicular axes.

of motions in the x, y, and z directions and each contributes $\frac{1}{2}kT$ to the internal energy of a molecule. It follows that the average translational kinetic energy of a molecule in a sample at a temperature T is $3 \times \frac{1}{2}kT$. The contribution to the molar internal energy (the energy per mole of molecules, U_m) is therefore

$$U_m(\text{translation}) = \frac{3}{2}\overbrace{N_A k}^{R}T = \frac{3}{2}RT$$

which evaluates to 3.72 kJ·mol^{-1} at 25 °C. A linear molecule, such as carbon dioxide, can rotate about two axes perpendicular to the line of atoms, and so it has two rotational modes of motion. Its average rotational energy is therefore $2 \times \frac{1}{2}kT = kT$, and the contribution to the molar internal energy is N_A times this value:

$$U_m(\text{rotation, linear}) = \overbrace{N_A k}^{R}T = RT$$

or 2.48 kJ·mol^{-1} at 25 °C. A nonlinear molecule can rotate around three mutually perpendicular axes, so the rotational contribution to the molar internal energy is

$$U_m(\text{rotation, nonlinear}) = \frac{3}{2}N_A kT = \frac{3}{2}RT$$

These contributions are summarized in Fig. 4B.5.

Self-test 4B.3A Estimate the contribution of motion to the molar internal energy of water vapor at 25 °C.

[**Answer:** 7.44 kJ·mol^{-1}]

Self-test 4B.3B Estimate the contribution of motion to the molar internal energy of benzene vapor at 100 °C.

The equipartition theorem makes it easy to estimate the change in internal energy when the temperature of a sample of monatomic ideal gas is changed. Thus, because the contribution to the molar internal energy of a monatomic ideal gas (such as argon) that arises from its motion is $\frac{3}{2}RT$, you can conclude that if the gas is heated through ΔT, then the change in its molar internal energy, ΔU_m, is $\Delta U_m = \frac{3}{2}R\Delta T$. For instance, if the gas is heated from 20. °C to 100. °C (so $\Delta T = +80.$ K), then its molar internal energy increases by $\frac{3}{2} \times (8.3145 \text{ J·K}^{-1}\text{·mol}^{-1}) \times (80.\text{ K}) = 1.0$ kJ·mol^{-1}.

Heat makes use of disorderly molecular motion in the surroundings; work makes use of orderly motion. The equipartition theorem can be used to estimate the translational and rotational contributions to the internal energy of an ideal gas. The vibrational contribution to the energy is insignificant at ordinary temperatures.

What have you learned in this Topic?

You have learned that the first law of thermodynamics states that the internal energy of an isolated system is constant. You have met the concept of a state function, which enables you to calculate the change in internal energy by selecting any convenient route between initial and final states. You now understand the distinction between heat and work at a molecular level and can use the equipartition theorem to estimate contributions to the internal energy.

The skills you have mastered are the ability to:

☐ **1.** State, and explain the implications of, the first law of thermodynamics (Sections 4B.1 and 4B.2).

☐ **2.** Recognize which functions of a system are state functions (Section 4B.2).

☐ **3.** Calculate the change in internal energy due to heat and work (Self-Test 4B.1).

☐ **4.** Calculate the change in internal energy along different paths (Example 4B.1).

☐ **5.** Use the equipartition theorem to estimate contributions to the internal energy of a gas Self-Test 4B.3).

Topic 4B Exercises

4B.1 A gas sample in a cylinder is supplied with 524 kJ of energy as heat. At the same time, a piston compresses the gas, doing 340 kJ of work. What is the change in internal energy of the gas during this process?

4B.2 A gas sample in a piston assembly expands, doing 171 kJ of work on its surroundings at the same time that 242 kJ of energy is supplied to the gas as heat. (a) What is the change in internal energy of the gas during this process? (b) Will the pressure of the gas be higher or lower when these changes are completed?

4B.3 The internal energy of a system increased by 982 J when it was supplied with 492 J of energy as heat. (a) Was work done by or on the system? (b) How much work was done?

4B.4 (a) Calculate the value of w for a system that absorbs 164 kJ of heat in a process for which the change in internal energy is $+152$ kJ. (b) Is work done on or by the system during this process?

4B.5 An ideal gas in a cylinder was placed in a heater and gained 5.50 kJ of energy as heat. If the cylinder increased in volume from 345 mL to 1846 mL against an atmospheric pressure of 750. Torr during this process, what is the change in internal energy of the gas in the cylinder?

4B.6 An electric heater rated at 100. W (1 W = 1 J·s^{-1}) operates for 20.0 min to heat an ideal gas in a cylinder. At the same time, the gas expands from 1.00 L to 6.00 L against a constant atmospheric pressure of 0.876 atm. What is the change in internal energy of the gas?

4B.7 In a combustion chamber, the total internal energy change produced from the burning of a fuel is -2573 kJ. The cooling system that surrounds the chamber absorbs 947 kJ as heat. How much work can be done by the fuel in the chamber?

4B.8 A laboratory animal exercised on a treadmill that was connected to a weight of mass 275 g through a pulley. The work done by the animal raised the weight through 1.01 m. At the same time, the animal lost 7.8 J of energy as heat. (a) Disregarding any other losses and treating the animal as a closed system, what was the change in internal energy of the animal? (b) What would the change be if the experiment took place in a space station with zero gravity rather than on Earth?

4B.9 In an adiabatic process, no energy is transferred as heat. Indicate whether each of the following statements about an adiabatic process in a closed system is always true, always false, or true in certain conditions (specify the conditions): (a) $\Delta U = 0$; (b) $q = 0$; (c) $q < 0$; (d) $\Delta U = q$; (e) $\Delta U = w$.

4B.10 Indicate whether each of the following statements about a process in a closed system with a constant volume is always true, always false, or true in certain conditions (specify the conditions): (a) $\Delta U = 0$; (b) $w = 0$; (c) $w < 0$; (d) $\Delta U = q$; (e) $\Delta U = w$.

4B.11 Each of the pictures below shows a molecular view of a system undergoing a change at constant temperature. In each case, indicate whether heat is absorbed or released by the system and whether expansion work is done on or by the system. Predict the signs of q and w for the process.

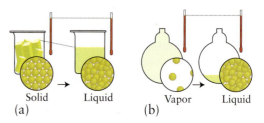

Solid   Liquid   Vapor   Liquid
(a)   (b)

4B.12 Each of the pictures below shows a molecular view of a system undergoing a change. In each case, indicate whether heat is absorbed or given off by the system and whether expansion work is done on or by the system. Predict the signs of q and w for the process.

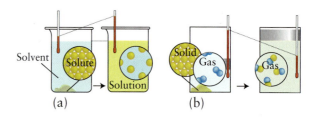

(a)   (b)

4B.13 Calculate the work for each of the following processes beginning with a gas sample in a piston assembly with $T = 305$ K, $P = 1.79$ atm, and $V = 4.29$ L: (a) irreversible expansion against a constant external pressure of 1.00 atm to a final volume of 6.52 L; (b) isothermal, reversible expansion to a final volume of 6.52 L.

4B.14 A sample of gas in a cylinder of volume 3.42 L at 298 K and 2.57 atm expands to 7.39 L by two different pathways. Path A is an isothermal, reversible expansion. Path B has two steps. In the first step, the gas is cooled at constant volume to 1.19 atm. In the second step, the gas is heated and allowed to expand against a constant external pressure of 1.19 atm until the final volume is 7.39 L. Calculate the work for each path.

4B.15 Estimate the contribution of motion to the molar internal energy of $CH_4(g)$ at 25 °C. Ignore vibrations.

4B.16 Estimate the contribution of motion to the molar internal energy of $N_2(g)$ at 25 °C. Ignore vibrations.

Topic 4C Enthalpy

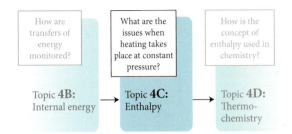

Why Do You Need to Know This Material? Enthalpy is used throughout chemistry when discussing changes at constant pressure, and it is the foundation of thermochemistry.

What Do You Need to Know Already? This Topic assumes that you are familiar with internal energy and heat capacity (Topics 4A and 4B). It supposes that you are familiar with the equipartition theorem (Topic 4B).

If a system has rigid walls, its volume remains the same even though other changes may take place, so it does no expansion work. If you also know that no nonexpansion work (such as electrical work) is done, then the change in the internal energy of the system is equal to the energy supplied to it as heat ($\Delta U = q_V$, where the V signifies a process at constant volume). However, in chemistry most chemical reactions take place in containers open to the atmosphere and therefore take place at a constant pressure of about 1 atm. Such systems are free to expand or contract. If a gas is evolved, the expanding gas needs to drive back the surrounding atmosphere to make room for itself. Work is done, even though there is no actual piston. It would be very useful if there were a state function that keeps track of energy changes at constant pressure by taking into account automatically losses of energy as expansion work during heat transfer.

4C.1 Heat Transfers at Constant Pressure

The state function that keeps track of losses of energy as expansion work during heat transfer at constant pressure (and the gain of energy if the process is compression) is called the **enthalpy,** H. Enthalpy is defined as

$$H = U + PV \tag{1}$$

where U, P, and V are the internal energy, pressure, and volume of the system. You can see that enthalpy is a state function by noting that U (from the first law), P, and V are all state functions, and so $H = U + PV$ must be a state function too. It can be shown that it follows from this definition that *a change in the enthalpy of a system is equal to the heat released or absorbed at constant pressure.*

How Is That Done?

Consider a process at constant pressure for which the change in internal energy is ΔU and the change in volume is ΔV. It then follows from the definition of enthalpy in Eq. 1 that the change in enthalpy is

$$\text{At constant pressure: } \Delta H = \Delta U + P\Delta V \tag{2}$$

Now write $\Delta U = q + w$, where q is the energy supplied to the system as heat and w is the energy supplied as work:

$$\Delta H = q + w + P\Delta V$$

Next, suppose that the system can do no work other than expansion work. In that case, Eq. 3 of Topic 4A is used to calculate the work ($w = -P_{ex}\Delta V$), and so

$$\Delta H = q - P_{ex}\Delta V + P\Delta V$$

Finally, because the system is open to the atmosphere or in a vessel that can adjust its size, the pressure of the system is the same as the external pressure; so $P_{ex} = P$, and the last two terms cancel to leave $\Delta H = q$.

You have seen that for a system that can do work only by expansion against a constant pressure,

$$\text{At constant pressure, and with no nonexpansion work: } \Delta H = q \tag{3}$$

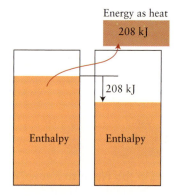

FIGURE 4C.1 The enthalpy of a system is like a measure of the height of water in a reservoir. When an exothermic reaction releases 208 kJ of heat at constant pressure, the "reservoir" falls by 208 kJ and $\Delta H = -208$ kJ.

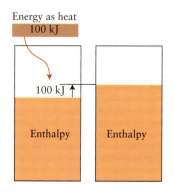

FIGURE 4C.2 If an endothermic reaction absorbs 100 kJ of heat at constant pressure, the height of the enthalpy "reservoir" rises by 100 kJ and $\Delta H = +100$ kJ.

This equation is often written $\Delta H = q_P$, where the subscript denotes constant pressure. Its implication is that, because chemical reactions typically take place at constant pressure in vessels open to the atmosphere, the heat that they release or require can be equated to the change in enthalpy of the system. It follows that if a reaction is studied in a calorimeter that is open to the atmosphere, then the observed temperature rise can be used as a measure of the enthalpy change that accompanies the reaction. For instance, if a reaction releases 1.25 kJ of heat in this kind of calorimeter, then $\Delta H = q_P = -1.25$ kJ.

When energy is transferred to a constant-pressure system as heat, the enthalpy of the system *increases* by that amount. When energy leaves a constant-pressure system as heat, the enthalpy of the system *decreases* by that amount. For example, the formation of zinc iodide from its elements, $Zn(s) + I_2(s) \rightarrow ZnI_2(s)$, is an exothermic reaction that (at constant pressure) releases 208 kJ of heat to the surroundings for each mole of ZnI_2 formed. Therefore, $\Delta H = -208$ kJ, because the enthalpy of the reaction mixture decreases by 208 kJ in this reaction (**FIG. 4C.1**). An endothermic process absorbs heat, and so when ammonium nitrate dissolves in water, the enthalpy of the system increases (**FIG. 4C.2**). That is, at constant pressure:

Exothermic reactions: $\Delta H < 0$
Endothermic reactions: $\Delta H > 0$

> **Self-test 4C.1A** In an exothermic reaction at constant pressure, 50. kJ of energy left the system as heat and 20. kJ of energy left the system as expansion work. What are the values of (a) ΔH and (b) ΔU for this process?
>
> [***Answer:*** (a) $-50.$ kJ; (b) $-70.$ kJ]
>
> **Self-test 4C.1B** In an endothermic reaction at constant pressure, 30. kJ of energy entered the system as heat. The products took up less volume than the reactants, and 40. kJ of energy entered the system as work as the outside atmosphere pressed down on it. What are the values of (a) ΔH and (b) ΔU for this process?

Enthalpy is a state function. The change in enthalpy of a system is equal to the heat supplied to the system at constant pressure. For an endothermic process, $\Delta H > 0$; for an exothermic process, $\Delta H < 0$.

4C.2 Heat Capacities at Constant Volume and Constant Pressure

As explained in Topic 4A, the heat capacity C is the constant of proportionality between the heat supplied to a system and the temperature rise that results ($q = C\Delta T$). However, the rise in temperature and therefore the heat capacity depend on the conditions under which the heating takes place because, at constant pressure, some of the heat is used to do expansion work as well as to raise the temperature of the system. The definition of heat capacity needs to be more precise.

Provided no nonexpansion work is done and no other changes take place, heat transferred at constant volume can be identified with the change in internal energy, $\Delta U = q$. This equality can be combined with $C = q/\Delta T$ to give the **heat capacity at constant volume,** C_V, as

$$C_V = \frac{\Delta U}{\Delta T} \tag{4a}$$

Similarly, because heat transferred at constant pressure can be identified with the change in enthalpy, ΔH, the **heat capacity at constant pressure,** C_P, is

$$C_P = \frac{\Delta H}{\Delta T} \tag{4b}$$

The corresponding molar heat capacities are these quantities divided by the amount of substance and are denoted $C_{V,m}$ and $C_{P,m}$.

THINKING POINT

Which would you expect to be larger for a given substance, $C_{V,m}$ or $C_{P,m}$?

The constant-volume and constant-pressure heat capacities of a solid substance are similar; the same is true of a liquid but not of a gas. The difference is a reflection of the fact that gases expand much more than solids and liquids when they are heated, so more energy is lost as work when a gas is heated at constant pressure than when a solid or a liquid is heated. It is possible to set up a simple quantitative relation between C_P and C_V for an ideal gas.

How Is That Done?

For an ideal gas, PV in the definition of enthalpy, $H = U + PV$, can be replaced by nRT, and so $H = U + nRT$. When a sample of an ideal gas is heated, the enthalpy, internal energy, and temperature all change and it follows that

$$\Delta H = \Delta U + nR\Delta T$$

The heat capacity at constant pressure can therefore be expressed as

$$C_P = \frac{\Delta H}{\Delta T} = \frac{\Delta U + nR\Delta T}{\Delta T} = \frac{\Delta U}{\Delta T} + nR = C_V + nR$$

To obtain the relation between the two molar heat capacities, divide this expression by n.

The calculation shows that the two molar heat capacities of an ideal gas are related by

$$C_{P,m} = C_{V,m} + R \tag{5}$$

As an example, the molar constant-volume heat capacity of argon is $12.8 \ \mathrm{J \cdot K^{-1} \cdot mol^{-1}}$, and so the corresponding molar constant-pressure value is $12.8 + 8.3 \ \mathrm{J \cdot K^{-1} \cdot mol^{-1}} = 21.1 \ \mathrm{J \cdot K^{-1} \cdot mol^{-1}}$, a difference of 65%. The heat capacity at constant pressure is greater than that at constant volume because at constant pressure not all the heat supplied is used to raise the temperature: some returns to the surroundings as expansion work and $C = q/\Delta T$ is larger (because ΔT is smaller) than at constant volume (when all the energy is used to increase the temperature of the system).

> *The molar heat capacity of an ideal gas at constant pressure is greater than that at constant volume; the two quantities are related by Eq. 5.*

4C.3 The Molecular Origin of the Heat Capacities of Gases

How heat capacities depend on molecular properties can be explored by using the equipartition theorem (Topic 4B). That theorem implies that the molar internal energy of a monatomic ideal gas at a temperature T is $U_m = \frac{3}{2}RT$ and that the change in molar internal energy when the temperature is changed by ΔT is $\Delta U_m = \frac{3}{2}R\Delta T$. It follows that the molar heat capacity at constant volume is

$$\text{For a monatomic gas: } C_{V,m} = \frac{\Delta U_m}{\Delta T} = \frac{\frac{3}{2}R\Delta T}{\Delta T} = \frac{3}{2}R$$

or about $12.5 \ \mathrm{J \cdot K^{-1} \cdot mol^{-1}}$, in agreement with the value found experimentally. From Eq. 5 ($C_{P,m} = C_{V,m} + R$), the molar heat capacity of an ideal gas at constant pressure is

$$\text{For a monatomic gas: } C_{P,m} = \frac{3}{2}R + R = \frac{5}{2}R$$

Note that for a monatomic ideal gas, both C_P and C_V are independent of temperature and pressure.

The molar heat capacities of gases composed of molecules (as distinct from atoms) are higher than those of monatomic gases because the molecules can store energy as rotational kinetic energy as well as translational kinetic energy. The rotational motion of linear molecules contributes another RT to the molar internal energy (Topic 4B). For a given change in temperature,

$$\text{For a linear molecule: } C_{V,m} = \frac{\Delta U_m}{\Delta T} = \frac{\frac{3}{2}R\Delta T + R\Delta T}{\Delta T} = \frac{5}{2}R$$

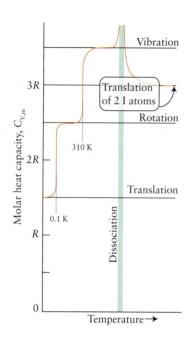

FIGURE 4C.3 Variation in the molar heat capacity of iodine vapor at constant volume. The translational motion of gaseous molecules contributes to the heat capacity even at very low temperatures. As the temperature rises past 0.05 K, rotation makes a significant contribution, but vibrations of the molecule contribute only at high temperatures (above about 310 K; for most other molecules, the temperature must be much higher). When the molecules dissociate, the heat capacity becomes very large during dissociation, but then settles down to a value characteristic of 2 mol I atoms undergoing only translational motion.

For nonlinear molecules, the rotational contribution is $\frac{3}{2}R\Delta T$ for a total (rotational + translational) contribution of $3RT$. In summary, for ideal gases:

	Atoms	Linear molecules	Nonlinear molecules
$C_{V,m}$	$\frac{3}{2}R$	$\frac{5}{2}R$	$3R$
$C_{P,m}$	$\frac{5}{2}R$	$\frac{7}{2}R$	$4R$

(In each case, $C_{P,m}$ has been calculated from $C_{P,m} = C_{V,m} + R$.) Note that the molar heat capacity increases with molecular complexity. The molar heat capacity of nonlinear molecules is higher than that of linear molecules because nonlinear molecules can rotate about three rather than only two axes.

The graph in **FIG. 4C.3** shows how $C_{V,m}$ for iodine vapor, $I_2(g)$, varies with temperature. At very low temperatures, $C_{V,m} = \frac{3}{2}R$ because the molecules move about without rotating, but it rises to $\frac{5}{2}R$ as molecular rotation takes place. At still higher temperatures, molecular vibrations start to absorb energy and the heat capacity rises toward $\frac{7}{2}R$. At 298 K, the experimental value is equivalent to $3.4R$.

EXAMPLE 4C.1   Calculating the enthalpy changes when heating an ideal gas

When balloons carrying people first took to the skies in the eighteenth century, they stimulated a lot of interest in the properties of gases. Calculate the final temperature and the change in enthalpy when 500. J of energy is transferred as heat to 0.900 mol $O_2(g)$ at 298 K and 1.00 atm at (a) constant pressure; (b) constant volume. Treat the gas as ideal.

ANTICIPATE You should expect the temperature to rise more as a result of heating at constant volume than at constant pressure, because at constant pressure some of the energy is used to do expansion work. That, in turn, suggests that the increase in enthalpy might be greater at constant volume than at constant pressure.

PLAN Oxygen is a linear molecule, and its heat capacities can be estimated from the equipartition theorem; then, use $q = C\Delta T$, with $C = nC_{V,m}$ or $nC_{P,m}$ for the changes at constant volume and constant pressure, respectively, to find the changes in temperature. The enthalpy change at constant pressure is equal to the heat supplied. At constant volume, find the enthalpy change by calculating ΔU and then converting it to ΔH by using $\Delta H = \Delta U + nR\Delta T$.

What should you assume? Assume that oxygen behaves as an ideal gas and that there is no vibrational contribution to the heat capacity.

SOLVE

From $C_{V,m} = \frac{5}{2}R$ and $C_{P,m} = C_{V,m} + R$,

$$C_{V,m} = \tfrac{5}{2}(8.3145\ \text{J·K}^{-1}\text{·mol}^{-1}) = 20.79\ \text{J·K}^{-1}\text{·mol}^{-1}$$

$$C_{P,m} = \tfrac{7}{2}(8.3145\ \text{J·K}^{-1}\text{·mol}^{-1}) = 29.10\ \text{J·K}^{-1}\text{·mol}^{-1}$$

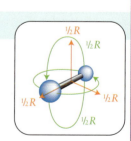

(a) From $\Delta T = q/nC_{P,m}$,

$$\Delta T = \frac{500.\ \text{J}}{(0.900\ \text{mol}) \times (29.10\ \text{J·K}^{-1}\text{·mol}^{-1})} = +19.1\ \text{K}$$

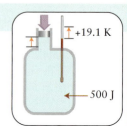

Find the final temperature.

$$T = 298 + 19.1 \text{ K} = 317 \text{ K, or } 44 \,°\text{C}$$

From $\Delta H = q_P$,

$$\Delta H = +500. \text{ J}$$

(b) To find ΔH for heating at constant volume, find the final temperature that is achieved at constant volume:

From $\Delta T = q/nC_{V,m}$,

$$\Delta T = \frac{500. \text{ J}}{(0.900 \text{ mol}) \times (20.79 \text{ J·K}^{-1}\text{·mol}^{-1})} = +26.7 \text{ K}$$

The final temperature is

$$T = 298 + 26.7 \text{ K} = 325 \text{ K, or } 52 \,°\text{C}$$

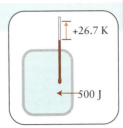

When 500. J is transferred at constant volume,

$$\Delta U = q = +500. \text{ J}$$

Now convert the change to a change in enthalpy by using $\Delta H = \Delta U + nR\Delta T$:

$$\Delta H = 500. \text{ J} + (0.900 \text{ mol}) \times (8.3145 \text{ J·K}^{-1}\text{·mol}^{-1}) \times (26.7 \text{ K}) = +700. \text{ J}$$

EVALUATE The increase in temperature and the change in enthalpy are smaller at constant pressure than at constant volume, as anticipated.

Self-test 4C.2A Calculate the final temperature and the change in enthalpy when 500. J of energy is transferred as heat to 0.900 mol Ne(g) at 298 K and 1.00 atm (a) at constant pressure; (b) at constant volume. Treat the gas as ideal.

[*Answer:* (a) 325 K, 500. J; (b) 343 K, 837 J]

Self-test 4C.2B Calculate the final temperature and the change in internal energy when 1.20 kJ of energy is transferred as heat to 1.00 mol H_2(g) at 298 K and 1.00 atm (a) at constant volume; (b) at constant pressure. Treat the gas as ideal.

Related Exercises 4C.3, 4C.4

Rotation requires energy and leads to higher heat capacities for complex molecules; the equipartition theorem can be used to estimate the molar heat capacities of gas-phase molecules.

4C.4 The Enthalpy of Physical Change

The molecules in a solid or liquid are held together by intermolecular attractions. A phase change in which the attractions between molecules are reduced, such as melting or vaporization, requires energy and is therefore endothermic. Phase changes that increase molecular contact, such as condensation or freezing, are exothermic because energy is released when molecules come closer together and can interact with one another more strongly. When a phase transition takes place at constant pressure, as is most common, the heat transfer accompanying the phase change is equal to the change in enthalpy of the substance.

At a given temperature, the vapor phase of a substance has a higher energy and therefore a higher enthalpy than the liquid phase. The difference in molar enthalpy between the vapor and the liquid states is called the **enthalpy of vaporization, ΔH_{vap}**:

$$\Delta H_{vap} = H_m(\text{vapor}) - H_m(\text{liquid}) \tag{6}$$

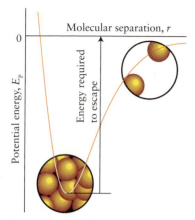

FIGURE 4C.4 The potential energies of molecules decrease as they approach one another and experience intermolecular attractions, and then rise again as the molecules are pressed closely together. The average intermolecular distance in a liquid is given by the position of the energy minimum. To escape from the liquid, the energy of a molecule must be raised from the bottom of the well to at least the energy of the horizontal part of the curve on the right.

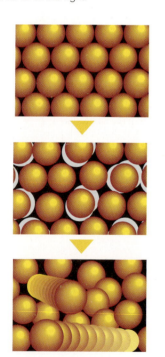

FIGURE 4C.5 Melting (fusion) is an endothermic process. Molecules in a solid sample vibrate about ordered positions (top) but, as molecules acquire more energy, they begin to be able to struggle past their neighbors (center). Finally, the solid changes completely into a liquid with disordered, mobile molecules (bottom).

TABLE 4C.1   Standard Enthalpies of Physical Change*

Substance	Formula	Freezing point, T_f/K	$\Delta H_{fus}°$/ (kJ·mol^{-1})	Boiling point, T_b/K	$\Delta H_{vap}°$/ (kJ·mol^{-1})
acetone	CH_3COCH_3	177.8	5.72	329.4	29.1
ammonia	NH_3	195.4	5.65	239.7	23.4
argon	Ar	83.8	1.2	87.3	6.5
benzene	C_6H_6	278.6	10.59	353.2	30.8
ethanol	C_2H_5OH	158.7	4.60	351.5	43.5
helium	He	3.5	0.021	4.22	0.084
mercury	Hg	234.3	2.292	629.7	59.3
methane	CH_4	90.7	0.94	111.7	8.2
methanol	CH_3OH	175.2	3.16	337.8	35.3
water	H_2O	273.2	6.01	373.2	40.7
					(44.0 at 25 °C)

*Values correspond to the temperature of the phase change. The superscript ° signifies that the change takes place at 1 bar and that the substance is pure (that is, the values are for standard states; see Topic 4D).

The enthalpy of vaporization of most substances changes little with temperature. For water at its boiling point, 100 °C, $\Delta H_{vap} = 40.7$ kJ·mol^{-1}; at 25 °C, $\Delta H_{vap} = 44.0$ kJ·mol^{-1}. The latter value means that to vaporize 1.00 mol H_2O(l), corresponding to 18.02 g of water, at 25 °C and constant pressure 44.0 kJ of energy must be supplied as heat.

Self-test 4C.3A A sample of benzene, C_6H_6, was brought to 80 °C, its normal boiling point. The heating was continued until an additional 15.4 kJ had been supplied, resulting in 39.1 g of boiling benzene becoming vaporized. What is the enthalpy of vaporization of benzene at its boiling point?

[*Answer:* 30.8 kJ·mol^{-1}]

Self-test 4C.3B A sample of ethanol, C_2H_5OH, of mass 23 g, was heated to its boiling point. It was found that an additional 22 kJ was required to vaporize all the ethanol. What is the enthalpy of vaporization of ethanol at its boiling point?

TABLE 4C.1 lists the enthalpies of vaporization of various substances. All enthalpies of vaporization are positive, and so it is conventional to report them without their sign. Notice that compounds with strong intermolecular forces, such as hydrogen bonds, tend to have the highest enthalpies of vaporization. That correlation is easy to explain, because the enthalpy of vaporization is a measure of the energy needed to separate molecules from their attractions in the liquid state into a free state in the vapor. In plots of the potential energy arising from intermolecular interactions, like that shown in **FIG. 4C.4**, the enthalpy of the substance in the liquid state, where molecular interactions are strong, is related to the depth of the bottom of the well in the curve. The enthalpy of the vapor state, in which the interactions are almost insignificant, corresponds to the horizontal continuation of the curve to the right. A substance with a high molar enthalpy of vaporization has a deep intermolecular potential well, indicating strong intermolecular attractions.

The molar enthalpy change that accompanies melting (more formally, fusion) is called the **enthalpy of fusion**, ΔH_{fus}, of the substance:

$$\Delta H_{fus} = H_m(\text{liquid}) - H_m(\text{solid}) \tag{7}$$

Melting, with only one known exception (helium), is endothermic, and so all enthalpies of fusion (with the exception of that special case) are positive and are reported without their sign (see Table 4C.1). The enthalpy of fusion of water at 0 °C is 6.01 kJ·mol^{-1}: to melt 1.0 mol H_2O(s) (18 g of ice) at 0 °C, 6.01 kJ of energy must be supplied as heat. Vaporizing the same amount of water takes much more energy (more than 40 kJ). When

FIGURE 4C.6 The enthalpy change for a reverse process has the same value, but the opposite sign, as the enthalpy change for the forward process at the same temperature.

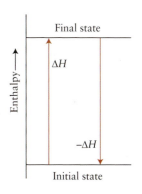

a solid melts, the molecules can move more freely, but because they stay nearly as close together in the liquid as in the solid, the interactions between them remain nearly as strong (**FIG. 4C.5**).

The **enthalpy of freezing** is the change in molar enthalpy of a liquid when it solidifies. Because enthalpy is a state property, the enthalpy of a sample of water must be the same after being frozen and then melted as it was before it was frozen. Therefore, the enthalpy of freezing of a substance is the negative of its enthalpy of fusion. For water at $0\,°C$, the enthalpy of freezing is $-6.01\ kJ\cdot mol^{-1}$, because 6.01 kJ of heat is *released* when 1 mol $H_2O(l)$ freezes. In general, to obtain the enthalpy change for the reverse of any process, just take the negative of the enthalpy change for the original process:

$$\Delta H_{\text{reverse pocess}} = -\Delta H_{\text{forward process}} \qquad (8)$$

This relation is illustrated in **FIG. 4C.6**.

Sublimation is the direct conversion of a solid into its vapor. Frost disappears on a cold, dry morning as the ice sublimes directly into water vapor. Solid carbon dioxide also sublimes, which is why it is called "dry ice." Each winter on Mars, solid carbon dioxide is deposited as polar frost, which sublimes when the feeble summer arrives (**FIG. 4C.7**). The **enthalpy of sublimation,** ΔH_{sub}, is the molar enthalpy change when a solid sublimes:

$$\Delta H_{\text{sub}} = H_m(\text{vapor}) - H_m(\text{solid}) \qquad (9)$$

Because enthalpy is a state function, the enthalpy of sublimation of a substance is the same whether the transition takes place in one step, directly from solid to gas, or we imagine it as taking place in two steps, first from solid to liquid and then from liquid to gas. The enthalpy of sublimation of a substance must therefore be equal to the sum of the enthalpies of fusion and vaporization, provided that they are measured at the same temperature (**FIG. 4C.8**):

$$\Delta H_{\text{sub}} = \Delta H_{\text{fus}} + \Delta H_{\text{vap}} \qquad (10)$$

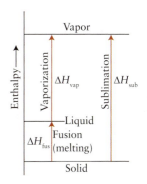

FIGURE 4C.7 The polar ice caps on Mars extend and recede with the seasons. They are mostly solid carbon dioxide and form by direct conversion of the gas into a solid. They disappear by sublimation. Although some water ice is also present in the polar caps, the temperature on Mars never becomes high enough to melt it. On Mars, ice is just another rock. (*NASA/JPL–Caltech/ MSSS.*)

Self-test 4C.4A The enthalpy of fusion of sodium metal is $2.6\ kJ\cdot mol^{-1}$ at $25\,°C$, and the enthalpy of sublimation of solid sodium at that temperature is $101\ kJ\cdot mol^{-1}$. What is the enthalpy of vaporization of sodium at $25\,°C$?

[*Answer:* $98\ kJ\cdot mol^{-1}$]

Self-test 4C.4B The enthalpy of vaporization of methanol is $38\ kJ\cdot mol^{-1}$ at $25\,°C$, and the enthalpy of fusion is $3\ kJ\cdot mol^{-1}$ at the same temperature. What is the enthalpy of sublimation of methanol at that temperature?

The enthalpy change for a reverse reaction is the negative of the enthalpy change for the forward reaction. Enthalpy changes can be added to obtain the value for an overall process.

4C.5 Heating Curves

The enthalpies of fusion and vaporization affect the appearance of the *heating curve* of a substance. A **heating curve** is a graph showing the variation in the temperature of a sample as it is heated at a constant rate at constant pressure and therefore at a constant rate of increase in enthalpy (**BOX 4C.1**).

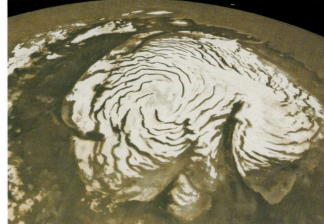

FIGURE 4C.8 Because enthalpy is a state property, the enthalpy of sublimation can be expressed as the sum of the enthalpies of fusion and vaporization measured at the same temperature.

Box 4C.1 HOW DO WE KNOW . . . THE SHAPE OF A HEATING CURVE?

One of two related techniques is normally used to generate a heating curve. In *differential thermal analysis* (DTA), equal masses of a sample and a reference material that will not undergo any phase changes, such as Al_2O_3 (which melts at a very high temperature), are inserted into two separate sample wells in a large steel block that acts as a heat sink (see the first illustration). Because the steel block is so large, it is possible to heat the sample and reference at the same slow and accurately monitored rate. Thermocouples are placed in each well and in the block itself to monitor temperature changes. Then the block is gradually heated and the temperatures of the sample and the reference are compared. An electrical signal is generated if the temperature of the sample stops rising while that of the reference continues to rise. Such an event signals an endothermic process in the sample, such as a phase change. The output of a DTA analysis is a *thermogram,* which shows the temperatures of phase changes as heat absorption peaks at the transition temperatures.

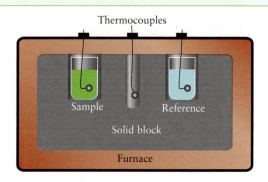

In differential thermal analysis, a sample and reference material are heated at the same rate in the same large metal heat sink. Changes in the heat capacity of the sample are measured by noting differences in temperature between the sample and the reference material.

In *differential scanning calorimetry* (DSC), higher precision can be obtained and heat capacities can be measured. The apparatus is similar to that for a DTA analysis, with the primary difference being that the sample and reference are in separate heat sinks that are heated by individual heaters (see the second illustration). The temperatures of the two samples are kept the same by differential heating. Even slight temperature differences between the reference and the sample will trigger a relay that sends more or less power to the sample to maintain the constant temperature. If the heat capacity of the sample is higher than the heat capacity of the reference, energy must be supplied at a greater rate to the sample cell. If a phase transition occurs in the sample, a great deal of energy must be supplied to the sample

until the phase transition is complete and the temperature begins to rise again.

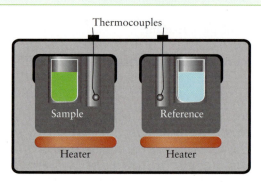

In a differential scanning calorimeter, a sample and reference material are heated in separate, but identical, metal heat sinks. The temperatures of the sample and reference material are kept the same by varying the power supplied to the two heaters. The output is the difference in power as a function of heat added.

The output of a differential scanning calorimeter is a measure of the power (the rate of energy supply) supplied to the sample cell. The thermogram in the third illustration shows a downward surge that signals a phase change. The thermogram does not look much like a heating curve, but it contains all the necessary information and is easily transformed into the familiar shape.

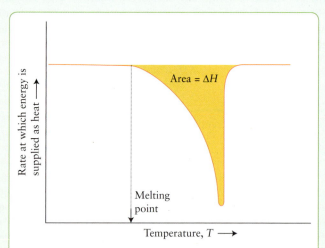

A thermogram from a differential scanning calorimeter. The downward surge indicates an endothermic phase change in the sample (fusion in this example) and the temperature at which the downward surge occurs indicates the melting point.

Consider what happens when a sample of very cold ice is heated. As shown in **FIG. 4C.9**, at first its temperature rises steadily. Although the molecules are still locked in a solid mass, they are oscillating more and more vigorously around their mean positions. However, once the temperature has reached the melting point, the molecules have enough energy to move past one another. At this temperature, all the added energy is used to

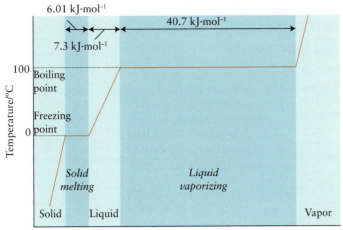

FIGURE 4C.9 The heating curve of water. The temperature of the solid rises as heat is supplied. At the melting point, the temperature remains constant, because all the heat is being used to melt the sample. When enough heat has been supplied to melt all the solid, the temperature begins to rise again. The heating curve for the liquid is not as steep as for the solid because the liquid has the higher heat capacity.

overcome the attractive forces between molecules. Thus, although heating continues, the temperature remains constant at the melting point until all the ice has melted. Only then does the temperature rise again, and the rise continues right up to the boiling point. At the boiling point, the temperature rise again comes to a halt. Now the water molecules have enough energy to escape into the vapor state, and all the heat supplied is used to form the vapor. After the sample has evaporated and heating continues, the temperature of the vapor rises again.

Most pure substances have heating curves similar to that in Fig. 4C.9. However, if the temperature rise is rapid, boiling might not begin until a few degrees over the boiling point. Once boiling begins, the temperature falls back to that of the boiling point. This phenomenon is called *superheating*. Similarly, when a sample is cooled rapidly, the temperature might fall below the freezing point briefly before freezing begins in a process called *supercooling*.

The steeper the slope of a heating curve, the lower is the heat capacity. In the heating curve for water, for instance, the heating curves of the solid and vapor phases have steeper slopes than the liquid; this observation implies that the liquid has a greater heat capacity than either the solid or the gas phase. The high heat capacity of liquid water is due largely to the extensive hydrogen-bonding network that survives in the liquid. The molecules are linked to one another by intermolecular hydrogen bonds, and the vibrations of these relatively weak bonds can take up energy more readily than can the stiff chemical bonds between atoms.

> *The temperature of a sample is constant at its melting and boiling points, even though heat is being supplied. The slope of a heating curve is steeper for a phase with a low heat capacity than for one with a high heat capacity.*

What have you learned in this Topic?

You have learned that when a system is at constant pressure, it is convenient to focus on the changes in the enthalpy, which takes expansion work into account. Changes of enthalpy are equal to the heat released or absorbed at constant pressure. You have seen how to relate the enthalpy changes of forward and reverse processes, and how to calculate the enthalpy change for a composite process. You now know how the enthalpy changes accompanying physical processes are reported and can explain their relative values.

The skills you have mastered are the ability to:

☐ **1.** Define the state function enthalpy (Section 4C.1).

☐ **2.** Interpret a change in enthalpy as the heat transferred at constant pressure.

☐ **3.** Relate the heat capacities at constant pressure and volume for an ideal gas (Section 4C.2).

☐ **4.** Determine the enthalpy change when heating an ideal gas (Example 4C.1).

☐ **5.** Define the enthalpies of vaporization, fusion, and sublimation (Section 4C.4).

☐ **6.** Interpret the heating curve of a substance (Section 4C.5).

Topic 4C Exercises

4C.1 Which gaseous compound do you expect to have the higher molar heat capacity, NO or NO_2? Why?

4C.2 Explain why the heat capacities of methane and ethane differ from the values expected for an ideal monatomic gas and from each other. The values of $C_{P,m}$ are 35.31 $J \cdot K^{-1} \cdot mol^{-1}$ for CH_4 and 52.63 $J \cdot K^{-1} \cdot mol^{-1}$ for C_2H_6.

4C.3 Calculate the final temperature and the change in enthalpy when 765 J of energy is transferred as heat to 0.820 mol Kr(g) at 298 K and 1.00 atm (a) at constant pressure; (b) at constant volume. Treat the gas as ideal.

4C.4 Calculate the final temperature and the change in enthalpy when 1.15 kJ of energy is transferred as heat to 0.640 mol Ne(g) at 298 K and 1.00 atm (a) at constant pressure; (b) at constant volume. Treat the gas as ideal.

4C.5 Predict the contribution of each type of molecular motion to the heat capacity $C_{V,m}$ and their total for each of the following atoms and molecules: (a) HCN; (b) C_2H_6; (c) Ar; (d) HBr. Ignore vibrations.

4C.6 Predict the contribution of each type of molecular motion to the heat capacity $C_{V,m}$ and their total for each of the following molecules: (a) NO; (b) NH_3; (c) HClO; (d) SO_2. Ignore vibrations.

4C.7 (a) At its boiling point, the vaporization of 0.579 mol CH_4(l) requires 4.76 kJ of heat. What is the enthalpy of vaporization of methane? (b) An electric heater was immersed in a flask of boiling ethanol, C_2H_5OH, and 22.45 g of ethanol was vaporized when 21.2 kJ of energy was supplied. What is the enthalpy of vaporization of ethanol?

4C.8 (a) When 25.23 g of methanol, CH_3OH, froze, 4.01 kJ of heat was released. What is the enthalpy of fusion of methanol? (b) A sample of benzene was vaporized at 25 °C. When 37.5 kJ of heat was supplied, 95 g of the liquid benzene vaporized. What is the enthalpy of vaporization of benzene at 25 °C?

4C.9 (a) Calculate the heat that must be supplied to a copper kettle of mass 500.0 g containing 400.0 g of water to raise its temperature from 22.0 °C to the boiling point of water, 100.0 °C. (b) What percentage of the heat is used to raise the temperature of the water? (See Table 4A.2.)

4C.10 (a) Calculate the heat that must be supplied to a stainless steel vessel of mass 500.0 g containing 400.0 g of water to raise its

temperature from 22.0 °C to the boiling point of water, 100.0 °C. (b) What percentage of the heat is used to raise the temperature of the water? (c) Compare these answers with those of Exercise 4C.9. (See Table 4A.2.)

4C.11 How much heat is needed to convert 80.0 g of ice at 0.0 °C into liquid water at 20.0 °C (see Tables 4A.2 and 4C.1)?

4C.12 If you start with 276 g of liquid water at 25. °C, how much heat must be supplied to convert all the liquid into vapor at 100. °C (see Tables 4A.2 and 4C.1)?

4C.13 An ice cube of mass 50.0 g at 0.0 °C is added to a glass containing 400.0 g of water at 45.0 °C. What is the final temperature of the system (see Tables 4A.2 and 4C.1)? Assume that no heat is lost to the surroundings.

4C.14 When 25.0 g of a metal at 90.0 °C is added to 50.0 g of water at 25.0 °C, the temperature of the water rises to 29.8 °C. What is the specific heat capacity of the metal (see Table 4A.2)?

4C.15 The following data were collected for a new compound used in cosmetics: $\Delta H_{fus} = 10.0$ $kJ \cdot mol^{-1}$, $\Delta H_{vap} = 20.0$ $kJ \cdot mol^{-1}$; heat capacities: 30 $J \cdot mol^{-1}$ for the solid; 60 $J \cdot mol^{-1}$ for the liquid; 30 $J \cdot mol^{-1}$ for the gas. Which heating curve below best matches the data for this compound?

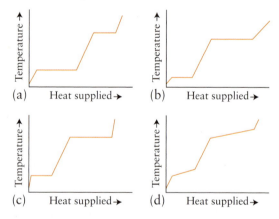

4C.16 Use the following information to construct a heating curve for bromine from −7.2 °C to 70.0 °C. The molar heat capacity of liquid bromine is 75.69 $J \cdot K^{-1} \cdot mol^{-1}$ and that of bromine vapor is 36.02 $J \cdot K^{-1} \cdot mol^{-1}$. The enthalpy of vaporization of liquid bromine is 30.91 $kJ \cdot mol^{-1}$. Bromine melts at −7.2 °C and boils at 58.78 °C.

Topic 4D Thermochemistry

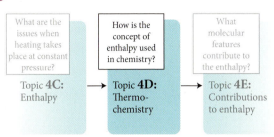

The same principles used to discuss the enthalpy changes accompanying physical processes (Topic 4C) apply to chemical changes. Enthalpies of chemical change are important in many areas of chemistry, such as the selection of materials that make good fuels, the design of chemical plants, and the study of biochemical processes. It is often important to know the ability of a reaction to release heat (as when a fuel burns), and **thermochemistry** is the study of the heat requirements of chemical reactions.

4D.1 Reaction Enthalpy

Every chemical reaction is accompanied by a change in energy, which is commonly released or absorbed as heat. For example, complete reaction with oxygen is called **combustion**, as in the combustion of methane, $CH_4(g) + 2 O_2(g) \rightarrow CO_2(g) + 2 H_2O(l)$, the major component of natural gas. Calorimetry shows that burning 1.00 mol $CH_4(g)$ at 298 K and 1 bar releases 890. kJ of energy as heat. This value is reported by writing

$$CH_4(g) + 2 O_2(g) \longrightarrow CO_2(g) + 2 H_2O(l) \qquad \Delta H = -890. \text{ kJ} \qquad \textbf{(A)}$$

This entire expression is called a **thermochemical equation,** and consists of a chemical equation together with a statement of the **reaction enthalpy** (or "enthalpy of reaction"), the corresponding enthalpy change. The stoichiometric coefficients indicate the number of moles of each reactant that gives the reported change in enthalpy. Therefore, in this case, the enthalpy change is that resulting from the complete reaction of 1 mol $CH_4(g)$ and 2 mol $O_2(g)$. If the same reaction is written with all the coefficients multiplied by 2, then the change in enthalpy will be twice as great because the equation now represents the burning of twice as much methane:

$$2 CH_4(g) + 4 O_2(g) \longrightarrow 2 CO_2(g) + 4 H_2O(l) \qquad \Delta H = -1780. \text{ kJ}$$

Note that, although burning takes place at high temperatures, the value of ΔH is the difference in enthalpies of the products and reactants, both measured at 298 K.

The first law of thermodynamics implies that, because enthalpy is a state function, the enthalpy change for the reverse of a process (such as a chemical reaction) is the negative of the enthalpy change of the forward process. For the reverse of reaction A, for instance,

$$CO_2(g) + 2 H_2O(l) \longrightarrow CH_4(g) + 2 O_2(g) \qquad \Delta H = +890. \text{ kJ}$$

Why Do You Need to Know This Material? Thermochemistry is one of the principal applications of thermodynamics to chemistry, because it enables the heat output (and requirements) of reactions to be discussed.

What Do You Need to Know Already? You need to understand the concept of enthalpy as a state function and the heat capacity at constant pressure (Topic 4C). You need to know how a calorimeter is used to measure the heat output of a reaction (Topic 4A).

EXAMPLE 4D.1 Determining a reaction enthalpy from experimental data

Benzene is added to gasoline to increase the octane number. When 0.113 g of benzene, C_6H_6, burns in excess oxygen in a calibrated constant-pressure calorimeter with a heat capacity of 551 J·(°C)$^{-1}$, the temperature of the calorimeter rises by 8.60 °C. Write the thermochemical equation for the reaction $2 C_6H_6(l) + 15 O_2(g) \rightarrow 12 CO_2(g) + 6 H_2O(l)$.

ANTICIPATE Because the temperature rises, the reaction is exothermic; therefore you should expect ΔH to be negative. All combustion reactions are exothermic.

PLAN The heat released by the reaction at constant pressure is calculated from the temperature change multiplied by the heat capacity of the calorimeter. Use the molar mass of one species (benzene) to convert the heat released into the reaction enthalpy corresponding to the thermochemical equation by noting the amount of C_6H_6 that reacts in the experiment (from

$n = m/M$, where m is the mass of benzene and M is its molar mass) and then scaling the heat released to 2 mol C_6H_6. The molar mass of benzene is 78.12 g·mol^{-1}.

What should you assume? Assume that all the heat released by the reaction is absorbed by the calorimeter, with none lost to the rest of the surroundings.

SOLVE

Find the heat transferred to the calorimeter from $q_{cal} = C_{cal}\Delta T$.

$$q_{cal} = \left[551 \text{ J·}(°C)^{-1}\right] \times (8.60 \text{ °C}) = 551 \text{ J} \times 8.60 = +4.74... \text{ kJ}$$

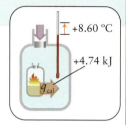

Calculate the amount of C_6H_6 that reacts from $n = m/M$.

$$n = \frac{0.113 \text{ g}}{78.12 \text{ g·mol}^{-1}} = \frac{0.113}{78.12} \text{ mol} = 1.45... \times 10^{-3} \text{ mol}$$

Calculate ΔH for 2 mol C_6H_6 by multiplying the heat output $q = -q_{cal}$ by (2 mol)/(1.45... × 10^{-3} mol). The change in enthalpy is negative because the reaction is exothermic.

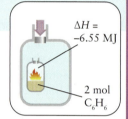

$$\Delta H = \overbrace{\frac{2 \text{ mol}}{1.45... \times 10^{-3}... \text{ mol}}}^{\text{adjust to 2 mol}} \times \overbrace{(-4.74... \text{ kJ})}^{q} = -6.55 \times 10^6 \text{ J} = -6.55 \text{ MJ}$$

The thermochemical equation is therefore

$$2 \text{ C}_6\text{H}_6(\text{l}) + 15 \text{ O}_2(\text{g}) \longrightarrow 12 \text{ CO}_2(\text{g}) + 6 \text{ H}_2\text{O}(\text{l}) \qquad \Delta H = -6.55 \text{ MJ}$$

EVALUATE   As expected, the enthalpy is negative, indicating an exothermic reaction.

Self-test 4D.1A   When 0.231 g of phosphorus reacts with chlorine to form phosphorus trichloride, PCl_3, in a constant-pressure calorimeter of heat capacity 216 J·(°C)$^{-1}$, the temperature of the calorimeter rises by 11.06 °C. Write the thermochemical equation for the reaction.

[**Answer:** 2 P(s) + 3 Cl$_2$(g) → 2 PCl$_3$(l), $\Delta H = -641$ kJ]

Self-test 4D.1B   When 0.338 g of pentane, C_5H_{12}, burns in an excess of oxygen to form carbon dioxide and liquid water in the same calorimeter as that used in Self-Test 4D.1A, the temperature rises by 76.7 °C. Write the thermochemical equation for the reaction.

Related Exercises   4D.3, 4D.4

A thermochemical equation is a statement of a chemical equation and the corresponding reaction enthalpy, the enthalpy change for the stoichiometric amounts of substances in the chemical equation.

4D.2 The Relation Between ΔH and ΔU

A constant-pressure calorimeter and a constant-volume calorimeter measure changes in different state functions: at constant volume, the heat transfer is interpreted as ΔU; at constant pressure, it is interpreted as ΔH (Topic 4C). It is sometimes necessary to convert the measured value of ΔU into ΔH. For example, it is easy to measure the heat released by the combustion of glucose in a constant-volume bomb calorimeter, but to use that information in assessing energy changes in metabolism, which take place at constant pressure, the enthalpy of reaction is needed.

For reactions in which no gas is generated or consumed, little expansion work is done as the reaction proceeds and the difference between ΔH and ΔU is negligible; so $\Delta H \approx \Delta U$. However, if a gas is formed in the reaction, so much expansion work is done to make room for the gaseous products that the difference can be significant. The ideal gas law can be used to relate the values of ΔH and ΔU for gases that behave ideally.

How Is That Done?

Start from the definition $H = U + PV$. Suppose that the amount of ideal gas reactant molecules is $n_1(g)$. Because for an ideal gas $PV = nRT$, the initial enthalpy is

$$H_1 = U_1 + PV_1 = U_1 + n_1(g)RT$$

After the reaction is complete, the amount of ideal gas product molecules is $n_2(g)$. The enthalpy is then

$$H_2 = U_2 + PV_2 = U_2 + n_2(g)RT$$

The difference is

$$\overbrace{H_2 - H_1}^{\Delta H} = \overbrace{U_2 - U_1}^{\Delta U} + \overbrace{\{n_2(g) - n_1(g)\}}^{\Delta n_{gas}} RT$$

and therefore

$$\Delta H = \Delta U + \Delta n_{gas} RT$$

You have seen that

$$\Delta H = \Delta U + \Delta n_{gas} RT \qquad (1)$$

where $\Delta n_{gas} = n_2(g) - n_1(g)$ is the change in the amount of gas molecules in the reaction (positive for net gas formation, negative for net gas consumption). Notice that ΔH is less negative (more positive) than ΔU for reactions that generate gases: less energy can be obtained from a reaction as heat at constant pressure than at constant volume, because the system must use some energy to expand to make room for the reaction products. For reactions with no change in the amount of gas molecules, the two quantities are negligibly different.

EXAMPLE 4D.2   Relating the enthalpy change and internal energy change for a chemical reaction

Glucose is the primary sugar in your bloodstream, supplying energy to your body. Dieticians therefore may need to investigate the enthalpy changes that accompany its reactions. A constant-volume bomb calorimeter was used to measure the heat generated by the combustion of 1.000 mol glucose molecules in the reaction $C_6H_{12}O_6(s) + 6\,O_2(g) \rightarrow 6\,CO_2(g) + 6\,H_2O(g)$, and found it to be 2559 kJ at 298 K; so $\Delta U = -2559$ kJ. What is the change in enthalpy for the same reaction?

ANTICIPATE Because there is a net formation of gas (from 6 mol to 12 mol), you should expect ΔH to be less negative than ΔU.

What should you assume? Assume that each gas is ideal.

SOLVE

From $\Delta n_{gas} = n_2(g) - n_1(g)$,

$$\Delta n_{gas} = (12 - 6)\ \text{mol} = +6\ \text{mol}$$

From $\Delta H = \Delta U + \Delta n_{gas} RT$,

$$\Delta H = -2559\ \text{kJ} + [(6\ \text{mol}) \times (8.3145 \times 10^{-3}\ \text{kJ·K}^{-1}\text{·mol}^{-1}) \times (298\text{K})]$$
$$= -2559\ \text{kJ} + 14.9\ \text{kJ} = -2544\ \text{kJ}$$

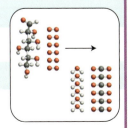

EVALUATE As anticipated, the reaction enthalpy is less negative than the reaction internal energy.

Self-test 4D.2A The thermochemical equation for the combustion of cyclohexane, C_6H_{12}, is $C_6H_{12}(l) + 9\,O_2(g) \rightarrow 6\,CO_2(g) + 6\,H_2O(l)$, $\Delta H = -3920$ kJ at 298 K. What is the change in internal energy for the combustion of 1.00 mol $C_6H_{12}(l)$ at 298 K?

[**Answer:** -3.91×10^3 kJ]

Self-test 4D.2B The reaction $4\,Al(s) + 3\,O_2(g) \rightarrow 2\,Al_2O_3(s)$ was investigated as part of a study on using aluminum as a rocket fuel (**FIG. 4D.1**). It was found that 1.000 mol Al produced 3378 kJ of heat in this reaction under constant-pressure conditions at 1000. °C. What is the change in internal energy for the combustion of 1.000 mol Al at 1000. °C?

Related Exercises 4D.7, 4D.8

FIGURE 4D.1 A technician inspects the solid propellant in a booster rocket, which is made up of powdered aluminum mixed with an oxidizing agent in a polymer base. (*George Shelton/NASA.*)

The reaction enthalpy is less negative (more positive) than the reaction internal energy for reactions that generate gases; for reactions with no change in the amount of gas, the two quantities are almost the same.

4D.3 Standard Reaction Enthalpies

Because the heat released or absorbed by a reaction depends on the physical states of the reactants and products, when calculating the reaction enthalpy the state of each substance must be specified. For example, when describing the combustion of ethene, two different thermochemical equations can be written for two different sets of products:

$$C_2H_4(g) + 3\,O_2(g) \longrightarrow 2\,CO_2(g) + 2\,H_2O(g) \qquad \Delta H = -1323 \text{ kJ} \qquad \textbf{(B)}$$
$$C_2H_4(g) + 3\,O_2(g) \longrightarrow 2\,CO_2(g) + 2\,H_2O(l) \qquad \Delta H = -1411 \text{ kJ} \qquad \textbf{(C)}$$

In the first reaction, the water is produced as a vapor; in the second, it is produced as a liquid. The heat generated is different in each case. Table 4C.1 shows that the enthalpy of water vapor is 44 kJ·mol^{-1} higher than that of liquid water at 25 °C. As a result, an additional 88 kJ (for 2 mol H_2O) remains stored in the system if water vapor is formed (**FIG. 4D.2**). If the 2 mol $H_2O(g)$ subsequently condenses, that 88 kJ is released as heat.

An enthalpy of reaction also depends on the conditions (such as the pressure). Unless otherwise stated, all the tables in this book list data for reactions in which each reactant and product is in its **standard state,** its pure form at exactly 1 bar. The standard state of liquid water is pure water at 1 bar. The standard state of ice is pure ice at 1 bar. A solute is in its standard state when its concentration is 1 mol·L^{-1}. The standard value of a property X (that is, the value of X for the standard state of the substance) is denoted $X°$.

The **standard reaction enthalpy** (or "standard enthalpy of reaction"), $\Delta H°$, is the reaction enthalpy when reactants in their standard states change into products in their standard states. For example, for reaction C, the value $\Delta H° = -1411$ kJ signifies that the heat output is 1411 kJ when 1 mol $C_2H_4(g)$ as pure ethene gas at 1 bar is allowed to react with pure oxygen gas at 1 bar, giving pure carbon dioxide gas and pure liquid water, both at 1 bar (**FIG. 4D.3**). Reaction enthalpies do not change very much with pressure over the small ranges normally encountered, so the standard value also gives a good indication of the change in enthalpy for pressures near 1 bar, such as at 1 atm.

You may see some tables reporting data at 1 atm, the former standard. The small change in standard pressure makes a negligible difference to most numerical values and, except in very precise work, it is normally safe to use data compiled for 1 atm instead of 1 bar.

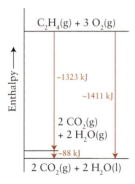

FIGURE 4D.2 The enthalpy changes for the reactions in which 1 mol $C_2H_4(g)$ burns to give carbon dioxide and water in either the gaseous (left) or the liquid (right) state. The difference in enthalpy is equal to 88 kJ, the enthalpy of vaporization of 2 mol $H_2O(l)$.

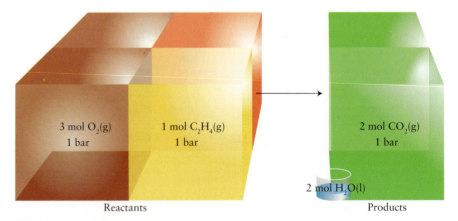

FIGURE 4D.3 The standard reaction enthalpy is the difference in enthalpy between the pure products and the pure reactants, each at 1 bar and the specified temperature (which is commonly but not necessarily 298 K). The scheme here is for the combustion of ethene to carbon dioxide and liquid water. Note that changes in the amount of gas dominate the change in volume in the course of the reaction.

Most thermochemical data are reported for 25 °C (more precisely, for 298.15 K). Temperature is not part of the definition of standard states: a standard state can be defined for any temperature; 298.15 K is simply the most common temperature used in tables of data. All reaction enthalpies used in this text are for 298.15 K unless another temperature is indicated.

A special type of reaction that plays an important role in thermodynamics, as well as in the everyday world, is combustion. A knowledge of the heat that can be obtained from the burning of a fuel is important for evaluating sources of energy (BOX 4D.1). The **standard enthalpy of combustion,** ΔH_c°, is the change in enthalpy per mole of a substance that is burned in a combustion reaction under standard conditions (TABLE 4D.1 and Appendix 2A). The products of the combustion of an organic compound are carbon dioxide gas and liquid water; any nitrogen present is released as N_2, unless other products are specified (such as $NO(g)$ and $NO_2(g)$). For practical applications it is necessary to take into account the mass of fuel that a mobile vehicle needs to carry or the volume it occupies. To take into account the burden of the mass of fuel, it is common to use the **specific enthalpy,** the standard enthalpy of combustion of a sample of the fuel divided by the mass of the sample. When the volume occupied by the fuel is important, its **enthalpy density,** the enthalpy of combustion of a sample divided by the volume of the sample, is used.

Box 4D.1 WHAT HAS THIS TO DO WITH . . . THE ENVIRONMENT?

ALTERNATIVE FUELS

Our complex modern life style was made possible by the discovery and refining of fossil fuels, fuels that are the result of the decay of organic matter laid down millions of years ago. The natural gas that heats our homes, the gasoline that powers our automobiles, and the coal, gas, and oil that provide much of our electrical power are fossil fuels. Vast reserves of petroleum, the source of liquid hydrocarbon fuels such as gasoline and coal, exist in many areas of the world. However, although they are large, these reserves are limited and essentially irreplaceable. Yet we are using them up rapidly. We need to develop *renewable* fuels, fuels that are replenished every year by the Sun.

Alternative and sustainable power-generation methods, such as hydroelectric power, wind power, geothermal power, solar power, and alternative fuels are being sought to reduce the demand for fossil fuels. Four of the most promising alternative fuels are hydrogen, ethanol, methane, and biodiesel. Hydrogen is extracted from ocean water by electrolysis or produced through steam reforming of hydrocarbons, as in this series of reactions:

$$CH_4(g) + H_2O(g) \longrightarrow CO(g) + 3 H_2(g)$$
$$CO(g) + H_2O(g) \longrightarrow CO_2(g) + H_2(g)$$

The bulk of industrial hydrogen production relies on hydrocarbons in petroleum and coal, but those fuels are not renewable. Ethanol is a renewable fuel obtained by fermenting *biomass,* the name given to plant materials that can be burned or reacted to produce fuels. Methane is a renewable fuel generated by bacterial digestion of wastes from sources such as sewage and agriculture. The use of hydrogen as a fuel is discussed in Topic 8B. Here we look at ethanol, methane, and biodiesel.

Ethanol, CH_3CH_2OH, is produced from the biological fermentation of the starches in grains, mainly corn. It is used as an additive in gasoline or as "E85," which is 85% ethanol and 15% gasoline by volume. Ethanol currently makes up about 10% by volume of gasoline in the United States, thereby reducing pollution as well as the use of petroleum. The 2007 Energy Independence and Security Act

These tanks at a water treatment facility are used to generate a mixture of methane and carbon dioxide by anaerobic digestion of sewage. The methane produced provides much of the power needed to run the facility. (© Maximilian/Prisma/agefotostock.)

requires that by 2022 the annual usage of renewable fuels increase to 36 billion gallons (1.4×10^{11} L, about 23% of the total annual volume of liquid fuels used in the United States). In 2013, 13 billion gallons (5.0×10^{10} L) of ethanol were produced in the United States and 23 billion gallons (8.9×10^{10} L) were produced worldwide.

The oxygen atom in the ethanol molecule reduces emissions of carbon monoxide and hydrocarbons by helping to ensure complete combustion. However, because ethanol is already partly oxygenated, it can provide less energy per liter; currently, mileage for a car burning E85 is about 15% lower than for one burning pure gasoline. One bushel of corn (about 30 L) can produce nearly 10 L of ethanol. A problem with ethanol as a fuel is that the sugars and starches fermented to produce it are expensive. However, the cellulose in straw and cornstalks left behind as stubble when grains are harvested is now attracting attention. Cellulose is the structural material in plants (see Topic 11E). It is made up of simple sugars, just as starches are, but the bacteria that ferment starches cannot digest cellulose. Research is now being conducted on enzymes that break down cellulose into sugars that can be digested. This process would greatly increase the biomass available for fuel production, because straw, wood, grasses, and in fact nearly any plant materials, could be used for fuel. It would also partially avoid the problem of the production of fuel being in competition with the production of food.

Methane, CH_4, is found in underground reserves as the primary component of natural gas, but it is also obtained from biological materials. The "digestion" of the biomaterials by bacteria is anaerobic, which means that it takes place in the absence of oxygen. Currently, many sewage treatment plants have anaerobic digesters that produce the methane used to operate the plants. To generate methane by anaerobic digestion on a large scale, additional materials, such as sugars from the enzymatic breakdown of biomass, would need to be used. Methane is less useful than ethanol as a transportation fuel because it is a gas. However, it can be used wherever natural gas is used.

Biodiesel is the term used for diesel fuel that comes from renewable, biological sources, such as algae or vegetable oil. Even used oils, such as those discarded from restaurant deep fryers, can be filtered and processed into biodiesel. Diesel engines are more efficient than gasoline engines, as the fuel has a high energy density (enthalpy of combustion per liter). A problem with biodiesel is that it is more viscous than traditional diesel fuel and can solidify at low temperatures.

These alternative fuels do produce carbon dioxide when burned and thus contribute to the greenhouse effect and global warming (Box 8B.1). However, they can be renewed every year, as long as the Sun shines and produces green plants.

Related Exercises 4.26–4.28, 4.33, 4.35, 4.63

For Further Reading Alternative Fuels & Advanced Vehicles Data Center, http://www.afdc.energy.gov/ (U.S. Department of Energy, accessed 2015). A Student's Guide to Alternative Fuel Vehicles, http://www.energyquest.ca.gov/transportation

This photobioreactor is used to study the growth of a new strain of algae from which liquid fuels can be extracted. Algae are efficient oil producers, and large arrays of algae tanks can produce more fuel per acre than other biofuels such as corn. *(Patrick Corkery/NREL.)*

(California Energy Commission, accessed 2015). National Renewable Energy Laboratory Biomass Research, http://www.nrel.gov/biomass (accessed 2015). Renewable and Alternative Energy Fuels, http://www.eia.doe.gov/renewable (U.S. Energy Information Administration, accessed 2015).

TABLE 4D.1   Standard Enthalpies of Combustion at 25 °C*

Substance	Formula	ΔH_c°/ (kJ·mol^{-1})	Specific enthalpy/ (kJ·g^{-1})	Enthalpy density[†]/ (kJ·L^{-1})
benzene	$C_6H_6(l)$	−3268	41.8	3.7×10^4
carbon	C(s, graphite)	−394	32.8	7.4×10^4
ethanol	$C_2H_5OH(l)$	−1368	29.7	2.3×10^4
ethyne (acetylene)	$C_2H_2(g)$	−1300.	49.9	53
glucose	$C_6H_{12}O_6(s)$	−2808	15.59	2.4×10^4
hydrogen	$H_2(g)$	−286	142	12
methane	$CH_4(g)$	−890.	55	36
octane	$C_8H_{18}(l)$	−5471	48	3.4×10^4
propane	$C_3H_8(g)$	−2220.	50.35	91
urea	$CO(NH_2)_2(s)$	−632	10.52	1.4×10^4

*In a combustion reaction, carbon is converted into carbon dioxide gas, hydrogen into liquid water, and nitrogen into nitrogen gas. More values are given in Appendix 2A.
[†]At 1 atm.

Once the standard reaction enthalpy is known, the enthalpy change, and therefore the heat released or required at constant pressure (from $\Delta H^\circ = q_P$, Topic 4C), can be calculated for any amount, mass, or volume of reactant consumed or product formed under standard conditions, even if that reaction cannot actually be carried out. To do so, a stoichiometry calculation is carried out like those described in *Fundamentals* L, but with heat treated as though it is a reactant or a product. Thus, instead of using a stoichiometric relation such as 1 mol $CH_4(g) \simeq$ 1 mol $CO_2(g)$ to calculate the amount of product in Reaction A, use 1 mol $CH_4(g) \simeq$ 890. kJ to calculate its heat output under standard conditions.

EXAMPLE 4D.3   Calculating the heat output of a fuel

Butane is a volatile liquid fuel that can be used in camping stoves. If you are planning a wilderness trek, you might need to assess how much butane you need to transport. How much butane should a backpacker carry to boil 1 L of water? Calculate the mass of butane that you would need to burn to obtain 350. kJ of heat, which is just enough energy to heat 1 L of water from 17 °C to boiling at sea level (if all heat losses are ignored). The thermochemical equation is

$$2\,C_4H_{10}(g) + 13\,O_2(g) \longrightarrow 8\,CO_2(g) + 10\,H_2O(l) \quad \Delta H° = -5756 \text{ kJ}$$

ANTICIPATE From experience, you might suspect that only a few grams of fuel will be required.

PLAN The first step is to convert heat output required into moles of fuel molecules by using the thermochemical equation. Then use the molar mass of the fuel to convert from moles of fuel molecules into a mass in grams.

What should you assume? Assume that no heat is lost to the surroundings, so that all the heat generated by the reaction is available to boil the water. In practice, of course, much of the heat is lost to the surroundings through inefficient transfer of energy from the burning gas to the water.

SOLVE

Note the relation of the enthalpy change to the amount of fuel molecules from the thermochemical equation.

$$5756 \text{ kJ} \simeq 2 \text{ mol } C_4H_{10}$$

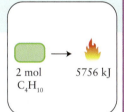

Convert heat output into moles of fuel molecules.

$$n(C_4H_{10}) = (350.\text{ kJ}) \times \frac{2 \text{ mol } C_4H_{10}}{5756 \text{ kJ}}$$

$$= \frac{350. \times 2}{5756} \text{ mol } C_4H_{10} = 0.122\ldots \text{ mol } C_4H_{10}$$

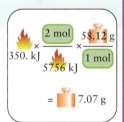

Use $m = nM$ and the molar mass of butane, 58.12 $g \cdot mol^{-1}$, to find the mass of reactant.

$$m(C_4H_{10}) = (0.122\ldots \text{ mol } C_4H_{10}) \times (58.12)g \cdot mol^{-1}$$
$$= 7.07 \text{ g } C_4H_{10}$$

EVALUATE The mass needed, just over 7 g, is consistent with the anticipation that only a few grams need be carried.

Self-test 4D.3A The thermochemical equation for the combustion of propane is

$$C_3H_8(g) + 5\,O_2(g) \longrightarrow 3\,CO_2(g) + 4\,H_2O(l) \quad \Delta H° = -2220.\text{ kJ}$$

What mass of propane must be burned to supply 350. kJ as heat? Would it be easier to pack propane rather than butane?

[***Answer:*** 6.95 g. Yes, slightly less propane would be needed.]

Self-test 4D.3B Ethanol trapped in a gel is another common camping fuel. What mass of ethanol must be burned to supply 350. kJ as heat?

$$C_2H_5OH(l) + 3\,O_2(g) \longrightarrow 2\,CO_2(g) + 3\,H_2O(l) \quad \Delta H° = -1368 \text{ kJ}$$

Related Exercises 4D.11, 4D.12

Standard reaction enthalpies refer to reactions in which the reactants and products are in their standard state, the pure form at 1 bar; they are usually reported for a temperature of 298.15 K. The heat absorbed or released by a reaction can be treated as a reactant or a product in a stoichiometric calculation.

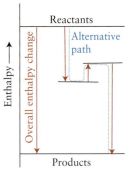

FIGURE 4D.4 If an overall reaction can be broken down into a series of steps, then the corresponding overall reaction enthalpy is the sum of the reaction enthalpies of those steps. None of the steps need be a reaction that can actually be carried out.

4D.4 Combining Reaction Enthalpies: Hess's Law

Enthalpy is a state function; therefore, the value of ΔH is independent of the path between given initial and final states. This approach is illustrated in Topic 4C, where the enthalpy change for an overall physical process (sublimation) is expressed as the sum of the enthalpy changes for a series of two individual steps (fusion and vaporization). The same rule applied to chemical reactions is known as **Hess's law:**

> The overall reaction enthalpy is the sum of the reaction enthalpies of the steps into which the reaction can be divided.

Hess's law applies even if the intermediate reactions or the overall reaction cannot actually be carried out; that is, they may be "hypothetical." Provided that the equation for each step balances and the individual equations add up to the equation for the reaction of interest, a reaction enthalpy can be calculated from any convenient sequence of reactions (**FIG. 4D.4**).

As an example of Hess's law, consider the oxidation of carbon as graphite, denoted C(gr), to carbon dioxide:

$$C(gr) + O_2(g) \longrightarrow CO_2(g) \qquad \textbf{(D)}$$

This reaction can be thought of as the outcome of two steps. One step is the oxidation of carbon to carbon monoxide,

$$C(gr) + \tfrac{1}{2}O_2(g) \longrightarrow CO(g) \qquad \Delta H = -110.5 \text{ kJ} \qquad \textbf{(E)}$$

The second step is the oxidation of carbon monoxide to carbon dioxide,

$$CO(g) + \tfrac{1}{2}O_2(g) \longrightarrow CO_2(g) \qquad \Delta H = -283.0 \text{ kJ} \qquad \textbf{(F)}$$

This two-step process is an example of a **reaction sequence,** a series of reactions in which the products of one reaction take part as reactants in another reaction. The equation for the overall reaction, the net outcome of the sequence, is the sum of the equations for the intermediate steps:

$$C(gr) + \tfrac{1}{2}O_2(g) \longrightarrow CO(g) \qquad \Delta H = -110.5 \text{ kJ} \qquad \textbf{(E)}$$

$$CO(gr) + \tfrac{1}{2}O_2(g) \longrightarrow CO_2(g) \qquad \Delta H = -283.0 \text{ kJ} \qquad \textbf{(F)}$$

$$\overline{C(gr) + O_2(g) \longrightarrow CO_2(g) \qquad \Delta H = -393.5 \text{ kJ} \quad (D) = (E) + (F)}$$

The same procedure is used to predict the enthalpies of reactions that cannot be measured directly in the laboratory. The procedure is described in **Toolbox 4D.1.**

Toolbox 4D.1 HOW TO USE HESS'S LAW

CONCEPTUAL BASIS

Because enthalpy is a state function, the enthalpy change of a system depends only on its initial and final states. Therefore, a reaction can be carried out in one step or, if that is not convenient, treated as the sum of several steps; the reaction enthalpy is the same in each case.

PROCEDURE

To use Hess's law to find the standard enthalpy of a given reaction, find a sequence of reactions with known reaction enthalpies that adds up to the reaction of interest. Different strategies may be used. Here is one that works in many cases.

Step 1 Select one of the reactants in the overall reaction and write down a chemical equation in which it also appears as a reactant.

Step 2 Select one of the products in the overall reaction and write down a chemical equation in which it also appears as a product. Add this equation to the equation written in step 1 and cancel species that appear on both sides of the equation.

Step 3 Cancel unwanted species in the sum obtained in step 2 by adding an equation that has the same substance or substances on the opposite side of the arrow.

Step 4 Once the sequence is complete, combine the standard reaction enthalpies.

In each step, it might be necessary to reverse the equation or multiply it by a factor. Recall that, if a chemical equation is reversed, then the sign of the reaction enthalpy is changed. If the stoichiometric coefficients need to be multiplied by a factor, then the reaction enthalpy must be multiplied by the same factor.

This procedure is illustrated in Example 4D.4.

EXAMPLE 4D.4 Using Hess's law

Propane, C_3H_8, is a gas used as a fuel for outdoor grills and alternative-fuel vehicles. The enthalpy change for the synthesis of propane from its elements in their standard states is difficult to measure directly, but if you are interested in assessing the thermodynamic properties of its reactions, you need to know its value. Calculate the standard enthalpy of the reaction $3 C(gr) + 4 H_2(g) \rightarrow C_3H_8(g)$ from the following experimental data:

(a) $C_3H_8(g) + 5 O_2(g) \longrightarrow 3 CO_2(g) + 4 H_2O(l)$ $\quad \Delta H° = -2220.$ kJ

(b) $C(gr) + O_2(g) \longrightarrow CO_2(g)$ $\quad \Delta H° = -394$ kJ

(c) $H_2(g) + \frac{1}{2}O_2(g) \longrightarrow H_2O(l)$ $\quad \Delta H° = -286$ kJ

PLAN Use the procedure in Toolbox 4D.1 to combine the chemical equations in a way that yields the required overall equation.

SOLVE

Step 1 For a reactant, choose carbon (graphite, denoted gr), then find an equation in which it appears as a reactant. Select (b) and multiply it through by 3 because the stoichiometric coefficient in the desired equation is 3.

$$3 C(gr) + 3 O_2(g) \longrightarrow 3 CO_2(g) \quad \Delta H° = 3 \times (-394 \text{ kJ}) = -1182 \text{ kJ}$$

Step 2 For a product, select propane; then find an equation in which it appears. Because propane is a product in the desired equation, equation (a) must be reversed, at the same time changing the sign of its reaction enthalpy.

$$3 CO_2(g) + 4 H_2O_2(l) \longrightarrow C_3H_8(g) + 5 O_2(g) \quad \Delta H° = +2220. \text{ kJ}$$

Add the two preceding equations and their standard enthalpies.

$3 C(gr) + 3 O_2(g) + 3 CO_2(g) + 4 H_2O(l) \longrightarrow C_3H_8(g) + 5 O_2(g) + 3 CO_2(g)$

$$\Delta H° = (-1182 + 2200) \text{ kJ} = +1038 \text{ kJ}$$

Simplify the equation by canceling species that appear both as products and as reactants.

$$3 C(gr) + 4 H_2O(l) \longrightarrow C_3H_8(g) + 2 O_2(g) \quad \Delta H° = +1038 \text{ kJ}$$

Step 3 To cancel the unwanted H_2O and O_2, multiply equation (c) by 4.

$$4 H_2(g) + 2 O_2(g) \longrightarrow 4 H_2O(l) \quad \Delta H° = 4 \times (-286 \text{ kJ}) = -1144 \text{ kJ}$$

Add the equations in steps 2 and 3 and combine the reaction enthalpies.

$3 C(gr) + 4 H_2(g) + 4 H_2O(l) + 2 O_2(g) \longrightarrow C_3H_8(g) + 2 O_2(g) + 4 H_2O(l)$

$$\Delta H° = 1038 + (-1144) \text{ kJ} = -106 \text{ kJ}$$

Step 4 Simplify the equation.

$$3 C(gr) + 4 H_2(g) \longrightarrow C_3H_8(g) \quad \Delta H° = -106 \text{ kJ}$$

Self-test 4D.4A Gasoline, which contains octane, may burn to carbon monoxide if the air supply is restricted. Calculate the standard reaction enthalpy for the incomplete combustion of liquid octane to carbon monoxide gas and liquid water from the standard reaction enthalpies for the combustions of octane and carbon monoxide:

$2 C_8H_{18}(l) + 25 O_2(g) \longrightarrow 16 CO_2(g) + 18 H_2O(l) \quad \Delta H° = -10\,942$ kJ

$2 CO(g) + O_2(g) \longrightarrow 2 CO_2(g) \quad \Delta H° = -566.0$ kJ

[*Answer:* $2 C_8H_{18}(l) + 17 O_2(g) \rightarrow 16 CO(g) + 18 H_2O(l)$,
$\Delta H° = -6414$ kJ]

Self-test 4D.4B Methanol is a clean-burning liquid fuel proposed as a replacement for gasoline. Suppose it could be produced by the controlled reaction of the oxygen in air with methane. Find the standard reaction enthalpy for the formation of 1 mol CH_3OH from methane and oxygen, given the following information:

$CH_4(g) + H_2O(g) \longrightarrow CO(g) + 3 H_2(g) \quad \Delta H° = +206.10$ kJ

$2 H_2(g) + CO(g) \longrightarrow CH_3OH(l) \quad \Delta H° = -128.33$ kJ

$2 H_2(g) + O_2(g) \longrightarrow 2 H_2O(g) \quad \Delta H° = -483.64$ kJ

Related Exercises 4D.13–4D.20

According to Hess's law, thermochemical equations for the individual steps of a reaction sequence may be combined to give the thermochemical equation for the overall reaction.

4D.5 Standard Enthalpies of Formation

There are millions of possible reactions, and it is impractical to list every one with its standard reaction enthalpy. However, chemists have devised an ingenious alternative. First, they report the "standard enthalpies of formation" of substances. Then they combine these quantities to obtain the standard enthalpy of any reaction.

The **standard enthalpy of formation**, ΔH_f°, of a substance is the standard reaction enthalpy per mole of formula units for the formation of a substance from its elements in their *most stable form*, as in the reaction for the formation of ethanol:

$$2\,C(gr) + 3\,H_2(g) + \tfrac{1}{2}O_2(g) \longrightarrow C_2H_5OH(l) \qquad \Delta H^\circ = -277.69\ kJ$$

where C(gr) denotes graphite, the most stable form of carbon at normal temperatures. The chemical equation that corresponds to the standard enthalpy of formation of a substance has the substance as the sole product with a stoichiometric coefficient of 1 (implying the formation of 1 mol of the substance). Sometimes, as here, fractional coefficients are required for the reactants. Because standard enthalpies of formation are expressed in kilojoules per mole of the substance of interest, in this case $\Delta H_f^\circ(C_2H_5OH, l) = -277.69$ kJ·mol^{-1}. Note also how the enthalpy change is labeled so that the species and its state (liquid, in this case) are specified unambiguously.

A Note on Good Practice: You should always be alert to the difference between a quantity *per mole* of molecules and the same quantity *for* or *of* a mole of molecules. Standard enthalpies of formation are expressed per mole of molecules, as in -277.69 kJ·mol^{-1}; the standard enthalpy of forming 1 mol $C_2H_5OH(l) = -277.69$ kJ. The point might seem picky, but it will help you to keep units straight.

It follows from the definition that the standard enthalpy of formation of an element in its most stable form is zero. For instance, the standard enthalpy of formation of C(gr) is zero because C(gr) → C(gr) is a "null reaction" (that is, nothing changes). In this case $\Delta H_f^\circ(C, gr) = 0$. However, the enthalpy of formation of an element in a form other than its most stable one is nonzero. For example, the conversion of carbon from graphite (its most stable form) into diamond is endothermic:

$$C(gr) \longrightarrow C(diamond) \qquad \Delta H^\circ = +1.9\ kJ$$

The standard enthalpy of formation of diamond is therefore reported as $\Delta H_f^\circ(C, diamond) = +1.9$ kJ·mol^{-1}. Values for a selection of other substances are listed in **TABLE 4D.2** and Appendix 2A.

Phosphorus is an exception: white phosphorus is used because it is much easier to obtain pure than the other, more stable allotropes.

TABLE 4D.2   Standard Enthalpies of Formation at 25 °C*

Substance	Formula	$\Delta H_f^\circ/$ (kJ·mol^{-1})	Substance	Formula	$\Delta H_f^\circ/$ (kJ·mol^{-1})
Inorganic compounds			**Organic compounds**		
ammonia	$NH_3(g)$	-46.11	benzene	$C_6H_6(l)$	$+49.0$
carbon dioxide	$CO_2(g)$	-393.51	ethanol	$C_2H_5OH(l)$	-277.69
carbon monoxide	$CO(g)$	-110.53	ethyne (acetylene)	$C_2H_2(g)$	$+226.73$
dinitrogen tetroxide	$N_2O_4(g)$	$+9.16$			
hydrogen chloride	$HCl(g)$	-92.31	glucose	$C_6H_{12}O_6(s)$	-1268
hydrogen fluoride	$HF(g)$	-271.1	methane	$CH_4(g)$	-74.81
nitrogen dioxide	$NO_2(g)$	$+33.18$			
nitric oxide	$NO(g)$	$+90.25$			
sodium chloride	$NaCl(s)$	-411.15			
water	$H_2O(l)$	-285.83			
	$H_2O(g)$	-241.82			

*A much longer list is given in Appendix 2A.

To see how to combine standard enthalpies of formation to obtain a standard reaction enthalpy, imagine that to carry out the reaction the reactants are first converted into the elements in their most stable form, then the products are formed from those elements. The standard reaction enthalpy of the first step is the negative of the standard enthalpies of formation of all the reactants (the reactants are being "unformed" into their elements), weighted by the amount present:

$$\text{"Unforming" of reactants: } \Delta H° = -\overbrace{\sum n\Delta H_f°(\text{reactants})}^{\text{sum over reactants}}$$

In this expression, the n are the stoichiometric coefficients in the chemical equation and the symbol Σ (sigma) means "form the sum of the following quantities." The standard enthalpy of the second step is the sum of the standard enthalpies of formation of all the products, again weighted by the amount present:

$$\text{Formation of products: } \Delta H° = \overbrace{\sum n\Delta H_f°(\text{products})}^{\text{sum over products}}$$

The sum of these two totals (taking into account the reversal of sign for the reactants) is the standard enthalpy of the overall reaction (**FIG. 4D.5**):

$$\Delta H° = \sum n\Delta H_f°(\text{products}) - \sum n\Delta H_f°(\text{reactants}) \tag{2}$$

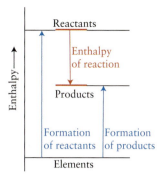

FIGURE 4D.5 The reaction enthalpy can be constructed from standard enthalpies of formation by imagining forming the reactants and products from their elements. Then the reaction enthalpy is the difference between the standard enthalpies of formation of the products and the reactants.

EXAMPLE 4D.5 Using standard enthalpies of formation to calculate a standard enthalpy of reaction

Amino acids are the building blocks of proteins, which have long chainlike molecules. They are oxidized in the body to urea, carbon dioxide, and liquid water. Is this reaction a source of heat for the body? Use the information in Appendix 2A to predict the standard reaction enthalpy for the oxidation of the simplest amino acid, glycine (NH_2CH_2COOH), a solid, to solid urea (H_2NCONH_2), carbon dioxide gas, and liquid water:

$$2\,NH_2CH_2COOH(s) + 3\,O_2(g) \longrightarrow H_2NCONH_2(s) + 3\,CO_2(g) + 3\,H_2O(l)$$

ANTICIPATE You should expect a strongly negative value, because all combustions are exothermic and this oxidation is like an incomplete combustion.

PLAN First, add together the standard enthalpies of formation of the products, multiplying each value by the appropriate number of moles from the balanced equation. Remember that the standard enthalpy of formation of an element in its most stable form is zero. Then, calculate the total standard enthalpy of formation of the reactants in the same way and use Eq. 2 to calculate the standard reaction enthalpy.

SOLVE

Calculate the total enthalpy of formation of the products from the data in Appendix 2A.

$$\Delta H_f°(H_2NCONH_2, s) = -333.51 \text{ kJ·mol}^{-1}$$
$$\Delta H_f°(CO_2, g) = -393.51 \text{ kJ·mol}^{-1}$$
$$\Delta H_f°(H_2O, l) = -285.83 \text{ kJ·mol}^{-1}$$

$$\sum n\Delta H_f°(\text{products}) = \overbrace{(1 \text{ mol})}^{H_2NCONH_2} \times (-333.51 \text{ kJ·mol}^{-1}) + \overbrace{(3 \text{ mol})}^{CO_2} \times (-393.51 \text{ kJ·mol}^{-1})$$
$$+ \overbrace{(3 \text{ mol})}^{H_2O} \times (-285.83 \text{ kJ·mol}^{-1}) = -2371.53 \text{ kJ}$$

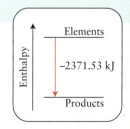

Calculate the total enthalpy of formation of the reactants.

$$\Delta H_f°(NH_2CH_2COOH, s) = -532.9 \text{ kJ·mol}^{-1}$$
$$\Delta H_f°(O_2, g) = 0$$

$$\sum n\Delta H_f°(\text{reactants}) = \overbrace{(2 \text{ mol})}^{NH_2CH_2COOH} \times (-532.9 \text{ kJ·mol}^{-1}) + \overbrace{(3 \text{ mol})}^{O_2} \times (0)$$
$$= \{2(-532.9) + 0\} = -1065.8 \text{ kJ}$$

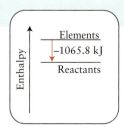

From $\Delta H° = \sum n\Delta H_f°(\text{products}) - \sum n\Delta H_f°(\text{reactants})$,

$$\Delta H° = -2371.53 - (-1065.8)\text{ kJ} = -1305.7\text{ kJ}$$

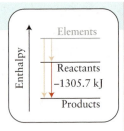

EVALUATE The thermochemical equation is therefore

$$2\text{ NH}_2\text{CH}_2\text{COOH(s)} + 3\text{ O}_2\text{(g)} \longrightarrow \text{H}_2\text{NCONH}_2\text{(s)} + 3\text{ CO}_2\text{(g)} + 3\text{ H}_2\text{O(l)} \qquad \Delta H° = -1305.7\text{ kJ}$$

As anticipated, the reaction is strongly exothermic. Because reactions in the body take place in aqueous solution, this value is not the same as the enthalpy change for the reaction in the body, but the two values are similar. Therefore, the oxidation of glycine is a potential source of energy in the body.

> **A Note on Good Practice:** Enthalpies of formation are expressed in kilojoules per mole and enthalpies of reaction in kilojoules for the reaction as written. Note how the stoichiometric coefficients are interpreted as numbers of moles, and that an unwritten coefficient of 1 for urea is included as 1 mol in the calculation.

Self-test 4D.5A Calculate the standard enthalpy of combustion of glucose from the standard enthalpies of formation in Table 4D.2 and Appendix 2A.

[***Answer:*** -2808 kJ·mol^{-1}]

Self-test 4D.5B You have an inspiration: maybe diamonds would make a great fuel! Calculate the standard enthalpy of combustion of diamonds from the information in Appendix 2A.

Related Exercises 4D.21, 4D.22, 4D.24

Standard enthalpies of formation are commonly determined from combustion data by using Eq. 2, as the following example illustrates.

EXAMPLE 4D.6 Using the enthalpy of combustion to calculate an enthalpy of formation

The information in Table 4D.2 and Appendix 2A must be determined from experimental data, but because some reactions cannot be carried out directly, chemists who compile these types of tables commonly use enthalpies of combustion. Use the information in Table 4D.2 and the enthalpy of combustion of propane gas to calculate the enthalpy of formation of propane, a gas distributed as LPG (for propane, $\Delta H_c° = -2220.$ kJ·mol^{-1}).

ANTICIPATE The enthalpy of formation for methane in Table 4D.2 is -74.81 kJ·mol^{-1}. Since propane contains more carbon and hydrogen atoms than methane, you should anticipate that the enthalpy of formation is more negative than -74.81 kJ·mol^{-1}.

PLAN Use Eq. 2 and the procedure set out in Example 4D.5, but solve for the standard enthalpy of formation of propane.

SOLVE

Write the thermochemical equation for the combustion of 1 mol $\text{C}_3\text{H}_8\text{(g)}$

$$\text{C}_3\text{H}_8\text{(g)} + 5\text{ O}_2\text{(g)} \longrightarrow 3\text{ CO}_2\text{(g)} + 4\text{ H}_2\text{O(l)} \qquad \Delta H° = -2220.\text{ kJ}$$

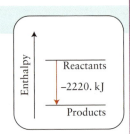

Find the total enthalpy of formation of the products.

$$\Delta H_f°(\text{CO}_2, \text{g}) = -393.51\text{ kJ·mol}^{-1}$$
$$\Delta H_f°(\text{HO}_2, \text{l}) = -285.83\text{ kJ·mol}^{-1}$$

$$\sum n\Delta H_f°(\text{products}) = \overbrace{(3\text{ mol})}^{\text{CO}_2} \times (-393.51\text{ kJ·mol}^{-1}) + \overbrace{(4\text{ mol})}^{\text{H}_2\text{O}} \times (-285.83\text{ kJ·mol}^{-1})$$
$$= -2323.85\text{ kJ}$$

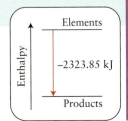

Find the total enthalpy of formation of the reactants.

$$\Delta H_f°(O_2, g) = 0$$

$$\sum n\Delta H_f°(\text{reactants}) = \underbrace{(1 \text{ mol})}_{C_3H_8} \times \{\Delta H_f°(C_3H_8, g)\} + \underbrace{(5 \text{ mol})}_{O_2} \times (0)$$

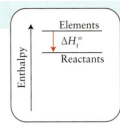

From $\Delta H_r° = \sum n\Delta H_f°(\text{products}) - \sum n\Delta H_f°(\text{reactants})$,

$$-2220. \text{ kJ} = -2323.85 \text{ kJ} - (1 \text{ mol})\, \Delta H_f°(C_3H_8, g)$$

Solve for $\Delta H_f°(C_3H_8, g)$.

$$\Delta H_f°(C_3H_8, g) = \frac{-2323.85 \text{ kJ} - (-2220.) \text{ kJ}}{1 \text{ mol}} = -104 \text{ kJ·mol}^{-1}$$

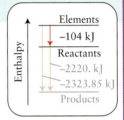

EVALUATE The enthalpy of formation for propane is more negative than that for methane, as anticipated.

Self-test 4D.6A Calculate the standard enthalpy of formation of ethyne, the fuel used in oxyacetylene welding torches, from the information in Table 4D.2 and given that $\Delta H_c°$ for ethyne is $-1300.$ kJ·mol^{-1}.

[**Answer:** $+227$ kJ·mol^{-1}]

Self-test 4D.6B Calculate the standard enthalpy of formation of urea, $CO(NH_2)_2$, a by-product of the metabolism of proteins, from the information in Table 4D.2 and given that $\Delta H_c°$ for urea is -632 kJ·mol^{-1}.

Related Exercises 4D.23, 4.13(a)

Standard enthalpies of formation can be combined to obtain the standard enthalpy of any reaction.

4D.6 The Variation of Reaction Enthalpy with Temperature

In some cases, you might know the reaction enthalpy for one temperature but require it for another temperature. For instance, the temperature of the human body is about 37 °C, but the data in Appendix 2A are for 25 °C. How can you find out whether an increase of 12 °C makes much difference to the reaction enthalpy of a metabolic process?

The enthalpies of both reactants and products increase with temperature. If the total enthalpy of the reactants increases more than that of the products when the temperature is raised, then the reaction enthalpy of an exothermic reaction will become more negative (**FIG. 4D.6**). On the other hand, if the enthalpy of the products increases more than that of the reactants, then the reaction enthalpy will become less negative. The increase in enthalpy of a substance when the temperature is raised depends on its heat capacity at constant pressure (Eq. 4b of Topic 4C, $C_P = \Delta H/\Delta T$), and it is easy to deduce **Kirchhoff's law** (see Exercise 4D.29), that

$$\Delta H°(T_2) = \Delta H°(T_1) + (T_2 - T_1) \times \Delta C_P \tag{3}$$

where ΔC_P is the difference between the constant-pressure heat capacities of the products and reactants:

$$\Delta C_P = \sum nC_{P,m}(\text{products}) - \sum nC_{P,m}(\text{reactants}) \tag{4}$$

The individual values can be found in Appendix 2A. Because the difference between $\Delta H°(T_2)$ and $\Delta H°(T_1)$ depends on the *difference* in the heat capacities of the reactants and products—a difference that is normally small—in most cases, the reaction enthalpy

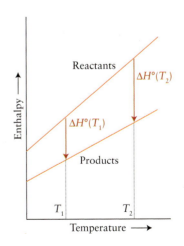

FIGURE 4D.6 If the heat capacity of the reactants is larger than that of the products, the enthalpy of the reactants will increase more sharply with increasing temperature. If the reaction is exothermic, the reaction enthalpy will become more negative, as shown here.

depends only weakly on the temperature and, for small ranges of temperature, it can be treated as constant.

EXAMPLE 4D.7   Predicting the reaction enthalpy at a different temperature

The standard enthalpy of reaction of $N_2(g) + 3 H_2(g) \rightarrow 2 NH_3(g)$ is $-92.22 \text{ kJ·mol}^{-1}$ at 298 K. The industrial synthesis takes place at 450. °C. What is the standard reaction enthalpy at the latter temperature?

ANTICIPATE There are more reactant molecules than product molecules, so you should suspect that the heat capacity of the products is less than that of the reactants and, therefore, that ΔC_P is negative. If that is so, an increase in temperature will make ΔH more negative and therefore the reaction more exothermic.

PLAN Calculate the difference in heat capacities of the reactants and products using Eq. 4, then use Eq. 3 to solve for $\Delta H°(T_2)$.

What should you assume? Assume that the heat capacities of reactants and products are constant over the temperature range of interest.

SOLVE

Calculate the difference in heat capacities.

$$\Delta C_P = \overbrace{(2 \text{ mol})C_{P,m}(NH_3, g)}^{products} - \overbrace{\{(1 \text{ mol})C_{P,m}(N_2, g) + (3 \text{ mol})C_{P,m}(H_2, g)\}}^{reactants}$$
$$= (2 \text{ mol}) \times (35.06 \text{ J·K}^{-1}\text{·mol}^{-1}) - \{(1 \text{ mol}) \times (29.12 \text{ J·K}^{-1}\text{·mol}^{-1})$$
$$+ (3 \text{ mol}) \times (28.82 \text{ J·K}^{-1}\text{·mol}^{-1})\}$$
$$= -45.46 \text{ J·K}^{-1}$$

Evaluate $T_2 - T_1$.

$$T_2 - T_1 = (450 + 273 \text{ K}) - 298 \text{ K} = 425 \text{ K}$$

Calculate the enthalpy change at the final temperature from $\Delta H°(T_2) = \Delta H°(T_1) + (T_2 - T_1) \times \Delta C_P$.

$$\Delta H°(450 \text{ °C}) = -92.22 \text{ kJ} + (425 \text{ K}) \times (-45.46 \text{ J·K}^{-1})$$
$$= -92.22 \text{ kJ} - 1.932 \times 10^4 \text{ J}$$
$$= -92.22 \text{ kJ} - 19.32 \text{ kJ} = -111.54 \text{ kJ}$$

EVALUATE The individual contributions of the changes in enthalpy of the reactants and products are depicted in **FIG. 4D.7**. As anticipated, ΔC_P is negative and the reaction is more exothermic at the higher temperature.

Self-test 4D.7A The reaction enthalpy of $4 Al(s) + 3 O_2(g) \rightarrow 2 Al_2O_3(s)$ is -3351 kJ at 298 K. Estimate its value at 1000. °C.

[***Answer:*** -3378 kJ]

Self-test 4D.7B The standard enthalpy of formation of ammonium nitrate is $-365.56 \text{ kJ·mol}^{-1}$ at 298 K. Estimate its value at 250. °C.

Related Exercises 4D.25–4D.28

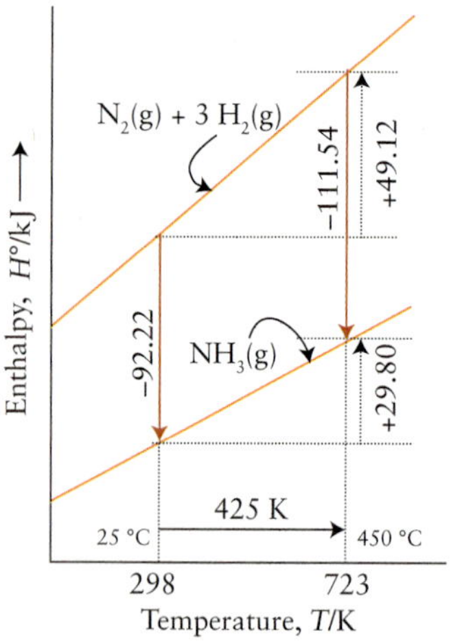

FIGURE 4D.7 A depiction of the individual contributions to the change in standard reaction enthalpy for the reaction in Example 4D.7.

The temperature variation of the standard reaction enthalpy is given by Kirchhoff's law, Eq. 3, in terms of the difference in heat capacities at constant pressure between the products and the reactants.

What have you learned in this Topic?

You have seen how to distinguish enthalpy and internal energy, report enthalpy changes in the form of a thermochemical equation, and specify the standard state of a substance. You now know how to predict the heat output from the stoichiometry of a reaction and how to use Hess's law to combine standard reaction enthalpies. You have also learned how to define a standard enthalpy of formation, use its value to calculate the standard enthalpy of any reaction, and predict how the value is affected by a change in temperature.

The skills you have mastered are the ability to:

☐ **1.** Write and use a thermochemical equation (Section 4D.1).

☐ **2.** Use experimental data to write a thermochemical equation (Example 4D.1).

☐ **3.** Convert between ΔH and ΔU for a reaction (Example 4D.2).

☐ **4.** Define the standard state of a substance.

☐ **5.** Predict the heat output of a reaction from its stoichiometry (Example 4D.3).

☐ **6.** Calculate an overall reaction enthalpy from the enthalpies of the reactions in a reaction sequence by using Hess's law (Toolbox 4D.1 and Example 4D.4).

☐ **7.** Use standard enthalpies of formation to calculate the standard enthalpy of a reaction, and vice versa (Examples 4D.5 and 4D.6).

☐ **8.** Use Kirchhoff's law to predict the variation of reaction enthalpy with temperature (Example 4D.7).

Topic 4D Exercises

4D.1 Carbon disulfide can be prepared from coke (an impure form of carbon) and elemental sulfur:

$$4\,C(s) + S_8(s) \longrightarrow 4\,CS_2(l) \qquad \Delta H° = +358.8 \text{ kJ}$$

How much heat is absorbed in the reaction of 1.25 mol S_8 at constant pressure? (b) Calculate the heat absorbed in the reaction of 197 g of carbon with an excess of sulfur. (c) If the heat absorbed in the reaction was 415 kJ, how much CS_2 was produced?

4D.2 Calculate the heat generated by a reaction mixture of 8.4 L of sulfur dioxide at 1.00 atm and 273 K and 10.0 g of oxygen in the reaction

$$2\,SO_2(g) + O_2(g) \longrightarrow 2SO_3(g) \qquad \Delta H° = -198 \text{ kJ}$$

4D.3 The reaction of 1.40 g of carbon monoxide with excess water vapor to produce carbon dioxide and hydrogen gases in a bomb calorimeter causes the temperature of the calorimeter assembly to rise from 22.113 °C to 22.799 °C. The calorimeter assembly is known to have a total heat capacity (calorimeter constant) of 3.00 kJ·(°C)$^{-1}$. (a) Write a balanced equation for the reaction. (b) Calculate the internal energy change, ΔU, for the reaction of 1.00 mol CO(g).

4D.4 Benzoic acid has a known internal energy of combustion (-3251 kJ·mol^{-1}). When a bomb calorimeter was calibrated by burning 0.825 g of benzoic acid in oxygen, the temperature rose 1.94 °C. A sample of the sugar D-ribose ($C_5H_{10}O_5$) of mass 0.727 g was then burned in excess oxygen in the same calorimeter, forming gaseous CO_2 and liquid H_2O. The temperature of the calorimeter rose from 21.81 °C to 22.72 °C. (a) Write a balanced equation for the combustion. (b) Calculate the internal energy change, ΔU, for the combustion of 1.00 mol ribose molecules.

4D.5 For a certain reaction at constant pressure, $\Delta H = -15$ kJ, and 22 kJ of expansion work is done on the system by compressing it into a smaller volume. What is ΔU for this process?

4D.6 For a certain reaction at constant pressure, $\Delta U = -68$ kJ, and 25 kJ of work is done on the system by compressing it into a smaller volume. What is ΔH for this process?

4D.7 Oxygen difluoride is a colorless, very poisonous gas that reacts rapidly and exothermically with water vapor to produce O_2 and HF:

$$OF_2(g) + H_2O(g) \longrightarrow O_2(g) + 2\,HF(g) \qquad \Delta H = -318 \text{ kJ}$$

What is the change in internal energy for the reaction of 1.00 mol OF_2?

4D.8 Ethanol is a renewable and clean-burning component of gasoline:

$$2\,C_2H_5OH(l) + 6\,O_2(g) \longrightarrow 4\,CO_2(g) + 6\,H_2O(l)$$
$$\Delta H = -1368 \text{ kJ}$$

What is the change in internal energy for the reaction of 1.00 mol $C_2H_5OH(l)$?

4D.9 The enthalpy of formation of trinitrotoluene (TNT) is -67 kJ·mol^{-1}, and the density of TNT is 1.65 g·cm^{-3}. In principle, it could be used as a rocket fuel, with the gases resulting from its decomposition streaming out of the rocket to give the required thrust. In practice, of course, it would be extremely dangerous as a fuel because it is sensitive to shock. Explore its potential as a rocket fuel by calculating its enthalpy density (enthalpy change per liter) for the reaction

$$4\,C_7H_5N_3O_6(s) + 21\,O_2(g) \longrightarrow 28\,CO_2(g) + 10\,H_2O(g) + 6\,N_2(g)$$

4D.10 A natural gas mixture is burned in a furnace at a power-generating station at a rate of 7.6 mol per minute. (a) If the fuel consists of 9.3 kmol CH_4, 3.1 kmol C_2H_6, 0.40 kmol C_3H_8, and 0.20 kmol C_4H_{10}, what mass of $CO_2(g)$ is produced per minute? (b) How much heat is released per minute?

4D.11 The oxidation of nitrogen in the hot exhaust of jet engines and automobiles occurs by the reaction

$$N_2(g) + O_2(g) \longrightarrow 2\,NO(g) \qquad \Delta H° = +180.6 \text{ kJ}$$

How much heat is absorbed in the formation of 1.55 mol NO? (b) How much heat is absorbed in the oxidation of 5.45 L of nitrogen measured at 1.00 atm and 273 K? (c) When the oxidation of N_2 to

NO was completed in a bomb calorimeter, the heat absorbed was measured as 492 J. What mass of nitrogen gas was oxidized?

4D.12 The combustion of octane is expressed by the thermochemical equation

$$C_8H_{18}(l) + \tfrac{25}{2}O_2(g) \longrightarrow 8\,CO_2(g) + 9\,H_2O(l) \qquad \Delta H^\circ = -5471\text{ kJ}$$

Estimate the mass of octane that would need to be burned to produce enough heat to raise the temperature of the air in a 12 ft × 12 ft × 8.0 ft room from 40. °F to 78 °F on a mild winter's day. Use the normal composition of air to determine its density and assume a pressure of 1.00 atm. (b) How much heat will be evolved from the combustion of 1.0 gallon of gasoline (assumed to be exclusively octane)? The density of octane is 0.70 g·mL^{-1}.

4D.13 Barium metal is produced by the reaction of aluminum metal with barium oxide. From the standard reaction enthalpies,

$$2\,Ba(s) + O_2(g) \longrightarrow 2\,BaO(s) \qquad \Delta H^\circ = -1107\text{ kJ}$$
$$2\,Al(s) + \tfrac{3}{2}O_2(g) \longrightarrow Al_2O_3(s) \qquad \Delta H^\circ = -1676\text{ kJ}$$

calculate the reaction enthalpy for the production of metallic barium in the reaction

$$3\,BaO(s) + 2\,Al(s) \xrightarrow{\Delta} Al_2O_3(s) + 3\,Ba(s)$$

4D.14 In the manufacture of nitric acid by the oxidation of ammonia, the first product is nitric oxide, which is then oxidized to nitrogen dioxide. From the standard reaction enthalpies,

$$N_2(g) + O_2(g) \longrightarrow 2\,NO(g) \qquad \Delta H^\circ = +180.5\text{ kJ}$$
$$N_2(g) + 2\,O_2(g) \longrightarrow 2\,NO_2(g) \qquad \Delta H^\circ = +66.4\text{ kJ}$$

calculate the standard reaction enthalpy for the oxidation of nitric oxide to nitrogen dioxide:

$$2\,NO(g) + O_2(g) \longrightarrow 2\,NO_2(g)$$

4D.15 Determine the reaction enthalpy for the hydrogenation of ethyne to ethane, $C_2H_2(g) + 2\,H_2(g) \to C_2H_6(g)$, from the following data: $\Delta H_c^\circ(C_2H_2,\,g) = -1300.\text{ kJ·mol}^{-1}$, $\Delta H_c^\circ(C_2H_6,\,g) = -1560.\text{ kJ·mol}^{-1}$, $\Delta H_c^\circ(H_2,\,g) = -286\text{ kJ·mol}^{-1}$.

4D.16 Determine the enthalpy for the partial combustion of methane to carbon monoxide, $2\,CH_4(g) + 3\,O_2(g) \to 2\,CO(g) + 4\,H_2O(l)$, from $\Delta H_c^\circ(CH_4,\,g) = -890.\text{ kJ·mol}^{-1}$ and $\Delta H_c^\circ(CO,\,g) = -283.0\text{ kJ·mol}^{-1}$.

4D.17 Use the data in Appendix 2A to calculate the standard reaction enthalpy for the reaction of pure nitric acid with hydrazine:

$$4\,HNO_3(l) + 5\,N_2H_4(l) \longrightarrow 7\,N_2(g) + 12\,H_2O(l)$$

4D.18 Use the data in Appendix 2A to calculate the standard reaction enthalpy for the reaction of magnesium carbonate with hydrochloric acid:

$$MgCO_3(s) + 2\,HCl(aq) \longrightarrow MgCl_2(aq) + H_2O(l) + CO_2(g)$$

4D.19 Calculate the reaction enthalpy for the synthesis of hydrogen bromide gas, $H_2(g) + Br_2(l) \to 2\,HBr(g)$, from the following data:

$$NH_3(g) + HBr(g) \longrightarrow NH_4Br(s) \qquad \Delta H^\circ = -188.32\text{ kJ}$$
$$N_2(g) + 3\,H_2(g) \longrightarrow 2\,NH_3(g) \qquad \Delta H^\circ = -92.22\text{ kJ}$$
$$N_2(g) + 4\,H_2(g) + Br_2(l) \longrightarrow 2\,NH_4Br(s) \qquad \Delta H^\circ = -541.66\text{ kJ}$$

4D.20 Calculate the reaction enthalpy for the formation of anhydrous aluminum bromide, $2\,Al(s) + 3\,Br_2(l) \to 2\,AlBr_3(s)$, from the following data:

$$2\,Al(s) + 6\,HBr(aq) \longrightarrow 2\,AlBr_3(aq) + 3\,H_2(g) \quad \Delta H^\circ = -1061\text{ kJ}$$
$$HBr(g) \longrightarrow HBr(aq) \qquad\qquad \Delta H^\circ = -81.15\text{ kJ}$$
$$H_2(g) + Br_2(l) \longrightarrow 2\,HBr(g) \qquad\qquad \Delta H^\circ = -72.80\text{ kJ}$$
$$AlBr_3(s) \longrightarrow AlBr_3(aq) \qquad\qquad \Delta H^\circ = -368\text{ kJ}$$

4D.21 Use standard enthalpies of formation from Appendix 2A to calculate the standard reaction enthalpy for each of the following reactions:
(a) the final stage in the production of nitric acid:

$$3\,NO_2(g) + H_2O(l) \longrightarrow 2\,HNO_3(aq) + NO(g)$$

(b) the industrial synthesis of boron trifluoride:

$$B_2O_3(s) + 3\,CaF_2(s) \longrightarrow 2\,BF_3(g) + 3\,CaO(s)$$

(c) the formation of a sulfide by the action of hydrogen sulfide on an aqueous solution of a base:

$$H_2S(aq) + 2\,KOH(aq) \longrightarrow K_2S(aq) + 2\,H_2O(l)$$

4D.22 Use standard enthalpies of formation from Appendix 2A to calculate the standard reaction enthalpy for each of the following reactions:
(a) the removal of hydrogen sulfide from natural gas:

$$2\,H_2S(g) + SO_2(g) \longrightarrow 3\,S(s) + 2\,H_2O(l)$$

(b) the oxidation of ammonia:

$$4\,NH_3(g) + 5\,O_2(g) \longrightarrow 4\,NO(g) + 6\,H_2O(g)$$

(c) the formation of phosphorous acid:

$$P_4O_6(s) + 6\,H_2O(l) \longrightarrow 4\,H_3PO_3(aq)$$

4D.23 Calculate the standard enthalpy of formation of dinitrogen pentoxide from the following data:

$$2\,NO(g) + O_2(g) \longrightarrow 2\,NO_2(g) \qquad \Delta H^\circ = -114.1\text{ kJ}$$
$$4\,NO_2(g) + O_2(g) \longrightarrow 2\,N_2O_5(g) \qquad \Delta H^\circ = -110.2\text{ kJ}$$

and from the standard enthalpy of formation of nitric oxide, NO (see Appendix 2A).

4D.24 An important reaction that takes place in the atmosphere is $NO_2(g) \to NO(g) + O(g)$, which is brought about by sunlight. How much energy must be supplied by the Sun to cause it? Calculate the standard enthalpy of the reaction from the following information:

$$O_2(g) \longrightarrow 2\,O(g) \qquad\qquad \Delta H^\circ = +498.4\text{ kJ}$$
$$NO(g) + O_3(g) \longrightarrow NO_2(g) + O_2(g) \qquad \Delta H^\circ = -200.\text{ kJ}$$

and from additional information in Appendix 2A.

4D.25 The enthalpy of combustion of 1.00 mol $CH_4(g)$ to carbon dioxide and water vapor is -802 kJ at 298 K. Calculate the enthalpy of this combustion at the temperature of a Bunsen burner flame at 1200. K.

4D.26 The amino acid glutamine (Gln) is produced in the body in an enzyme-catalyzed reaction of the amino acid glutamate (Glu) with the ammonium ion:

$$C_5H_8NO_4(aq) + NH_4^+(aq) \longrightarrow C_5H_{10}N_2O_3(aq) + H_2O(l)$$
$$\Delta H^\circ = +21.8\text{ kJ at 298 K}$$

Calculate the enthalpy of this reaction at 80.0 °C by using data in Appendix 2A and this information: $C_{P,m}(\text{Gln, aq}) = 187.0$ $J\cdot K^{-1}\cdot mol^{-1}$, $C_{P,m}(\text{Glu, aq}) = 177.0 \ J\cdot K^{-1}\cdot mol^{-1}$, $C_{P,m}(\text{NH}_4^+, \text{aq}) = 79.9 \ J\cdot K^{-1}\cdot mol^{-1}$.

4D.27 (a) From the data in Appendix 2A, calculate the enthalpy of vaporization of benzene (C_6H_6) at 298.2 K. The standard enthalpy of formation of gaseous benzene is $+82.93 \ kJ\cdot mol^{-1}$. (b) Given that, for liquid benzene, $C_{P,m} = 136.1 \ J\cdot K^{-1}\cdot mol^{-1}$ and that, for gaseous benzene, $C_{P,m} = 81.67 \ J\cdot K^{-1}\cdot mol^{-1}$, calculate the enthalpy of vaporization of benzene at its boiling point (353.2 K). (c) Compare the value obtained in part (b) with that found in Table 4C.1. What is the source of the difference between these numbers?

4D.28 (a) From the data in Appendix 2A, calculate the enthalpy required to vaporize 1 mol $CH_3OH(l)$ at 298.2 K. (b) Given that the molar heat capacity, $C_{P,m}$, of liquid methanol is $81.6 \ J\cdot K^{-1}\cdot mol^{-1}$ and that of gaseous methanol is $43.89 \ J\cdot K^{-1}\cdot mol^{-1}$, calculate the enthalpy of vaporization of methanol at its boiling point (64.7 °C). (c) Compare the value obtained in part (b) with that found in Table 4C.1. What is the source of difference between these values?

4D.29 Derive Kirchhoff's law for a reaction of the form $A + 2\ B \rightarrow 3\ C + D$ by considering the change in molar enthalpy of each substance when the temperature is changed from T_1 to T_2.

4D.30 Use the molar constant-volume heat capacities for gases given in Topic 4C (as multiples of R) to estimate the change in reaction enthalpy of $N_2(g) + 3\ H_2(g) \rightarrow 2\ NH_3(g)$ when the temperature is increased from 300. K to 500. K. Ignore the vibrational contributions to heat capacity. Is the reaction more or less exothermic at the higher temperature?

Topic 4E Contributions to Enthalpy

4E.1 Ion Formation

4E.2 The Born–Haber Cycle

4E.3 Bond Enthalpies

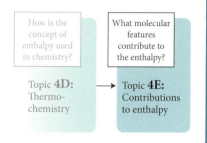

Why Do You Need to Know This Material? Your understanding of thermodynamics is enriched by relating thermodynamic properties, such as reaction enthalpy, to atomic and molecular properties.

What Do You Need to Know Already? You need to understand the concept of enthalpy as a state function (Topic 4C). This Topic draws on the discussion of ionization and electron gain (Topic 1F) and bond strength (Topic 2D).

The energies in the Topic 1F illustrations can be regarded as changes in internal energy at $T = 0$, and refer to the change in energy when only the ground state is occupied initially. At $T > 0$, there is a spread of populations over the available states, and the change in internal energy takes that spread into account. In practice, there is little numerical difference between the two quantities as almost all the population is in the ground state at normal temperatures.

One aspect of science that you will encounter often is that a deeper understanding of matter is obtained by seeing how bulk properties emerge from the behavior of individual atoms and molecules. Thermodynamics deals with bulk properties, but your insight into it is expanded considerably by understanding the origin of those properties in atomic terms.

4E.1 Ion Formation

Topic 1F introduces the concepts of ionization energy, I, and electron affinity, E_{ea}. Two closely related properties are used in thermodynamics. The **enthalpy of ionization,** ΔH_{ion}, is the change in standard enthalpy per mole of atoms for the loss of an electron. For any element X:

$$\text{Ionization: } X(g) \longrightarrow X^+(g) + e^-(g), \quad \Delta H_{ion}$$

The **enthalpy of electron gain,** ΔH_{eg}, is the analogous quantity for the gain of electrons.

$$\text{Electron gain: } X(g) + e^-(g) \longrightarrow X^-(g), \quad \Delta H_{eg}$$

Note that the electron-gain enthalpy and electron affinity (as defined in Topic 1F) have opposite signs. Thus, the electron-gain enthalpy is negative if electron gain releases energy (as for any exothermic process).

Enthalpies of ionization and electron gain are numerically very similar (apart from the change in sign between E_{ea} and ΔH_{eg}) to the corresponding energy changes—they differ by a few kilojoules per mole—and unless high precision is needed, it is usually appropriate to use the energies listed in Figs. 1F.8 and 1F.12 and Appendix 2. Trends in their values match trends in ionization energy and electron affinity, as set out in Topic 1F. Thus, alkali metal atoms have low (positive) enthalpies of ionization and halogen atoms have strongly negative enthalpies of electron gain.

The enthalpies of ionization and electron gain are the thermodynamic versions of ionization energy and electron affinity.

4E.2 The Born–Haber Cycle

For a given ionic solid, the difference in molar enthalpy between the solid and a gas of widely separated ions is called the **lattice enthalpy** of the solid, ΔH_L:

$$\Delta H_L = H_m(\text{ions, g}) - H_m(\text{solid}) \tag{1}$$

The lattice enthalpy can be identified with the heat required to vaporize the solid to widely separated gaseous ions at constant pressure. The greater the lattice enthalpy, the greater is the heat required. The discussion in Topic 2A referred to the lattice *energy*. The lattice *enthalpy* differs from the lattice energy by only a few kilojoules per mole and can be interpreted in a similar way.

Topic 2A explains how the energy changes accompanying the formation of an ionic solid can be estimated on the basis of a model—the "ionic model"—in which the principal contribution to the lattice energy is the Coulomb interaction between ions. However,

TABLE 4E.1 Lattice Enthalpies at 25 °C, $\Delta H_L/(kJ \cdot mol^{-1})$

Halides

LiF	1046	LiCl	861	LiBr	818	LiI	759
NaF	929	NaCl	787	NaBr	751	NaI	700.
KF	826	KCl	717	KBr	689	KI	645
AgF	971	AgCl	916	AgBr	903	AgI	887
$BeCl_2$	3017	$MgCl_2$	2524	$CaCl_2$	2260.	$SrCl_2$	2153
		MgF_2	2961	$CaBr_2$	1984		

Oxides

MgO	3850.	CaO	3461	SrO	3283	BaO	3114

Sulfides

MgS	3406	CaS	3119	SrS	2974	BaS	2832

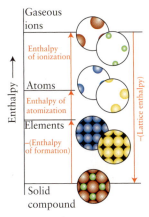

FIGURE 4E.1 In a Born–Haber cycle, select a sequence of steps that starts and ends at the same point (the solid compound, for instance). The lattice enthalpy is the enthalpy change accompanying the reverse of the step in which the solid is formed from a gas of ions. The sum of enthalpy changes around the complete cycle is zero because enthalpy is a state function.

a model can give only an estimate, and the model might be wrong: it is important to be able to measure the lattice enthalpy, not just estimate it. If the measured and calculated lattice energies are similar, then it is reasonably safe to conclude that the ionic model is reliable for the particular substance. If the two energies are markedly different, then the ionic model of that substance must be improved or even discarded.

The lattice enthalpy of a solid cannot normally be measured directly. However, because enthalpy is a state function, it can be obtained indirectly by combining other measurements. The procedure uses a **Born–Haber cycle,** a closed path of steps, one of which is the formation of a solid lattice from the gaseous ions. In the cycle the bulk elements are broken apart into atoms, the atoms are ionized, the gaseous ions form the ionic solid, and finally the elements are formed again from the ionic solid (**FIG. 4E.1**). Only the lattice enthalpy, the negative of the enthalpy of the step in which the ionic solid is formed from the gaseous ions, is unknown. The sum of the enthalpy changes around a complete Born–Haber cycle is zero, because the enthalpy of the system must be the same at the start and finish. **TABLE 4E.1** lists some lattice enthalpies found in this way.

EXAMPLE 4E.1   Using a Born–Haber cycle to calculate a lattice enthalpy

Potassium chloride is used as a pet-safe deicer. When investigating such applications, it is useful to understand the thermodynamic properties of the materials you are using. Devise and use a Born–Haber cycle to calculate the lattice enthalpy of potassium chloride.

ANTICIPATE A glance at Table 4E.1 shows that typical lattice enthalpies of ionic solids are close to 1000 $kJ \cdot mol^{-1}$, so you should expect a similar result for KCl.

PLAN Find the enthalpy change for each step, starting with the pure elements: atomize them to form gaseous atoms (steps a and b), ionize the atoms to form gaseous ions (steps c and d), allow the ions to form an ionic solid (this is the negative of the lattice enthalpy, $-\Delta H_L$), and then convert the solid back into the pure elements (step e). Steps c and d require ionization enthalpies for cation formation and electron-gain enthalpies for anion formation. Provided you are not seeking high precision, use ionization energies and electron affinities, being careful with the sign of the latter. Solve for the lattice enthalpy, ΔH_L, on the basis that the sum of the enthalpy changes for the complete cycle is zero.

SOLVE
The following steps are displayed in **FIG. 4E.2**.

(a) Find $\Delta H_f(K, g)$ in Appendix 2A.

$$K(s) \longrightarrow K(g) \qquad +89 \ kJ \cdot mol^{-1}$$

(b) Find $\Delta H_f(Cl, g)$ in Appendix 2A.

$$\tfrac{1}{2} Cl_2(g) \longrightarrow Cl(g) \qquad +122 \ kJ \cdot mol^{-1}$$

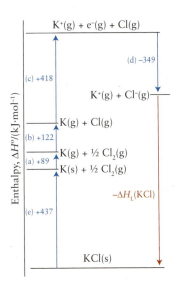

FIGURE 4E.2 The Born–Haber cycle used to determine the lattice enthalpy of potassium chloride (Example 4E.1). The enthalpy changes are in kilojoules per mole.

(c) Find the ionization energy of K in Appendix 2D.

$$K(g) \longrightarrow K^+(g) + e^-(g) \qquad +418 \text{ kJ·mol}^{-1}$$

(d) Write the electron-gain enthalpy of Cl as the negative of the electron affinity (Appendix 2D).

$$Cl(g) + e^-(g) \longrightarrow Cl^-(g) \qquad -349 \text{ kJ·mol}^{-1}$$

(e) Use Appendix 2A to write $-\Delta H_f(KCl)$.

$$KCl(s) \longrightarrow K(s) + \tfrac{1}{2}Cl_2(g) \qquad -(-437 \text{ kJ·mol}^{-1})$$

Set up the cycle, with a sum of zero.

$$\overbrace{\{89 + 122}^{(a + b)} + \overbrace{418 - 349}^{(c + d)} - \Delta H_L - \overbrace{(-437)\}}^{(e)} \text{ kJ·mol}^{-1} = 0$$

Solve for ΔH_L, the enthalpy of the reaction $KCl(s) \to K^+(g) + Cl^-(g)$.

$$\Delta H_L = (89 + 122 + 418 - 349 + 437) \text{ kJ·mol}^{-1} = +717 \text{ kJ·mol}^{-1}$$

EVALUATE   The lattice enthalpy of potassium chloride is 717 kJ·mol^{-1}, in line with the order of magnitude expected.

Self-test 4E.1A Calculate the lattice enthalpy of calcium chloride, $CaCl_2$, by using the data in Appendices 2A and 2D.

[*Answer:* 2259 kJ·mol^{-1}]

Self-test 4E.1B   Calculate the lattice enthalpy of magnesium bromide, $MgBr_2$, by using the data in Appendices 2A and 2D.

Related Exercises   4E.1–4E.3

The strength of interaction between ions in a solid is measured by the lattice enthalpy, which can be determined by using a Born–Haber cycle.

4E.3 Bond Enthalpies

The strength of a chemical bond is measured by the **bond enthalpy**, ΔH_B, the difference between the standard molar enthalpies of a molecule, X—Y (for instance, H_3C—OH), and its fragments X and Y (such as ·CH_3 and ·OH) in the gas phase:

$$\Delta H_B(X—Y) = \{H_m°(X, g) + H_m°(Y, g)\} - H_m°(X—Y, g) \tag{2}$$

Whereas a lattice enthalpy is equal to the heat required (at constant pressure) to break an ionic substance into gaseous ions, a bond enthalpy is the heat required to break a specific type of covalent bond at constant pressure. For example, the bond enthalpy of H_2 is derived from the thermochemical equation

$$H_2(g) \longrightarrow 2 H(g) \qquad \Delta H° = +436 \text{ kJ}$$

and is $\Delta H_B(H—H) = 436$ kJ·mol^{-1}. All bond enthalpies are positive because heat must be supplied to break a bond. In other words,

Bond breaking is always endothermic, and bond formation is always exothermic.

TABLE 4E.2 lists the bond enthalpies of some diatomic molecules.
A bond dissociation energy (Topic 2D) is the *energy* required to break the bond at $T = 0$; a bond enthalpy is the change in *standard enthalpy* at the temperature of interest (typically 298 K). The two quantities typically differ by only a few kilojoules per mole. Unless high precision is required, it is safe to use bond energies and bond enthalpies interchangeably. Because the numerical values are so similar, trends in bond enthalpy mirror those in bond energy (Topic 2D).

TABLE 4E.2	Bond Enthalpies of Diatomic Molecules
Molecule	$\Delta H_B/(\text{kJ·mol}^{-1})$
H_2	436
N_2	944
O_2	496
CO	1074
F_2	158
Cl_2	242
Br_2	193
I_2	151
HF	565
HCl	431
HBr	366
HI	299

TABLE 4E.3 Mean Bond Enthalpies

Bond	$\Delta H_B/(\text{kJ·mol}^{-1})$	Bond	$\Delta H_B/(\text{kJ·mol}^{-1})$
C—H	412	C—I	238
C—C	348	N—H	388
C=C	612	N—N	163
C⋯C*	518	N=N	409
C≡C	837	N—O	210.
C—O	360	N=O	630.
C=O	743	N—F	270.
C—N	305	N—Cl	200.
C—F	484	O—H	463
C—Cl	338	O—O	157
C—Br	276		

*In benzene.

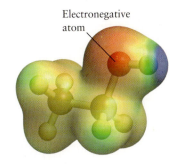

FIGURE 4E.3 An electrostatic potential diagram of ethanol, C_2H_5OH. An electronegative atom (here, the O atom) can pull electrons toward itself from more distant parts of the molecule, as shown by the red tint over the O atom. Hence, it can influence the strengths of bonds even between atoms to which it is not directly attached.

In a polyatomic molecule, all the atoms in the molecule exert a pull—through their electronegativities (Topic 2D)—on all the electrons in the molecule (**FIG. 4E.3**). As a result, the strength of a bond between a given pair of atoms varies to a small extent from one compound to another. For example, the O—H bond enthalpy in water, HO—H (492 kJ·mol^{-1}), is different from that of the same bond in methanol, CH$_3$O—H (437 kJ·mol^{-1}). However, these variations in bond enthalpy are not very great, and so the mean bond enthalpy, which is also denoted ΔH_B, is a guide to the strength of a bond in any molecule containing the bond (**TABLE 4E.3**).

Reaction enthalpies can be estimated by using mean bond enthalpies to determine the total energy required to break the reactant bonds and form the product bonds. In practice, only the bonds that change are treated. Because bond enthalpies refer to gaseous substances, to use the tabulated values, all substances must be gases or be converted into the gas phase.

The modern procedure for estimating a reaction enthalpy is to use commercial software to calculate the enthalpies of formation of the reactants and products and then to take the difference.

EXAMPLE 4E.2 Using mean bond enthalpies to estimate the enthalpy of a reaction

Mean bond enthalpies can be used to estimate the enthalpy of reaction when precise data are not available. Estimate the enthalpy of the reaction between liquid bromine and gaseous propene to form liquid 1,2-dibromopropane. The enthalpy of vaporization of Br$_2$ is 29.96 kJ·mol^{-1}, and that of CH$_3$CHBrCH$_2$Br is 35.61 kJ·mol^{-1}. The reaction is

$$Br_2(l) + CH_3CH{=}CH_2(g) \longrightarrow CH_3CHBrCH_2Br(l)$$

ANTICIPATE Two bonds must break in the reactants (Br—Br and C=C), whereas three bonds form in the product (C—C and two C—Br). Because more bonds are forming than are breaking, you should anticipate that the reaction is likely to be exothermic. Energy is required to vaporize the bromine, but it may be balanced by the energy released when the product condenses. However, the C=C bond has a high bond enthalpy, so you should expect the reaction to be only mildly exothermic.

PLAN Decide which bonds are broken and which bonds are formed. Use the mean bond enthalpies in Table 4E.3 to estimate the change in enthalpy when the reactant bonds break and the change in enthalpy when the new product bonds form. For diatomic molecules, use the information in Table 4E.2 for the specific molecule. Then, add the enthalpy change required to break the reactant bonds (a positive value) to the enthalpy change that occurs when the product bonds form (a negative value). Finally, because bond enthalpies are for gaseous substances, include the appropriate enthalpies of vaporization.

What should you assume? Assume that the mean bond enthalpies in the table are close to the actual bond enthalpies in the reactants and products.

SOLVE

Reactants: Break 1 mol C=C bonds in CH$_3$CH=CH$_2$ (average value 612 kJ·mol^{-1}) and 1 mol Br—Br bonds in Br$_2$ (193 kJ·mol^{-1}).

$$\Delta H° = \overbrace{612\ \text{kJ}}^{\text{C=C}} + \overbrace{193\ \text{kJ}}^{\text{Br—Br}} = +805\ \text{kJ}$$

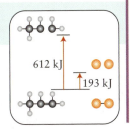

Products: Form 1 mol C—C bonds (average value 348 kJ·mol⁻¹) and 2 mol C—Br bonds (average value 276 kJ·mol⁻¹). The enthalpy change when the product bonds form is the negative of this sum.

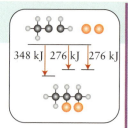

$$\Delta H° = -\{\overbrace{348 \text{ kJ}}^{\text{C—C}} + (2 \times \overbrace{276 \text{ kJ}}^{\text{C—Br}})\} = -900. \text{ kJ}$$

The overall enthalpy change is the sum of the bond enthalpy changes.

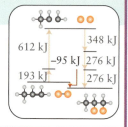

$$\Delta H° = 805 + (-900.) \text{ kJ} = -95 \text{ kJ}$$

The bromine must be vaporized, so the enthalpy of vaporization of 1 mol $Br_2(l)$ must be added to this result. When the product condenses to a liquid, heat is given off, so the enthalpy of vaporization of 1 mol $C_3H_6Br_2(l)$ must be subtracted.

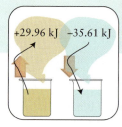

$$\Delta H° = -95 \text{ kJ} + \overbrace{29.96 \text{ kJ}}^{\Delta H_{vap}(Br_2)} - \overbrace{35.61 \text{ kJ}}^{\Delta H_{vap}(C_3H_6Br_2)} = -101 \text{ kJ}$$

The thermochemical equation is

$$Br_2(l) + CH_3CH{=}CH_2(g) \longrightarrow CH_3CHBrCH_2Br(l) \qquad \Delta H° = -101 \text{ kJ}$$

EVALUATE The reaction is exothermic, as suspected. Even though more bonds are formed than are broken, because the bond enthalpy of the C=C bond is large, the overall reaction enthalpy is small.

Self-test 4E.2A Use bond enthalpies to estimate the standard enthalpy of the reaction

$$CCl_3CHCl_2(g) + 2 HF(g) \longrightarrow CCl_3CHF_2(g) + 2 HCl(g)$$

[*Answer:* −24 kJ·mol⁻¹]

Self-test 4E.2B Use bond enthalpies to estimate the standard enthalpy of the reaction in which 1.00 mol of gaseous CH_4 reacts with gaseous F_2 to form gaseous CH_2F_2 and HF.

Related Exercises 4E.5–4E.7

A Note on Good Practice: The use of mean bond enthalpies requires caution, because actual bond enthalpies may differ considerably from mean values.

A mean bond enthalpy is the average molar enthalpy change accompanying the dissociation of a given type of bond.

What have you learned in this Topic?

You have seen how to relate some enthalpy changes to changes taking places at a molecular level, including ionization, electron gain, and bond dissociation. You have also learned how to use a Born–Haber cycle to infer the value of the lattice enthalpy from other data.

The skills you have mastered are the ability to:

☐ **1.** Define and use the ionization enthalpy and electron-gain enthalpy (Section 4E.1).

☐ **2.** Calculate a lattice enthalpy by using a Born–Haber cycle (Example 4E.1).

☐ **3.** Use mean bond enthalpies to estimate the standard enthalpy of a reaction (Example 4E.2).

Topic 4E Exercises

4E.1 Use the information in Fig. 1F.12, Appendix 2A, and Appendix 2D and the following data to calculate the lattice enthalpy of Na_2O: $\Delta H_f^\circ(Na_2O) = -409$ kJ·mol^{-1}; $\Delta H_f^\circ(O, g) = +249$ kJ·mol^{-1}.

4E.2 Use the information in Fig. 1F.12, Appendix 2A, and Appendix 2D and the following data to calculate the lattice enthalpy of $AlBr_3$: $\Delta H_f^\circ(Al, g) = +326$ kJ·mol^{-1}.

4E.3 Complete the following table (all values are in kilojoules per mole).

Compound, MX	ΔH_f°, M(g)	ΔH_{ion}, M	ΔH_f°, X(g)	ΔH_{eg}, X	ΔH_L, MX	ΔH_f°, MX(s)
(a) NaCl	108	494	122	−349	787	?
(b) KBr	89	418	97	−325	?	−394
(c) RbF	?	402	79	−328	774	−558

4E.4 By assuming that the lattice enthalpy of $NaCl_2$ is the same as that of $MgCl_2$, use enthalpy arguments based on data in Appendix 2A, Appendix 2D, and Fig. 1F.12 to explain why $NaCl_2$ is an unlikely compound.

4E.5 Use the bond enthalpies in Tables 4E.2 and 4E.3 to estimate the reaction enthalpy for
(a) $3\ C_2H_2(g) \rightarrow C_6H_6(g)$
(b) $CH_4(g) + 4\ Cl_2(g) \rightarrow CCl_4(g) + 4\ HCl(g)$
(c) $CH_4(g) + CCl_4(g) \rightarrow CHCl_3(g) + CH_3Cl(g)$

4E.6 Use the data in Tables 4E.2 and 4E.3 to estimate the reaction enthalpy for
(a) $HCl(g) + F_2(g) \rightarrow HF(g) + ClF(g)$, given that $\Delta H_B(Cl-F) = -256$ kJ·mol^{-1}
(b) $C_2H_4(g) + HCl(g) \rightarrow CH_3CH_2Cl(g)$
(c) $C_2H_4(g) + H_2(g) \rightarrow CH_3CH_3(g)$

4E.7 Use the data in Tables 4E.2 and 4E.3 to estimate the reaction enthalpy for
(a) $N_2(g) + 3\ F_2(g) \rightarrow 2\ NF_3(g)$
(b) $CH_3CHCH_2(g) + H_2O(g) \rightarrow CH_3CH(OH)CH_3(g)$
(c) $CH_4(g) + Cl_2(g) \rightarrow CH_3Cl(g) + HCl(g)$

4E.8 The bond enthalpy in NO is 632 kJ·mol^{-1} and that of each N—O bond in NO_2 is 469 kJ·mol^{-1}. Using Lewis structures and the data in Table 4E.3, explain (a) the difference in bond enthalpies between the two molecules; (b) the fact that the bond enthalpies of the two bonds in NO_2 are the same.

4E.9 Benzene is more stable and less reactive than would be predicted from its Kekulé structures. Use the data in Table 4E.3 to calculate the lowering in molar energy when resonance is allowed between the Kekulé structures of benzene.

4E.10 Investigate whether the replacement of a carbon–carbon double bond by single bonds is energetically favored by using Tables 4E.2 and 4E.3 to calculate the reaction enthalpy for the conversion of ethene, C_2H_4, to ethane, C_2H_6. The reaction is $H_2C{=}CH_2(g) + H_2(g) \rightarrow CH_3-CH_3(g)$.

Topic 4F Entropy

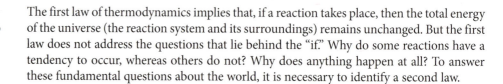

What are the basic concepts related to energy?	What determines the direction of natural change?	What molecular features contribute to the entropy?
Topic **4A:** Work and heat	Topic **4F:** Entropy	Topic **4G:** The molecular interpretation of entropy

Why Do You Need to Know This Material? The second law of thermodynamics introduces the concept of entropy, which is the key to understanding why one chemical reaction has a natural tendency to occur but another one does not.

What Do You Need to Know Already? The discussion draws on concepts related to the first law of thermodynamics, particularly enthalpy (Topic 4C), work of expansion of an ideal gas (Topic 4A), and heat capacity (Topic 4A). You also need to be aware of the meaning of reversible change (Topic 4A).

The first law of thermodynamics implies that, if a reaction takes place, then the total energy of the universe (the reaction system and its surroundings) remains unchanged. But the first law does not address the questions that lie behind the "if." Why do some reactions have a tendency to occur, whereas others do not? Why does anything happen at all? To answer these fundamental questions about the world, it is necessary to identify a second law.

4F.1 Spontaneous Change

A **spontaneous change** is a change that has a tendency to occur without needing to be driven by an external influence. A simple example is the cooling of a block of hot metal to the temperature of its surroundings (**FIG. 4F.1**). The reverse change, a block of metal spontaneously growing hotter than its surroundings, has never been observed. The expansion of a gas into a vacuum is spontaneous (**FIG. 4F.2**); a gas has no tendency to contract spontaneously into one part of its container.

A spontaneous change need not be fast. Molasses has a spontaneous tendency to flow out of an overturned can, but at low temperatures that flow may be very slow. Hydrogen and oxygen have a tendency to react to form water—the reaction is spontaneous in the thermodynamic sense—but a mixture of the two gases can be kept safely for centuries, provided it is not ignited with a spark. Diamonds have a natural tendency to turn into graphite, but diamonds last unchanged for countless years—on the human scale, diamonds are, in practice, forever. A spontaneous process has a natural *tendency* to occur; it does not necessarily take place rapidly or even occur at all. Whenever considering the thermodynamics of change, it is important to remember that only the *tendency* of a process to occur is being explored. Whether that tendency is actually realized in practice depends on its rate: rates are outside the reach of thermodynamics and are considered in Focus 7.

Changes can also be made to happen in their "unnatural" direction. For example, an electric current can be forced through a block of metal to heat it to a temperature higher than that of its surroundings. A gas can be driven into a smaller volume by pushing in a piston. In every case, though, to bring about a nonspontaneous change in a system, a way has to be contrived of *forcing* it to happen by applying an influence from outside the system. In short, *a nonspontaneous change can be brought about only by doing work.*

> *A process is spontaneous if it has a tendency to occur without being driven by an external influence; spontaneous changes need not be fast.*

4F.2 Entropy and Disorder

What is the pattern common to all spontaneous changes? Consider two scenarios. First, a hot block of metal in a cold room cools as the energy of its vigorously vibrating atoms spreads into the surroundings (the air and apparatus around it). The vigorously moving atoms of the metal collide with the slower atoms and molecules of the surroundings,

Spontaneous | Not spontaneous

FIGURE 4F.1 A red-hot block of metal (top) cools spontaneously to the temperature of its surroundings, the cool air around it (bottom). The reverse process, in which a block at the same temperature as its surroundings becomes hotter spontaneously, does not occur. *(W. H. Freeman photos by Ken Karp.)*

transferring some of their energy in the collisions. The reverse change is very improbable, because it would require energy to migrate from the cooler surroundings and concentrate in a small hot block of metal. Such a process would require that collisions of the less vigorously moving atoms of the surroundings with the more vigorously moving atoms of the metal would cause the latter to move even more vigorously. Second, the randomly moving molecules of a gas spread out all over their container; it is very unlikely that their random motion will bring them all simultaneously back into one corner. The pattern starting to emerge is that *energy and matter tend to disperse in a disorderly manner.*

In the language of thermodynamics, this simple idea is expressed in terms of the **entropy,** *S*, a measure of disorder. *Low entropy means little disorder; high entropy means great disorder.* It follows that the pattern can be expressed as the **second law of thermodynamics:**

> The entropy of an isolated system increases in the course of any spontaneous change.

Thus, the spontaneous cooling of hot metal is accompanied by an increase in entropy as energy spreads into the cooler surroundings. The "isolated system" in this case is taken to be the block of hot metal and its immediate surroundings. Likewise, the expansion of a gas is accompanied by an increase in entropy as the molecules spread through the container. The natural progression of a system and its surroundings (which together make up "the universe") is from order to disorder, from organized to random, from lower to higher entropy.

Provided the temperature is constant, a *change* in the entropy of a system can be calculated from the following expression:

$$\Delta S = \frac{q_{rev}}{T} \tag{1}$$

where ΔS is the entropy change of the system, *q* is the energy transferred as heat, and *T* is the (absolute) temperature at which the transfer takes place. The subscript "rev" on *q* signifies that the energy must be transferred reversibly in the sense described in Topic 4A. For a reversible transfer of energy as heat, the temperatures of the surroundings and the system must be only infinitesimally different and both must be constant. With heat measured in joules and temperature in kelvins, the change in entropy (and entropy itself) is measured in joules per kelvin ($J \cdot K^{-1}$).

What Does This Equation Tell You?
If a lot of energy is transferred as heat (large q_{rev}), a lot of disorder is created in the system and you should expect a correspondingly large increase in entropy. For a given transfer of energy, you should expect a greater change in disorder when the temperature is low than when it is high. The arriving energy stirs up the molecules of a cool system, which have little thermal motion, more noticeably than those of a hot system, in which the molecules are already moving vigorously. (Think of how sneezing in a quiet library will attract attention, but sneezing in a noisy street may pass unnoticed.)

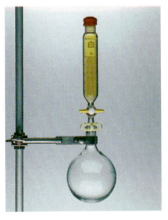

Spontaneous   Not spontaneous

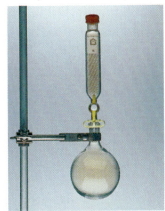

FIGURE 4F.2 A gas fills its container spontaneously. A glass cylinder containing the brown gas nitrogen dioxide (upper piece of glassware in the top illustration) is attached to an evacuated flask. When the stopcock between them is opened, the gas spontaneously fills both upper and lower vessels (bottom illustration). The reverse process, in which the gas now in both vessels spontaneously collects back in the upper vessel, does not occur. (*W. H. Freeman photos by Ken Karp.*)

EXAMPLE 4F.1   Calculating the change in entropy when a system is heated

Bullfrogs are cold-blooded animals, which means that they release any excess heat generated through metabolism into their environment. A sedentary bullfrog being studied in a large laboratory aquarium kept at a constant temperature of 25 °C transferred 100. J of energy reversibly to the water at 25 °C. What is the change in entropy of the water?

ANTICIPATE Because the water is absorbing energy as heat, you should expect its entropy to increase.

PLAN Use Eq. 1 to calculate the entropy change.

What should you assume? Assume that the aquarium is so large that its temperature remains virtually constant as the heat is transferred: Eq. 1 applies only at constant temperature.

SOLVE

Convert temperature into kelvins:

$$T = (273.15 + 25)\ K = 298\ K$$

From $\Delta S = q_{rev}/T$,

$$\Delta S = \frac{100.\ J}{298\ K} = +0.336\ J \cdot K^{-1}$$

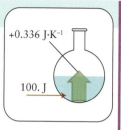

+0.336 J·K⁻¹

100. J

EVALUATE As predicted, the entropy of the water increases as a result of the flow of heat into it.

A Note on Good Practice: Note the + sign on the answer; always show the sign explicitly for the change in a quantity, even if it is positive.

Self-test 4F.1A Calculate the change in entropy of a large block of ice when 50. J of energy is removed reversibly from it as heat at 0 °C in a freezer.

[**Answer:** $-0.18\ J \cdot K^{-1}$]

Self-test 4F.1B Calculate the change in entropy of a large vat of molten copper when 50. J of energy is removed reversibly from it as heat at 1100. °C.

Related Exercises 4F.3, 4F.4

A very important characteristic of entropy that is not immediately obvious from Eq. 1 but can be proved by using thermodynamics is that *entropy is a state function*. This property is consistent with it being a measure of disorder, because the disorder of a system depends only on its current state and is independent of how that state was achieved.

Because entropy is a state function, the change in entropy of a system is independent of the path between its initial and final states. This independence means that, to calculate the entropy difference between a pair of states joined by an *irreversible* path, a *reversible* path should be found between the same two states and then Eq. 1 used for that path. For example, suppose an ideal gas undergoes free (irreversible) expansion at constant temperature. To calculate the change in entropy, imagine that the gas undergoes reversible, isothermal expansion between the same initial and final volumes, calculate the heat absorbed in this process, and use it in Eq. 1. Because entropy is a state function, the change in entropy of the gas calculated for this reversible path is also the change in its entropy for free expansion between the same two states.

> *Entropy is a measure of disorder; according to the second law of thermodynamics, the entropy of an isolated system increases in any spontaneous process. Entropy is a state function.*

4F.3 Entropy and Volume

Entropy is expected to increase when a given amount of matter spreads into a greater volume or is mixed with another substance. These processes disperse the molecules of the substance over a greater volume and increase the **positional disorder,** the disorder related to the locations of the molecules. The entropy change accompanying the isothermal expansion of an ideal gas illustrates this point and is calculated by using concepts from the first law.

How Is That Done?

To find out how the entropy of an ideal gas depends on its volume during an isothermal expansion (or compression), note that T is constant, so Eq. 1 can be used directly. Because $\Delta U = q + w$ and $\Delta U = 0$ for the isothermal expansion of an ideal gas (Topic 4B), $q = -w$: the energy the system loses by doing expansion work is replaced by an influx of energy as heat, so the internal energy remains the same. The same relation applies if the change is carried out reversibly: $q_{rev} = -w_{rev}$. Therefore, to find q_{rev}, calculate the work done when an ideal gas expands reversibly and isothermally and then change the sign. To calculate that work, use Eq. 4 from Topic 4A ($w_{rev} = -RT \ln(V_2/V_1)$). It follows that

$$\Delta S = \frac{q_{rev}}{T} = \frac{-w_{rev}}{T} = \frac{\overbrace{nRT \ln(V_2/V_1)}^{-w_{rev}}}{T} = nR \ln\frac{V_2}{V_1}$$

FIGURE 4F.3 The change in entropy as a sample of an ideal gas expands at constant temperature. Here $\Delta S/nR$ is plotted against the ratio of the final to the initial volume. The entropy increases logarithmically with volume.

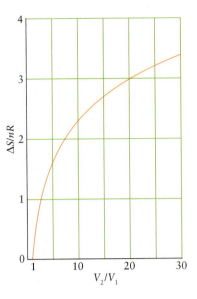

The calculation shows that the change in entropy of an ideal gas when it expands isothermally from a volume V_1 to a volume V_2 is

$$\Delta S = nR \ln \frac{V_2}{V_1} \qquad (2)$$

where n is the amount of gas molecules (in moles) and R is the gas constant (in joules per kelvin per mole). When the final volume is greater than the initial volume ($V_2 > V_1$), the change in entropy is positive, corresponding to an increase in entropy, as expected (**FIG. 4F.3**).

Although this change in entropy has been calculated for a reversible path, because entropy is a state function, Eq. 2 is also the change in entropy of the gas when it expands *irreversibly* between the same two states at any constant temperature. Do not jump to the conclusion, though, that there is no difference between reversible and irreversible processes: at this point only the change in entropy of the system has been considered and changes in the surroundings have not been taken into account (see Topic 4I).

EXAMPLE 4F.2   Calculating the change in entropy when an ideal gas expands isothermally

Meteorologists need to understand how the thermodynamic properties of air are affected by different conditions. They might begin by studying nitrogen, the principal component of air. What is the change in entropy of the gas when 1.00 mol $N_2(g)$ expands isothermally from 22.0 L to 44.0 L?

ANTICIPATE Because the greater volume provides more locations into which the molecules of the gas can spread, you should expect its entropy to increase.

PLAN Use Eq. 2 to calculate the entropy change for an isothermal expansion when the initial and final volumes are known.

What should you assume? Assume that nitrogen behaves as an ideal gas.

SOLVE

From $\Delta S = nR \ln (V_2/V_1)$,

$$\Delta S = (1.00 \text{ mol}) \times (8.3145 \text{ J·K}^{-1}\text{·mol}^{-1}) \times \ln \frac{44.0 \text{ L}}{22.0 \text{ L}}$$

$$= +5.76 \text{ J·K}^{-1}$$

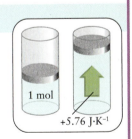

EVALUATE As expected, the entropy increases as the space available to the gas expands.

   A Note on Good Practice: Note that the entropy change *for* or *of* 1 mol of a substance is reported differently from the entropy change *per* mole: the units of the former are joules per kelvin ($J·K^{-1}$), whereas those of the latter are joules per kelvin per mole ($J·K^{-1}\text{·mol}^{-1}$).

Self-test 4F.2A Calculate the change in molar entropy of an ideal gas when it is compressed isothermally to one-third its initial volume.

[***Answer:*** -9.13 $J·K^{-1}\text{·mol}^{-1}$]

Self-test 4F.2B Calculate the change in molar entropy of carbon dioxide that is allowed to expand isothermally to 10. times its initial volume (treat carbon dioxide as an ideal gas).

Related Exercises 4F.5, 4F.6

The entropy change accompanying the isothermal compression or expansion of an ideal gas can be expressed in terms of its initial and final pressures. To do so, the ideal gas law—specifically, Boyle's law—is used to express the ratio of volumes in Eq. 2 in terms of the ratio of the initial and final pressures. Because pressure is inversely proportional to volume (Boyle's law), at constant temperature $V_2/V_1 = P_1/P_2$. Therefore,

$$\Delta S = nR \ln\frac{P_1}{P_2} \tag{3}$$

Self-test 4F.3A Calculate the change in entropy when the pressure of 1.50 mol Ne(g) is decreased isothermally from 20.00 bar to 5.00 bar. Assume ideal behavior.

[***Answer:*** $+17.3$ J·K^{-1}]

Self-test 4F.3B Calculate the change in entropy when the pressure of 70.9 g of chlorine gas is increased isothermally from 3.00 kPa to 24.00 kPa. Assume ideal behavior.

4F.4 Entropy and Temperature

Disorder is expected to increase when a system is heated because the supply of energy increases the thermal motion of the molecules. Heating increases the **thermal disorder,** the disorder arising from the thermal motion of the molecules. Equation 1 can be adapted to calculate the change in entropy when the temperature of a system is changed.

How Is That Done?

To calculate the entropy change in a system due to a temperature change, first note that Eq. 1 itself applies only when the temperature remains constant as heat is supplied to a system. Except in special cases, that is true only for infinitesimal transfers of heat, so the heating must be broken down into an infinite number of infinitesimal steps, with each step taking place at a constant but slightly different temperature, and then the infinitesimal entropy changes for all the steps are added together.

For an infinitesimal reversible transfer dq_{rev} at a temperature T, the increase in entropy is also infinitesimal; instead of Eq. 1, write

$$dS = \frac{dq_{rev}}{T}$$

The energy supplied as heat is related to the resulting increase in temperature, dT, by the heat capacity, C, of the system:

$$dq_{rev} = C dT$$

(This equation is the infinitesimal form of Eq. 5b of Topic 4A, $q = C\Delta T$, which also applies to reversible changes.) The combination of these two equations gives

$$dS = \frac{C dT}{T}$$

Now suppose that the temperature of a sample is increased from T_1 to T_2. The overall change in entropy is the sum (integral) of all these infinitesimal changes:

$$\Delta S = \int_{T_1}^{T_2} \frac{C dT}{T}$$

Provided the heat capacity is independent of temperature in the temperature range of interest, C can be taken outside the integral, which gives

$$\Delta S = C \int_{T_1}^{T_2} \frac{dT}{T} = C \ln\frac{T_2}{T_1}$$

The integral was evaluated by using

$$\int \frac{dx}{x} = \ln x + \text{constant}$$

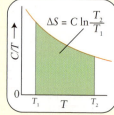

An important result from calculus (as mentioned in Topic 4A) is that a definite integral of a function between two limits is equal to the area under the graph of the function and lying between the two limits. In this case, a change in entropy when a substance is heated is equal to the area under the plot of C/T against T and lying between the initial and final temperatures.

FIGURE 4F.4 The change in entropy as a sample is heated for a system with a constant heat capacity (C) in the range of interest. Here $\Delta S/C$ is plotted against the ratio of the final temperature to the initial temperature. The entropy increases logarithmically with temperature.

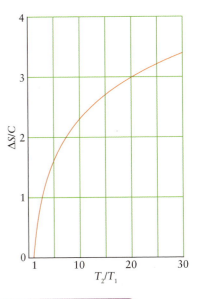

The calculation has shown that, provided the heat capacity can be treated as constant in the temperature range of interest, the entropy change that occurs when a system is heated from T_1 to T_2 is given by

$$\Delta S = C \ln \frac{T_2}{T_1} \qquad (4)$$

where C is the heat capacity of the system (C_V if the volume is constant, C_P if the pressure is constant). The dependence of the change in entropy on the ratio of final and initial temperatures is plotted in **FIG. 4F.4**.

What Does This Equation Tell You? If T_2 is higher than T_1, then $T_2/T_1 > 1$, the logarithm of this ratio is positive, and therefore ΔS is positive too, corresponding to the expected increase in entropy as the temperature is raised. The greater the heat capacity of the substance, the greater is the increase in entropy for a given change in temperature.

EXAMPLE 4F.3   Calculating the entropy change due to an increase in temperature

If you were an engineer in a power plant you might need to study the effects of heating on the air in a turbine. For some purposes, you can use nitrogen gas to model air. A sample of nitrogen gas of volume 20.0 L at 5.00 kPa is heated from 20. °C to 400. °C at constant volume. What is the change in the entropy of the nitrogen? The molar heat capacity of nitrogen at constant volume, $C_{V,m}$, is 20.81 J·K^{-1}·mol^{-1}.

ANTICIPATE Because the thermal disorder in a system increases as the temperature rises, you should expect a positive entropy change.

PLAN To use Eq. 4, first convert the temperature into kelvins and find the amount (in moles) of gas molecules by using the ideal gas law in the form $n = PV/RT$. Because the data are liters and kilopascals, use R expressed in those units. Then use Eq. 4, with the heat capacity at constant volume, $C_V = nC_{V,m}$. As always, avoid rounding errors by delaying the numerical calculation to the last possible stage.

What should you assume? Assume that nitrogen is an ideal gas (for the calculation of n) and that C_V is constant over the temperature range.

SOLVE

Convert temperatures into kelvins.

$$T_1 = 20. + 273.15 \text{ K} = 293 \text{ K}$$
$$T_2 = 400. + 273.15 \text{ K} = 673 \text{ K}$$

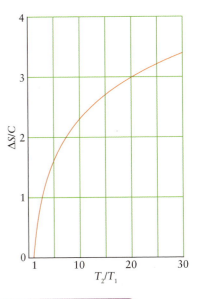

Find the amount of N_2 from $PV = nRT$ in the form $n = P_1V_1/RT_1$.

$$n = \frac{(5.00 \text{ kPa}) \times (20.0 \text{ L})}{(8.3145 \text{ L·kPa·K}^{-1}\text{·mol}^{-1}) \times (293 \text{ K})}$$

$$= \frac{5.00 \times 20.0}{8.3145 \times 293}\text{mol} = 0.0410\dots \text{ mol}$$

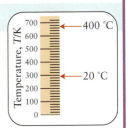

Calculate the change in entropy from $\Delta S = C \ln(T_2/T_1)$ with $C = nC_{V,m}$.

$$\Delta S = 0.0410\dots \text{ mol} \times (20.81 \text{ J·K}^{-1}\text{·mol}^{-1}) \times \ln\frac{673 \text{ K}}{293 \text{ K}}$$

$$= +0.710 \text{ J·K}^{-1}$$

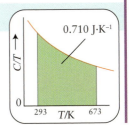

EVALUATE The change is positive (an increase), as expected.

Self-test 4F.4A The temperature of 1.00 mol He(g) is increased from 25 °C to 300. °C at constant volume. What is the change in the entropy of the helium? Assume ideal behavior and use the relation $C_{V,m} = \frac{3}{2}R$ from Topic 4C.

[***Answer:*** +8.15 J·K^{-1}]

Self-test 4F.4B The temperature of 5.5 g of stainless steel is increased from 20. °C to 100. °C. What is the change in the entropy of the stainless steel? The specific heat capacity of stainless steel is 0.51 J·(°C)$^{-1}$·g^{-1}.

Related Exercises 4F.7, 4F.8

The strategy of identifying a reversible path between the initial and final states can be applied to any change, even when more than one variable changes.

EXAMPLE 4F.4   Calculating the change in entropy when both temperature and volume change

In the production of iron in a blast furnace, compressed air and oxygen are driven into molten iron ore. Engineers must understand the thermodynamic aspects of each part of the process so that it can be optimized, including changes to the entropy of oxygen. In one experiment, 1.00 mol O_2(g) was compressed suddenly (and irreversibly) from 5.00 L to 1.00 L by driving in a piston, and in the process its temperature was increased from 20.0 °C to 25.2 °C. What is the change in entropy of the gas?

ANTICIPATE The entropy will decrease when a gas is compressed, but increase when its temperature rises. Because the temperature rise is small, you should expect an overall decrease, but you must do the calculation to be sure.

PLAN Because entropy is a state function, calculate the change in entropy by choosing a reversible path that results in the same final state. In this case, two variables are changed, so consider the following two hypothetical steps:

Step 1 Reversible isothermal compression at the initial temperature from the initial volume to the final volume (use Eq. 2).

Step 2 An increase in temperature of the gas at constant final volume to the final temperature (use C_V in Eq. 4).

What should you assume? Assume that oxygen is an ideal gas and that C_V is constant over the temperature range.

SOLVE

Step 1 From Eq. 2, $\Delta S = nR \ln(V_2/V_1)$,

$$\Delta S = (1.00 \text{ mol}) \times (8.3145 \text{ J·K}^{-1}\text{·mol}^{-1}) \times \ln\frac{1.00 \text{ L}}{5.00 \text{ L}}$$

$$= -13.4 \text{ J·K}^{-1}$$

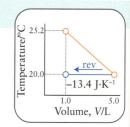

Step 2 From $\Delta S = C_V \ln(T_2/T_1)$ with $C_V = nC_{V,m}$,

$$\Delta S = (1.00 \text{ mol}) \times (20.79 \text{ J·K}^{-1}\text{·mol}^{-1}) \times \ln\frac{298.4 \text{ K}}{293.2 \text{ K}}$$

$$= +0.36 \text{ J·K}^{-1}$$

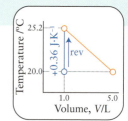

Add the entropy changes for the two steps, $\Delta S = \Delta S(\text{Step 1}) + \Delta S(\text{Step 2})$.

$$\Delta S = (-13.4 + 0.36) \text{ J·K}^{-1} = -13.0 \text{ J·K}^{-1}$$

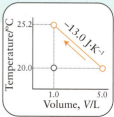

EVALUATE As expected, the decrease in entropy due to the fivefold compression is much greater than that due to the small rise in temperature.

The entropy of a system increases when its temperature increases and when its volume increases.

4F.5 Entropy and Physical State

When a solid melts, its molecules become more disordered as they form a liquid; thus, you should expect its entropy to increase. Similarly, you should expect an even greater increase in entropy when a liquid vaporizes, because its molecules then occupy a much greater volume and their motion is highly chaotic. To use Eq. 1 to calculate the entropy change for a substance undergoing a transition from one phase to another at its transition temperature, you need to note three facts:

1. At the transition temperature (such as the boiling point, if the transition is vaporization), the temperature of the substance remains constant as heat is supplied.

All the energy supplied is used to drive the phase transition, such as the conversion of liquid into vapor, rather than to raise the temperature. The T in the denominator of Eq. 1 is therefore a constant and may be set equal to the transition temperature (in kelvins). The temperature at which a solid melts when the pressure is 1 atm is the **normal melting point**. The temperature at which a liquid boils when the pressure is 1 atm is the **normal boiling point**. When the pressure is 1 bar, these transition temperatures are called the **standard melting point**, T_f (the f stands for fusion, another term for melting) and the **standard boiling point**, T_b, respectively. In practice, there is little difference between the normal and standard values.

2. At the temperature of a phase transition, the transfer of heat is reversible.

Provided the external pressure is fixed (at 1 atm, for instance), raising the temperature of the surroundings an infinitesimal amount results in complete vaporization, and lowering the temperature causes complete condensation.

3. Because the transition takes place at constant pressure (for instance, 1 atm), the heat supplied is equal to the change in enthalpy of the substance (Topic 4C).

It follows that q_{rev} in the expression for the entropy change can be replaced by ΔH for the phase change.

The **entropy of vaporization,** ΔS_{vap} is the change in entropy per mole of molecules when a substance changes from a liquid into a vapor. The heat required per mole to vaporize the liquid at constant pressure is equal to the enthalpy of vaporization (ΔH_{vap}, Topic 4C). It then follows from Eq. 1, by setting $q_{rev} = \Delta H_{vap}$, that

$$\text{At the boiling temperature: } \Delta S_{vap} = \frac{\Delta H_{vap}}{T_b} \tag{5}$$

where ΔH_{vap} is the enthalpy of vaporization *at the boiling point*. The **standard entropy of vaporization,** ΔS_{vap} is obtained when the liquid and the vapor are in their standard states (both pure, both at 1 bar) and the boiling temperature is for 1 bar (not 1 atm). Once again, it is important to remember that this value is the change in molar entropy *at the temperature of the boiling point*, and might be quite different from its value at another temperature. Because all standard entropies of vaporization are positive, they are normally reported without their positive sign.

TABLE 4F.1   Standard Entropy of Vaporization at the Normal Boiling Point*

Liquid	T_b/K	$\Delta S_{vap}°/(J \cdot K^{-1} \cdot mol^{-1})$
acetone	329.4	88.3
ammonia	239.7	97.6
argon	87.3	74
benzene	353.2	87.2
ethanol	351.5	124
helium	4.22	20.
mercury	629.7	94.2
methane	111.7	73
methanol	337.8	105
water	373.2	109

*The normal boiling point is the boiling temperature at 1 atm.

What Does This Equation Tell You? If a substance has very strong intermolecular forces, a lot of energy will be required to vaporize it and ΔS_{vap} will be large. However, strong intermolecular forces also increase the boiling temperature, which is in the denominator. Therefore, these effects may balance and the entropy of vaporization may be similar for many substances.

Self-test 4F.6A Calculate the standard entropy of vaporization of argon at its normal boiling point (see Table 4C.1).

[***Answer:*** 74 $J \cdot K^{-1} \cdot mol^{-1}$]

Self-test 4F.6B Calculate the standard entropy of vaporization of water at its normal boiling point (see Table 4C.1).

TABLE 4F.1 lists the standard entropies of vaporization of a number of liquids at their boiling points. These and other data show a striking pattern: many values are close to 85 $J \cdot K^{-1} \cdot mol^{-1}$. This observation is called **Trouton's rule.** The explanation of Trouton's rule is that approximately the same increase in positional disorder occurs when any liquid is converted into vapor, and so the change in entropy can be expected to be much the same in each case. If a liquid does not obey Trouton's rule, it is probably because its molecules have a more orderly arrangement in the liquid than is typical of most liquids, and so there is a greater increase in disorder and therefore a higher entropy of vaporization when such a liquid vaporizes. The molecules in a liquid are relatively highly ordered if there are extensive intermolecular interactions. Liquids such as water that can take part in extensive hydrogen bonding typically have entropies of vaporization greater than 85 $J \cdot K^{-1} \cdot mol^{-1}$.

THINKING POINT

Mercury atoms do not take part in hydrogen bonding. Why, then, does liquid mercury have such a high entropy of vaporization?

Self-test 4F.7A Use Trouton's rule to estimate the standard enthalpy of vaporization of liquid bromine, which boils at 59 °C.

[***Answer:*** 28 $kJ \cdot mol^{-1}$ (experimentally, 29.45 $kJ \cdot mol^{-1}$)]

Self-test 4F.7B Use Trouton's rule to estimate the standard enthalpy of vaporization of diethyl ether, $C_4H_{10}O$, which boils at 34.5 °C.

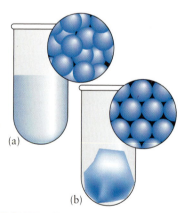

(a)

(b)

FIGURE 4F.5 A representation of the arrangement of molecules in (a) a liquid and (b) a solid. When the liquid freezes, the molecules lie in an ordered array; consequently, there is a decrease in the disorder of the system (a drop in entropy). The entropy rises again when the solid melts.

A smaller increase in entropy occurs when solids melt than when liquids vaporize, because a liquid is only slightly more disordered than a solid (**FIG. 4F.5**). By applying the same argument used for vaporization to the standard entropy of fusion of a substance at its melting (or freezing) point,

$$\text{At the freezing temperature: } \Delta S_{fus}° = \frac{\Delta H_{fus}°}{T_f} \qquad (6)$$

Here, $\Delta H_{fus}°$ is the standard enthalpy of fusion at the melting point and T_f is the temperature at the melting point. Almost all entropies of fusion are positive (there is a single known exception: solid helium-3), and so they are normally reported without their positive sign.

> **Self-test 4F.8A** Calculate the standard entropy of fusion of mercury at its freezing point (see Table 4C.1).
>
> [**Answer:** 9.782 J·K^{-1}·mol^{-1}]
>
> **Self-test 4F.8B** Calculate the standard entropy of fusion of benzene at its melting point (see Table 4C.1).

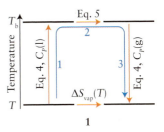

Equations 5 and 6 give the entropy change *at the transition temperature*. To find the entropy of transition at another temperature, the calculation has to be broken down into three steps (**1**). For example, to find the entropy of vaporization of water at 25 °C and 1 bar:

1. Heat the liquid from 25 °C to its boiling point, 100. °C,
2. Allow it to vaporize.
3. Cool the vapor back to 25 °C.

Because entropy is a state function, the changes in entropy for each of the individual steps can be added together to obtain the entropy of vaporization at 25 °C.

EXAMPLE 4F.5   Calculating the entropy of vaporization at temperatures other than the boiling point

Suppose you were investigating the correlation between intermolecular forces and the way in which molecules cohere to each other in a liquid and decided that you can gain some insight by comparing the entropies of vaporization of several liquids. Calculate the entropy of vaporization of acetone at 296 K with an external pressure of 1 bar. The molar heat capacity of liquid acetone is 127 J·K^{-1}·mol^{-1}, its boiling point is 329.4 K, and its enthalpy of vaporization is 29.1 kJ·mol^{-1}.

PLAN

Step 1 Use Eq. 4 to calculate the entropy change accompanying the heating of acetone from the "initial" temperature of 296 K to the "final" temperature of 329.4 K. The heat capacity of liquid acetone is approximately constant over this range.

Step 2 Use Eq. 5 to calculate the entropy of vaporization or look up the value in Table 4F.1.

Step 3 Use Eq. 4 to calculate the entropy change accompanying the cooling of acetone vapor from the new "initial" temperature of 329.4 K back to the new "final" temperature of 296 K (a negative value). The heat capacity of acetone vapor may be estimated from the equipartition theorem (Topic 4B) as $C_{P,m} = 4R$.

Step 4 Sum the three entropy changes to give $\Delta S_{vap}°(296 \text{ K})$.

What should you assume? Assume that acetone vapor is an ideal gas and that only translations and rotations contribute to the gas-phase heat capacity.

SOLVE

Step 1 Heat the liquid acetone at constant pressure. From $\Delta S = C_P \ln(T_2/T_1)$.

$$\Delta S = (127 \text{ J·K}^{-1}\text{·mol}^{-1}) \ln\frac{329.4 \text{ K}}{296 \text{ K}} = +13.5 \text{ J·K}^{-1}\text{·mol}^{-1}$$

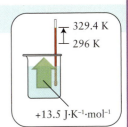

Step 2 Vaporize the acetone at its boiling point. From Table 4F.1 or Eq. 5,

$$\Delta S_{vap}° = \frac{29\,100 \text{ J·mol}^{-1}}{329.4} = +88.3 \text{ J·K}^{-1}\text{·mol}^{-1}$$

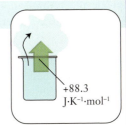

Step 3 Cool the acetone vapor. From $\Delta S = C_P \ln(T_2/T_1)$ and $C_{P,m} = 4R$,

$$\Delta S = \overbrace{(33.26\ \text{J·K}^{-1}\text{·mol}^{-1})}^{4R} \ln\frac{296\ \text{K}}{329.4\ \text{K}} = -3.54\ \text{J·K}^{-1}\text{·mol}^{-1}$$

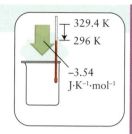

329.4 K
296 K

−3.54 J·K⁻¹·mol⁻¹

Step 4 Sum the entropies calculated in the previous steps.

$$\Delta S_{\text{vap}}°(296\ \text{K}) = (13.5 + 88.3 - 3.54)\ \text{J·K}^{-1}\text{·mol}^{-1}$$
$$= +98.3\ \text{J·K}^{-1}\text{·mol}^{-1}$$

Self-test 4F.9A Calculate the standard entropy of vaporization of ethanol, C_2H_5OH, at 285.0 K, given that the molar heat capacity at constant pressure of ethanol vapor is 78.3 J·K⁻¹·mol⁻¹ in this range (Tables 4A.2 and 4C.1).

[*Answer:* 140. J·K⁻¹·mol⁻¹]

Self-test 4F.9B Calculate the standard entropy of vaporization of benzene, C_6H_6, at 276.0 K, given that the molar heat capacity at constant pressure of benzene vapor is 82.4 J·K⁻¹·mol⁻¹ in this range (Tables 4A.2 and 4C.1).

Related Exercises 4F.17, 4F.18

The entropy of a substance increases when it melts and when it vaporizes.

What have you learned in this Topic?

You now know what "spontaneous" means in thermodynamics. You have learned that the entropy is a measure of the degree of disorder of an isolated system and that it increases in any spontaneous process. You have also learned that the entropy of a substance increases as its volume or temperature increases and when it melts or vaporizes.

The skills you have mastered are the ability to:

☐ **1.** Calculate the entropy change for reversible heat transfer (Example 4F.1).

☐ **2.** Evaluate the entropy change for the isothermal expansion or compression of an ideal gas (Examples 4F.2 and 4F.4).

☐ **3.** Calculate the entropy change when the temperature of a substance is changed (Examples 4F.3 and 4F.4).

☐ **4.** Calculate the standard entropy of a phase change at the transition temperature and at a temperature other than the transition temperature (Example 4F.5).

Topic 4F Exercises

4F.1 A human body generates heat at the rate of about 100. W (1 W = 1 J·s⁻¹). (a) At what rate does your body heat generate entropy in your surroundings, taken to be at 20. °C? (b) How much entropy do you generate each day? (c) Would the entropy generated be greater or less if you were in a room kept at 30. °C? Explain your answer.

4F.2 An aquarium heater is rated at 362 W (1 W = 1 J·s⁻¹). (a) At what rate does it generate entropy in a large aquarium maintained at 27 °C? (b) How much entropy does it generate in the course of a day? (c) Would the entropy generated be greater or less if the aquarium were maintained at 22 °C? Explain your answer.

4F.3 (a) Calculate the change in entropy of a block of copper at 25 °C that absorbs 65 J of energy from a heater. (b) If the block of copper is at 100. °C and it absorbs 65 J of energy from the heater, what is its entropy change? (c) Explain any difference in entropy change.

4F.4 (a) Calculate the change in entropy of 255 g of water at 0.0 °C when it absorbs 326 J of energy from a heater. (b) If the 1.0 L of water is at 99 °C, what is its entropy change? (c) Explain any difference in entropy change.

4F.5 Calculate the entropy change associated with the isothermal expansion of 5.25 mol of ideal gas atoms from 24.252 L to 34.058 L.

4F.6 Calculate the entropy change associated with the isothermal compression of 0.720 mol of ideal gas atoms from 24.32 L to 3.90 L.

4F.7 Assuming that the heat capacity of an ideal gas is independent of temperature, calculate the entropy change associated with raising the temperature of 1.00 mol of ideal gas atoms reversibly from 37.6 °C to 157.9 °C at (a) constant pressure and (b) constant volume.

4F.8 Assuming that the heat capacity of an ideal gas is independent of temperature, calculate the entropy change associated with lowering the temperature of 4.10 mol of ideal gas atoms from

225.71 °C to −12.50 °C at (a) constant pressure and (b) constant volume.

4F.9 Calculate the change in entropy when the pressure of 1.50 mol Ne(g) is decreased isothermally from 15.0 atm to 0.500 atm. Assume ideal behavior.

4F.10 Calculate the change in entropy when the pressure of 70.9 g of methane gas is increased isothermally from 7.00 kPa to 350.0 kPa. Assume ideal behavior.

4F.11 During the test of an internal combustion engine, 3.00 L of nitrogen gas at 18.5 °C was compressed suddenly (and irreversibly) to 0.500 L by driving in a piston. In the process, the temperature of the gas increased to 28.1 °C. Assume ideal behavior. What is the change in entropy of the gas?

4F.12 Calculate the change in entropy when the pressure of 5.75 g of helium gas is decreased from 320.0 kPa to 40.0 kPa while the temperature decreases from 423 K to 273 K. Assume ideal behavior.

4F.13 Use data in Table 4C.1 or Appendix 2A to calculate the entropy change for (a) the freezing of 1.00 mol $H_2O(l)$ at 0.00 °C; (b) the vaporization of 50.0 g of ethanol, C_2H_5OH, at 351.5 K.

4F.14 Use data in Table 4C.1 to calculate the entropy change for (a) the vaporization of 2.40 mol $H_2O(l)$ at 100. °C and 1 atm; (b) the freezing of 4.50 g of ethanol, C_2H_5OH, at 158.7 K.

4F.15 (a) Using Trouton's rule, estimate the boiling point of dimethyl ether, CH_3OCH_3, given that $\Delta H_{vap}° = 21.51$ kJ·mol^{-1}. (b) Using standard reference sources available in your library or on the Internet, find the actual boiling point of dimethyl ether and compare this value with the value obtained by using Trouton's rule. Explain any difference.

4F.16 (a) Using Trouton's rule, estimate the boiling point of methylamine, CH_3NH_2, given that $\Delta H_{vap}° = 25.60$ kJ·mol^{-1}. (b) Using standard reference sources available in your library or on the Internet, find the actual boiling point of methylamine and compare this value with the value obtained by using Trouton's rule. Explain any difference.

4F.17 Calculate the standard entropy of vaporization of water at 85 °C, given that its standard entropy of vaporization at 100. °C is 109.0 J·K^{-1}·mol^{-1} and the molar heat capacities at constant pressure of liquid water and water vapor are 75.3 J·K^{-1}·mol^{-1} and 33.6 J·K^{-1}·mol^{-1}, respectively, in this range.

4F.18 Calculate the standard entropy of vaporization of ammonia at 210.0 K, given that the molar heat capacities at constant pressure of liquid ammonia and ammonia vapor are 80.8 J·K^{-1}·mol^{-1} and 35.1 J·K^{-1}·mol^{-1}, respectively, in this range (see Table 4C.1).

Topic 4G The Molecular Interpretation of Entropy

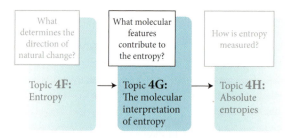

> What determines the direction of natural change?
>
> What molecular features contribute to the entropy?
>
> How is entropy measured?
>
> Topic **4F:** Entropy → Topic **4G:** The molecular interpretation of entropy → Topic **4H:** Absolute entropies

Why Do You Need to Know This Material? An important skill to develop is the ability to relate bulk thermodynamic properties to the properties of the molecules forming the sample.

What Do You Need to Know Already? You need to be aware of the thermodynamic definition of entropy (Topic 4F) and the quantum mechanical properties of a particle in a box (Topic 1C).

FIGURE 4G.1 Ludwig Boltzmann (1844–1906). (© *bilwissedition Ltd. & Co. KG/Alamy.*)

The set of hypothetical replications of a system corresponding to the same conditions is called an *ensemble.*

In Topic 4F, entropy is introduced as a measure of molecular disorder. What exactly does that mean and can the concept be expressed more precisely? In 1877 the Austrian physicist Ludwig Boltzmann (**FIG. 4G.1**) proposed a definition of entropy that deepened scientists' insight into the meaning of entropy at the molecular level. Moreover, his formula provides a way to calculate the entropy itself rather than, as in the thermodynamic definition ($\Delta S = q_{rev}/T$), being able to deduce only changes in its value.

4G.1 The Boltzmann Formula

The **Boltzmann formula** for the entropy is

$$S = k \ln W \tag{1}$$

where k is what is now called **Boltzmann's constant,** $k = 1.381 \times 10^{-23}$ J·K^{-1}. This constant also appears in Topic 4B in connection with the equipartition theorem, where it is pointed out that it is related to the gas constant by $R = kN_A$. The quantity W is the *number of ways that the atoms or molecules in the sample can be arranged with the same total energy.* Each arrangement of the molecules in a sample is a different state called a **microstate,** so W is equal to the number of different microstates that correspond to the same energy. A microstate typically lasts only for an instant; so, when the bulk properties of a system are measured, a time average is being taken over the many microstates that the system has occupied during the measurement. The entropy calculated from the Boltzmann formula is called the **statistical entropy.**

To understand Eq. 1, you need to imagine a large number of replicas of the system that is being investigated, with each copy having the same total energy. Although the energy of each replica is the same, individual molecules may be distributed differently over the available energy levels. Now imagine that all the replicas are placed in a box, and you select them blindly, one by one. If there is only one way of achieving the energy (for instance, all the molecules are in exactly the same energy level), then $W = 1$. In this case, every replica that you select will have exactly the same distribution of molecules and will be in exactly the same microstate. Such a system has zero entropy ($\ln 1 = 0$) and zero disorder, because each time you make a selection you find the system in an identical microstate. You can be absolutely certain that the selection will be the same in each case. However, if there is more than one way of arranging the molecules in the sample to achieve the same energy, then not every replica that you draw will be the same. For instance, if a certain energy can be achieved with 1000 different microstates ($W = 1000$), then there is only 1 chance in 1000 of drawing a *specific* one of those microstates. Because the disorder of the system is higher for this state than for that with $W = 1$, the entropy for this state is greater than zero.

EXAMPLE 4G.1   Calculating the statistical entropy

Sometimes scientists are surprised by apparently anomalous properties of the materials they study. When those anomalies are considered, they can reveal information about the structure of matter. Calculate the entropy of a tiny solid made up of four diatomic molecules of a compound such as carbon monoxide, CO, at $T = 0$ when (a) the four molecules have formed a perfectly ordered crystal in which all molecules are aligned with their C atoms on the left (top left image in **FIG. 4G.2**) and (b) the four molecules lie in random orientations (but parallel, as in any of the images in Fig. 4G.2).

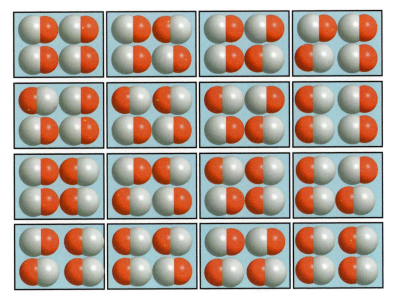

FIGURE 4G.2 A tiny sample of solid carbon monoxide consisting of four molecules. Each box represents a different arrangement of the four molecules. When there is only one way of arranging the four molecules so that they all lie with the carbon atom on the left (as in the top left image), the entropy of the solid is zero. When all 16 ways of arranging four CO molecules are accessible, the entropy of the solid is greater than zero.

ANTICIPATE You should expect that at $T = 0$ the sample in (a) will have zero entropy, because there is no disorder in either location or energy—all the molecules are in their lowest possible energy state. The sample in (b) has greater disorder, so you should expect its entropy to be greater than 0.

PLAN To find the number of microstates for a system, first determine the number of orientations, O, that each molecule can adopt and that have the same energy. Since each molecule can adopt O orientations, the number of microstates is O taken N times, $O \times O \times O \ldots$, where N is the total number of molecules in the system. The number of microstates in a system of this kind is therefore $W = O^N$. Then use that value of W in the Boltzmann formula (Eq. 1) to calculate the entropy of the sample.

SOLVE (a) Because there is only one way of arranging the molecules in the perfect crystal, $W = 1$.

From $S = k \ln W$ and $W = 1$,

$$S = k \ln 1 = 0$$

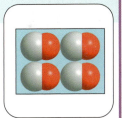

In this case, every arrangement is identical, and you can be sure that any selection results in the same microstate.
(b) Because each of the four molecules can take either of two orientations, the total number of ways of arranging them is 2 for each molecule, taken 4 times overall (that is, $O = 2$ and $N = 4$):

$$W = 2 \times 2 \times 2 \times 2 = 2^4$$

or 16 different possible arrangements corresponding to the same total energy. Now there is only 1 chance in 16 of selecting a given microstate, and so you can be less sure about the actual state of the system. The entropy of this tiny solid is therefore

From $S = k \ln W$ and $W = 2^4$,

$$S = k \ln 2^4 = (1.381 \times 10^{-23}\,\text{J·K}^{-1}) \times \ln 16$$
$$= 3.828 \times 10^{-23}\,\text{J·K}^{-1}$$

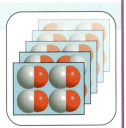

EVALUATE As expected, the entropy of the disordered solid is higher than that of the perfectly ordered solid.

Self-test 4G.1A Calculate the entropy of a sample of a solid in which the molecules may be found in any one of three orientations with the same energy at $T = 0$. Suppose there are 30 molecules in the sample.

[**Answer:** $4.5 \times 10^{-22}\,\text{J·K}^{-1}$]

Self-test 4G.1B Calculate the entropy of a sample of a solid in which it is supposed that a substituted benzene molecule, C_6H_5F, may be found in any one of six orientations with the same energy at $T = 0$. Suppose there is 1.0 mol of molecules in the sample.

Related Exercises 4G.1, 4G.2

For a more realistic sample size than that in Example 4G.1, one that contains 1.00 mol CO, corresponding to 6.02×10^{23} CO molecules, each of which could be oriented in either of two ways, there are $2^{6.02 \times 10^{23}}$ (an astronomically large number) different microstates, and a chance of only 1 in $2^{6.02 \times 10^{23}}$ of drawing a given microstate in a blind selection. The entropy at $T = 0$ is

$$\overset{\ln x^a = a \ln x}{S = k \ln 2^{6.02 \times 10^{23}} = k \times (6.02 \times 10^{23}) \times \ln 2}$$

$$= (1.381 \times 10^{-23} \, \text{J·K}^{-1}) \times (6.02 \times 10^{23}) \times \ln 2$$

$$= 5.76 \, \text{J·K}^{-1}$$

When the entropy of 1.00 mol CO(s) is actually measured at temperatures close to $T = 0$ (by using the technique in Topic 4G), the value found is 4.6 J·K^{-1}. This value—which is called the **residual entropy** of the sample, the entropy of a sample at $T = 0$ arising from positional disorder surviving at that temperature—is commonly regarded as close enough to 5.76 J·K^{-1} to suggest that in the crystal the molecules are indeed arranged nearly randomly. The reason for this randomness is that the electric dipole moment of a CO molecule is very small, and so there is little energy advantage in the molecules lying head to tail compared to head to head or tail to tail; thus the molecules lie in either orientation. For solid HCl, the same experimental measurements give $S \approx 0$, close to $T = 0$. This value indicates that the bigger dipole moments of the HCl molecules pull them into an orderly arrangement; so they lie strictly head to tail and there is no positional disorder at $T = 0$.

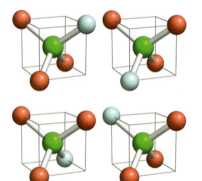

FIGURE 4G.3 The four possible orientations of a tetrahedral FClO$_3$ molecule in a solid.

EXAMPLE 4G.2   Using the Boltzmann formula to interpret a residual entropy

The entropy of a substance often gives a clue about its structure, including how its molecules are arranged in a solid. The entropy of 1.00 mol FClO$_3$(s) at $T = 0$ is 10.1 J·K^{-1}. Suggest an interpretation.

ANTICIPATE The observation that the entropy is nonzero at $T = 0$ suggests that the molecules are positionally disordered at that temperature.

PLAN From the shape of the molecule (which can be predicted by using VSEPR theory, Topic 2E), determine how many orientations, W, it is likely to be able to adopt in a crystal; then use the Boltzmann formula to see whether that number of orientations leads to the observed value of S.

SOLVE An FClO$_3$(s) molecule is tetrahedral, and—provided the energy of each orientation is almost the same—can take any of four orientations in the solid, with the F atom in a different position in each case (**FIG. 4G.3**). The total number of ways of arranging the molecules in a crystal containing N molecules with $O = 4$ is therefore

$$W = (4 \times 4 \times 4 \times \cdots \times 4)_{N \text{ times}} = 4^N$$

The entropy is

From $S = k \ln W$, $W = 4^N$, $N = 6.02 \times 10^{23}$, and $\ln x^a = a \ln x$

$$S = k \ln 4^{6.02 \times 10^{23}}$$

$$= (1.381 \times 10^{-23} \, \text{J·K}^{-1}) \times (6.02 \times 10^{23}) \times \ln 4$$

$$= 11.5 \, \text{J·K}^{-1}$$

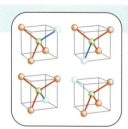

EVALUATE This value is reasonably close to the experimental value of 10.1 J·K^{-1}, which suggests that, at $T = 0$, the molecules lie nearly randomly in any of four possible orientations.

Self-test 4G.2A Explain the observation that the entropy of 1 mol N$_2$O(s) at $T = 0$ is 6 J·K^{-1}.
[**Answer:** In the crystal, the orientations NNO and ONN are equally likely.]

Self-test 4G.2B Suggest a reason why the entropy of ice is nonzero at $T = 0$; think about how the structure of ice is affected by the hydrogen bonds.

Related Exercises 4G.7, 4G.8

The Boltzmann formula relates the entropy of a substance to the number of arrangements of molecules that result in the same energy: the residual entropy is due to positional disorder at T = 0 when more than one orientation of a molecule is possible.

4G.2 The Equivalence of Statistical and Thermodynamic Entropies

The thermodynamic definition of entropy in Topic 4F ($\Delta S = q_{rev}/T$) and Boltzmann's formula for the statistical entropy ($S = k \ln W$) look very different. Moreover, the former comes from observations of the behavior of bulk matter and the latter by statistical analysis of molecular behavior. To verify that the two definitions are in fact consistent, it is necessary to show that the entropy changes predicted by each of them are the same. In the process of developing these ideas, you will also deepen your understanding of what is meant by "disorder."

One approach is to show that the Boltzmann formula predicts the same volume dependence of the entropy of an ideal gas as that obtained from the thermodynamic definition (Eq. 2 in Topic 4F, $\Delta S = nR \ln(V_2/V_1)$).

An ideal gas consists of a large number of molecules that can be viewed as occupying the energy levels characteristic of a particle in a box (Topic 1C). For simplicity, consider a one-dimensional box (**FIG. 4G.4a**), but the same considerations apply to a real three-dimensional container of any shape. At $T = 0$ (not shown), only the lowest energy level is occupied; so $W = 1$ and the entropy is zero. There is no "disorder," because which state each molecule occupies is known.

At any temperature $T > 0$ the molecules occupy large numbers of different energy levels, as shown in Fig. 4G.4. Because the molecules can now be found in a larger number of microstates, $W > 1$. Therefore, the entropy (which is proportional to $\ln W$) is greater than zero. Now the disorder has increased; consequently you can be less confident about precisely which state a given molecule occupies. For instance, in a tiny system with only two molecules, A and B, at $T = 0$ you can be confident that both molecules occupy the lowest energy level. However, if at a higher temperature a second state becomes available, you don't know whether it is molecule A that occupies the higher level, with B still in the ground state, or vice versa.

When the length of the box is increased at constant temperature (with $T > 0$), more energy levels become accessible to the molecules because the levels now lie closer together (Fig. 4G.4b). According to the discussion in Topic 1C, the separation of energy levels is proportional to $1/\text{length}^2$. Because the molecules are now distributed across more levels, you can be less sure about which energy level any given molecule occupies. Therefore, the value of W increases as the box is lengthened and, by the Boltzmann formula, the entropy increases, too. The same argument applies to a three-dimensional box: as the volume of the box increases, the number of accessible states increases, too. This relationship can be used to show that the thermodynamics expression for ΔS can be deduced from Eq. 1.

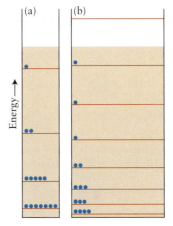

FIGURE 4G.4 (a) The energy levels of a particle in a box at $T > 0$. (b) The levels become closer together as the width of the box is increased. As a result, the number of levels accessible to the particles in the box at a given temperature increases, and the entropy of the system increases accordingly. The tinted band shows the range of thermally accessible levels. The change from part (a) to part (b) is a model of the isothermal expansion of an ideal gas. The total energy of the particles is the same in each case; only the distribution of energy levels differs.

How Is That Done?

To deduce $\Delta S = nR \ln(V_2/V_1)$ from the Boltzmann formula, start by supposing that the number of microstates available to a single molecule is proportional to the volume available to it and write $W = \text{constant} \times V$. For N molecules, the number of microstates is proportional to the Nth power of the volume:

$$W = \overbrace{(\text{constant} \times V) \times (\text{constant} \times V) \cdots}^{\text{N factors}}_{N \text{ times}} = (\text{constant} \times V)^N$$

Therefore, the change in the statistical entropy, $S = k \ln W$, when a sample expands isothermally from volume V_1 to a volume V_2 is

$$\Delta S = k \ln (\text{constant} \times V_2)^N - k \ln (\text{constant} \times V_1)^N$$

$$\overset{\ln x^a = a \ln x}{=} Nk \ln(\text{constant} \times V_2) - Nk \ln(\text{constant} \times V_1)$$

$$\overset{\ln x - \ln y = \ln(x/y)}{=} Nk \ln \frac{\text{constant} \times V_2}{\text{constant} \times V_1} \overset{\text{cancel}}{=} Nk \ln \frac{V_2}{V_1}$$

TABLE 4G.1   **Standard Molar Entropy of Water at Various Temperatures***

Phase	Temperature/ °C*	$S_m°/$ $(J·K^{-1}·mol^{-1})$
solid	−273 (0 K)	3.4
	0	43.2
liquid	0	65.2
	20	69.6
	50	75.3
	100	86.8
vapor	100	196.9
	200	204.1

*The "standard molar entropy" is the molar entropy of the pure substance, in this case water, at 1 bar.

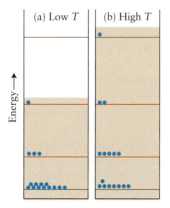

FIGURE 4G.5 More energy levels become accessible in a box of fixed width as the temperature is raised. The change from part (a) to part (b) is a model of the effect of heating an ideal gas at constant volume. The tinted band shows the range of thermally accessible levels. The average energy of the molecules also increases as the temperature is raised; that is, both internal energy and entropy increase with temperature.

This expression is identical to the thermodynamic expression because $N = nN_A$ (where N_A is Avogadro's constant) and $N_A k = R$, the gas constant; therefore,

$$\Delta S = \overbrace{Nk}^{nN_A} \ln \frac{V_2}{V_1} = \overbrace{nN_A k}^{R} \ln \frac{V_2}{V_1} = nR \ln \frac{V_2}{V_1}$$

In at least the case of the isothermal expansion of an ideal gas, the two definitions of entropy give identical results. Similarly, the Boltzmann formula predicts the increase in entropy of a substance as its temperature is raised (**TABLE 4G.1**). The same particle-in-a-box model of a gas can be used, but this reasoning also applies to liquids and solids, even though their energy levels are much more complicated. At low temperatures, the molecules of a gas can occupy only a few of the energy levels, so W is small and the entropy is low. As the temperature is raised, the molecules have access to larger numbers of energy levels (**FIG. 4G.5**); so W rises and the entropy increases, too.

The more general properties of the thermodynamic and statistical entropies are the same, too:

1.   Both are state functions.

Because the number of microstates available to the system depends only on its current state, not on its past history, W depends only on the current state of the system, and therefore the statistical entropy does, too.

2.   Both are extensive properties.

Doubling the number of molecules increases the number of microstates from W to W^2, and so the entropy changes from $k \ln W$ to $k \ln W^2$, or $2k \ln W$. Therefore, the statistical entropy, like the thermodynamic entropy, is an extensive property.

3.   Both increase in a spontaneous change.

In any irreversible change, the overall disorder of the system and its surroundings increases, which means that the number of microstates increases. If W increases, then so does $\ln W$, and the statistical entropy increases, too.

4.   Both increase with temperature.

When the temperature of the system increases, more microstates become accessible, and so the statistical entropy increases.

> *The equations used to calculate changes in the statistical entropy and the thermodynamic entropy lead to the same result.*

What have you learned in this Topic?

You have learned that there is a connection between the entropy and the distribution of molecules over the available energy levels of a system and that the connection is expressed by Boltzmann's formula. You have seen how his formula captures the volume dependence of the entropy of an ideal gas and have seen how it accounts for the increase in entropy as the temperature is raised. You also know that the statistical entropy has the same properties as the thermodynamic entropy. You now know that the entropy of a perfect crystal is zero at $T = 0$, and are familiar with the term "residual entropy."

The skills you have mastered are the ability to:

☐ **1.** Use the Boltzmann formula to calculate and interpret the entropy of an ideal gas (Example 4G.1).

☐ **2.** Predict the value of the residual entropy of simple substances (Example 4G.2).

Topic 4G Exercises

4G.1 Nanotechnologists have found ways to create and manipulate structures containing only a few molecules. However, orienting the molecules in specific ways to assemble such structures can be difficult. Calculate the entropy of a solid nanostructure made of 64 molecules in which the molecules (a) are all aligned in the same direction; (b) lie in any one of four orientations with the same energy.

4G.2 As in Exercise 4G.1, calculate the entropy of a solid nanostructure made of 16 molecules in which the molecules (a) are all aligned in the same direction; (b) lie in any one of three orientations with the same energy.

4G.3 Which would you expect to have a higher molar entropy at $T = 0$, single crystals of BF_3 or of COF_2? Why?

4G.4 On the basis of the structures of each of the following molecules, predict which ones would be most likely to have a residual entropy in their crystal forms at $T = 0$: (a) CO_2; (b) NO; (c) N_2O; (d) Cl_2.

4G.5 Considering positional disorder, would you expect a crystal of octahedral *cis*-MX_2Y_4 to have the same, higher, or lower residual entropy than the corresponding trans isomer? Explain your conclusion.

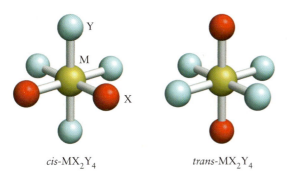

cis-MX_2Y_4 *trans*-MX_2Y_4

4G.6 There are three different substituted benzene compounds with the formula $C_6H_4F_2$. (a) Draw the structures of the three compounds. (b) Assume that the benzene rings pack similarly into their crystal lattices. If the positions of the H and F atoms are statistically disordered in the solid state, which isomer will have the *least* residual molar entropy?

4G.7 If tetrahedral SO_2F_2 adopts an orientationally disordered arrangement in its crystal form, what might its residual molar entropy be?

4G.8 What might you expect the residual molar entropy of trigonal pyramidal PH_2F to be if it were to adopt an orientationally disordered arrangement in its crystal form?

4G.9 Suppose that you create two tiny systems consisting of three atoms each, and each atom can accept energy in quanta of the same magnitude. (a) How many distinguishable arrangements are there of two quanta of energy distributed among the three atoms in one of these systems? (b) You now bring the two tiny systems together. How many distinguishable arrangements are there if the two quanta of energy are distributed among the six atoms? (c) In which direction will the energy quanta migrate starting from the initial arrangement?

4G.10 Suppose that you create two tiny systems consisting of four atoms each, and each atom can accept energy in quanta of the same magnitude. (a) How many distinguishable arrangements are there of two quanta of energy distributed among the four atoms in one of these systems? (b) You now bring the two tiny systems together. How many distinguishable arrangements are there if the two quanta of energy are distributed among the eight atoms? (c) Which state is more disordered, that in part (a) or that in part (b)?

Topic 4H Absolute Entropies

4H.1   Standard Molar Entropies

4H.2   Standard Reaction Entropies

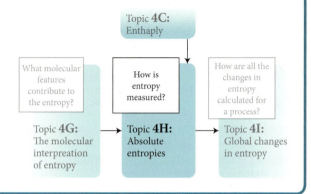

Why Do You Need to Know This Material?
One of the principal applications of thermodynamics is to decide whether a reaction is spontaneous; to do so, it is essential to know the contribution of the reactants and products to the entropy of the system.

What Do You Need to Know Already?
You need to know how to use the thermodynamic definition of entropy to calculate how it depends on the temperature (Topic 4F). This Topic also makes light use of the molecular interpretation of entropy (Topic 4G).

Entropy is a measure of disorder, and it is possible to imagine a perfectly orderly state of matter with neither positional nor, at $T = 0$, thermal disorder. Such a state represents a natural zero of entropy, a state of perfect order; therefore, it should be possible to set up an absolute scale of entropy. This idea is summarized by the **third law of thermodynamics:**

> The entropies of all perfect crystals approach zero as the absolute temperature approaches zero.

That is, $S \rightarrow 0$ as $T \rightarrow 0$. The "perfect crystal" part of this statement of the third law refers to a substance in which all the atoms are in a perfectly orderly array, and so there is no positional disorder. The $T \rightarrow 0$ part of the statement implies the absence of thermal motion—thermal disorder vanishes as the temperature approaches zero. As the temperature of a substance is raised from zero, more orientations become available to the molecules and their thermal disorder increases. Thus the entropy of any substance can be expected to be greater than zero above $T = 0$.

4H.1 Standard Molar Entropies

To measure the absolute entropy of a substance, you need to combine the thermodynamic definition, Eq. 1 of Topic 4F ($\Delta S = q_{rev}/T$), with the third law of thermodynamics. Because the third law implies that $S(0) = 0$,

$$S(T) = S(0) + \Delta S(\text{heating from 0 to } T) = \Delta S(\text{heating from 0 to } T)$$

The change in entropy upon heating is calculated from the heat capacity and the temperature, as explained in Topic 4F, where it is shown that if the heat capacity is constant throughout the temperature range of interest, then $\Delta S = nR \ln(T_2/T_1)$. However, the heat capacity is certainly not constant all the way down to $T = 0$, so a more general approach must be devised.

— **How Is That Done?** —

To take into account the possibility that the heat capacity changes with temperature, you need to refer to the expression derived in Topic 4F:

$$\Delta S(\text{heating from } T_1 \text{ to } T_2) = \int_{T_1}^{T_2} \frac{C dT}{T}$$

First, set $T_1 = 0$ and $T_2 = T$, the temperature of interest, in the limits of the integral. Heating commonly takes place at constant pressure, and so replace C by C_P. The expression then becomes

$$\Delta S(\text{heating from 0 to } T) = \int_0^T \frac{C_P dT}{T}$$

Therefore, by invoking the third law and setting $S(0) = 0$,

$$S(T) = \int_0^T \frac{C_P dT}{T}$$

As explained in Topic 4A, an integral of a function—in this case, the integral of C_P/T—is the area under the graph of the function between the two limits. Therefore, to measure the entropy of a substance, the heat capacity (typically the constant-pressure heat capacity) must be measured at all temperatures from $T = 0$ to the temperature of interest. Then the entropy of the substance is obtained by plotting C_P/T against T and measuring the area under the curve (FIG. 4H.1).

Note that C_P has been drawn as approaching zero as $T \rightarrow 0$. This feature is a general phenomenon and is explained by quantum mechanics. At low temperatures, the energy available is so small that there is not enough to stimulate transitions to higher energy states, so the sample cannot take up energy, and its "capacity for heat" is zero. This property of matter is one reason why the absolute zero of temperature has never been reached (BOX 4H.1).

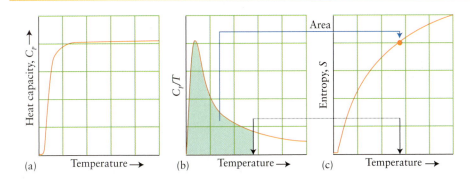

FIGURE 4H.1 The experimental determination of entropy. (a) The heat capacity (at constant pressure in this instance) of the substance is determined from close to absolute zero up to the temperature of interest. (b) The area under the plot of C_P/T against T is determined up to the temperature of interest. (c) This area is the entropy of the substance at that temperature.

Box 4H.1 FRONTIERS OF CHEMISTRY: THE QUEST FOR ABSOLUTE ZERO

The possibility of achieving absolute zero ($T = 0$) has intrigued scientists for many years. At absolute zero, it would be possible to study the properties of materials in their lowest allowed energy state. For instance, at very low temperatures an interference pattern of laser beams can be used to create an "optical lattice," a set of spatially symmetric potential energy wells in which atoms can be trapped to form perfect crystals. These arrays could be used to create tiny building blocks for very low-temperature nanoscale computers.

The third law of thermodynamics implies that it is not possible to reach absolute zero in a cyclical thermodynamic process with a finite number of steps (the version of the third law quoted in the text in terms of entropy is based on that observation). However, it may still be possible to reach absolute zero by using a noncyclic process, and record low temperatures have recently been reached by a variety of means.

One method to approach $T = 0$ is to alternate isothermal magnetization with adiabatic demagnetization. This approach takes advantage of the fact that, in a magnetic field, electrons with down spins ($\downarrow$) have slightly lower energies than electrons with up spins ($\uparrow$). If the up spins can be converted to down spins, the sample will have lower energy overall and therefore a lower temperature. If all the up spins could be converted to down spins, the temperature would reach absolute zero.

A cool paramagnetic sample that is not in a magnetic field has equal numbers of up and down electrons because the two spin states have the same energy (see Box 2G.1). This is true

whatever the temperature. So the first step is to cool the sample by conventional means (for instance, by putting it in contact with liquid helium, $T_b = 4.2$ K). Then a magnetic field is turned on. The down spin states now have lower energy than the up spin states, and so they are slightly more numerous. Call this state A. The sample in this state has a characteristic entropy that reflects the slightly greater number of down spins than up spins and the temperature of the sample. If there were even more down spins, the temperature of the sample would be reported as being lower. The task, therefore, is to turn some up spins into down spins. Two steps are used to achieve this conversion, and are summarized in the first figure.

In the first step, the magnetic field is increased. As a result, the energies of the two states diverge and the populations adjust by converting some up spins to down spins. This adjustment of populations takes place while the sample is in contact with its surroundings and occurs without change of temperature: it is the *isothermal magnetization step*. The entropy decreases in this step because, although the temperature remains the same, the spins have become more ordered. Call this state B.

In the second step, thermal contact with the surroundings is broken, so any subsequent change will be adiabatic—that is, not involve a transfer of energy as heat between the system and the surroundings. In any reversible adiabatic process the entropy does not change (because $\Delta S = q_{rev}/T$ and $q_{rev} = 0$). Under these adiabatic conditions, the magnetic field is reduced to the value it had in state A. However, although the magnetic field has been

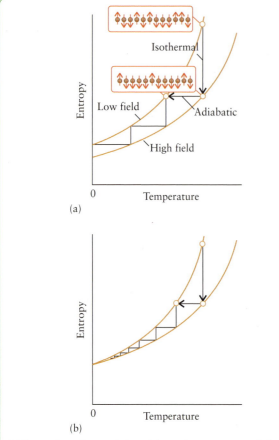

(a)

(b)

(a) The cooling of a sample by adiabatic demagnetization follows the steps shown here. The arrows in the inset represent the spin alignment of the electrons in the sample. The arrows in the graph show the direction the sample takes. First, the entropy decreases at constant temperature as a magnetic field is applied; then, as the sample is demagnetized in an adiabatic environment, its temperature decreases to maintain a constant entropy. If the two curves did not coincide, absolute zero could be reached. (b) However, the curves coincide before absolute zero can be reached.

reduced, the entropy is the same as in state B when a high magnetic field was present. This lower entropy corresponds to a lower temperature than the sample had initially: the sample has been cooled. The two steps are repeated until the temperature has been lowered as much as possible. However, because the two curves in the first illustration converge at $T = 0$, this cyclic process cannot achieve absolute zero itself. Sophisticated procedures of this kind have achieved temperatures as low as about 100 pK (1×10^{-10} K).

In a refinement of the process, nuclear spins are used instead of electron spins. Because nuclear magnetic moments are much weaker than electron magnetic moments, they do not interact so strongly with the surroundings and so are insulated from it more effectively. Very low temperatures can be reached using nuclear spins, which is called *nuclear adiabatic demagnetization*.

Other approaches have been used to reach very low temperatures. Because the average speed of atoms in a gas is proportional to the square root of the temperature, cooling can be achieved by slowing atoms down to very low speeds. In one such experiment, a group of 10 million rubidium atoms was slowed by arranging a beam of infrared radiation along the line of flight of the atoms in a beam traveling in the opposite direction. The impact of the infrared photons brought the atoms almost to a standstill. The atoms were then held in place by a magnetic field and cooled further by allowing the atoms with higher energies to escape.

At very low temperatures, atoms appear to lose their individual identities and, by a quantum mechanical process, form a "condensate," a new form of matter in which atoms merge to form a single large particle that behaves as a single atom. The second figure is a visualization of the condensate.

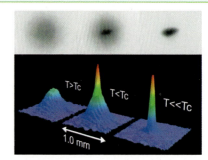

The sequence of images at the top shows how a cloud of atoms condenses as the temperature is decreased (left to right). The images below show temperature contour plots of the atom cloud at the three temperatures. *(CAL/NASA.)*

HOW MIGHT YOU CONTRIBUTE?

Methods for bringing large samples to very low temperatures need to be refined further. In addition, there are many unanswered questions about absolute zero that remain to be solved. For example, many substances become superconducting at very low temperatures. What would happen to conductivity at absolute zero? The U.S. National Aeronautics and Space Administration (NASA) is building a Cold Atom Laboratory on the International Space Station to study ultra-cold gases at temperatures lower than those possible to achieve on Earth, with the additional advantage of low gravity. This facility may open up more areas for research.

Related Exercises 4.53–4.56

For Further Reading P. Atkins, "The third law: The unattainability of zero," *Four Laws That Drive the Universe*, pp. 101–121, Oxford University Press, 2007. M. Lewenstein & A. Sanpera, "Probing quantum magnetism with cold atoms," *Science*, vol. 319, Jan. 18, 2008, pp. 292–293. Nova, "Absolute zero," http://www.pbs.org/wgbh/nova/zero. NASA Jet Propulsion Laboratory, "Cold Atom Laboratory: The coolest spot in the universe," http://coldatomlab.jpl.nasa.gov/.

TABLE 4H.1 Standard Molar Entropies at 25 °C*

Gases	$S_m°/(J \cdot K^{-1} \cdot mol^{-1})$	Liquids	$S_m°/(J \cdot K^{-1} \cdot mol^{-1})$	Solids	$S_m°/(J \cdot K^{-1} \cdot mol^{-1})$
ammonia, NH_3	192.4	benzene, C_6H_6	173.3	calcium oxide, CaO	39.8
carbon dioxide, CO_2	213.7	ethanol, C_2H_5OH	160.7	calcium carbonate, $CaCO_3^†$	92.9
hydrogen, H_2	130.7	water, H_2O	69.9	diamond, C	2.4
nitrogen, N_2	191.6			graphite, C	5.7
oxygen, O_2	205.1			lead, Pb	64.8

*Additional values are given in Appendix 2A.

†Calcite.

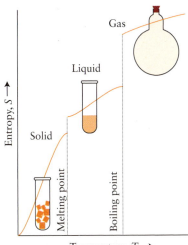

FIGURE 4H.2 The entropy of a solid increases as its temperature is raised. The entropy increases sharply when the solid melts to form the more disordered liquid and then gradually increases again up to the boiling point. A second, larger jump in entropy occurs when the liquid vaporizes. The shapes of the curves depend on the variation of the heat capacity with temperature.

You have seen that

$$S(T) = \text{area under the graph of } C_P/T \text{ against } T \text{ from 0 to the temperature of interest} \quad (1)$$

If a phase transition takes place between $T = 0$ and the temperature of interest, then the corresponding entropy of transition must be added by using Eq. 6 of Topic 4F ($\Delta S = \Delta H_{fus}/T_f$) with similar contributions from any other transitions (FIG. 4H.2). For instance, to measure the entropy of liquid water at 25 °C and 1 bar, it is necessary to measure the heat capacity of ice from $T = 0$ to $T = 273.15$ K, measure the entropy of fusion at that temperature from the enthalpy of fusion, and then measure the heat capacity of liquid water from $T = 273.15$ K to $T = 298.15$ K. TABLE 4H.1 gives selected values of the **standard molar entropy**, $S_m°$, the molar entropy of the pure substance at 1 bar, in this case at 298.15 K.

The molecular interpretation of entropy gives some insight into why some substances have high molar entropies and others have low molar entropies. For example, compare the molar entropy of diamond, 2.4 $J \cdot K^{-1} \cdot mol^{-1}$, with the much higher value for lead, 64.8 $J \cdot K^{-1} \cdot mol^{-1}$. The low entropy of diamond is what you should expect for a solid that has rigid bonds: at room temperature, its atoms are not able to jiggle around as much as the atoms of lead. Lead atoms are also much heavier than carbon atoms and have more thermally accessible vibrational energy levels.

The difference in molar entropy between two gases (compare H_2 and N_2 in Table 4H.1, for instance) can be understood in terms of the particle-in-a-box model and the fact that the greater the mass of the molecules, the closer together are the energy levels (FIG. 4H.3). You can also see that large, complex species have higher molar entropies than smaller, simpler ones in the same state of matter (compare $CaCO_3$ with CaO or C_2H_5OH with H_2O). Liquids have higher molar entropies than solids because the greater freedom of movement of the molecules in a liquid results in a less ordered state of matter. The molar entropies of gases, in which molecules occupy much larger volumes and undergo almost completely disordered motion, are substantially higher than those of the corresponding liquids.

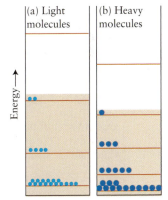

FIGURE 4H.3 The energy levels of a particle in a box are more widely spaced for (a) light molecules than for (b) heavy molecules. As a result, the number of thermally accessible levels, as shown by the tinted band, is greater for heavy molecules than for light molecules at the same temperature, and the entropy of the substance with heavy molecules is correspondingly greater.

Self-Test 4H.1A Which substance in each of the following pairs has the higher molar entropy: (a) $CO_2(g)$ at 25 °C and 1 bar or $CO_2(g)$ at 25 °C and 3 bar; (b) He(g) at 25 °C or He(g) at 100 °C in the same volume; (c) $Br_2(l)$ or $Br_2(g)$ at the same temperature? Explain your conclusions.

[*Answer:* (a) $CO_2(g)$ at 1 bar, because disorder increases with volume; (b) He(g) at 100 °C, because disorder increases with temperature; (c) $Br_2(g)$, because the vapor state has greater disorder than the liquid state.]

Self-Test 4H.1B Use the information in Table 4H.1 or Appendix 2A to determine which allotrope is the more ordered form and predict the sign of ΔS for each transition: (a) white tin changes into gray tin at 25 °C; (b) diamond changes into graphite at 25 °C.

Standard molar entropies increase as the complexity of a substance increases. The standard molar entropies of gases are higher than those of comparable solids and liquids at the same temperature.

4H.2 Standard Reaction Entropies

The entropy of a system changes as a result of a chemical reaction. In some cases, the sign of the entropy change of the system can be predicted without doing any calculation. For instance, a net increase in the amount of gas usually results in a positive change in entropy. Conversely, a net consumption of gas usually results in a negative change. Reactions that produce a large number of small molecules and the dissolving of a substance typically have positive entropies. However, for many reactions entropy changes are much more finely balanced and for these reactions numerical data must be used to calculate the sign of the entropy change of the system; numerical data must, of course, always be used to calculate actual numerical values.

To determine the change in entropy that accompanies a reaction, you need to know the molar entropies of all the substances taking part; then calculate the difference between the entropies of the products and those of the reactants. Specifically, the **standard reaction entropy**, $\Delta S°$, is the difference between the standard molar entropies of the products and those of the reactants, taking into account their stoichiometric coefficients. For example, the entropy change for the reaction $O_2(g) + 2\,H_2(g) \rightarrow 2\,H_2O(g)$ is

$$\Delta S° = (2\text{ mol}) \times S_m°(H_2O, g) - \{(1\text{ mol}) \times S_m°(O_2, g) + (2\text{ mol}) \times S_m°(H_2, g)\}$$

In general:

$$\Delta S° = \sum nS_m°(\text{products}) - \sum nS_m°(\text{reactants}) \tag{2}$$

The first term on the right is the total standard entropy of the products, and the second term is that of the reactants. In each case the standard molar entropy of a substance is multiplied by its amount (in moles) as given by the stoichiometric coefficient in the chemical equation. The standard molar entropies of many elements and compounds at 25 °C can be found in Appendix 2A.

EXAMPLE 4H.1   Calculating the standard reaction entropy

One of the most important industrial reactions is the conversion of nitrogen to ammonia. Because the efficiency of this reaction governs the economies of nations, it is necessary for chemical engineers to understand its thermodynamic properties. Calculate the standard reaction entropy of $N_2(g) + 3\,H_2(g) \rightarrow 2\,NH_3(g)$ at 25 °C.

ANTICIPATE You should expect a decrease in entropy, because there is a net reduction in the amount of gas molecules.

PLAN Use the chemical equation to write an expression for $\Delta S°$, as shown in Eq. 2, and then substitute values from Table 4H.1 or Appendix 2A.

SOLVE

From the chemical equation write

$$\Delta S° = \overbrace{(2\text{ mol}) \times S_m°(NH_3, g)}^{\text{product}} - \overbrace{\{(1\text{ mol}) \times S_m°(N_2, g) + (3\text{ mol}) \times S_m°(H_2, g)\}}^{\text{reactants}}$$

$$= (2\text{ mol}) \times (192.4\text{ J·K}^{-1}\text{·mol}^{-1})$$

$$\quad -\{(1\text{ mol}) \times (191.6\text{ J·K}^{-1}\text{·mol}^{-1}) + (3\text{ mol}) \times (130.7\text{ J·K}^{-1}\text{·mol}^{-1})\}$$

$$= 2(192.4\text{ J·K}^{-1}) - \{(191.6\text{ J·K}^{-1}) + 3(130.7\text{ J·K}^{-1})\}$$

$$= -198.9\text{ J·K}^{-1}$$

EVALUATE Because the value of $\Delta S°$ is negative, the product is less disordered than the reactants—in part because it occupies a smaller volume, as expected.

A Note on Good Practice: Avoid the error of setting the standard entropies of elements equal to zero, as you would for $\Delta H_f°$: the entropies to use are the absolute entropies for the given temperature and are zero only at $T = 0$.

Self-test 4H.2A Use data from Appendix 2A to calculate the standard reaction entropy of $N_2O_4(g) \rightarrow 2\ NO_2(g)$ at 25 °C.
[**Answer:** +175.83 J·K^{-1}]

Self-test 4H.2B Use data from Appendix 2A to calculate the standard reaction entropy of $C_2H_4(g) + H_2(g) \rightarrow C_2H_6(g)$ at 25 °C.

Related Exercises 4H.11, 4H.12

The standard reaction entropy is the difference between the standard molar entropy of the products and that of the reactants, weighted by the amounts of each species taking part in the reaction. When a gas is involved, it is positive (an increase in entropy) if there is a net production of gas in a reaction; it is negative (a decrease) if there is a net consumption of gas.

What have you learned in this Topic?

You have been introduced to the third law of thermodynamics and have seen how to use it to determine the absolute entropies of substances. You have also seen how to use those values to calculate the change in entropy of a system when a reaction takes place.

The skills you have mastered are the ability to:

☐ **1.** Use heat capacity data to determine the standard molar entropy of a substance (Section 4H.1).

☐ **2.** Calculate the standard reaction entropy from standard molar entropies (Example 4H.1).

Topic 4H Exercises

Exercises labeled $\int_{dx}^{C}$ **require calculus.**

4H.1 Which substance in each of the following pairs has the higher molar entropy at 298 K: (a) HBr(g) or HF(g); (b) NH$_3$(g) or Ne(g); (c) I$_2$(s) or I$_2$(l); (d) 1.0 mol Ar(g) at 1.00 atm or 1.0 mol Ar(g) at 2.00 atm?

4H.2 Which substance in each of the following pairs has the higher molar entropy? (Take the temperature to be 298 K unless otherwise specified.) (a) O$_2$(g) or O$_3$(g); (b) CH$_2$Br$_2$(g) or CH$_4$(g); (c) CaI$_2$(s) or CaI$_2$(aq); (d) O$_2$(g) at 278 K and 1.00 atm or O$_2$(g) at 278 K and 2.00 atm.

4H.3 List the following substances in order of increasing molar entropy at 298 K and 1 bar: H$_2$O(l), H$_2$O(g), H$_2$O(s), C(s, diamond). Explain your reasoning.

4H.4 List the following substances in order of increasing molar entropy at 298 K and 1 bar: NH$_3$(g), HF(g), H$_2$O(s), NH$_2$OH(g). Explain your reasoning.

4H.5 Which substance in each of the following pairs would you expect to have the higher standard molar entropy at 298 K and 1 bar? Explain your reasoning. (a) Iodine vapor or bromine vapor; (b) the two liquids cyclopentane and 1-pentene (see structures); (c) ethene (ethylene) or an equivalent mass of polyethylene, a substance formed by the polymerization of ethylene.

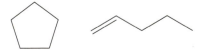

Cyclopentane, C$_5$H$_{10}$ 1-Pentene, C$_5$H$_{10}$

4H.6 Predict which of the hydrocarbons below has the greater standard molar entropy at 25 °C and 1 bar. Explain your reasoning.

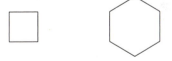

(a) Cyclobutane, C$_4$H$_8$ (a) Cyclohexane, C$_6$H$_{12}$

4H.7 Without performing any calculations, predict whether there is an increase or a decrease in entropy of the system for each of the following processes: (a) Cl$_2$(g) + H$_2$O(l) → HCl(aq) + HClO(aq); (b) Cu$_3$(PO$_4$)$_2$(s) → 3 Cu^{2+}(aq) + 2 PO$_4^{3-}$; (c) SO$_2$(g) + Br$_2$(g) + 2 H$_2$O(l) → H$_2$SO$_4$(aq) + 2 HBr(aq).

4H.8 Without performing any calculations, state whether the entropy of the system increases or decreases in each of the following processes: (a) the sublimation of dry ice, CO$_2$(s) → CO$_2$(g); (b) the formation of sulfurous acid from sulfur dioxide in the atmosphere, SO$_2$(g) + H$_2$O(l) → H$_2$SO$_3$(aq); (c) the melting of ice when salt is spread on sidewalks. Explain your reasoning.

4H.9 Container A is filled with 1.0 mol of the atoms of an ideal monatomic gas. Container B has 1.0 mol of atoms bound together as diatomic molecules that are not vibrationally active. Container C has 1.0 mol of atoms bound together as diatomic molecules that are vibrationally active. The containers all start at T_1 and the temperature is increased to T_2. Rank the containers in order of increasing change in entropy. Explain your reasoning.

4H.10 A closed vessel of volume 2.5 L contains a mixture of neon and fluorine. The total pressure is 3.32 atm at 0 °C. When the mixture is heated to 15 °C, the entropy of the mixture increases by

0.345 $J·K^{-1}$. What amount (in moles) of each substance (Ne and F_2) is present in the mixture?

4H.11 Use data in Table 4H.1 or Appendix 2A to calculate the standard reaction entropy for each of the following reactions at 25 °C. For each reaction, interpret the sign and magnitude of the reaction entropy. (a) The formation of 1.00 mol $H_2O(l)$ from the elements in their most stable state at 298 K and 1 bar. (b) The oxidation of 1.00 mol $CO(g)$ to carbon dioxide. (c) The decomposition of 1.00 mol calcite, $CaCO_3(s)$, to carbon dioxide gas and solid calcium oxide. (d) The decomposition of potassium chlorate: $4 KClO_3(s) \rightarrow 3 KClO_4(s) + KCl(s)$.

4H.12 Use data in Table 4H.1 or Appendix 2A to calculate the standard entropy change for each of the following reactions at 25 °C. For each reaction, interpret the sign and magnitude of the reaction entropy. (a) The preparation of manganese metal in the thermite reaction: $4 Al(s) + 3 MnO_2(s) \rightarrow 3 Mn(s) + 2 Al_2O_3(s)$.

(b) A reaction used to power rockets: $7 H_2O_2(l) + N_2H_4(l) \rightarrow 2 HNO_3(aq) + 8 H_2O(l)$. (c) The purification of silicon: $SiO_2(s) + 2 C(s) \rightarrow Si(s) + 2 CO(g)$. (d) The preparation of nitric oxide: $4 NH_3(g) + 5 O_2(g) \xrightarrow{1000\,°C,\,Pt} 4 NO(g) + 6 H_2O(g)$.

4H.13 The temperature dependence of the heat capacity of a substance is commonly written in the form $C_{P,\,m} = a + bT + c/T^2$, with a, b, and c constants. Obtain an expression for the entropy change when the substance is heated from T_1 to T_2. Evaluate this change for graphite, for which $a = 16.86\ J·K^{-1}·mol^{-1}$, $b = 4.77\ mJ·K^{-2}·mol^{-1}$, and $c = -8.54 \times 10^5\ J·K·mol^{-1}$, heated from 298 K to 400. K. What is the percentage error in assuming that the heat capacity is constant with its mean value in this range?

4H.14 At low temperatures, heat capacities of nonmetallic solids are proportional to T^3. Show that, near $T = 0$, the entropy of a substance is equal to one-third of its heat capacity at the same temperature.

Topic 4I Global Changes in Entropy

4I.1   The Surroundings

4I.2   The Overall Change in Entropy

4I.3   Equilibrium

Topic **4D:**
Thermo-chemistry

How is entropy measured?

How are all the changes in entropy calculated for a process?

Can the direction of natural change be related to changes in the system alone?

Topic **4H:**
Absolute entropies

Topic **4I:**
Global changes in entropy

Topic **4J:**
Gibbs free energy

Some processes seem to defy the second law. For example, water freezes to ice at low temperatures and cold packs for athletic injuries become ice cold, even on warm days, when the ammonium nitrate they contain dissolves in water inside the pack. Life itself appears to go against the second law. Every cell of a living organism is organized to an extraordinary extent. Thousands of different compounds, each one having a specific function to perform, move in the intricately organized dance called life. How can the molecules in our bodies form such highly organized, complex structures from slime, mud, and gas? Our existence seems at first sight to be a contradiction of the second law.

The dilemma is resolved by realizing that the second law refers to an *isolated* system. To interpret the second law correctly, any system must be treated as a part of a larger system that includes the surroundings of the system of interest. The process is spontaneous only if the *total* change in entropy, the sum of the changes in the system and the surroundings, is positive. The task, therefore, is to see how to calculate the change in entropy of the surroundings and then to combine that change with the change in entropy of the system.

Why Do You Need to Know This Material? To predict whether a process is spontaneous, you need to be able to assess the total entropy change of the system and its surroundings. The resulting discussion underlies the entire application of thermodynamics to chemical reactions.

What Do You Need to Know Already? This Topic draws on the thermodynamic definition of an entropy change (Topic 4F) and uses the expressions for the work of reversible isothermal expansion of an ideal gas (Topic 4A) and the entropy of reaction (Topic 4H).

4I.1 The Surroundings

The system of interest and its surroundings constitute the "isolated system" to which the second law refers (**FIG. 4I.1**). Only if the *total* entropy change,

$$\overbrace{\Delta S_{\text{tot}}}^{\substack{\text{total entropy} \\ \text{change}}} = \overbrace{\Delta S}^{\substack{\text{entropy change} \\ \text{of system}}} + \overbrace{\Delta S_{\text{surr}}}^{\substack{\text{entropy change} \\ \text{of surrondings}}} \tag{1}$$

is positive will the process be spontaneous. As is customary, properties of the system are written without subscripts; so here ΔS is the entropy change of the system and ΔS_{surr} that of the surroundings. A crucially important point is that a process in which ΔS is negative may actually be spontaneous, provided that the entropy of the surroundings increases so much that ΔS_{tot} is positive.

An example of the role of the surroundings in determining the spontaneous direction of a process is the freezing of water. The information in Table 4G.1 shows that the molar entropy of liquid water at 0 °C is 22.0 J·K^{-1}·mol^{-1} higher than that of ice at the same temperature. It follows that, when water freezes at 0 °C, its entropy *decreases* by 22.0 J·K^{-1}·mol^{-1}. Entropy changes for phase transitions do not vary much with temperature. Therefore, just below 0 °C, you can expect almost the same decrease. Yet you

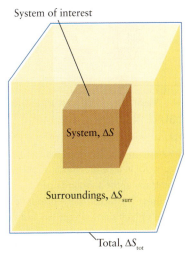

FIGURE 4I.1 The global isolated system that models events taking place in the world consists of a smaller system together with its immediate surroundings. The only events that can take place spontaneously in the global isolated system are those corresponding to an increase in the total entropy of both the smaller system and its immediate surroundings.

System of interest

System, ΔS

Surroundings, ΔS_{surr}

Total, ΔS_{tot}

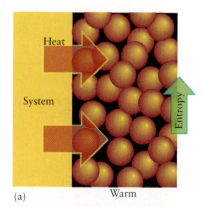

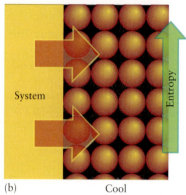

(a) Warm

(b) Cool

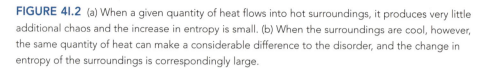

FIGURE 4I.2 (a) When a given quantity of heat flows into hot surroundings, it produces very little additional chaos and the increase in entropy is small. (b) When the surroundings are cool, however, the same quantity of heat can make a considerable difference to the disorder, and the change in entropy of the surroundings is correspondingly large.

know from everyday experience that the freezing of water is spontaneous below 0 °C. Clearly, the surroundings must be playing a deciding role: if their entropy increases by more than 22.0 J·K^{-1}·mol^{-1} when water freezes, then the total entropy change will be positive.

It is easy to predict the sign of the entropy change of the surroundings by noting whether the process is exothermic or endothermic. If the process is exothermic, then heat is released into the surroundings and their entropy increases ($\Delta S_{surr} > 0$). If the process is endothermic, heat leaves the surroundings, so their entropy decreases ($\Delta S_{surr} < 0$). To calculate the numerical value of the change, Eq. 1 of Topic 4F ($\Delta S = q_{rev}/T$) is used after noting the following points:

- Because the surroundings are always taken to be large, their temperature remains constant however much energy is transferred to or from them as heat, so the expression can always be used in the form $\Delta S_{surr} = q_{surr, rev}/T$.
- The heat that leaves the system enters the surroundings, $q_{surr} = -q$ (so if $q = -10$ kJ, meaning that 10 kJ has left the system, $q_{surr} = +10$ kJ, meaning that 10 kJ has entered the surroundings).
- For a system maintained at constant pressure the heat that leaves the system can be equated to the change in enthalpy of the system, so $q = \Delta H$, and therefore $q_{surr} = -\Delta H$.
- Because the surroundings are large, any transfer of heat into them will, from their perspective, be so tiny that it can be regarded as occurring reversibly, and hence $q_{surr, rev} = -\Delta H$.

It follows from these points that

$$\Delta S_{surr} = -\frac{\Delta H}{T} \text{ at constant temperature and pressure} \qquad (2)$$

Notice that, for a given enthalpy change of the system (that is, a given output of heat), the entropy of the surroundings increases more if their temperature is low than if it is high (**FIG. 4I.2**). The explanation is the "sneeze in the street" analogy mentioned in Topic 4F. Because ΔH is independent of path, Eq. 2 is applicable whether the process occurs reversibly or irreversibly.

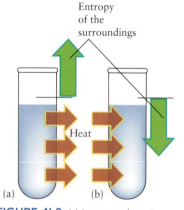

Entropy of the surroundings

Heat

(a)   (b)

FIGURE 4I.3 (a) In an exothermic process, heat escapes into the surroundings and increases their entropy. (b) In an endothermic process, the entropy of the surroundings decreases. The red arrows represent the transfer of heat between system and surroundings, and the green arrows indicate the entropy change of the surroundings.

EXAMPLE 4I.1   Calculating the change in entropy of the surroundings

One of the skills of a scientist is to be able to verify mathematically what is understood conceptually. For example, does the increase in entropy of the surroundings account for the fact that a liquid freezes when the temperature is lowered to below a certain temperature? Calculate the change in entropy of the surroundings per mole of Hg when mercury freezes at -49 °C; use $\Delta H_{fus}(Hg) = 2.292$ kJ·mol^{-1} at -49 °C.

ANTICIPATE The normal freezing point of mercury is -38 °C, so you should certainly expect it to freeze at a lower temperature. A contribution to this spontaneous freezing is the release of energy as heat to the surroundings, causing their entropy to increase (**FIG. 4I.3**).

PLAN To calculate the change in entropy of the surroundings when mercury freezes, write $\Delta H_{freeze} = -\Delta H_{fus}$ (freezing is the opposite of fusion) and convert the temperature to kelvins. The change in entropy of the system, the mercury itself, is -9.782 J·K^{-1}.

What should you assume? Assume that the temperature of the surroundings is constant.

SOLVE

$$\Delta H_{freeze} = -\Delta H_{fus} = -2.292 \text{ kJ·mol}^{-1}$$
$$T = (-49 + 273) \text{ K} = 224 \text{ K}$$

Then, from $\Delta S_{surr} = -\Delta H/T$,

$$\Delta S_{surr} = -\frac{(-2.292 \times 10^3 \, J \cdot mol^{-1})}{224 \, K}$$

$$= +10.2 \, J \cdot K^{-1} \cdot mol^{-1}$$

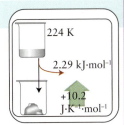

224 K

2.29 kJ·mol⁻¹

+10.2 J·K⁻¹·mol⁻¹

EVALUATE Note that the increase in entropy of the surroundings is positive and, as expected, greater than the decrease in entropy of the system, $-9.782 \, J \cdot K^{-1}$ (the value at $-49 \, °C$ is similar to that at $-39 \, °C$), so the overall change is positive and the freezing of mercury is spontaneous at $-49 \, °C$.

Self-test 41.1A Calculate the entropy change of the surroundings when 1.00 mol $H_2O(l)$ vaporizes at 90 °C and 1 bar. Take the enthalpy of vaporization of water as 40.7 kJ·mol⁻¹.

[*Answer:* $-112 \, J \cdot K^{-1}$]

Self-test 41.1B Calculate the entropy change of the surroundings when 2.00 mol $NH_3(g)$ is formed from the elements at 298 K.

Related Exercises 41.3, 41.4

The entropy change of the surroundings due to a process taking place at constant pressure and temperature is equal to $-\Delta H/T$, *where ΔH is the change in enthalpy of the system at the temperature T.*

41.2 The Overall Change in Entropy

As already emphasized, *to use the entropy to judge the direction of spontaneous change, you must consider the change in the entropy of the system plus the entropy change in the surroundings:*

If ΔS_{tot} is positive (an increase), the process is spontaneous.

If ΔS_{tot} is negative (a decrease), the reverse process is spontaneous.

If $\Delta S_{tot} = 0$, the process has no tendency to proceed in either direction.

EXAMPLE 41.2 Calculating the total change in entropy accompanying a reaction

The brilliant white glow of fireworks is due to the combustion of magnesium in air at a high temperature. But is that combustion spontaneous, in the thermodynamic sense, at ordinary temperatures? You can find out by calculating the total entropy change. Assess whether the combustion of magnesium, $2 \, Mg(s) + O_2(g) \rightarrow 2 \, MgO(s)$, is spontaneous at 25 °C under standard conditions, given $\Delta S° = -217 \, J \cdot K^{-1}$ and $\Delta H° = -1202 \, kJ$.

ANTICIPATE You can safely anticipate that all combustion reactions are spontaneous at normal temperatures.

PLAN Find the entropy change in the surroundings by using Eq. 2, then calculate the total entropy change by using Eq. 1.

SOLVE

Determine the change in entropy of the system (given in this case).

$$\Delta S° = -217 \, J \cdot K^{-1}$$

−217 J·K⁻¹

Determine the change in entropy of the surroundings from $\Delta S_{surr}° = -\Delta H°/T$.

$$\Delta S_{surr}° = -\frac{\overbrace{-1.202 \times 10^6 \, J}^{-1202 \, kJ}}{298 \, K}$$

$$= +4.03 \times 10^3 \, J \cdot K^{-1}$$

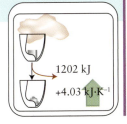

1202 kJ

+4.03 kJ·K⁻¹

Determine the total change in entropy from $\Delta S_{\text{tot}}° = \Delta S° + \Delta S_{\text{surr}}$.

$$\Delta S_{\text{tot}}° = \overbrace{-217 \text{ J·K}^{-1}}^{\text{system}} + \overbrace{(4.03 \times 10^3 \text{ J·K}^{-1})}^{\text{surroundings}}$$

$$= +3.81 \times 10^3 \text{ J·K}^{-1}$$

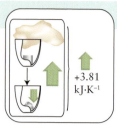

+3.81 kJ·K⁻¹

EVALUATE Because $\Delta S_{\text{tot}}°$ is positive, the reaction is spontaneous under standard conditions even though the entropy of the system decreases.

Self-test 4I.2A Is the formation of hydrogen fluoride from its elements spontaneous under standard conditions at 25 °C? For the reaction $H_2(g) + F_2(g) \rightarrow 2 HF(g)$, $\Delta H° = -542.2$ kJ and $\Delta S° = +14.1$ J·K⁻¹.

[**Answer:** $\Delta S_{\text{surr}} = +1819$ J·K⁻¹; therefore, $\Delta S_{\text{tot}} = +1833$ J·K⁻¹; spontaneous]

Self-test 4I.2B Is the formation of benzene from its elements spontaneous at 25 °C? For the reaction $6 C(s, \text{graphite}) + 3 H_2(g) \rightarrow C_6H_6(l)$, $\Delta H° = +441.0$ kJ and $\Delta S° = -253.18$ J·K⁻¹.

Related Exercises 4I.5–4I.8

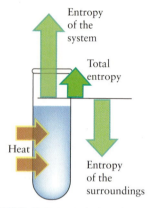

FIGURE 4I.4 An endothermic reaction is spontaneous only if the entropy of the system increases enough to overcome the decrease in entropy of the surroundings.

Spontaneous endothermic reactions were a puzzle for nineteenth-century chemists, who believed that reactions ran only in the direction of decreasing energy of the system. It seemed to them that, in endothermic reactions, reactants were rising spontaneously to higher energies, like a weight suddenly leaping up from the floor onto a table. However, the criterion for spontaneity is an *increase in the total entropy of the system and its surroundings*, not a decrease in the energy of a system. In an endothermic reaction, the entropy of the surroundings certainly decreases as energy flows as heat from them into the system. Nevertheless, there can still be an overall increase in entropy if the disorder of the system increases enough. Every endothermic reaction must be accompanied by increased disorder of the system if it is to be spontaneous (**FIG. 4I.4**): spontaneous endothermic reactions are driven by the dominating increase in disorder of the system.

By considering the total entropy change, it is possible to draw some far-reaching conclusions about processes going on in the universe. For instance, maximum work is achieved if expansion takes place reversibly, by matching the external pressure to the pressure of the system at every infinitesimal step (Topic 4A). That relation is always true: *a process produces maximum work if it takes place reversibly*. That is, w_{rev} is more negative (more energy leaves the system as work) than w_{irrev}. However, because the internal energy is a state function, ΔU is the same for any path between the same two states. Therefore, because $\Delta U = q + w$, it follows that q_{rev}, the heat absorbed along the reversible path, must be greater than q_{irrev}, the heat absorbed along any other path, because only then can the sums of q and w be the same for the two paths. If q_{rev} in the definition of entropy $\Delta S = q_{\text{rev}}/T$ is replaced by the smaller quantity q_{irrev}, then $\Delta S > q_{\text{irrev}}/T$. In general, the **Clausius inequality** is

$$\Delta S \geq \frac{q}{T} \tag{3}$$

with the equality applicable to a reversible process. For a fully isolated system like that in Fig. 4I.1, $q = 0$ for any process taking place inside the system. It then follows that

$$\Delta S \geq 0 \text{ for any process in an isolated system}$$

That is, *the entropy cannot decrease in an isolated system*. This is another statement of the second law of thermodynamics. It implies that, as a result of all the processes going on within and around us, the total entropy of the universe is steadily increasing.

Now consider an isolated system consisting of both the system of interest and its surroundings (again like that in Fig. 4I.1). For any spontaneous change in this isolated system, $\Delta S_{\text{tot}} > 0$. If for a particular hypothetical process it is calculated that $\Delta S_{\text{tot}} < 0$, then you can conclude that the reverse of that process is spontaneous.

Because entropy is a state function, the value of ΔS, the change in entropy of the system, is the same whether the process is reversible or irreversible. However, ΔS_{tot} is

TABLE 41.1   The Criteria for Spontaneity

ΔS	ΔS_{surr}	ΔS_{tot}	Character
+	+	+	spontaneous
−	−	−	not spontaneous; reverse change is spontaneous
+	−		spontaneous if ΔS dominates
−	+		spontaneous if ΔS_{surr} dominates

different for the two paths. For example, the isothermal expansion of an ideal gas always results in the change in the entropy of the system given by Eq. 2 of Topic 4F ($\Delta S = nR \ln(V_2/V_1)$). However, the change in entropy of the surroundings is different for reversible and irreversible paths because the surroundings are left in different states in each case (**FIG. 41.5**). **TABLE 41.1** lists the characteristics of reversible and irreversible processes.

EXAMPLE 41.3   Calculating the total entropy change for the expansion of an ideal gas

Engineers studying internal combustion engines, such as those in automobiles, and looking for ways to increase their efficiency, must understand how the entropy of gaseous mixtures is affected by different conditions. Calculate ΔS, ΔS_{surr}, and ΔS_{tot} for (a) the isothermal, reversible expansion and (b) the isothermal, free expansion of 1.00 mol of ideal gas molecules from 8.00 L to 20.00 L at 292 K. Explain any differences between the two paths.

ANTICIPATE (a) You should expect a positive increase in the entropy of the expanding gas itself but, because the expansion is reversible, no change in the total entropy. (b) For an irreversible expansion, you should expect the total entropy change to be positive.

PLAN Because entropy is a state function, the change in entropy of the system is the same regardless of the path between the two states. Thus, as stated in the text, Eq. 2 of Topic 4F ($\Delta S = nR \ln(V_2/V_1)$) can be used to calculate ΔS for both part (a) and part (b). For the entropy of the surroundings, find the heat transferred to the surroundings. In each case, combine the fact that $\Delta U = 0$ for an isothermal expansion of an ideal gas with $\Delta U = w + q$ and conclude that $q = -w$. Then use Eq. 4 in Topic 4A ($w = -nRT \ln(V_2/V_1)$) to calculate the work done in an isothermal, reversible expansion and Eq. 1 of this Topic to find the total entropy change.

SOLVE
(a) For the reversible path:

From $\Delta S = nR \ln(V_2/V_1)$,

$$\Delta S = (1.00 \text{ mol}) \times (8.3145 \text{ J·K}^{-1}\text{·mol}^{-1}) \times \ln \frac{20.0 \text{ L}}{8.00 \text{ L}}$$
$$= +7.6 \text{ J·K}^{-1}$$

The change in the entropy of the gas is positive, as expected. Because $\Delta U = 0$, $q = -w$. Therefore, because the energy flowing *into* the surroundings as heat is equal to that flowing *out* of the system,

From $q_{surr} = -q$, $q = -w$, and $w = -nRT \ln(V_2/V_1)$,

$$q_{surr} = -nRT \ln \frac{V_2}{V_1}$$

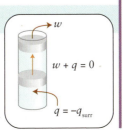

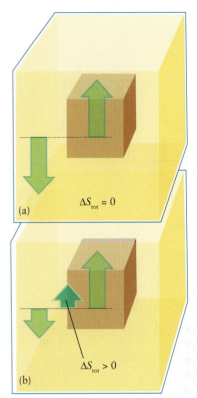

FIGURE 41.5 (a) When a process takes place reversibly—that is, when the system is at equilibrium with its surroundings—the change in entropy of the surroundings is the negative of the change in entropy of the system, and the overall change is zero. (b) For an irreversible change between the same two states of the system, the final state of the surroundings is different from the final state in part (a), and the change in the system is not canceled by the entropy change of the surroundings. Overall, there is an increase in entropy.

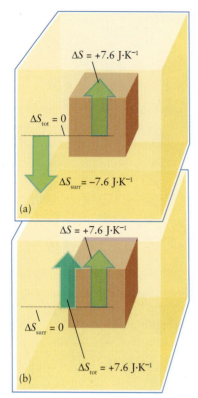

FIGURE 41.6 The changes in entropy and internal energy when an ideal gas undergoes (a) reversible and (b) irreversible changes between the same two states, as described in Example 41.3.

From $\Delta S_{surr} = q_{surr}/T$,

$$\Delta S_{surr} = -\frac{nRT}{T}\ln\frac{V_2}{V_1} = -nR\ln\frac{V_2}{V_1} = -\Delta S = -7.6\ \text{J·K}^{-1}$$

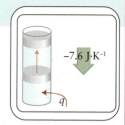

From $\Delta S_{tot} = \Delta S + \Delta S_{surr}$,

$$\Delta S_{tot} = 7.6\ \text{J·K}^{-1} - 7.6\ \text{J·K}^{-1} = 0$$

(b) For the irreversible process.

No work is done in free expansion (Topic 4A), so $w = 0$. Because $\Delta U = 0$, it follows that $q = 0$ and therefore

$$q_{surr} = 0$$

From $\Delta S_{surr} = q_{surr}/T$,

$$\Delta S_{surr} = 0$$

ΔS is the same as for the reversible path, at $+7.6\ \text{J·K}^{-1}$. The total change in entropy is therefore

$$\Delta S_{tot} = +7.6\ \text{J·K}^{-1}$$

EVALUATE   As expected, for the reversible process $\Delta S_{tot} = 0$ and for the irreversible process $\Delta S_{tot} > 0$. The total entropy changes for the two paths differ because the entropy of the surroundings is greater for the irreversible path than for the reversible path. The results are summarized in **FIG. 41.6.**

Self-test 41.3A   Determine ΔS, ΔS_{surr}, and ΔS_{tot} for (a) the reversible, isothermal expansion and (b) the isothermal free expansion of 1.00 mol of ideal gas molecules from 10.00 atm and 0.200 L to 1.00 atm and 2.00 L at 298 K.

[**Answer:** (a) $\Delta S = +19.1\ \text{J·K}^{-1}$; $\Delta S_{surr} = -19.1\ \text{J·K}^{-1}$; $\Delta S_{tot} = 0$;
(b) $\Delta S = +19.1\ \text{J·K}^{-1}$; $\Delta S_{surr} = 0$; $\Delta S_{tot} = +19.1\ \text{J·K}^{-1}$]

Self-test 41.3B   Determine ΔS, ΔS_{surr}, and ΔS_{tot} for the reversible, isothermal compression of 2.00 mol of ideal gas molecules from 1.00 atm and 4.00 L to 20.00 atm and 0.200 L at 298 K.

Related Exercises 41.9, 41.10

A process is spontaneous if it is accompanied by an increase in the total entropy of the system and the surroundings.

41.3 Equilibrium

Although this section provides only a brief introduction to equilibrium, the principles presented here are critically important because the tendency of reactions to proceed toward equilibrium is the basis of much of chemistry. These concepts are developed further in Focus 5.

A system at **equilibrium** has no tendency to change in either direction (forward or reverse). Such a system remains in its current state until it is disturbed by changing the conditions, such as raising the temperature, decreasing the volume, or adding more reactants. The equilibrium state important in chemistry is a **dynamic equilibrium,** in which forward and reverse processes still continue but at matching rates. For example, when a block of metal is at the same temperature as its surroundings, it is in thermal equilibrium with its surroundings (Topic 4A). As explained in that Topic, energy continues to flow as heat in both directions, but at the same rate; there is no *net* flow (**FIG. 41.7**). When a gas confined to a cylinder by a piston has the same pressure as that of the surroundings, the system is in **mechanical equilibrium** with the surroundings and the gas has no tendency to expand or contract. The pressure inside is causing the piston to move outward, but the pressure outside is causing the piston to move inward to the same extent, and there is no net change.

When a solid, such as ice, is in contact with its liquid form, such as water, at certain conditions of temperature and pressure (at 0 °C and 1 atm for water), the two states of matter are in dynamic equilibrium with each other, and there is no tendency for one form of matter to change into the other form. When solid and liquid water are at equilibrium, water molecules continually leave solid ice to form liquid water, and water molecules continually leave the liquid phase to form ice. However, there is no *net* change, because these processes occur at the same rate and so balance each other.

When a chemical reaction mixture reaches a certain composition, the reaction seems to come to a halt. A mixture of substances at **chemical equilibrium** has no tendency either to produce more products or to revert to reactants. At equilibrium, reactants are still forming products, but products are decaying into reactants at a matching rate and there is no *net* change of composition.

The common characteristic of any kind of dynamic equilibrium is the continuation of processes at the microscopic level but no *net* tendency for the system to change in either the forward or the reverse direction. That is, neither the forward nor the reverse process is spontaneous. Expressed thermodynamically,

$$\text{At equilibrium: } \Delta S_{tot} = 0 \tag{4}$$

The total entropy varies with the composition of a reaction mixture and, by looking for the composition at which $\Delta S_{tot} = 0$, it is possible to predict the equilibrium composition of the reaction. At this composition, the reaction has no tendency either to form more products or to decompose into reactants.

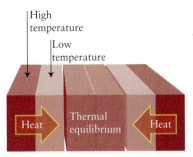

FIGURE 41.7 Energy flows as heat from a high-temperature region (dark red) to a low-temperature region (light red) through a thermally conducting interface. A system is in thermal equilibrium with its surroundings when the temperatures on both sides of the interface are the same (center). Because the systems are in contact, energy continues to flow as heat, but at the same rate in each direction.

> **Self-test 41.4A** Confirm that liquid water and water vapor are in equilibrium when the temperature is 100. °C and the pressure is 1 atm. Data are available in Tables 4C.1 and 4G.1.
> [***Answer:*** $\Delta S_{tot} = 0$ at 100. °C]
>
> **Self-test 41.4B** Confirm that liquid benzene and benzene vapor are in equilibrium at the normal boiling point of benzene, 80.1 °C, and 1 atm. The enthalpy of vaporization of benzene at its boiling point is 30.8 kJ·mol^{-1} and its entropy of vaporization is 87.2 J·K^{-1}·mol^{-1}.

The general criterion for equilibrium in thermodynamics is $\Delta S_{tot} = 0$.

What have you learned in this Topic?

You have learned that it is essential to take into account changes in both the system and its surroundings when using the entropy to assess whether a process is spontaneous, and you have seen how to calculate the entropy change of the surroundings from the enthalpy change of the system (at constant pressure). You have also seen that endothermic reactions are spontaneous, provided the entropy change of the system is large. Finally, you have had a first glimpse of how equilibria are discussed in chemistry.

The skills you have mastered are the ability to:

☐ **1.** Estimate the change in entropy of the surroundings due to heat transfer at constant pressure and temperature (Example 41.1).

☐ **2.** Calculate the total change in entropy for a process (Examples 41.2 and 41.3).

☐ **3.** State the Clausius inequality and use it to summarize the content of the second law of thermodynamics mathematically (Section 41.2).

Topic 4I Exercises

4I.1 What is the total entropy change accompanying a process in which 40.0 kJ of energy is transferred as heat from a large reservoir at 800. K to one at 200. K?

4I.2 What is the total entropy change accompanying a process in which 25.0 kJ of energy is transferred as heat from a large reservoir at 700. K to one at 320. K?

4I.3 The standard entropy of vaporization of benzene is approximately 85 J·K^{-1} mol^{-1} at its boiling point. (a) Estimate the standard enthalpy of vaporization of benzene at its boiling point of 80. °C. (b) What is the standard entropy change of the surroundings when 10. g of benzene, C_6H_6, vaporizes at its boiling point?

4I.4 The standard entropy of vaporization of acetone is approximately 85 J·K^{-1}·mol^{-1} at its boiling point. (a) Estimate the standard enthalpy of vaporization of acetone at its boiling point of 56.2 °C. (b) What is the entropy change of the surroundings when 10. g of acetone, CH_3COCH_3, condenses at its boiling point?

4I.5 Suppose that 50.0 g of water at 20.0 °C is mixed with 65.0 g of water at 50.0 °C at constant atmospheric pressure in a thermally insulated vessel. Calculate ΔS and ΔS_{tot} for the process.

4I.6 Suppose that 320.0 g of ethanol at 18.0 °C is mixed with 120.0 g of ethanol at 56.0 °C at constant atmospheric pressure in a thermally insulated vessel. Calculate ΔS and ΔS_{tot} for the process.

4I.7 Use the information in Table 4C.1 to calculate the changes in entropy of the surroundings and of the system for (a) the vaporization of 1.00 mol $CH_4(l)$ at its boiling point; (b) the melting of 1.00 mol $C_2H_5OH(s)$ at its melting point; (c) the freezing of 1.00 mol $C_2H_5OH(l)$ at its freezing point.

4I.8 Use the information in Table 4C.1 to calculate the changes in entropy of the surroundings and of the system for (a) the melting of 1.00 mol $NH_3(s)$ at its melting point; (b) the freezing of 1.00 mol $CH_3OH(l)$ at its freezing point; (c) the vaporization of 1.00 mol $H_2O(l)$ at its boiling point.

4I.9 Initially an ideal gas at 323 K occupies 1.67 L at 4.95 atm. The gas is allowed to expand to 7.33 L by two pathways: (a) isothermal, reversible expansion; (b) isothermal, irreversible free expansion. Calculate ΔS_{tot}, ΔS, and ΔS_{surr} for each pathway.

4I.10 Initially an ideal gas at 412 K occupies 12.62 L at 0.6789 atm. The gas is allowed to expand to 19.44 L by two pathways: (a) isothermal, reversible expansion and (b) isothermal, irreversible free expansion. Calculate ΔS_{tot}, ΔS, and ΔS_{surr} for each pathway.

4I.11 The following picture shows a molecular visualization of a system undergoing a spontaneous change. Account for the spontaneity of the process in terms of the entropy changes in the system and the surroundings. The thermometers show the temperature of the system.

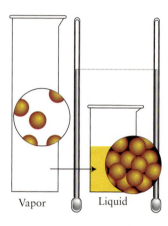

Vapor   Liquid

4I.12 The following picture shows a molecular visualization of a system undergoing a spontaneous change. Account for the spontaneity of the process in terms of the entropy changes in the system and the surroundings. The thermometers show the temperature of the system.

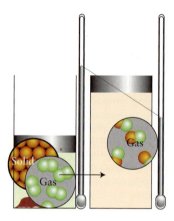

Topic 4J Gibbs Free Energy

Topic **4D:**
Thermo-
chemistry

How are all the changes in entropy calculated for a process?

Can the direction of natural change be related to changes in the system alone?

Topic **4I:**
Global changes in entropy

Topic **4J:**
Gibbs free energy

Why Do You Need to Know This Material? The Gibbs free energy is used throughout chemistry in the discussion of equilibria and electrochemistry.

What Do You Need to Know Already? You need to be aware of the definitions of enthalpy (Topic 4C) and entropy (Topic 4F) and to know that the direction of spontaneous change is indicated by the increase in total entropy (Topic 4I). You also need to be aware of the first law of thermodynamics (Topic 4B).

One of the problems with using the second law of thermodynamics to judge whether a reaction is spontaneous is that to assess the total entropy change, two quantities, the entropy change of the system and the entropy change of the surroundings, must be calculated and then added together. Some of this effort could be avoided if a single property combined the entropy calculations for the system and the surroundings. This simplification is achieved by introducing a new state function, the "Gibbs free energy," probably the single most widely used and useful quantity in the application of thermodynamics to chemistry. It is named for Josiah Willard Gibbs (**FIG. 4J.1**), the nineteenth-century American physicist who was responsible for turning thermodynamics from an abstract theory into a subject of great usefulness.

4J.1 Focusing on the System

The total entropy change, ΔS_{tot}, is the sum of the changes in the system, ΔS, and its surroundings, ΔS_{surr}, with $\Delta S_{tot} = \Delta S + \Delta S_{surr}$. For a process at constant temperature and pressure, the change in the entropy of the surroundings is given by Eq. 2 of Topic 4I ($\Delta S_{surr} = -\Delta H/T$). Therefore,

$$\Delta S_{tot} = \Delta S + \overbrace{\Delta S_{surr}}^{-\Delta H/T} = \Delta S - \frac{\Delta H}{T} \text{ at constant temperature and pressure} \quad (1)$$

This equation allows the total entropy change to be calculated from information about the system alone. The limitation is that it is valid only at constant temperature and pressure.

To take the next step, the **Gibbs free energy, G,** is introduced. It is defined as

$$G = H - TS \quad (2)$$

This quantity, which is commonly known as the *free energy* and more formally as the *Gibbs energy*, is defined solely in terms of state functions, so G itself is a state function. For a process occurring at constant temperature, the change in Gibbs free energy is

$$\Delta G = \Delta H - T\Delta S \text{ at constant temperature} \quad (3)$$

By comparing this expression rearranged into

$$\frac{\Delta G}{T} = \frac{\Delta H}{T} - \Delta S$$

with Eq. 1, where there is the additional constraint of constant pressure, we see that $\Delta G/T = -\Delta S_{tot}$ and therefore that

$$\Delta G = -T\Delta S_{tot} \text{ at constant temperature and pressure} \quad (4)$$

FIGURE 4J.1 Josiah Willard Gibbs (1839–1903). *(The New York Public Library/Art Resource, NY.)*

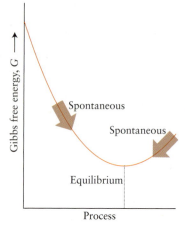

FIGURE 4J.2 At constant temperature and pressure, the direction of spontaneous change is toward lower Gibbs free energy. The horizontal axis represents the progress of the reaction or process. The equilibrium state of a system corresponds to the lowest point on the curve.

The minus sign in this equation means that, provided the temperature and pressure are constant, an increase in total entropy corresponds to a decrease in Gibbs free energy. Therefore (**FIG. 4J.2**),

> At constant temperature and pressure, the direction of spontaneous change is the direction of decreasing Gibbs free energy.

The great importance of the introduction of Gibbs free energy is that, *provided that the temperature and pressure are constant, the spontaneity of a process can be predicted solely in terms of the thermodynamic properties of the system.*

Equation 3 summarizes the factors that determine the direction of spontaneous change at constant temperature and pressure: for a spontaneous change, look for values of ΔH, ΔS, and T that result in a negative value of ΔG (**TABLE 4J.1**). One condition that may result in a negative ΔG is a large negative ΔH, such as that for a combustion reaction. A large negative ΔH corresponds to a large increase in the entropy of the surroundings. However, a negative ΔG can occur even if ΔH is positive (an endothermic reaction), provided that $T\Delta S$ is large and positive. In this case, the driving force of the reaction, the origin of its spontaneity, is the increase in entropy of the system.

Self-test 4J.1A Can a nonspontaneous process with a negative ΔS become spontaneous if the temperature is increased (assuming that ΔH and ΔS are both independent of temperature)?

[*Answer:* No]

Self-test 4J.1B Can a nonspontaneous process with a positive ΔS become spontaneous if the temperature is increased (assuming that ΔH and ΔS are both independent of temperature)?

TABLE 4J.1   Factors That Favor Spontaneity

ΔH	ΔS	Spontaneous?
−	+	yes, $\Delta G < 0$
−	−	yes, if $\lvert T\Delta S\rvert < \lvert\Delta H\rvert$, $\Delta G < 0$
+	+	yes, if $T\Delta S > \Delta H$, $\Delta G < 0$
+	−	no, $\Delta G > 0$

In Topic 4I it is established that the criterion of equilibrium is $\Delta S_{tot} = 0$. It follows from Eq. 4 that, for a process at constant temperature and pressure, the condition for equilibrium is

$$\Delta G = 0 \text{ at constant temperature and pressure} \qquad (5)$$

If $\Delta G = 0$ for such a process, then you know at once that the system is at equilibrium. For example, when ice and water are in equilibrium with each other at a particular temperature and pressure, the Gibbs free energy of 1 mol $H_2O(l)$ must be the same as the Gibbs free energy of 1 mol $H_2O(s)$. In other words, the molar Gibbs free energy, the Gibbs free energy per mole, of water in each phase is the same.

EXAMPLE 4J.1   Deciding whether a process is spontaneous

A good way to become familiar with thermodynamic processes is to start with a very simple system and consider how various changes might affect it. Calculate the change in molar Gibbs free energy, ΔG_m, for the process $H_2O(s) \rightarrow H_2O(l)$ at 1 atm and (a) 10. °C; (b) 0. °C. Decide for each temperature whether melting is spontaneous or not. Treat ΔH_{fus} and ΔS_{fus} as independent of temperature.

ANTICIPATE Because the solid and liquid phases are in equilibrium at the melting point, you should expect to find that $\Delta G = 0$ at 0. °C. Above that temperature the melting of the solid state is favored, so you should expect that ΔG will be negative at 10. °C.

PLAN Find the enthalpy of fusion of water in Table 4C.1 (the values there are for 1 atm and the transition temperature). The entropy of fusion of water can be determined from the values in Table 4G.1. Use Eq. 3 to calculate the change in Gibbs free energy.

What should you assume? Assume that ΔH_{fus} and ΔS_{fus} are constant over the temperature range of interest.

SOLVE The enthalpy of fusion is 6.01 kJ·mol^{-1} and, from Table 4G.1, the entropy of fusion is $S_m^{\circ}(l) - S_m^{\circ}(s) = (65.2 - 43.2)$ J·K^{-1}·mol^{-1} = 22.0 J·K^{-1}·mol^{-1}. These values are almost independent of temperature over the temperature range considered.

(a) At 10. °C

Convert temperature to kelvins:

$$T = (273.15 + 10.) \text{ K} = 283 \text{ K}$$

From $\Delta G_m = \Delta H_m - T\Delta S_m$,

$$\Delta G_m = \overbrace{6.0\ \text{kJ·mol}^{-1}}^{\Delta H_m} - (283\ \text{K}) \times \overbrace{(22.0\ \text{J·K}^{-1}\text{·mol}^{-1})}^{\Delta S_m}$$

$$= 6.0\ \text{kJ·mol}^{-1} - \underset{1\ \text{kJ}}{6.23 \times 10^3\ \text{J·mol}^{-1}}$$

$$= 6.0\ \text{kJ·mol}^{-1} - 6.23\ \text{kJ·mol}^{-1}$$

$$= -0.22\ \text{kJ·mol}^{-1}$$

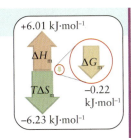

A Note on Good Practice: Be sure to convert joules to kilojoules before subtracting $T\Delta S$ from ΔH.

(b) At 0. °C

Convert temperature to kelvins:

$$T = (273.15 + 0.)\ \text{K} = 273\ \text{K}$$

From $\Delta G_m = \Delta H_m - T\Delta S_m$,

$$\Delta G_m = 6.01\ \text{kJ·mol}^{-1} - (273\ \text{K}) \times (22.0\ \text{J·K}^{-1}\text{·mol}^{-1})$$

$$= 6.01\ \text{kJ·mol}^{-1} - 6.01\ \text{kJ·mol}^{-1} = 0$$

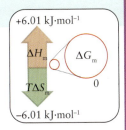

EVALUATE As expected, because the difference in molar Gibbs free energy is negative at 10. °C, melting is spontaneous at that temperature; but at 0. °C, ice and water are in equilibrium.

Self-test 4J.2A Calculate the change in molar Gibbs free energy for the process $H_2O(l) \rightarrow H_2O(g)$ at 1 atm and (a) 95 °C; (b) 105 °C. The enthalpy of vaporization at these temperatures is 40.7 kJ·mol^{-1}, and the entropy of vaporization is 109.1 J·K^{-1}·mol^{-1}. In each case, indicate whether vaporization would be spontaneous or not.

[**Answer:** (a) +0.6 kJ·mol^{-1}, not spontaneous; (b) −0.5 kJ·mol^{-1}, spontaneous]

Self-test 4J.2B Calculate the change in molar Gibbs free energy for the process $Hg(l) \rightarrow Hg(g)$ at 1 atm and (a) 350. °C; (b) 370. °C. The enthalpy of vaporization is 59.3 kJ·mol^{-1}, and the entropy of vaporization at these temperatures is 94.2 J·K^{-1}·mol^{-1}. In each case, indicate whether vaporization would be spontaneous or not.

Related Exercises 4J.3, 4J.4

The Gibbs free energy of a substance decreases (that is, it becomes less positive or more negative) as its temperature is raised at constant pressure. This conclusion follows from the definition $G = H - TS$ and the fact that the entropy of a pure substance is invariably positive; as T increases, TS increases too, and a larger quantity is subtracted from H. Another important conclusion is that, because the molar entropy of the gas phase of a substance is greater than that of its liquid phase, the Gibbs free energy decreases more sharply with temperature for the gas phase of a substance than for the liquid phase. Similarly, the Gibbs free energy of the liquid phase of a substance decreases more sharply than that of its solid phase (**FIG. 4J.3**).

You are now in a position to understand the thermodynamic origin of phase transitions. At low temperatures, the molar Gibbs free energy of the solid phase lies lowest, so there is a tendency for a liquid to freeze and thereby reduce its Gibbs free energy. Above a certain temperature, the molar Gibbs free energy of the liquid has fallen below that of the solid, and the substance has a tendency to melt. At an even higher temperature, the molar Gibbs free energy of the gas phase has fallen to below the liquid line, and the

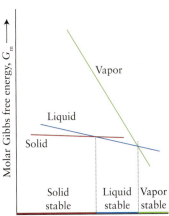

FIGURE 4J.3 The variation of the (molar) Gibbs free energy with temperature for three phases of a substance at a given pressure. The most stable phase is the phase with lowest molar Gibbs free energy. Note that, as the temperature is raised, the solid, liquid, and vapor phases in succession become the most stable.

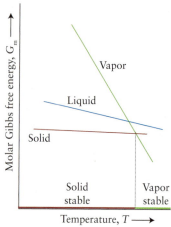

FIGURE 4J.4 For some substances and at certain pressures, the molar Gibbs free energy of the liquid phase might never lie lower than those of the other two phases. For such substances, the liquid is never the stable phase and, at constant pressure, the solid sublimes when the temperature is raised to the point of intersection of the solid and vapor lines.

Standard states are defined in Topic 4D: a standard state is the pure substance at 1 bar and the specified temperature. The standard state of a solute is at 1 bar and a concentration of 1 mol·L^{-1}.

substance has a tendency to vaporize. The temperature of each phase change corresponds to the point of intersection of the lines for the two phases, as can be seen in Fig. 4J.3.

The relative positions of the three lines in Fig. 4J.3 are different for each substance. One possibility—which depends on the strength of intermolecular interactions in the condensed phases—is for the liquid line to lie in the position shown in **FIG. 4J.4**. In this case, the liquid line is never the lowest, at any temperature. As soon as the temperature has been raised above the point corresponding to the intersection of the solid and gas lines, the direct transition of the solid to the vapor, sublimation, becomes spontaneous. This plot is the type expected for carbon dioxide, which sublimes at atmospheric pressure.

The change in Gibbs free energy for a process is a measure of the change in the total entropy of a system and its surroundings at constant temperature and pressure. Spontaneous processes at constant temperature and pressure are accompanied by a decrease in Gibbs free energy.

4J.2 Gibbs Free Energy of Reaction

The decrease in Gibbs free energy as a signpost of spontaneous change and $\Delta G = 0$ as a criterion of equilibrium are applicable to any kind of process, provided that it is occurring at constant temperature and pressure.

The thermodynamic function used as the criterion of spontaneity for a chemical reaction is the **Gibbs free energy of reaction,** ΔG (which is commonly referred to as the "reaction free energy"). This quantity is defined as the difference in molar Gibbs free energies, G_m, of the products and the reactants:

$$\Delta G = \sum nG_m(\text{products}) - \sum nG_m(\text{reactants}) \tag{6}$$

where the molar Gibbs free energy of each substance is multiplied by the amount (in moles) that appears in the chemical equation. For example, for the formation of ammonia, $N_2(g) + 3\,H_2(g) \rightarrow 2\,NH_3(g)$,

$$\Delta G = \{(2\text{ mol}) \times G_m(NH_3)\} - \{(1\text{ mol}) \times G_m(N_2) + (3\text{ mol}) \times G_m(H_2)\}$$

The molar Gibbs free energy of a substance in a mixture depends on what molecules it has as neighbors, so the molar Gibbs free energies of NH_3, N_2, and H_2 change as the reaction proceeds. For example, at an early stage of the reaction, an NH_3 molecule has mostly N_2 and H_2 molecules around it, but at a later stage of the reaction, most of its neighbors may be NH_3 molecules. Because the individual Gibbs free energies change as the reaction proceeds, the Gibbs free energy of reaction also changes. If $\Delta G < 0$ at a certain composition of the reaction mixture, then the forward reaction is spontaneous. If $\Delta G > 0$ at a certain composition, then the reverse reaction (the decomposition of ammonia in our example), is spontaneous.

The **standard Gibbs free energy of reaction,** $\Delta G°$, is defined like the Gibbs free energy of reaction but in terms of the *standard* molar Gibbs energies of the reactants and products:

$$\Delta G° = \sum nG_m°(\text{products}) - \sum nG_m°(\text{reactants}) \tag{7}$$

That is, the standard Gibbs free energy of reaction is the difference in Gibbs free energy between products in their standard states and reactants in their standard states (at the specified temperature). Because the *standard* state of a substance is its *pure* form at 1 bar, the *standard* Gibbs free energy of reaction is the difference in Gibbs free energy between the *pure* products and the *pure* reactants: it is a fixed quantity for a given reaction and does not vary as the reaction proceeds. Two points to remember are:

- $\Delta G°$ is fixed for a given reaction and temperature and so does not change as the reaction proceeds.
- ΔG depends on the composition of the reaction mixture and so it varies—and might even change sign—as the reaction proceeds.

Equations 6 and 7 are not very useful in practice because only *changes* in the Gibbs free energies of substances can be known, not their absolute values. However, the same technique as that used to find the standard reaction enthalpy in Topic 4D can be used, where each compound is assigned a standard enthalpy of formation, $\Delta H_f°$. Analogously, the **standard Gibbs free energy of formation,** $\Delta G_f°$ (the "standard free energy of formation"), of a substance is *the standard Gibbs free energy of reaction per mole for the formation of a compound from its elements in their most stable form* (**TABLE 4J.2**). For example, the standard Gibbs free energy of formation of hydrogen iodide gas at 25 °C is $\Delta G_f°(HI, g) = +1.70$ kJ·mol^{-1}. It is the standard Gibbs free energy per mole of HI for the reaction $\frac{1}{2} H_2(g) + \frac{1}{2} I_2(s) \rightarrow HI(g)$. It follows from the definition that the standard Gibbs free energies of formation of elements in their most stable forms are exactly zero; for example, $\Delta G_f°(I_2, s) = 0$ for the "null" reaction $I_2(s) \rightarrow I_2(s)$, because there is no change on going from reactants to products.

Standard Gibbs free energies of formation can be determined in various ways. One straightforward way is to combine standard enthalpy and entropy data from tables such as Tables 4D.2 and 4H.1. A list of values for several common substances is given in **TABLE 4J.3**, and a more extensive one appears in Appendix 2A.

TABLE 4J.2 **Examples of the Most Stable Forms of Elements**

Element	Most stable form at 25 °C and 1 bar
H_2, O_2, Cl_2, Xe	gas
Br_2, Hg	liquid
C	graphite
Na, Fe, I_2	solid

TABLE 4J.3 **Standard Gibbs Free Energies of Formation at 25 °C***

Gases	$\Delta G_f°/(kJ·mol^{-1})$	Liquids	$\Delta G_f°/(kJ·mol^{-1})$	Solids	$\Delta G_f°/(kJ·mol^{-1})$
NH_3	−16.45	benzene, C_6H_6	+124.3	$CaCO_3$†	−1128.8
CO_2	−394.4	ethanol, C_2H_5OH	−174.8	AgCl	−109.8
NO_2	+51.3	H_2O	−237.1		
H_2O	−228.6				

*Additional values are given in Appendix 2A.
†Calcite.

EXAMPLE 4J.2   Constructing a standard Gibbs free energy of formation from enthalpy and entropy data

Hydrogen iodide is a reactive compound that is sometimes used in the manufacture of methamphetamine. It can be produced by the direct reaction of the elements. Calculate the standard Gibbs free energy of formation of HI(g) at 25 °C from its standard molar entropy and standard enthalpy of formation.

PLAN Write the chemical equation for the formation of HI(g) and calculate the standard Gibbs free energy of reaction from $\Delta G° = \Delta H° - T\Delta S°$. It is best to write the equation with a stoichiometric coefficient of 1 for the compound of interest, because then $\Delta G°$ can be identified with $\Delta G_f°$. The standard enthalpy of formation is found in Appendix 2A. The standard reaction entropy is found as shown in Example 4H.1, by using the data from Table 4H.1 or Appendix 2A.

SOLVE The chemical equation is $\frac{1}{2} H_2(g) + \frac{1}{2} I_2(s) \rightarrow HI(g)$.

From data in Appendix 2A,

$$\Delta H° = (1 \text{ mol}) \times \Delta H_f°(HI, g) = +26.48 \text{ kJ}$$

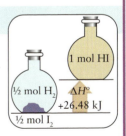

From Eq. 2 of Topic 4H,

$$\Delta S° = S_m°(HI, g) - \left\{ \frac{1}{2} S_m°(H_2, g) + \frac{1}{2} S_m°(I_2, s) \right\}$$
$$= \{(1 \text{ mol}) \times (206.6 \text{ J·K}^{-1}\text{·mol}^{-1})\}$$
$$- \left\{ \left(\frac{1}{2} \text{ mol}\right) \times (130.7 \text{ J·K}^{-1}\text{·mol}^{-1}) + \left(\frac{1}{2} \text{ mol}\right) \times (116.1 \text{ J·K}^{-1}\text{·mol}^{-1}) \right\}$$
$$= \left\{ 206.6 - \left(\frac{1}{2} \times 130.7 + \frac{1}{2} \times 116.1 \right) \right\} \text{ J·K}^{-1}$$
$$= +83.2 \text{ J·K}^{-1} = +0.0832 \text{ kJ·K}^{-1}$$

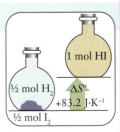

From $\Delta G° = \Delta H° - T\Delta S°$,

$$\Delta G° = (1 \text{ mol}) \times (26.48 \text{ kJ·K}^{-1}) - (298 \text{ K}) \times (0.0832 \text{ kJ·K}^{-1})$$
$$= +1.69 \text{ kJ}$$

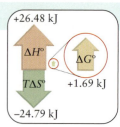

+26.48 kJ

$\Delta H°$ $\Delta G°$

$T\Delta S°$ +1.69 kJ

−24.79 kJ

EVALUATE The standard Gibbs free energy of formation of HI(g) is $+1.69 \text{ kJ·mol}^{-1}$, in good agreement with the value of $+1.70 \text{ kJ·mol}^{-1}$ quoted in the text. Note that, because this value is positive, the formation of pure HI from the elements at 1 bar and 25 °C is not spontaneous.

Self-test 4J.3A Calculate the standard Gibbs free energy of formation of NH_3(g) at 25 °C from the enthalpy of formation and the molar entropies of the species taking part in its formation.

[**Answer:** $-16.5 \text{ kJ·mol}^{-1}$]

Self-test 4J.3B Calculate the standard Gibbs free energy of formation of C_3H_6(g), cyclopropane, at 25 °C.

Related Exercises 4J.5–4J.8

The standard Gibbs free energy of formation at a given temperature is an indication of a compound's stability relative to its elements under standard conditions. The data in Appendix 2A are for 298.15 K, the conventional temperature for reporting thermodynamic data. If $\Delta G_f° < 0$ at a certain temperature, the pure compound has a lower Gibbs free energy than that of its pure elements and the elements are poised to change spontaneously into the compound at that temperature (**FIG. 4J.5**). The compound is then said to be "more stable" under standard conditions than the elements. If $\Delta G_f° > 0$, the Gibbs free energy of the compound is higher than that of its elements and the compound is poised to decompose spontaneously into the pure elements. In this case, the pure elements are said to be "more stable" than the pure compound. For example, the standard Gibbs free energy of formation of benzene is $+124 \text{ kJ·mol}^{-1}$ at 25 °C, and benzene is unstable with respect to its elements under standard conditions at 25 °C. Thus:

- A **thermodynamically stable compound** is a compound with a negative standard Gibbs free energy of formation (water is an example).
- A **thermodynamically unstable compound** is a compound with a positive standard Gibbs free energy of formation (benzene is an example).

The rates of reactions are discussed in Focus 7.

The word "labile" (see text below) is derived from the Latin word for "liable to slip."

The tendency to decompose may not be realized in practice, because it may be very slow. Benzene can, in fact, be kept indefinitely without decomposing at all. Substances that are thermodynamically unstable but survive for long periods are called **nonlabile** or even **inert**. For example, benzene is thermodynamically unstable but nonlabile. Substances that do decompose or react rapidly are called **labile**. Most radicals are labile. It is important to keep in mind the distinction between stability and lability:

- *Stable* and *unstable* are terms that refer to the thermodynamic tendency of a substance to decompose into its elements.
- *Labile*, *nonlabile*, and *inert* are terms that refer to the rate at which a thermodynamic tendency to react is realized.

Self-test 4J.4A Is glucose stable relative to its elements at 25 °C and under standard conditions?

[**Answer:** Yes; for glucose, $\Delta G_f° = -910. \text{ kJ·mol}^{-1}$, a negative value.]

Self-test 4J.4B Is methylamine, CH_3NH_2, stable relative to its elements at 25 °C and under standard conditions?

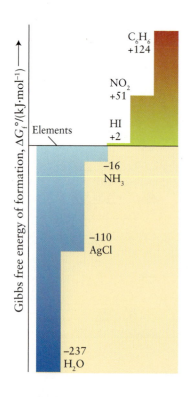

Gibbs free energy of formation, $\Delta G_f°/(\text{kJ·mol}^{-1})$

C_6H_6 +124

NO_2 +51

Elements

HI +2

−16 NH_3

−110 AgCl

−237 H_2O

FIGURE 4J.5 The standard Gibbs free energy of formation of a compound is defined as the standard reaction Gibbs free energy per mole of formula units of the compound when the compound is formed from its elements. It represents a "thermodynamic altitude" with respect to the elements at "sea level." The numerical values are in kilojoules per mole.

Just as standard enthalpies of formation can be combined to obtain standard reaction enthalpies, so standard Gibbs free energies of formation can be combined to obtain standard Gibbs free energies of reaction:

$$\Delta G° = \sum n\Delta G_f°(\text{products}) - \sum n\Delta G_f°(\text{reactants}) \qquad (8)$$

where, as usual, the n are the amounts of each substance as specified by the stoichiometric coefficients in the chemical equation.

EXAMPLE 4J.3　Calculating the standard Gibbs free energy of reaction

Ammonia survives indefinitely in air. An agronomist studying how long ammonia survives in soil might need to know whether it survives because its oxidation is not spontaneous under ordinary conditions or whether the oxidation is spontaneous but just very slow. Calculate the standard Gibbs free energy of the reaction $4 NH_3(g) + 5 O_2(g) \rightarrow 4 NO(g) + 6 H_2O(g)$ and decide whether the reaction is spontaneous under standard conditions at 25 °C.

PLAN Obtain the standard Gibbs free energies of formation from Appendix 2A, then use Eq. 8 to calculate the reaction Gibbs free energy.

SOLVE

From Appendix 2A and Eq. 8,

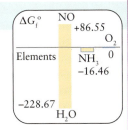

$$\Delta G° = \{(4 \text{ mol}) \times \Delta G_f°(\text{NO, g}) + (6 \text{ mol}) \times \Delta G_f°(\text{H}_2\text{O, g})\}$$
$$-\{(4 \text{ mol}) \times \Delta G_f°(\text{NH}_3\text{, g}) + (5 \text{ mol}) \times \Delta G_f°(\text{O}_2\text{, g})\}$$
$$= \{4(86.55) + 6(-228.57)\} - \{4(-16.45) + 0\} \text{ kJ}$$
$$= -959.42 \text{ kJ}$$

EVALUATE Because the standard reaction Gibbs free energy is negative, you can conclude that the combustion of ammonia is spontaneous at 25 °C under standard conditions. This spontaneous reaction is simply very slow under ordinary conditions.

Self-test 4J.5A Calculate the standard Gibbs free energy of the reaction $2 CO(g) + O_2(g) \rightarrow 2 CO_2(g)$ from standard Gibbs free energies of formation at 25 °C.

[***Answer:*** $\Delta G° = -514.38$ kJ]

Self-test 4J.5B Calculate the standard Gibbs free energy of the reaction $6 CO_2(g) + 6 H_2O(l) \rightarrow C_6H_{12}O_6(s, \text{glucose}) + 6 O_2(g)$ from standard Gibbs free energies of formation at 25 °C.

Related Exercises 4J.11–4J.12

The standard Gibbs free energy of formation of a substance is the standard Gibbs free energy of reaction per mole of compound when it is formed from its elements in their most stable forms. The sign of $\Delta G_f°$ indicates whether a compound is stable or unstable with respect to its elements. Standard Gibbs free energies of formation are used to calculate standard Gibbs free energies of reaction by using Eq. 8.

4J.3 The Gibbs Free Energy and Nonexpansion Work

The change in Gibbs free energy that accompanies a process is equal to the maximum *nonexpansion work* that the process can do at constant temperature and pressure. That is, the Gibbs free energy is a measure of the energy that is *free* to do nonexpansion work (hence the name "free energy"). As explained in Topic 4A, nonexpansion work, w_e, is any kind of work other than that due to expansion against an opposing pressure, so it includes electrical work and mechanical work (such as stretching a spring or carrying a weight up a slope). Nonexpansion work also includes the work of muscular activity, the work of linking amino acids together to form protein molecules, and the work of sending nerve signals through neurons, so a knowledge of changes in Gibbs free energy is central to an understanding of **bioenergetics,** the deployment and utilization of energy in living cells.

The subscript e in w_e stands for "extra."

The challenge is to derive a quantitative relation between the Gibbs free energy and the maximum nonexpansion work that a system can do.

How Is That Done?

To find the relation between the Gibbs free energy and the maximum nonexpansion work, start with Eq. 3 for an infinitesimal change (denoted d) in G at constant temperature:

$$dG = dH - TdS \quad \text{at constant temperature}$$

Then use Eq. 1 of Topic 4C ($H = U + PV$) to express the infinitesimal change in enthalpy at constant pressure in terms of the change in internal energy and volume:

$$dH = dU - PdV \quad \text{at constant pressure}$$

Next, substitute this expression into the preceding one:

$$dG = dU + PdV - TdS \quad \text{at constant temperature and pressure}$$

Now use Eq. 1 of Topic 4B for an infinitesimal change in internal energy ($dU = dw + dq$) and obtain

$$dG = dw + dq + PdV - TdS \quad \text{at constant temperature and pressure}$$

In order for a process to do the maximum possible work, it must occur reversibly. For a reversible change this equation becomes,

$$dG = dw_{rev} + dq_{rev} + PdV - TdS \quad \text{at constant temperature and pressure}$$

Now use the infinitesimal version of Eq. 1 of Topic 4F ($dS = dq_{rev}/T$) to replace dq_{rev} by TdS and cancel the two TdS terms:

$$dG = dw_{rev} + TdS + PdV - TdS$$
$$= dw_{rev} + PdV \quad \text{at constant temperature and pressure}$$

At this point, note that the system may do both expansion work and nonexpansion work:

$$dw_{rev} = dw_{rev,\,e} + dw_{rev,\,expansion}$$

Reversible expansion work (achieved by matching the external to the internal pressure) is given by the infinitesimal version of Eq. 3 of Topic 4A ($w_{expansion} = -P_{ex}\Delta V$, which becomes $dw_{expansion} = -P_{ex}dV$), and setting the external pressure equal to the pressure of the gas in the system at each stage of the expansion,

$$dw_{rev,\,expansion} = -PdV$$

It follows that

$$dw_{rev} = dw_{rev,\,e} - PdV$$

When this expression is substituted into the last line of the expression for dG ($dG = dw_{rev} + PdV$), the PdV terms cancel:

$$dG = dw_{rev,\,e} + PdV - PdV$$

leaving

$$dG = dw_{rev,\,e} \quad \text{at constant temperature and pressure}$$

Since $dw_{rev,\,e}$ is the *maximum* nonexpansion work that the system can do (because it is achieved reversibly), the final result is

$$dG = dw_{e,\,max} \quad \text{at constant temperature and pressure}$$

For a measurable change in the Gibbs free energy, the equation just derived becomes

$$\Delta G = w_{e,\,max} \quad \text{at constant temperature and pressure} \tag{9}$$

This important relation tells you that, if the change in Gibbs free energy of a process taking place at constant temperature and pressure is known, then you immediately know how much nonexpansion work it can do.

Equation 9 is also important in practice because it allows you to consider the energetics of biological processes quantitatively. For instance, for the oxidation of glucose, $C_6H_{12}O_6(s) + 6\,O_2(g) \rightarrow 6\,CO_2(g) + 6\,H_2O(l)$, the standard Gibbs free energy of reaction is -2879 kJ. Therefore, at 1 bar the maximum nonexpansion work obtainable from 1.000 mol $C_6H_{12}O_6(s)$, corresponding to 180.0 g of glucose, is 2879 kJ. Because about 17 kJ of work must be done to build 1 mol of peptide links (a link between amino acids) in a protein, the oxidation of 180 g of glucose can be used to build about (2879 kJ)/(17 kJ) = 170 moles of such links. In other words, the oxidation of one glucose molecule is needed to build about 170 peptide links. In practice, biosynthesis occurs indirectly, there are energy losses, and only about 10 such links can be built. A typical protein has several hundred peptide links, and so several glucose molecules must be sacrificed to build one protein molecule.

The change in Gibbs free energy for a process is equal to the maximum nonexpansion work that the system can do at constant temperature and pressure.

4J.4 The Effect of Temperature

The enthalpies of reactants and products depend on temperature, but the *difference* between their values changes only a little with temperature (this point is established in Topic 4D). The same is true of the entropies. As a result, the values of $\Delta H°$ and $\Delta S°$ do not change much with temperature. However, $\Delta G°$ does depend on temperature (remember the T in $\Delta G° = \Delta H° - T\Delta S°$) and might even change sign as the temperature is changed. There are four cases to consider (see **FIG. 4J.6**):

1. For an exothermic reaction ($\Delta H° < 0$) with a negative reaction entropy ($\Delta S° < 0$), $-T\Delta S°$ contributes a positive term to $\Delta G°$. At high temperatures $-T\Delta S°$ dominates $\Delta H°$ and $\Delta G°$ is positive (and the *reverse* reaction, the decomposition of pure products, is spontaneous). At lower temperatures $\Delta H°$ dominates $-T\Delta S°$, so then $\Delta G°$ is negative (and the formation of products is spontaneous) (Fig. 4J.6a). The temperature at which $\Delta G°$ changes sign is $T = \Delta H°/\Delta S°$.

2. For an endothermic reaction ($\Delta H° > 0$) with a positive reaction entropy ($\Delta S° > 0$), the reverse is true (Fig. 4J.6b). In this case, $\Delta G°$ is positive at low temperatures but may become negative when the temperature is raised to the point that $T\Delta S°$ becomes larger than $\Delta H°$. The formation of products from pure reactants becomes spontaneous when the temperature is high enough. As for the exothermic case, the temperature at which $\Delta G°$ changes sign is $T = \Delta H°/\Delta S°$.

3. For an endothermic reaction ($\Delta H° > 0$) with a negative reaction entropy ($\Delta S° < 0$), $\Delta G° > 0$ whatever the temperature, and the forward reaction is not spontaneous at any temperature because the entropies of both system and surroundings decrease during the process (Fig. 4J.6c).

4. For an exothermic reaction ($\Delta H° < 0$) with a positive reaction entropy ($\Delta S° > 0$), $\Delta G° < 0$ and the formation of products from pure reactants is spontaneous at all temperatures because the entropies of both system and surroundings increase during the process (Fig. 4J.6d).

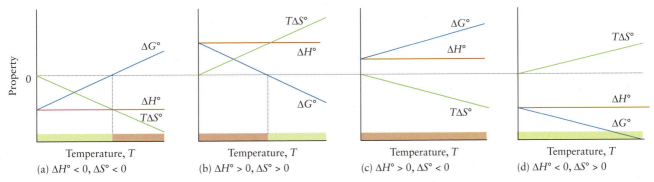

FIGURE 4J.6 The effect of an increase in temperature on the spontaneity of a reaction under standard conditions. In each case, "spontaneous" is taken to mean $\Delta G° < 0$ and "nonspontaneous" is taken to mean $\Delta G° > 0$. (a) An exothermic reaction with negative reaction entropy becomes spontaneous *below* the temperature marked by the vertical dotted line. (b) An endothermic reaction with a positive reaction entropy becomes spontaneous *above* the temperature marked by the vertical dotted line. (c) An endothermic reaction with negative reaction entropy is not spontaneous at any temperature. (d) An exothermic reaction with positive reaction entropy is spontaneous at all temperatures.

EXAMPLE 4J.4   Identifying the temperature at which an endothermic reaction becomes spontaneous

The production of steel from iron ore is endothermic. To reduce the heat that must be supplied, engineers need to find the lowest temperature at which the desired reactions are spontaneous. Estimate the temperature at which it is thermodynamically possible for carbon to reduce iron(III) oxide to iron under standard conditions by the endothermic reaction $2 Fe_2O_3(s) + 3 C(s) \rightarrow 4 Fe(s) + 3 CO_2(g)$.

ANTICIPATE   Because a gas is produced, $\Delta S° > 0$; because (as asserted) $\Delta H° > 0$ also, you should expect the reaction to become spontaneous at a high temperature.

PLAN   At low temperatures, $\Delta G° = \Delta H° - T\Delta S°$ and $\Delta H°$ is positive (the reaction is endothermic). If the temperature is increased there will come a point, at $T = \Delta H°/\Delta S°$, at which $\Delta G° = 0$. Above that temperature it is negative. Use data from Appendix 2A.

What should you assume?   Assume that $\Delta H°$ and $\Delta S°$ are constant over the temperature range considered.

SOLVE

From Eq. 2 of Topic 4D,

$$\Delta H° = (3 \text{ mol}) \times \Delta H_f°(CO_2, g) - (2 \text{ mol}) \times \Delta H_f°(Fe_2O_3, s)$$
$$= 3(-393.5) - 2(-824.2) \text{ kJ} = +467.9 \text{ kJ}$$

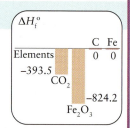

From Eq. 2 of Topic 4H,

$$\Delta S° = \{(4 \text{ mol}) \times S_m°(Fe, s) + (3 \text{ mol}) \times S_m°(CO_2, g)\}$$
$$- \{(2 \text{ mol}) \times S_m°(Fe_2O_3, s) + (3 \text{ mol}) \times S_m°(C, s)\}$$
$$= \{4(27.3) + 3(213.7)\} - \{2(87.4) + 3(5.7)\} \text{ J·K}^{-1} = +558.4 \text{ J·K}^{-1}$$

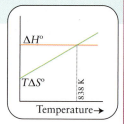

From $T = \Delta H°/\Delta S°$,

$$T = \frac{\overbrace{4.679 \times 10^5 \text{ J}}^{467.9 \text{ kJ}}}{558.4 \text{ J·K}^{-1}} = 838 \text{ K}$$

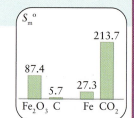

EVALUATE   As expected, because the reaction is endothermic, the minimum temperature at which reduction occurs at 1 bar is high, about 565 °C.

Self-test 4J.6A   What is the minimum temperature at which magnetite, Fe_3O_4, can be reduced to iron by using carbon (to produce CO_2)?

[*Answer:* 943 K]

Self-test 4J.6B   Estimate the temperature at which magnesium carbonate can be expected to decompose to magnesium oxide and carbon dioxide.

Related Exercises 4J.17, 4J.18

The Gibbs free energy increases with temperature for reactions with a negative $\Delta S°$ and decreases with temperature for reactions with a positive $\Delta S°$.

What have you learned in this Topic?

You have learned that the change in Gibbs free energy for a process is a measure of the change in the total entropy of a system and its surroundings at constant temperature and pressure. At constant temperature and pressure, spontaneous processes are accompanied by a decrease in Gibbs free energy. You also now know that the change in Gibbs free energy for a process is a measure of the maximum nonexpansion work that the system can do at constant temperature and pressure.

The skills you have mastered are the ability to:

☐ **1.** Use the change in Gibbs free energy to determine the spontaneity of a process at a given temperature (Example 4J.1).

☐ **2.** Calculate a standard Gibbs free energy of formation from enthalpy and entropy data (Example 4J.2).

☐ **3.** Calculate the standard Gibbs free energy of reaction from standard Gibbs free energies of formation (Example 4J.3).

☐ **4.** Predict the minimum temperature at which an endothermic reaction can occur spontaneously (Example 4J.4).

Topic 4J Exercises

4J.1 Why are so many exothermic reactions spontaneous?

4J.2 Explain how an endothermic reaction can be spontaneous.

4J.3 Calculate the change in molar Gibbs free energy for the process $NH_3(l) \rightarrow NH_3(g)$ at 1 atm and (a) $-15.0\,°C$; (b) $-45.\,°C$ (see Tables 4C.1 and 4F.1). In each case, indicate whether vaporization would be spontaneous.

4J.4 Calculate the change in molar Gibbs free energy for the process $CH_4(l) \rightarrow CH_4(g)$ at 1 atm and (a) $-140.0\,°C$; (b) $-180.\,°C$ (see Tables 4C.1 and 4F.1). In each case, indicate whether or not vaporization would be spontaneous.

4J.5 Write a balanced chemical equation for the formation reaction of (a) $NH_3(g)$; (b) $H_2O(g)$; (c) $CO(g)$; (d) $NO_2(g)$. For each reaction, determine $\Delta H°$, $\Delta S°$, and $\Delta G°$ from data in Appendix 2A.

4J.6 Write a balanced chemical equation for the formation reaction of (a) $HCl(g)$; (b) $C_6H_6(l)$; (c) $CuSO_4 \cdot 5H_2O(s)$; (d) $CaCO_3(s,$ calcite). For each reaction, determine $\Delta H°$, $\Delta S°$, and $\Delta G°$ from data in Appendix 2A.

4J.7 Calculate the standard reaction entropy, enthalpy, and Gibbs free energy for each of the following reactions from data in Appendix 2A:
(a) the decomposition of hydrogen peroxide:

$$2\,H_2O_2(l) \longrightarrow 2\,H_2O(l) + O_2(g)$$

(b) the preparation of hydrofluoric acid from fluorine and water:

$$2\,F_2(g) + 2\,H_2O(l) \longrightarrow 4\,HF(aq) + O_2(g)$$

4J.8 Calculate the standard reaction entropy, enthalpy, and Gibbs free energy for each of the following reactions from data in Appendix 2A:
(a) the production of "synthesis gas," a low-grade industrial fuel:

$$CH_4(g) + H_2O(g) \longrightarrow CO(g) + 3\,H_2(g)$$

(b) the thermal decomposition of ammonium nitrate:

$$NH_4NO_3(s) \longrightarrow N_2O(g) + 2\,H_2O(g)$$

4J.9 Calculate the standard enthalpy, entropy, and Gibbs free energy at 298 K for each of the following reactions by using data in Appendix 2A. For each case, confirm that the value obtained from the Gibbs free energies of formation is the same as that obtained by using the relation $\Delta G° = \Delta H° - T\Delta S°$.
(a) the oxidation of magnetite to hematite: $2\,Fe_3O_4(s) + O_2(g) \rightarrow 3\,Fe_2O_3(s)$
(b) the dissolution of CaF_2 in water: $CaF_2(s) \rightarrow CaF_2(aq)$
(c) the dimerization of NO_2: $2\,NO_2(g) \rightarrow N_2O_4(g)$

4J.10 Calculate the standard enthalpy, entropy, and Gibbs free energy for each of the following reactions at 298 K by using data in Appendix 2A. For each case, confirm that the value obtained from the Gibbs free energies of formation is the same as that obtained by using the relation $\Delta G° = \Delta H° - T\Delta S°$. Which of the reagents in parts (a)–(c) would you expect to be the most effective at removing water from a substance? Explain your reasoning.
(a) the hydration of copper sulfate: $CuSO_4(s) + 5\,H_2O(l) \rightarrow CuSO_4 \cdot 5\,H_2O(s)$
(b) the reaction of phosphorus(V) oxide with water: $P_4O_{10}(s) + 6\,H_2O(l) \rightarrow 4\,H_3PO_4(aq)$
(c) the reaction of $HNO_3(l)$ with water to form aqueous nitric acid: $HNO_3(l) \rightarrow HNO_3(aq)$

4J.11 Use the standard Gibbs free energies of formation in Appendix 2A to calculate $\Delta G°$ for each of the following reactions at 25 °C. Comment on the spontaneity of each reaction under standard conditions at 25 °C.
(a) $2\,SO_2(g) + O_2(g) \rightarrow 2\,SO_3(g)$
(b) $CaCO_3(s,$ calcite$) \rightarrow CaO(s) + CO_2(g)$
(c) $2\,C_8H_{18}(l) + 25\,O_2(g) \rightarrow 16\,CO_2(g) + 18\,H_2O(l)$

4J.12 Use the standard Gibbs free energies of formation in Appendix 2A to calculate $\Delta G°$ for each of the following reactions at 25 °C. Comment on the spontaneity of each reaction under standard conditions at 25 °C.
(a) $NH_4Cl(s) \rightarrow NH_3(g) + HCl(g)$
(b) $H_2(g) + D_2O(l) \rightarrow D_2(g) + H_2O(l)$
(c) $N_2(g) + NO_2(g) \rightarrow NO(g) + N_2O(g)$
(d) $2\,CH_3OH(g) + 3\,O_2(g) \rightarrow 2\,CO_2(g) + 4H_2O(l)$

4J.13 Determine which of the following compounds are stable with respect to decomposition into their elements under standard conditions at 25 °C (see Appendix 2A): (a) $PCl_5(g)$; (b) $HCN(g)$; (c) $NO(g)$; (d) $SO_2(g)$.

4J.14 Determine which of the following compounds are stable with respect to decomposition into their elements under standard conditions at 25 °C (see Appendix 2A): (a) $C_3H_6(g)$, cyclopropane; (b) $CaO(s)$; (c) $N_2O(g)$; (d) $HN_3(g)$.

4J.15 Which of the following compounds become less stable with respect to the elements as the temperature is raised: (a) $PCl_5(g)$; (b) $HCN(g)$; (c) $NO(g)$; (d) $SO_2(g)$?

4J.16 Which of the following compounds become less stable with respect to their elements as the temperature is raised: (a) $C_3H_6(g)$, cyclopropane; (b) $CaO(s)$; (c) $N_2O(g)$; (d) $HN_3(g)$?

4J.17 Assume that $\Delta H°$ and $\Delta S°$ are independent of temperature and use data in Appendix 2A to calculate $\Delta G°$ for each of the following reactions at 80. °C. Over what temperature range will each reaction be spontaneous under standard conditions?
(a) $B_2O_3(s) + 6\ HF(g) \longrightarrow 2\ BF_3(g) + 3\ H_2O(l)$
(b) $CaC_2(s) + 2\ HCl(aq) \longrightarrow CaCl_2(aq) + C_2H_2(g)$
(c) $C(s, graphite) \longrightarrow C(s, diamond)$

4J.18 Assume that $\Delta H°$ and $\Delta S°$ are independent of temperature and use data in Appendix 2A to calculate $\Delta G°$ for each of the following reactions at 250. °C. Over what temperature range will each reaction be spontaneous under standard conditions?
(a) $2\ NO(g) + O_2(g) \longrightarrow 2\ NO_2(g)$
(b) $CaO(s) + H_2O(l) \longrightarrow Ca(OH)_2(s)$
(c) $CaC_2(s) + 2\ H_2O(l) \longrightarrow Ca(OH)_2(aq) + C_2H_2(g)$

The following Example and Exercises draw on material from throughout Focus 4.

FOCUS 4 Online Cumulative Example

Your first experience with chemical transformations in a classroom might have been the reaction of baking soda ($NaHCO_3$) with vinegar ($CH_3COOH(aq)$) to produce carbon dioxide, water, and sodium acetate ($NaCH_3CO_2$). You wish to examine the thermodynamics of this reaction that you enjoyed as a child. A solution of baking soda is prepared by dissolving 15.9 g of $NaHCO_3$ in 50. mL of water. To this you add 200. mL of household vinegar, which is 0.833 M $CH_3COOH(aq)$. The reaction is carried out in an open container in a room in which the temperature is 22.1 °C and the atmospheric pressure is 1.00 atm.

(a) Use the data in Appendix 2A to write the thermochemical equation.

(b) Calculate the following quantities for the reaction: $\Delta H°$, $\Delta U°$, $\Delta S°$, and $\Delta G°$.

(c) What volume of $CO_2(g)$ is produced?

(d) Predict the values of w and q for this experiment and the corresponding changes in the internal energy and enthalpy (ΔU_{expt} and ΔH_{expt}).

 The Online Cumulative Example solution can be found at http://macmillanhighered.com/chemicalprinciples7e

FOCUS 4 Exercises

Exercises labeled $\int_{dx}^{C}$ require calculus.

4.1 How much heat is required to convert a block of ice of mass 42.30 g at −5.042 °C into water vapor at 150.35 °C?

4.2 A piece of stainless steel of mass 121.3 g was heated to 482 °C and quickly added to 37.55 g of ice at −23 °C in a well-insulated flask that was immediately sealed. (a) If there is no loss of energy to the surroundings, what will be the final temperature of this system? (b) What phases of water will be present and in what quantities when the system reaches its final temperature?

4.3 The heat capacity of liquid iodine is 80.7 $J·K^{-1}·mol^{-1}$, and its enthalpy of vaporization is 41.96 $kJ·mol^{-1}$ at its boiling point (184.3 °C). Using these facts and information in Appendix 2A, calculate the enthalpy of fusion of iodine at 25 °C.

4.4 A typical bathtub can hold 100. gallons of water. (a) Calculate the mass of natural gas that would need to be burned to heat the water for a tub of this size from 65 °F to 108 °F. Assume that the natural gas is pure methane, CH_4. (b) What volume of natural gas does this correspond to at 25 °C and 1.00 atm? See Table 4D.1.

4.5 In 1750, Joseph Black performed an experiment that eventually led to the discovery of enthalpies of fusion. He placed two samples of water, each of mass 150. g, at 0.00 °C (one ice and one liquid) in a room kept at a constant temperature of 5.00 °C. He then observed how long it took for each sample to reach its final temperature. The liquid sample reached 5.00 °C after 30.0 min. However, the ice took 10.5 h to reach 5.00 °C. He concluded that the difference in time that the two samples required to reach the same final temperature represented the difference in heat required to raise the temperatures of the samples. Use Black's data to calculate the enthalpy of fusion of ice in kilojoules per mole. Use the known heat capacity of liquid water.

4.6 (a) Calculate the work associated with the isothermal, reversible expansion of 1.000 mol of ideal gas molecules from 7.00 L to 15.50 L at 25.0 °C. (b) Calculate the work associated with the irreversible adiabatic expansion of the sample of gas described in part (a) against a constant atmospheric pressure of 760. Torr. (c) How will the temperature of the gas in part (b) compare with that in part (a) after the expansion?

4.7 (a) Calculate the work that must be done against the atmosphere for the expansion of the gaseous products in the combustion of 1.00 mol $C_6H_6(l)$ at 25 °C and 1.00 bar. (b) Using data in Appendix 2A, calculate the standard enthalpy of the reaction. (c) Calculate the change in internal energy, $\Delta U°$, of the system.

4.8 A system undergoes a two-step process. In step 1, it absorbs 60. J of heat at constant volume. In step 2, it releases 12 J of heat at 1.00 atm as it is returned to its original internal energy. Find the change in the volume of the system during the second step and identify it as an expansion or compression.

4.9 The high-temperature contribution of vibrational modes to the molar heat capacity of a solid at constant volume is R for each mode of vibrational motion. Hence, because each atom can vibrate in three dimensions, for a monatomic solid, the molar heat capacity at constant volume is approximately $3R$. (a) The specific heat capacity of a certain monatomic solid is 0.392 $J·K^{-1}·g^{-1}$. The chloride of this element (XCl_2) is 52.7% chlorine by mass. Identify the element. (b) This element crystallizes in a face-centered cubic unit cell and its atomic radius is 128 pm. What is the density of this solid?

4.10 Estimate the molar heat capacity (at constant volume) of sulfur dioxide gas. In addition to translational and rotational motion, there is vibrational motion. Each vibrational degree of freedom contributes R to the molar heat capacity at high temperature. The temperature needed for the vibrational modes to be accessible can be approximated by $\theta = h\nu_{vib}/k$, where k is Boltzmann's constant. The vibrational modes have frequencies 35 THz, 41 THz, and 16 THz (1 THz = 1×10^{12} Hz). (a) What is the high-temperature limit of the molar heat capacity at constant volume? (b) What is the molar heat capacity at constant volume at 1000. K? (c) What is the molar heat capacity at constant volume at room temperature?

4.11 Consider the hydrogenation of benzene to cyclohexane, which takes place by the step-by-step addition of two H atoms per step:

(1) $C_6H_6(l) + H_2(g) \longrightarrow C_6H_8(l)$ $\Delta H° = ?$
(2) $C_6H_8(l) + H_2(g) \longrightarrow C_6H_{10}(l)$ $\Delta H° = ?$
(3) $C_6H_{10}(l) + H_2(g) \longrightarrow C_6H_{12}(l)$ $\Delta H° = ?$

Draw Lewis structures for the products of the hydrogenation of benzene. If resonance is possible, show only one of the most important resonance structures. (b) Use bond enthalpies to estimate the enthalpy changes of each step and the total hydrogenation. Ignore the delocalization of electrons in this calculation and the fact that the compounds are liquids. (c) Use data from Appendix 2A to calculate the enthalpy of the complete hydrogenation of benzene to cyclohexane. (d) Compare the value obtained in part (c) with that obtained in part (b). Explain any difference.

4.12 Draw the Lewis structure for the hypothetical molecule N_6, consisting of a six-membered ring of nitrogen atoms. Using bond enthalpies, calculate the enthalpy of reaction for the decomposition of N_6 to $N_2(g)$. Do you expect N_6 to be a stable molecule?

4.13 Robert Curl, Richard Smalley, and Harold Kroto were awarded the Nobel Prize in Chemistry in 1996 for the discovery of the soccer-ball shaped molecule C_{60}. This molecule was the first of a new series of *molecular* allotropes of carbon. The enthalpy of combustion of C_{60} is $-25937 \text{ kJ·mol}^{-1}$, and its enthalpy of sublimation is $+233 \text{ kJ·mol}^{-1}$. There are 90 bonds in C_{60}, of which 60 are single bonds and 30 are double bonds. Like benzene, C_{60} has a set of multiple bonds for which resonance structures may be drawn. (a) Determine the enthalpy of formation of C_{60} from its enthalpy of combustion. (b) Calculate the expected enthalpy of formation of C_{60} from bond enthalpies, assuming the bonds to be isolated double and single bonds. (c) Is C_{60} more or less stable than predicted on the basis of the isolated-bond model? (d) Quantify the answer to part (c) by dividing the difference between the enthalpy of formation calculated from the combustion data and that obtained from the bond enthalpy calculation by 60 to obtain a per-carbon value. (e) How does the number in part (d) compare with the per-carbon resonance stabilization energy of benzene (the total resonance stabilization energy of benzene is approximately 150 kJ·mol^{-1})? (f) Why might these values differ? The enthalpy of atomization of C(gr) is $+717 \text{ kJ·mol}^{-1}$.

4.14 When sulfur burns, the product is normally SO_2, but SO_3 may also be formed under certain conditions. When 0.6192 g of sulfur was burned in the presence of oxygen in a bomb calorimeter with a heat capacity of 5.270 kJ·(°C)$^{-1}$, the temperature rose 1.140 °C. Assuming that all the sulfur was consumed in the reaction, what was the ratio of sulfur dioxide to sulfur trioxide produced?

4.15 Hydrochloric acid oxidizes zinc metal in a reaction that produces hydrogen gas and chloride ions. A piece of zinc metal of mass 8.5 g is dropped into an apparatus containing 800.0 mL of 0.500 M HCl(aq). If the initial temperature of the hydrochloric acid solution is 25 °C, what is the final temperature of this solution? Assume that the density and molar heat capacity of the hydrochloric acid solution are the same as those of water and that all the heat is used to raise the temperature of the solution.

4.16 Use only your knowledge of intermolecular forces to rank the following compounds in order of increasing enthalpy of vaporization of their liquid state: CH_4, H_2O, N_2, NaCl, C_6H_6, and H_2. Explain your ordering.

4.17 A technician carries out the reaction $2 SO_2(g) + O_2(g) \rightarrow 2 SO_3(g)$ at 25 °C and 1.00 atm in a cylinder fitted with a piston and maintained at constant pressure. Initially, 0.030 mol SO_2 and 0.030 mol O_2 are present in the cylinder. The technician then adds a catalyst to initiate the reaction. (a) Calculate the volume of the cylinder containing the reactant gases before reaction begins.

(b) What is the limiting reactant? (c) Assuming that the reaction goes to completion and that the temperature and pressure of the reaction remain constant, what is the final volume of the cylinder (include any excess reactant)? (d) How much work takes place, and is it done by the system or on the system? (e) How much enthalpy is exchanged, and does it leave or enter the system? (f) From your answers to parts (d) and (e), calculate the change in internal energy accompanying the reaction.

$\int_{dx}^{C}$ **4.18** Topic 4A explains how to calculate the work of reversible, isothermal expansion of an ideal gas. Now suppose that the reversible expansion is not isothermal and that the temperature decreases during expansion. (a) Derive an expression for the work when $T = T_{initial} - c(V - V_{initial})$, with c a positive constant. (b) Is the work in this case greater or smaller than that of isothermal expansion? Explain your conclusion.

4.19 Calculate the molar kinetic energy (in joules per mole) of Kr(g) at (a) 55.85 °C and (b) 54.85 °C. (c) The difference between the answers to parts (a) and (b) is the energy per mole that it takes to raise the temperature of Kr(g) by 1.00 °C. What is the value of the molar heat capacity of Kr(g)?

4.20 Calculate the molar kinetic energy (in joules per mole) of a sample of Ne(g) at (a) 25.00 °C and (b) 26.00 °C. (c) The difference between the answers to parts (a) and (b) is the energy per mole that it takes to raise the temperature of Ne(g) by 1 °C. What is the value of the molar heat capacity of Ne(g)?

4.21 According to current theories of biological evolution, complex amino and nucleic acids were produced from randomly occurring reactions of compounds thought to be present in the Earth's early atmosphere. These simple molecules then assembled into increasingly complex molecules, such as DNA and RNA. Is this process consistent with the second law of thermodynamics? Explain your answer.

4.22 The standard internal energy of formation, ΔU_f°, corresponds to the standard enthalpy of formation, but is measured at constant volume. From the data in Appendix A, determine ΔU_f° for (a) $H_2O(g)$; (b) $H_2O(l)$. Explain any differences between these values and ΔH_f° for the two phases of water.

4.23 In a microwave oven, radiation is absorbed by water in the food and the food is heated. How many photons of radiation of wavelength 4.50 mm are required to heat 350. g of water from 25.0 °C to 100.0 °C, assuming all their energy is used to raise the temperature?

4.24 Samples consisting of 1 mol $N_2(g)$ and 1 mol $CH_4(g)$ are in identical but separate containers, with initial temperatures of 500. K. Both gases gain 1200. J of heat at constant volume. Do the gases have the same final temperature? If not, which gas has the higher final temperature? Justify your reasoning.

4.25 A person of average mass burns about 15. kJ·min^{-1} playing tennis. Find the time such a person would have to spend playing tennis to burn up the energy provided by a 2.0-oz serving of cheese, which has a specific enthalpy of combustion of 17.0 kJ·g^{-1}.

4.26 "Synthesis gas" is a mixture of carbon monoxide, hydrogen, methane, and some noncombustible gases produced in the refining of petroleum. A certain synthesis gas is 40.0% by volume carbon monoxide, 25.0% hydrogen gas, 10.0% noncombustible gases, and the rest methane. What volume of this gas must be burned to raise the temperature of 5.5 L of water by 5 °C? Assume the gas is at 1.0 atm and 298 K and that all three gases are completely oxidized in the combustion.

4.27 Coal-fired steam engines get their power by using the heat from burning coal to boil water. Suppose that coal of density 1.5 g·cm^{-3} is carbon (it is, in fact, much more complicated, but this is a reasonable first approximation). The combustion of carbon is described by the equation

$$C(s) + O_2(g) \longrightarrow CO_2(g) \qquad \Delta H° = -394 \text{ kJ}$$

Calculate the heat produced when a lump of coal of size 7.0 cm × 6.0 cm × 5.0 cm is burned. (b) Estimate the mass of water that can be heated from 25 °C to 100. °C by burning this piece of coal.

4.28 A student rides a bicycle to class every day, a round trip of 10. miles that takes 30. min in each direction. The student burns 420 kJ·h^{-1} cycling. The same round trip in an automobile would require 0.40 gallons of gasoline. Assume that the student goes to class 150 days per year and that the enthalpy of combustion of gasoline can be approximated by that of octane, which has a density of 0.702 g·cm^{-3} (3.785 L = 1.000 gallon). What is the yearly energy requirement of this journey by (a) bicycle and (b) automobile?

4.29 Crude petroleum is often contaminated by poisonous hydrogen sulfide gas. The Claus process for the extraction of sulfur from petroleum has two steps:

$$2 \text{ H}_2\text{S}(g) + 3 \text{ O}_2(g) \longrightarrow 2 \text{ SO}_2(g) + 2 \text{ H}_2\text{O}(l)$$
$$2 \text{ H}_2\text{S}(g) + \text{SO}_2(g) \longrightarrow 3 \text{ S}(s) + 2 \text{ H}_2\text{O}(l)$$

Write a thermochemical equation for the overall reaction that does not contain SO$_2$. (b) What enthalpy change would be associated with the production of 60.0 kg of sulfur? (c) Would the reactor need to be cooled or heated to maintain a constant temperature?

4.30 A reaction being studied to regenerate oxygen from carbon dioxide on long spaceflights takes place in the following two steps:

$$CO_2(g) + 2 \text{ H}_2(g) \longrightarrow C(s) + 2 \text{ H}_2\text{O}(l)$$
$$2 \text{ H}_2\text{O}(l) \longrightarrow 2 \text{ H}_2(g) + O_2(g)$$

Write the thermochemical equation for the overall reaction. (b) What enthalpy change would be associated with the production of 32.0 L of oxygen at 0.82 atm and 300 K? (c) Would the reactor need to be cooled or heated to maintain a constant temperature?

4.31 Water gas is an inexpensive, low-grade fuel that can be made from coal. (a) Is the production of water gas exothermic or endothermic? The reaction is

$$C(s) + H_2O(g) \longrightarrow CO(g) + H_2(g)$$

(b) Calculate the enthalpy change for the production of 200. L of hydrogen at 500. Torr and 65 °C by this reaction.

4.32 The ABC cereal company is developing a new type of breakfast cereal to compete with a rival product that they call Brand X. You are asked to compare the energy content of the two cereals to see if the new ABC product is lower in calories, so you burn samples, each of mass 1.00 g, of the cereals in oxygen in a calorimeter with a heat capacity of 600. J·(°C)$^{-1}$. When the Brand X cereal sample burned, the temperature rose from 300.2 K to 309.0 K. When the ABC cereal sample burned, the temperature rose from 299.0 K to 307.5 K. (a) What is the heat output of each sample? (b) One serving of each cereal is 30.0 g. How would you label the

packages of the two cereals to indicate the fuel value per 30.0-g serving in joules and in nutritional Calories (kilocalories)?

4.33 An experimental automobile burns hydrogen for fuel. At the beginning of a test drive, the rigid 30.0-L tank was filled with hydrogen at 16.0 atm and 298 K. At the end of the drive, the temperature of the tank was still 298 K, but its pressure was 4.0 atm. (a) How many moles of H$_2$ were burned during the drive? (b) How much heat, in kilojoules, was given off by the combustion of that amount of hydrogen?

4.34 In hot, dry climates an inexpensive alternative to air conditioning is the swamp cooler. In this device, water continuously wets porous pads through which fans blow the hot air. The air is cooled as the water evaporates. Use the information in Tables 4A.2 and 4C.1 to determine how much water must be evaporated to cool the air in a room of dimensions 4.0 m × 5.0 m × 3.0 m by 20. °C. Assume that the enthalpy of vaporization of water is the same it is at 25 °C.

4.35 One step in the production of hydrogen as a fuel is the reaction of methane with water vapor:

$$CH_4(g) + H_2O(g) \xrightarrow{\text{Ni}} CO_2(g) + 3 \text{ H}_2(g) \qquad \Delta H = -318 \text{ kJ}$$

What is the change in internal energy for the production of 1.00 mol H$_2$?

4.36 (a) Before checking the numbers, which substance would you expect to have the higher standard molar entropy, CH$_3$COOH(l) or CH$_3$COOH(aq)? Having made this prediction, examine the values in Appendix 2A and explain your findings.

4.37 Under what conditions, if any, does the sign of each of the following quantities provide a criterion for assessing the spontaneity of a reaction? (a) $\Delta G°$; (b) $\Delta H°$; (c) $\Delta S°$; (d) ΔS_{tot}.

4.38 Three liquid samples with known masses are heated to their boiling points with the use of a heater rated at 500. W. Once their boiling points are reached, heating continues for 4.0 min and some of each sample is vaporized. After 4.0 min, the samples are cooled and the masses of the remaining liquids are determined. The process is performed at constant pressure. (a) Using the following data, calculate ΔS_{vap} and ΔH_{vap} for each sample. Assume that all the heat from the heater goes into the sample. (b) What do the values of ΔS_{vap} indicate about the relative degree of order in the liquids?

Liquid	Boiling temperature/°C	Initial mass/g	Final mass/g
C$_2$H$_5$OH	78.3	400.15	271.15
C$_4$H$_{10}$	0.0	398.05	74.95
CH$_3$OH	64.5	395.15	294.25

4.39 Consider the enthalpies of fusion and the melting points of the following elements: Pb, 5.10 kJ·mol^{-1}, 327 °C; Hg, 2.29 kJ·mol^{-1}, −39 °C; Na, 2.64 kJ·mol^{-1}, 98 °C. Given these data, determine whether a relation similar to Trouton's rule can be obtained for the entropy of fusion of the metallic elements.

4.40 Determine whether titanium dioxide can be reduced by carbon at 1000. K in each of the following reactions:

(a) TiO$_2$(s) + 2 C(s) $\longrightarrow$ Ti(s) + 2 CO(g)
(b) TiO$_2$(s) + C(s) $\longrightarrow$ Ti(s) + CO$_2$(g)

given that, at 1000. K, $\Delta G_f°(CO, g) = -200.$ kJ·mol^{-1}, $\Delta G_f°(CO_2, g) = -396$ kJ·mol^{-1}; and $\Delta G_f°(TiO_2, s) = -762$ kJ·mol^{-1}.

4.41 Which is the thermodynamically more stable iron oxide in air, $Fe_3O_4(s)$ or $Fe_2O_3(s)$? Justify your selection.

4.42 (a) Calculate the work that must be done at 298.15 K against the atmosphere at 1.00 bar for the production of $CO_2(g)$ and $H_2O(g)$ in the combustion of 0.825 mol $C_6H_6(l)$. (b) Calculate the change in the entropy of the system due to expansion of the product gases.

4.43 Hydrogen burns in an atmosphere of bromine gas to give hydrogen bromide gas. (a) What is the standard Gibbs free energy of the reaction $H_2(g) + Br_2(g) \rightarrow 2 HBr(g)$ at 298 K? (b) If 120. mL of H_2 gas at SATP combines with a stoichiometric amount of bromine and the resulting hydrogen bromide dissolves to form 150. mL of an aqueous solution, what is the molar concentration of the resulting hydrobromic acid?

4.44 Hydrogen reacts with nitrogen gas to form ammonia. (a) What is the standard Gibbs free energy of the reaction $3 H_2(g) + N_2(g) \rightarrow 2 NH_3(g)$ at 298 K? (b) If 50.1 L of H_2 gas at 1 bar and 298 K is added to 15.6 L of N_2 gas, also at 1 bar and 298 K, and the resulting ammonia dissolves to form 2.00 L of an aqueous solution, what amount of ammonia can be formed? (c) What is the molar concentration of the resulting aqueous ammonia solution?

4.45 Potassium nitrate dissolves readily in water, and its enthalpy of solution is $+34.9$ kJ·mol^{-1}. (a) Does the enthalpy of solution favor the dissolving process? (b) Is the entropy change of the system likely to be positive or negative when the salt dissolves? (c) Is the entropy change of the system primarily a result of changes in positional disorder or thermal disorder? (d) Is the entropy change of the surroundings primarily a result of changes in positional disorder or thermal disorder? (e) What is the driving force for the dissolution of KNO_3?

4.46 Explain why each of the following statements is false. (a) Reactions with negative Gibbs free energies of reaction occur spontaneously and rapidly. (b) Every sample of a pure element, regardless of its physical state, is assigned zero Gibbs free energy of formation. (c) An exothermic reaction producing more moles of gas molecules than are consumed has a positive standard reaction Gibbs free energy.

4.47 A common antiseptic used in first aid for cuts and scrapes is a 3% aqueous solution of hydrogen peroxide. The oxygen that bubbles out of the hydrogen peroxide as it is decomposed by enzymes in blood into oxygen and water helps to clean the wound. Two possible industrial routes for the synthesis of hydrogen peroxide are (i) $H_2(g) + O_2(g) \rightarrow H_2O_2(l)$, which uses either a metal such as palladium or the organic compound quinone as catalyst, and (ii) $2 H_2O(l) + O_2(g) \rightarrow 2 H_2O_2(l)$. (a) At 298 K and 1 atm, which method releases more energy per mole of O_2? (b) Which method has the more negative standard Gibbs free energy? (c) Once hydrogen peroxide has been used at home, can it be regenerated easily from oxygen and water?

4.48 Acetic acid, $CH_3COOH(l)$, could be produced from (a) the reaction of methanol with carbon monoxide; (b) the oxidation of ethanol; (c) the reaction of carbon dioxide with methane. Write balanced chemical equations for each process. Carry out a thermodynamic analysis of the three possibilities and decide which you would expect to be the easiest to accomplish.

4.49 Some entries for $S_m°$ in Appendix 2A are negative. What is common about these entries, and why would the entropy be negative?

4.50 Three isomeric alkenes have the formula C_4H_8 (see the following table). (a) Draw Lewis structures of these compounds. (b) Calculate $\Delta G°$, $\Delta H°$, and $\Delta S°$ for the three reactions that interconvert each pair of compounds. (c) Which isomer is the most stable? (d) Rank the isomers in order of decreasing $S_m°$.

Compound	$\Delta H_f°/(kJ·mol^{-1})$	$\Delta G_f°/(kJ·mol^{-1})$
2-methylpropene	-16.90	$+58.07$
cis-2-butene	-6.99	$+65.86$
trans-2-butene	-11.17	$+62.97$

4.51 Using values in Appendix 2A, calculate the standard Gibbs free energy for the vaporization of water at 25.0 °C, 100.0 °C, and 150.0 °C. (b) What should the value at 100.0 °C be? (c) Why is there a discrepancy?

4.52 Propose the argument that, for any liquid at atmospheric pressure (that is, a liquid that boils above room temperature when the external pressure is 1 atm), the numerical value of ΔH_{vap} in joules per mole is greater than the numerical value of ΔS_{vap} in joules per kelvin per mole. (Explain and justify each step and any assumptions.)

4.53 The molar entropy of electron spins in a magnetic field B is

$$S_m = R\left\{\frac{\Delta E/kT}{e^{\Delta E/kT} - 1} - \ln(1 - e^{-\Delta E/kT})\right\}$$

where $\Delta E = 2\mu_B B$ is the separation in energy of the two spin states in a magnetic field, and μ_B is the Bohr magneton, $\mu_B = 9.274 \times 10^{-24}$ J·T^{-1}. Plot this function against temperature for the following values of B: 0.1 T, 1 T, 10 T, and 100 T. (See Box 4H.1. Notice that the unit of magnetic induction, tesla, T, where 1 T $= 1$ kg·s^{-2}·A^{-1}, cancels.)

4.54 The populations p of the up and down spin states of electrons in a magnetic field B are given by

$$p_{down} = \frac{1}{1 + e^{-\Delta E/kT}} \quad \text{and} \quad p_{up} = \frac{e^{-\Delta E/kT}}{1 + e^{-\Delta E/kT}}$$

where $\Delta E = 2\mu_B B$ is the energy difference between the two spin states. (See Exercise 4.53.) Plot these two populations as a function of temperature for $B = 1$ T. (See Box 4H.1.)

4.55 Without doing any calculations, predict what temperature corresponds to equal populations of up and down spin states. (See Exercise 4.54.)

4.56 Suppose it could be arranged for there to be twice as many electrons with up spins than down spins in a sample in a magnetic field. Without doing any calculations, predict the sign of the temperature of such a sample. (See Exercise 4.54.)

4.57 It is helpful in understanding graphs of thermodynamic functions to interpret them in terms of molecular behavior. Consider the plot of the temperature dependence of the standard molar Gibbs free energy of the three phases of a substance in Fig. 4J.3. (a) Explain in terms of molecular behavior why the Gibbs free energy of each phase decreases with temperature. (b) Explain in terms of molecular behavior why the Gibbs free energy of the vapor phase decreases more rapidly with temperature than that of the solid or liquid phase.

4.58 Because state functions depend on only the current state of the system, when a system undergoes a series of processes that bring it back to the original state, a thermodynamic cycle has been completed and all the state functions have returned to their original value. However, properties that depend on the path may have changed. (a) Verify that there is no difference in the state function S for a process in which 1.00 mol of nitrogen molecules in a cylinder of volume 3.00 L at 302 K undergoes the following three steps: (i) cooling at constant volume until $T = 75.6$ K; (ii) heating at constant pressure until $T = 302$ K; (iii) compressing at constant temperature until $V = 3.00$ L. Calculate ΔU and ΔS for this entire cycle. (b) What are the values of q and w for the entire cycle? (c) What are ΔS_{surr} and ΔS_{total} for the cycle? If any values are nonzero, explain how this can be so, despite entropy being a state function. (d) Is the process spontaneous, nonspontaneous, or at equilibrium?

4.59 A technique used to overcome the unfavorable thermodynamics of one reaction is to "couple" that reaction to another process that is thermodynamically favored. For instance, the dehydrogenation of cyclohexane to form benzene and hydrogen gas is not spontaneous. Show that, if another molecule such as ethene is present to act as a hydrogen acceptor (that is, the ethene reacts with the hydrogen produced to form ethane), then the overall process is spontaneous.

4.60 Suppose that 200. J of energy is taken as heat from a hot source at 400. °C, passes through a turbine that converts some of the energy into work, and then releases the rest of the energy as heat into a cold sink at 20. °C. What is the maximum amount of work that can be produced by this engine if overall it is to operate

spontaneously? What is the efficiency of the engine, with work done divided by heat supplied expressed as a percentage? How could the efficiency be increased?

4.61 A scientist proposed the following two reactions to produce ethanol, a liquid fuel:

$$C_2H_4(g) + H_2O(g) \longrightarrow CH_3CH_2OH(l) \qquad \textbf{(A)}$$
$$C_2H_6(g) + H_2O(g) \longrightarrow CH_3CH_2OH(l) + H_2(g) \qquad \textbf{(B)}$$

Reaction B is preferred if it is spontaneous, because $C_2H_6(g)$ is a cheaper starting material than $C_2H_4(g)$. Assume standard-state conditions and determine if either reaction is thermodynamically spontaneous.

4.62 A rocket fuel would be useless if its oxidation were not spontaneous. Although rockets operate under conditions that are far from standard, an initial estimation of the potential of a rocket fuel might assess whether its oxidation at the high temperatures reached in a rocket is spontaneous. A chemist exploring potential fuels for use in space considered using vaporized aluminum chloride in a reaction for which the skeletal equation is

$$AlCl_3(g) + O_2(g) \longrightarrow Al_2O_3(s) + ClO(g)$$

Balance this equation. Then use the following data (which are for 2000 K) to decide whether the fuel is worth further investigation: ΔG_f° ($AlCl_3$, g) $= -467$ kJ·mol^{-1}, ΔG_f° (Al_2O_3, s) $= -1034$ kJ·mol^{-1}, ΔG_f° (ClO, g) $= +75$ kJ·mol^{-1}.

FOCUS 4 Cumulative Exercises

4.63 Petroleum-based fuels contribute to climate change, and alternative fuels are being sought (see Box 4D.1). Three compounds that could be produced biologically and used as fuels are methane, CH_4, which can be produced from the anaerobic digestion of sewage; dimethyl ether, $H_3C-O-CH_3$, a gas that can be produced from methanol and ethanol; and ethanol, CH_3CH_2OH, a liquid obtained from the fermentation of sugars.

(a) Draw the Lewis structure of each compound.

(b) Use bond enthalpies (and, for ethanol, its enthalpy of vaporization) to calculate the enthalpy of combustion of each fuel, assuming that they burn to produce gaseous CO_2 and gaseous H_2O. Explain any differences.

(c) Use the values for the enthalpies of combustion of organic compounds found in Appendix 2A to compare methane and ethanol with octane, a primary constituent of gasoline, as fuels by calculating the specific enthalpy (heat produced per gram) of each fuel. On the basis of this information, which would you choose as a fuel?

(d) What volume of methane gas at 10.00 atm and 298 K would you need to burn at constant pressure to produce the same amount of heat as 10.00 L of octane (the density of octane is 0.70 g·mL^{-1})?

(e) A problem with fuels containing carbon is that they produce carbon dioxide when they burn, and so a consideration governing the selection of a fuel could be the heat per mole of CO_2 produced. Calculate this quantity for methane,

ethanol, and octane. Which process produces more carbon dioxide in the environment for each kilojoule generated?

4.64 Vehicle air bags protect passengers by using a chemical reaction that generates gas rapidly. Such a reaction must be both spontaneous and explosively fast. A common reaction is the decomposition of sodium azide, NaN_3, to nitrogen gas and sodium metal.

(a) Write a balanced chemical equation for this reaction using the smallest whole-number coefficients.

(b) Predict the sign of the entropy of reaction without doing a calculation. Explain your reasoning.

(c) Determine the oxidation number of nitrogen in the azide ion and in nitrogen gas. Is nitrogen oxidized or reduced in the reaction?

(d) Use the data in Appendix 2A and the fact that, for sodium azide, $S_m^\circ = 96.9$ J·K^{-1}·mol^{-1}, to calculate ΔS° at 298 K for the decomposition of sodium azide.

(e) Use your result from part (d) and the fact that, for sodium azide, $\Delta H_f^\circ = +21.7$ kJ·mol^{-1}, to calculate ΔH° and ΔG° at 298 K for the decomposition of sodium azide.

(f) Is the reaction spontaneous at 298 K and a constant pressure of 1 bar?

(g) Can the reaction become nonspontaneous (at a constant pressure of 1 bar) if the temperature is changed? If so, must the temperature be raised or lowered?

INTERLUDE Free Energy and Life

The existence of living things may seem at first thought to be a contradiction of the second law of thermodynamics. Every cell of a living being is organized to an extraordinary extent. Thousands of different compounds, each one having a specific function to perform, move in the intricately choreographed dance we call life. We are examples of systems with very low entropy (in the sense that we are highly ordered). By considering the surroundings we can explain why the molecules in our body form a highly organized, complex structure, rather than slime, ooze, or gas.

Many biological reactions, such as the construction of a protein from amino acids or the construction of a DNA molecule, are not spontaneous and therefore must be driven by an external source of energy. That energy comes from sunlight and the chemicals in food that have stored solar energy (**FIG. 1**). When food is metabolized, the resulting exothermic reaction generates a lot of entropy in the surroundings, and if the reaction is coupled to a biochemical reaction that is nonspontaneous, then the *overall* change in entropy may be positive and the *overall* process spontaneous. In other words, *a reaction that produces a lot of entropy can drive a coupled nonspontaneous reaction forward.* In terms of Gibbs free energy, one biochemical process may be driven uphill in Gibbs free energy by another reaction that rolls downhill. Staying alive is very much like the effect of a heavy weight tied to another weight by a string that passes over a pulley (**FIG. 2**). The lighter weight could never fly up into the air on its own. However, when it is connected to a heavier weight falling downward on the other side of a pulley, the light weight can soar upward.

FIG. 1 Evidence for the conversion of carbon dioxide and water into sugars, a process driven by light, is found in the evolution of bubbles of oxygen gas from aquatic plants such as this aquatic fern, *Microsorum pteropus. (Leroy Laverman.)*

The hydrolysis of adenosine triphosphate, ATP (**1**), to adenosine diphosphate, ADP (**2**), is the reaction used most frequently by biological organisms to couple with and drive nonspontaneous reactions. The value of $\Delta G°$ for the hydrolysis of 1 mol ATP is about -30 kJ. To restore ADP back to ATP, which is accompanied by a Gibbs free-energy change of $+30$ kJ, an ADP molecule and a phosphate group must be linked by coupling them to another reaction for which the Gibbs free energy of reaction is more negative than -30 kJ. That is one reason why we have to eat. When we eat food containing glucose, we consume a fuel. If we were simply to burn glucose in an open container, it would do no work other than pushing back the atmosphere, and it would give off a lot of heat. However, in our bodies, the "combustion" is a

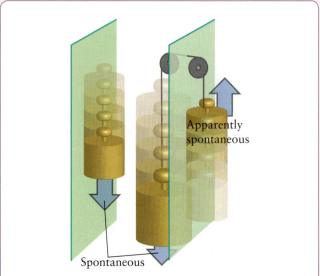

FIG. 2 A natural process can be represented as a falling weight (left). A weight rising spontaneously would be regarded as highly unusual until we recognize that it is actually part of a natural process overall (right). The natural fall of the heavier weight causes the apparently "unnatural" rise of the smaller weight.

highly controlled and sophisticated version of burning. In such a controlled reaction, the nonexpansion work that the process can do approaches 2500 kJ per mole of glucose molecules, which is enough to "recharge" about 80 mol ADP molecules.

When living organisms die, they no longer ingest the second-hand sunlight stored in molecules of carbohydrate, protein, and fat. Then the natural direction of change becomes dominant, and the intricate molecules that support life start to decompose. Living organisms are engaged in a constant battle to generate enough entropy in their surroundings to go on building and maintaining their elaborate interiors. As soon as they stop the battle, they stop generating that external entropy, and their bodies decay into the slime, ooze, and gas that in our lifetime we avoid becoming.

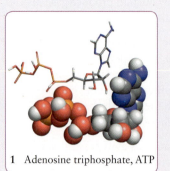

1 Adenosine triphosphate, ATP

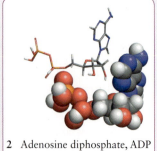

2 Adenosine diphosphate, ADP

EQUILIBRIUM

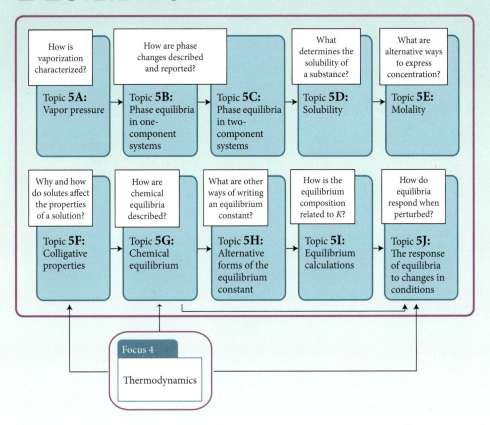

One of the most important aspects of chemistry is the tendency of chemical reactions to approach equilibrium. The equilibria encountered in chemistry are dynamic, which means that the forward and reverse processes continue, but at the same rate, so there is no net change. Equilibria are therefore living, responsive conditions, not dead, exhausted states.

The equilibria discussed in this Focus are both physical (as between different physical states) and chemical (as in chemical reactions). **TOPIC 5A** introduces a basic type of physical equilibrium, that between a liquid and its vapor. **TOPIC 5B** generalizes this discussion by introducing phase diagrams, which summarize the conditions under which each phase of a substance is the most stable and the equilibria between them. Because the conditions for equilibrium can be explained in different ways, the kinetic and thermodynamic perspectives on this discussion are presented in alternative parallel threads. **TOPIC 5C** considers physical equilibria in mixtures of two components, and introduces Raoult's law as well as some practical details about distillation. Thermodynamic aspects of solubility equilibria are treated in **TOPIC 5D**.

TOPIC 5E introduces "molality" and establishes its relation to molarity and other measures of concentration. **TOPIC 5F** discusses "colligative properties," properties of a solution that are affected by the amount of solute but not its identity, such as osmosis.

TOPIC 5G switches attention to chemical equilibria, introduces the concept of an "equilibrium constant," and shows, in alternative kinetic and thermodynamic threads, how it arises. The equilibrium constant can be expressed in a variety of ways, as described in **TOPIC 5H**, and its central importance for describing the composition of reaction mixtures at equilibrium is demonstrated in **TOPIC 5I**.

Finally, **TOPIC 5J** describes the dependence of the equilibrium constant on the conditions (once again, in parallel kinetic and thermodynamic threads) and shows how it is the key to controlling reactions and maximizing their yield.

Topic 5A Vapor Pressure

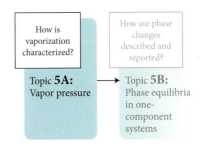

Why Do You Need to Know This Material? The process of vaporization is one of the most important types of phase transition because it reveals information about the forces between molecules and is used for the separation of substances.

What Do You Need to Know Already? The explanations in this Topic follow two threads, and which one you adopt depends on the course sequence. For the kinetic thread you need to be familiar with the concept of an activated process (Topic 7D); for the thermodynamic thread you need to know how the Gibbs free energy is used to describe equilibrium (Topic 4J). You also need to be familiar with intermolecular forces (Topic 3F).

A substance can exist in different *phases*, or physical forms. The phases of a substance include its solid, liquid, and gaseous forms. They might also include its different solid forms, such as the diamond and graphite phases of carbon. In one case—helium—two liquid phases are known to exist. The conversion of a substance from one phase into another, such as the melting of ice, the vaporization of water, and the conversion of graphite into diamond, is called a **phase transition.**

5A.1 The Origin of Vapor Pressure

A simple experiment can be used to show that, in a closed vessel, the liquid and vapor phases of a substance reach equilibrium with each other. First, set up a mercury barometer. The mercury inside the tube falls to a height proportional to the external atmospheric pressure, leaving it about 76 cm high at sea level. The space above the mercury is almost a vacuum (the trace of mercury vapor present is so small that it can be ignored). Now inject a tiny drop of water into the space above the mercury. All the water added evaporates and fills the space with vapor. The molecules in this vapor push the surface of the mercury down a few millimeters. The pressure exerted by the vapor—as measured by the change in the height of mercury—depends on the amount of water added. However, suppose enough water is added that a little liquid water remains on the surface of the mercury. The pressure of the vapor now reaches and remains at a constant value, no matter how much liquid water is present (**FIG. 5A.1**). You can conclude that, *at a fixed temperature, as long as some liquid is present, the vapor exerts a characteristic pressure regardless of the amount of liquid water present.* For example, at 20 °C, the mercury falls 18 mm, and so the pressure exerted by the water vapor is 18 Torr. The pressure of the water vapor is the same whether there is 0.1 mL or 1 mL of liquid water still present. This characteristic pressure is the *vapor pressure* of the liquid at the temperature of the experiment (**TABLE 5A.1**).

Liquids with high vapor pressures at ordinary temperatures are said to be **volatile.** Methanol (vapor pressure 98 Torr at 20 °C) is volatile; mercury (1.4 mTorr) is not. Solids also exert a vapor pressure, but their vapor pressures are usually very much lower than those of liquids because the molecules are held together more tightly in a solid than they are in a liquid. For instance, even at 1000 K the vapor pressure of iron is only 7×10^{-17} Torr, which is too low to sustain a mercury column even one atom high! Nevertheless, some pungent solids, such as menthol and iodine, do sublime (convert directly to the vapor) and can be detected by their odor. The vapor pressure of iodine, for instance, is 0.305 Torr at 25 °C.

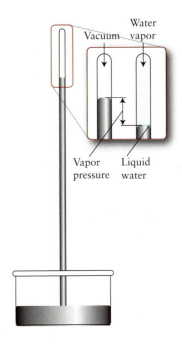

FIGURE 5A.1 The apparatus is a mercury barometer. In the close-up, the left-hand column shows the vacuum above the mercury and the right-hand column shows the effect of the addition of a small amount of water. At equilibrium, some of the water has evaporated and the vapor pressure exerted by the water on the mercury has lowered the height of the mercury column. The vapor pressure is the same however much water is present in the column.

TABLE 5A.1	Vapor Pressures at 25 °C
Substance	**Vapor pressure** P/**Torr**
benzene	94.6
ethanol	58.9
mercury	0.0017
methanol	122.7
toluene	29.1
water*	23.8

*For values at other temperatures, see Table 5A.2.

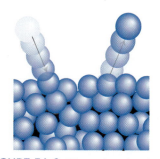

FIGURE 5A.2 When a liquid and its vapor are in dynamic equilibrium inside a closed container, the rate of condensation is equal to the rate of evaporation.

Equilibrium between the condensed and vapor phases can be described in two ways: either from the perspective of kinetics or from the perspective of thermodynamics.

How Is That Explained...

...using kinetics?

The *kinetic interpretation* of equilibrium is based on a comparison of competing rates, in this instance, the rates of evaporation and condensation. Vapor forms as molecules leave the surface of the liquid through evaporation. However, as the number of molecules in the vapor increases, more of them are available to condense, that is, to strike the surface of the liquid, stick to it, and become part of the liquid again. Eventually, the rate of molecules returning to the liquid matches the rate escaping (**FIG. 5A.2**). The vapor is now condensing as fast as the liquid is vaporizing, and so the equilibrium is *dynamic* in the sense that both the forward and reverse processes are still occurring but now their rates are equal. The **dynamic equilibrium** between liquid water and its vapor is denoted

$$H_2O(l) \rightleftharpoons H_2O(g)$$

Wherever the symbol $\rightleftharpoons$ appears, it means that the species on both sides of it are in dynamic equilibrium with each other. With this picture in mind, the **vapor pressure** of a liquid (or a solid) can be defined as the pressure exerted by its vapor when the vapor and the liquid (or the solid) are in dynamic equilibrium with each other.

...using thermodynamics?

In the *thermodynamic interpretation* of equilibrium, the condensed and vapor phases of a substance are in equilibrium, denoted

$$H_2O(l) \rightleftharpoons H_2O(g)$$

when there is no change in Gibbs free energy, $\Delta G = 0$ for the phase change process. In short, neither the forward nor the reverse process is spontaneous at equilibrium. The **vapor pressure** of a liquid (or a solid) is the pressure exerted by its vapor when the vapor and the liquid (or the solid) are in equilibrium with each other.

The vapor pressure of a substance is the pressure exerted by its vapor when the vapor is in dynamic equilibrium with the condensed phase. At equilibrium, the rate of vaporization is equal to the rate of condensation and neither vaporization nor condensation is spontaneous.

5A.2 Volatility and Intermolecular Forces

A high vapor pressure is expected when the molecules of a liquid are held together by weak intermolecular forces in the liquid; a low vapor pressure is expected when the intermolecular forces are strong. You should therefore expect compounds capable of forming hydrogen bonds (which are stronger than other types of intermolecular interactions) to be less volatile than molecular compounds of similar molar mass but that are not able to form hydrogen bonds.

The effect of hydrogen bonding can be seen by comparing dimethyl ether (**1**) and ethanol (**2**), which have the same molecular formula, C_2H_6O. Because the two compounds have the same numbers of electrons, they might be expected to have similar London interactions and therefore similar vapor pressures. However, an ethanol molecule has an —OH group and can form a hydrogen bond to another ethanol molecule. Ether molecules cannot form hydrogen bonds to one another, because all their hydrogen atoms are attached to

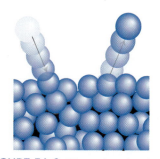

1 Dimethyl ether, C_2H_6O

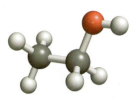

2 Ethanol, C_2H_6O

carbon atoms and the C—H bond is not very polar. The vapor pressure of ethanol at 295 K is 6.6 kPa, whereas that of dimethyl ether is 538 kPa. As a result of this difference, ethanol is a liquid at room temperature and pressure, whereas dimethyl ether is a gas.

THINKING POINT

Why does mercury have such a low vapor pressure at room temperature?

Self-test 5A.1A Which do you expect to have the higher vapor pressure at room temperature, tetrabromomethane, CBr_4, or tetrachloromethane, CCl_4? Give your reasons.

[*Answer:* CCl_4; weaker London forces]

Self-test 5A.1B Which do you expect to have the higher vapor pressure at 25 °C, CH_3CHO or $CH_3CH_2CH_3$?

The vapor pressure of a liquid at a given temperature is expected to be low if the forces acting between its molecules are strong.

TABLE 5A.2	Vapor Pressure of Water
Temperature/°C	Vapor pressure P/Torr
0	4.58
10	9.21
20	17.54
21	18.65
22	19.83
23	21.07
24	22.38
25	23.76
30	31.83
37*	47.08
40	55.34
60	149.44
80	355.26
100	760.00

*Human body temperature.

5A.3 The Variation of Vapor Pressure with Temperature

The vapor pressure of a liquid depends on how readily the molecules in the liquid can escape from the forces that hold them together. More energy to overcome these attractions is available at higher temperatures than at lower temperatures, and the vapor pressure of a liquid can therefore be expected to rise with increasing temperature. **TABLE 5A.2** shows the temperature dependence of the vapor pressure of water, and **FIG. 5A.3** shows how the vapor pressures of several liquids rise as the temperature increases.

Either kinetic arguments from Focus 7 *or* thermodynamic relations from Focus 4 can be used to find an expression for the temperature dependence of vapor pressure.

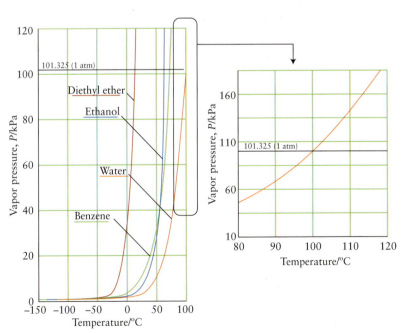

FIGURE 5A.3 The vapor pressures of liquids increase sharply with temperature, as shown here for diethyl ether (red), ethanol (blue), benzene (green), and water (orange). The normal boiling point is the temperature at which the vapor pressure is 1 atm (101.325 kPa). Notice that the curve for ethanol, which has a higher enthalpy of vaporization than benzene, rises more steeply than that of benzene, as predicted by the Clausius–Clapeyron equation (Eq. 1). The diagram on the right shows the vapor pressure of water close to its normal boiling point in more detail.

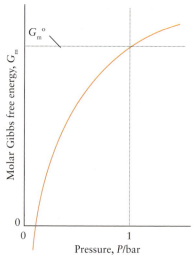

FIGURE 5A.4 The variation of the molar Gibbs free energy of an ideal gas with pressure. The Gibbs free energy has its standard value when the pressure of the gas is 1 bar. The value of the Gibbs free energy approaches minus infinity as the pressure falls to zero.

How Is That Done...

...using kinetics?

The escape of a molecule from the liquid is an activated process in the sense described in Topic 7D in connection with the temperature dependence of the rates of chemical reactions. Like that dependence, the rate of escape depends on the absolute temperature according to the Arrhenius equation:

$$\text{Rate of escape} = Ae^{-E_a/RT}$$

where E_a is the (molar) energy needed to remove molecules from the liquid to the vapor and A is a constant. Only a small error is introduced by setting $E_a = \Delta H_{vap}°$, the standard enthalpy of vaporization.

The rate of return of molecules to the liquid is proportional to the rate at which they strike its surface, which in turn is proportional to the pressure, P, of the vapor:

$$\text{Rate of return} = BP$$

where B is another constant. At equilibrium, the rates of escape and return are equal and P is the vapor pressure. Therefore

$$BP = Ae^{-\Delta H_{vap}°/RT}$$

If P_1 is the vapor pressure when the temperature is T_1 and P_2 is its value at T_2, it follows that

$$\frac{BP_2}{BP_1} = \frac{Ae^{-\Delta H_{vap}°/RT_2}}{Ae^{-\Delta H_{vap}°/RT_1}}$$

After canceling the blue Bs on the left and the blue As on the right,

$$\frac{P_2}{P_1} = \frac{e^{-\Delta H_{vap}°/RT_2}}{e^{-\Delta H_{vap}°/RT_1}}$$

use $e^x/e^y = e^{x-y}$
$$\stackrel{\sim}{=} e^{-(\Delta H_{vap}°/R)(1/T_2 - 1/T_1)}$$

Therefore, on taking logarithms and using $\ln e^x = x$,

$$\ln\frac{P_2}{P_1} = -\frac{\Delta H_{vap}°}{R}\left(\frac{1}{T_2} - \frac{1}{T_1}\right)$$

...using thermodynamics?

For an ideal gas at a pressure P

$$G_m(g, P) = G_m°(g) + RT\ln(P/P°)$$

where $P°$ is the standard pressure (1 bar), and $G_m°$ is the standard molar Gibbs free energy of the gas (its value at 1 bar), **FIG. 5A.4**. The Gibbs free energy of a liquid is almost independent of pressure, so $G_m(l,P) = G_m°(l)$. The Gibbs free energy of vaporization is $\Delta G_{vap} = G_m(g) - G_m(l)$. It follows that the Gibbs free energy of vaporization when a liquid vaporizes at a pressure P is

$$\Delta G_{vap} = \overbrace{G_m(g, P)}^{G_m°(g) + RT\ln(P/P°)} - \overbrace{G_m(l, P)}^{G_m°(l)}$$
$$= \{G_m°(g) + RT\ln(P/P°)\} - G_m°(l)$$

Therefore,

$$\Delta G_{vap} = \overbrace{G_m°(g) - G_m°(l)}^{\Delta G_{vap}°} + RT\ln(P/P°)$$
$$= \Delta G_{vap}° + RT\ln(P/P°)$$

At equilibrium, P is the vapor pressure and $\Delta G_{vap} = 0$, so

$$0 = \Delta G_{vap}° + RT\ln(P/P°)$$

It follows that

$$\ln(P/P°) = -\frac{\Delta G_{vap}°}{RT}$$

Now write $\Delta G_{vap}° = \Delta H_{vap}° - T\Delta S_{vap}°$, which gives

$$\ln(P/P°) = -\frac{\Delta G_{vap}°}{RT} = -\left(\frac{\Delta H_{vap}° - T\Delta S_{vap}°}{RT}\right)$$
$$= -\frac{\Delta H_{vap}°}{RT} + \frac{\Delta S_{vap}°}{R}$$

It follows that the vapor pressures P_1 and P_2 at the two temperatures T_1 and T_2 are related by

$$\ln\left(\frac{P_2}{P°}\right) - \ln\left(\frac{P_1}{P°}\right) = \overbrace{\left(-\frac{\Delta H_{vap}°}{RT_2} + \frac{\Delta S_{vap}°}{R}\right)}^{\ln(P_2/P°)}$$
$$- \overbrace{\left(-\frac{\Delta H_{vap}°}{RT_1} + \frac{\Delta S_{vap}°}{R}\right)}^{\ln(P_1/P°)}$$
$$= -\frac{\Delta H_{vap}°}{R}\left(\frac{1}{T_2} - \frac{1}{T_1}\right)$$

(The terms in blue cancel.) Finally, use the relation $\ln x - \ln y = \ln(x/y)$ to write the left-hand side of this expression as $\ln(P_2/P°) - \ln(P_1/P°) = \ln(P_2/P_1)$, and so obtain

$$\ln\frac{P_2}{P_1} = -\frac{\Delta H_{vap}°}{R}\left(\frac{1}{T_2} - \frac{1}{T_1}\right)$$

The outcome of this calculation by either route is the **Clausius–Clapeyron equation** for the vapor pressure of a liquid at two different temperatures:

$$\ln \frac{P_2}{P_1} = -\frac{\Delta H_{vap}{}^{\circ}}{R}\left(\frac{1}{T_2} - \frac{1}{T_1}\right) \tag{1}$$

What Does This Equation Tell You? When $T_2 > T_1$, the term in parentheses is negative. Then, because there is a minus sign on the right and the enthalpy of vaporization is positive, the right-hand side of the equation is positive. Consequently, $\ln(P_2/P_1)$ is positive too, which implies that P_2 is greater than P_1. In other words, the equation indicates that the vapor pressure increases with increasing temperature. Because $\Delta H_{vap}{}^{\circ}$ occurs in the numerator, the increase is greatest for substances with high enthalpies of vaporization (strong intermolecular interactions).

A simpler form of Eq. 1, which is commonly used to report the temperature dependence of the vapor pressure, is obtained first by writing $\ln(P_2/P_1) = \ln P_2 - \ln P_1$, and discarding the subscript 2, when it becomes

$$\ln P = \ln P_1 + \overbrace{\frac{\Delta H_{vap}{}^{\circ}}{RT_1}}^{A} - \overbrace{\frac{\Delta H_{vap}{}^{\circ}}{RT}}^{B/T}$$

This expression has the form

$$\ln P = A - \frac{B}{T}$$

where A and B are constants that depend on the identity of the substance. It follows that, for a given substance, a plot of $\ln P$ against $1/T$ should be a straight line with a slope determined by $B = \Delta H_{vap}{}^{\circ}/R$ (**FIG. 5A.5**).

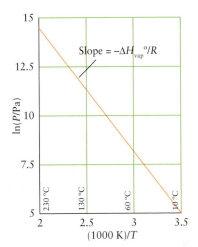

FIGURE 5A.5 A plot of the logarithm of the vapor pressure ($\ln P$) against the reciprocal of the temperature ($1/T$) is a straight line with a negative slope proportional to the enthalpy of vaporization of the liquid phase.

EXAMPLE 5A.1 Estimating the vapor pressure of a liquid from its value at a different temperature

Tetrachloromethane, CCl_4, which is now known to be carcinogenic, was once used as a dry-cleaning solvent and is a volatile liquid still used in the manufacture of coolants. The enthalpy of vaporization of CCl_4 is 33.05 kJ·mol^{-1}, and its vapor pressure at 57.8 °C is 405 Torr. What is the vapor pressure of tetrachloromethane at 25.0 °C?

ANTICIPATE Vapor pressure increases with temperature, so you should expect the vapor pressure of CCl_4 to be lower at 25.0 °C than at 57.8 °C and therefore less than 405 Torr.

PLAN Substitute the temperatures (in kelvins) and the enthalpy of vaporization (in joules per mole) into the Clausius–Clapeyron equation to find the ratio of vapor pressures. Then substitute the known vapor pressure to find the one required. Note that the vapor pressure P_1 corresponds to the temperature T_1.

What should you assume? Assume that $\Delta H_{vap}{}^{\circ}$ and $\Delta S_{vap}{}^{\circ}$ are constant over the temperature range of interest and that the vapor behaves as an ideal gas (so that the Clausius–Clapeyron equation can be used).

SOLVE Note that $\Delta H_{vap}{}^{\circ} = 33.05$ kJ·mol^{-1} corresponds to 3.305×10^4 J·mol^{-1}.

Convert temperatures to kelvins.

$$T_1 = 57.8 + 273.15 \text{ K} = 331.0 \text{ K}$$
$$T_2 = 25.0 + 273.15 \text{ K} = 298.2 \text{ K}$$

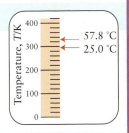

Using Eq. 1, $\ln(P_2/P_1) = (\Delta H_{vap}{}^{\circ}/R)(1/T_1 - 1/T_2)$,

$$\ln \frac{P_2}{P_1} = \frac{3.305 \times 10^4 \text{ J·mol}^{-1}}{8.3145 \text{ J·K}^{-1}\text{·mol}^{-1}}\left(\frac{1}{331.0 \text{ K}} - \frac{1}{298.2 \text{ K}}\right)$$

$$= \frac{3.305 \times 10^4}{8.3145}\left(\frac{1}{331.0} - \frac{1}{298.2}\right) = -1.33\ldots$$

Solve for P_2 (the pressure corresponding to the temperature T_2) by taking the exponential (e^x) of both sides and using $P_1 = 405$ Torr.

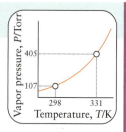

$$P_2 = \overbrace{(405 \text{ Torr})}^{P_1} \times e^{-1.33\ldots}$$
$$= 107 \text{ Torr}$$

EVALUATE   As expected, the vapor pressure at 25 °C, 107 Torr, is lower than its value at 57.8 °C (405 Torr).

A Note on Good Practice: Exponential functions are very sensitive to rounding errors, so it is important to carry out the numerical calculation in one step. A common error is to forget to express the enthalpy of vaporization in joules (not kilojoules) per mole, but keeping track of units will help you to avoid that mistake.

Self-test 5A.2A   The vapor pressure of water at 25 °C is 23.76 Torr and its standard enthalpy of vaporization at that temperature is 44.0 kJ·mol^{-1}. Estimate the vapor pressure of water at 35 °C.

[*Answer:* 42 Torr]

Self-test 5A.2B   The vapor pressure of benzene at 25 °C is 94.6 Torr and its standard enthalpy of vaporization is 30.8 kJ·mol^{-1}. Estimate the vapor pressure of benzene at 35 °C.

Related Exercises   5A.5, 5A.6

The vapor pressure of a liquid increases as temperature increases. The Clausius–Clapeyron equation gives the quantitative dependence of the vapor pressure on temperature.

5A.4 Boiling

Consider what happens when a liquid is heated in a container that is open to the atmosphere—water heated in a kettle is an example. When the temperature is raised to the point at which the vapor pressure is equal to the atmospheric pressure (for instance, when water is heated to 100 °C and the external pressure is 1 atm), vaporization occurs *through-out* the liquid, not just from its surface, and the liquid boils. At the boiling temperature any vapor formed can drive back the atmosphere and make room for itself. Thus, bubbles of vapor form in the liquid and rise rapidly to the surface. The **normal boiling point**, T_b, of a liquid is the temperature at which a liquid boils when the external pressure is 1 atm. To find the normal boiling points of the compounds in Fig. 5A.3, draw a horizontal line at $P = 1$ atm (101.325 kPa) and note the temperatures at which it intersects the curves.

THINKING POINT

Can a liquid boil in a rigid, sealed container?

Boiling occurs at a temperature higher than the normal boiling point when the external pressure is greater than 1 atm, as it is in a pressure cooker. In this case, a higher temperature is needed to raise the vapor pressure of the liquid to the higher pressure. Boiling occurs at a lower temperature when the external pressure is less than 1 atm, because now the vapor pressure matches the external pressure at a lower temperature. At the summit of Mt. Everest—where the pressure is about 253 Torr—water boils at only about 70 °C.

The lower the vapor pressure at room temperature, the higher is the boiling point. Therefore, a high normal boiling point is a sign of strong intermolecular forces.

EXAMPLE 5A.2   Estimating the boiling temperature of a liquid

Ethanol is produced for use as a fuel from corn and agricultural waste. Ethanol plant engineers need to know at what temperature ethanol boils at different pressures. The vapor pressure of ethanol at 34.9 °C is 13.3 kPa. Use the data in Table 4C.1 to estimate the boiling temperature of ethanol at 2.00 atm.

ANTICIPATE   In Table 4C.1, you can see that the normal boiling point of ethanol is 351.5 K (78.4 °C). At the higher pressure ethanol molecules must reach a higher temperature to break free and boil, so you should expect a boiling temperature greater than 351.5 K.

PLAN Use the Clausius–Clapeyron equation to find the temperature at which the vapor pressure has risen to 2.00 atm (203 kPa).

What should you assume? Assume that $\Delta H_{vap}°$ and $\Delta S_{vap}°$ are constant over the temperature range of interest, that the boiling point at 1 bar is approximately the same as at 1 atm, and that the vapor behaves as an ideal gas (so that the Clausius–Clapeyron equation can be used).

SOLVE From Table 4C.1, $\Delta H_{vap}° = 43.5$ kJ·mol^{-1} = 4.35×10^4 J·mol^{-1}.

Convert the given temperature to kelvins and set it equal to T_2.

$$T_2 = 34.9 + 273.15 \text{ K} = 308.0 \text{ K}$$

Rearrange the Clausius–Clapeyron equation to $1/T_1 = 1/T_2 + (R/\Delta H_{vap}°) \ln(P_2/P_1)$ and substitute the data with $P_1 = 203$ kPa and $P_2 = 13.3$ kPa.

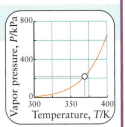

$$\frac{1}{T_1} = \frac{1}{308.0 \text{ K}} + \frac{\overbrace{8.3145 \text{ J·K}^{-1}\text{·mol}^{-1}}^{R}}{\underbrace{4.35 \times 10^4 \text{ J·mol}^{-1}}_{\Delta H_{vap}°}} \ln \frac{13.3 \text{ kPa}}{203 \text{ kPa}}$$

$$= \frac{1}{308.0 \text{ K}} + \frac{8.3145}{4.35 \times 10^4 \text{ K}} \ln \frac{13.3}{203}$$

$$= \frac{1}{308.0 \text{ K}} - \frac{1}{1919.62\ldots \text{ K}} = \frac{1}{367 \text{ K}}$$

Take the reciprocal of each side to obtain T_1.

$$T_1 = 367 \text{ K}$$

EVALUATE The calculated boiling point corresponds to 94 °C (greater than the normal boiling point of 78.4 °C, as anticipated).

Self-test 5A.3A The vapor pressure of acetone, C_3H_6O, at 7.7 °C is 13.3 kPa, and its enthalpy of vaporization is 29.1 kJ·mol^{-1}. Estimate the normal boiling point of acetone.

[**Answer:** 62.3 °C (actual: 56.2 °C)]

Self-test 5A.3B The vapor pressure of methanol, CH_3OH, at 49.9 °C is 400. Torr, and its enthalpy of vaporization is 35.3 kJ·mol^{-1}. Estimate the normal boiling point of methanol.

Related Exercises 5A.7, 5A.8

Boiling occurs when the vapor pressure of a liquid is equal to the external (atmospheric) pressure. Strong intermolecular forces usually lead to high normal boiling points.

What have you learned in this Topic?

You have been introduced to the concept of vapor pressure and have learned that the stronger the intermolecular forces in a substance, the lower its vapor pressure. You have seen how to predict the effect of temperature on vapor pressure and to interpret boiling points.

The skills you have mastered are the ability to:

☐ **1.** Explain the significance of dynamic equilibrium (Section 5A.1)

☐ **2.** Use the Clausius–Clapeyron equation to estimate the vapor pressure of a liquid (Example 5A.1).

☐ **3.** Estimate the boiling point of a liquid from vapor-pressure data (Example 5A.2).

Topic 5A Exercises

5A.1 Which do you expect to have the higher vapor pressure at room temperature, ammonia, NH_3, or phosphine, PH_3? Why?

5A.2 Which do you expect to have the higher vapor pressure at room temperature, octane, C_8H_{18}, or butane, C_4H_{10}? Why?

5A.3 Use the vapor-pressure curve in Fig. 5A.3 to estimate the boiling point of water when the atmospheric pressure is (a) 60. kPa; (b) 160. kPa.

5A.4 Use the vapor-pressure curve in Fig. 5A.3 to estimate the boiling point of benzene when the atmospheric pressure is (a) 50. kPa; (b) 80. kPa.

5A.5 Use data from Table 4C.1 to calculate the vapor pressure of methanol at 25.0 °C.

5A.6 Use data from Table 4C.1 to calculate the vapor pressure of mercury at 275 K.

5A.7 The vapor pressure of boron trichloride at −28 °C is 17.0 kPa and its enthalpy of vaporization is 23.77 kJ·mol⁻¹. What is the normal boiling point of boron trichloride?

5A.8 The vapor pressure of dimethyl ether at −58 °C is 18.1 kPa and its enthalpy of vaporization is 21.51 kJ·mol⁻¹. What is the normal boiling point of dimethyl ether?

5A.9 Arsine, AsH_3, is a highly toxic compound used in the electronics industry for the production of semiconductors. Its vapor pressure is 35 Torr at −111.95 °C and 253 Torr at −83.6 °C. Using these data, calculate (a) the standard enthalpy of vaporization;

(b) the standard entropy of vaporization; (c) the standard Gibbs free energy of vaporization; (d) the normal boiling point of arsine.

5A.10 The vapor pressure of chlorine dioxide, ClO_2, is 155 Torr at −22.75 °C and 485 Torr at 0.00 °C. Calculate (a) the standard enthalpy of vaporization; (b) the standard entropy of vaporization; (c) the standard Gibbs free energy of vaporization; (d) the normal boiling point of ClO_2.

5A.11 The normal boiling point of iodomethane, CH_3I, is 42.43 °C, and its vapor pressure at 0.00 °C is 140. Torr. Calculate (a) the standard enthalpy of vaporization of iodomethane; (b) the standard entropy of vaporization of iodomethane; (c) the vapor pressure of iodomethane at 25.0 °C.

5A.12 The normal boiling point of ethyl acetate, $CH_3COOC_2H_5$, used to remove nail polish, is 77.1 °C, and its vapor pressure at 16.2 °C is 10.0 kPa. Calculate (a) the standard enthalpy of vaporization of ethyl acetate; (b) the standard entropy of vaporization of ethyl acetate; (c) the vapor pressure of ethyl acetate at 30.0 °C.

Topic 5B Phase Equilibria in One-Component Systems

5B.1 One-Component Phase Diagrams
5B.2 Critical Properties

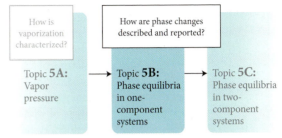

How is vaporization characterized?

How are phase changes described and reported?

Topic 5A: Vapor pressure → Topic 5B: Phase equilibria in one-component systems → Topic 5C: Phase equilibria in two-component systems

Why Do You Need to Know This Material? Phase diagrams are a convenient and widely used means of keeping track of the most stable phase of a substance at different conditions of temperature and pressure.

What Do You Need to Know Already? You need to be aware of vapor pressure, how equilibria are expressed in terms of the Gibbs free energy, and the dynamic nature of equilibria (Topic 5A).

Vaporization is one important type of phase transition (Topic 5A); freezing is another. A liquid solidifies (freezes) when its molecules have such low energies that they are unable to wriggle past their neighbors. In the solid, the molecules vibrate about their average positions but rarely move from place to place. The **freezing temperature,** the temperature at which the solid and liquid phases are in dynamic equilibrium, varies only slightly as the pressure is changed, and the **normal freezing point,** T_f, of a liquid is the temperature at which it freezes at 1 atm. In practice, a liquid sometimes does not freeze until the temperature is a few degrees below its freezing point, especially if cooling is rapid. A liquid that survives below its freezing point is said to be "supercooled." Melting, or **fusion,** is the opposite of freezing, when a solid turns into a liquid. The **normal melting point** of a solid is the same as the normal freezing point of the liquid and is also denoted T_f.

For most substances, the density of the solid phase is greater than that of the liquid phase, because the molecules pack together more closely in the solid. Applied pressure helps to hold the molecules together, and so a high temperature must be reached before the molecules of a solid can separate. As a result, most solids melt at higher temperatures when subjected to higher pressure. However, except at very high pressures, the effect of pressure on the freezing point is usually quite small. For example, iron at 1 atm melts at 1800 K, but it melts at only a few degrees higher when the pressure is a thousand times greater. At the center of the Earth, however, the pressure is high enough for iron to be solid despite the high temperatures there, so the Earth's inner core is believed to be solid.

At the melting point of ice, the molar volume of liquid water is less than that of ice. As a result, high pressure encourages the formation of the denser liquid and the melting point of ice decreases as the pressure is increased. This anomalous behavior is due to the hydrogen bonds in ice, which result in a very open structure (**FIG. 5B.1**). When ice melts, many of the hydrogen bonds collapse, thereby allowing the water molecules to pack together more closely.

The freezing points of most liquids increase with pressure. Water's hydrogen bonds make it anomalous: its melting point decreases with pressure.

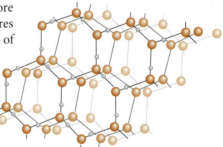

FIGURE 5B.1 The structure of ice; notice how the hydrogen bonds, which are strongest when the hydrogen atom is directly between two oxygen atoms, hold the water molecules apart from one another in a hexagonal array.

5B.1 One-Component Phase Diagrams

How do we keep track of the conditions under which each phase of a substance is stable? And how can we report the effect of pressure on a phase change? A **phase diagram** is a map showing which phase is the most stable at different pressures and temperatures. Phase diagrams are widely used to summarize the states of matter and are particularly useful when the sample is a mixture and its properties depend on the composition (mixtures are treated in Topic 5C).

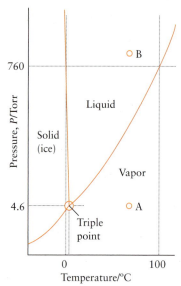

FIGURE 5B.2 The phase diagram of water (not to scale). The orange lines define the boundaries of the regions of pressure and temperature at which each phase is the most stable. Note that the freezing point decreases slightly with increasing pressure. The triple point is the point at which three phase boundaries meet. The letters A and B are referred to in Example 5B.1.

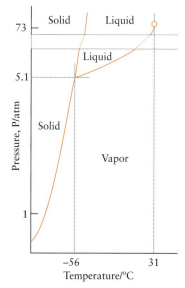

FIGURE 5B.3 The phase diagram of carbon dioxide (not to scale). The liquid can exist only at pressures above 5.1 atm. Note the slope of the boundary between the solid and liquid phases; it shows that the freezing point rises as pressure is applied. This characteristic implies that the liquid is less dense than the solid.

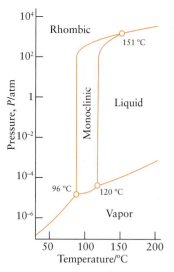

FIGURE 5B.4 The phase diagram of sulfur. Notice that there are two solid phases and three triple points. The pressure scale, which is logarithmic, covers a very wide range of values.

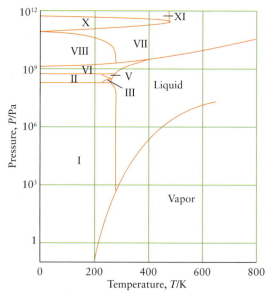

FIGURE 5B.5 The phase diagram of water drawn with a logarithmic scale for pressure, in order to show the different solid phases of water in the high-pressure region.

The phase diagram for water is shown in **FIG. 5B.2**, and that for carbon dioxide is shown in **FIG. 5B.3**. These graphs are examples of phase diagrams for a single substance and hence are known as "one-component phase diagrams." Any point in the region marked "solid" (or, specifically, "ice" for water) corresponds to conditions for which the solid phase of the substance is the most stable; similarly, the regions marked "liquid" and "vapor" (or "gas") indicate the conditions under which liquid and vapor, respectively, are the most stable phases. For example, the phase diagram for carbon dioxide shows that a sample of carbon dioxide at 10 °C and 2 atm will be a gas but, if the pressure is increased at constant temperature to 10 atm, then carbon dioxide will condense to a liquid. Sulfur has two solid phases (**FIG. 5B.4**), rhombic and monoclinic, corresponding to the two ways in which its crownlike S_8 molecules can stack together. The solid that forms when sulfur crystallizes depends on the temperature and pressure. Many substances have several solid phases; even water forms at least ten kinds of ice, depending on how the H_2O molecules fit together, but only one of them is stable at ordinary pressures (**FIG. 5B.5**).

The lines separating the regions in a phase diagram are called **phase boundaries.** Each point on a boundary between two regions represents a specific temperature and pressure at which the two neighboring phases coexist in dynamic equilibrium. If one of the phases is a vapor, the pressure corresponding to this equilibrium is the vapor pressure of the substance. Therefore, the liquid–vapor phase boundary shows how the vapor pressure of the liquid varies with temperature. For example, the point at 80 °C and 0.47 atm in the phase diagram for water lies on the phase boundary between liquid and vapor (**FIG. 5B.6**), which indicates that the vapor pressure of water at 80 °C is 0.47 atm. Similarly, the solid–vapor phase boundary shows how the vapor pressure of the solid varies with temperature.

The solid–liquid boundary, the almost vertical lines in Figs. 5B.2 and 5B.3, show the pressures and temperatures at which solid and liquid coexist in equilibrium. In other words, it shows how the melting temperature of the solid (or, equivalently, the freezing temperature of the liquid) varies with pressure. The steepness of the lines indicates that even large changes in pressure result in only small variations in melting temperature. The slope of the solid–liquid boundary depends on the relative densities of solid and liquid. A negative slope, such as that in the phase diagram for water, implies that the solid melts at a lower temperature as the pressure is raised. The physical reason is that the molecules can respond to the increased pressure by packing together more closely in the liquid, as remarked in the introduction to this Topic. Notice (most clearly from Fig. 5B.2) that as the pressure is increased on ice, it will eventually be converted to liquid. If the slope is positive, as it is for carbon dioxide (Fig. 5B.3), then as the pressure is increased, the temperature must be increased before the solid melts. In this case, the molecules favor the closer packing characteristic of the solid. If the pressure is increased on liquid carbon dioxide, it will solidify. This behavior can be summarized as follows:

- If the liquid is denser than the solid (as in water), then the melting point falls as the pressure is raised.
- If the solid is denser than the liquid (as in most materials), then the melting point rises as the pressure is raised.

> **Self-test 5B.1A** A given metal melts at 1650 K at 1 atm and at 1700 K at 100 atm. Predict which phase of this metal is more dense, the solid or liquid. Explain your conclusion.
>
> [*Answer:* The solid, because the solid–liquid boundary slopes up to the right, showing that the solid is the stable phase at higher pressures.]
>
> **Self-test 5B.1B** From the phase diagram for sulfur (Fig. 5B.4), predict which phase is more dense, liquid sulfur or monoclinic sulfur. Explain your conclusion.

A **triple point** is a point where three phase boundaries meet on a phase diagram. For water, the triple point for the solid, liquid, and vapor phases lies at 4.6 Torr and 0.01 °C (see Fig. 5B.2). At this triple point, all three phases (ice, liquid, and vapor) coexist in mutual dynamic equilibrium: solid is in equilibrium with liquid, liquid with vapor, and vapor with solid. The location of a triple point is a fixed property of a substance and cannot be altered by changing the conditions. The triple point of water is currently used to define the size of the kelvin: by definition, there are exactly 273.16 kelvins between absolute zero and the triple point of water. Because the normal freezing point of water is found to lie 0.01 K below the triple point, 0 °C (strictly, 0.00 °C) corresponds to 273.15 K.

Figure 5B.4 shows that sulfur can exist in any of four phases: two solid phases (rhombic and monoclinic sulfur), a liquid phase, and a vapor phase. There are three triple points in the diagram, where three different combinations of these phases simultaneously coexist: at 96 °C, rhombic solid, monoclinic solid, and vapor; at 120 °C, monoclinic solid, liquid, and vapor; at 151 °C and at a much higher pressure, monoclinic solid, rhombic solid, and liquid coexist in equilibrium. However, four phases in mutual equilibrium (such as the vapor, liquid, and rhombic and monoclinic solid forms of sulfur, all in mutual equilibrium) in a one-component system has never been observed, and thermodynamics can be used to prove that such a "quadruple point" cannot exist.

A phase diagram can be used to explain the changes that take place when the pressure or temperature of a substance is changed. Imagine that you have a sample of water in a cylinder fitted with a piston at low pressure. Suppose that the temperature is held constant at 50 °C, and that weights are placed on the piston to exert a pressure of 1.0 atm (**FIG. 5B.7**). Only liquid water is present. The piston presses on the surface of the liquid, as in Fig. 5B.7a. Now gradually reduce the pressure by removing some of the weights (Fig. 5B.7b). At first, nothing seems to happen. The high pressure is keeping all the water molecules in the liquid state, and the volume of a liquid changes very little with pressure. However, when so many weights have been removed that the pressure has fallen to 0.12 atm (93 Torr, the vapor pressure of water at 50 °C), vapor begins to appear (Fig. 5B.7c).

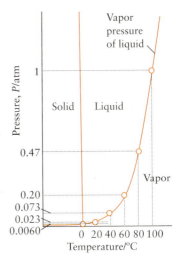

FIGURE 5B.6 The liquid–vapor boundary curve is a plot of the vapor pressure of the liquid (in this case, water) as a function of temperature. The liquid and its vapor are in equilibrium at each point on the curve. At each point on the solid–liquid boundary curve (for which the slope is slightly exaggerated), the solid and liquid are in equilibrium.

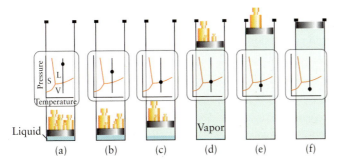

FIGURE 5B.7 The changes undergone by a liquid as its pressure is decreased at constant temperature. The dot on the vertical line in the phase diagram traces the path taken by the system, which is described in the text. The dark blue region in the container is the liquid and the light blue region above it is the vapor.

The sample is now at the vapor–liquid boundary on the phase diagram. The pressure remains constant so long as the liquid and vapor phases are both present at equilibrium and the temperature remains constant. You are free to pull the piston up by an arbitrary extent (Fig. 5B.7d), but enough water will evaporate to maintain the pressure at 0.12 atm. When you pull the piston out far enough, the liquid phase disappears (Fig. 5B.7e); you are now free to modify the pressure of the vapor at will (Fig. 5B.7f).

EXAMPLE 5B.1   Interpreting a phase diagram

Phase diagrams are very useful for predicting the changes that substances will undergo when the conditions are changed and are widely used in geology and metallurgy. Even a phase diagram of a simple, familiar substance can be revealing. Use the phase diagram in Fig. 5B.2 to describe the physical states and phase changes of water as the pressure on it is increased from 5 Torr to 800 Torr at 70 °C.

ANTICIPATE Because the pressure is increasing, you should expect the vapor to condense.

PLAN First, locate the initial and final conditions on the phase diagram. The region in which each of these points lies shows the stable phase of the sample under those conditions. If a point lies on one of the curves, then both phases are present in equilibrium.

SOLVE

From the phase diagram of water in Fig. 5B.2

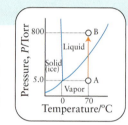

Although the phase diagram in Fig. 5B.2 is not to scale, the points can be located approximately. Point A is the starting point, at 5 Torr and 70 °C, so it lies in the vapor region. Increasing the pressure takes the vapor to the liquid–vapor phase boundary, at which point liquid begins to form. At this pressure, liquid and vapor are in equilibrium and the pressure remains constant until all the vapor has condensed. The pressure is increased further to 800 Torr, which takes it to point B, in the liquid region.

EVALUATE As expected, the vapor condenses.

Self-test 5B.2A The phase diagram for carbon dioxide is shown in Fig. 5B.3. Describe the physical states and phase changes of carbon dioxide as it is heated at 2 atm from −155 °C to 25 °C.

> [*Answer:* Solid CO_2 is heated until it begins to sublime at the solid–vapor boundary. The temperature remains constant until all the CO_2 has vaporized. The vapor is then heated to 25 °C.]

Self-test 5B.2B Describe what happens when liquid carbon dioxide in a container at 60 atm and 25 °C is released into a room at 1 atm and the same temperature.

Related Exercises 5B.1–5B.4

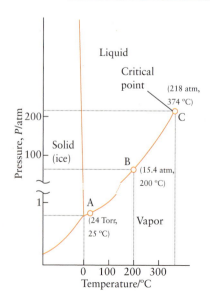

A phase diagram summarizes the regions of pressure and temperature at which each phase of a substance is most stable. The phase boundaries show the conditions under which two phases can coexist in dynamic equilibrium with each other. Three phases coexist in mutual equilibrium at a triple point.

5B.2 Critical Properties

A feature of the phase diagram in **FIG. 5B.8** is that the liquid–vapor boundary comes to an end at point C. To see what happens at that point, suppose that a vessel contains liquid water and water vapor at 25 °C and 24 Torr (the vapor pressure of water at 25 °C). The two phases are in equilibrium, and the system lies at point A on the liquid–vapor curve. Now raise the temperature, which moves the system from left to right along the phase boundary. At 100 °C, the vapor pressure is 760. Torr; at 200 °C, it has reached 11.7 kTorr

FIGURE 5B.8 A version of the phase diagram of water with more detail close to the critical point. Pressures are in atmospheres, except for point A.

(15.4 atm, point B). The liquid and vapor are still in dynamic equilibrium, but now the vapor is very dense because the pressure is so great.

When the temperature has been increased to 374 °C (point C), the vapor pressure has reached 218 atm—the container must be very strong! The density of the vapor is now so great that it is equal to that of the remaining liquid. At this stage, the surface separating the liquid from its vapor vanishes and a single uniform phase fills the container. Because a substance that fills any container that it occupies is by definition a gas, you have to conclude that this single uniform phase is a gas, despite its high density. Provided the temperature is at 374 °C or above, it is found that even if the pressure is increased by compressing the sample, no surface appears. That is, 374 °C is the **critical temperature**, T_c, of water, the temperature at and above which vapor cannot be condensed to liquid no matter how high the pressure. Similar considerations apply to other substances (**TABLE 5B.1**); for instance, the critical temperature of carbon dioxide is 31 °C (**FIG 5B.9**). The pressure corresponding to the end of the liquid–vapor phase boundary is called the **critical pressure**, P_c, of the substance. The critical pressure of water is 218 atm; that of carbon dioxide is 73 atm. The critical temperature and critical pressure jointly define the **critical point** of a substance.

A gas can be liquefied by applying pressure only if it is below its critical temperature (**FIG. 5B.10**). For example, carbon dioxide can be liquefied by applying pressure only if its temperature is lower than 31 °C. According to Table 5B.1, the critical temperature of oxygen is −118 °C, and so it cannot exist as a liquid at room temperature whatever the pressure.

A Note on Good Practice: Formally, a "vapor" is the gaseous phase of a substance that is below its critical temperature, so it can be condensed to a liquid by the application of pressure. A "gas" is a substance that is above its critical temperature and cannot be liquefied by pressure alone.

The dense fluid that exists above the critical temperature and pressure of a substance is called a **supercritical fluid.** It may be so dense that, although it is formally a gas, it is as dense as a liquid phase and can act as a solvent for liquids and solids. There is currently considerable interest in supercritical fluids as solvents for chemical reactions, as they are an important aspect of "green chemistry." In particular, the use of supercritical carbon dioxide avoids contamination with potentially harmful solvents and allows rapid extraction on account of the high mobility of the molecules through the

TABLE 5B.1	Critical Temperatures and Pressures of Selected Substances	
Substance	Critical temperature/ °C	Critical pressure P_c/atm
He	−268 (5.2 K)	2.3
Ne	−229	27
Ar	−123	48
Kr	−64	54
Xe	17	58
H_2	−240	13
O_2	−118	50
H_2O	374	218
N_2	−147	34
NH_3	132	111
CO_2	31	73
CH_4	−83	46
C_6H_6	289	49

FIGURE 5B.9 (a) At low temperatures the liquid and vapor phases of carbon dioxide in a sealed, constant-volume container are distinct. (b) As the temperature is raised, more of the liquid vaporizes; the density of the liquid decreases and that of the vapor increases. Near the critical temperature, T_c, the phases have nearly the same density. (c) At and above the critical temperature a single uniform phase, a supercritical fluid, fills the container. In these images, a small amount of a rhodium compound has been added to make the phases more distinct. (*Courtesy Professor Walter Leitner, RWTH Aachen University, and Dr. Nils Theyssen, Max-Planck-Institut für Kohlenforschung, Mülheim/Ruhr, Germany.*)

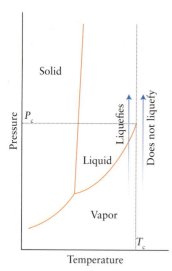

FIGURE 5B.10 A vapor may be liquefied when pressure is applied below its critical temperature. However, if the temperature is above the critical temperature (the vertical dotted line), an increase in pressure does not take the vapor into the liquid region. Above its critical temperature, at high pressure the substance becomes a supercritical fluid.

fluid. For example, because supercritical carbon dioxide can dissolve organic compounds, it is used to remove caffeine from coffee beans, to separate drugs from biological fluids for later analysis, and to extract fragrances from flowers and phytochemicals from herbs. Supercritical hydrocarbons are used to extract useful compounds from coal and fly ash, and a variety of other supercritical fluids are being investigated for recovering petroleum from oil-rich tar sands.

Self-test 5B.3A Identify trends in the data in Table 5B.1 that indicate the relationship of the strength of London forces to the critical temperature.

[*Answer:* Noble gases have higher critical temperatures as their atomic number increases, so critical temperature increases with the strength of London forces.]

Self-test 5B.3B Identify trends in the data in Table 5B.1 that indicate the effect of hydrogen bonding on the critical temperature.

The gaseous phase of a substance can be condensed to a liquid by the application of pressure only if the substance is below its critical temperature.

What have you learned in this Topic?

You have learned that a phase diagram depicts the regions of stability of each phase. You have also seen that phase boundaries show how transition temperatures change with pressure. You are now also familiar with the triple point and the critical point and can distinguish between a gas and a vapor.

The skills you have mastered are the ability to:

☐ **1.** Interpret a one-component phase diagram (Example 5B.1).
☐ **2.** Explain the significance of a triple point and the critical point (Section 5B.2).

Topic 5B Exercises

5B.1 Use Fig. 5B.2 to predict the phase of a sample of water under the following conditions: (a) 1 atm, 200 °C; (b) 100. atm, 50.0 °C; (c) 3 Torr, 10.0 °C.

5B.2 Use Fig. 5B.3 to predict the phase of a sample of CO_2 under the following conditions: (a) 6 atm, −80 °C; (b) 1 atm, −56 °C; (c) 80. atm, 25 °C; (d) 5.1 atm, −56 °C.

5B.3 The phase diagram for helium is shown here. (a) What is the maximum temperature at which superfluid helium-II can exist? (b) What is the minimum pressure at which solid helium can exist? (c) What is the normal boiling point of helium-I? (d) Can solid helium sublime?

5B.4 The phase diagram for carbon, shown here, indicates the extreme conditions that are needed to form diamonds from graphite. (a) At 2000 K, what is the minimum pressure needed before graphite changes into diamond? (b) What is the minimum temperature at which liquid carbon can exist at pressures below 10 000 atm? (c) What is the temperature and pressure of the diamond, liquid, and graphite triple point? (d) Are diamonds stable under normal conditions? If not, why is it that people can wear them without having to keep them under high pressure?

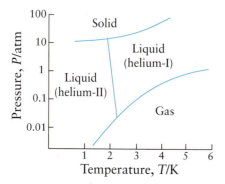

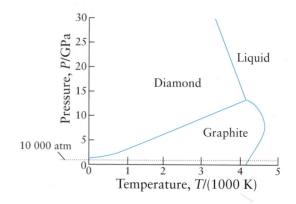

5B.5 Use the phase diagram for helium in Exercise 5B.3 (a) to describe the phases in equilibrium at each of helium's two triple points; (b) to decide which liquid phase is the more dense, helium-I or helium-II.

5B.6 Use the phase diagram for carbon in Exercise 5B.4 (a) to describe the phase transitions that carbon would undergo if the pressure on a sample is increased at a constant temperature of 2000 K from 100 atm to 1×10^6 atm; (b) to rank the diamond, graphite, and liquid phases of carbon in order of increasing density.

5B.7 Use the phase diagram for carbon dioxide (Fig. 5B.3) to predict what would happen to a sample of carbon dioxide gas at −50 °C and 1 atm if its pressure were suddenly increased to 73 atm at constant temperature. What would be the final physical state of the carbon dioxide?

5B.8 A new substance developed in a laboratory has the following properties: normal melting point, 83.7 °C; normal boiling point, 177 °C; triple point, 200. Torr and 38.6 °C. (a) Sketch the approximate phase diagram and label the solid, liquid, and gaseous phases and the solid–liquid, liquid–gas, and solid–gas phase boundaries. (b) Sketch an approximate heating curve for a sample at constant pressure, beginning at 500. Torr and 25 °C and ending at 200 °C.

Topic 5C Phase Equilibria in Two-Component Systems

5C.1   The Vapor Pressure of Mixtures
5C.2   Binary Liquid Mixtures
5C.3   Distillation
5C.4   Azeotropes

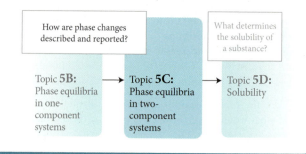

| How are phase changes described and reported? | | What determines the solubility of a substance? |

Topic **5B:**
Phase equilibria in one-component systems
→
Topic **5C:**
Phase equilibria in two-component systems
→
Topic **5D:**
Solubility

Why Do You Need to Know This Material? Many of the systems encountered in chemistry and the everyday world consist of two components, and chemists need to know how the properties of the mixture depend on the conditions and how to separate mixtures into their components.

What Do You Need to Know Already? This Topic draws on the discussion of vapor pressure in Topic 5A. It also uses mole fraction (which is introduced in Topic 3C) as a measure of composition.

Crude petroleum is a mixture of many compounds, which must be separated into the components of fuels and substances that serve as the feedstock of the chemical industry. Because the different vapor pressures of the components are used to separate them, it is essential to know how the total vapor pressure of a mixture depends on its composition.

5C.1 The Vapor Pressure of Mixtures

The French scientist François-Marie Raoult, who spent much of the later part of his life measuring vapor pressures of solutions and mixtures, discovered that *the vapor pressure of a liquid is proportional to its mole fraction.* This statement, which is called **Raoult's law,** is normally written

$$P_A = x_A P_A^\star \tag{1}$$

where P_A is the vapor pressure of the liquid A, x_A is its mole fraction (Topic 3C), and $P_A^\star$ is the vapor pressure of the pure liquid solvent (**FIG. 5C.1**). Equation 1 also applies to mixtures of volatile liquids (for instance, benzene and methylbenzene; **FIG. 5C.2**), in this case with A being benzene.

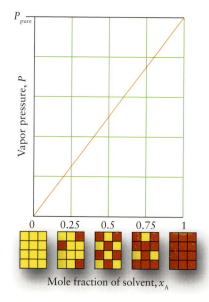

FIGURE 5C.1 According to Raoult's law the vapor pressure of a solvent in a solution is proportional to the mole fraction of the solvent molecules. The horizontal axis shows the mole fraction of solvent molecules (A, red squares) in the pure solute, three different solutions, and pure solvent, pictured below the graph. The vapor pressure of pure A is marked as P_{pure}. The solute in this solution (yellow squares) is not volatile, so it does not contribute to the vapor pressure.

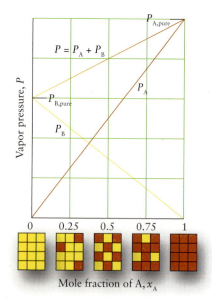

FIGURE 5C.2 The total vapor pressure of a mixture in which both components obey Raoult's law is the sum of the two partial vapor pressures (Dalton's law). The small rectangles below the graph represent the mole fraction of A (red squares).

What Does This Equation Tell You? The vapor pressure of the liquid is directly proportional to the mole fraction of its molecules in the solution. For instance, if 9 out of 10 of the molecules present in a solution are A molecules, then the vapor pressure of the solvent is nine-tenths the vapor pressure of pure A for the temperature of interest. Because in a mixture $x_A < 1$, provided Raoult's law is obeyed, a solute always lowers the vapor pressure of a solvent (**FIG. 5C.3**).

EXAMPLE 5C.1   Using Raoult's law

You are a chemical engineer at a processing plant for soft drinks. You have been asked to determine whether a particular solution will have a significantly different vapor pressure from pure water. Calculate the vapor pressure of water at 20 °C in a solution prepared by dissolving 10.00 g of sucrose, $C_{12}H_{22}O_{11}$, in 100.0 g of water.

ANTICIPATE Expect a lower vapor pressure when the solute is present. However, the vapor pressure is not lowered very much by a solute, so you should expect a value slightly below that of pure water, which Table 5A.2 gives as 17.54 Torr at 20 °C.

PLAN Calculate the mole fraction of the solvent (water) in the solution and then apply Raoult's law. To use Raoult's law, you need to know the vapor pressure of the pure solvent at the temperature of interest (Table 5A.1 or 5A.2).

SOLVE

Find the amount (in moles) of each species from the molar masses of solute and solvent and $n = m/M$:

$$\text{Amount of } C_{12}H_{22}O_{11} = \frac{10.00 \text{ g}}{342.3 \text{ g·mol}^{-1}} = \frac{10.00}{342.3} \text{ mol} = 0.02921\ldots \text{ mol}$$

$$\text{Amount of } H_2O = \frac{100.0 \text{ g}}{18.02 \text{ g·mol}^{-1}} = \frac{100.0}{18.02} \text{ mol} = 5.549\ldots \text{ mol}$$

Calculate the mole fraction of solvent from $x_{\text{solvent}} = n_{\text{solvent}}/(n_{\text{solute}} + n_{\text{solvent}})$:

$$x_{\text{water}} = \frac{\overbrace{5.549\ldots \text{ mol}}^{n_{H_2O}}}{\underbrace{(0.02921\ldots \text{ mol})}_{n_{C_{12}H_{22}O_{11}}} + \underbrace{(5.549\ldots \text{ mol})}_{n_{H_2O}}} = 0.995\ldots$$

Calculate the vapor pressure from $P_{\text{solvent}} = x_{\text{solvent}} P_{\text{solvent}}^{*}$ and the vapor pressure of pure solvent (Table 5A.2):

$$P_{\text{water}} = 0.995\ldots \times \overbrace{17.54 \text{ Torr}}^{P_{\text{water}}^{*}}$$

$$= 17.45 \text{ Torr}$$

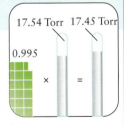

EVALUATE As expected, the vapor pressure is slightly less than that of pure water.

Self-test 5C.1A Calculate the vapor pressure of water at 90 °C for a solution prepared by dissolving 5.00 g of glucose ($C_6H_{12}O_6$) in 100. g of water. The vapor pressure of pure water at 90 °C is 524 Torr.

[*Answer:* 521 Torr]

Self-test 5C.1B Calculate the vapor pressure of ethanol in kilopascals at 19 °C for a solution prepared by dissolving 2.00 g of cinnamaldehyde, C_9H_8O, in 50.0 g of ethanol, C_2H_5OH. The vapor pressure of pure ethanol at that temperature is 5.3 kPa.

Related Exercises 5C.3, 5C.4

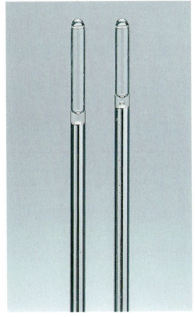

FIGURE 5C.3 The vapor pressure of a solvent is lowered by a nonvolatile solute. The barometer tube on the left has a small volume of pure water floating on the mercury. That on the right has a small volume of 10 M NaCl(aq), which has a lower vapor pressure than pure water. Note that the column on the right is depressed less by the vapor in the space above the mercury than the one on the left, showing that the vapor pressure is lower when solute is present. (*W. H. Freeman photo by Ken Karp.*)

The lowering of the vapor pressure of a solvent due to the presence of a solute can be explained either kinetically (in terms of rates) or thermodynamically.

How Is That Explained...

...using kinetics?

The presence of nonvolatile solute molecules blocks the escape of volatile solvent molecules from the surface of the solution, so their rate of escape is reduced. Solvent molecules that are already in the vapor phase return to the surface at a rate that depends on the pressure of the vapor, and they can stick to the surface wherever they land, on a solute molecule or on a solvent molecule (because all intermolecular interactions are the same in an ideal solution). Equilibrium is achieved when the return rate matches the now slower escape rate, which results in a lower vapor pressure than for the pure solvent.

...using thermodynamics?

At equilibrium, and in the absence of any solute, the molar Gibbs free energy of the vapor is equal to that of the pure liquid solvent (Topic 5A). A solute increases the disorder and therefore the entropy of the liquid phase. Because the entropy of the liquid phase is raised by a solute but the enthalpy is left unchanged, overall there is a decrease in the molar Gibbs free energy of the solvent. Because the Gibbs free energy of the solvent has been lowered, for the two phases to remain in equilibrium, it follows that the Gibbs free energy of the vapor must be lower too. Because the Gibbs free energy of a gas depends on its pressure, it follows that the vapor pressure also must be lower.

A hypothetical liquid mixture in which both volatile components obey Raoult's law at all concentrations, like that in Fig. 5C.2, is called an **ideal solution.** In an ideal solution, the interactions between the two types of molecules are the same as the interactions between each type of molecule in the pure state. That is, A–A, B–B, and A–B interactions are all the same in an ideal solution. Consequently, both types of molecule mingle freely with each other. Because the interactions are all the same, for an ideal solution, there is no release of energy as heat when the components mix: in thermodynamic terms, the "enthalpy of solution" is zero. Molecules of solutes that form nearly ideal solutions are often similar in composition and structure to the solvent molecules. For instance, toluene (methylbenzene), $C_6H_5CH_3$, forms nearly ideal solutions with benzene, C_6H_6. Real solutions do not obey Raoult's law at all concentrations; however, the lower the concentration of one component, the more closely they resemble ideal solutions. Like the ideal gas law (Topic 3B), Raoult's law is an example of a "limiting law," which in this case becomes increasingly valid as the concentration of one component approaches zero. A solution that does not obey Raoult's law over the full range of compositions is called a **real solution.** Real solutions approach ideal behavior at concentrations below about 0.1 mol·L^{-1}. Unless stated otherwise, all the mixtures discussed here are ideal.

The *solvent* in an electrolyte solution (such as salt in water) obeys Raoult's law provided the concentration of solute is very low, typically no more than 0.01 mol·L^{-1}. The greater departure from ideality in electrolyte solutions arises from the interactions between ions, which occur over a long distance and hence have a pronounced effect.

THINKING POINT

Laws are useful summaries of experience, but why are *limiting* laws useful?

The vapor pressure of a liquid is reduced by the presence of a solute; in an ideal solution of volatile liquids, the vapor pressure of each component is proportional to its mole fraction; in a solution of a nonvolatile solute, the vapor pressure of the solvent is proportional to its mole fraction.

5C.2 Binary Liquid Mixtures

A binary liquid mixture is a solution of two liquids. Because both liquids are volatile, both can contribute to the vapor pressure of the solution. Like one-component phase diagrams (Topic 5B), the phase diagrams of binary liquid mixtures show which phase is the more stable, but in addition to the temperature and pressure there is an additional variable, the composition. To keep matters simple, this discussion considers only pairs of liquids that are miscible in all proportions and supposes that the pressure is fixed at 1 atm. That leaves the temperature and composition as variables in the phase diagram (**FIG. 5C.4**). Only two phases will be considered: the liquid and the vapor. Because the pressure is fixed at 1 atm,

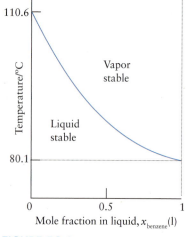

FIGURE 5C.4 A temperature–composition diagram for benzene and toluene. The curve shows how the boiling point of the mixture varies with its composition.

the boundary between them in the diagram is a plot of the boiling point against composition. The area above the line shows the temperatures at which the vapor is stable for any composition of the liquid mixture; the area below the line similarly shows the temperatures and compositions at which the liquid is the more stable phase.

The phase diagram can be adapted to carry more information. First, consider an ideal binary mixture of the volatile liquids A and B. You could think of A as benzene, and B as toluene, for example, because these two compounds have similar molecular structures and so form nearly ideal solutions. Because the mixture can be treated as ideal, each component has a vapor pressure given by Raoult's law:

$$P_A = x_A(l)P_A^* \qquad P_B = x_B(l)P_B^*$$

In these equations, $x_A(l)$ is the mole fraction of A in the liquid mixture and P_A^* is the vapor pressure of pure A; similarly, $x_B(l)$ is the mole fraction of B in the liquid and P_B^* is the vapor pressure of pure B (both vapor pressures depend on the temperature). According to Dalton's law (Topic 3C), the total pressure of the vapor, P, is the sum of these two partial pressures:

$$P = P_A + P_B = x_A(l)P_A^* + x_B(l)P_B^*$$

Therefore, the vapor pressure of an ideal mixture of two volatile liquids is intermediate in value between the vapor pressures of the two pure liquids. The total vapor pressure (the upper line in Fig. 5C.2) can be predicted if the individual vapor pressures at the temperature of interest and their mole fractions in the mixture are known. Similarly, the boiling point of the mixture of given composition can also be predicted by judging the temperature needed for the total vapor pressure to reach 1 atm.

EXAMPLE 5C.2 Predicting the vapor pressure of a mixture of two liquids

When you are designing a chemical plant, it is vitally important to know the pressures that will be encountered in reaction chambers and transfer piping. In some cases, that involves knowing the vapor pressures of the liquids present. What is the vapor pressure of each component at 25 °C and the total vapor pressure of a mixture of benzene and toluene in which one-third of the molecules are benzene (so $x_{benzene}(l) = \frac{1}{3}$ and $x_{toluene}(l) = \frac{2}{3}$)? The vapor pressures of pure benzene and toluene at 25 °C are 94.6 and 29.1 Torr, respectively.

ANTICIPATE The vapor pressure of the mixture will lie between the values of the pure liquids, closer to the more abundant component, which in this case is toluene. So, you should expect a value between 29 and 95 Torr, but closer to 29 Torr.

PLAN Use Raoult's law to calculate the vapor pressure of each component, and then use Dalton's law to calculate the total vapor pressure.

What should you assume? Assume that the liquid mixture and its vapor are ideal.

SOLVE

From Raoult's law in the form $P_A = x_A(l)P_A^*$ and $P_B = x_B(l)P_A^*$,

$$P_{benzene} = \frac{1}{3}(94.6 \text{ Torr}) = 31.5 \text{ Torr}$$

$$P_{toluene} = \frac{2}{3}(29.1 \text{ Torr}) = 19.4 \text{ Torr}$$

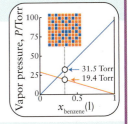

From Dalton's law, $P = P_A + P_B$,

$$P_{total} = 31.5 + 19.4 \text{ Torr} = 50.9 \text{ Torr}$$

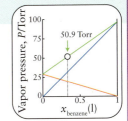

EVALUATE As expected, the vapor pressure of the mixture lies closer to that of toluene than that of benzene.

Self-test 5C.2A What is the total vapor pressure at 25 °C of a mixture of 3.00 mol C_6H_6 (benzene) and 2.00 mol $C_6H_5CH_3$ (toluene)?

[*Answer:* 68.4 Torr]

Self-test 5C.2B What is the total vapor pressure at 25 °C of a mixture of equal masses of benzene and toluene?

Related Exercises 5C.9, 5C.10

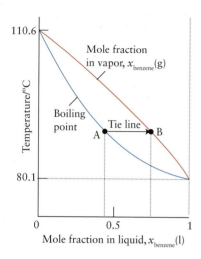

FIGURE 5C.5 A more elaborate temperature–composition diagram for benzene and toluene. The lower, blue curve shows how the boiling point of the mixture varies with composition. Horizontal tie lines connected to the upper, red curve give the composition of the vapor in equilibrium with the liquid at each boiling point. Thus, point B shows the vapor composition for a mixture that boils at point A.

Now consider the composition of the vapor. It can be expected to be richer than the liquid in the more volatile component (the component with the greater vapor pressure). Benzene, for instance, is more volatile than toluene, so the vapor in equilibrium with the liquid mixture can be expected to be richer in benzene than the liquid (**FIG. 5C.5**). If the composition of the vapor could be expressed in terms of the composition of the liquid, that expectation could be confirmed.

How is That Done?

First note that the definition of partial pressure ($P_A = x_A P$ for component A) and Dalton's law ($P = P_A + P_B$) allow the composition of the vapor of a mixture of liquids A and B to be expressed in terms of the partial pressures of the components:

$$x_A(g) = \frac{P_A}{P} = \frac{P_A}{P_A + P_B}$$

and likewise for $x_B(g)$. Next, express the vapor pressures of A and B in terms of the composition of the liquid by using Raoult's law:

$$x_A(g) = \frac{x_A(l)P_A^*}{x_A(l)P_A^* + x_B(l)P_B^*} \overset{x_B = 1 - x_A}{\cong} \frac{x_A(l)P_A^*}{x_A(l)P_A^* + (1 - x_A(l))P_B^*}$$

This expression relates the composition of the vapor (in terms of the mole fraction of A in the vapor) in a binary mixture to the composition of the liquid (in terms of the mole fraction of A in the liquid).

EXAMPLE 5C.3   Predicting the composition of the vapor in equilibrium with a binary liquid mixture

You are now a chemical engineer in the process of designing a plant to separate hydrocarbons obtained from crude oil. You need to keep track of the composition of the mixtures you are dealing with, both the vapors and the liquids. Find the mole fraction of benzene at 25 °C in the vapor of a solution of benzene in toluene in which one-third of the molecules in the liquid are benzene. See Example 5C.2 for data.

ANTICIPATE Benzene is more volatile than toluene, so you should expect the vapor to be richer in benzene than the liquid is.

SOLVE

From $x_A(g) = x_A(l)P_A^*/\{x_A(l)P_A^* + (1 - x_A(l))P_B^*\}$,

$$x_{benzene}(g) = \frac{\frac{1}{3} \times 94.6 \text{ Torr}}{\underbrace{\frac{1}{3} \times (94.6 \text{ Torr}) + (1 - \frac{1}{3}) \times (29.1 \text{ Torr})}_{x_{toluene}(l)}}$$

$$= 0.619$$

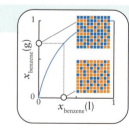

EVALUATE As expected, the vapor is richer in benzene than the liquid; in fact, the mole fraction of benzene in the vapor is nearly twice that in the liquid, as can be seen in Fig. 5C.5.

Self-test 5C.3A (a) Determine the vapor pressure at 25 °C of a solution of toluene in benzene in which the mole fraction of benzene is 0.900. (b) Calculate the mole fractions of benzene and toluene in the vapor.

[*Answer:* (a) 88.0 Torr; (b) 0.967 and 0.033]

Self-test 5C.3B (a) Determine the vapor pressure at 25 °C of a solution of benzene in toluene in which the mole fraction of benzene is 0.500. (b) Calculate the mole fractions of benzene and toluene in the vapor.

Related Exercises 5C.11, 5C.12

FIGURE 5C.6 Some of the steps that represent fractional distillation of a mixture of two volatile liquids (benzene and toluene). The original mixture boils at A and its vapor has composition B. After condensation of the vapor, the resulting liquid boils at C and the vapor has composition D, and so on.

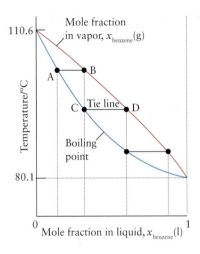

5C.3 Distillation

The upper curve in Fig. 5C.5 shows the composition of the vapor in equilibrium with the liquid mixture at the boiling point: it has been constructed using the same kind of calculation as in Example 5C.2 but at a number of different temperatures. To find the composition of the vapor in equilibrium with the boiling liquid, simply look along the **tie line,** the horizontal line at the boiling point, and see where it cuts through the upper curve. So, if a liquid mixture with the composition given by the vertical line through A in Fig. 5C.5 ($x_{benzene}(l) = 0.45$) is heated at a constant pressure of 1 atm, the mixture boils at the temperature corresponding to point A. At this temperature, the composition of the vapor in equilibrium with the liquid is given by point B ($x_{benzene}(g) = 0.73$).

When a mixture of benzene and toluene molecules with $x_{benzene}(l) = 0.20$ begins to boil (point A in **FIG. 5C.6**), the initial composition of the vapor formed is given by point B ($x_{benzene}(g) = 0.45$). If the vapor is cooled until it condenses, the first drop of condensed vapor, the **distillate,** will have the same composition as the vapor and is therefore richer in benzene than the original mixture; the liquid remaining in the pot is richer in toluene because more benzene than toluene has been lost. The separation is not very good: the distillate is still rich in toluene. However, if that drop of distillate is reheated, it will boil at the temperature represented by point C, and the vapor above the boiling solution will have the composition D ($x_{benzene}(g) = 0.73$), as indicated by the tie line. Note that the distillate from this second stage of distillation is even richer in benzene than the distillate from the first stage. If these steps of boiling, condensation, and boiling again, are repeated, a very tiny amount of nearly pure benzene would be obtained.

The process called **fractional distillation** uses a method of continuous redistillation to separate mixtures of liquids with similar boiling points, such as benzene and toluene: the mixture is heated and the rising vapor passes up through a tall column packed with material having a high surface area, such as glass beads (**FIG. 5C.7**). The vapor begins to condense on the beads near the bottom of the column. However, as heating continues, the vapor condenses and vaporizes over and over again as it rises; the liquid drips back into the boiling mixture. The vapor becomes richer in the component with the lower boiling point as it continues to rise through the column and finally passes out into the condenser. For example, in the case of benzene and toluene, the final distillate is nearly pure benzene, the more volatile of the components, whereas the liquid left in the pot is nearly pure toluene.

If the original sample consists of several volatile liquids, the component liquids appear in the distillate in succession in a series of **fractions,** or samples of the distillate that boil in specific temperature ranges. Giant fractionating columns are used in industry to separate complex mixtures such as crude petroleum (**FIG. 5C.8**). The volatile fractions are put to use as natural gas (boiling below 0 °C), gasoline (boiling in the range from 30 °C to 200 °C), and kerosene (boiling from 180 °C to 325 °C). Less volatile fractions are used as diesel fuel (boiling above 275 °C). The residue that remains behind after distillation is asphalt, which is used for surfacing roads.

> *The vapor of an ideal mixture of two volatile liquids is richer than the liquid in the more volatile component. Volatile liquids can be separated by fractional distillation.*

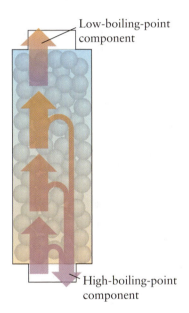

FIGURE 5C.7 A schematic illustration of the process of fractional distillation. The temperature in the fractionating column decreases with height. The condensations and reboilings illustrated in Fig. 5C.6 occur at increasing heights in the column. The less volatile component returns to the flask beneath the fractionating column, and the more volatile component escapes from the top, to be condensed and collected.

5C.4 Azeotropes

Most liquid mixtures are not ideal, and so their vapor pressures do not follow Raoult's law (**FIG. 5C.9**). In these cases, the temperature–composition curves are determined experimentally by analyzing the composition of the vapor.

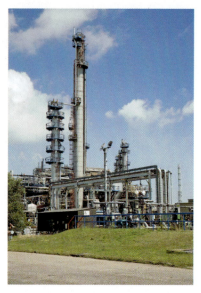

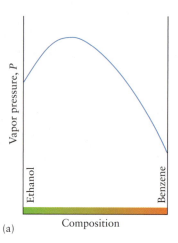

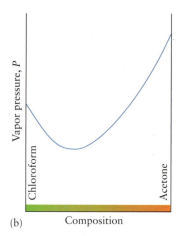

FIGURE 5C.9 A graphical illustration of the variation in the vapor pressures of (a) a mixture of ethanol and benzene and (b) a mixture of acetone and chloroform. Note that the mixture in (a) shows a vapor-pressure maximum and therefore displays a positive deviation from Raoult's law. The mixture in (b) shows a minimum and hence displays a negative deviation from Raoult's law.

The name "azeotrope" is formed from Greek words meaning "boiling without change."

Deviations from Raoult's law can be correlated with the **enthalpy of mixing,** ΔH_{mix}, the difference in molar enthalpy between the mixture and the pure components. The enthalpy of mixing of ethanol and benzene is positive—the mixing process is endothermic because the interactions between ethanol and benzene molecules are less favorable than between each type of molecule separately—and this mixture has a vapor pressure higher than that predicted by Raoult's law (a "positive deviation"). The enthalpy of mixing of acetone and chloroform is negative—the mixing process is exothermic because the interactions between acetone and chloroform molecules are stronger than those between molecules of chloroform or acetone separately—and this mixture has a vapor pressure lower than that predicted by the law (a "negative deviation").

Some deviations from Raoult's law can make it impossible to separate liquids completely by distillation. The temperature–composition diagram for mixtures of ethanol and benzene is shown in **FIG. 5C.10**. The lowest point on the boiling-point curve indicates the existence of a **minimum-boiling azeotrope.** Its components cannot be separated by distillation: in a fractional distillation, a mixture with the same azeotropic composition boils over first, not the more volatile pure liquid. The opposite behavior is found for a mixture of acetone in chloroform (**FIG. 5C.11**). This **maximum-boiling azeotrope** boils at a higher temperature than either constituent and is the last fraction to be collected, not the less volatile of the pure liquids.

Binary mixtures in which intermolecular forces are weaker in the solution than in the pure components have positive deviations from Raoult's law; some form minimum-boiling azeotropes. Mixtures in which intermolecular forces are stronger in the solution than in the pure components have negative deviations from Raoult's law; some form maximum-boiling azeotropes.

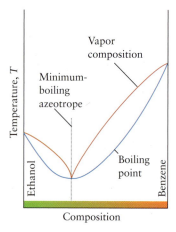

FIGURE 5C.10 The temperature–composition diagram of a minimum-boiling azeotrope (such as ethanol and benzene). When this mixture is fractionally distilled, the (more volatile) azeotropic mixture is obtained as the distillate.

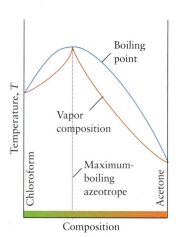

FIGURE 5C.11 The temperature–composition diagram showing a maximum-boiling azeotrope (such as acetone and chloroform). When this mixture is fractionally distilled, the (less volatile) azeotropic mixture is left in the flask.

What have you learned in this Topic?

You have learned how and why the presence of a nonvolatile solute lowers the vapor pressure. You have learned how to express the vapor pressure of a liquid in a mixture by using Raoult's law and how the total vapor pressure of a mixture varies with composition. You have been introduced to the concept of an ideal solution. You have also seen how the boiling point of a mixture of volatile liquids changes with composition and now understand the procedure of fractional distillation. You have been introduced to the concept of an azeotrope and have seen how its formation affects the process of distillation.

The skills you have mastered are the ability to:

☐ **1.** Calculate the vapor pressure of a solvent in a solution by using Raoult's law (Example 5C.1).

☐ **2.** Calculate the vapor pressure and vapor composition for a mixture of two liquids (Examples 5C.2 and 5C.3).

☐ **3.** Interpret a two-component phase diagram and discuss fractional distillation (Sections 5C.2 and 5C.3).

☐ **4.** Explain the existence and consequences of the formation of azeotropes (Section 5C.4).

Topic 5C Exercises

5C.1 Two beakers, one containing 0.010 M NaCl(aq) and the other containing pure water, are placed inside a bell jar and sealed. The beakers are left until the water vapor has come to equilibrium with any liquid in the container. The levels of the liquid in each beaker at the beginning of the experiment are the same, as pictured below. Draw the levels of the liquid in each beaker after equilibrium has been reached. Explain your reasoning.

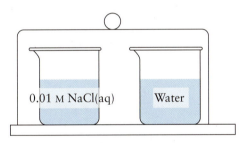

0.01 M NaCl(aq)   Water

5C.2 Two beakers, one containing 0.010 M NaCl(aq) and the other containing 0.010 M AlCl₃(aq), are placed inside a bell jar and sealed. The beakers are left until the water vapor has come to equilibrium with any liquid in the container. The levels of the liquid in each beaker at the beginning of the experiment are the same, as pictured below. Draw the levels of the liquid in each beaker after equilibrium has been reached. Explain your reasoning.

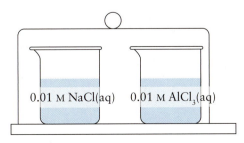

0.01 M NaCl(aq)   0.01 M AlCl₃(aq)

5C.3 Calculate the vapor pressure of the solvent in each of the following solutions. Use Table 5A.2 to find the vapor pressure of water in (a) an aqueous solution at 100 °C in which the mole fraction of sucrose is 0.100; (b) an aqueous solution at 100 °C in which the concentration of sucrose is 0.100 $mol \cdot L^{-1}$.

5C.4 What is the vapor pressure of the solvent in each of the following solutions: (a) the mole fraction of glucose is 0.0316 in an aqueous solution at 40 °C; (b) an aqueous solution at 23 °C is 0.0240 M $CO(NH_2)_2$, urea, a nonelectrolyte? Use the data in Table 5A.2 for the vapor pressure of water at various temperatures.

5C.5 The vapor pressure of benzene, C_6H_6, is 94.6 Torr at 25 °C. A nonvolatile compound was added to 0.300 mol $C_6H_6(l)$ at 25 °C and the vapor pressure of the benzene in the solution decreased to 75.0 Torr. What amount (in moles) of solute molecules was added to the benzene?

5C.6 The vapor pressure of benzene is 100.0 Torr at 26 °C. A nonvolatile compound was added to 0.400 mol $C_6H_6(l)$ at 26 °C and the vapor pressure of the benzene in the solution decreased to 68.0 Torr. What amount (in moles) of solute molecules was added to the benzene?

5C.7 When 8.05 g of an unknown compound X was dissolved in 100. g of benzene, the vapor pressure of the benzene decreased from 100.0 Torr to 94.8 Torr at 26 °C. What is (a) the mole fraction and what is (b) the molar mass of X?

5C.8 The normal boiling point of ethanol is 78.4 °C. When 9.15 g of a soluble nonelectrolyte was dissolved in 100. g of ethanol, the vapor pressure of the solution at that temperature was 7.40×10^2 Torr. (a) What are the mole fractions of ethanol and solute? (b) What is the molar mass of the solute?

5C.9 Benzene, C_6H_6, and toluene, $C_6H_5CH_3$, form an ideal solution. The vapor pressure of benzene is 94.6 Torr and that of toluene is 29.1 Torr at 25 °C. What is the vapor pressure of each component at 25 °C and what is the total vapor pressure of a mixture of 1.00 mol benzene and 0.400 mol toluene at 25 °C?

5C.10 Hexane, C_6H_{14}, and cyclohexane, C_6H_{12}, form an ideal solution. The vapor pressure of hexane is 151 Torr and that of cyclohexane is 98 Torr at 25.0 °C. What is the vapor pressure of each component at 25 °C, and what is the total vapor pressure of a mixture of 0.400 mol hexane and 0.200 mol cyclohexane at 25 °C?

5C.11 Benzene, C_6H_6, and toluene, $C_6H_5CH_3$, form an ideal solution. The vapor pressure of benzene is 94.6 Torr and that of toluene is 29.1 Torr at 25 °C. Calculate the total vapor pressure of each of the following solutions and the mole fraction of each substance in the vapor phase above those solutions at 25 °C: (a) 1.50 mol C_6H_6 mixed with 0.50 mol $C_6H_5CH_3$; (b) 15.0 g of benzene mixed with 64.3 g of toluene.

5C.12 Hexane, C_6H_{14}, and cyclohexane, C_6H_{12}, form an ideal solution. The vapor pressure of hexane is 151 Torr and that of cyclohexane is 98 Torr at 25.0 °C. Calculate the total vapor pressure of each of the following solutions and the mole fraction of each substance in the vapor phase above those solutions at 25 °C: (a) 0.25 mol C_6H_{14} mixed with 0.65 mol C_6H_{12}; (b) 10.0 g of hexane mixed with 10.0 g of cyclohexane.

5C.13 The vapor pressure of 1,1-dichloroethane, CH_3CHCl_2, is 228 Torr at 25 °C; at the same temperature, that of 1,1-dichlorotetrafluoroethane, CF_3CCl_2F, is 79 Torr. What mass of 1,1-dichloroethane must be mixed with 100.0 g of 1,1-dichlorotetrafluoroethane to give a solution with vapor pressure 157 Torr at 25 °C? Assume ideal behavior.

5C.14 The vapor pressure of butanone, $CH_3CH_2COCH_3$, is 100. Torr at 25 °C; at the same temperature that of propanone, CH_3COCH_3, is 222 Torr. What mass of propanone must be mixed with 350.0 g of butanone to give a solution with a vapor pressure of 135 Torr? Assume ideal behavior.

5C.15 Which of the following mixtures would you expect to show a positive deviation, a negative deviation, or no deviation (that is, form an ideal solution) from Raoult's law? Explain your conclusion. (a) methanol, CH_3OH, and ethanol, CH_3CH_2OH; (b) HF and H_2O; (c) hexane, C_6H_{14}, and H_2O.

5C.16 Which of the following mixtures would you expect to show a positive deviation, a negative deviation, or no deviation (that is, form an ideal solution) from Raoult's law? Explain your conclusion. (a) HBr and H_2O; (b) formic acid, HCOOH, and benzene; (c) cyclopentane, C_5H_{10}, and cyclohexane, C_6H_{12}.

Topic 5D Solubility

How are phase changes described and reported?

Topic **5C:** Phase equilibria in two-component systems → Topic **5D:** Solubility

What determines the solubility of a substance?

Why Do You Need to Know This Material? An understanding of the factors that affect the solubility of a substance is essential for the selection of appropriate solvents and for considering their environmental impact.

What Do You Need to Know Already? This Topic makes use of the concept of lattice enthalpy (Topic 4E) and the role of enthalpy and entropy in determining the value of the Gibbs free energy (Topic 4J).

The biggest liquid solutions on Earth are the oceans, which account for 1.4×10^{21} kg of the Earth's surface water. That mass amounts to nearly 2×10^8 t (1 t $= 10^3$ kg) of water for each inhabitant, more than enough to supply all the drinking water needed for a thirsty world. However, drinking seawater itself can be fatal, because of the high concentrations of dissolved salts, in particular Na^+ and Cl^- ions (**TABLE 5D.1**). Methods to remove the dissolved salts would provide more drinking water, but the nature of solutions must be understood for this is to be done efficiently. Many chemical reactions also take place in solution, and so it is important to understand why some substances dissolve but others do not.

5D.1 The Limits of Solubility

To set the scene for this Topic, imagine what happens when a crystal of glucose, $C_6H_{12}O_6$, is dropped into to some water. First, water molecules next to the surface of the crystal form hydrogen bonds to the glucose molecules. As a result, these glucose molecules are pulled toward the solution by water molecules but are held back by hydrogen bonds to other glucose molecules. When their interactions with water molecules are comparable to their interactions with the other glucose molecules, the glucose molecules can drift off into the solvent, surrounded by water molecules. A similar process takes place when an ionic solid dissolves. The polar water molecules hydrate the ions (that is, surround them in a closely held "solvent shell") and pry them away from the predominately attractive forces within the crystal lattice (**FIG. 5D.1**). Stirring and shaking speed the process because they bring more free water molecules to the surface of the solid and sweep the hydrated ions away.

If only a small amount—2 g, for instance—of glucose is added to 100 mL of water at room temperature, it all dissolves. However, if 200 g is added, some glucose remains undissolved (**FIG. 5D.2**). A solution is said to be **saturated** when the solvent has dissolved all the solute that it can and some undissolved solute remains. At this point, the concentration of solid solute in a saturated solution has reached its greatest value, and no more can dissolve. The **molar solubility,** *s,* of a substance is its molar concentration in a saturated solution. In other words, the molar solubility of a substance represents the limit of its ability to dissolve in a given quantity of solvent.

In a saturated solution, the solid solute still present continues to dissolve, but the rate at which

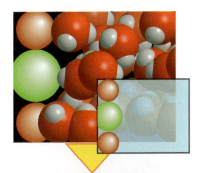
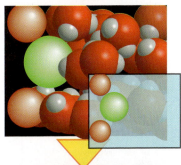
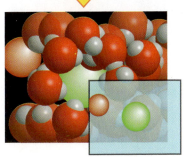

FIGURE 5D.1 The events that take place at the interface of a solid ionic solute and a solvent (water). Only the surface layer of ions is shown. When the ions at the surface of the solid become hydrated, they move off into the solution. The insets at the right show the ions alone.

TABLE 5D.1 **Principal Ions Found in Seawater**

Element	Principal form	Concentration $c/(g \cdot L^{-1})$
Cl	Cl^-	19.0
Na	Na^+	10.5
Mg	Mg^{2+}	1.35
S	SO_4^{2-}	0.89
Ca	Ca^{2+}	0.40
K	K^+	0.38
Br	Br^-	0.065
C	HCO_3^-, H_2CO_3, CO_2	0.028

FIGURE 5D.2 When a little glucose is stirred into 100 mL of water, all the glucose dissolves (left). However, when enough glucose is added, some undissolved glucose remains and the solution is saturated with glucose (right). *(W. H. Freeman photo by Ken Karp.)*

it dissolves exactly matches the rate at which the solute returns to the solid, so the concentration remains at its equilibrium value (**FIG. 5D.3**). In a saturated solution, the dissolved and undissolved solute are in dynamic equilibrium with each other, in the sense introduced in Topic 5A.

> *The molar solubility of a substance is its molar concentration in a saturated solution. A saturated solution is one in which the dissolved and undissolved solute are in dynamic equilibrium with each other and no more solute can dissolve.*

5D.2 The Like-Dissolves-Like Rule

An understanding of the interplay of forces that takes place when a solute dissolves can help to answer some practical questions. Suppose, for instance, that some oil has soiled a fabric. How should you select a good solvent to remove the oil? A good guide is the rule that *like dissolves like*. That is, a liquid consisting of polar molecules (a "polar liquid"), such as water, is generally the best solvent for ionic and polar compounds. Conversely, a liquid consisting of nonpolar molecules (a "nonpolar liquid"), including hexane and tetrachloroethene, $Cl_2C{=}CCl_2$, is a better solvent for nonpolar compounds, such as hydrocarbon oils, which are held together by London forces.

The "like-dissolves-like" rule can be understood by examining the attractions between solute and solvent molecules. In order for a substance to dissolve in a liquid solvent, the solute–solute attractions must be replaced by solute–solvent attractions, and dissolving can be expected to occur if the new interactions are similar to the original interactions (**FIG. 5D.4**). For example, when the main cohesive forces in a solute are hydrogen bonds, the solute is more likely to dissolve

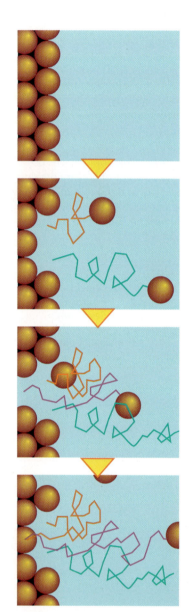

FIGURE 5D.3 The solute in a saturated solution is in dynamic equilibrium with the undissolved solute. If you could follow the solute particles (orange spheres), you would sometimes find them in solution and sometimes back within the solid. The colored lines represent the paths of individual solute particles. The solvent molecules are not shown.

Formation of cavities

Step 2

Enthalpy

Step 1

Separation of solute

Step 3

Cavities occupied by solute

Pure solvent and solute

Dissolution

Solution

FIGURE 5D.4 A representation of the changes in molecular interactions and energy associated with the formation of a dilute solution. In step 1, the solute molecules separate from one another. In step 2, some of the solvent molecules move apart and leave cavities in the solvent. In step 3, the solute molecules occupy the cavities in the solvent, a process that releases energy. The overall energy change for the process is the sum of the energy changes, shown by the red arrow. In the actual dissolving process, these steps do not occur sequentially.

in a hydrogen-bonding solvent than in other solvents. The molecules can slip into solution only if they can replace their strong solute–solute hydrogen bonds with strong solute–solvent hydrogen bonds. Glucose, for example, has hydrogen-bond-forming —OH groups and dissolves readily in water but not in hexane (**FIG. 5D.5**).

If the principal cohesive forces between solute molecules are London forces, then the best solvent is likely to be one that can mimic those forces. For example, a good solvent for nonpolar substances is the nonpolar liquid carbon disulfide, CS_2. It is a far better solvent than water for sulfur because solid sulfur is a molecular solid of S_8 molecules held together by London forces (**FIG. 5D.6**). The sulfur molecules cannot penetrate into the strongly hydrogen-bonded structure of water, because they cannot replace those bonds with interactions of similar strength.

The cleaning action of soaps and detergents relies on the like-dissolves-like rule. Soaps are the sodium salts of long-chain carboxylic acids, including sodium stearate (**1**). The anions of these acids have a polar carboxylate group ($—CO_2^-$), called the *head group,* at one end of a nonpolar hydrocarbon chain. The head group is **hydrophilic,** or water attracting, whereas the nonpolar hydrocarbon tails are **hydrophobic,** or water repelling. Because the hydrophilic head groups of the anions are able to dissolve in water and the hydrophobic hydrocarbon tails are able to dissolve in grease, soap is very effective at removing grease. The hydrocarbon tails sink into a blob of grease up to the head groups and the hydrophilic head groups remain on the surface of the blob. The casing of soap molecules, which is called a **micelle,** is soluble in water and so carries away the grease (**FIG. 5D.7**).

Soaps are made by heating sodium hydroxide with a fat such as coconut oil, olive oil, or animal fat, which contain *esters* formed between glycerol and long-chain carboxylic acids (Topic 11D). The sodium hydroxide attacks the esters and forms the soluble soap. In the case of animal fat, stearic acid forms the soap sodium stearate. Soaps, however, form a scum in hard water (water that contains Ca^{2+} and Mg^{2+} ions). The scum is an impure precipitate of almost insoluble calcium and magnesium stearates.

Modern commercial detergents are often sold as mixtures incorporating water softeners, such as sodium carbonate, and enzymes. Their most important component is a **surfactant,** or *surface-active agent.* Surfactant molecules are organic compounds with a structure and action similar to those of soap. Surfactants used in detergents typically contain sulfur atoms in their polar groups (**2**); they are superior to soaps because they do not form precipitates with the ions in hard water.

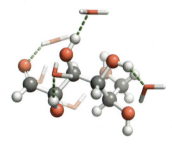

FIGURE 5D.5 Glucose ($C_6H_{12}O_6$) forms hydrogen bonds with water (dashed green lines).

1 Sodium stearate, $NaCH_3(CH_2)_{16}CO_2$

2 A typical surfactant ion

FIGURE 5D.6 Sulfur, which is a solid with nonpolar molecules, does not dissolve in water (left), but it does dissolve in carbon disulfide (right), with which the S_8 molecules have favorable London interactions. (©1989 *Chip Clark–Fundamental Photographs.*)

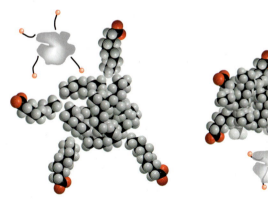

FIGURE 5D.7 On the left, the hydrocarbon tails of a soap or surfactant molecule begin to dissolve in grease. (Schematic representations of the molecules, with the grease depicted as a gray blob and the polar heads of the surfactant molecules shown in red, are shown next to the space-filling models.) The water-attracting head groups remain on the surface, where they can interact favorably with water (right). As more surfactant molecules dissolve in the grease, the whole blob, called a micelle, is suspended in water and is washed away.

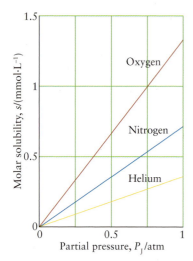

FIGURE 5D.8 The variation of the molar solubilities of oxygen, nitrogen, and helium gases with partial pressure. Note that the solubility of each gas doubles when its partial pressure is doubled.

TABLE 5D.2	Henry's Constants for Gases in Water at 20 °C
Gas	$k_H/(mol \cdot L^{-1} \cdot atm^{-1})$
air	7.9×10^{-4}
argon	1.5×10^{-3}
carbon dioxide	2.3×10^{-2}
helium	3.7×10^{-4}
hydrogen	8.5×10^{-4}
neon	5.0×10^{-4}
nitrogen	7.0×10^{-4}
oxygen	1.3×10^{-3}

A general guide to the suitability of a solvent is the rule that like dissolves like. Soaps and detergents contain surfactant molecules that have both hydrophobic and hydrophilic regions.

5D.3 Pressure and Gas Solubility

Almost all aquatic organisms rely on the presence of dissolved oxygen for respiration. Although oxygen is nonpolar, it is very slightly soluble in water and the extent to which it dissolves depends on its pressure above the surface of the liquid. The pressure of a gas arises from the impacts of its molecules. When a gas is present in a container together with a liquid, the gas molecules can burrow into the liquid like meteorites plunging into the ocean. Because the rate at which the gas molecules strike the liquid increases as the pressure of a gas increases, the solubility of the gas—its molar concentration when the dissolved gas is in dynamic equilibrium with the free gas—can be expected to increase as its pressure increases. If the gas above the liquid is a mixture (such as air), then the solubility of each component depends on that component's partial pressure (**FIG. 5D.8**).

The straight lines in Fig. 5D.8 show that *the solubility of a gas is directly proportional to its partial pressure, P*. This observation was first made in 1801 by the English chemist William Henry and is now known as **Henry's law**. The law is normally written

$$s = k_H P \tag{1}$$

The constant k_H, which is called **Henry's constant**, depends on the gas, the solvent, and the temperature (**TABLE 5D.2**).

EXAMPLE 5D.1   Estimating the solubility of a gas in a liquid

If you are an environmental scientist, you need to know the theoretical limits on the amount of oxygen that water can dissolve in order to monitor the capacity of natural waters to sustain life. Verify that the concentration of oxygen in lake water is normally adequate to sustain aquatic life, which requires an O_2 concentration of at least $0.13 \ mmol \cdot L^{-1}$. The partial pressure of oxygen is 0.21 atm at sea level.

ANTICIPATE You know that aquatic life is viable under normal circumstances, so you should expect to find the necessary concentration.

PLAN Use Eq. 1 to calculate the molar concentration of oxygen.

SOLVE

From $s = k_H P$,

$$s = (1.3 \times 10^{-3} \ mol \cdot L^{-1} \cdot atm^{-1}) \times (0.21 \ atm)$$
$$= 2.7 \times 10^{-4} \ mol \cdot L^{-1}$$

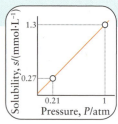

EVALUATE The molar concentration of O_2 is $0.27 \ mmol \cdot L^{-1}$, which is more than adequate to sustain life, as anticipated.

Self-test 5D.1A At the elevation of Bear Lake in Rocky Mountain National Park (2900 m), the partial pressure of oxygen is 0.14 atm. What is the molar solubility of oxygen in Bear Lake at 20 °C?
[*Answer:* $0.18 \ mmol \cdot L^{-1}$]

Self-test 5D.1B Use the information in Table 5D.2 to calculate the amount (in moles) of CO_2 that will dissolve in enough water to form 900. mL of solution at 20 °C when the partial pressure of CO_2 is 1.00 atm.

Related Exercises 5D.5, 5D.6

The solubility of a gas is proportional to its partial pressure, because an increase in pressure corresponds to an increase in the rate at which gas molecules strike the surface of the solvent.

5D.4 Temperature and Solubility

Most substances dissolve more quickly at higher temperatures than at low ones. However, that does not necessarily mean that they are more soluble—that is, reach a higher concentration of solute—at higher temperatures. In a number of cases, the solubility turns out to be lower at higher temperatures. It is important to distinguish the effect of temperature on the *rate* of a process from its effect on the final outcome.

Near room temperature, most gases become less soluble in water as the temperature is raised. The lower solubility of gases in warm water is responsible for the tiny bubbles that appear when cool water from the faucet is left to stand in a warm room. The bubbles consist of air that dissolved when the water was cooler; it comes out of solution as the temperature rises. In contrast, most ionic and molecular solids are more soluble in warm water than in cold (**FIG. 5D.9**). This characteristic is used in the laboratory to dissolve a substance and to grow crystals by letting a saturated-solution cool slowly. However, a few solids containing ions that are extensively hydrated in water, such as lithium carbonate, are less soluble at high temperatures than at low. A small number of compounds show a mixed behavior. For example, the solubility of sodium sulfate decahydrate increases up to 32 °C but then decreases as the temperature is raised further.

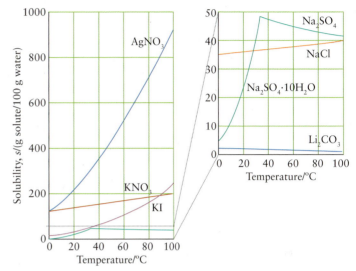

FIGURE 5D.9 The variation with temperature of the solubilities of six substances in water. The graph on the right is expanded vertically to show the variation in solubilities of four ionic compounds more clearly.

> *The rate of dissolving, but not necessarily the solubility of a substance, increases at higher temperatures. Most gases are less soluble in warm water than in cold water; solids show a more varied behavior.*

5D.5 The Thermodynamics of Dissolving

The change in molar enthalpy when a substance dissolves is called the **enthalpy of solution**, ΔH_{sol}. The change can be measured calorimetrically from the heat released or absorbed when the substance dissolves at constant pressure. However, because solute molecules interact with one another as well as with the solvent, to isolate the solute–solvent interactions it is necessary to focus on the **limiting enthalpy of solution,** the enthalpy change accompanying the formation of such dilute solutions that the solute–solute interactions are negligible (**TABLE 5D.3**). The data show that some solids, such as

TABLE 5D.3 Limiting Enthalpies of Solution, $\Delta H_{sol}/(kJ \cdot mol^{-1})$, at 25 °C*

Cation	Anion							
	fluoride	chloride	bromide	iodide	hydroxide	carbonate	sulfate	nitrate
lithium	+4.9	−37.0	−48.8	−63.3	−23.6	−18.2	−2.7	−29.8
sodium	+1.9	+3.9	−0.6	−7.5	−44.5	−26.7	+20.4	−2.4
potassium	−17.7	+17.2	+19.9	+20.3	−57.1	−30.9	+34.9	−23.8
ammonium	−1.2	+14.8	+16.0	+13.7	—	—	+25.7	+6.6
silver	−22.5	+65.5	+84.4	+112.2	—	+41.8	+22.6	+17.8
magnesium	−12.6	−160.0	−185.6	−213.2	+2.3	−25.3	−90.9	−91.2
calcium	+11.5	−81.3	−103.1	−119.7	−16.7	−13.1	−19.2	−18.0
aluminum	−27	−329	−368	−385	—	—	—	−350

*The limiting enthalpy of solution value for silver iodide, ΔH_{sol}, for example, is the entry found where the row labeled "silver" intersects the column labeled "iodide," and it is +112.2 kJ·mol^{-1}.

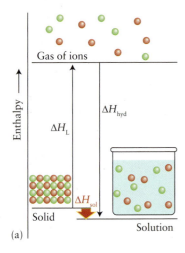

(a)

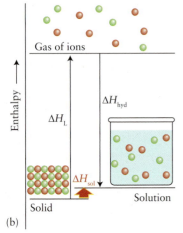

(b)

FIGURE 5D.10 The enthalpy of solution, ΔH_{sol}, is the sum of the enthalpy change required to separate the molecules or ions of the solute (the lattice enthalpy, ΔH_L; step 1 in Fig. 5D.4), and the enthalpy change accompanying their hydration (ΔH_{hyd}; steps 2 and 3 in Fig. 5D.4). The outcome is finely balanced: (a) in some cases, it is exothermic; (b) in others, it is endothermic. For gaseous solutes, the lattice enthalpy is zero because the molecules are already widely separated.

lithium chloride, dissolve exothermically (with a release of heat). Others, such as ammonium nitrate, dissolve endothermically (absorb heat).

To gain insight into the values in Table 5D.3, think of dissolving as a hypothetical two-step process (**FIG. 5D.10**). In the first step in the formation of an aqueous solution of a salt, imagine the ions separating from the solid to form a gas of ions. The change in molar enthalpy accompanying this highly endothermic step is the lattice enthalpy, ΔH_L, of the solid (Topic 2A and Table 4E.1). The lattice enthalpy of sodium chloride (787 kJ·mol^{-1}), for instance, is the molar enthalpy change for the process $NaCl(s) \rightarrow Na^+(g) + Cl^-(g)$. It is explained in Topic 2A that compounds formed from small, highly charged ions (such as Mg^{2+} and O^{2-}) cohere strongly and that a lot of energy is needed to break up the lattice. Such compounds have high lattice enthalpies. Compounds formed from large ions with low charges, such as potassium iodide, typically have weak attractive forces and correspondingly lower lattice enthalpies.

In the second hypothetical step, imagine the gaseous ions plunging into water and forming the final solution. The molar enthalpy of this step is called the **enthalpy of hydration**, ΔH_{hyd}, of the compound (**TABLE 5D.4**). Enthalpies of hydration are negative and comparable in value to the lattice enthalpies of the compounds. For sodium chloride, for instance, the enthalpy of hydration, the molar enthalpy change for the process $Na^+(g) + Cl^-(g) \rightarrow Na^+(aq) + Cl^-(aq)$ is -784 kJ·mol^{-1}. That is enough energy for 1 g of NaCl (as a gas of Na^+ and Cl^- ions) to raise the temperature of 100 mL of water by nearly 50 °C. Hydration is always exothermic for ionic compounds because of the formation of ion–dipole attractions between the water molecules and the ions. It is also exothermic for molecules that can form hydrogen bonds with water, such as sucrose, glucose, acetone, and ethanol.

Now bring the two steps of the dissolving process together and calculate the enthalpy change for the overall process: $\Delta H_{sol} = \Delta H_L + \Delta H_{hyd}$. When the data are included, the limiting enthalpy of solution of sodium chloride, the enthalpy change for the process $NaCl(s) \rightarrow Na^+(aq) + Cl^-(aq)$ is

$$\Delta H_{sol} = \overbrace{787 \text{ kJ·mol}^{-1}}^{\text{lattice enthalpy}} - \overbrace{784 \text{ kJ·mol}^{-1}}^{\text{enthalpy of hydration}} = \overbrace{+3 \text{ kJ·mol}^{-1}}^{\text{enthalpy of solution}}$$

Because the enthalpy of solution is positive, there is a net inflow of energy as heat when the solid dissolves. Sodium chloride therefore dissolves endothermically, but only to the extent of 3 kJ·mol^{-1}. As this example shows, the overall change in enthalpy depends on a very delicate balance between the lattice enthalpy and the enthalpy of hydration.

THINKING POINT

Why does sodium chloride dissolve in water spontaneously, even though the dissolving is endothermic?

TABLE 5D.4   Limiting Enthalpies of Hydration, $\Delta H_{hyd}/(\text{kJ·mol}^{-1})$, at 25 °C, of Some Halides*

Cation	Anion			
	F^-	Cl^-	Br^-	I^-
H^+	-1613	-1470	-1439	-1426
Li^+	-1041	-898	-867	-854
Na^+	-927	-784	-753	-740
K^+	-844	-701	-670	-657
Ag^+	-993	-850	-819	-806
Ca^{2+}	—	-2337	—	—

*The enthalpy of hydration for NaCl, ΔH_{hyd} for example, is the entry where the row labeled Na^+ intersects the column labeled Cl^-. The resulting value, -784 kJ·mol^{-1}, is for the process $Na^+(g) + Cl^-(g) \rightarrow Na^+(aq) + Cl^-(aq)$ when the resulting solution is very dilute.

High charge and small ionic radius both contribute to a high enthalpy of hydration. However, the same properties also contribute to high lattice enthalpy. It is therefore very difficult to make reliable predictions about solubility on the basis of ion charge and radius. The best that can be done is to use these properties to rationalize what is observed. With that limitation in mind, you can begin to understand the behavior of some everyday substances and the properties of some minerals. For example, nitrates have big, singly charged anions and hence low lattice enthalpies. Their hydration enthalpies, though, are quite large, because water can form hydrogen bonds with the nitrate anions. As a result, they are rarely found in mineral deposits because they are soluble in groundwater, the water that trickles through the ground and washes away soluble substances. Carbonate ions are about the same size as nitrate ions, but they are doubly charged. As a result, carbonates commonly have higher lattice enthalpies than do nitrates, and it is much harder to break the ions out of solids such as limestone (calcium carbonate). Hydrogen carbonate ions (bicarbonate ions, HCO_3^-) have a single charge, and their salts are more soluble than carbonates.

The difference in solubility between carbonates and hydrogen carbonates is responsible for the behavior of "hard water," which is water that contains dissolved calcium and magnesium salts. Hard water originates as rainwater, which dissolves carbon dioxide from the air and forms a very dilute solution of carbonic acid:

$$CO_2(g) + H_2O(l) \longrightarrow H_2CO_3(aq)$$

As the water runs over and through the ground, the carbonic acid reacts with the calcium carbonate of limestone or chalk and forms the more soluble hydrogen carbonate:

$$CaCO_3(s) + H_2CO_3(aq) \longrightarrow Ca^{2+}(aq) + 2\,HCO_3^-(aq)$$

These two reactions are reversed when water containing $Ca(HCO_3)_2$ is heated in a kettle or furnace:

$$Ca^{2+}(aq) + 2\,HCO_3^-(aq) \longrightarrow CaCO_3(s) + H_2O(l) + CO_2(g)$$

The carbon dioxide is driven off and the calcium carbonate is deposited as the hard, insoluble calcium carbonate deposit known as *scale*.

A negative enthalpy of solution implies that energy is released as heat when a substance dissolves. However, to judge whether dissolving is spontaneous at constant temperature and pressure, it is necessary to consider the change in Gibbs free energy, $\Delta G = \Delta H - T\Delta S$. In other words, changes in the entropy of the system must be considered, not just the change in enthalpy.

The disorder of a system typically increases when a solid dissolves (**FIG. 5D.11**). Therefore, in most cases, the entropy of the system can be expected to increase when a solution forms. Because $T\Delta S$ is positive, this increase in disorder makes a negative contribution to ΔG. If ΔH is negative, you can be confident that ΔG is negative overall. Therefore, you can expect most substances with negative enthalpies of solution to be soluble.

In some cases, the entropy of the system is lowered when the solution forms because the solvent molecules form cagelike structures around the solute molecules. As a result, ΔS is negative and the term $-T\Delta S$ makes a positive contribution to ΔG. Even if ΔH is negative, ΔG might be positive. In other words, even if energy is released to the surroundings, the increase in the entropy of the surroundings may not be enough to overcome the decrease in entropy of the system, which consists of the solute and solvent. In such a case the substance does not dissolve. For this reason hydrocarbons such as heptane are insoluble in water even though they have weakly negative enthalpies of solution.

If ΔH is positive, $\Delta G = \Delta H - T\Delta S$ can be negative only if $T\Delta S$ is positive and larger than ΔH. A substance with a strongly positive enthalpy of solution is likely to be insoluble, because the entropy of the surroundings may decrease so much as energy leaves them and enters the solution that dissolving corresponds to an overall decrease in disorder.

An understanding of the contributions to the Gibbs free energy of dissolving helps explain the temperature dependence of solubility. The Gibbs free energy becomes more negative and favorable to dissolving as the temperature increases only when ΔS is positive. For most ionic substances, dissolving does indeed result in an increase in the entropy of the system, and these substances are more soluble at higher temperatures (see Fig. 5D.9).

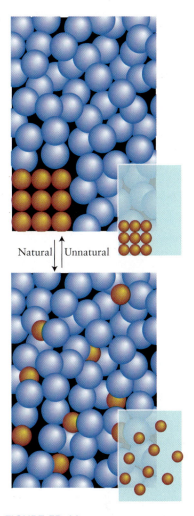

Natural | Unnatural

FIGURE 5D.11 The dissolution of a solid is a natural process that is usually accompanied by an increase in disorder. However, to decide whether the dissolution is spontaneous, the accompanying changes in the surroundings must be considered too. The insets show the solute particles alone.

Remember that ΔG is a measure of the overall change in entropy at constant temperature and pressure: ΔS is the change in entropy of the system and $-\Delta H/T$ is the change in entropy of the surroundings (Topic 4J).

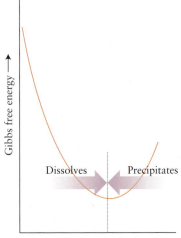

FIGURE 5D.12 At low concentrations of solute, dissolving is accompanied by a decrease in Gibbs free energy of the system, so it is spontaneous. At high concentrations, dissolving is accompanied by an increase in Gibbs free energy, so the reverse process, precipitation, is spontaneous. The concentration of a saturated solution corresponds to the state of lowest Gibbs free energy at the temperature of the experiment.

However, when a gas dissolves in a liquid, its molecules occupy much less space and there is a considerable reduction in the entropy of the system. In such a case, $-T\Delta S$ is positive. Consequently, an increase in temperature causes ΔG to rise and dissolving is less favorable.

Because ΔG for the dissolving of a solute depends on the concentration of the solute, even if ΔG is negative at low concentrations, it may become positive at high concentrations (**FIG. 5D.12**). A solute dissolves spontaneously only until $\Delta G = 0$. At that point, dissolved and undissolved solute are in equilibrium—the solution is saturated.

Enthalpies of solution in dilute solutions can be expressed as the sum of the lattice enthalpy and the enthalpy of hydration of the compound. Dissolving depends on the balance between the change in entropy of the solution and the change in entropy of the surroundings.

5D.6 Colloids

A **colloid** is a dispersion of small particles (from 1 nm to 1 μm in diameter) in a solvent. Colloidal particles are much larger than most molecules but are too small to be seen with an optical microscope. As a result, colloids have properties between those of a solution and those of a heterogeneous mixture. The small particles give the colloid a homogeneous appearance but are large enough to scatter light. The light scattering explains why milk is white, not transparent, and why searchlight and laser beams are more visible in fog, smoke, and clouds than in clear, dry air (**FIG. 5D.13**). Many foods are colloids, as are clay particles, smoke, and the fluids in living cells.

Colloids are classified according to the phases of their components (**TABLE 5D.5**). A colloid that is a suspension of solids in a liquid is called a **sol**, and a suspension of one liquid in another is called an **emulsion**. For example, muddy water is a sol in which tiny flakes of clay are dispersed in water; mayonnaise is an emulsion in which small droplets of water are suspended in vegetable oil. **Foam** is a suspension of a gas in a liquid or solid. Foam rubber, Styrofoam, soapsuds, and aerogels (insulating ceramic foams that have densities nearly as low as that of air; see the Interlude following Focus 3) are foams. Zeolites (Topic 7E) are a type of solid foam in which the openings in the solid are comparable in size to molecules.

A **solid emulsion** is a suspension of a liquid or solid phase in a solid. For example, opals are solid emulsions formed when partly hydrated silica fills the interstices between close-packed microspheres of silica aggregates. Gelatin desserts are a type of solid emulsion called a **gel**, which is soft but holds its shape. Photographic emulsions are gels that also contain solid colloidal particles of light-sensitive materials, such as silver bromide.

Aqueous colloids can be classified as hydrophilic or hydrophobic, depending on the strength of the molecular interactions between the suspended substance and water. Suspensions of fat in water (such as milk) and water in fat (such as mayonnaise and hand lotions) are hydrophobic colloids, because fat molecules have little attraction for water molecules. Gels and puddings are examples of hydrophilic colloids. The macromolecules

FIGURE 5D.13 Laser beams are invisible. However, they can be traced when they pass through smoky or misty environments because the light scatters from the particles suspended in the air. (© graeme hall snaps/Alamy.)

TABLE 5D.5   The Classification of Colloids*

Dispersed phase	Dispersion medium	Technical name	Examples
solid	gas	aerosol	smoke
liquid	gas	aerosol	hairspray, mist, fog
solid	liquid	sol or gel	printing ink, paint
liquid	liquid	emulsion	milk, mayonnaise
gas	liquid	foam	fire-extinguisher foam
solid	solid	solid dispersion	ruby glass (Au in glass); some alloys
liquid	solid	solid emulsion	bituminous road paving; ice cream
gas	solid	solid foam	insulating foam

*Based on R. J. Hunter, *Foundations of Colloid Science*, Vol. 1 (Oxford: Oxford University Press, 1987).

of the proteins in gelatin and the starch in pudding have many hydrophilic groups that attract water. The giant protein molecules in gelatin uncoil in hot water, and their numerous polar groups form hydrogen bonds with the water. When the mixture cools, the protein chains link together again, but now they have entwined to form a three-dimensional web that encloses many water molecules, as well as molecules of sugar, dye, and flavoring agents. The result is a gel: an open network of protein chains that holds the water in a flexible solid structure.

Many precipitates, such as $Fe(OH)_3$, form initially as colloidal suspensions. The tiny particles are kept from settling out by **Brownian motion,** the motion of small particles resulting from constant bombardment by solvent molecules. The sol is further stabilized by the adsorption of ions on the surfaces of the particles. The ions attract a layer of water molecules that prevents the particles from adhering to one another.

Biomimetic materials are materials that are modeled after naturally occurring materials. Materials such as gels or flexible polymers modeled after natural membranes and tissues are biomimetic materials with remarkable properties. Surfactant compounds called *phospholipids* are found in fats and form the membranes of living cells. These molecules are liquid crystals similar to soap molecules. Cell membranes consist of double layers of phospholipid molecules that line up with their hydrocarbon tails pointing into the membrane and their polar head groups forming the membrane surfaces (**FIG. 5D.14**). This structure separates the contents of the cell from intercellular fluid.

Colloids are suspensions of particles typically too small to be seen with a microscope but large enough to scatter light.

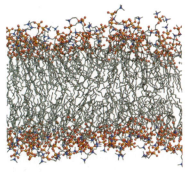

FIGURE 5D.14 A cross section of a section of the wall of a liposome, a tiny sac enclosed by a bilayer membrane formed from surfactant-like phospholipid molecules. *(Data from Biocomputing Group, University of Calgary.)*

 ANIMATION FIGURE 5D.14

What have you learned in this Topic?

You have learned that substances are likely to be most soluble in solvents that have similar intermolecular forces, with solubility depending on the balance of entropy changes in the system and surroundings. You are now aware of Henry's law and the effect of temperature on the solubilities of gases. You have seen the role of lattice enthalpy and hydration in determining whether dissolving is exothermic or endothermic and understand the role of entropy in the process of dissolving. You have also been introduced to the nature of colloidal suspensions and cell membranes.

The skills you have mastered are the ability to:

☐ **1.** Define the term solubility (Section 5D.1).

☐ **2.** Predict relative solubilities from molecular polarity (Section 5D.2).

☐ **3.** State and use Henry's law to calculate the solubility of a gas (Example 5D.3).

☐ **4.** Interpret enthalpies of solution in terms of lattice enthalpies and enthalpies of hydration (Section 5D.5).

☐ **5.** Discuss the thermodynamic aspects of dissolving in terms of contributions from the enthalpy and entropy (Section 5D.5).

☐ **6.** Identify the types of colloids and explain their properties (Section 5D.6).

Topic 5D Exercises

5D.1 Which would be the better solvent, water or benzene, for each of the following substances: (a) KCl; (b) CCl_4; (c) CH_3COOH?

5D.2 Which would be the better solvent, water or tetrachloromethane, for each of the following substances: (a) NH_3; (b) HNO_3; (c) N_2?

5D.3 The following groups are found in some organic molecules. Which are hydrophilic and which are hydrophobic: (a) —NH_2; (b) —CH_3; (c) —Br; (d) —COOH?

5D.4 The following groups are found in some organic molecules. Which are hydrophilic and which are hydrophobic: (a) —OH; (b) —CH_2CH_3; (c) —$CONH_2$; (d) —Cl?

5D.5 State the molar solubility in water of (a) O_2 at 50. kPa; (b) CO_2 at 500. Torr; (c) CO_2 at 0.10 atm. The temperature in each case is 20 °C, and the pressures are partial pressures of the gases. Use the information in Table 5D.2.

5D.6 Calculate the solubility in water (in milligrams per liter) of (a) air at 0.80 atm; (b) He at 0.80 atm; (c) He at 36 kPa. The temperature is 20 °C in each case, and the pressures are partial pressures of the gases. Use the information in Table 5D.2.

5D.7 The minimum mass concentration of oxygen required for fish life is 4 mg·L^{-1}. (a) Assume the density of lake water to be 1.00 g·mL^{-1}, and express this concentration in parts per million (which is equivalent to milligrams of O_2 per kilogram of water, mg·kg^{-1}). (b) What is the minimum partial pressure of O_2 that would supply the minimum mass concentration of oxygen in water to support fish life at 20 °C? (c) What is the minimum atmospheric pressure that would give this partial pressure, assuming that oxygen exerts about 21% of the atmospheric pressure? See Table 5D.2.

5D.8 The volume of blood in the body of a deep-sea diver is about 6.00 L. Blood cells make up about 55% of the blood volume, and the remaining 45% is the aqueous solution called plasma. What is the maximum volume of nitrogen measured at 1.00 atm and 37 °C that could dissolve in the diver's blood plasma at a depth of 93 m, where the pressure is 10.0 atm? (This is the volume that could come out of solution suddenly, causing the painful and dangerous condition called the bends, if the diver were to ascend too quickly.) Assume that Henry's constant for nitrogen at 37 °C (body temperature) is 5.8 × 10^{-4} mol·L^{-1}·atm^{-1}.

5D.9 The carbon dioxide gas dissolved in a sample of water in a partly filled, sealed container has reached equilibrium with its partial pressure in the air above the solution. Explain what happens to the solubility of the CO_2 if (a) the partial pressure of the CO_2 gas is doubled by the addition of more CO_2; (b) the total pressure of the gas above the liquid is doubled by the addition of nitrogen.

5D.10 Explain what happens to the solubility of the CO_2 in Exercise 5D.9 if (a) the partial pressure of $CO_2(g)$ is increased by compressing the gas to a third of its original volume; (b) the temperature is raised.

5D.11 A soft drink is made by dissolving CO_2 gas at 3.60 atm in a flavored solution and sealing the solution in aluminum cans at 20 °C. What amount (in moles) of CO_2 is contained in a can of the soft drink of volume 420. mL? At 20 °C the Henry's law constant for CO_2 is 2.3 × 10^{-2} mol·L^{-1}·atm^{-1}.

5D.12 A soft drink is made by dissolving CO_2 gas at 4.00 atm in a flavored solution and sealing the solution in aluminum cans at 20 °C. What amount (in moles) of CO_2 is contained in a can of the

soft drink of volume 360. mL? At 20 °C the Henry's law constant for CO_2 is 2.3 × 10^{-2} mol·L^{-1}·atm^{-1}

5D.13 Lithium sulfate dissolves exothermically in water. (a) Is the enthalpy of solution for Li_2SO_4 positive or negative? (b) Write the chemical equation for the dissolving process. (c) Which is larger for lithium sulfate, the lattice enthalpy or the enthalpy of hydration?

5D.14 The enthalpy of solution of ammonium nitrate in water is positive. (a) Does NH_4NO_3 dissolve endothermically or exothermically? (b) Write the chemical equation for the dissolving process. (c) Which is larger for NH_4NO_3, the lattice enthalpy or the enthalpy of hydration?

5D.15 Calculate the heat evolved or absorbed when 10.0 g of (a) NaCl; (b) NaI; (c) $AlCl_3$; (d) NH_4NO_3 is dissolved in 100. g of water. Assume that the enthalpies of solution in Table 5D.3 are applicable and that the specific heat capacity of the solution is 4.18 J·K^{-1}·g^{-1}.

5D.16 Determine the temperature change when 4.00 g of (a) KCl; (b) $MgBr_2$; (c) KNO_3; (d) NaOH is dissolved in 100. g of water. Assume that the specific heat capacity of the solution is 4.18 J·K^{-1}·g^{-1} and that the enthalpies of solution in Table 5D.3 are applicable.

5D.17 Distinguish between a foam and a sol. Give at least one example of each.

5D.18 Distinguish between an emulsion and a gel. Give at least one example of each.

5D.19 Some colloidal suspensions appear at first glance to be solutions. What quick, simple procedure could you use to distinguish colloids from solutions?

5D.20 Puddings contain large starch molecules that cause the mixture to thicken by a mechanism similar to that by which gelatin thickens. Which of the following suggestions is the best description of how puddings thicken? Explain your choice. (a) The starch molecules in pudding are insoluble in water and precipitate when mixed with water. (b) The strands of the starch molecules connect to each other by forming covalent bonds with each other. (c) The starch molecules form hydrogen bonds with water and encapsulate the water in a network. (d) The water molecules hydrate the starch molecules in pudding and the heat of hydration causes the starch molecules to decompose.

Topic 5E Molality

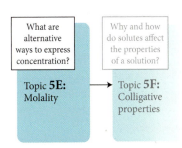

Topic **5E:** Molality → Topic **5F:** Colligative properties

Three measures of concentration are useful for the study of solutions. For a solute J:

- Molar concentration, commonly called molarity, and denoted c_J or $[J]$.
- Mole fraction, denoted x_J.
- Molality, denoted b_J.

In addition, particularly in environmental studies, the concentration of substances is often reported in parts per million (ppm) or parts per billion (ppb). For instance, if there are 25 pollutant molecules in an aqueous solution consisting of a million molecules, then it would be reported as present as 25 ppm. When the concentration is expressed in parts per million, it is important to indicate the units used in the calculation. Concentration expressed as parts per million by volume (microliters of solute per liter) is often denoted ppmv or ppm (v/v). When concentration is expressed as parts per million by mass (milligrams per kilograms, or micrograms per gram), it is denoted ppm (m/m).

Molar concentration (molarity) is introduced in *Fundamentals* G: it is the amount of solute molecules (in moles) divided by the volume of the solution (in liters):

$$c = \frac{n_{solute}}{V_{solution}} \quad (1)$$

It is commonly reported in moles per liter ($mol \cdot L^{-1}$). It is convenient to define its "standard" value as $c° = 1\ mol \cdot L^{-1}$.

Mole fraction and molality are important in the sense that they refer to the relative numbers of solute and solvent molecules. Mole fraction is introduced in Topic 4C, where it is defined as the ratio of the amount (in moles) of a species to the total amount of all the species present in a mixture:

$$x_J = \frac{n_J}{n} \qquad n = n_A + n_B + \dots \quad (2)$$

The **molality,** b_J, of a solute, the principal subject of this Topic, is the amount of solute species (in moles) in a solution divided by the mass of the solvent (in kilograms):

$$b = \frac{n_{solute}}{m_{solvent}} \quad (3)$$

A Note on Good Practice: The IUPAC *symbol* for molality is either m or b; we prefer the latter because it avoids confusion with m for mass (as in Eq. 3). The *units* of molality are commonly reported in moles of solute per kilogram of solvent ($mol \cdot kg^{-1}$); these units are often (but unofficially) denoted m (for example, a 1 m $NiSO_4(aq)$ solution) and read "molal." Because contexts differ, there is little chance of m for molality being confused with m for meter.

Like mole fraction but unlike molarity, the molality is independent of temperature. Note the emphasis on *solvent* in the definition of molality, but on *solution* in the definition of molarity. To prepare 1 m $NiSO_4(aq)$, dissolve 1 mol $NiSO_4$ in 1 kg of water (**FIG. 5E.1**). To prepare 1 M $NiSO_4(aq)$, dissolve 1 mol $NiSO_4$ in sufficient water to make 1 L of solution.

THINKING POINT

Why is molarity not independent of temperature?

Why Do You Need to Know This Material? It is sometimes necessary to emphasize the relative numbers of solute and solvent molecules in a solution: the molality provides a measure.

What Do You Need to Know Already? You need to be aware of the definitions of molar concentration (molarity, *Fundamentals* G) and mole fraction (Topic 3C).

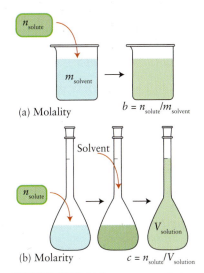

(a) Molality $\quad b = n_{solute}/m_{solvent}$

(b) Molarity $\quad c = n_{solute}/V_{solution}$

FIGURE 5E.1 (a) To prepare a solution of given molality, the appropriate amount of solute, $n_{solvent}$, is added to a known mass of solvent, $m_{solvent}$; then $b = n_{solute}/m_{solvent}$. (b) To prepare a solution of given molarity, the appropriate amount of solute, n_{solute}, is dissolved in a small amount of solvent. Then solvent is added until the volume of the solution is $V_{solution}$; then $c = n_{solute}/V_{solution}$.

EXAMPLE 5E.1 Calculating the molality of a solute

Suppose that you are a food scientist and are studying the contribution of various sugars to obesity in the hope of identifying one that, though very sweet, has a low calorie count. What is the molality of the sugar fructose ($C_6H_{12}O_6$, 180.15 g·mol^{-1}) in a solution prepared by dissolving 90.5 g of fructose in 250. g of water?

PLAN Convert the mass of fructose into its amount of $C_6H_{12}O_6$ in moles, express the mass of water in kilograms, and then divide the amount of solute by the mass of solvent.

SOLVE

From $b = n_{solute}/m_{solvent}$ and $n_{solute} = m_{solute}/M_{solute}$, so $b = (m_{solute}/M_{solute})/m_{solvent}$,

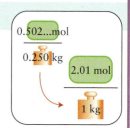

$$b(C_6H_{12}O_6) = \frac{\overbrace{(90.5\text{ g})/\overbrace{(180.15\text{ g·mol}^{-1})}^{M_{solute}}}^{n_{solute}}}{\underbrace{0.250\text{ kg}}_{m_{solvent}}} = 2.01\text{ mol·kg}^{-1}$$

You could report this concentration as 2.01 m $C_6H_{12}O_6$(aq).

Self-test 5E.1A Calculate the molality of $ZnCl_2$ in a solution prepared by dissolving 4.11 g of $ZnCl_2$ in 150. g of water.
[**Answer:** 0.201 mol·kg^{-1}]

Self-test 5E.1B Calculate the molality of $KClO_3$ in a solution prepared by dissolving 7.36 g of $KClO_3$ in 200. g of water.

Related Exercises 5E.1, 5E.2

Toolbox 5E.1 HOW TO USE THE MOLALITY

CONCEPTUAL BASIS

The molality of a solute is its concentration in moles per kilogram of solvent. Its value is independent of the temperature and is directly proportional to the relative numbers of solute and solvent molecules in the solution. To convert molarity to molality, note that the former is defined in terms of the volume of the solution; thus, you need the density of the solution to convert that overall volume to the mass of solvent present.

PROCEDURE

When working with molality, keep its definition ($b = n_{solute}/m_{solvent}$) in mind. The expressions derived here are summarized in **TABLE 5E.1**.

1. Calculating the mass of solute in a given mass of solvent from the molality

Step 1 Calculate the amount of solute molecules, n_{solute}, present in a given mass of solvent, $m_{solvent}$, by converting the mass of solvent from grams to kilograms, if necessary, and rearranging the equation defining molality (Eq. 3) into

$$n_{solute} = b \times m_{solvent}$$

Step 2 Use the molar mass of the solute, M_{solute}, to find the mass of the solute from its amount:

$$m_{solute} = n_{solute}M_{solute} = bm_{solvent}M_{solute}$$

A note on units: Molality is typically reported in moles per kilogram (mol·kg^{-1}). With $m_{solvent}$ in kilograms and M_{solute} in grams per mole (g·mol^{-1}), m_{solute} is obtained in grams: (mol·kg^{-1}) × kg × (g·mol^{-1}) = g.

2. Calculating the molality from a mole fraction

Step 1 Consider a solution composed of a total amount of molecules, n. If the mole fraction of the solute is x_{solute}, the amount of solute molecules present is

$$n_{solute} = x_{solute} \times n$$

Step 2 If there is only one solute, the mole fraction of solvent molecules is $1 - x_{solute}$. The amount of solvent molecules is then $n_{solvent} = (1 - x_{solute})n$. Convert this amount into mass by using the molar mass of the solvent, $M_{solvent}$

$$m_{solvent} = n_{solvent}M_{solvent} = \{(1 - x_{solute})n\}M_{solvent}$$

Step 3 From Eq.3,

$$b = \frac{x_{solute}n}{(1 - x_{solute})nM_{solvent}} = \frac{x_{solute}}{(1 - x_{solute})M_{solvent}}$$

A note on units: The mole fraction x_{solute} is a dimensionless number. Therefore b has the same units as $1/M_{solvent}$. Express $M_{solvent}$ in kilograms per mole (kg·mol^{-1}) by dividing its usual value (in grams per mole) by kg/g = 10^3 (so a molar mass of 55 g·mol^{-1} becomes 0.055 kg·mol^{-1}). Then the molality b is obtained in moles per kilogram (mol·kg^{-1}).

This expression can be rearranged to express the mole fraction in terms of the molality:

$$x_{solute} = \frac{bM_{solvent}}{1 + bM_{solvent}}$$

A note on units: With $M_{solvent}$ in kilograms per mole (kg·mol^{-1}) and molality b in moles per kilogram (mol·kg^{-1}), the product $bM_{solvent}$ is dimensionless: (mol·kg^{-1}) × (kg·mol^{-1}) = 1. Consequently, x_{solute} is also dimensionless.

Example 5E.2 shows how to carry out this conversion.

3. Calculating the molality, given the molarity

This conversion is more involved because the molality is defined in terms of the mass of *solvent* but the molarity is defined in terms of the volume of *solution*. To carry out the conversion you need to know the density of the solution.

Step 1 Calculate the total mass of a volume of solution $V_{solution}$ from the density, d, of the solution (not the solvent) by noting that the density is $d = m_{solution}/V_{solution}$

$$m_{solution} = d \times V_{solution}$$

Step 2 Use the definition of molarity, c, to calculate the amount of solute in the volume $V_{solution}$:

$$n_{solute} = c \times V_{solution}$$

Use the molar mass of the solute to convert this amount into the mass of solute present:

$$m_{solute} = n_{solute}M_{solute} = c \times V_{solution} \times M_{solute}$$

Step 3 Subtract the mass of solute (step 2) from the total mass (step 1) to find the mass of solvent in the solution,

$$m_{solvent} = m_{solution} - m_{solute} = dV_{solution} - cV_{solution}M_{solute}$$
$$= (d - cM_{solute})V_{solution}$$

Step 4 The molality, from Eq. 3 ($b = n_{solute}/m_{solvent}$), is then

$$b = \frac{cV_{solution}}{(d - cM_{solute})V_{solution}} = \frac{c}{d - cM_{solute}}$$

A note on units: With the molarity c expressed in moles per liter (mol·L^{-1}) and the volume of solution in liters (L), $cV_{solution}$ is obtained in moles: (mol·L^{-1}) × L = mol. With the molar mass in kilograms per mole (kg·mol^{-1}), the product cM_{solute} in the denominator is obtained in kilograms per liter: (mol·L^{-1}) × (kg·mol^{-1}) = kg·L^{-1}. The density d is typically expressed in grams per cubic centimeter, or equivalently grams per milliliter (g·mL^{-1}), which is numerically the same as kilograms per liter (because 1 g·mL^{-1} = 1 kg·L^{-1}). The term $d - cM_{solute}$ is therefore expressed in kilograms per liter (kg·L^{-1}) and the units of the product $(d - cM_{solute})V_{solution}$ are kilograms: (kg·L^{-1}) × L = kg. The value of b now turns out to be moles per kilogram (the numerator $cV_{solution}$ is in moles and the denominator is in kilograms, giving mol·kg^{-1}).

The last equation can be rearranged to give the molarity in terms of the molality:

$$c = \frac{bd}{1 + bM_{solute}}$$

A note on units: With the molality in moles per kilogram (mol·kg^{-1}) the product bd is obtained in moles per liter: (mol·kg^{-1}) × (kg·L^{-1}) = mol·L^{-1}. With the molar mass in kilograms per mole (kg·mol^{-1}), bM_{solute} in the denominator is dimensionless: (mol·kg^{-1}) × (kg·mol^{-1}) = 1. Consequently, c has the same units as bd, namely moles per liter (mol·L^{-1}).

This procedure is illustrated in Example 5E.3.

TABLE 5E.1   Relations Between Mole Fraction, Molarity, and Molality*

Conversion	Expression
amount of solute present from molality	$n_{solute} = bm_{solvent}$
mass of solute present from molality	$m_{solute} = bm_{solvent}M_{solute}$
molality from mole fraction	$b = \dfrac{x_{solute}}{(1 - x_{solute})M_{solvent}}$
mole fraction from molality	$x_{solute} = \dfrac{bM_{solvent}}{1 + bM_{solvent}}$
molality from molarity	$b = \dfrac{c}{d - cM_{solute}}$
molarity from molality	$c = \dfrac{bd}{1 + bM_{solute}}$

Symbols and units:
n: amount (mol); m: mass (kg).
b: molality (mol·kg^{-1}); c: molarity (molar concentration, mol·L^{-1}); x: mole fraction (unitless).
d: mass density of solution (g·mL^{-1} = g·cm^{-3} = kg·L^{-1}); M: molar mass (kg·mol^{-1}).

EXAMPLE 5E.2   Calculating a molality from a mole fraction

A colleague has been doing an experiment in which it was important to know the mole fractions of the components of a solution. You want to use the same solution, but you have in mind an experiment in which it is necessary to know the molality of the solute. What is the molality of benzene, C_6H_6, dissolved in toluene, $C_6H_5CH_3$, in a solution for which the mole fraction of benzene is 0.150?

PLAN Use procedure 2 in **Toolbox 5E.1**, taking note of the discussion of units.

SOLVE A molar mass of 92.13 g·mol^{-1} is equivalent to 0.092 13 kg·mol^{-1}.

From $b = x_{solute}/(1 - x_{solute})M_{solvent}$ with $M_{solvent}$ in kilograms per mole,

$$b = \frac{0.150}{(1 - 0.150) \times 0.092\,13 \text{ kg·mol}^{-1}} = 1.92 \text{ mol·kg}^{-1}$$

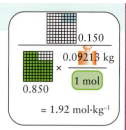

Self-test 5E.2A Calculate the molality of toluene dissolved in benzene, given that the mole fraction of toluene is 0.150. The molar mass of benzene is 78.11 g·mol^{-1}.

[**Answer:** 2.26 mol·kg^{-1}]

Self-test 5E.2B Calculate the molality of methanol in aqueous solution, given that the mole fraction of methanol is 0.250.

Related Exercises 5E.3, 5E.4

EXAMPLE 5E.3   Converting molarity into molality

You have made up a solution of known molarity but now realize that you need to know the molality instead. Find the molality of sucrose, $C_{12}H_{22}O_{11}$, in 1.06 M $C_{12}H_{22}O_{11}$(aq), which is known to have density 1.140 g·mL^{-1}.

ANTICIPATE The mass of 1 L of aqueous solution is close to 1 kg, so the numerical value of the molality can be expected to be close to that of the molarity, but with different units, of course.

PLAN Use procedure 3 in Toolbox 5E.1, taking note of the discussion of units.

SOLVE Note that a molar mass of 342.3 g·mol^{-1} is equivalent to 0.3423 kg·mol^{-1} and a density of 1.140 g·mL^{-1} is equivalent to 1.140 kg·L^{-1}.

From $b = c/(d - cM_{solute})$

$$b = \frac{1.06 \text{ mol·L}^{-1}}{1.140 \text{ kg·L}^{-1} - \underbrace{1.06 \text{ mol·L}^{-1} \times 0.3423 \text{ kg·mol}^{-1}}_{0.3628... \text{ kg·L}^{-1}}}$$

$$= \frac{1.06 \text{ mol·L}^{-1}}{0.7771... \text{ kg·L}^{-1}} = 1.36 \text{ mol·kg}^{-1}$$

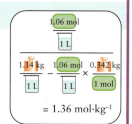

EVALUATE As expected, the numerical value of the molality of the sucrose, 1.36 (in mol·kg^{-1}), is close to the numerical value of its molarity, 1.06 (in mol·L^{-1}).

Self-test 5E.3A Battery acid is 4.27 M H_2SO_4(aq) and has a density of 1.25 g·cm^{-3}. What is the molality of H_2SO_4 in the solution?

[**Answer:** 5.14 m H_2SO_4(aq)]

Self-test 5E.3B The density of 1.83 M NaCl(aq) is 1.07 g·cm^{-3}. What is the molality of NaCl in the solution?

Related Exercises 5E.7, 5E.8

The molality of a solute in a solution is the amount (in moles) of solute divided by the mass of solvent used to prepare the solution.

What have you learned in this Topic?

You have learned how to use molality as a measure of concentration when it is appropriate to emphasize the relative numbers of solute and solvent molecules, and you have seen how to convert it to molarity and mole fraction.

The skills you have mastered are the ability to:

☐ **1.** Calculate the molality of a solute (Example 5E.1).

☐ **2.** Convert between molality and mole fraction or molarity (Toolbox 5E.1 and Examples 5E.2 and 5E.3).

Topic 5E Exercises

5E.1 Calculate (a) the molality of sodium chloride in a solution prepared by dissolving 25.0 g of NaCl in 500.0 g of water; (b) the mass (in grams) of NaOH that must be mixed with 345 g of water to prepare 0.18 m NaOH(aq); (c) the molality of urea, $CO(NH_2)_2$, in a solution prepared by dissolving 0.978 g of urea in 285 mL of water.

5E.2 Calculate (a) the molality of KOH in a solution prepared from 3.12 g of KOH and 67.0 g of water; (b) the mass (in grams) of ethylene glycol, HOC_2H_4OH, that should be added to 0.74 kg of water to prepare 0.28 m HOC_2H_4OH(aq); (c) the molality of an aqueous 3.68% by mass HCl solution.

5E.3 What is the molality of ethylene glycol, $C_2H_6O_2$, in an aqueous solution used for antifreeze, given that the mole fraction of ethylene glycol is 0.250?

5E.4 What is the molality of acetone, C_3H_6O, in an aqueous solution for which the mole fraction of acetone is 0.112?

5E.5 The density of a 5.00% by mass K_3PO_4 aqueous solution is 1.043 g·cm^{-3}. Determine (a) the molality; (b) the molarity of potassium phosphate in the solution.

5E.6 Calculate (a) the molality of 13.63 g of sucrose, $C_{12}H_{22}O_{11}$, dissolved in 612 mL of water; (b) the molality of CsCl in a 10.00% by mass aqueous solution; (c) the molality of acetone in an aqueous solution with a mole fraction for acetone of 0.197.

5E.7 The density of 14.8 M NH_3(aq) is 0.901 g·cm^{-3}. What is the molality of NH_3 in the solution?

5E.8 The density of 11.7 M $HClO_4$(aq) is 1.67 g·cm^{-3}. What is the molality of $HClO_4$ in the solution?

5E.9 Calculate (a) the molality of chloride ions in an aqueous solution of magnesium chloride in which $x_{MgCl_2} = 0.0120$; (b) the molality of 6.75 g of sodium hydroxide dissolved in 325 g of water; (c) the molality of 15.00 M HCl(aq) with a density of 1.0745 g·cm^{-3}.

5E.10 Calculate (a) the molality of chloride ions in an aqueous solution of iron(III) chloride for which $x_{FeCl_3} = 0.0312$; (b) the molality of hydroxide ions in a solution prepared from 3.24 g of barium hydroxide dissolved in 258 g of water; (c) the molality of 12.00 M NH_3(aq) with a density of 0.9519 g·cm^{-3}.

5E.11 (a) Calculate the mass of $CaCl_2 \cdot 6H_2O$ needed to prepare 0.125 m $CaCl_2$(aq) by using 500. g of water. (b) What mass of $NiSO_4 \cdot 6H_2O$ must be dissolved in 500. g of water to produce 0.22 m $NiSO_4$(aq)?

5E.12 The density of a 10.0% by mass H_2SO_4(aq) solution is 1.07 g·cm^{-3}. (a) What volume (in milliliters) of solution contains 8.37 g of H_2SO_4? (b) What is the molality of H_2SO_4 in the solution? (c) What mass (in grams) of H_2SO_4 is in 250. mL of the solution?

Topic 5F Colligative Properties

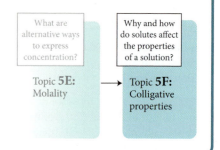

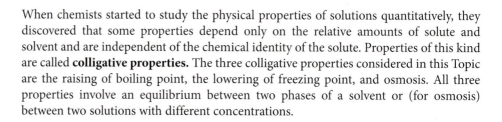

What are alternative ways to express concentration?

Why and how do solutes affect the properties of a solution?

Topic 5E: Molality → Topic 5F: Colligative properties

Why Do You Need to Know This Material? The effect of a solute on a solvent is an important step in formulating the thermodynamic description of physical and chemical equilibria. Moreover, in practice, it introduces the biologically, technologically, and analytically important property of osmosis.

What Do You Need to Know Already? This Topic makes use of the interpretation of entropy as a measure of disorder (Topic 4F) and the description of equilibria in terms of the Gibbs free energy. It uses the concept of molality (Topic 5E) and molarity (*Fundamentals* G).

The word colligative means "depending on the collection."

When chemists started to study the physical properties of solutions quantitatively, they discovered that some properties depend only on the relative amounts of solute and solvent and are independent of the chemical identity of the solute. Properties of this kind are called **colligative properties.** The three colligative properties considered in this Topic are the raising of boiling point, the lowering of freezing point, and osmosis. All three properties involve an equilibrium between two phases of a solvent or (for osmosis) between two solutions with different concentrations.

5F.1 Boiling-Point Elevation and Freezing-Point Depression

FIGURE 5F.1 shows how the molar Gibbs free energies of the liquid and vapor phases of a pure solvent vary with temperature. This graph is a plot of the equation $G_m = H_m - TS_m$, treating H_m and S_m as constants.

- At low temperatures, $G_m \approx H_m$, and so the line representing G_m of the vapor lies well above that for the liquid because the molar enthalpy of a vapor is considerably greater than that of a liquid.
- The slope of the line is $-S_m$. Because the molar entropy of the vapor is much greater than that of the liquid, the line for the vapor slopes down more steeply than that of the liquid.

The presence of a solute in the liquid phase of the solvent increases the entropy of the solvent and therefore (through $G_m = H_m - TS_m$) lowers its Gibbs free energy. As shown in Fig. 5F.1, the lines representing the molar Gibbs free energies of the liquid solution and the vapor intersect at a higher temperature than they do for the pure solvent. As a result, the boiling point is higher in the presence of the solute. This increase is called **boiling-point elevation.** The increase is usually quite small and is of little practical importance in science. A 0.1 M aqueous sucrose solution, for instance, boils at 100.05 °C.

FIGURE 5F.2 shows how the molar Gibbs free energies of the liquid and solid phases of a pure solvent vary with temperature. The explanation of the layout of the lines is similar to that for a liquid and its vapor, but the differences in position and slope are less pronounced:

- At low temperatures, $G_m \approx H_m$, and so the line representing G_m of the liquid lies above that for the solid (but not much above, because the enthalpies are not as different as for a liquid and a vapor).

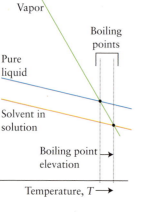

FIGURE 5F.1 The molar Gibbs free energy of a liquid and its vapor both decrease with increasing temperature, but that of the vapor decreases more sharply. The vapor is the stable phase at temperatures higher than the point of intersection of the two lines (the boiling point). When a nonvolatile solute is present, the molar Gibbs free energy of the solvent is lowered (an entropy effect), but that of the vapor is left unchanged. The point of intersection of the lines moves to a slightly higher temperature.

TABLE 5F.1   Boiling-Point and Freezing-Point Constants

Solvent	Freezing point/°C	k_f/ $(K \cdot kg \cdot mol^{-1})$	Boiling point/°C	k_b/ $(K \cdot kg \cdot mol^{-1})$
acetone	−95.35	2.40	56.2	1.71
benzene	5.5	5.12	80.1	2.53
camphor	179.8	39.7	204	5.61
carbon tetrachloride	−23	29.8	76.5	4.95
cyclohexane	6.5	20.1	80.7	2.79
naphthalene	80.5	6.94	217.7	5.80
phenol	43	7.27	182	3.04
water	0	1.86	100.0	0.51

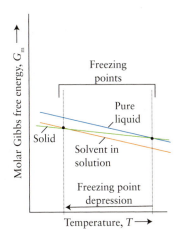

FIGURE 5F.2 The molar Gibbs free energy of a solid and its liquid phase both decrease with increasing temperature, but that of the liquid decreases slightly more steeply. The liquid is the stable phase at temperatures higher than the point of intersection of the two lines. When a solute is present, the molar Gibbs free energy of the solvent is lowered (an entropy effect), but that of the solid is left unchanged. The point of intersection of the lines moves to a lower temperature.

- Because the molar entropy of the liquid is greater than that of the solid, the line for the liquid slopes down more steeply than that of the solid (but not much more, because the entropies of these two phases are similar).

As can be seen from the illustration, the lines representing the molar Gibbs free energies of the liquid and solid phases of the solvent intersect at a lower temperature than they do for the pure solvent, and so the freezing point is lower in the presence of the solute. The resulting **freezing-point depression,** the lowering of the freezing point of the solvent caused by a solute, is more significant than the elevation of boiling point. For example, seawater freezes about 1 °C lower than fresh water. People living in regions with cold winters make use of the depression of the freezing point when they spread salt on icy walkways and roads. The salt lowers the freezing point of water as it produces a salt solution. In the laboratory, chemists make use of the effect to judge the purity of a solid compound: if impurities are present, the compound's melting point is lower than the accepted value.

The freezing-point depression for an ideal solution is proportional to the molality, b, of the solute. For a nonelectrolyte solution,

$$\text{Freezing-point depression: } \Delta T_f = k_f \times b \qquad \text{(1a)}$$

The constant k_f is called the **freezing-point constant** of the solvent; it is different for each solvent and must be determined experimentally (**TABLE 5F.1**). The effect is quite small; for instance, for a 0.1 m $C_{12}H_{22}O_{11}$(aq) (sucrose) solution,

$$\text{Freezing-point depression} = (1.86 \text{ K} \cdot \text{kg} \cdot \text{mol}^{-1}) \times (0.1 \text{ mol} \cdot \text{kg}^{-1}) = 0.2 \text{ K}$$

Because 0.2 K is the *depression* of the freezing point, the water in the solution freezes at −0.2 °C.

It is commonly said that antifreeze added to car engines is an example of the lowering of freezing point; it is, but it is not a colligative effect, as the concentration is far too high. The ethylene glycol molecules ($HOCH_2CH_2OH$, the common active component) lie between the water molecules and prevent their linking together to form ice.

Self-test 5F.1A Use the data in Table 5F.1 to determine at what temperature a solution of the analgesic codeine, $C_{18}H_{21}NO_3$, in benzene at a molality of 0.20 mol·kg^{-1} will freeze.

[*Answer:* 4.5 °C]

Self-test 5F.1B Use the data in Table 5F.1 to determine at what temperature a solution of the insecticide malathion, $C_{10}H_{19}O_6PS_2$, in camphor at a molality of 0.050 mol·kg^{-1} will freeze.

In an electrolyte solution, each formula unit contributes two or more ions. Sodium chloride, for instance, dissolves to give Na^+ and Cl^- ions, and both kinds of ions contribute to the depression of the freezing point. The cations and anions contribute nearly independently in very dilute solutions, and so the total solute molality is twice the molality of NaCl formula units. In place of Eq.1a,

$$\text{Freezing-point depression: } \Delta T_f = ik_f \times b \qquad \text{(1b)}$$

Here, i, the **van 't Hoff i factor,** is determined experimentally. In a very dilute solution (less than about 10^{-3} mol·L^{-1}), when all ions are independent, $i = 2$ for MX salts such as NaCl, $i = 3$ for MX$_2$ salts such as CaCl$_2$, and so on. For dilute nonelectrolyte solutions, $i = 1$. The i factor is so unreliable, however, that it is best to confine quantitative calculations of freezing-point depression to nonelectrolyte solutions. Even these solutions must be dilute enough to be approximately ideal.

The i factor can be used to help determine the extent to which a substance is dissociated into ions in solution. For example, in dilute solution, HCl has $i = 1$ in toluene and $i = 2$ in water. These values suggest that HCl retains its molecular form in toluene but is fully deprotonated in water (deprotonation, the loss of a proton from an acid, is described in *Fundamentals* J and more fully in Topic 6A). The strength of a weak acid in water (the extent to which it is deprotonated) can be estimated in this way. In an aqueous solution of a weak acid that is 5% deprotonated (5% of the acid molecules have given up their protons), each deprotonated molecule produces two ions and $i = 0.95 + (0.05 \times 2) = 1.05$.

Cryoscopy is the determination of the molar mass of a solute by measuring the depression of freezing point that it causes when dissolved in a solvent. Camphor has been used as the solvent for organic compounds because it has a large freezing-point constant; consequently, solutes depress its freezing point significantly. However, this procedure is now rarely used in modern laboratories because techniques such as mass spectrometry give far more reliable results. The procedure is described in Toolbox 5F.1 at the end of the next section.

Self-test 5F.2A What amount (in moles) of ions are present in a dilute solution containing 0.010 mol Na_2SO_4, assuming complete dissociation (separation of ions)? Estimate the i factor.
[*Answer:* 0.030 mol (0.020 mol Na^+ ions and 0.010 mol SO_4^{2-} ions); $i = 3$]

Self-test 5F.2B What amount (in moles) of ions are present in a solution containing 0.025 mol $CoCl_3$, assuming complete dissociation? Estimate the i factor.

The presence of a solute lowers the freezing point of a solvent; if the solute is nonvolatile, the boiling point is also raised. The freezing-point depression can be used to calculate the molar mass of the solute. If the solute is an electrolyte, the extent of its dissociation or (for an acid) deprotonation must also be taken into account.

5F.2 Osmosis

The name "osmosis" comes from the Greek word for "push."

Osmosis is the flow of solvent through a membrane into a more concentrated solution. The phenomenon can be demonstrated in the laboratory when a solution and the pure solvent are separated by a **semipermeable membrane,** a membrane that permits only certain types of molecules or ions to pass through (**FIG. 5F.3**). Cellulose acetate, for instance, allows water molecules to pass through it, but not solute molecules or ions with their bulky coating of hydrating water molecules. Initially, the heights of the solution and the pure solvent shown in the illustration are the same. However, the level of the solution inside the tube begins to rise as pure solvent passes through the membrane into the solution.

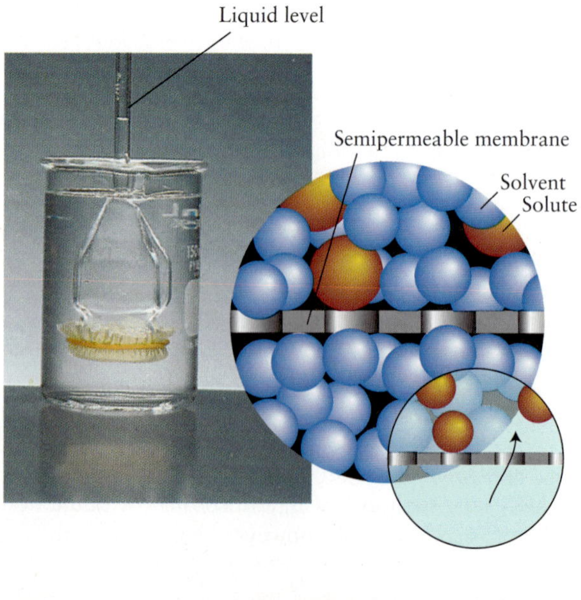

FIGURE 5F.3 An experiment to illustrate osmosis. Initially, the tube contained a sucrose solution and the beaker contained pure water; the initial heights of the two liquids were the same. At the stage shown here, water has passed into the solution through the membrane by osmosis, and the level of solution in the tube has risen above that of the pure water. The large inset shows the molecules of pure solvent (below the membrane) tending to join those in the solution (above the membrane) because the presence of solute molecules there has led to increased disorder. The small inset shows just the solute molecules; the arrow shows the direction of net flow of solvent molecules. (*W. H. Freeman photo by Ken Karp.*)

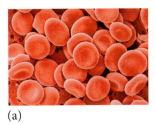
(a)

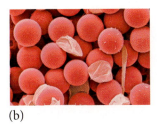
(b)

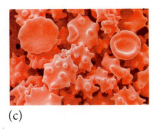
(c)

At equilibrium, the pressure exerted by the higher level of the column of solution is sufficiently great that the rate of flow of molecules through the membrane is the same in each direction, and so the net flow is zero. The pressure needed to stop the flow of solvent is called the **osmotic pressure,** Π (the Greek uppercase letter pi). The greater the osmotic pressure, the greater is the height of the solution needed to reduce the net flow to zero. When the net flow is zero, the solutions are said to be **isotonic** (have the same osmotic pressure).

Life depends on osmosis. Biological cell walls act as semipermeable membranes that allow water, small molecules, and hydrated ions to pass (**FIG. 5F.4**). However, they block the passage of the large enzymes and proteins that have been synthesized within the cell. The higher concentration of solutes within a cell compared to the solution outside the cell gives rise to an osmotic pressure, and water passes into the more concentrated solution in the interior of the cell, carrying small nutrient molecules with it. This influx of water also keeps the cell turgid (swollen). When the water supply is cut off, the turgidity is lost and the cell becomes dehydrated. In a plant, this dehydration results in wilting. Salted meat is preserved from bacterial attack by osmosis. In this case, the concentrated salt solution dehydrates—and kills—the bacteria by causing water to flow out of them. Osmotic pressure is an important consideration for designers of drug delivery systems that function automatically by sensing the body's needs (**BOX 5F.1**).

The pressure exerted by a vertical column of liquid is proportional to its height; see Eq. 2 in Topic 3A ($P = gdh$).

Box 5F.1 FRONTIERS OF CHEMISTRY: DRUG DELIVERY

The administration of drugs to ease disease and chronic, severe pain or to provide benefits such as hormone replacement therapy is difficult because drugs taken orally may lose much of their potency in the harsh conditions of the digestive system. In addition, they are distributed throughout the entire body, not just where they are needed, and side effects can be significant. Recently, however, techniques have been developed to deliver drugs gradually over time, to the exact location in the body where they are needed, and even at the time when they are needed.

Transdermal patches are applied to the skin. The drug is mixed with the adhesive for the patch, and so it lies next to the skin. In another style of patch, the drug is incorporated into a gel or solution reservoir separated from the skin by a permeable membrane that controls the rate of delivery. The skin can readily absorb many chemicals and so can absorb drugs such as nitroglycerin (for heart disease), morphine derivatives (for constant, severe pain), estrogen (for hormone replacement therapy), or nicotine (for easing symptoms that result when a patient stops smoking).

Implants provide a means of delivering drugs over a longer period of time at a controlled rate inside the body. Subcutaneous (under-the-skin) implants are used to provide appropriate doses of psychoactive medications, birth-control drugs, painkillers, and other medications that must be administered frequently. The implants last for as long as a month and can easily be replaced or renewed. When the location of the drug release is critical, implants can be placed deeper into the body. For example, implants can be introduced into the brain or spinal column to provide effective relief

from pain or to protect neurons from degeneration. The implant is contained inside a cylinder of porous foam through which the drug is released. Some implants contain living animal cells that have been genetically engineered to produce natural hormones or painkillers that are released to the body as they are produced. In other types of implants, membranes allow a gradual release of the drug.

Controlled-release drug delivery systems mimic nature. Phospholipids like those found in cell membranes (Topic 5C) assemble spontaneously into liquid-crystal structures in water (see Topic 3G); in these structures, sheets consisting of rows of molecules are lined up next to one another. The sheets can be coaxed into forming *liposomes*. Liposomes are similar to micelles

This implant containing live hamster cells was inserted into the spinal column of a patient for 17 weeks. After removal, the cells were still alive and secreting the hormone required to keep the patient healthy. (*Patrick Aebischer*)

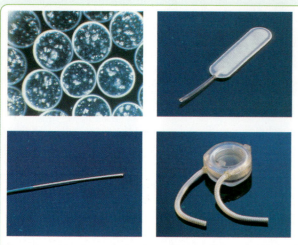

Examples of implants used to insert living cells into the body. The cells continually produce enzymes, hormones, or painkillers needed by the body. Often a long, thin plastic tail is attached as a tether to allow easy retrieval of the implant. (*Upper left: Patrick Aebischer. Upper right and bottom row: Sam Ogden Photography.*)

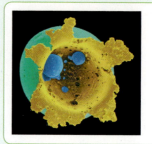

An electron microscope image of a drug capsule as it bursts open, revealing the tiny microcapsules inside. The image has been digitally colored. (*David McCarthy/ Science Source.*)

but are formed from a double layer of molecules, with polar heads forming each surface, like the membrane of a living cell. When a drug is present in the aqueous solution in which liposomes are being formed, some of the drug is encapsulated inside each liposome. The liposomes can then be injected into the body, where they stick only to certain types of cells—cancer cells, for instance. Compared with oral or intravenous medicine, a smaller dosage is required and side effects are greatly reduced.

Nanotechnology (Topic 3J) has led to very efficient versions of liposomes. Tiny hollow spheres only nanometers in diameter hold even tinier capsules of medicine. The spheres are made of silica covered with gold nanoparticles; when they are coated with antibodies, they attach to tumor cells. The spheres are sensitive to light of specific wavelengths, and when the light is applied, they either become hot and destroy the tumor, or burst, releasing the drugs within the capsules directly into the tumor.

Smart gels (see Box 8F.1) are being investigated for drug delivery in situations in which the drug dosage must be modified according to conditions in the body. For example, the amount of insulin that a nondiabetic person needs is delivered by the body according to the level of blood sugar. However, a diabetic person

must take insulin at specified times of the day, and the same amount each time. If the person's blood sugar is already low, a hypoglycemic reaction, and possibly coma, can result. An insulin delivery system that would be responsive to blood sugar levels and provide the correct dosage when it is needed is now being investigated. The system makes use of a smart gel in which molecules of insulin have been trapped. The gel incorporates into its structure molecules of phenylboronic acid, to which glucose (blood sugar) molecules adhere. If the blood sugar is high, more and more glucose molecules stick to the gel and cause it to swell. When the level of glucose rises above a certain concentration, the gel swells so much that it becomes porous, releasing insulin into the blood.

HOW MIGHT YOU CONTRIBUTE?

Both basic and applied research are needed for effective drug delivery systems. Basic research on the self-assembly of molecules may allow more innovative solutions in the future. Applied research may have more immediate benefits. For example, the optimal drug delivery system must be designed for each specific drug. Coatings for implants or nanospheres that are nontoxic and are similar in nature to body tissues need to be developed. Both the length of time during which a drug delivery system can remain active inside the body and the stability of the system need to be increased.

Related Exercises 5.31, 5.32

Further Reading O. C. Farokhzad and R. Langer, "Impact of nanotechnology on drug delivery," *Nano*, vol. 3, 2009, pp. 16–20. C. M. Henry, "Special delivery," *Chemical and Engineering News*, September 18, 2000, pp. 49–64. M. J. Lysaght and P. Aebischer, "Encapsulated cells as therapy," *Scientific American*, April 1999, pp. 76–82. S. Morrissey, "Nanotech meets medicine," *Chemical and Engineering News*, May 16, 2005, p. 30.

The thermodynamic origin of osmosis is that solvent tends to flow through a membrane until the molar Gibbs free energy of the solvent is the same on each side of it. A solute lowers the molar Gibbs free energy of the solution below that of the pure solvent (by increasing the entropy), and solvent therefore has a tendency to pass into the solution (**FIG. 5F.5**).

The same van 't Hoff responsible for the *i* factor showed that the osmotic pressure of a dilute solution is related to the molar concentration, *c*, of the solute in the solution:

$$\Pi = iRTc \qquad\qquad (3)$$

where *i* is the *i* factor, *R* is the gas constant, and *T* is the temperature. This expression is now known as the **van 't Hoff equation.** Notice that the osmotic pressure depends only on temperature and the total molar concentration of solute. It is independent of the identities of both the solute and the solvent. However, in an apparatus like that shown in Fig. 5F.3, the height to which the solvent is driven does depend on the identity of the solvent, because that height depends on its density (**FIG. 5F.6**).

You might be puzzled why the gas constant appears in the description of a property unrelated to gases. It is actually the Boltzmann constant, a more fundamental and widely applicable constant, in disguise: $R = N_A k$ (Topic 4G).

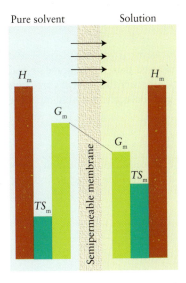

FIGURE 5F.5 On the left of the semipermeable membrane is the pure solvent with its characteristic molar enthalpy, entropy, and Gibbs free energy. On the right is the solution. The molar Gibbs free energy of the solvent is lower in the solution (an entropy effect), and so there is a spontaneous tendency for the solvent to flow into the solution.

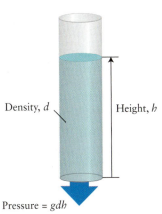

FIGURE 5F.6 The pressure at the base of a column of fluid is equal to the product of the acceleration of free fall, g, the density, d, of the liquid, and the height, h, of the column.

The van 't Hoff equation is used to determine the molar mass of a solute from osmotic pressure measurements. This technique, called **osmometry,** is used as explained in **Toolbox 5F.1.** Osmometry is very sensitive, even at low concentrations, and was once commonly used to determine very large molar masses, such as those of polymers and proteins.

Self-test 5F.3A What is the osmotic pressure of 0.0100 M KCl(aq) at 298 K? (Assume $i = 2$.)
[**Answer:** 0.489 atm]

Self-test 5F.3B What is the osmotic pressure of a 0.120 M sucrose solution at 298 K?

Toolbox 5F.1   HOW TO USE COLLIGATIVE PROPERTIES TO DETERMINE MOLAR MASS

CONCEPTUAL BASIS

The lowering of freezing point and the generation of osmotic pressure both depend on the total concentration of solute particles. Therefore, by using these colligative properties to determine the amount of solute present and knowing its mass, its molar mass can be inferred.

PROCEDURE

1. Cryoscopy

Step 1 Convert the observed freezing-point depression, ΔT_f, into solute molality b (Topic 5E) by writing Eq. 1b in the form

$$b = \frac{\Delta T_f}{i k_f}$$

Take the freezing-point constant from Table 5F.1.

Step 2 Calculate the amount of solute, n_{solute} (in moles), in the sample by multiplying the molality by the mass of solvent, $m_{solvent}$ (in kilograms):

$$n_{solute} = b \times m_{solvent}$$

Step 3 Determine the molar mass of the solute by dividing the given mass of solute, m_{solute} (in grams), by the calculated amount in moles (step 2).

$$M_{solute} = \frac{m_{solute}}{n_{solute}}$$

This procedure is illustrated in Example 5F.1.

2. Osmometry

Step 1 Convert the observed osmotic pressure into solute molar concentration (c) by writing Eq. 3 in the form

$$c = \frac{\Pi}{iRT}$$

In some cases, it may be necessary to calculate the osmotic pressure from the height, h, of the solution (in an apparatus like that in Fig. 5F.3) by using $\Pi = gdh$, where d is the density of the solution and g is the acceleration of free fall (see inside back cover).

Step 2 Because molar concentration is defined as $c = n_{solute}/V$, where n_{solute} is the amount of solute (in moles) and V is the volume of solution (in liters), calculate the amount of solute from

$$n_{solute} = cV$$

Step 3 Determine the molar mass of the solute by dividing the given mass of solute, m_{solute} (in grams), by its amount (step 2), as in the procedure for cryoscopy.

This procedure is illustrated in Example 5F.2.

EXAMPLE 5F.1   Determining molar mass cryoscopically

In modern laboratories, sophisticated instruments are used to determine molar mass. However, if you don't have access to such an instrument, you could still determine molar mass using nothing more complicated than a thermometer and a balance. The addition of 0.24 g of sulfur to 100. g of the solvent carbon tetrachloride lowers the solvent's freezing point by 0.28 °C. Sulfur is known to exist in molecular form. What is the molar mass and molecular formula of sulfur molecules?

ANTICIPATE   You might already be aware that sulfur commonly forms crownlike S_8 molecules.

PLAN   Use the procedure for cryoscopy in Toolbox 5F.1. Sulfur is a nonelectrolyte, so $i = 1$. Once you know the molar mass of the molecules, divide it by the molar mass of sulfur atoms (found in the list of elements inside the back cover) to find how many atoms are in each molecule.

SOLVE

Step 1 Convert the observed freezing-point depression, ΔT_f, into solute molality b by using $b = \Delta T_f / i k_f$ with $i = 1$:

$$b = \frac{0.28 \text{ K}}{29.8 \text{ K·kg·mol}^{-1}} = \frac{0.28}{29.8} \text{ mol·kg}^{-1} = 0.0093\ldots \text{ mol·kg}^{-1}$$

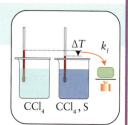

Step 2 Calculate the amount of solute, n_{solute} (in moles), in the sample by multiplying the molality by the mass of solvent, $m_{solvent}$ (in kilograms), $n_{solute} = b m_{solvent}$:

$$n(S_x) = (0.100 \text{ kg}) \times 0.0093\ldots \text{ mol·kg}^{-1} = 0.000\,93\ldots \text{ mol}$$

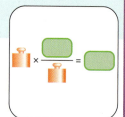

Step 3 Determine the molar mass of the solute by dividing the given mass of solute, m_{solute} (in grams), by the calculated amount in moles (step 2), $M_{solute} = m_{solute}/n_{solute}$:

$$M(S_x) = \frac{0.24 \text{ g}}{0.000\,93\ldots \text{ mol}} = 2.5\ldots \times 10^2 \text{ g·mol}^{-1}$$

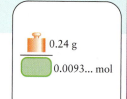

Use the molar mass of atomic sulfur (32.06 g·mol^{-1}) to find the value of x in the molecular formula S_x

$$x = \frac{2.5\ldots \times 10^2 \text{ g·mol}^{-1}}{32.06 \text{ g·mol}^{-1}} = 8.0$$

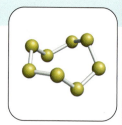

EVALUATE   Elemental sulfur is indeed composed of S_8 molecules.

Self-test 5F.4A   When 250. mg of eugenol, the molecular compound responsible for the odor of oil of cloves, was added to 100. g of camphor, it lowered the freezing point of camphor by 0.62 °C. Calculate the molar mass of eugenol.

[***Answer:*** $1.6 \times 10^2 \text{ g·mol}^{-1}$ (actual: 164.2 g·mol^{-1})]

Self-test 5F.4B   When 200. mg of linalool, a fragrant molecular compound found in cinnamon oil from Sri Lanka, was added to 100. g of camphor, it lowered the freezing point of camphor by 0.51 °C. What is the molar mass of linalool?

Related Exercises 5E.1–5E.2

EXAMPLE 5F.2   Using osmometry to determine molar mass

Osmometry has been widely used in the polymer industry because it is a sensitive technique for determining the huge molar masses of polymer molecules. Now imagine that you are a polymer chemist; you have devised a new way to make polyethylene and wish to know the molar mass of your new material. The osmotic pressure due to 2.20 g of polyethylene (PE) dissolved in enough benzene to produce 100.0 mL of solution was 1.10×10^{-2} atm at 25 °C. Calculate the molar mass of the polymer, which is a nonelectrolyte. The answer will be an average molar mass as not all the polymer molecules are the same length.

ANTICIPATE Because so many atoms are linked together in each polymer molecule, you should expect a high molar mass.

PLAN Use the procedure for osmometry in Toolbox 5E.1. Because polyethylene is a nonelectrolyte, $i = 1$. Use R in the units that match the data—in this case, liters and atmospheres.

SOLVE

Step 1 Convert the observed osmotic pressure into solute molar concentration, c, by writing Eq. 3 in the form $c = \Pi/iRT$ with $i = 1$:

$$c = \frac{1.10 \times 10^{-2}\,\text{atm}}{(0.0821\ \text{L·atm·K}^{-1}\text{·mol}^{-1}) \times (298\ \text{K})}$$

$$= \frac{1.10 \times 10^{-2}}{0.0821 \times 298}\ \text{mol·L}^{-1} = 4.50... \times 10^{-4}\ \text{mol·L}^{-1}$$

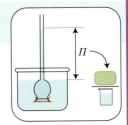

Step 2 Calculate the amount of solute molecules in the solution from $n_{\text{solute}} = c_{\text{solute}}V$.

$$n(\text{PE}) = (4.50... \times 10^{-4}\ \text{mol·L}^{-1}) \times (0.100\ \text{L})$$

$$= 4.50... \times 10^{-5}\ \text{mol}$$

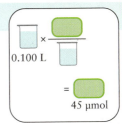

0.100 L

= 45 μmol

Step 3 Determine the molar mass of the solute by dividing the given mass of solute, m_{solute} (in grams), by its amount (step 2), $M_{\text{solute}} = m_{\text{solute}}/n_{\text{solute}}$:

$$M(\text{PE}) = \frac{2.20\ \text{g}}{4.50... \times 10^{-5}\ \text{mol}} = 4.89 \times 10^4\ \text{g·mol}^{-1}$$

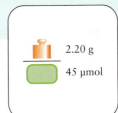

2.20 g

45 μmol

EVALUATE This molar mass is reported as 48.9 kg·mol^{-1}. As expected, the molar mass is very high.

Self-test 5F.5A The osmotic pressure of 3.0 g of polystyrene dissolved in enough benzene to produce 150. mL of solution was 1.21 kPa at 25 °C. Calculate the average molar mass of the sample of polystyrene.

[*Answer:* 41 kg·mol^{-1}]

Self-test 5F.5B The osmotic pressure of 1.50 g of polymethyl methacrylate dissolved in enough methylbenzene to produce 175 mL of solution was 2.11 kPa at 20 °C. Calculate the average molar mass of the sample of polymethyl methacrylate.

Related Exercises 5F.11–5F.14

In **reverse osmosis**, a pressure greater than the osmotic pressure is applied to the solution side of the semipermeable membrane. This application of pressure increases the rate at which solvent molecules leave the solution and thus reverses the flow of solvent, forcing it to flow from the solution to pure solvent. Reverse osmosis is used to remove salts from seawater to produce fresh water for drinking and irrigation. The water is almost literally squeezed out of the salt solution through the membrane. The technological challenge is to fabricate new membranes that are strong enough to withstand high pressures and do not easily become clogged. Commercial plants use cellulose acetate membranes at pressures as high as 70 atm.

THINKING POINT

Reverse osmosis is energy intensive. Why?

Osmosis is the flow of solvent through a semipermeable membrane into a solution; the osmotic pressure is proportional to the molar concentration of the solute. Osmometry is used to determine the molar masses of compounds with large molecules, such as polymers; reverse osmosis is used in water purification.

What have you learned in this Topic?

You have encountered the concept of colligative properties, including osmosis, which is used to determine the molar mass of compounds with large molecules, and have seen that the presence of a nonvolatile solute lowers the freezing point of a solvent and gives rise to an osmotic pressure. You are now able to account for these properties in thermodynamic terms, specifically in terms of the effect of the solute on the entropy of the solution.

The skills you have mastered are the ability to:

☐ **1.** Determine molar mass cryoscopically (Toolbox 5F.1 and Example 5F.1).

☐ **2.** Use osmometry to find the molar mass of a solute (Toolbox 5F.1 and Example 5F.2).

Topic 5F Exercises

5F.1 A solution consisting of a molecular substance of mass 1.14 g dissolved in 100. g of camphor freezes at 176.9 °C. What is the molar mass of the substance?

5F.2 When 1.78 g of a nonpolar solute was dissolved in 60.0 g of phenol, the latter's freezing point was lowered by 1.362 °C. Calculate the molar mass of the solute.

5F.3 The freezing point of a 1.00% by mass $NaCl(aq)$ is -0.593 °C. (a) Estimate the van 't Hoff i factor from the data. (b) Determine the total molality of all solute species. (c) Calculate the percentage dissociation of NaCl in this solution. (The molality calculated from the freezing-point depression is the sum of the molalities of the undissociated ion pairs, the Na^+ ions, and the Cl^- ions.)

5F.4 The freezing point of a 1.00% by mass $MgSO_4(aq)$ solution is -0.192 °C. (a) Estimate the van 't Hoff i factor from the data. (b) Determine the total molality of all solute species. (c) Calculate the percentage dissociation of $MgSO_4$ in this solution.

5F.5 Two unknown molecular compounds were being studied. A solution containing 5.00 g of compound A in 100. g of water froze at a lower temperature than a solution containing 5.00 g of compound B in 100. g of water. Which compound has the greater molar mass? Explain how you arrived at your answer.

5F.6 Two unknown compounds were being studied. Compound C is molecular and compound D is an ionic compound known to dissociate into ions completely in dilute aqueous solutions. A solution containing 0.30 g of compound C in 100. g of water froze at the same temperature as a solution containing 0.30 g of compound D in 100. g of water. Which compound has the greater molar mass? Explain how you arrived at your answer.

5F.7 Determine the freezing point of a 0.10 $mol \cdot kg^{-1}$ aqueous solution of a weak electrolyte that is 7.5% dissociated into two ions.

5F.8 A 0.124 m $CCl_3COOH(aq)$ solution has a freezing point of -0.423 °C. What is the percentage deprotonation of the acid?

5F.9 What is the osmotic pressure at 20 °C of (a) 0.010 M $C_{12}H_{22}O_{11}(aq)$; (b) 1.0 M HCl(aq); (c) 0.010 M $CaCl_2(aq)$? Assume complete dissociation of the $CaCl_2$.

5F.10 Which of the following solutions has the highest osmotic pressure at 50 °C: (a) 0.10 M KCl(aq); (b) 0.60 M $CO(NH_2)_2(aq)$; (c) 0.30 M $K_2SO_4(aq)$? Justify your answer by calculating the osmotic pressure of each solution.

5F.11 A sample of a polypeptide of mass 0.40 g dissolved in 1.0 L of an aqueous solution at 27 °C gave rise to an osmotic pressure of 3.74 Torr. What is the molar mass of the polypeptide?

5F.12 A solution prepared by adding 0.50 g of a polymer to 0.200 L of toluene (methylbenzene, a common solvent) showed an osmotic pressure of 0.582 Torr at 20 °C. What is the molar mass of the polymer?

5F.13 A polymer sample of mass 0.20 g dissolved in 0.100 L of toluene, gives rise to an osmotic pressure of 6.3 Torr at 20 °C. What is the molar mass of the polymer?

5F.14 Catalase, a liver enzyme, dissolves in water. A solution of volume 10.0 mL containing 0.166 g of catalase gives rise to an osmotic pressure of 1.2 Torr at 20 °C. What is the molar mass of catalase?

5F.15 Calculate the osmotic pressure at 20 °C of each of the following solutions, assuming complete dissociation for any ionic solutes: (a) 0.050 M $C_{12}H_{22}O_{11}(aq)$; (b) 0.0010 M NaCl(aq); (c) a saturated aqueous solution of AgCN of solubility 23 µg/100. g of water.

5F.16 Calculate the osmotic pressure at 20 °C of each of the following solutions, assuming complete dissociation of ionic compounds: (a) 4.5×10^{-3} M $C_6H_{12}O_6(aq)$; (b) 3.0×10^{-3} M $CaCl_2(aq)$; (c) 0.025 M $K_2SO_4(aq)$.

Topic 5G Chemical Equilibrium

How are chemical equilibria described?

What are other ways of writing an equilibrium constant?

Topic **5G:** Chemical equilibrium → Topic **5H:** Alternative forms of the equilibrium constant

Why Do You Need to Know This Material? Chemical equilibrium lies at the very heart of chemistry, as all chemical reactions tend toward equilibrium.

What Do You Need to Know Already? There are two alternative routes to an understanding of this Topic. To take the kinetics route, you need to know about rate laws and reaction mechanism (Focus 7). To take the thermodynamic route, you need to understand the concept of the Gibbs free energy of reaction (Topic 4J) and to be aware of the pressure-dependence of the Gibbs free energy (Topic 5A). The mathematical derivations depend on manipulations of logarithms (Appendix 1D).

The metal acts as a catalyst for these reactions, a substance that helps to make a reaction go faster (Topic 7E).

Like physical equilibria (Topic 5A), all chemical equilibria are dynamic. To say that chemical equilibrium is "dynamic" means that when a reaction has reached equilibrium, the forward and reverse reactions continue to take place, but with reactants being formed as fast as they are consumed. As a result, the composition of the reaction mixture remains constant. From the thermodynamic viewpoint, at equilibrium there is no tendency to form more reactants or more products.

5G.1 The Reversibility of Reactions

Some reactions, such as the explosive reaction of hydrogen and oxygen, appear to proceed to completion, but others seem to stop at an early stage. For example, consider the reaction that takes place when nitrogen and hydrogen are heated under pressure in the presence of a small amount of iron:

$$N_2(g) + 3 H_2(g) \xrightarrow{\text{Fe}} 2 NH_3(g) \qquad \textbf{(A)}$$

The reaction produces ammonia rapidly at first, but eventually the reaction seems to stop (**FIG. 5G.1**). As the graph shows, no matter how long you wait, no additional product forms. What actually happens when the formation of ammonia *appears* to stop is that the rate of the reverse reaction,

$$2 NH_3(g) \xrightarrow{\text{Fe}} N_2(g) + 3 H_2(g) \qquad \textbf{(B)}$$

increases as more ammonia is formed. The reaction reaches equilibrium when the ammonia decomposes as fast as it is being formed. This state of dynamic equilibrium is expressed,

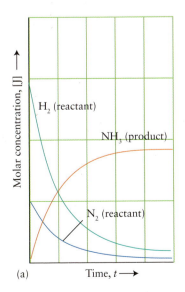

(a)

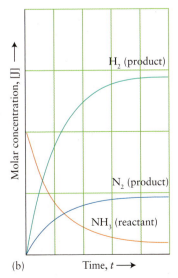

(b)

FIGURE 5G.1 (a) In the synthesis of ammonia, the concentrations of N_2 and H_2 decrease with time and that of NH_3 increases until they finally settle into values corresponding to a mixture in which all three are present and there is no further net change. (b) If the experiment is repeated with pure ammonia, it decomposes, and the composition settles down into a mixture of ammonia, nitrogen, and hydrogen. (The two graphs correspond to experiments at two different temperatures, and so they correspond to different equilibrium compositions.)

(a)

(b)

(c)

FIGURE 5G.2 (a) Methane burns in air with a steady flame, but, because matter is being added and removed, the reaction is not at equilibrium. (b) This sample of glucose in air is unchanging in composition, but it is not in equilibrium with its combustion products; the reaction proceeds far too slowly at room temperature. (c) Nitrogen dioxide (a brown gas) and dinitrogen tetroxide (a colorless volatile solid) are in equilibrium in these vessels. Note that when ice is added to change the temperature (right) the composition adjusts to a new value. *((a) SPL/Science Source; (b) W. H. Freeman photo by Ken Karp; (c) Leroy Laverman.)*

 LAB VIDEO FIGURE 5G.2

as in the discussion of physical equilibria (Topic 5A), by replacing the arrow in the chemical equation with equilibrium "harpoons":

$$N_2(g) + 3\,H_2(g) \rightleftharpoons 2\,NH_3(g) \qquad\qquad (C)$$

All chemical equilibria are dynamic equilibria. Dynamic equilibria are living equilibria in the sense that they respond to changes in temperature, pressure, and the addition or removal of even a small amount of reagent. A reaction that is simply not taking place (such as that in a mixture of hydrogen and oxygen at room temperature and pressure) has a composition that does not respond to small changes in the conditions.

THINKING POINT

Can you think of an experiment to show that a chemical equilibrium is dynamic? Think, perhaps, about using radioactive isotopes.

The characteristics of dynamic equilibria can be used to decide if a system is at equilibrium. Although all three systems in **FIG. 5G.2** appear at first glance to be unchanging, a closer look reveals the following:

- When methane, CH_4, burns with a steady flame to form carbon dioxide gas and water (Fig. 5G.2a), the combustion reaction is not at equilibrium because the composition is not constant (reactants continue to be added and the products spread away from the flame instead of reacting to form methane and oxygen again).
- The sample of glucose does not change (Fig. 5G.2b), even if it is left open to the atmosphere for a very long time. However, the glucose is not at equilibrium with the products of its combustion (carbon dioxide and water); it survives in air only because the rate of its combustion is immeasurably slow at room temperature.
- The gas-phase reaction (Fig. 5G.2c) is at equilibrium because additional experiments show that NO_2 is ceaselessly forming N_2O_4 and that N_2O_4 is decomposing into NO_2 at the same rate.

Chemical reactions reach a state of dynamic equilibrium in which the rates of forward and reverse reactions are equal and there is no net change in composition.

5G.2 Equilibrium and the Law of Mass Action

In 1864, the Norwegians Cato Guldberg (a mathematician) and Peter Waage (a chemist) discovered the mathematical relation that summarizes the composition of a reaction mixture at equilibrium. As an example of their approach, look at the data in **TABLE 5G.1** for the reaction between SO_2 and O_2:

$$2\,SO_2(g) + O_2(g) \rightleftharpoons 2\,SO_3(g) \qquad \text{(D)}$$

In each of these five experiments, a mixture with different initial compositions of the three gases was prepared and allowed to reach equilibrium at 1000. K. The compositions of the equilibrium mixtures and the total pressure P were then determined. At first, there seemed to be no pattern in the data. However, Guldberg and Waage noticed an extraordinary relation. They found (using modern notation) that the value of the quantity

$$K = \frac{(P_{SO_3}/P^\circ)^2}{(P_{SO_2}/P^\circ)^2(P_{O_2}/P^\circ)}$$

was nearly the same for every experiment, regardless of the initial composition. Here, P_J is the equilibrium partial pressure of gas J and $P^\circ = 1$ bar, the standard pressure. Note that K is unitless, because the units of P_J are canceled by the units of P° in each term. From now on, though, this expression will be written more simply as

$$K = \frac{(P_{SO_3})^2}{(P_{SO_2})^2 P_{O_2}}$$

with each P_J understood to be the numerical value of the partial pressure of J in bar.

Within experimental error and at a given temperature, Guldberg and Waage obtained the same value of K at equilibrium whatever the initial composition of the reaction mixture. This remarkable result shows that K is characteristic of the composition of the reaction mixture at equilibrium at a given temperature, and K is known as the **equilibrium constant** for the reaction. The **law of mass action** generalizes this result:

For the reaction

$$a\,A(g) + b\,B(g) \rightleftharpoons c\,C(g) + d\,D(g) \qquad \text{(E)}$$

between ideal gases, the constant

$$K = \frac{(P_C)^c(P_D)^d}{(P_A)^a(P_B)^b} \qquad \text{(1)}$$

is characteristic of the reaction (at a given temperature), with P_J denoting the numerical values of the partial pressures (in bar) at equilibrium. Note that the products appear in the numerator and the reactants in the denominator and that each partial pressure is raised to a power equal to the stoichiometric coefficient in the chemical equation.

TABLE 5G.1 Equilibrium Data and the Equilibrium Constant for the Reaction $2\,SO_2(g) + O_2(g) \rightleftharpoons 2\,SO_3(g)$ at 1000. K

P_{SO_2}/bar	P_{O_2}/bar	P_{SO_3}/bar	K^*
0.660	0.390	0.0840	0.0415
0.0380	0.220	0.00360	0.0409
0.110	0.110	0.00750	0.0423
0.950	0.880	0.180	0.0408
1.44	1.98	0.410	0.0409

*Average: 0.0413.

EXAMPLE 5G.1   Writing the expression for an equilibrium constant

The yearly production of ammonia is of the order of 1.3 Gt (1 Gt = 10^9 t), making this commodity one of the top ten chemicals produced globally. If you were studying the synthesis of ammonia, you would need to work with the equilibrium constant for the reaction. Write the equilibrium constant for the ammonia synthesis reaction, reaction C.

PLAN Write the equilibrium constant with the partial pressure of the product in the numerator, raised to a power equal to its stoichiometric coefficient in the balanced equation. Do the same for the reactants, but place their partial pressures in the denominator.

SOLVE

From $K = (P_C)^c (P_D)^d / (P_A)^a (P_B)^b$,

$$K = \frac{(P_{NH_3})^2}{P_{N_2}(P_{H_2})^3}$$

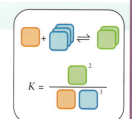

Each P_J in this expression should be interpreted as $P_J/P°$ (that is, as the numerical value of the pressure in bar).

Self-test 5G.1A Write the expression for the equilibrium constant for the reaction $4\,NH_3(g) + 5\,O_2(g) \rightleftharpoons 4\,NO(g) + 6\,H_2O(g)$.

[*Answer:* $K = (P_{NO})^4(P_{H_2O})^6/(P_{NH_3})^4(P_{O_2})^5$]

Self-test 5G.1B Write the expression for the equilibrium constant for $2\,H_2S(g) + 3\,O_2(g) \rightleftharpoons 2\,SO_2(g) + 2\,H_2O(g)$.

Related Exercises 5G.3, 5G.4

A different measure of concentration is used when writing expressions for the equilibrium constants of reactions that involve substances other than gases. Thus, for a substance J that forms an ideal solution, the partial pressure in the expression for K is replaced by the molar concentration [J] relative to its standard value, $c° = 1\ mol \cdot L^{-1}$. Although K should be written in terms of the dimensionless ratio $[J]/c°$, it is common practice to write K in terms of [J] alone and to interpret each [J] as the molar concentration with the units "$mol \cdot L^{-1}$" struck out. Pure liquids and solids do not appear in K. So, even though $CaCO_3(s)$ and $CaO(s)$ occur in the equilibrium

$$CaCO_3(s) \rightleftharpoons CaO(s) + CO_2(g) \tag{F}$$

neither appears in the equilibrium constant, which is $K = P_{CO_2}/P°$ (or, more simply, $K = P_{CO_2}$).

These empirical rules can be summarized by introducing the concept of the **activity,** a_J, of a substance J:

In this text, activities are simply quantities introduced to facilitate writing the expression for K. In advanced work, activities are used to take into account deviations from ideal behavior.

Substance	Activity	Simplified form
ideal gas	$a_J = P_J/P°$	$a_J = P_J$
solute in a dilute solution	$a_J = [J]/c°$	$a_J = [J]$
pure solid or liquid	$a_J = 1$	$a_J = 1$

Note that all activities are pure numbers and thus are unitless. When using the simplified form of the equilibrium constant, the activity is the numerical value of the pressure in bar or the numerical value of the molar concentration in moles per liter.

The use of activities provides a simple way to write a general expression for the equilibrium constant for any reaction:

In Eq. 2a the subscript "r" in n_r indicates that the values of n_r are unitless stoichiometric coefficients (see Section 5G.4).

$$K = \left\{ \frac{(\text{activities of products})^{n_r}}{(\text{activities of reactants})^{n_r}} \right\}_{\text{equilibrium}} \tag{2a}$$

More specifically, for a generalized version of reaction E, with phases not specified:

$$a \, A + b \, B \rightleftharpoons c \, C + d \, D \qquad K = \frac{(a_C)^c (a_D)^d}{(a_A)^a (a_B)^b} \qquad \textbf{(2b)}$$

Because activities are unitless, so too is K.

A Note on Good Practice: In some cases you will see an equilibrium constant denoted K_P to remind you that it is expressed in terms of partial pressures. However, the subscript P is unnecessary because, by definition, equilibrium constants for reactions involving gases are expressed in terms of partial pressures.

Chemical equilibria with reactants and products that are all in the same phase are called **homogeneous equilibria.** Equilibria C, D, and E are homogeneous. Equilibria in systems having more than one phase are called **heterogeneous equilibria.** Equilibrium F is heterogeneous; so too is the equilibrium between water vapor and liquid water in a closed system,

$$H_2O(l) \rightleftharpoons H_2O(g) \qquad \textbf{(G)}$$

In this "reaction," there is a gas phase and a liquid phase. Likewise, the equilibrium between a solid and its saturated solution is heterogeneous:

$$Ca(OH)_2(s) \rightleftharpoons Ca^{2+}(aq) + 2 \, OH^-(aq) \qquad \textbf{(H)}$$

The equilibrium constants for heterogeneous reactions are given by the general expression in Eq. 2, but in this case remember that the activity of a pure solid or liquid is 1. For instance, for reaction H,

$$K = \frac{a_{Ca^{2+}} (a_{OH^-})^2}{\underbrace{a_{Ca(OH)_2}}_{\text{1 for a pure solid}}} = [Ca^{2+}][OH^-]^2$$

> Remember that each [J] in the expression for K represents the molar concentration of J in moles per liter with the units struck out.

It is important to note that solid calcium hydroxide must be present for the equilibrium to exist, but it does not appear in the expression for the equilibrium constant.

Self-test 5G.2A Write the equilibrium constant for the reaction used in the purification of nickel, $Ni(s) + 4 \, CO(g) \rightleftharpoons Ni(CO)_4(g)$.

[***Answer:*** $K = P_{Ni(CO)_4}/(P_{CO})^4$]

Self-test 5G.2B Write the equilibrium constant K for $P_4(s) + 5 \, O_2(g) \rightleftharpoons P_4O_{10}(s)$.

Some reactions in solution involve the solvent as a reactant or product. When the solution is very dilute, the change in solvent concentration due to the reaction is insignificant. In such cases, the solvent is treated as a pure substance and ignored when writing K. In other words,

$$\text{For a nearly pure solvent, } a_{solvent} = 1$$

A final point is that, when a reaction involves fully dissociated ionic compounds in solution, the equilibrium constant should be written for the net ionic equation by using the activity for each type of ion. The concentrations of the spectator ions cancel and so do not appear in the equilibrium expression.

Self-test 5G.3A Write the equilibrium constant for the reaction $2 \, AgNO_3(aq) + 2 \, NaOH(aq) \rightleftharpoons Ag_2O(s) + 2 \, NaNO_3(aq) + H_2O(l)$. Remember to use the net ionic equation.

[***Answer:*** $K = 1/[Ag^+]^2[OH^-]^2$]

Self-test 5G.3B Write the equilibrium constant for the equilibrium $Zn(s) + 2 \, HCl(aq) \rightleftharpoons ZnCl_2(aq) + H_2(g)$.

TABLE 5G.2 Equilibrium Constants for Various Reactions

Reaction	T/K^*	K	$K_c^\dagger$
$H_2(g) + Cl_2(g) \rightleftharpoons 2\,HCl(g)$	300	4.0×10^{31}	4.0×10^{31}
	500	4.0×10^{18}	4.0×10^{18}
	1000	5.1×10^{8}	5.1×10^{8}
$H_2(g) + Br_2(g) \rightleftharpoons 2\,HBr(g)$	300	1.9×10^{17}	1.9×10^{17}
	500	1.3×10^{10}	1.3×10^{10}
	1000	3.8×10^{4}	3.8×10^{4}
$H_2(g) + I_2(g) \rightleftharpoons 2\,HI(g)$	298	794	794
	500	160	160
	700	54	54
$2\,BrCl(g) \rightleftharpoons Br_2(g) + Cl_2(g)$	300	377	377
	500	32	32
	1000	5	5
$2\,HD(g) \rightleftharpoons H_2(g) + D_2(g)$	100	0.52	0.52
	500	0.28	0.28
	1000	0.26	0.26
$F_2(g) \rightleftharpoons 2\,F(g)$	500	3.0×10^{-11}	7.3×10^{-13}
	1000	1.0×10^{-2}	1.2×10^{-4}
	1200	0.27	2.7×10^{-3}
$Cl_2(g) \rightleftharpoons 2\,Cl(g)$	1000	1.0×10^{-5}	1.2×10^{-7}
	1200	1.7×10^{-3}	1.7×10^{-5}
$Br_2(g) \rightleftharpoons 2\,Br(g)$	1000	3.4×10^{-5}	4.1×10^{-7}
	1200	1.7×10^{-3}	1.7×10^{-5}
$I_2(g) \rightleftharpoons 2\,I(g)$	800	2.1×10^{-3}	3.1×10^{-5}
	1000	0.26	3.1×10^{-3}
	1200	6.8	6.8×10^{-2}
$N_2(g) + 3\,H_2(g) \rightleftharpoons 2\,NH_3(g)$	298	6.8×10^{5}	4.2×10^{8}
	400	41	4.5×10^{4}
	500	3.6×10^{-2}	62
$2\,SO_2(g) + O_2(g) \rightleftharpoons 2\,SO_3(g)$	298	4.0×10^{24}	9.9×10^{25}
	500	2.5×10^{10}	1.0×10^{12}
	700	3.0×10^{4}	1.7×10^{6}
$N_2O_4(g) \rightleftharpoons 2\,NO_2(g)$	298	0.15	6.1×10^{23}
	400	47.9	1.44
	500	1.7×10^{3}	41

*Three significant figures.
$^\dagger K_c$ is the equilibrium constant in terms of molar concentrations of gases (Topic 5H).

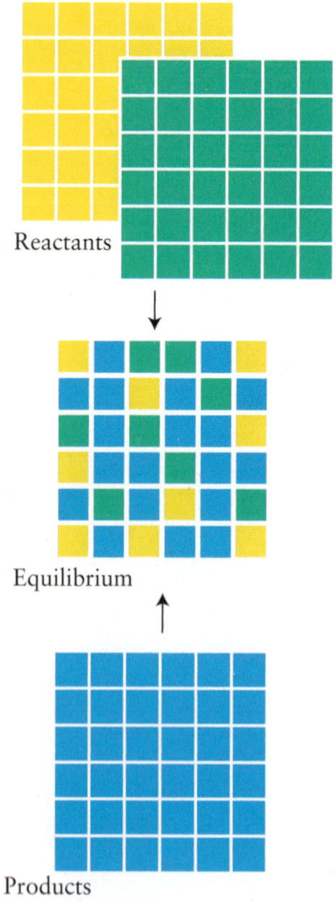

Reactants

Equilibrium

Products

FIGURE 5G.3 Regardless of whether a reaction begins with pure reactants or pure products, a reaction mixture always tends toward a mixture of reactants and products that has a composition in accord with the equilibrium constant for the reaction at the temperature of the experiment.

Each reaction has its own characteristic equilibrium constant, with a value that can be changed only by varying the temperature (**TABLE 5G.2**). The extraordinary empirical result, which is justified in the next two sections, is that, *regardless of the initial composition of a reaction mixture, the composition tends to adjust until the activities give rise to the characteristic value of K for the reaction and temperature* (**FIG. 5G.3**).

The equilibrium composition of a reaction mixture is described by the equilibrium constant, which is equal to the activities of the products (raised to powers equal to their stoichiometric coefficients in the balanced chemical equation for the reaction) divided by the activities of the reactants (also raised to powers equal to their stoichiometric coefficients).

5G.3 The Origin of Equilibrium Constants

Why do equilibrium constants have the form they do? There are two approaches to the answer, one kinetic and the other thermodynamic.

How Is That Explained . . .

. . . using kinetics?

The fact that a reaction reaches equilibrium when the forward and reverse rates of reaction become equal suggests that the form of the equilibrium constant ought to be related to the rate constants of those processes. A simple case to confirm this approach is a reaction of the form A + B $\rightleftharpoons$ P which is known to be second-order in the forward direction and first-order in the reverse direction:

$$A + B \longrightarrow P \quad Rate = k_r[A][B]$$
$$P \longrightarrow A + B \quad Rate = k_r'[P]$$

where k_r and k_r' are the rate constants for the forward and reverse reactions (Topic 7A). The rates are equal when $k_r[A][B] = k_r'[P]$, which can be rearranged into

$$\frac{[P]}{[A][B]} = \frac{k_r}{k_r'}$$

The ratio of rate constants is itself a constant, and the form of the ratio of concentrations on the left will be recognized as the form implied by the law of mass action. That is, the equilibrium constant is related to the two rate constants by

$$K = \frac{k_r}{k_r'}$$

and its form (as a ratio of concentrations) is such as to ensure that the forward and reverse reactions are occurring at the same rate.

A problem with this approach is that, as stressed in Topic 7A, except in special cases a rate law cannot be written down simply by examining the form of the chemical equation for the reaction. In contrast, the expression for the equilibrium constant can be written in that way. To resolve this paradox, it is necessary to think about the mechanism of the reaction. Then, provided each step is at equilibrium and the elementary chemical equations of all the steps add together to give the equation for the overall reaction, it turns out that the expression for the equilibrium constant *can* be written down from the overall equation. The details are set out in Topic 7C.

. . . using thermodynamics?

The thermodynamic explanation of the form of equilibrium constants is based on the Gibbs free energy (Topic 4J) and the criterion of equilibrium that at constant temperature and pressure $\Delta G = 0$ (**FIG. 5G.4**). The discussion that follows in Section 5G.4 is based on the following relation between the composition and the molar Gibbs free energy:

$$G_m(J) = G_m°(J) + RT \ln a_J$$

where a_J is the activity of J in the mixture at any stage of the reaction. That discussion goes on to establish that the "reaction Gibbs energy," the difference between the molar free energies of the products and reactants at any stage of the reaction is

$$\Delta G_r = \Delta G_r° + RT \ln Q$$

with the **reaction quotient** Q defined as

$$Q = \frac{(a_C)^c(a_D)^d}{(a_A)^a(a_B)^b}$$

At equilibrium, $\Delta G_r = 0$, and the activities have their equilibrium values, so

$$0 = \Delta G_r° + RT \ln Q_{equilibrium}$$

But the value of $Q_{equilibrium}$ is in fact the equilibrium constant, K, of the reaction. That is, in the thermodynamic approach, K has the form that guarantees that the Gibbs free energy of the reaction is zero, and the composition has no tendency to change in either direction.

The subscript "r" on ΔG_r indicates that the "molar" convention is being used and that the units of ΔG_r are kilojoules per mole (kJ·mol^{-1}, see the following section).

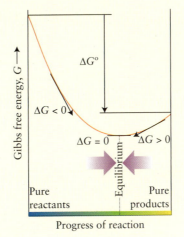

FIGURE 5G.4 The variation of Gibbs free energy of a reaction mixture with composition. A reaction mixture has a spontaneous tendency to change in the direction of decreasing Gibbs free energy. Note that ΔG is the slope of the line at each composition, whereas $\Delta G°$ is the difference between the standard Gibbs free energies of the pure reactants and those of the pure products.

5G.4 The Thermodynamic Description of Equilibrium

The thermodynamic explanation of the form of equilibrium constants is a very important application of thermodynamics. The explanation is based on the Gibbs free energy (Topic 4J) and the criterion of equilibrium that at constant temperature and pressure $\Delta G = 0$

(Fig. 5G.4). The value of ΔG at a particular stage of the reaction is the difference in the molar Gibbs free energies of the products and the reactants *at the partial pressures or concentrations that they have at that stage*, weighted by the stoichiometric coefficients interpreted as amounts in moles:

$$\Delta G = \sum nG_m(\text{products}) - \sum nG_m(\text{reactants}) \qquad \text{Units: kilojoules} \qquad \textbf{(3a)}$$

(This is Eq. 6 of Topic 4J.) It will sometimes be useful to interpret the n that appear in this equation as pure numbers (rather than amounts in moles). To signal that this "molar" convention ("molar" because the units of ΔG then become kilojoules per mole), is being used, a subscript r will be attached to ΔG and n (r for reaction) and write

$$\Delta G_r = \sum n_r G_m(\text{products}) - \sum n_r G_m(\text{reactants}) \qquad \text{Units: kilojoules per mole} \qquad \textbf{(3b)}$$

For instance, if $n = 2$ mol for a substance, then $n_r = 2$. Exactly when and why this convention is necessary is explained later in the "How Is That Done" section.

The molar Gibbs free energies G_m of each reactant and product changes during the course of the reaction because, when only reactants are present, each molecule is surrounded by reactant molecules, but, as products form, the environment of each molecule changes too. Because all the G_m change, ΔG changes as the reaction progresses. It is shown in Topic 5A that the molar Gibbs free energy of an ideal gas J is related to its partial pressure, P_J, by

$$G_m(J) = G_m°(J) + RT \ln \frac{P_J}{P°} \qquad \textbf{(4a)}$$

Thermodynamic arguments (which are not reproduced here) show that a similar expression applies to solutes and pure substances. In each case, the molar Gibbs free energy of a substance J can be written as

$$G_m(J) = G_m°(J) + RT \ln a_J \qquad \textbf{(4b)}$$

with the activity defined in Section 5G.2. The value of ΔG at any stage of a reaction can be expressed in terms of the composition of the reaction mixture at that stage.

How Is That Done?

To keep the units straight, use the "molar" convention for this calculation. To find an expression for ΔG_r for the reaction $a\,A + b\,B \rightleftharpoons c\,C + d\,D$, substitute Eq. 4b for each substance into Eq. 3b:

$$\Delta G_r = \overbrace{\{cG_m(C) + dG_m(D)\}}^{\text{products}} - \overbrace{\{aG_m(A) + bG_m(B)\}}^{\text{reactants}}$$

$$= \{c[G_m°(C) + RT \ln a_C] + d[G_m°(D) + RT \ln a_D]\}$$
$$- \{a[G_m°(A) + RT \ln a_A] + b[G_m°(B) + RT \ln a_B]\}$$

$$= \overbrace{\{cG_m°(C) + dG_m°(D)\} - \{aG_m°(A) + bG_m°(B)\}}^{\Delta G_r°}$$
$$+ RT\{(c \ln a_C + d \ln a_D) - (a \ln a_A + b \ln a_B)\}$$

As indicated, the combination of the first four terms in the final equation is the standard Gibbs free energy of reaction, $\Delta G_r°$:

$$\Delta G_r° = \{cG_m°(C) + dG_m°(D)\} - \{aG_m°(A) + bG_m°(B)\}$$

Therefore,

$$\Delta G_r = \Delta G_r° + RT\{(c \ln a_C + d \ln a_D) - (a \ln a_A + b \ln a_B)\}$$

Now tidy up the four logarithmic terms:

$$(c \ln a_C + d \ln a_D) - (a \ln a_A + b \ln a_B)$$

$$\overset{y \ln x = \ln x^y}{=} (\ln a_C{}^c + \ln a_D{}^d) - (\ln a_A{}^a + \ln a_B{}^b)$$

$$\overset{\ln x + \ln y = \ln xy}{=} \ln a_C{}^c a_D{}^d - \ln a_A{}^a a_B{}^b$$

$$\overset{\ln x - \ln y = \ln(x/y)}{=} \ln \frac{(a_C)^c (a_D)^d}{(a_A)^a (a_B)^b}$$

Bring this work together and obtain

$$\Delta G_r = \Delta G_r^\circ + RT \ln \frac{(a_C)^c (a_D)^d}{(a_A)^a (a_B)^b}$$

A Note on Good Practice: Notice that by using the "molar" convention, the units match: RT is a molar energy (in joules per mole), and so too are the two Gibbs free energy terms. You should always use the molar convention when RT appears in an equation and is not multiplied by an amount in moles.

The expression just derived may be written as

$$\Delta G_r = \Delta G_r^\circ + RT \ln Q \qquad (5)$$

with the **reaction quotient** Q defined as

$$Q = \frac{(a_C)^c (a_D)^d}{(a_A)^a (a_B)^b} \qquad (6)$$

Equations 5 and 6 show how the Gibbs free energy of reaction varies with the activities (the partial pressures of gases or molar concentrations of solutes) of the reactants and products. The expression for Q has the same form as the expression for K, but the activities refer to *any* stage of the reaction.

EXAMPLE 5G.2 Calculating the Gibbs free energy of reaction from the reaction quotient

The oxidation of SO_2 to SO_3 is one of the reactions involved in the formation of acid rain. If you want to predict the spontaneous direction of the reaction for a specific mixture of the gases, you need to calculate the reaction quotient under those conditions. The standard Gibbs free energy of reaction for $2\ SO_2(g) + O_2(g) \rightarrow 2\ SO_3(g)$ is $\Delta G_r^\circ = -141.74\ kJ \cdot mol^{-1}$ at 25.00 °C. (a) What is the Gibbs free energy of reaction when the partial pressure of each gas is 100. bar? (b) What is the spontaneous direction of the reaction under these conditions?

PLAN Calculate the reaction quotient and substitute it and the standard Gibbs free energy of reaction into Eq. 5. If $\Delta G_r < 0$, the forward reaction is spontaneous at the given temperature and composition. If $\Delta G_r > 0$, the reverse reaction is spontaneous at the given temperature and composition. If $\Delta G_r = 0$, there is no tendency to react in either direction; the reaction is at equilibrium. At 298.15 K, $RT = 2.479\ kJ \cdot mol^{-1}$.

SOLVE

(a) From $Q = (a_{SO_3})^2 / (a_{SO_2})^2 (a_{O_2}) = (P_{SO_3})^2 / (P_{SO_2})^2 (P_{O_2})$,

$$Q = \frac{(100.)^2}{(100.)^2 \times (100.)} = 1.00 \times 10^{-2}$$

From $\Delta G_r = \Delta G_r^\circ + RT \ln Q$,

$$\Delta G_r = \overbrace{-141.74\ kJ \cdot mol^{-1}}^{\Delta G_r^\circ} + \overbrace{(2.479\ kJ \cdot mol^{-1})}^{RT} \ln\overbrace{(1.00 \times 10^{-2})}^{Q}$$
$$= -153.16\ kJ \cdot mol^{-1}$$

(b) Because the Gibbs free energy of reaction is negative, the formation of products is spontaneous at this composition and temperature.

Self-test 5G.4A The standard Gibbs free energy of reaction for $H_2(g) + I_2(g) \rightarrow 2\ HI(g)$ is $\Delta G_r^\circ = -21.1\ kJ \cdot mol^{-1}$ at 500. K (at which temperature, $RT = 4.16\ kJ \cdot mol^{-1}$). What is the value of ΔG_r at 500. K when the partial pressures of the gases are $P_{H_2} = 1.5$ bar, $P_{I_2} = 0.88$ bar, and $P_{HI} = 0.065$ bar? What is the spontaneous direction of the reaction?

[*Answer:* $-45\ kJ \cdot mol^{-1}$; toward products]

Self-test 5G.4B The standard Gibbs free energy of reaction for $N_2O_4(g) \rightarrow 2\ NO_2(g)$ is $\Delta G_r^\circ = +4.73\ kJ \cdot mol^{-1}$ at 298 K. What is the value of ΔG_r when the partial pressures of the gases are $P_{N_2O_4} = 0.80$ bar and $P_{N_2O_2} = 2.10$ bar? What is the spontaneous direction of the reaction?

Related Exercises 5G.13–5G.16

You have now come to the most important point in this Topic. At equilibrium, the activities (the partial pressures or molar concentrations) of all the substances taking part in the reaction have their equilibrium values. At this point, the expression for Q (in which the activities now have their equilibrium values) has become the equilibrium constant, K, of the reaction. That is, at equilibrium, $Q = K$. Thermodynamics has explained the puzzling form of K: it is a direct consequence of Eq. 4b, which shows how the Gibbs free energy of a substance depends on its composition, and K is simply the value of Q when all the species have their equilibrium values.

Now take another important step. You know that $\Delta G_r = 0$ at equilibrium, and you have just seen that $Q = K$ at equilibrium. It follows from Eq. 5 that, at equilibrium,

$$0 = \Delta G_r^\circ + RT \ln K$$

and therefore that

$$\Delta G_r^\circ = -RT \ln K \tag{7}$$

This fundamentally important equation links thermodynamic quantities—which are widely available from tables of thermodynamic data—and the composition of a system at equilibrium. Note that:

- If ΔG_r° is negative, then $\ln K$ must be positive and therefore $K > 1$; products are favored at equilibrium.
- If ΔG_r° is positive, then $\ln K$ must be negative and therefore $K < 1$; reactants are favored at equilibrium.

A Note on Good Practice: Always write Eq. 7 with the standard-state symbol. Note too that, for the units to match on both sides (joules per mole), and as signaled by the presence of RT not multiplied by an amount in moles, the "molar" convention (subscript r) for the standard Gibbs free energy is being used.

THINKING POINT

A catalyst provides a lower-energy pathway between reactants and products. Will adding a catalyst to a reaction change the equilibrium constant?

EXAMPLE 5G.3   Predicting the value of K from the standard Gibbs free energy of reaction

Hydrogen iodide, HI, is used as a reagent in organic chemistry to transform primary alcohols into alkyl iodides. Suppose you are a chemist using HI; you would need to understand the equilibrium behavior of HI to maximize your yield of products. At 25.00 °C, the standard Gibbs free energy of reaction for $\frac{1}{2}H_2(g) + \frac{1}{2}I_2(s) \rightarrow HI(g)$ is $+1.70$ kJ·mol^{-1}; calculate the equilibrium constant for this reaction.

ANTICIPATE Because the standard Gibbs free energy of the reaction is positive, you should expect the equilibrium constant to be less than 1.

PLAN Use Eq. 7 with the temperature in kelvins.

SOLVE

From $\Delta G_r^\circ = -RT \ln K$ in the form $\ln K = -\Delta G_r^\circ/RT$, first converting the units of ΔG_r° to joules per mole and temperature to kelvins,

$$\ln K = -\frac{1.70 \times 10^3 \text{ J·mol}^{-1}}{(8.3145 \text{ J·K}^{-1}\text{·mol}^{-1}) \times (298.15 \text{ K})}$$

$$= -\frac{1.70 \times 10^3}{8.3145 \times 298.15} = -0.685\ldots$$

Taking the inverse logarithm ($e^{\ln x} = x$),

$$K = e^{-0.685\ldots} = 0.50$$

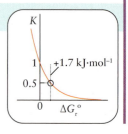

EVALUATE As expected, the equilibrium constant is less than 1.

> **A Note on Good Practice:** Exponential functions (inverse natural logarithms, e^x) are very sensitive to the value of x, so carry out all the arithmetic in one step to avoid rounding errors.

Self-test 5G.5A Use the thermodynamic data in Appendix 2A to calculate K from the value of $\Delta G_r°$ for the reaction $N_2O_4(g) \rightarrow 2\,NO_2(g)$ at 298 K.

[**Answer:** $K = 0.15$]

Self-test 5G.5B Use the thermodynamic data in Appendix 2A to calculate K from the value of $\Delta G_r°$ for the reaction $2\,NO(g) + O_2(g) \rightarrow 2\,NO_2(g)$ at 298 K.

Related Exercises 5G.21, 5G.22

You can now begin to acquire some insight into why some reactions have large equilibrium constants and others have small ones. It follows from $\Delta G_r° = \Delta H_r° - T\Delta S_r°$ and $\Delta G_r° = -RT \ln K$ in the form $\ln K = -\Delta G_r°/RT$ that

$$\ln K = -\frac{\Delta G_r°}{RT} = -\frac{\Delta H_r°}{RT} + \frac{\Delta S_r°}{R}$$

On taking inverse logarithms of both sides and using $e^{x+y} = e^x e^y$, this relation becomes

$$K = e^{-\Delta H_r°/RT + \Delta S_r°/R} = e^{-\Delta H_r°/RT}\, e^{\Delta S_r°/R} \qquad (8)$$

You can now see that K can be expected to be small if $\Delta H_r°$ is positive (because e^{-x} is small if x is positive). An endothermic reaction is therefore likely to have $K < 1$ and is unlikely to form much product. Only if $\Delta S_r°$ is large and positive, so the factor $e^{\Delta S_r°/R}$ is large, can you expect $K > 1$ for an endothermic reaction. Conversely, if a reaction is strongly exothermic, then $\Delta H_r°$ is large and negative, therefore you can expect $K > 1$ and the products to be favored. In other words, you can expect strongly exothermic reactions to go to completion.

> *The reaction quotient, Q, has the same form as K, the equilibrium constant, except that Q uses the activities evaluated at an arbitrary stage of the reaction. The equilibrium constant is related to the standard Gibbs free energy of reaction by $\Delta G_r° = -RT \ln K$.*

What have you learned in this Topic?

You have learned that reactions tend to proceed until they reach a composition corresponding to minimum Gibbs free energy. From a kinetic viewpoint, you have seen that an equilibrium constant represents the condition for all the steps in a reaction mechanism being at equilibrium. From a thermodynamic viewpoint, you have seen that the equilibrium constant is the value of the reaction quotient when the composition corresponds to equilibrium. You have encountered the term "activity" as a succinct way of writing reaction quotients and equilibrium constants.

The skills you have mastered are the ability to:

☐ **1.** Distinguish homogeneous and heterogeneous equilibria and write equilibrium constants for both types of reactions from a balanced equation (Example 5G.1).

☐ **2.** Relate the Gibbs free energy of reaction to the composition of the reaction mixture (Example 5G.2).

☐ **3.** Calculate an equilibrium constant from a standard Gibbs free energy (Example 5G.3).

Topic 5G Exercises

5G.1 State whether the following statements are true or false. If false, explain why.
(a) A reaction stops when equilibrium is reached.
(b) An equilibrium reaction is not affected by increasing the concentrations of products.
(c) If one starts with a higher pressure of reactant, the equilibrium constant will be larger.

(d) If one starts with higher concentrations of reactants, the equilibrium concentrations of the products will be larger.

5G.2 Determine whether the following statements are true or false. If false, explain why.
(a) In an equilibrium reaction, the reverse reaction begins only when all reactants have been converted to products.

(b) The equilibrium concentrations will be the same whether one starts with pure reactants or pure products.

(c) The rates of the forward and reverse reactions are the same at equilibrium.

(d) If the Gibbs free energy is greater than the standard Gibbs free energy of reaction, the reaction proceeds forward to equilibrium.

5G.3 Write the expression for K for each of the following reactions:
(a) $2\ C_2H_4(g) + O_2(g) + 4\ HCl(g) \rightleftharpoons 2\ C_2H_4Cl_2(g) + 2\ H_2O(g)$
(b) $4\ NH_3(g) + 6\ NO(g) \rightleftharpoons 7\ N_2(g) + 6\ H_2O(g)$

5G.4 Write the expression for K for each of the following reactions:
(a) $Br_2(g) + 3\ F_2(g) \rightleftharpoons 2\ BrF_3(g)$
(b) $4\ NH_3(g) + 3\ O_2(g) \rightleftharpoons 2\ N_2(g) + 6\ H_2O(g)$

5G.5 The following flasks show the dissociation of a diatomic molecule, X_2, over time. (a) Which flask represents the point in time at which the reaction has reached equilibrium? (b) What percentage of the X_2 molecules has decomposed at equilibrium? (c) Assuming that the initial pressure of X_2 was 0.10 bar, calculate the value of K for the decomposition.

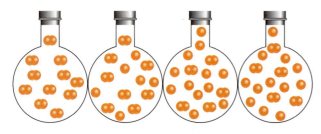

5G.6 The flask below contains atoms of A (red) and B (yellow). They react as follows $2\ A(g) + B(g) \rightarrow A_2B(g)$, with $K = 0.25$. Draw a picture of the flask and its contents after the reaction has reached equilibrium.

5G.7 Balance the following equations using the smallest whole-number coefficients, then write the expression for K for each reaction:
(a) $CH_4(g) + O_2(g) \rightleftharpoons CO_2(g) + H_2O(g)$
(b) $I_2(g) + F_2(g) \rightleftharpoons IF_5(g)$
(c) $NO_2(g) + F_2(g) \rightleftharpoons FNO_2(g)$

5G.8 Balance the following equations using the smallest whole-number coefficients, then write the expression for K for each reaction:
(a) $CH_4(g) + Cl_2(g) \rightleftharpoons CH_2Cl_2(g) + HCl(g)$
(b) $NH_3(g) + ClF_3(g) \rightleftharpoons HF(g) + N_2(g) + Cl_2(g)$
(c) $N_2(g) + O_2(g) \rightleftharpoons N_2O_5(g)$

5G.9 A sample of ozone, O_3, amounting to 0.10 mol, is placed in a sealed container of volume 1.0 L and the reaction $2\ O_3(g) \rightarrow 3\ O_2(g)$ is allowed to reach equilibrium. Then 0.50 mol O_3 is placed in a second container of volume 1.0 L at the same temperature and allowed to reach equilibrium. Without doing any calculations, predict which of the following will be different in the two containers at equilibrium. Which will be the same? (a) Amount of O_2; (b) partial pressure of O_2; (c) the ratio P_{O_2}/P_{O_3}; (d) the ratio $(P_{O_2})^3/(P_{O_3})^2$; (e) the ratio $(P_{O_3})^2/(P_{O_2})^3$. Explain each of your answers.

5G.10 A mixture consisting of 0.10 mol $H_2(g)$ and 0.10 mol $Br_2(g)$ is placed in a container of volume 2.0 L. The reaction $H_2(g) + Br_2(g) \rightarrow 2\ HBr(g)$ is allowed to come to equilibrium. Then 0.20 mol HBr is placed into a second sealed container of volume 2.0 L at the same temperature and allowed to reach equilibrium with H_2 and Br_2. Which of the following will be different in the two containers at equilibrium? Which will be the same? (a) Amount of Br_2; (b) partial pressure of H_2; (c) the ratio $P_{HBr}/P_{H_2}P_{Br_2}$; (d) the ratio P_{HBr}/P_{Br_2}; (e) the ratio $(P_{HBr})^2/P_{H_2}P_{Br_2}$; (f) the total pressure in the container. Explain each of your answers.

5G.11 Write the reaction quotient Q for
(a) $2\ BCl_3(g) + 2\ Hg(l) \rightarrow B_2Cl_4(s) + Hg_2Cl_2(s)$
(b) $P_4S_{10}(s) + 16\ H_2O(l) \rightarrow 4\ H_3PO_4(aq) + 10\ H_2S(aq)$
(c) $Br_2(g) + 3\ F_2(g) \rightarrow 2\ BrF_3(g)$

5G.12 Write the reaction quotient Q for
(a) $NCl_3(g) + 3\ H_2O(l) \rightarrow NH_3(g) + 3\ HClO(aq)$
(b) $P_4(s) + 3\ KOH(aq) + 3\ H_2O(l) \rightarrow PH_3(aq) + 3\ KH_2PO_2(aq)$
(c) $CO_3^{2-}(aq) + 2\ H_3O^+(aq) \rightarrow CO_2(g) + 3\ H_2O(l)$

5G.13 (a) Calculate the reaction Gibbs free energy of $I_2(g) \rightarrow 2\ I(g)$ at 1200. K ($K = 6.8$) when the partial pressures of I_2 and I are 0.13 bar and 0.98 bar, respectively. (b) Indicate whether this reaction mixture is likely to form reactants, is likely to form products, or is at equilibrium.

5G.14 Calculate the reaction Gibbs free energy of $PCl_3(g) + Cl_2(g) \rightarrow PCl_5(g)$ at 230 °C when the partial pressures of PCl_3, Cl_2, and PCl_5 are 0.35 bar, 0.45 bar, and 1.02 bar, respectively. What is the spontaneous direction of change, given that $K = 49$ at 230 °C?

5G.15 (a) Calculate the reaction Gibbs free energy of $N_2(g) + 3\ H_2(g) \rightarrow 2\ NH_3(g)$ when the partial pressures of N_2, H_2, and NH_3 are 4.2 bar, 1.8 bar, and 21 bar, respectively, and the temperature is 400. K. For this reaction, $K = 41$ at 400. K. (b) Indicate whether this reaction mixture is likely to form reactants, is likely to form products, or is at equilibrium.

5G.16 (a) Calculate the reaction Gibbs free energy of $H_2(g) + I_2(g) \rightarrow 2\ HI(g)$ at 700. K when the partial pressures of H_2, I_2, and HI are 0.35 bar, 0.18 bar, and 2.85 bar, respectively. For this reaction, $K = 54$ at 700. K. (b) Indicate whether this reaction mixture is likely to form reactants, is likely to form products, or is at equilibrium.

5G.17 Depict the progress of the reaction graphically (as in Fig. 5G.1) for the reaction in Exercise 5G.13.

5G.18 Depict the progress of the reaction graphically (as in Fig. 5G.1) for the reaction in Exercise 5G.13 if the starting partial pressures of I_2 and I are 0.75 bar and 0.12 bar, respectively.

5G.19 Calculate the standard Gibbs free energy of each of the following reactions:
(a) $I_2(g) \rightleftharpoons 2\,I(g)$, $K = 6.8$ at 1200. K
(b) $Ag_2CrO_4(s) \rightleftharpoons 2\,Ag^+(aq) + CrO_4^{2-}(aq)$,
$$K = 1.1 \times 10^{-12} \text{ at } 298 \text{ K}$$

5G.20 Calculate the standard Gibbs free energy for each of the following reactions:
(a) $H_2(g) + I_2(g) \rightleftharpoons 2\,HI(g)$, $K = 54$ at 700. K
(b) $CCl_3COOH(aq) + H_2O(l) \rightleftharpoons$
$$CCl_3CO_2^-(aq) + H_3O^+(aq), \quad K = 0.30 \text{ at } 298 \text{ K}$$

5G.21 Calculate the equilibrium constant at 25 °C for each of the following reactions, by using data in Appendix 2A:
(a) the combustion of hydrogen: $2\,H_2(g) + O_2(g) \rightleftharpoons 2\,H_2O(g)$
(b) the oxidation of carbon monoxide: $2\,CO(g) + O_2(g) \rightleftharpoons 2\,CO_2(g)$

(c) the decomposition of limestone: $CaCO_3(s) \rightleftharpoons CaO(s) + CO_2(g)$

5G.22 Calculate the equilibrium constant at 25 °C for each of the following reactions, by using data in Appendix 2A:
(a) the synthesis of trichloromethane (chloroform) from natural gas (methane). $\Delta G_f^\circ(CH_3Cl, g) = 48.5 \text{ kJ·mol}^{-1}$.

$$CH_4(g) + Cl_2(g) \rightleftharpoons CH_3Cl(g) + HCl(g)$$

(b) the hydrogenation of acetylene to ethane:

$$C_2H_2(g) + 2\,H_2(g) \rightleftharpoons C_2H_6(g)$$

(c) the final step in the industrial production of nitric acid:

$$3\,NO_2(g) + H_2O(l) \rightleftharpoons 2\,HNO_3(aq) + NO(g)$$

(d) the reaction of hydrazine and oxygen in a rocket:

$$N_2H_4(l) + O_2(g) \rightleftharpoons N_2(g) + 2\,H_2O(l)$$

Topic 5H Alternative Forms of the Equilibrium Constant

5H.1 Multiples of the Chemical Equation

5H.2 Composite Equations

5H.3 Molar Concentrations of Gases

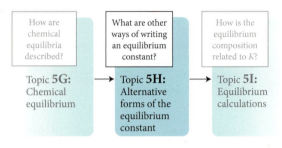

How are chemical equilibria described?

What are other ways of writing an equilibrium constant?

How is the equilibrium composition related to K?

Topic **5G:** Chemical equilibrium → Topic **5H:** Alternative forms of the equilibrium constant → Topic **5I:** Equilibrium calculations

---

Why Do You Need to Know This Material? A chemical reaction can be expressed by different chemical equations, and the equilibrium constants change accordingly. The equilibrium constant of a gas-phase reaction is expressed in terms of partial pressures; however, in some cases, it is necessary to know the molar concentrations of the gases instead.

What Do You Need to Know Already? You need to know how to write the expression for an equilibrium constant in terms of activities (Topic 5G) and how to use the ideal gas law (Topic 3B) to relate partial pressure to molar concentration.

The dynamic equilibrium achieved in the synthesis of ammonia can be expressed in a variety of ways, such as $N_2(g) + 3 H_2(g) \rightleftharpoons 2 NH_3(g)$ or $2 NH_3(g) \rightleftharpoons N_2(g) + 3 H_2(g)$ and each version will give rise to a different value of K. There may be a good reason to choose one version over another, and then it is necessary to convert a tabulated value for one version into a value for the version required. Another issue arises when molar concentrations of gases are used, because the thermodynamic procedure for expressing an equilibrium constant specifies that K is written in terms of the partial pressure of any gas that occurs in the reaction, but practical considerations often mean that the molar concentration of the gas is needed. How then is the equilibrium constant expressed and related to the thermodynamic version?

5H.1 Multiples of the Chemical Equation

The powers to which the activities are raised in the expression for an equilibrium constant must match the stoichiometric coefficients in the chemical equation, which is normally written with the smallest whole numbers for coefficients. Therefore, if the stoichiometric coefficients in a chemical equation are multiplied through by a factor, the equilibrium constant is adjusted accordingly. For example, at 500 K,

$$H_2(g) + I_2(g) \rightleftharpoons 2 HI(g) \qquad K_1 = \frac{(P_{HI})^2}{P_{H_2}P_{I_2}} = 160$$

If the chemical equation is multiplied through by 2, the equilibrium constant becomes

$$2 H_2(g) + 2 I_2(g) \rightleftharpoons 4 HI(g) \qquad K_2 = \frac{(P_{HI})^4}{(P_{H_2})^2(P_{I_2})^2} = K_1^2 = 160^2 = 2.96 \times 10^4$$

- In general, if a chemical equation is multiplied through by a factor N, K is raised to the Nth power.

Now suppose the original equation for the reaction is reversed:

$$2 HI(g) \rightleftharpoons H_2(g) + I_2(g)$$

This equation still describes the same equilibrium, but the equilibrium constant becomes

$$K_3 = \frac{P_{H_2}P_{I_2}}{(P_{HI})^2} = \frac{1}{K_1} = \frac{1}{160} = 0.0063$$

- In general the equilibrium constant for an equilibrium written in one direction is the reciprocal ($1/K$) of the equilibrium constant for the chemical equation written in the opposite direction.

These relations are summarized in **TABLE 5H.1**.

A Note on Good Practice: As these examples have shown, it is important to specify the chemical equation to which the equilibrium constant applies.

The expression for an equilibrium constant must reflect the manner in which the chemical equation is written, as summarized in Table 5H.1.

TABLE 5H.1	Relations Between Equilibrium Constants*
Chemical equation	**Equilibrium constant**
$a A + b B \rightleftharpoons c C + d D$	K_1
$c C + d D \rightleftharpoons a A + b B$	$K_2 = 1/K_1 = K_1^{-1}$
$N(a A + b B \rightleftharpoons c C + d D)$	$K_3 = K_1^N$

*For a reaction that can be expressed as the sum of other reactions, the equilibrium constant is the product of the equilibrium constants of the component reactions. Thus, for $A \rightleftharpoons B$ (K_1) and $B \rightleftharpoons C$ (K_2), then for $A \rightleftharpoons C$, $K = K_1 K_2$.

5H.2 Composite Equations

In some cases a chemical equation can be expressed as the sum of two or more chemical equations. For example, consider the three gas-phase reactions

$$2\,P(g) + 3\,Cl_2(g) \rightleftharpoons 2\,PCl_3(g) \qquad K_1 = \frac{(P_{PCl_3})^2}{(P_P)^2(P_{Cl_2})^3}$$

$$PCl_3(g) + Cl_2(g) \rightleftharpoons PCl_5(g) \qquad K_2 = \frac{P_{PCl_5}}{P_{PCl_3}P_{Cl_2}}$$

$$2\,P(g) + 5\,Cl_2(g) \rightleftharpoons 2\,PCl_5(g) \qquad K_3 = \frac{(P_{PCl_5})^2}{(P_P)^2(P_{Cl_2})^5}$$

The third reaction is the following sum of the first two reactions (with the second multiplied by a factor of 2):

$$2\,P(g) + 3\,Cl_2(g) \rightleftharpoons 2\,PCl_3(g)$$
$$\underline{2\,PCl_3(g) + 2\,Cl_2(g) \rightleftharpoons 2\,PCl_5(g)}$$
$$2\,P(g) + 5\,Cl_2(g) \rightleftharpoons 2\,PCl_5(g)$$

Remember, when combining chemical equations, species that appear as both reactants and products are canceled.

and the equilibrium constant, K_3, of the overall reaction can be written as the product of the equilibrium constants for the two reactions which, added together, give the overall reaction:

$$K_3 = \frac{(P_{PCl_5})^2}{(P_P)^2(P_{Cl_2})^5} = \overbrace{\frac{(P_{PCl_3})^2}{(P_P)^2(P_{Cl_2})^3}}^{K_1} \times \overbrace{\frac{(P_{PCl_5})^2}{(P_{PCl_3})^2(P_{Cl_2})^2}}^{K_2^{\,2}} = K_1K_2^2$$

Self-test 5H.1A At 500. K, K for $H_2(g) + D_2(g) \rightleftharpoons 2\,HD(g)$ is 3.6. What is the value of K for $4\,HD(g) \rightleftharpoons 2\,H_2(g) + 2\,D_2(g)$?

[**Answer:** 0.077]

Self-test 5H.1B At a certain temperature, K for $F_2(g) \rightleftharpoons 2\,F(g)$ is 7.3×10^{-13}. What is the value of K for $\frac{1}{2}F_2(g) \rightleftharpoons F(g)$?

The equilibrium constant for an overall reaction is the product of the equilibrium constants for its component reactions.

5H.3 Molar Concentrations of Gases

As established in Topic 5G, the equilibrium constant is defined in terms of activities, and the activities are interpreted in terms of the partial pressures of gases or molar concentrations of solutes:

$$a\,A + b\,B \rightleftharpoons c\,C + d\,D \qquad K = \frac{(a_C)^c(a_D)^d}{(a_A)^a(a_B)^b} \tag{1}$$

Gases *always* appear in K as the numerical values of their partial pressures in bar, and solutes in a condensed phase always appear as the numerical values of their molar concentrations in moles per liter. Often, however, especially in areas such as kinetics and atmospheric chemistry, gas-phase equilibria need to be discussed in terms of molar concentrations rather than partial pressures. To do so, the equilibrium constant K_c is introduced as

$$K_c = \frac{[C]^c[D]^d}{[A]^a[B]^b} \tag{2}$$

with each molar concentration raised to a power equal to the stoichiometric coefficient of the species in the chemical equation. To keep the notation clear, as in Topic 5G, $[J]/c^\circ$

has been replaced by [J], which represents the numerical value of the molar concentration of the gas J. For example, for the ammonia synthesis equilibrium,

$$N_2(g) + 3 H_2(g) \rightleftharpoons 2 NH_3(g) \qquad K_c = \frac{[NH_3]^2}{[N_2][H_2]^3} \tag{3}$$

You are free to choose either K or K_c to report the equilibrium constant of a reaction. However, it is important to remember that calculations of an equilibrium constant from thermodynamic tables of data (standard Gibbs free energies of formation, for instance) give K, not K_c. In some cases, you need to know K_c after you have calculated K from thermodynamic data, and so you need to be able to convert between these two constants.

How Is That Done?

The overall strategy for finding the relation between K and K_c is to replace the partial pressures that appear in K by the molar concentrations and thereby generate K_c. For this calculation, activities are written out in full as P_J/P° and $[J]/c^\circ$ in order to keep track of units, with $P^\circ = 1$ bar and $c^\circ = 1$ mol·L^{-1}.

The starting point is to assume that the gases are ideal and then write out the full form of Eq. 1:

$$K = \frac{(P_C/P^\circ)^c(P_D/P^\circ)^d}{(P_A/P^\circ)^a(P_B/P^\circ)^b}$$

The molar concentration of a gas J is $[J] = n_J/V$. For an ideal gas, the ideal gas law, $P_J V = n_J RT$, can be rearranged to show concentration explicitly:

$$P_J = \frac{n_J RT}{V} = RT \times \overbrace{\left(\frac{n_J}{V}\right)}^{[J]} = RT[J]$$

When this expression is substituted for each gas in the expression for K, it becomes

$$K = \frac{(RT[C]/P^\circ)^c(RT[D]/P^\circ)^d}{(RT[A]/P^\circ)^a(RT[B]/P^\circ)^b} = \left(\frac{RT}{P^\circ}\right)^{(c+d)-(a+b)}\frac{[C]^c[D]^d}{[A]^a[B]^b}$$

At this point recognize that K_c, Eq. 2, in fully expressed form (with the c° shown) is

$$K_c = \frac{([C]/c^\circ)^c([D]/c^\circ)^d}{([A]/c^\circ)^a([B]/c^\circ)^b} = \frac{(c^\circ)^{a+b}[C]^c[D]^d}{(c^\circ)^{c+d}[A]^a[B]^b} = (c^\circ)^{(a+b)-(c+d)}\frac{[C]^c[D]^d}{[A]^a[B]^b}$$

and therefore

$$\frac{[C]^c[D]^d}{[A]^a[B]^b} = K_c(c^\circ)^{(c+d)-(a+b)}$$

When this expression is substituted into the expression for K, the result is

$$K = \left(\frac{RT}{P^\circ}\right)^{(c+d)-(a+b)} \times K_c(c^\circ)^{(c+d)-(a+b)}$$

$$= \left(\frac{c^\circ RT}{P^\circ}\right)^{(c+d)-(a+b)} K_c$$

A useful way to remember the general form of the expression that has just been derived is to write it as

$$K = \left(\frac{c^\circ RT}{P^\circ}\right)^{\Delta n_r} K_c \tag{4a}$$

where Δn_r is the difference of (unitless) stoichiometric coefficients for the *gas-phase* species in the chemical equation, calculated as $\Delta n_r = n_{r,\text{products}} - n_{r,\text{reactants}}$ (so $\Delta n_r = 2 - (1+3) = -2$ for the ammonia synthesis reaction in Eq. 3 and $\Delta n_r = 1$ for $H_2O(l) \rightleftharpoons H_2O(g)$. If no gases take part in the reaction or if there are the same numbers of gas molecules on each side of the chemical equation, then $\Delta n_r = 0$ and $K = K_c$. The same relation holds between

Q and Q_c, the reaction quotient in terms of concentrations. Equation 4a is commonly written more succinctly as

$$K = (RT)^{\Delta n_r} K_c \qquad \textbf{(4b)}$$

but the full version clarifies the units and should be used in calculations. Some values of K_c are listed in Table 5G.2.

EXAMPLE 5H.1   Converting between K and K_c

Suppose you are a scientist studying the reactions of SO_2 and O_2. If you wish to use gas molar concentrations, you must first convert the equilibrium constant K to K_c. At 400 °C, the equilibrium constant K for $2\,SO_2(g) + O_2(g) \rightleftharpoons 2\,SO_3(g)$ is 3.1×10^4. What is the value of K_c at this temperature?

PLAN Because $P° = 1$ bar and $c° = 1$ mol·L^{-1}, it is sensible to use R expressed in bar and liters, $R = 8.3145 \times 10^{-2}$ L·bar·K^{-1}·mol^{-1}, and to note that

$$\frac{P°}{Rc°} = \frac{1 \text{ bar}}{(8.3145 \times 10^{-2} \text{L·bar·K}^{-1}\text{mol}^{-1}) \times (1 \text{ mol·L}^{-1})} = 12.03 \text{ K}$$

Then, Eq. 4a can be written as

$$K = \left(\frac{T}{12.03 \text{ K}}\right)^{\Delta n_r} K_c$$

To use this equation, identify the value of Δn_r for the reaction, convert the temperature to the Kelvin scale, and rearrange it to solve for K_c.

What should you assume? Assume that the gases are ideal.

SOLVE

From the chemical equation and reaction conditions,

$$\Delta n_r = 2 - (2 + 1) = -1 \text{ and } T = 400. + 273.15 \text{ K} = 673 \text{ K}$$

From $K = (T/12.03 \text{ K})^{\Delta n_r}K_c$ in the form $K_c = (T/12.03 \text{ K})^{-\Delta n_r}K$,

$$K_c = \left(\frac{673 \text{ K}}{12.03 \text{ K}}\right)^{\overset{1}{\overbrace{-(-1)}}} \times (3.1 \times 10^4) = 1.7 \times 10^6$$

Self-test 5H.2A The equilibrium constant for the ammonia synthesis is $K = 41$ at 127 °C. What is the value of K_c at that temperature?

[**Answer:** $\Delta n_r = -2$; so $K_c = 4.5 \times 10^4$]

Self-test 5H.2B At 127 °C, the equilibrium constant for $N_2O_4(g) \rightleftharpoons 2\,NO_2(g)$ is $K = 47.9$. What is the value of K_c at that temperature?

Related Exercises 5H.5, 5H.6

For thermodynamic calculations, gas-phase equilibria are expressed in terms of K; for practical calculations, however, they may be expressed in terms of molar concentrations by using Eq. 4.

What have you learned in this Topic?

You have learned how to relate the expression for the equilibrium constant to the way in which the chemical equation is written. You also know how to convert an equilibrium constant for a gas-phase reaction into an expression in terms of molar concentrations.

The skills you have mastered are the ability to:

☐ **1.** Calculate the effect on K of reversing a chemical equation, multiplying it by a factor, or combining it with another equation (Table 5H.1).

☐ **2.** Convert between K and K_c (Example 5H.1).

Topic 5H Exercises

5H.1 For the reaction $N_2(g) + 3H_2(g) \rightleftharpoons 2NH_3(g)$ at 400. K, $K = 41$. Find the value of K for each of the following reactions at the same temperature:

(a) $2NH_3(g) \rightleftharpoons N_2(g) + 3H_2(g)$

(b) $\frac{1}{2}N_2(g) + \frac{3}{2}H_2(g) \rightleftharpoons NH_3(g)$

(c) $2N_2(g) + 6H_2(g) \rightleftharpoons 4NH_3(g)$

5H.2 The equilibrium constant for the reaction $2NO(g) + O_2(g) \rightleftharpoons 2NO_2(g)$ is $K = 2.5 \times 10^{10}$ at 500. K. Find the value of K for each of the following reactions at the same temperature.

$\frac{1}{2}NO(g) + \frac{1}{4}O_2(g) \rightleftharpoons \frac{1}{2}NO_2(g)$

$4NO_2(g) \rightleftharpoons 4NO(g) + 2O_2(g)$

$6NO(g) + 3O_2(g) \rightleftharpoons 6NO_2(g)$

5H.3 Use the information in Table 5G.2 to determine the value of K at 300 K for the reaction $2BrCl(g) + H_2(g) \rightleftharpoons Br_2(g) + 2HCl(g)$.

5H.4 Use the information in Table 5G.2 to determine the value of K at 500 K for the reaction $2NH_3(g) + 3I_2(g) \rightleftharpoons N_2(g) + 6HI(g)$.

5H.5 Evaluate K_c for each of the following equilibria from the value of K:

(a) $2NOCl(g) \rightleftharpoons 2NO(g) + Cl_2(g)$, $K = 1.8 \times 10^{-2}$ at 500 K

(b) $CaCO_3(s) \rightleftharpoons CaO(s) + CO_2(g)$, $K = 167$ at 1073 K

5H.6 Evaluate K_c for each of the following equilibria from the value of K:

(a) $2SO_2(g) + O_2(g) \rightleftharpoons 2SO_3(g)$, $K = 3.4$ at 1000. K

(b) $NH_4HS(s) \rightleftharpoons NH_3(g) + H_2S(g)$, $K = 9.4 \times 10^{-2}$ at 24 °C

Topic 5I Equilibrium Calculations

5I.1 The Extent of Reaction

5I.2 The Direction of Reaction

5I.3 Calculations with Equilibrium Constants

What are other ways of writing an equilibrium constant?

Topic 5H: Alternative forms of the equilibrium constant

How is the equilibrium composition related to K?

Topic 5I: Equilibrium calculations

The equilibrium constant summarizes the composition of a reaction mixture that has reached equilibrium. It can be used to predict the individual equilibrium partial pressures or concentrations of reactants and products, given the starting conditions. There are two stages in this discussion: first, to understand the qualitative significance of the size of the equilibrium constant; second, to use it quantitatively to predict actual concentrations or partial pressures in the equilibrium mixture.

5I.1 The Extent of Reaction

As in Topic 5G, the significance of the equilibrium constant can be understood from either a kinetic or a thermodynamic viewpoint.

How Is That Explained...

...using kinetics?

- K is large when the forward rate constant is greater than the reverse rate constant; that is, $K \gg 1$ when $k_r \gg k_r'$.
- K is small when the reverse rate constant is greater than the forward rate constant; that is, $K \ll 1$ when $k_r' \gg k_r$.

...using thermodynamics?

- K is large when ΔG_r° for a reaction is strongly negative.
- K is small when ΔG_r° for a reaction is strongly positive.

When K is large, the reaction goes nearly to completion before it reaches equilibrium and the equilibrium reaction mixture consists mainly of products. When K is small, equilibrium is reached after very little reaction has taken place. For instance, consider the reaction

$$H_2(g) + Cl_2(g) \rightleftharpoons 2\,HCl(g) \qquad K = \frac{(P_{HCl})^2}{P_{H_2}P_{Cl_2}}$$

Experiment shows that $K = 4.0 \times 10^{18}$ at 500. K. Such a large value for K implies that when the system reaches equilibrium most of the reactants have been converted into HCl. In fact, this reaction essentially goes to completion. Now consider the equilibrium

$$N_2(g) + O_2(g) \rightleftharpoons 2\,NO(g) \qquad K = \frac{(P_{NO})^2}{P_{N_2}P_{O_2}}$$

Experiment gives $K = 3.4 \times 10^{-21}$ at 800. K. The very small value of K implies that equilibrium is reached when only a small amount of product has formed. The reactants N_2 and O_2 remain the dominant species in the system, even at equilibrium.

These remarks can be summarized as follows for chemical equations written with the smallest whole-number stoichiometric coefficients (**FIG. 5I.1**):

- Large values of K (larger than about 10^3): equilibrium favors the products.
- Intermediate values of K (approximately in the range 10^{-3} to 10^3): neither reactants nor products are strongly favored at equilibrium.
- Small values of K (smaller than about 10^{-3}): equilibrium favors the reactants.

Why Do You Need to Know This Material? Knowledge of the equilibrium constant for a reaction enables you to predict the composition of the reaction mixture at equilibrium. This ability is useful throughout chemistry, especially in the study of acids and bases and other reactions in solution.

What Do You Need to Know Already? You need to know how to write the equilibrium constant for a given reaction (Topics 5G and 5H) and how the equilibrium constant is related to the Gibbs free energy of reaction (Topic 5G). The calculations make use of reaction stoichiometry (*Fundamentals* L).

Because equilibrium constants must be raised to a power when a chemical equation is multiplied by a factor, and therefore change in magnitude according to how the equation is written (Topic 5H), the rules set out here are only general guidelines.

Reactants ▢ Products ▢

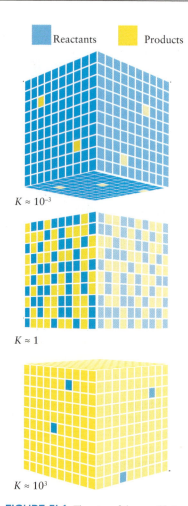

$K \approx 10^{-3}$

$K \approx 1$

$K \approx 10^3$

FIGURE 5I.1 The size of the equilibrium constant indicates whether the reactants or the products are favored at equilibrium. In this diagram, blue cubes represent the reactants and yellow cubes represent the products. Note that reactants are favored when K is small (top), products are favored when K is large (bottom), and reactants and products are in equal abundance when $K = 1$.

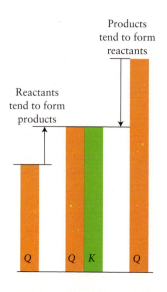

Products tend to form reactants

Reactants tend to form products

Q Q K Q

EXAMPLE 5I.1   Determining an equilibrium composition

Hydrogen chloride, HCl, can be manufactured commercially by reacting H_2 and Cl_2 in a graphite-lined furnace. Suppose you are a chemical engineer working with this reaction. You will need to know the expected equilibrium composition in order to control the reaction. In an equilibrium mixture of HCl, Cl_2, and H_2, the partial pressure of H_2 is 4.2 mPa and that of Cl_2 is 8.3 mPa. What is the partial pressure of HCl at 500. K in bar, given $K = 4.0 \times 10^{18}$ for $H_2(g) + Cl_2(g) \rightleftharpoons 2\,HCl(g)$?

ANTICIPATE Because the equilibrium constant is large, you should expect a high equilibrium partial pressure for the product, HCl.

PLAN At equilibrium, the partial pressures of the reactants and products (in bar) satisfy the expression for K. Therefore, rearrange the expression to give the one unknown concentration and substitute the data.

SOLVE First convert the units of partial pressures to bar using 1 bar = 10^5 Pa: $P_{H_2} = 4.2 \times 10^{-8}$ bar, $P_{Cl_2} = 8.3 \times 10^{-8}$ bar. Write the expression for the equilibrium constant from the chemical equation given.

From $K = (P_{HCl})^2/P_{H_2}P_{Cl_2}$,

$$P_{HCl} = (KP_{H_2}P_{Cl_2})^{1/2} = \{(4.0 \times 10^{18})(4.2 \times 10^{-8})(8.3 \times 10^{-8})\}^{1/2} = 1.2 \times 10^2$$

With units restored, this partial pressure is 1.2×10^2 bar, or 0.12 kbar.

EVALUATE As expected, at equilibrium, the partial pressure of product in the system, 120 bar, is overwhelming relative to the tiny partial pressures of the reactants.

Self-test 5I.1A Suppose that the equilibrium partial pressures of H_2 and Cl_2 are both 1.0 μPa. What is the equilibrium partial pressure of HCl at 500. K, given $K = 4.0 \times 10^{18}$?

[**Answer:** P_{HCl} = 20. mbar]

Self-test 5I.1B Suppose that the equilibrium partial pressures of N_2 and O_2 in the reaction $N_2(g) + O_2(g) \rightleftharpoons 2\,NO(g)$ at 800. K are both 52 kPa. What is the partial pressure (in pascals) of NO at equilibrium if $K = 3.4 \times 10^{-21}$ at 800. K?

Related Exercises 5I.1–5I.6

If K is large, products are favored at equilibrium (the equilibrium "lies to the right"); if K is small, reactants are favored (the equilibrium "lies to the left").

5I.2 The Direction of Reaction

Now suppose that a system of interest is not at equilibrium. For example, you might be given the concentrations of reactants and products for some arbitrary stage of a reaction, but want to know whether it will produce more products or more reactants as it moves toward equilibrium:

- If $Q < K$, the concentrations or partial pressures of the products are too low relative to those of the reactants for equilibrium. Hence, the reaction has a tendency to proceed toward products.
- If $Q = K$, the mixture has its equilibrium composition and has no tendency to change in either direction.
- If $Q > K$, the reverse reaction is spontaneous and the products tend to decompose into the reactants.

This pattern is summarized in **FIG. 5I.2** and can be interpreted either kinetically or thermodynamically.

FIGURE 5I.2 The relative sizes of the reaction quotient Q and the equilibrium constant K indicate the direction in which a reaction mixture tends to change. The arrows show that when $Q < K$, reactants form products (left), and when $Q > K$, products form reactants (right). There is no tendency to change once the reaction quotient has become equal to the equilibrium constant.

 ANIMATION FIGURE 5I.2

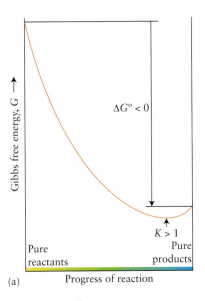

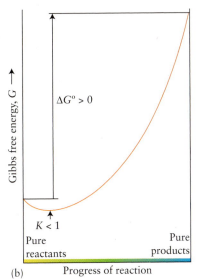

FIGURE 51.3 (a) A reaction that has the potential to go to completion ($K > 1$) is one in which the minimum of the Gibbs free-energy curve (the position of equilibrium) lies close to pure products. (b) A reaction that has little tendency to form products ($K < 1$) is one in which the minimum of the Gibbs free-energy curve lies close to pure reactants.

How Is That Explained...

...using kinetics?

As an example consider the simple reaction of $A \rightleftharpoons B$.

- If $k_r[A] > k_r'[B]$, the forward rate of reaction is faster than the reverse rate of reaction and the mixture proceeds toward products.
- If $k_r[A] = k_r'[B]$, the forward rate of reaction is equal to the reverse rate of reaction and the mixture has no tendency to change in either direction.
- If $k_r[A] < k_r'[B]$, the reverse rate of reaction is faster than the forward rate of reaction and the mixture proceeds toward reactants.

...using thermodynamics?

A plot of the Gibbs free energy of a reaction mixture against its changing composition (**FIG. 51.3**) reveals that the reaction tends to proceed toward the composition at the lowest point of the curve, because that is the direction of decreasing Gibbs free energy. The composition at the lowest point of the curve—the point of minimum Gibbs free energy—corresponds to equilibrium. For a system at equilibrium, any change leads to a reaction mixture with a greater Gibbs free energy, and so neither the forward nor the reverse reaction is spontaneous. When the minimum value of the Gibbs free energy lies very close to pure products, the equilibrium composition strongly favors products and "goes to completion" (Fig. 51.3a). When the Gibbs free-energy minimum lies very close to the pure reactants, the equilibrium composition strongly favors the reactants and the reaction "does not go" (Fig. 51.3b).

A Note on Good Practice: Note that the criterion of spontaneity is ΔG, not $\Delta G°$. Whether a reaction is spontaneous depends on the stage it has reached, so it is better to say that $K > 1$ for a reaction with a negative $\Delta G_r°$ rather than that it is spontaneous. However, for reactions with very large equilibrium constants, it is very unlikely that the mixture of reagents prepared in the laboratory will correspond to $Q > K$, and it is common to refer to such reactions as "spontaneous."

EXAMPLE 51.2   Predicting the direction of reaction

Suppose you are a chemical engineer working in a plant that produces hydrogen iodide and are exploring the efficiency of the production process. If you know the value of K, for any given composition you can predict in which direction the reaction has a tendency to go. A mixture of hydrogen and iodine, each at 55 kPa, and hydrogen iodide, at 78 kPa, was introduced into a container heated to 783 K. At this temperature, $K = 46$ for $H_2(g) + I_2(g) \rightleftharpoons 2 HI(g)$. Predict whether HI has a tendency to form or to decompose into $H_2(g)$ and $I_2(g)$.

PLAN Calculate Q and compare it with K. If $Q > K$, the products have a tendency to decompose into reactants until their concentrations match K. The opposite is true if $Q < K$; in that case, the reactants have a tendency to form more products.

SOLVE Substitute the data, noting that 1 kPa is equivalent to 0.01 bar, in the reaction quotient.

From $Q = (P_{HI})^2/P_{H_2}P_{I_2}$,

$$Q = \frac{(0.78)^2}{0.55 \times 0.55} = 2.0$$

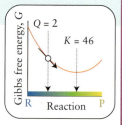

Because $Q < K$, you can conclude that the reaction will tend to form more product and consume reactants.

Self-test 5I.2A A mixture of H_2, N_2, and NH_3 with partial pressures 22 kPa, 44 kPa, and 18 kPa, respectively, was prepared and heated to 500. K, at which temperature $K = 3.6 \times 10^{-2}$ for reaction C. Decide whether ammonia tends to form or decompose.

[*Answer:* $Q = 6.9 > K$; tends to decompose]

Self-test 5I.2B For the reaction $N_2O_4(g) \rightleftharpoons 2 NO_2(g)$ at 298 K, $K = 0.15$. A mixture of N_2O_4 and NO_2 with initial partial pressures of 2.4 and 1.2 bar, respectively, was prepared at 298 K. Which compound will tend to increase its partial pressure?

Related Exercises 5I.7–5I.12

A reaction has a tendency to form products if $Q < K$ and to form reactants if $Q > K$.

5I.3 Calculations with Equilibrium Constants

The equilibrium constant of a reaction contains information about the equilibrium composition at the given temperature. However, in many cases, only the initial composition of the reaction mixture is known and you need to predict the equilibrium composition. If you know the value of K, the equilibrium composition can usually be inferred by using the reaction stoichiometry. The easiest way to proceed is to construct an **equilibrium table,** a table showing the initial composition, the changes needed to reach equilibrium in terms of some unknown quantity x, and the final equilibrium composition. The procedure is summarized in **Toolbox 5I.1** and illustrated in the examples that follow.

Toolbox 5I.1   HOW TO SET UP AND USE AN EQUILIBRIUM TABLE

CONCEPTUAL BASIS

The composition of a reaction mixture tends to adjust until the molar concentrations of solutes or partial pressures of gases ensure that $Q = K$. A change in the abundance of any one component is linked to changes in the others by the reaction stoichiometry.

PROCEDURE

Step 1 Write the equilibrium expression for the balanced chemical equation. Then set up an equilibrium table as shown here, with columns labeled by the species taking part in the reaction. If K_c is given for a reaction involving gases (Topic 5H), then proceed in the same way, but with molar concentrations for all the species.

The first line in the table shows how the reaction system was prepared before reaction began. As always, use molar concentrations (in moles per liter) and partial pressures (in bar), but strike out the units. Omit pure solids and liquids, as they do not appear in the equilibrium expression.

	Species 1	Species 2	Species 3	...
initial composition				...
change in composition				...
equilibrium composition				...

Step 2 In the second row, write the changes in the composition that are needed for the reaction to reach equilibrium.

When the changes are not known, write one of them as x or a multiple of x and then use the stoichiometry of the balanced equation to express the other changes in terms of that x.

Step 3 In the third row, write the equilibrium compositions in terms of x by adding the change in composition (from step 2) to the initial value for each substance (from step 1).

Although a change in composition may be positive (an increase) or negative (a decrease), the equilibrium value of each concentration or partial pressure itself must be positive.

Step 4 Use the equilibrium constant to determine the value of x, the unknown change in molar concentration or partial pressure, and find the equilibrium values for all species.

> **A Note on Good Practice:** A good habit to adopt is to check the answer by substituting the equilibrium composition into the expression for K.

An approximation technique can greatly simplify calculations when the change in composition (x) is less than about 5% of the initial value. To use it, assume that x is negligible when added to or subtracted from a number. Thus, all expressions like $A + x$ or $A - 2x$ can be replaced by A. When x occurs on its own (not added to or subtracted from another number), it is left unchanged. So, an expression such as $(0.1 - 2x)^2x$ simplifies to $(0.1)^2x$, provided that $2x \ll 0.1$ (specifically, if $2x < 0.005$). At the end of the calculation, it is important to verify that the

calculated value of x is indeed smaller than 5% of the initial values. If it is not, then the equation must be solved without making an approximation.

> **A Note on Good Practice:** Now that computational methods and calculators capable of solving equations are essentially universal, the approximation technique is largely unnecessary except for developing equations or when software is unavailable.

The approximation procedure is illustrated in Example 5I.3.

In some cases, the full equation for x is a quadratic equation of the form

$$ax^2 + bx + c = 0$$

When the approximation procedure is invalid, use the exact solutions of this equation:

$$x = \frac{-b \pm \sqrt{b^2 - 4ac}}{2a}$$

You have to decide which of the two solutions given by this expression is valid (the one with the $+$ sign or the one with the $-$ sign in front of the square root) by seeing which solution is chemically possible.

See Example 5I.4 for an illustration of this procedure.

For some reactions, the equation for x in terms of K may be a higher-order polynomial. If an approximation is not valid, use a graphing calculator or mathematical software to find the roots of the equation.

EXAMPLE 5I.3   Calculating the equilibrium composition by approximation

Dinitrogen oxide, N_2O, colloquially called "laughing gas," was first used as an anesthetic in dentistry in 1844. Suppose that you are a chemist attempting to prepare N_2O from N_2 and O_2; you might want to know the expected equilibrium composition. You plan to transfer a mixture of 0.482 mol N_2 and 0.933 mol O_2 to a reaction vessel of volume 10.0 L, where it will form N_2O at 800. K; at this temperature, $K = 3.2 \times 10^{-28}$ for the reaction $2\,N_2(g) + O_2(g) \rightleftharpoons 2\,N_2O(g)$. Calculate the partial pressures of the gases in the equilibrium mixture.

ANTICIPATE Because the equilibrium constant is so small, you should expect that equilibrium will be reached when only a very small amount of product has formed.

PLAN Use the procedure in Toolbox 5I.1. Because data are in liters and pressures are expected in bar, use the value of R that matches these units (namely, $L \cdot bar \cdot K^{-1} \cdot mol^{-1}$).

What should you assume? Assume that all the gases are ideal and, because the value of K is so small, that the amount of product formed will be so small that the change in partial pressures of the reactants is negligible.

SOLVE

First find the partial pressures of the reactants. From $P_J = n_J RT/V$,

$$P_{N_2} = \frac{(0.482 \text{ mol}) \times (8.3145 \times 10^{-2} \text{ L} \cdot \text{bar} \cdot \text{K}^{-1} \text{mol}^{-1}) \times (800. \text{ K})}{10.0 \text{ L}}$$

$$= 3.21 \text{ bar}$$

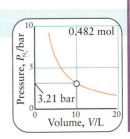

From $P_J = n_J RT/V$,

$$P_{O_2} = \frac{(0.933 \text{ mol}) \times (8.3145 \times 10^{-2} \text{ L·bar·K}^{-1}\text{mol}^{-1}) \times (800. \text{ K})}{10.0 \text{ L}}$$

$$= 6.21 \text{ bar}$$

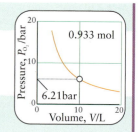

Step 1 Write the equilibrium constant and set up the equilibrium table. Fill in the first row with the initial pressures of reactants.

$$2 \text{ N}_2(g) + \text{O}_2(g) \rightleftharpoons 2 \text{ N}_2\text{O}(g) \qquad K = \frac{(P_{N_2O})^2}{(P_{N_2})^2(P_{O_2})}$$

	N$_2$	O$_2$	N$_2$O
initial pressure	3.21	6.21	0
change in pressure			
equilibrium pressure			

Step 2 Write the changes in composition required to reach equilibrium and fill out row 2 of the equilibrium table. The stoichiometry of the reaction implies that, if the partial pressure of O$_2$ decreases by x, then the partial pressure of N$_2$ decreases by $2x$ and that of N$_2$O increases by $2x$.

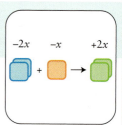

	N$_2$	O$_2$	N$_2$O
initial pressure	3.21	6.21	0
change in pressure	$-2x$	$-x$	$+2x$
equilibrium pressure			

Step 3 Add the values in the first two rows and write the equilibrium compositions in terms of x.

	N$_2$	O$_2$	N$_2$O
initial pressure	3.21	6.21	0
change in pressure	$-2x$	$-x$	$+x$
equilibrium pressure	$3.21 - 2x$	$6.21 - x$	$+2x$

Step 4 Insert the values in line 3 of the table into the expression for K. Make the approximation that $x \ll 3.21$ and 6.21, then solve the equilibrium expression for x.

$$K = \frac{(2x)^2}{(3.21 - 2x)^2(6.21 - x)} \approx \frac{(2x)^2}{(3.21)^2(6.21)}$$

$$x \approx \left\{ \frac{(3.21)^2 \times (6.21) \times (3.2 \times 10^{-28})}{4} \right\}^{1/2} = 7.2 \times 10^{-14}$$

The value of $2x$ is indeed very small compared with 3.21 (far smaller than 5% of it), and so the approximation is valid. You can conclude that, at equilibrium,

From $P_{N_2} = 3.21 - 2x$ bar,

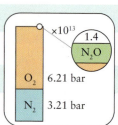

$$P_{N_2} = 3.21 \text{ bar}$$

From $P_{O_2} = 6.21 - x$ bar,

$$P_{O_2} = 6.21 \text{ bar}$$

From $P_{N_2O} = 2x$ bar,

$$P_{N_2O} = 1.4 \times 10^{-13} \text{ bar}$$

EVALUATE As expected, very little product has formed, and the initial pressures of reactants have remained virtually unchanged. When substituted into the full expression for the equilibrium constant, these values give $K = 3.1 \times 10^{-28}$, in good agreement with the experimental value.

Self-test 51.3A The initial partial pressures of nitrogen and hydrogen in a rigid, sealed vessel are 0.010 and 0.020 bar, respectively. The mixture is heated to a temperature at which $K = 0.11$ for $N_2(g) + 3 H_2(g) \rightleftharpoons 2 NH_3(g)$. What are the equilibrium partial pressures of each substance in the reaction mixture?

[**Answer:** N_2, 0.010 bar; H_2, 0.020 bar; NH_3, 9.4×10^{-5} bar]

Self-test 51.3B Hydrogen chloride gas is added to a reaction vessel containing solid iodine until its partial pressure reaches 0.012 bar. At the temperature of the experiment, $K = 3.5 \times 10^{-32}$ for $2 HCl(g) + I_2(s) \rightleftharpoons 2 HI(g) + Cl_2(g)$. Assume that some I_2 remains at equilibrium. What are the equilibrium partial pressures of each gaseous substance in the reaction mixture?

Related Exercises 51.15, 51.17, 51.29, 51.30

EXAMPLE 51.4 Calculating the equilibrium composition by using a quadratic equation

Phosphorus pentachloride, PCl_5, is used to convert alcohols (such as CH_3CH_2OH) to alkyl chlorides (such as CH_3CH_2Cl). If you were an industrial chemist, you might be asked to prepare some PCl_5 by the reaction of PCl_3 and Cl_2. At elevated temperatures, however, PCl_5 decomposes to these starting materials, so it will be important to estimate the equilibrium composition of the reaction mixture. Suppose that you place 3.12 g of PCl_5 in a reaction vessel of volume 500. mL and allow the sample to reach equilibrium with its decomposition products PCl_3 and Cl_2 at 250 °C, when $K = 78.3$ for the reaction $PCl_5(g) \rightleftharpoons PCl_3(g) + Cl_2(g)$. Find the composition of the equilibrium mixture.

ANTICIPATE Because the equilibrium constant is neither very small nor very large, you should expect that the partial pressures of reactants and products at equilibrium will be similar to each other.

PLAN Use the general procedure set out in Toolbox 51.1. Because data are in liters and pressures are expected in bar, use the value of R that matches these units (namely, $L \cdot bar \cdot K^{-1} \cdot mol^{-1}$).

SOLVE First, calculate the initial partial pressure of PCl_5 (the initial partial pressures of PCl_3 and Cl_2 are zero).

From $n_J = m_J/M_J$, the amount of PCl_5 added is:

$$n_{PCl_5} = \frac{3.12 \text{ g}}{208.24 \text{ g} \cdot mol^{-1}} = \frac{3.12}{208.24} \text{ mol} = 0.0150\ldots \text{ mol}$$

From $P_J = n_J RT/V$, the initial partial pressure of PCl_5 is therefore

$$P_{PCl_5} = (0.0150\ldots \text{ mol}) \times \frac{(8.3145 \times 10^{-2} \text{ L} \cdot bar \cdot K^{-1} \cdot mol^{-1}) \times (523 \text{ K})}{0.500 \text{ L}}$$

$$= 1.30 \text{ bar}$$

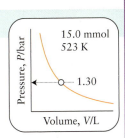

Step 1 Write the equilibrium expression and set up the equilibrium table. Fill in the first row with the initial pressures of reactants.

$$PCl_5(g) \rightleftharpoons PCl_3(g) + Cl_2(g) \qquad K = \frac{P_{PCl_3}P_{Cl_2}}{P_{PCl_5}}$$

	PCl₅	PCl₃	Cl₂
initial pressure	1.30	0	0
change in pressure			
equilibrium pressure			

Step 2 Write the changes in composition required to reach equilibrium and fill out row 2 of the equilibrium table. The stoichiometry of the reaction implies that, if the partial pressure of PCl_5 decreases by x, then the partial pressures of PCl_3 and Cl_2 each increase by x.

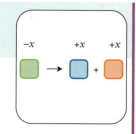

	PCl_5	PCl_3	Cl_2
initial pressure	1.30	0	0
change in pressure	$-x$	$+x$	$+x$
equilibrium pressure			

Step 3 Add the values in the first two rows and write the equilibrium values.

	PCl_5	PCl_3	Cl_2
initial pressure	1.30	0	0
change in pressure	$-x$	$+x$	$+x$
equilibrium pressure	$1.30 - x$	$+x$	$+x$

Step 4 Substitute the equilibrium values into the equilibrium expression.

$$K = \frac{x^2}{1.30 - x}$$

Rearrange the equation:

$$x^2 + Kx - 1.30K = 0$$

Substitute the value of K:

$$x^2 + 78.3x - 102 = 0$$

Solve for x by using the quadratic formula

$$x = \{-b \pm (b^2 - 4ac)^{1/2}\}/2a$$
$$= \frac{-78.3 \pm \{(78.3)^2 - 4 \times (-102)\}^{1/2}}{2}$$
$$= -79.6 \text{ or } 1.28$$

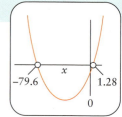

Because the partial pressures must be positive and because x is the partial pressure of PCl_3, select 1.28 as the solution. It follows that, at equilibrium,

From $P_{PCl_5} = 1.30 - x$ bar, $P_{PCl_5} = 1.30 - 1.28$ bar $= 0.02$ bar
From $P_{PCl_3} = x$ bar, $P_{PCl_3} = 1.28$ bar
From $P_{Cl_2} = x$ bar, $P_{Cl_2} = 1.28$ bar

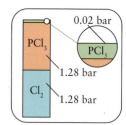

EVALUATE As expected, the equilibrium concentrations of reactants and products have similar values. **FIGURE 5I.4** illustrates how the reaction approaches equilibrium. Substituting these values into the equilibrium expression gives $K = 82$, in reasonable agreement with the reported value.

Self-test 5I.4A Bromine monochloride, BrCl, decomposes into bromine and chlorine and reaches the equilibrium $2\,BrCl(g) \rightleftharpoons Br_2(g) + Cl_2(g)$, for which $K = 32$ at 500. K. If initially pure BrCl is present at a concentration of 3.30 mbar, what is its partial pressure in the mixture at equilibrium?

[**Answer:** 0.3 mbar]

Self-test 5I.4B Chlorine and fluorine react at 2500. K to produce ClF and reach the equilibrium $Cl_2(g) + F_2(g) \rightleftharpoons 2\,ClF(g)$ with $K = 20$. If a gaseous mixture with $P_{Cl_2} = 0.200$ bar, $P_{F_2} = 0.100$ bar, and $P_{ClF} = 0.100$ bar is allowed to come to equilibrium at 2500. K, what is the partial pressure of ClF in the equilibrium mixture?

Related Exercises 5I.19, 5I.20, 5I.25, 5I.26

FIGURE 51.4 The approach of the composition of the reaction mixture to equilibrium when PCl_5 decomposes in a closed container. Note that the curves for Cl_2 and PCl_3 are superimposed because they increase by the same amount.

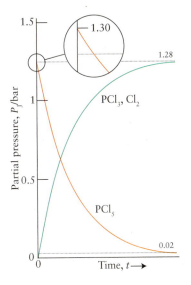

To calculate the equilibrium composition of a reaction mixture, set up an equilibrium table in terms of changes in the partial pressures or concentrations of reactants and products, express the equilibrium constant in terms of those changes, and solve the resulting equation.

What have you learned in this Topic?

You can now predict in which the direction a reaction will tend to go by comparing the value of the reaction quotient, Q with the equilibrium constant K. You have also learned how to determine individual concentrations or partial pressures from starting conditions by using the equilibrium constant and setting up an equilibrium table.

The skills you have mastered are the ability to:

☐ **1.** Calculate an equilibrium concentration from an equilibrium composition (Example 5I.1).

☐ **2.** Predict the direction of a reaction given the composition of a reaction mixture and the value of K (Example 5I.2).

☐ **3.** Use an equilibrium table to carry out equilibrium calculations (Toolbox 5I.1 and Examples 5I.3 and 5I.4).

☐ **4.** Make approximations when appropriate, and judge their validity (Toolbox 5I.1 and Example 5I.3).

Topic 5I Exercises

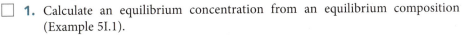

5I.1 At 500. K, the equilibrium constant for the reaction $Cl_2(g) + Br_2(g) \rightleftharpoons 2\,BrCl(g)$ is $K_c = 0.031$. If the equilibrium composition is 0.495 mol·L^{-1} Cl_2 and 0.145 mol·L^{-1} BrCl, what is the equilibrium molar concentration of Br_2?

5I.2 The equilibrium constant for the reaction $PCl_3(g) + Cl_2(g) \rightleftharpoons PCl_5(g)$ is $K = 3.5 \times 10^4$ at 760 °C. At equilibrium, the partial pressure of PCl_5 was 2.4×10^2 bar and that of PCl_3 was 8.32 bar. What was the equilibrium partial pressure of Cl_2?

5I.3 In a gas-phase equilibrium mixture of H_2, I_2, and HI at 500. K, $[HI] = 2.21 \times 10^{-3}$ mol·L^{-1} and $[I_2] = 1.46 \times 10^{-3}$ mol·L^{-1}. Given the value of the equilibrium constant in Table 5G.2, calculate the equilibrium molar concentration of H_2.

5I.4 In a gas-phase equilibrium mixture of H_2, Cl_2, and HCl at 1000. K, $[HCl] = 0.352$ mmol·L^{-1} and $[Cl_2] = 7.21$ mmol·L^{-1}. Use the information in Table 5G.2 to calculate the equilibrium molar concentration of H_2.

5I.5 In a gas-phase equilibrium mixture of PCl_5, PCl_3, and Cl_2 at 500. K, $P_{PCl_5} = 1.18$ bar, $P_{Cl_2} = 5.43$ bar. What is the partial pressure of PCl_3, given that $K = 25$ for the reaction $PCl_5(g) \rightleftharpoons PCl_3(g) + Cl_2(g)$?

5I.6 In a gas-phase equilibrium mixture of $SbCl_5$, $SbCl_3$, and Cl_2 at 500. K, $P_{SbCl_5} = 0.072$ bar and $P_{SbCl_3} = 5.02$ mbar. Calculate the equilibrium partial pressure of Cl_2, given that $K = 3.5 \times 10^{-4}$ for the reaction $SbCl_5(g) \rightleftharpoons SbCl_3(g) + Cl_2(g)$.

5I.7 If $Q = 1.0$ for the reaction $N_2(g) + O_2(g) \rightarrow 2\,NO(g)$ at 25 °C, will the reaction have a tendency to form products or reactants, or will it be at equilibrium?

5I.8 If $Q = 1.0 \times 10^{32}$ for the reaction $C(s) + O_2(g) \rightarrow CO_2(g)$ at 25 °C, will the reaction have a tendency to form products or reactants, or will it be at equilibrium?

5I.9 For the reaction $H_2(g) + I_2(g) \rightleftharpoons 2\,HI(g)$, $K = 160.$ at 500. K. An analysis of a reaction mixture at 500. K showed that it had the composition $P_{H_2} = 0.20$ bar, $P_{I_2} = 0.10$ bar, and $P_{HI} = 0.10$ bar. (a) Calculate the reaction quotient. (b) Is the reaction mixture at equilibrium? (c) If not, is there a tendency to form more reactants or more products?

5I.10 Analysis of a reaction mixture showed that it had the composition 0.520 mol·L^{-1} N_2, 0.485 mol·L^{-1} H_2, and 0.102 mol·L^{-1} NH_3 at 800. K, at which temperature $K_c = 0.278$ for $N_2(g) + 3\,H_2(g) \rightleftharpoons 2\,NH_3(g)$. (a) Calculate the reaction quotient Q_c. (b) Is the reaction mixture at equilibrium? (c) If not, is there a tendency to form more reactants or more products?

5I.11 A reaction vessel of volume 0.500 L at 700. K contains 1.20 mmol $SO_2(g)$, 0.50 mmol $O_2(g)$, and 0.10 mmol $SO_3(g)$. At 700. K, $K_c = 1.7 \times 10^6$ for the equilibrium $2\,SO_2(g) + O_2(g) \rightleftharpoons 2\,SO_3(g)$. (a) Calculate the reaction quotient Q_c. (b) Will more $SO_3(g)$ tend to form?

5I.12 Given that $K_c = 62$ for the reaction $N_2(g) + 3\,H_2(g) \rightleftharpoons 2\,NH_3(g)$ at 500. K, calculate whether more ammonia will tend to form when a mixture of composition 1.25 mmol·L^{-1} N_2, 4.58 mmol·L^{-1} H_2, and 0.875 mmol·L^{-1} NH_3 is present in a container at 500. K.

5I.13 (a) In an experiment, 2.0 mmol $Cl_2(g)$ was sealed into a reaction vessel of volume 2.0 L and heated to 1000. K to study its dissociation into Cl atoms. Use the information in Table 5G.2 to calculate the equilibrium composition of the mixture. (b) If 2.0 mmol F_2 was placed into the reaction vessel instead of the chlorine, what would be its equilibrium composition at 1000. K? (c) Use your results from parts (a) and (b) to determine which is thermodynamically more stable relative to its atoms at 1000. K, Cl_2 or F_2.

5I.14 (a) In an experiment, 5.0 mmol $Cl_2(g)$ was sealed into a reaction vessel of volume 2.0 L and heated to 1200. K, and the dissociation equilibrium was established. What is the equilibrium composition of the mixture? Use the information in Table 5G.2. (b) If 5.0 mol Br_2 is placed into the reaction vessel instead of the chlorine, what would be the equilibrium composition at 1200. K? (c) Use your results from parts (a) and (b) to determine which is thermodynamically more stable relative to its atoms at 1200. K, Cl_2, or Br_2.

5I.15 When solid NH_4HS and 0.400 mol $NH_3(g)$ were placed in a vessel of volume 2.0 L at 24 °C, the equilibrium $NH_4HS(s) \rightleftharpoons NH_3(g) + H_2S(g)$, for which $K_c = 1.6 \times 10^{-4}$, was reached. What are the equilibrium concentrations of NH_3 and H_2S?

5I.16 (a) When solid $NaHCO_3$ was placed in a rigid container of volume 2.50 L and heated to 160. °C, the equilibrium $2\,NaHCO_3(s) \rightleftharpoons Na_2CO_3(s) + CO_2(g) + H_2O(g)$ was reached. At equilibrium, some of the starting material remains and the *total* pressure in the container is 7.68 bar. What is the value of K? (b) The reaction is repeated, but this time the container initially contains not only the solid reactant but also CO_2, with a partial pressure of 1.00 bar. What are the equilibrium concentrations of CO_2 and H_2O for the new experiment?

5I.17 The equilibrium constant K_c for the reaction $N_2(g) + O_2(g) \rightleftharpoons 2\,NO(g)$ at 1200 °C is 1.00×10^{-5}. Calculate the equilibrium molar concentrations of NO, N_2, and O_2 in a reaction vessel of volume 1.00 L that initially held 0.114 mol N_2 and 0.114 mol O_2.

5I.18 The equilibrium constant K_c for the reaction $N_2(g) + O_2(g) \rightleftharpoons 2\,NO(g)$ at 1200 °C is 1.00×10^{-5}. Calculate the equilibrium molar concentrations of NO, N_2, and O_2 in a reaction vessel of volume 10.00 L that initially held 0.312 mol N_2 and 0.407 mol O_2.

5I.19 A reaction mixture that consisted of 0.400 mol H_2 and 1.60 mol I_2 was introduced into a flask of volume 3.00 L and heated. At equilibrium, 60.0% of the hydrogen gas had reacted. What is the equilibrium constant K for the reaction $H_2(g) + I_2(g) \rightleftharpoons 2\,HI(g)$ at this temperature?

5I.20 A reaction mixture that consisted of 0.20 mol N_2 and 0.20 mol H_2 was introduced into a reactor of volume 25.0 L and heated. At equilibrium, 5.0% of the nitrogen gas had reacted. What is the value of the equilibrium constant K_c for the reaction $N_2(g) + 3\,H_2(g) \rightleftharpoons 2\,NH_3(g)$ at this temperature?

5I.21 The equilibrium constant K_c for the reaction $2\,CO(g) + O_2(g) \rightleftharpoons 2\,CO_2(g)$ is 0.66 at 2000 °C. If 0.28 g of CO and 0.032 g of $O_2(g)$ are placed in a reaction vessel of volume 2.0 L and heated to 2000 °C, what will the equilibrium composition of the system be? (You may wish to use graphing software or a calculator to solve the cubic equation.)

5I.22 In the Haber process for ammonia synthesis, $K = 0.036$ for $N_2(g) + 3\,H_2(g) \rightleftharpoons 2\,NH_3(g)$ at 500. K. If a reactor of volume 2.0 L is charged with 1.42 bar of N_2 and 2.87 bar of H_2, what will the equilibrium partial pressures in the mixture?

5I.23 A reaction mixture consisting of 2.00 mol CO and 3.00 mol H_2 is placed in a reaction vessel of volume 10.0 L and heated to 1200. K. At equilibrium, 0.478 mol CH_4 was present in the system. Determine the value of K_c for the reaction $CO(g) + 3\,H_2(g) \rightleftharpoons CH_4(g) + H_2O(g)$ at 1200. K.

5I.24 A mixture consisting of 1.000 mol $H_2O(g)$ and 1.000 mol $CO(g)$ is placed in a reaction vessel of volume 10.00 L at 800. K. At equilibrium, 0.665 mol $CO_2(g)$ is present as a result of the reaction $CO(g) + H_2O(g) \rightleftharpoons CO_2(g) + H_2(g)$. What are (a) the equilibrium concentrations for all substances and (b) the value of K_c at 800. K?

5I.25 A reaction mixture is prepared by mixing 0.100 mol SO_2, 0.200 mol NO_2, 0.100 mol NO, and 0.150 mol SO_3 in a reaction vessel of volume 5.00 L. The reaction $SO_2(g) + NO_2(g) \rightleftharpoons NO(g) + SO_3(g)$ is allowed to reach equilibrium at 460 °C, when $K_c = 85.0$. What is the equilibrium concentration of each substance?

5I.26 In an experiment, 0.100 mol H_2S is placed in a reaction vessel of volume 10.0 L and heated to 1132 °C. At equilibrium, 0.0285 mol H_2 is present. Calculate the value of K_c for the reaction $2\,H_2S(g) \rightleftharpoons 2\,H_2(g) + S_2(g)$ at 1132 °C.

5I.27 The equilibrium constant $K_c = 0.56$ for the reaction $PCl_3(g) + Cl_2(g) \rightleftharpoons PCl_5(g)$ at 250 °C. Upon analysis, 1.50 mol PCl_5, 3.00 mol PCl_3, and 0.500 mol Cl_2 were found to be present in a reaction vessel of volume 0.500 L at 250 °C. (a) Is the reaction at equilibrium? (b) If not, in which direction does it tend to proceed? (c) What is the equilibrium composition of the reaction system?

5I.28 Suppose that 0.724 mol PCl_5 is placed in a reaction vessel of volume 0.500 L. What is the concentration of each substance when the reaction $PCl_5(g) \rightleftharpoons PCl_3(g) + Cl_2(g)$ has reached equilibrium at 250 °C (when $K_c = 1.80$)?

5I.29 At 25 °C, $K = 3.2 \times 10^{-34}$ for the reaction $2\,HCl(g) \rightleftharpoons H_2(g) + Cl_2(g)$. If a reaction vessel of volume 1.0 L is filled with HCl at 0.22 bar, what are the equilibrium partial pressures of HCl, H_2, and Cl_2?

5I.30 If 4.00 L of $HCl(g)$ at 1.00 bar and 273 K and 26.0 g of $I_2(s)$ are transferred to a reaction vessel of volume 12.0 L and heated to 25 °C, what will the equilibrium concentrations of HCl, HI, and Cl_2 be? $K_c = 1.6 \times 10^{-34}$ at 25°C for $2\,HCl(g) + I_2(s) \rightleftharpoons 2\,HI(g) + Cl_2(g)$.

5I.31 A reaction vessel of volume 3.00 L is filled with 0.342 mol $CO(g)$, 0.215 mol $H_2(g)$, and 0.125 mol $CH_3OH(g)$. Equilibrium is reached in the presence of a zinc oxide–chromium(III) oxide catalyst and, at 300 °C, $K_c = 1.1 \times 10^{-2}$ for the reaction $CO(g) + 2\,H_2(g) \rightleftharpoons CH_3OH(g)$. (a) As the reaction approaches equilibrium, will the concentration of CH_3OH increase, decrease, or remain unchanged? (b) What is the equilibrium composition of the mixture? (You may wish to use graphing or mathematical software to solve the cubic equation.)

5I.32 For the reaction $2\,NH_3(g) \rightleftharpoons N_2(g) + 3\,H_2(g)$, $K_c = 0.395$ at 350 °C. A sample of NH_3 of mass 25.6 g is placed in a reaction vessel of volume 5.00 L and heated to 350 °C. What are the equilibrium concentrations of NH_3, N_2, and H_2?

51.33 A sample of ammonium carbamate, $NH_4(NH_2CO_2)$, of mass 25.0 g was placed in an evacuated flask of volume 0.250 L and kept at 25 °C. At equilibrium, 17.4 mg of CO_2 was present. What is the value of K_c for the decomposition of ammonium carbamate into ammonia and carbon dioxide? The reaction is

$$NH_4(NH_2CO_2)(s) \rightleftharpoons 2NH_3(g) + CO_2(g).$$

51.34 Carbon monoxide and water vapor, each at 200. Torr, were introduced into a container of volume 0.250 L. When the mixture reached equilibrium at 700 °C, the partial pressure of $CO_2(g)$ was 88 Torr. Calculate the value of K for the equilibrium

$$CO(g) + H_2O(g) \rightleftharpoons CO_2(g) + H_2(g)$$

51.35 Consider the reaction $2\,NO(g) \rightleftharpoons N_2(g) + O_2(g)$. If the initial partial pressure of $NO(g)$ is 1.0 bar, and p is the equilibrium partial pressure of $N_2(g)$ in bar, what is the correct equilibrium relation? (a) $K = p^2/(1.0 - p)$; (b) $K = p^2$; (c) $K = p^2/(1.0 - 2p)^2$; (d) $K = 4p^3/(1.0 - 2p)^2$; (e) $K = 2p/(1.0 - p)^2$.

51.36 Consider the reaction $2\,NO_2(g) \rightleftharpoons 2\,NO(g) + O_2(g)$. If the initial molar concentration of $NO_2(g)$ is 0.030 mol·L^{-1}, and c is the equilibrium molar concentration of $O_2(g)$ in moles per liter, which of the following expressions is the correct equilibrium relation? (a) $K_c = c^3$; (b) $K_c = 2c^2/(0.030 - 2c)^2$; (c) $K_c = 4c^3/(0.030 - 2c)^2$; (d) $K_c = c^2/(0.030 - c)$; (e) $K_c = 2c/(0.030 - c)^2$.

Topic 5J The Response of Equilibria to Changes in Conditions

How are chemical equilibria described?

How do equilibria respond when perturbed?

Topic 5G: Chemical equilibrium → Topic 5J: The response of equilibria to changes in conditions

Why Do You Need to Know This Material? To design an efficient industrial or laboratory process you need to know how the conditions affect the equilibrium composition of a reaction mixture.

What Do You Need to Know Already? This Topic makes use of the equilibrium calculations described in Topic 5I. If you use the kinetic arguments, you need the information about the Arrhenius equation in Topic 7D; for the thermodynamic approach, you need the relation between equilibrium constants and standard reaction Gibbs free energies in Topic 5G.

Early in the twentieth century, the looming prospect of World War I created a desperate need for nitrogen compounds. Eventually, the German chemist Fritz Haber (**FIG. 5J.1**), in collaboration with the German chemical engineer Carl Bosch (**FIG. 5J.2**), found an economical way to harvest the nitrogen of the air. Haber heated nitrogen and hydrogen under pressure in the presence of iron:

$$N_2(g) + 3\,H_2(g) \overset{Fe}{\rightleftharpoons} 2\,NH_3(g) \tag{A}$$

The metal acts as a catalyst, a substance that helps to make a reaction go faster (Topic 7E).

The reaction proceeds until equilibrium is reached, typically with only a low concentration of ammonia. Haber sought ways to increase the amount of product formed by taking advantage of the fact that, because chemical equilibria are dynamic, they respond to changes in the conditions.

5J.1 Adding and Removing Reagents

How the composition of a reaction mixture at equilibrium tends to shift when the conditions are changed can be anticipated by using a principle identified by the French chemist Henri Le Chatelier (**FIG. 5J.3**):

> **Le Chatelier's principle:** When a stress is applied to a system in dynamic equilibrium, the equilibrium tends to adjust to minimize the effect of the stress.

This empirical (observation based) principle is no more than a rule of thumb. It provides neither an explanation nor a quantitative prediction. However, as it is developed, you will

FIGURE 5J.1 Fritz Haber (1868–1934). *(ullstein bild/The Granger Collection, NYC.)*

FIGURE 5J.2 Carl Bosch (1874–1940). *(© DIZ Muenchen GmbH, Sueddeutsche Zeitung Photo/Alamy.)*

FIGURE 5J.3 Henri Le Chatelier (1850–1936). *(Academie des Sciences, Paris, France/Archives Charmet/ Bridgeman Images.)*

FIGURE 5J.4 These graphs show the changes in composition that can be expected when additional hydrogen and then ammonia are added to an equilibrium mixture of nitrogen, hydrogen, and ammonia. Note that the addition of hydrogen results in the formation of ammonia, whereas the addition of ammonia results in the decomposition of some of the added ammonia as reactants are formed. In each case, the mixture settles into a composition in accord with the equilibrium constant of the reaction.

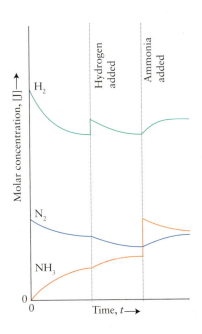

see its underlying kinetic and thermodynamic explanations and the powerful *quantitative* conclusions that can be made.

Suppose that the ammonia synthesis reaction, reaction A, has reached equilibrium. Now suppose that more hydrogen gas is pumped in. According to Le Chatelier's principle, the increase in the concentration of hydrogen molecules will tend to be minimized by the reaction of hydrogen with nitrogen. As a result, additional ammonia will be formed. If, instead of hydrogen, some ammonia is added, then the reaction will tend instead to form reactants at the expense of the added ammonia (**FIG. 5J.4**).

THINKING POINT

Suppose one of the products of a reaction at equilibrium is a pure solid. How will the equilibrium be affected if some of the solid is removed? If all the solid is removed?

The response of a system at equilibrium to the addition or removal of a substance can be explained by considering the relative sizes of Q and K. When reactants or products are added, only Q changes in response, while K, a characteristic of the reaction, remains constant. At equilibrium $Q = K$, so if the value of Q is perturbed, it will always tend to become equal to K again because that direction of change corresponds to a decrease in Gibbs free energy (**FIG. 5J.5**):

- When reactants are added to the equilibrium mixture, the reactant concentrations in the denominator of Q increase and so Q temporarily falls below K. Because $Q < K$, the reaction mixture responds by forming products and consuming reactants until $Q = K$ again. That is, when reactants are added to a system at equilibrium, it responds by converting reactants to products.
- When products are added to the equilibrium mixture, Q temporarily rises above K, because products appear in the numerator. Now, because $Q > K$, the reaction mixture responds by forming reactants at the expense of products until $Q = K$ again. That is, when products are added to a system at equilibrium, it responds by converting products to reactants.

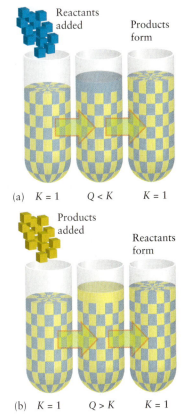

FIGURE 5J.5 (a) When a reactant (blue) is added to a reaction mixture at equilibrium, $Q < K$ and products have a tendency to form. (b) When a product (yellow) is added instead, $Q > K$ and reactants tend to be formed. For this reaction, we have used $K = 1$ for the equilibrium blue $\rightleftharpoons$ yellow.

ANIMATION FIGURE 5J.5

EXAMPLE 5J.1 Predicting the effect of adding or removing reactants and products

Nitric oxide, NO, is an intermediate in the production of nitric acid. It is produced commercially by the controlled oxidation of ammonia. Suppose you are considering how to increase the amount of NO produced each day in the equilibrium $4\,NH_3(g) + 5\,O_2(g) \rightleftharpoons 4\,NO(g) + 6\,H_2O(g)$. Predict the effect on each equilibrium concentration of (a) the removal of NO; (b) the addition of NH_3; (c) the addition of H_2O.

ANTICIPATE Because the removal of a product and the addition of a reactant cause a shift toward the formation of product, you should expect (a) and (b) to shift the equilibrium toward products, whereas (c) will shift the reaction toward reactants.

PLAN Consider how each change will affect the value of Q and what change is then needed to re-establish equilibrium.

SOLVE

For this reaction

$$4\,NH_3(g) + 5\,O_2(g) \rightleftharpoons 4\,NO(g) + 6\,H_2O(g) \qquad Q = \frac{(P_{NO})^4 (P_{H_2O})^6}{(P_{NH_3})^4 (P_{O_2})^5}$$

(a) The removal of NO (a product) from the equilibrium mixture decreases Q below K, so the equilibrium adjusts as more products are formed at the expense of reactants.
(b) When NH_3 is added to the system at equilibrium, Q drops below K, so, once again, the equilibrium adjusts as more products are formed at the expense of reactants.
(c) The addition of H_2O raises Q above K, so reactants form at the expense of products.

EVALUATE As anticipated, the reaction tends to shift toward products for (a) and (b) and toward reactants for (c).

Self-test 5J.1A Consider the equilibrium $SO_3(g) + NO(g) \rightleftharpoons SO_2(g) + NO_2(g)$. Predict the effect on the equilibrium of (a) the addition of NO; (b) the addition of NO_2; (c) the removal of SO_2.

[*Answer:* The equilibrium tends to shift toward (a) products; (b) reactants; (c) products.]

Self-test 5J.1B Consider the equilibrium $CO(g) + 2 H_2(g) \rightleftharpoons CH_3OH(g)$. Predict the effect on the equilibrium of (a) the addition of H_2; (b) the removal of CH_3OH; (c) the removal of CO.

Related Exercises 5J.1–5J.4

Le Chatelier's principle suggests a good way of ensuring that a reaction goes on generating a substance: simply remove products as they are formed. In its continuing hunt for equilibrium, the reaction goes on generating additional product. Industrial processes are rarely allowed to reach equilibrium for just this reason. In the commercial synthesis of ammonia, for instance, ammonia is continuously removed by circulating the equilibrium mixture through a refrigeration unit in which only the ammonia condenses. Therefore, the hydrogen and nitrogen continue to react to form additional product.

EXAMPLE 5J.2   Calculating the equilibrium composition after addition of a reagent

Suppose you are designing a chemical plant that is providing phosphorus compounds to other industries and you need to explore the equilibrium properties for the reaction of $PCl_5(g) \rightleftharpoons PCl_3(g) + Cl_2(g)$. The reaction has reached equilibrium at 250 °C (the equilibrium partial pressures of the components are $P_{PCl_5} = 0.02$ bar, $P_{PCl_3} = 1.28$ bar, $P_{Cl_2} = 1.28$ bar, and $K = 78.3$). You add 0.0100 mol $Cl_2(g)$ to the equilibrium mixture in the container (of volume 500. mL); then the system is once again allowed to reach equilibrium. Use this information to calculate the new composition of the equilibrium mixture.

ANTICIPATE Because a product has been added, you should expect the reaction to respond by forming reactants at the expense of products.

PLAN The general procedure is like that set out in Toolbox 5I.1, except that the direction of reaction might not be toward products. Write the expression for the equilibrium constant, and then set up an equilibrium table. In this case, use as the initial partial pressures those found immediately after addition of the reagent but before the reaction has responded.

SOLVE The 0.0100 mol $Cl_2(g)$ added to the container corresponds to an additional partial pressure of chlorine.

From $P_J = n_J RT/V$,

$$P_{Cl_2} = \frac{(0.0100 \text{ mol}) \times (8.3145 \times 10^{-2} \text{ L·bar·K}^{-1}\text{mol}^{-1}) \times (523 \text{ K})}{0.500 \text{ L}}$$

$$= 0.870 \text{ bar}$$

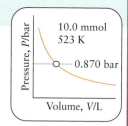

The total partial pressure of chlorine immediately after the addition of the chlorine gas is therefore $1.28 + 0.870$ bar $= 2.15$ bar. Construct the following table, with all partial pressures in bar, noting a decrease for partial pressures of the products and an increase for the partial pressure of the reactant. The reaction is

$$PCl_5(g) \rightleftharpoons PCl_3(g) + Cl_2(g) \qquad K = \frac{P_{PCl_3}P_{Cl_2}}{P_{PCl_5}}$$

	PCl_5	PCl_3	Cl_2
Step 1 initial partial pressure	0.02	1.28	2.15
Step 2 change in partial pressure	$+x$	$-x$	$-x$
Step 3 equilibrium partial pressure	$0.02 + x$	$1.28 - x$	$2.15 - x$

Then, from the expression for K:

$$K = \frac{(1.28 - x) \times (2.15 - x)}{0.02 + x} = \frac{2.75 - 3.43x + x^2}{0.02 + x}$$

Rearrange the equation with $K = 78.3$.

$$78.3 = \frac{2.75 - 3.43x + x^2}{0.02 + x}$$

$$78.3 \times (0.02 + x) = 2.75 - 3.43x + x^2$$

$$1.57 + 78.3x = 2.75 - 3.43x + x^2$$

$$x^2 - 81.7x + 1.18 = 0$$

Find the values of x,

Solve by using the quadratic formula (or software).

$$x = 81.7\ldots \text{ and } 0.0144\ldots$$

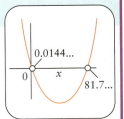

Select $x = 0.0144\ldots$ (because the partial pressures must be positive). It follows that, at equilibrium,

From $P_{PCl_5} = 0.02 + x$ bar, $P_{PCl_5} = 0.02 + 0.01$ bar $= 0.03$ bar

From $P_{PCl_3} = 1.28 - x$ bar, $P_{PCl_3} = 1.28 - 0.01$ bar $= 1.27$ bar

From $P_{Cl_2} = 2.15 - x$ bar, $P_{Cl_2} = 2.15 - 0.01$ bar $= 2.14$ bar

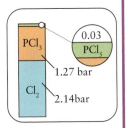

EVALUATE As anticipated, the partial pressure of reactant has increased and the partial pressures of the products have decreased from their initial values (**FIG. 5J.6**).

Self-test 5J.2A Suppose that the equilibrium mixture in Example 5I.3 is perturbed by adding 3.00 mol N_2(g) and then the system is once again allowed to reach equilibrium. Use this information and data from Example 5I.3 to calculate the new composition of the equilibrium mixture.

[*Answer:* 23.2 bar N_2; 6.21 bar O_2; 1.0×10^{-12} bar N_2O]

Self-test 5J.2B Nitrogen, hydrogen, and ammonia gases have reached equilibrium in a container of volume 1.00 L at 298 K. Equilibrium partial pressures are 0.080 atm, 0.050 atm, and 2.60 atm for nitrogen, hydrogen, and ammonia, respectively. For reaction A, $K = 6.8 \times 10^5$. Calculate the new equilibrium partial pressures if half the NH_3 is removed from the container and equilibrium is re-established.

Related Exercises 5J.7, 5J.8

When the equilibrium composition is perturbed by adding or removing a reactant or product, reaction tends to occur in the direction that restores the value of Q to that of the constant K.

5J.2 Compressing a Reaction Mixture

A gas-phase equilibrium may respond to compression—a reduction in volume—of the reaction vessel. According to Le Chatelier's principle, the composition will tend to change in a way that minimizes the resulting increase in pressure. For instance, in the dissociation of gaseous I_2 to form I atoms, I_2(g) $\rightleftharpoons$ 2 I(g), 1 mol of gaseous reactant molecules produces 2 mol of gaseous product atoms. The forward reaction increases the number of particles in the container and hence the total pressure of the system, and the reverse reaction decreases it. It follows that, when the mixture is compressed, the equilibrium composition will tend to shift in favor of the reactant, I_2, because that response minimizes the increase in pressure (**FIG. 5J.7**). Expansion of the system results in the opposite response, a tendency for I_2 to dissociate into free atoms. In the formation of ammonia, reaction A, 2 mol of gas molecules are produced from 4 mol of gas molecules. Therefore, Haber realized that, to increase the yield of ammonia, he needed to carry out the synthesis with highly compressed gases. The actual industrial process uses pressures of at least 250 atm (**FIG. 5J.8**).

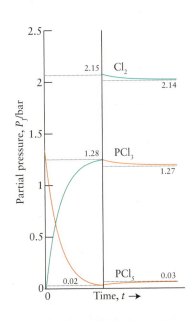

FIGURE 5J.6 The response of the equilibrium mixture illustrated in Fig. 5I.4 to the addition of chlorine. The curves for Cl_2 and PCl_3 are superimposed until the additional Cl_2 has been added.

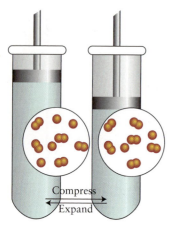

FIGURE 5J.7 Le Chatelier's principle predicts that, when a reaction at equilibrium is compressed, the number of molecules in the gas phase will tend to decrease. This diagram illustrates the effect of compression and expansion on the dissociation equilibrium of a diatomic molecule. Note the increase in the relative concentration of diatomic molecules as the system is compressed and the decrease when the system expands.

FIGURE 5J.8 A high-pressure vessel used for the catalytic synthesis of ammonia. The vessel must be able to withstand internal pressures of greater than 250 atm. *(ZD GONG/EPA/ Newscom)*

EXAMPLE 5J.3 Predicting the effect of compression on an equilibrium

Nitrogen tetroxide, N_2O_4, is a rocket propellant that can be prepared by the dimerization of nitrogen dioxide, NO_2. If you were a chemical engineer producing N_2O_4, you would need to know whether to use a high or a low pressure for the synthesis. Predict the effect of compression on the equilibrium composition of the reaction mixtures in which the equilibria (a) $2\,NO_2(g) \rightleftharpoons N_2O_4(g)$ and (b) $N_2(g) + O_2(g) \rightleftharpoons 2\,NO(g)$ have been established.

PLAN Reaction will take place in the direction that reduces the increase in pressure, and fewer gas molecules generate a lower pressure. Therefore, compare the number of reactant gas molecules with the number of product gas molecules.

SOLVE (a) In the forward reaction, two NO_2 molecules combine to form one N_2O_4 molecule. Hence, compression favors the formation of N_2O_4. (b) Because neither direction corresponds to a reduction in the number of gas-phase molecules, compressing the mixture should have no effect on the composition of the equilibrium mixture. (In practice, there will be a small effect due to the nonideality of the gases.)

Self-test 5J.3A Predict the effect of compression on the equilibrium composition of the reaction $CH_4(g) + H_2O(g) \rightleftharpoons CO(g) + 3\,H_2(g)$.

[*Answer:* Reactants favored]

Self-test 5J.3B Predict the effect of compression on the equilibrium composition of the reaction $CO_2(g) + H_2O(l) \rightleftharpoons H_2CO_3(aq)$.

Related Exercises 5J.9, 5J.10

The effect of compression on an equilibrium mixture can be explained by showing that compressing a system will change each partial pressure that occurs in the expression for K while leaving K itself unchanged.

How Is That Done?

Suppose you want to discover the effect of compression on the equilibrium $2\,NO_2(g) \rightleftharpoons N_2O_4(g)$. Write the equilibrium constant in its complete form (to be careful with units) as

$$K = \frac{P_{N_2O_4}/P^\circ}{(P_{NO_2}/P^\circ)^2}$$

Next, because the focus must be on the volume of the system, to consider compression express K in terms of the volume by writing $P_J = n_J RT/V$ for each substance:

$$K = \frac{n_{N_2O_4}RT/VP^\circ}{(n_{NO_2}RT/VP^\circ)^2} = \frac{n_{N_2O_4}}{(n_{NO_2})^2} \times \frac{P^\circ}{RT} \times V$$

Now, because P°/RT is a constant at constant temperature, for K to remain constant when the volume (V) of the system is reduced, the ratio $n_{N_2O_4}/(n_{NO_2})^2$ must increase. That is, the amount of NO_2 must decrease and the amount of N_2O_4 must increase. Therefore, as the volume of the system is decreased, the equilibrium shifts to a smaller total number of gas molecules. Conversely, if the system is expanded, more NO_2 would be produced and the equilibrium would shift in the direction of a larger total number of gas molecules.

Suppose that the total pressure inside a reaction vessel is increased by pumping in argon or some other inert gas at constant volume. The reacting gases continue to occupy the same volume, and so their individual molar concentrations and partial pressures remain unchanged despite the presence of an inert gas. In this case, therefore, provided that the gases can be regarded as ideal, the equilibrium composition is unaffected even though the total pressure has increased.

Compression of a reaction mixture at equilibrium tends to drive the reaction in the direction that reduces the number of gas-phase molecules; increasing the pressure by introducing an inert gas has no effect on the equilibrium composition.

5J.3 Temperature and Equilibrium

The equilibrium constant of a reaction depends on the temperature. Two experimental observations summarize this dependence. It is found that for exothermic (heat releasing) reactions, when the temperature is raised the composition of the equilibrium mixture shifts in favor of reactants (K decreases) but the opposite is true for endothermic (heat absorbing) reactions (K increases).

Le Chatelier's principle is consistent with these observations. Because the composition shifts toward reactants in an exothermic reaction, less heat is released, which can be regarded as helping to offset the rise in temperature. Similarly, because the composition shifts toward products in an endothermic reaction, more heat is absorbed and that helps to offset the increase in temperature.

An example is the decomposition of carbonates. A reaction such as $CaCO_3(s) \rightarrow CaO(s) + CO_2(g)$ is strongly endothermic, and an appreciable partial pressure of carbon dioxide is present at equilibrium only if the temperature is high. For instance, at 800 °C, the partial pressure is 0.22 atm at equilibrium. If the heating takes place in an open container, this partial pressure is never reached, because equilibrium is never reached. The gas drifts away, and the calcium carbonate decomposes completely, leaving a solid residue of CaO. However, if the surroundings are already so rich in carbon dioxide that its partial pressure exceeds 0.22 atm, then virtually no decomposition occurs: for every CO_2 molecule that is formed, one reacts with CaO and is so converted back into carbonate. This dynamic process is probably what happens on the hot surface of Venus (FIG. 5J.9), where the partial pressure of carbon dioxide is about 87 atm. This high pressure has led to speculation that the planet's surface is rich in carbonates, despite its high temperature (about 500 °C).

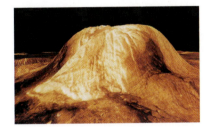

FIGURE 5J.9 A radar image of the surface of Venus. Although the rocks are very hot, the partial pressure of carbon dioxide in the atmosphere is so great that carbonates may be abundant. *(NASA/JPL)*

EXAMPLE 5J.4   Predicting the effect of temperature on an equilibrium

One step in the manufacture of sulfuric acid is the formation of sulfur trioxide by the combustion of SO_2 with O_2 in the presence of a vanadium(V) oxide catalyst. Suppose you are working out how to increase the equilibrium composition of sulfur trioxide. Predict how the equilibrium composition for the sulfur trioxide synthesis will tend to change when the temperature is raised.

ANTICIPATE Combustion reactions are all exothermic, so you should expect the equilibrium of this to shift in favor of reactants when the temperature is raised.

PLAN Verify that the reaction is exothermic. To find the standard reaction enthalpy, use the standard enthalpies of formation given in Appendix 2A.

SOLVE

The chemical equation is

$$2\,SO_2(g) + O_2(g) \xrightarrow{V_2O_5} 2\,SO_3(g)$$

The standard reaction enthalpy of the forward reaction is

$$\Delta H° = (2\ \text{mol}) \times \Delta H_f°(SO_3, g) - \{(2\ \text{mol}) \times \Delta H_f°(SO_2, g) + (1\ \text{mol}) \times \underbrace{\Delta H_f°(O_2, g)}_{\Delta H_f° = 0}\}$$
$$= 2(-395.75\ \text{kJ}) - 2(-296.83\ \text{kJ}) = -197.78\ \text{kJ}$$

EVALUATE Because the formation of SO_3 is exothermic, raising the temperature of the equilibrium mixture favors the decomposition of SO_3 to SO_2 and O_2; as a consequence, the pressures of SO_2 and O_2 will increase and that of SO_3 will decrease, as anticipated.

Self-test 5J.4A Predict the effect of raising the temperature on the equilibrium composition of the reaction $N_2O_4(g) \rightleftharpoons 2\ NO_2(g)$. See Appendix 2A for data.

[*Answer:* The pressure of NO_2 will increase.]

Self-test 5J.4B Predict the effect of lowering the temperature on the equilibrium composition of the reaction $2\ CO(g) + O_2(g) \rightleftharpoons 2\ CO_2(g)$. See Appendix 2A for data.

Related Exercises 5J.11, 5J.12

The effect of temperature on the equilibrium composition arises from the dependence of the equilibrium constant on the temperature. As with other aspects of equilibrium, there are kinetic and thermodynamic ways to understand the dependence.

How Is That Explained...

...using kinetics?

The forward and reverse rate constants in Eq. 3 of Topic 7C ($K = k_r/k_r{}'$) depend on temperature according to the Arrhenius equation. From $K = k_r/k_r{}'$ and the Arrhenius equations for the two rate constants

$$K = \frac{A e^{-E_a/RT}}{A' e^{-E_a'/RT}} \overset{e^x/e^y = e^{x-y}}{\cong} \frac{A}{A'} e^{-(E_a - E_a')/RT}$$

The difference between the forward and reverse activation energies can be identified with the reaction enthalpy (**FIG. 5J.10**), so

$$K = \text{constant} \times e^{-\Delta H_r^\circ/RT},$$

so

$$\ln xy = \ln x + \ln y,$$
$$\ln e^x = x$$
$$\ln K \overset{\frown}{=} \ln(\text{constant}) - \frac{\Delta H_r^\circ}{RT}$$

For the two temperatures T_1 and T_2, when the equilibrium constants are K_1 and K_2 and any temperature dependence of the reaction enthalpy is ignored, it follows that

$$\ln K_1 - \ln K_2 = \left\{ \ln(\text{constant}) - \frac{\Delta H_r^\circ}{RT_1} \right\}$$
$$- \left\{ \ln(\text{constant}) - \frac{\Delta H_r^\circ}{RT_2} \right\}$$
$$\ln K_1 - \ln K_2 = -\frac{\Delta H_r^\circ}{R} \left\{ \frac{1}{T_1} - \frac{1}{T_2} \right\}$$

...using thermodynamics?

The relations between the equilibrium constants K_1 and K_2 and the standard Gibbs free energies of reaction at two temperatures T_1 and T_2 is

$$\Delta G_{r,1}^\circ = -RT_1 \ln K_1$$
$$\Delta G_{r,2}^\circ = -RT_2 \ln K_2$$

These two expressions rearrange to

$$\ln K_1 = -\frac{\Delta G_{r,1}^\circ}{RT_1} \qquad \ln K_2 = -\frac{\Delta G_{r,2}^\circ}{RT_2}$$

Subtraction of the second from the first gives

$$\ln K_1 - \ln K_2 = -\frac{1}{R} \left\{ \frac{\Delta G_{r,1}^\circ}{T_1} - \frac{\Delta G_{r,2}^\circ}{T_2} \right\}$$

At this point, introduce the definition of ΔG_r° in terms of ΔH_r° and ΔS_r°:

$$\Delta G_{r,1}^\circ = \Delta H_{r,1}^\circ - T_1 \Delta S_{r,1}^\circ$$
$$\Delta G_{r,2}^\circ = \Delta H_{r,2}^\circ - T_2 \Delta S_{r,2}^\circ$$

which gives

$$\ln K_1 - \ln K_2 = -\frac{1}{R} \times$$

$$\left\{ \frac{\overbrace{\Delta H_{r,1}^\circ - T_1 \Delta S_{r,1}^\circ}^{\Delta G_{r,1}^\circ}}{T_1} - \frac{\overbrace{\Delta H_{r,2}^\circ - T_2 \Delta S_{r,2}^\circ}^{\Delta G_{r,2}^\circ}}{T_2} \right\}$$

$$= -\frac{1}{R} \left\{ \frac{\Delta H_{r,1}^\circ}{T_1} - \frac{\Delta H_{r,2}^\circ}{T_2} - \Delta S_{r,1}^\circ + \Delta S_{r,2}^\circ \right\}$$

It is usually reasonable to assume that ΔH_r° and ΔS_r° are both approximately independent of temperature over the range of temperatures of interest. When that approximation is made, the reaction entropies cancel, leaving

$$\ln K_1 - \ln K_2 = -\frac{\Delta H_r^\circ}{R} \left\{ \frac{1}{T_1} - \frac{1}{T_2} \right\}$$

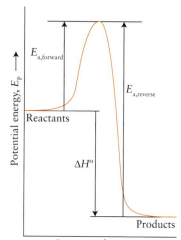

FIGURE 5J.10 The difference between the forward and reverse activation energies is the difference in (molar) energies between the products and reactants; to a good approximation, this difference is equal to the standard reaction enthalpy.

The expression just derived, from either approach, is a quantitative version of Le Chatelier's principle for the effect of temperature. It is normally rearranged into the **van 't Hoff equation** by multiplying through by -1 and then using $\ln a - \ln b = \ln(a/b)$:

$$\ln \frac{K_2}{K_1} = \frac{\Delta H_r^{\circ}}{R}\left\{\frac{1}{T_1} - \frac{1}{T_2}\right\} \tag{1}$$

In this expression, K_1 is the equilibrium constant when the temperature is T_1 and K_2 is the equilibrium constant when the temperature is T_2.

What Does This Equation Tell You? If the reaction is endothermic, then ΔH_r° is positive. If $T_2 > T_1$, then $1/T_2 < 1/T_1$ and the term in braces is also positive. Therefore, $\ln(K_2/K_1)$ is positive, which implies that $K_2/K_1 > 1$ and therefore that $K_2 > K_1$. In other words, an increase in temperature favors the formation of product if the reaction is endothermic. The opposite effect is predicted for an exothermic reaction, because ΔH_r° is then negative. Therefore, the van 't Hoff equation accounts for Le Chatelier's principle for the effect of temperature on an equilibrium.

This equation is sometimes called the *van 't Hoff isochore*, to distinguish it from van 't Hoff's osmotic pressure equation (Topic 5F). An *isochore* is the plot of an equation for a constant-volume process.

EXAMPLE 5J.5 Predicting the value of an equilibrium constant at a different temperature

Most reactions proceed faster at higher temperatures, and many industrial processes are carried out at high temperatures. However, for exothermic reactions increasing the temperature reduces the equilibrium constant and thus the reaction yield. Suppose you are studying how to increase the yield of the ammonia synthesis reaction and decide to investigate the effect of raising the temperature. The equilibrium constant K for the synthesis of ammonia (reaction A) is 6.8×10^5 at 298 K. Predict its value at 400. K.

ANTICIPATE The synthesis of ammonia is known to be exothermic, so you should expect the equilibrium constant to be smaller at the higher temperature.

PLAN To use the van 't Hoff equation, you need the standard reaction enthalpy, which can be calculated from the standard enthalpies of formation in Appendix 2A. Equation 1 requires you to use the "molar" convention.

What should you assume? Assume that the gases are ideal and that the reaction enthalpy is constant over the temperature range of interest.

SOLVE The standard reaction enthalpy for the forward reaction in A is

$$\Delta H_r^{\circ} = 2\Delta H_f^{\circ}(NH_3, g) = 2(-46.11 \text{ kJ·mol}^{-1})$$
$$= -92.22 \text{ kJ·mol}^{-1} = -9.222 \times 10^4 \text{ J·mol}^{-1}$$

From $\ln(K_2/K_1) = (\Delta H_r^{\circ}/R)\{(1/T_1) - (1/T_2)\}$,

$$\ln \frac{K_2}{K_1} = \frac{-9.222 \times 10^4 \text{ J·mol}^{-1}}{8.3145 \text{ J·K}^{-1}\text{·mol}^{-1}} \times \left(\frac{1}{298\text{K}} - \frac{1}{400.\text{K}}\right) = -9.49\ldots$$

Take antilogarithms (e^x):

$$K_2 = K_1 e^{-9.49\ldots} = (6.8 \times 10^5) \times e^{-9.49\ldots} = 51$$

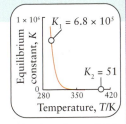

EVALUATE The equilibrium constant at 400. K is smaller, as expected, and is close to the experimental value of 41 in Table 5G.2. It is not exactly the same because ΔH_r° changes slightly over the temperature range in question. This conclusion shows why Haber needed to use a catalyst to accelerate a reaction instead of raising the temperature: raising the temperature would yield less ammonia at equilibrium.

Self-test 5J.5A The equilibrium constant K for $2 SO_3(g) \rightleftharpoons 2 SO_2(g) + O_2(g)$ is 2.5×10^{-25} at 298 K. Predict its value at 500. K.

[***Answer:*** 2.5×10^{-11}]

Self-test 5J.5B The equilibrium constant K for $PCl_5(g) \rightleftharpoons PCl_3(g) + Cl_2(g)$ 78.3 at 523 K. Predict its value at 800. K.

Related Exercises 5J.15, 5J.16

One word of warning: when using the van 't Hoff equation for reactions involving gases, the equilibrium constants must be K, not K_c. If you want a new value for K_c for a gas-phase reaction, you must convert from K_c into K at the initial temperature (by using Eq. 4b of Topic 5H, $K = (RT)^{\Delta n_r} K_c$), then use the van 't Hoff equation to calculate the value of K at the new temperature, and finally convert that K into the new K_c at the new temperature.

> *Raising the temperature of an exothermic reaction lowers the value of K; raising the temperature of an endothermic reaction increases the value of K. The van 't Hoff equation expresses the effect quantitatively.*

What have you learned in this Topic?

You have learned that because reactions at equilibrium are dynamic, they respond to changes in conditions such as addition of reactants or products and changes in temperature. You can now use Le Chatelier's principle to predict the effect of a change in conditions and make quantitative predictions by using the van 't Hoff equation for the effect of temperature.

The skills you have mastered are the ability to:

☐ **1.** Use Le Chatelier's principle to predict how the equilibrium composition of a reaction mixture is affected by adding or removing reagents (Examples 5J.1 and 5J.2), compressing or expanding the mixture (Example 5J.3), or changing the temperature (Example 5J.4).

☐ **2.** Predict the value of K at one temperature from its value at another temperature (Example 5J.5).

Topic 5J Exercises

5J.1 Consider the equilibrium $CO(g) + H_2O(g) \rightleftharpoons CO_2(g) + H_2(g)$. (a) If the partial pressure of CO_2 is increased, what happens to the partial pressure of H_2? (b) If the partial pressure of CO is decreased, what happens to the partial pressure of CO_2? (c) If the concentration of CO is increased, what happens to the concentration of H_2? (d) If the concentration of H_2O is decreased, what happens to the equilibrium constant for the reaction?

5J.2 Consider the equilibrium $CH_4(g) + 4 I_2(s) \rightleftharpoons CI_4(g) + 4 HI(g)$. (a) If the partial pressure of CH_4 is increased, what happens to the partial pressure of CI_4? (b) If the partial pressure of CI_4 is decreased, what happens to the amount of I_2? (c) If the concentration of HI is increased, what happens to the equilibrium constant for the reaction? (d) If the amount of I_2 is increased, what happens to the concentration of CI_4?

5J.3 The four gases NH_3, O_2, NO, and H_2O are mixed in a reaction vessel and allowed to reach equilibrium in the reaction $4 NH_3(g) + 5 O_2(g) \rightleftharpoons 4 NO(g) + 6 H_2O(g)$. Certain changes (see the following table) are then made to this mixture. Considering each change separately, state the effect (increase, decrease, or no change) that the change has on the original equilibrium values of the quantity in the second column (or K, if that is specified). The temperature and volume are constant.

Change	Quantity
(a) add NO	amount of H_2O
(b) add NO	amount of O_2
(c) remove H_2O	amount of NO
(d) remove O_2	amount of NH_3
(e) add NH_3	K
(f) remove NO	amount of NH_3
(g) add NH_3	amount of O_2

5J.4 The four substances HCl, I_2, HI, and Cl_2 are mixed in a reaction vessel and allowed to reach equilibrium in the reaction $2 HCl(g) + I_2(s) \rightleftharpoons 2 HI(g) + Cl_2(g)$. Certain changes (which are specified in the first column in the following table) are then made to this mixture. Considering each change separately, state the effect (increase, decrease, or no change) that the change has on the original equilibrium value of the quantity in the second column (or K, if that is specified). The temperature and volume are constant.

Change	Quantity
(a) add HCl	amount of HI
(b) add I_2	amount of Cl_2
(c) remove HI	amount of Cl_2
(d) remove Cl_2	amount of HCl
(e) add HCl	K
(f) remove HCl	amount of I_2
(g) add I_2	K

5J.5 State whether reactants or products will be favored by an increase in the total pressure (resulting from compression) on each of the following equilibria. If there is no change, explain why that is so.

(a) $2 O_3(g) \rightleftharpoons 3 O_2(g)$

(b) $H_2O(g) + C(s) \rightleftharpoons H_2(g) + CO(g)$

(c) $4 NH_3(g) + 5 O_2(g) \rightleftharpoons 4 NO(g) + 6 H_2O(g)$

(d) $2 HD(g) \rightleftharpoons H_2(g) + D_2(g)$

(e) $Cl_2(g) \rightleftharpoons 2 Cl(g)$

5J.6 State what happens to the concentration of the indicated substance when the total pressure on each of the following equilibria is increased (by compression):

(a) $NO_2(g)$ in $2 Pb(NO_3)_2(s) \rightleftharpoons 2 PbO(s) + 4 NO_2(g) + O_2(g)$

(b) $NO(g)$ in $3 NO_2(g) + H_2O(l) \rightleftharpoons 2 HNO_3(aq) + NO(g)$

(c) $HI(g)$ in $2 HCl(g) + I_2(s) \rightleftharpoons 2 HI(g) + Cl_2(g)$

(d) $SO_2(g)$ in $2 SO_2(g) + O_2(g) \rightleftharpoons 2 SO_3(g)$

(e) $NO_2(g)$ in $NO(g) + O_2(g) \rightleftharpoons 2 NO_2(g)$

5J.7 A reactor for the production of ammonia by the Haber process is found to be at equilibrium with $P_{N_2} = 3.11$ bar, $P_{H_2} = 1.64$ bar, and $P_{NH_3} = 23.72$ bar. If the partial pressure of N_2 is increased by 1.57 bar, what will be the partial pressure of each gas once equilibrium is re-established?

5J.8 In a laboratory studying the extraction of iron metal from iron ore, the following reaction was carried out at 1270 K in a reaction vessel of volume 10.0 L: $FeO(s) + CO(g) \rightleftharpoons Fe(s) + CO_2(g)$. At equilibrium the partial pressure of CO was 4.24 bar and that of CO_2 was 1.71 bar. The pressure of the CO_2 was reduced to 0.43 bar by reacting some of it with NaOH and the system was allowed to reach equilibrium again. What will be the partial pressure of each gas once equilibrium is re-established?

5J.9 Consider the equilibrium $3 NH_3(g) + 5 O_2(g) \rightleftharpoons 4 NO(g) + 6 H_2O(g)$. (a) What happens to the partial pressure of NH_3 when the partial pressure of NO is increased? (b) Does the partial pressure of O_2 decrease when the partial pressure of NH_3 is decreased?

5J.10 Consider the equilibrium $2 SO_2(g) + O_2(g) \rightleftharpoons 2 SO_3(g)$. (a) What happens to the partial pressure of SO_3 when the partial pressure of SO_2 is decreased? (b) If the partial pressure of SO_2 is increased, what happens to the partial pressure of O_2?

5J.11 Predict whether each of the following equilibria will shift toward products or reactants with a temperature increase:

(a) $N_2O_4(g) \rightleftharpoons 2 NO_2(g)$, $\Delta H° = +57$ kJ

(b) $X_2(g) \rightleftharpoons 2 X(g)$, where X is a halogen

(c) $Ni(s) + 4 CO(g) \rightleftharpoons Ni(CO)_4(g)$, $\Delta H° = -161$ kJ

(d) $CO_2(g) + 2 NH_3(g) \rightleftharpoons$
$$CO(NH_2)_2(s) + H_2O(g), \Delta H° = -90 kJ$$

5J.12 Predict whether each of the following equilibria will shift toward products or reactants with a temperature increase:

(a) $CH_4(g) + H_2O(g) \rightleftharpoons CO(g) + 3 H_2(g)$, $\Delta H° = +206$ kJ

(b) $CO(g) + H_2O(g) \rightleftharpoons CO_2(g) + H_2(g)$, $\Delta H° = -41$ kJ

(c) $2 SO_2(g) + O_2(g) \rightleftharpoons 2 SO_3(g)$, $\Delta H° = -198$ kJ

5J.13 A gaseous mixture consisting of 2.23 mmol N_2 and 6.69 mmol H_2 in a 500.-mL container was heated to 600. K and allowed to reach equilibrium. Will more ammonia be formed if that equilibrium mixture is then heated to 700. K? For $N_2(g) + 3 H_2(g) \rightleftharpoons 2 NH_3(g)$, $K = 1.7 \times 10^{-3}$ at 600. K and 7.8×10^{-5} at 700. K.

5J.14 A gaseous mixture consisting of 1.1 mmol SO_2 and 2.2 mmol O_2 in a 250-mL container was heated to 500. K and allowed to reach equilibrium. Will more sulfur trioxide be formed if that equilibrium mixture is cooled to 298 K? For the reaction $2 SO_2(g) + O_2(g) \rightleftharpoons 2 SO_3(g)$, $K = 2.5 \times 10^{10}$ at 500. K and 4.0×10^{24} at 298 K.

5J.15 Calculate the equilibrium constant at 25 °C and at 150 °C for each of the following reactions, using data available in Appendix 2A:

(a) $NH_4Cl(s) \rightleftharpoons NH_3(g) + HCl(g)$

(b) $H_2(g) + D_2O(l) \rightleftharpoons D_2(g) + H_2O(l)$

5J.16 Calculate the equilibrium constant at 25 °C and at 100 °C for each of the following reactions, using data available in Appendix 2A:

(a) $2 CuO(s) \rightleftharpoons 2 Cu(s) + O_2(g)$

(b) $C_2H_4(g) + H_2(g) \rightleftharpoons C_2H_6(g)$

5J.17 By combining the relation for K_c in terms of K with the van 't Hoff equation, find the analog of the van 't Hoff equation for K_c.

5J.18 The vaporization of a liquid can be treated as a special case of an equilibrium. How does the vapor pressure of a liquid vary with temperature? *Hint:* Devise a version of the van 't Hoff equation that applies to vapor pressure by first writing the equilibrium constant K for vaporization.

The following Example and Exercises draw on material from throughout Focus 5.

 Online Cumulative Example

Ethanol (CH_3CH_2OH) is added to gasoline to improve combustion and reduce air pollution. It is produced industrially by the hydration of ethene (C_2H_4) at high temperatures and pressures. You are a chemical engineer and wish to explore the thermodynamics and equilibrium properties of the reaction $C_2H_4(g) + H_2O(g) \rightleftharpoons CH_3CH_2OH(g)$.

(a) From the data in Appendix 2A, write the thermochemical equation for the industrial synthesis of ethanol.

(b) From the data in Appendix 2A, (i) determine $\Delta G°$ for this system at 25 °C and (ii) find the equilibrium constant at

300 °C. (iii) Would an increase in temperature favor reactants or products?

(c) A reactor is charged with 60 bar of $C_2H_4(g)$ and 40 bar of $H_2O(g)$. Determine the equilibrium pressures of $C_2H_4(g)$, $H_2O(g)$, and $CH_3CH_2OH(g)$ at 300 °C.

(d) The equilibrium mixture from part (c) is cooled and the water and ethanol condense. The excess ethene is allowed to escape. (i) What is the vapor pressure of the resulting liquid mixture at 25 °C? (ii) What is the mole fraction of ethanol in the vapor in equilibrium with the condensed mixture at 25 °C?

⟫ The Online Cumulative Example solution can be found at http://macmillanhighered.com/chemicalprinciples7e

FOCUS 5 Exercises

5.1 Are the water molecules oriented in the same way or differently around the cations and anions when sodium chloride dissolves? Explain your answer.

5.2 When sodium chloride dissolves in water, can a chloride ion be removed from the crystal by one water molecule, or are several required to remove it? Explain your answer.

5.3 Complete the following statements about the effect of intermolecular forces on the physical properties of a substance. (a) The higher the boiling point of a liquid, the (stronger, weaker) are its intermolecular forces. (b) Substances with strong intermolecular forces have (high, low) vapor pressures at room temperature. (c) Substances with strong intermolecular forces typically have (high, low) surface tensions. (d) The higher the vapor pressure of a liquid at a given temperature, the (stronger, weaker) are its intermolecular forces. (e) Because nitrogen, N_2, has (strong, weak) intermolecular forces, it has a (high, low) critical temperature. (f) Substances with high vapor pressures at room temperature have correspondingly (high, low) boiling points. (g) Because water has a high boiling point, it must have (strong, weak) intermolecular forces and a correspondingly (high, low) enthalpy of vaporization.

5.4 Hydrogen peroxide, H_2O_2, is a syrupy liquid with a vapor pressure lower than that of water and a boiling point of 152 °C. Account for the differences between these properties and those of water.

5.5 Explain the effect that an increase in temperature has on each of the following properties: (a) viscosity; (b) surface tension; (c) vapor pressure; (d) evaporation rate.

5.6 Explain how the vapor pressure of a liquid is affected by each of the following changes in conditions: (a) an increase in temperature; (b) an increase in surface area of the liquid; (c) an increase in volume above the liquid; (d) the addition of air to the volume above the liquid.

5.7 You have two beakers: one is filled with tetrachloromethane and the other with water. You also have two compounds, butane

($CH_3CH_2CH_2CH_3$) and calcium chloride. (a) In which liquid will butane dissolve? Sketch the local environment of the solute in the solution. (b) In which solvent will calcium chloride dissolve? Sketch the local environment of the solute in the solution.

5.8 Use the phase diagram for compound X below to answer these questions: (a) Is X a solid, liquid, or gas at normal room temperatures? (b) What is the normal melting point of X? (c) What is the vapor pressure of liquid X at −50 °C? (d) What is the vapor pressure of solid X at −100 °C?

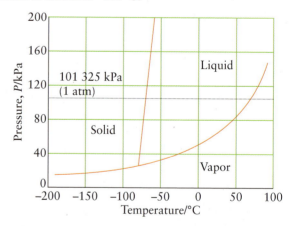

5.9 When a molar mass is determined from freezing-point depression, it is possible to make each of the following errors (among others). In each case, predict whether the error would cause the reported molar mass to be greater or less than the actual molar mass. (a) There was dust on the balance, causing the mass of solute to appear greater than it actually was. (b) The water was measured by volume, assuming a density of 1.00 g·cm⁻³, but the water was warmer and therefore less dense than assumed. (c) The thermometer was not calibrated accurately, and so the temperature of the freezing point was actually 0.5 °C higher than recorded. (d) The solution was not stirred sufficiently, and so not all the solute dissolved.

5.10 The freezing point of benzene is 5.53 °C. Suppose that 10.0 g of an organic compound used as a component of mothballs is dissolved in 80.0 g of benzene. The freezing point of the solution is 1.20 °C. (a) What is an approximate molar mass of the organic compound? (b) An elemental analysis of that substance indicated that the empirical formula is C_3H_2Cl. What is its molecular formula? (c) Using the atomic molar masses from the periodic table, calculate a more accurate molar mass of the compound.

5.11 The height of a column of liquid that can be supported by a given pressure is inversely proportional to its density. An aqueous solution of 0.010 g of a protein in 10. mL of water at 20 °C shows a rise of 5.22 cm in the apparatus shown in Fig. 5F.3. Assume the density of the solution to be 0.998 g·cm^{-3} and the density of mercury to be 13.6 g·cm^{-3}. (a) What is the molar mass of the protein? (b) What is the freezing point of the solution? (c) Which colligative property is best for measuring the molar mass of these large molecules? Give reasons for your answer.

5.12 A 0.060 M $C_6H_{12}O_6$(aq) solution (glucose) is separated from a 0.040 M $CO(NH_2)_2$(aq) solution (urea) by a semipermeable membrane at 25 °C. For both compounds, $i = 1$. (a) Which solution has the higher osmotic pressure? (b) Which solution becomes more dilute with the passage of H_2O molecules through the membrane? (c) To which solution should an external pressure be applied to maintain an equilibrium flow of H_2O molecules across the membrane? (d) What external pressure (in atm) should be applied in part (c)?

5.13 (a) From data in Appendix 2A, derive a numerical form of the Clausius–Clapeyron equation for methanol. (b) Use the equation to plot the appropriate quantities that should give a straight-line relation between vapor pressure and temperature. (c) Estimate the vapor pressure of methanol at 0.0 °C. (d) Estimate the normal boiling point of methanol.

5.14 (a) From data in Appendix 2A and Table 4C.1, derive a numerical form of the Clausius–Clapeyron equation for benzene. (b) Use the equation to plot the appropriate quantities that should give a straight-line relation between vapor pressure and temperature. (c) Estimate the boiling point of benzene when the external pressure is 0.655 atm. (d) Calculate $S_m°$ for gaseous benzene.

5.15 Colligative properties can be sources of insight into not only the properties of solutions, but also the properties of the solute. For example, acetic acid, CH_3COOH, behaves differently in two different solvents. (a) The freezing point of a 5.00% by mass aqueous acetic acid solution is −1.72 °C. What is the molar mass of the solute? Explain any discrepancy between the experimental and the expected molar mass. (b) The freezing-point depression associated with a 5.00% by mass solution of acetic acid in benzene is 2.32 °C. What is the experimental molar mass of the solute in benzene? What can you conclude about the nature of acetic acid in benzene?

5.16 Standard practice in chemical laboratories is to distill high-boiling-point substances under reduced pressure. Trichloroacetic acid has a standard enthalpy of vaporization of 57.8 kJ·mol^{-1} and a standard entropy of vaporization of 124 J·K^{-1}·mol^{-1}. Use this information to determine the pressure that is needed to distill trichloroacetic acid at 100. °C.

5.17 The temperature-dependence of the vapor pressure of phosphoryl chloride difluoride ($OPClF_2$) has been measured:

Temperature, T/K	190.	228	250.	273
Vapor pressure, P/Torr	3.2	68	240.	672

Plot $\ln P$ against $1/T$ (this plot is best created with the aid of a computer or a graphing calculator that can calculate a linear least-squares fit to the data). (b) From the plot (or a linear equation derived from it) in part (a), determine the standard enthalpy of vaporization of $OPClF_2$; (c) the standard entropy of vaporization of $OPClF_2$; and (d) the normal boiling point of $OPClF_2$. (e) If the pressure of a sample of $OPClF_2$ is reduced to 15 Torr, at what temperature will the sample boil?

5.18 From literature sources, find the critical pressures and temperatures of methane, methylamine (CH_3NH_2), ammonia, and tetrafluoromethane. Discuss the suitability of using each of these solvents for a supercritical extraction at room temperature in an autoclave that can withstand pressures of 100. atm.

5.19 From literature sources, find the critical temperatures for the gaseous hydrocarbons methane, ethane, propane, and butane. Explain the trends observed.

5.20 Consider an apparatus in which A and B are two 1.00-L flasks joined by a stopcock C. The volume of the stopcock is negligible. Initially, A and B are evacuated, the stopcock C is closed, and 1.50 g of diethyl ether, $C_2H_5OC_2H_5$, is introduced into flask A. The vapor pressure of diethyl ether is 57 Torr at −45 °C, 185 Torr at 0 °C, 534 Torr at 25 °C, and negligible below −86 °C. (a) If the stopcock is left closed and the flask is brought to equilibrium at −45 °C, what will be the pressure of diethyl ether in flask A? (b) If the temperature is raised to 25 °C, what will be the pressure of diethyl ether in the flask? (c) If the temperature of the assembly is returned to −45 °C and the stopcock C is opened, what will be the pressure of diethyl ether in the apparatus? (d) If flask A is maintained at −45 °C and flask B is cooled with liquid nitrogen (boiling point, −196 °C) with the stopcock open, what changes will take place in the apparatus? Assume ideal behavior.

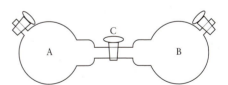

5.21 The apparatus in Exercise 5.20 is again evacuated. Then 35.0 g of chloroform, $CHCl_3$, is placed in flask A and 35.0 g of acetone, CH_3COCH_3, is placed in flask B. The system is allowed to come to equilibrium at 25 °C with stopcock C closed. The vapor pressures of chloroform and acetone at 25 °C are 195 Torr and 222 Torr, respectively. (a) What is the pressure in each flask at equilibrium? (b) The stopcock is now opened. What will be the final composition of the gas and liquid phases in each flask once the system reaches equilibrium, assuming ideal behavior? (c) Acetone and chloroform solutions show negative deviations from Raoult's law. How will this behavior affect the answers given in part (b)?

5.22 Pentane is a liquid with a vapor pressure of 512 Torr at 25 °C; at the same temperature, the vapor pressure of hexane is only 151 Torr. What composition must the liquid phase have if the gas-phase composition is to have equal partial pressures of pentane and hexane?

5.23 Using a spreadsheet program or graphing calculator, plot $\ln P$ against $1/T$ on the same set of axes for $\Delta H_{vap} = 15, 20., 25,$ and $30.$ kJ·mol^{-1} for the equation $\ln P = \Delta H_{vap}/RT$. Is the vapor pressure of a liquid more sensitive to changes in temperature if ΔH_{vap} is small or large? Explain your conclusion.

5.24 The combustion analysis of L-carnitine, an organic compound thought to build muscle strength, reported its composition as 52.16% C, 9.38% H, 8.69% N, and 29.78% O. The osmotic pressure of 100.00 mL of solution containing 0.322 g of L-carnitine in methanol was found to be 0.501 atm at 32 °C. Assuming that L-carnitine does not ionize in methanol, determine (a) the molar mass; (b) the molecular formula of L-carnitine.

5.25 Intravenous medications are often administered in 5.0% glucose, $C_6H_{12}O_6$(aq), by mass. What is the osmotic pressure of such solutions at 37 °C (body temperature)? Assume that the density of the solution is 1.0 g·mL^{-1}.

5.26 When sulfuric acid is added to water, so much energy is released as heat that the solution may boil. Would you expect this solution to show a deviation from Raoult's law? If so, which kind? Explain your reasoning.

5.27 Relative humidity at a particular temperature is defined as

$$\text{Relative humidity} = \frac{\text{partial pressure of water}}{\text{vapor pressure of water}} \times 100\%$$

The vapor pressure of water at various temperatures is given in Table 5A.2. (a) What is the relative humidity at 30 °C when the partial pressure of water is 25.0 Torr? (b) Explain what would be observed if the temperature of the air were to fall to 25 °C.

5.28 Interpret the following verse from Coleridge's *Rime of the Ancient Mariner:*

> Water, water, every where,
> And all the boards did shrink;
> Water, water every where,
> Nor any drop to drink.

5.29 A dextrose/saline aqueous solution that doctors commonly use to replace fluids in the body contains 1.75 g·L^{-1} NaCl and 40.0 g·L^{-1} dextrose ($C_6H_{12}O_6$). (a) What is the total molar concentration of all solutes in this solution? (b) What is the osmotic pressure of the solution at 25 °C? Assume total dissociation of the NaCl.

5.30 When 0.10 g of insulin is dissolved in 0.200 L of water, the osmotic pressure is 2.30 Torr at 20 °C. What is the molar mass of insulin?

5.31 An aqueous solution of the sugar mannitol ($C_6H_{12}O_6$) with a concentration of 180 mg·mL^{-1} is commonly used in veterinary medicine as an osmotic diuretic, which helps to remove water from living cells by osmosis. (a) What is the molarity of mannitol in the solution? (b) What is the osmotic pressure of the solution at 25 °C?

5.32 Human blood has an osmotic pressure relative to water of approximately 7.7 atm at body temperature (37 °C). In a hospital, intravenous glucose ($C_6H_{12}O_6$) solutions are often given. If a technician must mix 500. mL of a glucose solution for a patient, what mass of glucose should be used?

5.33 Dissociation of a diatomic molecule, $X_2(g) \rightleftharpoons 2\,X(g)$, occurs at 500 K. The equilibrium state of the reaction is shown in **1** and the equilibrium state in the same container after a change

has occurred is shown in **2**. Which of the following changes will produce the composition shown? (a) Increasing the temperature. (b) Adding X atoms. (c) Decreasing the volume. (d) Adding a catalyst. Explain your selections.

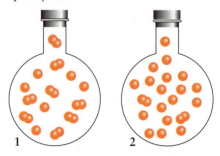

5.34 At 25 °C, $K = 47.9$ for $N_2O_4(g) \rightleftharpoons 2\,NO_2(g)$. (a) If 0.180 mol N_2O_4 and 0.0020 mol NO_2 are placed in a reaction vessel of volume 20.0 L and the reaction is allowed to reach equilibrium, what are the equilibrium concentrations of N_2O_4 and NO_2? (b) An additional 0.0020 mol NO_2 is added to the flask. How will this change affect the concentration of N_2O_4? (c) Justify your conclusion by calculating the new equilibrium concentrations of NO_2 and N_2O_4.

5.35 The following plot shows how the partial pressures of reactant and products vary with time for the decomposition of compound A into compounds B and C. All three compounds are gases. Use this plot to do the following: (a) Write a balanced chemical equation for the reaction. (b) Calculate the equilibrium constant for the reaction.

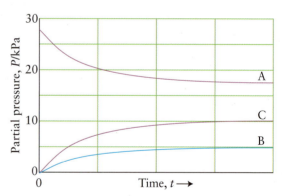

5.36 The following plot shows a system composed of the gaseous compounds A and B in a rigid, constant-volume flask. The system was initially at equilibrium, then a change occurred. (a) Describe the change that occurred and how it affected the system. (b) Write the chemical equation for the reaction that occurred. (c) Calculate the value of K_c for that reaction.

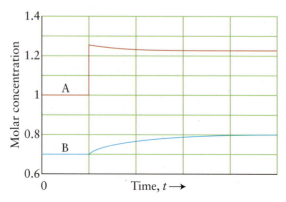

5.37 At 500 °C, $K_c = 0.061$ for $N_2(g) + 3 H_2(g) \rightleftharpoons 2 NH_3(g)$. If analysis shows that the composition of the reaction mixture at 500 °C is 3.00 $mol \cdot L^{-1}$ N_2, 2.00 $mol \cdot L^{-1}$ H_2, and 0.500 $mol \cdot L^{-1}$ NH_3, is the reaction at equilibrium? If not, in which direction does the reaction tend to proceed to reach equilibrium?

5.38 At 2500. K, the equilibrium constant is $K_c = 20.$ for the reaction $Cl_2(g) + F_2(g) \rightleftharpoons 2 ClF(g)$. An analysis of a reaction vessel at 2500. K revealed the presence of 0.18 $mol \cdot L^{-1}$ Cl_2, 0.31 $mol \cdot L^{-1}$ F_2, and 0.92 $mol \cdot L^{-1}$ ClF. Will ClF tend to form or to decompose as the reaction proceeds toward equilibrium?

5.39 In an experiment, 0.020 mol NO_2 was introduced into a flask of volume 1.00 L and the reaction $2 NO_2(g) \rightleftharpoons N_2O_4(g)$ was allowed to come to equilibrium at 298 K. (a) Using information in Table 5E.2, calculate the equilibrium concentrations of the two gases. (b) The volume of the flask is reduced to half its original value. Calculate the new equilibrium concentrations of the gases.

5.40 In an experiment, 0.100 mol SO_3 was introduced into a flask of volume 2.00 L and the reaction $2 SO_2(g) + O_2(g) \rightleftharpoons 2 SO_3(g)$ was allowed to come to equilibrium at 700 K. (a) Using information in Table 5G.2, calculate the equilibrium concentrations of the three gases. (b) The volume of the flask is increased to 6.00 L. Calculate the new equilibrium concentrations of the gases.

5.41 Let α be the fraction of PCl_5 molecules that has decomposed to PCl_3 and Cl_2 in the reaction $PCl_5(g) \rightleftharpoons PCl_3(g) + Cl_2(g)$ in a constant-volume container; then the amount of PCl_5 at equilibrium is $n(1 - \alpha)$, where n is the amount present initially. Derive an equation for K in terms of α and the total pressure P, and solve it for α in terms of P. Calculate the fraction decomposed at 556 K, at which temperature $K = 4.96$, and the total pressure is (a) 0.50 bar; (b) 1.00 bar.

5.42 The three compounds methylpropene, cis-2-butene, and trans-2-butene are isomers with the formula C_4H_8, with $\Delta G_f^\circ = +58.07, +65.86$, and $+62.97$ $kJ \cdot mol^{-1}$, respectively. In the presence of a suitable catalyst, these three compounds interconvert to give an equilibrium gaseous mixture. What will be the percentage of each isomer present at 25 °C once equilibrium is established?

5.43 (a) What is the standard Gibbs free energy of the reaction $CO(g) + H_2O(g) \rightarrow CO_2(g) + H_2(g)$ when $K = 1.00$? (b) From data available in Appendix 2A, estimate the temperature at which $K = 1.00$. (c) At this temperature, a cylinder is filled with CO(g) at 10.00 bar, $H_2O(g)$ at 10.00 bar, $H_2(g)$ at 5.00 bar, and $CO_2(g)$ at 5.00 bar. What will be the partial pressure of each of these gases when the system reaches equilibrium? (d) If the cylinder were filled instead with CO(g) at 6.00 bar, $H_2O(g)$ at 4.00 bar, $H_2(g)$ at 5.00 bar, and $CO_2(g)$ at 10.00 bar, what would the partial pressures be at equilibrium?

5.44 Consider the equilibrium $A(g) \rightleftharpoons 2 B(g) + 3 C(g)$ at 25 °C. When A is loaded into a cylinder at 10.0 atm and the system is allowed to come to equilibrium, the final pressure is found to be 20.04 atm. What is ΔG_r° for this reaction?

5.45 (a) Calculate the standard Gibbs free energies of formation of the halogen atoms X(g) at 1000. K from data available in Table 5G.2. (b) Show how these data correlate with the X—X bond strength by plotting the standard Gibbs free energy of formation of the atoms against the bond dissociation energy and atomic number. Rationalize any trends you observe.

5.46 The gas phosphine, PH_3, decomposes by the reaction $2 PH_3(g) \rightarrow 2 P(s) + 3 H_2(g)$. In an experiment, pure phosphine was placed in a rigid, sealed flask of volume 1.00 L at 0.64 bar and 298 K. After equilibrium was attained, the total pressure in the flask was found to be 0.93 bar. (a) Calculate the equilibrium partial pressures of H_2 and PH_3. (b) Calculate the mass (in grams) of P produced once equilibrium was reached. (c) Calculate K for this reaction.

5.47 (a) Calculate K at 25 °C for the reaction $Br_2(g) \rightleftharpoons 2 Br(g)$ from the thermodynamic data provided in Appendix 2A. (b) What is the vapor pressure of liquid bromine? (c) What is the partial pressure of Br(g) above the liquid in a bottle of bromine at 25 °C? (d) A student wishes to add 0.0100 mol Br_2 to a reaction and will do so by filling an evacuated flask with Br_2 vapor from a reservoir that contains only bromine liquid in equilibrium with its vapor. The flask will be sealed and then transferred to the reaction vessel. What volume container should the student use to deliver 0.010 mol $Br_2(g)$ at 25 °C?

5.48 A reaction vessel is filled with $Cl_2(g)$ at 1.00 bar and $Br_2(g)$ at 1.00 bar, which are allowed to react at 1000. K to form BrCl(g) according to the equation $Br_2(g) + Cl_2(g) \rightleftharpoons 2 BrCl(g)$, $K = 0.2$. Construct a plot of the reaction Gibbs free energy as a function of partial pressure of BrCl as the reaction approaches equilibrium.

5.49 The reaction $N_2O_4 \rightarrow 2 NO_2$ is allowed to reach equilibrium in chloroform solution at 25 °C. The equilibrium concentrations are 0.405 $mol \cdot L^{-1}$ N_2O_4 and 2.13 $mol \cdot L^{-1}$ NO_2. (a) Calculate K for the reaction. (b) An additional 1.00 mol NO_2 is added to 1.00 L of the solution and the system is allowed to reach equilibrium again at the same temperature. Use Le Chatelier's principle to predict the direction of change (increase, decrease, or no change) for N_2O_4, NO_2, and K after the addition of NO_2. (c) Calculate the final equilibrium concentrations after the addition of NO_2 and confirm that your predictions in part (b) were valid. If they do not agree, check your procedure and repeat it if necessary.

5.50 Using data from Table 5G.2 and standard graphing software, determine the standard enthalpy and entropy of the reaction $N_2O_4(g) \rightarrow 2 NO_2(g)$ and estimate the N—N bond enthalpy in N_2O_4. How does this value compare with the mean N—N bond enthalpy in Table 4E.3?

5.51 Estimate the vapor pressure of heavy water, D_2O, and of normal water at 25 °C by using data in Appendix 2A. How do these values compare with each other? Using your knowledge of intermolecular forces and any quantum effects, such as the zero-point energy of intermolecular vibrations, explain the reason for the difference observed.

5.52 Cyclohexane (C) and methylcyclopentane (M) are isomers with the chemical formula C_6H_{12}. The equilibrium constant for the rearrangement $C \rightleftharpoons M$ in solution is 0.140 at 25 °C. (a) A solution of 0.0200 $mol \cdot L^{-1}$ cyclohexane and 0.100 $mol \cdot L^{-1}$ methylcyclopentane is prepared. Is the system at equilibrium? If not, will it form more reactants or more products? (b) What are the concentrations of cyclohexane and methylcyclopentane at equilibrium? (c) If the temperature is raised to 50 °C, the concentration of cyclohexane becomes 0.100 $mol \cdot L^{-1}$ when equilibrium is re-established. Calculate the new equilibrium constant. (d) Is the reaction exothermic or endothermic at 25 °C? Explain your conclusion.

 5.53 Use the Living Graph "Variation of the Equilibrium Constant" on the Web site for this book to construct a plot from 250 K to 350 K for reactions with standard reaction Gibbs free energies of $+11$ kJ·mol^{-1} to $+15$ kJ·mol^{-1} in increments of 1 kJ·mol^{-1}. Enter free energies in joules. Which equilibrium constant is most sensitive to changes in temperature?

5.54 (a) Use a spreadsheet program or graphing software and data at 1000. K and 1200. K from Table 5G.2 to plot the expression for the temperature dependence of ln K given by the dissociation of the diatomic halogens into the atoms, $X_2(g) \rightleftharpoons 2\ X(g)$. (b) From the graphs, determine the enthalpies and entropies of dissociation. (c) Use these data to calculate the standard molar entropies of the gaseous halogen atoms $X(g)$.

5.55 A reaction used in the production of gaseous fuels from coal, which is mainly carbon, is $C(s) + H_2O(g) \rightleftharpoons CO(g) + H_2(g)$. (a) Evaluate K At 900 K, given that the standard Gibbs free energies of formation of $CO(g)$ and $H_2O(g)$ at 900 K are -191.28 kJ·mol^{-1} and -198.08 kJ·mol^{-1}, respectively. (b) A sample of graphite of mass 5.20 kg and 125 g of water were placed into a 10.0-L container and heated to 900 K. What are the equilibrium concentrations?

5.56 The volume of the container used for the reaction in Exercise 5.55 was compressed to 5.00 L and equilibrium was re-established. (a) Without doing a calculation, predict whether the equilibrium concentration of H_2 will have increased or decreased and explain how you decided. (b) Calculate the new equilibrium concentrations and evaluate your prediction. (c) To maximize the production of H_2, should the reaction be run at low or high pressure? At low or high temperature? Justify your answers.

5.57 The two air pollutants SO_3 and NO can react as follows: $SO_3(g) + NO(g) \rightarrow SO_2(g) + NO_2(g)$. (a) Predict the effect of the following changes to the amount of NO_2 when the reaction has come to equilibrium in a stainless steel bulb equipped with entrances for chemicals: (i) the amount of NO is increased; (ii) the SO_2 is removed by condensation; (iii) the pressure is tripled by pumping in helium. (b) Given that at a certain temperature $K = 6.0 \times 10^3$, calculate the amount (in moles) of NO that must be added to a 1.00-L vessel containing 0.245 mol $SO_3(g)$ to form 0.240 mol $SO_2(g)$ at equilibrium.

5.58 The distribution of Na^+ ions across a typical biological membrane is 10. mmol·L^{-1} inside the cell and 140 mmol·L^{-1} outside the cell. At equilibrium the concentrations would be equal, but in a living cell the ions are not at equilibrium. What is the difference in Gibbs free energy of Na^+ ions across the membrane at 37 °C (normal body temperature)? The concentration differential must be maintained by coupling to reactions that have at least that difference of Gibbs free energy.

5.59 The *Claus process*, which is used to remove sulfur found as sulfur dioxide in petroleum, is based on the reaction 2 $H_2S(g)$ + $SO_2(g) \rightleftharpoons 3\ S(s) + 2\ H_2O(g)$. (a) Use data from Appendix 2A to determine the equilibrium constant of this reaction at 25 °C. (b) Various changes (see the following table) are then made to this mixture. Considering each change separately, state the effect (increase, decrease, or no change) that the change has on the original equilibrium value of the quantity in the second column (or K, if that is specified). The temperature and volume are constant unless otherwise specified.

Change	Quantity
(a) add S	amount of H_2O
(b) add H_2S	amount of SO_2
(c) remove H_2O	amount of SO_2
(d) remove SO_2	amount of S
(e) add SO_2	K
(f) decrease volume	amount of SO_2
(g) increase temperature	amount of SO_2

5.60 To generate the starting material for a polymer that is used to make water bottles, hydrogen is removed from the ethane in natural gas to produce ethene in the catalyzed reaction $C_2H_6(g) \rightarrow H_2(g) + C_2H_4(g)$. Use the information in Appendix 2A to calculate the equilibrium constant for the reaction at 298 K. (a) If the reaction is begun by adding the catalyst to a flask containing C_2H_6 at 40.0 bar, what will be the partial pressure of the C_2H_4 at equilibrium? (b) Identify three steps the manufacturer can take to increase the yield of product.

5.61 The overall photosynthesis reaction is 6 $CO_2(g)$ +6 $H_2O(l) \rightarrow C_6H_{12}O_6(aq) + 6\ O_2(g)$, and $\Delta H° = +2802$ kJ. Suppose that the reaction is at equilibrium. State the effect that each of the following changes will have on the equilibrium composition: tends to shift toward the formation of reactants, tends to shift toward the formation of products, or has no effect. (a) The partial pressure of O_2 is increased. (b) The system is compressed. (c) The amount of CO_2 is increased. (d) The temperature is increased. (e) Some of the $C_6H_{12}O_6$ is removed. (f) Water is added. (g) The partial pressure of CO_2 is decreased.

5.62 Adenosine triphosphate (ATP) is a compound that provides energy for biochemical reactions in the body when it undergoes hydrolysis. For the hydrolysis of ATP at 37 °C (normal body temperature), $\Delta H_r° = -20.$ kJ·mol^{-1} and $\Delta S_r° = +34$ J·K^{-1}·mol^{-1}. Assuming that these quantities are independent of temperature, calculate the temperature at which the equilibrium constant for the hydrolysis of ATP becomes greater than 1.

FOCUS 5   Cumulative Exercises

5.63 Reactions between gases in the atmosphere are not at equilibrium, but for a thorough understanding of them we need to study both the rates at which they take place and their behavior under equilibrium conditions.

(a) The equilibrium for the depletion of ozone in the stratosphere is summarized by the equation 2 $O_3(g) \rightleftharpoons 3\ O_2(g)$.

From values in Appendix 2A, determine the standard Gibbs free energy and the standard entropy for the reaction.

(b) What is the equilibrium constant of the reaction in part (a) at 25 °C? What is the significance of your answer for ozone depletion?

(c) A reaction that destroys ozone in the stratosphere is $O_3(g) + O(g) \rightleftharpoons 2\,O_2(g)$. Calculate the value of the equilibrium constant for this reaction at 25 °C, given that at that temperature the reaction is catalyzed (accelerated) by NO_2 molecules in a two-step process:

$$NO_2(g) + O(g) \rightleftharpoons NO(g) + O_2(g) \quad K = 7 \times 10^{103}$$

$$NO(g) + O_3(g) \rightleftharpoons NO_2(g) + O_2(g) \quad K = 5.8 \times 10^{-34}$$

(d) Use your answer to part (c) to find the standard Gibbs free energy of formation of O atoms.

(e) The temperature dependence of the equilibrium constant of the reaction $N_2(g) + O_2(g) \rightleftharpoons 2\,NO(g)$, which makes an important contribution to the concentration of atmospheric nitrogen oxides, can be expressed as $\ln K = 2.5 - (21\,700\ \text{K})/T$. What is the standard enthalpy of the forward reaction at 298 K? Will this reaction proceed further at the very low temperatures of the stratosphere or at the very high temperatures in an internal combustion engine?

(f) An equimolar mixture of N_2 and O_2 was heated to a certain temperature until the reaction in part (e) came to equilibrium. The equilibrium reaction mixture was found to contain an equal number of moles of each reactant and product. At what temperature was the reaction carried out?

(g) An equimolar mixture of N_2 and O_2 with a total pressure of 4.00 bar was allowed to come to equilibrium in the reaction in part (e) at 1200. K. What will be the partial pressure of each reactant and product at equilibrium?

5.64 Sports drinks provide water to the body in the form of an isotonic solution (one having the same osmotic pressure as human blood). These drinks contain electrolytes such as NaCl and KCl as well as sugar and flavoring. One of the main flavoring agents in sport drinks is citric acid (**1**).

1 Citric acid

(a) Indicate the hybridization of each C atom in citric acid.

(b) Can citric acid take part in hydrogen bonding?

(c) Predict from a consideration of intermolecular forces whether citric acid is a gas, liquid, or solid at 25 °C and whether it is soluble in water.

(d) *Normal saline solution* is an isotonic solution (Topic 5F) containing 0.9% NaCl by mass in water. Assuming complete dissociation of the NaCl, what is the total molar concentration of all solutes in an isotonic solution? Assume a density of $1.00\ \text{g}\cdot\text{cm}^{-3}$ for the solution.

(e) If you decide to make up 500.0 mL of a sports drink with 1.0 g of NaCl and glucose, what mass of glucose do you need to add to the NaCl and water to make the solution isotonic (see part d)? Assume a density of $1.00\ \text{g}\cdot\text{cm}^{-3}$ for the solution.

(f) A paramedic treating injuries in a remote area has 300.0 mL of a 1.00% by mass solution of boric acid, $B(OH)_3$, that needs to be made isotonic (assume that the density is $1.00\ \text{g}\cdot\text{cm}^{-3}$). What mass of NaCl should be added? Assume that the NaCl is completely dissociated in the solution and that the volume of the solution does not change upon addition of the NaCl. Take into account the 0.007% deprotonation of boric acid.

INTERLUDE Homeostasis

Because a living being is not a closed system, true equilibrium in living systems can be reached only for very fast reactions such as those between acids and bases. In general, however, human bodies maintain nearly constant values of properties, such as temperature and levels of certain chemicals in the blood. The body maintains a beneficial environment through the process of **homeostasis,** the maintenance of constant internal conditions. Homeostasis is not a true equilibrium because there are usually slight variations above and below the desired point. However, living organisms respond to changes in conditions in the same way as a system in chemical equilibrium and conform to Le Chatelier's principle.

An important homeostatic biological process that involves chemical equilibria is the transport of oxygen. Most of the oxygen in the blood is carried by hemoglobin (Hb). When blood flows through lung tissues, about 98% of the hemoglobin molecules pick up oxygen molecules and a small amount of additional oxygen simply dissolves in the blood plasma (the solution in which blood cells are suspended). However, when blood enters the small blood vessels called capillaries in muscle tissue far from the lungs, the hemoglobin molecules are surrounded by tissues that are depleted in oxygen and the level of oxygen in blood plasma drops. The equilibrium $Hb(aq) + O_2(aq) \rightleftharpoons HbO_2(aq)$ is shifted toward reactants as some of the hemoglobin molecules release their oxygen molecules to re-establish the equilibrium compositions.

FIGURE 1 shows how O_2 uptake by hemoglobin and myoglobin (Mb), the oxygen storage protein, varies with the partial pressure of oxygen. The shape of the Hb saturation curve means that Hb can load O_2 almost as fully in the lungs as Mb, and unload it more completely than Mb in different regions of the organism. In the lungs, where $P_{O_2} \approx 105$ Torr, 98% of the Hb molecules have taken up O_2, resulting in almost complete saturation. In resting muscular tissue, the concentration of O_2 corresponds to a partial pressure of about 40 Torr, at which value 75% of the Hb molecules are saturated with oxygen. In this condition, sufficient oxygen is still available should a sudden surge of activity take place. If the local partial pressure of oxygen falls to 20 Torr, the fraction of Hb molecules saturated falls to about 10%.

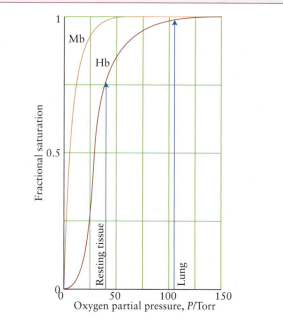

FIG. 1 The variation of the extent of saturation of myoglobin (Mb) and hemoglobin (Hb) with the partial pressure of oxygen. The different shapes of the curves account for the different biological functions of the two proteins.

Note that the steepest part of the curve falls in the range of typical tissue oxygen partial pressure. Myoglobin, on the other hand, begins to release O_2 only when P_{O_2} has fallen below about 20 Torr, and so it acts as a reserve to be drawn on only when the Hb oxygen has been used up.

Overloading the body's coping mechanisms can lead to failure to maintain homeostasis. The result can be sudden and even fatal illness. Mountain climbers encounter low oxygen conditions at high altitudes and, if they climb vigorously, their lungs may not be able to provide sufficient oxygen to maintain homeostasis. For this reason, mountain climbers spend some time in high-altitude base camps before climbing, so that their bodies can adjust by producing more hemoglobin molecules.

REACTIONS

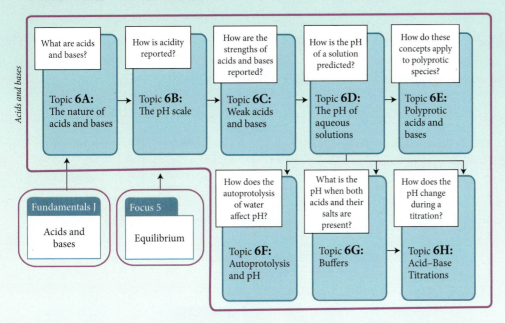

One of the most important types of reactions in chemistry is that between acids and bases. **TOPIC 6A** explores these reactions and introduces the modern view that their reaction is essentially the transfer of a proton (the nucleus of a hydrogen atom) from one species to another. A central aspect of an aqueous solution of an acid or base is therefore the concentration of protons. This concentration is reported as the pH of the solution, as explained in **TOPIC 6B**.

TOPIC 6C explains that there is a dynamic equilibrium between the protonated and unprotonated forms of most acids and bases, and **TOPIC 6D** explains how to calculate the pH of a solution in the presence of this equilibrium. A complication is that some acids can donate more than one proton; how to treat these "polyprotic acids" is described in **TOPIC 6E**. A further complication is that water itself acts as a very weak acid and as a very weak base. This character is important when the concentration of added acid or base is very low, and **TOPIC 6F** describes how to take it into account.

One very important goal of chemistry is to stabilize solutions against changes of pH. **TOPIC 6G** explains how these "buffer solutions" work. Another aspect of the reactions of acids and bases is their role in the common analytical technique of "titration." **TOPIC 6H** explains how the pH changes in the course of a titration and underlies the choice of indicators for monitoring its progress.

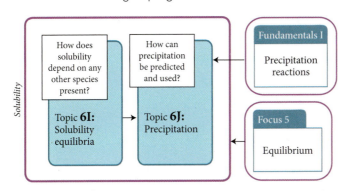

When certain pairs of solutions of soluble salts are mixed, one product might be insoluble. These "precipitation reactions" are described in **TOPIC 6I** from the viewpoint of the equilibrium that exists between dissolved and undissolved material. Predicting when precipitation reactions will occur and how these principles can be used in the analysis of mixed solutions is discussed in **TOPIC 6J**.

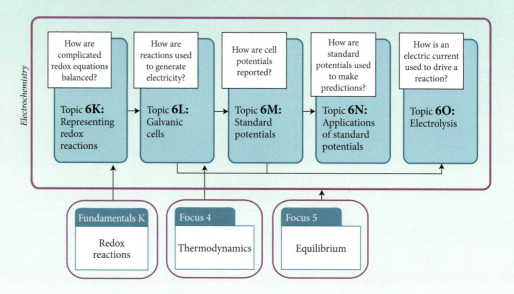

"Redox reactions" involve the reduction of one species and the oxidation of another and are the outcome of the transfer of electrons. **TOPIC 6K** describes a general procedure for balancing redox equations and introduces the concept of a "half-reaction." One major aspect of redox reactions is that it is possible to separate spatially the oxidation (electron loss) and reduction (electron gain) steps. **TOPIC 6L** shows how this separation is achieved in a "galvanic cell" and used to generate electric currents. As explained in **TOPIC 6M**, the reaction at each electrode of a galvanic cell can be regarded as making a characteristic contribution to the overall potential difference generated by the cell. **TOPIC 6N** explains how these electrode potentials are used to predict equilibrium constants and to determine the concentrations of dissolved species and extends the discussion to the hugely important concept of corrosion and its prevention.

A galvanic cell makes use of a chemical reaction to generate an electric current; the opposite is achieved in an "electrolytic cell," in which the application of a current forces an otherwise nonspontaneous reaction to occur. **TOPIC 6O** explains the principles of electrolysis.

Topic 6A The Nature of Acids and Bases

What are acids and bases?

How is acidity reported?

Topic **6A:**
The nature of acids and bases → Topic **6B:**
The pH scale

Why Do You Need to Know This Material? Acids and bases are a central aspect of chemistry and play a major role in all its applications, including those in biology, industry, and the environment.

What Do You Need to Know Already? You need to be familiar with the concept of acid and base as introduced in *Fundamentals* J and be able to write and interpret an equilibrium constant for a given reaction (Topic 5G). You need to be familiar with the concept of a coordinate covalent bond (Topic 2C).

When chemists see a pattern in the reactions of certain substances, they seek to define a class of substance that characterizes the pattern. Then, when a substance is found to belong to that class, they immediately know a great deal about its behavior. Classification of this kind opens the door to understanding and reduces the need to memorize the properties of individual substances. The reactions of the substances called "acids" and "bases" are an excellent illustration of this approach. The pattern in these reactions was first identified in aqueous solutions, and it led to the "Arrhenius definitions" of acids and bases (*Fundamentals* J). However, because similar reactions were found to take place in nonaqueous solutions and even in the absence of solvent, chemists realized that the original definitions could be replaced by more general definitions.

6A.1 Brønsted–Lowry Acids and Bases

In 1923, the Danish chemist Johannes Brønsted proposed that

> An **acid** is a proton donor.

> A **base** is a proton acceptor.

The term *proton* in these definitions refers to the hydrogen ion, H^+. An acid is therefore a species containing an **acidic hydrogen atom,** a hydrogen atom that can be transferred as its nucleus, a proton, to another species acting as a base. The same definitions were proposed independently by the English chemist Thomas Lowry, and the theory based on them is called the **Brønsted–Lowry theory** of acids and bases. A proton donor is commonly referred to as a **Brønsted acid** and a proton acceptor as a **Brønsted base.** Whenever an "acid" or a "base" is referred to in this Focus, it means a Brønsted acid or a Brønsted base.

A substance can act as an acid only if a base is present to accept its acidic protons. An acid does not simply release its acidic proton; the proton is *transferred* to the base through direct contact. For example, in the gas phase an HCl molecule remains intact. However, when HCl dissolves in water, each HCl molecule immediately transfers an H^+ ion to a neighboring H_2O molecule, which is acting as a base (**FIG. 6A.1**):

$$\overset{\text{acid}}{\overbrace{HCl(aq)}} + \overset{\text{base}}{\overbrace{H_2O(l)}} \longrightarrow H_3O^+(aq) + Cl^-(aq)$$

This process is a **proton transfer reaction,** a reaction in which a proton is transferred from one species to another. The HCl molecule is said to be **deprotonated.** Because at equilibrium almost all the HCl molecules in the solution have donated their protons to water, HCl is classified as a "strong acid." In this case, the proton transfer reaction essentially goes to completion. The H_3O^+ ion is called the *hydronium ion.* It is strongly hydrated in solution, and there is some evidence that a better representation of the species is $H_9O_4^+$ (or even larger clusters of water molecules attached to a proton). For simplicity, a hydrogen ion in water is sometimes represented as $H^+(aq)$, but you must remember that H^+ does not exist by itself in water and that $H_3O^+(aq)$ is a better representation because it indicates that a Brønsted base (H_2O) has accepted a proton.

Another example of an acid is hydrogen cyanide, HCN, which can transfer its proton to water when it dissolves to form the solution known as hydrocyanic acid, HCN(aq).

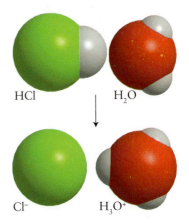

HCl H_2O

Cl^- H_3O^+

FIGURE 6A.1 The upper image shows that when an HCl molecule dissolves in water, a hydrogen bond forms between the H atom of HCl (the acid) and the O atom of a neighboring H_2O molecule (the base). The lower image shows the result: the nucleus of the hydrogen atom is pulled out of the HCl molecule to become part of a hydronium ion.

 **ANIMATION FIGURE 6A.1**

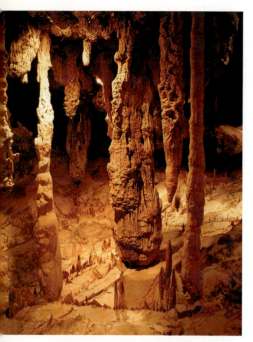

FIGURE 6A.2 Stalactites hang from the roof of a cave and stalagmites grow from the floor. Both are made of insoluble calcium carbonate formed from the soluble hydrogen carbonate ions in groundwater. *(Reinhard Dirscherl/WaterFrame/Getty Images.)*

A coordinate covalent bond is a bond in which both bonding electrons come from the same atom (Topic 2C).

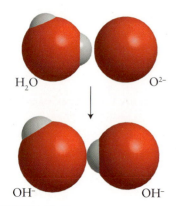

H_2O O^{2-}

OH^- OH^-

FIGURE 6A.3 When an oxide ion is present in water, it exerts such a strong attraction on the nucleus of a hydrogen atom in a neighboring water molecule that the hydrogen ion is pulled out of the molecule as a proton. As a result, the oxide ion forms two hydroxide ions.

However, only a small fraction of the HCN molecules donate their protons, and so, as described in *Fundamentals* J, it is classified as a "weak acid" in water. The proton transfer reaction is written as an equilibrium:

$$HCN(aq) + H_2O(l) \rightleftharpoons H_3O^+(aq) + CN^-(aq)$$

Like all chemical equilibria, this equilibrium is dynamic and you should think of protons as ceaselessly exchanging between HCN and H_2O molecules, with a constant but low concentration of CN^- and H_3O^+ ions. The proton transfer reaction of a strong acid such as HCl in water is also dynamic, but the equilibrium lies so strongly in favor of products that it is represented by only its forward reaction, with a single arrow.

In *Fundamentals* J, an Arrhenius acid is defined as a compound that produces hydronium ions in water and an Arrhenius base is a compound that produces hydroxide ions in water. The Brønsted definition is more general because it includes the possibility that an ion is an acid (an option not allowed by the Arrhenius definition). For instance, a hydrogen carbonate ion, HCO_3^-, one of the species present in natural waters, can act as an acid and donate a proton to an H_2O molecule (**FIG. 6A.2**):

$$HCO_3^-(aq) + H_2O(l) \rightleftharpoons H_3O^+(aq) + CO_3^{2-}(aq)$$

The distinction between strong and weak acids can be summarized as follows:

Strong acid: Almost all the molecules are deprotonated in solution.

Weak acid: Only a small fraction of the molecules or ions are deprotonated in solution.

The strength of an acid depends on the solvent, and an acid that is strong in water may be weak in another solvent or vice versa (Topic 6C). However, because almost all reactions in living tissues and most reactions in laboratories take place in water, unless otherwise specified, the solvent throughout this Focus is water.

A Brønsted base possesses a lone pair of electrons to which a proton can bond. For example, an oxide ion, O^{2-}, is a Brønsted base. When CaO dissolves in water, the strong electric field of the small, highly charged O^{2-} ion pulls a proton away from a neighboring H_2O molecule (**FIG. 6A.3**). In the process, a coordinate covalent bond forms between the proton and a lone pair of electrons on the oxide ion. By accepting a proton, the oxide ion has become **protonated.** Almost every oxide ion present accepts a proton from water, and so O^{2-} is an example of a "strong base" in water, a species that is fully protonated. That is, the following reaction goes essentially to completion:

$$O^{2-}(aq) + H_2O(l) \longrightarrow 2\,OH^-(aq)$$

Another example of a Brønsted base is ammonia. When an NH_3 molecule is present in water, the lone pair of electrons on the N atom can accept a proton from H_2O:

$$NH_3(aq) + H_2O(l) \rightleftharpoons NH_4^+(aq) + OH^-(aq)$$

Because an NH_3 molecule is electrically neutral, it has much less proton-pulling power than the oxide ion. As a result, only a very small proportion of the NH_3 molecules are converted into ammonium ions, NH_4^+ (**FIG. 6A.4**). Ammonia is therefore a "weak base." All amines, organic derivatives of ammonia, such as methylamine, CH_3NH_2, are weak bases in water. Because the proton transfer equilibrium in an aqueous solution of ammonia is dynamic, the protons are ceaselessly exchanging between NH_3 and H_2O molecules with a constant, very low concentration of NH_4^+ and OH^- ions. The proton transfer to the strong base O^{2-} is also dynamic, but the equilibrium lies so strongly in favor of products that, as for a strong acid, it is represented by its forward reaction with a single arrow.

The distinction between strong and weak bases can be summarized as follows:

Strong base: Almost all the molecules or ions are protonated in solution.

Weak base: Only a small fraction of the molecules or ions are protonated in solution.

As for acids, the strength of a base depends on the solvent: a base that is strong in water may be weak in another solvent and vice versa. The common strong bases in aqueous solution are listed in *Fundamentals* Table J.1.

A Note on Good Practice: The oxides and hydroxides of the alkali and alkaline earth metals are not Brønsted bases: the oxide and hydroxide *ions* they contain are the bases (the cations are spectator ions). However, for convenience, chemists typically refer to the compounds themselves as bases.

The product formed from the acid molecule when it transfers a proton to water can itself accept a proton from water and thus be classified as a base. For example, the CN^- ion produced when HCN is deprotonated can accept a proton from a neighboring H_2O molecule and form HCN again. Therefore, according to the Brønsted definition, CN^- is a base; it is called the "conjugate base" of the acid HCN. In general, a **conjugate base** of an acid is the species remaining when the acid donates a proton:

$$\text{Acid} \xrightarrow{\text{donates } H^+} \text{conjugate base}$$

Because HCN is the acid that forms when a proton is transferred to a cyanide ion, HCN is the "conjugate acid" of the base CN^-. In general, a **conjugate acid** of a base is the species formed when the base accepts a proton:

$$\text{Base} \xrightarrow{\text{accepts } H^+} \text{conjugate acid}$$

EXAMPLE 6A.1   Writing the formulas of conjugate acids and bases

Suppose you need to predict the products of a given reaction between an acid and a base. To do that you need to be able to write the formulas of the conjugate base and conjugate acid formed. Write the formulas of (a) the conjugate base of HCO_3^- and (b) the conjugate acid of O^{2-}.

PLAN Remove a proton (an H^+ ion) to form a conjugate base and add a proton to form a conjugate acid.

SOLVE

The conjugate base of an acid has one fewer H^+ ion than the acid; the conjugate acid of a base has one more H^+ ion than the base.

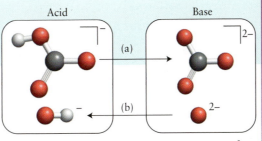

(a) The conjugate base of HCO_3^- is CO_3^{2-}.
(b) The conjugate acid of O^{2-} is OH^-.

Self-test 6A.1A What is (a) the conjugate acid of OH^-; (b) the conjugate base of HPO_4^{2-}?
[**Answer:** (a) H_2O; (b) PO_4^{3-}]

Self-test 6A.1B What is (a) the conjugate acid of H_2O; (b) the conjugate base of NH_3?

Related Exercises 6A.1–6A.6

The Brønsted definitions of acids and bases also apply to species in nonaqueous solvents and even to gas-phase reactions. For example, when acetic acid is dissolved in liquid ammonia, proton transfer takes place and the following equilibrium is reached:

$$CH_3COOH(am) + NH_3(l) \rightleftharpoons CH_3CO_2^-(am) + NH_4^+(am)$$

(The label "am" indicates a species dissolved in liquid ammonia.) An example of proton transfer in the gas phase is the reaction of hydrogen chloride and ammonia gases. They produce the fine powder of ammonium chloride often seen coating surfaces in chemical laboratories (**FIG. 6A.5**):

$$HCl(g) + NH_3(g) \longrightarrow NH_4Cl(s)$$

A Brønsted acid is a proton donor and a Brønsted base is a proton acceptor. The conjugate base of an acid is the base formed when the acid has donated a proton. The conjugate acid of a base is the acid that forms when the base has accepted a proton. A strong acid is treated as fully deprotonated in solution; a weak acid is only partially deprotonated in solution. A strong base is treated as completely protonated in solution; a weak base is only partially protonated in solution.

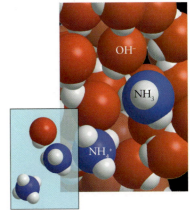

FIGURE 6A.4 In this molecular portrayal of the structure of a solution of ammonia in water at equilibrium, NH_3 molecules are still present because only a small percentage of them have been protonated by transfer of hydrogen ions from water. In a typical solution, only about 1 in 100 NH_3 molecules is protonated. The inset at the left shows only the solute species.

ANIMATION FIGURE 6A.4

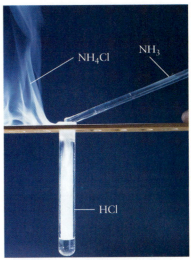

FIGURE 6A.5 The white powder is ammonium chloride formed by gaseous ammonia coming from the slanted tube and hydrogen chloride gas escaping from concentrated hydrochloric acid in the test tube. (*Andrew Lambert Photography/Science Source.*)

6A.2 Lewis Acids and Bases

The concepts of acids and bases have a much wider significance than the transfer of protons at the core of the Brønsted–Lowry theory. Even more substances can be classified as acids or bases under the definitions developed by the American chemist G. N. Lewis:

A **Lewis acid** is an electron pair acceptor.

A **Lewis base** is an electron pair donor.

When a Lewis base donates an electron pair to a Lewis acid, the two species share the pair and become joined by a coordinate covalent bond. A proton (H^+) is an electron pair acceptor. It is therefore a Lewis acid because it can attach to ("accept") a lone pair of electrons on a Lewis base. In other words, a Brønsted acid is a *supplier* of one particular Lewis acid, a proton.

The Lewis theory is more general than the Brønsted–Lowry theory. For instance, metal atoms and ions can act as Lewis acids, as in the formation of $Ni(CO)_4$ from nickel atoms (the Lewis acid) and carbon monoxide (the Lewis base), but they are not Brønsted acids. However, every Brønsted base is a special kind of Lewis base, one that can use a lone pair of electrons to form a coordinate covalent bond to a proton. For instance, an oxide ion is a Lewis base. It forms a coordinate covalent bond to a proton, a Lewis acid, by supplying both electrons for the bond:

The curved arrows show the direction in which a pair of bonding electrons moves and the red H indicates the proton that is transferred. Similarly, when the Lewis base ammonia, NH_3, dissolves in water, some of the molecules accept protons from water molecules:

An important point to note is that the entities regarded as acids and bases are different in each theory. In the Lewis theory, the proton is an acid; in the Brønsted theory, the species that *supplies* the proton is the acid. In both the Lewis and Brønsted theories, the species that accepts a proton is a base; in the Arrhenius theory, the compound that *supplies* the proton acceptor is the base (**FIG. 6A.6**).

FIGURE 6A.6 The left-hand column illustrates the action of acids (the species in the rounded red-edged rectangles) in the Arrhenius, Brønsted, and Lewis definitions. The right-hand column shows the action of the corresponding bases (the species in the rounded blue-edged rectangles). In each case, the small white circles represent hydrogen ions. The dark green disc represents an accompanying ion and the gray shapes other atoms. The Arrhenius definition includes any accompanying ion, whereas the Brønsted definition may apply either to a compound or to an ion. Only the Arrhenius definition requires the presence of water (the blue background).

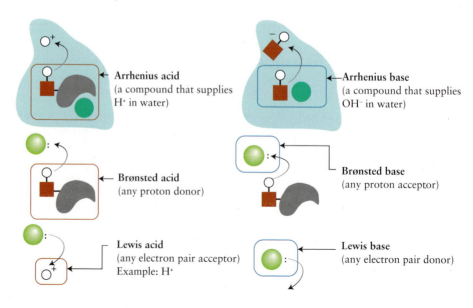

Many nonmetal oxides are Lewis acids that react with water to give Brønsted acids. An example is the reaction of CO_2 with water:

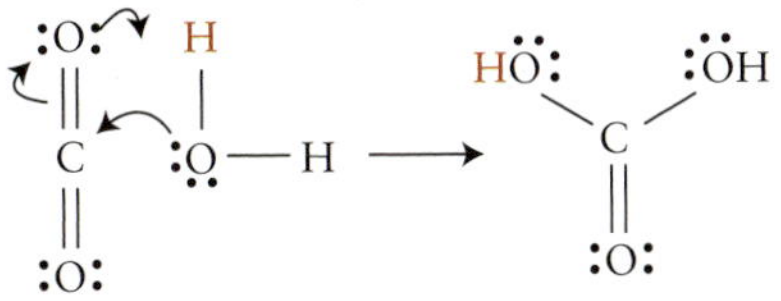

In this reaction, the C atom of CO_2, the Lewis acid, accepts an electron pair from the O atom of a water molecule, the Lewis base, and a proton migrates from an H_2O oxygen atom to a CO_2 oxygen atom. The product, an H_2CO_3 molecule, is a Brønsted acid.

Self-test 6A.2A (a) Identify the Brønsted acids and bases in both reactants and products in the proton transfer equilibrium $HNO_2(aq) + HPO_4^{2-}(aq) \rightleftharpoons NO_2^-(aq) + H_2PO_4^-(aq)$. (b) Which species (not necessarily shown explicitly) are Lewis acids and which are Lewis bases?
[*Answer:* (a) Brønsted acids, HNO_2, $H_2PO_4^-$; Brønsted bases, HPO_4^{2-} and NO_2^-;
(b) Lewis acid, H^+; Lewis bases, HPO_4^{2-} and NO_2^-]

Self-test 6A.2B (a) Identify the Brønsted acids and bases in both reactants and products in the proton transfer equilibrium $HCO_3^-(aq) + NH_4^+(aq) \rightleftharpoons H_2CO_3(aq) + NH_3(aq)$. (b) Which species (not necessarily shown explicitly) are Lewis acids and which are Lewis bases?

Because proton transfer plays a very special role in chemistry, the Brønsted definitions are central to most of the Topics in this Focus. The Lewis definitions, however, play an important role in the chemistry of d-metal ions (Topics 9C and 9D).

A Lewis acid is an electron pair acceptor; a Lewis base is an electron pair donor.
A proton acts as a Lewis acid when it attaches to a lone pair provided by a Lewis base.

6A.3 Acidic, Basic, and Amphoteric Oxides

An **acidic oxide** is an oxide that reacts with water to form a solution of a Brønsted acid. An example is CO_2, which forms H_2CO_3. Acidic oxides are *molecular* compounds, such as CO_2, that act as Lewis acids. Carbon dioxide, for instance, reacts with the OH^- present in aqueous sodium hydroxide:

$$2\,NaOH(aq) + CO_2(g) \longrightarrow Na_2CO_3(aq) + H_2O(l)$$

You can glimpse the underlying complexity of this apparently simple reaction by noting that it involves the attack on CO_2 by an OH^- ion acting as a Lewis base and proton transfer to another OH^- ion acting as a Brønsted base (although these processes are being shown together here, they do not necessarily take place in one step):

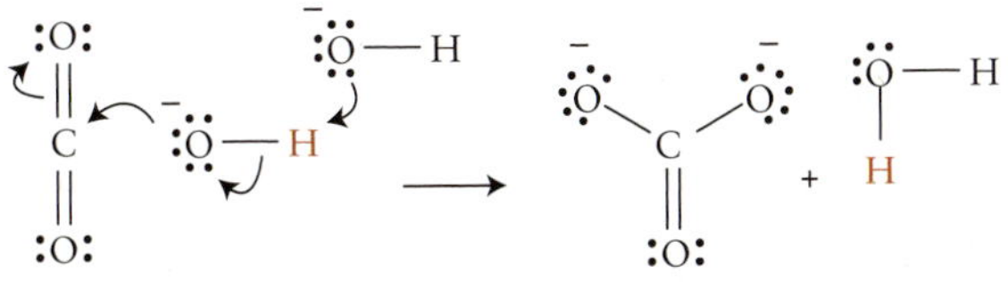

The white crust often seen on pellets of sodium hydroxide is a mixture of sodium carbonate formed in this way and of sodium hydrogen carbonate formed in a similar reaction:

$$NaOH(aq) + CO_2(g) \longrightarrow NaHCO_3(aq)$$

A **basic oxide** is an oxide that accepts protons from water to form a solution of hydroxide ions, as in the reaction

$$CaO(s) + H_2O(l) \longrightarrow Ca(OH)_2(aq)$$

Basic oxides are *ionic* compounds that can react with acids to give a salt and water. For instance, magnesium oxide, a basic oxide, reacts with hydrochloric acid:

$$MgO(s) + 2\,HCl(aq) \longrightarrow MgCl_2(aq) + H_2O(l)$$

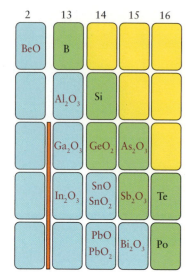

FIGURE 6A.7 The elements in and close to the diagonal line of metalloids typically form amphoteric oxides (indicated by the red lettering).

Square brackets in a chemical formula indicate that the species within the brackets is a complex ion, e.g., $[Al(OH)_4]^-$ (see Topic 9C). Don't confuse them with the notation for molar concentration.

"Amphi" derives from the Greek word for both.

In this reaction, the base O^{2-} accepts two protons from the hydronium ions present in the hydrochloric acid solution.

Metals typically form basic oxides and nonmetals typically form acidic oxides, but what about the elements that lie on the diagonal border between the metals and nonmetals? Along this frontier, from beryllium to polonium, metallic character blends into nonmetallic character and the oxides of these elements have both acidic and basic character (**FIG. 6A.7**). Substances that react with both acids and bases are classified as **amphoteric,** from the Greek word for "both." For example, aluminum oxide, Al_2O_3, is amphoteric. It reacts with acids,

$$Al_2O_3(s) + 6\,HCl(aq) \longrightarrow 2\,AlCl_3(aq) + 3\,H_2O(l)$$

and with bases,

$$2\,NaOH(aq) + Al_2O_3(s) + 3\,H_2O(l) \longrightarrow 2\,Na[Al(OH)_4](aq)$$

The product of the second reaction is sodium aluminate, which contains the aluminate ion, $Al(OH)_4^-$. Other main-group elements that form amphoteric oxides are shown in Fig. 6A.7. The acidic, amphoteric, or basic character of the oxides of the d-block metals depends on their oxidation state (**FIG. 6A.8**; also see Topic 9A).

Metals form basic oxides, nonmetals form acidic oxides; the elements on a diagonal line from beryllium toward polonium and several d-block metals form amphoteric oxides.

6A.4 Proton Exchange Between Water Molecules

An important implication of the Brønsted definitions of acids and bases is that the same substance may be able to act as both an acid and a base. For example, you have seen that a water molecule accepts a proton from an acid molecule (such as HCl or HCN) to form an H_3O^+ ion. So water is a base. However, a water molecule can donate a proton to a base (such as O^{2-} or NH_3) and become an OH^- ion. So water is also an acid. Water is therefore classified as **amphiprotic,** meaning that an H_2O molecule can act both as a proton donor and as a proton acceptor.

A Note on Good Practice: Distinguish between *amphoteric* and *amphiprotic*. Aluminum oxide is amphoteric (it reacts with both acids and bases), but it has no hydrogen atoms to donate as protons, and so is not amphiprotic.

Proton transfer between water molecules occurs even in pure water with one molecule acting as a proton donor and a neighboring molecule acting as a base:

$$\overbrace{H_2O(l)}^{\text{base}} + \overbrace{H_2O(l)}^{\text{acid}} \rightleftharpoons H_3O^+(aq) + OH^-(aq) \tag{A}$$

In more detail, the forward reaction, with the curved arrows showing how the electrons migrate and the hydrogen ion transferred indicated in red, is

The reaction is very fast in both directions, and so is always at equilibrium in water and in aqueous solutions. In every glass of water, protons from the hydrogen atoms are ceaselessly

FIGURE 6A.8 Certain elements of the d-block form amphoteric oxides, in some cases in oxidation states intermediate in their range (as shown here for certain Period 4 elements). Each stack of small squares represents an oxidation state.

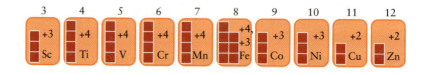

migrating between the molecules. This type of reaction, in which one molecule transfers a proton to another molecule of the same kind, is called **autoprotolysis** (**FIG. 6A.9**).

The equilibrium constant for reaction A is

$$K = \frac{a_{H_3O^+} a_{OH^-}}{(a_{H_2O})^2}$$

In dilute aqueous solutions (the only ones considered in this Focus), the solvent, water, is very nearly pure, and so its activity may be taken to be 1. The resulting expression is called the **autoprotolysis constant** of water and is written K_w:

$$K_w = a_{H_3O^+} a_{OH^-} \tag{1a}$$

As discussed in Topic 5G, the activity of a solute J in a dilute solution is approximately equal to the molar concentration relative to the standard molar concentration, $[J]/c^\circ$, with $c^\circ = 1 \text{ mol·L}^{-1}$, and so a practical form of this expression is

$$K_w = [H_3O^+][OH^-] \tag{1b}$$

where, as in Focus 5, the appearance of the expression is simplified by replacing $[J]/c^\circ$ by $[J]$, and interpreting it as the value of the molar concentration in moles per liter with the units struck out.

In pure water at 25 °C, the molar concentrations of H_3O^+ and OH^- are equal (the liquid is electrically neutral overall) and are known by experiment to be $1.0 \times 10^{-7} \text{ mol·L}^{-1}$. Therefore, at 25 °C (the only temperature we consider, unless otherwise stated),

$$K_w = (1.0 \times 10^{-7}) \times (1.0 \times 10^{-7}) = 1.0 \times 1.0^{-14}$$

The concentrations of H_3O^+ and OH^- are very low in pure water, which explains why pure water is such a poor conductor of electricity. To imagine the very tiny extent of autoprotolysis, think of each letter in this book as a water molecule. You would need to search through more than 50 books to find one ionized water molecule.

THINKING POINT

The autoprotolysis reaction is endothermic; do you expect K_w to increase or decrease with increasing temperature?

It is important to remember that K_w is fundamentally no different from the equilibrium constants discussed in Focus 5. *Because K_w is an equilibrium constant, the product of the concentrations of H_3O^+ and OH^- ions in any aqueous solution is always equal to K_w.* When the concentration of H_3O^+ ions is increased by adding acid, the concentration of OH^- ions immediately decreases to preserve the value of K_w. Alternatively, when the concentration of OH^- ions is increased by adding base, the concentration of H_3O^+ ions decreases correspondingly. The autoprotolysis equilibrium links the concentrations of H_3O^+ and OH^- ions rather like a seesaw: when one goes up, the other must go down (**FIG. 6A.10**).

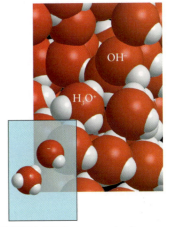

FIGURE 6A.9 As a result of autoprotolysis, pure water consists of hydronium ions and hydroxide ions as well as water molecules. The concentration of ions that results from autoprotolysis is only about 10^{-7} mol·L^{-1}, and so only about 1 molecule in 200 million is ionized. The inset at the left shows only the ions.

 **ANIMATION FIGURE 6A.9**

K_w is also widely called the autoionization constant and sometimes the ion product constant of water.

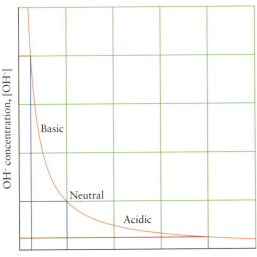

H$_3$O$^+$ concentration, [H$_3$O$^+$]

FIGURE 6A.10 The product of the concentrations of hydronium and hydroxide ions in water is a constant. If the concentration of one type of ion increases, the other must decrease to keep the product of the ion concentrations constant.

 **ANIMATION FIGURE 6A.10**

EXAMPLE 6A.2   Calculating the concentrations of ions in a solution of a metal hydroxide

Barium hydroxide is a base that is sometimes used for titrating acids. When using it, you need to know the actual concentration of hydroxide ion in the solution. What are the molar concentrations of H_3O^+ and OH^- in 0.0030 M $Ba(OH)_2$(aq) at 25 °C?

ANTICIPATE Because the compound provides OH^- ions, these ions will be present in overwhelming concentration, but nevertheless there must also be a tiny concentration of H_3O^+ ions to maintain the value of K_w.

PLAN Most hydroxides of Groups 1 and 2 can be treated as fully dissociated into ions in aqueous solution. Decide from the chemical formula how many OH^- ions are provided by each formula unit and calculate the concentrations of these ions in the solution. To find the concentration of H_3O^+ ions, use the water autoprotolysis constant $K_w = [H_3O^+][OH^-]$.

SOLVE

Decide whether the compound is fully dissociated in solution.

Because barium is an alkaline earth metal, $Ba(OH)_2$ dissociates almost completely in water to provide Ba^{2+} and OH^- ions.

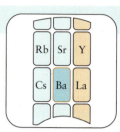

Find the mole ratio of hydroxide ion to barium hydroxide.

$$Ba(OH)_2(s) \longrightarrow Ba^{2+}(aq) + 2\,OH^-(aq); \quad 1\ mol\ Ba(OH)_2 \mathrel{\widehat{=}} 2\ mol\ OH^-$$

Calculate the hydroxide ion concentration from the solute concentration (here, with the units in place).

$$[OH^-] = 2 \times 0.0030\ mol \cdot L^{-1} = 0.0060\ mol \cdot L^{-1}$$

Use K_w in the form $[H_3O^+] = K_w/[OH^-]$ to find the concentration of H_3O^+ ions.

$$[H_3O^+] = \frac{K_w}{[OH^-]} = \frac{1.0 \times 10^{-14}}{0.0060} = 1.7 \times 10^{-12}$$

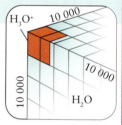

EVALUATE The concentration of H_3O^+ ions is 1.7 pmol·L^{-1} (1 pmol = 10^{-12} mol), which is very small, as expected; but it is not zero.

Self-test 6A.3A Estimate the concentrations of (a) H_3O^+ and (b) OH^- at 25 °C in 6.0×10^{-5} M HI(aq).

[*Answer:* (a) 60. µmol·L^{-1}; (b) 0.17 nmol·L^{-1}]

Self-test 6A.3B Estimate the concentrations of (a) H_3O^+ and (b) OH^- at 25 °C in 2.2×10^{-3} M NaOH(aq).

Related Exercises 6A.19, 6A.20, 6A.23, 6A.24

In aqueous solutions, the concentrations of H_3O^+ and OH^- ions are related by the autoprotolysis equilibrium; if one concentration is increased, then the other must decrease to maintain the value of K_w.

What have you learned in this Topic?

You have learned two definitions of acids and bases: the Lewis definition is more general, but the Brønsted–Lowry definition is more useful for aqueous solutions. Strong acids and bases are almost completely deprotonated or protonated, respectively, whereas weak acids and bases are not. You have seen how acids and bases form conjugate pairs and have encountered the concepts of autoprotolysis and amphoterism. You know that in water the concentrations of hydronium ions and hydroxide ions are always such as to satisfy the autoprotolysis constant.

The skills you have mastered are the ability to:

☐ **1.** Identify acid and base conjugate pairs (Example 6A.1).

☐ **2.** Recognize Lewis acid–base reactions (Section 6A.2).

☐ **3.** Use the autoprotolysis constant of water, K_w, to relate the concentrations of H_3O^+ and OH^- ions in aqueous solutions of acids and bases (Example 6A.2).

Topic 6A Exercises

6A.1 Write the formulas of the conjugate acids of (a) CH_3NH_2, methylamine; (b) NH_2NH_2, hydrazine; (c) HCO_3^-; and the conjugate bases of (d) HCO_3^-; (e) C_6H_5OH, phenol; (f) CH_3COOH.

6A.2 Write the formulas of the conjugate acids of (a) $C_2O_4^{2-}$, the oxalate ion; (b) $C_6H_5NH_2$, aniline; (c) NH_2OH, hydroxylamine; and the conjugate bases of (d) H_2O_2, hydrogen peroxide; (e) HNO_2; (f) $HCrO_4^-$.

6A.3 Write the chemical equations for the proton transfer equilibria of the following acids in aqueous solution and identify the conjugate acid–base pairs in each case: (a) H_2SO_4; (b) $C_6H_5NH_3^+$, anilinium ion; (c) $H_2PO_4^-$; (d) $HCOOH$, formic acid; (e) $NH_2NH_3^+$, hydrazinium ion.

6A.4 Write the chemical equations for the proton transfer equilibria of the following bases in aqueous solution and identify the conjugate acid–base pairs in each case: (a) CN^-; (b) NH_2NH_2, hydrazine; (c) CO_3^{2-}; (d) HPO_4^{2-}; (e) $CO(NH_2)_2$, urea.

6A.5 Identify (a) the Brønsted acid and base in the following reaction, and (b) the conjugate base and acid formed:

$$HNO_3(aq) + HPO_4^{2-}(aq) \longrightarrow NO_3^-(aq) + H_2PO_4^-(aq)$$

6A.6 Identify (a) the Brønsted acid and base in the following reaction, and (b) the conjugate base and acid formed:

$$HSO_3^-(aq) + CH_3NH_3^+(aq) \longrightarrow H_2SO_3(aq) + CH_3NH_2(aq)$$

6A.7 Below are molecular models of two oxoacids. Write the name of each acid and then draw the model of its conjugate base. (Red = O, white = H, green = Cl, and blue = N.)

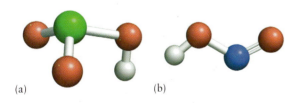

(a)   (b)

6A.8 Below are molecular models of two oxoacids. Write the name of each acid and then draw the model of its conjugate base. (Red = O, white = H, green = Cl, and blue = N.)

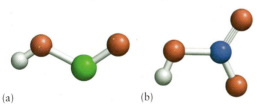

(a)   (b)

6A.9 Which of the following can be classified as reactions between Brønsted acids and bases? For those that can be so classified, identify the acid and the base. (*Hint:* It might help to write the net ionic equations.)

(a) $NH_4I(aq) + H_2O(l) \rightarrow NH_3(aq) + H_3O^+(aq) + I^-(aq)$

(b) $NH_4I(s) \xrightarrow{\Delta} NH_3(g) + HI(g)$

(c) $CH_3COOH(aq) + NH_3(aq) \rightarrow CH_3CONH_2(aq) + H_2O(l)$

(d) $NH_4I(am) + KNH_2(am) \rightarrow KI(am) + 2\,NH_3(l)$, where "am" indicates that liquid ammonia is the solvent

6A.10 Which of the following reactions can be classified as reactions between Brønsted acids and bases? For those that can be so classified, identify the acid and the base. (*Hint:* It might help to write the net ionic equations.)

(a) $KOH(aq) + CH_3I(aq) \rightarrow CH_3OH(aq) + KI(aq)$

(b) $AgNO_3(aq) + HCl(aq) \rightarrow AgCl(s) + HNO_3(aq)$

(c) $2\,NaHCO_3(am) + 2\,NH_3(l) \rightarrow Na_2CO_3(s) + (NH_4)_2CO_3(am)$, where "am" indicates that liquid ammonia is the solvent

(d) $H_2S(aq) + Na_2S(s) \rightarrow 2\,NaHS(aq)$

6A.11 Write the chemical equations of the two proton transfer equilibria that demonstrate the amphiprotic character of (a) HCO_3^-; (b) HPO_4^{2-}. Identify the conjugate acid–base pairs in each case.

6A.12 Write the chemical equations of the two proton transfer equilibria that demonstrate the amphiprotic character of (a) $H_2PO_3^-$; (b) NH_3. Identify the conjugate acid–base pairs in each case.

6A.13 Draw the Lewis structure or symbol for each of the following species and identify each one as a Lewis acid or Lewis base: (a) NH_3; (b) BF_3; (c) Ag^+; (d) F^-; (e) H^-.

6A.14 Draw the Lewis structure or symbol for each of the following species and identify each one as a Lewis acid or Lewis base: (a) SO_2; (b) I^-; (c) CH_3S^- (the C atom is the central atom); (d) NH_2^-; (e) NO_2.

6A.15 Draw the Lewis structure or symbol of each reactant, identify the Lewis acid and the Lewis base, and then draw the Lewis structure of the product (a complex) for the following Lewis acid–base reactions:

$$PF_5 + F^- \longrightarrow$$
$$Cl^- + SO_2 \longrightarrow$$

6A.16 Draw the Lewis structure or symbol of each reactant, identify the Lewis acid and the Lewis base, and then draw the Lewis structure of the product (a complex) for the following Lewis acid–base reactions:

$$F^- + BrF_3 \longrightarrow$$
$$FeCl_3 + Cl^- \longrightarrow$$

6A.17 State whether the following oxides are acidic, basic, or amphoteric: (a) BaO; (b) SO_3; (c) As_2O_3; (d) Bi_2O_3.

6A.18 State whether the following oxides are acidic, basic, or amphoteric: (a) SO_2; (b) CaO; (c) P_4O_{10}; (d) TeO_2.

6A.19 Calculate the molar concentration of OH^- in solutions with the following molar concentrations of H_3O^+: (a) 0.020 mol·L^{-1}; (b) 1.0×10^{-5} mol·L^{-1}; (c) 3.1 mol·L^{-1}.

6A.20 Calculate the molar concentration of H_3O^+ in solutions with the following molar concentrations of OH^-: (a) 0.0021 mol·L^{-1}; (b) 3.4×10^{-3} mol·L^{-1}; (c) 7.60 mmol·L^{-1}

6A.21 The value of K_w for water at body temperature (37 °C) is 2.1×10^{-14}. (a) What is the molar concentration of H_3O^+ ions at 37 °C? (b) What is the molar concentration of OH^- in neutral water at 37 °C?

6A.22 The molar concentration of H_3O^+ ions at the freezing point of water is 3.9×10^{-8} mol·L^{-1}. Calculate K_w at 0 °C.

6A.23 Calculate the molar concentration of $Ba(OH)_2(aq)$ and the molar concentrations of Ba^{2+}, OH^-, and H_3O^+ in an aqueous solution that contains 0.43 g of $Ba(OH)_2$ in 0.100 L of solution.

6A.24 Calculate the molar concentration of $KNH_2(aq)$ and the molar concentrations of K^+, NH_2^-, OH^-, and H_3O^+ in an aqueous solution that contains 0.40 g of KNH_2 in 0.450 L of solution.

Topic 6B The pH Scale

6B.1   The Interpretation of pH

6B.2   The pOH of Solutions

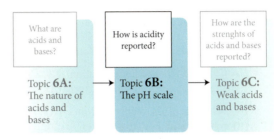

What are acids and bases?

How is acidity reported?

How are the strenghts of acids and bases reported?

Topic **6A:**
The nature of acids and bases

→ Topic **6B:**
The pH scale

→ Topic **6C:**
Weak acids and bases

One difficulty with describing the concentrations of acids and bases quantitatively is that the concentration of H_3O^+ ions can vary over many orders of magnitude: in some solutions, it is higher than 1 mol·L^{-1}, and in others it is lower than 10^{-14} mol·L^{-1}. Chemists avoid the awkwardness of working with such a wide range of values by reporting the hydronium ion concentration in terms of the **pH** of the solution, the negative logarithm (to the base 10) of the hydronium ion activity:

$$pH = -\log a_{H_3O^+} \tag{1a}$$

where (for solutions so dilute that they may be treated as ideal) $a_{H_3O^+} = [H_3O^+]/c°$. As in Topic 5G, the appearance of this expression is simplified by interpreting $[H_3O^+]$ as the molar concentration of H_3O^+ in moles per liter with the units struck out, and writing

$$pH = -\log[H_3O^+] \tag{1b}$$

For example, the pH of pure water, in which the concentration of H_3O^+ ions is 1.0×10^{-7} mol·L^{-1}. at 25 °C, is

$$pH = -\log(1.0 \times 10^{-7}) = 7.00$$

THINKING POINT

Do you expect the pH of pure water to increase or decrease with increasing temperature?

A Note on Good Practice: The number of digits following the decimal point in a pH value is equal to the number of significant figures in the corresponding molar concentration, because the digits preceding the decimal point simply report the power of 10 in the data (as in log $10^5 = 5$).

6B.1 The Interpretation of pH

The negative sign in the definition of pH means that the higher the concentration of H_3O^+ ions, the lower is the pH. For example, if the concentration of H_3O^+ is 1×10^{-7} mol·L^{-1}, the pH is 7.0, but if the concentration is increased to 1×10^{-6} mol·L^{-1}, the pH drops to 6.0. As this example shows, a change of one pH unit means that the concentration of H_3O^+ ions has changed by a factor of 10. It is important to remember that (at 25 °C):

- The pH of a basic solution is greater than 7.
- The pH of a neutral solution, such as pure water, is equal to 7.
- The pH of an acidic solution is less than 7.

Most solutions used in chemistry have a pH ranging from 0 to 14, but values outside this range are possible.

THINKING POINT

Can a pH be negative? If so, what would that signify?

Why Do You Need to Know This Material? The pH scale of hydronium ion concentration is used throughout chemistry, biology, medicine, and industry, and it is essential to know its definition and significance.

What Do You Need to Know Already? You need to understand the concepts of acid and base as introduced in Topic 6A and the properties of logarithms (Appendix 1D). You also need to be familiar with the significance of the autoprotolysis constant (Topic 6A).

The pH scale was introduced by the Danish chemist Søren Sørensen in 1909 in the course of his work on quality control in the brewing of beer. The negative logarithm is used so that most pH values are positive numbers.

EXAMPLE 6B.1   Calculating a pH from a concentration

You are working in a medical laboratory monitoring the recovery of patients in intensive care. The pH of their blood must be carefully monitored and controlled because even small deviations from normal levels can be fatal. What is the pH of (a) human blood, in which the concentration of H_3O^+ ions is 4.0×10^{-8} mol·L^{-1}; (b) 0.020 M HCl(aq); (c) 0.040 M KOH(aq)?

ANTICIPATE The concentration of H_3O^+ ions in blood is lower than in pure water, so you should expect pH > 7; in HCl(aq), an acid, you should expect pH < 7, and in KOH(aq), a base, pH > 7.

PLAN The pH is calculated from Eq. 1b. For strong acids, the molar concentration of H_3O^+ is equal to the molar concentration of the acid. For strong bases, first find the concentration of OH$^-$, then convert that concentration into $[H_3O^+]$ by using $[H_3O^+][OH^-] = K_w$ in the form $[H_3O^+] = K_w/[OH^-]$.

What should you assume? Assume that any strong acid (HCl here) is fully deprotonated in solution and any ionic compound (KOH here) is fully dissociated in solution.

SOLVE

(a) From pH = $-\log [H_3O^+]$,

$$pH = -\log(4.0 \times 10^{-8}) = 7.40$$

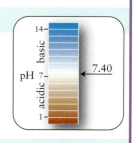

(b) HCl is a strong acid, so it is treated as completely deprotonated in water.

$$[H_3O^+] = [HCl] = 0.020 \text{ mol·L}^{-1}$$

From pH = $-\log [H_3O^+]$,

$$pH = -\log 0.020 = 1.70$$

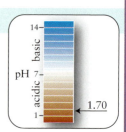

(c) Because KOH is assumed to dissociate completely in solution each formula unit provides one OH$^-$ ion,

$$[OH^-] = [KOH] = 0.040 \text{ mol·L}^{-1}$$

Find $[H_3O^+]$ from $[H_3O^+][OH^-] = K_w$ in the form $[H_3O^+] = K_w/[OH^-]$.

$$[H_3O^+] = \frac{K_w}{[OH^-]} = \frac{1.0 \times 10^{-14}}{0.040} = 2.5... \times 10^{-13}$$

From pH = $-\log [H_3O^+]$,

$$pH = -\log(2.5... \times 10^{-13}) = 12.60$$

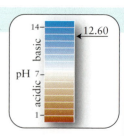

EVALUATE The calculated pH values are in line with what was anticipated.

Self-test 6B.1A Calculate the pH of (a) household ammonia, in which the OH$^-$ concentration is about 3×10^{-3} mol·L^{-1}; (b) 6.0×10^{-5} M HClO$_4$(aq).

[**Answer:** (a) 11.5; (b) 4.22]

Self-test 6B.1B Calculate the pH of 0.077 M NaOH(aq).

Related Exercises: 6B.3, 6B.4

The approximate value of the pH of an aqueous solution can be determined very quickly by using a strip of *universal indicator paper,* which turns different colors at different pH values. More precise measurements are made with a "pH meter" (**FIG. 6B.1**). This instrument consists of a voltmeter connected to two electrodes that dip into the solution. The difference in electrical potential between the electrodes is proportional to the hydronium ion activity (as explained in Topic 6L), so once the scale on the meter has been calibrated, the pH can be read directly.

To convert a pH value into a concentration of H_3O^+ ions, reverse the sign of the pH and then take its antilogarithm:

$$[H_3O^+] = 10^{-pH} \text{ mol·L}^{-1} \tag{2}$$

In many pH meters both electrodes are integrated into a single unit. Electrodes of this type are called *combination electrodes.*

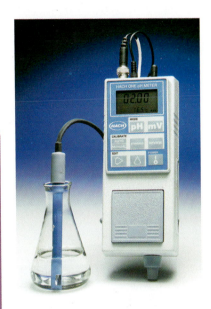

EXAMPLE 6B.2   Calculating the hydronium ion concentration from the pH

Sometimes the pH must be converted into the hydronium ion concentration. The quickest way to find the hydronium ion concentration in a solution is to use a pH meter to measure the pH and then calculate $[H_3O^+]$ from the pH. What is the molar concentration of hydronium ions in a solution with pH = 4.83?

ANTICIPATE Because the pH is less than 7, the pH of pure water at 25 °C, the solution must be acidic and you should expect a concentration of H_3O^+ ions greater than 10^{-7} mol·L^{-1}.

PLAN To calculate the precise value from Eq. 2, first change the sign of the pH and then take its antilogarithm.

SOLVE

From $[H_3O^+] = 10^{-pH} \text{ mol·L}^{-1}$,

$$[H_3O^+] = 10^{-4.83} \text{ mol·L}^{-1} = 1.5 \times 10^{-5} \text{ mol·L}^{-1}$$

EVALUATE As expected, the concentration of H_3O^+ ions is higher than in pure water, by a factor of more than 100.

Self-test 6B.2A The pH of stomach fluids is about 1.7. What is the molar concentration of H_3O^+ ions in the stomach?

[**Answer:** 2×10^{-2} mol·L^{-1}]

Self-test 6B.2B The pH of pancreatic fluids, which help to digest food once it has left the stomach, is about 8.2. What is the approximate molar concentration of H_3O^+ ions in pancreatic fluids?

Related Exercises 6B.7, 6B.8

FIGURE 6B.1 A pH meter is a voltmeter that is used to measure the pH electrochemically. (*Charles D. Winters/Science Source.*)

FIGURE 6B.2 shows the results of measuring the pH of a selection of liquids and beverages. Fresh lemon juice has a pH of 2.2, corresponding to an H_3O^+ concentration of 6 mmol·L^{-1} (with 1 mmol = 10^{-3} mol). Natural (unpolluted) rain, with acidity due largely to dissolved carbon dioxide, typically has a pH of about 5.7. In the United States, the Environmental Protection Agency (EPA) defines aqueous waste as "corrosive" if its pH is either lower than or equal to 2 (highly acidic) or higher than 11.5 (highly basic).

The pH scale is used to report the molar concentration of H_3O^+ ions: pH = −log $[H_3O^+]$; pH > 7 denotes a basic solution, pH < 7 an acidic solution; a neutral solution has pH = 7.

6B.2 The pOH of Solutions

Many quantitative expressions relating to acids and bases are greatly simplified when logarithms are used. The quantity pX is a generalization of pH:

$$pX = -\log X \tag{3}$$

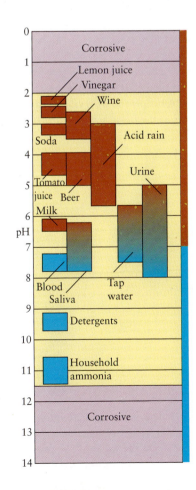

FIGURE 6B.2 Typical pH values of some common aqueous solutions. The purple-colored regions indicate the pH ranges for liquids regarded as corrosive.

For example, **pOH** is defined as

$$pOH = -\log a_{OH^-} \tag{4a}$$

which, for the same reasons as those for pH, is normally simplified to

$$pOH = -\log [OH^-] \tag{4b}$$

The pOH scale is convenient for reporting the concentration of OH^- ions in solution. For example, in pure water, where $[OH^-] = 1.0 \times 10^{-7}$ mol·L^{-1}, the pOH is 7.00. Similarly, by pK_w is meant

$$pK_w = -\log K_w = -\log(1.0 \times 10^{-14}) = 14.00 \text{ (at 25 °C)}$$

The values of pH and pOH for a given aqueous solution are related. To find that relation, start with the expression for the autoprotolysis constant of water written as $[H_3O^+] \times [OH^-] = K_w$. Then take logarithms of both sides:

$$\log([H_3O^+][OH^-]) = \log K_w$$

Now use $\log ab = \log a + \log b$ to obtain

$$\log [H_3O^+] + \log [OH^-] = \log K_w$$

Multiplication of both sides of the equation by -1 gives

$$\overbrace{-\log [H_3O^+]}^{pH} + \overbrace{(-\log [OH^-])}^{pOH} = \overbrace{-\log K_w}^{pK_w}$$

which is the same as

$$pH + pOH = pK_w \tag{5a}$$

Because $pK_w = 14.00$ at 25 °C, at that temperature

$$pH + pOH = 14.00 \tag{5b}$$

Equation 5 shows that the pH and pOH of a solution have complementary values: if one increases, the other decreases such that their sum remains constant (**FIG. 6B.3**).

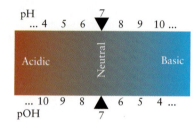

FIGURE 6B.3 The numbers along the top of the rectangle are values of pH for a range of aqueous solutions at 25 °C. Directly below each pH value is the pOH value of the same solution. Notice that the sum of the pH and pOH values for a given solution is always 14. Most values of pH and pOH lie in the range from 0 to 14, but pH and pOH values can lie outside this range and even be negative.

Self-test 6B.3A At 25 °C the pH of stomach fluids is about 1.7. What is the pOH of the fluids?

[*Answer:* 12.3]

Self-test 6B.3B At 25 °C the pOH of the solution of a common detergent is 9.4. What is the pH of this solution?

The pH and pOH of a solution are related by the expression pH + pOH = pK$_w$.

What have you learned in this Topic?

You have learned that the pH scale is convenient for expressing the wide range of hydronium ion concentrations found in aqueous solutions. You have seen that in aqueous solution, the pH and pOH have a seesaw relationship: if one goes up, the other goes down.

The skills you have mastered are the ability to:

☐ **1.** Determine the pH of a solution from a known concentration of acid or base (Example 6B.1).

☐ **2.** Find the molar concentration of hydronium ions from a known pH value (Example 6B.2).

☐ **3.** Interconvert between pH and pOH for aqueous solutions (Section 6B.2).

Topic 6B Exercises

6B.1 The molar concentration of HCl in hydrochloric acid is reduced to 12% of its initial value by dilution. What is the difference in the pH values of the two solutions?

6B.2 The molar concentration of $Ca(OH)_2$ in an aqueous solution is reduced to 5.2% of its initial value by dilution. What is the difference in the pH values of the two solutions?

6B.3 A careless laboratory technician wants to prepare 200.0 mL of a 0.025 M HCl(aq) solution but uses a volumetric flask of volume 250.0 mL by mistake. (a) What would the pH of the desired solution have been? (b) What will be the actual pH of the solution as prepared?

6B.4 A careless laboratory technician prepares 300.0 mL of 0.0175 M KOH(aq) and pipets 25.0 mL of the solution into a beaker. The beaker is allowed to stand in a warm place for two days before use, during which time some of the water evaporates and the volume is reduced to 18.0 mL. (a) What would the pH of the solution as initially prepared have been? (b) What will be the actual pH of the solution after the evaporation?

6B.5 Calculate the pH and pOH of each of the following aqueous solutions of a strong acid or base: (a) 0.0146 M HNO_3(aq); (b) 0.11 M HCl(aq); (c) 0.0092 M $Ba(OH)_2$(aq); (d) 2.00 mL of 0.175 M KOH(aq) after dilution to 0.500 L; (e) 13.6 mg of NaOH dissolved in 0.350 L of solution; (f) 75.0 mL of 3.5×10^{-4} M HBr(aq) after dilution to 0.500 L.

6B.6 Calculate the pH and pOH of each of the following aqueous solutions of strong acid or base: (a) 0.0356 M HI(aq); (b) 0.0725 M HCl(aq); (c) 3.46×10^{-3} M $Ba(OH)_2$(aq); (d) 10.9 mg of KOH dissolved in 10.0 mL of solution; (e) 10.0 mL of 5.00 M NaOH(aq) after dilution to 2.50 L; (f) 5.0 mL of 3.5×10^{-4} M $HClO_4$(aq) after dilution to 25.0 mL.

6B.7 The pH of several solutions was measured in the research laboratories of a food company; convert each of the following pH values into the molar concentration of H_3O^+ ions: (a) 3.3 (the pH of sour orange juice); (b) 6.7 (the pH of a saliva sample); (c) 4.4 (the pH of beer); (d) 5.3 (the pH of a coffee sample).

6B.8 The pH of several solutions was measured in a hospital laboratory; convert each of the following pH values into the molar concentration of H_3O^+ ions: (a) 4.8 (the pH of a urine sample); (b) 0.7 (the pH of a sample of fluid extracted from a stomach);

(c) 7.4 (the pH of blood); (d) 8.1 (the pH of exocrine pancreatic secretions).

6B.9 (a) Complete the following table. (b) Rank the four solutions in order of increasing acidity.

	$[H_3O^+]$	$[OH^-]$	pH	pOH
(i)	1.50 mol·L^{-1}			
(ii)		1.50 mol·L^{-1}		
(iii)			0.75	
(iv)				0.75

6B.10 (a) Complete the following table. (b) Rank the four solutions in order of increasing acidity.

	$[H_3O^+]$	$[OH^-]$	pH	pOH
(i)	0.50 mol·L^{-1}			
(ii)		0.50 mol·L^{-1}		
(iii)			−0.10	
(iv)				−0.10

6B.11 A student added solid Na_2O to a volumetric flask of volume 200.0 mL, which was then filled with water, resulting in 200.0 mL of NaOH solution. Then 5.00 mL of the solution was transferred to another volumetric flask and diluted to 500.0 mL. The pH of the diluted solution is 13.25. (a) What is the molar concentration of hydroxide ions in (i) the diluted solution, (ii) the original solution? (b) What mass of Na_2O was added to the first flask?

6B.12 A student added solid K_2O to a volumetric flask of volume 500.0 mL, which was then filled with water, resulting in 500.0 mL of KOH solution. Then 10.0 mL of the solution was transferred to another volumetric flask and diluted to 300.0 mL. The pH of the diluted solution is 14.12. (a) What is the molar concentration of hydroxide ion in (i) the diluted solution, (ii) the original solution? (b) What mass of K_2O was added to the first flask?

Topic 6C Weak Acids and Bases

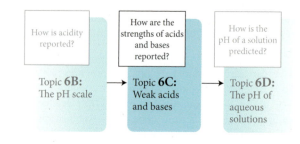

Material? Because weak acids and bases are common reagents you need to know how to identify them and, to develop your understanding of matter, to understand their relative strengths.

What Do You Need to Know Already? You need to be familiar with the concepts of strong acids and bases (Topic 6A) and pH (Topic 6B), and be able to work with equilibrium expressions (Topic 5G) and the autoprotolysis equilibrium of water (Topic 6A). This Topic draws on the effect of electronegativity on bond polarity and strength (Topic 2D).

Lemon juice, which contains citric acid, is often added to fish dishes to help eliminate the smell of some of these amines; it works by forming the less volatile salt.

Solutions of different acids having the same concentration might not have the same pH. For instance, the pH of 0.10 M $CH_3COOH(aq)$ is close to 3, but that of 0.10 M $HCl(aq)$ is close to 1. That is, the concentration of H_3O^+ ions in 0.10 M $CH_3COOH(aq)$ is *lower* than in 0.10 M $HCl(aq)$. Similarly, the concentration of OH^- ions is found to be lower in 0.10 M $NH_3(aq)$ than in 0.10 M $NaOH(aq)$. The explanation must be that, in water, CH_3COOH is not fully deprotonated and NH_3 is not fully protonated. Therefore, acetic acid and ammonia are, respectively, a weak acid and a weak base. The incomplete deprotonation of CH_3COOH explains why solutions of HCl and CH_3COOH of the same concentration react with the same metal at different rates (**FIG. 6C.1**).

Most acids and bases that exist in nature are weak. For example, the natural acidity of river water is due to the presence of weak acids such as carbonic acid (H_2CO_3, from dissolved CO_2), hydrogen phosphate ions, HPO_4^{2-}, dihydrogen phosphate ions, $H_2PO_4^-$ (from fertilizer runoff), and other carboxylic acids arising from the degradation of plant tissues. Similarly, most naturally occurring bases are weak. Some arise from the decomposition (in the absence of air) of compounds containing nitrogen. For example, the odor of dead fish is due to amines, which are weak bases.

6C.1 Acidity and Basicity Constants

When considering the composition of a solution of a weak acid in water, you should think of a solution that contains

- the acid molecules or ions;
- low concentrations of H_3O^+ ions and the conjugate base of the acid; and
- a very, very low concentration of OH^- ions that maintain the autoprotolysis equilibrium.

All these species are in ceaseless dynamic equilibrium. Similarly, for a solution of a weak base, you should consider

FIGURE 6C.1 The same mass of magnesium metal has been added to solutions of HCl, a strong acid (left), and CH_3COOH, a weak acid (right). Although the acid solutions were prepared with the same concentrations, the rate of hydrogen evolution, which depends on the concentration of hydronium ions, is much greater in the strong acid. (*W. H. Freeman photos by Ken Karp.*)

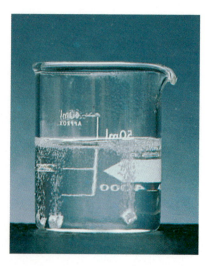

FIGURE 6C.2 In a solution of a weak acid, only some of the acidic hydrogen atoms are present as hydronium ions (the red sphere), and the solution contains a high proportion of the original acid molecules (HA, gray spheres). The green sphere represents the conjugate base of the acid and the blue spheres are water molecules. The inset at the left shows only the solute species.

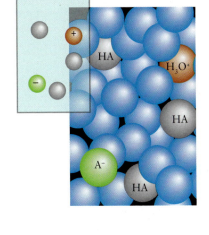

ANIMATION FIGURE 6C.2

- the base molecules or ions;
- low concentrations of OH^- ions and the conjugate acid of the base; and
- a very, very low concentration of H_3O^+ ions that maintain the autoprotolysis equilibrium.

An indicator of the strength of an acid or base is the size of the equilibrium constant for proton transfer to or from the solvent. For example, for acetic acid in water,

$$CH_3COOH(aq) + H_2O(l) \rightleftharpoons H_3O^+(aq) + CH_3CO_2^-(aq) \qquad \textbf{(A)}$$

the equilibrium constant is

$$K = \frac{a_{H_3O^+} a_{CH_3CO_2^-}}{a_{CH_3COOH} a_{H_2O}}$$

Because the only solutions considered in this text are dilute and the water is almost pure, the activity of H_2O can be set equal to 1. The resulting expression is called the **acidity constant**, K_a, of acetic acid. On making the further approximation of replacing the activities of the solute species by the numerical values of their molar concentrations, this expression becomes

K_a is also widely called the acid ionization constant or acid dissociation constant.

$$K_a = \underbrace{\frac{\overbrace{a_{H_3O^+} a_{CH_3CO_2^-}}^{[H_3O^+][CH_3CO_2^-]}}{\underbrace{a_{CH_3COOH}}_{[CH_3COOH]} \underbrace{a_{H_2O}}_{1}}}_{} = \frac{[H_3O^+][CH_3CO_2^-]}{[CH_3COOH]}$$

The experimental value of K_a for acetic acid at 25 °C is 1.8×10^{-5}. This small value implies that only a small proportion of CH_3COOH molecules donate their protons when dissolved in water. About 99 of 100 CH_3COOH molecules remain intact in 1 M $CH_3COOH(aq)$, but the actual value depends on the concentration of acid, (Topic 6D). This value is typical of weak acids in water (**FIG. 6C.2**). In general, the acidity constant of an acid HA is

$$HA(aq) + H_2O(l) \rightleftharpoons H_3O^+(aq) + A^-(aq) \qquad K_a = \frac{[H_3O^+][A^-]}{[HA]} \qquad \textbf{(1)}$$

TABLE 6C.1 lists the acidity constants of some weak acids in aqueous solution.

TABLE 6C.1 Acidity Constants at 25 °C*					
Acid	K_a	pK_a	**Acid**	K_a	pK_a
trichloroacetic acid, CCl_3COOH	3.0×10^{-1}	0.52	formic acid, HCOOH	1.8×10^{-4}	3.75
benzene sulfonic acid, $C_6H_5SO_3H$	2.0×10^{-1}	0.70	benzoic acid, C_6H_5COOH	6.5×10^{-5}	4.19
iodic acid, HIO_3	1.7×10^{-1}	0.77	acetic acid, CH_3COOH	1.8×10^{-5}	4.75
sulfurous acid, H_2SO_3	1.5×10^{-2}	1.81	carbonic acid, H_2CO_3	4.3×10^{-7}	6.37
chlorous acid, $HClO_2$	1.0×10^{-2}	2.00	hypochlorous acid, HClO	3.0×10^{-8}	7.53
phosphoric acid, H_3PO_4	7.6×10^{-3}	2.12	hypobromous acid, HBrO	2.0×10^{-9}	8.69
chloroacetic acid, $CH_2ClCOOH$	1.4×10^{-3}	2.85	boric acid, $B(OH)_3$†	7.2×10^{-10}	9.14
lactic acid, $CH_3CH(OH)COOH$	8.4×10^{-4}	3.08	hydrocyanic acid, HCN	4.9×10^{-10}	9.31
nitrous acid, HNO_2	4.3×10^{-4}	3.37	phenol, C_6H_5OH	1.3×10^{-10}	9.89
hydrofluoric acid, HF	3.5×10^{-4}	3.45	hypoiodous acid, HIO	2.3×10^{-11}	10.64

*The values for K_a listed here have been calculated from pK_a values with more significant figures than shown so as to minimize rounding errors. Values for polyprotic acids—those capable of donating more than one proton—refer to the first deprotonation.
†The proton transfer equilibrium is $B(OH)_3(aq) + 2 H_2O(l) \rightleftharpoons H_3O^+(aq) + B(OH)_4^-(aq)$

TABLE 6C.2 Basicity Constants at 25 °C*

Base	K_b	pK_b	Base	K_b	pK_b
urea, $CO(NH_2)_2$	1.3×10^{-14}	13.90	ammonia, NH_3	1.8×10^{-5}	4.75
aniline, $C_6H_5NH_2$	4.3×10^{-10}	9.37	trimethylamine, $(CH_3)_3N$	6.5×10^{-5}	4.19
pyridine, C_5H_5N	1.8×10^{-9}	8.75	methylamine, CH_3NH_2	3.6×10^{-4}	3.44
hydroxylamine, NH_2OH	1.1×10^{-8}	7.97	dimethylamine, $(CH_3)_2NH$	5.4×10^{-4}	3.27
nicotine, $C_{10}H_{14}N_2$	1.0×10^{-6}	5.98	ethylamine, $C_2H_5NH_2$	6.5×10^{-4}	3.19
morphine, $C_{17}H_{19}O_3N$	1.6×10^{-6}	5.79	triethylamine, $(C_2H_5)_3N$	1.0×10^{-3}	2.99
hydrazine, NH_2NH_2	1.7×10^{-6}	5.77			

*The values for K_b listed here have been calculated from pK_b values with more significant figures than shown so as to minimize rounding errors.

For the proton transfer of a base such as ammonia in water the equilibrium is

$$NH_3(aq) + H_2O(l) \rightleftharpoons NH_4^+(aq) + OH^-(aq) \qquad \textbf{(B)}$$

and the equilibrium constant is

$$K = \frac{a_{NH_4^+} a_{OH^-}}{a_{NH_3} a_{H_2O}}$$

In dilute solutions, the water is almost pure and its activity can be set equal to 1, to give the **basicity constant,** K_b. On making the further approximation of replacing the activities of the solute species by the numerical values of their molar concentrations, the basicity constant expression for ammonia becomes

K_b is also widely called the base ionization constant.

$$K_b = \frac{\overbrace{a_{NH_4^+} a_{OH^-}}^{[NH_4^+][OH^-]}}{\underbrace{a_{NH_3}}_{[NH_3]} \underbrace{a_{H_2O}}_{1}} = \frac{[NH_4^+][OH^-]}{[NH_3]}$$

The experimental value of K_b for ammonia in water at 25 °C is 1.8×10^{-5}. This small value implies that normally only a small proportion of the NH_3 molecules are present as NH_4^+. Equilibrium calculations of the kind described in Topic 6D show that only about 1 in 100 molecules is protonated in a typical solution (**FIG. 6C.3**).

In general, the basicity constant for a base B in water is

$$B(aq) + H_2O(l) \rightleftharpoons HB^+(aq) + OH^-(aq) \qquad K_b = \frac{[HB^+][OH^-]}{[B]} \qquad \textbf{(2)}$$

The value of K_b indicates how far the reaction proceeds to the right. The smaller the value of K_b the weaker is the ability of the base to accept a proton. **TABLE 6C.2** lists the basicity constants of some weak bases in aqueous solution.

Acidity and basicity constants are commonly reported as their negative logarithms, through the definitions

$$pK_a = -\log K_a \qquad pK_b = -\log K_b \qquad \textbf{(3)}$$

When thinking about the strengths of acids and bases, you should note that:

• The weaker the acid, the smaller the value of K_a and the greater the value of pK_a.

For example, the pK_a of trichloroacetic acid is 0.5, whereas that of acetic acid, a much weaker acid, is nearly 5. Similar remarks apply to bases:

• The weaker the base, the smaller the value of K_b and the greater the value of pK_b.

Values of pK_a and pK_b are included in Tables 6C.1 and 6C.2.

The proton-donating strength of an acid is measured by its acidity constant; the proton-accepting strength of a base is measured by its basicity constant. The smaller the constants, the weaker are the respective strengths. The larger the value of pK, the weaker is the acid or base.

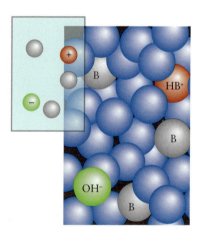

FIGURE 6C.3 In a solution of a weak base, only a small proportion of the base molecules (B, represented here by gray spheres) have accepted protons from water molecules (the blue spheres) to form HB^+ ions (the red sphere) and OH^- ions (green sphere). The inset at the left shows only the solute species.

ANIMATION FIGURE 6C.3

6C.2 The Conjugate Seesaw

Hydrochloric acid is classified as a strong acid because it is almost completely deprotonated in water. It follows that its conjugate base, Cl^-, must be a very, very weak proton acceptor (weaker, in fact, than H_2O). In contrast, acetic acid is a weak acid. Its conjugate base, the acetate ion, $CH_3CO_2^-$, must be a relatively good proton acceptor that readily forms CH_3COOH molecules when added to water. Similarly, because methylamine, CH_3NH_2, is a stronger base than ammonia (see Table 6C.2), the conjugate acid of methylamine—the methylammonium ion, $CH_3NH_3^+$—must be a weaker proton donor (and therefore a weaker acid) than NH_4^+. In general:

- the stronger the acid, the weaker is its conjugate base
- the stronger the base, the weaker is its conjugate acid

To express the relative strengths of an acid and its conjugate base (a "conjugate acid–base pair"), consider the ammonia proton transfer equilibrium, reaction B, for which the basicity constant is $K_b = [NH_4^+][OH^-]/[NH_3]$. The proton transfer equilibrium of ammonia's conjugate acid, NH_4^+, in water is

$$NH_4^+(aq) + H_2O(l) \rightleftharpoons H_3O^+(aq) + NH_3(aq) \qquad K_a = \frac{[H_3O^+][NH_3]}{[NH_4^+]}$$

Multiplication of the two equilibrium constants for the conjugate acid–base pair, K_a for NH_4^+ and K_b for NH_3, gives

$$\underbrace{\frac{[H_3O^+][NH_3]}{[NH_4^+]}}_{K_a} \times \underbrace{\frac{[NH_4^+][OH^-]}{[NH_3]}}_{K_b} = [H_3O^+][OH^-]$$

The product on the right is the autoprotolysis constant K_w (Topic 6A), so

$$K_a \times K_b = K_w \qquad (4a)$$

Equation 4a can be expressed in another way by taking logarithms of both sides of the equation:

$$\overset{\log xy = \log x + \log y}{\log(K_a \times K_b)} = \log K_a + \log K_b = \log K_w$$

Multiplication throughout by -1 turns this expression into

$$\underbrace{-\log K_a}_{pK_a} + \underbrace{(-\log K_b)}_{pK_b} = \underbrace{-\log K_w}_{pK_w}$$

and therefore into

$$pK_a + pK_b = pK_w \qquad (4b)$$

with $pK_w = 14.00$ at 25 °C. These expressions apply only to a conjugate acid–base pair, with K_a the acidity constant of the acid and K_b the basicity constant of its conjugate base.

Self-test 6C.1A Write the chemical formula for the conjugate acid of the base pyridine, C_5H_5N, and calculate its pK_a from the pK_b for pyridine.

[**Answer:** $C_5H_5NH^+$, 5.25]

Self-test 6C.1B Write the chemical formula for the conjugate base of the acid HIO_3 and calculate its pK_b from the pK_a for HIO_3.

Equation 4—in either form—confirms the seesaw relation between the strengths of acids and of their conjugate bases. Because K_w has a constant value at a given temperature, Eq. 4a implies that if an acid has a large K_a, then its conjugate base must have a small K_b. Similarly, if a base has a large K_b, then its conjugate acid must have a small K_a. Equation 4b implies that if the pK_a of an acid is large, then the pK_b of its conjugate base is small, and vice versa. This reciprocal relation is summarized in **FIG. 6C.4** and **TABLE 6C.3**. For example, because pK_b for ammonia in water is 4.75 at 25 °C, the pK_a of NH_4^+ is

$$pK_a = pK_w - pK_b = 14.00 - 4.75 = 9.25$$

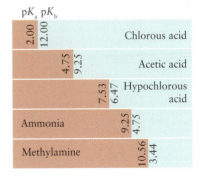

FIGURE 6C.4 The sum of the pK_a of an acid (pink) and the pK_b of its conjugate base (blue) is constant and equal to pK_w, which is 14.00 at 25 °C.

TABLE 6C.3   Conjugate Acid–Base Pairs Arranged by Strength

pK_a	Acid name	Acid formula		Base formula	Base name	pK_b
	Strong acid				*Very weak base*	
	hydroiodic acid	HI		I^-	iodide ion	
	perchloric acid	$HClO_4$		ClO_4^-	perchlorate ion	
	hydrobromic acid	HBr		Br^-	bromide ion	
	hydrochloric acid	HCl		Cl^-	chloride ion	
	sulfuric acid	H_2SO_4		HSO_4^-	hydrogen sulfate ion	
	chloric acid	$HClO_3$		ClO_3^-	chlorate ion	
	nitric acid	HNO_3		NO_3^-	nitrate ion	
	hydronium ion	H_3O^+		H_2O	*water*	
1.92	hydrogen sulfate ion	HSO_4^-		SO_4^{2-}	sulfate ion	12.08
3.37	nitrous acid	HNO_2		NO_2^-	nitrite ion	10.63
3.45	hydrofluoric acid	HF		F^-	fluoride ion	10.55
4.75	acetic acid	CH_3COOH		$CH_3CO_2^-$	acetate ion	9.25
6.37	carbonic acid	H_2CO_3		HCO_3^-	hydrogen carbonate ion	7.63
6.89	hydrosulfuric acid	H_2S		HS^-	hydrogen sulfide ion	7.11
9.25	ammonium ion	NH_4^+		NH_3	ammonia	4.75
9.31	hydrocyanic acid	HCN		CN^-	cyanide ion	4.69
10.25	hydrogen carbonate ion	HCO_3^-		CO_3^{2-}	carbonate ion	3.75
10.56	methylammonium ion	$CH_3NH_3^+$		CH_3NH_2	methylamine	3.44
	water	H_2O		OH^-	*hydroxide ion*	
	ammonia	NH_3		NH_2^-	amide ion	
	hydrogen	H_2		H^-	hydride ion	
	methane	CH_4		CH_3^-	methide ion	
	hydroxide ion	OH^-		O^{2-}	oxide ion	
	Very weak acid				**Strong base**	

This value shows that NH_4^+ is a weaker acid than boric acid ($pK_a = 9.14$) but stronger than hydrocyanic acid (HCN, $pK_a = 9.31$).

Although proton transfer in a solution of a strong acid is also an equilibrium, the proton-donating power of a strong acid, HA, is so much greater than that of H_3O^+ in the reverse reaction that proton transfer to water goes almost to completion. As a result, the solution can be treated as containing only H_3O^+ ions and A^- ions; there are virtually no HA molecules left. In other words, the only acid species present in an aqueous solution of a strong acid, other than the H_2O molecules, is the H_3O^+ ion. Because all strong acids in water behave as though they were solutions of the acid H_3O^+, strong acids are said to be **leveled** in water to the strength of the acid H_3O^+.

Sulfuric acid is a special case, because loss of its first acidic hydrogen leaves a conjugate base that is itself a weak acid, the HSO_4^- ion.

EXAMPLE 6C.1   Deciding which of two species is the stronger acid or base

Suppose you are conducting research relating acid and base strength to molecular structure. You might begin by comparing compounds with known acidity and basicity constants. Decide which member of each of the following pairs is the stronger acid or base in water: (a) acid: HF or HIO_3; (b) base: NO_2^- or CN^-.

PLAN   You need to compare the K_a of each weak acid. For anions, identify the conjugate weak acid and compare the K_a values.

SOLVE

Compare the relevant K_a and K_b (or pK_a and pK_b) values in Tables 6C.1 and 6C.2.

(a) Because $K_a(HIO_3) > K_a(HF)$, or $pK_a(HIO_3) < pK_a(HF)$, HIO_3 is a stronger acid than HF.

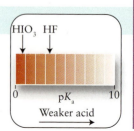

(b) Because $K_a(HNO_2) > K_a(HCN)$, or $pK_a(HNO_2) < pK_a(HCN)$, HCN is a weaker acid than HNO_2. Because the weaker acid has the stronger conjugate base, CN^- is the stronger base.

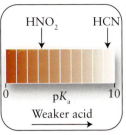

Self-test 6C.2A Use Tables 6C.1 and 6C.2 to decide which species in each of the following pairs is the stronger acid or base: (a) acid, HF or HIO; (b) base, $C_6H_5CO_2^-$ or $CH_2ClCO_2^-$; (c) base, $C_6H_5NH_2$ or $(CH_3)_3N$; (d) acid, $C_6H_5NH_3^+$ or $(CH_3)_3NH^+$.

[*Answer:* Stronger acids: (a) HF; (d) $C_6H_5NH_3^+$. Stronger bases: (b) $C_6H_5CO_2^-$; (c) $(CH_3)_3N$.]

Self-test 6C.2B Use Tables 6C.1 and 6C.2 to decide which species in each of the following pairs is the stronger acid or base: (a) base, C_5H_5N or NH_2NH_2; (b) acid, $C_5H_5NH^+$ or $NH_2NH_3^+$; (c) acid, HIO_3 or $HClO_2$; (d) base, ClO_2^- or HSO_3^-.

Related Exercises 6C.7–6C.14, 6C.17, 6C.18

The stronger the acid, the weaker is its conjugate base; the stronger the base, the weaker is its conjugate acid.

6C.3 Molecular Structure and Acid Strength

Chemists commonly interpret trends in the properties of compounds by considering the structures of their molecules. However, there are two reasons why the relative strengths of acids and bases are difficult to predict from molecular structure. First, K_a and K_b are equilibrium constants, and so they are related to the Gibbs free energy (Topic 4J, $\Delta G = \Delta H - T\Delta S$) of the proton transfer reaction (by Eq. 7 of Topic 5G, $\Delta G_r^\circ = -RT \ln K$). Their values therefore depend on considerations of entropy as well as enthalpy. Second, the solvent plays a significant role in proton transfer reactions, and so acid strength cannot be expected to be related solely to the acid molecule itself. However, although absolute values are difficult to predict, trends among series of compounds with similar structures and in the same solvent (normally water) can be expected. Because acid strength in aqueous solutions depends on the breaking of the H—A bond and the formation of an $H_2O—H^+$ bond, you might suspect that one factor in determining strength is the ease with which these bonds can be broken and formed.

Here is an opportunity for you to come up with a theory.

Although ionization energies and electron affinities are energy changes, not enthalpy changes, they differ by only a small amount and in fact these small differences cancel.

THINKING POINT

Do you think that the entropy of the system will increase or decrease when an acid molecule is deprotonated in water?

First, consider *binary acids,* acids composed of hydrogen and one other element, such as HCl and H_2S (in general, HA). Because the enthalpy change for a process does not depend on the path, the enthalpy change for the transfer of a proton from HA to H_2O in solution can be regarded as the outcome of the following hypothetical sequence (**FIG. 6C.5**):

Step	Reaction	Enthalpy change
Removal of HA from solution:	$HA(aq) \rightarrow HA(g)$	$-\Delta H_{solv}(HA)$
Dissociation of gaseous HA:	$HA(g) \rightarrow H(g) + A(g)$	$\Delta H_B(H—A)$
Ionization of H:	$H(g) \rightarrow H^+(g) + e^-(g)$	$I(H)$
Electron attachment to A:	$A(g) + e^-(g) \rightarrow A^-(g)$	$-E_{ea}(A)$
Hydration of H^+:	$H^+(g) + H_2O(l) \rightarrow H_3O^+(aq)$	$\Delta H_{hyd}(H^+)$
Hydration of A^-:	$A^-(g) \rightarrow A^-(aq)$	$\Delta H_{hyd}(A^-)$

The enthalpy change for the overall process, $HA(aq) + H_2O(l) \rightarrow H_3O^+(aq) + A^-(aq)$, is therefore the sum of these contributions:

$$\Delta H = -\Delta H_{solv}(HA) + \Delta H_B(H—A) + I(H) - E_{ea}(A) + \Delta H_{hyd}(H^+) + \Delta H_{hyd}(A^-)$$

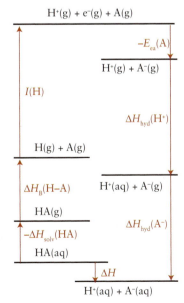

FIGURE 6C.5 The thermodynamic cycle for the analysis of contributions to the strengths of binary acids in water.

TABLE 6C.4   Contributions to the Enthalpy of Proton Transfer of Binary Acids*

Acid	$\Delta H_{solv}(HA)$	$\Delta H_B(H{-}A)$	$I(H)$	$E_{ea}(A)$	$\Delta H_{hyd}(H^+)$	$\Delta H_{hyd}(A^-)$	ΔH
NH_3	-34	453	1312	71	-1103	-500	$+125$
H_2O	-41	492	1312	178	-1103	-520	$+44$
HF	-49	565	1312	328	-1103	-510	-15
HCl	$?\ -35$	431	1312	349	-1103	-367	-41
HBr	$?\ -35$	366	1312	325	-1103	-336	-51
HI	$?\ -35$	299	1312	295	-1103	-291	-43

*(All values in kilojoules per mole ($kJ \cdot mol^{-1}$). Grayed-out values are contributions that are constant for all the acids. The colors are explained in the text. The values preceded with ? are estimates. Note that NH_3 is functioning as an acid (a proton donor) in this context.

TABLE 6C.4 lists the values of each of these contributions and the overall value of ΔH for a number of binary acids, including H_2O (the value for its removal from solution, $-\Delta H_{solv}$, corresponds to its vaporization). The task is to identify the contributions that account for the trends in acid strength. Bear in mind that any discrepancies between what is identified and what is observed might be due to the neglect of entropy considerations. With that caution in mind, take the view that as the enthalpy of proton transfer (the value of ΔH in the table) becomes more negative (proton transfer becomes more exothermic), the acid strength of HA increases.

First, consider trends across Period 2 (as represented by NH_3, H_2O, and HF acting as acids). The acid strength increases in the order $NH_3 < H_2O < HF$, and you can see from Table 6C.4 (final column in the top three rows) that ΔH follows this trend, becoming progressively more negative. The dominant contribution to this trend appears to be electron affinity (see the blue numbers in the table). One way to understand the role of electron affinity is to note that a high electron affinity is associated with a high electronegativity, and in turn the electronegativity controls bond polarity (Topic 2D):

- A high electron affinity of A suggests a high electronegativity and, therefore, a strongly polar H—A bond, $^{\delta+}H{-}A^{\delta-}$.
- The greater the partial positive charge on H, the stronger the $H_2O{\cdots}H{-}A$ hydrogen bond.
- The stronger the hydrogen bond, the more readily will HA transfer its proton to H_2O.

That is:

- *The greater the electron affinity (and therefore the electronegativity) of A, the stronger is the acid.*

For instance, the electronegativity difference is 0.8 for the N—H bond and 1.8 for F—H (see Fig. 2D.2); therefore, the F—H bond is markedly more polar than the N—H bond. This difference is consistent with the observation that HF is an acid in water, but NH_3 is not.

Now consider the relative strengths of binary acids within the same group. In Group 17 the acid strengths are in the order $HF < HCl < HBr < HI$. The values of ΔH (the bottom four rows in Table 6C.4) almost follow this trend. Hydrogen iodide is out of step: the difference might be explained by also taking entropy changes into account, but they are hard to assess. The dominant contribution to the relative strengths of acids down a group appears to be the trend in bond enthalpies, which decrease down a group (see the red numbers in the table). The enthalpies of hydration of A^- also change a lot, but hydration gets *less* exothermic upon descending the group, so the trend in bond enthalpies must dominate even that trend.

Acid strengths of binary acids across a period correlate with electron affinities; acid strengths down a group correlate with bond strength.

6C.4 The Strengths of Oxoacids and Carboxylic Acids

The effect of structure on acidity is illustrated by considering the oxoacids. The high polarity of the O—H bond is one reason that the proton of an —OH group in an oxoacid molecule is acidic. This factor can be seen in the case of phosphorous acid, H_3PO_3, which has the structure (HO)$_2$PHO (**1**): it can donate the protons from its two —OH groups but not the proton attached directly to the phosphorus atom. The difference in behavior can be traced to the much lower electronegativity of phosphorus (2.2) relative to that of oxygen (3.4).

1 Phosphorous acid, H_3PO_3

Consider a family of oxoacids in which the number of O atoms is constant, as in the hypohalous acids HClO, HBrO, and HIO (with structures A—O—H). By referring to **TABLE 6C.5**, you see that

- *the greater the electronegativity of the halogen, the stronger is the oxoacid.*

A partial explanation of this trend is that electrons are withdrawn slightly from the O—H bond as the electronegativity of A increases. As these bonding electrons are pulled toward the central atom, the O—H bond becomes more polar, and so (as in the discussion of binary acids) the molecule becomes a stronger acid. The high electronegativity of A also weakens the conjugate base by making the electrons on the O in A—O$^-$ less accessible to an incoming proton.

Now consider a family of oxoacids in which the number of oxygen atoms varies, as in the chlorine oxoacids HClO, $HClO_2$, $HClO_3$, and $HClO_4$ (with structures of the form O$_n$A—O—H) or in the sulfur oxoacids H_2SO_3 and H_2SO_4 (with structures of the form O$_n$A—(O—H)$_2$). When you refer to **TABLE 6C.6**, you see that

- *the greater the number of oxygen atoms attached to the central atom, the stronger is the acid.*

Another way to express this trend is to note that because the oxidation number of A increases as the number of O atoms increases,

- *the greater the oxidation number of the central atom, the stronger is the acid.*

This is the key to understanding the trend, because the greater the oxidation number of A, the greater is its electron-withdrawing power and the weaker is the O—H bond.

The effect of the number of O atoms on the strengths of organic acids is similar. For example, alcohols are organic compounds in which an —OH group is attached to a

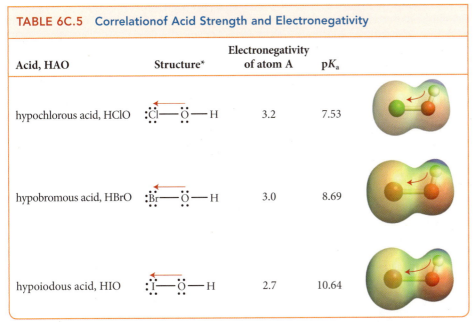

TABLE 6C.5   Correlation of Acid Strength and Electronegativity

Acid, HAO	Structure*	Electronegativity of atom A	pK_a	
hypochlorous acid, HClO	:Cl̈—Ö—H	3.2	7.53	
hypobromous acid, HBrO	:Br̈—Ö—H	3.0	8.69	
hypoiodous acid, HIO	:Ï—Ö—H	2.7	10.64	

*The red arrows indicate the direction of the shift of electron density away from the O—H bond.

TABLE 6C.6   Correlation of Acid Strength and Oxidation Number

Acid	Structure*	Oxidation number of Cl atom	pK_a	
hypochlorous acid, HClO	:Cl—O—H	+1	7.53	
chlorous acid, HClO_2	:Cl—O—H	+3	2.00	
chloric acid, HClO_3	:Cl—O—H	+5	strong	
perchloric acid, HClO_4	O=Cl—O—H	+7	strong	

*The red arrows indicate the direction of the shift of electron density away from the O—H bond. The Lewis structures shown are the ones with the most favorable formal charges, but it is unlikely that the bond orders are as high as these structures suggest.

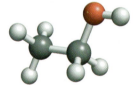

2 Ethanol, CH_3CH_2OH

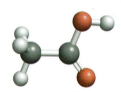

3 Acetic acid, CH_3COOH

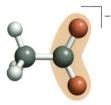

4 Acetate ion, $CH_3CO_2^-$

carbon atom (*Fundamentals* D), as in ethanol (**2**). Carboxylic acids, on the other hand, have two O atoms attached to the same carbon atom: one is a doubly bonded terminal O atom and the other is the O atom of an —OH group, as in acetic acid (**3**). Although carboxylic acids are weak acids, they are much stronger acids than alcohols, partly as a result of the electron-withdrawing power of the second O atom. In fact, alcohols have such weak proton-donating power that usually they are not regarded as oxoacids.

The strength of a carboxylic acid is also increased relative to that of an alcohol by electron delocalization in the conjugate base. The second O atom of the carboxyl group, —COOH, provides an additional electronegative atom over which the negative charge of the conjugate base can spread. This electron delocalization stabilizes the carboxylate anion, —CO_2^- (**4**). Moreover, because the charge is spread over several atoms, it is less effective at attracting a proton. A carboxylate ion is therefore a much weaker base than the conjugate base of an alcohol (for example, the ethoxide ion, $CH_3CH_2O^-$).

A Note on Good Practice: The formula of a carboxylic acid is written RCOOH, because the two O atoms are distinct (one is part of an —OH group); the formula of a carboxylate ion, however, is written RCO_2^- because the two O atoms are equivalent.

The strengths of carboxylic acids also vary with the total electron-withdrawing power of the atoms bonded to the carboxyl group. Because chlorine is more electronegative than hydrogen (3.2 and 2.2, respectively), the —CCl_3 group in trichloroacetic acid is more electron withdrawing than the —CH_3 group bonded to —COOH in acetic acid. Therefore, CCl_3COOH is expected to be a stronger acid than CH_3COOH. In agreement with this prediction, the pK_a of acetic acid is 4.75, whereas that of trichloroacetic acid is 0.52.

EXAMPLE 6C.2   Predicting relative acid strength from molecular structure

If you are working in a research laboratory preparing a series of compounds, you might be faced with having to make a choice between acids based on their strengths. Can you decide which to use based only on their formulas? Predict from their molecular structures which acid in each of the following pairs is the stronger: (a) H_2S and H_2Se; (b) H_2SO_4 and H_2SO_3; (c) H_2SO_4 and H_3PO_4.

PLAN Identify the relevant trends in the summary in **TABLE 6C.7**.

SOLVE

(a) Sulfur and selenium are in the same group, and so the H—Se bond is expected to be weaker than the H—S bond.

Thus, H_2Se can be expected to be the stronger acid.

(b) H_2SO_4 has the greater number of O atoms bonded to the S atom and the oxidation number of sulfur is +6, whereas in H_2SO_3 the sulfur has an oxidation number of only +4.

Thus, H_2SO_4 is expected to be the stronger acid.

(c) Both acids have four O atoms bonded to the central atom, but the electronegativity of sulfur is greater than that of phosphorus.

Thus, H_2SO_4 is expected to be the stronger acid.

Self-test 6C.3A For each of the following pairs, predict which acid is stronger: (a) H_2S and HCl; (b) HNO_2 and HNO_3; (c) H_2SO_3 and $HClO_3$.

[**Answer:** (a) HCl; (b) HNO_3; (c) $HClO_3$]

Self-test 6C.3B List the following carboxylic acids in order of increasing strength: $CHCl_2COOH$, CH_3COOH, and $CH_2ClCOOH$.

Related Exercises 6C.19–6C.20

TABLE 6C.7   Correlations of Molecular Structure and Acid Strength*

Acid type	Trend	
binary	The more polar the H—A bond, the stronger the acid. *This effect is dominant for acids of the same period.*	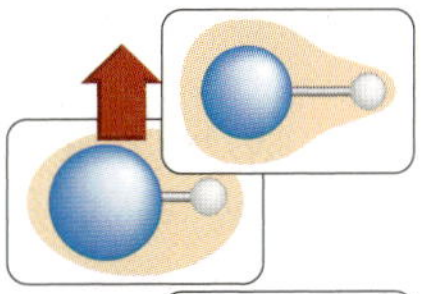
	The weaker the H—A bond, the stronger the acid. *This effect is dominant for acids of the same group.*	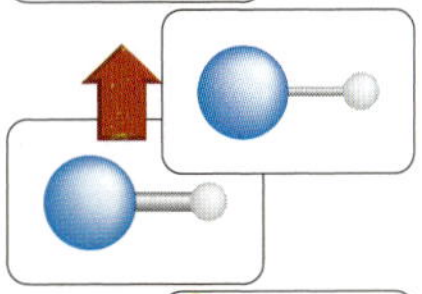
oxoacid	The greater the number of O atoms attached to the central atom (the greater the oxidation number of the central atom), the stronger the acid.	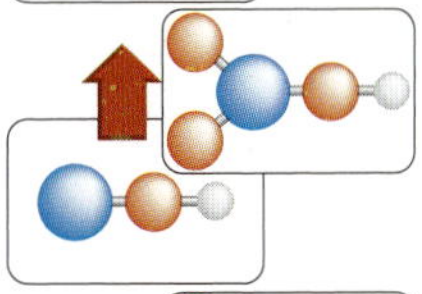
	For the same number of O atoms attached to the central atom, then the greater the electronegativity of the central atom, the stronger the acid.	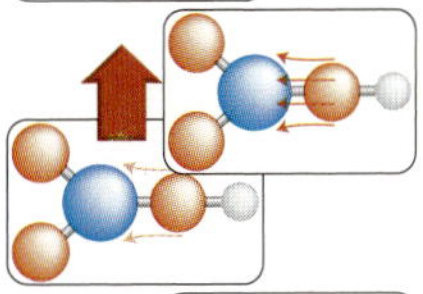
carboxylic	The greater the electronegativities of the groups attached to the carboxyl group, the stronger the acid.	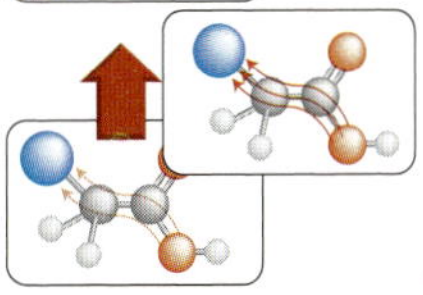

*In each diagram, the arrows indicate the corresponding increase in acid strength.

The greater the number of oxygen atoms and the more electronegative the atoms present in the molecules of an acid, the stronger is the acid. These trends are summarized in Table 6C.7.

What have you learned in this Topic?

You have learned that the acidity and basicity constants express the strength of an acid or base quantitatively. You have also learned that as the strength of an acid increases, the strength of its conjugate base decreases. You have seen that in a series of structurally related compounds as the electron-withdrawing power of groups attached to an —H or —OH increases, so does the strength of the acid.

The skills you have mastered are the ability to:

☐ **1.** Calculate pK_b from pK_a and vice-versa (Self-Test 6C.1).

☐ **2.** Use K_a and K_b values to identify the relative strengths of acids and bases (Example 6C.1).

☐ **3.** Predict trends in acid strength based on molecular structure (Example 6C.2).

Topic 6C Exercises

6C.1 Write (a) the chemical equation for the proton transfer equilibrium in water and the corresponding expression for K_a and (b) the chemical equation for the proton transfer equilibrium of the conjugate base and the corresponding expression for K_b for each of the following weak acids: (i) $HClO_2$; (ii) HCN; (iii) C_6H_5OH.

6C.2 Write (a) the chemical equation for the proton transfer equilibrium in water and the corresponding expression for K_b and (b) the chemical equation for the proton transfer equilibrium of the conjugate acid and the corresponding expression for K_a for each of the following weak acids: (i) $(CH_3)_2NH$, dimethylamine; (ii) $C_{14}H_{10}N_2$, nicotine; (iii) $C_6H_5NH_2$, aniline.

6C.3 (a) Give the K_a value of each of the following acids: (i) phosphoric acid, H_3PO_4, pK_a = 2.12; (ii) phosphorous acid, H_3PO_3, pK_a = 2.00; (iii) selenous acid, H_2SeO_3, pK_a = 2.46; (iv) hydrogen selenate ion, $HSeO_4^-$, pK_a = 1.92. (b) List the acids in order of increasing strength.

6C.4 (a) Give the pK_b values of the following bases: (i) ammonia, NH_3, K_b = 1.8 × 10⁻⁵; (ii) deuterated ammonia, ND_3, K_b = 1.1 × 10⁻⁵; (iii) hydrazine, NH_2NH_2, K_b = 1.7 × 10⁻⁶; (iv) hydroxylamine, NH_2OH, K_b = 1.1 × 10⁻⁸. (b) List the bases in order of increasing strength.

6C.5 Write the chemical formula for the conjugate base of formic acid, HCOOH and calculate its pK_b from the pK_a of formic acid (see Table 6C.1).

6C.6 Write the chemical formula for the conjugate acid of triethylamine, $(C_2H_5)_3N$, and calculate its pK_a from the pK_b of triethylamine (see Table 6C.2).

6C.7 Use data from Tables 6C.1 and 6C.2 to place the following acids in order of increasing strength: HNO_2, $HClO_2$, $^+NH_3OH$, $(CH_3)_2NH_2^+$.

6C.8 Use data from Tables 6C.1 and 6C.2 to place the following acids in order of increasing strength: HCOOH, $(CH_3)_3NH^+$, $N_2H_5^+$, HF.

6C.9 Use data from Tables 6C.1 and 6C.2 to place the following bases in order of increasing strength: F^-, NH_3, $CH_3CO_2^-$, C_5H_5N (pyridine).

6C.10 Use data from Tables 6C.1 and 6C.2 to place the following bases in order of increasing strength: $C_{10}H_{14}N_2$ (nicotine), ClO^-, $(CH_3)_3N$, HSO_3^-.

6C.11 The values of K_a for phenol and 2,4,6-trichlorophenol are 1.3 × 10⁻¹⁰ and 1.0 × 10⁻⁶, respectively. Which is the stronger acid? Account for the difference in strength.

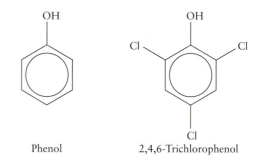

Phenol   2,4,6-Trichlorophenol

6C.12 The value of pK_b for aniline is 9.37 and that for 4-chloroaniline is 9.85. Which is the stronger base? Account for the difference in strength.

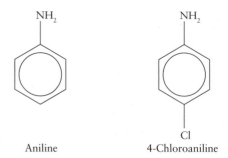

Aniline   4-Chloroaniline

6C.13 Arrange the following bases in order of increasing strength on the basis of the pK_a values of their conjugate acids, which are given in parentheses: (a) ammonia (9.26); (b) methylamine (10.56); (c) ethylamine (10.81); (d) aniline (4.63; see Exercise 6C.12). Is there a simple pattern of strengths?

6C.14 Arrange the following bases in order of increasing strength on the basis of the pK_a values of their conjugate acids, which are given in parentheses: (a) aniline (4.63; see Exercise 6C.12); (b) 2-hydroxyaniline (4.72); (c) 3-hydroxyaniline (4.17); (d) 4-hydroxyaniline (5.47). Is there a simple pattern of strengths?

2-Hydroxyaniline   3-Hydroxyaniline   4-Hydroxyaniline

6C.15 The pK_a of HIO(aq), hypoiodous acid, is 10.64 and that of HIO_3(aq), iodic acid, is 0.77. Account for the difference in strength.

6C.16 The pK_a of HClO(aq), hypochlorous acid, is 7.53 and that of HBrO(aq), hypobromous acid, is 8.69. Account for the difference in strength.

6C.17 Which is the stronger base, the hypobromite ion, BrO^-, or morphine, $C_{17}H_{19}O_3N$? Justify your answer.

6C.18 Which is the stronger acid, hydrocyanic acid, HCN, or the ammonium ion, NH_4^-? Justify your answer.

6C.19 Decide which acid in each of the following pairs is the stronger and explain why: (a) HF or HCl; (b) HClO or $HClO_2$; (c) $HBrO_2$ or $HClO_2$; (d) $HClO_4$ or H_3PO_4; (e) HNO_3 or HNO_2; (f) H_2CO_3 or H_2GeO_3.

6C.20 Decide which acid in each of the following pairs is the stronger and explain why: (a) H_3AsO_4 or H_3PO_4; (b) $HBrO_3$ or HBrO; (c) H_3PO_4 or H_3PO_3; (d) H_2Te or H_2Se; (e) H_2S or HCl; (f) HClO or HIO.

6C.21 Suggest an explanation for the different strengths of (a) acetic acid and trichloroacetic acid; (b) acetic acid and formic acid.

6C.22 Suggest an explanation for the different strengths of (a) ammonia and methylamine; (b) hydrazine and hydroxylamine.

Topic 6D The pH of Aqueous Solutions

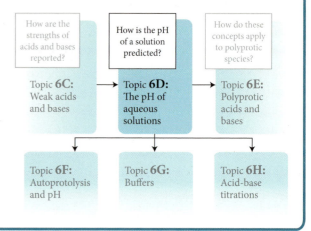

Proton transfer is a very rapid process: as soon as an acid or base is present in aqueous solution, a dynamic equilibrium is established. It follows that the composition of the solution, and specifically the concentration of hydronium ions expressed as pH, can be calculated by the techniques explained in Topics 5I and 6B.

6D.1 Solutions of Weak Acids

The **initial concentration** (or *formal concentration*) of an acid is its molar concentration as prepared, as if no acid molecules have donated any protons. For a strong acid HA, the molar concentration of H_3O^+ ions in solution is the same as the initial concentration of the strong acid, because almost all the HA molecules are deprotonated. However, to find the concentration of H_3O^+ ions in a solution of a weak acid HA, the equilibrium

$$HA(aq) + H_2O(l) \rightleftharpoons H_3O^+(aq) + A^-(aq) \qquad \textbf{(A)}$$

between the acid HA, its conjugate base A^-, and water, with $K_a = [H_3O^+][A^-]/[HA]$, must be taken into account. The pH can be expected to be somewhat higher (that is, the H_3O^+ concentration lower) than it would be for a strong acid with the same initial concentration. To find the pH of a solution of a weak acid, you can use an equilibrium table like those introduced in Topic 5I, as set out in **Toolbox 6D.1**. The calculation summarized in the Toolbox also allows the prediction of the **percentage deprotonation,** the percentage of HA molecules that are deprotonated in the solution:

$$\text{Percentage deprotonated} = \frac{\text{concentration of } A^-}{\text{initial concentration of HA}} \times 100\% \qquad \textbf{(1a)}$$

The equality $[H_3O^+] = [A^-]$, which follows from the stoichiometric relation 1 mol $A^- \mathbin{\widehat{=}}$ 1 mol H_3O^+ for the deprotonation reaction then implies that

$$\text{Percentage deprotonated} = \frac{[H_3O^+]}{[HA]_{\text{initial}}} \times 100\% \qquad \textbf{(1b)}$$

A small percentage of deprotonated molecules would indicate that the acid HA is very weak.

THINKING POINT

How might the extent of deprotonation of a weak acid be affected by its concentration?

Toolbox 6D.1 HOW TO CALCULATE THE pH OF A SOLUTION OF A WEAK ACID

CONCEPTUAL BASIS

Because the proton transfer equilibrium is established as soon as a weak acid is dissolved in water, the concentrations of acid, hydronium ion, and conjugate base of the acid must jointly correspond to the acidity constant of the acid. Any of these quantities can be calculated by setting up an equilibrium table like that in Toolbox 5I.1.

PROCEDURE

Step 1 Write the chemical equation and the expression for K_a for the proton transfer equilibrium. Set up a table with columns labeled by the acid (HA), H_3O^+, and the conjugate base of the acid (A^-). In the first row below the headings, show the initial concentration of each species. For this step, assume that no acid molecules have been deprotonated.

In the second row, write the changes in the concentration needed for the reaction to reach equilibrium. Assume that the concentration of the acid decreases by x $mol \cdot L^{-1}$ as a result of deprotonation. The reaction stoichiometry gives us the other changes in terms of x. In the third row, write the equilibrium concentration for each substance by adding the change in its concentration (line 2) to its initial concentration (line 1).

	Acid, HA	H_3O^+	Conjugate base, A^-
initial concentration	$[HA]_{initial}$	0	0
change in concentration	$-x$	$+x$	$+x$
equilibrium concentration	$[HA]_{initial} - x$	x	x

Although a *change* in concentration may be positive (an increase) or negative (a decrease), the value of the concentration itself must always be positive.

Step 2 Substitute the equilibrium concentrations into the expression for K_a.

Step 3 Solve for the value of x, which (from line 3) gives $[H_3O^+]$.

The calculation of x can often be simplified, as shown in Toolbox 5I.1, by ignoring changes of less than 5% of the initial concentration of the acid. A simple way to anticipate that the approximation can be used is to compare the values of K_a and the initial concentration of weak acid. If the numerical value of $[HA]_{initial}$ is at least two orders of magnitude greater than the value of K_a (is more than 10^2 times as large), then the approximation is probably valid. However, at the end of the calculation, check that x is consistent with the approximation, by calculating the percentage of acid deprotonated. If this percentage is greater than about 5%, then the exact expression for K_a must be solved for x. An exact calculation requires solving a quadratic equation. If the pH is greater than 6 (but less than 7), the acid is so dilute or so weak that the autoprotolysis of water contributes significantly to the pH. In such cases, use the procedures described in Topic 6F, which take autoprotolysis into account. The contribution of the autoprotolysis of water in an acidic solution can be ignored only when the calculated H_3O^+ concentration is substantially (about 10 times) higher than 10^{-7} $mol \cdot L^{-1}$, corresponding to a pH of 6 or less.

Step 4 Calculate the pH from $pH = -\log [H_3O^+]$.

Although pH should be calculated to the number of significant figures appropriate to the data, the answers are often considerably less reliable than that. One reason for this poor reliability is that interactions between the ions in solution are not taken into account.

This procedure is illustrated in Examples 6D.1 and 6D.2.

EXAMPLE 6D.1   Calculating the pH and percentage deprotonation of a weak acid

Acetic acid is a weak acid commonly found in both laboratory and household, but to what extent have its molecules actually been deprotonated? Calculate the pH and percentage deprotonation of CH_3COOH molecules in 0.080 M $CH_3COOH(aq)$, given that K_a for acetic acid is 1.8×10^{-5}.

ANTICIPATE Because the solution is that of an acid, expect pH < 7. Because the acid is weak, you should expect only a small percentage deprotonation.

PLAN Following the procedure in Toolbox 6D.1, write the proton transfer equilibrium and construct the equilibrium table with concentrations in moles per liter.

What should you assume? You can make two assumptions, but they need to be verified at the end of the calculation. (1) Deprotonation is so slight that the equilibrium concentration of the acid is approximately the same as its initial concentration. (2) The autoprotolysis of water does not contribute significantly to the pH.

SOLVE

Step 1 The proton transfer equilibrium and the corresponding equilibrium table are:

$$CH_3COOH(aq) + H_2O(l) \rightleftharpoons H_3O^+(aq) + CH_3CO_2^-(aq) \qquad K_a = \frac{[H_3O^+][CH_3CO_2^-]}{[CH_3COOH]}$$

	CH_3COOH	H_3O^+	$CH_3CO_2^-$
initial concentration	0.080	0	0
change in concentration	$-x$	$+x$	$+x$
equilibrium concentration	$0.080 - x$	x	x

Step 2 Substitute the equilibrium concentrations into the expression for K_a.

$$K_a = 1.8 \times 10^{-5} = \frac{x \times x}{0.080 - x}$$

Now assume that $x \ll 0.080$ and replace $0.080 - x$ by 0.080.

$$1.8 \times 10^{-5} \approx \frac{x^2}{0.080}$$

Step 3 Solve for x:

$$x \approx \sqrt{0.080 \times (1.8 \times 10^{-5})} = \pm 1.2 \times 10^{-3}$$

Select the positive root because x is also a concentration (it is equal to $[H_3O^+]$ in line 3 of the equilibrium table).

Step 4 From $x = [H_3O^+]$ and $pH = -\log [H_3O^+]$;

$$pH \approx -\log(1.2 \times 10^{-3}) = 2.92$$

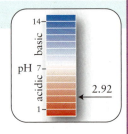

From Eq. 1b, Percentage deprotonated = $([H_3O^+]/[HA]_{initial}) \times 100\%$, with $[HA]_{initial} = 0.080$,

$$\text{Percentage deprotonated} = \frac{1.2 \times 10^{-3}}{0.080} \times 100\% = 1.5\%$$

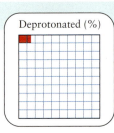

Deprotonated (%)

EVALUATE As anticipated, the pH is less than 7 and the percentage deprotonated is small. The red area in the grid in the final picture represents the percentage of the acid molecules that are deprotonated. Because x is less than 5% of 0.080, the approximation that $x \ll 0.080$ is valid. The assumption that the autoprotolysis of water can be ignored is also valid because $pH < 6$.

Self-test 6D.1A Calculate the pH and percentage deprotonation of 0.50 M aqueous lactic acid. See Table 6C.1 for K_a. Be sure to check any approximation to see whether it is valid.

[**Answer:** 1.69; 4.1%]

Self-test 6D.1B Calculate the pH and percentage deprotonation of 0.22 M aqueous chloroacetic acid. Be sure to check any approximation to see whether it is valid.

Related Exercises 6D.1, 6D.2

EXAMPLE 6D.2   Calculating the K_a and pK_a of a weak acid from the pH

Although there are extensive tables available for the pK_a of weak acids, you might be dealing with an unknown acid or a known acid at an unlisted temperature. You could then use a procedure like this to determine the K_a and pK_a. The pH of a 0.010 M aqueous solution of a certain carboxylic acid is 2.95. What are its K_a and pK_a?

ANTICIPATE All carboxylic acids are weak acids; therefore, expect $K_a \ll 1$.

PLAN Calculate the hydronium ion concentration from the pH and then calculate the value of K_a from the equilibrium concentration of the acid and the hydronium ion concentration.

What should you assume? As in Example 6D.1, because $pH < 6$, assume that the autoprotolysis of water does not contribute significantly to the pH.

SOLVE The equilibrium is

$$HA(aq) + H_2O(l) \rightleftharpoons H_3O^+(aq) + A^-(aq) \qquad K_a = \frac{[H_3O^+][A^-]}{[HA]}$$

From $[H_3O^+] = 10^{-pH}$,

$$[H_3O^+] = 10^{-2.95} \text{ mol·L}^{-1} = 0.0011... \text{ mol·L}^{-1}$$

Formulate the stoichiometric relations between equilibrium concentrations.

$$[H_3O^+] = [A^-] \qquad [HA] = [HA]_{\text{initial}} - [H_3O^+]$$

Use K_a in the form $K_a = [H_3O^+]^2/([HA]_{\text{initial}} - [H_3O^+])$ and substitute the data.

$$K_a = \frac{(0.0011...)^2}{0.010 - 0.0011...} = 1.4... \times 10^{-4}$$

From $pK_a = -\log K_a$,

$$pK_a = -\log 1.4... \times 10^{-4} = 3.85$$

EVALUATE As expected, $K_a \ll 1$. (In this case the acid is mandelic acid, $C_6H_5CH(OH)COOH$, an antiseptic.)

Self-test 6D.2A The pH of a 0.20 M aqueous solution of crotonic acid, C_3H_5COOH, which is used in medicinal research and in the manufacture of synthetic vitamin A, is 2.69. What are the K_a and pK_a of crotonic acid?

[*Answer:* 2.1×10^{-5}, 4.68]

Self-test 6D.2B The pH of a 0.50 M aqueous solution of the metabolic intermediate homogentisic acid is 2.35. What are the K_a and pK_a of homogentisic acid, $C_7H_5(OH)_2COOH$?

Related Exercises 6D.3, 6D.4

To calculate the pH and percentage deprotonation of a solution of a weak acid, set up an equilibrium table and determine the H_3O^+ concentration by using the acidity constant.

6D.2 Solutions of Weak Bases

Just as for weak acids, when a weak base, B, is dissolved in water the proton transfer equilibrium is established very rapidly:

$$B(aq) + H_2O(aq) \rightleftharpoons HB^+(aq) + OH^-(aq) \qquad \textbf{(B)}$$

with $K_b = [HB^+][OH^-]/[B]$. It is often useful to know the **percentage protonated,** the percentage of base molecules that have been protonated:

$$\text{Percentage protonated} = \frac{\text{concentration of } HB^+}{\text{initial concentration of B}} \times 100\%$$

$$= \frac{[HB^+]}{[B]_{\text{initial}}} \times 100\% \qquad \textbf{(2)}$$

Here $[B]_{\text{initial}}$ is the initial (or formal) concentration of base, its molar concentration based on the assumption that no protonation has occurred. The procedure for determining the pH of a solution of a weak base and its percentage protonation is analogous to that for weak acids and is set out in **Toolbox 6D.2**.

Toolbox 6D.2 HOW TO CALCULATE THE pH OF A SOLUTION OF A WEAK BASE

CONCEPTUAL BASIS

Proton transfer equilibrium is established as soon as a weak base is dissolved in water, and so the hydroxide ion concentration can be calculated from the initial concentration of the base and the value of its basicity constant. Because the hydroxide ions are in equilibrium with the hydronium ions, the values of pOH and pK_w can be used to calculate the pH.

PROCEDURE

Step 1 Write the chemical equation and the expression for K_b for the proton transfer equilibrium. Set up a table with columns labeled by the base (B), the conjugate acid of the base (HB^+), and OH^-. In the first row below the headings, show the initial concentration of each species. For this step, assume that no base molecules have been protonated.

Write the concentration changes that are needed for protonation of B to reach equilibrium. Assume that the concentration of the base decreases by x mol·L^{-1} as a result of protonation. The reaction stoichiometry gives the other changes in terms of x.

Write the equilibrium concentration for each substance by adding the change in its concentration (line 2) to its initial concentration (line 1).

	B	**HB$^+$**	**OH$^-$**
initial concentration	$[B]_{initial}$	0	0
change in concentration	$-x$	$+x$	$+x$
equilibrium concentration	$[B]_{initial} - x$	x	x

Although a change in concentration may be positive (an increase) or negative (a decrease), the concentration itself must be positive.

Step 2 Substitute the equilibrium concentrations into the expression for K_b.

Step 3 Use the value of K_b to calculate the value of x.

The calculation of x can often be simplified, as explained in Toolbox 6D.1. In this case the approximation can be used when the numerical value of $[B]_{initial}$ is at least 10^2 times greater than K_b. If the pH is below 8 (but higher than 7), the base is so dilute or so weak that the autoprotolysis of water contributes significantly to the pH. In such cases, use the procedures described in Topic 6F.

Step 4 Determine the pOH of the solution from pOH $= -\log [OH^-] = -\log x$.

Step 5 Calculate the pH from the pOH by using Eq. 5 in Topic 6B, pH $+$ pOH $= pK_w$ (use $pK_w = 14.00$ at 25 °C).

This procedure is illustrated in Example 6D.3.

As in Toolbox 6D.1, although the pH is calculated to the number of significant figures appropriate to the data, the answers are often considerably less reliable than that.

EXAMPLE 6D.3   Calculating the pH and percentage protonation of a weak base

Methylamine, CH_3NH_2, is a weak base used as a building block for some pharmaceuticals. If you prepare an aqueous solution of methylamine for use in a synthesis, you might need to know its pH to avoid unwanted side reactions. Calculate the pH and percentage protonation of a 0.20 M aqueous solution of methylamine at 25 °C. The K_b for CH_3NH_2 is 3.6×10^{-4}.

ANTICIPATE   Because methylamine is a weak base (like all amines), you should expect pH > 7 and only a small percentage protonation.

PLAN   Proceed as in Toolbox 6D.2.

What should you assume?   As in Example 6D.1, you can make two assumptions initially, but verify them at the end of the calculation. (1) The base is so weak that its equilibrium is approximately the same as its initial concentration. (2) The autoprotolysis of water does not contribute significantly to the pH.

SOLVE

Step 1   Write the proton transfer equilibrium and the corresponding equilibrium table, with all concentrations in moles per liter:

$$H_2O(l) + CH_3NH_2(aq) \rightleftharpoons CH_3NH_3^+(aq) + OH^-(aq) \qquad K_b = \frac{[CH_3NH_3^+][OH^-]}{[CH_3NH_2]}$$

	CH$_3$NH$_2$	**CH$_3$NH$_3^+$**	**OH$^-$**
initial concentration	0.20	0	0
change in concentration	$-x$	$+x$	$+x$
equilibrium concentration	$0.20 - x$	x	x

Step 2   Substitute the equilibrium concentrations into the expression for K_b.

$$K_b = \frac{x \times x}{0.20 - x}$$

Because x is likely to be small in the sense $x \ll 0.20$, replace $0.20 - x$ by 0.20.

$$K_b \approx \frac{x^2}{0.20}$$

Step 3   Use the value of $K_b = 3.6 \times 10^{-4}$ to calculate the value of x.

$$x \approx \sqrt{0.20 \times (3.6 \times 10^{-4})} = \pm 8.5 \times 10^{-3}$$

Take the positive root because x also appears in the equilibrium table as a concentration (in row 3).

Step 4   Determine the pOH of the solution. From pOH $= -\log [OH^-]$, with $[OH^-] = x$,

$$pOH \approx -\log(8.5 \times 10^{-3}) = 2.07$$

Step 5 Calculate the pH from $pH = pK_w - pOH$,

$$pH \approx 14.00 - 2.07 = 11.93$$

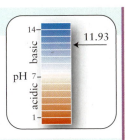

From Eq. 2, with $[HB^+] = x$ and $[B]_{initial} = 0.20$,

$$\text{Percentage protonated} = \frac{8.5 \times 10^{-3}}{0.20} \times 100\% = 4.2\%$$

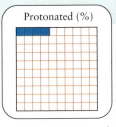

Protonated (%)

EVALUATE As anticipated, the pH is greater than 7 and the percentage protonated, 4.2%, is small. Because $x = 8.5 \times 10^{-3}$, the assumption that $x \ll 0.20$ is valid. Because the pH is greater than 8, the assumption that this equilibrium dominates the pH and that autoprotolysis can be ignored is also valid.

Self-test 6D.3A Estimate the pH and percentage of base that has been protonated in 0.15 M $NH_2OH(aq)$, aqueous hydroxylamine.

[*Answer:* 9.61; 0.027%]

Self-test 6D.3B Estimate the pH and percentage of base that has been protonated in 0.012 M $C_{10}H_{14}N_2(aq)$, nicotine.

Related Exercises 6D.5, 6D.6

To calculate the pH of a solution of a weak base, set up an equilibrium table to calculate pOH from the value of K_b and convert that pOH into pH by using $pH + pOH = 14.00$.

6D.3 The pH of Salt Solutions

A salt is produced by the neutralization of a base by an acid (*Fundamentals* J). However, if the pH of a solution of a salt is measured, it is not in general found to have the "neutral" value (pH = 7). For instance, if 0.3 M NaOH(aq) is neutralized with 0.3 M $CH_3COOH(aq)$, the resulting solution of sodium acetate has pH = 9.0. How can this be? The Brønsted–Lowry theory provides the explanation. According to this theory, an ion may be an acid or a base. The acetate ion, $CH_3CO_2^-$, for instance, is a base, and the ammonium ion, NH_4^+, is an acid. The pH of a solution of a salt depends on the relative acidity and basicity of its ions.

TABLE 6D.1 lists some cations and their acidities in water. They fall into four general categories.

- All cations that are the conjugate acids of weak bases produce acidic aqueous solutions.

Conjugate acids of weak bases, such as NH_4^+, act as proton donors, and so can be expected to form acidic solutions.

- Small, highly charged metal cations that can act as Lewis acids in water, such as Al^{3+} and Fe^{3+}, produce acidic solutions, even though the cations themselves have no protons to donate (**FIG. 6D.1**).

The protons come from the water molecules that hydrate these metal cations in solution (**FIG. 6D.2**). The water molecules act as Lewis bases and share electrons with the metal cations in coordinate covalent bonds (Topic 2C). This partial loss of electrons weakens the O—H bonds and allows one or more protons to be lost from the water molecules. Small, highly charged cations exert the greatest pull on the electrons, weaken the O—H bonds the most, and so form the most acidic solutions.

FIGURE 6D.1 These four solutions show that hydrated cations can be significantly acidic. The flasks contain, from left to right, pure water, 0.1 M $Al_2(SO_4)_3(aq)$, 0.1 M $Ti_2(SO_4)_3(aq)$, and 0.1 M $CH_3COOH(aq)$. All four tubes contain a few drops of universal indicator, which changes color from green in a neutral solution through yellow to red with increasing acidity. The superimposed numbers are the pH values of each solution. (*W. H. Freeman photo by Ken Karp.*)

The metal ions do have protons, of course, as components of their nuclei, but they are bound by strong nuclear forces and cannot be lost in a chemical process.

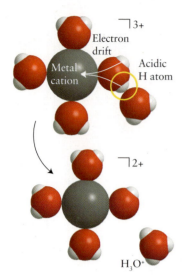

FIGURE 6D.2 In water, metal cations (here, any 3+ ion, such as Al^{3+}) are hydrated by water molecules that can act as Brønsted acids. Although, for clarity, only four water molecules are shown, a metal cation typically has six H_2O molecules attached to it. The acidity of the ion arises from the transfer of a hydrogen ion from one of the hydrating water molecules to a water molecule in the solution.

TABLE 6D.1   Acidic Character and K_a Values of Common Cations in Water*

Character	Examples	K_a	pK_a
Acidic			
conjugate acids of weak bases	anilinium ion, $C_6H_5NH_3^+$	2.3×10^{-5}	4.64
	pyridinium ion, $C_5H_5NH^+$	5.6×10^{-6}	5.24
	ammonium ion, NH_4^+	5.6×10^{-10}	9.25
	methylammonium ion, $CH_3NH_3^+$	2.8×10^{-11}	10.56
small, highly charged metal cations	Fe^{3+} as $Fe(H_2O)_6^{3+}$	3.5×10^{-3}	2.46
	Cr^{3+} as $Cr(H_2O)_6^{3+}$	1.3×10^{-4}	3.89
	Al^{3+} as $Al(H_2O)_6^{3+}$	1.4×10^{-5}	4.85
	Cu^{2+} as $Cu(H_2O)_6^{2+}$	3.2×10^{-8}	7.49
	Ni^{2+} as $Ni(H_2O)_6^{2+}$	9.3×10^{-10}	9.03
	Fe^{2+} as $Fe(H_2O)_6^{2+}$	8×10^{-11}	10.1
Neutral Group 1 and 2 cations metal cations with charge +1	Li^+, Na^+, K^+, Mg^{2+}, Ca^{2+} Ag^+		
Basic	none		

*As in Table 6C.1, the experimental pK_a values have more significant figures than shown here, and the K_a values have been calculated from these more precise data.

- Cations of Group 1 and 2 elements, as well as those of charge +1 from other groups, are such weak Lewis acids that the hydrated ions do not act as Brønsted acids.

These metal cations are too large or have too low a charge to have an appreciable polarizing effect on the hydrating water molecules that surround them, and so the water molecules do not readily release their protons. These cations are sometimes called "neutral cations," because they have so little effect on the pH.

- No cations are basic.

Cations cannot readily accept a proton, because the positive charge of the cation repels the positive charge of the incoming proton.

TABLE 6D.2 summarizes the properties of common anions in solution.

- Very few anions that contain hydrogen produce acidic solutions.

It is difficult for a positively charged proton to leave a negatively charged anion. The few anions that do act as (weak) acids include $H_2PO_4^-$ and HSO_4^-.

- All anions that are the conjugate bases of weak acids produce basic solutions.

For example, formic acid, HCOOH, the acid in ant venom, is a weak acid, and so the formate ion acts as a base in water:

$$H_2O(l) + HCO_2^-(aq) \rightleftharpoons HCOOH(aq) + OH^-(aq)$$

Formate ions and the other ions listed in the last row of Table 6D.2 act as bases in water.

TABLE 6D.2   Acidic and Basic Character of Common Anions in Water

Character	Examples
Acidic very few	HSO_4^-, $H_2PO_4^-$
Neutral conjugate bases of strong acids	Cl^-, Br^-, I^-, NO_3^-, ClO_4^-
Basic conjugate bases of weak acids	F^-, O^{2-}, OH^-, S^{2-}, HS^-, CN^-, CO_3^{2-}, PO_4^{3-}, NO_2^-, $CH_3CO_2^-$, other carboxylate ions

- The anions of strong acids—which include Cl^-, Br^-, I^-, NO_3^-, and ClO_4^-—are such weak bases that they have no significant effect on the pH of a solution.

These anions are considered to be "neutral" in water.

To decide whether the solution of a salt will be acidic, basic, or neutral, both the cation and the anion must be considered. First examine the anion to see whether it is the conjugate base of a weak acid. If the anion is neither acidic nor basic, examine the cation to see whether it is an acidic metal ion or the conjugate acid of a weak base. If one ion is an acid and the other a base, as in NH_4F, then the pH is affected by the reactions of both ions with water and both equilibria must be considered (Topic 6F).

Self-test 6D.4A Use Tables 6D.1 and 6D.2 to decide whether aqueous solutions of the salts (a) $Ba(NO_2)_2$, (b) $CrCl_3$, and (c) NH_4NO_3 are acidic, neutral, or basic.

                                     [**Answer:** (a) Basic; (b) acidic; (c) acidic]

Self-test 6D.4B Decide whether aqueous solutions of (a) Na_2CO_3, (b) $AlCl_3$, and (c) KNO_3 are acidic, neutral, or basic.

To calculate the pH of a salt solution, use the equilibrium table procedure described in Toolboxes 6D.1 and 6D.2—according to the Brønsted theory, an acidic cation is simply another weak acid and a basic anion is simply another weak base. However, often the K_a or K_b for the acidic or basic ion must first be calculated. Examples 6D.4 and 6D.5 illustrate the procedure.

EXAMPLE 6D.4   Calculating the pH of a salt solution with an acidic cation

You are working in the emergency room of a hospital where a patient suffering from influenza has developed metabolic alkalosis, a condition in which the pH of the blood is too high. You have available a stock solution of ammonium chloride, which is used to lower the pH of the blood of patients suffering from alkalosis, but you need to know its pH. Estimate the pH of 0.15 M NH_4Cl(aq) at 25 °C.

ANTICIPATE Because NH_4^+ is a weak acid and Cl^- is neutral, you should expect pH < 7.

PLAN Treat the solution as that of a weak acid, using an equilibrium table as in Toolbox 6D.1 to calculate the composition and hence the pH. First, write the chemical equation for proton transfer to water and the expression for K_a. Obtain the value of K_a from K_b for the conjugate base by using Eq. 4a in Topic 6C ($K_a = K_w/K_b$). The initial concentration of the acidic cation is equal to the concentration of the cation that the salt would produce if it retained all its acidic protons.

What should you assume? Assume that (1) the extent of deprotonation is so small that the change in concentration of NH_4^+ is insignificant, and (2) the autoprotolysis of water does not affect the pH significantly. Check these assumptions at the end of the calculation.

SOLVE
Step 1 The equilibrium for the acid NH_4^+ is

$$NH_4^+(aq) + H_2O(l) \rightleftharpoons H_3O^+(aq) + NH_3(aq) \qquad K_a = \frac{[H_3O^+][NH_3]}{[NH_4^+]}$$

From Table 6C.2, $K_b = 1.8 \times 10^{-5}$ for NH_3. Now construct the following equilibrium table, with all concentrations in moles per liter:

	NH_4^+	H_3O^+	NH_3
initial concentration	0.15	0	0
change in concentration	$-x$	$+x$	$+x$
equilibrium concentration	$0.15 - x$	x	x

From $K_a = K_w/K_b$,

$$K_a = \frac{1.0 \times 10^{-14}}{1.8 \times 10^{-5}} = 5.5... \times 10^{-10}$$

Step 2 Substitute the equilibrium concentrations into the expression for K_a and assume that $x \ll 0.15$:

$$K_a = \frac{x \times x}{0.15 - x} \approx \frac{x^2}{0.15}$$

Step 3 Set $K_a = 5.5... \times 10^{-10}$ and solve for x:

$$x \approx \sqrt{0.15 \times (5.5... \times 10^{-10})} = \pm 9.1... \times 10^{-6}$$

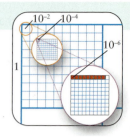

Because $x = [H_3O^+]$, which is a concentration, select the positive root.

Step 4 From $pH = -\log [H_3O^+]$ with $[H_3O^+] = x$,

$$pH \approx -\log(9.1... \times 10^{-6}) = 5.04$$

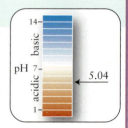

EVALUATE As anticipated, the pH is less than 7. The approximation that x is less than 5% of 0.15 is valid (by a large margin). The H_3O^+ concentration (9.1 μmol·L⁻¹) is substantially larger than that generated by the autoprotolysis of water (0.10 μmol·L⁻¹), and so ignoring the latter contribution is valid (but only just).

Self-test 6D.5A Estimate the pH of 0.10 M $CH_3NH_3Cl(aq)$, aqueous methylammonium chloride; the cation is $CH_3NH_3^+$.
[*Answer:* 5.78]

Self-test 6D.5B Estimate the pH of 0.10 M $NH_4NO_3(aq)$.

Related Exercises 6D.15, 6D.16, 6D.19, 6D.20

EXAMPLE 6D.5　Calculating the pH of a salt solution with a basic anion

Calcium acetate, $Ca(CH_3CO_2)_2(aq)$, is used to treat patients with a kidney disease that results in high levels of phosphate ions in the blood. The calcium binds to the phosphates so that they can be excreted. If you are using calcium acetate for this purpose, it is important to know the pH of the solution to avoid complications in the treatment. Estimate the pH of 0.15 M $Ca(CH_3CO_2)_2(aq)$ at 25 °C.

ANTICIPATE The $CH_3CO_2^-$ ion is the conjugate base of a weak acid; so the solution will be basic and you should expect pH > 7.

PLAN Use the procedure in Toolbox 6D.2, taking the initial concentration of base from the concentration of added salt. Calculate K_b for the basic anion from K_a for its conjugate acid. Convert pOH into pH by using Eq. 5b of Topic 6B (pH + pOH = 14.00).

What should you assume? As in Example 6D.2, you can make two assumptions initially: (1) because the protonation of the weak base is so small, the concentration of acetate ions retains its initial value; (2) the autoprotolysis of water does not affect the pH significantly. Be sure to verify these assumptions at the end of the calculation.

SOLVE
Step 1 The proton transfer equilibrium for the base $CH_3CO_2^-$ is

$$H_2O(l) + CH_3CO_2^-(aq) \rightleftharpoons CH_3COOH(aq) + OH^-(aq) \qquad K_b = \frac{[CH_3COOH][OH^-]}{[CH_3CO_2^-]}$$

The initial concentration of $CH_3CO_2^-$ is 2×0.15 mol·L⁻¹ = 0.30 mol·L⁻¹, because each formula unit of salt provides two $CH_3CO_2^-$ ions. Table 6C.1 gives the K_a of CH_3COOH as 1.8×10^{-5}. First, construct the equilibrium table:

	$CH_3CO_2^-$	CH_3COOH	OH^-
initial concentration	0.30	0	0
change in concentration	$-x$	$+x$	$+x$
equilibrium concentration	$0.30 - x$	x	x

Find the K_b of the $CH_3CO_2^-$ ion from $K_b = K_w/K_a$.

$$K_b = \frac{1.0 \times 10^{-14}}{1.8 \times 10^{-5}} = 5.5... \times 10^{-10}$$

Step 2 Substitute the equilibrium concentrations into the expression for K_b and assume that $x \ll 0.30$:

$$K_b = \frac{x \times x}{0.30 - x} \approx \frac{x^2}{0.30}$$

Step 3 Set $K_b = 5.5... \times 10^{-10}$ and solve for x

$$x \approx \sqrt{0.30 \times (5.5... \times 10^{-10})} = \pm 1.29... \times 10^{-5}$$

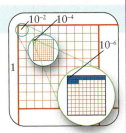

Because $x = [OH^-]$ is also a concentration, select the positive root.

Step 4 From $pOH = -\log [OH^-]$ with $[OH^-] = x$,

$$pOH \approx -\log(1.29... \times 10^{-5}) = 4.89$$

Step 5 From $pH = pK_w - pOH$,

$$pH \approx 14.00 - 4.89 = 9.11$$

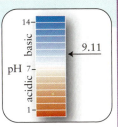

EVALUATE As expected, the pH is greater than 7. Because $x = 1.29... \times 10^{-5}$, the assumption that $x \ll 0.30$ is valid. Because the pH is outside the "danger zone" that falls between 7 and 8 for a base in water, the neglect of the contribution of the autoprotolysis of water is valid.

Self-test 6D.6A Estimate the pH of 0.10 M $KC_6H_5CO_2(aq)$, potassium benzoate; see Table 6C.1 for data.

[*Answer:* 8.59]

Self-test 6D.6B Estimate the pH of 0.020 M KF(aq); see Table 6C.1 for data.

Related Exercises 6D.17, 6D.18

Salts that contain the conjugate acids of weak bases produce acidic aqueous solutions; so do salts that contain small, highly charged metal cations. Salts that contain the conjugate bases of weak acids produce basic aqueous solutions.

What have you learned in this Topic?

You have learned that the pH of an aqueous solution can be calculated from a known concentration of acid or base and the acidity or basicity constants. Conversely, you have seen how to determine the equilibrium constant by measuring the pH of a solution of known concentration of acid or base. You have also learned how the pH of salt solutions depends on the acidic or basic characteristics of the ions in solution and have seen how to calculate the pH given the initial concentration of an acid, base, or salt.

Skills that you have mastered are the ability to:

☐ **1.** Calculate the pH of solutions of weak acids and bases and determine the percentage of deprotonation and protonation respectively (Examples 6D.1 and 6D.3).

☐ **2.** Use a measured value of pH in a solution of known concentration to determine an acidity constant (Example 6D.2).

☐ **3.** Predict whether a salt solution will be acidic, basic, or neutral (Self-Test 6D.4A).

☐ **4.** Calculate the pH of salt solutions with acidic or basic ions (Examples 6D.4 and 6D.5).

Topic 6D Exercises

Refer to Tables 6C.1 and 6C.2 for K_a and K_b. Assume 25 °C throughout.

6D.1 Calculate the pH, pOH, and percentage deprotonation of each of the following aqueous solutions: (a) 0.20 M $CH_3COOH(aq)$; (b) 0.20 M $CCl_3COOH(aq)$; (c) 0.20 M $HCOOH(aq)$. (d) Explain any differences in pH on the basis of molecular structure.

6D.2 Muscles produce lactic acid during exercise. Calculate the pH, pOH, and percentage deprotonation of the following aqueous solutions of lactic acid, $CH_3CH(OH)COOH$: (a) 0.11 M; (b) 3.7×10^{-3} M; (c) 8.2×10^{-5} M.

6D.3 (a) When the pH of 0.10 M $HClO_2(aq)$ was measured, it was found to be 1.2. What are the values of K_a and pK_a of chlorous acid? (b) The pH of a 0.10 M propylamine, $C_3H_7NH_2$, aqueous solution was measured as 11.86. What are the values of K_b and pK_b of propylamine?

6D.4 (a) The pH of 0.015 M $HNO_2(aq)$ was measured as 2.63. What are the values of K_a and pK_a of nitrous acid? (b) The pH of 0.10 M $C_4H_9NH_2(aq)$, butylamine, was measured as 12.04. What are the values of K_b and pK_b of butylamine?

6D.5 Calculate the pH, pOH, and percentage protonation of solute in each of the following aqueous solutions: (a) 0.057 M $NH_3(aq)$; (b) 0.162 M $NH_2OH(aq)$; (c) 0.35 M $(CH_3)_3N(aq)$; (d) 0.0073 M codeine, given that the pK_a of its conjugate acid is 8.21.

6D.6 Calculate the pOH, pH, and percentage protonation of solute in each of the following aqueous solutions: (a) 0.082 M $C_5H_5N(aq)$, pyridine; (b) 0.0103 M $C_{10}H_{14}N_2(aq)$, nicotine; (c) 0.060 M aqueous quinine, given that the pK_a of its conjugate acid is 8.52; (d) 0.045 M aqueous strychnine, given that the K_a of its conjugate acid is 5.49×10^{-9}.

6D.7 Find the initial concentration of the weak acid or base in each of the following aqueous solutions: (a) a solution of HClO with pH = 4.60; (b) a solution of hydrazine, NH_2NH_2, with pH = 10.20.

6D.8 Find the initial concentration of the weak acid or base in each of the following aqueous solutions: (a) a solution of HClO with pH = 5.4; (b) a solution of pyridine, C_5H_5N, with pH = 8.8.

6D.9 The percentage deprotonation of benzoic acid in a 0.110 M solution is 2.4%. What is the pH of the solution and the K_a of benzoic acid?

6D.10 An aqueous solution is 35.0% by mass methylamine (CH_3NH_2); its density is 0.85 g·cm^{-3}. (a) Draw the Lewis structures of a methylamine molecule and its conjugate acid. (b) If 80.0 mL of this solution is diluted to 300.0 mL, what would be the pH of the resulting solution?

6D.11 Decide whether an aqueous solution of each of the following salts has a pH equal to, greater than, or less than 7. If pH > 7 or pH < 7, write a chemical equation to justify your answer. (a) NH_4Br; (b) Na_2CO_3; (c) KF; (d) KBr; (e) $AlCl_3$; (f) $Cu(NO_3)_2$.

6D.12 Decide whether an aqueous solution of each of the following salts has a pH equal to, greater than, or less than 7. If pH > 7 or pH < 7, write a chemical equation to justify your answer. (a) $K_2C_2O_4$, potassium oxalate; (b) $Ca(NO_3)_2$; (c) CH_3NH_3Cl, methylamine hydrochloride; (d) K_3PO_4; (e) $FeCl_3$; (f) C_5H_5NHCl, pyridinium chloride.

6D.13 Rank the following solutions in order of increasing pH: (a) 1.0×10^{-5} M HCl(aq); (b) 0.20 M $CH_3NH_3Cl(aq)$; (c) 0.20 M $CH_3COOH(aq)$; (d) 0.20 M $C_6H_5NH_2(aq)$. Justify your ranking.

6D.14 Rank the following solutions in order of increasing pH: (a) 1.0×10^{-5} M NaOH(aq); (b) 0.20 M $NaNO_2(aq)$; (c) 0.20 M $NH_3(aq)$; (d) 0.20 M NaCN(aq). Justify your ranking.

6D.15 Calculate the pH of (a) 0.19 M $NH_4Cl(aq)$; (b) 0.055 M $AlCl_3(aq)$.

6D.16 Calculate the pH of (a) 0.15 M $CH_3NH_3Cl(aq)$; (b) 0.063 M $FeCl_3(aq)$.

6D.17 Calculate the pH of (a) 0.63 M $NaCH_3CO_2(aq)$; (b) 0.65 M KCN(aq).

6D.18 Calculate the pH of (a) 0.015 M $Na_2SO_3(aq)$; (b) 0.086 M NaF(aq).

6D.19 A sample of CH_3NH_3Cl of mass 15.5 g is dissolved in water to make 450. mL of solution. What is the pH of the solution?

6D.20 A sample of $C_6H_5NH_3Cl$ of mass 7.8 g is dissolved in water to make 350. mL of solution. What is the percentage deprotonation of the cations?

6D.21 During the analysis of an unknown acid HA, a 0.010 M solution of the sodium salt of the acid was found to have a pH of 10.35. Use Table 6C.1 to identify the acid.

6D.22 During the analysis of an unknown weak base B, a 0.10 M solution of the nitrate salt of the base was found to have a pH of 3.13. Use Table 6C.2 to identify the base.

Topic 6E Polyprotic Acids and Bases

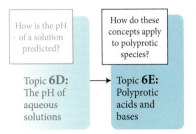

Why Do You Need to Know This Material? Many acids with biological and environmental importance are able to donate more than one proton. When working with such systems, you need to know how to calculate the pH and evaluate the concentration of all the ions in solution.

What Do You Need to Know Already? You need to be familiar with the properties of weak acids and bases (Topic 6C) and with equilibrium calculations related to them (Topic 6D).

A **polyprotic acid** is a compound that can donate more than one proton. Many common acids are polyprotic, including sulfuric acid, H_2SO_4, and carbonic acid, H_2CO_3, each of which can donate two protons, and phosphoric acid, H_3PO_4, which can donate three protons. Polyprotic acids play a critical role in biological systems, because many enzymes can be regarded as polyprotic acids that carry out their vital functions by donating one proton after another. A **polyprotic base** is a species that can accept more than one proton. Examples include the CO_3^{2-} and SO_3^{2-} anions, both of which can accept two protons, and the PO_4^{3-} anion, which can accept three protons.

THINKING POINT

Do successive deprotonations of a polyprotic acid result in successively stronger or weaker acids?

6E.1 The pH of a Polyprotic Acid Solution

Carbonic acid is an important natural component of the environment that forms whenever carbon dioxide dissolves in water. In fact, the oceans provide one of the critical mechanisms for maintaining a constant concentration of carbon dioxide in the atmosphere. Carbonic acid takes part in two successive proton transfer equilibria:

$$H_2CO_3(aq) + H_2O(l) \rightleftharpoons H_3O^+(aq) + HCO_3^-(aq) \qquad K_{a1} = 4.3 \times 10^{-7}$$

$$HCO_3^-(aq) + H_2O(l) \rightleftharpoons H_3O^+(aq) + CO_3^{2-}(aq) \qquad K_{a2} = 5.6 \times 10^{-11}$$

The conjugate base of H_2CO_3 in the first equilibrium, the ion HCO_3^-, acts as an acid in the second equilibrium. That ion produces, in turn, its own conjugate base, CO_3^{2-}.

Protons are donated successively by polyprotic acids, with the acidity constant decreasing significantly, usually by a factor of about 10^3 or more, with each proton lost: $K_{a1} \gg K_{a2} \gg K_{a3} \dots$ (**TABLE 6E.1**). The decrease can be traced to the attraction between opposite charges: it is harder to separate a positively charged proton from a negatively charged ion (such as HCO_3^-) than from the original uncharged molecule (H_2CO_3). Sulfuric acid, for example, is a strong acid, but its conjugate base, the hydrogen sulfate ion, HSO_4^-, is a weak acid.

Sulfuric acid is the only common polyprotic acid for which the first deprotonation can be considered complete. The second deprotonation adds to the H_3O^+ concentration slightly,

TABLE 6E.1   Acidity Constants of Polyprotic Acids at 25 °C

Acid	K_{a1}	pK_{a1}	K_{a2}	pK_{a2}	K_{a3}	pK_{a3}
sulfuric acid, H_2SO_4	strong		1.2×10^{-2}	1.92		
oxalic acid, $(COOH)_2$	5.9×10^{-2}	1.23	6.5×10^{-5}	4.19		
sulfurous acid, H_2SO_3	1.5×10^{-2}	1.81	1.2×10^{-7}	6.91		
phosphorous acid, H_3PO_3	1.0×10^{-2}	2.00	2.6×10^{-7}	6.59		
phosphoric acid, H_3PO_4	7.6×10^{-3}	2.12	6.2×10^{-8}	7.21	2.1×10^{-13}	12.68
tartaric acid, $C_2H_4O_2(COOH)_2$	6.0×10^{-4}	3.22	1.5×10^{-5}	4.82		
carbonic acid, H_2CO_3	4.3×10^{-7}	6.37	5.6×10^{-11}	10.25		
hydrosulfuric acid, H_2S	1.3×10^{-7}	6.89	7.1×10^{-15}	14.15		

and so the overall pH of a solution of the acid is slightly lower than that due to the first deprotonation alone. For example, in 0.010 M $H_2SO_4(aq)$ the first deprotonation is complete:

$$H_2SO_4(aq) + H_2O(l) \longrightarrow H_3O^+(aq) + HSO_4^-(aq)$$

and results in an H_3O^+ concentration equal to the initial concentration of the acid, 0.010 mol·L^{-1}, which corresponds to pH = 2.0. However, the conjugate base, HSO_4^-, is amphiprotic and also contributes protons to the solution. Therefore, the second proton transfer equilibrium needs to be taken into account:

$$HSO_4^-(aq) + H_2O(l) \rightleftharpoons H_3O^+(aq) + SO_4^{2-}(aq) \qquad K_{a2} = 0.012$$

The value of the equilibrium hydronium ion concentration is obtained from

$$\underbrace{K_{a2}}_{0.012} = \frac{[H_3O^+][SO_4^{2-}]}{\underbrace{[HSO_4^-]}_{0.010-x}} = \frac{(0.010 + x) \times x}{0.010 - x}$$

with the annotations $0.010+x$ over $[H_3O^+]$ and x over $[SO_4^{2-}]$.

Then, solving by using the quadratic formula gives $x = 4.3 \times 10^{-3}$, $[H_3O^+] = (0.010 + x)$ mol·L^{-1} = 0.014 mol·L^{-1}, and pH = 1.9, which is slightly lower than the pH = 2.0 that was calculated on the basis of the first deprotonation alone.

Self-test 6E.1A Estimate the pH of 0.050 M $H_2SO_4(aq)$.

[*Answer:* 1.23]

Self-test 6E.1B Estimate the pH of 0.10 M $H_2SO_4(aq)$.

Except for sulfuric acid (and a few other rare cases), to calculate the pH of a polyprotic acid, use K_{a1} and take only the first deprotonation into account; that is, treat the acid as a monoprotic weak acid (see Toolbox 6D.1). Subsequent deprotonations do take place, but provided K_{a2} is less than about $K_{a1}/1000$, they do not affect the pH significantly and can be ignored.

Estimate the pH of a polyprotic acid for which all deprotonations are weak by using only the first deprotonation equilibrium and assuming that further deprotonation is insignificant. An exception is sulfuric acid, the only common polyprotic acid that is a strong acid in its first deprotonation.

6E.2 Solutions of Salts of Polyprotic Acids

The conjugate base of a polyprotic acid is amphiprotic (Topic 6A): it can act as either an acid or a base because it can either donate its remaining acidic hydrogen atom as a proton or accept a proton and revert to the original acid. For example, a hydrogen sulfide ion, HS^-, in water acts as both an acid and a base:

$HS^-(aq) + H_2O(l) \rightleftharpoons H_3O^+(aq) + S^{2-}(aq)$ $K_{a2} = 7.1 \times 10^{-15}$; $pK_{a2} = 14.15$

$HS^-(aq) + H_2O(l) \rightleftharpoons H_2S(aq) + OH^-(aq)$ $K_{b1} = K_w/K_{a1} = 7.7 \times 10^{-8}$; $pK_{b1} = 7.11$

Because HS^- is amphiprotic, it is not immediately apparent whether an aqueous solution of NaHS will be acidic or basic. However, the pK_a and pK_b values of the HS^- ion can be used to conclude that:

- The pK_{a2} of H_2S (the pK_a of HS^-) is large (14.52), which implies that HS^- is a very weak acid; its conjugate base, S^{2-}, is therefore reasonably strong and its basic character is likely to dominate. As a result, pH > 7.
- The pK_{a1} of H_2S has an intermediate value (6.89), which implies that pK_{b1} of HS^- also has an intermediate value (7.11), indicating that this ion is a reasonably strong weak base; so again pH > 7.

This reasoning suggests that the pH will be high if both pK_{a1} and pK_{a2} are relatively large. Indeed, if some reasonable assumptions are made then in general

$$pH = \tfrac{1}{2}(pK_{a1} + pK_{a2}) \tag{1}$$

Margin notes:

If values of K_{a1} and K_{a2} for a polyprotic acid are close together, then the calculations are more complicated because both equilibria must be considered.

Exceptions are HSO_4^-, which is a very weak base, and HPO_3^{2-}, which does not act as an acid because its proton is not acidic (Topic 6C).

where, for an anion of formula HA^- (such as HS^-), K_{a1} is the first acidity constant of the parent acid H_2A (H_2S in this example) and K_{a2} is the second acidity constant of H_2A, which is the acidity constant of HA^- itself. This formula is reliable provided that $S \gg K_w/K_{a2}$ and $S \gg K_{a1}$ (and the analogous conditions for other salts), where S is the initial (that is, analytical or formal) concentration of the salt. If these criteria are not satisfied, a much more complicated expression must be used; it and its derivation, including the derivation of this simplified version, can be found on the website for this text.

EXAMPLE 6E.1 Estimating the pH of a solution of an amphiprotic salt

Salts are often used to create solutions with specific pH values. Suppose you need to prepare a salt solution with a pH of about 4.5 and have available sodium dihydrogen phosphate, NaH_2PO_4, and sodium citrate, $Na_2HC_6H_5O_7$. Which should you use? Estimate the pH of (a) 0.20 M NaH_2PO_4(aq); (b) 0.20 M $Na_2HC_6H_5O_7$(aq). For citric acid, $pK_{a2} = 5.95$ and $pK_{a3} = 6.39$.

ANTICIPATE Both salts have acidic anions, so you should expect the pH to be less than 7 but greater than the value for a 0.20 M strong acid solution, which is 0.7.

PLAN Verify that $S \gg K_w/K_{a2}$ and $S \gg K_{a1}$; if these conditions are satisfied, you can use Eq. 1. You need the pK_as of the appropriate acids. For a diprotic species H_2A^- (such as $H_2PO_4^-$), you need to consider the pK_as of H_3A (the pK_{a1} of H_3A) and H_2A^{2-} (the pK_{a2} of H_3A). For a monoprotic species HA (such as $HC_6H_5O_7^{2-}$), you need the pK_as of H_2A^- (the pK_{a2} of H_3A) and HA^{2-} (the pK_{a3} of H_3A). Use Eq. 1 in the form $pH = \frac{1}{2}(pK_{a2} + pK_{a3})$.

SOLVE
(a) For $H_2PO_4^-$, you need the pK_as of H_3PO_4 (the pK_{a1} of H_3PO_4) and $H_2PO_4^-$ (the pK_{a2} of H_3PO_4): $K_{a1} = 7.6 \times 10^{-3}$ and $K_{a2} = 6.2 \times 10^{-8}$, and so $pK_{a1} = 2.12$ and $pK_{a2} = 7.21$. Therefore, for the solution of $H_2PO_4^-$:

Check that $S \gg K_w/K_{a2}$ and $S \gg K_{a1}$.

$$S = 0.20, \quad \frac{K_w}{K_{a2}} = \frac{1.0 \times 10^{-14}}{6.2 \times 10^{-8}} = 1.6 \times 10^{-7}, \quad \text{so } S \gg K_w/K_{a2}$$

$$S = 0.20, \quad K_{a1} = 7.6 \times 10^{-3}, \quad \text{so } S \gg K_{a1}$$

Therefore, the use of Eq. 1 is valid.

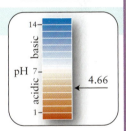

From $pH = \frac{1}{2}(pK_{a1} + pK_{a2})$,

$$pH = \frac{1}{2}(2.12 + 7.21) = 4.66$$

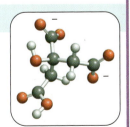

(b) For $HC_6H_5O_7^{2-}$, you need the pK_as of $H_2C_6H_5O_7^-$ (the pK_{a2} of citric acid, $H_3C_6H_5O_7$) and $HC_6H_5O_7^{2-}$ (the pK_{a3} of citric acid), $pK_{a2} = 5.95$ and $pK_{a3} = 6.39$. Therefore, for the solution of $HC_6H_5O_7^{2-}$:

Check that $S \gg K_w/K_{a3}$ and $S \gg K_{a2}$.

$$S = 0.20, \quad \frac{K_w}{K_{a3}} = \frac{1.0 \times 10^{-14}}{4.1 \times 10^{-7}} = 2.4 \times 10^{-8}, \quad \text{so } S \gg K_w/K_{a3}$$

$$S = 0.20, \quad K_{a2} = 1.1 \times 10^{-6}, \quad \text{so } S \gg K_{a2}$$

Therefore, the use of Eq. 1 is valid.

From $pH = \frac{1}{2}(pK_{a2} + pK_{a3})$,

$$pH = \frac{1}{2}(5.95 + 6.39) = 6.17$$

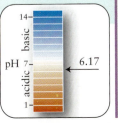

EVALUATE As expected, the pH values of both solutions are less than 7. The NaH_2PO_4 solution has a pH closer to the desired value of 4.5 and would be the better choice in this case.

Self-test 6E.2A Estimate the pH of 0.10 M $NaHCO_3$(aq).

[*Answer:* 8.31]

Self-test 6E.2B Estimate the pH of 0.50 M KH_2PO_4(aq).

Related Exercises 6E.5, 6E.6

Suppose you need to estimate the pH of an aqueous solution of the salt of a fully deprotonated polyprotic acid. The anion can be treated like the conjugate base of a monoprotic weak acid, provided K_{b2} is less than about $K_{b1}/1000$. Successive protonations do occur but to such a small extent they have negligible impact on the pH. An example is a solution of sodium sulfide, in which sulfide ions, S^{2-}, are present; another example is a solution of potassium phosphate, which contains PO_4^{3-} ions. In such a solution, the anion acts as a base: it accepts protons from water. For such a solution, it is necessary to use the techniques for calculating the pH of a basic anion illustrated in Example 6D.5. To find the value of K_b to work with in the calculation, use the K_a for the deprotonation that produces the ion being studied. For S^{2-}, use $K_{b2} = K_w/K_{a2}$ for H_2S; for PO_4^{3-}, use $K_{b3} = K_w/K_{a3}$ for H_3PO_4.

The pH of the aqueous solution of an amphiprotic salt can be estimated from the average of the pKₐs of the salt and its conjugate acid. The pH of a solution of a salt of the final conjugate base of a polyprotic acid is found from the reaction of the anion with water.

6E.3 The Concentrations of Solute Species

Environmental chemists studying the pollution caused by fertilizers in runoff from fields or mineralogists studying the formation of sedimentary rocks as groundwater trickles through rock formations need to know not only the pH but also the **speciation,** the concentrations of each of the ions present in the solution. For example, they may need to know the concentration of sulfite ions in a solution of sulfurous acid or the concentrations of phosphate and hydrogen phosphate ions in a solution of phosphoric acid. The calculations described in Toolbox 6D.1 give the pH—the hydrogen ion concentration— but for a polyprotic acid they do not give the concentrations of all the solute species in solution, which for H_3PO_4 includes H_3PO_4, $H_2PO_4^-$, HPO_4^{2-}, and PO_4^{3-}. To calculate them, it is necessary to take into account all the proton transfer equilibria in the solution.

To simplify the calculations, start by judging the relative concentration of each species in solution and identify terms that can be ignored. To make this judgment, use the general rule that concentrations of species present in the greater amount are not significantly affected by concentrations of species present in the smaller amount if the difference in concentrations is large. However, all assumptions should be checked and verified at the end of the calculation.

Toolbox 6E.1 HOW TO CALCULATE THE CONCENTRATIONS OF ALL SPECIES IN A POLYPROTIC ACID SOLUTION

CONCEPTUAL BASIS

Assume that the polyprotic acid, a weak acid for all its deprotonations, is the solute species present in largest amount. Assume that only the first deprotonation contributes significantly to $[H_3O^+]$ and that the autoprotolysis of water does not contribute significantly to $[H_3O^+]$ or $[OH^-]$.

PROCEDURE FOR A DIPROTIC ACID

Step 1 From the deprotonation equilibrium of the acid (H_2A), determine the concentrations of the conjugate base (HA^-) and H_3O^+, as illustrated in Example 6D.1.

Step 2 Find the concentration of A^{2-} from the second deprotonation equilibrium (that of HA^-) by substituting the concentrations of H_3O^+ and HA^- from step 1 into the expression for K_{a2}.

Step 3 Find the concentration of OH^- by dividing K_w by the concentration of H_3O^+.

PROCEDURE FOR A TRIPROTIC ACID

Step 1 From the deprotonation equilibrium of the acid (H_3A), determine the concentrations of the conjugate base (H_2A^-) and H_3O^+.

Step 2 Find the concentration of HA^{2-} from the second deprotonation equilibrium (that for deprotonation of H_2A^-) by substituting the concentrations of H_3O^+ and H_2A^- from step 1 into the expression for K_{a2}.

Step 3 Find the concentration of A^{3-} from the deprotonation equilibrium of HA^{2-} by substituting the concentrations of H_3O^+ and HA^{2-} from step 2 into the expression for K_{a3}.

The concentration of H_3O^+ is the same in all three calculations because only the first deprotonation makes a significant contribution to its value.

Step 4 Find the concentration of OH^- by dividing K_w by the concentration of H_3O^+.

This procedure is illustrated in Example 6E.2.

EXAMPLE 6E.2   Calculating the concentrations of all solute species in a polyprotic acid solution

You are an environmental chemist studying a local waterway and need to know the amounts of all forms of phosphate in the stream. You might begin by making up some solutions of phosphoric acid to use as standards. Calculate the concentrations of all solute species in 0.10 M $H_3PO_4(aq)$.

ANTICIPATE Because successive deprotonations result in ever weaker acids, you should expect the concentrations to be in the order $H_3PO_4 > H_2PO_4^- > HPO_4^{2-} > PO_4^{3-}$, with very, very little of the completely deprotonated species. Because the solution is acidic, you should also expect the concentration of OH^- to be very small.

PLAN Follow the procedure for a triprotic acid in **Toolbox 6E.1**.

What should you assume? Assume that only the first deprotonation affects the pH and that the autoprotolysis of water has no significant effect on the pH.

SOLVE

Step 1 The primary proton transfer equilibrium is

$$H_3PO_4(aq) + H_2O(l) \rightleftharpoons H_3O^+(aq) + H_2PO_4^-(aq)$$

and the first acidity constant, from Table 6E.1, is 7.6×10^{-3}. The equilibrium table, with the concentrations in moles per liter, is

	H_3PO_4	H_3O^+	$H_2PO_4^-$
initial concentration	0.10	0	0
change in concentration	$-x$	$+x$	$+x$
equilibrium concentration	$0.10 - x$	x	x

From $K_{a1} = [H_3O^+][H_2A^-]/[H_3A]$, assuming that $x \ll 0.10$, decide if it is valid to approximate the expression.

$$K_{a1} = \frac{x \times x}{0.10 - x} \approx \frac{x^2}{0.10}$$

with $K_{a1} = 7.6 \times 10^{-3}$ solves to $x \approx 0.028$, which is 28% of 0.10, and too large for the approximation to be acceptable.

Because the value of x is in fact larger than 5% of 0.10, the full quadratic expression, $K_{a1} = x^2/(0.10 - x)$, must be used, with $K_{a1} = 7.6 \times 10^{-3}$. Rearrange the equation to

$$x^2 + K_{a1}x - 0.10\,K_{a1} = 0$$

and substitute the value of K_{a1}:

$$\underbrace{1x^2}_{a} + \underbrace{(7.6 \times 10^{-3})x}_{b} \underbrace{-7.6 \times 10^{-4}}_{c} = 0$$

Solve for x with the quadratic formula $x = \{-b \pm \sqrt{b^2 - 4ac}\}/2a$:

$$x = \frac{-(7.6 \times 10^{-3}) \pm \sqrt{(7.6 \times 10^{-3})^2 - 4(1)(-7.6 \times 10^{-4})}}{2(1)}$$

$$x = 2.4 \times 10^{-2} \text{ or } -3.2 \times 10^{-2}$$

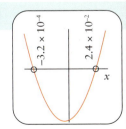

Reject the negative root because $x = [H_3O^+]$ is also a concentration.

$$[H_3O^+] = 2.4 \times 10^{-2} \text{ mol·L}^{-1}$$

From $[H_2PO_4^-] = [H_3O^+]$,

$$[H_2PO_4^-] = 2.4 \times 10^{-2} \text{ mol·L}^{-1}$$

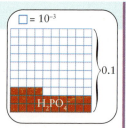

It follows from the equilibrium table that

$$[H_3PO_4] \approx 0.10 - 0.024 \text{ mol·L}^{-1} = 0.08 \text{ mol·L}^{-1}$$

(0.076 has been rounded to 0.08)

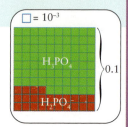

The illustrations represent the concentrations of species in the solution: red for $H_2PO_4^-$, green for H_3PO_4, and (as calculated below) yellow for HPO_4^{2-}.

Step 2 Now use $K_{a2} = 6.2 \times 10^{-8}$ to find the concentration of HPO_4^{2-}. Because $K_{a2} \ll K_{a1}$, it is safe to assume that the H_3O^+ concentration calculated in step 1 is unchanged by the second deprotonation. The proton transfer equilibrium is

$$H_2PO_4^-(aq) + H_2O(l) \rightleftharpoons H_3O^+(aq) + HPO_4^{2-}(aq)$$

Set up the equilibrium table with concentrations in moles per liter, using the results from step 1 for the concentrations of H_3O^+ and $H_2PO_4^-$:

	$H_2PO_4^-$	H_3O^+	HPO_4^{2-}
initial concentration (from step 1)	2.4×10^{-2}	2.4×10^{-2}	0
changes in concentration	$-x$	$+x$	$+x$
equilibrium concentration	$2.4 \times 10^{-2} - x$	$2.4 \times 10^{-2} + x$	x

From $K_{a2} = [H_3O^+][HA^{2-}]/[H_2A^-]$, assuming that $x \ll 2.4 \times 10^{-2}$.

$$K_{a2} = \frac{\overbrace{(2.4 \times 10^{-2} + x)}^{[HPO_4^{2-}]} \times \overbrace{x}^{\{H_3O^+\}}}{\underbrace{2.4 \times 10^{-2} - x}_{[H_2PO_4^-]}} \approx \frac{2.4 \times 10^{-2} \times x}{2.4 \times 10^{-2}}$$

and therefore $x \approx K_{a2}$ with $K_{a2} = 6.2 \times 10^{-8}$.

Check the assumption that $x \ll 2.4 \times 10^{-2}$.

$$6.2 \times 10^{-8} \ll 2.4 \times 10^{-2}$$

Therefore the assumption is valid.

From the final line of the equilibrium table

$$[HPO_4^{2-}] = x = 6.2 \times 10^{-8} \text{ mol·L}^{-1}$$
$$[H_2PO_4^-] = 2.4 \times 10^{-2} - x \approx 2.4 \times 10^{-2} \text{ mol·L}^{-1}$$
$$[H_3O^+] = 2.4 \times 10^{-2} + x \approx 2.4 \times 10^{-2} \text{ mol·L}^{-1}$$

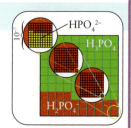

Step 3 The loss of the final proton from HPO_4^{2-} produces the phosphate ion, PO_4^{3-}:

$$HPO_4^{2-}(aq) + H_2O(l) \rightleftharpoons H_3O^+(aq) + PO_4^{3-}(aq)$$

The equilibrium constant is $K_{a3} = 2.1 \times 10^{-13}$, a very small value. Assume that the concentration of H_3O^+ calculated in step 1 and the concentration of HPO_4^{2-} calculated in step 2 are unaffected by the additional deprotonation. The equilibrium table is

	HPO_4^{2-}	H_3O^+	PO_4^{3-}
initial concentration (from step 2)	6.2×10^{-8}	2.4×10^{-2}	0
changes in concentration	$-x$	$+x$	$+x$
equilibrium concentration	$6.2 \times 10^{-8} - x$	$2.4 \times 10^{-2} + x$	x

From $K_{a3} = [H_3O^+][A^{3-}]/[HA^{2-}]$. Because K_{a3} is so small, assume that $x \ll 6.2 \times 10^{-8}$ and simplify the equation.

$$K_{a3} = \frac{(2.4 \times 10^{-2} + x) \times x}{6.2 \times 10^{-8} - x} \approx \frac{2.4 \times 10^{-2} \times x}{6.2 \times 10^{-8}}$$

Solve for x.

$$x \approx \frac{(6.2 \times 10^{-8})K_{a3}}{2.4 \times 10^{-2}} = \frac{(6.2 \times 10^{-8}) \times (2.1 \times 10^{-13})}{2.4 \times 10^{-2}}$$

$$= 5.4 \times 10^{-19}$$

From $[PO_4^{3-}] = x$,

$$[PO_4^{3-}] \approx 5.4 \times 10^{-19}\ mol \cdot L^{-1}$$

Step 4 From $[OH^-] = K_w/[H_3O^+]$,

$$[OH^-] = \frac{1.0 \times 10^{-14}}{2.4 \times 10^{-2}} = 4.2 \times 10^{-13}$$

At this point, the concentrations of all the solute species in 0.10 M H_3PO_4(aq) can be summarized as follows, in order of decreasing concentration:

Species:	H_3PO_4	H_3O^+	$H_2PO_4^-$	HPO_4^{2-}	OH^-	PO_4^{3-}
Concentration/(mol·L⁻¹):	0.08	2.4×10^{-2}	2.4×10^{-2}	6.2×10^{-8}	4.2×10^{-13}	5.4×10^{-19}

EVALUATE The second deprotonation proceeds only to a small extent. The HPO_4^{2-} ion is deprotonated only to a very tiny extent, too small to illustrate with a diagram, and the assumption that H_3O^+ is unaffected by the third deprotonation is justified.

Self-test 6E.3A Calculate the concentrations of all solute species in 0.20 M H_2S(aq).
[**Answer:** With concentrations in mol·L⁻¹: H_2S, 0.20; HS^-, 1.6×10^{-4}; H_3O^+, 1.6×10^{-4}; OH^-, 6.2×10^{-11}; S^{2-}, 7.1×10^{-15}]

Self-test 6E.3B Protonated glycine ($^+NH_3CH_2COOH$) is a diprotic amino acid with $K_{a1} = 4.5 \times 10^{-3}$ and $K_{a2} = 1.7 \times 10^{-10}$. Calculate the concentrations of all solute species in 0.50 M $NH_3CH_2COOHCl$(aq).

Related Exercises 6E.9–6E.12

The concentrations of all species in a solution of a polyprotic acid can be calculated by assuming that species present in smaller amounts do not affect the concentrations of species present in larger amounts.

6E.4 Composition and pH

In some applications, it is necessary to know how the concentrations of the ions present in a solution of a polyprotic acid vary with pH. This information is particularly important in the study of natural waters, such as rivers and lakes (**BOX 6E.1**). For example, if you were examining carbonic acid in rainwater, then, at low pH (when hydronium ions are abundant), you should expect the fully protonated species (H_2CO_3) to be dominant; at high pH (when hydroxide ions are abundant), you should expect the fully deprotonated species (CO_3^{2-}) to be dominant; at intermediate pH, the intermediate species (HCO_3^-) is expected to be dominant (**FIG. 6E.1**). These expectations can be verified quantitatively.

FIGURE 6E.1 The variation of the fractional composition of the species in carbonic acid with pH. Note that the more fully protonated species are dominant at lower pH and the more fully deprotonated species are dominant at higher pH.

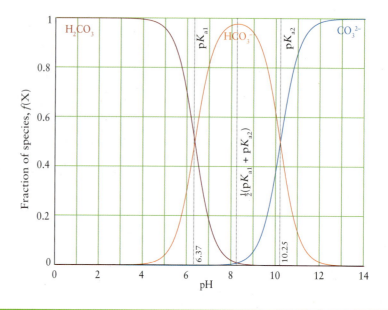

Box 6E.1 WHAT HAS THIS TO DO WITH . . . THE ENVIRONMENT?

ACID RAIN AND THE GENE POOL

The human impact on the environment affects many areas of our lives and the future of the planet. One example is the effect of acid rain on *biodiversity,* the diversity of living things. In the prairies that extend across the heartlands of North America and Asia, native plants have evolved that can survive even nitrogen-poor soil and drought. By studying prairie plants, scientists hope to breed food plants that will be hardy sources of food in times of drought. However, acid rain is making some of these plants extinct.

Acid rain is a regional phenomenon. The areas in different colors on the map shown here indicate average values of the pH of rain. Detailed analysis shows that the pH of rain decreases downwind (generally, east) of heavily populated areas. The low pH in heavily industrialized and populated areas is caused by the acidic oxides sulfur dioxide, SO_2, and the nitrogen oxides, NO and NO_2.

Rain unaffected by human activity contains mostly weak acids and has a pH of about 5.7. The primary acid present is carbonic acid, H_2CO_3, a weak acid that results when atmospheric carbon dioxide dissolves in water. The major pollutants in acid rain are strong acids that arise from human activities. Atmospheric nitrogen and oxygen can react to form NO, but the endothermic reaction is spontaneous only at the high temperatures of automobile internal combustion engines and electrical power stations:

$$N_2(g) + O_2(g) \rightleftharpoons 2\,NO(g)$$

Nitric oxide, NO, is not very soluble in water, but it is oxidized further in air to form nitrogen dioxide:

$$2\,NO(g) + O_2(g) \longrightarrow 2\,NO_2(g)$$

The NO_2 reacts with water, forming nitric acid and nitric oxide:

$$3\,NO_2(g) + 3\,H_2O(l) \longrightarrow 2\,H_3O^+(aq) + 2\,NO_3^-(aq) + NO(g)$$

The catalytic converters now used in automobiles can reduce NO to harmless N_2. They are required in many parts of the world for all new cars and trucks (Topic 7E).

Sulfur dioxide is produced as a byproduct of the burning of fossil fuels. It may combine with water directly to form sulfurous acid, a weak acid:

$$SO_2(g) + H_2O(l) \longrightarrow H_2SO_3(aq)$$

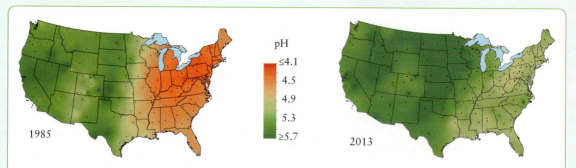

Precipitation over North America gradually becomes more acidic from west to east, especially in industrialized areas of the Northeast. This acid rain may be a result of the release of nitrogen and sulfur oxides into the atmosphere. The colors and numbers (see key) indicate pH measured at field laboratories in 1985 and 2013. Regulations reducing the amount of sulfur and nitrogen oxides in emissions from factories and automobiles have greatly reduced the impact of acid rain. Additional maps may be found on the National Atmospheric Deposition Program/National Trends Network website, http://nadp.sws.uiuc.edu/NTN/maps.aspx.

Eastern gama grass is a variety of a prairie plant that produces seeds rich in protein. It is the subject of sustainable agriculture research because it produces an abundance of seeds and yet is a perennial plant that is resistant to drought. *(Will & Deni McIntyre/ Science Source.)*

Alternatively, in the presence of particulate matter and aerosols, sulfur dioxide may react with atmospheric oxygen to form sulfur trioxide, which forms sulfuric acid, a strong acid, in water:

$$2\,SO_2(g) + O_2(g) \longrightarrow 2\,SO_3(g)$$
$$SO_3(g) + 2\,H_2O(l) \longrightarrow H_3O^+(aq) + HSO_4^-(aq)$$

Acid rain affects plants by changing the conditions in the soil. For example, nitric acid deposits nitrates, which fertilize the land. The nitrates allow fast-growing weeds such as quackgrass (*Elytrigia repens*) or couch grass (*Elymus repens*) to replace valuable prairie species. If these species were to become extinct, their genetic material would no longer be available for agricultural research.

Research on air pollution is complex. Forests and prairies cover vast areas, and the interplay of regional air pollutants is so subtle that it may take years to sort out all the environmental stresses. Controls being put in place are beginning to reduce the acidity of rain in North America and Europe. However, biologists have found that they are not enough for some sensitive environments. More stringent controls may be needed for us to maintain our quality of life without losing our precious heritage of native plants.

Related Exercise 6.83

Further Reading Environment Canada, "Acid rain" under "Pollution Issues," http://www.ec.gc.ca/air/. D. Malakoff, "Taking the sting out of acid rain," *Science,* vol. 330, November 12, 2010, pp. 910–911. U.S. Environmental Protection Agency, "Acid rain," http://www.epa.gov/acidrain. U.S. Geological Service, "Acid rain, atmospheric deposition, and precipitation chemistry," http://bqs .usgs.gov/acidrain.

How Is That Done?

To see how the concentrations of species present in a solution vary with pH, take the carbonic acid system as an example. Consider the following proton transfer equilibria:

$$H_2CO_3(aq) + H_2O(l) \rightleftharpoons H_3O^+(aq) + HCO_3^-(aq) \qquad K_{a1} = \frac{[H_3O^+][HCO_3^-]}{[H_2CO_3]}$$

$$HCO_3^-(aq) + H_2O(l) \rightleftharpoons H_3O^+(aq) + CO_3^{2-}(aq) \qquad K_{a2} = \frac{[H_3O^+][CO_3^{2-}]}{[H_2CO_3^-]}$$

Express the composition of the solution in terms of the fraction, $f(X)$, of each species X present, where X could be H_2CO_3, HCO_3^-, or CO_3^{2-}, and

$$f(X) = \frac{[X]}{[H_2CO_3] + [HCO_3^-] + [CO_3^{2-}]}$$

It will turn out to be helpful to express $f(X)$ in terms of the ratio of each species to the intermediate species, HCO_3^-. So divide the numerator and denominator by $[HCO_3^-]$ and get

$$f(X) = \frac{[X]/[HCO_3^-]}{[H_2CO_3]/[HCO_3^-] + 1 + [CO_3^{2-}]/[HCO_3^-]}$$

All three concentration ratios can be written in terms of the hydronium ion concentration. To do so, simply rearrange the expressions for the first and second acidity constants:

$$\frac{[H_2CO_3]}{[HCO_3^-]} = \frac{[H_3O^+]}{K_{a1}} \qquad \frac{[CO_3^{2-}]}{[HCO_3^-]} = \frac{K_{a2}}{[H_3O^+]}$$

then substitute them into the expression for $f(X)$ and rearrange it to obtain

$$f(H_2CO_3) = \frac{[H_3O^+]^2}{[H_3O^+]^2 + [H_3O^+]K_{a1} + K_{a1}K_{a2}}$$

$$f(HCO_3^-) = \frac{[H_3O^+]K_{a1}}{[H_3O^+]^2 + [H_3O^+]K_{a1} + K_{a1}K_{a2}}$$

$$f(CO_3^{2-}) = \frac{K_{a1}K_{a2}}{[H_3O^+]^2 + [H_3O^+]K_{a1} + K_{a1}K_{a2}}$$

FIGURE 6E.2 The variation of the fractional composition of the species in phosphoric acid with pH. As in Fig. 6E.1, the more fully protonated the species, the lower is the pH at which it is dominant.

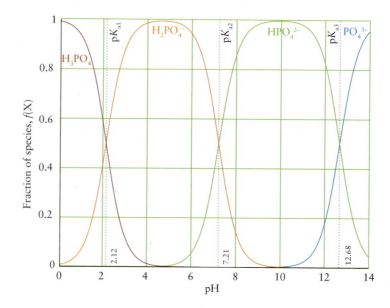

The expressions just derived give the fractions, f, of species in a solution of carbonic acid. They are easily generalized to any diprotic acid H_2A:

$$f(H_2A) = \frac{[H_3O^+]^2}{H} \qquad f(HA^-) = \frac{[H_3O^+]K_{a1}}{H} \qquad f(A^{2-}) = \frac{K_{a1}K_{a2}}{H} \qquad \text{(2a)}$$

where

$$H = [H_3O^+]^2 + [H_3O^+]K_{a1} + K_{a1}K_{a2} \qquad \text{(2b)}$$

What Do These Equations Tell You?

At high pH, the concentration of hydronium ions is very low; therefore the numerators in $f(H_2A)$ and $f(HA^-)$ are very small, and so these species are in very low abundance, as expected. At low pH, the concentration of hydronium ions is high; therefore the numerator in $f(H_2A)$ is large, and this species dominates.

The shapes of the curves predicted by Eq. 2 are shown for H_2CO_3 in Fig. 6E.1. You can see that $f(HCO_3^-) \approx 1$ at intermediate pH. The maximum value of $f(HCO_3^-)$ is at

$$\text{pH} = \tfrac{1}{2}(pK_{a1} + pK_{a2}) \qquad \text{(3)}$$

Equation 3 is the same as Eq. 1, but that is a coincidence: the derivations are different and the conditions for the validity of Eq. 1 do not apply to Eq. 3.

Note that the fully protonated form (H_2CO_3) is dominant when $\text{pH} < pK_{a1}$ and the fully deprotonated form (CO_3^{2-}) becomes dominant when $\text{pH} > pK_{a2}$. Similar calculations can be carried out for triprotic acids (**FIG. 6E.2**).

THINKING POINT

The expressions in Eq. 2 have a symmetry. Can you identify it and use it to write down the corresponding expressions for a triprotic acid?

The fraction of deprotonated species increases as the pH is increased, as summarized in Figs. 6E.1 and 6E.2.

What have you learned in this Topic?

You have learned that polyprotic acids can donate more than one proton and that successive deprotonations generally have much smaller acidity constants. You have seen that at low pH the fully protonated form of a weak acid dominates, whereas at high pH the fully deprotonated form is most abundant.

The skills you have mastered are the ability to:

☐ **1.** Calculate the pH of polyprotic acid solutions (Section 6E.1).

☐ **2.** Estimate the pH of an amphiprotic salt solution (Example 6E.1).

☐ **3.** Calculate the concentration of all ions in a polyprotic acid solution (Example 6E.2).

☐ **4.** Use fractional composition diagrams to determine the dominant form of a polyprotic acid system at a given pH (Section 6E.4).

Topic 6E Exercises

Refer to Table 6E.1 for K_a. Assume 25 °C throughout.

6E.1 Calculate the pH of 0.15 M H_2SO_4(aq) at 25 °C.

6E.2 Calculate the pH of 0.010 M H_2SeO_4(aq), given that K_{a1} is very large and $K_{a2} = 1.2 \times 10^{-2}$.

6E.3 Calculate the pH of each of the following solutions of diprotic acids at 25 °C, ignoring second deprotonations only when the approximation is justified: (a) 0.010 M H_2CO_3(aq); (b) 0.10 M $(COOH)_2$(aq); (c) 0.20 M H_2S(aq).

6E.4 Calculate the pH of each of the following solutions of diprotic acids at 25 °C, ignoring second deprotonations only when that approximation is justified: (a) 0.10 M H_2S(aq); (b) 0.15 M $H_2C_4H_4O_6$(aq), tartaric acid; (c) 1.1×10^{-3} M H_2TeO_4(aq), telluric acid, for which $K_{a1} = 2.1 \times 10^{-8}$ and $K_{a2} = 6.5 \times 10^{-12}$.

6E.5 Estimate the pH of (a) 0.15 M $NaHSO_3$(aq); (b) 0.050 M $NaHSO_3$(aq).

6E.6 Estimate the pH of (a) 0.0153 M $NaHCO_3$(aq); (b) 0.110 M $KHCO_3$(aq).

6E.7 Citric acid, which is extracted from citrus fruits and pineapples, undergoes three successive deprotonations with pK_a values of 3.14, 5.95, and 6.39. Estimate the pH of (a) a 0.15 M aqueous solution of the monosodium salt; (b) a 0.075 M aqueous solution of the disodium salt.

6E.8 Like sulfuric acid, a certain diprotic acid, H_2A, is a strong acid in its first deprotonation and a weak acid in its second deprotonation. A solution that is 0.015 M H_2A(aq) has a pH of 1.72. What is the value of K_{a2} for this acid?

6E.9 Calculate the molar concentrations of H_2CO_3, HCO_3^-, CO_3^{2-}, H_3O^+, and OH^- present in 0.0456 M H_2CO_3(aq).

6E.10 Calculate the molar concentrations of H_2SO_3, HSO_3^-, SO_3^{2-}, H_3O^+, and OH^- present in 0.125 M H_2SO_3(aq).

6E.11 Calculate the molar concentrations of H_2CO_3, HCO_3^-, CO_3^{2-}, H_3O^+, and OH^- present in 0.0456 M Na_2CO_3(aq).

6E.12 Calculate the molar concentrations of H_2SO_3, HSO_3^-, SO_3^{2-}, H_3O^+, and OH^- present in 0.170 M Na_2SO_3(aq).

6E.13 A large volume of 0.150 M H_2SO_3(aq) is treated with a strong base to adjust the pH to 5.50. Assume that the addition of the base, a solid, does not significantly affect the volume of the solution. Estimate the molar concentrations of H_2SO_3, HSO_3^-, and SO_3^{2-} present in the final solution.

6E.14 A large volume of 0.250 M H_2S(aq) is treated with a strong base to adjust the pH to 9.35. Assume that the addition of the base, a solid, does not significantly affect the volume of the solution. Estimate the molar concentrations of H_2S, HS^-, and S^{2-} present in the final solution.

6E.15 Calculate the pH of each of the following acid solutions at 25 °C; ignore second deprotonations only when that approximation is justified. (a) 1.0×10^{-4} M H_3BO_3(aq), boric acid acts as a monoprotic acid; (b) 0.015 M H_3PO_4(aq); (c) 0.10 M H_2SO_3(aq).

6E.16 Calculate the molar concentrations of $(COOH)_2$, $HOOCCO_2^-$, $(CO_2)_2^{2-}$, H_3O^+, and OH^- in 0.12 M $(COOH)_2$(aq), oxalic acid.

6E.17 Calculate the molar concentrations of all phosphate species in H_3PO_4(aq) at pH = 2.25, given that their *total* concentration is 15 mmol·L^{-1}.

6E.18 Ammonium acetate is prepared by the reaction of equal amounts of aqueous ammonia and acetic acid. Calculate the concentration of all solute species present in 0.100 M $NH_4CH_3CO_2$(aq).

Topic 6F Autoprotolysis and pH

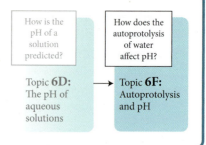

How is the pH of a solution predicted?

How does the autoprotolysis of water affect pH?

Topic **6D:** The pH of aqueous solutions → Topic **6F:** Autoprotolysis and pH

Why Do You Need to Know This Material? In very dilute solutions of acids and bases, autoprotolysis can contribute significantly to the pH. To make reliable calculations, you need to know how to recognize when autoprotolysis is significant and how to account for its contribution.

What Do You Need to Know Already? You need to be familiar with the concept of autoprotolysis (Topic 6B) and weak acid and base equilibrium calculations (Topic 6D).

If an ion is doubly charged, its concentration is multiplied by 2 in the charge-balance equation, and by 3 if it is triply charged. The charge-balance equation for an aqueous solution of $CaCl_2$ would therefore be $[Cl^-] = 2[Ca^{2+}]$.

Suppose you were asked to estimate the pH of 1.0×10^{-8} M HCl(aq). If you used the techniques of Example 6B.1 to calculate the pH from the concentration of the acid itself, you would find pH = 8.00. That value, though, is absurd, because it lies on the basic side of neutrality, whereas HCl is an acid! The error stems from there being two sources of hydronium ions, whereas only one has been taken into account. At very low acid concentrations, the supply of hydronium ions from the autoprotolysis of water is close to the supply provided by the very low concentration of HCl, and both sources must be taken into account.

6F.1 Very Dilute Solutions of Strong Acids and Bases

The contribution of autoprotolysis to pH must be taken into account when the concentration of strong acid or base is less than about 10^{-6} mol·L^{-1}. As an example, consider a solution of HCl, a strong acid. Other than water, the species present are H_3O^+, OH^-, and Cl^-. There are three unknown concentrations. To find them, three equations are needed.

One equation takes **charge balance** into account, the requirement that the solution must be electrically neutral overall. That is, the total charge of cations must equal that of anions. Because there is only one type of cation, H_3O^+, the concentration of H_3O^+ ions must equal the sum of the concentrations of the two types of anions, Cl^- and OH^-. The charge-balance relation $[H_3O^+] = [Cl^-] + [OH^-]$ then implies that

$$[OH^-] = [H_3O^+] - [Cl^-]$$

A second equation takes **material balance** into account, the requirement that all the added solute must be accounted for even though it is now present as ions. Because HCl is a strong acid, the concentration of Cl^- ions in the solution is equal to the concentration of the HCl added initially (assume that all the HCl molecules are deprotonated). If the numerical value of that initial concentration is denoted $[HCl]_{initial}$, the material-balance relation is

$$[Cl^-] = [HCl]_{initial}$$

When this relation is substituted into the charge balance expression the result is

$$[OH^-] = [H_3O^+] - [HCl]_{initial}$$

The third equation uses the autoprotolysis constant, K_w, to relate the concentration of OH^- ions to that of H_3O^+ ions

$$K_w = [H_3O^+][OH^-]$$

Now substitute the expression for $[OH^-]$ into this expression:

$$K_w = [H_3O^+](\overbrace{[H_3O^+] - [HCl]_{initial}}^{[OH^-]})$$
$$= [H_3O^+]^2 - [HCl]_{initial}[H_3O^+]$$

and rearrange it into a quadratic equation:

$$[H_3O^+]^2 - [HCl]_{initial}[H_3O^+] - K_w = 0 \qquad (1)$$

As will be shown in Example 6F.1, the quadratic formula (Toolbox 5I.1: for $ax^2 + bx + c = 0$, $x = \{-b \pm \sqrt{b^2 - 4ac}\}/2a$) with $x = [H_3O^+]$ can be used to solve this equation for the concentration of hydronium ions.

Now consider a very dilute solution of a strong base, such as NaOH. Apart from water, the species present in solution are Na^+, OH^-, and H_3O^+. As was done for HCl, three equations can be written down that relate the concentrations of these three ions by using charge balance, material balance, and the autoprotolysis constant. Because the cations present are hydronium ions and sodium ions, the charge-balance relation is

$$[OH^-] = [H_3O^+] + [Na^+]$$

Because NaOH is completely dissociated in solution, the concentration of sodium ions is the same as the initial concentration of NaOH, $[NaOH]_{initial}$. Therefore, the material-balance relation is $[Na^+] = [NaOH]_{initial}$. It follows that

$$[OH^-] = [H_3O^+] + [NaOH]_{initial}$$

The autoprotolysis constant now becomes

$$K_w = [H_3O^+]\overbrace{([H_3O^+] + [NaOH]_{initial})}^{[OH^-]}$$

This equation, after minor rearrangement, is also a quadratic equation for $[H_3O^+]$:

$$[H_3O^+]^2 + [NaOH]_{initial}[H_3O^+] - K_w = 0 \qquad (2)$$

which can be solved by using the quadratic formula.

EXAMPLE 6F.1 Calculating the pH of a very dilute aqueous solution of a strong acid

When chemists are using compounds as catalysts, sometimes only a very low concentration is needed. Suppose you are using 8.0×10^{-8} M HCl(aq) as a catalyst for an organic reaction and you are using a pH meter to gauge its acidity. What pH value should you expect?

ANTICIPATE Although the acid is very dilute, it is an acid, and you should expect a pH of slightly less than 7.

PLAN Let $[HCl]_{initial} = 8.0 \times 10^{-8}$ mol·L^{-1} and $[H_3O^+] = x$; substitute these values into Eq. 1 and solve for x.

SOLVE

From $[H_3O^+]^2 - [HCl]_{initial}[H_3O^+] - K_w = 0$, with $x = [H_3O^+]$,

$$(1) \overbrace{x^2}^{a} \overbrace{-(8.0 \times 10^{-8})x}^{b} \overbrace{-(1.0 \times 10^{-14})}^{c} = 0$$

From the quadratic formula, $x = \{-b \pm \sqrt{b^2 - 4ac}\}/2a$,

$$x = \frac{-(-8.0 \times 10^{-8}) \pm \sqrt{(8.0 \times 10^{-8})^2 - 4(1)(-1.0 \times 10^{-14})}}{2(1)}$$

$$= 1.5 \times 10^{-7} \text{ or } -6.8 \times 10^{-8}$$

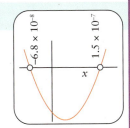

Reject the negative root because x is a concentration and use pH = $-\log x$.

$$pH = -\log(1.5 \times 10^{-7}) = 6.82$$

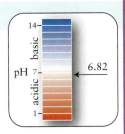

EVALUATE As expected, the pH is slightly less than 7.

Self-test 6F.1A What is the pH of 1.0×10^{-7} M HNO$_3$(aq)?

[*Answer:* 6.79]

Self-test 6F.1B What is the pH of 2.0×10^{-7} M NaOH(aq)?

Related Exercises 6F.1–6F.4

In very dilute solutions of strong acids and bases, the pH is significantly affected by the autoprotolysis of water. The pH is calculated by solving three simultaneous equations: the charge-balance equation, the material-balance equation, and the expression for K_w.

6F.2 Very Dilute Solutions of Weak Acids

Autoprotolysis also contributes to the pH of very dilute solutions of weak acids. In fact, some acids, such as hypoiodous acid, HIO, are so weak and undergo so little deprotonation that to find the pH of these solutions, the autoprotolysis of water must always be taken into account.

How Is That Done?

The calculation of pH for very dilute solutions of a weak acid HA is similar to that for strong acids. It is based on the fact that, apart from water, there are four species in solution—namely, HA, A^-, H_3O^+, and OH^-. Because there are four unknowns, four equations are needed to find their concentrations. Two of the relations are the autoprotolysis constant of water and the acidity constant of the acid HA:

$$K_w = [H_3O^+][OH^-] \qquad K_a = \frac{[H_3O^+][A^-]}{[HA]}$$

Charge balance implies that

$$[H_3O^+] = [OH^-] + [A^-]$$

Material balance provides the fourth equation: the total concentration of "A" groups (the acid itself and any conjugate base it forms; for instance, F^- ions if the acid added is HF) must be equal to the initial concentration of the acid:

$$[HA]_{initial} = [HA] + [A^-]$$

To find an expression for the concentration of hydronium ions in terms of the initial concentration of the acid, rearrange the charge-balance relation to express the concentration of A^- in terms of $[H_3O^+]$:

$$[A^-] = [H_3O^+] - [OH^-]$$

Then express $[OH^-]$ in terms of the hydronium ion concentration by using the autoprotolysis expression:

$$[A^-] = [H_3O^+] - \frac{K_w}{[H_3O^+]}$$

When this expression for $[A^-]$ is substituted into the material-balance equation in the form $[HA] = [HA]_{initial} - [A^-]$, that equation becomes

$$[HA] = [HA]_{initial} - \overbrace{\left([H_3O^+] - \frac{K_w}{[H_3O^+]}\right)}^{[A^-]}$$

Now substitute these expressions for [HA] and $[A^-]$ into K_a to obtain

$$K_a = \frac{[H_3O^+]\left([H_3O^+] - \dfrac{K_w}{[H_3O^+]}\right)}{[HA]_{initial} - [H_3O^+] + \dfrac{K_w}{[H_3O^+]}}$$

The resulting expression might look complicated, but experimental conditions often allow it to be simplified. For example, in many solutions of weak acids, $[H_3O^+] > 10^{-6}$ (that is, pH < 6). Under these conditions, $K_w/[H_3O^+] < 10^{-8}$ and this term can be ignored in both the numerator and the denominator. Then

$$K_a = \frac{[H_3O^+]^2}{[HA]_{initial} - [H_3O^+]}$$

Remember that $[H_3O^+]$ is really $[H_3O^+]/c°$.

which can be arranged into a quadratic equation for $[H_3O^+]$:

$$[H_3O^+]^2 + K_a[H_3O^+] - K_a[HA]_{initial} = 0 \qquad (3a)$$

However, when the acid is so dilute or so weak that $[H_3O^+] \leq 10^{-6}$ (that is, when the pH lies between 6 and 7), then the full expression for $[H_3O^+]$ must be solved after it has been rearranged into

$$[H_3O^+]^3 + K_a[H_3O^+]^2 - (K_w + K_a[HA]_{initial})[H_3O^+] - K_aK_w = 0 \qquad (3b)$$

To solve this cubic equation in $[H_3O^+]$, it is best to use a graphing calculator or mathematical software.

To see the connection with Eq. 3a, set $K_w = 0$ and then cancel one common factor of $[H_3O^+]$.

EXAMPLE 6F.2 Estimating the pH of a dilute aqueous solution of a weak acid when the autoprotolysis of water must be considered

Wine is a complex mixture of over 1000 compounds, many at very low concentrations. A large number of flavorful components are derivatives of phenol, C_6H_5OH, a weak acid. If you are an enologist studying wine chemistry, you may wish to know how dilute compounds such as phenol affect the acidity of wine. Use Eq. 3b to estimate the pH of a 1.0×10^{-4} M aqueous phenol solution.

ANTICIPATE Phenol is a weak acid, so you should expect pH < 7; but the pH will be only slightly less than 7 because the solution is so dilute.

PLAN Because Eq. 3b is complicated, first find the numerical factor of the third term, $K_w + K_a[HA]_{initial}$, and the value of the fourth term, K_aK_w. For simplicity, write $x = [H_3O^+]$.

SOLVE
With $x = [H_3O^+]$, Eq. 3b becomes

$$x^3 + K_ax^2 - (K_w + K_a[HA]_{initial})x - K_aK_w = 0$$

Find K_a for phenol in Table 6C.1.

$$K_a = 1.3 \times 10^{-10}$$

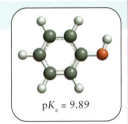

$pK_a = 9.89$

Evaluate $K_w + K_a[HA]_{initial}$.

$$K_w + K_a[HA]_{initial} = 1.0 \times 10^{-14} + (1.3 \times 10^{-10}) \times (1.0 \times 10^{-4})$$
$$= 2.3 \times 10^{-14}$$

Evaluate K_aK_w.

$$K_aK_w = (1.3 \times 10^{-10}) \times (1.0 \times 10^{-14}) = 1.3 \times 10^{-24}$$

Substitute the values into Eq. 3b with $[H_3O^+] = x$,

$$x^3 + (1.3 \times 10^{-10})x^2 - (2.3 \times 10^{-14})x - (1.3 \times 10^{-24}) = 0$$

To simplify the appearance of the coefficients (a purely cosmetic step), write $x = X \times 10^{-7}$ and divide the resulting equation through by 10^{-21}.

$$X^3 + 0.0013X^2 - 2.3X - 0.0013 = 0$$

Find the positive root.

The only positive root is $X = 1.516\ldots$; so $x = 1.516\ldots \times 10^{-7}$.

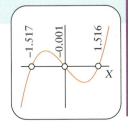

From pH $= -\log [H_3O^+]$ and $[H_3O^+] = x$,

$$pH = -\log(1.516\ldots \times 10^{-7}) = 6.82$$

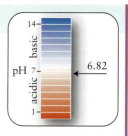

EVALUATE As expected, the pH is slightly less than 7.

Self-test 6F.2A Use Eq. 3b and the information in Table 6C.1 to estimate the pH of 2.0×10^{-4} M HCN(aq).

[*Answer*: 6.48]

Self-test 6F.2B Use Eq. 3b and the information in Table 6C.1 to estimate the pH of 1.0×10^{-2} M HIO(aq).

Related Exercises 6F.7–6F.10

In aqueous solutions of weak acids, the autoprotolysis of water must be taken into account if the hydronium ion concentration is less than 10^{-6} mol·L^{-1}. The expressions for K_w and K_a are combined with the equations for charge and material balance to find the pH.

What have you learned in this Topic?

You have learned that in very dilute aqueous solutions of acids or bases the autoprotolysis reaction of water contributes significantly to the hydronium ion concentration. You have also seen that for very weak acids, the contribution from autoprotolysis must be taken into account even when the acid solution is not dilute. You have seen how to use the charge- and material-balance expressions together with the relevant equilibrium expressions to solve for the hydronium ion concentration when the contribution from autoprotolysis is significant.

The skills you have mastered are the ability to:

☐ **1.** Calculate the pH of very dilute solutions of strong acids (Example 6F.1).

☐ **2.** Estimate the pH of solutions of weak acids when the autoprotolysis of water must be taken into account (Example 6F.2).

Topic 6F Exercises

6F.1 Calculate the pH of 6.55×10^{-7} M HClO$_4$(aq).

6F.2 Calculate the pH of 6.50×10^{-8} M HBr(aq).

6F.3 Calculate the pH of 9.78×10^{-8} M KOH(aq).

6F.4 Calculate the pH of 8.23×10^{-7} M NaNH$_2$(aq).

6F.5 The neglect of the autoprotolysis of water is invalid for $6 < pH < 8$. What is the lowest concentration of aqueous acetic acid that does not require you to take into account the autoprotolysis of water when calculating the pH of the solution to within ± 0.1?

6F.6 The neglect of the autoprotolysis of water is invalid for $6 < pH < 8$. At what concentration is it necessary to take into account the autoprotolysis of water when calculating the pH of an aqueous chloroacetic acid solution, ClCH$_2$COOH, to within ± 0.1? Account for any difference between this value and the answer to Exercise 6F.5.

For Exercises 6F.7–10, use a graphing calculator or mathematical software.

6F.7 Is the criterion $6 < pH < 8$ a good guide for when autoprotolysis can be ignored, or should it be modified to $5 < pH < 8$?

In Example 6D.4, the pH of 0.15 M NH$_4$Cl(aq) is found to be 5.04. However, the contribution to the pH from the autoprotolysis of water was ignored. Repeat the calculation of the pH of this solution, taking into account the autoprotolysis of water.

6F.8 Is the criterion $6 < pH < 8$ a good guide for when autoprotolysis can be ignored, or should it be modified to $6 < pH < 9.5$? In Example 6D.5 the pH of 0.15 M Ca(CH$_3$CO$_2$)$_2$(aq) is found to be 9.11. However, the contribution to the pH from the autoprotolysis of water was ignored. Repeat the calculation of the pH of this solution, taking into account the autoprotolysis of water.

6F.9 (a) Calculate the pH of 8.50×10^{-5} M and 7.37×10^{-6} M HCN(aq), ignoring the effect of the autoprotolysis of water. (b) Repeat the calculations, taking into account the autoprotolysis of water.

6F.10 (a) Calculate the pH of 1.89×10^{-5} M and 9.64×10^{-7} M HClO(aq), ignoring the effect of the autoprotolysis of water. (b) Repeat the calculations, taking into account the autoprotolysis of water.

Topic 6G Buffers

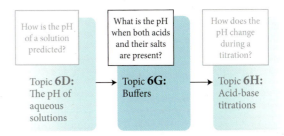

Why Do You Need to Know This Material? Buffer solutions are essential for controlling the pH of aqueous solutions used in industry and the fluids in living organisms.

What Do You Need to Know Already? You need to be familiar with the concepts of conjugate acids and bases (Topic 6C) and with equilibrium calculations involving weak acids and bases (Topic 6D).

The control of pH is crucial for the ability of organisms—including ourselves—to survive, because even minor drifts from the optimum value of the pH can cause enzymes to change their shape and cease to function. The information in this Topic is also used in industry to control the pH of reaction mixtures and to monitor natural waters. In medicine and biology, the information is used to control the conditions of cultures and biological cells and to maintain the proper pH of blood. In agriculture, it is used to maintain the soil at a pH optimal for plant growth. In the laboratory, it is used to interpret the change in pH of a solution during a titration.

6G.1 Buffer Action

The calculations in Topic 6D show how to estimate the pH of a solution of a weak acid or base. Suppose, however, that a salt of the acid or base is also present. How does that salt affect the pH of the solution? The crucial point to keep in mind throughout this Topic is that, according to the Brønsted–Lowry theory, the ions provided by a salt might themselves be acids or bases and so affect the pH.

To set the scene, suppose you have some dilute hydrochloric acid and add to it appreciable concentrations of sodium chloride, which contains the conjugate base of HCl, the Cl^- ion. Because HCl is a strong acid, its conjugate base is a very weak proton acceptor and its presence has no measurable effect on pH. The pH of 0.10 M HCl(aq) is about 1.0, even after 0.10 mol NaCl has been added to a liter of the solution.

Now suppose instead that the solution contains acetic acid to which sodium acetate has been added. Because $CH_3CO_2^-$, the conjugate base of CH_3COOH, is a weak base in water, its presence will increase the pH of the solution. Similarly, suppose ammonium chloride is added to a solution of ammonia. The NH_4^+ ion is a weak acid in water; therefore, its presence will lower the pH of the solution. As you will see, such "mixed solutions"—solutions that contain a weak acid or a weak base and one of its salts—provide a means of stabilizing the pH of aqueous solutions such as blood plasma, seawater, and reaction mixtures.

A **buffer** is a mixed solution in which the pH resists change when small amounts of strong acids or bases are added. A buffer consists of an aqueous solution of a weak acid and its conjugate base supplied as a salt, or a weak base and its conjugate acid supplied as a salt. Examples are a solution of acetic acid and sodium acetate and a solution of ammonia and ammonium chloride. Buffers are used to calibrate pH meters, to culture bacteria, and to control the pH of solutions in which chemical reactions are taking place. They are also administered intravenously to hospital patients. Human blood plasma is buffered to pH = 7.4; the ocean is buffered to about pH = 8.4 by a complex buffering process that depends on the presence of hydrogen carbonates and silicates.

When a drop of strong acid is added to pure water, the pH changes significantly. However, when the same amount of acid is added to a buffer, the pH hardly changes at all. To understand why not, consider the dynamic equilibrium between a weak acid and its

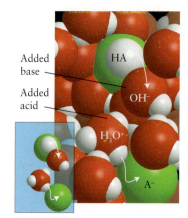

FIGURE 6G.1 A solution acts as a buffer if it contains a weak acid, HA, which donates protons when a base is added, and the conjugate base, A⁻, which accepts protons when an acid is added. In the inset, for clarity, water is represented by the blue background.

conjugate base in an aqueous solution that contains similar amounts of acid (CH_3COOH) and its salt ($NaCH_3CO_2$):

$$CH_3COOH(aq) + H_2O(l) \rightleftharpoons H_3O^+(aq) + CH_3CO_2^-(aq)$$

When a few drops of an acid are added to such a solution, the newly arrived H_3O^+ ions transfer protons to the $CH_3CO_2^-$ ions to form CH_3COOH and H_2O molecules (**FIG. 6G.1**). Because the added H_3O^+ ions are removed by the $CH_3CO_2^-$ ions, the pH remains almost unchanged, even if the added acid is strong. In effect, the acetate ions act as a "sink" for protons. If a small amount of base is added instead, the incoming OH^- ions remove protons from the CH_3COOH molecules to produce $CH_3CO_2^-$ ions and H_2O molecules. In this case, the acetic acid molecules act as a source of protons. Because the added OH^- ions are removed by the CH_3COOH molecules, the concentration of OH^- ions remains nearly unchanged. Consequently, the H_3O^+ concentration (and the pH) is also left nearly unchanged, even if the base is strong.

A similar effect occurs in a buffer solution containing similar amounts of base (NH_3) and its salt (NH_4Cl):

$$NH_3(aq) + H_2O(l) \rightleftharpoons NH_4^+(aq) + OH^-(aq)$$

When a few drops of a solution of base are added, the incoming OH^- ions remove protons from NH_4^+ ions to form NH_3 and H_2O molecules. If a few drops of an acid are added, the incoming protons attach to NH_3 molecules to make NH_4^+ ions and hence are removed from the solution. In each case, the pH is left almost unchanged, even if the added acid or base is strong.

THINKING POINT

Could an aqueous solution of glycine, $^-O_2CCH_2NH_3^+$, which contains both acidic and basic groups, act as a buffer?

A buffer is a mixture of a weak conjugate acid–base pair that stabilizes the pH of a solution by providing both a source and a sink for protons.

6G.2 Designing a Buffer

Suppose you need to make up a buffer with a particular pH. For instance, you might be culturing bacteria and need to maintain a precise pH to sustain their metabolism. To choose the most appropriate buffer system, you need to know the value of the pH at which a given buffer stabilizes a solution. A mixture of a weak acid and its salt will act as a buffer at pH < 7 (the acidic side of neutrality) and is known as an **acid buffer.** A mixture of a weak base and its salt will act as a buffer at pH > 7 (the basic side of neutrality) and is known as a **base buffer** (or "alkaline buffer"). To find the precise pH at which a mixed solution of given composition will act as a buffer, you need to carry out an equilibrium calculation that is very similar to those in Topic 6D, as shown in the following example. **TABLE 6G.1** summarizes the details of the compositions of acid and base buffers.

TABLE 6G.1   The Composition of Buffers

	Composition	Calculate pH from...	Examples	pK_a
Acid buffers (for pH < 7)				
acid (HA)	conjugate base (A^-) as the salt MA	$pK_a(HA)$	$CH_3COOH/CH_3CO_2^-$	4.75
			HNO_2/NO_2^-	3.37
			$HClO_2/ClO_2^-$	2.00
Base buffers (for pH > 7)				
base (B)	conjugate acid (HB^+) as the salt HBX	$pK_a(HB^+)$ by using $pK_a(HB^+) +$ $pK_b(B) = pK_w$	NH_4^+/NH_3	9.25
			$(CH_3)_3NH^+/(CH_3)_3N$	9.81
			$H_2PO_4^-/HPO_4^{2-}$	7.21

EXAMPLE 6G.1   Calculating the pH of a buffer solution

You are working in a microbiology laboratory culturing bacteria that require an acidic environment and you need to prepare a buffer to keep the culture at an appropriate pH. You prepare a buffer solution that is 0.040 M $NaCH_3CO_2$(aq) and 0.080 M CH_3COOH(aq) at 25 °C. What is the pH of the solution?

ANTICIPATE If the weak acid were present alone, you would expect pH < 7. Because some conjugate base has been added, you should expect the pH to be slightly higher than the acid-only value (which is estimated to be about 2.92 in Example 6D.1) but still less than 7.

PLAN First, identify the weak acid and its conjugate base. Then, write the proton transfer equilibrium between them and rearrange the expression for K_a to give $[H_3O^+]$. Finally, calculate the pH.

What should you assume? Assume that the protonation of acetate ions and the deprotonation of acetic acid molecules have both proceeded to such a small extent that the concentrations of both species are nearly the same as their initial (formal) values.

SOLVE The acid is CH_3COOH and its conjugate base is $CH_3CO_2^-$. The equilibrium to consider is

$$CH_3COOH(aq) + H_2O(l) \rightleftharpoons H_3O^+(aq) + CH_3CO_2^-(aq) \qquad K_a = \frac{[H_3O^+][CH_3CO_2^-]}{[CH_3COOH]}$$

From Table 6C.1, $pK_a = 4.75$ and $K_a = 1.8 \times 10^{-5}$.

Find the equilibrium concentration of H_3O^+ ions from the expression for K_a rearranged into

$$[H_3O^+] = K_a \times \frac{[CH_3COOH]}{[CH_3CO_2^-]}$$

Approximate the conjugate acid and base concentrations by their initial (formal) values.

$$[H_3O^+] \approx (1.8 \times 10^{-5}) \times \frac{0.080}{0.040} = 3.6... \times 10^{-5}$$

From pH $= -\log [H_3O^+]$,

$$pH \approx -\log(3.6... \times 10^{-5}) = 4.44$$

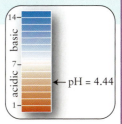

EVALUATE As expected, the solution is less acidic than it would be if only the acid were present (pH = 2.92) and acts as an acid buffer at pH ≈ 4.

Self-test 6G.1A Calculate the pH of a buffer solution that is 0.15 M HNO_2(aq) and 0.20 M $NaNO_2$(aq).

[***Answer:*** 3.49]

Self-test 6G.1B Calculate the pH of a buffer solution that is 0.040 M NH_4Cl(aq) and 0.030 M NH_3(aq).

Related Exercises 6G.7–6G.10

The point of using a buffer is to stabilize a solution against changes in pH when a strong base or a strong acid is added. The next example shows how to calculate the effect of added acid or base on the pH of an acid buffer.

EXAMPLE 6G.2   Calculating the pH change of a buffered solution

You continue to use the buffer that you prepared in Example 6G.1 but are worried about how adding sodium hydroxide to the buffer solution will affect the pH, which could upset your experiment. Suppose 1.2 g of sodium hydroxide (0.030 mol NaOH) is dissolved in 500. mL of the buffer solution described in Example 6G.1. Calculate the pH of the resulting solution and the change in pH.

ANTICIPATE Because the buffer solution contains a weak acid that will react with the strong base, you should expect only a small increase in the pH from the 4.44 of the initial solution.

PLAN Solve this problem in two stages. First, calculate the concentrations of acid and conjugate base by noting that the OH^- ions added to the buffer solution react with some of the acid of the buffer system, decreasing the amount of acid and correspondingly increasing the amount of conjugate base. Then rearrange the expression for K_a to obtain the pH of the solution, just as in Example 6G.1.

What should you assume? Assume that the equilibrium concentrations of the acid and its conjugate base remain nearly the same as their initial (formal) concentrations following neutralization and that the neutralization reaction goes to completion. Assume that the volume of the solution remains unchanged.

SOLVE The proton transfer equilibrium is

$$CH_3COOH(aq) + H_2O(l) \rightleftharpoons H_3O^+(aq) + CH_3CO_2^-(aq) \qquad K_a = \frac{[H_3O^+][CH_3CO_2^-]}{[CH_3COOH]}$$

Use the data in Example 6G.1, including $K_a = 1.8 \times 10^{-5}$. The OH^- from the added NaOH reacts with some of the CH_3COOH according to

$$CH_3COOH(aq) + OH^-(aq) \longrightarrow CH_3CO_2^-(aq) + H_2O(l)$$

Step 1 Find the new concentration of acid.

Find the initial amount of CH_3COOH in the solution from $n_J = V[J]$.

$$n(CH_3COOH)_{initial} = 0.500\ L \times 0.080\ mol \cdot L^{-1}$$
$$= 0.040\ mol$$

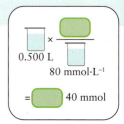

Calculate the amount of CH_3COOH that reacts by using 1 mol $CH_3COOH \simeq 1$ mol OH^-.

$$n(CH_3COOH)_{reacts} = 0.030\ mol\ OH^- \times \frac{1\ mol\ CH_3COOH}{1\ mol\ OH^-}$$
$$= 0.030\ mol\ CH_3COOH$$

Calculate the amount of CH_3COOH remaining from $n_{final} = n_{initial} - n_{reacts}$.

$$n(CH_3COOH)_{final} = 0.040 - 0.030\ mol = 0.010\ mol$$

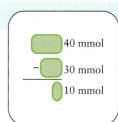

From $[J] = n_J/V$,

$$[CH_3COOH] = \frac{0.010\ mol}{0.500\ L} = 0.020\ mol \cdot L^{-1}$$

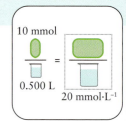

Step 2 Find the new concentration of the conjugate base.

Find the initial amount of $CH_3CO_2^-$ in the solution from $n_J = V[J]$.

$$n(CH_3CO_2^-)_{initial} = 0.500\ L \times 0.040\ mol \cdot L^{-1}$$
$$= 0.020\ mol$$

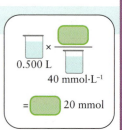

Add the increase in the amount of $CH_3CO_2^-$ due to the reaction.

$$n(CH_3CO_2^-)_{final} = 0.020 + 0.030\ mol = 0.050\ mol$$

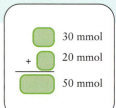

From $[J] = n_J/V$,

$$[CH_3CO_2^-] = \frac{0.050\ \text{mol}}{0.500\ \text{L}} = 0.10\ \text{mol·L}^{-1}$$

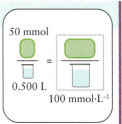

Step 3 Calculate the pH.

Calculate $[H_3O^+]$ from $[H_3O^+] = K_a[CH_3COOH]/[CH_3CO_2^-]$, approximating the concentrations of the acid and base by their values immediately following neutralization.

$$[H_3O^+] \approx (1.8 \times 10^{-5}) \times \frac{0.020}{0.10} = 3.6 \times 10^{-6}$$

From $pH = -\log[H_3O^+]$,

$$pH \approx -\log(3.6 \times 10^{-6}) = 5.44$$

The change in pH is $5.44 - 4.44 = 1.00$.

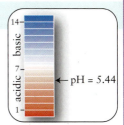

EVALUATE As expected, the pH of the solution changes only slightly, from about 4.4 to about 5.4.

Self-test 6G.2A Suppose that 0.0200 mol NaOH(s) is dissolved in 300. mL of the buffer solution of Example 6G.1. Calculate the pH of the resulting solution and the change in pH.

[**Answer:** 5.65, an increase of 1.21]

Self-test 6G.2B Suppose that 0.0100 mol HCl(g) is dissolved in 500. mL of the buffer solution of Example 6G.1. Calculate the pH of the resulting solution and the change in pH.

Related Exercises 6G.11, 6G.12

Buffers are often prepared with equal molar concentrations of the conjugate acid and base because then there are adequate supplies of the "source" and "sink" species that can stabilize the pH against change in either direction. The pH of these **equimolar** solutions, solutions with equal molar concentrations of the solutes, is very easy to predict. Consider the equilibrium

$$HA(aq) + H_2O(l) \rightleftharpoons H_3O^+(aq) + A^-(aq) \qquad K_a = \frac{[H_3O^+][A^-]}{[HA]} \qquad \textbf{(A)}$$

The values of [HA] and [A$^-$] that appear in K_a are the equilibrium concentrations of acid and base in the solution, not the concentrations added initially. However, a weak acid HA typically loses only a tiny fraction of its protons, and so [HA] is negligibly different from the concentration of the acid used to prepare the buffer, [HA]$_{\text{initial}}$. Likewise, only a tiny fraction of the weakly basic anions A$^-$ accept protons, and so [A$^-$] is negligibly different from the concentration of the base used to prepare the buffer, [A$^-$]$_{\text{initial}}$. Because the two initial concentrations are the same,

$$K_a \approx \frac{[H_3O^+][A^-]_{\text{initial}}}{[HA]_{\text{initial}}} \overset{\substack{\textit{Cancel}\\ \textit{blue}\\ \textit{terms}}}{=} [H_3O^+]$$

It follows by taking the negative logarithm of both sides that when [HA]$_{\text{initial}}$ = [A$^-$]$_{\text{initial}}$,

$$pH = pK_a \qquad \textbf{(1)}$$

This very simple result makes it easy to make an initial choice of a buffer: just select an acid that has a pK_a close to the desired pH and prepare an equimolar solution with its conjugate base.

Mixtures in which the conjugate acid and base have unequal concentrations—such as those in Examples 6G.1 and 6G.2—are buffers, but they may be less effective than those in which the concentrations are nearly equal. Table 6G.1 lists some typical buffer systems.

The pH at which a mixture acts as an acid buffer can be lowered by adding more of the weak acid. The pH can also be decreased by adding some strong acid to convert some of the conjugate base to weak acid. To raise the pH at which a mixture acts as an acid buffer, the concentration of the acid's conjugate base can be increased by adding more salt (which introduces more base A^-). Alternatively, some strong base could be added to convert some of the acid to the salt.

It is often helpful to make a quick estimate of the buffer pH by employing a form of the expression for K_a that gives the pH directly for any composition of the mixture. Thus, for the equilibrium in reaction A, rearrange the expression for K_a into

$$[H_3O^+] = K_a \times \frac{[HA]}{[A^-]}$$

from which it follows, by taking the negative logarithms of both sides, that

$$\underbrace{-\log[H_3O^+]}_{\text{pH}} = \underbrace{-\log K_a}_{\text{p}K_a} - \log\frac{[HA]}{[A^-]}$$

Then, from $\log x = -\log(1/x)$,

$$pH = pK_a - \log\frac{[HA]}{[A^-]} = pK_a + \log\frac{[A^-]}{[HA]}$$

As discussed earlier, $[HA]$ can be approximated by $[HA]_{initial}$ (which is written as $[acid]_{initial}$) and $[A^-]$ by $[A^-]_{initial}$ (which is written as $[base]_{initial}$); the result is the **Henderson–Hasselbalch equation:**

$$pH \approx pK_a + \log\frac{[base]_{initial}}{[acid]_{initial}} \qquad (2)$$

For an acetic acid/acetate buffer the expression takes the form

$$pH \approx pK_a(CH_3COOH) + \log\frac{[CH_3CO_2^-]_{initial}}{[CH_3COOH]_{initial}}$$

Equation 2 can also be used for a base buffer, with pK_a that of the conjugate acid of the base. For example, in the case of an ammonia buffer, the pK_a of NH_4^+ would be used and "base" identified with NH_3 and "acid" with NH_4^+. If only pK_b is available, pK_a is calculated by using Eq. 5b of Topic 6C ($pK_a + pK_b = pK_w$). Therefore, for an ammonia/ammonium buffer write

$$pH \approx pK_a(NH_4^+) + \log\frac{[NH_3]_{initial}}{[NH_4^+]_{initial}}$$

In practice, the Henderson–Hasselbalch equation is used to make rapid estimates of the pH of a mixed solution intended to be used as a buffer, and then the pH is adjusted to the precise value required by adding more acid or base and monitoring the pH of the solution with a pH meter.

EXAMPLE 6G.3   Selecting the composition for a buffer solution with a given pH

Carbonate and hydrogen carbonate (bicarbonate) ions contribute to buffering in a variety of natural systems. You are investigating their role in groundwater percolating through limestone hills into a recently discovered cave system, and need to understand how the pH of the water is controlled. Calculate the ratio of the molar concentrations of CO_3^{2-} and HCO_3^- ions required to achieve buffering at pH = 9.50. The pK_{a2} of H_2CO_3 is 10.25.

ANTICIPATE Because the desired pH is lower than the pK_{a2} of H_2CO_3, you need the logarithm in the Henderson–Hasselbalch equation to be negative. That will be the case when the ratio $[base]_{initial}/[acid]_{initial}$ is less than 1.

PLAN Rearrange the Henderson–Hasselbalch equation to solve for the ratio of the weak acid to its conjugate base.

What should you assume? As usual, assume that activities can be approximated by molar concentrations. You should also assume that the equilibrium concentrations of acid and its conjugate base can be approximated by their initial values.

SOLVE In the equilibrium $HCO_3^-(aq) + H_2O(l) \rightleftharpoons H_3O^+(aq) + CO_3^{2-}(aq)$ the acid is HCO_3^- and its conjugate base is CO_3^{2-}. Ignore any H_2CO_3 formed (see Topic 6E).

From $pH \approx pK_a + \log([base]_{initial}/[acid]_{initial})$,

$$\log\frac{[base]_{initial}}{[acid]_{initial}} \approx pH - pK_a$$

Use the initial concentration of CO_3^{2-} for the base and that of HCO_3^- for the acid and substitute the values of pH and pK_a.

$$\log\frac{[CO_3^{2-}]_{initial}}{[HCO_3^-]_{initial}} \approx 9.50 - 10.25 = -0.75$$

Now use $x = 10^{\log x}$:

$$\frac{[CO_3^{2-}]_{initial}}{[HCO_3^-]_{initial}} \approx 10^{-0.75} = 0.18$$

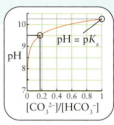

EVALUATE The solution acts as a buffer with a pH close to 9.50 if it is prepared by mixing the solutes in the ratio 0.18 mol CO_3^{2-} to 1.0 mol HCO_3^-. As expected, the ratio of concentrations is less than 1.

Self-test 6G.4A Calculate the ratio of the molar concentrations of acetate ions and acetic acid needed to buffer a solution at pH = 5.25. The pK_a of CH_3COOH is 4.75.

[***Answer: 3.16***]

Self-test 6G.4B Calculate the ratio of the molar concentrations of benzoate ions and benzoic acid (C_6H_5COOH) needed to buffer a solution at pH = 3.50. The pK_a of C_6H_5COOH is 4.19.

Related Exercises 6G.15, 6G.16

The pH of a buffer solution is close to the pK_a of the weak acid component when the acid and base have similar concentrations.

6G.3 Buffer Capacity

Just as a sponge can supply or hold only so much water, a buffer can supply or absorb only so many protons. Its proton sources and sinks become exhausted if too much strong acid or base is added to the solution. **Buffer capacity** is the maximum amount of acid or base that can be added before the buffer loses its ability to resist large changes in pH. A buffer with a high capacity is able to maintain its buffering action against the addition of strong acid or base better than can one with only a small capacity. The buffer is exhausted when most of the weak base has been converted into its conjugate acid or when most of the weak acid has been converted into its conjugate base. A concentrated buffer solution has a greater capacity than the same volume of a more dilute solution of the same buffer.

Buffer capacity also depends on the relative concentrations of weak acid and base. Broadly speaking, a buffer is found experimentally to have a high capacity for stabilizing against the addition of acid when the amount of weak base present is at least 10% of the amount of weak acid. Otherwise, the weak base is used up quickly as strong acid is added. Similarly, a buffer has a high capacity for stabilizing against the addition of base when the amount of weak acid present is at least 10% of the amount of weak base, because otherwise the weak acid is used up quickly as strong base is added.

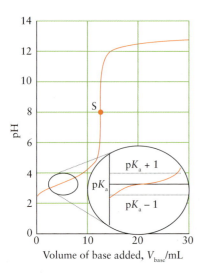

FIGURE 6G.2 This plot shows how the pH of a weak acid changes as a strong base is added. When the conjugate acid and base are present at similar concentrations, the curve shows that the pH changes very little as additional strong base or strong acid is added. As the inset shows, the pH lies between $pK_a \pm 1$ across the buffer region. S denotes the stoichiometric point (*Fundamentals* L).

These percentages can be used to express the effective range of buffer action in terms of the pH of the solution. The Henderson–Hasselbalch equation shows that, when the acid is 10 times as abundant as the base ([acid] = 10[base]), the pH of the solution is

$$pH = pK_a + \log\frac{[\text{base}]}{10[\text{base}]} = pK_a + \log\overset{-1}{\frac{1}{10}} = pK_a - 1 \qquad (3a)$$

Likewise, when the base is 10 times as abundant as the acid ([base] = 10[acid]), the pH is

$$pH = pK_a + \log\frac{10[\text{acid}]}{[\text{acid}]} = pK_a + \log\overset{+1}{10} = pK_a + 1 \qquad (3b)$$

Thus, the experimentally determined concentration range converts into a pH range of ± 1. That is, the buffer acts most effectively when the pH is within a range of ± 1 unit of pK_a (**FIG. 6G.2**). For instance, because the pK_a of $H_2PO_4^-$ is 7.21, a KH_2PO_4/K_2HPO_4 buffer can be expected to be most effective between about pH = 6.2 and pH = 8.2.

The composition of blood plasma, in which the concentration of HCO_3^- ions is about 20 times that of H_2CO_3, seems to be outside the range for optimum buffering. However, the principal waste products of living cells are carboxylic acids, such as lactic acid. Plasma, with its relatively high concentration of HCO_3^- ions, can absorb a significant surge of hydrogen ions from these carboxylic acids. The high proportion of HCO_3^- also helps us to withstand disturbances that lead to excess acid, such as disease and shock due to burns (**BOX 6G.1**).

Box 6G.1 WHAT HAS THIS TO DO WITH . . . STAYING ALIVE?

PHYSIOLOGICAL BUFFERS*

Buffer systems are so vital to the existence of living organisms that the most immediate threat to the survival of a person with severe injury or burns is a change in blood pH. Metabolic processes normally maintain the pH of human blood within a narrow range (7.35–7.45). To control blood pH, the body uses primarily the carbonic acid/hydrogen carbonate (bicarbonate) ion system. The normal ratio of HCO_3^- to H_2CO_3 in the blood is 20:1, with most of the carbonic acid in the form of dissolved CO_2. When the concentration of HCO_3^- increases further relative to that of H_2CO_3, blood pH rises. If the pH rises above the normal range, the condition is called *alkalosis*. Conversely, blood pH decreases when the ratio decreases; and when blood pH falls below the normal range, the condition is called *acidosis*. Because these conditions are life threatening and death can result within minutes, it is critical that the cause of any pH imbalance be identified and treated quickly.

The body maintains blood pH by two primary mechanisms: respiration and excretion. Carbonic acid concentration is controlled by respiration: as you exhale, you deplete your system of CO_2 and hence deplete it of H_2CO_3, too. This decrease in acid concentration raises the blood pH. Breathing faster and more deeply increases the amount of CO_2 exhaled and hence decreases the carbonic acid concentration in the blood, which in turn raises the blood pH. Hydrogen carbonate ion concentration is controlled by its rate of excretion in urine.

Respiratory acidosis results when decreased respiration resulting from conditions such as asthma, pneumonia, and emphysema or from inhaling smoke, raises the concentration of CO_2 in the blood and hence that of H_2CO_3. Respiratory acidosis is usually treated with a mechanical ventilator to assist the victim's breathing. The improved exhalation increases the excretion of CO_2 and raises blood pH. In many cases of asthma, chemicals can facilitate respiration by opening constricted bronchial passages.

Metabolic acidosis is caused by the release into the bloodstream of excessive amounts of lactic acid and other acidic byproducts of metabolism. These acids enter the bloodstream, react with hydrogen carbonate ion to produce H_2CO_3, and shift the ratio HCO_3^-/H_2CO_3 to a lower value. Heavy exercise, diabetes, and fasting can all produce metabolic acidosis. The normal response of the body is to increase the rate of breathing to eliminate some of the CO_2. Thus, you pant heavily when running uphill.

Metabolic acidosis can also result when a person is severely burned. Blood plasma leaks from the circulatory system into the injured area, producing edema (swelling) and reducing the blood volume. If the burned area is large, this loss of blood volume may be sufficient to reduce blood flow and oxygen supply to all the body's tissues. Lack of oxygen, in turn, causes the tissues to produce an excessive amount of lactic acid and leads to

---

*This box includes contributions from B. A. Pruitt, M.D., and A. D. Mason, M.D., U.S. Army Institute of Surgical Research.

metabolic acidosis. To minimize the decrease in pH, the injured person breathes harder to eliminate the excess CO_2. However, if blood volume drops below levels for which the body can compensate, a vicious circle ensues in which blood flow decreases still further, blood pressure falls, CO_2 excretion diminishes, and acidosis becomes more severe. People in this state are said to be in shock and will die if not treated promptly.

The dangers of shock are avoided or treated by intravenous infusion of large volumes of a salt-containing solution that is isotonic with blood (has the same osmotic pressure as blood, Topic 5F), usually one known as *lactated Ringer's solution*. The added liquid increases blood volume and blood flow, thereby improving oxygen delivery. The $[HCO_3^-]/[H_2CO_3]$ ratio then increases toward normal and allows the severely injured person to survive. Often, one of a paramedic's first steps in saving a life is to administer intravenous fluids.

Respiratory alkalosis is the rise in pH associated with excessive respiration. Hyperventilation, which can result from anxiety or high fever, is a common cause. The body may control blood pH during hyperventilation by fainting, which results in slower respiration. An intervention that may prevent fainting is to have a hyperventilating person breathe into a paper bag, which allows much of the respired CO_2 to be taken up again.

Metabolic alkalosis is the increase in pH resulting from illness or chemical ingestion. Repeated vomiting or the overuse of diuretics can cause metabolic alkalosis. Once again the body compensates, this time by decreasing the rate of respiration.

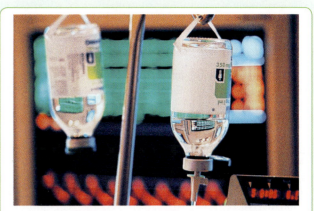

Patients who have suffered traumatic injuries must receive an immediate intravenous solution to combat the symptoms of shock and help maintain the pH of blood. *(Ben Edwards/The Image Bank/Getty Images.)*

Related Exercises 6.39 and 6.40

Further Reading J. C. Charles and R. L. Heilman, "Metabolic acidosis," *Hospital Physician,* March 2005, pp. 37–42. J. Squires, "Artificial blood," *Science,* vol. 295, Feb. 8, 2002, pp. 1002–1005. C. G. Morris and J. Low, "Metabolic acidosis in the critically ill: Part 2. Causes and treatment," *Anaesthesia,* vol. 63, 2008, pp. 396–411.

The capacity of a buffer is determined by its concentration and pH. A more concentrated buffer can react with more added acid or base than can a less concentrated one. A buffer solution is generally most effective when the pH is in the range $pK_a \pm 1$.

What have you learned in this Topic?

You have learned that solutions containing weak acids and their conjugate bases resist changes in pH when additional acid or base is added. You have also seen that a solution has optimal buffering ability when the conjugate acid and base have the same concentration. In addition you have seen that buffer solutions with higher concentrations can absorb more added acid or base than similar solutions of lower concentration.

The skills you have mastered are the ability to:

☐ **1.** Calculate the pH of a buffer solution (Example 6G.1).

☐ **2.** Calculate the change in pH of a buffer solution when acid or base has been added (Example 6G.2).

☐ **3.** Calculate the composition of a buffer solution for a desired pH (Example 6G.3).

☐ **4.** Select an appropriate buffer for a desired pH that effectively resists changes in pH (Sections 6G.2 and 6G.3).

Topic 6G Exercises

Refer to tables 6C.1, 6C.2, and 6E.1 for the values of K_a and K_b. Assume 25 °C throughout.

6G.1 Explain what happens to (a) the concentration of H_3O^+ ions in an acetic acid solution when solid sodium acetate is added; (b) the percentage deprotonation of benzoic acid in a benzoic acid solution when hydrochloric acid is added; (c) the pH of the solution when solid ammonium chloride is added to aqueous ammonia.

6G.2 Explain what happens to (a) the pH of a solution of phosphoric acid after the addition of solid sodium dihydrogen phosphate; (b) the percentage deprotonation of HCN in a hydrocyanic acid solution after the addition of hydrobromic acid; (c) the concentration of H_3O^+ ions when pyridinium chloride is added to an aqueous solution of the base pyridine.

6G.3 A solution of equal molar concentrations of glyceric acid and sodium glycerate was found to have pH = 3.52. (a) What are pK_a and K_a for glyceric acid? (b) What would the pH be if the concentration of acid was twice that of the salt?

6G.4 A solution of equal molar concentrations of the artificial sweetener saccharin and its sodium salt was found to have pH = 3.08. (a) What are the values of pK_a and K_a of saccharin? (b) What would the pH be if the concentration of acid was twice that of the salt?

6G.5 What is the concentration of hydronium ions in (a) a solution that is 0.075 M HCN(aq) and 0.060 M NaCN(aq); (b) a solution that is 0.20 M NH_2NH_2(aq) and 0.30 M NaCl(aq); (c) a solution that is 0.015 M HCN(aq) and 0.030 M NaCN(aq); (d) a solution that is 0.125 M NH_2NH_2(aq) and 0.125 M NH_2NH_3Br(aq)?

6G.6 What is the concentration of hydronium ions in (a) a solution that is 0.12 M HBrO(aq) and 0.160 M NaBrO(aq); (b) a solution that is 0.250 M CH_3NH_2(aq) and 0.150 M CH_3NH_3Cl(aq); (c) a solution that is 0.160 M HBrO(aq) and 0.320 M NaBrO(aq); (d) a solution that is 0.250 M CH_3NH_2(aq) and 0.250 M NaBr(aq)?

6G.7 Calculate the pH and pOH of (a) a solution that is 0.50 M $NaHSO_4$(aq) and 0.25 M Na_2SO_4(aq); (b) a solution that is 0.50 M $NaHSO_4$(aq) and 0.10 M Na_2SO_4(aq); (c) a solution that is 0.50 M $NaHSO_4$(aq) and 0.50 M Na_2SO_4(aq).

6G.8 Calculate the pH and pOH of (a) a solution that is 0.23 M Na_2HPO_4(aq) and 0.18 M Na_3PO_4(aq); (b) a solution that is 0.45 M Na_2HPO_4(aq) and 0.62 M Na_3PO_4(aq); (c) a solution that is 0.16 M Na_2HPO_4(aq) and 0.16 M Na_3PO_4(aq).

6G.9 Calculate the pH of the solution that results from mixing (a) 30.0 mL of 0.050 M HCN(aq) with 70.0 mL of 0.030 M NaCN(aq); (b) 40.0 mL of 0.030 M HCN(aq) with 60.0 mL of 0.050 M NaCN(aq); (c) 25.0 mL of 0.105 M HCN(aq) with 25.0 mL of 0.105 M NaCN(aq).

6G.10 Calculate the pH of the solution that results from mixing (a) 0.100 L of 0.050 M $(CH_3)_2NH$(aq) with 0.280 L of 0.040 M $(CH_3)_2NH_2Cl$(aq); (b) 45.0 mL of 0.015 M $(CH_3)_2NH$(aq) with 86.0 mL of 0.200 M $(CH_3)_2NH_2Cl$(aq); (c) 100.0 mL of 0.080 M $(CH_3)_2NH$(aq) with 40.0 mL of 0.075 M $(CH_3)_2NH_2Cl$(aq).

6G.11 A buffer solution of volume 100.0 mL is 0.100 M CH_3COOH(aq) and 0.100 M $NaCH_3CO_2$(aq). (a) What are the pH and the pH change resulting from the addition of 10.0 mL of 0.950 M NaOH(aq) to the buffer solution? (b) What are the pH and the pH change resulting from the addition of 20.0 mL of 0.100 M HNO_3(aq) to the initial buffer solution?

6G.12 A buffer solution of volume 100.0 mL is 0.140 M Na_2HPO_4(aq) and 0.120 M KH_2PO_4(aq). (a) What are the pH and the pH change resulting from the addition of 75.0 mL of 0.0100 M NaOH(aq) to the buffer solution? (b) What are the pH and the pH change resulting from the addition of 10.0 mL of 0.50 M HNO_3(aq) to the initial buffer solution?

6G.13 The pH of 0.40 M HF(aq) is 1.93. Calculate the change in pH when 0.356 g of sodium fluoride is added to 50.0 mL of the solution. Ignore any change in volume.

6G.14 The pH of 0.50 M HBrO(aq) is 4.50. Calculate the change in pH when 5.10 g of sodium hypobromite is added to 100. mL of the solution. Ignore any change in volume.

6G.15 Sodium hypochlorite, NaClO, is the active ingredient in many bleaches. Calculate the ratio of the concentrations of ClO^- and HClO in a bleach solution having a pH adjusted to 6.50 by the use of strong acid or strong base.

6G.16 Aspirin (acetylsalicylic acid, $K_a = 3.2 \times 10^{-4}$) is a product of the reaction of salicylic acid with acetic anhydride. Calculate the ratio of the concentrations of the acetylsalicylate ion to acetylsalicylic acid in a solution that has a pH adjusted to 4.13 by the use of strong acid or strong base.

6G.17 Predict the pH region in which each of the following buffers will be effective, assuming equal molar concentrations of the acid and its conjugate base: (a) sodium lactate and lactic acid; (b) sodium benzoate and benzoic acid; (c) potassium hydrogen phosphate and potassium phosphate; (d) potassium hydrogen phosphate and potassium dihydrogen phosphate; (e) hydroxylamine and hydroxylammonium chloride.

6G.18 Predict the pH region in which each of the following buffers will be effective, assuming equal molar concentrations of the acid and its conjugate base: (a) sodium nitrite and nitrous acid; (b) sodium formate and formic acid; (c) sodium carbonate and sodium hydrogen carbonate; (d) ammonia and ammonium chloride; (e) pyridine and pyridinium chloride.

6G.19 Suggest a conjugate acid–base system that would be an effective buffer at a pH close to (a) 2; (b) 7; (c) 3; (d) 12.

6G.20 Suggest a conjugate acid–base system that would be an effective buffer at a pH close to (a) 4; (b) 9; (c) 5; (d) 11.

6G.21 (a) What must be the ratio of the molar concentrations of CO_3^{2-} and HCO_3^- ions in a buffer solution having a pH of 11.0? (b) What mass of K_2CO_3 must be added to 1.00 L of 0.100 M $KHCO_3$(aq) to prepare a buffer solution with a pH of 11.0? (c) What mass of $KHCO_3$ must be added to 1.00 L of 0.100 M K_2CO_3(aq) to prepare a buffer solution with a pH of 11.0? (d) What volume of 0.200 M K_2CO_3(aq) must be added to 100. mL of 0.100 M $KHCO_3$(aq) to prepare a buffer solution with a pH of 11.0?

6G.22 (a) What must be the ratio of the molar concentrations of PO_4^{3-} and HPO_4^{2-} ions in a buffer solution having a pH of 12.0? (b) What mass of K_3PO_4 must be added to 1.00 L of 0.100 M K_2HPO_4(aq) to prepare a buffer solution with a pH of 12.0? (c) What mass of K_2HPO_4 must be added to 1.00 L of 0.100 M K_3PO_4(aq) to prepare a buffer solution with a pH of 12.0? (d) What volume of 0.150 M K_3PO_4(aq) must be added to 50.0 mL of 0.100 M K_2HPO_4(aq) to prepare a buffer solution with a pH of 12.0?

Topic 6H Acid–Base Titrations

6H.1 Strong Acid–Strong Base Titrations
6H.2 Strong Acid–Weak Base and Weak Acid–Strong Base Titrations
6H.3 Acid–Base Indicators
6H.4 Polyprotic Acid Titrations

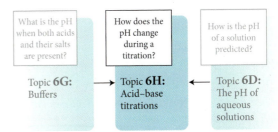

Topic **6G:** Buffers → Topic **6H:** Acid–base titrations ← Topic **6D:** The pH of aqueous solutions

Why Do You Need to Know This Material? The ability to calculate the pH during acid–base titrations and the knowledge of how to use indicators enables you to interpret the results when performing this widely used laboratory procedure.

What Do You Need to Know Already? You need to know how to calculate the pH of solutions of strong acids and bases (Topic 6B), weak acids and bases (Topic 6D), and buffers (Topic 6G). You also need to be familiar in Section 6H.4 with the proton transfer equilibria of polyprotic acids (Topic 6E). You should be familiar with the general procedure of titration set out in *Fundamentals* L.

The analytical technique of **titration** involves adding a solution, called the **titrant,** from a buret to a flask containing a sample called the **analyte.** For example, if an environmental chemist is monitoring acid mine drainage and needs to know the concentration of acid in the water, a sample of the effluent from the mine would be the analyte and a solution of base of known concentration would be the titrant. At the **stoichiometric point** in an acid–base titration, the amount of OH^- (or H_3O^+) added as titrant is equal to the amount of H_3O^+ (or OH^-) initially present in the analyte. The techniques in this Topic are used to identify the roles of different species in determining the pH of the analyte solution at any stage of a titration and to select the appropriate indicator for the titration.

6H.1 Strong Acid–Strong Base Titrations

When a strong acid is added to an aqueous solution of a strong base, a neutralization reaction occurs for which the net ionic equation is

$$H_3O^+(aq) + OH^-(aq) \longrightarrow 2 H_2O(l)$$

However, to ensure the correct stoichiometry when working with titrations, it is best to use the full chemical equation. For example, if hydrochloric acid is used to neutralize $Ca(OH)_2$, the fact that each formula unit of $Ca(OH)_2$ provides two OH^- ions must be taken into account:

$$2 HCl(aq) + Ca(OH)_2(aq) \longrightarrow CaCl_2(aq) + 2 H_2O(l)$$

A **pH curve** is a plot of the pH of the analyte solution against the volume of titrant added during a titration. The shape of the pH curve in **FIG. 6H.1** is typical of titrations in which a strong acid is added to a strong base. Initially, the pH falls slowly. Then, near the stoichiometric point, there is a sudden decrease in pH through 7. At this point, an indicator changes color or an automatic titrator responds electronically to the sudden change in pH. Titrations typically end at this point. However, if the titration is continued, the pH falls slowly toward the value of the acid itself as the dilution due to the original analyte solution becomes less and less important. How to find the values of the pH at any stage is set out in **Toolbox 6H.1**.

FIGURE 6H.2 shows a pH curve for a titration in which the analyte is a strong acid and the titrant is a strong base. This curve is the mirror image of the curve for the titration of a strong base with a strong acid.

The earlier term "equivalence point" is still widely used to refer to the stoichiometric point.

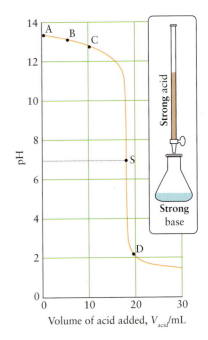

FIGURE 6H.1 The variation of pH during the titration of a strong base, 25.00 mL of 0.250 M NaOH(aq), with a strong acid, 0.340 M HCl(aq). The stoichiometric point (S) occurs at pH = 7. The other points on the pH curve are explained in Example 6H.1.

Toolbox 6H.1 HOW TO CALCULATE THE pH DURING A STRONG ACID–STRONG BASE TITRATION

CONCEPTUAL BASIS

The pH at any point in the titration of a strong acid with a strong base is governed by the major species in solution at that point.

PROCEDURE

First, use the reaction stoichiometry to find the amount of excess acid or base at the point of interest.

Step 1 Calculate the amount of H_3O^+ ions (if the analyte is a strong acid) or OH^- ions (if the analyte is a strong base) in the original analyte solution from $n_J = V[J]$, where J is H_3O^+ or OH^-.

Step 2 Calculate the amount of OH^- ions (if the titrant is a strong base) or H_3O^+ ions (if the titrant is a strong acid) in the volume of titrant added from $n_J = V[J]$, where J is OH^- or H_3O^+.

Step 3 To find the amount of analyte remaining after reaction, write the chemical equation for the neutralization and use the reaction stoichiometry to find the amount of H_3O^+ ions (or OH^- ions if the analyte is a strong base) that has reacted with the added titrant. Then subtract from the initial amount the amount of H_3O^+ or OH^- ions that has reacted.

Next, determine the concentration.

Step 4 Use the remaining amount of H_3O^+ (or OH^-) and the total volume of the combined solutions, $V = V_{analyte} + V_{titrant}$, to find the molar concentration of the unreacted H_3O^+ (or OH^-) ions in the solution from $[J] = n_J/V$.

Finally, calculate the pH.

Step 5 If acid is in excess, use $pH = -\log[H_3O^+]$. If base is in excess, find the pOH, and then convert pOH into pH by using the relation $pH + pOH = pK_w$.

This procedure is illustrated in Example 6H.1.

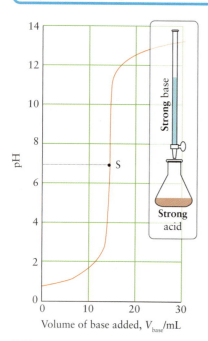

FIGURE 6H.2 The variation of pH during a typical titration of a strong acid (the analyte) with a strong base (the titrant). The stoichiometric point (S) occurs at pH = 7.

EXAMPLE 6H.1   Calculating points on the pH curve for a strong acid–strong base titration

You want to check the accuracy of a pH meter used for titrations and so calculate the expected pH at various points in a titration. The analyte consists of 25.00 mL of 0.250 M NaOH(aq), and the titrant is 0.340 M HCl(aq). Calculate (a) the pH of the original analyte solution and (b) the pH after the addition of 5.00 mL of the acid titrant.

ANTICIPATE You should expect the pH to decrease from its initial value as acid is added.

PLAN For part (a), determine the pOH of the solution and convert it to the pH. For part (b), follow the procedure outlined in Toolbox 6H.1.

What should you assume? Assume that so much acid or base is present that the autoprotolysis of water does not contribute significantly to the pH.

SOLVE (a) Initially, the pOH of the analyte is $pOH = -\log 0.250 = 0.602$, and so the pH of the solution is $pH = 14.00 - 0.602 = 13.40$. This is point A in Fig. 6H.1.
(b) Now follow the procedure in Toolbox 6H.1.

Step 1 Calculate the amount of OH^- ions in the original analyte solution from $n_J = V[J]$, where J is OH^-.

$$n(OH^-) = (25.00 \times 10^{-3}\,L) \times (0.250\ mol \cdot L^{-1})$$
$$= 6.25 \times 10^{-3}\,mol$$
$$= 6.25\ mmol$$

6.25 mmol OH^-

Step 2 Calculate the amount of H_3O^+ ions in the volume of titrant added from $n_J = V[J]$, where J is H_3O^+.

$$n(H_3O^+) = (5.00 \times 10^{-3}\,L) \times (0.340\ mol \cdot L^{-1})$$
$$= 1.70 \times 10^{-3}\,mol = 1.70\ mmol$$

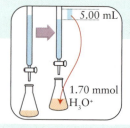

5.00 mL

1.70 mmol H_3O^+

Step 3 Write the chemical equation for the neutralization.

$$HCl(aq) + NaOH(aq) \longrightarrow NaCl(aq) + H_2O(l)$$

Use the reaction stoichiometry, 1 mol NaOH ≏ 1 mol HCl in the form 1 mol OH^- ≏ 1 mol H_3O^+ to find the amount of OH^- ions that has reacted with the added titrant:

$$n(OH^-) = 1.70\ mmol\ H_3O^+ \times \frac{1\ mol\ OH^-}{1\ mol\ H_3O^+} = 1.70\ mmol\ OH^-$$

To find the amount of OH⁻ ions that remains in the analyte solution, subtract the amount of OH⁻ ions that has reacted from the initial amount.

$$n(OH^-)_{final} = 6.25 - 1.70 \text{ mmol}$$
$$= 4.55 \text{ mmol}$$

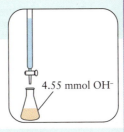

4.55 mmol OH⁻

Step 4 Use the remaining amount of OH⁻ and the total volume of the combined solutions, $V = V_{analyte} + V_{titrant}$, to find the molar concentration of the OH⁻ ions in the solution from $[J] = n_J/V$.

$$[OH^-] = \frac{4.55 \times 10^{-3} \text{ mol}}{(25.00 + 5.00) \times 10^{-3} \text{ L}} = 0.151... \text{ mol·L}^{-1}$$

Step 5 As base is in excess, find the pOH, and then convert pOH into pH by using the relation $pH = pK_w - pOH$.

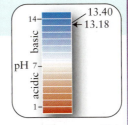

$$pOH = -\log(0.151...) = 0.82$$
$$pH = 14.00 - 0.82 = 13.18$$

This is point B in Fig. 6H.1.

EVALUATE Note that the pH has fallen, as expected, but only by a very small amount. This small change is consistent with the shallow slope of the pH curve at the start of the titration.

Self-test 6H.1A What is the pH of the solution that results from the addition of a further 5.00 mL of the HCl(aq) titrant to the analyte?

[**Answer:** 12.91, point C]

Self-test 6H.1B What is the pH of the solution that results from the addition of yet another 2.00 mL of titrant to the analyte?

Related Exercises 6H.3–6H.6

It is commonly observed that the pH changes abruptly close to the stoichiometric point in an acid–base titration. Suppose the stoichiometric point in the titration described in Example 6H.1 is reached and then an additional 1.00 mL of HCl(aq) is added. To find out by how much the pH changes, work through the steps in Toolbox 6H.1 as in Example 6H.1, except that now the acid is in excess. In this way it is found that after the addition, the pH has fallen to 2.1 (point D in Fig. 6H.1). This point is well below the pH (of 7) at the stoichiometric point, although only 1 mL more acid has been added.

In the titration of a strong acid with a strong base or of a strong base with a strong acid, the pH changes slowly initially, changes rapidly through pH = 7 at the stoichiometric point, and then changes slowly again.

6H.2 Strong Acid–Weak Base and Weak Acid–Strong Base Titrations

In many titrations, one solution—either the analyte or the titrant—contains a weak acid or base and the other solution contains a strong base or acid. For example, if you want to know the concentration of formic acid, the weak acid found in ant venom (**1**), you could titrate it with sodium hydroxide, a strong base. Alternatively, to find the concentration of ammonia, a weak base, in a solution, you might titrate it with hydrochloric acid, a strong acid. **FIGURES 6H.3** and **6H.4** show the different pH curves that are found experimentally for these two types of titrations. Notice that the stoichiometric point does not occur at pH = 7. Moreover, although the pH changes reasonably sharply near the stoichiometric point, it does not change as abruptly as it does in a strong acid–strong base titration. Weak acids are not normally titrated with weak bases because the stoichiometric point is too difficult to locate.

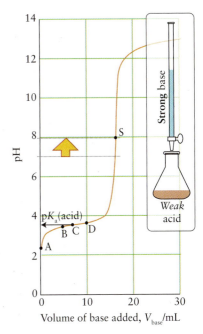

FIGURE 6H.3 The pH curve for the titration of a weak acid with a strong base: 25.00 mL of 0.100 M HCOOH(aq) with 0.150 M NaOH(aq). The stoichiometric point (S) occurs at pH > 7 because the anion HCO_2^- is a base. At point C, pH = pK_a. The other points on the curve are explained in Section 6H.2 and in Example 6H.3.

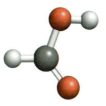

1 Formic acid, HCOOH

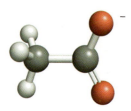

2 Acetate ion, $CH_3CO_2^-$

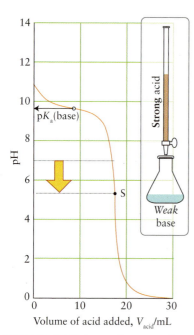

FIGURE 6H.4 A typical pH curve for the titration of a weak base with a strong acid. The stoichiometric point (S) occurs at pH < 7 because the salt formed by the neutralization reaction has an acidic cation. The pH after the addition of half of the volume of acid needed to reach the stoichiometric point S is equal to the pK_a of the conjugate acid of the base.

The pH at the stoichiometric point depends on the type of salt produced in the neutralization reaction. In the titration of acetic acid, CH_3OOH, with sodium hydroxide, the following reaction takes place:

$$CH_3OOH(aq) + NaOH(aq) \longrightarrow NaCH_3CO_2(aq) + H_2O(l)$$

At the stoichiometric point the solution consists of sodium acetate, $NaCH_3CO_2$, and water. Because Na^+ ions have virtually no effect on pH and the acetate ion, $CH_3CO_2^-$ (**2**), is a weak base, overall the solution is basic with pH > 7. The converse is true for the stoichiometric point of the titration of any weak acid with strong base. At the stoichiometric point of the titration of aqueous ammonia with hydrochloric acid, the solute is ammonium chloride. Because Cl^- ions have virtually no effect on the pH and NH_4^+ is an acid, expect pH < 7. The same is true for the stoichiometric point of the titration of any weak base with strong acid.

EXAMPLE 6H.2 Estimating the pH at the stoichiometric point of the titration of a weak acid with a strong base

Ant venom contains formic acid (HCOOH; "formica" is the Latin word for ant). Suppose you are at a pharmaceutical company working on a quick antidote and need to estimate the pH at the stoichiometric point when titrating a solution of formic acid. Estimate the pH at the stoichiometric point of the titration of 25.00 mL of 0.100 M HCOOH(aq) with 0.150 M NaOH(aq).

ANTICIPATE The formate ion produced in this titration is a base, so you should expect pH > 7.

PLAN To calculate the pH at the stoichiometric point, proceed as in Examples 6D.4 or 6D.5. Note that the amount of salt at the stoichiometric point is equal to the amount of titrant added and that the volume is the total volume of the combined analyte and titrant solutions. The K_b of a weak base is related to the K_a of its conjugate acid by $K_a \times K_b = K_w$; the K_a is listed in Table 6C.1.

What should you assume? Assume that the autoprotolysis of water has no significant effect on the pH and that because the formate ion is a very weak base, its concentration changes insignificantly when it is protonated by water.

SOLVE The salt present at the stoichiometric point is sodium formate, which provides basic formate ion. From Table 6C.1, $K_a = 1.8 \times 10^{-4}$ for formic acid; therefore, from Eq. 4 in Topic 6C, $K_b = K_w/K_a = 5.6 \times 10^{-11}$.

Find the initial amount of HCOOH in the analyte from $n_J = V[J]$.

$$n(HCOOH) = (2.500 \times 10^{-2} \text{ L}) \times (0.100 \text{ mol·L}^{-1})$$
$$= 2.50 \times 10^{-3} \text{ mol} = 2.50 \text{ mmol}$$

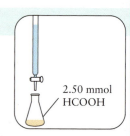

2.50 mmol
HCOOH

Find the amount of OH^- required to react with the HCOOH from the reaction stoichiometry, 1 mol $OH^- \widehat{=}$ 1 mol HCOOH.

$$n(OH^-) = (2.50 \times 10^{-3} \text{ mol HCOOH}) \times \frac{1 \text{ mol } OH^-}{1 \text{ mol HCOOH}}$$
$$= 2.50 \times 10^{-3} \text{ mol } OH^- = 2.50 \text{ mmol } OH^-$$

The amount of HCO_2^- in the solution at the stoichiometric point is the same as the amount of OH^- added.

$$n(HCO_2^-) = 2.50 \text{ mmol}$$

Find the volume of titrant containing this amount of OH^- ions from $V = n_J/[J]$.

$$V_{added} = \frac{2.5 \times 10^{-3} \text{ mol}}{0.150 \text{ mol·L}^{-1}}$$
$$= 1.67 \times 10^{-2} \text{ L} = 16.7 \text{ mL}$$

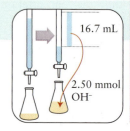

16.7 mL

2.50 mmol
OH^-

Find the total volume of the solution at the stoichiometric point from $V_{final} = V_{initial} + V_{added}$.

$$V_{final} = 25.00 + 16.7 \text{ mL} = 41.7 \text{ mL}$$

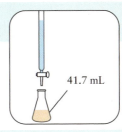

41.7 mL

Find the concentration of HCO_2^- ions at the stoichiometric point from $[J] = n_J/V_{final}$.

$$[HCO_2^-] = \frac{2.50 \times 10^{-3} \text{ mol}}{4.17 \times 10^{-2} \text{ L}} = 0.0600 \text{ mol·L}^{-1}$$

The pH of the solution can now be calculated as in Toolbox 6D.2.

Step 1 The equilibrium to consider is

$$HCO_2^-(aq) + H_2O(l) \rightleftharpoons HCOOH(aq) + OH^-(aq) \qquad K_b = \frac{[HCOOH][OH^-]}{[HCO_2^-]}$$

The equilibrium table, with all concentrations in moles per liter, is

	HCO_2^-	HCOOH	OH^-
initial concentration	0.0600	0	0
change in concentration	$-x$	$+x$	$+x$
equilibrium concentration	$0.0600 - x$	x	x

Step 2 From $K_b = [HCOOH][OH^-]/[HCO_2^-]$,

$$K_b = \frac{x \times x}{0.0600 - x}$$

Provided $x \ll 0.06$, the approximate form of this expression is:

$$K_b \approx \frac{x^2}{0.0600}$$

with $K_b = 5.6 \times 10^{-11}$.

Step 3 The solution is:

$$x \approx (0.0600 \times K_b)^{1/2} = (0.0600 \times 5.6 \times 10^{-11})^{1/2} = 1.8 \times 10^{-6}$$

It follows that the concentration of OH^- is 1.8 μmol·L^{-1}, which is about 18 times as great as the concentration of OH^- ions from the autoprotolysis of water (0.10 μmol·L^{-1}), and so ignoring the latter is reasonable. Because $x \ll 0.0600$, supposing that the initial concentration of formate ion is unchanged is also reasonable.

Step 4 Now write

$$pOH = -\log(1.8 \times 10^{-6}) = 5.74$$

Step 5 From $pH = pK_w - pOH$,

$$pH = 14.00 - 5.74 = 8.26 \text{ or about } 8.3$$

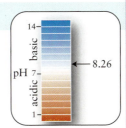

14—

basic

←— 8.26

pH 7—

acidic

1—

EVALUATE At the stoichiometric point, pH > 7, as expected.

Self-test 6H.2A Calculate the pH at the stoichiometric point of the titration of 25.00 mL of 0.010 M HClO(aq) with 0.020 M KOH(aq). See Table 6C.1 for K_a.

[*Answer:* 9.67]

Self-test 6H.2B Calculate the pH at the stoichiometric point of the titration of 25.00 mL of 0.020 M NH₃(aq) with 0.015 M HCl(aq). (For NH₄⁺, $K_a = 5.6 \times 10^{-10}$.)

Related Exercises 6H.9, 6H.10

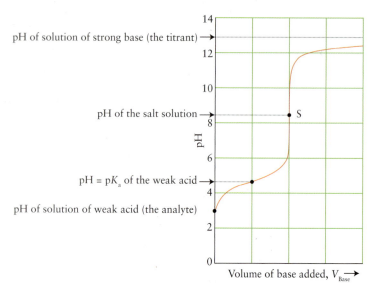

pH of solution of strong base (the titrant) →

pH of the salt solution →

pH = pK_a of the weak acid →

pH of solution of weak acid (the analyte)

S

Volume of base added, V_{Base} →

FIGURE 6H.5 The pK_a of an acid can be determined by carrying out a titration of the weak acid with a strong base and locating the pH of the solution after the addition of half the volume of base needed to reach the stoichiometric point S. The pH at that point is equal to the pK_a of the acid.

Now consider the overall shape of the pH curve. The slow change in pH about halfway to the stoichiometric point indicates that the solution acts as a buffer in that region (Topic 6G). At the halfway point of the titration, [HA] = [A⁻] and, as shown in Topic 6G, at that point pH = pK_a. The less steep slope of the curve near pH = pK_a illustrates the ability of a buffer solution to stabilize the pH. Moreover, you can now see how to determine pK_a: plot the pH curve during a titration, identify the pH halfway to the stoichiometric point, and set pK_a equal to that pH (**FIG. 6H.5**). To obtain the pK_b of a weak base, find pK_a of its conjugate acid in the same way but then find pK_b from pK_b = pK_w − pK_a. The values recorded in Tables 6C.1 and 6C.2 were obtained in this way.

After the stoichiometric point in the titration of a weak acid with a strong base, the pH depends only on the concentration of excess strong base. Note that in Fig. 6H.5 the pH well after the stoichiometric point approaches a constant value. For example, suppose you went on to add several liters of strong base from a giant buret. The presence of salt produced by the neutralization reaction would become negligible relative to the concentration of excess base. The pH would be that of the nearly pure titrant (the original base solution).

You have already seen how to estimate the pH of the initial analyte when only weak acid or weak base is present (point A in Fig. 6H.3, for instance), as well as the pH at the stoichiometric point (point S). Between these two points lie points corresponding to a mixed solution of some weak acid (or base) and some salt. The techniques described in **Toolbox 6H.2** and Example 6H.3 can then be used to account for the shape of the curve.

Toolbox 6H.2 HOW TO CALCULATE THE pH DURING A TITRATION OF A WEAK ACID OR A WEAK BASE

CONCEPTUAL BASIS

The pH is governed by the major solute species present in solution. As strong base is added to a solution of a weak acid, a salt of the conjugate base of the weak acid is formed. This salt affects the pH and needs to be taken into account. **TABLE 6H.1** outlines the regions encountered during a titration and the primary equilibrium to consider in each region.

PROCEDURE

The procedure is like that in Toolbox 6H.1, except that an additional step is required to calculate the pH from the proton transfer equilibrium. First use reaction stoichiometry to find the amount of excess acid or base. Begin by writing the chemical equation for the reaction, then:

Step 1 Calculate the amount of weak acid or base in the original analyte solution. Use $n_J = V_{analyte}[J]$.

Step 2 Calculate the amount of OH⁻ ions (or H_3O^+ ions if the titrant is an acid) in the volume of titrant added. Use $n_J = V_{titrant}[J]$.

Step 3 Use reaction stoichiometry to calculate the following amounts:

- Weak acid–strong base titration: the amount of conjugate base formed in the neutralization reaction, and the amount of weak acid remaining.

- Weak base–strong acid titration: the amount of conjugate acid formed in the neutralization reaction, and the amount of weak base remaining.

Calculate the concentrations.

Step 4 Find the "initial" molar concentrations of the conjugate acid and base in solution after neutralization, but before any proton transfer equilbrium with water is taken into account. Use [J] = n_J/V, where V is the total volume of the solution, $V = V_{analyte} + V_{titrant}$.

Calculate the pH.

Step 5 Use the expression for K_a or K_b to find the H_3O^+ concentration in a weak acid or the OH⁻ concentration in a weak base. Alternatively, if the concentrations of conjugate acid and base calculated in step 4 are both large relative to the concentration of hydronium ions, use them in the Henderson–Hasselbalch equation, Eq. 2 of Topic 6G, pH ≈ pK_a + log([base]$_{initial}$/[acid]$_{initial}$), to determine the pH. In each case, if the pH is less than 6 or greater than 8, assume that the autoprotolysis of water does not significantly affect the pH. If necessary, convert between K_a and K_b by using $K_a \times K_b = K_w$.

This procedure is illustrated in Example 6H.3.

TABLE 6H.1   Weak Acid and Weak Base Titration Equilibria

Point in titration	Primary species	Proton transfer equilibrium	Related Toolbox
1 Weak acid HA titrated with strong base			
initial	HA	$HA(aq) + H_2O(l) \rightleftharpoons H_3O^+(aq) + A^-(aq)$	6D.1
buffer region	HA, A^-	$HA(aq) + H_2O(l) \rightleftharpoons H_3O^+(aq) + A^-(aq)$	6H.2
stoichiometric point	A^-	$A^-(aq) + H_2O(l) \rightleftharpoons HA(aq) + OH^-(aq)$	6D.2
At the stoichiometric point, the solution is of a salt with a basic anion			
2 Weak base B titrated with strong acid			
initial	B	$B(aq) + H_2O(l) \rightleftharpoons HB^+(aq) + OH^-(aq)$	6D.2
buffer region	B, HB^+	$B(aq) + H_2O(l) \rightleftharpoons HB^+(aq) + OH^-(aq)$	6H.2
stoichiometric point	HB^+	$HB^+(aq) + H_2O(l) \rightleftharpoons H_3O^+(aq) + B(aq)$	6D.1
At the stoichiometric point, the solution is of a salt with an acidic anion			

EXAMPLE 6H.3   Calculating the pH before the stoichiometric point in a weak acid–strong base titration

As a part of your research program on formic acid, you need to titrate a solution of formic acid with sodium hydroxide solution and want to know what to expect. Calculate the pH of (a) 0.100 M HCOOH(aq) and (b) the solution resulting when 5.00 mL of 0.150 M NaOH(aq) is added to 25.00 mL of the acid. Use $K_a = 1.8 \times 10^{-4}$ for HCOOH.

ANTICIPATE When the strong base is added, some of the weak acid is neutralized, so you should expect the pH to rise from part (a) to part (b).

PLAN For part (a), use the procedure in Toolbox 6D.1. For part (b), find the pH using the procedure in Toolbox 6H.2.

What should you assume? Assume that the autoprotolysis of water does not contribute significantly to the pH and that the weak acid formic acid undergoes only a small degree of deprotonation.

SOLVE

(a) The proton transfer equilibrium is $HCOOH(aq) + H_2O(l) \rightleftharpoons H_3O^+(aq) + HCO_2^-(aq)$. From $[H_3O^+] \approx (K_a[HA]_{initial})^{1/2}$ and $pH = -\log[H_3O^+]$,

$$pH = -\log(1.8 \times 10^{-4} \times 0.100)^{1/2} = 2.37$$

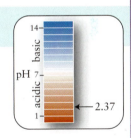

(b) The neutralization reaction is $HCOOH(aq) + OH^-(aq) \rightarrow HCO_2^-(aq) + H_2O(l)$.

Step 1 Calculate the amount of weak acid in the original analyte solution. Use $n_J = V_{analyte}[J]$.

$$n(HCOOH) = (2.500 \times 10^{-2}\,L) \times (0.100\,mol \cdot L^{-1})$$
$$= 2.50 \times 10^{-3}\,mol = 2.50\,mmol$$

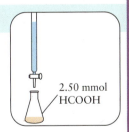

Step 2 Calculate the amount of OH^- in the volume of titrant added. Use $n_J = V_{titrant}[J]$.

$$n(OH^-) = (5.00 \times 10^{-3}\,L) \times (0.150\,mol \cdot L^{-1})$$
$$= 7.50 \times 10^{-4}\,mol = 0.750\,mmol$$

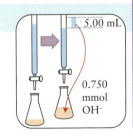

Step 3 Use the reaction stoichiometry, 1 mol OH$^-$ ≈ 1 mol HCO$_2^-$, to calculate the amount of conjugate base, HCO$_2^-$, formed in the neutralization reaction, and the amount of weak acid, HCOOH, remaining.

0.750 mmol OH$^-$ produces 0.750 mmol HCO$_2^-$ and leaves 2.50 − 0.750 mmol = 1.75 mmol HCOOH

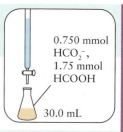

0.750 mmol HCO$_2^-$, 1.75 mmol HCOOH

30.0 mL

Step 4 Find the "initial" molar concentrations of the conjugate acid and base in solution after neutralization, but before any reaction with water is taken into account. Use [J] = n_J/V, where V is the total volume of the solution, $V = V_{analyte} + V_{titrant}$.

$$[\text{HCOOH}]_{initial} = \frac{1.75 \times 10^{-3}\ \text{mol}}{(25.00 + 5.00) \times 10^{-3}\ \text{L}} = 0.0583\ \text{mol·L}^{-1}$$

$$[\text{HCO}_2^-]_{initial} = \frac{7.50 \times 10^{-4}\ \text{mol}}{(25.00 + 5.00) \times 10^{-3}\ \text{L}} = 0.00250\ \text{mol·L}^{-1}$$

Step 5 Now evaluate the pH. The proton transfer equilibrium for HCOOH in water is

$$\text{HCOOH(aq)} + \text{H}_2\text{O(l)} \rightleftharpoons \text{H}_3\text{O}^+\text{(aq)} + \text{HCO}_2^-\text{(aq)}$$

Because the concentrations of conjugate acid and base calculated in step 4 are both large relative to the concentration of hydronium ions, it is appropriate to use them in the Henderson–Hasselbalch equation to determine the pH. From Table 6C.1, the pK_a for formic acid is 3.75.

From pH ≈ pK_a + log([HCO$_2^-$]$_{initial}$/[HCOOH]$_{initial}$),

$$\text{pH} \approx 3.75 + \log\frac{0.0250}{0.0583} = 3.38$$

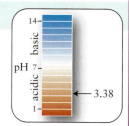

EVALUATE This pH corresponds to [H$_3$O$^+$] = 4.2 × 10^{-4} mol·L^{-1} point B in Fig. 6H.3, and, as expected, the contribution of autoprotolysis is negligible. As predicted, the pH of the mixed solution (3.38) is higher than that of the original acid (2.37).

Self-test 6H.3A Calculate the pH of the solution after the addition of another 5.00 mL of 0.150 M NaOH(aq).

[*Answer:* 3.93, point D in Fig. 6H.3]

Self-test 6H.3B Calculate the pH of the solution after the addition of yet another 5.00 mL of 0.150 M NaOH(aq).

Related Exercises 6H.11–6H.14, 6H.17, 6H.18

Figure 6H.5 summarizes the changes in pH of a solution during a titration of a weak acid by a strong base. Halfway to the stoichiometric point, the pH is equal to the pK$_a$ of the acid. The pH is greater than 7 at the stoichiometric point of the titration of a weak acid and strong base. The pH is less than 7 at the stoichiometric point of the titration of a weak base and strong acid.

6H.3 Acid–Base Indicators

A simple, reliable, and fast method of determining the pH of a solution and of monitoring a titration is with a pH meter, which uses a special electrode to measure H$_3$O$^+$ concentration (Topic 6N). An automatic titrator uses an integrated pH meter to monitor the pH of the analyte solution continuously (**FIG. 6H.6**). The stoichiometric point is detected by noting the characteristic rapid change in pH. Another common technique (but now largely superseded in commercial and research laboratories by the use of automatic titrators) is to use an indicator to detect the stoichiometric point. An **acid–base indicator** is a water-soluble dye

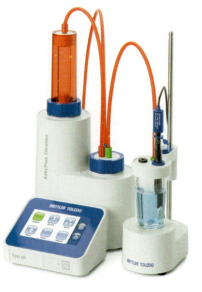

FIGURE 6H.6 A commercially available automatic titrator. The stoichiometric point of the titration is detected by the sudden change in pH that occurs in its vicinity; the pH is monitored electronically. (*Courtesy Mettler Toledo.*)

TABLE 6H.2 Indicator Color Changes*

Indicator	pK_{In}	pH range of color change	Color of acid form		Color of base form
thymol blue	1.7	1.2 to 2.8	red		yellow
methyl orange	3.4	3.2 to 4.4	red		yellow
bromophenol blue	3.9	3.0 to 4.6	yellow		blue
bromocresol green	4.7	3.8 to 5.4	yellow		blue
methyl red	5.0	4.8 to 6.0	red		yellow
litmus	6.5	5.0 to 8.0	red		blue
bromothymol blue	7.1	6.0 to 7.6	yellow		blue
phenol red	7.9	6.6 to 8.0	yellow		red
thymol blue	8.9	8.0 to 9.6	yellow		blue
phenolphthalein	9.4	8.2 to 10.0	colorless		pink
alizarin yellow R	11.2	10.1 to 12.0	yellow		red
alizarin	11.7	11.0 to 12.4	red		purple

*The colors of the acid and base forms are only a symbolic representation of the actual colors.

FIGURE 6H.7 The stoichiometric point of an acid–base titration may be detected by the color change of an indicator. Here we see the colors of solutions containing a few drops of phenolphthalein at (from left to right) pH of 7.0, 8.5, 9.4 (its end point), 9.8, and 13.0. At the end point, the concentrations of the conjugate acid and base forms of the indicator are equal. (©1991 Chip Clark–Fundamental Photographs.)

with a color that depends on the pH. The sudden change in pH that occurs at the stoichiometric point of a titration is signaled by a sharp change in color of the dye.

An acid–base indicator changes color with pH because it is a weak acid that has one color in its acid form (HIn, where In stands for indicator) and another color in its conjugate base form (In⁻). The color change results because the proton in HIn changes the structure of the molecule in such a way that the light absorption characteristics of HIn are different from those of In⁻. When the concentration of HIn is much greater than that of In⁻, the solution has the color of the acid form of the indicator. When the concentration of In⁻ is much greater than that of HIn, the solution has the color of the base form of the indicator.

Because it is a weak acid, an indicator takes part in a proton transfer equilibrium:

$$HIn(aq) + H_2O(l) \rightleftharpoons H_3O^+(aq) + In^-(aq) \qquad K_{In} = \frac{[H_3O^+][In^-]}{[HIn]}$$

The **end point** of an indicator is the point at which the concentrations of its acid and base forms are equal: $[HIn] = [In^-]$ and therefore $[H_3O^+]_{end\ point} = K_{In}$. That is, the color change occurs when

$$pH = pK_{In} \qquad (1)$$

The color starts to change perceptibly about 1 pH unit before pK_{In} and is effectively complete about 1 pH unit after pK_{In}. **TABLE 6H.2** gives the values of pK_{In} for a number of common indicators.

One common indicator is phenolphthalein (**FIG. 6H.7**). The acid form of this large molecule (**3**) is colorless; its conjugate base form (**4**) is pink. The structure of the base form of phenolphthalein allows electrons to be delocalized across all three of the benzenelike rings, and the increase in delocalization is part of the reason for the change in color. The pK_{In} of phenolphthalein is 9.4, and so the end point occurs in slightly basic solution. Litmus, another common indicator, has $pK_{In} = 6.5$; it is red for pH < 5 and blue for pH > 8.

There are many naturally occurring indicators. For instance, a single compound is responsible for the colors of red poppies and blue cornflowers: the pH of the sap is different in the two plants. The color of hydrangeas also depends on the acidity of their sap and can be controlled by modifying the acidity of the soil (**FIG. 6H.8**).

The *end point* is a property of the indicator; the *stoichiometric point* is a property of the chemical reaction taking place during the titration. It is important to select an

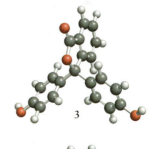

3

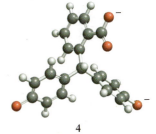

4

FIGURE 6H.8 The color of these hydrangeas depends on the acidity of the soil in which they are growing: acid soil results in blue flowers, and alkaline soil results in pink flowers. (© Darrell Gulin/Corbis.)

 **LAB VIDEO FIGURE 6H.8**

FIGURE 6H.9 Ideally, an indicator should have a sharp color change close to the stoichiometric point of the titration, which is at pH = 7 for a strong acid–strong base titration. However, the change in pH is so abrupt that phenolphthalein can be used. Phenolphthalein can also be used to detect the stoichiometric point of a weak acid–strong base titration, but methyl orange cannot. However, methyl orange can be used for a weak base–strong acid titration. Phenolphthalein would be inappropriate in this case, because its color change occurs well away from the stoichiometric point.

 **LAB VIDEO FIGURE 6H.9**

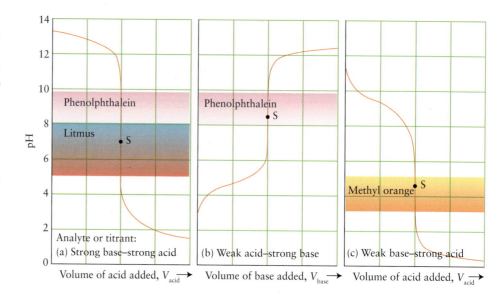

indicator with an end point close to the stoichiometric point of the titration (**FIG. 6H.9**). In practice, the pK_{In} of the indicator should be within about 1 pH unit of the stoichiometric point:

$$pK_{In} \approx pH(\text{at the stoichiometric point}) \pm 1$$

Phenolphthalein can be used for titrations with a stoichiometric point near pH = 9, such as a titration of a weak acid with a strong base. Methyl orange changes color between pH = 3.2 and pH = 4.4 and can be used in the titration of a weak base with a strong acid. Ideally, indicators for strong acid–strong base titrations should have end points close to pH = 7; however, in strong acid–strong base titrations, the pH changes rapidly over several pH units near the stoichiometric point, and even phenolphthalein can be used. Table 6H.2 includes the pH ranges over which several indicators can be used.

> *Acid–base indicators are weak acids that change color close to pH = pK_{In}; an indicator should be chosen so that its end point is close to the stoichiometric point of the titration.*

6H.4 Polyprotic Acid Titrations

Because many biological systems use polyprotic acids and their anions to control pH, it is important to be familiar with pH curves for polyprotic titrations and to be able to calculate the pH during such a titration. The titration of a polyprotic acid proceeds in the same way as that of a monoprotic acid, but there are as many stoichiometric points in the titration as there are acidic hydrogen atoms. It is essential to keep track of the major species in solution at each stage, as described in Topic 6E and summarized in Figs. 6E.1 and 6E.2.

Suppose you are titrating a solution of the dichloride salt of the amino acid histidine (**5**), $C_6H_{11}N_3O_2Cl_2(aq)$, with a solution of NaOH. The $C_6H_{11}N_3O_2^{2+}$ ion acts as a triprotic acid, which can be denoted H_3A^{2+}. The calculated pH curve, which matches the experimentally determined curve, is shown in **FIG. 6H.10**. Notice that there are three stoichiometric points (S_1, S_2, and S_3) and three buffer regions (B_1, B_2, and B_3). In pH calculations for these systems, assume that, as the hydroxide solution is added, initially NaOH reacts completely with the acid to form the diprotic conjugate base H_2A^+ (in this case, $C_6H_{10}N_3O_2^+$):

$$H_3A^{2+}(aq) + OH^-(aq) \longrightarrow H_2A^+(aq) + H_2O(l) \qquad \textbf{(A)}$$

At point B_1, the system is in the first buffer region and halfway to the first stoichiometric point, so pH = pK_{a1}. The pH in the first buffer region is determined by the proton transfer equilibrium between H_3A^{2+} ions and the H_2A^+ ions produced in the titration.

Once all the H_3A^{2+} ions have lost their first acidic protons, the system is at point S_1 and the primary species in solution are H_2A^+, Cl^- ions, and Na^+ ions. Point S_1 is the first stoichiometric point, and to reach it 1 mol NaOH needs to be supplied for each mole of H_3A^{2+}.

5   Histidine

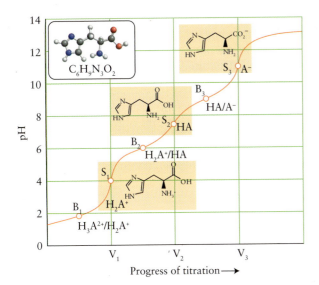

FIGURE 6H.10 The variation of the pH of the analyte solution during the titration of a triprotic acid (histidine). The major species present in solution at the three stoichiometric points (S_1, S_2, and S_3) and at points when half the titrant required to reach a stoichiometric point has been added (B_1, B_2, and B_3) are shown. The labels V_1, V_2, and V_3 denote the volumes of base required to reach the three stoichiometric points. Points A through F are explained in the text.

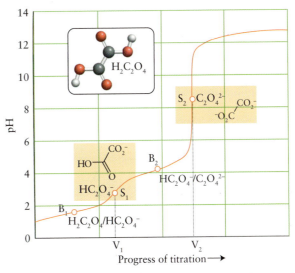

FIGURE 6H.11 The variation of the pH of the analyte solution during the titration of a diprotic acid (oxalic acid) and the major species present in solution at the two stoichiometric points (S_1 and S_2) and at points when half the titrant required to reach a stoichiometric point has been added (B_1 and B_2) are shown. The labels V_1 and V_2 denote the volumes of base required to reach the two stoichiometric points.

As more base is added, it reacts with the H_2A^+ ion to form its conjugate base, HA (in this case, $C_6H_9N_3O_2$):

$$H_2A^+(aq) + OH^-(aq) \longrightarrow HA(aq) + H_2O(l) \qquad \text{(B)}$$

At point B_2, the system is in the second buffer region and halfway to the second stoichiometric point, so pH = pK_{a2}. Addition of further base takes the titration to the second stoichiometric point, S_2. The primary species in solution are now HA molecules, Cl^- ions, and Na^+ ions. To reach the second stoichiometric point, an additional mole of NaOH is required for each mole of H_3A^{2+} ions originally present. A total of 2 mol NaOH for each mole of H_3A^{2+} ions has now been added.

Additional base reacts with HA to remove the last acidic proton and produce A^- (that is, $C_6H_8N_3O_2^-$):

$$HA(aq) + OH^-(aq) \longrightarrow A^-(aq) + H_2O(l) \qquad \text{(C)}$$

At point B_3, the system is in the third buffer region and halfway to the third stoichiometric point, so pH = pK_{a3}. When this reaction is complete, the primary species in solution are A^- ions, Cl^- ions, and Na^+ ions. To reach this stoichiometric point (labeled S_3 in the plot), another mole of OH^- must be added for each mole of H_3A^{2+} initially present as the salt. At this point, a total of 3 mol OH^- has been added for each mole of H_3A^{2+} that was originally present.

FIGURE 6H.11 shows the pH curve of a diprotic acid, such as oxalic acid, $H_2C_2O_4$. There are two stoichiometric points (S_1 and S_2) and two buffer regions (B_1 and B_2). The major species present in solution at each point are indicated. Note that it takes twice as much base to reach the second stoichiometric point as it does to reach the first.

Self-test 6H.4A What volume of 0.010 M NaOH(aq) is required to reach (a) the first stoichiometric point and (b) the second stoichiometric point in a titration of 25.00 mL of 0.010 M H_2SO_3(aq)?

[*Answer:* (a) 25 mL; (b) 50. mL]

Self-test 6H.4B What volume of 0.020 M NaOH(aq) is required to reach (a) the first stoichiometric point and (b) the second stoichiometric point in a titration of 30.00 mL of 0.010 M H_3PO_4(aq)?

The pH at any point in the titration of a polyprotic acid with a strong base can be predicted by using the reaction stoichiometry to recognize what stage has been reached in the titration. The principal solute species at that point are then identified together with the principal proton transfer equilibrium that determines the pH.

Self-test 6H.5A A sample of 0.200 M H_3PO_4(aq) of volume 25.0 mL was titrated with 0.100 M NaOH(aq). Identify the primary species in solution after the addition of the following volumes of NaOH solution: (a) 70.0 mL; (b) 100.0 mL.

[*Answer:* (a) Between the first and second stoichiometric points, Na^+ from the base, $H_2PO_4^-$, and HPO_4^{2-}; (b) at the second stoichiometric point, Na^+ and HPO_4^{2-}]

Self-test 6H.5B A sample of 0.100 M H_2S(aq) of volume 20.0 mL was titrated with 0.300 M NaOH(aq). Identify the primary species in solution after the addition of the following volumes of NaOH solution: (a) 5.0 mL; (b) 13.4 mL.

The titration of a polyprotic acid has a stoichiometric point corresponding to the removal of each acidic hydrogen atom. The pH of a solution of a polyprotic acid undergoing a titration is estimated by considering the primary species in solution and the proton transfer equilibrium that determines the pH.

What have you learned in this Topic?

You have learned how to apply the calculations introduced in Topic 6B to predict the pH along a titration curve for strong acid–base titrations. You have also seen how to apply the calculations from Topics 6C and 6D to predict the pH at each stage of strong acid–weak base and strong base–weak acid titrations. You now know how to choose an appropriate indicator for acid–base titrations. Last, you have learned how to predict the pH in the course of titrations of polyprotic acids.

The skills you have mastered are the ability to:

☐ **1.** Calculate the pH during a strong acid–strong base titration (Toolbox 6H.1 and Example 6H.1).

☐ **2.** Estimate the pH at the stoichiometric point for a strong base–weak acid titration (Example 6H.2).

☐ **3.** Calculate the pH during the titration of a weak acid or base (Toolbox 6H.2 and Example 6H.3).

☐ **4.** Select an appropriate indicator for an acid–base titration (Section 6H.3).

☐ **5.** Describe how the pH changes during the titration of a polyprotic acid with a strong base (Section 6H.4).

Topic 6H Exercises

Refer to Tables 6C.1, 6C.2, and 6E.1 for the values of K_a and K_b. Assume 25 °C throughout.

6H.1 (a) Sketch the titration curve for the titration of 5.00 mL 0.010 M NaOH(aq) with 0.0050 M HCl(aq), indicating the pH of the initial and final solutions and the pH at the stoichiometric point. What volume of titrant has been added at (b) the stoichiometric point; (c) the halfway point of the titration?

6H.2 (a) Sketch the titration curve for the titration of 5.00 mL 0.010 M HCl(aq) with 0.010 M $Ca(OH)_2$(aq), indicating the pH of the initial and final solutions and the pH at the stoichiometric point. What volume of titrant has been added at (b) the stoichiometric point; (c) the halfway point of the titration?

6H.3 Calculate the volume of 0.150 M HCl(aq) required to neutralize (a) one-half and (b) all the hydroxide ions in 25.0 mL of 0.110 M NaOH(aq). (c) What is the molar concentration of Na^+ ions at the stoichiometric point? (d) Calculate the pH of the solution after the addition of 20.0 mL of 0.150 M HCl(aq) to 25.0 mL of 0.110 M NaOH(aq).

6H.4 Calculate the volume of 0.120 M HCl(aq) required to neutralize (a) one-half and (b) all the hydroxide ions in 25.0 mL of 0.412 M KOH(aq). (c) What is the molar concentration of Cl^- ions at the stoichiometric point? (d) Calculate the pH of the solution after the addition of 30.0 mL of 0.120 M HCl(aq) to 15.0 mL of 0.310 M KOH(aq).

6H.5 Calculate the pH at each stage in the titration for the addition of 0.150 M HCl(aq) to 25.0 mL of 0.110 M NaOH(aq): (a) initially; (b) after the addition of 5.0 mL of acid; (c) after the addition of a further 5.0 mL; (d) at the stoichiometric point; (e) after the addition of 5.0 mL of acid beyond the stoichiometric point; (f) after the addition of 10.0 mL of acid beyond the stoichiometric point.

6H.6 Calculate the pH at each stage in the titration in which 0.116 M HCl(aq) is added to 25.0 mL of 0.215 M KOH(aq): (a) initially; (b) after the addition of 5.0 mL of acid; (c) after the addition of a further 5.0 mL; (d) at the stoichiometric point; (e) after the addition of 5.0 mL of acid beyond the stoichiometric point; (f) after the addition of 10.0 mL of acid beyond the stoichiometric point.

6H.7 Suppose that 1.436 g of impure sodium hydroxide is dissolved in $300.$ mL of aqueous solution and that 25.00 mL of this solution is titrated to the stoichiometric point with 34.20 mL of 0.0695 M HCl(aq). What is the (mass) percentage purity of the original sample?

6H.8 Suppose that 1.773 g of impure barium hydroxide is dissolved in enough water to produce $200.$ mL of solution and that 25.0 mL of this solution is titrated to the stoichiometric point with 13.1 mL of 0.0810 M HCl(aq). What is the (mass) percentage purity of the original sample?

6H.9 Benzoic acid, C_6H_5COOH, is used as a preservative in food and cosmetics because it is considered to be relatively safe. Suppose you are studying benzoic acid and need to predict the pH of a benzoic acid solution during a titration. Calculate the pH at the stoichiometric point of the titration of 25.00 mL of 0.120 M C_6H_5COOH(aq) with 0.0230 M NaOH(aq).

6H.10 Morphine, $C_{17}H_{19}O_3N$, is a potent painkiller. Suppose you are studying morphine and need to predict the pH of a morphine solution during a titration. Calculate the pH at the stoichiometric point of the titration of 30.00 mL of 0.0172 M $C_{17}H_{19}O_3N$(aq) with 0.0160 M HCl(aq).

6H.11 Suppose that 4.25 g of an unknown weak acid, HA, is dissolved in water. Titration of the solution with 0.350 M NaOH(aq) required 52.0 mL to reach the stoichiometric point. After the addition of 26.0 mL, the pH of the solution was found to be 3.82. (a) What is the molar mass of the acid? (b) What is the value of pK_a for the acid?

6H.12 Suppose that 0.483 g of an unknown weak acid, HA, is dissolved in water. Titration of the solution with 0.250 M NaOH(aq) required 42.0 mL to reach the stoichiometric point. After the addition of 21.0 mL, the pH of the solution was found to be 3.75. (a) What is the molar mass of the acid? (b) What is the value of pK_a for the acid? Identify the acid in Table 6C.1.

6H.13 Suppose that 25.0 mL of 0.10 M CH_3COOH(aq) is titrated with 0.10 M NaOH(aq). (a) What is the initial pH of the 0.10 M CH_3COOH(aq) solution? (b) What is the pH after the addition of 10.0 mL of 0.10 M NaOH(aq)? (c) What volume of 0.10 M NaOH(aq) is required to reach half way to the stoichiometric point? (d) Calculate the pH at that halfway point. (e) What volume

of 0.10 M NaOH(aq) is required to reach the stoichiometric point? (f) Calculate the pH at the stoichiometric point.

6H.14 Suppose that 30.0 mL of 0.12 M C_6H_5COOH(aq) is titrated with 0.20 M KOH(aq). (a) What is the initial pH of the 0.20 M C_6H_5COOH(aq)? (b) What is the pH after the addition of 5.00 mL of 0.20 M KOH(aq)? (c) What volume of 0.20 M KOH(aq) is required to reach half way to the stoichiometric point? (d) Calculate the pH at the halfway point. (e) What volume of 0.20 M KOH(aq) is required to reach the stoichiometric point? (f) Calculate the pH at the stoichiometric point.

6H.15 At what point in the titration of a weak base with a strong acid does the solution have its greatest buffer capacity?

6H.16 Why are indicators not used to identify the point at which a solution has its greatest buffer capacity?

6H.17 Suppose that 15.0 mL of 0.15 M NH_3(aq) is titrated with 0.10 M HCl(aq). (a) What is the initial pH of the 0.15 M NH_3(aq)? (b) What is the pH after the addition of 15.0 mL of 0.10 M HCl(aq)? (c) What volume of 0.10 M HCl(aq) is required to reach halfway to the stoichiometric point? (d) Calculate the pH at the halfway point. (e) What volume of 0.10 M HCl(aq) is required to reach the stoichiometric point? (f) Calculate the pH at the stoichiometric point. (g) Use Table 6H.2 to select an indicator for the titration.

6H.18 Suppose that 50.0 mL of 0.25 M CH_3NH_2(aq) is titrated with 0.35 M HCl(aq). (a) What is the initial pH of the 0.25 M CH_3NH_2(aq)? (b) What is the pH after the addition of 15.0 mL of 0.35 M HCl(aq)? (c) What volume of 0.35 M HCl(aq) is required to reach half way to the stoichiometric point? (d) Calculate the pH at the halfway point. (e) What volume of 0.35 M HCl(aq) is required to reach the stoichiometric point? (f) Calculate the pH at the stoichiometric point. (g) Use Table 6H.2 to select an indicator for the titration.

6H.19 Below is the titration curve for the neutralization of 25 mL of a monoprotic acid with a strong base. Answer the following questions about the reaction and explain your reasoning in each case. (a) Is the acid strong or weak? (b) What is the initial hydronium ion concentration of the acid? (c) What is K_a for the acid? (d) What is the initial concentration of the acid? (e) What is the concentration of base in the titrant? (f) Use Table 6H.2 to select an indicator for the titration.

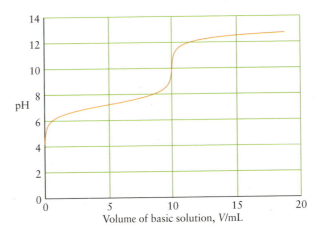

6H.20 Below is the titration curve for the neutralization of 25 mL of a base with a strong monoprotic acid. Answer the following questions about the reaction and explain your reasoning in each case. (a) Is the base strong or weak? (b) What is the initial hydroxide ion concentration of the base? (c) What is K_b for the base? (d) What is the initial concentration of the base? (e) What is the concentration of acid in the titrant? (f) Use Table 6H.2 to select an indicator for the titration.

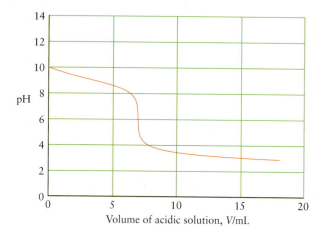

Volume of acidic solution, V/mL

6H.21 Which of the following indicators in Table 6H.2 could you use for a titration of 0.20 M $CH_3COOH(aq)$ with 0.20 M NaOH(aq): (a) methyl orange; (b) litmus; (c) thymol blue; (d) phenolphthalein? Explain your selections.

6H.22 Which of the following indicators in Table 6H.2 could you use for a titration of 0.20 M $NH_3(aq)$ with 0.20 M HCl(aq): (a) bromocresol green; (b) methyl red; (c) phenol red; (d) thymol blue? Explain your selections.

6H.23 Use Table 6H.2 to suggest suitable indicators for the titrations described in Exercises 6H.9 and 6H.14.

6H.24 Use Table 6H.2 to suggest suitable indicators for the titrations described in Exercises 6H.10 and 6H.12.

6H.25 What volume of 0.275 M KOH(aq) must be added to 75.0 mL of 0.137 M $H_3AsO_4(aq)$ to reach (a) the first stoichiometric point; (b) the second stoichiometric point; (c) the third stoichiometric point?

6H.26 What volume of 0.102 M NaOH(aq) must be added to 50.0 mL of 0.0510 M $H_2SO_3(aq)$ to reach (a) the first stoichiometric point; (b) the second stoichiometric point?

6H.27 What volume of 0.255 M $HNO_3(aq)$ must be added to 35.5 mL of 0.158 M $Na_2HPO_3(aq)$ to reach (a) the first stoichiometric point; (b) the second stoichiometric point?

6H.28 What volume of 0.650 M HCl(aq) must be added to 64.2 mL of 0.188 M $Na_3PO_4(aq)$ to reach (a) the first stoichiometric point; (b) the second stoichiometric point; (c) the third stoichiometric point?

6H.29 (a) Sketch the titration curve for the titration of 5.00 mL 0.010 M $H_2S_2O_3(aq)$ with 0.010 M KOH(aq), identifying the initial and final points, the stoichiometric points, and each of the halfway points. (b) What volume of KOH has been added at each stoichiometric point? (c) Determine the pH at each stoichiometric point. For $H_2S_2O_3(aq)$, $pK_{a1} = 0.6$ and $pK_{a2} = 1.74$.

6H.30 (a) Sketch the titration curve for the titration of 5.00 mL 0.010 M $H_3AsO_4(aq)$ with 0.010 M KOH(aq), identifying the initial and final points, the stoichiometric points, and each of the halfway points. (b) What volume of KOH has been added at each stoichiometric point? (c) Determine the pH at each stoichiometric point. For $H_3AsO_4(aq)$, $pK_{a1} = 2.25$ and $pK_{a2} = 6.77$.

6H.31 Suppose that 0.122 g of phosphorous acid, H_3PO_3, is dissolved in water and that the total volume of the solution is 50.0 mL. (a) Estimate the pH of this solution. (b) Estimate the pH of the solution that results when 5.00 mL of 0.175 M NaOH(aq) is added to the phosphorous acid solution. (c) Estimate the pH of the solution if an additional 5.00 mL of 0.175 M NaOH(aq) is added to the solution in part (b).

6H.32 Suppose that 0.242 g of oxalic acid, $(COOH)_2$, is dissolved in 50.0 mL of water. (a) Estimate the pH of this solution. (b) Estimate the pH of the solution that results when 15.0 mL of 0.150 M NaOH(aq) is added to the oxalic acid solution. (c) Estimate the pH of the solution that results if an additional 5.00 mL of NaOH(aq) solution is added to the solution from part (b).

6H.33 Estimate the pH of the solution that results when each of the following solutions is added to 50.0 mL of 0.275 M $Na_2HPO_4(aq)$: (a) 50.0 mL of 0.275 M HCl(aq); (b) 75.0 mL of 0.275 M HCl(aq); (c) 25.0 mL of 0.275 M HCl(aq).

6H.34 Estimate the pH of the solution that results when 75.0 mL of 0.0995 M $Na_2CO_3(aq)$ is mixed with (a) 25.0 mL of 0.130 M $HNO_3(aq)$; (b) 65.0 mL of 0.130 M $HNO_3(aq)$.

Topic 6I Solubility Equilibria

6I.1 The Solubility Product

6I.2 The Common-Ion Effect

6I.3 Complex Ion Formation

How does solubility depend on any other species present?

How can precipitation be predicted and used?

Topic **6I**: Solubility equilibria → Topic **6J**: Precipitation

Why Do You Need to Know This Material? Solubility equilibria form the basis for understanding chemical analysis. Many salts are only sparingly soluble and chemists need to know how to report and control their solubilities to understand their role in the environment and in laboratory procedures.

In a saturated solution, the dissolved solute is in equilibrium with undissolved solute. Its composition can therefore be treated as an example of chemical equilibrium. The ions adjust their concentrations to maintain the values of all relevant equilibrium constants. All the solutes treated in this Topic are ionic and are assumed to dissociate fully in solution.

6I.1 The Solubility Product

The equilibrium constant for the equilibrium between an ionic solid and its ions in solution is called the **solubility product,** K_{sp}. It is expressed in terms of the activities of the ions. For example, the solubility product for bismuth sulfide, Bi_2S_3, is

$$Bi_2S_3(s) \rightleftharpoons 2\,Bi^{3+}(aq) + 3\,S^{2-}(aq) \qquad K_{sp} = (a_{Bi^{3+}})^2(a_{S^{2-}})^3$$

Solid Bi_2S_3 does not appear in the expression for K_{sp} because it is a pure solid and its activity is 1 (Topic 5G). Because the concentrations of ions in a solution of a sparingly soluble salt are low, the activities can be approximated by the numerical values of the molar concentrations:

$$K_{sp} = [Bi^{3+}]^2[S^{2-}]^3$$

However, because ion–ion interactions are so strong, a solubility product expressed in this way is generally meaningful only for sparingly soluble salts. Another complication is that the separation of ions when a salt dissolves is rarely complete. A saturated solution of PbI_2, for instance, contains substantial concentrations of $Pb^{2+}I^-$ and $Pb^{2+}(I^-)_2$ ion clusters. Clusters of ions are most likely to form when one or both ions have a charge greater than ± 1. At best, the calculations in this Topic are only estimates.

One of the simplest ways to determine K_{sp} is to measure the **molar solubility** of the compound, the molar concentration of the compound in a saturated solution, but more advanced and accurate methods are also available. **TABLE 6I.1** gives some experimental values. In the following calculations, s is used to denote the numerical value of the molar solubility expressed in moles per liter; for example, if the molar solubility of a compound is 65 $\mu mol \cdot L^{-1}$ (that is, 6.5×10^{-5} mol·L^{-1}), then $s = 6.5 \times 10^{-5}$.

What Do You Need to Know Already? You need to be familiar with equilibrium calculations (Topic 5I) and how systems at equilibrium respond to changes in the conditions (Topic 5J).

K_{sp} is also called the *solubility product constant* and, most simply, the *solubility constant*.

Electrochemical methods for determining solubility products are discussed in Topic 6L.

EXAMPLE 6I.1   Determining the solubility product

You are an inorganic chemist with an interest in compounds of silver. One of your tasks is to compile a table of solubility products for its compounds from a list of their molar solubilities. The molar solubility of silver chromate, Ag_2CrO_4, is 65 $\mu mol \cdot L^{-1}$ at 25 °C (so $s = 6.5 \times 10^{-5}$). Estimate the value of K_{sp} for silver chromate at 25 °C.

ANTICIPATE Because the molar solubility is low, you should expect K_{sp} to be very small.

PLAN First, write the chemical equation for the equilibrium and the expression for the solubility product. Calculate the molar concentration of each type of ion formed by the salt from the molar solubility and the stoichiometric relations between the species.

What should you assume? Assume that the salt dissociates completely in water and that the anion is not protonated by water.

SOLVE

Write the chemical equation for the solubility equilibrium.

$$Ag_2CrO_4(s) \rightleftharpoons 2\,Ag^+(aq) + CrO_4{}^{2-}(aq)$$

TABLE 6I.1   Solubility Products at 25 °C

Compound	Formula	K_{sp}	Compound	Formula	K_{sp}
aluminum hydroxide	$Al(OH)_3$	1.0×10^{-33}	lead(II) fluoride	PbF_2	3.7×10^{-8}
antimony sulfide	Sb_2S_3	1.7×10^{-93}	iodate	$Pb(IO_3)_2$	2.6×10^{-13}
barium carbonate	$BaCO_3$	8.1×10^{-9}	iodide	PbI_2	1.4×10^{-8}
fluoride	BaF_2	1.7×10^{-6}	sulfate	$PbSO_4$	1.6×10^{-8}
sulfate	$BaSO_4$	1.1×10^{-10}	sulfide	PbS	8.8×10^{-29}
bismuth sulfide	Bi_2S_3	1.0×10^{-97}	magnesium ammonium phosphate	$MgNH_4PO_4$	2.5×10^{-13}
calcium carbonate	$CaCO_3$	8.7×10^{-9}	carbonate	$MgCO_3$	1.0×10^{-5}
fluoride	CaF_2	4.0×10^{-11}	fluoride	MgF_2	6.4×10^{-9}
hydroxide	$Ca(OH)_2$	5.5×10^{-6}	hydroxide	$Mg(OH)_2$	1.1×10^{-11}
sulfate	$CaSO_4$	2.4×10^{-5}	mercury(I) chloride	Hg_2Cl_2	2.6×10^{-18}
chromium(III) iodate	$Cr(IO_3)_3$	5.0×10^{-6}	iodide	Hg_2I_2	1.2×10^{-28}
copper(I) bromide	$CuBr$	4.2×10^{-8}	mercury(II) sulfide, black	HgS	1.6×10^{-52}
chloride	$CuCl$	1.0×10^{-6}	sulfide, red	HgS	1.4×10^{-53}
iodide	CuI	5.1×10^{-12}	nickel(II) hydroxide	$Ni(OH)_2$	6.5×10^{-18}
sulfide	Cu_2S	2.0×10^{-47}	silver bromide	$AgBr$	7.7×10^{-13}
copper(II) iodate	$Cu(IO_3)_2$	1.4×10^{-7}	carbonate	Ag_2CO_3	6.2×10^{-12}
oxalate	CuC_2O_4	2.9×10^{-8}	chloride	$AgCl$	1.6×10^{-10}
sulfide	CuS	1.3×10^{-36}	hydroxide	$AgOH$	1.5×10^{-8}
iron(II) hydroxide	$Fe(OH)_2$	1.6×10^{-14}	iodide	AgI	8×10^{-17}
sulfide	FeS	6.3×10^{-18}	sulfide	Ag_2S	6.3×10^{-51}
iron(III) hydroxide	$Fe(OH)_3$	2.0×10^{-39}	zinc hydroxide	$Zn(OH)_2$	2.0×10^{-17}
lead(II) bromide	$PbBr_2$	7.9×10^{-5}	sulfide	ZnS	1.6×10^{-24}
chloride	$PbCl_2$	1.6×10^{-5}			

Write the expression for the solubility product:

$$K_{sp} = [Ag^+]^2[CrO_4^{2-}]$$

From 2 mol $Ag^+ \mathrel{\hat=} 1$ mol Ag_2CrO_4

$$[Ag^+] = 2s$$

From 1 mol $CrO_4^{2-} \mathrel{\hat=} 1$ mol Ag_2CrO_4

$$[CrO_4^{2-}] = s$$

From $K_{sp} = [Ag^+]^2[CrO_4^{2-}] = (2s)^2(s) = 4s^3$, and $s = 6.5 \times 10^{-5}$,

$$K_{sp} = 4 \times (\overbrace{6.5 \times 10^{-5}}^{s})^3 = 1.1 \times 10^{-12}$$

EVALUATE   As expected, K_{sp} is very small.

Self-test 6I.1A   The molar solubility of lead(II) iodate, $Pb(IO_3)_2$, at 25 °C is 40. $\mu mol \cdot L^{-1}$. What is the value of K_{sp} for lead(II) iodate?

[**Answer:** 2.6×10^{-13}]

Self-test 6I.1B   The molar solubility of silver bromide, $AgBr$, at 25 °C is 0.88 $\mu mol \cdot L^{-1}$. What is the value of K_{sp} for silver bromide?

Related Exercises 6I.1–6I.4

EXAMPLE 6I.2   Estimating the molar solubility from the solubility product

Although molar solubilities are often of interest, you might find it difficult to locate the appropriate data. Solubility constants, however, are often easier to find, and can be converted to molar solubilities. According to Table 6I.1, $K_{sp} = 5.0 \times 10^{-6}$ for chromium(III) iodate in water at 25 °C. Estimate the molar solubility of the compound at 25 °C.

PLAN   Set up the solubility product in terms of the molar solubility, taking into account the stoichiometric relations implied by the chemical equation for the equilibrium, and then solve for the molar solubility.

What should you assume?   Assume that the salt dissociates completely in water and that the anion is not protonated by water.

SOLVE

Write the chemical equation for the solubility equilibrium.

$$Cr(IO_3)_3(s) \rightleftharpoons Cr^{3+}(aq) + 3\,IO_3^-(aq)$$

Write the expression for the solubility product.

$$K_{sp} = [Cr^{3+}][IO_3^-]^3$$

From 1 mol $Cr^{3+} \stackrel{\wedge}{=}$ 1 mol $Cr(IO_3)_3$,

$$[Cr^{3+}] = s$$

From 3 mol $IO_3^- \stackrel{\wedge}{=}$ 1 mol $Cr(IO_3)_3$,

$$[IO_3^-] = 3s$$

Write the expression for K_{sp} in terms of s.

$$K_{sp} = [Cr^{3+}][IO_3^-]^3 = s \times (3s)^3 = 27s^4$$

From $s = (K_{sp}/27)^{1/4}$,

$$s = \{(5.0 \times 10^{-6})/27\}^{1/4} = 0.021$$

The molar solubility of $Cr(IO_3)_3$ is therefore 0.021 mol·L^{-1}.

Self-test 61.2A The solubility product of silver sulfate, Ag_2SO_4, is 1.4×10^{-5}. Estimate the molar solubility of the salt.
[**Answer:** 15 mmol·L^{-1}]

Self-test 61.2B The solubility product of lead(II) fluoride, PbF_2, is 3.7×10^{-8}. Estimate the molar solubility of the salt.

Related Exercises 61.5, 61.6

The solubility product is the equilibrium constant for the equilibrium between an undissolved salt and its ions in a saturated solution.

A quick way to evaluate the fourth root is to take the square root twice in succession.

61.2 The Common-Ion Effect

The solubility of a sparingly soluble salt is reduced by the addition of another soluble salt that has an ion in common with the salt, such as the addition of a solution of a soluble chloride to a saturated solution of silver chloride (**FIG. 61.1**). This decrease in solubility is called the **common-ion effect**. Qualitatively, the effect can be understood in terms of Le Chatelier's principle (Topic 5J). The sparingly soluble salt responds to the added ions by coming out of solution; that is, its solubility decreases.

A quantitative understanding of the common-ion effect can be obtained by considering how a change in concentration of one of the ions affects the solubility product. Consider a saturated solution of silver chloride in water:

$$AgCl(s) \rightleftharpoons Ag^+(aq) + Cl^-(aq) \qquad K_{sp} = [Ag^+][Cl^-]$$

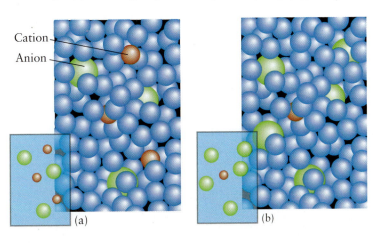

Cation
Anion

(a)   (b)

FIGURE 61.1 If the concentration of one of the ions of a slightly soluble salt is increased, the concentration of the other decreases to maintain a constant value of K_{sp}. (a) The cations (pink) and anions (green) in solution. (b) When more anions are added (together with their accompanying spectator ions, which are not shown), the concentration of cations decreases. The solubility of the original compound has been reduced by the presence of a common ion. In the insets, the blue background represents the solvent (water).

(a)   (b)

FIGURE 6I.2 (a) A saturated solution of zinc acetate in water. (b) When additional acetate ions are added as a single crystal of solid sodium acetate in the spatula shown in part (a), the solubility of the zinc acetate is reduced and more zinc acetate precipitates. (*W. H. Freeman photo by Ken Karp.*)

Experimentally, $K_{sp} = 1.6 \times 10^{-10}$ at 25 °C, and the molar solubility of AgCl in water is 13 μmol·L^{-1}. When sodium chloride is added to the solution, the concentration of Cl$^-$ ions increases. For the equilibrium constant to maintain its value, the concentration of Ag$^+$ ions must decrease. Because there is now less Ag$^+$ in solution, the solubility of AgCl must be lower in a solution of NaCl than it is in pure water. A similar effect occurs whenever two salts having a common ion are mixed (**FIG. 6I.2**).

It is difficult to predict the effect of a common ion reliably and accurately. Because ions interact with one another so strongly, simple equilibrium calculations are rarely valid: the activities of ions can differ markedly from their molar concentrations. However, as Example 6I.3 shows, estimates can be made.

THINKING POINT

Could the addition of a salt *without* a common ion affect the solubility of a sparingly soluble salt?

EXAMPLE 6I.3   Estimating the effect of a common ion on solubility

You know that silver chloride is not very soluble in water, but what do you know about its solubility in seawater? You decide to investigate the solubility of silver chloride in solutions with various concentrations of sodium chloride. Estimate the molar solubility of silver chloride in 10×10^{-4} M NaCl(aq) at 25 °C.

ANTICIPATE A common ion, Cl$^-$, is present, so you should expect the solubility of AgCl in NaCl(aq) to be lower than in pure water.

PLAN Set up the expression for the solubility product and solve for the concentration of silver ions.

What should you assume? Assume that the concentration of chloride ions from the AgCl is insignificant compared to the concentration of the chloride ions in the sodium chloride solution.

SOLVE For a given concentration of Cl$^-$ ions, the concentration of Ag$^+$ ions must satisfy K_{sp}.

From $K_{sp} = [\text{Ag}^+][\text{Cl}^-]$,

$$[\text{Ag}^+] = \frac{K_{sp}}{[\text{Cl}^-]}$$

It follows that silver chloride will dissolve in 1.0×10^{-4} M NaCl(aq), in which $[\text{Cl}^-] = 1.0 \times 10^{-4}$ mol·L^{-1}, until the concentration of Ag$^+$ ions is given by

$$[\text{Ag}^+] = \frac{1.6 \times 10^{-10}}{1.0 \times 10^{-4}} = 1.6 \times 10^{-6}$$

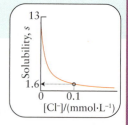

EVALUATE The concentration of Ag$^+$ ions, and hence the solubility of AgCl formula units, is 1.6 μmol·L^{-1}, which is about 10 times less than the solubility of AgCl in pure water.

Self-test 6I.3A What is the approximate molar solubility of calcium carbonate in 0.20 M CaCl$_2$(aq)?

[***Answer:*** 44 nmol·L^{-1}]

Self-test 6I.3B What is the approximate molar solubility of silver bromide in 0.10 M CaBr$_2$(aq)?

Related Exercises 6I.7–6I.10

The common-ion effect is the reduction in solubility of a sparingly soluble salt by the addition of a soluble salt that has an ion in common with it.

TABLE 61.2 Formation Constants of Complexes in Water at 25 °C

Equilibrium	K_f
$Ag^+(aq) + 2\,CN^-(aq) \rightleftharpoons Ag(CN)_2^-(aq)$	5.6×10^8
$Ag^+(aq) + 2\,NH_3(aq) \rightleftharpoons Ag(NH_3)_2^+(aq)$	1.6×10^7
$Au^+(aq) + 2\,CN^-(aq) \rightleftharpoons Au(CN)_2^-(aq)$	2.0×10^{38}
$Cu^{2+}(aq) + 4\,NH_3(aq) \rightleftharpoons Cu(NH_3)_4^{2+}(aq)$	1.2×10^{13}
$Hg^{2+}(aq) + 4\,Cl^-(aq) \rightleftharpoons HgCl_4^{2-}(aq)$	1.2×10^5
$Fe^{2+}(aq) + 6\,CN^-(aq) \rightleftharpoons Fe(CN)_6^{4-}(aq)$	7.7×10^{36}
$Ni^{2+}(aq) + 6\,NH_3(aq) \rightleftharpoons Ni(NH_3)_6^{2+}(aq)$	5.6×10^8

61.3 Complex Ion Formation

The solubility of a salt can be increased by finding a way to conceal the ions in solution, because then the dissolution process continues in a vain attempt to reach equilibrium. An ion can be concealed by making use of the fact that many metal cations are Lewis acids (Topic 6A). When a Lewis acid and a Lewis base react, they form a coordinate covalent bond (Topic 2C) and the product is called a **coordination complex.** An example is the formation of $Ag(NH_3)_2^+$, which occurs when an aqueous solution of the Lewis base ammonia is added to a solution of silver ions acting as a Lewis acid. The formation of the complex effectively removes some of the Ag^+ ions from solution. As a result, to preserve the value of K_{sp}, more silver chloride dissolves.

To treat the effect of complex formation on solubility quantitatively, note that both processes are at equilibrium:

$$AgCl(s) \rightleftharpoons Ag^+(aq) + Cl^-(aq) \qquad K_{sp} = [Ag^+][Cl^-] \qquad \textbf{(A)}$$

$$Ag^+(aq) + 2\,NH_3(aq) \rightleftharpoons Ag(NH_3)_2^+(aq) \qquad K_f = \frac{[Ag(NH_3)_2^+]}{[Ag^+][NH_3]^2} \qquad \textbf{(B)}$$

and that the sum of these two equilibria is the solubility equilibrium in the presence of complex formation. The resulting reaction and its equilibrium constant are:

$$AgCl(s) + 2\,NH_3(aq) \rightleftharpoons Ag(NH_3)_2^+(aq) + Cl^-(aq) \qquad K = K_{sp} \times K_f \qquad \textbf{(C)}$$

The equilibrium constant for the formation of a complex ion is called the **formation constant,** K_f (**TABLE 61.2**).

EXAMPLE 61.4   Calculating molar solubility in the presence of complex formation

Silver emulsion photographic film is now largely obsolete for amateur photography, but it is still used in a variety of medical and technical applications. You are working on the improvement of a particularly sensitive emulsion and need to investigate the solubility of silver chloride in various media. Calculate the molar solubility of silver chloride in 0.10 M $NH_3(aq)$.

ANTICIPATE Because silver ions form a complex with ammonia, you should expect the solubility of silver chloride to be greater in the ammonia solution than in water.

What should you assume? Assume that the reaction of ammonia with water does not significantly affect this equilibrium.

PLAN Because the solubility equilibrium, reaction C, is the sum of the equations for the solubility and complex formation equilibria, the equilibrium constant for the dissolution of AgCl in aqueous ammonia is the product of the equilibrium constants for reactions A and B (Topic 5H). Set up an equilibrium table for reaction C and solve for the equilibrium concentrations.

SOLVE The equilibrium to consider and the corresponding equilibrium table (with all concentrations in moles per liter) are

$$AgCl(s) + 2\,NH_3(aq) \rightleftharpoons Ag(NH_3)_2^+(aq) + Cl^-(aq) \qquad K = K_{sp} \times K_f = \frac{[Ag(NH_3)_2^+][Cl^-]}{[NH_3]^2}$$

	NH_3	$Ag(NH_3)_2^+$	Cl^-
initial concentration	0.10	0	0
change in concentration	$-2x$	$+x$	$+x$
equilibrium concentration	$0.10 - 2x$	x	x

Now determine the value of K and substitute the information from the table.

From $K = K_{sp} \times K_f$,

$$K = (1.6 \times 10^{-10}) \times (1.6 \times 10^7) = 2.6 \times 10^{-3}$$

Form $K = [Ag(NH_3)_2^+][Cl^-]/[NH_3]^2$

$$K = \frac{x \times x}{(0.10 - 2x)^2} = \left(\frac{x}{0.10 - 2x}\right)^2$$

Take the square root of both sides.

$$\frac{x}{0.10 - 2x} = K^{1/2}$$

Solve for x.

$$x = (0.10 - 2x)K^{1/2}$$
$$(1 + 2K^{1/2})x = 0.10K^{1/2}$$
$$x = \frac{0.10K^{1/2}}{1 + 2K^{1/2}} = \frac{0.10 \times (2.6 \times 10^{-3})^{1/2}}{1 + 2(2.6 \times 10^{-3})^{1/2}}$$
$$= 4.6 \times 10^{-3}$$

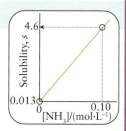

EVALUATE From line 3 in the equilibrium table, $x = [Ag(NH_3)_2^+] = 4.6 \times 10^{-3}$. Hence, the molar solubility of silver chloride in 0.10 M NH_3(aq) is 4.6 mmol·L^{-1}, more than 100 times the molar solubility of silver chloride in pure water (13 μmol·L^{-1}).

Self-test 6I.4A Use data from Tables 6I.1 and 6I.2 to calculate the molar solubility of silver bromide in 1.0 M NH_3(aq).

[*Answer:* 3.5 mmol·L^{-1}]

Self-test 6I.4B Use data from Tables 6I.1 and 6I.2 to calculate the molar solubility of copper(II) sulfide in 1.2 M NH_3(aq).

Related Exercises 6I.11, 6I.12

The solubility of a salt increases if the salt can form a complex ion with other species in the solution.

What have you learned in this Topic?

You have learned how to use equilibrium concepts to estimate the solubilities of sparingly soluble salts. Specifically, you have seen how the addition of a common ion can induce precipitation of another ion and, by use of solubility product constants, how to make predictions about which species will precipitate in a mixed solution.

The skills you have mastered are the ability to:

☐ **1.** Calculate the K_{sp} value for a sparingly soluble salt from its molar solubility (Example 6I.1).

☐ **2.** Determine the molar solubility of a salt from its solubility constant (Example 6I.2).

☐ **3.** Estimate the solubility of a salt in the presence of a common ion (Example 6I.3).

☐ **4.** Calculate the molar solubility of an ion in the presence of complex formation (Example 6I.4).

Topic 6I Exercises

Solubility products are listed in Table 6I.1. Unless otherwise stated, ignore any proton transfer reactions. Assume 25 °C throughout.

6I.1 Determine K_{sp} for each of the following sparingly soluble compounds, given their molar solubilities: (a) AgBr, 8.8×10^{-7} mol·L^{-1}; (b) PbCrO$_4$, 1.3×10^{-7} mol·L^{-1}; (c) Ba(OH)$_2$, 0.11 mol·L^{-1}; (d) MgF$_2$, 1.2×10^{-3} mol·L^{-1}.

6I.2 Determine K_{sp} for each of the following sparingly soluble compounds, given their molar solubilities: (a) AgI, 9.1×10^{-9} mol·L^{-1}; (b) Ca(OH)$_2$, 0.011 mol·L^{-1}; (c) Ag$_3$PO$_4$, 2.7×10^{-6} mol·L^{-1}; (d) Hg$_2$Cl$_2$, 5.2×10^{-7} mol·L^{-1}.

6I.3 The concentration of CrO$_4^{2-}$ in a saturated Tl$_2$CrO$_4$ solution is 6.3×10^{-5} mol·L^{-1}. What is the K_{sp} of Tl$_2$CrO$_4$?

61.4 The molar solubility of silver sulfite, Ag_2SO_3, is 1.55×10^{-5} mol·L^{-1}. What is the K_{sp} of silver sulfite?

61.5 Calculate the molar solubility in water of (a) BiI_3 ($K_{sp} = 7.71 \times 10^{-19}$); (b) CuCl; (c) $CaCO_3$.

61.6 Calculate the molar solubility in water of (a) $PbBr_2$; (b) Ag_2CO_3; (c) $Fe(OH)_2$.

61.7 Calculate the molar solubility of each substance in its respective solution: (a) silver chloride in 0.20 M NaCl(aq); (b) mercury(I) chloride in 0.150 M NaCl(aq); (c) lead(II) chloride in 0.025 M $CaCl_2$(aq); (d) iron(II) hydroxide in 2.5×10^{-3} M $FeCl_2$(aq).

61.8 Calculate the molar solubility of each substance in its respective solution: (a) silver iodide in 0.020 M NaI(aq); (b) calcium carbonate in 2.3×10^{-4} M Na_2CO_3(aq); (c) lead(II) fluoride in 0.21 M NaF(aq); (d) nickel(II) hydroxide in 0.450 M $NiSO_4$(aq).

61.9 Calculate the molar solubility of each of the following sparingly soluble substances in its respective solution: aluminum hydroxide at (a) pH = 7.0; (b) pH = 4.5; zinc hydroxide at (c) pH = 7.0; (d) pH = 6.0.

61.10 Calculate the molar solubility of each of the following sparingly soluble compounds in its respective solution: iron(III) hydroxide at (a) pH = 11.0; (b) pH = 3.0; iron(II) hydroxide at (c) pH = 8.0; (d) pH = 6.0.

61.11 Calculate the molar solubility of silver bromide in 0.10 M KCN(aq).

61.12 Precipitated silver chloride dissolves in ammonia solutions as a result of the formation of $Ag(NH_3)_2^+$. What is the molar solubility of silver chloride in 1.0 M NH_3(aq)?

Topic 6J Precipitation

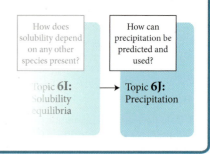

Why Do You Need to Know This Material? The ability to predict when a precipitate will form allows you to identify conditions that favor or hinder precipitation and provides a basis for one of the major applications of chemistry, the identification of substances present in a mixture.

What Do You Need to Know Already? You need to be familiar with solubility equilibria (Topic 6I) and equilibrium calculations (Topic 5I), and how systems at equilibrium respond to changes in the conditions (Topic 5J).

Precipitation, the formation of a solid (typically as a finely divided powder) when two solutions of soluble salts are mixed, is the result of a reaction in which one product is insoluble. This kind of reaction can be used to prepare ionic compounds and has important practical applications in municipal wastewater treatment, the extraction of minerals from seawater, the formation and loss of bones and teeth, and the global carbon cycle.

6J.1 Predicting Precipitation

How can you predict if a precipitate will form when you mix two solutions? If the concentrations of the ions in the solutions are known, the outcome can be predicted by comparing the values of Q, the reaction quotient, and K, the equilibrium constant, as discussed in Topic 5G. In this case, the equilibrium constant is the solubility product, K_{sp}, and the reaction quotient is denoted Q_{sp}. When the concentrations of the ions are high, Q_{sp} is greater than K_{sp}, precipitation is spontaneous, and the value of Q_{sp} adjusts until it is equal to K_{sp} (**FIG. 6J.1**).

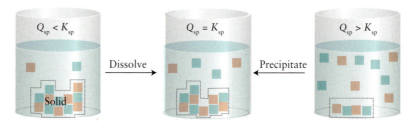

FIGURE 6J.1 The relative magnitudes of the solubility quotient, Q_{sp}, and the solubility product, K_{sp}, are used to decide whether a salt will dissolve (left) or precipitate (right). When the concentrations of the ions in solution are low (left), Q_{sp} is smaller than K_{sp}; when the ion concentrations are high (right), Q_{sp} is larger than K_{sp}.

EXAMPLE 6J.1   Predicting whether a precipitate will form when two solutions are mixed

You are an analytical chemist and are working out a procedure for precipitating lead from a solution of lead nitrate. You know that lead(II) iodide is insoluble, so you decide to use potassium iodide solution. Will the lead precipitate from the solutions that you have available? Will lead(II) iodide precipitate if you mix equal volumes of 0.2 M $Pb(NO_3)_2$(aq) and KI(aq) at 25 °C?

ANTICIPATE Because the concentrations of the Pb^{2+} and I^- ions are high and the solubility product of PbI_2 is low, you should expect precipitation to occur.

PLAN First calculate the new molar concentrations of the ions in the mixed solution, before reaction occurs. Then compare the value of the reaction quotient, Q_{sp} to K_{sp} for PbI_2. Remember that the final volume of the solution is the total volume of the mixture, so the molar concentrations to use in Q_{sp} must be adjusted accordingly.

What should you assume? Assume that the lead(II) iodide is fully dissociated (in the sense that its ions have separated) in aqueous solution and that activities can be replaced by molar concentrations.

SOLVE Table 6I.1 gives $K_{sp} = 1.4 \times 10^{-8}$ for PbI_2 at 25 °C.

Write the chemical equation and its equilibrium constant.

$$PbI_2(s) \rightleftharpoons Pb^{2+}(aq) + 2\,I^-(aq) \qquad K_{sp} = [Pb^{2+}][I^-]^2$$

Calculate the new molar concentrations of the ions, noting that the volume the ions occupy has doubled.

$$Pb^{2+}(aq): \tfrac{1}{2}(0.2 \text{ mol·L}^{-1}) = 0.1 \text{ mol·L}^{-1}$$

$$I^{-}(aq): \tfrac{1}{2}(0.2 \text{ mol·L}^{-1}) = 0.1 \text{ mol·L}^{-1}$$

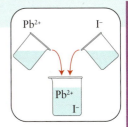

Form $Q_{sp} = [Pb^{2+}][I^{-}]^{2}$ and compare its value with that of K_{sp}:

$$Q_{sp} = 0.1 \times (0.1)^{2} = 1 \times 10^{-3}$$

$$\underbrace{1 \times 10^{-3}}_{Q_{sp}} \gg \underbrace{1.4 \times 10^{-8}}_{K_{sp}}$$

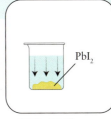

EVALUATE The value of Q_{sp} is considerably higher than K_{sp}, and so, as predicted, a precipitate forms (**FIG. 6J.2**).

Self-test 6J.1A Does a precipitate of silver chloride form when 200. mL of 1.0×10^{-4} M $AgNO_3$(aq) and 900. mL of 1.0×10^{-6} M KCl(aq) are mixed? Assume complete dissociation.

[**Answer:** No ($Q_{sp} = 1.5 \times 10^{-11}$; $Q_{sp} < K_{sp}$)]

Self-Test 6J.1B Does a precipitate of barium fluoride form when 100. mL of 1.0×10^{-3} M $Ba(NO_3)_2$(aq) is mixed with 200. mL of 1.0×10^{-3} M KF(aq)? Ignore possible protonation of F^{-}.

Related Exercises 6J.1−6J.6

A salt precipitates if Q_{sp} is greater than K_{sp}.

6J.2 Selective Precipitation

It is sometimes possible to separate different cations from a solution containing a mixture of ions by adding a soluble salt containing an anion with which they form salts with very different solubilities. For example, seawater is a mixture of many different ions. It is possible to precipitate magnesium ions from seawater by adding hydroxide ions. However, other cations are also present in seawater. Their individual concentrations and the relative solubilities of their hydroxides govern which will precipitate first if a certain amount of hydroxide is added. Optimum separation of two compounds is achieved when Q_{sp} for one species exceeds its K_{sp}, but Q_{sp} for the second species is significantly less than its K_{sp}. Example 6J.2 illustrates a strategy for predicting the order of precipitation.

EXAMPLE 6J.2 Predicting the order of precipitation

You have a sample of seawater collected off Hawaii, where there is a lot of volcanic activity. As part of an oceanographic study, you are going to analyze it to see if its composition is markedly different from a sample taken farther out in the ocean. The sample of seawater contains, among other solutes, the following concentrations of soluble cations: 0.050 mol·L^{-1} Mg^{2+}(aq) and 0.010 mol·L^{-1} Ca^{2+}(aq). (a) Use the information in Table 6I.1 to determine the order in which each ion precipitates as solid NaOH is added and give the molar concentration of OH^{-} when precipitation of each begins. Assume that there is no volume change upon addition of the NaOH and that the temperature is 25 °C. (b) If the first compound to precipitate is $X(OH)_2$, calculate the concentration of X^{2+} ions remaining in solution when the second ion precipitates.

ANTICIPATE In (a), because K_{sp} for $Mg(OH)_2$ is so much smaller than K_{sp} for $Ca(OH)_2$ and the formulas are analogous, you should expect $Mg(OH)_2$ to precipitate first. In (b), because much of X^{2+} has precipitated as $X(OH)_2$ and is no longer in solution, you should expect its concentration to be very small when the second ion precipitates.

PLAN (a) Use the common-ion effect to calculate the value of $[OH^{-}]$ required for each salt to precipitate by writing the expression for K_{sp} for each salt and then substituting the data

FIGURE 6J.2 When a few drops of lead(II) nitrate solution are added to a solution of potassium iodide, yellow lead(II) iodide immediately precipitates. (©1989 Chip Clark–Fundamental Photographs.)

LAB VIDEO FIGURE 6J.2

provided. (b) Calculate the remaining concentration of the first cation to precipitate by substituting the value of $[OH^-]$ into K_{sp} for that hydroxide.

SOLVE (a) Write the chemical equation and K_{sp} for the dissolving of $Ca(OH)_2$:

$$Ca(OH)_2(s) \rightleftharpoons Ca^{2+}(aq) + 2\,OH^-(aq) \qquad K_{sp} = [Ca^{2+}][OH^-]^2$$

From Table 6I.1, $K_{sp} = 5.5 \times 10^{-6}$. Then:

Find $[OH^-]$ from $K_{sp} = [Ca^{2+}][OH^-]^2$ in the form $[OH^-] = (K_{sp}/[Ca^{2+}])^{1/2}$.

$$[OH^-] = \left(\frac{5.5 \times 10^{-6}}{0.010}\right)^{1/2} = 0.023$$

corresponding to 23 mmol·L^{-1}.

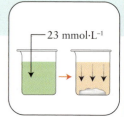

Write the chemical equation and K_{sp} for the solubility equilibrium of $Mg(OH)_2$.

$$Mg(OH)_2(s) \rightleftharpoons Mg^{2+}(aq) + 2\,OH^-(aq) \qquad K_{sp} = [Mg^{2+}][OH^-]^2$$

From Table 6I.1, $K_{sp} = 1.1 \times 10^{-11}$.

Find $[OH^-]$ from $K_{sp} = [Mg^{2+}][OH^-]^2$ in the form $[OH^-] = (K_{sp}/[Mg^{2+}])^{1/2}$.

$$[OH^-] = \left(\frac{1.1 \times 10^{-11}}{0.050}\right)^{1/2} = 1.5 \times 10^{-5}$$

corresponding to 15 μmol·L^{-1}.

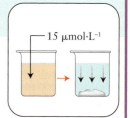

EVALUATE As predicted, the hydroxides are expected to precipitate in the order $Mg(OH)_2$ at 15 μmol·L^{-1} $OH^-(aq)$ and then $Ca(OH)_2$ at 23 mmol·L^{-1} $OH^-(aq)$.

(b) Find the concentration of magnesium ions when $[OH^-] = 0.023$ mol·L^{-1}.

Find $[Mg^{2+}]$ from $K_{sp} = [Mg^{2+}][OH^-]^2$ in the form $[Mg^{2+}] = K_{sp}/[OH^-]^2$.

$$[Mg^{2+}] = \frac{1.1 \times 10^{-11}}{(0.023)^2} = 2.1 \times 10^{-8}$$

corresponding to 21 nmol·L^{-1}.

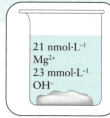

EVALUATE The concentration of magnesium ions remaining when $Ca(OH)_2$ begins to precipitate is indeed very small.

Self-test 6J.2A Potassium carbonate is added to a solution containing 0.030 mol·L^{-1} $Mg^{2+}(aq)$ and 0.0010 mol·L^{-1} $Ca^{2+}(aq)$. (a) Use the information in Table 6I.1 to determine the order in which each ion precipitates as the concentration of K_2CO_3 is increased and give the concentration of CO_3^{2-} when precipitation of each begins. (b) Calculate the concentration of the first ion to precipitate that remains in solution when the second ion precipitates.

[*Answer:* (a) $CaCO_3$ precipitates first, at 8.7 μmol·L^{-1} CO_3^{2-}, then $MgCO_3$ at 0.33 mmol·L^{-1} CO_3^{2-}; (b) 26 μmol·L^{-1} Ca^{2+}]

Self-test 6J.2B Chloride ions are added to a solution containing 0.020 mol·L^{-1} $Pb(NO_3)_2(aq)$ and 0.0010 mol·L^{-1} $AgNO_3(aq)$. (a) Use the information in Table 6I.1 to determine the order in which each cation precipitates as the concentration of chloride ions is increased, and give the concentration of Cl^- when precipitation of each begins. (b) Calculate the concentration of the first ion to precipitate that remains in solution when the second ion precipitates.

Related Exercises 6J.7, 6J.8, 6J.11, 6J.12

A mixture of ions in solution can be separated by adding oppositely charged ions with which they form salts with very different solubilities.

6J.3 Dissolving Precipitates

When a precipitate has been formed during the qualitative analysis of the ions present in a solution, it may be necessary to dissolve the precipitate again to identify the cation or anion. Several strategies are available.

One strategy is to remove one of the ions from the solubility equilibrium so that the precipitate will continue to dissolve, in a fruitless chase for equilibrium. Suppose, for example, that a solid hydroxide such as iron(III) hydroxide is in equilibrium with its ions in solution:

$$Fe(OH)_3(s) \rightleftharpoons Fe^{3+}(aq) + 3\,OH^-(aq)$$

To dissolve more of the solid, acid can be added. The H_3O^+ ions supplied by the added acid remove the OH^- ions by converting them into water, and more $Fe(OH)_3$ dissolves.

THINKING POINT

Silver ions can be brought into solution from solid Ag_2O by adding HNO_3, but not by adding HCl. Why cannot HCl be used?

Many carbonate, sulfite, and sulfide precipitates can be dissolved by the addition of acid, because the anions react with the acid to form a gas that bubbles out of solution. For example, in a saturated solution of zinc carbonate, solid $ZnCO_3$ is in equilibrium with its ions:

$$ZnCO_3(s) \rightleftharpoons Zn^{2+}(aq) + CO_3^{2-}(aq)$$

The CO_3^{2-} ions react with acid to form CO_2:

$$CO_3^{2-}(aq) + 2\,HNO_3(aq) \rightleftharpoons CO_2(g) + H_2O(l) + 2\,NO_3^-(aq)$$

The dissolution of carbonates by acid is an undesired result of acid rain, which has damaged the appearance of many historic marble and limestone monuments—marble and limestone are mineral forms of calcium carbonate (**FIG. 6J.3**).

An alternative procedure for removing an ion from solution is to change the identity of the ion by changing its oxidation state. The metal ions in very insoluble heavy metal sulfides can be dissolved by oxidizing the sulfide ion to elemental sulfur. For example, copper(II) sulfide, CuS, takes part in the equilibrium

$$CuS(s) \rightleftharpoons Cu^{2+}(aq) + S^{2-}(aq)$$

However, when nitric acid is added, the sulfide ions are oxidized to elemental sulfur:

$$3\,S^{2-}(aq) + 8\,HNO_3(aq) \rightleftharpoons 3\,S(s) + 2\,NO(g) + 4\,H_2O(l) + 6\,NO_3^-(aq)$$

This complicated oxidation removes the sulfide ions from the equilibrium, and Cu^{2+} ions dissolve as $Cu(NO_3)_2$.

Some precipitates dissolve when the temperature is changed because the solubility constant depends on temperature. This strategy is used to purify precipitates in a process called *recrystallization*. The mixture is heated to dissolve the solid and filtered to remove insoluble impurities. The solid is then allowed to re-form as the solution cools and is removed from the solution in a second filtration. As discussed in Section 6I.3, complex ion formation can also be used to dissolve metal ions.

The solubility of a solid can be increased by removing one of its ions from solution; acid can be used to dissolve a hydroxide, sulfide, sulfite, or carbonate precipitate; and nitric acid can be used to oxidize metal sulfides to sulfur and a soluble salt. Some solids can be dissolved by changing the temperature or by forming a complex ion.

6J.4 Qualitative Analysis

Complex formation, selective precipitation, and control of the pH of a solution all play important roles in the qualitative analysis of the ions present in aqueous solutions (*Fundamentals* I). There are many different schemes of analysis, but they all follow the same general principles. The following discussion illustrates a simple procedure for the identification of five cations by following the steps that might be used in the laboratory.

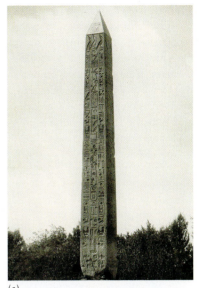

(a)

(b)

FIGURE 6J.3 The state of the carving on Cleopatra's Needle has deteriorated as a result of the action of acid rain: (a) after 3500 years in the Egyptian desert; (b) after a further 100 years in Central Park, New York City. (*Photo (a) ©SSPL/ The Image Works. Photo (b) Dr. Marli Miller/Getty Images.*)

FIGURE 6J.4 Part of a simple qualitative analysis scheme used to separate certain cations. In the first step, three cations are separated as insoluble chlorides. In the second step, cations that form highly insoluble sulfides are removed by precipitation at a low pH, and, in the third step, the remaining cations are precipitated as the sulfides at a higher pH.

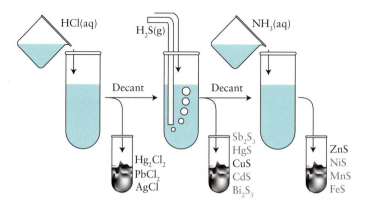

Suppose a solution contains lead(II), mercury(I), silver, copper(II), and zinc ions. The method is outlined in **FIG. 6J.4**, which includes additional cations, and is illustrated in **FIG. 6J.5**. Most chlorides are soluble; so, when hydrochloric acid is added to a mixture of salts, only certain chlorides precipitate (see Table 6I.1). Silver and mercury(I) chlorides have such small values of K_{sp} that, even with low concentrations of Cl^- ions, the chlorides precipitate. Lead(II) chloride, which is slightly soluble, will precipitate if the chloride ion concentration is high enough. The hydronium ions provided by the acid play no role in this step; they simply accompany the chloride ions. At this point, the precipitate can be separated from the solution by using a centrifuge to compact the solid and then decanting (pouring off) the solution. The solution now contains copper(II) and zinc ions, whereas the solid consists of $PbCl_2$, Hg_2Cl_2, and $AgCl$.

Because $PbCl_2$ is slightly soluble, if the precipitate is rinsed with hot water, the lead(II) chloride dissolves and the solution is separated from the precipitate. If sodium chromate is then added to that solution, the presence of lead can be verified because any lead(II) ions present will precipitate as yellow lead(II) chromate:

$$Pb^{2+}(aq) + CrO_4{}^{2-}(aq) \longrightarrow PbCrO_4(s)$$

At this point, the silver and mercury(I) chlorides remain as precipitates. To separate the Ag^+ and $Hg_2{}^{2+}$ ions, aqueous ammonia is added to the solid mixture. The silver chloride precipitate dissolves as the soluble complex ion $Ag(NH_3)_2{}^+$ forms:

$$Ag^+(aq) + 2\,NH_3(aq) \longrightarrow Ag(NH_3)_2{}^+(aq)$$

and mercury(I) reacts with ammonia to form a gray solid consisting of a mixture of mercury(II) ions precipitated as white $HgNH_2Cl(s)$ and black metallic mercury (**FIG. 6J.6**):

$$Hg_2Cl_2(s) + 2\,NH_3(aq) \longrightarrow Hg(l) + HgNH_2Cl(s) + NH_4{}^+(aq) + Cl^-(aq)$$

Now, any $Hg_2{}^{2+}$ has precipitated and any Ag^+ present is in solution as the complex ion. The solution is separated from the solid, and the presence of silver ion in the solution is

FIGURE 6J.5 Steps in the analysis of five cations by selective precipitation. (a) The original solution contains Pb^{2+}, $Hg_2{}^{2+}$, Ag^+, Cu^{2+}, and Zn^{2+} ions (left). Addition of HCl precipitates AgCl, Hg_2Cl_2, and $PbCl_2$, which can be removed by decanting or filtration (right), as shown in Fig. 6J.6. (b) Addition of H_2S to the solution remaining from the first step (left) precipitates CuS, which can be removed (right). (c) Making the solution from the second step (left) basic by adding ammonia precipitates ZnS (right). (*W. H. Freeman photos by Ken Karp.*)

LAB VIDEO FIGURE 6J.5

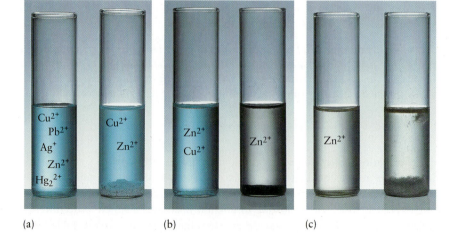

(a)   (b)   (c)

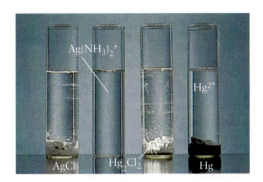

FIGURE 6J.6 When ammonia is added to a silver chloride precipitate, the precipitate dissolves. However, when ammonia is added to a precipitate of mercury(I) chloride, mercury metal and mercury(II) ions are formed in a redox reaction and the mass turns gray. Left to right: silver chloride in water, silver chloride in aqueous ammonia, mercury(I) chloride in water, and mercury(I) chloride in aqueous ammonia. (*W. H. Freeman photo by Ken Karp.*)

verified by addition of nitric acid. This acid pulls the ammonia out of the complex as NH_4^+, allowing white silver chloride to precipitate

$$Ag(NH_3)_2^+(aq) + Cl^-(aq) + 2\,H_3O^+(aq) \longrightarrow AgCl(s) + 2\,NH_4^+(aq) + 2\,H_2O(l)$$

Sulfides with widely different solubilities and solubility products can be selectively precipitated by adding S^{2-} ions to the solution removed from the chlorides in the first step (see Fig. 6J.4). Some metal sulfides (such as CuS, HgS, and Sb_2S_3) have very small solubility products and precipitate if there is the merest trace of S^{2-} ions in the solution. Such a very low concentration of S^{2-} ions is achieved by adding hydrogen sulfide, H_2S, to an acidified solution. A higher hydronium ion concentration shifts the equilibrium

$$H_2S(aq) + 2\,H_2O(l) \rightleftharpoons 2\,H_3O^+(aq) + S^{2-}(aq)$$

to the left, which ensures that almost all the H_2S is in its fully protonated form and hence that very little S^{2-} is present. Nevertheless, even that very low concentration will result in the precipitation of highly insoluble solids if the appropriate cations are present.

To confirm the presence of Zn^{2+} ions in the solution remaining after the first two steps, add H_2S followed by ammonia. The base removes the hydronium ion from the H_2S equilibrium, which shifts the equilibrium in favor of S^{2-} ions. The higher concentration of S^{2-} ions increases the Q_{sp} values of any remaining metal sulfides, such as ZnS or MnS, above their K_{sp} values, and they precipitate.

Qualitative analysis involves the separation and identification of ions by selective precipitation, complex formation, and the control of pH.

What have you learned in this Topic?

You have seen how to use solubility product constants to estimate whether a precipitate of a salt will form and to make predictions about the order of precipitation in a mixed solution. You have also learned how to use equilibrium concepts in qualitative analysis.

The skills you have mastered are the ability to:

☐ **1.** Predict whether a precipitate will form when two salt solutions are mixed. (Example 6J.1)

☐ **2.** Predict the order of precipitation when a common ion is added to a solution that contains multiple types of ions (Example 6J.2).

☐ **3.** Explain how to use solubility differences to identify the cations in a mixture of salts (Section 6J.4).

Topic 6J Exercises

Solubility products are listed in Table 6I.1. Assume 25 °C throughout.

6J.1 (a) What concentration (in moles per liter) of Ag^+ ions is required for the formation of a precipitate in 1.0×10^{-5} M NaCl(aq)? (b) What mass (in micrograms) of solid $AgNO_3$ needs to be added for the onset of precipitation in 100. mL of the solution in part (a)?

6J.2 Iodide ions can be used to precipitate lead(II) ions from 0.010 M $Pb(NO_3)_2(aq)$. (a) What minimum iodide ion concentration is required for the onset of PbI_2 precipitation? (b) What mass (in grams) of KI must be added to 25.0 mL of the $Pb(NO_3)_2(aq)$ solution for PbI_2 to form?

6J.3 Estimate the pH required for the onset of precipitation of $Ni(OH)_2$ from (a) 0.060 M $NiSO_4(aq)$; (b) 0.030 M $NiSO_4(aq)$.

6J.4 Decide whether a precipitate will form when each of the following pairs of solutions are mixed: (a) 5.0 mL of 0.10 M $K_2CO_3(aq)$ and 1.00 L of 0.010 M $AgNO_3(aq)$; (b) 3.3 mL of 1.0 M $HCl(aq)$, 4.9 mL of 0.0030 M $AgNO_3(aq)$, and enough water to dilute the solution to 50.0 mL. For the purpose of these calculations, ignore any reaction of the anion with water.

6J.5 Suppose that there are typically 20 average-sized drops in 1.0 mL of an aqueous solution. Will a precipitate form when 1 drop of 0.010 M $NaCl(aq)$ is added to 10.0 mL of (a) 0.0040 M $AgNO_3(aq)$; (b) 0.0040 M $Pb(NO_3)_2(aq)$?

6J.6 Assume 20 drops per milliliter. Will a precipitate form if (a) 7 drops of 0.0029 M $K_2CO_3(aq)$ are added to 25.0 mL of 0.0018 M $CaCl_2(aq)$; (b) 10 drops of 0.010 M $Na_2CO_3(aq)$ are added to 10.0 mL of 0.0040 M $AgNO_3(aq)$? For the purpose of these calculations, ignore any reaction of the anion with water.

6J.7 The concentrations of magnesium, calcium, and nickel(II) ions in an aqueous solution are 0.0010 mol·L^{-1}. (a) In what order do they precipitate when solid KOH is added? (b) Determine the pH at which each salt precipitates.

6J.8 Suppose that two hydroxides, MOH and M′$(OH)_2$, both have $K_{sp} = 1.0 \times 10^{-12}$ and that initially both cations are present accompanied by nitrate ions in a solution at concentrations of 0.0010 mol·L^{-1}. Which hydroxide precipitates first, and at what pH, when solid NaOH is added?

6J.9 You wish to separate magnesium ions and barium ions by selective precipitation. Which anion, fluoride or carbonate, would be the better choice for achieving this precipitation? Why?

6J.10 You wish to separate barium ions from calcium ions by selective precipitation. Which anion, fluoride or carbonate, would be the better choice for achieving this precipitation? Why?

6J.11 In the process of separating Pb^{2+} ions from Cu^{2+} ions as sparingly soluble iodates, what is the Pb^{2+} concentration when Cu^{2+} just begins to precipitate as sodium iodate is added to a solution that is initially 0.0010 M $Pb(NO_3)_2(aq)$ and 0.0010 M $Cu(NO_3)_2(aq)$?

6J.12 A chemist attempts to separate barium ions from lead ions by using the sulfate ion as a precipitating agent. (a) What sulfate ion concentrations are required for the precipitation of $BaSO_4$ and $PbSO_4$ from a solution containing 0.010 M $Ba^{2+}(aq)$ and 0.010 M $Pb^{2+}(aq)$? (b) What is the concentration of barium ions when the lead(II) sulfate begins to precipitate?

6J.13 Consider the two equilibria

$$CaF_2(s) \rightleftharpoons Ca^2(aq) + 2\,F^-(aq) \qquad K_{sp} = 4.0 \times 10^{-11}$$
$$F^-(aq) + H_2O(l) \rightleftharpoons HF(aq) + OH^-(aq)$$
$$K_b = 2.9 \times 10^{-11}$$

(a) Write the chemical equation for the overall equilibrium and calculate the corresponding equilibrium constant. (b) Estimate the solubility of CaF_2 at (i) pH = 7.0; (ii) pH = 3.0.

6J.14 Consider the two equilibria

$$BaF_2(s) \rightleftharpoons Ba^{2+}(aq) + 2\,F^-(aq) \qquad K_{sp} = 1.7 \times 10^{-6}$$
$$F^-(aq) + H_2O(l) \rightleftharpoons HF(aq) + OH^-(aq)$$
$$K_b = 2.9 \times 10^{-11}$$

(a) Write the chemical equation for the overall equilibrium and calculate the corresponding equilibrium constant. (b) Estimate the solubility of BaF_2 at (i) pH = 7.0; (ii) pH = 5.0.

6J.15 You find a bottle of a pure silver halide that could be AgCl or AgI. Develop a simple chemical test that would allow you to distinguish which compound was in the bottle.

6J.16 Which of the following compounds, if either, will dissolve in 1.00 M $HNO_3(aq)$: (a) $Bi_2S_3(s)$; (b) $FeS(s)$? Justify your answer by giving an appropriate calculation.

6J.17 A metal alloy sample is believed to contain silver, bismuth, and nickel. Explain how it could be determined qualitatively that all three of these metals are present.

6J.18 Zinc(II) readily forms the complex ion $Zn(OH)_4^{2-}$. Explain how this fact can be used to distinguish a solution of $ZnCl_2$ from $MgCl_2$.

Topic 6K Representing Redox Reactions

6K.1   Half-Reactions

6K.2   Balancing Redox Equations

How are complicated redox equations balanced?

How are reactions used to generate electricity??

Topic **6K:** Representing redox reactions → Topic **6L:** Galvanic cells

An important class of chemical reactions involves changes in the oxidation states of reactants. As explained in *Fundamentals* K, oxidation is the loss of one or more electrons from a reactant and reduction is the gain of one or more electrons. The joint process of electron loss and gain constitutes a "redox reaction." Redox reactions account for a wide range of chemical transformations, including the combustion of organic materials and the extraction of metals from ores.

One feature distinguishes redox reactions sharply from the proton transfer reactions characteristic of acids and bases (Topic 6A). Electrons are so intimately involved in bonding that when they migrate between species they often drag atoms, and even whole groups of atoms, with them. As a result, the chemical equations of redox reactions are often complicated because they commonly involve changes of atom partnerships as well as electron transfer.

6K.1 Half-Reactions

A **half-reaction** is the oxidation or reduction part of a reaction considered alone. An oxidation half-reaction shows the removal of electrons from a species that is being oxidized. For example to show just the oxidation of zinc in the reaction of zinc with silver ions,

$$Zn(s) + 2\,Ag^+(aq) \longrightarrow Zn^{2+}(aq) + 2\,Ag(s)$$

write:

$$Zn(s) \longrightarrow Zn^{2+}(s) + 2\,e^-$$

It is very important to understand that an oxidation half-reaction is merely a *conceptual* way of reporting an oxidation: the electrons are never actually free. In an equation for an oxidation half-reaction, the electrons released always appear on the right of the arrow. Their state is not given, because they are regarded as being in transit and as not having a definite physical state. The reduced and oxidized species in a half-reaction jointly form a **redox couple.** In this example, the redox couple consists of Zn^{2+} and Zn and is denoted Zn^{2+}/Zn. A redox couple has the form Ox/Red, where Ox is the oxidized form of the species and Red is the reduced form.

> **A Note on Good Practice:** Distinguish a conceptual half-reaction from an actual ionization, where the electron is removed and which is written, for example, as $Na(g) \rightarrow Na^+(g) + e^-(g)$, with the state of the electron specified.

Now consider reduction. To show the addition of electrons to a species, the corresponding half-reaction is written for electron gain. For example, to show the reduction of Ag^+ ions to Ag metal, write

$$Ag^+(aq) + e^- \longrightarrow Ag(s)$$

This half-reaction, too, is conceptual: as before, the electrons are not actually free and their state is not specified. In the equation for a reduction half-reaction, the electrons gained always appear on the left of the arrow. In this example, the redox couple is Ag^+/Ag.

Oxidation and reduction half-reactions always occur in combination; an oxidation cannot proceed without a corresponding reduction.

Half-reactions express the two contributions (oxidation and reduction) to an overall redox reaction.

Why Do You Need to Know This Material? Redox reactions constitute a major class of reactions with roles in analytical, synthetic, and biological chemistry. As a first step in understanding and using redox reactions, you need to be able to express them as "half-reactions" and to balance their often complex chemical equations.

What Do You Need to Know Already? You should be familiar with the basic concepts of redox reactions and oxidation numbers (*Fundamentals* K).

6K.2 Balancing Redox Equations

In redox reactions, it is conventional to write H^+ rather than H_3O^+ because that simplifies the equations and proton transfer is not the central focus in this context.

Balancing the chemical equation for a redox reaction by inspection can be a real challenge in some instances, especially for one taking place in aqueous solution, when water may participate and H_2O and either H^+ (in acidic solutions) or OH^- (in basic solutions) must be included. In such cases, it is easier to simplify the equation by separating it into its reduction and oxidation half-reactions, balance the half-reactions separately, and then add them to obtain the balanced equation for the overall reaction. When adding the equations for half-reactions, the number of electrons released by oxidation must be matched with the number used in reduction, because electrons are neither created nor destroyed in chemical reactions. The procedure is outlined in **Toolbox 6K.1** and illustrated in Examples 6K.1 and 6K.2.

Toolbox 6K.1 HOW TO BALANCE COMPLICATED REDOX EQUATIONS

CONCEPTUAL BASIS

When balancing redox equations, express the gain of electrons (reduction) and the loss of electrons (oxidation) as separate half-reactions, balance both atoms and charge in each of the two half-reactions, then combine the two half reactions. In the final step, the number of electrons released in the oxidation must equal the number used in the reduction.

PROCEDURE

In general, first balance the half-reactions separately, then combine them.

Step 1 Identify the species being oxidized and the species being reduced from the changes in their oxidation numbers.

Step 2 Write the two skeletal (unbalanced) equations for the oxidation and reduction half-reactions.

Step 3 Balance all elements in the half-reactions except O and H.

Step 4 (a) In acidic solution, balance O by using H_2O and then balance H by using H^+.

(b) In basic solution, balance O by using H_2O; then balance H by adding H_2O to the side of each half-reaction that needs H and adding OH^- to the other side. When ... OH^- ... → ... H_2O ... is

added to a half-reaction, one H atom is effectively added to the right. When ... H_2O .... → ... OH^- ... is added, one H atom is effectively added to the left. Note that in basic solution one H_2O molecule (along with one OH^- ion on the other side of the arrow) is added for each H atom needed. If necessary, cancel like species (typically H_2O) on opposite sides of the arrow.

Step 5 Balance the electric charges by adding electrons to the left for reductions and to the right for oxidations. Check that the number of electrons lost or gained in each half-reaction corresponds to the change in oxidation number of the element oxidized or reduced in that half-reaction.

Step 6 If necessary, multiply each entire half-reaction by the factor required to give equal numbers of electrons in the two half-reactions, and then add the two equations and include physical states. In some cases it is possible to simplify the half-reactions before they are combined by canceling species that appear on both sides of the arrow.

Step 7 Simplify the appearance of the complete equation by canceling species that appear on both sides of the arrow and check to make sure that charges as well as numbers of atoms balance.

Examples 6K.1 and 6K.2 illustrate this procedure.

EXAMPLE 6K.1 Balancing a redox equation in acidic solution

Permanganate ions are powerful oxidizing agents used in water treatment facilities to remove metals, such as iron, and toxic and malodorous chemicals, such as H_2S. If you are using permanganate solutions, you need to know the exact concentrations. Titration with oxalic acid is commonly used to determine the concentration of solutions of MnO_4^-, so you need to know the balanced redox equation. Permanganate ions, MnO_4^-, oxidize oxalic acid, $H_2C_2O_4$, in acidic aqueous solution. The skeletal equation (with states added) for the reaction is

$$MnO_4^-(aq) + H_2C_2O_4(aq) \longrightarrow Mn^{2+}(aq) + CO_2(g)$$

Write the balanced net ionic equation for this reaction.

PLAN To balance the equation, work through the procedure for an acidic solution set out in Toolbox 6K.1.

SOLVE

Reduction half-reaction

Step 1 Identify the species being reduced.

The oxidation number of Mn decreases from +7 to +2, so the Mn in MnO_4^- is reduced.

$$\overset{+7}{Mn}O_4^- \longrightarrow \overset{+2}{Mn}^{2+}$$

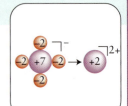

Step 2 Write the skeletal equation for this reduction.

$$MnO_4^- \longrightarrow Mn^{2+}$$

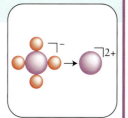

Step 3 Balance all elements except H and O.

$$MnO_4^- \longrightarrow Mn^{2+}$$

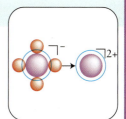

Step 4 Balance the O atoms by adding H_2O on the right.

$$MnO_4^- \longrightarrow Mn^{2+} + 4\,H_2O$$

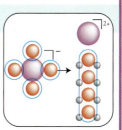

Balance the H atoms by adding H^+ on the left.

$$MnO_4^- + 8\,H^+ \longrightarrow Mn^{2+} + 4\,H_2O$$

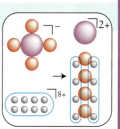

Step 5 Balance the net charges by adding electrons.

Net charge on the left is $+7$ and on the right $+2$; 5 electrons are needed on the left to lower the net charge from $+7$ to $+2$.

$$\overbrace{MnO_4^- + 8\,H^+ + 5\,e^-}^{\text{net charge} = +2} \longrightarrow \overbrace{Mn^{2+} + 4\,H_2O}^{\text{net charge} = +2}$$

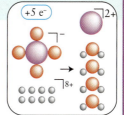

Note that the number of electrons transferred (5) corresponds to the change in oxidation number of Mn (from $+7$ to $+2$).

Oxidation half-reaction

Step 1 Identify the species being oxidized.

The oxidation number of carbon increases from $+3$ to $+4$, so the C atoms in oxalic acid are oxidized.

$$\overset{+3}{C_2}H_2O_4 \longrightarrow \overset{+4}{C}O_2$$

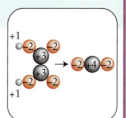

Step 2 Write the skeletal equation for the species being oxidized.

$$H_2C_2O_4 \longrightarrow CO_2$$

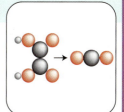

Step 3 Balance all elements except H and O.

$$H_2C_2O_4 \longrightarrow 2\ CO_2$$

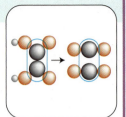

Step 4 Balance the O atoms by adding H_2O (none required).

$$H_2C_2O_4 \longrightarrow 2\ CO_2$$

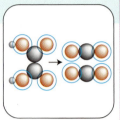

Balance the H atoms by adding H^+ on the right.

$$H_2C_2O_4 \longrightarrow 2\ CO_2 + 2\ H^+$$

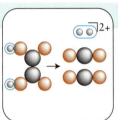

Step 5 Balance the net charges by adding electrons.

The net charge on the left is 0, and on the right $+2$; 2 electrons are needed on the right to lower the charge from $+2$ to 0.

$$\underbrace{H_2C_2O_4}_{\text{charge}\ =\ 0} \longrightarrow \underbrace{2\ CO_2 + 2\ H^+ + 2\ e^-}_{\text{net charge}\ =\ 0}$$

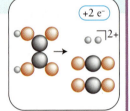

Note that the number of electrons transferred (2) corresponds to the change in oxidation number from 3 to 4 of two C atoms.

Now bring the two half-reactions together.

Step 6 First, match the numbers of electrons in each half-reaction.

Because 5 electrons are gained in one half-reaction and 2 are lost in the other, multiply the reduction half-reaction by 2 and the oxidation half-reaction by 5.

$$2\ MnO_4^- + 16\ H^+ + 10\ e^- \longrightarrow 2\ Mn^{2+} + 8\ H_2O$$
$$5\ H_2C_2O_4 \longrightarrow 10\ CO_2 + 10\ H^+ + 10\ e^-$$

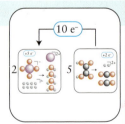

Add the equations and cancel the 10 electrons on each side of the arrow.

$$2\ MnO_4^- + 5\ H_2C_2O_4 + 16\ H^+ \longrightarrow 2\ Mn^{2+} + 8\ H_2O + 10\ CO_2 + 10\ H^+$$

Cancel 10 H^+ ions on each side and include physical states.

$$2\ MnO_4^-(aq) + 5\ H_2C_2O_4(aq) + 6\ H^+(aq) \longrightarrow 2\ Mn^{2+}(aq) + 8\ H_2O(l) + 10\ CO_2(g)$$

Note that all elements and charges now balance.

Self-test 6K.1A Copper reacts with dilute nitric acid to form copper(II) nitrate and the gas nitric oxide, NO. Write the balanced net ionic equation for the reaction.

[*Answer:* $3\ Cu(s) + 2\ NO_3^-(aq) + 8\ H^+(aq) \rightarrow 3\ Cu^{2+}(aq) + 2\ NO(g) + 4\ H_2O(l)$]

Self-test 6K.1B Acidified potassium permanganate solution reacts with sulfurous acid, $H_2SO_3(aq)$, to form sulfuric acid and manganese(II) ions. Write the balanced net ionic equation for the reaction. In acidic aqueous solution, H_2SO_3 is present as electrically neutral molecules and sulfuric acid is present as HSO_4^- ions.

Related Exercises 6K.1–6K.4, 6K.7

EXAMPLE 6K.2   Balancing a redox equation in basic solution

You are working in an analytical laboratory and have been asked to use the permanganate solution you prepared in Example 6K.1 to determine the concentration of bromide ions in a sample of groundwater sent to you by an environmental agency. The products of the reaction between bromide ions and permanganate ions, MnO_4^-, in basic aqueous solution are solid manganese(IV) oxide, MnO_2, and bromate ions, BrO_3^-. Balance the net ionic equation for the reaction.

PLAN Work through the procedure for a basic solution set out in Toolbox 6K.1.

Reduction half-reaction

Step 1 Identify the species being reduced.

The oxidation number of Mn changes from $+7$ in MnO_4^- to $+4$ in MnO_2, so the Mn in MnO_4^- is reduced.

$$\overset{+7}{MnO_4^-} \longrightarrow \overset{+4}{MnO_2}$$

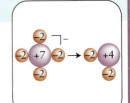

Step 2 Write the skeletal equation for reduction.

$$MnO_4^- \longrightarrow MnO_2$$

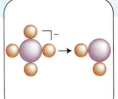

Step 3 The Mn atoms are balanced.

$$MnO_4^- \longrightarrow MnO_2$$

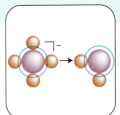

Step 4 Balance the O atoms by adding H_2O on the right.

$$MnO_4^- \longrightarrow MnO_2 + 2\,H_2O$$

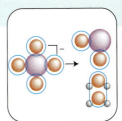

Four H atoms are needed on the left side. Balance the H atoms by adding 4 H_2O molecules to the left side and 4 OH^- ions to the right.

$$MnO_4^- + 4\,H_2O \longrightarrow MnO_2 + 2\,H_2O + 4\,OH^-$$
$$\uparrow\text{\_\_\_\_\_net 4 H atoms on left\_\_\_\_\_}\uparrow$$

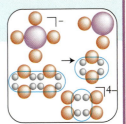

Cancel like species on opposite sides of the arrow (in this case, 2 H_2O).

$$MnO_4^- + 2\,H_2O \longrightarrow MnO_2 + 4\,OH^-$$

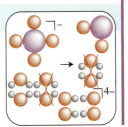

Step 5 Balance charge by adding electrons.

Net charge on the left is −1, and on the right it is −4; 3 electrons are needed on the left to lower the net charge from −1 to −4.

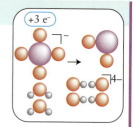

$$\overbrace{MnO_4^- + 2\,H_2O + 3\,e^-}^{\text{net charge} = -4} \longrightarrow \overbrace{MnO_2 + 4\,OH^-}^{\text{net charge} = -4}$$

Note that the number of electrons transferred (3) corresponds to the change in oxidation number of Mn from 7 to 4.

Oxidation half-reaction

Step 1 Identify the species being oxidized.

The oxidation number of Br increases from −1 in Br^- to +5 in BrO_3^-, and so Br^- is oxidized.

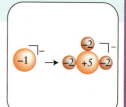

$$\overset{-1}{Br^-} \longrightarrow \overset{+5}{BrO_3^-}$$

Step 2 Write the skeletal equation for this oxidation.

$$Br^- \longrightarrow BrO_3^-$$

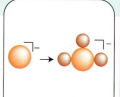

Step 3 The Br atoms are balanced.

$$Br^- \longrightarrow BrO_3^-$$

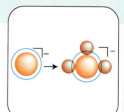

Step 4 Balance the O atoms by adding H_2O on the left.

$$Br^- + 3\,H_2O \longrightarrow BrO_3^-$$

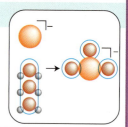

Six H atoms are needed on the right. Balance the H atoms by adding 6 H_2O molecules to the right and 6 OH^- ions to the left.

$$Br^- + 3\,H_2O + 6\,OH^- \longrightarrow BrO_3^- + 6\,H_2O$$
↑—net 6 H atoms on right—↑

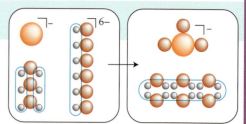

Cancel like species on opposite sides of the arrow (in this case, 3 H_2O).

$$Br^- + 6\,OH^- \longrightarrow BrO_3^- + 3\,H_2O$$

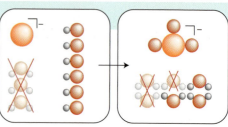

Step 5 Balance charge by adding electrons.

Net charge on the left is −7, and on the right it is −1; 6 electrons are needed on the right to lower the net charge from −1 to −7.

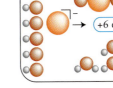

$$\overbrace{Br^- + 6\,OH^-}^{\text{net charge} = -7} \longrightarrow \overbrace{BrO_3^- + 3\,H_2O + 6\,e^-}^{\text{net charge} = -7}$$

Note that the number of electrons transferred (6) corresponds to the change in oxidation number of Br from −1 to +5.

Now bring the two half-reactions together.

Step 6 Match the numbers of electrons in each half-reaction.

Because 6 electrons are lost and 3 gained in the half-reactions as written, the reduction half-reaction must be multiplied by 2 for the numbers of electrons to match:

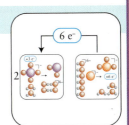

$$2\,MnO_4^- + 4\,H_2O + 6\,e^- \longrightarrow 2\,MnO_2 + 8\,OH^-$$

The oxidation half-reaction remains unchanged.

$$Br^- + 6\,OH^- \longrightarrow BrO_3^- + 3\,H_2O + 6\,e^-$$

Add the equations and cancel electrons.

$$2\,MnO_4^- + Br^- + 6\,OH^- + 4\,H_2O \longrightarrow 2\,MnO_2 + BrO_3^- + 8\,OH^- + 3\,H_2O$$

Step 7 Cancel 3 H_2O and 6 OH^- on each side and then include the physical states.

$$2\,MnO_4^-(aq) + Br^-(aq) + H_2O(l) \longrightarrow 2\,MnO_2(s) + BrO_3^-(aq) + 2\,OH^-(aq)$$

Self-test 6K.2A An alkaline (basic) solution of hypochlorite ions reacts with solid chromium(III) hydroxide to produce aqueous chromate ions and chloride ions. Write the net ionic equation for the reaction.

[***Answer:*** $2\,Cr(OH)_3(s) + 4\,OH^-(aq) + 3\,ClO^-(aq) \rightarrow 2\,CrO_4^{2-}(aq) + 5\,H_2O(l) + 3\,Cl^-(aq)$]

Self-test 6K.2B When iodide ions react with iodate ions in basic aqueous solution, triiodide ions, I_3^-, are formed. Write the net ionic equation for the reaction. (Note that the same product is obtained in each half-reaction.)

Related Exercises 6K.5, 6K.6, 6K.8

The chemical equation for a reduction half-reaction is added to the equation for an oxidation half-reaction to form the balanced chemical equation for the overall redox reaction.

What have you learned in this Topic?

You have learned how to recognize and write half-reactions for redox reactions. You have also learned a procedure for balancing redox equations in acidic and basic aqueous solutions.

The skills you have mastered are the ability to:

☐ **1.** Express oxidations and reductions as half-reactions.

☐ **2.** Balance a redox equation in acidic solution (Toolbox 6K.1 and Example 6K.1).

☐ **3.** Balance a redox equation in basic solution (Toolbox 6K.1 and Example 6K.2).

Topic 6K Exercises

6K.1 The following redox reaction is used in acidic solution in the Breathalyzer test to determine the level of alcohol in blood:

$$H^+(aq) + Cr_2O_7^{2-}(aq) + C_2H_5OH(aq) \longrightarrow$$
$$Cr^{3+}(aq) + C_2H_4O(aq) + H_2O(l)$$

Identify the elements undergoing oxidation or reduction and indicate their initial and final oxidation numbers. (b) Write and balance the oxidation half-reaction. (c) Write and balance

the reduction half-reaction. (d) Combine the half-reactions to produce a balanced redox equation.

6K.2 The following redox reaction is used in acidic solution to prepare orthotelluric acid:

$$Te(s) + ClO_3^-(aq) + H_2O(l) \longrightarrow H_6TeO_6(aq) + Cl_2(g)$$

Identify the elements undergoing oxidation or reduction and indicate their initial and final oxidation numbers. (b) Write and balance the oxidation half-reaction. (c) Write and balance the reduction half-reaction. (d) Combine the half-reactions to produce a balanced redox equation.

6K.3 Balance each of the following skeletal equations by using oxidation and reduction half-reactions. All the reactions take place in acidic solution. Identify the oxidizing agent and reducing agent in each reaction.
(a) Reaction of thiosulfate ion with chlorine gas:
$$Cl_2(g) + S_2O_3^{2-}(aq) \longrightarrow Cl^-(aq) + SO_4^{2-}(aq)$$
(b) Reaction of permanganate ion with sulfurous acid:
$$MnO_4^-(aq) + H_2SO_3(aq) \longrightarrow Mn^{2+}(aq) + HSO_4^-(aq)$$
(c) Reaction of hydrosulfuric acid with chlorine:
$$H_2S(aq) + Cl_2(g) \longrightarrow S(s) + Cl^-(aq)$$
(d) Reaction of chlorine in water:
$$Cl_2(g) \longrightarrow HClO(aq) + Cl_2(g)$$

6K.4 Balance each of the following skeletal equations by using oxidation and reduction half-reactions. All the reactions take place in acidic solution. Identify the oxidizing agent and reducing agent in each reaction.
(a) Reaction of the selenite ion with chlorate ion:
$$SeO_3^{2-}(aq) + ClO_3^-(aq) \longrightarrow SeO_4^{2-}(aq) + Cl_2(g)$$
(b) Formation of propanone (acetone), which is used in nail polish remover, from isopropanol (rubbing alcohol) by the action of dichromate ion:
$$C_3H_7OH(aq) + Cr_2O_7^{2-}(aq) \longrightarrow Cr^{3+}(aq) + C_3H_6O(aq)$$
(c) Reaction of gold with selenic acid:
$$Au(s) + SeO_4^{2-}(aq) \longrightarrow Au^{3+}(aq) + SeO_3^{2-}(aq)$$
(d) Preparation of stibine, SbH_3, from antimonic acid:
$$H_3SbO_4(aq) + Zn(s) \longrightarrow SbH_3(aq) + Zn^{2+}(aq)$$

6K.5 Balance each of the following skeletal equations by using oxidation and reduction half-reactions. All the reactions take place in basic solution. Identify the oxidizing agent and reducing agent in each reaction.

(a) Reaction of ozone with bromide ions:
$$O_3(aq) + Br^-(aq) \longrightarrow O_2(g) + BrO_3^-(aq)$$
(b) Reaction of bromine with itself (disproportionation) in aqueous solution:
$$Br_2(l) \longrightarrow BrO_3^-(aq) + Br^-(aq)$$
(c) Formation of chromate ions from chromium(III) ions:
$$Cr^{3+}(aq) + MnO_2(s) \longrightarrow Mn^{2+}(aq) + CrO_4^{2-}(aq)$$
(d) Formation of phosphine, PH_3, a poisonous gas with the odor of decaying fish:
$$P_4(s) \longrightarrow H_2PO_2^-(aq) + PH_3(aq)$$

6K.6 Balance each of the following skeletal equations by using oxidation and reduction half-reactions. All the reactions take place in basic solution. Identify the oxidizing agent and reducing agent in each reaction.
(a) Production of chlorite ions from dichlorine heptoxide by reaction with hydrogen peroxide solution:
$$Cl_2O_7(g) + H_2O_2(aq) \longrightarrow ClO_2^-(aq) + O_2(g)$$
(b) Reaction of permanganate ions with sulfide ions:
$$MnO_4^-(aq) + S^{2-}(aq) \longrightarrow S(s) + MnO_2(s)$$
(c) Reaction of hydrazine with chlorate ions:
$$N_2H_4(g) + ClO_3^-(aq) \longrightarrow NO(g) + Cl^-(aq)$$
(d) Reaction of plumbate ions and hypochlorite ions:
$$Pb(OH)_4^{2-}(aq) + ClO^-(aq) \longrightarrow PbO_2(s) + Cl^-(aq)$$

6K.7 The compound P_4S_3 is oxidized by nitrate ions in acid solution to give phosphoric acid, sulfate ions, and nitric oxide, NO. Write the balanced equation for each half-reaction and the overall equation for the reaction.

6K.8 Iron(II) hydrogen phosphite, $FeHPO_3$, is oxidized by hypochlorite ions in basic solution. The products are chloride ions, phosphate ions, and iron(III) hydroxide. Write the balanced equation for each half-reaction and the overall equation for the reaction.

Topic 6L Galvanic Cells

6L.1   The Structure of Galvanic Cells
6L.2   Cell Potential and Reaction Gibbs Free Energy
6L.3   The Notation for Cells

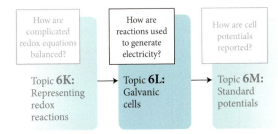

How are complicated redox equations balanced?

How are reactions used to generate electricity?

How are cell potentials reported?

Topic **6K:** Representing redox reactions → Topic **6L:** Galvanic cells → Topic **6M:** Standard potentials

The nature of electricity was unknown until the late eighteenth century, when the Italian scientist Luigi Galvani discovered that by touching the muscles of dead animals, mainly frogs, with rods bearing an electric charge, he could make their muscles twitch. He believed that electricity came from the muscles themselves. However, at the end of that century another Italian scientist, Alessandro Volta, suggested that the electricity was generated as a result of the muscles lying between two different metals when touched by the rod. He proved that the electricity came from metals by constructing a tower of alternating disks of different metals in layers separated by paper strips soaked in a solution of sodium chloride (**FIG. 6L.1**). This apparatus, a "voltaic pile," was the first device for producing an electric current, a simple battery.

6L.1 The Structure of Galvanic Cells

An **electrochemical cell** is a device in which an electric current—a flow of electrons through a circuit—is either produced by a spontaneous chemical reaction or used to bring about a nonspontaneous reaction. A **galvanic cell** is an electrochemical cell in which a spontaneous chemical reaction is used to generate an electric current. Technically, a **battery** is a collection of galvanic cells connected in series, so the voltage that it produces—its ability to push an electric current through a circuit—is the sum of the voltages of each cell.

To see how a spontaneous reaction can be used to generate an electric current, consider the redox reaction between zinc metal and copper(II) ions:

$$Zn(s) + Cu^{2+}(aq) \longrightarrow Zn^{2+}(aq) + Cu(s) \qquad \textbf{(A)}$$

When a piece of zinc metal is placed in an aqueous copper(II) sulfate solution, a layer of metallic copper begins to deposit on the surface of the zinc (as in Fig. K.5 of *Fundamentals* K). If the reaction could be observed at the atomic level, you would see that electrons are transferred from the Zn atoms to adjacent Cu^{2+} ions in the solution. These electrons reduce the Cu^{2+} ions to Cu atoms, which stick to the surface of the zinc or form a finely divided solid deposit in the beaker. The piece of zinc slowly disappears and the solution loses its blue color as the Zn atoms give up electrons and form colorless Zn^{2+} ions that drift off into the solution, replacing the blue Cu^{2+} ions.

Now suppose the reactants are separated but there is a pathway for the electrons to travel from the zinc metal to the copper(II) ions. As the reaction proceeds, electrons pass from the species being oxidized to the species being reduced. A galvanic cell makes use of this effect. It consists of two **electrodes,** or metallic conductors in contact with (but separated by) an **electrolyte,** an ionically conducting medium, inside the cell. In an ionic conductor, an electric current is carried by the movement of ions. The electrolyte is typically an aqueous solution of an ionic compound. Oxidation takes place at one electrode as the species being oxidized releases electrons

Why Do You Need to Know This Material? Electrochemistry is a crucial part of much of today's technology, including power generation for portable devices and vehicles. It also underlies analytical techniques and procedures for measuring thermodynamic properties.

What Do You Need to Know Already? You need to be familiar with redox reactions (*Fundamentals* K) and how to balance their equations (Topic 6K). The central connection between electrochemistry and thermodynamics is the Gibbs free energy and its relation to the maximum nonexpansion work that a reaction can do (Topic 4J).

Galvanic cells are also known as *voltaic cells.*

The formal term for "voltage" is "potential difference," measured in volts: $1 \text{ V} = 1 \text{ J}\cdot\text{C}^{-1}$ (see Section 6L.2).

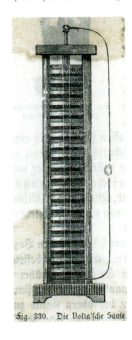

FIGURE 6L.1 Volta used a stack of alternating disks of silver and zinc separated by paper soaked in salt water to produce the first sustained electric current. (© *Bettmann/CORBIS.*)

The term "electrolyte" was first introduced in *Fundamentals* I, to refer to the solute. In the discussion of electrochemical cells, the term is commonly used to refer to the ionically conducting medium, which can be a liquid or a solid.

FIGURE 6L.2 In an electrochemical cell, a reaction takes place in two separate regions. Oxidation takes place at one electrode (the anode), and the electrons released travel through the external circuit to the other electrode, the cathode, where they cause reduction. The circuit is completed by the ions that carry the electric charge through the solution.

ANIMATION FIGURE 6L.2

A similar search for reliable power sources continues to this day, but now to power our mobile devices.

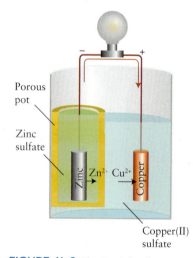

FIGURE 6L.3 The Daniell cell consists of copper and zinc electrodes dipping into solutions of copper(II) sulfate and zinc sulfate, respectively. The two solutions make contact through the porous barrier, which allows ions to pass through and complete the electric circuit.

into the metallic conductor and then into the external circuit. Reduction takes place at the other electrode, where the species that is undergoing reduction collects electrons from the metallic conductor connected to the external circuit (**FIG. 6L.2**). The overall chemical reaction can be thought of as pushing electrons on to one electrode from the oxidation process taking place there and pulling them off the other electrode from the reduction taking place at it. This push–pull process sets up a flow of electrons in the external circuit joining the two electrodes, and that current can be used to do electrical work.

The electrode at which oxidation takes place is called the **anode.** The electrode at which reduction takes place is called the **cathode.** Electrons are released by the oxidation half-reaction at the anode, travel through the external circuit, and re-enter the cell at the cathode, where they are used in the reduction half-reaction. A commercial galvanic cell has its cathode marked with a $+$ sign and its anode with a $-$ sign. Think of the $+$ sign as indicating the electrode at which the electrons enter and "add to" the cell and the $-$ sign as indicating the electrode at which the electrons leave the cell.

The *Daniell cell* is an early example of a galvanic cell that makes use of the oxidation of copper by zinc ions, as in reaction A. The British chemist John Daniell invented it in 1836, when the growth of telegraphy created an urgent need for a cheap, reliable, steady source of electric current. Daniell set up the arrangement shown in **FIG. 6L.3**, in which the two reactants are separated: the zinc metal is immersed in a solution of zinc sulfate and the copper electrode is immersed in a solution of copper(II) sulfate. For the electrons to travel from Zn atoms to Cu^{2+} ions and bring about the spontaneous reaction, they must pass from the zinc metal through the wire that serves as the external circuit, then through the copper electrode into the copper(II) solution. The Cu^{2+} ions are converted into Cu atoms at the cathode by the reduction half-reaction $Cu^{2+}(aq) + 2\,e^- \rightarrow Cu(s)$. At the same time, zinc atoms are converted into Zn^{2+} ions at the anode by the oxidation half-reaction $Zn(s) \rightarrow Zn^{2+}(aq) + 2\,e^-$. As Cu^{2+} ions are reduced, the solution at the cathode becomes negatively charged and the solution at the anode begins to develop a positive charge as additional Zn^{2+} ions enter the solution. To prevent this charge buildup, which would quickly stop the flow of electrons, the two solutions are in contact through a porous wall; ions provided by the electrolyte solutions move between the two compartments and complete the electrical circuit.

The electrodes in the Daniell cell are made of the metals involved in the reaction. However, not all electrode reactions include a conducting solid as a reactant or product. For example, to use the reduction $2\,H^+(aq) + 2\,e^- \rightarrow H_2(g)$ at an electrode, a chemically inert metallic conductor, such as an unreactive metal or graphite, must carry the electrons into or out of the electrode compartment. Platinum is customarily used as the electrode for hydrogen: the gas is bubbled over the metal immersed in a solution that contains hydrogen ions. This arrangement is called a *hydrogen electrode*. The entire compartment with metal conductor and electrolyte solution is commonly referred to as "the electrode" or, more formally, as a **half-cell.**

In a galvanic cell, a spontaneous chemical reaction draws electrons into the cell through the cathode, the site of reduction, and releases them at the anode, the site of oxidation.

6L.2 Cell Potential and Reaction Gibbs Free Energy

A reaction with a lot of pushing-and-pulling power generates a high potential difference (colloquially, a "high voltage"). A reaction with little pushing-and-pulling power generates only a small potential difference (a "low voltage"). An exhausted battery is a cell in which the reaction is at equilibrium; it has lost its power to move electrons and has a potential difference of zero. The SI unit of electric charge is the **coulomb.** One coulomb is defined as the magnitude of the charge delivered by a current of one ampere (1 A) flowing for one second: $1\ C = 1\ A\cdot s$. The SI unit of potential (and potential difference) is the **volt** (V). A volt is defined so that a charge of one coulomb (1 C) falling through a potential difference of one volt (1 V) releases one joule (1 J) of energy: $1\ C\cdot V = 1\ J$.

To express the ability of a cell to generate a potential difference quantitatively, it is helpful to note two points. First, electrical potential is analogous to gravitational potential. The maximum work that a falling weight can do is equal to its mass times the difference in gravitational potential. Similarly, the maximum work that an electron can do is equal to its charge times the difference in electrical potential through which it falls. Therefore, by calculating the work that can be done by an electron as it migrates between electrodes, it should be possible to infer the potential difference between them. Second, electrical work is a type of nonexpansion work because it involves moving electrons rather than changing the volume of the system. It is established in Topic 4J that, at constant temperature and pressure, the maximum nonexpansion work that a system can do is equal to the change in the Gibbs free energy. Thus, it should be possible to relate the potential difference caused by a reaction (an electrical property) to the Gibbs free energy of the reaction (a thermodynamic property).

How Is That Done?

The change in Gibbs free energy is equal to the maximum nonexpansion work that a reaction can do at constant pressure and temperature (Topic 4J):

$$\Delta G = w_{e,\,max}$$

The work done when an amount, n, of electrons (in moles) travels through a potential difference $\Delta \mathcal{V}$ is their total charge times the potential difference (Table 4A.1). The charge of one electron is $-e$; the charge per mole of electrons is $-eN_A$, where N_A is Avogadro's constant. Therefore, the total charge is $-neN_A$ and the work done is

$$w_e = \text{total charge} \times \text{potential difference} = (-neN_A) \times \Delta \mathcal{V}$$

(The "max" subscript on w_e will be restored shortly once the conditions for measuring $\Delta \mathcal{V}$ have been specified more precisely.) This expression is normally written in terms of **Faraday's constant**, F, the magnitude of the charge per mole of electrons (the product of the elementary charge e and Avogadro's constant N_A):

$$F = eN_A = (1.602\ 176\ldots \times 10^{-19}\ C) \times \{6.022\ 141\ldots \times 10^{23}\ (\text{mol e}^-)^{-1}\}$$
$$= 9.648\ 533\ldots \times 10^4\ C\cdot(\text{mol e}^-)^{-1}$$

Faraday's constant is normally abbreviated to $F = 9.6485 \times 10^4\ C\cdot mol^{-1}$ (or $96.485\ kC\cdot mol^{-1}$). Then

$$w_e = -nF\Delta \mathcal{V}$$

where n is the amount of electrons being transferred. Now restore the "max" subscript. Provided the cell is working *reversibly* (as explained in Topic 4A, in a reversible process the pushing force of the system is balanced against an equal opposing force), it produces a potential difference $\Delta \mathcal{V}_{rev}$ and the maximum amount of work, $w_{e,max}$. In this case, therefore, w_e can be identified with $w_{e,max}$ and hence with ΔG. Then

$$\Delta G = -nF\Delta \mathcal{V}_{rev}$$

The **cell potential**, E_{cell}, is the potential difference associated with a galvanic cell that is working reversibly. It follows by identifying E_{cell} with $\Delta \mathcal{V}_{rev}$ in the expression just derived that

$$\Delta G = -nFE_{cell} \tag{1a}$$

A cell works reversibly when its pushing power is balanced against an external matching source of potential difference. In practice, that means using a voltmeter with such a high resistance that the potential difference is measured without drawing any current. A *working cell,* one actually producing current, such as the cell in a media player, will produce a potential difference smaller than that predicted by this expression.

Equation 1a relates the thermodynamic information from Focuses 4 and 5 to the electrochemical information developed in this Focus. The units of ΔG are joules (or kilojoules),

The cell potential is still widely called the *electromotive force,* emf, of the cell, but it is not actually a force, so this term is becoming obsolete.

with a value that depends not only on E_{cell}, but also on the amount n (in moles) of electrons transferred in the reaction. Thus, in reaction A, $n = 2$ mol. As in the discussion of the relation between Gibbs free energy and equilibrium constants (Topic 5F), you will sometimes need to use this relation in its "molar" form, with n_r interpreted as a pure number (so, for reaction A, $n_r = 2$):

$$\Delta G_r = -n_r F E_{cell} \tag{1b}$$

The subscript "r" signals that the molar convention is being used (Topic 5G).

In this case, the units of ΔG_r are joules (or kilojoules) per mole.

Both forms of Eq.1 show how the potential difference produced in a galvanic cell working reversibly, E_{cell}, provides an experimental criterion of spontaneity of the reaction taking place in it. If the potential difference is positive, then the reaction Gibbs free energy at the current composition of the cell (for instance, as given by the concentration of reactants and products in the electrolyte) is negative, and the cell reaction has a spontaneous tendency to form products. If the potential difference is negative, then the reverse of the cell reaction is spontaneous, and the cell reaction has a spontaneous tendency to form reactants.

EXAMPLE 6L.1   Calculating the reaction Gibbs free energy

Suppose that you are in an engineering competition in which commercial power sources are prohibited. You might decide to use a Daniell cell to power a model electric car. You will need to know the potential of the cell corresponding to the concentrations of reagents you intend to use. The potential difference generated by the Daniell cell in which solutions of 1.00 mol·L^{-1} zinc and 0.01 mol·L^{-1} copper ions are used is $+1.04$ V. What is the reaction Gibbs free energy for this cell under these conditions?

ANTICIPATE   Because the cell generates a positive electric potential, the reaction is spontaneous under the conditions of the experiment, so you should expect to find that the Gibbs free energy of reaction is negative.

PLAN   Use Eq. 1a to determine the reaction Gibbs free energy for reaction A from the cell potential. Identify the value of n from the balanced equation (reaction A). In more complicated cases, you might need to refer to the two matched half-reactions.

From reaction A, $n = 2$ mol. Then, from $\Delta G = -nF E_{cell}$,

$$\Delta G = -(2 \text{ mol}) \times (9.6485 \times 10^4 \text{ C·mol}^{-1}) \times (1.04 \text{ V})$$

$$= -2.01 \times 10^5 \text{ C·V} = -201 \text{ kJ}$$

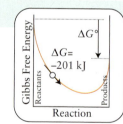

EVALUATE   Because the Gibbs free energy of reaction A under these conditions is -201 kJ, the reaction is spontaneous in the forward direction *for this composition of the cell*, as anticipated.

Self-test 6L.1A   The reaction taking place in a nicad (nickel–cadmium) cell is $Cd(s) + 2\ Ni(OH)_3(s) \rightarrow Cd(OH)_2(s) + 2\ Ni(OH)_2(s)$, and the cell potential when fully charged is $+1.25$ V. What is the reaction Gibbs free energy?

[**Answer:** -241 kJ]

Self-test 6L.1B   The reaction taking place in the silver cell used in some cameras and wristwatches is $Ag_2O(s) + Zn(s) \rightarrow 2\ Ag(s) + ZnO(s)$, and the cell potential when new is $+1.6$ V. What is the reaction Gibbs free energy?

Related Exercises 6L.1, 6L.2

The **standard cell potential**, $E_{cell}°$ is defined through

$$\Delta G° = -nF E_{cell}° \tag{2}$$

where $\Delta G°$, the standard Gibbs free energy of the reaction (Topic 4J), is defined as the difference of the molar Gibbs free energies of the products and reactants all in their standard states. As explained in that Topic, standard conditions are:

- all gases at 1 bar
- all participating solutes at 1 mol·L^{-1}
- all liquids and solids pure

More precisely, all the solutes should be at unit activity, not unit molarity. Activities differ appreciably from molarities in electrolyte solution because ions interact over long distances. However, this complication is ignored here. In some cases, it is possible to construct a cell that generates its standard potential. A Daniell cell in which the copper half-cell contains 1 M $CuSO_4(aq)$ and a pure copper electrode and the zinc half-cell contains 1 M $ZnSO_4(aq)$ and a pure zinc electrode generates its standard potential. However, in most cases the cell cannot be constructed with the reactants and products so neatly separated and, in general, it is best to think of $E_{cell}°$ as the value of $\Delta G°$ expressed as a potential difference in volts from Eq. 2 in the form $E_{cell}° = -\Delta G°/nF$.

It is important to understand the difference between $\Delta G°$ and ΔG (and therefore between $E_{cell}°$ and E_{cell}). The former is the difference in Gibbs free energies (appropriately weighted by the stoichiometric coefficients) between the isolated products and reactants *in their standard states*. The latter is the difference (again appropriately weighted) between the Gibbs free energies of the products and reactants that they have at an intermediate stage of the reaction, when they are surrounded by molecules that reflect the current composition of the mixture. Thus, whereas $\Delta G°$ has a fixed value characteristic of the reaction, ΔG changes as the reaction proceeds. Similarly $E_{cell}°$ has a fixed value characteristic of a reaction, whereas E_{cell} changes as the reaction proceeds.

When the chemical equation for a reaction is multiplied by a factor, ΔG (and $\Delta G°$) is increased by that factor, but E_{cell} (and $E_{cell}°$) remain the same. To see why that is so, note that when all the stoichiometric coefficients are multiplied by 2, the value of ΔG doubles. However, multiplying all the coefficients by 2 also doubles the value of n, and so $E_{cell} = -\Delta G/nF$ remains the same. That is, although the reaction Gibbs free energy (and its standard value) change when the chemical equation is multiplied by a factor, E_{cell} (and $E_{cell}°$) do not change:

	$\Delta G°$	$E_{cell}°$
$Zn(s) + Cu^{2+}(aq) \longrightarrow Zn^{2+}(aq) + Cu(s)$	-212 kJ	$+1.10$ V
$2 Zn(s) + 2 Cu^{2+}(aq) \longrightarrow 2 Zn^{2+}(aq) + 2 Cu(s)$	-424 kJ	$+1.10$ V

A practical implication of this conclusion is that the cell potential is independent of the size of the cell. To get a higher potential than predicted by Eq. 1, you would have to construct a battery by connecting cells in series; the potential is then the sum of the potentials of the individual cells (see the Interlude following Focus 6 for some examples).

> *Cell potential and reaction Gibbs free energy are related by Eq. 1 ($\Delta G = -nFE_{cell}$), and their standard values are related by Eq. 2 ($\Delta G° = -nFE_{cell}°$). The magnitude of the cell potential is independent of how the chemical equation is written.*

6L.3 The Notation for Cells

Chemists use a special notation to specify the structure of electrode compartments in a galvanic cell. The two electrodes in the Daniell cell, for instance, are denoted $Zn(s)|Zn^{2+}(aq)$ and $Cu^{2+}(aq)|Cu(s)$. Each vertical line represents an interface between phases—in this case, between solid metal and ions in solution in the order reactant|product.

The structure of a cell is expressed in a symbolic **cell diagram,** by using the conventions specified by IUPAC and used by chemists throughout the world. The diagram for the Daniell cell, for instance, is

$$Zn(s)|Zn^{2+}(aq)|Cu^{2+}(aq)|Cu(s)$$

In the Daniell cell, zinc sulfate and copper(II) sulfate solutions meet inside the porous barrier to complete the circuit. However, when different ions mingle together, they can affect the cell potential. To keep solutions from mixing, chemists use a "salt bridge" to join the two electrode compartments and complete the electrical circuit. A **salt bridge** typically consists of a gel containing a concentrated aqueous salt solution in an inverted U-tube (**FIG. 6L.4**). The bridge allows a flow of ions, and so it completes the electrical circuit, but the ions are chosen so that they do not affect the cell reaction (often KCl is

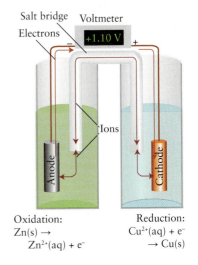

Salt bridge Voltmeter
Electrons +1.10 V

Ions

Anode   Cathode

Oxidation: Reduction:
$Zn(s) \rightarrow$ $Cu^{2+}(aq) + e^-$
$Zn^{2+}(aq) + e^-$ $\rightarrow Cu(s)$

FIGURE 6L.4 In a galvanic cell, electrons produced by oxidation at the anode ($-$) travel through the external circuit and re-enter the cell at the cathode ($+$), where they cause reduction. The circuit is completed inside the cell by migration of ions through the salt bridge. When the cell potential is measured, no current actually flows: the voltmeter measures the *tendency* of the electrons to flow from one electrode to the other.

used). In a cell diagram, a salt bridge is shown by a double vertical line ($\|$), and so the arrangement in Fig. 6L.4 is denoted

$$Zn(s)|Zn^{2+}(aq)\|Cu^{2+}(aq)|Cu(s)$$

THINKING POINT

Are there any salts that you definitely would not want to use in a salt bridge for this cell?

Any inert metallic component of an electrode is written as the outermost component of that electrode in the cell diagram. For example, a hydrogen electrode constructed with platinum is denoted $H^+(aq)|H_2(g)|Pt(s)$ when it is the right-hand electrode in a cell diagram and $Pt(s)|H_2(g)|H^+(aq)$ when it is the left-hand electrode. An electrode consisting of a platinum wire dipping into a solution of iron(II) and iron(III) ions is denoted either $Fe^{3+}(aq),Fe^{2+}(aq)|Pt(s)$ or $Pt(s)|Fe^{2+}(aq),Fe^{3+}(aq)$. In this case, the oxidized and reduced species are both in the same phase, and so a comma rather than a line is used to separate them. Pairs of ions in solution are normally written in the order Ox,Red.

> When it is important to emphasize the spatial arrangement of an electrode, the order may reflect that arrangement, as in $Cl^-(aq)|Cl_2(g)|Pt(s)$.

Self-test 6L.2A Write the diagram for a cell with a hydrogen electrode on the left and an iron(III)/iron(II) electrode on the right. The two electrode compartments are connected by a salt bridge, and platinum is used as the conductor at each electrode.

[**Answer**: $Pt(s)|H_2(g)|H^+(aq)\|Fe^{3+}(aq),Fe^{2+}(aq)|Pt(s)$]

Self-test 6L.2B Write the diagram for a cell that has an electrode consisting of a manganese wire dipping into a solution of manganese(II) ions on the left, a salt bridge, and a copper(II)/copper(I) electrode on the right with a platinum wire.

> Formerly, cell potentials were measured with a device called a potentiometer, but electronic devices are more reliable and easier to interpret.

Cell potential is measured with an electronic voltmeter (**FIG. 6L.5**). The cathode (the site of reduction) is identified by noting which terminal of the cell is positive. If the cathode is the electrode drawn on the right of the cell diagram, then, by convention, the cell potential is reported as positive, as in

electrons, e^-

Anode(−)   Cathode(+)
$$Zn(s)|Zn^{2+}(aq)\|Cu^{2+}(aq)|Cu(s)$$ $E_{cell} = +1$ V, for a given composition

In this case, the electrons can be thought of as tending to travel through the external circuit from the left of the cell as written (the anode) to the right (the cathode). However, if the cathode is found to be the electrode drawn on the left of the cell diagram, then the cell potential is reported as negative, as in

electrons, e^-

Cathode(+)   Anode(−)
$$Cu(s)|Cu^{2+}(aq)\|Zn^{2+}(aq)|Zn(s)$$ $E_{cell} = -1$ V for a given composition

FIGURE 6L.5 The cell potential is measured with an electronic voltmeter, a device designed to draw negligible current so that the composition of the cell does not change during the measurement. The display shows a positive value when the + terminal of the meter is connected to the cathode of the galvanic cell, which is also a + terminal. The salt bridge completes the electric circuit within the cell.
(W. H. Freeman photo by Ken Karp.)

LAB VIDEO FIGURE 6L.5

In summary, the sign of the cell potential reported in conjunction with a cell diagram is the same as the sign of the right-hand electrode in the diagram:

- A positive cell potential indicates that the right-hand electrode in the cell diagram is the cathode (the site of reduction, where electrons enter, "add to," the cell).
- A negative cell potential indicates that the right-hand electrode in the cell diagram is the anode (the site of oxidation, where electrons leave the cell).

A particular cell diagram corresponds to a specific layout of the corresponding cell reaction. *Solely for the purpose of writing the cell reaction corresponding to a given cell diagram,* the right-hand electrode in the diagram is supposed to be the site of reduction (the cathode) and the left-hand electrode is supposed to be the site of oxidation (the anode). The two half-reactions are then written as a reduction and an oxidation, respectively. Thus, for the cell written as

$$Zn(s)|Zn^{2+}(aq)\|Cu^{2+}(aq)|Cu(s)$$

and at a particular composition

Left (L)	Right (R)
$Zn(s) \longrightarrow Zn^{2+}(aq) + 2\,e^-$ (oxidation)	$Cu^{2+}(aq) + 2\,e^- \longrightarrow Cu(s)$ (reduction)
Overall (R + L): $Zn(s) + Cu^{2+}(aq) \longrightarrow Zn^{2+}(aq) + Cu(s)$	$E_{cell} = +1V$

Because $E_{cell} > 0$, and therefore $\Delta G < 0$ for this reaction, the cell reaction as written is spontaneous when the electrolytes have that particular composition. If instead the cell diagram is written

$$Cu(s)|Cu^{2+}(aq)\|Zn^{2+}(aq)|Zn(s)$$

and the composition is the same as before,

Left (L)	Right (R)
$Cu^{2+}(aq) + 2\,e^- \longrightarrow Cu(s)$ (oxidation)	$Zn(s) \longrightarrow Zn^{2+}(aq) + 2\,e^-$ (reduction)
Overall (L + R): $Cu(s) + Zn^{2+}(aq) \longrightarrow Cu^{2+}(aq) + Zn(s)$	$E_{cell} = -1V$

Because $E_{cell} < 0$, and therefore $\Delta G > 0$, the reverse of the cell reaction as written is spontaneous under these conditions.

The general procedure for writing the chemical equation for the reaction corresponding to a given cell diagram is set out in **Toolbox 6L.1.**

Toolbox 6L.1 HOW TO WRITE A CELL REACTION CORRESPONDING TO A CELL DIAGRAM

CONCEPTUAL BASIS

A cell diagram corresponds to a specific cell reaction in which the right-hand electrode in the cell diagram is treated as the site of reduction and the left-hand electrode is treated as the site of oxidation. The sign of the cell potential then distinguishes whether the resulting reaction is spontaneous in the direction written ($E_{cell} > 0$) or whether the reverse reaction is spontaneous ($E_{cell} < 0$) under the stated conditions.

PROCEDURE

Step 1 Write the half-reaction for the electrode on the right of the cell diagram as a reduction (remember: Right for Reduction).

Step 2 Write the half-reaction for the electrode on the left of the cell diagram as an oxidation.

Step 3 Multiply one or both equations by a factor if necessary to match the number of electrons in each half-reaction and then add the two equations.

If the cell potential is positive, then the reaction is spontaneous as written. If the cell potential is negative, then the reverse reaction is spontaneous.

This procedure is illustrated in Example 6L.2.

EXAMPLE 6L.2 Writing a cell reaction

The concentration of mercury, a toxic heavy metal pollutant, in aqueous solution depends in part on the redox properties of its compounds. Suppose you are studying the properties of mercury. You might need to construct an electrochemical cell and write the chemical equation for the cell reaction. Write the reaction for the cell $Pt(s)|H_2(g)|HCl(aq)|Hg_2Cl_2(s)|Hg(l)$.

PLAN Follow the procedure set out in Toolbox 6L.1.

SOLVE

Step 1 Write the equation for the reduction at the right-hand electrode.

$$Hg_2Cl_2(s) + 2\,e^- \longrightarrow 2\,Hg(l) + 2\,Cl^-(aq)$$

$Cl^-|Hg_2Cl_2|Hg$

Step 2 Write the equation for the oxidation at the left-hand electrode.

$$\tfrac{1}{2}\,H_2(g) \longrightarrow H^+(aq) + e^-$$

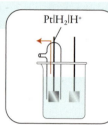

$Pt|H_2|H^+$

Step 3 To match the numbers of electrons, multiply the oxidation half-reaction by 2.

$$H_2(g) \longrightarrow 2\,H^+(aq) + 2\,e^-$$

Add the half-reactions together.

$$Hg_2Cl_2(s) + H_2(g) \longrightarrow 2\,Hg(l) + 2\,Cl^-(aq) + 2\,H^+(aq)$$

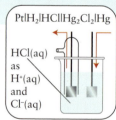

$Pt|H_2|HCl|Hg_2Cl_2|Hg$

$HCl(aq)$
as
$H^+(aq)$
and
$Cl^-(aq)$

If the concentrations in the cell are such that it is reported as having a positive potential (that is, the mercury/mercury(I) chloride electrode is found to be positive), then mercury(I) ions are being reduced and the reaction is spontaneous as written. If the concentrations are such that the potential is found to be negative (that is, the hydrogen electrode is found to be positive), then the reverse of the reaction as written would be spontaneous.

Self-test 6L.3A (a) Write the chemical equation for the reaction corresponding to the cell $Pt(s)|H_2(g)|H^+(aq)||Co^{3+}(aq), Co^{2+}(aq)|Pt(s)$. (b) Given that the cell potential is positive, is the cell reaction spontaneous as written?
[***Answer***: (a) $H_2(g) + 2\,Co^{3+}(aq) \rightarrow 2\,H^+(aq) + 2\,Co^{2+}(aq)$; (b) yes]

Self-test 6L.3B (a) Write the chemical equation for the reaction corresponding to the cell $Hg(l)|Hg_2Cl_2(s)|HCl(aq)||Hg_2(NO_3)_2(aq)|Hg(l)$. (b) Given that the cell potential is positive, is the cell reaction spontaneous as written?

Related Exercises 6L.3–6L.8

A final point to note might be surprising at first sight: the cell reaction need not be a redox reaction: all that is necessary is that the reaction can be expressed as a combination of reduction half-reactions taking place at the two electrodes. For example, the cell reaction for $Pt|H_2(g, P_L)|HCl(aq)|H_2(g, P_R)|Pt$ is simply $H_2(g, P_L) \rightarrow H_2(g, P_R)$, corresponding to the spontaneous expansion of the gas (provided $P_L > P_R$). The reduction half-reaction at each electrode is $2\,H^+(aq) + 2\,e^- \rightarrow H_2(g)$. The cell reaction might even be a precipitation reaction. For example, the cell reaction for $Ag(s)|AgCl(s)||Ag^+(aq)|Ag(s)$ is as follows:

R: $Ag^+(aq) + e^- \rightarrow Ag(s)$
L: $Ag(s) + Cl^-(aq) \rightarrow AgCl(s) + e^-$
R+L: $Ag^+(aq) + Cl^-(aq) \rightarrow AgCl(s)$

and corresponds to the precipitation of silver choride.

An electrode is designated by representing the interfaces between phases by a vertical line. A cell diagram depicts the physical arrangement of species and interfaces, with any salt bridge denoted by a double vertical line. The sign with which the cell potential is reported is the same as the measured sign of the right-hand electrode in the cell diagram. A positive cell potential indicates that the cell reaction is spontaneous as written for the given conditions.

What have you learned in this Topic?

You have learned how galvanic cells are constructed and how to relate the cell potential to the Gibbs free energy of a reaction. You have seen that the standard Gibbs free energy of a reaction can be expressed as a potential difference (in volts). You have also seen how to represent galvanic cells symbolically as a cell diagram.

The skills you have mastered are the ability to:

☐ **1.** Calculate the Gibbs free energy change for a reaction from a cell potential (Example 6L.1).

☐ **2.** Express the standard Gibbs free energy of a reaction as a potential difference.

☐ **3.** Depict a galvanic cell as a cell diagram.

☐ **4.** Write a reaction for a given cell diagram (Toolbox 6L.1 and Example 6L.2).

Topic 6L Exercises

6L.1 Calculate the standard reaction Gibbs free energy for the following cell reactions:
(a) $2\,Ce^{4+}(aq) + 3\,I^-(aq) \to 2\,Ce^{3+}(aq) + I_3^-(aq)$,
$E_{cell}° = +1.08\,V$
(b) $6\,Fe^{3+}(aq) + 2\,Cr^{3+}(aq) + 7\,H_2O(l) \to$
$\quad 6\,Fe^{2+}(aq) + Cr_2O_7^{2-}(aq) + 14\,H^+(aq),\ E_{cell}° = -1.29\,V$

6L.2 Calculate the standard reaction Gibbs free energy for the following cell reactions:
(a) $3\,Cr^{3+}(aq) + Bi(s) \to 3\,Cr^{2+}(aq) + Bi^{3+}(aq),\ E_{cell}° = -0.61\,V$
(b) $Mg(s) + 2\,H_2O(l) \to Mg^{2+}(aq) + H_2(g) + 2\,OH^-(aq)$,
$E_{cell}° = +2.36\,V$

6L.3 Write the half-reactions and the balanced equation for the cell reaction for each of the following galvanic cells:
(a) $Ni(s)|Ni^{2+}(aq)\|Ag^+(aq)|Ag(s)$
(b) $C(gr)|H_2(g)|H^+(aq)\|Cl^-(aq)|Cl_2(g)|Pt(s)$
(c) $Cu(s)|Cu^{2+}(aq)\|Ce^{4+}(aq),Ce^{3+}(aq)|Pt(s)$
(d) $Pt(s)|O_2(g)|H^+(aq)\|OH^-(aq)|O_2(g)|Pt(s)$
(e) $Pt(s)|Sn^{4+}(aq),Sn^{2+}(aq)\|Cl^-(aq)|Hg_2Cl_2(s)|Hg(l)$

6L.4 Write the half-reactions and the balanced equation for the cell reaction for each of the following galvanic cells:
(a) $Zn(s)|Zn^{2+}(aq)\|Au^{3+}(aq)|Au(s)$
(b) $Fe(s)|Fe^{2+}(aq)\|Fe^{3+}(aq)|Fe(s)$
(c) $Hg(l)|Hg_2Cl_2(s)|Cl^-(aq)\|Cl^-(aq)|CuCl(s)|Cu(s)$
(d) $Ag(s)|AgBr(s)|Br^-(aq)\|Cl^-(aq)|AgCl(s)|Ag(s)$
(e) $Pb(s)|Pb^{2+}(aq)\|MnO_4^-(aq),Mn^{2+}(aq),H^+(aq)|Pt(s)$

6L.5 Write the half-reactions, the balanced equation for the cell reaction, and the cell diagram for each of the following skeletal equations:
(a) $Ni^{2+}(aq) + Zn(s) \to Ni(s) + Zn^{2+}(aq)$
(b) $Ce^{4+}(aq) + I^-(aq) \to I_2(s) + Ce^{3+}(aq)$
(c) $Cl_2(g) + H_2(g) \to HCl(aq)$
(d) $Au^+(aq) \to Au(s) + Au^{3+}(aq)$

6L.6 Write the half-reactions, the balanced equation for the cell reaction, and the cell diagram for each of the following skeletal equations:
(a) $Mn(s) + Ti^{2+}(aq) \to Mn^{2+}(aq) + Ti(s)$
(b) $Fe^{3+}(aq) + H_2(g) \to Fe^{2+}(aq) + H^+(aq)$
(c) $Cu^+(aq) \to Cu(s) + Cu^{2+}(aq)$
(d) $MnO_4^-(aq) + H^+(aq) + Cl^-(aq) \to$
$\quad Cl_2(g) + Mn^{2+}(aq) + H_2O(l)$

6L.7 Write the half-reactions and devise a galvanic cell (write a cell diagram) to study each of the following reactions:
(a) $AgBr(s) \rightleftharpoons Ag^+(aq) + Br^-(aq)$, a solubility equilibrium
(b) $H^+(aq) + OH^-(aq) \to H_2O(l)$, the Brønsted neutralization reaction
(c) $Cd(s) + 2\,Ni(OH)_3(s) \to Cd(OH)_2(s) + 2\,Ni(OH)_2(s)$, the reaction in the nickel–cadmium cell

6L.8 Write balanced half-reactions and devise a galvanic cell (write a cell diagram) to study each of the following reactions:
(a) $Pb(NO_3)_2(aq) + K_2SO_4(aq) \to PbSO_4(s) + 2\,KNO_3(aq)$, a precipitation reaction
(b) $OH^-(aq, concentrated) \to OH^-(aq, dilute)$
(c) $Na(s) + S(s) \to Na^+(l) + S^{2-}(l)$, the reaction in a sodium–sulfur cell with a molten electrolyte

6L.9 (a) Write balanced half-reactions for the redox reaction of an acidified solution of potassium permanganate and iron(II) chloride. (b) Write the balanced equation for the cell reaction and devise a galvanic cell to study the reaction (write its cell diagram).

6L.10 (a) Write balanced half-reactions for the redox reaction between sodium perchlorate and copper(I) nitrate in an acidic solution. (b) Write the balanced equation for the cell reaction and devise a galvanic cell to study the reaction (write its cell diagram).

Topic 6M Standard Potentials

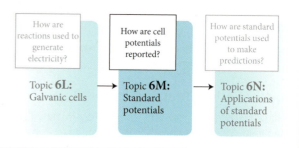

The term "electrode" in the discussion of electrochemical cells commonly refers to the entire half-cell of which the actual electrode, the metallic conductor, is a component and at which the potential is measured.

Standard potentials are also called _standard electrode potentials_. Because they are always written for reduction half-reactions, they are also sometimes called _standard reduction potentials_.

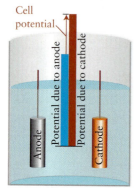

Thousands of galvanic cells can be devised and studied. Each cell contains two electrodes and, instead of having to list the potential of each possible cell, it is much more efficient to consider the contribution to the potential made by each half-cell.

6M.1 The Definition of Standard Potential

Under standard conditions (all solutes present at 1 mol·L^{-1}; all gases at 1 bar), each half-cell can be thought of as making a characteristic contribution to the cell potential called the **standard potential**, $E°$, of the electrode or of the corresponding redox couple. Each standard potential is a measure of the electron-pulling power of the reaction occurring at the electrode. In a galvanic cell, the reactions at each electrode pull in opposite directions, so the overall pulling power of the cell, the cell's standard potential, $E_{cell}°$, is the difference between the standard potentials of the two electrodes (**FIG. 6M.1**). That difference is always written as

$$E_{cell}° = E°(\text{electrode on right of cell diagram})$$
$$- E°(\text{electrode on left of cell diagram}) \tag{1a}$$

or, more succinctly,

$$E_{cell}° = E_R° - E_L° \tag{1b}$$

If $E_{cell}° > 0$, then the corresponding cell reaction is spontaneous under standard conditions (that is, as explained in Topic 5G, $K > 1$ for the reaction), and the electrode on the right of the cell diagram serves as the cathode. For example, for the cell

$$\text{Fe(s)}|\text{Fe}^{2+}(\text{aq})\|\text{Ag}^+(\text{aq})|\text{Ag(s)}$$

$$\text{corresponding to} \quad 2\,\text{Ag}^+(\text{aq}) + \text{Fe(s)} \longrightarrow 2\,\text{Ag(s)} + \text{Fe}^{2+}(\text{aq})$$

write

$$E_{cell}° = \overbrace{E°(\text{Ag}^+/\text{Ag})}^{E_R°} - \overbrace{E°(\text{Fe}^{2+}/\text{Fe})}^{E_L°}$$

and find (as explained later) $E_{cell}° = +1.24 \text{ V}$ at 25 °C. Because $E_{cell}° > 0$, the cell reaction has $K > 1$, with products dominant at equilibrium, and iron metal (in the couple Fe^{2+}/Fe) can reduce silver ions. If you had written the cell in the opposite order,

$$\text{Ag(s)}|\text{Ag}^+(\text{aq})\|\text{Fe}^{2+}(\text{aq})|\text{Fe(s)}$$

$$\text{corresponding to} \quad 2\,\text{Ag(s)} + \text{Fe}^{2+}(\text{aq}) \longrightarrow \text{Ag}^+(\text{aq}) + \text{Fe(s)}$$

you would have written

$$E_{cell}° = \overbrace{E°(\text{Fe}^{2+}/\text{Fe})}^{E_R°} - \overbrace{E°(\text{Ag}^+/\text{Ag})}^{E_L°}$$

FIGURE 6M.1 The cell potential can be thought of as the difference between the two potentials produced by the reactions occurring at the two electrodes.

and would have found $E_{cell}° = -1.24$ V. For the chemical equation written in this way, $K < 1$, with reactants dominant at equilibrium. However, the conclusion is the same: iron has a tendency to reduce silver.

A Note on Good Practice: Although $E_{cell}° > 0$ is often said to signify a spontaneous reaction, it actually signifies that the reaction is spontaneous only when the reactants and products all have their standard values. At other compositions, the reverse reaction might be spontaneous. It is much better to regard $E_{cell}° > 0$ as signifying that $K > 1$ for the reaction and $E_{cell}° < 0$ as signifying that $K < 1$, because the equilibrium constant is a fixed characteristic of the reaction. Whether the forward reaction is spontaneous then depends on the relative sizes of Q and K, as explained in Topic 5G.

A problem with compiling a list of standard potentials is that only the overall cell potential can be measured, not the contribution of a single half-cell. A voltmeter placed between the two electrodes of a galvanic cell measures the difference of their potentials, not the individual values. To provide numerical values for individual standard potentials, the standard potential of one particular electrode, the hydrogen electrode, is defined as zero at all temperatures:

$$2 H^+(g) + 2 e^- \longrightarrow H_2(g) \qquad E° = 0$$

In redox couple notation, where the couple denotes the reaction taking place at the metallic conductor, $E°(H^+/H_2) = 0$ at all temperatures. A hydrogen electrode in its standard state, with hydrogen gas at 1 bar and the hydrogen ions present at 1 mol·L^{-1} (strictly, unit activity), is called a **standard hydrogen electrode** (SHE). The standard hydrogen electrode is then used to define the standard potentials of all other electrodes:

The standard potential of an electrode is the standard potential of a cell (including the sign) in which the electrode lies on the right in the cell diagram and a hydrogen electrode lies on the left.

For example, for the cell

$$Pt(s)|H_2(g)|H^+(aq)\|Cu^{2+}(aq)|Cu(s)$$

the magnitude of the standard cell potential is found to be 0.34 V with the copper electrode acting as the cathode, and so $E_{cell}° = +0.34$ V. Because the hydrogen electrode contributes zero to the standard cell potential, that potential is attributed entirely to the copper electrode, and consequently

$$Cu^{2+}(aq) + 2 e^- \longrightarrow Cu(s) \qquad E°(Cu^{2+}/Cu) = +0.34 \text{ V}$$

The standard potential of an electrode is a measure of the tendency of the associated half-reaction to occur relative to the reduction of H^+ ions. For example, because the cell reaction

$$Cu^{2+}(aq) + H_2(g) \longrightarrow Cu(s) + 2 H^+(aq)$$

has $K > 1$ (because $E_{cell}° > 0$), the oxidizing ability of $Cu^{2+}(aq)$, as represented by the reduction half-reaction $Cu^{2+}(aq) + 2 e^- \rightarrow Cu(s)$, is greater than the oxidizing ability of $H^+(aq)$, as represented by $2 H^+(aq) + 2 e^- \rightarrow H_2(g)$. Consequently, Cu^{2+} ions can be reduced to metallic copper by hydrogen gas (in the sense that $K > 1$ for the reaction).

In general, the more positive the standard potential, the more strongly oxidizing is the oxidizing member of the redox couple and the more likely that member will be to undergo reduction itself (**FIG. 6M.2**).

Now consider the cell

$$Pt(s)|H_2(g)|H^+(aq)\|Zn^{2+}(aq)|Zn(s)$$

and the corresponding cell reaction

$$Zn^{2+}(aq) + H_2(g) \longrightarrow Zn(s) + 2 H^+(aq)$$

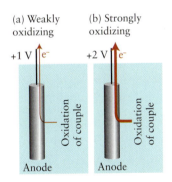

(a) Weakly oxidizing   (b) Strongly oxidizing

+1 V | e$^-$ +2 V | e$^-$

Oxidation of couple   Oxidation of couple

Anode   Anode

FIGURE 6M.2 (a) A redox couple at an electrode that has a small positive potential has only a small tendency to be reduced and so has a weak electron-pulling power (is a weak acceptor of electrons) relative to hydrogen ions; hence it is a weak oxidizing agent. (b) A redox couple at an electrode with a high positive potential has strong pulling power (is a strong acceptor of electrons) and is a strong oxidizing agent.

ANIMATION FIGURE 6M.2

The magnitude of the standard cell potential is 0.76 V, but in this case the hydrogen electrode (on the left) is found to be the cathode; therefore, the standard cell potential is reported as -0.76 V. Because the entire potential is attributed to the zinc electrode, write

$$Zn^{2+}(aq) + 2\,e^- \longrightarrow Zn(s) \qquad E°(Zn^{2+}/Zn) = -0.76 \text{ V}$$

The negative standard potential means that the Zn^{2+}/Zn electrode is the anode in a cell with H^+/H_2 as the other electrode and, therefore, that the reverse of the cell reaction, specifically,

$$Zn(s) + 2\,H^+(aq) \longrightarrow Zn^{2+}(aq) + H_2(g)$$

has $E_{cell}° > 0$ and therefore $K > 1$. The conclusion now is that the reducing ability of $Zn(s)$ in the half-reaction $Zn(s) \rightarrow Zn^{2+}(aq) + 2\,e^-$ is greater than the reducing ability of $H_2(g)$ in the half-reaction $H_2(g) \rightarrow 2\,H^+(aq) + 2\,e^-$. Consequently, zinc metal can reduce H^+ ions in acidic solution to hydrogen gas under standard conditions.

A common method for generating small quantities of $H_2(g)$ in teaching laboratories is to add $Zn(s)$ to aqueous solutions of HCl.

In Appendix 2B, standard potentials are listed both by numerical value and alphabetically, to make it easy to find the standard potential that you want.

In general, the more negative the potential, the more strongly reducing is the redox couple (**FIG. 6M.3**).

TABLE 6M.1 lists a number of standard potentials measured at 25 °C (the only temperature considered here); a longer list can be found in Appendix 2B. The standard potentials vary in a complicated way through the periodic table (**FIG. 6M.4**). However, the most negative—the most strongly reducing redox couples—are usually found toward the left of the periodic table, and the most positive—the most strongly oxidizing redox couples—are found toward the upper right.

Self-test 6M.1A Which is the stronger reducing agent for species in aqueous solution, lead or aluminum under standard conditions? (a) Evaluate the standard potential of the appropriate cell; (b) write the net ionic equation for the spontaneous reaction; (c) state your answer to the question.

[**Answer:** (a) $+1.53$ V; (b) $3\,Pb^{2+}(aq) + 2\,Al(s) \rightarrow 3\,Pb(s) + 2\,Al^{3+}(aq)$; (c) Al]

Self-test 6M.1B Which is the stronger oxidizing agent for species in aqueous solution, Cu^{2+} or Ag^+, under standard conditions? (a) Evaluate the standard potential of the appropriate cell; (b) write the net ionic equation for the corresponding cell reaction; (c) state your answer to the question.

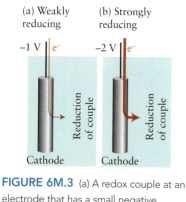

(a) Weakly reducing   **(b) Strongly reducing**

FIGURE 6M.3 (a) A redox couple at an electrode that has a small negative potential has only a small tendency to be oxidized and so has poor pushing power (is a weak donor of electrons) relative to hydrogen; hence it is a weak reducing agent. (b) A redox couple at an electrode with a *large* negative potential has strong pushing power (is a strong donor of electrons) and is a strong reducing agent.

ANIMATION FIGURE 6M.3

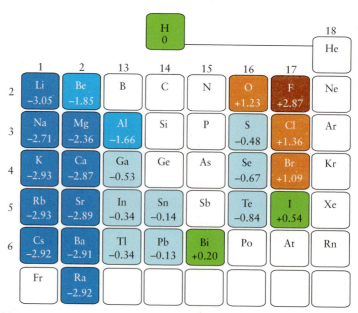

FIGURE 6M.4 The variation of standard potentials through the main groups of the periodic table. Note that the most negative values are in the s block and that the most positive values are close to fluorine.

TABLE 6M.1   Standard Potentials at 25 °C*

Species	Reduction half-reaction	$E°/V$
Oxidized form is strongly oxidizing		
F_2/F^-	$F_2(g) + 2\,e^- \longrightarrow 2\,F^-(aq)$	+2.87
Au^+/Au	$Au^+(aq) + e^- \longrightarrow Au(s)$	+1.69
Ce^{4+}/Ce^{3+}	$Ce^{4+}(aq) + e^- \longrightarrow Ce^{3+}(aq)$	+1.61
$MnO_4^-,H^+/Mn^{2+},H_2O$	$MnO_4^-(aq) + 8\,H^+(aq) + 5\,e^- \longrightarrow Mn^{2+}(aq) + 4\,H_2O\,(l)$	+1.51
Cl_2/Cl^-	$Cl_2(g) + 2\,e^- \longrightarrow 2\,Cl^-(aq)$	+1.36
$Cr_2O_7^{2-},H^+/Cr^{3+},H_2O$	$Cr_2O_7^{2-}(aq) + 14\,H^+(aq) + 6\,e^- \longrightarrow 2\,Cr^{3+}(aq) + 7\,H_2O(l)$	+1.33
$O_2,H^+/H_2O$	$O_2(g) + 4\,H^+(aq) + 4\,e^- \longrightarrow 2\,H_2O(l)$	+1.23; +0.82 at pH = 7
Br_2/Br^-	$Br_2(l) + 2\,e^- \longrightarrow 2\,Br^-(aq)$	+1.09
$NO_3^-,H^+/NO,H_2O$	$NO_3^-(aq) + 4\,H^+(aq) + 3\,e^- \longrightarrow NO(g) + 2\,H_2O(l)$	+0.96
Ag^+/Ag	$Ag^+(aq) + e^- \longrightarrow Ag(s)$	+0.80
Fe^{3+}/Fe^{2+}	$Fe^{3+}(aq) + e^- \longrightarrow Fe^{2+}(aq)$	+0.77
I_2/I^-	$I_2(s) + 2\,e^- \longrightarrow 2\,I^-(aq)$	+0.54
$O_2,H_2O/OH^-$	$O_2(g) + 2\,H_2O(l) + 4\,e^- \longrightarrow 4\,OH^-(aq)$	+0.40; +0.82 at pH = 7
Cu^{2+}/Cu	$Cu^{2+}(aq) + 2\,e^- \longrightarrow Cu(s)$	+0.34
$AgCl/Ag,Cl^-$	$AgCl(s) + e^- \longrightarrow Ag(s) + Cl^-(aq)$	+0.22
H^+/H_2	$2\,H^+(aq) + 2\,e^- \longrightarrow H_2(g)$	0, by definition
Fe^{3+}/Fe	$Fe^{3+}(aq) + 3\,e^- \longrightarrow Fe(s)$	−0.04
$O_2,H_2O/HO_2^-,OH^-$	$O_2(g) + H_2O(l) + 2\,e^- \longrightarrow HO_2^-(aq) + OH^-(aq)$	−0.08
Pb^{2+}/Pb	$Pb^{2+}(aq) + 2\,e^- \longrightarrow Pb(s)$	−0.13
Sn^{2+}/Sn	$Sn^{2+}(aq) + 2\,e^- \longrightarrow Sn(s)$	−0.14
Fe^{2+}/Fe	$Fe^{2+}(aq) + 2\,e^- \longrightarrow Fe(s)$	−0.44
Zn^{2+}/Zn	$Zn^{2+}(aq) + 2\,e^- \longrightarrow Zn(s)$	−0.76
$H_2O/H_2,OH^-$	$2\,H_2O(l) + 2\,e^- \longrightarrow H_2(g) + 2\,OH^-(aq)$	−0.83; −0.42 at pH = 7
Al^{3+}/Al	$Al^{3+}(aq) + 3\,e^- \longrightarrow Al(s)$	−1.66
Mg^{2+}/Mg	$Mg^{2+}(aq) + 2\,e^- \longrightarrow Mg(s)$	−2.36
Na^+/Na	$Na^+(aq) + e^- \longrightarrow Na(s)$	−2.71
K^+/K	$K^+(aq) + e^- \longrightarrow K(s)$	−2.93
Li^+/Li	$Li^+(aq) + e^- \longrightarrow Li(s)$	−3.05
Reduced form is strongly reducing		

*For a more extensive table, see Appendix 2B.

EXAMPLE 6M.1   Determining the standard potential of an electrode

You are working in the research laboratory of a company developing new forms of batteries for installation in satellites. As a part of your investigation, you have decided to study various combinations of metal electrodes. The standard potential of a Zn^{2+}/Zn electrode is −0.76 V, and the standard potential of the cell $Zn(s)|Zn^{2+}(aq)\|Sn^{4+}(aq),Sn^{2+}(aq)|Pt(s)$ is +0.91 V. What is the standard potential of the Sn^{4+}/Sn^{2+} electrode?

ANTICIPATE Because the cell potential is positive, the right-hand electrode (tin) is the cathode, the site of reduction. Therefore the zinc couple is more strongly reducing than the tin couple, and it has the more negative value. In other words, expect the tin couple to be less negative (more positive) than −0.76 V.

PLAN To determine the standard potential of an electrode from the potential of a standard cell in which the other electrode has a known standard potential, use Eq. 1 rearranged into $E_R° = E_{cell}° + E_L°$.

SOLVE

From $E_R° = E_{cell}° + E_L°$:

$$E°(Sn^{4+}/Sn^{2+}) = E_{cell}° + E°(Zn^{2+}/Zn)$$
$$= 0.91\text{ V} - 0.76\text{ V} = +0.15\text{ V}$$

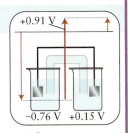

EVALUATE As expected, the Sn^{4+}/Sn^{2+} electrode has the less negative (more positive) standard potential.

Self-test 6M.2A The standard potential of an Ag^+/Ag electrode is $+0.80$ V, and the standard potential of the cell $Pt(s)|I_2(s)|I^-(aq)\|Ag^+(aq)|Ag(s)$ is $+0.26$ V. What is the standard potential of the I_2/I^- electrode?

[*Answer:* $+0.54$ V]

Self-test 6M.2B The standard potential of the Fe^{2+}/Fe electrode is -0.44 V and the standard potential of the cell $Fe(s)|Fe^{2+}(aq)\|Pb^{2+}(aq)|Pb(s)$ is $+0.31$ V. What is the standard potential of the Pb^{2+}/Pb electrode?

Related Exercises   6M.1, 6M.2

Tables of data do not always contain the standard potential required for a particular calculation but they might contain closely related values for the same element. For instance, you might require $E°(Ce^{4+}/Ce)$, whereas you are given only the values for $E°(Ce^{3+}/Ce)$ and $E°(Ce^{4+}/Ce^{3+})$. In such cases, when different numbers of electrons are involved in the half-reactions (here, 4, 3, and 1, respectively), the standard potentials cannot be added or subtracted directly. Instead, the values of $\Delta G°$ (which are additive) must be calculated for each half-reaction and then combined into the $\Delta G°$ for the desired half-reaction, which in turn is converted into the corresponding standard potential by using Eq. 2 of Topic 6L, $\Delta G° = -nFE_{cell}°$. The steps are:

- Find $E°$ values for the related half-reactions.
- Convert $E°$ values to $\Delta G°$.
- Combine the half-reactions and determine $\Delta G°$ for the desired reaction.
- Convert $\Delta G°$ into $E_{cell}°$ for the desired reaction.

EXAMPLE 6M.2   Calculating the standard potential of a redox couple from two related couples

You continue exploring different combinations of electrodes in pursuit of better batteries. You know that the Ce^{4+}/Ce^{3+} couple has been used in combination with Zn^{2+}/Zn to develop new power sources with high potentials and a large storage capacity. Use the information in Appendix 2B to determine $E°(Ce^{4+}/Ce)$, for which the reduction half-reaction is

$$Ce^{4+}(aq) + 4\,e^- \longrightarrow Ce(s) \tag{A}$$

PLAN Use the alphabetical listing in Appendix 2B to find half-reactions that can be combined to give the desired half-reaction. Combine these half-reactions and their standard Gibbs free energies of reaction. Use Eq. 1 of Topic 6L ($\Delta G° = -nFE°$) to obtain the standard potentials and then simplify the resulting expressions. Because the constant F is used in the second and third steps (as set out in the text), it cancels, and you do not need to insert its numerical value.

SOLVE From the data in Appendix 2B,

$$Ce^{3+}(aq) + 3\,e^- \longrightarrow Ce(s) \qquad E° = -2.48\ V \tag{B}$$
$$Ce^{4+}(aq) + e^- \longrightarrow Ce^{3+}(aq) \qquad E° = +1.61\ V \tag{C}$$

Convert the $E°$ values to $\Delta G°$ using Eq. 2 of Topic 6L, $\Delta G° = -nFE°$

(B) $Ce^{3+}(aq) + 3\,e^- \longrightarrow Ce(s)$
$$\Delta G° = -(3\ mol) \times F \times (-2.48\ V) = +7.44F\ V\cdot mol$$

(C) $Ce^{4+}(aq) + e^- \longrightarrow Ce^{3+}(aq)$
$$\Delta G° = -(1\ mol) \times F \times (+1.61\ V) = -1.61F\ V\cdot mol$$

When reactions B and C are added, the result is reaction A.

Add the reaction Gibbs free energies for reactions B and C to get the reaction Gibbs free energy for reaction A.

$$\Delta G° = \overbrace{7.44F\ V\cdot mol}^{\Delta G°(B)} + \overbrace{(-1.61F\ V\cdot mol)}^{\Delta G°(C)} = +5.83F\ V\cdot mol$$

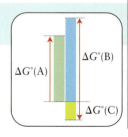

From Eq. 1 of Topic 6L in the form, $E° = -\Delta G°/nF$ with $n = 4$ mol:

$$E° = -\frac{5.83F \text{ V·mol}}{(4 \text{ mol}) \times F} = -1.46 \text{ V}$$

Note that this value is not the same as the sum of the potentials for half-reactions B and C (-0.87 V).

Self-test 6M.3A Use the data in Appendix 2B to calculate $E°(Au^{3+}/Au^{+})$.

[***Answer:*** $+1.26$ V]

Self-test 6M.3B Use the data in Appendix 2B to calculate $E°(Mn^{3+}/Mn)$.

Related Exercises 6M.9, 6M.10

The standard potential of an electrode is the standard potential of a cell in which the electrode lies on the right in the cell diagram and a hydrogen electrode lies on the left. A couple with negative standard potential has a thermodynamic tendency to reduce hydrogen ions in solution; a couple with a positive standard potential has a tendency to be reduced by hydrogen gas.

6M.2 The Electrochemical Series

Only a redox couple with a negative standard potential can reduce hydrogen ions under standard conditions (that is, have $K > 1$ for the reduction of hydrogen ions). A couple with a positive standard potential, such as Au^{3+}/Au, cannot reduce hydrogen ions under standard conditions in the sense that such a reaction would have $K < 1$, and commonly $K \ll 1$ for the reduction of hydrogen ions (**FIG. 6M.5**).

When Table 6M.1 is viewed as a table of relative strengths of oxidizing and reducing agents, it is called the **electrochemical series**. The species on the left of each half-reaction in Table 6M.1 are potentially oxidizing agents. They can themselves be reduced. Species on the right of the half-reactions are potentially reducing agents. An oxidized species in the list (on the left of the equation) has a tendency (in the sense that $K > 1$) to oxidize a reduced species that lies below it. For example, Cu^{2+} ions oxidize zinc metal. A reduced species (on the right of the equation) has a tendency to reduce an oxidized species that lies above it. For example, zinc metal reduces H^{+} ions to hydrogen gas.

The higher the position of a species on the left side of an equation in Table 6M.1, the greater is its oxidizing strength. For example, F_2 is a strong oxidizing agent, whereas Li^{+} is a very, very poor oxidizing agent. It also follows that, the lower the standard potential, the greater is the reducing strength of the reduced species on the right-hand side of a half-reaction in Table 6M.1. For example, lithium metal is the strongest reducing agent in the table.

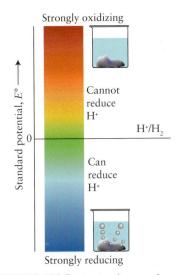

FIGURE 6M.5 The significance of standard potentials. Only redox couples with negative standard potentials (and hence lying below hydrogen in the electrochemical series) can reduce hydrogen ions under standard conditions. The reducing power increases as the standard potential decreases.

Self-test 6M.4A Can lead produce zinc metal from aqueous zinc sulfate under standard conditions?

[***Answer:*** No, because lead lies above zinc in Table 6M.1]

Self-test 6M.4B Can chlorine gas oxidize water to oxygen gas under standard conditions in basic solution?

The oxidizing and reducing power of a redox couple determines its position in the electrochemical series. The strongest oxidizing agents are at the top left of the series; the strongest reducing agents are at the bottom right of the series.

What have you learned in this Topic?

You have learned that standard potentials are reported by setting the standard potential of a hydrogen electrode equal to zero. You have also seen how to calculate the standard cell potential from tables of standard potentials and how to combine standard potentials to predict the value for a related couple.

The skills you have mastered are the ability to:

☐ **1.** Calculate a standard cell potential from two standard potentials (Self-Test 6M.1).

☐ **2.** Calculate a standard electrode potential from a standard cell potential (Example 6M.1).

☐ **3.** Calculate the standard cell potential from related redox couples (Example 6M.2).

☐ **4.** Use tables of standard potentials to assess the relative reducing and oxidizing strengths of redox couples.

Topic 6M Exercises

6M.1 A student was given a standard $Cu(s)|Cu^{2+}(aq)$ half-cell and another half-cell containing an unknown metal M in 1.00 M $M(NO_3)_2(aq)$ and formed the cell $M(s)|M^+(aq)||Cu^{2+}(aq)|Cu(s)$. The cell potential was found to be -0.689 V. What is the value of $E°(M^{2+}/M)$?

6M.2 A student was given a standard $Fe(s)|Fe^{2+}(aq)$ half-cell and another half-cell containing an unknown metal M in 1.00 M $MNO_3(aq)$ and formed the cell $M(s)|M^+(aq)||Fe^{2+}(aq)|Fe(s)$. At 25 °C, $E_{cell} = +1.24$ V. The reaction was allowed to continue overnight and the two electrodes were weighed. The iron electrode was found to be lighter and the unknown metal electrode was heavier. What is the value of $E°(M^+/M)$?

6M.3 Predict the standard potential of each of the following galvanic cells:
(a) $Pt(s)|Cr^{3+}(aq),Cr^{2+}(aq)||Cu^{2+}(aq)|Cu(s)$
(b) $Ag(s)|AgI(s)|I^-(aq)||Cl^-(aq)|AgCl(s)|Ag(s)$
(c) $Hg(l)|Hg_2Cl_2(s)|Cl^-(aq)||Hg_2^{2+}(aq)|Hg(l)$
(d) $C(gr)|Sn^{4+}(aq),Sn^{2+}(aq)||Pb^{4+}(aq),Pb^{2+}(aq)|Pt(s)$

6M.4 Predict the standard potential of each of the following galvanic cells:
(a) $Pt(s)|Fe^{3+}(aq),Fe^{2+}(aq)||Ag^+(aq)|Ag(s)$
(b) $U(s)|U^{3+}aq||V^{2+}(aq)|V(s)$
(c) $Sn(s)|Sn^{2+}(aq)||Sn^{4+}(aq),Sn^{2+}(aq)|Pt(s)$
(d) $Cu(s)|Cu^{2+}(aq)||Au^+(aq)|Au(s)$

6M.5 For each reaction that is spontaneous under standard conditions (that is, $K > 1$), write a cell diagram, determine the standard cell potential, and calculate $\Delta G°$ for the reaction:
(a) $2 NO_3^-(aq) + 8 H^+(aq) + 6 Hg(l) \rightarrow$
$\qquad 3 Hg_2^{2+}(aq) + 2 NO(g) + 4 H_2O(l)$
(b) $2 Hg^{2+}(aq) + 2 Br^-(aq) \rightarrow Hg_2^{2+}(aq) + Br_2(l)$
(c) $Cr_2O_7^{2-}(aq) + 14 H^+(aq) + 6 Pu^{3+}(aq) \rightarrow$
$\qquad 6 Pu^{4+}(aq) + 2 Cr^{3+}(aq) + 7 H_2O(l)$

6M.6 Predict the standard cell potential and calculate the standard reaction Gibbs free energy for galvanic cells having the following cell reactions:
(a) $3 Zn(s) + 2 Br^{3+}(aq) \rightarrow 3 Zn^{2+}(aq) + 2 Bi(s)$
(b) $2 H_2(g) + O_2(g) \rightarrow 2 H_2O(l)$ in acidic solution
(c) $2 H_2(g) + O_2(g) \rightarrow 2 H_2O(l)$ in basic solution
(d) $3 Au^+(aq) \rightarrow 2 Au(s) + Au^{3+}(aq)$

6M.7 Arrange the following metals in order of increasing strength as reducing agents for species in aqueous solution: (a) Cu, Zn, Cr, Fe; (b) Li, Na, K, Mg; (c) U, V, Ti, Al; (d) Ni, Sn, Au, Ag.

6M.8 Arrange the following species in order of increasing strength as oxidizing agents for species in aqueous solution: (a) Co^{2+}, Cl_2, Ce^{4+}, In^{3+}; (b) NO_3^-, ClO_4^-, HBrO, $Cr_2O_7^{2-}$, all in acidic solution; (c) O_2, O_3, HClO, and HBrO, all in acidic solution; (d) O_2, O_3, ClO^-, BrO^-, all in basic solution.

6M.9 Use the data in Appendix 2B to calculate $E°(U^{4+}/U)$.

6M.10 Use the data in Appendix 2B to calculate $E°(Ti^{3+}/Ti)$.

6M.11 Suppose that each of the following pairs of redox couples is combined to form a galvanic cell that generates a current under standard conditions. Identify the oxidizing agent and the reducing agent, write a cell diagram, and calculate the standard cell potential from the standard potentials of the electrodes. (a) Co^{2+}/Co and Ti^{3+}/Ti^{2+}; (b) La^{3+}/La and U^{3+}/U; (c) H^+/H_2 and Fe^{3+}/Fe^{2+}; (d) $O_3/O_2,OH^-$ and Ag^+/Ag.

6M.12 Suppose that each of the following pairs of redox couples is combined to form a galvanic cell that generates a current under standard conditions. Identify the oxidizing agent and the reducing agent, write a cell diagram, and calculate the standard cell potential from the standard potentials of the electrodes. (a) Pt^{2+}/Pt and $AgF/Ag,F^-$; (b) Cr^{3+}/Cr^{2+} and I_3^-/I^-; (c) H^+/H_2 and Ni^{2+}/Ni; (d) $O_3/O_2,OH^-$ and $O_3,H^+/O_2$.

6M.13 Identify the reactions with $K > 1$ in the following list and, for each such reaction, identify the oxidizing agent and calculate the standard cell potential.
(a) $Cl_2(g) + 2 Br^-(aq) \rightarrow 2 Cl^-(aq) + Br_2(l)$
(b) $MnO_4^-(aq) + 8 H^+(aq) + 5 Ce^{3+}(aq) \rightarrow$
$\qquad 5 Ce^{4+}(aq) + Mn^{2+}(aq) + 4 H_2O(l)$
(c) $2 Pb^{2+}(aq) \rightarrow Pb(s) + Pb^{4+}(aq)$
(d) $2 NO_3^-(aq) + 4 H^+(aq) + Zn(s) \rightarrow$
$\qquad Zn^{2+}(aq) + 2 NO_2(g) + 2 H_2O(l)$

6M.14 Identify the reactions with $K > 1$ among the following reactions and, for each such reaction, write balanced reduction and oxidation half-reactions. For those reactions, show that $K > 1$ by calculating the standard Gibbs free energy of the reaction. Use the smallest whole-number coefficients to balance the equations.
(a) $Mg^{2+}(aq) + Cu(s) \rightarrow ?$
(b) $Al(s) + Pb^{2+}(aq) \rightarrow ?$
(c) $Hg_2^{2+}(aq) + Ce^{3+}(aq) \rightarrow ?$
(d) $Zn(s) + Sn^{2+}(aq) \rightarrow ?$
(e) $O_2(g) + H^+(aq) + Hg(l) \rightarrow ?$

Topic 6N Applications of Standard Potentials

6N.1 Standard Potentials and Equilibrium Constants

6N.2 The Nernst Equation

6N.3 Ion-Selective Electrodes

6N.4 Corrosion

How are cell potentials reported?

How are standard potentials used to make predictions?

Topic 6M: Standard potentials → Topic 6N: Applications of standard potentials

Why Do You Need to Know This Material? Measurements of the potentials of galvanic cells are used to determine equilibrium constants and concentrations of dissolved ions and to monitor pH. An understanding of such cells gives insight into how to slow the corrosion of metals.

What Do You Need to Know Already? You need to be familiar with redox reactions (Topic 6K), galvanic cells (Topic 6L), standard potentials (Topic 6M), and the description of chemical equilibria in terms of the Gibbs free energy (Topic 5G).

In addition to supplying power for portable devices, galvanic cells have a wide variety of applications. For example, in chemistry they can be used to determine equilibrium constants and in medicine to monitor the concentration of ions such as Na^+, K^+, and Ca^{2+} in blood.

6N.1 Standard Potentials and Equilibrium Constants

One of the most useful applications of standard potentials is in the prediction of equilibrium constants from electrochemical data. As discussed in Topic 5G, the standard reaction Gibbs free energy, ΔG_r° (the "r" signifies the molar convention) is related to the equilibrium constant of the reaction by $\Delta G_r^\circ = -RT \ln K$. In Topic 6L, it is shown that the standard reaction Gibbs free energy is related to the standard potential of a galvanic cell by $\Delta G_r^\circ = -n_r F E_{cell}^\circ$, with n_r a pure number. The two equations are combined to give

$$n_r F E_{cell}^\circ = RT \ln K \qquad \text{(1a)}$$

This equation can be rearranged to express the equilibrium constant in terms of the standard cell potential:

$$\ln K = \frac{n_r F E_{cell}^\circ}{RT} \qquad \text{(1b)}$$

A Note on Good Practice: Because Eq. 1 uses the molar convention (n_r is a pure number), the units are kept straight: $F E_{cell}^\circ$ has the units joules per mole, as does RT, so the ratio $F E_{cell}^\circ / RT$ is a pure number; with n_r a pure number, the right-hand side is a pure number too (as it must be, if it is to be equal to a logarithm).

Because the magnitude of K increases exponentially with E_{cell}°,

- a reaction with a large positive E_{cell}° has $K \gg 1$
- a reaction with a large negative E_{cell}° has $K \ll 1$

THINKING POINT

What is the value of K for a reaction that has $E_{cell}^\circ = 0$?

The fact that E_{cell}° can be calculated by combining standard potentials implies that an equilibrium constant can be predicted for any reaction that can be expressed as two half-reactions. The reaction does not need to be a redox reaction (this point is emphasized at the end of Topic 6L). **Toolbox 6N.1** summarizes the steps and Example 6N.1 shows the steps in action.

Toolbox 6N.1 HOW TO CALCULATE EQUILIBRIUM CONSTANTS FROM ELECTROCHEMICAL DATA

CONCEPTUAL BASIS

The logarithm of the equilibrium constant of a reaction is proportional to the standard potential of the corresponding cell. A cell reaction with a large positive potential can be expected to have a strong tendency to take place and, therefore, to produce a high proportion of products at equilibrium. Therefore, expect $K > 1$ when $E_{cell}° > 0$ (and often $K \gg 1$). The opposite is true for a cell reaction with a negative standard potential.

PROCEDURE

The procedure for calculating an equilibrium constant from electrochemical data is as follows.

Step 1 Write the balanced equation for the reaction. Then find two reduction half-reactions in Table 6M.1 or Appendix 2B that combine to give that equation. Reverse one of the half-reactions and then add the two equations.

Step 2 Identify the numerical (unitless) value of n_r from the change in oxidation numbers or by examining the half-reactions (after multiplication by appropriate factors) for the number of electrons transferred in the balanced equation.

Step 3 To obtain $E_{cell}°$, subtract the standard potential of the electrode with the half-reaction that was reversed (to correspond to oxidation) from the standard potential of the unchanged electrode (which corresponds to reduction). In cell notation, $E_{cell}° = E_R° - E_L°$.

Step 4 Use the relation $\ln K = n_r F E_{cell}°/RT$ to calculate the value of K (with n_r a pure number).

At 25.00 °C (298.15 K), $RT/F = 0.025\ 693$ V; so, at that temperature,

$$\ln K = \frac{n_r E_{cell}°}{0.025\ 693\ \text{V}}$$

This procedure is illustrated in Example 6N.1.

EXAMPLE 6N.1   Calculating the equilibrium constant for a reaction

For reactions in which the equilibrium constant is very large or very small, it can be difficult to measure the concentration of all species in solution in order to determine K. An alternative method is to measure the cell potential for a reaction and then to use Eq. 1b to obtain the equilibrium constant. Calculate the equilibrium constant at 25.00 °C for the reaction $AgCl(s) \rightarrow Ag^+(aq) + Cl^-(aq)$. The equilibrium constant for this reaction is actually the solubility product, $K_{sp} = [Ag^+][Cl^-]$, for silver chloride (Topic 6I).

ANTICIPATE Because silver chloride is almost insoluble, you should expect K_{sp} to be very small (and $E_{cell}°$ therefore to be negative).

PLAN Follow the procedure in Toolbox 6N.1.

SOLVE

Step 1 Find the two reduction half-reactions required for the cell reaction and their standard potentials.

$$\text{R: } AgCl(s) + e^- \longrightarrow Ag(s) + Cl^-(aq) \qquad E° = +0.22\ \text{V}$$
$$\text{L: } Ag^+(aq) + e^- \longrightarrow Ag(s) \qquad E° = +0.80\ \text{V}$$

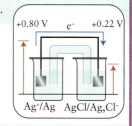

Reverse the second half-reaction.

$$\text{Oxidation: } Ag(s) \longrightarrow Ag^+(aq) + e^-$$

Add this equation to the reduction half-reaction and cancel species that appear on both sides of the equation (in blue, here).

$$AgCl(s) + Ag(s) + e^- \longrightarrow Ag^+(aq) + Ag(s) + Cl^-(aq) + e^-$$
$$\text{becomes: } AgCl(s) \longrightarrow Ag^+(aq) + Cl^-(aq)$$

Step 2 From the half-reactions, note the stoichiometric coefficient for the number of electrons transferred.

$$n_r = 1$$

Step 3 Find $E_{cell}°$ from $E_{cell}° = E_R° - E_L°$.

$$E_{cell}° = 0.22\ \text{V} - 0.80\ \text{V} = -0.58\ \text{V}$$

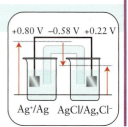

From $\ln K = n_rFE_{cell}°/RT = n_rE_{cell}°/(RT/F)$ with K identified as K_{sp} and $RT/F = 0.025\ 693$ V at 298 K,

$$\ln K_{sp} = \frac{(1) \times (-0.58\ \text{V})}{\underbrace{0.025\ 693\ \text{V}}_{RT/F}} = -\frac{0.58}{0.025\ 693} = -22.5...$$

Take the natural antilogarithm (e^x) of $\ln K_{sp}$.

$$K_{sp} = e^{-22.5...} = 1.6 \times 10^{-10}$$

EVALUATE The value of K_{sp} is indeed very small; it is also the same as that listed in Table 6I.1. Many of the solubility products listed in tables were determined from cell potential measurements and calculations like this one.

Self-test 6N.1A Use Appendix 2B to calculate the solubility product of mercury(I) chloride, Hg_2Cl_2.

[*Answer:* 2.6×10^{-18}]

Self-test 6N.1B Use the tables in Appendix 2B to calculate the solubility product of cadmium hydroxide, $Cd(OH)_2$.

Related Exercises 6N.1, 6N.2, 6N.11, 6N.12

The equilibrium constant of a reaction can be calculated from standard potentials by combining the equations for the half-reactions to give the cell reaction of interest and determining the standard potential of the corresponding cell.

6N.2 The Nernst Equation

As a reaction proceeds toward equilibrium, the concentrations of its reactants and products change and ΔG approaches zero. Therefore, as reactants are consumed in a working electrochemical cell, the cell potential also decreases until finally it reaches zero. A "dead" battery is one in which the cell reaction has reached equilibrium. At equilibrium, a cell generates zero potential difference across its electrodes, and the reaction can no longer do work. To describe this behavior quantitatively, it is necessary to find how the cell potential varies with the concentrations of species in the cell.

How Is That Done?

To establish how the cell potential depends on concentration, first note that Topic 5G provides an expression for the relation between the reaction Gibbs free energy and the composition:

$$\Delta G_r = \Delta G_r° + RT \ln Q$$

where Q is the reaction quotient for the cell reaction (Eq. 5 of Topic 5G). Because $\Delta G_r = -n_rFE_{cell}$ and $\Delta G_r° = -n_rFE_{cell}°$, it follows that

$$-n_rFE_{cell} = -n_rFE_{cell}° + RT \ln Q$$

At this point, divide through by $-n_rF$ to get an expression for E_{cell} in terms of Q, which is given below.

The equation for the concentration dependence of the cell potential that has just been derived,

$$E_{cell} = E_{cell}° - \frac{RT}{n_rF} \ln Q \tag{2a}$$

(with n_r a pure number) is called the **Nernst equation,** for the German electrochemist Walther Nernst, who first derived it. At 298.15 K, $RT/F = 0.025\ 693$ V; so at that temperature the Nernst equation takes the form

$$E_{cell} = E_{cell}° - \frac{0.025\ 693\ \text{V}}{n_r} \ln Q \tag{2b}$$

It is sometimes convenient to use this equation with common logarithms, in which case make use of the relation $\ln x = \ln 10 \times \log x = 2.303 \log x$. At 298.15 K,

$$E_{cell} = E_{cell}° - \frac{RT \ln 10}{n_rF} \log Q = E_{cell}° - \frac{0.059\ 160\ \text{V}}{n_r} \log Q \tag{2c}$$

The Nernst equation is widely used to estimate the potentials of cells under nonstandard conditions. In biology it is used, among other things, to estimate the potential difference across biological cell membranes, such as those of neurons.

THINKING POINT

When might the use of common logarithms in the Nernst equation be useful?

EXAMPLE 6N.2   Using the Nernst equation to predict a cell potential

As in Example 6L.1, you are planning to use a Daniell cell to power a model electric car. However, you find that you do not have standard solutions available. You have only dilute solutions, and you need to know whether they will be adequate to power the car. Calculate the potential of a Daniell cell at 25 °C in which the concentration of Zn^{2+} ions is 0.10 mol·L^{-1} and that of the Cu^{2+} ions is 0.0010 mol·L^{-1}.

PLAN Write the balanced equation for the cell reaction and the corresponding expression for Q, and note the value of n_r. Then determine $E_{cell}°$ from the standard potentials in Table 6M.1 or Appendix 2B. Evaluate Q for the stated conditions. Calculate the cell potential by substituting these values into the Nernst equation, Eq. 2b.

SOLVE The Daniell cell and the corresponding cell reaction are

$$Zn(s)|Zn^{2+}(aq)||Cu^{2+}(aq)|Cu(s) \qquad Cu^{2+}(aq) + Zn(s) \longrightarrow Zn^{2+}(aq) + Cu(s)$$

Set up the reaction quotient.

$$Q = \frac{a_{Zn^{2+}}}{a_{Cu^{2+}}} \approx \frac{[Zn^{2+}]}{[Cu^{2+}]} = \frac{0.10}{0.0010}$$

From the balanced equation, note the value of n_r.

$$n_r = 2$$

Determine the value of $E_{cell}° = E_R° - E_L°$.

$$E_{cell}° = 0.34 - (-0.76)\ V = +1.10\ V$$

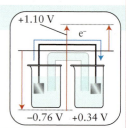

From $E_{cell} = E_{cell}° - (RT/n_rF)\ \ln Q$,

$$E_{cell} = 1.10\ V - \frac{0.025\ 693\ V}{2}\ \ln\frac{0.10}{0.0010}$$

$$= 1.10\ V - 0.059\ V = +1.04\ V$$

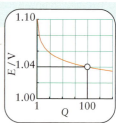

Self-test 6N.2A Calculate the potential of the cell $Zn(s)|Zn^{2+}(aq,\ 1.50\ mol·L^{-1})||Fe^{2+}(aq,\ 0.10\ mol·L^{-1})|Fe(s)$.

[**Answer:** +0.29 V]

Self-test 6N.2B Calculate the potential of the cell $Ag(s)|Ag^+(aq,\ 0.0010\ mol·L^{-1})||Ag^+(aq,\ 0.010\ mol·L^{-1})|Ag(s)$.

Related Exercises 6N.3, 6N.4

An important application of the Nernst equation is the measurement of concentration. In a **concentration cell,** the two half-cells are identical except for their concentrations. For such a cell, there is no driving force for change when the two concentrations are the same (which is the case when they are in their standard states), and so $E_{cell}° = 0$. Therefore, at 25 °C the potential corresponding to the cell reaction is related to Q by $E_{cell} = -(0.025\ 693\ V/n_r) \times \ln Q$. For example, a concentration cell having two Ag^+/Ag electrodes is

$$Ag(s)|Ag^+(aq,\ L)||Ag^+(aq,\ R)|Ag(s) \qquad Ag^+(aq,\ R) \longrightarrow Ag^+(aq,\ L)$$

The cell reaction has $n_r = 1$ and $Q = [Ag^+]_L/[Ag^+]_R$. If the concentration of Ag^+ in the right-hand electrode is 1 mol·L^{-1}, Q is equal to $[Ag^+]_L$, and the Nernst equation is

$$E_{cell} = (0.025\ 693\ V)\ \ln[Ag^+]_L$$

Therefore, by measuring E_{cell}, the concentration of Ag^+ in the left-hand electrode compartment can be determined. If the concentration of Ag^+ ions in the left-hand electrode is less than that in the right, then $E_{cell} > 0$ for the cell as specified and the right-hand electrode will be found to be the cathode.

EXAMPLE 6N.3   Using the Nernst equation to find a concentration

Measurement of ion concentrations with electrodes is used in a wide range of applications. Suppose you are interested in removing silver ions from exhausted electroplating solutions by precipitating the silver as silver chloride. To test the efficiency of your process, you decide to build an electrochemical cell to measure Ag^+ concentrations. Each electrode compartment of your galvanic cell contains a silver electrode and 10.0 mL of 0.10 M $AgNO_3(aq)$; they are connected by a salt bridge. You now add 10.0 mL of 0.10 M $NaCl(aq)$ to the left-hand electrode compartment. Almost all the silver precipitates as silver chloride, but a little remains in solution as a saturated solution of AgCl. The measured cell potential is $E_{cell} = +0.42$ V. What is the concentration of Ag^+ in the saturated solution?

ANTICIPATE Because silver chloride is very insoluble, you should expect a very low concentration.

PLAN Use the Nernst equation, Eq. 2, to find the concentration of Ag^+ in the compartment with the precipitate. The standard cell potential is 0 (in their standard states the half-cells are identical). At 25.00 °C, $RT/F = 0.025\ 693$ V.

SOLVE The cell and the corresponding cell reaction are

$$Ag(s)|Ag^+(aq, L)||Ag^+(aq, R)|Ag(s) \qquad Ag^+(aq, R) \longrightarrow Ag^+(aq, L)$$

with $[Ag^+(aq)]_L$ the unknown concentration and $[Ag^+(aq)]_R = 0.10\ mol \cdot L^{-1}$.

Set up the reaction quotient, Q.

$$Q = \frac{[Ag^+]_L}{[Ag^+]_R} = \frac{[Ag^+]_L}{0.10}$$

From the balanced equation, note the value of n_r.

$$n_r = 1$$

From $E_{cell} = E_{cell}° - (RT/n_rF) \ln Q$ rearranged into $\ln Q = -E_{cell}/(RT/n_rF)$,

$$\ln Q = \frac{-0.42\ V}{0.025\ 693\ V} = -16.3\ldots$$

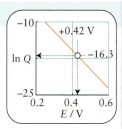

From $Q = e^{\ln Q}$,

$$Q = e^{-16.3\ldots}$$

From $[Ag^+]_L = Q[Ag^+]_R$,

$$[Ag^+]_L = e^{-16.3\ldots} \times 0.10 = 7.9 \times 10^{-9}$$

EVALUATE The concentration of Ag^+ ions in the saturated solution is 7.9 nmol·L^{-1}; as expected, this is a very low value.

Self-test 6N.3A Calculate the molar concentration of Y^{3+} in a saturated solution of YF_3 by using a cell constructed with two yttrium electrodes. The electrolyte in one compartment is 1.0 M $Y(NO_3)_3(aq)$. In the other compartment you have prepared a saturated solution of YF_3. The measured cell potential is +0.34 V at 298 K.

[**Answer:** $5.7 \times 10^{-18}\ mol \cdot L^{-1}$]

Self-test 6N.3B Calculate the potential of a cell constructed with two silver electrodes. The electrolyte in one compartment is 1.0 M $AgNO_3(aq)$. In the other compartment, NaOH has been added to a $AgNO_3$ solution until the pH = 12.5 at 298 K.

Related Exercises 6N.15, 6N.16

The dependence of cell potential on composition is expressed by the Nernst equation, Eq. 2.

6N.3 Ion-Selective Electrodes

An important application of the Nernst equation is the measurement of pH (and, through pH, acidity constants, Topic 6C). The pH of a solution can be measured electrochemically with a **pH meter.** The instrument makes use of a cell in which one electrode is sensitive to the H_3O^+ concentration and another electrode with a fixed potential serves as a reference. An electrode sensitive to the concentration of a particular ion is called an **ion-selective electrode.**

One combination of electrodes that can be used to determine pH is a hydrogen electrode connected through a salt bridge to a *calomel electrode.* The reduction half-reaction for the calomel electrode is

$$Hg_2Cl_2(s) + 2\,e^- \longrightarrow 2\,Hg(l) + 2\,Cl^-(aq) \qquad E^\circ = +0.27\ V$$

The overall cell reaction is

$$Hg_2Cl_2(s) + H_2(g) \longrightarrow 2\,H^+(aq) + 2\,Hg(l) + 2\,Cl^-(aq) \qquad Q = \frac{[H^+]^2[Cl^-]^2}{P_{H_2}}$$

Provided that the pressure of hydrogen is 1 bar, the reaction quotient can be written $Q = [H^+]^2[Cl^-]^2$. To find the concentration of hydrogen ions, write the Nernst equation:

$$E_{cell} = E_{cell}^\circ - \frac{RT}{2F}\ln([H^+]^2[Cl^-]^2)$$

Then apply $\ln(ab) = \ln a + \ln b$,

$$E_{cell} = E_{cell}^\circ - \frac{RT}{2F}\ln[Cl^-]^2 - \frac{RT}{2F}\ln[H^+]^2$$

Now use $\ln a^2 = 2\ln a$ to obtain

$$E_{cell} = \overbrace{E_{cell}^\circ - \frac{RT}{F}\ln[Cl^-]}^{E_{cell}'} - \frac{RT}{F}\ln[H^+]$$

The Cl^- concentration of a calomel electrode is fixed at the time of manufacture by saturating the solution with KCl, and so $[Cl^-]$ is a constant. As indicated, the first two terms on the right can be combined into a single constant, $E_{cell}' = E_{cell}^\circ - (RT/F)\ln[Cl^-]$:

$$E_{cell} = E_{cell}' - \frac{RT}{F}\ln[H^+]$$

Finally, because $\ln x = \ln 10 \times \log x$, and $(RT/F)\ln 10 = (0.025\ 693\ V) \times 2.303 = 0.0592$ V at 25 °C,

$$E_{cell} = E_{cell}' - \frac{RT}{F}\ln 10 \times \overbrace{\log[H^+]}^{-pH}$$
$$= E_{cell}' + (0.0592\ V) \times pH$$

Therefore, by measuring the cell potential, E_{cell}, the pH of the solution can be determined. The value of E_{cell}' is established by calibrating the cell, which requires measuring E_{cell} for a solution of known pH.

A glass electrode, a thin-walled glass bulb containing an electrolyte, is much easier to use than a hydrogen electrode and has a potential that varies linearly with the pH of the solution outside the glass bulb (**FIG. 6N.1**). Often there is a reference electrode built into the probe that makes contact with the test solution through a miniature salt bridge. A pH meter therefore usually has only one probe, called a "combination electrode," which forms a complete electrochemical cell once it is dipped into a solution. Today, the most commonly used reference electrode in pH meters is the saturated Ag/AgCl electrode,

$$Ag(s)|AgCl(s)|Cl^-(aq, sat.\ KCl)\| \qquad E_{sat} = +0.197\ V$$

Calomel is the common name for mercury(I) chloride, Hg_2Cl_2. Note that the compound contains the diatomic cation Hg_2^{2+}.

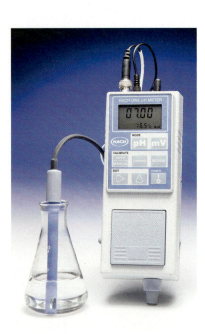

FIGURE 6N.1 A glass electrode in a protective plastic sleeve (left) is used to measure pH. *(Charles D. Winters/Science Source.)*

Note that the cell potential for this half reaction under standard conditions is +0.22 V.

FIGURE 6N.2 Iron nails stored in oxygen-free water (left) do not rust, because the oxidizing power of water itself is weak. When oxygen is present (as a result of air dissolving in the water, right), oxidation is thermodynamically spontaneous and rust soon forms. *(W. H. Freeman photo by Ken Karp.)*

The meter is calibrated with buffer solutions of known pH, and the measured cell potential is then automatically converted into the pH of the solution, which is displayed.

Commercially available electrodes used in pX meters are sensitive to other ions, such as Na^+, Ca^{2+}, NH_4^+, CN^-, or S^{2-}. They are used to monitor ion concentrations in blood, in industrial processes, and in pollution control.

The pH of a solution and the concentrations of ions can be measured by using an electrode that responds selectively to only one species of ion.

6N.4 Corrosion

Electrochemical cells play important roles in both the purification and the preservation of metals. Redox reactions are used throughout the chemical industry to extract metals from their ores. However, redox reactions also corrode the artifacts that industry produces. What redox reactions achieve, redox reactions can destroy.

Corrosion is the unwanted oxidation of a metal. The main culprit in corrosion is water. One half-reaction that must be taken into account is

$$2 H_2O(l) + 2 e^- \longrightarrow H_2(g) + 2 OH^-(aq) \qquad E° = -0.83 \text{ V}$$

This standard potential is for an OH^- concentration of 1 mol·L^{-1}, which corresponds to pH = 14, a strongly basic solution. However, from the Nernst equation, at pH = 7, the potential of this couple is $E = -0.42$ V. Any metal with a standard potential more negative than -0.42 V can therefore reduce water at pH = 7: in other words, at this pH, any such metal can be oxidized by water. Because $E° = -0.44$ V for $Fe^{2+}(aq) + 2 e^- \rightarrow Fe(s)$, iron has only a very slight tendency to be oxidized by water at pH = 7. For this reason, iron can be used for pipes in water supply systems and can be stored indefinitely in oxygen-free water without rusting (**FIG. 6N.2**).

The rusting of iron in the environment occurs when iron is exposed to damp air, with both oxygen and water present. Then the half-reaction

$$O_2(g) + 4 H^+(aq) + 4 e^- \longrightarrow 2 H_2O(l) \qquad E° = +1.23 \text{ V}$$

must be taken into account. The potential of this half-reaction at pH = 7 and $P_{O_2} = 0.2$ bar is $+0.81$ V, which lies well above the value for iron. Hence, iron can reduce oxygen in aqueous solution at pH = 7. In other words, oxygen and water can join forces to oxidize iron metal to iron(II) ions. They can also subsequently work together to oxidize the iron(II) ions to iron(III), because $E° = +0.77$ V for $Fe^{3+}(aq) + e^- \rightarrow Fe^{2+}(aq)$.

A drop of water on the surface of iron can act as the electrolyte for corrosion in a tiny electrochemical cell (**FIG. 6N.3**). At the edge of the drop, dissolved oxygen oxidizes the iron in the process

$$2 Fe(s) \longrightarrow 2 Fe^{2+}(aq) + 4 e^-$$

$$\underline{O_2(g) + 4 H^+(aq) + 4 e^- \longrightarrow 2 H_2O(l)}$$

$$\text{Overall: } 2 Fe(s) + O_2(g) + 4 H^+(aq) \longrightarrow 2 Fe^{2+}(aq) + 2 H_2O(l) \qquad \textbf{(A)}$$

The electrons withdrawn from the metal by this oxidation can be restored from another part of the conducting metal—in particular, from iron lying beneath the oxygen-poor region in the center of the drop. The iron atoms there give up their electrons to the iron atoms at the edge of the drop, form Fe^{2+} ions, and drift away into the surrounding water. This process results in the formation of tiny pits in the surface of the iron. The Fe^{2+} ions are then oxidized further to Fe^{3+} by the dissolved oxygen:

$$2 Fe^{2+}(aq) \longrightarrow 2 Fe^{3+}(aq) + 2 e^-$$

$$\underline{\tfrac{1}{2} O_2(g) + 2 H^+(aq) + 2 e^- \longrightarrow 2 H_2O(l)}$$

$$\text{Overall: } 2 Fe^{2+}(aq) + \tfrac{1}{2} O_2(g) + 2 H^+(aq) \longrightarrow 2 Fe^{3+}(aq) + H_2O(l) \qquad \textbf{(B)}$$

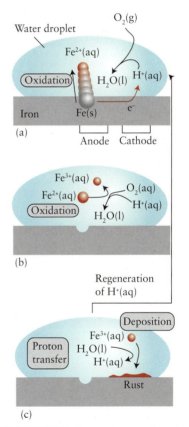

FIGURE 6N.3 The mechanism of rust formation in a drop of water. (a) Oxidation of the iron occurs at a location out of contact with the oxygen of the air. The surface of the metal acts as an anode in a tiny galvanic cell, with the metal at the outer edge of the drop serving as the cathode. (b) Further oxidation of Fe^{2+} results in the formation of Fe^{3+} ions. (c) Protons are removed from H_2O as oxide ions combine with Fe^{3+} ions to deposit as rust. These protons are recycled.

FIGURE 6N.4 Steel girders are galvanized by immersion in a bath of molten zinc. *(the palms/ Shutterstock.)*

These Fe^{3+} ions then precipitate as a hydrated iron(III) oxide, $Fe_2O_3 \cdot H_2O$, a brown, insoluble substance, which in its impure form is called rust. The oxide ions can be regarded as coming from deprotonation of water molecules and as immediately forming the hydrated solid by precipitation with the Fe^{3+} ions produced in reaction B. The net outcome is

$$4 H_2O(l) + 2 Fe^{3+}(aq) \longrightarrow 6 H^+(aq) + Fe_2O_3 \cdot H_2O(s) \qquad \textbf{(C)}$$

This step restores the $H^+(aq)$ ions needed for reaction A, and so hydrogen ions function as a catalyst. The removal of Fe^{3+} ions from solution drives the reaction forward. The overall process is the sum of reactions A, B, and C:

$$2 Fe(s) + \tfrac{3}{2} O_2(g) + H_2O(l) \longrightarrow Fe_2O_3 \cdot H_2O(s)$$

Water is more highly conducting when it has dissolved ions, and the formation of rust is then accelerated. That is one reason why the salt air of coastal cities and salt used for de-icing highways is so damaging to exposed metal.

Because corrosion is an electrochemical process, knowledge of redox reactions can be used to combat it. The simplest way to prevent corrosion is to protect the surface of the metal from exposure to air and water by painting. A method that achieves even greater protection is to **galvanize** the metal, coating it with an unbroken film of zinc (**FIG. 6N.4**). Zinc lies below iron in the electrochemical series; so, if a scratch exposes the metal beneath, the more strongly reducing zinc releases electrons to the iron. As a result, the zinc, not the iron, is oxidized. The zinc itself survives exposure on the unbroken surface because it is **passivated,** protected from further reaction, by a protective oxide. In general, the oxide of any metal that takes up more space than does the metal that it replaces acts as a **protective oxide,** an oxide that protects the metal from further oxidation. Zinc and chromium both form low-density protective oxides that can protect iron from oxidation. Aluminum is passivated by a thin layer of alumina, Al_2O_3, which forms when the metal is exposed to the environment. *Anodized aluminum* has a thick layer of aluminum oxide that has been formed electrolytically (Topic 6O) and is often dyed to give a variety of colors (**FIG. 6N.5**).

It is not possible to galvanize large metal structures, such as ships, underground pipelines, gasoline storage tanks, and bridges, but **cathodic protection,** the electrochemical protection of a metal object by connecting it to a more strongly reducing metal, can still be used. For example, a block of a more strongly reducing metal than iron, typically zinc or magnesium, can be buried in moist soil and connected to an underground pipeline (**FIG. 6N.6**). The block of magnesium is oxidized preferentially and supplies electrons to the iron for the reduction of oxygen. The block, which is called a **sacrificial anode,** protects the pipeline and is inexpensive to replace. For similar reasons, automobiles generally have negative ground systems as part of their electrical circuitry, which means that the

FIGURE 6N.5 Aluminum is protected from oxidation by electrodeposition of alumina, Al_2O_3, on the surface. The alumina can be dyed to give a wide range of colors. *(©1993 Paul Silverman– Fundamental Photographs.)*

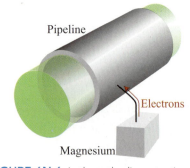

FIGURE 6N.6 In the cathodic protection of a buried pipeline or other large metal construction, the artifact is connected to a number of buried blocks of metal, such as magnesium or zinc. The sacrificial anodes (the magnesium block in this illustration) supply electrons to the pipeline (the cathode of the cell), thereby preserving it from oxidation.

body of the car is connected to the anode of the battery. The decay of the anode in the battery is the sacrifice that helps preserve the vehicle itself.

G A common means of protecting the steel bodies of automobiles and trucks is through **cationic electrodeposition** coatings. In this process, a corrosion-resistant primer is deposited electrochemically on to the automobile body, which serves as the cathode for the process. For many years, lead(IV) oxide was used to provide this corrosion protection. However, lead is toxic, and concerns about environmental pollution stimulated research into alternative metals for cationic electrodeposition coatings. Eventually it was discovered that yttrium oxide also offers corrosion resistance and, in fact, is twice as effective as lead. In addition, yttrium is nontoxic and its oxide is a ceramic. Thus, the oxide is insoluble in water and is not spread into the environment through waterways.

Self-test 6N.4A Which of the following procedures helps to prevent the corrosion of an iron rod in water: (a) decreasing the concentration of oxygen in the water; (b) painting the rod?

[**Answer:** Both]

Self-test 6N.4B Which of the following elements can act as a sacrificial anode for iron: (a) copper; (b) aluminum; (c) tin?

The corrosion of iron is accelerated by the presence of oxygen, moisture, and salt. Corrosion can be inhibited by coating the surface with paint or zinc or by using cathodic protection.

What have you learned in this Topic?

You have learned how galvanic cells can be used to measure concentrations of ions in solution and to determine equilibrium constants. You have also seen through the Nernst equation how the cell potential changes with its composition as expressed by the reaction quotient, Q. You have seen how the unwanted oxidation of metals called corrosion occurs and have met some ways by which it can be prevented.

The skills you have mastered are the ability to:

☐ **1.** Determine an equilibrium constant from a measured cell potential and from tables of standard potentials (Toolbox 6N.1 and Example 6N.1).

☐ **2.** Predict a cell potential for nonstandard conditions using the Nernst equation (Example 6N.2).

☐ **3.** Use the Nernst equation and a measured cell potential to determine the concentration of an ion in solution (Example 6N.3).

☐ **4.** Explain the process of corrosion and describe means of inhibiting it (Section 6N.4).

Topic 6N Exercises

Assume a temperature of 25 °C (298 K) unless instructed otherwise.

6N.1 Calculate the equilibrium constants for the following reactions:
(a) $Mn(s) + Ti^{2+}(aq) \rightleftharpoons Mn^{2+}(aq) + Ti(s)$
(b) $In^{3+}(aq) + U^{3+}(aq) \rightleftharpoons In^{2+}(aq) + U^{4+}(aq)$

6N.2 Calculate the equilibrium constants for the following reactions:
(a) $2\,Fe^{3+}(aq) + H_2(g) \rightleftharpoons 2\,Fe^{2+}(aq) + 2\,H^+(aq)$
(b) $2\,Cr(s) + O_2(g) + 2\,H_2O(l) \rightleftharpoons 2\,Cr^{2+}(aq) + 4\,OH^-(aq)$

6N.3 Predict the potential of each of the following cells:
(a) $Pt(s)|H_2(g, 1.0\ bar)|HCl(aq, 0.075\ mol \cdot L^{-1})||HCl(aq, 1.0\ mol \cdot L^{-1})|$ $H_2(g, 1.0\ bar)|Pt(s)$
(b) $Zn(s)|Zn^{2+}(aq, 0.37\ mol \cdot L^{-1})||Ni^{2+}(aq, 0.059\ mol \cdot L^{-1})|Ni(s)$
(c) $Pt(s)|Cl_2(g, 250\ Torr)|HCl(aq, 1.0\ mol \cdot L^{-1})||HCl(aq, 0.85\ mol \cdot L^{-1})|$ $H_2(g, 125\ Torr)|Pt(s)$

(d) $Sn(s)|Sn^{2+}(aq, 0.277\ mol \cdot L^{-1})||Sn^{4+}(aq, 0.867\ mol \cdot L^{-1})$, $Sn^{2+}(aq, 0.55\ mol \cdot L^{-1})|Pt(s)$

6N.4 Predict the potential of each of the following cells:
(a) $Cr(s)|Cr^{3+}(aq, 0.37\ mol \cdot L^{-1})||Pb^{2+}(aq, 9.5 \times 10^{-3}\ mol \cdot L^{-1})|Pb(s)$
(b) $Pt(s)|H_2(g, 2.0\ bar)|H^+(pH = 3.5)||Cl^-(aq, 0.75\ mol \cdot L^{-1})|$ $Hg_2Cl_2(s)|Hg(l)$
(c) $C(gr)|Sn^{4+}(aq, 0.059\ mol \cdot L^{-1}), Sn^{2+}(aq, 0.059\ mol \cdot L^{-1})||$
(d) $Fe^{3+}(aq, 0.15\ mol \cdot L^{-1}), Fe^{2+}(aq, 0.15\ mol \cdot L^{-1})|Pt(s)$
(e) $Ag(s)|AgI(s)|I^-(aq, 0.025\ mol \cdot L^{-1})||Cl^-(aq, 0.67\ mol \cdot L^{-1})|$ $AgCl(s)|Ag(s)$

6N.5 Evaluate the unknown quantity in each of the following cells:
(a) $Pt(s)|H_2(g, 1.0\ bar)|H^+(pH = ?)||Cl^-(aq, 1.0\ mol \cdot L^{-1})|$ $Hg_2Cl_2(s)|Hg(l)$, $E_{cell} = +0.33$ V.
(b) $C(gr)|Cl_2(g, 1.0\ bar)|Cl^-(aq, ?)||MnO_4^-(aq, 0.010\ mol \cdot L^{-1})$, $H^+(pH = 4.0), Mn^{2+}(aq, 0.10\ mol \cdot L^{-1})|Pt(s)$, $E_{cell} = -0.30$ V.

6N.6 Evaluate the unknown quantity in each of the following cells:
(a) $Pt(s)|H_2(g, 1.0 \text{ bar})|H^+(pH = ?)||Cl^-(aq, 1.0 \text{ mol·L}^{-1})|AgCl(s)|Ag(s)$, $E_{cell} = +0.47$ V.
(b) $Pb(s)|Pb^{2+}(aq, ?)||Ni^{2+}(aq, 0.20 \text{ mol·L}^{-1})|Ni(s)$, $E_{cell} = +0.045$ V.

6N.7 Calculate E_{cell} for each of the following concentration cells:
(a) $Cu(s)|Cu^{2+}(aq, 0.0010 \text{ mol·L}^{-1})||Cu^{2+}(aq, 0.010 \text{ mol·L}^{-1})|Cu(s)$
(b) $Pt(s)|H_2(g, 1 \text{ bar})|H^+(aq, pH = 4.0)||H^+(aq, pH = 3.0)|H_2(g, 1 \text{ bar})|Pt(s)$

6N.8 Calculate the unknown concentration of the ion in each of the following cells:
(a) $Pb(s)|Pb^{2+}(aq, ?)||Pb^{2+}(aq, 0.10 \text{ mol·L}^{-1})|Pb(s)$, $E_{cell} = +0.083$ V.
(b) $Pt(s)|Fe^{3+}(aq, 0.10 \text{ mol·L}^{-1}), Fe^{2+}(aq, 1.0 \text{ mol·L}^{-1})||Fe^{3+}(aq, ?), Fe^{2+}(aq, 0.0010 \text{ mol·L}^{-1})|Pt(s)$, $E_{cell} = +0.14$ V.

6N.9 A tin electrode in 0.015 M $Sn(NO_3)_2$(aq) is connected to a hydrogen electrode in which the pressure of H_2 is 1.0 bar. If the cell potential is 0.061 V at 25 °C, what is the pH of the electrolyte at the hydrogen electrode?

6N.10 A lead electrode in 0.020 M $Pb(NO_3)_2$(aq) is connected to a hydrogen electrode in which the pressure of H_2 is 1.0 bar. If the cell potential is 0.078 V at 25 °C, what is the pH of the electrolyte at the hydrogen electrode?

6N.11 (a) Use data from Appendix 2B to calculate the solubility product of Hg_2Cl_2. (b) Compare this number with the value listed in Table 6I.1 and comment on any difference.

6N.12 (a) The standard potential of the reduction of Ag_2CrO_4 to $Ag(s)$ and chromate ions is +0.446 V. Write the balanced half-reaction for the reduction of silver chromate. (b) Using the data from part (a) and Appendix 2B, calculate the solubility product of $Ag_2CrO_4(s)$.

6N.13 Calculate the reaction quotient, Q, for the following cell reactions, given the measured values of the cell potential. Balance the chemical equations by using the smallest whole-number coefficients.
(a) $Pt(s)|Sn^{4+}(aq), Sn^{2+}(aq)||Pb^{4+}(aq), Pb^{2+}(aq)|C(gr)$,
$E_{cell} = +1.33$ V.
(b) $Pt(s)|O_2(g)|H^+(aq)||Cr_2O_7^{2-}(aq), H^+(aq), Cr^{3+}(aq)|Pt(s)$,
$E_{cell} = +0.10$ V.

6N.14 Calculate the reaction quotient, Q, for the following cell reactions, given the measured values of the cell potential. Balance the chemical equations by using the smallest whole-number coefficients.
(a) $Ag(s)|Ag^+(aq)||ClO_4^-(aq), H^+(aq), ClO_3^-(aq)|Pt(s)$,
$E_{cell} = +0.40$ V.
(b) $C(gr)|Cl_2(g)|Cl^-(aq)||Au^{3+}(aq)|Au(s)$, $E_{cell} = 0.00$ V.

6N.15 Calculate the potential of a cell constructed with two nickel electrodes. The electrolyte in one compartment is 1.0 M $Ni(NO_3)_2$(aq).

In the other compartment, NaOH has been added to a $Ni(NO_3)_2$ solution until the pH = 11.0 at 298 K. See Table 6I.1.

6N.16 A cell was constructed with two lead electrodes. The electrolyte in one compartment is 1.0 M $Pb(NO_3)_2$(aq). In the other compartment, NaI has been added to a $Pb(NO_3)_2$ solution until a yellow precipitate forms and the concentration of I^- ions is 0.050 mol·L^{-1}. The potential of this cell is +0.155 V at 298 K. (a) Calculate the concentration of Pb^+ ions in the second compartment. (b) Calculate the solubility product of PbI_2.

6N.17 Consider the cell $Ag(s)|Ag^+(aq, 5.0 \text{ mmol·L}^{-1})||Ag^+(aq, 0.15 \text{ mol·L}^{-1})|Ag(s)$. Can this cell do work? If so, what is the maximum work that it can perform (per mole of Ag)?

6N.18 Consider the cell $Ag(s)|Ag^+(aq, 3.6 \text{ mmol·L}^{-1})||Pb^{2+}(aq, 0.25 \text{ mol·L}^{-1})|Pb(s)$. (a) Can this cell do work? If so, what is the maximum work that it can perform (per mole of Pb)? (b) What is the value of ΔH for the cell reaction (use standard enthalpies of formation), and the value of ΔS?

6N.19 Suppose the reference electrode for Table 6M.1 were the standard calomel electrode, Hg_2Cl_2/Hg, $Cl^-([Cl^-] = 1.00 \text{ mol·L}^{-1})$, with its $E°$ set equal to 0. Under this system, what would be the potential for (a) the standard hydrogen electrode; (b) the standard Cu^{2+}/Cu redox couple?

6N.20 Consider the question posed in Exercise 6N.19 except that a saturated calomel electrode (the solution is saturated with KCl instead of having $[Cl^-] = 1.00 \text{ mol·L}^{-1}$) is used in place of the standard calomel electrode. How will this replacement change your answers to Exercise 6N.19? The solubility of KCl is 35 g·(100 mL H_2O)$^{-1}$.

6N.21 (a) What is the approximate chemical formula of rust? (b) What is the oxidizing agent in the formation of rust? (c) How does the presence of salt accelerate the rusting process?

6N.22 (a) What is the electrolyte solution in the formation of rust? (b) How are steel (iron) objects protected by galvanizing and by sacrificial anodes? (c) Suggest two metals that could be used in place of zinc for galvanizing iron.

6N.23 (a) Suggest two metals that could be used for the cathodic protection of a titanium pipeline. (b) What factors other than relative positions in the electrochemical series need to be considered in practice? (c) Often copper piping is connected to iron pipes in household plumbing systems. What is a possible effect of the copper on the iron pipes?

6N.24 (a) Can aluminum be used for the cathodic protection of a steel underground storage container? (b) Which of the metals zinc, silver, copper, and magnesium cannot be used as a sacrificial anode in the protection of a buried iron pipeline? Explain your answer. (c) What is the electrolyte solution for the cathodic protection of an underground pipeline by a sacrificial anode?

Topic 60 Electrolysis

60.1 Electrolytic Cells

60.2 The Products of Electrolysis

60.3 Applications of Electrolysis

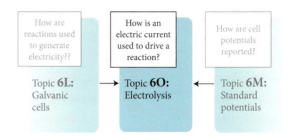

Redox reactions that have a positive Gibbs free energy of reaction are not spontaneous, but an electric current can be used to drive them forward. For example, fluorine cannot be isolated by any common chemical reaction. It was not until 1886 that the French chemist Henri Moissan found a way to force the formation of fluorine by passing an electric current through an anhydrous molten mixture of potassium fluoride and hydrogen fluoride. Fluorine is still prepared commercially by the same process.

60.1 Electrolytic Cells

An electrochemical cell in which electrolysis takes place is called an **electrolytic cell.** The arrangement of components in electrolytic cells is different from that in galvanic cells. Typically, the two electrodes share the same compartment, there is only one electrolyte, and concentrations and pressures are far from standard. As in all electrochemical cells, the ions present carry the current through the electrolyte. For example, when copper metal is refined electrolytically, the anode is impure copper, the cathode is pure copper, and the electrolyte is an aqueous solution of $CuSO_4$. As the Cu^{2+} ions in solution migrate toward the cathode, where they are reduced and deposited as Cu atoms, more Cu^{2+} ions are produced by oxidation of copper metal at the anode.

THINKING POINT

What metal impurities cannot be removed from copper during refining by electrolysis?

FIGURE 60.1 shows schematically the layout of an electrolytic cell used for the commercial production of magnesium metal from molten magnesium chloride (the *Dow process*). As in a galvanic cell, oxidation takes place at the anode and reduction takes place at the cathode, and electrons complete the circuit by traveling through an external wire. From anode to cathode, cations move through the electrolyte toward the cathode, and anions move toward the anode. Unlike a galvanic cell, however, in which the current is generated spontaneously, in an electrolytic cell a current must be supplied by an external electrical power source for reaction to occur. The result is to force oxidation at one electrode and reduction at the other. The following half-reactions are made to take place in the Dow process:

$$\text{Anode reaction: } 2\,Cl^-(\text{melt}) \longrightarrow Cl_2(g) + 2\,e^-$$

$$\text{Cathode reaction: } Mg^{2+}(\text{melt}) + 2\,e^- \longrightarrow Mg(l)$$

where "melt" signifies the molten salt, which serves as the electrolyte. A rechargeable battery functions as a galvanic cell when it is doing work and as an electrolytic cell when it is being charged.

To drive a reaction in a nonspontaneous direction, the external supply must generate a potential difference greater than the potential difference that would be produced by the reverse reaction. For example, because

$$2\,H_2(g) + O_2(g) \longrightarrow 2\,H_2O(l) \qquad E_{cell} = +1.23 \text{ V at pH} = 7, \text{ spontaneous}$$

Why Do You Need to Know This Material?
Electrolysis is used to produce metals from their salts and to purify impure metals. To treat electrolysis quantitatively, chemists need to know the current and time required to produce a given amount of material.

What Do You Need to Know Already?
You need to be familiar with cell potentials (Topics 6L and 6M) and stoichiometric relations (*Fundamentals* L).

The anode (the site of oxidation) of an electrolytic cell is labeled + and the cathode (the site of reduction) is labeled −, the opposite of a galvanic cell. The + sign indicates that the electrode is pulling electrons away from the species undergoing oxidation.

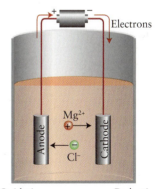

Electrons

Mg²⁺

Anode   Cathode

Cl⁻

Oxidation:
2 Cl⁻(aq) → Cl₂(g) + 2 e⁻

Reduction:
Mg²⁺(melt) + 2 e⁻ → Mg(s)

FIGURE 60.1 A schematic representation of the electrolytic cell used in the Dow process for magnesium. The electrolyte is molten magnesium chloride. Chloride ions are oxidized to chlorine gas at the anode and magnesium ions are reduced to magnesium metal at the cathode.

to achieve the nonspontaneous reverse reaction,

$$2\,H_2O(l) \longrightarrow 2\,H_2(g) + O_2(g) \qquad E_{cell} = -1.23\ \text{V at pH} = 7, \text{nonspontaneous}$$

under standard conditions (other than for the H^+ concentration, which is $10^{-7}\ \text{mol·L}^{-1}$ in pure water at 298 K), at least 1.23 V must be applied from the external source to overcome the reaction's natural pushing power in the opposite direction. In practice, the applied potential difference must usually be substantially greater than the cell potential to reverse a spontaneous cell reaction and achieve a significant rate of product formation. The additional potential difference, which varies with the type of electrode, is called the **overpotential.** For platinum electrodes, the overpotential for the production of hydrogen and oxygen from water is about 0.6 V, so about 1.8 V (that is, 0.6 + 1.23 V) is actually required to electrolyze water when platinum electrodes are used. Hydrogen is a clean-burning energy source, and much contemporary research on electrochemical cells involves attempts to reduce the overpotential and hence to increase the efficiency of electrolytic processes such as those that produce hydrogen.

When carrying out electrolysis in solution, the possibility that other species present might be oxidized or reduced by the electric current must be considered. For example, suppose that you want to electrolyze water to produce hydrogen and oxygen. To reverse the half-reaction

$$O_2(g) + 4\,H^+(aq) + 4\,e^- \longrightarrow 2\,H_2O(l) \qquad E = +0.82\ \text{V at pH} = 7$$

and bring about the oxidation of water, an applied potential difference is required of at least 0.82 V.

Pure water does not carry a current because the concentration of ions (H_3O^+ and OH^-) in it is very low. Therefore, an ionic solute with ions that are less easily oxidized or reduced than water must be added. Suppose the added salt is sodium chloride. When Cl^- ions are present at 1 mol·L^{-1} in water, is it possible that they, and not the water, will be oxidized? From Table 6M.1, the standard potential for the reduction of chlorine is +1.36 V:

$$Cl_2(g) + 2\,e^- \longrightarrow 2\,Cl^-(aq) \qquad E^\circ = +1.36\ \text{V}$$

To reverse this reaction and oxidize chloride ions, you would have to supply at least 1.36 V. Because only 0.82 V is needed to force the oxidation of water but 1.36 V is needed to force the oxidation of Cl^-, it appears that oxygen should be the product at the anode. However, the overpotential for oxygen production can be very high, and in practice chlorine also might be produced. At the cathode, the half-reaction

$$2\,H^+(aq) + 2\,e^- \longrightarrow H_2(g) \qquad E = +0.41\ \text{V at pH} = 7$$

is required. Hydrogen, rather than sodium metal, will be produced at the cathode, because the potential required to reduce sodium ions is significantly higher (+2.71 V).

EXAMPLE 60.1   Deciding which species will be produced at an electrode

Iodine is an essential element for proper thyroid health. Seawater contains trace amounts of iodide, and you are exploring electrolysis of solutions of iodide as a means to purify the element. You need to know whether iodide or water will be oxidized at lower potentials. Suppose that an aqueous solution with pH = 7 and containing I^- ions at 1 mol·L^{-1} is being electrolyzed. Will O_2 or I_2 be produced at the anode?

PLAN Decide which oxidation requires the lower potential. Then, provided the overpotentials are similar, that redox couple will be preferentially oxidized.

SOLVE

Find the standard potential of $I_2(s)$ from Table 6M.1 or Appendix 2B.

$$I_2(s) + 2\,e^- \longrightarrow 2\,I^-(aq) \qquad E^\circ = +0.54\ \text{V}$$

Find the standard (but at pH = 7) potential of $O_2(g)$ from Table 6M.1 or Appendix 2B.

$$O_2(g) + 4\,H^+(aq) + 4\,e^- \longrightarrow 2\,H_2O(l) \qquad E = +0.82\ \text{V at pH} = 7$$

From the potentials, it is clear that at least 0.54 V is needed to oxidize I^- and at least 0.82 V is needed to oxidize water under standard conditions (at pH = 7).

Therefore, provided that the overpotentials are similar, I^- ions are expected to be oxidized in preference to water.

Self-test 60.1A Predict the products resulting from the electrolysis of 1 M $AgNO_3$(aq).

[*Answer:* Cathode, Ag; anode, O_2]

Self-test 60.1B Predict the products resulting from the electrolysis of 1 M NaBr(aq).

Related Exercises 60.3, 60.4

The potential supplied to an electrolytic cell must be at least as great as that of the cell reaction to be reversed. If there is more than one reducible species in solution, the species with the greater potential for reduction is preferentially reduced. The same principle applies to oxidation.

60.2 The Products of Electrolysis

The calculation of the amount of product formed in an electrolysis is based on observations made by Michael Faraday (**FIG. 60.2**) and summarized—in more modern language than he used—as follows:

> **Faraday's law of electrolysis:** The amount of product formed or reactant consumed by an electric current is stoichiometrically equivalent to the amount of electrons supplied.

For instance, copper is refined electrolytically by using an impure form of copper metal called "blister copper" as the anode in an electrolytic cell (**FIG. 60.3**). The current supply drives the oxidation of the blister copper to copper(II) ions, which are then reduced to pure copper metal at the cathode in the reaction Cu^{2+}(aq) + 2 e^- → Cu(s). To calculate the amount of Cu produced by using a given amount of electrons, write the mole ratio from this half-reaction, 2 mol e^- ≏ 1 mol Cu, and then convert the amount of electrons to an amount of Cu atoms. For example, if 4.0 mol e^- had been supplied:

$$\text{Amount of Cu(mol)} = (4.0 \text{ mol } e^-) \times \frac{1 \text{ mol Cu}}{2 \text{ mol } e^-} = 2.0 \text{ mol Cu}$$

FIGURE 60.2 Michael Faraday (1791–1867). (© Hi-Story/Alamy.)

Self-test 60.2A What amount (in moles) of Al(s) can be produced from Al_2O_3 if 4.5 mol e^- is supplied?

[*Answer:* 1.5 mol Al]

Self-test 60.2B What amount (in moles) of Cr(s) can be produced from CrO_3 if 12.0 mol e^- is supplied?

In this context, Q is the charge supplied: don't confuse it with the reaction quotient Q!

The quantity, Q, of electricity passed through the electrolysis cell is measured in coulombs. It is determined by measuring the current, I, and the time, t, for which the current flows and is calculated from

$$\text{Charge supplied (C)} = \text{current (A)} \times \text{time (s)} \quad \text{or} \quad Q = It \qquad (1)$$

For example, because 1 A·s = 1 C, if 2.00 A is passed for 125 s, the charge supplied to the cell is

$$Q = (2.00 \text{ A}) \times (125 \text{ s}) = 250. \text{ A·s} = 250. \text{ C}$$

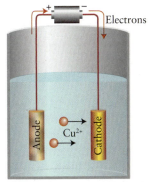

FIGURE 60.3 A schematic representation of the electrolytic process for refining copper. The anode is impure copper. The Cu^{2+} ions produced by oxidation of the anode migrate to the cathode, where they are reduced to pure copper metal. A similar arrangement is used for electroplating objects.

Oxidation:
Cu(s) →
Cu^{2+}(aq) + 2 e^-

Reduction:
Cu^{2+}(aq) + 2 e^-
→ Cu(s)

LAB VIDEO FIGURE 60.3

To determine the amount of electrons supplied by a given charge, use Faraday's constant, F, the magnitude of the charge per mole of electrons (Topic 6L). Because the charge supplied is $Q = nF$, where n is the amount of electrons (in moles), it follows that

$$n = \frac{Q}{F} = \frac{It}{F} \qquad (2)$$

Thus, if the current and the time for which it flows are known, the amount of electrons supplied can be determined. The mole ratio from the electrode reaction can then be used to convert the amount of electrons supplied into amount of product (see **1**).

Toolbox 60.1 HOW TO PREDICT THE RESULT OF ELECTROLYSIS

CONCEPTUAL BASIS

The number of electrons required to reduce a given amount of a species is related to the stoichiometric coefficients in the reduction half-reaction. The same is true of oxidation. Therefore, a stoichiometric relation can be set up between the reduced or oxidized species and the amount of electrons required to carry out the reaction. The amount of electrons required is calculated from the current and the length of time for which the current flows.

PROCEDURE

To determine the amount of product that can be produced:

Step 1 Identify the stoichiometric relation between electrons and the species of interest from the applicable half-reaction.

Step 2 Calculate the amount (in moles) of electrons supplied from Eq. 2, $n = It/F$. Use the stoichiometric relation from step 1

to convert n into the amount of substance. If required, use the molar mass to convert the amount of substance into mass (or molar volume to convert into volume).

This procedure is illustrated in Example 60.2.

To determine the time required for a given amount of product to be produced:

Step 1 Identify the stoichiometric relation between electrons and the species of interest from the applicable half-reaction.

Step 2 If required, use the molar mass to convert mass into amount (in moles). Use the stoichiometric relation from step 1 to convert amount of substance into amount of electrons passed, n (in moles).

Step 3 Substitute n, the current, and Faraday's constant into Eq. 2 rearranged to $t = Fn/I$ and solve for time.

This procedure is illustrated in Example 60.3.

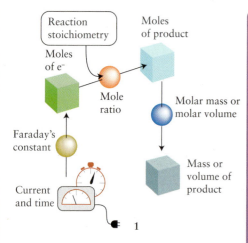

1

EXAMPLE 60.2   Calculating the amount of product formed by electrolysis

One of the largest uses of electricity is in the production of aluminum by electrolysis of its oxide dissolved in molten cryolite (Na_3AlF_6). As an engineer, you might need to predict how much aluminum can be produced by this process. Find the mass of aluminum that can be produced in 1.00 day in an electrolytic cell operating continuously at 1.00×10^5 A. The cryolite does not react.

ANTICIPATE In this industrial-scale process, with a high current acting for a long time, you should expect to generate many kilograms of aluminum.

PLAN Use the first procedure in **Toolbox 60.1**.

SOLVE

Step 1 Write the half-reaction for the reduction and find the amount of electrons required to reduce 1 mol Al^{3+} ions to the metal.

$$Al^{3+}(melt) + 3\,e^- \longrightarrow Al(l) \qquad 3 \text{ mol } e^- \mathrel{\hat{=}} 1 \text{ mol Al}$$

Step 2 From $m = nM$ and $n = It/F$, using $M(Al) = 26.98 \text{ g·mol}^{-1}$, 3600 s = 1 h, and 24 h = 1 d, and the stoichiometric relation between Al and the electrons,

$$m(Al) = n(e^-) \times \frac{1 \text{ mol Al}}{3 \text{ mol } e^-} \times \frac{26.98 \text{ g Al}}{1 \text{ mol Al}}$$

$$= \frac{\overbrace{(1.00 \times 10^5 \text{ C·s}^{-1})}^{I} \times \overbrace{(24.0 \times 3600 \text{ s})}^{t}}{\underbrace{9.65 \times 10^4 \text{ C·(mol } e^-)^{-1}}_{F}} \times \frac{1 \text{ mol Al}}{3 \text{ mol } e^-} \times \frac{26.98 \text{ g Al}}{1 \text{ mol Al}}$$

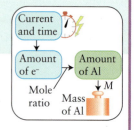

$$= 8.05 \times 10^5 \text{ g Al}$$

EVALUATE As expected, a large mass of aluminum (805 kg) is produced. The fact that the production of 1 mol Al requires 3 mol e⁻ accounts for the very high consumption of electricity that is characteristic of aluminum-production plants.

Self-test 60.3A Determine the mass (in grams) of magnesium metal that can be obtained from molten magnesium chloride by using a current of 7.30 A for 2.11 h. What volume of chlorine gas at 25 °C and 1.00 atm will be produced at the anode?

[**Answer:** 6.98 g; 7.03 L]

Self-test 60.3B What mass of chromium metal can be obtained from a 1 M solution of CrO_3 in dilute sulfuric acid by using a current of 6.20 A for 6.00 h?

Related Exercises 60.5, 60.6

EXAMPLE 60.3 Calculating the time required to produce a given mass of product

You are working in a plant that produces copper for use in electrical circuits by electrolysis of acidic solutions of $CuSO_4(aq)$. A customer has placed an order for a small quantity of high-purity copper, and you need to tell the customer how long it will take to produce the metal. How many hours are required to plate 25.00 g of copper metal from 1.00 M $CuSO_4(aq)$ by using a current of 3.00 A?

PLAN Use the second procedure in Toolbox 60.1.

SOLVE

Step 1 Find the stoichiometric relation between electrons and the species of interest from the half-reaction for the electrolysis.

$$Cu^{2+}(aq) + 2\,e^- \longrightarrow Cu(s) \qquad 2\;mol\;e^- \simeq 1\;mol\;Cu$$

Step 2 To find $n(e^-)$, convert mass of Cu into amount of Cu, and amount of Cu into amount of e⁻, with all amounts in moles.

$$n(e^-) = (25.00\;\text{g Cu}) \times \frac{1\;\text{mol Cu}}{63.54\;\text{g Cu}} \times \frac{2\;\text{mol e}^-}{1\;\text{mol Cu}} = \frac{25.00 \times 2}{63.54}\;\text{mol e}^-$$

$$= 0.786\ldots\;\text{mol e}^-$$

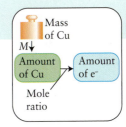

Step 3 Use $t = Fn/I$ and convert seconds to hours.

$$t = \frac{9.6485 \times 10^4\;\text{C·(mol e}^-)^{-1}}{3.00\;\text{C·s}^{-1}} \times 0.786\ldots\;\text{mol e}^- \times \frac{1\;\text{h}}{3600\;\text{s}} = 7.0\;\text{h}$$

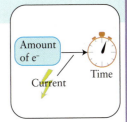

Self-test 60.4A Determine the time, in hours, required to electroplate 7.00 g of magnesium metal from molten magnesium chloride by using a current of 7.30 A.

[**Answer:** 2.12 h]

Self-test 60.4B How many hours are required to plate 12.00 g of chromium metal from a 1 M solution of CrO_3 in dilute sulfuric acid by using a current of 6.20 A?

Related Exercises 60.7, 60.8

The amount of product in an electrolysis reaction is calculated from the stoichiometry of the half-reaction, the current, and the time for which the current flows.

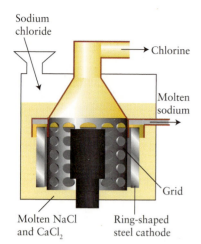

FIGURE 60.4 In the Downs process, molten sodium chloride is electrolyzed with a graphite anode (at which the Cl^- ions are oxidized to chlorine) and a steel cathode (at which the Na^+ ions are reduced to sodium). The sodium and chlorine are kept apart by the hood surrounding the anode. Calcium chloride is present to lower the melting point of sodium chloride to an economical temperature.

FIGURE 60.5 Chromium plating lends decorative flair as well as electrochemical protection to the steel of this motorcycle. Large quantities of electricity are needed for chromium plating because six electrons are required to produce each atom of chromium from chromium(VI) oxide. (© *tbkmedia.de/Alamy.*)

60.3 Applications of Electrolysis

The refining of copper and the electrolytic extraction of aluminum, magnesium, and fluorine has already been described. Another important industrial application of electrolysis is the production of sodium metal by the *Downs process*, the electrolysis of molten rock salt (**FIG. 60.4**):

$$\text{Cathode reaction: } 2\,Na^+(\text{melt}) + 2\,e^- \longrightarrow 2\,Na(l)$$
$$\text{Anode reaction: } 2\,Cl^-(\text{melt}) \longrightarrow Cl_2(g) + 2\,e^-$$

Sodium chloride is plentiful as the mineral rock salt, but the solid does not conduct electricity because the ions are locked into place: it must be molten for the ions to migrate and electrolysis to occur. The electrodes in the cell are made of inert materials such as carbon, and the cell is designed to keep the sodium and chlorine produced out of contact with each other and away from air. In a modification of the Downs process, the electrolyte is an aqueous solution of sodium chloride (see Topic 8C). The products of this *chloralkali process* are chlorine and aqueous sodium hydroxide.

THINKING POINT

In the Downs process, $CaCl_2$ is added to NaCl to reduce the latter's melting point. Why, then, is calcium metal not produced at the cathode?

Electroplating is the electrolytic deposition of a thin film of metal on an object (**FIG. 60.5**). The object to be electroplated (either metal or graphite-coated plastic) constitutes the cathode, and the electrolyte is an aqueous solution of a salt of the plating metal. Metal is deposited on the cathode by reduction of ions in the electrolyte solution. These cations are supplied either by the added salt or from oxidation of the anode, which is made of the plating metal.

> *Electrolysis is used industrially to extract metals from their salts; to prepare chlorine, fluorine, and sodium hydroxide; to refine copper; and in electroplating.*

What have you learned in this Topic?

You have learned that nonspontaneous redox reactions may be made to occur if a current is passed through an electrolytic cell and that the potential difference required must be greater than the cell potential that would be generated by the reverse reaction. You have also seen that Faraday's law in the form of Eq. 2 can be used to relate the amount of product formed to the time of electrolysis and the current that flowed.

The skills you have mastered are the ability to:

☐ **1.** Predict the products of electrolysis (Example 6O.1).

☐ **2.** Calculate the amount of product formed during an electrolysis (Toolbox 6O.1 and Example 6O.2).

☐ **3.** Determine the time required to form a given amount of product by electrolysis (Toolbox 6O.1 and Example 6O.3).

Topic 6O Exercises

For the exercises in this topic, base your answers on the potentials listed in Table 6M.1 or Appendix 2B, with the exception of the reduction and oxidation of water at pH = 7:

$$2\,H_2O(l) + 2\,e^- \longrightarrow H_2(g) + 2\,OH^-(aq),$$
$$E = -0.42 \text{ V at pH} = 7$$

$$O_2(g) + 4\,H^+(aq) + 4\,e^- \longrightarrow 2\,H_2O(l),$$
$$E = +0.82 \text{ V at pH} = 7$$

Ignore other factors such as overpotential.

6O.1 A 1.0 M $NiSO_4(aq)$ solution was electrolyzed by using inert electrodes. Write (a) the cathode reaction; (b) the anode reaction. (c) With no overpotential at the electrodes, what is the minimum potential that must be supplied to the cell for the onset of electrolysis?

6O.2 A 1.0 M KBr(aq) solution was electrolyzed by using inert electrodes. Write (a) the cathode reaction; (b) the anode reaction. (c) With no overpotential at the electrodes, what is the minimum potential that must be supplied to the cell for the onset of electrolysis?

6O.3 Aqueous solutions of (a) Mn^{2+}; (b) Al^{3+}; (c) Ni^{2+}; (d) Au^{3+} with concentrations of 1.0 mol·L^{-1} are electrolyzed at pH = 7. For each solution, determine whether the metal ion or water will be reduced at the cathode.

6O.4 The anode of an electrolytic cell was constructed from (a) Cr; (b) Pt; (c) Cu; (d) Ni. For each case, determine whether oxidation of the electrode or of water will take place at the anode when the electrolyte consists of a 1.0 M solution of the oxidized metal ions at pH = 7.

6O.5 A total charge of 4.5 kC is passed through an electrolytic cell. Determine the quantity of substance produced in each case: (a) the mass (in grams) of bismuth metal from a bismuth nitrate solution; (b) the volume (in liters at 273 K and 1.00 atm) of hydrogen gas from a sulfuric acid solution; (c) the mass of cobalt (in grams) from a cobalt(III) chloride solution.

6O.6 A total charge of 67.2 kC is passed through an electrolytic cell. Determine the amount of substance produced in each case: (a) the mass (in grams) of silver metal from a silver nitrate solution; (b) the volume (in liters at 273 K and 1.00 atm) of chlorine gas from a brine solution (concentrated aqueous sodium chloride solution); (c) the mass of copper (in grams) from a copper(II) chloride solution.

6O.7 (a) How much time is required to electroplate 1.50 g of silver from a silver nitrate solution by using a current of 13.6 mA? (b) If the same current is used for the same length of time, what mass of copper can be electroplated from a copper(II) sulfate solution?

6O.8 What current is required to electroplate 6.66 μg of gold in 30.0 min from a gold(III) chloride aqueous solution? (b) How much time is required to electroplate 6.66 μg of chromium from a potassium dichromate solution by using a current of 100 mA?

6O.9 (a) What current is required to produce 8.2 g of chromium metal from chromium(VI) oxide in 24 h? (b) What current is required to produce 8.2 g of sodium metal from molten sodium chloride in the same period?

6O.10 (a) When a current of 324 mA is used for 15 h, what volume (measured in liters at 298 K and 1.0 atm) of fluorine gas can be produced from a molten mixture of potassium and hydrogen fluorides? (b) With the same current and time period, how many liters of oxygen gas at 298 K and 1.0 atm can be produced from the electrolysis of water?

6O.11 When a ruthenium chloride solution was electrolyzed for 500 s with a 120-mA current, 31.0 mg of ruthenium was deposited. What is the oxidation number of ruthenium in the ruthenium chloride?

6O.12 A sample of manganese of mass 4.9 g was produced from a manganese nitrate aqueous solution when a current of 350 mA was passed for 13.7 h. What is the oxidation number of manganese in the manganese nitrate?

6O.13 Copper from 200.0 mL of a solution of copper(II) sulfate is plated on to the cathode of an electrolytic cell. (a) Hydronium ions are generated at one of the electrodes. Is this electrode the anode or cathode? (b) How many moles of H_3O^+ are generated if a current of 0.120 A is applied to the cell for 30.0 h? (c) If the pH of the solution was initially 7.0, what will be the pH of the solution after the electrolysis? Assume no change in the volume of solution.

6O.14 Thomas Edison was faced with the problem of measuring the electricity that each of his customers had used. His first solution was to use a zinc "coulometer," an electrolytic cell in which the quantity of electricity is determined by measuring the mass of zinc deposited. Only some of the current used by the customer passed through the coulometer. (a) What mass of zinc would be deposited in 1 month (of 31 days) if 1.0 mA of current passed through the cell continuously? (b) An alternative solution to this problem is to collect the hydrogen produced by electrolysis and measure its volume. What volume would be collected at 298 K and 1.00 bar under the same conditions? (c) Which method would be more practical?

6O.15 Suppose that 2.69 g of a silver salt (AgX) is dissolved in 550 mL of water. With a current of 3.5 A, 395.0 s was needed to plate out all the silver. (a) What is the mass percentage of silver in the salt? (b) What is the formula of the salt?

6O.16 Three electrolytic cells containing solutions of $CuNO_3$, $Sn(NO_3)_2$, and $Fe(NO_3)_3$, respectively, are connected in series. A current of 3.5 A is passed through the cells until 6.10 g of copper has been deposited in the first cell. (a) What masses of tin and iron are deposited? (b) For how long did the current flow?

The following Example and Exercises draw on material from throughout Focus 6.

FOCUS 6 Online Cumulative Example

You are a geologist studying the erosion of minerals and its impact on the pH of natural waters and have started to examine the properties of fluorite (calcium fluoride, CaF_2). Recalling that fluoride is a weak base, you construct an electrochemical cell to help determine the concentration of dissolved species in a saturated CaF_2 solution. A saturated solution of CaF_2 is placed in the left compartment of a galvanic cell along with a platinum electrode. In the other compartment, you insert a second platinum electrode in a solution of 0.10 M HCl(aq). The two compartments are connected through a salt bridge

and the measured cell potential is $E_{cell} = +0.365$ V at 25 °C. Note that during the measurement very little current flows and the potential is due to the difference in H_3O^+ concentrations in the right and left compartments. No other electrochemical reactions occur.

Determine the molar concentration of all solute species in the CaF_2 solution. From Table 6C.1, $K_a = 3.5 \times 10^{-4}$ for HF, and from Table 6I.1, $K_{sp} = 4.0 \times 10^{-11}$ for CaF_2. The cell and the corresponding cell reaction are

$$Pt(s)|H^+(aq, L)\|H^+(aq, R)|Pt(s)$$
$$H^+(aq, R) \longrightarrow H^+(aq, L)$$

 The Online Cumulative Example solution can be found at http://macmillanhighered.com/chemicalprinciples7e

FOCUS 6 Exercises

The values of K_a and K_b for weak acids and bases may be found in Tables 6C.1 and 6C.2.

6.1 The images below represent the solutes in the solutions of three acids (water molecules are not shown, hydrogen atoms and hydronium ions are represented by small gray spheres, conjugate bases by large colored spheres). (a) Which acid is a strong acid? (b) Which acid has the strongest conjugate base? (c) Which acid has the highest pK_a? Explain each of your answers.

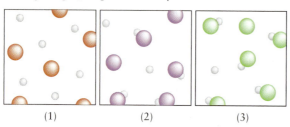

(1)   (2)   (3)

6.2 The images below represent the solutes in the solutions of three salts (water molecules are not shown, hydrogen atoms and hydronium ions are represented by small gray spheres, hydroxide ions by red and gray spheres, cations by pink spheres, and anions by green spheres). (a) Which salt has a cation that is the conjugate acid of a weak base? (b) Which salt has an anion that is the conjugate base of a weak acid? (c) Which salt has an anion that is the conjugate base of a strong acid? Explain each of your answers.

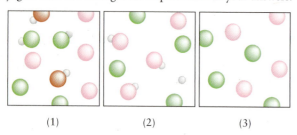

(1)   (2)   (3)

6.3 Estimate the enthalpy of deprotonation of formic acid at 25 °C, given that $K_a = 1.765 \times 10^{-4}$ at 20 °C and 1.768×10^{-4} at 30 °C.

6.4 How many hydronium ions are present at any moment in 100. mL of pure water at 25 °C?

6.5 Hydrogen peroxide, H_2O_2, reacts with sulfur trioxide to form peroxomonosulfuric acid, H_2SO_5, in a Lewis acid–base reaction. (a) Write the chemical equation for the reaction. (b) Draw the Lewis structures of the reactants and product (in the product, one —OH group in sulfuric acid is replaced by an —OOH group). (c) Identify the Lewis acid and Lewis base.

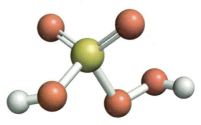

6.6 Dinitrogen monoxide, N_2O, reacts with water to form hyponitrous acid, $H_2N_2O_2(aq)$, in a Lewis acid–base reaction. (a) Write the chemical equation for the reaction. (b) Draw the Lewis structures of N_2O and $H_2N_2O_2$ (the atoms are attached in the order HONNOH). (c) Identify the Lewis acid and Lewis base.

6.7 A combustion analysis of 1.200 g of an anhydrous sodium salt gave 0.942 g of CO_2, 0.0964 g of H_2O, and 0.246 g of Na. The molar mass of the salt is 112.02 g·mol^{-1}. (a) What is the chemical formula of the salt? (b) The salt contains carboxylate groups (—CO_2^-), and the carbon atoms are bonded together. Draw the Lewis structure of the anion. (c) Next, 1.50 g of this sodium salt was dissolved in water and diluted to 50.0 mL. Identify the dissolved substance. Is it an acid, a base, or is it amphiprotic? Calculate the pH of the solution.

6.8 Many reactions that take place in water have analogous reactions in liquid ammonia (normal boiling point, −33 °C). (a) Write the chemical equation for the autoprotolysis of NH_3. (b) Write the formulas of the acid and base species that result from the autoprotolysis of liquid ammonia. (c) The autoprotolysis constant, K_{am}, of liquid ammonia has the value 1×10^{-33} at $-35\ °C$. What is the value of pK_{am} at that temperature? (d) What is the molar concentration of NH_4^+ ions in liquid ammonia? (e) Evaluate pNH_4 and pNH_2, which are the analogs of pH and pOH, in liquid ammonia at $-35\ °C$. (f) Derive the relation between pNH_4, pNH_2, and pK_{am}.

6.9 Acetic acid is used as a solvent for some reactions between acids and bases. (a) Nitrous acid and carbonic acids are both weak acids in water. Will either of them act as a strong acid in acetic acid? Explain your answer. (b) Will ammonia act as a strong or weak base in acetic acid? Explain your answer.

6.10 Decide on the basis of the information in Table 6C.3 whether carbonic acid is a strong or weak acid in liquid ammonia solvent. Explain your answer.

6.11 Write the equilibrium constant for the following reaction and calculate the value of K at 298 K for the reaction $HNO_2(aq) + NH_3(aq) \rightleftharpoons NH_4^+(aq) + NO_2^-(aq)$ using the data in Tables 6C.1 and 6C.2.

6.12 Write the equilibrium constant for the reaction $HIO_3(aq) + NH_2NH_2(aq) \rightleftharpoons NH_2NH_3^+(aq) + IO_3^-(aq)$ and calculate the value of K at 298 K using the data in Tables 6C.1 and 6C.2.

6.13 Draw the Lewis structure of boric acid, $B(OH)_3$. (a) Is resonance important for its description? (b) The proton transfer equilibrium for boric acid is given in a footnote to Table 6C.1. In that reaction does boric acid act as a Lewis acid, a Lewis base, or neither? Justify your answer by using Lewis structures of boric acid and its conjugate base.

6.14 Topic 6C discusses the relationship between molecular structure and the strengths of acids. The same ideas can be applied to bases. (a) Explain the relative strengths of the Brønsted bases OH^-, NH_2^-, and CH_3^- (see Table 6C.3). (b) Explain why NH_3 is a weak base in water, but PH_3 forms essentially neutral solutions. (c) If you were ranking the species in (a) or (b) as Lewis bases, would your rankings be the same or different? Explain your reasoning.

6.15 Use the thermodynamic data in Appendix 2A to calculate the acidity constant of $HF(aq)$.

6.16 The structure below shows a hydrated d-metal ion. Draw the structure of the conjugate base of this complex.

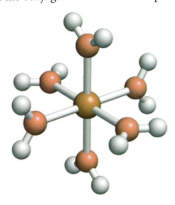

6.17 The autoprotolysis constant, K_{hw}, for heavy water, D_2O, at $25\ °C$ is 1.35×10^{-15}. (a) Write the chemical equation for the autoprotolysis (more precisely, the autodeuterolysis, because a deuteron is

being transferred) of D_2O. (b) Evaluate pK_{hw} for D_2O at $25\ °C$. (c) Calculate the molar concentrations of D_3O^+ and OD^- in pure heavy water at $25\ °C$. (d) Evaluate the pD and pOD of heavy water at $25\ °C$. (e) Find the relation between pD, pOD, and pK_{hw}.

6.18 The pK_{hw} for the autodeuterolysis of heavy water, D_2O (the analog of autoprotolysis, but involving the transfer of a deuteron), is 13.83 at 30. $°C$. Assuming $\Delta H_r^°$ for this reaction to be independent of temperature and using information from Exercise 6.17, calculate $\Delta S_r^°$ for the autodeuterolysis reaction. Suggest an interpretation of the sign. Suggest a reason why the autodeuterolysis constant of heavy water differs from the autoprotolysis constant of ordinary water.

6.19 Hemoglobin (Hb) molecules in blood carry O_2 molecules from the lungs, where the concentration of oxygen is high, to the tissues where it is low (see the Interlude following Focus 5). In the tissues the equilibrium $H_3O^+(aq) + HbO_2^-(aq) \rightleftharpoons HHb(aq) + H_2O(l) + O_2(aq)$ releases oxygen. When muscles work hard, they produce lactic acid as a byproduct. (a) What effect will the lactic acid have on the concentration of HbO_2^-? (b) When the hemoglobin returns to the lungs, where oxygen concentration is high, how does the concentration of HbO_2^- change?

6.20 Is the osmotic pressure of 0.10 M $H_2SO_4(aq)$ the same as, less than, or greater than that of 0.10 M $HCl(aq)$? Calculate the osmotic pressure of each solution to support your conclusion.

6.21 The two strands of the nucleic acid DNA are held together by hydrogen bonding between four organic bases. The structure of one of these bases, thymine, is shown below. (a) How many protons can this base accept? (b) Draw the structure of each conjugate acid that can be formed. (c) Mark with an asterisk any structure that can show amphiprotic behavior in aqueous solution.

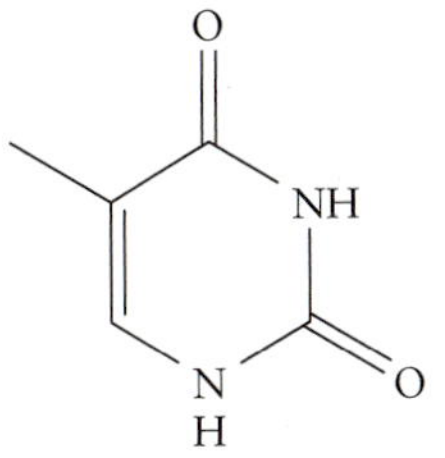

Thymine

6.22 The two strands of the nucleic acid DNA are held together by hydrogen bonding between four organic bases. The structure of one of these bases, cytosine, is shown below. (a) How many protons can this base accept? (b) Draw the structure of each conjugate acid that can be formed. (c) Mark with an asterisk any structure that can show amphiprotic behavior in aqueous solution.

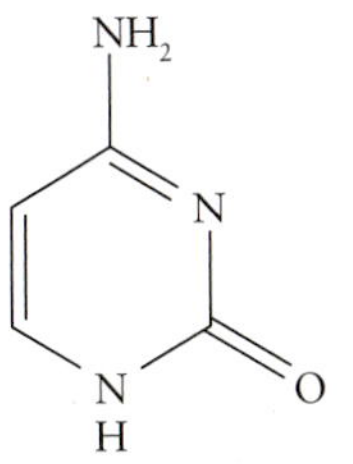

Cytosine

6.23 A buffer solution containing equal amounts of acetic acid and sodium acetate is prepared. What molar concentration of the buffer must be prepared to prevent a change in the pH by more than 0.20 after the addition of 1.00 mL of 6.00 M HCl(aq) to 100.0 mL of the buffer solution?

6.24 You require 0.150 L of a buffer solution with pH = 3.00. On the shelf is a bottle of trichloracetic acid/sodium trichloracetate buffer with pH = 2.95. The label also says [trichloracetate ion] = 0.200 mol·L^{-1}. What mass of which substance (trichloracetic acid or sodium trichloracetate) should you add to 0.150 L of the buffer to obtain the desired pH?

6.25 Malonic acid, $HOOCCH_2COOH$, a diprotic acid with $pK_{a1} = 2.8$ and $pK_{a2} = 5.7$, is titrated with KOH(aq). (a) What is the pH when $[HOOCCH_2COOH] = [HOOCCH_2CO_2^-]$? (b) What is the pH when $[HOOCCH_2CO_2^-] = [^-O_2CCH_2CO_2^-]$? (c) Which is the predominant species at pH = 4.2?

6.26 A species that can accept two protons is classified as dibasic. The dibasic molecule 1,2-ethanediamine, $H_2NC_2H_4NH_2$, which has $pK_{b1} = 3.19$ and $pK_{b2} = 6.44$, is titrated with HCl(aq). (a) What is the pH when $[H_2NC_2H_4NH_2] = [H_2NC_2H_4NH_3^+]$? (b) What is the pH when $[H_2NC_2H_4NH_3^+] = [^+H_3NC_2H_4NH_3^+]$? (c) Which is the predominant species at pH = 4.8?

6.27 A buffer solution is prepared by adding 55.0 mL of 0.15 M HNO$_3$(aq) to 45.0 mL of 0.65 M NaC$_6$H$_5$CO$_2$(aq). Determine the solubility of PbF$_2$ in this buffer solution.

6.28 A sample of 0.150 M Na$_2$CO$_3$(aq) of volume 25.0 mL is titrated with 0.100 M HCl(aq). What is the pH of the solution at each stoichiometric point in the titration?

6.29 In a "precipitation titration," the concentration of an ion is measured as it forms a precipitate. The concentration of CO$_3^{2-}$ ions in a sample of volume 25.0 mL was determined by titrating with 0.110 M AgNO$_3$(aq). Before the stoichiometric point, the Ag$^+$ ions react immediately with the CO$_3^{2-}$ ions, but after the stoichiometric point the concentration of Ag$^+$ ion increases rapidly. (a) The stoichiometric point is reached after the addition of 36.2 mL of the AgNO$_3$(aq). What is the concentration of CO$_3^{2-}$ ions in the sample? (b) The concentration of Ag$^+$ ions is followed by using a special electrode and measuring pAg (that is, $-\log[Ag^+]$). Sketch the plot of pAg against volume of AgNO$_3$(aq) and determine the value of pAg at the stoichiometric point.

6.30 What volume (in liters) of a saturated mercury(II) sulfide, HgS, solution contains an average of one mercury(II) ion, Hg^{2+}?

6.31 Two friends go to an all-you-can eat restaurant but eat too much and get heartburn. Both return to their rooms and look for a remedy. One friend takes two tablets, each containing 750 mg CaCO$_3$, and the second friend takes 3 teaspoons of milk of magnesia, which contains 400 mg MgO per teaspoon. What will be the pH of each friend's stomach after taking the medication, assuming a stomach acid volume of 100. mL and a concentration of HCl that is 0.10 mol·L^{-1}?

6.32 Consider the equilibria

$$ZnS(s) \rightleftharpoons Zn^{2+}(aq) + S^{2-}(aq)$$
$$S^{2-}(aq) + H_2O(l) \rightleftharpoons HS^-(aq) + OH^-(aq)$$
$$HS^-(aq) + H_2O(l) \rightleftharpoons H_2S(aq) + OH^-(aq)$$

Write the chemical equation for the overall equilibrium and determine the corresponding equilibrium constant. (b) Evaluate the

solubility of ZnS in a saturated H$_2$S solution, 0.1 M H$_2$S(aq), adjusted to pH = 7.0. (c) Evaluate the solubility of ZnS in a saturated H$_2$S solution, 0.1 M H$_2$S(aq), adjusted to pH = 10.0.

6.33 Use data available in the tables and appendixes to calculate the standard Gibbs free energy of formation of PbF$_2$(s).

6.34 Silver iodide is very insoluble in water. A common method for increasing its solubility is to increase the temperature of the solution containing the solid. Estimate the solubility of AgI at 85 °C.

6.35 A buffer solution of volume 300.0 mL is 0.200 M CH$_3$COOH(aq) and 0.300 M NaCH$_3$CO$_2$(aq). (a) What is the initial pH of this solution? (b) What mass of NaOH would have to be dissolved in this solution to bring the pH to 6.0?

6.36 A buffer solution of volume 250.0 mL is 0.300 M NH$_3$(aq) and 0.400 M NH$_4$Cl(aq). (a) What is the initial pH of this solution? (c) What volume of HCl gas at 2.00 atm and 25 °C would have to be dissolved in this solution to bring the pH to 8.0?

6.37 Novocaine, which is used by dentists as a local anesthetic, is a weak base with $pK_b = 5.05$. Blood is buffered to pH = 7.4. What is the ratio of the concentration of novocaine to that of its conjugate acid in the bloodstream?

6.38 To simulate blood conditions, a phosphate buffer system with a pH = 7.40 is desired. What mass of Na$_2$HPO$_4$ must be added to 0.500 L of 0.10 M NaH$_2$PO$_4$(aq) to prepare such a buffer?

6.39 The main buffer in the blood consists primarily of hydrogen carbonate ions (HCO$_3^-$) and H$_3$O$^+$ ions in equilibrium with water and CO$_2$:

$$H_3O^+(aq) + HCO_3^-(aq) \rightleftharpoons 2 H_2O(l) + CO_2(aq)$$
$$K = 7.9 \times 10^{-7}$$

This reaction assumes that all H$_2$CO$_3$ produced decomposes completely to CO$_2$ and H$_2$O. Suppose that 1.0 L of blood is removed from the body and brought to pH = 6.1. (a) If the concentration of HCO$_3^-$ is 5.5 μmol·L^{-1}, calculate the amount (in moles) of CO$_2$ present in the solution at this pH. (b) Calculate the change in pH that occurs when 0.65 μmol H$_3$O$^+$ is added to this sample of blood at this pH (that is, pH = 6.1). See Box 6G.1.

6.40 The pH of the blood is maintained by a buffering system consisting primarily of hydrogen carbonate ion (HCO$_3^-$) and H$_3$O$^+$ in equilibrium with water and CO$_2$:

$$H_3O^+(aq) + HCO_3^-(aq) \rightleftharpoons 2 H_2O(l) + CO_2(g)$$

During exercise, CO$_2$ is produced at a rapid rate in muscle tissue. (a) How does exercise affect the pH of blood? (b) Hyperventilation (rapid and deep breathing) can occur during intense exertion. How does hyperventilation affect the pH of the blood? (c) The normal first-aid treatment for hyperventilation is to have the patient breathe into a paper bag. Explain briefly why this treatment works and tell what effect the paper-bag treatment has on the pH of the blood. See Box 6G.1.

6.41 Fluoridation of city drinking water results in a fluoride ion concentration of approximately 5 × 10^{-5} mol·L^{-1}. Suppose you are using a water filter that adds calcium to the water. Will CaF$_2$ precipitate in water in which the Ca^{2+} ion concentration is 2 × 10^{-4} mol·L^{-1}?

6.42 A solution is prepared by dissolving 1 mol each of Cu(NO$_3$)$_2$, Ni(NO$_3$)$_2$, and AgNO$_3$ in 1 L of water. Using only data from

Appendix 2B, identify the metals (if any) that, when added to these solutions, (a) will leave the Ni^{2+} ions unaffected but will cause Cu and Ag to plate out of solution; (b) will leave the Ni^{2+} and Cu^{2+} ions in solution but will cause Ag to plate out of solution; (c) will leave all three metal ions in solution; (d) will leave Ni^{2+} and Ag^+ ions in solution but will cause Cu to plate out of solution.

6.43 Indicate for each of the following statements whether it applies to $E_{cell}°$, to E_{cell}, to both, or to neither: (a) decreases as the cell reaction progresses; (b) changes with temperature; (c) doubles when the coefficients of the equation are doubled; (d) can be calculated from K; (e) is a measure of how far the cell reaction is from equilibrium. Justify your answers.

6.44 State how the oxidizing strength of each of the following oxidizing agents would be affected by raising the pH (stronger, weaker, or no change): (a) Br_2; (b) MnO_4^-; (c) NO_3^-; (d) ClO_4^-; (e) Cu^{2+}. Justify your answers.

6.45 Volta discovered that when he used different metals in his "pile," some combinations had a stronger effect than others. From that information he constructed an electrochemical series. How would Volta have ordered the following metals, if he put the most strongly reducing metal first: Fe, Ag, Au, Zn, Cu, Ni, Co, Al?

6.46 Arrange the following metals in order of increasing strength as reducing agents: U, V, Ti, Ni, Sn, Cr, Rb.

6.47 A galvanic cell has the following cell reaction: $M(s) + 2 Zn^{2+}(aq) \rightarrow 2 Zn(s) + M^{4+}(aq)$. The standard potential of the cell is $+0.16$ V. What is the standard potential of the M^{4+}/M redox couple?

6.48 Using data in Appendix 2B, calculate the standard potential for the half-reaction $Ti^{4+}(aq) + 4 e^- \rightarrow Ti(s)$.

6.49 K_{sp} for $Cu(IO_3)_2$ is 1.4×10^{-7}. Using this value and data in Appendix 2B, calculate $E°$ for the half-reaction $Cu(IO_3)_2(s) + 2 e^- \rightarrow Cu(s) + 2 IO_3^-(aq)$.

6.50 K_{sp} for $Ni(OH)_2$ is 6.5×10^{-18}. Use this value and data from Appendix 2B to calculate $E°$ for the half-reaction $Ni(OH)_2(s) + 2 e^- \rightarrow Ni(s) + 2 OH^-(aq)$.

6.51 A galvanic cell functions only when the electrical circuit is complete. In the external circuit the current is carried by the flow of electrons through a metal wire. Explain how the current is carried through the cell itself.

6.52 A technical handbook contains tables of thermodynamic quantities for common reactions. If you want to know whether a certain cell reaction has a positive standard potential, which of the following properties would give you that information directly (on inspection)? Which would not? Explain your answer. (a) $\Delta G°$; (b) $\Delta H°$; (c) $\Delta S°$; (d) $\Delta U°$; (e) K.

6.53 (a) If you were to construct a concentration cell in which one half-cell contains 1.0 M $CrCl_3(aq)$ and the other half-cell contains 0.0010 M $CrCl_3(aq)$, and both electrodes were chromium, at which electrode would reduction be spontaneous? How will each of the following changes affect the cell potential? Justify your answers. (b) Adding 100 mL pure water to the anode compartment. (c) Adding 100 mL of 1.0 M $NaOH(aq)$ to the cathode compartment ($Cr(OH)_3$ is insoluble). (d) Increasing the mass of the chromium electrode in the anode compartment.

6.54 Dental amalgam, a solid solution of silver and tin in mercury, was used for filling tooth cavities. Two of the reduction half-reactions that the filling can undergo are

$$3 Hg_2^{2+}(aq) + 4 Ag(s) + 6 e^- \longrightarrow 2 Ag_2Hg_3(s) \qquad E° = +0.85 \text{ V}$$
$$Sn^{2+}(aq) + 3 Ag(s) + 2 e^- \longrightarrow Ag_3Sn(s) \qquad E° = -0.05 \text{ V}$$

Suggest a reason why, if you accidentally bite on a piece of aluminum foil with a tooth containing a silver filling, you may feel pain. Write a balanced chemical equation to support your suggestion.

6.55 Suppose that 25.0 mL of a solution of Ag^+ ions of unknown concentration is titrated with 0.015 M $KI(aq)$ at 25 °C. A silver electrode is immersed in this solution, and its potential is measured relative to a standard hydrogen electrode. A total of 16.7 mL of $KI(aq)$ was required to reach the stoichiometric point, when the potential was 0.325 V. (a) What is the molar concentration of Ag^+ in the solution? (b) Evaluate K_{sp} for AgI from these data.

6.56 Suppose that 35.0 mL of 0.012 M $Cu^+(aq)$ is titrated with 0.010 M $KBr(aq)$ at 25 °C. A copper electrode is immersed in this solution, and its potential is measured relative to a standard hydrogen electrode. What volume of the KBr solution must be added to reach the stoichiometric point, and what will the potential be at that point? $K_{sp}(CuBr) = 5.2 \times 10^{-9}$.

6.57 Use the data in Appendix 2B and the fact that, for the half-reaction $F_2(g) + 2 H^+(aq) + 2 e^- \rightarrow 2 HF(aq)$, $E° = +3.03$ V, to calculate the value of K_a for HF.

6.58 The following items are obtained from a stockroom for the construction of a galvanic cell: two 250-mL beakers and a salt bridge, a voltmeter with attached wires and clips, 200 mL of 0.0080 M $CrCl_3(aq)$, 200 mL of 0.12 M $CuSO_4(aq)$, a piece of copper wire, and a chrome-plated piece of metal. (a) Describe the construction of the galvanic cell. (b) Write the reduction half-reactions. (c) Write the overall cell reaction. (d) Write the cell diagram for the galvanic cell. (e) What is the expected cell potential?

6.59 (a) By considering the dependence of the Gibbs free energy of reaction on potential and on temperature, derive an equation for the temperature dependence of $E_{cell}°$. (b) Use your equation to predict the standard potential for the formation of water from hydrogen and oxygen in a fuel cell at 80 °C. Assume that $\Delta H°$ and $\Delta S°$ are independent of temperature.

6.60 (a) What is the standard cell potential ($E_{cell}°$) for the reaction below at 298 K? (b) What is the standard cell potential for the reaction at 335 K? (c) What is the cell potential for the reaction at 335 K when $[Zn^{2+}] = 2.0 \times 10^{-4}$ mol·L^{-1} and $[Pb^{2+}] = 1.0$ mol·L^{-1}? (See Exercise 6.59.)

$$Pb^{2+}(aq) + Zn(s) \longrightarrow Zn^{2+}(aq) + Pb(s)$$

6.61 In a neuron (a nerve cell), the concentration of K^+ ions inside the cell is about 20–30 times as great as that outside. What potential difference between the inside and the outside of the cell would you expect to measure if the difference is due only to the imbalance of potassium ions?

6.62 (a) The potential of the cell $Zn(s)|Zn^{2+}(aq, ?)||Pb^{2+}(aq, 0.10$ mol·$L^{-1})|Pb(s)$ is $+0.661$ V. What is the molar concentration of Zn^{2+} ions? (b) Write an equation showing how the Zn^{2+} ion concentration varies with the cell potential, assuming that all other aspects of the cell remain constant.

6.63 When a pH meter was calibrated with a boric acid–borate buffer with a pH of 9.40, the cell potential was +0.060 V. When the buffer was replaced with a solution of unknown hydronium ion concentration, the cell potential was +0.22 V. What is the pH of the solution?

6.64 What is the standard potential for the reduction of oxygen to water in (a) an acidic solution and (b) a basic solution? (c) Is MnO_4^- more likely to be reduced to MnO_4^{2-} in an acidic or a basic oxygenated solution (a solution saturated with oxygen gas at 1 atm)? Explain your conclusion.

6.65 What range (in volts) does a voltmeter need to have to measure pH in the range of 1 to 14 at 25 °C if the voltage is zero when pH = 7?

6.66 The entropy change of a cell reaction can be determined from the change of the cell potential with temperature. (a) Show that $\Delta S° = nF(E_{cell,2}° - E_{cell,1}°)/(T_2 - T_1)$. Assume that $\Delta S°$ and $\Delta H°$ are constant over the temperature range considered. (b) Calculate $\Delta S°$ and $\Delta H°$ for the cell reaction $Hg_2Cl_2(s) + H_2(g) \rightarrow 2\ Hg(l) + 2\ H^+(aq) + 2\ Cl^-(aq)$, given that $E° = +0.2699$ V at 293 K and +0.2669 V at 303 K.

6.67 A silver concentration cell is constructed with the electrolyte at both electrodes being initially 0.10 M $AgNO_3(aq)$ at 25 °C. The electrolyte at one electrode is diluted by a factor of 10 five times in succession and the cell potential measured each time. (a) Plot the potential of this cell against ln $[Ag^+]_{anode}$. (b) Calculate the value of the slope of the line. To what term in the Nernst equation does this value correspond? Is the value you determined from the plot consistent with the value you would calculate from the values in that term? If it is not consistent, calculate your percentage error. (c) What is the value of the y-intercept? To what term in the Nernst equation does this value correspond?

6.68 Consider the electroplating of a metal +1 cation from a solution of unknown concentration according to the half-reaction $M^+(aq) + e^- \rightarrow M(s)$, with a standard potential $E°$. When the half-cell is connected to an appropriate oxidation half-cell and current is passed through it, the M^+ cation begins plating out at E_1. To what value (E_2) must the applied potential be adjusted, relative to E_1, if 99.99% of the metal is to be removed from the solution?

6.69 Use only the data in Appendix 2B to calculate the acidity constant of HClO in water.

6.70 The magnitudes of the standard potentials of two metals M and X were determined to be

(1) $M^+(aq) + e^- \longrightarrow M(s)$ $|E°| = 0.25$ V
(2) $X^{2+}(aq) + 2\ e^- \longrightarrow X(s)$ $|E°| = 0.65$ V

When the two electrodes are connected, current flows from M to X in the external circuit. When the electrode corresponding to half-reaction 1 is connected to the standard hydrogen electrode (SHE), current flows from M to the SHE. (a) What are the signs of $E°$ of the two half-reactions? (b) What is the standard cell potential for the cell constructed from these two electrodes?

6.71 An aqueous solution of Na_2SO_4 was electrolyzed for 30.0 min; 25.0 mL of oxygen was collected at the anode over water at 22 °C and a total pressure of 722 Torr. Determine the current that was used to produce the gas. See Table 5A.2 for the vapor pressure of water.

See the Interlude following Focus 6 for Exercises 6.72–6.79.

6.72 A photoelectrochemical cell is an electrochemical cell that uses light to carry out an electrochemical reaction. The silicon electrodes in a photoelectrochemical cell being considered for the production of hydrogen react with water:

$$SiO_2(s) + 4\ H^+(aq) + 4\ e^- \longrightarrow Si(s) + 2\ H_2O(l)$$
$$E° = -0.84\ V$$

Calculate the standard cell potential for the reaction between silicon and water in a cell that also produces hydrogen from water and write the balanced equation for the cell reaction.

6.73 The "aluminum–air fuel cell" is used as a reserve battery in remote locations. In this cell, aluminum reacts with the oxygen in air in basic solution. (a) Write the oxidation and reduction half-reactions for this cell. (b) Calculate the standard cell potential.

6.74 What is (a) the electrolyte and (b) the oxidizing agent during discharge in a lead–acid battery? (c) Write the reaction that takes place at the cathode during the charging of the lead–acid battery.

6.75 (a) Write the cell reaction for the lead–acid battery. (b) Explain how each of the following change in a lead–acid battery during discharge: pH; amount of PbO_2; total amount of lead in the battery.

6.76 (a) Why are lead–antimony grids used as electrodes in the lead–acid battery rather than smooth plates? (b) What is the reducing agent in the lead–acid battery? (c) The lead–acid cell potential is about 2 V. How, then, does a car battery produce 12 V for its electrical system?

6.77 What is (a) the electrolyte and (b) the oxidizing agent in the mercury cell shown here? (c) Write the overall cell reaction for a mercury cell.

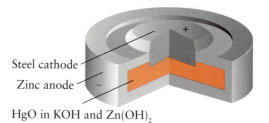

Steel cathode
Zinc anode
HgO in KOH and $Zn(OH)_2$

6.78 A fuel cell in which hydrogen reacts with nitrogen instead of oxygen is proposed. (a) Write the chemical equation for the reaction in water, which produces aqueous ammonia. (b) What would be the maximum free energy output of the cell for the consumption of 28.0 kg nitrogen? (c) Is this type of fuel cell thermodynamically feasible?

6.79 The body functions as a kind of fuel cell that uses oxygen from the air to oxidize glucose:

$$C_6H_{12}O_6(aq) + 6\ O_2(g) \longrightarrow CO_2(g) + 6\ H_2O(l)$$

During normal activity, a person uses the equivalent of about 10 MJ of energy a day. Assume that this value represents ΔG and estimate the average current through your body in the course of a day, assuming that all the energy that we use arises from the reduction of O_2 in the glucose oxidation reaction.

FOCUS 6 Cumulative Exercises

6.80 Soluble nontoxic salts such as $Fe_2(SO_4)_3$ are often used during water purification to remove soluble toxic contaminants, because they form gelatinous hydroxides that encapsulate the contaminants and allow them to be filtered from the water.

(a) Calculate the molar solubility in water of $Fe(OH)_3$ at 25 °C.

(b) What is the concentration of hydroxide ions in a saturated solution of $Fe(OH)_3$? Calculate the pH of the solution.

(c) Discuss whether your result in part (b) is reasonable for a solution of a basic hydroxide. Explain what assumptions were made in your calculations and evaluate the validity of the assumptions.

(d) A simplified equation for the reaction of Fe^{3+} ions with water is $Fe^{3+}(aq) + 6 H_2O(l) \rightleftharpoons Fe(OH)_3(s) + 3 H_3O^+(aq)$. Use the data in Table 6I.1 and K_w to calculate the equilibrium constant for this reaction.

(e) If 10.0 g of $Fe_2(SO_4)_3$ is dissolved in enough water to make up 1.00 L of aqueous solution and the pH of the solution is raised to 8.00 by addition of NaOH, what mass of solid $Fe(OH)_3$ will form?

(f) To test for the ability of $Fe_2(SO_4)_3$ to remove chloride ions from water, a standard aqueous solution containing 24.72 g of NaCl in 1.000 L of solution was prepared. A sample of the NaCl solution of volume 25.00 mL was then combined with the mixture described in part (e) and stirred. The $Fe(OH)_3$ precipitate containing encapsulated chloride ion was removed by filtration, then dissolved in acid. An aqueous solution of $AgNO_3$ was then added to the resulting solution and the solid AgCl formed filtered and dried. The mass of AgCl was 0.604 g. What percentage of the chloride ion in the sample had been removed from the solution?

6.81 Many important biological reactions involve electron transfer. Because the pH of bodily fluids is close to 7, the "biological standard potential" of an electrode, E^*, is measured at pH = 7.

(a) Calculate the biological standard potential for (i) the reduction of hydrogen ions to hydrogen gas; (ii) the reduction of nitrate ions to NO gas.

(b) Calculate the biological standard potential E^* for the reduction of the biomolecule NAD^+ to NADH in aqueous solution. The reduction half-reaction under thermodynamic standard conditions is $NAD^+(aq) + H^+(aq) + 2 e^- \rightarrow NADH(aq)$, with $E° = -0.099$ V.

(c) The pyruvate ion, $CH_3COCO_2^-$, is formed during the metabolism of glucose in the body. The ion has a chain of three carbon atoms. The central carbon atom has a double bond to a terminal oxygen atom, and one of the end carbon atoms is bonded to two oxygen atoms in a carboxylate group. Draw the Lewis structure of the pyruvate ion and assign a hybridization scheme to each carbon atom.

(d) The lactate ion has a similar structure to the pyruvate ion, except that the central carbon is now attached to an —OH group: $CH_3CH(OH)CO_2^-$. Draw the Lewis structure of the lactate ion and assign a hybridization scheme to the central carbon atom.

(e) During exercise, the pyruvate ion is converted to lactate ion in the body by coupling to the half-reaction for NADH given in part (b). For the half-reaction pyruvate $+ 2 H^+ + 2 e^- \rightarrow$ lactate, $E^* = -0.190$ V. Write the cell reaction for the spontaneous reaction that occurs between these two biological couples and calculate E^* and $E°$ for the overall reaction.

(f) Calculate the standard Gibbs free energy of reaction for the overall reaction in part (e).

(g) Calculate the equilibrium constant at 25 °C for the overall reaction in part (e).

INTERLUDE Practical Cells

An important application of galvanic cells is their use as the portable power sources commonly called "batteries." An ideal battery should be inexpensive, portable, safe to use, and environmentally benign. It should also maintain a potential difference that is stable with the passage of time (**TABLE 1**). Both the mass and the volume of a battery are critical parameters. The electrolyte in a battery uses as little water as possible, both to reduce leakage of the electrolyte and to keep the mass low. Much of the research on batteries deals with raising their specific energy (the energy that it can generate divided by the mass of the battery), typically expressed in kilowatt-hours per kilogram, $kW \cdot h \cdot kg^{-1}$.*

A **primary cell** is a galvanic cell with the reactants sealed inside at the time of manufacture. It cannot be recharged; when it runs down, it is discarded. A *dry cell* is the primary cell used in most common applications, such as remote controls and flashlights (**FIG. 1**). Its familiar cylindrical zinc container serves as the anode; in the center is the cathode, a carbon rod. The interior of the container is lined with paper that serves as the porous barrier. The electrolyte is a moist paste of ammonium chloride, manganese(IV) oxide, finely granulated carbon, and an inert filler, usually starch. The ammonia provided by the ammonium ions forms the complex ion $Zn(NH_3)_4^{2+}$ with the Zn^{2+} ions and prevents their buildup and a consequent reduction of the potential as the cell is discharged.

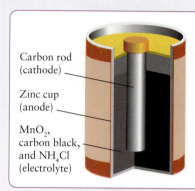

Carbon rod (cathode)
Zinc cup (anode)
MnO_2, carbon black, and NH_4Cl (electrolyte)

FIG. 1 A commercial dry cell. The dry cell is also called the Leclanché cell, for Georges Leclanché, the French engineer who invented it in about 1866. The electrolyte is a moist paste.

Two primary cells that provide a more stable and longer-lasting potential than the dry cell are the alkaline cell and the silver cell. An *alkaline cell* is similar to a dry cell but uses an alkaline electrolyte, with which the zinc electrode does not readily react when the battery is not in use. As a result, alkaline cells have longer lives than dry cells. They are used in smoke detectors and backup power supplies. A *silver cell* has a cathode made of solid Ag_2O and Ag. The relatively high potential of a silver cell, with its solid reactants and products, is maintained with great reliability over long periods of time, and the cell can be manufactured in

TABLE 1 Reactions in Commercial Batteries[†]

Primary cells

dry
$Zn(s)|ZnCl_2(aq), NH_4Cl(aq)|MnO(OH)(s)|MnO_2(s)|graphite,$ 1.5 V
Anode: $Zn(s) \longrightarrow Zn^{2+}(aq) + 2\,e^-$
    followed by $Zn^{2+}(aq) + 4\,NH_3(aq) \longrightarrow [Zn(NH_3)_4]^{2+}(aq)$
Cathode: $MnO_2(s) + H_2O(l) + e^- \longrightarrow MnO(OH)(s) + OH^-(aq)$
    followed by $NH_4^+(aq) + OH^-(aq) \longrightarrow H_2O(l) + NH_3(aq)$

alkaline
$Zn(s)|ZnO(s)|OH^-(aq)|Mn(OH)_2(s)|MnO_2(s)|graphite,$ 1.5 V
Anode: $Zn(s) + 2\,OH^-(aq) \longrightarrow ZnO(s) + H_2O(l) + 2\,e^-$
Cathode: $MnO_2(s) + 2\,H_2O(l) + 2\,e^- \longrightarrow Mn(OH)_2(s) + 2\,OH^-(aq)$

silver
$Zn(s)|ZnO(s)|KOH(aq)|Ag_2O(s)|Ag(s)|steel,$ 1.6 V
Anode: $Zn(s) + 2\,OH^-(aq) \longrightarrow ZnO(s) + H_2O(l) + 2\,e^-$
Cathode: $Ag_2O(s) + H_2O(l) + 2\,e^- \longrightarrow 2\,Ag(s) + 2\,OH^-(aq)$

Secondary cells

lead–acid
$Pb(s)|PbSO_4(s)|H^+(aq), HSO_4^-(aq)|PbO_2(s)|PbSO_4(s)|Pb(s),$ 2 V
Anode: $Pb(s) + HSO_4^-(aq) \longrightarrow PbSO_4(s) + H^+(aq) + 2\,e^-$
Cathode: $PbO_2(s) + 3\,H^-(aq) + HSO_4^-(aq) + 2\,e^- \longrightarrow PbSO_4(s) + 2\,H_2O(l)$

nicad
$Cd(s)|Cd(OH)_2(s)|KOH(aq)|Ni(OH)_3(s)|Ni(OH)_2(s)|Ni(s),$ 1.25 V
Anode: $Cd(s) + 2\,OH^-(aq) \longrightarrow Cd(OH)_2(s) + 2\,e^-$
Cathode: $2\,Ni(OH)_3(s) + 2\,e^- \longrightarrow 2\,Ni(OH)_2(s) + 2\,OH^-(aq)$

NiMH
$M(s)|MH(s)|KOH(aq)|NiOOH(s)|Ni(OH)_2(s)|Ni(s),$ 1.2 V
Anode: $MH(s)^{††} + OH^-(aq) \longrightarrow M(s) + H_2O(l) + e^-$
Cathode: $NiOOH(s) + H_2O(l) + e^- \longrightarrow Ni(OH)_2(s) + OH^-$

sodium–sulfur
$Na(l)|Na^+(ceramic\ electrolyte), S^{2-}(ceramic\ electrolyte)|S_8(l),$ 2.2 V
Anode: $2\,Na(l) \longrightarrow 2\,Na^+(electrolyte) + 2\,e^-$
Cathode: $S_8(l) + 16\,e^- \longrightarrow 8\,S^{2-}(electrolyte)$

[†]Cell notation is described in Topic 6L.
[††]The metal in a nickel–metal hydride battery is usually a complex alloy of several metals, such as Cr, Ni, Co, V, Ti, Fe, and Zr.

*1 $kW \cdot h = (10^3\ J \cdot s^{-1}) \times (3600\ s) = 3.6 \times 10^6\ J = 3.6$ MJ exactly.

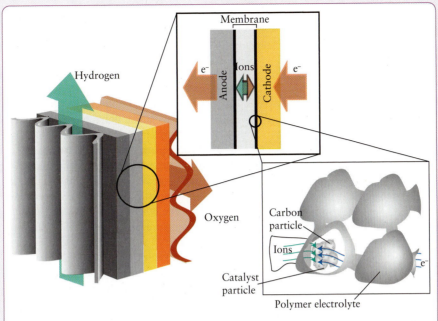

FIG. 2 A proton exchange membrane is used in both hydrogen and methanol fuel cells. It allows protons but not electrons to travel through it: the protons flow through the porous membrane to the cathode, where they combine with oxygen to form water, while the electrons flow through an external circuit. Several layers of cells are combined to generate the power required.

very small sizes. These features make it desirable for medical implants such as pacemakers, for hearing aids, and for cameras.

A **fuel cell** is like a primary cell, generating electricity directly from a chemical reaction, as in a battery, but uses reactants that are supplied continuously. Francis Bacon, a British scientist and engineer, developed an idea proposed by Sir William Grove in 1839 to produce a fuel cell that could generate 5 kW. A fuel cell that runs on hydrogen and oxygen was installed on the space shuttle (**FIG. 2**). An advantage of this fuel cell is that the only product of the cell reaction, water, can be used for life support.

In a simple version of a fuel cell, a fuel such as hydrogen gas is passed over a platinum electrode, oxygen is passed over the other, similar electrode, and the electrolyte is aqueous potassium hydroxide. A porous membrane that is permeable to ions separates the two electrode compartments. Many varieties of fuel cells are possible, and in some the electrolyte is a solid polymer membrane or a ceramic. Three of the most promising fuel cells are the alkali fuel cell, the phosphoric acid fuel cell, and the methanol fuel cell.

The hydrogen–oxygen cell that was used in the space shuttle is called an alkali fuel cell because it has an alkaline electrolyte:

Anode: $\quad 2\,H_2(g) + 4\,OH^-(aq) \rightarrow 4\,H_2O(l) + 4\,e^-$

Electrolyte: $\quad KOH(aq)$

Cathode: $\quad O_2(g) + 4\,e^- + 2\,H_2O(l) \rightarrow 4\,OH^-(aq)$

Although its cost prohibits its use in many applications, the alkali fuel cell is the primary fuel cell used in the aerospace industry.

An acid electrolyte may also be used, as in a phosphoric acid fuel cell:

Anode: $\quad 2\,H_2(g) \rightarrow 4\,H^+(aq) + 4\,e^-$

Electrolyte: $\quad H_3PO_4(aq)$

Cathode: $\quad O_2(g) + 4\,H^+(aq) + 4\,e^- \rightarrow 2\,H_2O(l)$

This fuel cell has shown promise for combined heat and power systems (CHP systems). In such systems, the waste heat is used to heat buildings or to do work. Efficiency in a CHP plant can reach 80%.

Although hydrogen gas is an attractive fuel, it has disadvantages for mobile applications: it is difficult to store and dangerous to handle. One possibility for portable fuel cells is to store the hydrogen in nanotubes made of carbon, silicon, or compounds such as WS_2 or TiO_2. Hydrogen molecules are readily adsorbed on the surfaces of these materials and nanotubes have very large surface areas. Carbon nanofibers in herringbone patterns have been shown to store huge amounts of hydrogen and to result in energy densities twice that of gasoline. Another option is the use of organometallic materials or inorganic hydrides, such as sodium aluminum hydride, $NaAlH_4$, doped with titanium. Despite the difficulties of storing hydrogen, many cities have deployed hydrogen fuel cell buses for mass transit (**FIG. 3**). An attractive fuel to use in fuel cells is methanol, which is easy to handle and is rich in hydrogen atoms:

Anode: $\quad CH_3OH(l) + 6\,OH^-(aq) \rightarrow 5\,H_2O(l) + CO_2(g) + 6\,e^-$

Electrolyte: polymeric materials

Cathode: $\quad O_2(g) + 4\,e^- + 2\,H_2O(l) \rightarrow 4\,OH^-(aq)$

The development of efficient and cost-effective methanol fuel cells for consumer applications is a focus of current research in fuel cell technology.

An exciting emerging technology is the biofuel cell. A biofuel cell is like a conventional fuel cell; however, in place of a platinum catalyst, it uses enzymes or even whole organisms. The electricity is extracted through organic molecules that can support the transfer of electrons. One application is as the power

FIG. 3 This bus operating in Reykjavik, Iceland is powered by a hydrogen fuel cell with a proton exchange membrane. Its operation is pollution free because the only product of the combustion is water. (*Martin Bond/Science Source*.)

source for medical implants, such as pacemakers, perhaps by using the glucose present in the bloodstream as the fuel.

Secondary cells are galvanic cells that must be charged before they can be used; this type of cell is normally rechargeable. The batteries used in portable computers and automobiles are secondary cells. In the charging process, an external source of electricity reverses the spontaneous cell reaction and creates a nonequilibrium mixture of reactants. After charging, the cell can again produce electricity.

The *lead–acid cell* of an automobile battery is a secondary cell that contains several grids that act as electrodes (**FIG. 4**). It has a low specific energy but, because the total surface area of these grids is large, the battery can generate large currents for short periods, such as the time needed for starting an engine. The electrodes are initially a hard lead–antimony alloy covered with a paste of lead(II) sulfate. The electrolyte is dilute sulfuric acid. During the first charging, some of the lead(II) sulfate is reduced to lead on one of the electrodes; that electrode will act as the anode during discharge. Simultaneously, during charging, lead(II) sulfate is oxidized to lead(IV) oxide on the electrode that will act

as the cathode during discharge. The lead–acid cell, which has a potential of 2 V, is used in a series of six cells to provide a 12 V source of power for starting the engine in most vehicles.

Hybrid vehicles make use of the rechargeable nickel–metal hydride (NiMH) cell to supplement energy provided by burning gasoline. In this type of battery, hydrogen is stored in the form of a metal hydride, using a heterogeneous alloy of several metals, commonly including titanium, vanadium, chromium, and nickel. The advantages include low mass, high energy density (the energy that it can generate divided by the volume of the cell), long shelf life, high current load capability, rapid charging, and good capacity (long time between charges). Because the materials are nontoxic, disposal of the batteries does not generate environmental problems.

The *lithium-ion* cell is used in laptop computers and many media players because it can be recharged many times. This type of battery has an electrolyte consisting of polypropylene oxide or polyethylene oxide mixed with molten lithium salts that are then allowed to cool. The resulting rubbery materials serve as good conductors of Li^+ ions. The low mass density of lithium gives it the highest available energy density, and lithium's very negative electrode potential provides a cell potential as high as 4 V.

A *sodium–sulfur* cell is one of the more startling batteries (**FIG. 5**). It has liquid reactants (sodium and sulfur) and a solid electrolyte (a porous aluminum oxide ceramic), it must operate at a temperature of about 320 °C, and it is highly dangerous in case of breakage. Because sodium has a low density, these cells have a very high specific energy. Their most common application is to power electric vehicles. Once the vehicle is operating, the heat generated by the battery is sufficient to maintain the temperature.

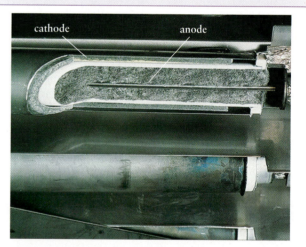

FIG. 5 A sodium–sulfur battery used in an electric vehicle. (*Takeshi Takahara/Science Source*.)

Related Exercises 6.72–6.79

Further Reading Breakthrough Technologies Institute, "The Online Fuel Cell Information Center," http://www.fuelcells.org/. S. Ritter, "Sunny forecast for fuel cells," *Chemical and Engineering News*, vol. 86, August 4, 2008, p. 7. D. Castelvecchi, "Is your phone out of juice? Biological fuel cell turns drinks into power," *Science News*, vol. 171, March 31, 2007, p. 197. S. M. Kwan and K. L. Yeung, "Zeolite micro fuel cell," *Chemical Communications*, 2008, p. 3631.

FIG. 4 A typical lead–acid battery consists of six cells in series and produces about 12 V.

KINETICS

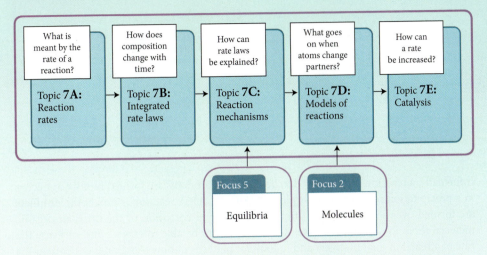

Thermodynamics (Focus 4) is used to predict the spontaneous direction of chemical change and the extent of reaction at equilibrium but says nothing about how quickly the reaction approaches equilibrium. Some spontaneous reactions—such as the decomposition of benzene into carbon and hydrogen—do not seem to proceed at all, whereas other reactions—such as proton transfer reactions—reach equilibrium very rapidly. This Focus examines the rates of reactions, including the details of how reactions proceed, what determines their rates, and how to control those rates. These aspects of chemical reactions constitute the field of "chemical kinetics."

TOPIC 7A introduces the concept of reaction rate and how it can be expressed in terms of the concentrations of the reactants (and sometimes products) involved in a reaction. The existence of these expressions, which are known as "rate laws," allows reactions to be classified according to their kinetic behavior. A rate law is expressed in terms of a "rate constant," a parameter that characterizes the rate of a given reaction. TOPIC 7B describes methods by which rate constants are determined experimentally and shows how this information is used to make predictions about how the concentrations of reactants and products change over time.

Rate laws are important because they provide a clue to how reactions take place at a molecular level. In particular, as shown in TOPIC 7C, they provide criteria for judging whether a "reaction mechanism," a suggested sequence of steps by which the overall reaction takes place, is acceptable. In a similar way, TOPIC 7D shows how the determination of the value of the rate constant and how it changes with temperature can be used to build models of the intimate details of how individual reaction events take place when bonds break and atoms exchange partners. These details point toward, in TOPIC 7E, an understanding of how "catalysts" function, and how their biological analogs, enzymes, act in organisms.

Topic 7A Reaction Rates

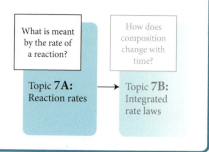

What is meant by the rate of a reaction?

How does composition change with time?

Topic **7A**: Reaction rates → Topic **7B**: Integrated rate laws

Why Do You Need to Know This Material? You need to know how to describe the rates of chemical reactions to be able to predict how quickly products are formed or reactants consumed.

What Do You Need to Know Already? You need to be familiar with units of concentration (*Fundamentals* G) and the ideal gas law (Topic 3B).

Informally, a reaction is considered to be fast if the products are formed rapidly, as occurs in a precipitation reaction or an explosion (**FIG. 7A.1**). A reaction is slow if the products are formed over a long period of time, as happens in corrosion or the decay of organic material (**FIG. 7A.2**). In each case, it is important to be able to express and measure the rate of a reaction quantitatively and to look for patterns in how those rates depend on the conditions. Once those patterns have been established, they can be used to discover details about how reactions take place at an atomic level and how yields can be modified.

7A.1 Concentration and Reaction Rate

In everyday life, a rate is defined as the change in a property divided by the time that it takes for that change to take place. For instance, the speed of an automobile, the rate of change of its position, is defined as distance traveled divided by the time taken. The *average speed* at some stage of the journey is obtained by dividing the distance traveled in an interval by the time interval. The *instantaneous speed* is obtained by reading the speedometer at some specific point on the journey. In chemistry, rates are expressed similarly. The **reaction rate** is defined, like the average speed of a car, as the change in concentration of one of the reactants or products at a selected stage of the reaction divided by the time interval over which the change takes place. Because the rate may change as time passes, the **average reaction rate** in a particular interval is defined as the change in molar concentration of a reactant R, $\Delta[R] = [R]_{t_2} - [R]_{t_1}$, divided by the time interval $\Delta t = t_2 - t_1$:

$$\text{Average rate of consumption of R} = -\frac{\Delta[R]}{\Delta t} \tag{1a}$$

FIGURE 7A.1 Reactions proceed at widely different rates. Some, such as this explosion at a military display in Nizhniy Tagil, Russia, are very fast. The gases are suddenly produced form the shockwave of the explosion. (*Sergei Butorin/Shutterstock.*)

FIGURE 7A.2 Some reactions are very slow, as in the gradual buildup of corrosion on the prow of the *Titanic* on the cold floor of the Atlantic Ocean. (*Emary Kristof/National Geographic Creative.*)

Because reactants are used up in a reaction, the concentration of R decreases as time passes, so $\Delta[R]$ is negative. The minus sign in Eq. 1a is included to ensure that the rate is positive, which is the normal convention in chemical kinetics. If instead the concentration of a product P is monitored, the average rate is expressed as

$$\text{Average rate of formation of P} = \frac{\Delta[P]}{\Delta t} \qquad \textbf{(1b)}$$

In this expression, $\Delta[P]$ is the change in molar concentration of P during the interval Δt; it is a positive quantity because products are formed as time goes on.

> **A Note on Good Practice:** Unlike in the discussion and formulation of equilibrium constants, in chemical kinetics the square brackets denote molar concentration with the units moles per liter ($mol \cdot L^{-1}$) retained.

EXAMPLE 7A.1   Calculating an average reaction rate

Suppose you need highly pure HI. You could prepare it by reacting hydrogen and iodine directly by the reaction $H_2(g) + I_2(g) \rightarrow 2\,HI(g)$, provided that the reaction is sufficiently fast. You might conduct an experiment to study the reaction rate to check if the preparation of HI in this process is fast enough. In an interval of 100. s, the concentration of HI increased from $3.50\ mmol \cdot L^{-1}$ to $4.00\ mmol \cdot L^{-1}$. What was the average rate of this reaction?

PLAN   Substitute the data into Eq. 1b.

SOLVE

From Average rate of formation of $P = \Delta[P]/\Delta t$,

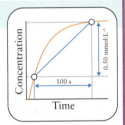

$$\text{Average rate of formation of HI} = \frac{(4.00 - 3.50)(mmol\ HI) \cdot L^{-1}}{100.s}$$

$$= 5.0 \times 10^{-3}(mmol\ HI) \cdot L^{-1} \cdot s^{-1}$$

$$= 5.0\ (\mu mol\ HI) \cdot L^{-1} \cdot s^{-1}$$

In the plot, the slope of the straight blue line gives the average rate.

> **A Note on Good Practice:** For the sake of clarity, it is good practice to choose multiples of units that minimize the powers of 10 shown explicitly. In this case, because 10^{-3} mmol = 1 μmol, the rate can be reported as $5.0\ (\mu mol\ HI) \cdot L^{-1} \cdot s^{-1}$.

Self-test 7A.1A   When the reaction $H_2(g) + I_2(g) \rightarrow 2\,HI(g)$ was carried out at a high temperature, the concentration of HI increased from $4.20\ mmol \cdot L^{-1}$ to $6.00\ mmol \cdot L^{-1}$ in 200. s. What was the average reaction rate?

[**Answer:** $9.00\ (\mu mol\ HI) \cdot L^{-1} \cdot s^{-1}$]

Self-test 7A.1B   Hemoglobin (Hb) carries oxygen through our bodies by forming a complex with it: $Hb(aq) + O_2(aq) \rightarrow HbO_2(aq)$. In a solution of hemoglobin exposed to oxygen, the concentration of hemoglobin fell from $1.2\ nmol \cdot L^{-1}$ (1 nmol = 10^{-9} mol) to $0.80\ nmol \cdot L^{-1}$ in 0.10 μs. What was the average rate at which hemoglobin reacted with oxygen in that solution, in millimoles per liter per microsecond?

Related Exercises 7A.3, 7A.4

In Example 7A.1, the reaction rate was reported in the units micromoles per liter per second ($\mu mol \cdot L^{-1} \cdot s^{-1}$), but other units for time (such as minutes or even hours) are commonly encountered for slower reactions. Note, too, that, when reporting a reaction rate, you must specify the species to which the rate refers, because species are consumed or produced at rates related to the stoichiometry of the reaction. For example, in the reaction in Example 7A.1, two HI molecules are produced from one H_2 molecule, and so the rate of consumption of H_2 is half the rate of formation of HI; therefore

$$\frac{\Delta[H_2]}{\Delta t} = -\frac{1}{2}\frac{\Delta[HI]}{\Delta t}$$

To avoid the ambiguity associated with several ways of reporting a reaction rate, a single unique average rate of a reaction can be reported without specifying the species. The

unique average rate of the reaction $aA + bB \rightarrow cC + dD$ is any of the following four equal quantities:

$$\text{Unique average reaction rate} = -\frac{1}{a}\frac{\Delta[A]}{\Delta t} = -\frac{1}{b}\frac{\Delta[B]}{\Delta t} = \frac{1}{c}\frac{\Delta[C]}{\Delta t} = \frac{1}{d}\frac{\Delta[D]}{\Delta t} \quad \textbf{(2)}$$

Division by the stoichiometric coefficients takes care of the stoichiometric relations between the reactants and products. There is no need to specify the species when reporting the unique average reaction rate because the value of the rate is the same, regardless of which species is indicated. However, the unique average rate does depend on the coefficients used in the balanced equation, and so the chemical equation should still be specified when reporting the unique rate.

THINKING POINT

By what factor does the unique average reaction rate change if the coefficients in a chemical equation are doubled?

Self-test 7A.2A The average rate of the reaction $N_2(g) + 3 H_2(g) \rightarrow 2 NH_3(g)$ over a certain period is reported as 1.15 (mmol NH_3)·L^{-1}·h^{-1}. (a) What is the average rate over the same period in terms of the consumption of H_2? (b) What is the unique average rate?

[*Answer:* (a) 1.72 (mmol H_2)·L^{-1}·h^{-1}; (b) 0.575 mmol·L^{-1}·h^{-1}]

Self-test 7A.2B Consider the reaction in Example 7A.1. What is (a) the average rate of consumption of H_2 in the same reaction and (b) the unique average rate, both over the same period?

The experimental technique used to measure a rate of reaction depends on how rapidly the reaction takes place. Special experimental techniques are used when the reaction is so fast that it is over in seconds or less. Two main considerations are critical for the study of fast reactions. One is to initiate the reaction at a very precise time. The other is to monitor the concentration at precise instants after the reaction has been initiated. Reactions that are initiated by mixing the reagents can be studied by the **stopped-flow technique,** in which solutions of the reactants are forced into a mixing chamber very rapidly and, within a few milliseconds, the formation of products or loss of reactants is monitored (**FIG. 7A.3**). This procedure is commonly used to study biologically important processes such as protein folding and enzyme reactions.

Some reactions can be initiated by a flash of light. Because lasers can produce very short pulses at precisely known times, they are often used to study the rates of such reactions. **Spectrometry**—the determination of concentrations by measuring the absorption of light by species (see *Major Technique 2* on the website of this book)—can respond very quickly to concentration changes and is often used in combination with a laser flash to study very fast reactions. For instance, suppose you are studying the effect of a chlorofluorocarbon on the concentration of ozone, a blue gas. You could use a spectrometer to monitor the light absorbed by the ozone and calculate the molar concentration of O_3 molecules from the intensity of absorption. By using pulsed lasers, chemists can study reactions that are complete in less than a picosecond (1 ps = 10^{-12} s). The newest techniques can even monitor processes that are complete after a few femtoseconds (1 fs = 10^{-15} s, **BOX 7A.1**), and chemists are now investigating processes on an attosecond time scale (1 as = 10^{-18} s). On that time scale, atoms seem hardly to be moving and are caught red-handed in the act of reaction.

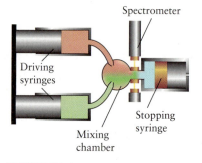

Spectrometer

Driving syringes

Mixing chamber

Stopping syringe

FIGURE 7A.3 In a stopped-flow experiment, the driving syringes on the left push reactant solutions into the mixing chamber, and the stopping syringe on the right stops the flow. The progress of the reaction is then monitored spectroscopically as a function of time.

The average rate of a reaction is the change in concentration of a species divided by the time over which the change takes place; the unique average rate is the average rate divided by the stoichiometric coefficient of the species monitored. Spectroscopic techniques are widely used to study reaction rates, particularly for fast reactions.

Box 7A.1 HOW DO WE KNOW . . . WHAT HAPPENS TO ATOMS DURING A REACTION?

The events that happen to an atom in a chemical reaction are on a time scale of approximately 1 femtosecond (1 fs = 10^{-15} s), the time that it takes for a bond to stretch or bend and, perhaps, break. If you could follow atoms on that time scale, you could make a movie of the changes in molecules as they take part in a chemical reaction. The field of *femtochemistry*, the study of very fast chemical processes, is bringing scientists closer to realizing that dream. Lasers can emit very intense but short pulses of electromagnetic radiation, and so they can be used to study processes on very short time scales. Recent developments have pushed the time scale of observations toward the attosecond region (1 at = 10^{-18} s), where even the motion of electrons is frozen.

So far, the technique has been applied only to very simple reactions. For example, it is possible to watch the ion pair Na^+I^- decompose in the gas phase into separate Na and I atoms. At the start, the sodium ion and iodide ion are held together by the coulombic attraction of opposite charges. The pair is then struck by a femtosecond pulse of radiation from a laser. That pulse excites an electron from the I^- ion on to the Na^+ ion, thereby creating an NaI molecule in which the two atoms are held together by a covalent bond. The molecule so formed has a lot of energy, and the bond length varies as the atoms swing in and out. At this point, a second femtosecond pulse is fired at the molecule. The radiation in the second pulse has a frequency that can be absorbed by the molecule only when the atoms have a particular separation. If the pulse is absorbed, the atoms of the vibrating molecule have that specific internuclear separation.

The illustration shows a typical result. The absorption reaches a maximum whenever the Na—I bond length returns to the length to which the second pulse is tuned. The peaks show

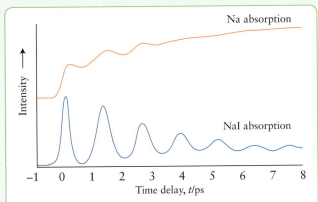

The femtosecond spectrum of the gas-phase NaI molecule as it dissociates into the separate atoms. A peak in the lower (blue) plot is observed whenever the bond distance in NaI reaches a certain value. The upper (orange) plot shows the formation of Na atoms as they escape from the NaI molecule.

that the sodium atom moves away from the iodine atom (corresponding to the dips in the curve), only to be recaptured (at the maxima) again. The separation of the maxima is about 1.3 ps, and so the Na atom takes that long to swing out and be recaptured by the I atom. The peaks progressively decrease in intensity, showing that some Na atoms escape from their I atom partners on each swing. It takes about 10 swings outward before an Na atom can be sure of escaping. When sodium bromide is studied by these methods, the sodium atom escapes after about one swing, showing that an Na atom can escape more readily from a Br atom than from an I atom.

7A.2 The Instantaneous Rate of Reaction

Like the speed of a car, it is often more important to know the instantaneous rate of a reaction, not the average rate over a long interval. Most reactions slow down as the reactants are used up. To determine the reaction rate at a given instant during the course of the reaction, two concentration measurements must be made and these measurements must be as close together in time as possible. When two points on the curve are brought successively closer together, the line joining them approaches being the "tangent" to the curve, a straight line that touches the curve at a point and indicates its slope at that point. The slope of the tangent to the plot of concentration against time at the time of interest is the rate at a single instant (**FIG. 7A.4**). This slope, which changes in the course of the reaction, is called the **instantaneous rate** of the reaction at the time of interest (**FIG. 7A.5**).

From now on, whenever a reaction rate is referred to, it will always mean an instantaneous rate. The definitions in Eqs. 1 and 2 can easily be adapted to refer to the instantaneous rate of a reaction.

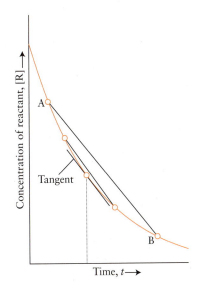

FIGURE 7A.4 The rate of reaction is the change in concentration of a reactant (or product) divided by the time interval over which the change occurs (the slope of the line AB, for instance). The *instantaneous rate* is the slope of the tangent to the curve at the time of interest.

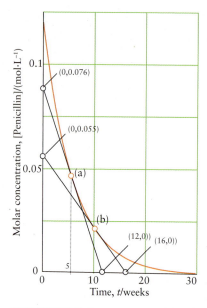

FIGURE 7A.5 Determination of the rate of deterioration of penicillin during storage at two different times. Note that the instantaneous rate (the slope of the tangent to the curve) at 5 weeks (a) is greater than the rate at 10 weeks, when less penicillin is present (b).

How Is That Done?

To set up expressions for the instantaneous rate of a reaction, consider the time interval Δt to be so brief that t and $t + \Delta t$ are very close together. The concentration of a reactant or product is determined at each of those times and the average rate is calculated from Eq. 1. Then the interval is decreased still further and the calculation is repeated. Now imagine continuing the process until the interval Δt has become infinitely small (denoted dt) and the change in molar concentration of a reactant R has also become infinitely small (denoted $d[R]$). Then the instantaneous rate is calculated:

$$\text{Rate of consumption of R} = -\frac{d[R]}{dt}$$

For a product P, write

$$\text{Rate of formation of P} = \frac{d[P]}{dt}$$

The "differential coefficients," the terms $d[R]/dt$ and $d[P]/dt$, are the mathematical expressions for the slope of the tangent drawn to a curve at the time of interest. Similarly, the unique instantaneous rate of a reaction is defined as in Eq. 2, but with differential coefficients in place of $\Delta[R]/\Delta t$ and $\Delta[P]/\Delta t$:

$$\text{Reaction rate} = -\frac{1}{a}\frac{d[A]}{dt} = -\frac{1}{b}\frac{d[B]}{dt} = \frac{1}{c}\frac{d[C]}{dt} = \frac{1}{d}\frac{d[D]}{dt}$$

Because it is difficult to draw a tangent accurately by eye, it is better to use a computer to analyze graphs of concentration against time. A superior method—which is described in Topic 7B—is to report rates by using a procedure that, although based on these definitions, avoids the use of tangents altogether.

The instantaneous reaction rate is the slope of a tangent drawn to the graph of concentration as a function of time; for most reactions, the instantaneous rate decreases as the reaction proceeds.

7A.3 Rate Laws and Reaction Order

Patterns in reaction rate data can often be identified by examining the **initial rate** of reaction, the instantaneous rate at the start of a reaction (**FIG. 7A.6**). The advantage of examining the initial rate is that the presence of products later in the reaction may affect the rate; the interpretation of the rate is then quite complicated.

To appreciate how initial rates are measured, suppose different amounts of solid dinitrogen pentoxide, N_2O_5, are measured into five flasks of the same volume, all the flasks are immersed in a water bath at 65 °C to vaporize the solid completely, and then spectrometry is used to monitor the concentration of N_2O_5 remaining in each flask as it decomposes:

$$2\,N_2O_5(g) \longrightarrow 4\,NO_2(g) + O_2(g) \tag{A}$$

Each flask has a different initial concentration of N_2O_5. The initial rate of reaction in each flask can be determined by plotting the concentration as a function of time for each flask and drawing the tangent to each curve at $t = 0$ (the black lines in Fig. 7A.6). Higher initial rates of decomposition of the vapor—steeper tangents—are found in the flasks with higher initial concentrations of N_2O_5. One way to find the pattern in the data is to plot the initial rate against concentration and examine the type of curve obtained. In this case, the plot of the initial rate against the initial concentration of N_2O_5 is a straight line, which indicates that the initial rate is proportional to the initial concentration (**FIG. 7A.7**):

$$\text{Initial rate of consumption of } N_2O_5 \propto [N_2O_5]_{\text{initial}}$$

By introducing a constant k_r, this proportionality can be written as an equality:

$$\text{Initial rate of consumption of } N_2O_5 = k_r \times [N_2O_5]_{\text{initial}}$$

The constant k_r is called the **rate constant** for the reaction and is characteristic of the reaction (in the sense that different reactions have different rate constants) and the temperature at which the reaction takes place. The experimental value of k_r for this reaction at 65 °C, the slope of the straight line in Fig. 7A.7, is $5.2 \times 10^{-3}\ \text{s}^{-1}$.

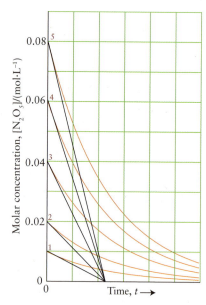

FIGURE 7A.6 The orange curves show how the concentration of N_2O_5 changes with time for five different initial concentrations. The initial rate of consumption of N_2O_5 can be determined by drawing a tangent (black line) to each curve at the start of the reaction.

FIGURE 7A.7 This graph was obtained by plotting the five initial rates from Fig. 7A.6 against the initial concentration of N_2O_5. The initial rate is directly proportional to the initial concentration. This graph also illustrates how the value of the rate constant k_r can be determined by calculating the slope of the straight line from two points.

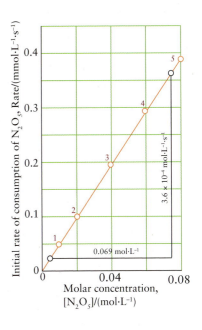

The initial rate of reaction A is proportional to the initial concentration of N_2O_5. If the reaction rate in one of the flasks is monitored as the reaction proceeds, then as the concentration of N_2O_5 falls, the rate would be found to fall too. More specifically, the rate at any instant is directly proportional to the concentration of N_2O_5 remaining at that instant, with the same constant of proportionality, k_r. It then follows that at *any* stage of the reaction

$$\text{Rate of consumption of } N_2O_5 = k_r[N_2O_5]$$

where $[N_2O_5]$ is the molar concentration of N_2O_5 at any instant. This equation is an example of a **rate law,** an expression for the instantaneous reaction rate in terms of the concentration of a reactant at any instant. Each reaction has its own characteristic rate law and rate constant, k_r (**TABLE 7A.1**). Rate laws may include product as well as reactant concentrations.

Other reactions have rate laws that may depend differently on concentration. Similar measurements for the reaction

$$2\,NO_2(g) \longrightarrow 2\,NO(g) + O_2(g) \qquad\qquad \textbf{(B)}$$

do not give a straight line when the rate is plotted against the concentration of NO_2 (**FIG. 7A.8a**). However, a plot of the rate against the *square* of the concentration of NO_2 is linear (Fig. 7A.8b). This result shows that the rate is proportional to the square of the concentration and therefore that the rate at any stage can be written

$$\text{Rate of consumption of } NO_2 = k_r[NO_2]^2$$

From the slope of the straight line in Fig. 7A.8b, $k_r = 0.54$ L·mol^{-1}·s^{-1} at 300. °C.

TABLE 7A.1   Rate Laws and Rate Constants

Reaction	Rate law*	Temperature, T/K†	Rate constant
Gas phase			
$H_2 + I_2 \longrightarrow 2\,HI$	$k_r[H_2][I_2]$	500	4.3×10^{-7} L·mol^{-1}·s^{-1}
		600	4.4×10^{-4}
		700	6.3×10^{-2}
		800	2.6
$2\,HI \longrightarrow H_2 + I_2$	$k_r[HI]^2$	500	6.4×10^{-9} L·mol^{-1}·s^{-1}
		600	9.7×10^{-6}
		700	1.8×10^{-3}
		800	9.7×10^{-2}
$2\,N_2O_5 \longrightarrow 4\,NO_2 + O_2$	$k_r[N_2O_5]$	298	3.7×10^{-5} s^{-1}
		318	5.1×10^{-4}
		328	1.7×10^{-3}
		338	5.2×10^{-3}
$2\,N_2O \longrightarrow 2\,N_2 + O_2$	$k_r[N_2O]$	1000	0.76 s^{-1}
		1050	3.4
$2\,NO_2 \longrightarrow 2\,NO + O_2$	$k_r[NO_2]^2$	573	0.54 L·mol^{-1}·s^{-1}
$C_2H_6 \longrightarrow 2\,CH_3$	$k_r[C_2H_6]$	973	5.5×10^{-4} s^{-1}
cyclopropane $\rightarrow$ propene	$k_r[\text{cyclopropane}]$	773	6.7×10^{-4} s^{-1}
Aqueous solution			
$H_3O^+ + OH^- \longrightarrow 2\,H_2O$	$k_r[H_3O^+][OH^-]$	298	1.5×10^{11} L·mol^{-1}·s^{-1}
$CH_3Br + OH^- \longrightarrow CH_3OH + Br^-$	$k_r[CH_3Br][OH^-]$	298	2.8×10^{-4} L·mol^{-1}·s^{-1}
$C_{12}H_{22}O_{11} + H_2O \longrightarrow 2\,C_6H_{12}O_6$	$k_r[C_{12}H_{22}O_{11}][H^+]$	298	1.8×10^{-4} L·mol^{-1}·s^{-1}

*For the unique instantaneous rate.
†Three significant figures.

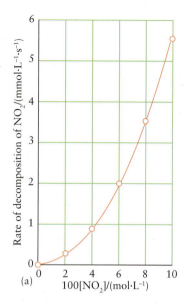

(a)

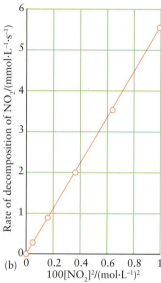

(b)

FIGURE 7A.8 (a) When the rates of disappearance of NO_2 are plotted against its concentration, a straight line is not obtained. (b) However, a straight line is obtained when the rates are plotted against the square of the concentration, indicating that the rate is directly proportional to the square of the concentration.

The rate laws for the decomposition reactions of N_2O_5 and NO_2 are different, but each has the form

$$\text{Rate} = \text{constant} \times [\text{concentration}]^a \qquad (3)$$

with $a = 1$ for the decomposition of N_2O_5 and $a = 2$ for the decomposition of NO_2. The decomposition of N_2O_5 is an example of a **first-order reaction,** because its rate is proportional to the *first* power of the concentration (that is, $a = 1$). The decomposition of NO_2 is an example of a **second-order reaction,** because its rate is proportional to the *second* power of the concentration (that is, $a = 2$). Doubling the concentration of a reactant in a first-order reaction doubles the reaction rate. Doubling the concentration of the reactant in any second-order reaction increases the reaction rate by a factor of $2^2 = 4$.

Most of the reactions considered in Focus 7 are either first or second order in each reactant, but some reactions have other orders (different values of a in Eq. 3). For example, ammonia decomposes into nitrogen and hydrogen on a hot platinum wire:

$$2\,NH_3(g) \longrightarrow N_2(g) + 3\,H_2(g) \qquad (\mathbf{C})$$

Experiments show that the decomposition takes place at a constant rate until all the ammonia has been consumed (**FIG. 7A.9**). Its rate law is therefore

$$\text{Rate of consumption of } NH_3 = k_r$$

That is, the rate is independent of the concentration of ammonia, so long as any is present. This decomposition is an example of a **zeroth-order reaction,** a reaction for which the rate is independent of concentration of the reactant (provided some is present).

The rate laws for the three most common orders of reaction are

Order in A	Rate law
0	$\text{Rate} = k_r$
1	$\text{Rate} = k_r[A]$
2	$\text{Rate} = k_r[A]^2$

Unless otherwise specified, the "Rate" in each expression is the instantaneous rate and $[A]$ is the concentration of the reactant at the instant of interest. A very important point is the following:

The rate law for a reaction is determined experimentally and cannot in general be inferred from the chemical equation for the reaction.

Zeroth-order reactions are so called because $\text{Rate} = k_r \times (\text{concentration})^0 = k_r$, a constant independent of concentration.

FIGURE 7A.9 (a) The concentration of the reactant in a zeroth-order reaction falls at a constant rate until the reactant is exhausted. (b) The rate of a zeroth-order reaction is independent of the concentration of the reactant and remains constant until all the reactant has been consumed, when the rate falls abruptly to zero.

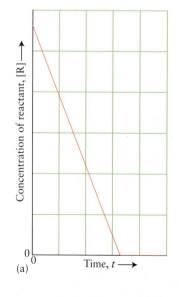

(a)

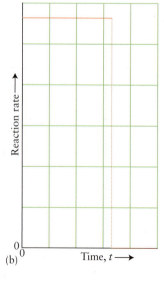

(b)

For instance, both the decomposition of N_2O_5, reaction A, and that of NO_2, reaction B, have a stoichiometric coefficient of 2 for the reactant, but one reaction is first order and the other is second order.

Many reactions have rate laws that depend on the concentrations of more than one reactant. An example is the redox reaction between persulfate ions and iodide ions:

$$S_2O_8^{2-}(aq) + 3\,I^-(aq) \longrightarrow 2\,SO_4^{2-}(aq) + I_3^-(aq) \qquad \textbf{(D)}$$

The rate law of this reaction is found to be

$$\text{Rate of consumption of } S_2O_8^{2-} = k_r[S_2O_8^{2-}][I^-]$$

The reaction is said to be first order with respect to $S_2O_8^{2-}$ (or "in" $S_2O_8^{2-}$) and first order in I^-. Doubling either the $S_2O_8^{2-}$ ion concentration or the I^- ion concentration doubles the reaction rate. Doubling both concentrations quadruples the reaction rate. The *overall* order of this reaction is the sum of the two orders, or 2. In general, if

$$\text{Rate} = k_r[A]^a[B]^b \ldots \qquad \textbf{(4)}$$

then the **overall order** is the sum of the powers $a + b \ldots$.

The units of k_r depend on the overall order of the reaction and ensure that $k_r \times$ (concentration)^a has the same units as the rate, namely, concentration/time. Thus, when the concentration is expressed in moles per liter and the rate is expressed in $\text{mol·L}^{-1}\text{·s}^{-1}$, the units of k_r are as follows:

Overall order:	1	2	3
Units of k_r:	s^{-1}	$\text{L·mol}^{-1}\text{·s}^{-1}$	$\text{L}^2\text{·mol}^{-2}\text{·s}^{-1}$

and so on. If the concentrations are expressed as partial pressures in kilopascals and the rate is reported in kPa·s^{-1}, then the units of k_r are:

Overall order:	1	2	3
Units of k_r:	s^{-1}	$\text{kPa}^{-1}\text{·s}^{-1}$	$\text{kPa}^{-2}\text{·s}^{-1}$

For an ideal gas, $PV = nRT$ implies that $n/V = P/RT$, so that n/V is proportional to the pressure. Because n/V is concentration (in mol·L^{-1}), concentration is proportional to pressure and therefore pressure can be used as a measure of concentration.

THINKING POINT

What would be the units of k_r for an overall order of $\frac{3}{2}$ if the concentrations were expressed in grams per milliliter, g·mL^{-1}?

Self-test 7A.3A   When the NO concentration is doubled, the rate of the reaction $2\,NO(g) + O_2(g) \rightarrow 2\,NO_2(g)$ increases by a factor of 4. When both the O_2 and the NO concentrations are doubled, the rate increases by a factor of 8. What are (a) the reaction order with respect to each reactant, (b) the overall order of the reaction, and (c) the units of k_r if the rate is expressed in moles per liter per second?

[*Answer:* (a) Second order in NO, first order in O_2; (b) Third order overall; (c) $\text{L}^2\text{·mol}^{-2}\text{·s}^{-1}$]

Self-test 7A.3B   When the concentration of 2-bromo-2-methylpropane, C_4H_9Br, is doubled, the rate of the reaction $C_4H_9Br(aq) + OH^-(aq) \rightarrow C_4H_9OH(aq) + Br^-(aq)$ increases by a factor of 2. When both the C_4H_9Br and the OH^- concentrations are doubled, the rate increase is the same, a factor of 2. What are (a) the reaction order with respect to each reactant, (b) the overall order of the reaction, and (c) the units of k_r if the rate is expressed in moles per liter per second?

The rate laws of reactions are empirical expressions, expressions established experimentally, and it is not surprising that some orders are not simple positive whole numbers. For instance, orders can be negative, as in (concentration)$^{-1}$, corresponding to order -1. Because $[A]^{-1} = 1/[A]$, a negative order implies that the concentration appears in the denominator of the rate law. Increasing the concentration of that species, usually a product, slows down the reaction because the species participates in a reverse reaction. An example is the decomposition of ozone, O_3, in the upper atmosphere:

$$2\,O_3(g) \longrightarrow 3\,O_2(g) \qquad \textbf{(E)}$$

The experimentally determined rate law for this reaction is

$$\text{Rate} = k_r\frac{[O_3]^2}{[O_2]} = k_r[O_3]^2[O_2]^{-1}$$

Note that a rate law may depend on the concentrations of products as well as those of reactants.

This rate law implies that the reaction is slower in regions of the atmosphere where O_2 molecules are abundant than where they are less abundant. Topic 7C shows how this rate law is an important clue to how the reaction takes place.

Some reactions have fractional orders. For example, the oxidation of sulfur dioxide to sulfur trioxide in the presence of platinum,

$$2\,SO_2(g) + O_2(g) \xrightarrow{\text{Pt}} 2\,SO_3(g)$$

is found to have the rate law

$$\text{Rate} = k_r\frac{[SO_2]}{[SO_3]^{1/2}} = k_r[SO_2][SO_3]^{-1/2}$$

and an overall order of $1 - \frac{1}{2} = \frac{1}{2}$. The presence of $[SO_3]$ in the denominator means that the reaction slows down as the concentration of product builds up. Once again, this rate law is a clue to how the reaction occurs.

All the reactions considered in this Topic are homogeneous and for all but zeroth-order reactions the rate depends on the concentration of one or more reactants. To increase the rate, the concentration of a reactant can be increased. Similarly, the rate of a heterogeneous reaction can be increased by increasing the surface area of a solid reactant (**FIG 7A.10**).

FIGURE 7A.10 Iron pots and pans can be heated in a flame without catching fire. However, a powder of finely divided iron filings oxidizes rapidly in air to form Fe_2O_3, because the powder presents a much greater surface area for reaction. (*W. H. Freeman photo by Ken Karp.*)

EXAMPLE 7A.2   Determining reaction orders and rate laws from experimental data

If you pursue a career in inorganic or physical chemistry, you may one day be studying the rate of the reaction of bromate ion with bromide ion. Suppose you conduct four experiments to discover how the initial rate of consumption of BrO_3^- ions varies as the concentrations of the reactants are changed in the reaction,

$$BrO_3^-(aq) + 5\,Br^-(aq) + 6\,H_3O^+(aq) \longrightarrow 3\,Br_2(aq) + 9\,H_2O(l)$$

(a) Use the experimental data in the following table to determine the order of the reaction with respect to each reactant and the overall order. (b) Write the rate law for the reaction and determine the value of k_r.

Experiment	Initial concentration, $[J]/(\text{mol·L}^{-1})$			Initial rate/ $((\text{mmol } BrO_3^-)\cdot\text{L}^{-1}\cdot\text{s}^{-1})$
	BrO_3^-	Br^-	H_3O^+	
1	0.10	0.10	0.10	1.2
2	0.20	0.10	0.10	2.4
3	0.10	0.30	0.10	3.5
4	0.20	0.10	0.15	5.5

PLAN Suppose that the concentration of a substance A is increased but no other concentrations change. The generic rate law, $\text{Rate} = k_r[A]^a[B]^b$, tells you that as the concentration of A increases by the factor f, the rate increases by f^a. To isolate the effect of each substance, compare experiments that differ in the concentration of only one substance at a time.

SOLVE

(a) *Order in BrO_3^-*: Compare experiments 1 and 2.

In experiments 1 and 2, the concentration of BrO_3^- is doubled ($f = 2$) but the other concentrations are held constant. Because the rate also doubles, $f^a = (2)^a = 2$. Therefore, $a = 1$ and the reaction is first order in BrO_3^-.

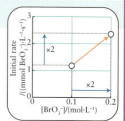

Order in Br^-: Compare experiments 1 and 3.

In experiments 1 and 3, when the other concentrations are held constant but the concentration of Br^- is changed by a factor of 3.0 ($f = 3.0$), the rate changes by a factor of $3.5/1.2 = 2.9$. Allowing for experimental error, $f^b = (3.0)^b = 3$, and so $b = 1$ and the reaction is first order in Br^-.

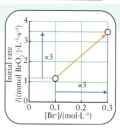

Order in H_3O^+: Compare experiments 2 and 4.

When the concentration of hydronium ions is increased from experiment 2 to experiment 4 by a factor of 1.5 ($f = 1.5$), the rate increases by a factor of $5.5/2.4 = 2.3$ when all other concentrations are held constant. Therefore, $f^c = (1.5)^c = 2.3$. To solve the relation $1.5^c = 2.3$ (and, in general, $f^c = x$), take logarithms of both sides.

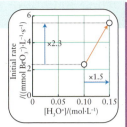

From $f^c = x$, $\ln f^c = c \ln f = \ln x$, and hence $c = (\ln x)/(\ln f)$,

$$c = \frac{\ln 2.3}{\ln 1.5} = 2.0$$

The reaction is second order in H_3O^+. The result can be checked by verifying that $(1.5)^2 = 2.3$. The overall order is $1 + 1 + 2 = 4$.

(b) The rate law is

$$\text{Rate of consumption of } BrO_3^- = k_r[BrO_3^-][Br^-][H_3O^+]^2$$

Find k_r by substituting the values from one of the experiments into the rate law and solving for k_r. For example, in experiment 4, note that the rate of reaction is $5.5 \text{ mmol·L}^{-1}\text{·s}^{-1} = 5.5 \times 10^{-3} \text{ mol·L}^{-1}\text{·s}^{-1}$; then substitute the concentrations into the rate law and solve for k_r.

From $k_r = (\text{Rate of consumption of } BrO_3^-)/[BrO_3^-][Br^-][H_3O^+]^2$,

$$k_r = \frac{5.5 \times 10^{-3} \text{ mol·L}^{-1}\text{·s}^{-1}}{(0.20 \text{ mol·L}^{-1}) \times (0.10 \text{ mol·L}^{-1}) \times (0.15 \text{ mol·L}^{-1})^2} = 12 \text{ L}^3\text{·mol}^{-3}\text{·s}^{-1}$$

EVALUATE The mean value of k_r calculated from the four experiments is $12 \text{ L}^3\text{·mol}^{-3}\text{·s}^{-1}$.

Self-test 7A.4A The reaction $2 \text{ NO(g)} + O_2(g) \rightarrow 2 \text{ NO}_2(g)$ occurs when automobile exhaust releases NO into the atmosphere. Write the rate law for the consumption of NO and determine the value of k_r, given the following data:

Experiment	Initial concentration, $[J]/(\text{mol·L}^{-1})$		Initial rate / $((\text{mmol NO})\text{·L}^{-1}\text{·s}^{-1})$
	NO	**O_2**	
1	0.012	0.020	0.102
2	0.024	0.020	0.408
3	0.024	0.040	0.816

[*Answer:* Rate of consumption of $NO = k_r[NO]^2[O_2]$; with the use of experiment 1, $k_r = 3.5 \times 10^4 \text{ L}^2\text{·mol}^{-2}\text{·s}^{-1}$]

Self-test 7A.4B Carbonyl chloride, $COCl_2$ (phosgene), is a highly toxic gas used to synthesize many organic compounds. Use the following data to write the rate law and determine the value of k_r for the reaction used to produce carbonyl chloride, $CO(g) + Cl_2(g) \rightarrow COCl_2(g)$, at a certain temperature:

Experiment	Initial concentration, $[J]/(\text{mol·L}^{-1})$		Initial rate/ $((\text{mmol } COCl_2)\text{·L}^{-1}\text{·s}^{-1})$
	CO	**Cl_2**	
1	0.12	0.20	0.121
2	0.24	0.20	0.241
3	0.24	0.40	0.682

Hint: One of the orders is not an integer.

Related Exercises 7A.13–7A.18

The order of a reaction is the power to which the concentration of the species is raised in the rate law; the overall order is the sum of the individual orders.

What have you learned in this Topic?

You have learned how to define a rate of reaction for the formation of products or consumption of reactants. You have also seen how to define a unique rate for a given reaction. From experimental data, you have seen how rates of reaction are determined and how those rates are used to write the rate law for the reaction.

The skills you have mastered are the ability to:

☐ **1.** Write the unique reaction rate for a chemical reaction (Section 7A.1).

☐ **2.** Calculate the average reaction rate from experimental data (Example 7A.1).

☐ **3.** Use experimental data to determine reaction orders and write the rate law for a reaction (Example 7A.2).

Topic 7A Exercises

All rates are unique reaction rates unless otherwise stated. Rate constants are listed in Table 7A.1.

7A.1 Complete the following statements relating to the production of ammonia by the Haber process, for which the overall reaction is $N_2(g) + 3 H_2(g) \rightarrow 2 NH_3(g)$. (a) The rate of consumption of N_2 is _____ times the rate of consumption of H_2. (b) The rate of formation of NH_3 is _____ times the rate of consumption of H_2. (c) The rate of formation of NH_3 is _____ times the rate of consumption of N_2.

7A.2 Complete the following statements for the reaction $6 Li(s) + N_2(g) \rightarrow 2 Li_3N(s)$. (a) The rate of consumption of N_2 is _____ times the rate of formation of Li_3N. (b) The rate of formation of Li_3N is _____ times the rate of consumption of Li. (c) The rate of consumption of N_2 is _____ times the rate of consumption of Li.

7A.3 Ethene is a component of natural gas, and its combustion has been thoroughly studied. At a certain temperature and pressure, the unique rate of the combustion reaction $C_2H_4(g) + 3 O_2(g) \rightarrow 2 CO_2(g) + 2 H_2O(g)$ is $0.44 \text{ mol·L}^{-1}\text{·s}^{-1}$. (a) What is the rate at which oxygen reacts? (b) What is the rate of formation of water?

7A.4 The "iodine clock reaction" is a popular chemical demonstration. As part of that demonstration, the I_3^- ion is generated in the reaction $S_2O_8^{2-}(aq) + 3 I^-(aq) \rightarrow 2 SO_4^{2-}(aq) + I_3^-(aq)$. In one trial, the unique rate of reaction was $4.5 \text{ μmol·L}^{-1}\text{·s}^{-1}$. (a) What was the rate of reaction of iodide ion? (b) What was the rate of formation of sulfate ions?

7A.5 The decomposition of gaseous hydrogen iodide, $2 HI(g) \rightarrow H_2(g) + I_2(g)$, gives the data shown here for 700. K.

Time, t/s	0.	1000.	2000.	3000.	4000.	5000.
$[HI]/(\text{mmol·L}^{-1})$	10.0	4.4	2.8	2.1	1.6	1.3

(a) Use a graphing calculator or standard graphing software to plot the concentration of HI as a function of time. (b) Estimate the rate of decomposition of HI at each time. (c) Plot the concentrations of H_2 and I_2 as a function of time on the same graph.

7A.6 The decomposition of gaseous dinitrogen pentoxide in the reaction $2 N_2O_5(g) \rightarrow 4 NO_2(g) + O_2(g)$ gives the data shown here at 298 K. (a) Using a graphing calculator or standard graphing software, plot the concentration of N_2O_5 as a function of time.

(b) Estimate the rate of decomposition of N_2O_5 at each time. (c) Plot the concentrations of NO_2 and O_2 as a function of time on the same graph.

Time, t/s	0.	1.11	2.22	3.33	4.44
$[N_2O_5]/(\text{mmol·L}^{-1})$	2.15	1.88	1.64	1.43	1.25

7A.7 Express the units for rate constants when the concentrations are in moles per liter and time is in seconds for (a) zeroth-order reactions; (b) first-order reactions; (c) second-order reactions.

7A.8 Rate laws for gas-phase reactions can also be expressed in terms of partial pressures, for instance, as $\text{Rate} = k_r P_J$ for a first-order reaction of a gas J. What are the units for the rate constants when partial pressures are expressed in Torr and time is expressed in seconds for (a) zeroth-order reactions; (b) first-order reactions; (c) second-order reactions?

7A.9 Dinitrogen pentoxide, N_2O_5, decomposes by a first-order reaction. What is the initial rate of decomposition of N_2O_5 when 3.45 g of N_2O_5 is confined in a container of volume 0.750 L and heated to 65 °C? For this reaction, $k_r = 5.2 \times 10^{-3} \text{ s}^{-1}$ in the rate law (for the rate of decomposition of N_2O_5).

7A.10 Ethane, C_2H_6, dissociates into methyl radicals by a first-order reaction at 700 °C. If 820. mg of ethane is confined to a reaction vessel of volume 2.00 L and heated to 700 °C, what is the initial rate of ethane decomposition if $k_r = 5.5 \times 10^{-4} \text{ s}^{-1}$ in the rate law (for the rate of dissociation of C_2H_6)?

7A.11 When 0.52 g of H_2 and 0.19 g of I_2 are confined to a reaction vessel of volume 750. mL and heated to 700. K, they react by a second-order process (first order in each reactant), with $k_r = 0.063 \text{ L·mol}^{-1}\text{·s}^{-1}$ in the rate law (for the rate of formation of HI). (a) What is the initial reaction rate? (b) By what factor does the reaction rate increase if the concentration of H_2 present in the mixture is doubled?

7A.12 When 510. mg of NO_2 is confined to a reaction vessel of volume 180. mL and heated to 300 °C, it decomposes by a second-order process. In the rate law for the decomposition of NO_2, $k_r = 0.54 \text{ L·mol}^{-1}\text{·s}^{-1}$. (a) What is the initial reaction rate? (b) How does the reaction rate change (and by what factor) if the mass of NO_2 present in the container is increased to 820. mg?

7A.13 In the reaction $CH_3Br(aq) + OH^-(aq) \rightarrow CH_3OH(aq) + Br^-(aq)$, when the OH^- concentration alone was doubled, the rate doubled; when the CH_3Br concentration alone was increased by a factor of 1.2, the rate increased by a factor of 1.2. Write the rate law for the reaction.

7A.14 In the reaction $4\,Fe^{2+}(aq) + O_2(g) + 4\,H_3O^+(aq) \rightarrow 4\,Fe^{3+}(aq) + 6\,H_2O(l)$, when the Fe^{2+} concentration alone was doubled, the rate increased by a factor of 8; when both the Fe^{2+} and the O_2 concentrations were increased by a factor of 2, the rate increased by a factor of 16. When the concentrations of all three reactants were doubled, the rate increased by a factor of 32. What is the rate law for the reaction?

7A.15 The following rate data were collected for the reaction $2\,A(g) + 2\,B(g) + C(g) \rightarrow 3\,G(g) + 4\,F(g)$:

	Initial concentration, $[J]_{initial}/(mmol \cdot L^{-1})$			Initial rate/
Experiment	A	B	C	$((mmol\ G) \cdot L^{-1} \cdot s^{-1})$
1	10.	100.	700.	2.0
2	20.	100.	300.	4.0
3	20.	200.	200.	16
4	10.	100.	400.	2.0
5	4.62	0.177	12.4	?

(a) What is the order for each reactant and the overall order of the reaction? (b) Write the rate law for the reaction. (c) Determine the reaction rate constant. (d) Predict the initial rate for Experiment 5.

7A.16 The following kinetic data were obtained for the reaction $A(g) + 2\,B(g) \rightarrow product$.

	Initial concentration, $[J]_{initial}/(mmol \cdot L^{-1})$		Initial rate/
Experiment	A	B	$(mmol \cdot L^{-1} \cdot s^{-1})$
1	0.60	0.30	12.6
2	0.20	0.30	1.4
3	0.60	0.10	4.2
4	0.17	0.25	?

What is the order with respect to each reactant, and the overall order of the reaction? (b) Write the rate law for the reaction. (c) From the data, determine the value of the rate constant. (d) Use the data to predict the reaction rate for Experiment 4.

7A.17 The following data were obtained for the reaction $A + B + C \rightarrow products$:

	Initial concentration, $[J]_{initial}/(mmol \cdot L^{-1})$			Initial rate/
Experiment	A	B	C	$((mmol) \cdot L^{-1} \cdot s^{-1})$
1	1.25	1.25	1.25	8.7
2	2.5	1.25	1.25	17.4
3	1.25	3.02	1.25	50.8
4	1.25	3.02	3.75	457
5	3.01	1.00	1.15	?

(a) Write the rate law for the reaction. (b) What is the order of the reaction? (c) Determine the value of the rate constant. (d) Use the data to predict the reaction rate for Experiment 5.

7A.18 The following kinetic data were obtained for the reaction $3\,A(g) + B(g) \rightarrow product$:

	Initial concentration, $[J]_{initial}/(mmol \cdot L^{-1})$		Initial rate/
Experiment	A	B	$(mol \cdot L^{-1} \cdot s^{-1})$
1	1.72	2.44	0.68
2	3.44	2.44	5.44
3	1.72	0.10	2.8×10^{-2}
4	2.91	1.33	?

(a) What is the order with respect to each reactant, and the overall order of the reaction? (b) Write the rate law for the reaction. (c) From the data, determine the value of the rate constant. (d) Use the data to predict the reaction rate for Experiment 4.

Topic 7B Integrated Rate Laws

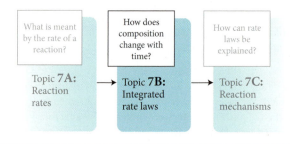

Why Do You Need to Know This Material?

The rate law of a reaction can be used to predict how reactant or product concentrations change with time. It also provides an accurate way to determine the rate constant of the reaction.

What Do You Need to Know Already?

You need to be familiar with the concepts of a rate law and reaction order (Topic 7A). The mathematical techniques used in this Topic are reviewed in Appendix 1E and 1F.

It is often useful to know how the concentration of a reactant or product varies with time. For example, how long will it take for a pollutant to decay? How much of the alternative fuel methanol can be produced in an hour from coal? How much of a sample of penicillin will be left after 6 months? These questions can be answered by using formulas derived from the experimental rate laws for the reactions. An **integrated rate law** gives the concentration of reactants or products at any time after the start of the reaction. Finding the integrated rate law from a rate law is very much like calculating the distance that a car will have traveled from its speed at each moment of the journey.

The integrated rate law for a zeroth-order reaction is easy to find. Because the rate is constant (at k_r), the difference in concentration of a reactant from its initial value, $[A]_0$, is proportional to the time for which the reaction is in progress, and

$$[A]_0 - [A] = k_r t \quad \text{or} \quad [A] = [A]_0 - k_r t$$

As shown in Fig. 7A.9a, a plot of concentration against time is a straight line of slope $-k_r$; the reaction comes to an end when $t = [A]_0/k_r$, because then all the reactant has been consumed ($[A] = 0$).

7B.1 First-Order Integrated Rate Laws

The aim here is to find the integrated rate law for a first-order reaction in the form of an expression for the concentration of a reactant A at a time t, denoted $[A]_t$, given that the initial molar concentration of A is $[A]_0$.

How Is That Done?

To find the concentration of reactant A in a first-order reaction at any time after it has begun, first write the rate law for the consumption of A as

$$\text{Rate of consumption of A} = -\frac{d[A]}{dt} = k_r[A]$$

Because an instantaneous rate is a derivative of concentration with respect to time, the techniques of integral calculus can be used to find how $[A]$ depends on time. First, divide both sides by $[A]$ and multiply through by $-dt$:

$$\frac{d[A]}{[A]} = -k_r dt$$

Now integrate both sides between the limits $t = 0$ (when $[A] = [A]_0$) and the time of interest, t (when $[A] = [A]_t$):

$$\int_{[A]_0}^{[A]_t} \frac{d[A]}{[A]} = -k_r \overbrace{\int_0^t dt}^{t} = -k_r t$$

To evaluate the integral on the left, use the standard form

$$\int \frac{dx}{x} = \ln x + \text{constant}$$

and obtain

$$\int_{[A]_0}^{[A]_t}\frac{d[A]}{[A]} = (\ln[A]_t + \text{constant}) - (\ln[A]_0 + \text{constant})$$

cancel the constants $\ln a - \ln b = \ln(a/b)$

$$\overset{\wedge}{=} \ln[A]_t - \ln[A]_0 \overset{\wedge}{=} \ln\frac{[A]_t}{[A]_0}$$

Therefore,

$$\ln\frac{[A]_t}{[A]_0} = -k_r t$$

Now take (natural) antilogarithms of both sides and obtain $[A]_t/[A]_0 = e^{-k_r t}$, and therefore

$$[A]_t = [A]_0 e^{-k_r t}$$

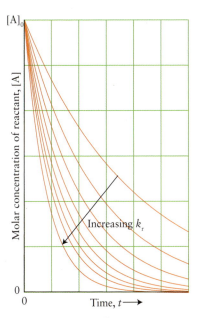

FIGURE 7B.1 The time-dependence of the concentration of a reactant in a first-order reaction is an exponential decay, as shown here. The larger the rate constant, the faster is the decay from the same initial concentration.

The two equations that have been derived,

$$\ln\frac{[A]_t}{[A]_0} = -k_r t \tag{1a}$$

$$[A]_t = [A]_0 e^{-k_r t} \tag{1b}$$

are two forms of the integrated rate law for a first-order reaction. The variation of concentration with time predicted by Eq. 1b is shown in **FIG. 7B.1**. This behavior is called **exponential decay** because the concentration of A is an exponential function of time. Initially the concentration changes rapidly, but it changes more slowly as time goes on and the reactant is used up.

EXAMPLE 7B.1 Calculating a concentration from a first-order integrated rate law

In polar stratospheric clouds, nitrogen can be found as N_2O_5, which takes part in the ozone cycle that protects life on Earth. However, N_2O_5 decomposes over time. If you were studying the role of the atmosphere in climate change, you might need to know how much N_2O_5 remains after a given period of time. What concentration of N_2O_5 remains 10.0 min (600. s) after the start of its decomposition at 65 °C in the reaction $2\,N_2O_5(g) \rightarrow 4\,NO_2(g) + O_2(g)$ when its initial concentration was $0.040\ \text{mol·L}^{-1}$? See Table 7A.1 for the rate law.

ANTICIPATE Because the reactant decays with time, you should expect a concentration that is lower than the initial value, but to assess the extent of the change you must do the calculation.

PLAN First identify the order of the reaction. If the reaction is first order in the specified reactant, use the exponential form of the first-order rate law (Eq. 1b) to find the new concentration of the reactant.

SOLVE

The reaction and its rate law, from Table 7A.1, are

$$2\,N_2O_5(g) \longrightarrow 4\,NO_2(g) + O_2(g)$$
$$\text{Rate of decomposition of } N_2O_5 = k_r[N_2O_5]$$

with $k_r = 5.2 \times 10^{-3}\ \text{s}^{-1}$. Therefore, the reaction is first order in N_2O_5 and Eq. 1b can be used.

From $[A]_t = [A]_0 e^{-k_r t}$,

$$[N_2O_5]_t = (0.040\ \text{mol·L}^{-1}) \times e^{-(5.2\times10^{-3}\text{s}^{-1})\times(600.\,\text{s})} = 0.0018\ \text{mol·L}^{-1}$$

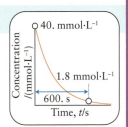

EVALUATE You can conclude that after 600. s, the concentration of N_2O_5 will have fallen from its initial value of $0.040\ \text{mol·L}^{-1}$ to $0.0018\ \text{mol·L}^{-1}$ (1.8 mmol·L^{-1}).

Self-test 7B.1A The rate law for the reaction $2 N_2O(g) \rightarrow 2 N_2(g) + O_2(g)$ is Rate of decomposition of $N_2O = k_r[N_2O]$. Calculate the concentration of N_2O remaining after the reaction has continued at 780 °C for 100. ms, if the initial concentration of N_2O is 0.20 mol·L^{-1} and $k_r = 3.4$ s^{-1}.

[**Answer:** 0.14 mol·L^{-1}]

Self-test 7B.1B Calculate the concentration of cyclopropane, C_3H_6 (**1**), remaining after the conversion into its isomer propene (**2**): $C_3H_6(g) \rightarrow CH_3$—$CH$=$CH_2(g)$ has continued at 773 K for 200. s, if the initial concentration of C_3H_6 is 0.100 mol·L^{-1} and $k_r = 6.7 \times 10^{-4}$ s^{-1}. The rate law is Rate of conversion of $C_3H_6 = k_r[C_3H_6]$.

Related Exercises 7B.1, 7B.2

An important application of an integrated rate law is to confirm that a reaction is in fact first order and to provide a way to measure its rate constant without drawing tangents to curves. Equation 1a can be written in the form of an equation for a straight line (see Appendix 1E; y = intercept + slope × x):

In Topic 7A, it is mentioned that there is a better way of determining rate constants than trying to draw tangents to curves: this is it.

$$\underbrace{\ln [A]_t}_{y} = \underbrace{\ln [A]_0}_{\text{intercept}} \underbrace{- k_r}_{\text{slope}} \underbrace{t}_{x} \tag{2}$$

Therefore, for a first-order process, a plot of $\ln[A]_t$ against t, should be a straight line with slope $-k_r$ and intercept (at $t = 0$) of $\ln [A]_0$.

EXAMPLE 7B.2   Measuring a rate constant

Many organic compounds can isomerize (turn into another compound with the same molecular formula) when heated. Suppose you are an organic chemist studying cyclopropane. You find that when cyclopropane (C_3H_6, **1**) is heated to 500. °C (773 K), it converts into an isomer, propene (**2**). You collect the following data, which show the concentration of cyclopropane at a series of times after the start of the reaction. Confirm that the reaction is first order in C_3H_6 and calculate the rate constant.

1 Cyclopropane, C_3H_6

Time, t/min	0	5	10.	15
$[C_3H_6]_t$/(mol·L^{-1})	1.50×10^{-3}	1.24×10^{-3}	1.00×10^{-3}	0.83×10^{-3}

PLAN Plot the natural logarithm of the reactant concentration against t. If a straight line is obtained, the reaction is first order and the slope of the graph is $-k_r$. If the plot is not linear, then the reaction is not first order in the specified reactant. Use the linear regression methods in the Living Graph section of the textbook Web site, a spreadsheet program, or a graphing calculator to plot the data and determine the slope of the graph.

2 Propene, C_3H_6

SOLVE

For the graphical procedure, begin by setting up the following table:

Time, t/min	0	5	10.	15
$\ln [C_3H_6]_t$	-6.50	-6.69	-6.91	-7.09

Plot the data.

The points are plotted in **FIG. 7B.2**. The graph is a straight line, confirming that the reaction is first order in cyclopropane. When points A and B on the plot are used, the slope of the line is

$$\text{Slope} = \frac{(-7.02) - (-6.56)}{(13.3 - 1.7) \text{ min}} = -0.040 \text{ min}^{-1}$$

Therefore, because $k_r = -\text{slope}$, $k_r = 0.040$ min^{-1}.

EVALUATE The value found for k_r is equivalent to 6.7×10^{-4} s^{-1}, as in Table 7A.1.

A Note on Good Practice: Be sure to include the appropriate units for k_r.

Self-test 7B.2A Data on the decomposition of N_2O_5 at 25 °C are

Time, t/min	0	200.	400.	600.	800.	1000.
$[N_2O_5]_t$/(mmol·L^{-1})	15.0	9.6	6.2	4.0	2.5	1.6

Confirm that the reaction is first order and find the value of k_r for the reaction $N_2O_5(g) \rightarrow$ products.

[**Answer:** The plot of ln $[N_2O_5]$ against time is linear, so the reaction is first order; $k_r = 2.2 \times 10^{-3}$ min^{-1}.]

Self-test 7B.2B Azomethane, $CH_3N_2CH_3$, decomposes to ethane and nitrogen gas in the reaction $CH_3N_2CH_3(g) \rightarrow CH_3CH_3(g) + N_2(g)$. The reaction was followed at 460. K by measuring the partial pressure of azomethane over time:

Time, t/s	0	1000.	2000.	3000.	4000.
Pressure, $P_{CH_3N_2CH_3}$/Torr	8.20×10^{-2}	5.72×10^{-2}	3.99×10^{-2}	2.78×10^{-2}	1.94×10^{-2}

Confirm that the reaction is first order of the form Rate $= k_rP$, where P is the partial pressure of azomethane, and find the value of k_r.

Related Exercises 7B.11, 7B.12

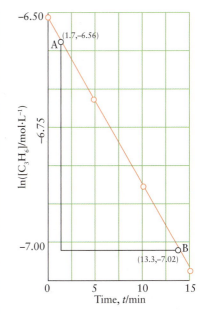

FIGURE 7B.2 A first-order reaction can be identified by plotting the natural logarithm of the reactant concentration against time. The graph is linear if the reaction is first order. The slope of the line, which is calculated for the system in Example 7B.2, is equal to the negative of the rate constant.

LIVING GRAPH FIGURE 7B.2

EXAMPLE 7B.3 Predicting how long it will take for a concentration to change by a specified amount

As in Example 7B.1, you are studying the decomposition of N_2O_5, but now you need to know how long it takes for a given amount of N_2O_5 to decompose. A sample of N_2O_5 is allowed to decompose by the reaction $2 N_2O_5(g) \rightarrow 4 NO_2(g) + O_2(g)$. How long will it take for the concentration of N_2O_5 to decrease from 20. mmol·L^{-1} to 2.0 mmol·L^{-1} at 65 °C? Use the data in Table 7A.1.

PLAN First, identify the order of the reaction in N_2O_5 by referring to Table 7A.1. If the reaction is first order, rearrange Eq. 1a into an equation for t in terms of the given concentrations. Then substitute the numerical value of the rate constant and the concentration data and evaluate t.

SOLVE The reaction is first order with $k_r = 5.2 \times 10^{-3}$ s^{-1} at 65 °C (338 K).

From $\ln([A]_t/[A]_0) = -k_rt$ rearranged into $t = (1/k_r) \ln([A]_0/[A]_t)$,

$$t = \frac{1}{5.2 \times 10^{-3}s^{-1}} \ln \frac{20. \text{ mmol·L}^{-1}}{2.0 \text{ mmol·L}^{-1}} = 4.4 \times 10^2 \text{ s}$$

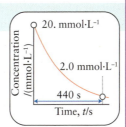

Self-test 7B.3A How long does it take for the concentration of A to decrease to 1.0% of its initial value in a first-order reaction of the form A $\rightarrow$ products with $k_r = 1.0$ s^{-1}?

[**Answer:** 4.6 s]

Self-test 7B.3B Cyclopropane isomerizes to propene in a first-order process. How long does it take for the concentration of cyclopropane to decrease from 1.0 mol·L^{-1} to 0.0050 mol·L^{-1} at 500. °C? Use the data in Table 7A.1.

Related Exercises 7B.8, 7B.9, 7B.17

Some rate laws depend on the concentration of more than one species. For instance, oxidation of iodide ions by persulfate ions (reaction D of Topic 7A)

$$S_2O_8^{2-}(aq) + 3 I^-(aq) \rightarrow 2 SO_4^{2-}(aq) + I_3^-(aq)$$

follows the rate law

$$\text{Rate of consumption of } I^- = k_r[S_2O_8^{2-}][I^-]$$

In such cases, the analysis can be simplified by ensuring that one species remains effectively at the same concentration throughout the reaction and so can be treated as a constant. One way to do this is to start with persulfate ions in such a high concentration that their concentration barely changes in the course of the reaction. For instance, the concentration of persulfate ions might be 100 times as great as the concentration of iodide ions; so, even when all the iodide ions have been oxidized, the persulfate concentration is almost the same as it was at the beginning of the reaction. Because $[S_2O_8^{2-}]$ is virtually constant, the rate law can be written

$$\text{Rate of consumption of } I^- = k_r'[I^-]$$

where $k_r' = k_r[S_2O_8^{2-}]$, another constant. The actual second-order reaction has been turned into a **pseudo-first-order reaction,** a reaction that is effectively first order. The rate law for a pseudo-first-order reaction is much easier to analyze than the true rate law because it depends on the concentration of only one substance, and the variation of iodide ion concentration with time can be treated in the same way as a true first-order reaction. Then, by knowing the concentration of $S_2O_8^{2-}$ ions used in the experiment, the value of the overall rate constant can be deduced from $k_r = k_r'/[S_2O_8^{2-}]$.

In a first-order reaction, the concentration of reactant decays exponentially with time. To verify that a reaction is first order, plot the natural logarithm of its concentration against time and expect a straight line; the slope of the straight line is $-k_r$.

7B.2 Half-Lives for First-Order Reactions

The **half-life,** $t_{1/2}$, of a reactant is the time needed for its concentration to fall to one-half its initial value. Knowing the half-lives of pollutants such as chlorofluorocarbons allows their environmental impact to be assessed. If their half-lives are short, they may not survive long enough to reach the stratosphere, where they can destroy ozone. Half-lives are also important in planning storage systems for radioactive materials, because the decay of radioactive nuclei is a first-order process (Topic 10B).

You already know that the higher the value of k_r, the more rapid is the consumption of a reactant. Therefore, you should be able to deduce a relation for a first-order reaction that shows that the greater the rate constant, the shorter is the half-life of a substance.

How Is That Done?

To find the relation between the rate constant and the half-life of a reactant in a first-order reaction, begin by rearranging Eq. 1a into an expression for the time needed to reach a given concentration:

From $\ln([A]_t/[A]_0) = -k_r t$ and $\ln(1/x) = -\ln x$

$$t = \frac{1}{k_r}\ln\frac{[A]_0}{[A]_t}$$

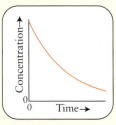

Now set $t = t_{1/2}$ and $[A]_t = \frac{1}{2}[A]_0$

$$t_{1/2} = \frac{1}{k_r}\ln\frac{[A]_0}{\frac{1}{2}[A]_0} = \frac{1}{k_r}\ln 2$$

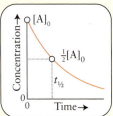

The initial concentration has canceled.

The derivation has shown that for the rate law Rate of consumption of A = $k_r[A]$,

$$t_{1/2} = \frac{\ln 2}{k_r} \quad (3)$$

As anticipated, the greater the value of the rate constant k_r, the shorter is the half-life of the reactant (**FIG. 7B.3**). Note that the half-life of a substance that decays by a first-order reaction depends on only the rate constant, not on the concentration. Therefore, the half-life has the same value at all stages of the reaction: whatever the reactant concentration might be at any stage of a first-order reaction, it takes the same time ($t_{1/2}$) for that concentration to fall to half its value.

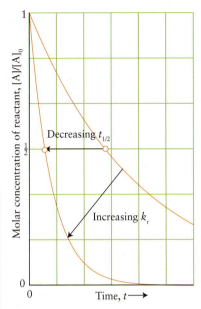

FIGURE 7B.3 The time-dependence of the concentration of the reactant in two first-order reactions plotted on the same graph. When the first-order rate constant is large, the half-life of the reactant is short because the exponential decay of the concentration of the reactant is then fast.

EXAMPLE 7B.4   Using a half-life to calculate the amount of reactant remaining

In 1989, a teenager in Ohio was poisoned by breathing vapors from spilled mercury. The mercury level in his urine, which is proportional to its concentration in his body, was found to be 1.54 mg·L^{-1}. Suppose you are working in a toxicology laboratory. You need to know from a urine analysis how much mercury a person has ingested and decide to use this case as a standard. Mercury(II) is eliminated from the body by a first-order process and has a half-life in the body of 6 days (6.0 d). What would be the concentration of mercury(II) in the urine of the teenager in milligrams per liter after 30. d if therapeutic measures are not taken?

ANTICIPATE   Thirty days is five half-lives, so you should expect that most of the mercury will have been eliminated.

PLAN   The level of mercury(II) in the urine can be predicted by using the integrated first-order rate law, Eq. 1b. To use this equation, you need the rate constant. Therefore, start by calculating the rate constant from the half-life (Eq. 3) and substitute the result into Eq. 1b.

SOLVE

From $t_{1/2} = (\ln 2)/k_r$ in the form $k_r = (\ln 2)/t_{1/2}$,

$$k_r = \frac{\ln 2}{6.0 \text{ d}} = \frac{\ln 2}{6.0}\text{d}^{-1} = 0.115\ldots \text{ d}^{-1}$$

From $[A]_t = [A]_0 e^{-k_r t}$

$$[A]_t = (1.54 \text{ mg·L}^{-1})e^{-(0.115\ldots \text{ d}^{-1}) \times (30. \text{ d})} = 0.048 \text{ mg·L}^{-1}$$

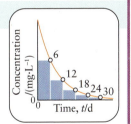

EVALUATE   The concentration of mercury in the urine is therefore 0.048 mg·L^{-1}. As expected, this is a very small value.

A Note on Good Practice: Note that, because exponential functions e^x are so sensitive to the value of x, avoid rounding errors by leaving the numerical calculation to a single final step.

Self-test 7B.4A   In 1972, grain treated with methyl mercury was released for human consumption in Iraq, resulting in 459 deaths. The half-life of methyl mercury in body tissues is 70. d. How many days are required for the amount of methyl mercury to drop to 10.% of the original value after ingestion?

[*Answer:* 230 d]

Self-test 7B.4B   Soil at the Rocky Flats Nuclear Processing Facility in Colorado was found to be contaminated with radioactive plutonium-239, which has a half-life of 24 ka (2.4×10^4 years). The soil was loaded into drums for storage. How many years must pass before the radioactivity drops to 20.% of its initial value?

Related Exercises 7B.15, 7B.16

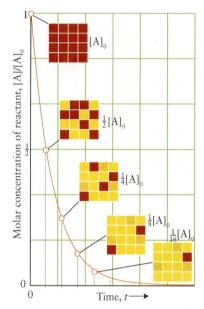

Molar concentration of reactant, $[A]/[A]_0$

$[A]_0$

$\frac{1}{2}[A]_0$

$\frac{1}{4}[A]_0$

$\frac{1}{8}[A]_0$
$\frac{1}{16}[A]_0$

Time, $t \longrightarrow$

FIGURE 7B.4 For first-order reactions, the half-life is the same whatever the concentration at the start of the chosen period. Therefore, it takes one half-life to fall to half the initial concentration, two half-lives to fall to one-fourth the initial concentration, three half-lives to fall to one-eighth, and so on. The boxes portray the composition of the reaction mixture at the end of each half-life; the red squares represent the reactant A and the yellow squares represent the product.

The concentration of the reactant does not appear in Eq. 3: for a first-order reaction, the half-life of a substance is independent of its initial concentration. It follows that the concentration of A can be used as its "initial" concentration at any stage of the reaction: if at some stage the concentration of A happens to be $[A]$, then after a further time $t_{1/2}$, the concentration of A will have fallen to $\frac{1}{2}[A]$; after a further $t_{1/2}$ that $\frac{1}{2}[A]$ will have fallen to $\frac{1}{4}[A]$; and so on (**FIG. 7B.4**). In general, at any stage of a first-order reaction, the concentration remaining after n half-lives is $(\frac{1}{2})^n[A]$. For example, in Example 7B.4, because 30 days corresponds to 5 half-lives, after that interval $[A]_t = (\frac{1}{2})^5[A]_0$, or $[A]_0/32$, which evaluates to 3%, the same as the result obtained in the example.

Self-test 7B.5A Calculate (a) the number of half-lives and (b) the time required for the concentration of N_2O to fall to one-eighth of its initial value in a first-order decomposition at 1000. K. Consult Table 7A.1 for the rate constant.

[*Answer:* (a) 3 half-lives; (b) 2.7 s]

Self-test 7B.5B Calculate (a) the number of half-lives and (b) the time required for the concentration of C_2H_6 to fall to one-sixteenth of its initial value as it dissociates into CH_3 radicals at 973 K. Consult Table 7A.1 for the rate constant.

The half-life of a reactant that decays in a first-order reaction is characteristic of the reaction and independent of the initial concentration. It is inversely proportional to the rate constant of the reaction.

7B.3 Second-Order Integrated Rate Laws

As with first-order reactions, it is useful to be able to predict how the concentration of a reactant or product changes with time in second-order reactions. To make these predictions you need to derive the integrated form of the rate law

$$\text{Rate of consumption of A} = k_r[A]^2$$

How Is That Done?

To obtain the integrated rate law for a second-order reaction, begin by writing the rate law as

$$-\frac{d[A]}{dt} = k_r[A]^2$$

After division by $[A]^2$ and multiplication by $-dt$, this equation becomes

$$\frac{d[A]}{[A]^2} = -k_r dt$$

To integrate this equation, use the same limits as in the first-order case:

$$\int_{[A]_0}^{[A]_t} \frac{d[A]}{[A]^2} = -k_r \overbrace{\int_0^t dt}^{t} = -kt$$

This time you need the integral

$$\int \frac{dx}{x^2} = -\frac{1}{x} + \text{constant}$$

to write the integral on the left as

$$\int_{[A]_0}^{[A]_t} \frac{d[A]}{[A]^2} = \left(-\frac{1}{[A]_t} + \text{constant}\right) - \left(-\frac{1}{[A]_0} + \text{constant}\right)$$

$$= \frac{1}{[A]_0} - \frac{1}{[A]_t}$$

Therefore

$$\frac{1}{[A]_t} - \frac{1}{[A]_0} = k_r t$$

which can be rearranged into

$$[A]_t = \frac{[A]_0}{1 + k_r t [A]_0}$$

The two equations that have been derived are

$$\frac{1}{[A]_t} - \frac{1}{[A]_0} = k_r t \qquad \textbf{(4a)}$$

$$[A]_t = \frac{[A]_0}{1 + k_r t [A]_0} \qquad \textbf{(4b)}$$

Equation 4b is plotted in **FIG. 7B.5**. It shows that the concentration of the reactant decreases rapidly at first but then changes more slowly than it would for a first-order reaction with the same initial rate. This slowing down of second-order reactions has important environmental consequences: because many pollutants degrade by second-order reactions, they remain at low concentration in the environment for long periods. Equation 4a can be written in the form of an equation for a straight line (y = intercept + slope x):

$$\overbrace{\frac{1}{[A]_t}}^{y} = \overbrace{\frac{1}{[A]_0}}^{\text{intercept}} + \overbrace{k_r}^{\text{slope}}\;\overbrace{t}^{x} \qquad \textbf{(4c)}$$

Therefore, to determine whether a reaction is second order in a reactant, plot the inverse of the concentration against the time and see whether a straight line is obtained. If the line is straight, then the reaction is second order and the slope of the line is equal to k_r (**FIG. 7B.6**).

The half-life of a reactant in a second-order reaction is found by setting $t = t_{1/2}$ and $[A]_t = \frac{1}{2}[A]_0$ in Eq. 4, and then solving for $t_{1/2}$. The resulting expression,

$$t_{1/2} = \frac{1}{k_r [A]_0}$$

shows that the half-life of the reactant in a second-order reaction is inversely proportional to the concentration of reactant. The half-life increases as the reaction proceeds and reactant concentration decreases. Because of this variation, the half-life is not very useful for describing reactions with second-order kinetics.

As you have seen for first- and second-order rate laws, each integrated rate law can be rearranged into an equation that, when plotted, gives a straight line; the rate constant

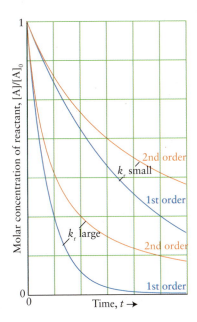

FIGURE 7B.5 The time dependence of the concentration of a reactant in a second-order reaction. The larger the rate constant, k_r, the greater is the dependence of the rate on the concentration of the reactant. The lower blue lines are the curves for first-order reactions with the same initial rates as for the corresponding second-order reactions. Note how the concentrations for second-order reactions fall away much less rapidly at longer times than those for first-order reactions do.

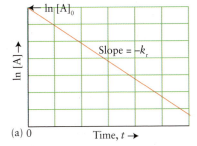
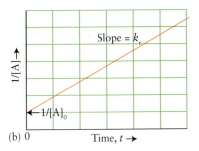

FIGURE 7B.6 The plots that allow the determination of reaction order. (a) If a plot of ln [A] against time is a straight line, the reaction is first order in [A]. (b) If a plot of 1/[A] against time is a straight line, the reaction is second order in [A].

TABLE 7B.1   **Integrated Rate Laws, Rate Law Plots, and Half-Lives**

	Order of reaction		
	0	**1**	**2**
Rate law	Rate $= k_r$	Rate $= k_r[A]$	Rate $= k_r[A]^2$
Integrated rate law	$[A]_t = -k_r t + [A]_0$	$[A]_t = [A]_0 e^{-k_r t}$	$[A]_t = \dfrac{[A]_0}{1 + k_r t[A]_0}$ $\dfrac{1}{[A]_t} = \dfrac{1}{[A]_0} + k_r t$
Plot to determine order	plot of $[A]_t$ vs t, starting at $[A]_0$, slope $-k_r$	plot of $\ln [A]_t$ vs t, starting at $\ln [A]_0$, slope $-k_r$	plot of $1/[A]_t$ vs t, starting at $1/[A]_0$, slope k_r
Slope of the line plotted	$-k_r$	$-k_r$	k_r
Half-life	$t_{1/2} = \dfrac{[A]_0}{2k_r}$	$t_{1/2} = \dfrac{\ln 2}{k_r} \approx \dfrac{0.693}{k_r}$	$t_{1/2} = \dfrac{1}{k_r[A]_0}$
	(not used)		(not used)

can then be obtained from the slope of the plot. **TABLE 7B.1** summarizes the relationships to use.

A second-order reaction has a long tail of low concentration at long reaction times. The half-life of a second-order reaction is inversely proportional to the concentration of the reactant.

What have you learned in this Topic?

You have learned that the integrated form of a rate law gives you an expression that allows you to predict how the concentrations of reactants and products change with time. You have also seen how experimental data can be treated to give straight-line plots in which the slopes are proportional to the rate constant, k_r. For first-order reactions, you have learned that the half-life depends only on the rate constant for the particular reaction and is independent of the concentration of the reactant.

The skills you have mastered are the ability to:

☐ **1.** Predict the concentration of a reactant or product at a given time after a reaction has been initiated (Example 7B.1).

☐ **2.** Plot experimental data to determine the rate constant for a reaction (Example 7B.2).

☐ **3.** Predict how long it will take for the concentration of a reactant or product to change by a given value (Example 7B.3).

☐ **4.** Use the half-life of a reaction to predict the amount of reactant remaining after a given time period (Example 7B.4).

☐ **5.** Plot experimental data to determine the rate constant for a second-order reaction (Section 7B.3).

Topic 7B Exercises

Exercises labeled $\int_{dx}^{C}$ require calculus.

All rates are unique reaction rates unless otherwise stated.

7B.1 Beta blockers are drugs that are used to manage hypertension. It is important for doctors to know how rapidly a beta blocker is eliminated from the body. A certain beta blocker is eliminated in a first-order process with a rate constant of $7.6 \times 10^{-3} \, min^{-1}$ at normal body temperature ($37 \, ^{\circ}C$). A patient is given 20. mg of the drug. What mass of the drug remains in the body 5.0 h after administration?

7B.2 In the brewing of beer, ethanal, which smells like green apples, is an intermediate in the formation of ethanol. Ethanal decomposes in the following first-order reaction: $CH_3CHO(g) \rightarrow CH_4(g) + CO(g)$. At a certain temperature, the rate constant for the decomposition is $1.5 \times 10^{-3} \, s^{-1}$. What concentration of ethanal, which had an initial concentration of $0.100 \, mol \cdot L^{-1}$, remains 40.0 min after the start of its decomposition at this temperature?

7B.3 Determine the rate constant for each of the following first-order reactions, in each case expressed for the rate of loss of A: (a) $A \rightarrow B$, given that the concentration of A decreases to one-half its initial value in 1000. s; (b) $A \rightarrow B$, given that the concentration of A decreases from $0.67 \, mol \cdot L^{-1}$ to $0.53 \, mol \cdot L^{-1}$ in 25 s; (c) $2 \, A \rightarrow B + C$, given that $[A]_0 = 0.153 \, mol \cdot L^{-1}$ and that after 115 s the concentration of B rises to $0.034 \, mol \cdot L^{-1}$.

7B.4 Determine the rate constant for each of the following first-order reactions: (a) $A \rightarrow B + C$, given that the concentration of A decreases to one-fourth its initial value in 125 min; (b) $2 \, A \rightarrow D + E$, given that $[A]_0 = 0.0421 \, mol \cdot L^{-1}$ and that after 63 s the concentration of D increases to $0.00132 \, mol \cdot L^{-1}$; (c) $3 \, A \rightarrow F + G$, given that $[A]_0 = 0.080 \, mol \cdot L^{-1}$ and that after 11.4 min the concentration of F rises to $0.015 \, mol \cdot L^{-1}$. In each case, write the rate law for the rate of loss of A.

7B.5 Dinitrogen pentoxide, N_2O_5, decomposes by first-order kinetics with a rate constant of $3.7 \times 10^{-5} \, s^{-1}$ at 298 K. (a) What is the half-life (in hours) of N_2O_5 at 298 K? (b) If $[N_2O_5]_0 = 0.0567 \, mol \cdot L^{-1}$, what will be the concentration of N_2O_5 after 3.5 h? (c) How much time (in minutes) will elapse before the N_2O_5 concentration decreases from $0.0567 \, mol \cdot L^{-1}$ to $0.0135 \, mol \cdot L^{-1}$?

7B.6 Dinitrogen pentoxide, N_2O_5, decomposes by first-order kinetics with a rate constant of $0.15 \, s^{-1}$ at 353 K. (a) What is the half-life (in seconds) of N_2O_5 at 353 K? (b) If $[N_2O_5]_0 = 0.0567 \, mol \cdot L^{-1}$, what will be the concentration of N_2O_5 after 2.0 s? (c) How much time (in minutes) will elapse before the N_2O_5 concentration decreases from $0.0567 \, mol \cdot L^{-1}$ to $0.0135 \, mol \cdot L^{-1}$?

7B.7 Substance A decomposes in a first-order reaction and its half life is 355 s. How much time must elapse for the concentration of A to decrease to (a) one-eighth of its initial concentration; (b) one-fourth of its initial concentration; (c) 15% of its initial concentration; (d) one-ninth of its initial concentration?

7B.8 The first-order rate constant for the photodissociation of A is $1.24 \times 10^{-3} \, min^{-1}$. Calculate the time needed for the concentration of A to decrease to (a) 25 % of its initial concentration; (b) one-sixth of its initial concentration.

7B.9 For the first-order reaction $A \rightarrow 3 \, B + C$, when $[A]_0 = 0.015 \, mol \cdot L^{-1}$, the concentration of B increases to $0.018 \, mol \cdot L^{-1}$ in 3.0 min. (a) What is the rate constant for the reaction expressed as the rate of loss of A? (b) How much more time would be needed for the concentration of B to increase to $0.030 \, mol \cdot L^{-1}$?

7B.10 Pyruvic acid is an intermediate in the fermentation of grains. During fermentation the enzyme pyruvate decarboxylase causes the pyruvate ion to release carbon dioxide. In one experiment, an aqueous solution of pyruvate ions of volume 200 mL at an initial concentration of $3.23 \, mmol \cdot L^{-1}$ is placed in a sealed, rigid flask of volume 500 mL at 293 K. Because the concentration of the enzyme was kept constant, the reaction was pseudo-first order in pyruvate ion. The elimination of CO_2 by the reaction was monitored by measuring the partial pressure of the CO_2 gas. The pressure of the gas was found to rise from 0 to 100. Pa in 522 s. What is the rate constant of the pseudo-first-order reaction?

7B.11 The following data were collected for the reaction $2 \, HI(g) \rightarrow H_2(g) + I_2(g)$ at 580 K. (a) Use a graphing calculator or software to plot the data in an appropriate fashion to determine the order of the reaction. (b) From the graph, determine the rate constant for (i) the rate law for the loss of HI and (ii) the unique rate law.

Time, t/s	0	1000.	2000.	3000.	4000.
$[HI]/(mol \cdot L^{-1})$	1.0	0.11	0.061	0.041	0.031

7B.12 The following data were collected for the reaction $H_2(g) + I_2(g) \rightarrow 2 \, HI(g)$ at 780 K. (a) Using a graphing calculator or software, plot the data in an appropriate fashion to determine the order of the reaction. (b) From the graph, determine the rate constant for the consumption of I_2.

Time, t/s	0	1.0	2.0	3.0	4.0
$[I_2]/(mmol \cdot L^{-1})$	1.00	0.43	0.27	0.20	0.16

7B.13 The half-life of A in a second-order reaction is 50.5 s when $[A]_0 = 0.84 \, mol \cdot L^{-1}$. Calculate the time needed for the concentration of A to decrease to (a) one-sixteenth; (b) one-fourth; (c) one-fifth of its original value.

7B.14 Determine the rate constant for each of the following second-order reactions: (a) $2 \, A \rightarrow B + 2 \, C$, given that the concentration of A decreases from $2.50 \, mmol \cdot L^{-1}$ to $1.25 \, mmol \cdot L^{-1}$ in 100 s; (b) $A \rightarrow C + 2 \, D$, given that $[A]_0 = 0.300 \, mol \cdot L^{-1}$ and that the concentration of C increases to $0.010 \, mol \cdot L^{-1}$ in 200. s. In each case, express your results in terms of the rate law for the loss of A.

7B.15 Sulfuryl chloride, SO_2Cl_2, decomposes by first-order kinetics, and $k_r = 2.81 \times 10^{-3} \, min^{-1}$ at a certain temperature. (a) Determine the half-life for the reaction. (b) Determine the time needed for the concentration of SO_2Cl_2 to decrease to 10% of its initial concentration. (c) If 14.0 g of SO_2Cl_2 is sealed in a reaction vessel of volume 2500. L and heated to the specified temperature, what mass will remain after 1.5 h?

7B.16 Ethane, C_2H_6, forms $\cdot CH_3$ radicals at 700. $^{\circ}C$ in a first-order reaction, for which $k_r = 1.98 \, h^{-1}$. (a) What is the half-life for the reaction? (b) Calculate the time needed for the amount of ethane to fall from 2.26×10^{-3} mol to 1.45×10^{-4} mol in a reaction vessel of volume 500. mL at 700. °C. (c) How much of a sample of ethane with mass of 5.44 mg placed in a reaction vessel of 500. mL volume at 700. °C will remain after 42 min?

7B.17 Calculate the time required for each of the following second-order reactions to take place: (a) $2A \rightarrow B + C$, for the concentration of A to decrease from 0.10 mol·L^{-1} to 0.080 mol·L^{-1}, given that $k_r = 0.015$ L·mol^{-1}·min^{-1} for the rate law expressed in terms of the loss of A; (b) $A \rightarrow 2B + C$, when $[A]_0 = 0.15$ mol·L^{-1}, for the concentration of B to increase to 0.19 mol·L^{-1}, given that $k_r = 0.0035$ L·mol^{-1}·min^{-1} in the rate law for the loss of A.

7B.18 The second-order rate constant for the decomposition of NO_2 (to NO and O_2) at 573 K is 0.54 L·mol^{-1}·s^{-1}. Calculate the time for an initial NO_2 concentration of 0.20 mol·L^{-1} to decrease to (a) one-half; (b) one-sixteenth; (c) one-ninth of its initial concentration.

7B.19 Suppose a reaction has the form $aA \rightarrow$ products and the rate of the reaction is written Rate $= k_r[A]$. Derive expressions for the concentration of A at a time t and for the half-life of A in terms of a and k_r.

7B.20 Suppose a reaction has the form $aA \rightarrow$ products and the rate of the reaction is written Rate $= k_r[A]^2$. Derive an expression for the concentration of A at a time t in terms of a and k_r.

7B.21 Derive an expression for the half-life of a reactant A that decays by a third-order reaction with rate constant k_r.

7B.22 Derive an expression for the half-life of a reactant A that decays by an nth-order reaction (with $n > 1$) with rate constant k_r.

Topic 7C Reaction Mechanisms

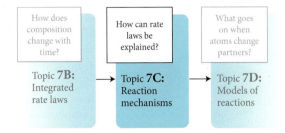

A knowledge of how a reaction takes place at the molecular level provides answers to many important questions. For example, what controls the rate of formation of the DNA double helix from its individual strands? What molecular events convert ozone into oxygen or turn a mixture of fuel and air into carbon dioxide and water when it ignites in an engine? In this Topic you will see how to use the empirical (experimentally determined) rate law for a reaction as a window onto these events.

7C.1 Elementary Reactions

All but the simplest reactions are the outcome of several, and sometimes many, steps called **elementary reactions.** Each elementary reaction describes a distinct event in the progress of a reaction, often a collision of particles. To describe how a reaction takes place, chemists propose a **reaction mechanism,** a sequence of elementary reactions or steps that they believe take place as reactants are transformed into products.

Although several different mechanisms might be proposed for a reaction, rate measurements can be used to eliminate some of them. For example, in the decomposition of ozone, $2 O_3(g) \rightarrow 3 O_2(g)$, the reaction might be supposed to take place in one of two different ways:

One-step mechanism: Two O_3 molecules collide and rearrange into three O_2 molecules (**FIG. 7C.1**):

$$O_3 + O_3 \longrightarrow O_2 + O_2 + O_2$$

Two-step mechanism: In the first step, an O_3 molecule is energized by solar radiation and dissociates into an O atom and an O_2 molecule. In the second step, the O atom attacks another O_3 molecule to produce two more O_2 molecules (**FIG. 7C.2**):

Step 1 $O_3 \xrightarrow{h\nu} O_2 + O$

Step 2 $O + O_3 \longrightarrow O_2 + O_2$

The free O atom in the second mechanism is a **reaction intermediate,** a species that plays a role in a reaction but does not appear in the chemical equation for the overall reaction; it is produced in one step but is used up in a later step. The two equations for the elementary reactions add together to give the equation for the overall reaction. That is true of any proposed mechanism: the sum of the elementary reactions must be the same as the chemical equation for the overall reaction.

> **A Note on Good Practice:** The chemical equations for elementary reactions are written without state symbols. They differ from the overall chemical equation, which summarizes bulk behavior, because they show how *individual* atoms and molecules take part in the reaction. Stoichiometric coefficients are not used for elementary reactions. Instead, to emphasize that a specific process involving individual molecules is being depicted, the formula of a species appears as many times as required, as in $O_2 + O_2 + O_2$, not $3 O_2$.

Why Do You Need to Know This Material? The mechanism of a reaction can give insight into how reactions occur and suggest conditions to optimize a process. Chemists often look for ways to increase the rates of chemical reactions or minimize unwanted products.

What Do You Need to Know Already? You need to be familiar with the concept of a rate law (Topic 7A) and the equilibrium constants of reactions (Topic 5G).

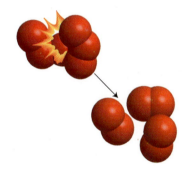

FIGURE 7C.1 A representation of a proposed one-step mechanism for the decomposition of ozone in the atmosphere. This reaction takes place in a single bimolecular collision.

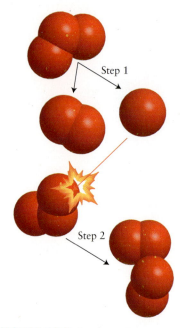

FIGURE 7C.2 An alternative, two-step mechanism for the decomposition of ozone. In the first step, an energized ozone molecule shakes off an oxygen atom. In the second step, that oxygen atom attacks another ozone molecule.

Elementary reactions are classified according to their **molecularity,** the number of reactant molecules, atoms, or ions taking part in a specified elementary reaction. The equation for the first step in the two-step mechanism for the decomposition of ozone ($O_3 \rightarrow O_2 + O$) is an example of a **unimolecular reaction,** because only one reactant molecule participates. In this case, an ozone molecule can be pictured as acquiring energy from sunlight and vibrating so vigorously that it shakes itself apart. In the second step of the two-step mechanism, the O atom produced by the dissociation of O_3 goes on to attack another O_3 molecule ($O + O_3 \rightarrow O_2 + O_2$). This elementary reaction is an example of a **bimolecular reaction,** because two reactant species come together to react. The molecularity of a unimolecular reaction is 1 and that of a bimolecular reaction is 2.

Any forward reaction that can take place is also accompanied, in principle at least, by the corresponding reverse reaction. Therefore, the unimolecular decay of O_3 in step 1 of the two-step mechanism is accompanied by the formation reaction

$$O_2 + O \longrightarrow O_3$$

The reverse of the reaction in step 1 is bimolecular. Similarly, the bimolecular attack of O on O_3 in step 2 of the two-step mechanism is accompanied by its bimolecular reverse reaction,

$$O_2 + O_2 \longrightarrow O + O_3$$

These two equations add up to give the reverse of the overall equation, $3\,O_2(g) \rightarrow 2\,O_3(g)$. The forward and reverse reactions jointly provide a mechanism for reaching dynamic equilibrium between the reactants and the products in the overall process.

The reverse of the single elementary step in the one-step mechanism is

$$O_2 + O_2 + O_2 \longrightarrow O_3 + O_3$$

This step is an example of a **termolecular reaction,** an elementary reaction requiring the simultaneous collision of three molecules. Termolecular reactions are uncommon because it is very unlikely that three molecules will collide simultaneously with one another under normal conditions.

Self-test 7C.1A What is the molecularity of (a) $C_2N_2 \rightarrow CN + CN$; (b) $NO_2 + NO_2 \rightarrow NO + NO_3$?

[*Answer:* (a) Unimolecular; (b) bimolecular]

Self-test 7C.1B What is the molecularity of (a) $C_2H_5Br + OH^- \rightarrow C_2H_5OH + Br^-$; (b) $Br_2 \rightarrow Br + Br$?

Many reactions take place by a series of steps called elementary reactions. The molecularity of an elementary reaction is the number of reactant particles that take part in the step.

7C.2 The Rate Laws of Elementary Reactions

To verify that a proposed reaction mechanism agrees with experimental data, chemists construct the overall rate law implied by the mechanism and check to see whether it is consistent with the experimentally determined rate law. However, although the constructed rate law and the experimental rate law may be the same, the proposed mechanism may still be incorrect because some other mechanism might also lead to the same rate law. Kinetic information can only support a proposed mechanism; it can never prove that a mechanism is correct. The acceptance of a suggested mechanism is more like the process of proof in an ideal court of law than a proof in mathematics, with evidence being assembled to give a convincing, consistent picture.

To construct an overall rate law from a mechanism, the rate law for each elementary reaction is identified and then combined. Because an elementary reaction shows how that step of the reaction occurs, its rate law (but *not* the rate law for the overall reaction) can be written from its chemical equation, with each power of concentration ($[A]$, $[A]^2$, etc.) in the rate law being the same as the number of reactant species of a given type participating in the step. For instance, in a unimolecular elementary reaction, the reactant simply shakes itself apart, so the rate is proportional to the number of reactant molecules. In a

TABLE 7C.1   The Rate Laws of Elementary Reactions

Molecularity	Elementary reaction		Rate law*
1	$A \longrightarrow$	products	rate $= k_r[A]$
2	$A + B \longrightarrow$	products	rate $= k_r[A][B]$
	$A + A \longrightarrow$	products	rate $= k_r[A]^2$
3	$A + B + C \longrightarrow$	products	rate $= k_r[A][B][C]$
	$A + A + B \longrightarrow$	products	rate $= k_r[A]^2[B]$
	$A + A + A \longrightarrow$	products	rate $= k_r[A]^3$

* When it is necessary to distinguish the steps in a mechanism, replace k_r by k_1, k_2, etc.

bimolecular elementary reaction the rate depends on the frequency of collisions of two molecules, so it is proportional to the square of the concentration (this point is developed in Topic 7D). That is, a unimolecular reaction has a first-order rate law and a bimolecular reaction has a second-order rate law, as summarized in **TABLE 7C.1**.

To see how this works in practice, consider the rate laws for the steps in a mechanism proposed for the gas-phase oxidation of NO to NO_2. Its overall rate law has been determined experimentally:

$$2\,NO(g) + O_2(g) \longrightarrow 2\,NO_2(g) \qquad \text{Rate of formation of } NO_2 = k_r[NO]^2[O_2]$$

The following mechanism has been proposed:

Step 1 A fast bimolecular dimerization and its reverse:

Forward: $NO + NO \rightarrow N_2O_2$ Rate of formation of $N_2O_2 = k_1[NO]^2$
Reverse: $N_2O_2 \rightarrow NO + NO$ Rate of consumption of $N_2O_2 = k_1'[N_2O_2]$

Step 2 A slow bimolecular reaction, in which an O_2 molecule collides with the dimer, and its reverse:

Forward: $O_2 + N_2O_2 \rightarrow NO_2 + NO_2$ Rate of consumption of $N_2O_2 = k_2[N_2O_2][O_2]$
Reverse: $NO_2 + NO_2 \rightarrow O_2 + N_2O_2$ Rate of formation of $N_2O_2 = k_2'[NO_2]^2$

As may be verified, the sum of the two forward reactions is the observed overall reaction, and the sum of the two reverse reactions is the reverse of the overall reaction.

A Note on Good Practice: Use k_r (or k_1, k_2, ...) for a forward reaction and k_r' (or k_1', k_2', ...) for the accompanying reverse reaction.

The rate law of an elementary unimolecular reaction is first-order; that of a bimolecular elementary reaction is second-order.

7C.3 Combining Elementary Rate Laws

To evaluate the plausibility of the proposed mechanism for the gas-phase oxidation of NO to NO_2, the overall rate law implied by the rate laws for the elementary reactions must be constructed and compared with experiment. At this stage it is normally necessary to introduce one or more assumptions about the relative rates of the elementary reactions. These assumptions are often guided by additional experimental information. For instance, experiments show that the rate of the reaction $NO_2 + NO_2 \rightarrow O_2 + N_2O_2$ is so slow that its contribution to the overall rate law can be ignored.

The first step is to identify all elementary reactions that result in the formation or consumption of the overall product (NO_2) and write the equation for its net rate of formation. The **net rate of formation** of a species is the sum of the rates of all elementary reactions resulting in its formation minus the rates of any elementary reactions that lead to its consumption. In this case, NO_2 is formed in the forward reaction in step 2; its reverse reaction is so slow that it can be ignored. Therefore,

$$\text{Rate of formation of } NO_2 = 2k_2[N_2O_2][O_2]$$

The factor of 2 appears in the rate law because two NO_2 molecules are formed from each N_2O_2 molecule consumed. This expression is not yet an acceptable rate law for the overall reaction because it includes the concentration of an intermediate, N_2O_2, and intermediates do not appear in the overall rate law for a reaction.

According to the mechanism, N_2O_2 is formed in the forward reaction in step 1, removed in the reverse reaction, and removed in the forward reaction of step 2. Therefore its net rate of formation is

$$\text{Net rate of formation of } N_2O_2 = \overbrace{k_1[NO]^2}^{\text{Step 1}} - \overbrace{k_1'[N_2O_2]}^{\text{Step 1, reverse}} - \overbrace{k_2[N_2O_2][O_2]}^{\text{Step 2}}$$

To proceed, in the **steady-state approximation,** the assumption is made that the concentrations of the intermediates remain low and do not change significantly in the course of the reaction. The justification for this approximation is that the intermediate is so reactive that it reacts as soon as it is formed. Because the concentration of the intermediate is constant, its net rate of formation is zero, and this equation becomes

$$k_1[NO]^2 - k_1'[N_2O_2] - k_2[N_2O_2][O_2] = 0$$

This equation can be rearranged to find the concentration of N_2O_2:

$$[N_2O_2] = \frac{k_1[NO]^2}{k_1' + k_2[O_2]}$$

which is then substituted into the rate law:

$$\text{Rate of formation of } NO_2 = 2k_2[N_2O_2][O_2] = \frac{2k_1k_2[NO]^2[O_2]}{k_1' + k_2[O_2]} \tag{1}$$

This overall rate law is not the same as the experimental one (Rate of formation of $NO_2 = k_r[NO]^2[O_2]$). It has been stressed that a reaction mechanism is plausible only if its predictions are in line with experimental results; so should the proposed mechanism be discarded? Before doing so, it is always wise to explore whether under certain conditions the predictions do in fact agree with experimental data. In this case, the rate of the forward reaction in step 2 is very slow relative to the reverse reaction of step 1, so $k_1'[N_2O_2] \gg k_2[N_2O_2][O_2]$, which implies that $k_1' \gg k_2[O_2]$ when $[N_2O_2]$ is canceled on each side of the inequality. The term $k_2[O_2]$ in the denominator of Eq. 1 can now be ignored and Eq. 1 simplifies to

$$\text{Rate of formation of } NO_2 = \frac{2k_1k_2}{k_1'}[NO]^2[O_2] \tag{2a}$$

which agrees with the experimentally determined rate law with

$$k_r = \frac{2k_1k_2}{k_1'} \tag{2b}$$

With these assumptions about the relative rates of the elementary steps, the proposed mechanism is consistent with experiment. A further check would be to determine the experimental values of k_1, k_2, and k_1' (if that is possible) and to verify that their combination is consistent with the experimental value of k_r.

EXAMPLE 7C.1   Setting up an overall rate law from a proposed mechanism

The decomposition of ozone in the stratosphere is an issue of great concern because stratospheric ozone protects life on Earth. Suppose you are studying the mechanism of ozone decomposition. The following rate law has been determined for the decomposition of ozone discussed at the beginning of this Topic,

$$2\,O_3(g) \longrightarrow 3\,O_2(g) \qquad \text{Rate of decomposition of } O_3 = k_r\frac{[O_3]^2}{[O_2]}$$

The following mechanism has been proposed:

Step 1 $O_3 \rightleftarrows O_2 + O$
Step 2 $O + O_3 \rightleftarrows O_2 + O_2$

Measurements of the rates of the elementary forward reactions show that the slow step is the forward reaction in the second step, the attack of O on O_3. Its reverse, $O_2 + O_2 \rightarrow O + O_3$, is so slow that it can be ignored. Derive the rate law implied by the mechanism, and confirm that it matches the observed rate law.

A Note on Good Practice: Be careful to distinguish the sign $\rightleftharpoons$ (paired regular arrows), which signifies only that both forward and reverse reactions may occur, from the sign $\rightleftharpoons$ (paired single-barbed arrows), which signifies that the reactions are at equilibrium.

PLAN Write the rate laws for the elementary reactions and combine them into the overall rate law. If necessary, use the steady-state approximation for any intermediates.

SOLVE The rate laws for the elementary reactions are

Step 1 Forward: $O_3 \longrightarrow O_2 + O$ Rate of consumption of $O_3 = k_1[O_3]$ (fast)
     Reverse: $O_2 + O \longrightarrow O_3$ Rate of formation of $O_3 = k_1'[O_2][O]$ (fast)
Step 2 Forward: $O + O_3 \longrightarrow O_2 + O_2$ Rate of consumption of $O_3 = k_2[O][O_3]$ (slow)
     Reverse: $O_2 + O_2 \longrightarrow O + O_3$ Rate of formation of $O_3 = k_2'[O_2]^2$ (very slow, ignore)

Write the rate law for the net rate of decomposition of O_3.

$$\text{Net rate of decomposition of } O_3 = k_1[O_3] - k_1'[O_2][O] + k_2[O][O_3]$$

Because O is an intermediate, the O atom concentration must be eliminated from this expression. Therefore, write out the net rate of formation of O atoms and use the steady-state approximation to set that net rate equal to zero.

$$\text{Rate of formation of } O = k_1[O_3] - k_1'[O_2][O] - k_2[O][O_3] = 0$$

This equation rearranges to

$$[O] = \frac{k_1[O_3]}{k_1'[O_2] + k_2[O_3]}$$

Substitution of this expression into the rate law for the net rate of decomposition of ozone gives

$$\text{Net rate of decomposition of } O_3 = k_1[O_3] - \overbrace{\frac{k_1 k_1'[O_2][O_3]}{k_1'[O_2] + k_2[O_3]}}^{k_1'[O_2][O]} + \overbrace{\frac{k_1 k_2[O_3]^2}{k_1'[O_2] + k_2[O_3]}}^{k_2[O][O_3]}$$

Multiply the first term by $(k_1'[O_2] + k_2[O_3])/(k_1'[O_2] + k_2[O_3])$

$$= \frac{k_1 k_1'[O_2][O_3] + k_1 k_2[O_3]^2}{k_1'[O_2] + k_2[O_3]} - \frac{k_1 k_1'[O_2][O_3]}{k_1'[O_2] + k_2[O_3]} + \frac{k_1 k_2[O_3]^2}{k_1'[O_2] + k_2[O_3]}$$

$$= \frac{k_1 k_1'[O_2][O_3] + k_1 k_2[O_3]^2 - k_1 k_1'[O_2][O_3] + k_1 k_2[O_3]^2}{k_1'[O_2] + k_2[O_3]}$$

Canceling the terms in blue gives

$$\text{Net rate of decomposition of } O_3 = \frac{2 k_1 k_2[O_3]^2}{k_1'[O_2] + k_2[O_3]}$$

Because step 2 is slow, $k_2[O][O_3] \ll k_1'[O_2][O]$, or equivalently by canceling the [O], $k_2[O_3] \ll k_1'[O_2]$, this expression simplifies to

$$\text{Net rate of decomposition of } O_3 = \frac{2 k_1 k_2[O_3]^2}{k_1'[O_2]} = k_r \frac{[O_3]^2}{[O_2]} \quad \text{with } k_r = \frac{2 k_1 k_2}{k_1'}$$

EVALUATE The final rate law has exactly the same form as that of the observed rate law. Because the mechanism is consistent with both the balanced chemical equation and the experimental rate law, you can consider it plausible.

Self-test 7C.2A Consider the following mechanism for the formation of a DNA double helix from its strands A and B:

$$A + B \rightleftharpoons \text{unstable helix} \quad (\text{fast, } k_1 \text{ and } k_1' \text{ both large})$$

$$\text{Unstable helix} \longrightarrow \text{stable double helix} \quad (\text{slow, } k_2 \text{ very small})$$

Derive the rate equation for the formation of the stable double helix and express the rate constant of the overall reaction in terms of the rate constants of the individual steps.

[**Answer:** Rate $= k_r[A][B]$, $k_r = k_1 k_2/k_1'$]

Self-test 7C.2B The proposed two-step mechanism for a reaction is $H_2A + B \rightarrow HB^+ + HA^-$ and its reverse, both of which are fast, followed by $HA^- + B \rightarrow HB^+ + A^{2-}$, which is slow. Find the rate law with HA^- treated as the intermediate and write the equation for the overall reaction.

Related Exercises 7C.5–7C.8

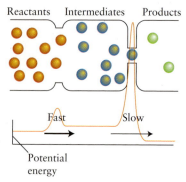

FIGURE 7C.3 The rate-determining step in a reaction is an elementary reaction that governs the rate at which products are formed in a multistep series of reactions. The narrow neck representing the rate-determining step in the drawing is like a ferry that cannot handle the busy traffic arriving on a highway. The superimposed "reaction profile" shows the energy requirements of each step. The step that requires the most energy is the slowest (Topic 7D).

In some cases, there is a quicker way to combine the rate laws for the elementary steps of a mechanism. It may be that the rate at which the intermediate goes on to form product is very slow relative to its rates of formation and decomposition back to reactants. If that is so, you can suppose that the reactants and the intermediate quickly reach their equilibrium concentrations, with the slow loss of intermediate having an insignificant effect on those concentrations. That is, the reactants and the intermediate establish a **pre-equilibrium,** in which the intermediate is formed and sustained in a rapid formation reaction and its reverse. For instance, if it is supposed that the first step in the NO oxidation gives rise to a pre-equilibrium, then the concentrations of the reactants and intermediate are related by an equilibrium constant expressed in terms of concentrations, K_c:

$$NO + NO \rightleftharpoons N_2O_2 \qquad K_c = \frac{[N_2O_2]}{[NO]^2}$$

This equilibrium implies that $[N_2O_2] = K_c[NO]^2$. The rate of formation of NO_2 (from the beginning of Section 7C.3) is therefore

$$\text{Rate of formation of } NO_2 = 2k_2[N_2O_2][O_2] = 2k_2K_c[NO]^2[O_2]$$

A Note on Good Practice: As stated in Topic 5G, equilibrium constants are unitless. To keep the units consistent, the equilibrium expression should formally be written as $K_c = ([N_2O_2]/c^\circ)/([NO]/c^\circ)^2$, with $c^\circ = 1 \text{ mol·L}^{-1}$. Then, $[N_2O_2] = K_c[NO]^2/c^\circ$ and the rate of formation of NO_2 becomes $2k_2[N_2O_2][O_2] = (2k_2K_c/c^\circ)[NO]^2[O_2]$. The rate of formation of NO_2 then has the proper units of $\text{mol·L}^{-1}\text{·s}^{-1}$. Here, c° is omitted for clarity.

As before, the mechanism gives rise to an overall third-order rate law, in agreement with experiment, with $k_r = k_2K_c$. Although this procedure is much simpler than the steady-state approach, it is less flexible: it is more difficult to extend to more complex mechanisms, and the conditions under which the approximation is valid are harder to establish.

The slowest elementary step in a sequence of reactions that governs the overall rate of formation of products—in the reaction between O_2 and the intermediate N_2O_2 in step 2—is called the **rate-determining step** of the reaction (**FIG. 7C.3**). A rate-determining step is like a slow ferry on the route between two cities. The rate at which the traffic arrives at its destination is governed by the rate at which the vehicles are ferried across the river, because that part of the journey is much slower than any of the others. Steps that follow the rate-determining step take place as soon as the intermediate has been formed and have a negligible effect on the overall rate. Therefore, these steps can be ignored when writing the overall rate law from the mechanism. In some reactions there may be more than one possible route to products (Topic 7E). In that case, the path with the lowest energy requirements governs the overall rate of reaction (**FIG. 7C.4**).

In simple cases, the rate-determining step can be identified by considering the experimental rate law. For example, the reaction

$$(CH_3)_3 CBr(sol) + OH^-(sol) \longrightarrow (CH_3)_3 COH(sol) + Br^-(sol)$$

takes place in an organic solvent ("sol") with the rate law Rate of formation of products = $k_r[(CH_3)_3CBr]$. An acceptable two-step mechanism for the reaction might be

$$(CH_3)_3CBr \longrightarrow (CH_3)_3C^+ + Br^-$$
$$(CH_3)_3C^+ + OH^- \longrightarrow (CH_3)_3COH$$

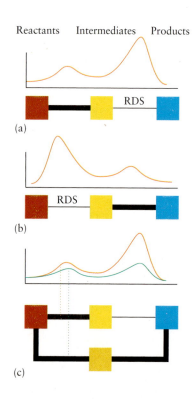

(a)

(b)

(c)

FIGURE 7C.4 (a) If the rate-determining step (RDS) is the second step, the rate law for that step determines the rate law for the overall reaction. The orange curve shows the "reaction profile" for such a mechanism, with a lot of energy required for the slow step. The rate law derived from such a mechanism takes into account steps that precede the RDS. (b) If the rate-determining step is the first step, the rate law for that step must match the rate law for the overall reaction. Later steps do not affect the rate or the rate law. (c) If two routes to the product are possible, the faster one (in this case, the lower one) determines the rate of the reaction; in the mechanism forming the upper route, the slow step (thinner line) is not an RDS.

In this mechanism, the rate law for the first step is Rate = $k_r[(CH_3)_3CBr]$, the same as the experimental rate law. Because the reaction is zeroth order in OH^-, the hydroxide ion concentration does not affect the rate and the first step can be assumed to be the slow step. As soon as $(CH_3)_3C^+$ is formed, it is immediately converted into the alcohol, $(CH_3)_3COH$.

A rate law is often derived from a proposed mechanism by imposing the steady-state approximation or assuming that there is a pre-equilibrium. To be plausible, a proposed mechanism must be consistent with the experimental rate law and the chemical equation for the overall reaction.

7C.4 Rates and Equilibrium

At equilibrium, the rates of the forward and reverse overall reactions and the rates of each pair of individual forward and reverse elementary reactions in a given step of the mechanism are equal. Because the rates depend on rate constants and concentrations, it should be possible to find a relation between rate constants for elementary reactions and the equilibrium constant for the overall reaction.

How Is That Done?

To deduce the relation between rate constants and equilibrium constants, note that the equilibrium constant for a chemical reaction in solution that has the form $A + B \rightleftharpoons C + D$ is

$$K = \frac{[C][D]}{[A][B]}$$

Suppose that experiments show that both the forward reaction and the reverse reaction are elementary second-order reactions. In that case their rate laws are

$$A + B \longrightarrow C + D \qquad Rate = k_r[A][B]$$
$$C + D \longrightarrow A + B \qquad Rate = k_r'[C][D]$$

At equilibrium, these two rates are equal, and so

$$k_r[A][B] = k_r'[C][D]$$

It follows that, at equilibrium,

$$\frac{[C][D]}{[A][B]} = \frac{k_r}{k_r'}$$

Comparison of this expression with the expression for the equilibrium constant then shows that

$$K = \frac{k_r}{k_r'}$$

If the reaction involves gas-phase species and the rate law is expressed in terms of concentrations, use K_c instead of K.

The calculation shows that for a reaction $A + B \rightleftharpoons C + D$ that proceeds in a single bimolecular step in each direction the equilibrium constant for the overall reaction is related to the rate constants of the forward and reverse elementary reactions by

$$K = \frac{k_r}{k_r'} \qquad (3)$$

This conclusion provides another way of deciding (apart from considerations of the Gibbs free energy of the reaction), when to expect a large equilibrium constant: $K \gg 1$ (and products are favored) when k_r for the forward direction is much larger than k_r' for the reverse direction. In this case, the fast forward reaction builds up a high concentration of products before reaching equilibrium (FIG. 7C.5). In contrast, $K \ll 1$ (and reactants are favored) when k_r is much smaller than k_r'. Now the reverse reaction consumes the products rapidly, and so their concentrations are very low.

Equation 3 applies when the reaction takes place by a single step in each direction. If a reaction has a complex mechanism in which the elementary reactions have rate

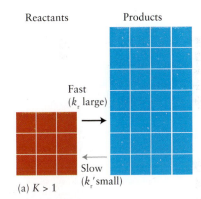

Reactants · Products

Fast (k_r large)

Slow (k_r' small)

(a) $K > 1$

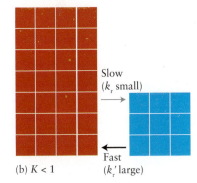

Slow (k_r small)

Fast (k_r' large)

(b) $K < 1$

FIGURE 7C.5 The equilibrium constant for a reaction is equal to the ratio of the rate constants for the forward and reverse reactions. (a) When the forward rate constant (k_r) is large compared to the reverse rate constant, the forward rate matches the reverse rate when the reaction has neared completion and the concentration of reactants is low. (b) Conversely, if the reverse rate constant (k_r') is larger than the forward rate constant, then the forward and reverse rates are equal when little reaction has taken place and the concentration of products is low.

constants k_1, k_2, ... and the reverse elementary reactions have rate constants k_1', k_2', ... then, by an argument similar to that for the single-step reaction, the overall equilibrium constant is related to the rate constants of the individual reaction steps as follows:

$$K = \frac{k_1}{k_1'} \times \frac{k_2}{k_2'} \times \cdots \qquad (4)$$

The equilibrium constant for an elementary reaction is equal to the ratio of the forward and reverse rate constants of the reaction or, for multistep reactions, the ratio of the product of the forward rate constants to the product of the reverse rate constants.

7C.5 Chain Reactions

Some reactions, such as some explosions and, more benignly, the polymerization reactions used to make plastics, take place by a "chain reaction." In a **chain reaction,** a highly reactive intermediate reacts to produce another highly reactive intermediate, which reacts to produce another, and so on (**FIG. 7C.6**). In many cases, the reaction intermediate—which in this context is called a **chain carrier**—is a radical (a species with an unpaired electron, Topic 2C), and the reaction is called a "radical chain reaction." In a **radical chain reaction,** one radical reacts with a molecule to produce another radical, that radical goes on to attack another molecule to produce yet another radical, and so on.

The formation of HBr in the reaction

$$H_2(g) + Br_2(g) \longrightarrow 2\,HBr(g)$$

occurs by a chain reaction. The chain carriers are hydrogen atoms (H·) and bromine atoms (Br·). The first step in any chain reaction is **initiation,** the formation of chain carriers from a reactant. Often heat (denoted Δ) or light (denoted $h\nu$) is used to generate the chain carriers:

$$Br_2 \xrightarrow{\Delta \text{ or } h\nu} Br\cdot + Br\cdot$$

Once chain carriers have been formed, the chain can propagate in a reaction in which one carrier reacts with a reactant molecule to produce another carrier. The elementary reactions for the **propagation** of the chain are

$$Br\cdot + H_2 \longrightarrow HBr + H\cdot$$
$$H\cdot + Br_2 \longrightarrow HBr + Br\cdot$$

The chain carriers—radicals here—produced in these reactions can go on to attack other reactant molecules (H$_2$ and Br$_2$), thereby allowing the chain to continue. The elementary reaction that ends the chain, a process called **termination,** takes place when chain carriers combine to form products. Two examples of termination reactions are

$$Br\cdot + Br\cdot \longrightarrow Br_2$$
$$H\cdot + Br\cdot \longrightarrow HBr$$

Explosions can be expected when **chain branching** occurs—that is, when more than one chain carrier is formed in a propagation step. The characteristic pop that occurs when a mixture of hydrogen and oxygen is ignited is a consequence of chain branching. The two gases combine in a radical chain reaction in which the initiation step may be the formation of hydrogen atoms:

$$\text{Initiation: } H_2 \xrightarrow{\Delta} H\cdot + H\cdot$$

After the reaction has been initiated, two new radicals are formed when one hydrogen atom attacks an oxygen molecule:

$$\text{Branching: } H\cdot + O_2 \longrightarrow HO\cdot + \cdot O\cdot$$

Two radicals are also produced when the oxygen atom attacks a hydrogen molecule:

$$\text{Branching: } \cdot O\cdot + H_2 \longrightarrow HO\cdot + H\cdot$$

As a result of these branching processes, the chain produces an increasingly larger number of radicals that can take part in even more branching steps. The reaction rate increases rapidly, and an explosion typical of many combustion reactions might occur (**FIG. 7C.7**).

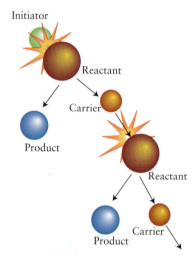

FIGURE 7C.6 In a chain reaction, the product of one step in a reaction is a highly reactive reactant in a subsequent step, which in turn produces reactive species that can take part in other reaction steps.

FIGURE 7C.7 This flame front was caught during the fast combustion—almost a miniature explosion—that occurs inside an internal combustion engine every time a spark plug ignites the mixture of fuel and air. (*W. H. Freeman photo by Ken Karp.*)

Chain reactions begin with the initiation of a reactive intermediate that propagates the chain and terminate when radicals combine. Branching chain reactions can be explosively fast.

What have you learned in this Topic?

You have learned that chemical reactions occur in a series of one or more elementary steps and that together these steps form a proposed mechanism for a reaction. You have seen that to verify the plausibility of a proposed mechanism, you must write the overall rate law for the elementary steps proposed and confirm that it is consistent with the experimentally determined rate law. You have also seen that the equilibrium constant is equal to the ratio of the forward and reverse rate constants of reaction. Last, you have learned that some reactions produce highly reactive intermediates that react to form additional reactive intermediates in chain reactions.

The skills you have mastered are the ability to:

☐ **1.** Identify the molecularity of elementary reactions in a proposed mechanism (Section 7C.1).

☐ **2.** Deduce the overall rate law given a series of elementary reactions in a proposed mechanism (Example 7C.1).

☐ **3.** Describe and use the steady-state and pre-equilibrium approximations (Section 7C.3).

☐ **4.** Identify the rate-determining step of a reaction (Section 7C.3).

☐ **5.** Show how the equilibrium constant is related to the forward and reverse rate constants of the elementary reactions contributing to an overall reaction (Section 7C.4).

☐ **6.** Identify initiation, propagation, and termination steps in a radical chain reaction mechanism (Section 7C.5).

Topic 7C Exercises

Exercises labeled require calculus.

All rates are unique reaction rates unless otherwise stated.

7C.1 Each of the following is an elementary reaction. Write its rate law and state its molecularity. (a) $NO + NO \rightarrow N_2O_2$; (b) $Cl_2 \rightarrow Cl + Cl$.

7C.2 Each of the following is an elementary reaction. Write its rate law and state its molecularity. (a) $OH + NO_2 + N_2 \rightarrow HNO_3 + N_2$ (N_2 takes part in the collision, but is unchanged chemically); (b) $ClO^- + H_2O \rightarrow HClO + OH^-$.

7C.3 Write the overall reaction for the mechanism proposed below and identify any reaction intermediates.

Step 1 $AC + B \longrightarrow AB + C$
Step 2 $AC + AB \longrightarrow A_2B + C$

7C.4 Write the overall reaction for the mechanism proposed below and identify any reaction intermediates.

Step 1 $C_4H_9Br \longrightarrow C_4H_9^+ + Br^-$
Step 2 $C_4H_9^+ + H_2O \longrightarrow C_4H_9OH_2^+$
Step 3 $C_4H_9OH_2^+ + H_2O \longrightarrow C_4H_9OH + H_3O^+$

7C.5 The following mechanism has been proposed for the gas-phase reaction between HBr and NO_2:

Step 1 $HBr + NO_2 \longrightarrow HBrO + NO$ (slow)
Step 2 $HBr + HBrO \longrightarrow H_2O + Br_2$ (fast)

(a) Write the overall reaction. (b) Write the rate law for each step and indicate its molecularity. (c) What is the reaction intermediate?

7C.6 A reaction was believed to occur by the following mechanism.

Step 1 $A_2 \longrightarrow A + A$
Step 2 $A + A + B \longrightarrow A_2B$
Step 3 $A_2B + C \longrightarrow A_2 + BC$

(a) Write the overall reaction. (b) Write the rate law for each step and indicate its molecularity. (c) What are the reaction intermediates? (d) A catalyst is a substance that accelerates the rate of a reaction and is regenerated in the process. Which substance is functioning as a catalyst in the reaction?

7C.7 The following mechanism has been proposed for the reaction between nitric oxide and bromine:

Step 1 $NO + Br_2 \longrightarrow NOBr_2$ (slow)
Step 2 $NOBr_2 + NO \longrightarrow NOBr + NOBr$ (fast)

Write the rate law for the formation of NOBr implied by this mechanism.

7C.8 The mechanism proposed for the oxidation of iodide ion by the hypochlorite ion in aqueous solution is as follows:

Step 1 $ClO^- + H_2O \rightleftharpoons HClO + OH^-$ and its reverse (both fast, equilibrium)
Step 2 $I^- + HClO \longrightarrow HIO + Cl^-$ (slow)
Step 3 $HIO + OH^- \longrightarrow IO^- + H_2O$ (fast)

Write the rate law for the formation of HIO implied by this mechanism.

7C.9 Three mechanisms for the reaction $NO_2(g) + CO(g) \rightarrow CO_2(g) + NO(g)$ have been proposed:

(I)   **Step 1** $NO_2 + CO \longrightarrow CO_2 + NO$

(II) **Step 1** $NO_2 + NO_2 \longrightarrow NO + NO_3$ (slow)
      Step 2 $NO_3 + CO \longrightarrow NO_2 + CO_2$ (fast)

(III) **Step 1** $NO_2 + NO_2 \rightleftharpoons NO + NO_3$ and its reverse
      (both fast, equilibrium)
      Step 2 $NO_3 + CO \longrightarrow NO_2 + CO_2$ (slow)

Which mechanism agrees with the following rate law: Rate $= k_r[NO_2]^2$? Explain your reasoning.

7C.10 When the rate of the reaction $2\,NO(g) + O_2(g) \rightarrow 2\,NO_2(g)$ was studied, the rate was found to double when the O_2 concentration alone was doubled but to quadruple when the NO concentration alone was doubled. Which of the following mechanisms accounts for these observations? Explain your reasoning.

(I)   **Step 1** $NO + O_2 \rightleftharpoons NO_3$ and its reverse
      (both fast, equilibrium)
      Step 2 $NO + NO_3 \longrightarrow NO_2 + NO_2$ (slow)

(II) **Step 1** $NO + NO \longrightarrow N_2O_2$ (slow)
      Step 2 $O_2 + N_2O_2 \longrightarrow N_2O_4$ (fast)
      Step 3 $N_2O_4 \longrightarrow NO_2 + NO_2$ (fast)

7C.11 Indicate whether each of the following statements is true or false. If a statement is false, explain why. (a) For a reaction with a very large equilibrium constant, the rate constant of the forward reaction is much larger than the rate constant of the reverse reaction. (b) At equilibrium, the rate constants of the forward and reverse reactions are equal. (c) Increasing the concentration of a reactant increases the rate of a reaction by increasing the rate constant in the forward direction.

7C.12 Indicate whether each of the following statements is true or false. If a statement is false, explain why. (a) The equilibrium constant for a reaction equals the rate constant for the forward reaction divided by the rate constant for the reverse reaction. (b) In a reaction that is a series of equilibrium steps, the overall equilibrium constant is equal to the product of all the forward rate constants divided by the product of all the reverse rate constants. (c) Increasing the concentration of a product increases the rate of the reverse reaction, and so the rate of the forward reaction must then increase, too.

7C.13 Consider the reaction $A \rightleftharpoons B$, which is first order in each direction with rate constants k_r and k_r'. Initially, only A is present. Show that the concentrations approach their equilibrium values at a rate that depends on k_r and k_r'.

7C.14 Repeat Exercise 7C.13 for the same reaction taken to be second order in each direction.

Topic 7D Models of Reactions

7D.1 The Effect of Temperature

7D.2 Collision Theory

7D.3 Transition State Theory

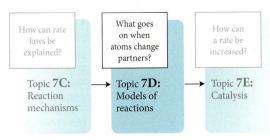

How can rate laws be explained?

What goes on when atoms change partners?

How can a rate be increased?

Topic 7C: Reaction mechanisms → Topic 7D: Models of reactions → Topic 7E: Catalysis

Rate laws and rate constants are windows onto the molecular processes of chemical change. Although a reaction mechanism might have been established experimentally (Topic 7C), there remain the questions of why the rate constants for the individual steps have the values they do and why they vary with temperature.

7D.1 The Effect of Temperature

The rates of chemical reactions depend on temperature. The *qualitative* observation is that most reactions go faster as the temperature is raised (**FIG. 7D.1**). An increase of 10 °C from room temperature typically doubles the rate of reaction of organic species in solution. That is one reason why foods are cooked: heating accelerates reactions that lead to the breakdown of cell walls and the decomposition of proteins. Foods are refrigerated to slow down the natural chemical reactions that lead to their decomposition.

The temperature dependence of the rate constant summarizes how a reaction rate varies with temperature. If the rate constant increases with temperature, then for a given concentration of reactants the reaction goes faster at the higher temperature. In the late nineteenth century, the Swedish chemist Svante Arrhenius examined quantitatively the effect of temperature on rates of reactions and found that the plot of the logarithm of the rate constant ($\ln k_r$) against the inverse of the absolute temperature ($1/T$) is a straight line. In other words, he established that

$$\ln k_r = \text{intercept} + \text{slope} \times \frac{1}{T}$$

The intercept is denoted $\ln A$ and, for reasons that will become clear, the slope is denoted $-E_a/R$, where R is the gas constant. With this notation, the empirical **Arrhenius equation** is

$$\ln k_r = \ln A - \frac{E_a}{RT} \tag{1a}$$

An alternative form of this expression, which is obtained by taking antilogarithms of both sides, is

$$k_r = Ae^{-E_a/RT} \tag{1b}$$

The two constants, A and E_a, are known as the **Arrhenius parameters** for the reaction and are found from experiment; A is called the **pre-exponential factor,** and E_a is the **activation energy.** Both A and E_a are nearly independent of temperature but have values that depend on the reaction being studied. The Arrhenius equation applies to reactions of any order.

FIGURE 7D.1 Reaction rates almost always increase with temperature. The beaker on the left contains magnesium in cold water and that on the right contains magnesium in hot water. Phenolphthalein indicator has been added to show the formation of an alkaline solution as magnesium reacts. (*W. H. Freeman photo by Ken Karp.*)

Why Do You Need to Know This Material? Chemists use information about the temperature dependence of rate constants to build models of atomic events. Models of reactions can lend insight into the nature of the temperature dependence of not only reaction rate constants but also of equilibrium constants.

What Do You Need to Know Already? You should be familiar with the kinetic molecular theory of gases (Topic 3D), rate laws (Topic 7A), and the relationship of rate constants to equilibrium constants (Topic 7C).

The use of the gas constant does not mean that the Arrhenius equation applies only to gases; it applies to a wide range of reactions in solution, too.

The units of the pre-exponential factor are the same as those of the rate constant. Thus the logarithm terms in Eq. 1a are unitless because the expression can be rearranged into the form $\ln(k_r/A) = -E_a/RT$ and the units cancel.

621

EXAMPLE 7D.1   Measuring an activation energy

An important reaction in organic chemistry is that of organic halides with hydroxide ion to form alcohols. You are interested in learning about the impact of temperature on the rate of this kind of reaction, so you decide to determine the Arrhenius parameters for one such reaction. The rate constant for the second-order reaction between bromoethane and hydroxide ions in water, $C_2H_5Br(aq) + OH^-(aq) \rightarrow C_2H_5OH(aq) + Br^-(aq)$, was measured at several temperatures, with the results shown here:

Temperature/°C	25	30.	35	40.	45	50.
Rate constant, $k_r/(L \cdot mol^{-1} \cdot s^{-1})$	8.8×10^{-5}	1.6×10^{-4}	2.8×10^{-4}	5.0×10^{-4}	8.5×10^{-4}	1.4×10^{-3}

Determine the activation energy of the reaction.

ANTICIPATE Activation energies of organic reactions are typically between 10 and 100 $kJ \cdot mol^{-1}$, so you should expect a value within that range.

PLAN Activation energies are found from the Arrhenius equation (Eq. 1). Plot $\ln k_r$ against $1/T$, with T in kelvins. Because the slope is equal to $-E_a/R$, to find the activation energy multiply the slope of the graph by $-R$ with $R = 8.3145 \ J \cdot K^{-1} \cdot mol^{-1}$. A spreadsheet, curve-fitting program, or graphing calculator is very useful for this type of calculation.

SOLVE The table to use for drawing the graph is

Temperature/°C	Temperature, T/K	$1/(T/K)$	Rate constant, $k_r/(L \cdot mol^{-1} \cdot s^{-1})$	$\ln k_r$
25	298	3.35×10^{-3}	8.8×10^{-5}	-9.34
30.	303	3.30×10^{-3}	1.6×10^{-4}	-8.74
35	308	3.25×10^{-3}	2.8×10^{-4}	-8.18
40.	313	3.19×10^{-3}	5.0×10^{-4}	-7.60
45	318	3.14×10^{-3}	8.5×10^{-4}	-7.07
50.	323	3.10×10^{-3}	1.4×10^{-3}	-6.57

Calculate the slope of the data plotted in **FIG. 7D.2** from two points on the graph or by using the Living Graph on the textbook Web site or other suitable linear regression tool.

$$\text{Slope} = -1.1 \times 10^4 \ K$$

Because the slope is equal to $-E_a/R$, from $E_a = -R \times \text{slope}$,

$$E_a = -(8.3145 \ J \cdot K^{-1} \cdot mol^{-1}) \times (-1.1 \times 10^4 \ K)$$
$$= 8.9 \times 10^4 \ J \cdot mol^{-1}$$

EVALUATE The calculated value corresponds to 89 $kJ \cdot mol^{-1}$, which is within the anticipated range.

Self-test 7D.1A The rate constant for the second-order gas-phase reaction $HO(g) + H_2(g) \rightarrow H_2O(g) + H(g)$ varies with temperature as shown here:

Temperature/°C	100.	200.	300.	400.
Rate constant, $k/(L \cdot mol^{-1} \cdot s^{-1})$	1.1×10^{-9}	1.8×10^{-8}	1.2×10^{-7}	4.4×10^{-7}

Determine the activation energy.

[**Answer:** 42 $kJ \cdot mol^{-1}$]

Self-test 7D.1B The rate of a reaction increased from 3.00 $mol \cdot L^{-1} \cdot s^{-1}$ to 4.35 $mol \cdot L^{-1} \cdot s^{-1}$ when the temperature was raised from 18 °C to 30. °C. What is the activation energy of the reaction?

Related Exercises 7D.3, 7D.4

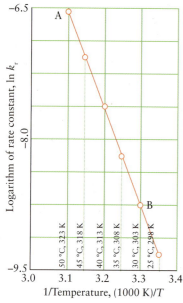

FIGURE 7D.2 An Arrhenius plot is a graph of $\ln k_r$ against $1/T$. If, as here, the line is straight, then the reaction is said to show Arrhenius behavior in the temperature range studied. This plot has been constructed from the data in Example 7D.1.

Reactions that give a straight line when $\ln k_r$ is plotted against $1/T$ are said to show **Arrhenius behavior.** A wide range of reactions show Arrhenius behavior. Even fireflies flash more quickly on hot nights than on cold nights, and their rate of flashing is

TABLE 7D.1 Arrhenius Parameters

Reaction	A	$E_a/(\text{kJ·mol}^{-1})$
First order, gas phase		
cyclopropane $\longrightarrow$ propene	$1.6 \times 10^{15}\ \text{s}^{-1}$	272
$CH_3NC \longrightarrow CH_3CN$	$4.0 \times 10^{13}\ \text{s}^{-1}$	160
$C_2H_6 \longrightarrow 2\ CH_3$	$2.5 \times 10^{17}\ \text{s}^{-1}$	384
$N_2O \longrightarrow N_2 + O$	$8.0 \times 10^{11}\ \text{s}^{-1}$	250
$2\ N_2O_5 \longrightarrow 4\ NO_2 + O_2$	$4.0 \times 10^{13}\ \text{s}^{-1}$	103
Second order, gas phase		
$O + N_2 \longrightarrow NO + N$	$1 \times 10^{11}\ \text{L·mol}^{-1}\text{·s}^{-1}$	315
$OH + H_2 \longrightarrow H_2O + H$	$8 \times 10^{10}\ \text{L·mol}^{-1}\text{·s}^{-1}$	42
$2\ CH_3 \longrightarrow C_2H_6$	$2 \times 10^{10}\ \text{L·mol}^{-1}\text{·s}^{-1}$	0
Second order, in aqueous solution		
$C_2H_5Br + OH^- \longrightarrow C_2H_5OH + Br^-$	$4.3 \times 10^{11}\ \text{L·mol}^{-1}\text{·s}^{-1}$	90
$CO_2 + OH^- \longrightarrow HCO_3^-$	$1.5 \times 10^{10}\ \text{L·mol}^{-1}\text{·s}^{-1}$	38
$C_{12}H_{22}O_{11} + H_2O \longrightarrow 2\ C_6H_{12}O_6$	$1.5 \times 10^{15}\ \text{L·mol}^{-1}\text{·s}^{-1}$	108

Arrhenius-like over a narrow range of temperatures. This observation suggests that the biochemical reactions responsible for the flashing have rate constants that increase with temperature in accord with Eq. 1. Some Arrhenius parameters are listed in **TABLE 7D.1**.

Because the slope of an Arrhenius plot is proportional to E_a, it follows that *the higher the activation energy, the stronger is the temperature dependence of the rate constant*. Reactions with low activation energies (about 10 kJ·mol^{-1}, with not very steep Arrhenius plots) have rates that increase only slightly with temperature. Reactions with high activation energies (above approximately 60 kJ·mol^{-1}, with steep Arrhenius plots) have rates that depend strongly on the temperature (**FIG. 7D.3**).

The Arrhenius equation is used to predict the value of a rate constant at one temperature from its value at another temperature.

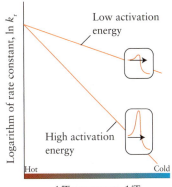

FIGURE 7D.3 The dependence of the rate constant on temperature for two reactions with different activation energies. The higher the activation energy, the more strongly the rate constant depends on temperature. The small insets show the activation barriers for the two reactions.

--- **How Is That Done?** ---

The Arrhenius equations for the two temperatures, which are denoted T_1 and T_2, when the rate constant for the reaction has the values k_{r1} and k_{r2}, respectively, are

$$\text{At temperature } T_1: \ln k_{r1} = \ln A - \frac{E_a}{RT_1}$$

$$\text{At temperature } T_2: \ln k_{r2} = \ln A - \frac{E_a}{RT_2}$$

Eliminate $\ln A$ by subtracting the first equation from the second:

$$\underbrace{\ln k_{r2} - \ln k_{r1}}_{\ln(k_{r2}/k_{r1})} = -\frac{E_a}{RT_2} + \frac{E_a}{RT_1}$$

The expression that has been derived can be rearranged into

$$\ln \frac{k_{r2}}{k_{r1}} = \frac{E_a}{R}\left(\frac{1}{T_1} - \frac{1}{T_2}\right) \tag{2}$$

This expression has the following interpretation:

- When $T_2 > T_1$, the right-hand side is positive, so $\ln(k_{r2}/k_{r1})$ is positive, which means that $k_{r2} > k_{r1}$. That is, the rate constant increases with temperature.
- For fixed values of T_1 and T_2, $\ln(k_{r2}/k_{r1})$ is large when E_a is large; so the increase in rate constant is large for reactions with a high activation energy.

EXAMPLE 7D.2   Using the activation energy to predict a rate constant

The hydrolysis of sucrose is a part of the digestive process. Suppose you want to investigate how strongly the rate depends on body temperature and decide to calculate the rate constant for the hydrolysis of sucrose at 35.0 °C, given that $k_r = 1.0$ mL·mol^{-1}·s^{-1} at 37.0 °C (normal body temperature) and that the activation energy of the reaction is 108 kJ·mol^{-1}. What is the rate constant for the hydrolysis of sucrose at 35.0 °C?

ANTICIPATE You should expect a smaller rate constant at the lower temperature.

PLAN Use Eq. 2 with $T_1 = 310.0$ K and $T_2 = 308.0$ K and express R in kilojoules per kelvin per mole (to cancel the units of E_a).

SOLVE

From $\ln(k_{r2}/k_{r1}) = (E_a/R)(1/T_1 - 1/T_2)$,

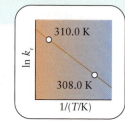

$$\ln\frac{k_{r2}}{k_{r1}} = \frac{108 \text{ kJ·mol}^{-1}}{8.3145 \times 10^{-3} \text{ kJ·K}^{-1}\text{·mol}^{-1}} \times \left(\frac{1}{310.0 \text{ K}} - \frac{1}{308.0 \text{ K}}\right)$$

$$= -0.27\ldots$$

From $x = e^{\ln x}$,

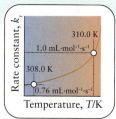

$$\frac{k_{r2}}{k_{r1}} = e^{-0.27\ldots}$$

Finally, because $k_{r1} = 1.0$ mL·mol^{-1}·s^{-1},

$$k_{r2} = (1.0 \text{ mL·mol}^{-1}\text{·s}^{-1}) \times e^{-0.27\ldots} = 0.76 \text{ mL·mol}^{-1}\text{·s}^{-1} \text{ at } 35 \text{ °C}$$

EVALUATE As expected, the rate constant at the lower temperature is lower. The high activation energy of the reaction means that its rate is very sensitive to temperature.

Self-test 7D.2A The rate constant for the second-order reaction between CH_3CH_2Br and OH^- in water is 0.28 mL·mol^{-1}·s^{-1} at 35.0 °C. What is the value of the rate constant at 50.0 °C? See Table 7D.1 for data.

[*Answer:* 1.4 mL·mol^{-1}·s^{-1}]

Self-test 7D.2B The rate constant for the first-order isomerization of cyclopropane, C_3H_6, to propene, $CH_3CH\!=\!CH_2$, is 6.7×10^{-4} s^{-1} at 500. °C. What is its value at 300. °C? See Table 7D.1 for data.

Related Exercises 7D.5, 7D.6

An Arrhenius plot of ln k_r against 1/T is used to determine the Arrhenius parameters of a reaction; a high activation energy signifies a high sensitivity of the rate constant to changes in temperature.

7D.2 Collision Theory

Any model of how reactions take place at a molecular level must account for the temperature dependence of rate constants, as expressed by the Arrhenius equation; it should also reveal the significance of the Arrhenius parameters A and E_a. Reactions in the gas phase are conceptually simpler than those in solution, and so they are considered first.

First, assume that a reaction can take place only if reactant molecules meet. In a gas, that meeting is a "collision," and so a model based on that picture is called the **collision theory** of reactions. In this model, molecules are considered to behave like defective billiard balls: they bounce apart if they collide at low speed, but they might smash into pieces when the impact is more energetic. Likewise, if two molecules collide with less than a certain kinetic energy, they simply bounce apart. If they meet with more than that kinetic energy, reactant bonds can break and new bonds can form, resulting in products (**FIG. 7D.4**). The minimum kinetic energy needed for reaction is denoted E_{min}.

To set up a quantitative theory based on this qualitative picture, it is necessary to know the frequency at which molecules collide and the fraction of those collisions that have at least the energy E_{min} required for reaction. The **collision frequency**, the number of collisions per second, between A and B molecules in a gas at a temperature T can be calculated from the kinetic model of a gas (Topic 3D):

$$\text{Collision frequency} = \sigma \bar{v}_{rel} N_A^2 [A][B] \qquad (3)$$

in which N_A is Avogadro's constant and σ (sigma) is the **collision cross section**, the area that a molecule presents as a target during a collision. The bigger the collision cross section, the greater is the collision frequency, because big molecules are easier targets than small molecules. The quantity $\bar{v}_{rel}$ is the **mean relative speed**, the mean speed at which the molecules approach each other in a gas. The mean relative speed is calculated by multiplying each possible speed by the fraction of molecules that have that speed and then adding all the products together. When the temperature is T and the molar masses are M_A and M_B, this mean relative speed turns out to be

$$\bar{v}_{rel} = \left(\frac{8RT}{\pi M}\right)^{1/2} \qquad M = \frac{M_A M_B}{M_A + M_B} \qquad (4)$$

The collision frequency is greater the higher the relative speeds of the molecules, and therefore the higher the temperature.

Although the mean relative speed of the molecules increases with temperature, and the collision frequency therefore increases too, Eq. 4 shows that it increases only as the square root of the temperature. This dependence is far too weak to account for observations. If you used Eq. 4 to predict the temperature dependence of reaction rates, you would conclude that an increase in temperature of 10 °C at about room temperature (from 273 K to 283 K) increases the collision frequency by a factor of only 1.02, whereas experiments show that many reaction rates double over that range. Therefore, another factor must be affecting the rate.

That factor is the fraction of molecules that collide with a kinetic energy equal to or greater than a certain minimum energy, E_{min}, for only these energetic collisions can result in reaction. Because kinetic energy is proportional to the square of the speed, this fraction can be obtained from the Maxwell distribution of speeds (Topic 3D). **FIGURE 7D.5** shows the kind of result expected. As indicated for a specific reaction by the shaded area under the blue curve, at a low temperature very few molecules have enough kinetic energy to react. At higher temperatures, a much larger fraction of molecules can react, as represented by the shaded area under the red curve. This fraction must be incorporated in the model.

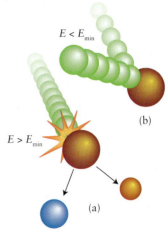

FIGURE 7D.4 (a) In the collision theory of chemical reactions, reaction may take place only when two molecules collide with a kinetic energy at least equal to a minimum value, E_{min} (which later will be identified with the activation energy). (b) Otherwise, they simply bounce apart.

How Is That Done?

To find how the fraction of molecules that collides with at least an energy ε_{min} affects the rate of an elementary reaction, it is necessary to know that, at a temperature T, the fraction of collisions with at least the energy ε_{min} is equal to $e^{-\varepsilon_{min}/kT}$, where k is Boltzmann's constant. This factor is normally expressed in terms of the molar energy $E_{min} = N_A \varepsilon_{min}$ and the gas constant $R = N_A k$, as $e^{-E_{min}/RT}$. That result comes from an expression known as the *Boltzmann distribution*, which is not derived here. The rate of reaction is the product of this factor and the collision frequency:

Rate of reaction = collision frequency × fraction with sufficient energy

$$= \overbrace{\sigma \bar{v}_{rel} N_A^2 [A][B]}^{\substack{\text{collision} \\ \text{frequency}}} \times \overbrace{e^{-E_{min}/RT}}^{\substack{\text{fraction with} \\ \text{energy} \geq E_{min}}}$$

The rate law of an elementary reaction that depends on collisions of A with B is Rate = $k_r[A][B]$, where k_r is the rate constant. Therefore, the expression for the rate constant is

$$k_r = \frac{\text{rate of reaction}}{[A][B]} = \frac{\sigma \bar{v}_{rel} N_A^2 [A][B] \times e^{-E_{min}/RT}}{[A][B]} = \sigma \bar{v}_{rel} N_A^2 e^{-E_{min}/RT}$$

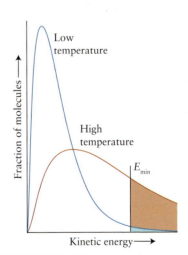

FIGURE 7D.5 The fraction of molecules that collide with a kinetic energy that is at least equal to a certain minimum value, E_{min} (which later is identified with the activation energy, E_a), is given by the shaded areas under each curve. Note that this fraction increases rapidly as the temperature is raised.

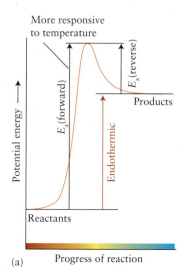

(a)

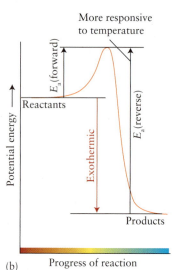

(b)

FIGURE 7D.6 (a) The activation energy for an endothermic reaction is larger in the forward direction than in the reverse, and so the rate of the forward reaction is more sensitive to temperature and the equilibrium shifts toward products as the temperature is raised. (b) The opposite is true for an exothermic reaction, in which case the reverse reaction is more sensitive to temperature and the equilibrium shifts toward reactants as the temperature is raised.

The calculation shows that, according to collision theory, the rate constant is

$$k_r = \sigma \bar{v}_{rel} N_A^2 e^{-E_{min}/RT} \qquad (5)$$

This exponential temperature dependence for k_r is much stronger than the weak temperature dependence of the collision frequency itself. By comparing Eq. 5 with Eq. 1b, the term $\sigma \bar{v}_{rel} N_A^2$ can be identified as the pre-exponential factor, A, and E_{min} can be identified as the activation energy, E_a. That is:

- The pre-exponential factor, A, is a measure of the rate at which molecules collide.
- The activation energy, E_a, is the minimum (molar) kinetic energy required for a collision to result in reaction.

Why some thermodynamically spontaneous reactions do not take place at a measurable rate now becomes clear: they have such high activation energies that hardly any of the collisions between reactants result in reaction. Although the molecules are colliding with one another billions of times a second, a mixture of hydrogen and oxygen can survive for years. The activation energy for the production of radicals is very high, and no radicals are formed until an energizing spark or flame is brought into contact with the mixture.

The dependence of the rate constant on temperature, its sensitivity to the activation energy, and the fact that the equilibrium constant is equal to the ratio of the forward and reverse rate constants (Eq. 3 of Topic 7C, $K_c = k_r/k_r'$), jointly provide a kinetic explanation for the temperature dependence of the equilibrium constant. To see what is involved, a **reaction profile,** a diagram that represents the energy changes that occur during a reaction, is constructed (**FIG. 7D.6**). Then, with that diagram in mind and remembering that a high activation energy indicates a high sensitivity of the rate constant to temperature, the following conclusions can be drawn:

- If the reaction is endothermic in the forward direction, the activation energy is higher for the forward direction than for the reverse direction.

The higher activation energy means that the rate constant of the forward reaction depends more strongly on temperature than does the rate constant of the reverse reaction. Therefore, when the temperature is raised, the rate constant for the forward reaction increases more than that of the reverse reaction. As a result, $K_c = k_r/k_r'$ will increase and formation of products will become favored.

- If the reaction is exothermic in the forward direction, the activation energy is lower for the forward direction than for the reverse direction.

In this case the lower activation energy for the forward reaction means that its rate constant depends less strongly on temperature than does the rate constant of the reverse reaction. Therefore, when the temperature is raised, $K_c = k_r/k_r'$ will decrease and formation of products will become less favored. Both conclusions are in accord with Le Chatelier's principle (Topic 5J).

Whenever a model has been developed, it should always be checked for consistency with experimental results. In the present case, the results of careful experiments show that the collision model of reactions is not complete because the experimental rate constant is normally smaller than that predicted by collision theory. The model can be improved by realizing that the relative *orientation* of the molecules when they collide also might affect the reaction rate. For example, the results of experiments of the kind described in **BOX 7D.1** have shown that, in the gas-phase reaction of chlorine atoms with HI molecules, HI + Cl → HCl + I, the Cl atom reacts with the HI molecule only if it approaches from a favorable direction (**FIG. 7D.7**). A dependence on relative orientation is called the **steric requirement**

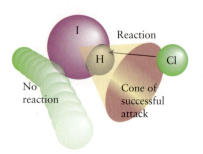

FIGURE 7D.7 Whether a reaction takes place when two species collide in the gas phase depends on their relative orientations. In the reaction between a Cl atom and an HI molecule, for example, only those collisions in which the Cl atom approaches the HI molecule from a direction that lies inside the cone indicated here lead to reaction, even though the energy of collisions in other relative orientations may exceed the activation energy.

Box 7D.1 HOW DO YOU KNOW . . . WHAT HAPPENS DURING A MOLECULAR COLLISION?

When molecules collide with sufficient energy, old bonds give way to new ones and atoms in one molecule become part of another. Is there any way of finding out *experimentally* what is happening at this climactic stage of a reaction?

Molecular beams give the necessary insight. A molecular beam consists of a stream of molecules moving in the same direction with nearly the same speed. A beam may be directed at a gaseous sample or into the path of a second beam, consisting of molecules of a second reactant. The molecules may react when the beams collide; the experimenters can then detect the products of the collision and the direction at which the products emerge from the collision. They also use spectroscopic techniques to determine the vibrational and rotational excitation of the products formed in the collision of the beams.

By repeating the experiment with molecules having different speeds and different states of rotational or vibrational excitation, chemists can learn more about the collision itself. For example, it has been found that, in the reaction between a Cl atom and an HI molecule, the best direction of attack is within a cone of half-angle 30° surrounding the H atom (see Fig. 7D.7).

In a "sticky" collision, the reactant molecules orbit around each other for one revolution or more. As a result, the product molecules emerge in random directions because any "memory" of the approach direction is lost. However, a rotation takes time—about 1 ps. If the reaction is over before that, the product molecules will emerge in a specific direction that depends on the direction of the collision. In the collision of K and I_2, for example, most of the products are thrown off in the forward direction. This observation is consistent with the "harpoon mechanism" that has been proposed for this reaction. In this mechanism, an electron jumps across from the K atom to the I_2 molecule when they are quite far apart, resulting in an I_2^- ion and a K^+ ion. The K^+ ion then extracts one of the I atoms as an I^- ion. The electron acts like a harpoon, the electrostatic attraction being the line attached to the harpoon, and I_2 a whale. Because the crucial "harpooning" occurs at large distances and there is no actual collision, the products are thrown off in approximately the same direction as the reactants were traveling initially.

The reaction between a K atom and a CH_3I molecule takes place by a different mechanism. A collision leads to reaction

only if the two reactants approach each other very closely. In this mechanism, the K atom effectively bumps into a brick wall, and the KI product bounces out in the backward direction.

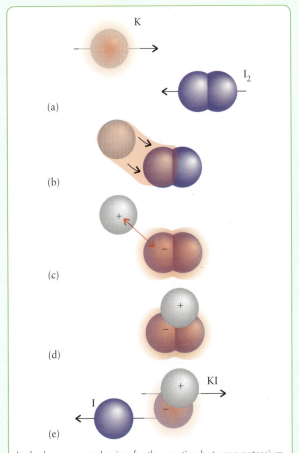

In the harpoon mechanism for the reaction between potassium and iodine to form potassium iodide, as a K atom approaches an I_2 molecule (a), an electron passes from the K atom to the I_2 molecule (b). The charge difference now tethers the two ions together (c and d) until an I^- ion separates and leaves with the K^+ ion (e).

of the reaction. It is normally taken into account by introducing an empirical factor, P, called the **steric factor,** and changing Eq. 5 to

$$k_r = \overbrace{P}^{\substack{\text{steric} \\ \text{requirement}}} \times \overbrace{\sigma \overline{v}_{rel} N_A^2}^{\substack{\text{collision} \\ \text{factor}}} \times \overbrace{e^{-E_{min}/RT}}^{\substack{\text{energy} \\ \text{requirement}}} \qquad (6)$$

TABLE 7D.2 lists some values of P. All the values shown are less than 1 because the steric requirement reduces the probability of reaction. For collisions between complex

TABLE 7D.2 The Steric Factor, *P*

Reaction	P
$NOCl + NOCl \longrightarrow NO + NO + Cl_2$	0.16
$NO_2 + NO_2 \longrightarrow NO + NO + O_2$	5.0×10^{-2}
$ClO + ClO \longrightarrow Cl_2 + O_2$	2.5×10^{-3}
$H_2 + C_2H_4 \longrightarrow C_2H_6$	1.7×10^{-6}

species, the steric requirement may be severe, and P is very small. In such cases, the reaction rate is considerably less than the rate at which high-energy collisions occur.

THINKING POINT

Why might the harpoon mechanism described in Box 7D.1 result in a value of P greater than 1?

According to the collision theory of gas-phase reactions, a reaction takes place only if the reactant molecules collide with a kinetic energy of at least the activation energy, and they do so in the correct relative orientation.

7D.3 Transition State Theory

Transition state theory is also known as *activated complex theory*, ACT.

Although collision theory applies to gas-phase reactions, some of its concepts can be extended to explain why the Arrhenius equation also applies to reactions in solution. There, molecules do not speed through space and collide, but jostle through the solvent and stay in one another's vicinity for relatively long periods. The more general theory that accounts both for this behavior (and for reactions in gases) is called **transition state theory.** This theory improves on collision theory by suggesting a way of calculating the rate constant even when steric requirements are significant.

In transition state theory two molecules are imagined as approaching each other and being distorted as they come close enough to affect each other. In the gas phase, that meeting and distortion are equivalent to the "collision" of collision theory. In solution, the approach is a jostling zigzag walk among solvent molecules, and the distortion might not take place until after the two reactant molecules have met and received a particularly vigorous kick from the solvent molecules around them (**FIG. 7D.8**). In either case, the collision or the kick does not immediately tear the molecules apart. Instead, the encounter results in the formation of an **activated complex,** an arrangement of the two molecules that can either go on to form products or fall apart again into the unchanged reactants.

In the activated complex, the original bonds have lengthened and weakened, and the new bonds are only partly formed. For example, in the proton transfer reaction between the weak acid HCN and water, the activated complex can be pictured as consisting of an HCN molecule with its hydrogen atom in the process of forming a hydrogen bond to the oxygen atom of a water molecule and poised midway between the two molecules. At this point, the hydrogen atom could re-form HCN or its proton leave as H_3O^+.

In transition state theory, the activation energy is a measure of the energy of the activated complex relative to that of the reactants. The reaction profile in **FIG. 7D.9** shows how the energy of the reaction mixture changes as the reactants meet, form the activated complex, and then go on to form products. A reaction profile shows the potential energy

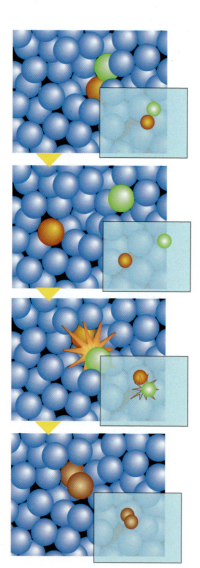

FIGURE 7D.8 This sequence of images shows the reactant molecules in solution as they meet, then either move apart or acquire enough energy by impacts from the solvent molecules to form an activated complex, which may go on to form products.

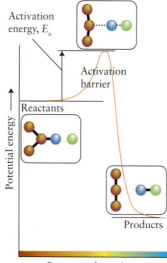

FIGURE 7D.9 A reaction profile for an exothermic reaction. In the transition state theory of reaction rates, it is supposed that the potential energy (the energy due to position) increases as the reactant molecules approach each other and form an activated complex, then reaches a maximum at the transition state. It then decreases as the atoms rearrange into the bonding pattern characteristic of the products, and these products separate. Only molecules with enough energy can cross the barrier and react to form products.

of the reactants and products, their total energy arising from their relative location, not their relative speed. Consider what happens when reactants approach each other with a certain kinetic energy. As the reactants approach, they lose kinetic energy as they "climb" up the left side of the barrier (that is, as their potential energies increase due to the repulsion resulting from their closer approach and the distortions of their bonds). If the initial kinetic energy of the reactants is less than E_a, they cannot climb to the top of the potential barrier, and they "roll" back down on the left and separate again. If their initial kinetic energy is at least E_a, they can form the activated complex, pass over the top of the barrier, a specific arrangement of atoms known as a **transition state,** and roll down the other side, where they separate as products.

A **potential energy surface** can help to depict the energy changes in the course of a reaction as a function of the locations of the atoms. In this three-dimensional plot, the z-axis is a measure of the total potential energy of the reactants and products and the x- and y-axes represent interatomic distances. For example, the plot in **FIG. 7D.10** shows the potential energy changes that occur in the attack of a bromine atom on a hydrogen molecule and the reverse process, the attack of a hydrogen atom on an HBr molecule:

$$H_2 + Br\cdot \rightleftarrows HBr + H\cdot$$

The low-energy locations corresponding to the reactants and products are separated by a barrier over which there is a path of minimum potential energy and which the kinetic energy of the approaching molecules must overcome. The actual path of the encounter depends on the total energies of the particles, but insight into the reaction process can be obtained by examining potential energy changes alone. For example, suppose the H—H bond remained the same length as the Br atom approached. That would take the system to point A, a state of very high potential energy. In fact, at normal temperatures, the colliding species might not have enough kinetic energy to reach this point. The path with the lowest potential energy is the one up the foot of the valley, over the "saddle point" (the saddle-shaped region) at the top of the pass, and down the foot of the valley on the other side of the pass. Only the lowest-energy path is available: the H—H bond must lengthen as the new H—Br bond begins to form.

As in collision theory, the rate of the reaction depends on the rate at which reactants can form the activated complex and pass through the transition state at the top of the barrier. The resulting expression for the rate constant is very similar to the one given in Eq. 3, and so this more general theory also accounts for the form of the Arrhenius equation and the observed dependence of the reaction rate on temperature.

> *In transition state theory, a reaction takes place only if two molecules acquire enough energy, perhaps from the surrounding solvent, to form an activated complex and cross through a transition state at the top of an energy barrier.*

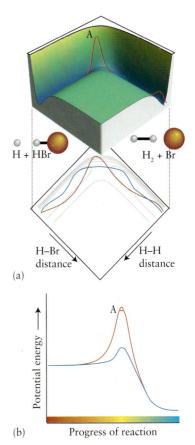

FIGURE 7D.10 (a) The contours of a potential energy surface for the reaction between a hydrogen molecule and a bromine atom (on the right). The path of lowest potential energy (blue) is up one valley, across the pass—the saddle-shaped high point—and down the floor of the other valley to the products (on the left). The path shown in red would take the atoms to very high potential energies, point A. (b) The reaction profiles for the high energy path (red) and lowest energy path (blue).

What have you learned in this Topic?

You have learned that the temperature dependence of reaction rates is described empirically by the Arrhenius equation. You have also seen that models of reactions in the gas phase based on collision theory can be used to deduce the temperature dependence of reaction rates and by introducing an empirical steric factor they can predict rate constants. You have also seen that transition state theory can extend the concepts of collision theory and account for the temperature dependence of reactions occurring in solution as well as the temperature dependence of equilibrium constants.

The skills you have mastered are the ability to:

☐ **1.** Determine the activation energy from the experimental temperature dependence of reaction rate constants (Example 7D.1).

☐ **2.** Predict the rate constant for a reaction at a new temperature if the activation energy and rate constant at one temperature are known (Example 7D.2).

☐ **3.** Discuss the Arrhenius parameters, A and E_a, in terms of models of reactions (Sections 7D.2 and 7D.3).

Topic 7D Exercises

7D.1 The rate constant of the first-order reaction $2\ N_2O(g) \rightarrow 2\ N_2(g) + O_2(g)$ is $0.76\ s^{-1}$ at 1000. K and $0.87\ s^{-1}$ at 1030. K. Calculate the activation energy of the reaction.

7D.2 The rate constant of the second-order reaction $2\ HI(g) \rightarrow H_2(g) + I_2(g)$ is $2.4 \times 10^{-6}\ L\cdot mol^{-1}\cdot s^{-1}$ at 575 K and $6.0 \times 10^{-5}\ L\cdot mol^{-1}\cdot s^{-1}$ at 630. K. Calculate the activation energy of the reaction.

7D.3 (a) Use a graphing calculator or standard graphing software to make an Arrhenius plot of the data shown here for the conversion of cyclopropane into propene and calculate the activation energy for the reaction. (b) What is the value of the rate constant at 600 °C?

T/K	750.	800.	850.	900.
k_r/s^{-1}	1.8×10^{-4}	2.7×10^{-3}	3.0×10^{-2}	0.26

7D.4 (a) Use a graphing calculator or standard graphing software to make an Arrhenius plot of the data shown here for the decomposition of iodoethane into ethene and hydrogen iodide, $C_2H_5I(g) \rightarrow C_2H_4(g) + HI(g)$, and determine the activation energy for the reaction. (b) What is the value of the rate constant at 400 °C?

T/K	660	680	720	760
k_r/s^{-1}	7.2×10^{-4}	2.2×10^{-3}	1.7×10^{-2}	0.11

7D.5 The rate constant of the reaction between CO_2 and OH^- in aqueous solution to give the HCO_3^- ion is $1.5 \times 10^{10}\ L\cdot mol^{-1}\cdot s^{-1}$ at 25 °C. Determine the rate constant at human body temperature (37 °C), given that the activation energy for the reaction is $38\ kJ\cdot mol^{-1}$.

7D.6 Ethane, C_2H_6, dissociates into methyl radicals at 700. °C with a rate constant $k_r = 5.5 \times 10^{-4}\ s^{-1}$. Determine the rate constant at 870. °C, given that the activation energy of the reaction is $384\ kJ\cdot mol^{-1}$.

7D.7 For the reversible, one-step reaction $A + A \rightleftharpoons B + C$, the forward rate constant for the formation of B is $265\ L\cdot mol^{-1}\cdot min^{-1}$ and the rate constant for the reverse reaction is $392\ L\cdot mol^{-1}\cdot min^{-1}$. The activation energy for the forward reaction is $39.7\ kJ\cdot mol^{-1}$ and that of the reverse reaction is $25.4\ kJ\cdot mol^{-1}$. (a) What is the equilibrium constant for the reaction? (b) Is the reaction exothermic or endothermic? (c) What will be the effect of raising the temperature on the rate constants and the equilibrium constant?

7D.8 For the reversible, one-step reaction $A + B \rightleftharpoons C + D$ the forward rate constant is $52.4\ L\cdot mol^{-1}\cdot h^{-1}$ and the rate constant for the reverse reaction is $32.1\ L\cdot mol^{-1}\cdot h^{-1}$. The activation energy was found to be $35.2\ kJ\cdot mol^{-1}$ for the forward reaction and $44.0\ kJ\cdot mol^{-1}$ for the reverse reaction. (a) What is the equilibrium constant for the reaction? (b) Is the reaction exothermic or endothermic? (c) What will be the effect of raising the temperature on the rate constants and the equilibrium constant?

Topic 7E Catalysis

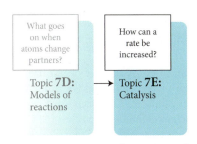

The rates of many reactions increase if the concentration of reactants is increased or the temperature is raised (Topics 7A and 7D). Another way to increase the rate of a reaction is to use a **catalyst,** a substance that increases the rate without being consumed in the reaction (**FIG. 7E.1**). The name comes from the Greek words meaning "breaking down by coming together." In many cases, only a small amount of catalyst is necessary, because it is not consumed but acts over and over again. This is why small amounts of chlorofluorocarbons can have such a devastating effect on the ozone layer in the stratosphere—they break down into radicals that catalyze the destruction of ozone (**BOX 7E.1**).

7E.1 How Catalysts Work

A catalyst speeds up a reaction by providing an alternative pathway—a different reaction mechanism—between reactants and products. This new pathway has a lower activation energy than the original pathway (**FIG. 7E.2**). At the same temperature, a greater fraction of reactant molecules can cross the lower barrier of the catalyzed path and convert into products than when no catalyst is present. Although the reaction takes place more quickly, a catalyst has no effect on the equilibrium composition. Both forward and reverse reactions are accelerated on the catalyzed path, leaving the equilibrium constant unchanged.

A **homogeneous catalyst** is a catalyst that is in the same phase as the reactants. For reactants that are gases, a homogeneous catalyst is also a gas. If the reactants are in a liquid solution, a homogeneous catalyst is dissolved in the solution. Dissolved bromine is a homogeneous liquid-phase catalyst for the decomposition of aqueous hydrogen peroxide:

$$2\,H_2O_2(aq) \xrightarrow{\;Br_2\;} 2\,H_2O(l) + O_2(g)$$

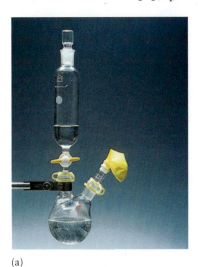

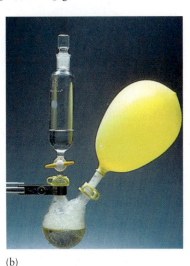

(a) (b)

FIGURE 7E.1 A small amount of catalyst—in this case, potassium iodide in aqueous solution—can accelerate the decomposition of hydrogen peroxide to water and oxygen. (a) The very slow inflation of the balloon when no catalyst is present. (b) Its rapid inflation when a catalyst is present. (*W. H. Freeman photos by Ken Karp.*)

Why Do You Need to Know This Material? Almost the whole of chemical industry depends on the existence and development of catalysts. Without them, the fertilizers used for food production and the polymers used for everyday objects would be difficult or impossible to manufacture economically on large scales.

What Do You Need to Know Already? You need to be familiar with rate laws (Topic 7A), reaction mechanisms (Topic 7C), and the concept of activation energy (Topic 7D).

The Chinese characters for catalyst, which translate as "marriage broker," capture the sense quite well.

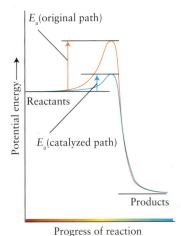

FIGURE 7E.2 A catalyst provides a new reaction pathway with a lower activation energy, thereby allowing more reactant molecules to cross the barrier and form products. The reverse reaction is also accelerated, so the equilibrium composition is not affected.

Box 7E.1 WHAT HAS THIS TO DO WITH . . . THE ENVIRONMENT?

PROTECTING THE OZONE LAYER

Every year our planet is bombarded with enough energy from the Sun to destroy all life. Only the ozone in the stratosphere protects us from that onslaught. The ozone, though, is threatened by modern lifestyles. Chemicals used as refrigerants and propellants, such as chlorofluorocarbons (CFCs), and the nitrogen oxides in jet exhausts have been found to create holes in Earth's protective ozone layer, particularly over Antarctica. Because they act as catalysts, even small amounts of these chemicals can cause large changes in the vast reaches of the stratosphere.

Ozone forms in the stratosphere in two steps. First, O_2 molecules or other oxygen-containing compounds are broken apart into atoms by sunlight, a process called *photodissociation*:

$$O_2 \xrightarrow{\text{sunlight, } \lambda < 340 \text{ nm}} O + O$$

Then the O atoms, which are reactive radicals with two unpaired electrons, react with the more abundant O_2 molecules to form ozone. The ozone molecules are created in such a high-energy state that their vibrational motions would quickly tear them apart unless another molecule, such as O_2 or N_2, collides with them first. The other molecule, indicated as M, carries off some of the energy:

$$O + O_2 \longrightarrow O_3{}^*$$
$$O_3{}^* + M \longrightarrow O_3 + M$$

where * denotes a high-energy state. The net reaction, the sum of these two elementary reactions (after multiplying them by 2) and the reaction presented earlier, is $3\,O_2 \rightarrow 2\,O_3$. Some of the ozone is decomposed by ultraviolet radiation:

$$O_3 \xrightarrow{\text{UV, } \lambda < 340 \text{ nm}} O + O_2$$

The oxygen atom produced in this step can react with oxygen molecules to produce more ozone, and so the ozone concentration

Seasonal variation in the mean area of the ozone hole over Antarctica for 2014.

in the stratosphere normally remains constant, with seasonal variations. Because the decomposition of ozone absorbs ultraviolet radiation, ozone helps to shield the Earth from radiation damage.

Warnings of the possibility that *anthropogenic* (human-generated) chemicals can threaten the ozone in the stratosphere began to appear in 1970 and 1971, when Paul Crutzen determined experimentally that NO and NO_2 molecules catalyze the destruction of ozone. Nitrogen oxides are produced naturally in the troposphere by lightning and combustion in automobile and airplane engines. However, only N_2O is sufficiently unreactive to travel up to the stratosphere, where it is converted into NO_2.

Mario Molina and Sherwood Rowland used Crutzen's work and other data in 1974 to build a model of the stratosphere that explained how chlorofluorocarbons could threaten the ozone layer.* In 1985, ozone levels over Antarctica were indeed found

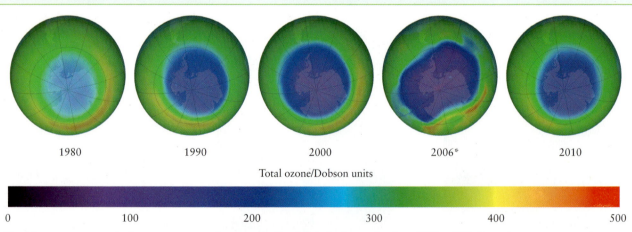

Monthly average stratospheric ozone concentrations over the South Pole for September from 1980 to 2010. Normal ozone concentration at temperate latitudes is about 350 DU (green). (1 dobson unit, 1 DU, corresponds to a layer of ozone that would be 10 μm thick at 0 °C and 1 atm, so 350 DU corresponds to a layer of thickness 3.50 mm.) The largest observed hole in the ozone layer was recorded on September 24, 2006 (for additional maps see http://ozonewatch.gsfc.nasa.gov/) (NASA).

---

*Crutzen, Molina, and Rowland were awarded the 1995 Nobel Prize in chemistry "for their work in atmospheric chemistry, particularly concerning the formation and decomposition of ozone."

to be decreasing and had dropped to the lowest ever observed; by the year 2000, the ozone hole extended to Chile. These losses are now known to be global in extent, and it has been postulated that they may be contributing to global warming in the Southern Hemisphere.

Susan Solomon and James Anderson showed that CFCs produce chlorine atoms and chlorine oxide under the conditions present in the ozone layer and identified the CFCs emanating from everyday objects, such as cans of hair spray, refrigerators, and air conditioners, as primary culprits in the destruction of stratospheric ozone. The CFC molecules are sufficiently unreactive to survive attack by hydroxyl radicals, $\cdot OH$, in the troposphere and (like N_2O) rise up into the stratosphere, where they are exposed to ultraviolet radiation from the Sun. They readily dissociate in the presence of this radiation and form chlorine atoms, which destroy ozone by various mechanisms, one of which is

$$Cl\cdot + O_3 \longrightarrow ClO\cdot + O_2$$
$$ClO\cdot + \cdot O\cdot \longrightarrow Cl\cdot + O_2$$

The O atoms are produced when ozone is decomposed by ultraviolet light, as described previously. Notice that the net reaction, $O_3 + O \rightarrow O_2 + O_2$, does not involve chlorine. Chlorine atoms act as continuously regenerated catalysts, and so even a low concentration can do a lot of damage.

Most nations signed the Montreal Protocol of 1987 and the amendments added in 1992, which required the more dangerous CFCs to be phased out by 1996. The concentration of ozone in the stratosphere is measured by NASA's Total Ozone Mapping Spectrometer (TOMS). The TOMS data show that the levels of compounds that deplete ozone are decreasing and that the ozone hole has been shrinking, although in 2008 that trend reversed and it began to grow again. However, it is thought that if the Montreal Protocol continues to be observed and there are no major volcanic eruptions (which release dust that accelerates the destruction of ozone), the ozone levels in the stratosphere may return to protective levels.

Related Exercises 7.29–7.31

Further Reading C. Baird and M. Cann, "Stratospheric chemistry: the ozone layer" and "The ozone holes," *Environmental Chemistry*, 5th ed. (New York: W. H. Freeman and Company, 2012). F. S. Rowland, "Stratospheric ozone depletion," *Philosophical Transactions of the Royal Society B*, vol. 361, 2006, pp. 769–790. J. Shanklin, "Reflections on the ozone hole," *Nature*, vol. 465, 2010, pp. 34–35. NASA, "Ozone Hole Watch," http://ozonewatch.gsfc.nasa.gov/.

In the absence of bromine or another catalyst, a solution of hydrogen peroxide can be stored for a long time at room temperature; however, bubbles of oxygen form as soon as a drop of bromine is added. Bromine's role in this reaction is believed to be its reduction to Br^- in one step, followed by oxidation back to Br_2 in a second step. The overall equations for each step (not the elementary reactions, which are numerous in each case and not fully identified here) are

$$Br_2(aq) + H_2O_2(aq) \longrightarrow 2\,Br^-(aq) + 2\,H^+(aq) + O_2(g)$$
$$2\,Br^-(aq) + H_2O_2(aq) + 2\,H^+(aq) \longrightarrow Br_2(aq) + 2\,H_2O(l)$$

When the two equations are added, both the catalyst, Br_2, and the intermediate, Br^-, cancel, leaving the overall equation as $2\,H_2O_2(aq) \rightarrow 2\,H_2O(l) + O_2(g)$. Hence, although Br_2 molecules have participated in the reaction, they are not consumed and act over and over again.

Although a catalyst does not appear in the balanced equation for a reaction, the concentration of a homogeneous catalyst might appear in the rate law. For example, the reaction between the triiodide ion and the azide ion is very slow unless a catalyst such as carbon disulfide is present:

$$I_3^-(aq) + 2\,N_3^-(aq) \xrightarrow{CS_2} 3\,I^-(aq) + 3\,N_2(g)$$

The rate-determining step of this reaction is the very first step, in which a reactive intermediate is formed:

$$CS_2 + N_3^- \longrightarrow S_2CN_3^- \quad \text{(slow)}$$

The intermediate then reacts rapidly with triiodide ion in a series of fast elementary reactions that can be summarized as

$$2\,S_2CN_3^- + I_3^- \longrightarrow 2\,CS_2 + 3\,N_2 + 3\,I^-$$

The rate law derived from this mechanism is the same as the experimental rate law:

$$\text{Rate of consumption of } I_3^- = k_r[CS_2][N_3^-]$$

Notice that the rate law is first order in the catalyst, carbon disulfide, but zeroth order in triiodide ion.

A **heterogeneous catalyst** is a catalyst present in a phase different from that of the reactants. The most common heterogeneous catalysts are finely divided or porous solids used in gas-phase or liquid-phase reactions. They are finely divided or porous so that they will provide a large surface area for the elementary reactions that provide the catalytic pathway.

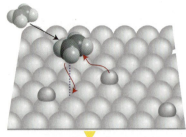

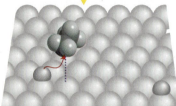

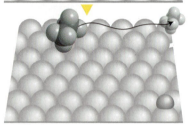

FIGURE 7E.3 The reaction between ethene, CH$_2$=CH$_2$, and hydrogen on a catalytic metal surface that has adsorbed some hydrogen molecules, which have dissociated and stuck to the surface as hydrogen atoms. In this sequence of images, the ethene molecule approaches the metal surface. Next, after the ethene molecule sticks to the surface, it meets a hydrogen atom and forms a bond. At this stage, the ·CH$_2$CH$_3$ radical is attached to the surface by one of its carbon atoms until it meets another hydrogen atom; then ethane is formed and escapes from the surface.

One example is the iron catalyst used in the Haber process for ammonia (Topic 5J). Another is finely divided nickel, which is used in the hydrogenation of ethene:

$$H_2C{=}CH_2(g) + H_2(g) \xrightarrow{\text{Ni}} H_3C{-}CH_3(g)$$

The reactant is adsorbed on the catalyst's surface. As a reactant molecule attaches to the surface of the catalyst, its bonds are weakened and the reaction can proceed more quickly because the bonds are more easily broken (**FIG. 7E.3**).

EXAMPLE 7E.1 Determining the effect of a catalyst on a reaction rate

You are working for a manufacturer of cleaning supplies and are investigating the use of hydrogen peroxide, H$_2$O$_2$, as a stain removal agent in laundry detergent. However, the water available in the market test region contains an iron oxide component and you want to know if this will interfere with the use of hydrogen peroxide. The activation energy for the decomposition of hydrogen peroxide is 75.3 kJ·mol^{-1}. In the presence of an iron oxide catalyst the activation energy for decomposition was found to be 32.8 kJ·mol^{-1}. By what factor does the rate of decomposition increase at 25 °C in the presence of the catalyst if all other factors are equal?

ANTICIPATE Because the activation energy for the catalyzed reaction is much lower, you should expect a large increase in the decomposition rate.

PLAN Because the rate constant is related to the activation energy according to Eq. 1b of Topic 7D ($k_r = Ae^{-E_a/RT}$), write an expression for the ratio of the rate constants at the two temperatures, k_{r2}/k_{r1}, and solve the expression for k_{r2} in terms of k_{r1}.

What should you assume? Assume that the pre-exponential factor for decomposition with and without catalyst is the same.

SOLVE

From the ratio of rate constants given by Eq. 1b of Topic 7D, $k_r = Ae^{-E_a/RT}$

$$\frac{k_{r2}}{k_{r1}} = \frac{Ae^{-E_{a2}/RT}}{Ae^{-E_{a1}/RT}} \overset{e^x/e^y = e^{x-y}}{=\joinrel=} e^{-(E_{a2}-E_{a1})/RT}$$

Then, with

$$\frac{E_{a2}-E_{a1}}{RT} = \frac{(32.8 - 75.3) \times 10^3 \,\overset{\text{kJ}}{\text{J·mol}^{-1}}}{(8.3145 \,\text{J·K}^{-1}\text{·mol}^{-1}) \times (298 \,\text{K})} = -17.15\ldots$$

The ratio of rate constants is

$$\frac{k_{r2}}{k_{r1}} = e^{-(-17.15\ldots)} = e^{17.15\ldots} = 2.8 \times 10^7$$

Now express k_{r2} in terms of k_{r1}

$$k_{r2} = 2.8 \times 10^7 \, k_{r1}$$

EVALUATE In the presence of the catalyst the rate constant for decomposition of hydrogen peroxide is much greater, by a factor of nearly 30 million, as anticipated.

Self-test 7E.1A A reaction rate increases by a factor of 1000. in the presence of a catalyst at 25 °C. The activation energy of the original pathway is 98 kJ·mol^{-1}. What is the activation energy of the new pathway, all other factors being equal? In practice, the new pathway also has a different pre-exponential factor.

[*Answer:* 81 kJ·mol^{-1}]

Self-test 7E.1B A reaction rate increases by a factor of 500. in the presence of a catalyst at 37 °C. The activation energy of the original pathway is 106 kJ·mol^{-1}. What is the activation energy of the new pathway, all other factors being equal? In practice, the new pathway also has a different pre-exponential factor.

Related Exercises 7E.3, 7E.4

Catalysts participate in reactions but are not themselves consumed; they provide a reaction pathway with a lower activation energy. Catalysts are classified as either homogeneous or heterogeneous.

7E.2 Industrial Catalysts

The catalytic converters of automobiles use heterogeneous catalysts to bring about the complete and rapid combustion of unburned fuel (**FIG. 7E.4**). The mixture of gases leaving an engine includes not only carbon dioxide and water, but also carbon monoxide, unburned hydrocarbons, and the nitrogen oxides collectively referred to as NO_x. Air pollution is decreased if the carbon compounds are oxidized to carbon dioxide and the NO_x reduced, by another catalyst, to nitrogen. The challenge is to find a catalyst—or a mixture of catalysts—that will accelerate both the oxidation and the reduction reactions and be active when the car is first started and the engine is cool.

Microporous catalysts are heterogeneous catalysts used in catalytic converters and for many other specialized applications, because of their very large surface areas and reaction specificity. **Zeolites,** for example, are microporous aluminosilicates (Topic 3I) with three-dimensional structures riddled with hexagonal channels connected by tunnels (**FIG. 7E.5**). In catalytic converters, they trap nitrogen oxides so that they can be reduced to harmless nitrogen gas. Catalysts with a different formulation trap incompletely burned hydrocarbons and oxidize them to carbon dioxide. The enclosed nature of the active sites in zeolites gives them a special advantage over other heterogeneous catalysts, because an intermediate can be held in place inside the channels until the products form. Moreover, the channels allow products to grow only to a particular size. One successful application of zeolites is the catalyst ZSM-5, which is used to convert methanol to gasoline. The pores of the zeolite are just large enough to allow product hydrocarbons consisting of about eight carbon atoms to escape, so the chains do not grow too long.

Catalysts can be **poisoned,** or inactivated. A common cause of such poisoning is the adsorption of a molecule so tightly to the catalyst that it seals the surface of the catalyst against further reaction. Some heavy metals, especially lead, are very potent poisons for heterogeneous catalysts, which is one reason why lead-free gasoline must be used in engines fitted with catalytic converters.

Microporous catalysts are heterogenous catalysts such as zeolites that have large surface areas.

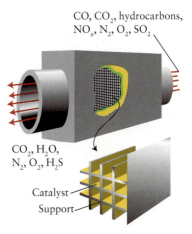

CO, CO_2, hydrocarbons, NO_x, N_2, O_2, SO_2

CO_2, H_2O, N_2, O_2, H_2S

Catalyst
Support

FIGURE 7E.4 The catalytic converter of an automobile is made from a mixture of catalysts bonded to a honeycomb ceramic support through which the exhaust gases flow.

7E.3 Living Catalysts: Enzymes

Living cells contain thousands of different kinds of homogenous catalysts, each of which is necessary to life. Many of these catalysts are proteins called *enzymes,* large molecules with a pocketlike active site, where reaction takes place (**FIG. 7E.6**). The **substrate,** the molecule on which the enzyme acts, fits into the pocket as a key fits into a lock (**FIG. 7E.7**). However, unlike an ordinary lock, a protein molecule distorts slightly as the substrate molecule approaches, and its ability to undergo the correct distortion also determines whether the "key" will fit. This refinement of the original lock-and-key model is known as the **induced-fit mechanism** of enzyme action.

Once in the active site, the substrate undergoes reaction. The product is then released for use in the next stage, which is controlled by another enzyme, and the original enzyme molecule is free to receive the next substrate molecule. One example of an enzyme is amylase, which is present in your mouth. The amylase in saliva helps to break down starches in food into the more easily digested glucose. If you chew a cracker long enough, you can notice the increased sweetness.

The kinetics of enzyme reactions were first studied by the German chemists Leonor Michaelis and Maud Menten in the early part of the twentieth century. They found that, when the concentration of substrate is low, the rate of an enzyme-catalyzed reaction increases with the concentration of the substrate, as shown in the plot in **FIG. 7E.8**. However, when the concentration of substrate is high, the reaction rate depends only on the concentration of the enzyme. In the **Michaelis–Menten mechanism** of enzyme reaction, the enzyme E binds reversibly to the substrate S to give the bound enzyme–substrate complex, ES:

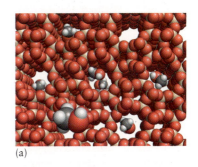

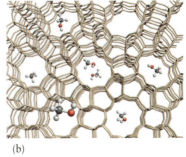

FIGURE 7E.5 (a) The structure of the ZSM-5 zeolite catalyst. Reactants diffuse through the channels, which are narrow enough to hold intermediates in positions favorable for reaction. (b) In this rendering, only the zeolite bonds are shown, so that the structure can be seen more clearly.

$$E + S \rightleftharpoons ES$$

forward: second order   Rate = $k_1[E][S]$
reverse: first order   Rate = $k_1'[ES]$

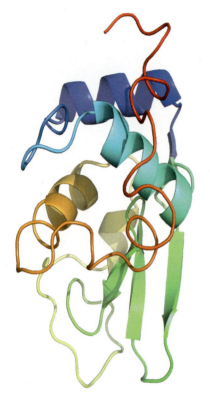

FIGURE 7E.6 The lysozyme molecule is a typical enzyme molecule. Lysozyme is present in a number of places in the body, including tears and the mucus in the nose. One of its functions is to attack the cell walls of bacteria and destroy them. This "ribbon" representation shows only the general arrangement of the atoms, to emphasize the overall shape of the molecule; the ribbon actually consists of amino acids linked together (Topic 11E).

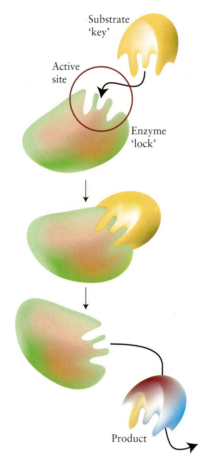

FIGURE 7E.7 In the lock-and-key model of enzyme action, the correct substrate is recognized by its ability to fit into the active site like a key into a lock. In a refinement of this model, the enzyme changes its shape slightly as the key enters.

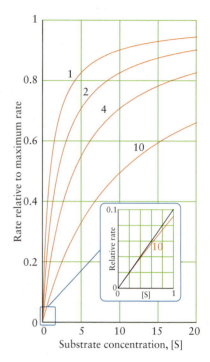

FIGURE 7E.8 A plot of the rate of an enzyme-catalyzed reaction (relative to its maximum value, $k_2[E]_0$, when S is in very high concentration) as a function of concentration of substrate for various values of K_M. At low substrate concentrations, the rate of reaction is directly proportional to the substrate concentration (as indicated by the extrapolated black line for $K_M = 10$). At high substrate concentrations, the rate becomes constant at $k_2[E]_0$ once the enzyme molecules are "saturated" with substrate. The units of [S] are the same as those of K_M.

The complex decays with first-order kinetics, releasing the enzyme to act again:

$$\text{ES} \longrightarrow \text{E} + \text{product} \qquad \text{forward: first order} \qquad \text{Rate} = k_2[\text{ES}]$$

When the overall rate law is worked out (see Exercise 7.14), it is found that

$$\text{Rate of formation of product} = \frac{k_2[\text{E}]_0[\text{S}]}{K_M + [\text{S}]} \tag{1a}$$

where $[\text{E}]_0$ is the total concentration of enzyme (bound plus unbound) and the **Michaelis constant,** K_M, is

$$K_M = \frac{k_1' + k_2}{k_1} \tag{1b}$$

When the rate given by Eq. 1a is plotted against the concentration of substrate, the resulting curve is exactly like the one observed experimentally.

One form of biological poisoning mirrors the effect of lead on a catalytic converter. The activity of an enzyme is destroyed if an alien substrate attaches too strongly to the enzyme's active site, because then the site is blocked and made unavailable to the true substrate (**FIG. 7E.9**). As a result, the chain of biochemical

FIGURE 7E.9 (a) An enzyme poison (represented by the purple sphere) can act by attaching so strongly to the active site that it blocks the site, thereby taking the enzyme out of action. (b) Alternatively, the poison molecule may attach elsewhere, so distorting the enzyme molecule and its active site that the substrate no longer fits.

reactions in the cell stops, and the cell dies. Nerve gases act by blocking the enzyme-controlled reactions that allow impulses to travel through nerves. Arsenic, that favorite of fictional poisoners, acts in a similar way. After ingestion as As(V) in the form of arsenate ions (AsO_4^{3-}), it is reduced to As(III), which binds to enzymes and inhibits their action. However, not all enzyme poisoning is detrimental. Cyclooxygenase enzymes are responsible for producing prostaglandins and thromboxanes, which lead to inflammation. In patients with chronic arthritis, these enzymes are overactive, leading to painful inflammation in joints. Aspirin (acetyl salicylic acid, **1**) reduces inflammation by reacting irreversibly with cyclooxygenases to halt their catalytic activity.

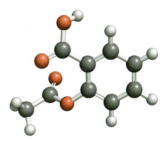

1   Acetyl salicylic acid (aspirin)

Enzymes are biological catalysts that function by modifying substrate molecules to promote reaction.

What Have You Learned In This Topic?

You have learned that catalysts can increase reactions rates by providing a lower activation energy pathway to form products. You have seen that although catalysts do not appear in the balanced equation for a reaction they do appear in the rate law. You have also learned how enzymes function as catalysts by providing active sites for substrates.

The skills you have mastered are the ability to:

☐ **1.** Discuss how a catalyst increases the rate of a reaction (Section 7E.1).

☐ **2.** Calculate the increase in rate constant when the activation energies in the presence and absence of a catalyst are known (Example 7E.1).

☐ **3.** Differentiate a homogeneous from a heterogeneous catalyst (Sections 7E.1 and 7E.2).

☐ **4.** Explain what is meant by catalyst poisoning (Section 7E.2 and 7E.3).

☐ **5.** Describe the induced fit mechanism of enzyme action (Section 7E.3).

Topic 7E Exercises

7E.1 How does a catalyst affect (a) the rate of the reverse reaction; (b) the value of ΔH_r° for the reaction?

7E.2 How does a homogeneous catalyst affect (a) the rate law; (b) the equilibrium constant of a reaction?

7E.3 The presence of a catalyst provides a reaction pathway in which the activation energy of a certain reaction is reduced from 125 kJ·mol^{-1} to 75 kJ·mol^{-1}. (a) By what factor does the rate of the reaction increase at 298 K, all other factors being equal? (b) By what factor would the rate change if the reaction were carried out at 350. K instead?

7E.4 The presence of a catalyst provides a reaction pathway in which the activation energy of a certain reaction is reduced from 75 kJ·mol^{-1} to 52 kJ·mol^{-1}. (a) By what factor does the rate of the reaction increase at 200. K, all other factors being equal? (b) By what factor would the rate change if the reaction were carried out at 300. K instead?

7E.5 The hydrolysis of an organic nitrile, a compound containing a —C≡N group, in basic solution, is proposed to proceed by the following mechanism. Write a complete balanced equation for the overall reaction, list any intermediates, and identify the catalyst in this reaction.

Step 1   R—C≡N + OH$^-$ ⟶ R—C$\underset{\backslash OH}{\overset{/\!/\,N^-}{}}$

Step 2   R—C$\underset{\backslash OH}{\overset{/\!/\,N^-}{}}$ + H$_2$O ⟶ R—C$\underset{\backslash OH}{\overset{/\!/\,NH}{}}$ + OH$^-$

Step 3   R—C$\underset{\backslash OH}{\overset{/\!/\,NH}{}}$ ⟶ R—C$\underset{\backslash\!\backslash O}{\overset{/\,NH_2}{}}$

7E.6 Consider the following mechanism for the hydrolysis of ethyl acetate. Write a complete balanced equation for the overall reaction, list any intermediates, and identify the catalyst in this reaction.

Step 1 $H_3C - C \overset{O}{\underset{O-CH_2-CH_3}{\parallel}}$ + OH⁻   ⟶

$H_3C - C \overset{O}{\underset{OH}{\parallel}}$ + CH₃CH₂O⁻

Step 2   CH₃CH₂O⁻ + H₂O ⟶ CH₃CH₂OH + OH⁻

7E.7 Decide which of the following statements about catalysts are true. If the statement is false, explain why. (a) In an equilibrium process, a catalyst increases the rate of the forward reaction but leaves the rate of the reverse reaction unchanged. (b) A catalyst is not consumed in the course of a reaction. (c) The pathway for a reaction is the same in the presence of a catalyst as in its absence, but the rate constants are decreased in both the forward and the reverse directions. (d) A catalyst must be carefully chosen to shift the equilibrium toward the products.

7E.8 Decide which of the following statements about catalysts are true. If the statement is false, explain why. (a) A heterogeneous catalyst works by binding one or more of the molecules undergoing reaction to the surface of the catalyst. (b) Enzymes are naturally occurring proteins that serve as catalysts in biological systems. (c) The equilibrium constant for a reaction is greater in the presence of a catalyst. (d) A catalyst changes the pathway of a reaction in such a way that the reaction becomes more exothermic.

7E.9 The Michaelis–Menten rate equation for enzyme reactions is typically written as the rate of formation of product (Eq. 1a). This equation implies that 1/rate (where rate is the rate of formation of product) depends linearly on the inverse of the substrate concentration [S]. This relation allows K_M to be determined. Derive this equation and sketch 1/rate against 1/[S]. Label the axes, the y-intercept, and the slope with their corresponding expressions in terms of K_M and [S].

7E.10 The Michaelis constant (K_M) is an index of the stability of an enzyme–substrate complex. Does a high Michaelis constant indicate a long-lived or short-lived enzyme–substrate complex? Explain your reasoning.

The following Example and Exercises draw on material from throughout Focus 7.

FOCUS 7 Online Cumulative Example

You are working in a water treatment facility and want to measure the concentration of iron(II) ions in the water supply. Iron(II) reacts with 1,10-phenanthroline (phen, **1**) to form the

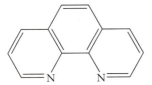

1 1,10-Phenanthroline

deep red complex ion ferroin, $Fe(phen)_3^{2+}$, which can be used to determine its concentration by spectrophotometry. However, in acidic solutions the complex decomposes in the following overall reaction,

$$Fe(phen)_3^{2+}(aq) + 3\,H_3O^+(aq) \longrightarrow$$
$$Fe^{2+}(aq) + 3\,Hphen^+(aq) + 3\,H_2O(l)$$

To know how fast the concentration of the complex changes at different temperatures, you first need to determine the activation energy of the reaction.

(a) The data in the following table were collected at 40 °C. Use these data to determine the reaction order with respect to $Fe(phen)_3^{2+}$ and H_3O^+. (b) Identify the rate law for the reaction and determine the value of the rate constant k_r at 40 °C.

Experiment	$[Fe(phen)_3^{2+}]/$ $(mol\cdot L^{-1})$	$[H_3O^+]/$ $(mol\cdot L^{-1})$	Initial rate/ $(mol\cdot L^{-1}\cdot s^{-1})$
1	7.5×10^{-3}	0.5	9.0×10^{-6}
2	7.5×10^{-3}	0.05	9.0×10^{-6}
3	3.75×10^{-2}	0.05	4.5×10^{-5}

(c) Use the rate constant from part (b) and the data in the following table to determine the activation energy for this reaction. (d) How long would it take for the concentration of $Fe(phen)_3^{2+}$ to decrease by half at 25 °C?

Temperature/°C	50	60	70
Rate constant, k_r/s^{-1}	5.4×10^{-3}	2.2×10^{-2}	8.5×10^{-2}

 The Online Cumulative Example solution can be found at http://macmillanhighered.com/chemicalprinciples7e

FOCUS 7 Exercises

Exercises labeled $\int_{dx}^{C}$ require calculus.
All rates are unique reaction rates unless otherwise stated.

7.1 In some reactions, two or more different products can be formed. If the product formed by the fastest reaction predominates, the reaction is considered to be under "kinetic control." If the predominant product is the most thermodynamically stable, the reaction is considered to be under "thermodynamic control." In the reaction of HBr with the reactive intermediate $CH_3CH{=}CHCH_2^+$, at low temperatures the predominant product is $CH_3CHBrCH{=}CH_2$, but at high temperatures, the predominant product is $CH_3CH{=}CHCH_2Br$. (a) Which product is formed by the reaction with the larger activation energy? (b) Does kinetic control predominate at low or high temperatures? Explain your answers.

7.2 An organic compound A can decompose by either of two kinetically controlled pathways to form product B or C (see Exercise 7.1). The activation energy for the formation of B is greater than that for the formation of C. Will the ratio [B]/[C] increase or decrease as the temperature is increased? Explain your answer.

7.3 The equilibrium constant for the binding of a substrate to the active site of an enzyme was found to be 326 at 310 K. At the same temperature, the rate constant for the second-order binding reaction is $7.4 \times 10^7\ L\cdot mol^{-1}\cdot s^{-1}$. What is the rate constant for the loss of unreacted substrate from the active site (the reverse of the binding reaction)?

7.4 Compounds A and B both decompose by first-order reactions. At 398 K, the rate constant for the decomposition of A is $3.6 \times 10^{-5}\ s^{-1}$. Separate containers of A and B were prepared, with initial concentrations of 0.120 (mol A)$\cdot L^{-1}$ and 0.240 (mol B)$\cdot L^{-1}$. After 5.0 h, it was found that the concentration of A was 3.0 times the concentration of B. (a) What was the concentration of A at that time? (b) What is the rate constant for the decomposition of B at 398 K?

7.5 The first-order decomposition of compound X, a gas, is carried out and the data are represented in the following pictures. The green spheres represent the compound; the decomposition products are not shown. The times at which the images were taken are shown below each flask. (a) Determine the half-life of the reaction. (b) Draw the appearance of the molecular image at 8.0 s.

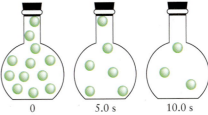

| 0 | 5.0 s | 10.0 s |

7.6 Determine the molecularity of the following symbolic elementary reactions:
(a) $\bigcirc + \bullet{-}\bullet \longrightarrow \bigcirc{-}\bullet + \bullet$
(b) $\bigcirc{-}\bigcirc \longrightarrow \bigcirc + \bigcirc$
(c) $\bigcirc + \bigcirc + \bullet \longrightarrow \bullet + \bigcirc{-}\bigcirc$

639

7.7 The following rate laws were each derived from an elementary reaction. In each case, write the chemical equation for the reaction, state its molecularity, and propose a structure for the activated complex:
(a) Rate = $k_r[CH_3CHO]$ (Products are CH_3 and CHO.)
(b) Rate = $k_r[I]^2[Ar]$ (Products are I_2 and Ar; the role of the Ar is to remove energy as the products form.)
(c) Rate = $k_r[O_2][NO]$ (Products are NO_2 and O.)

7.8 The following mechanism has been proposed for the formation of hydrazine in the overall reaction, $N_2(g) + 2 H_2(g) \rightarrow N_2H_4(g)$:

Step 1 $N_2 + H_2 \longrightarrow N_2H_2$
Step 2 $H_2 + N_2H_2 \longrightarrow N_2H_4$

The rate law for the overall reaction is Rate = $k_r[N_2][H_2]^2$. Which step is the slow step? Show your work.

7.9 The hydrolysis of sucrose ($C_{12}H_{22}O_{11}$) produces fructose and glucose: $C_{12}H_{22}O_{11}(aq) + H_2O(l) \rightarrow C_6H_{12}O_6(glucose, aq) + C_6H_{12}O_6$ (fructose, aq). Two mechanisms are proposed for this reaction:

(i) **Step 1** $C_{12}H_{22}O_{11} \longrightarrow C_6H_{12}O_6 + C_6H_{10}O_5$ (slow)
    Step 2 $C_6H_{10}O_5 + H_2O \longrightarrow C_6H_{12}O_6$ (fast)
(ii) $C_{12}H_{22}O_{11} + H_2O \longrightarrow C_6H_{12}O_6 + C_6H_{12}O_6$ (slow)

Under what conditions can these two mechanisms be distinguished by using kinetic data?

7.10 Some organic compounds containing the C=O group can react with themselves in a process known as *aldol condensation*. The mechanism for this reaction in acidic solution is shown here. Write the overall reaction, identify any intermediates, and determine the role of the hydrogen ion.

Step 1

Step 2

Step 3

Step 4

7.11 The rate law of the reaction $2 NO(g) + 2 H_2(g) \rightarrow N_2(g) + 2 H_2O(g)$ is Rate = $k_r[NO]^2[H_2]$, and the mechanism that has been proposed is

Step 1 $NO + NO \longrightarrow N_2O_2$
Step 2 $N_2O_2 + H_2 \longrightarrow N_2O + H_2O$
Step 3 $N_2O + H_2 \longrightarrow N_2 + H_2O$

(a) Which step in the mechanism is likely to be rate determining? Explain your answer. (b) Sketch a reaction profile for the overall reaction, which is known to be exothermic. Label the activation energies of each step and the overall reaction enthalpy.

7.12 (a) Use a graphing calculator or graphing software to calculate the activation energy for the acid hydrolysis of sucrose to give glucose and fructose (see Exercise 7.9) from an Arrhenius plot of the data below. (b) Calculate the rate constant at 37 °C (human body temperature). (c) From data in Appendix 2A, calculate the enthalpy change for this reaction, assuming that the solvation enthalpies of the sugars are negligible. Draw an energy profile for the overall process.

Temperature/°C	24	28	32	36	40.
k_r/s^{-1}	4.8×10^{-3}	7.8×10^{-3}	13×10^{-3}	$20. \times 10^{-3}$	32×10^{-3}

7.13 The decomposition of A has the rate law Rate = $k_r[A]^a$. Show that for this reaction the ratio $t_{1/2}/t_{3/4}$, where $t_{1/2}$ is the half-life and $t_{3/4}$ is the time for the concentration of A to decrease to three-fourths of its initial concentration, can be written in terms of a alone and can therefore be used to make a quick assessment of the order of the reaction in A.

7.14 (a) From the following mechanism, derive Eq. 1a in Topic 7E, which Michaelis and Menten proposed to represent the rate of formation of products in an enzyme-catalyzed reaction. (b) Show that the rate is independent of substrate concentration at high concentrations of substrate.

$$E + S \rightleftharpoons ES \qquad k_1, k_1'$$
$$ES \longrightarrow E + P \qquad k_2$$

where E is the free enzyme, S is the substrate, ES is the enzyme–substrate complex, and P is the product. Note that the steady-state concentration of free enzyme will be equal to the initial concentration of the enzyme less the amount of enzyme that is present in the enzyme–substrate complex: $[E] = [E]_0 - [ES]$.

7.15 For the enzyme-catalyzed conversion of a certain substrate, $K_M = 0.038$ mol·L^{-1} at 25 °C. When the substrate concentration is 0.156 mol·L^{-1}, the rate of the reaction is 1.21 mmol·L^{-1}s^{-1}. The maximum rate of the conversion reaction is reached at high substrate concentrations (see Exercise 7.14). Calculate the maximum rate of this enzyme-catalyzed reaction.

7.16 A gas composed of molecules of diameter 0.50 nm takes part in a chemical reaction at 300. K and 1.0 atm with another gas (present in large excess) consisting of molecules of about the same size and mass to form a gas-phase product at 300. K. The activation energy for the reaction is 25 kJ·mol^{-1}. Use collision theory to calculate the ratio of the reaction rate at 320. K to that at 300. K.

7.17 Refer to the illustration below for the reaction A → D. (a) How many steps does this reaction have? (b) Which is the rate-determining step in this reaction? (c) Which step is the fastest? (d) How many intermediates must form in the reaction? (e) A catalyst is added that accelerates the third step only. What effect, if any, will the catalyst have on the rate of the overall reaction?

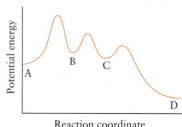

Reaction coordinate

7.18 The following schematic reaction profile is for the reaction A → D. (a) Is the overall reaction exothermic or endothermic? Explain your answer. (b) How many intermediates are there? Identify them. (c) Identify each activated complex and reaction intermediate. (d) Which step is rate determining? Explain your answer. (e) Which step is the fastest? Explain your answer.

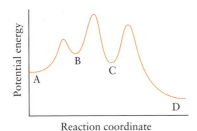

Reaction coordinate

7.19 The half-life of a substance taking part in a third-order reaction A → products is inversely proportional to the square of the initial concentration of A. How can this half-life be used to predict the time needed for the concentration to fall to (a) one-half; (b) one-fourth; (c) one-sixteenth of its initial value?

7.20 Suppose that a pollutant is entering the environment at a steady rate R and that, once there, its concentration decays by a first-order reaction. Derive an expression for (a) the concentration of the pollutant at equilibrium in terms of R and (b) the half-life of the pollutant species when $R = 0$.

7.21 Which of the following plots will be linear? (a) [A] against time for a reaction that is first order in A; (b) [A] against time for a reaction that is zeroth order in A; (c) ln [A] against time for a reaction that is first order in A; (d) 1/[A] against time for a reaction that is second order in A; (e) k_r against temperature; (f) initial rate against [A] for a reaction that is first order in A; (g) half-life against [A] for a reaction that is zeroth order in A; (h) half-life against [A] for a reaction that is second order in A.

7.22 The pre-equilibrium and the steady-state approximations are two different approaches to deriving a rate law from a proposed mechanism. For the following mechanism, determine the rate law (a) by the pre-equilibrium approximation and (b) by the steady-state approximation. (c) Under what conditions do the two methods give the same answer? (d) What will the rate laws become at high concentrations of Br⁻?

$$CH_3OH + H^+ \rightleftharpoons CH_3OH_2^+ \quad \text{(fast equilibrium)}$$
$$CH_3OH_2^+ + Br^- \longrightarrow CH_3Br + H_2O \quad \text{(slow)}$$

7.23 (a) What is the overall reaction for the following mechanism?

$$ClO^- + H_2O \rightleftharpoons HClO + OH^- \quad \text{(fast equilibrium)}$$
$$HClO + I^- \longrightarrow HIO + Cl^- \quad \text{(very slow)}$$
$$HIO + OH^- \rightleftharpoons IO^- + H_2O \quad \text{(fast equilibrium)}$$

(b) Write the rate law based on this mechanism. (c) How will the reaction rate depend on the pH of the solution? (d) How would the rate law differ if the reactions were carried out in an organic solvent?

7.24 The second-order reactions in Table 7D.1 show wide variations in activation energy. In an activated complex, the reactant bonds are lengthened while the product bonds are beginning to form. Consider what bonds need to be stretched to form the acti-

vated complex for each reaction and use bond enthalpies (Topics 2D and 4E) to explain the differences in activation energies.

7.25 Vision depends on the protein rhodopsin, which absorbs light in the retina of the eye in a reaction in which one form, metarhodopsin I, is converted to another, metarhodopsin II. The half-life of this reaction in cattle eyes is 600 μs at 37 °C but 1 s at 0 °C, whereas in frog eyes the same process has a half-life that differs by a factor of only 6 over the same temperature range. Suggest an explanation and speculate on the survival advantages that this difference provides the frog.

7.26 Raw milk sours in about 4 h at 28 °C but in about 48 h in a refrigerator at 5 °C. What is the activation energy for the souring of milk?

7.27 To prepare a dog of mass 1.5 kg for surgery, 150 mg of the anesthetic phenobarbitol is administered intravenously. The reaction in which the anesthetic is metabolized (decomposed in the body) is first order in phenobarbitol and has a half-life of 4.5 h. After about 2 h, the drug begins to lose its effect. However, the surgical procedure requires more time than had been anticipated. What mass of phenobarbitol must be re-injected to restore the original level of the anesthetic in the dog?

7.28 Models of population growth are analogous to chemical reaction rate equations. In the model developed by Malthus in 1798, the rate of change of the population N of Earth is $dN/dt = \text{births} - \text{deaths}$. The numbers of births and deaths are proportional to the population, with proportionality constants b and d. Derive the integrated rate law for population change. How well does it fit the approximate data for the population of Earth over time given below?

Year	1750	1825	1922	1960	1974	1987	2000
$N/10^9$	0.5	1	2	3	4	5	6

7.29 The following mechanism has been suggested to explain the contribution of chlorofluorocarbons to the destruction of the ozone layer:

Step 1 $O_3 + Cl \longrightarrow ClO + O_2$
Step 2 $ClO + O \longrightarrow Cl + O_2$

(a) What is the reaction intermediate, and what is the catalyst? (b) Identify the radicals in the mechanism. (c) Identify the steps as initiating, propagating, or terminating. (d) Write a chain-terminating step for the reaction. (See Box 7E.1.)

7.30 The contribution to the destruction of the ozone layer caused by high-flying aircraft has been attributed to the following mechanism:

Step 1 $O_3 + NO \longrightarrow NO_2 + O_2$
Step 2 $NO_2 + O \longrightarrow NO + O_2$

(a) Write the overall reaction. (b) Write the rate law for each step and indicate its molecularity. (c) What is the reaction intermediate? (d) A catalyst is a substance that accelerates the rate of a reaction and is regenerated in the process. What is the catalyst in the reaction? (See Box 7E.1.)

7.31 The rate constant of the reaction O(g) + N₂(g) → NO(g) + N(g), which takes place in the stratosphere, is 9.7×10^{10} L·mol⁻¹·s⁻¹ at 800. °C. The activation energy of the reaction is 315 kJ·mol⁻¹. What is the rate constant at 700. °C? (See Box 7E.1.)

FOCUS 7 Cumulative Exercise

7.32 Cyanomethane, commonly known as acetonitrile, CH_3CN, is a toxic volatile liquid which is used as a solvent to purify steroids and to extract fatty acids from fish oils. Acetonitrile can be synthesized from methyl isonitrile by the isomerization reaction $CH_3NC(g) \rightarrow CH_3CN(g)$.

(a) Draw the Lewis structures of methyl isonitrile and cyanomethane; assign a hybridization scheme to each C atom and indicate whether each molecule is polar or nonpolar.

(b) Estimate $\Delta H°$ for the isomerization reaction by using the mean bond enthalpies in Table 4E.3. Which isomer has the lower enthalpy of formation?

(c) The isomerization reaction obeys the first-order rate law, Rate = $k_r[CH_3NC]$, in the presence of argon. The activation energy for this reaction is 161 kJ·mol^{-1}, and the rate constant at 500. K is 6.6×10^{-4} s^{-1}. Calculate the rate constant at 300. K and the time (in seconds) needed for the concentration of CH_3NC to decrease to 75% of its initial value at 300. K.

(d) Draw a reaction profile for the isomerization reaction.

(e) Calculate the temperature at which the concentration of CH_3NC will decrease to 75% of its original concentration in 1.0 h.

(f) What purpose does the excess argon serve?

(g) At low concentrations of Ar, the rate is no longer first order in CH_3NC; suggest an explanation.

THE MAIN-GROUP ELEMENTS

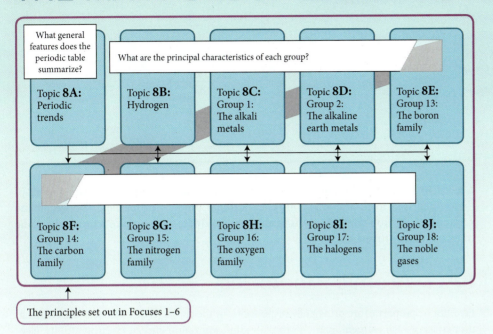

What general features does the periodic table summarize?

What are the principal characteristics of each group?

Topic 8A: Periodic trends

Topic 8B: Hydrogen

Topic 8C: Group 1: The alkali metals

Topic 8D: Group 2: The alkaline earth metals

Topic 8E: Group 13: The boron family

Topic 8F: Group 14: The carbon family

Topic 8G: Group 15: The nitrogen family

Topic 8H: Group 16: The oxygen family

Topic 8I: Group 17: The halogens

Topic 8J: Group 18: The noble gases

The principles set out in Focuses 1–6

FOCUS 8

The modern age requires materials for medicine, transportation, and communication that were not even imagined 100 years ago. To conceive and develop such materials scientists and engineers need a thorough knowledge of the elements and their compounds. Focus 8 surveys the properties and significant applications of the main-group elements and their compounds in relation to their locations in the periodic table. The chemistry of the elements and their compounds is very much the heart of chemistry, where chemical principles are applied to the structures of compounds and the reactions they undergo.

To prepare for this journey through the periodic table and the extraordinary variety of the properties of the elements, their chemical "personalities," **TOPIC 8A** first reviews the trends in properties discussed in Focuses 1 and 2, which set out the principles of chemistry. **TOPIC 8B** describes the unique element hydrogen and **TOPICS 8C** through 8J summarize the major periodic trends exhibited by the other elements of the main groups.

The opening page of each Topic in this Focus is different from those of the other Topics in this book. Because the purpose of each one is similar, except for Topic 8A the opening questions ("Why do you need to know this material?" and "What do you need to know already?") are omitted. In each case the material is important to know because periodic trends are often the key to what material should be used or what chemical reaction to expect from a substance. Similarly, there are no summarizing checklists at the ends of the Topics, for they would largely have to reproduce the material treated in each Topic.

Topic 8A Periodic Trends

8A.1 Atomic Properties

8A.2 Bonding Trends

8A.3 Trends Exhibited by Hydrides and Oxides

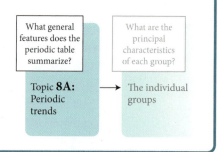

What general features does the periodic table summarize?

What are the principal characteristics of each group?

Topic **8A**: Periodic trends → The individual groups

Why Do You Need to Know This Material? The trends in periodic properties demonstrate how the elements are related to one another and suggest the properties an element and its compounds are likely to have.

What Do You Need to Know Already? This Topic draws on periodic properties (Topics 1E and 1F) and chemical bonding (Topics 2A through 2D).

The electron configuration of an element controls the number of bonds that an atom of the element can form and affects its chemical and physical properties. The systematic variation of electron configurations through the periodic table accounts for the periodic trends in these properties.

8A.1 Atomic Properties

Five atomic properties are principally responsible for the characteristic physical and chemical properties of each element, namely atomic radius, ionization energy, electron affinity, electronegativity, and polarizability. All five properties are related to the interaction of charged particles as described by Coulomb's law (Topic 1E) and therefore to trends in the effective nuclear charge experienced by the valence electrons and their distance from the nucleus.

The valence electrons of an atom experience increasing effective nuclear charge on going across a period from left to right (Topic 1E and Fig. 1F.3). The attraction of a valence electron to the nucleus weakens sharply when electrons enter a new shell that is farther from the nucleus and surrounds an electron-rich core. The valence electrons of elements at the upper right of the periodic table, near fluorine, are shielded least from the nuclear charge; the valence electrons of the elements at the lower left, near cesium, are shielded most (Topic 1E). As a result:

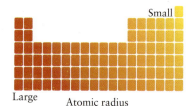

Small

Large   Atomic radius

FIGURE 8A.1 Atomic radius tends to decrease from left to right across a period and to increase down a group. This diagram and those in Figs. 8A.2 through 8A.4 are highly schematic representations of periodic trends, with the density of the red tint representing the value of the property: the denser the tint, the higher the value.

- Atomic radii typically decrease from left to right across a period and increase down a group (**FIG. 8A.1**; see also Figs. 1F.4 and 1F.5).

As the effective nuclear charge experienced by the valence electrons increases across a period, the electrons are pulled closer to the nucleus, thereby decreasing the atomic radius. Down a group the valence electrons are progressively farther from the nucleus because they occupy shells with higher principal quantum numbers, so atomic radii increase down a group. Ionic radii follow similar periodic trends (see Fig. 1F.6).

- First ionization energies typically increase from left to right across a period and decrease down a group (**FIG. 8A.2**; see also Figs. 1F.8 and 1F.9).

The increasing effective nuclear charge across a period grips the electrons more tightly and hence increases the ionization energy. The decrease in ionization energy down a group shows that it is easier to remove valence electrons from shells that are farther from the nucleus and are more effectively shielded by a greater number of core electrons.

- The highest electron affinities are found at the top right of the periodic table (see Fig. 1F.12; the trends are too ill-defined to depict simply).

The electron affinity of an element is a measure of the energy released when an anion is formed from a neutral atom. Except for the noble gases, elements near fluorine tend to have the highest electron affinities, and so are present as anions in compounds with metallic elements. The increasing effective nuclear charge across a period accounts for the trends in electron affinity: more energy is released when electrons attach to atoms with high effective nuclear charge.

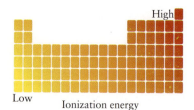

High

Low   Ionization energy

FIGURE 8A.2 Ionization energy tends to increase from left to right across a period and to decrease down a group.

- Electronegativities typically increase from left to right across a period and decrease down a group (**FIG. 8A.3**; see also Fig. 2D.2).

The greater the effective nuclear charge and the closer the electrons are to the nucleus, the stronger is the pull on the electrons in a bond. The electronegativity of an element—a measure of the tendency of an atom to attract electrons to itself when it is part of a compound—is a useful guide to the type of bond that the element is likely to form. When there is a large difference in electronegativity between two elements, their atoms form largely ionic bonds with each other; when the difference is small, the bonds are largely covalent.

- Polarizabilities typically decrease from left to right along a period and increase down a group (**FIG. 8A.4**).

Polarizability measures the ease with which an electron cloud can be distorted by charges on neighboring species and is greatest for the electron-rich, heavier atoms of a group and for negatively charged ions, which are also electron rich (Topic 2D). On the other hand, high *polarizing power*—the ability to distort the electron cloud of a neighboring atom or ion—is commonly associated with small size and high positive charge. A bond between a highly polarizable atom or ion (such as iodine or the iodide ion) and an atom or ion of high polarizing power (such as beryllium) is likely to have considerable covalent character.

THINKING POINT

Can you identify some of the physical properties that are associated with ionic and covalent character?

Self-test 8A.1A Which of the elements carbon, aluminum, and germanium has atoms with the greatest polarizability?

[**Answer:** Germanium]

Self-test 8A.1B Which of the elements oxygen, gallium, and tellurium has the greatest electron affinity?

Atomic radii and polarizabilities decrease from left to right across a period and increase down a group; ionization energies increase across a period and decrease down a group; electron affinities and electronegativities are highest near fluorine.

8A.2 Bonding Trends

The number of bonds an element can form (its "valence") and their types are related to its position in the periodic table. It is usually possible to predict the valence of an element in Period 2 from the number of electrons in the valence shell and the octet rule. For example, carbon, with four valence electrons, commonly forms four bonds; oxygen, with six valence electrons and therefore needing two more to complete its octet, typically forms two bonds.

On account of its small size, the element at the head of a group often has features that distinguish it from its **congeners,** the other members of the group. For example, elements in Period 3 and later have access to empty d-orbitals and can use them to expand their valence shells. Elements at the bottom and toward the left of the p-block also exhibit variable valence and can form compounds in which they have an oxidation number 2 less than their group number suggests (the inert-pair effect, Topic 1F). The three points to keep in mind about bonding trends are therefore:

- Period 2 elements typically obey the octet rule.
- Period 3 and later elements can expand their valence shells.
- Elements in Periods 5 and 6 of the p-block show variable valence (the inert-pair effect).

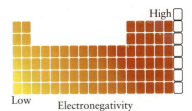

FIGURE 8A.3 Electronegativity generally increases from left to right across a period and decreases down a group.

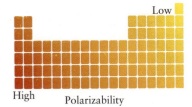

FIGURE 8A.4 The polarizability of atoms tends to decrease from left to right across a period and to increase down a group.

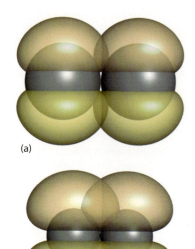

(a)

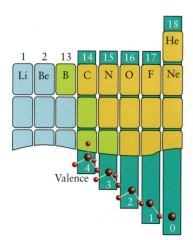

(b)

FIGURE 8A.5 (a) The p-orbitals of Period 3 and later elements are held apart by the cores of the atoms (shown in gray) and have very little overlap with each other. (b) In contrast, the atoms of elements in Period 2 are small; consequently, their p-orbitals can overlap effectively with each other and with those of elements in later periods.

The radius of an atom helps to determine how many other atoms can bond to it. One reason that small atoms typically have low valences, and another reason for the distinctiveness of Period 2 elements compared to their congeners, is that so few other atoms can pack around them. Nitrogen, for instance, never forms pentahalides, but phosphorus does (Topic 2C). With few exceptions, only Period 2 elements form multiple bonds with themselves or other elements in the same period, because only they are small enough for their p-orbitals to have substantial π overlap (**FIG. 8A.5**).

Valence and oxidation state are directly related to the valence-shell electron configuration of a group.

8A.3 Trends Exhibited by Hydrides and Oxides

Periodic trends in the chemical properties of the main-group elements become apparent when the binary compounds that they form with specific elements are compared. All the main-group elements, with the exception of the noble gases and, possibly, indium and thallium, form binary compounds with hydrogen, so hydrides can be used to illustrate periodic trends in bonding.

The formulas of **hydrides** of main-group elements are related directly to the group number and reveal the typical valences of the elements (**FIG. 8A.6**):

Group	14	15	16	17
Element	C	N	O	F
Hydride	CH_4	NH_3	H_2O	HF
Valence	4	3	2	1

The nature of a binary hydride is related to the characteristics of the element bonded to hydrogen (**FIG. 8A.7**). Most Group 1 and 2 elements form ionic compounds with hydrogen in which the latter is present as a hydride ion, H^-. These ionic compounds are called **saline hydrides** (or "saltlike hydrides"). They are formed by all members of the s-block, with the exception of beryllium, and are made by heating the metal in hydrogen:

$$2 K(s) + H_2(g) \longrightarrow 2 KH(s)$$

The saline hydrides are white, high-melting-point solids with crystal structures that resemble those of the corresponding halides. The alkali metal hydrides, for instance, have the rock-salt structure (Fig. 3H.28).

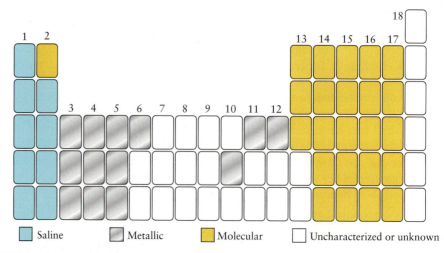

FIGURE 8A.6 The chemical formulas of the hydrides of the elements of the main groups display the valence characteristic of each group.

Saline   Metallic   Molecular   Uncharacterized or unknown

FIGURE 8A.7 The different classes of binary hydrogen compounds and their distribution throughout the periodic table.

The **metallic hydrides** are black, powdery, electrically conducting solids formed by heating certain of the d-block metals in hydrogen (**FIG. 8A.8**):

$$2\,Cu(s) + H_2(g) \longrightarrow 2\,CuH(s)$$

Because the metallic hydrides release their hydrogen (as H_2 gas) when they are heated or treated with acid, they are being investigated as possible means of storing and transporting hydrogen.

Nonmetals form covalent **molecular hydrides,** which consist of discrete molecules. These compounds are volatile, and many are Brønsted acids. Some are gases—for example, ammonia, the hydrogen halides (HF, HCl, HBr, HI), and the lighter hydrocarbons, such as methane and ethane. Liquid molecular hydrides include water and hydrocarbons such as octane and benzene.

Broadly speaking (that is, there are exceptions), the pattern of hydrides to note is

	s-block	d-block	p-block
Type	saline	metallic	molecular

All the main-group elements except the noble gases react with oxygen. Like hydrides, oxides reveal periodic trends in the chemical properties of the elements. The trend in bonding type is from soluble ionic oxides on the left of the periodic table through insoluble, high-melting-point oxides on the left of the p-block to low-melting-point and often gaseous molecular oxides on the right. Metallic elements with low ionization energies commonly form basic ionic oxides. Elements with intermediate ionization energies, such as beryllium, boron, aluminum, and the metalloids, form *amphoteric* oxides (oxides that react with both acids and bases) that do not react with or dissolve in water but do dissolve in both acidic and basic solutions.

Many oxides of nonmetals, such as CO_2, NO, and SO_3, are gaseous molecular compounds. Most can act as Lewis acids, because the electronegative oxygen atoms withdraw electrons from the central atom, enabling it to act as an electron pair acceptor (Topic 6A). Oxides of nonmetals that react with water typically form acidic solutions and hence are called **acid anhydrides.** The familiar laboratory acids HNO_3 and H_2SO_4, for instance, are derived from the anhydrides N_2O_5 and SO_3, respectively. Even oxides that do not react with water can be regarded as the *formal* anhydrides of acids. A **formal anhydride** of an acid is the molecule obtained by striking out the elements of water (H, H, and O) from the molecular formula of the acid. Carbon monoxide, for instance, is the formal anhydride of formic acid, HCOOH, although CO does not react with cold water to form the acid.

The general pattern of oxides to keep in mind is

	s-block	d-block	p-block
Type	ionic	ionic	covalent
Acid–base character	basic	basic to amphoteric	amphoteric to acidic

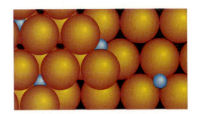

FIGURE 8A.8 In a metallic hydride, the tiny hydrogen atoms (the small spheres) occupy gaps—called interstices—between the larger metal atoms (the large spheres).

Xenon forms an oxide, but by an indirect route.

Self-test 8A.2A An element ("E") in Period 4 forms a molecular hydride with the formula HE. Identify the element.

[*Answer:* Bromine]

Self-test 8A.2B An element ("E") in Period 3 forms an amphoteric oxide with the formula E_2O_3. Identify the element.

Binary hydrides are classified as saline, metallic, or molecular. Oxides of metals tend to be ionic and to form basic solutions in water. Oxides of nonmetals are molecular, and many are the anhydrides of acids.

What have you learned in this Topic?

You have seen that the properties of elements are related to their locations in the periodic table and that trends in the properties of an element and the types of chemical bonds it is likely to form can be predicted from its position. Hydrides and oxides of main-group elements can be used to illustrate the periodicity of chemical properties.

The skills you have mastered are the ability to:

☐ **1.** Predict and explain trends in the properties of the main-group elements and in the formulas of their oxides and hydrides.

☐ **2.** Identify the type of hydride a given element is likely to form.

☐ **3.** Explain the differences in the properties of metal oxides and nonmetal oxides.

☐ **4.** Identify acid anhydrides and write the formulas for their corresponding acids.

Topic 8A Exercises

These exercises also serve as a review of principles covered in Topics 1E, 1F, and 2D.

8A.1 State which atom of each of the following pairs is larger: (a) fluorine, nitrogen; (b) potassium, calcium; (c) gallium, arsenic; (d) chlorine, iodine.

8A.2 State which atom of each of the following pairs is larger: (a) tellurium, tin; (b) silicon, lead; (c) calcium, rubidium; (d) germanium, oxygen.

8A.3 State which element of each of the following pairs is more electronegative: (a) sulfur, phosphorus; (b) selenium, tellurium; (c) sodium, cesium; (d) silicon, oxygen.

8A.4 State which element of each of the following pairs is less electronegative: (a) calcium, barium; (b) gallium, arsenic; (c) tellurium, sulfur; (d) tin, germanium.

8A.5 Arrange the following elements in order of increasing first ionization energy: oxygen, tellurium, selenium.

8A.6 Arrange the following atoms in order of increasing first ionization energy: boron, thallium, gallium.

8A.7 (a) Which element has the greater electron affinity: bromine or chlorine? (b) Explain your answer.

8A.8 (a) Which element has the greater electron affinity: bromine or selenium? (b) Explain your answer.

8A.9 (a) Which of the following species has the greatest polarizability: chloride ions, bromine atoms, bromide ions? (b) Explain your answer.

8A.10 (a) Which of the following species has the greatest polarizing power: sodium ions, magnesium atoms, aluminum ions? (b) Explain your answer.

8A.11 Which bond distance is longer: (a) the Li–Cl distance in lithium chloride or the K–Cl distance in potassium chloride; (b) the K–O distance in potassium oxide or the Ca–O distance in calcium oxide?

8A.12 Which of the following bonds do you expect to be longer: (a) the Br—O bond in BrO^- or in BrO_2^-; (b) the C—H bond in CH_4 or the Si—H bond in SiH_4?

8A.13 Write the balanced chemical equation for the reaction between potassium and hydrogen.

8A.14 Write the balanced chemical equation for the reaction between calcium and hydrogen.

8A.15 Classify each of the following hydrides as saline, molecular, or metallic: (a) LiH; (b) NH_3; (c) HBr; (d) UH_3.

8A.16 Classify each of the following hydrides as saline, molecular, or metallic: (a) B_2H_6; (b) SiH_4; (c) CaH_2; (d) PdH_x, $x < 1$.

8A.17 For each of the following oxides, state whether the compound is acidic, basic, or amphoteric. (a) NO_2; (b) Al_2O_3; (c) $B(OH)_3$; (d) MgO.

8A.18 For each of the following oxides, state whether the compound is acidic, basic, or amphoteric. (a) CuO; (b) P_2O_3; (c) ClO_2; (d) GeO_2.

8A.19 Give the formula for the formal anhydride of each of the following acids: (a) H_2CO_3; (b) $B(OH)_3$.

8A.20 Give the formula for the acid corresponding to each of the following formal anhydrides: (a) N_2O_5; (b) P_4O_{10}; (c) SeO_3.

Topic 8B Hydrogen

8B.1 The Element

8B.2 Compounds of Hydrogen

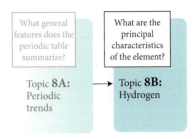

Hydrogen is widely considered to be the fuel of our future, because it burns cleanly and it is abundant (in compounds) on Earth. It occupies a unique place in the periodic table. Although a hydrogen atom has the same valence electron configuration as atoms of the Group 1 elements, ns^1, and forms $+1$ ions, the element has few other similarities to the alkali metals. Hydrogen is a nonmetal that resembles the halogens: it needs only one electron to complete its valence electron configuration, it can form -1 ions, and it exists as diatomic molecules (H_2). However, the chemical properties of hydrogen are very different from those of the halogens. Because it does not fit clearly into either group of elements, in this book it is not assigned to a group.

Be aware that many periodic tables assign hydrogen either to Group 1 or (less commonly) to both Group 1 and Group 17.

8B.1 The Element

Hydrogen is the most abundant element in the universe, accounting for 89% of all atoms. Hydrogen atoms were formed in the first few seconds after the Big Bang, the event that marked the beginning of the universe. However, there is little free hydrogen on Earth because H_2 molecules, being very light, move at such high average speeds that they escape from the Earth's gravity. **FIGURE 8B.1** summarizes the abundances of the elements in three settings: the universe as a whole, the Earth's crust, and the human body.

THINKING POINT

What are some of the factors that account for the different abundances in these sources?

Most of the Earth's hydrogen is present as water, either in the oceans or trapped inside minerals and clays. Hydrogen is also found in the hydrocarbons that make up the **fossil fuels:** coal, petroleum, and natural gas. It takes energy to release hydrogen from these compounds and one of the challenges to achieving its potential as a fuel is to produce the gas by a means that takes less energy than is released when it burns.

Because water is its only combustion product, hydrogen burns without polluting the air or contributing significantly to the greenhouse effect (**BOX 8B.1**). Petroleum, coal, and natural gas are becoming increasingly rare, but there is enough water in the oceans to generate all the hydrogen fuel Earth will ever need. Hydrogen is obtained from water by electrolysis, but that process requires electricity generated elsewhere. Chemists are currently seeking ways of using sunlight instead of other fuels to drive the **water-splitting reaction,** the photochemical decomposition of water into its elements:

Water-splitting reaction: $2\,H_2O(l) \xrightarrow{\text{light}} 2\,H_2(g) + O_2(g)$ $\Delta G° = +474\text{ kJ}$

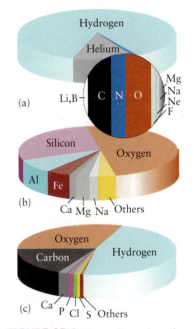

FIGURE 8B.1 These charts show the relative abundances of the principal elements in (a) the universe (the "cosmic abundances"); (b) the crust of the Earth; and (c) the human body.

So many vehicles now operate on hydrogen fuel cells (see the Interlude following Focus 6) that hydrogen refueling stations have opened in many cities.

Self-test 8B.1A Suppose you drove the water-splitting reaction forward by coupling it to a hydroelectric plant. What mass of the water driving the turbines would need to fall through 10. m to provide enough energy to produce 1.0 mol H_2? *Hint:* Use mgh to calculate the energy when a mass m falls through a distance h.

[*Answer:* 2.4 t]

Self-test 8B.1B (a) What amount (in moles) of ultraviolet photons is needed to produce 1 mol H_2 in the same reaction? Take $\lambda = 250$ nm. (b) If a source generates 1.0×10^{14} photons in 1 s, for how long would the water need to be irradiated?

649

Box 8B.1 WHAT HAS THIS TO DO WITH . . . THE ENVIRONMENT?

THE GREENHOUSE EFFECT

The Sun acts like a black body (Topic 1B), emitting radiation with a peak in intensity at 500 nm, in the visible region of the spectrum. About 55% of the solar radiation that falls on the Earth is reflected away or used in natural processes. The remaining 45% is converted into thermal motion (heat). Because the Earth also behaves a little like a black body, it radiates energy. Because the temperature of Earth is lower than that of the Sun, most of this energy escapes as infrared radiation with wavelengths between 4 and 50 μm.

The *greenhouse effect* is the trapping of this infrared radiation by certain gases in the atmosphere. This effect warms the Earth, as if the entire planet were enclosed in a huge greenhouse*. Oxygen and nitrogen, which make up roughly 99% of the atmosphere, do not absorb infrared radiation. However, water vapor and CO_2 do. Even though these two gases make up only about 1% of the atmosphere, they trap enough radiation to raise the temperature of the Earth by 33 °C. Without this naturally occurring greenhouse effect, the average surface temperature of the Earth would be well below the freezing point of water.

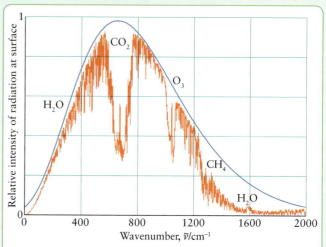

The intensity of infrared radiation at various wavenumbers that would be lost from Earth in the absence of greenhouse gases is shown by the smooth line. The jagged line is the intensity of the radiation actually emitted.

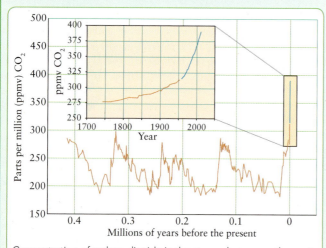

Concentration of carbon dioxide in the atmosphere, over the past 400 000 years. The orange lines show atmospheric CO_2 concentrations, which are determined from ice core samples collected in Antarctica. The blue line shows atmospheric CO_2 concentrations measured at the Mauna Loa observatory in Hawaii.

The major greenhouse gases include water vapor, carbon dioxide, methane, dinitrogen oxide (nitrous oxide), ozone, and certain chlorofluorocarbons. Water is the most important greenhouse gas. It absorbs strongly near 6.3 μm and at wavelengths longer than 12 μm, as shown on the plot. Atmospheric carbon dioxide absorbs about half the infrared radiation with wavelengths of 14–16 μm.

The concentration of water vapor in the atmosphere is thought to have remained steady over time, but concentrations of some other greenhouse gases are rising. From the year 1000 or

earlier until about 1750, the CO_2 concentration in the atmosphere remained fairly stable at about 280 ± 10 parts per million by volume (ppmv). Since then, the CO_2 concentration has increased to about 400 ppmv in 2014 (see the graph on the left). The concentration of methane, CH_4, has more than doubled during this time and is now at its highest level in 160 000 years. Studies of air pockets in ice cores taken from Antarctica show that changes in the concentrations of both atmospheric carbon dioxide and methane over the past 160 000 years correlate well with changes in the global surface temperature. The rising concentrations of carbon dioxide and methane are therefore of great concern.

Where is the additional CO_2 coming from? Human activities are largely responsible. Some is generated when limestone, $CaCO_3$, is heated and decomposed in cement making (see Topic 3I). Large amounts of CO_2 are also released into the atmosphere by deforestation, which involves burning large areas of brush and trees. However, most of it comes from the burning of fossil fuels, which began on a large scale after 1850 and increased by about 100% from 1970 to 2014. The additional methane is coming mainly from the petroleum industry and from agriculture.

Until 2000 the temperature of the surface of the Earth rose about 0.2 °C per decade (see the third graph). Although the temperature rise has slowed, if current trends in population growth and energy use continue, by the middle of the twenty-first century the concentration of CO_2 in the atmosphere will be about twice its value prior to the Industrial Revolution.

What are the likely consequences of this doubling of the CO_2 concentration? The Intergovernmental Panel on Climate Change (IPCC) estimated in 2014 that, by the year 2100, the Earth will undergo an increase in temperature of 3.7 to 4.8 °C, with a rise in sea level of 0.5 to 0.9 m. A rise of 4.8 °C may not sound like much. However, the temperature during the last ice age was only 6 °C lower than at present. Furthermore, the *rate* of temperature change is likely to be faster than at any time in the

---

*The actual insulation mechanism is different in a greenhouse. The glass not only inhibits the escape of infrared radiation, it also keeps the warm air inside.

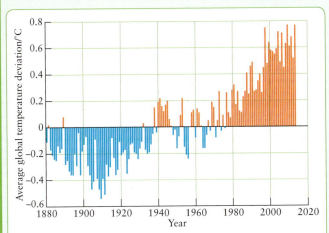

0.8
0.6
0.4
0.2
0
−0.2
−0.4
−0.6

Average global temperature deviation/°C

1880 1900 1920 1940 1960 1980 2000 2020

Year

The deviation of the annual global temperature from 1880 to 2014 with respect to the nineteenth century global average temperature. For further information see http://www.ncdc.noaa.gov/cag/time-series/global.

and held constant at that level, the concentration of CO_2 in the atmosphere would continue to increase at about 1.5 ppmv per year for the next century. Second, to maintain CO_2 at its current concentration of 400 ppmv, we would have to reduce fossil fuel consumption by about 50% immediately.

Alternatives to fossil fuels, such as hydrogen, are explored in Box 4D.1. Coal, which is mostly carbon, can be converted into fuels with a lower proportion of carbon. Its conversion into methane, CH_4, for instance, would reduce CO_2 emissions per unit of energy. We can also work with nature by accelerating the uptake of carbon by the natural processes of the carbon cycle. For example, one proposed solution is to pump CO_2 exhaust deep into the ocean, where it would dissolve to form carbonic acid and bicarbonate ions. Carbon dioxide can also be removed from power plant exhaust gases by passing the exhaust through an aqueous slurry of calcium silicate to produce harmless solid products:

$$2\,CO_2(g) + H_2O(l) + CaSiO_3(s) \longrightarrow$$
$$SiO_2(s) + Ca(HCO_3)_2(s)$$

Related Exercises 8.25–8.27

Further Reading P. Cox & C. Jones, "Climate change: Illuminating the modern dance of climate and CO_2," *Science*, vol. 321, 2008, pp. 1642–1644. Intergovernmental Panel on Climate Change, *Climate Change 2014: Mitigation of Climate Change*, IPCC, 2014, http://mitigation2014.org/report. S. K. Ritter, S. Solomon, G.-K. Plattner, R. Knutti, & P. Friedlingstein, "Irreversible climate change due to carbon dioxide emissions," *Proceedings of the National Academy of Sciences*, vol. 106, 2009, pp. 1704–1709.

past 10 000 years. Rapid climate changes may bring about droughts and storms and have detrimental effects on many of the Earth's ecosystems.

Computer projections of atmospheric CO_2 concentration for the next 200 years predict escalating increases in CO_2 concentration. Only about half the CO_2 released by humans is absorbed by Earth's natural systems. The other half increases the CO_2 concentration in the atmosphere by about 1.5 ppmv per year. Two conclusions can be drawn from these facts. First, even if CO_2 emissions were reduced to the amount emitted in 1990

At present, most commercial hydrogen is obtained as a by-product of petroleum refining in a sequence of two catalyzed reactions. The first is a **re-forming reaction,** in which a hydrocarbon and steam are converted into carbon monoxide and hydrogen over a nickel catalyst:

Re-forming reaction: $CH_4(g) + H_2O(g) \xrightarrow{\text{Ni}} CO(g) + 3\,H_2(g)$

The mixture of products, called **synthesis gas,** is also the starting point for the manufacture of many other compounds, including methanol. The re-forming reaction is followed by the **shift reaction,** in which the carbon monoxide in the synthesis gas reacts with more water:

Shift reaction: $CO(g) + H_2O(g) \xrightarrow{\text{Fe/Cu}} CO_2(g) + H_2(g)$

Hydrogen is prepared in small amounts in the laboratory by reducing hydrogen ions from a strong acid (such as hydrochloric acid) with a metal that has a negative standard potential, such as zinc:

$$Zn(s) + 2\,H^+(aq) \longrightarrow Zn^{2+}(aq) + H_2(g)$$

Hydrogen is a colorless, odorless, tasteless gas (**TABLE 8B.1**). Because H_2 molecules are small and nonpolar, they can attract each other only by very weak London forces. As a result, hydrogen does not condense to a liquid until it is cooled to a very low temperature (20. K at 1 atm). One striking physical property of liquid hydrogen is its very low density (0.070 g·cm^{-3}), which is less than one-tenth that of water (**FIG. 8B.2**). This low density makes hydrogen a very lightweight fuel. Hydrogen has the highest specific enthalpy of any known fuel (the highest enthalpy of combustion per gram), and so liquid hydrogen was used with liquid oxygen to power the space shuttle's main rocket engines.

Each year, about half the 0.3 Mt (3×10^8 kg) of hydrogen used in industry is converted on site into ammonia by the Haber process (Topic 5J). Through the reactions of

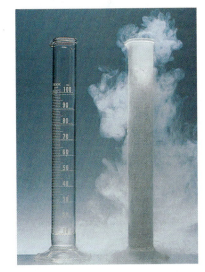

FIGURE 8B.2 The two measuring cylinders contain the same mass of liquid. The cylinder on the left holds about 10 mL of water, that on the right liquid hydrogen at −253 °C, which is one-tenth as dense and consequently occupies about 100 mL. (*W. H. Freeman photo by Ken Karp.*)

TABLE 8B.1   Physical Properties of Hydrogen

Valence configuration: $1s^1$
Normal form*: colorless, odorless gas

Z	Name	Symbol	Molar mass/ $(g \cdot mol^{-1})$	Abundance/%	Melting point/°C	Boiling point/°C	Density†/ $(g \cdot L^{-1})$
1	hydrogen	H	1.008	99.98	−259 (14 K)	−253 (20 K)	0.089
1	deuterium	^{2}H or D	2.014	0.02	−254 (19 K)	−249 (24 K)	0.18
1	tritium	^{3}H or T	3.016	radioactive	−252 (21 K)	−248 (25 K)	0.27

*Normal form means the state and appearance of the element at 25 °C and 1 atm.
†At 25 °C and 1 atm.

ammonia, hydrogen finds its way into numerous other important nitrogen compounds such as hydrazine and sodium amide (Topic 8G).

Hydrogen is produced as a byproduct of the refining of fossil fuels and by electrolysis of water. It has a low density and weak intermolecular forces.

8B.2 Compounds of Hydrogen

Hydrogen is unusual because it can form both a cation (H^+), so resembling the alkali metals, and an anion (H^-), so resembling the halogens. Moreover, its intermediate electronegativity (2.2 on the Pauling scale) means that it typically forms covalent bonds when in combination with nonmetals and metalloids. Because hydrogen forms compounds with so many elements (**TABLE 8B.2**; also see Topic 8A), most of the compounds of hydrogen are discussed in sections on the other elements.

The hydride ion, H^-, is large, with a radius of 154 pm (**1**), lying in size between the fluoride and chloride ions. Because of the large radius of this two-electron ion, the single positive charge of the nucleus has little control over the electrons, making the ion highly polarizable and contributing to covalent character in its bonds to cations. The two electrons in the H^- ion are so weakly held that they are easily lost. Consequently, ionic hydrides are very powerful reducing agents, with $E°(H_2/H^-) = -2.25$ V. This value is similar to the standard potential $E°(Na^+/Na) = -2.71$ V, and, like sodium metal, hydride ions react with water as soon as they come into contact with it:

$$NaH(s) + H_2O(l) \longrightarrow NaOH(aq) + H_2(g)$$

Because this reaction produces hydrogen, saline hydrides are potentially useful as transportable sources of hydrogen fuel.

In compounds with N—H, O—H, and F—H bonds, in which hydrogen is attached to a highly electronegative atom, the H atom can participate in *hydrogen bonding* (Topic 3F). A hydrogen bond is about 5% as strong as a covalent bond between the same types of atoms but about ten times stronger than other intermolecular interactions. For example, the O—H bond enthalpy is 463 $kJ \cdot mol^{-1}$ whereas the O—H···O hydrogen bond enthalpy is about 20 $kJ \cdot mol^{-1}$. There are several models of hydrogen bonding; the

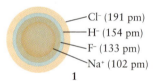

Cl⁻ (191 pm)
H⁻ (154 pm)
F⁻ (133 pm)
Na⁺ (102 pm)

1

TABLE 8B.2   Chemical Properties of Hydrogen

Reactant	Reaction with hydrogen
Group 1 metals (M)	$2\,M(s) + H_2(g) \longrightarrow 2\,MH(s)$
Group 2 metals (M, not Be or Mg)	$M(s) + H_2(g) \longrightarrow MH_2(s)$
some d-block metals (M)	$2\,M(s) + x\,H_2(g) \longrightarrow 2\,MH_x(s)$
oxygen	$O_2(g) + 2\,H_2(g) \longrightarrow 2\,H_2O(l)$
nitrogen	$N_2(g) + 3\,H_2(g) \longrightarrow 2\,NH_3(g)$
halogen (X_2)	$X_2(g, l, s) + H_2(g) \longrightarrow 2\,HX(g)$

simplest is as a coulombic interaction between the partial positive charge on a hydrogen atom and the partial negative charge of an atom in another molecule, as in $O—H^{\delta+}\cdots^{\delta-}O$.

THINKING POINT

Deuterium, ^2H, differs from ^1H in the mass of its nucleus. Can you think of how that difference might affect its properties and those of its compounds?

Self-test 8B.2A What are some properties of hydrogen that argue against its classification as a Group 17 element?

[***Answer:*** Hydrogen has no p-electrons and has a very low electron affinity.]

Self-test 8B.2B What are some properties of hydrogen that argue against its classification as a Group 1 element?

The hydride ion has a large radius and is highly polarizable. Ionic hydrides are powerful reducing agents. Hydrogen can form hydrogen bonds to lone pairs of electrons on highly electronegative elements.

Topic 8B Exercises

8B.1 Is there any chemical support for the view that hydrogen should be classified as a member of Group 1? Give evidence that supports this view.

8B.2 Is there any chemical support for the view that hydrogen should be classified as a member of Group 17? Give evidence that supports this view.

8B.3 Write a balanced chemical equation for (a) the hydrogenation of ethyne (acetylene, C_2H_2) to ethene (C_2H_4) by hydrogen (give the oxidation numbers of the carbon atoms in the reactant and product); (b) the shift reaction; (c) the reaction of barium hydride with water.

8B.4 Write a balanced chemical equation for (a) the reaction between sodium hydride and water; (b) the formation of synthesis gas; (c) the hydrogenation of ethene, $H_2C{=}CH_2$, and give the oxidation numbers of the carbon atoms in the reactant and product; (d) the reaction of magnesium with hydrochloric acid.

8B.5 Identify the products and write a balanced equation for the reaction of hydrogen with (a) chlorine; (b) sodium; (c) phosphorus; (d) copper.

8B.6 Identify the products and write a balanced equation for the reaction of hydrogen with (a) nitrogen; (b) fluorine; (c) cesium; (d) copper(II) ions.

8B.7 The enthalpy of dissociation of hydrogen bonds, ΔH_{HBond}, is a measure of their strength. Explain the trend seen in the data for the following pure substances, which were measured in the gas phase:

Substance	NH_3	H_2O	HF
$\Delta H_{HBond}/(kJ \cdot mol^{-1})$	17	25	29

8B.8 Methanoic acid (formic acid), HCOOH, forms dimers in the gas phase. Propose a reason for this behavior.

Topic 8C Group 1: The Alkali Metals

8C.1 The Group 1 Elements
8C.2 Compounds of Lithium, Sodium, and Potassium

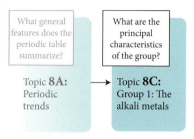

The members of Group 1 are called the *alkali metals*. The chemical properties of these elements are unique and strikingly similar. Nevertheless, there are differences, and the subtlety of some of these differences is the basis of the most subtle property of matter: consciousness. Your thinking, which relies on the transmission of signals along neurons, is achieved by the concerted action of sodium and potassium ions and their carefully regulated migration across membranes. So, even to learn about sodium and potassium, you have to make use of them in your brain.

The valence electron configuration of the alkali metals is ns^1, where n is the period number. Their physical and chemical properties are dominated by the ease with which the single valence electron can be removed (**TABLE 8C.1**).

TABLE 8C.1 The Group 1 Elements

Common name: alkali metals
Valence configuration: ns^1
Normal form*: soft, silver-gray metals

Z	Name	Symbol	Molar mass/ (g·mol^{-1})	Melting point/°C	Boiling point/°C	Density/ (g·cm^{-3})
3	lithium	Li	6.94	181	1347	0.53
11	sodium	Na	22.99	98	883	0.97
19	potassium	K	39.10	64	774	0.86
37	rubidium	Rb	85.47	39	688	1.53
55	cesium	Cs	132.91	28	678	1.87
87	francium	Fr	(223)	27	677	—

*Normal form means the state and appearance of the element at 25 °C and 1 atm.

8C.1 The Group 1 Elements

All the Group 1 elements are soft, lustrous, low-density metals (**FIG. 8C.1**). Lithium, sodium, and potassium are the only metals that are less dense than water. As explained in Topic 3H, in a metallic solid the cations are bound together by a "sea" of electrons that lies between them. Because the valence shell consists of a single electron, bonding in the metals is weak, leading to low melting and boiling points that decrease down the group (**FIG. 8C.2**). Cesium, which melts at 28 °C, is barely a solid at room temperature. Lithium is the hardest alkali metal, but it is softer than lead. Francium is a rare, intensely radioactive element about which little is known.

The alkali metals are the most violently reactive of all the metals and the most difficult to extract. They are too easily oxidized to be found in the free state in nature and cannot be extracted from their compounds by ordinary chemical reducing agents. The metals are obtained by electrolysis of their molten salts, as in the electrolytic Downs process (Topic 6O) or, in the case of potassium, by exposing molten potassium chloride to sodium vapor:

$$KCl(l) + Na(g) \xrightarrow{750\,°C} NaCl(s) + K(g)$$

(a)   (b)   (c)   (d)

FIGURE 8C.1 The alkali metals of Group 1: (a) lithium; (b) sodium; (c) potassium; (d) rubidium and cesium. Francium has never been isolated in visible quantities. The first three elements were freshly cut and you can see how rapidly they have been corroded in moist air; rubidium and cesium are even more reactive and have to be stored (and photographed) in sealed, airless containers. (*W. H. Freeman photos by Ken Karp.*)

Although the equilibrium constant for this reaction is not particularly favorable, the reaction runs to the right because potassium is more volatile than sodium: the potassium vapor is driven off by the heat and condensed in a cooled collecting vessel.

Lithium metal had few uses until after World War II, when thermonuclear weapons were developed (Topic 10C). This application has had an effect on the molar mass of lithium. Because only lithium-6 can be used in these weapons, the proportion of lithium-7 and, as a result, the molar mass of commercially available lithium has increased. A growing application of lithium is in rechargeable lithium–ion batteries. Because lithium has the most negative standard potential of all the elements, it can produce a high potential when it is used in a galvanic cell. Furthermore, because lithium has such a low density, lithium–ion batteries are light.

The first ionization energies of the alkali metals are low, and so in their compounds they exist as singly charged cations, such as Na^+. As a result, most of their compounds are ionic. The standard potentials of the alkali metals are all negative. However, $E°(M^+/M)$ does not vary as smoothly as the ionization energy, because the lattice energy of the solid and the hydration energy of the ions play a role in determining $E°(M^+/M)$; entropy effects also contribute.

Because their standard potentials are so strongly negative, all the alkali metals are strong reducing agents that react with water:

$$2\,Na(s) + 2\,H_2O(l) \longrightarrow 2\,NaOH(aq) + H_2(g)$$

The vigor of this reaction increases down the group (**FIG. 8C.3**). The reaction of water with sodium and potassium is vigorous enough to ignite the hydrogen that is formed; with rubidium and cesium it is dangerously explosive. Rubidium and cesium are more dense than water, and so they sink and react beneath the surface; the rapidly evolved hydrogen gas forms a shock wave that can shatter the vessel. Lithium is the least active of the Group 1 metals in the reaction with water, mainly because of the greater strength with which its atoms are bound together, but molten lithium is one of the most active metals known.

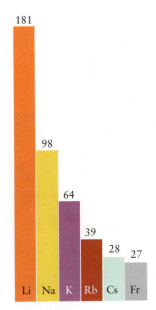

FIGURE 8C.2 The melting points of the alkali metals decrease down the group. The numerical values shown here are degrees Celsius.

(a)   (b)   (c)

FIGURE 8C.3 The alkali metals react with water, producing gaseous hydrogen and a solution of the alkali metal hydroxide. (a) Lithium reacts slowly. (b) Sodium reacts so vigorously that the heat released melts the unreacted metal and ignites the hydrogen. (c) Potassium reacts even more vigorously. (©1994 *Richard Megna–Fundamental Photographs.*)

FIGURE 8C.4 The blue swirls in this flask of liquid ammonia formed when small pieces of sodium were added and dissolved. (©1994 Richard Megna–Fundamental Photographs.)

FIGURE 8C.5 Although the alkali metals form a mixture of products when they react with oxygen, lithium gives mainly the oxide (left), sodium the very pale yellow peroxide (center), and potassium the yellow superoxide (right). (W. H. Freeman photo by Ken Karp.)

The alkali metals also release their valence electrons when they dissolve in liquid ammonia, but the outcome is different. Instead of reducing the ammonia, the electrons occupy cavities formed by groups of NH_3 molecules and give ink-blue *metal–ammonia solutions* (**FIG. 8C.4**). These solutions of solvated electrons (and cations of the metal) are often used to reduce organic compounds. As the metal concentration is increased, the blue gives way to a metallic bronze, and the solutions begin to conduct electricity like liquid metals.

THINKING POINT

Suppose you stir one of these bronze solutions. Would you expect its electrical conductivity to change?

All alkali metals react directly with almost all nonmetals except the noble gases. However, only lithium reacts with nitrogen, which it reduces to the nitride ion (N^{3-}):

$$6\,Li(s) + N_2(g) \longrightarrow 2\,Li_3N(s)$$

The principal product of the reaction of the alkali metals with oxygen varies systematically down the group (**FIG. 8C.5**). Lithium forms mainly the oxide, Li_2O. Sodium forms predominantly the very pale yellow sodium peroxide, Na_2O_2, which contains the peroxide ion O_2^{2-}, and potassium forms mainly the superoxide, KO_2, which contains the superoxide ion, O_2^-.

Self-test 8C.1A Write the chemical equation for the reaction between lithium and oxygen.
[*Answer:* $4\,Li(s) + O_2(g) \rightarrow 2\,Li_2O(s)$]

Self-test 8C.1B Write the chemical equation for the reaction between potassium and oxygen.

The alkali metals are highly reactive metals and so they are usually found as singly charged cations. They react with water with increasing vigor down the group.

8C.2 Compounds of Lithium, Sodium, and Potassium

Lithium is typical of an element at the head of its group in that it differs significantly from its congeners. The differences stem in part from the small size of the Li^+ cation, which gives it a strong polarizing power and hence a tendency to form bonds with significant covalent character. A further consequence of the small size of a Li^+ ion is strong ion–dipole interactions, with the result that many lithium salts form hydrates in which H_2O molecules are clustered around the Li^+ ion.

Lithium compounds are used in ceramics, lubricants, and medicine. Small daily doses of lithium carbonate are an effective treatment for bipolar (manic-depressive) disorder, but its action is not fully understood. Lithium soaps—the lithium salts of long-chain carboxylic acids—are used as thickeners in lubricating greases for high-temperature applications because they have higher melting points than more conventional sodium and potassium soaps.

THINKING POINT

Why do lithium soaps have higher melting points than sodium soaps?

Two factors that make sodium compounds important are their low cost and their high solubility in water. Sodium chloride is readily mined as rock salt, which is a deposit of sodium chloride left as ancient oceans evaporated; it is also obtained from the evaporation of brine (salt water) from present-day seas and salt lakes (**FIG. 8C.6**). Sodium

FIGURE 8C.6 An evaporation pond. The blue color is due to a dye added to the brine to increase heat absorption and hence speed up evaporation. (Pete McBride/National Geographic Creative.)

chloride is used in large quantities for the electrolytic production of chlorine and sodium hydroxide from brine.

Sodium hydroxide, NaOH, is a soft, waxy, white, corrosive solid that is sold commercially as "lye." It is an important industrial chemical because it is an inexpensive starting material for the production of other sodium salts. The amount of electricity used by industry to electrolyze brine to produce NaOH in the *chloralkali process* is second only to the amount used to extract aluminum from its ores. The process produces chlorine and hydrogen gases as well as aqueous sodium hydroxide (**FIG. 8C.7**). The net ionic equation for the reaction is

$$2\ Cl^-(aq) + 2\ H_2O(l) \xrightarrow{\text{electrolysis}} Cl_2(g) + 2\ OH^-(aq) + H_2(g)$$

Sodium hydrogen carbonate, $NaHCO_3$ (sodium bicarbonate), is commonly called *bicarbonate of soda* or *baking soda*. The rising action of baking soda in batter depends on the reaction of a weak acid, HA, with the hydrogen carbonate ions:

$$HCO_3^-(aq) + HA(aq) \longrightarrow A^-(aq) + H_2O(l) + CO_2(g)$$

The release of gas causes the batter to rise. The weak acids are provided by the recipe, generally in the form of lactic acid from sour milk or buttermilk, citric acid from lemons, or the acetic acid in vinegar. *Baking powder* contains a solid weak acid (often a mixture of monocalcium phosphate and sodium aluminum sulfate) as well as the hydrogen carbonate, and carbon dioxide is released when water is added.

Sodium carbonate decahydrate, $Na_2CO_3 \cdot 10H_2O$, was once widely used as *washing soda*. It is still sometimes added to water to precipitate Mg^{2+} and Ca^{2+} ions as carbonates, as in

$$Ca^{2+}(aq) + CO_3^{2-}(aq) \longrightarrow CaCO_3(s)$$

and to provide an alkaline environment that helps to remove grease from fabrics and dishes. Anhydrous sodium carbonate, or *soda ash,* is used in large amounts in the glass industry as a source of sodium oxide, into which it decomposes when heated (see the Interlude following Focus 3).

THINKING POINT

Why is calcium carbonate much less soluble than sodium carbonate?

Potassium compounds have many similarities to sodium compounds. The principal mineral sources of potassium are *carnallite*, $KCl \cdot MgCl_2 \cdot 6H_2O$, and *sylvite*, KCl, which is incorporated directly into some fertilizers as a source of essential potassium. Potassium compounds are generally more expensive than the corresponding sodium compounds, but in some applications their advantages outweigh their expense. For example, potassium nitrate, KNO_3, releases oxygen when heated, in the reaction

$$2\ KNO_3(s) \xrightarrow{\Delta} 2\ KNO_2(s) + O_2(g)$$

and is used to facilitate the ignition of matches. It is less hygroscopic (water absorbing) than the corresponding sodium compounds, because the potassium cation is larger and is less strongly hydrated by H_2O molecules.

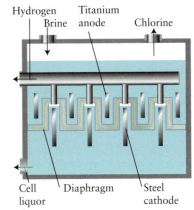

FIGURE 8C.7 A diaphragm cell for the electrolytic production of sodium hydroxide from brine (aqueous sodium chloride solution), represented by the blue color. The diaphragm prevents the chlorine produced at the titanium anodes from mixing with the hydrogen and the sodium hydroxide formed at the steel cathodes. The liquid (cell liquor) is drawn off and the water is partly evaporated. The unconverted sodium chloride crystallizes, leaving the sodium hydroxide dissolved in the cell liquor.

Self-test 8C.2A Use Table 6M.1 to determine the minimum potential difference that must be applied under standard conditions to carry out the chloralkali process.

[*Answer:* 2.2 V]

Self-test 8C.2B Like KO_2, cesium superoxide, CsO_2, can be used to remove exhaled CO_2 and generate oxygen from water. Explain why KO_2 is preferred for this purpose on spacecraft.

Lithium resembles magnesium, and its compounds show covalent character. Sodium compounds are soluble in water, plentiful, and inexpensive. Potassium compounds are generally less hygroscopic than sodium compounds.

Topic 8C Exercises

8C.1 Explain why lithium differs from its congeners in its chemical and physical properties. Give two examples to support your explanation.

8C.2 Explain why K^+ ions move more rapidly through water than Li^+ ions.

8C.3 (a) Write the valence electron configurations of the alkali metal atoms. (b) Explain why the alkali metals are strong reducing agents in terms of electron configurations, ionization energies, and the hydration of their ions.

8C.4 Explain why LiI and CsF are highly soluble in water, but LiF and CsI are only slightly soluble.

8C.5 Write the chemical equation for the reaction between (a) sodium and oxygen; (b) lithium and nitrogen; (c) sodium and water; (d) potassium superoxide and water.

8C.6 Write the chemical equation for the reaction between (a) cesium and oxygen (cesium reacts with oxygen in the same way as potassium); (b) sodium oxide and water; (c) lithium and hydrochloric acid; (d) cesium and iodine.

Topic 8D Group 2: The Alkaline Earth Metals

8D.1 The Group 2 Elements
8D.2 Compounds of Beryllium, Magnesium, and Calcium

What general features does the periodic table summarize?

What are the principal characteristics of the group?

Topic **8A:** Periodic trends → Topic **8D:** Group 2: The alkaline earth metals

The Group 2 metals include magnesium, which is the doorway to life because it is found in every chlorophyll molecule, and calcium, which builds your bones as well as being present in the concrete of buildings. Calcium, strontium, and barium are called the *alkaline earth metals* because their "earths"—the old name for oxides—are basic (alkaline). The name alkaline earth metal is often extended to all the members of Group 2 (**TABLE 8D.1**). The valence electron configuration of the atoms of the Group 2 elements is ns^2. The second ionization energy is low enough to be recovered from the lattice enthalpy of the compounds they form (**FIG. 8D.1**). Hence, the Group 2 elements occur with an oxidation number of $+2$, as the cation M^{2+} in all their compounds.

TABLE 8D.1 The Group 2 Elements

Common name: alkaline earth metals (Ca, Sr, Ba)
Valence configuration: ns^2
Normal form*: soft, silver-gray metals

Z	Name	Symbol	Molar mass/ (g·mol^{-1})	Melting point/°C	Boiling point/°C	Density/ (g·cm^{-3})
4	beryllium	Be	9.01	1285	2470	1.85
12	magnesium	Mg	24.31	650	1100	1.74
20	calcium	Ca	40.08	840	1490	1.53
38	strontium	Sr	87.62	770	1380	2.58
56	barium	Ba	137.33	710	1640	3.59
88	radium	Ra	(226)	700	1500	5.00

Normal form means the state and appearance of the element at 25 °C and 1 atm.

8D.1 The Group 2 Elements

Group 2 metals have many characteristics in common with Group 1 metals, but there are some important differences. All the Group 2 elements are too reactive to occur in the free state in nature (**FIG. 8D.2**). Instead, they are generally found in compounds as doubly charged cations. The element beryllium occurs mainly as the mineral *beryl*, $3BeO\cdot Al_2O_3\cdot 6SiO_2$, sometimes in crystals so big that they weigh several tons. The gemstone *emerald* is a form of beryl; its green color is caused by Cr^{3+} ions present as impurities (**FIG. 8D.3**). Magnesium occurs in seawater and as the mineral *dolomite*, $CaCO_3\cdot MgCO_3$. Calcium also occurs as $CaCO_3$ in *limestone, calcite,* and *chalk* (Topic 3I).

(a) Beryllium The element beryllium, Be, is obtained by electrolytic reduction of molten beryllium chloride. The element's low density makes it useful for the construction of components of missiles and satellites. Beryllium is also used as windows for x-ray tubes: because Be atoms have so few electrons, thin sheets of the metal are transparent to x-rays. Beryllium is added in small amounts to copper: these hard, electrically conducting beryllium–copper alloys are used to make nonsparking tools and in the electronics industry to form tiny nonmagnetic parts that resist deformation and corrosion.

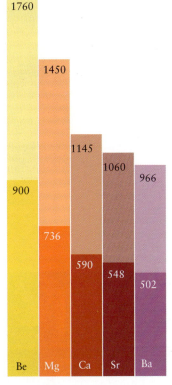

FIGURE 8D.1 The first (front row) and second (back row) ionization energies (in kilojoules per mole) of the Group 2 elements. Although the second ionization energies are larger than the first, they are not enormous, and both valence electrons are lost from each atom in all the ionic compounds of these elements.

(a)

(b)

(c)

(d)

(e)

FIGURE 8D.2 The Group 2 elements: (a) beryllium; (b) magnesium; (c) calcium; (d) strontium; and (e) barium. The four central elements (magnesium through barium) were discovered by Humphry Davy in a single year (1808). The two outer elements were discovered later: beryllium in 1828 (by Friedrich Wöhler) and radium (which is not shown here) in 1898 (by Pierre and Marie Curie). (*W. H. Freeman photos by Ken Karp.*)

FIGURE 8D.3 An emerald is a crystal of beryl with some Cr^{3+} ions, which are responsible for the green color. (*M. Claye/Science Source.*)

(b) Magnesium   Metallic magnesium, Mg, is produced by either chemical or electrolytic reduction of its compounds. In chemical reduction, first magnesium oxide is obtained from the thermal decomposition of dolomite. Then ferrosilicon, an alloy of iron and silicon, is used to reduce the MgO to the metal at about 1200 °C. At this temperature, the magnesium produced is immediately vaporized and carried away. The electrolytic method uses seawater as its principal raw material: magnesium hydroxide is precipitated by adding slaked lime, $Ca(OH)_2$, the precipitate is filtered off and treated with hydrochloric acid to produce magnesium chloride, and the dried molten salt is electrolyzed.

Magnesium is a silver-white metal that is protected from extensive oxidation in air by a film of white oxide, which gives the metal a dull gray coat. Its density is only two-thirds that of aluminum and the pure metal is very soft. However, its alloys have great strength and are used in applications requiring lightness and toughness—in airplanes, for instance. The use of magnesium alloys in automobiles increased by a factor of 5 between 1989 and 1995 as manufacturers sought to reduce the weight of vehicles. However, don't expect to be driving a magnesium car soon: magnesium is more expensive than steel and more difficult to work.

Magnesium burns vigorously in air with a brilliant white flame, partly because it reacts with nitrogen and carbon dioxide as well as with oxygen. Because the reaction is accelerated when burning magnesium is sprayed with water or exposed to carbon dioxide, neither water nor CO_2 fire extinguishers should ever be used on a magnesium fire. They will only make the fire worse!

(c) Calcium, strontium, and barium   The true alkaline earth metals—calcium (Ca), strontium (Sr), and barium (Ba)—are obtained either by electrolysis or by reduction with aluminum in a version of the thermite process (see Fig. 4A.8):

$$3\,BaO(s) + 2\,Al(s) \xrightarrow{\Delta} 2\,Al_2O_3(s) + 3\,Ba(s)$$

As in Group 1, reactions of the Group 2 metals with oxygen and water increase in vigor going down the group. Beryllium, magnesium, calcium, and strontium are partly protected in air by a surface layer of oxide.

All the Group 2 elements with the exception of beryllium react with water; for example,

$$Ca(s) + 2\,H_2O(l) \longrightarrow Ca^{2+}(aq) + 2\,OH^-(aq) + H_2(g)$$

Beryllium does not react with water, even when it is red hot: its protective oxide film survives even at high temperatures. Magnesium reacts with hot water (see Fig. 7D.1), and calcium reacts with cold water (**FIG. 8D.4**).

The alkaline earth metals can be detected in burning compounds by the colors that they give to flames. Calcium burns orange-red, strontium crimson, and barium yellow-green.

FIGURE 8D.4 Calcium reacts gently with cold water to produce hydrogen and calcium hydroxide, $Ca(OH)_2$. (*Andrew Lambert Photography/Science Source.*)

Fireworks are often made from their salts (typically nitrates and chlorates, because the anions then provide an additional supply of oxygen) together with magnesium powder.

Beryllium shows a hint of nonmetallic character, but the other Group 2 elements are all reactive metals. The vigor of reaction with water and oxygen increases down the group.

8D.2 Compounds of Beryllium, Magnesium, and Calcium

Apart from a tendency toward nonmetallic character in beryllium, Group 2 elements have all the chemical characteristics of metals, such as forming basic oxides and hydroxides.

(a) Beryllium   Beryllium resembles its diagonal neighbor aluminum in its chemical properties. It is the least metallic element of the group, and many of its compounds have properties commonly attributed to covalent bonding. Beryllium is amphoteric and reacts with both acids and alkalis. Like aluminum, beryllium reacts with water in the presence of sodium hydroxide; the products are the *beryllate ion*, $Be(OH)_4^{2-}$, and hydrogen:

$$Be(s) + 2 OH^-(aq) + 2 H_2O(l) \longrightarrow Be(OH)_4^{2-}(aq) + H_2(g)$$

Beryllium compounds are very toxic and must be handled with great caution. Their properties are dominated by the highly polarizing character of the Be^{2+} ion and its small size. The strong polarizing power results in moderately covalent compounds, and its small size limits to four the number of groups that can attach to the ion. These two features together are responsible for the prominence of the tetrahedral BeX_4 unit (**1**), like that in the beryllate ion. A tetrahedral unit is also found in the solid chloride (**2**) and hydride. Beryllium hydride was once thought to consist of chains of BeH_2 groups, but it is now known to have a network structure (**FIG. 8D.5**). The chloride is made by the action of chlorine on the oxide in the presence of carbon:

$$BeO(s) + C(s) + Cl_2(g) \xrightarrow{600-800\,°C} BeCl_2(g) + CO(g)$$

The Be atoms in $BeCl_2$ act as Lewis acids and accept electron pairs from the Cl atoms of the neighboring linear $BeCl_2$ groups, forming a chain of tetrahedral $BeCl_4$ units in the solid.

(b) Magnesium   Magnesium has more pronounced metallic properties than beryllium; its compounds are primarily ionic, with some covalent character. Magnesium oxide, MgO, is formed when magnesium burns in air, but the product is contaminated by magnesium nitride. To prepare the pure oxide, the hydroxide or the carbonate is heated. Magnesium oxide dissolves only very slightly in water. One of its most striking properties is that it is *refractory* (able to withstand high temperatures); it melts at 2800 °C. This high stability can be traced to the small ionic radii of the Mg^{2+} and O^{2-} ions and hence to their very strong electrostatic interaction with each other. The oxide has two other useful characteristics: it conducts heat very well, and it conducts electricity poorly. All three properties make it useful as an insulator in electric heaters.

Magnesium hydroxide, $Mg(OH)_2$, is a base. It is not very soluble in water but forms instead a white colloidal suspension, a mist of small particles dispersed through a liquid (Topic 5G), which is known as *milk of magnesia* and is used as a stomach antacid. Because this base is not very soluble, it is not absorbed from the stomach but remains to act on whatever acids are present. Hydroxides have an advantage over hydrogen carbonates (which also are used as antacids) in that their neutralization does not lead to the formation of carbon dioxide and its inconvenient consequence, belching. Milk of magnesia

1 BeX_4 unit

2   Beryllium chloride, $BeCl_2$

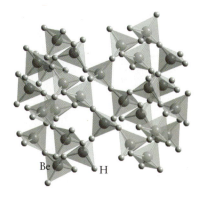

FIGURE 8D.5 The network structure of beryllium hydride is based on tetrahedral BeX_4 units.

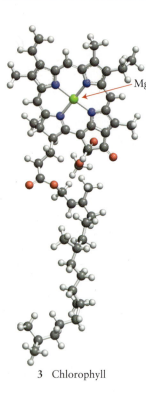

3   Chlorophyll

must be used sparingly as an antacid, because the salt produced by the neutralization of milk of magnesia in the stomach is magnesium chloride, which can act as a purgative. Magnesium sulfate, as *Epsom salts*, $MgSO_4 \cdot 7H_2O$, is another common purgative. Because the doubly charged ions become hydrated, they inhibit the absorption of water from the intestine. The resulting increased flow of water through the intestine triggers the mechanism that results in defecation.

Arguably the most important compound of magnesium is chlorophyll. This green organic compound consists of large molecules that capture light from the Sun and channel its energy into photosynthesis. One function of the Mg^{2+} ion, which lies just above the plane of a ring of nitrogen and carbon atoms (3), appears to be to keep the ring rigid. This rigidity helps to ensure that the energy captured from the incoming photon is not lost as heat before it has been used to bring about a chemical reaction. Magnesium also plays an important role in energy generation in living cells. For example, it is involved in the contraction of muscles.

(c) Calcium   The most common compound of calcium is calcium carbonate, $CaCO_3$, which occurs naturally in a variety of forms such as chalk and limestone (Topic 3I).

Calcium carbonate decomposes to calcium oxide, CaO, or *quicklime*, when heated:

$$CaCO_3(s) \xrightarrow{\Delta} CaO(s) + CO_2(g)$$

Quicklime is produced in enormous quantities throughout the world. About 40% of this output is used in metallurgy. In ironmaking (Topic 9B), it is used as a Lewis base; its O^{2-} ion reacts with silica, SiO_2, impurities in the ore to form a liquid *slag*:

$$CaO(s) + SiO_2 \xrightarrow{\Delta} CaSiO_3(l)$$

Calcium oxide is called quicklime because it reacts so exothermically and rapidly with water:

$$CaO(s) + H_2O(l) \longrightarrow Ca^{2+}(aq) + 2\,OH^-(aq)$$

The product, calcium hydroxide, is commonly known as *slaked lime* because, as calcium hydroxide, the thirst of lime for water has been quenched (slaked). Slaked lime is the form in which lime is normally sold, because quicklime can set fire to moist wood and paper. In fact, the wooden boats that were once used to transport quicklime sometimes caught fire in the heat of reaction when water seeped into their holds. Slaked lime is used as an inexpensive base in industry and to adjust the pH of soils in agriculture. Perhaps surprisingly, it is also used to *remove* Ca^{2+} ions from hard water containing $Ca(HCO_3)_2$. Its role there is to convert HCO_3^- into CO_3^{2-} by providing OH^- ions:

$$HCO_3^-(aq) + OH^-(aq) \longrightarrow CO_3^{2-}(aq) + H_2O(l)$$

As the concentration of CO_3^{2-} ions rises, Ca^{2+} ions precipitate in the reaction

$$Ca^{2+}(aq) + CO_3^{2-}(aq) \longrightarrow CaCO_3(s)$$

This reaction removes the Ca^{2+} ions that were present initially and those that were added as lime; overall, the Ca^{2+} ion concentration is reduced.

THINKING POINT

Why is calcium hydrogen carbonate (bicarbonate) more soluble than calcium carbonate?

Teeth are generally denser than bone and are protected by a hard enamel coating. Tooth enamel is a *hydroxyapatite*, $Ca_5(PO_4)_3OH$. Tooth decay begins when acids attack the enamel:

$$Ca_5(PO_4)_3OH(s) + 4\,H_3O^+(aq) \longrightarrow 5\,Ca^{2+}(aq) + 3\,HPO_4^{2-}(aq) + 5\,H_2O(l)$$

The principal agents of tooth decay are the carboxylic acids produced when bacteria act on the remains of food. A more resistant coating forms when the OH^- ions in the apatite are replaced by F^- ions. The resulting mineral is called *fluorapatite*:

$$Ca_5(PO_4)_3OH(s) + F^-(aq) \longrightarrow Ca_5(PO_4)_3F(s) + OH^-(aq)$$

The addition of fluoride ions to domestic water supplies (in the form of NaF) is now widespread and has resulted in a dramatic decrease in dental cavities. Fluoridated toothpastes, containing either tin(II) fluoride or sodium monofluorophosphate (MFP, Na_2FPO_3), are also recommended to strengthen tooth enamel.

Self-test 8D.2A Explain why beryllium compounds have covalent characteristics.
   [**Answer:** The small size and high charge on the beryllium ion make it highly polarizing.]

Self-test 8D.2B Is the reaction between CaO and SiO_2 a redox reaction or a Lewis acid–base reaction? If it is redox, identify the oxidizing and reducing agents. If it is Lewis acid–base, identify the acid and the base.

Beryllium compounds have a pronounced covalent character, and the structural unit is commonly tetrahedral. The small size of the magnesium cation results in a thermally stable oxide with low solubility in water. Calcium compounds are common in structural materials because the small, highly charged Ca^{2+} ion results in rigid structures.

Topic 8D Exercises

8D.1 Write the chemical equation for the reaction between magnesium metal and hot water.

8D.2 Write the chemical equation for the reaction between strontium metal and hydrogen gas.

8D.3 The compounds CaO and BaO are sometimes used as drying agents for organic solvents such as pyridine, C_5H_5N. The drying agent and the product of the drying reaction are both insoluble in the organic solvent. (a) Write the balanced chemical equation corresponding to the reaction that dries the solvent. (b) For CaO, estimate the standard Gibbs free energy of the drying reaction at 25 °C.

8D.4 Aluminum and beryllium have a diagonal relationship. Write the chemical equations for the reaction of aluminum with aqueous sodium hydroxide to that of beryllium with aqueous sodium hydroxide.

8D.5 Predict the products of each of the following reactions and then balance each equation:
(a) $Mg(OH)_2 + HCl(aq) \rightarrow$
(b) $Ca(s) + H_2O(l) \rightarrow$
(c) $BaCO_3(s) \xrightarrow{\Delta}$

8D.6 Predict the products of each of the following reactions and then balance each equation:
(a) $Mg(s) + Br_2(l) \rightarrow$
(b) $BaO(s) + Al(s) \rightarrow$
(c) $CaO(s) + SiO_2(s) \rightarrow$

8D.7 (a) Draw the Lewis structure of $BeCl_2$. (b) Predict the Cl—Be—Cl bond angle. (c) What hybrid orbitals are used in the bonding in $BeCl_2$? (d) Why does $MgCl_2$ not have the same structure?

8D.8 In the gas phase, $BeCl_2$ forms a dimer by forming chlorine-atom bridges like those in the $AlCl_3$ dimer. Draw the Lewis structure of the $BeCl_2$ dimer and assign formal charges.

Topic 8E Group 13: The Boron Family

8E.1   The Group 13 Elements

8E.2   Group 13 Oxides, Halides, and Nitrides

8E.3   Boranes, Borohydrides, and Borides

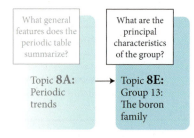

Group 13 is the first group of the p block. Here, close to the center of the periodic table, elements have subtle properties, because neither gaining nor losing electrons is energetically favorable. Members of Group 13 have an ns^2np^1 electron configuration (**TABLE 8E.1**) and so have a maximum oxidation number of $+3$. The oxidation numbers of B and Al are $+3$ in almost all their compounds. However, the heavier elements in the group are more likely to keep their s-electrons (the inert-pair effect, Topic 1F); so the oxidation number $+1$ becomes increasingly important down the group, and thallium(I) compounds are as common as thallium(III) compounds.

8E.1 The Group 13 Elements

This Topic focuses on the two most important members of the group, boron and aluminum.

(a) Boron   The element boron, B, forms perhaps the most extraordinary structures of all the elements. It has a high ionization energy and is a metalloid that forms covalent bonds, like its diagonal neighbor silicon. However, because it has only three electrons in its valence shell and has a small atomic radius, it tends to form compounds that have incomplete octets (Topic 2C) or are electron deficient. These unusual bonding characteristics have made boron an essential element of modern technology and, in particular, nanotechnology.

Boron is mined as *borax* and *kernite,* $Na_2B_4O_7 \cdot xH_2O$, with $x = 10$ and 4, respectively. Large deposits from ancient hot springs are found in volcanic regions, such as the Mojave Desert region of California. In the extraction process, the ore is converted into boron oxide with acid and then reduced with magnesium to an impure brown, amorphous form of boron:

$$B_2O_3(s) + 3\,Mg(s) \xrightarrow{\Delta} 2\,B(s) + 3\,MgO(s)$$

Here, perhaps, is an opportunity for another young chemist like Hall to transform the production of boron as Hall did for aluminum (described later in this Topic).

A product of higher purity is obtained by reducing a volatile boron compound, such as BCl_3 or BBr_3, with hydrogen on a heated tantalum filament:

$$2\,BBr_3(g) + 3\,H_2(g) \xrightarrow{\Delta,\,Ta} 2\,B(s) + 6\,HBr(g)$$

Because of the expense, boron production remains quite low despite the element's desirable properties of hardness and low density.

TABLE 8E.1   The Group 13 Elements

Valence configuration: ns^2np^1

Z	Name	Symbol	Molar mass/ (g·mol^{-1})	Melting point/°C	Boiling point/°C	Density/ (g·cm^{-3})	Normal form*
5	boron	B	10.81	2300	3931	2.47	powdery brown metalloid
13	aluminum	Al	26.98	660	2467	2.70	silver-white metal
31	gallium	Ga	69.72	30	2403	5.91	silver metal
49	indium	In	114.82	156	2080	7.29	silver-white metal
81	thallium	Tl	204.38	304	1457	11.87	soft metal

*Normal form means the state and appearance of the element at 25 °C and 1 atm.

Elemental boron exists in a variety of allotropic forms in which each atom strives to find ways to share eight electrons despite being so small and having only three electrons to contribute. It is most commonly found as either a gray-black nonmetallic, high-melting-point solid or a dark brown powder with an icosahedral (20-faced) structure based on clusters of 12 atoms (**1**). Because of the three-dimensional network formed by these bonds, boron is very hard and is chemically unreactive. When boron fibers are incorporated into plastics, the result is a very tough composite material that is stiffer than steel yet lighter than aluminum; this material is used in aircraft, missiles, and body armor (Topic 11E).

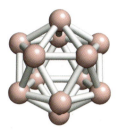

1 B_{12}

(b) Aluminum   The element aluminum, Al, is the most abundant metallic element in the Earth's crust and, after oxygen and silicon, the third most abundant element (see Fig. 8B.1). However, the aluminum content in most minerals is low, and the commercial source of aluminum, *bauxite,* is a hydrated, impure oxide, $Al_2O_3 \cdot xH_2O$, where x can range from 1 to 3. Bauxite ore, which is red from the iron oxides that it contains (**FIG. 8E.1**), is processed to obtain alumina, Al_2O_3, in the *Bayer process*. In this process, the ore is first treated with aqueous sodium hydroxide, which dissolves the amphoteric alumina as the aluminate ion, $Al(OH)_4^-$. Carbon dioxide is then bubbled through the solution to remove OH^- ions as HCO_3^- and to convert some of the aluminate ions into aluminum hydroxide, which precipitates. The aluminum hydroxide is removed and dehydrated to the oxide by heating to 1200 °C.

FIGURE 8E.1 The iron oxides in bauxite ore give it a red tint. (*Michael James/Science Source.*)

Obtaining aluminum metal from the oxide was a challenge for early scientists and engineers. When it was first isolated, aluminum was a rare and expensive metal. During the nineteenth century it so symbolized modern technology that the Washington Monument was given an expensive aluminum tip. That rarity and expense were transformed by electrochemistry. Aluminum metal is now obtained on a huge scale by the *Hall process*. In 1886, Charles Hall discovered that, by mixing the mineral cryolite, Na_3AlF_6, with alumina, he got a mixture that melts at a much more economical temperature, 950 °C, instead of the 2050 °C of pure alumina. The melt is electrolyzed in a cell that uses graphite (or carbonized petroleum) anodes and a carbonized steel-lined vat that serves as the cathode (**FIG. 8E.2**). The electrolysis half-reactions are

Napoleon reserved his aluminum plates for his special guests; the rest had to make do with gold.

Cathode reaction: $Al^{3+}(melt) + 3\,e^- \longrightarrow Al(l)$

Anode reaction: $2\,O^{2-}(melt) + C(s, gr) \longrightarrow CO_2(g) + 4\,e^-$

Overall reaction: $4\,Al^{3+}(melt) + 6\,O^{2-}(melt) + 3\,C(s, gr) \longrightarrow 4\,Al(l) + 3\,CO_2(g)$

Note that the carbon electrode is consumed in the reaction. It follows from the reaction stoichiometry that a current of 1 A must flow for 80 h to produce 1 mol Al (27 g of aluminum, about enough for two soft-drink cans). The very high energy consumption can be greatly reduced by recycling, which requires less than 5% of the electricity needed to extract aluminum from bauxite. Although a typical beverage can contains less than 14 g of aluminum, the energy thrown away when an aluminum can is discarded is equivalent to burning the amount of gasoline that would fill half the can.

Aluminum has a low density; it is a strong metal and an excellent electrical conductor. Although it is strongly reducing and therefore easily oxidized, aluminum is resistant to corrosion because its surface is passivated in air by a stable oxide film (Topic 6N). The thickness of the oxide layer can be increased by making aluminum the anode of an electrolytic cell; the result is called *anodized aluminum*.

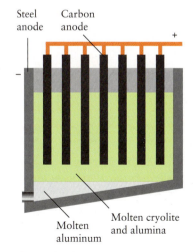

Aluminum's low density, wide availability, and corrosion resistance make it ideal for construction and for the aerospace industry. Aluminum is a soft metal, and so it is usually alloyed with copper and silicon for greater strength. Its lightness and good electrical conductivity have also led to its use for overhead power lines, and its negative electrode potential has led to its use in fuel cells. Perhaps one day your automobile will not only be made of aluminum but fueled by it too.

Aluminum is amphoteric, reacting both with nonoxidizing acids (such as hydrochloric acid) to form aluminum ions,

$$2\,Al(s) + 6\,H^+(aq) \longrightarrow 2\,Al^{3+}(aq) + 3\,H_2(g)$$

and with hot aqueous alkali to form aluminate ions,

$$2\,Al(s) + 2\,OH^-(aq) + 6\,H_2O(l) \longrightarrow 2\,Al(OH)_4^-(aq) + 3\,H_2(g)$$

FIGURE 8E.2 In the Hall process, aluminum oxide is dissolved in molten cryolite and the mixture is electrolyzed in a cell with carbon anodes and a steel cathode. The molten aluminum flows out of the bottom of the cell.

Steel anode   Carbon anode   Molten aluminum   Molten cryolite and alumina

An oxidizing acid is an oxoacid in which the anion is an oxidizing agent; examples are HNO_3 and $HClO_3$.

Gallium, which is produced as a byproduct in the Bayer process, has important uses in the electronics industry. It is a common doping agent for semiconductors, and some of its compounds, such as GaAs, are used in light-emitting diodes to convert electricity into light. Thallium, at the bottom of Group 13, is a dangerously poisonous heavy metal.

Boron is a hard metalloid with pronounced nonmetallic properties. Aluminum is a light, strong, amphoteric, reactive metallic element with a surface that becomes passivated when exposed to air.

8E.2 Group 13 Oxides, Halides, and Nitrides

Boron, a metalloid with largely nonmetallic properties, has acidic oxides. Aluminum, its metallic neighbor, has amphoteric oxides (like its diagonal neighbor in Group 2, beryllium).

(a) Boron Boric acid, $B(OH)_3$, is a toxic white solid that melts at 171 °C. Because the boron atom in $B(OH)_3$ has an incomplete octet, it can act as a Lewis acid and form a bond by accepting a lone pair of electrons from an H_2O molecule acting as a Lewis base:

$$(OH)_3B + :OH_2 \longrightarrow (OH)_3BOH_2$$

The compound so formed is a weak *mono*protic acid:

$$(OH)_3BOH_2(aq) + H_2O(l) \rightleftharpoons H_3O^+(aq) + B(OH)_4^-(aq) \qquad pK_a = 9.14$$

The major use of boric acid is as the starting material for its anhydride, boron oxide, B_2O_3. Because it melts (at 450 °C) to a liquid that dissolves many metal oxides, boron oxide (often as the acid) is used as a *flux*, a substance that cleans metals as they are soldered or welded. Boron oxide is also used to make fiberglass and borosilicate glass, a glass with a very low thermal expansion, such as Pyrex (see the Interlude following Focus 3).

THINKING POINT

Why does borosilicate glass expand less than other glasses when it is heated?

The boron halides are useful industrial catalysts that are made either by direct reaction of the elements at a high temperature or from boron oxide. The most important is boron trifluoride, BF_3, which is produced by the reaction between boron oxide, calcium fluoride, and sulfuric acid:

$$B_2O_3(s) + 3\,CaF_2(s) + 3\,H_2SO_4(l) \xrightarrow{\Delta} 2\,BF_3(g) + 3\,CaSO_4(s) + 3\,H_2O(l)$$

Boron trichloride, BCl_3, which is also widely used as a catalyst, is produced commercially by the action of chlorine gas on the oxide in the presence of carbon:

$$B_2O_3(s) + 3\,C(s) + 3\,Cl_2(g) \xrightarrow{500\,°C} 2\,BCl_3(g) + 3\,CO(g)$$

The B atom has an incomplete octet in all its trihalides. The compounds consist of trigonal planar molecules with an empty 2p-orbital perpendicular to the molecular plane. The empty orbital allows the molecules to act as Lewis acids, which accounts for the catalytic action of BF_3 and BCl_3.

When boron oxide, B_2O_3, is heated to 900 °C in ammonia, boron nitride, BN, is formed as a fluffy, slippery powder:

$$B_2O_3(s) + 2\,NH_3(g) \xrightarrow{\Delta} 2\,BN(s) + 3\,H_2O(g)$$

Its structure resembles that of graphite, but the latter's flat planes of carbon hexagons are replaced in boron nitride by planes of hexagons of alternating B and N atoms (**FIG. 8E.3**).

Although it is isoelectronic with graphite, boron nitride is white and has much lower electrical conductivity than graphite because the electrons are localized on the nitrogen atoms. Under high pressure, boron nitride is converted into a very hard, diamondlike crystalline form called Borazon (a proprietary name). In recent years, boron nitride nanotubes similar to those formed by carbon have been synthesized, and they have been found to be semiconducting (Topic 3J).

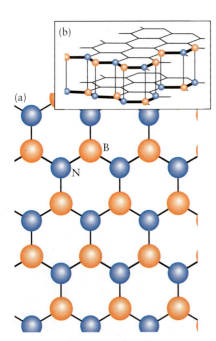

FIGURE 8E.3 (a) The structure of hexagonal boron nitride, BN, resembles that of graphite, consisting of flat planes of hexagons of alternating B and N atoms (in place of C atoms) but, as shown for two adjacent layers in part (b), the planes are stacked differently, with each B atom directly over an N atom and vice versa (compare with Fig. 3H.8).

FIGURE 8E.4 Some of the impure forms of α-alumina are prized as gems. (a) Ruby is alumina with Cr^{3+} in place of some Al^{3+} ions. (b) Sapphire is alumina with Fe^{3+} and Ti^{4+} impurities. (c) Topaz is alumina with Fe^{3+} impurities. (*Part (a) Jacana/Science Source; part (b) ©Boltin Picture Library/Bridgeman Images; part (c) Roberto de Gugliemo/Science Source.*)

(a)

(b)

(b) Aluminum Aluminum oxide, Al_2O_3, is known almost universally as *alumina*. It exists with a variety of crystal structures, many of which form important ceramic materials (see the Interlude following Focus 3). Some impure forms of alumina are beautiful, rare, and highly prized (**FIG. 8E.4**). A less dense and more reactive form of the oxide is γ-alumina. This form absorbs water and is used as the stationary phase in chromatography.

γ-Alumina is produced by heating aluminum hydroxide. It is moderately reactive and is amphoteric, dissolving readily in bases to produce the aluminate ion and in acids to produce the hydrated Al^{3+} ion:

$$Al_2O_3(s) + 2\,OH^-(aq) + 3\,H_2O(l) \longrightarrow 2\,Al(OH)_4^-(aq)$$
$$Al_2O_3(s) + 6\,H_3O^+(aq) + 3\,H_2O(l) \longrightarrow 2\,Al(H_2O)_6^{3+}(aq)$$

One of the most important aluminum salts prepared by the action of an acid on alumina is aluminum sulfate, $Al_2(SO_4)_3$:

$$Al_2O_3(s) + 3\,H_2SO_4(aq) \longrightarrow Al_2(SO_4)_3(aq) + 3\,H_2O(l)$$

Aluminum sulfate is called *papermaker's alum* and is used in the paper industry to coagulate cellulose fibers into a hard, nonabsorbent surface. True alums (from which aluminum takes its name) are mixed sulfates of formula $M^+M'^{3+}(SO_4)_2 \cdot 12H_2O$. They include potassium alum, $KAl(SO_4)_2 \cdot 12H_2O$ (which is used in water and sewage treatment), and ammonium alum, $NH_4Al(SO_4)_2 \cdot 12H_2O$ (which is used for pickling cucumbers and as the Brønsted acid in baking powders).

Sodium aluminate, $NaAl(OH)_4$, is used along with aluminum sulfate in water purification. When mixed with aluminate ions, the acidic hydrated Al^{3+} cation from the aluminum sulfate produces aluminum hydroxide:

$$Al^{3+}(aq) + 3\,Al(OH)_4^-(aq) \longrightarrow 4\,Al(OH)_3(s)$$

The aluminum hydroxide is formed as a fluffy, gelatinous network that entraps impurities as it settles, and this precipitate can be removed by filtration (**FIG. 8E.5**).

Aluminum chloride, $AlCl_3$, a major industrial catalyst, is made by the action of chlorine on aluminum or on alumina in the presence of carbon:

$$2\,Al(s) + 3\,Cl_2(g) \longrightarrow 2\,AlCl_3(s)$$
$$Al_2O_3(s) + 3\,C(s) + 3\,Cl_2(g) \longrightarrow 2\,AlCl_3(s) + 3\,CO(g)$$

Aluminum chloride is an ionic solid in which each Al^{3+} ion is coordinated to six Cl^- ions. However, it sublimes at 192 °C to a vapor of Al_2Cl_6 molecules (**2**). In this molecule, the Al atom of one $AlCl_3$ planar fragment acts as a Lewis acid and accepts an electron pair from a Cl atom of the other $AlCl_3$ fragment, which is acting as a Lewis base.

(c)

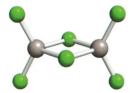

2 Aluminum chloride dimer, Al_2Cl_6

FIGURE 8E.5 Aluminum hydroxide, $Al(OH)_3$, forms as a white, fluffy precipitate. The fluffy form of the solid captures impurities and is used in the purification of water. (*W. H. Freeman photo by Ken Karp.*)

Boron oxide is an acid anhydride. Aluminum shows some nonmetallic character in that its oxide is amphoteric. Boron and aluminum halides have incomplete octets and act as Lewis acids. Boron nitride has structures related to those formed by carbon.

8E.3 Boranes, Borohydrides, and Borides

Boron forms a remarkable series of binary compounds with hydrogen—the *boranes*. These compounds include diborane, B_2H_6, and more complex compounds such as decaborane, $B_{10}H_{14}$. Anionic versions of these compounds, the *borohydrides,* are known; the most important is $BH_4{}^-$ as sodium borohydride, $NaBH_4$.

Sodium borohydride is a very useful reducing agent. At pH = 14 (strongly alkaline conditions), the potential of the half-reaction

$$H_2BO_3{}^-(aq) + 5\,H_2O(l) + 8\,e^- \longrightarrow BH_4{}^-(aq) + 8\,OH^-(aq)$$

is -1.24 V. Because this potential is well below that of the Ni^{2+}/Ni couple (-0.23 V), borohydride ions can reduce Ni^{2+} ions to metallic nickel. This reduction is the basis of the "chemical plating" of nickel. The advantage of this chemical plating over electroplating is that the item being plated does not have to be an electrical conductor.

THINKING POINT

Do you have any possessions that may have been plated in this way?

The boranes are an extensive series of binary compounds of boron and hydrogen, somewhat analogous to the hydrocarbons. The starting point for borane production is the reaction (in an organic solvent) of sodium borohydride with boron trifluoride:

$$4\,BF_3 + 3\,BH_4{}^- \longrightarrow 3\,BF_4{}^- + 2\,B_2H_6$$

The product B_2H_6 is diborane (**3**), a colorless gas that bursts into flame in air. On contact with water, it immediately reduces the hydrogen in the water:

$$B_2H_6(g) + 6\,H_2O(l) \longrightarrow 2\,B(OH)_3(aq) + 6\,H_2(g)$$

When diborane is heated to a high temperature, it decomposes into hydrogen and pure boron:

$$B_2H_6(g) \xrightarrow{\Delta} 2\,B(s) + 3\,H_2(g)$$

This sequence of reactions is a useful route to the pure element, but more complex boranes form when the heating is less severe. When diborane is heated to 100 °C, for instance, it forms decaborane, $B_{10}H_{14}$, a solid that melts at 100 °C. Decaborane is stable in air, is oxidized by water only slowly, and is an example of the general rule that heavier boranes are less flammable than boranes of low molar mass.

The boranes are electron-deficient compounds (Topic 2G): valid Lewis structures cannot be drawn for them, because too few electrons are available. For instance, there are 8 atoms in diborane, so there must be at least 7 bonds; however, there are only 12 valence electrons, and so no more than 6 electron pair bonds can be formed. Molecular orbital theory readily overcomes this difficulty: the 6 electron pairs are regarded as delocalized over the entire molecule and their bonding power is shared by several atoms. In diborane, for instance, a single electron pair is delocalized over a B—H—B unit. It binds all 3 atoms together with bond order of $\frac{1}{2}$ for each of the B—H bridging bonds. The molecule has two such bridging **three-center bonds** (**4**).

THINKING POINT

Why are three-center bonds characteristic of boron but not other elements?

3 Diborane, B_2H_6

4 A three-center bond

Numerous borides of the metals and nonmetals are known. Their formulas are typically unrelated to their location in the periodic table and include AlB_2, CaB_6, $B_{13}C_2$, $B_{12}S_2$, Ti_3B_4, TiB, and TiB_2. In some metal borides, the boron atoms are found at the centers of clusters of metal atoms. More commonly, the boron atoms form extended structures such as zigzag chains, branched chains, or networks of hexagonal rings of boron atoms, as in MgB_2, which is a superconductor below 39 K (Topic 3J). The hardness and thermal stability related to their network structures make some borides suitable for rocket nozzles and turbine blades.

Self-test 8E.2A Compare the reactions for the formation of BF_3 and BCl_3. Which reaction is a redox reaction?

[**Answer:** The formation of BCl_3]

Self-test 8E.2B What is the oxidation number of boron in (a) $NaBH_4$; (b) $H_2BO_3^-$?

The boranes are an extensive series of highly reactive electron-deficient binary compounds of boron and hydrogen. Borohydrides are useful reducing agents. The boron atoms in borides form extended structures.

Topic 8E Exercises

8E.1 Write a balanced chemical equation for the industrial preparation of aluminum from its oxide.

8E.2 Write a balanced chemical equation for the industrial preparation of impure boron.

8E.3 Complete and balance the following equations:
(a) $B_2O_3(s) + Mg(l) \rightarrow$
(b) $Al(s) + Cl_2(g) \rightarrow$
(c) $Al(s) + O_2(g) \rightarrow$

8E.4 Complete and balance the following equations:
(a) $Al_2O_3(s) + OH^-(aq) \rightarrow$
(b) $Al_2O_3(s) + H_3O^+(aq) + H_2O(l) \rightarrow$
(c) $B(s) + NH_3(g) \rightarrow$

8E.5 Suggest a Lewis structure for B_4H_{10} and deduce the formal charges on the atoms. *Hint:* There are four B—H—B bridges.

8E.6 The diatomic molecule BF can be obtained by the reaction between BF_3 and B at a high temperature and low pressure. (a) Identify the electron configuration of the molecule in terms of the occupied molecular orbitals and calculate the bond order. (b) CO is isoelectronic with BF. How do the molecular orbitals in the two molecules differ?

8E.7 Many gallium compounds have structures that are similar to those of the corresponding aluminum and boron compounds. Draw the Lewis structure and describe the shape of $GaBr_4^-$.

8E.8 Many gallium compounds have structures that are similar to those of the corresponding aluminum and boron compounds. Draw the Lewis structure and describe the shape of Ga_2Cl_6.

Topic 8F Group 14: The Carbon Family

8F.1 The Group 14 Elements

8F.2 Oxides of Carbon and Silicon

8F.3 Other Important Group 14 Compounds

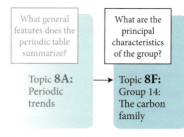

What general features does the periodic table summarize?

What are the principal characteristics of the group?

Topic **8A:** Periodic trends → Topic **8F:** Group 14: The carbon family

FIGURE 8F.1 The elements of Group 14. Back row, from left to right: silicon; tin. Front row: carbon (as graphite); germanium; lead. (©1984 Chip Clark–Fundamental Photographs.)

Carbon is central to life and natural intelligence. Silicon and germanium are central to electronic technology and artificial intelligence (**FIG. 8F.1**). The unique properties of Group 14 elements make both types of intelligence possible. Carbon, at the head of the group, forms so many compounds that it has its own branch of chemistry, organic chemistry (Focus 11).

The valence electron configuration is ns^2np^2 for all members of Group 14. All four electrons are approximately equally available for bonding in the lighter elements, and carbon and silicon are characterized by their ability to form four covalent bonds. However, as in Group 13, the heavier elements display the inert-pair effect (Topic 1F) and consequently the most common oxidation number for lead is +2.

8F.1 The Group 14 Elements

The Group 14 elements show increasing metallic character down the group (**TABLE 8F.1**). Carbon has definite nonmetallic properties and the oxides of carbon and silicon are acidic. Germanium is a typical metalloid in that it exhibits metallic or nonmetallic properties according to the other element present in the compound. Tin and, even more so, lead have definite metallic properties. However, even though tin is classified as a metal, it is not far from the metalloids in the periodic table, and it does have some amphoteric properties. For example, tin reacts with both hot concentrated hydrochloric acid and hot alkali:

$$Sn(s) + 2 H_3O^+(aq) \longrightarrow Sn^{2+}(aq) + H_2(g) + 2 H_2O(l)$$
$$Sn(s) + 2 OH^-(aq) + 2 H_2O(l) \longrightarrow Sn(OH)_4^{2-}(aq) + H_2(g)$$

(a) Carbon Because carbon, C, stands at the head of its group, it is expected to differ from the other members of the group. Some of the differences between carbon and silicon stem from the smaller atomic radius of carbon, which explains the wide occurrence of C=C and C=O double bonds relative to the rarity of Si=Si and Si=O double bonds. Silicon atoms are too large for the side-by-side overlap of p-orbitals necessary for π-bonds to form

TABLE 8F.1 The Group 14 Elements

Valence configuration: ns^2np^2

Z	Name	Symbol	Molar mass/ (g·mol⁻¹)	Melting point/°C	Boiling point/°C	Density/ (g·cm⁻³)	Normal form*
6	carbon	C	12.01	3370s†	—	1.9–2.3	black nonmetal (graphite)
						3.2–3.5	transparent nonmetal (diamond)
							brown nonmetal (fullerite)
14	silicon	Si	28.09	1410	2620	2.33	gray metalloid
32	germanium	Ge	72.61	937	2830	5.32	gray-white metalloid
50	tin	Sn	118.71	232	2720	7.29	white lustrous metal
82	lead	Pb	207.2	328	1760	11.34	blue-white lustrous metal

*Normal form means the state and appearance of the element at 25 °C and 1 atm.
† The letter s denotes that the element sublimes.

between them, except in specially contrived arrangements in which other groups of atoms either hold the two Si atoms closely together or protect the fragile double bond from attack by preventing the close approach of reagents, as in tetramesityldisilene (**1**). Carbon dioxide, which consists of discrete O=C=O molecules, is a gas that we exhale. Silicon dioxide (silica), which consists of networks of —O—Si—O— groups, is a mineral that we stand on.

Silicon compounds can also act as Lewis acids, whereas carbon compounds typically cannot. Because a silicon atom is bigger than a carbon atom and can expand its valence shell by using its d-orbitals, it can accommodate the lone pair of an attacking Lewis base. A carbon atom is smaller and has no available d-orbitals, so in general it cannot act as a Lewis acid. An exception to this behavior is when the carbon atom has multiple bonds, because then a π-bond can give way to the formation of a σ-bond, as in the reaction of CO_2 with H_2O (Topic 6A).

Solid carbon exists as graphite, diamond, and other allotropes such as the fullerenes, which have structures related to that of graphite. Graphite is the thermodynamically most stable of these allotropes under ordinary conditions. The single sheets of carbon atoms in a hexagonal honeycomb array that make up graphite are known as *graphene* (**2**) and are currently of great interest because of their electrical, optical, and mechanical properties. Despite its existence as extremely thin monolayers, graphene is highly opaque, has a thermal conductivity higher than that of diamond, and is one of the strongest materials known. Graphene and its chemical modifications have potential not just in nanoscience for the creation of miniaturized electronic circuits, but also in applications ranging from the (almost) frivolous room-temperature distillation of vodka to the seriously useful, such as antibacterial action and solar cells.

Coke is an impure form of carbon obtained from the solid residue remaining after the destructive distillation of coal; because it is inexpensive, it is used extensively in steel production (Topic 9B). *Soot* and *carbon black* contain very small crystals of graphite and other forms of carbon. *Carbon black*, which is produced by heating gaseous hydrocarbons to nearly 1000 °C in the absence of air, is used for reinforcing rubber, for pigments, and for printing inks, such as the ink on a page of a book. *Activated carbon,* which is also called *activated charcoal,* consists of granules of microcrystalline carbon. It is produced by heating waste organic matter in the absence of air and then processing it to increase the porosity. The very high specific surface area (as much as about 2000 $m^2 \cdot g^{-1}$, so 1 g has twice the area of a tennis court and its surrounds) of the porous carbon enables it to remove organic impurities from liquids and gases by adsorption. It is used in air purifiers, gas masks, and aquarium water filters. On a larger scale, activated carbon is used in water purification plants to remove organic compounds from drinking water.

In diamond, each carbon atom is sp^3 hybridized and linked tetrahedrally to its four neighbors, with all electrons in C—C σ-bonds (see Fig. 3H.6). In contrast, graphite consists of planar sheets of sp^2 hybridized carbon atoms in a hexagonal network (see Fig. 3H.8 and the description of graphite in Topic 3H). The graphite found in nature is the result of changes in ancient organic remains. Pure graphite is produced in industry by passing a high electric current through rods of coke.

Chemists were greatly surprised when soccer-ball-shaped carbon molecules were first identified in 1985, particularly because they might be even more abundant than graphite and diamond! The C_{60} molecule (**3**) is named buckminsterfullerene after the American architect R. Buckminster Fuller, whose geodesic domes it resembles. Within two years, scientists had succeeded in making crystals of buckminsterfullerene; the solid samples are called *fullerite* (**FIG. 8F.2**). The discovery of this molecule and others with similar structures, such as C_{70}, opened up the prospect of a whole new field of chemistry. For instance, the interior of a C_{60} molecule is big enough to hold an atom of another element, and chemists are now busily preparing a whole new periodic table of these "shrink-wrapped" atoms.

1 Tetramesityldisilene

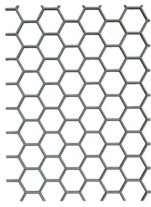

2 Graphene

The 2010 Nobel Prize for physics was awarded to Andre Geim and Konstantin Novoselov of the University of Manchester, England, "for ground-breaking experiments regarding the two-dimensional material graphene."

3 Buckminsterfullerene, C_{60}

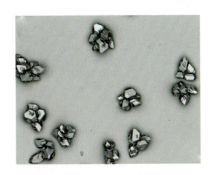

FIGURE 8F.2 These crystals are fullerite, in which buckminsterfullerene molecules are packed together in a close-packed lattice. (*Republished with permission of Takeyama, Y. et al, "Ionic liquid-mediated epitaxy of high-quality C60 crystallites in a vacuum" CrystEngComm, 2012, 14, 4939–4945. Permission conveyed through Copyright Clearance Center, Inc.*)

The *fullerenes* are members of the family of molecules resembling (and including) buckminsterfullerene. They are formed in smoky flames and by red giants (stars with low surface temperatures and large diameters), and so the universe might contain huge amounts of them. Graphite and diamond are network solids that are insoluble in all liquid solvents except some liquid metals. However, the fullerenes, which are molecular, can be dissolved by suitable solvents (such as benzene); buckminsterfullerene itself forms a red-brown solution. Fullerite currently has few uses, but some of the compounds of the fullerenes have great promise. For example, K_3C_{60} is a superconductor below 18 K, and other fullerene compounds appear to be active against cancer and diseases such as HIV-AIDS.

Carbon nanotubes consist of concentric tubes with walls like graphene sheets rolled into cylinders. These tiny structures form strong, conducting fibers with a large surface area. As a consequence, they have unusually interesting and promising properties that have become a major thrust of nanotechnology research. Topic 3J describes some of the applications of nanomaterials produced when carbon and silicon are coaxed into desired configurations, and **BOX 8F.1** describes materials that result when molecules of carbon and silicon compounds assemble themselves into organized structures.

(b) Silicon   The metalloid silicon, Si, is the second most abundant element in the Earth's crust. It occurs widely in rocks as *silicates,* compounds containing the silicate ion, SiO_3^{2-}, and as the silica, SiO_2, of sand (Topic 3I). Pure silicon is obtained from *quartzite,* a granular form of *quartz* (another solid phase of SiO_2), by reduction with high-purity carbon in an electric arc furnace:

$$SiO_2(s) + 2\,C(s) \xrightarrow{\Delta} Si(s) + 2\,CO(g)$$

The crude product is exposed to chlorine to form silicon tetrachloride, which is then distilled and reduced with hydrogen to a purer form of the element:

$$SiCl_4(l) + 2\,H_2(g) \longrightarrow Si(s) + 4\,HCl(g)$$

Silicon is used in semiconductors after purification. In one process, a large single crystal is grown by pulling a solid rod of the element slowly from the melt. The silicon is then purified by **zone refining,** in which a hot, molten zone is dragged from one end of the rod to the other, collecting impurities as it goes (**FIG. 8F.3**). The result is "ultrapure" silicon, which has fewer than one impurity atom per billion Si atoms. An alternative technique is the decomposition of silane, SiH_4, by an electric discharge. This method produces an amorphous form of silicon with a significant hydrogen content. *Amorphous silicon* is used in photovoltaic devices, which produce electricity from sunlight.

(c) Germanium, tin, and lead   Germanium, Ge, is recovered from the flue dust of industrial plants processing zinc ores (in which it occurs as an impurity). It is used mainly in the semiconductor industry to create very fast integrated circuits.

Tin, Sn, and lead, Pb, are obtained very easily from their ores and have been known since antiquity. Tin occurs chiefly as the mineral *cassiterite,* SnO_2, and is obtained from it by reduction with carbon at 1200 °C:

$$SnO_2(s) + C(s) \xrightarrow{1200\,°C} Sn(l) + CO_2(g)$$

The principal lead ore is *galena,* PbS. It is roasted in air, which converts it into PbO, and this oxide is then reduced with carbon monoxide:

$$2\,PbS(s) + 3\,O_2(g) \xrightarrow{\Delta} 2\,PbO(s) + 2\,SO_2(g)$$
$$PbO(s) + CO(g) \longrightarrow Pb(s) + CO_2(g)$$

Tin is expensive and not very strong, but it is resistant to corrosion. Its main use is in tinplating, which accounts for about 40% of its consumption. Tin is also used in alloys such as bronze and pewter (Topic 3I).

Lead's durability (its chemical inertness) and malleability make it useful in the construction industry. The inertness of lead under normal conditions can be traced to the passivation of its surface by oxides, chlorides, and sulfates (Topic 6N). Another important property of lead is its high density, which makes it useful as a radiation shield because its numerous electrons absorb high-energy radiation. Lead was once commonly used in leaded gasoline (as tetraethyl lead, $Pb(CH_2CH_3)_4$), but lead is a toxic heavy metal and that

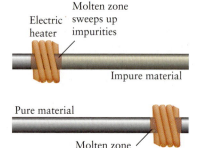

Electric heater — Molten zone sweeps up impurities

Impure material

Pure material

Molten zone that has collected impurities

FIGURE 8F.3 In the technique of zone refining, a molten zone is passed repeatedly from one end of the solid sample to the other. The impurities collect in the molten zone and move along the solid with the heater, leaving a pure substance behind.

BOX 8F.1   FRONTIERS OF CHEMISTRY: SELF-ASSEMBLING MATERIALS

A major concern of scientists working with nanomaterials is how to coax molecules into organizing themselves into the desired nanoscale arrangements. Many biological processes operate at the nanoscale level. Proteins, for example, fold up into the shape that optimizes their function (Topic 11E), cell membranes form spontaneously when certain molecules called phospholipids encounter water (Topic 5D), and the DNA molecules of our genes replicate themselves every time a cell divides. Although the expression has a variety of definitions, the common meaning of *self-assembly of molecules* is the spontaneous formation of organized structures from individual parts.

Self-assembly can be static or dynamic. In *static self-assembly*, the structure formed is stable and the process does not readily reverse. Two examples are the folding of polypeptide chains into a protein molecule and the formation of the double helix of DNA. *Dynamic self-assembly* involves interactions that dissipate energy and can easily be reversed. Two examples are oscillating chemical reactions and convection patterns. On the macroscopic scale, the coordinated motion of a school of fish or a flock of migrating birds is regarded as an example of dynamic self-assembly. However, the most promising area of self-assembly is in the intermediate realm between molecular and macroscopic, where it may become possible to design nanoscale sensors and machines to carry out specific tasks. Designing materials that can assemble themselves is a strategy that is beginning to result in materials that can respond to stimuli and act in seemingly intelligent ways.

In the Belousov–Zhabotinskii reaction, beautiful regular patterns form spontaneously as the result of the oscillating concentrations of reactants and products due to competing reactions. *(Ted Kinsman/Science Source.)*

You might be using a self-assembled material the next time you use in-line skates. One type of these skates contains a "smart gel" that is a liquid at normal temperatures but assembles itself into a firm, rubbery gel when exposed to body temperature. The gel fills the space between the lining of the skate boot and the sides and, once the skate has been secured, it sets to conform to the exact shape of the foot. Such a gel generally consists of an aqueous suspension of organic molecules with long chains containing different regions, some of which have strong intermolecular forces and some of which have weak intermolecular forces. The temperature determines how the chains are coiled. At low temperatures, the regions with strong intermolecular forces are turned inward, and so the molecules remain in liquid solution. However, at higher temperatures, those regions are turned outward, where they can attract other chains to form a flexible, but firm, network.

Soft gels expand and contract as their structures change in response to electrical signals and are being investigated for use in artificial limbs that would respond and feel like real ones. One material being studied for use in artificial muscle contains a mixture of

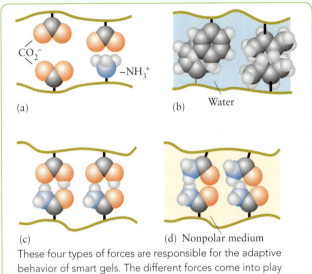

These four types of forces are responsible for the adaptive behavior of smart gels. The different forces come into play when the network of polymer chains composing a gel is disturbed. (a) Charged ionic regions can attract or repel each other. (b) Nonpolar hydrophobic regions exclude water. (c) Hydrogen bonds may form from one chain to another. (d) Dipole–dipole interactions can attract or repel chains.

polymers, silicone oil (a polymer with a —(—O—Si—O—Si—)$_n$— backbone and hydrocarbon side chains), and salts. When exposed to an electric field, the molecules of the soft gel rearrange themselves so that the material contracts and stiffens. If struck, the stiffened material can break but, upon softening, the gel is re-formed. The transition between gel and solid state is therefore reversible.

Another recent development is the creation, by self-assembly, of molecules that act as tiny motors when illuminated with light or energized by the catalyzed oxidation of glucose. These miniature motors can even rotate in a specified direction rather than clockwise or counterclockwise at random. Molecular motors that can be put into reverse by the addition of a base (to remove a crucial proton) have also been made.

HOW MIGHT YOU CONTRIBUTE?

The properties of self-assembled materials need to be studied and categorized. For example, how do molecules recognize one another in a mixture? In addition, strategies to promote self-assembly need to be designed. Simple two-dimensional assemblies have been created, and these techniques need to be expanded to three dimensions. Applications are emerging in nanotechnology, in the fields of microelectronics and robotics, and in medicine, where biosensor arrays have shown high selectivity for individual toxins and bioagents.

Related Exercise 8.29

Further Reading R. Dagani, "Intelligent gels," *Chemical and Engineering News*, June 9, 1997, pp. 26–27. J. S. Moore and M. L. Kraft, "Synchronized self-assembly," *Science*, vol. 320, May 2, 2008, pp. 620–621. G. Whitesides and B. Grzybowski, "Self-assembly at all scales," *Science*, vol. 295, March 29, 2002, pp. 2418–2421. "Innovative imaging technique clarifies molecular self-assembly," May 5, 2014, http://phys.org/news/2014-05-imaging-technique-molecular-self-assembly.html.

usage has been banned in most countries as concern for the amount of lead in the environment has increased. The main use of lead today is for the electrodes of rechargeable storage batteries (see the Interlude following Focus 6).

Metallic character increases significantly down Group 14. Carbon is the only member of Group 14 that commonly forms multiple bonds with itself; singly bonded silicon atoms can act as Lewis acids because a silicon atom can expand its valence shell. Carbon has an important series of allotropes: diamond, graphite, and the fullerenes.

8F.2 Oxides of Carbon and Silicon

The oxides of carbon and silicon are hugely important in industry as the starting materials for a large number of important materials such as cement. However, these oxides have widely different properties: carbon's oxides are gases; silicon's are solids.

THINKING POINT

What accounts for the difference in properties of carbon and silicon oxides?

(a) Carbon Carbon dioxide, CO_2, is formed when organic matter burns in a plentiful supply of air and during animal respiration. There is widespread and well-founded concern that an increase in atmospheric carbon dioxide due to the combustion of fossil fuels is contributing to global warming (Box 8B.1).

Carbon dioxide is the acid anhydride of carbonic acid, H_2CO_3, which forms when the gas dissolves in water. However, not all the dissolved molecules react to form the acid, and a solution of carbon dioxide in water is an equilibrium mixture of CO_2, H_2CO_3, HCO_3^-, and a very small amount of CO_3^{2-}. Carbonated beverages are made by using high partial pressures of CO_2 to produce high concentrations of carbon dioxide in water. When the partial pressure of the CO_2 is reduced by removing a bottle cap or the seal, the equilibrium $H_2CO_3(aq) \rightleftharpoons CO_2(g) + H_2O(l)$ shifts from H_2CO_3 toward CO_2 and the liquid effervesces.

Carbon monoxide, CO, is produced when carbon or organic compounds burn in a limited supply of air, as happens in cigarettes, badly tuned automobile engines, and charcoal fires. It is produced commercially as a component of synthesis gas by the re-forming reaction (Topic 8B). Carbon monoxide is the formal anhydride of formic acid, HCOOH, and the gas can be produced in the laboratory by the dehydration of formic acid with hot, concentrated sulfuric acid:

$$HCOOH(l) \xrightarrow{150\ °C,\ H_2SO_4} CO(g) + H_2O(l)$$

Although the reverse of this reaction cannot be carried out directly, carbon monoxide does react with hydroxide ions in hot alkali to produce formate ions:

$$CO(g) + OH^-(aq) \longrightarrow HCO_2^-(aq)$$

Carbon monoxide is a colorless, odorless, flammable, almost insoluble, very toxic gas that condenses to a colorless liquid at -90 °C. It is not very reactive, largely because its bond enthalpy ($1074\ kJ·mol^{-1}$) is higher than that of any other molecule. However, it is a Lewis base, and the lone pair on the carbon atom forms covalent bonds with d-block atoms and ions. Carbon monoxide is also a Lewis acid, because its empty antibonding π-orbitals can accept electron density from a metal (**FIG. 8F.4**). This dual character makes carbon monoxide very useful for forming complexes, and numerous metal carbonyls are known (Topic 9B). Complex formation is also responsible for carbon monoxide's toxicity: it attaches more strongly than oxygen to the iron in hemoglobin and prevents it from accepting oxygen from the air in the lungs. As a result, the victim suffocates (Box 9C.1).

Because it can be further oxidized, carbon monoxide is a reducing agent. It is used in the production of a number of metals such as lead and iron (Topic 9B).

(b) Silicon *Silica*, SiO_2, is a hard, rigid network solid that is insoluble in water. It occurs naturally as quartz and as sand, which consists of small fragments of quartz, usually colored golden brown by iron oxide impurities. Some precious and semiprecious stones are

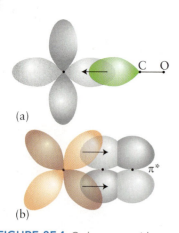

FIGURE 8F.4 Carbon monoxide can bind to a d-metal atom in two ways: (a) by using the lone pair on the C atom to form a σ-bond or (b) by using one of its empty antibonding π-orbitals to accept electrons donated from a d-orbital on the metal atom.

impure silica (**FIG. 8F.5**). *Flint* is silica colored black by carbon impurities. The structures and uses of some of the silicates are described in Topic 3I.

Metasilicic acid, H_2SiO_3, and *orthosilicic acid*, H_4SiO_4, are weak acids. However, when a solution of sodium orthosilicate is acidified, instead of H_4SiO_4, a gelatinous precipitate of silica is produced:

$$4 H_3O^+(aq) + SiO_4^{4-}(aq) + x H_2O(l) \longrightarrow SiO_2(s)\cdot xH_2O(gel) + 6 H_2O(l)$$

After it is washed, dried, and granulated, this *silica gel* has a very high specific surface area (about 700 $m^2\cdot g^{-1}$) and is useful as a drying agent, a support for catalysts, a packing for chromatography columns, and a thermal insulator. Packets of silica gel are often included in the packaging of moisture-sensitive electronic items.

FIGURE 8F.5 Some impure forms of silica that are semiprecious stones: amethyst (left), in which the color is due to Fe^{3+} impurities; agate (center); and onyx (right). (*W. H. Freeman photo by Ken Karp.*)

> **Self-test 8F.1A** What is (a) the oxidation number and (b) the formal charge of carbon in CO (use the Lewis structure $:C\equiv O:$)?
>
> [*Answer:* (a) +2; (b) −1.]
>
> **Self-test 8F.1B** What type of reaction is the reaction of carbon monoxide with hydroxide ion?

Carbon has two important oxides: carbon dioxide and carbon monoxide. The former is the acid anhydride of carbonic acid, the parent acid of the hydrogen carbonates and the carbonates. Silica is a hard network material. The silicic acids are weak acids.

8F.3 Other Important Group 14 Compounds

Carbon is the only Group 14 element that forms both monatomic and polyatomic anions. There are three classes of carbides: saline carbides (saltlike carbides), covalent carbides, and interstitial carbides.

The *saline carbides* are formed most commonly from the metals of Groups 1 and 2, aluminum, and a few other metals. The s-block metals form saline carbides when their oxides are heated with carbon. The anions present in saline carbides are either C_2^{2-} or C^{4-}. All the C^{4-} carbides, which are called *methides*, produce methane and the corresponding hydroxide in water:

$$Al_4C_3(s) + 12 H_2O(l) \longrightarrow 4 Al(OH)_3(s) + 3 CH_4(g)$$

This reaction shows that the methide ion is a very strong Brønsted base. The species C_2^{2-} is the *acetylide ion*, and the carbides that contain it are called *acetylides*. The acetylide ion is also a strong Brønsted base, and acetylides react with water to produce ethyne (acetylene) and the corresponding hydroxide. Calcium carbide, CaC_2, is the most common saline carbide. It was once used in miners' lanterns, in which water from a reservoir was allowed to drip slowly on to chunks of calcium carbide. The acetylene produced was then burned for the light it created.

The *covalent carbides* include silicon carbide, SiC, which is sold as carborundum:

$$SiO_2(s) + 3 C(s) \xrightarrow{2000\ ^\circ C} SiC(s) + 2 CO(g)$$

Pure silicon carbide is colorless, but iron impurities normally impart an almost black color to the crystals. Carborundum is an excellent abrasive because it is very hard, with a diamondlike structure that fractures into pieces with sharp edges (**FIG. 8F.6**).

The interstitial carbides are compounds formed by the direct reaction of a d-block metal and carbon at temperatures above 2000 °C. As in steel, the C atoms pin the metal atoms together into a rigid structure, resulting in very hard substances with melting points often well above 3000 °C. Tungsten carbide, WC, is used for the cutting surfaces of drills.

All Group 14 elements form liquid molecular tetrachlorides. The least stable is $PbCl_4$, which decomposes to solid $PbCl_2$ when it is warmed to about 50 °C. Carbon tetrachloride, CCl_4 (tetrachloromethane), was widely used as an industrial solvent; however, now

FIGURE 8F.6 Carborundum crystals, showing the sharp fractured edges that give the substance its abrasive power. (*Chip Clark/Fundamental Photographs, NYC.*)

Now that their environmental effects have been recognized, the use of chlorofluorocarbons has been greatly reduced (see Box 7E.1).

that it is known to be carcinogenic, it is used primarily as the starting point for the manufacture of chlorofluorocarbons. Carbon tetrachloride is formed by the action of chlorine on methane:

$$CH_4(g) + 4 Cl_2(g) \xrightarrow{\Delta} CCl_4(g, l \text{ when cooled}) + 4 HCl(g)$$

Silicon reacts directly with chlorine to form silicon tetrachloride, $SiCl_4$. This compound differs strikingly from CCl_4 in that it reacts readily with water as a Lewis acid, accepting a lone pair of electrons from H_2O:

$$SiCl_4(l) + 2 H_2O(l) \longrightarrow SiO_2(s) + 4 HCl(aq)$$

Because carbon so readily forms bonds to itself, there are huge numbers of hydrocarbons (Focus 11). Silicon forms a much smaller number of compounds with hydrogen, called the *silanes*. The simplest silane is silane itself, SiH_4, the analog of methane.

Because it absorbs water rapidly, silane is used as a water scavenger and in adhesives to promote bonding in humid environments. Silane is much more reactive than methane and bursts into flame on contact with air. Although it survives in pure water, it forms SiO_2 when a trace of alkali is present:

$$SiH_4(g) + 2 H_2O(l) \xrightarrow{OH^-} SiO_2(s) + 4 H_2(g)$$

THINKING POINT

Is the existence of alien life forms based on silicon plausible?

Self-test 8F.2A Both carbide ions, C_2^{2-} and C^{4-}, react as bases with water. Predict which carbide ion is the stronger base. Explain your answer.

[*Answer:* The C^{4-} ion, because it has a higher negative charge]

Self-test 8F.2B Explain the observation that SiH_4 reacts with water containing OH^- ions but CH_4 does not.

Carbon forms ionic carbides with the metals of Groups 1 and 2. Covalent and interstitial carbides are very hard. Silicon compounds are more reactive than carbon compounds; they can act as Lewis acids.

Topic 8F Exercises

8F.1 Describe the sources of silicon and write balanced chemical equations for the three steps in the industrial preparation of silicon.

8F.2 Suggest a reason why the size of the silicon atom does not permit a silicon analog of the graphite structure.

8F.3 Identify the oxidation number of germanium in the following compounds and ions: (a) GeO_4^{4-}; (b) $K_4Ge_4Te_{10}$; (c) Ca_3GeO_5.

8F.4 Identify the oxidation number of tin in the following compounds and ions: (a) $Sn(OH)_6^{2-}$; (b) $SnHPO_3$; (c) $NaSn_2F_5$.

8F.5 Balance the following skeletal equations and classify them as acid–base or redox:
(a) $MgC_2(s) + H_2O(l) \rightarrow C_2H_2(g) + Mg(OH)_2(s)$
(b) $Pb(NO)_2(s) \xrightarrow{\Delta} PbO(s) + NO_2(g) + O_2(g)$

8F.6 Balance the following skeletal equations and classify them as acid–base or redox:
(a) $CH_4(g) + S_8(s) \rightarrow CS_2(l) + H_2S(g)$
(b) $Sn(s) + KOH(aq) + H_2O(l) \rightarrow K_2Sn(OH)_6(aq) + H_2(g)$

8F.7 Calculate the values of $\Delta H_r°$, $\Delta S_r°$, and $\Delta G_r°$ for the production of high-purity silicon by the reaction $SiO_2(s) + 2 C(s, \text{graphite}) \rightarrow Si(s) + 2 CO(g)$ at 25 °C and estimate the temperature at which the equilibrium constant becomes greater than 1.

8F.8 Calculate the values of $\Delta H_r°$, $\Delta S_r°$, and $\Delta G_r°$ for the reaction $2 CO(g) + O_2(g) \rightarrow 2 CO_2(g)$ at 25 °C and estimate the temperature at which the equilibrium constant becomes less than 1.

Topic 8G Group 15: The Nitrogen Family

What general features does the periodic table summarize?

What are the principal characteristics of the group?

Topic **8A:** Periodic trends → Topic **8G:** Group 15: The nitrogen family

Atoms of the Group 15 elements have the valence electron configuration ns^2np^3 (TABLE 8G.1). The chemical and physical properties of the elements vary sharply in this group, from the almost unreactive gas nitrogen, through the soft nonmetal phosphorus, which is so reactive that it ignites in air, to the important semiconducting materials arsenic and antimony and the largely metallic bismuth (FIG. 8G.1). The range of oxidation numbers of Group 15 elements is from −3 to +5, but only nitrogen and phosphorus display the whole range.

FIGURE 8G.1 The elements of Group 15. Back row, from left to right: liquid nitrogen, red phosphorus, arsenic. Front row, from left to right: antimony and bismuth. (©1984 Chip Clark–Fundamental Photographs.)

8G.1 The Group 15 Elements

The metallic character of the Group 15 elements increases down the group, but the only element considered to be metallic is bismuth, at the bottom of the group.

(a) Nitrogen   Although nitrogen, N, is rare in the Earth's crust, elemental nitrogen is the principal component of our atmosphere (76% by mass). Pure nitrogen gas is obtained by the fractional distillation of liquid air. Air is cooled to below −196 °C by repeated expansion and compression in a refrigerator like that described in Topic 3E. The liquid mixture is then warmed, and the nitrogen (b.p. −196 °C) boils off. Nitrogen gas produced industrially is used primarily as a raw material for the synthesis of ammonia; the liquid is used as a coolant.

Plants need nitrogen to grow. However, they cannot use N_2 directly, because the strength of the N≡N bond (944 kJ·mol^{-1}) makes nitrogen gas almost as inert as the noble gases. To be available for organisms, nitrogen must first be "fixed," or combined with other elements into more useful compounds. Once it is fixed, nitrogen can be converted into compounds for use as medicines, fertilizers, explosives, and plastics. Lightning converts some atmospheric nitrogen into its oxides, which dissolve in rain and are then washed into the soil. Some bacteria also fix nitrogen in nodules on the roots of clover, beans, peas, alfalfa, and other legumes (FIG. 8G.2). An intensely active field of research is the search for catalysts that can mimic these bacteria and fix nitrogen at ordinary

Lavoisier named the element *azote*, which means "lifeless." Ironically, we now know that there would be no life as we know it without nitrogen.

TABLE 8G.1 The Group 15 Elements

Valence configuration: ns^2np^3

Z	Name	Symbol	Molar mass/ (g·mol^{-1})	Melting point/°C	Boiling point/°C	Density/ (g·cm^{-3})	Normal form*
7	nitrogen	N	14.01	−210	−196	1.04†	colorless gas
15	phosphorus	P	30.97	44	280	1.82	white or red nonmetal
33	arsenic	As	74.92	613s‡	—	5.78	gray metalloid
51	antimony	Sb	121.76	631	1750	6.69	blue-white lustrous metalloid
83	bismuth	Bi	208.98	271	1650	8.90	white-pink metal

Normal form means the state and appearance of the element at 25 °C and 1 atm.

†For the liquid at its boiling point.

‡The letter s denotes that the element sublimes.

FIGURE 8G.2 The bacteria that inhabit these nodules on the roots of a pea plant are responsible for fixing atmospheric nitrogen and making it available to the plant. (*Hugh Spencer/ Science Source.*)

1 Dinitrogen trioxide, N_2O_3

The name phosphorus means "light bringer."

2 Phosphorus(III) oxide, P_4O_6

3 Phosphorus, P_4

FIGURE 8G.3 The minerals (from left to right) orpiment, As_2S_3; stibnite, Sb_2S_3; and realgar, As_4S_4; are all ores that have been used as sources of Group 15 elements. (©1984 *Chip Clark– Fundamental Photographs.*)

temperatures. At present, the Haber synthesis of ammonia is the main industrial route for fixing nitrogen, but it requires costly high pressures (Topic 5J).

Nitrogen differs in many ways from its congeners. Because of its high electronegativity ($\chi = 3.0$, about the same as for chlorine), nitrogen is the only element in Group 15 that forms hydrides capable of hydrogen bonding. Because its atoms are small, nitrogen can form multiple bonds with other Period 2 atoms by using its p-orbitals.

Nitrogen is found with a wide range of oxidation numbers: compounds are known for each whole-number oxidation number from -3 (in NH_3) to $+5$ (in nitric acid and the nitrates). It also occurs with fractional oxidation numbers, such as $-\frac{1}{3}$ in the azide ion, N_3^-.

(b) Phosphorus   The atomic radius of phosphorus is nearly 50% larger than that of nitrogen, and so two phosphorus atoms are too big to approach each other closely enough for their 3p-orbitals to overlap and form π-bonds. Thus, whereas nitrogen can form multiply bonded structures such as N_2O_3 (**1**), phosphorus forms additional single bonds, as in P_4O_6 (**2**). The size of its atoms and the availability of 3d-orbitals enable phosphorus to form as many as six bonds (as in PCl_6^-), whereas nitrogen can form only four (Topic 2C).

Phosphorus is obtained from the *apatites,* which are mineral forms of calcium phosphate, $Ca_3(PO_4)_2$. The rocks are heated in an electric furnace with carbon and sand:

$$2\,Ca_3(PO_4)_2(s) + 6\,SiO_2(s) + 10\,C(s) \xrightarrow{\Delta} P_4(g) + 6\,CaSiO_3(l) + 10\,CO(g)$$

The phosphorus vapor condenses as *white phosphorus,* a soft, white, poisonous molecular solid consisting of tetrahedral P_4 molecules (**3**). This allotrope is highly reactive, in part because of the strain associated with the 60° angles between the bonds. It is very dangerous to handle because it bursts into flame on contact with air and can cause severe burns. White phosphorus is normally stored under water. It changes into *red phosphorus* when it is heated in the absence of air. Red phosphorus is less reactive than the white allotrope; however, it can be ignited by friction and is used in the striking surfaces of matchbooks. The friction created by rubbing a match across the surface ignites the phosphorus, which, in turn, lights the highly flammable material in the match head. Red phosphorus is thought to consist of chains of linked P_4 tetrahedra.

THINKING POINT

What is the molecular origin of friction?

Vapor rising from white phosphorus in moist air glows with a yellow-green light. The oxides produced by the reaction of phosphorus with the oxygen in air are formed in electronically excited states, and light is given off as electrons fall back into the ground state in the process called chemiluminescence (Topic 3J).

(c) Arsenic, antimony, and bismuth   Arsenic, As, and antimony, Sb, are metalloids. They have been known in the pure state since ancient times because they are easily obtained from their ores (**FIG. 8G.3**). In the elemental state, they are used primarily in the semiconductor industry and in the lead alloys used as electrodes in storage batteries. Gallium arsenide is used in lasers, including the lasers used in CD players.

Metallic bismuth, Bi, with its large, weakly bonded atoms, has a low melting point and is used in alloys that serve as fire detectors in sprinkler systems: the alloy melts when a fire breaks out nearby, and the sprinkler system is activated. Like ice, solid bismuth is less dense than the liquid. As a result, molten bismuth does not shrink when it solidifies in molds, and so it is used to make low-temperature castings.

Nitrogen is unreactive as an element, largely because of its strong triple bond. White phosphorus is highly reactive. The differences in properties between the nonmetals nitrogen and phosphorus can be traced to the larger atomic radius and lower electronegativity of phosphorus and the availability of d-orbitals in its valence shell. Metallic character increases down the group.

8G.2 Compounds with Hydrogen and the Halogens

The compounds of the Group 15 elements with hydrogen and the halogens have a wide range of important applications: ammonia, for instance, is of crucial importance in agriculture. The differences between the compounds of nitrogen and phosphorus also illustrate the importance of the atomic radius of the element in controlling the types of compounds it can form.

(a) Nitrogen   By far the most important hydrogen compound of a Group 15 element is ammonia, NH_3, which is prepared in huge amounts (over 140 Mt worldwide in 2015) by the Haber process (Topic 5J). Small quantities of ammonia are present naturally in the atmosphere as a result of the bacterial decomposition of organic matter in the absence of air. This decomposition typically occurs in lake and riverbeds, in swamps, and in cattle feedlots.

Ammonia is a pungent, toxic gas that condenses to a colorless liquid at $-33\ °C$. The liquid resembles water in its physical properties, including its ability to act as a solvent for a wide range of substances. Because the dipole moment of the NH_3 molecule (1.47 D) is lower than that of the H_2O molecule (1.85 D), salts with strong ionic character, such as KCl, cannot dissolve in ammonia. Salts with polarizable anions tend to be more soluble in ammonia than are salts with greater ionic character. For example, iodides are more soluble than chlorides in ammonia. Liquid ammonia undergoes much less extensive autoprotolysis than water:

$$2\,NH_3(am) \rightleftharpoons NH_4^+(am) + NH_2^-(am) \qquad K_{am} = [NH_4^+][NH_2^-] = 1 \times 10^{-33} \text{ at } -35\ °C$$

Bases that would be fully protonated in water, such as the cyclopentadienide anion, $C_5H_5^-$, behave as very weak bases in ammonia.

Ammonia is very soluble in water because the NH_3 molecules can form hydrogen bonds to H_2O molecules. Ammonia is a weak Brønsted base in water; it is also a reasonably strong Lewis base, particularly toward d-block elements. For example, it reacts with $Cu^{2+}(aq)$ ions to give a deep-blue complex (**FIG. 8G.4**):

$$Cu^{2+}(aq) + 4\,NH_3(aq) \longrightarrow Cu(NH_3)_4^{2+}(aq)$$

Ammonium salts decompose when heated:

$$(NH_4)_2CO_3(s) \xrightarrow{\Delta} 2\,NH_3(g) + CO_2(g) + H_2O(g)$$

The pungent smell of decomposing ammonium carbonate once made it an effective "smelling salt," a stimulant used to revive people who had fainted.

The ammonium cation of an ammonium salt may be oxidized by an anion that has oxidizing character, such as a nitrate. The products depend on the temperature of the reaction:

$$NH_4NO_3(s) \xrightarrow{250\ °C} N_2O(g) + 2\,H_2O(g)$$

$$2\,NH_4NO_3(s) \xrightarrow{>300\ °C} 2\,N_2(g) + O_2(g) + 4\,H_2O(g)$$

The explosive violence of the second reaction is the reason that ammonium nitrate is used as a component of dynamite. Ammonium nitrate has a high nitrogen content (33.5% by mass) and is highly soluble in water. These characteristics make it attractive as a fertilizer, which is its principal use.

Hydrazine, NH_2NH_2, is an oily, colorless liquid. It is prepared by the gentle oxidation of ammonia with alkaline hypochlorite solution:

$$2\,NH_3(aq) + ClO^-(aq) \xrightarrow{\text{aqueous alkali}} N_2H_4(aq) + Cl^-(aq) + H_2O(l)$$

Its physical properties are very similar to those of water; for instance, its melting point is $1.5\ °C$ and its boiling point is $113\ °C$. However, its chemical properties are very different. It is dangerously explosive and is normally stored and used in aqueous solution.

Nitrogen has oxidation number $+3$ in the highly reactive nitrogen halides. Nitrogen trifluoride, NF_3, is the most stable halide; it does not react with water. However, NCl_3 does

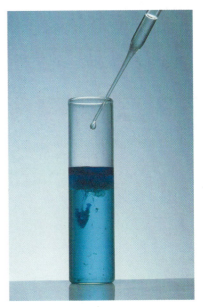

FIGURE 8G.4 When aqueous ammonia is added to a copper(II) sulfate solution, first a light-blue precipitate of $Cu(OH)_2$ forms (the cloudy region at the top, which appears dark because it is backlit). The precipitate disappears when more ammonia is added to form the dark blue complex $Cu(NH_3)_4^{2+}$ by a Lewis acid–base reaction. (*W. H. Freeman photo by Ken Karp.*)

The pungency of heated ammonium chloride was known in antiquity to the Ammonians, the worshippers of the Egyptian god Ammon.

Like many of the reactions in this book, heating ammonium nitrate is extremely hazardous. Don't try it.

FIGURE 8G.5 Nitrogen triiodide decomposes explosively with the slightest touch. (*Charles D. Winters/ Science Source.*)

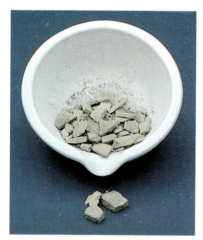

FIGURE 8G.6 This sample of magnesium nitride was formed by burning magnesium in an atmosphere of nitrogen. When magnesium burns in air, the product is the oxide as well as the nitride. (*©1986 Chip Clark– Fundamental Photographs.*)

react with water, forming ammonia and hypochlorous acid. Nitrogen triiodide, NI_3, which is known only in combination with ammonia as the "ammoniate" $NI_3 \cdot NH_3$ (the analog of a hydrate), is so unstable that it decomposes explosively when touched, even lightly. It can actually be set off by touching it with a feather (**FIG 8G.5**).

Nitrides are solids that contain the nitride ion, N^{3-}. Nitrides are stable only for small cations such as lithium or magnesium. Boron nitride, BN, is an important ceramic discussed in Topic 8F. Magnesium nitride, Mg_3N_2, is formed together with the oxide when magnesium is burned in air (**FIG. 8G.6**):

$$3\,Mg(s) + N_2(g) \xrightarrow{\Delta} Mg_3N_2(s)$$

Magnesium nitride, like all nitrides, reacts with water to produce ammonia and the corresponding hydroxide:

$$Mg_3N_2(s) + 6\,H_2O(l) \longrightarrow 3\,Mg(OH)_2(s) + 2\,NH_3(g)$$

In this reaction, the nitride ion acts as a strong base, accepting protons from water to form ammonia.

The azide ion is a highly reactive polyatomic anion of nitrogen, N_3^-. Its most common salt, sodium azide, NaN_3, is prepared from dinitrogen oxide and molten sodium amide:

$$N_2O(g) + 2\,NaNH_2(l) \xrightarrow{175\,°C} NaN_3(s) + NaOH(l) + NH_3(g)$$

Sodium azide, like most azide salts, is shock sensitive. It is used in automobile air bags, where it decomposes to elemental sodium and nitrogen when detonated:

$$2\,NaN_3(s) \longrightarrow 2\,Na(s) + 3\,N_2(g)$$

The azide ion is a weak base and accepts a proton to form its conjugate acid, hydrazoic acid, HN_3. Hydrazoic acid is a weak acid similar in strength to acetic acid.

(b) Phosphorus   The hydrogen compounds of other members of Group 15 are much less stable than ammonia and decrease in stability down the group. Phosphine, PH_3, is a poisonous gas that smells faintly of garlic and bursts into flame in air if it is slightly impure. It is much less soluble than ammonia in water because PH_3 cannot form hydrogen bonds to water. Aqueous solutions of phosphine are neutral, because the electronegativity of phosphorus is so low that the lone pair in PH_3 is spread over the hydrogen atoms as well as the phosphorus atom; consequently, the molecule has only a very weak tendency to accept a proton ($pK_b = 27.4$). Because PH_3 is the very weak parent acid of the strong Brønsted base P^{3-}, phosphine can be prepared by protonating phosphide ions with a Brønsted acid. Even water is a sufficiently strong proton donor:

$$2\,P^{3-}(s) + 6\,H_2O(l) \longrightarrow 2\,PH_3(g) + 6\,OH^-(aq)$$

Phosphorus trichloride, PCl_3, and phosphorus pentachloride, PCl_5, are the two most important halides of phosphorus. The former is prepared by direct chlorination of phosphorus. Phosphorus trichloride, a liquid, is a major intermediate for the production of pesticides, oil additives, and flame retardants. Phosphorus pentachloride, a solid, is made by allowing phosphorus trichloride to react with more chlorine (see Fig. 2C.1).

A typical reaction of the nonmetal halides is their reaction with water to give oxoacids, without a change in oxidation number:

$$PCl_3(l) + 3\,H_2O(l) \longrightarrow H_3PO_3(s) + 3\,HCl(g)$$

This reaction is an example of a **hydrolysis reaction,** a reaction with water in which new element–oxygen bonds are formed. Another example is the violent and dangerous reaction of PCl_5 (phosphorus in the oxidation state $+5$) with water to produce phosphoric acid, H_3PO_4 (also phosphorus in the oxidation state $+5$):

$$PCl_5(s) + 4\,H_2O(l) \longrightarrow H_3PO_4(l) + 5\,HCl(g)$$

An interesting feature of phosphorus pentachloride is that it is an ionic solid of tetrahedral PCl_4^+ cations and octahedral PCl_6^- anions, but it vaporizes to a gas of trigonal bipyramidal PCl_5 molecules (Topic 2C). Phosphorus pentabromide is also molecular in the vapor and ionic as the solid; in the solid, however, the anions are simply Br^- ions, presumably because six bulky Br atoms simply do not fit around a central P atom.

THINKING POINT

Size is important. Can you think of some other ways in which the size of an atom influences the atomic, physical, and chemical properties of an element?

Self-test 8G.1A Explain why the nitrogen trihalides become less stable as the molar mass of the halogen increases.

[*Answer:* The N atom is small; as the molar mass of the halogen increases, fewer can fit easily around it.]

Self-test 8G.1B (a) Draw the Lewis structure for the azide ion and assign formal charges to the atoms. (b) You will find it possible to write a number of Lewis structures. Which is likely to make the biggest contribution to the resonance? (c) Predict the shape of the ion and its polarity.

The important compounds of nitrogen with hydrogen are ammonia, hydrazine, and hydrazoic acid, the parent of the shock-sensitive azides. Phosphine forms neutral solutions in water; hydrolysis of phosphorus halides produces oxoacids without change in oxidation number.

8G.3 Nitrogen Oxides and Oxoacids

Nitrogen forms several oxides, with oxidation numbers ranging from $+1$ to $+5$. All nitrogen oxides are acidic oxides, and some are the acid anhydrides of the nitrogen oxoacids (TABLE 8G.2). In atmospheric chemistry, where the oxides play an important two-edged role in both maintaining and polluting the atmosphere, they are referred to collectively as NO_x (read "nox")

Dinitrogen oxide, N_2O (oxidation number $+1$), is commonly called nitrous oxide. It is formed by heating ammonium nitrate gently:

$$NH_4NO_3(s) \xrightarrow{250\,°C} N_2O(g) + 2\,H_2O(g)$$

Because it is tasteless, unreactive, nontoxic in small amounts, and soluble in fats, N_2O is sometimes used as a foaming agent and propellant for whipped cream.

Nitrogen monoxide, NO (oxidation number $+2$), is also known as nitric oxide or nitrogen oxide. It is a colorless gas prepared industrially by the catalytic oxidation of ammonia:

$$4\,NH_3(g) + 5\,O_2(g) \xrightarrow{1000\,°C,\ Pt} 4\,NO(g) + 6\,H_2O(g)$$

In the laboratory, nitrogen monoxide can be prepared by reducing a nitrite with a mild reducing agent such as I^-:

$$2\,NO_2^-(aq) + 2\,I^-(aq) + 4\,H^+(aq) \longrightarrow 2\,NO(g) + I_2(aq) + 2\,H_2O(l)$$

THINKING POINT

Why can a strong reducing agent not be used to prepare NO?

TABLE 8G.2 Nitrogen Oxides and Oxoacids

Oxidation number	Oxide formula	Oxide name	Oxoacid formula	Oxoacid name
5	N_2O_5	dinitrogen pentoxide	HNO_3	nitric acid
4	NO_2*	nitrogen dioxide	—	
	N_2O_4	dinitrogen tetroxide	—	
3	N_2O_3	dinitrogen trioxide	HNO_2	nitrous acid
2	NO	nitrogen monoxide, nitric oxide	—	
1	N_2O	dinitrogen monoxide, nitrous oxide	$H_2N_2O_2$	hyponitrous acid

*$2\,NO_2 \rightleftharpoons N_2O_4$.

FIGURE 8G.7 Dinitrogen trioxide, N_2O_3, condenses to a deep-blue liquid that freezes at −100 °C to a pale-blue solid, as shown here. On standing, it turns green as a result of partial decomposition into nitrogen dioxide, a yellow-brown gas (inset). (*Leroy Laverman.*)

Nitrogen monoxide is rapidly oxidized to nitrogen dioxide upon exposure to air, a reaction that contributes to acid rain (Box 6E.1):

$$2\,NO(g) + O_2(g) \longrightarrow 2\,NO_2(g)$$

Nitrogen monoxide plays both harmful and beneficial roles in our lives. The conversion of atmospheric nitrogen into NO in hot airplane and automobile engines contributes to the problem of acid rain and the formation of smog, as well as to the destruction of the ozone layer (Box 7E.1). However, in small amounts in the body, nitrogen monoxide acts as a neurotransmitter, helps to dilate blood vessels, and participates in other physiological processes, such as immune response. It is a subtle neurotransmitter because it is highly mobile, on account of its small size, but it is quickly destroyed, on account of being a radical.

Nitrogen dioxide, NO_2 (oxidation number +4), is a choking, poisonous, brown gas that contributes to the color and odor of smog. The molecule has an odd number of electrons, and in the gas phase it exists in equilibrium with its colorless dimer N_2O_4. Only the dimer exists in the solid phase, and so the brown gas condenses to a colorless solid. When it dissolves in water, NO_2 disproportionates into nitric acid (oxidation number +5) and nitrogen oxide (oxidation number +2):

$$3\,NO_2(g) + H_2O(l) \longrightarrow 2\,HNO_3(aq) + NO(g)$$

Nitrogen dioxide in the atmosphere undergoes the same reaction and contributes to the formation of acid rain. It also initiates a complex sequence of smog-forming photochemical reactions.

The blue gas dinitrogen trioxide, N_2O_3 (**FIG. 8G.7, 1**), in which the oxidation number of nitrogen is +3, is the anhydride of nitrous acid, HNO_2, and forms that acid when it dissolves in water:

$$N_2O_3(g) + H_2O(l) \longrightarrow 2\,HNO_2(aq)$$

Nitrous acid, the parent acid of the nitrites, has not been isolated in pure form but is widely used in aqueous solution. Nitrites are produced by the reduction of nitrates with hot metal:

$$KNO_3(s) + Pb(s) \xrightarrow{350\,°C} KNO_2(s) + PbO(s)$$

Most nitrites are soluble in water and mildly toxic. Despite their toxicity, nitrites are used in the curing of meat products because they retard bacterial growth. During the curing process, nitrites react with acid to generate nitric oxide, forming a pink nitrosyl complex with myoglobin that inhibits the oxidation of blood (a reaction that would otherwise turn the meat brown). The NO–myoglobin complexes are responsible for the pink color of ham, sausages, and other cured meat.

Nitric acid, HNO_3 (oxidation number +5) is used extensively in the production of fertilizers and explosives. It is made by the three-step *Ostwald process:*

Step 1 Oxidation of ammonia:

$$4\,\overset{-3}{N}H_3(g) + 5\,O_2(g) \xrightarrow{350\,°C,\,5\,atm,\,Pt/Rh} 4\,\overset{+2}{N}O(g) + 6\,H_2O(g)$$

Step 2 Oxidation of nitrogen oxide:

$$2\,\overset{+2}{N}O(g) + O_2(g) \longrightarrow 2\,\overset{+4}{N}O_2(g)$$

Step 3 Disproportionation in water:

$$3\,\overset{+4}{N}O_2(g) + H_2O(l) \longrightarrow 2\,H\overset{+5}{N}O_3(ag) + \overset{+2}{N}O(g)$$

Nitric acid, a colorless liquid that boils at 83 °C, is normally used in aqueous solution. Concentrated nitric acid is often pale yellow as a result of partial decomposition of the acid to NO_2. Because nitrogen has its highest oxidation number (+5) in HNO_3, nitric acid is an oxidizing agent as well as an acid.

Nitrogen forms oxides with each of its integer oxidation numbers from +1 to +5; the properties of the oxides and oxoacids are related to the oxidation number of nitrogen in the compound.

8G.4 Phosphorus Oxides and Oxoacids

The oxoacids and oxoanions of phosphorus are among the chemicals manufactured in the greatest amounts. Phosphate fertilizer production consumes two-thirds of all the sulfuric acid produced in the United States.

The structures of the phosphorus oxides are based on the tetrahedral PO_4 unit, which is similar to the structural unit in the oxides of its neighbor silicon (Topic 3I). White phosphorus burns in a limited supply of air to form phosphorus(III) oxide, P_4O_6 (**2**):

$$P_4(s, white) + 3\,O_2(g) \longrightarrow P_4O_6(s)$$

The molecules are tetrahedral, like P_4, but an O atom lies between each pair of P atoms. Phosphorus(III) oxide is the anhydride of phosphorous acid, H_3PO_3 (**4**), and is converted into the acid by cold water:

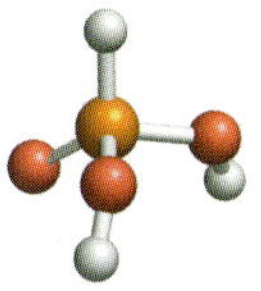

$$P_4O_6(s) + 6\,H_2O(l) \longrightarrow 4\,H_3PO_3(aq)$$

Although its formula suggests that it should be a triprotic acid, H_3PO_3 is in fact diprotic because one of the H atoms is attached directly to the P atom and the P—H bond is nonpolar (Topic 6C).

When phosphorus burns in an ample supply of air, it forms phosphorus(V) oxide, P_4O_{10} (**5**). This white solid reacts so vigorously with water that it is widely used in the laboratory as a drying agent. Phosphorus(V) oxide is the anhydride of phosphoric acid, H_3PO_4 (**6**):

$$P_4O_{10}(s) + 6\,H_2O(l) \longrightarrow 4\,H_3PO_4(aq)$$

Phosphoric acid is used primarily for the production of fertilizers, in processed food to increase acidity, and in detergents. Many soft drinks owe their tart taste to the presence of low concentrations of phosphoric acid. Pure phosphoric acid, H_3PO_4, is a colorless solid with a melting point of 42 °C; but in the laboratory it is normally a syrupy liquid because of the water that it has absorbed. In fact, phosphoric acid is usually purchased as 85% H_3PO_4. Its high viscosity can be traced to extensive hydrogen bonding. Although phosphorus has a high oxidation number (+5), the acid shows appreciable oxidizing power only at temperatures above about 350 °C; it may be used where nitric acid and sulfuric acid would be too strongly oxidizing.

Phosphoric acid is the parent of the phosphates, which are of great commercial importance. Phosphate rock is mined in huge quantities in Florida and Morocco. After being crushed, it is treated with sulfuric acid to give a mixture of sulfates and phosphates called *superphosphate*, a major fertilizer:

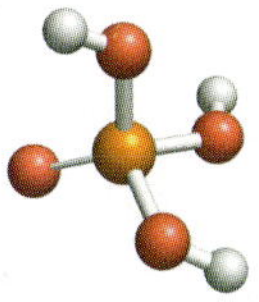

$$Ca_3(PO_4)_2(s) + 2\,H_2SO_4(l) \longrightarrow 2\,CaSO_4(s) + Ca(H_2PO_4)_2(s)$$

When phosphoric acid is heated, it undergoes a **condensation reaction,** a reaction in which two molecules combine together with the elimination of a small molecule, normally that of water:

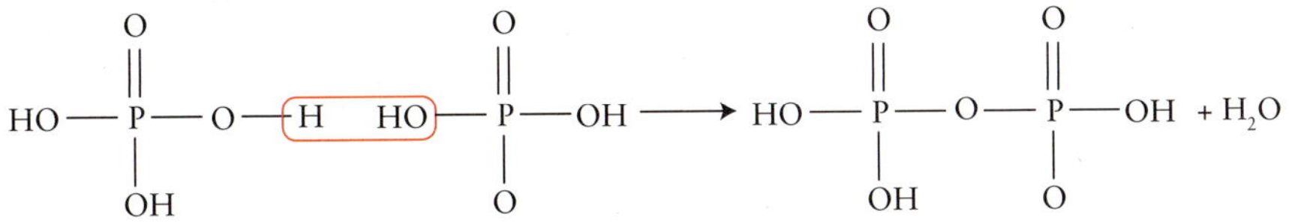

The product, $H_4P_2O_7$, is pyrophosphoric acid. Further heating gives even more complicated products that have chains and rings of PO_4 groups. The products are called polyphosphoric acids. Polyphosphoric acids are far from being of only academic interest: they empower our actions and thoughts. The most important polyphosphate is adenosine triphosphate, ATP (**7**), which is found in every living cell. The triphosphate part of this molecule is a chain of three phosphate groups that are converted into adenosine diphosphate, ADP, in the body:

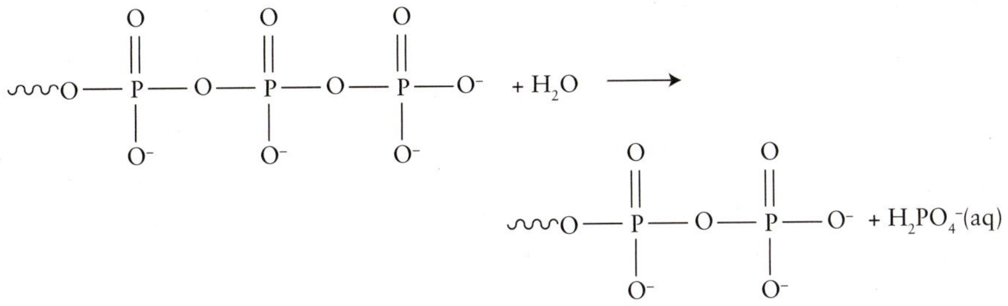

(where the wiggly line indicates the rest of the molecule). Some energy is required to break the O—H bond in water and the relatively weak O—P bond in ATP, but more

4 Phosphorous acid, H_3PO_3

5 Phosphorus(V) oxide, P_4O_{10}

6 Phosphoric acid, H_3PO_4

7 Adenosine triphosphate, ATP

energy is released when the new O—H and O—P bonds form in the product ($\Delta G° = -30$ kJ at pH = 7). This energy is used to power energy-demanding processes in cells.

Self-test 8G.2A How does (a) the acidity of a nitrogen oxoacid and (b) its strength as an oxidizing agent change as the oxidation number of N increases from +1 to +5?

[**Answer:** (a) Increases; (b) increases]

Self-test 8G.2B What is the molarity of phosphoric acid in 85% (by mass) H_3PO_4(aq), which has a density of 1.7 g·mL^{-1}?

The oxides of phosphorus have structures based on the tetrahedral PO$_4$ unit; P$_4$O$_6$ and P$_4$O$_{10}$ are the anhydrides of phosphorous acid and phosphoric acid, respectively. Polyphosphates are extended structures used (as ATP) by living cells to store and transfer energy.

Topic 8G Exercises

8G.1 Nitrogen can be found in compounds with oxidation numbers ranging from −3 to +5. For each of the possible integral oxidation numbers, give one example of a nitrogen compound or ion.

8G.2 Give the oxidation number of phosphorus in (a) white phosphorus; (b) red phosphorus; (c) Ca_5P_8; (d) PH_4I; (e) dihydrogen phosphate ion, $H_2PO_4^-$; (f) P_5H_5; (g) P_2O_5.

8G.3 Urea, $CO(NH_2)_2$, reacts with water to form ammonium carbonate. Write the chemical equation and calculate the mass of ammonium carbonate that can be obtained from 4.0 kg of urea.

8G.4 (a) Nitrous acid reacts with hydrazine in acidic solution to form hydrazoic acid, HN_3. Write the chemical equation and determine the mass of hydrazoic acid that can be produced from 15.0 g of hydrazine. (b) Suggest a method for preparing sodium azide, NaN_3. (c) Is the production of hydrazoic acid an oxidation or a reduction of hydrazine?

8G.5 Lead azide, $Pb(N_3)_2$, is used as a detonator. (a) What volume of nitrogen at STP (1 atm, 0 °C) does 1.5 g of lead azide produce when it decomposes into lead metal and nitrogen gas? (b) Would 1.5 g of mercury(II) azide, $Hg(N_3)_2$, which is also used

as a detonator, produce a larger or smaller volume, given that its decomposition products are elemental mercury and nitrogen gas? (c) Metal azides in general are potent explosives. Why?

8G.6 Sodium azide is used to inflate protective air bags in automobiles. What mass of solid sodium azide is needed to provide 65.0 L of N_2(g) at 1.2 atm and 25 °C?

8G.7 The common acid anhydrides of nitrogen are N_2O, N_2O_3, and N_2O_5. Write the formulas of their corresponding acids and chemical equations for the formation of the acids by the reaction (in one case, hypothetical) of the anhydrides with water.

8G.8 The common acid anhydrides of phosphorus are P_4O_6 and P_4O_{10}. Write the formulas of their corresponding acids and chemical equations for the formation of the acids by the reaction of the anhydrides with water.

8G.9 The normal boiling point of NH_3 is −33 °C and that of NF_3, which has a greater molar mass, is −129 °C. Explain this difference.

8G.10 Explain the observations that NH_3 is a weak Brønsted base in water, whereas NF_3 is not.

Topic 8H Group 16: The Oxygen Family

8H.1 The Group 16 Elements

8H.2 Compounds with Hydrogen

8H.3 Sulfur Oxides and Oxoacids

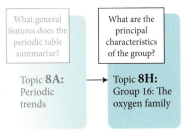

What general features does the periodic table summarize?

What are the principal characteristics of the group?

Topic **8A:** Periodic trends → Topic **8H:** Group 16: The oxygen family

The elements become increasingly nonmetallic toward the right-hand side of the periodic table; by Group 16, even polonium, at the foot of the group, is best regarded as a metalloid (**TABLE 8H.1**). Here, close to the far right of the periodic table, the electron configurations are only a few electrons short of a closed-shell, noble-gas configuration, and the effective nuclear charge is high. As a result, when the elements in Group 16 form compounds with other nonmetals, they do so by sharing electrons to form covalent bonds. The atoms have the valence electron configuration ns^2np^4, and so they need only two more electrons to complete their valence shells.

8H.1 The Group 16 Elements

The members of Group 16 are collectively called the **chalcogens,** from the Greek words for "copper" and "source," but more broadly interpreted as "ore giver," as the elements are commonly found in combination with metals as ores. Electronegativities decrease down Group 16 (**FIG. 8H.1**) and atomic and ionic radii increase (**FIG. 8H.2**).

(a) Oxygen Oxygen is the most abundant element in the Earth's crust; the free element accounts for 23% of the mass of the atmosphere. Oxygen is much more reactive than nitrogen, the other major component of our atmosphere. The combustion of all living organisms in oxygen is thermodynamically spontaneous; however, people do not burst into flame at normal temperatures, because combustion has a high activation energy.

Oxygen normally occurs as a colorless, tasteless, odorless gas of O_2 molecules; the gas condenses to a pale-blue liquid at $-183\,°C$ (**FIG. 8H.3**). Although O_2 has an even number of electrons, two of them are unpaired, making the molecule paramagnetic; in other words, O_2 behaves like a tiny magnet and is attracted into a magnetic field (Topic 3G).

More than 20 Mt (1 Mt = 10^9 kg) of liquid oxygen is produced each year in the United States (about 80 kg per inhabitant) by fractional distillation of liquid air. The biggest consumer of oxygen is the steel industry, which needs about 1 t of oxygen to produce

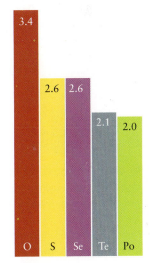

FIGURE 8H.1 The electronegativities of the Group 16 elements decrease down the group.

TABLE 8H.1 The Group 16 Elements

Common name: chalcogens
Valence configuration: ns^2np^4

Z	Name	Symbol	Molar mass/ (g·mol⁻¹)	Melting point/°C	Boiling point/°C	Density/ (g·cm⁻³)	Normal form*
8	oxygen	O	16.00	−218	−183	1.14†	colorless paramagnetic gas (O_2)
				−192	−112	1.35†	blue gas (ozone, O_3)
16	sulfur	S	32.06	115	445	2.09	yellow nonmetallic solid (S_8)
34	selenium	Se	78.96	220	685	4.79	gray nonmetallic solid
52	tellurium	Te	127.60	450	990	6.25	silver-white metalloid
84	polonium‡	Po	(209)	254	960	9.40	gray metalloid

Normal form means the appearance and state of the element at 25 °C and 1 atm.
†For the liquid at its boiling point.
‡Radioactive.

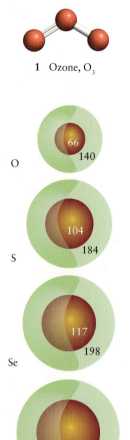

1 Ozone, O_3

FIGURE 8H.2 The atomic and ionic radii of the Group 16 elements increase steadily down the group. The values shown are in picometers, and each anion (shown green) is substantially larger than its neutral parent atom (brown).

1 t of steel. In steelmaking, oxygen is blown into molten iron to oxidize any impurities, particularly carbon (Topic 9B). Elemental oxygen is also used for welding (to create a very hot flame in oxyacetylene torches) and in medicine. Physicians administer oxygen to relieve strain on the heart and lungs and as a stimulant.

Ozone, O_3 (**1**), is an allotrope of oxygen formed in the stratosphere by the effect of solar radiation on O_2 molecules. Its total abundance in the atmosphere is equivalent to a layer that, at the ordinary conditions of 25 °C and 1 bar, would cover the Earth to a thickness of only 3 mm, yet its presence in the stratosphere is vital to the maintenance of life on Earth (Box 7E.1). Ozone can be made in the laboratory by passing an electric discharge through oxygen. It is a blue gas that condenses at -112 °C to an explosive blue liquid that looks like ink (**FIG. 8H.4**). Its pungent smell can often be detected near electrical equipment and after lightning. Ozone is also present in smog, where it is produced by the reaction of oxygen molecules with oxygen atoms:

$$O + O_2 \longrightarrow O_3$$

The oxygen atoms are produced by the photochemical decomposition of NO_2, which is generated from emission products of automobile engines:

$$NO_2 \xrightarrow{\text{UV radiation}} NO + O$$

(b) Sulfur   The differences between oxygen and sulfur, S, are similar to those between nitrogen and phosphorus, and for similar reasons: the atomic radius of sulfur is 58% larger than that of oxygen, and sulfur has a lower first ionization energy and electronegativity. The bonds that sulfur forms to hydrogen are much less polar than the bonds between oxygen and hydrogen. Consequently, S—H bonds do not take part in hydrogen bonding; as a result, H_2S is a gas, whereas H_2O is a liquid, despite the smaller number of electrons and hence weaker London forces. Sulfur also has weaker tendencies to form multiple bonds, tending instead to form additional single bonds to as many as six other atoms by using its d-orbitals (or, perhaps, simply its greater size).

Sulfur has a striking ability to **catenate**, or form chains of atoms. Oxygen's ability to form chains is very limited, with H_2O_2, O_3, the anions O_2^-, O_2^{2-}, and O_3^-, and other peroxides the only examples. Sulfur's ability is much more pronounced. It appears, for instance, in the existence of S_8 rings, their fragments, and the long strands of "plastic sulfur" that form when sulfur is heated to about 200 °C and suddenly cooled (**FIG 8H.5**). The —S—S— links that connect different parts of the chains of amino acids in proteins are another example of catenation. These "disulfide links" contribute to the shapes of proteins, including the keratin of your hair; thus, sulfur helps to keep you alive and, perhaps, curly haired (Topic 11E).

FIGURE 8H.3 Liquid oxygen is pale blue. (The gas itself is colorless.) The inset shows that the gas consists of diatomic molecules known formally as dioxygen. (©1984 Chip Clark–Fundamental Photographs.)

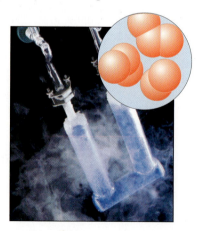

FIGURE 8H.4 Ozone is a blue gas that condenses to a dark-blue, highly unstable liquid. The inset shows that the gas consists of triatomic molecules. (Ross Chapple.)

FIGURE 8H.5 Plastic sulfur forms when molten sulfur is rapidly cooled in a beaker of water. (©1992 Richard Megna–Fundamental Photographs.)

FIGURE 8H.6 A collection of sulfide ores. From left to right: galena, PbS; cinnabar, HgS; pyrite, FeS₂; sphalerite, ZnS. Pyrite has a lustrous golden color and has frequently been mistaken for gold; hence, it is also known as fool's gold. Gold and fool's gold are readily distinguished by their densities. (©1985 Chip Clark–Fundamental Photographs.)

Sulfur is widely distributed as sulfide ores, which include *galena*, PbS; *cinnabar*, HgS; *iron pyrite*, FeS₂; and *sphalerite*, ZnS (**FIG. 8H.6**). Because these ores are so common, sulfur is a byproduct of the extraction of a number of metals, especially copper. Sulfur is also found as deposits of the native element (called *brimstone*), which are formed by bacterial action on H₂S. The low melting point of sulfur (115 °C) is utilized in the **Frasch process,** in which superheated water is used to melt solid sulfur underground and compressed air pushes the resulting slurry to the surface. Sulfur is also commonly found in petroleum, and extracting it chemically has been made inexpensive and safe by the use of heterogeneous catalysts, particularly zeolites (Topic 7E). One method used to remove sulfur in the form of H₂S from petroleum and natural gas is the **Claus process,** in which some of the H₂S is first oxidized to sulfur dioxide:

$$2\,H_2S(g) + 3\,O_2(g) \longrightarrow 2\,SO_2(g) + 2\,H_2O(l)$$

The SO₂ is then used to oxidize the remainder of the hydrogen sulfide:

$$2\,H_2S(g) + SO_2(g) \xrightarrow{300\,°C,\,Al_2O_3} 3\,S(s) + 2\,H_2O(l)$$

Sulfur is of major industrial importance. Most of the sulfur that is produced is used to make sulfuric acid, but an appreciable amount is used to vulcanize rubber (Topic 11E).

Elemental sulfur is a yellow, tasteless, almost odorless, insoluble, nonmetallic molecular solid of crownlike S₈ rings (**2**). The two common crystal forms of sulfur are monoclinic sulfur and rhombic sulfur (see Fig. 5B.4). The more stable form under normal conditions is rhombic sulfur, which forms beautiful yellow crystals (**FIG. 8H.7**). At low temperatures, sulfur vapor consists mainly of S₈ molecules. At temperatures above 720 °C, the vapor has a blue tint from the S₂ molecules that form (**FIG. 8H.8**). The latter are paramagnetic, like O₂.

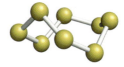

2 Sulfur, S₈

THINKING POINT

Why is sulfur a solid but oxygen a gas?

(a)

(b)

FIGURE 8H.7 One of the two most common forms of sulfur is the blocklike rhombic form (a). It differs from the needlelike monoclinic sulfur (b) in the manner in which the S₈ rings are stacked together. (Part (a) Chip Clark/ Fundamental Photographs, NYC. Part (b) sciencephotos/Alamy.)

FIGURE 8H.8 Hot sulfur vapor burns in the Kawah Ijen volcano with the typical blue light emitted by an electronically excited S₂ molecule. (© Olivier Grunewald.)

FIGURE 8H.9 Two of the Group 16 elements: selenium, which is a nonmetal, on the left; and tellurium, a metalloid, on the right. *(Chip Clark/ Fundamental Photographs, NYC.)*

(c) Selenium, tellurium, and polonium Selenium, Se, and tellurium, Te, occur in sulfide ores; they are also recovered from the anode sludge formed during the electrolytic refining of copper (Topic 9B). Both elements have several allotropes, the most stable consisting of long zigzag chains of atoms. Although these allotropes look like silver-white metals, they are poor electrical conductors (**FIG. 8H.9**). The conductivity of selenium is increased by exposure to light, and so it is used in solar cells, photoelectric devices, and photocopying machines. Selenium also occurs as a deep-red solid consisting of Se_8 molecules.

Polonium, Po, is a radioactive, low-melting metalloid. It is a useful source of α particles (helium-4 nuclei; they are described in more detail in Topic 8J) and is used in antistatic devices in textile mills: the α particles reduce static by counteracting the negative charges that tend to build up on the fast-moving fabric.

Metallic character increases down Group 16 as electronegativity decreases. Oxygen and sulfur occur naturally in the elemental state; sulfur forms chains and rings with itself, but oxygen does not.

8H.2 Compounds with Hydrogen

Hydrogen reacts with all the Group 16 elements, and by far the most important of these compounds is water, H_2O. Much of it might have come from outer space in the form of comets, which are like huge dirty snowballs. Another source of water is rocks; the water locked up in mineral hydrates is released when they melt deep in the Earth and emerge from volcanoes.

(a) Oxygen Water purification provides the safe drinking water that has allowed cities to grow. Municipal water supplies normally undergo several stages of purification (**FIG. 8H.10**). Raw water is aerated by bubbling air through it to remove foul-smelling dissolved gases such as H_2S, to oxidize some organic compounds to CO_2, and to add oxygen. Adding slaked lime, $Ca(OH)_2$, reduces acidity, precipitates Mg^{2+}, Fe^{3+}, Cu^{2+}, and other metal ions as hydroxides, and softens hard water (Topics 5D and 8D). After the lime is added, the water is pumped into a primary settling basin. Because the precipitates tend to form as a colloid (a very fine powder that remains suspended in the water, Topic 5D), either $Fe_2(SO_4)_3$ or alum (specifically, $Al_2(SO_4)_3 \cdot 18H_2O$) is added to coagulate and flocculate the precipitate so that it can be filtered. **Coagulation** is the irreversible aggregation of small particles into larger particles; **flocculation** is the reversible, loose aggregation of somewhat larger particles to form a fluffy gel. At this point the water is basic and carbon dioxide is often added to lower the pH, which promotes precipitation of the aluminum as $Al(OH)_3$ so that it can be removed by filtration (Topic 8E).

As the precipitate settles slowly in a secondary basin, it adsorbs any remaining suspended $CaCO_3$, bacteria, and other particles, such as dirt and algae. Precipitates from the primary and secondary basins are combined in a sludge lagoon for disposal. The water is then passed through a sand filter to remove any remaining suspended particles.

The pH of the water is checked again and made slightly basic to reduce acid corrosion of the pipes. At this point, a disinfectant, usually chlorine, is added. In the United States, the chlorine level is required to be greater than 1 g of Cl_2 per 1000 kg of water (that

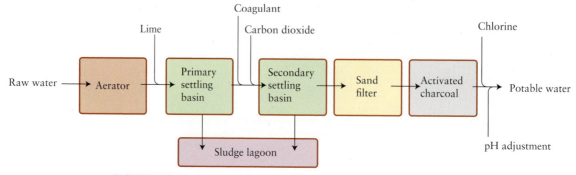

FIGURE 8H.10 Typical steps in the purification of drinking water.

is, 1 ppm by mass) at the point of consumption. In water, chlorine forms hypochlorous acid, which is highly toxic to bacteria:

$$Cl_2(g) + 2\,H_2O(l) \longrightarrow H_3O^+(aq) + Cl^-(aq) + HClO(aq)$$

Additional water purification steps, such as reverse osmosis, may be required (Topic 5F), depending on the source and condition of the water.

Because water is a common solvent, it seems to be merely a passive medium in which chemical reactions take place. However, water is a reactive compound, and an alien raised in a nonaqueous environment might consider it aggressively corrosive and be surprised at our survival. For instance, water is an oxidizing agent:

$$2\,H_2O(l) + 2\,e^- \longrightarrow 2\,OH^-(aq) + H_2(g) \qquad E = -0.42\ \text{V at pH} = 7$$

One example is its reaction with the alkali metals, as in

$$2\,Na(s) + 2\,H_2O(l) \longrightarrow 2\,NaOH(aq) + H_2(g)$$

However, unless the other reactant is a strong reducing agent, water acts as an oxidizing agent only at high temperatures, as in the re-forming reaction (Topic 8B).

Water is a very mild reducing agent:

$$4\,H^+(aq) + O_2(g) + 4\,e^- \longrightarrow 2\,H_2O(l) \qquad E = +0.82\ \text{V at pH} = 7$$

However, few substances besides fluorine are strong enough oxidizing agents to accept the electrons released in the reverse of this half-reaction.

Water is also a Lewis base, because an H_2O molecule can donate one of its lone pairs to a Lewis acid and form complexes such as $Fe(H_2O)_6^{3+}$. Water's ability to act as a Lewis base is also the origin of its ability to hydrolyze substances, such as phosphorus pentachloride (Topic 8G).

Hydrogen peroxide, H_2O_2 (**3**), is a very pale blue liquid that is appreciably denser than water (1.44 $g \cdot mL^{-1}$ at 25 °C) but similar in other physical properties: its melting point is −0.4 °C, and its boiling point is 152 °C. Chemically, though, hydrogen peroxide and water differ greatly. The presence of the second oxygen atom makes H_2O_2 a very weak acid ($pK_{a1} = 11.75$). Hydrogen peroxide is a stronger oxidizing agent than water. For example, H_2O_2 oxidizes Fe^{2+} and Mn^{2+} in acidic and basic solutions. It can also act as a reducing agent in the presence of more powerful oxidizing agents such as permanganate ions and chlorine (usually in basic solution).

Hydrogen peroxide is normally sold for industrial use as a 30% by mass aqueous solution. When it is used as a hair bleach (as a 6% solution), it acts by oxidizing the pigments in the hair. A 3% H_2O_2 aqueous solution is used as a mild antiseptic. Contact with blood catalyzes its disproportionation into water and oxygen gas, which cleanses the wound:

$$2\,H_2O_2(aq) \longrightarrow 2\,H_2O(l) + O_2(g)$$

Because it oxidizes unpleasant effluents without producing any harmful byproducts, H_2O_2 is also being used increasingly as an oxidizing agent to control pollution.

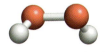

3 Hydrogen peroxide, H_2O_2

Self-test 8H.1A From the Lewis structures of the following molecules, determine which are paramagnetic and explain how you decided: (a) N_2O_3; (b) NO; (c) N_2O.
[***Answer:*** (b) NO is paramagnetic, because it has an unpaired electron.]

Self-test 8H.1B Write the half-reactions and the overall reaction for the oxidation of water by F_2 in acidic solution. Determine the standard potential and $\Delta G°$ for the overall reaction.

(b) Sulfur   Except for water, all the Group 16 binary compounds with hydrogen (the compounds H_2E, where E is a Group 16 element) are toxic gases with offensive odors. They are insidious poisons because they paralyze the olfactory nerve and, soon after exposure, the victim cannot smell them. Rotten eggs smell of hydrogen sulfide, H_2S, because egg proteins contain sulfur and give off the gas when they decompose. Another sign of the formation of sulfides in eggs is the pale green discoloration due to FeS sometimes seen in overcooked hard-boiled eggs where the white meets the yolk. Hydrogen sulfide,

FIGURE 8H.11 The blue stones in this ancient Egyptian ornament are lapis lazuli. This semiprecious stone is an aluminosilicate colored by S_2^- and S_3^- impurities. The blue color is due to S_3^- and the hint of green to S_2^-. (*Egyptian National Museum, Cairo, Egypt/The Bridgeman Art Library.*)

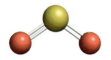

4   Sulfur dioxide, SO_2

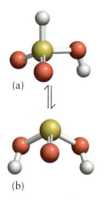

(a)

(b)

5   Sulfurous acid, H_2SO_3

6   Sulfur trioxide, SO_3

H_2S, is prepared either by protonation of the sulfide ion, a Brønsted base, in a reaction such as

$$FeS(s) + 2\,HCl(aq) \longrightarrow FeCl_2(aq) + H_2S(g)$$

or by direct reaction of the elements at 600 °C.

Hydrogen sulfide dissolves in water to give a solution of hydrosulfuric acid, a weak diprotic acid and the parent acid of the hydrogen sulfides (which contain the HS^- ion) and the sulfides (which contain the S^{2-} ion).

The sulfur analog of hydrogen peroxide also exists and is an example of a *polysulfane*, a catenated molecular compound of composition $HS-S_n-SH$, where n can take on values from 0 through 6. The polysulfide ions obtained from the polysulfanes include two ions found in the semiprecious stone lapis lazuli (**FIG. 8H.11**).

Water can act as a Lewis base, an oxidizing agent, and a weak reducing agent; hydrogen peroxide is a stronger oxidizing agent than water. Hydrogen sulfide is a weak acid; the polysulfanes exhibit the ability of sulfur to catenate.

8H.3 Sulfur Oxides and Oxoacids

Sulfur forms several oxides that in atmospheric chemistry are referred to collectively as SO_x (read "sox"). The most important oxides and oxoacids of sulfur are the dioxide and trioxide, which are the anhydrides of sulfurous and sulfuric acids, respectively. Sulfur burns in air to form sulfur dioxide, SO_2 (**4**), a colorless, choking, poisonous gas. About 70 Mt of sulfur dioxide is produced annually in nature from the decomposition of vegetation and from volcanic emissions. In addition, approximately 100 Mt of naturally occurring hydrogen sulfide is oxidized each year to the dioxide by atmospheric oxygen:

$$2\,H_2S(g) + 3\,O_2(g) \longrightarrow 2\,SO_2(g) + 2\,H_2O(g)$$

Industry and transport contribute another 150 Mt of SO_2, of which about 70% comes from oil and coal combustion—mainly in electricity-generating plants. Because, like many other countries, the United States and Canada have increased restrictions on emissions of sulfur oxides, emissions of SO_2 into the atmosphere in North America fell by about 54% between 1980 and 2007. The more stringent regulations that came into effect in the twenty-first century should allow for a further 50% reduction (Box 6E.1).

Sulfur dioxide is the anhydride of sulfurous acid, H_2SO_3, which is the parent acid of the hydrogen sulfites (or bisulfites) and the sulfites:

$$SO_2(g) + H_2O(l) \longrightarrow H_2SO_3(aq)$$

Sulfurous acid is an equilibrium mixture of two molecules (**5a** and **5b**); in (a) it resembles phosphorous acid, with one of the H atoms attached directly to the S atom. These molecules are also in equilibrium with molecules of SO_2, each of which is surrounded by a cage of water molecules. The evidence for this equilibrium is that crystals of composition $SO_2 \cdot xH_2O$, with x about 7, are obtained when the solution is cooled. Such substances, in which a molecule occupies a cage formed by other molecules, are called **clathrates.** Methane, carbon dioxide, and the noble gases also form clathrates with water.

Sulfur dioxide is easily liquefied under pressure and can therefore be used as a refrigerant. It is also a preservative for dried fruit and a bleach for textiles and flour, but its most important use is in the production of sulfuric acid.

The oxidation number of sulfur in sulfur dioxide and the sulfites is +4, an intermediate value in sulfur's range from −2 to +6. Hence, these compounds can act as either oxidizing agents or reducing agents. By far the most important reaction of sulfur dioxide is its slow oxidation to sulfur trioxide, SO_3 (**6**), in which the oxidation number of sulfur is +6:

$$2\,SO_2(g) + O_2(g) \longrightarrow 2\,SO_3(g)$$

At normal temperatures, sulfur trioxide is a volatile liquid (its boiling point is 45 °C), composed of trigonal planar SO_3 molecules. In the solid and to some extent in the liquid, these molecules form trimers (unions of three molecules) of composition S_3O_9 (7) as well as larger clusters.

Sulfuric acid, H_2SO_4, is produced commercially in the *contact process*, in which sulfur is first burned in oxygen and the SO_2 produced is oxidized to SO_3 over a V_2O_5 catalyst:

$$S(s) + O_2(g) \xrightarrow{1000\ °C} SO_2(g)$$

$$2\ SO_2(g) + O_2(g) \xrightarrow{500\ °C,\ V_2O_5} 2\ SO_3(g)$$

Because sulfur trioxide forms a corrosive acid mist with water vapor, it is absorbed instead in 98% concentrated sulfuric acid to give the dense, oily liquid called *oleum:*

$$SO_3(g) + H_2SO_4(l) \longrightarrow H_2S_2O_7(l)$$

Oleum is then converted into the acid by reaction with water:

$$H_2S_2O_7(l) + H_2O(l) \longrightarrow 2\ H_2SO_4(l)$$

Sulfuric acid is the inorganic chemical manufactured in the greatest amount worldwide, the annual production in the United States alone being more than 30 Mt. The low cost of sulfuric acid leads to its widespread use in industry, particularly for the production of fertilizers (Topic 8G).

Sulfuric acid is a colorless, corrosive, oily liquid that boils (and decomposes) at about 300 °C. It has three chemically important properties: it is a strong Brønsted acid, a dehydrating agent, and an oxidizing agent (**FIG. 8H.12**). Sulfuric acid is a strong acid in the sense that its first deprotonation is almost complete at normal concentrations in water. Its conjugate base HSO_4^- is a weak acid, with $pK_a = 1.92$ (Topic 6E). Sulfuric acid forms strong hydrogen bonds with water. The energy released when these bonds form can cause the mixture to boil violently. Sulfuric acid should always be diluted slowly and with great care, making sure to add the acid to the water, instead of the reverse, to avoid splashing concentrated acid.

The powerful dehydrating ability of sulfuric acid is seen when a little concentrated acid is poured on sucrose, $C_{12}H_{22}O_{11}$. A black, frothy mass of carbon forms as a result of the extraction of H_2O (**FIG. 8H.13**):

$$C_{12}H_{22}O_{11}(s) \longrightarrow 12\ C(s) + 11\ H_2O(l)$$

The froth is caused by the generation of the gases CO and CO_2 in side reactions.

Self-test 8H.2A What is the oxidation number of sulfur in (a) the dithionate ion, $S_2O_6{}^{2-}$; (b) the thiosulfate ion, $S_2O_3{}^{2-}$?

[***Answer:*** (a) +5; (b) +2]

Self-test 8H.2B What is the oxidation number of sulfur in (a) S_2Cl_2; (b) oleum, $H_2S_2O_7$?

Sulfur dioxide is the anhydride of sulfurous acid, and sulfur trioxide is the anhydride of sulfuric acid. Sulfuric acid is a strong acid, a dehydrating agent, and an oxidizing agent.

7 Sulfur trioxide trimer, S_3O_9

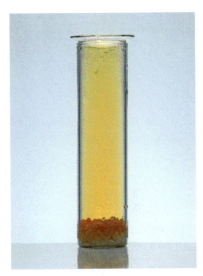

FIGURE 8H.12 Sulfuric acid is an oxidizing agent. When some concentrated acid is poured on to solid sodium bromide, NaBr, the bromide ions are oxidized to elemental bromine, which colors the mixture red-brown. *(W. H. Freeman photo by Ken Karp.)*

(a)   (b)   (c)

FIGURE 8H.13 Sulfuric acid is a dehydrating agent. When concentrated sulfuric acid is poured on to sucrose (a), the sucrose, a carbohydrate, is dehydrated (b), leaving a frothy black mass of carbon (c). *(Chip Clark/ Fundamental Photographs, NYC.)*

Topic 8H Exercises

8H.1 Write chemical equations for (a) the burning of lithium in oxygen; (b) the reaction of sodium metal with water; (c) the reaction of fluorine gas with water; (d) the oxidation of water at the anode of an electrolytic cell.

8H.2 Write chemical equations for the reaction of (a) sodium oxide and water; (b) sodium peroxide and water; (c) sulfur dioxide and water; (d) sulfur dioxide and oxygen, with the use of a vanadium pentoxide catalyst.

8H.3 Complete and balance each of the following equations:
(a) $H_2S(aq) + O_2(g) \rightarrow$
(b) $CaO(s) + H_2O(l) \rightarrow$
(c) $H_2S(g) + SO_2(g) \rightarrow$

8H.4 Complete and balance each of the following equations:
(a) $FeS(s) + HCl(aq) \rightarrow$
(b) $H_2(g) + S_8(s) \rightarrow$
(c) $Br^-(aq) + Cl_2(g) \rightarrow$

8H.5 Explain from structural considerations why the dipole moment of the NH_3 molecule is smaller than that of the H_2O molecule.

8H.6 Explain from structural considerations why water and hydrogen peroxide have similar physical properties but different chemical properties.

8H.7 (a) Draw the Lewis structure of H_2O_2 and predict the approximate H—O—O bond angle. (b) Which of the following ions will be oxidized by hydrogen peroxide in acidic solution: (i) Cu^+; (ii) Mn^{2+}; (iii) Ag^+; (iv) F^-?

8H.8 (a) Draw the Lewis structure for oleum, $(HO)_2OSOSO(OH)_2$, prepared by treating sulfuric acid with SO_3. (b) Determine the formal charges on the sulfur and oxygen atoms. (c) What is the oxidation number of sulfur in this compound?

8H.9 If 2.00 g of sodium peroxide is dissolved to form 200. mL of an aqueous solution, what will be the pH of the solution? For H_2O_2, $K_{a1} = 1.8 \times 10^{-12}$ and K_{a2} is negligible.

8H.10 What is the pH of 0.010 M NaHS(aq)?

8H.11 Describe the trend in acidity of the binary hydrogen compounds of the Group 16 elements and account for the trend in terms of bond strength.

8H.12 When lead(II) sulfide is treated with hydrogen peroxide, the possible products are either lead(II) sulfate or lead(IV) oxide and sulfur dioxide. (a) Write balanced equations for the two reactions. (b) Use data from Appendix 2A to decide which possibility is more likely.

Topic 8I Group 17: The Halogens

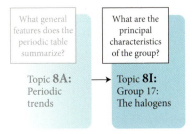

The unique properties of the halogens (**TABLE 8I.1**), the members of Group 17, can be traced to their valence configuration, ns^2np^5, which needs only one more electron to reach a closed-shell configuration. To complete the octet of valence electrons, in the elemental state all halogen atoms combine to form diatomic molecules, such as F_2 and I_2. With the exception of fluorine, the halogens can also lose valence electrons, and their oxidation numbers can range from -1 to $+7$.

8I.1 The Group 17 Elements

The halogens form a family showing the smooth trends in physical properties that are expected when London forces between molecules are the dominant intermolecular force. Because electronegativity decreases down the group (**FIG. 8I.1**) and atomic and ionic radii increase smoothly down the group (**FIG. 8I.2**), chemical properties show smooth trends as well, with the exception of some properties of fluorine.

(a) Fluorine   The first element of the group, fluorine, F, is the halogen of greatest abundance in the Earth's crust. It occurs widely in many minerals, including *fluorite (fluorspar)*, CaF_2; *cryolite*, Na_3AlF_6; and the *fluorapatites,* $Ca_5(PO_4)_3F$. Because fluorine is the most strongly oxidizing element ($E° = +2.87$ V), it cannot be obtained from its compounds by oxidation with another element. Fluorine is produced by electrolyzing an anhydrous molten mixture of potassium fluoride and hydrogen fluoride at about 75 °C with a carbon anode.

Fluorine itself is a highly reactive, almost colorless gas of F_2 molecules. Most of the fluorine produced by industry is used to make the volatile solid UF_6 used for processing nuclear fuel (Topic 10C). Much of the rest is used in the production of SF_6 for electrical equipment.

Fluorine has a number of peculiarities that stem from its high electronegativity, small size, and lack of available d-orbitals. It is the most electronegative element and has an oxidation number of -1 in all its compounds. Its high electronegativity and small size

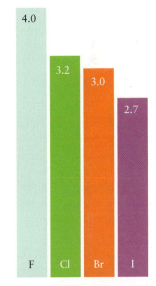

FIGURE 8I.1 The electronegativities of the halogens decrease steadily down the group.

TABLE 8I.1   The Group 17 Elements

Common name: halogens
Valence configuration: ns^2np^5

Z	Name	Symbol	Molar mass/ (g·mol⁻¹)	Melting point/°C	Boiling point/°C	Density/ (g·cm⁻³)	Normal form*
9	fluorine	F	19.00	-220	-188	1.51†	almost colorless gas
17	chlorine	Cl	35.45	-101	-34	1.66†	yellow-green gas
35	bromine	Br	79.90	27	59	3.12	red-brown liquid
53	iodine	I	126.90	114	184	4.95	purple-black nonmetallic solid
85	astatine‡	At	(210)	300	350	—	nonmetallic solid

*Normal form means the appearance and state of the element at 25 °C and 1 atm.
†For the liquid at its boiling point.
‡Radioactive.

693

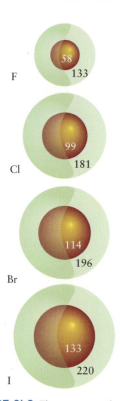

F — 58, 133

Cl — 99, 181

Br — 114, 196

I — 133, 220

FIGURE 8I.2 The atomic and ionic radii of the halogens increase steadily down the group as electrons occupy outer shells of the atoms. The values shown are in picometers. In all cases, the radii of the anions (represented by the green spheres) are larger than the atomic radii.

FIGURE 8I.3 Iron reacts vigorously and exothermically with chlorine to form anhydrous iron(III) chloride. (*Chip Clark/ Fundamental Photographs, NYC.*)

allow it to oxidize other elements to their highest oxidation states. The small size helps, because it allows several F atoms to pack around a central atom, as in IF_7.

Because the fluoride ion is so small, the lattice enthalpies of its ionic compounds tend to be high (see Table 4E.1). As a result, fluorides are less soluble than other halides. This difference in solubility is one of the reasons why the oceans are salty with chlorides rather than fluorides, even though fluorine is more abundant than chlorine in the Earth's crust. Chlorides are more readily dissolved and washed out to sea. There are some exceptions to this trend in solubilities, including AgF, which is soluble; the other silver halides are insoluble. The exception arises because the covalent character of the silver halides increases from AgCl to AgI as the anion becomes larger and more polarizable. Silver fluoride, which contains the small and almost unpolarizable fluoride ion, is freely soluble in water because it is predominantly ionic.

(b) Chlorine   The element chlorine, Cl, is one of the chemicals manufactured in the greatest amount worldwide. It is obtained from sodium chloride by electrolysis of molten rock salt or brine (recall Topics 6O and 8C). It is a pale yellow-green gas of Cl_2 molecules that condenses at $-34\ °C$. It reacts directly with nearly all the elements (notable exceptions are carbon, nitrogen, oxygen, and the noble gases). It is a strong oxidizing agent and oxidizes metals to high oxidation states; for example, anhydrous iron(III) chloride, not iron(II) chloride, is formed when chlorine reacts with iron (**FIG. 8I.3**):

$$2\ Fe(s) + 3\ Cl_2(g) \longrightarrow 2\ FeCl_3(s)$$

Chlorine is used in a number of industrial processes, including the manufacture of plastics, solvents, and pesticides. It is used as a bleach in the paper and textile industries and as a disinfectant in water treatment (Topic 8H).

(c) Bromine   The element bromine, Br, exists as a corrosive, red-brown fuming liquid of Br_2 molecules that has a penetrating odor. It is produced from brine wells by the oxidation of Br^- ions with elemental chlorine (see *Fundamentals* Fig. K.2):

$$2\ Br^-(aq) + Cl_2(g) \longrightarrow Br_2(l) + 2\ Cl^-(aq)$$

Air is bubbled through the solution to vaporize the bromine and drive it into a cooling apparatus where it can be collected.

Organic bromides are incorporated into textiles as fire retardants and are used as pesticides; inorganic bromides, particularly silver bromide, are used in photographic emulsions.

(d) Iodine   Iodine, I, occurs as iodide ions in brines and as an impurity in Chile saltpeter. It was once obtained from seaweed, which contains high concentrations accumulated from seawater: 2000 kg of seaweed produce about 1 kg of iodine. The best modern source is the brine from oil wells; the oil itself was produced by the decay of marine organisms that had accumulated the iodine while they were alive.

Like bromine, elemental iodine is produced by oxidation with chlorine:

$$Cl_2(g) + 2\ I^-(aq) \longrightarrow I_2(aq) + 2\ Cl^-(aq)$$

The blue-black lustrous solid sublimes easily and forms a purple vapor.

When iodine dissolves in organic solvents, it produces solutions having a variety of colors. These colors arise from the different interactions between the I_2 molecules and the solvent (**FIG. 8I.4**). The element is only slightly soluble in water, unless I^- ions are present, in which case the soluble brown triiodide ion, I_3^-, is formed. Iodine itself has few direct uses; however, when it is dissolved in alcohol, it is familiar as a mild oxidizing antiseptic. Because it is an essential trace element for living systems but scarce in inland areas, iodides are added to table salt (sold as "iodized salt") to prevent iodine deficiency.

Self-test 8I.1A (a) Identify the oxidation number of Cl in Cl_2O_7 and (b) write the chemical equation for its reaction with water.

[*Answer:* (a) +7; (b) $Cl_2O_7(aq) + H_2O(l) \rightarrow 2HClO_4(aq)$]

Self-test 8I.1B Explain the trends in the melting points and boiling points of the halogens.

FIGURE 81.4 Solutions of iodine in a variety of solvents. From left to right, the first three solvents are tetrachloromethane (carbon tetrachloride), water, and potassium iodide solution, in which the brown I_3^- ion forms. In the solution on the far right, some starch has been added to a solution of I_3^-. The intense blue color that results has led to the use of starch as an indicator for the presence of iodine. (©1990 Chip Clark, Fundamental Photographs.)

The halogens show smooth trends in chemical properties down the group; fluorine has some anomalous properties, such as its strength as an oxidizing agent and the lower solubilities of most fluorides.

81.2 Compounds of the Halogens

The uses of halogen compounds are widely varied: from nonstick cookware coated with the fluorine-containing polymer sold as Teflon, to organochlorine compounds used as pesticides and ammonium perchlorate used as rocket fuel.

(a) Interhalogens   The halogens form compounds among themselves. These **interhalogens** have the formulas XX', XX'_3, XX'_5, and XX'_7, where X is the heavier (and larger) of the two halogens. Only some of the possible combinations have been synthesized (**TABLE 81.2**). They are all prepared by direct reaction of the two halogens, the product formed being determined by the proportions of reactants used. For example,

$$Cl_2(g) + 3 F_2(g) \longrightarrow 2 ClF_3(g)$$
$$Cl_2(g) + 5 F_2(g) \longrightarrow 2 ClF_5(g)$$

The physical properties of the interhalogens are intermediate between those of their parent halogens. Trends in the chemistry of the interhalogen fluorides can be related to the decrease in bond dissociation energy as the central halogen atom becomes heavier. The fluorides of the heavier halogens are all very reactive: bromine trifluoride gas is so reactive that even asbestos burns in it.

(b) Hydrogen halides   The hydrogen halides, HX, can be prepared by the direct reaction of the elements:

$$H_2(g) + X_2(g) \longrightarrow 2 HX(g)$$

Fluorine reacts explosively by a radical chain reaction as soon as the gases are mixed. A mixture of hydrogen and chlorine also explodes when exposed to light. Bromine and iodine react with hydrogen much more slowly. A less hazardous laboratory source of the hydrogen halides is the action of a nonvolatile acid on a metal halide, as in

$$CaF_2(s) + 2 H_2SO_4(aq, conc) \longrightarrow Ca(HSO_4)_2(aq) + 2 HF(g)$$

Because Br^- and I^- are oxidized by sulfuric acid, phosphoric acid is used in the preparation of HBr and HI:

$$KI(s) + H_3PO_4(aq) \xrightarrow{\Delta} KH_2PO_4(aq) + HI(g)$$

All the hydrogen halides are colorless, pungent gases except hydrogen fluoride, which is a liquid at temperatures below 20 °C. Its low volatility is a sign of extensive hydrogen bonding, and short zigzag chains of hydrogen-bonded molecules, up to about $(HF)_5$, survive to some extent in the vapor. All the hydrogen halides dissolve in water to give acidic solutions. Hydrofluoric acid has the distinctive property of attacking glass and

TABLE 81.2	Known Interhalogens
Interhalogen	**Normal form***
XF_n	
ClF	colorless gas
ClF_3	colorless gas
ClF_5	colorless gas
BrF	pale-brown gas
BrF_3	pale-yellow liquid
BrF_5	colorless liquid
IF	unstable
IF_3	yellow solid
IF_5	colorless liquid
IF_7	colorless gas
XCl_n	
BrCl	red-brown gas
ICl	red solid
I_2Cl_6	yellow solid
XBr_n	
IBr	black solid

*Normal form means the appearance and state of the compound at 25 °C and 1 atm.

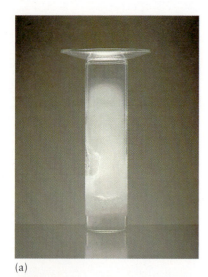

(a)

(b)

FIGURE 8I.5 When a mixture of hydrofluoric acid and ammonium fluoride is swirled inside a glass vial (a), reaction with the silica in the glass frosts the surface of the cover glass as well as the walls of the vial (b). (*W. H. Freeman photos by Ken Karp.*)

silica. The interiors of some lamp bulbs are frosted by the vapors from an aqueous solution of hydrofluoric acid and ammonium fluoride (**FIG. 8I.5**).

THINKING POINT

Why does hydrogen fluoride dissolve in water as a weak acid whereas the other hydrogen halides dissolve as strong acids?

Hydrogen fluoride is used to make fluorinated carbon compounds, such as Teflon (polytetrafluoroethylene) and the refrigerant R134a (CF_3CH_2F). Most fluoro-substituted hydrocarbons are relatively inert chemically: they are inert to oxidation by air, hot nitric acid, concentrated sulfuric acid, and other strong oxidizing agents. Consequently, Teflon is used to line chemical reactor vessels and to coat the surfaces of nonstick cookware. Because R134a does not contribute to global warming or the destruction of ozone in the stratosphere and is relatively unreactive and nontoxic, it is now the refrigerant most recommended for use in air conditioners in the United States.

(c) Halogen oxoacids and oxoanions The acid strengths and oxidizing abilities of the halogen oxoacids increase with the oxidation number of the halogen (Topic 6C). The hypohalous acids, HXO (halogen oxidation number +1), are prepared by direct reaction of the halogen with water. For example, chlorine gas disproportionates in water to produce hypochlorous acid and hydrochloric acid:

$$\overset{0}{Cl_2}(g) + H_2O(aq) \longrightarrow \overset{+1}{H}ClO(aq) + \overset{-1}{H}Cl(aq)$$

Hypohalite ions, XO⁻, are formed when a halogen is added to a basic aqueous solution. Sodium hypochlorite, NaClO, is produced from the electrolysis of brine when the electrolyte is rapidly stirred, and the chlorine gas produced at the anode reacts with the hydroxide ion generated at the cathode. The chlorine gas disproportionates to produce hypochlorite and chloride ions:

$$\overset{0}{Cl_2}(g) + 2\,OH^-(aq) \longrightarrow \overset{+1}{C}lO^-(aq) + \overset{-1}{C}l^-(aq) + H_2O(aq)$$

Because hypochlorites oxidize organic material, they are used in liquid household bleaches and as disinfectants. Their action as oxidizing agents stems partly from the decomposition of hypochlorous acid in solution:

$$2\,HClO(aq) \longrightarrow 2\,H^+(aq) + 2\,Cl^-(aq) + O_2(g)$$

The oxygen either bubbles out of the solution or attacks oxidizable material. Calcium hypochlorite is the main component of bleach powders and is used for purifying the water in home swimming pools.

Chlorate ions, ClO_3^- (chlorine oxidation number +5), form when chlorine reacts with hot concentrated aqueous alkali:

$$3\,Cl_2(g) + 6\,OH^-(aq) \overset{\Delta}{\longrightarrow} ClO_3^-(aq) + 5\,Cl^-(aq) + 3\,H_2O(l)$$

They decompose when heated, to an extent that depends on whether a catalyst is present:

$$4\,KClO_3(s) \overset{\Delta}{\longrightarrow} 3\,KClO_4(s) + KCl(s)$$
$$2\,KClO_3(s) \overset{\Delta,\ MnO_2}{\longrightarrow} 2\,KCl(s) + 3\,O_2(g)$$

The latter reaction is a convenient laboratory source of oxygen.

Chlorates are useful oxidizing agents. Potassium chlorate is used as an oxidant in fireworks and in matches. The heads of safety matches consist of a paste of potassium chlorate, antimony sulfide, and sulfur, with powdered glass to create friction when the match is struck; as mentioned in Topic 8G, the striking strip contains red phosphorus, which ignites the match head.

The principal use of sodium chlorate is as a source of chlorine dioxide, ClO_2. The chlorine in ClO_2 has oxidation number +4, and so the chlorate must be reduced to form it. Sulfur dioxide is a convenient reducing agent for this reaction:

$$2\,NaClO_3(aq) + SO_2(g) + H_2SO_4(aq, dilute) \longrightarrow 2\,NaHSO_4(aq) + 2\,ClO_2(g)$$

Chlorine dioxide has an odd number of electrons and is a paramagnetic yellow gas. Despite the environmental damage it creates, it is often used to bleach paper pulp, because it can oxidize the various pigments in the pulp without degrading the wood fibers.

The perchlorates, ClO_4^- (chlorine oxidation number $+7$), are prepared by electrolytic oxidation of aqueous chlorates:

$$ClO_3^-(aq) + H_2O(l) \longrightarrow ClO_4^-(aq) + 2\,H^+(aq) + 2\,e^-$$

Perchloric acid, $HClO_4$, is a colorless liquid and the strongest of all common acids. Because chlorine has its highest oxidation number, $+7$, in these compounds, they are powerful oxidizing agents; contact between perchloric acid and even a small amount of organic material can result in a dangerous explosion.

One spectacular example of the oxidizing ability of perchlorates was their use in the booster rockets of the space shuttles. The solid propellant consisted of aluminum powder (the fuel), ammonium perchlorate (the oxidizing agent as well as a fuel), and iron(III) oxide (the catalyst). These reactants were mixed into a liquid polymer, which set to a solid inside the rocket shell. A variety of products can form when the mixture is ignited. One of the reactions is

$$3\,NH_4ClO_4(s) + 3\,Al(s) \xrightarrow{Fe_2O_3} Al_2O_3(s) + AlCl_3(s) + 6\,H_2O(g) + 3\,NO(g)$$

The solid products formed thick clouds of white powder during liftoff (**FIG. 8I.6**)

FIGURE 8I.6 The white smoke emitted by the space shuttle booster rockets consisted of powdered aluminum oxide and aluminum chloride. (*NASA/Scott Andrews.*)

Self-test 8I.2A Which oxoacid of bromine is (a) the strongest acid? (b) the strongest oxidizing agent? Explain your answers.

[*Answer:* For both (a) and (b), $HBrO_4$, because it has the most O atoms]

Self-test 8I.2B How does pH affect the oxidizing power of $HClO_3$?

The interlalogens have properties intermediate between those of the constituent halogens. Nonmetals form covalent halides; metals typically form ionic halides. The oxoacids of chlorine are all oxidizing agents; both acidity and oxidizing strength of oxoacids increase as the oxidation number of the halogen increases.

Topic 8I Exercises

8I.1 Identify the oxidation number of the halogen in (a) hypoiodous acid; (b) ClO_2; (c) dichlorine heptoxide; (d) $NaIO_3$.

8I.2 Identify the oxidation number of the halogen atoms in (a) iodine heptafluoride; (b) sodium periodate; (c) hypobromous acid; (d) sodium chlorite.

8I.3 Write the balanced chemical equation for (a) the thermal decomposition of potassium chlorate without a catalyst; (b) the reaction of bromine with water; (c) the reaction between sodium chloride and concentrated sulfuric acid. (d) Identify each reaction as a Brønsted acid–base, Lewis acid–base, or redox reaction.

8I.4 Write the balanced chemical equation for the reaction of chlorine with water in (a) a neutral aqueous solution; (b) a dilute basic solution; (c) a concentrated basic solution. (d) Verify that each reaction is a disproportionation reaction.

8I.5 (a) Arrange the chlorine oxoacids in order of increasing oxidizing strength. (b) Suggest an interpretation of that order in terms of oxidation numbers.

8I.6 (a) Arrange the hypohalous acids in order of increasing acid strength. (b) Suggest an interpretation of that order in terms of electronegativities.

8I.7 Draw the Lewis structure of Cl_2O. Predict the shape of the Cl_2O molecule and estimate the Cl—O—Cl bond angle.

8I.8 Draw the Lewis structure of BrF_3. What is the hybridization of the bromine atom in the molecule?

8I.9 Suggest reasons why the following interhalogens are not stable: (a) ICl_5; (b) IF_2; (c) $ClBr_3$.

8I.10 Chlorine occurs as the oxoanions ClO^-, ClO_2^-, ClO_3^-, and ClO_4^- in many chemical stockrooms. However, fluorine forms no stable oxoanions. Explain this observation.

8I.11 The interhalogen IF_x can be made only by indirect routes. For example, xenon difluoride gas can react with iodine gas to produce IF_x and xenon gas. In one experiment, xenon difluoride is introduced into a rigid container until a pressure of 3.6 atm is reached. Iodine vapor is then introduced until the total pressure is 7.2 atm. Reaction is then allowed to proceed at constant temperature until completion by solidifying the IF_x as it is produced. The final pressure in the flask due to the xenon and excess iodine vapor is 6.0 atm. (a) What is the formula of the interhalogen? (b) Write the chemical equation for its formation.

8I.12 The interhalogen ClF_x has been used as a rocket fuel. It reacts with hydrazine to form the gases hydrogen fluoride, nitrogen, and chlorine. In one study of this reaction, ClF_x gas is introduced into a rigid container until a pressure of 1.2 atm is reached. Gaseous hydrazine is then introduced until the total pressure is 3.0 atm. Reaction is then allowed to proceed at constant temperature until completion. The final pressure in the flask is 6.0 atm, which includes 0.9 atm due to excess hydrazine. (a) What is the formula of the interhalogen? (b) Write the chemical equation for its reaction with hydrazine.

8I.13 Use data from Appendix 2B to decide whether chlorine gas will oxidize Mn^{2+} to form the permanganate ion in an acidic solution.

8I.14 (a) Use the data in Appendix 2B to decide which of ozone and fluorine is the stronger oxidizing agent in water. (b) Does your answer depend on whether the reaction is carried out in acidic or basic solution?

8I.15 The concentration of F^- ions can be measured by adding an excess of lead(II) chloride solution and weighing the lead(II) chlorofluoride (PbClF) precipitate. Calculate the molar concentration of F^- ions in 25.00 mL of a solution that gave a lead chlorofluoride precipitate of mass 0.765 g.

8I.16 Suppose 25.00 mL of an aqueous solution of iodine was titrated with 0.0250 M $Na_2S_2O_3(aq)$, with starch as the indicator. The blue color of the starch–iodine complex disappeared when 18.33 mL of the thiosulfate solution had been added. What was the molar concentration of I_2 in the original solution? The titration reaction is $I_2(aq) + 2 S_2O_3^{2-}(aq) \rightarrow 2 I^-(aq) + S_4O_6^{2-}(aq)$.

Topic 8J Group 18: The Noble Gases

8J.1   The Group 18 Elements

8J.2   Compounds of the Noble Gases

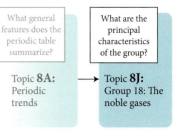

The elements in Group 18, the noble gases, get their group name from their very low reactivity (TABLE 8J.1). Experiments on the gases and, later, recognition of their closed-shell electron configurations (ns^2np^6) prompted the belief that these elements were chemically inert. Indeed, no compounds of the noble gases were known until 1962. That year, the British chemist Neil Bartlett synthesized the first noble-gas compound, xenon hexafluoroplatinate, $XePtF_6$. Soon after, chemists at Argonne National Laboratory made xenon tetrafluoride, XeF_4, from a high-temperature mixture of xenon and fluorine.

8J.1 The Group 18 Elements

All the Group 18 elements occur in the atmosphere as monatomic gases; together they make up about 1% of its mass. Argon is the third most abundant gas in the atmosphere after nitrogen and oxygen (discounting the variable amount of water vapor). All the noble gases except helium and radon are obtained by the fractional distillation of liquid air.

(a) Helium   The element helium, He, is the second most abundant element in the universe after hydrogen; it is rare on Earth because its atoms are so light that a large proportion of them reach high speeds and escape from the atmosphere; unlike hydrogen, they cannot be anchored within compounds. However, helium is found as a component of natural gases trapped under rock formations (notably in Texas), where it has collected as a result of the emission of α particles by radioactive elements. An α particle is a helium-4 nucleus ($^4He^{2+}$), and an atom of the element forms when the particle picks up two electrons from its surroundings.

Helium gas is twice as dense as hydrogen under the same conditions. Nevertheless, because its density is still very low and it is nonflammable, it is used to provide buoyancy in airships such as blimps. Helium is also used to dilute oxygen for use in hospitals and in deep-sea diving, to pressurize rocket fuels, as a coolant, and in helium–neon lasers. The element has the lowest boiling point of any substance (4.2 K), and it does not freeze to a solid at any temperature unless pressure is applied to hold the light, mobile atoms together. These properties and its chemical inertness make helium useful for **cryogenics,** the study

TABLE 8J.1   The Group 18 Elements

Common name: noble gases
Valence configuration: ns^2np^6
Normal form: colorless monatomic gas

Z	Name	Symbol	Molar mass/ (g·mol^{-1})	Melting point/°C	Boiling point/°C
2	helium	He	4.00	—	−269 (4.2 K)
10	neon	Ne	20.18	−249	−246
18	argon	Ar	39.95	−189	−186
36	krypton	Kr	83.80	−157	−153
54	xenon	Xe	131.29	−112	−108
86	radon*	Rn	(222)	−71	−62

*Radioactive.

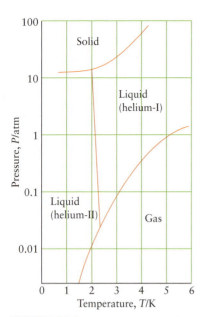

FIGURE 8J.1 The phase diagram for helium-4 shows the two liquid phases of helium. Helium-II, the low-temperature liquid phase, is a superfluid.

of matter at very low temperatures, such as those used for the study of superconductivity (Topic 3J). Helium is the only substance known to have more than one liquid phase (**FIG. 8J.1**). Below 2 K, liquid helium-II shows the remarkable property of **superfluidity**, the ability to flow without viscosity.

(b) Neon, argon, krypton, and xenon   Neon, Ne, which emits an orange-red glow when an electric discharge is passed through it, is widely used in advertising signs and displays (**FIG. 8J.2**). Argon, Ar, is used to provide an inert atmosphere for welding (to prevent oxidation) and to fill some types of light bulbs, where it conducts heat away from the filament. Krypton, Kr, gives an intense white light when an electric discharge is passed through it, and so it is used in airport runway lighting. Because krypton is produced by nuclear fission, its atmospheric abundance is one measure of worldwide nuclear activity. Xenon, Xe, is used in halogen lamps for automobile headlights and in high-speed photographic flash tubes; it is also being investigated as an anesthetic.

The noble gases are all found naturally as unreactive monatomic gases. Helium has two liquid phases; the lower-temperature liquid phase exhibits superfluidity.

8J.2 Compounds of the Noble Gases

The ionization energies of the noble gases are very high but decrease down the group (**FIG. 8J.3**). Xenon's ionization energy is low enough for electrons to be lost to very electronegative elements. No compounds of helium, neon, and argon exist, except under very special conditions, such as the capture of atoms of He and Ne inside a buckminsterfullerene cage. Krypton forms only one known stable neutral molecule, KrF_2. In 1988, a compound with a Kr—N bond was reported, but it is stable only below $-50\ ^\circ C$. This leaves xenon as the noble gas with the richest chemistry. It forms several compounds with fluorine and oxygen, and compounds with Xe—N and Xe—C bonds have been reported, such as $(C_6F_5)_2Xe$.

The starting point for the synthesis of xenon compounds is the preparation of xenon difluoride, XeF_2, and xenon tetrafluoride, XeF_4, by heating a mixture of the elements to 400 °C at 6 atm. At higher pressures, fluorination proceeds as far as xenon hexafluoride, XeF_6. All three fluorides are crystalline solids (**FIG. 8J.4**). In the gas phase, all are molecular

FIGURE 8J.2 The colors of this fluorescent lighting art are due to emission from noble-gas atoms. Neon is responsible for the red light; when it is mixed with a little argon, the color becomes blue-green. The yellow color is achieved by coating the inside of the glass with phosphors that give off yellow light when excited. (© Andrea Heselton/Alamy.)

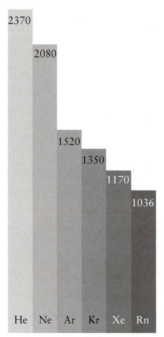

FIGURE 8J.3 The ionization energies of the noble gases decrease steadily down the group. The values shown are in kilojoules per mole.

FIGURE 8J.4 Crystals of xenon tetrafluoride, XeF_4. This compound was first prepared in 1962 by the reaction of xenon and fluorine at 6 atm and 400 °C. (Science Source.)

compounds. Solid xenon hexafluoride, however, is ionic, with a complex structure consisting of XeF_5^+ cations bridged by F^- anions.

The xenon fluorides are used as powerful fluorinating agents (reagents for attaching fluorine atoms to other substances). The tetrafluoride will even fluorinate platinum metal:

$$Pt(s) + XeF_4(s) \longrightarrow Xe(g) + PtF_4(s)$$

The xenon fluorides are used to prepare the xenon oxides and oxoacids and, in a series of disproportionations, to bring the oxidation number of xenon up to +8. First, xenon tetrafluoride is hydrolyzed to xenon trioxide, XeO_3 in a disproportionation reaction:

$$6\, XeF_4(s) + 12\, H_2O(l) \longrightarrow 2\, XeO_3(aq) + 4\, Xe(g) + 3\, O_2(g) + 24\, HF(aq)$$

Xenon trioxide, XeO_3 (**1**), is the anhydride of xenic acid, H_2XeO_4 (**2**). It reacts with aqueous alkali to form a hydrogen xenate ion, $HXeO_4^-$. This ion slowly disproportionates into xenon and the octahedral perxenate ion, XeO_6^{4-} (**3**), in which the oxidation number of xenon is +8. Aqueous perxenate solutions are yellow and are very powerful oxidizing agents as a result of the high oxidation number of xenon. When barium perxenate is treated with sulfuric acid, it is dehydrated to the anhydride of perxenic acid, xenon tetroxide, XeO_4 (**4**), an explosively unstable gas.

1 Xenon trioxide, XeO_3

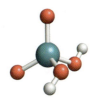

2 Xenic acid, H_2XeO_4

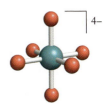

3 Perxenate ion, XeO_6^{4-}

Self-test 8J.1A Explain why the ease of compound formation increases down Group 18.
[***Answer:*** Ionization energy decreases down the group, and so it is easier for the heavier elements to share their electrons.]

Self-test 8J.1B Predict the shape of the $XeOF_4$ molecule.

4 Xenon tetroxide, XeO_4

Xenon is the only noble gas known to form an extensive series of compounds with fluorine and oxygen; xenon fluorides are powerful fluorinating agents, and xenon oxides are powerful oxidizing agents.

Topic 8J Exercises

8J.1 What are the sources for the production of helium and argon?

8J.2 What are the sources for the production of krypton and xenon?

8J.3 Identify the oxidation number of the noble-gas element in (a) KrF_2; (b) XeF_6; (c) KrF_4; (d) XeO_4^{2-}.

8J.4 Identify the oxidation number of the noble-gas element in (a) XeO_3; (b) XeO_6^{4-}; (c) XeF_2; (d) $HXeO_4^-$.

8J.5 Xenon tetrafluoride is a powerful oxidizing agent. In an acidic solution, it is reduced to xenon. Write the corresponding half-reaction.

8J.6 Complete and balance each of the following reactions:
(a) $XeF_6(s) + H_2O(l) \rightarrow XeO_3(aq) + HF(aq)$
(b) $Pt(s) + XeF_4(s) \rightarrow$
(c) $Kr(g) + F_2(g) \xrightarrow{\text{electric discharge}}$

8J.7 Predict the relative acid strengths of H_2XeO_4 and H_4XeO_6. Explain your conclusions.

8J.8 Predict the relative oxidizing strengths of H_2XeO_4 and H_4XeO_6. Explain your conclusions.

The following Example and Exercises draw on material from throughout Focus 8.

FOCUS 8 Online Cumulative Example

You are working for an environmental research laboratory and are interested in the energy costs and greenhouse gas emissions for the production of industrial raw materials, in particular, aluminum. The production of aluminum from ores requires 85 600 MJ of energy per tonne (1 t = 10^3 kg); 70% of which is used in the electrolysis of Al_2O_3 dissolved in cryolite. The overall chemical equation is

$$4\,Al^{3+}(melt) + 6\,O^{2-}(melt) + 3\,C(s, gr) \longrightarrow$$
$$4\,Al(l) + 3\,CO_2\,(g)$$

(a) How long will it take to produce 1.0 t of aluminum if an aluminum smelter is operating at 300. kA? (b) The Gibbs free energy required to drive the electrolysis reaction forward in an operating cell is 2390 kJ. What is the minimum potential difference required to drive the process forward in such a cell?

(c) Assume that all the 85 600 MJ is provided by the burning of coal. The average mass of CO_2 generated from coal (regarded as carbon, C) is 0.265 g of CO_2 for each kilojoule of energy produced. What mass of coal is consumed for each tonne of aluminum produced? (d) What volume of $CO_2(g)$ at 1.0 atm and 20 °C is formed for each tonne of aluminum? (e) In an alternative method for aluminum production, alumina, Al_2O_3, reacts with carbon at high temperatures to produce aluminum carbide, Al_4C_3, and carbon monoxide. In a second reaction, the aluminum carbide reacts with additional alumina to form aluminum metal and carbon monoxide. Balance the skeletal reactions below and indicate the oxidation numbers of all the elements.

$$Al_2O_3(s) + C(s) \longrightarrow Al_4C_3(s) + CO(g) \quad \triangle$$
$$Al_4C_3(s) + Al_2O_3(s) \longrightarrow Al(l) + CO(g) \quad \triangle$$

 The online Cumulative Example solution can be found at http://macmillanhighered.com/chemicalprinciples7e

FOCUS 8 Exercises

8.1 (a) Use graphing software to plot standard potential against atomic number for the elements of Groups 1 and 2 (refer to Appendix 2B for data). (b) What generalizations can be deduced from the graph?

8.2 Use graphing software and data from Appendix 2B to plot ionization energy against standard potential for the elements of Groups 1 and 2. What generalizations can be drawn from the graph?

8.3 Indicate whether the following statements about Group 14 elements are true or false. If false, explain what is wrong with the statement. (a) Only two of the elements are metals. (b) The oxides of all the elements are amphoteric. (c) The outer s- and p-orbitals become closer in energy on going down the group.

8.4 Indicate whether the following statements about Group 16 elements are true or false. If false, explain what is wrong with the statement. (a) Four of the elements are nonmetals. (b) H_2SeO_3 is a stronger acid than H_2SO_3. (c) The oxides of all the elements are acidic.

8.5 Describe evidence for the statement that hydrogen can act as both a reducing agent and an oxidizing agent. Give chemical equations to support your evidence.

8.6 What justification is there for regarding the ammonium ion as an analog of a Group 1 metal cation? Consider properties such as solubility, charge, and radius. The radius of NH_4^+ treated as a sphere is 151 pm.

8.7 Hydrogen burns in an atmosphere of bromine to give hydrogen bromide. If 135 mL of H_2 gas at 273 K and 1.00 atm combines with a stoichiometric amount of bromine and the resulting

hydrogen bromide dissolves to form 225 mL of an aqueous solution, what is the molar concentration of the resulting hydrobromic acid solution?

8.8 The saline hydrides all react rapidly with water. They also react similarly with liquid ammonia. (a) Write a balanced equation for the reaction of CaH_2 with liquid ammonia. (b) Would it be best to classify this reaction as redox, Brønsted acid–base, or Lewis acid–base? Explain your answer.

8.9 The azide ion forms many ionic and covalent compounds that are similar to those of the halides. (a) Write the Lewis formula for the azide ion and predict the N—N—N bond angle. (b) Compare the acidity of hydrazoic acid with those of the hydrohalic acids and explain any differences (for HN_3, $K_a = 1.7 \times 10^{-5}$). (c) Write the formulas of three ionic or covalent azides.

8.10 The standard enthalpy of formation of $SiCl_4(g)$ at 298 K is -662.75 kJ·mol^{-1} and its standard molar entropy is $+330.86$ J·K^{-1}·mol^{-1}. Calculate the temperature at which the reduction of $SiCl_4(g)$ with hydrogen gas to Si(s) and HCl(g) becomes spontaneous.

8.11 (a) Examine the structures of diborane, B_2H_6, and Al_2Cl_6. Compare the bonding in these two compounds. How are they similar? (b) What are the differences, if any, in the types of bonds formed? (c) What is the hybridization of the Group 13 element? (d) Describe the shapes of the molecules and the expected bond angles.

8.12 (a) Draw the Lewis structure and assign a hybridization scheme to the atoms in the P_4 molecule. (b) Explain why the bonds in this molecule are regarded as strained.

8.13 Like the atoms of elements, molecules have ionization energies. (a) Define the ionization energy of a molecule. (b) Predict which of these compounds has a higher ionization energy and justify your prediction: $SiCl_4$ or SiI_4.

8.14 Isoelectronic species have the same number of electrons. (a) Divide the following species into two isoelectronic groups: CN^-, N_2, NO_2^-, C_2^{2-}, O_3. (b) Which species in each group is likely to be (i) the strongest Lewis base; (ii) the strongest reducing agent?

8.15 Isoelectronic species have the same number of electrons. (a) Divide the following species into three isoelectronic groups: NH_3, NO, NO_2^+, N_2O, H_3O^+, O_2^+. (b) Which species in each group is likely to be (i) the strongest Lewis acid; (ii) the strongest oxidizing agent?

8.16 The ground state of O_2 has two unpaired π^* electrons with parallel spins. There are two known low-lying excited states of O_2. State A has the two π^* electrons with the spins antiparallel but in different orbitals. State B has the two π^* electrons paired in the same orbital. (a) The energies of the excited states are 94.72 $kJ \cdot mol^{-1}$ and 157.85 $kJ \cdot mol^{-1}$ above the ground state. Which state corresponds to which excitation energy? (b) What is the wavelength of the radiation absorbed in the transition from the ground state to the first excited state?

8.17 Thiosulfuric acid, $H_2S_2O_3$, has a structure similar to that of sulfuric acid, except that a sulfur atom has replaced one terminal oxygen atom. How would you expect the physical and chemical properties of thiosulfuric acid and sulfuric acid to differ?

8.18 The Topics on the main group elements in this Focus are organized by group. Discuss whether it would be helpful to organize the elements according to period and give examples of trends that could be displayed helpfully in that way.

8.19 Account for the observation that solubility in water generally increases from chloride to iodide for ionic halides with low covalent character (such as the potassium halides) but decreases from chloride to iodide for ionic halides in which the bonds are significantly covalent (such as the silver halides).

8.20 Account for the observation that melting and boiling points generally decrease from fluoride to iodide for ionic halides but increase from fluoride to iodide for molecular halides.

8.21 The nitrosonium ion, NO^+, is isoelectronic with N_2 and has two fewer electrons than O_2. The ion is stable and can be purchased as its hexafluorophosphate (PF_6^-) or tetrafluoroborate (BF_4^-) salt. (a) Draw its molecular orbital diagram and compare it with those of N_2 and O_2. (b) NO^+ is diamagnetic. Can you tell which orbital, the σ_{2p} or the π_{2p}, is higher in energy from this information?

8.22 (a) Draw the molecular orbital energy-level diagram for O_2. Using the diagram, determine the bond order and magnetic properties of O_2. (b) What molecular property of oxygen is explained by its molecular orbital diagram but not by its Lewis structure? (c) Describe the nature of the highest occupied molecular orbital—is it bonding, antibonding, or nonbonding? (d) Predict the bond order and magnetic properties of the peroxide ion and the superoxide ion on the basis of molecular orbital theory.

8.23 One method used to produce hydrogen as a fuel decomposes methanol in the reaction $CH_3OH(l) \rightarrow 2\,H_2(g) + CO(g)$. (a) What is the standard reaction enthalpy of this process? (b) What is the standard enthalpy of combustion of the hydrogen produced in the reaction? (c) What is the enthalpy change accompanying the combustion of 1.00 mol $CH_3OH(l)$? (d) Which process generates more heat at constant pressure, the direct combustion of methanol in part (c), or the two-step process (parts (a) and (b) combined)?

8.24 One method for preparing hydrogen as a fuel from methanol uses a two-step process with the overall reaction $CH_3OH(l) + H_2O(l) \rightarrow 3\,H_2(g) + CO_2(g)$. (a) What is the standard reaction enthalpy of this process? (b) What is the standard enthalpy of combustion of the hydrogen produced? (c) What is the enthalpy change accompanying the combustion of 1.00 mol $CH_3OH(l)$? (d) Which process generates more heat at constant pressure, the direct combustion of methanol in part (c) or the two-step process (parts (a) and (b) combined)?

8.25 Infrared radiation is absorbed only when there is a change in the dipole moment of the molecule as the molecule vibrates. Which of the following gases found in the atmosphere can absorb in the infrared region and so function as greenhouse gases: (a) CO; (b) NH_3; (c) CF_4; (d) O_3; (e) Ar? Explain your reasoning. See Box 8B.1.

8.26 Carbon dioxide absorbs infrared energy during bending or stretching motions that are accompanied by a change in dipole moment (from zero). Which of the transitions pictured in Fig. 2b of *Major Technique* 1 on the website of this book can absorb infrared radiation? Explain your reasoning. See Box 8B.1.

8.27 Methanol, CH_3OH, is a clean-burning liquid fuel being used as a replacement for gasoline. Calculate the theoretical yield in kilograms of CO_2 produced by the combustion of 1.00 L of methanol (of density 0.791 $g \cdot cm^{-3}$) and compare it with the 2.16 kg of CO_2 generated by the combustion of 1.00 L of octane. Which fuel contributes more CO_2 per liter to the atmosphere when burned? What other factors would you take into consideration when deciding which of the two fuels to use? See Box 8B.1.

8.28 The concentration of nitrate ion can be determined by a multistep process. A sample of water of volume 25.00 mL from a rural well contaminated with $NO_3^-(aq)$ was made basic and treated with an excess of zinc metal, which reduces the nitrate ion to ammonia. The evolved ammonia gas was passed into 50.00 mL of 2.50×10^{-3} M HCl(aq). The unreacted HCl(aq) was titrated to the stoichiometric point with 28.22 mL of 1.50×10^{-3} M NaOH(aq). (a) Write balanced chemical equations for the three reactions. (b) What is the molar concentration of nitrate ion in the well water?

8.29 (a) Consider the substances shown in parts (a), (c), and (d) in the second illustration in Box 8F.1. (b) In which of these three parts of the figure will the interactions be the strongest? (c) In which will they be the weakest?

FOCUS 8   Cumulative Exercises

8.30 The gas NO is released to the stratosphere by jet engines. Because it can contribute to the destruction of stratospheric ozone, its concentration is monitored closely. One monitoring technique measures the chemiluminescence given off by the reaction $NO(g) + O_3(g) \rightarrow NO_2^*(g) + O_2(g)$. The asterisk indicates that NO_2 is produced in an excited state. The excited NO_2 molecule gives off red light and infrared radiation as it returns to the ground state. Because both NO and NO_2 may be present in the air sample, part of the sample is exposed to a reducing agent, which converts all the NO_2 to NO. Two samples are then analyzed, first air from the unreduced original sample, which gives the concentration of NO, then air from the reduced sample. The concentration of NO_2 in the air is the difference between the two measurements. In one experiment, two samples were collected, one from a cloud and one from clear air. The following measurements were made at 9 km above the Pacific Ocean (ppt denotes parts per trillion, 1 part per 10^{12}, by molecules):

	Unreduced sample	Reduced sample
Cloud	860 ppt	1110 ppt
Clear air	480 ppt	740 ppt

(a) Determine the concentrations (in ppt) of NO and NO_2 in each air sample.

(b) In which part of the atmosphere is the concentration of NO greater?

(c) Explain the difference, taking into account other sources of NO in the atmosphere besides airplanes.

(d) Some of the NO molecules released by jet engines react with the hydroxyl radical, ·OH. Draw the Lewis structure of the most likely product and state its name.

(e) Is the product of the reaction in part (d) likely to be more stable than NO? Explain your reasoning.

(f) In the stratosphere, NO catalyzes the conversion of O_3 to O_2 in a two-step reaction with the intermediate NO_2. The overall equation for the reaction is $O_3(g) + O(g) \rightarrow 2\,O_2(g)$ and the rate law for the reaction is Rate $= k_r[NO][O_3]$. Write an acceptable two-step mechanism for the reaction, indicating which step is the slow step.

(g) At $-30\,°C$, the temperature of the air from which the samples were taken, the rate constant for the reaction in part (f) is 6×10^{-15} cm³·molecule⁻¹·s⁻¹. If the concentration of ozone in

the air from which the NO clear air sample was taken was 480 ppt and the total pressure of the air was 220 mbar, what was the rate at which ozone was being destroyed in the air at $-30\,°C$?

8.31 The tiny structures such as spheres and tubes formed by carbon atoms are the basis for a large part of the field of nanotechnology. Boron nitride forms similar structures.

(a) What is the hybridization of carbon atoms in carbon nanotubes and of boron and nitrogen atoms in boron nitride nanotubes?

(b) A simple carbon nanotube is made up of a sheet resembling a layer of graphene (similar to chicken wire) that has been curled up and bonded to itself. How many hexagons must be strung together (the hexagons denoted by the colored bonds in the following diagram) around the circumference of a nanotube with the "armchair" configuration to form a tube that is approximately 1.3 nm in diameter? The diagram shows the orientation of hexagons with respect to the curvature of the nanotube; a representative nanotube is shown for reference. The C—C bond distance in carbon nanotubes is 142 pm.

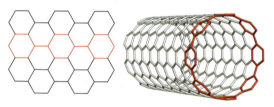

(c) Buckminsterfullerene, C_{60}, can be hydrogenated, but, a compound with the formula $C_{60}H_{60}$ has not yet been prepared. The most hydrogenated form of C_{60} known is $C_{60}H_{36}$. Why does hydrogenation stop at this point?

(d) Boron nitride can form nanotubes, but it does not form spheres that resemble buckminsterfullerene. Examine the structure of buckminsterfullerene and suggest a reason why BN cannot form spheres.

(e) In its cubic crystalline form the nitrogen atoms of BN form a face-centered cubic unit cell in which half the tetrahedral interstitial sites are occupied by B atoms (if all B and N atoms were replaced by C atoms, the diamond structure would result). Calculate the density of cubic BN if the edge length of the unit cell is 361.5 pm.

(f) The density of hexagonal BN is 2.29 g·cm⁻³. Which form of BN is favored at high pressures, hexagonal or cubic?

THE d-BLOCK ELEMENTS

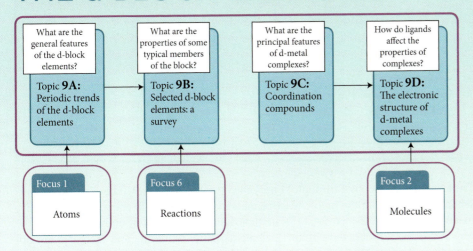

The d-block elements are present in virtually every aspect of modern life. Iron and copper helped civilization rise from the Stone Age and are still among our most important industrial metals. Other members of the block include the metals of new technologies, such as titanium for the aerospace industry and vanadium for catalysts in the petrochemical industry. The precious metals silver, platinum, and gold are prized as much for their appearance, rarity, and durability as for their usefulness. Compounds of cobalt, molybdenum, and zinc are found in vitamins and essential enzymes. Some compounds make life more interesting and colorful. The beautiful color of cobalt blue glass, the brilliant greens and blues of kiln-baked pottery, and many pigments used by artists make use of d-block compounds.

Trends in properties of the d-block metals are interpreted in terms of their locations in the Periodic Table in **TOPIC 9A**. These trends are illustrated in **TOPIC 9B** by looking more closely at some important representative elements from the block. A striking feature of the d-block elements is their ability to form a wide variety of complexes with characteristic shapes and colors that are described in **TOPIC 9C**. Finally, **TOPIC 9D** explores the origins of these characteristic colors and establishes a link between color and magnetism.

Topic 9A Periodic Trends of the d-Block Elements

9A.1 Trends in Physical Properties
9A.2 Trends in Chemical Properties

What are the general features of the d-block elements?

What are the properties of some typical members of the block?

Topic **9A:** Periodic trends of the d-block elements

→ Topic **9B:** Selected d-block elements: a survey

Why Do You Need to Know This Material? A knowledge of how physical and chemical properties of the d-block elements depend on location in the periodic table is key to understanding their similarities and differences.

What Do You Need to Know Already? This Topic draws on many of the principles introduced in the earlier Focuses, particularly electron configurations (Topic 1E).

Some periodic tables designate the elements La–Yb or La–Lu as the first row of the f-block. In this book, the elements Ce–Lu are designated as the first row of the f-block, the lanthanoids, which means "lanthanum-like."

The elements in Groups 3 through 11 are called the *transition metals* because they represent a transition from the highly reactive metals of the s-block to the much less reactive metals of the p-block (**FIG. 9A.1**). Note that the transition metals do not extend all the way across the d-block: the Group 12 elements (zinc, cadmium, and mercury) are not normally considered to be transition metals. Because their d-orbitals are full and except in rare instances do not participate in bonding, the Group 12 elements have properties that are more like those of main-group metals than those of transition metals. Just after the start of the third row of the d-block in Period 6, following lanthanum, electrons begin to occupy the seven 4f-orbitals, and the lanthanoids (the "rare earths," commonly known as the lanthanides) delay the completion of the period. These elements, together with the actinoids (commonly called the actinides), the analogous series in Period 7, are sometimes referred to as the *inner transition metals*.

9A.1 Trends in Physical Properties

The ground-state electron configurations of the atoms of d-block elements differ primarily in the occupation of the $(n-1)$d-orbitals. According to the rules of the building-up principle (Topic 1E), these orbitals are the last orbitals to be occupied. However, once they are occupied, they lie slightly lower in energy than the outer ns-orbitals. Because there are five d-orbitals in a given shell and each one can accommodate up to two electrons, there are 10 elements in each row of the d-block. Because the $(n-1)$d-orbitals of transition metals are inner orbitals, the physical properties of the elements tend to be similar.

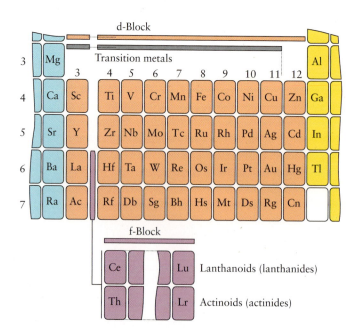

FIGURE 9A.1 The orange rectangles identify the elements in the d-block of the periodic table (Groups 3–12). Note that the f-block, consisting of the inner transition metals, intervenes in Periods 6 and 7 as indicated by the purple color. The transition metals are the elements in Groups 3–11.

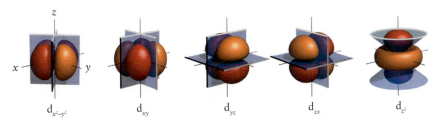

FIGURE 9A.2 The five d-orbitals, with their nodal planes indicated.

All the d-block elements are metals. Most of these so-called "d-metals" are good electrical conductors. In fact, at room temperature, silver is the best electrical conductor of all metals. Most of the d-metals are malleable, ductile, lustrous, and silver-white in color. Generally, their melting and boiling points are higher than those of the main-group elements. There are a few notable exceptions: copper is red-brown, gold is yellow, and mercury has such a low melting point that it is a liquid at room temperature.

The shapes of the d-orbitals (**FIG. 9A.2**, which repeats Fig. 1D.10) affect the properties of the d-block elements in two ways:

- The lobes of two d-orbitals on the same atom occupy markedly different regions of space. As a result, electrons in different d-orbitals are relatively far apart and repel one another only weakly.
- The electron density in d-orbitals is low near the nucleus (in part because the high angular momentum of a d-electron flings it away from the nucleus), and so they are not very effective at shielding other electrons from the positive nuclear charge.

One consequence of these two characteristics is the trend in atomic radii of the d-block metals, which gradually decrease across a period and then rise again (**FIG. 9A.3**). Nuclear charge and number of d-electrons both increase from left to right across each row. Because the repulsion between d-electrons is weak, initially the increasing nuclear charge can draw them inward, and so the atoms become smaller. Farther across the block, there are so many d-electrons that electron–electron repulsion increases more rapidly than the attraction from the nuclear charge, and the radii begin to increase again. Because these attractions and repulsions are finely balanced, the range of d-metal atomic radii is small. In fact, some of the atoms of one d-metal can easily replace atoms of another d-metal in a crystal lattice (**FIG. 9A.4**). The d-metals can therefore form a wide range of alloys (Topic 3I).

The atomic radii of the second row of d-metals (Period 5) are typically greater than those in the first row (Period 4). The atomic radii in the third row (Period 6), however, are about the same as those in the second row and smaller than expected. This effect is due to the **lanthanide contraction,** the decrease in radius along the first row of the f-block (**FIG. 9A.5**). This decrease is due to the increasing nuclear charge along the period coupled with the poor shielding ability of f-electrons. When the d-block resumes (at hafnium), the atomic radius has fallen from 188 pm for lanthanum to 156 pm for hafnium.

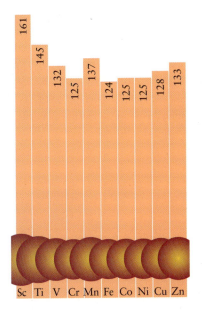

FIGURE 9A.3 The atomic radii (in picometers) of the elements of the first row of the d-block.

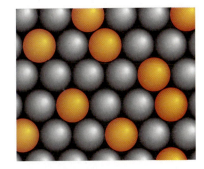

FIGURE 9A.4 Because the atomic radii of the d-block elements are so similar, the atoms of one element can replace the atoms of another element with minor modification of the atomic locations; consequently, d-block metals form a wide range of alloys.

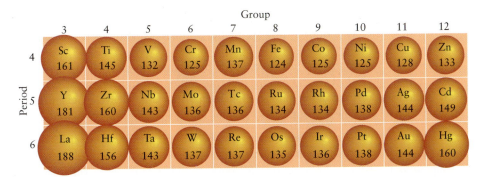

FIGURE 9A.5 The atomic radii of the d-block elements (in picometers). Notice the similarity of all the values and, in particular, the close similarity between the second and the third rows as a result of the lanthanide contraction.

FIGURE 9A.6 The densities (in grams per centimeter cubed, g·cm⁻³) of the d-metals at 25 °C. The lanthanide contraction has a pronounced effect on the densities of the elements in Period 6 (front row in this illustration), which are among the densest of all the elements.

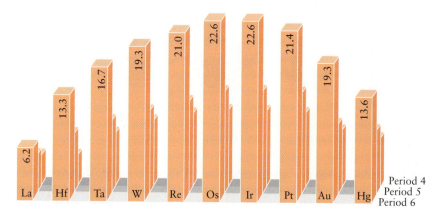

A technologically important effect of the lanthanide contraction is the high density of the Period 6 elements (**FIG. 9A.6**). The atomic radii of these elements are comparable to those of the Period 5 elements, but their atomic masses are about twice as great, so more mass is packed into the same volume. A block of iridium, for example, contains about as many atoms as a block of rhodium of the same volume, but its density is nearly twice as great as that of rhodium. In fact, iridium is one of the two densest elements; its neighbor osmium is the other. Another effect of the contraction is the low reactivity—the "nobility"—of gold and platinum. Due to the poor shielding offered by d-electrons, the valence electrons of gold and platinum are relatively close to the nucleus, causing them to be tightly bound and not readily available for chemical reactions.

The atomic radii of the d-block metals are similar but tend to decrease across a series. The lanthanide contraction accounts for the smaller-than-expected atomic radii and higher densities of the d-block elements in Period 6.

9A.2 Trends in Chemical Properties

The d-block elements lose their valence s-electrons when they form compounds. Most of them can also lose a variable number of d-electrons and exist in a variety of oxidation states. The only elements of the block that do not use their d-electrons in compound formation are the members of Group 12. Zinc and cadmium lose only their s-electrons; mercury loses a d-electron only in rare instances. The ability to exist in different oxidation states is responsible for many of the special chemical properties of the transition metals and plays a role in the action of many biomolecules.

Most d-block elements form compounds with more than one oxidation number. The distribution of oxidation numbers looks daunting at first sight (**FIG. 9A.7**), but some patterns become apparent when you examine the illustration:

- Elements close to the center of each row have the widest range of oxidation numbers; manganese, at the center of its row, is found with seven oxidation numbers.
- Except for mercury, the elements at the ends of each row of the d-block occur with only one oxidation number other than 0.
- All the other elements of each row are found with at least two positive oxidation numbers.
- Elements in the second and third rows of the block are more likely to reach higher oxidation numbers than are those in the first row. Notice that ruthenium and osmium are found with every oxidation number available to them, and that even gold and mercury, which lie near the end of the block, can be found with three oxidation numbers.

The pattern of oxidation numbers underlies trends in the chemical properties of the d-block elements. An element with a high oxidation number is easily reduced; therefore, the compound is likely to be a good oxidizing agent. For example, manganese has oxidation number +7 in the permanganate ion, MnO_4^-, and this ion is a good oxidizing agent in acidic solution:

$$MnO_4^-(aq) + 8 H^+(aq) + 5 e^- \longrightarrow Mn^{2+}(aq) + 4 H_2O(l) \qquad E° = +1.51 V$$

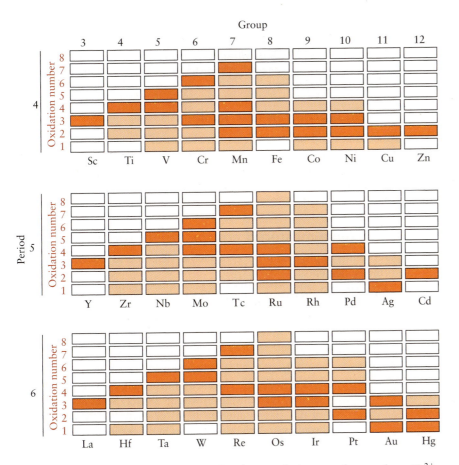

FIGURE 9A.7 The oxidation numbers of the d-block elements. The dark orange blocks mark the common oxidation numbers for each element; the light orange blocks mark the other known oxidation numbers.

Compounds that contain the element with a low oxidation number, such as Cr^{2+}, are often good reducing agents:

$$Cr^{3+}(aq) + e^- \longrightarrow Cr^{2+}(s) \qquad E° = -0.41\ V$$

The pattern of oxidation numbers correlates with the pattern of acid–base behavior of d-metal oxides. Although most d-metal oxides are basic, the oxides of a given element show a shift toward acidic character with increasing oxidation number, just as the oxoacids do (Topic 6C). The family of chromium oxides is a good example:

CrO	+2	basic
Cr_2O_3	+3	amphoteric
CrO_3	+6	acidic

Chromium(VI) oxide, CrO_3, is the anhydride of chromic acid, H_2CrO_4, the parent acid of the chromates.

The elements on the left of the d-block resemble the s-block metals in being much more difficult to extract from their ores than the metals on the right. Indeed, on moving from right to left, the elements are encountered in very approximately the order in which they were put to use as metals as civilization emerged. On the far right are copper and zinc, which jointly were responsible for the Bronze Age. As metal workers discovered how to achieve higher temperatures, they found out how to reduce iron oxides, and the Iron Age began. The metals at the left of the block (for example, titanium) require such extreme conditions for their extraction, including the use of other active metals or electrolysis, that they became widely available only in the twentieth century, when these techniques were developed on a commercial scale.

Self-test 9A.1A Predict trends in ionization energies of the d-block metals.
[*Answer:* Ionization energy increases from left to right across a row and decreases down a group.]

Self-test 9A.1B Six of the d-block metals in Period 4 form +1 ions. Predict trends in the radii of those ions.

The range of oxidation numbers of a d-block element increases toward the center of the block. Compounds in which the d-block element has a high oxidation number tend to be oxidizing; compounds in which the d-block element has a low oxidation number tend to be reducing. The acidic character of d-block oxides increases with the oxidation number of the element.

What have you learned in this Topic?

You have learned that the d-block elements have atoms of similar size and that the lanthanide contraction leads to smaller atomic radii and higher densities than expected for the elements in Period 6. You have also learned that the d-block elements can be found with a variety of oxidation numbers and that their oxidizing ability increases with oxidation number.

The skills you have mastered are the ability to:

☐ **1.** Explain trends in physical properties such as density and atomic radius among the d-block elements.

☐ **2.** Account for trends in chemical properties such as common oxidation numbers and acid–base character of the d-block metal oxides.

Topic 9A Exercises

You might find Appendix 2 helpful when answering these questions.

9A.1 Which members of the d-block, those at the left or at the right of the block, are likely to have the more strongly negative standard potentials? Explain your prediction.

9A.2 Identify five elements in the d-block that have positive standard potentials.

9A.3 Write the formulas for the oxoanions of the following elements in which the element is found with its highest oxidation number (see Fig. 9A.7). In each case, the charge of the oxoanion is given in parentheses. (a) Mn (-1); (b) Mo (-2).

9A.4 Write the formulas for the oxoanions of the following elements in which the element is found with its highest oxidation number (see Fig. 9A.7). In each case, the charge of the oxoanion is given in parentheses. (a) V (-3); (b) Ti (-4).

9A.5 Identify the element with the higher first ionization energy in each of the following pairs: (a) iron and nickel; (b) nickel and copper; (c) osmium and platinum; (d) nickel and palladium; (e) hafnium and tantalum.

9A.6 Identify the element with the higher first ionization energy in each of the following pairs: (a) manganese and cobalt; (b) manganese and rhenium; (c) chromium and zinc; (d) chromium and molybdenum; (e) palladium and platinum.

9A.7 The energies of the 3d-orbitals are nearly unchanged across the d-elements in Period 4. However, the 3d-orbitals in zinc atoms have significantly lower energies and those of gallium atoms are even lower in energy. Suggest an explanation for these differences.

9A.8 In some older forms of the periodic table, zinc, cadmium, and mercury are classified as belonging to "Group 2B," with the alkaline earth metals in "Group 2A." (a) What is the justification of this classification in terms of electronic structure? (b) How would you explain differences in properties between "Group 2B" and "Group 2A"?

9A.9 Explain why the density of mercury (13.55 $g \cdot cm^{-3}$) is significantly higher than that of cadmium (8.65 $g \cdot cm^{-3}$), whereas the density of cadmium is only slightly greater than that of zinc (7.14 $g \cdot cm^{-3}$).

9A.10 Explain why the density of vanadium (6.11 $g \cdot cm^{-3}$) is significantly less than that of chromium (7.19 $g \cdot cm^{-3}$). Both vanadium and chromium crystallize in a body-centered cubic lattice.

9A.11 (a) Describe the trend in the stability of oxidation states moving down a group in the d-block (for example, from chromium to molybdenum to tungsten). (b) How does this trend compare with the trend in the stabilities of oxidation states observed for the p-block elements when moving down a group?

9A.12 Which oxoanion, MnO_4^- or ReO_4^-, is expected to be the stronger oxidizing agent? Explain your choice.

9A.13 Which of the elements vanadium, chromium, and manganese is most likely to form an oxide with the formula MO_3? Explain your decision.

9A.14 Which of the elements scandium, molybdenum, and copper is most likely to form a chloride with the formula MCl_4? Explain your decision.

Topic 9B Selected d-Block Elements: A Survey

9B.1   Scandium Through Nickel

9B.2   Groups 11 and 12

What are the general features of the d-block elements?

What are the properties of some typical members of the block?

Topic **9A:** Periodic trends of the d-block elements → Topic **9B:** Selected d-block elements: a survey

Why Do You Need to Know This Material? The d-block metals are the workhorse elements of the periodic table and play a role in all branches of chemistry and industry.

What Do You Need to Know Already? You should keep in mind the general trends summarized in Topic 9A.

Although the physical properties of the d-block elements are similar (Topic 9A), the chemical properties of these elements are so diverse that it is impossible to summarize them briefly. However, it is possible to discover some of the major trends within the d-block by considering the properties of certain representative elements, particularly those in the first row of the block.

9B.1 Scandium Through Nickel

TABLE 9B.1 summarizes the physical properties of the Period 4 transition metals from scandium through nickel. Notice the similarities in their melting and boiling points, but the gradual increase in density.

(a) Scandium   The element scandium, Sc, which was first isolated in 1937, is a reactive metal that reacts with water about as vigorously as calcium does. It has few uses and is not thought to be essential to life. The small, highly charged Sc^{3+} ion is strongly hydrated in water (like Al^{3+}), and the resulting $[Sc(OH_2)_6]^{3+}$ complex is about as strong a Brønsted acid as acetic acid.

Square brackets are commonly used to indicate the presence of a d-metal complex.

> **A Note on Good Practice:** When water is part of a complex, its formula is written OH_2 to emphasize that the water molecule links to the metal atom through the O atom.

(b) Titanium   The strong, light metal titanium, Ti, is used where these properties are critical—in widely diverse applications that include jet engines, bicycle frames, and dental fixtures such as partial plates. Although titanium is relatively reactive, unlike scandium it is resistant to corrosion because it has a protective layer of oxide on its surface. The principal sources of the metal are the ores *ilmenite*, $FeTiO_3$, and *rutile*, TiO_2.

Titanium requires strong reducing agents for extraction from its ores. It was not widely exploited commercially until demand from the aerospace industry increased in the latter part of the twentieth century. The metal is obtained by first treating the ores with chlorine in the presence of coke (impure carbon obtained by heating coal in the

TABLE 9B.1   Physical Properties of the d-Block Elements Scandium Through Nickel

Z	Name	Symbol	Valence electron configuration	Melting point/°C	Boiling point/°C	Density/ (g·cm^{-3})
21	scandium	Sc	$3d^14s^2$	1540	2800	2.99
22	titanium	Ti	$3d^24s^2$	1660	3300	4.55
23	vanadium	V	$3d^34s^2$	1920	3400	6.11
24	chromium	Cr	$3d^54s^1$	1860	2600	7.19
25	manganese	Mn	$3d^54s^2$	1250	2120	7.47
26	iron	Fe	$3d^64s^2$	1540	2760	7.87
27	cobalt	Co	$3d^74s^2$	1494	2900	8.80
28	nickel	Ni	$3d^84s^2$	1455	2150	8.91

absence of air) to form titanium(IV) chloride; the volatile chloride is then reduced by passing it through liquid magnesium:

$$TiCl_4(g) + 2\,Mg(l) \xrightarrow{700\,°C} Ti(s) + 2\,MgCl_2(s)$$

The most common oxidation state of titanium is $+4$, in which the atom has lost both its 4s-electrons and its two 3d-electrons. Its most important compound is titanium(IV) oxide, TiO_2, which is almost universally known as titanium dioxide. This oxide is a brilliantly white (when finely powdered), nontoxic, stable solid used as the white pigment in paints and paper. It is used in photovoltaic cells to generate an electric current in the presence of solar radiation. Titanium also forms a series of oxoanions called *titanates*, which are prepared by heating TiO_2 with a stoichiometric amount of the oxide or carbonate of a second metal.

THINKING POINT

Why is titanium dioxide powder white?

(c) Vanadium The soft silver-gray metal vanadium, V, is produced by reducing its oxide or chloride. For example, vanadium(V) oxide is reduced by calcium:

$$V_2O_5(s) + 5\,Ca(l) \xrightarrow{\Delta} 2\,V(s) + 5\,CaO(s)$$

Vanadium is used to make tough steels for automobile and truck springs. Because it is not economical to add the pure metal to iron, a **ferroalloy** of the metal, an alloy with iron and carbon, that is less expensive to produce, is used instead.

Vanadium(V) oxide, V_2O_5, commonly known as vanadium pentoxide, is the most important compound of vanadium. This orange-yellow solid is used as an oxidizing catalyst (a catalyst that accelerates oxidation) in the contact process for the manufacture of sulfuric acid (Topic 8H). The wide range of colors of vanadium compounds, including the blue of the vanadyl ion, VO^{2+} (**FIG. 9B.1**), has led to their use in glazes in the ceramics industry.

(d) Chromium The name of chromium, Cr, comes from the Greek word for "color" and was inspired by its colorful compounds. Chromium is a bright, lustrous, corrosion-resistant metal. It is obtained from the black mineral *chromite*, $FeCr_2O_4$, by reduction with carbon in an electric arc furnace:

$$FeCr_2O_4(s) + 4\,C(s) \xrightarrow{\Delta} Fe(l) + 2\,Cr(l) + 4\,CO(g)$$

A less abundant source of chromium is the green mineral *chrome ochre*, Cr_2O_3, which is reduced to the metal by aluminum in the thermite process:

$$Cr_2O_3(s) + 2\,Al(s) \xrightarrow{\Delta} Al_2O_3(s) + 2\,Cr(l)$$

Chromium metal is important in metallurgy, because it is used to make stainless steel and for chromium plating (Topic 6O). Chromium(IV) oxide, CrO_2, a brown-black solid, is a ferromagnetic material that is being investigated for applications in spin transport electronics, or *spintronics*, in which information is stored as nuclear spin orientation.

FIGURE 9B.1 Many vanadium compounds form vividly colored solutions in water. They are also used in pottery glazes. The blue colors here are due to the vanadyl ion, VO^{2+}. (*W. H. Freeman photo by Ken Karp.*)

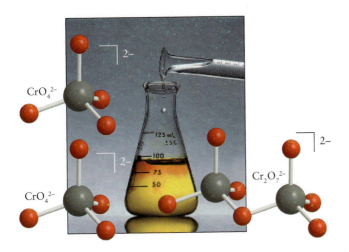

Sodium chromate, Na_2CrO_4, a yellow solid, is the starting material for the preparation of most other chromium compounds, including fungicides, pigments, and ceramic glazes. The chromate ion changes into the orange dichromate ion, $Cr_2O_7^{2-}$, in the presence of acid (**FIG. 9B.2**):

$$2\,CrO_4^{2-}(aq) + 2\,H^+(aq) \longrightarrow Cr_2O_7^{2-}(aq) + H_2O(l)$$

In the laboratory, acidified solutions of dichromates, in which the oxidation number of chromium is +6, are useful oxidizing agents:

$$Cr_2O_7^{2-}(aq) + 14\,H^+(aq) + 6\,e^- \longrightarrow 2\,Cr^{3+}(aq) + 7\,H_2O(l) \qquad E° = +1.33\,V$$

(e) Manganese   The gray metal manganese, Mn, resembles iron. It is much less resistant to corrosion than chromium and becomes coated with a thin brown oxide layer when exposed to air. The metal is rarely used alone, but it is an important component of alloys. Sulfur is an impurity in iron: it forms FeS and weakens the material. When manganese is added to iron as ferromanganese, it reacts preferentially with sulfur to form MnS. Both FeS and MnS collect in the cracks between the grains of steel, but whereas FeS melts at a low temperature and allows the cracks to spread, MnS does not. Manganese also increases iron's hardness, toughness, and resistance to abrasion. Another useful alloy is *manganese bronze* (39% by mass Zn, 1% Mn, a small amount of iron and aluminum, and the rest copper), which is very resistant to corrosion and is used for the propellers of ships. Manganese is also alloyed with aluminum to increase the stiffness of beverage cans, allowing them to be manufactured with thinner walls.

Manganese is currently obtained by the thermite process from *pyrolusite*, a mineral form of manganese dioxide:

$$3\,MnO_2(s) + 4\,Al(s) \xrightarrow{\Delta} 3\,Mn(l) + 2\,Al_2O_3(s)$$

Manganese lies near the center of its row (in Group 7) and occurs with a wide variety of oxidation numbers. The most stable state has oxidation number +2, but +4, +7, and, to a lesser extent, +3 are common in manganese compounds. Its commercially most important compound is manganese(IV) oxide, MnO_2, commonly called manganese dioxide. This compound is a brown-black solid used in dry cells, as a decolorizer to conceal the green tint of glass, and as the starting point for the production of other manganese compounds.

Potassium permanganate is a strong oxidizing agent in acidic solution and is used to oxidize organic compounds and as a mild disinfectant. Its usefulness stems not only from its thermodynamic tendency to oxidize other species but also from its ability to act by a variety of mechanisms; as a result, it is likely to be able to find a path with low activation energy and act rapidly.

(f) Iron   The most widely used of all the d-metals, iron, Fe, is the most abundant element on Earth and the second most abundant metal in the Earth's crust (after aluminum). Its principal ores are the oxides *hematite*, Fe_2O_3, and *magnetite*, Fe_3O_4, which are used to

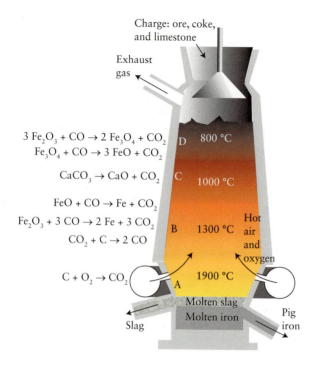

Exhaust
gas

$3 Fe_2O_3 + CO \rightarrow 2 Fe_3O_4 + CO_2$
$Fe_3O_4 + CO \rightarrow 3 FeO + CO_2$

D   800 °C

$CaCO_3 \rightarrow CaO + CO_2$

C   1000 °C

$FeO + CO \rightarrow Fe + CO_2$

$Fe_2O_3 + 3 CO \rightarrow 2 Fe + 3 CO_2$

$CO_2 + C \rightarrow 2 CO$

B   1300 °C   Hot
air
and
oxygen

$C + O_2 \rightarrow CO_2$

A   1900 °C

Molten slag
Molten iron   Pig
iron

Slag

FIGURE 9B.3 The reduction of iron ore takes place in a blast furnace containing a mixture of the ore with coke and limestone. Different reactions take place in different zones (labeled A through D) when the blast of air and oxygen is admitted. The ore, an oxide, is reduced to the metal by reduction with carbon monoxide produced in the furnace.

make steel. The sulfide mineral *pyrite*, FeS_2 (see Fig. 8H.6), is widely available, but it is not used in steelmaking because the sulfur is difficult to remove.

Before about 1400 CE, iron was available primarily as *cast iron*, which has a high carbon content, is hard, has a high melting point, and is brittle. Cast iron is still widely used for objects that are exposed to little mechanical and thermal shock, such as ornamental railings, engine blocks, and transmission housings. Steel was probably first prepared by heating iron in a charcoal fire, which results in a lower carbon content than cast iron, and thanks to its hardness, strength, and corrosion resistance, it is the primary form in which iron is used today.

The production of steel begins with iron ore, generally found as the oxide Fe_2O_3, which is reduced through a series of redox and Lewis acid–base reactions. A mixture of iron ore, coke, and limestone (calcium carbonate) is fed continuously into the top of a blast furnace (**FIG. 9B.3**), which is approximately 40 m high. Each kilogram of iron produced requires about 1.75 kg of ore, 0.75 kg of coke, and 0.25 kg of limestone. In the heat of the furnace, limestone decomposes to calcium oxide (lime) and carbon dioxide. The calcium oxide, which contains the Lewis base O^{2-}, helps to remove the acidic and amphoteric oxide impurities from the ore:

$$CaO(s) + SiO_2(s) \xrightarrow{\Delta} CaSiO_3(l)$$

$$CaO(s) + Al_2O_3(s) \xrightarrow{\Delta} Ca(AlO_2)_2(l)$$

$$6 CaO(s) + P_4O_{10}(s) \xrightarrow{\Delta} 2 Ca_3(PO_4)_2(l)$$

The mixture of products, which is known as *slag*, is molten at the temperatures in the furnace and floats on the denser molten iron. It is drawn off and used to make rocklike material for the construction industry.

Molten iron is produced through a series of reactions in the four main temperature zones of the furnace. At the bottom, in Zone A, preheated air is blown into the furnace under pressure, and the coke is oxidized to heat the furnace to 1900 °C and provide carbon in the form of carbon dioxide. Higher up, the iron is reduced in stages to the metal, which melts and flows from Zone C to Zone A. Although the melting point of pure iron is 1540 °C, iron mixed with 4% carbon melts at about 1015 °C. As the carbon dioxide moves up through the furnace to Zone B, it reacts with some of the carbon provided by the coke, producing carbon monoxide. This reaction is endothermic and lowers the temperature to 1300 °C. The carbon monoxide produced in this reaction rises to Zones C and D, where it reduces iron ore in a series of reactions, some of which are shown in Fig. 9B.3. The iron runs out of the bottom of the furnace as molten pig iron.

The pig iron produced in blast furnaces must be processed further to produce steel. The first stage is to lower the carbon content of the iron and to remove the remaining impurities. In the *basic oxygen process*, oxygen and powdered limestone are forced through the molten metal. In the second stage, steels are produced by adding the appropriate metals, often as *ferroalloys* (mixtures of the metal with iron, such as ferrovanadium, an alloy of iron and vanadium), to the molten iron. The steel that results is an alloy of 2% or less carbon in iron. The higher the carbon content, the harder and more brittle is the steel. Low-carbon steel (less than 0.15% carbon) is so soft and ductile that it is used to make wire, but high-carbon steel (0.61 to 1.5% carbon) is hard enough to make knives and drill bits. *Stainless steels* are highly corrosion resistant; they are typically about 15% chromium by mass (**TABLE 9B.2**).

Iron is quite reactive and corrodes in moist air. It does not react with oxidizing acids such as HNO_3, which form a protective oxide film, but reacts with nonoxidizing acids, evolving hydrogen and forming iron(II) salts. The colors of these salts vary from pale yellow to dark green-brown. Iron(II) salts are readily oxidized to iron(III) salts. The oxidation is slow in acidic solution but rapid in basic solution, where insoluble iron(III)

TABLE 9B.2   The Composition of Different Steels

Element blended into iron	Typical amount (%)	Effect
manganese	0.5 to 1.0	increases strength and hardness but lowers ductility
	13	increases wear resistance
nickel	<5	increases strength and shock resistance
	>5	increases corrosion resistance (stainless) and hardness
chromium	variable	increases hardness and wear resistance
	>12	increases corrosion resistance (stainless)
vanadium	variable	increases hardness
tungsten	<20	increases hardness, especially at high temperatures

hydroxide, $Fe(OH)_3$, is precipitated. Although $[Fe(OH_2)_6]^{3+}$ ions are pale purple and Fe^{3+} ions give amethyst its purple color, the colors of aqueous solutions of iron(III) salts are dominated by the conjugate base of $[Fe(OH_2)_6]^{3+}$, the yellow $[FeOH(OH_2)_5]^{2+}$ ion:

$$[Fe(OH_2)_6]^{3+}(aq) + H_2O(l) \rightleftharpoons H_3O^+(aq) + [FeOH(OH_2)_5]^{2+}(aq)$$

Like some other d-block metals, such as nickel, iron can form compounds in which its oxidation number is zero. For example, when iron is heated in carbon monoxide, it reacts to form *iron pentacarbonyl*, $Fe(CO)_5$, a yellow molecular liquid that boils at 103 °C.

A healthy adult human body contains about 3 to 4 g of iron, mostly as hemoglobin. Because about 1 mg is lost daily (in sweat, feces, and hair), and women lose about 20 mg in each menstrual cycle, iron must be ingested daily to maintain the balance.

THINKING POINT

Which common food sources provide iron in the diet?

(g) Cobalt Ores containing cobalt, Co, are often found in association with copper(II) sulfide. Cobalt is a silver-gray metal and is used mainly for alloying with iron. *Alnico steel*, an alloy of aluminum, nickel, and cobalt with iron, is used to make permanent magnets, such as those in loudspeakers. Cobalt steels are hard enough to be used as surgical instruments, drill bits, and lathe tools. The color of cobalt glass is due to a blue pigment that forms when cobalt(II) oxide is heated with silica and alumina.

(h) Nickel The element nickel, Ni, is also used in alloys. It is a hard, silver-white metal used mainly for the production of stainless steel and for alloying with copper to produce *cupronickels*, the alloys used for nickel coins (which are about 25% Ni and 75% Cu). Nickel is also used in nickel–cadmium (NiCad) batteries and as a catalyst, especially for the addition of hydrogen to organic compounds, as in the hydrogenation of vegetable oils (Topic 11B).

About 70% of the Western world's supply of nickel comes from iron and nickel sulfide ores that were brought close to the surface nearly 2 billion years ago by the violent impact of a huge meteorite at Sudbury, Ontario (**FIG. 9B.4**). The ore is first *roasted* (heated in air) to form nickel(II) oxide, which is reduced to the metal either electrolytically or by reaction with hydrogen gas in the first step of the *Mond process*:

$$NiO(s) + H_2(g) \xrightarrow{\Delta} Ni(s) + H_2O(g)$$

The impure nickel is then refined by first exposing it to carbon monoxide, with which it forms nickel tetracarbonyl, $Ni(CO)_4$:

$$Ni(s) + 4\,CO(g) \longrightarrow Ni(CO)_4(g)$$

Nickel tetracarbonyl is a volatile, poisonous liquid that boils at 43 °C and can therefore be removed from impurities. Nickel metal is then obtained by heating the pure nickel tetracarbonyl to about 200 °C, at which temperature it decomposes.

Nickel's most common oxidation number is +2, and the green color of aqueous solutions of nickel salts is due to the presence of $[Ni(OH_2)_6]^{2+}$ ions.

The slightly yellow hue of cupronickels is removed by the addition of small amounts of cobalt.

FIGURE 9B.4 The mine at Sudbury, Ontario, supplies most of the Western world's supply of nickel. (©NHPA/ SuperStock.)

FIGURE 9B.5 Three important copper ores (from left to right): chalcopyrite, $CuFeS_2$; malachite, $CuCO_3 \cdot Cu(OH)_2$; and chalcocite, Cu_2S. (©1984 Chip Clark–Fundamental Photographs.)

FIGURE 9B.6 In this industrial-scale copper refinery, the molten impure copper produced by smelting is poured into molds. Next, the copper will be purified by electrolysis. (Joel Sartore/ National Geographic Creative.)

FIGURE 9B.7 Copper corrodes in air to form an attractive pale green layer of basic copper carbonate. This patina, or incrustation, passivates the surface, which helps to protect it from further corrosion. (© Boltin Picture Library/ Bridgeman Images.)

Self-test 9B.1A Which oxide, Fe_2O_3 or Fe_3O_4, would you expect to be more acidic in water? Explain your reasoning.

[*Answer:* Fe_2O_3, because Fe has the higher oxidation number in that compound.]

Self-test 9B.1B (a) What type of reaction is the formation of nickel tetracarbonyl, Lewis acid–base or redox? (b) If Lewis acid–base, identify the acid and the base. If redox, identify the oxidizing agent and the reducing agent.

The Period 4 d-block elements titanium through nickel are obtained chemically from their ores, with the ease of reduction increasing to the right in the periodic table. They have many industrial uses, particularly as alloys.

9B.2 Groups 11 and 12

The elements close to the right-hand edge of the d-block have full d-orbitals. Group 11 contains the **coinage metals**—copper, silver, and gold—which have $(n-1)d^{10}ns^1$ valence electron configurations (**TABLE 9B.3**). Group 12 contains zinc, cadmium, and mercury, with valence configurations $(n-1)d^{10}ns^2$. The low reactivity of the coinage metals is due partly to the poor shielding abilities of the d-electrons and hence the tight grip that the nucleus can exert on the outermost electron. This effect is enhanced in Period 6 by the lanthanide contraction, which helps to account for the inertness of gold.

(a) Copper   The metal copper, Cu, is sufficiently unreactive for some to be found "native," as the metal, but most is produced from its sulfides, particularly the ore *chalcopyrite*, $CuFeS_2$ (**FIG. 9B.5**).

Processes for extracting metals from their ores are generally classified as **pyrometallurgical**, when high temperatures are used (**FIG. 9B.6**), or **hydrometallurgical**, when aqueous solutions are used. Copper is extracted by both methods.

The impure copper from either process is refined electrolytically: it is made into anodes and plated onto cathodes of pure copper. Other metals may be present in the impure copper, and those with highly positive electrode potentials also are reduced. The rare metals—most notably, platinum, silver, and gold—obtained from the anode sludge are sold to recover much of the cost of the electricity used in the electrolysis.

Copper alloys such as brass and bronze, which are harder and more resistant to corrosion than is copper, are important construction materials. Copper corrodes in moist air in the presence of oxygen and carbon dioxide:

$$2\,Cu(s) + H_2O(l) + O_2(g) + CO_2(g) \xrightarrow{\Delta} Cu_2(OH)_2CO_3(s)$$

The pale green product is called *basic copper carbonate* and is responsible for the green patina of copper and bronze objects (**FIG. 9B.7**). The patina adheres to the surface, protects the metal from further oxidation, and has a pleasing appearance.

Like all the coinage metals, copper forms compounds with oxidation number +1. However, in water, copper(I) salts disproportionate into metallic copper and copper(II) ions. The latter exist as pale blue $[Cu(OH_2)_6]^{2+}$ ions in water.

Copper is essential in animal metabolism. In mammals, copper-bearing enzymes are necessary for healthy nerves and connective tissue. In some animals, such as the octopus

Z	Name	Symbol	Valence electron configuration	Melting point/°C	Boiling point/°C	Density/ $(g \cdot cm^{-3})$
TABLE 9B.3	**Physical Properties of Elements in Groups 11 and 12**					
29	copper	Cu	$3d^{10}4s^1$	1083	2567	8.93
47	silver	Ag	$4d^{10}5s^1$	962	2212	10.50
79	gold	Au	$5d^{10}6s^1$	1064	2807	19.28
30	zinc	Zn	$3d^{10}4s^2$	420	907	7.14
48	cadmium	Cd	$4d^{10}5s^2$	321	765	8.65
80	mercury	Hg	$5d^{10}6s^2$	−39	357	13.55

and certain arthropods, it transports oxygen through the blood, a role performed by iron in mammals. As a result, the blood of these animals is green rather than red.

(b) Silver   Some silver, Ag, is found in nature as the metal, but most is obtained as a by-product of the refining of copper and lead, and a considerable amount was once recycled through the photographic industry. Silver has a positive standard potential, and so it does not reduce $H^+(aq)$ to hydrogen. Silver reacts readily with sulfur and sulfur compounds, producing the familiar black tarnish on silver dishes and cutlery.

In almost all its compounds, silver has oxidation number +1 and Ag(I) does not disproportionate in aqueous solution. Apart from silver nitrate, $AgNO_3$, silver fluoride, AgF, and a few others, silver salts are generally only sparingly soluble in water. Silver nitrate is the most important compound of silver and the starting point for the manufacture of silver halides.

(c) Gold   The noble metal gold, Au, is so inert that most of it is found in nature as the metal itself. Pure gold is classified as "24-carat gold." Its alloys with silver and copper, which differ in hardness and hue, are classified according to the proportion of gold that they contain (**FIG. 9B.8**). For example, 10- and 14-carat golds contain, respectively, $\frac{10}{24}$ and $\frac{14}{24}$ parts by mass of gold. Gold is a highly malleable metal: 1 g of gold can be worked into a leaf covering an area of about 1 m^2 or pulled out into a wire more than 2 km long.

Gold is too noble to react even with strong oxidizing agents such as nitric acid. Both the gold couples

$$Au^+(aq) + e^- \longrightarrow Au(s) \qquad E° = +1.69 \text{ V}$$
$$Au^{3+}(aq) + 3\,e^- \longrightarrow Au(s) \qquad E° = +1.40 \text{ V}$$

lie above H^+/H_2 and $NO_3^-, H^+/NO, H_2O$:

$$NO_3^-(aq) + 4\,H^+(aq) + 3\,e^- \longrightarrow NO(g) + 2\,H_2O(l) \qquad E° = +0.96 \text{ V}$$

However, gold does react with *aqua regia,* a mixture of concentrated nitric and hydrochloric acids, because the complex ion $[AuCl_4]^-$ forms:

$$Au(s) + 6\,H^+(aq) + 3\,NO_3^-(aq) + 4\,Cl^-(aq) \longrightarrow$$
$$[AuCl_4]^-(aq) + 3\,NO_2(g) + 3\,H_2O(l)$$

Despite the unfavorable equilibrium constant for the formation of Au^{3+} from gold, the reaction proceeds because any Au^{3+} ions formed are immediately complexed by Cl^- ions and removed from the equilibrium. In a process that is widely used in the refining of the metal, gold also reacts with sodium cyanide in an aerated aqueous solution to form the complex ion $[Au(CN)_2]^-$:

$$4\,Au(s) + 8\,NaCN(aq) + O_2(aq) + 2\,H_2O(l) \longrightarrow 4\,Na[Au(CN)_2](aq) + 4\,NaOH(aq)$$

(d) Zinc   The metal zinc, Zn, is found mainly as its sulfide, ZnS, in *sphalerite,* often in association with lead ores. The metal is extracted by roasting and then smelting with coke:

$$2\,ZnS(s) + 3\,O_2(g) \xrightarrow{\Delta} 2\,ZnO(s) + 2\,SO_2(g)$$

$$ZnO(s) + C(s) \xrightarrow{\Delta} Zn(l) + CO(g)$$

Zinc is used mainly for galvanizing iron (Topic 6N). Like copper, it is protected by a hard film of basic carbonate, $Zn_2(OH)_2CO_3$, which forms on contact with air.

Zinc and cadmium are both silvery, reactive metals that are similar to each other but differ sharply from mercury. Zinc is amphoteric (like its main-group neighbor aluminum). It reacts with acids to form Zn^{2+} ions and with alkalis to form the zincate ion, $[Zn(OH)_4]^{2-}$:

$$Zn(s) + 2\,OH^-(aq) + 2\,H_2O(l) \longrightarrow [Zn(OH)_4]^{2-}(aq) + H_2(g)$$

Galvanized containers should therefore not be used for transporting alkalis. Cadmium, which is lower down the group and is more metallic, has a more basic oxide.

(e) Cadmium   Like zinc, cadmium, Cd, has an oxidation number of +2 in all its compounds. However, the biological effects of the two metals are very different. Zinc is an essential element for human health. It is present in many enzymes and plays a role in the expression of DNA and in growth. Zinc is toxic only in very high amounts. However,

FIGURE 9B.8 The color of commercial gold depends on its composition. Left to right: 8-carat gold, 14-carat gold, white gold, 18-carat gold, and 24-carat gold. White gold consists of 6 parts Au and 18 parts Ag by mass. *(Field Museum of Natural History, Chicago/Getty Images.)*

cadmium is a deadly poison that disrupts metabolism by substituting for other essential metals in the body such as zinc and calcium, leading to soft bones and to kidney and lung disorders.

(f) Mercury   Mercury, Hg, occurs mainly as HgS in the mineral *cinnabar*, from which it is obtained by roasting in air:

$$HgS(s) + O_2(g) \xrightarrow{\Delta} Hg(g) + SO_2(g)$$

The volatile metal is separated by distillation and condensed. Mercury is the only metallic element that is liquid at room temperature (gallium and cesium are liquids on warm days). It has a long liquid range, from its melting point of $-39\ °C$ to its boiling point of $357\ °C$, and so it is well suited for its use in thermometers, silent electrical switches, and high-vacuum pumps.

Because mercury lies above hydrogen in the electrochemical series, it is not oxidized by hydrogen ions. However, it does react with nitric acid:

$$3\,Hg(l) + 8\,H^+(aq) + 2\,NO_3^-(aq) \longrightarrow 3\,Hg^{2+}(aq) + 2\,NO(g) + 4\,H_2O(l)$$

In nearly all its compounds, mercury has the oxidation number $+1$ or $+2$. Its compounds with oxidation number $+1$ are unusual in that the mercury(I) cation is the covalently bonded diatomic ion $(Hg—Hg)^{2+}$, written Hg_2^{2+}.

Compounds containing mercury, particularly its organic compounds, are acutely poisonous. Mercury vapor is an insidious poison because its effect is cumulative. Frequent exposure to low levels of mercury vapor can allow mercury to accumulate in the body. The effects include impaired neurological function, hearing loss, and other ailments.

Self-test 9B.2A   Use standard Gibbs free energies of formation to calculate $\Delta G°$ at 298 K for the reaction $CuS(s) + O_2(g) \rightarrow Cu(s) + SO_2(g)$. ($\Delta G_f°(CuS, s) = -49.0\ kJ\cdot mol^{-1}$.)

[Answer: $\Delta G° = -251.2\ kJ$]

Self-test 9B.2B   Calculate $E_{cell}°$ for a cell powered by the reaction of mercury metal and nitric acid to form aqueous mercury(I) and gaseous NO.

Metals in Groups 11 and 12 are easily reduced from their compounds and have low reactivity as a result of poor shielding of the nuclear charge by the d-electrons. Copper is extracted from its ores by either pyrometallurgical or hydrometallurgical processes.

What have you learned in this Topic?

You have encountered some of the principal uses of Period 4 d-block elements and have seen how they are produced commercially. You have also become acquainted with some of the physical and chemical properties of the d-block elements and with how and why they vary across the periodic table.

The skills you have mastered are the ability to:

☐ **1.** Describe and write balanced equations for the principal reactions used to produce the elements in the first row (Period 4) of the d-block and in Groups 11 and 12.

☐ **2.** Describe the names, properties, and reactions of some of the principal compounds of the elements in the first row of the d-block.

☐ **3.** Describe the operation of a blast furnace and explain how steel is made.

Topic 9B Exercises

9B.1 Predict the major products of each of the following reactions and then balance the equations:

(a) $TiCl_4(s) + Mg(s) \xrightarrow{\Delta}$

(b) $CoCO_3(s) + HNO_3(aq) \longrightarrow$

(c) $V_2O_5(s) + Ca(l) \xrightarrow{\Delta}$

9B.2 Predict the major products of each of the following reactions and then balance the equations:

(a) $FeCr_2O_4(s) + C(s) \xrightarrow{\Delta}$

(b) $CrO_4^{2-}(s) + H_3O^+(aq) \longrightarrow$

(c) $MnO_2(s) + Al(s) \xrightarrow{\Delta}$

9B.3 Give the systematic name and chemical formula of the principal component of (a) rutile; (b) hematite; (c) pyrolusite; (d) chromite.

9B.4 Give the systematic name and chemical formula of the principal component of (a) pyrite; (b) chalcopyrite; (c) basic copper carbonate; (d) cinnabar.

9B.5 What is the oxidation number of (a) Ti in $BaTiO_3$; (b) Zn in $Zn_2(OH)_2CO_3$?

9B.6 What is the oxidation number of (a) V in VO^{2+}; (b) Zn in $[Zn(OH)_4]^{2-}$?

9B.7 (a) What reducing agent is used in the production of iron from its ore? (b) Write chemical equations for the production of iron in a blast furnace. (c) What is the major impurity in the product of the blast furnace?

9B.8 (a) Why is limestone added to a blast furnace? (b) Write chemical equations that show its reactions in the furnace.

9B.9 Write the chemical equation that describes each of the following processes: (a) solid V_2O_5 reacts with acid to form the VO^{2+} ion; (b) solid V_2O_5 reacts with base to form the VO_4^{3-} ion.

9B.10 Write the chemical equation that describes each of the following processes: (a) the production of chromium by the thermite reaction; (b) the corrosion of copper metal by carbon dioxide in moist air; (c) the purification of nickel by using carbon monoxide.

9B.11 By considering electron configurations, explain why gold and silver are less reactive than copper.

9B.12 By considering electron configurations, suggest a reason why iron(III) is readily prepared from iron(II) but the conversion of nickel(II) and cobalt(II) into nickel(III) and cobalt(III) is much more difficult.

9B.13 Use Appendix 2B to determine whether an acidic sodium dichromate solution can oxidize (a) bromide ions to bromine and (b) silver(I) ions to silver(II) ions under standard conditions.

9B.14 Use Appendix 2B to determine whether an acidic potassium permanganate solution can oxidize (a) chloride ions to chlorine and (b) mercury metal to mercury(I) ions under standard conditions.

9B.15 (a) Explain why a chromium(III) salt produces an acidic solution in water. (b) Explain why the slow addition of hydroxide ions to a solution containing chromium(III) ions first produces a gelatinous precipitate that subsequently dissolves with further addition of hydroxide ions. Write chemical equations showing these aspects of the behavior of chromium(III) ions.

9B.16 Some of the properties of manganese differ markedly from those of its neighbors. For example, at constant pressure it takes 400 kJ (2 sf) to atomize 1.0 mol Cr(s) and 420 kJ to atomize 1.0 mol Fe(s), but only 280 kJ to atomize 1.0 mol Mn(s). Propose an explanation, using the electron configurations of the gaseous atoms, for the lower enthalpy of atomization of manganese.

Topic 9C Coordination Compounds

What are the principal features of d-metal complexes?	How do ligands affect the properties of complexes?
Topic **9C**: Coordination compounds	→ Topic **9D**: The electronic structure of d-metal complexes

Why Do You Need to Know This Material? One of the most striking properties of the d-block elements is their ability to form coordination compounds, which are used widely in chemistry, medicine, and industry.

What Do You Need to Know Already? This Topic makes use of the electron configurations of atoms and ions (Topics 1E and 9A), and the classification of species as Lewis acids and bases (Topic 6A).

The formation of coordinate covalent bonds is described in Topics 2C and 6A.

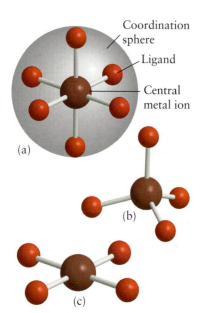

(a)

(b)

(c)

FIGURE 9C.1 (a) Almost all six-coordinate complexes are octahedral. Four-coordinate complexes are either (b) tetrahedral or (c) square planar.

The name "ligand" comes from the Latin word meaning "to bind."

Many of the d-block elements form characteristically colored solutions in water. For example, although solid copper(II) chloride is brown and copper(II) bromide is black, their aqueous solutions are both light blue. The blue color is due to the hydrated copper(II) ions, $Cu(OH_2)_6^{2+}$, that form when the solids dissolve. As the formula suggests, these hydrated ions have a specific composition. They can be regarded as the outcome of a reaction in which the water molecules act as Lewis bases (electron pair donors, Topic 6A) and the Cu^{2+} ion acts as a Lewis acid (an electron pair acceptor). This type of Lewis acid–base reaction is characteristic of many cations of d-block elements. The main-group elements behave similarly, but the d-block elements display the property to a striking and characteristic extent.

The hydrated ion $Cu(OH_2)_6^{2+}$ is an example of a **complex,** a species consisting of a central metal atom or ion to which a number of molecules or ions are attached by coordinate covalent bonds. The chemical formula of a complex ion (but not of a neutral complex) is commonly enclosed within square brackets, so this ion would be denoted $[Cu(OH_2)_6]^{2+}$. A **coordination compound** is an electrically neutral compound in which at least one of the ions present is a complex. However, the terms *coordination compound* (the overall neutral compound) and *complex* (one or more of the ions or neutral species present in the compound) are often used interchangeably. Coordination compounds include complexes in which the central metal atom is electrically neutral, such as $Ni(CO)_4$, and ionic compounds, such as $K_4[Fe(CN)_6]$.

Much research focuses on the structures, properties, and uses of the complexes formed between d-metal ions acting as Lewis acids and a variety of Lewis bases, partly because they participate in many biological reactions. Hemoglobin and vitamin B_{12}, for example, are both complexes—the former of iron and the latter of cobalt (BOX 9C.1). Complexes of the d-metals are often brightly colored and magnetic and are used in chemistry for analysis, to dissolve ions, in the electroplating of metals, and for catalysis. They are also the target of current research in solar-energy conversion, in atmospheric nitrogen fixation, and in pharmaceuticals.

9C.1 Coordination Complexes

The Lewis bases attached to the central metal atom or ion in a d-metal complex are known as **ligands;** they can be either ions or molecules. An example of an ionic ligand is the cyanide ion. In the hexacyanoferrate(II) ion, $[Fe(CN)_6]^{4-}$, the CN^- ions provide the electron pairs that form bonds to the Lewis acid Fe^{2+}. In the neutral complex $Ni(CO)_4$, the Ni atom acts as the Lewis acid and the ligands are the CO molecules, acting as Lewis bases.

Each ligand in a complex typically has at least one lone pair of electrons with which it bonds to the central atom or ion by forming a coordinate covalent bond. The ligands are said to **coordinate** to the metal when they form the complex in this way. These ligands make up the **coordination sphere** of the central ion. The number of points at which ligands are attached to the central metal atom is called the **coordination number** of the complex (FIG. 9C.1). The coordination number is 4 in $Ni(CO)_4$ and 6 in $[Fe(CN)_6]^{4-}$.

Coordination sphere

Ligand

Central metal ion

Box 9C.1 WHAT HAS THIS TO DO WITH...STAYING ALIVE?

WHY WE NEED TO EAT d-METALS

Some of the critical enzymes in our cells are *metalloproteins*, large organic molecules made up of long chains of amino acids that also include at least one metal atom. Biochemists study these metalloproteins intensely because they control life and protect against disease. The d-block metal complexes in metalloproteins catalyze redox reactions, form components of membrane, muscle, skin, and bone, catalyze acid–base reactions, and control the flow of energy and oxygen.

Hemoglobin and myoglobin, in which an iron(II) ion lies at the center of a heme group, are the most familiar metalloproteins. They serve as the oxygen transport and storage mechanism in mammalian systems. The points of attachment of the ligand to the central Fe atom are the four nitrogen atoms from amine groups in the planar heme. A nitrogen atom on the amino acid histidine (see Table 11E.3) serves as a fifth point of connection in a modified square pyramidal shape about the Fe atom. The oxygen molecule acts as an additional ligand, attaching directly to the Fe atom and producing a modified octahedral shape (see the figure below).

Cobalt is a d-metal needed to prevent pernicious anemia and some kinds of mental illness. It is an essential part of a coenzyme required for the activity of vitamin B_{12} (which is also called cobalamin) and gives the vitamin its red color. The cobalt atom is found in an octahedral complex in which five of the ligands are attached through nitrogen atoms from organic amine groups and one ligand attaches through a —CH_2— group.

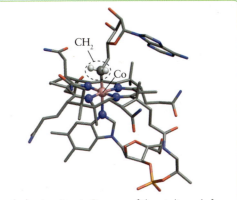

In cobalamin, vitamin B_{12}, one of the six ligands forming an octahedral structure about a cobalt atom is an organic molecule that is attached through a carbon—cobalt bond. The bond is weak and easily broken.

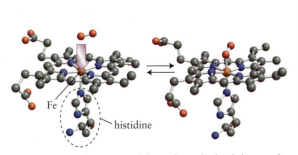

Molecular oxygen is transported throughout the body by attaching to iron(II) atoms in the heme group of hemoglobin molecules; it is also stored in the heme group of myoglobin. The iron(II) atom lies at the center of the base of a square pyramidal complex formed by nitrogen atoms.

ANIMATION BOX 9C.1

Cobalamin is the only biomolecule known to have a metal–carbon bond. The ease with which this bond is broken and the ability of the cobalt ion to change from one oxidation state to another are responsible for the importance of cobalamin as a biological catalyst.

Zinc enzymes play several important roles in metabolism, including the expression of our genes, the digestion of food, the storage of insulin, and the building of collagen. In fact, zinc has so many functions in our systems that it has been called a "master hormone." Its concentration in our bodies is about the same as that of iron.

Other d-metals are also vital to health. For example, chromium(III) plays a role in the regulation of glucose metabolism and copper(I) is an essential nutrient for healthy cells. However, it is important to note that only trace amounts of d-metals are needed in human nutrition; large amounts of any of them can be toxic.

Related Exercise 9.25

Further Reading J. J. R. Fraústo da Silva and R. J. P. Williams, *The Biological Chemistry of the Elements: The Inorganic Chemistry of Life* (Oxford: Oxford University Press, 1991). "Biological inorganic chemistry," Chapter 26 in M. Weller, T. Overton, J. Rourke, and F. Armstrong, *Inorganic Chemistry*, 7th edition (Oxford: Oxford University Press, 2014). "Metalloproteins," *Nature*, vol. 460, no. 7257, pp. 813–862, 2009; available at http://www.nature.com/nature/supplements/insights/metalloproteins/

Because water is a Lewis base, it forms complexes with most d-block ions when they dissolve in it. Aqueous solutions of d-metal ions are usually solutions of their H_2O complexes: $Fe^{2+}(aq)$, for instance, is actually $[Fe(OH_2)_6]^{2+}$. Many complexes are prepared simply by mixing aqueous solutions of a d-metal ion and the appropriate Lewis base (**FIG. 9C.2**); for example,

$$[Fe(OH_2)_6]^{2+}(aq) + 6\,CN^-(aq) \longrightarrow [Fe(CN)_6]^{4-}(aq) + 6\,H_2O(l)$$

FIGURE 9C.2 When potassium cyanide is added to a solution of iron(II) sulfate, the cyanide ions replace the H_2O ligands of the $[Fe(OH_2)_6]^{2+}$ complex (left) and produce a new complex, the hexacyanoferrate(II) ion, $[Fe(CN)_6]^{4-}$ (right). The blue color is due to the polymeric compound called Prussian blue, which forms from the cyanoferrate ion. *(W. H. Freeman photo by Ken Karp.)*

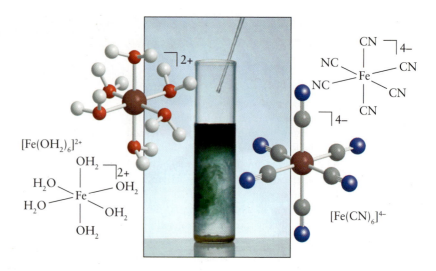

This is an example of a **substitution reaction**, a reaction in which one Lewis base takes the place of another. Here the CN^- ions drive out H_2O molecules from the coordination sphere of the $[Fe(OH_2)_6]^{2+}$ complex and take their place. Replacement is less complete when certain other ions, such as Cl^-, are added to an iron(II) solution:

$$[Fe(OH_2)_6]^{2+}(aq) + Cl^-(aq) \longrightarrow [FeCl(OH_2)_5]^+(aq) + H_2O(l)$$

Because the color of a d-metal complex depends on the identity of the ligands as well as that of the metal, impressive color changes often accompany substitution reactions (**FIG. 9C.3**). These reactions often occur quickly because many coordination complexes are labile (short-lived). A labile complex in aqueous solution, such as $[Cu(OH_2)_6]^{2+}(aq)$, rapidly exchanges water ligands with the water molecules of the solvent, during which a small amount of the five-coordinate metal complex is formed:

$$[Cu(OH_2)_6]^{2+}(aq) \underset{\text{fast}}{\overset{\text{fast}}{\rightleftharpoons}} [Cu(OH_2)_5]^{2+}(aq) + H_2O(l)$$

The five-coordinate complex immediately reacts with any Lewis base that is present. The equilibrium for the formation of a coordination complex is described by its formation constant, K_f (Topic 6I). For example, the formation of $[Cu(NH_3)_4]^{2+}(aq)$ by reaction of $[Cu(OH_2)_6]^{2+}(aq)$ with $NH_3(aq)$ is

pale blue $\qquad\qquad$ intense, deep blue

$$\overbrace{[Cu(OH_2)_6]^{2+}(aq)} + 4 NH_3(aq) \rightleftharpoons \overbrace{[Cu(NH_3)_4]^{2+}(aq)} + 6 H_2O(l)$$

$$K_f = \frac{[[Cu(NH_3)^4]^{2+}]}{[[Cu(OH_2)_6]^{2+}][NH_3]^4}$$

with $K_f = 1.2 \times 10^{13}$ at 25 °C. This high value indicates that the Cu—N bond in $[Cu(NH_3)_4]^{2+}$ is much stronger than the Cu—O bond in $[Cu(OH_2)_6]^{2+}$. The formation constants for other coordination complexes in aqueous solution are listed in Table 6I.2.

The names of coordination compounds can become awesomely long, because the identity and number of each type of ligand must be included. In most cases, chemists avoid the problem by using the chemical formula rather than the name itself. For instance, it is much easier to refer to $[FeCl(OH_2)_5]^+$ than to pentaaquachloridoiron(II) ion, its formal name. However, names are sometimes needed, and they can be constructed and interpreted, in simple cases at least, by using the rules set out in **Toolbox 9C.1**. **TABLE 9C.1** gives the names of common ligands and their abbreviations, which are used in the formulas of complexes. The rules have recently been changed, but as the older names are still in wide use, both are given.

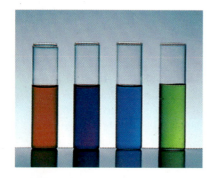

FIGURE 9C.3 Some of the highly colored d-metal complexes. Left to right: aqueous solutions of $[Fe(SCN)(OH_2)_5]^{2+}$, $[Co(SCN)_4(OH_2)_2]^{2-}$, $[Cu(NH_3)_4(OH_2)_2]^{2+}$, and $[CuBr_4]^{2-}$. *(W. H. Freeman photo by Ken Karp.)*

Toolbox 9C.1 HOW TO NAME d-METAL COMPLEXES AND COORDINATION COMPOUNDS

CONCEPTUAL BASIS

d-Metal complexes are identified by giving the names and numbers of the individual ligands. Because some names can be quite long, interpreting them is rather like eating a large bun: nibble it bite by bite, don't try to swallow it in one gulp. The rules used here are consistent with the latest (2005) IUPAC recommendations. Further information on naming complexes can be found at http://old.iupac.org/publications/books/seriestitles/nomenclature.html (referred to as the "Red Book").

PROCEDURE

The following rules are adequate for most common complexes; more elaborate rules are needed if the complex contains more than one metal atom. Some rules apply to naming the complex (formally, the "coordination entity") and others to writing its chemical formula.

A. Writing the formula

1. Write the chemical symbol of the central atom (usually a metal atom) first, followed by symbols for the ligands, then enclose the entire collection of symbols between square brackets and add the overall charge.

2. Write the chemical symbols of ligands in alphabetical order; when different ligands contain the same element, those denoted by a single letter (e.g., O) take precedence over symbols represented by two or more letters (e.g., OH). If a particular point needs to be emphasized, then the order may be varied. Water as a ligand is denoted OH_2, to emphasize that the O links to the metal atom.

   Examples: $[FeCl(OH_2)_5]^+$ $[Fe(NH_3)_5(OH_2)]^{3+}$

3. To avoid any ambiguity, the linking atom can be underlined.

   Examples: $[Fe(\underline{N}CS)(OH_2)_5]^{2+}$ $[Fe(NC\underline{S})(OH_2)_5]^{2+}$

B. Naming the complex

1. Name the ligands first, and then the metal atom or ion. A Roman numeral in parentheses denotes the oxidation number of the central metal ion.

2. Neutral ligands, such as $H_2NCH_2CH_2NH_2$ (ethylenediamine), have the same name as the molecule, except for H_2O (aqua), NH_3 (ammine), CO (carbonyl), and NO (nitrosyl).

3. Anionic ligands end in -o; for anions that end in -ide (such as chloride), -ate (such as sulfate), and -ite (such as nitrite), change the endings as follows:

   -ide ⟶ -ido   -ate ⟶ -ato   -ite ⟶ -ito

   Examples: chlorido, sulfato, and nitrito

4. Greek prefixes indicate the number of each type of ligand in the complex ion:

2	3	4	5	6	...
di-	tri-	tetra-	penta-	hexa-	...

   If the ligand already contains a Greek prefix (such as the di- in ethylenediamine) or if it is polydentate (able to attach at more than one binding site simultaneously), then the following prefixes are used instead:

2	3	4	...
bis-	tris-	tetrakis-	...

5. Ligands are named in alphabetical order, ignoring any Greek prefix.

   $[FeCl(OH_2)_5]^+$ pentaaquachloridoiron(II) ion
   $[Cr(Cl)_2(NH_3)_4]^+$ tetraamminedichloridochromium(III) ion

   (Note that in some cases the order of ligands in the name is not the same as the order in the formula.)

6. If there is an ambiguity in identifying which atom is linked to the metal atom, then κE is added to the name in parentheses, where E denotes the connecting atom (and κ is kappa):

   $[Fe(\underline{N}CS)(OH_2)_5]^{2+}$ pentaaquathiocyanato(κN)iron(III) ion
   $[Fe(NC\underline{S})(OH_2)_5]^{2+}$ pentaaquathiocyanato(κS)iron(III) ion

7. If the complex has an overall negative charge (an anionic complex), the suffix -ate is added to the stem of the metal's name. If the symbol of the metal originates from a Latin name (as listed in Appendix 2D), then the Latin stem is used. For example, the symbol for iron is Fe, from the Latin *ferrum*. Therefore, any anionic complex of iron ends with -ferrate followed by the oxidation number of the metal in Roman numerals:

   $[Fe(CN)_6]^{4-}$ hexacyanidoferrate(II) ion
   $[Ni(CN)_4]^{2-}$ tetracyanidonickelate(II) ion

8. The name of a coordination compound (as distinct from a complex cation or anion) is built in the same way as that of a simple compound, with the cation named before the anion:

   $NH_4[PtCl_3(NH_3)]$
           ammonium amminetrichloridoplatinate(II)

   $[Cr(NH_3)_4(OH)_2]Br$
           tetraamminedihydroxidochromium(III) bromide

This procedure is illustrated in Example 9C.1.

TABLE 9C.1   Common Ligands

Formula*	Name
Neutral ligands	
OH_2	aqua
NH_3	ammine
NO	nitrosyl
CO	carbonyl
$NH_2CH_2CH_2NH_2$	ethylenediamine (en)[†]
$NH_2CH_2CH_2NHCH_2CH_2NH_2$	diethylenetriamine (dien)[‡]
Anionic ligands	
F^-	fluorido
Cl^-	chlorido
Br^-	bromido
I^-	iodido
OH^-	hydroxido
O^{2-}	oxido
$\underline{C}N^-$	cyanido-κC
$C\underline{N}^-$	isocyano, cyanido-κN
$\underline{N}CS^-$	isothionato, thiocyanato-κN
$NC\underline{S}^-$	thiocyanato-κS
NO_2^- as $\underline{O}NO^-$	nitrito-κO
NO_2^- as $\underline{N}O_2^-$	nitro, nitrito-κN
CO_3^{2-} as $\underline{O}CO_2^{2-}$	carbonato-κO
$C_2O_4^{2-}$ as $^-O_2CCO_2^-$	oxalato (ox)[†]
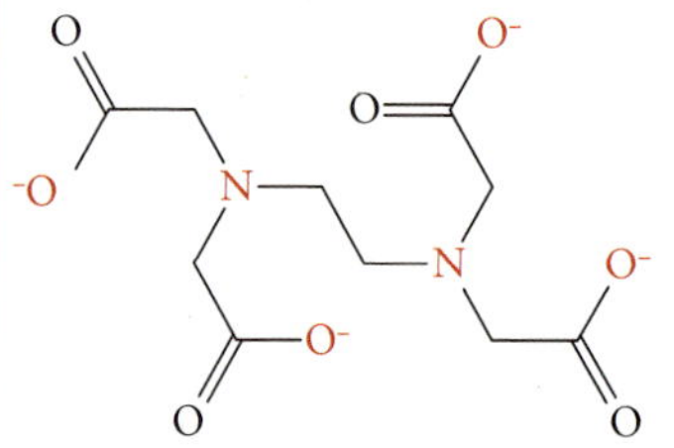	ethylenediaminetetraacetato (edta)[§]
SO_4^{2-} as $\underline{O}SO_3^{2-}$	sulfato

*Ligand atoms that bond to the metal atom are underlined in ambiguous cases.
[†]Bidentate (attaches to two sites).
[‡]Tridentate (attaches to three sites).
[§]Hexadentate (attaches to six sites).

EXAMPLE 9C.1   Naming complexes and coordination compounds

You are working in the chemical stockroom of a university, and a teaching assistant requests two inorganic reagents. You want to make sure you deliver the correct materials, so you need to compare both the formulas and names from the request with the labels on the bottles. (a) Name the coordination compound $[Co(NH_3)_3(OH_2)_3]_2(SO_4)_3$. (b) Write the formula of sodium dichloridobis(oxalato)platinate(IV).

PLAN Apply the rules in Toolbox 9C.1.

SOLVE

(a) There are three SO_4^{2-} ions for every two complex ions.

The complex cation must have a charge of $+3$: $[Co(NH_3)_3(OH_2)_3]^{3+}$.

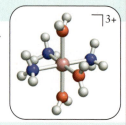

All the ligands are neutral.

Co is present as cobalt(III).

There are three NH_3 molecules (ammine) and three H_2O molecules (aqua); ammine precedes aqua alphabetically.

The cation is triamminetriaquacobalt(III), and the compound is triamminetriaquacobalt(III) sulfate.

(b) Two Cl^- ligands and two $C_2O_4^{2-}$ ions are attached to Pt^{4+}.

The charge on the complex is -2.

From Table 9C.1, the symbol for oxalate is ox; Cl precedes ox alphabetically; the use of "bis" indicates that two oxalato ligands are present; "bis" is used instead of "di" because oxalate is polydentate (in this case, bidentate, with two points of connection).

The complex anion is $[PtCl_2(ox)_2]^{2-}$.

The compound is $Na_2[PtCl_2(ox)_2]$.

Self-test 9C.1A (a) Name the compound $[Fe(OH)(OH_2)_5]Cl_2$. (b) Write the formula of potassium diaquabis(oxalato)chromate(II).

[***Answer:*** (a) Pentaaquahydroxidoiron(III) chloride;
(b) $K_2[Cr(OH_2)_2(ox)_2]$]

Self-test 9C.1B (a) Name the compound $[CoBr(NH_3)_5]SO_4$. (b) Write the formula of tetraamminediaquachromium(III) bromide.

Related Exercises 9C.1–9C.4

A complex is formed between a Lewis acid (the metal atom or ion) and a number of Lewis bases (the ligands).

9C.2 The Shapes of Complexes

The richness of coordination chemistry is enhanced by the variety of shapes that complexes can adopt. The most common complexes have coordination number 6. Almost all these species have their ligands at the vertices of a regular octahedron, with the metal ion at the center, and are called **octahedral complexes** (1). An example of an octahedral complex is the hexacyanidoferrate(II) ion, $[Fe(CN)_6]^{4-}$.

The next most common coordination number is 4. Two shapes are typically found for this coordination number. In a **tetrahedral complex,** the four ligands are found at the vertices of a tetrahedron, as in the tetrachloridocobaltate(II) ion, $[CoCl_4]^{2-}$ (2). An alternative arrangement, most notably for atoms and ions with d^8 electron configurations such as Pt^{2+} and Au^{3+}, is for the ligands to lie at the corners of a square, giving a **square planar complex** (3).

Many other shapes are possible for complexes. The simplest are linear, with coordination number 2. An example is dimethylmercury(0), $Hg(CH_3)_2$ (4), which is a toxic compound formed by bacterial action on aqueous solutions of Hg^{2+} ions. Coordination numbers as high as 12 are found for members of the f-block, but they are rare in the d-block. One interesting type of d-metal compound in which there are 10 links between the ligands and the central metal ion is ferrocene, dicyclopentadienyliron(0), $Fe(C_5H_5)_2$ (5). Ferrocene is an aptly named "sandwich compound," with the two planar cyclopentadienyl ligands the "bread" and the metal atom the "filling." The formal name for a sandwich compound is a **metallocene.**

Complexes of molybdenum and tungsten with eight ligands are known. These complexes have antiprismatic (6) or dodecahedral shapes (7). However, complexes with more than six ligands are rare.

Some ligands are **polydentate** ("many toothed") and can occupy more than one binding site simultaneously. For example, each end of the *bidentate* (that is, two-toothed) ethylenedi-amine molecule, $NH_2CH_2CH_2NH_2$ (8), has a nitrogen atom with a lone pair of electrons. This ligand is widely used in coordination chemistry and in formulas is abbreviated to en, as

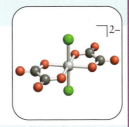

1 An octahedral complex

2 A tetrahedral complex

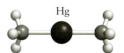

3 A square planar complex

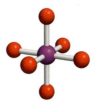

Hg

4 Dimethyl mercury(0)

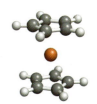

5 Ferrocene, $Fe(C_5H_5)_2$

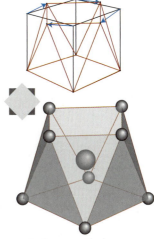

6 Square antiprism

7 Dodecahedral complex

8 Ethylenediamine, $NH_2CH_2CH_2NH_2$

9 $[Co(en)_3]^{3+}$

Isomerism is also important in organic chemistry (Topic 11A).

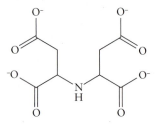

10 Ethylenediaminetetraacetic acid

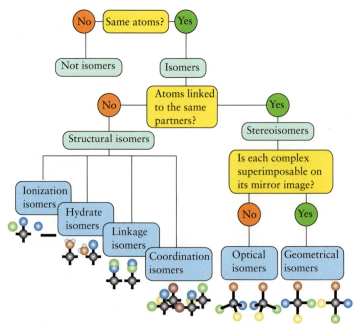

11 An EDTA complex

12 Iminodisuccinate ion

in tris (ethylenediamine)cobalt(III), $[Co(en)_3]^{3+}$ (**9**). The metal ion in $[Co(en)_3]^{3+}$ lies at the center of the three ligands as though pinched by three molecular claws. It is an example of a **chelate** (from the Greek word for "claw"), a complex containing one or more ligands that form a ring of atoms that includes the central metal atom. There are few hexadentate ligands, but a common example is the ethylenediaminetetraacetate ion, edta (the fully protonated acid is shown as (**10**); the red arrows show the points of attachment). This ligand forms complexes with many metal ions, including Pb^{2+} (**11**); hence, it is used as an antidote to lead poisoning.

Chelating ligands are quite common in nature. Mosses and lichens secrete chelating ligands to extract essential metal ions from the rocks on which they dwell. Chelate formation also lies behind the body's strategy of producing a fever when infected by bacteria. The higher temperature kills bacteria by reducing their ability to synthesize a particular iron-chelating ligand.

The production of some chelates releases toxic chemicals such as cyanides to the environment. However, new types of chelates that **sequester** d-block metals, bind with them and remove them from solution, are helping to solve some of society's trickiest environmental problems. For example, the chelating agent sodium iminodisuccinate, which contains the hexadentate iminodisuccinate ion (**12**), can scavenge metal ions from wastewater and serves as a nontoxic additive to detergents. It is rapidly degraded to nontoxic products in the environment. Other environmentally friendly chelates accelerate the action of hydrogen peroxide, and the combination is replacing chlorine bleaches in the production of paper, greatly reducing the release of toxic pollutants to the environment.

Complexes with coordination number 6 are typically octahedral; those with coordination number 4 are either tetrahedral or square planar. Polydentate ligands can form chelates.

9C.3 Isomers

Many complexes and coordination compounds exist as **isomers,** compounds that contain the same numbers of the same atoms but in different arrangements. For example, the ions shown in (**13a**) and (**13b**) differ only in the positions of the Cl^- ligands, but they are distinct species, because they have different physical and chemical properties. Isomerism is of more than academic interest: for example, anticancer drugs based on complexes of platinum are active only if they are the correct isomer. The complex needs to have a particular shape to interact with DNA molecules.

FIGURE 9C.4 summarizes the types of isomerism found in coordination complexes. The two major classes of isomers are **structural isomers,** in which the atoms are

FIGURE 9C.4 The various types of isomerism in coordination compounds.

connected to different partners, and **stereoisomers,** in which the atoms have the same partners but are arranged differently in space. Structural isomers of coordination compounds are subdivided into ionization, hydrate, linkage, and coordination isomers.

Ionization isomers differ by the exchange of a ligand with an anion or a neutral molecule outside the coordination sphere. For instance, $[CoBr(NH_3)_5]SO_4$ and $[Co(NH_3)_5SO_4]Br$ are ionization isomers because the Br^- ion is a ligand of the cobalt in the former but an accompanying anion in the latter. The isomers can be distinguished by their chemical properties, because an ion inside the complex is not available for reaction. Thus, the addition of a barium salt will result in the precipitation of barium sulfate from a solution of $[CoBr(NH_3)_5]SO_4$, but not from a solution of $[Co(NH_3)_5SO_4]Br$.

Hydrate isomers differ by the exchange of an H_2O molecule with another ligand in the coordination sphere (**FIG. 9C.5**). For example, the solid hexahydrate of chromium(III) chloride, $CrCl_3 \cdot 6H_2O$, may be any of the three compounds $[Cr(OH_2)_6]Cl_3$, $[CrCl(OH_2)_5]Cl_2 \cdot H_2O$, or $[CrCl_2(OH_2)_4]Cl \cdot 2H_2O$. Hydrate isomers can often be distinguished by the stoichiometry of reactions in which the ion is exchanged with water. For example, 2 mol AgCl can be precipitated from 1 mol $[CrCl(OH_2)_5]Cl_2 \cdot H_2O$, but only 1 mol AgCl can be produced from 1 mol $[CrCl_2(OH_2)_4]Cl \cdot 2H_2O$.

> **Self-test 9C.2A** When excess silver nitrate is added to 0.0010 mol $CrCl_3 \cdot 6H_2O$ in aqueous solution, 0.0010 mol AgCl is formed. Which hydrate isomer is present?
> [***Answer:*** $[CrCl_2(OH_2)_4]Cl \cdot 2H_2O$]
>
> **Self-test 9C.2B** When excess silver nitrate is added to 0.0010 mol $CrCl_3 \cdot 6H_2O$ in aqueous solution, 0.0030 mol AgCl is formed. Which hydrate isomer is present?

Linkage isomers differ in the identity of the atom used by a given ligand to attach to the metal ion (**FIG. 9C.6**). Common ligands that show linkage isomerism are SCN^- versus NCS^-, NO_2^- versus ONO^-, and CN^- versus NC^-, where the coordinating atom is written first in each pair. For example, NO_2^- can form $[CoCl(NH_3)_4(NO_2)]^+$ and $[CoCl(NH_3)_4(ONO)]^+$. In the current system of nomenclature, the atom through which the ligand is coordinated is underlined, so these two complexes would be denoted $[CoCl(NH_3)_4(\underline{N}O_2)]^+$ and $[CoCl(NH_3)_4(N\underline{O}_2)]^+$, respectively. The name used to specify the ligand is different in each case. For instance, nitro (modern name: nitrito-κN) signifies that the ligand is linked through the N atom and nitrito (modern name: nitrito-κO) that it is linked through an O atom. Table 9C.1 lists the names to use for **ambidentate ligands,** ligands that can attach through atoms of different elements.

Coordination isomers differ by the exchange of one or more ligands between a cationic complex and an anionic complex (**FIG. 9C.7**). Thus, $[Cr(NH_3)_6][Fe(CN)_6]$ and $[Fe(NH_3)_6][Cr(CN)_6]$ are coordination isomers.

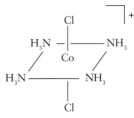

(a) *trans*-$[CoCl_2(NH_3)_4]^+$

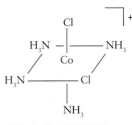

(b) *cis*-$[CoCl_2(NH_3)_4]^+$

13

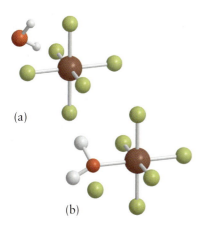

FIGURE 9C.5 Hydrate isomers. In part (a), the water molecule is simply part of the surrounding solvent; in part (b), the water molecule is present in the coordination sphere and a ligand (green sphere) is now present in the solution.

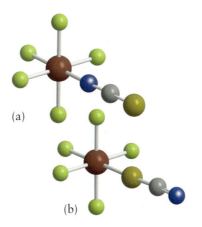

FIGURE 9C.6 Linkage isomers. In part (a) the ligand (here NCS^-) is attached through its N atom, but in part (b) it is attached through its S atom.

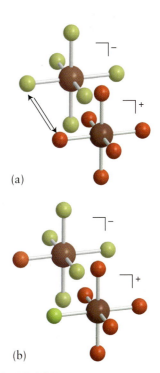

FIGURE 9C.7 The compounds in parts (a) and (b) are coordination isomers. In these compounds, a ligand has been exchanged between the cationic and anionic complexes.

(a) *cis*-PtCl$_2$(NH$_3$)$_2$

(b) *trans*-PtCl$_2$(NH$_3$)$_2$

14

Self-test 9C.3A Identify the type of isomer represented by each of the following pairs: (a) [Cu(NH$_3$)$_4$][PtCl$_4$] and [Pt(NH$_3$)$_4$][CuCl$_4$]; (b) [Cr(NH$_3$)$_4$(OH)$_2$]Br and [CrBr(NH$_3$)$_4$(OH)]OH.
[***Answer:*** (a) Coordination; (b) ionization]

Self-test 9C.3B Identify the type of isomer represented by each of the following pairs: (a) [Co(N̲CS)(NH$_3$)$_5$]Cl$_2$ and [Co(NC̲S)(NH$_3$)$_5$]Cl$_2$; (b) [CrCl(OH$_2$)$_5$]Cl$_2$·H$_2$O and [CrCl$_2$(OH$_2$)$_4$]Cl·2H$_2$O.

Although they are built from the same numbers and kinds of atoms, *structural isomers* have different chemical formulas, because the formulas show how the atoms are grouped in or outside the coordination sphere. *Stereoisomers,* on the other hand, have the same formulas, because their atoms have the same partners in the coordination spheres; only the spatial arrangement of the ligands differs. There are two types of stereoisomerism, geometrical and optical.

In **geometrical isomers,** atoms are bonded to the same neighbors but have different locations relative to each other, as in (**13a**) and (**13b**): the complex with the Cl$^-$ ligands on opposite sides of the central atom is called the *trans isomer*, and the complex with the ligands on the same side is called the *cis isomer*. Geometrical isomers can occur for square planar and octahedral complexes but not for tetrahedral complexes because, in the latter, any pair of vertices is equivalent to any other pair. The chemical and physiological properties of geometrical isomers can differ greatly. For example, *cis*-PtCl$_2$(NH$_3$)$_2$ (**14a**) is used for chemotherapy treatment of cancer patients, but *trans*-PtCl$_2$(NH$_3$)$_2$ (**14b**) is therapeutically inactive. **Optical isomers** are nonsuperimposable mirror images of each other (**FIG. 9C.8**). Both geometrical and optical isomerism can occur in an octahedral complex, as in [CoCl$_2$(en)$_2$]$^+$: the trans isomer (**15a**) is green, and the two alternative cis isomers (**15b**) and (**15c**), which are optical isomers of one another, are violet. Optical isomers can also occur whenever four different groups form a tetrahedral complex, but not if they form a square-planar complex.

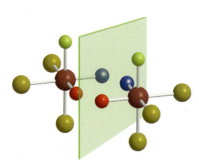

FIGURE 9C.8 Optical isomers. The two complexes are each other's mirror image; no matter how we rotate them, one complex cannot be superimposed on the other.

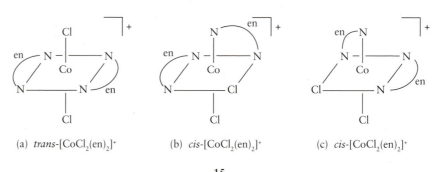

(a) *trans*-[CoCl$_2$(en)$_2$]$^+$ (b) *cis*-[CoCl$_2$(en)$_2$]$^+$ (c) *cis*-[CoCl$_2$(en)$_2$]$^+$

15

A **chiral** complex is one that is not identical to its mirror image and so cannot be superimposed on it. Thus, all optical isomers are chiral. The cis isomers of [CoCl$_2$(en)$_2$]$^+$ are chiral, and a chiral complex and its mirror image form a pair of **enantiomers.** The trans isomer is superimposable on its mirror image; complexes with this property are called **achiral.** Enantiomers differ in one physical property: chiral molecules display **optical activity,** the ability to rotate the plane of polarization of light (**BOX 9C.2**). In ordinary light, the plane of wave motion lies at random orientations around the direction of travel. In plane-polarized light, the wave lies in a single plane (**FIG. 9C.9**). Plane-polarized light can be prepared by passing ordinary light through a special filter, such as the material used to make polarized sunglasses. One enantiomer of a chiral complex

The name enantiomer comes from the Greek words meaning "both parts."

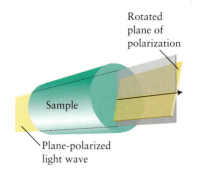

Rotated plane of polarization

Sample

Plane-polarized light wave

FIGURE 9C.9 Plane-polarized light consists of radiation in which all the wave motion lies in one plane (as represented by the plane on the left). When such light passes through a solution of an optically active substance, the plane of the polarization is rotated through a characteristic angle that depends on the identity and concentration of the solute and the length of the path through the sample (right).

Box 9C.2 HOW DO WE KNOW . . . THAT A SUBSTANCE IS OPTICALLY ACTIVE?

The electric field of plane-polarized light oscillates in a single plane. It can be prepared by passing ordinary, nonpolarized light through a polarizer, which contains a material that allows the light to pass only if the electric field is aligned in a certain direction.

An optically active substance, such as a chiral complex, rotates the plane of polarization of a beam of light by an angle that depends on the substance, its concentration, and the length of the sample cell. The light is passed through a sample cell about 10 cm long. To detect chirality, a solution of the chiral complex is placed in the cell. When the light emerges from the far end of the cell, the angle of its plane of polarization may have rotated from its original angle. To determine the angle, the light is passed through an analyzer that contains another polarizing filter. The filter is rotated until the intensity of the light that has passed through the polarizer, sample, and filter reaches its maximum. The angle of the plane of polarization is determined from the angle through which the filter was rotated to achieve this maximum setting. If the sample is not optically active, the light is not rotated by the sample and the maximum intensity is observed at an angle of 0°. The sample is optically active if the angle of rotation is different from 0°. The actual value depends

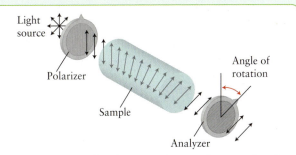

A polarimeter is used to determine the optical activity of a substance by measuring the angle through which plane-polarized light is rotated by a sample.

on the identity of the complex, its concentration, the wavelength of the light, and the length of the sample cell.

The determination of the angle of rotation is called *polarimetry.* In some cases, it can help a chemist follow a reaction. For example, if a reaction destroys the chirality of a complex, then the angle of optical rotation decreases with time as the concentration of the complex falls.

rotates the plane of polarization clockwise; its mirror-image partner rotates it by the same amount the other way. Achiral complexes are not optically active: they do not rotate the plane of polarization of polarized light.

Some complexes are synthesized in the laboratory as **racemic mixtures,** or mixtures of enantiomers in equal proportions. Because enantiomers rotate the plane of polarization of light in opposite directions, a racemic sample is not optically active.

EXAMPLE 9C.2 Identifying optical isomerism

New pharmaceuticals must be enantiomerically pure for use in human medicine. If you work for a large biotechnology company, you will need to be able to recognize chiral sites in complex molecules. Which of the following complexes are chiral, and which form enantiomeric pairs?

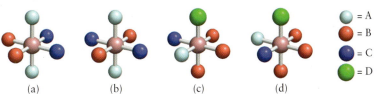

(a) (b) (c) (d)

= A
= B
= C
= D

PLAN Draw the mirror image of each complex and mentally rotate it; judge whether any rotation will allow the mirror image to be superimposed on the original molecule. If not, then the complex is chiral. Determine which complexes form enantiomeric pairs by finding pairs in which the two complexes are the nonsuperimposable mirror images of each other. If imagining the three-dimensional structure is difficult, build simple paper models of the complexes.

SOLVE The mirror image of each complex is shown on the right of each pair.

(a) Rotating the mirror image about A—A gives a structure identical to the original, and so superimposable on it.

Not chiral

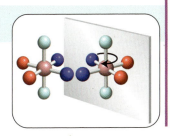

(b) The mirror image is superimposable on the original by rotation about A—A.

Not chiral

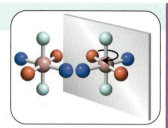

(c) No rotation will allow the complex to be superimposed on its mirror image.

Chiral

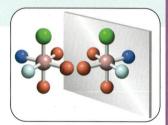

(d) No rotation will allow the complex to be superimposed on its mirror image.

Chiral

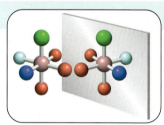

When the mirror image of (c) is rotated by 180° around the vertical B—D axis, it becomes the complex (d).

(c) and (d) are a pair of enantiomers.

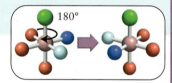

Self-test 9C.4A Repeat Example 9C.2 for the following complexes:

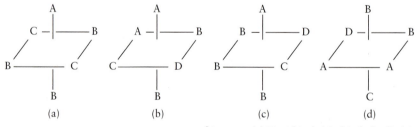

[*Answer:* (a) Not chiral; (c) chiral; (b, d) chiral and enantiomeric]

Self-test 9C.4B Repeat Example 9C.2 for the following complexes:

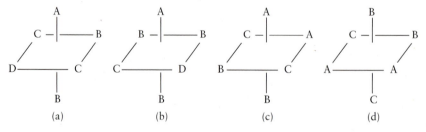

Related Exercises 9C.17, 9C.18.

The varieties of isomerism are summarized in Fig. 9C.4. Enantiomeric pairs of optical isomers rotate the plane of polarization of light in opposite directions.

What have you learned in this Topic?

You now know how to name d-metal coordination compounds and have seen how the three-dimensional arrangement of ligands around a central metal atom can result in structural isomers and stereoisomers.

The skills you have mastered are the ability to:

☐ **1.** Name and write formulas for d-metal complexes (Toolbox 9C.1 and Example 9C.1).

☐ **2.** Identify pairs of ionization, linkage, hydrate, coordination, geometrical, and optical isomers (Self-tests 9C.2 and 9C.3 and Example 9C.2).

Topic 9C Exercises

9C.1 Name each of the following complex ions and identify the oxidation number of the metal: (a) $[Fe(CN)_6]^{4-}$; (b) $[Co(NH_3)_6]^{3+}$; (c) $[Co(CN)_5(OH_2)]^{2-}$; (d) $[Co(NH_3)_5(SO_4)]^+$.

9C.2 Name each of the following complex ions and identify the oxidation number of the metal: (a) $[CrCl_3(NH_3)_2(OH_2)]^+$; (b) $[Rh(en)_3]^{3+}$; (c) $[Fe(Br)_4(ox)]^{3-}$; (d) $[Ni(OH)(OH_2)_5]^{2+}$.

9C.3 Use the information in Table 9C.1 to write the formula for each of the following coordination compounds:
(a) potassium hexacyanidochromate(III)
(b) pentaamminesulfatocobalt(III) chloride
(c) tetraamminediaquacobalt(III) bromide
(d) sodium bisoxalato(diaqua)ferrate(III)

9C.4 Use the information in Table 9C.1 to write the formula for each of the following coordination compounds:
(a) triamminediaquabromidocobalt(II) hydroxide
(b) dichloridobisethylenediaminecobalt(III) bromide
(c) sodium triamminetrichloridonickelate(II)
(d) barium tris(oxalato)ferrate(III)
(e) diaquadichloridoplatinum(IV) iodide

9C.5 Which of the following ligands can be polydentate? If the ligand can be polydentate, give the maximum number of places on the ligand that can bind simultaneously to a single metal center: (a) $HN(CH_2CH_2NH_2)_2$; (b) CO_3^{2-}; (c) H_2O; (d) oxalate.

9C.6 Which of the following ligands can be polydentate? If the ligand can be polydentate, give the maximum number of places on the ligand that can bind simultaneously to a single metal center: (a) chloride ion; (b) cyanide ion; (c) ethylenediaminetetraacetate; (d) $N(CH_2CH_2NH_2)_3$.

9C.7 Which of the following isomers of diaminobenzene can form chelating complexes? Explain your reasoning.

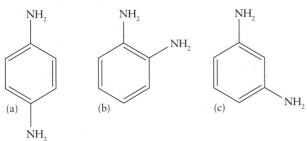

9C.8 Which of the following ligands do you expect to form chelating complexes? Explain your reasoning.

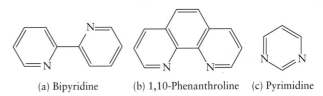

(a) Bipyridine   (b) 1,10-Phenanthroline   (c) Pyrimidine

9C.9 With the help of Table 9C.1, determine the coordination number of the metal ion in each of the following complexes: (a) $[NiCl_4]^{2-}$; (b) $[Ag(NH_3)_2]^+$; (c) $[PtCl_2(en)_2]^{2+}$; (d) $[Cr(edta)]^-$.

9C.10 With the help of Table 9C.1, determine the coordination number of the metal ion in each of the following complexes: (a) $PtBr_2(NH_3)_2$; (b) $[Ni(en)_2I_2]^+$; (c) $[Co(ox)_3]^{3-}$; (d) $[Mn(CO)_5]^-$.

9C.11 Identify the type of structural isomerism that exists in each of the following pairs of compounds:
(a) $[Co(NH_3)_5(\underline{N}O_2)]Br_2$ and $[Co(NH_3)_5(\underline{O}NO)]Br_2$
(b) $[Pt(NH_3)_4(SO_4)](OH)_2$ and $[Pt(NH_3)_4(OH)_2]SO_4$
(c) $[CoCl(NCS)(NH_3)_4]Cl$ and $[CoCl(\underline{N}CS)(NH_3)_4]Cl$
(d) $[CrCl(NH_3)_5]Br$ and $[CrBr(NH_3)_5]Cl$

9C.12 Identify the type of structural isomerism that exists in each of the following pairs of compounds or ions:
(a) $[Pt(OH_2)_4][PtCl_6]$ and $[PtCl_2(OH_2)_4][PtCl_4]$
(b) $[Cr(en)_3][Co(ox)_3]$ and $[Co(en)_3][Cr(ox)_3]$
(c) $[Fe(C\underline{N})(OH)_5]^{3-}$ and $[Fe(\underline{C}N)(OH)_5]^{3-}$
(d) $[CoBr_2(NH_3)_4]Br \cdot H_2O$ and $[CoBr(NH_3)_4(OH_2)]Br_2$

9C.13 Which of the following coordination compounds can have cis and trans isomers? If such isomerism exists, draw the two structures and name the compound: (a) $[CoCl_2(NH_3)_4]$ $Cl \cdot H_2O$; (b) $[CoCl(NH_3)_5]Br$; (c) $PtCl_2(NH_3)_2$, a square planar complex.

9C.14 Which of the following complexes can have cis and trans isomers? If such isomerism exists, draw the two structures and label the ions as cis or trans: (a) $[Fe(OH)_2(OH_2)_4]^+$; (b) $[RuBr_2(NH_3)_4]^{2+}$; (c) $[Co(NH_3)_3(OH_2)_3]^{3+}$.

9C.15 How many isomers are possible for $[Cr(NH_3)_5(NO_2)]Cl_2$? Consider all types of isomerism and draw each isomer.

9C.16 How many isomers are possible for $[CoCl(NCS)(OH_2)_4]$ $Cl \cdot H_2O$? Consider all types of isomerism and draw each isomer.

9C.17 Is either of the following complexes chiral? If both complexes are chiral, do they form an enantiomeric pair?

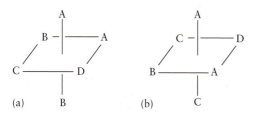

(a)   (b)

9C.18 Is either of the following complexes chiral? If both complexes are chiral, do they form an enantiomeric pair?

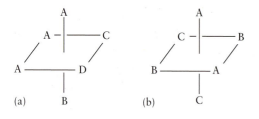

(a)   (b)

9C.19 Ethylenediamine (en) forms a chelate complex with Co^{3+}, $[Co(en)_3]^{3+}$. Are isomers of this complex possible and, if so, what type of isomers are they?

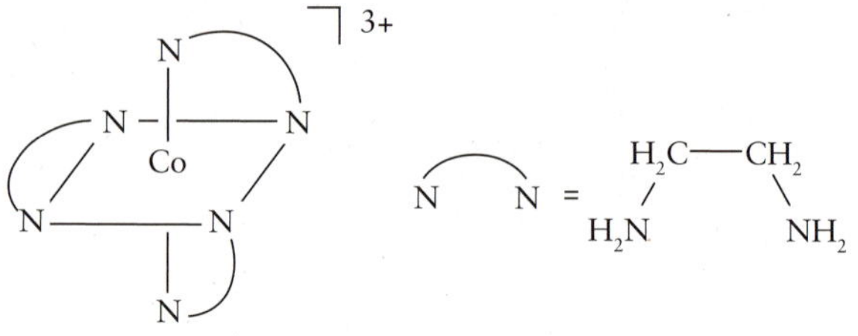

$[Co(en)_3]^{3+}$

9C.20 One structure of the octahedral complex $FeCl_2(NH_3)_3SCN$ is shown below. Draw all possible isomers of this complex.

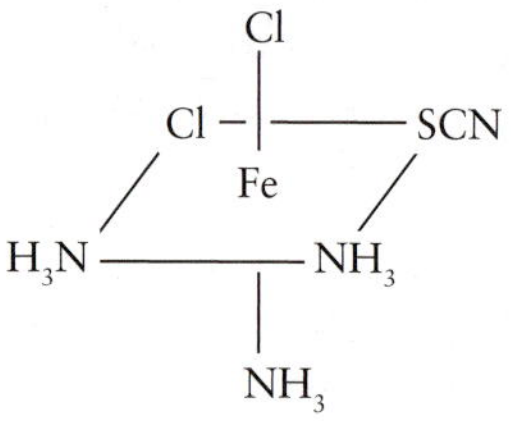

Topic 9D The Electronic Structure of d-Metal Complexes

9D.1 Crystal Field Theory

9D.2 The Spectrochemical Series

9D.3 The Colors of Complexes

9D.4 Magnetic Properties of Complexes

9D.5 Ligand Field Theory

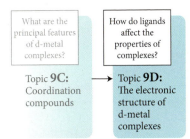

What are the principal features of d-metal complexes?

How do ligands affect the properties of complexes?

Topic **9C:** Coordination compounds → Topic **9D:** The electronic structure of d-metal complexes

Crystal field theory was first devised to explain the colors of solids, particularly ruby, which owes its color to Cr^{3+} ions, and then adapted to individual complexes. Crystal field theory is simple to apply and leads directly to useful predictions. However, it does not account for all the properties of complexes. A more sophisticated approach, *ligand field theory*, is based on molecular orbital theory and provides more detailed explanations.

9D.1 Crystal Field Theory

Crystal field theory takes a very simple view of the environment of the central metal atom (or ion): it supposes that each ligand can be represented by a point negative charge. These negative charges represent the ligand lone pairs directed toward the central metal atom (**FIG. 9D.1**). Because the metal atom at the center of a complex is usually a positively charged ion, the negative charges representing the ligands are attracted to it. In most cases, however, there are still d-electrons on the central metal ion, and the point charges representing the ligands interact with each electron to different extents that depend on the orientation and shape of the d-orbital that it occupies. Crystal field theory explains how these differences account for the optical and magnetic properties of the complex.

As an example, consider an octahedral d^1 complex, such as one containing a Ti^{3+} ion. In a free Ti^{3+} ion, all five 3d-orbitals have the same energy, and the d-electron is equally likely to occupy any one of them. However, when a Ti^{3+} ion is dissolved in water, six H_2O molecules surround it and form a $[Ti(OH_2)_6]^{3+}$ complex. The six point charges representing the ligands lie on opposite sides of the central metal ion along the x-, y-, and z-axes. **FIGURE 9D.2** shows that three of the orbitals (d_{xy}, d_{yz}, and d_{zx}) have their lobes directed between the point charges. In an octahedral complex these three d-orbitals are called **t_{2g}-orbitals.** The other

Why Do You Need to Know This Material? Two of the striking features of the complexes formed by the d-block elements are their wide range of colors and their magnetic properties. Both features, and the relation between them, are explained by two alternative theories of their electronic structure.

What Do You Need to Know Already? This Topic makes use of the electron configurations of atoms and ions (Topics 1E and 2A). Molecular orbital theory (Topic 2G) plays an important role in the final section.

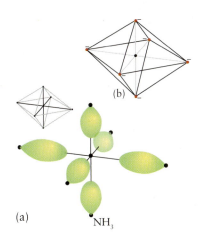

(a) NH_3

(b)

FIGURE 9D.1 In the crystal field theory of complexes, the lone pairs of electrons that serve as the Lewis base sites on the ligands (a) are treated as equivalent to point negative charges (b).

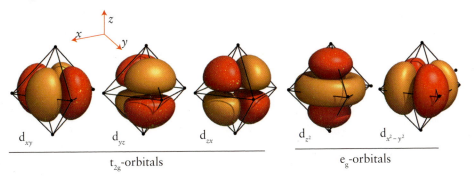

d_{xy} d_{yz} d_{zx} d_{z^2} $d_{x^2-y^2}$

t_{2g}-orbitals e_g-orbitals

FIGURE 9D.2 In an octahedral complex with a central d-metal atom or ion, a d_{xy}-orbital is directed between the ligand sites, and an electron that occupies it has a relatively low energy. The same lowering of energy occurs for d_{yz}- and d_{zx}-orbitals. A d_{z^2}-orbital points directly toward two ligands, and an electron that occupies it has a relatively high energy. The same rise in energy occurs for a $d_{x^2-y^2}$-electron.

ANIMATION FIGURE 9D.2

733

The labels t_{2g} and e_g are derived from group theory, the mathematical theory of symmetry. The letter g indicates that the orbital does not change sign when you imagine starting from any point, passing through the nucleus, and ending at the corresponding point on the other side of the nucleus.

Ligand field splitting is also called "crystal field splitting."

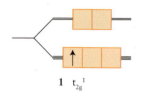

1 $t_{2g}^{\ 1}$

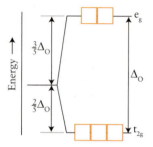

FIGURE 9D.3 The energy levels of the d-orbitals in an octahedral complex with the ligand field splitting Δ_O. The horizontal line on the left represents the average energy of the d-orbitals once the complex has formed; the lines on the right show the modification of their energies due to their different interactions with the ligands. Each orbital (represented by a box) can hold two electrons.

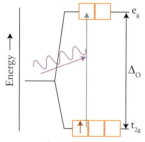

FIGURE 9D.4 When a complex is exposed to light with the appropriate frequency, an electron can be excited to a higher-energy orbital and a photon of that light is absorbed.

The subscript "g" is not used to label the orbitals in a tetrahedral complex, because there is no center of symmetry (which is also called an "inversion center").

two d-orbitals (d_{z^2} and $d_{x^2-y^2}$), have lobes that point directly toward the point charges and are called **e_g-orbitals.** Because the negative point charges representing the ligands repel electrons, the energy of the single d-electron is increased when the complex forms, but it is raised more if it occupies an e_g-orbital than if it occupies a t_{2g}-orbital. Typically, the difference in energies of electrons in the t_{2g}- and the e_g-orbitals accounts only for about 10% of that energy, but it plays a major role in the optical and magnetic properties of the complex.

The energy separation between the two sets of orbitals is called the **ligand field splitting**, Δ_O (the O denotes octahedral). The three t_{2g}-orbitals lie at an energy that is $\frac{2}{5}\Delta_O$ below the average d-orbital energy in the complex, and the two e_g-orbitals lie at an energy $\frac{3}{5}\Delta_O$ above the average (**FIG. 9D.3**). Because the t_{2g}-orbitals have the lower energy, in the ground state of the $[Ti(OH_2)_6]^{3+}$ complex the electron occupies one of them in preference to an e_g-orbital and hence the lowest-energy electron configuration of the complex is $t_{2g}^{\ 1}$. This ground-state configuration is represented by the box diagram in (**1**).

The single d-electron of an octahedral $[Ti(OH_2)_6]^{3+}$ complex can be excited from the t_{2g}-orbital into an e_g-orbital if it absorbs a photon of energy Δ_O (**FIG. 9D.4**). The greater the splitting, the shorter is the wavelength of the light that is absorbed by the complex. Therefore, the wavelength of the electromagnetic radiation absorbed by a complex can be used to determine the ligand field splitting.

EXAMPLE 9D.1 Determining the ligand field splitting

You are a scientist investigating complexes of titanium for use as photoactive coatings for self-cleaning windows. These coatings can oxidize organic materials (dirt) upon exposure to light, leaving a clean surface. The excited-state energy must be large to be useful in this application, and you need to calculate the excited-state energy for one of your promising materials. The complex $[Ti(OH_2)_6]^{3+}$ absorbs light of wavelength 510. nm. What is the ligand field splitting in the complex in kilojoules per mole ($kJ \cdot mol^{-1}$)?

PLAN Because a photon has energy $h\nu$, where h is Planck's constant and ν (nu) is the frequency of the radiation, it can be absorbed if $h\nu = \Delta_O$. The wavelength, λ (lambda), of light is related to the frequency by Eq. 1 of Topic 1A ($\lambda = c/\nu$, where c is the speed of light). Therefore, the wavelength of light absorbed and the ligand field splitting are related by $\Delta_O = hc/\lambda$. To report the ligand field splitting as a molar energy, multiply this expression by Avogadro's constant: $\Delta_O = N_A hc/\lambda$.

SOLVE Because the wavelength absorbed is 510. nm (corresponding to 5.10×10^{-7} m), it follows that the ligand field splitting is

From $\Delta_O = N_A hc/\lambda$,

$$\Delta_O = \frac{\overbrace{6.022 \times 10^{23}\ mol^{-1}}^{N_A} \times \overbrace{6.626 \times 10^{-34}\ J \cdot s}^{h} \times \overbrace{2.998 \times 10^{8}\ m \cdot s^{-1}}^{c}}{\underbrace{5.10 \times 10^{-7}\ m}_{\lambda}}$$

$$= 2.35 \times 10^{5}\ J \cdot mol^{-1} = 235\ kJ \cdot mol^{-1}$$

235 kJ·mol⁻¹

510 nm

Self-test 9D.1A The complex $[Fe(OH_2)_6]^{3+}$ absorbs light of wavelength 700. nm. What is the value (in kilojoules per mole) of the ligand field splitting?

[**Answer:** 171 kJ·mol⁻¹]

Self-test 9D.1B The complex $[Fe(CN)_6]^{4-}$ absorbs light of wavelength 305 nm. What is the value (in kilojoules per mole) of the ligand field splitting?

Related Exercises 9D.5, 9D.6

The relative energies of the d-orbitals are different in complexes with different shapes. For example, in a tetrahedral complex, the three t_2-orbitals point more directly at the ligands than the two e-orbitals do. As a result, in a tetrahedral complex, the t_2-orbitals have a higher energy than the e-orbitals (**FIG. 9D.5**). The ligand field splitting, Δ_T (where the T denotes tetrahedral), is generally smaller than in octahedral complexes, in part because there are fewer repelling ligands.

FIGURE 9D.5 The energy levels of the d-orbitals in a tetrahedral complex with the ligand field splitting Δ_T. Each box (that is, orbital) can hold two electrons.

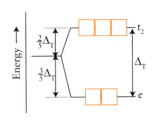

THINKING POINT

Into what groups do you think the d-orbitals are split in a square planar complex?

In octahedral complexes, the e_g-orbitals (d_{z^2} and $d_{x^2-y^2}$), lie higher in energy than the t_{2g}-orbitals (d_{xy}, d_{yz}, and d_{zx}). The opposite is true in a tetrahedral complex, for which the ligand field splitting is smaller.

9D.2 The Spectrochemical Series

Different ligands affect the d-orbitals of a given metal atom or ion to different degrees and thus produce different values of the ligand field splitting. For example, the ligand field splitting is much greater in $[Fe(CN)_6]^{4-}$ than it is in $[Fe(OH_2)_6]^{2+}$. The relative strengths of the splitting produced by a given ligand are much the same regardless of the identity of the d-metal in the complex. Ligands can therefore be arranged in a **spectrochemical series** according to the relative magnitudes of the ligand field splittings that they produce (**FIG. 9D.6**). Ligands below the horizontal line in Fig. 9D.6 produce only a small ligand field splitting, and so they are called **weak-field ligands;** ligands above the line produce a larger splitting and are called **strong-field ligands.** A CN^- ion is therefore a strong-field ligand, whereas an H_2O molecule is a weak-field ligand.

Knowledge of relative ligand strengths allows the color of a complex ion and its magnetism to be explained. Because all five d-orbitals of an isolated metal atom or ion have the same energy, electrons occupy each orbital separately (Hund's rule, Topic 1E) until five electrons have been accommodated. However, when the atom is part of a complex, the difference in energies between the t_{2g}- and e_g-orbitals affects the order in which the orbitals are occupied. The replacement of one ligand by another provides chemical control over color, because the ligands control the energy difference of the t_{2g}- and e_g-orbitals. The substitution of weak-field ligands for strong-field ligands (or vice versa) also acts like a chemical switch for turning paramagnetism on and off, because the value of the ligand field splitting determines which of the d-orbitals are occupied and thus how many of the electrons are paired.

First, consider the metal atom or ion at the center of an octahedral complex. The energies of its d-orbitals are split by the ligands as shown in Fig. 9D.3. The three t_{2g}-orbitals all have the same energy and lie below the two e_g-orbitals. The single d-electron of a d^1 complex occupies one of the t_{2g}-orbitals, and so the ground-state configuration is t_{2g}^1 (as in **1**). The two d-electrons of a d^2 complex occupy separate t_{2g}-orbitals and give rise to the configuration t_{2g}^2 (**2**). Similarly, a d^3 complex will have the ground-state configuration t_{2g}^3 (**3**). According to Hund's rule, all these electrons have parallel spins, because that arrangement corresponds to the lowest energy.

A d^4 octahedral complex presents a problem. The fourth electron could enter a t_{2g}-orbital, resulting in a t_{2g}^4 configuration. However, to do so, it would have to enter an orbital that is already half full and experience a strong repulsion from the electron already there (**4**). To avoid this repulsion, it could occupy an empty e_g-orbital to give a $t_{2g}^3 e_g^1$ configuration (**5**), but now it experiences a strong repulsion from the ligands. Which configuration has the lower energy depends on the ligands present. If Δ_O is large (as it is for strong-field ligands), the energy difference between the t_{2g}- and e_g-orbitals will be large and the configuration t_{2g}^4 will have a lower energy than $t_{2g}^3 e_g^1$. If Δ_O is small (as it is for weak-field ligands), $t_{2g}^3 e_g^1$ will be the lower energy configuration and the one adopted by the complex.

A Note on Good Practice: Note that a configuration with a single electron in an orbital is written with a superscript 1, as in $t_{2g}^3 e_g^1$, not $t_{2g}^3 e_g$.

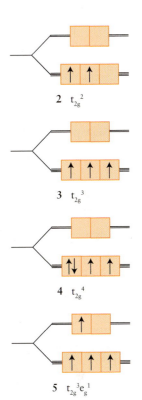

2   t_{2g}^2

3   t_{2g}^3

4   t_{2g}^4

5   $t_{2g}^3 e_g^1$

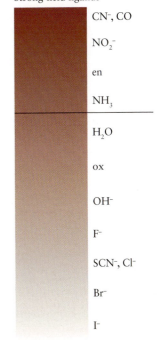

Strong-field ligands

CN^-, CO

NO_2^-

en

NH_3

H_2O

ox

OH^-

F^-

SCN^-, Cl^-

Br^-

I^-

Weak-field ligands

FIGURE 9D.6 The spectrochemical series. Strong-field ligands give rise to a large splitting between the t_{2g}- and e_g-orbitals, whereas weak-field ligands give rise to only a small splitting. The horizontal line marks the approximate frontier between the two kinds of ligands. The increasing intensity of the color represents the increasing strength of the ligand field.

EXAMPLE 9D.2   Predicting the electron configuration of a complex

Suppose you are working in a laboratory synthesizing complexes of Mn and Fe. You have created a new ligand and need to know how strong- or weak-field ligands might affect the properties of your new complex. Predict the electron configuration of an octahedral d^5 complex with (a) strong-field ligands and (b) weak-field ligands, and state the number of unpaired electrons in each case.

ANTICIPATE (a) For strong-field ligands, the d-orbitals are widely separated in energy, and so you should expect the electrons to pair in the lower-energy orbitals, resulting in fewer unpaired electrons. (b) For weak-field ligands, the d-orbitals are all close together in energy, so you should expect the electrons to be distributed across all of them with the maximum number of unpaired electrons.

PLAN Add electrons to the orbitals in accordance with the building-up principle, to obtain the lowest energy configuration.

SOLVE

(a) Because Δ_O is large, all five electrons enter the t_{2g}-orbitals and four electrons must pair.

t_{2g}^5; 1 unpaired electron

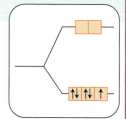

(b) Because Δ_O is small, the five electrons occupy all five orbitals and can do so without pairing.

$t_{2g}^3 e_g^2$; 5 unpaired electrons

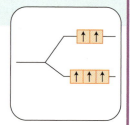

EVALUATE As expected, the number of unpaired electrons is greater for weak-field ligands than it is for strong-field ligands.

Self-test 9D.2A Predict the electron configurations of and the number of unpaired electrons found in an octahedral d^6 complex with (a) strong-field ligands and (b) weak-field ligands.

[***Answer:*** (a) t_{2g}^6, 0; (b) $t_{2g}^4 e_g^2$, 4]

Self-test 9D.2B Predict the electron configurations of and the number of unpaired electrons found in an octahedral d^7 complex with (a) strong-field ligands and (b) weak-field ligands.

Related Exercises 9D.7, 9D.8

TABLE 9D.1 lists the configurations for d^1 through d^{10} octahedral complexes, including the alternative configurations for d^4 through d^7 octahedral complexes. A d^N complex with the maximum number of unpaired spins is called a **high-spin complex.** High-spin complexes are expected for weak-field ligands because the electrons can easily occupy both the t_{2g}- and the e_g-orbitals, and the greatest number of electrons can then have parallel spins. A d^N complex with the minimum number of unpaired spins is called a **low-spin complex.** Strong-field ligands lead to a low-spin complex, because the energy required to reach the e_g-orbitals is large, and so electrons enter the t_{2g}-orbitals until they are completely full, even though they have to pair their spins. The spectrochemical series therefore allows you to make the following predictions:

- If the ligands are strong field, expect a low-spin complex.
- If the ligands are weak field, expect a high-spin complex.

Tetrahedral complexes are almost always high spin. Because ligand field splittings are smaller for tetrahedral than for octahedral complexes, even if the ligands are classified as strong-field ligands for octahedral complexes, the splitting is so small in the corresponding tetrahedral complex that the t_2-orbitals are energetically accessible.

TABLE 9D.1   The Electron Configurations of d^N Complexes

Number of d-electrons, d^N	Configuration		
	Octahedral complexes		Tetrahedral complexes
d^1	t_{2g}^1		e^1
d^2	t_{2g}^2		e^2
d^3	t_{2g}^3		$e^2 t_2^1$
	low spin	**high spin**	
d^4	t_{2g}^4	$t_{2g}^3 e_g^1$	$e^2 t_2^2$
d^5	t_{2g}^5	$t_{2g}^3 e_g^2$	$e^2 t_2^3$
d^6	t_{2g}^6	$t_{2g}^4 e_g^2$	$e^3 t_2^3$
d^7	$t_{2g}^6 e_g^1$	$t_{2g}^5 e_g^2$	$e^4 t_2^3$
d^8		$t_{2g}^6 e_g^2$	$e^4 t_2^4$
d^9		$t_{2g}^6 e_g^3$	$e^4 t_2^5$
d^{10}		$t_{2g}^6 e_g^4$	$e^4 t_2^6$

The electron configurations of d-block metal atoms and ions in complexes are obtained by applying the building-up principle to the d-orbitals, taking into account the strength of the ligand field splitting. Relative field-splitting strengths are summarized by the spectrochemical series.

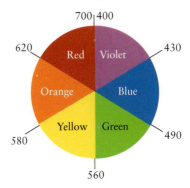

FIGURE 9D.7 In a color wheel, the color of light absorbed is opposite the color perceived. For example, a complex that absorbs orange light appears blue to the eye.

9D.3 The Colors of Complexes

White light is a mixture of all wavelengths of electromagnetic radiation from about 400 nm (violet) to about 700 nm (red). When some of these wavelengths are removed from a beam of white light when it passes through a sample, the emerging light is no longer white. For example, if red light is absorbed from white light, then the light that remains appears green. Conversely, if green is removed, then the light appears red. Red and green are said to be each other's **complementary color**—each is the color that white light appears when the other is removed (**FIG. 9D.7**).

The color wheel in the illustration can be used to suggest the wavelength range over which a complex has significant absorption (not necessarily maximum absorption). For example, if a substance looks blue (as does a copper(II) sulfate solution, for instance), then it is absorbing orange (580–620 nm) light. Conversely, from the wavelength (and therefore the color) of the light that a substance absorbs, it is possible to predict the color of the substance by noting the complementary color on the color wheel. For example, because $[Ti(OH_2)_6]^{3+}$ absorbs 510 nm light, which is yellow-green light, the complex looks violet (**FIG. 9D.8**).

FIGURE 9D.8 Because $[Ti(OH_2)_6]^{3+}$ absorbs yellow-green light (of wavelengths close to 510 nm), it looks violet in white light. *(©1994 Richard Megna–Fundamental Photographs.)*

THINKING POINT

What color would you expect a complex that absorbs violet and blue light to appear?

Because weak-field ligands generate small splittings, the complexes that they form absorb low-energy, long-wavelength radiation. The long wavelengths correspond to red light, and so these complexes exhibit colors near green. Because strong-field ligands result in large splittings, the complexes that they form should absorb high-energy, short-wavelength radiation, corresponding to the violet end of the visible spectrum. Such complexes can therefore be expected to have colors near orange and yellow (**FIG. 9D.9**).

The colors described so far arise from **d–d transitions,** in which an electron is excited from one d-orbital into another. Another source of color in d-metal complexes is the **charge-transfer transition,** in which an electron is excited from a ligand onto the metal atom or vice versa. Charge-transfer transitions are often very intense and are the most common cause of the familiar colors of d-metal complexes, such as the deep purple of permanganate ions, MnO_4^- (**FIG. 9D.10**).

The absorption of visible light by d-metal complexes can be used to measure their concentrations by using a spectrophotometer (see *Major Technique* 2 on the website of this book).

FIGURE 9D.9 The effect on the color of the complex of substituting ligands with increasing ligand field strengths in octahedral nickel(II) complexes in aqueous solution (from left to right: $[Ni(OH_2)_6^{2+}]$, $[Ni(NH_3)_6^{2+}]$, and $[Ni(en)_3^{2+}]$). *(©1994 Richard Megna–Fundamental Photographs.)*

Absorbance is sometimes called "optical density," and an older name for the molar absorption coefficient is "extinction coefficient."

At a given wavelength, the **absorbance**, A, of a solution is defined as the common (base 10) logarithm of the ratio of the incident intensity of light, I_0, to the intensity of the light transmitted through the sample, I (**FIG. 9D.11**):

$$A = \log\left(\frac{I_0}{I}\right) \tag{1}$$

The solution is contained in a transparent rectangular tube called a "cuvette." The absorbance is proportional to the path length through the solution, L, and the molar concentration of the complex, c (that is, $A \propto Lc$). The coefficient of proportionality is written ε (epsilon) and is called the **molar absorption coefficient:**

$$A = \varepsilon L c \tag{2}$$

This relation is often written in terms of the intensities themselves by substituting the definition of A and taking antilogarithms (10^x in this case) of both sides, as

$$I = I_0 10^{-\varepsilon L c} \tag{3}$$

This form of the relation is called **Beer's law.** It shows that the transmitted intensity falls off sharply with path length: doubling the path length results in a 100-fold reduction in transmitted intensity. The molar absorption coefficient is characteristic of the compound and the wavelength of the incident light.

According to Eq. 2, a plot of absorbance versus the molar concentration of the complex is a straight line with a slope εL, so ε can be determined for the wavelength being used. With ε known, the measured absorbance of a solution can be used to determine the concentration of a species in a sample. This technique, called **spectrophotometry**, is widely used in analytical chemistry.

FIGURE 9D.10 In a ligand-to-metal charge-transfer transition, an energetically excited electron migrates from a ligand to the central metal ion. This type of transition is responsible for the intense purple of the permanganate ion, MnO_4^-. *(Richard Megna/ Fundamental Photographs.)*

EXAMPLE 9D.3   Determining the formation constant for a d-metal complex

You are analyzing water samples from a local stream and want to use the intense red color of $Fe(SCN)^{2+}$ to measure the concentration of Fe^{3+}. You need to know the formation constant (Topic 6I) for this complex to determine the concentration of Fe^{3+}.

$$Fe^{3+}(aq) + SCN^-(aq) \longrightarrow Fe(SCN)^{2+}(aq) \qquad K_f = \frac{[Fe(SCN)^{2+}]}{[Fe^{3+}][SCN^-]}$$

The molar absorption coefficient, ε, for $Fe(SCN)^{2+}$ at 457 nm is 4.8×10^3 L·mol^{-1}·cm^{-1}. A solution is prepared with initial concentrations of Fe^{3+} and SCN^- of 0.30 and 0.20 mmol·L^{-1}, respectively. The resulting red solution has an absorbance at 457 nm of 0.0474 in a cuvette with path length 1.00 cm. Determine the formation constant, K_f, for $Fe(SCN)^{2+}$.

PLAN Set up an equilibrium table as described in Toolbox 5H.1, then use Beer's law to determine the equilibrium concentration of $Fe(SCN)^{2+}$. Solve for the equilibrium concentrations of Fe^{3+} and SCN^- and substitute into the equilibrium expression to determine K_f.

SOLVE

First, set up an equilibrium table with all concentrations in millimoles per liter.

	Fe^{3+}	SCN^-	$Fe(SCN)^{2+}$
Initial concentration	0.30	0.20	0
Change in concentration	$-x$	$-x$	$+x$
Equilibrium concentration	$0.30 - x$	$0.20 - x$	x

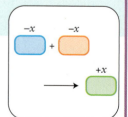

To determine the equilibrium concentration of $Fe(SCN)^{2+}$, use Eq. 2 in the form $c = A/L\varepsilon$,

$$c = \frac{0.0474}{(1.00 \text{ cm}) \times (4.8 \times 10^3 \text{ mol}^{-1}\text{·L·cm}^{-1})}$$

$$= 9.87\ldots \times 10^{-6} \text{ mol·L}^{-1} = 0.009\,87\ldots \text{ mmol·L}^{-1}$$

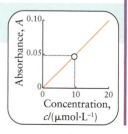

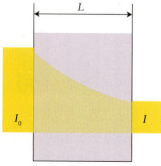

FIGURE 9D.11 The absorbance of a sample with path length L is $A = \log(I_0/I)$ where I_0 and I are the incident and transmitted intensities of light, respectively.

This value, from the equilibrium table, is also the value of x (from the third column, bottom row in the table).

From the bottom row of the equilibrium table it follows that

$$[Fe^{3+}] = 0.30 - 0.009\,87\ldots \text{ mmol·L}^{-1} = 0.29 \text{ mmol·L}^{-1}$$
$$= 2.9 \times 10^{-4} \text{ mol·L}^{-1}$$
$$[SCN^-] = 0.20 - 0.009\,87\ldots \text{ mmol·L}^{-1} = 0.19 \text{ mmol·L}^{-1}$$
$$= 1.9 \times 10^{-4} \text{ mol·L}^-$$
$$[Fe(SCN)^{2+}] = 0.009\,87\ldots \text{ mmol·L}^{-1} = 9.87\ldots \times 10^{-6} \text{ mol·L}^{-1}$$

Substitute the numerical values of the equilibrium concentrations into the equilibrium expression and evaluate K_f.

$$K_f = \frac{9.87\ldots \times 10^{-6}}{(2.9 \times 10^{-4}) \times (1.9 \times 10^{-4})} = 180$$

Self-test 9D.3A The concentrations of solutions of the purple permanganate ion are frequently determined spectrophotometrically. If a cell of path length 1.00 cm containing a solution of $KMnO_4$ has an absorbance of 0.398 at 525 nm, calculate the concentration of MnO_4^- given that the molar absorption coefficient at 525 nm is 2455 L·mol^{-1}·cm^{-1}.

[**Answer:** $[MnO_4^-] = 0.162$ mmol·L^{-1}]

Self-test 9D.3B Calculate the molar absorption coefficient of human oxyhemoglobin if a 4.15-μmol·L^{-1} solution placed in a 1.00-cm cuvette has an absorbance of 0.531 at 415 nm.

Related Exercises 9D.11, 9D.12

Transitions between d-orbitals or between the ligands and the metal atom in complexes give rise to color; the wavelength of d−d transitions can be correlated with the magnitude of the ligand field splitting. The absorbance of a compound in solution is proportional to its molar concentration. Beer's law can be used to determine the concentration of solutes.

9D.4 Magnetic Properties of Complexes

A substance with unpaired electrons is paramagnetic and is pulled into a magnetic field (Box 2G.2). A substance without unpaired electrons is diamagnetic and is pushed out of a magnetic field. Many d-metal complexes have unpaired d-electrons and are therefore paramagnetic. Because a high-spin d^N complex has more unpaired electrons than does a low-spin d^N complex, a high-spin complex is more strongly paramagnetic than a low-spin complex with the same number of d-electrons. Whether a complex is high spin or low spin depends on the ligands present. The d^4 through d^7 complexes of strong-field ligands have large energy gaps and so tend to be low spin and therefore only weakly paramagnetic or diamagnetic (**FIG. 9D.12**). The d^4 through d^7 complexes of weak-field ligands have only a small energy gap between the t_{2g}- and e_g-orbitals and therefore tend to be high spin and strongly paramagnetic.

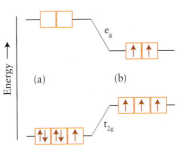

FIGURE 9D.12 (a) A strong-field ligand is likely to lead to a low-spin complex (in this case, the configuration is that of Fe^{3+}). (b) Substituting weak-field ligands is likely to result in a high-spin complex.

EXAMPLE 9D.4   Predicting the magnetic properties of a complex

You are an engineer studying materials for computer hard-drive fabrication and need to predict the magnetic properties of the iron complexes you are investigating. Compare the magnetic properties of (a) $[Fe(OH_2)_6]^{2+}$; (b) $[Fe(CN)_6]^{4-}$.

ANTICIPATE Because H_2O is a weak-field ligand, you should expect the maximum number of unpaired electrons; hence, it is likely to be high spin and thus paramagnetic. The CN^- ion is a strong-field ion and so its complexes are likely to be low spin (and possibly diamagnetic).

PLAN To predict the magnetic properties of a complex, you need to know both the number of d-electrons on the central atom or ion and the position of the ligands in the spectrochemical series. Decide from the spectrochemical series whether the ligands are weak field or strong field. Then judge whether the complex is high spin or low spin.

SOLVE

(a) Identify the number of d-electrons.

The Fe^{2+} ion is a d^6 ion.

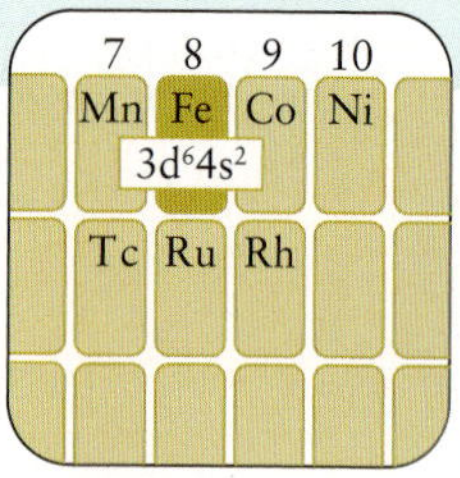

7	8	9	10
Mn	Fe	Co	Ni
	$3d^6 4s^2$		
Tc	Ru	Rh	

Classify the ligand strength.

H_2O is a weak-field ligand.

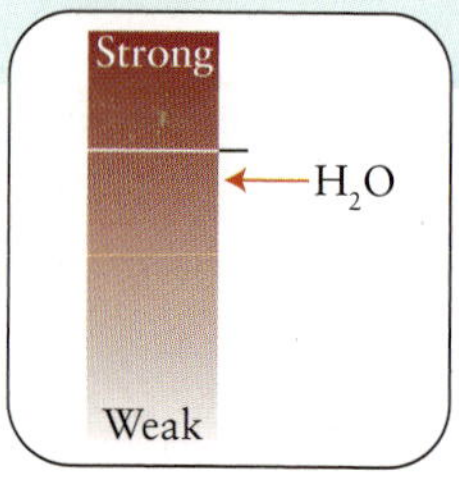

Strong

← H_2O

Weak

For a weak-field ligand, predict a high-spin $t_{2g}^4 e_g^2$ configuration.

$[Fe(OH_2)_6]^{2+}$ will have four unpaired electrons; therefore, paramagnetic.

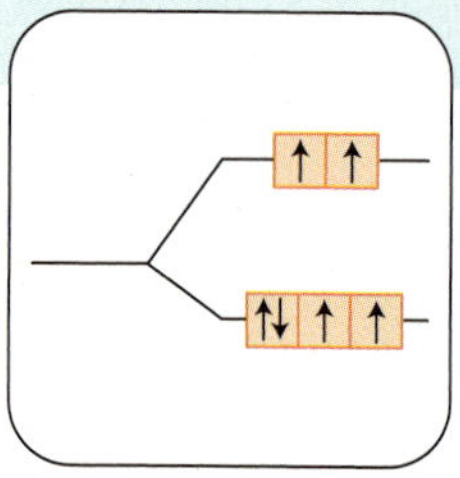

(b) Classify the strength of the ligand.

CN^- is a strong-field ligand.

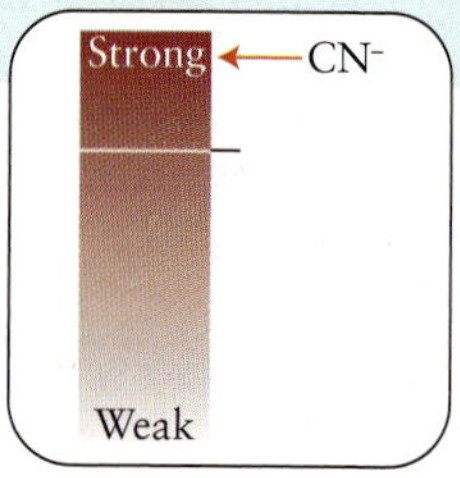

Strong ← CN^-

Weak

As noted above, the Fe^{2+} ion is a d^6 ion. For a strong-field ligand, predict a low-spin t_{2g}^6 configuration.

$[Fe(CN)_6]^{4-}$ will have no unpaired electrons; therefore, diamagnetic.

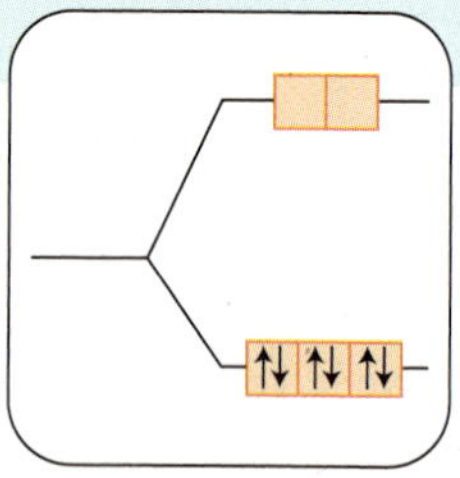

EVALUATE As expected, the aqua complex is paramagnetic and the cyanido complex diamagnetic.

Self-test 9D.4A What change in magnetic properties can be expected when NO_2^- ligands in an octahedral complex are replaced by Cl^- ligands in (a) a d^6 complex; (b) a d^3 complex?

[**Answer:** (a) The complex becomes paramagnetic; (b) there is no change in magnetic properties.]

Self-test 9D.4B Compare the magnetic properties of $[Ni(en)_3]^{2+}$ with those of $[Ni(OH_2)_6]^{2+}$.

Related Exercises 9D.13, 9D.14, 9D.21, 9D.22

The magnetic properties of a complex depend on the magnitude of the ligand field splitting. Strong-field ligands tend to form low-spin, weakly paramagnetic complexes; weak-field ligands tend to form high-spin, strongly paramagnetic complexes.

9D.5 Ligand Field Theory

Crystal field theory is based on a very primitive model of bonding. For example, ligands are not point charges: they are actual molecules or ions. The theory also leaves several questions unanswered. Why, for instance, is the electrically neutral molecule CO a strong-field ligand but the negatively charged ion Cl^- a weak-field ligand?

To improve models of bonding in complexes, chemists have turned to molecular orbital theory (Topic 2G). **Ligand field theory** describes bonding in complexes in terms of molecular orbitals built from the metal atom d-orbitals and ligand orbitals. In contrast to crystal field theory, which models the structure of the complex in terms of point charges, ligand field theory assumes, more realistically, that ligands are attached to the central metal atom or ion by covalent bonds. Much of the description of crystal field theory can be transferred into ligand field theory; the principal difference is the origin of the ligand field splitting.

To describe the electronic structure of a complex, molecular orbitals are built from the available atomic orbitals in the complex, just as if it were a molecule. For example, consider an octahedral complex of a d-metal in Period 4, such as iron, cobalt, or copper. All nine 4s-, 4p-, and 3d-orbitals of the central metal ion must be considered, because they have similar energies. To simplify the discussion initially, consider only one atomic orbital on each of the ligands. For instance, for a Cl^- ligand, use the Cl3p-orbital directed toward the metal atom; for an NH_3 ligand, use the sp^3 lone-pair orbital of the nitrogen atom. The six orbitals provided by the six ligands in an octahedral complex are represented by the tear-shaped lobes in **FIG. 9D.13**. Each of these orbitals has cylindrical symmetry around the metal–ligand axis, and so each can form a σ-orbital.

There are nine valence orbitals on the metal atom and six on the ligands, giving 15 in all. Fifteen molecular orbitals can therefore be constructed. It turns out (as explained shortly) that six are bonding, three are nonbonding, and six are antibonding. The energies of all 15 are displayed in **FIG. 9D.14**, together with the labels that they are commonly given. Notice in Fig. 9D.14 that the t_{2g}-orbitals on the metal atom have no partners among the ligands: there are no ligand orbitals to match them. Therefore, these three orbitals are nonbonding orbitals in the complex, whereas the e_g-orbitals are antibonding.

Now the building-up principle is used to work out the ground-state electron configuration, as in the discussion of diatomic molecules (Topic 2G). First, count the available electrons. In a d^N complex, N electrons are supplied by the metal; each ligand orbital (a Lewis base) supplies two electrons, and so 12 electrons are supplied by the ligands, giving $12 + N$ in all. The first 12 electrons fill the six bonding orbitals. That leaves N electrons to be accommodated in the nonbonding and antibonding orbitals. At this point, notice that the next available orbitals (those in the upper, red box in Fig. 9D.14) lie in exactly the same pattern as that encountered in crystal field theory. The only difference is that, in ligand field theory, the three t_{2g}-orbitals are recognized as nonbonding orbitals and the two e_g-orbitals as antibonding between the metal and the ligands. That is,

- The ligand field splitting can be identified with the energy separation between nonbonding and antibonding orbitals built from the d-orbitals.

The remaining four orbitals are high-energy antibonding orbitals that are ordinarily unavailable to the electrons.

From this point on, the analysis is the same as in crystal field theory. The order of filling these two sets of orbitals is exactly the same as in crystal field theory, and so is the discussion of optical and magnetic properties. If the ligand field splitting is large, then the t_{2g}-orbitals are filled first, and a low-spin complex is expected. If the ligand field splitting is small, then the e_g-orbitals are occupied before spin-pairing begins in the t_{2g}-orbitals, giving rise to a high-spin complex.

Although ligand field theory puts the discussion of bonding on a firmer basis, it has not yet explained all the puzzling features of crystal field theory. In particular, why is CO a strong-field ligand? Why is Cl^- a weak-field ligand despite its negative charge?

The model can be refined to accommodate these points by considering the effects of other ligand orbitals. The molecular orbitals just constructed considered only ligand orbitals that pointed directly at the central metal atom and formed σ-bonding and

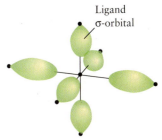

FIGURE 9D.13 The tear-shaped objects are representations of the six-ligand atomic orbitals that are used to build the molecular orbitals of an octahedral complex in ligand field theory. They might represent s- or p-orbitals on the ligands or hybrids of the two.

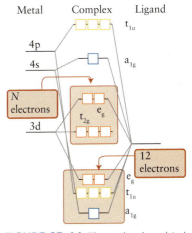

FIGURE 9D.14 The molecular orbital energy-level diagram for an octahedral complex. The 12 electrons provided by the six ligands fill the lowest six orbitals, which are all bonding orbitals. The N d-electrons provided by the central metal atom or ion are accommodated in the orbitals inside the upper box. The ligand field splitting is the energy separation of the nonbonding (t_{2g}) and antibonding (e_g) orbitals in the box.

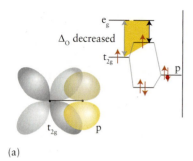

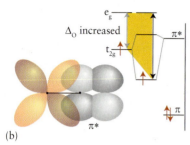

FIGURE 9D.15 The effect of π-bonding on ligand field splitting. (a) In this case, the occupied p-orbital of the ligand is close in energy to the metal t_{2g}-orbitals and they overlap to form bonding and antibonding combinations. The ligand field splitting is reduced. (b) In this case, the unoccupied antibonding π*-orbital of the ligand is close in energy to the metal t_{2g}-orbitals and they overlap to form bonding and antibonding combinations. In this case, the ligand field splitting is increased.

antibonding combinations. Ligands also have orbitals perpendicular to the metal–ligand axis, which can contribute to bonding and antibonding π-orbitals. As **FIG. 9D.15** shows, a p-orbital on the ligand perpendicular to the axis of the metal–ligand bond can overlap with one of the t_{2g}-orbitals of the metal atom to produce two new molecular orbitals, one bonding and one antibonding. The resulting bonding orbital lies below the energy of the original nonbonding t_{2g}-orbitals; the antibonding orbital lies above them.

If the ligand is Cl^- the Cl3p-orbital used to build the metal–ligand π-orbital is full. It provides two electrons, which occupy a new molecular orbital, the bonding metal–ligand combination shown in Fig. 9D.15a. The N d-electrons provided by the metal must therefore occupy the *antibonding* metal–ligand π-orbital. Because this molecular orbital is higher in energy than the original nonbonding t_{2g}-orbitals from which it is formed, the ligand field splitting is decreased by π-bonding. Therefore, Cl^- can be a weak-field ligand despite its negative charge.

Now suppose that the ligand is CO. The orbital that overlaps with the metal t_{2g}-orbitals in this case is either the full bonding π-orbital or the empty antibonding π*-orbital of the CO molecule. It turns out that the latter orbital is closer in energy to the metal orbitals, and so it plays the dominant role in bond formation to the metal, as in Fig. 9D.15b. There are no electrons from the ligand to accommodate, because its π*-orbital is empty. The N d-electrons therefore enter the *bonding* metal–ligand orbital. Because this new molecular orbital is lower in energy than the original t_{2g}-orbitals, the ligand field splitting is increased by π-bonding and CO is a strong-field ligand despite being electrically neutral.

According to ligand field theory, the ligand field splitting is the energy separation between nonbonding and antibonding molecular orbitals built principally from d-orbitals. When π-bonding is possible, the ligand field splitting is decreased if the ligand supplies π-electrons and is increased if the ligand does not supply π-electrons.

What have you learned in this Topic?

You have seen how crystal field theory, and the more sophisticated ligand field theory, can be used to account for the electron configurations of d-metal complexes and their optical and magnetic properties. You have learned that strong field ligands lead to low-spin complexes and that weak field ligands lead to high-spin complexes. You have also seen how to determine the formation constant of a d-metal complex.

The skills you have mastered are the ability to:

☐ **1.** Calculate the ligand field splitting from the wavelength of light absorbed by a complex (Example 9D.1).

☐ **2.** Use the spectrochemical series to predict the effect of a ligand on the color, electron configuration, and magnetic properties of a d-metal complex (Examples 9D.2 and 9D.4).

☐ **3.** Determine the formation constant of a d-metal complex (Example 9D.3).

☐ **4.** Describe bonding in d-metal complexes in terms of ligand field theory and account for the classification of ligands as strong field or weak field. (Section 9D.5).

Topic 9D Exercises

9D.1 Identify the number of valence electrons (including d-electrons) present in each of the following metal ions: (a) Ti^{2+}; (b) Tc^{2+}; (c) Ir^+; (d) Ag^+; (e) Y^{3+}; (f) Zn^{2+}.

9D.2 Identify the number of valence electrons (including d-electrons) present in each of the following metal ions: (a) Co^{2+}; (b) Mo^{4+}; (c) Ru^{4+}; (d) Pt^{2+}; (e) Os^{3+}; (f) V^{3+}.

9D.3 Draw an orbital energy-level diagram (like those in Figs. 9D.3 and 9D.5) showing the configuration of d-electrons on the metal ion in each of the following complexes: (a) $[Co(NH_3)_6]^{3+}$; (b) $[NiCl_4]^{2-}$ (tetrahedral); (c) $[Fe(OH_2)_6]^{3+}$; (d) $[Fe(CN)_6]^{3-}$. Predict the number of unpaired electrons for each complex.

9D.4 Draw an orbital energy-level diagram (like those in Figs. 9D.3 and 9D.5) showing the configuration of d-electrons on the metal ion in each of the following complexes: (a) $[Zn(OH_2)_6]^{2+}$; (b) $[CoCl_4]^{2-}$ (tetrahedral); (c) $[Co(CN)_6]^{3-}$; (d) $[CoF_6]^{3-}$. Predict the number of unpaired electrons for each complex.

9D.5 Solutions of the $[V(OH_2)_6]^{3+}$ ion are green and absorb light of wavelength 560 nm. What is the ligand field splitting in the complex in kilojoules per mole?

9D.6 Solutions of the $[V(OH_2)_6]^{2+}$ ion are lilac in color and absorb light of wavelength 806 nm. What is the ligand field splitting in the complex in kilojoules per mole?

9D.7 The complexes (a) $[Co(en)_3]^{3+}$ and (b) $[Mn(CN)_6]^{3-}$ have low-spin electron configurations. Predict the electron configurations and the number of unpaired electrons, if any, in each complex.

9D.8 The complexes (a) $[RhF_6]^{3-}$ and (b) $[Ni(ox)_3]^{4-}$ have high-spin electron configurations. Predict the electron configurations and the number of unpaired electrons, if any, in each complex.

9D.9 Explain the difference between a weak-field ligand and a strong-field ligand. What measurements can be used to classify them as such?

9D.10 Describe the changes that may take place in a compound's properties when weak-field ligands are replaced by strong-field ligands.

9D.11 The total concentration of cis- and trans-$[CoCl(en)_2SCN]^+$ in a solution can be studied at 540 nm, where both isomers have $\varepsilon = 175$ L·mol^{-1}·cm^{-1}. In one experiment the absorbance of a solution of the isomers in a cuvette with a path length of 1.00 cm was 0.262 at 540 nm. What was the total concentration of $[CoCl(en)_2SCN]^+$ in the solution?

9D.12 Iron in blood serum can be measured by reducing it to Fe^{2+} and reacting it with ferrozine to form $[Fe(ferrozine)_3]^{4-}$, a purple complex that has a maximum absorbance at 562 nm. The molar absorption coefficient, ε, for the complex at 562 nm is 2.79×10^4 L·mol^{-1}·cm^{-1}. A solution of the complex ion had an absorbance at 562 nm of 0.703 in a cuvette with path length 2.00 cm. Determine the molar concentration of $[Fe(ferrozine)_3]^{4-}$ in the solution.

9D.13 When the paramagnetic $[Fe(CN)_6]^{3-}$ ion is reduced to $[Fe(CN)_6]^{4-}$, the ion becomes diamagnetic. However, when the paramagnetic $[FeCl_4]^-$ ion is reduced to $[FeCl_4]^{2-}$, the ion remains paramagnetic. Explain these observations.

9D.14 When the paramagnetic $[Co(CN)_6]^{4-}$ ion is oxidized to $[Co(CN)_6]^{3-}$, the ion becomes diamagnetic. However, when the paramagnetic $[Co(ox)_3]^{4-}$ is oxidized to $[Co(ox)_3]^{3-}$, the ion remains paramagnetic. Explain these observations.

9D.15 Of the two complexes (a) $[CoF_6]^{3-}$ and (b) $[Co(en)_3]^{3+}$, one appears yellow and the other appears blue. Match the complex to the color and explain your choice.

9D.16 Which of the complexes $[Cu(OH_2)_6]^{2+}$ or $[CuBr_6]^{4-}$ absorbs at the longer wavelength?

9D.17 The complex $[Ni(NH_3)_6]^{2+}$ has a ligand field splitting of 209 kJ·mol^{-1} and forms a purple solution. What is the wavelength and color of the absorbed light?

9D.18 The complex $[TiCl_6]^{3-}$ has a ligand field splitting of 160. kJ·mol^{-1} and forms an orange solution. What is the wavelength and color of the absorbed light?

9D.19 The complex $[Co(CN)_6]^{3-}$ is pale yellow. (a) How many unpaired electrons are present in the complex? (b) If ammonia molecules replace the cyanide ions as ligands, will the wavelength of the radiation absorbed be longer or shorter?

9D.20 In aqueous solution, water competes effectively with bromide ions for coordination to Cu^{2+} ions. The hexaaquacopper(II) ion is the predominant species in solution. However, in the presence of a large concentration of bromide ions, the solution becomes deep violet. This violet color is due to the presence of the tetrabromidocuprate(II) ions, which are tetrahedral. This process is reversible, and so the solution becomes light blue again upon dilution with water. (a) Write the formulas of the two complex ions of copper(II) that form. (b) Is the change in color from violet to blue upon dilution expected? Explain your reasoning.

9D.21 Suggest a reason why Zn^{2+}(aq) ions are colorless. Would you expect zinc compounds to be paramagnetic? Explain your answer.

9D.22 Suggest a reason why copper(II) compounds are often colored but copper(I) compounds are colorless. Which oxidation number results in paramagnetic compounds?

9D.23 (a) Estimate the ligand field splitting for (i) $[CrCl_6]^{3-}$ ($\lambda_{max} = 740.$ nm), (ii) $[Cr(NH_3)_6]^{3+}$ ($\lambda_{max} = 460.$ nm), and (iii) $[Cr(OH_2)_6]^{3+}$ ($\lambda_{max} = 575$ nm), where λ_{max} is the wavelength of the most intensely absorbed light. (b) Arrange the ligands in order of increasing ligand field strength.

9D.24 (a) Estimate the ligand field splitting for (i) $[CoF_6]^{3-}$ ($\lambda_{max} = 700.$ nm), (ii) $[Co(NH_3)_6]^{3+}$ ($\lambda_{max} = 435$ nm), and (iii) $[Co(OH_2)_6]^{3+}$ ($\lambda_{max} = 540.$ nm), where λ_{max} is the wavelength of the most intensely absorbed light. (b) Arrange the ligands in order of increasing ligand field strength.

9D.25 Which d-orbitals on the metal ion are used to form σ-bonds between metal ions and ligands in an octahedral complex?

9D.26 Which d-orbitals on the metal ion are used to form π-bonds between metal ions and ligands in an octahedral complex?

9D.27 Ligands that can interact with a metal center by forming a π-bond are commonly called π-acids and π-bases. The definitions of π-acid and π-base are similar to those used for Lewis acidity. A π-acid is a ligand that can accept electrons by using π-orbitals, and a π-base donates electrons by using π-orbitals. (a) On the basis of these definitions, describe each of the following ligands as either a π-acid, a π-base, or neither: (i) CN^-; (ii) Cl^-; (iii) H_2O; (iv) en. (b) Place them in order of increasing ligand field splitting.

9D.28 (a) Describe each of the following ligands as either a π-acid, a π-base, or neither (see Exercise 9D.27): (i) NH_3; (ii) ox; (iii) F^-; (iv) CO. (b) Place them in order of increasing ligand field splitting.

9D.29 Is the best description of the t_{2g}-orbitals in $[Co(OH_2)_6]^{3+}$ bonding, antibonding, or nonbonding? Explain how you reached your conclusion.

9D.30 Is the best description of the t_{2g}-orbitals in $[Fe(CN)_6]^{3-}$ bonding, antibonding, or nonbonding? Explain how you reached your conclusion.

9D.31 Is the best description of the e_g-orbitals in $[Fe(OH_2)_6]^{3+}$ bonding, antibonding, or nonbonding? Explain how you reached your conclusion.

9D.32 Is the best description of the t_{2g}-orbitals in $[CoF_6]^{3-}$ bonding, antibonding, or nonbonding? Explain how you reached your conclusion.

9D.33 Using concepts of ligand field theory, explain why water is a weaker field ligand than ammonia.

9D.34 Using concepts of ligand field theory, explain why ethylenediamine is a weaker field ligand than CO.

The following Example and Exercises draw on material from throughout Focus 9.

FOCUS 9 Online Cumulative Example

You are an inorganic chemist interested in the impact of various ligands on the properties of transition metal complexes. To explore changes systematically, you have prepared a series of octahedrally coordinated cobalt(III) complexes of the form $[Co(NH_3)_5X]^{n+}$, where X = Cl^-, NH_3, H_2O, $\underline{N}O_2^-$, and $N\underline{O}_2^-$. Aqueous solutions of each compound were prepared and the wavelength of maximum absorbance, λ_{max}, is given below.

Complex	λ_{max}/nm
$[CoCl(NH_3)_5]^{2+}$	530
$[Co(NH_3)_6]^{3+}$	475
$[Co(NH_3)_5OH_2]^{3+}$	495
$[Co(NH_3)_5\underline{O}NO]^{2+}$	485
$[Co(NH_3)_5\underline{N}O_2]^{2+}$	460

(a) Name the $[Co(NH_3)_5\underline{N}O_2]^{2+}$ ion. (b) Calculate the ligand field splitting energy for $[Co(NH_3)_5OH_2]^{2+}$. (c) Predict the color of an aqueous solution of $[Co(NH_3)_5\underline{O}NO]^{2+}$. (d) From the information provided, place the ligands (Cl^-, NH_3, H_2O, $\underline{N}O_2^-$, and $N\underline{O}_2^-$) in order of increasing ligand field splitting. (e) Draw the crystal field splitting diagram for $[Co(NH_3)_6]^{3+}$ and indicate if the complex is high or low spin and if it is diamagnetic or paramagnetic. (f) Which ions in this study may have a significant impact on the pH of an aqueous solution?

 The Online Cumulative Example solution can be found at http://macmillanhighered.com/chemicalprinciples7e

FOCUS 9 Exercises

9.1 The compound $Cr(OH)_3$ is very insoluble in water; therefore, electrochemical methods must be used to determine its K_{sp}. Given that the reduction of $Cr(OH)_3(s)$ to $Cr(s)$ and hydroxide ions has a standard potential of -1.34 V, calculate the solubility product of $Cr(OH)_3$.

9.2 The complex ion $[Ni(NH_3)_6]^{2+}$ forms in a solution containing 0.16 mol·L^{-1} $NH_3(aq)$ and 0.015 mol·L^{-1} $Ni^{2+}(aq)$. If the formation constant of $[Ni(NH_3)_6]^{2+}$ is 1.0×10^9, what are the equilibrium molar concentrations?

9.3 (a) Draw all the possible isomers of the square planar complex $PtBrCl(NH_3)_2$ and name each isomer. (b) How can the existence of these isomers be used to show that the complex is square planar rather than tetrahedral?

9.4 Draw the structures of all possible isomeric forms of $CrBrClI(NH_3)_3$. Which isomers are chiral?

9.5 Suggest a chemical test for distinguishing between (a) $[Ni(SO_4)(en)_2]Cl_2$ and $[NiCl_2(en)_2]SO_4$; (b) $[NiI_2(en)_2]Cl_2$ and $[NiCl_2(en)_2]I_2$.

9.6 Trigonal bipyramidal complexes in which a metal ion is surrounded by five ligands are rarer than octahedral or tetrahedral complexes, but many are known. Will trigonal bipyramidal compounds of the formula MX_3Y_2 exhibit isomerism? If so, what types of isomerism are possible?

9.7 (a) Sketch the orbital energy-level diagrams for $[MnCl_6]^{4-}$ and $[Mn(CN)_6]^{4-}$. (b) How many unpaired electrons are present in each complex? (c) Which complex absorbs the longer wavelengths of incident electromagnetic radiation? Explain your reasoning.

9.8 (a) Which types of ligands are π-acid ligands in general (see Exercise 9D.27), strong field or weak field? (b) Which types of ligands are π-base ligands in general, strong field or weak field?

9.9 The trigonal prismatic structure shown in the illustration was once proposed for the complex $CoCl_2(NH_3)_4$. Use the fact that only two isomers of the complex are known to rule out the prismatic structure.

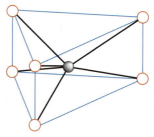

9.10 A hexagonal planar structure was once proposed for the $CrCl_3(NH_3)_3$ complex. Use the fact that only two isomers of the complex are known to rule out the hexagonal planar structure.

9.11 Before the structures of octahedral complexes were established, various means were used to explain the fact that d-metal ions could bind to a greater number of ligands than expected on the basis of their charges. For example, Co^{3+} can bind to six ligands, not just three. An early theory attempted to explain this behavior by postulating that once three ligands had attached to a metal ion of charge $+3$, the others bonded to the attached ligands, forming chains of ligands. Thus, the compound $[Co(NH_3)_6]Cl_3$, which we now know to have an octahedral structure, would have been described as $Co(NH_3-NH_3-Cl)_3$. Show how the chain theory is not consistent with at least two properties of coordination compounds.

9.12 How many chelate rings are present in (a) $[Ru(ox)_3]^{3-}$; (b) $[Fe(trien)]^{3+}$ (trien is triethylenetetramine); (c) $[Cu(dien)_2]^{2+}$?

trien

dien

9.13 Suggest the form that the orbital energy-level diagram would take for a square planar complex with the ligands in the *xy* plane, and discuss how the building-up principle applies. *Hint:* The d_{z^2}-orbital has a greater electron density in the *xy*-plane than the d_{zx}- or d_{yz}-orbitals but less than the d_{xy}-orbital.

9.14 Nickel(II) complexes with weak-field ligands such as the bromide ion have been found to be octahedral, but nickel(II) complexes with strong-field ligands have been found to be square planar. Explain these findings. (See Exercise 9.13.)

9.15 In a nickel-containing enzyme, various groups of atoms in the enzyme form a complex with the metal, which was found to have oxidation number $+2$ and to have no unpaired electrons. What is the most probable geometry of the Ni^{2+} complex: (a) octahedral; (b) tetrahedral; (c) square planar (see Exercise 9.13)? Justify your answer by drawing the orbital energy-level diagram of the ion.

9.16 Is there a correlation between the ligand field strength of the halide ions F^-, Cl^-, Br^-, and I^- and the electronegativity of the halogen? If so, can this correlation be explained by ligand field theory? Justify your answer.

9.17 A solid pink salt has the formula $CoCl_3 \cdot 5NH_3 \cdot H_2O$. When $AgNO_3$ is added to the pink solution containing 0.0010 mol $CoCl_3 \cdot 5NH_3 \cdot H_2O$, 0.43 g of $AgCl$ precipitates. Heating the salt produces a purple solid with the formula $CoCl_3 \cdot 5NH_3$. (a) Write the correct formula and name of the pink salt. (b) When $AgNO_3$ is added to a solution containing 0.0010 mol $CoCl_3 \cdot 5NH_3$, a purple salt, 0.29 g of $AgCl$ precipitates. Write the correct formula and name of the purple salt.

9.18 Two chemists prepared a complex and determined its formula, which they wrote as $[CrCl_3NH_3] \cdot 2H_2O$. However, when they dissolved 2.11 g of the compound in water and added an excess of silver nitrate, 2.87 g of $AgCl$ precipitated and they realized that the formula was incorrect. Write the correct formula of the compound and draw its structure, including all possible isomers.

9.19 Molecular models of the iridium complex $IrCl(CO)$ $[P(C_6H_5)_3]_2$ (ChemSpider ID 21106488) can be found on the Internet. (a) View the three-dimensional structure and determine the coordination geometry of the iridium atom. (b) What is the

oxidation number of the iridium atom? (c) Are isomers of this compound possible? If so, draw them. Label any that are optical isomers.

9.20 *cis*-Platin is an anticancer drug with a three-dimensional structure that can be viewed on the Internet. (a) What are the formula and systematic name for the compound *cis*-platin? (b) Draw any isomers that are possible for this compound. Label any isomers that are optically active. (c) What is the coordination geometry of the platinum atom?

9.21 The concentration of Fe^{2+} ions in an acid solution can be determined by a redox titration with either $KMnO_4$ or $K_2Cr_2O_7$. The reduction products of these reactions are Mn^{2+} and Cr^{3+}, and in each case the iron is oxidized to Fe^{3+}. In one titration of an acidified Fe^{2+} solution, 25.20 mL of 0.0210 M $K_2Cr_2O_7(aq)$ was required for complete reaction. If the titration had been carried out with 0.0420 M $KMnO_4(aq)$, what volume of the permanganate solution would have been required for complete reaction?

9.22 Ligand field theory predicts that different types of metal ions will form more stable complexes with certain types of ligands. From your understanding of ligand field theory, predict what types of ligands (weak field or strong field) would form the more stable complexes with the early d-metals in their highest oxidation states. Likewise, predict the types of ligands that would form the more stable complexes of the late d-block metals (those to the right of the block) in their lowest oxidation states. Explain the reasoning for your choices.

9.23 An analytical chemistry team analyzed a green mineral salt found in the glaze of an ancient pot. When they heated the mineral, it gave off a colorless gas that turned limewater milky white. When they dissolved the mineral in sulfuric acid, the same colorless gas was released and a blue solution formed. Suggest a possible formula for the compound and justify your conclusion.

9.24 An analytical chemist analyzed a yellow mineral salt found in the pigments used by a Renaissance artist. When hydrochloric acid was added, the mineral dissolved and formed an orange solution. The original salt was dissolved in water to form a yellow solution. When barium chloride was added to the solution, a yellow solid formed. The solid was filtered and the water evaporated from the filtrate. The colorless chloride remaining was subjected to a flame test. It gave a bright yellow color to the flame. Suggest a possible formula for the compound and justify your conclusion.

9.25 Vanadium is used in steelmaking because it forms V_4C_3, which increases the strength of the steel and its resistance to wear. It is formed in the steel by adding V_2O_5 to iron ore as it is reduced by carbon during the refining process. (a) What is the likely oxidation state of vanadium in V_4C_3? (b) Write a balanced equation for the reaction of V_2O_5 with carbon to form V_4C_3 and CO_2.

FOCUS 9 Cumulative Exercise

9.26 Hemoglobin contains one heme group per subunit. The heme group is an Fe^{2+} complex that is coordinated to the four N atoms in a porphyrin ligand in a square planar arrangement and to an N atom on the histidine residue (see Table 11E.3) on the subunit. The hemoglobin molecule transports O_2 through the body by using a bond between O_2 and the Fe^{2+} ion at the center of the heme group. The Fe^{2+} ion can be viewed as having octahedral coordination, with one position unoccupied and available for bonding to oxygen. (See Box 9C.1.)

(a) Draw the structure of the heme group with the sixth site not coordinated.

(b) The deoxygenated heme is a high-spin complex of the Fe^{2+} ion. When the oxygen molecule binds to the Fe^{2+} ion as the sixth ligand, the resulting octahedral complex is low spin. Predict the number of unpaired electrons in (i) deoxygenated and (ii) oxygenated heme.

(c) Other compounds can bond to the iron atom, displacing oxygen. Identify which of the following species cannot bond to the iron in a heme group and explain your reasoning: CO; Cl^-; BF_3; NO_2^-.

(d) Hemoglobin in both its oxygenated (HbO_2) and deoxygenated (Hb) forms helps to maintain the pH of blood at an optimal level. Hemoglobin has several acidic protons, but the most important deprotonation is the first. At 25 °C, for the deoxygenated form, $HbH(aq) + H_2O(l) \rightleftharpoons Hb^-(aq) + H_3O^+(aq)$

and $pK_{a1} = 6.62$. For the oxygenated form, $HbO_2H(aq) + H_2O(l) \rightleftharpoons HbO_2^-(aq) + H_3O^+(aq)$ and $pK_{a1} = 8.18$. Calculate the percentage of each form of hemoglobin that is deprotonated at the pH of blood, 7.4.

(e) Use your answer to part (d) to determine how the oxygenation of hemoglobin affects the pH of blood. That is, will the pH increase, decrease, or stay the same as more hemoglobin molecules become oxygenated? Explain your reasoning.

NUCLEAR CHEMISTRY

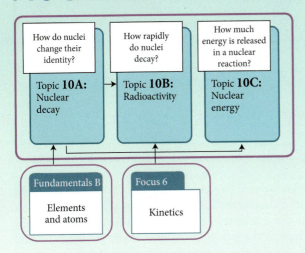

In much of chemistry, the nucleus is an inert passenger, carrying with it the electrons that are responsible for physical and chemical change. However, a nucleus does have an internal structure and may undergo change. The forces that bind nucleons together are so great that nuclear changes can release huge amounts of energy. Chemists can exploit these changes and are deeply involved in the consequences of using them to generate electrical power.

TOPIC 10A describes the types of radiation that are emitted when nucleons adjust their relative positions within nuclei or when the nucleus emits particles. This radiation was originally a great puzzle, but it has given considerable insight into the composition of nuclei and has found extensive use in medicine and energy production. TOPIC 10B explains how the intensity of emitted radiation is reported, how it varies with time, and how that information is used in practical applications. TOPIC 10C takes the story further by assessing the energy locked up in nuclei and released when they undergo fission and fusion. It concludes with an outline of how that energy is captured and how chemists are solving the resulting environmental problems.

Topic 10A Nuclear Decay

How do nuclei change their identity?

How rapidly do nuclei decay?

Topic **10A**: Nuclear decay → Topic **10B**: Radioactivity

Why Do You Need to Know This Material? Many of the problems associated with nuclear power can be solved with chemistry, and you need to know how nuclei change if you are to contribute to their solution.

What Do You Need to Know Already? This Topic builds on the brief description of the atomic nucleus and subatomic particles in *Fundamentals* B.

Atomic nuclei are extraordinary particles. They contain all the protons in the atom crammed together in a tiny volume, despite their positive charges (**FIG. 10A.1**). Most nuclei survive indefinitely despite the immense repulsive forces between their protons because the neutrons also contribute to the "strong force" that binds the nucleons (the protons and neutrons) together. In some nuclei, though, the repulsions that protons exert on one another overcome that strong force. Fragments of the nucleus are then ejected, and the nucleus is said to "decay."

10A.1 The Evidence for Spontaneous Nuclear Decay

In 1896, the French scientist Henri Becquerel happened to store a lump of uranium oxide in a drawer that contained some photographic plates (**FIG. 10A.2**). He was astonished to find that the plates had become darkened in certain areas, even though they were covered with an opaque material. Becquerel realized that the uranium compound must give off some kind of radiation. Marie Sklodowska Curie (**FIG. 10A.3**), a young Polish doctoral student, showed that the radiation, which she called **radioactivity,** was emitted by uranium regardless of the compound in which it was found. She concluded that the source must be the uranium atoms themselves. Together with her husband, Pierre, she went on to show that thorium, radium, and polonium are also radioactive.

The origin of radioactivity was initially a mystery because the existence of the atomic nucleus was unknown at the time. However, in 1898, Ernest Rutherford took the first step to discover its origin when he identified three different types of radioactivity by observing the effect of electric fields on radioactive emissions (**FIG. 10A.4**). Rutherford called the three types α (alpha), β (beta), and γ (gamma) radiation.

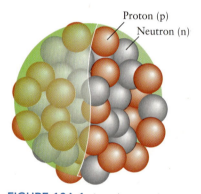

Proton (p)

Neutron (n)

FIGURE 10A.1 A nucleus can be represented as a cluster of tightly bonded protons (red) and neutrons (gray). The diameter of a nucleus is about 10 fm (1 fm = 10^{-15} m).

Marie Curie's work with radioactivity won her two Nobel Prizes, one shared with her husband, the French physicist Pierre Curie, and Becquerel.

FIGURE 10A.2 Henri Becquerel discovered radioactivity when he noticed that an unexposed photographic plate left near some uranium oxide became fogged. This photograph shows one of his original plates annotated with his record of the event. *(Granger, NYC)*

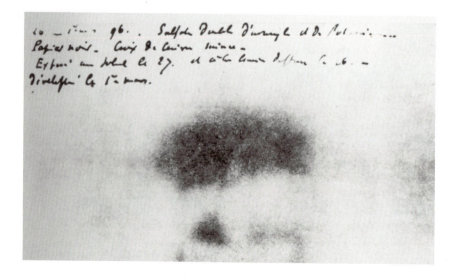

FIGURE 10A.3 Marie Sklodowska Curie (1867–1934). (Granger, NYC)

FIGURE 10A.4 The effects of an electric field on nuclear radiation. The direction of deflection shows that α particles are positively charged, β particles are negatively charged, and γ-rays are uncharged.

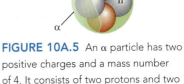

FIGURE 10A.5 An α particle has two positive charges and a mass number of 4. It consists of two protons and two neutrons and is the same as the nucleus of a helium-4 atom.

When Rutherford allowed the radiation to pass between two electrically charged electrodes, he found that one type was attracted to the negatively charged electrode. He proposed that the radiation attracted to the negative electrode consists of positively charged particles, which he called **α particles.** Once Rutherford had identified the atomic nucleus (in 1908, *Fundamentals* B), he realized that an α particle must be a helium nucleus, He^{2+}. An α particle is denoted $_2^4\alpha$, or simply α. You can think of it as a tightly bound cluster of two protons and two neutrons (**FIG. 10A.5**).

Rutherford found that a second type of radiation was attracted to the positively charged electrode. He proposed that this type of radiation consists of a stream of negatively charged particles. By measuring the charge and mass of these particles, he showed that they are electrons. The rapidly moving electrons emitted by nuclei are called **β particles** and denoted β⁻. Because a β particle has no protons or neutrons, its mass number is 0 and it can be written $_{-1}^{0}e$.

A Note on Good Practice: An electron emitted or captured by a nucleus does not have an atomic number, so its charge (−1) is written as a subscript preceding the symbol. You will see later that this is a convenient bookkeeping device.

The third common type of radiation that Rutherford identified, **γ radiation,** is not affected by an electrical field. Like visible light, γ radiation is electromagnetic radiation but of much higher frequency—greater than about 10^{20} Hz and corresponding to wavelengths less than about 1 pm. It can be regarded as a stream of very-high-energy photons, each photon being emitted by a single nucleus as that nucleus discards energy. The frequencies, ν, of γ-rays are related to the energy discarded by the nucleus, ΔE, by $\nu = \Delta E/h$ (Topic 1A). The frequency is very high because the energy difference between the excited and the ground nuclear states is very large. Both α and β radiation are often accompanied by γ radiation: the new nucleus may be formed with its nucleons in a high-energy arrangement, and a γ-ray photon is emitted when nucleons settle down into a state of lower energy (**FIG. 10A.6**).

It is important to distinguish between β emission (electrons emitted from the nucleus) and ionization (electrons removed from the valence shell).

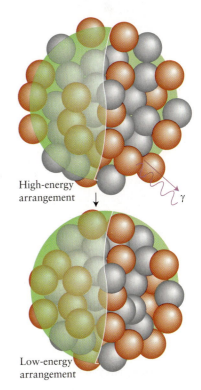

High-energy arrangement

Low-energy arrangement

FIGURE 10A.6 After a nucleus decays, the nucleons remaining in the nucleus may be left in a high-energy state, as shown by the expanded arrangement in the upper part of the illustration. As the nucleons adjust to a lower-energy arrangement (bottom), the excess energy is released as a γ-ray photon.

TABLE 10A.1   Nuclear Radiation

Type	Degree of penetration	Speed†	Particle‡	Mass number	Charge	Example
α	not penetrating but damaging	10% of c	helium-4 nucleus, $^4_2\text{He}^{2+}$, $^4_2\alpha$, α	4	+2	$^{226}_{88}\text{Ra} \longrightarrow {}^{222}_{86}\text{Rn} + \alpha$ (Fig.10A.7)
β	moderately penetrating	<90% of c	electron, $^{\ 0}_{-1}\text{e}$, β⁻, β, e⁻	0	−1	$^3_1\text{H} \longrightarrow {}^3_2\text{He} + {}^{\ 0}_{-1}\text{e}$ (Fig. 10A.8)
electron capture§	—	—	electron, $^{\ 0}_{-1}\text{e}$, e⁻	0	−1	$^{44}_{22}\text{Ti} + {}^{\ 0}_{-1}\text{e} \longrightarrow {}^{44}_{21}\text{Sc}$ (Fig. 10A.9)
γ	very penetrating; often accompanies other radiation	c	photon, γ	0	0	$^{60}_{27}\text{Co}^\star \longrightarrow {}^{60}_{27}\text{Co} + \gamma^{\|}$ (Fig. 10A.6)
β⁺	moderately penetrating	<90% of c	positron, $^0_{+1}\text{e}$, β⁺	0	+1	$^{22}_{11}\text{Na} \longrightarrow {}^{22}_{10}\text{Ne} + {}^0_{+1}\text{e}$ (Fig.10A.10)
p	moderate or low penetration	10% of c	proton, $^1_1\text{H}^+$, ^1_1p, p	1	+1	$^{53}_{27}\text{Co} \longrightarrow {}^{52}_{26}\text{Fe} + {}^1_1\text{p}$
n	very penetrating	<10% of c	neutron, ^1_0n, n	1	0	$^{137}_{53}\text{I} \longrightarrow {}^{136}_{53}\text{I} + {}^1_0\text{n}$

†c is the speed of light.
‡Alternative symbols are given for the particles; often it is sufficient to use the simplest (the one on the right).
§Electron capture is not nuclear radiation but is included for completeness.
‖An energetically excited state of a nucleus is usually denoted by an asterisk (*).

Since Rutherford's work, scientists have identified other types of nuclear radiation. Some consist of rapidly moving particles, such as neutrons or protons. Others consist of rapidly moving **antiparticles,** particles with a mass equal to that of one of the subatomic particles but with an opposite charge. For example, the **positron** has the same mass as an electron but a positive charge; it is denoted β⁺ or $^0_{+1}\text{e}$. When an antiparticle encounters its corresponding particle, both particles are annihilated and completely converted into energy. In **electron capture,** an electron in an atomic orbital is captured by the nucleus and a proton is converted into a neutron. **TABLE 10A.1** summarizes the properties of particles commonly found in nuclear radiation.

THINKING POINT

How would you write the symbol for an antiproton? What name would you give it?

The most common types of radiation emitted by radioactive nuclei are α particles (the nuclei of helium atoms), β particles (fast electrons ejected from the nucleus), and γ-rays (high-frequency electromagnetic radiation).

10A.2 Nuclear Reactions

The discoveries of Becquerel, Curie, and Rutherford and Rutherford's later development of the nuclear model of the atom showed that radioactivity is produced by **nuclear decay,** the partial breakup of a nucleus. A change in the composition of a nucleus is called a **nuclear reaction.** A specific atom with a given atomic number (Z, the number of protons in its nucleus) and mass number (A, the total number of nucleons) is called a **nuclide.** Thus, ^1H, ^2H, and ^{16}O are three different nuclides, the first two being isotopes of the same element. Nuclei that change their structure spontaneously and emit radiation are called **radioactive.** Often the result is a different nuclide.

Nuclear reactions differ in some important ways from chemical reactions. First, different isotopes of the same element undergo essentially the same chemical reactions, but their nuclei undergo very different nuclear reactions. Second, some nuclear reactions, such as those in which α or β particles are emitted from the nucleus, leave behind a nucleus with a different number of protons. The product, which is called the **daughter nucleus** (**FIG. 10A.7** and **FIG. 10A.8**), is therefore the nucleus of an atom of a different

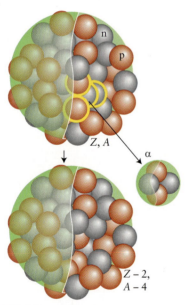

FIGURE 10A.7 When a nucleus ejects an α particle, the atomic number of the nuclide decreases by 2 and the mass number decreases by 4. The nucleons ejected from the upper nucleus are indicated by the gold boundary.

The term "nuclide" originally denoted the nucleus of an atom, but it now denotes the entire atom.

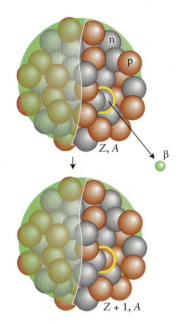

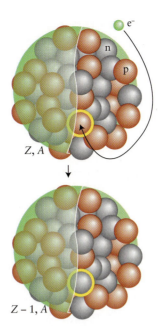

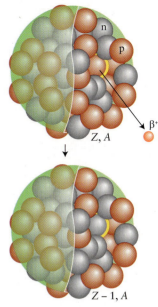

FIGURE 10A.8 When a nucleus ejects a β particle, the atomic number of the nuclide increases by 1 and the mass number remains unchanged. The neutron that we can regard as the source of the electron is indicated by the gold boundary in the upper nucleus.

FIGURE 10A.9 In electron capture, a nucleus captures one of the surrounding electrons. The effect is to convert a proton (outlined in gold, above) into a neutron (outlined in gold, below). As a result, the atomic number of the nuclide decreases by 1 but the mass number remains the same.

FIGURE 10A.10 In positron (β⁺) emission, the nucleus ejects a positron. The effect is to convert a proton into a neutron. As a result, the atomic number of the nuclide decreases by 1 but the mass number remains the same.

element. For example, when a sodium-24 nucleus emits a β particle, a magnesium-24 nucleus is formed. In this case, a **nuclear transmutation,** the conversion of one element into another, has taken place. Another important difference between nuclear and chemical reactions is that energy changes are very much greater for nuclear reactions than for chemical reactions. For example, the combustion of 1.0 g of methane produces about 52 kJ of energy as heat. In contrast, a nuclear reaction of 1.0 g of uranium-235 produces about 8.2×10^7 kJ of energy, more than a million times as much.

To predict the identity of a daughter nucleus, you need to note how the atomic number and mass number change when the parent nucleus ejects a particle. For example, when a radium-226 nucleus, with $Z = 88$, undergoes α decay, it emits an α particle, which has a nuclear charge of +2 and a mass number of 4. Because both total mass number and total nuclear charge are conserved in a nuclear reaction, the fragment remaining must be a nucleus of atomic number 86 (radon) and mass number 222; so the daughter nucleus is radon-222:

$$^{226}_{88}\text{Ra} \longrightarrow \, ^{222}_{86}\text{Rn} + \, ^4_2\alpha$$

The expression for these changes is called a **nuclear equation.** The following example shows how to use nuclear equations to identify daughter nuclei.

EXAMPLE 10A.1   Predicting the outcome of electron capture and positron emission

When radioactive nuclides are used in medicine and industry, it is important to know what daughter nuclide to expect, as that nuclide might also be radioactive. Suppose you are working in a medical nuclear research laboratory and are studying calcium-41 and oxygen-15. What nuclide is produced when (a) calcium-41 undergoes electron capture; (b) oxygen-15 undergoes positron emission?

ANTICIPATE   (a) In electron capture, a proton is turned into a neutron and, although there is no change in mass number, the atomic number is reduced by 1 (**FIG. 10A.9**). (b) A positron has the same tiny mass as an electron but a single positive charge. Positron emission can be thought of as the positive charge emitted by a proton as it is converted into a neutron. As a result, the atomic number decreases by 1 but there is no change in mass number (**FIG. 10A.10**).

PLAN Write the nuclear equation for each reaction, representing the daughter nuclide as E, with atomic number Z and mass number A. Then find Z and A from the requirement that both mass number and atomic number are conserved in a nuclear reaction.

SOLVE

(a) Write the equation for the nuclear reaction.

$$^{41}_{20}\text{Ca} + ^{\ 0}_{-1}\text{e} \longrightarrow ^{A}_{Z}\text{E}$$

Express mass and charge conservation.

$$41 + 0 = A \quad \text{or} \quad A = 41, \text{unchanged}$$
$$20 - 1 = Z \quad \text{or} \quad Z = 19$$

Identify the element.

$$Z = 19 \text{ corresponds to K}$$

Write the nuclear equation.

$$^{41}_{20}\text{Ca} + ^{\ 0}_{-1}\text{e} \longrightarrow ^{41}_{19}\text{K}$$

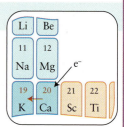

(b) Write the equation for the nuclear reaction.

$$^{15}_{8}\text{O} \longrightarrow ^{A}_{Z}\text{E} + ^{\ 0}_{+1}\text{e}$$

Express mass and charge conservation.

$$15 = A + 0 \quad \text{or} \quad A = 15, \text{unchanged}$$
$$8 = Z - 1 \quad \text{or} \quad Z = 7$$

Identify the element.

$$Z = 7 \text{ corresponds to N}$$

Write the nuclear equation.

$$^{15}_{8}\text{O} \longrightarrow ^{15}_{7}\text{N} + ^{\ 0}_{+1}\text{e}$$

EVALUATE As expected, in both electron capture and positron emission a proton has been converted into a neutron, so there was a decrease in atomic number but no change in mass.

Self-test 10A.1A Identify the nuclide produced and write the nuclear equation for (a) electron capture by beryllium-7; (b) positron emission by sodium-22.

[*Answer:* (a) $^{7}_{4}\text{Be} + ^{\ 0}_{-1}\text{e} \rightarrow ^{7}_{3}\text{Li}$; (b) $^{22}_{11}\text{Na} \rightarrow ^{22}_{10}\text{Ne} + ^{\ 0}_{+1}\text{e}$]

Self-test 10A.1B Identify the nuclide produced and write the nuclear equation for (a) electron capture by iron-55; (b) positron emission by carbon-11.

Related Exercises 10A.5–10A.8.

Nuclear reactions may result in the formation of different elements. The transmutation of a nucleus can be predicted by noting the atomic numbers and the mass numbers in the nuclear equation for the process.

10A.3 The Pattern of Nuclear Stability

The nuclei of some elements are stable, but others decay the moment they are formed. It would be useful to know if there is a pattern to the stabilities and instabilities of nuclei, for then you could make predictions about the modes of nuclear decay. One clue is that elements with even atomic numbers are consistently more abundant than neighboring elements with odd atomic numbers. This difference can be seen in **FIG. 10A.11**, which is a plot of the cosmic abundance of the elements against atomic number. The same pattern occurs on Earth. Of the eight elements present as 1% or more of the mass of the Earth, only one, aluminum, has an odd atomic number.

Nuclei with even numbers of both protons and neutrons are more stable than those with any other combination. The least stable nuclei are nuclei with odd numbers of both protons and neutrons (**FIG. 10A.12**). Nuclei are more likely to be stable if they are built from certain numbers of either kind of nucleons. These numbers—namely, 2, 8, 20, 50, 82, 114, 126, and 184—are called **magic numbers.** For example, there are ten stable isotopes of tin ($Z = 50$), the most of any element, but only two stable isotopes of its neighbor, antimony ($Z = 51$). The α particle itself is a "doubly magic" nucleus, with two protons and two neutrons. This pattern of nuclear stability is similar to the pattern of electron stability in atoms: the noble gas atoms have 2, 10, 18, 36, 54, and 86 electrons around their nuclei.

FIGURE 10A.13 is a plot of mass number against atomic number for known nuclides. Stable nuclei are found in a **band of stability** surrounded by a **sea of instability,** the region of unstable nuclides that decay spontaneously with the emission of radiation. For atomic numbers up to about 20, the stable nuclides have approximately equal numbers of neutrons and protons, and so A is close to $2Z$. For higher atomic numbers, all common nuclides—both stable and unstable—have more neutrons than protons, and so $A > 2Z$.

The increase in the ratio of neutrons to protons with increasing atomic number can be explained by considering the role of the neutrons in helping to overcome the electrical

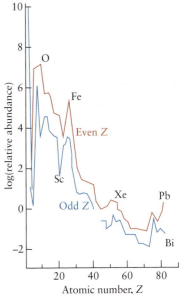

FIGURE 10A.11 The variation in cosmic nuclear abundance with atomic number. Note that elements with even atomic numbers (red line) are consistently more abundant than neighboring elements with odd atomic numbers (blue line).

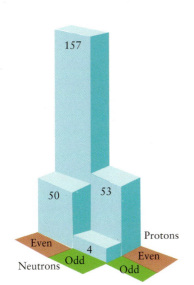

FIGURE 10A.12 The numbers of stable nuclides having even or odd numbers of neutrons and protons. With the exception of hydrogen, stable nuclides are much more likely to have even numbers of both protons and neutrons. Only four stable nuclides have odd numbers of both protons and neutrons.

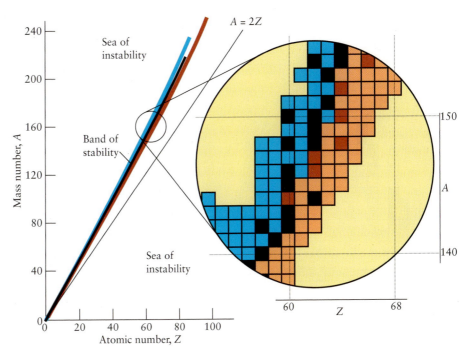

FIGURE 10A.13 The manner in which nuclear stability depends on the atomic number and the mass number. Nuclides along the narrow black band (the band of stability) are generally stable. Nuclides in the blue region are likely to emit a β particle, and those in the red region are likely to emit an α particle. Nuclei in the orange region are likely either to emit positrons or to undergo electron capture. The straight line indicates the position that nuclides would have if the numbers of neutrons and protons were equal ($A = 2Z$). The inset shows a magnified view of the diagram near $Z = 60$.

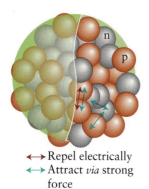

←→ Repel electrically
←→ Attract *via* strong
force

FIGURE 10A.14 The protons in a nucleus repel one another electrically, but the strong force, which acts between all nucleons, holds the nucleus together.

repulsions between protons. The strong force that holds the protons and neutrons together in a nucleus is powerful enough to overcome the electrostatic repulsion between the protons, but it can act over only a very short distance—about the diameter of a nucleus (**FIG. 10A.14**). Because neutrons are uncharged, they can contribute to the strong force without adding to the electrostatic repulsion. A lot of neutrons are needed to overcome the mutual repulsion of the protons in a nucleus of high atomic number, which explains why the band of stability curves upward.

Nuclei with certain even numbers of protons and of neutrons are the most stable.

10A.4 Predicting the Type of Nuclear Decay

Figure 10A.13 can be used to predict the type of disintegration that a radioactive nuclide is likely to undergo. Nuclei that lie above the band of stability are neutron rich. **Neutron-rich nuclei** are nuclei with a high proportion of neutrons. These nuclei tend to decay in such a way that the final n/p ratio is closer to that found in the band of stability. For example, a $^{14}_{6}C$ nucleus can reach a lower energy by ejecting a β particle, which reduces the n/p ratio as a result of the conversion of a neutron into a proton (**FIG. 10A.15**):

$$^{14}_{6}C \longrightarrow {}^{14}_{7}N + {}^{0}_{-1}e$$

Proton-rich nuclei have a low proportion of neutrons and lie below the band of stability. These isotopes tend to decay in such a way that the atomic number is reduced. For example, proton-rich $^{29}_{15}P$ is radioactive and decays by emitting a positron, a process that converts a proton into a neutron and raises the final n/p ratio:

$$^{29}_{15}P \longrightarrow {}^{29}_{14}S + {}^{0}_{+1}e$$

As shown in Example 10A.1, electron capture and proton emission also decrease the proton count of proton-rich nuclides.

Very few nuclides with $Z < 60$ emit α particles. All nuclei with $Z > 82$ are unstable and decay mainly by α-particle emission. They must discard protons to reduce

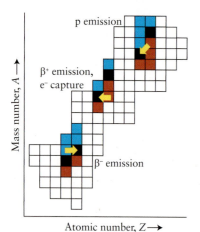

FIGURE 10A.15 Three different ways of reaching the band of stability (black). Nuclei that are neutron rich (blue region) tend to convert neutrons into protons by β emission; nuclei that are proton rich (red) tend to reach stability (black) by emitting a positron, capturing an electron, or emitting a proton.

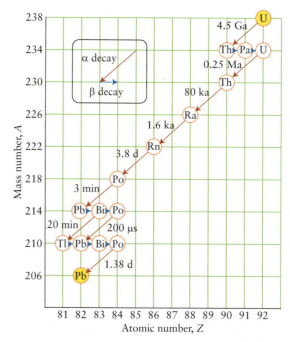

FIGURE 10A.16 The uranium-238 decay series. The times are the half-lives of the nuclides (see Topic 10B). The unit a, for annum, is the SI abbreviation for year.

their atomic number, and they generally need to lose neutrons, too. These nuclei decay in a step-by-step manner and give rise to a **radioactive series,** a characteristic sequence of nuclides (**FIG. 10A.16**). First, one α particle is ejected, then another α particle or a β particle is ejected, and so on, until a stable nucleus, such as an isotope of lead (with the magic atomic number 82) is formed. For example, the uranium-238 series ends at lead-206, the uranium-235 series ends at lead-207, and the thorium-232 series ends at lead-208.

Bismuth ($Z = 83$) was thought to have a stable isotope, but in 2005 even that isotope was found to decay very slowly.

Self-test 10A.2A Which of the following processes, (a) electron capture, (b) proton emission, (c) β⁻ emission, (d) β⁺ emission, might a $^{145}_{64}\mathrm{Gd}$ nucleus undergo to begin to reach stability? Refer to Figs. 10A.13 and 10A.15.

[*Answer:* a, b, d]

Self-test 10A.2B Which of the set of processes listed in Self-Test 10A.2A might a $^{148}_{58}\mathrm{Ce}$ nucleus undergo to begin to reach stability?

The pattern of nuclear stability can be used to predict the likely mode of radioactive decay: neutron-rich nuclei tend to reduce their neutron count; proton-rich nuclei tend to reduce their proton count. In general, only heavy nuclides emit α particles.

10A.5 Nucleosynthesis

The process of forming elements is **nucleosynthesis.** This process occurs naturally. Hydrogen and helium were produced in the Big Bang; all other naturally occurring elements are descended from these two, as a result of nuclear reactions taking place either in stars or in space. Some elements—among them technetium and promethium—are found in only trace amounts on Earth. Although these elements were made in stars, their short lifetimes did not allow them to survive long enough to contribute to the formation of our planet. However, nuclides that are too unstable to be found on Earth can be made by artificial techniques, and scientists have added about 2700 different nuclides to the approximately 270 that occur naturally.

To overcome the energy barriers to nuclear synthesis, particles must collide vigorously with one another (**FIG. 10A.17**), as they do in the hot interiors of stars. Therefore, to make elements artificially, it is necessary to simulate the conditions found inside a star. If it is traveling fast enough, a proton, an α particle, or another positively charged nucleus has enough kinetic energy to overcome the electrostatic repulsion from the nucleus. The incoming particle penetrates into the nucleus, where it is captured by the strong force. The high speeds required can be achieved in a particle accelerator.

The transmutation of elements—the conversion of one element into another, particularly of lead into gold—was the dream of the alchemists and one of the roots of modern chemistry. However, the alchemists had access only to chemical techniques, which proved ineffective because they involved energy changes far too weak to force nucleons into nuclei. Transmutation has now been recognized in nature and achieved in the laboratory, but by using methods undreamed of by the early alchemists. Rutherford achieved the first artificial nuclear transmutation in 1919. He bombarded nitrogen-14 nuclei with high-speed α particles. The products of the transmutation were oxygen-17 and a proton:

$$^{14}_{7}\mathrm{N} + {}^{4}_{2}\alpha \longrightarrow {}^{17}_{8}\mathrm{O} + {}^{1}_{1}\mathrm{p}$$

A large number of nuclides have been synthesized on Earth. For instance, technetium was prepared (as technetium-97) for the first time on Earth in 1937 by the reaction between molybdenum and deuterium nuclei:

$$^{97}_{42}\mathrm{Mo} + {}^{2}_{1}\mathrm{H} \longrightarrow {}^{97}_{43}\mathrm{Tc} + 2\,{}^{1}_{0}\mathrm{n}$$

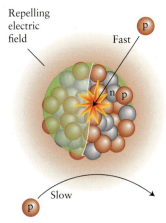

FIGURE 10A.17 When a positively charged particle approaches a nucleus, it is repelled strongly. However, a very fast particle can reach the nucleus before the repulsion turns it aside, and a nuclear reaction may result.

Such transmutation processes are commonly written simply as

$$^{97}_{42}\text{Mo}(d, 2n)^{97}_{43}\text{Tc}$$

where d denotes a deuteron, $^2\text{H}^+$. The general form of representation is:

Target (incoming particle, ejected particle) product

with the number of each particle included. Technetium is now moderately abundant because it accumulates in the decay products of nuclear power plants. Another isotope, technetium-99, has pharmaceutical applications, particularly for bone scans (BOX 10A.1).

Box 10A.1 WHAT HAS THIS TO DO WITH . . . STAYING ALIVE?

NUCLEAR MEDICINE

Nuclear chemistry has transformed medical diagnosis, treatment, and research. Radioactive tracers are used to measure organ function, sodium-24 to monitor blood flow, and strontium-87 to study bone growth. However, the most dramatic effect of radioisotopes on diagnosis has been in the field of imaging. Technetium-99m (the m denotes a "metastable," reasonably long-lived, state) is the most widely used radioactive nuclide in medicine, especially for bone scans. This isotope is highly active and gives off γ-rays that pass through the body. The γ-rays cause much less damage than α particles would, and the isotope is so short-lived that the risk to the patient is minimal.

Cobalt-60 is used for the *gamma knife*, a technique that is actually knifeless and that can kill tumor cells in locations such as the brain when surgery is not possible. In gamma-knife therapy, about 200 γ-rays are focused from different angles onto a tumor with a known shape and location. Each γ-ray by itself has a low amplitude and so causes little or no damage to the tissues through which it passes. However, at the intersection of 200 of these energetic rays, the radiation is very strong and sufficiently powerful to destroy tumor cells. The process is painless and fast, and the patient can remain awake through the treatment.

Positron emission tomography (PET) makes use of a short-lived positron emitter such as fluorine-18 to image human tissue with a degree of detail not possible with x-rays. It has been used extensively to study brain function (see illustrations) and in medical diagnosis. For example, when the hormone estrogen is labeled with fluorine-18 and injected into a cancer patient, the fluorine-bearing compound is preferentially absorbed by the tumor. The positrons given off by the fluorine atoms are quickly annihilated when they meet electrons. The resulting γ-rays are detected by a scanner that moves slowly over the part of the body containing the tumor. The growth of the tumor can be estimated quickly and accurately with this technique. A PET imaging facility must be located near a cyclotron so that the positron emitters can be incorporated into the desired compounds as soon as they are created.

Several kinds of cancer therapy use radiation to destroy malignant cells. *Boron neutron-capture therapy* is unusual in that boron-10, the isotope injected, is not radioactive. However, when boron-10 is bombarded with neutrons, it gives off highly destructive α particles. In boron neutron-capture therapy, the boron-10 is incorporated into a compound that is absorbed preferentially by tumors. The patient is then exposed to brief periods of neutron bombardment. As soon as the bombardment ceases, the boron-10 stops generating α particles.

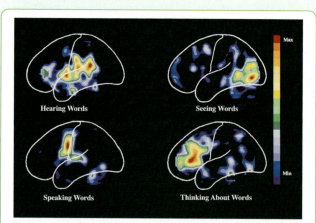

These four PET scans show how blood flow to different parts of the brain is affected by various activities. In this case, an oxygen isotope that is taken up by the hemoglobin in blood is used as a source of positrons. (*M.E. Raichle, Mallinckrodt Institute of Radiology, Washington University School of Medicine.*)

Related Exercises 10.15–10.19

Further Reading P. Barry, "Unintended consequences of cancer therapies," *Science News*, vol. 171, p. 334, May 26, 2007. V. Marx, "Molecular imaging," *Chemical and Engineering News*, vol. 83(30), 2005, pp. 25–34. G. Miller, "Neuroscience: A better view of brain disorders," *Science*, vol. 313, September 8, 2006, pp. 1376–1379.

This patient is about to undergo a PET scan of brain function. (*Marcus E. Raichle and Washington University School of Medicine.*)

THINKING POINT

Why is it impractical to transmute lead into gold?

A neutron can get close to a target nucleus more easily than a proton can. Because a neutron has no charge and hence is not repelled by the nuclear charge, it need not be accelerated to such high speeds. An example of **neutron-induced transmutation** is the formation of cobalt-60, which is used in the radiation treatment of cancer. The three-step process starts from iron-58. First, iron-59 is produced:

$$^{58}_{26}\text{Fe} + {}^{1}_{0}\text{n} \longrightarrow {}^{59}_{26}\text{Fe}$$

The second step is the β decay of iron-59 to cobalt-59:

$$^{59}_{26}\text{Fe} \longrightarrow {}^{59}_{27}\text{Co} + {}^{0}_{-1}\text{e}$$

In the final step, cobalt-59 absorbs another neutron from the incident neutron beam and is converted into cobalt-60:

$$^{59}_{27}\text{Co} + {}^{1}_{0}\text{n} \longrightarrow {}^{60}_{27}\text{Co}$$

The overall reaction is

$$^{58}_{26}\text{Fe} + 2\,{}^{1}_{0}\text{n} \longrightarrow {}^{60}_{27}\text{Co} + {}^{0}_{-1}\text{e} \quad \text{or} \quad {}^{58}_{26}\text{Fe}(2\text{n}, \beta^-){}^{60}_{27}\text{Co}$$

Self-test 10A.3A Complete the following nuclear reactions: (a) ? + ${}^{4}_{2}\alpha \rightarrow {}^{243}_{96}\text{Cm} + {}^{1}_{0}\text{n}$; (b) ${}^{242}_{96}\text{Cm} + {}^{4}_{2}\alpha \rightarrow {}^{245}_{98}\text{Cf} + ?$.

[**Answer:** (a) ${}^{240}_{94}\text{Pu}$; (b) ${}^{1}_{0}\text{n}$]

Self-test 10A.3B Complete the following nuclear reactions: (a) ${}^{250}_{98}\text{Cf} + ? \rightarrow {}^{257}_{103}\text{Lr} + 4\,{}^{1}_{0}\text{n}$; (b) ? + ${}^{12}_{6}\text{C} \rightarrow {}^{254}_{102}\text{No} + 4\,{}^{1}_{0}\text{n}$.

The **transuranium elements** are the elements following uranium in the periodic table. The elements from rutherfordium (Rf, $Z = 104$) through meitnerium (Mt, $Z = 109$) were formally named in 1997. The **transmeitnerium elements,** the elements beyond meitnerium (including hypothetical nuclides that have not yet been made) are named systematically, at least until they have been identified and there is international agreement on a permanent name. Their systematic names use the prefixes in **TABLE 10A.2**, which identify their atomic numbers, with the ending *-ium*. Thus, element 110 was known as ununnilium until it was named darmstadtium (Ds) in 2003.

New elements and isotopes of known elements are made by nucleosynthesis; the repulsive electrical forces of like-charged particles are overcome when very fast particles collide.

What have you learned in this Topic?

You have learned about the properties of different types of radioactive decay and the relative stabilities of nuclei, and have seen how to write nuclear equations.

The skills you have mastered are the ability to:

☐ **1.** Write, complete, and balance nuclear equations (Example 10A.1).

☐ **2.** Distinguish α, β, and γ radiation by their response to an electric field (Section 10A.1).

☐ **3.** Use the band of stability to predict the types of decay that a given radioactive nucleus is likely to undergo (Self-Test 10A.2).

☐ **4.** Write the nuclear reactions for nuclear transformation processes (Section 10A.5).

TABLE 10A.2 Notation for the Systematic Nomenclature of Elements*

Digit	Prefix	Abbreviation
0	nil	n
1	un	u
2	bi	b
3	tri	t
4	quad	q
5	pent	p
6	hex	h
7	sep	s
8	oct	o
9	enn	e

*For instance, element 123 would be named unbitriium, Ubt.

Topic 10A Exercises

10A.1 When the nucleons rearrange in the following daughter nuclei, the energy changes by the amount shown and a γ-ray is emitted. Determine the frequency and wavelength of the γ-ray in each case: (a) nickel-60 in an excited state, 1.33 MeV; (b) arsenic-80, 1.64 MeV; (c) iron-59, 1.10 MeV. (1 MeV = 1.602×10^{-13} J.)

10A.2 When the nucleons rearrange in the following daughter nuclei, the energy changes by the amount shown and a γ-ray is emitted. Determine the frequency and wavelength of the γ-ray emitted in the decay of each of following nuclides: (a) iron-53, 3.04 MeV; (b) vanadium-52, 1.43 MeV; (c) scandium-44, 0.27 MeV. (1 MeV = 1.602×10^{-13} J.)

10A.3 *Isotones* are nuclides that have the same number of neutrons. Which isotopes of argon and calcium are isotones of potassium-40?

10A.4 Which isotopes of krypton and selenium are isotones (see Exercise 10A.3) of bromine-80?

10A.5 Write the balanced nuclear equation for each of the following decays: (a) β decay of boron-12; (b) α decay of bismuth-214; (c) β decay of technetium-98; (d) α decay of radium-226.

10A.6 Write the balanced nuclear equation for each of the following processes: (a) α decay of francium-221; (b) α decay of radon-212; (c) β decay of actinium-228; (d) electron capture by protactinium-230.

10A.7 Write the balanced nuclear equation for each of the following processes: (a) positron emission by indium-109; (b) positron decay of magnesium-23; (c) electron capture by lead-202; (d) electron capture by arsenic-76.

10A.8 Write the balanced nuclear equation for each of the following processes: (a) β^+ decay of germanium-66; (b) electron capture by iridium-190; (c) β^+ decay of iodine-124; (d) electron capture by niobium-92.

10A.9 Determine the particle emitted and write the balanced nuclear equation for each of the following nuclear transformations: (a) sodium-24 to magnesium-24; (b) ^{128}Sn to ^{128}Sb; (c) lanthanum-140 to barium-140; (d) ^{228}Th to ^{224}Ra.

10A.10 Determine the particle emitted and write the balanced nuclear equation for each of the following nuclear transformations: (a) gadolinium-148 to samarium-144; (b) fluorine-17 to oxygen-17; (c) silver-112 to cadmium-112; (d) plutonium-238 to uranium-234.

10A.11 Complete the following equations for nuclear reactions:
(a) ^{11}B + ? → 2 n + ^{13}N
(b) ? + D → n + ^{36}Ar
(c) ^{96}Mo + D → ? + ^{97}Tc
(d) ^{45}Sc + n → α + ?

10A.12 Complete the following equations for nuclear decay reactions:
(a) ^{12}O → 2 p + ?
(b) ^{17}C → ? + n + β^-
(c) ^{148}Ba → ^{147}La + ? + n
(d) ^{18}Ne → β^+ + ?

10A.13 The following nuclides lie outside the band of stability. Predict whether each is most likely to undergo β decay, β^+ decay, or α decay, and identify the daughter nucleus: (a) copper-68; (b) cadmium-103.

10A.14 The following nuclides lie outside the band of stability. Predict whether each is most likely to undergo β decay, β^+ decay, or α decay, and identify the daughter nucleus: (a) copper-60; (b) xenon-140.

10A.15 Identify the daughter nuclides in each step of the radioactive decay of uranium-235, if the string of particle emissions is α, β, α, β, α, α, α, β, α, β, α. Write a balanced nuclear equation for each step.

10A.16 Neptunium-237 undergoes an α, β, α, α, β, α, α, α, β, α, β sequence of radioactive decays. Write a balanced nuclear equation for each step.

10A.17 Complete the following nuclear equations:
(a) $^{14}_{7}$N + ? → $^{17}_{8}$O + $^{1}_{1}$p
(b) ? + $^{1}_{0}$n → $^{249}_{97}$Bk + $^{0}_{-1}$e
(c) $^{243}_{95}$Am + $^{1}_{0}$n → $^{244}_{96}$Cm + ? + γ
(d) $^{13}_{6}$C + $^{1}_{0}$n → ? + γ

10A.18 Complete the following nuclear equations:
(a) ? + $^{1}_{1}$p → $^{21}_{11}$Na + γ
(b) $^{1}_{1}$H + $^{1}_{1}$p → $^{2}_{1}$H + ?
(c) $^{15}_{7}$N + $^{1}_{1}$p → $^{12}_{6}$C + ?
(d) $^{20}_{10}$Ne + ? → $^{24}_{12}$Mg + γ

10A.19 Complete the following equations for nuclear transmutation:
(a) $^{20}_{10}$Ne + $^{4}_{2}$α → ? + $^{16}_{8}$O
(b) $^{20}_{10}$Ne + $^{20}_{10}$Ne → ? + $^{16}_{8}$O + γ
(c) $^{44}_{20}$Ca + ? → γ + $^{48}_{22}$Ti
(d) $^{27}_{13}$Al + $^{2}_{1}$H → ? + $^{28}_{13}$Al

10A.20 Complete the following equations for nuclear transmutation:
(a) ? + γ → $^{0}_{-1}$e + $^{20}_{10}$Ne
(b) $^{44}_{22}$Ti + $^{0}_{-1}$e → $^{0}_{1}$e + ?
(c) $^{241}_{95}$Am + ? → 4 $^{1}_{0}$n + $^{248}_{100}$Fm
(d) ? + $^{1}_{0}$n → $^{0}_{-1}$e + $^{244}_{96}$Cm

10A.21 One explanation for the existence of elements heavier than iron is the rapid neutron-capture process (r-process). In the r-process proposed to occur in supernovae, many high-speed neutrons smash into an iron nucleus. Some of these neutrons are captured, resulting in a highly unstable nucleus that eventually decays. In each decay step, a neutron is converted into a proton. Write the overall nuclear equation for the absorption of six neutrons by an iron-56 nucleus and their decay to six protons.

10A.22 It has been claimed that the neon present in meteorites was generated by nuclear reactions. One such reaction occurs when a silicon-28 atom is bombarded by cosmic-ray protons. When one of these energetic protons is absorbed, the result is an isotope of neon, three protons, a neutron, and an α particle. Identify the isotope of neon formed and write the nuclear equation for the process.

10A.23 Write a nuclear equation for each of the following processes: (a) oxygen-17 produced by α-particle bombardment of nitrogen-14; (b) americium-241 produced by neutron bombardment of plutonium-239.

10A.24 Write a nuclear equation for each of the following transformations: (a) ^{257}Rf produced by the bombardment of californium-245 with carbon-12 nuclei; (b) the first synthesis of ^{266}Mt by the bombardment of bismuth-209 with iron-58 nuclei. Given that the first decay of meitnerium is by α emission, what is the daughter nucleus?

10A.25 What will be the systematic name and atomic symbol given to (a) element 126; (b) element 136; (c) element 200?

10A.26 What will be the systematic name and atomic symbol given to (a) element 118; (b) element 127; (c) element 202?

Topic 10B Radioactivity

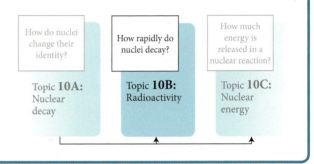

Why Do You Need to Know This Material? When handling materials with radioactive nuclei, it is essential to understand their modes of decay and how rapidly they decay. The rate of decay is the basis of a method used to determine the age of ancient objects.

What Do You Need to Know Already? You need to know the origins and classification of radiation from nuclei (Topic 10A) and the kinetics of first-order decay (Topic 7B).

Nuclear radiation is sometimes called **ionizing radiation,** because it is energetic enough to eject electrons from atoms. Hospitals use nuclear radiation to kill unwanted tissue, such as cancerous cells (see Box 10A.1). Yet the same powerful effects that help to diagnose and cure illness can also damage healthy tissue. The damage depends on the intensity of the source, the type of radiation, and the length of exposure. The three principal types of nuclear radiation have different abilities to penetrate matter and cause damage (TABLE 10B.1).

10B.1 The Biological Effects of Radiation

Heavy, highly charged α particles interact so strongly with matter that they slow down, capture electrons from surrounding matter, and change into bulky helium atoms before traveling very far. They penetrate only the first layer of skin and can be stopped by glass, by clothing, and even by a sheet of paper. The surface layer of dead skin absorbs most α radiation, where it can do little harm. However, α particles can be extremely dangerous if inhaled or ingested. The energy of their impact can knock atoms out of molecules, which can lead to serious illness and death. For example, plutonium, considered to be one of the most toxic radioactive materials, is an α emitter and can be handled safely with minimal shielding. However, it is easily oxidized to Pu^{4+}, which has chemical properties similar to those of Fe^{3+}. As a result, plutonium can take the place of iron in the body and become absorbed into bone, where it destroys the body's ability to produce red blood cells. The result is radiation sickness, cancer, and death.

Next in penetrating power is β radiation. These fast electrons can penetrate about 1 cm into flesh before their electrostatic interactions with the electrons and nuclei of molecules bring them to a standstill.

Most penetrating of all is γ radiation. The high-energy γ-ray photons can pass through buildings and bodies and can cause damage by ionizing the molecules in their path. Protein molecules and DNA that have been damaged in this way can no longer function, and the results can be radiation sickness and cancer. Intense sources of γ-rays must be surrounded by shields built from lead bricks or thick concrete to absorb this penetrating radiation.

The **absorbed dose** of radiation is the energy deposited in a sample (in particular, the human body) when it is exposed to radiation. The SI unit of absorbed dose is the **gray,** Gy, which corresponds to an energy deposit of $1\ J \cdot kg^{-1}$. The first unit used for reporting dose was the **radiation absorbed dose** (rad), the amount of radiation that deposits 10^{-2} J of

TABLE 10B.1	Shielding Requirements of α, β, and γ Radiation	
Radiation	**Relative penetrating power**	**Shielding required**
α	1	paper, skin
β	100	3 mm aluminum
γ	10 000	concrete, lead

energy per kilogram of tissue, and so 1 rad $= 10^{-2}$ Gy. A dose of 1 rad corresponds to a 65-kg person absorbing a total of 0.65 J, which is not very much energy: it is enough to boil only 0.2 mg of water. However, the energy of a particle of nuclear radiation is highly localized, like the impact of a subatomic bullet. As a result, the incoming particles can break individual bonds as they collide with molecules in their path.

The extent of radiation damage to living tissue depends on the type of radiation and the type of tissue. The **relative biological effectiveness,** Q, must also be included when assessing the damage that a given dose of each type of radiation may cause. For β and γ radiation, Q is set arbitrarily at about 1; however, for α radiation, Q is close to 20. A dose of 1 Gy of γ radiation causes about the same amount of damage as 1 Gy of β radiation, but 1 Gy of α particles is about 20 times as damaging (even though these particles are the least penetrating). The precise figures depend on the total dose, the rate at which the dose accumulates, and the type of tissue, but these values are typical.

The **dose equivalent** is the actual dose modified to take into account the different destructive powers of the various types of radiation in combination with various types of tissue. It is obtained by multiplying the actual dose (in gray) by the value of Q for the radiation type. The result is expressed in the SI unit called a **sievert** (Sv):

$$\text{Dose equivalent(Sv)} = Q \times \text{absorbed dose(Gy)} \qquad\qquad (1)$$

The former (non-SI) unit of dose equivalent was the **roentgen equivalent man** (rem), which was defined in the same way as the sievert but with the absorbed dose in rad; thus, 1 rem $= 10^{-2}$ Sv.

A dose of 0.3 Gy (30 rad) of γ radiation corresponds to a dose equivalent of 0.3 Sv (30 rem), enough to cause a reduction in the number of white blood cells (the cells that fight infection), but 0.3 Gy of α radiation corresponds to 6 Sv (600 rem), which is enough to cause death if an α emitter is ingested or inhaled. A typical average annual dose equivalent that you receive from natural sources, called **background radiation,** is about 2 mSv·a^{-1} (where a, for annum, is the SI abbreviation for year), but this figure varies, depending on your life style and where you live. About 20% of background radiation comes from your own body. About 30% comes from cosmic rays (a mix of γ-rays and high-energy subatomic particles from outer space) that continuously bombard the Earth, and 40% comes from radon seeping out of the ground. The remaining 10% is a result largely of medical diagnoses (a typical chest x-ray gives a dose equivalent of about 0.07 mSv). Emissions from nuclear power plants and other nuclear facilities contribute about 0.1% in countries where they are widely used.

The principal source of radioactivity in the human body is potassium-40. About 35 000 potassium-40 nuclei disintegrated in your body while you were reading this sentence.

Human exposure to radiation is monitored by reporting the absorbed dose and the dose equivalent; the latter takes into account the effects of different types of radiation on tissues.

10B.2 Measuring the Rate of Nuclear Decay

Geiger counters make use of the ionization of a gas, usually argon, when it is exposed to nuclear radiation, and **scintillation counters** measure radiation by counting the flashes of light that are generated when radiation strikes a substance called a phosphor. Both are used to measure the rate at which a radioactive nucleus decays (**BOX 10B.1**). Each click of a Geiger counter or flash of a phosphor in a scintillation counter indicates that one nuclear disintegration has been detected. The **activity** of a sample is the number of nuclear disintegrations in a given time interval divided by the length of the interval. The SI unit of activity is the **becquerel** (Bq): 1 Bq is equal to one nuclear disintegration per second. Another commonly used (non-SI) unit of radioactivity is the curie (Ci). It is equal to 3.7×10^{10} nuclear disintegrations per second, the radioactive output of 1 g of radium-226. Because the curie is a very large unit, most activities are expressed in millicuries (mCi) or microcuries (μCi). **TABLE 10B.2** summarizes these units.

The equation for the decay of a nucleus (parent nucleus $\rightarrow$ daughter nucleus + radiation) has exactly the same form as a unimolecular elementary reaction (Topic 7B),

Box 10B.1 HOW DO WE KNOW . . . HOW RADIOACTIVE A MATERIAL IS?

The ability of nuclear radiation to eject electrons from atoms and ions can be used to measure its intensity. Henri Becquerel first gauged the intensity of radiation by determining the degree to which it blackened a photographic film. The blackening results from the same redox processes as those used in photographic film, such as

$$Ag^+ + Br^- \xrightarrow{h\nu} Ag + Br$$

except that the initial oxidation of the bromide ions is caused by nuclear radiation instead of light. Becquerel's technique is still used in film *dosimeters*, which monitor the cumulative exposure of workers to radiation and are worn as a badge.

Thermoluminescent dosimeters contain a material such as lithium fluoride. Any incident radiation knocks electrons out of the fluoride ions. The electrons migrate away from the fluorine atoms but remain trapped in the crystal. When the crystal is heated, the electrons fall back on to the fluorine atoms and give off the difference in energy as light. The dose of radiation received is determined by the intensity of the light. Dosimeters can be used over a period of time ranging from one day to several weeks, because the excited electrons accumulate with continued exposure, allowing a long-term dosage to be evaluated. Heating the dosimeter during readout releases the stored energy and so thermoluminescent dosimeters are reusable.

A *Geiger counter* monitors radiation by detecting the ionization of a low-pressure gas, as shown in the illustration. The radiation ionizes atoms of the gas inside a cylinder and allows a brief flow of current between the electrodes. The resulting electrical signal can be recorded directly or converted into an audible click. The frequency of the clicks indicates the intensity of the radiation. A limitation of Geiger counters is that they do not respond well to γ-rays. Only about 1% of the γ-ray photons are detected, whereas all the β particles incident on the counter are detected. Because the efficiency of a Geiger counter depends on the size of the tube, a counter used to monitor a wide range of activities usually contains two tubes of different sizes.

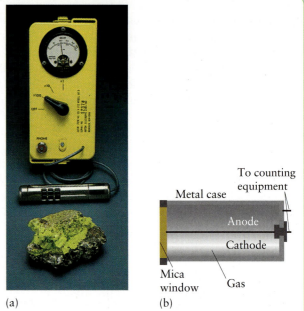

(a) A Geiger counter with a piece of uranium ore. (b) The detector in a Geiger counter contains a gas (often either argon and a little ethanol vapor or neon with some bromine vapor) in a cylinder with a high potential difference (500–1200 V) between the central wire and the walls. When radiation ionizes the gas, the ions allow current to flow briefly, resulting in the characteristic click. (Photo ©1989 Chip Clark–Fundamental Photographs.)

A *scintillation counter* makes use of the fact that phosphorescent substances such as sodium iodide and zinc sulfide give a flash of light—a scintillation—when exposed to radiation. The counter also contains a photomultiplier tube, which converts light into an electrical signal. The intensity of the radiation is determined from the strength of the electronic signal.

with an unstable nucleus taking the place of a reactant molecule. This type of decay is expected for a process that does not depend on any external factors but only on the intrinsic instability of the nucleus. The rate of nuclear decay depends only on the identity of the isotope, not on its chemical form or temperature.

TABLE 10B.2   Radiation Units*

Property	Unit name	Symbol	Definition
activity	becquerel	Bq	1 disintegration per second
	curie	Ci	3.7×10^{10} disintegrations per second
absorbed dose	gray	Gy	$1 \ J \cdot kg^{-1}$
	radiation absorbed dose	rad	$10^{-2} \ J \cdot kg^{-1}$
dose equivalent	sievert	Sv	$Q \times$ absorbed dose[†]
	roentgen equivalent man	rem	$Q \times$ absorbed dose[†]

*Non-SI units are in red.
[†]Q is the relative biological effectiveness of the radiation. Normally, $Q \approx 1 \ Sv \cdot Gy^{-1}$ for γ, β, and most other radiation, but $Q \approx 20 \ Sv \cdot Gy^{-1}$ for α radiation and fast neutrons. A further factor of 5 (that is, 5Q) is used for bone under certain circumstances.

As in a unimolecular chemical reaction, the rate law for nuclear decay is first order. That is, the relation between the rate of decay and the number N of radioactive nuclei present is given by the **law of radioactive decay:**

$$\text{Activity} = \text{rate of decay} = k \times N \qquad (2)$$

In this context, k is called the **decay constant.** The law implies that the activity of a radioactive sample is proportional to the number of atoms in the sample. As discussed in Topic 7B, a first-order rate law implies an exponential decay. It follows that the number N of radioactive nuclei remaining after a time t is given by

$$N = N_0 e^{-kt} \qquad (3)$$

where N_0 is the number of radioactive nuclei present initially (at $t = 0$); this expression is plotted in **FIG. 10B.1**.

Radioactive decay is normally discussed in terms of the **half-life**, $t_{1/2}$, the time needed for half the initial number of nuclei to disintegrate. Just as in Topic 7B, $t_{1/2}$ can be related to k (the analog of the first-order rate constant k_r) by setting $N = \frac{1}{2}N_0$ and $t = t_{1/2}$ in Eq. 3:

$$t_{1/2} = \frac{\ln 2}{k} \qquad (4)$$

This equation shows that, the larger the value of k, the shorter is the half-life of the nuclide. Nuclides with short half-lives are less stable than nuclides with long half-lives. They are more likely to decay in a given period of time and are "hotter" (more intensely radioactive) than nuclides with long half-lives.

THINKING POINT

Might there be advantages in nuclear reactors producing intensely radioactive waste?

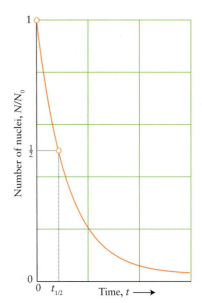

FIGURE 10B.1 The exponential decay of the activity of a sample shows that the number of radioactive nuclei in the sample also decays exponentially with time. The curve is characterized by the half-life, $t_{1/2}$.

EXAMPLE 10B.1   Using the law of radioactive decay

One of the reasons why thermonuclear weapons have to be serviced regularly is the nuclear decay of the tritium, ^3H, that they contain. Suppose you are monitoring the decay of tritium. What mass of a tritium sample initially of mass 1.00 g will remain after 5.0 a (1 a = 1 year)? The decay constant of tritium is 0.0564 a^{-1}.

ANTICIPATE The half-life of tritium (**TABLE 10B.3**) is 12.3 years, and so you should expect that more than half of the sample will remain after only 5 years.

PLAN The total mass of isotope in a sample is proportional to the number of nuclei of that isotope that the sample contains; therefore, the time dependence of the mass of a radioactive isotope follows the same radioactive decay law as the number of nuclides in the sample. That is, because $m \propto N$, in place of Eq. 3 you can write $m = m_0 e^{-kt}$, where m is the total mass of the radioactive isotope at time t and m_0 is its initial mass.

SOLVE

From $m = m_0 e^{-kt}$,

$$m = (1.00 \text{ g}) \times e^{-(0.0564 \text{ a}^{-1}) \times (5.0 \text{ a})} = 0.75 \text{ g}$$

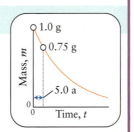

TABLE 10B.3   Half-Lives of Radioactive Isotopes*

Nuclide	Half-life, $t_{1/2}$
tritium	12.3 a
carbon-14	5.73 ka
carbon-15	2.4 s
potassium-40	1.26 Ga
cobalt-60	5.26 a
strontium-90	28.1 a
iodine-131	8.05 d
cesium-137	30.17 a
radium-226	1.60 ka
uranium-235	0.71 Ga
uranium-238	4.5 Ga
fermium-244	3.3 ms

*d = day; a = year.

EVALUATE As expected, more than half of the sample, 0.75 g, remains after 5.0 years.

Self-test 10B.1A The decay constant for fermium-254 is 210 s^{-1}. What mass of the isotope will be present if a sample of mass 1.00 μg is kept for 10. ms?

[***Answer:*** 0.12 mg]

Self-test 10B.1B The decay constant for the nuclide neptunium-237 is 3.3×10^{-7} a^{-1}. What mass of the isotope will be present if a sample of mass 5.0 μg survives for 1.0 Ma (1.0 million years)?

Related Exercises 10B.5, 10B.6

The half-lives of radioactive nuclides span a very wide range (Table 10B.1). Consider strontium-90, for which the half-life is 28.1 a. This nuclide is present in nuclear **fallout,** the fine dust that settles from clouds of airborne particles after the explosion of a nuclear bomb and that may also be present in the accidental release of radioactive materials into the air. Because it is chemically very similar to calcium, strontium may accompany that element through the environment and become incorporated into bones; once there, it continues to emit radiation for many years. About 10 half-lives (for strontium-90, 281 years) must pass before the activity of a sample has fallen to 1/1000 of its initial value. Iodine-131 is a radioisotope that was released in the accidental fire at the 1986 Chernobyl nuclear power plant and in the explosion at the Fukushima Daiichi nuclear plant after the tsunami in Japan in 2011. It has a half-life of only 8.05 d, but it accumulates in the thyroid gland. Supplemental iodine pills were issued to people near these disasters so that their thyroids would already be saturated with iodine, thus reducing the amount of radioactive iodine that could be absorbed. Despite these precautions, several cases of thyroid cancer were linked to iodine-131 exposure from the Chernobyl accident. Plutonium-239 has a half-life of 24 ka (24 000 years). Consequently, very-long-term storage facilities are required for plutonium waste, and land contaminated with plutonium cannot be inhabited again for thousands of years without expensive remediation efforts.

The constant half-life of a nuclide is used to determine the ages of archaeological artifacts. In **isotopic dating,** the activity of the radioactive isotopes that they contain is measured. Isotopes used for dating objects include uranium-238, potassium-40, and tritium (^{3}H). However, the most important example is **radiocarbon dating,** which uses the β decay of carbon-14, for which the half-life is 5730 a.

Carbon-12 is the principal isotope of carbon, but a small proportion of carbon-14 is present in all living organisms. Its nuclei are produced when nitrogen nuclei in the atmosphere are bombarded by neutrons formed in the collisions of cosmic rays with other nuclei:

$$^{14}_{7}\text{N} + ^{1}_{0}\text{n} \longrightarrow ^{14}_{6}\text{C} + ^{1}_{1}\text{p}$$

Carbon-14 atoms are produced in the atmosphere at an almost constant rate and, as a result, the proportion of carbon-14 to carbon-12 atoms in the atmosphere is nearly constant over time. Carbon-14 atoms enter living organisms as $^{14}CO_2$ through photosynthesis and digestion. They leave the organisms by the normal processes of excretion and respiration and also because the nuclei decay at a steady rate. As a result, all living organisms have a fixed ratio (of about 1 to 10^{12}) of carbon-14 atoms to carbon-12 atoms, and 1.0 g of naturally occurring carbon has an activity of 15 disintegrations per minute.

When an organism dies, it no longer exchanges carbon with its surroundings. However, carbon-14 nuclei already inside the organism continue to decay with a constant half-life, and so the ratio of carbon-14 to carbon-12 decreases. The ratio observed in a sample of dead tissue can therefore be used to estimate the time since death.

In the technique developed by Willard Libby in Chicago in the late 1940s, the proportion of carbon-14 in a sample is determined by monitoring the β radiation from CO_2 obtained by burning the sample. In the modern version of the technique, which requires only a few milligrams of sample, the carbon atoms are converted into C^- ions by bombardment of the sample with cesium atoms. The C^- ions are then accelerated with electric fields, and the carbon isotopes are separated and counted with a mass spectrometer (**FIG. 10B.2**).

Nuclear testing has increased the amount of carbon-14 in the air, and sensitive radiocarbon dating techniques take this increase into account.

FIGURE 10B.2 In a carbon-14 dating measurement, a mass spectrometer is used to determine the proportion of carbon-14 nuclei in the sample relative to the number of carbon-12 nuclei. (*Enrico Sacchetti/Science Source.*)

EXAMPLE 10B.2   Interpreting carbon-14 dating

Suppose you are working in a carbon dating laboratory and are examining a piece of wood found in an archaeological site in Arizona. You find that a carbon sample of mass 1.00 g from the piece of wood undergoes 7.90×10^3 carbon-14 disintegrations in a period of 20.0 h. In the same period, 1.00 g of carbon from a modern source you are using as a standard undergoes 1.84×10^4 disintegrations. Calculate the age of the sample, given that the half-life of ^{14}C is 5.73 ka.

ANTICIPATE Because the half-life of ^{14}C is 5.73 ka and the activity of the sample has fallen to less than half that of the modern sample, you should expect that it is more than 5730 years old.

PLAN First, rearrange Eq. 3 to give an expression for the time and then express k in terms of $t_{1/2}$ by using Eq. 4. The carbon-14 activity of the modern sample can be used to represent the *original* activity in the ancient sample. Therefore, N/N_0 is equal to the ratio of the number of disintegrations in the ancient and modern samples.

SOLVE

From $N = N_0 e^{-kt}$,

$$t = \frac{1}{k} \ln\left(\frac{N}{N_0}\right)$$

From $t_{1/2} = (\ln 2)/k$,

$$t = -\frac{t_{1/2}}{\ln 2} \ln\left(\frac{N}{N_0}\right)$$

Substitute the data.

$$t = -\frac{5.73 \text{ ka}}{\ln 2} \times \ln\left(\frac{7900}{18\,400}\right) = 6.99 \text{ ka}$$

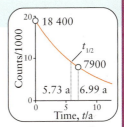

EVALUATE As expected, the sample is more than 5730 years old. You can conclude that nearly 7000 years have elapsed since the piece of wood was part of a living tree.

Self-test 10B.2A A sample of carbon of mass 250. mg from wood found in a tomb in Israel underwent 2480 carbon-14 disintegrations in 20. h. Estimate the time since death, assuming the same activity for a modern sample as in Example 10B.2.

[*Answer:* 5.1 ka]

Self-test 10B.2B A sample of carbon of mass 1.00 g from scrolls found near the Dead Sea underwent 1.4×10^4 carbon-14 disintegrations in 20. h. Estimate the approximate time since the sheepskins composing the scrolls were removed from the sheep, assuming the same activity for a modern sample as in Example 10B.2.

Related Exercises 10B.7–10B.10

The law of radioactive decay implies that the number of radioactive nuclei decreases exponentially with time with a characteristic half-life. Radioactive isotopes are used to determine the ages of objects.

10B.3 Uses of Radioisotopes

Radioisotopes are radioactive isotopes. They are used not only to cure disease (as described in Box 10A.1), but also to preserve food, to trace mechanisms of reactions, and to power equipment in spacecraft.

Radioactive **tracers** are isotopes that are used to track how substances change and move. For example, a sample of sugar can be **labeled** with carbon-14; that is, some carbon-12 atoms in sugar molecules are replaced by carbon-14 atoms, which can be detected by radiation counters. In this way, the progress through the body of even a tiny amount of sugar molecules, too small an amount to be detected by other means, can be monitored. Chemists and biochemists use tracers to help determine the mechanisms of reactions. For example, if water containing oxygen-18 is used in photosynthesis, the oxygen produced is oxygen-18 (red):

$$6 \, CO_2(g) + 6 \, H_2O(l) \longrightarrow C_6H_{12}O_6(s, \text{glucose}) + 3 \, O_2(g) + 3 \, O_2(g)$$

This result shows that the oxygen produced in photosynthesis comes from the water molecules, not the carbon dioxide molecules.

Radioisotopes have important commercial applications. For example, americium-241 is used in smoke detectors. Its role is to ionize any smoke particles, which then allow a current to flow and set off the alarm. Exposure to radiation is also used to sterilize food and inhibit the sprouting of potatoes. Radioisotopes that give off a lot of energy as heat are used to provide power in remote locations, where refueling of generators is not possible. The instruments in unmanned spacecraft, such as *Voyager 2*, which has now left the solar system, are powered by radiation from isotopes with long half-lives, such as plutonium.

THINKING POINT

Why are some people opposed to the irradiation of food?

Isotopes are also used to study the environment. Just as carbon-14 is used to date organic materials, geologists can determine the age of very old substances such as rocks by measuring the abundance in rocks of radioisotopes with longer half-lives. Uranium-238 ($t_{1/2} = 4.5$ Ga, 1 Ga $= 10^9$ years) and potassium-40 ($t_{1/2} = 1.26$ Ga) are used to date very old rocks. For example, potassium-40 decays by electron capture to form argon-40. The rock is placed under vacuum and crushed, and a mass spectrometer is used to measure the amount of argon gas that escapes. This technique was used to determine the age of rocks collected on the surface of the Moon; they were found to be 3.5–4.0 billion years old, about the same age as the oldest rocks on Earth.

Radioisotopes are used as long-lasting power sources, to study the environment, and to track movement. They are used in biology to trace metabolic pathways, in chemistry to reveal reaction mechanisms, and in geology to determine the ages of rocks.

What have you learned in this Topic?

You have learned that radioactive decay follows first-order kinetics and so has a constant half-life that is characteristic of each isotope.

The skills you have mastered are the ability to:

☐ **1.** Describe the relative penetrating powers of the three main types of radiation (Section 10B.1).

☐ **2.** Predict the amount of a radioactive sample that will remain after a given time period, given the decay constant or half-life of the sample (Example 10B.1).

☐ **3.** Use the half-life of an isotope to determine the age of an object (Example 10B.2).

Topic 10B Exercises

Exercises labeled $\int_{dx}^C$ require calculus.

Note that the SI symbol for 1 year is 1 a and that it takes the usual prefixes, as in 1 ka $= 10^3$ a and 1 Ga $= 10^9$ a.

10B.1 Evaluate the decay constant for (a) tritium, $t_{1/2} = 12.3$ a; (b) lithium-8, $t_{1/2} = 0.84$ s; (c) nitrogen-13, $t_{1/2} = 10.0$ min.

10B.2 Evaluate the half-life of (a) potassium-40, $k = 5.3 \times 10^{-10}$ a^{-1}; (b) cobalt-60, $k = 0.132$ a^{-1}; (c) nobelium-255, $k = 3.85 \times 10^{-3}$ s^{-1}.

10B.3 The activity of a sample of a radioisotope was found to be 2150 disintegrations per minute. After 6.0 h, the activity was found to be 1324 disintegrations per minute. What is the half-life of the radioisotope?

10B.4 A sample of pure cobalt-60 has an activity of 1 μCi. (a) How many atoms of cobalt-60 are present in the sample? (b) What is the mass in grams of the sample?

10B.5 (a) What percentage of a carbon-14 sample remains after 3.00 ka? (b) Determine the percentage of a tritium sample that remains after 12.0 a.

10B.6 (a) What percentage of a strontium-90 sample remains after 8.5 a? (b) Determine the percentage of an iodine-131 sample that remains after 6.0 d.

10B.7 Potassium-40, which is presumed to have existed at the formation of the Earth, is used for dating minerals. If three-fifths of the original potassium-40 exists in a rock, how old is the rock?

10B.8 A piece of wood, found in an archaeological dig, has a carbon-14 activity that is 62% of the current carbon-14 activity. How old is the piece of wood?

10B.9 A sample of carbon of mass 250. mg from a piece of cloth excavated from an ancient tomb in Nubia undergoes 1.50×10^3 disintegrations in 10.0 h. If a current sample of carbon of mass 1.00 g shows 921 disintegrations per hour, how old is the piece of cloth?

10B.10 A current sample of carbon of mass 1.00 g shows 921 disintegrations per hour. If 1.00 g of charcoal from an archaeological dig in a limestone cave in Slovenia shows 5.50×10^3 disintegrations in 24.0 h, what is the age of the charcoal sample?

10B.11 Deoxyglucose labeled with fluorine-18 is commonly used in PET scans to locate tumors. Fluorine-18 has a half-life of 109 min. How long will it take for the level of fluorine-18 in the body to drop to 10% of its initial value?

10B.12 Technetium-99m (the m signifies a "metastable," or moderately stable, species) is generated in nuclear reactors and shipped to hospitals for use in medical imaging. The radioisotope has a half-life of 6.01 h. If a sample of technetium-99m of mass 165 mg is shipped from a nuclear reactor to a hospital 125 km away in a truck that averages $50.0 \ km \cdot h^{-1}$, what mass of technetium-99m will remain when it arrives at the hospital?

10B.13 A sample of mass 1.40 g containing radioactive cobalt was kept for 2.50 a, at which time it was found to contain 0.266 g of ^{60}Co. The half-life of ^{60}Co is 5.27 a. What percentage (by mass) of the original sample was ^{60}Co?

10B.14 A radioactive sample contains 3.25×10^{18} atoms of a nuclide that decays at a rate of 3.4×10^{13} disintegrations per 15 min. (a) What percentage of the nuclide will have decayed after 150 d? (b) How many atoms of the nuclide will remain in the sample? (c) What is the half-life of the nuclide?

10B.15 A radioactive isotope X with a half-life of 27.4 d decays into another radioactive isotope Y with a half-life of 18.7 d, which decays into the stable isotope Z. Set up and solve the rate laws for the amounts of the three nuclides as a function of time, and plot your results as a graph.

10B.16 Suppose that the nuclide Y in Exercise 10B.15 is needed for medical research and that 2.00 g of nuclide X was supplied at $t = 0$. At what time will Y be most abundant in the sample?

10B.17 A chemist is studying the mechanism of the following hydrolysis reaction of the organic ester methyl acetate: $CH_3COOCH_3 + H_2O \rightarrow CH_3COOH + CH_3OH$. The chemist wants to decide if the O atom in the product methanol comes from the methyl acetate or the water. Propose an experiment using isotopes that would allow the chemist to determine the origin of the oxygen atom.

10B.18 The blood plasma volume in a cardiology patient was measured by injecting 5.0 mL of aqueous human serum albumin labeled with ^{125}I ($t_{1/2} = 59.4$ d). The activity of the sample was 5.1 µCi. After 20. min, blood was withdrawn from the patient and centrifuged to obtain the plasma. The activity of 10.0 mL of plasma was found to be 11 nCi. Report the blood plasma volume of the patient.

10B.19 Measurements of the tritium levels in aqueous solutions can be used to determine their age. The activity of the tritium in a bottle of wine was found to be 8.3% of that of a sample of fresh grape juice from the same region where the wine was bottled. How old is the wine?

10B.20 A nutrient solution containing sulfur-35, which has a half-life of 88 d, is being used to study chemical reactions in which sulfur is used by bacteria. The sample has an activity of 10.0 Ci. What mass of sulfur-35 is present in the nutrient solution? The molar mass of sulfur-35 is $35.0 \ g \cdot mol^{-1}$.

Topic 10C Nuclear Energy

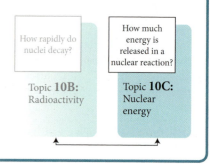

How rapidly do nuclei decay?

How much energy is released in a nuclear reaction?

Topic **10B**: Radioactivity

Topic **10C**: Nuclear energy

Why Do You Need to Know This Material? One of the most pressing problems of the age is the generation of electric power. Nuclear energy is one source, but brings with it problems that chemists can help to resolve.

What Do You Need to Know Already? You need to be aware of the various modes of nuclear decay (Topic 10A).

When nuclei decay, the nucleons adopt new, lower-energy arrangements. The technological problem confronted by the nuclear industry is how to capture that energy difference and then use it to generate electrical power. The associated problem is that the material that remains after the fuel has been spent is highly radioactive and needs to be isolated from the environment. Both problems can be addressed with chemistry.

10C.1 Mass–Energy Conversion

The energy released when a radioactive nucleus decays can be calculated by comparing the masses of the nuclear reactants and products. Einstein's theory of relativity implies that the mass of an object is a measure of its energy content: the greater the mass of an object, the greater its energy. Specifically, the total energy, E, and the mass, m, are related by Einstein's famous equation

$$E = mc^2 \tag{1}$$

where c is the speed of light (3.00×10^8 m·s^{-1}). It follows from this relation that loss of energy is always accompanied by loss of mass.

The mass loss that always accompanies the loss of energy is normally far too small to detect. Even in a strongly exothermic chemical reaction, such as one that releases 10^3 kJ of energy, the masses of the products and the reactants differ by only 10^{-8} g. In a nuclear reaction, the energy changes are very large, the mass loss is measurable, and the energy released can be calculated from the observed change in mass.

A nucleus can be regarded as the result of bringing nucleons (protons and neutrons) together. The **nuclear binding energy,** E_{bind}, is the energy *released* in that process. All binding energies are positive, which means that a nucleus has a lower energy than its constituent nucleons; the greater the binding energy, the lower the energy of the nucleus.

Einstein's equation can be used to calculate the nuclear binding energy from the difference in mass, Δm, between the nucleus and the separated nucleons. For example, iron-56 has 26 protons, each of mass m_p, and 30 neutrons, each of mass m_n. The difference in mass between the nucleus and the separate nucleons is

$$\Delta m = m(\text{nucleus}) - \sum m(\text{nucleons}) = m(^{56}_{26}\text{Fe nucleus}) - (26m_p + 30m_n)$$

Masses of bare nuclei are not readily available and are normally replaced by the mass of the nuclide. That mass includes the total mass of all the Z electrons in the atom of atomic number Z, but that can be canceled by replacing the Z protons by Z hydrogen atoms with their Z electrons. There is a small discrepancy in the binding energy calculated in this way, because the contributions of the electron energies are slightly different in the atom and in the hydrogen atoms, but this contribution is so small that it can be ignored except in the most detailed calculations. Thus, in the present example,

$$\Delta m = m(^{56}_{26}\text{Fe atom}) - (26m_H + 30m_n)$$

where m_H is the mass of a hydrogen atom. The binding energy is then calculated from the difference in mass:

$$E_{\text{bind}} = -\Delta m \times c^2 \tag{2}$$

A nuclide is an atom of specified atomic number and mass number, including the electrons (Topic 10A).

The negative sign is present because a loss of mass ($\Delta m < 0$) corresponds to a positive binding energy. Binding energies are commonly reported in electronvolts (eV) or, more specifically, millions of electronvolts (1 MeV $= 10^6$ eV):

$$1 \text{ eV} = 1.602\ 18 \times 10^{-19} \text{ J}$$

Because the masses of nuclides are so small, they are normally reported as a multiple of the atomic mass constant: $m_u = 1.660\ 54 \times 10^{-27}$ kg. The atomic mass constant is defined as exactly 1/12 the mass of one atom of carbon-12.

EXAMPLE 10C.1 Calculating the nuclear binding energy

If you were a scientist working on the development of nuclear fusion reactions you would have to know how much energy is stored in a nucleus as its binding energy. Calculate the nuclear binding energy in electronvolts of a helium-4 nucleus, given the following masses: ^{4}He, $4.0026m_u$; ^{1}H, $1.0078m_u$; n, $1.0087m_u$, with m_u the atomic mass constant ($1.660\ 54 \times 10^{-27}$ kg).

ANTICIPATE Because nuclear reactions can release huge amounts of energy, you should expect a large value for the nuclear binding energy.

PLAN Write the nuclear equation for the formation of the nuclide from hydrogen atoms and neutrons, and calculate the difference in masses between the products and the reactants (H atoms are used instead of protons to account for the mass of the electrons in the He atom); convert the result from a multiple of the atomic mass constant into kilograms. Then, use the Einstein relation to calculate the energy corresponding to this loss of mass and convert the units to electronvolts.

SOLVE

Write the nuclear equation.

$$2\ ^1\text{H} + 2\ ^1\text{n} \longrightarrow\ ^4\text{He}$$

Calculate the change in mass.

$$\Delta m = m(^4\text{He}) - (2m_\text{H} + 2m_\text{n})$$
$$= 4.0026m_u - \{2(1.0078) + 2(1.0087)\}m_u$$
$$= -0.0304m_u$$

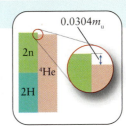

Express the change in mass in kilograms.

$$\Delta m = -0.0304 \times 1.6605 \times 10^{-27} \text{ kg} = -5.04\ldots \times 10^{-29} \text{ kg}$$

Calculate the binding energy from $E_\text{bind} = -\Delta m \times c^2$,

$$E_\text{bind} = -(-5.04\ldots \times 10^{-29} \text{ kg}) \times (3.00 \times 10^8 \text{ m}\cdot\text{s}^{-1})^2$$
$$= 4.54\ldots \times 10^{-12} \text{ kg}\cdot\text{m}^2\cdot\text{s}^{-2} = 4.54\ldots \times 10^{-12} \text{ J}$$

Convert the binding energy to millions of electronvolts:

$$E_\text{bind} = 4.54\ldots \times 10^{-12} \text{ J} \times \frac{1 \text{ eV}}{1.602 \times 10^{-19} \text{ J}} \times \frac{1 \text{ MeV}}{10^6 \text{ eV}} = 28.4 \text{ MeV}$$

EVALUATE The value of the binding energy shows that 4.54 pJ (1 pJ $= 10^{-12}$ J) or 28.4 MeV is released when one ^{4}He nucleus forms from its nucleons. Although this may seem only a small energy, on the atomic scale it is very large, as expected. The total binding energy of 1 mol He atoms is 2.73×10^{12} J, or 2.73 TJ.

Self-test 10C.1A Calculate the binding energy of a carbon-12 nucleus in electronvolts.

[***Answer:*** 92.3 MeV]

Self-test 10C.1B Calculate the binding energy of a uranium-235 nucleus in electronvolts. The mass of one uranium-235 atom is $235.0439m_u$.

Related Exercises 10C.3, 10C.4

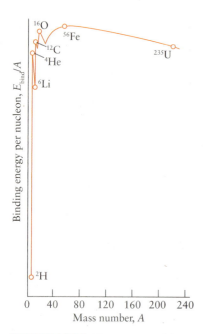

FIGURE 10C.1 The variation of the nuclear binding energy per nucleon. The maximum binding energy per nucleon occurs near iron and nickel. Their nuclei have the lowest energies of all because their nucleons are the most tightly bound.

FIGURE 10C.1 shows the binding energy per nucleon, E_{bind}/A, for the elements. The graph shows that nucleons are bonded together most strongly in the elements near iron and nickel. This high binding energy is one of the reasons why iron and nickel are so abundant in meteorites and on a rocky planet such as Earth. Nuclei of light atoms become more stable when they "fuse" together and heavy nuclei release energy when they undergo "fission" and split into lighter nuclei.

Nuclear binding energies are determined by applying Einstein's formula to the mass difference between the nucleus and its components. Iron and nickel have the highest binding energy per nucleon.

10C.2 The Extraction of Nuclear Energy

In a nuclear **fission** process, a large nucleus is broken into fragments. In a nuclear **fusion** process, a larger nucleus is formed from smaller nuclei. In 1938, Lise Meitner, Otto Hahn, and Fritz Strassmann realized that, by bombarding heavy atoms such as uranium with neutrons, they could "split" the atoms into smaller fragments in fission reactions, releasing huge amounts of energy. The energy that would be released can be estimated by using Einstein's equation. Specifically, the energy change in a fission or fusion process is related to the difference in binding energies of the final and initial nuclei and therefore to their masses.

EXAMPLE 10C.2   Calculating the energy released during fission

Suppose you are a member of a scientific team in a nuclear power plant. You would need to be able to predict how much energy to expect from given fission reactions. When uranium-235 nuclei are bombarded with neutrons, they can split apart in a variety of ways, like glass balls that shatter into pieces of different sizes. In one process, uranium-235 forms barium-142 and krypton-92:

$$^{235}_{92}U + ^{1}_{0}n \longrightarrow ^{142}_{56}Ba + ^{92}_{36}Kr + 2\,^{1}_{0}n$$

Calculate the energy (in joules) released when 1.0 g of uranium-235 undergoes this fission reaction. The masses of the particles are $^{235}_{92}U$, $235.04m_u$; $^{142}_{56}Ba$, $141.92m_u$; $^{92}_{36}Kr$, $91.92m_u$; n, $1.0087m_u$.

ANTICIPATE Fission processes are used to produce energy; therefore, you should expect that a large amount of energy will be released.

PLAN If you know the mass loss, Δm, you can calculate the energy released by one uranium nucleus by using Einstein's equation in the form $\Delta E = \Delta m \times c^2$. To do this, calculate the total mass of the particles on each side of the nuclear equation, take the difference, and substitute the mass difference into this relation. Then determine the number of nuclei in the sample from $N = m(\text{sample})/m(\text{atom})$ and, finally, multiply the energy released from the fission of one nucleus by that number to find the energy released by the sample.

SOLVE

Calculate the total mass of products.

$$m(\text{products}) = m(\text{Ba}) + m(\text{Kr}) + 2m_n$$
$$= \{141.92 + 91.92 + 2(1.0087)\}m_u$$
$$= 235.86m_u$$

Calculate the total mass of reactants.

$$m(\text{reactants}) = m(\text{U}) + m(\text{n})$$
$$= 235.04m_u + 1.0087m_u = 236.05m_u$$

Calculate the change in mass.

$$\Delta m = 235.86m_u - 236.05m_u = -0.19m_u$$

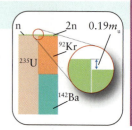

Notice that the neutrons are not canceled even though they appear on both sides of the equation. Like equations for elementary chemical reactions, nuclear equations show the specific process.

Express this change in kilograms.

$$\Delta m = -0.19 \times 1.6605 \times 10^{-27}\,\text{kg} = -3.1\ldots \times 10^{-28}\,\text{kg}$$

Calculate the change in energy for the fission of one nucleus from $\Delta E = \Delta m \times c^2$.

$$\Delta E = (-3.1\ldots \times 10^{-28}\,\text{kg}) \times (3.00 \times 10^8\,\text{m·s}^{-1})^2$$
$$= (-3.1 \times 10^{-28}) \times (3.00 \times 10^8)^2\,\text{J} = -2.8\ldots \times 10^{-11}\,\text{J}$$

Find the number of atoms, N, in the sample from $N = m(\text{sample})/m(\text{atom})$.

$$N = \frac{1.0 \times 10^{-3}\,\text{kg}}{235.04 m_\text{u}} = \frac{1.0 \times 10^{-3}\,\text{kg}}{235.04 \times (1.6605 \times 10^{-27}\,\text{kg})} = 2.56\ldots \times 10^{21}$$

Calculate the total energy change of the sample from $\Delta E(\text{total}) = N\Delta E$.

$$\Delta E(\text{total}) = (2.56\ldots \times 10^{21}) \times (-2.8\ldots \times 10^{-11}\,\text{J}) = -7.3 \times 10^{10}\,\text{J or 73 GJ}$$

EVALUATE The energy released is large, as expected; 73 GJ is 1.3 million times as much energy as would be produced by burning 1.0 g of methane, the main component of natural gas.

Self-test 10C.2A Another mode in which uranium-235 can undergo fission is

$$^{235}_{92}\text{U} + ^1_0\text{n} \longrightarrow ^{135}_{52}\text{Te} + ^{100}_{40}\text{Zr} + ^1_0\text{n}$$

Calculate the energy released when 1.0 g of uranium-235 undergoes fission in this way. The masses needed are $^{235}_{92}\text{U}$, $235.04 m_\text{u}$; n, $1.0087 m_\text{u}$; $^{135}_{52}\text{Te}$, $134.92 m_\text{u}$; $^{100}_{40}\text{Zr}$, $99.92 m_\text{u}$.

[*Answer:* 77 GJ]

Self-test 10C.2B One mode of induced fission of uranium-235 is

$$^{235}_{92}\text{U} + ^1_0\text{n} \longrightarrow ^{138}_{56}\text{Ba} + ^{86}_{36}\text{Kr} + 12\,^1_0\text{n}$$

How much energy is released when 1.0 g of uranium-235 undergoes fission in this manner? The additional masses needed are $^{138}_{56}\text{Ba}$, $137.91 m_\text{u}$; $^{86}_{36}\text{Kr}$, $85.91 m_\text{u}$.

Related Exercises 10C.5, 10C.6, 10C.9, 10C.10

Spontaneous nuclear fission takes place when the natural oscillations of a heavy nucleus cause it to break into two nuclei of similar mass. An example is the spontaneous disintegration of americium-244 into iodine and molybdenum:

$$^{244}_{95}\text{Am} \longrightarrow ^{134}_{53}\text{I} + ^{107}_{42}\text{Mo} + 3\,^1_0\text{n}$$

Fission does not occur in precisely the same way in every instance. For example, more than 200 isotopes of 35 different elements have been identified among the fission products of uranium-235, with most products having mass numbers close to 90 or 130 (**FIG. 10C.2**).

Induced nuclear fission is fission caused by bombarding a heavy nucleus with neutrons (**FIG. 10C.3**). The nucleus breaks into two fragments when struck by a projectile. Nuclei that can undergo induced fission are called **fissionable.** For most nuclei, fission takes place only if the impinging neutrons travel so rapidly that they can smash into the nucleus and drive it apart with the shock of impact; uranium-238 undergoes fission in this way. **Fissile nuclei,** however, are nuclei that can be nudged into breaking apart even by slow neutrons. They include uranium-235, uranium-233, and plutonium-239—the fuels of nuclear power plants.

Once nuclear fission has been induced, it can continue, even if the supply of neutrons from outside is discontinued, provided that the fission produces more neutrons. Such self-sustaining fission takes place in uranium-235, which undergoes numerous fission processes, including

$$^{235}_{92}\text{U} + ^1_0\text{n} \longrightarrow ^{141}_{56}\text{Ba} + ^{92}_{36}\text{Kr} + 3\,^1_0\text{n}$$

If the three product neutrons strike three other fissile nuclei, then after the next round of fission there will be nine neutrons, which can induce fission in nine more nuclei. The neutrons are chain carriers in a branched chain nuclear reaction.

Neutrons produced in a chain reaction are moving very fast, and most escape into the surroundings without colliding with another fissionable nucleus. However, if a large

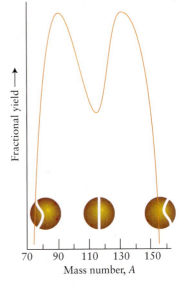

FIGURE 10C.2 The fission yield of uranium-235. Note that the majority of fission products lie in the regions close to $A = 90$ and 130 and that relatively few nuclides corresponding to symmetrical fission (A close to 117) are formed.

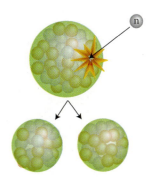

FIGURE 10C.3 In induced nuclear fission, the impact of an incoming neutron causes the nucleus to break apart.

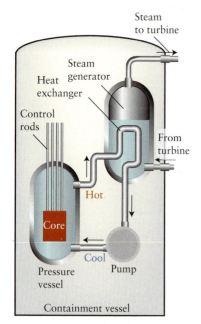

FIGURE 10C.4 A schematic representation of one type of nuclear reactor in which water acts as a moderator for the nuclear reaction. In this pressurized water reactor (PWR), the coolant is water under pressure. The fission reactions produce heat, which boils water in the steam generator; the resulting steam turns the turbines that generate electricity.

enough number of uranium nuclei are present in the sample, enough neutrons can be captured to sustain the chain reaction. In that case, there is a **critical mass,** a mass of fissionable material above which so few neutrons escape from the sample that the fission chain reaction is sustained. If a sample is **supercritical,** with a mass in excess of the critical value, then the reaction is not only self-sustaining but difficult to control and may result in an explosion. The critical mass for a solid sphere of pure plutonium of normal density is about 15 kg, about the size of a grapefruit.

Explosive fission cannot occur in a nuclear reactor because the fuel is not dense enough. Instead, reactors sustain a much slower, controlled chain reaction by making efficient use of a limited supply of neutrons and slowing the neutrons down. The fuel is shaped into long rods and inserted into a **moderator,** a material that slows down the neutrons as they pass between fuel rods; the slower neutrons have a greater probability of collision with a nucleus (**FIG. 10C.4**). The first moderator used was graphite. Heavy water, D_2O, is also an effective neutron moderator, but light-water reactors (LWRs), the most common type of nuclear reactor in the United States, use ordinary water as a moderator.

If the rate of the chain reaction exceeds a certain level, the reactor will become too hot and begin to melt. Control rods—rods made from neutron-absorbing elements, such as boron or cadmium—are inserted between the fuel rods to control the number of available neutrons and the rate of nuclear reaction.

Although fission reactors do not generate chemical pollution, they do produce highly hazardous radioactive waste. However, another kind of nuclear reaction being investigated for power generation is **nuclear fusion,** which is essentially free of long-lived radioactive waste products, and its abundant fuel is readily extracted from seawater. The reaction is the fusion of hydrogen nuclei to form helium nuclei.

Because nuclear binding energy increases on going from light elements such as hydrogen to heavier elements, energy is released when small nuclei fuse together. The strong electrical repulsion between protons makes it difficult for them to approach each other closely, but nuclei of the heavier isotopes of hydrogen fuse together more readily because the additional neutrons contribute to the strong force. To achieve the high kinetic energies needed for successful collisions, fusion reactors need to operate at temperatures above 10^8 K. One method for achieving a controlled release of nuclear energy is to heat a **plasma,** or ionized gas, by passing an electric current through it.

> *Nuclear energy can be extracted by arranging for a nuclear chain reaction to take place in a critical mass of fissionable material. Nuclear fusion makes use of the energy released when light nuclei fuse together to form heavier nuclei.*

10C.3 The Chemistry of Nuclear Power

Chemistry is used in the preparation of uranium, the recovery of important fission products, and the safe disposal or utilization of nuclear waste. The most important mineral source of uranium is *pitchblende*, UO_2 (**FIG. 10C.5**), much of which is obtained (in the United States) from strip mines in New Mexico and Wyoming. The uranium obtained by reduction of the ore is **enriched,** that is, the abundance of a specific isotope, in this case uranium-235, is increased. The natural abundance of uranium-235 is about 0.7%; for use in a nuclear reactor, this fraction must be increased to about 3%. For a nuclear weapon, the enrichment must be much greater.

The enrichment procedure uses the small mass difference between the hexafluorides of uranium-235 and uranium-238 to separate them. The first procedure to be developed converts the uranium into uranium hexafluoride, UF_6, which can be vaporized readily. The different effusion rates of the two isotopic fluorides are then used to separate them. From Graham's law of effusion (rate of effusion $\propto 1/(\text{molar mass})^{1/2}$; Topic 3D), the rates of effusion of $^{235}UF_6$ (molar mass, 349.0 g·mol^{-1}) and $^{238}UF_6$ (molar mass, 352.1 g·mol^{-1}) should be in the ratio

FIGURE 10C.5 Pitchblende is a common uranium ore. It is a variety of uranite, UO_2. (*Chip Clark/Fundamental Photographs, NYC.*)

$$\frac{\text{Rate of effusion of } ^{235}UF_6}{\text{Rate of effusion of } ^{238}UF_6} = \sqrt{\frac{352.1}{349.0}} = 1.004$$

FIGURE 10C.6 Containers of high-level waste products, including cesium-137 and strontium-90, glow under a protective layer of water. If the canisters were unshielded, the radiation that they emit would be great enough to cause death within about 4 s. The blue glow is due to Cherenkov radiation that results when charged particles travel faster than the speed of light in that medium (water in this case). (*Earl Roberge/Science Source.*)

This ratio is so close to 1 that the vapor must be allowed to effuse repeatedly through porous barriers consisting of screens with large numbers of minute holes. In practice, it is allowed to do so thousands of times.

Because the effusion process is technically demanding and uses a lot of energy, scientists and engineers continue to look for alternative enrichment procedures. One of these approaches uses a centrifuge that rotates samples of uranium hexafluoride vapor at very high speed. This rotation causes the heavier $^{238}UF_6$ molecules to be thrown outward and collected as a solid on the outer parts of the rotor, leaving a high proportion of $^{235}UF_6$ closer to the axis of the rotor, from where it can be removed.

Spent nuclear fuel remains radioactive and consists of a mixture of uranium and fission products. Nuclear reactor waste can be processed and some of it reused, but the percentage processed depends on the price of uranium. When the price is high, as it is in the early twenty-first century, much of the nuclear waste is processed for reuse.

The processing of nuclear waste is complex. Any remaining uranium-235 must be recovered, any plutonium produced must be extracted, and the largely useless but radioactive fission products must be stored safely (**FIG. 10C.6**). The highly radioactive fission (HRF) products from used nuclear fuel rods must be stored until their level of radioactivity is no longer dangerous (about 10 half-lives). Generally they are buried underground, but even burial of radioactive wastes is not without problems. Metal storage drums can corrode and allow liquid radioactive waste to seep into aquifers that supply drinking water (**FIG. 10C.7**). Leakage can be minimized by incorporating the HRF products into a glass—a solid, complex network of silicon and oxygen atoms—in a process called *vitrification*. Most of the fission products are oxides of the type that form one of the components of glass: they are network formers; that is, they promote the formation of a relatively disorderly Si—O network rather than inducing crystallization into an orderly array of atoms. Crystallization is dangerous because crystalline regions fracture easily, exposing the incorporated radioactive materials to moisture. Water could dissolve them and carry them away from the storage area. An alternative is to incorporate radioactive waste into hard ceramic materials (see the Interlude following Focus 3). An example is Synroc, a fracture-resistant, titanate-based ceramic that can accept radioactive waste products into its lattice.

> *Uranium is extracted by a series of reactions that lead to uranium hexafluoride; the isotopes are then separated by a variety of procedures. Some radioactive waste is currently converted into glass or ceramic materials for storage underground.*

FIGURE 10C.7 This 35-year-old drum of radioactive waste has corroded and leaked radioactive materials into the soil. The drum was located in one of the nuclear waste disposal sites at the U.S. Department of Energy's Hanford, Washington, nuclear manufacturing and research facility. Several storage sites at this facility became seriously contaminated and had to be cleaned and reconfigured for more stable storage. (*U.S. Department of Energy.*)

What have you learned in this Topic?

You have learned that the energy bound up in a nucleus is expressed as the nuclear binding energy and calculated from the difference of masses of the nucleus and its nucleons (and in practice, from the mass of the nuclide). You have also seen how that energy may be released and have seen a little of the contributions that chemists make to the problem of storing nuclear waste.

The skills you have mastered are the ability to:

☐ **1.** Calculate the nuclear binding energy from mass information (Example 10C.1).

☐ **2.** Calculate the energy released in nuclear fission (Example 10C.2).

Topic 10C Exercises

10C.1 The Sun emits radiant energy at the rate of 3.9×10^{26} J·s^{-1}. What is the rate of mass loss (in kilograms per second) of the Sun?

10C.2 (a) For the fusion reaction 6 D $\rightarrow$ 2 ^{4}He + 2 ^{1}H + 2 n, 3×10^8 kJ of energy is released by a certain sample of deuterium, D (D denotes ^{2}H). What is the mass loss (in grams) for the reaction? (b) What was the mass of deuterium converted? The molar mass of deuterium is 2.014 g·mol^{-1}.

10C.3 Calculate the binding energy per nucleon (J·nucleon^{-1}) for (a) ^{62}Ni, 61.928 346m_u; (b) ^{239}Pu, 239.0522m_u; (c) ^{2}H, 2.0141m_u; (d) ^{3}H, 3.016 05m_u. (e) Which nuclide is the most stable?

10C.4 Calculate the binding energy per nucleon (J·nucleon^{-1}) for (a) ^{98}Mo, 97.9055m_u; (b) ^{151}Eu, 150.9196m_u; (c) ^{56}Fe, 55.9349m_u; (d) ^{232}Th, 232.0382m_u. (e) Which nuclide is the most stable?

10C.5 Calculate the energy released per gram of starting material in the fusion reaction represented by each of the following equations:
(a) $D + D \rightarrow {}^3He + n$ (D, 2.0141m_u; ^{3}He, 3.0160m_u)
(b) $^3He + D \rightarrow {}^4He + {}^1H$ (^{1}H, 1.0078m_u; ^{4}He, 4.0026m_u)
(c) $^7Li + {}^1H \rightarrow 2\,{}^4He$ (^{7}Li, 7.0160m_u)
(d) $D + T \rightarrow {}^4He + n$ (T, 3.0160m_u); D denotes ^{2}H and T denotes ^{3}H.

10C.6 Calculate the energy released per gram of starting material in the nuclear reaction represented by each of the following equations:
(a) $^7Li + {}^1H \rightarrow n + {}^7Be$ (^{1}H, 0.0078m_u; ^{7}Li, 7.0160m_u; ^{7}Be, 7.0169m_u)
(b) $^{59}Co + D \rightarrow {}^1H + {}^{60}Co$ (^{59}Co, 58.9332m_u; ^{60}Co, 59.9529m_u; D, 2.0141m_u)
(c) $^{40}K + \beta \rightarrow {}^{40}Ar$ (^{40}K, 39.9640m_u; ^{40}Ar, 39.9624m_u; β, 0.0005m_u)
(d) $^{10}B + n \rightarrow {}^4He + {}^7Li$ (^{10}B, 10.0129m_u; ^{4}He, 4.0026m_u)

10C.7 Sodium-24 (23.990 96m_u) decays to magnesium-24 (23.985 04m_u). (a) Write a nuclear equation for the decay.

(b) Determine the change in energy that accompanies the decay.
(c) Calculate the change in the binding energy per nucleon.

10C.8 (a) How much energy is emitted in each α decay of plutonium-234? (^{234}Pu, 234.0433m_u; ^{230}U, 230.0339m_u). (b) The half-life of plutonium-234 is 8.8 h. How much heat is released by the α decay of a sample of plutonium-234 of mass 1.00 μg in a 24-h period?

10C.9 One fission reaction that takes place in nuclear reactors is

$$^{235}_{92}U + {}^1_0n \longrightarrow {}^{139}_{56}Ba + {}^{94}_{36}Kr + 3\,{}^1_0n$$

(a) Calculate the energy released (in joules) when 5.0 g of uranium-235 undergoes this reaction. The masses of the isotopes are $^{235}_{92}U$, 235.04m_u; $^{139}_{56}Ba$, 138.91m_u; $^{94}_{36}Kr$, 93.93m_u; n, 1.0087m_u. (b) Calculate the mass of coal that would have to be burned to release the same amount of energy. Assume that the coal consists entirely of graphite.

10C.10 One fission reaction that takes place in nuclear reactors is

$$^{235}_{92}U + {}^1_0n \longrightarrow {}^{140}_{54}Xe + {}^{92}_{38}Sr + 4\,{}^1_0n$$

(a) Calculate the energy released (in joules) when 5.0 g of uranium-235 undergoes this reaction. The masses of the isotopes are $^{235}_{92}U$, 235.04m_u; $^{140}_{54}Xe$, 139.92m_u; $^{92}_{38}Sr$, 91.91m_u; n, 1.0087m_u. (b) Calculate the mass of coal that would have to be burned to release the same amount of energy. Assume that the coal consists entirely of graphite.

The following Example and Exercises draw on material from throughout Focus 10.

FOCUS 10 Online Cumulative Example

Targeted alpha therapy (TAT) is an emerging method of cancer treatment. The procedure involves administering a drug that contains a nuclide that is an α emitter linked to a bioactive molecule that is absorbed preferentially in tumor tissues. Because α particles are highly damaging to soft tissues but do not penetrate very far, cell damage occurs only in a small localized region. You are developing new TAT drugs and are considering using astatine-211 as an α particle source and need to understand its decay characteristics. (a) Predict the identity of the daughter nuclide formed when ^{211}At undergoes α decay and write the balanced nuclear reaction. (b) The energy released when one ^{211}At nucleus decays is 5.97 MeV. Predict the mass of the daughter nuclide, given that the mass of ^{211}At is $210.9875 m_u$ and that of ^{4}He is $4.0026 m_u$. (c) The half-life of ^{211}At is 7.2 hours. How long will a patient have to wait before 95.0% of an administered drug containing ^{211}At has undergone nuclear decay?

 The online Cumulative Example solution can be found at http://macmillanhighered.com/chemicalprinciples7e

FOCUS 10 Exercises

Note that the SI symbol for 1 year is 1 a and that it takes the usual prefixes, as in 1 ka = 10^3 a and 1 Ga = 10^9 a.

10.1 State whether the following statements are true or false. If false, explain why. (a) The dose equivalent is lower than the actual dose of radiation because it takes into account the different effects of different types of radiation. (b) Exposure to 1×10^8 Bq of radiation would be much more hazardous than exposure to 10 Ci of radiation. (c) Spontaneous radioactive decay follows first-order kinetics.

10.2 State whether the following statements are true or false. If false, explain why. (a) Fissile nuclei can undergo fission when struck with slow neutrons, whereas fast neutrons are required to split fissionable nuclei. (b) For fusion to occur, the colliding particles must have high relative kinetic energy. (c) The larger the binding energy per nucleon, the more stable is the nucleus.

10.3 (a) How many radon-222 nuclei ($t_{1/2}$ = 3.82 d) decay per minute to produce an activity of 4.0 pCi? (b) A bathroom in the basement of a home measures 2.0 m $\times$ 3.0 m $\times$ 2.5 m. If the activity of radon-222 in the room is 4.0 pCi·L^{-1}, how many nuclei decay during a shower lasting 5.0 min?

10.4 Tritium undergoes β decay, and the emitted β particle has an energy of 0.0186 MeV (1 MeV = 1.602×10^{-13} J). If a sample of tissue of mass 1.0 g absorbs 10% of the decay products of 1.0 mg of tritium, what dose equivalent does the tissue absorb?

10.5 It is found that 20. μmol ^{222}Rn ($t_{1/2}$ = 3.82 d) has seeped into a closed basement with a volume of 2.0×10^3 m^3. (a) What is the initial activity of the radon in picocuries per liter (pCi·L^{-1})? (b) How many atoms of ^{222}Rn will remain after 1 day (24 h)? (c) How long will it take for the radon to decay to below the EPA-recommended level of 4 pCi·L^{-1}?

10.6 A radioactive sample contains ^{32}P (half-life, 14.28 d), ^{33}S (half-life, 87.2 d), and ^{59}Fe (half-life, 44.6 d). After 90 d, a sample that originally weighed 8.00 g contains 0.0254 g of ^{32}P, 1.466 g of ^{33}S, and 0.744 g of ^{59}Fe. What was the percentage composition (by mass) of the original sample?

10.7 Uranium-238 decays through a series of α and β emissions to lead-206, with an overall half-life for the entire process of 4.5 Ga. How old is a uranium-bearing ore that is found to have a ^{238}U/^{206}Pb ratio of (a) 1.00; (b) 1.25?

10.8 Suppose a planet on which life is based on silicon instead of carbon has been discovered. However, a meteor strike destroyed most of the life on the planet. To determine how long ago the meteor strike occurred, the activity of silicon-32, which has a half-life of 1.6×10^2 a, was measured in samples from the fossil remains of life forms killed in the meteor strike and in existing life forms. Activity in the fossil samples was found to be 0.015% of the activity in the living samples. How long ago did the meteor strike occur?

10.9 Technetium-99m is produced by a sequence of reactions in which molybdenum-98 is bombarded with neutrons to form molybdenum-99, which undergoes β decay to technetium-99m. (a) Write the balanced nuclear equations for this sequence. (b) Compare the neutron-to-proton ratio of the final daughter product with that of molybdenum-99. Which is closer to the band of stability?

10.10 Actinium-225 decays by successive emission of three α particles. (a) Write the nuclear equations for the three decay processes. (b) Compare the neutron-to-proton ratio of the final daughter product with that of actinium-225. Which is closer to the band of stability?

10.11 What volume of helium at 1.0 atm and 298 K will be collected if 2.5 g of ^{222}Rn is stored in a container able to expand to maintain constant pressure for 23 d? (^{222}Rn decays to ^{218}Po with a half-life of 3.824 d.)

10.12 A nuclear waste storage facility is being proposed for your locality and you have been asked to prepare a recommendation for how highly radioactive fission products should be processed and stored. In your recommendation, discuss the benefits and drawbacks of at least three modes of nuclear waste storage.

10.13 The radioactivity from a sample of Na$_2$^{14}CO$_3$ was measured by noting the time required for the count of disintegrations to reach 8000. Five times were recorded: 21.25, 23.46, 20.97, 22.54, and 23.01 min. Background radiation was measured by noting the time required for 500 counts to be registered. Three times were recorded: 5.26, 5.12, and 4.95 min. What is the average level of radioactivity in the $Na_2^{14}CO_3$ sample, corrected for background radiation, in (a) disintegrations per minute; (b) microcuries?

10.14 (a) Use a standard graphing program to plot the fraction of ^{14}C remaining in an archaeological sample 40.0 ka old as a function of time. For convenience, use intervals of 1000 a. (b) Plot the natural logarithm of the fraction of ^{14}C remaining against time. (c) After what period of time will less than 1.00% of the original ^{14}C in the sample remain?

10.15 Radiopharmaceuticals have one of two general functions: (1) they may be used to *detect* or *image* biological problems such as tumors and (2) they may be used to *treat* an illness. Which type of radiation (α, β, or γ) would be the most suitable for (a) detection and (b) therapy? Justify your selections. (c) From standard literature or online sources, find at least two radionuclides that have been used for imaging body tissues. (d) What are the half-lives of these radionuclides?

10.16 Radioactive metal ions that have short half-lives are being intensely studied as pharmaceuticals. The strategy is to attach a well-designed ligand to the metal ion so that the complex very selectively aggregates in one particular type of body tissue. What properties of the ligand are important to the design of effective radiopharmaceutical therapeutic agents?

10.17 Sodium-24 is used for monitoring blood circulation. (a) If 2.0 mg of sodium-24 has an activity of 17.3 Ci, what is its decay constant and its half-life? (b) What mass of the sodium-24 sample remains after 2.0 d? The mass of a sodium-24 atom is $24m_u$.

10.18 A positron has the same mass as that of an electron, but the opposite charge. When a positron emitted in a PET scan encounters an electron, annihilation occurs in the body: electromagnetic energy is produced and no matter remains. How much energy (in joules) is produced in the encounter? See Box 10A.1.

10.19 Do nuclei that are positron emitters lie above or below the band of stability? Which of the following isotopes might be suitable for PET scans? Explain your reasoning and write the equation for the decay: (a) ^{18}O; (b) ^{13}N; (c) ^{11}C; (d) ^{20}F; (e) ^{15}O. See Box 10A.1.

10.20 The biological half-life of a radioisotope is the time required for the body to excrete half of the radioisotope. The *effective half-life* is the time required for the amount of a radioisotope in the body to be reduced to half its original amount, as a result of both the decay of the radioisotope and its excretion. Sulfur-35 ($t_{1/2} = 87.4$ d) is used in cancer research. The biological half-life of sulfur-35 in the human body is 90. d. What is the effective half-life of sulfur-35?

10.21 Barium-140 ($t_{1/2} = 12.8$ d) released in the fire at the Chernobyl nuclear plant has been found in some agricultural products in the region. The biological half-life of barium-140 in the human body is 65 d. What is the effective half-life (see Exercise 10.20) of barium-140?

FOCUS 10 Cumulative Exercise

10.22 A person with pernicious anemia lacks "intrinsic factor," a compound required for the absorption of vitamin B_{12} and its storage in the liver. The diagnosis is confirmed with the Schilling test. In this test the patient is given a small dose of vitamin B_{12} labeled with radioactive ^{57}Co or ^{58}Co, followed by a saturating dose of unlabeled B_{12}, which releases the stored B_{12}. If the patient has intrinsic factor, a 24-hour urine sample will contain 13–15% of the labeled B_{12}. If intrinsic factor is absent, less than 6% will be excreted. The patient is then given intrinsic factor and the test is repeated for comparison. In one administration of the Schilling test, the patient was given a capsule containing 0.5 μCi of $^{58}CoB_{12}$, followed by 1.0 mg of unlabeled B_{12}. Then 1200 mL of urine was collected over the next 24 hours. A sample of the urine of volume 3.0 mL gave 83 cpm (counts per minute) and a standard sample of volume 3.0 mL containing 0.4 nCi per mL gave 910 cpm. The test was repeated one week later with the administration of 30. mg of intrinsic factor. The sample of urine from the second test gave 120 cpm. The half-life of ^{58}Co is 72 days.

(a) Calculate the percentage of ^{58}Co excreted with and without the intrinsic factor. Assume that the first test did not contaminate the second test.

(b) What would be the activity of the $^{58}CoB_{12}$ in the original capsule if it were stored for 7 days?

(c) If the biological half-life of B_{12} is 180 days, what is the effective half-life of $^{58}CoB_{12}$ in the body? See Exercise 10.20.

(d) Use the effective half-life of $^{58}CoB_{12}$ to determine what fraction of the counts in the second test would be due to the dose given in the first test.

(e) The radioisotope ^{58}Co decays to another radioisotope, ^{59}Fe. The total mass of iron in the patient was 2.5 g. If all the ^{59}Fe produced in the first test were incorporated into hemoglobin, what percentage of the total iron would be ^{59}Fe the day after the first test? Ignore the decay of ^{59}Fe.

ORGANIC CHEMISTRY

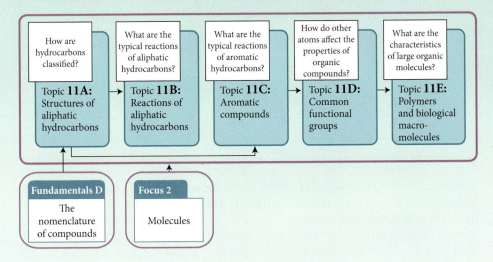

How are hydrocarbons classified?

Topic **11A:** Structures of aliphatic hydrocarbons

What are the typical reactions of aliphatic hydrocarbons?

Topic **11B:** Reactions of aliphatic hydrocarbons

What are the typical reactions of aromatic hydrocarbons?

Topic **11C:** Aromatic compounds

How do other atoms affect the properties of organic compounds?

Topic **11D:** Common functional groups

What are the characteristics of large organic molecules?

Topic **11E:** Polymers and biological macro-molecules

Fundamentals D

The nomenclature of compounds

Focus 2

Molecules

Carbon is the basis of many materials that are essential to modern technology and even to life itself. Understanding the compounds that carbon forms and their reactions is therefore important not only to our technological and medical advances but also to our very survival. Carbon forms such a huge variety of compounds that an entire field of chemistry, *organic chemistry*, is devoted to its study. Carbon atoms are so adaptable because they can string together to form chains and rings of endless variety. This versatility allows the element to form thousands of complicated biomolecules that contribute to the structure and function of every organism on Earth.

The first three Topics deal with "hydrocarbons," compounds built from only carbon and hydrogen. **TOPIC 11A** deals with the structures of the so-called aliphatic compounds, **TOPIC 11B** with their reactions; **TOPIC 11C** describes both the structure and the reactions of a second group of compounds, the so-called "aromatic" compounds.

The final two Topics deal with organic compounds containing atoms of elements other than carbon and hydrogen. **TOPIC 11D** introduces the structures and properties of some of the common groups of these atoms and introduces some of the characteristic mechanisms by which they enliven the hydrocarbons. One major role of these groups of atoms is to provide ways to link molecules together into long chains, and **TOPIC 11E** examines how they are used to create the polymers that are so characteristic of the modern world. As in so many spheres, nature has preceded chemists' explorations. The final part of this Topic shows the role of organic compounds in nature: how they sustain us, feed us, and reproduce us.

Topic 11A Structures of Aliphatic Hydrocarbons

11A.1 Types of Aliphatic Hydrocarbons

11A.2 Isomers

11A.3 Physical Properties of Alkanes and Alkenes

How are hydrocarbons classified?

What are the typical reactions of aliphatic hydrocarbons?

Topic 11A: Structures of aliphatic hydrocarbons → Topic 11B: Reactions of aliphatic hydrocarbons

Why Do You Need to Know This Material? All life on Earth is based on compounds of carbon; so are fuel, food, and clothing. They are the foundation of the petrochemical industry and are used in the flexible, strong polymeric and composite materials that make modern communication and transportation possible.

What Do You Need to Know Already? This Topic draws on the introduction to organic formulas and nomenclature in *Fundamentals* C and D, the structure of molecules (Focus 2), intermolecular forces (Topic 3F), reaction enthalpy (Topic 4D), reaction mechanisms (Topic 7C), and the concept of isomers (Topic 9C).

A hydrocarbon is a compound of carbon and hydrogen. There are two major types of hydrocarbons: aliphatic and aromatic. **Aliphatic hydrocarbons** have no benzenelike aromatic rings; **aromatic hydrocarbons** have at least one. The compound shown as (**1**) is aliphatic; compound (**2**) is aromatic. More complex molecules are said to have an "aromatic region" consisting of benzenelike rings and an "aliphatic region" containing chains of carbon atoms. Aliphatic hydrocarbons can be regarded as the basic framework for many organic compounds, not just the hydrocarbons themselves.

1 Pentane, C_5H_{12}

2 Ethylbenzene, $C_6H_5CH_2CH_3$

11A.1 Types of Aliphatic Hydrocarbons

Aliphatic hydrocarbons are divided into two broad classes according to the type of bonding between their carbon atoms. A **saturated hydrocarbon** is an aliphatic hydrocarbon with no multiple carbon–carbon bonds; an **unsaturated hydrocarbon** has one or more double or triple carbon–carbon bonds. More hydrogen atoms can be added to compounds in which there are multiple bonds, but compounds with only single bonds are considered to be "saturated" with hydrogen. Compound (**3**) is saturated; compounds (**4**) and (**5**) are unsaturated.

Because many organic molecules are very complicated, chemists have developed a simple way to represent their structures. It is often sufficient to give a **condensed structural formula,** which shows how the atoms are grouped together (see *Fundamentals* C). For instance, $CH_3CH_2CH_2CH_3$ is the condensed structural formula for butane, and $CH_3CH(CH_3)CH_3$ is the formula for methylpropane. The parentheses around one of the CH_3 groups show that it is attached to the carbon atom on its left (or, if the formula begins with the parenthetical group, the carbon atom on its right). When several groups of atoms are repeated, they can be collected together; so the formula of butane could also be written as $CH_3(CH_2)_2CH_3$ and that of methylpropane as $(CH_3)_3CH$.

An even simpler representation of a hydrocarbon is as a "line structure" (introduced in *Fundamentals* C), in which a chain of carbon atoms is shown as a zigzag line. The end of each short line in the zigzag represents a carbon atom. Because carbon nearly always has a valence of 4 in organic compounds, it is not necessary to show H atoms: just fill in the correct number of hydrogen atoms mentally, as has been done for methylbutane (**6**), isoprene (**7**), and propyne (**8**).

3 Hexane, C_6H_{14}

4 2-Hexene, C_6H_{12}

5 3-Hexene, C_6H_{12}

6 Methylbutane, $(CH_3)_2CHCH_2CH_3$

7 Isoprene, $CH_2=C(CH_3)CH=CH_2$

8 Propyne, $CH_3C≡CH$

A benzene ring is often represented either by a hexagon of alternating single and double bonds or as a circle inside a hexagon (Topic 2B); in either case only one H atom is attached to each C atom.

Self-test 11A.1A Draw (a) the line structure of aspirin (**9a**) and (b) the structural formula for $CH_3(CH_2)_2C(CH_3)_2CH_2C(CH_3)_3$.

[***Answer:*** (a) (**9b**); (b) (**10**)]

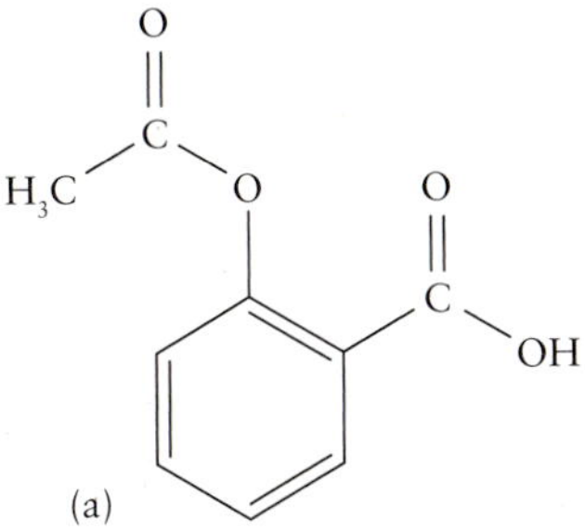

(a) 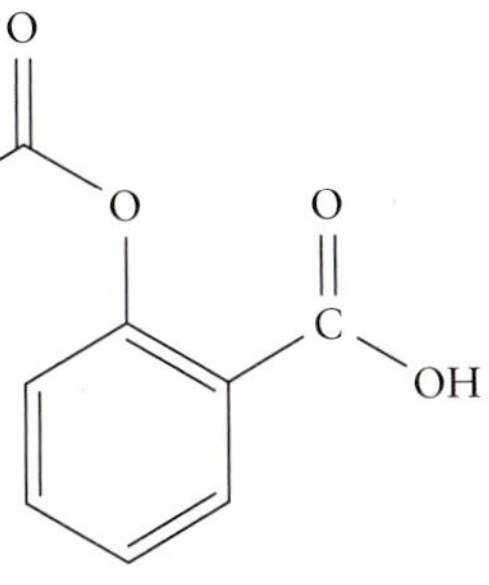 (b)

9 Acetylsalicylic acid (aspirin) **10**

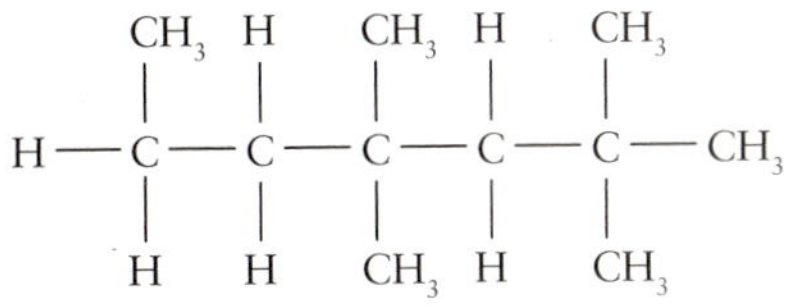

Self-test 11A.1B Write (a) the structural formula of carvone (**11a**) and (b) the condensed structural formula for (**11b**).

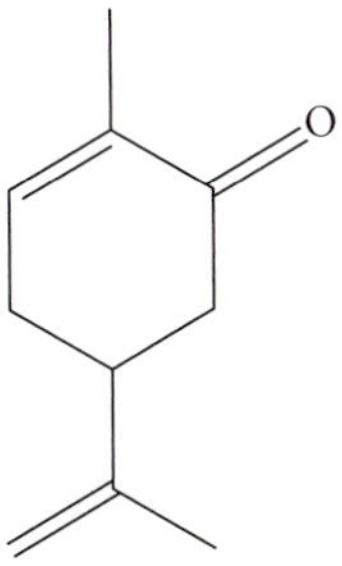

11a Carvone 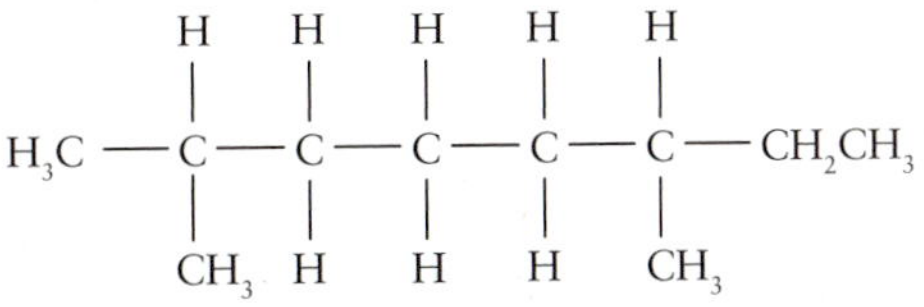 **11b**

Saturated hydrocarbons are called **alkanes.** Each carbon atom in an alkane has four single bonds in a tetrahedral arrangement, with sp^3 hybridization (Topic 2F). The simplest alkane is methane, CH_4 (**12**). The formulas of other alkanes can be thought of as derived from CH_4 by inserting CH_2 groups between pairs of atoms. Although alkane structures are usually drawn in two dimensions as flat structures with 90° angles, as in (**11b**), it is important to remember that they consist of tetrahedral arrangements of bonds at each carbon atom. Moreover, because all the C—C bonds are single, the different parts of an alkane molecule can rotate relative to each other. In liquids and gases, the chains of atoms in an alkane molecule are in constant motion, at some instants rolled up into a ball (**13**) and at others stretched out into a zigzag (**14**).

To name an alkane in which the carbon atoms form a single chain, a prefix denoting the number of carbon atoms is combined with the suffix *-ane* (**TABLE 11A.1**). For example, CH_3—CH_3 (more simply, CH_3CH_3) is ethane and CH_3—CH_2—CH_3 (that is, $CH_3CH_2CH_3$) is propane. Cyclopropane, C_3H_6 (**15**), and cyclohexane, C_6H_{12} (**16**), are **cycloalkanes,** alkanes that contain rings of carbon atoms.

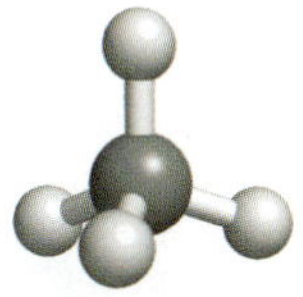

12 Methane, CH_4

13 Decane, $C_{10}H_{22}$

14 Decane, $C_{10}H_{22}$

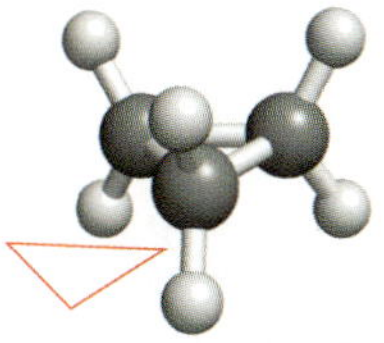

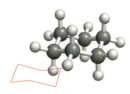

15 Cyclopropane, C_3H_6 **16 Cyclohexane, C_6H_{12}**

17   Butane, C_4H_{10}

18   Methylpropane, C_4H_{10}

The name isomer comes from the Greek words for "equal parts," suggesting that isomers are built from the same kit of parts.

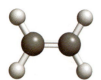

19   Ethene, C_2H_4

20   Ethyne, C_2H_2

TABLE 11A.1   Alkane Nomenclature*

Number of carbon atoms	Formula	Name of alkane	Name of alkyl group	Formula
1	CH_4	methane	methyl	$CH_3—$
2	CH_3CH_3	ethane	ethyl	$CH_3CH_2—$
3	$CH_3CH_2CH_3$	propane	propyl	$CH_3CH_2CH_2—$
4	$CH_3(CH_2)_2CH_3$	butane	butyl	$CH_3(CH_2)_2CH_2—$
5	$CH_3(CH_2)_3CH_3$	pentane	pentyl	$CH_3(CH_2)_3CH_2—$
6	$CH_3(CH_2)_4CH_3$	hexane	hexyl	$CH_3(CH_2)_4CH_2—$
7	$CH_3(CH_2)_5CH_3$	heptane	heptyl	$CH_3(CH_2)_5CH_2—$
8	$CH_3(CH_2)_6CH_3$	octane	octyl	$CH_3(CH_2)_6CH_2—$
9	$CH_3(CH_2)_7CH_3$	nonane	nonyl	$CH_3(CH_2)_7CH_2—$
10	$CH_3(CH_2)_8CH_3$	decane	decyl	$CH_3(CH_2)_8CH_2—$
11	$CH_3(CH_2)_9CH_3$	undecane	undecyl	$CH_3(CH_2)_9CH_2—$
12	$CH_3(CH_2)_{10}CH_3$	dodecane	dodecyl	$CH_3(CH_2)_{10}CH_2—$

*Greek prefixes are used for alkanes and alkyl groups with chains of more than 12 carbon atoms.

Another reason for the variety of compounds that carbon can form is that the same atoms can bond together in different arrangements. Thus, four carbon atoms can link together in a chain to form butane (**17**) or in a Y-shape to form methylpropane (**18**). Different compounds with the same molecular formula are called **isomers** (Topic 9C). Thus, butane and methylpropane are isomers with the same molecular formula, C_4H_{10}.

THINKING POINT

How many other scientific terms with the prefix iso- do you know?

The simplest unsaturated hydrocarbon with a double bond is ethene, C_2H_4 or $H_2C=CH_2$, which is commonly called ethylene (**19**). Ethene is the parent of the **alkenes,** a series of compounds with formulas derived from $H_2C=CH_2$ by inserting CH_2 groups. For example, the next member of the family is propene, $H—CH_2—CH=CH_2$ (more simply, $CH_3CH=CH_2$). The name of an alkene is the same as the name of the corresponding alkane, except that it ends in -*ene*. The location of the double bond is specified by numbering the carbon atoms in the chain and writing the lower of the two numbers of the carbon atoms joined by the double bond. Thus, $CH_3CH_2CH=CH_2$ is 1-butene and $CH_3CH=CHCH_3$ is 2-butene (**Toolbox 11A.1**). The term alkene also includes hydrocarbons with more than one double bond, as in $CH_2=CH—CH=CH_2$, 1,3-butadiene.

Alkynes are hydrocarbons that have at least one carbon–carbon triple bond. The simplest is ethyne, $HC≡CH$, which is commonly called acetylene (**20**). Alkynes are named like alkenes but with the suffix -*yne*.

Toolbox 11A.1   HOW TO NAME ALIPHATIC HYDROCARBONS

CONCEPTUAL BASIS

Because the types of carbon–carbon bonds present in a hydrocarbon molecule typically dominate its properties, an aliphatic hydrocarbon is first classified as an alkane, alkene, or alkyne. Then the longest chain of carbon atoms is used to form the "root" of the name. Other hydrocarbon groups attached to the longest chain are named as side chains.

PROCEDURE

The following rules for naming hydrocarbons have been adopted by the International Union for Pure and Applied Chemistry (IUPAC). For additional information on naming

hydrocarbons, see *Section A* in the IUPAC "Blue Book" (http://www.acdlabs.com/iupac/nomenclature).

Alkanes

Step 1 Count carbon atoms in the longest chain.

The names of the first 12 straight-chain alkanes are given in Table 11A.1; they all end in -*ane*.

Step 2 Identify and count substituents.

Name a hydrocarbon side chain as a substituent by changing the ending -*ane* to -*yl* (as in the last two columns in Table 11A.1).

Example: $CH_3CH_2—$ is the ethyl group.

The names of branched-chain hydrocarbons and hydrocarbon derivatives are based on the name of the *longest* continuous carbon chain in the molecule (which might not be shown in a horizontal line).

Example: H_3C — CH — CH_3
CH_2 — CH_3 is methylbutane.

A cyclic (ring) hydrocarbon is designated by the prefix *cyclo-*.

Example: CH_2
H_2C — CH_2 is cyclopropane.

Step 3 Number the backbone carbon atoms from the end that gives the lowest numbers to the substituents.

To indicate the position of a branch or substituent, the carbon atoms in the longest chain are numbered consecutively from one end to the other, starting at the end that will give the *lower* numbers to the substituents.

Examples:

$CH_3CH_2CH_2CHCH_2CH_3$
CH_2CH_3

3-Ethylhexane

$CH_3C(CH_3)_2CH_2CH_3$

2,2-Dimethylbutane

Step 4 Indicate how many of each substituent are in the molecule by the appropriate prefix.

The prefixes *di-, tri-, tetra-, penta-, hexa-,* and so on, indicate how many of each substituent are in the molecule. Numbers set off by hyphens specify to which carbon atoms the groups are attached.

Examples:

2,2,3-Trimethylbutane

1-Ethyl-2-methylcyclopentane

Step 5 List the substituents in alphabetical order (disregarding the Greek prefixes) and attach them to the root name. When numbering substituents in equivalent positions, the one listed first is assigned the lower number, as in 1-ethyl-2-methylcyclopentane (not 2-ethyl-1-methylcyclopentane).

The names of substituents other than hydrocarbon groups are discussed more fully in Topic 11D (see Toolbox 11D.1).

Example 11A.1 illustrates how to name alkanes.

Alkenes and Alkynes

Double bonds in hydrocarbons are indicated by changing the suffix *-ane* to *-ene*, and triple bonds by changing the suffix to *-yne*. The position of the multiple bond is given by the number of the first (lower-numbered) carbon atom involved in the multiple bond. If more than one multiple bond of the same type is present, the number of these bonds is indicated by a Greek prefix. Then follow the rules for naming alkanes with Steps 2 and 3 replaced by:

Step 2 Identify and count substituents and multiple bonds.

Step 3 Number the backbone carbon atoms starting at one end so as to give the lowest number to the multiple bond.

Examples:

H_3C — CH_2 — $C \equiv CH$
1-Pentyne

H_3C — CH_2 — $CH = CH$ — CH_3
2-Pentene

$H_2C = CH$ — CH_2 — $CH = CH_2$
1,4-Pentadiene

When numbering atoms in the chain, the lowest numbers are given preferentially to (a) groups of atoms named by suffixes (see Toolbox 11D.1), (b) double bonds, (c) triple bonds, and (d) groups named by prefixes.

Example 11A.2 illustrates how to name alkenes.

EXAMPLE 11A.1   Naming alkanes and cycloalkanes

You will often come across labels on pharmaceuticals and in other contexts that are expressed in IUPAC names, and it is a part of your scientific training to be able to interpret them. (a) Name the compound shown as (**21**) and (b) write the structural formula of 2-ethyl-1, 1-dimethylcyclohexane.

PLAN Use the procedure in Toolbox 11A.1. For part (b), interpret the name by identifying the root of the name and then attaching the substituents to the specified locations.

SOLVE

(a) Count carbon atoms in the longest chain.

The longest carbon chain (in red) of (**21**) has five carbon atoms. The molecule is a substituted pentane.

21

Identify and count substituents.

There are three methyl groups (CH₃—) on the longest chain. The molecule is a trimethylpentane.

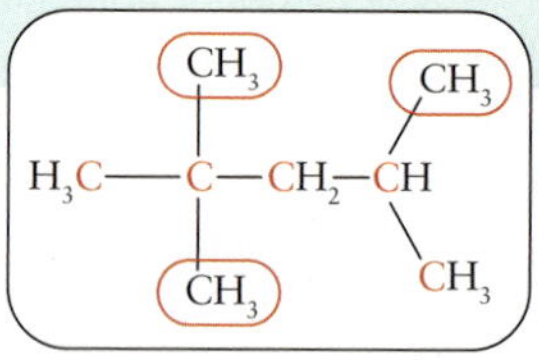

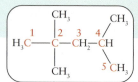

Number the backbone carbon atoms, starting at one end, so as to give the lowest numbers to the substituents.

2,2,4-trimethylpentane

(b) Draw the longest chain of carbon atoms first.

"Cyclohexane" implies that the molecule has a ring of six carbon atoms.

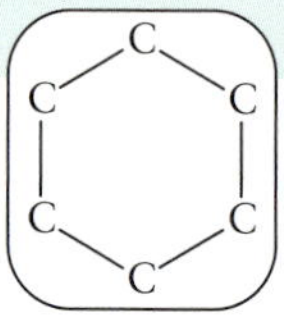

Number the carbon atoms and add the substituents according to the number given in the name.

Add two methyl groups to one carbon atom, which becomes carbon atom 1. Then add an ethyl group to carbon atom 2.

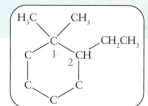

Add hydrogen atoms as needed to give each carbon atom a valence of 4.

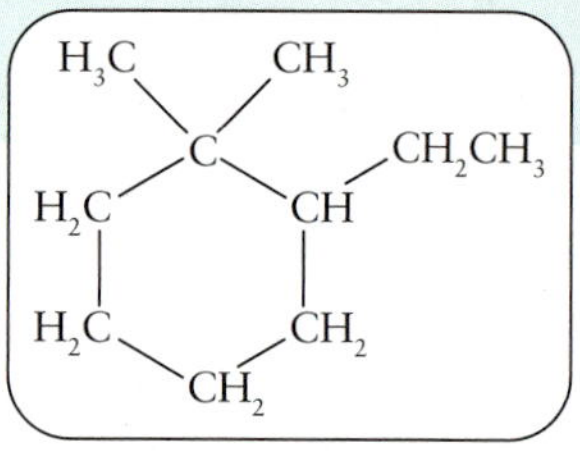

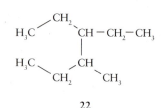

22

Self-test 11A.2A (a) Name the compound shown as (**22**) and (b) write the structural formula of 5-ethyl-2, 2-dimethyloctane.

[**Answer:** (a) 3-Ethyl-4-methylhexane; (b) (**23**)]

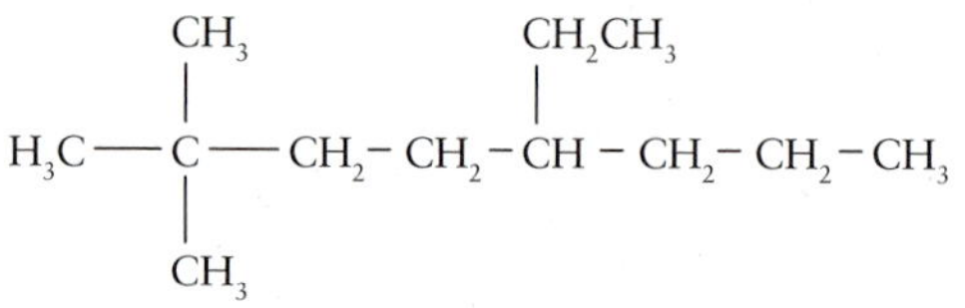

$$H_3C - C - CH_2 - CH_2 - CH - CH_2 - CH_2 - CH_3$$

23

Self-test 11A.2B (a) Name the compound $(CH_3)_2CHCH_2CH(CH_2CH_3)_2$ and (b) write the structural formula of 3,3,5-triethylheptane.

Related Exercises 11A.3–11A.8

EXAMPLE 11A.2   Naming alkenes

Learning organic chemistry is very much (in part, at least) like learning a language and being able to interpret the words (the names of compounds) that you encounter. (a) Name the alkene $CH_3CH_2CH=CH_2$ and (b) write the condensed structural formula for 5-methyl-1,3-hexadiene.

PLAN (a) Use the procedure in Toolbox 11A.1. (b) Interpret the name by identifying the backbone and the locations of the double bonds and substituents.

SOLVE

(a) *Step 1* Count carbon atoms in the longest chain.

The longest chain (in red) is four, indicating the root but-.

Step 2 Identify and count substituents and multiple bonds.

There are no substituents, but there is a double bond, so the suffix is -ene.

Step 3 Number the backbone carbon atoms, from the end that gives the lowest numbers to the location of the double bond.

$CH_3CH_2CH{=}CH_2$ is 1-butene (not 3-butene).

(b) 5-Methyl-1,3-hexadiene has two double bonds, at C atoms 1 and 3, with a methyl group on C atom 5: $CH_2{=}CHCH{=}CHCH(CH_3)CH_3$.

Self-test 11A.3A (a) Name the alkene $(CH_3)_2CHCH{=}CH_2$ and (b) write the condensed structural formula for 2-methylpropene.

[***Answer:*** (a) 3-Methyl-1-butene; (b) $CH_2{=}C(CH_3)_2$]

Self-test 11A.3B (a) Name the alkene $(CH_3CH_2)_2CHCH{=}CHCH_3$ and (b) write the structural formula for cyclopropene.

Related Exercises 11A.9–11A.14

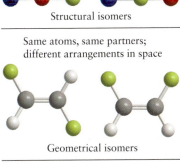

FIGURE 11A.1 A summary of the types of isomerism in organic compounds.

Saturated hydrocarbons have only single bonds; unsaturated hydrocarbons have one or more multiple bonds. Alkanes are saturated hydrocarbons. Alkenes and alkynes are unsaturated hydrocarbons: the former have carbon–carbon double bonds and the latter have triple bonds.

11A.2 Isomers

FIGURE 11A.1 summarizes the types of isomerism that are found in organic compounds. Molecules that are **structural isomers** are built from the same atoms, but the atoms are connected differently; that is, the molecules have a different **connectivity**. For example, a CH_2 group can be inserted into the C_3H_8 molecule in two different ways to give two different compounds with the formula C_4H_{10}, one butane (**24**) and the other methylpropane (**25**). Although the CH_2 group could be inserted in other places, the free rotation about single C—C bonds in hydrocarbons allows the resulting molecules to be twisted into one or the other of these two isomers. Both compounds are gases, but butane condenses at $-1\,°C$, whereas methylpropane condenses at $-12\,°C$.

Two molecules that differ only by rotation about one or more bonds might look different on paper, but they are not isomers of each other; they are different **conformations** of the same molecule. Example 11A.3 illustrates how to tell whether two molecules are different isomers or different conformations of the same isomer.

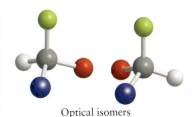

24 Butane, C_4H_{10}

25 Methylpropane, C_4H_{10}

EXAMPLE 11A.3   Writing the formulas of structural isomers

Suppose you have synthesized a new compound. Even if the combustion analysis agrees with the empirical formula of the compound you had intended to prepare, you might have made an isomer of the desired compound instead. Therefore, it is useful to be able to identify all possible isomers. Draw two-dimensional structural formulas for all the isomeric alkanes of formula C_5H_{12}.

PLAN Isomers cannot be changed into one another simply by rotating either the entire structure or parts of the structure on the page. One approach is to insert CH_2 groups into different parts of the formulas of the two C_4H_8 molecules given earlier (as **24** and **25**), then discard formulas that repeat those already obtained. It is often easier to identify different conformations of the same isomer by building molecular models that allow rotation about single bonds.

SOLVE

From butane, you can form

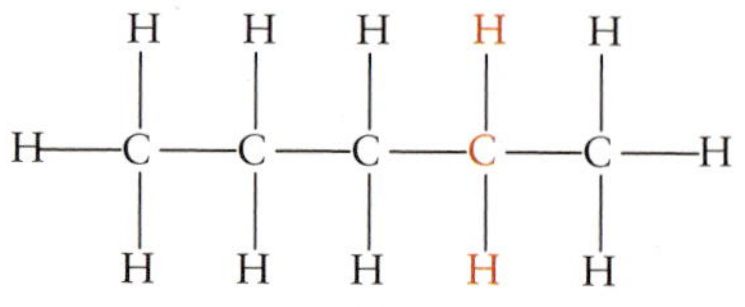

(a)

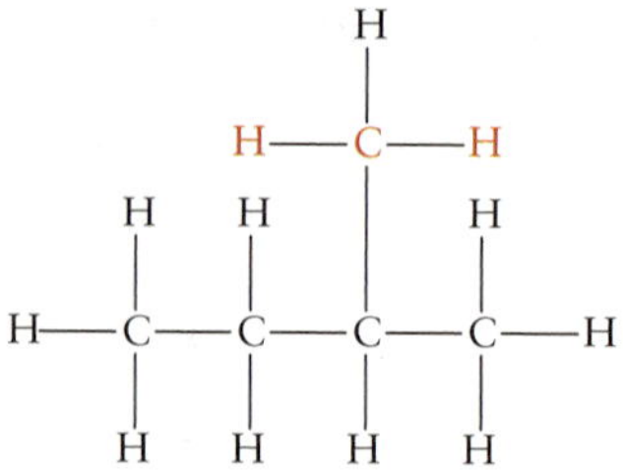

(b)

From methylpropane, you can form

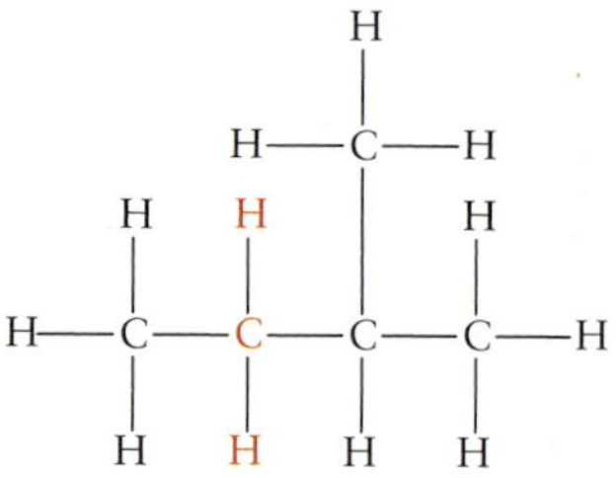

(c)

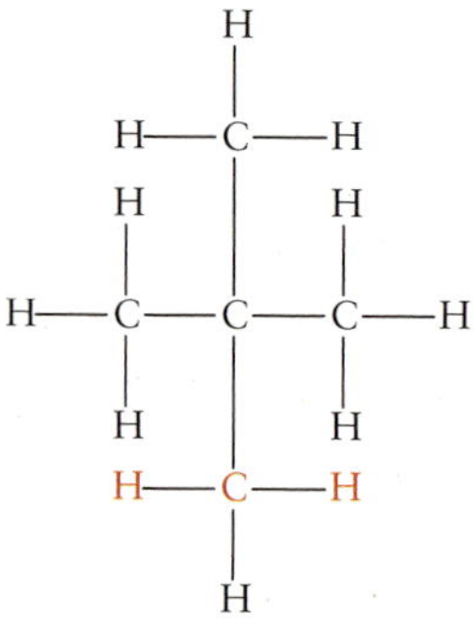

(d)

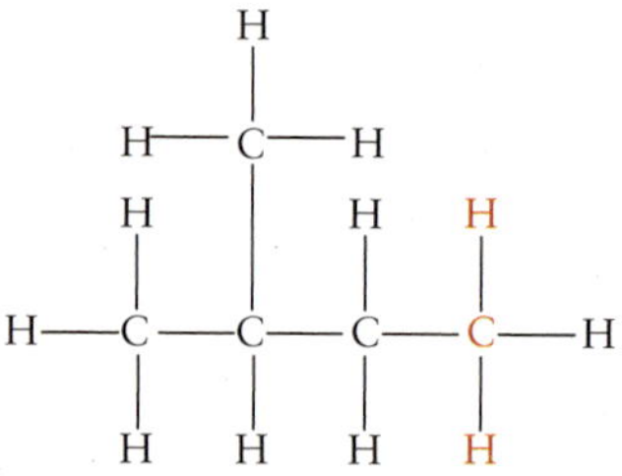

(e)

Molecules (b) and (c) are identical. The atoms of molecule (e) are joined together in the same arrangement as that in (b) and (c), but the structure looks different because it has been rotated 180° and twisted; so (b), (c), and (e) are the same. There are therefore only three distinct isomers with formula C_5H_{12}: (a) $CH_3(CH_2)_3CH_3$, (b) $CH_3CH_2CH(CH_3)_2$, and (d) $C(CH_3)_4$.

Self-test 11A.4A Write the condensed structural formulas for the five isomeric alkanes of molecular formula C_6H_{14}.
  [**Answer:** $CH_3(CH_2)_4CH_3$; $CH_3(CH_2)_2CH(CH_3)_2$; $CH_3CH_2CH(CH_3)CH_2CH_3$; $CH_3CH_2C(CH_3)_3$; $(CH_3)_2CHCH(CH_3)_2$]

Self-test 11A.4B Halogen atoms can substitute for hydrogen atoms in hydrocarbons. Write the condensed structural formulas for the four isomers with the molecular formula C_4H_9Br.

Related Exercises 11A.21, 11A.22

In **stereoisomers,** the molecules have the same connectivity but the atoms are arranged differently in space. One class of stereoisomers consists of **geometrical isomers,** in which atoms have different arrangements on either side of a double bond or above and below the ring of a cycloalkane or cycloalkene (**FIG. 11A.2**). Geometrical isomers of organic molecules are distinguished by the prefixes *cis-* and *trans-*. For example, in the upper part of the illustration there are two different 2-butenes: in the cis isomer, both methyl groups are on the same side of the double bond; in the trans isomer, the methyl groups are across the double bond. Geometrical isomers have the same molecular formula and structural formula but different properties.

Self-test 11A.5A Identify (**26a**) and (**26b**) as cis or trans.
            [**Answer:** (**26a**) is *trans*-2-pentene; (**26b**) is *cis*-2-pentene]

Self-test 11A.5B Identify (**27a**) and (**27b**) as cis or trans.

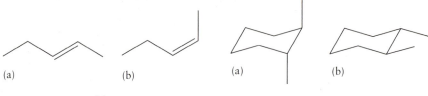

(a)   (b)   (a)   (b)

26   27

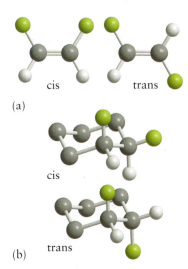

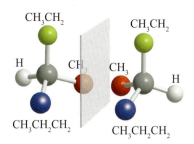

FIGURE 11A.2 Two pairs of geometrical isomers. In geometrical isomerism, two groups take different positions across a double bond, as in part (a), or different positions above and below a ring, as in part (b). Note that the neighbors of each atom in a pair of isomers are the same, but the arrangement of the atoms in space is different. (The compound in part (a) is 2-butene when the green sphere is CH₃ and the white sphere is H.)

FIGURE 11A.3 The molecule of 3-methylhexane on the right is the mirror image of the one on the left. Each group is represented by a sphere of a different color. The molecule on the left cannot be superimposed on the one on the right, and so these two molecules are distinct optical isomers and enantiomers of each other.

Another type of stereoisomerism is optical isomerism. **Optical isomers** are two compounds with molecules that are nonsuperimposable mirror images of each. To understand this definition, consider 3-methylhexane, $CH_3CH_2CH(CH_3)CH_2CH_2CH_3$, and its mirror image (**FIG. 11A.3**). No matter how you twist and turn the molecules, you cannot superimpose the mirror-image molecule on the original molecule. It is like trying to superimpose your right hand on your left hand. A **chiral molecule,** such as 3-methylhexane, is a molecule that is not identical to its mirror image. A chiral molecule and its mirror image form a pair of **enantiomers,** or mirror-image isomers. Although they have the same composition, the two enantiomers are, in fact, two distinct compounds. In organic compounds, optical isomers occur whenever four different groups are attached to the same carbon atom, which is then called a "chiral carbon atom." The alkane 3-methylpentane (**28**) does not have a chiral carbon atom. It is an example of an **achiral molecule,** a molecule that can be superimposed on its mirror image, just as you could superimpose a cube on its mirror image.

Enantiomers have identical chemical properties except when they react with other chiral compounds. Because many biochemical substances are chiral, one consequence of this difference in reactivity is that enantiomers may have different odors and pharmacological activities. To be effective, the molecule has to fit into a cavity, or slot, of a certain shape, either in an odor receptor in the nose or in an enzyme. Only one member of the enantiomeric pair may be able to fit. For example, R-carvone (**29a**) is a primary flavor in oil of caraway seeds, whereas its mirror image s-carvone (**29b**) is a primary flavor in oil of spearmint. The chiral carbon atom is marked with an asterisk (*) in each structure. The line structure of carvone (**11a**) is the same for both isomers: only a three-dimensional representation can distinguish them.

Enantiomers differ in one physical property: chiral molecules display optical activity, the ability to rotate the plane of polarization of light (Topic 9C). If a chiral molecule rotates the plane of polarization clockwise, then its mirror-image partner rotates it through the same angle in the opposite direction.

Often organic compounds synthesized in the laboratory are "racemic mixtures," or mixtures of enantiomers in equal proportions (Topic 9C). In contrast, reactions in living cells commonly lead to only one enantiomer. It is a remarkable and currently unexplained feature of nature that almost all naturally occurring amino acids in animals have the same handedness.

The name "chiral" comes from the Greek word for hand; "enantiomer" is formed from the Greek words for "both parts."

28   3-Methylpentane, C_6H_{14}

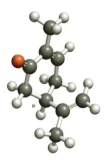

29a   R-Carvone

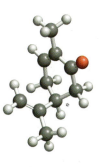

29b   s-Carvone

EXAMPLE 11A.4   Deciding whether a compound is chiral

Many pharmaceuticals are highly specific in the sense that they are active only if they have the correct chirality. In some cases, a compound with the wrong chirality might be lethal. It is vitally important to be able to distinguish molecules with different chiralities. A bromoalkane is formed from an alkane when a bromine atom replaces a hydrogen atom. Decide whether the bromoalkanes (a) $CH_3CH_2CHBrCH_3$ and (b) $CH_3CHBrCH_3$ are chiral.

PLAN Identify carbon atoms as chiral if they have four different groups attached.

SOLVE

Draw the compounds. Mark carbon atoms as chiral (*) if they have four different groups attached.

(a) The C atom marked * is attached to four different groups and hence is chiral; therefore, the molecule is not superimposable on its mirror image. The molecule is chiral.

(b) No atom is chiral. The molecule is superimposable on its mirror image. The molecule is achiral. The blue rectangle is a plane of symmetry. Each half of the molecule is the mirror image of the other half.

(a) $CH_3CH_2CHBrCH_3$ (b) $CH_3CHBrCH_3$

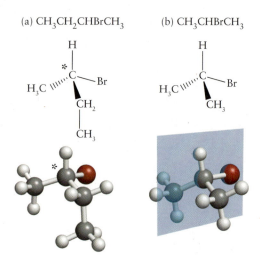

A Note on Good Practice: Dashed and solid wedge-shaped bonds are commonly used when displaying organic structures to convey a sense of the three-dimensional shapes. The dashed wedge-shaped bonds go into the page and the solid wedge-shaped bonds come toward you. The thin solid lines are in planes parallel to the plane of the paper.

Self-test 11A.6A A chlorofluorocarbon molecule contains Cl and F atoms as well as C and H. Which of the following chlorofluorocarbons is chiral: (a) CH_3CF_2Cl; (b) CH_3CHFCl; (c) CH_2FCl?
[*Answer:* (b)]

Self-test 11A.6B In an alcohol, an —OH group is attached to a C atom. Which of the following alcohols is chiral: (a) $CH_3CH_2CH_2CH_2OH$; (b) $CH_3CH(OH)CH_3$; (c) $CH_3CH(OH)CH_2CH_3$?

Exercises 11A.27, 11A.28

Structural isomers have identical molecular formulas, but their atoms are linked to different neighbors. Geometrical isomers have the same molecular and structural formulas but different arrangements in space. Molecules with four different groups attached to a single carbon atom are chiral; they are optical isomers.

11A.3 Physical Properties of Alkanes and Alkenes

The electronegativities of carbon and hydrogen (2.55 and 2.20, respectively) are so similar and rotation about bonds so free that hydrocarbon molecules are best regarded as non-polar. The dominant interaction between alkane molecules is therefore the London force (Topic 3F). Because the strength of this interaction increases with the number of electrons in the molecule, the alkanes in petroleum, their major source, become less volatile with increasing molar mass and therefore are separable by fractional distillation (**FIG. 11A.4**). The lightest members, methane through butane, are gases at room temperature. Pentane is a volatile liquid, and hexane through undecane ($C_{11}H_{24}$) are moderately volatile liquids that are present in gasoline, which is discussed further in the Interlude following Focus 11. All alkanes are insoluble in water. Because their densities are lower than that of water, they float on the surface of water. Even a small oil spill at sea can form an organic layer on the surface of the ocean and spread over a huge area.

THINKING POINT

How could you estimate the area that 1 L of gasoline might spread over on water?

A double bond consists of a σ-bond and a π-bond (Topic 2F). Each carbon atom in a double bond is sp^2 hybridized and uses the three hybrid orbitals to form three σ-bonds. The unhybridized p-orbitals on the two carbon atoms overlap with each other and form

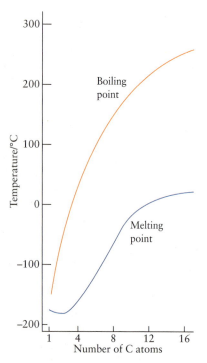

FIGURE 11A.4 The melting and boiling points of the unbranched alkanes from CH_4 to $C_{16}H_{34}$.

FIGURE 11A.5 The π-bond (represented by the yellow electron clouds) in an alkene molecule makes the molecule resistant to twisting around a double bond. Consequently, all six atoms (the two C atoms that form the bond and the four atoms attached to them) lie in the same plane.

 ANIMATION FIGURE 11A.5

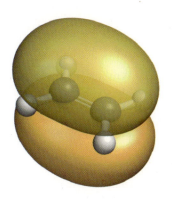

a π-bond. The C=C group and all four atoms attached to it lie in the same plane and are locked into that arrangement by the resistance to twisting of the π-bond (**FIG. 11A.5**). Because alkene molecules cannot roll up into a ball as compactly as alkanes or rotate into favorable positions, they cannot pack together as closely as alkanes; so alkenes have lower melting points than alkanes of similar molar mass.

The carbon–carbon double bond in alkenes is more reactive than carbon–carbon single bonds and gives alkenes their characteristic properties (Topic 11B).

> *The strength of the London forces between alkane molecules increases as the molar mass of the molecules increases. Alkenes have a double bond consisting of a σ-bond and a π-bond.*

Topic 2F describes the origin of the resistance to rotation of the C=C bond.

What have you learned in this Topic?

You have learned that aliphatic hydrocarbons are classified as alkanes, alkenes, and alkynes. You now know how to name hydrocarbons and how to identify different kinds of isomers, including geometrical and optical isomers. You have seen how their structures affect their physical properties.

The skills you have mastered are the ability to:

☐ **1.** Distinguish alkanes, alkenes, and alkynes by differences in bonding and structure.

☐ **2.** Name simple hydrocarbons (Toolbox 11A.1 and Examples 11A.1 and 11A.2).

☐ **3.** Identify two molecules, given their structural formulas, as structural, geometrical, or optical isomers (Section 11A.2).

☐ **4.** Write the formulas of isomeric molecules (Example 11A.3).

☐ **5.** Judge whether a compound is chiral (Example 11A.4).

☐ **6.** Describe general trends in the physical properties of alkanes (Section 11A.3).

Topic 11A Exercises

11A.1 Draw line structures of the following molecules and identify each as an alkane, alkene, or alkyne: (a) CH_3CCCH_3; (b) $CH_3CH_2CH_2CH_3$; (c) $CH_2CHCH_2CH_3$; (d) $CH_3CHCHCH_2CCCH_3$; (e) $CH_2CHCH_2CHCH_2$.

11A.2 Give the molecular formula and identify each of the following molecules as being an alkane, alkene, or alkyne:

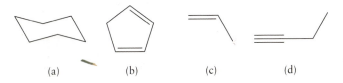

(a)   (b)   (c)   (d)

11A.3 Name each of the following unbranched alkanes: (a) C_3H_8; (b) C_4H_{10}; (c) C_7H_{16}; (d) $C_{10}H_{22}$.

11A.4 Name each of the following unbranched alkanes: (a) C_9H_{20}; (b) C_6H_{14}; (c) $C_{11}H_{24}$; (d) $C_{12}H_{26}$.

11A.5 Name the following substituents: (a) CH_3-; (b) $CH_3(CH_2)_3CH_2-$; (c) $CH_3CH_2CH_2-$; (d) $CH_3CH_2CH_2CH_2CH_2CH_2-$.

11A.6 Name the following substituents: (a) $CH_3(CH_2)_7CH_2-$; (b) $CH_3CH_2CH_2CH_2CH_2-$; (c) CH_3CH_2-; (d) $CH_3CH_2CH_2CH_2CH_2CH_2CH_2-$.

11A.7 Give the systematic name of (a) $CH_3CH_2CH_3$; (b) CH_3CH_3; (c) $CH_3(CH_2)_3CH_3$; (d) $(CH_3)_2CHCH(CH_3)_2$.

11A.8 Give the systematic name of (a) $CH_3CH_2CH(CH_3)CH_2CH_3$; (b) $CH_3CH(CH_2CH_3)CH(CH_3)_2$; (c) $(CH_3)_3C(CH_2)_3CH(CH_3)_2$; (d) $(CH_3)_3CC(CH_3)_3$.

11A.9 Give the systematic name of (a) $CH_3CH=CHCH(CH_3)_2$; (b) $CH_3CH_2CH(CH_3)C(CH_3)_3$.

11A.10 Give the systematic name of (a) $CH_2=CHCH_2CH(C_6H_5)(CH_2)_4CH_3$; (b) $(CH_3)_2CHCH(CH_3)CHClC≡CCH_3$.

11A.11 Write the shortened (condensed) structural formula of (a) 3-methyl-1-pentene; (b) 4-ethyl-3,3-dimethylheptane; (c) 5,5-dimethyl-1-hexyne; (d) 3-ethyl-2,4-dimethylpentane.

11A.12 Write the shortened (condensed) structural formula of (a) 4-ethyl-2, 2-dimethylhexane; (b) 3-ethyl-4-methyl-1-pentene; (c) *cis*-4-ethyl-3-heptene; (d) *trans*-4-methyl-2-hexene.

11A.13 Write the structural formula of (a) 4,4-dimethylnonane; (b) 4-propyl-5,5-diethyl-1-decyne; (c) 2,2,4-trimethylpentane; (d) *trans*-3-hexene.

11A.14 Write the structural formula of (a) 4-ethyl-2,3,6-trimethyloctane; (b) *cis*-4-ethyl-2-hexene; (c) 1-ethyl-2,3-dimethylcyclopentane; (d) 5-ethyl-1-heptene.

11A.15 Draw line structures to represent each of the following molecules: (a) nonane, $CH_3(CH_2)_7CH_3$; (b) cyclopropane, C_3H_6; (c) cyclohexene, C_6H_{10}.

11A.16 Draw line structures to represent each of the following species:(a) 2,2,3,3-tetramethylhexane, $CH_3C(CH_3)_2C(CH_3)_2CH_2CH_2CH_3$; (b) the trityl cation, $(C_6H_5)_3C^+$; (c) butadiene, $CH_2CHCHCH_2$.

11A.17 Identify the type and number of bonds on carbon atom 2 in (a) pentane; (b) 2-pentene; (c) 2-pentyne.

11A.18 Predict the geometry and hybridization of the orbitals used in bonding on carbon atom 2 in (a) pentane; (b) 2-pentene; (c) 2-pentyne.

11A.19 Draw the structures of *cis*-1,2-dichloropropene and *trans*-1,2-dichloropropene. Which of these molecules is polar?

11A.20 Draw the structure of 1,3-pentadiene. Use valence-bond and molecular orbital pictures to describe the bonding for the σ-framework and π-orbitals, respectively.

11A.21 Write the structural formulas for and name (a) at least 10 *alkene* isomers having the formula C_6H_{12}; (b) at least 10 *cycloalkane* isomers having the formula C_6H_{12}.

11A.22 Write the structural formulas for and name all the isomers (including geometrical isomers) of the alkenes (a) C_4H_8; (b) C_5H_{10}.

11A.23 Identify each of the following pairs as structural isomers, geometrical isomers, or not isomers: (a) butane and cyclobutane; (b) cyclopentane and pentene;

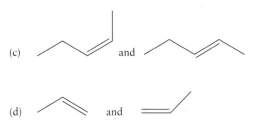

11A.24 Identify each of the following pairs as structural isomers, geometrical isomers, or not isomers: (a) 1-chlorohexane and chlorocyclohexane;

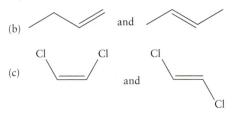

11A.25 A branched hydrocarbon C_4H_{10} reacts with chlorine in the presence of light to give two branched structural isomers with the formula C_4H_9Cl. Write the structural formulas of (a) the hydrocarbon; (b) the isomeric products.

11A.26 A branched hydrocarbon C_6H_{14} reacts with chlorine in the presence of light to give only two structural isomers with the formula $C_6H_{13}Cl$. Write the structural formulas of (a) the hydrocarbon; (b) the two isomeric products.

11A.27 Indicate which of the following molecules are optical isomers and identify the chiral carbon atoms in those that are: (a) $CH_3CHBrCH_2CH_3$; (b) $CH_3CH_2CHCl_2$; (c) 1-bromo-2-chloropropane; (d) 1,2-dichloropentane.

11A.28 Indicate which of the following molecules are optical isomers and identify the chiral carbon atoms in those that are: (a) $CH_3CHBrCH_2Br$; (b) $CH_3CH_2CHClCH_2CH_3$; (c) 2-bromo-2-methylpropane; (d) 2,3-dimethylpentane.

Topic 11B Reactions of Aliphatic Hydrocarbons

11B.1 Alkane Substitution Reactions

11B.2 Synthesis of Alkenes and Alkynes

11B.3 Electrophilic Addition

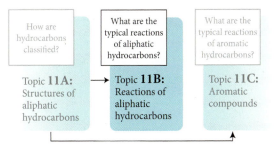

| How are hydrocarbons classified? | What are the typical reactions of aliphatic hydrocarbons? | What are the typical reactions of aromatic hydrocarbons? |

Topic **11A:** Structures of aliphatic hydrocarbons → Topic **11B:** Reactions of aliphatic hydrocarbons   Topic **11C:** Aromatic compounds

The chemical properties of hydrocarbons can be understood in terms of the mechanisms of their reactions, the details of the steps that turn one compound into another. Knowledge of the mechanisms of hydrocarbon reactions is critical for developing synthetic routes leading to new compounds. The reactions of alkanes differ markedly from those of alkenes and alkynes.

11B.1 Alkane Substitution Reactions

The alkanes were once called the *paraffins*, from Latin words meaning "little affinity." As this name suggests, they are not very reactive. One reason for their resistance to chemical onslaught is thermodynamic: the C—C and C—H bonds are strong (their mean bond enthalpies are 348 kJ·mol^{-1} and 412 kJ·mol^{-1}, respectively), and so there is little energy advantage in replacing them with most other bonds. The most notable exceptions are C=O (743 kJ·mol^{-1}), C—O (360 kJ·mol^{-1}), and C—F (484 kJ·mol^{-1}) bonds. The alkanes are commonly used as fuels, because their combustion to carbon dioxide and water is highly exothermic (Topic 4D):

$$CH_4(g) + 2\,O_2(g) \longrightarrow CO_2(g) + 2\,H_2O(g) \qquad \Delta H° = -890 \text{ kJ}$$

In this reaction, the strong carbon–hydrogen bonds are replaced by the even stronger O—H bonds (463 kJ·mol^{-1}), and the oxygen–oxygen bond (496 kJ·mol^{-1}) is replaced by two very strong C=O bonds. The excess energy is given off as heat.

Alkanes are used as the raw materials for the synthesis of many more reactive compounds. Starting with alkanes obtained from the refining of petroleum, organic chemists introduce reactive groups of atoms into the molecules in the process called **functionalization.** The functionalization of alkanes can be achieved by a **substitution reaction,** a reaction in which an atom or group of atoms replaces an atom (in alkanes, a hydrogen atom) in the original molecule (**FIG. 11B.1**). An example of a substitution reaction is that between methane and chlorine. A mixture of these two gases survives indefinitely in the dark but, when exposed to ultraviolet radiation or heated to more than 300 °C, the gases react explosively:

$$CH_4(g) + Cl_2(g) \xrightarrow{\text{light or heat}} CH_3Cl(g) + HCl(g)$$

The reaction takes place by a radical chain mechanism (Topic 7C). Chloromethane (CH$_3$Cl) is only one of the products; dichloromethane (CH$_2$Cl$_2$), trichloromethane (CHCl$_3$), and tetrachloromethane (CCl$_4$) also form, especially at high concentrations of chlorine.

Alkane substitution takes place by a radical chain mechanism.

11B.2 Synthesis of Alkenes and Alkynes

Most alkenes and alkynes used in industry are produced during the refining of petroleum. One of the first refining steps is a reaction that uses a catalyst to convert some of the abundant alkanes into the more reactive alkenes:

$$CH_3CH_3(g) \xrightarrow{Cr_2O_3} CH_2{=}CH_2(g) + H_2(g)$$

Why Do You Need to Know This Material? Reactions of aliphatic hydrocarbons are important for the refining of fuels and for the synthesis of new compounds.

What Do You Need to Know Already? This Topic builds on the description of aliphatic compounds in Topic 11A and draws on the concept of bond enthalpy (Topic 4E) and the notion of reaction mechanism (Topic 7C).

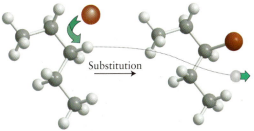

Substitution

FIGURE 11B.1 In an alkane substitution reaction, an incoming atom or group of atoms (represented by the red sphere) replaces a hydrogen atom in the alkane molecule.

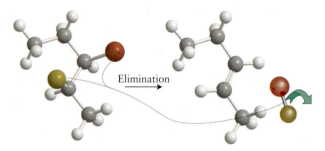

FIGURE 11B.2 In an elimination reaction, two atoms or groups (the red and gold spheres) attached to neighboring carbon atoms are eliminated from the molecule, leaving a double bond between the two carbon atoms.

ANIMATION FIGURE 11B.2

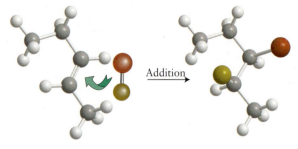

FIGURE 11B.3 In an addition reaction, the atoms provided by an incoming molecule (the red and gold spheres) form bonds to the carbon atoms originally joined by a multiple bond.

ANIMATION FIGURE 11B.3

This step is an example of an **elimination reaction,** a reaction in which two groups or atoms on neighboring carbon atoms are removed from a molecule, leaving a multiple bond (**FIG. 11B.2**).

Another route commonly used in the laboratory to produce alkenes is the **dehydrohalogenation** of haloalkanes, the removal of a hydrogen atom and a halogen atom from neighboring carbon atoms:

$$CH_3CH_2CHBrCH_3 + CH_3CH_2O^- \xrightarrow{\text{ethanol at 70 °C}}$$
$$CH_3CH{=}CHCH_3 + CH_3CH_2OH + Br^-$$

The states of reactants and products are often not given for organic reactions, because the reaction may take place at the surface of a catalyst or it may take place in a nonaqueous solvent, as here. The reaction is another example of an elimination reaction and is carried out in hot ethanol, with sodium ethoxide, $NaCH_3CH_2O$ (the ionic compound $Na^+CH_3CH_2O^-$), as the reagent. Some $CH_3CH_2CH{=}CH_2$ is also formed in this reaction.

The double bonds in alkenes can be generated by elimination reactions.

11B.3 Electrophilic Addition

Although a double carbon–carbon bond is stronger than a single carbon–carbon bond, the carbon–carbon π-bond is weaker than the σ-bond. The overlap responsible for the formation of the π-bond is less extensive than that responsible for the formation of the σ-bond, and the enhanced electron density does not lie directly between the two nuclei (Topic 2F). A consequence of this relative weakness is the characteristic chemical reaction of an alkene, an **addition reaction,** in which atoms supplied by the reactant form σ-bonds to the two atoms originally joined by the double bond (**FIG. 11B.3**). In the process, the π-bond is lost but the carbon–carbon σ-bond survives. An example is **halogenation,** the addition of two halogen atoms at a double bond, as in the formation of 1,2-dichloroethane:

$$CH_2{=}CH_2 + Cl_2 \longrightarrow CH_2Cl{-}CH_2Cl$$

The addition of hydrogen chloride to give chloroethane is an example of a **hydrohalogenation** reaction:

$$CH_2{=}CH_2 + HCl \longrightarrow CH_3{-}CH_2Cl$$

The high electron density in the region of the double bond makes alkenes susceptible to addition reactions (**FIG. 11B.4**). Because electrons are negatively charged, this region represents an accumulation of negative charge that can attract a positively charged reactant. A reactant that is attracted to a region of high electron density is called an **electrophile.** The mechanism of alkene addition is by electrophilic attack on the carbon atoms that make up the double bond. An electrophile may be a positively charged species or it may be a species that has a partial positive charge or can acquire one in the course of the reaction.

An example is the bromination of ethene. When ethene (or any other alkene) is bubbled through a solution of bromine, the solution is decolorized as the red bromine

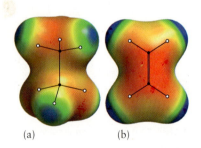

(a)   (b)

FIGURE 11B.4 (a) The electrostatic potential diagram for an ethane molecule: the blue regions show where the positive charge of the nuclei outweighs the negative charge of the electrons, and the red regions show where the opposite is true. (b) The electron distribution in an ethene molecule shows the negative region of charge associated with the accumulation of electrons in the region of the double bond.

ANIMATION FIGURE 11B.4

reacts to form dibromoethane (this reaction is illustrated in Fig. 2B.1). Bromine molecules are polarizable; as a Br_2 molecule approaches the high electron density of an alkene double bond, a partial positive charge is induced on the Br atom closer to the double bond (**FIG. 11B.5**). This separation of charges means that a Br_2 molecule can act as an electrophile. As it moves in for the attack, the partially positively charged Br atom becomes more and more like Br^+, and its partner becomes more like Br^-. The bond between the two Br atoms snaps, and the Br^- ion forms a bridge between the two carbon atoms of the alkene, giving a cyclic "bromonium ion":

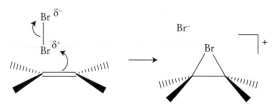

Almost immediately, a Br^- ion swoops in for the attack, attracted by the positive charge of the bromonium ion. It forms a bond to one carbon atom, and the bond from that carbon atom to the bromine atom already present breaks as the bromine atom moves onto the second carbon atom, giving 1,2-dibromoethane:

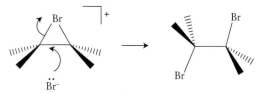

With the use of a solid-state catalyst, hydrogen can be added to carbon–carbon double bonds in a **hydrogenation** reaction:

$$H_2(g) + \cdots C{=}C\cdots \xrightarrow{\text{catalyst}} \cdots CH{-}CH\cdots$$

This reaction is used in the food industry to convert vegetable oils into shortening (**FIG. 11B.6**) and to make some peanut butters less runny at room temperature. Oil and solid fat molecules both have long hydrocarbon chains, but oils have more double bonds. Because double bonds resist twisting, oil molecules do not pack together well, and so the oil is liquid at ordinary temperatures. When some of the double bonds are replaced by single bonds, the chains become much more flexible, and so the molecules pack together better and form a solid.

Alkynes also undergo addition reactions and can be converted either to alkenes or to alkanes, depending on the conditions and the stoichiometric ratios of reactants.

Self-test 11B.1A Write the condensed structural formula of the product of the addition of hydrogen to 2-butene: $CH_3CH{=}CHCH_3 + H_2 \rightarrow$ product.

[***Answer:*** $CH_3CH_2CH_2CH_3$]

Self-test 11B.2B Write the condensed structural formula of the compound formed by the addition of hydrogen chloride to 2-butene.

The mechanism of addition to alkenes and alkynes is electrophilic attack.

What have you learned in this Topic?

You have learned that alkanes are not very reactive but that the multiple bonds in alkenes and alkynes are sites of addition reactions.

The skills you have mastered are the ability to:

- [ ] **1.** Distinguish alkanes from alkenes and alkynes by differences in reactivity.
- [ ] **2.** Predict the products of given elimination, addition, and substitution reactions (Sections 11B.2 and 11B.3).

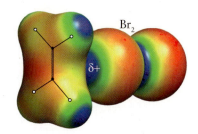

FIGURE 11B.5 As a bromine molecule approaches a double bond in an alkene, the atom closer to the ethene molecule acquires a partial positive charge. The computation that produced this image was carried out for the point at which the bromine molecule is so close to the double bond that a carbon–bromine bond is starting to form.

Curved arrows like the ones here are often used to illustrate organic reaction mechanisms. They show the direction in which electron pairs move as they form new bonds.

You can identify which foods' ingredients have undergone this reaction by looking for "hydrogenated" or "partially hydrogenated" oils on the packaging.

FIGURE 11B.6 When the runny oil (top) is hydrogenated, it is converted into a solid fat (bottom). Hydrogen adds to the carbon–carbon double bonds, converting them into single bonds. The resulting more flexible molecules can pack together more closely and form a solid. (*W. H. Freeman photo by Ken Karp.*)

Topic 11B Exercises

11B.1 Use the data in Appendix 2A to write balanced equations and calculate the heat released when (a) 1.00 mol and (b) 1.00 g of each of the following compounds is burned in excess oxygen: propane, butane, and pentane. Is there a trend in the amount of heat released per mole of molecules or per gram of compound? If so, what is it?

11B.2 Write the balanced chemical equation for the fluorination of methane to difluoromethane. Use bond enthalpies (Tables 4E.2 and 4E.3) to estimate the enthalpy of this reaction. The corresponding reaction using chlorine is much less exothermic. To what can this difference be attributed?

11B.3 How many different products containing two carbon atoms are possible in the reaction of chlorine with ethane? Do any of the products exist as optical isomers?

11B.4 How many different products that retain the three-membered ring are possible in the reaction of chlorine with cyclopropane? Do any of the products exist as stereoisomers?

11B.5 Two structural isomers can result when hydrogen bromide reacts with 2-pentene. (a) Write their structural formulas. (b) What name is given to this type of reaction?

11B.6 (a) Write a balanced equation for the production of 2,3-dichlorohexane from 2-hexyne. (b) What name is given to this type of reaction?

11B.7 (a) Write a balanced equation for the reaction of bromocyclohexane with sodium ethoxide in ethanol. (b) Draw structural formulas of the cyclic reactant and product. (c) What name is given to this type of reaction?

11B.8 Draw line structures of the possible products for the reaction of sodium ethoxide with (a) 1-bromobutane; (b) 2-bromobutane.

11B.9 Use bond enthalpies (Tables 4E.2 and 4E.3) to estimate the reaction enthalpies for the halogenation of ethene by chlorine, bromine, and iodine. What trend, if any, exists in these values?

11B.10 Use bond enthalpies (Tables 4E.2 and 4E.3) to estimate the reaction enthalpies for the hydrohalogenation of ethene by HX, where X = Cl, Br, I. What trend, if any, exists in these values?

Topic 11C Aromatic Compounds

11C.1 Nomenclature

11C.2 Electrophilic Substitution

| How are hydrocarbons classified? | What are the typical reactions of aromatic hydrocarbons? | How do other atoms affect the properties of organic compounds? |

Topic **11A:** Structures of aliphatic hydrocarbons → Topic **11C:** Aromatic compounds → Topic **11D:** Common functional groups

Why Do You Need to Know This Material? Aromatic compounds are so central to organic chemistry that you need to be familiar with their nomenclature and typical reactions.

What Do You Need to Know Already? This Topic draws on the introduction to organic hydrocarbons in Topic 11A and the structure of molecules (Focus 2).

Aromatic compounds are important in industry as solvents and building blocks for polymers. In biology, they serve as components of some amino acids and contribute to the structure of DNA. Their name reflects the distinctive odors many of them have. They all contain an aromatic ring, usually the six-membered ring of benzene (Topic 2B). An abundant source of aromatic hydrocarbons is coal, which is a very complex mixture of compounds, many of which consist of extensive networks containing aromatic rings (Interlude following Focus 11).

11C.1 Nomenclature

Aromatic compounds are formally called **arenes.** The parent compound is benzene itself, C_6H_6 (**1**). When the benzene ring is named as a substituent, it is called the *phenyl* group, as in 2-phenylbutane, $CH_3CH(C_6H_5)CH_2CH_3$. In general, aromatic hydrocarbon groups are called *aryl* groups. Arenes include the fused-ring analogs of benzene, such as naphthalene, $C_{10}H_8$ (**2**), and anthracene, $C_{14}H_{10}$ (**3**), obtained by the distillation of coal.

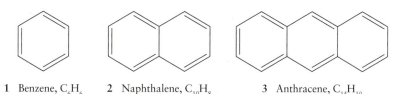

1 Benzene, C_6H_6 **2** Naphthalene, $C_{10}H_8$ **3** Anthracene, $C_{14}H_{10}$

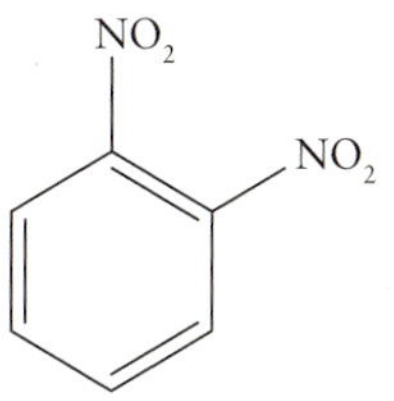

In an older but still widely used system of nomenclature, when there is a substituent in location 1 of a benzene ring, locations 2, 3, and 4 are denoted *ortho-* (abbreviated *o-*), *meta-* (*m-*), and *para-* (*p-*), respectively. Thus, (**4**) is *ortho*-dinitrobenzene. In systematic nomenclature, the locations of substituents on the aromatic ring are identified by numbering the carbon atoms from 1 to 6 around the ring, selecting the direction that corresponds to the lower substituent numbers. In this system, (**4**) is 1,2-dinitrobenzene ($-NO_2$ is the nitro group). However, (**5**) is 2,4,6-trinitrophenol because the compound C_6H_5OH is phenol, and so the carbon atom attached to the $-OH$ group is numbered 1.

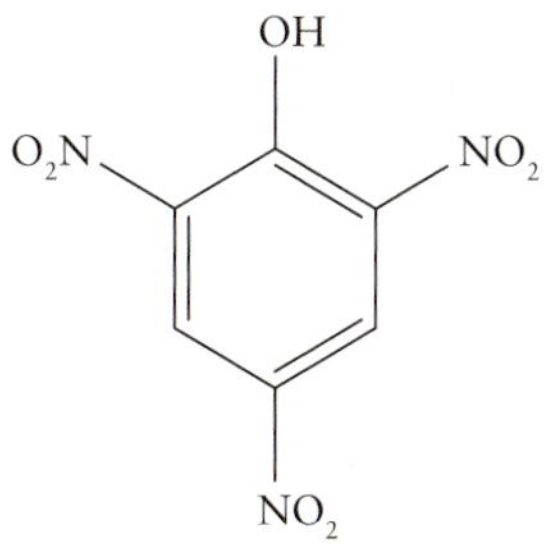

4 1,2-Dinitrobenzene

5 2,4,6-Trinitrophenol

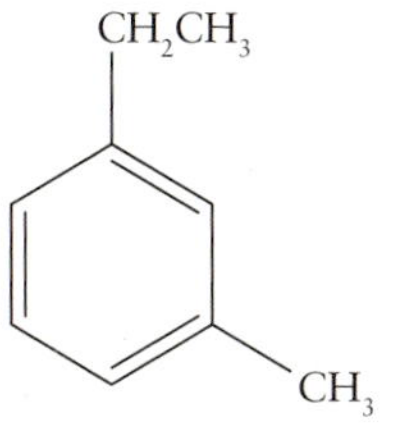

6

EXAMPLE 11C.1   Naming an aromatic compound

Many of the compounds found in commercial preparations are listed by their formal names, and it is essential to be able to interpret them in terms of their structures. Many of these compounds are aromatic or have aromatic regions. Name (a) compound (**6**) and the three dimethylbenzenes (xylenes), in which the second methyl group is on carbon atom (b) 2; (c) 3; (d) 4.

PLAN Name the compounds by counting around the ring in the direction that gives the smallest numbers to the substituents.

7 Benzaldehyde, C_6H_5CHO

8 Cinnamaldehyde

SOLVE
(a) 1-Ethyl-3-methylbenzene
(b) 1,2-Dimethylbenzene
(*o*-xylene)
(c) 1,3-Dimethylbenzene
(*m*-xylene)
(d) 1,4-Dimethylbenzene
(*p*-xylene)

(a) (b) (c) (d)

Self-test 11C.1A Name compound (**7**).

[**Answer:** 1-Ethyl-3-propylbenzene]

Self-test 11C.1B Name compound (**8**).

Related Exercises 11C.1–11C.4

Aromatic compounds are named by giving the substituents on the benzene ring the lowest numbers. When the benzene ring is itself named as a substituent, it is called a phenyl group.

11C.2 Electrophilic Substitution

Arenes are unsaturated but, unlike the alkenes, they are not very reactive despite their double bonds. Whereas alkenes commonly take part in addition reactions, arenes undergo predominantly *substitution* reactions, with the π-bonds of the ring left intact. For example, bromine immediately adds to a double bond of an alkene but reacts with benzene only in the presence of a catalyst—typically, iron(III) bromide—and it does not affect the bonding in the ring. Instead, one of the bromine atoms replaces a hydrogen atom to give bromobenzene, C_6H_5Br:

$$C_6H_6 + Br_2 \xrightarrow{FeBr_3} C_6H_5Br + HBr$$

The mechanism of substitution on an electron-rich benzene ring is **electrophilic substitution**, electrophilic attack on an atom and the replacement of one atom by another or by a group of atoms. That substitution rather than addition occurs can be traced to the stability of the delocalized π-electrons in the ring. Delocalization gives the electrons such low energy—that is, they are bound so tightly—that they are unavailable for forming new σ-bonds (Topics 2B and 2G).

The bromination of benzene illustrates the difference between electrophilic addition to alkenes and electrophilic substitution of arenes. First, to achieve the bromination of benzene, it is necessary to use a catalyst such as iron(III) bromide. The catalyst acts as a Lewis acid, binding to the bromine molecule (a Lewis base) and ensuring that the outer bromine atom has a pronounced partial positive charge:

$$:Br\!-\!Br: + FeBr_3 \longrightarrow {}^{\delta+}:Br\!-\!BrFeBr_3{}^{\delta-}$$

(For simplicity, only one lone pair is shown on each of the first two bromine atoms.) The outer bromine atom of the complex is now primed to act as a strong electrophile (**FIG. 11C.1**).

Electrophilic substitution begins like electrophilic addition, with an attack on a region of high electron density to form a positively charged intermediate:

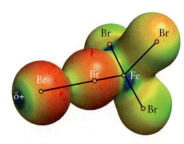

FIGURE 11C.1 The catalyst $FeBr_3$ acts by forming a complex with a bromine molecule. Because the iron atom withdraws electrons, the bromine atom not directly attached to the iron atom acquires a partial positive charge (the blue region on the left). This partial charge enhances the ability of the bromine molecule to act as an electrophile.

However, because the delocalized π-electrons constitute such a stable arrangement, the hydrogen atom shown can be pulled from the ring as H$^+$ by a bromine atom in the FeBr$_4^-$ complex. Delocalization is regained and the products are C$_6$H$_5$Br and HBr:

The iron(III) bromide catalyst is released in this step and is free to activate another bromine molecule.

One of the most extensively studied examples of electrophilic substitution is the nitration of benzene. A mixture of nitric acid and concentrated sulfuric acid converts benzene slowly into nitrobenzene. The actual nitrating agent is the electrophile NO$_2^+$ (the nitronium ion, ONO$^+$), a linear triatomic ion:

$$HNO_3 + H_2SO_4 \longrightarrow NO_2^+ + HSO_4^- + H_2O$$

The accepted mechanism of the reaction is

In the second step, the hydrogen ion is pulled out of the ring by the HSO$_4^-$ ion acting as a Brønsted base. As in the bromination reaction, the restoration of the delocalization of the π-electrons facilitates the removal of the hydrogen ion.

Certain groups attached to an aromatic ring can donate electrons into its delocalized molecular orbitals. Each of these electron-donating substituents has an electronegative atom with at least one lone pair of electrons attached directly to the aromatic ring. Examples include —NH$_2$ and —OH. Electrophilic substitution of benzene is much faster when an electron-donating substituent is present. For example, the nitration of phenol, C$_6$H$_5$OH, proceeds so quickly that it requires no catalyst. Moreover, when the products are analyzed, the only products are found to be 2-nitrophenol (*ortho*-nitrophenol, **9**) and 4-nitrophenol (*para*-nitrophenol, **10**).

Why is the meta position such an unattractive location for substitution, and why does phenol react much faster than benzene? An electrophile is attracted to regions of high electron density. Therefore, to account for the fast reaction of phenol, the electron density must be greater in the ring when the electron-donating —OH substituent is present. To account for the dominance of the ortho and para products, electron density must be relatively high at the ortho and para positions. A molecular orbital calculation, such as those described in *Major Technique* 5 on the website of this book, shows that there is indeed a higher concentration of electrons in the ring in phenol, especially at the ortho and para positions, than in benzene, largely because the O atom has lone pairs of electrons that can participate in π-bonding with the carbon atoms (**FIG. 11C.2**).

Long before their theories were supported by computations, organic chemists found a way to use resonance structures to explain the product distribution in electrophilic substitution. Thus, the Lewis structure for phenol is regarded as a resonance hybrid of the following structures:

The curved arrows show how one resonance structure relates to another. Notice that the formal negative charge is located on the ortho and para positions, exactly where reaction takes

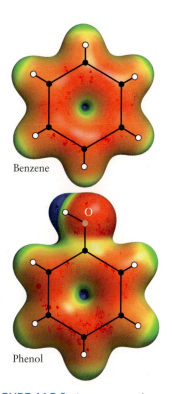

9 Vanillin   **10** *p*-Nitrophenol

Benzene

Phenol

FIGURE 11C.2 The presence of an —OH group in phenol alters the distribution of electrons in the benzene ring. The blue regions denote relatively positively charged parts of the molecule and the green, yellow, and red regions denote progressively more negatively charged regions. Note how the electron-rich yellow regions spread over the carbon and hydrogen atoms more in phenol than in benzene, so these atoms are more susceptible to electrophilic attack. A more detailed analysis shows that the ortho and para carbon atoms have the greatest accumulation of negative charge.

place most quickly. Other ortho- and para-directing groups include —NH_2, —Cl, and —Br. All have an atom with a lone pair of electrons next to the ring, and all accelerate reaction.

THINKING POINT

Can you suggest a molecular orbital explanation for this directing effect?

You have seen that an —OH group can accelerate a reaction. Are there substituents that can slow down the electrophilic substitution of benzene? One way to reduce the electron density in the benzene ring and to make it less attractive to electrophiles is to substitute for a hydrogen atom on the benzene ring a highly electronegative atom or group that can withdraw some of the electron density from the benzene ring. Alternatively, a substituent that removes electrons by resonance can be used. A number of substituents, such as the carboxyl group (—COOH), have both effects. For instance, the nitration of benzoic acid, C_6H_5COOH, is found to be much slower than the nitration of benzene; moreover, most of the product is the meta nitro compound:

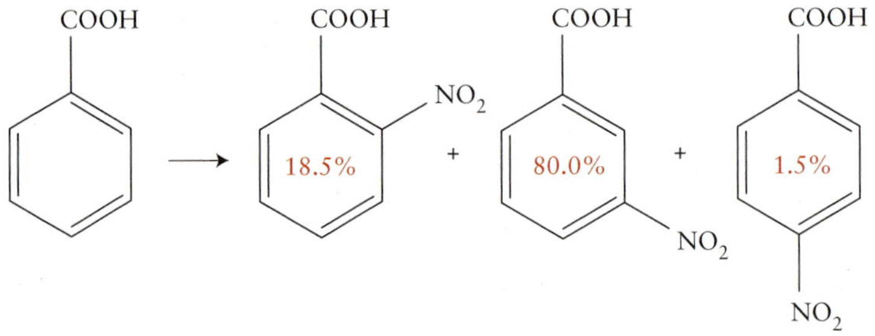

The electronegative O atoms of the carboxylic acid group withdraw electrons from the whole ring, thereby reducing its overall electron density. Moreover, resonance preferentially removes electrons from the ortho and para positions. To focus on the essentials, only the lone pairs of electrons involved in resonance are shown:

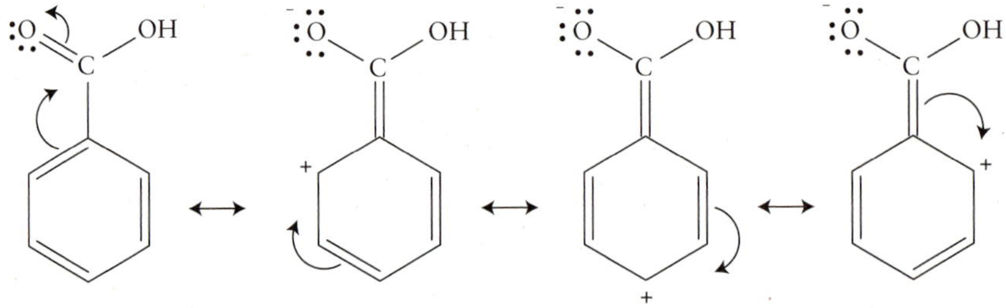

As a result of both effects, the reaction rate is lowered, especially at the ortho and para positions, and the meta position is left as the most likely place for attack. Other meta-directing electron-withdrawing substituents include —NO_2, —CF_3, and —C≡N. Notice that none of these substituents has a lone pair of electrons on the atom next to the ring.

Aromatic rings are much less reactive than their double-bond character would suggest; they commonly undergo substitution rather than addition. Electrophilic substitution of benzene with electron-donating substituents is accelerated and takes place at the ortho and para positions preferentially. Electrophilic substitution of benzene with electron-withdrawing substituents takes place at a reduced rate and primarily at the meta positions.

What have you learned in this Topic?

You have learned that aromatic hydrocarbons contain a benzenelike aromatic ring that enhances the stability of the compound and is resistant to addition. You have seen that a typical reaction of aromatic compounds is electrophilic substitution, with the preservation of the aromatic ring.

The skills you have mastered are the ability to:

☐ **1.** Name simple aromatic compounds (Example 11C.1).

☐ **2.** Predict the products of electrophilic substitution reactions (Section 11C.2).

☐ **3.** Explain why ortho- and para-directing groups accelerate electrophilic substitution on the benzene ring and identify such groups (Section 11C.2).

Topic 11C Exercises

11C.1 Name the following compounds:

(a)

(b)

11C.2 Name the following compounds:

(a)

(b)

11C.3 Draw the structural formula of (a) methylbenzene, more commonly known as toluene; (b) *p*-chlorotoluene; (c) 1,3-dimethylbenzene; (d) 4-chloromethylbenzene.

11C.4 Draw the structural formula of (a) *p*-xylene (1,4-dimethylbenzene); (b) 1,2-dibromobenzene; (c) 3-phenylpropene; (d) 2-ethyl-1,4-dimethylbenzene.

11C.5 (a) Draw all the isomeric dichloromethylbenzenes. (b) Name each one and indicate which are polar and which are nonpolar.

11C.6 (a) Draw all the isomeric diaminodichlorobenzenes. (b) Name each one and indicate which are polar and which are nonpolar.

11C.7 Draw resonance structures of cyanobenzene (C_6H_5CN) that show how it functions as a meta-directing substituent.

11C.8 One can prepare bromonitrobenzenes either by nitrating bromobenzene or by brominating nitrobenzene. Will these two reactions give the same product (or product distribution)? If not, how will the products differ?

11C.9 How many different compounds can be produced when naphthalene, $C_{10}H_8$ (**2**), undergoes electrophilic aromatic substitution by a single electrophile (denoted E)? Draw line structures to represent them.

11C.10 The structure of benzaldehyde, the main flavoring agent in bitter almond oil, is

Do you expect the aldehyde functional group (CHO) to act as a meta-directing or as an ortho, para-directing group? Explain your conclusion.

11C.11 Give the molecular formulas corresponding to the following line structures and identify each as an alkane, an alkene, an alkyne, or an aromatic hydrocarbon.

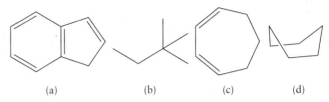

(a)   (b)   (c)   (d)

11C.12 Give the molecular formulas corresponding to the following line structures and identify each as an alkane, an alkene, an alkyne, or an aromatic hydrocarbon:

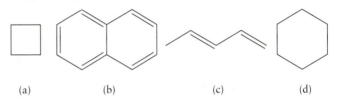

(a)   (b)   (c)   (d)

11C.13 Give the molecular formulas corresponding to the following line structures and identify each as an alkane, an alkene, an alkyne, or an aromatic hydrocarbon:

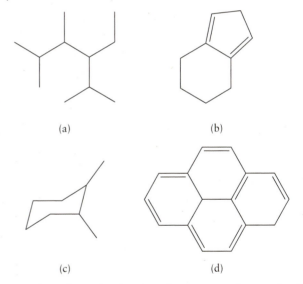

(a)   (b)

(c)   (d)

11C.14 Give the molecular formulas corresponding to the following line structures and identify each as an alkane, an alkene, an alkyne, or an aromatic hydrocarbon:

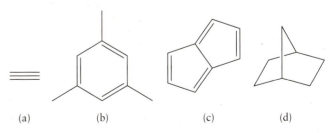

(a)   (b)   (c)   (d)

Topic 11D Common Functional Groups

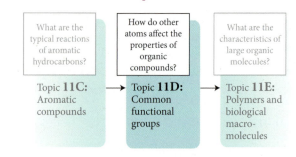

Why Do You Need to Know This Material? The great diversity of organic molecules found in living systems results from the presence of groups of atoms with characteristic functions. Once you are familiar with the properties of these groups, you will begin to be able to predict the properties of other organic molecules, even those of giant biomolecules and synthetic polymeric materials.

What Do You Need to Know Already? This Topic builds on Topics 11A through 11C and requires the same preparation, as well as the concepts of acid–base and redox reactions (*Fundamentals* J and K). It also makes use of the concepts of molecular polarity (Topic 2E) and solubility (Topic 5D).

The immense variety of organic compounds can be intimidating at first sight. Fortunately, organic compounds can be understood in terms of a relatively small number of **functional groups,** which are small groups of atoms that exhibit characteristic properties. Whereas hydrocarbons are built only from carbon and hydrogen, functional groups may include atoms of other elements.

Functional groups are either attached to the carbon backbone of a molecule or form part of that chain. Examples are the chlorine atom in chloroethane, CH_3CH_2Cl, and the —OH group in ethanol, CH_3CH_2OH. Carbon–carbon multiple bonds, such as the double bond in 2-butene, $CH_3CH=CHCH_3$, are also often considered functional groups. **TABLE 11D.1** lists the most common functional groups. Double and triple carbon–carbon bonds are considered in Topics 11A and 11B.

TABLE 11D.1 Common Functional Groups

Group	Class of compound	Group	Class of compound
—X	halide (X = F, Cl, Br, or I)	—COOH	carboxylic acid
—OH	alcohol, phenol	—COOR	ester
—O—	ether	—N<	amine
—CHO	aldehyde	—CO—N<	amide
—CO—	ketone		

11D.1 Haloalkanes

The **haloalkanes** (which are also called alkyl halides) are alkanes in which at least one hydrogen atom has been replaced by a halogen atom. Although they have important uses, many haloalkanes are highly toxic and a threat to the environment. The haloalkane 1,2-dichlorofluoroethane, $CHClFCH_2Cl$, is an example of a hydrochlorofluorocarbon (HCFC), one of the compounds held responsible for the depletion of the ozone layer (see Box 7E.1). Many pesticides are aromatic compounds with several halogen atoms.

Carbon–halogen bonds are polar (**1**). The carbon atom to which the halogen atom is attached is partially positive, making it susceptible to **nucleophilic substitution,** in which a nucleophile replaces the halogen atom. A **nucleophile** is a reactant that seeks out centers of positive charge in a molecule. Two examples are the hydroxide ion, OH^- and the water molecule, H_2O. In each case a lone pair on the oxygen atom is attracted to regions of partial positive charge; the hydroxide ion is the stronger nucleophile because of its negative charge. Water acts as a nucleophile in a "hydrolysis reaction," a reaction with water in which a carbon–element bond is replaced by a carbon–oxygen bond. For example, bromomethane undergoes hydrolysis by hydroxide ions in aqueous ethanol to methanol and bromide ions:

$$CH_3Br + OH^- \longrightarrow CH_3OH + Br^-$$

Haloalkanes are alkanes in which at least one hydrogen atom has been replaced by a halogen atom; they undergo nucleophilic substitution.

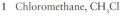

1 Chloromethane, CH_3Cl

11D.2 Alcohols

The **hydroxyl group** is an —OH group covalently bonded to a carbon atom. An **alcohol** is an organic compound that contains a hydroxyl group not connected directly to a benzene ring or to a >C=O group. One of the best-known organic compounds is ethanol, CH_3CH_2OH, which is also called *ethyl alcohol* and *grain alcohol*. The —OH group in an alcohol molecule is polar and alcohols can give up their hydroxyl protons in certain solvents (**FIG. 11D.1**), but their conjugate bases are so strong that they are not acids in water.

There are three common methods for naming alcohols:

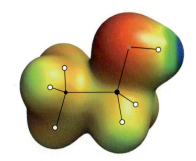

FIGURE 11D.1 The charge distribution in an ethanol molecule. Red denotes the region of net negative charge around the oxygen atom and blue denotes regions of net positive charge. Because the molecule is polar and can form hydrogen bonds, it is highly soluble in water.

- Add the suffix *-ol* to the stem of the parent hydrocarbon, as in methanol and ethanol. When the location of the —OH group needs to be specified (to avoid ambiguity), the number of the carbon atom to which it is attached is given, as in 1-propanol for $CH_3CH_2CH_2OH$ and 2-propanol for $CH_3CH(OH)CH_3$.
- Name the parent hydrocarbon as a group and attach the name alcohol, as in methyl alcohol, CH_3OH, and ethyl alcohol, CH_3CH_2OH.
- Name the —OH group as a substituent, in which case the name *hydroxy* is used, as in 2-hydroxybutane for $CH_3CH(OH)CH_2CH_3$.

This book normally uses the first of these three methods.

Alcohols are divided into three classes according to the number of organic groups attached to the carbon atom connected to the —OH group:

2-Propanol is often known by its common name, isopropyl alcohol.

Type of alcohol	Structure	Example	
primary	RCH_2—OH	ethanol, CH_3CH_2OH	
secondary	R_2CH—OH	2-butanol, $CH_3CH(OH)CH_2CH_3$	
tertiary	R_3C—OH	2-methyl-2-propanol, $(CH_3)_3COH$	

The common names of 2-methyl-2-propanol are tertiary-butanol and tertiary-butyl alcohol (usually shortened to *tert*-butanol and *tert*-butyl alcohol or even *t*-butanol and *t*-butyl alcohol).

Each R represents an organic group, such as a methyl or ethyl group; they need not all be the same.

Methanol is usually prepared industrially from synthesis gas, which is obtained from coal (Topic 8B):

$$CO(g) + 2 H_2(g) \xrightarrow{\text{catalyst, 250 °C, 50-100 atm}} CH_3OH(g)$$

Ethanol is produced in large quantities throughout the world by the fermentation of carbohydrates. It is also prepared by the hydration of ethene in an addition reaction:

$$CH_2=CH_2(g) + H_2O(g) \xrightarrow{\text{catalyst, 300 °C}} CH_3CH_2OH(g)$$

The laboratory synthesis of alcohols is by nucleophilic substitution of haloalkanes, such as

$$CH_3CHClCH_3 + OH^- \longrightarrow CH_3CH(OH)CH_3 + Cl^-$$

Ethylene glycol, or 1,2-ethanediol, $HO(CH_2)_2OH$ (**2**), is an example of a **diol,** which is a compound with two hydroxyl groups. Ethylene glycol is used as a component of antifreeze mixtures and in the manufacture of some synthetic fibers.

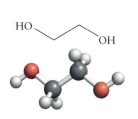

2 1,2-Ethanediol, $HO(CH_2)_2OH$

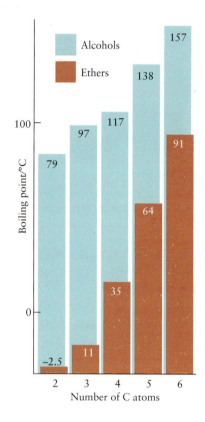

FIGURE 11D.2 The boiling points of ethers are lower than those of isomeric alcohols because there is hydrogen bonding in alcohols but not in ethers. All the molecules represented here are unbranched.

Alcohols have much lower vapor pressures than hydrocarbons with approximately the same molar mass; consequently, alcohols with low molar masses are liquids. For example, ethanol is a liquid at room temperature, but butane, which has a higher molar mass, is a gas. The relatively low volatility of alcohols is a sign of the strength of hydrogen bonds. The ability of alcohols to form hydrogen bonds also accounts for the solubility in water of alcohols with low molar mass.

The formulas of alcohols are derived from that of water by replacing one of the hydrogen atoms with an organic group. Like water, they form intermolecular hydrogen bonds.

11D.3 Ethers

An **ether** is an organic compound of the form R—O—R, where R is an alkyl group (the two R groups need not be the same). You can think of an ether as an HOH molecule in which both H atoms have been replaced by alkyl groups:

H—O—H	CH_3CH_2—O—H	CH_3CH_2—O—CH_2CH_3
Water	Ethanol (an alcohol)	Diethyl ether (an ether)

Ethers are more volatile than alcohols of the same molar mass because their molecules do not form hydrogen bonds to one another (**FIG. 11D.2**). They can form hydrogen bonds with water molecules, but they are less soluble in water than alcohols because they can only accept protons, not donate them, and as a result they cannot form as many hydrogen bonds to water molecules as alcohols do. Because ethers are not very reactive and not very polar, they are useful solvents for other organic compounds. However, ethers are dangerously flammable; diethyl ether is easily ignited and must be used with great care.

Ethers are not very reactive. They are more volatile than alcohols with similar molar masses because their molecules cannot form hydrogen bonds with one another.

11D.4 Phenols

In a **phenol,** a hydroxyl group is attached directly to an aromatic ring. The parent compound, phenol itself, C_6H_5OH (**3**), is a white, crystalline, molecular solid. Many substituted phenols occur naturally, some being responsible for the fragrances of plants. They are often components of *essential oils,* the oils that can be distilled from flowers and leaves. Thymol (**4**), for instance, is the active ingredient of oil of thyme, and eugenol (**5**) provides most of the scent and flavor of oil of cloves.

Phenols differ from alcohols in that they are weak acids. Similar to the resonance in phenol itself (Topic 11C), the resonance in the anion

3 Phenol, C_6H_5OH

4 Thymol

5 Eugenol

delocalizes the negative charge of the conjugate base of phenol and hence stabilizes the anion. As a result, $C_6H_5O^-$ is a weaker conjugate base than the corresponding conjugate base of an alcohol, such as $CH_3CH_2O^-$, the *ethoxide ion,* which is formed from ethanol.

Consequently, phenol, C_6H_5OH, is a stronger acid than ethanol, CH_3CH_2OH, and phenols that are otherwise insoluble in water are soluble in basic solutions. However, even a single —CH_2— group can insulate the O atom from the benzene ring; for example, phenylmethanol, $C_6H_5CH_2OH$ (benzyl alcohol, **6**), is an alcohol, not a phenol.

Phenols are weak acids as a result of delocalization and stabilization of the conjugate base.

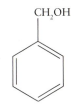

6 Benzyl alcohol, phenylmethanol, $C_6H_5CH_2OH$

11D.5 Aldehydes and Ketones

The **carbonyl group**, $C{=}O$, occurs in two closely related families of compounds:

- **Aldehydes** are compounds of the form $\overset{R}{\underset{H}{C}}{=}O$

- **Ketones** are compounds of the form $\overset{R}{\underset{R}{C}}{=}O$

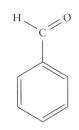

7 Benzaldehyde, C_6H_5CHO

In an aldehyde, the carbonyl group is found at the end of the carbon chain, whereas in a ketone it is at an intermediate position. The R groups may be either aliphatic or aromatic, and the two R groups of a ketone need not be the same. In condensed structural formulas, the carbonyl group of a ketone is written —CO—, as in CH_3COCH_3, propanone (acetone), a common laboratory solvent. The carbonyl group in an aldehyde is normally written —CHO, as in HCHO (formaldehyde), the simplest aldehyde. *Formalin,* the liquid used to preserve biological specimens, is an aqueous solution of formaldehyde. Wood smoke contains formaldehyde, and formaldehyde's destructive effect on bacteria is one reason why smoking food helps to preserve it.

The systematic names of aldehydes are obtained by replacing the ending -*e* by -*al,* as in methanal for HCHO and ethanal for CH_3CHO. Note that the carbon atom of the carbonyl group is included in the count of carbon atoms when determining the alkane from which the aldehyde is derived. Ketones are given the suffix -*one,* as in propanone, CH_3COCH_3. To avoid ambiguity, a number is used to denote the carbon atom that has become the carbonyl group; thus, $CH_3CH_2CH_2COCH_3$ is 2-pentanone and $CH_3CH_2COCH_2CH_3$ is 3-pentanone.

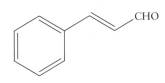

8 Cinnamaldehyde

Aldehydes occur naturally in essential oils and contribute to the flavors of fruits and the odors of plants. Benzaldehyde, C_6H_5CHO (**7**), contributes to the characteristic aroma of cherries and almonds. Cinnamaldehyde (**8**) is found in cinnamon, and vanilla extract contains vanillin (**9**), which is present in oil of vanilla. Ketones can also be fragrant. For example, carvone (structure **11a** in Topic 11A) is the essential oil of spearmint.

Formaldehyde is prepared industrially by the catalytic oxidation of methanol:

$$2\,CH_3OH(g) + O_2(g) \xrightarrow{600\,°C,\,Ag} 2\,HCHO(g) + 2\,H_2O(g)$$

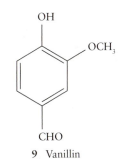

9 Vanillin

Further oxidation of the product to a carboxylic acid (Section 11D.6) can be avoided by using a mild oxidizing agent. There is less risk of further oxidation for ketones than for aldehydes, because a C—C bond would have to be broken for a carboxylic acid to form.

Dichromate oxidation of secondary alcohols produces ketones in good yield, with little additional oxidation. For example, $CH_3CH_2CH(OH)CH_3$ can be oxidized to $CH_3CH_2COCH_3$. The difference between the ease of oxidation of aldehydes and that of ketones is used to distinguish them. Aldehydes can reduce silver ions to form a silver mirror—a coating of silver on test-tube walls—in the *Tollens test,* which uses a solution of Ag^+ ions in aqueous ammonia (**FIG. 11D.3**):

FIGURE 11D.3 An aldehyde (left) produces a silver mirror with Tollens reagent, but a ketone (right) does not, because it is less easily oxidized than an aldehyde. (*W. H. Freeman photo by Ken Karp.*)

- Aldehydes: $CH_3CH_2CHO + Ag^+$ (the Tollens test) gives CH_3CH_2COOH and Ag(s)
- Ketones: $CH_3COCH_3 + Ag^+$ (the Tollens test) gives no reaction

Aldehydes and ketones can be prepared by the oxidation of alcohols. Aldehydes can be more easily oxidized than ketones can.

LAB VIDEO FIGURE 11D.3

11D.6 Carboxylic Acids

The **carboxyl group**, — C (=O)(OH) , normally abbreviated —COOH, is the functional group found in **carboxylic acids,** which are weak acids of the form R—COOH. The strengths of carboxylic acids are related to their structures, as described in Topic 6C. The simplest carboxylic acid is formic acid, HCOOH, the acid in ant venom. Another common carboxylic acid is acetic acid, CH_3COOH, the acid of vinegar. It forms when the ethanol in wine is oxidized by air:

$$CH_3CH_2OH \xrightarrow{O_2} H_3C-C(=O)(OH) + H_2O$$

Carboxylic acids are named systematically by replacing the -*e* of the parent hydrocarbon by the suffix -*oic acid;* the carbon atom of the carboxyl group is included in the count of atoms to determine the parent hydrocarbon molecule. Thus, formic acid is formally methanoic acid, and acetic acid is ethanoic acid.

Carboxylic acids can be prepared by oxidizing primary alcohols and aldehydes with a strong oxidizing agent, such as acidified aqueous potassium permanganate solution. In some processes of industrial importance, an alkyl group can be oxidized directly to a carboxyl group.

Carboxylic acids contain the acidic carboxyl group.

11D.7 Esters

10 Ethyl acetate, $CH_3COOC_2H_5$

The name "ester" comes from a shortening of the German words *Essig* (vinegar) and *Aether* (ether). At the time the term was coined much less was known about organic chemistry!

The product of the reaction between a carboxylic acid and an alcohol is called an **ester** (**10**). Acetic acid and ethanol, for example, react when heated to about 100 °C in the presence of a strong acid. The products of this **esterification** are ethyl acetate and water:

$$H_3C-C(=O)(OH) + H-O-CH_2CH_3 \longrightarrow H_3C-C(=O)(O-CH_2CH_3) + H_2O$$

Many esters have fragrant odors and contribute to the flavors of fruits. For example, benzyl acetate, $CH_3COOCH_2C_6H_5$, is an active component of oil of jasmine. Other naturally occurring esters include fats and oils. For example, the animal fat tristearin (**11**), which is a component of beef fat, is an ester formed from glycerol and stearic acid.

11 Tristearin, $C_{57}H_{110}O_6$

Ester formation is an example of a **condensation reaction,** in which two molecules combine to form a larger one and a small molecule is eliminated (**FIG. 11D.4**). The reaction is catalyzed by a small amount of strong acid, such as sulfuric acid. In an esterification of a carboxylic acid and an alcohol, the eliminated molecule is H_2O.

Condensation reactions provide a solution to a problem resulting from the partial hydrogenation of oils (Topic 11B). During the hydrogenation process, the remaining double bonds are converted from their naturally occurring cis forms to the trans isomer. These fats, which are called "trans fats," can contribute to health problems such as arteriosclerosis. *Transesterification* is a technique in which enzymes are used to promote switching of the long carboxylic acid chains from one alcohol location to another in a *fatty acid,* a compound with a carboxylic acid group at the end of a long hydrocarbon chain. The resulting fat has the desired higher melting point yet is still unsaturated and has no trans fats. The new process also reduces the need for cooling water and does not use toxic or flammable solvents.

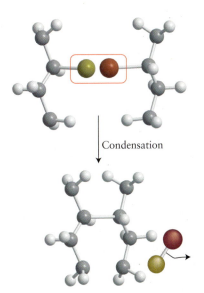

Condensation

FIGURE 11D.4 In a condensation reaction, two molecules are linked as a result of removing two atoms or groups (the red and yellow spheres) as a small molecule.

> **Self-test 11D.1A** (a) Write the condensed structural formula of the ester formed from the reaction between propanoic acid, CH_3CH_2COOH, and methanol, CH_3OH. (b) Write the condensed structural formulas of the acid and alcohol that react to form pentyl ethanoate, $CH_3COOC_5H_{11}$, a contributor to the flavor of bananas.
> [*Answer:* (a) $CH_3CH_2COOCH_3$; (b) CH_3COOH and $CH_3(CH_2)_3CH_2OH$]
>
> **Self-test 11D.1B** (a) Write the condensed structural formula of the ester formed from the reaction between formic acid, $HCOOH$, and ethanol, CH_3CH_2OH. (b) Write the condensed structural formulas of the acid and alcohol that react to form methyl butanoate, $CH_3(CH_2)_2COOCH_3$, a contributor to the flavor of apples.

Alcohols condense with carboxylic acids to form esters.

11D.8 Amines, Amino Acids, and Amides

An **amine** is a compound with a formula derived from NH_3 in which various numbers of H atoms are replaced by organic groups, which may be either aliphatic or aromatic. Amines are classified as primary, secondary, or tertiary according to the number of R groups on the nitrogen atom:

Type of amine	Structure	Example
primary	RNH_2	methylamine, CH_3NH_2
secondary	R_2NH	dimethylamine, $(CH_3)_2NH$
tertiary	R_3N	trimethylamine, $(CH_3)_3N$

Each R represents an organic group, such as a methyl or ethyl group; they need not all be the same. In each case, the N atom is sp^3 hybridized, with one lone pair of electrons and three σ-bonds. A *quaternary ammonium ion* is a tetrahedral ion of the form R_4N^+, where the R groups may all be different (and include up to three H atoms). For example, the tetramethylammonium ion, $(CH_3)_4N^+$, and the trimethylammonium ion, $(CH_3)_3NH^+$,

FIGURE 11D.5 The variation of the fractional composition of a solution of a typical amino acid with pH (here the amino acid is alanine, R = CH₃). The vertical dotted lines are labeled with the pK values for the three proton transfer equilibria involved (for $^+H_3NCHRCOOH$, $^+H_3NCHRCO_2^-$, and $H_2NCHRCOOH$, from left to right. Notice that the concentration of the molecular form is extremely low at all pH values; its concentration had to be multiplied by a factor of 10^8 for it to be visible on the graph. Amino acids are present almost entirely in ionic form in aqueous solution.

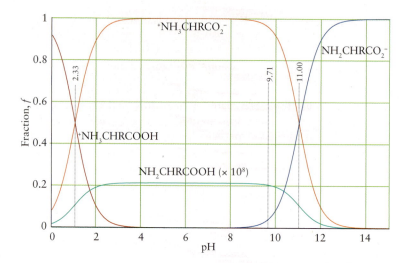

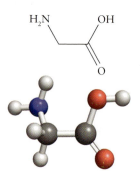

12   Glycine, NH_2CH_2COOH

Zwitter is the German word for hybrid or hermaphrodite, a blend of male and female.

are quaternary ammonium ions. The **amino group,** the parent functional group of amines, is —NH₂.

Amines are widespread in nature. Many of them have a pungent and often unpleasant odor. Because proteins are organic polymers containing nitrogen, amines are present in the decomposing remains of living matter and, together with sulfur compounds, are responsible for the stench of decaying flesh. The common names of two diamines—putrescine, $NH_2(CH_2)_4NH_2$, and cadaverine, $NH_2(CH_2)_5NH_2$—speak for themselves. Like ammonia, amines are weak bases (Topic 6C), but quaternary ammonium ions with at least one H atom attached to the N atom are generally acidic.

An **amino acid** is a carboxylic acid that contains an amino group as well as a carboxyl group. The simplest example is glycine, NH_2CH_2COOH (**12**). Notice that an amino acid has both a basic group (—NH₂) and an acidic group (—COOH) in the same molecule. In aqueous solution with pH close to 7, amino acids are present as the **zwitterion,** such as $^+H_3NCH_2CO_2^-$, in which the amine group is protonated and the carboxyl group is deprotonated. There are four possible forms of an amino acid, which differ in the extent of protonation of the two functional groups. In very acidic solution, alanine exists as $^+H_3NCH(CH_3)$ COOH, but, as base is added, the proton on the N is removed first, leaving $NH_2CH(CH_3)$ COOH (mainly in its zwitterion form), and then the proton from the carboxylic acid, to form $NH_2CH(CH_3)CO_2^-$. The concentrations of the various species in a solution of an amino acid can be calculated by the same methods used for polyprotic acids in Topic 6E. **FIGURE 11D.5** shows how the concentrations of alanine species present in a solution vary with pH. Notice that the molecular form of alanine is almost nonexistent at any pH: its concentration had to be multiplied by a factor of 100 million for it to be visible on the plot.

By far the most important amino acids are the α-amino acids, in which the —NH₂ group is attached to the carbon atom next to the carboxyl group, as in glycine. Like alcohols, amines condense with carboxylic acids:

$$H_3C-\overset{O}{\overset{\|}{C}}\overset{\diagdown}{\underset{\boxed{OH \quad H}}{}}-\overset{CH_2CH_3}{\underset{NH}{|}} \longrightarrow H_3C-\overset{O}{\overset{\|}{C}}\overset{\diagup}{\underset{HN-CH_2CH_3}{}} \quad + H_2O$$

The product is an **amide.** When the reactant is a primary amine, RNH₂, the product is a molecule of the form R—(CO)—NHR. Many amides have N—H bonds that can take part in hydrogen bonding, and so the intermolecular forces between their molecules are strong.

The mechanism of amide formation is a source of insight into the properties of carboxylic acids and amines. Initially, an amine might be expected to act as a base and simply accept a proton from the carboxylic acid. Indeed, that does happen, and a quaternary ammonium salt is formed when the reagents are mixed in the absence of a solvent. For example,

$$CH_3COH + CH_3NH_2 \longrightarrow CH_3CO_2^- + CH_3NH_3^+$$

However, upon heating to about 200 °C, a thermodynamically more favorable reaction takes place. The proton transfer is reversed, and the amine acts as a nucleophile as it attacks the carbon atom of the carboxyl group in a condensation reaction:

Self-test 11D.2A Predict whether an ester or a primary amine of the same molar mass is likely to have the higher boiling point and explain why.

[**Answer:** The amine; it can form hydrogen bonds with the —NH_2 group, but the ester group cannot form hydrogen bonds in the pure state.]

Self-test 11D.2B What are the hybridization schemes of the C and N atoms in formamide, $HCONH_2$?

Toolbox 11D.1 HOW TO NAME SIMPLE COMPOUNDS WITH FUNCTIONAL GROUPS

CONCEPTUAL BASIS

In general, the names of compounds containing functional groups follow the same conventions and numbering system as those used in the names of hydrocarbons (Toolbox 11A.1), with the ending of the name changed to indicate the functional group. The aim here, as there, is to be succinct but unambiguous.

PROCEDURE

The functional group has priority in numbering over hydrocarbon side chains.

Alcohols

To form the systematic name, identify the parent hydrocarbon and replace the ending -e by -ol. The location of the hydroxyl group is denoted by numbering the C atoms of the backbone, starting at the end of the chain that results in the lowest number for the —OH group location. When —OH is named as a substituent, it is called *hydroxy*. For example, $CH_3CH(OH)CH_2CH_2CH_3$ is 2-pentanol or 2-hydroxypentane.

Aldehydes and ketones

For aldehydes, identify the parent hydrocarbon: include the C of —CHO in the count of carbon atoms. Then change the final -e of the hydrocarbon name to -al. The C in the —CHO group is always carbon 1, at the end of a carbon chain, and is not explicitly numbered. For ketones, change the -e of the parent

hydrocarbon to -one and number the chain in the order that gives the carbonyl group the lower number. Thus, $CH_3CH_2CH_2COCH_3$ is 2-pentanone.

Carboxylic acids

Change the -e of the parent hydrocarbon to -oic acid. To identify the parent acid, include the C atom of the —COOH group when counting carbon atoms. Thus, $CH_3CH_2CH_2COOH$ is butanoic acid.

Esters

Change the -anol of the alcohol to -yl and the -oic acid of the parent acid to -oate. Thus, CH_3COOCH_3 (from methanol and ethanoic acid) is methyl ethanoate.

Amines

Amines are named systematically by specifying the groups attached to the nitrogen atom in alphabetical order (disregarding prefixes), followed by the suffix -amine. Amines with two amino groups are called *diamines*. The —NH_2 group is called *amino*- when it is a substituent. Thus, $(CH_3CH_2)_2NCH_3$ is diethylmethylamine and $CH_3CH(NH_2)CH_3$ is 2-aminopropane.

Halides

Name the halogen atom as a substituent by changing the -ine part of its name to -o, as in 2-chlorobutane, $CH_3CHClCH_2CH_3$.

These rules are illustrated in Example 11D.1.

EXAMPLE 11D.1 Naming compounds with functional groups

You are working in a pharmaceutical research lab and need to retrieve reagents from storage. The compounds are arranged alphabetically, so you need to determine the names from the structural formulas given in the synthetic procedures you are testing. Name the compounds (a) $CH_3CH(CH_3)CH(OH)CH_3$; (b) $CH_3CHClCH_2COCH_3$; and (c) $(CH_3CH_2)_2NCH_2CH_2CH_3$.

PLAN Follow the procedures in **Toolbox 11D.1**.

SOLVE

(a) Count the number of carbon atoms in the longest chain and replace the ending of the name of that chain with -ol:

The alcohol is 3-methyl-2-butanol.

$$CH_3-\underset{4}{C}H-\underset{3}{C}H-\underset{2}{C}H-\underset{1}{C}H_3$$
with CH₃ on C3 and OH on C2

(b) The compound is both a ketone and a haloalkane. Number the hydrocarbon chain in the direction that gives the ketone group the lower number. The chain has five carbon atoms; the ketone group is on the second carbon atom from one end, and the chlorine atom is on the fourth:

The compound is 4-chloro-2-pentanone.

$$\underset{5}{C}H_3-\underset{4}{C}H-\underset{3}{C}H_2-\underset{2}{C}-\underset{1}{C}H_3$$
with Cl on C4 and O double bond on C2

(c) The compound $(CH_3CH_2)_2NCH_2CH_2CH_3$ has two ethyl groups and one propyl group attached to a nitrogen atom. Give the ethyl groups the prefix *di-* to indicate their number and list the groups in alphabetical order.

The compound is diethylpropylamine.

H_3C-CH_2 and H_3C-CH_2 both attached to N, with N—$CH_2-CH_2-CH_3$

Self-test 11D.3A Name (a) $CH_3CH(CH_2CH_2OH)CH_3$; (b) $CH_3CH(CHO)CH_2CH_3$; (c) $(C_6H_5)_3N$.

[**Answer:** (a) 3-methyl-1-butanol; (b) 2-methylbutanal; (c) triphenylamine]

Self-test 11D.3B Name (a) $CH_3CH_2CH(OH)CH_2CH_3$; (b) $CH_3CH_2COCH_2CH_3$; (c) $CH_3CH_2NHCH_3$.

Related Exercises 11D.5, 11D.6, 11D.11, 11D.12

Amines are derived from ammonia by the replacement of hydrogen atoms with organic groups. Amides result from the condensation of amines with carboxylic acids. Amines and many amides take part in hydrogen bonding.

What have you learned in this Topic?

You have seen that the study of organic chemistry is simplified by recognizing that there is a small set of functional groups that exhibit characteristic properties in any compound in which they occur. You have been introduced to several functional groups and have seen how to name the compounds that contain them.

The skills you have mastered are the ability to:

☐ **1.** Recognize a simple haloalkane, alcohol, ether, phenol, aldehyde, ketone, carboxylic acid, amine, amide, or ester, given a molecular structure.

☐ **2.** Predict the oxidation products of aldehydes and ketones (Section 11D.5).

☐ **3.** Write the structural formula of an ester or an amide formed from the condensation reaction of a given carboxylic acid with a given alcohol or amine (Sections 11D.7 and 11D.8).

☐ **4.** Predict the influence of hydrogen bonding on the physical properties of organic compounds (Self-Test 11D.2).

☐ **5.** Name compounds with simple functional groups (Toolbox 11D.1 and Example 11D.1).

Topic 11D Exercises

11D.1 Write the general formula of each of the following types of compounds, using R to denote an organic group: (a) amine; (b) alcohol; (c) carboxylic acid; (d) aldehyde.

11D.2 Write the general formula of each of the following types of compounds, using R to denote an organic group: (a) ether; (b) ketone; (c) ester; (d) amide.

11D.3 Identify each type of compound: (a) R—O—R; (b) R—CO—R; (c) R—NH₂; (d) R—COOR.

11D.4 Identify each type of compound: (a) R—CHO; (b) R—COOH; (c) R—CONHR; (d) R—OH.

11D.5 Name the following compounds: (a) $CH_3CHClCH_3$; (b) $CH_3CH_2C(CH_3)ClCH_2CHClCH_3$; (c) CH_3Cl_3; (d) CH_2Cl_2.

11D.6 Name the following compounds: (a) $CH_3CCl_2CH_3$; (b) $CH_2CHCH_2CH_2Br$; (c) $CH_3CCCH_2CH(OH)CH_3$; (d) $CH_3OCH_2CH_2CH_2CH_3$.

11D.7 Write the formulas of the following compounds and state whether each one is a primary, secondary, or tertiary alcohol or a phenol: (a) 1-chloro-2-hydroxybenzene; (b) 2-methyl-3-pentanol; (c) 2,4-dimethyl-1-hexanol; (d) 2-methyl-2-butanol.

11D.8 Write the formulas of the following compounds and state whether each one is a primary, secondary, or tertiary alcohol or a phenol: (a) 1-hexanol; (b) 1-phenyl-1-ethanol; (c) 2,4-dimethylphenol; (d) 3-ethyl-2,2-dimethyl-3-pentanol.

11D.9 Write the formula of (a) methyl propyl ether; (b) ethyl butyl ether; (c) dipropyl ether.

11D.10 Write the formula of (a) methyl cyclopropyl ether; (b) dibutyl ether; (c) ethyl pentyl ether.

11D.11 Name each compound:
(a) $CH_3CH_2CH_2OCH_2CH_2CH_2CH_3$; (b) $C_6H_5OCH_3$;
(c) $CH_3CH_2CH_2OCH_2CH_2CH_2CH_2CH_3$.

11D.12 Name each compound: (a) $CH_3(CH_2)_3CH_2OCH_3$;
(b) $CH_3CH_2OCH_2CH_2CH_2CH_3$; (c) $CH_3CH_2CH_2OCH_2CH_2CH_3$.

11D.13 Identify each compound as an aldehyde or a ketone and give its systematic name: (a) CH_3CHO; (b) CH_3COCH_3;
(c) $(CH_3CH_2)_2CO$.

11D.14 Identify each compound as an aldehyde or a ketone and give its systematic name: (a) $CH_3CH_2CH(CH_3)CHO$;
(b)

(c) $(CH_3CH_2)_2CHCH_2COCH_2CH_3$

11D.15 Write the structural formula of (a) butanal;
(b) 3-hexanone; (c) 2-heptanone.

11D.16 Write the structural formula of (a) 2-ethyl-2-methylpentanal; (b) 3,5-dihydroxy-4-octanone;
(c) 4,5-dimethyl-3-hexanone.

11D.17 Give the systematic name of (a) CH_3COOH;
(b) $CH_3CH_2CH_2COOH$; (c) $CH_2(NH_2)COOH$.

11D.18 Give the systematic name of (a) $CH_3CH(CH_3)CH_2COOH$;
(b) CH_2ClCH_2COOH; (c) $CH_3(CH_2)_7COOH$.

11D.19 Draw the structure of (a) benzoic acid, C_6H_5COOH;
(b) 2-chloro-3-methylpentanoic acid; (c) hexanoic acid;
(d) propenoic acid.

11D.20 Draw the structure of (a) 2-methylpropanoic acid;
(b) 2,2-dichlorobutanoic acid; (c) 2,2,2-trifluoroethanoic acid;
(d) 4,4-dimethylpentanoic acid.

11D.21 Give the systematic name of each of the following amines: (a) CH_3NH_2; (b) $(CH_3CH_2)_2NH$; (c) $o\text{-}CH_3C_6H_4NH_2$.

11D.22 Give the systematic name of each of the following amines: (a) $CH_3CH_2CH_2NH_2$; (b) $(CH_3CH_2)_4N^+$;
(c) $p\text{-}ClC_6H_4NH_2$.

11D.23 Write the structural formula of each of the following amines: (a) o-methylphenylamine; (b) triethylamine; (c) tetramethylammonium ion.

11D.24 Write the structural formula of each of the following amines: (a) methylpropylamine; (b) dimethylamine;
(c) m-methylphenylamine.

11D.25 Which of the following molecules or ions may function as a nucleophile in a nucleophilic substitution reaction: (a) NH_3;
(b) CO_2; (c) Br^-; (d) SiH_4?

11D.26 Which of the following molecules or ions may function as a nucleophile in a nucleophilic substitution reaction: (a) OH^-;
(b) NH_4^+; (c) NH_2^-; (d) H_2O?

11D.27 Suggest an alcohol that could be used for the preparation of each of the following compounds and indicate how the reaction would be carried out: (a) ethanal; (b) 2-octanone;
(c) 5-methyloctanal.

11D.28 Suggest an alcohol that could be used for the preparation of each of the following compounds and indicate how the reaction would be carried out: (a) propanal; (b) 2-pentanone;
(c) 5-ethyl-3-nonanone.

11D.29 Draw the structure of the principal product formed from each pair of reagents in condensation reactions: (a) butanoic acid with 2-propanol; (b) ethanoic acid with 1-pentanol; (c) hexanoic acid with methylethyl amine; (d) ethanoic acid with propylamine.

11D.30 Draw the structure of the principal product formed from each pair of reagents in condensation reactions: (a) propanoic acid with 2-methylpropanol; (b) ethanoic acid with cyclohexanol;
(c) butanoic acid with dimethylamine; (d) 2-methyl-pentanoic acid with ethylamine.

11D.31 Classify each of the following reactions as an addition reaction, a nucleophilic substitution reaction, an electrophilic substitution reaction, or a condensation reaction: (a) the reaction of 1-butene with chlorine in the absence of light; (b) the polymerization of the amino acid glycine; (c) the hydrogenation of 1-butyne; (d) the polymerization of styrene, $CH_2CHC_6H_5$; (e) the reaction of methylamine with butanoic acid. (See Topic 11E.)

11D.32 Classify each of the following reactions as an addition reaction, a nucleophilic substitution reaction, an electrophilic substitution reaction, or a condensation reaction: (a) the reaction of terephthalic acid with 1,2-ethanediol; (b) the reaction of 3-chlorohexane with concentrated sodium hydroxide; (c) the reaction of water with 2-iodo-2-methylpropane; (d) the reaction of propanoic acid with ethanol; (e) the reaction of toluene with bromine in the presence of $FeBr_3$.

11D.33 You are given samples of propanal, 2-propanone, and ethanoic acid. Describe how you would use chemical tests including acid–base indicators to distinguish among the three compounds.

11D.34 You are given samples of 1-propanol, pentane, and ethanoic acid. Describe how you would use chemical tests such as aqueous solubility and acid–base indicators, to distinguish among the three compounds.

11D.35 Rank the following acids in order of increasing acid strength: $ClCH_2COOH$, Cl_3CCOOH, CH_3COOH, and CH_3CH_2COOH. Justify your answer.

11D.36 Rank methylamine, dimethylamine, and diethylamine in order of increasing base strength. Explain your rankings in relation to their molecular structure.

Topic 11E Polymers and Biological Macromolecules

How do other atoms affect the properties of organic compounds?

Topic **11D:** Common functional groups → Topic **11E:** Polymers and biological macromolecules

What are the characteristics of large organic molecules?

Why Do You Need to Know This Material? The development of polymeric materials has changed society. Knowledge of their structure and properties will allow you to understand their role in the modern world. Moreover, biological organisms function by using analogous molecules, and biology cannot be understood at a molecular level without knowing their structures and reactions.

What Do You Need to Know Already? This Topic draws on the introduction to functional groups in Topic 11D, intermolecular forces (Topic 3F), reaction mechanisms (Topic 7C), and isomers (Topic 11A).

The chains of carbon atoms in organic compounds can reach enormous lengths and form **macromolecules,** which are molecules containing hundreds and sometimes thousands of atoms. **Polymers,** such as polypropylene and polytetrafluoroethylene (sold as Teflon), are macromolecular compounds consisting of chains or networks of small repeating units. Although polymers can be large and complex, their properties can be understood once the functional groups they contain are known.

Polymers are made by two main types of reactions, *addition reactions* and *condensation reactions*. Which type of reaction takes place depends on the functional groups present in the starting materials. Most of the raw material for polymers comes from petroleum, but some polymers are made from agricultural products such as corn and soybeans. Biological macromolecules include the proteins that act as enzymes or contribute to the structures of organisms. They also include carbohydrates and molecules such as DNA and RNA, which control inheritance.

11E.1 Addition Polymerization

Alkenes can react with themselves to form long chains in a process called **addition polymerization.** For example, an ethene molecule may form a bond to another ethene molecule; another ethene molecule may add to that, and so on, forming a long hydrocarbon chain. The original alkene, such as ethene, is a small molecule called a **monomer.** Each monomer becomes one **repeating unit,** the structure that repeats over and over to produce the polymer chain. The product, the chain of covalently linked repeating units, is the polymer. The simplest addition polymer is polyethylene, $—(CH_2CH_2)_n—$, which is made by polymerizing ethene and so consists of long chains of thousands of repeating $—CH_2CH_2—$ units.

The plastics industry has developed polymers from a number of monomers with the formula $CHX{=}CH_2$, where X is a single atom (such as the Cl in vinyl chloride, $CHCl{=}CH_2$) or a group of atoms (such as the CH_3 in propene). These substituted ethenes are used to make polymers of formula $—(CHXCH_2)_n—$, including polyvinyl chloride (PVC), $—(CHClCH_2)_n—$, and polypropylene, $—(CH(CH_3)CH_2)_n—$ (**TABLE 11E.1**). They differ in appearance, rigidity, transparency, and resistance to weathering.

A widely used synthetic procedure is **radical polymerization,** polymerization by a radical chain reaction (Topic 7C). In a typical procedure, a monomer (such as ethene) is compressed to about 1000 atm and heated to 100 °C in the presence of a small amount of an organic peroxide (a compound of formula R—O—O—R, where R is an organic group). The reaction is initiated by dissociation of the O—O bond, giving two radicals:

$$R—O—O—R \longrightarrow R—O\cdot + \cdot O—R$$

Once it has been initiated, the chain reaction can propagate as the radicals attack the monomer molecules $CHX{=}CH_2$ (with X = H for ethene itself) and form a new, highly reactive radical:

TABLE 11E.1   Common Addition Polymers

Monomer name	Formula	Polymer formula	Common name
ethene*	$CH_2{=}CH_2$	$-(CH_2-CH_2)_n-$	polyethylene
vinyl chloride	$CHCl{=}CH_2$	$-(CHCl-CH_2)_n-$	polyvinyl chloride
styrene	$CH(C_6H_5){=}CH_2$	$-(CH(C_6H_5)-CH_2)_n-$	polystyrene
acrylonitrile	$CH(CN){=}CH_2$	$-(CH(CN)-CH_2)_n-$	Orlon, Acrilan
propene*	$CH(CH_3){=}CH_2$	$-(CH(CH_3)-CH_2)_n-$	polypropylene
methyl methacrylate	$CH_3OOCC(CH_3){=}CH_2$	(see structure below)	Plexiglas, Lucite
tetrafluoroethene*	$CF_2{=}CF_2$	$-(CF_2-CF_2)_n-$	Teflon, PTFE[†]

$$\left(\!\!\begin{array}{c} CH_3 \\ | \\ -C-CH_2- \\ | \\ O{=}C \\ \diagdown OCH_3 \end{array}\!\!\right)_n$$

*The suffix -ene is replaced by -ylene in the common names of these compounds, hence the names of the corresponding polymers.

[†]PTFE, polytetrafluoroethylene.

This radical attacks another monomer molecule, and the chain grows longer:

$$R-O-CH_2-\underset{X}{\overset{H}{C}}\cdot \;+\; H_2C{=}\underset{X}{\overset{H}{C}} \longrightarrow R-O-CH_2-\underset{X}{\overset{H}{C}}-CH_2-\underset{X}{\overset{H}{C}}\cdot$$

The reaction continues until all the monomer molecules have been used up or until it terminates when pairs of chains have linked together into single nonradical species. The product consists of macromolecules with many repeating units. Polyethylene (with X = H), for instance, consists of long chains of formula $-(CH_2CH_2)_n-$, in which n can reach many thousands. Many addition polymer molecules also have a number of branches, which are generated as new chains sprout at intermediate points along the "backbone" chain.

Strong, tough polymers have chains that pack together well. One problem with early attempts to make polypropylene was that the orientations of the H and CH_3 groups on each C atom were random, thereby preventing the chains from packing well. The resulting material was amorphous, sticky, and nearly useless. Now, however, the stereochemistry of the chains can be controlled by using a **Ziegler–Natta catalyst,** a catalyst consisting of the compounds titanium tetrachloride, $TiCl_4$, and triethylaluminum, $(CH_3CH_2)_3Al$. A polymer in which each unit or pair of repeating units has the same relative orientation is described as being **stereoregular;** the stereoregularity arises from the manner in which the chains grow on the catalyst (**FIG. 11E.1**). The chains of stereoregular polymers produced by Ziegler–Natta catalysts pack together well to form highly crystalline, dense materials (**FIG. 11E.2**).

Rubber is a polymer of isoprene (**1**). Natural rubber is obtained from the bark of the rubber tree as a milky white liquid, which is called *latex* (**FIG. 11E.3**) and consists of a suspension of rubber particles in water. The rubber itself is a soft white solid that becomes even softer when warm. It is used for pencil erasers and was once used as crepe rubber for the soles of shoes.

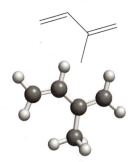

1   Isoprene

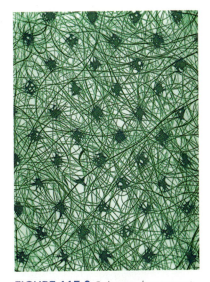

FIGURE 11E.2 Polypropylene carpeting gets its strength from the regularity of the structure of its chains. The dark patches are regions in which the fibers have been heat bonded together. (SPL/Science Source.)

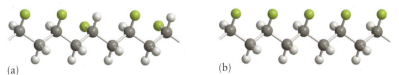

(a)   (b)

FIGURE 11E.1 (a) A polymer in which the substituents lie on random sides of the chain. (b) A stereoregular polymer produced by using a Ziegler–Natta catalyst. In this case, the substituents all lie on the same side of the chain.

The name "gutta percha" comes from the Malay words for "perca sap," perca being the name of the tree from which it is obtained.

FIGURE 11E.3 Collecting latex from a rubber tree in Malaysia, a principal producer of latex. (© *Imagestate Media Partners Limited–Impact Photos/Alamy.*)

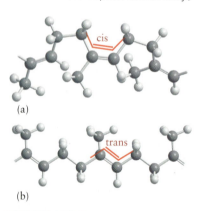

FIGURE 11E.4 (a) In natural rubber, the isoprene units are polymerized to be all cis. (b) The harder material, gutta-percha, is the all-trans polymer.

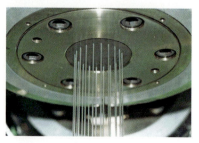

FIGURE 11E.5 Synthetic fibers are made by extruding liquid polymer from small holes in an industrial version of a spider's spinneret. (*Courtesy Sunline Co., Ltd.*)

Chemists were unable to synthesize rubber for a long time, even though they knew it was a polymer of isoprene. The enzymes in the rubber tree produce a stereoregular polymer in which all the links between monomers are in a cis arrangement (**FIG. 11E.4**); straightforward radical polymerization, however, produces a random mixture of cis and trans links and a sticky, useless product. The stereoregular polymer was achieved with a Ziegler–Natta catalyst, and almost pure, rubbery *cis*-polyisoprene can now be produced. *trans*-Polyisoprene, in which all the links are trans, is the hard, naturally occurring material *gutta-percha*, once used inside golf balls and still used to fill root canals in teeth.

Alkenes undergo addition polymerization. When a Ziegler–Natta catalyst is used, the polymer is stereoregular and has a high density.

11E.2 Condensation Polymerization

In **condensation polymers,** the monomers are linked together by condensation reactions, like those used to form ester or amide links. Polymers formed by linking together monomers that have carboxylic acid groups with those that have alcohol groups are called **polyesters.** Polymers of this type are widely used to make artificial fibers. A typical polyester is Dacron, or Terylene, a polymer produced from the esterification of terephthalic acid with ethylene glycol (1,2-ethanediol, $HOCH_2CH_2OH$); its technical name is polyethylene terephthalate, or PETE. The first condensation is

A new ethylene glycol molecule can condense with the carboxyl group on one end of the product, and another terephthalic acid molecule can condense with the hydroxyl group on the other end. As a result, the polymer grows at both ends and becomes

In the radical addition polymerization of alkenes, side chains can grow out from the main chain. However, in condensation polymerization, growth can occur only at the functional groups, and so chain branching is much less likely. As a result, polyester molecules make good fibers, because the unbranched chains can be made to lie side by side when the heated product is stretched and forced through small holes (**FIG. 11E.5**). The fibers produced can then be spun into yarn (**FIG. 11E.6**). Polyesters can also be molded and used in surgical implants, such as artificial hearts, or made into thin films for use in packaging and adhesive tape.

FIGURE 11E.6 A scanning electron micrograph of Dacron polyester and cotton fibers in a blended shirt fabric. Compare the smooth cylinders of polyester (colored orange) with the irregular surface of the cotton (colored green). The smooth polyester fibers resist wrinkles, and the irregular cotton fibers produce a more comfortable and absorbent texture. (*Andrew Syred/Science Photo Library/ Science Source.*)

Condensation polymerization of amines with carboxylic acids leads to the **polyamides,** substances more commonly known as *nylons.* A common polyamide is nylon-66, which is a polymer of 1,6-diaminohexane, $H_2N(CH_2)_6NH_2$, and adipic acid, $HOOC(CH_2)_4COOH$. The 66 in the name indicates that there are six carbon atoms in each monomer.

For condensation polymerization, it is necessary to have two functional groups on each monomer and to mix stoichiometric amounts of the reactants. In polyamide production, the starting materials first form "nylon salt" by proton transfer:

$$HOOC(CH_2)_4COOH + H_2N(CH_2)_6NH_2 \longrightarrow {}^-O_2C(CH_2)_4CO_2^- + {}^+H_3N(CH_2)_6NH_3^+$$

At this point, the excess acid or amine can be removed. Then, when the nylon salt is heated, the condensation begins, just as in the preparation of simple amides. The first step is

FIGURE 11E.7 A rather crude nylon fiber can be made by dissolving the salt of an amine in water and dissolving the acid in a layer of hexane, which floats on the water. The polymer forms at the interface of the two layers, and a long string can be pulled out slowly. (©1986 Chip Clark–Fundamental Photographs.)

⚡ LAB VIDEO FIGURE 11E.7

The amide grows at both ends by further condensations (**FIG. 11E.7**), and the final product is

The long polyamide (nylon) chains can be spun into fibers (like polyesters) or molded. The $N—H\cdots O=C$ hydrogen bonding that takes place between neighboring chains accounts for much of the strength of nylon fibers (**FIG. 11E.8**).

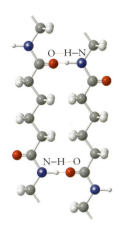

FIGURE 11E.8 The strength of nylon fibers is an indication of the strength of the hydrogen bonds between neighboring polyamide chains, shown as red dotted lines.

⚡ ANIMATION FIGURE 11E.8

EXAMPLE 11E.1   Identifying the formulas of polymers and monomers

Recognizing the monomer units in polymers is important for understanding how the polymers are prepared. Suppose you are studying the properties of various polymer materials and want to relate their properties to the building blocks used in their synthesis. Write the formulas of (a) the monomers of Kevlar, a strong fiber used to make bulletproof vests,

and (b) two repeating units of the polymer that is formed when peroxides are added to $CH_3CH_2CH=CH_2$ at a high temperature and pressure.

PLAN (a) Look at the backbone of the polymer, the long chain to which the other groups are attached. If the atoms are all carbon atoms, then the compound is an addition polymer. If ester groups are present in the backbone, then the polymer is a polyester and the monomers will be an acid and an alcohol. If the backbone contains amide groups, then the polymer is a polyamide and the monomers will be an acid and an amine. (b) If the monomer is an alkene or alkyne, then the monomers will add to one another; a π-bond will be replaced by new σ-bonds between the monomers. If the monomers consist of an acid and an alcohol or amine, then a condensation polymer forms with the loss of a molecule of water.

SOLVE

(a) Amide groups are present in the backbone, and so the polymer is a polyamide. Because the amide groups face in opposite directions, there are two different monomers, one with two acid groups and the other with two amine groups. Cleave each amide group apart and add a molecule of water to each amide link.

(b) The monomer is an alkene; so it forms an addition polymer. Replace each π-bond by an additional σ-bond to an adjacent monomer:

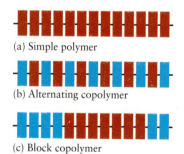

2 Lactic acid, $CH_3CH(OH)COOH$

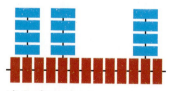

3 Poly(methyl methacrylate)

Self-test 11E.1A (a) Write the formula for the monomer of the polymer sold as Teflon, $—(CF_2CF_2)_n—$. (b) The polymer of lactic acid (**2**) is a biodegradable polymer made from renewable resources. It is used in surgical sutures that dissolve in the body. Write the formula for a repeating unit of this polymer.

[*Answer:* (a) $CF_2{=}CF_2$; (b) $(OCH(CH_3)CO)_n$]

Self-test 11E.1B Write the formula for (a) the monomer of poly(methyl methacrylate), used in contact lenses (**3**); (b) two repeating units of polyalanine, the polymer of the amino acid alanine, $CH_3CH(NH_2)COOH$.

Related Exercises 11E.1–11E.4

Condensation polymers are typically formed by the reaction of a carboxylic acid with an alcohol to form a polyester or with an amine to form a polyamide.

11E.3 Copolymers and Composite Materials

Copolymers are polymers made up of more than one type of repeating unit (**FIG. 11E.9**). One example is nylon-66, in which the repeating units are formed from 1,6-diaminohexane, $H_2N(CH_2)_6NH_2$, and adipic acid, $HOOC(CH_2)_4COOH$. They form an **alternating copolymer,** in which acid and amine monomers alternate.

(a) Simple polymer

(b) Alternating copolymer

(c) Block copolymer

(d) Graft copolymer

FIGURE 11E.9 The classification of copolymers. (a) A simple polymer made from a single monomer represented by the red rectangles. (b) An alternating co-polymer of two monomers, represented by the red and blue rectangles. (c) A block copolymer. (d) A graft copolymer.

In a **block copolymer,** a long segment made from one monomer is followed by a segment formed from the other monomer. One example is the block copolymer formed from styrene and butadiene. Pure polystyrene is a transparent, brittle material that is easily broken; polybutadiene is a synthetic rubber that is very resilient, but soft and opaque. A block copolymer of the two monomers produces *high-impact polystyrene,* a material that is a durable, strong, yet transparent plastic. A different formulation of the two polymers produces *styrene–butadiene rubber* (SBR), which is used mainly for automobile tires and running shoes, but also in chewing gum.

In a **random copolymer,** different monomers are linked in no particular order. A **graft copolymer** consists of long chains of one monomer with shorter chains of the other monomer attached as side groups. For example, the polymer used to make hard contact lenses is a nonpolar hydrocarbon that repels water. The polymer used to make soft contact lenses is a graft copolymer that has a backbone of nonpolar monomers but side groups of a different water-absorbing monomer. So much water is absorbed by the side chains that 50% of the volume of the contact lens is water, which makes it pliable, soft, and more comfortable to wear than a hard contact lens.

A **composite material** consists of two or more substances that have been combined into a heterogeneous material while retaining their individual characteristics. Seashells are natural composites that owe their strength to a tough organic matrix and their hardness to the calcium carbonate crystals imbedded in the matrix (**FIG. 11E.10**). Some lightweight composites, such as the graphite composite used for tennis rackets in which graphite fibers are imbedded in a polymer matrix, can have three times the strength-to-density ratio of steel (**FIG. 11E.11**). A composite material containing ceramic flakes in a polylactic acid polymer is being used to provide instant mends for broken bones. The material is injected as a paste into the cracked bone, where it sets at body temperature into a scaffold that acts like bone tissue and promotes healing as new bone cells form around it.

> Self-test 11E.2A Use Fig. 11E.9 to identify the type of copolymer formed by monomers A and B: —AAAABBBBB—.
>
> [***Answer:*** Block copolymer]
>
> Self-test 11E.2B Identify the type of copolymer formed by monomers A and B: —ABABABAB—.

Copolymers and composite materials combine the advantages of more than one component material.

11E.4 Physical Properties of Polymers

A polymer can be designed to have the appropriate properties for an application. The first consideration is chain length. Because synthetic polymers consist of molecules with different lengths, they do not have definite molar masses. You can speak only of the *average* molar mass and the *average* chain length of a polymer. Because of the variability in chain length, polymers do not have definite melting points; rather, they soften gradually as the temperature is raised. The viscosity of a polymer, its ability to flow when molten (Topic 3G), depends on its chain length. The longer the chains, the more tangled together they may be, and hence the slower their flow.

The mechanical strength of a polymer increases as the strength of the interactions between chains increases. Therefore, the longer the chains, the greater is the strength of the polymer. For chains of the same length, stronger intermolecular forces result in greater mechanical strength. The nature of the groups of atoms attached to or that form a part of the backbone of a polymer affects the strength of intermolecular forces and contributes to the mechanical strength of the polymer. For example, nylon is a polyamide and so has a series of repeating —NH— and —CO— groups that are hydrophilic and can take part in hydrogen bonding. Consequently, nylon is a strong polymer. It is also

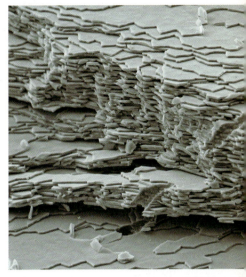

FIGURE 11E.10 A photomicrograph of the cross section of the mother-of-pearl lining a mollusk shell. The composite material making up mother-of-pearl consists of flat crystals of calcium carbonate embedded in a tough, flexible organic matrix that resists cracking. *(Eye of Science/Science Source.)*

FIGURE 11E.11 The body of this electric car is constructed of composite materials that are stronger than steel, but much lighter, resulting in higher performance. *(© Drive Images/Alamy.)*

FIGURE 11E.12 The two samples of polyethylene, a polymer of the hydrocarbon ethene, in the test tube were produced by different processes. The floating (low-density) sample was produced by high-pressure polymerization and the chains contain many branches. The sample at the bottom (high-density) was produced with a special catalyst that allowed it to pack together uniformly with a minimum of branching. As the insets show, the latter has a higher density because the polymer chains pack together better. *(W. H. Freeman photo by Ken Karp.)*

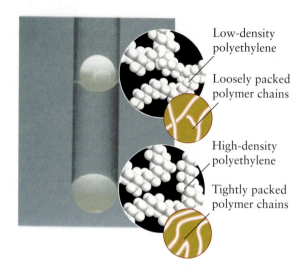

Low-density polyethylene

Loosely packed polymer chains

High-density polyethylene

Tightly packed polymer chains

(a)

(b)

FIGURE 11E.13 (a) High-density polyethylene body armor protects law enforcement personnel without restricting movement because it is soft and flexible. (b) Kevlar fibers have very large tensile strength and can also be woven into fibers for use in personal armor. *(Photo (a) Tom Vickers/Splash/Newscom; photo (b) Sinclair Stammers/Science Source.)*

hygroscopic (absorbs water), because water molecules are attracted to these polar groups. In contrast, polyethylene is a hydrocarbon that contains only C—C and C—H bonds, which are hydrophobic. As a result, whereas polyethylene repels water, water can penetrate through thin nylon fabrics because H_2O molecules can migrate by forming and breaking hydrogen bonds to the polymer molecules.

Chain-packing arrangements that maximize intermolecular contact result in greater strength as well as greater density. Long, unbranched chains can line up next to one another like uncooked spaghetti and form crystalline regions that result in strong, dense materials. Branched polymer chains cannot fit together as closely and form weaker, less dense materials (**FIG. 11E.12**). A soft, lightweight body armor has been developed by creating arrays of long polyethylene chains that are closely aligned in the same direction, giving rise to very strong intermolecular forces. This body armor is reported to be 15 times stronger than steel, yet it has such a low density that it floats on water. It is also soft and flexible, so it is comfortable to wear (**FIG. 11E.13**).

The **elasticity** of a polymer is its ability to return to its original shape after being stretched. Natural rubber has long hydrocarbon chains with low elasticity and is easily softened by heating. However, the "vulcanization" of rubber increases its elasticity. In vulcanization, rubber is heated with sulfur. The sulfur atoms form cross-links between the polyisoprene chains and produce a three-dimensional network of atoms (**FIG. 11E.14**). Because the chains are covalently linked together, vulcanized rubber does not soften as much as natural rubber when the temperature is raised. Vulcanized rubber is also much more resistant to deformation when stretched, because the cross-links pull it back. Polymeric materials that return to their original shapes after stretching are called **elastomers.** However, extensive cross-linking can produce a rigid network that resists stretching. For example,

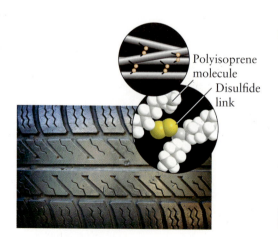

Polyisoprene molecule

Disulfide link

FIGURE 11E.14 The gray cylinders in the small inset represent the long chains of rubber molecules, and the beaded yellow strings represent disulfide (—S—S—) links that are introduced when the rubber is vulcanized, or heated with sulfur. These cross-links increase the resilience of the rubber and make it more useful than natural rubber. Automobile tires are made of vulcanized rubber and a number of additives, including carbon. *(© adam korzeniewski/Alamy.)*

ANIMATION FIGURE 11E.14

TABLE 11E.2 Recycling Codes

Recycling code	Polymer	Recycling code	Polymer
1 ♲ PETE	polyethylene terephthalate	5 ♲ PP	polypropylene
2 ♲ HDPE	high-density polyethylene	6 ♲ PS	polystyrene
3 ♲ PVC	polyvinyl chloride	7 ♲ OTHER	other
4 ♲ LDPE	low-density polyethylene		

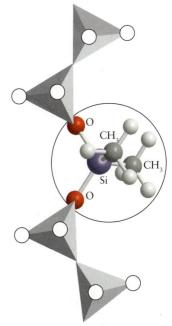

FIGURE 11E.15 A typical silicone structure. The hydrocarbon groups give the substance a water-repelling quality. Note the similarity of this structure to that of the purely inorganic pyroxenes in Fig. 31.8.

high concentrations of sulfur result in very extensive cross-linking and the hard material called *ebonite,* which is used to make fountain pens and bowling balls.

Plastics can be distinguished by their reaction to heat. A **thermoplastic polymer** is one that can be softened again after it has been molded; a **thermosetting polymer** is one that takes on a permanent shape in a mold and does not soften when heated. Many thermoplastic materials are made of addition polymers and can be recycled by melting and reprocessing. Examples are polyethylene and polyethylene terephthalate (**TABLE 11E.2**). Thermosetting plastics are used where heat resistance is important. For example, the vulcanized rubber in vehicle tires and the urea–formaldehyde foam used in the manufacture of plywood are thermosetting plastics.

Silicones are synthetic polymeric materials that are based on silicon instead of carbon. They consist of long —O—Si—O—Si— chains with the two remaining bonding positions on each Si atom occupied by organic groups, such as the methyl group, —CH_3 (**FIG. 11E.15**). Silicones are used to waterproof fabrics because their oxygen atoms attach to the fabric, leaving the hydrophobic (water-repelling) methyl groups like tiny, inside-out umbrellas sticking up out of the fabric's surface. Silicone-based soft materials have a wide variety of medical uses such as implants and drug delivery. They are also used in the aerospace and electronics industries as heat-resistant adhesives and insulators.

Because they are molecular compounds, most polymers are electrical insulators. However, polymers that have alternating double bonds along the chain can be used to conduct electricity (**BOX 11E.1**).

Box 11E.1 FRONTIERS OF CHEMISTRY: CONDUCTING POLYMERS*

One day, if you want to check your e-mail in a remote location, you might unroll a thin plastic sheet with a tiny microprocessor embedded in it. When you activate the microprocessor, your messages will appear and you might reply by touching it, writing on it with a special pen, or speaking to it. The remarkable material of this thin and flexible computer already exists: one form of it was discovered by accident in the early 1970s, when a chemist who was polymerizing ethyne (acetylene) added a thousand times too much catalyst. Instead of a synthetic rubber, he made a thin, flexible film. It looked like metal foil (see the photograph), and—very much like a metal—it conducted electricity.*

Metals conduct electricity because their valence electrons move easily from atom to atom. Most covalently bonded solids do not conduct electricity, because their valence electrons are locked into individual bonds and are not free to move. Exceptions such as graphite and carbon nanotubes have delocalized π-bonds

This organic photovoltaic cell makes use of a flexible conducting organic polymer that converts energy from the Sun into electricity. (*Patrick Landmann/Science Source.*)

in connected aromatic rings through which electrons can move freely because there are empty molecular orbitals close in energy to occupied orbitals (Topics 3H and 3J). However, a disadvantage is that commercial graphite is fragile and brittle.

Conducting polymers provide an exciting alternative. They are organic compounds that do not rust and that have low densities. They can be molded or drawn into shells, fibers, or thin plastic sheets, but they can still act like metallic conductors. They can be made to glow with almost any color and to change conductivity with conditions. Imagine cases of food labeled with polymer tags that change in conductivity if the cases are left unrefrigerated for too long.

All conducting polymers have a common feature: a long chain of sp^2 hybridized carbon atoms, often with nitrogen or sulfur atoms included in the chains. Polyacetylene, the first conducting polymer, is also the simplest, consisting of thousands of —CH=CH— units:

The alternation of double and single bonds means that each C atom has an unhybridized p-orbital that can overlap with the p-orbital on either side. This arrangement allows electrons to be delocalized along the entire chain like a one-dimensional version of graphite.

One conducting polymer, polypyrrole,

has been used in smart windows, which darken from a transparent yellow-green to a nearly opaque blue-black in bright sunlight. Fibers of polypyrrole are also woven into radar camouflage cloth that absorbs microwaves. Because it does not reflect microwaves back to their source, the cloth appears on radar as a patch of empty space.

Polyaniline, which has the structure

is being used in flexible coaxial cable; rechargeable, flat, button-like batteries; and laminated, rolled films that could be used as flexible computer or television displays. Thin films of poly-*p*-phenylenevinylene, PPV,

give off light when exposed to an electric field, a process called *electroluminescence*. By varying the composition of the polymer, scientists have coaxed it to glow in a wide range of colors. Such multicolor organic light-emitting diode (OLED) displays can be as bright as fluorescent ones.

HOW MIGHT YOU CONTRIBUTE?

All-plastic transistors and other electronic components can be miniaturized to an amazing degree, raising the possibility of nanoscale organic microprocessors and computers that could survive highly corrosive conditions, such as those in the body or at marine research sites. Because conducting polymers can also be designed to change shape with the level of electrical current, they could serve as artificial muscles, biosensors, and neural probes. For these goals to be achieved, the characteristics of these polymers must be known and their responses to various conditions have to be studied.

Related Exercises 11.49, 11.50

Further Reading S. Günes, H. Neugebauer, and N. S. Sariciftci, "Conjugated polymer-based organic solar cells," *Chemical Reviews*, vol. 107, pp. 1324–1338, 2007. D. W. Hatchett and M. Josowicz, "Composites of intrinsically conducting polymers as sensing nanomaterials," *Chemical Reviews*, vol. 108, pp. 746–769, 2008. T. A. Skotheim and J. R. Reynolds, *Handbook of Conducting Polymers*, 3rd ed., CRC Press, 2007. Scholz, F., ed., *Conducting Polymers: A New Era in Electrochemistry*, Springer, 2008. Information on the 2000 Nobel Prize, at http://nobelprize.org/nobel_prizes/chemistry/laureates/2000/adv.html.

*The 2000 Nobel Prize in chemistry was awarded to A. J. Heeger, A. G. MacDiarmid, and H. Shirakawa for their discovery of conducting polymers.

Polymers melt over a range of temperatures, and polymers consisting of long chains tend to have high viscosities. Polymer strength increases with increasing chain length and the extent of crystallization. Thermoplastic polymers are recyclable.

11E.5 Proteins

At one level, life can be regarded as a collection of hugely complex reactions taking place between organic compounds in oddly shaped containers. Many of these organic compounds are polymers, including the cellulose of wood, natural fibers such as cotton and silk, the proteins and carbohydrates in our food, and the nucleic acids of our genes.

Protein molecules are condensation copolymers of up to 20 different naturally occurring amino acids that differ in their side chains (**TABLE 11E.3**). Human bodies can synthesize 11 amino acids in sufficient amounts for our needs. However, not all the proteins necessary for life can be produced and the other nine must be present in the diet; they are known as the **essential amino acids.**

A molecule formed by the condensation of two or more amino acids is called a **peptide.** An example is the combination of glycine and alanine, denoted Gly-Ala:

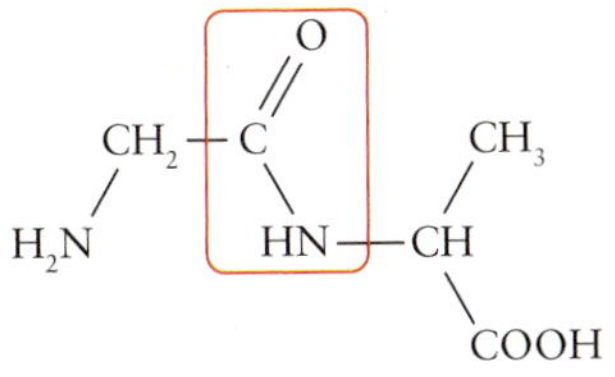

The —CO—NH— link shown in the red box is called a **peptide bond,** and each monomer used to form a peptide is called a **residue.** A typical protein is a polypeptide chain of more than a hundred residues joined through peptide bonds and arranged in a strict order. When only a few amino acid residues are present, the molecule is called an **oligopeptide.** The artificial sweetening agent aspartame is a type of oligopeptide called a **dipeptide,** because it has two residues.

The **primary structure** of a protein is the sequence of residues in the peptide chain. Aspartame is made from the methyl ester of phenylalanine, denoted PME, and

TABLE 11E.3 The Naturally Occurring Amino Acids, R—CH(NH₂)COOH

R	Name	Abbreviation	R	Name	Abbreviation
—H	glycine	Gly		arginine	Arg
—CH₃	alanine	Ala			
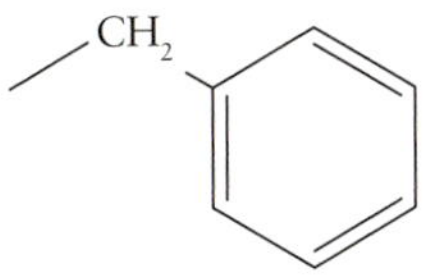	phenylalanine*	Phe			
—CH(CH₃)₂	valine*	Val		histidine*	His
—CH₂CH(CH₃)₂	leucine*	Leu			
—CH(CH₃)CH₂CH₃	isoleucine*	Ile			
—CH₂OH	serine	Ser			
—CH(OH)CH₃	threonine*	Thr			
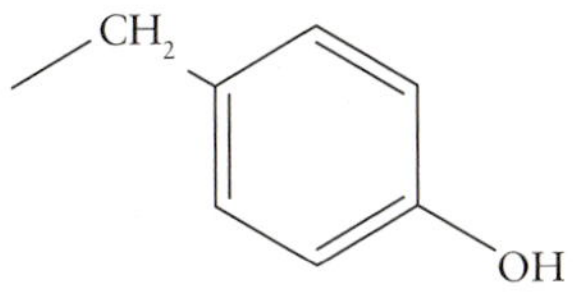	tyrosine	Tyr		tryptophan*	Trp
—CH₂COOH	aspartic acid	Asp	—CH₂CONH₂	asparagine	Asn
—CH₂CH₂COOH	glutamic acid	Glu	—CH₂CH₂CONH₂	glutamine	Gln
—CH₂SH	cysteine	Cys			
—CH₂CH₂SCH₃	methionine*	Met		proline†	Pro
—CH₂(CH₂)₃NH₂	lysine*	Lys			

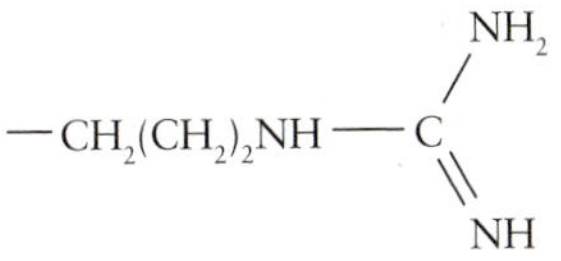

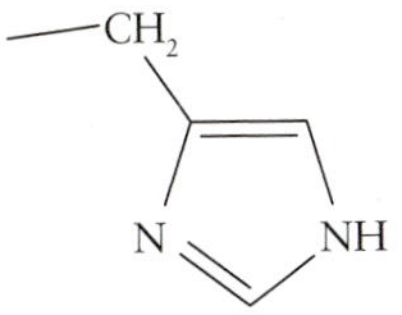

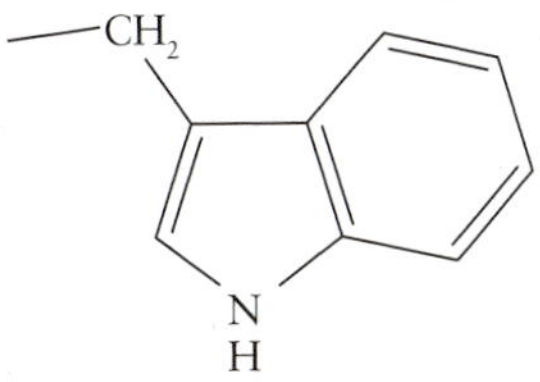

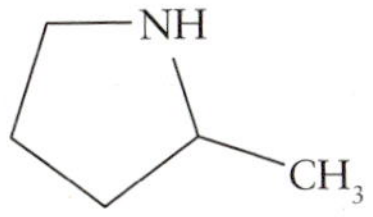

*Essential amino acids for humans.
†The entire amino acid is shown.

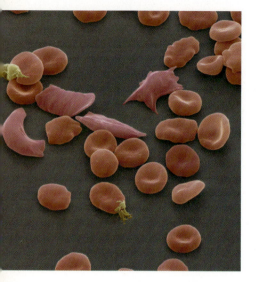

FIGURE 11E.16 Sickle-shaped red blood cells form when only one amino acid (glutamic acid) in a polypeptide chain is replaced by another amino acid (valine). These cells are less able to take up oxygen than normal cells. (*Eye of Science/Science Source.*)

aspartic acid (Asp), and so its primary structure is [PME-Asp]. Three fragments of the primary structure of human hemoglobin are

Leu-Ser-Pro-Ala-Lys-Thr-Asn-Val-Lys-...

...-Val-Lys-Gly-Trp-Ala-Ala-...

...-Ser-Thr-Val-Leu-Thr-Ser-Lys-Ser-Lys-Try-Arg

The determination of the primary structure of a protein is a very demanding analytical task, but, thanks to automated procedures, many of these structures are now known. Any modification of the primary structure of a protein—the replacement of one amino acid residue by another—may lead to a congenital disease. Even one wrong amino acid in the chain can disrupt the normal function of the molecule (**FIG. 11E.16**).

The **secondary structure** of a protein is the shape adopted by the polypeptide chain—in particular, how it coils or forms sheets. The order of the amino acids in the chain controls the secondary structure, because their intermolecular forces hold the chains together at specific points. The most common secondary structure in animal proteins is the **α helix**, a helical conformation of a polypeptide chain held in place by hydrogen bonds between residues (**FIG. 11E.17**). One alternative secondary structure is the **β sheet**, which is characteristic of the protein known as silk. In silk, protein molecules lie side by side to form nearly flat sheets. The molecules of many other proteins consist of alternating α-helical and β-sheet regions (**FIG. 11E.18**).

The **tertiary structure** of a protein is the shape into which its secondary structure is folded as a result of interactions between residues. The globular form of each chain in hemoglobin is an example. One important type of link responsible for tertiary structure is the covalent **disulfide link,** —S—S—, between residues containing sulfur. Other links are formed by the various types of intermolecular forces. In most cases, a given protein will fold into a precise three-dimensional conformation determined by the locations of the hydrophobic and hydrophilic groups on the chain (**FIG. 11E.19**). Sometimes, how-

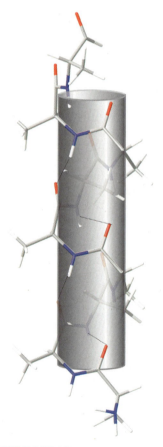

FIGURE 11E.17 A representation of part of an α helix, one of the secondary structures adopted by polypeptide chains. The cylinder encloses the "backbone" of the polypeptide chain, and the side groups project outward from it. The thin lines represent the hydrogen bonds that maintain the helical shape.

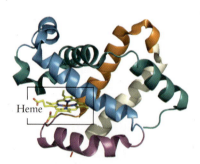

FIGURE 11E.18 One of the four polypeptide chains that make up the human hemoglobin molecule. The chains consist of alternating regions of α helices and β sheets. The α-helix regions are represented by colored helices. The oxygen molecules that we inhale attach to the iron atom at the center of the heme and are carried through the bloodstream.

 ANIMATION FIGURE 11E.18

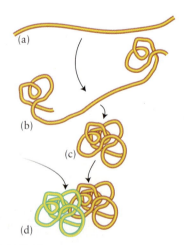

FIGURE 11E.19 These structures show how a protein first forms α helices and β sheets and then how the coils and sheets fold together to form the shape of a protein. Finally, if the protein has a quaternary structure, the protein subunits stack together. (a) Newly formed polypeptide; (b) intermediate; (c) subunit; (d) mature (in this case, dimeric) protein.

FIGURE 11E.20 The amylose molecule, one component of starch, is a polysaccharide, a polymer of glucose. It consists of glucose units linked together to give a structure like the one shown but with a moderate degree of branching.

🔁 **ANIMATION FIGURE 11E.20**

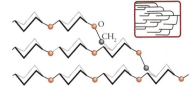

FIGURE 11E.21 The amylopectin molecule is another component of starch. It has a more highly branched structure than amylose, as emphasized in the inset.

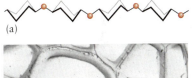

4 D-Glucose, $C_6H_{12}O_6$

5 Fructose, $C_6H_{12}O_6$

ever, the proteins do not fold properly. When this incorrect folding happens in the body it can lead to illnesses such as Alzheimer's disease. In Alzheimer's disease, deposits of proteins that have misfolded and can no longer function properly inhibit the activity of the brain. The so-called "prion diseases," such as variant Creutzfeldt–Jakob disease ("mad cow disease"), are also due to protein misfolding. If scientists can solve the puzzle of why proteins misfold and discover how to correct that, then diseases now considered untreatable can be cured.

Proteins may also have a **quaternary structure,** in which neighboring polypeptide units stack together in a specific arrangement. The hemoglobin molecule, for example, has a quaternary structure of four polypeptide units, one of which is shown in Fig. 11E.18.

The loss of structure by a protein is called **denaturation.** This structural change may be a loss of quaternary, tertiary, or secondary structure; it may also be degradation of the primary structure by cleavage of the peptide bonds. Even mild heating can cause irreversible denaturation. When you cook an egg, the protein called albumen denatures into a white mass. The "permanent waving" of hair, which consists primarily of long α helices of the protein keratin, is a result of partial denaturation and then construction of new links to give a more fashionable appearance.

Proteins are polymers made of amino acid units. The primary structure of a polypeptide is the sequence of amino acid residues; secondary structure is the formation of helices and sheets; tertiary structure is the folding into a compact unit; quaternary structure is the packing of individual protein units together.

11E.6 Carbohydrates

The **carbohydrates** are so called because many have the empirical formula CH_2O, which suggests a hydrate of carbon. They include starches, cellulose, and sugars such as glucose, $C_6H_{12}O_6$ (**4**), which contains an aldehyde group, and fructose (fruit sugar), a structural isomer of glucose that is a ketone (**5**). Carbohydrates have many —OH groups, and so they may also be regarded as alcohols. The presence of these —OH groups allows them to form numerous hydrogen bonds with one another and with water.

Polysaccharides are polymers of glucose. They include starch, which can be digested by humans, and cellulose, which cannot. Starch is made up of two components: amylose and amylopectin. Amylose, which makes up about 20% to 25% of most starches, consists of chains made up of several thousand glucose units linked together (**FIG. 11E.20**). Amylopectin is also made up of glucose chains (**FIG. 11E.21**), but its chains are linked into a branched structure and its molecules are much larger. Each molecule consists of about a million glucose units.

Cellulose is the structural material of plants. It is a polymer with the same monomer (glucose) that starch has, but the units in cellulose are linked differently, forming flat, ribbonlike strands (**FIG. 11E.22**). Hydrogen bonds between these strands lock them

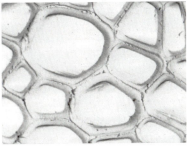

(a)

(b)

FIGURE 11E.22 (a) Cellulose is yet another polysaccharide constructed from glucose units. The units in cellulose link together in such a way that they form long, flat ribbons, which can produce a fibrous material through hydrogen bonding. (b) These long tubes of cellulose (shown in cross section) form the structural material of an aspen tree. *(Photo permission of the Institute of Paper Science and Technology.)*

🔁 **ANIMATION FIGURE 11E.22**

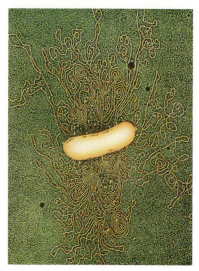

FIGURE 11E.23 A DNA molecule is very large, even in bacteria. In this micrograph, a DNA molecule has spilled out through the damaged cell wall of a bacterium. (*Science Source*)

together into a rigid structure that for us (but not for termites) is indigestible. Cellulose is the most abundant organic chemical in the world, and billions of tons of it are produced annually by photosynthesis. In research on alternative fuels, enzymes are used to break down the cellulose in waste biomass into glucose, which is then fermented to make ethanol for use as a fuel (see Box 4D.1).

> *Carbohydrates include sugars, starches, and cellulose. Glucose is an alcohol and an aldehyde that polymerizes to form starch and cellulose.*

11E.7 Nucleic Acids

The nucleus of every living cell contains at least one deoxyribonucleic acid (DNA) molecule to control the production of proteins and to carry genetic information from one generation of cells to the next. Human DNA molecules are immense: if a molecule of DNA could be extracted without damage from a cell nucleus and drawn out straight from its highly coiled natural shape, it would be about 2 m long (**FIG. 11E.23**). The ribonucleic acid (RNA) molecule is closely related to DNA. One of its functions is to carry information stored by DNA to a region of the cell where that information can be used in protein synthesis.

DNA is a polymer with a backbone built of repeating units derived from the sugar ribose (**6**). For DNA, the ribose molecule has been modified by removing the oxygen atom at carbon atom 2, the second carbon atom clockwise from the ether oxygen atom in the five-membered ring. Therefore, the repeating unit—the monomer—is called deoxyribose (**7**).

Attached by a covalent bond to carbon atom 1 of the deoxyribose ring is an amine (and therefore a base), which may be adenine, A (**8**); guanine, G (**9**); cytosine, C (**10**); or thymine, T (**11**). In RNA, uracil, U (**12**), replaces thymine. The base bonds to carbon atom 1 of deoxyribose through the nitrogen of the —NH— group (printed in red); the compound so formed is called a **nucleoside**. All nucleosides have a similar structure, summarized in (**13**); the lens-shaped object represents the attached amine.

The DNA monomers are each completed by a phosphate group, $—O—PO_3^{2-}$, covalently bonded to carbon atom 5 of the ribose unit to give a compound called a **nucleotide** (**14**). Because there are four possible nucleoside monomers (one for each base), there are four possible nucleotides in each type of nucleic acid.

Molecules of DNA and RNA are **polynucleotides,** polymeric species built from nucleotide units. Polymerization takes place when the phosphate group of one nucleotide (which is the conjugate base of an organic phosphoric acid) condenses with the —OH group on carbon atom 3 of the ribose unit of another nucleotide, thereby forming an ester link and releasing a molecule of water. As this condensation continues, it results in a structure like that shown in **FIG. 11E.24**, a compound known as a **nucleic acid**. The DNA molecule itself is a double helix in which two long nucleic acid strands are wound around each other.

The ability of DNA to replicate lies in its double-helical structure. There is a precise correspondence between the bases in the two strands. Adenine in one strand always forms two hydrogen bonds to thymine in the other, and guanine always forms

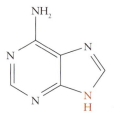

6 Ribose, $C_5H_{10}O_5$

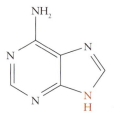

7 Deoxyribose, $C_5H_{10}O_4$

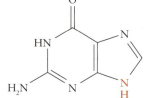

8 Adenine

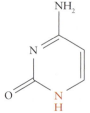

9 Guanine

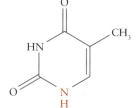

10 Cytosine

11 Thymine

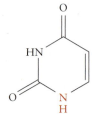

12 Uracil

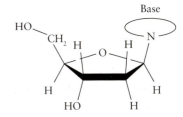

13 A nucleoside

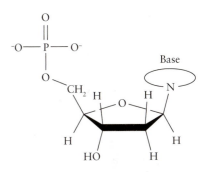

14 A nucleotide

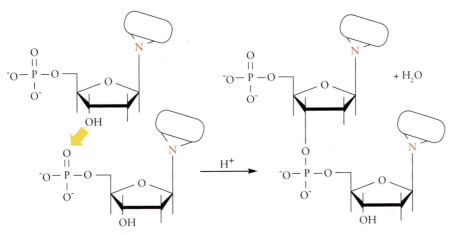

FIGURE 11E.24 The condensation of nucleotides that leads to the formation of a nucleic acid, a polynucleotide.

three hydrogen bonds to cytosine; so, across the helix, the base pairs are always AT and GC (**FIG. 11E.25**). Any other combination would not be held together as well. During replication of the DNA, the hydrogen bonds, which are relatively weak compared with the covalent bonds in the strands, can be broken by an enzyme while the strands themselves remain intact. Nucleotides in the cellular fluid then attach in the appropriate places on each strand to form the double helices of two new molecules of DNA.

As well as replication—the production of copies of itself for reproduction and cell division—DNA governs the production of proteins by serving as the template during the synthesis of molecules of RNA. The RNA molecules, with U in place of T, are considered to be "messenger" molecules, because they carry information about segments of the genetic message out of the cell nucleus to the locations where protein synthesis takes place. In this way, the chemical reactions of functional groups, and in a broader sense, the principles of chemistry, bring matter to life.

> *Nucleic acids are copolymers of four nucleotides joined by phosphate ester links. The nucleotide sequence stores all genetic information.*

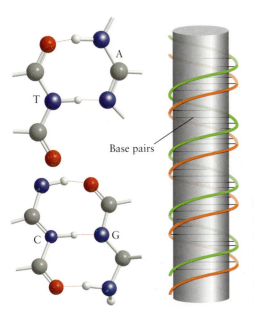

FIGURE 11E.25 The bases in the DNA double helix fit together by virtue of the hydrogen bonds that they can form (red dotted lines), as shown on the left. Once formed, the AT and GC pairs are almost identical in size and shape. As a result, the turns of the helix shown on the right are regular and consistent.

What have you learned in this Topic?

You have been introduced to some of the reactions that are used to produce polymers such as plastics and have seen how the reactions of functional groups account for the structures of biologically important polymers such as proteins and DNA.

The skills you have mastered are the ability to:

☐ **1.** Predict the type of polymer that a monomer can form and identify monomers, given the repeating unit of a polymer (Example 11E.1).

☐ **2.** Distinguish the various types of copolymers (Self-test 11E.2).

☐ **3.** Explain how the physical properties of polymers are related to their structures (Section 11E.4).

☐ **4.** Describe the composition of proteins and distinguish their primary, secondary, tertiary, and quaternary structures (Section 11E.5).

☐ **5.** Describe the composition of carbohydrates (Section 11E.6).

☐ **6.** Describe the structures and functions of nucleic acids (Section 11E.7).

Topic 11E Exercises

11E.1 Sketch three repeating units of the polymer formed from (a) $CH_2=C(CH_3)_2$; (b) $CH_2=CHCN$; (c) isoprene,

11E.2 Sketch three repeating units of the polymer formed from (a) 1,1-dichloroethene; (b) 1-phenyl-2-methylethene; (c) $CH_3CH=CH_2CF_3$.

11E.3 A polyamide has the repeating unit $-(COC_6H_{12}CONHC_6H_4NH)_n-$. Identify the monomers of the polyamide.

11E.4 A polyester has the repeating unit $-(OCH_2C_6H_4COOCH_2C_6H_4CO)_n-$. Identify the monomers of the polyester.

11E.5 Write the structural formulas of the monomers of each of the following polymers, for which one repeating unit is shown: (a) polyvinylchloride (PVC), $-(CHClCH_2)_n-$; (b) Kel-F, $-(CFClCF_2)_n-$.

11E.6 Write the structural formulas of the monomers of each of the following polymers, for which one repeating unit is shown: (a) a polymer used to make carpets, $-OC(CH_3)_2CO)_n-$; (b) $-CH(CH_3)CH_2)_n-$; (c) a polypeptide, $-(NHCH_2CO)_n-$.

11E.7 Draw the structural formula of two units of the polymer formed from (a) the reaction of oxalic acid (ethanedioic acid), HOOCCOOH, with 1,4-diaminobutane, $H_2NCH_2CH_2CH_2CH_2NH_2$; (b) the polymerization of the amino acid alanine (2-aminopropanoic acid).

11E.8 Draw the structural formula of two units of the polymer formed from (a) the reaction of terephthalic acid with 1,2-diaminoethane, $H_2NCH_2CH_2NH_2$; (b) the polymerization of 4-hydroxybenzoic acid.

Terephthalic acid   4-Hydroxybenzoic acid

11E.9 Identify the type of copolymer formed by monomers A and B: —BBBBAA—.

11E.10 Identify the type of copolymer formed by monomers A and B: —AABABBAA —.

11E.11 Why do polymers not have definite molar masses? How does the fact that polymers have average molar masses affect their melting points?

11E.12 Rank the following polymers according to increasing value as fibers: polyesters, polyamides, polyalkenes. Explain your reasoning.

11E.13 The polymer polylactic acid (see structure (**2**)) was used for a while to replace polyethylene terephthalate (PETE, see Section 11E.2) in the bags containing certain kinds of snack foods. However, polylactic acid is stiffer than PETE and so the bags made loud crackling sounds when they were handled. The problem was solved by using a special adhesive between the layers of the bags. Compare the structures of polylactic acid and PETE and suggest a reason why polylactic acid chains are more rigid than those of PETE.

11E.14 Polyacrylamide is used in coatings for wooden floors. It is made by addition polymerization of the monomer $H_2C=CHCONH_2$. (a) Draw three repeating units of polyacrylamide. (b) How do the side chains in polyacrylamide contribute to its hardness and strength?

11E.15 How does average molar mass (which is proportional to the chain length) affect each of the following polymer characteristics: (a) softening point; (b) viscosity; (c) strength?

11E.16 How does the polarity of the side chains in a polymer affect each of the following characteristics of the polymer: (a) softening point; (b) viscosity; (c) strength?

11E.17 Describe how the linearity of the polymer chain affects polymer strength.

11E.18 Describe how cross-linking affects the elasticity and rigidity of a polymer.

11E.19 (a) Draw the structure of the peptide bond that links the amino acids in proteins. (b) Identify the functional group formed. (c) Identify the type of polymer formed (addition or condensation).

11E.20 (a) Draw the structure of the link between glucose units that creates amylose. (b) Identify the functional group formed. (c) Identify the type of polymer formed (addition or condensation).

11E.21 Name the amino acids in Table 11E.3 that contain side groups capable of forming hydrogen bonds. This interaction contributes to the tertiary structure of a protein.

11E.22 Name the amino acids in Table 11E.3 that contain non-polar side groups. These groups contribute to the tertiary structure of a protein by preventing contact with water.

11E.23 Draw the structure of the peptide formed from the reaction of the acid group of tyrosine with the amine group of glycine.

11E.24 Draw the line structure of the peptide formed from the reaction of the acid group of aspartic acid with the amine group of phenylalanine. *Note*: The acid side chain of aspartic acid is not involved in this reaction. The resulting peptide is related to the artificial sweetener aspartame.

11E.25 Identify (a) the functional groups and (b) the chiral carbon atoms in the mannose molecule shown here.

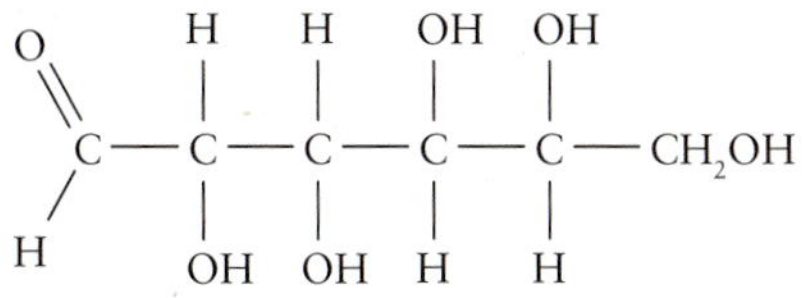

11E.26 Identify (a) the functional groups and (b) the chiral carbon atoms in the histidine molecule shown here.

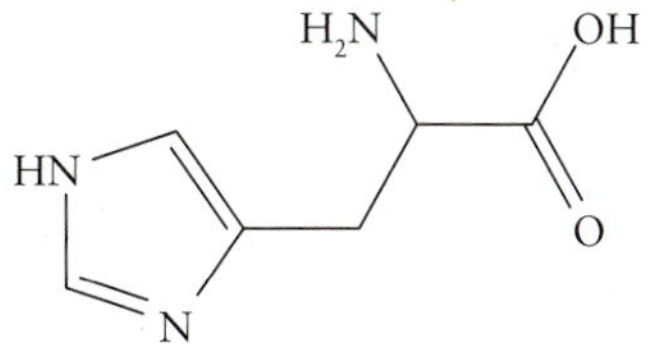

11E.27 Write the complementary nucleic acid sequence that would pair with each of the following DNA sequences: (a) CATGAGTTA; (b) TGAATTGCA.

11E.28 Write the complementary nucleic acid sequence that would pair with each of the following DNA sequences: (a) ATTAGATCAT; (b) GACTAGGATCT.

The following Example and Exercises draw on material from throughout Focus 11.

FOCUS 11 Online Cumulative Example

Metabolism in mammals produces damaging byproducts, such as hydrogen peroxide, superoxide ions, and oxygen-containing radicals, which are collectively known as "reactive oxygen species." Glutathione (GSH) is an important tripeptide that serves as a potent antioxidant. The thiol (—SH) group acts as a target for the oxidizing agents, loses a hydrogen atom, and forms a disulfide bond (—S—S—) with another GSH molecule. You are investigating ways to protect against oxidative stress (Box 2C.1) and want to know more about the chemistry of this essential compound.

(a) Draw the structural formula of glutathione from its line structure (**1**).

(b) Identify the three amino acids from which glutathione is derived. (*Hint*: One peptide bond is formed with the side-chain of an amino acid.)

(c) Identify and label any chiral carbon atoms in the structure of GSH. How many enantiomeric pairs are possible?

(d) The pK_a values for glutathione are $pK_{a1} = 2.12$ and $pK_{a2} = 3.59$ for successive deprotonation of the two COOH groups, $pK_{a3} = 8.75$ for the NH_2 group, and $pK_{a4} = 9.65$ for the SH group. Predict the predominant form of GSH at the physiological pH of 7.4.

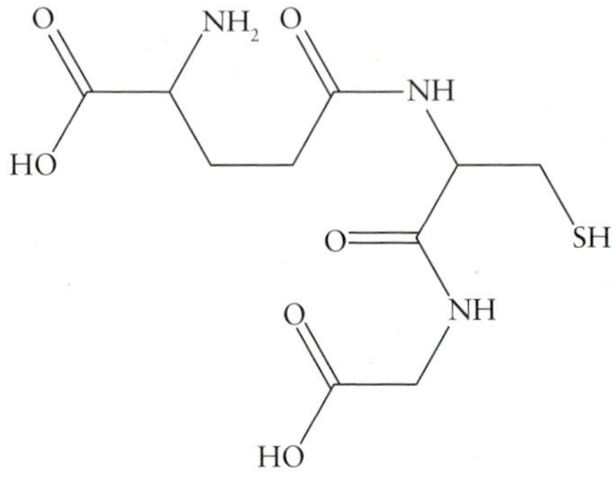

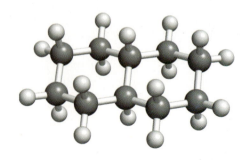

1   Glutathione (GSH)

 The online Cumulative Example solution can be found at http://macmillanhighered.com/chemicalprinciples7e

FOCUS 11 Exercises

11.1 Why do branched-chain alkanes have lower melting points and boiling points than unbranched alkanes with the same number of carbon atoms?

11.2 Arrange the following compounds in order of increasing boiling point: $CH_3CH_2CH_2CH_2CH_3$, $CH_3CH_2CH(CH_3)_2$, $C(CH_3)_4$. Explain your reasoning.

11.3 Classify each of the following reactions as addition or substitution and write its chemical equation: (a) chlorine reacts with methane when exposed to light; (b) bromine reacts with ethene in the absence of light.

11.4 Classify each of the following reactions as addition or substitution and write its chemical equation: (a) hydrogen reacts with 2-pentene in the presence of a nickel catalyst; (b) hydrogen chloride reacts with propene.

11.5 The dehydrohalogenation of haloalkanes to produce alkenes is always carried out in a nonaqueous solvent, usually ethanol. Give two reasons why water cannot be used as the solvent for the reaction.

11.6 View the animation of the alkene addition mechanism associated with Fig. 11B.3 on the website for this book. (a) In the addition of HCl to propene an intermediate is formed. Is that intermediate positively or negatively charged? How is the charge eliminated? (b) Suggest a two-step mechanism for the reaction.

11.7 Give the systematic name of each of the following compounds. If geometrical isomers are possible, write the names of each one: (a) $CH_2{=}C(CH_3)_2$; (b) $CH_3CH{=}C(CH_3)CH_2CH_3$; (c) $HC{\equiv}CCH_2CH_2CH_2CH_3$; (d) $CH_3CH_2C{\equiv}CCH_2CH_3$; (e) $CH_3CH_2CH_2C{\equiv}CCH_3$.

11.8 Give the systematic name of each of the following compounds. If geometrical isomers are possible, write the names of each one: (a) $CH_3CH_2CH{=}CHCH_3$; (b) $CH_3C(CH_3){=}CHCH_3$; (c) $(CH_3)_2C{=}CHCH_2CH_3$; (d) $HC{\equiv}CCH_2CH(CH_2CH_2CH_3)CH_3$; (e) $CH_3CH_2C{\equiv}CCH(CH_3)CH_2CH_3$.

11.9 The structure of decalin is shown below. (a) From an examination of this structure, determine its chemical formula. (b) From what aromatic hydrocarbon may decalin be produced by complete hydrogenation? (c) Are isomers possible for decalin? If so, draw the appropriate line structures.

11.10 Nonsystematic names for organic compounds may still be found in the chemical literature and chemical supply catalogs, so it is important to be familiar with them as well as with the IUPAC names. (a) Give the systematic name of (i) isobutane and (ii) isopentane. (b) Formulate a rule for using the prefix *iso-* and predict the structure of isohexane.

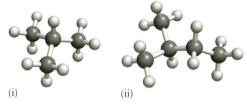

(i)   (ii)

11.11 A carbocation is a positively charged organic ion. The first stable carbocation, the pentamethylcyclopentadienyl cation, was reported in 2002 (*Angewandte Chemie International Edition*, vol. 41, p. 1429). This cation has a charge of +1. (a) Draw the molecular structure of this ion and identify which carbon atom bears most of the positive charge. (b) How many resonance structures are possible? (c) How many π-electrons are present in the ion?

11.12 The chemists who synthesized the pentamethylcyclopentadienyl cation (see Exercise 11.11) found that its π-electrons are found in one double bond and a π-orbital that is delocalized across the three other carbon atoms. (a) Draw a structure for the ion that illustrates this bonding. (b) On which carbon atom would you place the positive charge?

11.13 The following names of organic molecules are incorrect. Draw line structures and identify the correct systematic name for (a) 4-methyl-3-propylheptane; (b) 4,6-dimethyloctane; (c) 2,2-dimethyl-4-propylhexane; (d) 2,2-dimethyl-3-ethylhexane.

11.14 Alkanes undergo substitution of hydrogen atoms when treated with halogens. Bromination of which of the following compounds could give rise to chiral monosubstituted products: (a) ethane; (b) propane; (c) butane; (d) pentane? Include in your answer the names of the chiral compounds that would be formed.

11.15 A hydrocarbon with the formula C_3H_6 reacts with bromine only in the presence of light, producing C_3H_5Br. What is the name of this hydrocarbon?

11.16 A hydrocarbon is 90% carbon by mass and 10% hydrogen by mass and has a molar mass of 40 g·mol^{-1}. It decolorizes bromine water, and 1.46 g of the hydrocarbon reacts with 1.60 L of hydrogen (measured at STP) in the presence of a nickel catalyst. Write the molecular formula of the hydrocarbon and the structural formulas of two possible isomers.

11.17 The compound 1-bromo-4-nitrobenzene (*p*-bromonitrobenzene) can be brominated. What do you expect to be the major product and why? Draw appropriate structures to support your answer.

11.18 Consider the nitration of the compound

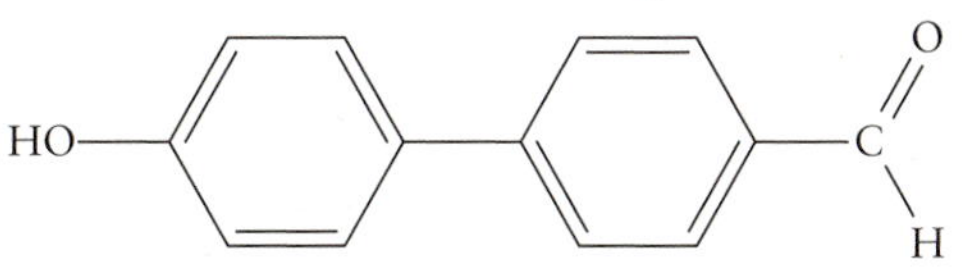

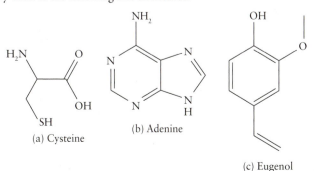

If the reaction can be controlled so that one NO_2 group replaces one H atom of the molecule, where do you expect the nitro group to end up in the product?

11.19 (a) Draw the structure of 3,4,6-trimethyl-1-heptene. (b) Identify the chiral carbon atoms in the structure with stars. (c) Are cis and trans isomers possible for this molecule?

11.20 Trinitrotoluene, TNT, is a well-known explosive. (a) Use the following structure to identify the systematic name of TNT. (b) TNT is made by nitrating toluene (methylbenzene) with a mixture of concentrated nitric and sulfuric acids. Explain why this particular trisubstituted isomer is the one that is formed in the nitration.

11.21 Conjugated polyenes are hydrocarbons with alternating single and double bonds. They are commonly used as dyes because they absorb in the visible range. Two such molecules have the formulas C_6H_8 and C_8H_{10}. Of the two, which has its maximum absorption at the longer wavelength? Justify your answer. See *Major Technique* 2 on the website of this book.

11.22 (a) How many liters of hydrogen at 1.00 atm and 298 K are needed to hydrogenate (i) 1.00 mol C_6H_{10}, cyclohexene; (ii) 1.00 mol C_6H_6, benzene, completely? (b) Estimate the reaction enthalpy of each hydrogenation from the average bond enthalpies in Tables 4E.2 and 4E.3. (c) A Kekulé structure of benzene suggests that its enthalpy of hydrogenation should be three times that of cyclohexene. Do your calculations support this implication? Explain any differences.

11.23 Write the chemical formula of the compound represented by each of the following line structures:

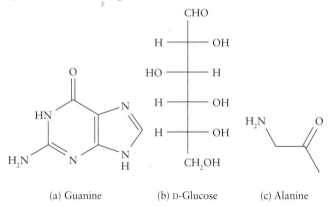

(a) Guanine   (b) D-Glucose   (c) Alanine

11.24 Write the chemical formula of the compound represented by each of the following line structures:

NH$_2$

H$_2$N

OH

SH
(a) Cysteine

(b) Adenine

OH

O

(c) Eugenol

11.25 Identify all the functional groups in each of the following compounds:

(a) vanillin, the compound responsible for vanilla flavor,

(b) carvone, the compound responsible for spearmint flavor,

(c) caffeine, the stimulant in coffee, tea, and cola drinks,

11.26 Identify all the functional groups in each of the following compounds:

(a) zingerone, the pungent, hot component of ginger,

(b) Tylenol (acetaminophen), an analgesic,

(c) procaine, a local anesthetic,

11.27 Identify the chiral carbon atoms in each of the following compounds:

(a) camphor, used in cooling salves,

(b) testosterone, a male sex hormone,

11.28 Identify the chiral carbon atoms in each of the following compounds:

(a) menthol, the flavor of peppermint,

(b) estradiol, a female sex hormone,

11.29 Identify the hybridization of each carbon and nitrogen atom in guanine (see Exercise 11.23a).

11.30 Identify the hybridization of each carbon and nitrogen atom in caffeine (see Exercise 11.25c).

11.31 The structures of the following molecules can be found on the website for this book. Draw the structure of each and identify its chiral carbon atoms: (a) cocaine; (b) aflatoxin B2, a toxin and carcinogen that occurs naturally in peanuts as a byproduct of the growth of fungi.

11.32 The structures of the following molecules can be found on the website for this book. Draw the line structure of each and identify its chiral carbon atoms: (a) cephalosporin C, toxic to penicillin-resistant staphylococci; (b) thromboxane A2 ($C_{20}H_{32}O_5$), a substance that promotes the clotting of blood.

11.33 (a) Write the structural formulas of diethyl ether and 1-butanol (note that they are isomers). (b) The boiling point of 1-butanol is 117 °C, higher than that of diethyl ether (35 °C), yet the solubility of both compounds in water is about 8 g per 100 mL. Account for these observations.

11.34 Consider the following organic molecules, which have approximately the same molar masses but might contain different functional groups: $CH_3CH_2CH_2CH_2CH_2CH_3$, $CH_3CH_2CH_2CH_2CHCH_2$, $CH_3CH_2CH_2CH_2CH_2OH$, $CH_3CH_2CH_2CH_2CHO$, $CH_3CH_2CH_2COOH$, $CH_3CH_2COOCH_3$, $CH_3CH_2COCH_2CH_3$, $CH_3CH_2CH_2CH_2OCH_3$.

(a) Draw a Lewis structure for each molecule, name it, and classify it by any functional group that is present. (b) Which molecules are isomers of each other? Are any chiral? If so, which ones? (c) For each molecule, list the types of intermolecular forces that are present. (d) Use your answers to parts (a) and (b) to predict the order of boiling points, from lowest to highest.

11.35 Write the structural formula of the product of (a) the reaction of glycerol (1,2,3-trihydroxypropane) with stearic acid, $CH_3(CH_2)_{16}COOH$, to produce a saturated fat; (b) the oxidation of 4-hydroxybenzyl alcohol by sodium dichromate in an acidic organic solvent.

11.36 Write the condensed structural formulas of the principal products of the reaction that takes place when (a) ethylene glycol, 1,2-ethanediol, is heated with stearic acid, $CH_3(CH_2)_{16}COOH$; (b) ethanol is heated with oxalic acid, $HOOCCOOH$; (c) 1-butanol is heated with propanoic acid.

11.37 A segment of a protein is analyzed and found to contain the amino acid sequence Glu-Leu-Asp. Draw the Lewis structure of this segment, showing the peptide bonds.

11.38 The pK_a values for phenol, o-nitrophenol, m-nitrophenol, and p-nitrophenol are 9.89, 7.17, 8.28, and 7.15, respectively. Explain the origin of these differences in pK_a.

11.39 Acrylic resins are polymeric materials used to make warm yet lightweight garments. The osmotic pressure of a solution prepared by dissolving 47.7 g of an acrylic resin in enough water to make 500. mL of solution is 0.325 atm at 25 °C. (a) What is the average molar mass of the polymer? (b) How many monomers constitute the "average" molecule? The repeating unit of this acrylic resin has the formula $-CH_2CH(CN)-$. (c) What would be the vapor pressure of the solution given that the vapor pressure of pure water at 25 °C is 0.0313 atm? (Assume that the density of the solution is 1.00 g·cm^{-3}.) (d) Which approach (osmometry or the lowering of vapor pressure) would you prefer for the determination of very high molar masses such as those of acrylic resins? Why?

11.40 The average molar mass of a sample of polypropylene was determined by measuring the osmotic pressure of 500. mL of a solution of 3.16 g of polypropylene in benzene. A pressure of 0.0112 atm was observed at 25 °C. (a) What is the average molar mass of the polymer? (b) How many propene monomer units with formula $-CH(CH_3)CH_2-$ were combined on average to create each chain? (c) If this sample consists only of linear chains and the carbon–carbon bond lengths in the polymer are close to their average value, what is the average chain length?

11.41 (a) Explain the differences among the primary, secondary, tertiary, and quaternary structures of a protein. (b) Identify the forces holding each structure together as covalent bonds or primarily intermolecular forces.

11.42 Haloalkanes may react with hydroxide ions, undergoing nucleophilic displacement of the halide ion to form an alcohol. A complication of such reactions is competition from elimination reactions rather than substitution. (a) Predict the possible products from the reaction of 2-bromopentane with sodium hydroxide. (b) What can be done to favor the substitution reaction over the elimination pathway or vice versa?

11.43 The protonated form of glycine ($^+H_3NCH_2COOH$) has $K_{a1} = 4.47 \times 10^{-3}$, and $K_{a2} = 1.66 \times 10^{-10}$. (a) Write the chemical equations for the proton transfer equilibria. (b) What is the dominant form of glycine in solution at pH = 2, pH = 5, and pH = 12 (Topic 6E)?

11.44 In the amide group, rotation is restricted about the C—N bond, so that the C, N, and O atoms in that group are normally in the same plane. This rigidity is partly responsible for the secondary structure of proteins. The amino acid glycine can form a dipeptide, composed of two glycine monomers. (a) Draw the Lewis structure of the glycine dipeptide. (b) Explain how resonance structures can account for the restricted rotation and draw a second resonance structure for the dipeptide.

11.45 Explain the process of condensation polymerization. How might the polymer obtained from benzene-1,2-dicarboxylic acid and ethylene glycol differ from Dacron?

11.46 The average molar mass of a hydrogen-bonded pair of nucleotides in a DNA molecule is 625 g·mol^{-1}. Each successive pair is found at a distance of 340 pm along the chain. If the total length of one strand of a DNA molecule is 0.299 m, what is the molar mass of the molecule?

11.47 Propane torches are used for minor home repairs, but welders must use torches that burn acetylene (ethyne). (a) Write chemical equations for the combustion of propane and ethyne. (b) Calculate the enthalpy of combustion for each fuel per gram and per mole. (c) Use your data in part (b) to explain why welders do not use propane torches.

11.48 Polyphosphazenes, which are used as heat-resistant polymers in the aerospace industry and as flexible scaffolds for bone regeneration, are inorganic polymers with the repeating unit $(-PR_2=N-)_n$, in which R stands for side chains such as $-CH_3$ groups. How would you expect the properties of the polyphosphazene with methyl side chains to differ from those of the corresponding silicone polymer (see Fig. 11E.15)? Explain your reasoning.

11.49 The monomer of the conducting polymer polyaniline is the compound aniline (aminobenzene). (a) Draw the structural formula of the aniline monomer. (b) What is the hybridization of the N atom in (i) aniline; (ii) polyaniline? (c) Indicate the locations of lone pairs, if any, in polyaniline. (d) Do the N atoms help to carry the current? Explain your reasoning. See Box 11E.1.

11.50 Fibers of the conducting polymer polypyrrole are woven into radar camouflage cloth. Because it absorbs microwaves, rather than reflecting them back to their source, the cloth appears to be a patch of empty space on radar. (a) What is the hybridization of the N atom in polypyrrole? (b) Explain why polypyrrole absorbs microwave radiation, but small organic molecules do not. See Box 11E.1 for the structure of polypyrrole.

11.51 Pheromones are commonly called sex attractants, although they have other signaling functions, too. A pheromone of the queen bee is *trans*-CH$_3$CO(CH$_2$)$_5$CH=CHCOOH. (a) Write the structural formula of the pheromone. (b) Identify and name the functional groups in the molecule.

The following six exercises are based on the information in the Interlude following Focus 11.

11.52 Why are hydrocarbons with between one and four carbon atoms not suitable for gasoline?

11.53 What are the major problems associated with using coal as a fuel?

11.54 Propene and butane can be combined by the process called alkylation to form a straight-chain hydrocarbon used in gasoline. Write a chemical equation for the reaction.

11.55 The compound C$_{18}$H$_{38}$ is a component of fuel oil that can be converted by a cracking process into two compounds suitable for use in gasoline. The compounds have the formulas C$_n$H$_{2n+2}$ and C$_n$H$_{2n}$, with the value of *n* being the same in each. Each compound has a straight chain of carbon atoms, with no branches. Draw possible structural formulas for the two compounds and name them.

11.56 *Aromatization* is a process used to improve the quality of gasoline by converting aliphatic hydrocarbons into aromatic ones. (a) When heptane is converted into toluene, a byproduct is formed. Does this byproduct make another valuable fuel, or is it a pollutant? (b) Write a balanced chemical equation for the conversion of heptane into toluene.

11.57 Isomerization is a process used to improve the quality of gasoline by converting straight-chain hydrocarbons into branched ones. If octane is isomerized into a branched pentane, several structural isomers are possible. (a) Draw the structures of these isomers. (b) One of these isomers has a chiral carbon atom. Mark that atom with a star.

FOCUS 11   Cumulative Exercises

11.58 A gaseous compound used to make both bubble gum and automobile tires was analyzed to determine its properties and toxicity.

(a) When 0.108 g of the compound was analyzed by combustion analysis, 0.352 g of CO$_2$ and 0.109 g of H$_2$O were produced. What is the empirical formula of the compound?

(b) The molar mass of the compound was found to be 54.09 g·mol^{-1}. What is the molecular formula of the compound?

(c) Draw the structural formulas of at least six possible structural isomers of the compound.

(d) When the compound is exposed to hydrogen gas over a catalyst, the molar mass of the hydrogenated product is 58.12 g·mol^{-1}. Which of the isomers in part (c) does this information help to eliminate?

(e) A flask of volume 2.45 L was filled with the compound at 25 °C and 1.0 atm. Hydrogen bromide gas was pumped into the flask until the total pressure in the flask reached 5.0 atm. Reaction then proceeded to completion because the single product was removed as fast as it was formed. After completion of the reaction, the pressure of the excess HBr in the flask was 2.0 atm. What is the molecular formula of the product?

(f) The product in (e) has three structural isomers, two of which are optical isomers. Draw the structural formulas of the original compound and each of the product isomers. Mark each chiral carbon atom with a star.

(g) Assign a hybridization scheme to each C atom in the original compound.

(h) Write the name of the original compound.

11.59 Waste reduction is an important goal of the green chemistry movement. In many chemical syntheses in industry, not all the atoms required for the reaction appear in the product. Some end up in byproducts and are wasted. "Atom economy" is the use of as few atoms as possible to reach an end product and is calculated as a percentage, by using atom economy = (mass of desired product obtained)/(mass of all reactants consumed) × 100%.

(a) Consider the following synthesis of CH$_3$CH=CHCH$_3$:

$$CH_3CH_2CHBrCH_3 + CH_3CH_2O^- \xrightarrow{\text{ethanol}}$$
$$CH_3CH=CHCH_3 + CH_3CH_2OH + Br^-$$

Identify the type of reaction (substitution, elimination, addition).

(b) Name each organic reactant and product.

(c) Does the CH$_3$CH$_2$O$^-$ ion function as a nucleophile, an electrophile, or neither?

(d) Calculate the atom economy of the reaction, assuming 100% yield.

(e) An alternative synthesis of CH$_3$CH=CHCH$_3$ is

$$CH_3CH_2CHBrCH_3 + CH_3O^- \xrightarrow{\text{methanol}}$$
$$CH_3CH=CHCH_3 + CH_3OH + Br^-$$

Calculate the atom economy for the reaction, assuming 100% yield.

(f) Another alternative synthesis of CH$_3$CH=CHCH$_3$ is

$$CH_3CH_2CHBrCH_3 + CH_3S^- \xrightarrow{\text{methanethiol}}$$
$$CH_3CH=CHCH_3 + CH_3SH + Br^-$$

Calculate the atom economy for the reaction, assuming 100% yield.

(g) Which of the three syntheses produces the lowest mass of waste? Which produces the highest?

(h) Assume that you carry out each of the three syntheses, starting with 50.0 g of CH$_3$CH$_2$CHBrCH$_3$ and with the second reagent in excess in each case. Your yields of CH$_3$CH=CHCH$_3$ for the three reactions are as follows: (a) 16.2 g; (e) 15.4 g; (f) 13.1 g. Calculate the percentage yield and experimental atom economy for each reaction.

(i) Which reaction would you recommend to a manufacturer? Explain your reasoning.

INTERLUDE Technology: Fuels

The primary sources of hydrocarbons are the fossil fuels petroleum and coal. Aliphatic hydrocarbons are obtained primarily from petroleum, which is a mixture of aliphatic and aromatic hydrocarbons, together with some organic compounds containing sulfur and nitrogen (**FIG. 1**). Coal is another major source of aromatic hydrocarbons.

FIG. 1 Because fossil fuel reserves are limited, they must be extracted from wherever they are found. This platform is used to extract petroleum from beneath the ocean. The natural gas accompanying it cannot easily be transported, and so it is burned off. (*Doug Menuez/Photodisc/Media Bakery.*)

Gasoline

The hydrocarbons in petroleum are separated by fractional distillation (see the table).* *Kerosene*, a fuel used in jet and diesel engines, contains a number of alkanes having formulas in the range C_{10} to C_{16}. Lubricating oils are mixtures in the range C_{17} to C_{22}. The heavier members of the series include the *paraffin waxes* and *asphalt*. However, the primary use of petroleum is for gasoline production, and the gasoline fraction (from C_5 to C_{11} hydrocarbons) is too small to meet the demand. In addition, the straight-chain alkanes make poor gasoline. Therefore, petroleum is refined to increase both the quantity and the quality of gasoline.

Hydrocarbon Constituents of Petroleum

Hydrocarbons	Boiling range/°C	Fraction
C_1 to C_4	−160 to 0	natural gas and propane
C_5 to C_{11}	30 to 200	gasoline
C_{12} to C_{16}	180 to 400	kerosene, fuel oil
C_{17} to C_{22}	350 and above	lubricants
C_{23} to C_{34}	low-melting-point solids	paraffin wax
C_{35} upward	soft solids	asphalt

*Fractional distillation is discussed in Topic 5C.

The quantity of gasoline that can be obtained from petroleum is increased by *cracking*, or breaking down long hydrocarbon chains, and by *alkylation*, or combining small molecules to make larger ones. In cracking, the less volatile fractions are exposed to high temperatures in the presence of a catalyst, often a modified zeolite (Topic 7E). For example, fuel oil can be converted into a mixture of octene and octane isomers:

$$C_{16}H_{34} \xrightarrow{\Delta, \text{ catalyst}} C_8H_{16} + C_8H_{18}$$

Alkylation also requires a catalyst to achieve the desired chain length. Octane, for instance, can be synthesized from a mixture of butane and butene:

$$C_4H_{10} + C_4H_8 \xrightarrow{\text{catalyst}} C_8H_{18}$$

The quality of gasoline, which determines how smoothly it burns, is measured by the *octane rating*. For example, the straight-chain molecule octane, $CH_3(CH_2)_6CH_3$, burns so poorly that it has an octane rating of −19, but its isomer 2,2,4-trimethylpentane, which is commonly called isooctane, has an octane rating of 100. The octane rating is improved by increasing the branching of the molecules and by introducing unsaturation and rings. In *isomerization*, straight-chain hydrocarbons are converted into their branched-chain isomers. For example:

$$CH_3(CH_2)_6CH_3 \xrightarrow{AlCl_3} (CH_3)_3CCH_2CH(CH_3)_2$$

Aromatization is the conversion of an alkane into an arene:

$$CH_3(CH_2)_5CH_3 \xrightarrow{AlCl_3, \ Cr_2O_3} CH_3C_6H_5 + 4 H_2$$

The product of this reaction, toluene (methylbenzene), has an octane rating of 120.

The quality of gasoline is also improved by the addition of ethanol, which has an octane rating of 120. The use of ethanol helps to reduce the demand for petroleum. Unlike petroleum, of which there are limited reserves on Earth, ethanol is a renewable fuel that can be regenerated every year (see Box 4D.1).

Coal

As reserves of petroleum decline worldwide, interest in making better use of coal is growing. The thought of an automobile that runs on coal is strange, but the use of gasoline substitutes derived from coal is a real possibility. Unfortunately, the increased usage of coal raises environmental concerns. Coal contains a much lower ratio of hydrogen to carbon than petroleum and is harder to purify, to work, and to transport. Although both coal and petroleum contribute to the greenhouse effect, the use of coal is potentially more damaging. When it burns, coal also releases a large amount of pollution in the form of particulate matter (primarily ash) and sulfur and nitrogen oxides. Much of the research on coal is aimed at converting it into more useful fuels.

Coal contains many aromatic rings (**FIG. 2**) and is the end product of the decay of swamp vegetation in *anaerobic* conditions (with a very low concentration of oxygen). Oxygen and hydrogen are gradually lost through the *coalification* process. As coalification proceeds, hydrogen is released and the aromatic structures increase.

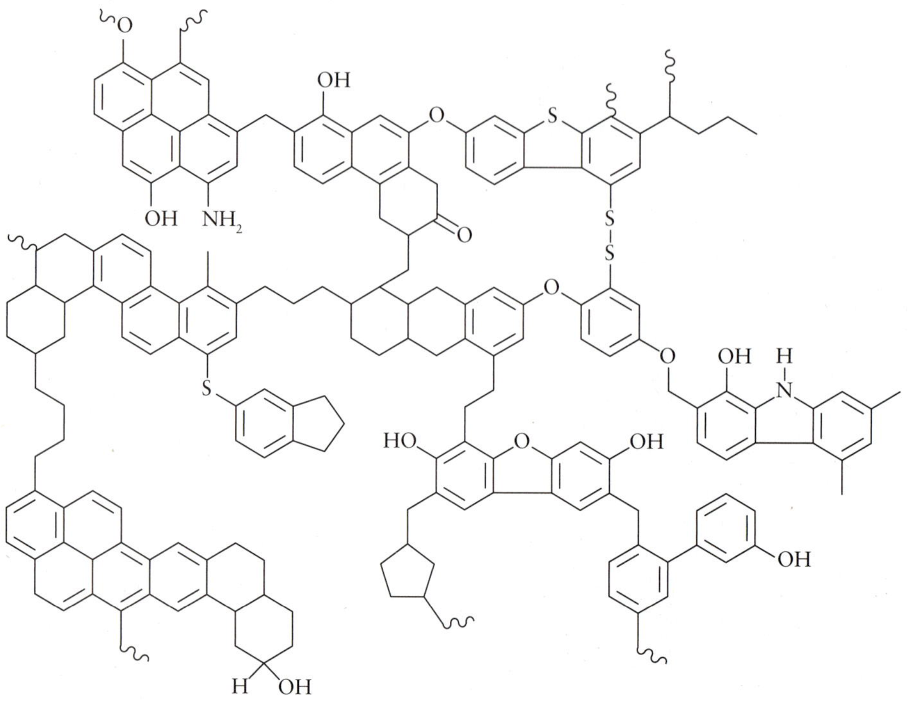

FIG. 2 A highly schematic representation of a part of the structure of coal. When coal is heated in the absence of oxygen, the structure breaks up and a complex mixture of products—many of them aromatic—is obtained.

When coal is destructively distilled—heated in the absence of oxygen so that it decomposes and vaporizes—its sheetlike molecules break up, and the fragments include aromatic hydrocarbons and their derivatives. *Coal gas,* which is given off first, contains carbon monoxide, hydrogen, methane, and small amounts of other gases. The complex liquid mixture that remains is called *coal tar.* A large number of pharmaceuticals, dyes, and fertilizers come from coal tar. Benzene is the raw material for many plastics, detergents, and pesticides. Naphthalene is used to make synthetic indigo (the dye in blue jeans), ammonia is used as fertilizer, and pitch, which contains the heaviest tars, is used for waterproofing and rustproofing. Coal gas is also the starting point for a number of alternative fuels, such as methane and hydrogen and liquid fuels such as methanol. These fuels are highly desired because they burn cleanly, producing little air pollution.

SYMBOLS, UNITS, AND MATHEMATICAL TECHNIQUES

1A SYMBOLS

Each physical quantity is represented by an italic or oblique Greek symbol (thus, m for mass, not m; Π for osmotic pressure, not Π). Table 1 lists most of the symbols used in this textbook together with their units (see also Appendix 1B). The symbols may be modified by attaching subscripts, as set out in Table 2. Fundamental constants are not included in the lists but can be found inside the back cover of the book. The symbols for mathematical constants are upright (thus, π not π, e, not e). All labels are upright (thus, s-orbital, not *s*-orbital; π-electron, not *π*-electron).

TABLE 1 Common Symbols and Units		
Symbol	**Physical quantity**	**SI unit**
α (alpha)	polarizability	$C^2 \cdot m^2 \cdot J^{-1}$
γ (gamma)	surface tension	$N \cdot m^{-1}$
δ (delta)	chemical shift	—
ε (epsilon)	molecular energy	J
θ (theta)	colatitude	degree (°), rad
λ (lambda)	wavelength	m
μ (mu)	dipole moment	$C \cdot m$
ν (nu)	frequency	Hz
Π (pi)	osmotic pressure	Pa
σ (sigma)	cross section	m^2
ϕ (phi)	azimuth	degree (°), rad
χ (chi)	electronegativity	—
ψ (psi)	wavefunction	$m^{-n/2}$ (in n dimensions)
a	activity	—
	van der Waals parameter	$L^2 \cdot bar \cdot mol^{-2}$
	unit-cell parameter	m
A	area	m^2
	mass number	—
	Madelung constant	—
b	van der Waals parameter	$L \cdot mol^{-1}$
	molality	$mol \cdot kg^{-1}$, "m"
B	second virial coefficient	$L \cdot mol^{-1}$
C	heat capacity	$J \cdot K^{-1}$
	third virial coefficient	$L^2 \cdot mol^{-2}$
c	molar concentration, molarity	$mol \cdot L^{-1}$, "M"
d	density	$kg \cdot m^{-3}$ ($g \cdot cm^{-3}$)
	length of unit-cell diagonal	m
E	energy	J
	electrode potential	V, $(J \cdot C^{-1})$
E_a	activation energy	$J \cdot mol^{-1}$ ($kJ \cdot mol^{-1}$)
E_{bind}	nuclear binding energy	J
E_{cell}	cell potential	V, $(J \cdot C^{-1})$
E_{ea}	electron affinity	$J \cdot mol^{-1}$ ($kJ \cdot mol^{-1}$)
E_k	kinetic energy	J
E_p	potential energy	J
e	elementary charge	C

(continued)

TABLE 1   Common Symbols and Units (*continued*)

Symbol	Physical quantity	SI unit
F	force	N
G	Gibbs free energy	J
H	enthalpy	J
h	height	m
I	ionization energy	$J \cdot mol^{-1}$ ($kJ \cdot mol^{-1}$)
	electric current	A ($C \cdot s^{-1}$)
i	i factor	—
[J]	molarity, molar concentration	$mol \cdot L^{-1}$, "M"
k_r	rate constant	(depends on order)
k	decay constant	s^{-1}
k_b	boiling-point constant	$K \cdot kg \cdot mol^{-1}$
k_f	freezing-point constant	$K \cdot kg \cdot mol^{-1}$
k_H	Henry's law constant	$mol \cdot L^{-1} \cdot atm^{-1}$
K	equilibrium constant	—
K_a	acidity constant	—
K_b	basicity constant	—
K_c	equilibrium constant	—
K_f	formation constant	—
K_M	Michaelis constant	$mol \cdot L^{-1}$
K_P	equilibrium constant	—
K_{sp}	solubility product	—
K_w	water autoprotolysis constant	—
l, L	length	m
m	mass	kg
M	molar mass	$kg \cdot mol^{-1}$ ($g \cdot mol^{-1}$)
N	number of entities	—
n	amount of substance	mol
p	linear momentum	$kg \cdot m \cdot s^{-1}$
P	pressure	Pa
P_J	partial pressure	Pa
q	heat	J
Q	electric charge	C
Q	reaction quotient	—
	relative biological effectiveness	—
r	radius	m
R	radial wavefunction	$m^{-3/2}$
S	entropy	$J \cdot K^{-1}$
s	dimensionless molar solubility	—
t	time	s
$t_{1/2}$	half-life	s
T	absolute temperature*	K
U	internal energy	J
v	velocity	$m \cdot s^{-1}$
V	volume	m^3, L
$\mathcal{V}$	electric potential	V ($J \cdot C^{-1}$)
w	work	J
x_J	mole fraction	—
Y	angular wavefunction	—
Z	compression factor	—
	atomic number	—

*Throughout this text, T denotes an absolute temperature

TABLE 2   Subscripts for Symbols

Subscript	Meaning	Example (units)
a	acid	acidity constant, K_a
b	base	basicity constant, K_b
	boiling	boiling temperature, T_b (K)
B	bond	bond enthalpy, ΔH_B ($kJ \cdot mol^{-1}$)
bind	binding	binding energy, E_{bind} (eV)

(continued)

TABLE 2 Subscripts for Symbols *(continued)*

Subscript	Meaning	Example (units)
c	concentration	equilibrium constant, K_c
	combustion	enthalpy of combustion, ΔH_c (kJ·mol^{-1})
	critical	critical temperature, T_c (K)
e	nonexpansion (extra) work	electrical work, w_e (J)
f	formation	enthalpy of formation, ΔH_f (kJ·mol^{-1})
		formation constant, K_f
	freezing	freezing temperature, T_f (K)
fus	fusion	enthalpy of fusion, ΔH_{fus} (kJ·mol^{-1})
H	Henry	Henry's law constant, k_H
In	indicator	indicator constant, K_{In}
k	kinetic	kinetic energy, E_k (J)
L	lattice	lattice enthalpy, ΔH_L (kJ·mol^{-1})
m	molar	molar volume, $V_m = V/n$ (L·mol^{-1})
M	Michaelis	Michaelis constant, K_M
mix	mixing	enthalpy of mixing, ΔH_{mix} (kJ·mol^{-1})
p	potential	potential energy, E_p (J)
P	constant pressure	heat capacity at constant pressure, C_P (J·K^{-1})
r	reaction	reaction enthalpy, ΔH_r (kJ·mol^{-1})
s	specific	specific heat capacity, $C_s = C/m$ (J·K^{-1}·g^{-1})
sol	solution	enthalpy of solution, ΔH_{sol} (kJ·mol^{-1})
sp	solubility product	solubility product, K_{sp}
sub	sublimation	enthalpy of sublimation, ΔH_{sub} (kJ·mol^{-1})
surr	surroundings	entropy of surroundings, S_{surr} (J·K^{-1})
tot	total	total entropy, S_{tot} (J·K^{-1})
V	constant volume	heat capacity at constant volume, C_V (J·K^{-1})
vap	vaporization	enthalpy of vaporization, ΔH_{vap} (kJ·mol^{-1})
w	water	water autoprotolysis constant, K_w
0	initial	initial concentration, $[A]_0$
	ground state	wavefunction, ψ_0

1B UNITS AND UNIT CONVERSIONS

Each physical quantity is reported as a multiple of a defined unit:

Physical quantity = numerical value × unit

For instance, a length may be expressed as a multiple of the unit of length: 1 meter or 1 m; as in $l = 2.0 \times 1$ m = 2.0 m. All units are denoted by roman letters, such as m for meter and s for second.

Units are treated like algebraic quantities that may be multiplied and divided. Thus, the preceding equation may be expressed as

Physical quantity/unit = numerical value

For instance, to report a length of 2.0 m you could write $l/m = 2.0$.

The **Système International** (SI) is the internationally accepted form of the metric system. It defines seven **base units** in terms of which all physical quantities can be expressed:

meter, m The meter, the unit of length, is the length of the path traveled by light during a time interval of 1/299 792 458 of a second.

kilogram, kg The kilogram, the unit of mass, is the mass of a standard cylinder maintained at a laboratory in France.

second, s The second, the unit of time, is 9 192 631 770 periods of a certain spectroscopic transition in a cesium-133 atom.

ampere, A The ampere, the unit of electric current, is defined in terms of the force exerted between two parallel wires carrying the current.

kelvin, K The kelvin, the unit of temperature, is 1/273.16 of the absolute temperature of the triple point of water.

mole, mol The mole, the unit of chemical amount, is the amount of substance that contains as many specified entities as there are atoms in exactly 12 g of carbon-12.

candela, cd The candela, the unit of luminous intensity, is defined in terms of a carefully specified source. We do not use the candela in this book.

In 2012 it was agreed internationally to replace the definition of the kilogram in terms of the prototype cylinder by a more subtle definition in terms of the fundamental constants; the new definition has been agreed but not yet (in 2016) implemented. New definitions of the mole and the kelvin have also been agreed, but not yet implemented. The redefined base units are expected to be adopted by the end of 2018.

Any unit may be modified by one of the prefixes given in Table 3, which denote multiplication or division by a power

TABLE 3 Typical SI Prefixes

Prefix:	deca-	kilo-	mega-	giga-	tera-	peta-			
Abbreviation:	da	k	M	G	T	P			
Factor:	10	10^3	10^6	10^9	10^{12}	10^{15}			
Prefix:	deci-	centi-	milli-	micro-	nano-	pico-	femto-	atto-	zepto-
Abbreviation:	d	c	m	μ (mu)	n	p	f	a	z
Factor:	10^{-1}	10^{-2}	10^{-3}	10^{-6}	10^{-9}	10^{-12}	10^{-15}	10^{-18}	10^{-21}

of 10 of the unit. Thus, 1 mm = 10^{-3} m and 1 MK = 10^6 K. Note that all the prefixes are upright (Roman), not italic.

Derived units are combinations of the base units. Table 4 lists some derived units. The names of units derived from the names of people all begin with a lowercase letter, but the initial letter of their abbreviation is uppercase (thus, joule and its symbol J).

TABLE 4 Derived Units with Special Names

Physical quantity	Name of unit	Abbreviation	Definition
absorbed dose	gray	Gy	$J \cdot kg^{-1}$
dose equivalent	sievert	Sv	$J \cdot kg^{-1}$
electric charge	coulomb	C	$A \cdot s$
electric potential	volt	V	$J \cdot C^{-1}$
energy	joule	J	$N \cdot m$, $kg \cdot m^2 \cdot s^{-2}$
force	newton	N	$kg \cdot m \cdot s^{-2}$
frequency	hertz	Hz	s^{-1}
power	watt	W	$J \cdot s^{-1}$
pressure	pascal	Pa	$N \cdot m^{-2}$, $kg \cdot m^{-1} \cdot s^{-2}$
volume	liter	L	dm^3

It is often necessary to convert units from another system (for instance, calories for energy and inches for length) into SI units. Table 5 lists some common conversions.

TABLE 5 Relations Between Units

Physical quantity	Common unit	Abbreviation	SI equivalent*
mass	pound	lb	**0.453 592 37 kg**
	tonne	t	**10^3 kg (1 Mg)**
	ton (short, U.S.)	ton	907.184 74 kg
	ton (long, U.K.)	ton	1016.046 kg
length	inch	in.	**2.54 cm**
	foot	ft	**30.48 cm**
volume	U.S. quart	qt	**0.946 3525 L**
	U.S. gallon	gal	**3.785 41 L**
	Imperial quart	qt	**1.136 5225 L**
	Imperial gallon	gal	**4.546 09 L**
time	minute	min	**60 s**
	hour	h	**3600 s**
energy	calorie (thermochemical)	cal	**4.184 J**
	electronvolt	eV	$1.602\ 177 \times 10^{-19}$ J
	kilowatt–hour	kWh	**3.6×10^6 J**
	liter–atmosphere	L·atm	**101.325 J**
pressure	torr	Torr	133.322 Pa
	atmosphere	atm	**101 325 Pa (760 Torr)**
	bar	bar	**10^5 Pa**
	pounds/square inch	psi	6894.76 Pa
power	horsepower	hp	**745.7 W**
dipole moment	debye	D	$3.335\ 64 \times 10^{-30}$ C·m

*Values in boldface type are exact.

As explained in *Fundamentals* A, to convert between units, use a **conversion factor** of the form

$$\text{Conversion factor} = \frac{\text{units required}}{\text{units given}}$$

When using a conversion factor, the units are treated just like algebraic quantities: they are multiplied or canceled in the normal way.

The conversion of temperatures is carried out slightly differently. Because the Fahrenheit degree (°F) is smaller than a Celsius degree by a factor of $\frac{5}{9}$ (because there are 180 Fahrenheit degrees between the freezing point and boiling point of water but only 100 Celsius degrees between the same two points) and because 0 °C coincides with 32 °F, use

$$\text{Temperature (°F)} = \left\{ \tfrac{9}{5} \times \text{temperature (°C)} \right\} + 32$$

(The 32 is exact.) For example, to convert 37 °C (blood temperature) into degrees Fahrenheit, write

$$\text{Temperature (°F)} = \left\{ \tfrac{9}{5} \times 37 \right\} + 32 = 99$$

and the temperature is reported as 99 °F. A more sophisticated way of expressing the same relation is to write

$$\text{Temperature/°F} = \left\{ \tfrac{9}{5} \times \text{temperature/°C} \right\} + 32$$

In this expression, the temperature units are treated like numbers and canceled when it is appropriate. The same conversion then becomes

$$\text{Temperature/°F} = \left\{ \tfrac{9}{5} \times (37\,°C)/°C \right\} + 32$$
$$= \left\{ \tfrac{9}{5} \times 37 \right\} + 32 = 99$$

and multiplication through by °F gives

$$\text{Temperature} = 99\ °F$$

The corresponding expression for conversion between the Celsius and Kelvin scales is

$$\text{Temperature/°C} = \text{temperature/K} - 273.15$$

(The 273.15 is exact.) Note that the size of the degree Celsius is the same as that of the kelvin, so a property with a value reported as $100\ \text{J·(°C)}^{-1}$ can be interpreted as $100\ \text{J·K}^{-1}$.

1C SCIENTIFIC NOTATION

In **scientific notation,** a number is written as $A \times 10^a$. Here A is a decimal number with one nonzero digit in front of the decimal point and a is a whole number. For example, 333 is written 3.33×10^2 in scientific notation, because $10^2 = 10 \times 10 = 100$:

$$333 = 3.33 \times 100 = 3.33 \times 10^2$$

Numbers between 0 and 1 are expressed in the same way but with a negative power of 10; they have the form $A \times 10^{-a}$, with $10^{-1} = 0.1$, and so on. Thus, 0.0333 in decimal notation is 3.33×10^{-2} because

$$10^{-2} = \frac{1}{10} \times \frac{1}{10} = \frac{1}{100}$$

and therefore

$$0.033 = 3.33 \times \frac{1}{100} = 3.33 \times 10^{-2}$$

In each case the number of zeros following the decimal point is one less than the number (disregarding the sign) to which 10 is raised. Thus, 10^{-5} is written as a decimal point followed by $5 - 1 = 4$ zeros and then a 1:

$$10^{-5} = 10^{-1} \times 10^{-1} \times 10^{-1} \times 10^{-1} \times 10^{-1}$$
$$= 0.000\,01$$

Note the space separating groups of three digits, which is used to make numbers easier to interpret. However, if that grouping results in a single remaining digit, then it joins the preceding group (thus 0.1234, not 0.123 4; 0.123 4567, not 0.123 456 7).

The digits in a reported measurement are called the **significant figures.** There are two significant figures (written 2 sf) in 1.2 cm^3 and 3 sf in 1.78 g. *Fundamentals* A describes how to find the number of significant figures in a measurement.

Some zeros are legitimately measured digits, but other zeros serve only to mark the place of the decimal point. Trailing zeros (the last ones after a decimal point), as in 22.0 mL, are significant, because they were measured. Thus, 22.0 mL has 3 sf. The "captive" zero in 80.1 kg is a measured digit, and so 80.1 kg has 3 sf. However, the leading digits in 0.0025 g are not significant; they are only placeholders used to indicate powers of 10, not measured numbers. That they are only placeholders can be seen by reporting the mass as 2.5×10^{-3} g, which has 2 sf.

The results of measurements, which are always uncertain, are distinguished from the results of *counting*, which are *exact*. For example, the report "12 eggs" means that there are exactly 12 eggs present, not a number somewhere between 11.5 and 12.5.

Some ambiguity arises with whole numbers ending in zero. Does a length reported as 400 m have 3 sf (4.00×10^2), 2 sf (4.0×10^2), or only 1 sf (4×10^2)? In such cases, the use of scientific notation removes any ambiguity. If it is not convenient to use scientific notation, a final decimal point can be used to indicate that every digit to the left of the decimal is significant. Thus, 400 m is ambiguous and cannot be taken to have more than 1 sf unless other information is given. However, 400. m unambiguously has 3 sf. The final decimal point is rarely used in the everyday world (thus, "the speed limit is 50 mph" is ambiguous in science but not in law), but we adopt it throughout this text.

Different rounding-off rules are needed for addition (and its reverse, subtraction) and multiplication (and its reverse, division). In both procedures, the answers need to be rounded off to the correct number of significant figures.

Rounding off In calculations, round *up* if the last digit is above 5 and round *down* if it is below 5. For numbers ending in 5, always round to the nearest even number. For example, 2.35 rounds to 2.4 and 2.65 rounds to 2.6. In a calculation with multiple steps, round off only in the final step; if possible, carry all digits in the memory of the calculator until that stage. In this text intermediate results are reported using . . . (as in 22.0/7.0 = 3.142 . . .) to denote a value that has not yet been rounded.

Addition and subtraction When adding or subtracting, make sure that the number of decimal places in the result is the same as the *smallest number of decimal places* in the data. For example, 0.10 g + 0.024 g = 0.12 g.

Multiplication and division When multiplying or dividing, make sure that the number of significant figures in the result is the same as the *smallest number of significant figures* in the data. For example, (8.62 g)/(2.0 cm^3) = 4.3 g·cm^{-3}.

Integers and exact numbers In multiplication or division by an integer or an exact number, the uncertainty of the result is determined by the measured value. Some unit conversion factors are defined exactly, even though they are not whole numbers. For example, 1 in. is defined as *exactly* 2.54 cm and the 273.15 in the conversion between Celsius and Kelvin temperatures is exact; so 100.000 °C converts into 373.150 K.

Logarithms and exponentials The mantissa of a common logarithm (the digits following the decimal point, see Appendix 1D) has the same number of significant figures as the original number. Thus, log 2.45 = 0.389. A common antilogarithm of a number has the same number of significant figures as the mantissa (see below) of the original number. Thus, $10^{0.389} = 2.45$ and $10^{12.389} = 2.45 \times 10^{12}$. There is no simple rule for assessing the correct number of significant figures when natural logarithms are used: one way is to convert natural logarithms into common logarithms and then to use the rules just specified.

1D EXPONENTS AND LOGARITHMS

To multiply numbers in scientific notation, multiply the decimal parts of the numbers and add the powers of 10:

$$(A \times 10^a) \times (B \times 10^b) = (A \times B) \times 10^{a+b}$$

An example is

$$(1.23 \times 10^2) \times (4.56 \times 10^3) = 1.23 \times 4.56 \times 10^{2+3}$$
$$= 5.61 \times 10^5$$

(We are supposing that the initial factors here and below are measurements with 3 sf.) This rule also applies if the powers of 10 are negative:

$$(1.23 \times 10^{-2}) \times (4.56 \times 10^{-3}) = 1.23 \times 4.56 \times 10^{-2-3}$$
$$= 5.61 \times 10^{-5}$$

The results of such calculations are then adjusted so that one digit precedes the decimal point:

$$(4.56 \times 10^{-3}) \times (7.65 \times 10^6) = 34.88 \times 10^3$$
$$= 3.488 \times 10^4$$

When dividing two numbers in scientific notation, divide the decimal parts of the numbers and subtract the powers of 10:

$$\frac{A \times 10^a}{B \times 10^b} = \frac{A}{B} \times 10^{a-b}$$

An example is

$$\frac{4.31 \times 10^5}{9.87 \times 10^{-8}} = \frac{4.31}{9.87} \times 10^{5-(-8)} = 0.437 \times 10^{13}$$
$$= 4.37 \times 10^{12}$$

Before adding and subtracting numbers in scientific notation, rewrite the numbers as decimal numbers multiplied by the same power of 10:

$$1.00 \times 10^3 + 2.00 \times 10^2 = 1.00 \times 10^3 + 0.200 \times 10^3$$
$$= 1.20 \times 10^3$$

When raising a number in scientific notation to a particular power, raise the decimal part of the number to the power and multiply the power of 10 by the power:

$$(A \times 10^a)^b = A^b \times 10^{a \times b}$$

For example, 2.88×10^4 raised to the third power is

$$(2.88 \times 10^4)^3 = 2.88^3 \times (10^4)^3 = 2.88^3 \times 10^{3 \times 4}$$
$$= 23.9 \times 10^{12} = 2.39 \times 10^{13}$$

The rule follows from the fact that

$$(10^4)^3 = 10^4 \times 10^4 \times 10^4 = 10^{4+4+4} = 10^{3 \times 4}$$

The **common logarithm** of a number x, denoted log x, is the power to which 10 must be raised to equal x. Thus, the logarithm of 100 is 2, written log 100 = 2, because $10^2 = 100$. The logarithm of 1.5×10^2 is 2.18 because

$$10^{2.18} = 10^{0.18+2} = 10^{0.18} \times 10^2 = 1.5 \times 10^2$$

The number to the left of the decimal point in the logarithm (the 2 in log(1.5×10^2) = 2.18) is called the **characteristic** of the logarithm: it is the power of 10 in the original number (the power 2 in 1.5×10^2). The decimal fraction (the numbers to the right of the decimal point, such as 0.18 in our example) is called the **mantissa** (from the Latin word for "makeweight"). It is the logarithm of the decimal number written with one nonzero digit to the left of the decimal point (the 1.5 in the example).

The distinction between the characteristic and the mantissa is important when deciding how many significant figures to retain in a calculation that includes logarithms (as in the calculation of pH). Just as the power of 10 in a decimal number indicates only the location of the decimal point and plays no role in the determination of significant figures, so the characteristic of a logarithm is not included in the count of significant figures in a logarithm (see Appendix 1C). The number of significant figures in the mantissa is equal to the number of significant figures in the decimal number.

The **common antilogarithm** of a number x is the number that has x as its common logarithm. In practice, the common antilogarithm of x is simply another name for 10^x, and so the common antilogarithm of 2 is $10^2 = 100$ and that of 2.18 is

$$10^{2.18} = 10^{0.18+2} = 10^{0.18} \times 10^2 = 1.5 \times 10^2$$

The logarithm of a number greater than 1 is positive, and the logarithm of a number less than 1 (but greater than 0) is negative. For any number x,

$$\text{If } x > 1, \log x > 0$$
$$\text{If } x = 1, \log x = 0$$
$$\text{If } x < 1, \log x < 0$$

Logarithms are not defined either for 0 or for negative numbers.

The **natural logarithm** of a number x, denoted ln x, is the power to which the number e = 2.718 . . . must be raised to equal x. Thus, ln 10.0 = 2.303, signifying that $e^{2.303} = 10.0$. The value of e may seem a peculiar choice, but it occurs naturally in a number of mathematical expressions, and its use simplifies many formulas. Common and natural logarithms are related by the expression

$$\ln x = \ln 10 \times \log x$$

In practice, a convenient approximation is

$$\ln x \approx 2.303 \times \log x$$

The **natural antilogarithm** of x is normally called the **exponential** of e; it is the value of e raised to the power x. Thus, the natural antilogarithm of 2.303 is $e^{2.303} = 10.0$.

The following relations between logarithms are useful. Written here mainly for common logarithms, they also apply to natural logarithms.

Relation	Example
$\log 10^x = x$	$\log 10^{-7} = -7$
$\ln e^x = x$	$\ln e^{-kt} = -kt$
$\log x + \log y = \log xy$	$\log[\text{Ag}^+] + \log[\text{Cl}^-] =$
	$\log[\text{Ag}^+][\text{Cl}^-]$
$\log x - \log y = \log(x/y)$	$\log A_0 - \log A = \log(A_0/A)$
$x \log y = \log y^x$	$2\log[\text{H}^+] = \log([\text{H}^+]^2)$
$\log(1/x) = -\log x$	$\log(1/[\text{H}^+]) = -\log[\text{H}^+]$

Logarithms are useful for solving expressions of the form

$$a^x = b$$

for the unknown x. (This type of calculation can arise in the study of chemical kinetics to determine the order of a reaction.) We take logarithms of both sides,

$$\log a^x = \log b$$

and, from a relation given in the preceding table, write the expression as

$$x \log a = \log b$$

Therefore,

$$x = \frac{\log b}{\log a}$$

1E EQUATIONS AND GRAPHS

A **quadratic equation** is an equation of the form

$$ax^2 + bx + c = 0$$

The two **roots** of the equation (the solutions) are given by the expression

$$x = \frac{-b \pm \sqrt{b^2 - 4ac}}{2a}$$

The roots can also be determined graphically (by using a graphing calculator, for instance) by noting where the graph of $y(x) = ax^2 + bx + c$ against x passes through $y = 0$ (Fig. 1). When a quadratic equation arises in connection with a chemical calculation, we accept only the root that leads to a physically plausible

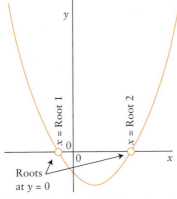

FIGURE 1 A graph of a function of the form $y(x) = ax^2 + bx + c$ passes through $y = 0$ at two points, which are the two roots of the quadratic equation $ax^2 + bx + c = 0$.

result. For example, if x is a concentration, then it must be a positive number, and a negative root can be ignored.

A calculation may result in a cubic equation:

$$ax^3 + bx^2 + cx + d = 0$$

Cubic equations are often very tedious to solve exactly, and so it is better to use mathematical software or a graphing calculator, and to identify the locations where the graph of $y(x)$ against x passes through $y = 0$ (Fig. 2).

Experimental data can often be analyzed most effectively from a graph. In many cases, the best procedure is to find a way of plotting the data as a straight line. It is easier to identify deviations of data points from a straight line than deviations from a curve. Moreover, it is also easy to calculate the slope of a straight line, to **extrapolate** (extend) a straight line beyond the range of the data, and to **interpolate** between the data points (that is, find a value between two measured values).

The formula of a straight-line graph of y (the vertical axis) plotted against x (the horizontal axis) is

$$y = mx + b$$

Here b is the **intercept** of the line with the y-axis (Fig. 3), the value of y when $x = 0$. The **slope** of the graph, its gradient, is m. The slope can be calculated by choosing two points, x_1 and x_2, and their corresponding values on the y-axis, y_1 and y_2, and substituting the values into the formula

$$m = \frac{y_2 - y_1}{x_2 - x_1}$$

Because b is the intercept and m is the slope, the equation of the straight line is equivalent to

$$y = (\text{slope} \times x) + \text{intercept}$$

The modern recommended procedure is to plot dimensionless values against dimensionless values. Thus, to plot the volume, V, of a gas (in cubic centimeters, cm^3) against pressure, P (in pascals, Pa), the values of V/cm^3 are plotted against $P/$Pa. A consequence of this approach is that both the slope and the intercept are pure numbers. However, you will also come

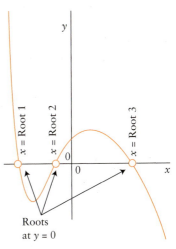

FIGURE 2 A graph of a function of the form $y(x) = ax^3 + bx^2 + cx + d$ passes through $y = 0$ at three points, which are the three roots of the cubic equation $ax^3 + bx^2 + cx + d = 0$.

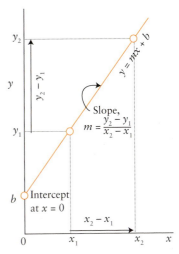

FIGURE 3 The straight line $y(x) = mx + b$; its intercept with the vertical axis at $x = 0$ is b and its slope is m.

across the common practice of labeling the axes of a graph as V (cm^3) and P (Pa).

1F CALCULUS

Differential calculus is the part of mathematics that deals with the slopes of curves and with infinitesimal quantities. The slope of a curve at a point can be calculated by considering the straight line joining two points x and $x + \delta x$, where δx is small. The slope of this line is

$$\text{Slope} = \frac{y(x + \delta x) - y(x)}{\delta x}$$

In differential calculus, the slope of the curve is found by letting the separation of the points become infinitesimally small. The **first derivative** of the function y with respect to x is then defined as

$$\frac{dy}{dx} = \lim_{\delta x \to 0} \frac{y(x + \delta x) - y(x)}{\delta x}$$

where "lim" means the limit of whatever follows—in this case, the value of the expression as δx is allowed to approach zero. For example, if $y(x) = x^2$,

$$\frac{dy}{dx} = \lim_{\delta x \to 0} \frac{(x + \delta x)^2 - x^2}{\delta x}$$
$$= \lim_{\delta x \to 0} \frac{x^2 + 2x\delta x + (\delta x)^2 - x^2}{\delta x}$$
$$= \lim_{\delta x \to 0} \frac{2x\delta x + (\delta x)^2}{\delta x} = \lim_{\delta x \to 0} (2x + \delta x) = 2x$$

Therefore, the slope of the graph of the function $y = x^2$ at any point x is $2x$. The same procedure can be applied to other functions. However, in practice, it is usually more convenient to consult tables of first derivatives that have already been worked out. A selection of common functions and their first derivatives is given here.

Function, $y(x)$	Derivative, dy/dx
x^n	nx^{n-1}
$\ln x$	$1/x$
e^{ax}	ae^{ax}
$\sin ax$	$a \cos ax$
$\cos ax$	$-a \sin ax$

Integral calculus provides a way to determine the original function, given its first derivative. Thus, if we know that the first derivative is $2x$, then the integral calculus establishes that the function itself is $y = x^2 + $ constant. The constant is included because when we differentiate $x^2 + $ constant, we get $2x$ regardless of the value of the constant. Formally:

$$\int (2x)\, dx = x^2 + \text{constant}$$

It follows that the functions in the left-hand column of the preceding table are the integrals (to within a constant) of the functions in the right-hand column. More formally, they are the **indefinite integrals** of the function ("indefinite" on account of the presence of the unknown constant), in contrast with the "definite" integrals described next. Tables of indefinite integrals may be consulted for more complex examples, and mathematical software or a graphing calculator can be used to evaluate them.

An integral has a further important interpretation: the integral of a function evaluated between two points a and b, which is denoted $\int_a^b \ldots$ is the *area* beneath the graph of the function between the two points (Fig. 4). For example, the area beneath the curve $y(x) = \sin x$ between $x = 0$ and $x = \pi$ is

$$\text{Area} = \int_0^{\pi} \sin x\, dx = \left(\int \sin x\, dx \right)_{\text{at } b} - \left(\int \sin x\, dx \right)_{\text{at } a}$$

$$= \left(\overset{-1}{-\cos x} + \text{constant} \right)_{\text{at } \pi} - \left(\overset{1}{-\cos x} + \text{constant} \right)_{\text{at } 0}$$

$$= 1 + 1 = 2$$

An integral with limits attached, as in this example, is called a **definite integral** (because the unknown constant has canceled).

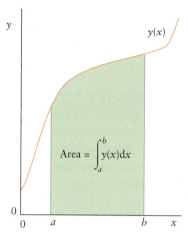

FIGURE 4 The definite integral of the function $y(x)$ between $x = a$ and $x = b$ is equal to the area bounded by the curve, the x-axis, and the two vertical lines through a and b.

EXPERIMENTAL DATA

2A THERMODYNAMIC DATA AT 25 °C

Inorganic Substances

Substance	Molar mass, $M/(\text{g·mol}^{-1})$	Enthalpy of formation, $\Delta H_f°/(\text{kJ·mol}^{-1})$	Gibbs free energy of formation, $\Delta G_f°/(\text{kJ·mol}^{-1})$	Molar heat capacity, $C_{P,m}/(\text{J·K}^{-1}\text{·mol}^{-1})$	Molar entropy,* $S_m°/(\text{J·K}^{-1}\text{·mol}^{-1})$
Aluminum					
Al(s)	26.98	0	0	24.35	28.33
Al^{3+}(aq)	26.98	−524.7	−481.2	—	−321.7
$Al(OH)_3$(s)	78.00	−1276	—	—	—
Al_2O_3(s)	101.96	−1675.7	−1582.35	79.04	50.92
$AlCl_3$(s)	133.33	−704.2	−628.8	91.84	110.67
$AlBr_3$(s)	266.68	−527.2	—	100.6	180.2
Antimony					
Sb(s)	121.76	0	0	25.23	45.69
SbH_3(g)	124.78	+145.11	+147.75	41.05	232.78
$SbCl_3$(g)	228.11	−313.8	−301.2	76.69	337.80
$SbCl_5$(g)	299.01	−394.34	−334.29	121.13	401.94
Arsenic					
As(s), gray	74.92	0	0	24.64	35.1
AsO_4^{3-}(aq)	138.92	−888.14	−648.41	—	−162.8
As_2S_3(s)	246.05	−169.0	−168.6	116.3	163.6
Barium					
Ba(s)	137.33	0	0	28.07	62.8
Ba^{2+}(aq)	137.33	−537.64	−560.77	—	+9.6
BaO(s)	153.33	−553.5	−525.1	47.78	70.42
$BaCO_3$(s)	197.34	−1216.3	−1137.6	85.35	112.1
$BaCO_3$(aq)	197.34	−1214.78	−1088.59	—	−47.3
Boron					
B(s)	10.81	0	0	11.09	5.86
BF_3(g)	67.81	−1137.0	−1120.3	50.46	254.12
B_2O_3(s)	69.62	−1272.8	−1193.7	62.93	53.97
Bromine					
Br_2(l)	159.80	0	0	75.69	152.23
Br_2(g)	159.80	+30.91	+3.11	36.02	245.46
Br(g)	79.90	+111.88	+82.40	20.79	175.02
Br^-(aq)	79.90	−121.55	−103.96	—	+82.4
HBr(g)	80.91	−36.40	−53.45	29.14	198.70
Calcium					
Ca(s)	40.08	0	0	25.31	41.42
Ca(g)	40.08	+178.2	+144.3	20.79	154.88
Ca^{2+}(aq)	40.08	−542.83	−553.58	—	−53.1

(continued)

Inorganic Substances (*continued*)

Substance	Molar mass, $M/(\text{g·mol}^{-1})$	Enthalpy of formation, $\Delta H_f°/(\text{kJ·mol}^{-1})$	Gibbs free energy of formation, $\Delta G_f°/(\text{kJ·mol}^{-1})$	Molar heat capacity, $C_{P,m}/(\text{J·K}^{-1}\text{·mol}^{-1})$	Molar entropy,* $S_m°/(\text{J·K}^{-1}\text{·mol}^{-1})$
CaO(s)	56.08	−635.09	−604.03	42.80	39.75
CaC$_2$(s)	64.10	−59.8	−64.9	62.72	69.96
Ca(OH)$_2$(s)	74.10	−986.09	−898.49	87.49	83.39
Ca(OH)$_2$(aq)	74.10	−1002.82	−868.07	—	−74.5
CaF$_2$(s)	78.08	−1219.6	−1167.3	67.03	68.87
CaF$_2$(aq)	78.08	−1208.09	−1111.15	—	−80.8
CaCO$_3$(s), calcite	100.09	−1206.9	−1128.8	81.88	92.9
CaCO$_3$(s), aragonite	100.09	−1207.1	−1127.8	81.25	88.7
CaCO$_3$(aq)	100.09	−1219.97	−1081.39	—	−110.0
CaCl$_2$(s)	110.98	−795.8	−748.1	72.59	104.6
CaCl$_2$(aq)	110.98	−877.1	−816.0	—	59.8
CaSO$_4$(s)	136.14	−1434.11	−1321.79	99.66	106.7
CaSO$_4$(aq)	136.14	−1452.10	−1298.10	—	−33.1
CaBr$_2$(s)	199.88	−682.8	−663.6	72.59	130

Carbon (for organic compounds, see the next table)

Substance					
C(s), graphite	12.01	0	0	8.53	5.740
C(s), diamond	12.01	+1.895	+2.900	6.11	2.377
C(g)	12.01	+716.68	+671.26	20.84	158.10
HCN(g)	27.03	+135.1	+124.7	35.86	201.78
HCN(l)	27.03	+108.87	+124.97	70.63	112.84
HCN(aq)	27.03	+107.1	+119.7	—	124.7
CO(g)	28.01	−110.53	−137.17	29.14	197.67
CO$_2$(g)	44.01	−393.51	−394.36	37.11	213.74
CO$_3{}^{2-}$(aq)	60.01	−677.14	−527.81	—	−56.9
CS$_2$(l)	76.15	+89.70	+65.27	75.7	151.34
CCl$_4$(l)	153.81	−135.44	−65.21	131.75	216.40

Cerium

Substance					
Ce(s)	140.12	0	0	26.94	72.0
Ce^{3+}(aq)	140.12	−696.2	−672.0	—	−205
Ce^{4+}(aq)	140.12	−537.2	−503.8	—	−301

Chlorine

Substance					
Cl$_2$(g)	70.90	0	0	33.91	223.07
Cl(g)	35.45	+121.68	+105.68	21.84	165.20
Cl$^-$(aq)	35.45	−167.16	−131.23	—	+56.5
HCl(g)	36.46	−92.31	−95.30	29.12	186.91
HCl(aq)	36.46	−167.16	−131.23	—	56.5

Copper

Substance					
Cu(s)	63.55	0	0	24.44	33.15
Cu$^+$(aq)	63.55	+71.67	+49.98	—	+40.6
Cu^{2+}(aq)	63.55	+64.77	+65.49	—	−99.6
CuO(s)	79.55	−157.3	−129.7	42.30	42.63
Cu$_2$O(s)	143.10	−168.6	−146.0	63.64	93.14
CuSO$_4$(s)	159.61	−771.36	−661.8	100.0	109
CuSO$_4$·5H$_2$O(s)	249.69	−2279.7	−1879.7	280	300.4

Deuterium

Substance					
D$_2$(g)	4.028	0	0	29.20	144.96
D$_2$O(g)	20.028	−249.20	−234.54	34.27	198.34
D$_2$O(l)	20.028	−294.60	−243.44	34.27	75.94

Substance	Molar mass, $M/(\text{g·mol}^{-1})$	Enthalpy of formation, $\Delta H_f°/(\text{kJ·mol}^{-1})$	Gibbs free energy of formation, $\Delta G_f°/(\text{kJ·mol}^{-1})$	Molar heat capacity, $C_{P,m}/(\text{J·K}^{-1}\text{·mol}^{-1})$	Molar entropy,* $S_m°/(\text{J·K}^{-1}\text{·mol}^{-1})$
Fluorine					
$F_2(g)$	38.00	0	0	31.30	202.78
$F^-(aq)$	19.00	−332.63	−278.79	—	−13.8
$HF(g)$	20.01	−271.1	−273.2	29.13	173.78
$HF(aq)$	20.01	−330.08	−296.82	—	88.7
Hydrogen (see also Deuterium)					
$H_2(g)$	2.0158	0	0	28.82	130.68
$H(g)$	1.0079	+217.97	+203.25	20.78	114.71
$H^+(aq)$	1.0079	0	0	0	0
$H_2O(l)$	18.02	−285.83	−237.13	75.29	69.91
$H_2O(g)$	18.02	−241.82	−228.57	33.58	188.83
$H_3O^+(aq)$	19.02	−285.83	−237.13	75.29	+69.91
$H_2O_2(l)$	34.02	−187.78	−120.35	89.1	109.6
$H_2O_2(aq)$	34.02	−191.17	−134.03	—	143.9
Iodine					
$I_2(s)$	253.80	0	0	54.44	116.14
$I_2(g)$	253.80	+62.44	+19.33	36.90	260.69
$I^-(aq)$	126.90	−55.19	−51.57	—	+111.3
$HI(g)$	127.91	+26.48	+1.70	29.16	206.59
Iron					
$Fe(s)$	55.84	0	0	25.10	27.28
$Fe^{2+}(aq)$	55.84	−89.1	−78.90	—	−137.7
$Fe^{3+}(aq)$	55.84	−48.5	−4.7	—	−315.9
$FeS(s, \alpha)$	87.90	−100.0	−100.4	50.54	60.29
$FeS(aq)$	87.90	—	+6.9	—	—
$FeS_2(s)$	119.96	−178.2	−166.9	62.17	52.93
$Fe_2O_3(s)$, hematite	159.68	−824.2	−742.2	103.85	87.40
$Fe_3O_4(s)$, magnetite	231.52	−1118.4	−1015.4	143.43	146.4
Lead					
$Pb(s)$	207.2	0	0	26.44	64.81
$Pb^{2+}(aq)$	207.2	−1.7	−24.43	—	+10.5
$PbO_2(s)$	239.2	−277.4	−217.33	64.64	68.6
$PbSO_4(s)$	303.3	−919.94	−813.14	103.21	148.57
$PbBr_2(s)$	367.0	−278.7	−261.92	80.12	161.5
$PbBr_2(aq)$	367.0	−244.8	−232.34	—	175.3
Magnesium					
$Mg(s)$	24.31	0	0	24.89	32.68
$Mg(g)$	24.31	+147.70	−113.10	20.79	148.65
$Mg^{2+}(aq)$	24.31	−466.85	−454.8	—	−138.1
$MgO(s)$	40.31	−601.70	−569.43	37.15	26.94
$MgCO_3(s)$	84.32	−1095.8	−1012.1	75.52	65.7
$MgCl_2(s)$	95.21	−641.8	—	—	—
$MgBr_2(s)$	184.11	−524.3	−503.8	—	117.2
Manganese					
Mn	54.94	0	0	26.3	32.01
MnO_2	86.94	−520.0	−465.1	54.1	53.05
Mercury					
$Hg(l)$	200.59	0	0	27.98	76.02
$Hg(g)$	200.59	+61.32	+31.82	20.79	174.96
$HgO(s)$	216.59	−90.83	−58.54	44.06	70.29
$Hg_2Cl_2(s)$	472.08	−265.22	−210.75	102	192.5

(continued)

Inorganic Substances (*continued*)

Substance	Molar mass, $M/(g \cdot mol^{-1})$	Enthalpy of formation, $\Delta H_f°/(kJ \cdot mol^{-1})$	Gibbs free energy of formation, $\Delta G_f°/(kJ \cdot mol^{-1})$	Molar heat capacity, $C_{P,m}/(J \cdot K^{-1} \cdot mol^{-1})$	Molar entropy,* $S_m°/(J \cdot K^{-1} \cdot mol^{-1})$
Nitrogen					
$N_2(g)$	28.02	0	0	29.12	191.61
$NH_3(g)$	17.03	−46.11	−16.45	35.06	192.45
$NH_3(aq)$	17.03	−80.29	−26.50	—	111.3
$NH_4^+(aq)$	18.04	−132.51	−79.31	—	+113.4
$NO(g)$	30.01	+90.25	+86.55	29.84	210.76
$N_2H_4(l)$	32.05	+50.63	+149.34	139.3	121.21
$NH_2OH(s)$	33.03	−114.2	—	—	—
$HN_3(g)$	43.04	+294.1	+328.1	98.87	238.97
$N_2O(g)$	44.02	+82.05	+104.20	38.45	219.85
$NO_2(g)$	46.01	+33.18	+51.31	37.20	240.06
$NH_4Cl(s)$	53.49	−314.43	−202.87	—	94.6
$NO_3^-(aq)$	62.02	−205.0	−108.74	—	+146.4
$HNO_3(l)$	63.02	−174.10	−80.71	109.87	155.60
$HNO_3(aq)$	63.02	−207.36	−111.25	—	146.4
$NH_4NO_3(s)$	80.05	−365.56	−183.87	84.1	151.08
$N_2O_4(g)$	92.02	+9.16	+97.89	77.28	304.29
$NH_4ClO_4(s)$	117.49	−295.31	−88.75	—	186.2
Oxygen					
$O_2(g)$	32.00	0	0	29.36	205.14
$OH^-(aq)$	17.01	−229.99	−157.24	—	−10.75
$O_3(g)$	48.00	+142.7	+163.2	39.29	238.93
Phosphorus					
$P(s)$, white	30.97	0	0	23.84	41.09
$PH_3(g)$	33.99	+5.4	+13.4	37.11	210.23
$H_3PO_3(aq)$	81.99	−964.8	—	—	—
$H_3PO_4(l)$	97.99	−1266.9	—	—	—
$H_3PO_4(aq)$	97.99	−1288.34	−1142.54	—	158.2
$P_4(g)$	123.88	+58.91	+24.44	67.15	279.98
$PCl_3(l)$	137.32	−319.7	−272.3	—	217.18
$PCl_3(g)$	137.32	−287.0	−267.8	71.84	311.78
$PCl_5(g)$	208.22	−374.9	−305.0	112.8	364.6
$PCl_5(s)$	208.22	−443.5	—	—	—
$P_4O_6(s)$	219.88	−1640	—	—	—
$P_4O_{10}(s)$	283.88	−2984.0	−2697.0	—	228.86
Potassium					
$K(s)$	39.10	0	0	29.58	64.18
$K(g)$	39.10	+89.24	+60.59	20.79	160.34
$K^+(aq)$	39.10	−252.38	−283.27	—	+102.5
$KOH(s)$	56.11	−424.76	−379.08	64.9	78.9
$KOH(aq)$	56.11	−482.37	−440.50	—	91.6
$KF(s)$	58.10	−567.27	−537.75	49.04	66.57
$KCl(s)$	74.55	−436.75	−409.14	51.30	82.59
$K_2S(s)$	110.26	−380.7	−364.0	—	105
$K_2S(aq)$	110.26	−471.5	−480.7	—	190.4
$KBr(s)$	119.00	−393.80	−380.66	52.30	95.90
$KClO_3(s)$	122.55	−397.73	−296.25	100.25	143.1
$KClO_4(s)$	138.55	−432.75	−303.09	112.38	151.0
$KI(s)$	166.00	−327.90	−324.89	52.93	106.32
Silicon					
$Si(s)$	28.09	0	0	20.00	18.83
$SiO_2(s, \alpha)$	60.09	−910.94	−856.64	44.43	41.84

Substance	Molar mass, $M/(\text{g·mol}^{-1})$	Enthalpy of formation, $\Delta H_f^\circ/(\text{kJ·mol}^{-1})$	Gibbs free energy of formation, $\Delta G_f^\circ/(\text{kJ·mol}^{-1})$	Molar heat capacity, $C_{P,m}/(\text{J·K}^{-1}\text{·mol}^{-1})$	Molar entropy,* $S_m^\circ/(\text{J·K}^{-1}\text{·mol}^{-1})$
Silver					
Ag(s)	107.87	0	0	25.35	42.55
Ag$^+$(aq)	107.87	+105.58	+77.11	—	+72.68
AgCl(s)	143.32	−127.07	−109.79	50.79	96.2
AgCl(aq)	143.32	−61.58	−54.12	—	129.3
AgNO$_3$(s)	169.88	−124.39	−33.41	93.05	140.92
AgBr(s)	187.77	−100.37	−96.90	52.38	107.1
AgBr(aq)	187.77	−15.98	−26.86	—	155.2
Ag$_2$O(s)	231.74	−31.05	−11.20	65.86	121.3
AgI(s)	234.77	−61.84	−66.19	56.82	115.5
AgI(aq)	234.77	+50.38	+25.52	—	184.1
Sodium					
Na(s)	22.99	0	0	28.24	51.21
Na(g)	22.99	+107.32	+76.76	20.79	153.71
Na$^+$(aq)	22.99	−240.12	−261.91	—	+59.0
NaOH(s)	40.00	−425.61	−379.49	59.54	64.46
NaOH(aq)	40.00	−470.11	−419.15	—	48.1
NaCl(s)	58.44	−411.15	−384.14	50.50	72.13
NaBr(s)	102.89	−361.06	−348.98	51.38	86.82
NaI(s)	149.89	−287.78	−286.06	52.09	98.53
Sulfur					
S(s), rhombic	32.06	0	0	22.64	31.80
S(s), monoclinic	32.06	+0.33	+0.1	23.6	32.6
S^{2-}(aq)	32.06	+33.1	+85.8	—	−14.6
H$_2$S(g)	34.08	−20.63	−33.56	34.23	205.79
H$_2$S(aq)	34.08	−39.7	−27.83	—	121
SO$_2$(g)	64.06	−296.83	−300.19	39.87	248.22
SO$_3$(g)	80.06	−395.72	−371.06	50.67	256.76
SO$_4^{2-}$(aq)	96.06	−909.27	−744.53	—	+20.1
HSO$_4^-$(aq)	97.07	−887.34	−755.91	—	+131.8
H$_2$SO$_4$(l)	98.08	−813.99	−690.00	138.9	156.90
SF$_6$(g)	146.06	−1209	−1105.3	97.28	291.82
Tin					
Sn(s), white	118.71	0	0	26.99	51.55
Sn(s), gray	118.71	−2.09	+0.13	25.77	44.14
SnO(s)	134.71	−285.8	−256.9	44.31	56.5
SnO$_2$(s)	150.71	−580.7	−519.6	52.59	52.3
Zinc					
Zn(s)	65.41	0	0	25.40	41.63
Zn^{2+}(aq)	65.41	−153.89	−147.06	—	−112.1
ZnO(s)	81.41	−348.28	−318.30	40.25	43.64

*The standard entropies of individual ions in solution are determined by setting the standard entropy of H$^+$ in water equal to 0 and then defining the standard entropies of all other ions relative to this value; hence a negative standard entropy is one that is lower than the standard entropy of H$^+$ in water. All *absolute* entropies are positive, and no sign need be given; all entropies of ions are relative to that of H$^+$ and are listed here with a sign (either + or −).

Organic Compounds

Substance	Molar mass, M/(g·mol^{-1})	Enthalpy of combustion, $\Delta H_c°$/ (kJ·mol^{-1})	Enthalpy of formation, $\Delta H_f°$/ (kJ·mol^{-1})	Gibbs free energy of formation, $\Delta G_f°$/ (kJ·mol^{-1})	Molar heat capacity, $C_{P,m}$/ (J·K^{-1}·mol^{-1})	Molar entropy, $S_m°$/ (J·K^{-1}·mol^{-1})
Hydrocarbons						
CH$_4$(g), methane	16.04	−890	−74.81	−50.72	35.69	186.26
C$_2$H$_2$(g), ethyne (acetylene)	26.04	−1300	+226.73	+209.20	43.93	200.94
C$_2$H$_4$(g), ethene (ethylene)	28.05	−1411	+52.26	+68.15	43.56	219.56
C$_2$H$_6$(g), ethane	30.07	−1560	−84.68	−32.82	52.63	229.60
C$_3$H$_6$(g), propene (propylene)	42.08	−2058	+20.42	+62.78	63.89	266.6
C$_3$H$_6$(g), cyclopropane	42.08	−2091	+53.30	+104.45	55.94	237.4
C$_3$H$_8$(g), propane	44.09	−2220	−103.85	−23.49	73.5	270.2
C$_4$H$_{10}$(g), butane	58.12	−2878	−126.15	−17.03	97.45	310.1
C$_5$H$_{12}$(g), pentane	72.14	−3537	−146.44	−8.20	120.2	349
C$_6$H$_6$(l), benzene	78.11	−3268	+49.0	+124.3	136.1	173.3
C$_6$H$_6$(g)	78.11	−3302	+82.9	+129.72	81.67	269.31
C$_6$H$_{12}$(l), cyclohexane	84.15	−3920	−156.4	+26.7	156.5	204.4
C$_6$H$_{12}$(g)	84.15	−3953	—	—	—	—
C$_7$H$_8$(l), toluene	92.13	−3910	+12.0	+113.8	—	221.0
C$_7$H$_8$(g)	92.13	−3953	+50.0	+122.0	103.6	320.7
C$_8$H$_{18}$(l), octane	114.22	−5471	−249.9	+6.4	—	358
Alcohols and phenols						
CH$_3$OH(l), methanol	32.04	−726	−238.86	−166.27	81.6	126.8
CH$_3$OH(g)	32.04	−764	−200.66	−161.96	43.89	239.81
C$_2$H$_5$OH(l), ethanol	46.07	−1368	−277.69	−174.78	111.46	160.7
C$_2$H$_5$OH(g)	46.07	−1409	−235.10	−168.49	65.44	282.70
C$_6$H$_5$OH(s), phenol	94.11	−3054	−164.6	−50.42	—	144.0
Carboxylic acids						
HCOOH(l), formic acid	46.02	−255	−424.72	−361.35	99.04	128.95
CH$_3$COOH(l), acetic acid	60.05	−875	−484.5	−389.9	124.3	159.8
CH$_3$COOH(aq)	60.05	—	−485.76	−396.46	—	178.7
CH$_3$CO$_2$$^-$(aq)	59.04	—	−486.0	−396.30	—	+86.6
(COOH)$_2$(s), oxalic acid	90.04	−254	−827.2	−697.9	117	120
C$_6$H$_5$COOH(s), benzoic acid	122.12	−3227	−385.1	−245.3	146.8	167.6
Aldehydes and ketones						
HCHO(g), methanal (formaldehyde)	30.03	−571	−108.57	−102.53	35.40	218.77
CH$_3$CHO(l), ethanal (acetaldehyde)	44.05	−1166	−192.30	−128.12	—	160.2
CH$_3$CHO(g)	44.05	−1192	−166.19	−128.86	57.3	250.3
CH$_3$COCH$_3$(l), propanone (acetone)	58.08	−1790	−248.1	−155.4	124.7	200

Substance	Molar mass, $M/(g \cdot mol^{-1})$	Enthalpy of combustion, $\Delta H_c^\circ/$ $(kJ \cdot mol^{-1})$	Enthalpy of formation, $\Delta H_f^\circ/$ $(kJ \cdot mol^{-1})$	Gibbs free energy of formation, $\Delta G_f^\circ/$ $(kJ \cdot mol^{-1})$	Molar heat capacity, $C_{P,m}/$ $(J \cdot K^{-1} \cdot mol^{-1})$	Molar entropy, $S_m^\circ/$ $(J \cdot K^{-1} \cdot mol^{-1})$
Sugars						
$C_6H_{12}O_6(s)$, glucose	180.15	−2808	−1268	−910	—	212
$C_6H_{12}O_6(aq)$	180.15	—	—	−917	—	—
$C_6H_{12}O_6(s)$, fructose	180.15	−2810	−1266	—	—	—
$C_{12}H_{22}O_{11}(s)$, sucrose	342.29	−5645	−2222	−1545	—	360
Nitrogen compounds						
$CH_3NH_2(g)$, methylamine	31.06	−1085	−22.97	+32.16	53.1	243.41
$CO(NH_2)_2(s)$, urea	60.06	−632	−333.51	−197.33	93.14	104.60
$NH_2CH_2COOH(s)$, glycine	75.07	−969	−532.9	−373.4	99.2	103.51
$C_6H_5NH_2(l)$, aniline	93.13	−3393	+31.6	+149.1	—	191.3

2B STANDARD POTENTIALS AT 25 °C

Potentials in Electrochemical Order

Reduction half–reaction	$E°/V$	Reduction half–reaction	$E°/V$
Strongly oxidizing		$NO_3^- + H_2O + 2\,e^- \rightarrow NO_2^- + 2\,OH^-$	+0.01
$H_4XeO_6 + 2\,H^+ + 2\,e^- \rightarrow XeO_3 + 3\,H_2O$	+3.0	$Ti^{4+} + e^- \rightarrow Ti^{3+}$	0.00
$F_2 + 2\,e^- \rightarrow 2\,F^-$	+2.87	$2\,H^+ + 2\,e^- \rightarrow H_2$	0, by definition
$O_3 + 2\,H^+ + 2\,e^- \rightarrow O_2 + H_2O$	+2.07	$Fe^{3+} + 3\,e^- \rightarrow Fe$	−0.04
$S_2O_8^{2-} + 2\,e^- \rightarrow 2\,SO_4^{2-}$	+2.05	$O_2 + H_2O + 2\,e^- \rightarrow HO_2^- + OH^-$	−0.08
$Ag^{2+} + e^- \rightarrow Ag^+$	+1.98	$Pb^{2+} + 2\,e^- \rightarrow Pb$	−0.13
$Co^{3+} + e^- \rightarrow Co^{2+}$	+1.81	$In^+ + e^- \rightarrow In$	−0.14
$H_2O_2 + 2\,H^+ + 2\,e^- \rightarrow 2\,H_2O$	+1.78	$Sn^{2+} + 2\,e^- \rightarrow Sn$	−0.14
$Au^+ + e^- \rightarrow Au$	+1.69	$AgI + e^- \rightarrow Ag + I^-$	−0.15
$Pb^{4+} + 2\,e^- \rightarrow Pb^{2+}$	+1.67	$Ni^{2+} + 2\,e^- \rightarrow Ni$	−0.23
$2\,HClO + 2\,H^+ + 2\,e^- \rightarrow Cl_2 + 2\,H_2O$	+1.63	$V^{3+} + e^- \rightarrow V^{2+}$	−0.26
$Ce^{4+} + e^- \rightarrow Ce^{3+}$	+1.61	$Co^{2+} + 2\,e^- \rightarrow Co$	−0.28
$2\,HBrO + 2\,H^+ + 2\,e^- \rightarrow Br_2 + 2\,H_2O$	+1.60	$In^{3+} + 3\,e^- \rightarrow In$	−0.34
$MnO_4^- + 8\,H^+ + 5\,e^- \rightarrow Mn^{2+} + 4\,H_2O$	+1.51	$Tl^+ + e^- \rightarrow Tl$	−0.34
$Mn^{3+} + e^- \rightarrow Mn^{2+}$	+1.51	$PbSO_4 + 2\,e^- \rightarrow Pb + SO_4^{2-}$	−0.36
$Au^{3+} + 3\,e^- \rightarrow Au$	+1.40	$Ti^{3+} + e^- \rightarrow Ti^{2+}$	−0.37
$Cl_2 + 2\,e^- \rightarrow 2\,Cl^-$	+1.36	$In^{2+} + e^- \rightarrow In^+$	−0.40
$Cr_2O_7^{2-} + 14\,H^+ + 6\,e^- \rightarrow 2\,Cr^{3+} + 7\,H_2O$	+1.33	$Cd^{2+} + 2\,e^- \rightarrow Cd$	−0.40
$O_3 + H_2O + 2\,e^- \rightarrow O_2 + 2\,OH^-$	+1.24	$Cr^{3+} + e^- \rightarrow Cr^{2+}$	−0.41
$O_2 + 4\,H^+ + 4\,e^- \rightarrow 2\,H_2O$	+1.23	$Fe^{2+} + 2\,e^- \rightarrow Fe$	−0.44
$MnO_2 + 4\,H^+ + 2\,e^- \rightarrow Mn^{2+} + 2\,H_2O$	+1.23	$In^{3+} + 2\,e^- \rightarrow In^+$	−0.44
$ClO_4^- + 2\,H^+ + 2\,e^- \rightarrow ClO_3^- + H_2O$	+1.23	$S + 2\,e^- \rightarrow S^{2-}$	−0.48
$Pt^{2+} + 2\,e^- \rightarrow Pt$	+1.20	$In^{3+} + e^- \rightarrow In^{2+}$	−0.49
$Br_2 + 2\,e^- \rightarrow 2\,Br^-$	+1.09	$Ga^+ + e^- \rightarrow Ga$	−0.53
$Pu^{4+} + e^- \rightarrow Pu^{3+}$	+0.97	$O_2 + e^- \rightarrow O_2^-$	−0.56
$NO_3^- + 4\,H^+ + 3\,e^- \rightarrow NO + 2\,H_2O$	+0.96	$U^{4+} + e^- \rightarrow U^{3+}$	−0.61
$2\,Hg^{2+} + 2\,e^- \rightarrow Hg_2^{2+}$	+0.92	$Se + 2\,e^- \rightarrow Se^{2-}$	−0.67
$ClO^- + H_2O + 2\,e^- \rightarrow Cl^- + 2\,OH^-$	+0.89	$Cr^{3+} + 3\,e^- \rightarrow Cr$	−0.74
$Hg^{2+} + 2\,e^- \rightarrow Hg$	+0.85	$Zn^{2+} + 2\,e^- \rightarrow Zn$	−0.76
$NO_3^- + 2\,H^+ + e^- \rightarrow NO_2 + H_2O$	+0.80	$Cd(OH)_2 + 2\,e^- \rightarrow Cd + 2\,OH^-$	−0.81
$Ag^+ + e^- \rightarrow Ag$	+0.80	$2\,H_2O + 2\,e^- \rightarrow H_2 + 2\,OH^-$	−0.83
$Hg_2^{2+} + 2\,e^- \rightarrow 2\,Hg$	+0.79	$Te + 2\,e^- \rightarrow Te^{2-}$	−0.84
$AgF + e^- \rightarrow Ag + F^-$	+0.78	$Cr^{2+} + 2\,e^- \rightarrow Cr$	−0.91
$Fe^{3+} + e^- \rightarrow Fe^{2+}$	+0.77	$Mn^{2+} + 2\,e^- \rightarrow Mn$	−1.18
$BrO^- + H_2O + 2\,e^- \rightarrow Br^- + 2\,OH^-$	+0.76	$V^{2+} + 2\,e^- \rightarrow V$	−1.19
$MnO_4^{2-} + 2\,H_2O + 2\,e^- \rightarrow MnO_2 + 4\,OH^-$	+0.60	$Ti^{2+} + 2\,e^- \rightarrow Ti$	−1.63
$MnO_4^- + e^- \rightarrow MnO_4^{2-}$	+0.56	$Al^{3+} + 3\,e^- \rightarrow Al$	−1.66
$I_2 + 2\,e^- \rightarrow 2\,I^-$	+0.54	$U^{3+} + 3\,e^- \rightarrow U$	−1.79
$I_3^- + 2\,e^- \rightarrow 3\,I^-$	+0.53	$Be^{2+} + 2\,e^- \rightarrow Be$	−1.85
$Cu^+ + e^- \rightarrow Cu$	+0.52	$Mg^{2+} + 2\,e^- \rightarrow Mg$	−2.36
$Ni(OH)_3 + e^- \rightarrow Ni(OH)_2 + OH^-$	+0.49	$Ce^{3+} + 3\,e^- \rightarrow Ce$	−2.48
$O_2 + 2\,H_2O + 4\,e^- \rightarrow 4\,OH^-$	+0.40	$La^{3+} + 3\,e^- \rightarrow La$	−2.52
$ClO_4^- + H_2O + 2\,e^- \rightarrow ClO_3^- + 2\,OH^-$	+0.36	$Na^+ + e^- \rightarrow Na$	−2.71
$Cu^{2+} + 2\,e^- \rightarrow Cu$	+0.34	$Ca^{2+} + 2\,e^- \rightarrow Ca$	−2.87
$Hg_2Cl_2 + 2\,e^- \rightarrow 2\,Hg + 2\,Cl^-$	+0.27	$Sr^{2+} + 2\,e^- \rightarrow Sr$	−2.89
$AgCl + e^- \rightarrow Ag + Cl^-$	+0.22	$Ba^{2+} + 2\,e^- \rightarrow Ba$	−2.91
$Bi^{3+} + 3\,e^- \rightarrow Bi$	+0.20	$Ra^{2+} + 2\,e^- \rightarrow Ra$	−2.92
$SO_4^{2-} + 4\,H^+ + 2\,e^- \rightarrow H_2SO_3 + H_2O$	+0.17	$Cs^+ + e^- \rightarrow Cs$	−2.92
$Cu^{2+} + e^- \rightarrow Cu^+$	+0.15	$Rb^+ + e^- \rightarrow Rb$	−2.93
$Sn^{4+} + 2\,e^- \rightarrow Sn^{2+}$	+0.15	$K^+ + e^- \rightarrow K$	−2.93
$AgBr + e^- \rightarrow Ag + Br^-$	+0.07	$Li^+ + e^- \rightarrow Li$	−3.05
		Strongly reducing	

Reduction half–reaction	$E°/V$	Reduction half–reaction	$E°/V$
$Ag^+ + e^- \rightarrow Ag$	$+0.80$	$In^{2+} + e^- \rightarrow In^+$	-0.40
$Ag^{2+} + e^- \rightarrow Ag^+$	$+1.98$	$In^{3+} + e^- \rightarrow In^{2+}$	-0.49
$AgBr + e^- \rightarrow Ag + Br^-$	$+0.07$	$In^{3+} + 2 e^- \rightarrow In^+$	-0.44
$AgCl + e^- \rightarrow Ag + Cl^-$	$+0.22$	$In^{3+} + 3 e^- \rightarrow In$	-0.34
$AgF + e^- \rightarrow Ag + F^-$	$+0.78$	$K^+ + e^- \rightarrow K$	-2.93
$AgI + e^- \rightarrow Ag + I^-$	-0.15	$La^{3+} + 3 e^- \rightarrow La$	-2.52
$Al^{3+} + 3 e^- \rightarrow Al$	-1.66	$Li^+ + e^- \rightarrow Li$	-3.05
$Au^+ + e^- \rightarrow Au$	$+1.69$	$Mg^{2+} + 2 e^- \rightarrow Mg$	-2.36
$Au^{3+} + 3 e^- \rightarrow Au$	$+1.40$	$Mn^{2+} + 2 e^- \rightarrow Mn$	-1.18
$Ba^{2+} + 2 e^- \rightarrow Ba$	-2.91	$Mn^{3+} + e^- \rightarrow Mn^{2+}$	$+1.51$
$Be^{2+} + 2 e^- \rightarrow Be$	-1.85	$MnO_2 + 4 H^+ + 2 e^- \rightarrow Mn^{2+} + 2 H_2O$	$+1.23$
$Bi^{3+} + 3 e^- \rightarrow Bi$	$+0.20$	$MnO_4^- + e^- \rightarrow MnO_4^{2-}$	$+0.56$
$Br_2 + 2 e^- \rightarrow 2 Br^-$	$+1.09$	$MnO_4^- + 8 H^+ + 5 e^- \rightarrow Mn^{2+} + 4 H_2O$	$+1.51$
$BrO^- + H_2O + 2 e^- \rightarrow Br^- + 2 OH^-$	$+0.76$	$MnO_4^{2-} + 2 H_2O + 2 e^- \rightarrow MnO_2 + 4 OH^-$	$+0.60$
$Ca^{2+} + 2 e^- \rightarrow Ca$	-2.87	$NO_3^- + 2 H^+ + e^- \rightarrow NO_2 + H_2O$	$+0.80$
$Cd^{2+} + 2 e^- \rightarrow Cd$	-0.40	$NO_3^- + 4 H^+ + 3 e^- \rightarrow NO + 2 H_2O$	$+0.96$
$Cd(OH)_2 + 2 e^- \rightarrow Cd + 2 OH^-$	-0.81	$NO_3^- + H_2O + 2 e^- \rightarrow NO_2^- + 2 OH^-$	$+0.01$
$Ce^{3+} + 3 e^- \rightarrow Ce$	-2.48	$Na^+ + e^- \rightarrow Na$	-2.71
$Ce^{4+} + e^- \rightarrow Ce^{3+}$	$+1.61$	$Ni^{2+} + 2 e^- \rightarrow Ni$	-0.23
$Cl_2 + 2 e^- \rightarrow 2 Cl^-$	$+1.36$	$Ni(OH)_3 + e^- \rightarrow Ni(OH)_2 + OH^-$	$+0.49$
$ClO^- + H_2O + 2 e^- \rightarrow Cl^- + 2 OH^-$	$+0.89$	$O_2 + e^- \rightarrow O_2^-$	-0.56
$ClO_4^- + 2 H^+ + 2 e^- \rightarrow ClO_3^- + H_2O$	$+1.23$	$O_2 + 4 H^+ + 4 e^- \rightarrow 2 H_2O$	$+1.23$
$ClO_4^- + H_2O + 2 e^- \rightarrow ClO_3^- + 2 OH^-$	$+0.36$	$O_2 + H_2O + 2 e^- \rightarrow HO_2^- + OH^-$	-0.08
$Co^{2+} + 2 e^- \rightarrow Co$	-0.28	$O_2 + 2 H_2O + 4 e^- \rightarrow 4 OH^-$	$+0.40$
$Co^{3+} + e^- \rightarrow Co^{2+}$	$+1.81$	$O_3 + 2 H^+ + 2 e^- \rightarrow O_2 + H_2O$	$+2.07$
$Cr^{2+} + 2 e^- \rightarrow Cr$	-0.91	$O_3 + H_2O + 2 e^- \rightarrow O_2 + 2 OH^-$	$+1.24$
$Cr_2O_7^{2-} + 14 H^+ + 6 e^- \rightarrow 2 Cr^{3+} + 7 H_2O$	$+1.33$	$Pb^{2+} + 2 e^- \rightarrow Pb$	-0.13
$Cr^{3+} + 3 e^- \rightarrow Cr$	-0.74	$Pb^{4+} + 2 e^- \rightarrow Pb^{2+}$	$+1.67$
$Cr^{3+} + e^- \rightarrow Cr^{2+}$	-0.41	$PbSO_4 + 2 e^- \rightarrow Pb + SO_4^{2-}$	-0.36
$Cs^+ + e^- \rightarrow Cs$	-2.92	$Pt^{2+} + 2 e^- \rightarrow Pt$	$+1.20$
$Cu^+ + e^- \rightarrow Cu$	$+0.52$	$Pu^{4+} + e^- \rightarrow Pu^{3+}$	$+0.97$
$Cu^{2+} + 2 e^- \rightarrow Cu$	$+0.34$	$Ra^{2+} + 2 e^- \rightarrow Ra$	-2.92
$Cu^{2+} + e^- \rightarrow Cu^+$	$+0.15$	$Rb^+ + e^- \rightarrow Rb$	-2.93
$F_2 + 2 e^- \rightarrow 2 F^-$	$+2.87$	$S + 2 e^- \rightarrow S^{2-}$	-0.48
$Fe^{2+} + 2 e^- \rightarrow Fe$	-0.44	$SO_4^{2-} + 4 H^+ + 2 e^- \rightarrow H_2SO_3 + H_2O$	$+0.17$
$Fe^{3+} + 3 e^- \rightarrow Fe$	-0.04	$S_2O_8^{2-} + 2 e^- \rightarrow 2 SO_4^{2-}$	$+2.05$
$Fe^{3+} + e^- \rightarrow Fe^{2+}$	$+0.77$	$Se + 2 e^- \rightarrow Se^{2-}$	-0.67
$Ga^+ + e^- \rightarrow Ga$	-0.53	$Sn^{2+} + 2 e^- \rightarrow Sn$	-0.14
$2 H^+ + 2 e^- \rightarrow H_2$	$\mathbf{0,}$ by definition	$Sn^{4+} + 2 e^- \rightarrow Sn^{2+}$	$+0.15$
$2 HBrO + 2 H^+ + 2 e^- \rightarrow Br_2 + 2 H_2O$	$+1.60$	$Sr^{2+} + 2 e^- \rightarrow Sr$	-2.89
$2 HClO + 2 H^+ + 2 e^- \rightarrow Cl_2 + 2 H_2O$	$+1.63$	$Te + 2 e^- \rightarrow Te^{2-}$	-0.84
$2 H_2O + 2 e^- \rightarrow H_2 + 2 OH^-$	-0.83	$Ti^{2+} + 2 e^- \rightarrow Ti$	-1.63
$H_2O_2 + 2 H^+ + 2 e^- \rightarrow 2 H_2O$	$+1.78$	$Ti^{3+} + e^- \rightarrow Ti^{2+}$	-0.37
$H_4XeO_6 + 2 H^+ + 2 e^- \rightarrow XeO_3 + 3 H_2O$	$+3.0$	$Ti^{4+} + e^- \rightarrow Ti^{3+}$	0.00
$Hg_2^{2+} + 2 e^- \rightarrow 2 Hg$	$+0.79$	$Tl^+ + e^- \rightarrow Tl$	-0.34
$Hg^{2+} + 2 e^- \rightarrow Hg$	$+0.85$	$U^{3+} + 3 e^- \rightarrow U$	-1.79
$2 Hg^{2+} + 2 e^- \rightarrow Hg_2^{2+}$	$+0.92$	$U^{4+} + e^- \rightarrow U^{3+}$	-0.61
$Hg_2Cl_2 + 2 e^- \rightarrow 2 Hg + 2 Cl^-$	$+0.27$	$V^{2+} + 2 e^- \rightarrow V$	-1.19
$I_2 + 2 e^- \rightarrow 2 I^-$	$+0.54$	$V^{3+} + e^- \rightarrow V^{2+}$	-0.26
$I_3^- + 2 e^- \rightarrow 3 I^-$	$+0.53$	$Zn^{2+} + 2 e^- \rightarrow Zn$	-0.76
$In^+ + e^- \rightarrow In$	-0.14		

2C GROUND-STATE ELECTRON CONFIGURATIONS*

Z	Symbol	Configuration	Z	Symbol	Configuration
1	H	$1s^1$	58	Ce	$[Xe]4f^15d^16s^2$
2	He	$1s^2$	59	Pr	$[Xe]4f^36s^2$
3	Li	$[He]2s^1$	60	Nd	$[Xe]4f^46s^2$
4	Be	$[He]2s^2$	61	Pm	$[Xe]4f^56s^2$
5	B	$[He]2s^22p^1$	62	Sm	$[Xe]4f^66s^2$
6	C	$[He]2s^22p^2$	63	Eu	$[Xe]4f^76s^2$
7	N	$[He]2s^22p^3$	64	Gd	$[Xe]4f^75d^16s^2$
8	O	$[He]2s^22p^4$	65	Tb	$[Xe]4f^96s^2$
9	F	$[He]2s^22p^5$	66	Dy	$[Xe]4f^{10}6s^2$
10	Ne	$[He]2s^22p^6$	67	Ho	$[Xe]4f^{11}6s^2$
11	Na	$[Ne]3s^1$	68	Er	$[Xe]4f^{12}6s^2$
12	Mg	$[Ne]3s^2$	69	Tm	$[Xe]4f^{13}6s^2$
13	Al	$[Ne]3s^23p^1$	70	Yb	$[Xe]4f^{14}6s^2$
14	Si	$[Ne]3s^23p^2$	71	Lu	$[Xe]4f^{14}5d^16s^2$
15	P	$[Ne]3s^23p^3$	72	Hf	$[Xe]4f^{14}5d^26s^2$
16	S	$[Ne]3s^23p^4$	73	Ta	$[Xe]4f^{14}5d^36s^2$
17	Cl	$[Ne]3s^23p^5$	74	W	$[Xe]4f^{14}5d^46s^2$
18	Ar	$[Ne]3s^23p^6$	75	Re	$[Xe]4f^{14}5d^56s^2$
19	K	$[Ar]4s^1$	76	Os	$[Xe]4f^{14}5d^66s^2$
20	Ca	$[Ar]4s^2$	77	Ir	$[Xe]4f^{14}5d^76s^2$
21	Sc	$[Ar]3d^14s^2$	78	Pt	$[Xe]4f^{14}5d^96s^1$
22	Ti	$[Ar]3d^24s^2$	79	Au	$[Xe]4f^{14}5d^{10}6s^1$
23	V	$[Ar]3d^34s^2$	80	Hg	$[Xe]4f^{14}5d^{10}6s^2$
24	Cr	$[Ar]3d^54s^1$	81	Tl	$[Xe]4f^{14}5d^{10}6s^26p^1$
25	Mn	$[Ar]3d^54s^2$	82	Pb	$[Xe]4f^{14}5d^{10}6s^26p^2$
26	Fe	$[Ar]3d^64s^2$	83	Bi	$[Xe]4f^{14}5d^{10}6s^26p^3$
27	Co	$[Ar]3d^74s^2$	84	Po	$[Xe]4f^{14}5d^{10}6s^26p^4$
28	Ni	$[Ar]3d^84s^2$	85	At	$[Xe]4f^{14}5d^{10}6s^26p^5$
29	Cu	$[Ar]3d^{10}4s^1$	86	Rn	$[Xe]4f^{14}5d^{10}6s^26p^6$
30	Zn	$[Ar]3d^{10}4s^2$	87	Fr	$[Rn]7s^1$
31	Ga	$[Ar]3d^{10}4s^24p^1$	88	Ra	$[Rn]7s^2$
32	Ge	$[Ar]3d^{10}4s^24p^2$	89	Ac	$[Rn]6d^17s^2$
33	As	$[Ar]3d^{10}4s^24p^3$	90	Th	$[Rn]6d^27s^2$
34	Se	$[Ar]3d^{10}4s^24p^4$	91	Pa	$[Rn]5f^26d^17s^2$
35	Br	$[Ar]3d^{10}4s^24p^5$	92	U	$[Rn]5f^36d^17s^2$
36	Kr	$[Ar]3d^{10}4s^24p^6$	93	Np	$[Rn]5f^46d^17s^2$
37	Rb	$[Kr]5s^1$	94	Pu	$[Rn]5f^67s^2$
38	Sr	$[Kr]5s^2$	95	Am	$[Rn]5f^77s^2$
39	Y	$[Kr]4d^15s^2$	96	Cm	$[Rn]5f^76d^17s^2$
40	Zr	$[Kr]4d^25s^2$	97	Bk	$[Rn]5f^97s^2$
41	Nb	$[Kr]4d^45s^1$	98	Cf	$[Rn]5f^{10}7s^2$
42	Mo	$[Kr]4d^55s^1$	99	Es	$[Rn]5f^{11}7s^2$
43	Tc	$[Kr]4d^55s^2$	100	Fm	$[Rn]5f^{12}7s^2$
44	Ru	$[Kr]4d^75s^1$	101	Md	$[Rn]5f^{13}7s^2$
45	Rh	$[Kr]4d^85s^1$	102	No	$[Rn]5f^{14}7s^2$
46	Pd	$[Kr]4d^{10}$	103	Lr	$[Rn]5f^{14}6d^17s^2$
47	Ag	$[Kr]4d^{10}5s^1$	104	Rf	$[Rn]5f^{14}6d^27s^2$ (?)
48	Cd	$[Kr]4d^{10}5s^2$	105	Db	$[Rn]5f^{14}6d^37s^2$ (?)
49	In	$[Kr]4d^{10}5s^25p^1$	106	Sg	$[Rn]5f^{14}6d^47s^2$ (?)
50	Sn	$[Kr]4d^{10}5s^25p^2$	107	Bh	$[Rn]5f^{14}6d^57s^2$ (?)
51	Sb	$[Kr]4d^{10}5s^25p^3$	108	Hs	$[Rn]5f^{14}6d^67s^2$ (?)
52	Te	$[Kr]4d^{10}5s^25p^4$	109	Mt	$[Rn]5f^{14}6d^77s^2$ (?)
53	I	$[Kr]4d^{10}5s^25p^5$	110	Ds	$[Rn]5f^{14}6d^87s^2$ (?)
54	Xe	$[Kr]4d^{10}5s^25p^6$	111	Rg	$[Rn]5f^{14}6d^{10}7s^1$ (?)
55	Cs	$[Xe]6s^1$	112	Cn	$[Rn]5f^{14}6d^{10}7s^2$ (?)
56	Ba	$[Xe]6s^2$	114	Fl	$[Rn]5f^{14}6d^{10}7s^2\,7p^2$(?)
57	La	$[Xe]5d^16s^2$	116	Lv	$[Rn]5f^{14}6d^{10}7s^2\,7p^4$(?)

*The electron configurations followed by a question mark are speculative.

2D THE ELEMENTS

Element	Symbol	Atomic Number Z	Molar mass $M/(\text{g·mol}^{-1})$	Normal state	Density $d/(\text{g·cm}^{-3})$	Melting point/°C	Boiling point/°C	Ionization energies $I/(\text{kJ·mol}^{-1})$	Electron affinity $E_{ea}/(\text{kJ·mol}^{-1})$	Electronegativity χ	Principal oxidation numbers	Atomic radius r/pm	Ionic radius r/pm
actinium (Greek *aktis*, ray)	Ac	89	(227)	s, m	10.07	1050	3200	499, 1170, 1900	—	1.1	+3	188	118(3+)
aluminium (from alum, salts of the form $KAl(SO_4)_2 \cdot 12H_2O$)	Al	13	26.98	s, m	2.70	660	2467	577, 1817, 2744	+43	1.6	+3	143	54(3+)
americium (the Americas)	Am	95	(243)	s, m	13.67	990	2600	578	—	1.3	+3	173	107(3+)
antimony (from the Greek *anti-monos*, not alone; Latin *stibium*)	Sb	51	121.76	s, md	6.69	631	1750	834, 1794, 2443	+103	2.1	−3, +3, +5	141	89(3+)
argon (Greek *argos*, inactive)	Ar	18	39.95	g, nm	1.66‡	−189	−186	1520	<0	—	0	174	—
arsenic (Greek *arsenikos*, male)	As	33	74.92	s, md	5.78	613§	—	947, 1798	+78	2.2	−3, +3, +5	125	222(3−)
astatine (Greek *astatos*, unstable)	At	85	(210)	s, nm	—	300	350	1037, 1600	+270	2.0	−1	—	227(1−)
barium (Greek *barys*, heavy)	Ba	56	137.33	s, m	3.59	710	1640	502, 965	+14	0.89	+2	217	135(2+)
berkelium (Berkeley, California)	Bk	97	(247)	s, m	14.79	986	—	601	—	1.3	+3	—	87(4+)
beryllium (from the mineral beryl, $Be_3Al_2SiO_{18}$)	Be	4	9.01	s, m	1.85	1285	2470	900, 1757	<0	1.6	+2	113	34(2+)
bismuth (German *weisse masse*, white mass)	Bi	83	208.98	s, m	8.90	271	1650	703, 1610, 2466	+91	2	+3, +5	155	96(3+)
bohrium (Niels Bohr)	Bh	107	(264)	—	—	—	—	660	—	—	+5	128#	83(5+)#
boron (Arabic *buraq*, borax, $Na_2B_4O_7 \cdot 10H_2O$; *bor*(ax) + (carb)*on*)	B	5	10.81	s, md	2.47	2300	3931	799, 2427, 3660	+27	2	+3	88	23(3+)
bromine (Greek *bromos*, stench)	Br	35	79.90	l, nm	3.12	−7	59	1140, 2104	+325	3	−1, +1, +3, +4, +5, +7	114	196(1−)
cadmium (Greek *Cadmus*, founder of Thebes)	Cd	48	112.41	s, m	8.65	321	765	868, 1631	<0	1.7	+2	149	103(2+)
calcium (Latin *calx*, lime)	Ca	20	40.08	s, m	1.53	840	1490	590, 1145, 4910	+2	1.3	+2	197	100(2+)
californium (California)	Cf	98	(251)	s, m	—	—	—	608	—	1.3	+3	169	117(2+)
carbon (Latin *carbo*, coal or charcoal)	C	6	12.01	s, nm	2.27	3700§	—	1090, 2352, 4620	+122	2.6	−4, −1, +2, +4	77	260(4−)

(continued)

Element	Symbol	Atomic Number, Z	Molar mass, M/(g·mol⁻¹)	Normal† state	Density, d/(g·cm⁻³)	Melting Point/°C	Boiling Point/°C	Ionization energies, I/(kJ·mol⁻¹)	Electron affinity, E_{ea}/(kJ·mol⁻¹)	Electronegativity, χ	Principal oxidation numbers	Atomic radius, r/pm	Ionic radius, r/pm
cerium (the asteroid Ceres, discovered 2 days earlier)	Ce	58	140.12	s, m	6.71	800	3000	527, 1047, 1949	<50	1.1	+3, +4	183	107(3+)
cesium (Latin *caesius*, sky blue)	Cs	55	132.91	s, m	1.87	28	678	376, 2420	+46	0.79	+1	265	167(1+)
chlorine (Greek *chloros*, yellowish green)	Cl	17	35.45	g, nm	1.66‡	−101	239	1255, 2297	+349	3.2	−1, +1, +3, +4, +5, +6, +7	99	181(1−)
chromium (Greek *chroma*, color)	Cr	24	52.00	s, m	7.19	1860	2600	653, 1592, 2987	+64	1.7	+2, +3	125	84(2+)
cobalt (German *Kobold*, evil spirit; Greek *kobalos*, goblin)	Co	27	58.93	s, m	8.80	1494	2900	760, 1646, 3232	+64	1.9	+3, +6	125	64(3+)
copernicium (Nicolaus Copernicus)	Cn	112	(285)	—	—	—	—	—	—	—	—	—	—
copper (Latin *cuprum*, from Cyprus)	Cu	29	63.55	s, m	8.93	1083	2567	785, 1958, 3554	+118	1.9	+1, +2	128	72(2+)
curium (Marie Curie)	Cm	96	(247)	s, m	13.30	1340	—	581	—	1.3	+3	174	99(3+)
darmstadtium (Darmstadt, a town in Germany)	Ds	110	—	—	—	—	—	—	—	—	—	—	—
dubnium (Dubna)	Db	105	(262)	s, m	29	—	—	640	—	—	+5	139#	68(5+)#
dysprosium (Greek *dysprositos*, hard to get at)	Dy	66	162.50	s, m	8.53	1410	2600	572, 1126, 2200	—	1.2	+3	177	91(3+)
einsteinium (Albert Einstein)	Es	99	(252)	s, m	—	—	—	619	<50	1.3	+3	203	98(3+)
erbium (Ytterby, a town in Sweden)	Er	68	167.26	s, m	9.04	1520	2600	589, 1151, 2194	<50	1.2	+3	176	89(3+)
europium (Europe)	Eu	63	151.96	s, m	5.25	820	1450	547, 1085, 2404	<50	—	+3	204	98(3+)
fermium (Enrico Fermi, an Italian physicist)	Fm	100	(257)	s, m	—	—	—	627	—	1.3	+3	—	91(3+)
flerovium (Georgy Flyorov, a Russian physicist)	Fl	114	(298)	—	—	—	—	—	—	—	—	—	—
fluorine (Latin *fluere*, to flow)	F	9	19.00	g, nm	1.51‡	−220	−188	1680, 3374	+328	4.0	−1	58	133(1−)
francium (France)	Fr	87	(223)	s, m	—	27	677	400	+44	0.7	+1	270	180(1+)
gadolinium (Johann Gadolin, a Finnish chemist)	Gd	64	157.25	s, m	7.87	1310	3000	592, 1167, 1990	<50	1.2	+2, +3	180	97(3+)

Element (derivation)	Symbol	Z	Molar mass	State	Density	m.p.	b.p.	Ionization energy	Electron affinity	Electronegativity	Oxidation states	Atomic radius	Ionic radius
gallium (Latin *Gallia*, France)	Ga	31	69.72	s, m	5.91	30	2403	577, 1979, 2963	+29	1.6	+1, +3	122	62(3+)
germanium (Latin *Germania*, Germany)	Ge	32	72.64	s, md	5.32	937	2830	784, 1557, 3302	+116	2.0	+2, +4	122	90(2+)
gold (Anglo–Saxon *gold*; Latin *aurum*, gold)	Au	79	196.97	s, m	19.28	1064	2807	890, 1980	+223	2.5	+1, +3	144	91(3+)
hafnium (Latin *Hafnia*, Copenhagen)	Hf	72	178.49	s, m	13.28	2230	5300	642, 1440, 2250	0	1.3	+4	156	84(3+)
hassium (Hesse, the German state)	Hs	108	(277)	—	—	—	—	750	—	—	+3	126[#]	80(4+)[#]
helium (Greek *helios*, the sun)	He	2	4.00	g, nm	0.12‡	—	−269	2370, 5250	<0	—	0	128	—
holmium (Latin *Holmia*, Stockholm)	Ho	67	164.93	s, m	8.8	1470	2300	581, 1139	<50	1.2	+3	177	89(3+)
hydrogen (Greek *hydro* + *genes*, water–forming)	H	1	1.0079	g, nm	0.070‡	−259	−253	1310	+73	2.2	−1, +1	30	154(1−)
indium (from the bright indigo line in its spectrum)	In	49	114.82	s, m	7.29	156	2080	556, 1821	+29	1.8	+1, +3	163	80(3+)
iodine (Greek *ioeidēs*, violet)	I	53	126.90	s, nm	4.95	114	184	1008, 1846	+295	2.7	−1, +1, +3, +5, +7	133	220(1−)
iridium (Greek and Latin *iris*, rainbow)	Ir	77	192.22	s, m	22.56	2447	4550	880	+151	2.2	+3, +4	136	75(3+)
iron (Anglo–Saxon *iron*; Latin *ferrum*)	Fe	26	55.84	s, m	7.87	1540	2760	759, 1561, 2957	+16	1.8	+2, +3	124	82(2+)
krypton (Greek *kryptos*, hidden)	Kr	36	83.80	g, nm	3.00‡	−157	−153	1350, 2350	<0	—	+2	189	169(1+)
lanthanum (Greek *lanthanein*, to lie hidden)	La	57	138.91	s, m	6.17	920	3450	538, 1067, 1850	+50	1.1	+3	188	122(3+)
lawrencium (Ernest Lawrence, an American physicist)	Lr	103	(262)	s, m	—	—	—	—	—	1.3	+3	—	88(3+)
lead (Anglo–Saxon *lead*; Latin *plumbum*)	Pb	82	207.2	s, m	11.34	328	1760	716, 1450	+35	2.3	+2, +4	175	132(2+)
lithium (Greek *lithos*, stone)	Li	3	6.94	s, m	0.53	181	1347	519, 7298	+60	1.0	+1	152	76(1+)
livermorium (Lawrence Livermore National Laboratory)	Lv	116	(293)	—	—	—	—	—	—	—	—	—	—
lutetium (*Lutetia*, ancient name of Paris)	Lu	71	174.97	s, m	9.84	1700	3400	524, 1340, 2022	<50	1.3	+3	173	85(3+)
magnesium (Magnesia, a district in Thessaly, Greece)	Mg	12	24.31	s, m	1.74	650	1100	736, 1451	<0	1.3	+2	160	72(2+)
manganese (Greek and Latin *magnes*, magnet)	Mn	25	54.94	s, m	7.47	1250	2120	717, 1509	<0	1.6	+2, +3, +4, +7	137	91(2+)
meitnerium (Lise Meitner)	Mt	109	(268)	—	—	—	—	840	—	—	+2	—	83(2+)
mendelevium (Dmitri Mendeleev)	Md	101	(258)	—	—	—	—	635	—	1.3	+3	—	90(3+)

(continued)

Element	Symbol	Atomic Number, Z	Molar mass, $M/(\text{g}\cdot\text{mol}^{-1})$	Normal* state	Density, $d/(\text{g}\cdot\text{cm}^{-3})$	Melting Point/°C	Boiling Point/°C	Ionization energies, $I/(\text{kJ}\cdot\text{mol}^{-1})$	Electron affinity, $E_{ea}/(\text{kJ}\cdot\text{mol}^{-1})$	Electronegativity, χ	Principal oxidation numbers	Atomic radius, r/pm	Ionic radius, r/pm
mercury (the planet Mercury; Latin *hydrargyrum*, liquid silver)	Hg	80	200.59	l, m	13.55	−39	357	1007, 1810	−18	2	+1, +2	160	112(2+)
molybdenum (Greek *molybdos*, lead)	Mo	42	95.94	s, m	10.22	2620	4830	685, 1558, 2621	+72	2.2	+4, +5, +6	136	92(2+)
neodymium (Greek *neos + didymos*, new twin)	Nd	60	144.24	s, m	7.00	1024	3100	530, 1035	<0	1.1	+3	182	104(3+)
neon (Greek *neos*, new)	Ne	10	20.18	g, nm	1.44‡	−249	−246	2080, 3952	0	—	0	—	—
neptunium (the planet Neptune)	Np	93	(237)	s, m	20.45	640	—	597	—	1.4	+5	150	88(5+)
nickel (German *Nickel*, Old Nick, Satan)	Ni	28	58.69	s, m	8.91	1455	2732	737, 1753	+156	1.9	+2, +3	125	78(2+)
niobium (Niobe, daughter of Tantalus; see tantalum)	Nb	41	92.91	s, m	8.57	2425	5000	664, 1382	+86	1.6	+5	143	69(5+)
nitrogen (Greek *nitron + genes*, soda-forming)	N	7	14.01	g, nm	1.04‡	−210	−196	1400, 2856	−7	3.0	−3, +3, +5	75	171(3−)
nobelium (Alfred Nobel, the founder of the Nobel prizes)	No	102	(259)	s, m	—	—	—	642	—	1.3	+2	—	113(2+)
osmium (Greek *osme*, a smell)	Os	76	190.23	s, m	22.58	3030	5000	840	+106	2.2	+3, +4	135	81(3+)
oxygen (Greek *oxys + genes*, acid forming)	O	8	16.00	g, nm	1.14‡	−218	−183	1310, 3388	+141, −844	3.4	−2	66	140(2−)
palladium (the asteroid Pallas, discovered at about the same time)	Pd	46	106.42	s, m	12.00	1554	3000	805, 1875	+54	2.2	+2, +4	138	86(2+)
phosphorus (Greek *phosphoros*, light bearing)	P	15	30.97	s, nm	1.82	44	280	1011, 1903, 2912	+72	2.2	−3, +3, +5	110	212(3−)
platinum (Spanish *plata*, silver)	Pt	78	195.08	s, m	21.45	1772	3720	870, 1791	+205	2.3	+2, +4	138	85(2+)
plutonium (the planetlike Pluto)	Pu	94	(244)	s, m	19.81	640	3200	585	—	1.3	+3, +4	151	108(3+)
polonium (Poland)	Po	84	(209)	s, md	9.40	254	960	812	+174	2.0	+2, +4	167	65(4+)
potassium (from potash; Latin *kalium* and Arabic *qali*, alkali)	K	19	39.10	s, m	0.86	64	774	418, 3051	+48	0.82	+1	227	138(1+)

Element (etymology)	Symbol	Z	Atomic mass	State	Density	m.p.	b.p.	Ionization	Electron affinity	Electroneg.	Oxidation states	Atomic radius	Ionic radius
praseodymium (Greek *prasios* + *didymos*, green twin)	Pr	59	140.91	s, m	6.78	935	3000	523, 1018	<50	1.1	+3	183	106(3+)
promethium (Prometheus, the Greek god)	Pm	61	(145)	s, m	7.22	1168	3300	536, 1052	<50	—	+3	181	106(3+)
protactinium (Greek *protos* + *aktis*, first ray)	Pa	91	231.04	s, m	15.37	1200	4000	568	—	1.5	+5	161	89(5+)
radium (Latin *radius*, ray)	Ra	88	(226)	s, m	5.00	700	1500	509, 979	—	0.9	+2	223	152(2+)
radon (from radium)	Rn	86	(222)	g, nm	4.40‡	−71	−62	1036, 1930	<0	—	+2	—	—
rhenium (Latin *Rhenus*, Rhine)	Re	75	186.21	s, m	21.02	3180	5600	760, 1260	+14	1.9	+4, +7	137	72(4+)
rhodium (Greek *rhodon*, rose; its aqueous solutions are often rose-colored)	Rh	45	102.9	s, m	12.42	1963	3700	720, 1744	+110	2.3	+3	134	75(3+)
roentgenium (W. Roentgen, discoverer of x-rays)	Rg	111	—	—	—	—	—	—	—	—	—	—	—
rubidium (Latin *rubidus*, deep red, "flushed")	Rb	37	85.47	s, m	1.53	39	688	402, 2632	+47	0.82	+1	248	152(1+)
ruthenium (Latin *Ruthenia*, Russia)	Ru	44	101.07	s, m	12.36	2310	4100	711, 1617	+101	2.2	+2, +3, +4	134	77(3+)
rutherfordium (Ernest Rutherford)	Rf	104	(261)	—	—	—	—	490	—	—	+4	150#	67(4+)#
samarium (from samarskite, a mineral)	Sm	62	150.36	s, m	7.54	1060	1600	543, 1068	<50	1.2	+3	180	100(3+)
scandium (Latin *Scandia*, Scandinavia)	Sc	21	44.96	s, m	2.99	1540	2800	631, 1235	+18	1.4	+3	161	83(3+)
seaborgium (Glenn Seaborg)	Sg	106	(266)	—	—	—	—	730	—	—	+6	132#	86(5+)#
selenium (Greek *selēnē*, the moon)	Se	34	78.96	s, nm	4.79	220	685	941, 2044	+195	2.6	−2, +4, +6	117	198(2−)
silicon (Latin *silex*, flint)	Si	14	28.09	s, md	2.33	1410	2620	786, 1577	+134	1.9	+4	117	26(4+)
silver (Anglo-Saxon *seolfor*; Latin *argentum*)	Ag	47	107.87	s, m	10.50	962	2212	731, 2073	+126	1.9	+1	144	113(1+)
sodium (English *soda*; Latin *natrium*)	Na	11	22.99	s, m	0.97	98	883	494, 4562	+53	0.93	+1	154	102(1+)
strontium (Strontian, Scotland)	Sr	38	87.62	s, m	2.58	770	1380	548, 1064	+5	0.95	+2	215	118(2+)
sulfur (Sanskrit *sulvere*)	S	16	32.06	s, nm	2.09	115	445	1000, 2251	+200, −532	2.6	−2, +4, +6	104	184(2−)
tantalum (Tantalos, Greek mythological figure)	Ta	73	180.95	s, m	16.65	3000	5400	761	+14	1.5	+5	143	72(3+)
technetium (Greek *technētos*, artificial)	Tc	43	(98)	s, m	11.50	2200	4600	702, 1472	+96	1.9	+4, +7	136	72(4+)
tellurium (Latin *tellus*, earth)	Te	52	127.60	s, md	6.25	450	990	870, 1775	+190	2.1	−2, +4	137	221(2−)
terbium (Ytterby, a town in Sweden)	Tb	65	158.93	s, m	8.27	1360	2500	565, 1112	<50	—	+3	178	97(3+)

(continued)

Element	Symbol	Atomic Number, Z	Molar mass, M/(g·mol⁻¹)	Normal[†] state	Density, d/(g·cm⁻³)	Melting Point/°C	Boiling Point/°C	Ionization energies, I/(kJ·mol⁻¹)	Electron affinity, E_{ea}/(kJ·mol⁻¹)	Electronegativity, χ	Principal oxidation numbers	Atomic radius, r/pm	Ionic radius, r/pm[∥]
thallium (Greek *thallos*, a green shoot)	Tl	81	204.38	s, m	11.87	304	1457	590, 1971	+19	2.0	+1, +3	170	105(3+)
thorium (Thor, Norse god)	Th	90	232.04	s, m	11.73	1700	4500	587, 1110	—	1.3	+4	180	99(4+)
thulium (Thule, early name for Scandinavia)	Tm	69	168.93	s, m	9.33	1550	2000	597, 1163	<50	1.2	+3	175	94(3+)
tin (Anglo-Saxon *tin*; Latin *stannum*)	Sn	50	118.71	s, m	7.29	232	2720	707, 1412	+116	2.0	+2, +4	141	93(2+)
titanium (Titans, Greek mythological figures, sons of the Earth)	Ti	22	47.87	s, m	4.55	1660	3300	658, 1310	+7.6	1.5	+4	145	69(4+)
tungsten (Swedish *tung* + *sten*, heavy stone; from wolframite)	W	74	183.84	s, m	19.30	3387	5420	770	+79	2.4	+5, +6	137	62(6+)
uranium (the planet Uranus)	U	92	238.03	s, m	18.95	1135	4000	584, 1420	—	1.4	+6	154	80(6+)
vanadium (Vanadis, Scandinavian mythological figure)	V	23	50.94	s, m	6.11	1920	3400	650, 1414	+51	1.6	+4, +5	132	61(4+)
xenon (Greek *xenos*, stranger)	Xe	54	131.29	g, nm	3.56‡	−112	−108	1170, 2046	<0	2.6	+2, +4, +6	218	190(1+)
ytterbium (Ytterby, a town in Sweden)	Yb	70	173.04	s, m	6.97	824	1500	603, 1176	<50	—	+3	194	86(3+)
yttrium (Ytterby, a town in Sweden)	Y	39	88.91	s, m	4.48	1510	3300	616, 1181	+30	1.2	+3	181	106(3+)
zinc (Anglo-Saxon *zinc*)	Zn	30	65.41	s, m	7.14	420	907	906, 1733	+9	1.6	+2	133	83(2+)
zirconium (Arabic *zargun*, gold color)	Zr	40	91.22	s, m	6.51	1850	4400	660, 1267	+41	1.3	+4	160	87(4+)

*Parentheses around molar mass indicate the most stable isotope of a radioactive element.
[†]The normal state is the state of the element at normal temperature and pressure (20 °C and 1 atm). s denotes solid, l, liquid, and g, gas; m denotes metal, nm, nonmetal, and md, metalloid.
‡The density quoted is for the liquid.
§The solid sublimes.
∥Charge in parentheses.
#Atomic and ionic radii are estimated.

3A THE NOMENCLATURE OF POLYATOMIC IONS

Charge number	Chemical formula	Name	Oxidation number of central element	Charge number	Chemical formula	Name	Oxidation number of central element
+2	Hg_2^{2+}	mercury(I)	+1		O_3^-	ozonide	$-\frac{1}{3}$
	UO_2^{2+}	uranyl	+6		OH^-	hydroxide	$-2(O)$
	VO^{2+}	vanadyl	+4		SCN^-	thiocyanate	—
+1	NH_4^+	ammonium	−3	−2	C_2^{2-}	carbide (acetylide)	−1
	PH_4^+	phosphonium	−3				
−1	$CH_3CO_2^-$	acetate (ethanoate)	0(C)		CO_3^{2-}	carbonate	+4
					$C_2O_4^{2-}$	oxalate	+3
	HCO_2^-	formate (methanoate)	+2(C)		CrO_4^{2-}	chromate	+6
					$Cr_2O_7^{2-}$	dichromate	+6
	CN^-	cyanide	+2(C), −3(N)		O_2^{2-}	peroxide	−1
	ClO_4^-	perchlorate*	+7		S_2^{2-}	disulfide	−1
	ClO_3^-	chlorate*	+5		SiO_3^{2-}	metasilicate	+4
	ClO_2^-	chlorite*	+3		SO_4^{2-}	sulfate	+6
	ClO^-	hypochlorite*	+1(Cl)		SO_3^{2-}	sulfite	+4
	MnO_4^-	permanganate	+7		$S_2O_3^{2-}$	thiosulfate	+2
	NO_3^-	nitrate	+5		AsO_4^{3-}	arsenate	+5
	NO_2^-	nitrite	+3	−3	BO_3^{3-}	borate	+3
	N_3^-	azide	$-\frac{1}{3}$		PO_4^{3-}	phosphate	+5

*These names are representative of the halogen oxoanions.

When a hydrogen ion bonds to a −2 or −3 anion, add "hydrogen" before the name of the anion. For example, HSO_3^- is hydrogen sulfite (or hydrogensulfite). If two hydrogen ions bond to a −3 anion, add "dihydrogen" before the name of the anion. For example, $H_2PO_4^-$ is dihydrogen phosphate (or dihydrogenphosphate).

Oxoacids and Oxoanions

The names of oxoanions and their parent acids can be determined by noting the oxidation number of the central atom and then referring to the table shown here. For example, the nitrogen in $N_2O_2^{2-}$ has an oxidation number of +1; because nitrogen belongs to Group 15, the ion is a hyponitrite ion.

Group number					
14	15	16	17	Oxoanion	Oxoacid
—	—	—	+7	per . . . ate	per . . . ic acid
+4	+5	+6	+5	. . . ate	. . . ic acid
—	+3	+4	+3	. . . ite	. . . ous acid
—	+1	+2	+1	hypo . . . ite	hypo . . . ous acid

3B COMMON NAMES OF CHEMICALS

Many chemicals are often referred to by their common names, sometimes as a result of their use over hundreds of years and sometimes because they appear on the labels of consumer products, such as detergents, beverages, and antacids. The following names are just a few that have found their way into the language of everyday life.

Common name	Formula	Chemical name
baking soda	$NaHCO_3$	sodium hydrogen carbonate (sodium bicarbonate)
bleach, laundry	$NaClO$	sodium hypochlorite
borax	$Na_2B_4O_7 \cdot 10H_2O$	sodium tetraborate decahydrate
brimstone	S_8	sulfur
calamine	$ZnCO_3$	zinc carbonate
chalk	$CaCO_3$	calcium carbonate
Epsom salts	$MgSO_4 \cdot 7H_2O$	magnesium sulfate heptahydrate
fool's gold, pyrite	FeS_2	iron(II) disulfide
gypsum	$CaSO_4 \cdot 2H_2O$	calcium sulfate dihydrate
lime (quicklime)	CaO	calcium oxide
lime (slaked lime)	$Ca(OH)_2$	calcium hydroxide
limestone	$CaCO_3$	calcium carbonate
lye, caustic soda	$NaOH$	sodium hydroxide
marble	$CaCO_3$	calcium carbonate
milk of magnesia	$Mg(OH)_2$	magnesium hydroxide
plaster of Paris	$CaSO_4 \cdot \frac{1}{2}H_2O$	calcium sulfate hemihydrate
potash*	K_2CO_3	potassium carbonate
quartz	SiO_2	silcon dioxide
table salt	$NaCl$	sodium chloride
vinegar	CH_3COOH	acetic acid (ethanoic acid)
washing soda	$Na_2CO_3 \cdot 10H_2O$	sodium carbonate decahydrate

*Potash also refers collectively to K_2CO_3, KOH, K_2SO_4, KCl, and KNO_3.

3C TRADITIONAL NAMES OF SOME COMMON CATIONS WITH VARIABLE CHARGE NUMBERS

Modern nomenclature includes the oxidation number of elements with variable oxidation states in the names of their compounds, as in cobalt(II) chloride. However, the traditional nomenclature, in which the suffixes −ous and −ic are used, is still encountered. The table translates from one system into the other for some common elements.

Element	Cation	Old-style name	Modern name
cobalt	Co^{2+}	cobaltous	cobalt(II)
	Co^{3+}	cobaltic	cobalt(III)
copper	Cu^+	cuprous	copper(I)
	Cu^{2+}	cupric	copper(II)
iron	Fe^{2+}	ferrous	iron(II)
	Fe^{3+}	ferric	iron(III)
lead	Pb^{2+}	plumbous	lead(II)
	Pb^{4+}	plumbic	lead(IV)
manganese	Mn^{2+}	manganous	manganese(II)
	Mn^{3+}	manganic	manganese(III)
mercury	Hg_2^{2+}	mercurous	mercury(I)
	Hg^{2+}	mercuric	mercury(II)
tin	Sn^{2+}	stannous	tin(II)
	Sn^{4+}	stannic	tin(IV)

***ab initio* method** The calculation of molecular structure by solving the *Schrödinger equation* numerically. Compare with *semiempirical method*.

absolute zero ($T = 0$; that is, 0 on the *Kelvin scale*) The lowest possible temperature ($-273.15\,°C$).

absorb (1) To accept one substance into and throughout the bulk of another substance. Compare with *adsorb*. (2) To remove energy from a beam of radiation.

absorbance (A) A measure of the extent of absorption of radiation by a sample: $A = \log(I_0/I)$.

absorbed dose (of radiation) The energy deposited in a given mass of sample when it is exposed to radiation (particularly but not exclusively nuclear radiation). Absorbed dose is measured in *rad* or *gray*.

absorption spectrum The wavelength dependence of the absorption of a sample, determined by measuring the extent to which the sample absorbs electromagnetic radiation as the wavelength is varied over a range.

abundance (of an isotope) The percentage (in terms of the numbers of atoms) of the isotope present in a sample of the element. See also *natural abundance*.

acceleration The rate of change of velocity (either its direction or its magnitude).

acceleration of free fall (g) The acceleration experienced by a body owing to the gravitational field at the surface of the Earth.

accuracy Freedom from systematic error. Compare with *precision*.

accurate measurement A measurement that has small systematic error and gives a result close to the accepted value of the property.

achiral Not chiral: identical with its mirror image. See also *chiral*.

acid See *Arrhenius acid*; *Brønsted acid*; *Lewis acid*. Used alone, "acid" normally means a Brønsted acid.

acid anhydride A compound that forms an oxoacid when it reacts with water. See also *formal anhydride*. *Example:* SO_3, the anhydride of sulfuric acid.

acid buffer See *buffer*.

acid ionization (dissociation) constant (K_a) See *acidity constant*.

acid–base indicator See *indicator*.

acid–base titration See *titration*.

acidic hydrogen atom A hydrogen atom (more exactly, the proton of that hydrogen atom) that can be donated to a base.

acidic ion An ion that acts as a Brønsted acid. *Examples:* NH_4^+; $[Al(OH_2)_6]^{3+}$.

acidic oxide An oxide that reacts with water to give an acid; the oxides of nonmetallic elements generally are acidic oxides. *Examples:* CO_2; SO_3.

acidic solution A solution with pH < 7.

acidity The strength of the tendency to donate a proton.

acidity constant (K_a) The equilibrium constant for proton transfer to water; for an acid HA, $K_a = [H_3O^+][A^-]/[HA]$ at equilibrium.

actinide Former (and still very common) term for *actinoid*.

actinoid A member of the second row of the f-block of the periodic table (thorium through nobelium).

activated complex An unstable combination of reactant molecules that can either go on to form products or fall apart into the unchanged reactants.

activated complex theory See *transition state theory*.

activation energy (E_a) (1) The minimum energy needed for reaction. (2) The height of the activation barrier. (3) An empirical parameter that describes the temperature dependence of the rate constant of a reaction.

activity (1) In thermodynamics, a_J, the effective concentration or pressure of a species J expressed as the partial pressure or concentration of the species relative to its standard value. (2) In radioactivity, the number of nuclear disintegrations per second.

addition polymerization The polymerization, usually of alkenes, by an addition reaction propagated by radical or ionic intermediates. See *polymerization*.

addition reaction A chemical reaction in which atoms or groups bond to two atoms joined by a multiple bond. The product of the reaction is a single molecule that contains all the reactant atoms. *Example:* $CH_3CH{=}CH_2 + HBr \rightarrow CH_3CH_2CH_2Br$.

adhesion Binding to a surface.

adhesive forces Forces that bind a substance to a surface.

adiabatic Not permitting or accompanied by the passage of energy as heat. *Example:* adiabatic wall.

adsorb To bind a substance to a surface; the surface *adsorbs* the substance. Distinguish from *absorb*.

aerosol A fine mist of solid particles or droplets of liquid suspended in a gas.

alcohol An organic molecule containing an —OH group attached to a carbon atom that is not part of a carbonyl group or an aromatic ring. Alcohols are classified as *primary*, *secondary*, or *tertiary* according to the number of carbon atoms attached to the C—OH carbon atom. *Examples:* CH_3CH_2OH (primary); $(CH_3)_2CHOH$ (secondary); $(CH_3)_3COH$ (tertiary).

aldehyde An organic compound containing the —CHO group. *Examples:* CH_3CHO, ethanal (acetaldehyde); C_6H_5CHO, benzaldehyde.

aliphatic hydrocarbon A hydrocarbon that does not have benzene rings in its structure.

alkali An aqueous solution of a strong base. *Example:* aqueous NaOH.

alkali metal   A member of Group 1 of the periodic table (the lithium family).

alkaline earth metal   Calcium, strontium, and barium; more informally, a member of Group 2 of the periodic table (the beryllium family).

alkaline solution   An aqueous solution with pH > 7.

alkane   (1) A hydrocarbon with no carbon–carbon multiple bonds. (2) A saturated hydrocarbon. (3) A member of a series of hydrocarbons derived from methane by the repetitive insertion of $-CH_2-$ groups; alkanes have the molecular formula C_nH_{2n+2}. *Examples:* CH_4; CH_3CH_3; $CH_3(CH_2)_6CH_3$.

alkene   (1) A hydrocarbon with at least one carbon–carbon double bond. (2) A member of a series of hydrocarbons derived from ethene by the repetitive insertion of $-CH_2-$ groups; alkenes with one double bond have the molecular formula C_nH_{2n}. *Examples:* $CH_2{=}CH_2$; $CH_3CH{=}CH_2$; $CH_3CH{=}CHCH_2CH_3$.

alkyne   (1) A hydrocarbon with at least one carbon–carbon triple bond. (2) A member of a series of hydrocarbons derived from ethyne by the repetitive insertion of $-CH_2-$ groups; alkynes with one triple bond have the molecular formula C_nH_{2n-2}. *Examples:* $CH{\equiv}CH$; $CH_3C{\equiv}CCH_3$.

allotropes   Alternative forms of an element that differ in the way in which the atoms are linked. *Examples:* O_2 and O_3; white and gray tin.

alloy   A mixture of two or more metals formed by melting, mixing, and then cooling. A *substitutional alloy* is an alloy in which atoms of one metal are substituted for atoms of another metal. An *interstitial alloy* is an alloy in which atoms of one metal lie in the gaps in the lattice formed by atoms of another metal. A *homogeneous alloy* is an alloy in which the atoms of the elements are distributed uniformly. A *heterogeneous alloy* is an alloy that consists of (micro) crystalline phases with different compositions.

alpha (α) decay   Nuclear decay due to *α-particle* emission.

alpha (α) helix   One type of secondary structure adopted by a polypeptide chain, in the form of a right-handed helix.

alpha (α) particle   Positively charged, subatomic particle emitted from some radioactive nuclei; nucleus of a helium atom ($_2^4He^{2+}$).

alternating copolymer   See *copolymer*.

ambidentate ligand   A ligand that can coordinate to a metal atom by using atoms of different elements. *Example:* SCN^-, which can coordinate through S or N.

amide   An organic compound formed by the reaction of an amine and a carboxylic acid in which the acidic $-OH$ group has been replaced by an amino group or a substituted amino group. An amide contains the group $-CONR_2$. *Example:* CH_3CONH_2, acetamide.

amine   A compound derived from ammonia by replacing various numbers of H atoms with organic groups; the number of hydrogen atoms replaced determines the classification as *primary*, *secondary*, or *tertiary*. *Examples:* CH_3NH_2 (primary); $(CH_3)_2NH$ (secondary); $(CH_3)_3N$ (tertiary). See also *quaternary ammonium ion*.

amino acid   A carboxylic acid that also contains an amino group. The *essential amino acids* are amino acids that must be ingested as a part of the diet. *Example:* NH_2CH_2COOH, glycine. See Table 11E.3.

amino group   The functional group $-NH_2$, characteristic of *amines*.

ammonium cation   The cation NH_4^+.

amorphous solid   A solid in which the atoms, ions, or molecules lie in a random jumble with no long-range order. *Examples:* glass; butter. Compare with *crystalline solid*.

amount of substance (n)   The number of entities in a sample divided by Avogadro's constant. Also referred to as *chemical amount*. See *mole*.

ampere (A)   The SI unit of electric current. See also Appendix 1B.

amphiprotic   Having the ability both to donate and to accept protons. See *amphoteric*. *Examples:* H_2O; HCO_3^-.

amphoteric   Having the ability to react with both acids and bases. See *amphiprotic*. *Examples:* Al; Al_2O_3.

amplitude   The height of a mathematical function above 0. On a graph depicting a wave, the height of the wave above the center line.

analysis   See *chemical analysis*.

analyte   The solution of unknown concentration in a titration. Normally, the analyte is in the flask, not the buret.

angular wavefunction ($Y(\theta,\phi)$)   The angular part of a wavefunction, particularly the angular component of the wavefunctions of the hydrogen atom; the probability amplitude of an electron as a function of orientation around the nucleus.

anhydride   See *acid anhydride*.

anhydrous   Lacking water. *Example:* $CuSO_4$, the anhydrous form of copper(II) sulfate. Compare with *hydrate*.

anion   A negatively charged ion. *Examples:* F^-; SO_4^{2-}

anisotropic   Depending on orientation.

anode   The electrode at which oxidation takes place.

antibonding orbital   A molecular orbital that, when occupied, contributes to an overall raising of the energy of a molecule.

antiferromagnetic material   A substance in which electron spins on neighboring atoms are locked into an antiparallel arrangement over large regions. *Example:* manganese.

antilogarithm   If the logarithm to the base B is x, then the antilogarithm of x is B^x. The *common antilogarithm* of x is 10^x. The *natural antilogarithm* of x is the exponential e^x.

antioxidant   A substance that reacts with radicals and so prevents the oxidation of another substance.

antiparticle   A particle with the same mass as a subatomic particle but with opposite charge. *Example:* positron, the antiparticle of an electron.

aqueous solution   A solution in which the solvent is water.

arene   An aromatic hydrocarbon.

aromatic compound   An organic compound that includes a benzene ring as part of its structure. *Examples:* C_6H_6 (benzene); C_6H_5Cl (chlorobenzene); $C_{10}H_8$ (naphthalene).

aromatic hydrocarbon   See *aromatic compound* (the more general term).

Arrhenius acid   A compound that contains hydrogen and releases hydrogen ions (H^+) in water. *Examples:* HCl; CH_3COOH; but not CH_4.

Arrhenius base   A compound that produces hydroxide ions (OH^-) in water. *Examples:* NaOH; NH_3; but not Na, because it is not a compound.

Arrhenius behavior   A reaction shows Arrhenius behavior if a plot of ln k_r against $1/T$ is a straight line. See *Arrhenius equation*.

Arrhenius equation   The equation $\ln k_r = \ln A - E_a/RT$ for the commonly observed temperature dependence of a rate constant k_r. An *Arrhenius plot* is a graph of ln k_r against $1/T$.

Arrhenius parameters   The *pre-exponential factor A* (also called the *frequency factor*) and the *activation energy E_a*. See also *Arrhenius equation*.

aryl group   An aromatic group. *Example:* —C_6H_5, phenyl.

atmosphere   (1) The layer of gases surrounding a planet (specifically, the air for planet Earth). (2) A unit of pressure (1 atm = 1.013 25 × 10^5 Pa exactly).

atom   (1)The smallest particle of an element that has the chemical properties of that element. (2) An electrically neutral species consisting of a nucleus and its surrounding electrons.

atomic hypothesis   The proposal advanced by John Dalton that matter is composed of atoms.

atomic mass constant (m_u, formerly amu)   One-twelfth the mass of one atom of carbon-12.

atomic nucleus   See *nucleus*.

atomic number (Z)   The number of protons in the nucleus of an atom; this number determines the identity of the element and the number of electrons in the neutral atom.

atomic orbital   A region of space in which there is a high probability of finding an electron in an atom. An *s-orbital* is a spherical region; a *p-orbital* has two lobes, on opposite sides of the nucleus; a *d-orbital* typically has four lobes, with the nucleus at the center; an *f-orbital* has a more complicated arrangement of lobes.

atomic radius   Half the distance between the centers of neighboring atoms in a solid or a homonuclear molecule.

atomic structure   The arrangement of electrons around the nucleus of an atom.

atomic weight   The numerical value of the *molar mass* of an element.

Aufbau principle   See *building-up principle*.

autoionization   See *autoprotolysis*.

autoprotolysis   A reaction in which a proton is transferred between two molecules of the same substance. The products are the conjugate acid and conjugate base of the substance. *Example:* 2 $H_2O(l) \rightleftharpoons H_3O^+(aq) + OH^-(aq)$.

autoprotolysis constant   The equilibrium constant for an autoprotolysis reaction. *Example:* for water, K_w, with $K_w = [H_3O^+][OH^-]$.

average bond enthalpy ($\Delta H_B(A—B)$)   See *mean bond enthalpy*.

average reaction rate   The reaction rate calculated by measuring the change in concentration of a reactant or product over a finite time interval (and hence an average of the changing rate within that interval). See also *unique average reaction rate*.

Avogadro's constant   The number of objects per mole of objects ($N_A = 6.022\ 14 × 10^{23}\ \text{mol}^{-1}$). *Avogadro's number* is the number of objects in 1 mol of objects (that is, the dimensionless number 6.022 14 × 10^{23}).

Avogadro's principle   The volume of a sample of gas at a given temperature and pressure is proportional to the amount of gas molecules in the sample: $V \propto n$.

axial bond   A bond that is perpendicular to the molecular plane in a bipyramidal molecule.

axial lone pair   A lone pair lying on the axis of a bipyramidal molecule.

azeotrope   A mixture of liquids that boils without change of composition. A *minimum-boiling azeotrope* has a boiling point lower than that of either component; a *maximum-boiling azeotrope* has a boiling point higher than that of either component.

azimuthal quantum number (l)   See *orbital angular momentum quantum number*.

background radiation   The average nuclear radiation to which the Earth's inhabitants are exposed daily.

balanced equation   See *chemical equation*.

ball-and-stick model   A depiction of a molecule in which atoms are represented by balls and bonds are represented by sticks.

Balmer series   A family of spectral lines (some of which lie in the visible region) in the spectrum of atomic hydrogen.

band gap   A range of energies for which there are no orbitals in a solid. Unless stated otherwise, the band gap refers to the gap between the valence band and the conduction band.

band of stability   A region of a plot of mass number against atomic number corresponding to the existence of stable nuclei.

bar   A unit of pressure: 1 bar = 10^5 Pa.

barometer   An instrument for measuring the atmospheric pressure.

base   See *Arrhenius base; Brønsted base; Lewis base*. Used alone, "base" normally means a Brønsted base.

base buffer   See *buffer*.

base ionization constant   See *basicity constant*.

base pair   Two specific nucleotides that link one complementary strand of a DNA molecule to the other by means of hydrogen bonding: adenine pairs with thymine and guanine pairs with cytosine.

base units   The units of measurement in the International System (SI) in terms of which all other units are defined. *Examples: kilogram* for mass; *meter* for length; *second* for time; *kelvin* for temperature; *ampere* for electric current.

basic ion   An ion that acts as a Brønsted base. *Example:* $CH_3CO_2^-$.

basic oxide   An oxide that is a Brønsted base. The oxides of metallic elements are generally basic. *Examples:* Na_2O; MgO.

basic solution   A solution with pH > 7.

basicity constant (K_b)   The equilibrium constant for proton transfer from water to a base; for a base B, $K_b = [BH^+][OH^-]/[B]$.

battery   A collection of galvanic cells joined in series; the voltage that the battery produces is the sum of the voltages of each cell.

becquerel (Bq)   The SI unit of radioactivity (one disintegration per second).

Beer's law   The absorbance of electromagnetic radiation by a sample is proportional to the molar concentration of the absorbing species and the length of the sample through which the radiation passes.

beta (β) decay   Nuclear decay due to *β-particle* emission.

beta (β) particle   A fast electron emitted from a nucleus in a radioactive decay.

beta (β) sheet One type of planar secondary structure adopted by a polypeptide, in the form of a pleated sheet.

bimolecular reaction An elementary reaction in which two molecules, atoms, or ions come together and form a product. *Example:* $O + O_3 \rightarrow O_2 + O_2$.

binary Consisting of two components, as in *binary mixture* and *binary (ionic* or *molecular) compound. Examples:* acetone and water (a binary mixture); HCl, $CaCl_2$, C_6H_6 (binary compounds; $CaCl_2$ is ionic, HCl and C_6H_6 are molecular).

binary compound A compound consisting of atoms of two different elements (see *binary*).

bioenergetics The deployment and utilization of energy in living cells.

biomass The organic material of the planet produced annually by photosynthesis.

biomimetic material A material modeled after a naturally occurring material.

biradical A species with two unpaired electrons. *Example:* $\cdot CH_2CH_2CH_2\cdot$.

black body An object that absorbs and emits all frequencies of radiation without favor.

black-body radiation The electromagnetic radiation emitted by a *black body*.

block (s-block, p-block, d-block, f-block) The region of the periodic table containing elements for which, according to the building-up principle, the corresponding subshell is currently being filled.

block copolymer See *copolymer*.

body-centered cubic structure (bcc) A crystal structure with a unit cell in which a central atom lies at the center of a cube formed by eight others.

Bohr frequency condition The relation between the change in energy of an atom or molecule and the frequency of radiation emitted or absorbed: $\Delta E = h\nu$.

Bohr radius (a_0) In an early model of the hydrogen atom, the radius of the lowest energy orbit; now a specific combination of fundamental constants ($a_0 = 4\pi\varepsilon_0\hbar^2/m_ee^2 = 52.9$ pm) in the description of the wavefunctions of hydrogen.

boiling Rapid vaporization taking place throughout a liquid. See *boiling temperature*.

boiling point (b.p.) See *boiling temperature; normal boiling point*.

boiling-point constant (k_b) The constant of proportionality between the boiling-point elevation and the molality of a solute.

boiling-point elevation The increase in normal boiling point of a solvent caused by the presence of a solute (a *colligative property*).

boiling temperature (1) The temperature at which a liquid boils. (2) The temperature at which a liquid is in equilibrium with its vapor at the pressure of the surroundings; vaporization then occurs throughout the liquid, not only at the liquid's surface.

Boltzmann formula (for the entropy) The formula $S = k \ln W$, where k is *Boltzmann's constant* and W is the number of atomic arrangements that correspond to the same energy.

Boltzmann's constant (k) A fundamental constant; $k = 1.380\ 65 \times 10^{-23}$ J·K^{-1}. Note that $R = N_Ak$.

bond A link between atoms. See also *covalent bond; double bond; ionic bond; triple bond*.

bond angle In an A—B—C molecule or part of a molecule, the angles between the B—A and B—C bonds.

bond enthalpy ($\Delta H_B(X-Y)$) The enthalpy change accompanying the dissociation of a bond. *Example:* $H_2(g) \rightarrow 2\ H(g)$, $\Delta H_B(H-H) = +436$ kJ·mol^{-1}.

bond length The distance between the centers of two atoms joined by a bond.

bond order The number of electron pair bonds that link a specific pair of atoms.

bonding orbital A molecular orbital that, when occupied, results in an overall lowering of the energy of a molecule.

Born–Haber cycle A closed series of reactions used to express the enthalpy of formation of an ionic solid in terms of contributions that include the lattice enthalpy.

Born interpretation The interpretation of the square of the wavefunction, ψ, of a particle as the probability density for finding the particle in a region of space.

Born–Mayer equation The formula for the minimum energy of an ionic solid.

boundary condition A constraint on the value of the wavefunction of a particle.

boundary surface The surface showing the region of space within which there is about 90% probability of finding an electron when it occupies a specific orbital in an atom or molecule.

Boyle's law At constant temperature, and for a given sample of gas, the volume is inversely proportional to the pressure: $P \propto 1/V$.

Bragg's equation An equation relating the angle of diffraction of x-rays to the spacing of layers of atoms in a crystal ($\lambda = 2d \sin \theta$).

branched alkane An alkane with hydrocarbon side chains.

branching Description of a step in a chain reaction in which more than one chain carrier is formed in a propagation step. *Example:* $\cdot O\cdot + H_2 \rightarrow \cdot OH + \cdot H$. See also *propagation*.

Bravais lattices The 14 basic patterns of unit cells from which a crystal can be built.

Brønsted acid A proton donor (a source of hydrogen ions, H$^+$). *Examples:* HCl; CH_3COOH; HCO_3^-; NH_4^+.

Brønsted base A proton acceptor (a species to which hydrogen ions, H$^+$, can bond). *Examples:* OH$^-$; Cl$^-$; $CH_3CO_2^-$; HCO_3^-; NH_3.

Brønsted–Lowry definition A definition of acids and bases in terms of the ability of molecules and ions to participate in proton transfer. See *Brønsted–Lowry theory*.

Brønsted–Lowry theory A theory of acids and bases involving proton transfer from one species to another. See also *Brønsted acid* and *Brønsted base*.

Brownian motion The ceaseless jittering motion of colloidal particles caused by the impact of solvent molecules.

buffer A solution that resists any change in pH when small amounts of acid or base are added. An *acid buffer* stabilizes solutions at pH < 7 and a *base buffer* stabilizes solutions at pH > 7. *Examples:* a solution containing CH_3COOH and $CH_3CO_2^-$; (acid buffer); a solution containing NH_3 and NH_4^+ (base buffer).

buffer capacity An indication of the amount of acid or base that can be added before a buffer loses its ability to resist the change in pH.

building-up principle   The procedure for arriving at the ground-state electron configurations of atoms and molecules.

bulk matter   Matter composed of large numbers of atoms. See *bulk property.*

bulk property   A property that depends on the collective behavior of large numbers of atoms. *Examples:* melting point; vapor pressure; internal energy.

buret   A narrow, graduated tube fitted with a stopcock, used to measure the volume of liquid delivered into another vessel.

calibration   Interpretation of an observation by comparison with known information.

calorie (cal)   A unit of energy. The unit is now defined in terms of the joule by 1 cal = 4.184 J exactly. The *nutritional calorie* is 1 kcal.

calorimeter   An apparatus used to determine the heat released or absorbed in a process by measuring the temperature change.

candela (cd)   The SI unit of luminous intensity. See also Appendix 1B.

capillary action   The rise of liquids up narrow tubes.

carbohydrate   A compound of general formula $C_m(H_2O)_n$, although small deviations from this general formula are often encountered. Carbohydrates include celluloses, starches, and sugars. *Examples:* $C_6H_{12}O_6$, glucose; $C_{12}H_{22}O_{11}$, sucrose.

carbonyl group   A >CO group in an inorganic or organic compound.

carboxyl group   The functional group —COOH. See *carboxylic acid.*

carboxylate ion   The deprotonated form of a *carboxylic acid.* *Examples:* $CH_3CO_2^-$, acetate ion; $C_6H_5CO_2^-$, benzoate ion.

carboxylic acid   An organic compound containing the carboxyl group, —COOH. *Examples:* CH_3COOH, acetic acid; C_6H_5COOH, benzoic acid.

carrier   See *chain carrier.*

catalyst   A substance that increases the rate of a reaction without being consumed in the reaction. A catalyst is *homogeneous* if it is present in the same phase as the reactants and *heterogeneous* if it is in a different phase from the reactants. *Examples:* homogeneous, Br^- (aq) for the decomposition of H_2O_2(aq); heterogeneous, Pt in the Ostwald process.

catenate   To form chains or rings of atoms. *Examples:* O_3; S_8.

cathode   The electrode at which reduction takes place.

cathodic protection   Protection of a metal object by connecting it to a more strongly reducing metal.

cation   A positively charged ion. *Examples:* Na^+; NH_4^+; Al^{3+}.

cationic electrodeposition   The electrochemical deposition of a corrosion-resistant protective coating onto a metal.

cell diagram   A description of an electrochemical cell that corresponds to a given cell reaction. *Example:* $Zn(s)|Zn^{2+}(aq)||Cu^{2+}(aq)|Cu(s)$.

cell potential (E_{cell})   (1) The potential difference between the electrodes of an electrochemical cell when it is producing no current. (2) An indication of the tendency of a reaction in an electrochemical cell to occur spontaneously.

Celsius scale   A temperature scale on which the freezing point of water is at 0 degrees and its normal boiling point is at 100 degrees. Units on this scale are degrees Celsius, °C.

ceramic   (1) A solid obtained by the action of heat on clay. (2) A noncrystalline inorganic solid usually containing oxides, borides, and carbides.

cesium chloride structure   A crystal structure the same as that of solid cesium chloride.

chain branching   A propagation step in a chain reaction when more than one chain carrier is formed.

chain carrier   An intermediate in a chain reaction.

chain reaction   A reaction that is propagated when an intermediate reacts to produce another intermediate in a series of elementary reactions. *Example:* $Br\cdot + H_2 \rightarrow HBr + H\cdot$ followed by $H\cdot + Br_2 \rightarrow HBr + Br\cdot$.

chalcogens   Oxygen, sulfur, selenium, and tellurium in Group 16 of the periodic table.

change of state   The change of a substance from one of its physical states to another of its physical states. *Example:* melting, solid → liquid.

characteristic (of a logarithm)   The number preceding the decimal point.

charge   A measure of the strength with which a particle can interact with an electric field.

charge balance   The requirement that, because a solution is neutral overall, the concentration of positive charge due to cations must equal the concentration of negative charge due to anions.

charge-transfer transition   A transition in which an electron is excited from the ligands of a complex to the metal atom or vice versa.

Charles's law   The volume of a given sample of gas at constant pressure is directly proportional to its absolute temperature: $V \propto T$.

chelate   A complex containing at least one polydentate ligand that forms a ring of atoms including the central metal atom. *Example:* $[Co(en)_3]^{3+}$.

chemical analysis   The determination of the chemical composition of a sample. See also *qualitative; quantitative.*

chemical bond   See *bond.*

chemical change   The conversion of one or more substances into different substances.

chemical element   See *element.*

chemical energy   The energy available from a chemical reaction. *Example:* the energy released in the combustion of a fuel.

chemical equation   A statement in terms of chemical formulas summarizing the qualitative information about the chemical changes taking place in a reaction and the quantitative information that atoms are neither created nor destroyed in a chemical reaction. In a *balanced equation* (commonly called a "chemical equation"), the same number of atoms of each element appears on both sides of the equation.

chemical equilibrium   A dynamic equilibrium between reactants and products in a chemical reaction.

chemical formula   A collection of chemical symbols and subscripts that shows the composition of a substance. See also *condensed structural formula; empirical formula; molecular formula; structural formula.*

chemical kinetics   The study of the rates of chemical reactions and the steps by which they take place.

chemical nomenclature   The systematic naming of compounds.

chemical plating   The deposition of a metal surface on an object by making use of a chemical reduction reaction.

chemical property   The ability of a substance to participate in a chemical reaction.

chemical reaction   A chemical change in which one substance responds to the presence of another, to a change of temperature, or to some other influence.

chemical symbol   The abbreviation of the name of an element.

chemiluminescence   The emission of light by products formed in energetically excited states during a chemical reaction.

chemistry   The branch of science concerned with the study of matter and the changes that matter can undergo.

chiral (molecule or complex)   Not able to be superimposed on its own mirror image. *Examples:* $CH_3CH(NH_2)COOH$; $CHBrClF$; $[Co(en)_3]^{3+}$.

chloralkali process   The production of chlorine and sodium hydroxide by the electrolysis of aqueous sodium chloride.

cholesteric phase   A liquid crystal phase in which layers of parallel molecules are twisted relative to one another in such a way that the orientations of the molecules form a helical structure.

chromatogram   A record of the signal from the detector (or the paper record) obtained in a chromatographic analysis of a mixture.

chromatography   A separation technique that relies on the ability of different phases to adsorb substances to different extents.

classical mechanics   The laws of motion proposed by Isaac Newton in which particles travel in definite paths in response to forces.

clathrate   A structure in which a molecule of one substance sits in a cage made up of molecules of another substance, typically water. *Example:* SO_2 in water.

Claus process   A process for obtaining sulfur from the H_2S in oil wells by the oxidation of H_2S with SO_2; the latter is formed by the oxidation of H_2S with oxygen.

Clausius inequality   The relation $\Delta S \geq q/T$.

Clausius–Clapeyron equation   An equation that gives the quantitative dependence of the vapor pressure of a substance on the temperature.

closed shell (or subshell)   A shell (or subshell) containing the maximum number of electrons allowed by the exclusion principle. *Example:* the neonlike core $1s^2 2s^2 2p^6$.

closed system   See *system.*

close-packed structure   A crystal structure in which atoms occupy the smallest total volume with the least empty space. *Examples:* hexagonal close packing and cubic close packing of identical spheres.

coagulation   The formation of aggregates from colloidal particles.

cohesion   The act or state in which the particles of a substance stick to one another.

cohesive forces   The forces that bind the molecules of a substance together to form a bulk material and are responsible for condensation.

coinage metals   The elements copper, silver, and gold.

colligative property   A property that depends only on the relative number of solute and solvent particles present in a solution and not on the chemical identity of the solute. *Examples:* elevation of boiling point; depression of freezing point; osmosis.

collision cross section   The area that a molecule presents as a target during a collision.

collision frequency   The number of collisions per second between the molecules of two reactants in a gas.

collision theory   The theory of elementary gas-phase bimolecular reactions in which molecules are assumed to react only if they collide with a characteristic minimum kinetic energy.

colloid (or colloidal suspension)   A dispersion of tiny particles with diameters between 1 nm and 1 μm in a gas, liquid, or solid. *Example:* milk.

combined gas law   A combination of Boyle's law and Charles's law that allows the pressure, volume, or temperature of a sample of an ideal gas to be predicted after a change in state. $P_1V_1/n_1T_1 = P_2V_2/n_2T_2$.

combustion   A reaction in which an element or compound burns in oxygen. *Example:* $CH_4(g) + 2 O_2(g) \rightarrow CO_2(g) + 2 H_2O(l)$.

combustion analysis   The determination of the composition of a sample by the measurement of the masses of the products of its combustion.

common antilogarithm   See *antilogarithm.*

common logarithm   See *logarithm.*

common name   An informal name for a compound that may give little or no clue to the compound's composition. *Examples:* water; aspirin; acetic acid.

common-ion effect   Reduction of the solubility of one salt by the presence of another salt with one ion in common. *Example:* the lower solubility of AgCl in NaCl(aq) than in pure water.

competing reaction   A reaction taking place at the same time as the reaction of interest and using some of the same reactants, but forming different products.

complementarity   The impossibility of knowing the position of a particle with arbitrarily great precision if its linear momentum is known precisely.

complementary color   The color that white light becomes when one of the colors present in it is removed.

complete ionic equation   A balanced chemical equation expressed in terms of the cations and anions present in solution. *Example:* $Ag^+(aq) + NO_3^-(aq) + Na^+(aq) + Cl^-(aq) \rightarrow AgCl(s) + NO_3^-(aq) + Na^+(aq)$.

complex   (1) The combination of a Lewis acid and a Lewis base linked by a coordinate covalent bond. (2) A species consisting of several ligands (the Lewis bases) that have an independent existence bonded to a single central metal atom or ion (the Lewis acid). *Examples:* (1) $H_3N—BF_3$; (2) $[Fe(OH_2)_6]^{3+}$; $[PtCl_4]^-$.

composite material   A synthetic material composed of a polymer and one or more other substances that have been solidified together.

compound   (1) A specific combination of elements that can be separated into its elements by chemical techniques but not physical techniques. (2) A substance consisting of atoms of two or more elements in a definite, unchangeable ratio.

compressible   Able to be compressed into a smaller volume.

compression   The act of reducing the volume of a sample.

compression factor (Z)   The ratio of the actual molar volume of a gas to the molar volume of an ideal gas under the same conditions.

concentration   The quantity of a substance in a given volume. See also *molar concentration*.

concentration cell   A galvanic cell in which the electrodes have the same composition but are at different concentrations.

condensation   The formation of a liquid or solid phase from the gas phase of the substance.

condensation polymer   A polymer formed by a chain of condensation reactions. *Examples:* polyesters; polyamides (nylon).

condensation reaction   A reaction in which two molecules combine to form a larger one and a small molecule is eliminated. *Example:* $CH_3COOH + C_2H_5OH \rightarrow CH_3COOC_2H_5 + H_2O$.

condensed phase   A solid or liquid phase; not a gas.

condensed structural formula   A compact version of the structural formula, showing how the atoms are grouped together. *Example:* $CH_3CH(CH_3)CH_3$ for methylpropane.

conduction band   An incompletely occupied band of energy levels in a solid.

configuration   See *electron configuration*.

conformations   Molecular shapes that can be interchanged by rotation about bonds, without bond breakage and re-formation.

congeners   Elements in the same group of the periodic table.

conjugate acid   The Brønsted acid formed when a Brønsted base has accepted a proton. *Example:* NH_4^+ is the conjugate acid of NH_3.

conjugate acid–base pair   A Brønsted acid and its conjugate base. *Examples:* HCl and Cl^-; NH_4^+ and NH_3.

conjugate base   The Brønsted base formed when a Brønsted acid has donated a proton. *Example:* NH_3 is the conjugate base of NH_4^+.

conjugated double bonds   A sequence of alternating single and double bonds, as in —C=C—C=C—.

connectivity (of atoms in a molecule)   The pattern in which the atoms in a molecule are bonded to one another.

constructive interference   Interference that results in an increased amplitude of a wave. Compare with *destructive interference*.

conversion factor   A factor that is used to convert a measurement from one unit into another.

Cooper pair   A pair of electrons that can travel together almost freely through a crystal lattice and give rise to superconductivity.

coordinate   Use of a lone pair to form a coordinate covalent bond. *Examples:* $F_3B + :NH_3 \rightarrow F_3B—NH_3$; $Ni + 4\,CO \rightarrow Ni(CO)_4$.

coordinate covalent bond   A bond formed between a Lewis base and a Lewis acid by sharing an electron pair originally belonging to the Lewis base. See *coordinate*.

coordination complex   The product of the reaction of a Lewis acid and a Lewis base to form a coordinate covalent bond. See also *coordination compound*.

coordination compound   A neutral complex or an ionic compound in which at least one of the ions is a complex. *Examples:* $Ni(CO)_4$; $K_3[Fe(CN)_6]$.

coordination isomers   Isomers that differ by the exchange of one or more ligands between a cationic complex and an anionic complex.

coordination number   (1) The number of nearest neighbors of an atom in a solid. (2) For ionic solids, the coordination number of an ion is the number of nearest neighbors of opposite charge. (3) For complexes, the number of points at which ligands are attached to the central metal ion.

coordination sphere   The ligands attached directly to the central ion in a complex.

copolymer   A polymer formed from a mixture of different monomers. In *random copolymers*, the sequence of monomers has no particular order; in *alternating copolymers*, two monomers alternate; in block *copolymers*, regions of one monomer alternate with regions of another; in *graft copolymers*, chains of one monomer are attached to a backbone chain of a second monomer.

core   The inner closed shells of an atom.

core electrons   The electrons that belong to an atom's core.

corrosion   The unwanted reaction of a material that results in the dissolution or consumption of the material. *Example:* the unwanted oxidation of a metal.

corrosive   (1) A reagent that can cause corrosion. (2) Having a high reactivity, such as the reactivity of a strong oxidizing agent or a concentrated acid or base.

coulomb (C)   The unit of electric charge. One coulomb is the magnitude of the charge delivered by a current of one ampere flowing for one second (1 C = 1 A·s).

Coulomb potential energy   The potential energy of an electric charge in the vicinity of another electric charge; the potential energy is inversely proportional to the separation of the charges.

coulombic attraction   The attraction between opposite electric charges.

couple   See *redox couple*.

covalent bond   A pair of electrons shared between two atoms.

covalent radius   The contribution of an atom to the length of a covalent bond.

cracking   The process of converting petroleum fractions into smaller molecules with more double bonds. *Example:* $CH_3(CH_2)_6CH_3 \rightarrow CH_3(CH_2)_3CH_3 + CH_3CH=CH_2$.

critical mass   The mass of fissionable material above which so few neutrons escape from a sample of nuclear fuel that the fission chain reaction is sustained; a greater mass is *supercritical* and a smaller mass is *subcritical*.

critical point   The point in a phase diagram at the critical pressure and critical temperature.

critical pressure (P_c)   The vapor pressure of a liquid at its critical temperature.

critical temperature (T_c)   The temperature at and above which a substance cannot exist as a liquid.

cryogenics   The study of matter at very low temperatures.

cryoscopy   The measurement of molar mass by using the depression of freezing point.

crystal face   A flat plane forming a side of a crystal.

crystal field   The electrostatic influence of the ligands (modeled as point negative charges) on the central ion of a complex. *Crystal field theory* is a rationalization of the optical, magnetic, and thermodynamic properties of complexes in terms of the crystal field of their ligands.

crystalline solid   A solid in which the atoms, ions, or molecules lie in an orderly array. *Examples:* NaCl; diamond; graphite. Compare with *amorphous solid*.

crystallization   The process in which a solute comes out of solution as crystals.

cubic close-packed structure (ccp)   A close-packed structure with an ABCABC . . . pattern of layers.

curie (Ci)   A unit of activity (for radioactivity).

current (I)   The rate of supply of charge; current is measured in *amperes* (A), with $1\ A = 1\ C \cdot s^{-1}$.

cycle   (1) In thermodynamics, a sequence of changes that begins and ends at the same state. (2) In spectroscopy, one complete reversal of the direction of the electromagnetic field and its return to the original direction.

cycloalkane   A saturated aliphatic hydrocarbon in which the carbon atoms form a ring. *Example:* C_6H_{12}, cyclohexane.

Dalton's law of partial pressures   The total pressure of a mixture of gases is the sum of the partial pressures of its components.

data   The information provided or obtained from experiments.

daughter nucleus   A nucleus that is the product of a nuclear decay.

d–d transition   A transition in which an electron is excited from one d-orbital to another.

de Broglie relation   The proposal that every particle has wavelike properties and that its wavelength, λ, is related to its momentum by $\lambda = h/p$, with $p = mv$.

debye (D)   The unit commonly used to report electric dipole moments: $1\ D = 3.336 \times 10^{-30}\ C \cdot m$.

decant   To pour off a liquid from on top of another, denser liquid or from a solid.

decay constant (k)   The rate constant for radioactive decay.

decomposition   A reaction in which a substance is broken down into simpler substances; *thermal decomposition* is decomposition brought about by heat. *Example:*
$$CaCO_3(s) \xrightarrow{\Delta} CaO(s) + CO_2(g).$$

definite integral   An *integral* with limits attached. See also Appendix 1F.

degenerate   Having the same energy. *Example:* atomic orbitals in the same subshell.

dehydrating agent   A reagent that removes water or the elements of water from a compound. *Example:* H_2SO_4.

dehydrogenation   The removal of a hydrogen atom from each of two neighboring carbon atoms, resulting in the formation of a carbon–carbon multiple bond.

dehydrohalogenation   The removal of a hydrogen atom and a halogen atom from neighboring carbon atoms in a haloalkane.

delocalized   Spread over a region. In particular, *delocalized electrons* are electrons that spread over several atoms in a molecule.

delta (Δ, in a chemical equation)   A symbol that signifies that the reaction takes place at elevated temperatures.

delta X (ΔX)   The difference between the final and initial values of a property, $\Delta X = X_{final} - X_{initial}$. *Examples:* ΔT; ΔE.

denaturation   The loss of structure of a large molecule, such as a protein.

density (d)   The mass of a sample of a substance divided by its volume: $d = m/V$.

density isosurface   A graphic image that represents a molecular structure as a surface and shows the distribution of electrons in a molecule; the surface corresponds to locations with the same electron density.

deposition   The condensation of a vapor directly to a solid. Deposition is the reverse of *sublimation*.

deprotonation   Loss of a proton from a Brønsted acid. *Example:*
$NH_4^+(aq) + H_2O(l) \rightarrow H_3O^+(aq) + NH_3(aq)$.

derived unit   A combination of base units. *Examples:* centimeters cubed (cm^3); joules ($kg \cdot m^2 \cdot s^{-2}$).

descriptive chemistry   The description of the preparation, properties, and applications of the elements and their compounds.

destructive interference   Interference that results in a reduced amplitude of a wave. Compare with *constructive interference*.

deuteron   The nucleus of a deuterium atom, $^2H^+$, consisting of a proton and a neutron.

diagonal relationship   A similarity in properties between diagonal neighbors in the periodic table, especially for main-group elements in Periods 2 and 3 at the left-hand side of the table. *Examples:* Li and Mg; Be and Al.

diamagnetic (substance)   A substance that tends to be pushed out of a magnetic field, consisting of atoms, ions, or molecules with no unpaired electrons. *Examples:* most common substances.

diamine   An organic compound that contains two —NH_2 groups.

diathermic   Permitting the passage of energy as heat. *Example:* diathermic walls.

diatomic ion   An ion that consists of two atoms with a net charge.

diatomic molecule   A molecule that consists of two atoms. *Examples:* H_2; CO.

differential calculus   The part of mathematics that deals with the slopes of curves and with infinitesimal quantities. See also Appendix 1F.

diffraction   The deflection of waves and the resulting interference caused by an object in their path. See also *x-ray diffraction*.

diffraction pattern   The pattern of bright spots against a dark background resulting from diffraction.

diffusion   The spreading of one substance through another substance.

dilute   (1) verb: To reduce the concentration of a solute by adding more solvent. (2) adjective: Describes a solution in which the solute has a low concentration.

dimer   The union of two identical molecules. *Example:* Al_2Cl_6 formed from two $AlCl_3$ molecules.

diol   An organic compound with two —OH groups.

dipeptide   A *peptide* formed by the condensation of two amino acids.

dipole   See *electric dipole; instantaneous dipole moment*.

dipole–dipole interaction   The interaction between two electric dipoles: like partial charges repel and opposite partial charges attract.

dipole–induced-dipole interaction   The interaction between an electric dipole and the instantaneous dipole that it induces in a nonpolar molecule.

dipole moment   See *electric dipole moment*.

diprotic   An acid with two acidic hydrogen atoms. See also *polyprotic acid or base*.

disaccharide   A carbohydrate molecule that is composed of two saccharide units. *Example:* $C_{12}H_{22}O_{11}$, sucrose.

dispersion   (1) The spatial separation of light into its component colors (as by a prism). (2) See *suspension*.

dispersion force   See *London force*.

disproportionation   A redox reaction in which a single element is simultaneously oxidized and reduced. *Example:* $2\,Cu^+(aq) \rightarrow Cu(s) + Cu^{2+}(aq)$.

dissociation   (1) The breaking of a bond. (2) The separation of ions that occurs when an ionic solid dissolves.

dissociation constant   See *acidity constant*.

dissociation energy (D)   The energy required to separate bonded atoms.

distillate   A liquid obtained by distillation.

distillation   The separation of the components of a mixture by making use of their different volatilities.

distribution (of molecular speeds)   The fraction of gas molecules moving at each speed at any instant.

disulfide link   An —S—S— link that contributes to the secondary and tertiary structures of polypeptides.

domain   A region of a metal in which the electron spins of the atoms are aligned, resulting in *ferromagnetism*.

doping   The addition of a known, small amount of a second substance to an otherwise pure solid substance.

d-orbital   See *atomic orbital*.

dose equivalent   The actual dose of radiation experienced by a sample, modified to take into account the *relative biological effectiveness* of the radiation. The dose equivalent is measured in sievert (and formerly rem).

double bond   (1) Two electron pairs shared by neighboring atoms. (2) One σ-bond and one π-bond between neighboring atoms.

drying agent   A substance that absorbs water and thus maintains a dry atmosphere. *Example:* phosphorus(V) oxide.

ductility   The ability to be drawn out into a wire (as for a metal).

duplet   The $1s^2$ electron pair of the heliumlike electron configuration.

dynamic equilibrium   The condition in which a forward process and its reverse are taking place simultaneously at equal rates. *Examples:* vaporizing and condensing; chemical reactions at equilibrium.

effective nuclear charge (Z_{eff})   The net nuclear charge after taking into account the shielding caused by other electrons in the atom.

effervesce   To bubble out of solution as a gas.

effusion   The escape of a substance (particularly a gas) through a small hole.

elasticity   The ability to return to the original shape after distortion.

elastomer   An elastic polymer. *Example:* rubber (polyisoprene).

electric conductivity   A measure of the ability of a substance or solution to conduct electricity.

electric current   See *current*.

electric dipole   A positive charge next to an equal but opposite negative charge.

electric dipole moment (μ)   A measure of the magnitude of an electric dipole (commonly in debye).

electric field   A region of influence that affects charged particles.

electrical conduction   The conduction of electric charge through matter. See also *electronic conductor; ionic conduction*.

electrochemical cell   A system consisting of two electrodes in contact with an electrolyte. A *galvanic cell* (*voltaic cell*) is an electrochemical cell used to produce electricity, and an *electrolytic cell* is an electrochemical cell in which an electric current is used to cause chemical change.

electrochemical series   Redox couples arranged in order of oxidizing and reducing strengths; usually arranged with strong oxidizing agents at the top of the list and strong reducing agents at the bottom.

electrochemistry   The branch of chemistry that deals with the use of chemical reactions to produce electricity, the relative strengths of oxidizing and reducing agents, and the use of electricity to produce chemical change.

electrode   A metallic conductor that makes contact with an electrolyte in an electrochemical cell. A *half-cell* is sometimes also referred to as an electrode.

electrolysis   A process in which a chemical change is produced by passing an electric current through a liquid.

electrolyte   (1) An ionically conducting medium. (2) A substance that dissolves to give an electrically conducting solution. A *strong electrolyte* is a substance that is fully ionized in solution. A *weak electrolyte* is a molecular substance that is only partially ionized in solution. A *nonelectrolyte* does not ionize in solution. *Examples:* NaCl is a strong electrolyte; CH_3COOH is a weak electrolyte; $C_6H_{12}O_6$ is a nonelectrolyte.

electrolyte solution   A solution of an electrolyte.

electrolytic cell   See *electrochemical cell*.

electromagnetic field   The region of influence generated by accelerated charged particles.

electromagnetic radiation   A wave of oscillating electric and magnetic fields; includes light, x-rays, and γ-rays.

electromotive force (emf)   See *cell potential*.

electron (e^-)   A negatively charged subatomic particle found outside the nucleus of an atom.

electron affinity (E_{ea})   The energy released when an electron is added to a gas-phase atom or monatomic ion.

electron arrangement (VSEPR model)   The three-dimensional geometry of the arrangement of bonds and lone pairs about a central atom in a molecule or ion.

electron capture   The capture by a nucleus of one of its own atom's s-electrons.

electron configuration   The occupancy of orbitals in an atom or molecule. *Example:* N, $1s^2 2s^2 2p^3$.

electron-deficient compound   A compound with too few valence electrons for it to be assigned a valid Lewis structure. *Example:* B_2H_6.

electronegative element   An element with a high electronegativity. *Examples:* O; F.

electronegativity (χ, chi)   The ability of an atom to attract electrons to itself when it is part of a compound.

electronic conductor A substance that conducts electricity by the transfer of electrons.

electronic structure The details of the distribution of the electrons that surround the nuclei in atoms and molecules.

electronvolt (eV) A unit of energy; the change in potential energy of an electron when it moves through a potential difference of 1 V; $1 \text{ eV} = 1.602\ 18 \times 10^{-19}$ J.

electrophile A reactant that is attracted to a region of high electron density. *Examples:* Br_2; NO_2^+.

electrophilic substitution Substitution that takes place as a result of an attack by an *electrophile*. *Example:* nitration of benzene.

electroplating The deposition of a thin film of metal on an object by electrolysis.

electropositive element An element in the electrochemical series with strong reducing power *Examples:* Cs; Mg.

electrostatic potential surface A molecular structure in which the net charge is calculated at each point of the density isosurface and depicted by different colors; an "elpot" surface.

element (1) A substance that cannot be separated into simpler components by chemical techniques. (2) A substance consisting of atoms of the same atomic number. *Examples:* hydrogen; gold; uranium.

elementary reaction An individual step in a proposed reaction mechanism.

elimination reaction A reaction in which two groups or atoms on neighboring carbon atoms are removed from a molecule, thereby leaving a multiple bond between the carbon atoms. *Example:* $CH_3CHBrCH_3 + OH^- \rightarrow CH_3CH{=}CH_2 + H_2O + Br^-$.

empirical Determined by experiment.

empirical formula A chemical formula that shows the relative numbers of atoms of each element in a compound by using the simplest whole-number subscripts. *Examples:* P_2O_5; CH for benzene.

emulsion A suspension of droplets of one liquid dispersed throughout another liquid.

enantiomers A pair of optical isomers that are mirror images of, but not superimposable on, each other.

end point The stage in a titration at which enough titrant has been added to bring the indicator to a color halfway between its initial and its final colors.

endothermic process A process that absorbs heat ($\Delta H > 0$). *Examples:* vaporization; $N_2O_4(g) \rightarrow 2\ NO_2(g)$.

energy (E) The capacity of a system to do work or supply heat. *Kinetic energy* is the energy due to motion, and *potential energy* is the energy arising from position. The total energy is the sum of the kinetic and potential energies.

energy level A permitted value of the energy in a quantized system such as an atom or a molecule.

enrich In nuclear chemistry, to increase the abundance of a specific isotope.

ensemble A collection of hypothetical replications of a system.

enthalpy (H) A state property; $H = U + PV$. A change in enthalpy is equal to the heat transferred at constant pressure.

enthalpy density (of a fuel) The enthalpy of combustion per liter (without the negative sign).

enthalpy of freezing The enthalpy change per mole accompanying freezing; the negative of the *enthalpy of fusion*.

enthalpy of fusion (ΔH_{fus}) The enthalpy change per mole accompanying fusion (melting).

enthalpy of hydration (ΔH_{hyd}) The enthalpy change accompanying the hydration of gas-phase ions.

enthalpy of ionization (ΔH_{ion}) The molar enthalpy change accompanying the loss of electrons from a gas-phase atom, ion, or molecule.

enthalpy of melting (ΔH_{melt}) See *enthalpy of fusion*.

enthalpy of mixing (ΔH_{mix}) The change in enthalpy when two fluids (liquids or gases) mix.

enthalpy of solution (ΔH_{sol}) The change in enthalpy when a substance dissolves. The *limiting enthalpy of solution* is the enthalpy of solution for the formation of an infinitely dilute solution.

enthalpy of sublimation (ΔH_{sub}) The enthalpy change per mole accompanying sublimation (the direct conversion of a solid into a vapor).

enthalpy of vaporization (ΔH_{vap}) The enthalpy change per mole accompanying vaporization (the conversion of a substance from the liquid state into the vapor state).

entropy (S) (1) A measure of the disorder of a system. (2) A change in entropy is equal to the heat supplied reversibly to a system divided by the temperature at which the transfer takes place.

entropy of vaporization (ΔS_{vap}) The entropy change per mole accompanying vaporization (the conversion of a substance from the liquid state into the vapor state).

enzyme A biological catalyst.

e-orbital One of the orbitals d_{z^2} or $d_{x^2-y^2}$ in an octahedral or tetrahedral complex. In an octahedral complex, the orbitals are designated e_g.

equation of state A mathematical expression relating the pressure, volume, temperature, and amount of substance present in a sample. *Example:* ideal gas law, $PV = nRT$.

equatorial bond A bond perpendicular to the axis of a molecule (particularly trigonal bipyramidal and octahedral molecules).

equatorial lone pair A *lone pair* in the plane perpendicular to the molecular axis.

equilibrium See *chemical equilibrium; dynamic equilibrium*.

equilibrium constant (K) An expression characteristic of the equilibrium composition of the reaction mixture, with a form given by the law of mass action. *Example:* $N_2(g) + 3\ H_2(g) \rightleftharpoons 2\ NH_3(g)$, $K = (P_{NH_3})^2/P_{N_2}(P_{H_2})^3$.

equilibrium table A table used to calculate the composition of a reaction mixture at equilibrium, given the initial composition. The columns are headed by the species and the rows are, successively, the initial composition, the change to reach equilibrium, and the equilibrium composition.

equimolar Having the same molar concentration or the same number of moles.

equipartition theorem The average energy of each quadratic contribution to the energy of a molecule in a sample at a temperature T is equal to $\frac{1}{2}kT$ (where k is *Boltzmann's constant*).

equivalence point See *stoichiometric point*.

essential amino acid An amino acid that is an essential component of the diet because it cannot be synthesized in the body.

ester   The product (other than water) of the reaction between a carboxylic acid and an alcohol and having the formula RCOOR′. *Example:* $CH_3COOC_2H_5$, ethyl acetate.

esterification   The formation of an *ester*.

ether   An organic compound of the form R—O—R′. *Examples:* $CH_3OC_2H_5$, ethyl methyl ether; $C_2H_5OC_2H_5$, diethyl ether.

eutectic   A mixture that melts at the lowest temperature and does so without change of composition.

evaporate   To vaporize completely.

excited state   A state other than the state of lowest energy.

exclusion principle   No more than two electrons can occupy any given orbital; and, when two electrons do occupy one orbital, their spins must be paired.

exothermic process   A process that releases heat ($\Delta H < 0$). *Examples:* freezing; $N_2(g) + 3\ H_2(g) \rightarrow 2\ NH_3(g)$.

expanded valence shell   A valence shell containing more than eight electrons. Also called an *expanded octet*. *Examples:* the valence shells of P and S in PCl_5 and SF_6.

expansion work   See *work*.

experiment   A test carried out under carefully controlled conditions.

exponential   The exponential of x is the natural antilogarithm of x, namely, e^x.

exponential decay   A variation with time of the form e^{-kt}. *Example:* $[A] = [A]_0 e^{-k,t}$

extensive property   A physical property that depends on the size of the sample. *Examples:* volume; internal energy; enthalpy; entropy.

extrapolate   To extend a graph outside the region covered by the data.

extrinsic semiconductor   A material in which semiconduction arises from the presence of a low concentration of a dopant. *Example:* Arsenic added to very pure silicon.

face   See *crystal face*.

face-centered cubic structure (fcc)   A crystal structure built from a cubic unit cell in which there is an atom at the center of each face and one at each corner.

Fahrenheit scale   A temperature scale on which the freezing point of water is at 32 degrees and the normal boiling point is at 212 degrees. Units on this scale are degrees Fahrenheit, °F.

fallout   The fine dust that settles from clouds of airborne particles after a nuclear explosion.

Faraday's constant (F)   The magnitude of the charge per mole of electrons; $F = N_A e = 96.485\ kC \cdot mol^{-1}$.

Faraday's law of electrolysis   The amount of product formed by an electric current is chemically equivalent to the amount of electrons supplied.

fat   An ester of glycerol and carboxylic acids with long hydrocarbon chains; fats act as long-term energy storage in living systems.

fatty acid   A carboxylic acid with a long hydrocarbon chain. *Example:* $CH_3(CH_2)_{16}COOH$, stearic acid.

ferrimagnetic material   A material in which the electron spins on neighboring atoms are different and locked together in an antiferromagnetic arrangement.

ferroalloy   An alloy of a metal with iron and, often, carbon. *Example:* ferrovanadium.

ferrofluid   A magnetic liquid that is a suspension of a finely powdered magnetic material such as magnetite, Fe_3O_4, in a viscous, oily liquid (such as mineral oil) that contains a detergent.

ferromagnetic material   See *ferromagnetism*.

ferromagnetism   The ability of some substances to be permanently magnetized; the electron spins of neighboring atoms are aligned. *Examples:* iron; magnetite, Fe_3O_4.

ferrous alloy   An alloy based on iron and often including various other d-block metals. *Examples:* The varieties of steels.

field   An influence spreading over a region of space. *Examples:* an *electric field* from a charge; a *magnetic field* from a magnet or a moving charge.

filtration   The separation of a heterogeneous mixture of a solid and liquid by passing the mixture through a fine mesh.

first derivative (dy/dx)   A measure of the slope of a curve. See also Appendix 1F.

first ionization energy (I_1)   The minimum energy required to remove an electron from an atom. See *ionization energy*.

first law of thermodynamics   The internal energy of an isolated system is constant.

first-order reaction   A reaction in which the rate is proportional to the first power of the concentration of a substance.

fissile nucleus   A nucleus having the ability to undergo fission induced by slow neutrons. *Example:* ^{235}U is fissile.

fission (nuclear)   The breakup of a nucleus into two smaller nuclei of similar mass; fission may be *spontaneous* or *induced* (particularly by the impact of neutrons).

fissionable   Having the ability to undergo induced nuclear fission.

fix (nitrogen)   Convert elemental nitrogen to its compounds, particularly ammonia.

flocculation   The reversible aggregation of colloidal particles into larger particles that can be filtered.

fluorescence   The emission of light from molecules excited by radiation of higher frequency.

foam   (1) A frothy collection of bubbles formed by a liquid. (2) A type of *colloid* formed by a gas of tiny bubbles in a liquid or solid.

f-orbital   See *atomic orbital*.

force (F)   An influence that changes the state of motion of an object. *Examples:* an electrostatic force from an electric charge; a mechanical force from an impact.

formal anhydride   An *acid anhydride* that does not necessarily react with water to form the corresponding acid. *Example:* CO, the formal anhydride of formic acid, HCOOH.

formal charge   (1) The electric charge of an atom in a molecule assigned on the assumption that the bonding is nonpolar covalent. (2) Formal charge (FC) = number of valence electrons in the free atom − (number of lone-pair electrons + $\frac{1}{2}$ × number of shared electrons).

formation constant (K_f)   The equilibrium constant for complex formation. The *overall formation constant* is the product of *step-by-step formation constants*. The inverse of the formation constant ($1/K_f$) is called the *stability constant*.

formula unit   The group of ions that matches the formula of the smallest unit of an ionic compound. *Example:* NaCl, one Na^+ ion and one Cl^- ion.

formula weight   The numerical value of the *molar mass* of an ionic compound.

fossil fuel The partly decomposed remains of vegetable and marine life (mainly coal, oil, and natural gas).

fraction Samples of distillate obtained in different ranges of boiling temperatures.

fractional distillation Separation of the components of a liquid mixture by repeated distillation, making use of their differing volatilities.

Frasch process A process for mining sulfur that uses superheated water to melt the sulfur and compressed air to force it to the surface.

free energy See *Gibbs free energy*.

free expansion Expansion against zero opposing pressure.

freezing-point constant (k_f) The constant of proportionality between the freezing-point depression and the solute molality.

freezing-point depression The lowering of the freezing point of a solvent caused by the presence of a solute (a colligative property).

freezing temperature The temperature at which a liquid freezes.

frequency (of radiation) (ν, nu) The number of cycles (repeats of the waveform) per second (unit: *hertz*, Hz).

fuel cell A primary electrochemical cell in which the reactants are supplied continuously from outside while the cell is in use.

functional group A group of atoms that brings a characteristic set of chemical properties to an organic molecule. *Examples:* —OH; —Br; —COOH.

functionalization The introduction of functional groups into an alkane molecule.

fundamental charge (e) The magnitude of the charge of an electron.

fusion (1) Melting. (2) The merging of nuclei to form the nucleus of a heavier element.

galvanic cell See *electrochemical cell*.

galvanize Coat a metal with an unbroken film of zinc.

gamma (γ) **radiation** Very-high-frequency, short-wavelength electromagnetic radiation emitted by nuclei.

gas A fluid form of matter that fills the container that it occupies and can easily be compressed into a much smaller volume. (A gas differs from a vapor in that a gas is a substance at a higher temperature than its critical temperature; a vapor is a gaseous form of matter at a temperature below its critical temperature.)

gas constant (R) (1) The constant that appears in the ideal gas law. See inside the back cover of the book for values. (2) $R = N_A k$, where k is *Boltzmann's constant*.

gauge pressure The pressure inside a container less than that outside the container.

Geiger counter A device that is used to detect and measure radioactivity by relying on ionization caused by incident radiation.

gel A soft, solid colloid that typically consists of a liquid trapped within a solid network.

geometrical isomers Stereoisomers that differ in the spatial arrangement of the atoms. Geometrical isomers exhibit cis–trans isomerism.

Gibbs free energy ($G = H - TS$) The energy of a system that is free to do work at constant temperature and pressure. The direction of spontaneous change at constant pressure and temperature is the direction of decreasing Gibbs free energy.

Gibbs free energy of reaction The difference in molar Gibbs free energies of the products and reactants, weighted by the stoichiometric coefficients in the chemical equation.

glass An ionic solid with an amorphous structure resembling that of a liquid.

glass electrode A thin-walled glass bulb containing an electrolyte solution and a metallic contact; used for measuring pH.

graft copolymer See *copolymer*.

Graham's law of effusion The rate of effusion of a gas is inversely proportional to the square root of its molar mass.

gravimetric analysis An analytical method using measurements of mass.

gray (Gy) The SI unit of *absorbed dose*; 1 Gy corresponds to an energy deposit of 1 J·kg^{-1}. See also *rad*.

green chemistry The practice of chemistry that conserves resources and minimizes impact on the environment.

greenhouse effect The blocking by some atmospheric gases (notably carbon dioxide) of the radiation of heat from the surface of the Earth back into space, leading to the possibility of a worldwide rise in temperature.

greenhouse gas A gas that contributes to the *greenhouse effect*.

ground state The state of lowest energy.

group A vertical column in the periodic table.

Haber process (Haber–Bosch process) The catalyzed synthesis of ammonia at high pressure.

half-cell One compartment of an electrochemical cell consisting of an electrode and an electrolyte.

half-life ($t_{1/2}$) (1) In chemical kinetics, the time needed for the concentration of a substance to fall to half its initial value. (2) In radioactivity, the time needed for half the initial number of radioactive nuclei to disintegrate.

half-reaction A hypothetical oxidation or reduction reaction showing either electron loss or electron gain. *Examples:* $Na(s) \rightarrow Na^+(aq) + e^-$; $Cl_2(g) + 2\,e^- \rightarrow 2\,Cl^-(aq)$.

halide ion An anion formed from a halogen atom. *Examples:* F^-; I^-.

Hall process (Hall–Hérault process) The production of aluminum by the electrolysis of aluminum oxide dissolved in molten cryolite.

haloalkane An alkane with a halogen substituent. *Example:* CH_3Cl, chloromethane.

halogen A member of Group 17.

halogenation The incorporation of a halogen into a compound (particularly, into an organic compound).

hamiltonian The operator H in the Schrödinger equation; $H\psi = E\psi$.

hard matter Solid matter that can withstand strong forces without deforming.

hard water Water that contains dissolved calcium and magnesium salts.

heat (q) The energy that is transferred as the result of a temperature difference between a system and its surroundings.

heat capacity (C) The ratio of heat supplied to the temperature rise produced. The *heat capacity at constant pressure*, C_P, and the *heat capacity at constant volume*, C_V, are normally distinguished. See also *molar heat capacity; specific heat capacity*.

heating The act of transferring energy as heat.

heating curve A graph of the variation of the temperature of a sample as it is heated at a constant rate.

Heisenberg uncertainty principle If the location of a particle is known to within an uncertainty Δx, then the linear momentum parallel to the x-axis can be known only to within an uncertainty Δp, where $\Delta p \Delta x \geq \hbar/2$.

Henderson–Hasselbalch equation An approximate equation for estimating the pH of a solution containing a conjugate acid and base. See also Topic 6G.

Henry's constant The constant k_H that appears in *Henry's law*.

Henry's law The solubility of a gas in a liquid is proportional to its partial pressure above the liquid: solubility = $k_H \times$ *partial pressure*.

hertz (Hz) The SI unit of frequency: 1 Hz is one complete cycle per second; $1 \text{ Hz} = 1 \text{ s}^{-1}$.

Hess's law A reaction enthalpy is the sum of the enthalpies of any sequence of reactions (at the same temperature and pressure) into which the overall reaction can be divided.

heterogeneous alloy See *alloy*.

heterogeneous catalyst See *catalyst*.

heterogeneous equilibrium An equilibrium in which at least one substance is in a different phase from the others. *Example:*. $AgCl(s) \rightleftharpoons Ag^+(aq) + Cl^-(aq)$

heterogeneous mixture A mixture in which the individual components, although mixed together, lie in distinct regions that can be distinguished with an optical microscope. *Example:* a mixture of sand and sugar.

heteronuclear diatomic molecule A molecule consisting of two atoms of different elements. *Examples:* HCl; CO.

hexagonal close-packed structure (hcp) A close-packed structure with an ABABAB . . . pattern of layers.

high-boiling azeotrope See *azeotrope*.

highest occupied molecular orbital (HOMO) The highest-energy molecular orbital in the ground state of a molecule occupied by at least one electron.

high-spin complex A d^n complex with the maximum number of unpaired electron spins.

high-temperature superconductor A material that becomes superconducting at temperatures well above the transition temperature for the first generation of superconductors, typically 100 K and above.

homeostasis The maintenance of constant physiological conditions.

homogeneous alloy See *alloy*.

homogeneous catalyst See *catalyst*.

homogeneous equilibrium A chemical equilibrium in which all the substances taking part are in the same phase. *Example:* $H_2(g) + I_2(g) \rightleftharpoons 2 HI(g)$.

homogeneous mixture A mixture in which the individual components are uniformly mixed, even on a molecular scale. *Examples:* air; solutions.

homonuclear diatomic molecule A molecule consisting of two atoms of the same element. *Examples:* H_2; N_2.

Hund's rule If more than one orbital in a subshell is available, add electrons with parallel spins to different orbitals of that subshell.

hybrid orbital A mixed orbital formed by blending together atomic orbitals on the same atom. *Example:* an sp^3 hybrid orbital.

hybridization The formation of *hybrid orbitals*.

hydrate A solid compound containing H_2O molecules. *Example:* $CuSO_4 \cdot 5H_2O$.

hydrate isomers Isomers that differ by an exchange of an H_2O molecule and a ligand in the coordination sphere.

hydrated Having water molecules attached. See *hydration*.

hydration (1) of ions: The attachment of water molecules to a central ion. (2) of organic compounds: The addition of water across a multiple bond (H to one carbon atom, OH to the other). *Example:* $CH_2{=}CH_2 + H_2O \rightarrow CH_3CH_2OH$.

hydride A binary compound of a metal or metalloid with hydrogen; the term is often extended to include all binary compounds of hydrogen. A *saline* or *saltlike hydride* is a compound of hydrogen and a strongly electropositive metal; a *molecular hydride* is a compound of hydrogen and a nonmetal; a *metallic hydride* is a compound of certain d-block metals and hydrogen.

hydrocarbon A binary compound of carbon and hydrogen. *Examples:* CH_4; C_6H_6.

hydrogen bond A link formed by a hydrogen atom lying between two strongly electronegative atoms (O, N, or F). The electronegative atoms may be located on different molecules or in different regions of the same molecule.

hydrogenation The addition of hydrogen to multiple bonds. *Example:* $CH_3CH{=}CH_2 + H_2 \rightarrow CH_3CH_2CH_3$.

hydrohalogenation The addition of a hydrogen halide to an alkene to form a haloalkane. *Example:* $CH_3CH{=}CH_2 + HCl \rightarrow CH_3CHClCH_3$.

hydrolysis reaction The reaction of water with a substance, resulting in the formation of a new element–oxygen bond. *Example:* $PCl_5(s) + 4 H_2O(l) \rightarrow H_3PO_4(aq) + 5 HCl(aq)$.

hydrolyze To undergo hydrolysis. See also *hydrolysis reaction*.

hydrometallurgical process The extraction of metals by reduction of their ions in aqueous solution. *Example:* $Cu^{2+}(aq) + Fe(s) \rightarrow Cu(s) + Fe^{2+}(aq)$.

hydronium ion The ion H_3O^+.

hydrophilic Water attracting. *Example:* hydroxyl groups are hydrophilic.

hydrophobic Water repelling. *Example:* hydrocarbon chains are hydrophobic.

hydroxyl group An —OH group in an organic compound.

hypervalent compound A compound that contains an atom with more atoms attached than are allowed by the octet rule. *Example:* SF_6.

hypothesis A suggestion put forward to account for a series of observations. *Example:* Dalton's atomic hypothesis.

i **factor** The factor that takes into account the existence of ions in an electrolyte solution, particularly for the interpretation of colligative properties. It indicates the number of particles formed from one formula unit of the solute. *Example:* $i \approx 2$ for very dilute NaCl(aq).

ideal gas A gas that satisfies the ideal gas law and is described by the *kinetic model*.

ideal gas law ($PV = nRT$) All gases obey the law more and more closely as the pressure is reduced to very low values.

ideal solution A solution that obeys *Raoult's law* at any concentration; all solutions behave ideally as the concentration approaches zero. *Example:* benzene and toluene form an almost ideal system.

incandescence Light emitted by a hot body.

incomplete octet A valence shell of an atom that has fewer than eight electrons. *Example:* the valence shell of B in BF_3.

indefinite integral An *integral* without limits attached. See also Appendix 1F.

indicator A substance that changes color when it goes from its acidic to its basic form (an *acid–base indicator*) or from its oxidized to its reduced form (a *redox indicator*).

induced dipole moment An electric dipole moment produced in a polarizable molecule by an electric field.

induced nuclear fission See *fission*.

induced-fit mechanism A model of the action of an enzyme in which the enzyme molecule adjusts its shape to accommodate the incoming substrate molecule. A modification of the *lock-and-key mechanism* of enzyme action.

inert (1) Unreactive. (2) Thermodynamically unstable but surviving for long periods (*nonlabile*).

inert-pair effect The observation that an element displays a valence lower than would be expected from its group number. An *inert pair* is a pair of valence-shell s-electrons that are tightly bound to the atom and might not participate in bond formation.

infrared radiation Electromagnetic radiation with a lower frequency (longer wavelength) than that of red light but a higher frequency (shorter wavelength) than microwave radiation.

initial concentration (of a weak acid or base) The concentration as prepared, as if no deprotonation or protonation had taken place.

initial rate The rate at the start of the reaction when products are present in concentrations too low to affect the rate.

initiation The formation of reactive intermediates that serve as chain carriers from a reactant at the start of a chain reaction. *Example:* $Br_2 \rightarrow Br\cdot + Br\cdot$.

inner transition metal A member of the f-block of the periodic table (the *lanthanoids* and *actinoids*).

inorganic chemistry The study of the elements other than carbon and their compounds.

inorganic compound A compound that is not organic. See also *organic compound*.

insoluble substance A substance that does not dissolve in a specified solvent. When the solvent is not specified, water is generally meant.

instantaneous dipole moment A dipole moment that arises from a transient redistribution of charge and is responsible for the *London force*.

instantaneous rate The slope of the tangent of a graph of concentration against time.

insulator (electrical) A substance that does not conduct electricity. *Examples:* nonmetallic elements; molecular solids.

integral (1) The sum of infinitesimal quantities. (2) The inverse of a derivative in the sense that the integral of the first derivative of a function is the original function. See *definite integral, indefinite integral*.

integral calculus The part of mathematics that deals with the combination of infinitesimal quantities and the areas under curves. See also Appendix 1F.

integrated rate law An expression for the concentration of a reactant or product in terms of the time, obtained from the rate law of the reaction. *Example:* $[A] = [A]_0 e^{-k,t}$

intensity The brightness of electromagnetic radiation. The intensity of a wave of electromagnetic radiation is proportional to the square of its *amplitude*.

intensive property A physical property of a substance that is independent of the size of the sample. *Examples:* density; molar volume; temperature.

intercept (of a graph) The value at which the line cuts the specified (usually vertical) axis. See also Appendix 1E.

interference Interaction between waves, leading to a greater amplitude (*constructive interference*) or to a smaller one (*destructive interference*).

interhalogen A binary compound of two halogens. *Example:* IF_3.

intermediate See *reaction intermediate*.

intermolecular Between molecules.

intermolecular forces The forces of attraction and repulsion between molecules. *Examples:* hydrogen bonding; dipole–dipole force; London force. See also *van der Waals interactions*.

internal energy (U) In thermodynamics, the total energy of a system.

International System See *SI*.

internuclear axis The straight line between the nuclei of two bonded atoms.

interpolate To find a value between two measured values.

interstice A hole or gap in a crystal lattice.

interstitial alloy See *alloy*.

intramolecular Within a molecule.

intrinsic semiconductor A pure substance in which an empty conduction band lies close in energy to a full valence band.

ion An electrically charged atom or group of atoms. *Examples:* Al^{3+}; SO_4^{2-}. See also *anion; cation*.

ion–dipole interaction The attraction between an ion and the opposite partial charge of the electric dipole of a polar molecule.

ion exchange The exchange of one type of ion in solution for another.

ion pair A cation and anion in close proximity.

ionic bond The attraction between the opposite charges of cations and anions.

ionic character The extent to which ionic structures contribute to the resonance of a molecule or ion.

ionic compound A compound that consists of ions. *Examples:* $NaCl$; KNO_3.

ionic conduction Electrical conduction in which the charge is carried by ions. See also *electrical conduction; electronic conductor*.

ionic equation See *complete ionic equation*.

ionic liquid An ionic compound that is liquid at ordinary temperatures because one of its ions is a relatively large, organic ion. Ionic liquids are used as nontoxic, nonvolatile solvents.

ionic model   The description of bonding in terms of ions.

ionic radius   The contribution of an ion to the distance between neighboring ions in a solid ionic compound.

ionic solid   A solid built from cations and anions. *Examples:* $NaCl$; KNO_3.

ionization   (1) of atoms and molecules: Conversion into ions by the transfer of electrons. *Example:* $K(g) \rightarrow K^+(g) + e^-(g)$. (2) of acids and bases: See *protonation* and *deprotonation*.

ionization constant   See *acidity constant*.

ionization energy (I)   The minimum energy required to remove an electron from the ground state of a gaseous atom, molecule, or ion. (See *first ionization energy*). The *second ionization energy* is the ionization energy for removal of a second electron, and so on.

ionization isomers   Isomers that differ by the exchange of a ligand with an anion or neutral molecule outside the coordination sphere.

ionizing radiation   High-energy radiation (typically but not necessarily nuclear radiation) that can cause ionization.

ion-selective electrode   An electrode sensitive to the concentration of a particular ion.

irreversible process   A process that is not reversed by an infinitesimal change in a variable.

isoelectronic   Having the same number of atoms and the same number of valence electrons. *Examples:* F^- and Ne; SO_2 and O_3; CN^- and CO.

isolated system   See *system*.

isomer   One of two or more compounds that contain the same number of the same atoms in different arrangements. In *structural isomers*, the atoms have different partners or lie in a different order; in *stereoisomers*, the atoms have the same partners but are in different arrangements in space. *Optical isomers* are related like an object and its mirror image; they are types of stereoisomers. *Examples:* CH_3OCH_3 and CH_3CH_2OH (structural isomers); *cis*- and *trans*-2-butene (stereoisomers).

isomerization   A reaction in which a compound is converted into one of its isomers. *Example: cis*-butene $\rightarrow$ *trans*-butene.

isotherm   A line on a graph depicting the variation of a property at constant temperature.

isothermal change   A change that occurs at constant temperature.

isotonic   Having the same osmotic pressure.

isotope   One of two or more atoms that have the same atomic number but different mass numbers. *Example:* ^1H, ^2H, and ^3H are all isotopes of hydrogen.

isotopic abundance   See *abundance*.

isotopic dating   The determination of the age of objects by measuring the activity of the radioactive isotopes that they contain, particularly ^{14}C.

isotopic label   See *tracer*.

isotropic   Not depending on orientation.

joule (J)   The SI unit of energy ($1\ J = 1\ kg \cdot m^2 \cdot s^{-2}$).

Joule–Thomson effect   The cooling of a gas as it expands.

Kekulé structures   Two Lewis structures of benzene, consisting of alternating single and double bonds.

kelvin (K)   The SI unit of temperature. See also Appendix 1B.

Kelvin scale   A fundamental scale of temperature on which the triple point of water lies at 273.16 K and the lowest attainable temperature is at 0. The unit on the Kelvin scale is the *kelvin*, K.

ketone   An organic compound containing a carbonyl group between two carbon atoms, having the form R–CO–R′. *Example:* CH_3–CO–CH_2CH_3, butanone.

kilogram (kg)   The SI unit of mass. See also Appendix 1B.

kinetic energy (E_k)   The energy of a particle due to its motion. Kinetic energy may be *translational* (arising from motion through space), *rotational* (arising from rotation about a center of mass), or *vibrational* (arising from the oscillating motion of the atoms in a molecule). *Example:* the translational kinetic energy of a particle of mass m and speed v is $\frac{1}{2}mv^2$.

kinetic model   A model of the properties of an ideal gas in which pointlike molecules are in continuous random motion in straight lines until collisions occur between them.

kinetic molecular theory (KMT)   The mathematical version of the *kinetic model* of gases.

Kirchhoff's law   The relation between the standard reaction enthalpies at two temperatures in terms of the temperature difference and the difference in heat capacities (at constant pressure) of the products and reactants.

labeled   The replacement of an atom in a compound with a radioisotope of the same element in order to allow very small amounts of the compound to be detected.

labile   Refers to species that survive only for short periods.

lanthanide   Former (and still widely used) term for *lanthanoid*.

lanthanide contraction   The reduction of atomic radius of the elements following the lanthanides below the values that would be expected by extrapolation of the trend down a group (and arising from the poor shielding ability of f-electrons).

lanthanoid   A member of the first row of the f-block (cerium through ytterbium).

lattice   An orderly array of atoms, molecules, or ions in a crystal.

lattice energy   The difference between the potential energy of ions in a crystal and that of the same widely separated ions in a gas.

lattice enthalpy   The standard enthalpy change for the conversion of an ionic solid into a gas of ions.

law   A summary of an extensive series of observations.

law of conservation of energy   Energy can be neither created nor destroyed.

law of conservation of mass   Matter (and specifically atoms) is neither created nor destroyed in a chemical reaction.

law of constant composition   A compound has the same composition whatever its source.

law of mass action   For an equilibrium of the form $a\,A + b\,B \rightleftharpoons c\,C + d\,D$, the ratio $a_C^{\,c}a_D^{\,d}/a_A^{\,a}a_B^{\,b}$ evaluated at equilibrium is equal to a constant K, which has a specific value for a given chemical equation and temperature.

law of partial pressures   See *Dalton's law of partial pressures*.

law of radioactive decay   The rate of decay is proportional to the number of radioactive nuclides in the sample.

LCAO and LCAO-MO   See *linear combination of atomic orbitals*.

Le Chatelier's principle   When a stress is applied to a system in dynamic equilibrium, the equilibrium adjusts to minimize the effect of the stress. *Example:* a reaction at equilibrium tends to proceed in the endothermic direction when the temperature is raised.

leveling   The observation that strong acids all have the same strength in water, and all behave as though they were solutions of H_3O^+ ions.

Lewis acid   An electron pair acceptor. *Examples:* H^+; Fe^{3+}; BF_3.

Lewis base   An electron pair donor. *Examples:* OH^-; H_2O; NH_3.

Lewis formula (for an ionic compound)   A representation of the structure of an ionic compound showing the formula unit of ions in terms of their Lewis diagrams.

Lewis structure   A diagram showing how electron pairs are shared between atoms in a molecule.

Lewis symbol (for atoms and ions)   The chemical symbol of an element, with a dot for each valence electron.

ligand   A group attached to the central metal ion in a complex; a *polydentate ligand* occupies more than one binding site. See also *ambidentate ligand.*

ligand field splitting (Δ)   The energy separation of the e- and t-orbitals in a complex induced by the presence of ligands.

ligand field theory   The theory of bonding in d-metal complexes, a more complete version of *crystal field theory.* See also *crystal field.*

light   See *visible radiation.*

limiting enthalpy of solution   See *enthalpy of solution.*

limiting law   A law that is accurately obeyed only at the limit of a property, such as when a property (the pressure of a gas, for example) is made very small.

limiting reactant   The reactant that governs the theoretical yield of product in a given reaction.

line structure   A representation of the structure of an organic molecule in which lines represent bonds; carbon atoms and the hydrogen atoms attached to them are not usually shown explicitly.

linear combination of atomic orbitals (LCAO)   A sum of atomic orbitals to form a molecular orbital (an LCAO-MO).

linear momentum (p)   The product of mass and velocity.

linkage isomers   Isomers that differ in the identity of the atom that a ligand uses to attach to a metal ion.

lipid   A naturally occurring organic compound that dissolves in hydrocarbons but not in water. *Examples:* fats; steroids; terpenes; the molecules that form cell membranes.

liquid   A fluid form of matter that has a well-defined surface and takes the shape of the part of the container that it occupies.

liquid crystal   A substance that flows like a liquid but has molecules that lie in a moderately orderly array. Liquid crystals may be *nematic, smectic,* or *cholesteric,* depending on the arrangement of the molecules.

lock-and-key mechanism   A model of enzyme action in which the enzyme is thought of as a lock and its substrate as a matching key.

logarithm   If a number x is written as B^y, then y is the logarithm of x to the base B. For *common logarithms* (denoted log x), $B = 10$. For *natural logarithms* (denoted ln x), $B = e$. See also Appendix 1D.

London (dispersion) interaction   The interaction between instantaneous electric dipoles on neighboring molecules.

lone pair   A pair of valence electrons that is not taking part in bonding.

long period   A period of the periodic table with more than eight members.

long-range order   An orderly arrangement of atoms or molecules that is repeated over long distances.

lowest unoccupied molecular orbital (LUMO)   The lowest-energy molecular orbital that is unoccupied in the ground state.

low-spin complex   A d^n complex with the minimum number of unpaired electron spins.

luminescence   The emission of light from a process, other than incandescence, that results from the formation of an excited state.

Lyman series   A series of lines in the spectrum of atomic hydrogen in which the transitions are to orbitals with $n = 1$.

lyotropic liquid crystal   A liquid crystal that results from the action of a solvent on a solute.

macromolecules   Very large molecules consisting of hundreds of atoms.

macroscopic level   The level at which visible objects can be observed directly.

Madelung constant (A)   A number that appears in the expression for the lattice energy and depends on the type of crystal lattice.

magic numbers   The numbers of protons or neutrons that correlate with enhanced nuclear stability. *Examples:* 2, 8, 20, 50, 82, and 126.

magnetic field   A region of influence that affects the motion of moving charged particles.

magnetic quantum number (m_l)   The quantum number that identifies the individual orbitals of a subshell of an atom and determines their orientation in space.

main group   Any one of the groups forming the s- and p-blocks of the periodic table (Groups 1, 2, and 13 through 18).

malleability   The ability to be deformed by being struck by a hammer (as a metal).

manometer   An instrument used for measuring the pressure of a gas confined inside a container.

mantissa (of a logarithm)   The numbers to the right of the decimal point.

many-electron atom   An atom with more than one electron.

mass (m)   A measure of the quantity of matter in a sample.

mass number (A)   The total number of nucleons (protons plus neutrons) in the nucleus of an atom. *Example:* ^{14}C, with mass number 14, has 14 nucleons (6 protons and 8 neutrons).

mass percentage composition   The mass of a substance present in a sample, expressed as a percentage of the total mass of the sample.

mass spectrometer   An instrument used in *mass spectrometry.*

mass spectrometry   Technique for measuring the masses and abundances of atoms and molecules by passing a beam of ions through a magnetic field.

mass spectrum   The display of the relative number of particles with each specified mass; the output generated by the detector of a mass spectrometer. See also *mass spectrometry.*

material balance   The requirement that the sum of the concentrations of all forms of a solute in a solution be equal to the initial concentration of the solute. *Example:* the sum of the concentrations of HCN and CN$^-$ in an aqueous solution of HCN is equal to the initial concentration of HCN.

materials science   The study of the chemical structures, compositions, and properties of materials.

matter   Anything that has mass and takes up space.

maximum-boiling azeotrope   See *azeotrope*.

Maxwell distribution of speeds   The formula for calculating the fraction of molecules that move at any given speed in a gas at a specified temperature.

mean bond enthalpy (ΔH_B)   The mean of A—B bond enthalpies for a number of different molecules containing the A—B bond. See also *bond enthalpy*.

mean free path   The average distance that a molecule travels between collisions.

mean relative speed   The mean speed at which two molecules approach each other in a gas.

mechanical equilibrium   The state in which the pressure of a system is equal to that of its surroundings.

mechanism   See *reaction mechanism*.

melting temperature   The temperature at which a substance melts. See *normal melting point* and *standard melting point*.

meniscus   The curved surface that a liquid forms in a narrow tube.

mesophase   A state of matter showing some of the properties of both a liquid and a solid (a liquid crystal).

metal   (1) A substance that conducts electricity, has a metallic luster, is malleable and ductile, forms cations, and has basic oxides. (2) A metal consists of cations held together by a sea of electrons. *Examples:* iron; copper; uranium.

metallic bond   The form of bonding characteristic of metals in which cations are held together by a sea of electrons.

metallic conductor   An electronic conductor with a resistance that increases as the temperature is raised.

metallic hydride   See *hydride*.

metallic solid   See *metal*.

metallocene   A compound in which a metal atom lies between two cyclic ligands, thus resembling a sandwich. *Example:* ferrocene (dicyclopentadienyliron(0), $Fe(C_5H_5)_2$).

metalloid   An element that has the physical appearance and properties of a metal but behaves chemically like a nonmetal. *Examples:* arsenic; polonium.

meter (m)   The SI unit of length. See also Appendix 1B.

micelle   A compact, often nearly spherical cluster of oriented detergent (surfactant) molecules.

Michaelis constant (K_M)   A constant in the rate law for the *Michaelis–Menten mechanism*.

Michaelis–Menten mechanism   A model of enzyme catalysis in which the enzyme and its substrate reach a rapid *pre-equilibrium* with the bound substrate–enzyme complex.

microporous catalyst   A catalyst with an open, porous structure. *Example:* a *zeolite*.

microscopic level   A level of description that refers to the very small, such as atoms.

microstate   A permissible instantaneous arrangement of molecules in a sample (in the context of statistical thermodynamics and the statistical definition of entropy).

microwaves   Electromagnetic radiation with wavelengths close to 1 cm.

millimeter of mercury (mmHg)   The pressure exerted by a column of mercury of height 1 mm (at 15 °C and in a standard gravitational field).

minerals   Substances that are mined; more generally, inorganic substances.

minimum-boiling azeotrope   See *azeotrope*.

mixture   A type of matter that consists of more than one substance and can be separated into its components by making use of the different physical properties of the substances present.

model   A simplified description of nature.

moderator   A substance that slows neutrons. *Examples:* graphite; heavy water.

molality (b_J)   The amount of solute (in moles) divided by the mass of solvent (in kilograms).

molar   Refers to the quantity per mole. *Examples: molar mass*, the mass per mole; *molar volume*, the volume per mole. (*Molar concentration* and some related quantities are exceptions.)

molar absorption coefficient   The constant of proportionality between the absorbance of a sample and the product of its molar concentration and path length. See also *Beer's law*.

molar concentration ([J], c_J)   The amount (in moles) of solute divided by the volume of the solution (in liters).

molar heat capacity (C_m)   The heat capacity per mole of substance.

molar mass (M)   (1) The mass per mole of atoms of an element. (2) The mass per mole of molecules of a molecular compound. (3) The mass per mole of formula units of an ionic compound.

molar solubility (s)   The numerical value of the molar concentration of a saturated solution of a substance.

molar volume (V_m)   The volume of a sample divided by the amount (in moles) of atoms, molecules, or formula units that it contains.

molarity ([J], c_J)   The informal name for *molar concentration*.

mole (mol)   The SI unit of chemical amount. See also Appendix 1B.

mole fraction (x)   The amount of particles (molecules, atoms, or ions) of a substance in a mixture expressed as a fraction of the total amount of particles in the mixture.

mole ratio   The stoichiometric relation between two species in a chemical reaction written as a conversion factor. *Example:* (2 mol H_2)/(1 mol O_2) in the reaction 2 H_2(g) + O_2(g) → 2 H_2O(l).

molecular biology   The study of the functions of living organisms in terms of their molecular composition.

molecular compound   A compound that consists of molecules. *Examples:* water; sulfur hexafluoride; benzoic acid.

molecular formula   A combination of chemical symbols and subscripts showing the actual numbers of atoms of each element present in a molecule. *Examples:* H_2O; SF_6; C_6H_5COOH.

molecular hydride   See *hydride*.

molecular orbital   A wavefunction that spreads throughout a molecule and gives the probability (through its square) that an electron will be found at each location.

molecular orbital energy-level diagram A portrayal of the relative energies of the molecular orbitals in a molecule.

molecular orbital theory The description of molecular structure in which electrons occupy orbitals that spread throughout a molecule.

molecular potential energy curve A plot of the energy of a molecule against internuclear distance, with the nuclei stationary.

molecular solid A solid consisting of a collection of individual molecules held together by intermolecular forces. *Examples:* glucose; aspirin; sulfur.

molecular weight The numerical value of the *molar mass* of a molecular compound.

molecularity The number of reactant molecules (or free atoms) taking part in an *elementary reaction*. See also *bimolecular reaction; termolecular reaction; unimolecular reaction.*

molecule (1) The smallest particle of a compound that possesses the chemical properties of the compound. (2) A definite, distinct, electrically neutral group of bonded atoms. *Examples:* H_2; NH_3; CH_3COOH.

momentum (p) See *linear momentum.*

monatomic ion An ion formed from a single atom. *Examples:* Na^+; Cl^-.

monomer A small molecule from which a polymer is formed. *Examples:* $CH_2{=}CH_2$ for polyethylene; $NH_2(CH_2)_6NH_2$ for nylon.

monoprotic acid A Brønsted acid with one acidic hydrogen atom. *Example:* CH_3COOH.

monosaccharide An individual unit from which carbohydrates are considered to be composed. *Example:* $C_6H_{12}O_6$, glucose.

multiple bond A double or triple bond between two atoms.

nanomaterials Materials composed of particles that range in size from about 1 to 100 nm and can be manufactured and manipulated at the molecular level.

nanoscience The study of materials at the nanometer scale. These materials are larger than single atoms but too small to exhibit most bulk properties.

nanotechnology The study and manipulation of matter at the atomic level (nanometer scale).

natural abundance (of an isotope) The abundance of an isotope in a sample of a naturally occurring material.

natural antilogarithm See *exponential.*

natural logarithm See *logarithm.*

natural product An organic substance that occurs naturally in the environment.

naturally occurring Found in nature without needing to be synthesized.

negative deviation (from Raoult's law) The tendency of a nonideal solution to have a vapor pressure lower than that predicted by *Raoult's law.*

nematic phase A liquid-crystal phase in which rod-shaped molecules have axes arranged parallel to each other but staggered with respect to one another in other directions.

Nernst equation The equation expressing the potential of an electrochemical cell in terms of the concentrations of the reagents taking part in the cell reaction; $E_{cell} = E_{cell}° - (RT/n_rF) \ln Q$.

net ionic equation The equation showing the net change in a chemical reaction, obtained by canceling the spectator ions in a complete ionic equation. *Example:* $Ag^+(aq) + Cl^-(aq) \rightarrow AgCl(s)$.

network solid A solid consisting of atoms linked together covalently throughout its extent. *Examples:* diamond; silica.

neutralization reaction The reaction of an acid with a base to form a salt. *Example:* $HCl(aq) + NaOH(aq) \rightarrow NaCl(aq) + H_2O(l)$.

neutron (n) An electrically neutral subatomic particle found in the nucleus of an atom; it has approximately the same mass as a proton.

neutron-induced transmutation The conversion of one nucleus into another by the impact of a neutron. *Example:* $^{58}_{26}Fe + 2\,^1_0n \rightarrow\,^{60}_{27}Co +\,^0_{-1}e$.

neutron-rich nucleus A nucleus with such a high proportion of neutrons that it lies above the *band of stability.*

noble gas A member of Group 18 of the periodic table.

nodal plane A plane on which an electron in an atom or molecule will not be found.

node A point or surface on which a wavefunction passes through zero.

nomenclature See *chemical nomenclature.*

nonaqueous solution A solution in which the solvent is not water. *Example:* sulfur in carbon disulfide.

nonbonding orbital A valence-shell atomic orbital that has not been used to form a bond to another atom.

nonelectrolyte A substance that dissolves to give a solution that does not conduct electricity. *Example:* sucrose.

nonelectrolyte solution A solution of a nonelectrolyte.

nonexpansion work (w_e) See *work.*

nonferrous alloy An alloy based on metals other than iron. *Examples:* brass, bronze.

nonideal solution A solution that does not obey *Raoult's law.* Compare with *ideal solution.*

nonlabile Refers to a species that survives for a long period, even if it is thermodynamically unstable.

nonmetal A substance that does not conduct electricity and is neither malleable nor ductile. *Examples:* all gases; phosphorus; sodium chloride.

nonpolar bond (1) A covalent bond between two atoms that have zero partial charges. (2) A covalent bond between two atoms with the same or nearly the same electronegativity.

nonpolar molecule A molecule with zero electric dipole moment.

normal boiling point (T_b) (1) The boiling temperature when the pressure is 1 atm. (2) The temperature at which the vapor pressure of a liquid is 1 atm.

normal form The form of a substance under typical everyday conditions (for instance, close to 1 atm, 25 °C).

normal freezing point (T_f) The temperature at which a liquid freezes at 1 atm.

normal melting point (T_f) The melting point of a substance at a pressure of 1 atm.

NO_x An oxide, or mixture of oxides, of nitrogen, typically in atmospheric chemistry.

ns-orbital An atomic orbital with principal quantum number n and $l = 0$.

n-type semiconductor   See *semiconductor*.

nuclear atom   The structure of the atom proposed by Rutherford: a central, small, very dense, positively charged nucleus surrounded by electrons.

nuclear binding energy (E_{bind})   The energy released when Z protons and $A - Z$ neutrons come together to form a nucleus with atomic number Z and mass number A. The greater the binding energy per nucleon, the lower is the energy of the nucleus.

nuclear chemistry   The study of the chemical consequences of nuclear reactions.

nuclear decay   The spontaneous partial breakup of a nucleus (including its fission). Nuclear decay is also referred to as *nuclear disintegration*. *Example*: $^{226}_{88}Ra \rightarrow {}^{222}_{86}Rn + {}^{4}_{2}\alpha$

nuclear equation   A summary of the changes in a nuclear reaction, written in a form resembling a chemical equation.

nuclear fission   See *fission*.

nuclear fusion   See *fusion*.

nuclear model   A model of the atom in which the electrons surround a minute central nucleus.

nuclear reaction   A change that a nucleus undergoes (such as a nuclear transmutation).

nuclear reactor   A device for achieving controlled self-sustaining nuclear fission.

nuclear transmutation   The conversion of one element into another. *Example*: $^{12}_{6}C + {}^{4}_{2}\alpha \rightarrow {}^{16}_{8}O + \gamma$.

nucleic acid   (1) The product of a condensation of nucleotides. (2) A molecule containing an organism's genetic information.

nucleon   A proton or a neutron; thus, either of the two principal components of a nucleus.

nucleophile   A reactant that seeks out centers of positive charge in a molecule. *Examples*: H_2O, OH^-.

nucleophilic substitition   Substitution that results from attack by a nucleophile. *Examples*: hydrolysis of haloalkanes, $CH_3Br + H_2O \rightarrow CH_3OH + HBr$.

nucleoside   A combination of an organic base and a ribose or deoxyribose molecule.

nucleosynthesis   The formation of an element.

nucleotide   A nucleoside with a phosphate group attached to the carbohydrate ring; one of the units from which nucleic acids are made.

nucleus   The small, positively charged particle at the center of an atom that is responsible for most of its mass.

nuclide   An atom of specified atomic number and mass number. *Examples*: $^{1}_{1}H$; $^{16}_{8}O$

nutritional calorie (Cal)   A unit used in food science to measure the caloric value of food; 1 Cal = 1 kcal.

occupy   Have the characteristics of the wavefunction of a specified state; to be in a specified state.

octahedral complex   A complex in which six ligands are arranged at the corners of a regular octahedron, with a metal atom or ion at the center. *Example*: $[Fe(CN)_6]^{4-}$.

octahedral hole   A cavity in a (usually close packed) crystal lattice formed from six spheres at the corners of a regular octahedron.

octet   An ns^2np^6 valence electron configuration.

octet rule   When atoms form bonds, they proceed as far as possible toward completing their octets by sharing electron pairs.

oligopeptide   A short chain of amino acids connected by amide (peptide) bonds.

one-component phase diagram   See *phase diagram*.

open system   See *system*.

optical activity   The ability of a substance to rotate the plane of polarized light passing through it.

optical isomers   Isomers that are related like an object and its mirror image. Optical isomerism is the existence of optical isomers. Optical isomerism is a type of stereoisomerism.

orbital   See *atomic orbital* and *molecular orbital*.

orbital angular momentum   A measure of the rate of rotation.

orbital angular momentum quantum number (l)   The quantum number that specifies the subshell of a given shell in an atom and determines the shapes of the orbitals in the subshell; $l = 0$, $1, 2, \ldots, n - 1$. *Examples*: $l = 0$ for the s-subshell; $l = 1$ for the p-subshell. (The quantum number l also specifies the magnitude of the angular momentum of the electron around the nucleus.)

order of reaction   The power to which the concentration of a single substance is raised in a rate law. *Example*: if the rate = $k_r[SO_2][SO_3]^{-1/2}$, then the reaction is first order in SO_2, and of order $-\frac{1}{2}$ in SO_3. See *first-order reaction, second-order reaction, zeroth-order reaction*.

ore   The natural mineral source of a metal. *Example*: Fe_2O_3, hematite, an iron ore.

organic chemistry   The branch of chemistry that deals with *organic compounds*.

organic compound   A compound containing the element carbon and usually hydrogen. (The carbonates are normally excluded.)

organometallic compound   A compound that contains a metal–carbon bond. *Example*: $Ni(CO)_4$. (Complexes of CN^- are normally excluded.)

oscillating   Varying in a periodic manner with time.

osmometry   The determination of the molar mass of a solute from osmotic pressure measurements.

osmosis   The tendency of a solvent to flow through a semipermeable membrane into a more concentrated solution (a colligative property).

osmotic pressure (Π, pi)   The pressure needed to stop the flow of solvent through a semipermeable membrane. See also *osmosis*.

overall formation constant   See *formation constant*.

overall order   The sum of the powers to which individual concentrations are raised in the rate law of a reaction. *Example*: If rate = $k_r[SO_2][SO_3]^{-1/2}$, then the overall order is $\frac{1}{2}$.

overall reaction   The net outcome of a sequence of reactions.

overlap   The merging of orbitals belonging to different atoms of a molecule.

overpotential   The additional potential difference that must be applied beyond the cell potential to cause appreciable electrolysis.

oxidant   See *oxidizing agent*.

oxidation   (1) Combination with oxygen. (2) A reaction in which an atom, ion, or molecule loses an electron. (3) A half-reaction in which the oxidation number of an element is increased. *Examples*: (1, 2) $2 Mg(s) + O_2(g) \rightarrow 2 MgO(s)$; (2, 3) $Mg(s) \rightarrow Mg^{2+}(s) + 2 e^-$.

oxidation number The effective charge on an atom in a compound, calculated according to a set of rules (see Toolbox K.1). An increase in oxidation number corresponds to oxidation; a decrease corresponds to reduction.

oxidation–reduction reaction See *redox reaction*.

oxidation state The actual condition of a species with a specified oxidation number.

oxidizing agent A species that removes electrons from a species being oxidized (and is itself reduced) in a redox reaction. *Examples*: O_2; O_3; MnO_4^-; Fe^{3+}.

oxoacid An acid that contains oxygen. *Examples*: H_2CO_3; HNO_3; HNO_2; $HClO$.

oxoanion An anion of an oxoacid. *Examples*: HCO_3^-; CO_3^{2-}.

paired electrons Two electrons with opposed spins ($\uparrow\downarrow$).

parallel spins Electrons with spins aligned in the same direction ($\uparrow\uparrow$).

paramagnetic substance See *paramagnetism*.

paramagnetism The tendency to be pulled into a magnetic field; a paramagnetic substance is composed of atoms or molecules with unpaired electrons. *Examples*: O_2; $[Fe(CN)_6]^{3-}$.

parent nucleus In a nuclear reaction, the nucleus that undergoes disintegration or transmutation.

partial charge A charge arising from small shifts in the distributions of electrons. A partial charge can be either positive ($\delta+$) or negative ($\delta-$).

partial pressure (P_J) The pressure that a gas (J) in a mixture would exert if it alone occupied the container.

particle in a box A particle confined between rigid walls.

parts per million (ppm) (1) The ratio of the mass of a solute to the mass of the solution, multiplied by 10^6. (2) The *mass percentage composition* multiplied by 10^4. (*Parts per billion*, ppb, the mass ratio multiplied by 10^9, also may be used.)

pascal (Pa) The SI unit of pressure: $1 Pa = 1 kg \cdot m^{-1} \cdot s^{-2}$. See also Appendix 1B.

passivation Protection from further reaction by a surface film. *Example*: aluminum in air.

Pauli exclusion principle See *exclusion principle*.

p-electron An electron in a p-orbital.

penetration The possibility that an s-electron may be found inside the inner shells of an atom and hence close to the nucleus.

peptide A molecule formed by a condensation reaction between amino acids; often described in terms of the number of units, for example, *dipeptide, oligopeptide, polypeptide*.

peptide bond The —CONH— group.

percentage composition See *mass percentage composition*; *volume percentage composition*.

percentage deprotonation The fraction of a weak acid, expressed as a percentage, that is present as its conjugate base in a solution.

percentage ionization The fraction of molecules of a substance, expressed as a percentage, that is present as ions.

percentage protonated The fraction of a base, expressed as a percentage, that is present as its conjugate acid in a solution.

percentage yield The percentage of the theoretical yield of a product achieved in practice.

period A horizontal row in the periodic table; the number of the period is equal to the principal quantum number of the valence shell of the atoms.

periodic law The recognition of the periodicity of the properties of the elements.

periodic table A chart in which the elements are arranged in order of atomic number and divided into groups and periods in a manner that shows the relationships between the properties of the elements.

pH The negative logarithm of the molar concentration of hydronium ions in a solution: $pH = -\log [H_3O^+]$. A $pH < 7$ indicates an acidic solution; $pH = 7$, a neutral solution; and $pH > 7$, a basic solution.

pH curve A graph of the pH of a reaction mixture against volume of titrant added in an acid–base titration.

pH meter An electronic device used to measure the pH of a solution.

phase A specific physical state of matter. A substance may exist in solid, liquid, and gaseous phases and, in certain cases, in more than one solid or liquid phase. *Examples*: white and gray tin are two solid phases of tin; ice, liquid, and vapor are three phases of water.

phase boundary (1) A line separating two areas in a phase diagram; the points on a phase boundary represent the conditions at which the two adjoining phases are in dynamic equilibrium. (2) The surface between two phases.

phase diagram A summary in graphical form of the conditions of temperature and pressure at which the various solid, liquid, and gaseous phases of a substance exist. A *one-component phase diagram* is a phase diagram for a single substance.

phase transition The conversion of a substance from one phase into another phase. *Examples*: vaporization; white tin $\rightarrow$ gray tin.

phenol An organic compound in which a hydroxyl group is attached directly to an aromatic ring (Ar—OH). *Example*: C_6H_5OH, phenol.

phosphor A phosphorescent material that emits light after excitation to higher energy states.

phosphorescence Long-lasting *luminescence*.

photochemical reaction A reaction caused by light. *Example*: $H_2(g) + Cl_2(g) \xrightarrow{h\nu} 2 HCl(g)$.

photoelectric effect The emission of electrons from the surface of a metal when electromagnetic radiation strikes it.

photon A particle-like packet of electromagnetic radiation. The energy of a photon of frequency ν is $E = h\nu$.

physical change A change in which the identity of the substance does not change and only its physical properties are different. *Example*: freezing.

physical chemistry The study of the principles of chemistry.

physical equilibrium A state in which two or more phases of a substance coexist without a tendency to change. Example: ice and water at $0\,°C$ and 1 atm.

physical property A characteristic observed or measured without changing the identity of the substance.

physical state The condition of being a solid, a liquid, or a gas at a particular temperature.

pi (π) bond   A bond formed by the side-to-side overlap of two p-orbitals.

pi (π) orbital   A molecular orbital that has one nodal plane cutting through the internuclear axis.

piezoelectric   Having the property of becoming electrically charged when mechanically distorted. *Example*: $BaTiO_3$.

pipet   A narrow tube, sometimes with a central bulb, calibrated to deliver a specified volume.

pK_a and pK_b   The negative logarithms of the acidity and basicity constants: $pK = -\log K$. The larger the value of pK_a or pK_b, the weaker is the acid or base, respectively.

Planck's constant (h)   A fundamental constant of nature with the value 6.626×10^{-34} J·s.

plasma   (1) An ionized gas. (2) In biology, the colorless component of blood in which the red and white blood cells are dispersed.

p–n junction   An interface between a p-type and an n-type semiconductor.

pOH   The negative logarithm of the hydroxide ion molarity in a solution; $pOH = -\log [OH^-]$.

poison   Verb: To inactivate a catalyst.

polar covalent bond   A covalent bond between atoms that have partial electric charges. *Examples*: H—Cl; O—S.

polar molecule   A molecule with a nonzero electric dipole moment. *Examples*: HCl; NH_3.

polarizability (α)   A measure of the ease with which the electron cloud of a molecule can be distorted.

polarizable   An easily polarized species. See *polarize*.

polarize   To distort the electron cloud of an atom, ion, or molecule.

polarized light   Plane-polarized light is light in which the wave motion occurs in a single plane.

polarizing power   The ability of an ion to polarize a neighboring atom or ion.

polyamide   A polymer in which the monomers are linked by amide bonds formed by condensation.

polyatomic ion   An ion in which more than two atoms are linked by covalent bonds. *Examples*: NH_4^+; NO_3^-; SiF_6^{2-}.

polyatomic molecule   A molecule that consists of more than two atoms. *Examples*: O_3; $C_{12}H_{22}O_{11}$.

polydentate ligand   A ligand that can attach at several binding sites.

polyester   A polymer in which the monomers are linked by ester groups formed by condensation polymerization.

polymer   A substance with large molecules consisting of chains of covalently linked repeating units formed from small molecules known as *monomers*. *Examples*: polyethylene; nylon. See also *copolymer*.

polymerization   The formation of a *polymer* from its *monomers*.

polynucleotide   A polymer built from nucleotide units. *Examples*: DNA; RNA.

polypeptide   A polymer formed by the condensation of amino acids.

polyprotic acid   A molecule that can donate more than one proton. (A polyprotic acid is sometimes called a polybasic acid.) *Example*: H_3PO_4, triprotic acid.

polyprotic base   A molecule that can accept more than one proton. *Example*: N_2H_4.

polysaccharide   A chain of many saccharide units, such as glucose, linked together. *Examples*: cellulose; amylose.

p-orbital   See *atomic orbital*.

positional disorder   Disorderly locations of molecules; a contribution to the entropy.

positive ion   An ion formed by loss of one or more electrons from an atom or molecule; a *cation*.

positron   A fundamental particle with the same mass as an electron (β^+) but with opposite charge.

positron emission   A mode of radioactive decay in which a nucleus emits a *positron*.

potential difference   An electrical potential difference between two points is a measure of the work that must be done to move an electric charge from one point to the other. Potential difference is measured in volts, V, and is commonly called *voltage*.

potential energy (E_p)   The energy arising from position. *Example*: the Coulomb potential energy of a charge is inversely proportional to its distance from another charge.

potential energy surface   A surface showing the variation of the potential energy with the relative location of the atoms in a polyatomic cluster of atoms (as in a collision between a diatomic molecule and an atom).

power   The rate of supply of energy. The SI unit of power is the watt, W ($1\ W = 1\ J·s^{-1}$). See also Appendix 1B.

precipitate   The solid formed in a *precipitation reaction*.

precipitation   The process in which a solute comes out of solution rapidly as a finely divided powder, called a *precipitate*.

precipitation reaction   A reaction in which a *precipitate* is formed when two solutions are mixed. *Example*: $KBr(aq) + AgNO_3(aq) \rightarrow KNO_3(aq) + AgBr(s)$.

precise measurements   (1) Measurements with a large number of significant figures. (2) A series of measurements with small random error and hence in close mutual agreement. See *precision*.

precision   Freedom from random error. Compare with *accuracy*.

pre-equilibrium condition   A pre-equilibrium arises (or is assumed) when an intermediate is formed in a rapid equilibrium reaction prior to a slow step in a reaction mechanism.

pre-exponential factor (A)   The constant obtained from the y intercept in an Arrhenius plot.

pressure (P)   Force divided by the area to which it is applied.

primary alcohol   See *alcohol*.

primary cell   A *galvanic cell* that produces electricity from chemicals sealed within it at the time of manufacture. It cannot be recharged.

primary pollutant   A pollutant introduced directly into the environment. *Example*: SO_2.

primary structure   The sequence of amino acids in the polypeptide chain of a protein.

primitive cubic structure   A structure in which the unit cell consists of spheres (representing atoms or ions) at the corners of a cube.

principal quantum number (n)   The quantum number that specifies the energy of an electron in a hydrogen atom and labels the shells of the atom.

probability density (of a particle)   A function that, when multiplied by the volume of the region, gives the probability that the particle will be found in that region of space. See also *Born interpretation.*

product   A species formed in a chemical reaction.

promotion (of an electron)   The conceptual excitation of an electron to an orbital of higher energy in the description of bond formation.

propagation   A series of steps in a chain reaction in which one chain carrier reacts with a reactant molecule to produce another carrier. See also *chain reaction.*

property   A characteristic of matter. *Examples:* vapor pressure; color; density; temperature. See also *chemical property; physical property.*

protective oxide   An oxide that protects a metal from further oxidation. *Example:* aluminum oxide.

proton (p)   A positively charged subatomic particle found in the nucleus of an atom.

proton emission   A nuclear decay process in which a proton is emitted. In proton emission, the mass and charge numbers of the nucleus both decrease by 1.

proton transfer equilibrium   The equilibrium involving the transfer of a hydrogen ion between an acid and a base.

proton transfer reaction   See *proton transfer equilibrium.*

protonation   Proton transfer to a Brønsted base. *Example:* $H_3O^+(aq) + HS^-(s) \rightarrow H_2S(g) + H_2O(l)$.

proton-rich nucleus   A nucleus that has a low proportion of neutrons and lies below the *band of stability.*

pseudo-first-order reaction   A reaction with a rate law that is effectively first order because all but one species have virtually constant concentrations.

p-type semiconductor   See *semiconductor.*

pX   The quantity $-\log X$. *Example:* $pOH = -\log [OH^-]$.

pyrometallurgical process   The extraction of metals by using reactions at high temperatures. *Example:* $Fe_2O_3(s) + 3CO(g) \xrightarrow{\Delta} 2Fe(l) + 3CO_2(g)$.

quadratic equation   An equation of the form $ax^2 + bx + c = 0$. See also Appendix 1E.

qualitative   A non-numerical description of the properties of a substance, system, or process. See *qualitative analysis.*

qualitative analysis   The identification of the substances present in a sample.

quanta   The plural of *quantum.*

quantitative   A numerical description of the properties of a substance, system, or process. See *quantitative analysis.*

quantitative analysis   The determination of the amount of each substance present in a sample.

quantization   The restriction of a property to certain values. *Examples:* the quantization of energy and angular momentum.

quantum   A packet of energy.

quantum dot   A small three-dimensional cluster of semiconducting materials. *Example:* 10 to 10^5 atoms of Cd and Se (as cadmium selenide, CdSe).

quantum mechanics   The description of matter that takes into account the wave–particle duality of matter and the fact that the energy of an object may be changed only in discrete steps.

quantum number   An integer (sometimes, a half-integer) that labels a wavefunction and specifies the value of a property. *Example:* principal quantum number, n.

quaternary ammonium ion   An ion of the form NR_4^+, where R denotes hydrogen or an alkyl group (the four groups may be different).

quaternary structure   The manner in which neighboring polypeptide units stack together to form a protein molecule.

racemic mixture   A mixture containing equal amounts of two enantiomers.

rad   A (non-SI) unit of *absorbed dose* of radiation (see *radiation absorbed dose*); 1 rad corresponds to an energy deposit of 10^{-2} J·kg^{-1}. See also *gray.*

radial distribution function   The function that gives the probability that an electron in an atom will be found at a particular radius regardless of the direction.

radial wavefunction $(R(r))$ The radial part of a wavefunction, particularly the radial component of the wavefunctions of the hydrogen atom; the probability amplitude of an electron as a function of distance from the nucleus.

radiation absorbed dose (rad)   The amount of radiation that deposits 10^{-2} J of energy per kilogram of tissue.

radical   An atom, molecule, or ion with at least one unpaired electron. *Examples:* ·NO; ·O·; ·CH$_3$.

radical chain reaction   A chain reaction propagated by radicals.

radical polymerization   A polymerization procedure that utilizes a radical chain reaction.

radioactive   A nucleus is radioactive if it can change its structure spontaneously and emit radiation.

radioactive series   A step-by-step nuclear decay path in which α and β particles are successively ejected and which terminates at a stable nuclide (often an isotope of lead).

radioactivity   The spontaneous emission of radiation by nuclei. Such nuclei are radioactive.

radiocarbon dating   *Isotopic dating* based specifically on the use of carbon-14.

radioisotope   A radioactive isotope.

radius ratio   The ratio of the radius of the smaller ion in an ionic solid to the radius of the larger ion. The radius ratio controls which crystal structure is adopted by a simple ionic solid.

random copolymer   See *copolymer.*

random error   An error that varies randomly from measurement to measurement, sometimes giving a high value and sometimes a low one.

Raoult's law   The vapor pressure of a liquid solution of a nonvolatile solute is directly proportional to the mole fraction of the solvent in the solution: $P = x_{solvent}P_{pure}$, where P_{pure} is the vapor pressure of the pure solvent.

rate constant (k_r) The constant of proportionality in a rate law.

rate-determining step   The elementary reaction that governs the rate of the overall reaction. *Example:* the step $O + O_3 \rightarrow O_2 + O_2$ in the decomposition of ozone.

rate law   An equation expressing the instantaneous reaction rate in terms of the concentrations, at any instant, of the substances taking part in the reaction. *Example:* rate $= k_r[NO_2]^2$.

rate of reaction   See *instantaneous rate; reaction rate.*

reactant   A species acting as a starting material in a chemical reaction; a reagent taking part in a specified reaction.

reaction enthalpy (ΔH)   The change of enthalpy for the reaction as expressed by the chemical equation, with stoichiometric coefficients interpreted as the amounts in moles. *Example:* $CH_4(g) + 2\,O_2(g) \rightarrow CO_2(g) + 2\,H_2O(l)$, $\Delta H = -890$ kJ.

reaction Gibbs free energy (ΔG)   See *Gibbs free energy of reaction.*

reaction intermediate   A species that is produced and consumed during a reaction but does not appear in the overall chemical equation.

reaction mechanism   The pathway that is proposed for an overall reaction and accounts for the experimental rate law.

reaction order   See *order of reaction.*

reaction profile   The variation in potential energy as two reactants meet, form an activated complex, and separate as products.

reaction quotient (Q)   The ratio of the activities of the products to those of the reactants, each raised to a power equal to the stoichiometric coefficient (as in the definition of the equilibrium constant, but at an arbitrary stage of a reaction). *Example:* for $N_2(g) + 3\,H_2(g) \rightarrow 2\,NH_3(g)$, $Q = (P_{NH_3})^2/P_{N_2}(P_{H_2})^3$

reaction rate   The unique rate of a chemical reaction calculated by dividing the change in concentration of a substance by the interval during which the change takes place and by taking into account the stoichiometric coefficient of the substance. See also *unique average reaction rate.*

reaction sequence   A series of reactions in which the products of one reaction take part as reactants in the next. *Example:* $2\,C(s) + O_2(g) \rightarrow 2\,CO(g)$, followed by $2\,CO(g) + O_2(g) \rightarrow 2\,CO_2(g)$.

reaction stoichiometry   The quantitative relation between the amounts of reactants consumed and those of products formed in chemical reactions as expressed by the balanced chemical equation for the reaction.

reagent   A substance or a solution that reacts with other substances.

real gas   An actual gas; a gas that differs from an ideal gas in its behavior.

recrystallization   Purification by repeated dissolving and crystallization.

redox couple   The oxidized and reduced forms of a substance taking part in a reduction or oxidation half-reaction. The notation is oxidized species/reduced species. *Example:* H^+/H_2.

redox indicator   See *indicator.*

redox reaction   A reaction in which oxidation and reduction take place. *Example:* $S(s) + 3\,F_2(g) \rightarrow SF_6(g)$.

redox titration   See *titration.*

reducing agent   The species that supplies electrons to a substance being reduced (and is itself oxidized) in a redox reaction. *Examples:* H_2; H_2S; SO_3^{2-}.

reductant   See *reducing agent.*

reduction   (1) The removal of oxygen from, or the addition of hydrogen to, a compound. (2) A reaction in which an atom, ion, or molecule gains an electron. (3) A half-reaction in which the oxidation number of an element is decreased. *Example:* $Cl_2(g) + 2\,e^- \rightarrow 2\,Cl^-$ (aq).

re-forming reaction   A reaction in which a hydrocarbon is converted into carbon monoxide and hydrogen over a nickel catalyst.

refractory   Able to withstand high temperatures.

relative biological effectiveness (Q)   A factor used when assessing the damage caused by a given dose of radiation.

rem   See *roentgen equivalent man.*

repeating unit   The combination of atoms in a polymer that repeats over and over again in the polymer chain.

residual entropy   The nonzero entropy at $T = 0$ in certain systems, which is due to a surviving disorder in the orientation of molecules.

residue (biochemical)   An amino acid in a polypeptide chain.

resistance (electrical)   A measure of the ability of matter to conduct electricity: the lower the resistance, the better it conducts.

resonance   A blending of Lewis structures into a single composite, hybrid structure. *Example:* $\ddot{O}{=}S{-}\ddot{O}: \longleftrightarrow :\ddot{O}{-}S{=}\ddot{O}$.

resonance hybrid   The composite structure that results from resonance.

reverse osmosis   The passage of solvent out of a solution when a pressure greater than the osmotic pressure is applied on the solution side of a semipermeable membrane.

reversible isothermal expansion   Expansion at constant temperature against an external pressure matched to the pressure of the system.

reversible process   A process that can be reversed by an infinitesimal change in a variable.

rock-salt structure   A crystal structure the same as that of a mineral form of sodium chloride.

roentgen equivalent man (rem)   The (non-SI) unit for reporting *dose equivalent.* See also *sievert.*

root mean square speed (v_{rms})   The square root of the average value of the squares of the speeds of the molecules in a sample.

roots (of an equation)   The solutions of the equation $f(x) = 0$. See also Appendix 1E.

rotational kinetic energy   See *kinetic energy.*

rounding off   The adjustment of a numerical result to the correct number of significant figures.

Rydberg constant ($\mathcal{R}$)   The constant in the formula for the frequencies of the lines in the spectrum of atomic hydrogen; $\mathcal{R} = 3.290 \times 10^{15}$ Hz.

sacrificial anode   A metal electrode that is allowed to decay in order to protect a metallic artifact. See *cathodic protection.*

saline hydride   See *hydride.*

salt   (1) An ionic compound. (2) The ionic product of the reaction between an acid and a base. *Examples:* NaCl; K_2SO_4.

salt bridge   An inverted U-shaped tube containing a concentrated salt (potassium chloride or potassium nitrate) in a gel that acts as an electrolyte and provides a conducting path between two compartments of an electrochemical cell.

sample   A representative part of a whole.

saturated   Unable to take up further material.

saturated hydrocarbon   A hydrocarbon with no carbon–carbon multiple bonds. *Example:* CH_3CH_3.

saturated solution   A solution in which the dissolved and undissolved solute are in dynamic equilibrium. See *saturated.*

Schrödinger equation An equation for calculating the wavefunction of a particle, especially for an electron in an atom or molecule. See also *wavefunction*.

scientific method A set of procedures employed to develop a scientific understanding of nature.

scientific notation The expression of numbers in the form $n.nnn\ldots \times 10^a$.

scintillation counter A device for detecting and measuring radioactivity that makes use of the fact that certain substances give a flash of light when they are exposed to radiation.

sea of instability A region in a graph of mass number against atomic number corresponding to unstable nuclei that decay spontaneously with the emission of radiation. See also *band of stability*.

second (s) The SI unit of time. See also Appendix 1B.

second derivative (d^2y/dx^2) A measure of the curvature of a function. See also Appendix 1F.

second ionization energy (I_2) The energy needed to remove an electron from a singly charged gas-phase cation. *Example:* $Cu^+(g) \rightarrow Cu^{2+}(g) + e^-(g)$, $I_2 = 1958\ kJ\cdot mol^{-1}$.

second law of thermodynamics A spontaneous change is accompanied by an increase in the total entropy of the system and its surroundings.

second virial coefficient (B) See *virial equation*.

secondary alcohol See alcohol.

secondary cell A *galvanic cell* that must be charged (or recharged) by using an externally supplied current before it can be used.

secondary pollutant A pollutant formed by the chemical reaction of another species in the environment. *Example:* SO_3 from the oxidation of SO_2.

secondary structure The manner in which a polypeptide chain is coiled. *Examples:* α helix; β sheet.

second-order reaction (1) A reaction with a rate that is proportional to the square of the molar concentration of a reactant. (2) A reaction with an overall order of 2.

selective precipitation The precipitation of one compound in the presence of other, more soluble compounds.

semiconductor An *electronic conductor* with a resistance that decreases as the temperature is raised. In an *n-type semiconductor*, the current is carried by electrons in a largely empty band; in a *p-type semiconductor*, the conduction is a result of electrons missing from otherwise filled bands.

semi-empirical method The calculation of molecular structure that draws on experimental information to simplify the procedure. Compare with *ab initio method*.

semipermeable membrane A membrane that allows only certain types of molecules or ions to pass.

sequester Form a complex between a cation and a bulky molecule or ion that enwraps the central ion. *Example:* Ca^{2+} with $O_3POPO_2OPO_3{}^{3-}$.

series (in spectroscopy) A family of spectral lines arising from transitions that have one state in common. *Example:* the Balmer series in the spectrum of atomic hydrogen.

shell All the orbitals of a given principal quantum number. *Example:* the single 2s- and three 2p-orbitals of the shell with $n = 2$.

shielding The repulsion experienced by an electron in an atom that arises from the other electrons present and opposes the attraction exerted by the nucleus.

shift reaction A reaction between carbon monoxide and water: $CO(g) + H_2O(g) \rightarrow CO_2(g) + H_2(g)$; the reaction is used in the manufacture of hydrogen.

short-range order Atoms or molecules lying in a regular arrangement that does not extend very far past their nearest neighbors.

SI (*Système International*) The International System of units; a collection of definitions of units and symbols and their deployment. It is an extension and rationalization of the metric system. See also Appendix 1B.

side chain A hydrocarbon substituent on a hydrocarbon chain.

sievert (Sv) The SI unit of dose equivalent: $1\ Sv = 1\ J\cdot kg^{-1}$.

sigma (σ) **bond** Two electrons in a cylindrically symmetrical cloud between two atoms.

sigma (σ) **orbital** A molecular orbital that has no nodal plane containing the internuclear axis.

significant figures (sf, in a measurement) The digits in the measurement, up to and including the first uncertain digit in scientific notation. *Example:* 0.0260 mL (that is, 2.60×10^{-2} mL), a measurement with three significant figures (3 sf). See also Appendix 1C.

single bond A shared electron pair.

skeletal equation An unbalanced equation that summarizes the qualitative information about the reaction. *Example:* $H_2 + O_2 \rightarrow H_2O$.

slope (of graph) The gradient of a graph. See also Appendix 1E.

smectic phase A liquid-crystal phase in which molecules lie parallel to one another and form layers.

soft matter Matter that deforms readily when subjected to an applied force.

sol A colloidal dispersion of solid particles in a liquid.

solid A rigid form of matter that maintains the same shape whatever the shape of its container.

solid electrolyte A solid ionic conductor.

solid emulsion A colloidal dispersion of a liquid in a solid. *Example:* butter, an emulsion of water in butterfat.

solid solution A solid homogeneous mixture of two or more substances.

solubility The concentration of a saturated solution of a substance.

solubility constant See *solubility product*.

solubility product (K_{sp}) The product of ionic molar concentrations in a saturated solution; the dissolution equilibrium constant. *Example:* $Hg_2Cl_2(s) \rightleftharpoons Hg_2{}^{2+}(aq) + 2\ Cl^-(aq)$, $K_{sp} = [Hg_2{}^{2+}][Cl^-]^2$.

solubility rules A summary of the solubility pattern of a range of common compounds in water. See also *Fundamentals* Table I.1.

soluble substance A substance that dissolves to a significant extent in a specified solvent. When the solvent is not specified, water is generally meant.

solute A dissolved substance.

solution A homogeneous mixture. See also *solute; solvent*.

solvated   Surrounded by and linked to solvent molecules. *Hydration* is the special case in which the solvent is water.

solvent   (1) The most abundant component of a solution. (2) The component of a solution in which the other components are considered to be dissolved.

solvent extraction   A process for separating a mixture of substances that makes use of their differing solubilities in various solvents.

s-orbital   See *atomic orbital*.

space-filling model   A depiction of a molecule in which atoms are represented by spheres that indicate the space occupied by each atom.

sp^3d^n hybrid orbital   A hybrid orbital formed from one s-orbital, three p-orbitals, and n d-orbitals.

speciation   The concentrations of each of the ions present in a solution of a polyprotic acid.

species   In this text, an atom, ion, or molecule.

specific enthalpy (of a fuel)   The enthalpy of combustion per gram (without the negative sign).

specific heat capacity   The heat capacity of a sample divided by its mass.

spectator ion   An ion that is present but remains unchanged during a reaction. *Examples:* Na^+ and NO_3^- in $NaCl(aq) + AgNO_3(aq) \rightarrow NaNO_3(aq) + AgCl(s)$.

spectral line   Radiation of a single wavelength emitted or absorbed by an atom or molecule.

spectrochemical series   Ligands ordered according to the strength of the ligand field splitting that they produce.

spectrometer   An instrument for recording the spectrum of a sample.

spectrophotometer   An instrument for measuring and recording electronically the intensity of radiation passing through a sample as the wavelength of the radiation is changed and hence recording the spectrum of the sample.

spectrophotometry   An analytical technique in which the absorbance of a solution is measured to determine the nature or concentration of a substance. See *spectrophotometer*.

spectroscopy   The analysis of the electromagnetic radiation emitted, absorbed, or scattered by substances.

spectrum   The frequencies or wavelengths of the electromagnetic radiation emitted, absorbed, or scattered by substances.

speed   The magnitude of the velocity; the rate of change of position.

sphalerite structure   See *zinc-blende structure*.

spherical polar coordinates   The coordinates of a point expressed in terms of the radius r, the colatitude θ, and the azimuth ϕ.

spherically symmetrical   Independent of orientation about a central point.

spin magnetic quantum number (m_s)   The quantum number that distinguishes the two spin states of an electron: $m_s = +\frac{1}{2}(\uparrow)$ and $m_s = -\frac{1}{2}(\downarrow)$.

spin   The intrinsic angular momentum of an electron; the spin cannot be eliminated and can occur in only two senses, denoted $\uparrow$ and $\downarrow$ or α and β.

sp^n hybrid orbital   A hybrid orbital constructed from an s-orbital and n p-orbitals. There are two *sp hybrid orbitals*, three *sp^2 hybrid orbitals*, and four *sp^3 hybrid orbitals*.

spontaneous change   A natural change, one that has a tendency to occur without needing to be driven by an external influence. *Examples:* a gas expanding into a vacuum; a hot object cooling; methane burning.

spontaneous nuclear fission   See *fission*.

square planar complex   A complex in which four ligands lie at the corners of a square with the metal atom at the center.

stability constant   See *formation constant*.

stable   See *thermodynamically stable compound*.

standard ambient temperature and pressure (SATP)   25 °C (298.15 K) and 1 bar.

standard boiling point (T_b)   The temperature at which a liquid boils when the pressure is 1 bar.

standard cell potential ($E_{cell}°$)   The *cell potential* when the concentration of each solute taking part in the cell reaction is 1 mol·L^{-1} (strictly, unit activity) and all the gases are at 1 bar. The standard cell potential of a galvanic cell is the difference between its two standard potentials: $E_{cell}° = E_R° - E_L°$, where R and L denote the right and left electrodes in the specification of the cell.

standard emf   See *standard cell potential*.

standard enthalpy of combustion ($\Delta H_c°$)   The change of enthalpy per mole of substance when it burns (reacts with oxygen) completely under standard conditions.

standard enthalpy of formation ($\Delta H_f°$)   The standard reaction enthalpy per mole of compound for the compound's synthesis from its elements in their most stable form at 1 bar and the specified temperature.

standard entropy of fusion ($\Delta S_{fus}°$)   The standard entropy change per mole accompanying *fusion*.

standard entropy of vaporization ($\Delta S_{vap}°$)   The standard entropy change per mole accompanying *vaporization*.

standard Gibbs free energy of formation ($\Delta G_f°$)   The standard reaction Gibbs free energy per mole for the formation of a compound from its elements in their most stable form.

standard Gibbs free energy of fusion ($\Delta G_{fus}°$)   The standard Gibbs free energy change per mole accompanying *fusion*.

standard Gibbs free energy of reaction ($\Delta G°$)   The Gibbs free energy of reaction under standard conditions.

standard Gibbs free energy of vaporization ($\Delta G_{vap}°$)   The standard Gibbs free energy change per mole accompanying *vaporization*.

standard hydrogen electrode (SHE)   A hydrogen electrode that is in its standard state (hydrogen ions at concentration 1 mol·L^{-1} (strictly, unit activity) and hydrogen pressure 1 bar) and is defined as having $E° = 0$ at all temperatures.

standard melting point (T_f)   The temperature at which a solid melts when the pressure is 1 bar.

standard molar entropy ($S_m°$)   The entropy per mole of a pure substance at 1 bar.

standard potential ($E°$)   (1) The contribution of an electrode to the standard potential of a cell. (2) The standard potential of a cell when the left-hand electrode is a standard hydrogen electrode and the right-hand electrode is the electrode of interest.

standard pressure ($P°$)   A pressure of 1 bar exactly.

standard reaction enthalpy ($\Delta H°$)   The reaction enthalpy under standard conditions.

standard reaction entropy ($\Delta S°$)   The reaction entropy under standard conditions.

standard state   The pure form of a substance at 1 bar; for a solute, at a concentration of 1 mol·L^{-1}.

standard temperature and pressure (STP)   0 °C (273.15 K) and 1 atm (101.325 kPa).

state function   A property of a substance that is independent of how the sample was prepared. *Examples:* pressure; enthalpy; entropy; color.

state of matter   The physical condition of a sample. The most common states of a pure substance are solid, liquid, or gas (vapor).

state property   See *state function*.

state symbol   A symbol (abbreviation) denoting the state of a species. *Examples:* s (solid); l (liquid); g (gas); aq (aqueous solution).

statistical entropy   The entropy calculated from statistical thermodynamics; $S = k \ln W$.

statistical thermodynamics   The interpretation of the laws of thermodynamics in terms of the behavior of large numbers of atoms and molecules.

steady-state approximation   The assumption that the net rate of formation of reaction intermediates is 0.

Stefan–Boltzmann law   The total intensity of radiation emitted by a heated black body is proportional to the fourth power of the absolute temperature.

stereoisomers   Isomers in which atoms have the same partners arranged differently in space.

stereoregular polymer   A polymer in which each unit or pair of repeating units has the same relative orientation.

steric factor (P)   An empirical factor that takes into account the steric requirement of a reaction.

steric requirement   A constraint on an elementary reaction in which the successful collision of two molecules depends on their relative orientation.

stick structure   See *line structure*.

stock solution   A solution stored in concentrated form.

stoichiometric coefficients   The numbers multiplying chemical formulas in a chemical equation. *Examples:* 1, 1, and 2 in $H_2 + Br_2 \rightarrow 2\ HBr$.

stoichiometric point   The stage in a titration when exactly the right volume of solution needed to complete the reaction has been added.

stoichiometric proportions   Reactants in the same proportions as their coefficients in the chemical equation. *Example:* equal amounts of H_2 and Br_2 in the formation of HBr.

stoichiometric relation   An expression that equates the relative amounts of reactants and products that participate in a reaction. *Example:* 1 mol $H_2 \mathrel{\hat{=}} 2$ mol HBr for $H_2 + Br_2 \rightarrow 2\ HBr$.

stoichiometry   See *reaction stoichiometry*.

stopped-flow technique   A procedure for observing fast reactions, involving the spectrometric analysis of a reaction mixture immediately after the rapid injection of reactants into a mixing chamber.

STP   See *standard temperature and pressure*.

strong acids and bases   Acids and bases that are fully deprotonated or protonated, respectively, in solution. *Examples:* HCl, $HClO_4$ (strong acids); NaOH, $Ca(OH)_2$ (strong bases).

strong electrolyte   See *electrolyte*.

strong force   A short-range but very strong force that acts between nucleons and binds them together to form a nucleus.

strong-field ligand   A ligand that produces a large *ligand field splitting* and lies above H_2O in the spectrochemical series.

structural formula   A chemical formula that shows how atoms in a compound are attached to one another. See also *condensed structural formula*.

structural isomers   Isomers in which the atoms have different partners.

subatomic particle   A particle smaller than an atom. *Examples:* electron; proton; neutron.

subcritical   Having a mass less than the critical mass.

sublimation   The direct conversion of a solid into a vapor without first forming a liquid.

sublimation vapor pressure   The vapor pressure of a solid.

subshell   All the atomic orbitals of a given shell of an atom that have the same value of the quantum number l. *Example:* the five 3d-orbitals of an atom.

substance   A single, pure type of matter; either a compound or an element.

substituent   An atom or group that has replaced a hydrogen atom in an organic molecule.

substitution reaction   (1) A reaction in which an atom (or a group of atoms) replaces an atom in the original molecule. (2) In complexes, a reaction in which one Lewis base expels another and takes its place. *Examples:* (1) $C_6H_5OH + Br_2 \rightarrow BrC_6H_4OH + HBr$; (2) $[Fe(OH_2)_6]^{3+}(aq) + 6\ CN^-(aq) \rightarrow [Fe(CN)_6]^{3-}(aq) + 6\ H_2O(l)$.

substitutional alloy   See *alloy*.

substrate   The chemical species on which an enzyme acts.

superconductor   An *electronic conductor* that conducts electricity with zero resistance. See also *high-temperature superconductor*.

supercooled   Refers to a liquid cooled to below its freezing point but not yet frozen.

supercritical   Having a mass greater than the *critical mass*.

supercritical fluid   A fluid phase of a substance above its *critical temperature* and *critical pressure*.

superfluidity   The ability to flow without *viscosity*.

superimpose   The combination of atomic orbitals to form a molecular orbital.

surface-active agent   See *surfactant*.

surface tension (γ)   The tendency of molecules at the surface of a liquid to be pulled inward, resulting in a smooth surface.

surfactant   A substance that accumulates at the surface of a solution and affects the surface tension of the solvent; a component of detergents. *Example:* the stearate ion of soaps.

surroundings   The region outside a system, where observations are made.

suspension   A mist of small particles in a fluid medium.

sustainable development   The economical utilization and renewal of resources coupled with hazardous waste reduction and concern for the environment.

symbolic level   The discussion of chemical phenomena in terms of chemical symbols and mathematical equations.

synthesis   A reaction in which a substance is formed from simpler starting materials. *Example:* $N_2(g) + 3\ H_2(g) \rightarrow 2\ NH_3(g)$.

synthesis gas   A mixture of carbon monoxide and hydrogen produced by the catalyzed reaction of a hydrocarbon and water.

system   The object of study, usually a reaction vessel and its contents. An *open system* can exchange both matter and energy with the surroundings. A *closed system* has a fixed amount of matter but can exchange energy with the surroundings. An *isolated system* has no contact with its surroundings.

systematic error   An error that persists in a series of measurements and does not average out. See also *accuracy*.

systematic name   The name of a compound that reveals which elements are present (and, in its most complete form, how the atoms are arranged). *Example:* methylbenzene is the systematic name for toluene.

Système International d'Unités   See *SI*.

temperature (*T*)   (1) How hot or cold a sample is. (2) The intensive property that determines the direction in which heat will flow between two objects in contact.

temperature–composition diagram   A *phase diagram* showing how the normal boiling point of a liquid mixture varies with composition.

termination   A step in a *chain reaction* in which chain carriers combine to form products. *Example:* Br· + Br· → Br$_2$.

termolecular reaction   An *elementary reaction* in which three species collide simultaneously.

tertiary alcohol   See *alcohol*.

tertiary structure   The shape into which the α-helical and β-sheet sections of a polypeptide are twisted as a result of interactions between peptide groups lying in different parts of the primary structure.

tetrahedral complex   A complex in which four ligands lie at the corners of a regular tetrahedron with a metal atom at the center. *Example:* [Cu(NH$_3$)$_4$]$^{2+}$.

tetrahedral hole   A cavity in a (usually close-packed) crystal structure formed by one sphere lying in the dip formed by three others.

theoretical yield   The maximum quantity of product that can be obtained, according to the reaction stoichiometry, from a given quantity of a specified reactant.

theory   A collection of ideas and concepts used to account for a scientific law.

thermal decomposition   See *decomposition*.

thermal disorder   The disorder arising from the thermal motion of molecules.

thermal energy   The sum of the potential and kinetic energies due to thermal motion.

thermal equilibrium   The state in which a system is at the same temperature as its surroundings.

thermal motion   The random, chaotic motion of atoms.

thermochemical equation   An expression consisting of the balanced chemical equation and the corresponding reaction enthalpy.

thermochemistry   The study of the heat released or absorbed by chemical reactions; a branch of thermodynamics.

thermodynamically stable compound   (1) A compound with no thermodynamic tendency to decompose into its elements. (2) A compound with a negative Gibbs free energy of formation.

thermodynamically unstable compound   (1) A compound with a thermodynamic tendency to decompose into its elements. (2) A compound with a positive Gibbs free energy of formation.

thermodynamics   The study of the transformations of energy from one form to another. See also *first law of thermodynamics; second law of thermodynamics, third law of thermodynamics*.

thermonuclear explosion   An explosion resulting from uncontrolled nuclear fusion.

thermoplastic polymer   A polymer that can be softened by heating after it has been molded.

thermosetting polymer   A polymer that takes on a permanent shape in a mold and does not soften upon heating.

thermotropic liquid crystal   A liquid crystal prepared by melting the solid phase.

third law of thermodynamics   The entropies of all perfect crystals are the same at the absolute zero of temperature.

third virial coefficient   See *virial equation*.

three-center bond   A chemical bond in which a hydrogen atom lies between two other atoms (typically boron atoms) and one electron pair binds all three atoms together.

tie line (in a temperature–composition phase diagram)   A line joining the point indicating the boiling point of a mixture of given composition with the corresponding composition of the vapor at that temperature.

titrant   The solution of known concentration added from a buret in a titration.

titration   The analysis of composition by measuring the volume of one solution needed to react with a given volume of another solution. In an *acid–base titration*, an acid is titrated with a base; in a *redox titration*, an oxidizing agent is titrated with a reducing agent.

t-orbital   One of the orbitals d$_{xy}$, d$_{yz}$, and d$_{zx}$ in an octahedral or tetrahedral complex. In an octahedral complex, these orbitals are designated t$_{2g}$, and in a tetrahedral complex, t$_2$.

torr (symbol: Torr)   A unit of pressure: 760 Torr = 1 atm exactly.

total energy   See *energy*.

tracer   An isotope that can be tracked from compound to compound in the course of a sequence of reactions.

trajectory   The path of a particle on which location and linear momentum are specified at each instant.

transition   A change of state. (1) In thermodynamics, a change of physical state. (2) In spectroscopy, a change of quantum state.

transition metal   An element that belongs to Groups 3 through 11. *Examples:* vanadium; iron; gold.

transition state   An unstable arrangement of atoms that can either form products or revert to reactants. See *activated complex*.

transition state theory   A theory of reaction rates in which the reactants form an activated complex.

translational kinetic energy   See *kinetic energy*.

transmeitnerium elements   The elements beyond meitnerium; those with $Z > 109$.

transmutation   See *nuclear transmutation*.

transuranium elements   The elements beyond uranium; those with $Z > 92$.

triboluminescence Luminescence that results from mechanical shock to a crystal.

triple bond (1) Three electron pairs shared by two neighboring atoms. (2) One σ-bond and two π-bonds between neighboring atoms.

triple point The point where three phase boundaries meet in a phase diagram. Under the conditions represented by the triple point, all three adjoining phases coexist in dynamic equilibrium.

triprotic See *polyprotic acid or base*.

Trouton's rule The empirical observation that the entropy of vaporization at the boiling point (the enthalpy of vaporization divided by the boiling temperature) is approximately $85 \text{ J·K}^{-1}\text{·mol}^{-1}$ for many liquids.

ultraviolet catastrophe The classical prediction that any black body at any temperature should emit intense ultraviolet radiation.

ultraviolet radiation Electromagnetic radiation with a higher frequency (shorter wavelength) than that of violet light.

unbranched alkane An alkane with no side chains, in which all the carbon atoms lie in a linear chain.

uncertainty principle See *Heisenberg uncertainty principle*.

unimolecular reaction An elementary reaction in which a single reactant molecule changes into products. *Example:* $O_3 \rightarrow O_2 + O$.

unique average reaction rate The rate of change in the concentration of a reactant or product divided by its stoichiometric coefficient in the balanced equation. All unique average rates are reported as positive quantities. See also *average reaction rate*.

unit See *base units*.

unit cell The smallest unit that, when stacked together repeatedly without any gaps, can reproduce an entire crystal.

universe (in thermodynamics) The system and its surroundings.

unsaturated hydrocarbon A hydrocarbon with at least one carbon–carbon multiple bond. *Examples:* $CH_2{=}CH_2$; C_6H_6.

valence The number of bonds that an atom can form.

valence band In the theory of solids, a band of energy levels that is fully occupied by electrons.

valence electrons The electrons that belong to the valence shell.

valence shell The outermost shell of an atom. *Example:* the $n = 2$ shell of Period 2 atoms.

valence-bond theory The description of bond formation in terms of the pairing of spins in the atomic orbitals of neighboring atoms.

valence-shell electron-pair repulsion model (VSEPR model) A model for predicting the shapes of molecules, using the fact that electron pairs repel one another.

van der Waals equation An approximate equation of state for a real gas in which two parameters represent the effects of intermolecular forces.

van der Waals interactions Intermolecular interactions that depend on the inverse sixth power of the separation. See *intermolecular forces*.

van der Waals parameters The experimentally determined coefficients that appear in the van der Waals equation and are unique for each real gas. The parameter a is an indication of the strength of attractive intermolecular forces, and the parameter b is an indication of the strength of repulsive intermolecular forces. See also *van der Waals equation*.

van der Waals radius Half the distance between the centers of nonbonded, touching atoms in a solid.

van 't Hoff equation (1) The equation for the osmotic pressure in terms of the molar concentration, $\Pi = i[J]RT$. (2) An equation that shows how the equilibrium constant varies with temperature.

van 't Hoff *i* factor See *i factor*.

vapor The gaseous phase of a substance (specifically, of a substance that is a liquid or solid at the temperature in question). See also *gas*.

vapor pressure The pressure exerted by the vapor of a liquid (or a solid) when the vapor and the liquid (or solid) are in dynamic equilibrium.

vaporization The formation of a gas or a vapor from a liquid.

vapor-phase synthesis A process in which a substance is vaporized and then condensed or mixed with a reactant and the condensed product forms tiny crystals.

variable covalence The ability of an element to form different numbers of covalent bonds. *Example:* S in SO_2 and SO_3.

variable valence The ability of an element to form ions with different charges. *Example:* In^+ and In^{3+}.

velocity The rate of change of position.

vibrational kinetic energy See *kinetic energy*.

virial equation An equation of state expressed in powers of $1/V_m$: specifically, $PV = nRT(1 + B/V_m + C/V_m^2 + \ldots)$, where B is the *second virial coefficient* and C is the *third virial coefficient*.

viscosity The resistance of a fluid (a gas or a liquid) to flow: the higher the viscosity, the slower is the flow.

visible light See *visible radiation*.

visible radiation Electromagnetic radiation that can be detected by the human eye, with wavelengths in the range from 700 nm to 400 nm. Visible radiation is also called *visible light* or simply *light*.

volatile Having a high vapor pressure at ordinary temperatures. A substance is typically regarded as volatile if its boiling point is below 100 °C.

volatility The readiness with which a substance vaporizes. See also *volatile*.

volt (V) The SI unit of electrical potential. See also Appendix 1B.

voltaic cell See *electrochemical cell*.

volume (V) The amount of space that a sample occupies.

volumetric analysis An analytical method using measurement of volume, as in a *titration*.

volumetric flask A flask calibrated to contain a specified volume.

water autoprotolysis constant (K_w) The equilibrium constant for the autoprotolysis (autoionization) of water, $2\,H_2O(l) \rightleftharpoons H_3O^+(aq) + OH^-(aq)$, $K_w = [H_3O^+][OH^-]$.

water of hydration See *hydration*.

water-splitting reaction The photochemical decomposition of water into hydrogen and oxygen.

wavefunction (ψ) A solution of the Schrödinger equation; the probability amplitude.

wavelength (λ) The peak-to-peak distance of a wave.

wave–particle duality   The combined wavelike and particle-like character of both radiation and matter.

weak acids and bases   Acids and bases that are only incompletely deprotonated or protonated, respectively (commonly, in aqueous solution) at normal concentrations. *Examples:* HF, CH_3COOH (weak acids); NH_3, CH_3NH_2 (weak bases).

weak electrolyte   See *electrolyte*.

weak-field ligand   A ligand that produces a small *ligand field splitting* and that lies below NH_3 in the *spectrochemical series*.

Wien's law   The wavelength corresponding to the maximum in the radiation emitted by a heated *black body* is inversely proportional to the absolute temperature.

work (w)   The energy expended during the act of moving an object against an opposing force. In *expansion work*, the system expands against an opposing pressure; *nonexpansion work* is work that does not arise from a change in volume.

work function (Φ)   The energy required to remove an electron from a metal.

x-ray   Electromagnetic radiation with wavelengths from about 10 pm to about 1000 pm.

x-ray diffraction   The analysis of crystal structures by studying the interference pattern in a beam of x-rays.

yield   See *percentage yield; theoretical yield.*

zeolite   A microporous aluminosilicate.

zeroth-order reaction   A reaction with a rate that is independent of the concentration of the reactant. *Example:* the catalyzed decomposition of ammonia.

zero-point energy   The lowest possible energy of a system. *Example:* $E = h^2/8mL^2$ for a particle of mass m in a box of length L.

Ziegler–Natta catalyst   A stereospecific catalyst for polymerization reactions, consisting of titanium tetrachloride and triethylaluminum.

zinc-blende structure   A crystal structure in which the cations occupy half the tetrahedral holes in a nearly close-packed cubic lattice of anions; also known as *sphalerite structure*.

zone refining   A method for purifying a solid by repeatedly passing a molten zone along the length of a sample.

zwitterion   A form of an amino acid in which in which the amino group is protonated and the carboxyl group is deprotonated. *Example:* $^+H_3NCH_2CO_2{}^-$.

SELF-TESTS B

ANSWERS

Fundamentals

A.1B 250. g × 1.000 lb/453.6 g × 16 oz/1 lb = 8.82 oz

A.2B 9.81 m·s^{-2} × 1 km/10^3 m × (3600 s/1 h)^2 = 1.27 × 10^5 km·h^{-2}

A.3B $V = m/d$ = (10.0 g)/(0.176 85 g·L^{-1}) = 56.5 L

A.4B $E_k = mv^2/2 = {}^1\!/_2$ × (1.5 kg) × (3.0 m·s^{-1})^2 = 6.8 J

A.5B $E_k = mgh$ = (0.350 kg) × (9.81 m·s^{-2}) × (443 m) × (10^{-3} kJ/J) = 1.52 kJ

B.1B number of Au atoms = m(sample)/m(one atom) = (0.0123 kg)/(3.27 × 10^{-25} kg) = 3.76 × 10^{22} Au atoms

B.2B (a) 8, 8, 8; (b) 92, 144, 92

B.3B (a) Sn; (b) Na; (c) iodine; (d) yttrium

C.1B (a) Potassium is a Group 1 metal. Cation, +1, so K$^+$. (b) Sulfur is a nonmetal in Group 16. Anion, 16 − 18 = −2, so S^{2-}.

C.2B (a) Li$_3$N; (b) SrBr$_2$

D.1B (a) dihydrogen arsenate; (b) ClO$_3^-$

D.2B (a) gold(III) chloride; (b) calcium sulfide; (c) manganese(III) oxide

D.3B (a) phosphorus trichloride; (b) sulfur trioxide; (c) hydrobromic acid

D.4B (a) Cs$_2$S·4H$_2$O; (b) Mn$_2$O$_7$; (c) HCN; (d) S$_2$Cl$_2$

D.5B (a) pentane; (b) carboxylic acid

E.1B atoms of H = (3.14 mol H$_2$O) × (2 mol H)/(1 mol H$_2$O) × (6.022 × 10^{23} atoms/mol) = 3.78 × 10^{24} H atoms

E.2B (a) $n = m/M$ = (5.4 × 10^3 g)/(26.98 g·mol^{-1}) = 2.0 × 10^2 mol; (b) $N = N_A × n$ = (6.022 × 10^{23} Al atoms/mol) × (2.0 × 10^2 mol) = 1.2 × 10^{26} Al atoms

E.3B copper-63: (62.94 g·mol^{-1}) × 0.6917 = 43.536 g·mol^{-1}; copper-65: (64.93 g·mol^{-1}) × 0.3083 = 20.018 g·mol^{-1}; 43.536 g·mol^{-1} + 20.018 g·mol^{-1} = 63.55 g·mol^{-1}

E.4B (a) phenol: 6 C, 6 H, 1 O; 6(12.01 g·mol^{-1}) + 6(1.008 g·mol^{-1}) + (16.00 g·mol^{-1}) = 94.11 g·mol^{-1}; (b) Na$_2$CO$_3$·10H$_2$O: 2 Na, 1 C, 13 O, 20 H; 2(22.99 g·mol^{-1}) + (12.01 g·mol^{-1}) + 13(16.00 g·mol^{-1}) + 20(1.008 g·mol^{-1}) = 286.15 g·mol^{-1}

E.5B Ca(OH)$_2$: 1 Ca, 2 O, 2 H; (40.08 g·mol^{-1}) + 2(16.00 g·mol^{-1}) + 2(1.008 g·mol^{-1}) = 74.10 g·mol^{-1}; (1.00 × 10^3 g lime)/(74.10 g·mol^{-1}) = 13.5 mol formula units of lime

E.6B CH$_3$COOH: 2 C, 4 H, 2 O; 2(12.01 g·mol^{-1}) + 4(1.008 g·mol^{-1}) + 2(16.00 g·mol^{-1}) = 60.05 g·mol^{-1}; (60.05 g·mol^{-1})(1.5 mol) = 90. g

F.1B % C = (6.61 g/7.50 g) × 100% = 88.1%; % H = (0.89 g/7.50 g) × 100% = 11.9%

F.2B AgNO$_3$: (107.87 g·mol^{-1}) + (14.01 g·mol^{-1}) + 3(16.00 g·mol^{-1}) = 169.88 g·mol^{-1}; % Ag = (107.87 g·mol^{-1})/(169.88 g·mol^{-1}) × 100% = 63.498%

F.3B n(O) = (18.59 g)/(16.00 g·mol^{-1}) = 1.162 mol O; n(S) = (37.25 g)/(32.07 g·mol^{-1}) = 1.162 mol; n(F) = (44.16 g)/(19.00 g·mol^{-1}) = 2.324 mol. 1:1:2 ratio, empirical formula is SOF$_2$.

F.4B M(CHO$_2$) = 45.012 g·mol^{-1}. (90.0 g·mol^{-1})/(45.012 g·mol^{-1}) = 2.00; 2 × (CHO$_2$) = C$_2$H$_2$O$_4$

G.1B M(Na$_2$SO$_4$) = 142.05 g·mol^{-1}. (15.5 g)/(142.05 g·mol^{-1}) = 0.109 mol; (0.109 mol)/(0.350 L) = 0.312 м Na$_2$SO$_4$(aq)

G.2B (0.125 mol·L^{-1}) × (0.05000 L) = 0.00625 mol oxalic acid. M(oxalic acid) = 90.036 g·mol^{-1}. (0.00625 mol) × (90.036 g·mol^{-1}) = 0.563 g oxalic acid

G.3B (2.55 × 10^{-3} mol HCl)/(0.358 mol HCl/L) = 7.12 × 10^{-3} L = 7.12 mL

G.4B $V_{initial} = (c_{final} × V_{final})/c_{initial}$ = (1.59 × 10^{-5} mol·L^{-1}) × (0.02500 L)/(0.152 mol·L^{-1}) = 2.62 × 10^{-3} mL

H.1B Mg$_3$N$_2$(s) + 4 H$_2$SO$_4$(aq) → 3 MgSO$_4$(aq) + (NH$_4$)$_2$SO$_4$(aq)

I.1B (a) molecular compound and not an acid, so a nonelectrolyte, does not conduct electricity; (b) ionic compound, so a strong electrolyte, conducts electricity

I.2B 3 Hg$_2^{2+}$(aq) + 2 PO$_4^{3-}$(aq) → (Hg$_2$)$_3$(PO$_4$)$_2$(s)

I.3B SrCl$_2$ and Na$_2$SO$_4$; Sr^{2+}(aq) + SO$_4^{2-}$(aq) → SrSO$_4$(s)

J.1B (a) is neither an acid nor a base; (b) and (c) are acids; (d) *supplies* the base OH$^-$

J.2B 3 Ca(OH)$_2$(aq) + 2 H$_3$PO$_4$(aq) → Ca$_3$(PO$_4$)$_2$(s) + 6 H$_2$O

K.1B Cu$^+$(aq) is oxidized to Cu^{2+}, I$_2$(s) is reduced to I$^-$.

C1

K.2B (a) $x + 3(-2) = -2$; $x = +4$ for S; (b) $x + 2(-2) = -1$; $x = +3$ for N; (c) $x + 1 + 3(-2) = 0$; $x = +5$ for Cl

K.3B (a) $4(-2) + x = 0$, $x = +4$; (b) $3(-2) + x = -1$, $x = +5$

K.4B H_2SO_4 is the oxidizing agent (S is reduced from $+6$ to $+4$); NaI is the reducing agent (I is oxidized from -1 to $+5$).

K.5B $2\,Ce^{4+}(aq) + 2\,I^-(aq) \rightarrow 2\,Ce^{3+}(aq) + I_2(s)$

L.1B $(2\text{ mol Fe})/(1\text{ mol Fe}_2O_3) \times 25\text{ mol Fe}_2O_3 = 50.\text{ mol Fe}$

L.2B $2\text{ mol CO}_2/1\text{ mol CaSiO}_3$; $\text{mol CO}_2 = (3.00 \times 10^2\text{ g})/(44.01\text{ g·mol}^{-1}) = 6.82\text{ mol}$; $(1\text{ mol CaSiO}_3/2\text{ mol CO}_2) \times (6.82\text{ mol CO}_2) = 3.41\text{ mol CaSiO}_3$; $(3.41\text{ mol CaSiO}_3) \times (116.17\text{ g·mol}^{-1}\text{ CaSiO}_3) = 396\text{ g CaSiO}_3$

L.3B $2\,KOH + H_2SO_4 \rightarrow K_2SO_4 + 2\,H_2O$; $2\text{ mol KOH} \mathrel{\hat=} 1\text{ mol H}_2SO_4$; $(0.255\text{ mol KOH·L}^{-1}) \times (0.025\text{ L}) = 6.375 \times 10^{-3}\text{ mol KOH}$. $(6.375 \times 10^{-3}\text{ mol KOH}) \times (1\text{ mol H}_2SO_4)/(2\text{ mol KOH}) = 3.19 \times 10^{-3}\text{ mol H}_2SO_4$; $(3.19 \times 10^{-3}\text{ mol H}_2SO_4)/(0.016\,45\text{ L}) = 0.194\text{ M H}_2SO_4(aq)$

L.4B $(0.100 \times 0.028\,15)\text{ mol KMnO}_4 \times (5\text{ mol As}_4O_6)/(8\text{ mol KMnO}_4) \times 395.28\text{ g·mol}^{-1} = 6.96 \times 10^{-2}\text{ g As}_4O_6$

M.1B $(15\text{ kg Fe}_2O_3)/159.69\text{ g·mol}^{-1}) \times (2\text{ mol Fe})/(1\text{ mol Fe}_2O_3) \times (55.85\text{ g·mol}^{-1}) = 10.5\text{ kg Fe}$; $8.8\text{ kg}/10.5\text{ kg} \times 100\% = 84\%$ yield

M.2B $2\,NH_3 + CO_2 \rightarrow OC(NH_2)_2 + H_2O$; $n(NH_3) = (14.5 \times 10^3\text{ g})/(17.034\text{ g·mol}^{-1}) = 851\text{ mol NH}_3$; $n(CO_2) = (22.1 \times 10^3\text{ g})/(44.01\text{ g·mol}^{-1}) = 502\text{ mol CO}_2$; $2\text{ mol NH}_3 \mathrel{\hat=} 1\text{ mol CO}_2$; (a) NH_3 is the limiting reactant. $(851\text{ mol NH}_3/2) < (502\text{ mol CO}_2)$. (b) $2\text{ mol NH}_3/1\text{ mol urea}$. 426 mol, or 25.6 kg of urea can be produced. (c) $(502 - 426)\text{ mol} = 76\text{ mol CO}_2$ excess $= 3.3\text{ kg CO}_2$

M.3B There is 0.61 mol NO_2 and 1.0 mol H_2O. $1\text{ mol H}_2O \mathrel{\hat=} 3\text{ mol NO}_2$; therefore, there is not enough NO_2, so NO_2 is the limiting reactant. 22 g, or 0.35 mol HNO_3, were produced. Theoretical yield is $(0.61\text{ mol NO}_2) \times (2\text{ mol HNO}_3)/(3\text{ mol NO}_2) = 0.407\text{ mol HNO}_3$. Percentage yield $= (0.35\text{ mol})/(0.407\text{ mol}) \times 100\% = 86\%$.

M.4B The sample contains 0.0118 mol C (0.142 g C) and 0.0105 mol H (0.0106 g H). Mass of O $= 0.236 - (0.142 + 0.0105)\text{g} = 0.0834\text{ g O}$ (0.00521 mol O). The C:H:O mole ratios are $0.0118:0.0105:0.005\,21$, or $2.26:2.02:1$. Multiplying these numbers by 4 gives $9:8:4$ and the empirical formula $C_9H_8O_4$.

Focus 1

1A.1B $\lambda = c/\nu = (2.998 \times 10^8\text{ m·s}^{-1})/(98.4 \times 10^6\text{ Hz}) = 3.05\text{ m}$

1A.2B $\nu = \mathcal{R}(1/2^2 - 1/5^2) = 21\mathcal{R}/100$; $\lambda = c/\nu = 100c/21$; $\mathcal{R} = (100 \times 2.998 \times 10^8\text{ m·s}^{-1})/(21 \times 3.29 \times 10^{15}\text{ s}^{-1}) = 434\text{ nm}$; violet line

1B.1B $T = \text{constant}/\lambda_{max} = (2.9 \times 10^{-3}\text{ m·K})/(700. \times 10^{-9}\text{ m}) = 4.1 \times 10^3\text{ K}$

1B.2B $E = h\nu = (6.626 \times 10^{-34}\text{ J·s}) \times (4.8 \times 10^{14}\text{ Hz}) = 3.2 \times 10^{-19}\text{ J}$

1B.3B (a) $E_k = 1/2 \times (9.109 \times 10^{-31}\text{ kg}) \times (7.85 \times 10^5\text{ m·s}^{-1})^2 = 2.81 \times 10^{-19}\text{ J}$; (b) $3.63\text{ eV} \times (1.602 \times 10^{-19}\text{ J·eV}^{-1}) = 5.82 \times 10^{-19}\text{ J}$, $\lambda = [(3.00 \times 10^8\text{ m·s}^{-1}) \times (6.626 \times 10^{-34}\text{ J·s})]/(5.82 \times 10^{-19}\text{ J}) = 342\text{ nm}$

1B.4B $\lambda = h/m\nu = (6.626 \times 10^{-34}\text{ J·s})/(0.0050\text{ kg} \times 2 \times 331\text{ m·s}^{-1}) = 2.0 \times 10^{-34}\text{ m}$

1B.5B $\Delta\nu = \hbar/2m\Delta x = (1.054\,57 \times 10^{-34}\text{ J})/(2 \times 2.0\text{ t} \times 10^3\text{ kg·t}^{-1} \times 1\text{ m}) = 3 \times 10^{-38}\text{ m·s}^{-1}$. No, the uncertainty is too small.

1C.1B $E_3 - E_2 = 5h^2/8m_eL^2 = h\nu$; $\nu = 5h/8m_eL^2$; $\lambda = c/\nu = 8m_ecL^2/5h = [8 \times (9.109\,39 \times 10^{-31}\text{ kg}) \times (2.998 \times 10^8\text{ m·s}^{-1}) \times (1.50 \times 10^{-10}\text{ m})^2]/(5 \times 6.626 \times 10^{-34}\text{ J·s}) = 14.8 \times 10^{-9}\text{ m}$, or 14.8 nm

1D.1B $\text{ratio} = (e^{-6a_0/a_0}/\pi a_0^3)/(1/\pi a_0^3) = e^{-6} = 0.0025$

1D.2B $3p$

1E.1B $1s^2 2s^2 2p^6 3s^2 3p^1$ or $[\text{Ne}]3s^2 3p^1$

1E.2B $[\text{Ar}]3d^{10} 4s^2 4p^3$

1F.1B (a) $r(Ca^{2+}) < r(K^+)$; (b) $r(Cl^-) < r(S^{2-})$

1F.2B In the third ionization of Be, the electron is removed from the noble-gas core; however, in the third ionization of B, the electron is removed from the valence shell. Core electrons are closer to the nucleus, so greater amounts of energy are required to remove them.

1F.3B In fluorine (Group 17), an additional electron fills the single vacancy in the valence shell; the shell now has the noble-gas configuration of neon and is complete. In neon, an additional electron would have to enter a new shell, where it would be farther from the attraction of the nucleus.

Focus 2

2A.1B (a) $[\text{Ar}]3d^5$; (b) $[\text{Xe}]4f^{14}5d^{10}$

2A.2B I^-, $[\text{Kr}]4d^{10}5s^2 5p^6$

2A.3B $:\!\ddot{\text{Br}}\!:^-$ Mg^{2+} $:\!\ddot{\text{Br}}\!:^-$

2A.4B KCl, because Cl^- has a smaller radius than Br^-

2B.1B $H\!-\!\ddot{\text{Br}}\!:$; H has no lone pairs, Br has 3 lone pairs.

2B.2B $H\!-\!\underset{\underset{H}{|}}{\ddot{\text{N}}}\!-\!H$

2B.3B $H\!-\!\underset{\underset{H}{|}}{\ddot{\text{N}}}\!-\!\underset{\underset{H}{|}}{\ddot{\text{N}}}\!-\!H$

2B.4B $\left[:\!\ddot{\text{O}}\!-\!\ddot{\text{N}}\!=\!\ddot{\text{O}}\right]^- \longleftrightarrow \left[\ddot{\text{O}}\!=\!\ddot{\text{N}}\!-\!\ddot{\text{O}}\!:\right]^-$

2B.5B $:\!\ddot{\text{F}}\!-\!\ddot{\text{O}}\!-\!\ddot{\text{F}}\!:$
$\quad\;\; 0 \quad\;\; 0 \quad\;\; 0$

2C.1B $:\!\ddot{\text{O}}\!-\!\text{N}\!=\!\ddot{\text{O}}$

2C.2B $:\!\ddot{\text{I}}\!-\!\ddot{\text{I}}\!-\!\ddot{\text{I}}\!:^-$ 10 electrons

2C.3B $\ddot{O}=\ddot{O}-\ddot{\underset{..}{O}}:$
$\quad\quad\quad\; 0 \quad\; +1 \quad -1$

2D.1B (a) CO_2

2D.2B CaS

2E.1B linear

2E.2B (a) trigonal planar; (b) angular

2E.3B (a) AX_2E_2; (b) tetrahedral; (c) angular

2E.4B square planar

2E.5B (a) nonpolar; (b) polar

2F.1B (a) 3 σ, no π; (b) 2 σ, 2 π

2F.2B Three σ bonds formed from two C2sp hybrids: one bond between the two C atoms and two connecting each C atom to an H atom in a linear arrangement; two π bonds, one between the two C2p_x-orbitals and the other between the two C2p_y-orbitals.

2F.3B (a) octahedral; (b) square planar; (c) sp^{3}d^2

2F.4B Carbon atom of CH_3 group is sp^3 hybridized and forms four σ bonds at 109.5°. The other two carbon atoms are both sp^2 hybridized and each forms three σ bonds and one π bond; bond angles are about 120°.

2G.1B O_2^+: $\sigma_{2s}^2\sigma_{2s}^{*2}\sigma_{2p}^2\pi_{2p}^4\pi_{2p}^{*1}$; BO = (8 − 3)/2 = 2.5

2G.2B CN^-: $1\sigma^2 2\sigma^{*2} 1\pi^4 3\sigma^2$

Focus 3

3A.1B $h = P/dg = (1.01 \times 10^5 \text{ kg·m}^{-1}\text{·s}^{-2})/[(998 \text{ kg·m}^{-3}) \times (9.80665 \text{ m·s}^{-2})] = 10.3$ m

3A.2B $P = (10. \text{ cmHg}) \times (10 \text{ mmHg})/(1 \text{ cmHg}) \times (1.01325 \times 10^5 \text{ Pa})/(760 \text{ mmHg}) = 1.3 \times 10^4$ Pa

3A.3B (630. Torr) $\times$ (133.3 Pa/1 Torr) = 8.40×10^4 Pa or 84.0 kPa

3B.1B $V_2 = P_1V_1/P_2 = (1.00 \text{ bar}) \times (750. \text{ L})/(5.00 \text{ bar}) = 150.$ L

3B.2B $P_2 = P_1T_2/T_1 = (760. \text{ mmHg}) \times (573 \text{ K})/(293 \text{ K}) = 1.49 \times 10^3$ mmHg

3B.3B $P_2 = P_1n_2/n_1 = (1.20 \text{ atm}) \times (300. \text{ mol})/(200. \text{ mol}) = 1.80$ atm

3B.4B $V/\text{min} = (n/\text{min}) \times RT/P = (1.00 \text{ mol/min}) \times (8.206 \times 10^{-2} \text{ L·atm·K}^{-1}\text{·mol}^{-1}) \times (300. \text{ K})/(1.00 \text{ atm}) = 24.6 \text{ L·min}^{-1}$

3B.5B $V_2 = P_1V_1/P_2 = (1.00 \text{ atm}) \times (80. \text{ cm}^3)/(3.20 \text{ atm}) = 25 \text{ cm}^3$

3B.6B $P_2 = P_1V_1T_2/V_2T_1 = [(1.00 \text{ atm}) \times (250. \text{ L}) \times (243 \text{ K})]/[(800. \text{ L}) \times (293 \text{ K})] = 0.259$ atm

3B.7B $n = (1 \text{ mol He}/4.003 \text{ g He}) \times (2.0 \text{ g He}) = 0.50 \text{ mol He}$; $V = nRT/P = (0.50 \text{ mol}) \times (24.47 \text{ L mol}^{-1}) = 12$ L

3B.8B $M = dRT/P = (1.04 \text{ g·L}^{-1}) \times (62.364 \text{ L·Torr·K}^{-1}\text{·mol}^{-1}) \times (450. \text{ K})/(200. \text{ Torr}) = 146 \text{ g·mol}^{-1}$

3C.1B $2 \text{ H}_2\text{O}(l) \rightarrow 2 \text{ H}_2(g) + \text{O}_2(g)$; (2 mol H_2/3 mol gas molecules) $\times$ (720. Torr) = 480. Torr H_2; (1 mol O_2/3 mol gas molecules) $\times$ (720. Torr) = 240. Torr O_2

3C.2B $n(O_2) = (141.2 \text{ g O}_2)/(32.00 \text{ g·mol}^{-1}) = 4.412 \text{ mol O}_2$; $n(\text{Ne}) = (335.0 \text{ g Ne})/(20.18 \text{ g·mol}^{-1}) = 16.60 \text{ mol Ne}$; $P_{O_2} = (4.412 \text{ mol O}_2/21.01 \text{ mol total}) \times (50.0 \text{ atm}) = 10.5$ atm

3C.3B $2 \text{ H}_2(g) + \text{O}_2(g) \rightarrow 2 \text{ H}_2\text{O}(l)$, and so 2 mol $H_2O \hat{=} 1$ mol O_2; $n(O_2) = PV/RT = [(1.00 \text{ atm}) \times (100.0 \text{ L})]/[(8.206 \times 10^{-2} \text{ L·atm·K}^{-1}\text{·mol}^{-1}) \times (298 \text{ K})] = 4.09 \text{ mol O}_2$. $n(H_2O) = 2(4.09 \text{ mol O}_2) = 8.18 \text{ mol H}_2\text{O}$; $m(H_2O) = (8.18 \text{ mol H}_2\text{O}) \times (18.02 \text{ g·mol}^{-1}) = 147 \text{ g H}_2\text{O}$

3D.1B (10. s) $\times$ $[(16.04 \text{ g·mol}^{-1})/(4.00 \text{ g·mol}^{-1})]^{1/2}$ = 20. s

3D.2B $v_{rms} = (3RT/M)^{1/2} = [3 \times (8.3145 \text{ J·K}^{-1}\text{·mol}^{-1}) \times (298 \text{ K})/(16.04 \times 10^{-3} \text{ kg·mol}^{-1})]^{1/2} = 681 \text{ m·s}^{-1}$

3E.1B $P = [nRT/(V - nb)] - an^2/V^2 = [\{20. \text{ mol CO}_2 \times (8.3145 \times 10^{-2} \text{ L·bar·K}^{-1}\text{·mol}^{-1}) \times (293 \text{ K})\}/\{100. - (20. \text{ mol} \times 4.29 \times 10^{-2} \text{ L·mol}^{-1})\}] - \{3.658 \text{ L}^2\text{·bar·mol}^{-2} \times (20. \text{ mol})^2\}/(100.)^2 = 4.8$ bar

3F.1B 1,1-dichloroethane, because it has a dipole moment

3F.2B Unlike CF_4, CHF_3 has a net dipole moment and, thus, higher boiling point, even though one might expect the CF_4 molecule (with more electrons) to exhibit stronger London interactions.

3F.3B (a) CH_3OH and (c) HClO

3H.1B (a) $\{(710 \text{ nm} \times 1 \text{ cm}/10^7 \text{ nm}) \times (10 \text{ cm}) \times (0.5 \text{ mm} \times 1 \text{ cm}/10 \text{ mm})\} \times (2.27 \text{ g·cm}^{-3}) = 8 \times 10^{-5} \text{ g}$; $(8 \times 10^{-5} \text{ g})/(12.01 \text{ g·mol}^{-1}) = 7 \times 10^{-6} \text{ mol}$; (b) $(1.3 \times 10^{-6} \text{ mol}) \times (6.022 \times 10^{23} \text{ atoms·mol}^{-1}) = 4 \times 10^{18}$ atoms

3H.2B 8(1/8) + 2(1/2) + 2(1) = 4 atoms

3H.3B $d_{fcc} = (4M)/(8^{3/2}N_Ar^3) = (4 \times 55.85 \text{ g·mol}^{-1})/\{8^{3/2} \times (6.022 \times 10^{23} \text{ mol}^{-1}) \times (1.24 \times 10^{-8} \text{ cm})^3\} = 8.60 \text{ g·cm}^{-3}$; $d_{bcc} = (3^{3/2}M)/(32N_Ar^3) = (3^{3/2} \times 55.85 \text{ g·mol}^{-1})/\{32 \times (6.022 \times 10^{23} \text{ mol}^{-1}) \times (1.24 \times 10^{-8} \text{ cm})^3\} = 7.90 \text{ g·cm}^{-3}$. The observed density is closer to that predicted for a body-centered cubic structure.

3H.4B $\rho = (100 \text{ pm})/(184 \text{ pm}) = 0.54$, rock-salt structure

3H.5B $\rho = (167 \text{ pm})/(220 \text{ pm}) = 0.76$, cesium-chloride structure;

$$d = \frac{M}{N_A\left(\dfrac{b}{3^{1/2}}\right)^3} = \frac{(132.91 + 126.90)\text{g·mol}^{-1}}{(6.022 \times 10^{23} \text{ mol}^{-1})\left(\dfrac{7.74 \times 10^{-8} \text{ cm}}{3^{1/2}}\right)^3}$$

$$= 4.83 \text{ g·cm}^{-3}$$

3I.1B $d(\text{brass})/d(\text{Cu})$

$$= \frac{0.5000 \times 65.41 \text{ g·mol}^{-1} + 0.5000 \times 63.55 \text{ g·mol}^{-1}}{63.55 \text{ g·mol}^{-1}}$$

$$= 1.015$$

3I.2B $x_{Ag} = 0.37$, $x_{Ni} = 0.63$ (from Figure 3I.5). In 1.00 mol of the mixture there are (0.37 mol)(107.87 g·mol^{-1}) = 40. g Ag, and $(0.63 \text{ mol})(58.69 \text{ g·mol}^{-1}) = 37 \text{ g Ni}$. Total mass = $(40. + 37) \text{ g} = 77 \text{ g}$. Mass% (Ag) = $(40./77) \times 100\% = 52\%$; mass% (Ni) = $(37/77) \times 100\% = 48\%$

3I.3B $1.0 \text{ cm}^3 \text{ diamond} \times \dfrac{3.51 \text{ g diamond}}{1 \text{ cm}^3 \text{ diamond}} \times \dfrac{1 \text{ cm}^3 \text{ graphite}}{2.27 \text{ g graphite}} =$

1.6 cm^3; $(1 + x)^3 = 1.6 \text{ cm}^3$, $x = +0.170 \text{ cm}$

3I.4B $1.0 \text{ kg hydrate} \times \dfrac{10^3 \text{ g hydrate}}{1 \text{ kg hydrate}} \times \dfrac{1 \text{ mol hydrate}}{228.33 \text{ g hydrate}} \times$

$\dfrac{18 \text{ mol water}}{6 \text{ mol hydrate}} \times \dfrac{18.02 \text{ g water}}{1 \text{ mol water}} \times \dfrac{1 \text{ kg water}}{10^3 \text{ g water}} = 0.24 \text{ kg}$

3J.1B $1.0 \text{ cm} \times \dfrac{10^{12} \text{ pm}}{10^2 \text{ cm}} \times \dfrac{1 \text{ Ca atom}}{(2 \times 197 \text{ pm})} \times$

$\dfrac{2 \text{ valence e}^-}{1 \text{ Ca atom}} \times \dfrac{1 \text{ orbital}}{2 \text{ valence e}^-} = 2.5 \times 10^7$

3J.2B p-type

3J.3B In chemiluminescence, light is emitted as the result of molecules being excited during a chemical reaction. In phosphorescence, light is emitted long after the stimulus has ceased.

Focus 4

4A.1B $w = -P\Delta V = -(9.60 \text{ atm}) \times (2.2 \text{ L} - 0.22 \text{ L}) \times (101.325 \text{ J·L}^{-1} \cdot \text{atm}^{-1}) = -1926 \text{ J} = -1.9 \text{ kJ}$

4A.2B $w = -P\Delta V = -(1.00 \text{ atm}) \times (4.00 \text{ L} - 2.00 \text{ L}) \times 101.325 \text{ J·L}^{-1} \cdot \text{atm}^{-1} = -202 \text{ J}$; $w = -nRT \ln(V_{\text{final}}/V_{\text{initial}}) = -(1.00 \text{ mol}) \times (8.3145 \text{ J·K}^{-1} \cdot \text{mol}^{-1}) \times (303 \text{ K}) \times \ln(4.00 \text{ L}/2.00 \text{ L}) = -1.75 \text{ kJ}$; isothermal reversible expansion does more work

4A.3B $q = nC_m\Delta T = (3.00 \text{ mol}) \times (111 \text{ J·K}^{-1} \cdot \text{mol}^{-1}) \times (15.0 \text{ K}) = 5.0 \text{ kJ}$

4A.4B $C_{\text{cal}} = \dfrac{q_{\text{cal}}}{\Delta T} = \dfrac{4.16 \text{ kJ}}{3.24 \text{ °C}} = 1.28 \text{ kJ·(°C)}^{-1}$

4B.1B $w = \Delta U - q = -150 \text{ J} - (+300 \text{ J}) = -450 \text{ J}$; $w < 0$ (the system did work)

4B.2B $q = +1.00 \text{ kJ}$; $w = -(2.00 \text{ atm}) \times (3.00 \text{ L} - 1.00 \text{ L}) - (101.325 \text{ J·L}^{-1} \cdot \text{atm}^{-1}) = -405 \text{ J}$; $\Delta U = q + w = 1.00 \text{ kJ} + (-0.405 \text{ kJ}) = +0.60 \text{ kJ}$

4B.3B $U_m \text{ (motion)} = U_m\text{(translation)} + U_m\text{(rotation)} = 2 \times (3/2)(RT) = 3 \times (8.3145 \text{ J·K}^{-1} \cdot \text{mol}^{-1}) \times (298 \text{ K}) \times (1 \text{ kJ}/1000 \text{ J}) = 7.43 \text{ kJ·mol}^{-1}$.

4C.1B (a) $\Delta H = +30 \text{ kJ}$; (b) $\Delta U = q + w = 30 \text{ kJ} + 40 \text{ kJ} = +70 \text{ kJ}$

4C.2B (a) $\Delta T = \dfrac{q}{nC_{V,m}} =$

$\dfrac{1.20 \text{ kJ} \times \dfrac{1000 \text{ J}}{1 \text{ kJ}}}{1.00 \text{ mol} \times \left(\dfrac{5}{2} \times 8.3145 \text{ L·atm·mol}^{-1} \cdot \text{K}^{-1}\right)} = 57.7 \text{ K},$

$T_f = 298 \text{ K} + 57.7 \text{ K} = 356 \text{ K}$; $\Delta U = q + w = 1.20 \text{ kJ} + 0 = 1.20 \text{ kJ}$

(b) (step 1) constant volume, find final temperature:

$\Delta T = \dfrac{q}{nC_P} =$

$\dfrac{1.20 \text{ kJ} \times \dfrac{1000 \text{ J}}{1 \text{ kJ}}}{1.00 \text{ mol} \times \left(\dfrac{7}{2} \times 8.3145 \text{ L·atm·mol}^{-1} \cdot \text{K}^{-1}\right)} = 41.2 \text{ K};$

$T_f = 298 \text{ K} + 41.2 \text{ K} = 339 \text{ K}$; $\Delta U = q + w = q + 0$, so $\Delta U = q = nC_V\Delta T = (1 \text{ mol}) \times (5/2) \times (8.3145 \text{ J·mol}^{-1} \cdot \text{K}^{-1}) \times (41.2 \text{ K}) = 856 \text{ J}$; (step 2) isothermal, $\Delta T = 0$, so $\Delta U = 0$; Therefore, after both steps, $\Delta U = +856 \text{ J}$, $T_f = 339 \text{ K}$

4C.3B $\dfrac{22 \text{ kJ}}{23 \text{ g}} \times \dfrac{46.07 \text{ g}}{1 \text{ mol}} = 44 \text{ kJ·mol}^{-1}$

4C.4B $\Delta H_{\text{sub}} = \Delta H_{\text{vap}} + \Delta H_{\text{fus}} = (38 + 3) \text{ kJ·mol}^{-1} = 41 \text{ kJ·mol}^{-1}$

4D.1B $q_r = -q_{\text{cal}} = -(216 \text{ J·°C}^{-1})(76.7 \text{ °C}) = -1.66 \times 10^4 \text{ J};$

$\Delta H_r = \dfrac{-1.66 \times 10^4 \text{ J}}{0.338 \text{ g}} \times \dfrac{72.15 \text{ g}}{1 \text{ mol}} \times \dfrac{1 \text{ kJ}}{1000 \text{ J}}$

$= -3.54 \times 10^3 \text{ kJ};$

$C_5H_{12}(l) + 8 O_2(g) \rightarrow 5 CO_2(g) + 6 H_2O(l)$, $\Delta H = -3.54 \times 10^3 \text{ kJ}$

4D.2B $\Delta U = \Delta H - \Delta n_{\text{gas}}RT = -3378 \text{ kJ} - \left(-\dfrac{3}{4} \text{ mol}\right) \times \left(\dfrac{8.3145 \text{ J}}{\text{mol·K}}\right) \times 1273 \text{ K} \times \left(\dfrac{1 \text{ kJ}}{1000 \text{ J}}\right) = -3.37 \times 10^3 \text{ kJ}$

4D.3B $-350. \text{kJ} \times \dfrac{1 \text{ mol C}_2\text{H}_5\text{OH}}{-1368 \text{ kJ}} \times \dfrac{46.07 \text{ g C}_2\text{H}_5\text{OH}}{1 \text{ mol C}_2\text{H}_5\text{OH}} =$

$11.8 \text{ g C}_2\text{H}_5\text{OH}$

4D.4B $CH_4(g) + \dfrac{1}{2}O_2(g) \rightarrow CH_3OH(l)$, $\Delta H° = 206.10 \text{ kJ} + (-128.33 \text{ kJ}) + \dfrac{1}{2}(-483.64 \text{ kJ}) = -164.05 \text{ kJ}$

4D.5B $C(\text{diamond}) + O_2(g) \rightarrow CO_2(g)$; $\Delta H_r° = \Delta H_f°(CO_2) - \Delta H_f°(C, \text{diamond}) - \Delta H_f°(O_2, g) = -393.51 \text{ kJ·mol}^{-1} - (+1.895 \text{ kJ·mol}^{-1}) - 0 \text{ kJ·mol}^{-1} = -395.41 \text{ kJ·mol}^{-1}$

4D.6B $CO(NH_2)_2(s) + \dfrac{3}{2}O_2(g) \rightarrow CO_2(g) + 2 H_2O(l) + N_2(g)$; $\Delta H_r° = \Delta H_f°(CO_2) + 2\Delta H_f°(H_2O) - \Delta H_f°(CO(NH_2)_2)$; $-632 \text{ kJ} = -393.51 \text{ kJ} + 2(-285.83 \text{ kJ}) - \Delta H_f°(CO(NH_2)_2)$; $\Delta H_f°(CO(NH_2)_2) = -333 \text{ kJ·mol}^{-1}$

4D.7B $\Delta H_r(523 \text{ K}) = \Delta H_r(298 \text{ K}) + \Delta C_{P,r}\Delta T = -365.56 \text{ kJ·mol}^{-1}$

$+ \left[\left(84.1 - \dfrac{3}{2}(29.36) - 2(28.82) - 29.12\right) \text{J·K}^{-1} \cdot \text{mol}^{-1} \times \dfrac{1 \text{ kJ}}{1000 \text{ J}}\right](523 - 298)\text{K} = -376 \text{ kJ·mol}^{-1}$

4E.1B $[524.3 + 147.70 + 2(111.88) + 736 + 1451 - 2(325)] \text{ kJ} - \Delta H_L = 0$; $\Delta H_L = +2433 \text{ kJ}$

4E.2B $CH_4(g) + 2 F_2(g) \rightarrow CH_2F_2(g) + 2 HF(g)$; bonds broken [2(C—H), 2(F—F)]: $2(412 \text{ kJ·mol}^{-1}) + 2(158 \text{ kJ·mol}^{-1}) = 1140 \text{ kJ·mol}^{-1}$; bonds formed [2(C—F), 2(H—F)]: $2(484 \text{ kJ·mol}^{-1}) + 2(565 \text{ kJ·mol}^{-1}) = 2098 \text{ kJ·mol}^{-1}$; $\Delta H_r° = 1140 \text{ kJ·mol}^{-1} - 2098 \text{ kJ·mol}^{-1} = -958 \text{ kJ·mol}^{-1}$

4F.1B $\Delta S = -50. \text{ J}/1373 \text{ K} = -0.036 \text{ J·K}^{-1}$

4F.2B $\Delta S = nR \ln(V_2/V_1) = (8.3145 \text{ J·K}^{-1}\text{·mol}^{-1}) \ln(10/1) = +19 \text{ J·K}^{-1}\text{·mol}^{-1}$

4F.3B $\Delta S = nR \ln(P_1/P_2) = 70.9 \text{ g} \times (1 \text{ mol}/70.9 \text{ g}) \times (8.3145 \text{ J·mol}^{-1}\text{·K}^{-1}) \times \ln(3.00 \text{ kPa}/24.00 \text{ kPa}) = -17.3 \text{ J·K}^{-1}$

4F.4B $\Delta S = (5.5 \text{ g})(0.51 \text{ J·K}^{-1}\text{·g}^{-1}) \ln(373/293) = +0.68 \text{ J·K}^{-1}$

4F.5B (1) $\Delta S = (23.5 \text{ g})(1 \text{ mol}/32.00 \text{ g})(8.3145 \text{ J·K}^{-1}\text{·mol}^{-1}) \times \ln(2.00 \text{ kPa}/8.00 \text{ kPa}) = -8.46 \text{ J·K}^{-1}$; (2) $\Delta S = (23.5 \text{ g})(1 \text{ mol}/32.00 \text{ g})(20.786 \text{ J·K}^{-1}\text{·mol}^{-1}) \times \ln(360 \text{ K}/240 \text{ K}) = +6.19 \text{ J·K}^{-1}$; $\Delta S = -8.46 + 6.19 \text{ J·K}^{-1} = -2.27 \text{ J·K}^{-1}$

4F.6B $\Delta S_{vap}° = \dfrac{\Delta H_{vap}}{T_b} = \dfrac{40.7 \text{ kJ·mol}^{-1}}{373.2 \text{ K}} \times \dfrac{10^3 \text{ J}}{1 \text{ kJ}}$
$= 109 \text{ J·K}^{-1}\text{·mol}^{-1}$

4F.7B $\Delta H_{vap}° = (85 \text{ J·K}^{-1}\text{·mol}^{-1})(273.2 \text{ K} + 24.5 \text{ K}) \times (1 \text{ kJ}/10^3 \text{ J}) = 26.2 \text{ kJ·mol}^{-1}$

4F.8B $\Delta S_{fus}° = \Delta H_{fus}°/T_f = (10.59 \times 10^3 \text{ J·mol}^{-1})/(278.6 \text{ K}) = 38.01 \text{ J·K}^{-1}\text{·mol}^{-1}$

4F.9B (1) Heat: $\Delta S = (136 \text{ J·K}^{-1}\text{·mol}^{-1}) \ln\left(\dfrac{353.2}{276.0}\right) =$

$+33.5 \text{ J·K}^{-1}\text{·mol}^{-1}$; (2) Vaporize:

$\Delta S_{vap}° = \dfrac{30800 \text{ J·mol}^{-1}}{353.2 \text{ K}} = 87.2 \text{ J·K}^{-1}\text{·mol}^{-1}$;

(3) Cool: $\Delta S = (82.4 \text{ J·K}^{-1}\text{·mol}^{-1}) \ln\left(\dfrac{276.0}{353.2}\right) =$

$-20.3 \text{ J·K}^{-1}\text{·mol}^{-1}$; $\Delta S_{vap}°(296 \text{ K}); = (33.5 + 87.2 - 20.3) \text{ J·K}^{-1}\text{·mol}^{-1} = 100.4 \text{ J·K}^{-1}\text{·mol}^{-1}$

4G.1B $\Delta S = k \ln W = (1.38066 \times 10^{-23} \text{ J·K}^{-1})(1.0 \text{ mol})(6.022 \times 10^{23} \text{ mol}^{-1}) \ln(6) = +15 \text{ J·K}^{-1}$

4G.2B In ice, each O atom is surrounded by four H atoms: two of the H atoms are covalently bonded to the O atom; the other two H atoms, which belong to neighboring water molecules, are interacting with the central O atom through hydrogen bonds. Thus, more than one orientation is possible in the crystal and entropy will not be zero at $T = 0$.

4H.1B (a) $\Delta S = S_{gray} - S_{white} = (44.14 - 51.55) \text{ J·K}^{-1}\text{·mol}^{-1} = -7.41 \text{ J·K}^{-1}\text{·mol}^{-1}$; the gray form; (b) $\Delta S = S_{graphite} - S_{diamond} = (5.7 - 2.4) \text{ J·K}^{-1}\text{·mol}^{-1} = +3.3 \text{ J·K}^{-1}\text{·mol}^{-1}$; diamond

4H.2B $\Delta S_r° = (229.60 - 219.56 - 130.68) \text{ J·K}^{-1} = -120.64 \text{ J·K}^{-1}$

4I.1B $\Delta S_{surr} = -\Delta H/T = -(2.00 \text{ mol} \times -46.11 \text{ kJ·mol}^{-1})/298 \text{ K} \times (10^3 \text{ J}/1 \text{ kJ}) = +309 \text{ J·K}^{-1}$

4I.2B $\Delta S_{surr} = -\Delta H/T = -(49.0 \text{ kJ}/298 \text{ K}) \times (10^3 \text{ J}/1 \text{ kJ}) = -164 \text{ J·K}^{-1}$; $\Delta S_{tot} = -164 \text{ J·K}^{-1} + (-253.18 \text{ J·K}^{-1}) = -417 \text{ J·K}^{-1}$; no

4I.3B $\Delta S = nR \times \ln(V_2/V_1) = (2.00 \text{ mol}) \times (8.3145 \text{ J·K}^{-1}\text{·mol}^{-1}) \times \ln(0.200/4.00) = -49.8 \text{ J·K}^{-1}\text{·mol}^{-1}$; $\Delta S_{surr} = +49.8 \text{ J·K}^{-1}\text{·mol}^{-1}$; $\Delta S_{tot} = 0$

4I.4B $\Delta S_{surr} = -\Delta H_{vap}/T = -(30.8 \times 10^3 \text{ J·mol}^{-1})/353.2 \text{ K} = -87.2 \text{ J·K}^{-1}\text{·mol}^{-1}$; $\Delta S = +87.2 \text{ J·K}^{-1}$; $\Delta S_{tot} = \Delta S + \Delta S_{surr} = +87.2 \text{ J·K}^{-1} + (-87.2 \text{ J·K}^{-1}) = 0$ at 353.2 K

4J.1B Yes. $\Delta G = \Delta H - T\Delta S$. When $\Delta S > 0$, $T\Delta S > 0$. Hence, as temperature increases, $-T\Delta S$ becomes more negative; eventually $\Delta G < 0$ and the process becomes spontaneous.

4J.2B (a) $\Delta G_m = \Delta H_m - T\Delta S_m = 59.3 \text{ kJ·mol}^{-1} - (623 \text{ K}) \times (0.0942 \text{ kJ·K}^{-1}\text{·mol}^{-1}) = +0.6 \text{ kJ·mol}^{-1}$; vaporization is nonspontaneous; (b) $\Delta G_m = \Delta H_m - T\Delta S_m = 59.3 \text{ kJ·mol}^{-1} - (643 \text{ K})(0.0942 \text{ kJ·K}^{-1}\text{·mol}^{-1}) = -1.3 \text{ kJ·mol}^{-1}$; vaporization is spontaneous.

4J.3B $3 \text{ H}_2 \text{ (g)} + 3 \text{ C (s, graphite)} \rightarrow \text{C}_3\text{H}_6 \text{ (g)}$, $\Delta S_r = 237.4 \text{ J·K}^{-1}\text{·mol}^{-1} - [3(130.68 \text{ J·K}^{-1}\text{·mol}^{-1}) + 3(5.740 \text{ J·K}^{-1}\text{·mol}^{-1})] = -171.86 \text{ J·K}^{-1}\text{·mol}^{-1}$, $\Delta G_r = \Delta H_r - T\Delta S_r = +53.30 \text{ kJ·mol}^{-1} - (298 \text{ K})(-171.86 \text{ J·K}^{-1}\text{·mol}^{-1}) \times (1 \text{ kJ}/10^3 \text{ J}) = +104.5 \text{ kJ·mol}^{-1}$

4J.4B From Appendix 2A, $\Delta G_f°(\text{CH}_3\text{NH}_2, \text{g}) = +32.16 \text{ kJ·mol}^{-1}$ at 298 K. Because $\Delta G_f° > 0$, CH_3NH_2 is less stable than its elements at the stated conditions.

4J.5B $\Delta G = [-910 + (6 \text{ mol})(0)] \times [(6 \text{ mol})(-394.36) + (6 \text{ mol})(-237.13)] = +2879 \text{ kJ}$

4J.6B $\text{MgCO}_3(\text{s}) \rightarrow \text{MgO}(\text{s}) + \text{CO}_2(\text{g})$; $\Delta H° = -601.70 + (-393.51) - (-1095.8) = +100.6 \text{ kJ}$; $\Delta S° = 26.94 + 213.74 - 65.7 \text{ J·K}^{-1} = +175.0 \text{ J·K}^{-1}$;
$T = \dfrac{\Delta H°}{\Delta S°} = \dfrac{100.6 \text{ kJ}}{175.0 \text{ J·K}^{-1}} \times \dfrac{10^3 \text{ J}}{1 \text{ kJ}} = 574.9 \text{ K}$

Focus 5

5A.1B $\text{CH}_3\text{CH}_2\text{CH}_3$; molecules of the two substances have the same molar mass and thus the same number of electrons and comparable London forces. However, CH_3CHO is polar and also experiences dipole−dipole forces.

5A.2B $\ln\left(\dfrac{P_2}{94.6 \text{ Torr}}\right) = \dfrac{30.8 \times 10^3 \text{ J·mol}^{-1}}{8.3145 \text{ J·mol}^{-1}\text{·K}^{-1}}\left(\dfrac{1}{298 \text{ K}} - \dfrac{1}{308 \text{ K}}\right)$;
$P_2 = 142 \text{ Torr}$

5A.3B $\ln\left(\dfrac{760 \text{ Torr}}{400 \text{ Torr}}\right) = \dfrac{35.3 \times 10^3 \text{ J·mol}^{-1}}{8.3145 \text{ J·mol}^{-1}\text{·K}^{-1}}\left(\dfrac{1}{323 \text{ K}} - \dfrac{1}{T_2}\right)$;
$T_2 = 339 \text{ K}$

5B.1B The positive slope of the solid−liquid boundary shows that monoclinic sulfur is more dense than liquid sulfur over the temperature range in which monoclinic sulfur is stable; the solid is more stable at high pressure.

5B.2B Carbon dioxide is liquid at 60 atm and 25 °C. When it is released into a room at 1 atm and 25 °C, as the pressure decreases, the system reaches the liquid−vapor boundary, at which pressure the liquid is changed to vapor. The vaporization absorbs sufficient heat to cool the CO_2 to below its sublimation temperature at 1 atm. As a result, fine particles of solid CO_2 "snow" are produced.

5B.3B Critical temperature increases with the strength of intermolecular forces. For example, CH_4 cannot form hydrogen bonds; so it has a lower critical temperature than either NH_3 or H_2O, which can form hydrogen bonds.

5C.1B $n_{C_9H_8O} = 2.00 \text{ g} \times \dfrac{1 \text{ mol}}{132.16 \text{ g}} = 0.0151 \text{ mol}$;

$n_{C_2H_5OH} = 50.0 \text{ g} \times \dfrac{1 \text{ mol}}{46.07 \text{ g}} = 1.09 \text{ mol}$;

$x_{C_2H_5OH} = \dfrac{1.09 \text{ mol}}{(1.09 + 0.0151) \text{ mol}} = 0.986$;

$P = (0.986)(5.3 \text{ kPa}) = 5.2 \text{ kPa}$

5C.2B Equal masses; therefore, assume 50.00 g of each.

$n_{C_6H_6} = 50.00 \text{ g} \times \dfrac{1 \text{ mol}}{78.11 \text{ g}} = 0.6401 \text{ mol}$;

$n_{C_7H_8} = 50.00 \text{ g} \times \dfrac{1 \text{ mol}}{92.13 \text{ g}} = 0.5427 \text{ mol}$;

$x_{C_6H_6} = \dfrac{0.6401 \text{ mol}}{(0.6401 + 0.5427) \text{ mol}} = 0.5412$;

$x_{C_7H_8} = \dfrac{0.5427 \text{ mol}}{(0.6401 + 0.5427) \text{ mol}} = 0.4588$;

$P_{\text{total}} = (0.5412) \times (94.6 \text{ Torr}) + (0.4588) \times (29.1 \text{ Torr})$
$= 64.5 \text{ Torr}$

5C.3B (a) $P = (0.500)(94.6 \text{ Torr}) + (0.500)(29.1 \text{ Torr}) = 61.8 \text{ Torr}$;

(b) $x_{C_6H_6,\text{ vap}} = \dfrac{(0.500) \times (94.6 \text{ Torr})}{61.8 \text{ Torr}} = 0.765$; $x_{C_7H_8,\text{ vap}} = 1 - 0.765 = 0.235$

5D.1B $s = (2.3 \times 10^{-2} \text{ mol·L}^{-1}\text{·atm}^{-1})(1.00 \text{ atm}) = 2.3 \times 10^{-2}$ mol·L^{-1}; $n(CO_2) = 0.900 \text{ L} \times (2.3 \times 10^{-2} \text{ mol·L}^{-1}) = 0.021 \text{ mol}$

5E.1B $\text{molality} = \dfrac{7.36 \text{ g KClO}_3}{0.200 \text{ kg H}_2\text{O}} \times \dfrac{1 \text{ mol KClO}_3}{122.55 \text{ g KClO}_3} = 0.300 \text{ mol·kg}^{-1}$

5E.2B $x_{H_2O} = 1 - 0.250 = 0.750$; $0.750 \text{ mol H}_2\text{O} \times 18.02$ $\text{g·mol}^{-1} \times 1 \text{ kg}/10^3 \text{ g} = 0.0135 \text{ kg}$; $m(H_2O) = 0.0135 \text{ kg}$ H_2O; $\text{molality} = \dfrac{0.250 \text{ mol CH}_3\text{OH}}{0.0135 \text{ kg H}_2\text{O}} = 18.5 \text{ mol·kg}^{-1}$

5E.3B Assume a 1L solution; $m_{\text{NaCl}} = 1 \text{ L} \times \dfrac{1.83 \text{ mol NaCl}}{1 \text{ L soln}} \times$ $58.44 \text{ g·mol}^{-1} = 106.9 \text{ g NaCl}$; $m_{\text{solution}} = 1 \text{ L soln} \times$ $\dfrac{10^3 \text{ mL}}{1 \text{ L}} \times \dfrac{1.07 \text{ g}}{1 \text{ mL}} = 1.07 \times 10^3 \text{ g}$; $m_{H_2O} =$ $1070 \text{ g} - 107 \text{ g} = 9.6 \times 10^3 \text{ g}$; $1.83 \text{ mol}/0.96 \text{ kg} =$ 1.9 mol·kg^{-1}

5F.1B $\Delta T_f = k_f m = (39.7 \text{ K·kg·mol}^{-1})(0.050 \text{ mol·kg}^{-1}) =$ $1.99 \text{ K} = 1.99 \,^{\circ}\text{C}$; $T_f = 179.8 \,^{\circ}\text{C} - 1.99 \,^{\circ}\text{C} = 177.8 \,^{\circ}\text{C}$

5F.2B 0.100 mol ions (0.025 mol Co^{3+} ions and 0.075 mol Cl^- ions); $i = 4$

5F.3B $i = 1$ since sucrose is a nonelectrolyte and does not dissociate; $(1)(0.08206 \text{ L·atm·mol}^{-1}\text{·K}^{-1})(298 \text{ K})$ $(0.120 \text{ mol·L}^{-1}) = 2.93 \text{ atm}$

5F.4B $b = 0.51 \text{ K}/39.7 \text{ kg·K·mol}^{-1} = 0.0128 \text{ mol·kg}^{-1}$; $n(\text{linalool}) = 0.100 \text{ kg} \times (0.0128 \text{ mol}/1 \text{ kg}) = 1.28 \times$ 10^{-3} mol; $M = 0.200 \text{ g}/(1.28 \times 10^{-3} \text{ mol}) = 156 \text{ g·mol}^{-1}$

5F.5B $n = cV = \left(\dfrac{2.11 \text{ kPa}}{(1) \times (8.3145 \text{ L·kPa·K}^{-1}\text{·mol}^{-1}) \times (293 \text{ K})}\right) \times$

$0.175 \text{ L} = 1.52 \times 10^{-4} \text{ mol}$;

$M = \left(\dfrac{1.50 \text{ g}}{1.52 \times 10^{-4} \text{ mol}}\right) = 9.87 \times 10^3 \text{ g·mol}^{-1}$
$= 9.87 \text{ kg·mol}^{-1}$

5G.1B $K = (P_{SO_2})^2(P_{H_2O})^2/(P_{H_2S})^2(P_{O_2})^3$

5G.2B $K = 1/(P_{O_2})^5$

5G.3B $K = [ZnCl_2](P_{H_2})/[HCl]^2$

5G.4B $\Delta G_r = 4.73 \text{ kJ·mol}^{-1} + (8.3145 \text{ J·mol}^{-1}\text{·K}^{-1})(298 \text{ K}) \times$ $\ln\{(2.10)^2/0.80\} \times (1 \text{ kJ}/10^3 \text{ J}) = +8.96 \text{ kJ·mol}^{-1}$. Because $\Delta G > 0$, reaction proceeds toward reactants.

5G.5B $\Delta G^{\circ} = 2\Delta G_f^{\circ} (NO_2(g)) - [\Delta G_f^{\circ} (O_2(g)) + 2\Delta G_f^{\circ}$ $(NO(g))] = [2(51.31 - (0 + 2(86.55))] \text{ kJ·mol}^{-1} =$ $-70.48 \text{ kJ·mol}^{-1}$; $\ln K = -(-70.48 \times 10^3 \text{ J·mol}^{-1})/$ $(8.3145 \text{ kJ·K}^{-1}\text{·mol}^{-1} \times 298 \text{ K}) = 28.45$; $K = 2.3 \times 10^{12}$

5H.1B $K = (7.3 \times 10^{-13})^{1/2} = 8.5 \times 10^{-7}$

5H.2B $K_c = K(T/12.03 \text{ K})^{-\Delta n}$, $\Delta n = 2 - 1 = +1$; $K_c = (47.9) \times$ $(12.03 \text{ K}/400. \text{ K}) = 1.44$

5I.1B $52 \text{ kPa} \times (1\text{bar})/(10^2 \text{ kPa}) = 0.52 \text{ bar}$; $P_{NO} = [K \times P_{N_2} \times$ $P_{O_2}]^{1/2} = [(3.4 \times 10^{-21}) \times (0.52 \text{ bar}) \times (0.52 \text{ bar})]^{1/2} =$ $3.0 \times 10^{-11} \text{ bar}$, or $3.0 \times 10^{-6} \text{ Pa}$

5I.2B $Q = (1.2)^2/2.4 = 0.60$; $K = 0.15$, and so $Q > K$ and the partial pressure of N_2O_4 will increase.

5I.3B $K = (2x)^2(x)/(0.012 - 2x)^2 \approx 4x^3/(0.012)^2 = 3.5 \times$ 10^{-32}; $x = 1.1 \times 10^{-12}$. At equilibrium: $P_{HCl} = 0.012 \text{ bar}$; $P_{HI} = 2.2 \times 10^{-12} \text{ bar}$; $P_{Cl_2} = 1.1 \times 10^{-12} \text{ bar}$; some solid I_2 remains as well.

5I.4B $Q = (0.100)^2/(0.200)(0.100) = 0.5$, so $Q < K$; $K = 20. =$ $(0.100 + 2x)^2/[(0.200 - x)(0.100 - x)]$; $x = 0.0750 \text{ bar}$; $P_{ClF} = 0.100 + 2x = 0.250 \text{ bar}$

5J.1B The equilibrium tends to shift toward (a) products; (b) products; (c) reactants.

5J.2B $Q = (1.30)^2/(0.080)(0.050)^3 = 1.7 \times 10^5$; $Q < K$. $K =$ $(1.30 + 2x)^2/(0.080 - x)(0.050 - x)^3 = 6.8 \times 10^5$; $x =$ $5.86 \times 10^{-3} \text{ bar}$; 0.074 bar N_2; 1.31 bar NH_3; 0.032 bar H_2

5J.3B Compression affects gaseous species only. $CO_2(g)$ reacts to form an aqueous species. Therefore, compression favors the formation of $H_2CO_3(aq)$.

5J.4B $\Delta H_r^{\circ} = 2(-393.51 \text{ kJ·mol}^{-1}) - 2 (-110.53 \text{ kJ·mol}^{-1}) -$ $0 = -565.96 \text{ kJ·mol}^{-1}$. The reaction is exothermic; therefore, lowering the temperature shifts the reaction to the products. The pressure of CO_2 will increase.

5J.5B $\Delta H_r^\circ = 0 + (-287.0 \text{ kJ·mol}^{-1}) - (-374.9 \text{ kJ·mol}^{-1}) = +87.9 \text{ kJ·mol}^{-1}$;

$$\ln\left(\frac{K_2}{K_1}\right) = \frac{87.9 \times 10^3 \text{ J·mol}^{-1}}{8.3145 \text{ J·mol}^{-1}\text{·K}^{-1}}\left[\frac{1}{523 \text{ K}} - \frac{1}{800 \text{ K}}\right] = 6.99\ldots$$

$K_2 = K_1 e^{6.99\ldots} = (78.3)e^{6.99\ldots} = 8.6 \times 10^4$

Focus 6

6A.1B (a) H_3O^+ (it has one more H^+ than H_2O); (b) NH_2^- (it has one fewer H^+ than NH_3)

6A.2B (a) Brønsted acids: $NH_4^+(aq)$, $H_2CO_3(aq)$; Brønsted bases: $HCO_3^-(aq)$, $NH_3(aq)$; (b) Lewis acid: $H^+(aq)$; Lewis bases: $NH_3(aq)$, $HCO_3^-(aq)$

6A.3B (a) $[H_3O^+] = (1.0 \times 10^{-14})/(2.2 \times 10^{-3}) = 4.5 \times 10^{-12}$ mol·L^{-1}; (b) $[OH^-] = [\text{NaOH}]_{\text{initial}} = 2.2 \times 10^{-3}$ mol·L^{-1}

6B.1B $[OH^-] = [\text{NaOH}]_{\text{initial}} = 0.077$ mol·L^{-1}; $[H_3O^+] = (1.0 \times 10^{-14})/(0.077) = 1.30 \times 10^{-13}$ mol·L^{-1}; pH $= -\log(1.3 \times 10^{-13}) = 12.89$

6B.2B $[H_3O^+] = 10^{-8.2} = 6 \times 10^{-9}$ mol·L^{-1}

6B.3B pH $= 14.0 - 9.4 = 4.6$

6C.1B The conjugate base of HIO_3 is IO_3^-; $pK_b = pK_w - pK_a = 14.00 - 0.77 = 13.23$

6C.2B (a) $1.8 \times 10^{-9} = K_b(C_5H_5N) < K_b(NH_2NH_2) = 1.7 \times 10^{-6}$, so NH_2NH_2 is the stronger base. (b) From part a, C_5H_5N is the weaker base, so $C_5H_5NH^+$ is the stronger acid. (c) $1.7 \times 10^{-1} = K_a(HIO_3) > K_a(HClO_2) = 1.0 \times 10^{-2}$, so HIO_3 is the stronger acid. (d) $K_b(HSO_3^-) = (1.0 \times 10^{-14})/(1.5 \times 10^{-2}) = 6.7 \times 10^{-13}$; $K_b(ClO_2^-) = (1.0 \times 10^{-14})/(1.0 \times 10^{-2}) = 1.0 \times 10^{-12}$; $K_b(HSO_3^-) < K_b(ClO_2^-)$; therefore, ClO_2^- is the stronger base.

6C.3B $CH_3COOH < CH_2ClCOOH < CHCl_2COOH$ (The electronegativity of Cl is greater than that of H. Therefore, acidity increases as the number of chlorine atoms increases.)

6D.1B Check approximation: $K_a = 1.4 \times 10^{-3} = x^2/(0.22 - x)$, $x = 0.018, > 5\%$ of 0.22, so the quadratic equation is required: $x^2 + (1.4 \times 10^{-3})x - 3.1 \times 10^{-4} = 0$; $x = 1.7 \times 10^{-2}$ mol·L$^{-1} = [H_3O^+]$; pH $= 1.77$; percentage deprotonation $= [(1.7 \times 10^{-2})/0.22] \times 100\% = 7.7\%$

6D.2B $[H_3O^+] = 10^{-2.35} = 4.5 \times 10^{-3}$ mol·L^{-1}; $K_a = (4.5 \times 10^{-3})^2/\{0.50 - (4.5 \times 10^{-3})\} = 4.1 \times 10^{-5}$

6D.3B $K_b = 1.0 \times 10^{-6} = x^2/(0.012 - x) \approx x^2/.012$; $x = 1.1 \times 10^{-4} = [OH^-]$; pOH $= 3.96$; pH $= 14.00 - 3.96 = 10.04$; percentage protonated $= (1.1 \times 10^{-4})/(0.012) \times 100 = 0.92\%$

6D.4B (a) CO_3^{2-} is the conjugate base of the weak acid HCO_3^-; therefore, the solution is basic. (b) $K_a(\text{Al}(H_2O)_6^{3+}) = 1.4 \times 10^{-5}$; therefore, the solution is acidic. (c) K^+ is a "neutral cation" and NO_3^- is a conjugate base of a strong acid; therefore, the aqueous solution is neutral.

6D.5B $NH_4^+(aq) + H_2O(l) \rightleftharpoons NH_3(aq) + H_3O^+(aq)$; $K_b(NH_3) = 1.8 \times 10^{-5}$; $K_a(NH_4^+) = (1.0 \times 10^{-14})/(1.8 \times 10^{-5}) = 5.6 \times 10^{-10}$; $K_a = 5.6 \times 10^{-10} = x^2/(0.10 - x) \approx x^2/0.10$; $x = 7.5 \times 10^{-6}$; pH $= -\log(7.5 \times 10^{-6}) = 5.12$

6D.6B $F^-(aq) + H_2O(l) \rightleftharpoons HF(aq) + OH^-(aq)$; $K_a(HF) = 3.5 \times 10^{-4}$; $K_b(F^-) = (1.0 \times 10^{-14})/(3.5 \times 10^{-4}) = 2.9 \times 10^{-11}$; $K_b = 2.9 \times 10^{-11} = x^2/(0.020 - x) \approx x^2/0.020$; $x = 7.56 \times 10^{-7}$; pOH $= 6.12$; pH $= 14.00 - 6.12 = 7.88$

6E.1B $0.012 = (0.10 + x)(x)/(0.10 - x)$; $x^2 + 0.11x - 0.0012 = 0$; from the quadratic formula, $x = 0.0098$; $[H_3O^+] = 0.10 + 0.0098 = 0.11$; pH $= 0.96$

6E.2B $pK_{a1}(H_3PO_4) = 2.12$; $pK_{a2}(H_3PO_4) = 7.21$; pH $= \frac{1}{2}(2.12 + 7.21) = 4.66$

6E.3B $[Cl^-] = 0.50$ mol·L^{-1}; $^+NH_3CH_2COOH(aq) + H_2O(l) \rightleftharpoons {}^+NH_3CH_2CO_2^-(aq) + H_3O^+(aq)$, $K_{a1} = 4.5 \times 10^{-3}$ but $> 5\%$ deprotonation, so approximation not valid, $x^2 + (4.5 \times 10^{-3})x - 0.00225 = 0$; $x = 0.045 = [H_3O^+] = [^+NH_3CH_2CO_2^-]$; $[^+NH_3CH_2COOH] = (0.50 - 0.045)$ mol·L$^{-1} = 0.45$ mol·L^{-1}; $^+NH_3CH_2CO_2^-(aq) + H_2O(l) \rightleftharpoons NH_2CH_2CO_2^-(aq) + H_3O^+(aq)$, $K_{a2} = 1.7 \times 10^{-10} = (x)(0.045)/(0.045 - x)$, $x = 1.7 \times 10^{-10}$ mol·L$^{-1} = [NH_2CH_2CO_2^-]$ ($< 5\%$ ionization, so approximation is valid); $[OH^-] = K_w/[H_3O^+] = (1.0 \times 10^{-14})/(0.045) = 2.2 \times 10^{-13}$ mol·L^{-1}

6F.1B $K_w = [H_3O^+]([H_3O^+] + [\text{NaOH}]_{\text{initial}}) = x(x + [\text{NaOH}]_{\text{initial}})$, $x^2 + (2.0 \times 10^{-7})x - (1.0 \times 10^{-14})$, $x = 4.1 \times 10^{-8}$ mol·L^{-1}, pH $= -\log(4.1 \times 10^{-8}) = 7.39$

6F.2B $K_a(HIO) = 2.3 \times 10^{-11}$, $x^3 + (2.3 \times 10^{-11})x^2 - [1.0 \times 10^{-14} + (2.3 \times 10^{-11})(1.0 \times 10^{-2})]x - (2.3 \times 10^{-11})(1.0 \times 10^{-14}) = 0$, $x^3 + (2.3 \times 10^{-11})x^2 - (2.4 \times 10^{-13})x - (2.3 \times 10^{-25}) = 0$, $x = 4.9 \times 10^{-7} = [H_3O^+]$, pH $= -\log(4.9 \times 10^{-7}) = 6.31$

6G.1B $K_a = 5.6 \times 10^{-10}$; pH $= -\log(5.6 \times 10^{-10}) + \log(0.030/0.040) = 9.13$

6G.2B $n(CH_3CO_2^-)_{\text{final}} = [(0.500 \text{ L}) \times (0.040 \text{ mol·L}^{-1})] - 0.0100 \text{ mol} = 0.0100 \text{ mol}$, $[CH_3CO_2^-]_{\text{final}} = (0.0100 \text{ mol})/(0.500 \text{ L}) = 0.0200$ mol·L^{-1}; $n(CH_3COOH)_{\text{final}} = [(0.500 \text{ L}) \times (0.080 \text{ mol·L}^{-1})] + 0.0100 \text{ mol} = 0.0500 \text{ mol}$, $[CH_3COOH]_{\text{final}} = (0.0500 \text{ mol})/(0.500 \text{ L}) = 0.100$ mol·L^{-1}; pH $= 4.75 + \log(0.0200/0.100) = 4.05$; pH$_{\text{final}} -$ pH$_{\text{initial}} = 4.05 - 4.45 = -0.4$ (a decrease of 0.4)

6G.3B $(CH_3)_3NH^+/(CH_3)_3N$, because the $pK_a = 9.81$, which is close to 10

6G.4B pH $- pK_a = 3.50 - 4.19 = -0.69$; $([C_6H_5CO_2^-]/[C_6H_5COOH]) = 10^{-0.69} = 0.20{:}1$

6H.1B Amount of H_3O^+ added $= 0.012 \text{ L} \times (0.340 \text{ mol})/1 \text{ L} = 0.004\ 08$ mol; amount of OH^- remaining $= 0.006\ 25 - 0.004\ 08 \text{ mol} = 0.002\ 17$ mol, $[OH^-] = (0.00217 \text{ mol})/(0.0370 \text{ L}) = 0.0586$ mol·L^{-1}; pOH $= 1.232$; pH $= 14.00 - 1.232 = 12.77$

6H.2B initial amount of NH_3 = 0.025 L $\times$ (0.020 mol·L^{-1}) = 5.0 $\times$ 10^{-4} mol; volume of HCl added = (5.0 $\times$ 10^{-4} mol)/(0.015 mol·L^{-1}) = 0.0333 L; [NH_4^+] = (5.0 $\times$ 10^{-4} mol)/(0.0333 + 0.02500)L = 0.0086 mol·L^{-1}; K_a = [H_3O^+][NH_3]/[NH_4^+] = 5.6 $\times$ 10^{-10} = (x^2)/(0.0086 $-$ x) $\approx$ (x^2)/(0.0086); [H_3O^+] = x = 2.2 $\times$ 10^{-6}; pH = 5.66

6H.3B final volume of solution = 25.00 + 15.00 mL = 40.00 mL; amount of HCO_2^- formed = amount of OH^- added = (0.150 mol·L^{-1}) $\times$ (15.00 mL) = 2.25 mmol; amount of HCOOH remaining = 2.50 $-$ 2.25 mmol = 0.25 mmol; [HCOOH] = (0.25 $\times$ 10^{-3} mol)/(0.04000 L) = 0.062 mol·L^{-1}; [HCO_2^-] = (2.25 $\times$ 10^{-3} mol)/(0.04000 L) = 0.0562 mol·L^{-1}; pH = 3.75 + log (0.0562/0.0062) = 4.71

6H.4B (a) amount of H_3PO_4 = (0.030 L)(0.010 mol·L^{-1}) = 3.0 $\times$ 10^{-4} mol; at the first stoichiometric point, the initial amount of H_3PO_4 = amount of NaOH added; volume of NaOH = 3.0 $\times$ 10^{-4} mol/0.020 mol·L^{-1} = 0.015 L, or 15 mL. (b) 2 $\times$15 mL = 30. mL

6H.5B (0.020 L)(0.100 mol·L^{-1}) = 0.0020 mol H_2S; 0.0020 mol/(0.300 mol·L^{-1}) = 0.0067 L, or 6.7 mL NaOH for the first stoichiometric point and 2 $\times$ 6.7 mL = 13.4 mL for the second stoichiometric point; (a) before the first stoichiometric point; primary species present are Na^+, H_2S, and HS^-; (b) at the second stoichiometric point, primary species present are Na^+ and S^{2-} (also HS^- and OH^- because S^{2-} has a relatively large K_b)

6I.1B K_{sp} = [Ag^+][Br^-] = (8.8 $\times$ 10^{-7})^2 = 7.7 $\times$ 10^{-13}

6I.2B K_{sp} = [Pb^{2+}][F^-]^2 = (s)(2s)^2 = 4s^3; 3.7 $\times$ 10^{-8} = 4s^3; s = 2.1 $\times$ 10^{-3} mol·L^{-1}

6I.3B In 0.10 M $CaBr_2$(aq), [Br^-] = 2 $\times$ (0.10 mol·L^{-1}) = 0.20 mol·L^{-1}; s = [Ag^+] = K_{sp}/[Br^-] = (7.7 $\times$ 10^{-13})/(0.20) = 3.8 $\times$ 10^{-12} mol·L^{-1}

6I.4B CuS(s) + 4 NH_3(aq) $\rightleftharpoons$ $Cu(NH_3)_4^{2+}$(aq) + S^{2-}(aq); K = (1.3 $\times$ 10^{-36}) $\times$ (1.2 $\times$ 10^{13}) = 1.6 $\times$ 10^{-23}; K = [$Cu(NH_3)_4^{2+}$][S^{2-}]/[NH_3]^4 = 1.6 $\times$ 10^{-23} = x^2/(1.2 $-$ 4x)^4 $\approx$ x^2/(1.2)^4; x/(1.2)^2 = 4.0 $\times$ 10^{-12}; x = [S^{2-}] = 5.8 $\times$ 10^{-12} mol·L^{-1}

6J.1B [Ba^{2+}] = {(1.0 $\times$ 10^{-3} mol·L^{-1}) $\times$ (100 mL)}/300 mL = 3.3 $\times$ 10^{-4} mol·L^{-1}. [F^-] = {(1.0 $\times$ 10^{-3} mol·L^{-1}) $\times$ (200 mL)}/300 mL = 6.7 $\times$ 10^{-4} mol·L^{-1}. Q_{sp} = (3.3 $\times$ 10^{-4})(6.7 $\times$ 10^{-4})^2 = 1.5 $\times$ 10^{-10} < K_{sp} = 1.7 $\times$ 10^{-6}; therefore, no precipitate of BaF_2 forms.

6J.2B (a) For $PbCl_2$ to precipitate, [Cl^-] = (K_{sp}/[Pb^{2+}])$^{1/2}$ = (1.6 $\times$ 10^{-5}/0.020)$^{1/2}$ = 2.8 $\times$ 10^{-2} mol·L^{-1}. For AgCl to precipitate, [Cl^-] = (K_{sp}/[Ag^+]) = (1.6 $\times$ 10^{-10}/0.0010) = 1.6 $\times$ 10^{-7} mol·L^{-1}. Therefore, AgCl precipitates first at [Cl^-] = 1.6 $\times$ 10^{-7} mol·L^{-1}, then $PbCl_2$ precipitates at [Cl^-] = 2.8 $\times$ 10^{-2} mol·L^{-1}. (b) When the $PbCl_2$ precipitates, [Ag^+] = (K_{sp}/[Cl^-]) = (1.6 $\times$ 10^{-10})/(2.8 $\times$ 10^{-2}) = 5.7 $\times$ 10^{-9} mol·L^{-1}.

6K.1B reduction: (8 H^+ + MnO_4^- + 5 e^- $\rightarrow$ Mn^{2+} + 4 H_2O) $\times$ 2; oxidation: (H_2O + H_2SO_3 $\rightarrow$ HSO_4^- + 2 e^- + 3 H^+) $\times$ 5; net reaction: H^+(aq) + 2 MnO_4^- (aq) + 5 H_2SO_3(aq) $\rightarrow$ 2 Mn^{2+}(aq) + 5 HSO_4^-(aq) + 3 H_2O(l)

6K.2B oxidation: (3 I^- $\rightarrow$ I_3^- + 2 e^-) $\times$ 8; reduction: 9 H_2O + 16 e^- + 3 IO_3^- $\rightarrow$ I_3^- + 18 OH^-; net reaction: 3 H_2O(l) + IO_3^- (aq) + 8 I^-(aq) $\rightarrow$ 3 I_3^- (aq) + 6 OH^-(aq)

6L.1B Because Zn loses two electrons, n = 2 mol; ΔG = $-$(2 mol)(9.6485 $\times$ 10^4 C·mol^{-1})(1.6 V) = $-$3.09 $\times$ 10^5 C·V = $-$3.1 $\times$ 10^2 kJ.

6L.2B Mn(s)|Mn^{2+}(aq)||Cu^{2+}(aq),Cu^+(aq)|Pt(s)

6L.3B (a) left: 2 Hg(l) + 2 HCl(aq) $\rightarrow$ Hg_2Cl_2(s) + 2 e^-; right: 2 e^- + $Hg_2(NO_3)_2$(aq) $\rightarrow$ 2 Hg(l) + 2 NO_3^-(aq); 2 HCl(aq) + $Hg_2(NO_3)_2$(aq) $\rightarrow$ Hg_2Cl_2(s) + 2 HNO_3(aq). (b) yes

6M.1B (a) oxidation half-reaction: Cu^{2+} + 2 e^- $\rightarrow$ Cu, $E°$ = +0.34 V, reduction half-reaction: Ag^+ + e^- $\rightarrow$ Ag, $E°$ = +0.80 V; $E_{cell}°$ = 0.80 $-$ 0.34 = + 0.46 V. (b) Cu(s)|Cu^{2+}(aq)||Ag^+(aq)|Ag(s); Cu(s) + 2 Ag^{2+}(aq) $\rightarrow$ Cu^{2+}(aq) + 2 Ag(s); (c) Ag^+ is the stronger oxidizing agent.

6M.2B $E°$(Pb^{2+}/Pb) = $E_{cell}°$ + $E°$(Fe^{2+}/Fe) = 0.31 V + ($-$0.44 V) = $-$0.13 V

6M.3B Mn^{3+} + e^- $\rightarrow$ Mn^{2+}, $E_1°$ = +1.51 V; Mn^{2+} + 2e^- $\rightarrow$ Mn(s), $E_2°$ = $-$1.18 V; net half-reaction: Mn^{3+} + 3 e^- $\rightarrow$ Mn(s), $E_3°$ = [($n_1E_1°$ + $n_2E_2°$)/(n_3)] = [(1 mol)(1.51 V) + (2 mol)($-$1.18 V)]/(3 mol) = $-$0.28 V

6M.4B O_2 + 2 H_2O + 4e^- $\rightarrow$ 4 OH^-, $E°$ = +0.40 V; Cl_2 + 2 e^- $\rightarrow$ 2 Cl^-, $E°$ = +1.36 V. Yes, Cl_2(g) can oxidize H_2O to O_2(g) in basic solution under standard conditions because the Cl_2 reduction half-reaction has a more positive standard potential than the O_2 reduction half-reaction.

6N.1B $Cd(OH)_2$ + 2 e^- $\rightarrow$ Cd + 2 OH^-, $E°$ = $-$0.81 V; Cd^{2+} + 2 e^- $\rightarrow$ Cd, $E°$ = $-$0.40 V; net: $Cd(OH)_2$(s) $\rightarrow$ Cd^{2+}(aq) + 2 OH^- (aq), $E_{cell}°$ = $-$0.41 V; ln K_{sp} = (2)($-$0.41 V)/0.025 693 V = $-$31.92; K_{sp} = 1.4 $\times$ 10^{-14}

6N.2B cell reaction: Ag^+(aq, 0.010 mol·L^{-1}) $\rightarrow$ Ag^+(aq, 0.0010 mol·L^{-1}), $E_{cell}°$ = 0.0 V, n = 1; E_{cell} = 0.0 V $-$ (0.025693 V) $\times$ ln(0.0010/0.010) = +0.059 V

6N.3B At pH = 12.5, pOH = 1.5 and [OH^-] = 0.032 mol·L^{-1}; [Ag^+] = (K_{sp}/[OH^-]) = (1.5 $\times$ 10^{-8}/(0.032) = 4.7 $\times$ 10^{-7} mol·L^{-1}. E = $-$(0.025693 V/1) ln(4.7 $\times$ 10^{-7}/1.0) = 0.37 V

6N.4B (b) Aluminum. It has a standard potential ($-$1.66 V) lower than that of iron (+0.44 V), so it is more easily oxidized than iron.

6O.1B The reduction half-reactions are: Br_2(l) + 2 e^- $\rightarrow$ 2 Br^- (aq), $E°$ = +1.09 V; O_2(g) + 4 H^+(aq) + 4 e^- $\rightarrow$ 2H_2O(l), E = +0.82 V at pH = 7; 2 H^+(aq) + 2 e^- $\rightarrow$ H_2(g), $E°$ = 0.00 V. Product at the cathode: H_2; product at the anode: O_2 and Br_2. (The product should be H_2O. However, because of the high overpotential for oxygen, bromine may also be produced.)

6O.2B   12.0 mol e^- × (1 mol Cr)/(6 mol e^-) = 2 mol Cr

6O.3B $m_{Cr} = \dfrac{(6.20\ \text{C·s}^{-1})(6.00\ \text{h} \times 3600\ \text{s·h}^{-1})}{9.6485 \times 10^4\ \text{C·mol}^{-1}} \times$

$\dfrac{1\ \text{mol Cr}}{6\ \text{mol } e^-} \times \dfrac{52.00\ \text{g Cr}}{1\ \text{mol Cr}} = 12.0\ \text{g Cr}$

6O.4B   12.00 g Cr $\times \dfrac{1\ \text{mol Cr}}{52.00\ \text{g Cr}} \times \dfrac{6\ \text{mol } e^-}{1\ \text{mol Cr}} = 1.385\ \text{mol } e^-$;

$t = \dfrac{(1.385\ \text{mol } e^-)(9.6485 \times 10^4\ \text{C·mol}^{-1})}{6.20\ \text{C·s}^{-1}} \times$

$\dfrac{1\ \text{h}}{3600\ \text{s}} = 5.99\ \text{h}$

Focus 7

7A.1B   average rate of disappearance of Hb =

$\dfrac{-[(8.0 \times 10^{-7}) - (1.2 \times 10^{-6})](\text{mmol Hb})\text{·L}^{-1}}{0.10\ \mu s} =$

$4 \times 10^{-6}\ \text{mmol·L}^{-1}\text{·}\mu s^{-1}$

7A.2B   (a) $\frac{1}{2}[5.0 \times 10^{-3}\ (\text{mmol HI})\text{·L}^{-1}\text{·s}^{-1}] = 2.5 \times 10^{-3}\ (\text{mmol H}_2)\text{·L}^{-1}\text{·s}^{-1}$; (b) unique average rate = $\Delta[H_2]/\Delta t = -\frac{1}{2}(\Delta[HI]/\Delta t) = 2.5 \times 10^{-3}\ \text{mmol·L}^{-1}\text{·s}^{-1}$

7A.3B   (a) first order in C_4H_9Br; zero order in OH^-; (b) first order overall; (c) s^{-1}

7A.4B   rate = $k_r[CO]^m[Cl_2]^n$

$\dfrac{\text{rate (2)}}{\text{rate (1)}} = \dfrac{0.241}{0.121} = \left(\dfrac{0.24}{0.12}\right)^m\left(\dfrac{0.20}{0.20}\right)^n$; $2 = (2)^m$; $m = 1$;

$\dfrac{\text{rate (3)}}{\text{rate (2)}} = \dfrac{0.682}{0.241} = \left(\dfrac{0.24}{0.24}\right)^m\left(\dfrac{0.40}{0.20}\right)^m$; $2.8 = (2)^n$;

$n = \dfrac{\log 2.8}{\log 2} = 1.5$;

therefore, rate = $k_r[CO][Cl_2]^{3/2}$. From experiment 1,

$k_r = \dfrac{0.121\ \text{mol·L}^{-1}\text{·s}^{-1}}{(0.12\ \text{mol·L}^{-1})(0.20\ \text{mol·L}^{-1})^{3/2}}$

$= 11\ \text{L}^{3/2}\text{·mol}^{-3/2}\text{·s}^{-1}$

7B.1B $[C_3H_6]_t = [C_3H_6]_0\ e^{-k_r t} = (0.100\ \text{mol·L}^{-1})$
$[e^{-(6.7 \times 10^{-4}\text{·s}^{-1})(200\ \text{s})}] = 0.087\ \text{mol·L}^{-1}$

7B.2B   A plot of $\ln[CH_3N_2CH_3]$ against time is linear, showing that the reaction must be first order; $k_r = 3.60 \times 10^{-4}\ s^{-1}$.

7B.3B   Reaction is first order with $k_r = 6.7 \times 10^{-4}\ s^{-1}$ at 500° C;

$t = \dfrac{1}{k_r}\ln\dfrac{[C_3H_6]_0}{[C_3H_6]_t} = \dfrac{1}{6.7 \times 10^{-4}\ s^{-1}}\ln\left(\dfrac{1.0\ \text{mol·L}^{-1}}{0.0050\ \text{mol·L}^{-1}}\right)$

$= 7.9 \times 10^3\ s = 2.2\ h$

7B.4B $k_r = \dfrac{\ln 2}{2.4 \times 10^4\ y}$; $t = \dfrac{2.4 \times 10^4\ y}{\ln 2}\left[\ln\left(\dfrac{1.0}{0.20}\right)\right]$

$= 5.6 \times 10^4\ y$

7B.5B   (a) Four half-lives are required for the concentration to fall to one-sixteenth of its initial value: $(1/2)^4 = 1/16$; (b) The reaction is first order with $k_r = 5.5 \times 10^{-4}\ s^{-1}$ at 973 K. $t_{1/2} = (\ln 2)/k_r = (\ln 2)/(5.5 \times 10^{-4}\ s^{-1}) = 1.3 \times 10^3\ s = 21\ \text{min}$. $t = 4t_{1/2} = 4 \times 21\ \text{min} = 84\ \text{min}$.

7C.1B   (a) bimolecular (two reactants); (b) unimolecular (one reactant)

7C.2B   net rate$_{\text{disappearance of B}}$ = $k_1[H_2A][B] + k_2[HA^-][B] - k_1'[HA^-][BH^+]$; net rate$_{\text{formation of HA}^-}$ = $k_1[H_2A][B] - k_1'[HA^-][BH^+] - k_2[HA^-][B] = 0$; $[HA^-] = (k_1[H_2A][B])/(k_1'[BH^+] + k_2[B])$; substitute for $[HA^-]$: net rate$_{\text{disappearance of B}}$ =

$k_1[H_2A][B] + \dfrac{(k_2k_1[H_2A][B]^2)}{k_1'[BH^+] + k_2[B]} - \dfrac{k_1'k_1[H_2A][B][BH^+]}{k_1'[BH^+] + k_2[B]}$;

assume $k_2[B] \ll k_1'[BH^+]$; then net rate$_{\text{disappearance of B}}$
$= k_r[H_2A][B]^2[BH^+]^{-1}$; $k_r = (2k_2k_1)/k_1'$; $H_2A + 2\ B \rightarrow 2\ BH^+ + A^{2-}$

7D.1B $\ln\left(\dfrac{k_{r2}}{k_{r1}}\right) = \dfrac{E_a}{R}\left[\dfrac{1}{T_1} - \dfrac{1}{T_2}\right]$; $\ln\left(\dfrac{4.35}{3.00}\right) =$

$\dfrac{E_a}{8.3145\ \text{J·mol}^{-1}\text{·K}^{-1}}\left[\dfrac{1}{291\ K} - \dfrac{1}{303\ K}\right]$;

$E_a = 2.3 \times 10^4\ \text{J·mol}^{-1} = 23\ \text{kJ·mol}^{-1}$

7D.2B $\ln\left(\dfrac{k_{r2}}{k_{r1}}\right) = \dfrac{E_a}{R}\left[\dfrac{1}{T_1} - \dfrac{1}{T_2}\right]$; $\ln\left(\dfrac{k_{r2}}{k_{r1}}\right) =$

$\dfrac{2.72 \times 10^5\ \text{J·mol}^{-1}}{8.3145\ \text{J·mol}^{-1}\text{·K}^{-1}}\left[\dfrac{1}{773} - \dfrac{1}{573}\right] = -14.77$ $k_{r2}/k_{r1} =$

$e^{-14.77}$; $k_{r2}' = (6.7 \times 10^{-4}\ s^{-1})(e^{-14.77}) = 2.6 \times 10^{-10}\ s^{-1}$

7E.1B $\dfrac{k_{r2}}{k_{r1}} = e^{-(E_{a,\ cat} - E_a)/RT} = 500.$; $\ln 500. = -E_{a,cat}/RT + E_a/RT$;

$E_{a,cat} = E_a - RT(\ln 500.)$; $E_{a,cat} = 106\ \text{kJ·mol}^{-1} - (8.3145 \times 10^{-3}\ \text{kJ·K}^{-1}\text{·mol}^{-1})(310.\ K)(\ln 500.) = 90.\ \text{kJ·mol}^{-1}$

Focus 8

8A.1B   Oxygen. It is above both gallium and tellurium in the periodic table and to the right of gallium.

8A.2B   Aluminum forms an amphoteric oxide in which it has the oxidation state +3; therefore, aluminum is the element.

8B.1B   (a) Energy of 1.0 mol photons $= \left(\dfrac{hc}{\lambda}\right)N_A =$

$\left(\dfrac{6.626 \times 10^{-34}\ \text{J·s} \times 2.998 \times 10^8\ \text{m·s}^{-1}}{250 \times 10^{-9}\ \text{m}}\right) \times (6.022 \times 10^{23}\ \text{mol}^{-1}) =$

$4.78 \ldots \times 10^5\ \text{J·mol}^{-1}$. 1.0 mol H_2 requires $(474\ \text{kJ})/2 = 237\ \text{kJ}$ or $2.37 \times 10^5\ \text{J}$, thus $(2.37 \times 10^5\ \text{J})/(4.78 \ldots \times 10^5\ \text{J·mol}^{-1}) = 0.50\ \text{mol photons}$. (b) $(0.50\ \text{mol} \times 6.022 \times 10^{23}\ \text{mol}^{-1})/(1.0 \times 10^{14}\ \text{photons·s}^{-1}) = 3.0 \times 10^9\ s$, or 95 years.

8B.2B   Hydrogen is a nonmetal and a diatomic gas at room temperature. It has an intermediate electronegativity

(χ = 2.2), so it forms covalent bonds with nonmetals and forms anions in combination with metals. In contrast, Group 1 elements are solid metals that have low electronegativities and form cations in combination with nonmetals.

8C.1B $K(s) + O_2(g) \rightarrow KO_2(s)$

8C.2B The molar mass of potassium is smaller than that of cesium. Thus a smaller mass of oxide is required if KO_2 is used.

8D.1B $2\,Ba(s) + O_2(g) \rightarrow 2\,BaO(s)$

8D.2B Lewis acid−base reaction; CaO is the base and SiO_2 is the acid.

8E.1B The boron atom in $B(OH)_3$ has an incomplete octet and can accept a lone pair of electrons from a water molecule, which is acting as a Lewis base. The complex formed is a weak Brønsted acid in which an acidic proton can be lost from an H_2O molecule in the complex.

8E.2B (a) +3, because the oxidation number of H in this hydride is taken to be −1; (b) +3

8F.1B Lewis acid−base reaction; CO is the Lewis acid and OH^- is the Lewis base.

8F.2B SiH_4 is a Lewis acid that can react with the Lewis base OH^-. CH_4 is not a Lewis acid, because the C atom is much smaller than the Si atom and has no accessible d-orbitals to accommodate added electron pairs.

8G.1B (a) $\overset{-1}{\ddot{N}}=\overset{+1}{N}=\overset{-1}{\ddot{N}}$

(b) Other resonance structures violate the octet rule, so the one shown in (a) is the major resonance contributor. (c) Linear and nonpolar.

8G.2B Assume 1 L of solution; 1 L soln × (10^3 mL/1 L) × (1.7 g soln/1 mL soln) × (85 g H_3PO_4/100 g soln) × (1 mol H_3PO_4/97.99 g H_3PO_4) = 15 M H_3PO_4(aq)

8H.1B Reduction: $2\,(2\,e^- + F_2 \rightarrow 2\,F^-)$; oxidation: $2\,H_2O \rightarrow O_2 + 4\,H^+ + 4\,e^-$; overall: $2\,H_2O(l) + 2\,F_2(g) \rightarrow O_2(g) + 4\,H^+(aq) + 4\,F^-(aq)$; $E_{cell}° = 2.87 - 1.23\,V = +1.64\,V$; $\Delta G° = -(4\,mol)(96.485\,kC\cdot mol^{-1})(1.64\,V) = -633\,kJ$

8H.2B (a) +1; (b) +6

8I.1B The melting and boiling points of the halogens increase down the group because the London forces between their molecules become stronger.

8I.2B In the reduction of $HClO_3$, an acid is a reactant. Therefore, increasing the pH reduces the oxidizing ability of $HClO_3$.

8J.1B Square pyramidal

Focus 9

9A.1B Each metal atom will lose one 4s-electron. V^+, Mn^+, Co^+, and Ni^+ will have one electron in the 4s-orbital, but Cr^+ and Cu^+ will have no electrons in the 4s-orbital. Because their outermost valence electrons will be in

3d-orbitals and not the 4s-orbital, their radii should be smaller than those of the other cations.

9B.1B (a) Lewis acid-base. (b) Ni is the Lewis acid; CO is the Lewis base.

9B.2B Oxidation: $2\,Hg(l) \rightarrow Hg_2^{2+}(aq) + 2\,e^-$; reduction: $4\,H^+(aq) + NO_3^-(aq) + 3\,e^- \rightarrow NO(aq) + 2\,H_2O(l)$; overall: $6\,Hg(l) + 8\,H^+(aq) + 2\,NO_3^-(aq) \rightarrow 3\,Hg_2^{2+}(aq) + 2\,NO(aq) + 4\,H_2O(l)$; $E_{cell}° = 0.96\,V - 0.79\,V = +0.17\,V$

9C.1B (a) pentaaminebromidocobalt(II) sulfate; (b) $[Cr(NH_3)_4(OH_2)_2]Br_3$

9C.2B The ratio of $CrCl_3·6H_2O$ to Cl is 1:3; therefore, the isomer $[Cr(OH_2)_6]Cl_3$ is present.

9C.3B (a) linkage isomers; (b) hydrate isomers

9C.4B (a) not chiral; (b) chiral; (c) not chiral; (d) chiral; no entantiomeric pairs

9D.1B $\Delta_O = hc/\lambda = (6.022 \times 10^{23}\,mol^{-1})(6.626 \times 10^{-34}\,J\cdot s^{-1})(2.998 \times 10^8\,m\cdot s^{-1})/(305 \times 10^{-9}\,m) \times (1\,kJ/10^3\,J) = 392\,kJ\cdot mol^{-1}$

9D.2B (a) Large Δ_O, so $t_{2g}^6 e_g^1$ with one unpaired electron. (b) Small Δ_O, so $t_{2g}^5 e_g^2$ with three unpaired electrons.

9D.3B $\varepsilon = A/Lc = 0.531/(1.00\,cm)(4.15 \times 10^{-6}\,mol\cdot L^{-1}) = 1.28 \times 10^5\,L\cdot mol^{-1}\cdot cm^{-1}$.

9D.4B Because Ni^{2+} has eight d-electrons it is a d^8 complex, and there is no empty e_g orbital. Therefore, both complexes have a $t_{2g}^6 e_g^2$ configuration and are paramagnetic.

Focus 10

10A.1B (a) $^{55}_{26}Fe + ^{0}_{-1}e \rightarrow ^{A}_{Z}E$; $^{55}_{25}Mn$ is produced.

(b) $^{11}_{6}C \rightarrow ^{A}_{Z}E + ^{0}_{-1}e$; $^{11}_{5}B$ is produced.

10A.2B (c) Ce-148 is neutron rich (in the blue band) and therefore might undergo β emission, which converts a neutron into a proton, to reach stability.

10A.3B (a) mass number: $250 + A = 257 + 4(1)$, $A = 11$; atomic number: $98 + Z = 103 + 4(0)$, $Z = 5$; the missing nuclide is $^{11}_{5}B$; (b) mass number: $A + 12 = 254 + 4(1)$, $A = 246$; atomic number: $Z + 6 = 102 + 4(0)$, $Z = 96$; the missing nuclide is $^{246}_{96}Cm$.

10B.1B $m = m_0 e^{-kt} = (5.0\,\mu g)e^{-(3.3 \times 10^{-7}a^{-1})(1.0 \times 10^6\,a)} = 3.6\,\mu g$

10B.2B $t = -\left(\dfrac{5.73 \times 10^3\,a}{\ln 2}\right) \times \ln\left(\dfrac{1.4 \times 10^4}{18400}\right) = 2.3 \times 10^3\,a$

10C.1B $\Delta m = \{(235.0439m_u) - 92(1.0078m_u) - 143(1.0087m_u)\} \times (1.6605 \times 10^{-27}\,kg) = -3.1845 \times 10^{-27}\,kg$; $E_{bind} = -(-3.1845 \times 10^{-27}\,kg) \times (3.00 \times 10^8\,m\cdot s^{-1})^2 \times (1\,eV)/(1.602\,18 \times 10^{-19}\,J) = 1.79 \times 10^9\,eV$

10C.2B $\Delta m = \{137.91m_u + 85.91m_u + 12(1.0087m_u) - (235.04m_u + 1.0087m_u)\} \times 1.6605 \times 10^{-27}\,kg = -2.06 \times 10^{-28}\,kg$ per nucleus; $\Delta E = (1.0 \times 10^{-3}\,kg)/(235.0 \times 1.6605 \times 10^{-27}\,kg) \times (-2.06 \times 10^{-28}\,kg) \times (3.00 \times 10^8\,m\cdot s^{-1})^2 = -4.8 \times 10^{10}\,J$

Focus 11

11A.1B (a)

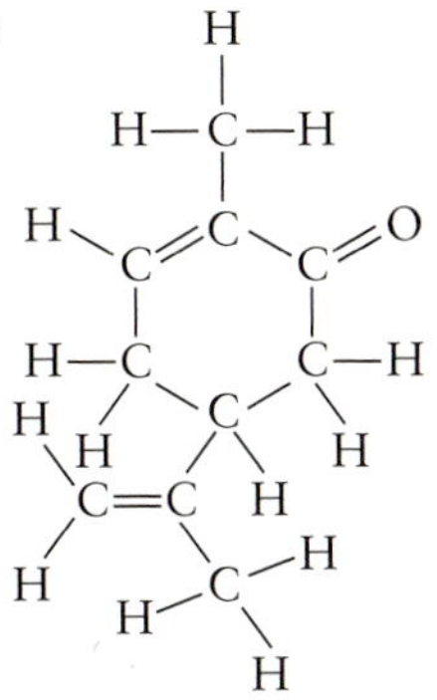

(b) $(CH_3)_2CH(CH_2)_3CH(CH_3)CH_2CH_3$

11A.2B (a) 4-ethyl-2-methylhexane;

(b)

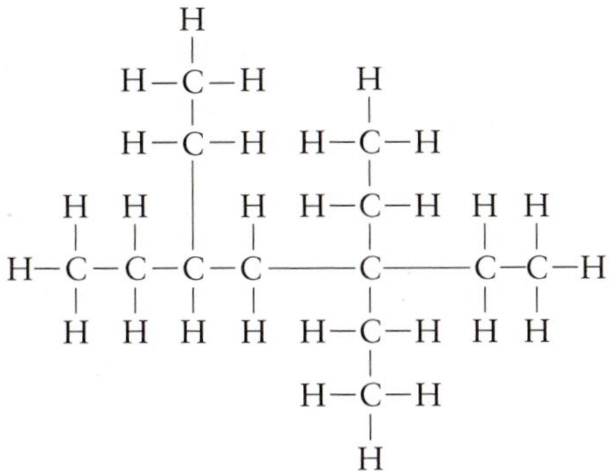

11A.3B (a) 4-ethyl-2-hexene;   (b)

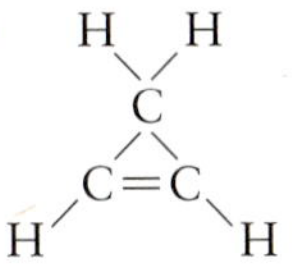

11A.4B $CH_3(CH_2)_2CH_2Br$; $CH_3CH_2CHBrCH_3$; $(CH_3)_2CHCH_2Br$; $(CH_3)_3CBr$

11A.5B 27a and 27b are both trans.

11A.6B (c) is chiral, because the second C atom is attached to four different groups.

11B.1B $CH_3CHClCH_2CH_3$

11C.1B 1-ethyl-3,5-dimethyl-2-propylbenzene

11D.1B (a) $HCOOCH_2CH_3$; (b) CH_3OH and $CH_3(CH_2)_2COOH$

11D.2B C: sp^2 hybridization; N: sp^3 hybridization

11D.3B (a) 3-pentanol; (b) 3-pentanone; (c) ethylmethylamine

11E.1B (a) $CH_2{=}C(CH_3)COOCH_3$; (b) $(-NHCH(CH_3)CO-NHCH(CH_3)CO-)_n$

11E.2B Alternating copolymer

ODD-NUMBERED EXERCISES

Fundamentals

A.1 (a) law; (b) hypothesis; (c) hypothesis; (d) hypothesis; (e) hypothesis

A.3 (a) chemical; (b) physical; (c) physical

A.5 The temperature of the injured camper and the evaporation and condensation of water are physical properties. The ignition of propane is a chemical change.

A.7 (a) physical; (b) chemical; (c) chemical

A.9 (a) intensive; (b) intensive; (c) intensive; (d) extensive

A.11 (a) 1 kilograin; (b) 1 centibatman; (c) 1 megamutchkin

A.13 236 mL

A.15 (a) 5.4×10^8 pm < (b) 1.3×10^9 pm

A.17 $d = 19.0$ g·cm^{-3}

A.19 $V = 0.0427$ cm^3

A.21 $d_{liquid} = 0.8589$ g·cm^{-3}

A.23 $V = 7.41$ cm^3. Since the area is 1 cm, the thickness must be 7.41 cm.

A.25 0.3423; keep 4 significant figures

A.27 0.989

A.29 (a) 4.82×10^3 pm; (b) 30.5 mm^3·s^{-1}; (c) 1.88×10^{-12} kg; (d) 2.66×10^3 kg·m^{-3}; (e) 0.044 mg·cm^{-3}

A.31 (a) $d = 1.72$ g·cm^{-3}; (b) $d = 1.7$ g·cm^{-3}

A.33 (a) Formula: °X $= 50 + 2 \times$ °C; (b) 94 °X

A.35 32 J

A.37 $E_{recovered} = 8.1 \times 10^2$ kJ; $h = 29$ m

A.39 $E = 6.0$ J

A.41 $g = \dfrac{Gm_E}{R_E^2}$

B.1 1.40×10^{22} atoms

B.3 (a) 5p, 6n, 5e; (b) 5p, 5n, 5e; (c) 15p, 16n, 15e; (d) 92p, 146n, 92e

B.5 (a) ^{194}Ir; (b) ^{22}Ne; (c) ^{51}V

B.7

Element	Symbol	Protons	Neutrons	Electrons	Mass number
Chlorine	^{36}Cl	17	19	17	36
Zinc	^{65}Zn	30	35	30	65
Calcium	^{40}Ca	20	20	20	40
Lanthanum	^{137}La	57	80	57	137

B.9 (a) They all have the same mass. (b) They have differing numbers of protons, neutrons, and electrons.

B.11 (a) 0.5359; (b) 0.4638; (c) 2.526×10^{-4}; (d) 535.9 kg

B.13 (a) Scandium is a Group 3 metal. (b) Strontium is a Group 2 metal. (c) Sulfur is a Group 16 nonmetal. (d) Antimony is a Group 15 metalloid.

B.15 (a) Sr, metal; (b) Xe, nonmetal; (c) Si, metalloid

B.17 (a) Alkali metal: none; (b) Transition metals: cadmium; (c) lanthanoid: cerium

B.19 (a) d-block; (b) p-block; (c) d-block; (d) s-block; (e) p-block; (f) d-block

B.21 (a) Pb; Group 14; Period 6; metal; (b) Cs; Group 1; Period 6; metal.

C.1 Container (a) holds a mixture (one is single compound and another is single element). Container (b) holds a single element.

C.3 The chemical formula of xanthophyll is $C_{40}H_{56}O_2$.

C.5 (a) $C_3H_7O_2N$; (b) C_2H_7N

C.7 (a) Cesium is a metal in Group 1; it will form Cs$^+$ ions. (b) Iodine is a nonmetal in Group 17 and will form I$^-$ ions. (c) Selenium is a Group 16 nonmetal and will form Se^{2-} ions. (d) Calcium is a Group 2 metal and will form Ca^{2+} ions.

C.9 (a) ^{10}Be^{2+} has 4 protons, 6 neutrons, and 2 electrons. (b) ^{17}O^{2-} has 8 protons, 9 neutrons, and 10 electrons. (c) ^{80}Br$^-$ has 35 protons, 45 neutrons, and 36 electrons. (d) ^{75}As^{3-} has 33 protons, 42 neutrons, and 36 electrons.

C.11 (a) ^{19}F$^-$; (b) ^{24}Mg^{2+}; (c) ^{128}Te^{2-}; (d) ^{86}Rb$^+$

C.13 (a) Al_2Te_3; (b) MgO; (c) Na_2S; (d) RbI

C.15 (a) Group 13; (b) aluminum, Al

C.17 (a) 0.542; (b) 0.458; (c) 2.49×10^{-4}; (d) 14 kg

C.19 (a) Na_2HPO_3; (b) $(NH_4)_2CO_3$; (c) +2; (d) +2

C.21 (a) HCl, molecular compound (in the gas phase); (b) S_8, element (molecular substance); (c) CoS, ionic compound; (d) Ar, element; (e) CS_2, molecular compound; (f) $SrBr_2$, ionic compound

D.1 (a) Bromite ion; (b) HSO_3^-

D.3 (a) $MnCl_2$; (b) $Ca_3(PO_4)_2$; (c) $Al_2(SO_3)_3$; (d) Mg_3N_2

D.5 (a) phosphorus pentafluoride; (b) iodine trifluoride; (c) oxygen difluoride; (d) diboron tetrachloride; (e) cobalt (II) sulfate heptahydrate; (f) mercury (II) bromide; (g) iron(III) hydrogen phosphate (or ferric hydrogen phosphate; or iron(III) biphosphate); (h) tungsten(V) oxide; (i) osmium (III) bromide

D.7 (a) calcium phosphate; (b) tin(IV) sulfide, stannic sulfide; (c) vanadium(V) oxide; (d) copper(I) oxide, cuprous oxide

D.9 (a) sulfur hexafluoride; (b) dinitrogen pentoxide; (c) nitrogen triiodide; (d) xenon tetrafluoride; (e) arsenic tribromide; (f) chlorine dioxide

D.11 (a) hydrochloric acid; (b) sulfuric acid; (c) nitric acid; (d) acetic acid; (e) sulfurous acid; (f) phosphoric acid

D.13 (a) $HClO_4$; (b) HClO; (c) HIO; (d) HF; (e) H_3PO_3; (f) HIO_4

D.15 (a) TiO_2; (b) $SiCl_4$; (c) CS_2; (d) SF_4; (e) Li_2S; (f) SbF_5; (g) N_2O_5; (h) IF_7

D.17 (a) ZnF_2; (b) $Ba(NO_3)_2$; (c) AgI; (d) Li_3N; (e) Cr_2S_3

D.19 (a) $BaCl_2$; (b) ionic

D.21 (a) sodium sulfite; (b) iron(III) oxide; (c) iron(II) oxide; (d) magnesium hydroxide; (e) nickel(II) sulfate hexahydrate; (f) phosphorus pentachloride;

(g) chromium(III) dihydrogen phosphate; (h) diarsenic trioxide; (i) ruthenium(II) chloride

D.23 (a) $CuCO_3$, copper(II) carbonate; (b) K_2SO_3, potassium sulfite; (c) LiCl, lithium chloride

D.25 (a) heptane; (b) propane; (c) pentane; (d) butane

D.27 (a) cobalt(III) oxide monohydrate; $Co_2O_3 \cdot H_2O$; (b) cobalt(II) hydroxide; $Co(OH)_2$

D.29 E = Si; SiH_4, silicon tetrahydride; Na_4Si, sodium silicide

D.31 (a) lithium aluminum hydride, ionic (with a molecular anion); (b) sodium hydride, ionic

D.33 (a) selenic acid; (b) sodium arsenate; (c) calcium tellurite; (d) barium arsenate; (e) antimonic acid; (f) nickel(III) selenate

D.35 (a) alcohol; (b) carboxylic acid; (c) haloalkane

E.1 1.73×10^{11} km

E.3 3 At atoms

E.5 (a) 1.2×10^{-14} mol; (b) 2.6×10^6 years;

E.7 3.5×10^{-15} mol C

E.9 (a) 1.38×10^{23} atoms of O; (b) 1.26×10^{22} formula units; (c) 0.146 mol

E.11 (a) 6.94 g·mol^{-1}; (b) 6.96 g·mol^{-1}

E.13 % ^{11}B = 73.8%; % ^{10}B = 26.2%

E.15 CaS; molar mass = 72.15 g·mol^{-1}

E.17 (a) 75 g of indium contains more moles of atoms than 80. g of tellurium. (b) 15.0 g of P has slightly more atoms than 15.0 g of S. (c) They have the same number of moles.

E.19 (a) 20.027 g·mol^{-1}; (b) 1.11 g·cm^{-3}; (c) The volume of the spherical tank $V = 9.05 \times 10^8$ cm^3; the volume calculated from the density data, $V = 9.01 \times 10^8$ cm^3. (d) The reported D_2O mass was not accurate. (e) The assumption in part (b) is reasonable.

E.21 (a) 0.0981 mol Al_2O_3; 5.91×10^{22} Al_2O_3 molecules; (b) 1.30×10^{-3} mol HF; 7.83×10^{20} HF molecules; (c) 4.56×10^{-5} mol H_2O_2; 2.75×10^{19} H_2O_2 molecules; (d) 6.94 mol glucose; 4.18×10^{24} glucose molecules; (e) 0.312 mol N; 1.88×10^{23} N atoms; 0.156 mol N_2; 9.39×10^{22} N_2 molecules

E.23 (a) 0.0134 mol Cu^{2+}; (b) 8.74×10^{-3} mol SO_3; (c) 430. mol F^-; (d) 0.0699 mol H_2O

E.25 (a) 4.52×10^{23} formula units; (b) 124 mg; (c) 3.036×10^{22} formula units

E.27 (a) 2.992×10^{-23} g; (b) 3.34×10^{25} H_2O molecules

E.29 (a) 0.0417 mol $CuCl_2 \cdot 4H_2O$; (b) 0.0834 mol Cl^-; (c) 1.00×10^{23} H_2O molecules; (d) 0.3099

E.31 (a) 1.6 kg of water; $109.28 per L H_2O; (b) $70.20

E.33 0.39%

F.1 (a) $C_{10}H_{16}O$; (b) 78.90%; 10.59%; 10.51%

F.3 (a) HNO_3; (b) O (oxygen)

F.5 $C_7H_{15}NO_3$; C, 52.15%; H, 9.3787%; N, 8.691%; O, 29.78%

F.7 (a) 143. g·mol^{-1}; (b) copper(I) oxide

F.9 atom ratio: 1 O : 2.67 C : 2.67 H; the formula is $C_8H_8O_3$

F.11 (a) Na_3AlF_6; (b) $KClO_3$; (c) NH_6PO_4 or $[NH_4][H_2PO_4]$, ammonium dihydrogen phosphate.

F.13 (a) PCl_5; (b) phosphorus pentachloride

F.15 $C_{16}H_{13}ClN_2O$

F.17 $Os_3C_{12}O_{12}$

F.19 $C_8H_{10}N_4O_2$

F.21 $C_{49}H_{78}N_6O_{12}$

F.23 ethene (85.63%) > heptane (83.91%) > propanol (59.96%)

F.25 (a) empirical formula: C_2H_3Cl, molecular formula: $C_4H_6Cl_2$; (b) empirical formula: CH_4N, molecular formula: $C_2H_8N_2$

F.27 45.1% $NaNO_3$

G.1 (a) false; (b) true; (c) false

G.3 (a) heterogeneous, decanting; (b) heterogeneous, dissolving followed by filtration and distillation; (c) homogeneous, distillation

G.5 (a) 13.5 mL; (b) 62.5 mL; (c) 5.92 mL

G.7 Measure 482.2 g of H_2O on a balance and pour it into a beaker. Then weigh 27.8 g of KNO_3 and mix it with the water until it totally dissolves.

G.9 15.2 g

G.11 16.2 mL

G.13 1.0×10^{-2} mol

G.15 (a) 4.51 mL; (b) 12.0 mL of 2.5 M NaOH solution are added to 48.0 mL of water.

G.17 (a) 8.0 g; (b) 12 g

G.19 (a) 0.067 57 M; (b) 0.0732 M

G.21 (a) 4.58×10^{-2} M; (b) 9.07×10^{-3} M

G.23 0.13 M Cl^-

G.25 No X remaining. No health benefits because there are no molecules of the active substance, X, left in the solution.

G.27 600. mL

G.29 It is satisfactory.

H.1 (a) You cannot add a different compound or element to the chemical equation that is not produced (or involved) in the chemical reaction. (b) $2\,Cu + SO_2 \rightarrow 2\,CuO + S$

H.3 $2\,SiH_4 + 4\,H_2O \rightarrow 2\,SiO_2 + 8\,H_2$

H.5 (a) $NaBH_4(s) + 2\,H_2O(l) \rightarrow NaBO_2(aq) + 4\,H_2(g)$ (b) $Mg(N_3)_2(s) + 2\,H_2O(l) \rightarrow Mg(OH)_2(aq) + 2\,HN_3(aq)$ (c) $2\,NaCl(aq) + SO_3(g) + H_2O(l) \rightarrow$ $Na_2SO_4(aq) + 2\,HCl(aq)$ (d) $4\,Fe_2P(s) + 18\,S(s) \rightarrow P_4S_{10}(s) + 8\,FeS(s)$

H.7 (a) $Ca(s) + 2\,H_2O(l) \rightarrow H_2(g) + Ca(OH)_2(aq)$ (b) $Na_2O(s) + H_2O(l) \rightarrow 2\,NaOH(aq)$ (c) $3\,Mg(s) + N_2(g) \rightarrow Mg_3N_2(s)$ (d) $4\,NH_3(g) + 7\,O_2(g) \rightarrow 6\,H_2O(g) + 4\,NO_2(g)$

H.9 (a) $3\,Pb(NO_3)_2(aq) + 2\,Na_3PO_4(aq) \rightarrow Pb_3(PO_4)_2(s) + 6\,NaNO_3(aq)$ (b) $Ag_2CO_3(aq) + 2\,NaBr(aq) \rightarrow 2\,AgBr(s) + Na_2CO_3(aq)$

H.11 (I) $3\,Fe_2O_3(s) + CO(g) \rightarrow 2\,Fe_3O_4(s) + CO_2(g)$ (II) $Fe_3O_4(s) + 4\,CO(g) \rightarrow 3\,Fe(s) + 4\,CO_2(g)$

H.13 (I) $N_2(g) + O_2(g) \rightarrow 2\,NO(g)$ (II) $2\,NO(g) + O_2(g) \rightarrow 2\,NO_2(g)$

H.15 $4\,HF(aq) + SiO_2(s) \rightarrow SiF_4(aq) + 2\,H_2O(l)$

H.17 $C_7H_{16}(l) + 11\,O_2(g) \rightarrow 7\,CO_2(g) + 8\,H_2O(g)$

H.19 $C_{14}H_{18}N_2O_5(s) + 16\,O_2(g) \rightarrow 14\,CO_2(g) + 9\,H_2O(l) + N_2(g)$

H.21 $2\,C_{10}H_{15}N(s) + 26\,O_2(g) \rightarrow 19\,CO_2(g) + 13\,H_2O(l) + CH_4N_2O(aq)$

H.23 (I) $H_2S(g) + 2\,NaOH(s) \rightarrow Na_2S(aq) + 2\,H_2O(l)$ (II) $4\,H_2S(g) + Na_2S(alc) \rightarrow Na_2S_5(alc) + 4\,H_2(g)$ (III) $2\,Na_2S_5(alc) + 9\,O_2(g) + 10\,H_2O(l) \rightarrow$ $2\,Na_2S_2O_3 \cdot 5H_2O(s) + 6\,SO_2(g)$

H.25 (a) first oxide: P_2O_5; second oxide: P_2O_3; (b) P_4O_{10} (phosphorus (V) oxide), P_4O_6 (phosphorus(III) oxide); (c) $P_4(s) + 3\,O_2(g) \rightarrow P_4O_6(s)$; $P_4(s) + 5\,O_2(g) \rightarrow P_4O_{10}(s)$

I.1 The picture would show a precipitate, $CaSO_4(s)$, at the bottom of the flask. Sodium and chloride ions, $NaCl(aq)$, would remain throughout the solution.

I.3 (a) CH_3OH, nonelectrolyte; (b) $BaCl_2$, strong electrolyte; (c) KF, strong electrolyte

I.5 (a) $3\,BaBr_2(aq) + 2\,Li_3PO_4(aq) \rightarrow$
$$Ba_3(PO_4)_2(s) + 6\,LiBr(aq)$$
$3\,Ba^{2+}(aq) + 6\,Br^-(aq) + 6\,Li^+(aq) + 2\,PO_4^{3-}(aq) \rightarrow$
$$Ba_3(PO_4)_2(s) + 6\,Li^+(aq) + 6\,Br^-(aq)$$
net ionic equation: $3\,Ba^{2+}(aq) + 2\,PO_4^{3-}(aq) \rightarrow$
$$Ba_3(PO_4)_2(s)$$
(b) $2\,NH_4Cl(aq) + Hg_2(NO_3)_2(aq) \rightarrow$
$$2\,NH_4NO_3(aq) + Hg_2Cl_2(s)$$
$2\,NH_4^+(aq) + 2\,Cl^-(aq) + Hg_2^{2+}(aq) + 2\,NO_3^-(aq) \rightarrow$
$$Hg_2Cl_2(s) + 2\,NH_4^+(aq) + 2\,NO_3^-(aq)$$
net ionic equation: $Hg_2^{2+}(aq) + 2\,Cl^-(aq) \rightarrow Hg_2Cl_2(s)$
(c) $2\,Co(NO_3)_3(aq) + 3\,Ca(OH)_2(aq) \rightarrow$
$$2\,Co(OH)_3(s) + 3\,Ca(NO_3)_2(aq)$$
$2\,Co^{3+}(aq) + 6\,NO_3^-(aq) + 3\,Ca^{2+}(aq) + 6\,OH^-(aq) \rightarrow$
$$2\,Co(OH)_3(s) + 3\,Ca^{2+}(aq) + 6\,NO_3^-(aq)$$
net ionic equation: $Co^{3+}(aq) + 3\,OH^-(aq) \rightarrow$
$$Co(OH)_3(s)$$

I.7 (a) soluble; (b) slightly soluble; (c) insoluble; (d) insoluble

I.9 (a) $Na^+(aq)$ and $I^-(aq)$;
(b) $Ag^+(aq)$ and $CO_3^{2-}(aq)$, Ag_2CO_3 is insoluble.
(c) $NH_4^+(aq)$ and $PO_4^{3-}(aq)$;
(d) $Fe^{2+}(aq)$ and $SO_4^{2-}(aq)$

I.11 (a) $Fe(OH)_3$, precipitate; (b) Ag_2CO_3, precipitate forms; (c) No precipitate will form because all possible products are soluble in water.

I.13 (a) net ionic equation: $Fe^{2+}(aq) + S^{2-}(aq) \rightarrow FeS(s)$; spectator ions: Na^+, Cl^-
(b) net ionic equation: $Pb^{2+}(aq) + 2\,I^-(aq) \rightarrow PbI_2(s)$; spectator ions: K^+, NO_3^-
(c) net ionic equation: $Ca^{2+}(aq) + SO_4^{2-}(aq) \rightarrow$
$$CaSO_4(s); \text{ spectator ions: } NO_3^-, K^+$$
(d) net ionic equation: $Pb^{2+}(aq) + CrO_4^{2-}(aq) \rightarrow$
$$PbCrO_4(s); \text{ spectator ions: } Na^+, NO_3^-$$
(e) net ionic equation: $Hg_2^{2+}(aq) + SO_4^{2-}(aq) \rightarrow$
$$Hg_2SO_4(s); \text{ spectator ions: } K^+, NO_3^-$$

I.15 (a) overall equation: $(NH_4)_2CrO_4(aq) + BaCl_2(aq) \rightarrow$
$$BaCrO_4(s) + 2\,NH_4Cl(aq)$$
complete ionic equation:
$2\,NH_4^+(aq) + CrO_4^{2-}(aq) + Ba^{2+}(aq) + 2\,Cl^-(aq) \rightarrow$
$$BaCrO_4(s) + 2\,NH_4^+(aq) + 2\,Cl^-(aq)$$
net ionic equation: $Ba^{2+}(aq) + CrO_4^{2-}(aq) \rightarrow$
$$BaCrO_4(s); \text{ spectator ions: } NH_4^+, Cl^-$$
(b) $CuSO_4(aq) + Na_2S(aq) \rightarrow CuS(s) + Na_2SO_4(aq)$
complete ionic equation:
$Cu^{2+}(aq) + SO_4^{2-}(aq) + 2\,Na^+(aq) + S^{2-}(aq) \rightarrow$
$$CuS(s) + 2\,Na^+(aq) + SO_4^{2-}(aq)$$
net ionic equation: $Cu^{2+}(aq) + S^{2-}(aq) \rightarrow CuS(s)$;
spectator ions: Na^+, SO_4^{2-}

(c) $3\,FeCl_2(aq) + 2\,(NH_4)_3PO_4(aq) \rightarrow$
$$Fe_3(PO_4)_2(s) + 6\,NH_4Cl(aq)$$
complete ionic equation:
$3\,Fe^{2+}(aq) + 6\,Cl^-(aq) + 6\,NH_4^+(aq) + 2\,PO_4^{3-}(aq) \rightarrow$
$$Fe_3(PO_4)_2(s) + 6\,NH_4^+(aq) + 6\,Cl^-(aq)$$
net ionic equation: $3\,Fe^{2+}(aq) + 2\,PO_4^{3-}(aq) \rightarrow$
$$Fe_3(PO_4)_2(s); \text{ spectator ions: } Cl^-, NH_4^+$$
(d) $K_2C_2O_4(aq) + Ca(NO_3)_2(aq) \rightarrow$
$$CaC_2O_4(s) + 2\,KNO_3(aq)$$
complete ionic equation:
$2\,K^+(aq) + C_2O_4^{2-}(aq) + Ca^{2+}(aq) + 2\,NO_3^-(aq) \rightarrow$
$$CaC_2O_4(s) + 2\,K^+(aq) + 2\,NO_3^-(aq)$$
net ionic equation: $Ca^{2+}(aq) + C_2O_4^{2-}(aq) \rightarrow CaC_2O_4(s)$;
spectator ions: K^+, NO_3^-
(e) $NiSO_4(aq) + Ba(NO_3)_2(aq) \rightarrow$
$$Ni(NO_3)_2(aq) + BaSO_4(s)$$
complete ionic equation:
$Ni^{2+}(aq) + SO_4^{2-}(aq) + Ba^{2+}(aq) + 2\,NO_3^-(aq) \rightarrow$
$$Ni^{2+}(aq) + 2\,NO_3^-(aq) + BaSO_4(s)$$
net ionic equation: $Ba^{2+}(aq) + SO_4^{2-}(aq) \rightarrow BaSO_4(s)$;
spectator ions: Ni^{2+}, NO_3^-

I.17 (a) $AgNO_3$ and Na_2CrO_4; (b) $CaCl_2$ and Na_2CO_3; (c) $Cd(ClO_4)_2$ and $(NH_4)_2S$

I.19 (a) Diluted H_2SO_4 solution will be used as reagent.
$Pb^{2+}(aq) + SO_4^{2-}(aq) \rightarrow PbSO_4(s)$
(b) H_2S solution will also be used as reagent.
$Mg^{2+}(aq) + S^{2-}(aq) \rightarrow MgS(s)$

I.21 (a) $2\,Ag^+(aq) + SO_4^{2-}(aq) \rightarrow Ag_2SO_4(s)$
(b) $Hg^{2+}(aq) + S^{2-}(aq) \rightarrow HgS(s)$
(c) $3\,Ca^{2+}(aq) + 2\,PO_4^{3-}(aq) \rightarrow Ca_3(PO_4)_2(s)$
(d) $AgNO_3$ and Na_2SO_4; Na^+, NO_3^-; $Hg(CH_3CO_2)_2$ and $Hg(CH_3CO_2)_2$ and Li_2S; Li^+, $CH_3CO_2^-$ $CaCl_2$ and $CaCl_2$ and K_3PO_4; K^+, Cl^-

I.23 white ppt. = $AgCl(s)$, Ag^+; no ppt. with H_2SO_4, no Ca^{2+}; black ppt. = ZnS, Zn^{2+}

I.25 (a) $2\,NaOH(aq) + Cu(NO_3)_2(aq) \rightarrow$
$$Cu(OH)_2(s) + 2\,NaNO_3(aq)$$
complete ionic equation:
$2\,Na^+(aq) + 2\,OH^-(aq) + Cu^{2+}(aq) + 2\,NO_3^-(aq) \rightarrow$
$$Cu(OH)_2(s) + 2\,Na^+(aq) + 2\,NO_3^-(aq)$$
net ionic equation:
$Cu^{2+}(aq) + 2\,OH^-(aq) \rightarrow Cu(OH)_2(s)$
(b) 0.0800 M

I.27 (a) $Ag^+(aq) + I^-(aq) \rightarrow AgI(s)$; (b) 1.01×10^{-2} M Ag^+

J.1 (a) base; (b) acid; (c) base; (d) acid; (e) base

J.3 CH_3COOH.

J.5 (a) overall equation:
$HF(aq) + NaOH(aq) \rightarrow NaF(aq) + H_2O(l)$
complete ionic equation:
$HF(aq) + Na^+(aq) + OH^-(aq) \rightarrow$
$$Na^+(aq) + F^-(aq) + H_2O(l)$$
net ionic equation:
$HF(aq) + OH^-(aq) \rightarrow F^-(aq) + H_2O(l)$
(b) overall equation:
$(CH_3)_3N(aq) + HNO_3(aq) \rightarrow (CH_3)_3NHNO_3(aq)$
complete ionic equation:
$(CH_3)_3N(aq) + H^+(aq) + NO_3^-(aq) \rightarrow$
$$(CH_3)_3NH^+(aq) + NO_3^-(aq)$$

net ionic equation:

$(CH_3)_3N(aq) + H^+(aq) \rightarrow (CH_3)_3NH^+(aq)$

(c) overall equation:

$LiOH(aq) + HI(aq) \rightarrow LiI(aq) + H_2O(l)$

complete ionic equation:

$Li^+(aq) + OH^-(aq) + H^+(aq) + I^-(aq) \rightarrow$
$$Li^+(aq) + I^-(aq) + H_2O(l)$$

net ionic equation: $OH^-(aq) + H^+(aq) \rightarrow 2\,H_2O(l)$

J.7 (a) $HBr(aq) + KOH(aq) \rightarrow KBr(aq) + H_2O(l)$

(b) $Zn(OH)_2(aq) + 2\,HNO_2(aq) \rightarrow$
$$Zn(NO_2)_2(aq) + 2\,H_2O(l)$$

(c) $Ca(OH)_2(aq) + 2\,HCN(aq) \rightarrow$
$$Ca(CN)_2(aq) + 2\,H_2O(l)$$

(d) $3\,KOH(aq) + H_3PO_4(aq) \rightarrow K_3PO_4(aq) + 3\,H_2O(l)$

J.9 (a) KCH_3CO_2, potassium acetate;

$CH_3COOH(aq) + K^+(aq) + OH^-(aq) \rightarrow$
$$K^+(aq) + CH_3CO_2^-(aq) + H_2O(l)$$

(b) $(NH_4)_3PO_4$, ammonium phosphate;

$3\,NH_3(aq) + 3\,H^+ + PO_4^{3-}(aq) \rightarrow 3\,NH_4^+(aq) + PO_4^{3-}(aq)$

(c) $Ca(BrO_2)_2$, calcium bromite;

$Ca^{2+}(aq) + 2\,OH^-(aq) + 2\,HBrO_2(aq) \rightarrow$
$$2\,H_2O(l) + Ca^{2+}(aq) + 2\,BrO_2^-(aq)$$

(d) Na_2S, sodium sulfide;

$2\,Na^+(aq) + 2\,OH^-(aq) + H_2S(aq) \rightarrow$
$$2\,H_2O(l) + 2\,Na^+(aq) + S^{2-}(aq)$$

J.11 (b)

J.133 (a) acid: $H_3O^+(aq)$; base: $CH_3NH_2(aq)$;   (b) acid: CH_3COOH; base: CH_3NH_2;   (c) acid: $HI(aq)$; base: $CaO(s)$

J.15 (a) CHO_2;   (b) $C_2H_2O_4$;   (c) $(COOH)_2(aq) + 2\,NaOH(aq) \rightarrow Na_2C_2O_4(aq) + 2\,H_2O(l)$;

net ionic equation: $(COOH)_2(aq) + 2\,OH^- \rightarrow C_2O_4^{2-}(aq) + 2\,H_2O(l)$

J.17 (a) $C_6H_5O^-(aq) + H_2O(l) \rightarrow C_6H_5OH(aq) + OH^-(aq)$

(b) $ClO^-(aq) + H_2O(l) \rightarrow HClO(aq) + OH^-(aq)$

(c) $C_5H_5NH^+(aq) + H_2O(l) \rightarrow C_5H_5N(aq) + H_3O^+(aq)$

(d) $NH_4^+(aq) + H_2O(l) \rightarrow NH_3(aq) + H_3O^+(aq)$

J.19 (a) $c(C_6H_5NH_3^+) = \left(\dfrac{40.0\text{ g }C_6H_5NH_3Cl}{210.0\text{ mL}}\right)\left(\dfrac{1000\text{ mL}}{1\text{ L}}\right)$
$$\times \left(\dfrac{1\text{ mol }C_6H_5NH_3Cl}{129.45\text{ g }C_6H_5NH_3Cl}\right)$$

$= 1.47\text{ M }C_6H_5NH_3Cl = 1.47\text{ M }C_6H_5NH_3^+$

(b) $\underset{\text{acid}}{C_6H_5NH_3^+} + \underset{\text{base}}{H_2O(l)} \rightarrow \underset{\substack{\text{conjugate}\\\text{base}}}{C_6H_5NH_2(aq)} + \underset{\substack{\text{conjugate}\\\text{acid}}}{H_3O^+}$

J.21 (a) $AsO_4^{3-}(aq) + H_2O(l) \rightarrow HAsO_4^{2-}(aq) + OH^-(aq)$;

$HAsO_4^{2-}(aq) + H_2O(l) \rightarrow H_2AsO_4^-(aq) + OH^-(aq)$;

$H_2AsO_4^-(aq) + H_2O(l) \rightarrow H_3AsO_4(aq) + OH^-(aq)$.

In each equation, H_2O is the acid.   (b) 0.505 mol Na^+

J.23 (a) $CO_2(g) + H_2O(l) \rightarrow H_2CO_3(aq)$ (carbonic acid);

(b) $SO_3(g) + H_2O(l) \rightarrow H_2SO_4(aq)$ (sulfuric acid)

K.1 (a) $+2$;   (b) $+2$;   (c) $+6$;   (d) $+4$;   (e) $+1$

K.3 (a) $+4$;   (b) $+4$;   (c) -2;   (d) $+5$;   (e) $+1$;   (f) 0

K.5 Fe^{2+} ions and copper will form. Iron is a reducing agent, so it reduces Cu^{2+} to copper metal. The iron is then oxidized to Fe^{2+}.

K.7 (a) Methanol $CH_3OH(aq)$ is oxidized, and $O_2(g)$ is reduced.   (b) Mo is reduced, and some of the S present

in $Na_2S(s)$ is oxidized. The sulfur present in $MoS_2(s)$ remains in the -2 oxidation state.   (c) $Tl^+(aq)$ is both oxidized and reduced.

K.9 (a) oxidizing agent: H^+ in $HCl(aq)$; reducing agent: $Zn(s)$;   (b) oxidizing agent: $SO_2(g)$; reducing agent: $H_2S(g)$;   (c) oxidizing agent: $B_2O_3(s)$; reducing agent: $Mg(s)$

K.11 $CO_2(g) + 4\,H_2(g) \rightarrow CH_4(g) + 2\,H_2O(l)$; oxidation–reduction reaction; CO_2 is oxidizing reagent; H_2 is reducing reagent

K.13 (a) $2\,NO_2(g) + O_3(g) \rightarrow N_2O_5(g) + O_2(g)$;

(b) $S_8(s) + 16\,Na(s) \rightarrow 8\,Na_2S(s)$;

(c) $2\,Cr^{2+}(aq) + Sn^{4+}(aq) \rightarrow 2\,Cr^{3+}(aq) + Sn^{2+}(aq)$;

(d) $2\,As(s) + 3\,Cl_2(g) \rightarrow 2\,AsCl_3(l)$

K.15 (a) oxidizing agent: $WO_3(s)$; reducing agent: $H_2(g)$;

(b) oxidizing agent: HCl; reducing agent: $Mg(s)$;

(c) oxidizing agent: $SnO_2(s)$; reducing agent: $C(s)$;

(d) oxidizing agent: $N_2O_4(g)$; reducing agent: $N_2H_4(g)$

K.17 (a) $Cl_2(g) + H_2O(l) \rightarrow HClO(aq) + HCl(aq)$;

oxidizing agent: $Cl_2(g)$; reducing agent: $Cl_2(g)$;

(b) $4\,NaClO_3(aq) + 2\,SO_2(g) + 2\,H_2SO_4(aq, \text{dilute}) \rightarrow$
$$4\,NaHSO_4(aq) + 4\,ClO_2(g);$$

oxidizing agent: $NaClO_3(aq)$; reducing agent: $SO_2(g)$;

(c) $2\,CuI(aq) \rightarrow 2\,Cu(s) + I_2(s)$;

oxidizing agent: $CuI(aq)$; reducing agent: $CuI(aq)$

K.19 (a) $Mg(s) + Cu^{2+}(aq) \rightarrow Mg^{2+}(aq) + Cu(s)$;

(b) $Fe^{2+}(aq) + Ce^{4+}(aq) \rightarrow Fe^{3+}(aq) + Ce^{3+}(aq)$;

(c) $H_2(g) + Cl_2(g) \rightarrow 2\,HCl(g)$;

(d) $4\,Fe(s) + 3\,O_2(g) \rightarrow 2\,Fe_2O_3(s)$

K.21 (a) $-\frac{1}{2}$;   (b) -1;   (c) -1;   (d) -1;   (e) $-\frac{1}{3}$

K.23 (a) need a reducing agent;   (b) need a reducing agent

K.25 (a) redox reaction: oxidizing agent: $I_2O_5(s)$; reducing agent: $CO(g)$;   (b) redox reaction: oxidizing agent: $I_2(aq)$; reducing agent: $S_2O_3^{2-}(aq)$;   (c) precipitation reaction: $Ag^+(aq) + Br^-(aq) \rightarrow AgBr(s)$;   (d) redox reaction: oxidizing agent: $UF_4(g)$; reducing agent: $Mg(s)$

L.1 0.050 mol Br_2 will be obtained.

L.3 (a) 8.6×10^{-5} mol H_2;   (b) 11.3 g Li_3N

L.5 (a) 507.1 g Al;   (b) 6.613×10^6 g Al_2O_3

L.7 (a) 505 g H_2O;   (b) 1.33×10^3 g O_2

L.9 4.3×10^3 g H_2O

L.11 0.482 g HCl

L.13 3.50×10^{-2} M $Ca(OH)_2$

L.15 (a) 0.271 M;   (b) 0.163 g NaOH

L.17 (a) 0.209 M;   (b) 0.329 g HNO_3 in solution

L.19 63.0 g·mol^{-1}

L.21 0.150 M

L.23 (a) $Na_2CO_3(aq) + 2\,HCl(aq) \rightarrow 2\,NaCl(aq) + H_2CO_3(aq)$;

(b) 12.6 M

L.25 0.28%

L.27 $I_3^- + SnCl_2(aq) + 2\,Cl^- \rightarrow 3\,I^- + SnCl_4(aq)$

L.29 (a) $S_2O_3^{2-}$ is both oxidized and reduced.   (b) 11.1 g $S_2O_3^{2-}$ present initially.

L.31 Pt

L.33 $x = 2$, $BaBr_2 + Cl_2 \rightarrow BaCl_2 + Br_2$

L.35 509 kg Fe

L.37 (a) Pipette 31 mL of 16 M HNO_3 into a 1.00 L volumetric flask that contains about 800 mL of H_2O. Dilute to the mark with H_2O. Shake the flask to mix the solution thoroughly.   (b) 2.5×10^2 mL

L.39 (a) empirical formula: SnO_2;   (b) tin(IV) oxide
L.41 (a) It will not affect the reported KOH concentration.
(b) It will make the reported KOH concentration too high.
(c) It will make the reported KOH concentration too high.
(d) It will make the reported KOH concentration too high.
M.1 76.6%
M.3 93.1% yield
M.5 (a) BrF_3;   (b) 12 mol ClO_2F and 2 mol Br_2 will be produced; 1 mol BrF_3 will remain.
M.7 (a) $B_2O_3(s) + 3\ Mg(s) \rightarrow 3\ MgO(s) + 2\ B(s)$;
(b) 3.71×10^4 g B can be produced.
M.9 (a) $Cu^{2+}(aq) + 2\ OH^-(aq) \rightarrow Cu(OH)_2(s)$;   (b) 2.44 g $Cu(OH)_2$
M.11 (a) O_2;   (b) 5.77 g P_4O_{10};   (c) 5.7 g P_4O_6 remains
M.13 The number of chlorine atoms is 6.
M.15 (a) $2\ Al(s) + 3\ Cl_2(g) \rightarrow 2\ AlCl_3(s)$;   (b) 671 g $AlCl_3$;
(c) 44.7%
M.17 81.2%
M.19 The empirical formula is: $C_4H_5N_2O$.
The molecular formula is: $C_8H_{10}N_4O_2$.
$2\ C_8H_{10}N_4O_2(s) + 19\ O_2(g) \rightarrow$
$$16\ CO_2(g) + 10\ H_2O(l) + 4\ N_2(g)$$
M.21 The empirical formula is $C_8H_{16}N_4O_3$.
M.23 (a) The solid is calcium phosphate, $Ca_3(PO_4)_2$.
(b) 130. g $Ca_3(PO_4)_2$
M.25 93.0%
M.27 (a) The empirical formula is: $C_{11}H_{14}O_3$.   (b) The molecular formula of the compound is: $C_{22}H_{28}O_6$.

Focus 1

1A.1 (a) Radiation may pass through a metal foil.   (b) The slower speed supports the particle model.   (c) The radiation model.   (d) The particle model.
1A.3 (a) no;   (b) no;   (c) yes;   (d) no
1A.5 microwaves < visible light < ultraviolet light < x-rays < γ-rays
1A.7 (a) 420 nm;   (b) 150 nm
1A.9

Frequency (2 s.f.)	Wavelength (2 s.f.)	Energy of photon (2 s.f.)	Event
8.7×10^{14} Hz	340 nm	5.8×10^{-19} J	suntan
5.0×10^{14} Hz	600 nm	3.3×10^{-19} J	reading
300 MHz	1 m	2×10^{-25} J	microwave popcorn
1.2×10^{17} Hz	2.5 nm	7.9×10^{-17} J	dental x-ray

1A.11 For the Lyman series, the lower energy level is $n = 1$; for the Balmer series, $n = 2$; for the Paschen series, $n = 3$; and for the Brackett series, $n = 4$.
1A.13 (a) 121 nm;   (b) Lyman series;   (c) This absorption lies in the ultraviolet region.
1A.15 The transition is $n_1 = 1$ to $n_2 = 3$.
1A.17 30.4 nm
1B.1 (a) false;   (b) true;   (c) false
1B.3 the photoelectric effect
1B.5 8.8237 pm

1B.7 (a) 3.37×10^{-19} J;   (b) 44.1 J;   (c) 203 kJ
1B.9 number of photons = 1.4×10^{20} photons
moles of photons = 2.3×10^{-4} mol photons
1B.11 3400 K
1B.13 $\lambda_{max} = 1.59 \times 10^{-6}$ m, or 1590 nm
1B.15 (a) 2.0×10^{-10} m;   (b) 1.66×10^{-17} J;   (c) 8.8 nm;
(d) x-ray
1B.17 The heavier person (80 kg) should have a shorter wavelength
1B.19 1.44 pm for both; the wavelength of a proton and a neutron are identical to 3 significant figures.
1B.21 1.1×10^{-34} m
1B.23 3.96×10^3 m·s^{-1}
1B.25 $\Delta v = 1.65 \times 10^5$ m·s^{-1}
1B.27 $\Delta x = 1.3 \times 10^{-36}$ m
1C.1 (a) 8.24 nm;   (b) 10.6 nm
1C.3 yes; $n_1 = 1, n_2 = 2, n_1 = 1, n_2 = 3, n_1 = 2, n_2 = 3$
1C.5 (a)

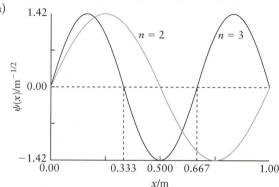

(b) For $n = 2$, there is one node at $x = 0.500$ m.
(c) For $n = 3$, there are two nodes, one at $x = 0.333$ m and 0.667 m.
(d) The number of nodes is equal to $n - 1$.
(e) For $n = 2$, a particle is most likely to be found at $x = 0.25$ m and $x = 0.75$ m.
(f) For $n = 3$, a particle is most likely to be found at $x = 0.17, 0.50,$ and 0.83 m.
1C.7 (a) Integrate over the "left half of the box" or from 0 to $\frac{1}{2}L$:
$$\int_0^{L/2} \psi^2 dx = \frac{2}{L}\int_0^{L/2}\left(\sin\frac{n\pi x}{L}\right)^2 dx$$
$$= \frac{2}{L}\left[\left(\frac{-1}{2n\pi}\cdot\cos\frac{n\pi x}{L}\cdot\sin\frac{n\pi x}{L} + \frac{x}{2}\right)\Bigg|_0^{L/2}\right]$$
given n is an integer:
$$= \frac{2}{L}\left[\left(\frac{L/2}{2}\right) - 0\right] = \frac{1}{2}$$
1D.1 (a) energy will increase;   (b) n increases;   (c) l increases;
(d) radius increases
1D.3 0.33
1D.5 With one electron in each p-orbital, the electron distribution is spherically symmetric.
1D.7 (a) The probability (P) is 32.3%.   (b) The probability (P) is 76.1%.
1D.9 (a)

1s   2p   3d

(b) A region in space where the wavefunction ψ passes through 0.
(c) The simplest s-orbital has 0 nodes, the simplest p-orbital has 1 nodal plane, and the simplest d-orbital has 2 nodal planes. (d) An f-orbital would be expected to have 3 nodal planes.

1D.11 (a) 1 orbital; (b) 5 orbitals; (c) 3 orbitals; (d) 7 orbitals

1D.13 (a) 7 values: 0, 1, 2, 3, 4, 5, 6; (b) 5 values: $-2, -1, 0, +1, +2$, (c) 3 values: $-1, 0, +1$; (d) 4 subshells: 4s, 4p, 4d, and 4f

1D.15 (a) $n = 6; l = 1$; (b) $n = 3; l = 2$; (c) $n = 2; l = 1$; (d) $n = 5; l = 3$

1D.17 (a) $-1, 0, +1$; (b) $-2, -1, 0, +1, +2$; (c) $-1, 0, +1$; (d) $-3, -2, -1, 0, +1, +2, +3$

1D.19 (a) 3 orbitals; (b) 5 orbitals; (c) 1 orbital; (d) 7 orbitals

1D.21 (a) 5d, five; (b) 1s, one; (c) 6f, seven; (d) 2p, three

1D.23 (a) 3; (b) 1; (c) 4; (d) 1

1D.25 (a) cannot exist; (b) exists; (c) cannot exist; (d) exists

1E.1 (a) energy increases; (b) n increases; (c) l increases; (d) radius increases. All of these are the same as in Exercise 1D.1.

1E.3 (a) $V(r) = \left(\dfrac{-3e^2}{4\pi\varepsilon_0}\right)\left(\dfrac{1}{r_2} + \dfrac{1}{r_2} + \dfrac{1}{r_3}\right) + \dfrac{e^2}{4\pi\varepsilon_0}\left(\dfrac{1}{r_{12}} + \dfrac{1}{r_{13}} + \dfrac{1}{r_{23}}\right)$
(b) The first term represents the coulombic attractions between the nucleus and each electron, and the second term represents the coulombic repulsions between each pair of electrons.

1E.5 (a) false; (b) true; (c) false; (d) true

1E.7 Only (d) is the configuration expected for a ground-state atom.

1E.9 (a) possible; (b) not possible; (c) not possible

1E.11 (a) sodium [Ne]3s^1
(b) silicon [Ne]3s^{2}3p^2
(c) chlorine [Ne]3s^{2}3p^5
(d) rubidium [Kr]5s^1

1E.13 (a) silver [Kr]4d^{10}5s^1
(b) beryllium [He]2s^2
(c) antimony [Kr]4d^{10}5s^{2}5p^3
(d) gallium [Ar]3d^{10}4s^{2}4p^1
(e) tungsten [Xe]4f^{14}5d^{4}6s^2
(f) iodine [Kr]4d^{10}5s^{2}5p^5

1E.15 (a) tellurium; (b) vanadium; (c) carbon; (d) thorium

1E.17 (a) 4p; (b) 4s; (c) 6s; (d) 6s

1E.19 (a) 5; (b) 11; (c) 5; (d) 20

1E.21 (a) 3; (b) 2; (c) 3; (d) 2

1E.23

Element	Electron configuration	Unpaired electrons
Ga	[Ar]3d^{10}4s^{2}4p^1	1
Ge	[Ar]3d^{10}4s^{2}4p^2	2
As	[Ar]3d^{10}4s^{2}4p^3	3
Se	[Ar]3d^{10}4s^{2}4p^4	2
Br	[Ar]3d^{10}4s^{2}4p^5	1

1E.25 (a) ns^1; (b) ns^2np^3; (c) $(n-1)d^5ns^2$; (d) $(n-1)d^{10}ns^1$

1F.1 (a) silicon (118 pm) > sulfur (104 pm) > chlorine (99 pm);
(b) titanium (147 pm) > chromium (129 pm) > cobalt (125 pm);
(c) mercury (155 pm) > cadmium (152 pm) > zinc (137 pm);
(d) bismuth (182 pm) > antimony (141 pm) > phosphorus (110 pm)

1F.3 $P^{3-} > S^{2-} > Cl^-$

1F.5 (a) Ca; (b) Na; (c) Na

1F.7 (a) oxygen (1310 kJ·mol^{-1}) > selenium (941 kJ·mol^{-1}) > tellurium (870 kJ·mol^{-1}); ionization energies decrease down a group. (b) gold (890 kJ·mol^{-1}) > osmium (840 kJ·mol^{-1}) > tantalum (761 kJ·mol^{-1}); ionization energies decrease from right to left in the periodic table. (c) lead (716 kJ·mol^{-1}) > barium (502 kJ·mol^{-1}) > cesium (376 kJ·mol^{-1}); ionization energies decrease from right to left in the periodic table.

1F.9 The first ionization energy for sulfur and phosphorus atoms are nearly the same due to greater electron–electron repulsions in S, making the energy of the outermost electrons higher than predicted. Once the first electron is removed, electrons are held tighter due to the smaller size of the S$^+$ ion, and this is reflected in the much greater second ionization energy of sulfur.

1F.11 (a) iodine; (b) they are equal; (c) sulfur; (d) they are equal

1F.13 (a) The tendency to form ions that are two units lower in charge than that expected based on group number.
(b) Because of the poor shielding ability of the d-electrons in heavy elements, there is enhanced ability of the s-electrons to penetrate to the nucleus and be bound tighter than expected.

1F.15 (a) A similarity in chemical properties between an element in the periodic table and one lying one period lower and one group to the right. (b) This is due to the similarity in size of the ions. (c) Al^{3+} and Ge^{4+} compounds show the diagonal relationship, as do Li$^+$ and Mg^{2+}.

1F.17 Only (b) Li and Mg exhibit a diagonal relationship.

1F.19 The ionization energies of the s-block metals are considerably lower, thus making it easier for them to lose electrons in chemical reactions.

1F.21 (a) metal; (b) nonmetal; (c) metal; (d) metalloid; (e) metalloid; (f) metal

1.1 397 nm

1.3 750 J

1.5 (a) $\displaystyle\int_0^L \left(\sin\dfrac{\pi\cdot x}{L}\right)\cdot\left(\sin\dfrac{2\pi\cdot x}{L}\right)dx$
$= \dfrac{L}{2\pi}\left(\sin\dfrac{\pi\cdot x}{L}\right) - \dfrac{L}{6\pi}\left(\sin\dfrac{3\pi\cdot x}{L}\right)\Big|_0^L = 0$

(b)
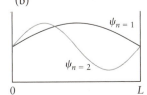
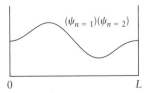

1.7 (a) 0; Because the integral is zero, one would not expect to observe a transition between the $n = 1$ and $n = 3$ states. (b) The intensity will increase as the length of the box increases.

1.9 A $2p_x$ orbital has two lobes, one that has a wavefunction with a positive sign and one with a wavefunction with a negative sign. Thus there is a $\frac{1}{2}$ or 0.50 probability that an electron excited to the $2p_x$ orbital would be found in the region of space for which the wavefunction has a positive sign.

1.11 (a) The observed values 75.7 eV (7.30 MJ·mol^{-1}) and 5.28 eV (0.519 MJ·mol^{-1}) correspond respectively to the second (7300 kJ·mol^{-1}) and first (519 kJ·mol^{-1}) ionization energies of Li ($1s^2 2s^1$). (b) The PES values observed 153 eV (14.8 MJ·mol^{-1}) and 933 eV (0.90 MJ·mol^{-1}) correspond respectively to the third (14800 kJ·mol^{-1}) and first (900 kJ·mol^{-1}) ionization energies of Be ($1s^2 2s^2$).

1.13 Oxygen is the first element encountered in which the p-electrons must be paired. This added electron–electron repulsion energy causes the ionization energy to be lower.

1.15 molar volume (cm^3·mol^{-1}) = molar mass (g·mol^{-1})/ density (g·cm^{-3})

Element	Molar volume	Element	Molar volume
Li	13	Na	24
Be	4.87	Mg	14.0
B	4.38	Al	9.99
C	5.29	Si	12.1
N	16	P	17.0
O	14.0	S	15.3
F	17.1	Cl	21.4
Ne	16.7	Ar	24.1

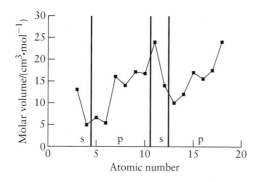

The molar volume roughly parallels atomic size (volume), which decreases as the s-subshell begins to fill and subsequently increases as the p-subshell fills.

1.17 (a) In copper it is energetically favorable for an electron to be promoted from the 4s-orbital to a 3d-orbital, giving a completely filled 3d-subshell. In the case of Cr, it is energetically favorable for an electron to be promoted from the 4s-orbital to a 3d-orbital to half-fill the 3d-subshell. (b) Nb, Mo, Ru, Rh, Pd, Ag, Pt, and Au. Of these, the explanation used for chromium and copper is valid for Mo, Pd, Ag, and Au. (c) Because the np-orbitals are so much lower in energy than the $(n + 1)$s-orbitals.

1.19 Based on (a) the element is probably a member of Group 4 (the titanium family). From (b) we know that it must be in Period 5 and thus the element is likely zirconium. However, rhodium, with its $4d^8$ configuration, is a second possibility.

1.21 (a) radius of neutral atom = 285 pm (approx. 20 pm larger than Cs); (b) radius of +1 ion = 194 pm (approx. 20 pm larger than Cs$^+$); (c) $I_1 = 356$ kJ·mol^{-1} (approx. 20 kJ less than Cs)

1.23 <u>**A**</u> = Na; <u>**B**</u> = Cl; <u>**C**</u> = Na$^+$; <u>**D**</u> = Cl$^-$

1.25 (a) $1s^2 2s^1$; (b) +1; (c) $1s^2 2s^2 2p^{12}$; therefore it should have $Z = 20$.

1.27 (a) 478 nm; (b) 7.12×10^{14} s^{-1}

1.29 Neon ($Z = 10$) would have to lose a 2p-electron to ionize. The longest detectable wavelength would be 3.64 nm, which is in the x-ray region.

1.31 (a) The violet GaN laser will provide enough energy to eject the electron. (b) 2.10×10^{-20} J.

1.33 (a) 9.35×10^{-18} J; (b) This energy corresponds to a wavelength of 2.13×10^{-8} m or 21.3 nm. This falls within the x-ray region. (c) If the chain is 10 carbons long, then there would be 9 C—C bonds. If a wavefunction extends over two adjacent carbon atoms then the minimum number of wavefunctions would be 9. (d) 4.25×10^{-19} J; (e) This energy corresponds to a wavelength of 4.69×10^{-7} m or 469 nm. This falls within the visible region; (f) 2.75×10^{-18} m^2. This will give $L = 1.658 \times 10^{-9}$ m = 1658 pm. Since each C—C bond is 139 pm long, this length corresponds to the chain of carbon atoms having twelve C—C.

Focus 2

2A.1 (a) 5; (b) 4; (c) 7; (d) 3

2A.3 (a) [Ar]; (b) [Ar]$3d^{10}4s^2$; (c) [Kr]$4d^5$; (d) [Ar]$3d^{10}4s^2$

2A.5 (a) [Ar]$3d^{10}$; (b) [Xe]$4f^{14}5d^{10}6s^2$; (c) [Ar]$3d^{10}$; (d) [Xe]$4f^{14}5d^{10}$

2A.7 (a) [Kr]$4d^{10}5s^2$; same; In$^+$ and Sn^{2+} lose 5p valence electrons; (b) none; (c) [Kr]$4d^{10}$; Pd

2A.9 (a) Co^{2+}; (b) Fe^{2+}; (c) Mo^{2+}; (d) Nb^{2+}

2A.11 (a) Co^{3+}; (b) Fe^{3+}; (c) Ru^{3+}; (d) Mo^{3+}

2A.13 (a) 4s; (b) 3p; (c) 3p; (d) 4s

2A.15 (a) −2; (b) −2; (c) +1; (d) +3; (e) +2

2A.17 (a) 3; (b) 6; (c) 6; (d) 2

2A.19 (a) [Kr]$4d^{10}5s^2$; no unpaired electrons; (b) [Kr]$4d^{10}$; no unpaired electrons; (c) [Xe]$4f^{14}5d^4$; four unpaired electrons; (d) [Kr]; no unpaired electrons; (e) [Ar]$3d^8$; two unpaired electrons

2A.21 (a) [Ar]; no unpaired electrons; (b) [Kr]$4d^{10}5s^2$; no unpaired electrons; (c) [Xe]; no unpaired electrons; (d) [Kr]$4d^{10}$; no unpaired electrons

2A.23 (a) Mg_3As_2; (b) In_2S_3; (c) AlH_3; (d) H_2Te; (e) BiF_3

2A.25 (a) :Cl:$^-$ Tl^{3+} :Cl:$^-$:Cl:$^-$

(b) :S:$^{2-}$ Al^{3+} :S:$^{2-}$ Al^{3+} :S:$^{2-}$

(c) Ba^{2+} :O:$^{2-}$

2A.27 (b) Ga^{3+}, O^{2-}

2A.29 Lattice energy is inversely proportional to the distance between the ions; thus the larger rubidium ion will have the lower lattice energy.

2B.1 (a) :Cl—C—Cl: with :Cl: top and :Cl: bottom (b) :Cl—C—Cl: with :O: double bonded on top (c) Ö=N—F̈:

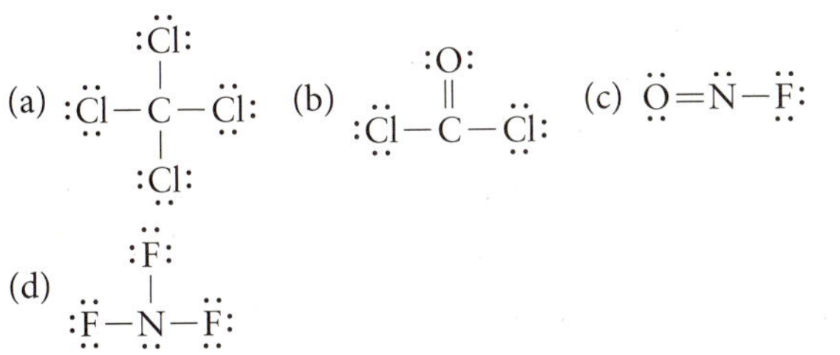

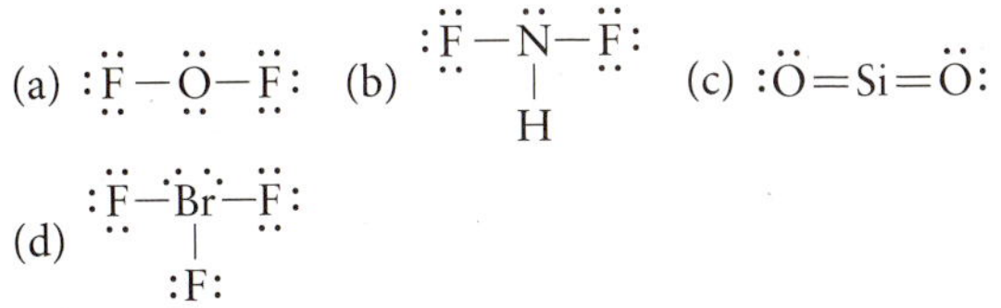

(d) :F̈—N—F̈: with :F: top and :F: bottom

2B.3 (a) :F̈—Ö—F̈: (b) :F̈—N—F̈: with H below N (c) :Ö=Si=Ö:

(d) :F̈—Br̈—F̈: with :F: below Br

2B.5 (a) [H—B—H]⁻ with H top and H bottom (b) [:Br̈—Ö:]⁻ (c) [H—N̈—H]⁻

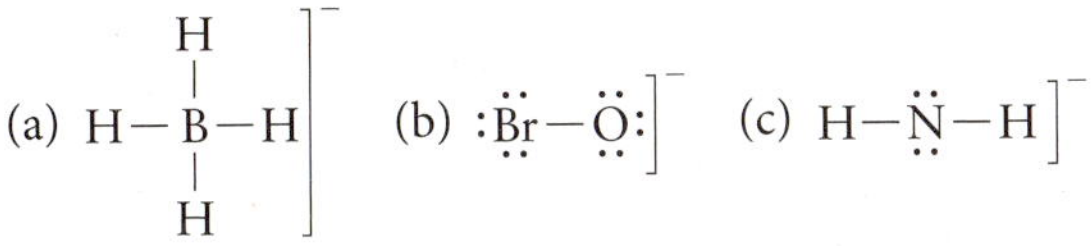

2B.7 E is phosphorus (P).

2B.9 (a) [H—N—H]⁺ with H top and H bottom :Cl̈:⁻ (b) K⁺ [:P̈:]³⁻ K⁺ with K⁺ above

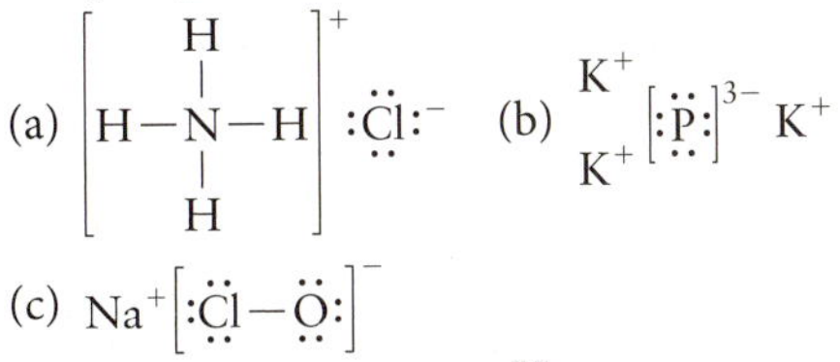

(c) Na⁺ [:Cl̈—Ö:]⁻

2B.11 (a) H—C—H with :O: double bonded on top (b) H—C—Ö—H with H top and H bottom

(c) H—N̈—C—C—Ö—H with H, H below N, and :O: double bonded above second C

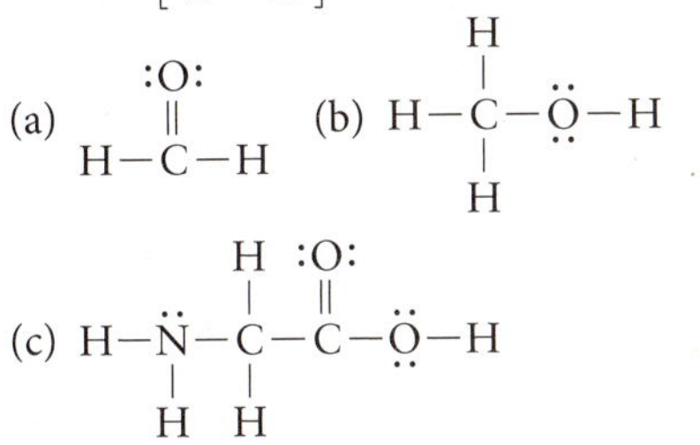

2B.13 The four possible resonance structures for anthracene are:

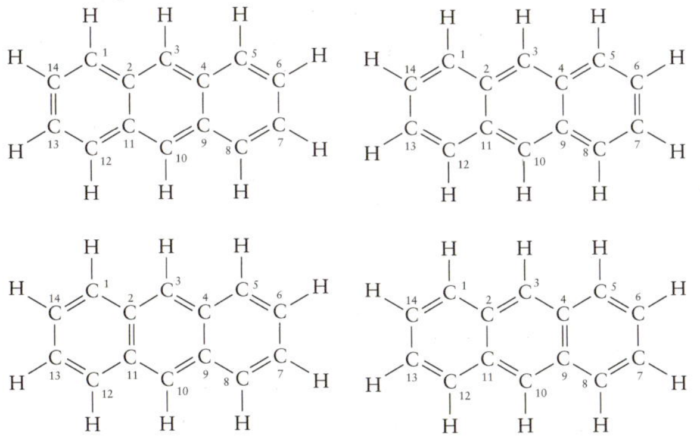

2B.15 Ö=N—Cl̈: with :O: on left ; :Ö—N—Cl̈: with :O: double bonded on top

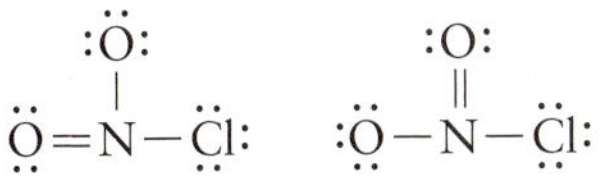

In both structures, the N has a formal charge of +1 and the singly bound O has a formal charge of −1. All other atoms have formal charge of 0.

2B.17 The two resonance structures for cyclobutadiene are shown below (the carbons have been numbered for clarity):

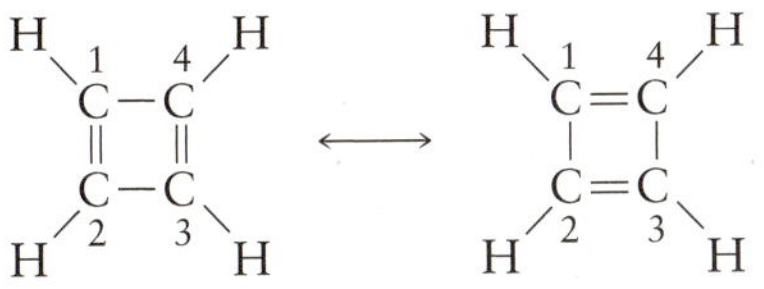

2B.19 (a) :N≡O:⁺ with charges 0 +1 (b) :N≡N: with charges 0 0 (c) :C≡O: with charges −1 +1

(d) [:C≡C:]²⁻ with charges −1 −1 (e) [:C≡N:]⁻ with charges −1 0

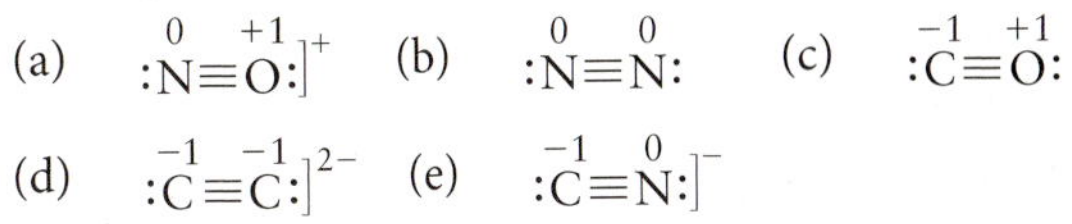

2B.21 Two possible structures for hypochlorous acid are:

H—Cl̈—Ö: (charges +1 −1)   H—Ö—Cl̈: (charges 0 0)

Based on formal charge, the structure on the right is the more likely.

2B.23 (a) Ö=Cl—Ö: with :O: H below, charges 0 0 0, lower energy ; :Ö—Cl—Ö: with :O: H below, charges −1 +2 0 and −1

(b) Ö=C=S̈ charges 0 0 0, lower energy ; :Ö—C≡S: charges −1 0 +1

(c) H—C≡N: charges 0 0, lower energy ; H—C=N̈ charges −1 +1

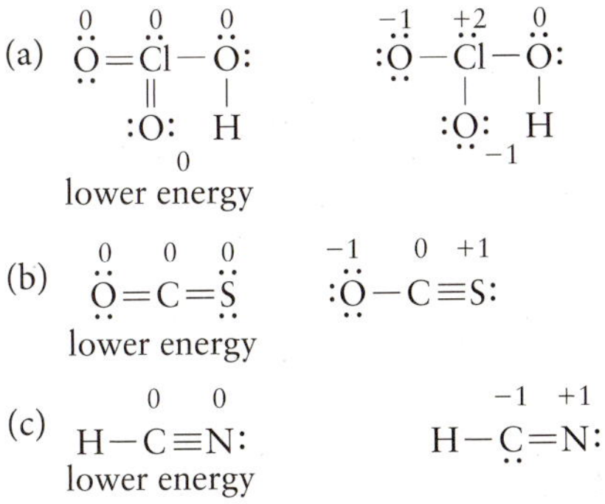

2C.1 Only (b) and (c) are radicals.

2C.3 (a) The periodate ion has one Lewis structure that obeys the octet rule:

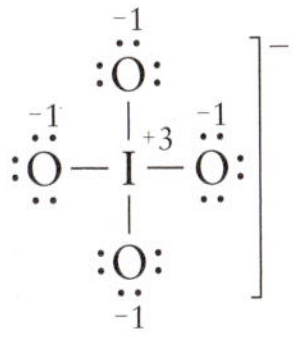

The formal charge at I can be reduced from +3 to 0 by including three double-bond contributions, thereby giving rise to four resonance forms.

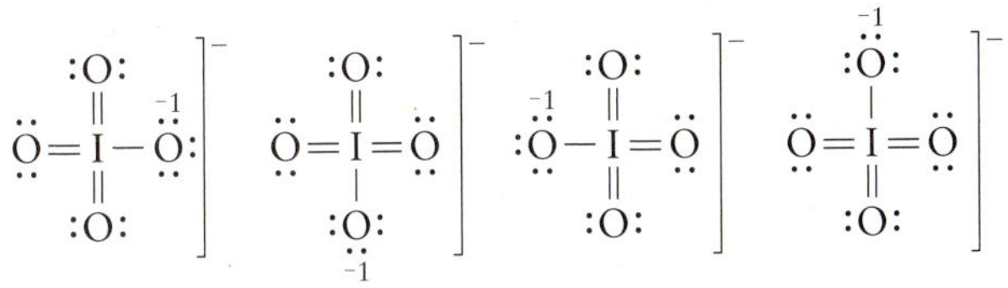

(b) The hydrogen phosphate ion has one Lewis structure that obeys the octet rule (the first structure shown below). Including one double bond to oxygen lowers the formal charge at P from +1 to 0. There are three resonance forms that include this contribution.

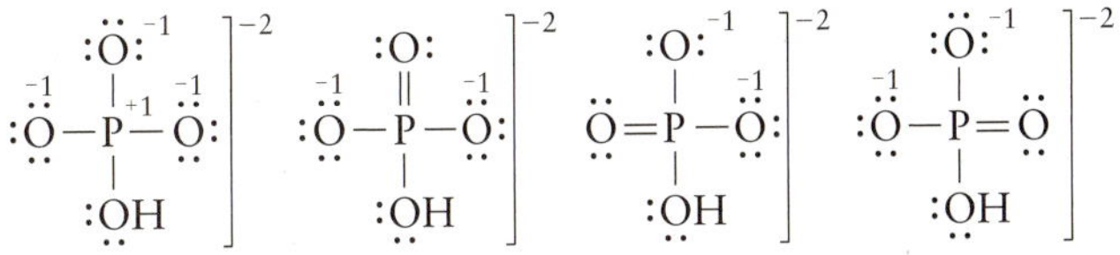

(c) There is one Lewis structure that obeys the octet rule shown below at the left. The formal charge at chlorine can be reduced to +1 by including one double bond contribution. The formal charge can be reduced to 0 if there are two double bond contributions. These contributions give rise to the following resonance structures.

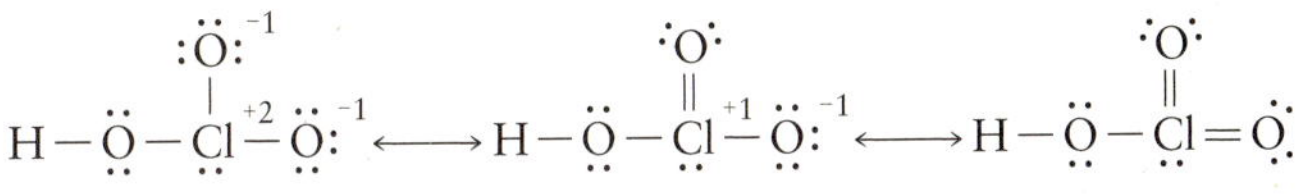

(d) The arsenate ion has one Lewis structure that obeys the octet rule.

$$\left[\begin{array}{c} \overset{\displaystyle :\!\ddot{\text{O}}\!:^{-1}}{\underset{\displaystyle {}^{-1}:\!\ddot{\text{O}}\!:}{\overset{-1}{:}\ddot{\text{O}}\!-\!\overset{+1}{\text{As}}\!-\!\ddot{\text{O}}\!:^{-1}}} \end{array}\right]^{3-}$$

Just as in part (a), including one double bond to oxygen lowers the formal charge at As from +1 to 0. There are four resonance forms that include this contribution.

$$\left[:\!\ddot{\text{O}}\!-\!\text{As}\!-\!\ddot{\text{O}}\!:\right]^{3-} \quad \left[\ddot{\text{O}}\!=\!\text{As}\!-\!\ddot{\text{O}}\!:\right]^{3-} \quad \left[:\!\ddot{\text{O}}\!-\!\text{As}\!-\!\ddot{\text{O}}\!:\right]^{3-} \quad \left[:\!\ddot{\text{O}}\!-\!\text{As}\!=\!\ddot{\text{O}}\right]^{3-}$$

2C.5 Only (a) is a radical.

(a) $:\!\dot{\ddot{\text{Cl}}}\!-\!\ddot{\text{O}}\!:$ (b) $:\!\ddot{\text{Cl}}\!-\!\ddot{\text{O}}\!-\!\ddot{\text{O}}\!-\!\ddot{\text{Cl}}\!:$

(c)
$$\ddot{\text{O}}\!=\!\text{N}\!-\!\ddot{\text{O}}\!-\!\ddot{\text{Cl}}\!: \quad (\text{with } :\!\text{O}:\ \text{above N})$$

2C.7 (a) $\left[:\!\ddot{\text{Cl}}\!-\!\text{I}\!-\!\ddot{\text{Cl}}\!:\right]^{+}$ I has 2 bonding pairs and 2 lone pairs

(b) $\left[\text{ICl}_4\right]^{-}$ I has 4 bonding pairs and 2 lone pairs

(c) ICl_3 I has 3 bonding pairs and 2 lone pairs

(d) ICl_5 I has 5 bonding pairs and 1 lone pair

2C.9 (a) SF_6, 12 electrons (b) XeF_4, 10 electrons (c) $[\text{AsF}_6]^{-}$, 12 electrons (d) TeCl_4, 10 electrons

2C.11 (a) $\text{F}\!-\!\text{Xe}\!-\!\text{F}$ with $=\!\text{O}$, 2 lone pairs (b) XeF_4, 2 lone pairs (c) XeO_2F_2, 1 lone pair

2C.13 (a) In BeCl_2, there are 4 electrons around the central beryllium:

$$:\!\ddot{\text{Cl}}\!-\!\text{Be}\!-\!\ddot{\text{Cl}}\!:$$

(b) In ClO_2 there is an odd number of electrons around the central chlorine:

$$:\!\ddot{\text{O}}\!-\!\dot{\ddot{\text{Cl}}}\!-\!\ddot{\text{O}}\!:$$

2C.15 (a)

$$\left[\overset{-1}{:}\!\ddot{\text{O}}\!-\!\overset{0}{\text{S}}\!-\!\ddot{\text{O}}\!:^{0} \right]^{-} \text{ with } \overset{\|}{:\!\ddot{\text{O}}\!:_{0}} \text{ and H} \qquad \text{lower energy}$$

$$\left[\overset{-1}{:}\!\ddot{\text{O}}\!-\!\overset{+1}{\text{S}}\!-\!\ddot{\text{O}}\!:^{0} \right]^{-} \text{ with } :\!\ddot{\text{O}}\!:^{-1} \text{ and H}$$

(b)

$$\left[\overset{0}{:}\!\text{O:}\quad\text{H} \atop \overset{-1}{:}\!\ddot{\text{O}}\!-\!\overset{+1}{\text{S}}\!-\!\ddot{\text{O}}\!:^{0}\right]^{-} \text{ with } {}^{-1}:\!\ddot{\text{O}}\!: \qquad \text{lower energy}$$

$$\left[{}^{-1}:\!\ddot{\text{O}}:\quad\text{H} \atop \overset{-1}{:}\!\ddot{\text{O}}\!-\!\overset{+2}{\text{S}}\!-\!\ddot{\text{O}}\!:^{0}\right]^{-} \text{ with } {}^{-1}:\!\ddot{\text{O}}\!:$$

2C.17 (a) The first structure is favored on the basis of formal charges.   (b) The first structure is preferred based on formal charge.

2D.1 In (1.78) < Sn (1.96) < Sb (2.05) < Se (2.55)

2D.3 BaBr_2 would have bonds that are primarily ionic.

2D.5 (a) HCl; (b) CF_4; (c) C—O bonds

2D.7 (a) KCl; (b) BaO

2D.9 $\text{Rb}^{+} < \text{Sr}^{2+} < \text{Be}^{2+}$; smaller, more highly charged cations have greater polarizing power.

2D.11 $\text{O}^{2-} < \text{N}^{3-} < \text{Cl}^{-} < \text{Br}^{-}$; the polarizability increases as the ion gets larger and less electronegative.

2D.13 (a) $\text{CO}_3^{2-} > \text{CO}_2 > \text{CO}$; (b) $\text{SO}_3^{2-} > \text{SO}_2 \approx \text{SO}_3$; (c) $\text{CH}_3\text{NH}_2 > \text{CH}_2\text{NH} > \text{HCN}$

2D.15 CF_4; shortest bond

2D.17 (a) ca. 127 pm; (b) the C—O bond: 127 ppm; the C—N bonds: 142 pm; (c) 172 pm; (d) 158 pm

2D.19 (a) 77 pm + 58 pm = 135 pm; (b) 111 pm + 58 pm = 169 pm; (c) 141 pm + 58 pm = 199 pm. Bond distance increases with atomic size going down Group 14.

2E.1 (a) must have lone pairs; (b) may have lone pairs

2E.3 (a) $:\!\text{N}\!\equiv\!\text{C}\!-\!\text{H}$ linear; (b) CHF_3 tetrahedral

2E.5 (a) angular; (b) slightly less than 120°

2E.7 (a) trigonal pyramidal; (b) The O—S—Cl angles are identical; (c) Slightly less than 109.5°

2E.9 (a) T-shaped; (b) slightly less than 90°

2E.11 (a) SCl_4 (b) ICl_3 (c) $[\text{IF}_4]^{-}$ (d) XeO_2

(a) seesaw, AX_4E; (b) T-shaped, AX_3E_2; (c) square planar, AX_4E_2; (d) trigonal pyramidal, AX_3E

2E.13 The Lewis structures are:

(a) $\left[\text{I}_3\right]^{-}$

(b) $:\!\ddot{\text{Cl}}\!-\!\text{P}\!-\!\ddot{\text{Cl}}\!:$ with $=\!\text{O}$ above and $:\!\ddot{\text{Cl}}\!:$ below

(c)

$$\begin{bmatrix} :\ddot{O}: \\ | \\ :\ddot{O}-I \\ | \\ :\ddot{O}: \end{bmatrix}^{-}$$

(d) $:N\equiv N-\ddot{O}:$

and

$\ddot{N}=N=\ddot{O}$

(resonance forms possible)

(a) linear, 180°, X_2E_3; (b) tetrahedral, 109.5°, AX_4; (c) trigonal pyramid, less than 109.5°, AX_3E; (d) linear, 180°, AX_2

2E.15 The Lewis structures are:

(a) $:\ddot{Cl}-\overset{\displaystyle :\ddot{F}:}{\underset{\displaystyle :\ddot{F}:}{C}}-\ddot{F}:$ (b) $:\ddot{Cl}-\overset{\displaystyle :\ddot{Cl}:}{\underset{\displaystyle :\ddot{Cl}:}{Te}}-\ddot{Cl}:$

(c) $:\ddot{F}-\overset{\displaystyle :O:}{\overset{\displaystyle \|}{C}}-\ddot{F}:$ (d) $\begin{bmatrix} H \\ | \\ H-C-H \\ | \\ H \end{bmatrix}^{-}$

(a) tetrahedral; 109.5°, AX_4; (b) seesaw, 90° and 120°, AX_4E; (c) trigonal planar, 120°, AX_3; (d) trigonal pyramidal, slightly less than 109.5°, AX_3E

2E.17 (a) slightly less than 120°; (b) 180°; (c) 180°; (d) slightly less then 109.5°

2E.19 (a) tetrahedral, 109.5°; (b) tetrahedral about the carbon atoms (109.5°); CBeC angle of 180°; (c) angular, H—B—H angle slightly less than 120°; (d) angular, slightly less than 120°

2E.21 (a) trigonal planar; 120°

$\overset{\displaystyle H}{\underset{\displaystyle H}{}}C=C\overset{\displaystyle H}{\underset{\displaystyle H}{}}$

(b) linear, 180°
$:N\equiv C-\ddot{Cl}:$

(c) tetrahedral, 109.5°

$:\ddot{Cl}-\overset{\displaystyle :O:}{\underset{\displaystyle :\ddot{Cl}:}{P}}-\ddot{Cl}:$ $:\ddot{Cl}-\overset{\displaystyle :O:}{\overset{\displaystyle \|}{\underset{\displaystyle :\ddot{Cl}:}{P}}}-\ddot{Cl}:$

(d) trigonal pyramidal; 107°

$\overset{\displaystyle H\ \ H}{H-\underset{..}{N}-\underset{..}{N}-H}$

2E.23 (a)

$:\ddot{Cl}-\overset{\displaystyle :\ddot{O}:}{\underset{\displaystyle :\ddot{Cl}:}{Sb}}-\ddot{Cl}:$

tetrahedral

(b)

$:\ddot{O}-\overset{\displaystyle :\ddot{O}:}{\underset{\displaystyle :\ddot{Cl}:}{S}}-\ddot{Cl}:$

tetrahedral

(c)

$\begin{bmatrix} :\ddot{O}: \\ | \\ :\ddot{O}-I-\ddot{F}: \\ | \\ :\ddot{F}: \end{bmatrix}^{-}$

seesaw

2E.25 The Lewis structures are:

(a) $:\ddot{Cl}-\overset{\displaystyle H}{\underset{\displaystyle :\ddot{Cl}:}{C}}-H$ (b) $:\ddot{Cl}-\overset{\displaystyle :\ddot{Cl}:}{\underset{\displaystyle :\ddot{Cl}:}{C}}-\ddot{Cl}:$

(c) $\ddot{S}=C=\ddot{S}$ (d) $\overset{\displaystyle :\ddot{F}:}{\underset{\displaystyle :\ddot{F}:}{\overset{\displaystyle :\ddot{F}\diagdown}{\diagup}}}\ddot{S}:$

Molecules (a) and (d) are polar; (b) and (c) are nonpolar.

2E.27 (a) pyridine: polar; (b) ethane: nonpolar; (c) trichloromethane: polar

2E.29 (a) **1** and **2** are polar; **3** is nonpolar. (b) **1** has the largest dipole moment.

2F.1 (a) sp^3 orbitals oriented toward corners of a tetrahedron (109.5° apart); (b) sp orbitals oriented directly opposite to each other (180° apart); (c) sp^3d^2 orbitals oriented toward the corners of an octahedron (90° and 180°); (d) sp^2 orbitals oriented toward the corners of an equilateral triangle (120°).

2F.3 (a) 2 σ-bonds; 0 π-bonds; (b) 2 σ-bonds; 1 π-bond; there is also a resonance structure that includes 2 σ-bonds. 2 π-bonds.

2F.5 (a) sp; (b) sp^2; (c) sp^3; (d) sp^3

2F.7 (a) sp^2; (b) sp^3; (c) sp^3d; (d) sp^3

2F.9 (a) sp^3; (b) sp^3d^2; (c) sp^3d; (d) sp^3

2F.11 (a) sp^3; (b) nonpolar

2F.13

$\overset{\displaystyle H}{\underset{\displaystyle H}{}}C=C\overset{\displaystyle H}{\underset{\displaystyle C}{}}$
$\quad\quad\quad\quad\quad\quad\ \ \|\!\|\!\|$
$\quad\quad\quad\quad\quad\quad\quad N.$

The CH_2 and CH are sp^2 hybridized, 120°; the C bonded to N is sp hybridized, 180°.

2F.15 As the s-character of a hybrid orbital increases, the bond angle increases.

2F.17 In formaldehyde, both the C and the O are sp^2 hybridized, the bond angle is 120°; the molecule has three sigma bonds and one pi bond.

2F.19 The two hybrid orbitals are $h_1 = s + p_x + p_y + p_z$ and $h_2 = s - p_x + p_y - p_z$. Therefore, to show these two orbitals are orthogonal, $\int h_1 h_2 \, d\tau = 0$

$\int h_1 h_2 \, d\tau = \int (s + p_x + p_y + p_z)(s - p_x + p_y - p_z) \, d\tau =$

$\int (s^2 - sp_x + sp_y - sp_z + sp_x - p_x^2 + p_x p_y - p_x p_z + sp_y - p_x p_y + p_y^2 - p_y p_z + sp_z - p_z p_x + p_z p_y - p_z^2) \, d\tau$

This integral of a sum may be written as a sum of integrals:

$\int s^2 \, d\tau - \int sp_x \, d\tau + \int sp_y \, d\tau - \int sp_z \, d\tau + \cdots$

Since the hydrogen wavefunctions are mutually orthogonal, integrals of a product of two different wavefunctions are zero. This simplifies to

$\int s^2 \, d\tau - \int p_x^2 \, d\tau + \int p_y^2 \, d\tau - \int p_z^2 \, d\tau = 1 - 1 + 1 - 1 = 0$

2F.21 The hybridization is $sp^{0.67}$.

2G.1 The molecular orbital diagrams are as follows (only the valence electrons are shown):

(a)

—	σ_{2p}^*
— —	π_{2p}^*
—	σ_{2p}
— —	π_{2p}
⇅	σ_{2s}^*
—	σ_{2s}

(b)

—	σ_{2p}^*
— —	π_{2p}^*
—	σ_{2p}
— —	π_{2p}
↑	σ_{2s}^*
—	σ_{2s}

(c)

—	σ_{2p}^*
— —	π_{2p}^*
—	σ_{2p}
↑ —	π_{2p}
⇅	σ_{2s}^*
—	σ_{2s}

(a) Li_2 $b = \frac{1}{2}(2) = 1$; diamagnetic, no unpaired electrons

(b) Li_2^+ $b = \frac{1}{2}(1) = \frac{1}{2}$; paramagnetic, one unpaired electron

(c) Li_2^- $b = \frac{1}{2}(2 - 1) = \frac{1}{2}$; paramagnetic, one unpaired electron

2G.3 (a) (1) $(\sigma_{2s})^2(\sigma_{2s}^*)^2(\sigma_{2p})^2(\pi_{2p})^4(\pi_{2p}^*)^4(\sigma_{2p}^*)^1$; (2) $(\sigma_{2s})^2(\sigma_{2s}^*)^2(\sigma_{2p})^2(\pi_{2p})^4(\pi_{2p}^*)^3$; (3) $(\sigma_{2s})^2(\sigma_{2s}^*)^2(\sigma_{2p})^2(\pi_{2p})^4(\pi_{2p}^*)^4(\sigma_{2p}^*)^2$; (b) (i) 0.5; (ii) 1.5; (iii) 0; (c) (i) and (ii); (d) σ for (i) and (iii), π for (ii).

2G.5 The charge on C_2^{n-} is -2 and the bond order is 3.

2G.7 HeH^- would not be expected to exist because its bond order is 0; $b = \frac{1}{2}(2 - 2) = 0$. HeH^+, on the other hand, would have a bond order of 1 and thus be lower in energy.

2G.9 (a) The energy level diagram for N_2 is as follows:

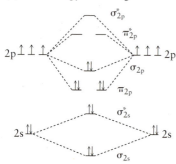

(b) The oxygen atom is more electronegative, which will make its orbitals lower in energy than those of N. Energy level diagram for NO^+:

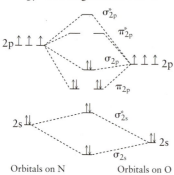

Orbitals on N   Orbitals on O

(c) Higher probability of being at O; O is more electronegative and its orbitals are lower in energy.

2G.11 (a) B_2 (6 valence electrons): $(\sigma_{2s})^2(\sigma_{2s}^*)^2(\pi_{2p})^2$, bond order = 1. (b) Be_2 (4 valence electrons): $(\sigma_{2s})^2(\sigma_{2s}^*)^2$, bond order = 0. (c) F_2 (14 valence electrons): $(\sigma_{2s})^2(\sigma_{2s}^*)^2(\sigma_{2p})^2(\pi_{2p})^4(\pi_{2p}^*)^4$, bond order = 1.

2G.13 CO $(\sigma_{2s})^2(\sigma_{2s}^*)^2(\pi_{2p})^4(\sigma_{2p})^2$; bond order = 3
CO^+ $(\sigma_{2s})^2(\sigma_{2s}^*)^2(\pi_{2p})^4(\sigma_{2p})^1$; bond order = 2.5
Due to the higher bond order for CO, it should form a stronger bond.

2G.15 (a)–(c) All are paramagnetic. B_2^- and B_2^+ each have one unpaired electron; B_2 has two unpaired electrons.

2G.17 (a) F_2 has a bond order of 1; F_2^- has a bond order of 1/2. F_2 will have the stronger bond. (b) B_2 has a bond order of 1; B_2^+ has a bond order of 1/2. B_2 will have the stronger bond.

2G.19 $C_2^+ = (\sigma_{2s})^2(\sigma_{2s}^*)^2(\pi_{2p})^3$; $C_2 = (\sigma_{2s})^2(\sigma_{2s}^*)^2(\pi_{2p})^4$; $C_2^- = (\sigma_{2s})^2(\sigma_{2s}^*)^2(\pi_{2p})^4(\sigma_{2p})^1$; C_2^- is expected to have the lowest ionization energy.

2G.21 $N = \sqrt{\dfrac{1}{2 + 2S}}$

2.1 Atoms for which no formal charge is shown are zero:

(a) [structures shown]

(b) $\left[\overset{+1}{\underset{..}{Br}} = \overset{..}{\underset{..}{O}} \right]^+$ (c) $\left[\overset{-1}{:C} \equiv \overset{-1}{C:} \right]^{2-}$

2.3 Iron(III) chloride would be expected to have the greater lattice energy.

2.5 [structures shown]

2.7 The first of the three resonance structures shown below is the most important Lewis structure (no two like charges are near one another):

$\overset{+1}{:N} \equiv N - \overset{-1}{\underset{..}{N}} - \overset{+1}{N} \equiv N:$ $\overset{-1}{\underset{..}{N}} = N = \overset{+1}{N} - \overset{+1}{N} \equiv N:$

$\overset{+1}{:N} \equiv N - \overset{+1}{\underset{..}{N}} = N = \overset{-1}{\underset{..}{N}}$

2.9 (a) $H-C \equiv C-H$ $H-C \equiv Si-H$

$H-Si \equiv Si-H$ $H-C \equiv N:$ $:N \equiv N:$

(b) The structure of benzene is shown in the box below:

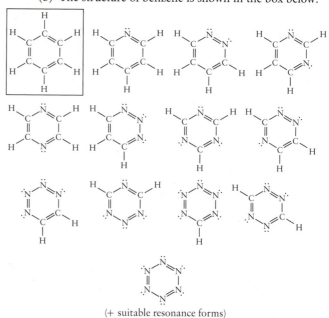

(+ suitable resonance forms)

2.11 (a)

$$\left[\begin{array}{c} \overset{\displaystyle :\!\ddot{O}\!:}{\underset{\displaystyle}{\|}} \\ H-C\overset{C}{\underset{C}{\diagdown}}\overset{C}{\underset{H}{\diagup}}H \\ \text{(ring of carbons with H's and bottom }:\ddot{O}:) \end{array}\right]^{2-}$$

A benzene-like ring: top carbon double-bonded to $:\ddot{O}:$, with H atoms on the four side carbons and bottom carbon bonded to $:\ddot{O}:$.

(b) All the atoms have formal charge 0 except the two oxygen atoms, which are −1. The negative charge is most likely to be concentrated at the oxygen atoms.   (c) The protons will bond to the oxygen atoms.

2.13 (a)

$$H-\ddot{\ddot{O}}-\ddot{N}=C=\ddot{O} \qquad H-\ddot{\ddot{O}}-N\overset{+1}{\equiv}C-\overset{-1}{\ddot{O}}\!:$$

$$H-\ddot{O}\overset{+1}{=}\ddot{N}-\overset{-1}{\ddot{C}}=\ddot{O}$$

(b)

$$\underset{H}{\overset{H}{\diagdown}}C=\ddot{S}=\ddot{O} \qquad \underset{H}{\overset{H}{\diagdown}}C=\overset{+1}{\ddot{S}}-\overset{-1}{\ddot{O}}\!:$$

(c)

$$\underset{H}{\overset{H}{\diagdown}}C=\overset{+1}{N}=\overset{-1}{\ddot{N}} \qquad \underset{H}{\overset{H}{\diagdown}}\overset{-1}{:}C-\overset{+1}{N}\equiv N:$$

(d)

$$\overset{+1}{\ddot{O}}=N=\overset{-1}{\ddot{N}} \qquad \ddot{O}=N-C\equiv N:$$

2.15 (a)

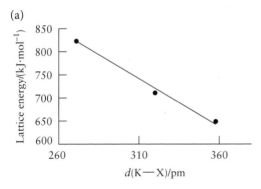

(b) From the graph, lattice energy $= -1.984\,d_{M-X} + 1356$. With $d_{K-Br} = 338$ pm, the lattice energy of KBr is about 693 kJ·mol^{-1}.   (c) The experimental value for the lattice energy for KBr is 689 kJ·mol^{-1} so the agreement is very good.

2.17

$$\begin{array}{c}
\text{R H H H R'} \\
\mid \ \ \mid \ \ \mid \ \ \mid \ \ \mid \\
H-C=C-\overset{.}{C}-C=C-H
\end{array}
\longleftrightarrow
\begin{array}{c}
\text{R H H H R'} \\
\mid \ \ \mid \ \ \mid \ \ \mid \ \ \mid \\
H-\overset{.}{C}-C=C-\overset{.}{C}=C-H
\end{array}$$

$$\searrow \qquad \begin{array}{c} \text{R H H H R'} \\ \mid \ \ \mid \ \ \mid \ \ \mid \ \ \mid \\ H-C=C-C=C-\overset{.}{C}-H \end{array} \qquad \nearrow$$

2.19 (a) I: Tl$_2$O$_3$; II: Tl$_2$O;   (b) +3; +1;   (c) [Xe]4f^{14}5d^{10}; [Xe]4f^{14}5d^{10}6s^2;   (d) compound II (the +3 ion is more polarizing)

2.21 (a)

$$\left[\begin{array}{c}:O:\\ \|\\ :\ddot{O}\underset{-1}{-}\overset{|+1}{Cl}=\ddot{O}\\ :\ddot{O}\!:_{-1}\end{array}\right]^{-} \left[\begin{array}{c}:O:\\ \|\\ \ddot{O}=\overset{|+1}{Cl}-\overset{-1}{\ddot{O}}\!:\\ :\ddot{O}\!:_{-1}\end{array}\right]^{-} \left[\begin{array}{c}\overset{-1}{:}\ddot{O}:\\ \|\\ \ddot{O}=\overset{\|+1}{Cl}-\underset{-1}{\ddot{O}}\\ :O:\end{array}\right]^{-} \left[\begin{array}{c}\overset{-1}{:}\ddot{O}:\\ \|\\ :\underset{-1}{\ddot{O}}-\overset{\|+1}{Cl}=\ddot{O}\\ :O:\end{array}\right]^{-}$$

$$\left[\begin{array}{c}:O:\\ \|\\ :\ddot{O}\underset{-1}{-}\overset{\|+1}{Cl}-\overset{-1}{\ddot{O}}\!:\\ :O:\end{array}\right]^{-} \left[\begin{array}{c}\overset{-1}{:}\ddot{O}:\\ |\\ \ddot{O}=\overset{|+1}{Cl}=\ddot{O}\\ :O:_{-1}\end{array}\right]^{-}$$

$$\left[\begin{array}{c}:O:\\ \|\\ \ddot{O}=Cl=\ddot{O}\\ :O:_{-1}\end{array}\right]^{-} \left[\begin{array}{c}:O:\\ \|\\ \ddot{O}=Cl-\ddot{O}\!:\\ :O:\end{array}\right]^{-} \left[\begin{array}{c}\overset{-1}{:}\ddot{O}:\\ |\\ \ddot{O}=Cl=\ddot{O}\\ :O:\end{array}\right]^{-} \left[\begin{array}{c}:O:\\ \|\\ :\ddot{O}\underset{-1}{-}Cl=\ddot{O}\\ :O:\end{array}\right]^{-}$$

$$\left[\begin{array}{c}:O:\\ \overset{-1}{:}\ddot{O}\overset{\|}{=}Cl=\ddot{O}\\ \|\\ :O:\end{array}\right]^{-} \left[\begin{array}{c}\overset{-1}{:}\ddot{O}:\\ |\\ :\ddot{O}-Cl-\ddot{O}\!:^{-1}\\ :O:_{-1}\overset{|+3}{}\end{array}\right]^{-}$$

$$\left[\begin{array}{c}\overset{-1}{:}\ddot{O}:\\ |\\ :\ddot{O}\underset{-1}{-}\overset{|+2}{Cl}=\ddot{O}\\ :O:_{-1}\end{array}\right]^{-} \left[\begin{array}{c}:O:\\ |\\ :\ddot{O}\underset{-1}{-}\overset{|+2}{Cl}-\ddot{O}\!:^{-1}\\ :O:_{-1}\end{array}\right]^{-} \left[\begin{array}{c}\overset{-1}{:}\ddot{O}:\\ |\\ \ddot{O}=\overset{|+2}{Cl}-\ddot{O}\!:^{-1}\\ :O:_{-1}\end{array}\right]^{-} \left[\begin{array}{c}\overset{-1}{:}\ddot{O}:\\ |\\ :\ddot{O}\underset{-1}{-}\overset{\|+2}{Cl}-\ddot{O}\!:^{-1}\\ :O:\end{array}\right]^{-}$$

The four structures with three double bonds (third row) and the one with four double bonds are the most plausible Lewis structures.   (b) The structure with four double bonds.   (c) +7; The structure with all single bonds fits this criterion best.   (d) Approaches (a) and (b) are consistent but approach (c) is not because oxidation numbers are assigned by assuming ionic bonding.

2.23 (a) The S—S bond has some double bond character. (b) and (c) The Lewis structures for the two possible S$_2$F$_2$ are:

$$\boxed{:\ddot{F}-\ddot{S}-\ddot{S}-\ddot{F}\!:} \longleftrightarrow :\ddot{F}-\overset{+1}{\ddot{S}}=\overset{-1}{\ddot{S}}-\ddot{F}\!: \quad \textit{isomer 1}$$
$$\text{favored}$$

$$\underset{:\ddot{F}:}{\overset{:\ddot{F}\overset{+1}{\diagdown}_{-1}}{S-\ddot{S}:}} \longleftrightarrow \boxed{\underset{:\ddot{F}:}{\overset{:\ddot{F}\diagdown}{S=\ddot{S}}}} \quad \textit{isomer 2}$$
$$\text{favored}$$

Because of resonance it is expected that the S—S bond length is between a single and a double bond in length.

2.25 (a) The CN bond in CH$_3$NH$_2$ would be expected to be longer.   (b) The PF bond in PF$_3$ would be expected to be longer.

2.27 (a) The Lewis structures are:

$$\left[\underset{H}{\overset{H}{\underset{|}{\overset{|}{H-C}}}}\right]^{+} \quad \underset{H}{\overset{H}{\underset{|}{\overset{|}{H-C-H}}}} \quad \left[\underset{H}{\overset{H}{\underset{|}{\overset{|}{H-C}}}}\!:\right]^{-} \quad \left[\underset{H}{\overset{H}{\underset{|}{\overset{|}{C}}}}\!:\right]^{2+} \quad \left[\underset{H}{\overset{H}{\underset{|}{\overset{|}{:C}}}}\!:\right]^{2-}$$

(b) None are radicals. (c) Based on the Lewis structure and VSEPR theory the order of increasing HCH bond angle is $CH_2^{2-} < CH_3^- < CH_4 < CH_2 < CH_3^+ < CH_2^{2+}$

2.29 (a) Empirical formula = CH_4O; this corresponds to the compound methanol (H_3C-OH). All bond angles around carbon are 109.5°; bond angles around oxygen are somewhat less than 109.5°; (b) Both carbon and oxygen are sp^3 hybridized. (c) The molecule is polar.

2.31
Bonding   Antibonding

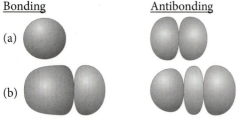

(a)

(b)

(c) The bonding and antibonding orbitals for HF appear different due to the fact that a p-orbital from the F atom is used to construct the orbitals.

2.33 (a) $CF^+ < CF < CF^-$; (b) CF^+ ion will be diamagnetic

2.35 The Lewis structure of borazine is:

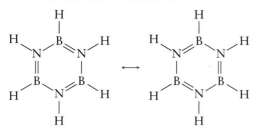

The orbitals at each B and N atom will be sp^2 hybridized.

2.37 The antibonding orbital is proportional to
$\psi \propto e^{-r/a_0} - e^{-r/a_0} = 0$.

2.39 (a)   σ   π   δ

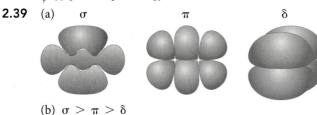

(b) $\sigma > \pi > \delta$

2.41 (a) Bond order changes from 3 to 2. (b) Bond order changes from 3 to 2. Both ions will be paramagnetic.

2.43 (a) The Lewis structure of benzyne is:

H
$\mid$
H$\diagdown$ sp^2
   sp^2 C
    C=$\qquad$C sp
$\mid\qquad\qquad\mid\mid\mid$
    C=C sp
H sp^2 C sp^2
   $\mid$
   H

(b) Benzyne would be highly reactive because the two carbon atoms that are sp hybridized are constrained to have a very strained structure compared to what their hybridization would like to adopt—namely a linear arrangement. Instead of 180° angles at these carbon atoms, the angles by necessity of being in a six-membered ring are constrained to be close to 120°.

2.45 (a–b) All atoms in this molecule have a formal charge of zero.

O sp^2
$\sigma(C2sp^2,C2sp^2)$
        $\sigma(C2sp^2,O2sp^2)$ and
        $\pi(C2p,O2p)$
C—H
$\sigma(C2sp^2,H1s)$ C=C $\sigma(C2sp^2,H1s)$
$\sigma(C2sp^2,C2sp^2)$ and
$\pi(C2p,C2p)$

2.47 (a)
$$\left[\begin{array}{c}:\!\ddot{F}\!:\quad^{+1}\quad:\!\ddot{F}\!:\\:\!\ddot{F}\diagdown\!\!\mid\!\!\diagup\ddot{F}\diagdown\!\!\mid\!\!\diagup\ddot{F}\!:\\\quad\ddot{Sb}\quad\quad\ddot{Sb}\\:\!\mid_{-1}\quad_{-1}\!\mid:\\:\!\ddot{F}\!:\quad\quad:\!\ddot{F}\!:\end{array}\right]^-$$

(b) Each Sb atom in this structure has a hybridization of sp^3d.

2.49
$$\left[\begin{array}{c}^{+1}\\:\!\ddot{Cl}\!:\\:\!\ddot{Cl}-Bi\quad\quad Bi-\ddot{Cl}\!:\\_{-2}\diagup\ddot{Cl}\diagdown_{-2}\\:\!\ddot{Cl}\!:\\^{+1}\end{array}\right]^{2-}$$

2.51
$$\left[\begin{array}{c}\quad\quad a\\\quad\quad:\!\ddot{I}\!:\\\quad b\diagup\\:\!\ddot{I}\diagdown\,c\diagup\ddot{I}\!:\\:\!\ddot{I}\diagdown\,\ddot{I}\!:\\\quad\quad d\diagdown\\\quad\quad:\!\ddot{I}\!:\\\quad\quad e\end{array}\right]^-$$

I_b and I_d have a formal charge of -1 and I_c has a formal charge of $+1$. I_b and I_d are sp^3d hybridized and I_c is sp^3 hybridized.

2.53 (a) bond order changes from 2 to 1.5; longer bond; paramagnetic; (b) bond order changes from 3 to 2.5; longer bond; paramagnetic; (c) bond order changes from 2 to 2.5; a shorter bond; paramagnetic.

2.55 (a) sp^3d^3; (b) sp^3d^3f; (c) sp^2d

2.57 All are σ-bonds, except the triple bond, which is one σ- and two π-bonds.

H$\quad$H$\quad$180°
  $sp^3$$\longrightarrow$C$\diagup$C$\equiv$N:
H$\quad$109.5°
          sp

2.59 (a) H$-\overset{H}{\underset{H}{C}}-\ddot{S}-$H (b) $\ddot{S}=C=\ddot{S}$ (c) $:\ddot{Cl}-\overset{H}{\underset{H}{C}}-\ddot{Cl}:$

2.61 (a) The Lewis structure for HOCO is:
H$-\ddot{O}-C=\ddot{O}$

(b) It is a radical because the C has an unpaired electron on it.

2.63 Angles a and c are expected to be approximately 120°. Angle b is expected to be around 109.5°.

2.65 (a) The Lewis structures of NO and NO_2 are:

$\dot{N}=\ddot{O}$ (best possible structure)

$:\overset{-1}{\ddot{O}}-\overset{+1}{\ddot{N}}=\ddot{O}\quad\longleftrightarrow\quad\ddot{O}=\overset{+1}{\ddot{N}}-\overset{-1}{\ddot{O}}:$

(equivalent resonance structures)

From Table 2D.2, the average bond dissociation energy of a N=O bond is 630 kJ·mol^{-1}, which is right in line with the Lewis structure of NO. The bond dissociation energy of each NO bond in NO$_2$ is 469 kJ·mol^{-1}, which is about halfway between a N—O double and an N—O single bond, suggesting that the resonance picture of NO$_2$ is a reasonable one. (b) An N—O single bond should have a bond length of 149 pm, whereas an N—O double bond should have a bond length of 120 pm (values are obtained by summing the respective covalent radii values from Figure 2D.11). From Table 2D.3, the length of an N—O triple bond can be estimated to be between 105 and 110 pm. Because NO itself has a bond length of 115 pm, this suggests that its actual bond order is somewhere between that of a double and a triple bond.

(c) (d)

(e) $N_2O_5(g) + H_2O(l) \rightarrow 2\ HNO_3(aq)$; nitric acid is produced

(f) $4.05\ g\ N_2O_5 \times \dfrac{1\ mol\ N_2O_5}{108.02\ g\ N_2O_5} \times \dfrac{2\ mol\ HNO_3}{1\ mol\ N_2O_5}$

$= 7.50 \times 10^{-2}\ mol\ HNO_3$

Molarity $= \dfrac{7.50 \times 10^{-2}\ mol\ HNO_3}{1.00\ L}$

$= 7.50 \times 10^{-2}\ mol·L^{-1}\ HNO_3$

(g) The oxidation numbers of nitrogen for the various nitrogen oxides are: NO: +2; NO$_2$: +4; N$_2$O$_3$: +3; and N$_2$O$_5$: +5. An oxidizing agent is a species that gains electrons; based on oxidation number, N$_2$O$_5$ should be the most potent oxidizing agent of the nitrogen oxides, as it possesses the most positive oxidation number for N of the group.

Focus 3

3A.1 (a) 8×10^9 Pa; (b) 80 kbar; (c) 6×10^7 Torr; (d) 1×10^6 lb·in^{-2}

3A.3 Pressure is the total weight of the air in a column divided by the area of the base of the column. At a higher altitude, the weight of air above a region is lower.

3A.5 (a) 86 Torr; (b) The side attached to the bulb will be higher. (c) 848 Torr

3A.7 924 cm

3A.9 2.9×10^3 lb

3B.1 (a) 22.4 in; (b) The pressure of gas in tube (1) is 41.85 inHg. The pressure of gas in tube (2) is 59.85 inHg.

3B.3 (a) 20. L; (b) When additional mole of gases is added into the same volume, the pressure will be doubled at each T in (a). The plot is

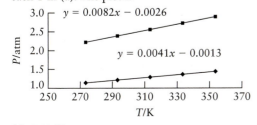

(c) 0.32 K

3B.5 (a) 1.06×10^3 kPa; (b) 4.06×10^2 m L; (c) $m = 2.0 \times 10^2$ g; (d) $n = 3.24 \times 10^5$ mol CH$_4$

3B.7 (a)

Volume, L	$(nR/V)/(\text{atm·K}^{-1})$
0.01	8.21
0.02	4.10
0.03	2.74
0.04	2.05
0.05	1.64

(b) The slope is equal to nR/V; (c) The intercept is equal to 0.00 for all the plots.

3B.9 (a) 1.5×10^3 kPa; (b) 4.5×10^3 Torr

3B.11 10.7 L

3B.13 The volume must be increased by 10.% to keep P and T constant.

3B.15 0.050 g

3B.17 (a) 11.8 mL; (b) 0.969 atm; (c) 199 K

3B.19 (a) 63.4 L·mol^{-1}; (b) 6.3 L·mol^{-1}

3B.21 18.1 L

3B.23 4.18 atm

3B.25 621 g

3B.27 2.52×10^{-3} mol

3B.29 (a) 3.6 m^3; (b) 1.8×10^2 m^3

3B.31 NH$_3$ < N$_2$ < N$_2$H$_4$

3B.33 615 K or 342 °C

3B.35 (a) 1.28 g·L^{-1}; (b) 3.90 g·L^{-1}

3B.37 (a) 70. g·mol^{-1}; (b) The compound is most likely CHF$_3$. (c) 2.9 g·L^{-1}

3B.39 $C_2H_2Cl_2$

3B.41 44.0 g·mol^{-1}

3C.1 (a) $x_{HCl} = 0.9$; $x_{benzene} = 1 - 0.9 = 0.1$; (b) $P_{HCl} = 0.72$ atm; $P_{benzene} = 0.08$ atm

3C.3 (a) The partial pressure of N$_2$(g) is 1.0 atm; the partial pressure of H$_2$ (g) is 2.0 atm. (b) The total pressure is 3.0 atm.

3C.5 $P_{O_2} = 2.0$ bar

3C.7 (a) 739.2 Torr; (b) $H_2O(l) \rightarrow H_2(g) + \frac{1}{2}O_2(g)$; (c) 0.142 g

3C.9 (a) 3.0×10^5 L; (b) 1.0×10^4 L

3C.11 (b)

3C.13 (a) 97.8 L; (b) 32.5 L; (c) 19.6 L

3C.15 (a) 4.25×10^{-3} g; (b) HCl left after the reaction; $P = 0.0757$ atm

3D.1 no

3D.3 $C_{10}H_{10}$

3D.5 (a) time = 154 s; (b) time = 123 s; (c) time = 33.0 s; (d) time = 186 s

3D.7 molar mass = 110. g·mol^{-1}; the molecular formula is C_8H_{12}

3D.9 (a) 627 m·s^{-1}; (b) 458 m·s^{-1}; (c) 378 m·s^{-1}

3D.11 241 m·s^{-1}

3D.13 271 K

3D.15 0.316

3D.17 (a) The most probable speed is the one that corresponds to the maximum on the distribution curve. (b) The percentage of molecules having the most probable speed decreases as the temperature is raised.

3D.19 $T = 16 \times 200. \, K = 3.20 \times 10^3 \, K$

3E.1 Hydrogen bonding is important in HF.

3E.3 (a) H_2 molecules have greater root mean square speed. (b) NH_3 will have stronger deviation from ideality.

3E.5 (a) 1.63 atm (ideal gas); 1.62 atm (real gas); (b) 48.9 atm (ideal gas); 39.0 atm (real gas); (c) 489 atm (ideal gas); 2.00×10^3 atm (real gas). At low pressures, the ideal gas law gives essentially the same values as the van der Waals equation, but at high pressures there are very significant differences.

3E.7

$a/(bar \cdot L^2 \cdot mol^{-2})$	Substance
17.813	CH_3CN
3.700	HCl
2.303	CH_4
0.208	Ne

3E.9 (a) 1.00 atm; (b) (i) 0.993 atm (ii) 1.09 atm

3E.11 Ammonia: $a = 4.169 \, L^2 \cdot atm \cdot mol^{-2}$; $b = 0.0371 \, L \cdot mol^{-1}$; Oxygen: $a = 1.364 \, L^2 \cdot atm \cdot mol^{-2}$; $b = 0.0319 \, L \cdot mol^{-1}$

Volume/L	$P_{ammonia}$/atm	P_{oxygen}/atm	P_{ideal}/atm
0.05	228.05	805.44	489.08
0.2	45.89	111.37	122.27
0.4	41.33	57.91	61.13
0.6	31.86	39.26	40.76
0.8	25.54	29.71	30.57
1	21.23	23.90	24.45

3E.13 (a) $3.95 \times 10^7 \, pm^3 \cdot atom^{-1}$; $r = 211$ pm; (b) $8.78 \times 10^6 \, pm^3$; (c) The difference in these values illustrates that there is no easy definition for the boundaries of an atom.

3F.1 (a) London forces, dipole–dipole, hydrogen bonding; (b) London forces; (c) London forces, dipole–dipole, hydrogen bonding; (d) London forces, dipole–dipole

3F.3 Only (b) CH_3Cl, (c) CH_2Cl_2, and (d) $CHCl_3$ will have dipole–dipole interactions. The molecules CH_4 and CCl_4 do not have permanent dipole moments.

3F.5 (a) NaCl; (b) butanol; (c) triiodomethane; (d) methanol

3F.7 (a) PF_3 and PBr_3 are both trigonal pyramidal, but PBr_3 has the greater number of electrons and should have the higher boiling point. (b) SO_2 and O_3 are bent and have dipole moments, but SO_2 has the higher boiling point. (c) BF_3 and BCl_3 are both trigonal planar. BCl_3 has the higher boiling point.

3F.9 The order is (b) dipole-induced dipole $\cong$ (c) dipole–dipole in the gas phase $<$ (e) dipole–dipole in the solid phase $<$ (a) ion–dipole $<$ (d) ion–ion.

3F.11 only HNO_2

3F.13 II, because the dipoles are aligned with oppositely charged ends closest to each other, thereby maximizing dipole–dipole attractions.

3F.15 AsF_3 is a polar molecule, whereas AsF_5 is not. So the dipole–dipole intermolecular forces in AsF_3 are stronger than the London (dispersion) forces in AsF_5, which results in a higher boiling point of AsF_3.

3F.17 Al^{3+} ion will attract water molecule more strongly than Be^{2+} ion.

3F.19 (a) Xenon is larger. (b) Hydrogen bonding in water causes the molecules to be held together more tightly than in diethyl ether. (c) Pentane is a linear molecule compared to dimethylpropane, which is a compact, spherical molecule. The compactness of the dimethylpropane gives it a lower surface area.

3F.21 $F = \dfrac{-dE_p}{dr} = \dfrac{-d}{dr}\left(\dfrac{1}{r^6}\right) = -\left(\dfrac{-6}{r^7}\right) \propto \dfrac{1}{r^7}$

3G.1 (a) As intermolecular forces increase, the boiling point will increase. (b) Viscosity is likely to increase with an increase in intermolecular forces. (c) Surface tension also increases with an increase in intermolecular forces.

3G.3 (a) cis-Dichloroethene has the greater intermolecular forces and the greater surface tension; (b) Benzene at 20 °C has a higher surface tension.

3G.5 C_6H_6 (nonpolar) $< C_6H_5SH$ (polar, but no hydrogen bonding) $< C_6H_5OH$ (polar and with hydrogen bonding).

3G.7 CH_4, -162 °C; CH_3CH_3, -88.5 °C; $(CH_3)_2CHCH_2CH_3$, 28 °C; $CH_3(CH_2)_3CH_3$, 36 °C; CH_3OH, 64.5 °C; CH_3CH_2OH, 78.3 °C; $CH_3CHOHCH_3$, 82.5 °C; C_5H_9OH (cyclic, but not aromatic), 140 °C; $C_6H_5CH_3OH$ (aromatic ring), 205 °C; $OHCH_2CHOHCH_2OH$, 290 °C

3G.9 Due to the silanol group (Si—OH) on the glass tube wall, water molecules that are close to the glass tube wall will interact strongly with the silanol group by forming H-bonds and climbing up the glass wall to form a concave meniscus. Water molecules do not have such interactions with a plastic tube wall due to the hydrophobic character of the plastics, so water molecules will interact with each other through H-bonding and form a convex meniscus.

3G.11 Water will rise to a higher level than ethanol.

3G.13 There are too many ways that these molecules can rotate and twist so that they do not remain rodlike.

3G.15 Because benzene is an isotropic solvent and its viscosity is the same in every direction.

3G.17 Substance (a), $C_5H_6N^+Cl^-$, would be a better choice as an ionic liquid solvent.

3H.1 (a) Glucose will be held in the solid by London forces, dipole–dipole interactions, and hydrogen bonds; benzophenone will be held in the solid by dipole–dipole interactions and London forces. (b) (m.p. = 48 °C) $<$ glucose (m.p. = 148−155 °C).

3H.3 (a) network; (b) ionic; (c) molecular; (d) molecular; (e) network

3H.5 Substance A: ionic; substance B: metallic; substance C: molecular solid

3H.7 (a) 2 atoms; (b) coordination number is 8; (c) 286 pm

3H.9 (a) $d = 2.72 \, g \cdot cm^{-3}$; (b) $d = 0.813 \, g \cdot cm^{-3}$

3H.11 (a) 138.7 pm; (b) 143.1 pm

3H.13 Rhodium metal is close-packed structure.

3H.15 (a) mass of one unit cell = 3.73×10^{-22} g; (b) 8 atoms per unit cell

3H.17 90.7%

3H.19 In diamond, carbon is sp^3 hybridized and forms a tetrahedral, three-dimensional network structure, which is extremely rigid. Graphite carbon is sp^2 hybridized and planar.

3H.21 (a) 1.1×10^{16} C atoms; (b) 18 nmol

3H.23 (a) Indium arsenide has a (4,4)-coordination. (b) The formula of indium arsenide is InAs.

3H.25 (a) one Cs^+ and one Cl^- in each unit cell and one formula unit, CsCl, per unit cell; (b) two titanium atoms and four oxygen atoms per unit cell and two formula units, TiO_2, per unit cell; (c) The Ti atoms are 6-coordinate and the O atoms are 3-coordinate.

3H.27 (a) Rhenium oxide has a (6,2)-coordination. (b) The formula of rhenium oxide is ReO_3.

3H.29 (a) cesium-chloride structure with (8,8) coordination; (b) rock-salt structure with (6,6) coordination; (c) rock-salt structure with (6, 6) coordination

3H.31 (a) $d = 3.36$ g·cm^{-3}; (b) $d = 4.80$ g·cm^{-3}

3H.33 (a) One instance of the smallest possible rectangular unit cell is:

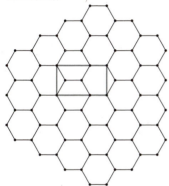

(b) There are four carbon atoms in each unit cell.
(c) The coordination number is 3 (C_3).

3H.35 3.897×10^{10} pm = 3.897 cm

3I.1 $d = 0.9703$ g·cm^{-3}

3I.3 Alloys are usually (1) harder and more brittle, and (2) poorer conductors of electricity than the metals from which they are made.

3I.5 (a) The alloy is interstitial because the atomic radius of nitrogen is much smaller (74 pm vs. 124 pm) than that of iron. (b) Nitriding will make iron harder and stronger, with a lower electrical conductivity.

3I.7 (a) 2.8 Cu per Ni; (b) The atom ratio will be 15 Sn : 1.2 Sb : 1 Cu.

3I.9 (a) It belongs to a hexagonal crystal system. (b) The total number of carbonate ions is six in the unit cell. The total number of calcium ions in the cell is six. The formula is calcite, $CaCO_3$.

3I.11 $V = 15.4$ cm^3

3I.13

$$\left[\begin{array}{c} \vdots\ddot{O}\text{:} \\ | \\ \text{:}\ddot{O}-Si-\ddot{O}\text{:} \\ | \\ \text{:}\ddot{O}\text{:} \end{array} \right]^{4-}$$

formal charge: Si = 0, O = −1; oxidation number: Si = +4, O = −2. This is an AX_4 VSEPR structure; therefore, the shape is tetrahedral.

3I.15 The $Si_2O_7^{6-}$ ion is built from two SiO_4^{4-} tetrahedral ions in which the silicate tetrahedrals share one O atom. (b) The pyroxenes, for example, jade, $NaAl(SiO_3)_2$, consist of chains of SiO_4 units in which two O atoms are shared by neighboring units.

3J.1 The conductivity of a semiconductor increases with temperature as increasing numbers of electrons are promoted into the conduction band, whereas the conductivity of a metal will decrease as the motion of atoms will slow down the migration of electrons.

3J.3 $n = 4.2 \times 10^6$

3J.5 (a) In and Ga; (b) P and Sb

3J.7 (a) In fluorescence, light absorbed by molecules is immediately emitted, whereas in phosphorescence, molecules remain in an excited state for a period of time before emitting the absorbed light. (b) Mechanistically, fluorescence involves the retention of the relative spin orientation of the excited electron, whereas phosphorescence involves changes of electron spin state so relaxation is a slow process.

3J.9 Paramagnetic elements have at least one unpaired electron.

Sc = $3d^14s^2 \rightarrow$ paramagnetic	Ti = $4s^23d^2 \rightarrow$ paramagnetic
V = $3d^34s^2 \rightarrow$ paramagnetic	Mn = $4s^23d^5 \rightarrow$ paramagnetic
Fe = $3d^64s^2 \rightarrow$ ferromagnetic	Co = $4s^23d^7 \rightarrow$ ferromagnetic
Ni = $4s^23d^8 \rightarrow$ ferromagnetic	Y = $4d^15s^2 \rightarrow$ paramagnetic
Zr = $4d^25s^2 \rightarrow$ paramagnetic	Nb = $4d^45s^1 \rightarrow$ paramagnetic
Mo = $4d^55s^1 \rightarrow$ paramagnetic	Tc = $4d^55s^2 \rightarrow$ paramagnetic
Ru = $4d^75s^1 \rightarrow$ paramagnetic	Rh = $4d^85s^1 \rightarrow$ paramagnetic
Lu = $6s^24f^{14}5d^1 \rightarrow$ paramagnetic	Hf = $6s^24f^{14}5d^2 \rightarrow$ paramagnetic
Ta= $6s^24f^{14}5d^3 \rightarrow$ paramagnetic	W = $6s^24f^{14}5d^4 \rightarrow$ paramagnetic
Re = $6s^2 4f^{14}5d^5 \rightarrow$ paramagnetic	Os = $6s^24f^{14}5d^6 \rightarrow$ paramagnetic;
Ir = $6s^24f^{14}5d^7 \rightarrow$ paramagnetic	Pt = $4f^{14}5d^96s^1 \rightarrow$ paramagnetic;
Au = $4f^{14}5d^{10}6s^1 \rightarrow$ paramagnetic	Zn = $3d^{10}4s^2 \rightarrow$ diamagnetic

More information can be found at: http://www.periodictable.com/Properties/A/MagneticType.html

3J.11 The compound is ferromagnetic below T_C because the magnetization is higher.

3J.13 (a) molarity of QDs = 4.5×10^{-6} M

3J.15 $E_{111} = \dfrac{h^2}{8m_eL^2}(1^2 + 1^2 + 1^2) = \dfrac{3h^2}{8m_eL^2}$

Similarly, the other two energy levels are determined to be

$E_{211} = \dfrac{h^2}{8m_eL^2}(2^2 + 1^2 + 1^2) = \dfrac{3h^2}{4m_eL^2}$

and

$E_{221} = \dfrac{h^2}{8m_eL^2}(2^2 + 2^2 + 1^2) = \dfrac{9h^2}{8m_eL^2}$

The 211, 121, and 112 levels are degenerate; that is $E_{211} = E_{121} = E_{112}$; and the 221, 122, and 212 levels are degenerate; that is $E_{221} = E_{122} = E_{212}$.

3.1 (a) P_{Ar} = 280. Torr; (b) P_{total} = 700. Torr

3.3 2.4×10^4 L

3.5 0.481 M

3.7 (a) $N_2O_4(g) \rightarrow 2 NO_2(g)$; (b) 2.33 atm; (c) 4.65 atm; (d) x_{NO_2} = 0.426; $x_{N_2O_4}$ = 0.574

3.9   (a) The molecular formula is N_2H_4.

(b)
$$H-\overset{\displaystyle H}{\underset{\displaystyle H}{N}}-\overset{\displaystyle H}{N}-H$$

(c) 4.2×10^{-4} mol

3.11   The ratio is not independent of temperature.

3.13 254 g·mol^{-1}; the formula is OsO_4

3.15   (a) A, 300 K; B, 500 K; C, 1000 K;   (b) $v_{rms} = 667$ m·s^{-1}

3.17   (a) $ClNO_2$;   (b) $ClNO_2$;

(c)
$$\overset{\displaystyle :\ddot{Cl}:}{\underset{}{\ddot{O}=N-\ddot{O}:}} \qquad \overset{\displaystyle :\ddot{Cl}:}{\underset{}{:\ddot{O}-N=\ddot{O}}}$$

(d) trigonal planar

3.19   (a) $V = 0.0153$ L;   (b) attractive forces dominate

3.21   (a) $P_{CO_2} = 0.556$ atm; $P_{N_2} = 0.834$ atm; $P_{H_2O} = 1.11$ atm.

(b) The partial pressure of $N_2(g)$ and $H_2O(g)$ will remain the same as those in part (a).

3.23   (a) 138 g NaN_3;   (b) $v_{rms} = 515$ m·s^{-1}

3.25   (a) The NI_3 Lewis structure is

The molecular shape is trigonal pyramidal and it is a polar molecule. It can participate in dipole–dipole interactions.

(b) The BI_3 Lewis structure is

The molecular shape is trigonal planar and it is a nonpolar molecule. It cannot participate in dipole–dipole interactions.

3.27   (a) Surface area $= 1.92 \times 10^6$ pm^2;   (b) Pentane has the larger surface area.   (c) The pentane should have the higher boiling point.

3.29 $r/\text{pm} = \{2.936 \times 10^5 \times M/(\text{g·mol}^{-1})/(d)\}^{1/3}$
where M is the atomic mass in g·mol^{-1} and r is the radius in pm. For the different gases we calculate the results given in the following table:

Gas	Density/ (g·cm^3)	Molar mass/ (g·mol^{-1})	Radius/pm
Neon	1.20	20.18	170
Argon	1.40	39.95	203
Krypton	2.16	83.80	225
Xenon	2.83	131.30	239
Radon	4.4	222	246

3.31 21.0 g·cm^{-3}

3.33   (a) The average oxidation state of Ti is $+2.36$.

(b) The fraction of $Ti^{2+} = 0.64$ and the fraction of $Ti^{3+} = 0.36$.

3.35   (a) true;   (b) false;   (c) true;   (d) false

3.37   (a)

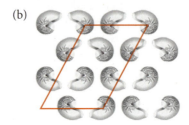

(b)

3.39   (a) $r = 50.2$ pm;   (b) K^+ must occupy octahedral holes. The percentage of holes filled is 75%.

3.41   The empirical formula is MA_2.

3.43   (a) For pure metals, the heat capacity per atom for each metal is very similar, which means that the Dulong–Petit law applies well here (a good approximation).   (b) The heat capacities per mole of atoms are different for different ionic crystals.   (c) Heat capacity per mole of atoms for $CuSO_4 = 16.67$ J·K^{-1}·mol^{-1}; heat capacity per mole of atoms for $PbSO_4 = 17.20$ J·K^{-1}·mol^{-1}.

3.45   (a) 52% of the space is occupied, so 48% of this unit cell would be empty.   (b) The percentage of empty space in an fcc unit cell is 26%, so the fcc cell is much more efficient at occupying the space available.

3.47   The Lewis structure of $(HF)_3$ chain is

$$H-\ddot{\ddot{F}}:\cdots H-\ddot{\ddot{F}}:\cdots H-\ddot{\ddot{F}}:$$

The bond angle is ~180°.

3.49 38.1 g Fe_2O_3 is the maximum mass that can be produced.

3.51   (a) They are classified as salts based on classical definition (M^{3+} and N^{3-}; they are called interstitial nitrides).   (b) The covalent character of MN is shown as the triple bond between nitrogen (N) and d-block metal (M);   (c) because Group 11 and 12 metals have their d-orbitals filled with electrons and do not have space to hold the electrons from N.

3.53   formula $= YBa_2Cu_3O_7$

3.55 $SiC(s) + 2\,NaOH(l) + 2\,O_2(g) \rightarrow$
$$Na_2SiO_3(s) + H_2O(g) + CO_2(g)$$

3.57   Plot (a) was obtained from an organic dye and plot (b) was obtained from a quantum dot suspension.

3.59   When semiconductors are at very low temperatures, the valence band is fully occupied and the conduction band is empty, so the conductivity is low. As the temperature increases in a vacuum, some electrons are promoted from the valence band to the conduction band and cause the conductivity to increase. However, when semiconductors are heated in oxygen, chemisorption will happen, and adsorbed oxygen on the surface will trap electrons, forming oxygen anions, which will decrease the conductivity.

3.61 (a) The oxidation number of phosphorus in $Li_7P_3S_{11}$ is +5.
(b) The oxidation number of titanium in $BaTiO_3$ is +4.

3.63 (a) crystalline phase; (b) crystalline phase;
(c) crystalline phase; (d) glassy phase

3.65 The oxidation number of silicon in the anion is +4.

3.67 Ionic fluorides react with water to liberate HF, which then reacts with the glass.

3.69 (a) $4 FeS_2(s) + 11 O_2(g) \rightarrow 2 Fe_2O_3(s) + 8 SO_2(g)$;
(b) 146 g Fe_2O_3; (c) 3.66 mol of SO_2 will dissolve in 5.00 L of water to form a solution that is (3.66 mol ÷ 5.00 L) = 0.732 M in H_2SO_3, according to the equation
$SO_2(g) + H_2O(l) \rightarrow H_2SO_3(aq)$; (d) 50 kg SO_2;
(e) 2×10^4 L; (f) (i) 0.787 atm; (ii) 1.03 g. (g) By using ideal gas law and van der Waals equation to calculate the pressures for different amounts of SO_2, the results are listed in the following table:

Moles of SO_2	P(ideal)/atm	P(vdw)/atm	Percentage deviation
0.100	2.4618	2.407213	2.267647
0.200	4.9236	4.705575	4.633339
0.300	7.3854	6.895579	7.103409
0.400	9.8472	8.97773	9.684746
0.500	12.309	10.95254	12.38485

(h) The percentage deviation is calculated and included in the table in (g). (i) Intermolecular attraction (a term) has larger effect on the pressure of SO_2. (j) Based on the data from the table in (g), when $P \approx 5$ atm, SO_2 becomes a "real" gas. To be exact, when the observed $P = 5.04$ atm, SO_2 becomes a "real" gas.

Focus 4

4A.1 (a) isolated; (b) closed; (c) isolated;
(d) open; (e) closed; (f) open

4A.3 (a) 28 J; (b) positive; (c) 8 J

4A.5 reversible expansion

4A.7 (a) 1.12×10^2 kJ; (b) 89%

4A.9 25 °C

4A.11 14.8 kJ·(°C)$^{-1}$

4A.13 −1.19 kJ

4B.1 864 kJ

4B.3 (a) on; (b) 90×10^2 J

4B.5 5.35 kJ

4B.7 −1626 kJ

4B.9 (a) true if no work is done; (b) always true;
(c) always false; (d) true only if $w = 0$; (e) always true

4B.11 (a) absorbed, by the system, q is positive, w is negative;
(b) released, on the system, q is negative, w is positive

4B.13 (a) −226 J; (b) −326 J

4B.15 7.44 kJ·mol^{-1}

4C.1 NO_2. The heat capacity increases with molecular complexity—as more atoms are present in the molecule, there are more possible bond vibrations that can absorb added energy.

4C.3 373 K

4C.5 (a) 5/2 R; (b) 3R; (c) 3/2 R; (d) 5/2 R

4C.7 (a) 8.22 kJ·mol^{-1} (b) 43.5 kJ·mol^{-1}

4C.9 90%

4C.11 33.4 kJ

4C.13 31 °C

4C.15 (c)

4D.1 (a) 448 kJ; (b) 1.47×10^3 kJ; (c) 352 g CS_2

4D.3 (a) $CO(g) + H_2O(g) \rightarrow CO_2(g) + H_2(g)$
(b) −41.2 kJ·mol^{-1}

4D.5 7 kJ

4D.7 −320. kJ

4D.9 23.9×10^3 kJ·L^{-1}

4D.11 (a) 140 kJ; (b) 43.9 kJ; (c) 0.0762 g

4D.13 −16 kJ

4D.15 −312 kJ·mol^{-1}

4D.17 −2986.71 kJ·mol^{-1}

4D.19 −72.80 kJ

4D.21 (a) −138.18 kJ; (b) 752.3 kJ; (c) 15.28 kJ

4D.23 11.3 kJ

4D.25 −793.11 kJ

4D.27 (a) 33.93 kJ·mol^{-1}; (b) 30.94 kJ·mol^{-1};
(c) The values are very similar (table value = 30.8 kJ·mol^{-1}). Heat capacities are not strictly constant with temperature.

4D.29 For the reaction: A + 2 B → 3 C + D the molar enthalpy of raction at temperature 2 is given by:
$$\Delta H_{r,2}° = H_{m,2}°(\text{products}) - H_{m,2}°(\text{reactants})$$
$$= 3H_{m,2}°(C) + H_{m,2}°(D) - H_{m,2}°(A) - 2H_{m,2}°(B)$$
$$= 3[H_{m,1}°(C) + C_{P,m}(C)(T_2 - T_1)] + [H_{m,1}°(D)$$
$$+ C_{P,m}(D)(T_2 - T_1)] - [H_{m,1}°(A) + C_{P,m}(A)$$
$$(T_2 - T_1)] - 2[H_{m,1}°(B) + C_{P,m}(B)(T_2 - T_1)]$$
$$= 3H_{m,1}°(C) + H_{m,1}°(D) - H_{m,1}°(A) - 2H_{m,1}°(B)$$
$$+ [3C_{P,m}(C) + C_{P,m}(D) - C_{P,m}(A)$$
$$- 2C_{P,m}(B)](T_2 - T_1)$$
$$= \Delta H_{r,1}° + [3C_{P,m}(C) + C_{P,m}(D) - C_{P,m}(A)$$
$$- 2C_{P,m}(B)](T_2 - T_1)$$
Finally, $\Delta H_{r,2}° = \Delta H_{r,1}° + \Delta C_{P,r}(T_2 - T_1)$, which is Kirchhoff's law.

4E.1 2564 kJ·mol^{-1}

4E.3 (a) −412 kJ·mol^{-1}; (b) 673 kJ·mol^{-1}; (c) 63 kJ·mol^{-1}

4E.5 (a) −597 kJ·mol^{-1}; (b) −460 kJ·mol^{-1}; (c) 0 kJ·mol^{-1}

4E.7 (a) −202 kJ·mol^{-1}; (b) −45 kJ·mol^{-1}; (c) −115 kJ·mol^{-1}

4E.9 approximately 228 kJ

4F.1 (a) 0.341 J·K^{-1}·s^{-1}; (b) 29.5 J·K^{-1}·day^{-1};
(c) Less, because in the equation $\Delta S_{surr} = \dfrac{-\Delta H}{T}$, if T is larger, ΔS_{surr} is smaller.

4F.3 (a) 0.22 J·K^{-1}; (b) 0.17 J·K^{-1}; (c) Since T is in the denominator the entropy change is smaller at higher temperatures.

4F.5 14.8 J·K^{-1}

4F.7 (a) 6.80 J·K^{-1}; (b) 4.08 J·K^{-1}

4F.9 42.4 J·K^{-1}

4F.11 −14.6 J·K^{-1}

4F.13 (a) −22.0 J·K^{-1}; (b) 134 J·K^{-1}

4F.15 (a) 253 K; (b) 248 K. The value 85 J·K^{-1}·mol^{-1} is an average value for the entropy of vaporization of organic liquids and therefore deviations are expected when using this average value for individual organic liquids.

4F.17 $111 \text{ J·K}^{-1}\text{·mol}^{-1}$

4G.1 (a) 0; (b) $1.22 \times 10^{-21} \text{ J·K}^{-1}$

4G.3 COF_2. COF_2 and BF_3 are both trigonal planar molecules, but it would be possible for COF_2 to be disordered with the fluorine and oxygen atoms occupying the same locations. Because all the groups attached to boron are identical, such disorder is not possible.

4G.5 Higher; for the cis compound there are 12 different orientations but for the trans compound there are only three different orientations.

4G.7 14.9 J·K^{-1}

4G.9 (a) $W = 3$; (b) $W = 12$; (c) Initially, one of the three-atom systems had two atoms in higher energy states. In part (b), the system will be at equilibrium when each three- atom system has one quantum of energy. Therefore energy will flow from the system with two quanta to the system with none.

4H.1 (a) $HBr(g)$; (b) $NH_3(g)$; (c) $I_2(l)$; (d) 1.0 mol $Ar(g)$ at 1.00 atm

4H.3 $C(s, \text{diamond}) < H_2O(s) < H_2O(l) < H_2O(g)$. Entropy will increase when going from a solid to a liquid to a gas. $C(s, \text{diamond})$ will have less entropy than solid water because water is a molecular substance held together in the solid phase by weak hydrogen bonds, and in $C(s, \text{diamond})$ the carbon is more rigidly held in place and will thus have less entropy.

4H.5 (a) Iodine is expected to have higher entropy due to its larger mass and consequently larger number of fundamental particles. (b) 1-Pentene is expected to have higher entropy due to its more flexible framework. (c) Ethene (or ethylene), because it is a gas. Also, for the same mass, a sample of ethene will be composed of many small molecules, whereas polyethylene will be made up of fewer but larger molecules.

4H.7 (a) decrease; (b) increase; (c) decrease

4H.9 $\Delta S_B < \Delta S_C < \Delta S_A$. The change in entropy for container A is greater than that for container B or C due to the greater number of particles. The change in entropy in container C is greater than that of container B because of the disorder due to the vibrational motion of the molecules in container C.

4H.11 (a) $-163.34 \text{ J·K}^{-1}\text{·mol}^{-1}$; the entropy change is negative because the number of moles of gas has decreased by 1.5. (b) $-86.5 \text{ J·K}^{-1}\text{·mol}^{-1}$; the entropy change is negative because the number of moles of gas has decreased by 0.5. (c) $160.6 \text{ J·K}^{-1}\text{·mol}^{-1}$; the entropy change is positive because the number of moles of gas has increased by 1. (d) $-36.8 \text{ J·K}^{-1}\text{·mol}^{-1}$; the 4 moles of solid products are more ordered than the 4 moles of solid reactants.

4H.13 $dS = \dfrac{dq_{\text{rev}}}{T} = \dfrac{C_{P,m}dT}{T}$ and, therefore, $\Delta S = \displaystyle\int_{T_1}^{T_2} \dfrac{C_{P,m}}{T} dT$.

$C(s, \text{graphite}) + \frac{1}{2}O_2(g) \rightarrow CO(g)$

If $C_{P,m} = a + bT + c/T^2$, then

$$\Delta S = \int_{T_1}^{T_2} \frac{a + bT + c/T^2}{T} dT$$

$$= \int_{T_1}^{T_2} \left(\frac{a}{T} + b + \frac{c}{T^3}\right) dT = \left(a \ln(T) + bT - \frac{c}{2T^2}\right)\Bigg|_{T_1}^{T_2}$$

$$= a \ln\left(\frac{T_2}{T_1}\right) + b(T_2 - T_1) - \frac{c}{2}\left(\frac{1}{T_2^2} - \frac{1}{T_1^2}\right)$$

$\Delta S(\text{true}) = 3.31 \text{ J·K}^{-1}\text{·mol}^{-1}$; $\Delta S(\text{mean}) = 3.41 \text{ J·K}^{-1}\text{·mol}^{-1}$; 3.0%

4I.1 150 J·K^{-1}

4I.3 (a) $30. \text{ kJ·mol}^{-1}$; (b) -11 J·K^{-1}

4I.5 $\Delta S_{\text{cold}} = 11.9 \text{ J·K}^{-1}$; $\Delta S_{\text{hot}} = -11.1 \text{ J·K}^{-1}$; $\Delta S_{\text{tot}} = 0.8 \text{ J·K}^{-1}$

4I.7 (a) $\Delta S_{\text{surr}} = -73 \text{ J·K}^{-1}$; $\Delta S_{\text{system}} = 73 \text{ J·K}^{-1}$; (b) $\Delta S_{\text{surr}} = -29.0 \text{ J·K}^{-1}$; $\Delta S_{\text{system}} = 29.0 \text{ J·K}^{-1}$; (c) $\Delta S_{\text{surr}} = 29.0 \text{ J·K}^{-1}$; $\Delta S_{\text{system}} = -29.0 \text{ J·K}^{-1}$;

4I.9 (a) $\Delta S_{\text{tot}} = 0 \text{ J·K}^{-1}$; $\Delta S_{\text{surr}} = -3.84 \text{ J·K}^{-1}$; $\Delta S = 3.84 \text{ J·}^{-1}$; (b) $\Delta S_{\text{tot}} = 3.84 \text{ J·K}^{-1}$; $\Delta S_{\text{surr}} = 0 \text{ J·K}^{-1}$; $\Delta S = 3.84 \text{ J·K}^{-1}$

4I.11 Spontaneous change means $\Delta S_{\text{tot}} > 0$. A vapor to liquid change is condensation and it releases heat. However, as shown in the diagram, the temperature of the system does not change. Therefore the released heat must have left the system to the surroundings. Since heat flows from hot to cold, the temperature of the surroundings (T_{surr}) must be lower than the temperature of the system (T_{sys}), and since $\Delta S_{\text{sys}} = \dfrac{q_{\text{rev}}}{T_{\text{sys}}}$ and $\Delta S_{\text{surr}} = \dfrac{q_{\text{rev}}}{T_{\text{surr}}}$, then $\Delta S_{\text{tot}} > 0$.

4J.1 Exothermic reactions tend to be spontaneous because the result is an increase in the entropy of the surroundings. Using the mathematical relationship $\Delta G_r = \Delta H_r - T\Delta S_r$, it is clear that if ΔH_r is large and negative compared to ΔS_r, then the reaction will generally be spontaneous.

4J.3 (a) -1.8 kJ·mol^{-1}; spontaneous; (b) 1.1 kJ·mol^{-1}; not spontaneous

4J.5 (a) $\frac{1}{2}N_2(g) + \frac{3}{2}H_2(g) \rightarrow NH_3(g)$; $\Delta H_r^\circ = -46.11 \text{ kJ·mol}^{-1}$; $\Delta S_r^\circ = -99.38 \text{ J·K}^{-1}\text{·mol}^{-1}$; $\Delta G_r^\circ = -16.49 \text{ kJ·mol}^{-1}$; (b) $H_2(g) + \frac{1}{2}O_2(g) \rightarrow H_2O(g)$; (b) $\Delta H_r^\circ = -241.82 \text{ kJ·mol}^{-1}$; $\Delta S_r^\circ = -44.42 \text{ J·K}^{-1}\text{·mol}^{-1}$; $\Delta G_r^\circ = -228.58 \text{ kJ·mol}^{-1}$; (c); (b) $\Delta H_r^\circ = -110.53 \text{ kJ·mol}^{-1}$; $\Delta S^\circ = 89.36 \text{ J·K}^{-1}\text{·mol}^{-1}$; $\Delta G^\circ = -137.2 \text{ kJ·mol}^{-1}$; (d) $\frac{1}{2}N_2(g) + O_2(g) \rightarrow NO_2(g)$; $\Delta H_r^\circ = 33.18 \text{ kJ·mol}^{-1}$; $\Delta S_r^\circ = -60.89 \text{ J·K}^{-1}\text{·mol}^{-1}$; $\Delta G_r^\circ = 51.33 \text{ kJ·mol}^{-1}$

4J.7 (a) $\Delta S_r^\circ = 125.8 \text{ J·K}^{-1}\text{·mol}^{-1}$; $\Delta H_r^\circ = -196.10 \text{ kJ·mol}^{-1}$; $\Delta G_r^\circ = -233.56 \text{ kJ·mol}^{-1}$; (b) $\Delta S_r^\circ = 14.6 \text{ J·K}^{-1}\text{·mol}^{-1}$; $\Delta H_r^\circ = -748.66 \text{ kJ·mol}^{-1}$; $\Delta G_r^\circ = -713.02 \text{ kJ·mol}^{-1}$

4J.9 (a) $\Delta H_r^\circ = -235.8 \text{ kJ·mol}^{-1}$; $\Delta S_r^\circ = -133.17 \text{ J·K}^{-1}\text{·mol}^{-1}$; $\Delta G_r^\circ = -195.8 \text{ kJ·mol}^{-1}$; using $\Delta G_r^\circ = \Delta H_r^\circ - T\Delta S_r^\circ = -196.1 \text{ kJ·mol}^{-1}$; (b) $\Delta H_r^\circ = 11.5 \text{ kJ·mol}^{-1}$; $\Delta S_r^\circ = -149.7 \text{ J·K}^{-1}\text{·mol}^{-1}$; $\Delta G_r^\circ = 56.2 \text{ kJ·mol}^{-1}$; using $\Delta G_r^\circ = \Delta H_r^\circ - T\Delta S_r^\circ = 56.1 \text{ kJ·mol}^{-1}$; (c) $\Delta H_r^\circ = -57.20 \text{ kJ·mol}^{-1}$; $\Delta S_r^\circ = -175.83 \text{ J·K}^{-1}\text{·mol}^{-1}$; $\Delta G_r^\circ = -4.73 \text{ kJ·mol}^{-1}$; using $\Delta G_r^\circ = \Delta H_r^\circ - T\Delta S_r^\circ = -4.80 \text{ kJ·mol}^{-1}$

4J.11 (a) $\Delta G_r^\circ = -141.74 \text{ kJ·mol}^{-1}$, spontaneous; (b) $\Delta G_r^\circ = 130.4 \text{ kJ·mol}^{-1}$, not spontaneous; (c) $\Delta G_r^\circ = -10590.9 \text{ kJ·mol}^{-1}$, spontaneous

4J.13 (a) and (d) are thermodynamically stable.

4J.15 (a) $-234.2 \text{ J·K}^{-1}\text{·mol}^{-1}$. The compound is less stable at higher temperatures. (b) $34.90 \text{ J·K}^{-1}\text{·mol}^{-1}$. $HCN(g)$ is more stable at higher T. (c) $12.38 \text{ J·K}^{-1}\text{·mol}^{-1}$. $NO(g)$ is more stable as T increases. (d) $11.28 \text{ J·K}^{-1}\text{·mol}^{-1}$. $SO(g)$ is more stable as T increases.

4J.17 (a) $\Delta G_r^\circ = -98.42 \text{ kJ·mol}^{-1}$, spontaneous below 612.9 K; (b) $\Delta G_r^\circ = -283.7 \text{ kJ·mol}^{-1}$, spontaneous at all temperatures; (c) $\Delta G_r^\circ = 3.082 \text{ kJ·mol}^{-1}$, nonspontaneous at all temperatures

4.1 132.0 kJ

4.3 12.71 kJ·mol^{-1}

4.5 7.53 kJ·mol^{-1}

4.7 (a) 3.72 kJ; (b) −3267.5 kJ; (c) −3263.8 kJ

4.9 (a) Cu; (b) 8.90 g·cm^{-3}

4.11 (a) (1)

(2)

(3)

(b) −372 kJ; (c) −205.4 kJ; (d) The hydrogenation of benzene is much less exothermic than predicted by bond enthalpy estimations. Part of this difference is due to the inherent inaccuracy of using average values, but the difference is so large that this cannot be the complete explanation. As may be expected, the resonance energy of benzene makes it more stable than would be expected by treating it as a set of three isolated double and three isolated single bonds. The difference in these two values [−205 kJ − (−372 kJ) = 167 kJ]] is a measure of how much more stable benzene is than the Kekulé structure would predict.

4.13 (a) 2326 kJ·mol^{-1}; (b) 3547 kJ·mol^{-1}; (c) more stable; (d) 20 kJ per carbon atom; (e) 25 kJ per carbon atom; (f) There is slightly less stabilization per carbon atom in C$_{60}$ than in benzene. This fits with expectations, as the C$_{60}$ molecule is forced by its geometry to be curved.

4.15 31 °C

4.17 (a) 1.5 L; (b) SO$_2$ (g); (c) 1.1 L; (d) 40 J, on the system; (e) −3.0 kJ, leaves the system; (f) −2960 J

4.19 (a) 4103.2 J·mol^{-1}; (b) 4090.7 J·mol^{-1}; (c) 12.5 kJ·mol^{-1}

4.21 According to the second law of thermodynamics, the formation of complex molecules from simpler precursors would not be spontaneous, because such processes create order from disorder. If there is an external input of energy, however, a more ordered system could be created.

4.23 2.49 × 10^{27} photons

4.25 64 min

4.27 (a) −1.0 × 10^4 kJ; (b) 3.2 × 10^4 g

4.29 (a) 2 H$_2$S(g) + O$_2$(g) → 2 S(s) + 2 H$_2$O(l); (b) −4.96 × 10^5 kJ; (c) The reactor would need to be cooled.

4.31 (a) endothermic; (b) 623 kJ

4.33 (a) 14.7 mol; (b) −4.20 × 10^3 kJ

4.35 −108 kJ

4.37 (a) $\Delta G° < 0$; (b) $\Delta H°$ unable to tell; (c) $\Delta S°$ unable to tell; (d) $\Delta S_{total} > 0$

4.39 According to Trouton's rule, the entropy of vaporization of an organic liquid is a constant of approximately 85 J·mol^{-1}·K^{-1}.

For Pb: $\Delta S_{fus}° = \dfrac{5100 \text{ J}}{600 \text{ K}} = 8.50 \text{ J·K}^{-1}$

For Hg: $\Delta S_{fus}° = \dfrac{2290 \text{ J}}{234 \text{ K}} = 9.79 \text{ J·K}^{-1}$

For Na: $\Delta S_{fus}° = \dfrac{2640 \text{ J}}{371 \text{ K}} = 7.12 \text{ J·K}^{-1}$

These numbers are reasonably close but clearly much smaller than the value associated with Trouton's rule.

4.41 Fe$_2$O$_3$ is thermodynamically more stable because $\Delta G_r°$ for the interconversion between Fe$_3$O$_4$ to Fe$_2$O$_3$ is negative.

4.43 (a) $\Delta G_r° = -110.0$ kJ·mol^{-1}; (b) 7.14 × 10^{-2} mol·L^{-1}

4.45 (a) no; (b) positive; (c) positional disorder; (d) thermal disorder; (e) dispersal of matter

4.47 (a) method (i); (b) method (i); (c) no

4.49 The entries all correspond to aqueous ions. The fact that they are negative is due to the reference point that has been established. Because ions cannot actually be separated and measured independently, a reference point that defines $S_m°(\text{H}^+, \text{aq}) = 0$ has been established. This definition is then used to calculate the standard entropies for the other ions. The fact that they are negative will arise in part because the solvated ion M(H$_2$O)$_x{}^{n+}$ will be more ordered than the isolated ion and solvent molecules (M^{n+} + x H$_2$O).

4.51 (a) 8.57 kJ at 298 K, −0.35 kJ at 373 K, −6.29 kJ at 423 K; (b) 0; (c) The discrepancy arises because the enthalpy and entropy values calculated from the tables are not rigorously constant with temperature.

4.53

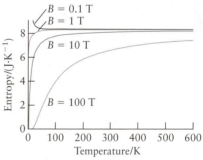

4.55 $T \rightarrow \infty$

4.57 (a) Because standard molar entropies increase with temperature (more translational, vibrational, and rotational motion), the $-T\Delta S_m°$ term becomes more negative at higher temperatures. (b) The increase in molar entropy depends on the relative sizes of the molar heat capacities. In some cases, the latter are larger for gases than for condensed phases, therefore, their $-T\Delta S_m°$ term is more negative.

4.59 Dehydrogenation of cyclohexane to benzene:

$$\text{C}_6\text{H}_{12}(l) \rightarrow \text{C}_6\text{H}_6(l) + 3 \text{ H}_2(g) \qquad \Delta G_r° = 97.6 \text{ kJ·mol}^{-1}$$

The reaction of ethene with hydrogen:

$$\text{C}_2\text{H}_4(g) + \text{H}_2(g) \rightarrow \text{C}_2\text{H}_6(g) \qquad \Delta G_r° = -100.97 \text{ kJ·mol}^{-1}$$

Combine these two reactions:

$$\text{C}_6\text{H}_{12}(l) \rightarrow \text{C}_6\text{H}_6(l) + 3 \text{ H}_2(g) \qquad \Delta G_r° = +97.6 \text{ kJ·mol}^{-1}$$
$$+3[\text{C}_2\text{H}_4(g) + \text{H}_2(g) \rightarrow \text{C}_2\text{H}_6(g)] \qquad \Delta G_r° = 3(-100.97 \text{ kJ·mol}^{-1})$$

$$\text{C}_6\text{H}_{12}(l) + 3 \text{ C}_2\text{H}_4(g) \rightarrow \text{C}_6\text{H}_6(l) + 3 \text{ C}_2\text{H}_6(g) \quad \Delta G_r° = -205.31 \text{ kJ·mol}^{-1}$$

The overall process is spontaneous.

4.61 Reaction A is spontaneous, but reaction B is not spontaneous.

4.63 (a)

(b) $CH_4(g) + 2 O_2(g) \rightarrow CO_2(g) + 2 H_2O(g)$

$\Delta H_c° = (4 \times 412) + (2 \times 496) + (2 \times -743)$
$\qquad + (4 \times -463)$
$\qquad = +2640 - 3338 = -698 \text{ kJ·mol}^{-1}$

$H_3C—O—CH_3(g) + 3 O_2(g) \rightarrow$
$\qquad\qquad\qquad\qquad 2 CO_2(g) + 3 H_2O(g)$

$\Delta H_c° = (2 \times 360) + (6 \times 412) + (3 \times 496)$
$\qquad + (4 \times -743) + (6 \times -463)$
$\qquad = -1070 \text{ kJ·mol}^{-1}$

Given: $CH_3CH_2OH(l) \rightarrow CH_3CH_2OH(g)$
$\qquad \Delta H_{vap}° = 43.5 \text{ kJ·mol}^{-1}$

$CH_3CH_2OH(g) + 3 O_2(g) \rightarrow 2 CO_2(g) + 3 H_2O(g)$

$\Delta H_c° = (360) + (463) + (348) + (5 \times 412)$
$\qquad + (3 \times 496) + (4 \times -743) + (6 \times -463)$
$\qquad = -1031 \text{ kJ·mol}^{-1}$

For the combustion of $CH_3CH_2OH(l)$, $\Delta H_c° = 43.5 - 1031 = -988 \text{ kJ·mol}^{-1}$.

The combustion of 1 mole of dimethyl ether releases the most heat.

(c) $\dfrac{-890 \text{ kJ·mol}^{-1}}{16.01 \text{ g·mol}^{-1}} = -55.6 \text{ kJ·g}^{-1}$

$\dfrac{-1368 \text{ kJ·mol}^{-1}}{46.02 \text{ g·mol}^{-1}} = -29.73 \text{ kJ·g}^{-1}$

$\dfrac{-5471 \text{ kJ·mol}^{-1}}{114.08 \text{ g·mol}^{-1}} = -47.96 \text{ kJ·g}^{-1}$

Methane, as it releases the most heat per gram. (d) 921 L; (e) methane gas, -890 kJ·mol^{-1} CO_2 (less CO_2); ethanol liquid, -684 kJ·mol^{-1} CO_2 (more CO_2); octane liquid, -684 kJ·mol^{-1} CO_2 (more CO_2)

Focus 5

5A.1 PH_3 will have the higher vapor pressure (no hydrogen bonding).

5A.3 (a) 87 °C; (b) 113 °C

5A.5 $P_{25.0\,°C} = 0.19$ atm or 1.5×10^2 Torr

5A.7 289 K, or 16 °C

5A.9 (a) $+28.3 \text{ kJ·mol}^{-1}$; (b) $91.6 \text{ J·K}^{-1}\text{·mol}^{-1}$; (c) $+1.0 \text{ kJ·mol}^{-1}$; (d) 309 K

5A.11 (a) $+28.6 \text{ kJ·mol}^{-1}$; (b) $90.6 \text{ J·K}^{-1}\text{·mol}^{-1}$; (c) 0.53 atm ($4.0 \times 10^2$ Torr)

5B.1 (a) vapor; (b) liquid; (c) vapor

5B.3 (a) 2.4 K; (b) about 10 atm; (c) 5.5 K; (d) no

5B.5 (a) Lower pressure triple point: liquid helium-I and -II are in equilibrium with helium gas; higher pressure triple point: liquid helium-I and -II are in equilibrium with solid helium. (b) helium-I

5B.7 solid

5C.1 The molar Gibbs free energy of the solvent in the NaCl solution will always be lower than the molar Gibbs free energy of pure water; if given enough time, all of the water from the "pure water" beaker will become part of the NaCl solution, leaving an empty beaker.

5C.3 (a) 0.900 atm (684 Torr); (b) 0.998 atm (758 Torr)

5C.5 7.80×10^{-2} mol

5C.7 (a) 0.052; (b) 115 g·mol^{-1}

5C.9 75.8 Torr

5C.11 (a) $P_{total} = 78.2$ Torr; vapor composition: $x_{benzene} = 0.91$; $x_{toluene} = 0.09$; (b) $P_{total} = 43.2$ Torr; vapor composition: $x_{benzene} = 0.473$; $x_{toluene} = 0.527$

5C.13 63 g

5C.15 (a) An ideal solution is predicted (intermolecular attractions in the mixture are similar to those in the component liquids). (b) A negative deviation is predicted (hydrogen bonding in the solution). (c) A positive deviation is predicted (weak intermolecular forces between molecules in the solution).

5D.1 (a) Water would be the best choice; (b) benzene; (c) water

5D.3 (a) hydrophilic; (b) hydrophobic; (c) hydrophobic; (d) hydrophilic

5D.5 (a) 6.4×10^{-4} M; (b) 1.5×10^{-2} M; (c) 2.3×10^{-3} M

5D.7 (a) 4 ppm; (b) 0.1 atm; (c) 0.5 atm

5D.9 (a) CO_2 concentration will double; (b) no change

5D.11 3.48×10^{-2} mol CO_2 (about 1.5 g)

5D.13 (a) negative; (b) $Li_2SO_4(s) \rightarrow 2 Li^+(aq) + SO_4^{2-}(aq) + \text{heat}$; (c) the enthalpy of hydration

5D.15 (a) $+0.67$ kJ; (b) -0.50 kJ; (c) -24.7 kJ; (d) $+0.82$ kJ

5D.17 A *foam* is a colloid that is a suspension of a gas in a liquid or a solid matrix, whereas a *sol* is a suspension of a solid in a liquid.

5D.19 Colloids will scatter light, whereas true solutions do not.

5E.1 (a) 0.856 m; (b) 2.5 g; (c) 0.0571 m

5E.3 18.5 m

5E.5 (a) 0.248 m; (b) 0.246 M

5E.7 22.8 m

5E.9 (a) 1.35 m; (b) 0.519 m; (c) 28.43 m

5E.11 (a) 13.9 g; (b) 29 g

5F.1 $1.6 \times 10^2 \text{ g·mol}^{-1}$

5F.3 (a) $i = 1.84$; (b) $0.318 \text{ mol·kg}^{-1}$; (c) 83.8% dissociation

5F.5 B has the greater molar mass (fewer particles in solution than A).

5F.7 -0.20 °C

5F.9 (a) 0.24 atm; (b) 48 atm; (c) 0.72 atm

5F.11 $2.0 \times 10^3 \text{ g·mol}^{-1}$

5F.13 $5.8 \times 10^3 \text{ g·mol}^{-1}$

5F.15 (a) 1.2 atm; (b) 0.048 atm; (c) 8.3×10^{-5} atm

5G.1 (a) false; (b) false; (c) false; (d) true

5G.3 (a) $K = \dfrac{(P_{C_2H_4Cl_2})^2 (P_{H_2O})^2}{(P_{C_2H_4})^2 (P_{O_2}) (P_{HCl})^4}$

(b) $K = \dfrac{(P_{N_2})^7 (P_{H_2O})^6}{(P_{NH_3})^4 (P_{NO})^6}$

5G.5 (a) Flask 3 represents the point of reaction equilibrium.
(b) 54.5%; (c) $K = 0.26$

5G.7 (a) $CH_4(g) + 2 O_2(g) \rightleftharpoons CO_2(g) + 2 H_2O(g)$

$$K = \frac{(P_{CO_2})(P_{H_2O})^2}{(P_{CH_4})(P_{O_2})^2}$$

(b) $I_2(g) + 5 F_2(g) \rightleftharpoons 2 IF_5(g)$ $K = \frac{(P_{IF_5})^2}{(P_{I_2})(P_{F_2})^5}$

(c) $2 NO_2(g) + F_2(g) \rightleftharpoons 2 FNO_2(g)$ $K = \frac{(P_{FNO_2})^2}{(P_{F_2})(P_{NO_2})^2}$

5G.9 (a) The number of moles of O_2 is larger in the second experiment. (b) The concentration of O_2 is larger in the second case. (c) Although $(P_{O_2})^3/(P_{O_3})^2$ is the same, $(P_{O_2})/(P_{O_3})$ will be different. (d) Because K_c is a constant, $(P_{O_2})^3/(P_{O_3})^2$ is the same. (e) Because $(P_{O_2})^3/(P_{O_3})^2$ is the same in (d), its reciprocal, $(P_{O_3})^2/(P_{O_2})^3$, must be the same.

5G.11 (a) $\frac{1}{P_{BCl_3}^2}$; (b) $[H_3PO_4]^4 [H_2S]^{10}$; (c) $\frac{P_{BrF_3}^2}{P_{Br_2} P_{F_2}^3}$

5G.13 $\Delta G_r = 8.3 \times 10^{-1}$ kJ·mol^{-1}. Because ΔG_r is positive, the reaction will be spontaneous to produce I_2.

5G.15 $\Delta G_r = -27$ kJ·mol^{-1}. Because ΔG_r is negative, the reaction will proceed to form products.

5G.17

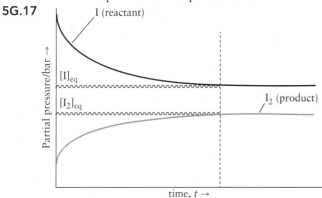

5G.19 (a) $\Delta G_r° = -19$ kJ·mol^{-1}; (b) $\Delta G_r° = +68$ kJ·mol^{-1}

5G.21 (a) $K = 1 \times 10^{80}$; (b) $K = 1 \times 10^{90}$; (c) $K = 1 \times 10^{-23}$

5H.1 (a) $K = 0.024$; (b) $K = 6.4$; (c) $K = 1.7 \times 10^3$

5H.3 $K = 1.5 \times 10^{34}$

5H.5 (a) $K_c = 4.3 \times 10^{-4}$; (b) $K_c = 1.87$

5I.1 $[Br_2] = 1.4$ mol·L^{-1}

5I.3 $[H_2] = 2.1 \times 10^{-5}$ mol·L^{-1}

5I.5 $P_{PCl_3} = 5.4$ bar

5I.7 $K = 4 \times 10^{-31}$. Because $Q > K$, the reaction will tend to proceed to produce reactants.

5I.9 (a) $Q = \frac{(0.10)^2}{(0.20)(0.10)} = 0.50$; (b) $Q \neq K$, therefore, the system is not at equilibrium. (c) Because $Q < K$, more products will be formed.

5I.11 (a) $Q_c = 6.9$. (b) Because $Q_c < K_c$, more products will tend to form, which will result in the formation of more SO_3.

5I.13 (a) The concentration of Cl_2 is essentially unchanged. The concentration of Cl atoms is $2 \times (5.5 \times 10^{-6}) = 1.1 \times 10^{-5}$ mol·L^{-1}. (b) The concentration of F_2 is 8×10^{-4} mol·L^{-1}. The concentration of F atoms is 3.2×10^{-4} mol·L^{-1}. (c) Cl_2 is more stable.

5I.15 $[NH_3] = 0.200$ mol·L^{-1}; $[H_2S] = 8 \times 10^{-4}$ mol·L^{-1}

5I.17 $[NO] = 3.6 \times 10^{-4}$ mol·L^{-1}; the concentrations of N_2 and O_2 remain essentially unchanged at 0.114 mol·L^{-1}.

5I.19 $K_c = 1.1$

5I.21 $[CO_2] = 8.6 \times 10^{-5}$ mol·L^{-1}; $[CO] = 4.9 \times 10^{-3}$ mol·L^{-1}; $[O_2] = 4.6 \times 10^{-4}$ mol·L^{-1}

5I.23 $K_c = 3.88$

5I.25 $[SO_2] = 0.0011$ mol·L^{-1}; $[NO_2] = 0.0211$ mol·L^{-1}; $[NO] = 0.0389$ mol·L^{-1}; $[SO_3] = 0.0489$ mol·L^{-1}

5I.27 (a) Because $Q \neq K$, the reaction is not at equilibrium. (b) Because $Q_c < K_c$, the reaction will proceed to form products. (c) $[PCl_5] = 3.07$ mol·L^{-1}; $[PCl_3] = 5.93$ mol·L^{-1}; $[Cl_2] = 0.93$ mol·L^{-1}

5I.29 $P_{HCl} = 0.22$ bar; $P_{H_2} = P_{Cl_2} = 3.9 \times 10^{-18}$ bar

5I.31 (a) Because $Q_c > K_c$, the reaction will proceed to produce more of the reactants; (b) $[CO] = 0.156$ mol·L^{-1}; $[H_2] = 0.155$ mol·L^{-1}; $[CH_3OH] = 4.1 \times 10^{-5}$ mol·L^{-1}

5I.33 $K_c = 1.58 \times 10^{-8}$

5I.35 (c) $K = p^2/(1.0 - 2p)^2$

5J.1 (a) The H_2 partial pressure will decrease. (b) The CO_2 partial pressure will decrease. (c) The H_2 concentration will increase. (d) The equilibrium constant for the reaction is unchanged because it is unaffected by any change in concentration.

5J.3 (a) The amount of water will decrease. (b) The amount of O_2 will increase. (c) The amount of NO will increase. (d) The amount of NH_3 will increase. (e) The equilibrium constant will not be affected. (f) The amount of NH_3 will decrease. (g) The amount of oxygen will decrease.

5J.5 (a) reactants; (b) reactants; (c) reactants; (d) no change; (e) reactants

5J.7 $P_{N_2} = 4.61$ bar, $P_{H_2} = 1.44$ bar, $P_{NH_3} = 23.85$ bar

5J.9 (a) The pressure of NH_3 will increase. (b) The pressure of O_2 will increase.

5J.11 (a) and (b) are endothermic and raising the temperature should favor the formation of products; (c) and (d) are exothermic and raising the temperature should favor the formation of reactants.

5J.13 Less ammonia will be present at higher temperature, assuming no other changes occur to the system

5J.15 (a) At 298 K, $K = 1 \times 10^{-16}$; at 423 K, $K = 1 \times 10^{-7}$; (b) At 298 K, $K = 7.8 \times 10^{-2}$; at 423 K, $K = 0.22$

5J.17 $\ln\left(\dfrac{K_{c2}}{K_{c1}}\right) = \dfrac{\Delta H_r°}{R}\left(\dfrac{1}{T_1} - \dfrac{1}{T_2}\right) - \Delta n_r \ln\left(\dfrac{T_2}{T_1}\right)$

5.1 Water is a polar molecule and as a result will orient itself differently around cations and anions, aligning its dipole in such a way as to present the most favorable interaction possible.

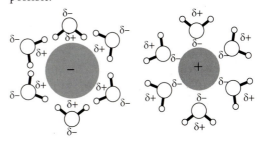

5.3 (a) stronger; (b) low; (c) high; (d) weaker;
(e) weak, low; (f) low; (g) strong, high

5.5 (a, b) decrease with increasing temperature;
(c, d) increase with increasing temperature

5.7 (a) Butane (a nonpolar molecule) will dissolve in the nonpolar solvent (tetrachloromethane).

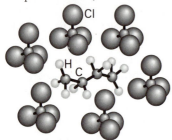

(b) Calcium chloride will dissolve in the polar solvent.

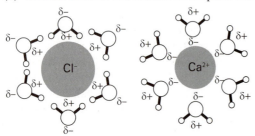

5.9 (a) M_{solute} appears greater than actual molar mass;
(b) M_{solute} appears lower than the actual molar mass;
(c) M_{solute} appears lower than the actual molar mass;
(d) M_{solute} appears greater than actual molar mass.

5.11 (a) $M_{protein} = 4.8 \times 10^3$ g·mol^{-1}; (b) freezing point is -3.9×10^{-4} °C; (c) The freezing point change is so small that it cannot be measured accurately, so osmotic pressure would be the preferred method.

5.13 (a) $\ln P = -\dfrac{4595\ K}{T} + 13.59$; (b) The relationship to plot is $\ln P$ versus $\dfrac{1}{T}$. (c) $P = 0.040$ atm or 30. Torr;
(d) $T = 338.1$ K

5.15 (a) 56.9 g·mol^{-1}; some must be dissociating;
(b) 116 g·mol^{-1}; acetic acid forms dimers in benzene.

5.17 (a) Linear regression provides the following:
$y = -3358.714x + 12.247$

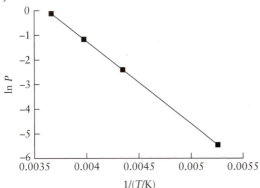

(b) 28 kJ·mol^{-1}; (c) 1.0×10^2 J·K^{-1}·mol^{-1};
(d) 2.7×10^2 K; (d) 2.1×10^2 K

5.19 The critical temperatures are CH$_4$: -82.1 °C; C$_2$H$_6$ = 32.2 °C; C$_3$H$_8$ = 96.8 °C; C$_4$H$_{10}$ = 152 °C. The critical temperatures increase due to the influence of the stronger London forces.

5.21 (a) Flask A will have a pressure of 195 Torr and flask B will have a pressure of 222 Torr. (b) For the liquid phase $x_{acetone} = 0.67$ and $x_{chloroform} = 0.33$. For the vapor phase $x_{acetone} = 0.70$ and $x_{chloroform} = 0.30$. (c) The solution shows negative deviation from Raoult's law.

5.23 The fact that ΔH_{vap} is small indicates that it takes little energy to volatilize the sample, which means that the intermolecular forces are weaker. Hence we expect the ratio P_2/P_1 in the Clausius–Clapeyron equation to be larger.

5.25 The osmotic pressure is 7.1 atm.

5.27 (a) 78.5%; (b) Water vapor in the air would condense as dew or fog.

5.29 (a) 0.222 M dextrose and 2.99×10^{-2} M NaCl; (b) 6.89 atm

5.31 (a) 1.0 M; (b) 24 atm

5.33 Increasing the temperature (a) would increase the formation of X(g).

5.35 (a) $2\,A(g) \rightarrow B(g) + 2\,C(g)$; (b) $K = 1.54 \times 10^{-2}$

5.37 Because $Q \neq K$, the system is not at equilibrium, and because $Q < K$, the reaction will proceed to produce more products.

5.39 (a) $[N_2O_4] = 0.0065$ mol·L^{-1}; $[NO_2] = 7.0 \times 10^{-3}$ mol·L^{-1};
(b) $[N_2O_4] = 0.015$ mol·L^{-1}; $[NO_2] = 0.010$ mol·L^{-1}

5.41 (a) At conditions $K = 4.96$ and $P = 0.50$ bar, $\alpha = 0.953$.
(b) At conditions $K = 4.96$ and $P = 1.00$ bar, $\alpha = 0.912$.

5.43 (a) If $K = 1$, then $\Delta G° = 0$; (b) $T = 978$ K (or 705 °C);
(c) All pressures are equal to 7.50 bar. (d) $P_{CO} = 7.04$ bar;
$P_{H_2O} = 5.04$ bar; $P_{CO_2(g)} = 8.96$ bar; $P_{H_2} = 3.96$ bar

5.45 (a)

Halogen	Bond Dissociation Energy $/\Delta G_f°$(kJ·mol^{-1})	$\Delta G_f°$ /(kJ·mol^{-1})
Fluorine	146	19.1
Chlorine	230	47.9
Bromine	181	42.8
Iodine	139	5.6

(b)

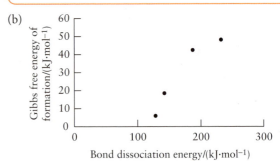

5.47 (a) $K = 4.5 \times 10^{-29}$; (b) The vapor pressure of bromine will be 0.285 bar or 0.289 atm.

(c) $P_{Br} = 3.6 \times 10^{-15}$ bar or 3.6×10^{-15} atm;
(d) $V = 0.846$ L or 846 mL

5.49 (a) $K = 2.77 \times 10^4$; (b) The amount of NO_2 will be greater than initially present, but less than the 3.13 mol·L^{-1} present immediately upon making the addition. K_c will not be affected. (c) At equilibrium, $[N_2O_4] = 0.64$ mol·L^{-1}; $[NO_2] = 2.67$ mol·L^{-1}.
$$\frac{[NO_2]^2}{[N_2O_4]} = \frac{(2.67)^2}{0.64} = 11.1 \approx K_C$$
These concentrations are consistent with the predictions in (b).

5.51 For H_2O, $P_{H_2O} = 0.032$ bar or 24 torr. For D_2O, $P_{D_2O} = 0.028$ bar or 21 torr.

5.53

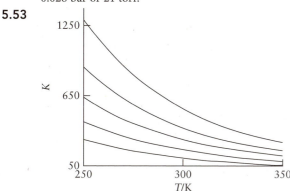

From the plot, it is clear that the larger the equilibrium constant, the more sensitive to the change of temperature.

5.55 (a) $K_c = 0.403$; Since K_c is quite small, there will not be a large amount of hydrogen gas produced when the reaction reaches equilibrium. (b) $[CO] = [H_2] = 0.364$ mol·L^{-1}; $[H_2O] = 0.330$ mol·L^{-1}.

5.57 (a) (i) More NO_2 will form. (ii) More NO_2 will form. (iii) There will be no change in the amount of NO_2.
(b) 0.242 moles of NO

5.59 (a) $K = 5.6 \times 10^{15}$
Part (b) The effect of each change to the equilibrium is:
(a) The amount of H_2O will not change. (b) The amount of SO_2 will decrease. (c) The amount of SO_2 should decrease. (d) The amount of S should decrease. (e) The equilibrium constant will be unaffected. (f) The amount of SO_2 should decrease. (g) The amount of SO_2 should increase.

5.61 (a) The equilibrium shifts toward the reactants side of the equation. (b) There will be little or no effect on compressing the system. (c) The reaction shifts toward the formation of products. (d) Raising the temperature will favor the formation of products. (e) The reaction shifts toward the formation of products. (f) Changing the amount of water will not affect the reaction. (g) Decreasing the partial pressure of a reactant (CO_2) will favor the production of more reactants.

5.63 (a) The hybridization of the carbon atoms in citric acid are:

(b) Yes; all four —OH groups present can participate in hydrogen bonding; the remaining three oxygens can accept hydrogen bonds. (c) Because citric acid can form hydrogen bonds with itself, it should have large intermolecular forces and should be a solid. The ability to both donate and accept hydrogen bonds should also make it soluble in water. (d) 100.0 g of a 0.9% NaCl solution will contain 0.9 g of NaCl ($M = 58.44$ g·mol^{-1}), which is 1.54×10^{-2} mol of NaCl. If NaCl completely dissociates, the amount of solute present is 3.08×10^{-2} mol of solute in 100.0 mL, making the concentration of normal saline 0.3 mol·L^{-1}. (e) To make a 500.0 mL sports drink isotonic, you need it to have a total molarity of 0.3 mol·L^{-1} (from part d). This means that the amount of solute needed is 0.3 mol·L^{-1} × 0.5000 L = 0.15 mol total solute. Addition of 1.0 g of NaCl to the 500-mL sports drink will account for 0.034 mol of solute (twice the moles of NaCl added), leaving 0.116 mol of glucose needed to achieve isotonicity. Because glucose has a molar mass of 180 g·mol^{-1}, this means that in addition to the 1.0 g of NaCl, 21 g of glucose needs to be added to the 500.0-mL sports drink. (f) 300.0 mL of a 1.00% boric acid = 300.0 g of solution (assuming the density of the solution is 1.00 g·cm^{-3}), meaning that there are 3.00 g of $B(OH)_3$ or 4.85×10^{-2} mol^{-1} of $B(OH)_3$ in the solution (because $M_{boric\ acid} = 61.81$ g·mol^{-1}. To be isotonic, 300.0 mL should have 0.3 mol·L^{-1} × (0.3000 L) = 0.09 mol total solute. Subtracting out the amount of boric acid present means that 0.04 mol of solute needed to be added to achieve isotonicity; because 1 mol of NaCl provides 2 mol of solute, 0.02 mol, or 1.17 g, of NaCl must be added to the boric acid solution.

Focus 6

6A.1 (a) $CH_3NH_3^+$; (b) $NH_2NH_3^+$; (c) H_2CO_3; (d) CO_3^{2-}; (e) $C_6H_5O^-$; (f) $CH_3CO_2^-$

6A.3 (a) H_2SO_4 and HSO_4^- form a conjugate acid–base pair in which H_2SO_4 is the acid and HSO_4^- is the base.
(b) $C_6H_5NH_3^+$ and $C_6H_5NH_2$ form a conjugate acid–base pair in which $C_6H_5NH_3^+$ is the acid and $C_6H_5NH_2$ is the base.
(c) $H_2PO_4^-$ and HPO_4^{2-} form a conjugate acid–base pair in which $H_2PO_4^-$ is the acid and HPO_4^{2-} is the base.
(d) HCOOH and HCO_2^- form a conjugate acid–base pair in which HCOOH is the acid and HCO_2^- is the base.
(e) $NH_2NH_3^+$ and NH_2NH_2 form a conjugate acid–base pair in which $NH_2NH_3^+$ is the acid and NH_2NH_2 is the base.

6A.5 (a) Brønsted acid: HNO_3; Brønsted base: HPO_4^{2-};
(b) conjugate base to HNO_3, NO_3^-; conjugate acid to HPO_4^{2-}, $H_2PO_4^-$

6A.7 (a) $HClO_3$ (chloric acid); conjugate base, ClO_3^-

(b) HNO$_2$ (nitrous acid); conjugate base, NO$_2^-$

6A.9 (a) proton transferred from NH$_4^+$ to H$_2$O, NH$_4^+$ (acid), H$_2$O (base); (b) proton transferred from NH$_4^+$ to I$^-$, NH$_4^+$ (acid), I$^-$ (base); (c) no proton transferred; (d) proton transferred from NH$_4^+$ to NH$_2^-$, NH$_4^+$ (acid), NH$_2^-$ (base)

6A.11 (a) HCO$_3^-$ as an acid: HCO$_3^-$(aq) + H$_2$O(l) $\rightleftharpoons$ H$_3$O$^+$(aq) + CO$_3^{2-}$(aq), HCO$_3^-$ (acid) and CO$_3^{2-}$ (base), H$_2$O (base) and H$_3$O$^+$ (acid); HCO$_3^-$ as a base: H$_2$O(l) + HCO$_3^-$(aq) $\rightleftharpoons$ H$_2$CO$_3$(aq) + OH$^-$(aq), HCO$_3^-$ (base) and H$_2$CO$_3$ (acid), H$_2$O (acid) and OH$^-$ (base); (b) HPO$_4^{2-}$ as an acid: HPO$_4^{2-}$(aq) + H$_2$O(l) $\rightleftharpoons$ H$_3$O$^+$(aq) + PO$_4^{3-}$(aq), HPO$_4^{2-}$ (acid) and PO$_4^{3-}$ (base), H$_2$O (base) and H$_3$O$^+$ (acid); HPO$_4^{2-}$ as a base: HPO$_4^{2-}$(aq) + H$_2$O(l) $\rightleftharpoons$ H$_2$PO$_4^-$(aq) + OH$^-$(aq), HPO$_4^{2-}$ (base) and H$_2$PO$_4^-$ (acid), H$_2$O (acid) and OH$^-$ (base)

6A.13 The Lewis structures of (a) to (e) are:

(a) Lewis base (b) Lewis acid (c) Lewis acid

(d) Lewis base (e) Lewis base

6A.15 (a)

Lewis acid   Lewis base   Product

(b)

Lewis acid   Lewis base   Product

6A.17 (a) basic; (b) acidic; (c) amphoteric; (d) amphoteric

6A.19 (a) [OH$^-$] = 5.0 × 10^{-13} mol·L^{-1}; (b) [OH$^-$] = 1.0 × 10^{-9} mol·L^{-1}; (c) [OH$^-$] = 3.2 × 10^{-12} mol·L^{-1}

6A.21 (a) [H$_3$O$^+$] = 1.4 × 10^{-7} mol·L^{-1} (b) [OH$^-$] = [H$_3$O$^+$] = 1.4 × 10^{-7} mol·L^{-1}

6A.23 [Ba(OH)$_2$]$_0$ = 2.5 × 10^{-2} mol·L^{-1} = [Ba^{2+}]; [OH$^-$] = 5.0 × 10^{-2} mol·L^{-1}; [H$_3$O$^+$] = 2.0 × 10^{-13} mol·L^{-1}

6B.1 ΔpH = 0.92

6B.3 (a) The pH of the desired solution = 1.6. (b) The actual pH = 1.7

6B.5 (a) pH = 1.84, pOH = 12.16; (b) pH = 0.96, pOH = 13.04; (c) pOH = 1.74, pH = 12.26; (d) pOH = 3.15, pH = 10.85; (e) pOH = 3.01, pH = 10.99; (f) pH = 4.28, pOH = 9.72

6B.7 (a) [H$_3$O$^+$] = 5 × 10^{-4} mol·L^{-1}; (b) [H$_3$O$^+$] = 2 × 10^{-7} mol·L^{-1}; (c) [H$_3$O$^+$] = 4 × 10^{-5} mol·L^{-1}; (d) [H$_3$O$^+$] = 5 × 10^{-6} mol·L^{-1}

6B.9 (a)

	[H$_3$O$^+$]/ (mol·L^{-1})	[OH$^-$]/ (mol·L^{-1})	pH	pOH
(i)	**1.50**	1.50 × 10^{-14}	0.176	13.824
(ii)	1.50 × 10^{-14}	**1.50**	13.824	0.176
(iii)	0.18	5.6 × 10^{-14}	**0.75**	13.25
(iv)	5.6 × 10^{-14}	0.18	13.25	**0.75**

(b) (ii) < (iv) < (iii) < (i)

6B.11 (a) (i) [OH$^-$] = 0.18 mol·L^{-1}; (ii) [OH$^-$] = 18 mol·L^{-1}; (b) 110 grams Na$_2$O

6C.1 (i) HClO$_2$
(a) HClO$_2$(aq) + H$_2$O(l) $\rightleftharpoons$ H$_3$O$^+$(aq) + ClO$_2^-$(aq)
$$K_a = \frac{[H_3O^+][ClO_2^-]}{[HClO_2]}$$
(b) ClO$_2^-$(aq) + H$_2$O(l) $\rightleftharpoons$ HClO$_2$(aq) + OH$^-$(aq)
$$K_b = \frac{[HClO_2][OH^-]}{[ClO_2^-]}$$
(ii) HCN
(a) HCN(aq) + H$_2$O(l) $\rightleftharpoons$ H$_3$O$^+$(aq) + CN$^-$(aq)
$$K_a = \frac{[H_3O^+][CN^-]}{[HCN]}$$
(b) CN$^-$(aq) + H$_2$O(l) $\rightleftharpoons$ HCN(aq) + OH$^-$(aq)
$$K_b = \frac{[HCN][OH^-]}{[CN^-]}$$
(iii) C$_6$H$_5$OH
(a) C$_6$H$_5$OH(aq) + H$_2$O(l) $\rightleftharpoons$ H$_3$O$^+$(aq) + C$_6$H$_5$O$^-$(aq)
$$K_a = \frac{[H_3O^+][C_6H_5O^-]}{[C_6H_5OH]}$$
(b) C$_6$H$_5$O$^-$(aq) + H$_2$O(l) $\rightleftharpoons$ C$_6$H$_5$OH(aq) + OH$^-$(aq)
$$K_b = \frac{[C_6H_5OH][OH^-]}{[C_6H_5O^-]}$$

6C.3 (a)

Acid	pK$_{a1}$	K$_{a1}$
(i) H$_3$PO$_4$	2.12	7.6 × 10^{-3}
(ii) H$_3$PO$_3$	2.00	1.0 × 10^{-2}
(iii) H$_2$SeO$_3$	2.46	3.5 × 10^{-3}
(iv) HSeO$_4^-$	1.92	1.2 × 10^{-2}

(b) H$_2$SeO$_3$ < H$_3$PO$_4$ < H$_3$PO$_3$ < HSeO$_4^-$

6C.5 HCO$_2^-$ pK$_b$ = pK$_w$ − pK$_a$ = 14.0 − 3.75 = 10.25

6C.7 (CH$_3$)$_2$NH$_2^+$ (14.00 − 3.27 = 10.73) < $^+$NH$_3$OH (14.00 − 7.97 = 6.03) < HNO$_2$ (3.37) < HClO$_2$ (2.00)

6C.9 F$^-$ (14.00 − 3.45 = 10.55) < CH$_3$CO$_2^-$ (14.00 − 4.75 = 9.25) < C$_5$H$_5$N (8.75) ≪ NH$_3$ (4.75)

6C.11 2,4,6-Trichlorophenol is the stronger acid.

6C.13 (1) arylamines < ammonia < alkylamines; (2) methyl < ethyl < etc.

6C.15 HIO_3 is the stronger acid, with the lower pK_a.

6C.17 Hypobromite ion is a stronger base.

6C.19 (a) HCl is stronger;   (b) $HClO_2$ is stronger;
(c) $HClO_2$ is stronger;   (d) $HClO_4$ is stronger;
(e) HNO_3 is stronger;   (f) H_2CO_3 is stronger

6C.21 (a) Trichloroacetic acid is the stronger acid.   (b) Formic acid is a slightly stronger acid than acetic acid.

6D.1 (a) pH = 2.72; pOH = 11.28; % deprotonation = 0.95%;
(b) pH = 0.85; pOH = 13.15; % deprotonation = 70%;
(c) pH = 2.22; pOH = 11.78; % deprotonation = 3.0%;
(d) Acidity increases when the hydrogen atoms in the methyl group of acetic acid are replaced by atoms that have a higher electronegativity.

6D.3 (a) $K_a = 0.09$; $pK_a = 1.0$;   (b) $K_b = 5.6 \times 10^{-4}$; $pK_b = 3.25$

6D.5 (a) pOH = 3.00; pH = 11.00; percentage protonation = 1.8%;
(b) pOH = 4.38; pH = 9.62; percentage protonation = 0.026%;
(c) pOH = 2.32; pH = 11.68; percentage protonation = 1.4%;
(d) pOH = 3.96; pH = 10.04; percentage protonation = 2.5%

6D.7 (a) $[HClO] = 0.021$ mol·L^{-1};
(b) $[NH_2NH_2] = 1.5 \times 10^{-2}$ mol·L^{-1}

6D.9 pH = 2.58; $K_a = 6.5 \times 10^{-5}$

6D.11 (a) less than 7;   (b) greater than 7;   (c) greater than 7;
(d) neutral;   (e) less than 7;   (f) less than 7

6D.13 The rank in increasing solution pH is   (c) < (a) < (b) < (d).

6D.15 (a) pH = 5.00;   (b) pH = 3.06

6D.17 (a) pH = 9.28;   (b) pH = 11.56

6D.19 pH = 5.42

6D.21 The formula of the acid is HBrO.

6E.1 pH = 0.80

6E.3 (a) pH = 4.18;   (b) pH = 1.28;   (c) pH = 3.80

6E.5 (a) pH = 4.37;   (b) pH = 4.37.

6E.7 (a) pH = 4.55;   (b) pH = 6.17

6E.9 $[H_2CO_3] = 0.0455$ mol·L^{-1};
$[H_3O^+] = [HCO_3^-] = 1.4 \times 10^{-4}$ mol·L^{-1};
$[CO_3^{2-}] = 5.6 \times 10^{-11}$ mol·L^{-1},
$[OH^-] = 7.1 \times 10^{-11}$ mol·L^{-1}.

6E.11 $[H_2CO_3] = 2.3 \times 10^{-8}$ mol·L^{-1};
$[OH^-] = [HCO_3^-] = 0.0028$ mol·L^{-1};
$[CO_3^{2-}] = 0.0428$ mol·L^{-1},
$[H_3O^+] = 3.6 \times 10^{-12}$ mol·L^{-1}.

6E.13 $[HSO_3^-] = 0.14$ mol·L^{-1}; $[H_2SO_3] = 3.2 \times 10^{-5}$ mol·L^{-1};
$[SO_3^{2-}] = 0.0054$ mol·L^{-1}

6E.15 (a) pH = 6.54;   (b) pH = 2.12;   (c) pH = 1.49

6E.17 $[H_3PO_4] = 6.4 \times 10^{-3}$ mol·L^{-1},
$[H_2PO_4^-] = 8.6 \times 10^{-3}$ mol·L^{-1},
$[HPO_4^{2-}] = 9.5 \times 10^{-8}$ mol·L^{-1},
$[PO_4^{3-}] = 3.5 \times 10^{-18}$ mol·L^{-1}.

6F.1 pH = 6.174

6F.3 pH = 7.205

6F.5 $[HA]_{initial} = 1.0 \times 10^{-6}$ mol·L^{-1}

6F.7 pH = 5.04; the pH does not change

6F.9 (a) pH = 6.69; pH = 7.22;   (b) pH = 6.64; pH = 6.92

6G.1 (a) When solid sodium acetate is added to an acetic acid solution, the concentration of H_3O^+ decreases.
(b) The percentage of benzoic acid that is deprotonated decreases.   (c) The concentration of OH^- decreases.

6G.3 (a) pH = pK_a = 3.52, $K_a = 3.02 \times 10^{-4}$;   (b) pH = 3.22

6G.5 (a) $[H_3O^+] = 6.1 \times 10^{-10}$ mol·L^{-1};
(b) $[H_3O^+] = 1.7 \times 10^{-11}$ mol·L^{-1};
(c) $[H_3O^+] = 2.5 \times 10^{-10}$ mol·L^{-1};
(d) $[H_3O^+] = 5.9 \times 10^{-9}$ mol·L^{-1}

6G.7 (a) pH = 1.62, pOH = 12.38;
(b) pH = 1.22, pOH = 12.78;
(c) pH = pK_a = 1.92, pOH = 12.08

6G.9 (a) pH = 9.46;   (b) pH = 9.71;
(c) pH = pK_a = 9.31

6G.11 (a) pH = 4.75 (initial pH); pH = 6.3 (after adding NaOH); ΔpH = 1.55;   (b) pH = 4.75 (initial pH); pH = 4.58 (after adding NaOH); ΔpH = −0.17

6G.13 ΔpH = 1.16

6G.15 $\dfrac{[ClO^-]}{[HClO]} = 9.3 \times 10^{-2}$

6G.17 (a) pK_a = 3.08; pH range 2–4;   (b) pK_a = 4.19; pH range 3–5;   (c) pK_{a3} = 12.68; pH range 11.5–13.5;
(d) pK_{a2} = 7.21; pH range 6–8;   (e) pK_b = 7.97, pK_a = 6.03; pH range 5–7

6G.19 (a) $HClO_2$ and $NaClO_2$, pK_a = 2.00;
(b) NaH_2PO_4 and Na_2HPO_4, pK_{a2} = 7.21;
(c) $CH_2ClCOOH$ and $NaCH_2ClCO_2$, pK_a = 2.85;
(d) Na_2HPO_4 and Na_3PO_4, pK_a = 12.68

6G.21 (a) $\dfrac{[CO_3^{2-}]}{[HCO_3^-]} = 5.6$;   (b) 77 g K_2CO_3;
(c) 1.8 g $KHCO_3$;   (d) 2.8×10^2 mL

6H.1 (a) The titration curve is (see example 6H.1 for calculations):

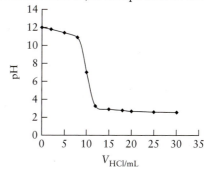

(b) 10.0 mL;   (c) 5.00 mL

6H.3 (a) 9.17×10^{-3} L HCl(aq);   (b) 0.0183 L;
(c) $[Na^+] = 0.0635$ mol·L^{-1};   (d) pH = 2.25

6H.5 (a) pOH = −log(0.110) = 0.959, pH = 14.00 − 0.959 = 13.04;   (b) pH = 12.82;   (c) pH = 12.55;
(d) pH = 7.00;   (e) pH = 1.80;   (f) pH = 1.55

6H.7 percent purity = 79.4%

6H.9 pH = 14 − 5.77 = 8.23

6H.11 (a) 234 g·mol^{-1};   (b) pK_a = 3.82

6H.13 (a) initial pH = 2.89;   (b) pH = 4.56;   (c) 25.0 mL NaOH is required to reach the stoichiometric point, and 12.5 mL NaOH is required to reach halfway to the stoichiometric point.;   (d) pH = pK_a = 4.75;   (e) 25.0 mL;
(f) pH = 8.72

6H.15 At a point when 50% of the weak base is neutralized with a strong acid.

6H.17 (a) initial pH = 11.20; (b) pH = 8.99; (c) 11. mL; (d) pH = 9.25; (e) 22 mL; (f) pH = 5.24; (g) Methyl red will be appropriate for this titration.

6H.19 (a) The acid is a weak acid; (b) $[H_3O^+] = 1 \times 10^{-5} \ mol \cdot L^{-1}$; (c) $K_a = 3 \times 10^{-8}$; (d) $[HA]_{initial} = 3 \times 10^{-3} \ mol \cdot L^{-1}$; (e) $[B] = 8 \times 10^{-3} \ M$; (f) Phenolphthalein will be an appropriate indicator.

6H.21 Phenolphthalein and thymol blue; the others would not.

6H.23 For both Exercise 6H.9 and Exercise 6H.14, thymol blue or phenolphthalein.

6H.25 (a) 0.0374 L or 37.4 mL; (b) 74.8 mL; (c) 112 mL

6H.27 (a) 0.0220 L or 22.0 mL; (b) 44.0 mL

6H.29 (a)

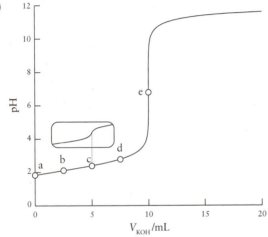

(b) first stoichiometric point: (c) = 5.0 mL; second stoichiometric point: (e) = 10. mL; (c) first stoichiometric point: pH = 2.39; second stoichiometric point: pH = 7.04

6H.31 (a) pH = 1.89; (b) pH = 2.42; (c) pH = 5.91

6H.33 (a) pH = 4.66; (b) pH = 2.80; (c) pH = 7.21

6I.1 (a) $K_{sp} = 7.7 \times 10^{-13}$; (b) $K_{sp} = 1.7 \times 10^{-14}$; (c) $K_{sp} = 5.3 \times 10^{-3}$; (d) $K_{sp} = 6.9 \times 10^{-9}$

6I.3 $K_{sp} = 1.0 \times 10^{-12}$

6I.5 (a) $S = 1.30 \times 10^{-5} \ mol \cdot L^{-1}$; (b) $S = 1.0 \times 10^{-3} \ mol \cdot L^{-1}$; (c) $S = 9.3 \times 10^{-5} \ mol \cdot L^{-1}$

6I.7 (a) $S = 8.0 \times 10^{-10} \ mol \cdot L^{-1}$; (b) $S = 1.2 \times 10^{-16} \ mol \cdot L^{-1}$; (c) $S = 4.6 \times 10^{-3} \ mol \cdot L^{-1}$; (d) $S = 1.3 \times 10^{-6} \ mol \cdot L^{-1}$

6I.9 (a) $S = 1.0 \times 10^{-12} \ mol \cdot L^{-1}$; (b) $S = 3.1 \times 10^{-5} \ mol \cdot L^{-1}$; (c) $S = 2.0 \times 10^{-3} \ mol \cdot L^{-1}$; (d) $S = 0.20 \ mol \cdot L^{-1}$

6I.11 $S = 2.0 \times 10^{-3} \ mol \cdot L^{-1}$

6J.1 (a) $[Ag^+] = 1.6 \times 10^{-5} \ mol \cdot L^{-1}$; (b) $2.7 \times 10^2 \ \mu g \ AgNO_3$

6J.3 (a) pH = 6.00; (b) pH = 6.18

6J.5 (a) will precipitate because $Q_{sp}(2 \times 10^{-7}) > K_{sp}(1.6 \times 10^{-10})$; (b) will not precipitate because $Q_{sp} (1 \times 10^{-11}) < K_{sp} (1.6 \times 10^{-5})$

6J.7 (a) the order of precipitation is $Ni(OH)_2$, $Mg(OH)_2$, $Ca(OH)_2$; (b) pH = 14.00 − 1.13 = 12.87

6J.9 CO_3^{2-} is the better choice of anion.

6J.11 $[Pb^{2+}] = 1.8 \times 10^{-9} \ mol \cdot L^{-1}$

6J.13 (a) $K = K_{sp} \cdot K_b^2 = 3.4 \times 10^{-32}$; (b) (i) $2.0 \times 10^{-4} \ mol \cdot L^{-1}$; (ii) $4 \times 10^{-4} \ mol \cdot L^{-1}$

6J.15 The two salts can be distinguished by their solubility in NH_3.

6J.17 In order to use qualitative analyses, the sample must first be dissolved. This can be accomplished by digesting the sample with concentrated HNO_3 and then diluting the resulting solution. Once the sample is dissolved and diluted, an aqueous solution containing chloride ions can be introduced. This should precipitate the Ag^+ as $AgCl$ but would leave the bismuth and nickel in solution, as long as the solution was acidic. The remaining solution can then be treated with H_2S. In acidic solution, Bi_2S_3 will precipitate but NiS will not. Once the Bi_2S_3 has been precipitated, the pH of the solution can be raised by addition of base. Once this is done, NiS should precipitate.

6K.1 (a) Cr is reduced from +6 to +3; C oxidized from −2 to −1; (b) $C_2H_5OH(aq) \rightarrow C_2H_4O(aq) + 2 H^+(aq) + 2 e^-$; (c) $Cr_2O_7^{2-}(aq) + 14 H^+(aq) + 6 e^- \rightarrow$ $2 Cr^{3+}(aq) + 7 H_2O(l)$; (d) $8 H^+(aq) + Cr_2O_7^{2-}(aq) + 3 C_2H_5 OH(aq) \rightarrow$ $2 Cr^{3+}(aq) + 3 C_2H_4 O(aq) + 7 H_2O(l)$

6K.3 (a) $4 Cl_2(g) + S_2O_3^{2-}(aq) + 5 H_2O(l) \rightarrow 8 Cl^-(aq) + 2 SO_4^{2-}(aq) +10 H^+(aq)$; Cl_2 is the oxidizing agent and $S_2O_3^{2-}$ is the reducing agent. (b) $2 MnO_4^-(aq) + H^+(aq) + 5 H_2SO_3(aq) \rightarrow 2 Mn^{2+}(aq) + 3 H_2O(l) + 5 HSO_4^-(aq)$; is the oxidizing agent and H_2SO_3 is the reducing agent. (c) $Cl_2(g) + H_2S(aq) \rightarrow 2 Cl^-(aq) + S(s) + 2 H^+(aq)$; is the oxidizing agent and H_2S is the reducing agent. (d) $Cl_2(g) + H_2O(l) \rightarrow HOCl(aq) + Cl^-(aq) + H^+(aq)$; Cl_2 is both the oxidizing and the reducing agent.

6K.5 (a) $3 O_3(g) + Br^-(aq) \rightarrow 3 O_2(g) + BrO_3^-(aq)$; is the oxidizing agent and Br^- is the reducing agent. (b) $3 Br_2(l) + 6 OH^-(aq) \rightarrow 5 Br^-(aq) + BrO_3^-(aq) + 3 H_2O(l)$; is both the oxidizing agent and the reducing agent. (c) $2 Cr^{3+}(aq) + 4 OH^-(aq) + 3 MnO_2(s) \rightarrow 2 CrO_4^{2-}(aq) + 2 H_2O(l) + 3 Mn^{2+}(aq)$; is the reducing agent and MnO_2 is the oxidizing agent.; (d) $P_4(s) + 3 OH^-(aq) + 3 H_2O(l) \rightarrow 3 H_2PO_2^-(aq) + PH_3(g)$; is both the oxidizing and the reducing agent.

6K.7 half reactions: $NO_3^-(aq) + 4 H^+(aq) + 3 e^- \rightarrow NO(aq) + 2 H_2O(l)$

$P_4S_3(aq) + 28 H_2O(l) \rightarrow$ $4 H_3PO_4(aq) + 3 SO_4^{2-}(aq) + 44 H^+(aq) + 38 e^-$; overall reaction: $3 P_4S_3(aq) + 38 NO_3^-(aq) + 20 H^+(aq) + 8 H_2O(l) \rightarrow 12 H_3PO_4(aq) + 9 SO_4^{2-}(aq) + 38 NO(g)$

6L.1 (a) $-2.08 \times 10^5 \ J. \ mol^{-1}$; (b) $7.47 \times 10^5 \ J. \ mol^{-1}$

6L.3 (a) anode: $Ni(s) \rightarrow Ni^{2+}(aq) + 2 e^-$, cathode: $Ag^+(aq) + e^- \rightarrow Ag(s)$; overall: $2 Ag^+(aq) + Ni(s) \rightarrow 2 Ag(s) + Ni^{2+}(aq)$; (b) anode: $H_2(g) \rightarrow 2 H^+(aq) + 2 e^-$; cathode: $Cl_2(g) + 2 e^- \rightarrow 2 Cl^-(aq)$; overall: $Cl_2(g) + H_2(g) \rightarrow 2 H^+(aq) + 2 Cl^-(aq)$; (c) anode: $Cu(s) \rightarrow Cu^{2+}(aq) + 2 e^-$; cathode: $Ce^{4+}(aq) + e^- \rightarrow Ce^{3+}(aq)$; overall: $2 Ce^{4+}(aq) + Cu(s) \rightarrow Cu^{2+}(aq) + 2 Ce^{3+}(aq)$; (d) anode: $2 H_2O(l) \rightarrow O_2(g) + 4 H^+(aq) + 4 e^-$; cathode: $O_2(g) + 2 H_2O(l) + 4 e^- \rightarrow 4 OH^-(aq)$;

overall: $H_2O(l) \rightarrow H^+(aq) + OH^-(aq)$;
(e) anode: $Sn^{2+}(aq) \rightarrow Sn^{4+}(aq) + 2\,e^-$;
cathode: $Hg_2Cl_2(s) + 2\,e^- \rightarrow 2\,Hg(l) + 2\,Cl^-(aq)$;
overall: $Sn^{2+}(aq) + Hg_2Cl_2(s) \rightarrow$
$$2\,Hg(l) + 2\,Cl^-(aq) + Sn^{4+}(aq)$$

6L.5 (a) anode: $Zn(s) \rightarrow Zn^{2+}(aq) + 2\,e^-$,
cathode: $Ni^{2+}(aq) + 2\,e^- \rightarrow Ni(s)$;
overall: $Ni^{2+}(aq) + Zn(s) \rightarrow Ni(s) + Zn^{2+}(aq)$;
$Zn(s)|Zn^{2+}(aq)\|Ni^{2+}(aq)|Ni(s)$;
(b) anode: $2\,I^-(aq) \rightarrow 2\,e^- + I_2(s)$,
cathode: $Ce^{4+}(aq) + e^- \rightarrow Ce^{3+}(aq)$;
overall: $2\,I^-(aq) + 2\,Ce^{4+}(aq) \rightarrow 2\,Ce^{3+}(aq) + I_2$ (s);
$Pt(s)|I^-(aq)|I_2(s)\|Ce^{4+}(aq),\ Ce^{3+}(aq)|Pt(s)$;
(c) anode: $H_2(g) \rightarrow 2\,H^+(aq) + 2\,e^-$,
cathode: $Cl_2(g) + 2\,e^- \rightarrow 2\,Cl^-(aq)$;
overall: $H_2(g) + Cl_2(g) \rightarrow 2\,HCl(aq)$;
$Pt(s)|H_2(g)|H^+(aq)\|Cl^-(aq)|Cl_2(g)|Pt(s)$;
(d) anode: $Au(s) \rightarrow Au^{3+}(aq) + 3\,e^-$,
cathode: $Au^+(aq) + e^- \rightarrow Au(s)$;
overall: $3\,Au^+(aq) \rightarrow 2\,Au(s) + Au^{3+}(aq)$,
$Au(s)|Au^{3+}(aq)\|Au^+(aq)|Au(s)$

6L.7 (a) anode: $Ag(s) + Br^-(aq) \rightarrow AgBr(s) + e^-$;
cathode: $Ag^+(aq) + e^- \rightarrow Ag(s)$;
$Ag(s)|AgBr(s)|Br^-(aq)\|Ag^+(aq)|Ag(s)$;
(b) anode: $4\,OH^-(aq) \rightarrow O_2(g) + 2\,H_2O(l) + 4\,e^-$;
cathode: $O_2(g) + 4\,H^+(aq) + 4\,e^- \rightarrow 2\,H_2O(l)$;
$Pt(s)|O_2(g)|OH^-(aq)\|H^+(aq)|O_2(g)|Pt(s)$;
(c) anode: $Cd(s) + 2\,OH^-(aq) \rightarrow Cd(OH)_2(s) + 2\,e^-$;
cathode: $Ni(OH)_3(s) + e^- \rightarrow Ni(OH)_2(s) + OH^-(aq)$;
$Cd(s)|Cd(OH)_2(s)|KOH(aq)\|Ni(OH)_3(s)|Ni(OH)_2(s)|Ni(s)$

6L.9 (a) anode: $Fe^{2+}(aq) \rightarrow Fe^{3+}(aq) + e^-$;
cathode: $MnO_4^-(aq) + 8\,H^+(aq) + 5\,e^- \rightarrow$
$$Mn^{2+}(aq) + 4\,H_2O(l);$$
(b) $MnO_4^-(aq) + 5\,Fe^{2+}(aq) + 8\,H^+(aq) \rightarrow$
$$Mn^{2+}(aq) + 5\,Fe^{3+}(aq) + 4\,H_2O(l)$$
$Pt(s)\ |\ Fe^{3+}(aq),\ Fe^{2+}(aq)\ \|\ H^+(aq),$
$$MnO_4^-(aq),\ Mn^{2+}(aq)\ |\ Pt(s)$$

6M.1 -0.349 V

6M.3 (a) $+0.75$ V; (b) $+0.37$ V; (c) $+0.52$ V; (d) $+1.52$ V

6M.5 (a) $E_{cell}° = +0.17$ V; $\Delta G_r° = -98$ kJ·mol^{-1};
$Hg(l)\ |\ Hg_2^{2+}(aq)\ \|\ NO_3^-(aq),\ H^+(aq)\ |\ O(g)\ |\ Pt(s)$;
(b) not spontaneous;
(c) $E_{cell}° = +0.36$ V; $\Delta G_r° = -208$ kJ·mol^{-1};
$Pt(s)\ |\ Pu^{3+}(aq),\ Pu^{4+}(aq)\ \|\ Cr_2O_7^{2-}(aq),$
$$Cr^{3+}(aq),\ H^+(aq)\ |\ Pt(s)$$

6M.7 (a) Cu < Fe < Zn < Cr; (b) Mg < Na < K < Li;
(c) V < Ti < Al < U; (d) Au < Ag < Sn < Ni

6M.9 -1.50 V

6M.11 (a) oxidizing agent: Co^{2+}; reducing agent: Ti^{2+};
$Pt(s)\ |\ Ti^{2+}(aq),\ Ti^{3+}(aq)\ |\ Co^{2+}(aq)\ |\ Co(s)$, $+0.09$ V;
(b) oxidizing agent: U^{3+}; reducing agent: La;
$La(s)\ |\ La^{3+}(aq)\ |\ U^{3+}(aq)\ |\ U(s)$, $+0.73$ V;
(c) oxidizing agent: Fe^{3+}; reducing agent: H_2;
$Pt(s)\ |\ H_2(g)\ |\ H^+(aq)\ |\ Fe^{2+}(aq),\ Fe^{3+}(aq)\ |\ Pt(s)$, $+0.77$ V;
(d) oxidizing agent: O_3; reducing agent: Ag;
$Ag(s)\ |\ Ag^+(aq)\ |\ OH^-(aq)\ |\ O_3(g),\ O_2(g)\ |\ Pt(s)$, $+0.44$ V

6M.13 (a) $Cl_2(g)$, $+0.27$ V; (b) and (c) do not favor products;
(d) $NO_3^-(aq)$, $+1.56$ V

6N.1 (a) 6×10^{-16}; (b) 1×10^4

6N.3 (a) $+0.067$ V; (b) $+0.51$ V; (c) -1.33 V; (d) $+0.31$ V

6N.5 (a) pH = 1.0; (b) $[Cl^-] = 10^{-1}$ mol·L^{-1}

6N.7 (a) $+0.030$ V; (b) $+0.06$ V

6N.9 pH = 2.25

6N.11 (a) $K_{sp} = 1.2 \times 10^{-17}$; (b) The calculated value is a
factor of 10 greater than the experimentally determined
(measured) value (1.3×10^{-18}).

6N.13 (a) $Q = 10^6$, $Pb^{4+}(aq) + Sn^{2+}(aq) \rightarrow Pb^{2+}(aq) + Sn^{4+}(aq)$;
(b) $Q = 1.0$, $2\,Cr_2O_7^{2-}(aq) + 16\,H^+(aq) \rightarrow 4\,Cr^{3+}(aq) +$
$8\,H_2O(l) + 3\,O_2(g)$

6N.15 $+0.3$ V

6N.17 yes; 8.4 kJ per mole of Ag

6N.19 (a) -0.27 V; (b) $+0.07$ V

6N.21 (a) $Fe_2O_3 \cdot H_2O$; (b) H_2O and O_2 jointly oxidize iron.
(c) Water is more highly conducting if it contains
dissolved ions, so the rate of rusting is increased.

6N.23 (a) aluminum or magnesium; (b) cost, availability, and
toxicity of products in the environment; (c) Fe could act as
the anode of an electrochemical cell if Cu^{2+} or Cu^+ were
present; therefore, it could be oxidized at the point of contact.
Water with dissolved ions would act as the electrolyte.

6O.1 (a) cathode: $Ni^{2+}(aq) + 2\,e^- \rightarrow Ni(s)$;
(b) anode: $2\,H_2O(l) \rightarrow O_2(g) + 4\,H^+(aq) + 4\,e^-$;
(c) $+1.46$ V

6O.3 (a) water; (b) water; (c) Ni^{2+}; (d) Al^{3+}

6O.5 (a) 3.3 g; (b) 0.54 L; (c) 0.94 g

6O.7 (a) 27 h; (b) 0.44 g

6O.9 (a) 1.1 A; (b) 0.40 A

6O.11 $+2$

6O.13 (a) anode; (b) 0.134 mol; (c) pH = 0.17

6O.15 (a) 57.4%; (b) AgBr

6.1 (a) Acid (1) is a strong acid; (b) Acid (3) has the
strongest base; (c) Acid (3) has the largest pK_a.

6.3 $\Delta H_r° = 1.3 \times 10^2$ J·mol^{-1}

6.5 (a) The reaction is: $H_2O_2(g) + SO_3(g) \rightarrow H_2SO_5(g)$
(b) H_2O_2, $H-\ddot{O}-\ddot{O}-H$
SO_3,

H_2SO_5,

(c) H_2O_2 = Lewis base; SO_3 = Lewis acid

6.7 (a) NaC_2HO_4
(b)

(c) The dissolved substance is sodium oxalate,
amphiprotic; pH = 2.7

6.9 (a) Nitrous acid will act like a strong acid.
(b) Ammonia will act like a strong base.

6.11 $K = \dfrac{K_a(HNO_2) \times K_b(NH_3)}{K_w}$; $K = 7.7 \times 10^5$

6.13 The Lewis structure of boric acid is

H—Ö:
    |
    B—Ö:
    |
H—Ö: H

Boric acid

$\left[\begin{array}{c}:\ddot{O}-H \\ | \\ H-\ddot{O}-B-\ddot{O}-H \\ | \\ H-\ddot{O}: \end{array}\right]^{-}$

Conjugate base

$B(OH)_3$ is a very weak acid because it does not have a conjugate system to delocalize the electrons on the oxygen to weaken the O—H bond. (b) In that reaction, boric acid acts as a Lewis acid.

6.15 $K = 6.9 \times 10^{-4}$

6.17 (a) $D_2O + D_2O \rightleftharpoons D_3O^+ + OD^-$; (b) $pK_{hw} = 14.870$;
(c) $[D_3O^+] = [OD^-] = 3.67 \times 10^{-8}\ mol \cdot L^{-1}$;
(d) $pD = 7.435 = pOD$; (e) $pD + pOD = pK_{hw} = 14.870$

6.19 (a) The H^+ in lactic acid can interact with HbO_2^- to produce HHb and the concentration of HbO_2^- will be lower in the tissues. (b) The concentration of HbO_2^- will increase.

6.21 (a) Two protons can be accepted.
(b)

(c) Each of the two nitro- groups will show amphiprotic behavior in aqueous solution (can either accept a proton and donate a proton).

6.23 The initial buffer solution must contain at least 0.026 mol CH_3COOH and 0.026 mol $NaCH_3CO_2$. The concentration of the initial solution will then be 0.260 M in both acetic acid and sodium acetate.

6.25 (a) $pH = pK_{a1} = 2.8$; (b) $pH = pK_{a2} = 5.7$; (c) At $pH = 4.2$, $HOOCCH_2CO_2^-$ is the predominant species.

6.27 2.2×10^{-3} M

6.29 (a) $[CO_3^{2-}] = 0.0796\ mol \cdot L^{-1}\ CO_3^{2-}$
(b)

At stoichiometric point, pAg = 3.94.

6.31 For Friend #1: If assuming all CO_2 will be escaped from stomach, the pH in the stomach will be 7.00; for Friend #2: pH = 13. 64.

6.33 $\Delta G_f^{\circ}(PbF_2, s) = -345.65\ kJ \cdot mol^{-1}$

6.35 (a) pH = 4.93; (b) mass of NaOH = 2.08 g NaOH

6.37 $[N]/[HN^+] = 10^{-1.55} = 2.8 \times 10^{-2}$

6.39 (a) $3 \times 10^{-18}\ mol\ CO_2(aq)$; (b) $\Delta pH = -0.3$

6.41 Because $Q_{sp}(CaF_2) < K_{sp}(CaF_2)$, there will be no CaF_2 precipitation at this condition.

6.43 (a) E_{cell}; (b) both; (c) neither; (d) E_{cell}°; (e) E

6.45 Al, Zn, Fe, Co, Ni, Cu, Ag, Au

6.47 -0.92 V

6.49 $+0.14$ V

6.51 A negatively charged electrolyte flows from the cathode to the anode.

6.53 (a) Reduction takes place at the electrode with the higher concentration, which would be the chromium electrode in contact with the 1.0 M $CrCl_3(aq)$. (b) Dilution of the anode concentration would increase the cell potential; (c) Formation of insoluble $Cr(OH)_3$ will decrease the Cr^{3+} concentration and therefore decrease the cell potential. (d) no effect

6.55 (a) 1.0×10^{-2} M; 8.5×10^{-17}

6.57 2.5×10^{-3}

6.59 (a) $E_2^{\circ} = E_1^{\circ} + \Delta S_r^{\circ}(T_2 - T_1)/n_r F$; (b) $+1.18$ V

6.61 0.08 V to 0.09 V

6.63 pH = 12

6.65 $+0.828$ V to -0.828 V

6.67 (a) A plot of cell potential, E_{cell}, versus $\ln [Ag^+]_{anode}$ would be a linear increase with positive slope; (b) slope = 0.025 693 V, corresponding to $\dfrac{RT}{n_r F}$, consistent;
(c) y-intercept is E_{cell}°, and for all concentration cells $E_{cell}^{\circ} = 0$.

6.69 4×10^{-9}

6.71 0.205 A

6.73 (a) oxidation/anode: $6\ OH^-(aq) + 2\ Al(s) \rightarrow 2\ Al(OH)_3(aq) + 6\ e^-$;
reduction/cathode: $3\ H_2O(l) + \frac{3}{2}O_2(g) + 6\ e^- \rightarrow 6\ OH^-(aq)$;
(b) $+2.06$ V

6.75 (a) $Pb(s) + PbO_2(s) + 2\ HSO_4^-(aq) + H^+(aq) \rightarrow PbSO_4(s) + 2\ H_2O(l)$;
(b) increases; decreases; stays the same

6.77 (a) $KOH(aq)/HgO(s)$; (b) $HgO(s)$;
(c) $HgO(s) + Zn(s) \rightarrow Hg(l) + ZnO(s)$

6.79 100 A

6.81 (a) (i) $E^* = 0 - 0.41\ V = -0.41$ V;
(ii) $E^* = +0.96\ V - 0.55\ V = +0.41$ V;
(b) $E^* = -0.099\ V - 0.207\ V = -0.31$ V
(c)   (d)

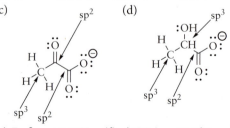

(e) $E_{cell}^{\circ} = +0.33V$; (f) $\Delta G_r^{\circ} = -63.7\ kJ \cdot mol^{-1} = -64\ kJ \cdot mol^{-1}$; (g) $K = +1.43 \times 10^{11}$

Focus 7

7A.1 (a) one-third; (b) two thirds; (c) two

7A.3 (a) $1.3 \text{ mol·L}^{-1}\text{·s}^{-1}$; (b) $0.88 \text{ mol·L}^{-1}\text{·s}^{-1}$

7A.5 (a) and (c); Note that the curves for $[I_2]$ and $[H_2]$ are identical and only the $[I_2]$ curve is shown.

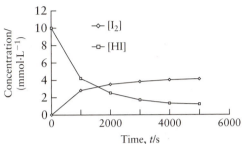

(b)

Time, t/s	Rate/(mmol·L^{-1}·s^{-1})
0	0.0060
1000	0.003
2000	0.00098
3000	0.00061
4000	0.00040
5000	0.00031

7A.7 (a) $\text{mol·L}^{-1}\text{·s}^{-1}$; (b) s^{-1}; (c) $\text{L·mol}^{-1}\text{·s}^{-1}$;

7A.9 $2.2 \times 10^{-4} \text{ (mol N}_2\text{O}_5)\text{·L}^{-1}\text{·s}^{-1}$

7A.11 (a) $2.2 \times 10^{-5} \text{ mol·L}^{-1}\text{·s}^{-1}$; (b) factor of 2

7A.13 rate $= k_r[CH_3Br][OH^-]$

7A.15 (a) first order with respect to A, second-order with B, zeroth-order with respect to C, third-order overall;
(b) rate $= k_r[A][B]^2$; (c) $k_r = 2.0 \times 10^{-5} \text{ L}^2\text{·mmol}^{-2}\text{·s}^{-1}$;
(d) $2.9 \times 10^{-6} \text{ mmol·L}^{-1}\text{·}^{-1}$

7A.17 (a) rate $= k_r[A][B]^2[C]^2$; (b) overall order $= 5$
(c) $k_r = 2.85 \times 10^{12} \text{ L}^4\text{·mmol}^4\text{·s}^{-1}$; (d) $1.13 \times 10^{-2} \text{ mmol·L}^{-1}\text{·s}^{-1}$

7B.1 2.0 mg

7B.3 (a) $k_r = 6.93 \times 10^{-4} \text{ s}^{-1}$; (b) $k_r = 9.4 \times 10^{-3} \text{ s}^{-1}$;
(c) $k_r = 5.1 \times 10^{-3}\text{s}^{-1}$

7B.5 (a) 5.2 h; (b) $3.5 \times 10^{-2} \text{ mol·L}^{-1}$; (c) 6.5×10^2 min

7B.7 (a) 1065 s; (b) 710 s; (c) 9.7×10^2 s; (d) 1.1×10^3 s

7B.9 (a) $k_r = 0.17 \text{ min}^{-1}$; (b) an additional 3.5 min

7B.11 (a)

(b) (i) $7.8 \times 10^{-3}\text{L·mol}^{-1}\text{·s}^{-1}$.(ii) $3.9 \times 10^{-3}\text{L·mol}^{-1}\text{·s}^{-1}$

7B.13 (a) 7.4×10^2 s; (b) 1.5×10^2 s; (c) 2.0×10^2 s

7B.15 (a) 247 min; (b) 819 min; (c) 10.9 g

7B.17 (a) 1.7×10^2 min; (b) 3.3×10^3 min

7B.19 $[A]_t = [A]_0 \, e^{-ak_r t}$; $t_{1/2} = \ln 2/ak_r$

7B.21 $t_{1/2} = 3/(2k_r[A]_0^2)$

7C.1 (a) rate $= k_r[NO]^2$; bimolecular; (b) rate $= k_r[Cl_2]$; unimolecular

7C.3 $2 \text{ AC} + B \rightarrow A_2B + 2 \text{ C}$; intermediate is AB

7C.5 (a) $2 \text{ HBr} + NO_2 \rightarrow NO + H_2O + Br_2$;
(b) Step 1: rate $= k_1[HBr][NO_2]$; bimolecular;
Step 2: rate $= k_2[HBr][HOBr]$; bimolecular;
(c) HOBr

7C.7 rate $= k_r[NO][Br_2]$

7C.9 If mechanism (I) were correct, the rate law would be rate $= k_r[NO_2][CO]$. But this expression does not agree with the experimental result and can be eliminated as a possibility. Mechanism (II) has rate $= k_r[NO_2]^2$ from the slow step. Step 2 does not influence the overall rate, but it is necessary to achieve the correct overall reaction; thus this mechanism agrees with the experimental data. Mechanism (III) is not correct, which can be seen from the rate expression for the slow step, rate $= k_r[NO_3][CO]$. [CO] cannot be eliminated from this expression to yield the experimental result, which does not contain [CO].

7C.11 (a) True; (b) False. At equilibrium, the *rates* of the forward and reverse reactions are equal, *not the rate constants*.
(c) False. Increasing the concentration of a reactant causes the rate to increase by providing more reacting molecules. It does not affect the rate constant of the reaction.

7C.13 The overall rate of formation of A is rate $= -k_r[A] + k_r'[B]$. The first term accounts for the forward reaction and is negative as this reaction reduces [A]. The second term, which is positive, accounts for the back reaction which increases [A]. Given the 1:1 stoichiometry of the reaction, if no B were present at the beginning of the reaction, [A] and [B] at any time are related by the equation $[A] + [B] = [A]_0$, where $[A]_0$ is the initial concentration of A. Therefore, the rate law may be written as
$$\frac{d[A]}{dt} = -k_r[A] + k_r'([A]_0 - [A]) = -(k_r + k_r')[A] + k_r'[A]_0$$
The solution of this first-order differential equation is
$$[A] = \frac{k_r' + k_r e^{-(k_r' + k_r)t}}{k_r' + k_r}[A]_0$$
As $t \rightarrow \infty$, the exponential term in the numerator goes to zero and the concentrations reach their equilibrium values given by
$$[A]_{eq} = \frac{k_r'[A]_0}{k_r' + k_r} \text{ and } [B]_{eq} = [A]_0 - [A]_\infty = \frac{k_r[A]_0}{k_r + k_r'}$$
taking the ratio of products over reactants we see that
$$\frac{[B]_{eq}}{[A]_{eq}} = \frac{k_r}{k_r'} = K$$
where K is the equilibrium constant for the reaction.

7D.1 39 kJ·mol^{-1}

7D.3

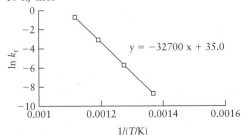

(a) $2.72 \times 10^2 \text{ kJ·mol}^{-1}$; (b) 0.088 s^{-1}

7D.5 2.7×10^{10} mol·L^{-1}·s^{-1}

7D.7 (a) 0.676; (b) endothermic; (c) Raising the temperature will increase the rate constant of the reaction with the higher activation barrier more than it will the rate constant of the reaction with the lower energy barrier. We expect the rate of the forward reaction to go up substantially more than for the reverse reaction in this case. k_r will increase more than k_r' and consequently the equilibrium constant K will increase.

7E.1 (a) In the presence of a catalyst, both the forward and reverse reaction rates will increase. (b) A catalyst will not affect the value of $\Delta H_r°$ for the reaction.

7E.3 (a) 6×10^8; (b) 3×10^7

7E.5 $RCN + H_2O \longrightarrow RCONH_2$; intermediates: $RC(=N^-)OH$, $RC(=NH)OH$; catalyst: OH^-

7E.7 (a) False. A catalyst increases the rate of both the forward and reverse reactions by providing a completely different pathway. (b) True, although a catalyst may be poisoned and lose activity. (c) False. There is a completely different pathway provided for the reaction in the presence of a catalyst. (d) False. The position of the equilibrium is unaffected by the presence of a catalyst.

7E.9 (a) To obtain the Michaelis–Menten rate equation, we will begin by employing the steady-state approximation, setting the rate of change in the concentration of the ES intermediate equal to zero:
$$\frac{d[ES]}{dt} = k_1[E][S] - k'_1[ES] - k_2[ES] = 0$$
Rearranging gives
$$[E][S] = \left(\frac{k_2 + k'_1}{k_1}\right)[ES] = K_M[ES].$$
The total bound and unbound enzyme concentration, $[E]_0$, is given by $[E]_0 = [E] + [ES]$, and, therefore, $[E] = [ES] - [E]_0$. Substituting this expression for $[E]$ into the preceding equation, we obtain $([ES] - [E]_0)[S] = K_M [ES]$
Rearranging to obtain $[ES]$ gives
$$[ES] = \frac{[E]_0[S]}{K_M + [S]}$$
From the mechanism, the rate of appearance of the product is given by rate $= k_2[ES]$. Substituting the preceding equation for $[ES]$, we obtain
$$\text{Rate} = \frac{k_2[E]_0[S]}{K_M + [S]},$$
the Michaelis–Menten rate equation, which can be rearranged to obtain
$$\frac{1}{\text{rate}} = \frac{K_M}{k_2[E]_0[S]} + \frac{1}{k_2[E]_0}.$$
If one plots $\dfrac{1}{\text{rate}}$ versus $\dfrac{1}{[S]}$, the slope will be $\dfrac{K_M}{k_2[E]_0}$ and the y-intercept will be $\dfrac{1}{k_2[E]_0}$.

(b)

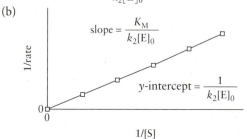

7.1 (a) $CH_3CH=CHCH_2Br$ because the positive charge on the reactive intermediate is on a primary carbon atom. This also agrees with the result that at high temperatures this is the predominant product as there is sufficient energy to overcome the larger activation energy.
(b) Kinetic control predominates at low temperatures. The reaction pathway with the lower activation energy will predominate at low temperatures because the lower activation energy barrier results in a larger rate constant and therefore a faster reaction.

7.3 2.3×10^5 L·mol^{-1}·s^{-1}

7.5 (a) $t_{1/2} = 5$ s; (b) There should be four molecules at $t = 8$ s.

7.7 (a) $CH_3CHO \rightarrow CH_3 + CHO$, unimolecular, $[CH_3\cdots CHO]^\ddagger$; (b) $2\,I \rightarrow I_2$, termolecular, $[I\cdots I\cdots Ar]^\ddagger$; (c) $O_2 + NO \rightarrow NO_2 + O$, bimolecular, $[O\cdots O\cdots NO]^\ddagger$

7.9 The anticipated rate for mechanism (i) is: rate $= k_r[C_{12}H_{22}O_{11}]$, while the expected rate for mechanism (ii) is: rate $= k_r[C_{12}H_{22}O_{11}][H_2O]$. The rate for mechanism (ii) will be pseudo-first-order in dilute solutions of sucrose because the concentration of water will not change. Therefore, in dilute solutions kinetic data cannot be used to distinguish between the two mechanisms. However, in a highly concentrated solution of sucrose, the concentration of water will change during the course of the reaction. As a result, if mechanism (ii) is correct the kinetics will display a first-order dependence on the concentration of H_2O while mechanism (i) predicts that the rate of the reaction is independent of $[H_2O]$.

7.11 (a) If step 2 is the slow step, if step 1 is a rapid equilibrium, and if step 3 is fast also, then our proposed rate law will be rate $= k_2[N_2O_2][H_2]$. Consider the equilibrium of step 1:
$$k_1[NO]^2 = k_1'[N_2O_2]$$
$$[N_2O_2] = \frac{k_1}{k_1'}[NO]^2$$
Substituting in our proposed rate law, we have
rate $= k_2 \frac{k_1}{k_1'}[NO]^2[H_2] = k[NO]^2[H_2]$ where $k = k_2\frac{k_1}{k_1'}$. The assumptions made above reproduce the observed rate law; therefore, step 2 is the slow step.
(b)

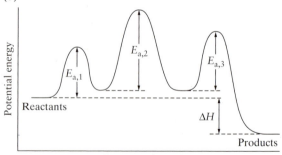

7.13 To get an expression for $t_{1/2}$ in terms of n, we need to evaluate an integral such as:
$$\int_{[A]_0}^{[A]} \frac{d[A]}{[A]^n} = -k_r \int_0^t dt = -k_r t$$
$$\frac{1}{n-1}\left(\frac{1}{[A]^{n-1}} - \frac{1}{[A]_0^{n-1}}\right) = k_r t$$
An expression for $t_{1/2}$ is then

$$\frac{1}{n-1}\left(\frac{2^{n-1}}{[A]_0^{n-1}} - \frac{1}{[A]_0^{n-1}}\right) = k_r t_{1/2}$$

$$\frac{1}{n-1}\left(\frac{2^{n-1} - 1}{[A]_0^{n-1}}\right) = k_r t_{1/2}$$

An expression for $t_{3/4}$ could be found by setting up $[A] = \frac{3}{4}[A]_0$:

$$\frac{1}{n-1}\left(\frac{4^{n-1}}{3^{n-1}[A]_0^{n-1}} - \frac{1}{[A]_0^{n-1}}\right) = k_r t_{3/4}$$

$$\frac{1}{n-1}\left(\frac{(4/3)^{n-1} - 1}{[A]_0^{n-1}}\right) = k_r t_{3/4}$$

The ratio $t_{1/2}/t_{3/4}$ is then

$$t_{1/2}/t_{3/4} = \left(\frac{2^{n-1} - 1}{(4/3)^{n-1} - 1}\right)$$

7.15 1.5×10^{-3} mol·L^{-1}·s^{-1}

7.17 (a) three steps; (b) first step; (c) third step; (d) two; (e) none

7.19 For a third-order reaction,

$$t_{1/2} \propto \frac{1}{[A]_0^2} \text{ or } t_{1/2} = \frac{\text{constant}}{[A]_0^2}$$

(a) The time necessary for the concentration to fall to one-half of the initial concentration is one half-life:

$$\text{first half-life} = t_1 = t_{1/2} = \frac{\text{constant}}{[A]_0^2}$$

(b) This time, $t_{1/4}$, is two half-lives, but because of different starting concentrations, the half-lives are not the same:

$$\text{second half-life} = t_2 = \frac{\text{constant}}{(\frac{1}{2}[A]_0)^2} = \frac{4(\text{constant})}{[A]_0^2} = 4t_1$$

$$\text{total time} = t_1 + t_2 = t_1 + 4t_1 = 5t_1 = t_{1/4}$$

(c) This time, $t_{1/16}$, is four half-lives; again, the half-lives are not the same:

$$\text{third half-life} = t_3 = \frac{\text{constant}}{(\frac{1}{4}[A]_0)^2} = \frac{16(\text{constant})}{[A]_0^2} = 16t_1$$

$$\text{fourth half-life} = t_4 = \frac{\text{constant}}{(\frac{1}{8}[A]_0)^2} = \frac{64(\text{constant})}{[A]_0^2} = 64t_1$$

$$\text{total time} = t_1 + t_2 + t_3 + t_4 = t_1 + 4t_1 + 16t_1 + 64t_1$$
$$= 85t_1 = t_{1/16}$$

If t_1 is known, the times $t_{1/4}$ and $t_{1/16}$ can be calculated easily.

7.21 The following plots are linear: (b), (c), (d), (f), (g)

7.23 (a) overall reaction: $ClO^- + I^- \rightarrow IO^-$;

(b) rate $= \dfrac{k_2 k_1}{k_1'} \dfrac{[ClO^-][I^-]}{[OH^-]}$;

(c) As the pH increases, the concentration of OH^- increases and the rate decreases. (d) If the reaction is carried out in an organic solvent, then H_2O is no longer the solvent and its concentration must be included in calculating the equilibrium concentration of HOCl:

$$\text{rate} = \frac{k_2 k_1}{k_1'} \frac{[ClO^-][I^-][H_2O]}{[OH^-]}$$

7.25 Frogs are poikilotherms, which means that they need to be able to function over a wide range of body core temperatures. This adaptation in the rhodopsin in frog eyes allows their vision to be constant despite temperature fluctuations.

7.27 40 mg must be re-injected.

7.29 (a) intermediate = ClO; catalyst = Cl;
(b) Cl, ClO, O, O$_2$; (c) step 1 and step 2 are propagating;
(d) Cl + Cl $\longrightarrow$ Cl$_2$

7.31 2.5×10^9 L·mol^{-1}·s^{-1}

Focus 8

8A.1 (a) nitrogen; (b) potassium; (c) gallium; (d) iodine

8A.3 (a) sulfur; (b) selenium; (c) sodium; (d) oxygen

8A.5 tellurium < selenium < oxygen

8A.7 (a) chlorine; (b) Chlorine has a higher effective nuclear charge.

8A.9 (a) bromide ion; (b) Bromide has larger size.

8A.11 (a) KCl; (b) K—O

8A.13 $2 K(s) + H_2(g) \rightarrow 2 KH(s)$

8A.15 (a) saline; (b) molecular; (c) molecular; (d) metallic

8A.17 (a) acidic; (b) amphoteric; (c) acidic; (d) basic

8A.19 (a) CO_2; (b) B_2O_3

8B.1 In the majority of its reactions, hydrogen acts as a reducing agent; that is, $H_2(g) \rightarrow 2 H^+(aq) + 2 e^-$, $E° = 0$. In these reactions, hydrogen resembles Group 1 elements, such as Na and K. Hydrogen also has a similar electron affinity to elements in Group 1. The electron affinity of H is $+73$ kJ·mol^{-1}, which is similar to that of Li, which is 60 kJ·mol^{-1}.

8B.3 (a) $C_2H_2(g) + H_2(g) \rightarrow H_2C{=}CH_2(g)$; oxidation number of C in $C_2H_2 = -1$; of C in $H_2C{=}CH_2 = -2$;
(b) $CO(g) + H_2O(g) \rightarrow CO_2(g) + H_2(g)$;
(c) $BaH_2(s) + 2 H_2O(l) \rightarrow Ba(OH)_2 + 2 H_2(g)$

8B.5 (a) $H_2(g) + Cl_2(g) \xrightarrow{\text{light}} 2 HCl(g)$; (b) $H_2(g) + 2 Na(l) \xrightarrow{\Delta} 2 NaH(s)$; (c) $P_4(s) + 6 H_2(g) \rightarrow 4 PH_3(g)$; (d) $2 Cu(s) + H_2(g) \rightarrow 2 CuH(s)$

8B.7 The reason for this trend is primarily the trend of the electronegativity of the central atom (N < O < F).

8C.1 This behavior is related to the small ionic radius of Li^+, 58 pm, which is closer to the ionic radius of Mg^{2+}, 72 pm, but substantially less than that of Na^+, 102 pm. Lithium is the only Group 1 element that reacts directly with nitrogen to form lithium nitride and with oxygen to form mainly the oxide.

8C.3 (a) ns^1; (b) Reducing agents give up one or more electrons. It is relatively easy to remove the one valence electron of the alkali metals because they all have low first ionization energies and the resulting cation has the electron configuration of a noble gas. Alkali metal ions are strongly hydrated; the stability generated by solvation makes them unreactive toward chemical reducing agents and therefore makes the ionic form highly favorable.

8C.5 (a) $4 Na(s) + O_2(g) \rightarrow 2 Na_2O(s)$; (b) $6 Li(s) + N_2(g) \xrightarrow{\Delta} 2 Li_3N(s)$; (c) $2 Na(s) + 2 H_2O(l) \rightarrow 2 NaOH(aq) + H_2(g)$; (d) $4 KO_2(s) + 2 H_2O(g) \rightarrow 4 KOH(s) + 3 O_2(g)$

8D.1 $Mg(s) + 2 H_2O(l) \rightarrow Mg(OH)_2 + H_2(g)$

8D.3 (a) $CaO(s) + H_2O(l) \rightarrow Ca(OH)_2(s)$;
(b) $\Delta G_r° = -57.33$ kJ·mol^{-1}

8D.5 (a) $Mg(OH)_2(s) + 2 HCl(aq) \rightarrow MgCl_2(aq) + 2 H_2O(l)$;
(b) $Ca(s) + 2 H_2O(l) \rightarrow Ca(OH)_2(aq) + H_2(g)$;
(c) $BaCO_3(s) \xrightarrow{\Delta} BaO(s) + CO_2(g)$

8D.7 (a) $:\!\ddot{Cl}\!-\!Be\!-\!\ddot{Cl}\!:$; (b) 180°; (c) sp; (d) $MgCl_2$ is ionic, $BeCl_2$ is a molecular compound; thus they will have different structures.

8E.1 $4\,Al^{3+}(melt) + 6\,O^{2-}(melt) + 3\,C(s,\,gr) \rightarrow 4\,Al(s) + 3\,CO_2(g)$

8E.3 (a) $B_2O_3(s) + 3\,Mg(l) \xrightarrow{\Delta} 2\,B(s) + 3\,MgO(s)$;
(b) $2\,Al(s) + 3\,Cl_2(g) \rightarrow 2\,AlCl_3(s)$; (c) $4\,Al(s) + 3\,O_2(g) \rightarrow 2\,Al_2O_3(s)$

8E.5

(structure of B_3H_6-type ring with H atoms)

8E.7 The Lewis structure of $GaBr_4{}^-$ is:

(Lewis structure of $GaBr_4^-$)

The shape of the $GaBr_4{}^-$ is tetrahedral.

8F.1 Silicon occurs widely in the Earth's crust in the form of silicates in rocks and as silicon dioxide in sand.
(1) $SiO_2(s) + 2\,C(s) \rightarrow Si(s,\,crude) + 2\,CO(g)$
(2) $Si(s,\,crude) + 2\,Cl_2(g) \rightarrow SiCl_4(l)$
(3) $SiCl_4(l) + 2\,H_2(g) \rightarrow Si(s,\,pure) + 4\,HCl(g)$

8F.3 (a) +4; (b) +4; (c) +4

8F.5 (a) $MgC_2(s) + 2\,H_2O(l) \rightarrow C_2H_2(g) + Mg(OH)_2(s)$ (acid–base);
(b) $2\,Pb(NO_3)_2(s) \rightarrow 2\,PbO(s) + 4\,NO_2(g) + O_2(g)$ (redox)

8F.7 $\Delta H_r° = +689.88\ kJ\cdot mol^{-1}$; $\Delta S_r° = +360.85\ J\cdot K^{-1}\cdot mol^{-1}$; $\Delta G_r° = +582.29\ kJ\cdot mol^{-1}$; $T = 1912\ K$

8G.1

−3	NH_3, Li_3N, $LiNH_2$, $NH_2{}^-$
−2	H_2NNH_2
−1	N_2H_2, NH_2OH
0	N_2
+1	N_2O, N_2F_2
+2	NO
+3	NF_3, $NO_2{}^-$, NO^+
+4	NO_2, N_2O_4
+5	HNO_3, $NO_3{}^-$, NO_2F

8G.3 $CO(NH_2)_2(aq) + 2\,H_2O(l) \rightarrow (NH_4)_2CO_3(aq)$; 6.4 kg $(NH_4)_2CO_3$

8G.5 (a) 0.35 L $N_2(g)$; (b) $Hg(N_3)_2$ would produce a larger volume. (c) The azide ion is thermodynamically unstable with respect to the production of $N_2(g)$.

8G.7 $N_2O/H_2N_2O_2$; $N_2O(g) + H_2O(l) \rightarrow H_2N_2O_2(aq)$
N_2O_3/HNO_2; $N_2O_3(g) + H_2O(l) \rightarrow 2\,HNO_2(aq)$
N_2O_5/HNO_3; $N_2O_5(g) + H_2O(l) \rightarrow 2\,HNO_3(aq)$

8G.9 Ammonia (NH_3) can undergo hydrogen bonding with itself.

8H.1 (a) $4\,Li(s) + O_2(g) \xrightarrow{\Delta} 2\,Li_2O(s)$;
(b) $2\,Na(s) + 2\,H_2O(l) \rightarrow 2\,NaOH(aq) + H_2(g)$;
(c) $2\,F_2(g) + 2\,H_2O(l) \rightarrow 4\,HF(aq) + O_2(g)$;
(d) $2\,H_2O(l) \rightarrow O_2(g) + 4\,H^+(aq) + 4\,e^-$

8H.3 (a) $2\,H_2S(g) + 3\,O_2(g) \xrightarrow{\Delta} 2\,SO_2(g) + 2\,H_2O(g)$;
(b) $CaO(s) + H_2O(l) \rightarrow Ca(OH)_2(aq)$;
(c) $2\,H_2S(g) + SO_2(g) \xrightarrow{300\,°C,\ Al_2O_3} 3\,S(s) + 2\,H_2O(l)$

8H.5 H_2O has two unshared electron pairs, whereas NH_3 only has one.

8H.7 (a) Hydrogen peroxide has the following Lewis structure:
$H\!-\!\ddot{O}\!-\!\ddot{O}\!-\!H$
Each O in H_2O_2 is an AX_2E_2 structure; therefore, the bond angle is predicted to be <109.5°. (b)–(e) The reduction potential of H_2O_2 is +1.78 V in acidic solution so only Cu^+ and Mn^{2+} will be oxidized.

8H.9 pH = 13.12

8H.11 The weaker the H—X bond, the stronger the acid. Therefore, the acid strengths are $H_2Te > H_2Se > H_2S > H_2O$

8I.1 (a) +1; (b) +4; (c) +7; (d) +5

8I.3 (a) $4\,KClO_3(l) \xrightarrow{\Delta} 3\,KClO_4(s) + KCl(s)$;
(b) $Br_2(l) + H_2O(l) \rightarrow HBrO(aq) + HBr(aq)$;
(c) $NaCl(s) + H_2SO_4(aq) \rightarrow NaHSO_4(aq) + HCl(g)$;
(d) Both (a) and (b) are redox reactions, whereas (c) is an acid–base reaction.

8I.5 (a) $HClO < HClO_2 < HClO_3 < HClO_4$; (b) The higher the oxidation number the stronger the oxidizing agent.

8I.7 $:\!\ddot{Cl}\!-\!\ddot{O}\!-\!\ddot{Cl}\!:$, AX_2E_2, angular, about 109°.

8I.9 (a) too many large chlorine atoms present on the central iodine; (b) IF_2 is a radical and highly reactive. (c) too many large atoms on a small central atom.

8I.11 (a) IF_3. (b) $3\,XeF_2(g) + I_2(g) \rightarrow 2\,IF_3(s) + 3\,Xe(g)$

8I.13 $Cl_2(g)$ will not oxidize Mn^{2+} to form $MnO_4{}^-$ in an acidic solution.

8I.15 0.117 $mol\cdot L^{-1}$

8J.1 Helium occurs as a component of natural gases found under certain rock formations. Argon is obtained by distillation of liquid air.

8J.3 (a) +2; (b) +6; (c) +4; (d) +6

8J.5 $XeF_4(aq) + 4\,H^+(aq) + 4\,e^- \rightarrow Xe(g) + 4\,HF(aq)$

8J.7 H_4XeO_6, because it has more O atoms bonded to Xe.

8.1 (a)

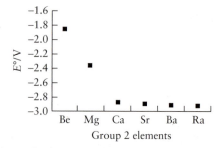

(b) For both groups, the trend in standard potentials with increasing atomic number is overall downward (they become more negative). This downward trend is because we expect that it is easier to remove electrons that are farther away from the nuclei.

8.3 (a) True; (b) False. The oxides of carbon are acidic, whereas those of tin and lead are basic. (c) true

8.5 In the majority of its reactions, hydrogen acts as a reducing agent. Examples are $2 H_2(g) + O_2(g) \rightarrow 2 H_2O(l)$ and various ore reduction processes, such as $NiO(s) + H_2(g) \xrightarrow{\Delta} Ni(s) + H_2O(g)$. With highly electropositive elements, such as the alkali earth metals, $H_2(g)$ acts as an oxidizing agent and forms metal hydrides; for example, $2 K(s) + H_2(g) \rightarrow 2 KH(s)$.

8.7 0.0538 M

8.9 (a) The structure of the azide anion is:

$$\left[:\ddot{N}=N=\ddot{N}: \right]^-$$ AX_2, linear 180°

(b) HCl, HBr, and HI are all strong acids. For the weak acid HF, $K_a = 3.5 \times 10^{-4}$, so HF is slightly more acidic than HN_3. The small size of the azide ion suggests that the H—N bond in HN_3 is similar in strength to that of the H—F; (c) ionic: NaN_3, $Pb(N_3)_2$, AgN_3; covalent: HN_3, $B(N_3)_3$, FN_3.

8.11 (a) Both have the same basic structure in how the atoms are arranged in space. (b) The bonding between the boron atoms and the bridging hydrogen atoms is electron deficient (a three-center, two-electron bond). The bonding is conventional (all bonds involve two atoms and two electrons). (c) The hybridization is sp^3 at the B and Al atoms. (d) The molecules are not planar. Bond angles in the ring are expected to be approximately 90°, whereas the angle between the terminal hydrogens and the Group 13 element is expected to be greater than 109.5°.

8.13 (a) The ionization energy of a molecule is the energy required to strip one electron out of a gaseous molecule. (b) SiI_4 has more electrons than $SiCl_4$; I is also less electronegative and much bigger (and therefore more polarizable) than Cl. As a result it should be easier to remove an electron from SiI_4, meaning that $SiCl_4$ should have the higher ionization energy.

8.15 (a) 10-electron species: NH_3 and H_3O^+; 15-electron species: NO and O_2^+; 22-electron species: N_2O^+ and NO_2^+. (b) strongest Lewis acids: H_3O^+, NO, and NO_2^+; (c) strongest oxidizing agents: H_3O^+, NO, and NO_2^+

8.17 The structures of thiosulfuric acid and sulfuric acid are

Due to the replacement of one of the doubly bonded oxygens in sulfuric acid with a sulfur atom, it should be expected that an aqueous solution of thiosulfuric acid should be slightly less acidic; also, the boiling point should also be expected to be slightly lower (due to reduced hydrogen bonding).

8.19 The solubility of the ionic halides is determined by a variety of factors, especially the lattice enthalpy. Lattice

enthalpies decrease from chloride to iodide, so water molecules can more readily separate the ions in the latter. Less ionic halides, such as the silver halides, generally have a much lower solubility, and the trend in solubility is the reverse of the more ionic halides, because the ions are not easily hydrated, making them less soluble.

8.21 (a) The molecular orbital diagram for NO^+ should have the oxygen orbitals slightly lower in energy than the nitrogen orbitals, because oxygen is more electronegative. This will cause the bonding to be more ionic than in either N_2 or O_2. There is an ambiguity, however, in that the MO diagram could be similar to either that of N_2 or that of O_2. There are consequently two possibilities for the orbital energy diagram:

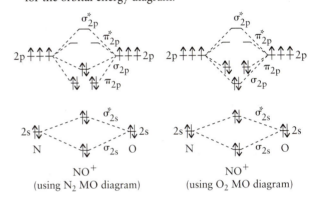

NO^+
(using N_2 MO diagram)

NO^+
(using O_2 MO diagram)

(b) The two orbital diagrams predict the same bond order (3) and the same magnetic properties (diamagnetic), and so these properties cannot be used to determine which diagram is the correct one.

8.23 (a) $\Delta H_r° = +128.33 \text{ kJ·mol}^{-1}$; (b) $\Delta H_c°(CH_3OH) = -726 \text{ kJ·mol}^{-1}$; (c) $-483.64 \text{ kJ·mol}^{-1}$. (d) The direct combustion of one mole of methanol will produce more heat (-726 kJ) than decomposing methanol followed by combusting the hydrogen gas formed.

8.25 Species (a), (b), (c), and (d) can all function as greenhouse gases, whereas (e) cannot. Any molecule other than a homonuclear diatomic can exhibit a changing dipole moment as it vibrates with certain vibrational modes. Because argon is monoatomic, it has no covalent bonds, no vibrational modes, and no dipole moment.

8.27 1.09 kg of CO_2 will be formed. This is about half the amount generated by combusting an equivalent volume of octane (2.16 kg per liter). However, one needs to consider how much energy is produced per liter of fuel and how the mass of carbon dioxide produced compares for a given amount of energy produced. Calculations yield the energy per liter of methanol to be $1.79 \times 10^4 \text{ kJ}$ and that for octane to be $3.37 \times 10^4 \text{ kJ}$. The combustion of octane produces almost twice as much energy per liter as methanol (octane/methanol = 1.88). For an equivalent amount of combustion energy, methanol produces 2.05 kg of CO_2, which is still slightly less than octane.

8.29 Ion–ion forces are among the strongest intermolecular interactions. Therefore, the interactions in (a) are the strongest shown. Hydrogen bonding is stronger than a dipole–dipole interaction, indicating that the interactions in (c) are stronger than those shown in (d).

8.31 (a) C: sp^2; B: sp^2; N: sp^2;   (b) 8 units are strung together around the circumference of the given nanotube. Because each unit contains 2 hexagons, this makes 16 hexagons per circumference.   (c) In C_{60}, the carbon atoms are sp^2 hybridized and are nearly planar. However, the curvature of the molecule introduces some strain at the carbon atoms, so that there is a tendency for some of the carbon atoms to undergo conversion to sp^3 hybridization. However, to make every carbon sp^3 hybridized would introduce much more strain on the carbon cage and, after a certain point, further addition of hydrogen becomes unfavorable.   (d) The spherical structures require the formation of five-member rings (see structure **3** in Topic 8F, C_{60}). Boron nitride cannot form these rings because they would require high-energy boron–boron or nitrogen–nitrogen bonds.   (e) $d = 3.491$ g·cm^{-3}; (f) We would expect the cubic form to be favored at high pressures.

Focus 9

9A.1 Elements at the left of the d-block tend to have strongly negative standard potentials; this can be attributed to their lower ionization energies.

9A.3 (a) MnO_4^-;   (b) MoO_4^{2-}

9A.5 (a) Fe; (b) Cu; (c) Pt; (d) Pd; (e) Ta

9A.7 With zinc, the last of the 3d-electrons is added; this closes the $n = 3$ shell, thereby significantly lowering the energy of the 3d-orbitals. With gallium, the increase in Z_{eff} causes the 3d-orbitals to be drawn down closer to the nucleus, thus lowering their energy even more.

9A.9 This is due to the lanthanide contraction.

9A.11 (a) Higher oxidation states become more stable upon going down a group.   (b) Higher oxidation states tend to be less stable as one descends a group.

9A.13 M has an oxidation number of +6, which is the most common one for Cr.

9B.1 (a) $TiCl_4(g) + 2\,Mg(l) \xrightarrow{\Delta} Ti(s) + 2\,MgCl_2(s)$;
(b) $CoCO_3(s) + HNO_3(aq) \rightarrow$
$\qquad Co^{2+}(aq) + HCO_3^-(aq) + NO_3^-(aq)$;
(c) $V_2O_5(s) + 5\,Ca(l) \xrightarrow{\Delta} 2\,V(s) + 5\,CaO(s)$

9B.3 (a) titanium(IV) oxide, TiO_2;
(b) iron(III) oxide, Fe_2O_3;
(c) manganese(IV) oxide, MnO_2;
(d) iron(II) chromite, $FeCr_2O_4$

9B.5 (a) Titanium has an oxidation number of +4.   (b) Zinc has an oxidation number of +2.

9B.7 (a) CO;   (b) In Zones D and C, $Fe_2O_3(s) + CO(g) \rightarrow$ $2\,FeO(s) + CO_2(g)$; in Zone B, $Fe_2O_3(s) + 3\,CO(g) \rightarrow$ $2\,Fe(s) + 3\,CO_2(g)$ and $FeO(s) + CO(g) \rightarrow$ $Fe(s) + CO_2(g)$;   (c) carbon

9B.9 (a) $V_2O_5(s) + 2\,H_3O^+(aq) \rightarrow 2\,VO_2^+(aq) + 3\,H_2O(l)$;
(b) $V_2O_5(s) + 6\,OH^-(aq) \rightarrow 2\,VO_4^{3-}(aq) + 3\,H_2O(l)$

9B.11 Cu, Ag, and Au have the electron configuration $(n-1)d^{10}ns^1$. As the value of n increases, d- and f-electrons become less effective at shielding, leading to a higher Z_{eff}, which makes it more difficult to oxidize the element.

9B.13 (a) Br^- will be oxidized to Br_2.   (b) No reaction will occur.

9B.15 (a) Cr^{3+} ions in water form the complex $[Cr(OH_2)_6]^{3+}(aq)$, which behaves as a Brønsted acid.   (b) The precipitate is $Cr(OH)_3$, which dissolves as the $Cr(OH)_4^-$ complex ion is formed:
$Cr^{3+}(aq) + 3\,OH^-(aq) \rightarrow Cr(OH)_3(s)$
$Cr(OH)_3(s) + OH^-(aq) \rightarrow Cr(OH)_4^-(aq)$

9C.1 (a) hexacyanoferrate(II) ion, +2;
(b) hexaamminecobalt(III) ion, +3;
(c) aquapentacyanocobaltate(III) ion, +3;
(d) pentaamminesulfatocobalt(III) ion, +3

9C.3 (a) $K_3[Cr(CN)_6]$; (b) $[Co(NH_3)_5(SO_4)]Cl$;
(c) $[Co(NH_3)_4(OH_2)_2]Br_3$; (d) $Na[Fe(OH_2)_2(C_2O_4)_2]$

9C.5 (a) tridentate ligand;   (b) mono- or bidentate ligand;
(c) monodentate ligand;   (d) bidentate ligand

9C.7 Only the molecule (b) can function as a chelating ligand. The MH_2 groups in (a) and (c) are arranged so that they would not be able to coordinate simultaneously to the same metal center.

9C.9 (a) 4; (b) 2; (c) 6 (en is bidentate);   (d) 6 (EDTA is hexadentate)

9C.11 (a) structural and linkage isomers;   (b) structural and ionization isomers;   (c) structural and linkage isomers;
(d) structural and ionization isomers

9C.13 (a) yes

trans-Tetraamminedichloridocobalt(III) chloride monohydrate

and

cis-Tetraamminedichloridocobalt(III) chloride monohydrate

(b) no;

(c) yes

cis-Diamminedichloridoplatinum(II)

and

trans-Diamminedichloridoplatinum(II)

9C.15 Three isomers are possible:

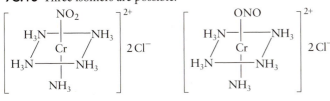

9C.17 (a) chiral; yes

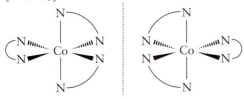

(b) no

9C.19 $[Co(en)_3]^{3+}$ can form enantiomers:

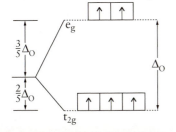

Nonsuperimposable mirror images

9D.1 (a) 2; (b) 5; (c) 8; (d) 10; (e) 0 (or 8); (f) 10

9D.3 (a) octahedral; strong-field ligand, 6 e⁻, no unpaired electrons

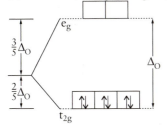

(b) tetrahedral: weak-field ligand, 8 e⁻, 2 unpaired electrons

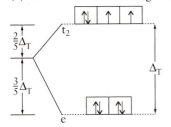

(c) octahedral: weak-field ligand, 5 e⁻, 5 unpaired electrons

(d) octahedral: strong-field ligand, 5 e⁻, 1 unpaired electron

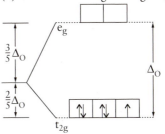

9D.5 $\Delta_O = 214$ kJ·mol⁻¹

9D.7 (a) no unpaired electrons; (b) 2 unpaired electrons

9D.9 Weak-field ligands do not interact strongly with the d-electrons in the metal ion and produce a small crystal field splitting of the d-electron energy states, whereas the opposite is true of strong-field ligands. Magnetic susceptibility (paramagnetism) establishes whether the associated ligand is weak-field or strong-field.

9D.11 1.50×10^{-3} mol·L⁻¹

9D.13 Due to the fact that CN⁻ is a strong field and Cl⁻ is a weak field ligand.

9D.15 (a) $[CoF_6]^{3-}$ is blue and (b) $[Co(en)_3]^{3+}$ is yellow. Because F⁻ is a weak-field ligand and en a strong-field ligand, the splitting between levels is less in (a) than in (b), so (a) will absorb light of longer wavelength than (b).

9D.17 573 nm; this wavelength is in the yellow region of the visible spectrum.

9D.19 (a) none; (b) longer

9D.21 The 3d-orbitals are filled so there can be no electronic transitions between the t and e levels; no visible light is absorbed producing colorless solutions. Zn compounds would be diamagnetic (no unpaired electrons).

9D.23 (a) (i) 162 kJ·mol⁻¹; (ii) 260 kJ·mol⁻¹; (iii) 208 kJ·mol⁻¹; (b) Cl⁻ < H₂O < NH₃

9D.25 The e_g set

9D.27 (a) (i) π-acid; (ii) π-base; (iii) π-base; (iv) neither; (b) Cl⁻ < H₂O < en < CN⁻

9D.29 Nonbonding or slightly antibonding. In a complex that forms only σ-bonds, this set of orbitals is nonbonding; if weak interactions with the p-orbitals on the ligands occur, they become slightly antibonding.

9D.31 Antibonding, due to interactions with ligand orbitals forming σ-bonds.

9D.33 The two lone pairs of electrons on water are used to form the σ-bond to the metal ion as well as a weak π-bond, causing the t_{2g} set of orbitals to move up in energy, making Δ_O smaller; ammonia does not have this extra lone pair of electrons and consequently cannot function as a π-donor ligand.

9.1 $K_{sp} = e^{-70.1} = 3.6 \times 10^{-31}$

9.3 (a) PtBrCl(NH₃)₂

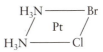

cis-Diamminebromidochloridoplatinum(II)

and

trans-Diamminebromidochloridoplatinum(II)

(b) If tetrahedral, there would be only one compound, not two.

9.5 (a) $[Ni(SO_4)(en)_2]Cl_2$ will give a precipitate of AgCl when $AgNO_3$ is added; the other will not.
(b) $[NiCl_2(en)_2]I_2$ will show free I_2 upon mild oxidation; the first will not.

9.7 (a) $[MnCl_6]^{4-}$; 5 e$^-$, Cl$^-$ is a weak-field ligand

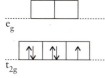

$[Mn(CN)_6]^{4-}$; 5 e$^-$, CN$^-$ is a strong-field ligand

(b) $[MnCl_6]^{4-}$: five; $[Mn(CN)_6]^{4-}$: one; (c) $[MnCl_6]^{4-}$; weak-field complexes absorb longer-wavelength light.

9.9 If the prismatic structure were true, there are four possible isomers, not two.

9.11 If chain theory were true, the chloride ions could not be precipitated as AgCl; also VSEPR theory would predict the ion to have a trigonal planar ligand arrangement, which would have d-orbital energies that are not the same as the octahedral arrangement and would lead to different spectroscopic and magnetic properties than observed.

9.13 Here shown for a d^8 ion (Pd^{2+}):

$$
\begin{array}{ll}
\text{—} & d_{x^2-y^2} \\
\uparrow\downarrow & d_{xy} \\
\uparrow\downarrow & d_{z^2} \\
d_{zx}\ \uparrow\downarrow \quad \uparrow\downarrow\ d_{yz} &
\end{array}
$$

E

9.15 To have no unpaired electrons, it would have to be square planar; both the octahedral and tetrahedral geometries require two unpaired electrons.

9.17 AgCl will only form from Cl$^-$ found outside the coordination sphere. (a) $[Co(NH_3)_5OH_2]Cl_3$, pentaammineaquacobalt (III) chloride;
(b) $[CoCl(NH_3)_5]Cl_2$, pentaamminechloridocobalt (III) chloride.

9.19 (a) square planar; (b) +1; (c) There are two possible isomers. Neither the cis nor the trans form is optically active.

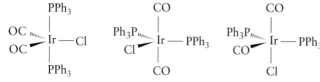

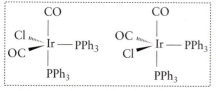

Nonsuperimposable mirror images
(enantiomers)

9.21 15.12 mL

9.23 basic copper carbonate $(Cu_2CO_3(OH)_2)$

9.25 (a) +3 is the most likely oxidation state of vanadium;
(b) $2\ V_2O_5(s) + 8\ C(s) \longrightarrow V_4C_3(s) + 5\ CO_2(g)$

Focus 10

10A.1 (a) $\nu = 3.22 \times 10^{20}$ Hz; $\lambda = 9.32 \times 10^{-13}$ m;
(b) $\nu = 3.97 \times 10^{20}$ Hz; $\lambda = 7.56 \times 10^{-13}$ m;
(c) $\nu = 2.66 \times 10^{20}$ Hz; $\lambda = 1.13 \times 10^{-12}$ m

10A.3 $^{39}_{18}$Ar and $^{41}_{20}$Ca are isotones of K-40.

10A.5 (a) $^{12}_{5}B \rightarrow\ ^{0}_{-1}e +\ ^{12}_{6}C$; (b) $^{214}_{83}Bi \rightarrow\ ^{4}_{2}\alpha +\ ^{210}_{81}Tl$;
(c) $^{98}_{43}Tc \rightarrow\ ^{0}_{-1}e +\ ^{98}_{44}Ru$; (d) $^{266}_{88}Ra \rightarrow\ ^{4}_{2}\alpha +\ ^{262}_{86}Rn$

10A.7 (a) $^{109}_{49}In \rightarrow\ ^{0}_{+1}e +\ ^{109}_{48}Cd$; (b) $^{23}_{12}Mg \rightarrow\ ^{0}_{+1}e +\ ^{23}_{11}Na$;
(c) $^{202}_{82}Pb +\ ^{0}_{-1}e \rightarrow\ ^{202}_{81}Tl$; (d) $^{76}_{33}As +\ ^{0}_{-1}e \rightarrow\ ^{76}_{32}Ge$

10A.9 (a) $^{24}_{11}Na \rightarrow\ ^{24}_{12}Mg +\ ^{0}_{-1}e$; a β particle is emitted.
(b) $^{128}_{50}Sn \rightarrow\ ^{128}_{51}Sb +\ ^{0}_{-1}e$; a β particle is emitted.
(c) $^{140}_{57}La \rightarrow\ ^{140}_{56}Ba +\ ^{0}_{+1}e$; a positron (β$^+$) is emitted.
(d) $^{228}_{90}Th \rightarrow\ ^{224}_{88}Ra +\ ^{4}_{2}\alpha$; an α particle is emitted.

10A.11 (a) $^{11}_{5}B +\ ^{4}_{2}\alpha \rightarrow 2\ ^{1}_{0}n +\ ^{13}_{7}N$;
(b) $^{35}_{17}Cl +\ ^{2}_{1}D \rightarrow\ ^{1}_{0}n +\ ^{36}_{18}Ar$;
(c) $^{96}_{42}Mo +\ ^{2}_{1}D \rightarrow\ ^{1}_{0}n +\ ^{97}_{43}Tc$;
(d) $^{45}_{21}Sc +\ ^{1}_{0}n \rightarrow\ ^{4}_{2}\alpha +\ ^{42}_{19}K$

10A.13 (a) $^{68}_{29}Cu \rightarrow\ ^{0}_{-1}e +\ ^{68}_{30}Zn$; (b) $^{103}_{48}Cd \rightarrow\ ^{0}_{+1}e +\ ^{103}_{47}Ag$

10A.15 α $^{235}_{92}U \rightarrow\ ^{4}_{2}\alpha +\ ^{231}_{90}Th$
β $^{231}_{90}Th \rightarrow\ ^{0}_{-1}e +\ ^{231}_{91}Pa$
α $^{231}_{91}Pa \rightarrow\ ^{4}_{2}\alpha +\ ^{227}_{89}Ac$
β $^{227}_{89}Ac \rightarrow\ ^{0}_{-1}e +\ ^{227}_{90}Th$
α $^{227}_{90}Th \rightarrow\ ^{4}_{2}\alpha +\ ^{223}_{88}Ra$
α $^{223}_{88}Ra \rightarrow\ ^{4}_{2}\alpha +\ ^{219}_{86}Rn$
α $^{219}_{86}Rn \rightarrow\ ^{4}_{2}\alpha +\ ^{215}_{84}Po$
β $^{215}_{84}Po \rightarrow\ ^{0}_{-1}e +\ ^{215}_{85}At$
α $^{215}_{85}At \rightarrow\ ^{4}_{2}\alpha +\ ^{211}_{83}Bi$
β $^{211}_{83}Bi \rightarrow\ ^{0}_{-1}e +\ ^{211}_{84}Po$
α $^{211}_{84}Po \rightarrow\ ^{4}_{2}\alpha +\ ^{207}_{82}Pb$

10A.17 (a) $^{14}_{7}N +\ ^{4}_{2}\alpha \rightarrow\ ^{17}_{8}O +\ ^{1}_{1}p$;
(b) $^{248}_{96}Cm +\ ^{1}_{0}n \rightarrow\ ^{249}_{97}Bk +\ ^{0}_{-1}e$;
(c) $^{243}_{95}Am +\ ^{1}_{0}n \rightarrow\ ^{244}_{96}Cm +\ ^{0}_{-1}e + \gamma$;
(d) $^{13}_{6}C +\ ^{1}_{0}n \rightarrow\ ^{14}_{6}C + \gamma$

10A.19 (a) $^{20}_{10}Ne +\ ^{4}_{2}\alpha \rightarrow\ ^{8}_{4}Be +\ ^{16}_{8}O$;
(b) $^{20}_{10}Ne +\ ^{20}_{10}Ne \rightarrow\ ^{16}_{8}O +\ ^{24}_{12}Mg$;
(c) $^{44}_{20}Ca +\ ^{4}_{2}\alpha \rightarrow \gamma +\ ^{48}_{22}Ti$;
(d) $^{27}_{13}Al +\ ^{2}_{1}H \rightarrow\ ^{1}_{1}p +\ ^{28}_{13}Al$

10A.21 $^{56}_{26}Fe + 6\ ^{1}_{0}n \rightarrow\ ^{62}_{26}Fe \rightarrow\ ^{62}_{32}Ge + 6\ ^{0}_{-1}e$

10A.23 (a) $^{14}_{7}N +\ ^{4}_{2}\alpha \rightarrow\ ^{17}_{8}O +\ ^{1}_{1}p$;
(b) $^{239}_{94}Pu + 2\ ^{1}_{0}n \rightarrow\ ^{241}_{95}Am +\ ^{0}_{-1}e$

10A.25 (a) unbihexium, Ubh; (b) untrihexium, Uth;
(c) binilnilium, Bnn

10B.1 (a) 5.64×10^{-2} a^{-1}; (b) 0.83 s^{-1}; (c) 0.0693 min^{-1}

10B.3 8.8 h

10B.5 (a) 69.6%; (b) 50.9%

10B.7 $x = 9.29 \times 10^8$ a

10B.9 3.54×10^3 years old

10B.11 400 min

10B.13 26.4%

10B.15 $[X] = [X]_0 e^{-k_1 t}$; $[Y] = \dfrac{k_1}{k_2 - k_1}(e^{-k_1 t} - e^{-k_2 t})[X]_0$;

because $[X] + [Y] + [Z] = [X]_0$ at all times, $[Z] = [X]_0 - ([X] + [Y])$, or

$[Z] = [X]_0 - \left([X]_0 e^{-k_1 t} + \dfrac{k_1}{k_2 - k_1}(e^{-k_1 t} - e^{-k_2 t})[X]_0\right) =$

$[X]_0\left(1 + \dfrac{k_1 e^{-k_2 t} - k_2 e^{-k_1 t}}{k_2 - k_1}\right)$

The values of the rate constants can be found from the half-lives:
$$k_1 = 0.0253 \ \mathrm{d}^{-1} \text{ and } k_2 = 0.0371 \ \mathrm{d}^{-1}$$
Using these constants and assuming $[X]_0 = 2.00$ g, the graph is

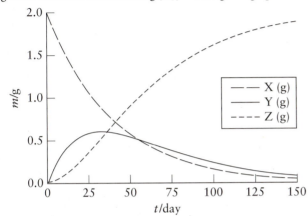

10B.17 Use $H_2{}^{18}O$, in the reaction, separate the products, and use a suitable technique such as vibrational spectroscopy or mass spectrometry to determine whether the product has incorporated the ^{18}O.

10B.19 44 years

10C.1 $-4.3 \times 10^9 \ \mathrm{kg \cdot s^{-1}}$ (mass is lost)

10C.3 (a) $1.41 \times 10^{-12} \ \mathrm{J \cdot nucleon^{-1}}$; (b) 1.21×10^{-12} $\mathrm{J \cdot nucleon^{-1}}$; (c) $1.8 \times 10^{-13} \ \mathrm{J \cdot nucleon^{-1}}$; (d) $4.3 \times 10^{-13} \ \mathrm{J \cdot nucleon^{-1}}$; (e) ^{62}Ni

10C.5 energy released: (a) $7.8 \times 10^{10} \ \mathrm{J \cdot g^{-1}}$; (b) 3.52×10^{11} $\mathrm{J \cdot g^{-1}}$; (c) $2.09 \times 10^{11} \ \mathrm{J \cdot g^{-1}}$; (d) $3.36 \times 10^{11} \ \mathrm{J \cdot g^{-1}}$

10C.7 (a) $^{24}_{11}Na \rightarrow {}^{24}_{12}Mg + {}^{0}_{-1}e$; (b) -8.85×10^{-13} J; (c) $-3.69 \times 10^{-14} \ \mathrm{J \cdot nucleon^{-1}}$

10C.9 (a) 3.5×10^8 kJ of energy released; (b) 1.1×10^4 kg of coal

10.1 (a) false; the dose equivalent is either equal to or larger than the actual dose, due to the Q factor; (b) false; 1.8×10^8 Bq = 0.0031 Ci, which is much smaller than 10 Ci; (c) true

10.3 (a) 9 dpm; (b) 7×10^5 decays

10.5 (a) $3.4 \times 10^8 \ \mathrm{pCi \cdot L^{-1}}$; (b) 1.0×10^{19} atoms; (c) 1×10^2 days

10.7 (a) 4.5 Ga; (b) 3.8 Ga

10.9 (a) $^{98}_{42}Mo + {}^{1}_{0}n \rightarrow {}^{99}_{42}Mo \rightarrow {}^{99m}_{43}Tc + {}^{0}_{-1}e$; (b) Tc-99m

10.11 0.27 L

10.13 (a) 262 dpm; (b) $1.18 \times 10^{-4} \ \mu$Ci

10.15 (a) γ radiation for diagnosis (least destructive and not stopped by body tissues); (b) α particles (most destructive); (c) and (d) Two examples are ^{131}I, 8d (used to image the thyroid) and ^{99m}Tc, 6 h (used for various body tissues).

10.17 (a) $k = 1.1 \times 10^{-3} \ \mathrm{d^{-1}}$; half-life = 617 d; (b) 1.0 mg

10.19 Nuclei that are positron emitters are proton-rich and lie below the band of stability and are suitable for PET scans.

(a) ^{18}O is neutron rich; not suitable; (b) ^{13}N is proton-rich, suitable for PET: $^{13}_{7}N \rightarrow {}^{13}_{6}C + {}^{0}_{1}e$; (c) ^{11}C is proton rich, will emit a positron; suitable for PET: $^{11}_{6}C \rightarrow {}^{11}_{5}B + {}^{0}_{1}e$; (d) ^{20}F is neutron rich, not suitable; (e) ^{15}O is proton-rich, so is suitable for PET: $^{15}_{8}O \rightarrow {}^{15}_{7}N + {}^{0}_{1}e$

10.21 To determine the effective half-life, we need to determine the effective rate constant, k_E. This constant is equal to the sum of the biological rate constant (k_B) and the radioactive decay rate constant (k_R), both of which can be obtained from the respective half-lives:

$$k_E = k_B + k_R = \frac{\ln 2}{65 \ \mathrm{d}} + \frac{\ln 2}{12.8 \ \mathrm{d}} = 6.5 \times 10^{-2} \ \mathrm{d^{-1}}$$

$$t_{1/2}(\text{effective}) = \frac{\ln 2}{6.5 \times 10^{-2} \ \mathrm{d^{-1}}} = 11 \ \mathrm{d}$$

Focus 11

11A.1 (a) alkyne
(b) alkane
(c) alkene
(d) alkene and alkyne
(e) alkene

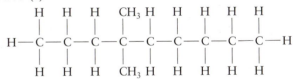

11A.3 (a) propane; (b) butane; (c) heptane; (d) decane

11A.5 (a) methyl; (b) pentyl; (c) propyl; (d) hexyl

11A.7 (a) propane; (b) ethane; (c) pentane; (d) 2,3-dimethylbutane

11A.9 (a) 4-methyl-2-pentene; (b) 2,2,3-trimethylpentane

11A.11 (a) $CH_2{=}CHCH(CH_3)CH_2CH_3$;
(b) $CH_3CH_2C(CH_3)_2CH(CH_2CH_3)(CH_2)_2CH_3$; (c) $HC{\equiv}C(CH_2)_2C(CH_3)_3$; (d) $CH_3CH(CH_3)CH(CH_2CH_3)CH(CH_3)_2$

11A.13 (a)

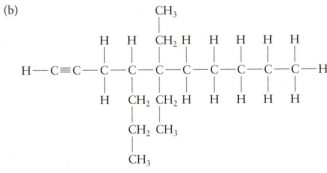

11A.15 (a)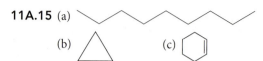

(b) 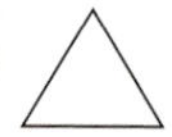 △   (c)

11A.17 (a) four σ-type single bonds;   (b) two σ-type single bonds and one double bond with a σ- and a π-bond;   (c) one σ-type single bond and one triple bond with a σ-bond and two π-bonds

11A.19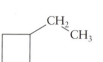

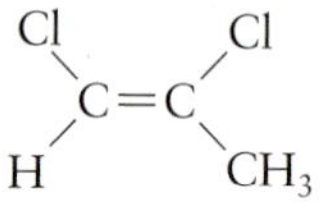

cis-1,2-Dichloropropene 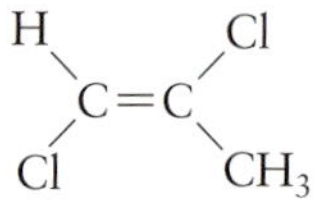   *trans*-1,2-Dichloropropene

cis-1,2-Dichloropropene is polar, although *trans*-1,2-dichloropropene is slightly polar also.

11A.21 (a) hexenes:

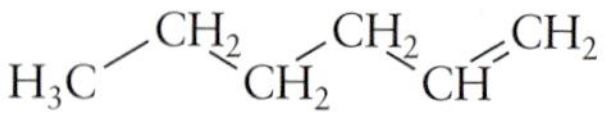

1-Hexene 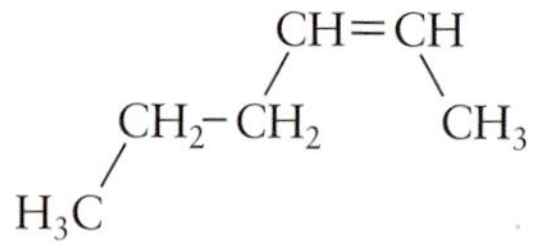   *cis*-2-Hexene

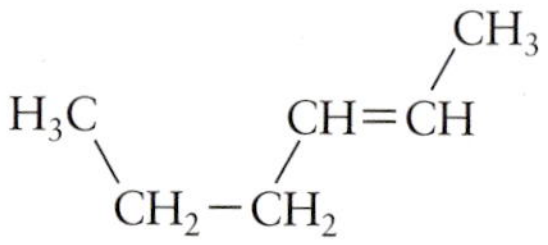

trans-2-Hexene 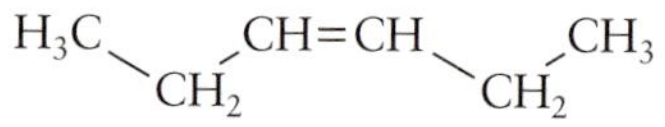   *cis*-3-Hexene

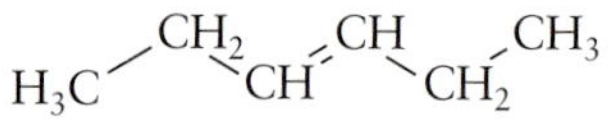

trans-3-Hexene

pentenes:

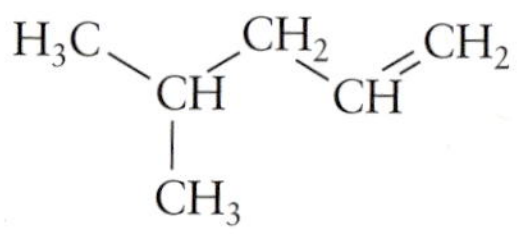

4-Methyl-1-pentene 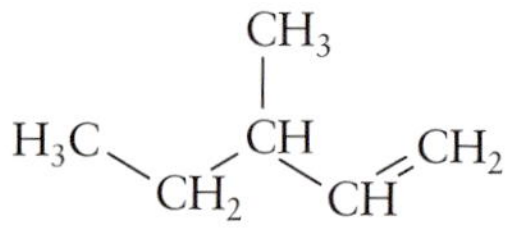   3-Methyl-1-pentene

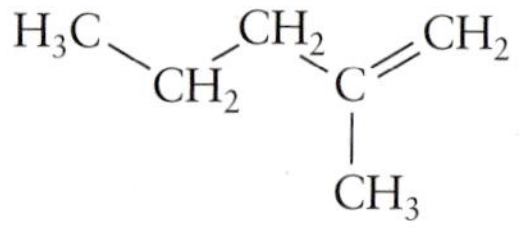

2-Methyl-1-pentene 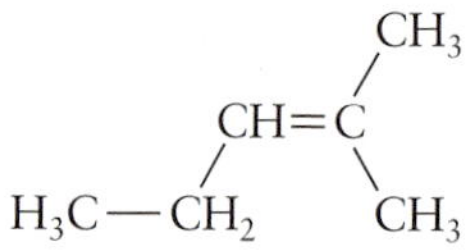   2-Methyl-2-pentene

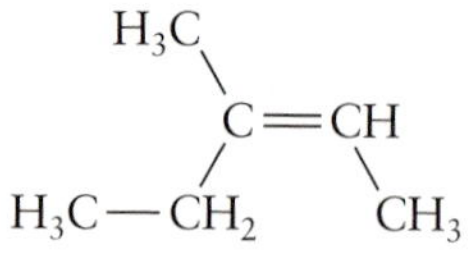

cis-3-Methyl-2-pentene (+ trans isomer) 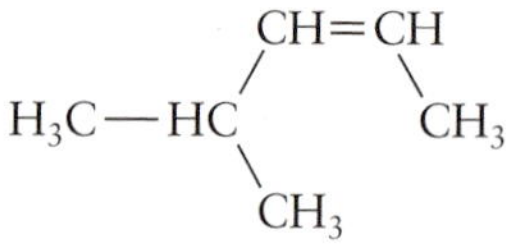   *cis*-4-Methyl-2-pentene (+ trans isomer)

butenes:

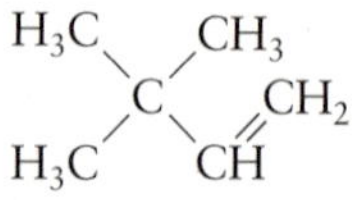

3,3-Dimethyl-1-butene 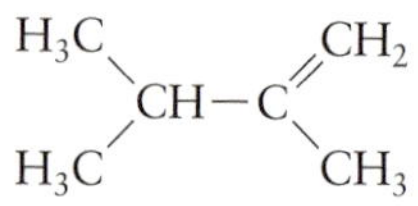   2,3-Dimethyl-1-butene

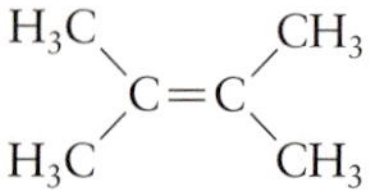

2,3-Dimethyl-2-butene

(b) cyclic molecules:

Cyclohexane 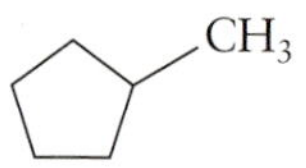   Methylcyclopentane

Ethylcyclobutane 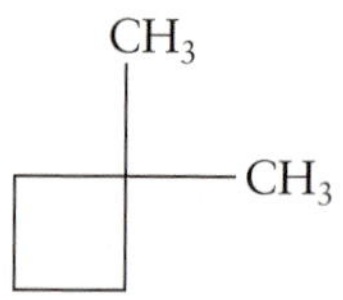   1,1-Dimethylcyclobutane

The following structures are drawn to emphasize the stereochemistry:

cis-1,2-Dimethylcyclobutane

trans-1,2-Dimethylcyclobutane (nonsuperimposable mirror images)

trans-1,3-Dimethylcyclobutane   *cis*-1,3-Dimethylcyclobutane

Propylcyclopropane   Isopropylcyclopropane or 2-cyclopropylpropane

1-Ethyl-1-methylcyclopropane

trans-1-Ethyl-2-methylcyclopropane (nonsuperimposable mirror images)

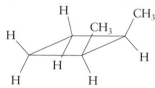
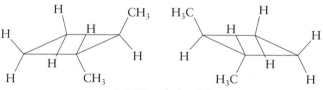
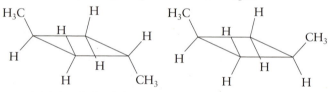
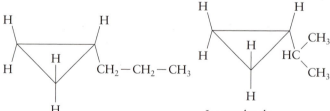
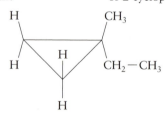
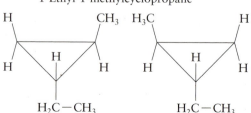

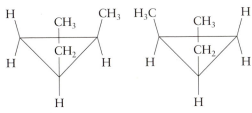

1,1,2-Trimethylcyclopropane
(nonsuperimposable mirror images)

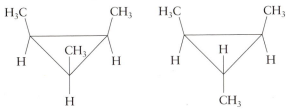

***cis*-1-Ethyl-2-methylcyclopropane**
(nonsuperimposable mirror images)

1,2,3-Trimethylcyclopropane **1,2,3-Trimethylcyclopropane**
(all cis isomer)   (cis–trans isomer)

11A.23 (a) not isomers;   (b) structural isomers;
(c) geometrical isomers;   (d) not isomers

11A.25 (a)

$$H_3C-\underset{\underset{CH_3}{|}}{\overset{\overset{H}{|}}{C}}-CH_3$$

(b) If only two isomeric products are formed and they are both branched, then the only possibilities are:

$$H_3C-\underset{\underset{CH_3}{|}}{\overset{\overset{Cl}{|}}{C}}-CH_3 \qquad H_2C-\underset{\underset{CH_3}{|}}{\overset{\overset{Cl\ H}{|\ |}}{C}}-CH_3$$

11A.27 An * designates a chiral carbon:

(a) optically active,

$$H_3C-\underset{\underset{Br}{|}}{\overset{\overset{H}{|}}{C^*}}-CH_2-CH_3$$

(b) not optically active,

$$Cl-\underset{\underset{Cl}{|}}{\overset{\overset{H}{|}}{C}}-\underset{\underset{H}{|}}{\overset{\overset{H}{|}}{C}}-CH_3$$

(c) optically active,

$$H-\underset{\underset{H}{|}}{\overset{\overset{Br}{|}}{C}}-\underset{\underset{H}{|}}{\overset{\overset{Cl}{|}}{C^*}}-CH_3$$

(d) optically active,

$$H-\underset{\underset{H}{|}}{\overset{\overset{Cl}{|}}{C}}-\underset{\underset{H}{|}}{\overset{\overset{Cl}{|}}{C^*}}-CH_2-CH_2-CH_3$$

11B.1 The balanced equations are

$$C_3H_8(g) + 5\,O_2(g) \rightarrow 3\,CO_2(g) + 4\,H_2O(l)$$
$$C_4H_{10}(g) + \tfrac{13}{2}\,O_2(g) \rightarrow 4\,CO_2(g) + 5\,H_2O(l)$$
$$C_5H_{12}(g) + 8\,O_2(g) \rightarrow 5\,CO_2(g) + 6\,H_2O(l)$$

The enthalpies of combustion that correspond to these reactions are listed in Appendix 2:

Compound	(a) Enthalpy of combustion/ $(kJ \cdot mol^{-1})$	(b) Heat released/ $(kJ \cdot g^{-1})$
Propane	-2220	50.3
Butane	-2878	49.5
Pentane	-3537	49.0

The enthalpy of combustion increases with molar mass as might be expected, because the number of moles of CO_2 and H_2O formed will increase as the number of carbon and hydrogen atoms in the compounds increases. The heat released per gram of these hydrocarbons is essentially the same, because the H-to-C ratio is similar in the three hydrocarbons.

11B.3 There are nine possible products:

$$H-\underset{\underset{H}{|}}{\overset{\overset{H}{|}}{C}}-\underset{\underset{H}{|}}{\overset{\overset{H}{|}}{C}}-Cl$$

One monochloro compound

$$Cl-\underset{\underset{H}{|}}{\overset{\overset{H}{|}}{C}}-\underset{\underset{H}{|}}{\overset{\overset{H}{|}}{C}}-Cl \qquad H-\underset{\underset{H}{|}}{\overset{\overset{H}{|}}{C}}-\underset{\underset{H}{|}}{\overset{\overset{Cl}{|}}{C}}-Cl$$

Two dichloro compounds

$$H-\underset{\underset{H}{|}}{\overset{\overset{H}{|}}{C}}-\underset{\underset{Cl}{|}}{\overset{\overset{Cl}{|}}{C}}-Cl \qquad Cl-\underset{\underset{H}{|}}{\overset{\overset{Cl}{|}}{C}}-\underset{\underset{H}{|}}{\overset{\overset{Cl}{|}}{C}}-Cl$$

Two trichloro compounds

$$Cl-\underset{\underset{H}{|}}{\overset{\overset{Cl}{|}}{C}}-\underset{\underset{H}{|}}{\overset{\overset{Cl}{|}}{C}}-Cl \qquad H-\underset{\underset{H}{|}}{\overset{\overset{Cl}{|}}{C}}-\underset{\underset{Cl}{|}}{\overset{\overset{Cl}{|}}{C}}-Cl$$

Two tetrachloro compounds

$$Cl-\underset{\underset{Cl}{|}}{\overset{\overset{H}{|}}{C}}-\underset{\underset{Cl}{|}}{\overset{\overset{Cl}{|}}{C}}-Cl \qquad Cl-\underset{\underset{Cl}{|}}{\overset{\overset{Cl}{|}}{C}}-\underset{\underset{Cl}{|}}{\overset{\overset{Cl}{|}}{C}}-Cl$$

One pentachloro compound   One hexachloro compound

None of these form optical isomers.

11B.5 (a)

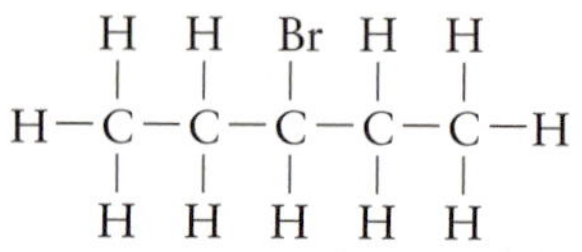

```
    H   H   Br  H   H
    |   |   |   |   |
H — C — C — C — C — C — H
    |   |   |   |   |
    H   H   H   H   H
```
3-Bromopentane

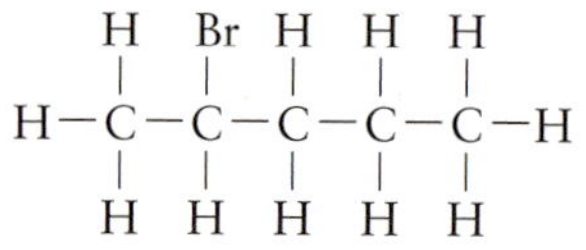

```
    H   Br  H   H   H
    |   |   |   |   |
H — C — C — C — C — C — H
    |   |   |   |   |
    H   H   H   H   H
```
2-Bromopentane

(b) addition reaction

11B.7 (a) $C_6H_{11}Br + NaOCH_2CH_3 \rightarrow C_6H_{10} + NaBr + HOCH_2CH_3$

(b)

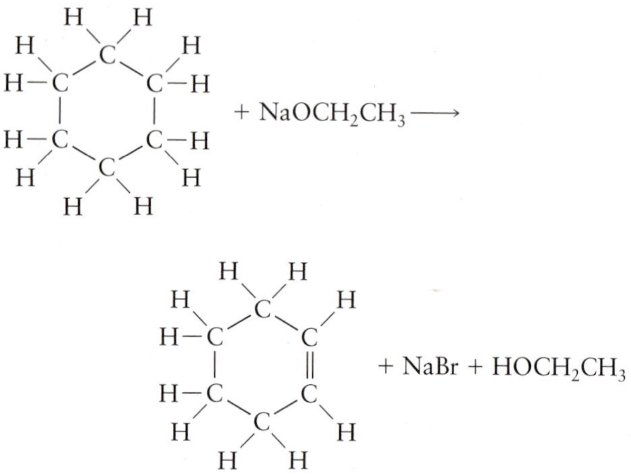

```
        H   H
   H     C      H
    \   /  \   /
  H — C      C — H
      |      |
  H — C      C — H
    /   \  /   \
   H     C      H
        H   H
```
+ NaOCH₂CH₃ ⟶

```
        H   H
   H     C      H
    \   /  \   /
  H — C      C — H
      |      ‖
  H — C      C
    /   \  /   \
   H     C      H
        H   H
```
+ NaBr + HOCH₂CH₃

(c) elimination reaction

11B.9 $C_2H_4 + X_2 \rightarrow C_2H_4X_2$; break one X—X bond and form two C—X bonds. Using bond enthalpies:

Halogen	Cl	Br	I
X—X bond breakage/ (kJ·mol⁻¹)	+242	+193	+151
C—X bond formation/ (kJ·mol⁻¹)	−2(338)	−2(276)	−2(238)
Total/(kJ·mol⁻¹)	−434	−359	−325

The reaction is less exothermic as the halogen becomes heavier. In general, the reactivity, and also the danger associated with use of the halogens in reactions, decreases as one descends the periodic table.

11C.1 (a) 1-ethyl-3-methylbenzene; (b) pentamethylbenzene (1,2,3,4,5-pentamethylbenzene is also correct, but, because there is only one possible pentamethylbenzene, the use of the numbers is not necessary)

11C.3 (a) CH₃ (b) CH₃

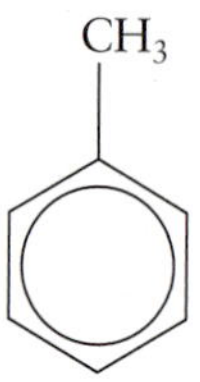

(benzene ring with CH₃)   (benzene ring with CH₃ at top and Cl at bottom)
                                              Cl

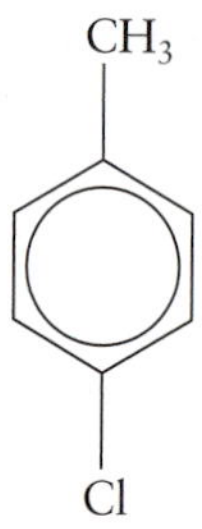

(c) CH₃

(benzene ring with CH₃ at top and CH₃ at lower right)
                CH₃

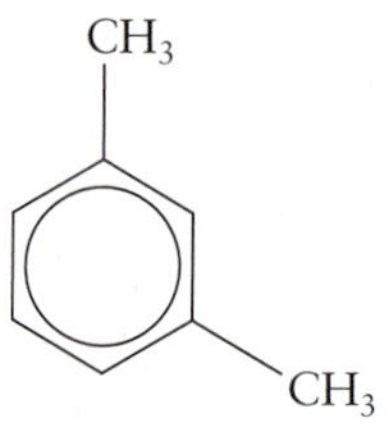

(d) CH₃

(benzene ring with CH₃ at top and Cl at bottom)
                Cl

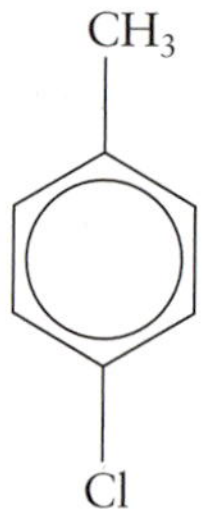

11C.5

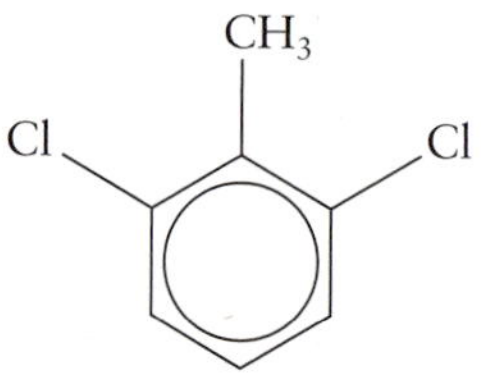

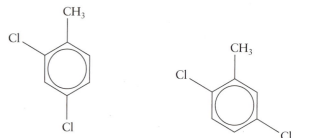

1,3-Dichloro-2-methylbenzene 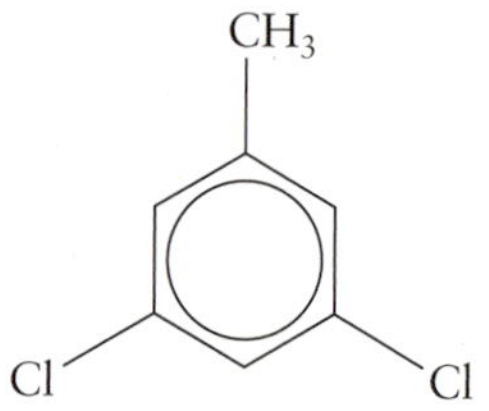   1,3-Dichloro-5-methylbenzene

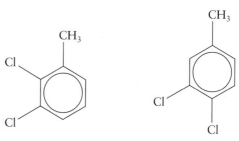

1,3-Dichloro-4-methylbenzene   1,4-Dichloro-2-methylbenzene

1,2-Dichloro-3-methylbenzene   1,2-Dichloro-4-methylbenzene

(b) All of these molecules are polar.

11C.7

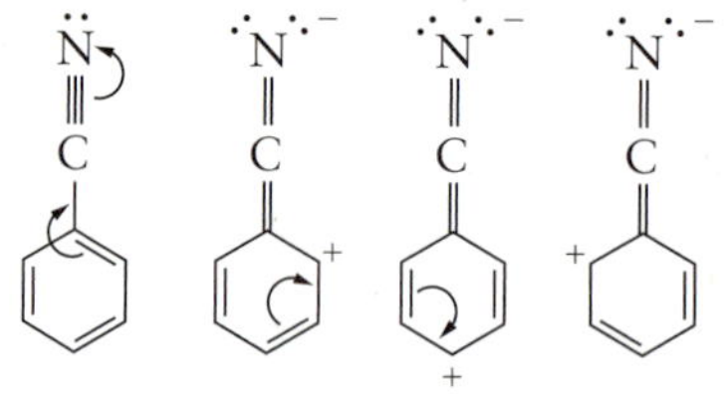

(four resonance structures of benzonitrile/related anion)

Electrophiles tend to avoid the ortho and para positions which have slight + charges in the resonance forms.

11C.9 Two compounds can be produced. Resonance makes positions 1, 4, 6, and 9 equivalent. It also makes positions 2, 3, 7, and 8 equivalent. Positions 5 and 10 are equivalent but have no H atom.

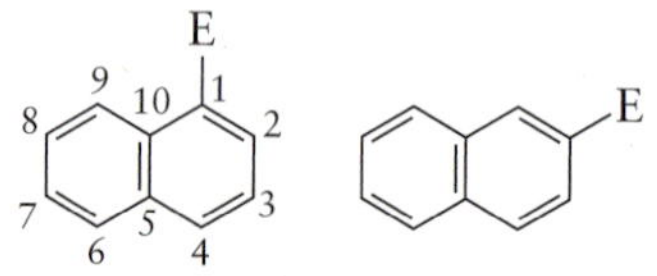

(naphthalene numbered 1–10 with E substituent, and second naphthalene with E substituent)

11C.11 (a) C_9H_8, aromatic hydrocarbon; (b) C_6H_{14}, alkane; (c) C_7H_{10}, alkene; (d) C_6H_{12}, alkane

11C.13 (a) $C_{11}H_{24}$, alkane; (b) C_9H_{12}, alkene; (c) C_8H_{16}, alkane; (d) $C_{16}H_{12}$, alkene

11D.1 (a) RNH_2, R_2NH, R_3N; (b) ROH; (c) RCOOH; (d) RCHO

11D.3 (a) ether; (b) ketone; (c) amine; (d) ester

11D.5 (a) 2-chloropropane; (b) 2,4-dichloro-4-methylhexane; (c) 1,1,1,-triiodoethane; (d) dichloromethane

11D.7 (a)

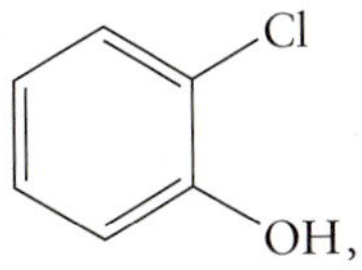

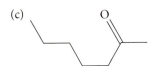

OH, phenol

(b) $CH_3CH(CH_3)CH(OH)CH_2CH_3$, secondary alcohol;

(c) $CH_3CH_2CH(CH_3)CH_2CH(CH_3)CH_2OH$, primary alcohol;

(d) $CH_3C(CH_3)(OH)CH_2CH_3$, tertiary alcohol

11D.9 (a) $CH_3OCH_2CH_2CH_3$;
(b) $CH_3CH_2CH_2CH_2OCH_2CH_3$;
(c) $CH_3CH_2CH_2OCH_2CH_2CH_3$

11D.11 (a) butyl propyl ether; (b) methyl phenyl ether; (c) pentyl propyl ether

11D.13 (a) aldehyde, ethanal; (b) ketone, propanone; (c) ketone, 3-pentanone

11D.15 (a)

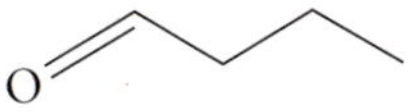

Butanal

(b)

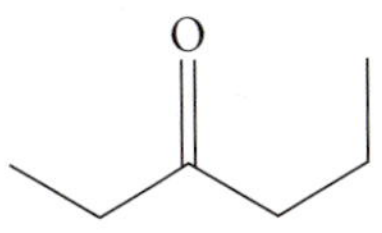

3-Hexanone

(c)

2-Heptanone

11D.17 (a) ethanoic acid; (b) butanoic acid; (c) 2-aminoethanoic acid

11D.19

(a) (b)

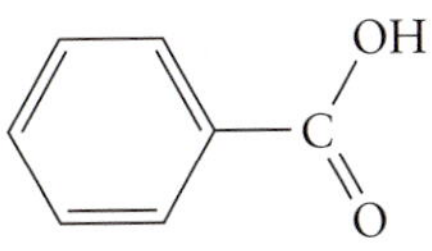

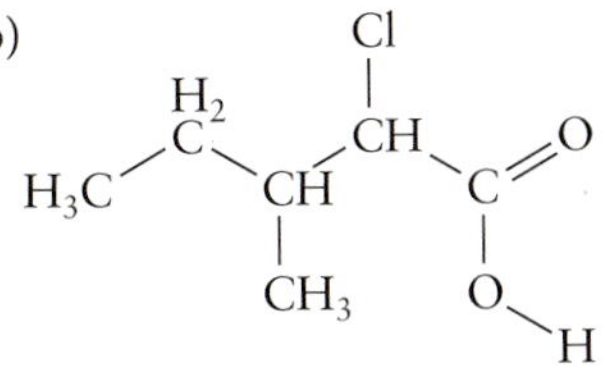

(c) 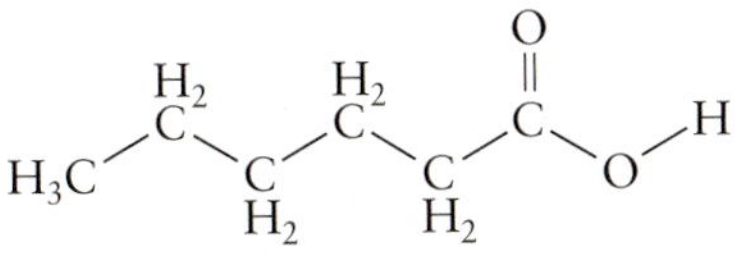 (d) $H_2C=CH$

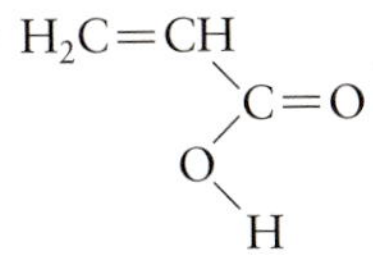

11D.21 (a) methylamine; (b) diethylamine;
(c) o-methylaniline, 2-methylaniline, o-methylphenylamine, or 1-amino-2-methylbenzene

11D.23 (a) (b)

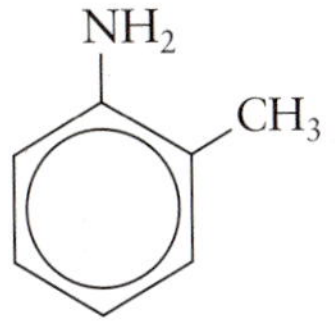

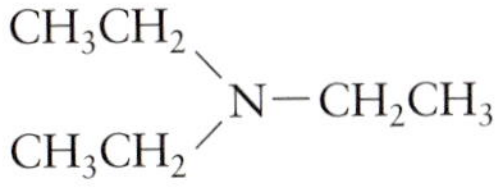

(c)

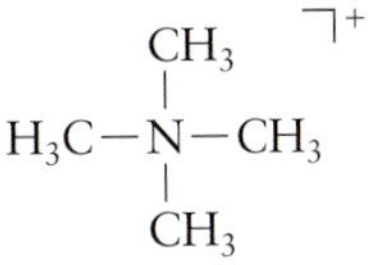

11D.25 (a) and (c)

11D.27 (a) ethanol, use an oxidizing agent such as the salt pyridinium chlorochromate (PCC), $C_5H_5NH[CrO_3Cl]$;
(b) 2-octanol, use an oxidizing agent such as acidified sodium dichromate, $Na_2Cr_2O_7$, or the salt pyridinium chlorochromate (PCC), $C_5H_5NH[CrO_3Cl]$;
(c) 5-methyl-1-octanol, use an oxidizing agent such as the salt pyridinium chlorochromate (PCC), $C_5H_5NH[CrO_3Cl]$

11D.29 (a)

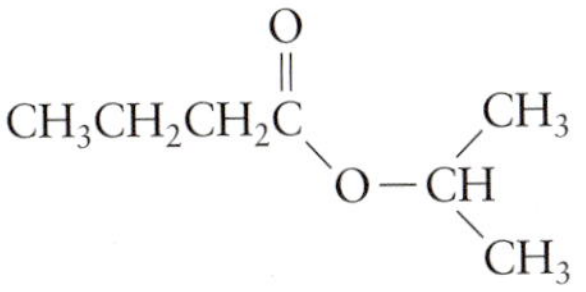

(b)

(c)

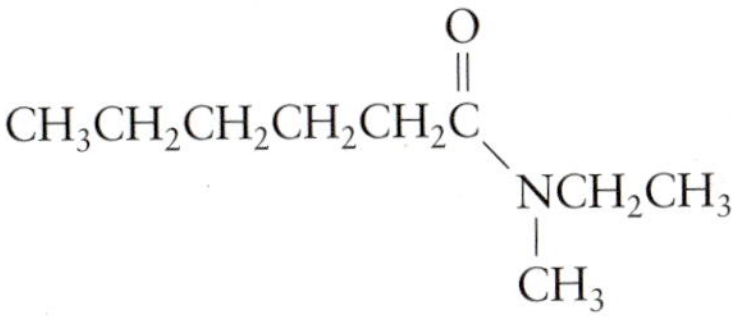

(d)

11D.31 (a) addition; (b) condensation; (c) addition; (d) addition; (e) condensation

11D.33 The following procedures can be used:
1. Dissolve the compounds in water and use an acid–base indicator to look for a color change.
2. $CH_3CH_2CHO \xrightarrow{\text{Tollens regent}} CH_3CH_2COOH + Ag(s)$
3. $CH_3COCH_3 \xrightarrow{\text{Tollens regent}}$ no reaction
Procedure 1 will distinguish ethanoic acid; procedures 2 and 3 will distinguish propanal from 2-propanone.

11D.35 $CH_3CH_2COOH < CH_3COOH < ClCH_2COOH < Cl_3CCOOH$. The greater the electronegativities of the groups attached to the carboxyl group, the stronger the acid. Propanoic acid is less acidic than acetic acid because alkyl groups are more electron-donating.

11E.1
(a) $-CH_2-C(CH_3)_2-CH_2-C(CH_3)_2-CH_2-C(CH_3)_2-$

(b) $-CH-CH_2-CH-CH_2-CH-CH_2-$
     | | |
     CN   CN   CN

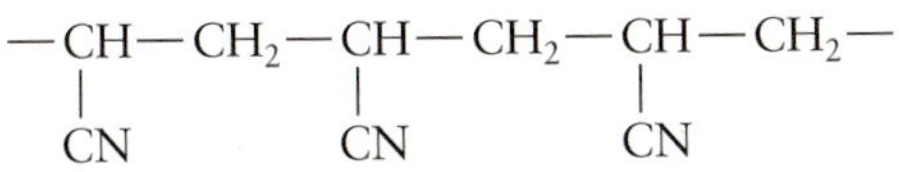

(c)

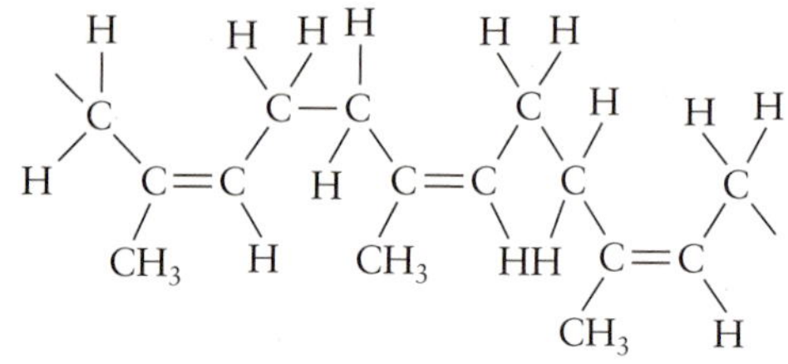

cis version

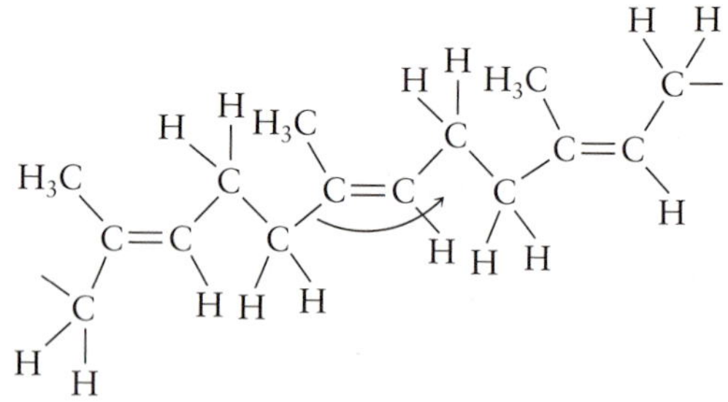

trans version

11E.3 HOOCC$_6$H$_{12}$COOH, NH$_2$C$_6$H$_4$NH$_2$

11E.5 (a) CHCl=CH$_2$; (b) CFCl=CF$_2$

11E.7 (a) —OCCONH(CH$_2$)$_4$NHCOCONH(CH$_2$)$_4$NH—;
(b) —OC—CH(CH$_3$)—NH—OC—CH(CH$_3$)—NH—

11E.9 block copolymer

11E.11 Polymers generally do not have definite molecular masses because there is no fixed point at which the chain-lengthening process will cease. The chain stops growing due to lack of nearby monomer units of the appropriate kind or a lack of properly oriented smaller polymeric aggregates. A polymer is, in a sense, not a pure compound, but rather a mixture of similar compounds of different chain length. There is no fixed molar mass, only an average molar mass. Because there is no one unique compound, there is no one unique melting point, rather a range of melting points.

11E.13 The smaller polylactic acid (PLC) chains pack closer together, making PLC materials stiffer than polyethylene terephthalate (PETE).

11E.15 Larger average molar mass corresponds to longer average chain length. Longer chain length allows for greater intertwining of the chains, making them more difficult to pull apart. This intertwining results in (a) higher softening points, (b) greater viscosity, and (c) greater mechanical strength.

11E.17 Highly linear, unbranched chains allow for maximum interaction between chains. The greater the intermolecular contact between chains, the stronger the forces between them, and the greater the strength of the material.

11E.19 (a) O
        ‖
      —C
         \
          NH—

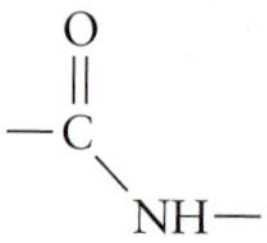

(b) amide; (c) condensation

11E.21 Serine, threonine, tyrosine, aspartic acid, glutamic acid, lysine, arginine, histidine, asparagine, and glutamine satisfy the criteria. Proline and tryptophan generally do not contribute through hydrogen bonding, because they are typically found in hydrophobic regions of proteins.

11E.23

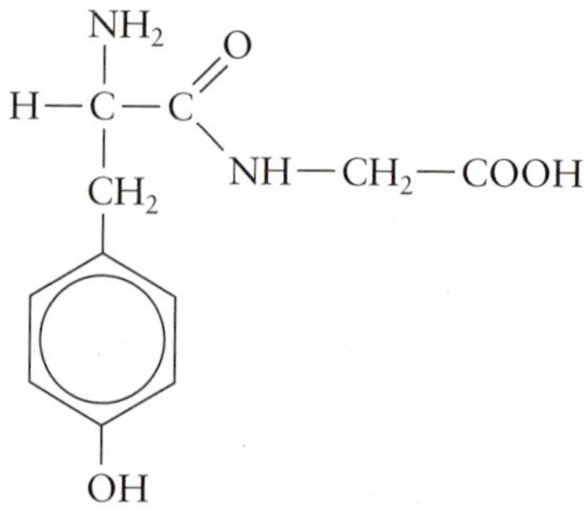

11E.25 (a) alcohols and aldehyde; (b) The chiral carbon atoms are marked with asterisks.

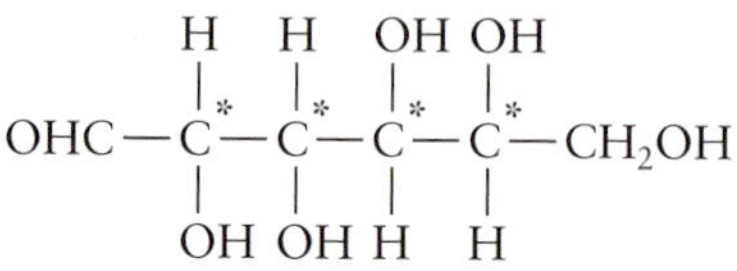

OHC—C—C—C—C—CH$_2$OH

11E.27 (a) GTACTCAAT; (b) ACTTAACGT

11.1 The difference can be traced to the weaker London forces that exist in branched molecules. Atoms in neighboring branched molecules cannot lie as close together as they can in unbranched isomers.

11.3 (a) substitution, CH$_4$ + Cl$_2$ → CH$_3$Cl + HCl
(b) addition, CH$_2$=CH$_2$ + Br$_2$ → CH$_2$Br—CH$_2$Br

11.5 Water is not used because the nonpolar reactants will not readily dissolve in a highly polar solvent such as water. Also, the ethoxide ion reacts with water.

11.7 (a) 2-methyl-1-propene, no geometrical isomers;
(b) *cis*-3-methyl-2-pentene, *trans*-3-methyl-2-pentene;
(c) 1-hexyne, no geometrical isomers; (d) 3-hexyne, no geometrical isomers; (e) 2-hexyne, no geometrical isomers.

11.9 (a) C$_{10}$H$_{18}$; (b) naphthalene, [structure], C$_{10}$H$_8$;
(c) yes. *Cis* and *trans* forms (relative to the C—C bond common to the two six-membered rings) are possible.

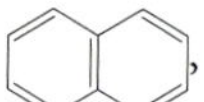

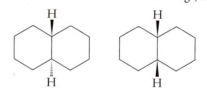

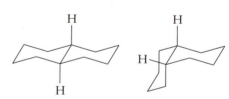

trans-Decalin   *cis*-Decalin

11.11 (a)

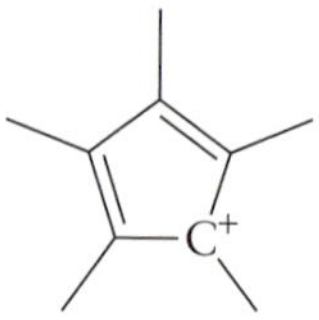

(b) five resonance structures (positive charge on any one of the five carbon atoms)
(c) four π-electrons

11.13 (a)

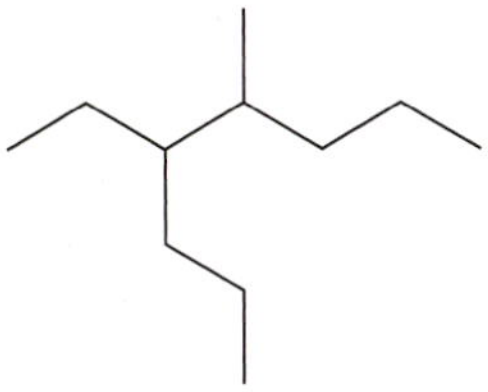

4-Ethyl-5-methyloctane

(b)

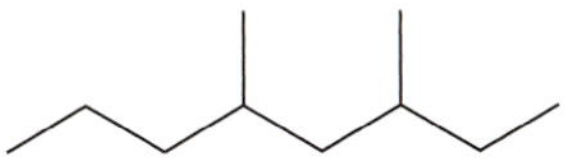

3,5-Dimethyloctane

(c)

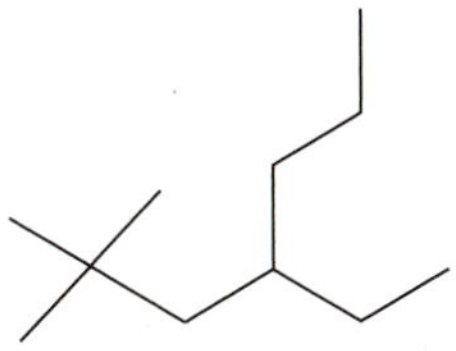

2,2-Dimethyl-4-ethylheptane

(d)

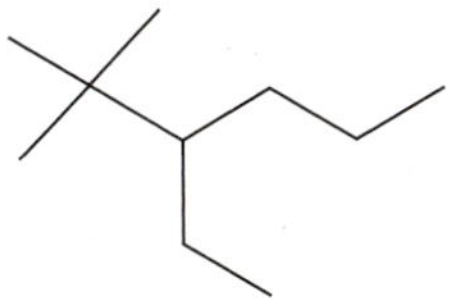

3-Ethyl-2,2-dimethylhexane

11.15 cyclopropane

11.17 The NO_2 group is a meta-directing group and the Br atom is an ortho, para-directing group. Because the position para to Br is already substituted with the NO_2 group, further bromination will not occur there. The resonance forms show that the bromine atom will activate the position ortho to it, as expected. The NO_2 group will deactivate the group ortho to itself, thus in essence enhancing the reactivity of the position meta to the NO_2 group. This position is ortho to the Br atom, so the effects of the Br and NO_2 groups reinforce each other. Bromination is thus expected to occur as shown:

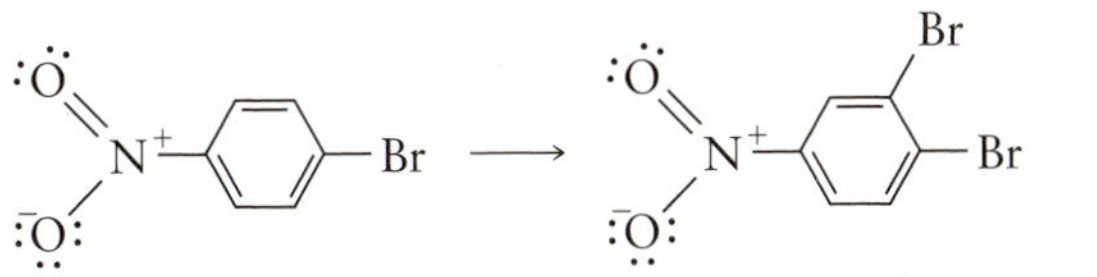

11.19 (a) and (b)

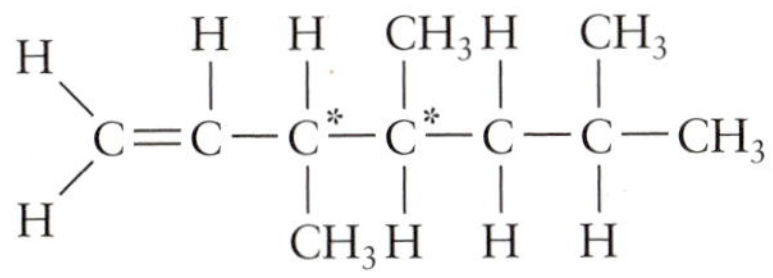

(c) No, there are no cis/trans isomers for this molecule.

11.21 C_8H_{10} will have an absorption maximum at a longer wavelength. Molecular orbital theory predicts that in conjugated hydrocarbons (molecules that contain a chain of carbon atoms with alternating single and double bonds) electrons become delocalized and are free to move up and down the chain of carbon atoms. Such electrons may be described using the one-dimensional "particle in a box" model introduced in Focus 1. According to this model, as the box to which electrons are confined lengthens, the quantized energy states available to the electrons get closer together. More electrons must be accommodated, but the LUMO-HOMO gap is smaller the greater the length of the chain. Therefore, lower energy photons, that is, photons with longer wavelengths, will be absorbed by the C_8H_{10} molecule because it provides a longer "box" than C_6H_8.

11.23 (a) $C_5H_5N_5O$; (b) $C_6H_{12}O_6$; (c) $C_3H_7NO_2$

11.25 (a) alcohol, ether, aldehyde, aromatic ring; (b) ketone, alkene; (c) amine, amide, alkene

11.27 An asterisk (*) denotes a chiral carbon atom.
(a)

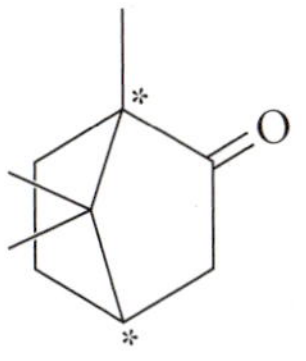

(b)

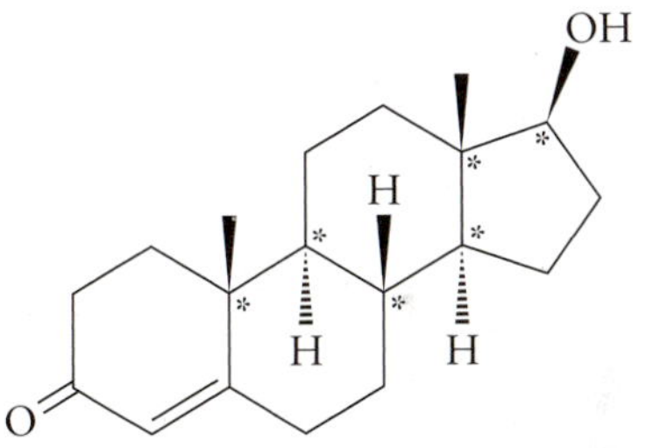

11.29

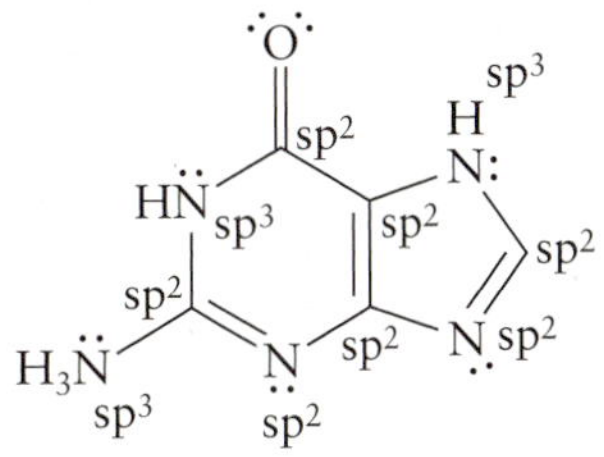

11.31 (a)

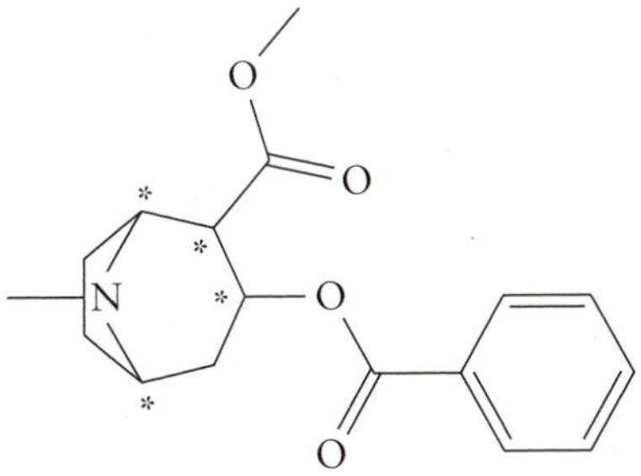

(b)

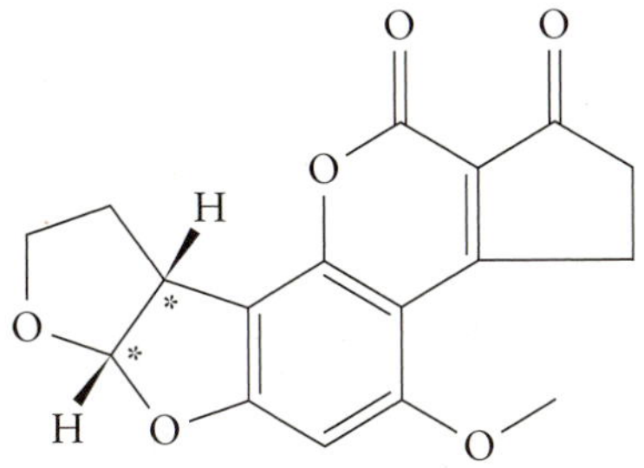

11.33 (a)

$$H-\overset{\overset{\displaystyle H}{|}}{\underset{\underset{\displaystyle H}{|}}{C}}-\overset{\overset{\displaystyle H}{|}}{\underset{\underset{\displaystyle H}{|}}{C}}-O-\overset{\overset{\displaystyle H}{|}}{\underset{\underset{\displaystyle H}{|}}{C}}-\overset{\overset{\displaystyle H}{|}}{\underset{\underset{\displaystyle H}{|}}{C}}-H$$

Diethyl ether

$$H-\overset{\overset{\displaystyle H}{|}}{\underset{\underset{\displaystyle H}{|}}{C}}-\overset{\overset{\displaystyle H}{|}}{\underset{\underset{\displaystyle H}{|}}{C}}-\overset{\overset{\displaystyle H}{|}}{\underset{\underset{\displaystyle H}{|}}{C}}-\overset{\overset{\displaystyle H}{|}}{\underset{\underset{\displaystyle H}{|}}{C}}-O-H$$

1-Butanol

(b) 1-Butanol can hydrogen bond with itself but diethyl ether cannot, so 1-butanol molecules are held together more strongly in the liquid. Therefore, 1-butanol has the higher boiling point. Both compounds can form hydrogen bonds with water and therefore have similar solubilities.

11.35 (a)

$$CH_2-O-\overset{\overset{\displaystyle O}{||}}{C}-(CH_2)_{16}CH_3$$
$$HC-O-\overset{\overset{\displaystyle O}{||}}{C}-(CH_2)_{16}CH_3$$
$$CH_2-O-\overset{\overset{\displaystyle O}{||}}{C}-(CH_2)_{16}CH_3$$

(b)

$$HO-\bigcirc-CHO$$

11.37

$$H_2\ddot{N}-\underset{\underset{\underset{\underset{\displaystyle :\underset{..}{O}H}{|}}{\underset{\displaystyle C=\ddot{O}}{|}}}{\underset{\displaystyle CH_2}{|}}}{\overset{\overset{\displaystyle :O:}{||}}{CH}}-\overset{\overset{\displaystyle :O:}{||}}{\underset{..}{C}}-H\ddot{N}-\underset{\underset{\underset{\displaystyle CH_3}{|}}{\underset{\displaystyle CH-CH_3}{|}}}{\overset{}{CH}}-\overset{\overset{\displaystyle :O:}{||}}{\underset{..}{C}}-H\ddot{N}-\underset{\underset{\underset{\displaystyle :\underset{..}{O}H}{|}}{\underset{\displaystyle C=\ddot{O}}{|}}}{\underset{\displaystyle CH_2}{|}}{\overset{}{CH}}-\overset{\overset{\displaystyle :O:}{||}}{C}-\ddot{O}H$$

11.39 (a) 7180 g·mol^{-1}; (b) 135 monomers; (c) 23.75 Torr; (d) Measuring the change in osmotic pressure proves to be a better method in this case. The osmotic pressure developed by the resulting polymer solution is readily measured, whereas the change in partial pressure of $H_2O(g)$ changes by less than 0.1% upon addition of the polymer.

11.41 (a) Primary structure is the sequence of amino acids along a protein chain. Secondary structure is the conformation of the protein, or the manner in which the chain is coiled or layered, as a result of interactions between amide and carboxyl groups. Tertiary structure is the shape into which sections of the proteins twist and intertwine, as a result of interactions between side groups of the amino acids in the protein. If the protein consists of several polypeptide units, then the manner in which the units stick together is the quaternary structure. (b) The primary structure is held together by covalent bonds. Intermolecular forces provide the major stabilizing force of the secondary structure. The tertiary and quaternary structures are maintained by a combination of London forces, hydrogen bonding, and sometimes ion–ion interactions.

11.43 (a) $^+H_3NCH_2COOH(aq) + H_2O(l) \rightarrow$
$\qquad\qquad ^+H_3NCH_2CO_2^-(aq) + H_3O^+(aq)$
$^+H_3NCH_2CO_2^-(aq) + H_2O(l) \rightarrow$
$\qquad\qquad H_2NCH_2CO_2^-(aq) + H_3O^+(aq)$

(b) $pK_{a1} = 2.35$; $pK_{a2} = 9.78$; pH = 2, $^+H_3NCH_2COOH$; pH = 5, $^+H_3NCH_2CO_2^-$; pH = 12, $H_2NCH_2CO_2^-$

11.45 Condensation polymerization involves the loss of a small molecule, often water or HCl, when monomers are combined. Dacron is more linear than the polymer obtained from benzene-1,2-dicarboxylic acid and ethylene glycol, so Dacron can be more readily spun into yarn.

11.47 (a) $CH_3CH_2CH_3(g) + 5\,O_2(g) \rightarrow 3\,CO_2(g) + 4\,H_2O(g)$
$CHCH + \frac{5}{2}O_2(g) \rightarrow 2\,CO_2(g) + H_2O(g)$

(b) propane, -2043.96 kJ·mol^{-1}, -46.35 kJ·g^{-1}; ethyne, -1255.57 kJ·mol^{-1}, -48.22 kJ·g^{-1};
(c) More heat is released per gram of ethyne, resulting in a hotter flame.

11.49 (a)

$$\overset{\displaystyle NH_2}{\bigcirc}$$

(b) (i) sp^3; (ii) sp^3; (c) Each N atom carries one lone pair of electrons. (d) Yes, the N atoms help to carry the current because the unhybridized p-orbital on each N atom is part of the extended π-conjugation (delocalized π-bonds) that allows electrons to move freely along the polymer.

11.51 (a)

$$CH_3-\overset{\overset{\displaystyle O}{||}}{C}-(CH_2)_5\,\,\,\underset{\displaystyle H}{\overset{\displaystyle H}{C=C}}\overset{\displaystyle \overset{\displaystyle O}{||}}{\underset{\displaystyle C}{}}\overset{}{OH}$$

(b)

$$\underset{}{\overset{}{C=O}} \qquad \text{Carbonyl group, ketone}$$

$$\underset{}{\overset{}{C=C}} \qquad \text{Alkene}$$

$$-C\overset{\displaystyle O}{\underset{\displaystyle OH}{}} \qquad \text{Carboxylic acid}$$

11.53 Coal is not a pure substance and, as a result, does not burn cleanly. Some types of coal produce considerable amounts of sulfur and nitrogen oxides, which contribute to air pollution. The burning of high-sulfur coal contributed very much to the environmental damage in many countries. This damage persists to this day. Coal is also not as easy to transport as gasoline because it is a solid rather than a liquid or gas. Liquids or gases can be placed in fuel tanks and pipelines.

11.55 1-nonene: $CH_2CHCH_2CH_2CH_2CH_2CH_2CH_2CH_3$
2-nonene: $CH_3CHCHCH_2CH_2CH_2CH_2CH_2CH_3$

11.57 (a) (b)

11.59 (a) elimination; (b) $CH_3CH_2CHBrCH_3$ is 2-bromobutane; $CH_3CH_2O^-$ is ethoxide; $CH_3CH=CHCH_3$ is 2-butene; CH_3CH_2OH is ethanol; Br^- is bromide; (c) neither, it is a base; (d) 30.8%; (e) 33.4%; (f) 30.4%; (g) The lowest mass of waste is synthesis (e). The highest mass of waste is synthesis (f). (h) The respective experimental yields for the three reactions are 79.4, 75.5, and 64.2%. The respective experimental atom economy values for the three reactions are 24.4, 25.1, and 19.5%. (i) Based on the highest mass of 2-butene produced (and the highest percentage yield) synthesis (a) looks the best. However, based on the experimental atom economy, synthesis (e) looks the best.

Note: Page numbers starting with A refer to Appendices; those starting with F refer to the *Fundamentals* section.

KEY EQUATIONS

1. General

Roots of the equation $ax^2 + bx + c = 0$:

$$x = \frac{-b \pm \sqrt{b^2 - 4ac}}{2a}$$

Kinetic energy of a particle of mass m and speed v:

$$E_k = \tfrac{1}{2}mv^2$$

Gravitational potential energy of a body of mass m at height h:

$$E_p = mgh$$

Coulomb potential energy of two charges Q_1 and Q_2 at a separation r in a vacuum:

$$E_p = Q_1 Q_2 / 4\pi\varepsilon_0 r$$

2. Structure and spectroscopy

Relation between the wavelength, λ, and frequency, ν, of electromagnetic radiation:

$$\lambda\nu = c$$

Energy of a photon of electromagnetic radiation of frequency ν:

$$E = h\nu$$

de Broglie relation:

$$\lambda = h/p$$

Heisenberg uncertainty principle:

$$\Delta p \Delta x \geq \tfrac{1}{2}\hbar$$

Energy of a particle of mass m in a one-dimensional box of length L:

$$E_n = n^2 h^2 / 8mL^2, n = 1, 2, \ldots$$

Bohr frequency condition:

$$h\nu = E_{upper} - E_{lower}$$

Energy levels of a hydrogenlike atom of atomic number Z:

$$E_n = -Z^2 h\mathcal{R}/n^2, n = 1, 2, \ldots$$

Formal charge:

$$FC = V - \left(L + \tfrac{1}{2}B\right)$$

3. Thermodynamics

Ideal gas law:

$$PV = nRT$$

Expansion work against constant external pressure:

$$w = -P_{ex}\Delta V$$

Work of reversible, isothermal expansion of an ideal gas from V_1 to V_2:

$$w = -nRT \ln(V_2/V_1)$$

First law of thermodynamics:

$$\Delta U = q + w$$

Definition of entropy change:

$$\Delta S = q_{rev}/T$$

Definition of enthalpy:

$$H = U + PV$$

Definition of Gibbs free energy:

$$G = H - TS$$

Change in Gibbs free energy at constant temperature:

$$\Delta G = \Delta H - T\Delta S$$

Relation between the constant-pressure and constant-volume molar heat capacities of an ideal gas:

$$C_{P,m} = C_{V,m} + R$$

Standard reaction enthalpy ($X = H$) and Gibbs free energy ($X = G$) from standard enthalpies and Gibbs free energies of formation:

$$\Delta X^\circ = \sum n\Delta X_f^\circ(\text{products}) - \sum n\Delta X_f^\circ(\text{reactants}), n \text{ in moles}$$

$$\Delta X_r^\circ = \sum n_r\Delta X_f^\circ(\text{products}) - \sum n_r\Delta X_f^\circ(\text{reactants}), n_r \text{ a pure number}$$

Standard reaction entropy:

$$\Delta S^\circ = \sum nS_m^\circ(\text{products}) - \sum nS_m^\circ(\text{reactants}), n \text{ in moles}$$

Kirchhoff's law:

$$\Delta H_2^\circ = \Delta H_1^\circ + \Delta C_P(T_2 - T_1)$$

Change in entropy when a substance of constant heat capacity, C, is heated from T_1 to T_2:

$$\Delta S = C \ln(T_2/T_1)$$

Change in entropy when an ideal gas expands isothermally from V_1 to V_2:

$$\Delta S = nR \ln(V_2/V_1)$$

Boltzmann's formula for the statistical entropy:

$$S = k \ln W$$

Entropy change of the surroundings for a process in a system with enthalpy change ΔH:

$$\Delta S_{surr} = -\Delta H/T$$

4. Equilibrium and electrochemistry

Definition of activity (for ideal systems):

For an ideal gas: $a_J = P_J/P^\circ, P^\circ = 1$ bar

For a solute in an ideal solution: $a_J = [J]/c^\circ, c^\circ = 1 \text{ mol·L}^{-1}$

For a pure liquid or solid: $a_J = 1$

Reaction quotient and equilibrium constant:

For the reaction $a\,A + b\,B \longrightarrow c\,C + d\,D, Q = a_C^c a_D^d / a_A^a a_B^b$

For the equilibrium $a\,A + b\,B \rightleftharpoons c\,C + d\,D$

$$K = (a_C^c a_D^d / a_A^a a_B^b)_{\text{equilibrium}}$$

Variation of Gibbs free energy of reaction with composition:

$$\Delta G_r = \Delta G_r^\circ + RT \ln Q$$

Relation between standard reaction Gibbs free energy and equilibrium constant:

$$\Delta G_r^\circ = -RT \ln K$$

van 't Hoff equation:

$$\ln \frac{K_2}{K_1} = \frac{\Delta H_r^\circ}{R}\left(\frac{1}{T_1} - \frac{1}{T_2}\right)$$

Relation between K and K_c:

$$K = (c^\circ RT/P^\circ)^{\Delta n_r} K_c \cdot P^\circ/Rc^\circ = 12.03 \text{ K}$$

Clausius–Clapeyron equation:

$$\ln \frac{P_2}{P_1} = \frac{\Delta H_{vap}^\circ}{R}\left(\frac{1}{T_1} - \frac{1}{T_2}\right)$$

Relation between Gibbs free energy and maximum nonexpansion work:

$$\Delta G = w_{e,max} \text{ at constant temperature and pressure}$$

Relation between pH and pOH:

$$\text{pH} + \text{pOH} = pK_w$$

Relation between acidity and basicity constants of a conjugate acid–base pair:

$$pK_a + pK_b = pK_w$$

Henderson–Hasselbalch equation:

$$\text{pH} = pK_a + \log([\text{base}]_{\text{initial}}/[\text{acid}]_{\text{initial}})$$

Relation between the Gibbs free energy of reaction and the cell potential:

$$\Delta G = -nFE_{cell}$$

Relation between the equilibrium constant for a cell reaction and the standard cell potential:

$$\ln K = n_r FE_{cell}^\circ/RT$$

Nernst equation:

$$E_{cell} = E_{cell}^\circ - (RT/n_r F) \ln Q$$

5. Kinetics

Average reaction rate:

$$\text{Rate of consumption of R} = -\frac{\Delta[\text{R}]}{\Delta t}$$

$$\text{Rate of formation of P} = \frac{\Delta[\text{P}]}{\Delta t}$$

Unique average rate for $a\,A + b\,B \longrightarrow c\,C + d\,D$:

$$\text{Unique average reaction rate} = -\frac{1}{a}\frac{\Delta[\text{A}]}{\Delta t} = -\frac{1}{b}\frac{\Delta[\text{B}]}{\Delta t} = \frac{1}{c}\frac{\Delta[\text{C}]}{\Delta t} = \frac{1}{d}\frac{\Delta[\text{D}]}{\Delta t}$$

Integrated rate laws:

For rate of disappearance of $A = k_r[\text{A}]$,

$$\ln \frac{[\text{A}]_t}{[\text{A}]_0} = -k_r t \quad [\text{A}]_t = [\text{A}]_0 e^{-k_r t}$$

For rate of disappearance of $A = k_r[\text{A}]^2$,

$$\frac{1}{[\text{A}]_t} - \frac{1}{[\text{A}]_0} = k_r t \quad [\text{A}]_t = \frac{[\text{A}]_0}{1 + [\text{A}]_0 k_r t} \quad \frac{1}{[\text{A}]_t} = k_r t + \frac{1}{[\text{A}]_0}$$

Half-life of a reactant in a first-order reaction:

$$t_{1/2} = (\ln 2)/k_r$$

Arrhenius equation:

$$\ln k_r = \ln A - E_a/RT$$

The rate constant at one temperature in terms of its value at another temperature:

$$\ln \frac{k_2}{k_1} = \frac{E_a}{R}\left(\frac{1}{T_1} - \frac{1}{T_2}\right)$$

The equilibrium constant in terms of the rate constants:

$$K = k_{forward}/k_{reverse}$$

THE ELEMENTS

Element	Symbol	Atomic number, Z	Molar mass,* $M/(\text{g·mol}^{-1})$	Element	Symbol	Atomic number, Z	Molar mass,* $M/(\text{g·mol}^{-1})$
Actinium	Ac	89	(227)	Manganese	Mn	25	54.94
Aluminum	Al	13	26.98	Meitnerium	Mt	109	(276)
Americium	Am	95	(243)	Mendelevium	Md	101	(258)
Antimony	Sb	51	121.76	Mercury	Hg	80	200.59
Argon	Ar	18	39.95	Molybdenum	Mo	42	95.94
Arsenic	As	33	74.92	Neodymium	Nd	60	144.24
Astatine	At	85	(210)	Neon	Ne	10	20.18
Barium	Ba	56	137.33	Neptunium	Np	93	(237)
Berkelium	Bk	97	(247)	Nickel	Ni	28	58.69
Beryllium	Be	4	9.01	Niobium	Nb	41	92.91
Bismuth	Bi	83	208.98	Nitrogen	N	7	14.01
Bohrium	Bh	107	(272)	Nobelium	No	102	(259)
Boron	B	5	10.81	Osmium	Os	76	190.23
Bromine	Br	35	79.90	Oxygen	O	8	16.00
Cadmium	Cd	48	112.41	Palladium	Pd	46	106.42
Calcium	Ca	20	40.08	Phosphorus	P	15	30.97
Californium	Cf	98	(251)	Platinum	Pt	78	195.08
Carbon	C	6	12.01	Plutonium	Pu	94	(244)
Cerium	Ce	58	140.12	Polonium	Po	84	(209)
Cesium	Cs	55	132.91	Potassium	K	19	39.10
Chlorine	Cl	17	35.45	Praseodymium	Pr	59	140.91
Chromium	Cr	24	52.00	Promethium	Pm	61	(145)
Cobalt	Co	27	58.93	Protactinium	Pa	91	231.04
Copernicium	Cn	112	(285)	Radium	Ra	88	(226)
Copper	Cu	29	63.55	Radon	Rn	86	(222)
Curium	Cm	96	(247)	Rhenium	Re	75	186.21
Darmstadtium	Ds	110	(281)	Rhodium	Rh	45	102.90
Dubnium	Db	105	(268)	Roentgenium	Rg	111	(280)
Dysprosium	Dy	66	162.50	Rubidium	Rb	37	85.47
Einsteinium	Es	99	(252)	Ruthenium	Ru	44	101.07
Erbium	Er	68	167.26	Rutherfordium	Rf	104	(265)
Europium	Eu	63	151.96	Samarium	Sm	62	150.36
Fermium	Fm	100	(257)	Scandium	Sc	21	44.96
Flerovium	Fl	114	(289)	Seaborgium	Sg	106	(266)
Fluorine	F	9	19.00	Selenium	Se	34	78.96
Francium	Fr	87	(223)	Silicon	Si	14	28.09
Gadolinium	Gd	64	157.25	Silver	Ag	47	107.87
Gallium	Ga	31	69.72	Sodium	Na	11	22.99
Germanium	Ge	32	72.64	Strontium	Sr	38	87.62
Gold	Au	79	196.97	Sulfur	S	16	32.06
Hafnium	Hf	72	178.49	Tantalum	Ta	73	180.95
Hassium	Hs	108	(270)	Technetium	Tc	43	(98)
Helium	He	2	4.00	Tellurium	Te	52	127.60
Holmium	Ho	67	164.93	Terbium	Tb	65	158.93
Hydrogen	H	1	1.0079	Thallium	Tl	81	204.38
Indium	In	49	114.82	Thorium	Th	90	232.04
Iodine	I	53	126.90	Thulium	Tm	69	168.93
Iridium	Ir	77	192.22	Tin	Sn	50	118.71
Iron	Fe	26	55.84	Titanium	Ti	22	47.87
Krypton	Kr	36	83.80	Tungsten	W	74	183.84
Lanthanum	La	57	138.91	Uranium	U	92	238.03
Lawrencium	Lr	103	(262)	Vanadium	V	23	50.94
Lead	Pb	82	207.2	Xenon	Xe	54	131.29
Lithium	Li	3	6.94	Ytterbium	Yb	70	173.04
Livermorium	Lv	116	(293)	Yttrium	Y	39	88.91
Lutetium	Lu	71	174.97	Zinc	Zn	30	65.41
Magnesium	Mg	12	24.31	Zirconium	Zr	40	91.22

*Parentheses around molar mass indicate the most stable isotope of a radioactive element.